T0136248

THE UNIVERSITY OF ARIZONA SPACE SCIENCE SERIES
RICHARD P. BINZEL, GENERAL EDITOR

Meteorites and the Early Solar System II
D. S. Lauretta and H. Y. McSween Jr.,
editors, 2006, 943 pages

Comets II
M. C. Festou, H. U. Keller, and H. A. Weaver,
editors, 2004, 745 pages

Asteroids III
William F. Bottke Jr., Alberto Cellino, Paolo Paolicchi,
and Richard P. Binzel, editors, 2002, 785 pages

TOM GEHRELS, GENERAL EDITOR

Origin of the Earth and Moon
R. M. Canup and K. Righter, editors, 2000, 555 pages

Protostars and Planets IV
Vincent Mannings, Alan P. Boss, and Sara S. Russell, editors, 2000, 1422 pages

Pluto and Charon
S. Alan Stern and David J. Tholen, editors, 1997, 728 pages

Venus II—Geology, Geophysics, Atmosphere, and Solar Wind Environment
S. W. Bougher, D. M. Hunten, and R. J. Phillips, editors, 1997, 1376 pages

Cosmic Winds and the Heliosphere
J. R. Jokipii, C. P. Sonett, and M. S. Giampapa, editors, 1997, 1013 pages

Neptune and Triton
Dale P. Cruikshank, editor, 1995, 1249 pages

Hazards Due to Comets and Asteroids
Tom Gehrels, editor, 1994, 1300 pages

Resources of Near-Earth Space
John S. Lewis, Mildred S. Matthews, and Mary L. Guerrieri, editors, 1993, 977 pages

Protostars and Planets III
Eugene H. Levy and Jonathan I. Lunine, editors, 1993, 1596 pages

Mars
Hugh H. Kieffer, Bruce M. Jakosky, Conway W. Snyder,
and Mildred S. Matthews, editors, 1992, 1498 pages

Solar Interior and Atmosphere
A. N. Cox, W. C. Livingston, and M. S. Matthews, editors, 1991, 1416 pages

The Sun in Time
C. P. Sonett, M. S. Giampapa, and M. S. Matthews, editors, 1991, 990 pages

Uranus
Jay T. Bergstralh, Ellis D. Miner, and Mildred S. Matthews, editors, 1991, 1076 pages

Asteroids II
Richard P. Binzel, Tom Gehrels, and Mildred S. Matthews, editors, 1989, 1258 pages

Origin and Evolution of Planetary and Satellite Atmospheres
S. K. Atreya, J. B. Pollack, and Mildred S. Matthews, editors, 1989, 1269 pages

Mercury
Faith Vilas, Clark R. Chapman, and Mildred S. Matthews, editors, 1988, 794 pages

Meteorites and the Early Solar System
John F. Kerridge and Mildred S. Matthews, editors, 1988, 1269 pages

The Galaxy and the Solar System
Roman Smoluchowski, John N. Bahcall, and Mildred S. Matthews, editors, 1986, 483 pages

Satellites
Joseph A. Burns and Mildred S. Matthews, editors, 1986, 1021 pages

Protostars and Planets II
David C. Black and Mildred S. Matthews, editors, 1985, 1293 pages

Planetary Rings
Richard Greenberg and André Brahic, editors, 1984, 784 pages

Saturn
Tom Gehrels and Mildred S. Matthews, editors, 1984, 968 pages

Venus
D. M. Hunten, L. Colin, T. M. Donahue, and V. I. Moroz, editors, 1983, 1143 pages

Satellites of Jupiter
David Morrison, editor, 1982, 972 pages

Comets
Laurel L. Wilkening, editor, 1982, 766 pages

Asteroids
Tom Gehrels, editor, 1979, 1181 pages

Protostars and Planets
Tom Gehrels, editor, 1978, 756 pages

Planetary Satellites
Joseph A. Burns, editor, 1977, 598 pages

Jupiter
Tom Gehrels, editor, 1976, 1254 pages

Planets, Stars and Nebulae, Studied with Photopolarimetry
Tom Gehrels, editor, 1974, 1133 pages

Meteorites and the Early Solar System II

Meteorites and the Early Solar System II

D. S. Lauretta
H. Y. McSween Jr.

Editors

With 88 collaborating authors

Foreword by Richard P. Binzel

Dedicated to Robert M. Walker and Alastair G. W. Cameron

THE UNIVERSITY OF ARIZONA PRESS
Tucson

in collaboration with

LUNAR AND PLANETARY INSTITUTE
Houston

About the front cover:

The cover painting illustrates the setting of planetesimals in the primordial solar nebula and shows the solar nebular disk, some of the first solid bodies, a dust-reddened Sun, and bipolar outflows emanating from the growing protostar, perpendicular to the solar nebula. This image also conveys the astrophysical setting of the nebula, a region of high-mass star formation, by showing a sky full of our placental star cluster, nebulosity, abundant proplyds, and shock fronts. *Painting by William K. Hartmann, Planetary Science Institute, Tucson, Arizona.*

About the back cover:

Fireball close to the southern celestial pole, photographed in November 2003 by a prototype desert fireball observatory. The initial network, comprising three observatories, began operations in December 2005. *Image courtesy of Phil A. Bland, Imperial College London and Natural History Museum of London.*

The University of Arizona Press
in collaboration with the Lunar and Planetary Institute

∞ This book is printed on acid-free, archival-quality paper.
Manufactured in the United States of America

11 10 09 08 07 06 6 5 4 3 2 1

Library of Congress Cataloging-in-Publication Data

Meteorites and the early solar system II / D. S. Lauretta,
H. Y. McSween, Jr., editors ; foreword by Richard P. Binzel.
p. cm. — (The University of Arizona space science series)
Includes bibliographical references and index.
ISBN-13: 978-0-8165-2562-1 (acid-free paper)
ISBN-10: 0-8165-2562-5 (acid-free paper)
1. Meteorites. 2. Solar system. I. Lauretta, D. S. (Dante S.),
1970— II. McSween, Harry Y. III. Series.
QB755.M4854 2006
523.5'1—dc22
2005036084

Contents

PART IV: THE FIRST NEBULAR EPOCH: GENESIS OF THE FIRST SOLAR SYSTEM MATERIALS

PART V: THE SECOND NEBULAR EPOCH: MATERIALS PROCESSING IN THE NEBULA

PART VI: THE ACCRETION EPOCH: FORMATION OF PLANETESIMALS

PART VII: THE PARENT-BODY EPOCH: A. ALTERATION AND METAMORPHISM

PART VIII: THE PARENT-BODY EPOCH: B. MELTING AND DIFFERENTIATION

PART IX: THE PLANETARY EPOCH: METEORITES AND THE EARTH

Collaborating Authors

Scientific Organizing Committee

The editors thank the following scientists for serving as associate editors and assisting in the planning stages of this book:

Alan P. Boss
Harold C. Connolly Jr.
Otto Eugster
Matthieu Gounelle
Alexander N. Krot
Laurie A. Leshin
Kevin D. McKeegan
Larry R. Nittler
Joseph A. Nuth III
Frank Podosek
Bo Reipurth
François Robert
Alan Rubin
Meenakshi Wadhwa

Acknowledgment of Reviewers

The editors gratefully acknowledge the following individuals, as well as several anonymous reviewers, for their time and effort in reviewing chapters for this volume:

Carl B. Agee
Conel M. O'D. Alexander
Yuri Amelin
John Beckett
Gretchen K. Benedix
Edwin A. Bergin
Phil Bland
Peter Bodenheimer
John Bradley
Thomas H. Burbine
Robin Canup
Pat Cassen
John Chambers
Fred J. Ciesla
Nicolas Dauphas
Andrew M. Davis
Jeremy S. Delaney
Jason P. Dworkin
Denton S. Ebel
Pascale Ehrenfreund
Ahmed El Goresy
Jon M. Friedrich
Michael J. Gaffey
Hans-Peter Gail
Brett Gladman
Joe Goldstein
Monica Grady
Jeff Grossman
Lawrence Grossman
Henning Haack
Nader Haghighipour
Ralph P. Harvey
Roger H. Hewins
Peter Hoppe
Gary R. Huss
Shigeru Ida
Rhian Jones
Timothy Jull
Noriko Kita
Jeffrey Klein
Thorsten Kleine
Alfred Kracher
Dante S. Lauretta
Gary E. Lofgren
Stanley G. Love
Guenter W. Lugmair
Kurt Marti
Tim McCoy
Harry Y. McSween Jr.
Bradley S. Meyer
Michael R. Meyer
Rolf Michel
M. Miyamoto
Alessandro Morbidelli
Carsten Münker
Yoshi Nakagawa
Larry R. Nittler
Joseph A. Nuth III
Ulrich Ott
Herbert Palme
Dimitri A. Papanastassiou
Julie Paque
Robert O. Pepin
Michail I. Petaev
Robert C. Reedy
Kevin Righter
François Robert
Alan Rubin
Sara Russell
Alex Ruzicka
Scott Sandford
Ed Scott
Steve Simon
Angela Speck
Tomasz Stepinski
Marilena Stimpfl
Timothy D. Swindle
Shogo Tachibana
Richard J. Walker
S. J. Weidenschilling
Michael K. Weisberg
Kees C. Welten
Rainer Wieler
Akira Yamaguchi
Qing-zhu Yin
Andrew N. Youdin
Ernst Zinner
Michael Zolensky

Foreword

"I would sooner believe Yankee professors would lie than stones would fall from the heavens" is the apocryphal quote attributed to Thomas Jefferson following the 1807 meteorite fall in Weston, Connecticut, USA. With certitude we know that this period marks a turning point after which a triumvirate of well-documented and analyzed falls (1795, Wold Cottage, Yorkshire, England; 1803, L'Aigle, Normandy, France; 1807, Weston) conclusively swayed acceptance of their extraterrestrial origin. Two hundred years later, *Meteorites and the Early Solar System II* provides a bicentennial benchmark for the field with the goal of serving as the foundation for ongoing advancement. Our modern view recognizes meteorites as messengers across both space and time. Meteorites lure and challenge us to trace their origins and unravel their recorded histories of early solar system and planet-building processes. Newly recognized pieces of the puzzle include micrometeorites, interplanetary dust particles, and presolar grains, where the latter are identifiable as samples from multiple stellar systems.

Herein, 88 authors convey the beauty and complexity of the mostly consistent, but sometimes conflicting, messages delivered to and decoded within our laboratories. The story of how these basic data constrain, challenge, break, and reshape our models for solar system history is science at its best. All Space Science Series authors are challenged to convey what we know, how we know it, and where we go from here. Our target audience is new researchers entering the field as well as current practitioners expanding and updating their horizons with eyes wide open for new cross-disciplinary connections.

Broad international interaction continues to be as important to meteoritics today as it was two centuries ago, as exemplified by authors from nine different countries contributing to this volume. Recovery of returned samples from the *Genesis* mission, the return of samples from the *Stardust* mission, and the *Hayabusa* (MUSES-C) mission all point to a future of international exploration and *in situ* sample collection. Being able to target the selection of samples, rather than (not ungraciously) being at the mercy of random delivery mechanisms, offers the potential for answering carefully focused science questions that we are only beginning to define.

The authors and editors of *Meteorites and the Early Solar System II* deserve our tremendous thanks for the future they are enabling through the creation of this volume. Less visible, but most especially deserving, are Renée Dotson and co-workers of the Lunar and Planetary Institute (LPI), who brought this volume to reality, literally page by page. The unflagging support and dedication for the ongoing success of the Space Science Series by LPI Director Dr. Stephen Mackwell and the staff of the University of Arizona Press is key for creating these volumes as gateways to the present and future of space science. That future rests most especially in the hands of new students, picking up this book for the first time. Welcome inside!

Richard P. Binzel
Space Science Series General Editor
Lexington, Massachusetts
March 2006

Preface

Meteorites and their accompanying visible phenomenon, meteors, have always fascinated humanity. Both were known to ancient peoples, although the connection between the two was not forged until the late eighteenth century. Aristotle described the origin of meteors in his *Meteorologica* as the result of exhalations from crevices in the Earth that ignited and scattered across the sky. This idea remained with humanity for centuries with Isaac Newton postulating that combustible material ascended into the atmosphere to ignite and produce lightning, thunder, and fiery meteors. After Benjamin Franklin's experiments proved that lightning was electrical in nature, this idea was almost immediately extended to meteors. By the mid-eighteenth century, many scientists believed that meteors and fireballs were manifestations of atmospheric electricity.

The fact that stones fell from the sky was known to ancient peoples and recorded in many legends and histories. Aristotle was forced to conclude that these stones were raised high in the air by the wind before falling back to Earth. During the sixteenth, seventeenth, and eighteenth centuries, natural scientists debated as to whether or not such stones could form in the atmosphere. After Franklin's electrical theory of lightning, the idea that meteorites originated *in situ* during a lightning strike became a leading explanation for the origin of meteorites.

The theory that material from outer space entered Earth's atmosphere, produced fireballs and meteors, and subsequently dropped meteorites did not become widely accepted by the scientific community until the end of the eighteenth century. This paradigm shift was the result of several factors, including the publication of Ernst Chladni's seminal work *On the Origin of the Mass of Iron Discovered by Pallas and Others Similar to It, and on Some Natural Phenomena Related to Them*; the publication of Edward Howard's groundbreaking chemical analyses of meteoritic stones and iron, which revealed the ubiquitous presence of nickel in these materials; and the astonishing number of witnessed meteorite falls that occurred during this time period, including the fall of many meteoritic stones at Sienna, Italy, in 1794, a witnessed fall in 1795 at Wold Cottage, England, and a shower of nearly three thousand stones in 1803 near L'Aigle in Normandy. This last event left no doubt that meteorites do indeed fall from the sky and represent material distinct in its composition and structure from rocks formed on Earth.

Since this time, the links between the field of meteoritics and planetary astronomy, astrophysics, analytical chemistry, and the study of our origins have strengthened immensely. It is now known that the composition, chemistry, and mineralogy of meteorites collectively provide evidence for a wide variety of chemical and physical processes, an understanding of which are crucial to describing the birth of solar systems in general, and ours in particular. This book represents a synthesis of the field of meteoritics as it stands at the beginning of the twenty-first century. The depth and breadth of this field is astonishing, making compilation of this volume both at times appear to be an insurmountable task and at others a humbling yet amazing learning experience. Unlike many other volumes in the Space Science Series, this book is not the end result of a conference on this topic. Instead, the content and structure of this book was carefully laid out almost from the time of its conception. In contrast to most books on meteorites, this volume is process-oriented and emphasizes the constraints that meteoritical investigations place on the nature, duration, and extent of the primary processes that occurred during the early stages of solar system formation. The book is arranged chronologically, with each section representing a different epoch in the formation and processing of planetary material. It is important to keep in mind that this timeline is not always linear but represents variations in space as well as time.

A key distinction of this book from the first *Meteorites and the Early Solar System* is the emphasis on the connection between astrophysics and meteoritics. When Henry Russell discovered the correlation between the elemental abundance in the solar photosphere and the bulk composition of chondritic meteorites, and Hans Suess and Harold Urey demonstrated that the relative abundances of atomic nuclei in chondritic meteorites are the result of nuclear processes in astrophysical environments, the role of meteoritics in unraveling the complex history of the presolar processes became apparent. However, it was Robert M. Walker who conjectured that grains that originated in stars could be found in meteorites and analyzed individually to provide new insights into basic astrophysical processes. With his colleagues at Washington University in St. Louis, he pushed the boundaries of analytical chemistry until such grains could be routinely identified and characterized and the field of presolar grain studies was born. In the past two decades this area of research rapidly matured and is now a cornerstone of modern meteoritics. A synthesis of this field is presented in the section of this book that covers "The Presolar Epoch."

Another major advance since the publication of the first *Meteorites and the Early Solar System* is the role of meteoritics in understanding the astrophysical processes of star and planet formation. One of the most influential researchers in this field was Alastair G. W. Cameron, a pioneer in the development of the theory of nucleosynthesis in stellar interiors based on the elemental abundances in chondritic meteorites. Throughout his career, Cameron emphasized the importance of communication between meteoriticists and astrophysicists in order to understand the complex history and astronomical setting of the early solar system. Today meteorite studies provide constraints on the proximity of supernovae and AGB stars to the solar nebula, the role of high-energy photon irradiation in establishing the chemistry of solar system materials, and the likelihood of solar system formation in a region of high-mass star formation similar to the Orion nebula. These fields are summarized in both "The Disk Formation Epoch" and "The First Nebular Epoch" sections of this book. The cross-talk between the astrophysical and meteoritic communities have greatly strengthened both fields. This complementarity is demonstrated throughout this volume in sections related to "Materials Processing in the Nebula" and "The Formation of Planetesimals."

Another key distinction of this work from its predecessor in the Space Science Series was the decision to extend the scope beyond the formation of chondritic planetesimals and tell the story of "Parent-Body Alteration and Metamorphism," culminating in "Melting and Differentiation." These disciplines are central to meteoritics and provide otherwise unavailable insights into the development of the terrestrial planets. The book ends with a series of chapters describing the connections between "Meteorites and the Earth." This section emphasizes the contributions of meteoritics to understanding the formation of our home planet and that of our planetary companion and neighbors. In addition, we describe the effects on meteorites from their residence on the surface of

Earth, both to provide a means to see through these processes back to the early history of the solar system and also to better understand the care that must be taken to preserve these treasures for future generations of scientists.

This book represents almost four years of planning, writing, coaxing, reviewing, rewriting, editing, and production. We expect that the result will be valuable both as a textbook for graduate courses on meteoritics and as a reference for meteoriticists and researchers in allied fields worldwide for many years to come. Its existence would not have been possible without the hard work of many individuals. We are indebted to the many authors and referees that worked hard to produce a comprehensive yet accessible volume. We also thank the members of our scientific organizing committee for believing in the original vision of this book and working hard to make it a reality. The editor of the Space Science Series, Richard P. Binzel, provided continued support of this vision, and his timely and effective cajoling were essential to completing this book. William K. Hartmann provided another modern masterpiece in planetary art for the cover. Phil A. Bland captured the stunning image of an incoming meteor that adorns the back cover. We thank Michael J. Drake for both financial and moral support throughout the long process of producing this book. Linda Mamassian of Mamassian Editorial Services did a thorough and meticulous job in compiling the index for this book. We thank Stephen Mackwell for providing the resources and support of the Lunar and Planetary Institute. The tireless efforts of Renée Dotson and the entire production staff at the Lunar and Planetary Institute were invaluable in coordinating the efforts of the editors, authors, and referees. Their pleasant demeanor and efficient competence are largely responsible for the professional appearance of the book you now hold in your hands. Finally, the editors thank our spouses for their continued love and support throughout the production of this volume.

Dante S. Lauretta and Harry Y. McSween Jr.

Part I:
Meteoritics Overview

Types of Extraterrestrial Material Available for Study

Monica M. Grady
The Natural History Museum
(now at The Open University)

Ian Wright
The Open University

There is a vast range of extraterrestrial materials available for study: meteorites (asteroidal, lunar, and martian, possibly cometary), micrometeorites and interplanetary dust particles (asteroidal and cometary), and Apollo and Luna samples. We review source objects for various materials. We relate meteorites, and components separated from them, to stages of solar system evolutionary history, from condensation of primordial solids through aggregation, alteration, differentiation, and brecciation. We also consider the presolar history of grains from interstellar and circumstellar environments. The hierarchy of discrete components is intimately intertwined with the parent object categories, as the parent objects all carry material from a variety of sources and processes. These materials can be studied with an enormous variety of analytical techniques and instrumentation. We also briefly discuss source objects that have the potential to yield material for analysis, but have not yet been recognized as having done so (Venus, Mercury), and the outward signs of planetary impacts (craters, tektites).

1. INTRODUCTION

There is a vast range of extraterrestrial material available for investigation. A brief summary of that which is most readily recognized includes meteorites (asteroidal, lunar and martian, plus possibly cometary), micrometeorites (asteroidal and cometary), and Apollo and Luna samples. The purpose of this introductory chapter is to outline the research potential that the material represents. A traditional view is that meteorites are subdivided into stones, stony-irons, and irons, and that these different groups yield information about early solar system and planetary history. And that is a useful description. But increasing numbers of meteorites available for laboratory analysis, coupled with high-precision instrumentation capable of measuring ever-more specific areas or zones within meteorites, have redefined the ways in which we interpret the data now available to us. We use meteorites, and components separated from them, to follow all stages of solar system evolutionary history, from condensation of primordial solids, through aggregation, alteration, differentiation, and brecciation. We can also trace the presolar history of subsets of grains. Interplanetary dust particles (IDPs) collected from the stratosphere are frequently highly porous, fragile aggregates (<10 μm in diameter) of anhydrous minerals believed to come from comets. We have no other material that, with any such certainty, could be taken as cometary. Hydrated IDPs, and the larger micrometeorites (<500 μm diameter) collected from polar meltwater, are asteroidal, and might sample regions of the asteroid belt not supplied to Earth as meteorites. We can use extraterrestrial material to place our solar system into the wider context of the astrophysical evolution of the local galactic neighborhood. In order to achieve this goal, extraterrestrial materials are investigated using a range of complementary techniques, both space- and groundbased. This chapter will provide a very brief overview of what extraterrestrial materials are available for study, what their source regions are, and the types of investigations applied to them. We also attempt to place the different extraterrestrial materials in the approximate context of their astrophysical evolution, as detailed in other chapters in this volume.

2. TOOLS FOR THE STUDY OF EXTRATERRESTRIAL MATERIALS

2.1. Remote Measurements

Much of the discussion in this volume is based on results from laboratory-based studies of meteorites; however, it is useful to summarize the other types of observations and measurements that can be applied to extraterrestrial materials. At the very least, it helps place meteorites in their correct scientific context within the exploration of the solar system and beyond. It also helps to provide a bridge between the different communities of scientists who study extraterrestrial objects.

2.1.1. Telescope instrumentation. The acquisition of images and spectral data from both space- and groundbased telescopes is an important complement to laboratory data. Images from the Hubble Space Telescope (HST) have by themselves served to revolutionize our understanding and appreciation of the universe that we inhabit. Solar system objects (comets, asteroids, and planets and their satellites) are regularly observed by telescopes. Observations of star formation and of the presence and structure of protoplanetary disks around young stars have informed models of grain growth, disk formation, and evolution (e.g., *Throop et al.,* 2001). Spectral measurements across a range of wavelengths, and perhaps most particularly in the infrared (first by the Infrared Astronomical Satellite, then the Infrared Space Observatory, and currently the Spitzer Space Telescope), have also brought about a fresh understanding of the properties of dust grains in different astrophysical sites. The identification of crystalline, rather than amorphous, silicates, plus recognition of silicate compositional variations in the dust around stars (e.g., *Molster et al.,* 2002), have strengthened links between astronomical and laboratory measurements. Instruments on groundbased telescopes have also contributed to understanding of our solar system evolution, with observations of planet-forming disks (e.g., *Greaves et al.,* 1998; *Wahhaj et al.,* 2003), discoveries of extrasolar planets (e.g., *Mayor and Queloz,* 1995), asteroid observations (e.g., *Bus et al.,* 2002), and measurement of micrometeoroid flux [e.g., the radar observations described by *Meisel et al.* (2002)]. As new generations of groundbased telescopes, with large collecting surfaces and adaptive optics, are developed, such contributions can only become increasingly frequent.

2.1.2. Orbiting (and flyby) spacecraft. In the 15 years since *Meteorites and the Early Solar System* was published (*Kerridge and Matthews,* 1988), we have become much more familiar with the surface landscapes of many of the bodies within the solar system as a result of the high-quality, high-resolution images taken by a variety of spacecraft. Magellan at Venus (e.g., *Saunders et al.,* 1992); Clementine and Lunar Prospector at the Moon (e.g., *McCallum,* 2001); Mars Global Surveyor (e.g., *Albee et al.,* 1998), Mars Express (e.g., *Chicarro,* 2004), and Mars Odyssey at Mars (e.g., *Saunders et al.,* 2004); Near Earth Asteroid Rendezvous (NEAR) at asteroid 433 Eros (e.g., *Veverka,* 2002); Galileo at Jupiter (e.g., *Young,* 1998); and Cassini at Saturn (e.g., *Porco et al.,* 2005) have all been instrumental in giving us new perspectives on our neighbors. Imagery has been complemented by radar, spectral, and magnetic surveys, allowing understanding of planetary structure, surface composition, and atmosphere. Dust fluxes in space have been measured by spacecraft on their way to other destinations, including Galileo (*Grün et al.,* 1992), Ulysses (*Grün et al.,* 1992, 1993), Cassini (*Altobelli et al.,* 2003), and Stardust (*Tuzzolino et al.,* 2003).

2.1.3. Landing spacecraft. Apart from the Moon, only Mars, Venus, Saturn's giant satellite Titan, and a single asteroid have had craft land on their surfaces. In the past 15 years, 11 spacecraft have been launched to Mars, 7 of which carried landers; unfortunately, only 3 (Pathfinder, Spirit, and Opportunity) landed successfully. Even so, we have made great strides in our comprehension of surface processes on the Red Planet. Landing on Venus is an even greater challenge than Mars, and nothing has been attempted since the Soviet lander, Venera 14, spent 60 minutes on the venusian surface in 1982 before failing (e.g., *Moroz,* 1983). The Huygens probe descended through Titan's thick and cloudy atmosphere, revealing tantalizing glimpses of a landscape apparently carved by rivers (possibly of methane) and icy boulders covering an icy surface (*Lebreton et al.,* 2005). The NEAR probe made a successful, controlled descent to the surface of 433 Eros, taking the closest images that we have of an asteroid surface (e.g., *Veverka et al.,* 2001). While not technically a "landing," the Galileo mission sent a probe down through Jupiter's atmosphere in 1995, sending back data on the composition and structure of the gas giant (e.g., *Young et al.,* 1996).

2.1.4. Sample return. The next stage, relating remote observations to laboratory analyses, has to be through direct sample return. Lunar samples (totaling almost 300 kg) from the Apollo and Luna missions have been available for analysis for over 30 years. Missions are in progress to collect material from additional sources, although at a much reduced weight level (picograms to micrograms). Genesis, a mission to collect samples of the solar wind, returned to Earth in September 2004; although the landing was less successful than planned, it is anticipated that almost all the mission science will be recoverable. Stardust, the mission to Comet Wild 2, will return cometary, interplanetary, and interstellar dust in 2006, and Hayabusa, the mission to asteroid Itokawa, will bring back a sample of the asteroid in 2007. Beyond these missions, both the European Space Agency and NASA are planning martian sample return missions within the next decade. The much lower masses of returned material required for analysis from post-Apollo missions are a result of major advances in the sensitivity, precision, and resolution of analytical instrumentation.

2.2. Direct Measurements

Direct analysis of an extraterrestrial sample begins with a preliminary low-resolution optical study. In other words, someone looks at it! While this is painfully obvious, it is a necessary first step, and serves to distinguish between metallic and stony objects. The presence of fusion crust shows the freshness, or otherwise, of a fall, while its texture (glossy or matte) can help distinguish between melted and unmelted stones, and its degree of completeness can indicate whether the meteorite is a complete stone, or part of one that broke on impact. The occurrence and extent of weathering rind, surficial evaporates, oxide veining, and rusting also give some idea of the terrestrial environment in which the meteorite has resided prior to collection.

Following preliminary examination, almost any and every analytical method known to scientists has been used to analyze extraterrestrial materials. Texture and mineralogy is determined by conventional thin-section techniques, after which structure and major, minor, trace-element, and isotopic chemistries are determined using a vast range of instrumentation. Techniques include optical and electron microscopy; electron, ion, proton, and X-ray probes; stable and radioactive isotope and organic mass spectrometry (using thermal, resonance, acceleration, or plasma ionization); and optical, infrared (IR), ultraviolet (UV), Mössbauer, nuclear

TABLE 1. Sources of extraterrestrial material.

Source	Material	Epoch
Sun	Meteorites; lunar material (Apollo, Luna, meteorites)	Planetary
Moon	Lunar material (Apollo, Luna, meteorites)	Planetary
Earth	Craters; tektites	Planetary
Mars	Meteorites	Planetary
Asteroids	Meteorites	
	Unmelted	Nebular
		Accretionary
		Parent-body epoch A (aqueous and thermal alteration)
	Melted	Parent-body epoch B (differentiation and core formation)
	Brecciated	Planetary
Comets	IDPs	Nebular
Extrasolar	Meteorites, IDPs	Presolar

magnetic resonance (NMR), and Raman spectroscopy — the list is almost endless. And these instruments are becoming increasingly precise, using ever smaller amounts of material, allowing subtle effects of zonation, overgrowths, and alteration to be traced.

3. PARENT OBJECTS

If all the observational and analytical techniques outlined above were brought into play, there would be very few extraterrestrial materials *not* available for study. However, we will confine our discussion to physical samples that can be examined in the laboratory. So, what sort of parent objects are we considering (Table 1)?

3.1. Asteroids

Most of the extraterrestrial materials that we study come from the asteroid belt, a region of space between about 2 and 4 AU from the Sun. The belt is not an homogeneous array of asteroids. Rather, the objects within it are from groups of varying composition — metallic and nonmetallic, melted and unmelted. There are many different classes based on asteroid composition (from reflectance data) — a recent compilation recognizes 26 separate groups (*Bus et al.,* 2002). Asteroids are not expelled at random from within the asteroid belt, but through a combination of gravitational, collisional, and thermal radiation effects, drift into specific regions, or resonances, from which they are subsequently ejected. The meteorites that fall on Earth, then, are an incomplete sample of the asteroid belt, probably sampling <2% of the cumulative surface area of the asteroids, albeit distributed over several broad regions (e.g., *Farinella et al.,* 1993; *Bottke et al.,* 2002). The asteroidal origin of meteorites is inferred from their ancient age, and deduced by observation. Two classes of observation link meteorites to asteroids. Spectral reflectance measurements of asteroidal and meteorite surfaces give a close match for asteroids and meteorites of a variety of classes [e.g., basaltic achondrites match closely to Vesta (*Burbine et al.,* 2001); carbonaceous chondrites match C-class asteroids (*Burbine et al.,* 2002)]. There are, however, paradoxes in this comparison, in that the largest group of stony meteorites, the ordinary chondrites, has no good spectral match with any common asteroidal class (*Binzel et al.,* 1998). Additionally, the asteroid for which we have the best surface composition, the S-class 433 Eros, might be an altered ordinary chondrite, or, less plausibly, a primitive achondrite (*Nittler et al.,* 2001). Space weathering has been proposed as an explanation for the dichotomy; further remote observations and an asteroid sample return mission are required for successful resolution of the puzzle (*McCoy et al.,* 2001).

The second method for linking meteorites with asteroids rests on observations of incoming fireballs and the ability to track their trajectory and thus calculate an orbit for the meteoroid (*Halliday et al.,* 1996). This has been undertaken successfully for about half a dozen specific meteorites, all of which have been shown to follow orbits that have their apogee within the asteroid belt. It is not possible to link specific meteorites with specific asteroids. Automatic camera networks have the potential to photograph incoming objects, allowing the subsequent recovery of newly fallen meteorites that would otherwise have become lost (e.g., *Bland et al.,* 1999).

3.2. Comets

There are two possible reservoirs of cometary material available for study, one that is widely accepted as cometary, the other more problematic. The former is the reservoir derived from interplanetary dust particles (IDPs) — numerous measurements using a variety of techniques have indicated that anhydrous, C-rich IDPs are likely to be cometary in nature (e.g., *Bradley and Brownlee,* 1986; *Thomas et al.,* 1993; *Bradley,* 2003). More problematic is whether there are any meteorites that might be derived from comets. Each time a comet makes its periodic visit to the inner solar system, more of the ice that comprises a substantial fraction of the cometary nucleus sublimes, releasing dust and gas to form a spectacular tail. Eventually, as more and more material is lost from the nucleus, the cometary orbit decays, and the comet falls into the Sun. Images from the Solar and Heliospheric Observatory (SOHO) satellite have shown that fragmentation and erosion of Sun-grazing comets are relatively common occurrences (*Sekanina,* 2003). However, it is conceivable that a cometary nucleus might fall to Earth and be collected as a meteorite (e.g., *Anders,* 1975). *Campins and Swindle* (1998) reviewed the likely features of cometary meteorites, and concluded that no single known meteorite group matched all the characteristics. They also discussed the possible presence of cometary material as xenoliths within unequilibrated meteorites, a suggestion that has been pursued by analysis of the Krymka unequilibrated ordinary chondrite (*Semenenko et al.,* 2003).

3.3. The Moon

The Moon has been a source of extraterrestrial materials for laboratory study ever since the return of samples from the Apollo 11 mission in 1969. Material collected by astronauts is available from the six Apollo missions (~382 kg) and the three robotic Luna probes that returned material: Luna 16, 20, and 24 (~0.3 kg). Much of the Apollo and Luna material is regolith soil of varying composition, containing a variety of clast types (e.g., *Korotev,* 1998a,b, 2001; *Papike et al.,* 1998). A new source of lunar material became available to the community in the early 1980s — that recovered on Earth as lunar meteorites. Since recognition of ALHA 81005 as a lunar meteorite in 1982 (*Score and Mason,* 1982; *Bogard,* 1983), many fragments of lunar meteorites have been found by meteorite recovery programs; the Sahara and Oman Deserts seem to be particularly fruitful search areas. Although the arrival of lunar meteorites is a haphazard process, with no specific region of the Moon being sampled in any systematic fashion, lunar meteorites have extended the range of lunar materials available for study (e.g., *Warren et al.,* 1989). They are frequently fragmental breccias, containing clasts of unusual basalts not previously identified in any catalog of lunar samples (e.g., *Korotev,* 1999). The most comprehensive and up-to-date list of lunar meteorites is maintained by R. L. Korotev at Washington University (*http://epsc.wustl.edu/admin/resources/meteorites/moon_meteorites_list.html*).

3.4. Mars

Unlike the Moon, no samples have (as yet) been returned directly from Mars, either by robotic probe or astronaut; since the earliest martian sample return mission is not set to launch until after 2010, we are likely to be closer to the production of *Meteorites and the Early Solar System III* than the current volume before any such samples do come back. We are, however, in receipt of a number of martian meteorites that sample a variety of igneous rock types (e.g., *Meyer,* 2003). A rationale for the martian origin of the rocks has been rehearsed many times (e.g., *McSween,* 1985), but a précis of the explanation is that the meteorites have crystallization ages younger than asteroids (e.g., *Nyquist et al.,* 2001), and that many of them contain pockets of melt glass in which martian atmospheric gases have been trapped (*Bogard and Johnson,* 1983). The assumption that these meteorites are from Mars is now widely accepted. The original three groups, shergottites, nakhlites, and Chassigny, have been supplemented with the orthopyroxenite ALH 84001. Shergottites have also been subdivided into basalts and lherzholites (*McSween,* 1994), with a recent suggestion that there might be a third subdivision, olivine-phyric shergottites (*Goodrich,* 2003). Although the martian meteorites are all igneous rocks, they have varying crystallization histories, and have been used to explore planetary processes on Mars. Petrogenetic analysis of the shergottites has allowed development of theories of basaltic fractionation and crystallization on Mars during magma genesis (e.g., *McCoy et al.,* 1992; *Wadhwa et al.,* 1994; *Borg et al.,* 2002; *Goodrich,* 2003). The recognition of complex assemblages of secondary minerals (halite with carbonates and clays) in nakhlites has led to interpretation of the scale and mode of fluid flow on the surface of Mars (e.g., *Bridges and Grady,* 1999, 2000). Although the "life on Mars" question continues to be debated quite actively, the identification of structures argued to be the fossilized remains of martian micro-organisms in ALH 84001 (*McKay et al.,* 1996) does not attract much support: *Treiman* (1998, 2003) presents a viewpoint that is held by many professional meteoriticists.

3.5. Nucleosynthetic Sites

Materials that originated in several different nucleosynthetic sites are to be found within primitive chondrites; they are present at levels of a few parts per billion to parts per million. The grains have generally been isolated through vigorous and lengthy acid-dissolution procedures (e.g., *Amari et al.,* 1994; *Lewis et al.,* 1994), thus they are very refractory materials. Because of the nature of the preparation procedures, any presolar silicates were destroyed during isolation of the refractory species; more recent developments in analytical techniques have allowed presolar silicates to be identified *in situ* in IDPs and meteorites (*Messenger et al.,* 2003; *Nagashima et al.,* 2004). The most abundant materials that have been identified as presolar include silicon carbide (SiC), graphite, diamond, and oxides. Less-abundant grains include nitrides. Specific types of grains and their characteristics are outlined below; here, we give a very brief summary of the parent objects from which they are likely to be derived. The presolar cloud has been the recipient of material from a multitude of sites; on the basis of their isotopic compositions, the grains may be assigned to specific astrophysical locations.

Nanodiamonds are generally assumed to be the most abundant of the presolar grains found in meteorites, present in concentrations of several hundred parts per million in CI1 and CM2 chondrites (e.g., *Huss and Lewis,* 1995; *Russell et al.,* 1996a). Although their precise origin is not fully understood, the nanodiamonds are presumed to have formed by chemical vapor deposition in the expanding shell of a type II supernova (*Clayton et al.,* 1995) with heavier elements produced by the r- and p-processes (e.g., *Ott,* 2003), then trapped by ion implantation (e.g., *Koscheev et al.,* 2001). Diamonds have also been observed in the dusty disks around stars (e.g., *Van Kerckhoven et al.,* 2002), implying possible circumstellar, as well as interstellar, origins. However, in contrast, there is some evidence, based on the relative abundances of nanodiamonds in meteorites and IDPs, that not all the nanodiamonds are presolar (*Dai et al.,* 2002).

Presolar graphite comprises ~2 ppm of CM2 chondrites (*Anders and Zinner,* 1993); transmission electron microscopy (TEM) and Raman spectroscopy of presolar graphite show that it exhibits a range of crystallinities, from poorly graphitized carbon to well-crystalline graphite (*Bernatowicz et al.,* 1991; *Zinner et al.,* 1995). The precise origins of presolar graphite are the least well understood of all the "exotic" grains. Possible sources for different groups of graphite include an ONe nova explosion (*Amari et al.,* 2001), He-burning in Wolf-Rayet stars or type II supernovae (*Anders and Zinner,* 1993; *Hoppe et al.,* 1995), or ion-molecule reactions in molecular clouds (*Zinner et al.,* 1995). Meteoritic SiC grains have also emanated from several different astrophysical environments, including low-mass AGB stars, J- and R-type carbon stars, and ejecta from type II supernovae (e.g., *Alexander,* 1993; *Anders and Zinner,* 1993; *Hoppe et al.,* 1996; *Zinner,* 1997). However, most SiC grains (~90% of the total) belong to a single population, thought to be synthesized in an expanding envelope around thermally pulsing low-mass asymptotic giant branch (AGB) stars (e.g., *Anders and Zinner,* 1993; *Hoppe et al.,* 1996).

As well as inorganic species, organic presolar grains are present, mostly in CM and CI chondrites (e.g., *Sephton and Gilmour,* 2000). They are presumed to be species produced by ion-molecule reactions on grain surfaces in the interstellar medium (*Ehrenfreund and Charnley,* 2000). Remnants of such interstellar material have also been identified in IDPs (*Messenger,* 2000).

One of the most exciting and philosophically rewarding results of separation and analysis of presolar grains from meteorites has been the dialogue that this has enabled between astronomers, astrophysicists, and meteoriticists, and the fresh understanding the results have given to nucleosynthetic processes during stellar cycling. Astrophysicists can now constrain their theoretical calculations using real data obtained on components from stars at different stages of their lifetimes. Likewise, results from presolar organic components can be used to constrain the thermodynamics of ion-molecule reactions in the interstellar medium.

3.6. The Sun

The Sun is a source of material for laboratory investigation, although perhaps only indirectly. Samples from the Sun are preserved, implanted within surfaces that have been exposed to the solar wind (SW) and to solar energetic particles (SEP). The composition of the present-day SW is well known from analysis of foils exposed at the lunar surface during the Apollo missions (*Geiss et al.,* 2004). The Moon has been a major source of SW-bearing materials. The lunar regolith has provided samples of both the current and ancient SW, and attempts have been made to use Apollo samples of varying antiquity and maturity to trace variation in SW through time (e.g., *Becker and Pepin,* 1989; *Wiens et al.,* 2004a,b). Meteorite regolith breccias have also contributed to understanding of the SW; gas-rich breccias record implanted SW and SEP (e.g., *Pedroni and Begemann,* 1994). More recently, the Genesis mission exposed a variety of pure collector materials to the SW for 2.5 years (*Burnett et al.,* 2003) in order to obtain sufficient material for precise elemental and isotopic compositional determination.

3.7. Additional Sources of Extraterrestrial Material?

Thus far, we have discussed source objects that we know have supplied extraterrestrial material to Earth. But there are other potential reservoirs of extraterrestrial specimens, not all of which are necessarily recognized by conventional wisdom as being bona fide sources. In this section we will consider, in an increasingly speculative fashion, what other materials might be, or might become, available for study.

3.7.1. Extraterrestrial "meteorites." The strict definition of a "meteorite" refers to a naturally occurring sample of extraterrestrial material that has fallen to Earth. Analogous specimens have been identified within samples collected from the Moon (e.g., *Anders et al.,* 1973; *McSween,* 1976; *Rubin,* 1997) and given the name of the locality from which they were collected. Pictures of the surface of Mars show landscapes strewn with rocks, and in January 2005, the Opportunity rover took images of an iron meteorite. This first meteorite to be recorded on Mars' surface has been given the rather uninformative name "Heat Shield Rock" (*Jet Propulsion Laboratory,* 2005). We do not know what proportion of martian surface rocks are likely to be samples of asteroids (or indeed the Moon or even Earth). The statistics of this particular problem have been considered by *Bland and Smith* (2000). Ordinary chondrites are the most numerous type of meteorite on Earth; it would be unfortunate if the first-ever sample returned from Mars by a space mission turned out to be an ordinary chondrite.

3.7.2. Space debris. Artificial space debris includes micrometer- to millimeter-sized particles from solid fuel rocket motors, flakes of paint, globules of sodium from coolant systems, fragments of craft destroyed by impacts or explosions, and frozen astronaut urine from early space missions (*Graham et al.,* 1999, 2001). As more satellites are launched into orbit, it is apparent that the flux of artificial space debris will increase, providing not just a greater hazard to orbiting craft and astronauts, but also interfering with the collection of natural micrometeoroids. On a larger scale, space debris also encompasses the hulks of satellites and rocket stages long since forgotten in terms of their primary missions. Because of the inherent dangers of space debris to future missions, communications satellites, and astronauts, the orbits of all pieces greater than a few centimeters in size are monitored (e.g., *Africano and Stansbery,* 2002). It has been argued that such objects should be preserved in space as part of our cultural heritage (e.g., *Gorman,* 2005). This would effectively raise the status of space debris to that of exoarchaeological artifact, the management of which should ensure its survival into the future.

3.7.3. Meteorites from Mercury and Venus. Other parent bodies that may come to be represented in the meteorite collection on Earth include Venus and Mercury. For many reasons, the likelihood that venusian meteorites are possible is somewhat remote (e.g., *Gladman et al.,* 1996). However, it is important to recall that in the late 1970s a number of dynamical arguments were used to counter the notion of martian meteorites. So we should remain open to the possibility of Venus as a source of meteorites (e.g., *Goodrich and Jones,* 1987). The prospects for the recognition of a mercurian meteorite have been outlined by *Love and Keil* (1995). A conclusion of this work, and that of *Gladman et al.* (1996), was that meteorites from Mercury were a distinct possibility. The potential identification of the first mercurian meteorite (*Palme,* 2002) was complete speculation based on the description of an unusual meteorite (*Ebihara et al.,* 2002).

3.7.4. Ice meteorites. Comets have impacted Earth throughout its history. Indeed, there is evidence (somewhat controversial) for accretion of relatively small comets by Earth (e.g., *Frank and Sigwarth,* 1997). It has been suggested that ice meteorites (rich in ammonia) might fall, and could be collected in Antarctica (e.g., *Bérczi and Lukács,* 1994), although this idea is not generally accepted, and it is hard to envisage how the mechanism might work.

3.7.5. Sedimentary meteorites. All the martian meteorites that we have in our collections are igneous. But given that fluid has played such a prominent role in shaping Mars' surface, it is possible that sedimentary rocks might be present — and hence might be ejected to Earth as sedimentary meteorites. The method by which we could identify such specimens is unclear: Experiments to simulate atmospheric entry and survival of sedimentary meteorites were inconclusive (*Brack et al.,* 2002). *Wright et al.* (1995) speculated that such meteorites could be collected from Antarctica. When such samples are eventually collected, group names based on composition have already been suggested ["amathosites" for sandstones and "calcarites" for limestones (*Cross,* 1947)], although these terms are so outdated that a more modern nomenclature would be much more appropriate.

3.7.6. Extraterrestrial artifacts. Throughout the ages, iron meteorites have been turned into artifacts, both for ceremonial and for utilitarian use (see examples quoted in *Buchwald,* 1975). There is, however, another aspect of this process of misidentification that needs to be considered, i.e., one that involves extraterrestrial artifacts. While this is a subject that demands skepticism, there are compilations of accounts of high-technology artifacts and unusual materials allegedly observed to fall from the sky (e.g., *Corliss,* 1977, 1978). It has been proposed that the possibility of extraterrestrial artifacts from pre-human layers of the geological record is "worth objective analysis" (*Arkhipov,* 1994, 1996). The objectives of the Search for Extraterrestrial Intelligence (SETI) are considered by most people to be reasonably worthwhile; the search for extraterrestrial artifacts on Earth is merely an extension of this philosophy.

4. STAGES IN METEORITE HISTORY AND EVOLUTION

4.1. Introduction

As of June 2005, there were around 31,000 recognized meteorites maintained in collections throughout the world (*Grady,* 2000; *Grossman,* 2000; *Grossman and Zipfel,* 2001; *Russell et al.,* 2002, 2003, 2004, 2005), so there is no shortage of material for study. However, of these, only 1200 have been observed to fall, and so are free from the highest levels of terrestrial contaminants that can obscure results from analysis. Meteorite classification is treated fully in the next chapter; here, we confine ourselves to a bare outline of different meteorite types, to give context to the description of meteoritic components. Thus asteroidal meteorites are now divided into two large groups, one that includes melted meteorites (irons and stony-irons and the stony achondrites), the other of which comprises all the unmelted stones (chondrites). It is the chondritic meteorites, with their primitive compositions, that have yielded the most insight to preplanetary processes, but only 63 (around 5%) of the classified meteorite falls are carbonaceous chondrites or unequilibrated ordinary chondrites. In contrast, melted asteroidal meteorites have played a prominent role in the comprehension of planetary differentiation and core formation (83 of the meteorite falls, or almost 7% of the total). Analyses of whole-rock meteorites and components separated from them have provided complementary datasets that enable complex meteorite histories to be deduced. The most cursory inspection of a meteorite shows it to be heterogeneous (Fig. 1). Even the earliest analyses of meteorites separated metallic phases from nonmetallic (*Howard,* 1802). There are, however, excellent reasons for continuing to perform chemical and isotopic analyses of whole-rock specimens. It is the bulk (chemical and isotopic) chemistry of a meteorite that enables recognition of relationships between meteorites, and therefore assignation to an individual parent body (although not to a precise parental source such as a specific asteroid). However, although bulk analytical data enable meteorite groups to be recognized, analysis of discrete components within a meteorite is required to give more detailed and specific accounts of formation histories. An individual meteorite can be separated into discrete components, each set of which may trace different processes. Discrete components can be ordered in a hierarchical fashion that approximates a chronology of processes experienced by the parent

Fig. 1. Thin section of the Vigarano chondrite, showing chondrules and CAIs (nebular) and interchondrule matrix (accretionary). Presolar grains are embedded within the specimen. Subtle differences in color and chondrule/matrix ratio indicate that the meteorite has been brecciated and reaggregated.

object (Table 2). The subdivisions within this section approximate the sections into which the rest of this volume is divided.

4.2. The Presolar Epoch

A volumetrically insignificant but scientifically critical component within chondrites is their complement of interstellar and circumstellar grains. The family of grains most recently analyzed in the laboratory is that of grains formed prior to the accretion of the solar nebula. Lumped together in the broad category of "presolar grains," these materials have a variety of origins. The grains thus pre-date the major chondritic components, and occur as several populations of grains with different grain sizes (e.g., *Anders and Zinner,* 1993; *Nittler,* 2003). The presence of the grains was first inferred in the late 1970s to early 1980s on the basis of the isotopic composition of noble gases (*Alaerts et al.,* 1980). The unusual isotopic signatures of the noble gases implied the existence of several different hosts; analyses of acid-resistant residues suggested that the hosts might be carbon-rich. In 1983, combustion of a set of residues yielded the first C and N isotopic compositions of the grains (*Lewis et*

TABLE 2. Extraterrestrial material available for analysis.

Epoch	Process	Meteorites		IDPs	Component
		Unmelted	Melted		
Presolar	Nucleosynthesis	X		X	Circumstellar and interstellar grains (SiC; graphite; nanodiamonds, etc.); GEMS
	Interstellar chemistry	X		X	Organic molecules
Nebular	Formation of primary solids	X		X	CAIs; chondrules; silicates; metal; sulfides
Accretionary	Formation of planetesimals	X			Chondrule rims; dark inclusions; interchondrule matrix; phyllosilicates(?)
Parent-Body A	Aqueous alteration	X			Secondary minerals (carbonates, magnetite, phyllosilicates, etc.)
	Thermal alteration	X			Silicates
Parent-Body B	Differentiation		X		Achondrites
	Core formation		X		Iron meteorites
Planetary	Early solar irradiation	X	X	X	Regolith breccias
	Long-term solar irradiation				Lunar samples (Apollo, Luna, meteorites)
	Cosmic-ray irradiation	X	X	X	Plus returned space hardware
	Lunar history		X		Lunar samples (Apollo, Luna, meteorites)
	Martian history		X		Martian meteorites
	Collision	X	X		Shock veins; impact melts; implanted species
	Impact				Craters, tektites

al., 1983; *Swart et al.,* 1983). Individual grains of all these types except diamonds are now analyzed routinely by ion microprobe. Individual crystallites of nanodiamonds are only ~3 nm across, thus isotopic measurement of discrete grains is not yet possible (and may never be possible), even with the most sensitive of techniques. The grains are characterized and classified according to the isotopic composition of their major elements (including C, Si, O, Mg) (e.g., *Hoppe et al.,* 1994, 1995). Gases trapped within the grains (including N, Ne, Xe) are measured by mass spectrometry (e.g., *Huss et al.,* 2003). The isotopic compositions of minor and trace elements associated with specific nucleosynthetic processes are also measured by ion microprobe (e.g., Ba, Sm) (*Ott and Begemann,* 1990; *Zinner et al.,* 1991) or resonance ionization mass spectrometry (e.g., Zr, Mo) (*Nicolussi et al.,* 1998), while the identification of subgrains within grains has been recognized through analytical transmission electron microscopy (e.g., *Bernatowicz et al.,* 1991), and can now be analyzed by ion microprobe (*Stadermann et al.,* 2003a). Meteorites are not the only objects that are host to presolar grains: Interplanetary dust particles contain species that were irradiated prior to accretion and might therefore be interstellar in origin (*Bradley,* 1994). These subcomponents, known as GEMS (glass with embedded metal and sulfides), contain silicates, a few of which have unusual O-isotopic compositions (*Messenger et al.,* 2003) indicative of a preserved circumstellar component; the search for other isotopically anomalous species within GEMS has not yet been successful (*Stadermann et al.,* 2003b). Interplanetary dust particles contain other presolar components, including circumstellar forsterite crystals (*Messenger et al.,* 2003) and organic compounds with large enrichments in D and N (e.g., *Keller et al.,* 2000). The organic species are similar to those observed in molecular clouds, and along with the presence of inorganic presolar grains imply that IDPs are the most isotopically primitive materials available for laboratory investigation.

Apart from the GEMS, the measurements outlined above have almost exclusively been undertaken on populations of grains produced after extensive demineralization of whole-rock primitive chondrites (e.g., *Amari et al.,* 1994, *Lewis et al.,* 1994). Less-destructive separation techniques are being developed (*Tizard et al.,* 2005), and analysis of individual presolar grains *in situ* within a meteorite is a goal that is coming into sight (*Nagashima et al.,* 2004). Automated mapping of thin sections by probe techniques show the most promise in this respect (e.g., *Messenger et al.,* 2003; *Mostefaoui et al.,* 2003), and can also be applied to components within other types of extraterrestrial materials.

The foregoing discussion focuses on inorganic (although still mainly carbon-bearing) populations of presolar grains. There is also a vast range of organic presolar components available for measurement. These components are present in the most primitive chondrites (CI, CM, and unequilibrated ordinary chondrites, or UOCs), as well as in IDPs. They are characterized by elevated D/H and $^{13}C/^{12}C$ ratios, and encompass a range of materials from simple aliphatic to complex aromatic (e.g., *Cronin and Chang,* 1993). Laboratory analysis of presolar organic materials tends to be the domain of organic chemists and their range of techniques, including pyrolysis and combustion gas chromatography–isotope ratio mass spectrometry (GC–IRMS).

4.3. First and Second Nebular Epochs

Components separated from extraterrestrial materials that trace primary processes are materials that are presumed to have formed in the solar nebula; in this volume, that period is divided into the first and second nebular epochs, reflecting first the genesis of the materials, then their processing into objects prior to accretion into planetesimals. CI chondrites are the meteorites with compositions closest to that of the solar photosphere, and so are often taken to be the most primitive, i.e., closest in composition to the dust from which the solar system aggregated. They have, however, suffered extensive exposure to fluids. Perhaps more useful for a better understanding of primordial materials are the CV3 and CO3 carbonaceous chondrites and UOCs. These are specimens that have chondritic compositions and are altered very little by secondary parent-body processing (*McSween,* 1979). They are not, as was originally thought, totally unaltered, but exhibit minimum levels of aqueous alteration and thermal metamorphism. They have distinct chondrules and CAIs, and thus provide readily recognizable components that can be identified as representative of material from the earliest nebular epochs. As well as primary silicates and oxides in CAIs and chondrules, components include metal and sulfides.

Mineralogy plus major-, minor-, and trace-element and isotopic chemistry of mineral grains allows inference of condensation sequences, nebular oxygen fugacity, and irradiation history, as well as thermal evolution. Isotopic composition of primary components has been particularly successful in delineating the heterogeneous nature of the presolar nebula. An absolute chronology for primary components has been outlined on the basis of high-precision Pb-Pb dating (e.g., *Amelin et al.,* 2002); the results provide a marker to which all other processes can be related. The presence of decay products from short-lived radionuclides has been used to infer relative chronologies for the different nonmetallic components in primitive chondrites (e.g., *Russell et al.,* 1996b; *Srinivasan et al.,* 1996), assuming production of the nuclides by an external source just prior to collapse of the nebular cloud (e.g., *Wadhwa and Russell,* 2000). However, scenarios involving spallation by neutrons close to an energetic young Sun have upset the chronological argument (*Shu et al.,* 1997). Distinction between the two competing hypotheses is still awaited and, we hope, almost resolved, as increasing instrumental precision allows measurement of ever-smaller quantities of decay products from radionuclide systems with shorter and shorter halflives (*McKeegan et al.,* 2000; *Chaussidon et al.,* 2002).

4.4. Accretion Epoch

The accretion epoch is the period of time during which primordial components aggregated into planetesimals. Nebular gas had started to dissipate, and the accretion disk had begun to accumulate into discrete planetesimals. During aggregation into parent bodies, and the subsequent lithification of those parents, there was a period of "intermediate alteration." Processing of primary components at this stage occurred through gas-solid exchange, during solid-solid collision, and possibly through fluid-solid interactions. Materials most likely to have been produced and processed during this epoch include interchondrule matrix (*Alexander,* 1995), accretionary rims on chondrules (e.g., *Metzler et al.,* 1992; *Bischoff,* 1998; *Vogel et al.,* 2003) and CAIs (e.g., *Krot et al.,* 2001; *Wark and Boynton,* 2001), and dark inclusions (e.g., *Weisberg and Prinz,* 1998; *Zolensky et al.,* 2003). There are arguments as to whether such components were produced during the final stages of the nebula prior to accumulation into parent bodies or were produced on the parents themselves (e.g., *Krot et al.,* 1995). Recent work has suggested that phyllosilicates, commonly assumed to have a parent-body origin, might also have been formed in the nebula (*Ciesla et al.,* 2003).

4.5. Parent-Body Epoch A: Thermal and Aqueous Alteration

Following on from aggregation into planetesimals, the first planetary epoch encompasses the time during which mild to moderate heating of the planetesimals took place, as well as collisions between planetesimals. As with the nebular and accretion epochs, most of the information for this period is obtained from components within chondritic meteorites. In theory, the recognition of secondary components in chondritic meteorites should be fairly straightforward. Melting of ice, and the subsequent reactions between solid and fluid on parent bodies, led to the formation of secondary minerals (e.g., *Grimm and McSween,* 1989) such as carbonates and magnetite, the alteration of primary silicates (olivine, pyroxene) to phyllosilicates, and oxidation of metal and sulfide grains (e.g., *Browning et al.,* 1996). Carbon-rich planetesimals, such as the parent objects of CM2 meteorites, had an additional suite of secondary alteration products resulting from hydrous pyrolysis of organic compounds (*Sephton et al.,* 2000). As heating continued, or reached higher levels in planetesimals with little or no ice, thermal alteration took over from aqueous alteration, and dehydration and thermal metamorphism produced a different crop of secondary minerals (e.g., *Nakamura et al.,* 2000). It has been recognized for many years that textural and mineralogical changes brought about by secondary alteration helps to subclassify chondrites into petrologic types (e.g., *Sears et al.,* 1980). Analysis of secondary minerals yields information on the extent of alteration, fluid composition, and alteration temperature. Radiogenic isotope systematics, such as I-Xe, of secondary minerals reveal the timing of fluid alteration (e.g., *Swindle,* 1998). Variations in O-isotope composition of different components (carbonate, magnetite, phyllosilicates) show the extent of fluid-solid interaction and constrain the fluid composition and alteration temperature (e.g., *Clayton,* 2003).

4.6. Parent-Body Epoch B: Differentiation and Core Formation

Continued thermal evolution of asteroidal parents eventually resulted in melting of the silicates, differentiation between metal and silicate, and segregation of metal into planetesimal cores; collision between asteroids also continued. All these processes can be traced through analysis of components from melted meteorites. These are a very broad category of meteorites that includes magmatic and nonmagmatic iron meteorites, pallasites, mesosiderites, and achondrites.

4.6.1. Differentiation. Trace-element and isotope variations in silicates from achondrites record the magmatic processes that they have experienced and allow associations of complementary igneous rocks to be recognized. So, for instance, it is clear that the howardites, eucrites, and diogenites form a sequence of rocks that sample different depths from a single asteroid class (e.g., *Mittlefehldt et al.,* 1998). An interesting development of recent years has been recognition of primitive achondrites as bridges between melted achondrites and unmelted chondrites; this recognition has been based on the analysis of mineral chemistry, as well as textural variations (e.g., *McCoy et al.,* 1997) and O-isotope relationships (*Clayton and Mayeda,* 1996). Radiogenic isotope systems, such as Mn-Cr, Fe-Ni, and Sm-Nd, record the onset of planetary differentiation and its duration (e.g., *Lugmair and Shukolyukov,* 1998; *Shukolyukov and Lugmair,* 1993; *Blichert-Toft et al.,* 2002).

4.6.2. Core formation. The most extensive planetary melting ultimately leads to metal extraction and core formation. To trace these processes, we have the magmatic iron meteorites and pallasites. The original subdivision of iron meteorites on the basis of texture has been replaced by classification into groups with distinct trace element chemistries. However, the two schemes, textural and chemical, give complementary information on cooling history as well as parent-body relationships (e.g., *Malvin et al.,* 1984; *Naryan and Goldstein,* 1985). Components within iron meteorites available for study include the major iron-nickel alloys (kamacite and taenite) plus minor components such as sulfides, phosphides, and graphite; many iron meteorites also contain silicates, allowing connections to be made with less-differentiated materials. Shock-produced phases, e.g., diamonds, record collisional history for iron meteorite parents. Radionuclide systems, such as Pd-Ag (*Chen and Wasserburg,* 1990), Re-Os (e.g., *Horan et al.,* 1998), and Hf-W (*Kleine et al.,* 2002; *Yin et al.,* 2002), are used to understand the differentiation and cooling histories of melted meteorites (*Halliday et al.,* 2001). The chemistry of silicate and metal fractions in pallasites give information about processes occurring within their source regions, although the location of this source region, at the core-mantle interface (e.g., *Wasson and Choi,* 2003; *Minowa and Ebihara,* 2002) or at a more shallow depth below a thick regolith (*Hsu,* 2003), is still debated.

4.7. Planetary Epoch

Parent-body surfaces (including the Moon and Mars, as well as asteroids) are subject to modification by external agents. Evolution of asteroidal parents continues even after the internal heat sources that drove alteration and metamorphism have cooled and the parent objects have solidified. The main agents of change are collision and irradiation, and both melted and unmelted meteorite groups are equally affected. Ejection from the asteroid belt, transition to Earth, and terrestrial history prior to collection are all events that can be traced through the study of meteorites.

4.7.1. Parent-body modification. Collision between asteroids leads to regolith formation, compaction, and lithification, and the signatures of these processes are found in brecciated meteorites. The type of materials that were produced during these tertiary processes are veins and pockets of shock-produced glass, minerals altered by remobilization of fluids (sulfidization), as well as the introduction of clasts or xenoliths of the impactor (*Keil et al.,* 1997). Investigation of clast variety and the composition of clasts in different hosts allows elucidation of the collisional history of the asteroid belt and the extent to which bodies have been broken up and reaggregated (e.g., *Scott,* 2002). Recognition of melted clasts in primitive meteorites is also a constraint on the chronology of parent-body formation (e.g., *Gilmour et al.,* 2000).

Implantation of species from the SW also modifies regolithic surfaces; this is particularly relevant for samples from the lunar surface, where implantation at different epochs has allowed changes in composition of the SW to be inferred, with all the implications that this has for solar evolution (e.g., *Kerridge et al.,* 1991). "Gas-rich" meteorites represent compacted regoliths from the surfaces of primitive asteroids, and record the effects of SW interaction from early in the history of the solar system. These same samples also experienced the effects of irradiation by cosmic rays (generally protons of 10–100 MeV energies, capable of penetrating rocks to depths of typically 50 cm). The result of cosmic-ray exposure is production of damage tracks (by relatively heavy ions) in the irradiated zone, along with the transmutation of elements by spallation, i.e., nuclear erosion resulting in the production of light elements from heavier target atoms. On the basis of measurements of spallogenic (or cosmogenic) isotopes made on inclusions separated from gas-rich meteorites, *Wieler et al.* (1989) concluded that irradiation times were on the order of millions to tens of millions of years (compared with a few hundred million years for lunar regolith samples).

Measurements of cosmogenic nuclides are also widely used to assess the length of time that meteorites have been exposed in space as small bodies, i.e., following their removal from an asteroidal (or planetary) parent body. Cosmic-ray-exposure age indicates the length of transit time between ejection from a parent object and its arrival on Earth. For example, on the basis of such measurements, *Schultz et al.* (1991) found that over half of all H-group chondrites had cosmic-ray-exposure ages of around 7 Ma, and inferred that the parent body of the H chondrites must have undergone a major disturbance about 7 m.y. ago.

4.7.2. Terrestrial history. The focus of most research on extraterrestrial materials is on preterrestrial components. In practically all cases, a meteoriticist would prefer to analyze only freshly recovered meteorite "falls," to ensure a minimum of terrestrial contamination. Realistically, though, this is almost impossible — the "fall" population is much smaller and more restricted than that of meteorite "finds," and accounts for less than 4% of all known meteorites (*Grady,* 2000; *Grossman,* 2000; *Grossman and Zipfel,* 2001; *Russell et al.,* 2002, 2003, 2004, 2005). In order to gain a full understanding of solar and presolar history, it is necessary to study the terrestrial histories of meteorites, to disentangle any effects that might result from terrestrial rather than preterrestrial alteration. Hence characterization of the type of weathering products (degree of rusting, formation of clay minerals, and secondary salts such as carbonates and sulfates) and their extent are essential steps that must be taken to ensure that the petrology, texture, and chemistry of the samples have not been compromised by their terrestrial sojourn. The terrestrial history of a meteorite, of course, can also be used to generate useful information about solar system history. Terrestrial age dating using several radiogenic systems (^{10}Be, ^{14}C, ^{36}Cl) allows pairing of meteorite "finds" (e.g., *Jull et al.,* 2000; *Nishiizumi et al.,* 2000). Population studies of meteorite "falls" and "finds" have been used to infer flux rates (e.g., *Bland et al.,* 1996), the breakup of asteroidal parents (e.g., *Schultz et al.,* 1991), and a possible change in population of the asteroid belt (e.g., *Dennison et al.,* 1986).

4.7.3. Craters and tektites. The final types of extraterrestrial materials considered here are not materials *per se*, but signatures left behind by extraterrestrial impactors. These signatures fall into two groups that record the terrestrial consequences of extraterrestrial objects. The first group comprises craters, their sizes and morphologies, and the changes manifest in the host rocks through impact. The presence of craters on planetary surfaces yields information about the age and activity of those surfaces (e.g., *Shoemaker,* 1998); crater-counting studies yield relative chronologies (e.g., *Hartmann and Neukum,* 2001), while crater-size distributions help interpretation of impactor fluxes (e.g., *Ivanov et al.,* 2002). Many craters have meteorites associated with them, almost all of which are iron meteorites. For craters where no macrometeorites occur, it is possible to derive the composition of the projectile from analysis of impact melt rocks; the technique also is applicable to lunar melt breccias (e.g. *Norman et al.,* 2002). The more minor group of samples that comes under this section is that of tektites, glasses formed during impact (e.g., *Koeberl,* 1986). They can be useful indicators of impact angle and the areal distribution of impact ejecta, and can also be used to date craters (e.g., *Deutsch and Schaerer,* 1994).

5. SUMMARY

We have attempted to outline the range of extraterrestrial materials available for study, and which form the basis for the observations and conclusions contained in the rest of this volume. We have categorized materials on the basis of their original parent body (asteroid, comet, etc.), and then tried to formulate a hierarchical description of discrete components in terms of the types of processes the material might have experienced (primary condensate, secondary alteration, etc.). The hierarchy of discrete components is intimately intertwined with the parent-object categories, as the parent objects all carry material from a variety of sources and processes. We have tried to relate them to the subjects of the chapters that follow this one. We have briefly discussed source objects that have the potential to yield material for analysis, but have not yet been recognized as having done so (Venus, Mercury), and the outward signs of planetary impacts (craters and tektites).

Acknowledgments. Financial support from the Particle Physics and Astronomy Research Council is gratefully acknowledged. Thanks go to J. Bradley, T. McCoy, and E. Scott for comments that helped to improve this manuscript. This is IARC manuscript number 2005-0618.

REFERENCES

Africano J. L. and Stansbery E. G. (2002) Orbital debris and NASA's measurement program. *Bull. Am. Astron. Soc., 34,* 754.

Alaerts L., Lewis R. S., Matsuda J., and Anders E. (1980) Isotopic anomalies of noble gases in meteorites and their origins — VI. Presolar components in the Murchison C2 chondrite. *Geochim. Cosmochim. Acta, 44,* 189–209.

Albee A. L., Palluconi F. D., and Arvidson R. E. (1998) Mars Global Surveyor mission: Overview and status. *Science, 279,* 1671.

Alexander C. M. O'D. (1993) Presolar SiC in chondrites: How variable and how many sources? *Geochim. Cosmochim. Acta, 57,* 2869–2888.

Alexander C. M. O'D. (1995) Trace element contents of chondrule rims and interchondrule matrix in ordinary chondrites. *Geochim. Cosmochim. Acta, 59,* 3247–3266.

Altobelli N., Kempf S., Landgraf M., Srama R., Dikarev V., Krüger H., Moragas-Klostermeyer G., and Grün E. (2003) Cassini between Venus and Earth: Detection of interstellar dust. *J. Geophys. Res., 108,* LIS7-1.

Amari S., Lewis R. S., and Anders E. (1994) Interstellar grains in meteorites. I — Isolation of SiC, graphite, and diamond; size distributions of SiC and graphite. *Geochim. Cosmochim. Acta, 58,* 459–470.

Amari S., Gao X., Nittler L. R., Zinner E. K., José J., Hernanz M., and Lewis R. S. (2001) Presolar grains from novae. *Astrophys. J., 551,* 1065–1072.

Amelin Y., Krot A. N., Hutcheon I. D., and Ulyanov A. A. (2002) Lead isotopic ages of chondrules and calcium-aluminium-rich inclusions. *Science, 297,* 1678–1683.

Anders E. (1975) Do stony meteorites come from comets. *Icarus, 24,* 363–371.

Anders E. and Zinner E. K. (1993) Interstellar grains in primitive meteorites: Diamond, silicon carbide and graphite. *Meteoritics, 28,* 490–514.

Anders E., Ganapathy R., Krähenbühl U., and Morgan J. W. (1973) Meteoritic material on the moon. *Moon, 8,* 3–24.

Arkhipov A. V. (1994) Galactic debris? *Spaceflight, 36,* 249.

Arkhipov A. V. (1996) On the possibility of extraterrestrial-artifact finds on Earth. *Observatory, 116,* 175–176.

Becker R. H. and Pepin R. O. (1989) Long-term changes in solar wind elemental and isotopic ratios — A comparison of two lunar ilmenites of different antiquities. *Geochim. Cosmochim. Acta, 53,* 1135–1166.

Bérczi Sz. and Lukács B. (1994) Icy meteorites in Antarctica? In *Proceedings of the 19th Symposium on Antarctic Meteorites,* pp. 185–188. National Institute of Polar Research, Tokyo.

Bernatowicz T. J., Amari S., Zinner E. K., and Lewis R. S. (1991) Interstellar grains within interstellar grains. *Astrophys. J. Lett., 373,* L73–L76.

Binzel R. P., Bus S. J., and Burbine T. H. (1998) Relating S-asteroids and ordinary chondrite meteorites: The new big picture. *Bull. Am. Astron. Soc., 30,* 1041.

Bischoff A. (1998) Aqueous alteration of carbonaceous chondrites: Evidence for preaccretionary alteration. A review. *Meteoritics & Planet. Sci., 33,* 1113–1122.

Bland P. A. and Smith T. B. (2000) Meteorite accumulations on Mars. *Icarus, 144,* 21–26.

Bland P. A., Bevan A. W. R., and Verveer A. (1999) A Nullarbor camera network (abstract). *Meteoritics & Planet. Sci., 34,* A12.

Bland P. A., Smith T. B., Jull A. J. T., Berry F. J., Bevan A. W. R., Cloudt S., and Pillinger C. T. (1996) The flux of meteorites to the Earth over the last 50,000 years. *Mon. Not. R. Astron. Soc., 283,* 551–565.

Blichert-Toft J., Boyet M., Télouk P., and Albarède F. (2002) ^{147}Sm-^{143}Nd and ^{176}Lu-^{176}Hf in eucrites and the differentiation of the HED parent body. *Earth Planet. Sci. Lett., 204,* 167–181.

Bogard D. (1983) A meteorite from the Moon: Editorial. *Geophys. Res. Lett., 10,* 773–774.

Bogard D. D. and Johnson P. (1983) Martian gases in an Antarctic meteorite? *Science, 221,* 651–654.

Borg L. E., Nyquist L. E., Wiesmann H., and Reese Y. (2002) Constraints on the petrogenesis of martian meteorites from the Rb-Sr and Sm-Nd isotopic systematics of the lherzolitic shergottites ALH77005 and LEW88516. *Geochim. Cosmochim. Acta, 66,* 2037–2053.

Bottke W. F. Jr, Vokrouhlický D., Rubincam D. P., and Brož M. (2002) The effect of Yarkovsky thermal forces on the dynamical evolution of asteroids and meteoroids. In *Asteroids III* (W. F. Bottke Jr. et al., eds.), pp. 653–667. Univ. of Arizona, Tucson.

Brack A., Baglioni P., Borruat G., Brandstätter F., Demets R., Edwards H. G. M., Genge M., Kurat G., Miller M. F., Newton E. M., Pillinger C. T., Roten C.-A., and Wäsch E. (2002) Do meteoroids of sedimentary origin survive terrestrial atmospheric entry? The ESA artificial meteorite experiment STONE. *Planet. Space Sci., 50,* 763–772.

Bradley J. P. (1994) Chemically anomalous, preaccretionally irradiated grains in interplanetary dust from comets. *Science, 265,* 925–929.

Bradley J. P. (2003) Interplanetary dust particles. In *Treatise on Geochemistry, Vol. 1: Meteorites, Comets, and Planets* (A. M. Davies, ed.), pp. 689–711. Elsevier, Oxford.

Bradley J. P. and Brownlee D. E. (1986) Cometary particles — thin sectioning and electron beam analysis. *Science, 231,* 1542–1544.

Bridges J. C. and Grady M. M. (1999) A halite-siderite-anhydrite-chlorapatite assemblage in Nakhla: Mineralogical evidence for evaporites on Mars. *Meteoritics & Planet. Sci., 34,* 407–416.

Bridges J. C. and Grady M. M. (2000) Evaporite mineral assemblages in the nakhlite (martian) meteorites. *Earth Planet. Sci. Lett., 176,* 267–279.

Browning L. B., McSween H. Y., and Zolensky M. E. (1996) Correlated alteration effects in CM carbonaceous chondrites. *Geochim. Cosmochim. Acta, 60,* 2621–2633.

Buchwald V. F. (1975) *Handbook of Iron Meteorites: Their History, Distribution, Composition and Structure.* Univ. of California, Berkeley.

Burbine T. H., McCoy T. J., Meibom A., Gladman B., and Keil K. (2002) Meteoritic parent bodies: Their number and identification. In *Asteroids III* (W. F. Bottke Jr. et al., eds.), pp. 653–667. Univ. of Arizona, Tucson.

Burbine T. H., Buchanan P. C., Binzel, R. P., Bus S. J., Hiroi T., Hinrichs J. L., Meibom A., and McCoy T. J. (2001) Vesta, vestoids, and the howardite, eucrite, diogenite group: Relationships and the origin of spectral differences. *Meteoritics & Planet. Sci., 36,* 761–781.

Burnett D. S., Barraclough B. L., Bennett R., Neugebauer M., Oldham L. P., Sasaki C. N., Sevilla D., Smith N., Stansbery E., Sweetnam D., and Wiens R. C. (2003) The Genesis Discovery mission: Return of solar matter to Earth. *Space Sci. Rev., 105,* 509–534.

Bus S. J., Vilas F., and Barucci M. A. (2002) Visible wavelength spectroscopy of asteroids. In *Asteroids III* (W. F. Bottke Jr. et al., eds.), pp. 169–182. Univ. of Arizona, Tucson.

Campins H. and Swindle T. D. (1998) Expected characteristics of cometary meteorites. *Meteoritics & Planet. Sci., 33,* 1201–1211.

Chaussidon M., Robert F., and McKeegan K. D. (2002) Incorporation of short-lived ^{7}Be in one CAI from the Allende meteorite (abstract). In *Lunar and Planetary Science XXXIII,* Abstract #1563. Lunar and Planetary Institute, Houston (CD-ROM).

Chen J. H. and Wasserburg G. J. (1990) The isotopic composition of Ag in meteorites and the presence of Pd-107 in protoplanets. *Geochim. Cosmochim. Acta, 54,* 1729–1743.

Chicarro A. F. (2004) The Mars Express mission — initial scientific results from orbit (abstract). In *Lunar and Planetary Science XXXV,* Abstract #2174. Lunar and Planetary Institute, Houston (CD-ROM).

Ciesla F. J., Lauretta D. S., Cohen B. A., and Hood L. L. (2003) A nebular origin for chondritic fine-grained phyllosilicates. *Science, 299,* 549–552.

Clayton D. D., Meyer B. S., Sanderson C. I., Russell S. S., and Pillinger C. T. (1995) Carbon and nitrogen isotopes in type II supernova diamonds. *Astrophys. J., 447,* 894–905.

Clayton R. N. (2003) Oxygen isotopes in the solar system. *Space Sci. Rev., 106,* 19–32.

Clayton R. N. and Mayeda T. K. (1996) Oxygen isotope studies of achondrites. *Geochim. Cosmochim. Acta, 60,* 1999–2017.

Corliss W. R. (1977) *Handbook of Unusual Natural Phenomena.* The Sourcebook Project, Maryland. 533 pp.

Corliss W. R. (1978) *Ancient Man: A Handbook Of Puzzling Artifacts.* The Sourcebook Project, Maryland. 774 pp.

Cronin J. R. and Chang S. (1993) Organic matter in meteorites: Molecular and isotopic analyses of the Murchison meteorite. In *The Chemistry of Life's Origins* (J. M. Greenberg, ed.), pp. 209–258. Kluwer, Dordrecht.

Cross F. C. (1947) Hypothetical meteorites of sedimentary origin. *Pop. Astron., 55,* 96–102.

Dai Z. R., Bradley J. P., Joswiak D. J., Brownlee D. E., Hill H. G. M., and Genge M. J. (2002) Possible in situ formation of meteoritic nanodiamonds in the early solar system. *Nature, 418,* 157–159.

Dennison J. E., Lingner D. W., and Lipschutz M. E. (1986) Antarctic and non-Antarctic meteorites form different populations. *Nature, 319,* 390–393.

Deutsch A. and Schaerer U. (1994) Dating terrestrial impact events. *Meteoritics, 29,* 301–322.

Ebihara M., Oura Y., Miura Y. N., Haramura H., Misawa K., Kojima H., and Nagao K. (2002) A new source of basaltic meteorites inferred from Northwest Africa 011. *Science, 296,* 334–336.

Ehrenfreund P. and Charnley S. B. (2000) Organic molecules in the interstellar medium, comets, and meteorites: A voyage from dark clouds to the early Earth. *Annu. Rev. Astron. Astrophys., 38,* 427–483.

Farinella P., Gonczi R., Froeschlé C., and Froeschlé C. (1993) The injection of asteroid fragments into resonances. *Icarus, 101,* 174–187.

Frank L. A. and Sigwarth J. B. (1997) Detection of atomic oxygen trails of small comets in the vicinity of the Earth. *Geophys. Res. Lett., 24,* 2431–2434.

Geiss J., Bühler F., Cerutti H., Eberhardt P., Filleux C., Meister J., and Signer P. (2004) The Apollo SWC experiment: Results, conclusions, consequences. *Space Sci. Rev., 110,* 307–335.

Gilmour J. D., Whitby J. A., Turner G., Bridges J. C., and Hutchison R. (2000) The iodine-xenon system in clasts and chondrules from ordinary chondrites: Implications for early solar system chronology. *Meteoritics & Planet. Sci., 35,* 445–456.

Gladman B. J., Burns J. A., Duncan M., Lee P., and Levison H. F. (1996) The exchange of impact ejecta between terrestrial planets. *Science, 271,* 1387–1392.

Goodrich C. A (2003) Petrogenesis of olivine-phyric shergottites Sayh al Uhaymir 005 and Elephant Moraine A79001 lithology A. *Geochim. Cosmochim. Acta, 67,* 3735–3772.

Goodrich C. A. and Jones J. H. (1987) Complex igneous activity on the ureilite parent body (abstract). In *Lunar and Planetary Science XVIII,* pp. 347–348. Lunar and Planetary Institute, Houston.

Gorman A. C. (2005) The cultural landscape of interplanetary orbit. *J. Social Archaeol., 5(1),* 85–107.

Grady M. M. (2000) *Catalogue of Meteorites, 5th edition.* Cambridge Univ., Cambridge. 689 pp.

Graham G. A., Kearsley A. T., Grady M. M., Wright I. P., Griffiths A. D., and McDonnell J. A. M. (1999) Hypervelocity impacts in low Earth orbit: Cosmic dust versus space debris. *Adv. Space Res., 23,* 95–100.

Graham G. A., Kearsley A. T., Drolshagen G., McBride N., Green S. F., and Wright I. P. (2001) Microparticle impacts upon HST solar cells. *Adv. Space Res., 28,* 1341–1346.

Greaves J. S., Holland W. S., Moriarty-Schieven G., Jenness T.,

Dent W. R. F., Zuckerman B., McCarthy C., Webb R. A., Butner H. M., Gear W. K., and Walker H. J. (1998) A dust ring around ε Eridani: Analog to the young solar system. *Astrophys. J. Lett., 506,* L133–L137.

Grimm R. E. and McSween H. Y. (1989) Water and the thermal evolution of carbonaceous chondrite parent bodies. *Icarus, 82,* 244–280.

Grossman J. N. (2000) Meteoritical Bulletin No. 84. *Meteoritics & Planet. Sci., 35,* A199–A225.

Grossman J. N. and Zipfel J. (2001) Meteoritical Bulletin No. 85. *Meteoritics & Planet. Sci., 36,* A293–A322.

Grün E., Baguhl M., Fechtig H., Hanner M. S., Kissel J., Lindblad B.-A., Linkert D., Linkert G., Mann I., and McDonnell J. A. M. (1992) Galileo and ULYSSES dust measurements — from Venus to Jupiter. *Geophys. Res. Lett., 19,* 1311–1314.

Grün E., Zook H. A., Baguhl M., Balogh A., Bame S. J., Fechtig H., Forsyth R., Hanner M. S., Horanyi M., Kissel J., Lindblad B.-A., Linkert D., Linkert G., Mann I., McDonnell J. A. M., Morfill G. E., Phillips J. L., Polanskey C., Schwehm G., Siddique N., Staubach P., Svestka J., and Taylor A. (1993) Discovery of Jovian dust streams and interstellar grains by the ULYSSES spacecraft. *Nature, 362,* 428–430.

Halliday A. N., Lee D., Porcelli D., Wiechert U., Schönbächler M., and Rehkämper M. (2001) The rates of accretion, core formation and volatile loss in the early solar system. *Philos. Trans. R. Soc. Lond., 359,* 2111–2135.

Halliday I., Griffin A. A., and Blackwell A. T. (1996) Detailed data for 259 fireballs from the Canadian camera network and inferences concerning the influx of large meteoroids. *Meteoritics & Planet. Sci., 31,* 185–217.

Hartmann W. K. and Neukum G. (2001) Cratering chronology and the evolution of Mars. *Space Sci. Rev., 96,* 165–194.

Hoppe P., Amari S., Zinner E., Ireland T., and Lewis R. S. (1994) Carbon, nitrogen, magnesium, silicon, and titanium isotopic compositions of single interstellar silicon carbide grains from the Murchison carbonaceous chondrite. *Astrophys. J., 430,* 870–890.

Hoppe P., Amari S., Zinner E., and Lewis R. S. (1995) Isotopic compositions of C, N, O, Mg, and Si, trace element abundances, and morphologies of single circumstellar graphite grains in four density fractions from the Murchison meteorite. *Geochim. Cosmochim. Acta, 59,* 4029–4056.

Hoppe P., Strebel R., Eberhardt P., Amari S., and Lewis R. S. (1996) Small SiC grains and a nitride grain of circumstellar origin from the Murchison meteorite: Implications for stellar evolution and nucleosynthesis. *Geochim. Cosmochim. Acta, 60,* 883–907.

Horan M. F., Smoliar M. I., and Walker R. J. (1998) W-182 and Re(187)-Os-187 systematics of iron meteorites — chronology for melting, differentiation, and crystallization in asteroids. *Geochim. Cosmochim. Acta, 62,* 545–554.

Howard E. C. (1802) Experiments and observations on certain stony and metalline substances which at different times are said to have fallen on the Earth; also on various kinds of native iron. *Philos. Trans. R. Soc. Lond., 92,* 168–212.

Hsu W. (2003) Minor element zoning and trace element geochemistry of pallasites. *Meteoritics & Planet. Sci., 38,* 1217–1241.

Huss G. R. and Lewis R. S. (1995) Presolar diamond, SiC and graphite in primitive chondrites: Abundances as a function of meteorite class and petrologic type. *Geochim. Cosmochim. Acta, 59,* 115–160.

Huss G. R., Meshik A. P., Smith J. B., and Hohenberg C. M. (2003) Presolar diamond, silicon carbide, and graphite in carbonaceous chondrites: Implications for thermal processing in the solar nebula. *Geochim. Cosmochim. Acta, 67,* 4823–4848.

Ivanov B. A., Neukum G., Bottke W. F., and Hartmann W. K. (2002) The comparison of size-frequency distributions of impact craters and asteroids and the planetary cratering rate. In *Asteroids III* (W. F. Bottke Jr. et al., eds.), pp. 89–101. Univ. of Arizona, Tucson.

Jet Propulsion Laboratory (2005) Opportunity rover finds an iron meteorite on Mars. Available on line at http://marsrovers.jpl.nasa.gov/newsroom/pressreleases/20050119a.html, January 19, 2005.

Jull A. J. T., Bland P. A., Klandrud S. E., McHargue L. R., Bevan A. W. R., Kring D. A., and Wlotzka F. (2000) Using ^{14}C and ^{14}C-^{10}Be for terrestrial ages of desert meteorites. In *Workshop on Extraterrestrial Materials from Cold and Hot Deserts* (L. Schultz et al., eds.), pp. 41–43. LPI Contribution No. 997, Lunar and Planetary Institute, Houston.

Keil K., Stoeffler D., Love S. G., and Scott E. R. D. (1997) Constraints on the role of impact heating and melting in asteroids. *Meteoritics & Planet. Sci., 32,* 349–363.

Keller L. P., Messenger S., and Bradley J. P. (2000) Analysis of a deuterium-rich interplanetary dust particle (IDP) and implications for presolar material in IDPs. *J. Geophys. Res., 105,* A10397–10402.

Kerridge J. F. and Matthews M. S., eds. (1988) *Meteorites and the Early Solar System.* Univ. of Arizona, Tucson. 1269 pp.

Kerridge J. F., Signer P., Wieler R., Becker R. H., and Pepin R. O. (1991) Long-term changes in composition of solar particles implanted in extraterrestrial materials. In *The Sun in Time* (C. P. Sonnett et al., eds.), pp. 389–412. Univ. of Arizona, Tucson.

Kleine T., Münker C., Mezger K., and Palme H. (2002) Rapid accretion and early core formation on asteroids and the terrestrial planets from Hf-W chronometry. *Nature, 418,* 952–955.

Koeberl C. (1986) Geochemistry of tektites and impact glasses. *Annu. Rev. Earth Planet. Sci., 14,* 323–350.

Korotev R. L. (1998a) Compositional variation in lunar regolith samples: Lateral. In *Workshop on New Views of the Moon: Integrated Remotely Sensed, Geophysical, and Sample Datasets* (B. L. Jolliff and G. Ryder, eds.), pp. 45–46. LPI Contribution No. 958, Lunar and Planetary Institute, Houston.

Korotev R. L. (1998b) Compositional variation in lunar regolith samples: Vertical. In *Workshop on New Views of the Moon: Integrated Remotely Sensed, Geophysical, and Sample Datasets* (B. L. Jolliff and G. Ryder, eds.), pp. 46–47. LPI Contribution No. 958, Lunar and Planetary Institute, Houston.

Korotev R. L. (1999) Lunar meteorites and implications for compositional remote sensing of the lunar surface. In *Workshop on New Views of the Moon II: Understanding the Moon Through the Integration of Diverse Datasets,* p. 36. LPI Contribution No. 980, Lunar and Planetary Institute, Houston.

Korotev R. L. (2001) On the systematics of lunar regolith compositions (abstract). In *Lunar and Planetary Science XXXII,* Abstract #1134. Lunar and Planetary Institute, Houston.

Koscheev A. P., Gromov M. D., Mohapatra R. K., and Ott U. (2001) History of trace gases in presolar diamonds inferred from ion-implantation experiments. *Nature, 412,* 615–617.

Krot A. N., Scott E. R. D., and Zolensky M. E. (1995) Mineralogical and chemical modification of components in CV3 chondrites: Nebular or asteroidal processing? *Meteoritics, 30,* 748–775.

Krot A. N., Ulyanov A. A., Meibom A., and Keil K. (2001)

Forsterite-rich accretionary rims around Ca, Al-rich inclusions from the reduced CV3 chondrite Efremovka. *Meteoritics & Planet. Sci., 36,* 611–628.

Lebreton J. P., Matson D. L., and the Huygens Team (2005) The Huygens mission at Titan: Results highlights (abstract). In *Lunar and Planetary Science XXXVI,* Abstract #2024. Lunar and Planetary Institute, Houston (CD-ROM).

Lewis R. S., Amari S., and Anders E. (1994) Interstellar grains in meteorites: II. SiC and its noble gases. *Geochim. Cosmochim. Acta, 58,* 471–494.

Lewis R. S., Anders E., Wright I. P., Norris S. J., and Pillinger C. T. (1983) Isotopically anomalous nitrogen in primitive meteorites. *Nature, 305,* 767–771.

Love S. G. and Keil K. (1995) Recognizing Mercurian meteorites. *Meteoritics, 30,* 269–278.

Lugmair G. W. and Shukolyukov A. (1998) Early solar system timescales according to ^{53}Mn-^{53}Cr systematics. *Geochim. Cosmochim. Acta, 62,* 2863–2886.

McCoy T. J., Taylor G. J., and Keil K. (1992) Zagami: Product of a two-stage magmatic history. *Geochim. Cosmochim. Acta, 56,* 3571–3582.

McCoy T. J., Keil K., Muenow D. W., and Wilson L. (1997) Partial melting and melt migration in the acapulcoite-lodranite parent body. *Geochim. Cosmochim. Acta, 61,* 639–650.

McCoy T. J., Burbine T. H., McFadden L. A., Starr R. D., Gaffey M. J., Nittler L. R., Evans L. G., Izenberg N., Lucey P., Trombka J. I., Bell J. F. III, Clark B. E., Clark P. E., Squyres S. W., Chapman C. R., Boynton W. V., and Veverka J. (2001) The composition of 433 Eros: A mineralogical-chemical synthesis. *Meteoritics & Planet. Sci., 36,* 1661–1672.

McCallum I. S. (2001) A new view of the Moon in light of data from Clementine and Prospector missions. *Earth Moon Planets, 85,* 253–269.

McKay D. S., Gibson E. K., Thomas-Keprta K. L., Vali H., Romanek C. S., Clemett S. J., Chiller X. D. F., Maechling C. R., and Zare R. N. (1996) Search for past life on Mars: Possible relic biogenic activity in martian meteorite ALH84001. *Science, 273,* 924–930.

McKeegan K. D., Chaussidon M., and Robert F. (2000) Incorporation of short-lived Be-10 in a calcium-aluminum-rich inclusion from the Allende meteorite. *Science, 289,* 1334–1337.

McSween H. Y. (1976) A new type of chondritic meteorite found in lunar soil. *Earth Planet. Sci. Lett., 31,* 193–199.

McSween H. Y. (1979) Are carbonaceous chondrites primitive or processed — A review. *Rev. Geophys., 17,* 1059–1078.

McSween H. Y. (1985) SNC meteorites: Clues to martian petrologic evolution? *Rev. Geophys., 23,* 391–416.

McSween H. Y. (1994) What we have learned about Mars from SNC meteorites. *Meteoritics, 29,* 757–779.

Malvin D. J., Wang D., and Wasson J. T. (1984) Chemical classification of iron meteorites. X — Multielement studies of 43 irons, resolution of group IIIE from IIIAB, and evaluation of Cu as a taxonomic parameter. *Geochim. Cosmochim. Acta, 48,* 785–804.

Mayor M. and Queloz D. (1995) A Jupiter-mass companion to a solar-type star. *Nature, 378,* 355–359.

Meisel D. D., Janches D., and Mathews J. D. (2002) The size distribution of Arecibo interstellar particles and its implications. *Astrophys. J., 579,* 895–904.

Messenger S. (2000) Identification of molecular-cloud material in interplanetary dust particles. *Nature, 404,* 968–971.

Messenger S., Keller L. P., Stadermann F. J., Walker R. M., and Zinner E. (2003) Samples of stars beyond the solar system: Silicate grains in interplanetary dust. *Science, 300,* 105–108.

Metzler K., Bischoff A., and Stöffler D. (1992) Accretionary dust mantles in CM chondrites — Evidence for solar nebula processes. *Geochim. Cosmochim. Acta, 56,* 2873–2897.

Meyer C. (2003) Mars Meteorite Compendium. Available on line at http://www-curator.jsc.nasa.gov/curator/antmet/mmc/mmc.htm.

Minowa H. and Ebihara M. (2002) Rare earth elements in pallasite olivines (abstract). In *Lunar and Planetary Science XXXIII,* Abstract #1386. Lunar and Planetary Institute, Houston (CD-ROM).

Mittlefehldt D. W., McCoy T. J., Goodrich C. A., and Kracher A. (1998) Non-chondritic meteorites from asteroidal bodies. In *Planetary Materials* (J. J. Papike, ed.), pp. 4-1 to 4-195. Reviews in Mineralogy, Vol. 36, Mineralogical Society of America.

Molster F. J., Waters L. B. F. M., and Tielens A. G. G. M. (2002) Crystalline silicate dust around evolved stars. II. The crystalline silicate complexes. *Astron. Astrophys., 382,* 222–240.

Moroz V. I. (1983) Summary of preliminary results of the Venera 13 and Venera 14 missions. In *Venus* (D. M. Hunten et al., eds.), pp. 45–68. Univ. of Arizona, Tucson.

Mostefaoui S., Hoppe P., Marhas K. K., and Gröner E. (2003) Search for *in situ* presolar oxygen-rich dust in meteorites. *Meteoritics & Planet. Sci., 38,* Abstract No. 5185.

Nagashima K., Krot A. N., and Yurimoto H. (2004) Stardust silicates from primitive meteorites. *Nature, 428,* 921–924.

Nakamura T., Kitajima F., and Takaoka N. (2000) Thermal metamorphism of CM carbonaceous chondrites deduced from phyllosilicate decomposition and trapped noble gas abundance. In *Proceedings of the 25th Antarctic Meteorite Symposium,* pp. 102–105. National Institute of Polar Research, Tokyo.

Narayan C. and Goldstein J. I. (1985) A major revision of iron meteorite cooling rates — an experimental study of the growth of the Widmanstätten pattern. *Geochim. Cosmochim. Acta, 49,* 397–410.

Nicolussi G. K., Pellin M. J., Lewis R. S., Davis A. M., Clayton R. N., and Amari S. (1998) Zirconium and molybdenum in individual circumstellar graphite grains: New isotopic data on the nucleosynthesis of heavy elements. *Astrophys. J., 504,* 492–499.

Nishiizumi K., Caffee M. W., and Welten K. C. (2000) Terrestrial ages of Antarctic meteorites: Update 1999. In *Workshop on Extraterrestrial Materials from Cold and Hot Deserts* (L. Schultz et al., eds.), pp. 64–68. LPI Contribution No. 997, Lunar and Planetary Institute, Houston.

Nittler L. R. (2003) Presolar stardust in meteorites: Recent advances and scientific frontiers. *Earth Planet. Sci. Lett., 209,* 259–273.

Nittler L. R., Starr R. D., Lim L., McCoy T. J., Burbine T. H., Reedy R. C., Trombka J. I., Gorenstein P., Squyres S. W., Boynton W. V., McClanahan T. P., Bhangoo J. S., Clark P. E., Murphy M. E., and Killen R. (2001) X-ray fluorescence measurements of the surface elemental composition of asteroid 433 Eros. *Meteoritics & Planet. Sci., 36,* 1673–1695.

Norman M. D., Bennett V. C., and Ryder G. (2002) Targeting the impactors: Siderophile element signatures of lunar impact melts from Serenitatis. *Earth Planet. Sci. Lett., 202,* 217–228.

Nyquist L. E., Bogard D. D., Shih C.-Y., Greshake A., Stöffler D., and Eugster O. (2001) Ages and geologic histories of martian meteorites. *Space Sci. Rev., 96,* 105–164.

Ott U. (2003) Isotopes of volatiles in pre-solar grains. *Space Sci. Rev., 106,* 33–48.

Ott U. and Begemann F. (1990) Discovery of s-process barium in the Murchison meteorite. *Astrophys. J. Lett., 353,* L57–L60.

Palme H. (2002) Planetary science: A new solar system basalt. *Science, 296,* 271–273.

Papike J. J., Ryder G., and Shearer C. K. (1998) Lunar samples. In *Planetary Materials* (J. J. Papike, ed.), pp. 5-1 to 5-234. Reviews in Mineralogy, Vol. 36, Mineralogical Society of America.

Pedroni A. and Begemann F. (1994) On unfractionated solar noble gases in the H3-6 meteorite Acfer 111. *Meteoritics, 29,* 632–642.

Porco C. C., Baker E., Barbara J., Beurle K., Brahic A., Burns J. A., Charnoz S., Cooper N., Dawson D. D., Del Genio A. D., Denk T., Dones L., Dyudina U., Evans M. W., Giese B., Grazier K., Helfenstein P., Ingersoll A. P., Jacobson R. A., Johnson T. V., McEwen A., Murray C. D., Neukum G., Owen W. M., Perry J., Roatsch T., Spitale J., Squyres S., Thomas P., Tiscareno M., Turtle E., Vasavada A. R., Veverka J., Wagner R., and West R. (2005) Cassini imaging science: Initial results on Saturn's rings and small satellites. *Science, 307,* 1243–1247.

Rubin A. E. (1997) The Hadley Rille enstatite chondrite and its agglutinate-like rim: Impact melting during accretion to the Moon. *Meteoritics & Planet. Sci., 32,* 135–141.

Russell S. S., Arden J. W., and Pillinger C. T. (1996a) A carbon and nitrogen isotope study of diamond from primitive chondrites. *Meteoritics & Planet. Sci., 31,* 343–355.

Russell S. S., Srinivasan G., Huss G. R., Wasserburg G. J., and MacPherson G. J. (1996b) Evidence for widespread ^{26}Al in the solar nebula and constraints for nebula timescales. *Science, 273,* 757–762.

Russell S. S., Zipfel J., Grossman J. N., and Grady M. M. (2002). Meteoritical Bulletin No. 86. *Meteoritics & Planet. Sci., 37,* A157–A184.

Russell S. S., Zipfel J., Folco L., Jones R., Grady M. M., McCoy T. J., and Grossman J. N. (2003) Meteoritical Bulletin No. 87. *Meteoritics & Planet. Sci., 38,* A189–A248.

Russell S. S., Folco L., Grady M. M., Zolensky M. E., Jones R., Righter K., Zipfel J., and Grossman J. N. (2004) Meteoritical Bulletin No. 88. *Meteoritics & Planet. Sci., 39,* A215–A272.

Russell S. S., Zolensky M. E., Righter K., Folco L., Jones R., Connolly H. C. Jr., Grady M. M., and Grossman J. N. (2005) Meteoritical Bulletin No. 89. *Meteoritics & Planet. Sci., 40,* in press.

Saunders R. S., Spear A. J., Allin P. C., Austin R. S., Berman A. L., Chandlee R. C., Clark J., deCharon A. V., De Jong E. M., Griffith D. G., Gunn J. M., Hensley S., Johnson W. T. K., Kirby C. E., Leung K. S., Lyons D. T., Michaels G. A., Miller J., Morris R. B., Morrison A. D., Piereson R. G., Scott J. F., Shaffer S. J., Slonski J. P., Stofan E. R., Thompson T. W., and Wall S. D. (1992) Magellan mission summary. *J. Geophys. Res., 97,* E13067–E13090.

Saunders R. S., Arvidson R. E., Badhwar G. D., Boynton W. V., Christensen P. R., Cucinotta F. A., Feldman W. C., Gibbs R. G., Kloss C., Landano M. R., Mase R. A., McSmith G. W., Meyer M. A., Mitrofanov I. G., Pace G. D., Plaut J. J., Sidney W. P., Spencer D. A., Thompson T. W., and Zeitlin C. J. (2004) 2001 Mars Odyssey mission summary. *Space Sci. Rev., 110,* 1–36.

Schultz L., Weber H. W., and Begemann F. (1991) Noble gases in H-chondrites and potential differences between Antarctic and non-Antarctic meteorites. *Geochim. Cosmochim. Acta, 55,* 59–66.

Score R. A. and Mason B. (1982) *Antarct. Met. Newslett., 5,* 1–2.

Scott E. R. D. (2002) Meteorite evidence for the accretion and collisional evolution of asteroids. In *Asteroids III* (W. F. Bottke Jr. et al., eds.), pp. 697–709. Univ. of Arizona, Tucson.

Sears D. W., Grossman J. N., Melcher C. L., Ross L. M., and Mills A. A. (1980) Measuring metamorphic history of unequilibrated ordinary chondrites. *Nature, 287,* 791–795.

Sekanina Z. (2003) Erosion model for the sungrazing comets observed with the Solar and Heliospheric Observatory. *Astrophys. J., 597,* 1237–1265.

Semenenko V. P., Jessberger E. K., Chaussidon M., Weber I., Wies C., and Stephan T. (2003) Carbonaceous xenoliths from the Krymka chondrite as probable cometary material. *Meteoritics & Planet. Sci., 38,* Abstract No. 5005.

Sephton M. A. and Gilmour I. (2000) Aromatic moieties in meteorites: Relics of interstellar grain processes? *Astrophys. J., 540,* 588–591.

Sephton M. A., Pillinger C. T., and Gilmour I. (2000) Aromatic moieties in meteoritic macromolecular materials: Analyses by hydrous pyrolysis and δ^{13}C of individual compounds. *Geochim. Cosmochim. Acta, 64,* 321–328.

Shoemaker E. M. (1998) Impact cratering through geologic time. *J. R. Astron. Soc. Canada, 92,* 297–309.

Shu F H., Shang H., Glassgold A. E., and Lee T. (1997) X-rays and fluctuating x-winds from protostars. *Science, 277,* 1475–1479.

Shukolyukov A. and Lugmair G. W. (1993) Iron-60 in eucrites. *Earth Planet. Sci. Lett., 119,* 154–166.

Srinivasan G., Sahijpal S., Ulyanov A. A., and Goswami J. N. (1996) Ion microprobe studies of Efremovka CAIs: II. Potassium isotope composition and ^{41}Ca in the early solar system. *Geochim. Cosmochim. Acta, 60,* 1823–1835.

Stadermann F. J., Bernatowicz T., Croat T. K., Zinner E., Messenger S., and Amari, S. (2003a) Titanium and oxygen isotopic compositions of sub-micrometer TiC crystals within presolar graphite (abstract). In *Lunar and Planetary Science XXXIV,* Abstract #1627. Lunar and Planetary Institute, Houston (CD-ROM).

Stadermann F. J. and Bradley J. P. (2003b) The isotopic nature of GEMS in interplanetary dust particles. *Meteoritics & Planet. Sci., 38,* Abstract No. 5236.

Swart P. K., Grady M. M., Pillinger C. T., Lewis R. S., and Anders E. (1983) Interstellar carbon in meteorites. *Science, 220,* 406–410.

Swindle T. D. (1998) Implications of I-Xe studies for the timing and location of secondary alteration. *Meteoritics & Planet. Sci., 33,* 1147–1155.

Thomas K. L., Blanford G. E., Keller L. P., Klock W. and McKay D. S. (1993) Carbon abundance and silicate mineralogy of anhydrous interplanetary dust particles. *Geochim Cosmochim. Acta, 57,* 1551–1566.

Throop H. B., Bally J., Esposito L. W., and McCaughrean M. J. (2001) Evidence for dust grain growth in young circumstellar disks. *Science, 292,* 1686–1689.

Tizard J., Lyon I., and Henkel T. (2005) The gentle separation of presolar SiC grains from meteorites. *Meteoritics & Planet. Sci., 40,* 335–342.

Treiman A. H. (1998) The history of Allan Hills revised: Multiple shock events. *Meteoritics & Planet. Sci., 33,* 753–764.

Treiman A. H. (2003) Submicron magnetite grains and carbon compounds in martian meteorite ALH84001: Inorganic, abiotic formation by shock and thermal metamorphism. *Astrobiology, 3,* 369–392.

Tuzzolino A. J., Economou T. E., McKibben R. B., Simpson J. A., McDonnell J. A. M., Burchell M. J., Vaughan B. A. M., Tsou P., Hanner M. S., Clark B. C., and Brownlee D. E. (2003) Dust Flux Monitor Instrument for the Stardust mission to comet Wild 2. *J. Geophys. Res., 108,* SRD5–1.

Van Kerckhoven C., Tielens A. G. G. M., and Waelkens C. (2002) Nanodiamonds around HD 97048 and Elias 1. *Astron. Astrophys., 384,* 568–584.

Veverka J. (2002) NEAR at Eros. *Icarus, 155,* 1–2.

Veverka J., Farquhar B., Robinson M., Thomas P., Murchie S., Harch A., Antreasian P. G., Chesley S. R., Miller J. K., Owen W. M., Williams B. G., Yeomans D., Dunham D., Heyler G., Holdridge M., Nelson R. L., Whittenburg K. E., Ray J. C., Carcich B., Cheng A., Chapman C., Bell J. F., Bell M., Bussey B., Clark B., Domingue D., Gaffey M. J., Hawkins E., Izenberg N., Joseph J., Kirk R., Lucey P., Malin M., McFadden L., Merline W. J., Peterson C., Prockter L., Warren J., and Wellnitz D. (2001) The landing of the NEAR-Shoemaker spacecraft on asteroid 433 Eros. *Nature, 413,* 390–393.

Vogel N., Wieler R., Bischoff A., and Baur H. (2003) Microdistribution of primordial Ne and Ar in fine-grained rims, matrices, and dark inclusions of unequilibrated chondrites — Clues on nebular processes. *Meteoritics & Planet. Sci., 38,* 1399–1418.

Wadhwa M. and Russell S. S. (2000) Timescales of accretion and differentiation in the early solar system: The meteoritic evidence. In *Protostars and Planets IV* (V. Mannings et al., eds.), pp. 995–1018. Univ. of Arizona, Tucson.

Wadhwa M., McSween H. Y. Jr., and Crozaz G. (1994) Petrogenesis of shergottite meteorites inferred from minor and trace element microdistributions. *Geochim. Cosmochim. Acta, 58,* 4213–4229.

Wahhaj Z., Koerner D. W., Ressler M. E., Werner M. W., Backman D. E., and Sargent A. I. (2003) The inner rings of β Pictoris. *Astrophys. J. Lett., 584,* L27–L31.

Wark D. and Boynton W. V. (2001) The formation of rims on calcium-aluminium-rich inclusions: Step I — flash heating. *Meteoritics & Planet. Sci., 36,* 1135–1166.

Warren P. H., Jerde E. A., and Kallemeyn G. W. (1989) Lunar meteorites — siderophile element contents, and implications for the composition and origin of the moon. *Earth Planet. Sci. Lett., 91,* 245–260.

Wasson J. T. and Choi B.-G. (2003) Main-group pallasites — chemical composition, relationship to IIIAB irons, and origin. *Geochim. Cosmochim. Acta, 67,* 3079–3096.

Weisberg M. K. and Prinz M. (1998) Fayalitic olivine in CV3 chondrite matrix and dark inclusions: A nebular origin. *Meteoritics & Planet. Sci., 33,* 1087–1099.

Wieler R., Graf T., Pedroni A., Signer P., Pellas P., Fieni C., Suter M., Vogt S., Clayton R. N., and Laul J. C. (1989) Exposure history of the regolithic chondrite Fayetteville. II — Solar-gas-free light inclusions. *Geochim. Cosmochim. Acta, 53,* 1449–1459.

Wiens R. C., Bochsler P., Burnett D. S., and Wimmer-Schweingruber R. F. (2004a) Solar and solar-wind isotopic compositions. *Earth Planet. Sci. Lett., 222,* 697–712.

Wiens R. C., Bochsler P., Burnett D. S., and Wimmer-Schweingruber R. F. (2004b) Erratum to "Solar and solar wind isotopic compositions." *Earth Planet. Sci. Lett., 226,* 549–565.

Wright I. P., Grady M. M., and Pillinger C. T. (1995) The acquisition of martian sedimentary rocks: For the time being, collection as meteorites from terrestrial desert areas represents the best hope. In *Workshop on Meteorites from Cold and Hot Deserts* (L. Schultz et al., eds.), pp. 77–78. LPI Tech. Rpt. 95-02, Lunar and Planetary Institute, Houston.

Yin Q., Jacobsen S. B., Yamashita K., Blichert-Toft J., Télouk P., and Albarède F. (2002) A short timescale for terrestrial planet formation from Hf-W chronometry of meteorites. *Nature, 418,* 949–952.

Young R. E. (1998) The Galileo probe mission to Jupiter: Science overview. *J. Geophys. Res., 103,* E22775–E22790.

Young R. E., Smith M. A., and Sobeck C. K. (1996) Galileo probe: In situ observations of Jupiter's atmosphere. *Science, 272,* 837–838.

Zinner E. K. (1997) Presolar material in meteorites: An overview. In *Astrophysical Implications of the Laboratory Study of Presolar Materials* (T. J. Bernatowicz and E. K. Zinner, eds.), pp. 3–26. American Institute of Physics Conference Proceedings 402.

Zinner E., Amari S., and Lewis R. S. (1991) S-process Ba, Nd, and Sm in presolar SiC from the Murchison meteorite. *Astrophys. J. Lett., 382,* L47–L50.

Zinner E. K., Amari S., Wopenka B., and Lewis R. S. (1995) Interstellar graphite in meteorites: Isotopic compositions and structural properties of single graphite grains from Murchison. *Meteoritics, 30,* 209–226.

Zolensky M., Nakamura K., Weisberg M. K., Prinz M., Nakamura T., Ohsumi K., Saitow A., Mukai M., and Gounelle M. (2003) A primitive dark inclusion with radiation-damaged silicates in the Ningqiang carbonaceous chondrite. *Meteoritics & Planet. Sci., 38,* 305–322.

Systematics and Evaluation of Meteorite Classification

Michael K. Weisberg
Kingsborough Community College of the City University of New York and American Museum of Natural History

Timothy J. McCoy
Smithsonian Institution

Alexander N. Krot
University of Hawai'i at Manoa

Classification of meteorites is largely based on their mineralogical and petrographic characteristics and their whole-rock chemical and O-isotopic compositions. According to the currently used classification scheme, meteorites are divided into chondrites, primitive achondrites, and achondrites. There are 15 chondrite groups, including 8 carbonaceous (CI, CM, CO, CV, CK, CR, CH, CB), 3 ordinary (H, L, LL), 2 enstatite (EH, EL), and R and K chondrites. Several chondrites cannot be assigned to the existing groups and may represent the first members of new groups. Some groups are subdivided into subgroups, which may have resulted from asteroidal processing of a single group of meteorites. Each chondrite group is considered to have sampled a separate parent body. Some chondrite groups and ungrouped chondrites show chemical and mineralogical similarities and are grouped together into clans. The significance of this higher order of classification remains poorly understood. The primitive achondrites include ureilites, acapulcoites, lodranites, winonaites, and silicate inclusions in IAB and IIICD irons and probably represent recrystallization or residues from a low-degree partial melting of chondritic materials. The genetic relationship between primitive achondrites and the existing groups of chondritic meteorites remains controversial. Achondrites resulted from a high degree of melting of chondrites and include asteroidal (angrites, aubrites, howardites-diogenites-eucrites, mesosiderites, 3 groups of pallasites, 15 groups of irons plus many ungrouped irons) and planetary (martian, lunar) meteorites.

1. INTRODUCTION

Meteorite classification is the basic framework from which meteoriticists and cosmochemists work and communicate. It is a process designed to group similar meteorites together and evolves with the collection of new data, discovery of new types of meteorites, and new ideas about relationships between meteorites. Although classification does not necessarily need to imply genetic relationship, it has provoked questions about the processes that resulted in the similarities and differences among the meteorite groups and individual members within groups. Additionally, the relationship between primitive (chondrites) and differentiated (achondrites) meteorites and the parent bodies from which meteorites are derived is an important issue in meteorite study. Each chondrite group, for example, is often interpreted to represent materials from a single parent body. Differences in the characteristics among the chondrite groups have provided information on the types of primitive materials that were available during planetary accretion and the systematic geochemical trends of planetary bodies with their increasing distance from the Sun (e.g., *Wasson,* 1985). The wide variety of meteorite types has given rise to thoughts on the number of planetary bodies (asteroids) represented in the meteorite record.

A variety of characteristics are used to classify meteorites (e.g., mineralogy, petrology, whole-rock chemistry, and O isotopes), but the groupings based on these data are not always consistent with each other. Additionally, a number of meteorites do not fit into the existing groups. In this chapter, we present the basic systematics of meteorite classification and describe meteorite classes, clans, groups, and subgroups to illustrate the diversity of solar system materials and the methods used to classify them. We also provide a critical evaluation of current meteorite classification. The reader is also directed to a number of contemporaneous writings on meteorite classification and properties that provide additional detail to the information and discussion offered here (e.g., *Brearley and Jones,* 1998; *Mittlefehldt et al.,* 1998; *Krot et al.,* 2004; *Mittlefehldt,* 2004).

2. HISTORICAL NOTE

The current classification scheme for meteorites had its beginnings in the 1860s with G. Rose's classification of the meteorite collection at the University Museum of Berlin and

Maskelyne's classification of the British Museum collection. Rose was the first to split stones into chondrites and nonchondrites. Maskelyne classified meteorites into siderites (irons), siderolites (stony irons), and aerolites (stones). This was followed by Tshermak's modification of the Rose classification in 1883 and Brezina's modification of the Tshermak-Rose classification scheme (see *Prior,* 1920). The first classification of meteorites based on chemical composition was by *Farrington* (1907), who chemically analyzed irons. Based on these earlier schemes, *Prior* (1920) developed a comprehensive meteorite classification scheme, introducing terms like mesosiderite and lodranite, that became the backbone of the current meteorite classification. Prior's system was later modified by *Mason* (1967).

The meteorite classification system that is currently used and discussed in this chapter is broadly based on *Prior* (1920) and *Mason* (1967) and its modifications over the last four decades by a number of modern studies of the petrologic, bulk chemical, and O-isotopic characteristics of meteorites, infused with continual discovery of new meteorite samples.

3. CLASSIFICATION METHODS

3.1. Terminology and Nomenclature

Meteorites are classified in a variety of ways to communicate their different attributes, which have important implications. Meteorites are considered *falls* if they can be associated with an observed fall event and *finds* if they cannot be connected to a recorded fall event. This is an important distinction because finds, depending on the time they spent on Earth, are more prone to chemical interaction with the terrestrial environment. Therefore, the chemical trends they record need to be carefully considered. Most of the meteorites in our collections are finds and many of these are materials recovered from private and nationally organized search efforts in hot (Africa, Australia) and cold (Antarctica) deserts.

Meteorites are given names based on the location in which they are found (e.g., the Allende meteorite is a fall from Pueblito de Allende, Mexico). Meteorites from desert regions, such as Antarctica, are given names and numbers [e.g., Allan Hills (ALH) 84001 is a meteorite find that was collected in Allan Hills, Antarctica, during the 1984–1985 field season and was the first meteorite described in that field season].

The terms *stones, irons,* and *stony irons* are useful but this terminology is coarse and only serves as an initial description of the material. A more meaningful approach is division into *chondrites* (undifferentiated meteorites) and *achondrites* (differentiated meteorites) (Fig. 1). The chondrites are typically defined as meteorites that contain small (1–2 mm) spheres called chondrules (Fig. 2a,b). This is a misconception, since some lack chondrules. More properly, chondrites are meteorites that have solar-like compositions (minus the highly volatile elements) and are derived from asteroids (and possibly comets) that did not experience planetary differentiation. The achondrites are igneous rock (melts, partial melts, melt residues) or breccias of igneous rock fragments from differentiated asteroids and planetary bodies (Mars, Moon). Some meteorites have achondritic (igneous or recrystallized) textures (Plates 1 and 2), but retain a primitive chemical affinity to their chondritic precursors. They are referred to as *primitive achondrites* because they are nonchondritic, but are thought to be closer to their primitive chondritic parent than other achondrites.

A number of variations on classification have been presented over the years. Like any natural history discipline, meteorite taxonomy is an attempt to group objects of similar type. At the lowest level (e.g., objects so similar that they must have originated from common materials, processes, and locations), general agreement exists as to groupings of meteorites. Disagreements center on higher-order systems that attempt to relate disparate materials through common processes (e.g., melted vs. unmelted), nebular reservoirs (e.g., O-isotopic composition or oxidation state), or parent bodies. The terms used for meteorite classification have evolved over time but they have not always been rigorously and consistently applied. Furthermore, the classification of chondrites and achondrites evolved independently, producing terminology for classification of achondrites and primitive achondrites that differs from that of chondrites. *Wasson* (1985) divided meteorites into chondrites and differentiated meteorites, and subdivided the differentiated meteorites into achondrites, stony irons, and stones. *Krot et al.* (2004) recently divided meteorites into chondrites and nonchondrites and subdivided the nonchondrites into achondrites, primitive achondrites, stony irons, and irons. *Kallemeyn et al.* (1996) recognized problems in dividing chondrites into classes and in distinguishing between class and clan affiliations in chondrites. They divided chondrites into two major classes, which they called carbonaceous and noncarbonaceous "super clans." The noncarbonaceous super clan included the O, R, and E chondrites. They then divided the super clans into clans.

The hierarchy of terms for chondrite classification used in this chapter include, in order, class, clan, group, and subgroup. A chondrite *class* consists of two or more groups sharing primary whole-rock chemical and O-isotopic properties (e.g., *Hutchison,* 2004). Chondrites within a class have broadly similar refractory lithophile-element abundances that are enriched, depleted, or equal to solar abundances (Fig. 3a,b), with CI chondrites defining our best estimate of solar composition (e.g., *Anders and Grevesse,* 1989), and their O-isotopic compositions plot in the same general region (above, on, or below the terrestrial fractionation line) on a three-isotope diagram (Plate 3a,b). The chondrites are divided into three major *classes — carbonaceous* (C), *ordinary* (O), and *enstatite* (E). The letter nomenclature is probably best since the terms "carbonaceous" and "ordinary" are historical terms that are not very meaningful. For example, "ordinary," originally applied because these are the most common meteorites, is an unfortunate term that

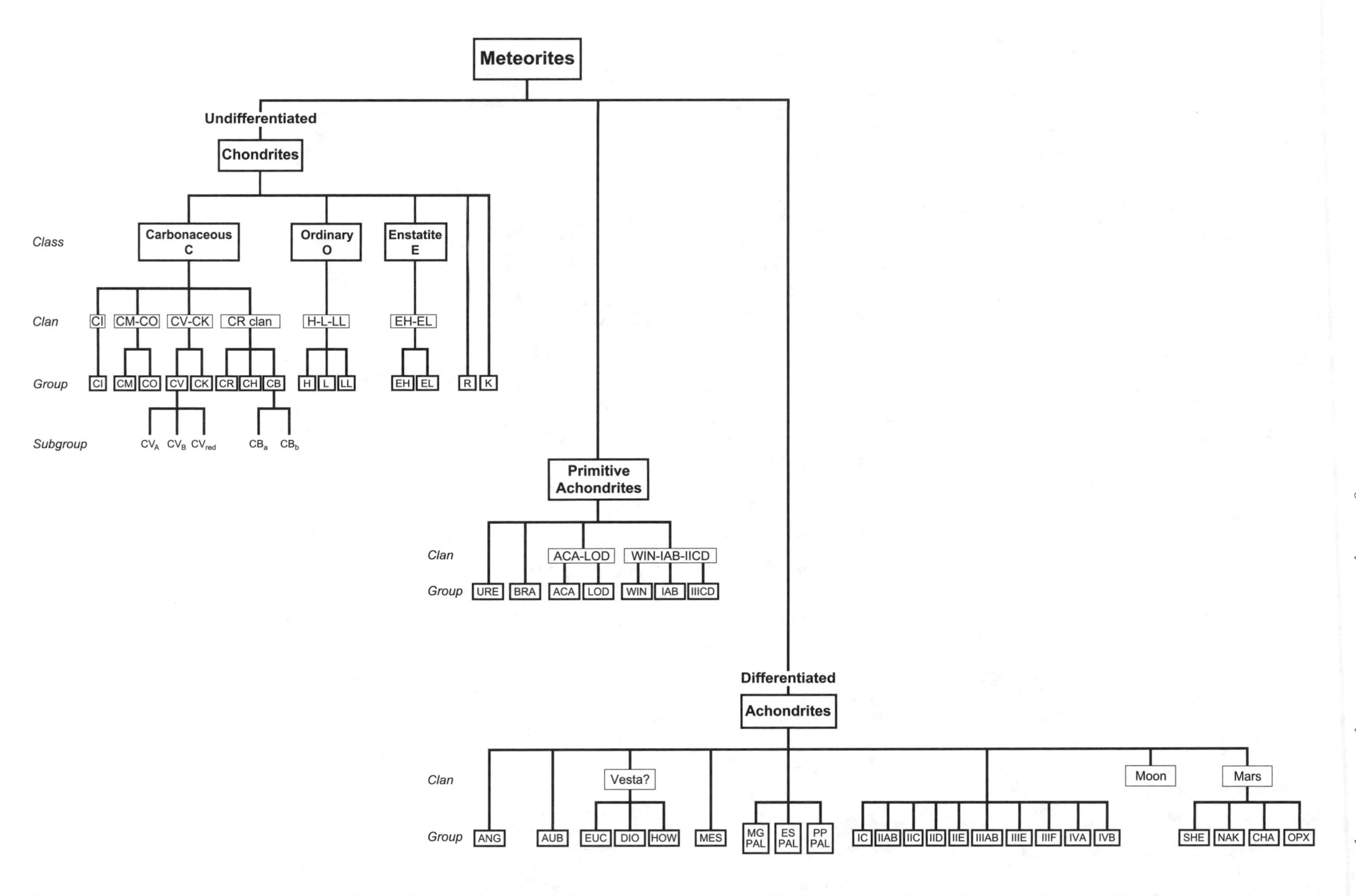

Fig. 1. Diagram expressing the systematics of meteorite classification and showing the major meteorite divisions, classes, clans, and groups and relationships among meteorite groups. URE — ureilite, ACA — acapulcoite, LOD — lodranite, ANG — angrite, AUB — aubrite, BRA — brachinite, WIN — winonaite, HED — howardite-eucrite-diogenite, MES — mesosiderite, MG PAL — main-group pallasite, ES PAL — Eagle Station pallasite, PP PAL — pyroxene pallasite, SHE — shergottite, NAK — nakhlite, CHA — chassignite, OPX — orthopyroxenite.

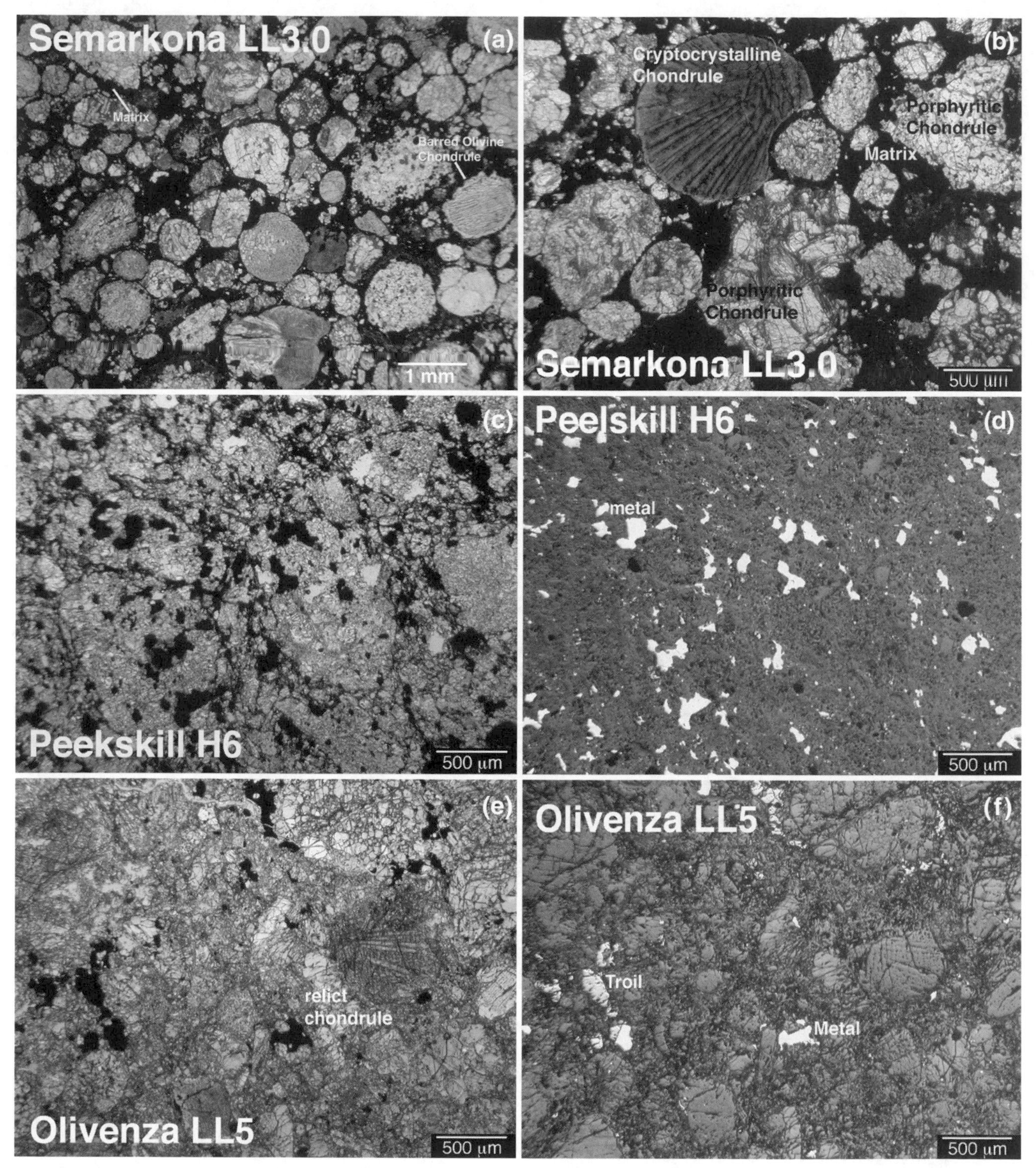

Fig. 2. Thin section, plane-polarized light (PL) and reflected light (RF) photomicrographs of representative examples of chondrites showing petrographic variations among chondrite groups. **(a)** The Semarkona L3.0 ordinary chondrite in PL showing a high density of chondrules with different textures (textural types) including barred and porphyritic chondrules, surrounded by opaque matrix. **(b)** The Semarkona chondrite in PL showing a large indented cryptocrystalline chondrule, porphyritic chondrules, and matrix. **(c)** The Peekskill H6 chondrite in PL showing a recrystallized ordinary chondrite in which the primary texture has been essentially destroyed and chondrule boundaries are not easily discernable. **(d)** Peekskill shown in RF showing about 7% metal. **(e)** The Olivenza LL5 ordinary chondrite in PL showing the texture of a recrystallized LL chondrite in which some relict chondrule are discernable, but chondrule boundaries are not as sharp as in Semarkona [**(a)**,**(b)**]. **(f)** Olivenza in RF showing a lower abundance of metal (~2 vol%) than in the Peekskill H6 chondrite.

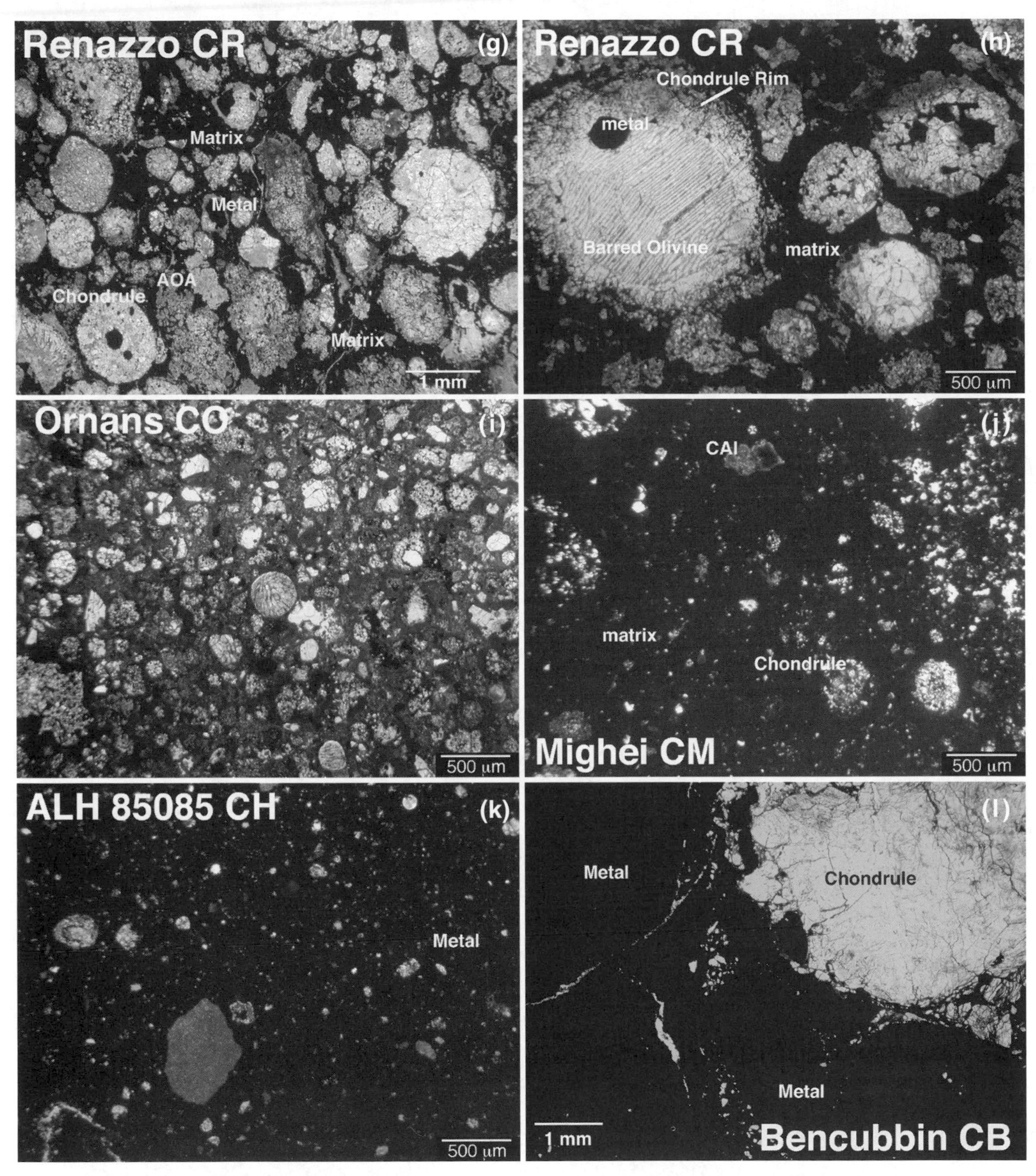

Fig. 2. (continued). **(g)** The Renazzo CR carbonaceous chondrite showing chondrules, opaque matrix, an amoeboid olivine aggregate (AOA) and a variety of other irregular-shaped objects. **(h)** Renazzo PL image showing a large (2 mm) barred olivine chondrule containing a metal bleb and a silicate rim. **(i)** The Ornans CO chondrite in PL showing the small (~200 μm) chondrules typical of CO chondrites and about 30% matrix. **(j)** PL photo of the Mighei CM chondrite showing the very high matrix abundance (up to 70 vol%) and small chondrules (100–200 μm) typical of CM chondrites. **(k)** The ALH 85085 CH chondrite in PL showing its microchondrules (most less than 100 μm in diameter) and fragments surrounded by Fe,Ni metal (~20 vol%). Most of the opaque areas are metal. **(l)** The Bencubbin CB chondrite in PL showing a large chondrule that is near 1 cm in diameter and opaque large Fe,Ni metal spheres.

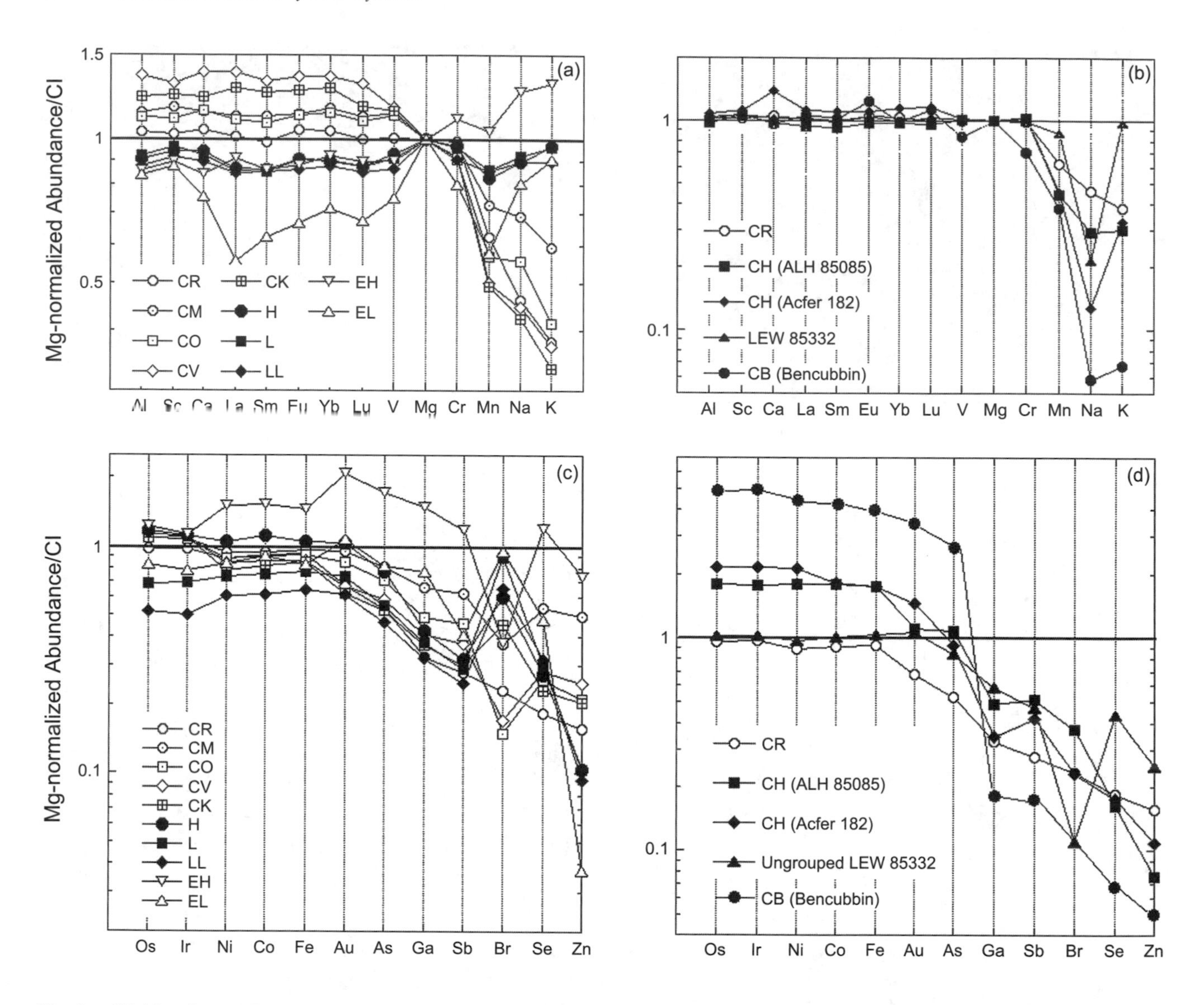

Fig. 3. "Spider diagram" showing the chemical differences among the various chondrite groups. Shown in this diagram are average whole-chondrite lithophile-element abundances in **(a)** various chondrite groups and **(b)** chondrite groups in the CR clan and siderophile-element abundances in **(c)** various chondrite groups and **(d)** chondrite groups in the CR clan. All data are normalized to Mg and CI chondrites. Elements are arranged according to their volatility. Data are from *Kallemeyn and Wasson* (1981, 1982, 1985), *Kallemeyn et al.* (1978, 1989, 1991, 1994, 1996), *Sears et al.* (1982), and *Wasson and Kallemeyn* (1988).

understates their importance; "carbonaceous" is essentially a misnomer since the carbon content of some C chondrites is very low.

The term *clan* is relatively new in chondrite classification and is used as a higher order of classification than group, but less inclusive than class. It was originally defined to encompass chondrites that have chemical, mineralogical, and isotopic similarities and were thought to have formed in the same local region of the solar nebula, i.e., in a narrow range of heliocentric distances (*Kallemeyn and Wasson,* 1981; *Kallemeyn et al.,* 1996). *Weisberg et al.* (1995a) used the term clan for a limited number of chondrites within a class that have chemical, mineralogical, and isotopic similarities suggesting a petrogenetic kinship, but have petrologic and/or bulk chemical characteristics that preclude a group relationship. A clan consists of chondrite groups that may have closely similar refractory lithophile-element abundances (Fig. 3b), plot on the same O-isotopic mixing line (Plate 3a,b), share an isotopic anomaly, or have some similarities in petrologic characteristics (e.g., chondrule size or primary mineral compositions). Four C-chondrite clans have been recognized: the CR clan, which includes the CR, CH, CB, and ungrouped chondrite Lewis Cliff (LEW) 85332; the CM-CO clan; the CV-CK clan; and the CI clan (*Weisberg et al.,* 1995a; *Kallemeyn et al.,* 1996). The significance of chondrite clans is not well understood and requires further research and evaluation. They may represent materials from the same local nebular reservoir that were sorted according to size, or they may represent materials that experienced different secondary histories such as brecciation, impact events, oxidation or reduction, and/or hydrothermal alteration.

Kallemeyn et al. (1996) questioned the usefulness of dividing chondrites into the C, O, and E classes. The H, L, and LL groups overlap in O-isotopic compositions and petrologic characteristics, indicating that they are closely related at the *clan* level. The same is true for EH and EL chondrite groups. As shown in Fig. 1, the O and E chondrite classes consist of only one clan.

Group is arguably the most basic and significant unit used in meteorite taxonomy. Group is commonly interpreted to indicate that the meteorites are from the same parent body. The term *group* is defined as a minimum of five unpaired chondrites of closely similar petrologic (chondrule size, modal abundances, and mineral compositions), whole-rock chemical and O-isotopic characteristics (Table 1, Fig. 3, Plate 3). Groups within the C, O, and E chondrite classes are each named differently, due to historical practice. Groups within the C-chondrite class are designated with the letter C and a second letter that is usually the first initial of the type specimen for that group. Eight C-chondrite groups are currently recognized. These include CI (Ivuna-like), CM (Mighei-like), CO (Ornans-like), CV (Vigarano-like), CK (Karoonda-like), CR (Renazzo-like), CB (Bencubbin-like), and CH (high Fe; ALH 85085-like). Groups within the O-chondrite class are named H, L, and LL based on the abundance of Fe and the ratio of metallic Fe (Fe^0) to oxidized Fe (FeO): H chondrites have the highest Fe and highest Fe^0/FeO ratios (H, 0.58; L, 0.29; LL, 0.11) among O chondrites and LL chondrites the lowest. In the equilibrated O chondrites (petrologic types 4–6), this is also reflected in their olivine [(in mol% Fa): H, 16–20; L, 23–26; LL, 27–32] and low-Ca pyroxene compositions. Based on mineralogy and bulk chemistry, the E chondrites are divided into EH (high Fe) and EL (low Fe) groups. The R (Rumurutti-like) chondrites have whole-chondrite and O-isotopic characteristics that are consistent with their affiliation to the O-chondrite class. However, their O-isotopic compositions plot considerably above the ordinary chondrites on the three-isotope diagram (Plate 3a,b), and therefore may be a separate class or O-chondrite clan.

For cases in which there are less than five members, the term *grouplet* has been applied. The K (Kakangari-like) meteorites may be the most significant grouplet, for they fall outside the properties of the C, O, and E chondrite classes. They have refractory-lithophile-element abundances closer to the O chondrites and O-isotopic compositions that plot below the terrestrial fractionation line, like the C chondrites (Plate 3).

Systematic petrologic differences have been recognized in members of some chondrite groups, which led to division of these groups into *subgroups*. For example, the CV chondrites can be divided into the oxidized and reduced subgroups, based on matrix/chondrule ratios, metal/magnetite ratios, and Ni content of metal and sulfides.

Some chondrites are mineralogically and/or chemically unique and defy classification into the existing chondrite groups; these are commonly called *ungrouped* chondrites. Among the most significant of these are Acfer 094, which appears to have largely escaped parent-body processing, and the unusual breccia Kaidun, which contains clasts of a variety of chondrite classes mixed on a millimeter scale. Meteorites within a group that have an atypical feature are often referred to *anomalous*.

The primitive achondrites have textures (Plate 1) suggesting that they were molten, partially molten, melt residues, or extensively recrystallized, but have a petrologic characteristic, whole-rock chemical compositions, or O-isotopic compositions that indicate a close affinity to the chondrites from which they formed (e.g., *Prinz et al.,* 1983) (see the discussions of each primitive achondrite group below). The nomenclature of primitive achondrites tends to follow the terminology first established by Rose for achondrites, applying the "-ite" suffix to a particularly well-known or well-characterized member of the groups (e.g., acapulcoites for Acapulco), although it can also include the terminology of iron meteorites. The emergence of primitive achondrites as a branch in the tree of classification is one of the major classification developments in the past several years. Although a useful new branch of the classification tree, the term "primitive achondrite" suffers from a lack of concise definition and consistent application. The definition of *Prinz et al.* (1983) is useful for recognizing primitive achondrites. We suggest that primitive achondrites are generally meteorites that exceeded their solidus temperature on the parent body — thus experiencing partial melting — but did not crystallize from a melt. Or, if they were molten, they are derived from parent bodies in which planetary differentiation did not achieve isotopic equilibrium. For the classic core groups of the primitive achondrites (acapulcoites, lodranites, winonaites, silicate-bearing IAB and IIICD irons), evidence as residues of partial melting is clear and classification as primitive achondrites is relatively consistent. It is noted, however, that the definition separates residues (e.g., lodranites) from partial melts (e.g., eucrites), even though both may have formed as a result of partial melting at similar parent-body temperatures, and could obscure relationships between future recoveries of basalts and residues from a common parent body. For other groups, the criterion of not crystallizing from a melt remains contentious. Scientific debate continues about the origin of ureilites and brachinites as residues or cumulates. In this chapter, we argue that brachinites and ureilites are properly included as primitive achondrites based on their whole chemical or O-isotopic compositions.

Achondrites include irons, stony irons, and stones that are from differentiated asteroids, Mars, and the Moon. Historically the term achondrite referred to stony meteorites only and irons and stony irons have been considered separately. However, most iron meteorites are thought to be the cores of differentiated asteroids; some stony irons have been interpreted to be mixtures of core and mantle materials, core and crustal materials, or products of impact from differentiated asteroids. Therefore, they are considered here as achondrites. The hierarchical terms (class, clan) used for chondrites have not been applied rigorously or systematically to achondrites. To help resolve this issue, we used the term *clan* for achondrites and primitive achondrites as defined above for the chondrites (Fig. 1). The term *petrogenetic as-*

TABLE 1. Average petrologic characteristics of the major chondrite groups.

	CI	CM	CO	CV	CK	CR	CH	CB	H	L	LL	EH	EL	R	K
Chondrule abundance (vol%)*	≪1	20[†]	48	45	45	50–60	70	20–40	60–80	60–80	60–80	60–80	60–80	>40	27
Matrix abundance (vol%)	>99	70[†]	34	40	40	30–50	5	≪1	10–15	10–15	10–15	2–15[‡]	2–15[‡]	36	73
CAI – AOA abundance (vol%)	≪1	5	13	10	10	0.5	0.1	≪1	≪1	≪1	≪1	≪1	≪1	0	≪1
Metal abundance (vol%)	0	0.1	1–5	0–5	0–5	5–8	20	60–80	8	4	2	10	10	0.1	7
Average chondrule diameter (mm)	NA	0.3	0.15	1.0	1.0	0.7	0.02	(0.2–1 cm)	0.3	0.7	0.9	0.2	0.6	0.4	0.6
Olivine composition (mol% Fa)	[††]	[††]	[††]	[††]	(<1–47)[§] 29–33	1–3[¶]	(<1–36)[§] 2[¶]	(2–3)[§] 3[¶]	(16–20)[§] 19.3[**]	(23–26)[§] 25.2[**]	(27–32)[§] 31.3[**]	0.4	0.4	38.0[**]	2.2

[*] Chondrule abundance includes lithic and mineral fragments.
[†] Highly variable.
[‡] The amount of matrix, if any, is not well established in E chondrites.
[§] Range of compositions.
[¶] Mode.
[**] Data for equilibrated varieties.
[††] Highly variable, unequilibrated.

NA = not applicable.

sociation has been used for achondrite groups with similar chemical and isotopic characteristics that suggest they are from the same or similar asteroids. As we discuss below, this term is vague and probably should be discarded. The term *group* has had a different application and significance for achondrites compared to its use for the chondrites. It is generally implied that members in a chondrite group are fragments of the same or similar asteroid. This is not the case within the achondrites. While some groups (e.g., howardites) fit the definition of having formed on a single parent body, other groups encompass materials that formed on multiple parent bodies. Pallasites, in particular, appear to have formed on multiple parent bodies, as evidenced by the main group, Eagle Station, and pyroxene pallasites and the ungrouped pallasite Milton. Brachinites may have formed on multiple parent bodies and aubrites include the anomalous aubrite Shallowater that likely requires a different parent body from other aubrites. Thus, terms like pallasite, mesosiderite, and brachinite are more like textural terms that describe meteorites that formed by a similar process, but are not necessarily from the same parent body. Similarly, meteorites thought to have formed on the same parent body (e.g., acapulcoites-lodranites; winonaites-IAB-IIICD irons; howardites-eucrites-diogenites; basaltic-anorthositic lunar meteorites; shergottites-nakhlites-chassignites-orthopyroxenites from Mars) sometimes occur within different groups. Thus, the term group may carry a very different connotation for achondrites than for chondrites. Clearly, it would be ideal to standardize the naming conventions, perhaps applying the same class-clan-group-subgroup-ungrouped nomenclature to achondrites, although the historical entrenchment of these terms would make acceptance of any new system very difficult. The naming of achondrite groups itself is historical and somewhat confusing, adding the "-ite" suffix to well-known meteorites (e.g., angrites for Angra dos Reis), early workers in meteoritics (e.g., howardites for early chemist Edmund Howard), and even Greek terms (e.g., eucrites from the Greek for "easily determinable" in reference to the coarse grains within these basalts).

The classification of irons is discussed later. The terms presented in Fig. 1 are the chemical classification of irons, which includes a roman numeral and a one- or two-letter designation. Currently, there are 13 recognized groups, although *Wasson and Kallemeyn* (2002) argued that IAB and IIICD form a single group termed the IAB complex. While the geochemical differences between the groups owe their existence to nebular processes, the geochemical trends within 10 of these groups are consistent with fractional crystallization, with each group forming in the core of a separate asteroid (*Scott,* 1972).

3.2. Primary Classification Parameters

The classification scheme outlined above is largely predicated on knowledge of a meteorite's petrologic characteristics, which include its texture, mineralogy, and mineral compositions; whole-rock chemical composition; and O-isotopic composition (Tables 1 and 2, Figs. 2 and 3, Plates 1–3). As we discuss below, the vast majority of meteorites are characterized based on their petrology alone, although knowledge of all three is a virtual prerequisite for defining a new group. Other characteristics can occasionally provide supporting data. Nitrogen and C abundances and isotopes, for example, have been used to distinguish between chondrite groups (e.g., *Kung and Clayton,* 1978; *Kerridge,* 1985). Similar cosmic-ray-exposure ages for meteorites that are petrologically, chemically, or isotopically similar can point to a single ejection event from a common parent body and strengthen evidence of a grouping and genetic linkage.

Among the three classification methods, whole-rock chemical composition has been in use the longest. This is due in large part to the ability of early workers to determine chemical composition, long before the polarizing microscope made mineral identification and separation practical. Chondrite compositions are usually expressed as Mg- or Si-normalized abundances relative to CI chondrites. One of the defining chemical characteristics of the chondrites is that their whole-rock chemical compositions are not fractionated relative to the solar photosphere, with the exception of the moderately volatile to highly volatile elements (Fig. 3). CI chondrites are considered the best match to solar abundances, whereas the other chondrite groups are defined by a narrow range of uniform enrichments or depletions in refractory elements and volatility controlled depletions in the moderately volatile lithophile and siderophile elements relative to CI chondrites (Fig. 3). Other bulk compositional criteria for resolving the chondrite groups involve plotting reduced Fe (metallic Fe and Fe as FeS) against oxidized Fe (Fe in silicate and oxide phases). Also used are compositional diagrams of elements of different volatility: e.g., Sb/Ni vs. Ir/Ni, Sc/Mg vs. Ir/Ni, and Al/Mn vs. Zn/Mn (e.g., *Kallemeyn et al.,* 1996). Bulk chemical composition is less commonly used as a primary classification criteria for stony achondrites, although it has been used to discriminate different igneous trends in the eucrites. For iron meteorites, the trace elements Ga, Ge, Ir, and Ni provide the basis for classification into chemical groups (Fig. 4).

Petrologic examination, including both textural and mineralogical studies in the polarizing microscope and mineralogical and mineral chemical studies using the electron microprobe or scanning electron microscope, have largely supplanted bulk chemical studies as the primary tool for classification of most meteorites, in large part because of their simplicity. A basic distinction between the chondrites and achondrites is texture. Chondrites generally have characteristic aggregational textures that sharply distinguish them from the igneous or recrystallized textured achondrites (Fig. 2). Most meteorites have unique textural and mineralogical properties that allow their classification by petrologic thin section analysis. In chondrites, these properties include chondrule size, chondrule to matrix ratios, abundance of metal, and mineral compositions (Table 1, Fig. 2). For achondrites, characteristic textures (e.g., igneous vs.

TABLE 2. Average petrologic characteristics of the major (asteroidal) achondrite and primitive achondrite groups.

	URE	ACA	LOD	ANG	AUB	BRA	WIN	HED	MES	PAL
Texture	coarse, some mosaicized	fine	coarse	medium to coarse	coarse/ brecciated	equigranular, with triple junctures	fine to medium	fine to coarse equigranular brecciated	breccias/ impact melt	coarse
Olivine composition Fa (mol%)	2–26*	4–13	3–13	11–66 (La 0.1–48)	<0.1	30–35	1–8	27–44	8–37	8–30
Low-Ca pyroxene Fs (mol%)	13–25	1–9	1–9		0.1–1.2	trace	1–9	14–79	23–59	—
Ca-pyroxene										
Fs (mol%)	13–32	46–50	46–50	12–50	0–0.2	10–13	2–4			
Wo (mol%)	2–16	43–46	43–46	50–55	40–46	38.7–47 (up to 5% TiO_2) (up to 12% Al_2O_3)	44–45	—	—	
Plagioclase An (mol%)	rare to absent	12–31	12–31	86–99.7	2–23	22–32	11–22	73–96	91–93	—
Metal	rare	present	present	rare	rare	minor†	present	rare	(stony-iron)	(stony-iron)
Other minerals present	augite graphite	troilite phosphates spinel graphite	troilite phosphates spinel graphite	spinel troilite phosphates oxides	variety of cubic suflides	sulfide oxides	troilite daubreelite schreibersite graphite	pigeonite silica chromite	pigeonite silica phosphates	phosphates troilite

* Olivine has reduction rims with near-end-member compositions. La-Ca end memeber.
† Tafassasset is a metal-rich (8 vol%) brachinite.

URE = ureilites, ACA = acapulcoite, LOD = lodranite, ANG = angrites, AUB = aubrite, BRA = brachinite, WIN = winonaites, HED = howardite = eucrite = diogenite, MES = mesosiderite, PAL = pallasite.

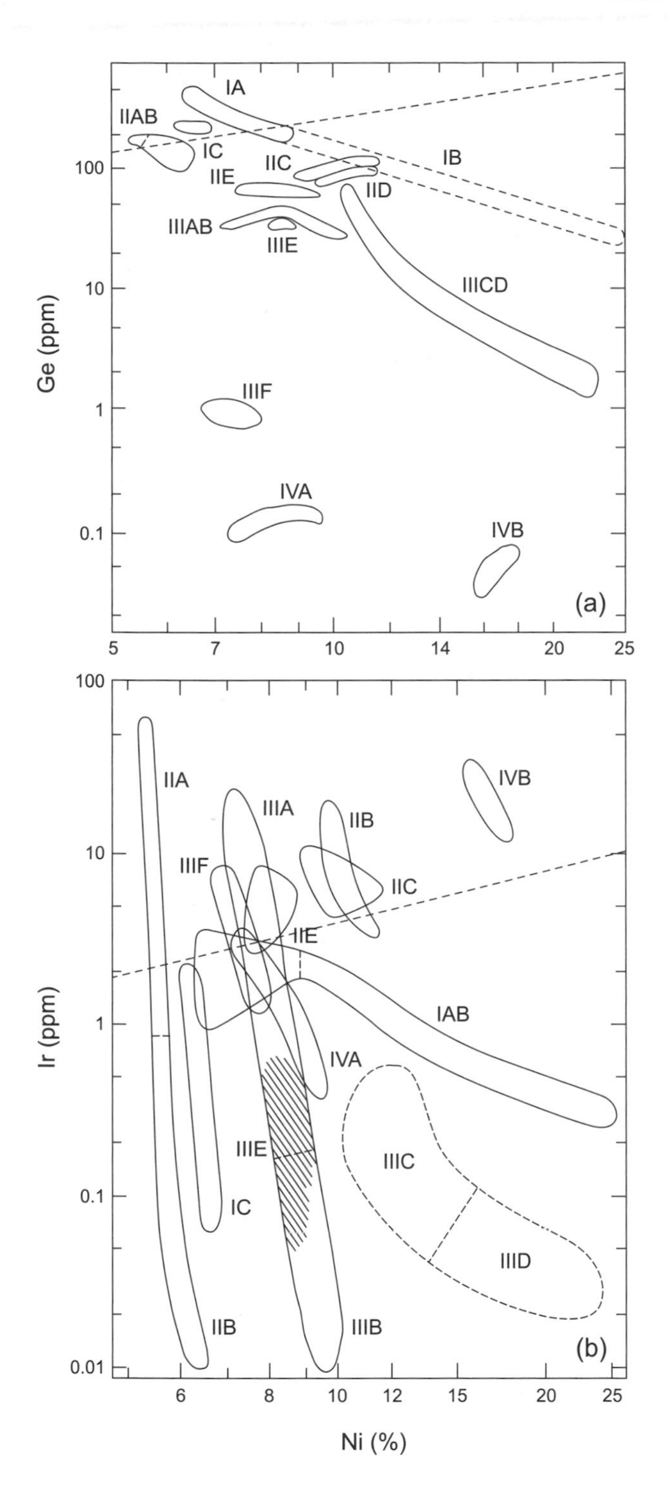

Fig. 4. Plots of **(a)** Ni vs. Ge and **(b)** Ni vs. Ir showing the fields for the 13 iron meteorite groups (from *Scott and Wasson,* 1975).

granoblastic), modal abundances of minerals, and mineral compositions are used to identify them by thin section examination (e.g., Table 2, Plates 1 and 2). As an example, eucrites are basalts that have igneous intergrowths dominated by plagioclase and clinopyroxene with an Fe/Mn ratio in the pyroxene of ~30. In general, petrologic studies and whole-rock chemical compositions tend to mirror one another, as the chemical composition tends to reflect the composition and abundance of the individual minerals and, at least in achondrites, the bulk chemical composition controls mineral crystallization.

Pioneered by R. Clayton and colleagues at the University of Chicago, the discovery that small anomalies in the abundance of the minor isotope ^{17}O could provide a unique fingerprint of nebular or parent-body origin opened a new way to determine relationships among meteorites not necessarily apparent from petrology or whole-rock chemical composition. Each meteorite group occupies a specific region on a three-isotope plot for O (Plate 3). Nominally anhydrous chondrites plot along lines of slope-1, suggesting that their O-isotopic trends result from mass-independent fractionation in the solar nebula. For example, the CV, CO, and CK chondrites plot along a slope-1 line commonly referred to as the C-chondrite anhydrous mineral (CCAM) mixing line. In contrast, hydrated chondrites (e.g., CM and CR chondrites) generally plot along shallower slopes (~0.7), indicative of modification by aqueous alteration. Like the highly differentiated and homogenized Earth, achondrites from largely molten bodies (e.g., Moon, Mars, Vesta) plot along slopes of 0.52, the result of mass fractionation imparted during melting. Some primitive achondrites (e.g., ureilites, acapulcoites-lodranites) exhibit greater deviation from this slope-0.52 line. Among these, the ureilites exhibit the most dramatic O-isotopic heterogeneity. Despite similar petrologic and bulk chemical characteristics and igneous textures suggesting a cumulate or partial melt residue origin, their whole-rock O-isotopic compositions plot along a mixing line that is an extension of the CCAM line (Plate 3), indicating that if ureilites are from a single asteroid, it was not internally equilibrated in terms of O isotopes during differentiation.

In the practical matter of actually placing a new meteorite within one of the classificational bins — or deciding that it is ungrouped or joins with others to form a new group — petrologic studies are the tool of choice. More than 98% of the world's known meteorites have been adequately classified from petrologic studies alone, although that number is dominated by equilibrated O chondrites from hot and cold deserts. Because many meteorite groups have characteristic textures and metal abundances (e.g., Fig. 2, Plates 1 and 2), a trained observer can make a good first estimate of classification with the unaided eye. Good examples include the unusual white fusion crust on aubrites or the presence of small, rare chondrules in black matrix of a CM chondrite. Petrologic studies typically involve preparation of a polished thin section to examine a variety of textural and mineralogical features, including the variety of minerals present and identifiable with the polarizing microscope, the presence or absence of chondrules, the texture, and mineral abundances. Although tentative classification can often be made by the trained observer in hand sample or thin section, mineral analyses are essential for definitive classification. Typically, olivine, pyroxene, and, sometimes, plagioclase are analyzed, since they are widespread in meteorites and have diagnostic mineral compositions between

TABLE 3. Summary of the criteria for classifying chondrites according to petrologic type, based on *Van Schmus and Wood* (1967).

Criterion	1	2	3	4	5	6	7
Homogeneity of olivine compositions	—	>5% mean deviations		≤5%	Homogeneous		
Structural state of low-Ca pyroxene	—	Predominantly monoclinic		>20% monoclinic	≤20% monoclinic	Orthorhombic	
Feldspar	—	Minor primary grains		Secondary <2-μm grains	Secondary 2–50-μm grains	Secondary >50-μm grains	
Chondrule glass	Altered or absent	Mostly altered, some preserved	Clear, isotropic	Devitrified	Absent		
Metal: Maximum Ni (wt%)	—	<20 taenite minor or absent	>20 kamacite and taenite in exsolution relationship				
Sulfides: Mean Ni (wt%)	—	>0.5	<0.5				
Matrix	Fine grained opaque	Mostly fine-grained opaque	Opaque to transparent	Transparent, recrystallized			
Chondrule-matrix integration	No chondrules	Sharp chondrule boundaries		Some chondrules can be discerned, fewer sharp edges		Chondrules poorly delineated	Primary textures destroyed
Carbon (wt%)	3–5	0.8–2.6	0.2–1	<0.2			
Water (wt%)	18–22	2–16	0.3–3	<1.5			

After *Van Schmus and Wood* (1967) with modifications by *Sears and Dodd* (1988), *Brearley and Jones* (1998), and this work.

the different groups. Mineral chemistries are typically determined with the electron microprobe or scanning electron microscope to determine mineral composition and degree of mineral heterogeneity. Among the more abundant meteorites from hot and cold deserts, older techniques, such as oil immersion analyses, are used to distinguish between critical ranges of mineral compositions and classify the abundant equilibrated O chondrites. In some cases, other minerals provide diagnostic compositions, such as the concentration of Si in the metal of E chondrites or the Fe/Mn ratio of pyroxene in lunar, martian, and eucritic basalts.

Petrologic studies alone have produced some spectacular misclassifications. For example, the basalt Elephant Moraine (EET) 87251 was originally thought to be a eucrite (asteroidal basalt) and the ALH 84001 pyroxenite was classified as a diogenite (asteroidal orthopyroxenite). Elephant Moraine 87251 is now known to be a lunar meteorite and ALH 84001 is martian. In the case of ALH 84001, this misclassification stemmed from a lack of appreciation of the petrologic diversity likely to be found on Mars. As a general rule, misclassifications are minimized if a group of competent workers with a range of specializations within the field are consulted about meteorites for which one or more petrologic features seem odd. In these cases, O-isotopic compositions tend to be sought. Chemical compositions are less commonly sought as a primary classificational tool, although they are essential for iron meteorites. It is important to emphasize that even some of the best-studied meteorites in the world cannot be confidently binned into the existing scheme, suggesting either transitional properties between groups or sampling of a previously unknown type of meteorite.

3.3. Secondary and Tertiary Properties

Chondrites are assigned a number according to petrologic type (e.g., *Van Schmus and Wood,* 1967; Table 3). Type 3.0 is thought to represent the most pristine materials, 3.1 to 6 indicate increasing degree of petrologic equilibration and recrystallization, and 2 to 1 represent increasing degree of hydrous alteration in chondrites. Type 7 has been used to indicate chondrites that have been completely recrystallized or melted, but some or most of these meteorites may be impact melted. Petrologic types can be assigned based on petrologic observation, mineral composition (Table 3), and thermoluminescence properties of the meteorite. However, the mineral criteria listed in Table 3 do not work equally well for all chondrite groups (*Huss et al.,* 2006). For example, one of the major criteria used to distinguish between type 3 and 4 is heterogeneity of olivine composition. This is difficult to apply to E chondrites because olivine is low in abundance. Also, the amount of Fe in silicate is low due to the reducing conditions under which the meteorites formed and is not a reflection of the degree of metamorphic equilibration.

In some cases, the petrologic type is reflected in whole-rock O-isotopic composition; e.g., in equilibrated (type 4–6) chondrites, whole-rock O-isotopic compositions become increasingly homogenous. The slope ~0.7 line of CM chondrites, most of which are (hydrated) petrologic type 2, on an O three-isotope plot has been attributed to closed-system hydration reactions with water that has heavier O than the silicates (*Clayton and Mayeda,* 1999).

Figure 5 shows the petrologic types that are represented in each of the chondrite groups. Some groups (e.g., H, L, LL, EH, EL) do not have heavily hydrated (petrologic types 2 and 1) members. Other groups (e.g., CI, CM, CR) lack less-altered (type 3) members. This is particularly unfortunate in the case of the CI chondrites, which chemically are among the most primitive solar system materials.

Thermoluminescence (TL) has been used to subdivide the petrologic type 3 chondrites into 3.0 to 3.9 subtypes (*Sears et al.,* 1980). This has been largely applied to O chondrites. Thermoluminescence sensitivity is largely dependent on the degree of crystallization of the chondrule mesostasis. Thus, we can recognize meteorites such as Semarkona (LL3.0) and Krymka (LL3.1), which are interpreted to be examples of the most pristine O chondrites. The TL data are supported by the high degree of disequilibrium in the mineral assemblage of chondrules and matrix and the abundance of glassy mesostasis in the chondrules of the low petrologic type (<3.5) chondrites (*Huss et al.,* 2006).

It can be argued that the current scheme for petrologic type needs extensive revision. For example, does it make sense to have a scheme where type 3 denotes the least-altered meteorites and alteration proceeds to lower and higher values depending on whether it is hydrous alteration or ther-

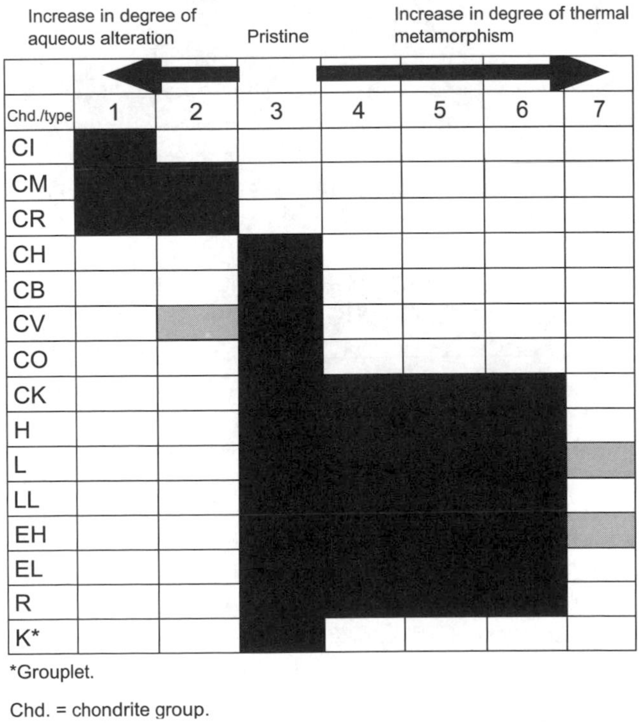

Increase in degree of aqueous alteration
Pristine
Increase in degree of thermal metamorphism
Chd./type 1 2 3 4 5 6 7
CI
CM
CR
CH
CB
CV
CO
CK
H
L
LL
EH
EL
R
K*
*Grouplet.
Chd. = chondrite group.

Fig. 5. Diagram showing the petrologic types for each chondrite group.

mal metamorphism? More importantly, there are meteorites that have been hydrously altered and have been later reheated during thermal metamorphism. There is no way to label such cases in the current scheme. One solution may be to have separate schemes for denoting degree of hydrous alteration and degree of metamorphism.

Chondrites are also assigned values that represent the degree of shock pressure that they experienced. These values range from S1 (unshocked, pressure <5 GPa) to S6 (very strongly shocked, pressures up to 90 GPa). The shock stages are assigned based on petrographic features of the silicate minerals olivine, pyroxene, and plagioclase and include undulatory extinction, fracturing, and mosaicism, as well as transformation of plagioclase into maskelynite (e.g., *Stöffler et al.,* 1991; *Rubin et al.,* 1997). Other indicators of high degrees of shock in chondrites include the presence of high-pressure minerals, such as majorite garnet and wadslyite, a high-pressure form of olivine (e.g., *Kimura et al.,* 2003). However, the degree of shock can vary within a meteorite on the scale of centimeters.

Finally, meteorites have been subjected to 4.5 b.y. of impact processing, producing myriad breccias. With rare exceptions, chondrite breccias are *genomict,* i.e., consist of clasts belonging to the same chondrite group. Relatively few chondrite breccias contain clasts belonging to different meteorite groups and are called *polymict.* This observation has led to the conclusion that each chondrite group represents a single parent body [see Table 1 in *Bischoff et al.* (2006) for nomenclature of chondrite breccias]. Some chondrites are termed genomict breccias (*Bischoff et al.,* 2006), indicating that they contain fragments of the same chondrite group, but different petrologic types. Many achondrites are brecciated. Most notable among these are the howardites, which are, by definition, breccias consisting of eucrite and diogenite components. It is interesting to note that no iron meteorites are true breccias (e.g., recemented fragments of previously formed and fragmented iron meteorites), perhaps pointing to the inability of impacts to relithify iron meteorite regolith.

4. MAJOR GROUPINGS AND THEIR SIGNIFICANCE

The chondrites are among the most primitive solar system materials available for laboratory study. They are composed of submillimeter- to centimeter-sized components, including chondrules, Ca-Al-rich inclusions (CAIs), amoeboid olivine aggregates (AOAs), Fe, Ni-metal, and fine-grained matrix (Fig. 2a–d). These components are thought to have formed as independent objects in the protoplanetary disk by high-temperature processes that included condensation and evaporation (e.g., *Ebel,* 2006; *Fedkin and Grossman,* 2006). As a result, chondritic components of primitive (unaltered and unmetamorphosed) chondrites preserved records of the physical-chemical properties of the protoplanetary disk region(s) where they formed, and are ground truth for astrophysical models of nebular evolution. Chondrules are spherical to semispherical objects, generally composed of mafic minerals and interpreted to have formed as molten or partly melted droplets from transient heating event(s) in the solar nebula. Chondrule textures are illustrated in Fig. 2 and in Fig. 1 from *Lauretta et al.* (2006). Calcium-aluminum-rich inclusions and AOAs are irregular to spherical objects having primary minerals that are composed of more refractory elements than those in the chondrules. The mineralogy of CAIs is discussed by *Beckett et al.* (2006). Surrounding the chondrules and inclusions is a fine-grained (generally <5 μm) silicate-rich matrix. Chondrite matrix is a separate component, not simply composed of pulverized chondrules and CAIs, and its relationship to the chondrules and CAIs in a chondrite is not well understood. Chondrite matrix is discussed by *Krot et al.* (2006) and *Brearley* (2006). Additionally, chondrites contain small (<5–10 μm) presolar grains of diamond, silicon carbide, graphite, silicon nitride, and oxides (e.g., *Anders and Zinner,* 1993; *Bernatowicz and Zinner,* 1997).

The achondrites include meteorites from asteroids, Mars, and the Moon. Primitive achondrites may represent melts, partial melts, or melt residues from local heating events on planetary bodies during the earliest stages of differentiation or possibly from local impact events. We include ureilites, brachinites, acapulcoites, lodranites, winonaites, and IAB and IIICD irons as primitive achondrites. Below we illustrate important characteristics of some meteorite clans, groups, and subgroups and the data used to establish the groupings. Further discussions of achondrites are given by *McCoy et al.* (2006); see *Chabot and Haack* (2006) for a discussion of irons. The descriptions below are meant to give the reader an overview of some of the distinguishing characteristics of the variety of meteorites available for study.

5. PROPERTIES OF CHONDRITES

5.1. Carbonaceous (C) Chondrites

Carbonaceous chondrites are characterized by whole-chondrite (Mg-normalized) refractory-lithophile-element abundances ≥1× CI (Fig. 3) and O-isotopic compositions that plot near or below the terrestrial fractionation (TF) line. They have a range of chondrule sizes, from the large rimmed chondrules in CV and CR chondrites to the small chondrules in CH chondrites (Table 1, Fig. 2). Although many C chondrites (e.g., CI, CM, CO, CV, CR) are matrix-rich, some (e.g., CH, CB) are matrix-poor.

5.1.1. CI (Ivuna-type) chondrite group/clan. The CI chondrites (seven members are known) are often considered the most primitive solar system materials, having bulk compositions close to that of the solar photosphere. They are altered, brecciated rocks devoid of chondrules and CAIs. All the CI chondrites are of petrologic type 1, indicating that they are heavily hydrated meteorites consisting of fine-grained, phyllosilicate-rich matrix with minor magnetite, sulfides, sulfates, and carbonates, and rare isolated olivine and pyroxene fragments (e.g., *McSween and Richardson,*

1977; *Endress and Bischoff,* 1993, 1996; *Endress et al.,* 1996; *Leshin et al.,* 1997). It is not clear if the CI chondrites ever contained chondrules and CAIs and are essentially all matrix component, or if the chondrules and CAIs were consumed and their primary chondrite textures destroyed during extensive aqueous alteration. It is noteworthy, and somewhat miraculous, that the CI chondrites have the most primitive chemical compositions (similar to that of the solar photosphere) despite their being hydrothermally altered, brecciated chondrites with their primary mineralogy entirely erased.

The CI chondrites have been assigned to a separate clan because there is nothing to link them to the chondrite groups in the other clans (*Kallemeyn and Wasson,* 1981).

5.1.2. CM-CO clan. The CM (Mighei-like) chondrites are the most abundant group of C chondrites. In addition, CM-like materials commonly occur as clasts in other chondrite groups and achondrites (e.g., *Zolensky et al.,* 1996), suggesting their wide distribution in the early solar system. They are characterized by relatively small (~300 μm) chondrules (Fig. 2j), CAIs, and AOAs that have been partially to completely replaced by phyllosilicates. Although most CM chondrites are of petrologic type 2 (CM2), the degree of aqueous alteration experienced by CM chondrites varies widely (e.g., *McSween,* 1979; *Browning et al.,* 1996). Some CM chondrites [Yamato (Y) 82042 and EET 83334] contain virtually no anhydrous silicates and have been classified as petrologic type 1 (CM1) (*Zolensky et al.,* 1997).

Bells is an example of an anomalous CM chondrite (*Rowe et al.,* 1994; *Brearley,* 1995; *Mittlefehldt,* 2002). Its matrix has a higher abundance of magnetite than typical CM chondrites, and the matrix phyllosilicates are more similar to those in CI chondrites. Bells has refractory-lithophile-element abundances like those of CI chondrites, but its moderately volatile- and volatile-lithophile-element abundances are like those of CM chondrites. Refractory-siderophile-element abundances are higher than those of CI chondrites, and Fe, Co, and Ni abundances are at CI levels (*Kallemeyn,* 1995; *Mittlefehldt,* 2002).

The CO (Ornans-like) chondrites have relatively small (~150 μm) chondrules (*Rubin,* 1989) and relatively high matrix abundances (30–45 vol%). They range in petrologic type from 3.0 to 3.7 (*McSween,* 1977a; *Scott and Jones,* 1990). Chondrules, CAIs, and AOAs in metamorphosed CO chondrites contain secondary minerals (e.g., nepheline, sodalite, ferrous olivine, hedenbergite, andradite, and ilmenite); these secondary minerals are virtually absent in the more pristine CO3.0 chondrites (e.g., *Rubin et al.,* 1985; *Keller and Buseck,* 1990a; *Tomeoka et al.,* 1992; *Rubin,* 1998; *Russell et al.,* 1998a). Whole-chondrite O-isotopic compositions of the CO chondrites plot along the CCAM line, overlapping with those of the CV chondrites (Plate 3).

Allan Hills A77307 is a CO3.0 chondrite and is commonly referred to as the most primitive CO chondrite (e.g., *Brearley,* 1993). However, it is significantly depleted in ^{18}O, has smaller chondrules and higher matrix content (~40–50 vol%) compared to other CO chondrites, and its bulk composition is more like that of CM than CO chondrites, making its classification uncertain (*Scott et al.,* 1981; *Kallemeyn and Wasson,* 1982; *Sears and Ross,* 1983; *Rubin,* 1989; *Greenwood et al.,* 2002).

Three characteristics have been used to link the CM and CO chondrites and support a CM-CO clan (*Kallemeyn and Wasson,* 1982): (1) chondrules are similar in size and anhydrous minerals have similar compositions; (2) refractory-lithophile-element abundances are similar; and (3) O-isotopic compositions of high-temperature minerals are similar.

5.1.3. CK-CV clan. The CV (Vigarano-like) chondrites have a high abundance of matrix, large (about 1 mm) chondrules, and a high abundance of large CAIs and AOAs (Table 1, Plate 4). All the CV chondrites are classified as type 3. They are a diverse group of meteorites divided into oxidized (CV_{ox}) and reduced (CV_{red}) subgroups, largely based on modal metal/magnetite ratios and Ni content of metal and sulfides (*McSween,* 1977b). Subsequently, *Weisberg et al.* (1997a) subdivided oxidized CVs into the Allende-like (CV_{oxA}) and Bali-like (CV_{oxB}) subgroups. Thus, three subgroups (CV_{red}, CV_{oxA}, CV_{oxB}) are now recognized.

Division of the CV chondrites is based mainly on their petrologic characteristics. Matrix/chondrule ratios increase in the order CV_{red} (0.5–0.6)-CV_{oxA} (0.6–0.7)-CV_{oxB} (0.7–1.2), whereas metal/magnetite ratios tend to decrease in the same order. The whole-rock O-isotopic compositions of the CV chondrites plot along the CCAM line, with the CV_{oxB} chondrites being slightly depleted in ^{16}O relative to the CV_{red} and CV_{oxA} chondrites. Although all the CV chondrites have been classified as type 3, the CV_{oxB} chondrites contain abundant (hydrated) phyllosilicates and can be classified as type 2 (e.g., *Tomeoka and Buseck,* 1990; *Keller and Buseck,* 1990b; *Keller et al.,* 1994). Recent studies have shown that there are significant mineralogical differences between the CV subgroups resulting from varying degrees of late-stage alteration (*Krot et al.,* 1998, 1995, 2003). Thus, these subgroups may reflect secondary characteristics superimposed onto the members of this meteorite group.

The CV_{oxB} chondrites experienced aqueous alteration and contain hydrous phyllosilicates, magnetite, Fe,Ni-sulfides, Fe,Ni-carbides, fayalite, Ca,Fe-pyroxenes (Fs_{10-50} Wo_{45-50}), and andradite. The CV_{oxB} matrices consist of very fine-grained (<1–2 μm) ferrous olivine (~Fa_{50}), concentrically zoned nodules of Ca,Fe-pyroxenes and andradite, coarse (>10 μm) grains of nearly pure fayalite ($Fa_{>90}$), and phyllosilicates (Fig. 6). The CV_{oxA} chondrites are more extensively altered than the CV_{oxB}, but contain very minor phyllosilicates. The major secondary minerals include ferrous olivine (Fa_{40-60}), Ca,Fe-pyroxenes, andradite, nepheline, sodalite, Fe,Ni-sulfides, magnetite, and Ni-rich metal. The CV_{oxA} matrices are coarser-grained than those in the CV_{oxB} chondrites and largely consist of lath-shaped ferrous olivine (~Fa_{50}), Ca, Fe-pyroxene ± andradite nodules, and nepheline (Fig. 6). Some oxidized CVs [e.g., Meteorite Hills (MET) 00430] are mineralogically intermediate between the CV_{oxB} and CV_{oxA} chondrites. For example, the matrix in MET 00430 contains ferrous olivine with grain sizes inter-

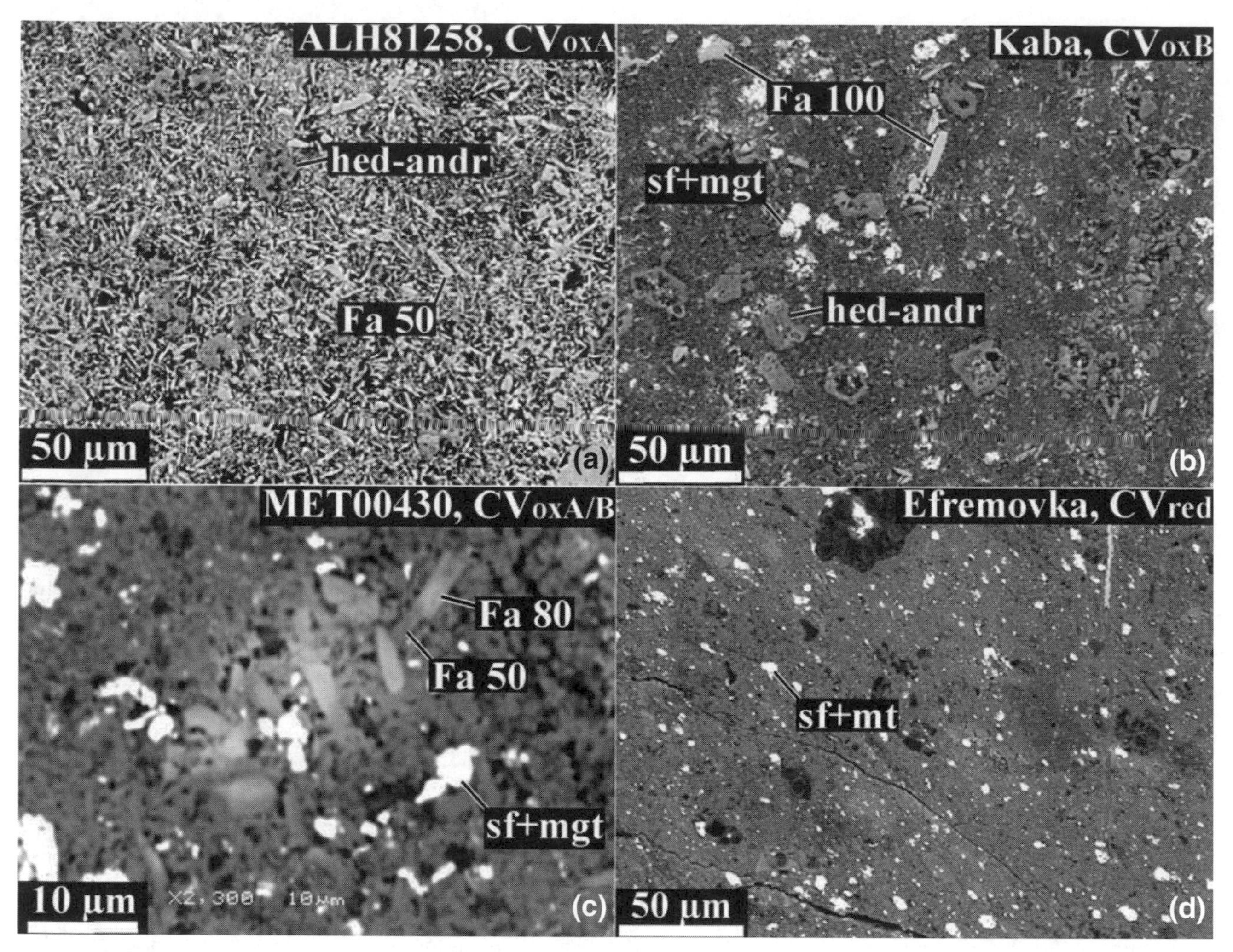

Fig. 6. Backscattered electron images of matrices in **(a)** ALH 81258 (CV_{oxA}), **(b)** Kaba (CV_{oxB}), **(c)** MET 00430 ($CV_{oxA/B}$), and **(d)** Efremovka (CV_{red}). Matrix in ALH 81258 contains lath-shaped, relatively coarse-grained and compositionally uniform (~Fa_{50}) ferrous olivine. Matrix in Kaba contains nearly coarse grains of nearly pure fayalite (~Fa_{100}), sulfide-magnetite grains (sf-mgt), and very fine-grained groundmass largely composed of ferrous olivine (~Fa_{50}). Matrix in MET 00430 contains fayalite grains showing inverse compositional zoning (Fa_{80-50}), sulfide-magnetite grains, and fine-grained, lath-shaped ferrous olivine (~Fa_{50}). Matrix in Efremovka is heavily compacted by shock metamorphism; it is largely composed of ferrous olivine (~Fa_{50}) and sulfide-Fe,Ni-metal (sf-mt) grains. All matrices contain Ca,Fe-pyroxenes-andradite (hed-andr) nodules; the Efremovka matrix has the lowest abundance of these nodules.

mediate between those in the matrices of the CV_{oxA} and CV_{oxB} chondrites and with inverse compositional zoning (Fa_{80-50}). The reduced CV chondrites Efremovka and Leoville experienced alteration similar to that of the CV_{oxA} meteorites, but of a smaller degree. Both meteorites virtually lack phyllosilicates and contain nepheline, sodalite, and Ca,Fe-pyroxenes; however, these minerals are much less abundant than in the CV_{oxA} meteorites. The CV chondrite Mokoia contains clasts of the CV_{oxA} and CV_{oxB} materials (*Krot et al.,* 1998). The CV chondrite breccia Vigarano con-tains the CV_{red}, CV_{oxB}, and CV_{oxA} materials (*Krot et al.,* 2000c). These observations may indicate that the CV subgroups represent different lithological varieties of the CV asteroidal body that experienced complex, multistage alteration.

The CK (Karoonda-like) chondrites contain a high abundance of matrix and large chondrules (700–1000 µm), most of which have porphyritic textures. Chondrules with glassy, cryptocrystalline, or barred olivine textures are rare. The CK chondrites are highly oxidized, as indicated by high fayalite contents of their olivine (Fa_{29-33}), nearly complete absence of Fe,Ni-metal, high Ni content in sulfides, and abundant magnetite with exsolution lamellae of ilmenite and spinel. The whole-rock O-isotopic compositions of the CK chondrites plot along the CCAM line, within the range of the CO and CV chondrites (Plate 3a). Their refractory-lithophile-element abundances (~1.2× CI) are between those of CV and CO chondrites with depletions in moderately volatile elements greater than those of CV and CO chondrites (*Kallemeyn et al.,* 1991). Most CK chondrites are of high (4–6) petrologic type.

The CK chondrites of low petrologic type (<4) show some chemical, mineralogical, and petrographic similarities to the CV chondrites (*Greenwood et al.,* 2005). For ex-

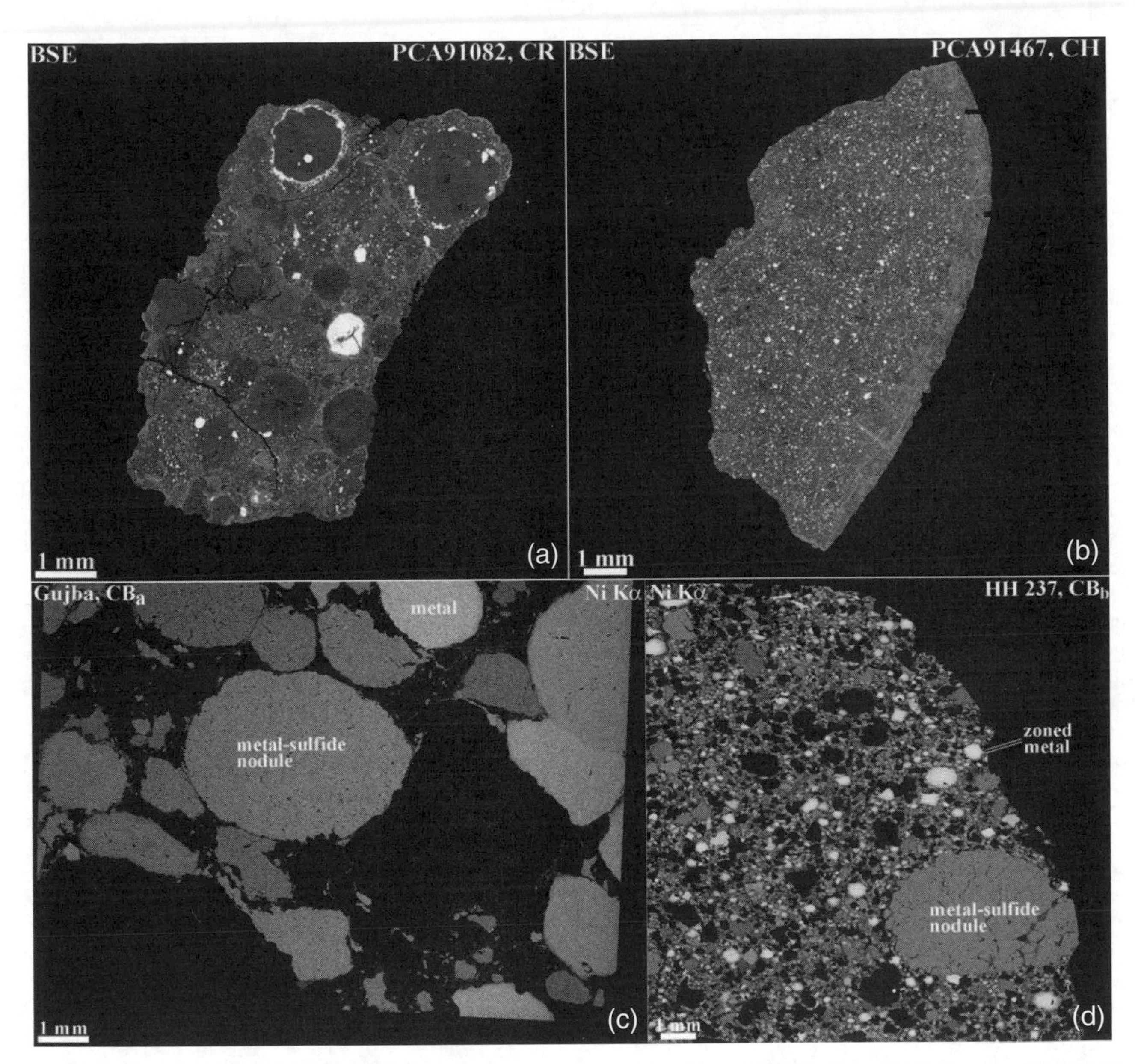

Fig. 7. Backscattered electron images [**(a)**,**(b)**] and elemental maps in Ni Kα X-rays [**(c)**,**(d)**] of the **(a)** CR chondrite PCA 91082, **(b)** CH chondrite PAT 91546, **(c)** CB_a chondrite Gujba, and **(d)** CB_b chondrite Hammadah al Hamra 237. In the CR chondrite PCA 91082 **(a)**, FeNi-metal occurs as nodules in chondrules, igneous rims around chondrules, and in the matrix. In the CH chondrite PAT 91546 **(b)**, FeNi-metal occurs largely as irregularly shaped grains outside. In the CB_a chondrite Gujba **(c)**, FeNi metal occurs as large nodules several millimeters in size. In the CB_b chondrite Hammadah al Hamra 237 **(d)**, metal occurs as both large nodules, like those in CBb chondrites, and as numerous smaller spherical to irregularly shaped nodules.

ample, the CK3 and the oxidized CV3 chondrites have a similar range of O-isotopic compositions on the CCAM line and their magnetites have similar compositions (e.g., *Greenwood et al.,* 2005). Ningqiang is an ungrouped C chondrite with characteristics that overlap CV and CK (*Rubin et al.,* 1988; *Kallemeyn et al.,* 1991; *Kallemeyn,* 1996; *Koeberl et al.,* 1987; *Mayeda et al.,* 1988; *Guimon et al.,* 1995). These observations support a close (clan) relationship between the CK and CV chondrites, which needs further study.

5.1.4. CR clan. Most CR (Renazzo-like) chondrites are of petrologic type 2 and are characterized by large (millimeter-sized), Fe,Ni-metal-rich, porphyritic, forsterite- or enstatite-rich (type I) chondrules, some of which are multilayered, with rims of metal and silicates (olivine, pyroxene, glassy mesostasis, ±silica phase or phyllosilicates in the more-altered chondrules) (*Weisberg et al.,* 1993) (Fig. 2g,h and Fig. 7a, Plate 5a). The abundances of CAIs and AOAs are relatively low. Grosvenor Mountains (GRO) 95577 contains no anhydrous silicates and is classified as a CR1 chondrite (*Weisberg and Prinz,* 2000). Matrix ranges from 40 to 70 vol% and is hydrated, containing phyllosilicates, magnetites, carbonates, and sulfides. Matrix-like fragments, commonly referred to as dark inclusions, are common in the CR chondrites. An intriguing aspect of the CR chondrites is that the Fe,Ni-metal in the matrix and chondrules shows a large compositional range with a positive Ni vs. Co trend and with a solar Ni/Co ratio, suggesting it is a highly primitive material. The whole-rock refractory-lithophile-element abundances of CR chondrites are close to solar, while the moderately volatile lithophiles are highly depleted (Fig. 3). On the O three-isotope diagram, the CR chondrites define a unique mixing line, of slope ~0.7, that differs from the CCAM line and the CM chondrite line (Plate 3a). The whole-rock N-isotopic compositions show large positive anomalies with $\delta^{15}N$ up to ~200‰ (*Weisberg et al.,* 1993), suggesting a heavy N carrier in the CR chondrites, the identity of which has not yet been resolved. The mineralogy, petrology, and chemistry of CR chondrites and their components are described in detail in several papers (*McSween,* 1977b; *Bischoff et al.,* 1993a; *Weisberg et al.,* 1993, 2004; *Kallemeyn et al.,* 1994; *Krot et al.,* 2000a, 2002a,b; *Aléon et al.,* 2002).

Al Rais is considered to be an anomalous CR chondrite due to a high matrix (and dark inclusions) component (>70 vol%), which is reflected in its whole-chondrite chemical and O-isotopic composition (*Weisberg et al.,* 1993). It may simply be a breccia that acquired a larger amount of the matrix (or dark inclusion) component than other CR chondrites.

The CH (ALH 85085-like) chondrites are characterized by small (<50–100 μm in diameter) chondrules, which are mostly cryptocrystalline in texture, and high abundance of Fe,Ni-metal (~20 vol%) (Fig. 2k and Fig. 7b, Plate 5b) with a large range in composition, positive Ni vs. Co trend, and a solar Ni/Co ratio. Interchondrule matrix is virtually absent; heavily hydrated, lithic, matrix-like clasts are present. CH chondrites have ~CI bulk refractory-lithophile-element abundances, but are highly depleted in volatile and moderately-volatile lithophile elements. Whole-rock O-isotopic compositions plot along the CR-mixing line; whole-rock Nisotopic compositions show large positive anomalies in $\delta^{15}N$, up to 1000‰ (*Wasson and Kallemeyn,* 1990; *Bischoff et al.,* 1993b).

The unusual characteristics of the CH chondrites, including the dominance of small cryptocrystalline chondrules, high abundance of metal, and lack of matrix led to the interpretation that the CH chondrites are "subchondritic" meteorites formed as a result of an asteroidal collision and are not true primary materials formed in the solar nebula (*Wasson and Kallemeyn,* 1990). However, based on the presence of CAIs uniformly enriched in ^{16}O, Fe,Ni-metal condensates, and pristine chondrules in CH chondrites, it was suggested that these meteorites are in fact pristine products of the solar nebula (*Grossman et al.,* 1988; *Scott,* 1988; *Weisberg et al.,* 1988; *Meibom et al.,* 1999; *Krot et al.,* 2000b). However, the origin of these meteorites is currently being studied.

The CB (Bencubbin-like) chondrites have characteristics that sharply distinguish them from other chondrites (*Weisberg et al.,* 2001); these include (1) very high metal abundances (60–80 vol%), (2) almost exclusively cryptocrystalline and barred textures of chondrules (Fig. 2l and Fig. 7c,d, Plate 5c,d,), (3) large depletion in moderately-volatile-lithophile elements, and (4) large enrichment in heavy N.

Based on their petrologic and chemical characteristics, the CB chondrites are divided into the CB_a (Gujba, Bencubbin, Weatherford) and CB_b (Hammadah al Hamra 237, QUE 94411) subgroups (*Weisberg et al.,* 2001, 2002). The CB_a chondrites contain ~60 vol% metal, chondrules are centimeter-sized, Fe,Ni-metal contains 5–8 wt% Ni, and $\delta^{15}N$ is up to ~1000‰. The CB_b contain >70 vol% metal, chondrules are millimeter-sized, Fe,Ni-metal contains 4–15% wt% Ni, and $\delta^{15}N$ is ~200‰.

The origin and significance of the CB chondrites have generated much controversy. They have been interpreted to be highly primitive nebular materials containing metal that condensed directly from the solar nebula (*Newsom and Drake,* 1979; *Weisberg et al.,* 1990, 2001; *Meibom et al.,* 1999, 2000a, 2001; *Campbell et al.,* 2001; *Petaev et al.,* 2001; *Krot et al.,* 2002b). Others argue that CB chondrites formed in a vapor cloud produced during an impact event on a chondritic planetesimal (*Kallemeyn et al.,* 1978; *Campbell et al.,* 2002). Like in CH chondrites, fine-grained matrix material is essentially absent, but rare matrix-like clasts are present in some CB chondrites (*Lovering,* 1962; *Mason and Nelen,* 1968; *Kallemeyn et al.,* 1978; *Newsom and Drake,* 1979; *Weisberg et al.,* 1990, 2001; *Rubin et al.,* 2001; *Krot et al.,* 2001a, 2002b; *Greshake et al.,* 2002). It has also been suggested that the CB_b chondrites are the primitive nebular-formed chondritic materials and the CB_a chondrites are products of impact onto a CB_b-like target material (*Campbell et al.,* 2002). An intriguing characteristic of CB chondrites is that they all contain impact melt areas between the metal and silicate chondrules, which could be a shock-melted matrix material (*Newsom and Drake,* 1979; *Weisberg et al.,* 1990, 2001; *Krot et al.,* 2002b; *Meibom et al.,* 2000b).

The CR, CH, and CB chondrite groups and the ungrouped meteorite LEW 85332 (discussed below) have grossly different petrographic characteristics (Plate 5, Fig. 7) as well as differences in bulk chemistry (see above). However, they all share a number of characteristics that suggest a close relationship, and as a result are grouped together into the CR clan (*Weisberg et al.,* 1995a, *Krot et al.,* 2002b). Members of the CR clan have the following similar characteristics: (1) They are all metal-rich, but metal abundances vary widely among them (Ta-ble 1); (2) Fe,Ni-metal has a broad compositional range and is characterized by an approximately solar Ni/Co ratio; (3) they are highly depleted in moderately-volatile lithophile elements, with the depletions correlating with increasing volatility (Fig. 3); (4) their bulk O-isotopic compositions plot on or near the CR-mixing line on a three-isotope diagram (Plate 3a); (5) their bulk N-isotopic compositions have large positive anomalies in $\delta^{15}N$ ranging from 200‰ in the CR chondrites to >1000‰ in the CB and CH chondrites; (6) most of their anhydrous mafic silicates are Mg-rich (Fa and Fs <4 mol%); and (7) they contain heavily-hydrated matrix and/or matrix lumps composed of serpentine, saponite, sulfides, framboidal and platelet magnetites, and carbonates.

5.2. Ordinary (O) Chondrites: H-L-LL Chondrite Clan and Related Meteorites

Ordinary chondrites include the H, L, and LL groups. The O chondrites are the most common materials in our meteorite collections, constituting more than 85% of observed falls. They are characterized by a high abundance of large (millimeter-sized) chondrules with various textures and mineral compositions. Matrix abundances (10–15 vol%) are generally lower than in C chondrites; CAIs and AOAs are very rare (Table 1). Ordinary chondrites show a wide range in petrologic types from 3 to 6 (Fig. 2a–f), with several less-altered (petrologic types 3.0–3.1) members. Some of the least-metamorphosed O chondrites (e.g., Semarkona and Bishunpur) show evidence for aqueous alteration (e.g., *Hutchison et al.,* 1987; *Alexander et al.,* 1989; *Sears and Weeks,* 1991; *Sears et al.,* 1980, 1991). Hydrous alteration took place mainly in the matrix and in some cases chondrule mesostases (*Hutchison et al.,* 1987). The degree of alteration varies, and Bishunpur appears to be less altered than Semarkona (e.g., *Brearley and Jones,* 1998). The whole-rock Mg-normalized refractory lithophile abundances are <1× CI (Fig. 3) and their O-isotopic compositions plot above the TF line, sharply distinguishing them from the C chondrites (Plate 3a,b). In general, the H, L, and LL chondrites have similar, overlapping petrologic characteristics and O-isotopic compositions that indicate they are closely related members of a clan.

A number of parameters are used to resolve the H, L, and LL groups. Small, systematic differences among the O-chondrite groups are observed for metal abundances and chondrule size (Table 1, Fig. 2). On a plot of Co concentration in kamacite vs. fayalite content in olivine, there is a hiatus between H and L, but no hiatus between L and LL chondrites (*Rubin,* 1990). Olivine composition is not a reliable indicator of group for unequilibrated O chondrites of low petrologic type (<3.5), but is successfully used for distinguishing the equilibrated varieties (Table 1). Siderophile-element abundances can also be used to distinguish H, L, and LL (*Kallemeyn et al.,* 1989; *Sears et al.,* 1991).

Several O chondrites (Bjurböle, Cynthiana, Knyahinya, Qidong, Xi Ujimgin) are intermediate between L and LL chondrites, possibly indicating formation on a separate parent body (*Kallemeyn et al.,* 1989). Based on their siderophile-element abundances and olivine and kamacite compositions, Tieschitz and Bremervörde are intermediate between H and L and have been classified as H/L chondrites (*Kallemeyn et al.,* 1989).

The chemical properties of the Burnwell, Willaroy, Suwahib (Buwah), Moorabie, Cerro los Calvos, and Wray (a) ordinary chondrites plot on extensions of the H-L-LL trends of O chondrites toward more reducing compositions and may represent a fourth group. These meteorites have lower fayalite contents in olivine (Fa_{13-15}), ferrosilite in orthopyroxene, Co in kamacite (0.30–0.45 wt%), and FeO in their whole-chondrite chemical compositions and their bulk O-isotopic compositions are heavier than those of equilibrated H chondrites. However, their abundances of refractory lithophile elements, Zn and other elements, and their matrix/chondrule abundance ratios are typical of O chondrites (*Wasson et al.,* 1993; *McCoy et al.,* 1994; *Russell et al.,* 1998b). *Wasson et al.* (1993) argue that the low-FeO chondrites were originally normal H or L chondrites that have been altered by metamorphism in a highly reducing regolith. *McCoy et al.* (1994) and *Russell et al.* (1998b) suggested that reduction did not play a role in establishing their mineral compositions and they are samples from different parent bodies than those of the H, L, and LL chondrites.

5.3. Enstatite (E) Chondrites: EH-EL Chondrite Clan and Related Meteorites

Enstatite chondrites formed under highly reducing nebular conditions, as recorded by their mineralogy and mineral chemistry (e.g., *Keil,* 1968). They contain Fe-poor silicates and Si-bearing metal, and elements that are generally lithophile in most meteorite groups (Mn, Mg, Ca, Na, K) can behave as chalcophile elements. Enstatite ($MgSiO_3$) is the major silicate mineral in the chondrules. Chondrule olivine rarely occurs in low petrologic types (EH3 and EL3) chondrites and is essentially absent in the higher petrologic types. Additionally, they contain a variety of unusual sulfide, metal, and nitride phases that are essentially absent in other chondrite groups. These include oldhamite (CaS), niningerite [(Mg,Fe,Mn)S], alabandite [(Mn,Fe)S], osbornite (TiN), sinoite (Si_2N_2O), Si-rich kamacite, daubreelite ($FeCr_2S_4$), caswellsilverite ($NaCrS_2$), and perryite [$(Ni,Fe)_x(Si,P)$]. They are the only chondrites with O-isotopic compositions that plot on the TF line and close to the O composition of Earth and the Moon (Plate 3) (*Clayton et al.,* 1984).

The E chondrites are divided into EH and EL groups. The composition of the (Mg,Mn,Fe)S phase clearly distinguishes the EH and EL chondrites. The EH chondrites are more reduced than the EL chondrites and contain niningerite and various alkali sulfides (e.g., caswellsilverite, djerfisherite), while alabandite is characteristic for EL chondrites (e.g., *Keil,* 1968; *Lin and El Goresy,* 2002). In addition, EH chondrites have higher Si (2–3 wt%) in the Fe,Ni-metal, whereas EL chondrites generally have less than 1.0 wt% Si in the metal. Two E chondrites, LEW 87223 and Y 793225, are ungrouped (*Grossman et al.,* 1993; *Weisberg et al,* 1995b; *Lin and Kimura,* 1998). Lin and Kimura showed that Y 793225 sulfide has an intermediate composition between niningerite and alabandite and *Kimura et al.* (2005) referred to Y 793225 as "EI," because its mineral compositions are intermediate between EH and EL chondrites.

The EH and EL chondrites range in petrologic type from EH3 to EH5 and from EL3 to EL6. EH3 chondrites differ from EL3 in having higher modal abundances of sulfides

(7–16 vol% vs. 7–10 vol%), lower abundances of enstatite (56–63 vol% vs. 64–66 vol%), and more Si-rich (1.6–4.9 wt% vs. 0.2–1.2 wt%) and Ni-poor (2–4.5 wt% vs. 3.6–8.7 wt%) metal compositions. Lewis Cliff 87223 differs from both EH3 and EL3 chondrites in having the highest abundance of metal (~17 vol%) and FeO-rich enstatite, and the lowest abundance (~3 vol%) of sulfides (*Weisberg et al.,* 1995b).

Some meteorites are classified as E-chondrite melt rocks (*McCoy et al.,* 1995; *Weisberg et al.,* 1997b; *Rubin and Scott,* 1997). They are complex meteorites that have homogeneous, in some cases coarse-grained (millimeter-sized enstatite), achondritic textures and are chemically and O-isotopically similar to the E chondrites. Some are metal-rich with as much as 25 vol% metal. These E-chondrite melt rocks have been interpreted to be the result of impact melting on the E-chondrite-like parent bodies (*McCoy et al.,* 1995; *Rubin and Scott,* 1997) or internally derived melts (*Olsen et al.,* 1977; *Weisberg et al,* 1997b; *Patzer et al.,* 2001). Itqiy is an E-chondrite melt rock that is depleted in plagioclase and has a fractionated REE pattern that suggests it may be the product of partial melting (*Patzer et al.,* 2001).

5.4. R (Rumuruti-like) Chondrites

The R chondrites (56 are known) have refractory-lithophile-element abundances and O-isotopic compositions that are consistent with placing within the O-chondrite class, but this has not yet been established. They are quite different from the H, L, and LL chondrites. They are highly oxidized meteorites characterized by NiO-bearing, FeO-rich olivine (Fa_{37-40}) and nearly complete absence of Fe,Ni-metal. They have a very high abundance of matrix (~50 vol%) similar to some C chondrites. Most R chondrites are metamorphosed (petrologic type >3.6) and are brecciated. Although the refractory and moderately volatile-lithophile-element abundances of the R chondrites are similar to those of O chondrites (~0.95× CI), absence of significant depletions in Mn and Na and enrichment in volatile elements such as Ga, Se, S, and Zn distinguish R chondrites from H, L, and LL chondrites. The R chondrites have whole-rock $\Delta^{17}O$ values higher than for any other chondrite group (Plate 3) (*Weisberg et al.,* 1991; *Bischoff et al.,* 1994; *Rubin and Kallemeyn,* 1994; *Schulze et al.,* 1994; *Kallemeyn et al.,* 1996; *Russell,* 1998).

5.5. Ungrouped Chondrites

There are a number of chondrites that do not fit into any of the existing chondrite groups. These meteorites are referred to as ungrouped. In many cases, they have characteristics that are intermediate between chondrite groups. They may be anomalous members of chondrite groups or the first representatives of new groups.

Acfer 094 is a type 3 C-chondrite breccia with mineral compositions, petrologic characteristics, N-isotopic, and O-isotopic similarities to the CM and CO chondrites (*Newton et al.,* 1995; *Greshake,* 1997). Its bulk chemical composition is similar to that of CM chondrites, and O-isotopic composition and matrix modal abundance are similar to those of CO chondrites. In contrast to CM chondrites, Acfer 094 shows no evidence for aqueous alteration and has a different C-isotopic composition.

Adelaide is an ungrouped, type 3 C chondrite with affinities to the CM-CO clan, but appears to have escaped the thermal metamorphism and aqueous alteration commonly observed in the CM and CO chondrites (*Fitzgerald and Jones,* 1977; *Davy et al.,* 1978; *Kallemeyn and Wasson,* 1982; *Hutcheon and Steele,* 1982; *Kerridge,* 1985; *Brearley,* 1991; *Huss and Hutcheon,* 1992; *Krot et al.,* 2001b,c).

Coolidge and Loongana 001 comprise a C-chondrite grouplet with chemical, O-isotopic, and petrographic characteristics similar to those of the CV chondrites (*Kallemeyn and Wasson,* 1982; *Scott and Taylor,* 1985; *Noguchi,* 1994; *Kallemeyn and Rubin,* 1995). However, both meteorites are of higher petrologic type, 3.8–4, and have smaller chondrules and lower volatile-element abundances than CV chondrites. They may be part of the CK-CV clan.

The K-chondrite (Kakangari-like) grouplet consists of two meteorites, Kakangari and LEW 87232, that are chemically, mineralogically, and isotopically distinct from the O, E, and C classes (*Srinivasan and Anders,* 1977; *Davis et al.,* 1977; *Zolensky et al.,* 1989; *Brearley,* 1989; *Weisberg et al.,* 1996a). Lea County 002 may also be classified as a K chondrite (*Prinz et al.,* 1991; *Weisberg et al.,* 1996a). The K chondrites have high matrix abundance (70–77 vol%) like some C chondrites, metal abundances (6–9 vol%) similar to H chondrites, and average olivine (Fa_2) and enstatite (Fs_4) compositions are Fe-poor, indicating an oxidation state intermediate between H and E chondrites. The matrix of K chondrites is mineralogically unique and composed largely of enstatite (Fs_3) that is compositionally similar to that in the chondrules.

The K chondrites have refractory-lithophile- and volatile-element abundances similar to those of the O chondrites. Chalcophile-element abundances are between those of H and E chondrites. Their whole-rock O-isotopic compositions plot below the TF line on the three-isotope diagram (Plate 3), near the CR, CB, and CH chondrites (CR mixing line). K chondrites are excellent candidates for being the first members of a fourth major class of chondrites.

Kaidun is a complex chondritic breccia that appears to consist mainly of C-chondrite material, but also contains clasts of E and O chondrites. Several C-chondrite groups may be present (*Ivanov,* 1989; *Clayton and Mayeda,* 1999; *Zolensky and Ivanov,* 2003).

Lewis Cliff 85332 is an ungrouped, metal-rich type 3 C-chondrite breccia with chemical, O-isotopic, and petrographic characteristics similar to those of the CR chondrites and is considered a member of the CR clan (*Rubin and Kallemeyn,* 1990; *Prinz et al.,* 1992; *Weisberg et al.,* 1995a; *Brearley,* 1997).

MacAlpine Hills (MAC) 87300 and MAC 88107 comprise a grouplet of type 2–3 C chondrites with bulk chemical compositions intermediate between those of CO and CM chondrites, and O-isotopic compositions similar to those of the CK chondrites (*Clayton and Mayeda,* 1999). Contrary to the CM chondrites, which have low natural or induced thermoluminescence (TL), MAC 87300 and MAC 88107 have high TL (*Sears et al.,* 1991). They experienced very minor aqueous alteration that resulted in formation of saponite, serpentine, magnetite, Ni-bearing sulfides, fayalite, and hedenbergite (*Krot et al.,* 2000c).

Ningqiang is an ungrouped C chondrite that shares many petrologic as well as O-isotopic and TL characteristics with CV and CK chondrites (*Rubin et al.,* 1988; *Kallemeyn et al.,* 1991; *Kallemeyn,* 1996; *Koeberl et al.,* 1987; *Mayeda et al.,* 1988; *Guimon et al.,* 1995). As a result, it has been classified as CV-anomalous (*Koeberl et al.,* 1987; *Rubin et al.,* 1988, *Weisberg et al.,* 1996b) and as CK-anomalous (*Kallemeyn et al.,* 1991). Whichever is correct, it is clearly a member of the CK-CV clan.

Belgica 7904, Y 86720, Y 82162, and Dhofar 225 have petrographic, mineralogical, and chemical affinities to the CM and CI chondrites but have distinct O-isotopic compositions and appear to have experienced aqueous alteration followed by reheating and dehydration (e.g., *Akai,* 1988; *Tomeoka et al.,* 1989; *Tomeoka,* 1990; *Paul and Lipschutz,* 1990; *Bischoff and Metzler,* 1991; *Ikeda,* 1992; *Ikeda and Prinz,* 1993; *Hiroi et al.,* 1993a, 1996; *Clayton and Mayeda,* 1999; *Ivanova et al.,* 2002).

Sahara 00182 (SAH 00182) is an ungrouped metal-rich C3 chondrite (*Grossman and Zipfel,* 2001; *Weisberg,* 2001). It contains large (up to 2 mm diameter), metal-rich, multilayered chondrules that are texturally similar to those in the CR chondrites. However, it differs from the CR chondrites in that its chondrules contain troilite (FeS), and the chondrules and matrix lack the (secondary alteration) hydrous phyllosilicates that are characteristic of many CR chondrites. Its whole-chondrite O-isotopic composition plots on the C3 (carbonaceous chondrite anhydrous mineral) mixing line (*Grossman and Zipfel,* 2001).

Tagish Lake is an ungrouped C chondrite that has similarities to CI and CM chondrites (*Brown et al.,* 2000; *Grady et al.,* 2002; *Mittlefehldt,* 2002; *Zolensky et al.,* 2002). It is a type 2, hydrously altered chondrite composed of two lithologies that have been altered to different degrees (*Zolensky et al.,* 2002). Its whole-rock refractory-lithophile-element abundances are similar to those of CM chondrites and its moderately volatile and volatile-lithophile-element abundances are intermediate between CI and CM (*Mittlefehldt,* 2002; *Friedrich et al.*, 2002). Siderophile and chalcophile elements are also similar to both CI and CM (*Mittlefehldt,* 2002; *Friedrich et al.*, 2002).

6. PROPERTIES OF PRIMITIVE ACHONDRITES

6.1. Ureilites

The ureilites are a major group of primitive achondrites with almost 200 known specimens; most are monomict and some are polymict breccias. They have textures, mineralogies, and lithophile-element chemistries that suggest they are highly fractionated rocks from an achondrite parent body, yet their O-isotopic compositions do not follow a mass-dependent fractionation trend characteristic of planetary differentiation. Instead, they plot along the CCAM line, suggesting a possible relationship to the CV chondrites (Plate 3) (e.g., *Clayton and Mayeda,* 1988).

Ureilites are olivine-pyroxene rocks with interstitial C (graphite and microdiamonds) mixed with metal, sulfides, and minor silicates. There are three major types of ureilites: (1) olivine-pigeonite, (2) olivine-orthopyroxene, and (3) polymict ureilites. Most ureilites are essentially devoid of feldspar with the exception of the polymict ureilites and some rare exceptions. The ureilites that have not been heavily modified by shock typically display large, elongate olivine and pyroxene grains (about 1 mm in size) that form triple junctures and have curved intergranular boundaries (Plate 1a). Some ureilites display a mineral fabric suggestive of crystal settling and/or compaction.

The presence of interstitial C adds to the challenge of understanding ureilite petrogenesis. It contains trapped noble gases in abundances similar to those of primitive chondrites. One would expect these gases to have been driven off if the ureilites formed during high-temperature igneous processes. Late-stage injection may provide a better explanation for the retention of the noble gases in the C. Another question concerning the C is the relationship of the diamonds to the graphite. Are the diamonds shock-formed or the result of chemical vapor deposition in the early solar nebula? One well-known characteristic of the ureilites is that olivine in contact with the graphite has experienced reduction of Fe. Evidence of the reduction reaction is visibly displayed in olivine grains having reduced rims that are composed of Fe-poor olivine speckled with tiny blebs of low-Ni metal.

The polymict ureilites (e.g., North Haig, Nilpena, EET 83309, EET 87720, Dar al Gani (DaG) 319, and DaG 164/165) are polymict breccias that contain a variety of rock (lithic) and mineral fragments. The clasts that they contain include monomict ureilite fragments, feldspar-bearing lithic clasts, isolated mineral fragments, chondrite and chondrule fragments, and dark chondritic inclusions (*Ikeda and Prinz,* 2000; *Ikeda et al.,* 2000; *Kita et al.,* 2004).

Because of their gross petrologic similarities to terrestrial ultramafic rocks, ureilites have been perceived to be products of internal magmatic differentiation. The olivine-augite-orthopyroxene ureilites appear to resemble magmatic cumulates (e.g., *Goodrich et al.,* 1987, 2001). The olivine-pigeonite ureilites have been interpreted to be partial melt residues (e.g., *Scott et al.,* 1993; *Warren and Kallemeyn,* 1992; *Goodrich et al.,* 2002). The missing basaltic component has been explained as being largely lost through explosive volcanism on the ureilite parent body(ies); it may have been preserved in feldspathic material in polymict ureilites [see *Goodrich* (1992) for a review of models for ureilite petrogenesis]. The observed variations in whole-rock O-isotopic compositions of ureilites may suggest that ureilites formed by melting of chondritic material in a number of isolated magma systems and not from a common magma source. These observations have led to other models for the origin of ureilites, including a planetesimal-scale collision model involving extraction of partial melt and recrystallization of the mafic minerals to form the ureilites (*Takeda,* 1987) and an origin as primitive aggregates of olivine condensates (*Kurat,* 1988). Most recently, ureilites have been interpreted to be mantle rock from a partially melted asteroid. The ureilite parent body (UPB) may have been stratified in magnesian content and pyroxene type and abundance as a result of the pressure dependence of C-redox reactions (*Goodrich et al.,* 2004). Lack of magmatic homogenization may have preserved the primitive O-isotopic signature of the ureilites.

6.2. Acapulcoite-Lodranite Clan

The acapulcoites and lodranites may be considered a primitive achondrite clan. The acapulcoites are fine-grained (150–230 μm), equigranular rocks with approximately chondritic abundances of olivine, pyroxene, plagioclase, metal, and troilite, while lodranites are coarse-grained (540–700 μm) olivine, pyroxene rocks depleted in troilite and plagioclase (Table 2, Plate 1b) (*Nagahara,* 1992; *McCoy et al.,* 1996; *Mittlefehldt et al.,* 1998). The whole-rock O-isotopic compositions of acapulcoites and lodranites are similar and they form a cluster with significant spread on the three-isotope diagram (Plate 3c,d). Rare relict chondrules have been reported in several acapulcoites (e.g., *Schultz et al.,* 1982; *McCoy et al.,* 1996), suggesting that they are close to their chondrite precursor and therefore designated primitive achondrite. Acapulcoites likely formed from very low degrees of partial melting at temperatures just above the Fe,Ni-FeS cotectic (~950°C), while lodranites are residues from higher degrees of partial melting and basaltic melt removal at temperatures of ~1250°C.

Elephant Moraine 84302 and Graves Nunataks (GRA) 95209 are texturally and mineralogically intermediate between acapulcoites and lodranites (*Takeda et al.,* 1994; *Mittlefehldt et al.,* 1996; *McCoy et al.,* 1997a,b; *Mittlefehldt and Lindstrom,* 1998; *McCoy and Carlson,* 1998). Lewis Cliff 86220 consists of an acapulcoite host with intruded Fe,Ni-FeS and gabbroic partial melts (*McCoy et al.,* 1997b). Frontier Mountain 93001 may be an impact melt from the acapulcoite-lodranite parent body (*Burroni and Folco,* 2003).

6.3. Brachinites

Brachinites are olivine-rich achondrites that show diverse petrology with some having differences in their bulk chemistry and O-isotopic compositions. The brachinites are medium- to coarse-grained (0.1–2.7 mm) equigranular olivine-rich (74–98%) rocks (Plate 1d) that contain minor augite (4–15%), plagioclase (0–10%), orthopyroxene (traces), chromite (0.5–2%), Fe-sulfides (3–7%), phosphates, and Fe,Ni-metal. Lithophile-element abundances in Brachina are close to chondritic and are unfractionated and thus brachinites are considered primitive achondrites. Other brachinites show depletions in Al, Ca, Rb, K, and Na (*Mittlefehldt et al.,* 1998). Siderophile-element abundances are variable in brachinites (~0.1–1× CI). The petrologic and geochemical differences among the different brachinites suggest that they may not have all formed by the same process or on the same parent body (*Mittlefehldt et al.,* 1998; *Mittlefehldt,* 2004).

Tafassasset and LEW 88763 are two meteorites that are petrologically and chemically similar to the brachinites, but Tafassasset is metal-rich (~8 vol%) and both Tafassaset and LEW 88763 have O-isotopic compositions that are far removed from the region in which most brachinites plot on the three-isotope diagram (*Nehru et al.,* 2003). Instead, they plot close to the CR chondrites on the three-isotope diagram (*Bourot-Denise et al.,* 2002). They may have formed by a process similar to that of the other brachinites but clearly represent a different nebular reservoir and parent body. The petrologic and chemical differences among the brachinites suggest that they should not be considered a meteorite group. They may need to be divided into a number of groups.

The origin of brachinites remains controversial. They may represent oxidized and recrystallized chondritic materials, partial melt residues, or igneous cumulates (*Warren and Kallemeyn,* 1989; *Nehru et al.,* 1996; *Mittlefehldt,* 2004). Mittlefehldt argued that the brachinites are igneous cumulate rocks from a differentiated asteroid and are therefore achondrites and not primitive achondrites.

6.4. Winonaites and IAB and IIICD Iron Clan

Winonaites have generally chondritic mineralogy and chemical composition, but achondritic, recrystallized textures (Plate 1e). They are fine- to medium-grained, mostly equigranular rocks, but some (Pontlyfni and Mount Morris) have regions interpreted to be relic chondrules based on their textures. The mineral compositions of winonaites are intermediate between those of E and H chondrites, and Fe,Ni-FeS veins that may represent the first partial melts of a chondritic precursor material are common (*Benedix et al.,* 1998).

Silicate inclusions in IAB and IIICD irons consist of variable amounts of low-Ca pyroxene, high-Ca pyroxene, olivine, plagioclase, troilite, graphite, phosphates, and Fe,Ni-metal, and minor amounts of daubreelite and chromite. They have similar mineral compositions to winonaites and the O-isotopic compositions of the IAB silicates are similar to those of the winonaites (Plate 3c,d) (*Kracher,* 1982; *Kallemeyn and Wasson,* 1985; *Palme et al.,* 1991; *Kimura et al.,* 1992; *McCoy et al.,* 1993; *Choi et al.,* 1995; *Yugami et al.,* 1997; *Takeda et al.,* 1997; *Clayton and Mayeda,* 1996; *Benedix et al.,* 1998, 2000). Based on these similarities, it was inferred that the silicate inclusions in IAB and winonaites formed on the same or similar parent body; it is less clear whether the IIICD irons are from the same parent body (*McCoy et al.,* 1993; *Choi et al.,* 1995; *Benedix et al.,* 1998, 2000).

7. Properties of Achondrites

Under the heading of achondrite, we include the metal-poor stony meteorites, as well as the stony irons and irons; each of these types contains several meteorite groups and ungrouped members. We apply the term "clan" (which has not been previously used for achondrites) to some achondrite groups.

7.1. Angrites

The angrites are a perplexing group of achondrites. They are medium- to coarse-grained (up to 2–3 mm) vesicular igneous rocks of generally basaltic composition but unusual mineralogies consisting of Ca-Al-Ti-rich pyroxene, Ca-rich olivine, and anorthitic plagioclase (Plate 2e) with accessory spinel, troilite, kirschteinite, whitlockite, titanomagnetite, and Fe,Ni metal (e.g., *McKay et al.,* 1988, 1990; *Yanai,* 1994; *Mittlefehldt et al.,* 1998). Interestingly, the type meteorite — Angra dos Reis — is unlike all other angrites. It is essentially a monomineralic rock consisting of ~90% Ca-Al-Ti-rich pyroxene (e.g., *Prinz et al.,* 1977). Oxygen-isotopic compositions of the angrites are similar to those of the HED meteorites, brachinites, and mesosiderites (Plate 3c,d). However, the unusual mineralogies and compositions of the angrites suggest that they are not related to any of these meteorite groups. Angrites are the most alkali-depleted basalts in the solar system and have low abundances of the moderately volatile element Ga. However, they are not notably depleted in the highly volatile elements Br, Se, Zn, In, and Cd compared to lunar basalts and basaltic eucrites. It has been suggested that the angrites formed from melts enriched in a CAI component (*Prinz et al.,* 1977). Partial melting experiments of a CV chondrite at IW + 1 (*Jurewicz et al.,* 1993) produced melt compositions similar to some bulk angrites, although olivine xenocrysts appear to be present in other angrites. It has recently been suggested (*Campbell and Humayun,* 2005) that the volatile-depleted IVB irons may have originated on the same parent body as the angrites. The origin of Angra dos Reis is uncertain.

7.2. Aubrites

The aubrites are highly reduced achondrites with close affinities to the E chondrites and are often referred to as enstatite achondrites. All the aubrites are breccias with the exception of Shallowater, which has an igneous texture (*Keil et al.,* 1989). They consist mostly (~75–95 vol%) of nearly FeO-free enstatite, with minor albitic plagioclase and nearly FeO-free diopside (Plate 2) and forsterite (e.g., *Watters and Prinz,* 1979). Other phases include accessory Si-bearing Fe,Ni metal (e.g., *Casanova et al.,* 1993) and a variety of sulfides similar to those in the E chondrites (*Watters and Prinz,* 1979; *Keil and Brett,* 1974; *Wheelock et al.,* 1994). The aubrite parent body likely formed by partial melting, removal of that melt, continued melting to very high temperatures (~1600°C), and crystallization, although Shallowater may require an impact mixing origin on a distinct parent body. The highly reduced state, O-isotopic composition, and presence of ferroan alabandite in the aubrites may link them to the EL chondrites. However, they are probably derived from different asteroidal parent bodies (*Keil,* 1989).

7.3. HED Clan: Howardites, Eucrites, Diogenites

Eucrites and diogenites are basalts and orthopyroxene cumulates, respectively, thought to have originated in near-surface flows or intrusives on a common asteroidal parent body. Some of their textural variations are displayed in Plate 2. Many HED meteorites are impact-produced monomict or polymict breccias. Howardites are polymict breccias containing eucritic and diogenitic components (*Fredriksson and Keil,* 1963; *Delaney et al.,* 1984). HED meteorites have similar whole-rock O-isotopic compositions (Plate 3) and similar Fe/Mn ratios in their pyroxenes, suggesting a close genetic relationship (e.g., *Papike,* 1998). Similarities in the mineralogical compositions of the HEDs and the surface mineralogy of asteroid 4 Vesta, as determined by ground-based visible and near-infrared spectroscopy, suggest that the HEDs are probably impact ejecta off this asteroid (*McCord et al.,* 1970; *Binzel and Xu,* 1993). Although classified as a eucrite, Northwest Africa (NWA) 011 differs in its O-isotopic composition and must have originated on a separate parent body (*Yamaguchi et al.,* 2002). Therefore, it should be considered an ungrouped achondrite. The finding of eucrites from additional parent bodies is reassuring, given the identification of core materials (iron meteorites) from dozens of differentiated asteroids.

7.4. Mesosiderites

Mesosiderites are breccias composed of approximately equal proportions of silicates and Fe,Ni-metal plus troilite. Their silicate fraction consists of mineral and lithic clasts in a fine-grained fragmental or igneous matrix (e.g., *Floran,* 1978; *Floran et al.,* 1978; *Hewins,* 1983; *Mittlefehldt et al.,* 1998). The lithic clasts are largely basalts, gabbros, and pyroxenites with minor amounts of dunite and rare anorthosite (*Scott et al.,* 2001). Mineral clasts consist of coarse-grained orthopyroxene, olivine, and plagioclase; Fe,Ni-metal in mesosiderites is mostly in the form of millimeter or submillimeter grains that are intimately mixed with similarly sized silicate grains.

Chaunskij is an unusual, highly metamorphosed, heavily shocked, metal-rich, cordierite-bearing mesosiderite (*Petaev et al.,* 1993).

7.5. Pallasites

Pallasites are essentially olivine (35–85 vol%)-metal (and troilite) assemblages. The term pallasite has been used to indicate a group of meteorites, but there are four separate kinds of pallasites that are distinguished by differences in their silicate mineralogy and composition, metal composition, and O-isotopic composition. These include (1) the main-group pallasites, (2) the Eagle Station grouplet, (3) the pyroxene-pallasite grouplet (e.g., *Boesenberg et al.,* 1995; *Mittlefehldt et al.,* 1998), and (4) the Milton ungrouped pallasite. The differences among the pallasite groups suggest that they originated on separate asteroidal bodies. Aside from textural similarities, the pallasites appear to be unrelated and therefore the term pallasite does not meet the definition of clan or group. The term pallasite is best thought of as a descriptive/textural term for olivine-metal rocks.

Although O-isotopic compositions of the main-group pallasites are similar to those of the HEDs (Plate 3c,d) and their metal is close in composition to that of Ni-rich IIIAB irons, no genetic relationship between these meteorite groups has been established.

The Eagle Station grouplet is mineralogically similar to the main-group pallasites, but has more ferroan and Ca-rich olivine (*Davis and Olsen,* 1991; *Buseck,* 1977) and different O-isotopic compositions. Their metal is close in composition to IIF irons and has higher Ni and Ir contents than those of the main-group pallasites (*Scott,* 1977). *Kracher et al.* (1980) suggested that Eagle Station grouplet pallasites and IIF irons formed in a close proximity in the solar nebula region but not on the same parent asteroid.

The pyroxene-pallasite grouplet consists of Vermillion and Y 8451 (*Boesenberg et al.,* 1995). These meteorites contain ~14–63 vol% olivine, 30–43 vol% metal, 0.7–3 vol% pyroxene, 0–1 vol% troilite, and minor whitlockite. The occurrence of millimeter-sized pyroxenes, their metal compositions, and their O-isotopic compositions distinguish them from the main-group and Eagle Station grouplet pallasites (*Hiroi et al.,* 1993b; *Boesenberg et al.,* 1995; *Yanai and Kojima,* 1995). One of the most perplexing characteristics of the pallasites is that the pyroxene pallasites fall on a long mixing line drawn from the main-group pallasites to the Eagle Station pallasites (Fig. 3) (*Boesenberg et al.,* 1995). The significance of this line, if any, is unclear.

7.6. Martian Meteorite Clan

The martian meteorites are crustal rocks from Mars. They have been called SNC (an acronym for the shergottites-nakhlites-chassignite groups) and are volcanic (basaltic shergottites and nakhlites) and plutonic (lherzolitic and olivine-phyric shergottites, Chassigny, ALH 84001) rocks. Their O-isotopic compositions differ from those of other meteorites and they define a unique slope-0.52 mass fractionation line (Plate 3), suggesting their origin from the same, differentiated parent body. The relatively young crystallization age of some of these meteorites (<1.3 Ga) suggests derivation from a planet-sized body. Most importantly, gases trapped in impact-produced glass inclusions and veins in some martian meteorites are compositionally and isotopically similar to the martian atmosphere, as determined by the Viking landers, demonstrating that these meteorites are from the planet Mars (e.g., *McSween and Treiman,* 1998; *Bogard and Johnson,* 1983; *Marti et al.,* 1995; *Nyquist et al.,* 2001; *McSween,* 2002). The acronym SNC does not adequately describe the current variety of martian meteorites in the world's collections. Recent new finds from Antarctica and hot deserts have brought the total number of SNC meteorites to ~35 and have yielded the new types of lherzolitic shergottites (lherzolites) and the famous orthopyroxenite ALH 84001.

The martian meteorites have crystallization ages of a few hundred million years, with the exception of ALH 84001, which has an age of ~4.5 Ga (e.g., *Nyquist et al.,* 2001). Allan Hills 84001 consists mainly of coarse, up to 6 mm long, orthopyroxene crystals (96%), with 2% chromite, 1% plagioclase (maskelynite), 0.15% phosphate, and accessory phases such as olivine, augite, pyrite, and carbonates (*Mittlefehldt,* 1994). One of the most intriguing aspects of this meteorite is that it has been postulated to contain nanofossils, as well as other evidence of life, of martian origin (*McKay et al.,* 1996).

7.7. Lunar Meteorite Clan

Lunar meteorites are classified by their texture, petrology, and chemistry. They include a wide range of rock types including brecciated and unbrecciated mare basalts, mixed mare/highlands breccias, highland regolith breccias, and highland impact-melt breccias [e.g, see *Papike et al.* (1998) for a more detailed description of the lunar meteorites]. The lunar meteorites share many common mineralogical (e.g., Fe/Mn ratios in pyroxenes), bulk chemical, and O-isotopic compositions with the Apollo and Luna samples. However, they almost certainly come from areas not sampled by these missions and some lunar meteorites (e.g., NWA 773), providing sample lithologies unknown from the mission samples.

7.8. Iron Groups and Ungrouped Irons

The classification of irons, perhaps more than any other group of meteorites, is historically deep. The well-known Widmanstätten structure was first recognized in the early nineteenth century by W. Thompson and A. von Widmanstätten. In the 1870s and 1880s, G. Tschermak introduced the structural classification of iron meteorites still used today, dividing them into octahedrites (O), hexahedrites (H), and ataxites [(D), for the German "dicht" (dense)]. He further subdivided the octahedrites by the width of their kamacite bands into Of ("fein," fine), Om ("mittlere," medium), and Og ("gros," large). The structural classification remains a useful tool for the characterization of irons.

In the 1950s, the modern chemical classification of iron meteorites emerged. *Goldberg et al.* (1951) first introduced the concept of grouping irons of similar composition using Ni and Ga. *Lovering et al.* (1957) and *Wasson and Kimber-*

lin (1967) expanded on this idea. It was the former who recognized four groups (labeled with Roman numerals I–IV) in order of decreasing Ga and Ge concentrations. These elements are the most volatile siderophile elements and tend to be strongly fractionated between groups (*Wai and Wasson,* 1977) by nebular processes. The widespread application of the chemical classification (e.g., Fig. 4) stems largely from the work of J. Wasson and coworkers, who have systematically analyzed and classified nearly every iron meteorite known over the last 30 years. These authors introduced letters to the Roman numerals (e.g., IA, IVB) and, with increasing numbers of analyses, recognized that some of these groups were not independent, leading to groups like IAB (a grouping of the former IA and IB). Currently, there are 13 recognized groups, although *Wasson and Kallemeyn* (2002) argued that IAB and IIICD form a single group termed the IAB complex. While the geochemical differences between the groups owe their existence to nebular processes, the geochemical trends within 10 of these groups are consistent with fractional crystallization, with each group forming in the core of a separate asteroid (*Scott,* 1972). The geochemical trends in IAB, IIICD, and IIE irons differ substantially and have been the subject of intense debate, with suggested origins including crystallization of a S-rich core (*Kracher,* 1982), partial asteroidal differentiation followed by impact mixing (*Benedix et al.,* 2000), and formation in isolated impact melt pools (*Wasson and Kallemeyn,* 2002). These groups are also noteworthy in containing silicate inclusions best described as primitive achondritic. In the case of IIE irons, silicates in some members (e.g., Netschaëvo) have strong similarities to H chondrites. The term "nonmagmatic" has been widely applied to IAB, IIICD, and IIE irons, stemming from the 1970s-era belief that these meteorites could not have formed from melts. This idea has been fully refuted and the term should be discarded.

Although the geochemical classification has largely supplanted the older structural classification, owing largely to the lack of practitioners of the latter, it is noteworthy that the two often correlate and the geochemical classification of many meteorites can be inferred from the structure, including the IVA irons (fine octahedrites), the IIAB irons (coarse octahedrites), and many IAB irons (medium octahedrites often with silicate inclusions). This overlap primarily reflects the control of geochemistry (particularly bulk Ni concentration) on the growth of the Widmanstätten structure.

Finally, about 15% of all iron meteorites do not fit into the well-described groups. This percentage increases to about 40% in the Antarctic meteorites (*Wasson,* 1990), where smaller sample masses may more closely reflect the true rate of incoming material. The origin of these meteorites is not well understood. Some may represent extreme fractional crystallization products of well-known groups and others could be impact melt products. However, some may represent the only sample of an asteroid. *Wasson* (1990) estimated that 30–50 asteroids may be represented in the ungrouped irons and it is significant to note that this may be as many as all other groups of meteorites combined (*Burbine et al.,* 2002).

8. SYNTHESIS

8.1. Alternative Scheme for Meteorite Classification

Comparison of the mineralogical-petrological properties with O-isotopic compositions of meteorites has revealed a number of interesting relationships. Some meteorites with differing petrologic and whole-chondrite chemical compositions have similar O-isotopic compositions. This has been interpreted to indicate derivation of these groups (or their components), which have grossly different petrographic features from the same local nebular reservoir or parent asteroid. For example, the CR, CH, and CB chondrites, which have very different chondrule sizes and textures and metal abundances, are considered part of the CR clan and may have formed from the same O-isotope reservoir. The howardites, eucrites, and diogenites (HEDs) are considered an association, largely because howardites are a physical mixture of eucritic and diogenitic clasts. Oxygen isotopes revealed a shared signature with main-group pallasites and IIIAB irons, suggesting a possible genetic relationship. Although this may indicate derivation from a common nebular reservoir, spectroscopic and dynamic studies that point to the intact, differentiated asteroid 4 Vesta as the parent of HEDs (*McCord et al.,* 1970) would require a separate parent body or bodies for main-group pallasites and IIIAB irons.

On the other hand, some meteorites with strong petrologic similarity, suggesting group affiliation, plot in different regions of the O-isotope diagram, suggesting derivation from different O reservoirs. For example, Tafassasset and LEW 88763 are petrologically and chemically similar to the brachinites, but their O-isotopic compositions are far removed from the region in which most brachinites plot on the O three-isotope diagram (*Nehru et al.,* 2003). Eagle Station pallasites, pyroxene pallasites, main-group pallasites, and the ungrouped pallasite Milton are petrologically similar, but their O-isotopic compositions differ dramatically (Plate 3), suggesting that they represent separate parent bodies and do not belong to the same group.

The disagreement between O-isotopic compositions and petrologic-chemical studies — both in terms of pointing to differences between petrologically similar materials and possibly between petrologically disparate materials — has produced most of the debate about the classificational structure over the last few decades. The classificational scheme illustrated in Fig. 1 stems largely from chemical-petrological studies, and perhaps only in the case of C chondrites and a few groups of achondrites (e.g., acapulcoite-lodranites vs. winonaites) do the O isotopes play a prominent role. It is possible to develop an entirely different classificational

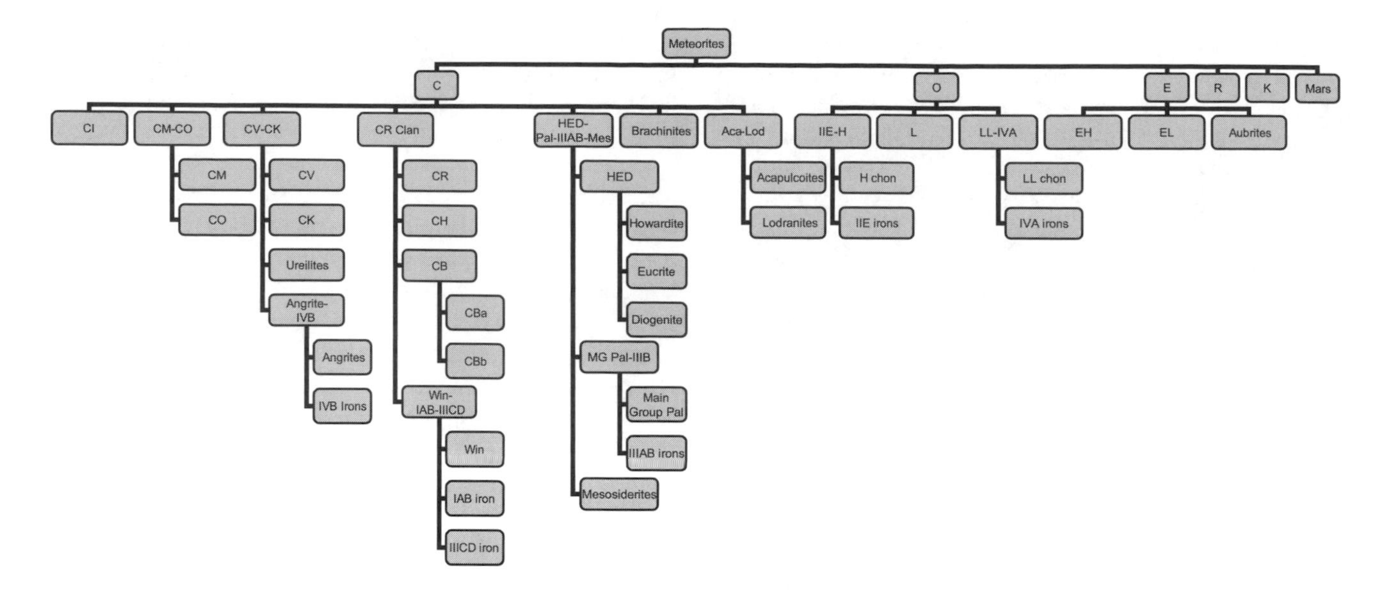

Fig. 8. Alternative classification scheme in which meteorites are linked by origins from common parent bodies or derivation from common nebular reservoirs. Some links are based on petrology, chemistry, and O-isotopic compositions (e.g., enstatite chondrites and aubrites), while in other cases, the link is based almost completely on O isotopes (e.g., IVA irons and LL chondrites). We note that this scheme is highly interpretive and is just one of a number of possibilities.

scheme and we illustrate one such scheme in Fig. 8. In such a scheme, meteorites are grouped to point out origins from common parent bodies or derivation from common nebular reservoirs or precursor materials. In this sense, we find groupings between chondrites and irons (e.g., IVA-LL, IIE-H), between achondrites and irons (e.g., angrites-IVB), and between chondrites and achondrites (CV-ureilites, EH-EL-aubrites). The problem with such a scheme is that it is highly interpretive. In some cases, links are based on petrology, chemistry, and O-isotopic compositions (e.g., E chondrites and aubrites), while in other cases the link rests almost completely on O isotopes (e.g., IVA irons and LL chondrites). In addition, it is unclear what, if anything, these links actually represent. In some cases, a common parent body may be indicated (e.g., angrites-IVB irons), while in others, the link may only reflect derivation of a particular achondrite or iron from a precursor similar, but not identical, to the known chondrite (e.g., EH-EL-aubrites). Such a diagram can provide some interesting insights, such as the apparent paucity of achondrites and irons apparently derived from the abundant O chondrites, while most achondrites and irons appear linked to C chondrites. Even in this case, care must be applied, since this may be essentially an artifact of defining meteorites below the terrestrial fractionation line as C chondrites. If such a simplistic link were applied to the terrestrial planets, Earth would be linked to E chondrites and aubrites, while Mars would derive from a yet-undiscovered chondrite more akin to O chondrites. Neither of these conclusions appears valid at this time. Given these uncertainties, and the ongoing debate about meteorite origins reflected in the remainder of this book, we find it unlikely that such a classification scheme will supplant the more traditional version (Fig. 1) in the foreseeable future, although knowledge of the possible links is a key piece of information for any worker new to the field.

8.2. The Future of Classification

The evolution of meteorite classification necessitates a continuing influx of new samples and continuing efforts to determine their petrologic characteristics and bulk chemical and O-isotopic compositions. While classification schemes will continue to evolve, in both the groups of meteorites and the techniques used to classify them, one of the ultimate goals of classification is to put meteorites into geologic context. In many respects, most classification is simply an effort to deduce the kind of information that the terrestrial field geologist knows when they collect their samples (where the samples came from and what their spatial relationship was to the rocks around it). In the next few decades, it is quite possible that missions to asteroids, Mars, the Moon, and comets will allow us to place meteorites in geologic context and, ultimately, the goal should be to say this meteorite came from this asteroid/type or asteroid/family of asteroids.

Acknowledgments. Reviews by M. Grady, J. Grossman, and D. Lauretta and editorial handling by L. Leshin are greatly appreciated and led to major improvements in this chapter. This work was supported by NASA Grants NAG5-11546 (M.K.W., P.I.), NAG5-13464 (T.J.M., P.I.), NAG5-10610, NAG5-12882 (A.N.K., P.I.), and NAG5-11591 (K. Keil, P.I.). This is Hawai'i Institute of Geophysics and Planetology Publication No. 1424 and School of Ocean and Earth Science and Technology Publication No. 6709.

REFERENCES

Aléon J., Krot A. N., and McKeegan K. D. (2002) Calcium-aluminum-rich inclusions and amoeboid olivine aggregates from the CR carbonaceous chondrites. *Meteoritics & Planet. Sci., 37,* 1729–1755.

Akai J. (1988) Incompletely transformed serpentine-type phyllosilicates in the matrix of Antarctic CM chondrites. *Geochim. Cosmochim. Acta, 52,* 1593–1599.

Alexander C. M. O'D., Hutchison R., and Barber D. J. (1989) Origin of chondrule rims and interchondrule matrices in unequilibrated ordinary chondrites. *Earth Planet. Sci. Lett., 95,* 187–207.

Anders E. and Grevesse N. (1989) Abundances of elements: Meteoritic and solar. *Geochim. Cosmochim. Acta, 53,* 197–214.

Anders E. and Zinner E. (1993) Interstellar grains in primitive meteorites: Diamond, silicon carbide, and graphite. *Meteoritics, 28,* 490–514.

Beckett J. R., Connolly H. C., and Ebel D. S. (2006) Chemical processes in calcium-aluminum-rich inclusions: A mostly CMAS view of melting and crystallization. In *Meteorites and the Early Solar System II* (D. S. Lauretta and H. Y. Mc-Sween Jr., eds.), this volume. Univ. of Arizona, Tucson.

Benedix G. K., McCoy T. J., Keil K., Bogard D. D., and Garrison D. H. (1998) A petrologic and isotopic study of winonaites: Evidence for early partial melting, brecciation, and metamorphism. *Geochim. Cosmochim. Acta, 62,* 2535–2553.

Benedix G. K., McCoy T. J., Keil K., and Love S. G. (2000) A petrologic study of the IAB iron meteorites: Constraints on the formation of the IAB-winonaite parent body. *Meteoritics & Planet. Sci., 35,* 1127–1141.

Bernatowicz T. J. and Zinner E. (1997) *Astrophysical Implications of the Laboratory Study of Presolar Materials.* American Institute of Physics, New York. 750 pp.

Binzel R.P. and Xu S. (1993) Chips off of asteroid 4 Vesta: Evidence for the parent body of basaltic achondrite meteorites. *Science, 260,* 186–191.

Bischoff A. and Metzler K. (1991) Mineralogy and petrography of the anomalous carbonaceous chondrites Yamato-86720, Yamato-82162, and Belgica-7904. *Proc. NIPR Symp. Antarct. Meteorites, 4,* 226–246.

Bischoff A., Palme H., Ash R. D., Clayton R. N., Schultz L., Herpers U., Stöffler D., Grady M. M., Pillinger C. T., Spettel B., Weber H., Grund T., Endreß M., and Weber D. (1993a) Paired Renazzo-type (CR) carbonaceous chondrites from the Sahara. *Geochim. Cosmochim. Acta, 57,* 1587–1603.

Bischoff A., Palme H., Schultz L., Weber D., Weber H., and Spettel B. (1993b) Acfer 182 and paired samples, an iron-rich carbonaceous chondrite: Similarities with ALH85085 and relationship to CR chondrites. *Geochim. Cosmochim. Acta, 57,* 2631–2648.

Bischoff A., Geiger T., Palme H., Spettel B., Schultz L., Scherer P., Loeken T., Bland P., Clayton R. N., Mayeda T. K., Herpers U., Meltzow B., Michel R., and Dittrich-Hannen B. (1994) Acfer 217; a new member of the Rumuruti chondrite group (R). *Meteoritics, 29,* 264–274.

Bischoff A., Scott E. R. D., Metzler K., and Goodrich C. A. (2006) Nature and origins of meteoritic breccias. In *Meteorites and the Early Solar System II* (D. S. Lauretta and H. Y. McSween Jr., eds.), this volume. Univ. of Arizona, Tucson.

Boesenberg J. S., Prinz M., Weisberg M. K., Davis A. M., Clayton R. N., Mayeda T. K., and Wasson J. T. (1995) Pyroxene-pallasites: A new pallasite grouplet (abstract). *Meteoritics, 30,* 488–489.

Bogard D. D. and Johnson P. (1983) Martian gases in an Antarctic meteorite. *Science, 221,* 651–654.

Bourot-Denise M., Zanda B., and Javoy M. (2002) Tafassesset: An equilibrated CR chondrite (abstract). In *Lunar and Planetary Science XXXIII,* Abstract #1611. Lunar and Planetary Institute, Houston (CD-ROM).

Bowman L. E., Spilde M. N., and Papike J. J. (1997) Automated energy dispersive spectrometer modal analysis applied to diogenites. *Meteoritics & Planet. Sci., 32,* 869–875.

Brearley A. J. (1989) Nature and origin of matrix in the unique type 3 chondrite, Kakangari. *Geochim. Cosmochim. Acta, 53,* 197–214.

Brearley A. J. (1991) Mineralogical and chemical studies of matrix in the Adelaide meteorite, a unique carbonaceous chondrite with affinities to ALH A77307 (CO3) (abstract). In *Lunar and Planetary Science XXII,* pp. 133–134. Lunar and Planetary Institute, Houston.

Brearley A. J. (1993) Matrix and fine-grained rims in the unequilibrated CO3 chondrite, ALHA77307: Origins and evidence for diverse, primitive nebular dust components. *Geochim. Cosmochim. Acta, 57,* 1521–1550.

Brearley A. J. (1995) Aqueous alteration and brecciation in Bells, an unusual, saponite-bearing, CM chondrite. *Geochim. Cosmochim. Acta, 59,* 2291–2317.

Brearley A. J. (1997) Phyllosilicates in the matrix of the unique carbonaceous chondrite, LEW 85332 and possible implications for the aqueous alteration of CI chondrites. *Meteoritics & Planet. Sci., 32,* 377–388.

Brearley A. J. (2006) The action of water. In *Meteorites and the Early Solar System II* (D. S. Lauretta and H. Y. McSween Jr., eds.), this volume. Univ. of Arizona, Tucson.

Brearley A. J. and Jones R. H. (1998) Chondritic meteorites. In *Planetary Materials* (J. J. Papike ed.), pp. 3-1 to 3-398. Reviews in Mineralogy, Vol. 36, Mineralogical Society of America.

Brown P. G., Hildebrand A. R., Zolensky M. E., Grady M., Clayton R. N., Mayeda T. K., Tagliaferri E., Spalding R., MacRae N. D., Hoffman E. L., Mittlefehldt D. W., Wacker J. F., Bird J. A., Campbell M. D., Carpenter R., Gingerich H., Glatiotis M., Greiner E., Mazur M. J., McCausland P. J., Plotkin H., and Mazur T. R. (2000) The fall, recovery, orbit, and composition of the Tagish Lake meteorite; a new type of carbonaceous chondrite. *Science, 290,* 320–325.

Browning L. B., McSween H. Y. Jr., and Zolensky M. E. (1996) Correlated alteration effects in CM carbonaceous chondrites. *Geochim. Cosmochim. Acta, 60,* 2621–2633.

Burbine T. H., McCoy T. J., Meibom A., Gladman B., and Keil K. (2002) Meteoritic parent bodies: Their number and identification. In *Asteroids III* (W. F. Bottke Jr. et al., eds.), pp. 653–667. Univ. of Arizona, Tucson.

Burroni A. and Folco L. (2003) Frontier Mountains 93001: An igneous-textured member of the Acapulco-Lodran clan. *Meteoritics & Planet. Sci., 38,* A98.

Buseck P. R. (1977) Pallasite meteorites mineralogy, petrology and geochemistry. *Geochim. Cosmochim. Acta, 41,* 711–740.

Campbell A. J. And Humayan M. (2005) The Tishomingo iron: Relationship to IVB irons, CR clan chondrites and angrites and implications for the origin of volatile-depleted iron meteorites (abstract). In *Lunar and Planetary Science XXXVI,* Abstract

#2062. Lunar and Planetary Institute, Houston (CD-ROM).

Campbell A. J., Humayun M., Krot A., and Keil K. (2001) Origin of zoned metal grains in the QUE 94411 chondrite. *Geochim. Cosmochim. Acta, 65,* 163–180.

Campbell A. J., Humayun M., and Weisberg M. K. (2002) Siderophile element constraints on the formation of metal in the metal-rich chondrites Bencubbin, Gujba and Weatherford. *Geochim. Cosmochim. Acta, 66,* 647–660.

Casanova I., Keil K., and Newsom H. E. (1993) Composition of metal in aubrites: Constraints on core formation. *Geochim. Cosmochim. Acta, 57,* 675–682.

Chabot N. L. and Haack H. (2006) Evolution of asteroidal cores. In *Meteorites and the Early Solar System II* (D. S. Lauretta and H. Y. McSween Jr., eds.), this volume. Univ. of Arizona, Tucson.

Choi B-G., Quyang X., and Wasson J. T. (1995) Classification and origin of IAB and IIICD iron meteorites. *Geochim. Cosmochim. Acta, 59,* 593–612.

Clayton R. N. and Mayeda Y. K. (1988) Formation of ureilites by nebular processes. *Geochim. Cosmochim. Acta, 52,* 1313–1318.

Clayton R. N. and Mayeda T. K. (1996) Oxygen-isotope studies of achondrites. *Geochim. Cosmochim. Acta, 60,* 1999–2018.

Clayton R. N. and Mayeda T. K. (1999) Oxygen isotope studies of carbonaceous chondrites. *Geochim. Cosmochim. Acta, 63,* 2089–2104.

Clayton R. N., Mayeda T. K., and Rubin A. E. (1984) Oxygen isotopic compositions of enstatite chondrites and aubrites. *Proc. Lunar Planet. Sci. Conf. 15th,* in *J. Geophys. Res., 89,* C245–C249.

Davis A. M. and Olsen E. J. (1991) Phosphates in pallasite meteorites as probes of mantle processes in small planetary bodies. *Nature, 353,* 637–640.

Davis A. M., Grossman L., and Ganapathy R. (1977) Yes, Kakangari is a unique chondrite. *Nature, 265,* 230–232.

Davy R., Whitehead S. G., and Pitt G. (1978) The Adelaide meteorite. *Meteoritics, 13,* 121–139.

Delaney J. S., Prinz M., and Takeda H. (1984) The polymict eucrites. *Proc. Lunar Planet. Sci. Conf. 15th,* in *J. Geophys. Res., 89,* C251–C288.

Ebel D. S. (2006) Condensation of rocky material in astrophysical environments. In *Meteorites and the Early Solar System II* (D. S. Lauretta and H. Y. McSween Jr., eds.), this volume. Univ. of Arizona, Tucson.

Endress M. and Bischoff A. (1993) Mineralogy, degree of brecciation, and aqueous alteration of CI chondrites Orgueil, Ivuna, and Alais (abstract). *Meteoritics, 28,* 345–346.

Endress M. and Bischoff A. (1996) Carbonates in CI chondrites: Clues to parent body evolution. *Geochim. Cosmochim. Acta, 60,* 489–507.

Endress M., Zinner E., and Bischoff A. (1996) Early aqueous activity on primitive meteorite parent bodies. *Nature, 379,* 701–703.

Farrington O. C. (1907) Analysis of iron meteorites, compiled and classified. *Field Columbian Museum Pub. 120, Geol. Ser. Vol. 3,* 59–110.

Fedkin A. V. and Grossman L. (2006) The fayalite content of chondritic olivine: Obstacle to understanding the condensation of rocky material. In *Meteorites and the Early Solar System II* (D. S. Lauretta and H. Y. McSween Jr., eds.), this volume. Univ. of Arizona, Tucson.

Fitzgerald M. J. and Jones J. B. (1977) Adekaide and Bench Crater — members of a new subgroup of the carbonaceous chondrites. *Meteoritics, 12,* 443–458.

Floran R. J. (1978) Silicate petrography, classification, and origin of the mesosiderites: Review and new observations. *Proc. Lunar Planet. Sci. Conf. 9th,* pp. 1053–1081.

Floran R. J., Caulfield J. B. D., Harlow G. E., and Prinz M. (1978) Impact origin for the Simondium, Pinnaroo, and Hainholz mesosiderites: Implications for impact processes beyond the Earth-Moon system. *Proc. Lunar Planet. Sci. Conf. 9th,* pp. 1083–1114.

Fredriksson K. and Keil K. (1963) The light-dark structure in the Pantar and Kapoeta stone meteorites. *Geochim. Cosmochim. Acta, 27,* 717–739.

Friedrich J. M., Wang M-S, and Lipschutz M. E. (2002) Comparison of the trace element composition of Tagish Lake with other primitive carbonaceous chondrites. *Meteoritics & Planet. Sci., 37,* 677–686.

Goldberg E., Uchiyama A., and Brown H. (1951) The distribution of nickel, cobalt, gallium, palladium and gold in iron meteorites. *Geochim. Cosmochim. Acta, 2,* 1–25.

Goodrich C. A. (1992) Ureilites: A critical review. *Meteoritics, 27,* 327–352.

Goodrich C. A., Jones J. H., and Berkley J. L. (1987) Origin and evolution of the ureilite parent magmas: Multi-stage igneous activity on a large parent body. *Geochim. Cosmochim. Acta, 51,* 2255–2273.

Goodrich C. A., Fioretti A. M., Tribaudino M., and Molin G. (2001) Primary trapped melt inclusions in olivine in the olivine-augite-orthopyroxene ureilite Hughes 009. *Geochim. Cosmochim. Acta, 65,* 621–652.

Goodrich C. A., Krot A. E., Scott E. R. D., Taylor G. J., Fioretti A. M., and Keil K. (2002) Formation and evolution of the ureilite parent body and its offspring (abstract). In *Lunar and Planetary Science XXXIII,* Abstract #1379. Lunar and Planetary Institute, Houston (CD-ROM).

Goodrich C. A., Scott E. R. D, Fioretti A. M. (2004) Ureilitic breccias: Clues to the petrologic structure and impact disruption of the ureilite parent body. *Chem. Erde, 64,* 283–327.

Grady M. M., Verchovsky A. B., Franchi I. A., Wright I. P., and Pillinger C. T. (2002) Light element geochemistry of the Tagish Lake CI2 chondrite: Comparison with CI1 and CM2 meteorites. *Meteoritics & Planet. Sci., 37,* 713–735.

Greenwood R. C., Franchi I. A., and Pillinger C. T. (2002) Oxygen isotopes in CO3 chondrites (abstract). In *Lunar and Planetary Science XXXIII,* Abstract #1609. Lunar and Planetary Institute, Houston (CD-ROM).

Greenwood R. C., Franchi I. A., Kearsley A. T., and Alard O. (2005) The relationship between CK and CV chondrites: A single parent body source (abstract). In *Lunar and Planetary Science XXXV,* Abstract #1664. Lunar and Planetary Institute, Houston (CD-ROM).

Greshake A. (1997) The primitive matrix components of the unique carbonaceous chondrite Acfer 094; a TEM study. *Geochim. Cosmochim. Acta, 61,* 437–452.

Greshake A., Krot A. N., Meibom A., Weisberg M. K., and Keil K. (2002) Heavily-hydrated matrix lumps in the CH and metal-rich chondrites QUE 94411 and Hammadah al Hamra 237. *Meteoritics & Planet. Sci., 37,* 281–294.

Grossman J. N. and Zipfel J. (2001) The Meteoritical Bulletin, No. 85. *Meteoritics & Planet. Sci., 36,* A293–A322.

Grossman J. N., Rubin A. E., and MacPherson (1988) ALH85085; a unique volatile-poor carbonaceous chondrite with possible implications for nebular fractionation processes. *Earth Planet. Sci. Lett., 91,* 33–54.

Grossman J. N., MacPherson G. J., and Crozaz G. (1993) LEW 87223: A unique E chondrite with possible links to H chondrites (abstract). *Meteoritics, 28,* 358.

Guimon R. K., Symes S. J. K., Sears D. W. G., and Benoit P. H. (1995) Chemical and physical studies of type 3 chondrites XII: The metamorphic history of CV chondrites and their components. *Meteoritics & Planet. Sci., 30,* 704–714.

Hewins R. H. (1983) Impact versus internal origins for mesosiderites. *Proc. Lunar Planet. Sci. Conf. 14th,* in *J. Geophys. Res., 88,* B257–B266.

Hiroi T., Pieters C. M., Zolensky M. E., and Lipschutz M. E. (1993a) Evidence of thermal metamorphism on the C, G, B and F asteroids. *Science, 261,* 1016–1018.

Hiroi T., Bell J. F., Takeda H., and Pieters C. M. (1993b) Spectral comparison between olivine-rich asteroids and pallasites. *Proc. NIPR Symp. Antarct. Meteorites, 6,* 234–245.

Hiroi T., Pieters C. M., Zolensky M. E., and Prinz M. (1996) Reflectance spectra (UV-3 µm) of heated Ivuna (CI) meteorite and newly identified thermally metamorphosed CM chondrites (abstract). In *Lunar and Planetary Science XXVII,* pp. 551–552. Lunar and Planetary Institute, Houston.

Huss G. R. and Hutcheon I. D. (1992) Abundant $^{26}Mg^*$ in Adelaide refractory inclusions (abstract). *Meteoritics, 27,* 236.

Huss G. R., Rubin A. E., and Grossman J. N. (2006) Thermal metamorphism in chondrites. In *Meteorites and the Early Solar System II* (D. S. Lauretta and H. Y. McSween Jr., eds.), this volume. Univ. of Arizona, Tucson.

Hutcheon I. D. and Steele I. M. (1982) Refractory inclusions in the Adelaide chondrite (abstract). In *Lunar and Planetary Science XXIII,* pp. 352–353. Lunar and Planetary Institute, Houston.

Hutchison R. (2004) *Meteorites: A Petrologic, Chemical and Isotopic Synthesis.* Cambridge Univ., Cambridge.

Hutchison R., Alexander C. M. O'D., and Barber D. J. (1987) The Semarkona meteorite: First recorded occurrence of smectite in an ordinary chondrite, and its implications. *Geochim. Cosmochim. Acta, 31,* 1103–1106.

Ikeda Y. (1992) An overview of the research consortium, "Antarctic carbonaceous chondrites with CI affinities, Yamato-86720, Yamato-82162, and Belgica-7904." *Proc. NIPR Symp. Antarct. Meteorites, 5,* 49–73.

Ikeda Y. and Prinz M. (1993) Petrologic study of the Belgica 7904 carbonaceous chondrite: Hydrous alteration, thermal metamorphism, and relationship to CM and CI chondrites. *Geochim. Cosmochim. Acta, 57,* 439–452.

Ikeda Y. and Prinz M. (2000) Magmatic inclusions and felsic clasts in the Dar al Gani 319 polymict ureilite. *Meteoritics & Planet. Sci., 36,* 481–500.

Ikeda Y., Prinz M., and Nehru C. E. (2000) Lithic and mineral clasts in the Dar al Gani (DAG) 319 polymict ureilite. *Antarct. Meteorit. Res., 13,* 177–221.

Ivanov A. V. (1989) The meteorite Kaidun: Composition and history of formation. *Geokhimiya, 2,* 259–266.

Ivanova M. A., Taylor L. A.., Clayton R. N., Mayeda T. K., Nazarov M. A., Brandstaetter F., and Kurat G. (2002) Dhofar 225 vs. the CM clan: Metamorphosed or new type of carbonaceous chondrite? (abstract). In *Lunar and Planetary Science XXXIII,* Abstract #1437. Lunar and Planetary Institute, Houston (CD-ROM).

Jurewicz A. J. G., Mittlefehldt D. W., and Jones J. H. (1993) Experimental partial melting of the Allende (CV) and Murchison (CM) chondrites and the origin of asteroidal basalts. *Geochim. Cosmochim. Acta, 57,* 2123–3139.

Kallemeyn G. W. (1995) Bells and Essebi: To be or not to be (CM) (abstract). *Meteoritics, 30,* 525–526.

Kallemeyn G. W. (1996) The classificational wanderings of the Ningqiang chondrite (abstract). In *Lunar and Planetary Science XXVII,* pp. 635–636. Lunar and Planetary Institute, Houston.

Kallemeyn G. W. and Rubin A. E. (1995) Coolidge and Loongana 001: A new carbonaceous chondrite grouplet. *Meteoritics, 30,* 20–27.

Kallemeyn G. W. and Wasson J. T. (1981) The compositional classification of chondrites: I. The carbonaceous chondrite groups. *Geochim. Cosmochim. Acta, 45,* 1217–1230.

Kallemeyn G. W. and Wasson J. T. (1982) The compositional classification of chondrites: III. Ungrouped carbonaceous chondrites. *Geochim. Cosmochim. Acta, 46,* 2217–2228.

Kallemeyn G. W. and Wasson J. T. (1985) The compositional classification of chondrites: IV. Ungrouped chondritic meteorites and clasts. *Geochim. Cosmochim. Acta, 49,* 261–270.

Kallemeyn G. W., Boynton W. V., Willis J., and Wasson J. T. (1978) Formation of the Bencubbin polymict meteoritic breccia. *Geochim. Cosmochim. Acta, 42,* 507–515.

Kallemeyn G. W., Rubin A. E., Wang D., and Wasson J. T. (1989) Ordinary chondrites: Bulk compositions, classification, lithophile-element fractionations, and composition-petrographic type relationships. *Geochim. Cosmochim. Acta, 53,* 2747–2767.

Kallemeyn G. W., Rubin A. E., and Wasson J. T. (1991) The compositional classification of chondrites: V. The Karoonda (CK) group of carbonaceous chondrites. *Geochim. Cosmochim. Acta, 55,* 881–892.

Kallemeyn G. W., Rubin A. E., and Wasson J. T. (1994) The compositional classification of chondrites: VI. The CR carbonaceous chondrite group. *Geochim. Cosmochim. Acta, 58,* 2873–2888.

Kallemeyn G. W., Rubin A. E., and Wasson J. T. (1996) The compositional classification of chondrites: VII. The R chondrite group. *Geochim. Cosmochim. Acta, 60,* 2243–2256.

Keil K. (1968) Mineralogical and chemical relationships among enstatite chondrites. *J. Geophys. Res., 73,* 6945–6976.

Keil K. (1989) Enstatite meteorites and their parent bodies. *Meteoritics, 24,* 195–208.

Keil K. and Brett R. (1974) Heideite, $(Fe,Cr)_{1+x}(Ti,Fe)_2S_4$, a new mineral in the Bustee enstatite achondrite. *Am. Mineral., 59,* 465–470.

Keil K., Ntaflos Th., Taylor G. J., Brearley A. J., Newsom H. E., and Romig A. D. Jr. (1989) The Shallowater aubrite: Evidence for origin by planetesimal impacts. *Geochim. Cosmochim. Acta, 53,* 3291–3307.

Keller L. P. and Buseck P. R. (1990a) Matrix mineralogy of the Lance CO3 carboncaeous chondrite. *Geochim. Cosmochim. Acta, 54,* 1155–1163.

Keller L. P. and Buseck P. R. (1990b) Aqueous alteration in the Kaba CV3 carbonaceous chondrite. *Geochim. Cosmochim. Acta, 54,* 2113–2120.

Keller L. P., Thomas K. L., Clayton R. N., Mayeda T. K., DeHart J. M., and McKay D. S. (1994) Aqueous alteration of the Bali CV3 chondrite: Evidence from mineralogy, mineral chemistry, and oxygen isotopic compositions. *Geochim. Cosmochim. Acta, 58,* 5589–5598.

Kerridge J. F. (1985) Carbon, hydrogen, and nitrogen in carbonaceous chondrites: Abundances and isotopic compositions in bulk samples. *Geochim. Cosmochim. Acta, 49,* 1707–1714.

Kimura M., Tsuchiayama A., Fukuoka T., and Iimura Y. (1992) Antarctic primitive achondrites, Yamato-74025, -75300, and -75305: Their mineralogy, thermal history, and the relevance to winonaites. *Proc. NIPR Symp. Antarct. Meteorites, 5,* 165–190.

Kimura M., Chen M., Yoshida Y., El Goresy A., and Ohtani E. (2003) Back transformation of high pressure phases in a shock melt vein of an H-chondrite during atmosphere passage: Implications for the survival of high-pressure phases after decompression. *Earth Planet. Sci. Lett., 217,* 141–150.

Kimura M., Weisberg M. K., Lin Y., Suzuki A., Ohtani E., and Okazaki R. (2005) Thermal history of enstatite chondrites from silica polymorphs. *Meteoritics & Planet. Sci., 40,* 855–868.

Kita N. T., Ikeda Y., Togashi S., Liu Y., Morishita Y., and Weisberg M. K. (2004) Origin of ureilites inferred from a SIMS oxygen isotopic and trace element study of clasts in the Dar al Gani 319 polymict ureilite. *Geochim. Cosmochim. Acta, 68,* 4213–4235.

Koeberl C., Ntaflos T., Kurat G., and Chai C. F. (1987) Petrology and geochemistry of the Ningqiang (CV3) chondrite (abstract). In *Lunar and Planetary Science XVIII,* pp. 499–500. Lunar and Planetary Institute, Houston.

Kracher A. (1982) Crystallization of a S-saturated melt, and the origin of the iron meteorite groups IAB and IIICD. *Geophys. Res. Lett., 9,* 412–415.

Kracher A., Willis J., and Wasson J. T. (1980) Chemical classification of iron meteorites; IX. A new group (IIF), revision of IAB and IIICD, and data on 57 additional irons. *Geochim. Cosmochim. Acta, 44,* 773–787.

Krot A. N., Scott E. R. D., and Zolensky M. E. (1995) Mineralogic and chemical variations among CV3 chondrites and their components: Nebular and asteroidal processing. *Meteoiritics, 30,* 748–775.

Krot A. N., Petaev M. I., Scott E. R. D., Choi B.-G., Zolensky M. E., and Keil K. (1998) Progressive alteration in CV3 chondrites: More evidence for asteroidal alteration. *Meteoritics & Planet. Sci., 33,* 1065–1085.

Krot A. N., Weisberg M. K., Petaev M. I., Keil K., and Scott E. R. D. (2000a) High-temperature condensation signatures in Type I chondrules from CR carbonaceous chondrites (abstract). In *Lunar and Planetary Science XXXI,* Abstract #1470. Lunar and Planetary Institute, Houston (CD-ROM).

Krot A. N., Meibom A., and Keil K. (2000b) Volatile-poor chondrules in CH carbonaceous chondrites: Formation at high ambient nebular temperature (abstract). In *Lunar and Planetary Science XXXI,* Abstract #1481. Lunar and Planetary Institute, Houston (CD-ROM).

Krot A. N., Meibom A., and Keil K. (2000c) A clast of Bali-like oxidized CV3 material in the reduced CV3 chondrite breccia Vigarano. *Meteoritics & Planet. Sci., 35,* 817–827.

Krot A. N., McKeegan K. D., Russell S. S., Meibom A., Weisberg M. K., Zipfel J., Krot T. V., Fagan T. J., and Keil K. (2001a) Refractory Ca,Al-rich inclusions and Al-diopside-rich chondrules in the metal-rich chondrites Hammadah al Hamra 237 and QUE 94411. *Meteoritics & Planet. Sci., 36,* 1189–1217.

Krot A. N., Huss G. R., and Hutcheon I. D. (2001b) Corundum-hibonite refractory inclusions from Adelaide: Condensation or crystallization from melt? (abstract). *Meteoritics & Planet. Sci., 36,* A105.

Krot A. N., Hutcheon I. D., and Huss G. R. (2001c) Aluminum-rich chondrules and associated refractory inclusions in the unique carbonaceous chondrite Adelaide (abstract). *Meteoritics & Planet. Sci., 36,* A105–A106.

Krot A. N., Aleon J., and McKeegan K. D. (2002a) Mineralogy, petrography and oxygen-isotopic compositions of Ca,Al-rich inclusions and amoeboid olivine aggregates in the CR carbonaceous chondrites (abstract). In *Lunar and Planetary Science XXXIII,* Abstract #1412. Lunar and Planetary Institute, Houston (CD-ROM).

Krot A. N., Meibom A., Weisberg M. K., and Keil K. (2002b) The CR chondrite clan: Implications for early solar system processes. *Meteoritics & Planet. Sci., 37,* 1451–1490.

Krot A. N., Petaev M. I., and Bland P. A. (2003) Growth of ferrous olivine in the oxidized CV chondrites during fluid-assisted thermal metamorphism (abstract). *Meteoritics & Planet. Sci., 38,* A73.

Krot A. N., Keil K., Goodrich C. A., Scott E. R. D., and Weisberg M. K. (2004) Classification of Meteorites. In *Treatise on Geochemistry, Vol. 1. Meteorites, Comets, and Planets* (A. M. Davis, ed.), pp. 83–128. Elsevier, Oxford.

Krot A. E., Hutcheon I. D., Brearley A. J., Pravdivtseva O. V., Pataev M. I., and Hohenberg C. M. (2006) Timescales and settings for alteration of chondritic meteorites. In *Meteorites and the Early Solar System II* (D. S. Lauretta and H. Y. McSween Jr., eds.), this volume. Univ. of Arizona, Tucson.

Kung C. C. and Clayton R. N. (1978) Nitrogen abundances and isotopic compositions in stony meteorites. *Earth Planet. Sci. Lett., 38,* 421–435.

Kurat G. (1988) Primitive meteorites: An attempt toward unification. *Philos. Trans. R. Soc. London, A325,* 459–482.

Lauretta D. S., Nagahara H., and Alexander C. M. O'D. (2006) Petrology and origin of ferromagnesian silicate chondrules. In *Meteorites and the Early Solar System II* (D. S. Lauretta and H. Y. McSween Jr., eds.), this volume. Univ. of Arizona, Tucson.

Leshin L. A., Rubin A. E., and McKeegan K. D. (1997) The oxygen isotopic composition of olivine and pyroxene from CI chondrites. *Geochim. Cosmochim. Acta, 61,* 835–845.

Lin Y. and El Goresy A. (2002) A comparative study of opaque phases in Qingzhen (EH3) and MacAlpine Hills 88136 (EL3): Representatives of EH and EL parent bodies. *Meteoritics & Planet. Sci., 37,* 577–599.

Lin Y. and Kimura M. (1998) Petrographic and mineralogical study of new EH melt rocks and a new enstatite chondrite grouplet. *Meteoritics & Planet. Sci., 33,* 501–511.

Lovering J. F. (1962) The evolution of the meteorites — Evidence for the co-existence of chondritic, achondritic and iron meteorites in a typical parent meteorite body. In *Researches on Meteorites* (C. B. Moore, ed.), pp. 179–197. Wiley, New York.

Lovering J. F., Nichiporuk W., Chodos A., and Brown H. (1957) The distribution of gallium, germanium, cobalt, chromium, and copper in iron and stony-iron meteorites in relation to Ni content and structure. *Geochim. Cosmochim. Acta, 11,* 263–278.

Marti K., Kim J. S., Thakur A. N., McCoy T. J., and Keil K. (1995) Signatures of the martian atmosphere in glass of the Zagami meteorite. *Science, 267,* 1981–1984.

Mason B. (1967) Meteorites. *Am. Sci., 55,* 429–455.

Mason B. and Nelen J. (1968) The Weatherford meteorite. *Geochim. Cosmochim Acta, 32,* 661–664.

Mayeda K., Clayton R. N., Krung D. A., and Davis A. M. (1988) Oxygen, silicon and magnesium isotopes in Ningqiang chondrules (abstract). *Meteoritics, 23,* 288.

McCoy T. J. and Carlson W. D. (1998) Opaque minerals in the GRA 95209 lodranite: A snapshot of metal segregation (abstract). In *Lunar and Planetary Science XXIX,* Abstract #1675. Lunar and Planetary Institute, Houston (CD-ROM).

McCoy T. J., Scott E. R. D., and Haack H. (1993) Genesis of the IIICD iron meteorites: Evidence from silicate-bearing inclusions. *Meteoritics, 28,* 552–560.

McCoy T. J., Keil K., Scott E. R. D., Benedix G. K., Ehlmann A. J., Mayeda T. K., and Clayton R. N. (1994) Low FeO ordinary chondrites: A nebular origin and a new chondrite parent body (abstract). In *Lunar and Planetary Science XXV,* pp. 865–866. Lunar and Planetary Institute, Houston.

McCoy T. J., Keil K., Bogard D. D., Garrison D. H., Casanova I., Lindstrom M. M., Brearley A. J., Kehm K., Nichols R. H. Jr., and Hohenberg C. M. (1995) Origin and history of impact-melt rocks of enstatite chondrite parentage. *Geochim. Cosmochim. Acta, 59,* 161–175.

McCoy T. J., Keil K., Clayton R. N., Mayeda T. K., Bogard D. D., Garrison D. H., Huss G. R., Hutcheon I. D., and Wieler R. (1996) A petrologic, chemical and isotopic study of Monument Draw and comparison with other acapulcoites: Evidence for formation by incipient partial melting. *Geochim. Cosmochim. Acta, 60,* 2681–2708.

McCoy T. J., Keil K., Clayton R. N., Mayeda T. K., Bogard D. D., Garrison D. H., and Wieler R. (1997a) A petrologic and isotopic study of lodranites: Evidence for early formation as partial melt residues from heterogeneous precursors. *Geochim. Cosmochim. Acta, 61,* 623–637.

McCoy T. J., Keil K., Muenow D. W., and Wilson L. (1997b) Partial melting and melt migration in the acapulcoite-lodranite parent body. *Geochim. Cosmochim. Acta, 61,* 639–650.

McCoy T. J., Mittlefehldt D. W., and Wilson L. (2006) Asteroid differentiation. In *Meteorites and the Early Solar System II* (D. S. Lauretta and H. Y. McSween Jr., eds.), this volume. Univ. of Arizona, Tucson.

McCord T. B., Adams J. B., and Johnson T. V. (1970) Asteroid Vesta: Spectral reflectivity and compositional implications. *Science, 168,* 1445–1447.

McKay D. S., Gibson E. K., Thomas-Keprta, Vali H., Romanek C. S., Clement S. J., Chillier X. D. F., Maechling C. R., and Zare R. N. (1996) Search for past life on Mars: Possible relic biogenic activity in martian meteorite ALH84001. *Science, 273,* 924–930.

McKay G., Lindstrom D., Yang S-R., and Wagstaff J. (1988) Petrology of a unique achondrite Lewis Cliff 86010 (abstract). In *Lunar and Planetary Science XIX,* pp. 762–763. Lunar and Planetary Institute, Houston.

McKay G., Crozaz G., Wagstaff J., Yang S-R., and Lundberg L. (1990) A petrographic, electron microprobe, and ion microprobe study of mini-angrite Lewis Cliff 87051 (abstract). In *Lunar and Planetary Science XXI,* pp. 771–772. Lunar and Planetary Institute, Houston.

McSween H. Y. Jr. (1977a) Carbonaceous chondrites of the Ornans type: A metamorphic sequence. *Geochim. Cosmochim. Acta, 41,* 479–491.

McSween H. Y. Jr. (1977b) Petrographic variations among carbonaceous chondrites of the Vigarano type. *Geochim. Cosmochim. Acta, 41,* 1777–1790.

McSween H. Y. Jr. (1979) Are carbonaceous chondrites primitive or processed? A review. *Rev. Geophys. Space Phys., 17,* 1059–1078.

McSween H. Y. Jr. (2002) The rocks of Mars, from far and near. *Meteoritics & Planet. Sci., 37,* 7–25.

McSween H. Y. Jr. and Richardson S. M. (1977) The composition of carbonaceous chondrite matrix. *Geochim. Cosmochim. Acta, 41,* 1145–1161.

McSween H. Y. Jr. and Treiman A. H. (1998) Martian meteorites. In *Planetary Materials* (J. J. Papike ed.), pp. 6-1 to 6-53. Reviews in Mineralogy, Vol. 36, Mineralogical Society of America.

Meibom A., Petaev M. I., Krot A. N., Wood J. A., and Keil K. (1999) Primitive Fe,Ni metal grains in CH carbonaceous chondrites formed by condensation from a gas of solar composition. *J. Geophys. Res., 104,* 22053–22059.

Meibom A., Desch S. J., Krot A. N., Cuzzi J. N., Petaev M. I., Wilson L., and Keil K. (2000a) Large scale thermal events in the solar nebula recorded in Fe, Ni metal condensates in primitive meteorites. *Science, 288,* 839–841.

Meibom A., Krot A. N., Righter K., Chabot N., and Keil K. (2000b) Metal/sulfide-ferrous silicate shock melts in QUE 94411 and Hammadah al Hamra 237: Remains of the missing matrix? (abstract). In *Lunar and Planetary Science XXXI,* Abstract #1420. Lunar and Planetary Institute, Houston (CD-ROM).

Meibom A., Petaev M. I., Krot A. N., Keil K., and Wood J. A. (2001) Growth mechanism and additional constraints on FeNi metal condensation in the solar nebula. *J. Geophys. Res., 106,* 32797–32801.

Mittlefehldt D. W. (1994) ALH84001, a cumulate orthopyroxenite member of the martian meteorite clan. *Meteoritics, 29,* 214–221.

Mittlefehldt D. W. (2002) Geochemistry of the ungrouped carbonaceous chondrite Tagish Lake, the anomalous CM chondrite Bells, and comparison with CI and CM chondrites. *Meteoritics & Planet. Sci., 37,* 703–712.

Mittlefehldt D. W. (2004) Achondrites. In *Treatise on Geochemistry, Vol. 1: Meteorites, Comets, and Planets* (A. M. Davis, ed.), pp. 291–324. Elsevier, Oxford.

Mittlefehldt D. W. and Lindstrom M. M. (1998) Petrology and geochemistry of lodranite GRA 95209 (abstract). *Meteoritics & Planet. Sci., 33,* A111.

Mittlefehldt D. W., Lindstron M. M., Bogard D. D., Garrison D. H., and Field S. W. (1996) Acapulco- and Lodran-like achondrites: Petrology, geochemistry, chronology and origin. *Geochim. Cosmochim. Acta, 60,* 867–882.

Mittlefehldt D. W., McCoy T. J., Goodrich C. A., and Kracher A. (1998) Non-chondritic meteorites from asteroidal bodies. In *Planetary Materials* (J. J. Papike, ed.), pp. 4-1 to 4-495. Reviews in Mineralogy, Vol. 36, Mineralogical Society of America.

Nagahara H. (1992) Yamato-8002: Partial melting residue on the "unique" chondrite parent body. *Proc. NIPR Symp. Antarct. Meteorites, 5,* 191–223.

Nehru C. E., Prinz M., Weisberg M. K., Ebihara M. E., Clayton R. E., and Mayeda T. K. (1996) Brachinites: A new primitive achondrite group (abstract). *Meteoritics, 27,* 267.

Nehru C. E., Weisberg M. K., Boesenberg J. S., and Kilgore M. (2003) Tafassasset: A metal-rich primitive achondrite with affinities to brachinites (abstract). In *Lunar and Planetary Science XXXIV,* Abstract #1370. Lunar and Planetary Institute, Houston (CD-ROM).

Newsom H. E. and Drake M. J. (1979) The origin of metal clasts in the Bencubbin meteoritic breccia. *Geochim. Cosmochim. Acta, 43,* 689–707.

Newton J., Bischoff A., Arden J. W., Franci I. A., Geiger T., Greshake A., and Pillinger C. T. (1995) Acfer 094, a uniquely primitive carbonaceous chondrite from the Sagara. *Meteoritics, 30,* 47–56.

Noguchi T. (1994) Petrology and mineralogy of the Coolidge meteorite (CV4). *Proc. NIPR Symp. Antarct. Meteorites, 7,* 42–72.

Nyquist L. E., Bogard D. D., Shih C-Y., Greshake A., Stöffler D., and Eugster O. (2001) Ages and geologic histories of martian meteorites. *Space Sci. Rev., 96,* 105–164.

Olsen E., Bunch T. E., Jarosewich E., Noonan A. F., and Huss G. I. (1977) Happy Canyon: A new type of enstatite chondrite. *Meteoritics, 12,* 109–123.

Palme H., Hutcheon I. D., Kennedy A. K., Sheng Y. J., and Spettel B. (1991) Trace element distributions in minerals from a silicate inclusion in the Caddo IAB-iron meteorite (abstract). In *Lunar and Planetary Science XXII,* pp. 1015–1016. Lunar and Planetary Institute, Houston.

Papike J. J. (1998) Comparative planetary mineralogy: Chemistry of melt-derived pyroxene, feldspar, and olivine. In *Planetary Materials* (J. J. Papike, ed.), pp. 7-1 to 7-11. Reviews in Mineralogy, Vol. 36, Mineralogical Society of America.

Papike J. J., Ryder G., and Shearer C. K. (1998) Lunar samples. In *Planetary Materials* (J. J. Papike, ed.), pp. 5-1 to 5-234. Reviews in Mineralogy, Vol. 36, Mineralogical Society of America.

Patzer A., Hill D. H., and Boynton W. V. (2001) Itqiy: A metal-rich enstatite meteorite with achondritic texture. *Meteoritics & Planet. Sci., 36,* 1495–1505.

Paul R. L. and Lipschutz M. E. (1990) Consortium study of labile trace elements in some Antarctic carbonaceous chondrites: Antarctic and non-Antarctic meteorite comparisons. *Proc. NIPR Symp. Antarct. Meteorites, 3,* 80–95.

Petaev M. I., Clarke R. S. Jr., Olsen E. J., Jarosewich E., Davis A. M., Steele I. M., Litschutz M. E., Wang M-S., Clayton R. N., Mayeda T. K., and Wood J. A. (1993) Chaunskij: The most highly metamorphosed, shock-modified and metal-rich mesosiderite (abstract). In *Lunar and Planetary Science XXIV,* pp. 1131–1132. Lunar and Planetary Institute, Houston.

Petaev M. I., Meibom A., Krot A. N., Wood J. A., and Keil K. (2001) The condensation origin of zoned metal grains in Queen Alexandra Range 94411: Implications for the formation of the Bencubbin-like chondrites. *Meteoritics & Planet. Sci., 36,* 93–106.

Prinz M., Keil K., Hlava P. F., Berkley J. L., Gomes C. B., and Curvello W. S. (1977) Studies of Brazilian meteorites, III. Origin and history of the Angra dos Reis achondrite. *Earth Planet. Sci. Lett., 35,* 317–330.

Prinz M., Nehru C. E., Delaney J. S., and Weisberg M. (1983) Silicates in IAB and IIICD irons, winonaites, lodranites and Brachina: A primitive and modified primitive group (abstract). In *Lunar and Planetary Science XIV,* pp. 616–617. Lunar and Planetary Institute, Houston.

Prinz M., Chatterjee N., Weisberg M. K., Clayton R. N., and Mayeda T. K. (1991) Lea County 002: A second Kakangari-type chondrite (abstract). In *Lunar and Planetary Science XXII,* pp. 1097–1098. Lunar and Planetary Institute, Houston.

Prinz M., Weisberg M. K., Brearley A., Grady M. M., Pillinger C. T., Clayton R. N., and Mayeda T. K. (1992) LEW85332; a C2 chondrite in the CR clan (abstract). *Meteoritics, 27,* 278–279.

Prior G. T. (1920) The classification of meteorites. *Mineral. Mag., 19,* 51–63.

Rowe M. W., Clayton R. N., and Mayeda T. K. (1994) Oxygen isotopes in separated components of CI and CM chondrites. *Geochim. Cosmochim. Acta, 58,* 5341–5347.

Rubin A. E. (1989) Size-frequency distributions of chondrules in CO3 chondrites. *Meteoritics, 24,* 179–189.

Rubin A. E. (1990) Kamacite and olivine in ordinary chondrites: Intergroup and intragroup relationships. *Geochim. Cosmochim. Acta, 54,* 1217–1232.

Rubin A. E. (1998) Correlated petrologic and geochemical characteristics of CO3 chondrites. *Meteoritics & Planet. Sci., 33,* 385–391.

Rubin A. E. and Kallemeyn G. W. (1990) Lewis Cliff 85332: A unique carbonaceous chondrite. *Meteoritics, 25,* 215–225.

Rubin A. E. and Kallemeyn G. W. (1994) Pecora Escarpment 91002; a member of the Rumuruti (R) chondrite group. *Meteoritics, 29,* 255–264.

Rubin A. E. and Scott E. R. D. (1997) Abee and related EH chondrite impact-melt breccias. *Geochim. Cosmcohim. Acta, 61,* 425–435.

Rubin A. E., James J. A., Keck B. D., Weeks K. S., Sears D. W. G., and Jarosewich E. (1985) The Colony meteorite and variations in CO3 chondrite properties. *Meteoritics, 20,* 175–197.

Rubin A. E., Wang D., Kallemeyn G. W., and Wasson J. T. (1988) The Ningqiang meteorite; classification and petrology of an anomalous CV chondrite. *Meteoritics, 23,* 13–23.

Rubin A. E., Keil K., and Scott E. R. D. (1997) Shock metamoprhism of enstatite chondrites. *Geochim. Cosmochim. Acta, 61,* 847–858.

Rubin A. E., Kallemeyn G. W., Wasson J. T., Clayton R. N., Mayeda T. K., Grady M. M., and Verchovky A. B. (2001) Cujba: A new Bencubbin-like meteorite fall from Nigeria (abstract). In *Lunar and Planetary Science XXXII,* Abstract #1779. Lunar and Planetary Institute, Houston (CD-ROM).

Russell S. S. (1998) A survey of calcium-aluminum-rich inclusions from Rumuritiite chondrites: Implications for relationships between meteorite groups (abstract). *Meteoritics & Planet. Sci., 33,* A131–A132.

Russell S. S., Huss G. R., Fahey A. J., Greenwood R. C., Hutchison R., and Wasserburg G. J. (1998a) An isotopic and petrologic study of calcium-aluminum-rich inclusions from CO3 chondrites. *Geochim. Cosmochim. Acta, 62,* 689–714.

Russell S. S., McCoy T. J., Jarosewich E., and Ash R. D. (1998b) The Burnwell, Kentucky, low iron oxide chondrite fall: Description, classification and origin. *Meteoritics & Planet. Sci., 33,* 853–856.

Schulze H., Bischoff A., Palme H., Spettel B., Dreibus G., and Otto J. (1994) Mineralogy and chemistry of Rumuruti: The first meteorite fall of the new R chondrite group. *Meteoritics & Planet. Sci., 29,* 275–286.

Shultz L., Palme, H., Spettel B., Weber H. W., Wänke H., Christophe Michel-Levy., and Lorin J. C. (1982) Allan Hills A77081 — An unusual stony meteorite. *Earth Planet. Sci. Lett., 61,* 23–31.

Scott E. R. D. (1972) Chemical fractionation in iron meteorites and its interpretation. *Geochim. Cosmochim. Acta, 36,* 1205–1236.

Scott E. R. D. (1977) Geochemical relationship between some pallasites and iron meteorites. *Mineral. Mag., 41,* 262–275.

Scott E. R. D. (1988) A new kind of primitive chondrite, Allan Hills 85085. *Earth Planet. Sci. Lett., 91,* 1–18.

Scott E. R. D. and Jones R. H. (1990) Disentangling nebular and asteroidal features of CO3 carbonaceous chondrites. *Geochim. Cosmochim. Acta, 54,* 2485–2502.

Scott E. R. D. and Taylor G. J. (1985) Petrology of types 4–6 carbonaceous chondrites. *Proc. Lunar Planet. Sci. Conf. 15th,* in *J. Geophys. Res., 90,* C699–C709.

Scott E. R. D. and Wasson J. T. (1975) Classification and properties of iron meteorites. *Rev. Geophys. Space Phys., 13,* 527–546.

Scott E. R. D., Taylor G. J., Maggiore P., Keil K., McKinley S. G., and McSween H. Y. Jr. (1981) Three CO3 chondrites from Antarctica; comparison of carbonaceous and ordinary type 3 chondrites (abstract). *Meteoritics, 16,* 385.

Scott E. R. D., Taylor G. J., and Keil K. (1993) Origin of ureilite meteorites and implications for planetary accretion. *Geophys. Res. Lett., 20,* 415–418.

Scott E. R. D., Haack H., and Love S. G. (2001) Formation of mesosiderites by fragmentation and reaccretion of a large differentiated asteroid. *Meteoritics & Planet. Sci., 36,* 869–881.

Sears D. W. G. and Dodd R. T. (1988) Overview and classification of meteorites. In *Meteorites and the Early Solar System* (J. F. Kerridge and M. S. Matthews, eds.), pp. 3–31. Univ. of Arizona, Tucson.

Sears D. W. G. and Ross M. (1983) Classification of the Allan Hills A77307 meteorite. *Meteoritics, 18,* 1–7.

Sears D. W. G. and Weeks K. S. (1991) Chemical and physical studies of type 3 chondrites: 2. Thermoluminescence of sixteen type 3 ordinary chondrites and relationships with oxygen isotopes. *Proc. Lunar Planet. Sci. Conf. 14th,* in *J. Geophys. Res., 88,* B301–B311.

Sears D. W. G., Grossman J. N., Melcher C. L., Ross L. M., and Mills A. A. (1980) Measuring the metamorphic history of unequilibrated ordinary chondrites. *Nature, 287,* 791–795.

Sears D. W. G., Kallemeyn G. W., and Wasson J. T. (1982) The compositional classification of chondrites: II. The enstatite chondrite groups. *Geochim. Cosmochim. Acta, 46,* 597–608.

Sears D. W. G., Hasan F. A., Batchelor J. D., and Lu J. (1991) Chemical and physical studies of type 3 chondrites; XI, Metamorphism, pairing, and brecciation of ordinary chondrites. *Proc. Lunar Planet. Sci., Vol. 22,* pp. 493–512.

Srinivasan B. and Anders E. (1977) Noble gases in the unique chondrite, Kakangari. *Meteoritics, 12,* 417–424.

Stöffler D., Keil K., and Scott E. R. D. (1991) Shock metamorphism in ordinary chondrites. *Geochim. Cosmochim. Acta, 55,* 3845–3867.

Takeda H. (1987) Mineralogy of Antarctic ureilites and a working hypothesis for their origin and evolution. *Earth Planet. Sci. Lett., 81,* 358–370.

Takeda H., Mori H., Hiroi T., and Saito J. (1994) Mineralogy of new Antarctic achondrites with affinity to Lodran and a model of their evolution in an asteroid. *Meteoritics, 29,* 830–842.

Takeda H., Yugami K., Bogard D. D., and Miyamoto M. (1997) Plagioclase-augite-rich gabbro in the Caddo County IAB iron and the missing basalts assoicated with iron meteorites. In *Lunar and Planetary Science XXVIII,* pp. 1409–1410. Lunar and Planetary Institute, Houston.

Tomeoka K. (1990) Mineralogy and petrology of Belgica-7904: A new kind of carbonaceous chondrite from Antarctica. *Proc. NIPR Symp. Antarct. Meteorites, 3,* 40–54.

Tomeoka K. and Buseck P. R. (1990) Phyllosilicates in the Mokoia CV carbonaceous chondrite: Evidence for aqueous alteration in an oxidizing condition. *Geochim. Cosmochim. Acta, 54,* 1745–1754.

Tomeoka K., Kojima H., and Yanai K. (1989) Yamato-86720: A CM carbonaceous chondrite having experienced extensive aqueous alteration and thermal metamoprhism. *Proc. NIPR Symp. Antarct. Meteorites, 2,* 55–74.

Tomeoka K., Nomura K., and Takeda H. (1992) Na-bearing Ca-Al-rich inclusions in the Yamato-791717 CO carbonaceous chondrite. *Meteoritics, 27,* 136–143.

Van Schmus W. R. and Wood J. A. (1967) A chemical-petrologic classification for the chondritic meteorites. *Geochim. Cosmochim. Acta, 31,* 747–765.

Wai C. M. and Wasson J. T. (1977) Nebular condensation of moderately volatile elements and their abundances in ordinary chondrites. *Earth. Planet. Sci. Lett., 36,* 1–13.

Warren P. H. and Kallemeyn G. W. (1989) Allan Hills 84025: The second brachinite, far more differentiated than Brachina, and an ultramafic achondritic clast from L chondrite Yamato 75097. *Proc. Lunar Planet. Sci. Conf. 19th,* pp. 475–486.

Warren P. H. and Kallemeyn G. W. (1992) Explosive volcanism and the graphite-oxygen fugacity buffer on the parent asteroid(s) of the ureilite meteorites. *Icarus, 100,* 110–126.

Wasson J. T. (1985) *Meteorites: Their Record of the Early Solar-System History.* Freeman, New York.

Wasson J. T. (1990) Ungrouped iron meteorites in Antarctica: Origin of anomalously high abundance. *Science, 249,* 900–902.

Wasson J. T. and Kallemeyn G. W. (1988) Composition of chondrites. *Philos. Trans. R. Soc. London, A325,* 535–544.

Wasson J. T. and Kallemeyn G. W. (1990) Allan Hills 85085: A subchondritic meteorite of mixed nebular and regolithic heritage. *Earth Planet. Sci. Lett., 101,* 148–161.

Wasson J. T. and Kallemeyn G. W. (2002) The IAB iron-meteorite complex: A group, five subgroups, numerous grouplets, closely related, mainly formed by crystal segregation in rapidly cooling melts. *Geochim. Cosmochim. Acta, 66,* 2445–2473.

Wasson J. T. and Kimberlin J. (1967) The chemical classification of iron meteorites; II, Irons and pallasites with germanium concentrations between 8 and 100 ppm. *Geochim. Cosmochim. Acta, 31,* 149–178.

Wasson J. T., Rubin A. E., and Kallemeyn G. W. (1993) Reduction during metamoprhism of four ordinary chondrites. *Geochim. Cosmochim. Acta, 57,* 1867–1869.

Watters T.R. and Prinz M. (1979) Aubrites: Their origin and relationship to enstatite chondrites. *Proc. Lunar Planet. Sci. Conf. 10th,* pp. 1073–1093.

Weisberg M. K. (2001) Sahara 00182, the first CR3 chondrite and formation of multi-layered chondrules (abstract). *Meteoritics & Planet. Sci., 36,* A222–A223.

Weisberg M. K. and Prinz M. (2000) The Grosvenor Mountains 95577 CR1 chondrite and hydration of the CR chondrites (abstract). *Meteoritics & Planet. Sci., 35,* A168.

Weisberg M. K., Prinz M., and Nehru C. E. (1988) Petrology of ALH85085: A chondrite with unique characteristics. *Earth Planet. Sci. Lett., 91,* 19–32.

Weisberg M. K., Prinz M., and Nehru C. E. (1990) The Bencubbin chondrite breccia and its relationship to CR chondrites and the ALH85085 chondrite. *Meteoritics, 25,* 269–279.

Weisberg M. K., Prinz M., Kojima H., Yanai K., Clayton R. N., and Mayeda T. K. (1991) The Carlisle Lakes-type chondrites; a new grouplet with high $\Delta^{17}O$ and evidence for nebular oxidation. *Geochim. Cosmochim. Acta, 55,* 2657–2669.

Weisberg M. K., Prinz M., Clayton R. N., and Mayeda T. K. (1993) The CR (Renazzo-type) carbonaceous chondrite group and its implications. *Geochim. Cosmochim. Acta, 57,* 1567–1586.

Weisberg M. K., Prinz M., Clayton R. N., Mayeda T. K., Grady M. M., and Pillinger C. T. (1995a) The CR chondrite clan.

Proc. NIPR Symp. Antarct. Meteorites, 8, 11–32.

Weisberg M. K., Boesenberg J. S., Kozhushko G., Prinz M., Clayton R. N., and Mayeda T. K. (1995b) EH3 and EL3 chondrites: A petrologic-oxygen isotopic study (abstract). In *Lunar and Planetary Science XXVI,* pp. 1481–1482. Lunar and Planetary Institute, Houston.

Weisberg M. K., Prinz M., Clayton R. N., Mayeda T. K., Grady M. M., Franchi I., Pillinger C. T., and Kallemeyn G. W. (1996a) The K (Kakangari) chondrite grouplet. *Geochim. Cosmochim. Acta, 60,* 4253–4263.

Weisberg M. K., Prinz M., Zolensky M. E., Clayton R. N., Mayeda T. K., and Ebihara M. (1996b) Ningqiang and its relationship to oxidized CV3 chondrites (abstract). *Meteoritics & Planet. Sci., 31,* 150–151.

Weisberg M. K., Prinz M., Clayton R. N., and Mayeda T. K. (1997a) CV3 chondrites; three subgroups, not two (abstract). *Meteoritics & Planet. Sci., 32,* 138–139.

Weisberg M. K., Prinz M., and Nehru C. E. (1997b) QUE94204: An EH-chondritic melt rock (abstract). In *Lunar and Planetary Science XXVIII,* pp. 1525–1526. Lunar and Planetary Institute, Houston.

Weisberg M. K., Prinz M., Clayton R. N., Mayeda T. K., Sugiura N., Zashu S., and Ebihara M. (2001) A new metal-rich chondrite grouplet. *Meteoritics & Planet. Sci., 36,* 401–418.

Weisberg M. K., Boesenberg J. S. and Ebel D. S. (2002) Gujba and the origin of the CB chondrites (abstract). In *Lunar and Planetary Science XXXIII,* Abstract #1551. Lunar and Planetary Institute, Houston (CD-ROM).

Weisberg M. K., Connolly H. C. Jr., and Ebel D. S. (2004) Petrology and origin of amoeboid olivine aggregates in CR chondrites. *Meteoritics & Planet. Sci., 39,* 1741–1753.

Wheelock M. M., Keil K., Floss C., Taylor G. J., and Crozaz G. (1994) REE geochemistry of oldhamite-dominated clasts from the Norton County aubrite: Igneous origin of oldhamite. *Geochim. Cosmochim. Acta, 58,* 449–458.

Yanai K. (1994) Angrite Asuka-881371: Preliminary examination of a unique meteorite in the Japanese collection of Antarctic meteorites. *Proc. NIPR Symp. Antarct. Meteorites, 7,* 30–41.

Yanai K. and Kojima H. (1995) Yamato-8451: A newly identified pyroxene-bearing pallasite. *Proc. NIPR Symp. Antarct. Meteorites, 8,* 1–10.

Yamaguchi A., Clayton R. N., Mayeda T. K., Ebihara M., Oura Y., Miura Y. N., Haramura H., Misawa K., Kojima H., and Nagao K. (2002) A new source of basaltic meteorites inferred from Northwest Africa 001. *Science, 296,* 334–336.

Yugami K., Takeda H., Kojima H., and Miyamoto M. (1997) Modal abundances of primitive achondrites and the endmember mineral assemblage of the differentiated trend (abstract). *NIPR Symp. Antarct. Meteorites, 22,* 220–222.

Zolensky M. E. and Ivanov A. (2003) The Kaidun microbreccia meteorite: A harvest from the inner and outer asteroid belt. *Chem. Erde, 63,* 185–246.

Zolensky M. E., Score R., Schutt J. W., Clayton R. N., and Mayeda T. K. (1989) Lea County 001, an H5 chondrite and Lea County 002, an ungrouped chondrite. *Meteoritics, 24,* 227–232.

Zolensky M. E., Weisberg M. K., Buchanan P. C., and Mittlefehldt D. W. (1996) Mineralogy of carbonaceous chondrite clasts in HED achondrites and the moon. *Meteoritics & Planet. Sci., 31,* 518–537.

Zolensky M. E., Mittlefehldt D. W., Lipschutz M. E., Wang M-S., Clayton R. N., Mayeda T. K., Grady M. M., Pillinger C. T., and Barber D. (1997) CM chondrites exhibit the complete petrologic range from type 2 to 1. *Geochim. Cosmochim. Acta, 61,* 5099–5115.

Zolensky M. E., Nakamura K., Gounelle M., Mikouchi T., Kasama T., Tachikawa O., and Tonui (2002) Mineralogy of Tagish Lake: An ungrouped type 2 carbonaceous chondrite. *Meteoritics & Planet. Sci., 37,* 737–761.

Recent Advances in Meteoritics and Cosmochemistry

Harry Y. McSween Jr.
University of Tennessee

Dante S. Lauretta
University of Arizona

Laurie A. Leshin
Arizona State University

Major advances in research on extraterrestrial materials (meteorites, micrometeorites, interplanetary dust particles, and presolar grains) since the first edition of *Meteorites and the Early Solar System* are highlighted. Topics include the importance of newly recognized meteorite groups; new discoveries about the presolar and nebular components of chondrites and interplanetary dust particles; improvements in early solar system chronology using short-lived radionuclides; new insights into cosmochemical abundances, fractionations, and nebular reservoirs; and advances in reading nebular and prenebular records through a haze of secondary parent body processes. Some significant new insights based on these data are that the early solar system experienced a bewildering variety of nucleosynthetic inputs, the nebula was a dynamic entity with transient heating events, and the system of planetesimals and planets evolved more rapidly than previously appreciated.

1. INTRODUCTION

The goal of the first edition of *Meteorites and the Early Solar System* (*Kerridge and Matthews,* 1988) was to interpret the meteoritic record in terms of constraints on theories of the origin and early evolution of the solar system. That is also the intent of this volume. Here we briefly highlight significant new discoveries in meteoritics and cosmochemistry since the publication of the first edition. The following chapters will explain in more detail how these breakthroughs have constrained our understanding of processes and events prior to and within the early solar system. The breakthroughs introduced here are assigned to epochs, following the general organization of the book.

We acknowledge that our choices of critical advances in meteoritics and cosmochemistry are subjective but, we believe, defensible. In citing literature, we have focused on original discovery papers or, in some cases, comprehensive reviews. To make this task tractable, we consider research on meteoritic samples *per se* and do not review the considerable progress made in providing geologic context through the identification and study of their asteroid or cometary parent bodies. A comprehensive treatment of advances in asteroid spectroscopy, spacecraft encounters with asteroids, and mechanisms for the delivery of meteorites from asteroids to Earth can be found in a recently published companion volume, *Asteroids III* (*Bottke et al.,* 2002). Likewise, the context and insights provided by new astrophysical theory and astronomical observations are not catalogued here, but will be considered in other chapters. Given that the present volume is not concerned with later planetary evolution, we omit martian and lunar meteorites from consideration, except where they bear on the timing of planet accretion and early differentiation.

2. EXTRATERRESTRIAL MATERIALS AVAILABLE FOR STUDY

2.1. Newly Recognized Chondrite Groups

Meteorite classification attempts to document differences and linkages among meteorites — a critical first step because variations and trends among members of a group can reveal the processes that produced them. In the first edition, 9 groups of chondrites, 4 groups of achondrites, 2 stony-iron groups, and 13 iron meteorite groups were recognized. Other anomalous meteorites (distinct individual specimens or groups having less than five members) were also known. Since that time, new meteorites found in Antarctica and hot deserts have considerably expanded our collections, and a recent compilation of meteorites (including ungrouped samples, clasts in breccias, and micrometeorites) suggested that perhaps ~135 asteroidal parent bodies may be represented (*Meibom and Clark,* 1999).

Several new classes of chondrites — one distinct chondrite group (the Rumuruti or R group) and four carbonaceous chondrite groups (CR, CH, CB, CK) — have been defined. The R chondrites (*Bischoff et al.,* 1993; *Rubin and Kallemeyn,* 1994; *Schulze et al.,* 1994; *Kallemeyn et al.,* 1996) are the most highly oxidized anhydrous chondrite group, containing no metal and high olivine fayalite contents, with fewer chondrules than ordinary chondrites

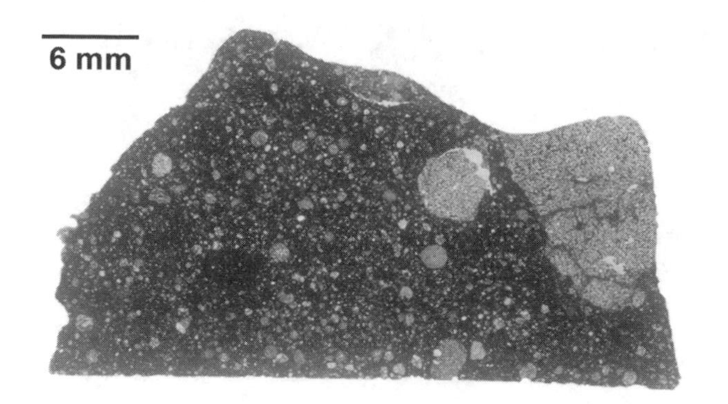

Fig. 1. Photomicrograph of the Dar al Gani 013 Rumaruti chondrite (width of section = 36 mm), showing the brecciated texture that is common in these meteorites. Large clasts at the upper center and right are impact melts, whereas other clasts are unmelted but show different degrees of thermal metamorphism. Courtesy of A. Bischoff, University of Münster.

(Fig. 1). The CR (*Weisberg et al.,* 1993), CH (*Grossman et al.,* 1988; *Scott,* 1988), and CB (*Weisberg et al.,* 2001) groups are among the most primitive types of chondrites (*Krot et al.,* 2002a). These chondrites have large amounts of metal, but are classified as carbonaceous based on chemistry and O isotopes. The CK group (*Kallemeyn et al.,* 1991) contains the most highly metamorphosed carbonaceous chondrites. However, *Greenwood et al.* (2003) have questioned whether CK and equilibrated CV chondrites are really distinct groups, and *Rubin et al.* (2003) suggested that the CB chondrite designation may be misleading because that group may have formed by nonnebular processes.

Several remarkable chondrites that have proved difficult to classify have been described as well. The newly recovered Tagish Lake meteorite (*Brown et al.,* 2000), a carbonaceous chondrite of uncertain affinity, contains volatile-element abundances straddling those of CI and CM chondrites (*Mittlefehldt,* 2002), along with chondrules and inclusions similar to those in CM chondrites (*Simon and Grossman,* 2003). However, Tagish Lake experienced a more pervasive alteration history than CMs. Anomalous ordinary chondrites of very low metamorphic grade and containing large quantities of refractory inclusions have also been described (*Kimura et al.,* 2002).

2.2. Newly Recognized Achondrite Groups

Primitive achondrites, which were not distinguished from achondrites (crystallized melts) in the first edition, are now generally recognized to be residues from partial melting (*Goodrich and Delaney,* 2000). The major-element compositions of most primitive achondrites are approximately chondritic, but elements residing in minerals with low melting points and incompatible trace elements that are concentrated in partial melts have been depleted in them. Some primitive achondrites have experienced higher degrees of melting, so that they no longer retain chondritic major-element compositions, and other primitive achondrites have experienced such low degrees of melting that they retain relict chondritic textures. However, we favor applying the classification to all residues from partial melting, regardless of the degree of melting. Iron/magnesium/manganese systematics provide an effective means of distinguishing primitive achondrites and (igneous) achondrites (*Goodrich and Delaney,* 2000). Newly classified groups of primitive achondrites include the acapulcoites-lodranites (Fig. 2) (*McCoy et al.,* 1996, 1997), brachinites (*Nehru et al.,* 1992), and winonaites (*Benedix et al.,* 1998). Ureilites, although not a newly defined group, are now also recognized as primitive achondrites, and the complementary basaltic component for these residues occurs as plagioclase-bearing clasts (*Kita et al.,* 2004).

Among the achondrites, angrites (*Mittlefehldt and Lindstrom,* 1990) constitute a newly defined group due to the addition of new meteorites. These basaltic meteorites are dominated by high abundances of refractory elements and low oxidation state. A few lunar and martian meteorites (all achondrites) were recognized prior to publication of the first edition, but not discussed in that volume; the numbers of these meteorites have markedly increased. The discovery that the NWA 011 basaltic achondrite (Fig. 3) has distinct Fe/Mn and O isotopes from otherwise similar eucrites (*Yamaguchi et al.,* 2002) demonstrates that multiple differentiated asteroids yielded basaltic meteorites.

3. THE PRESOLAR EPOCH

3.1. Presolar Grains

The study of minute stardust grains found in the fine-grained (matrix) portions of chondrites has burgeoned since the first edition, owing to the isolation of these grains in the laboratory and their analysis using the ion microprobe. The phases most intensely studied are diamond, silicon carbide, graphite, corundum, spinel, silicon nitride, and hibonite (*Anders and Zinner,* 1993; *Ott,* 1993; *Bernatowicz*

Fig. 2. Photomicrograph of Acapulco, a primitive achondrite (acapulcoite) (width of section = 4.7 mm), showing its thoroughly recrystallized texture. Courtesy of T. McCoy, Smithsonian Institution.

Fig. 3. Photomicrograph of Northwest Africa 011, a unique basaltic achondrite consisting mostly of large pigeonite grains in a groundmass of pigeonite and plagioclase (plain polarized light, width of field = 5.2 mm). Courtesy of A. Yamaguchi, National Institute of Polar Research.

and Zinner, 1997; *Zinner,* 1998; *Hoppe and Zinner,* 2000; *Nittler,* 2003). Titanium carbide and FeNi metal occur as minute inclusions within presolar graphite grains (Fig. 4). All these phases are refractory, and the carbon-bearing phases formed under reducing conditions. Presolar silicates have also now been recognized in both interplanetary dust particles (*Messenger et al.,* 2003) and chondrites (*Nguyen and Zinner,* 2004).

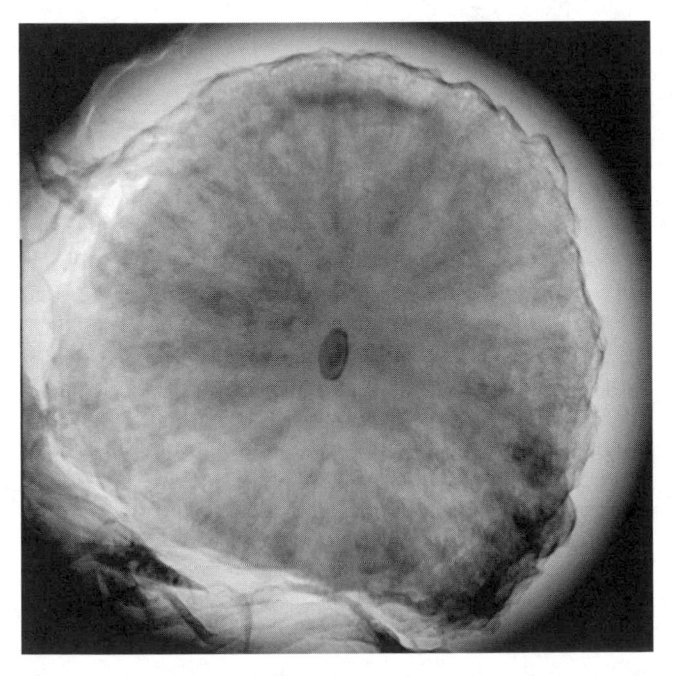

Fig. 4. Ultrathin section of presolar graphite, containing a 70-nm-long crystal of TiC that served as a nucleation center for condensation of graphite in the outflowing gas around an AGB star. The graphite structure is concentric layers, and the radial spokes are an electron diffraction effect. Courtesy of T. J. Bernatowicz, Washington University.

The exotic nature of presolar grains is demonstrated by their isotopic compositions, the diversity of which points toward formation in multiple stellar environments. For example, the carriers of the ^{22}Ne-rich components Ne-E(L) and Ne-E(H) (released at low and high temperatures respectively), which provided the first hint that chondrites contained interstellar materials, have now been identified as graphite (*Amari et al.,* 1990) and SiC (*Lewis et al.,* 1990), respectively. The morphologies and microstructures of presolar grains also reveal information on the conditions of dust formation in stellar regions (*Bernatowicz et al.,* 1996, 2003). Lack of sputtering effects and the possibility of organic surface coatings suggest mantles of ice may have protected SiC grains in the interstellar medium (*Bernatowicz et al.,* 2003). Presolar grains occur in the matrices of all chondrite groups in varying proportions (*Huss and Lewis,* 1995).

Isotopic analyses of interstellar grains have provided rigorous tests of models for nucleosynthesis and evolution in red giants, asymptotic giant branch (AGB) stars, supernovae, and novae (*Bernatowicz and Zinner,* 1997; *Zinner,* 1998; *Nittler,* 2003). These data provide stunning confirmations of predictions of the products of *s*-process nucleosynthesis, although no pure *r*-process pattern has yet been found in presolar grains. These grains also provide information on galactic chemical evolution. For example, isotopes of Si and Ti in silicon carbide do not conform to predictions for stellar nucleosynthesis and are thought to reflect ranges of initial composition of parent stars (*Alexander and Nittler,* 1999).

3.2. Interplanetary Dust Particles

Tiny interplanetary dust particles (IDPs), collected in the stratosphere by aircraft, were described in the first edition. Advances in microanalytical technology have allowed improved characterization of these very small meteoritic materials (*Rietmeijer,* 1998). Interplanetary dust particles sample, at least in part, parent bodies distinct from those of macrometeorites.

The first edition focused on the mineralogies, textures, and optical properties (from transmission spectra) of IDPs. More recent research indicates that most IDPs are more primitive than CI chondrites. *Bradley et al.* (1996) obtained the first reflectance spectra of anhydrous IDPs, for which there are no meteorite counterparts. They suggested that these IDPs are samples of primitive P or D asteroids in the outer main belt. Some anhydrous IDPs are probably from comets (*Brownlee et al.,* 1995).

Detailed studies reveal that anhydrous IDPs contain abundant (organic) molecular cloud material and silicate stardust (*Messenger,* 2000; *Messenger et al.,* 2003). Glass with embedded metal and sulfide (GEMS) (Fig. 5) are another interesting component of IDPs (*Bradley,* 1994). These grains have appreciable radiation histories, as expected if they once resided in the interstellar medium, so they may be the irradiated remnants of cosmically abundant presolar silicate grains. However, very few GEMS have anomalous

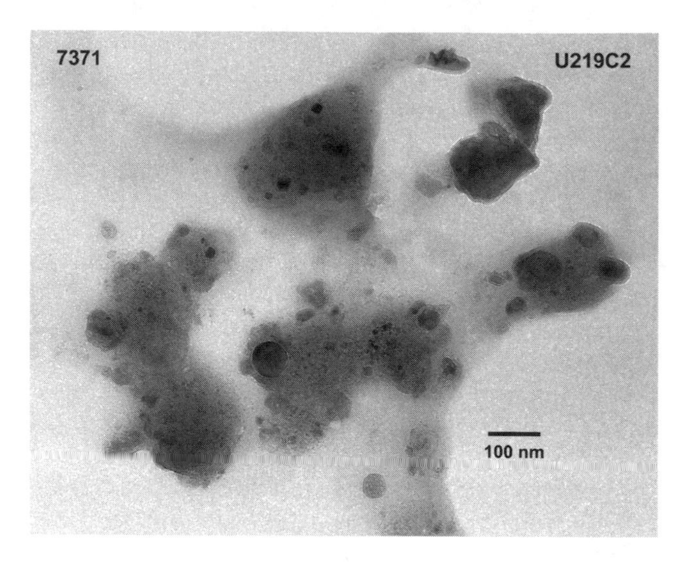

Fig. 5. Brightfield transmission electron micrograph of glass with embedded metal and sulfides (GEMS) within a thin section of an interplanetary dust particle. GEMS are composed of rounded, nanometer-sized kamacite and pyrrhotite crystals in Mg-rich silicate glass. The properties of GEMS appear to have been shaped by chronic exposure to ionizing radiation, and they resemble amorphous silicates in the interstellar medium. Courtesy of J. Bradley, Lawrence Livermore National Laboratory.

O isotopic compositions (*Messenger et al.,* 2003) of the magnitude of other presolar oxides (*Nittler et al.,* 1997).

3.3. Organic Compounds

The structure and stable isotopic composition of organic compounds in chondrites reflect complex interstellar, nebular, and planetary processes whose resolution may be confounded by terrestrial contamination (*Gilmour,* 2003). The first edition focused on abiotic synthesis mechanisms inferred for the most extensively studied organic compounds in CM carbonaceous chondrites.

Compound-specific stable-isotope measurement on meteoritic organic matter is a significant new development. The D/H enrichment of bulk organic matter, described in the first edition, has now been measured in amino and sulfonic acids (*Cooper et al.,* 1997) and suggests synthesis by ion-molecule reactions in the interstellar medium. Enrichment of ^{33}S in sulfonic acid is attributed to UV irradiation in an interstellar environment (*Cooper et al.,* 1997). Values of δ^{13}C in low-molecular-weight compounds become more positive as the amount of C in the molecule increases, whereas the opposite trend is observed in high-molecular-weight compounds (*Sephton et al.,* 1999, 2000). These are kinetic signatures of both bond formation (synthesis from simpler compounds) and bond destruction (cracking of higher homologs to form lower ones). Such trends may suggest that the carbon skeletons formed within icy mantles on interstellar grains, where radiation-induced reactions create and destroy organic matter.

A significant discovery concerning organic compounds is the determination that nonbiogenic amino acids are present as nonracemic mixtures in carbonaceous chondrites. Small but measurable excesses of the L-enantiomers of amino acids have been demonstrated to have a nonbiologic (*Cronin and Pizzarello,* 1997) and nonterrestrial (*Engel and Macko,* 1997) origin. How slightly optically active organic compounds would be produced in the interstellar medium is not understood, but selective destruction of one enantiomer by UV circularly polarized light from neutron stars in a presolar cloud has been invoked. Hydrocarbons with large H- and N-isotopic anomalies in anhydrous IDPs likely formed in molecular clouds (*Keller et al.,* 2004), although organic matter occurs in comparable abundance in hydrated IDPs (*Flynn et al.,* 2003).

4. THE DISK FORMATION EPOCH

4.1. Presolar Grain Survival and Chronology

Because of destructive oxidation reactions, the lifetimes of presolar SiC grains exposed to a hot nebula are short compared to nebula cooling timescales (*Mendybaev et al.,* 2002). Attempts to date presolar silicon carbide grains based on the production of spallogenic ^{21}Ne give surprisingly low ages of ~10–130 Ma (*Tang and Anders,* 1988). However, recoil experiments suggest that micrometer-sized SiC grains will have lost virtually all Ne produced in the interstellar medium (*Ott and Begemann,* 2000), casting doubt on these ages.

4.2. Nebular Elemental and Isotopic Abundances

Anders and Grevesse (1989), *Lodders* (2003), and *Palme and Jones* (2003) compiled new average solar system elemental and isotopic abundances for CI chondrites. These were compared with solar photosphere compositions, which have also been updated (*Palme and Jones,* 2003, and references therein). The solar O abundance has been revised downward by 50% and the C abundance by a smaller amount (*Allende Prieto et al.,* 2001), leading to a higher C/O ratio and thus somewhat more reducing nebular gas than envisioned in the first edition (*Allende Prieto et al.,* 2002). There have also been significant downward revisions in S, Se, and P, and new solar abundances of light elements support the previous conclusion that Li is depleted in the Sun.

The smoothness of heavy-element abundance patterns was formerly used to support the choice of CI chondrites in determining solar system abundances; however, some measured discontinuities are real (*Burnett et al.,* 1989) so this reasoning is flawed. The recognition that sulfate veins in Orgueil probably formed during the meteorite's residence on Earth (*Gounelle and Zolensky,* 2001) suggests open-system modification of soluble components like S. The Tagish Lake meteorite has been suggested to be a better proxy for solar system abundances than CI chondrites (*Brown et al.,* 2000; *Zolensky et al.,* 2002).

4.3. Isotopic Reservoirs

The slope of the O-isotope mixing line for bulk refractory inclusions (calcium-aluminum-rich inclusions, or CAIs) is ~0.95, but new microanalyses of inclusions suggest a slope of 1.00 (*Young and Russell,* 1998). The distinction is important for distinguishing between competing mechanisms for production of a ^{16}O-rich reservoir. One interpretation is that this reservoir represents a nucleosynthetic enrichment in the nebula, possibly inherited from a nearby supernova. An alternative hypothesis is mass-independent fractionation of ^{16}O (*Thiemens,* 1996), perhaps through self-shielding of CO in the nebula (*Clayton,* 2002). A 0.95 slope would rule out the possibility of mass-independent chemical fractionation of O isotopes in the nebula, but either mechanism is permitted by a 1.00 slope (*Ireland and Fegley,* 2000). Analyses of O isotopes in condensate olivines in amoeboid olivine aggregates (AOAs) suggest a ^{16}O-rich nebular gas (*Krot et al.,* 2002b), whereas chondrules generally formed in equilibrium with a more ^{16}O-poor gas. The ^{16}O-rich gas may have resulted from mass-independent fractionation or evaporation of dust-enhanced regions.

5. THE FIRST NEBULAR EPOCH

5.1. Nebular Chronology and Origin of Short-lived Radioisotopes

The first edition included seven chapters on chronology, reflecting the importance of understanding the temporal sequence and duration of early solar system events and processes. The most significant new advance in meteorite geochronology is the development of new high-resolution chronometers based on short-lived radionuclides. The six extinct radionuclides discussed in the first edition have now been augmented by meteorite dating techniques based on ^{10}Be-^{10}B, ^{41}Ca-^{41}K, ^{60}Fe-^{60}Ni, ^{146}Sm-^{142}Nd, ^{182}Hf-^{182}W, ^{92}Nb-^{92}Zr, ^{187}Re-^{187}Os, and ^{107}Pd-^{107}Ag (*McKeegan and Davis,* 2004). A major challenge in using these isotopic systems is in understanding how they were produced, which bears on what events are actually dated.

Significant advances have been made in the ^{53}Mn-^{53}Cr (*Lugmair and Shukolyukov,* 2001) system, which appears to be a useful chronometer despite evidence that ^{53}Mn was not distributed homogeneously in the nebula. This chronometer dates fractionation of Mn from Cr, due primarily to Mn volatility (*Palme,* 2001). However, ^{53}Mn data have found a variety of applications, such as constraining the timing of aqueous alteration to form carbonates in carbonaceous chondrites to <20 m.y. after accretion (*Endress et al.,* 1996).

Considerable attention has also been focused on ^{26}Al-^{26}Mg. The idea that ^{26}Al was homogenously distributed in the neb-ula (*MacPherson et al.,* 1995) has been challenged by argu-ments that it was produced by local irradiation in an X-wind region of the early Sun (*Gounelle et al.,* 2001). If refractory condensates incorporated this nuclide and were then dispersed, the canonical $^{26}Al/^{27}Al$ ratio would not apply to other nebular materials, perhaps explaining an apparent age difference of several million years (based on ^{26}Al) between CAIs and chondrules. However, this age difference is supported by ^{53}Mn and ^{129}I data (*Swindle et al.,* 1996) and very precise Pb-Pb ages (*Amelin et al.,* 2002) for CAIs and chondrules.

As noted in the first edition, short-lived radionuclide chronometers commonly show inconsistent timescales, a problem that has not yet been fully resolved. However, ^{129}I-^{129}Xe dates for unshocked meteorites can now be reconciled with ^{26}Al and ^{53}Mn chronologies (*Gilmour et al.,* 2000).

The age of the solar system is now based on a combination of long- and short-lived chronometers applied to refractory inclusions in chondrites. $^{235}U/^{238}U$-$^{207}Pb/^{206}Pb$ ages of 4566 Ma for Allende CAIs (*Allègre et al.,* 1995) and 4567 Ma for Efremovka CAIs (*Amelin et al.,* 2002) may be a lower age limit if the inclusions were altered. This age is slightly older than the 4559-Ma estimate in the first edition. *Lugmair and Shukolyukov* (2001) defined lower (4568 Ma) and upper (4571 Ma) age limits, based on constraints from ^{53}Mn-^{53}Cr, in agreement with the upper limit for the CAI ^{207}Pb-^{206}Pb age and with ^{26}Al and ^{129}I data for the ordinary chondrite Ste. Marguerite.

Unambiguous evidence of three especially important extinct radionuclides has been discovered since the first edition. Two of these bear directly on the question of the sources of short-lived nuclides, with apparently contradictory implications. The finding of live ^{10}Be in CAIs (*McKeegan et al.,* 2000) is important because this nuclide is not produced via nucleosythesis in stars and thus is generally interpreted to be evidence for local production of at least some short-lived isotopes by spallation reactions in the nebula. However, ^{10}Be and ^{26}Al are decoupled in many CAIs (*Marhas et al.,* 2002), suggesting that ^{26}Al was not a product of spallation. Evidence in meteorites for live ^{60}Fe (*Shukolyukov and Lugmair,* 1993; *Tachibana and Huss,* 2003), which is only formed through stellar nucleosynthesis, also contradicts the idea that all short-lived radioactivities were locally produced. These nuclides were probably produced in stars and imported within presolar grains (*MacPherson et al.,* 2003). Finally, with a half-life of 103 Ka, ^{41}Ca is the shortest-lived isotope yet confirmed to be present in chondrites (*Srinivasan et al.,* 1994). This age constrains the timescale from synthesis to injection of short-lived isotopes to be very short.

5.2. Nebular Condensates and Residues

The requirement for a hot nebula to allow vaporization and condensation, a popular idea in the first edition, has now been circumvented by models in which heating of the disk occurs close to the protosun and the materials are redistributed outward (the X-wind model). *Shu et al.* (1996) described a scenario in which various chondrite components might have been formed far from the positions of their accretion into chondrites. The existence of ^{10}Be in CAIs (*McKeegan et al.,* 2000), not formed during stellar nucleosyn-

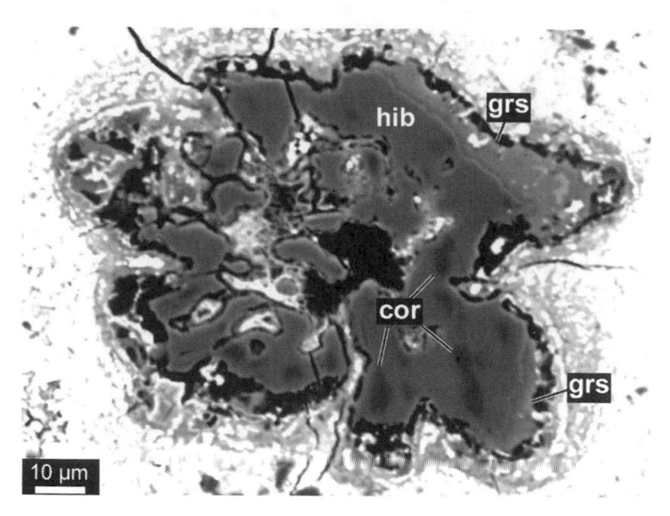

Fig. 6. Backscattered electron image of a CAI in the Adelaide carbonaceous chondrite. Corundum (cor) is replaced by hibonite (hib), which is in turn corroded by grossite (grs). The inferred crystallization sequence is consistent with the equilibrium condensation at a total pressure $\leq 10^{-5}$ bar and ~1000 CI dust/gas enrichment relative to solar composition. Courtesy of A. Krot, University of Hawai'i.

thesis, has been taken as support for the X-wind model. However, there are still many proponents of the hot nebula hypothesis, which is consistent with models and observations of the earliest era of disk formation when mass accretion rates were high (e.g., *Bell et al.,* 2000). Regardless of which model is advocated, it is clear that we envision a more dynamic nebula with transient heating events.

Refractory inclusions (CAIs) were described in the first edition as nebular condensates, commonly modified by melting, recrystallization, and alteration (Fig. 6). These ancient objects are now generally understood to be mixtures of condensate materials and refractory residues from evaporation (*Ireland and Fegley,* 2000). High-temperature condensation is required to produce the observed fractionations in refractory trace elements, whereas the widespread persistence of mass-dependent isotopic fractionation in refractory elements such as Mg and Si indicates evaporation (*Grossman et al.,* 2002). It is now recognized that the large CAIs in Allende have been extensively processed after accretion and are not representative of CAIs in other meteorite classes. Calcium-aluminum-rich inclusions in other chondrite groups have now been characterized (*MacPherson,* 2003).

Amoeboid olivine aggregates (AOAs) are the most common type of refractory inclusions in many types of carbonaceous chondrites. Detailed studies of AOAs, which provide a compositional link between CAIs and chondrules, indicate they are clumps of lower-temperature nebula condensates that originated in a ^{16}O-rich gaseous reservoir (*Krot et al.,* 2002b). Smooth, concentric zoning of some FeNi metal grains in primitive CH chondrites suggests their formation by gas-solid condensation (*Meibom et al.,* 1999). These grains (Fig. 7) may be the first documented metal condensates.

5.3. Nebular Chemical Fractionations

Cosmochemical behavior is dominated by element volatility, presumably reflecting either incomplete condensation or evaporation. The refractory elements are always assumed to be present in chondritic (CI) relative proportions; however, two cases have now been shown to violate this assumption. Refractory Re/Os ratios in carbonaceous chondrites are systematically lower than those in ordinary and enstatite chondrites (*Walker et al.,* 2002), and Nb/Ta ratios in chondrites are variable (*Weyer et al.,* 2003).

Moderately volatile elements are depleted in achondrite parent bodies and in planets, relative to chondrites. However, measured ^{41}K/^{39}K ratios among the terrestrial planets, the Moon, achondrites, and chondrites are uniform to within ~2‰ (*Humayun and Clayton,* 1995). This lack of K-isotope fractionation has been argued to preclude evaporation as an explanation for the observed volatile depletions. However, evaporation of nebular crystalline dust prior to its accretion into planetesimals is possible (*Young,* 2000).

6. THE SECOND NEBULAR EPOCH

6.1. Chondrules

Four chapters in the first edition were devoted to chondrules. New observational, experimental, and theoretical studies of chondrules have resulted in some changes, many of which were detailed in *Chondrules and the Protoplanetary Disk* (*Hewins et al.,* 1996). The presence of C as a reducing agent in chondrule precursors produces characteristic mineralogical features of chondrules (*Connolly et al.,* 1994). Melting intervals for chondrules, previously thought

Fig. 7. X-ray elemental map in Ni Kα of a zoned FeNi metal grain from the Hammadah al Hamra 237 carbonaceous chondrite. This metal grain is inferred to be a nebular condensate. Courtesy of A. Krot, University of Hawai'i.

to have lasted for hours, are now restricted to a few minutes, to allow retention of moderately volatile elements (*Hewins,* 1997). Moreover, volatile retention apparently requires that chondrules could not have formed in the canonical nebula, but must have formed by flash heating at higher pressures (*Yu and Hewins,* 1998). Many chondrules appear to have experienced multiple heating events (*Rubin and Krot,* 1996), and most were not completely melted. Correlations between the relative ages, based on ^{26}Al, and bulk compositions of chondrules (*Mostefaoui et al.,* 2002) suggest that Si and volatile elements evaporated from chondrule melts and then recondensed as precursors for the next generation of chondrules. Experiments constrain peak temperatures to 1770–2120 K, and chondrule cooling rates were decidedly nonlinear, very fast and then slow, consistent with shock melting (*Desch and Connolly,* 2002). The origin of chondrules remains enigmatic, but thermal processing of particles in nebular shock waves appears to meet most constraints imposed by chondrule properties (*Ciesla and Hood,* 2002; *Desch and Connolly,* 2002).

6.2. Metal and Sulfide

Metal grains in ordinary chondrites, suggested to be condensates in the first edition, have apparently been modified after accretion (*Zanda et al.,* 1994). Sulfides in chondrites are now understood to have formed by reaction of metal with the gaseous nebula. *Lauretta et al.* (1997) determined that corrosion of FeNi metal by H_2S produces Fe-bearing sulfides rapidly enough to occur within the lifetime of the nebula, and showed that sulfides produced in this way are morphologically similar to those in chondrite matrix.

6.3. Matrix

The mineralogy of chondrite matrices has become increasingly well characterized by application of transmission electron microscopy. Matrix is most easily studied in carbonaceous chondrites (*Buseck and Hua,* 1993), where it is most abundant. New descriptions of matrix in the most primitive carbonaceous chondrites reveal amorphous silicate material, forsterite, and enstatite, with minor carbonaceous, refractory, and presolar materials — apparently mixtures of nebular materials and interstellar grains. Hydrothermal alteration of such matrix has produced the matrices of most carbonaceous chondrites. The mineralogy of interstitial matrix and rims on chondrules in unequilibrated ordinary chondrites (*Brearley,* 1996) is more complex, consisting mostly of olivine, sometimes with amorphous silicate material, and pyroxenes, feldspars, metal, a variety of spinels and sulfides, whitlockite, and various alteration phases.

6.4. Organic Compounds

The Tagish Lake meteorite's fall and clean collection under subfreezing conditions make it a relatively uncontaminated sample. The organic components of Tagish Lake are dominated by soluble carboxyl and dicarboxyl compounds but few amino compounds. Its abundant aromatic compounds have a more restricted distribution than those in other carbonaceous chondrites (*Pizzarello et al.,* 2001), possibly reflecting catalytic processes that tend to be more selective than other synthetic pathways. Tiny, hollow organic globules (Fig. 8) in Tagish Lake (*Nakamura et al.,* 2002) are similar to material produced by laboratory simulation of UV photolysis of interstellar ice analogs.

Calculations support the idea that the soluble fraction of the organic compounds in chondrites formed during aqueous alteration of parent bodies (*Schulte and Shock,* 2004). Some progress has been made in characterizing the insoluble macromolecular organic matter in carbonaceous chondrites, cited as a daunting task in the first edition. Similarities among macromolecules in CI and CM chondrites suggest that essentially the same aromatic structural units were common to all (*Sephton et al.,* 1998, 2000). *Cody et al.* (2002) determined that this material in Murchison was 61–66% aromatic and that its aliphatic chains are highly branched.

6.5. Cosmogenic Nuclides

There is continuing disagreement over whether or not meteorites show noble gas isotopic evidence of irradiation by an energetic early Sun (*Woolum and Hohenberg,* 1993; *Wieler et al.,* 2000). As noted earlier, however, the discovery of the former presence of ^{10}Be in refractory inclusions (*McKeegan et al.,* 2000) attests to intense irradiation in the early solar system. The production by spallation of other short-lived radionuclides found in chondrites has also been suggested (*Gounelle et al.,* 2001); however, ^{10}Be and ^{26}Al are decoupled in many CAIs (*Marhas et al.,* 2002). Some short-lived nuclides appear to have been produced by spallation in the nebular, some by nucleosynthesis in other stars, and some possibly (but not necessarily) by both processes.

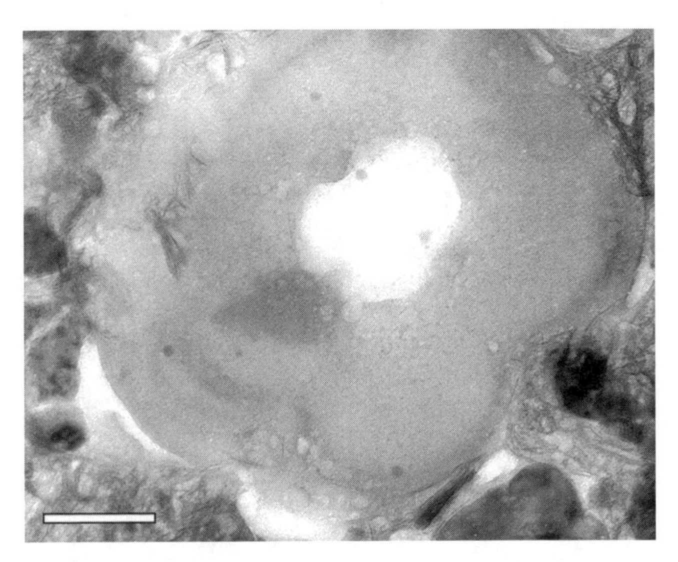

Fig. 8. TEM image of a hollow globule of organic matter (scale bar = 200 nm) in the Tagish Lake carbonaceous chondrite. Courtesy of K. Nakamura, Kobe University.

7. THE ACCRETION EPOCH

7.1. Mixing and Sorting of Nebular Components

Presolar grains constrain mixing between reservoirs in the nebula. The average isotopic compositions of carbon and silicon in SiC grains are the same among different meteorite groups, as would be expected if the nebula sampled a well-mixed reservoir of debris from various stars (*Huss and Lewis,* 1995; *Russell et al.,* 1997). The limited size distributions of chondrules are now explained as the result of aerodynamic sorting in turbulent nebula eddies (*Cuzzi et al.,* 1996; *Kuebler et al.,* 1999).

7.2. Sampling Other Accreted Materials

Calculations of IDP orbital evolution show that the atmospheric entry velocity of comet-derived dust is generally higher than that of asteroid-derived particles. Entry velocity controls the degree of atmospheric heating, which, in turn, affects the abundances of ^{4}He (*Nier and Schlutter,* 1993) and other volatile elements (especially zinc) (*Kehm et al.,* 2002), as well as mineralogy (*Greshake et al.,* 1998). These changes can potentially be used to identify IDPs of cometary origin.

Micrometeorites are larger than IDPs and are commonly melted partly or completely during atmospheric passage. The resulting cosmic spherules represent the peak in mass distribution of meteoritic materials accreted to Earth, so a comprehensive chemical study of them may provide a less-biased sampling of asteroid compositions than do meteorites. Most cosmic spherules have CM-chondrite-like compositions (*Brownlee et al.,* 1997).

8. THE PARENT BODY EPOCH

The nebular record in meteorites is commonly obscured, at least in part, by parent-body processes — thermal metamorphism, aqueous alteration, melting, impact brecciation, and shock metamorphism. The first edition devoted six chapters to these topics, but ambiguities remained. Significant advances in unraveling these geologic overprints in meteorites now enable a better understanding of which materials and properties are of nebular origin and which are not.

8.1. Chondritic Parent Bodies

All asteroids have been heated to some degree, probably by radioactive decay of short-lived nuclides such as ^{26}Al (*MacPherson et al.,* 1995; *Russell et al.,* 1996; *Srinivasan et al.,* 1999; *Kita et al.,* 2000). Thermal evolution models based on ^{26}Al heating provide a context for interpreting measureable properties in chondrites — equilibration temperatures, cooling rates, and radiometric ages (*McSween et al.,* 2002). These models are appropriate for bodies that develop concentric metamorphic zones, and the existence of such "onion-shell" asteroids has been demonstrated by measuring the times of cooling for ordinary chondrites metamorphosed at various peak temperatures (*Göpel et al.,* 1994; *Trieloff et al.,* 2003). Differences in ^{182}Hf-^{182}W and ^{235}U/^{238}U-^{207}Pb/^{206}Pb ages (equal to 2–12 Ma for H chondrites) have been used to estimate the interval of ordinary chondrite metamorphism (*Humayun and Campbell,* 2002).

The chondrite classification system based partly on thermal metamorphic grade was in wide use prior to the first edition but the action of fluids, indicated by changes in oxidation state during ordinary chondrite metamorphism (*McSween and Labotka,* 1993), had not been recognized. Scales quantifying the intensity of aqueous alteration (*Browning et al.,* 1996) and shock metamorphism (*Stöffler et al.,* 1991; *Scott et al.,* 1992; *Rubin et al.,* 1997) now allow effects accompanying these processes to be discerned. Noteworthy examples are destruction of presolar grains (*Huss,* 1990), the redistribution of minor elements in chondritic metal during metamorphism (*Zanda et al.,* 1994), and the synthesis of new organic compounds during aqueous alteration and heating (*Sephton et al.,* 2003; *Pearson et al.,* 2002). In the first edition, aqueous alteration was described only as a parent-body process, still a prominent view (*Krot et al.,* 1995). However, other observations suggest some carbonaceous chondrite components may have been altered before accretion, either in the nebula or within small, precursor planetesimals (*Metzler et al.,* 1992).

Halite-bearing clasts (Fig. 9) have been discovered in several recently fallen ordinary chondrite regolith breccias (*Zolensky et al.,* 1999). Irradiation effects show that these salts are preterrestrial, and fluid inclusions demonstrate that they formed by precipitation from aqueous liquids at low temperatures in asteroid regoliths (*Whitby et al.,* 2000).

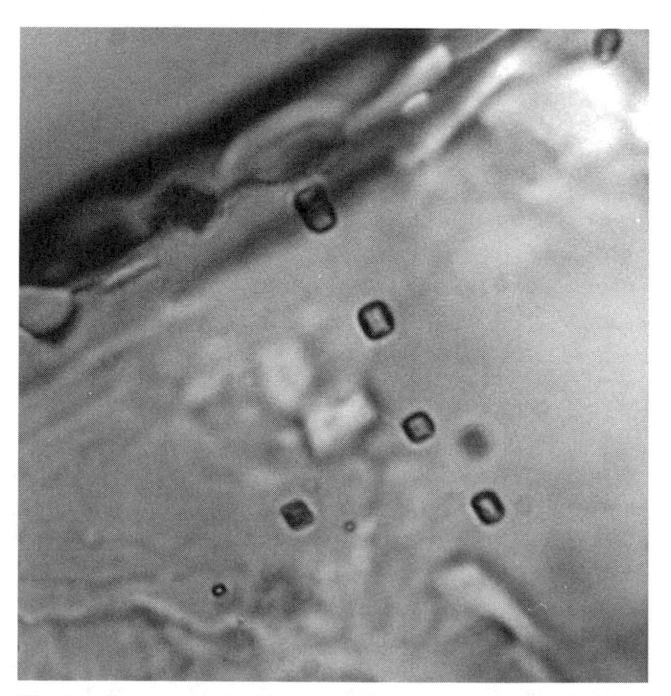

Fig. 9. Halite crystals (3–4 nm) in the Zag ordinary chondrite demonstrate the former presence of evaporating fluids in asteroids. Courtesy of M. Zolensky, NASA Johnson Space Center.

8.2. Differentiated Parent Bodies

A new development in understanding achondrite magmatism is evidence supporting that eucrites and related diogenites may have formed in a magma body of asteroidal scale (*Righter and Drake,* 1997; *Ruzicka et al.,* 1997). However, despite extensive petrographic and geochemical studies, there is still little consensus on major questions of eucrite petrogenesis (*Mittlefehldt et al.,* 1998). Angrites provide an interesting comparison with eucrites, mimicking the dichotomy of alkaline and tholeiitic basalts on Earth (*Mittlefehldt and Lindstrom,* 1990). Angrites and eucrites have similar O-isotopic compositions (*Clayton and Mayeda,* 1996) and experiments suggest they could have been produced by melting similar chondritic protoliths under different redox conditions. The paucity of meteoritic basalt groups relative to other differentiated meteorites might be explained by explosive volcanism (*Wilson and Keil,* 1991).

Ureilites, the most perplexing primitive achondrites, were formerly argued to be cumulates but are now generally viewed as smelting residues (*Goodrich,* 1992; *Walker and Grove,* 1993; *Singletary and Grove,* 2003). Supporting the interpretation that the ureilite parent body did not undergo extensive melting is the surprising finding that ureilites have primitive O-isotopic compositions that correlate with the Fe contents of olivine and pyroxene (*Clayton and Mayeda,* 1988).

New information on Fe-Ni phase equilibria, such as a revised Fe-Ni phase diagram (*Yang et al.,* 1996), has provided an improved understanding of the cooling histories of asteroid cores. An investigation of Ni-rich areas of widmanstätten patterns in iron meteorites (*Yang et al.,* 1997) indicated these are not taenite, as previously thought, but intergrowths of tetrataenite, martensite, and awaruite. This revelation constitutes a drastic revision of iron meteorite mineralogy, and its implications for Fe-Ni phase relations are not yet clear. *Hopfe and Goldstein* (2001) have significantly revised the metallographic cooling rate method.

The difficulties in segregating metallic liquids from host silicates have been clarified (*Taylor,* 1992). Numerical models of fractional crystallization in irons have become more sophisticated, and now explore the effects of imperfect mixing, liquid immiscibility, and trapping of S-rich liquid (*Haack and McCoy,* 2003, and references therein). Application of Re-Os dating has demonstrated that the cores representing all the various iron meteorite groups crystallized within 100 m.y. of each other (*Smoliar et al.,* 1996; *Shen et al.,* 1996).

9. THE PLANETARY EPOCH

9.1. Formation Chronology

The application of ^{182}Hf-^{182}W to iron meteorites and chondrites has constrained the early times of core formation in asteroids (*Lee and Halliday,* 1997). A revision in the W-isotopic composition for chondrites (*Kleine et al.,* 2002; *Schoenberg et al.,* 2002; *Yin et al.,* 2002) significantly lowered the formation ages of Earth (11–30 Ma) and Mars (<15 Ma) relative to the formation time of CAIs, suggesting rapid accretion and differentiation of the terrestrial planets.

9.2. Organic and Putative Biogenic Substances

Chyba and Sagan (1992) concluded that the dominant source of organic material to the early Earth was IDPs, which are ~10% organics by mass, because their gentle atmospheric deceleration could deliver organic compounds intact.

No review of advances in meteoritics or biogenic molecules would be complete without mentioning ALH 84001, originally misclassified as a diogenite but recognized as a unique martian meteorite by *Mittlefehldt* (1994). If for no other reason, ALH 84001 is important as the oldest (~4.5 Ga) planetary sample (*Nyquist et al.,* 2001). The proposal that this meteorite contains evidence of extraterrestrial life (*McKay et al.,* 1996) generated a firestorm of controversy that placed meteoritics on the front pages of newspapers. The meteoritics community responded with an outpouring of research on carbonates, nanophase magnetites, putative microfossils, and organic compounds. Although few planetary scientists have accepted the proffered evidence for this hypothesis, its lingering effects are a revitalized Mars exploration program focused on the search for life, as well as a significant increase in the price of meteorites. Scientific consequences include research on the formation and stability of polycyclic aromatic hydrocarbons, recognition of the difficulties in identifying potential mineralogic or isotopic biomarkers and fossils, discovery of novel mechanisms for carbonate formation, and acknowledgment that terrestrial micro-organisms eventually contaminate all meteorites (*Steele et al.,* 2000).

9.3. Implications of Cosmic-Ray Exposure Ages

The large number of exposure ages for different meteorite groups now available (*Wieler and Graf,* 2001; *Herzog,* 2003) permits new insights into the orbital mechanics of meteoroids. The pronounced age clumps for various groups suggest that a small number of collisions produced a large fraction of known meteorites. In the first edition, the reigning view was that Earth capture depended on impact injection of meteoroids into chaotic resonances. However, the calculations of *Gladman et al.* (1997) showed that transport through resonances took much less time than clocked by cosmic-ray exposure ages. The order-of-magnitude discrepancies have now been resolved by considering the role of the Yarkovsky effect (small accelerations produced by asymmetric solar heating) in slowly moving meteoroids into resonances (*Farinella et al.,* 1998).

Acknowledgments. Thoughtful reviews by H. Palme and E. Zinner have improved this chapter. We also thank various individuals who provided images, each identified in the respective captions.

REFERENCES

Alexander C. M. O'D. and Nittler L. R. (1999) The galactic chemical evolution of Si, Ti and O isotopes. *Astrophys. J., 519,* 222–235.

Allègre C. J., Gerard M., and Göpel C. (1995) The age of the Earth. *Geochim. Cosmochim. Acta, 59,* 1445–1456.

Allende Prieto C., Lambert D. L., and Asplund M. (2001) The forbidden abundance of oxygen in the sun. *Astrophys. J. Lett., 556,* L63–L66.

Allende Prieto C., Lambert D. L., and Asplaund M. (2002) A reappraisal of the solar photospheric C/O ratio. *Astrophys. J. Lett., 573,* L137–L140.

Amari S., Anders E., Virag A., and Zinner E. (1990) Interstellar graphite in meteorites. *Nature, 345,* 238–240.

Amelin Y., Krot A. N., Hutcheon I. D., and Ulyanov A. A. (2002) Lead isotopic ages of chondrules and calcium-aluminum-rich inclusions. *Science, 297,* 1678–1683.

Anders E. and Grevesse N. (1989) Abundances of the elements: Meteoritic and solar. *Geochim. Cosmochim. Acta, 53,* 197–214.

Anders E. and Zinner E. (1993) Interstellar grains in primitive meteorites: Diamond, silicon carbide and graphite. *Meteoritics, 28,* 490–514.

Bell K. R., Cassen P. M., Wasson J. T., and Woolum D. S. (2000) The FU Orionis phenomenon and solar nebular material. In *Protostars and Planets IV* (V. Manning et al., eds.), pp. 897–926. Univ. of Arizona, Tucson.

Benedix G. K., McCoy T. J., Keil K., Bogard D. D., and Garrison D. H. (1998) A petrologic and isotopic study of winonaites: Evidence for early partial melting, brecciation, and metamorphism. *Geochim. Cosmochim. Acta, 62,* 2535–2554.

Bernatowicz T. J. and Zinner E. K. (1997) *Astrophysical Implications of the Laboratory Study of Presolar Materials*. American Institute of Physics, New York. 750 pp.

Bernatowicz T. J., Cowsik R., Gibbons P. C., Lodders K., Fegley B., Amari S., and Lewis R. S. (1996) Constraints on stellar grain formation from presolar graphite in the Murchison meteorite. *Astrophys. J., 472,* 760–782.

Bernatowicz T. J., Messenger S., Pravdivtseva O., Swan P., and Walker R. M. (2003) Pristine presolar silicon carbide. *Geochim. Cosmochim. Acta, 67,* 4679–4691.

Bischoff A., Palme H., Schultz L., Weber D., Weber H. W., and Spettel B. (1993) Acfer 182 and paired samples, an iron-rich carbonaceous chondrite: Similarities with ALH85085 and its relationship to CR chondrites. *Geochim. Cosmochim. Acta, 57,* 2631–2648.

Bottke W. F. Jr., Cellino A., Paolicchi P., and Binzel R. P., eds. (2002) *Asteroids III.* Univ. of Arizona, Tucson. 785 pp.

Bradley J. P. (1994) Chemically anomalous, preaccretionally irradiated grains in interplanetary dust from comets. *Science, 265,* 925–929.

Bradley J. P., Keller L. P., Brownlee D. E., and Thomas K. L. (1996) Reflectance spectroscopy of interplanetary dust particles. *Meteoritics & Planet. Sci., 31,* 394–402.

Brearley A. J. (1996) The nature of matrix in unequilibrated chondritic meteorites and its possible relationship to chondrules. In *Chondrules and the Protoplanetary Disk* (R. H. Hewins et al., eds.), pp. 137–152. Cambridge Univ., Cambridge.

Brown P. G., Hildebrand A. R., Zolensky M. E., Grady M., Clayton R. N., Mayeda T. K., Taliaferri E., Spanding R., MacRae N. D., Hoffman E. L., Mittlefehldt D. W., Wacker J. F., Bird J. A., Campbell M. D., Carpenter R., Gingerich H., Glatiotis M., Greiner E., Mazur M. J., McCausland P. J., Plotkin H., and Mazur T. R. (2000) The fall, recovery, orbit, and composition of the Tagish Lake metorite: A new type of carbonaceous chondrite. *Science, 290,* 320–325.

Browning L. B., McSween H. Y., and Zolensky M. E. (1996) Correlated alteration effects in CM carbonaceous chondrites. *Geochim. Cosmochim. Acta, 60,* 2621–2633.

Brownlee D. E., Joswiak D. J., Schlutter D. J., Pepin R. O., Bradley J. P., and Love S. J. (1995) Identification of individual cometary IDPs by thermally stepped He release (abstract). In *Lunar and Planetary Science XXVI,* pp. 183–184. Lunar and Planetary Institute, Houston.

Brownlee D. E., Bates B., and Schramm L. (1997) The elemental composition of stony cosmic spherules. *Meteoritics & Planet. Sci., 32,* 157–175.

Burnett D. S., Woolum D. S., Benjamin T. M., Rogers P. X. Z., Duffy C. J., and Maggiore C. (1989) A test of the smoothness of the elemental abundances of carbonaceous chondrites. *Geochim. Cosmochim. Acta, 53,* 471–481.

Buseck P. R. and Hua X. (1993) Matrices of carbonaceous chondrite meteorites. *Annu. Rev. Earth Planet. Sci., 21,* 255–305.

Chyba C. F. and Sagan C. (1992) Endogenous production, exogenous delivery and impact-shock synthesis of organic molecules: An inventory for the origins of life. *Nature, 355,* 125–132.

Ciesla F. J. and Hood L. L. (2002) The nebular shock wave model for chondrule formation: Shock processing in a particle-gas suspension. *Icarus, 158,* 281–293.

Clayton R. N. (2002) Self-shielding in the solar nebula. *Nature, 415,* 860–861.

Clayton R. N. and Mayeda T. K. (1988) Formation of ureilites by nebular processes. *Geochim. Cosmochim. Acta, 52,* 1313–1318.

Clayton R. N. and Mayeda T. K. (1996) Oxygen isotope studies of achondrites. *Geochim. Cosmochim. Acta, 60,* 1999–2017.

Cody G. D., Alexander C. M. O'D., and Tera F. (2002) Solid-state (^{1}H and ^{13}C) nuclear magnetic resonanance spectroscopy of insoluble organic residue in the Murchison meteorite: A self-consistent quantitative analysis. *Geochim. Cosmochim. Acta, 66,* 1851–1865.

Connolly H. C., Hewins R. H., Ash R. D., Zanda B., Lofgren G. E., and Bourot-Denise M. (1994) Carbon and the formation of reduced chondrules. *Nature, 371,* 136–139.

Cooper G. W., Thiemens M. H., Jackson T.-L., and Chang S. (1997) Sulfur and hydrogen isotope anomalies in meteorite sulfonic acids. *Science, 277,* 1072–1074.

Cronin J. R. and Pizzarello S. (1997) Enantiomeric excesses in meteoritic amino acids. *Science, 275,* 951–955.

Cuzzi J. N., Dobrovolskis A. R., and Hogan R. C. (1996) Turbulence, chondrules, and planetesimals. In *Chondrules and the Protoplanetary Disk* (R. H. Hewins et al., eds.), pp. 35–43. Cambridge Univ., New York.

Desch S. J. and Connolly H. C. (2002) A model of the thermal processing of particles in solar nebula shocks: Application to the cooling rates of chondrules. *Meteoritics & Planet. Sci., 37,* 183–207.

Endress M., Zinner E., and Bischoff A. (1996) Early aqueous activity on primitive meteorite parent bodies. *Nature, 279,* 701–703.

Engel M. H. and Macko S. A. (1997) Isotopic evidence for extraterrestrial non-racemic amino acids in the Murchison meteorite. *Nature, 389,* 265–268.

Farinella P., Vokrouhlický D., and Hartmann W. (1998) Meteorite delivery via Yarkovsky orbital drift. *Icarus, 132,* 378–387.

Flynn G. J., Keller L. P., Feser M., Wirick S., and Jacobsen C. (2003) The origin of organic matter in the solar system: Evidence from the interplanetary dust particles. *Geochim. Cosmochim. Acta, 67,* 4791–4806.

Gilmour I. (2003) Structural and isotopic analysis of organic matter in carbonaceous chondrites. In *Treatise on Geochemistry, Vol. 1: Meteorites, Comets, and Planets* (A. M. Davis, ed.), pp. 269–290. Elsevier, Oxford.

Gilmour J. D., Whitby J. A., Turner G., Bridges J. C., and Hutchison R. (2000) The iodine-xenon system in clasts and chondrules from ordinary chondrites: Implications for early solar system chronology. *Meteoritics & Planet. Sci., 35,* 445–455.

Gladman B. J., Migliorini F., Morbidelli A., Zappala V., Michel P., Cellino A., Froeschle C., Levinson H. F., Bailey M., and Duncan M. (1997) Dynamical lifetimes of objects injected into asteroid belt resonances. *Science, 277,* 197–201.

Goodrich C. A. (1992) Ureilites: A critical review. *Meteoritics, 27,* 327–352.

Goodrich C. A. and Delaney J. S. (2000) Fe/Mg-Fe/Mn relations of meteorites and primary heterogeneity of primitive achondrite parent bodies. *Geochim. Cosmochim. Acta, 64,* 149–160.

Göpel C., Manhes G., and Allègre C. J. (1994) U-Pb systematics of phosphates from equilibrated ordinary chondrites. *Earth Planet. Sci. Lett., 121,* 153–171.

Gounelle M. and Zolensky M. E. (2001) A terrestrial origin for sulfate veins in CI1 chondrites. *Meteoritics & Planet. Sci., 36,* 1321–1329.

Gounelle M., Shu f. H., Shang H., Glassgold A. E., Rehm K. E., and Lee T. (2001) Extinct radioactivities and protosolar cosmic rays: Self-shielding and light elements. *Astrophys. J., 548,* 1051–1070.

Greenwood R. C., Kearsley A. T., and Franchi I. A. (2003) Are CK chondrites really a distinct group or just equilibrated CVs? (abstract). *Meteoritics & Planet. Sci., 38,* A96.

Greshake A., Klock W., Arndt P., Maetz M., Flynn G. J., Bajt S., and Bischoff A. (1998) Heating experiments simulating atmospheric entry heating of micrometeorites: Clues to their parent body sources. *Meteoritics & Planet. Sci., 33,* 267–290.

Grossman J. N., Rubin A. E., and MacPherson G. J. (1988) ALH 85085: A unique volatile-poor carbonaceous chondrite with possible implications for nebular fractionation processes. *Earth Planet. Sci. Lett., 91,* 33–54.

Grossman L., Ebel D. S., and Simon S. B. (2002) Formation of refractory inclusions by evaporation of condensate precursors. *Geochim. Cosmochim. Acta, 66,* 145–161.

Haack H. and McCoy T. J. (2003) Iron and stony-iron meteorites. In *Treatise on Geochemistry, Vol. 1: Meteorites, Comets, and Planets* (A. M. Davis, ed.), pp. 325–345. Elsevier, Oxford.

Herzog G. F. (2003) Cosmic-ray exposure ages of meteorites. In *Treatise on Geochemistry, Vol. 1: Meteorites, Comets, and Planets* (A. M. Davis, ed.), pp. 347–380. Elsevier, Oxford.

Hewins R. H. (1997) Chondrules. *Annu. Rev. Earth Planet. Sci., 25,* 61–83.

Hewins R. H., Jones R. H., and Scott E. R. D., eds. (1996) *Chondrules and the Protoplanetary Dis*k. Cambridge Univ., New York.

Hopfe W. D. and Goldstein J. I. (2001) The metallographic cooling rate method revised: Application to iron meteorites and mesosiderites. *Meteoritics & Planet. Sci., 36,* 135–154.

Hoppe P. and Zinner E. (2000) Presolar dust grains from meteorites and their stellar sources. *J. Geophys. Res., 105,* 10371–10385.

Humayun M. and Clayton R. N. (1995) Potassium isotope cosmochemistry: Genetic implications of volatile element depletion. *Geochim. Cosmochim. Acta, 59,* 2131–2148.

Humayun M. and Campbell A. J. (2002) The duration of ordinary chondrite metamorphism inferred from tungsten microdistribution in metal. *Earth Planet. Sci. Lett., 198,* 225–243.

Huss G. R. (1990) Ubiquitous interstellar diamond and SiC in primitive chondrites: Abundances reflect metamorphism. *Nature, 347,* 159–162.

Huss G. R. and Lewis R. S. (1995) Presolar diamond, SiC, and graphite in primitive chondrites: Abundances as a function of meteorite class and petrologic type. *Geochim. Cosmochim. Acta, 59,* 115–160.

Ireland T. R. and Fegley B. Jr. (2000) The solar system's earliest chemistry: Systematics of refractory inclusions. *Intern. Geol. Rev., 42,* 865–894.

Kallemeyn G. W., Rubin A. E., and Wasson J. T. (1991) The compositional classification of chondrites: V. The Karoonda (CK) group of carbonaceous chondrites. *Geochim. Cosmochim. Acta, 55,* 881–892.

Kallemeyn G. W., Rubin A. E., and Wasson J. T. (1996) The compositional classification of chondrites: VII. The R chondrite group. *Geochim. Cosmochim. Acta, 60,* 2243–2256.

Kehm K., Flynn G. J., Sutton S. R., and Hohenberg C. M. (2002) Combined noble gas and trace element measurements on individual stratospheric interplanetary dust particles. *Meteoritics & Planet. Sci., 37,* 1323–1335.

Kerridge J. F. and Matthews M. S., eds. (1988) *Meteorites and the Early Solar System.* Univ. of Arizona, Tucson. 1268 pp.

Keller L. P., Messenger S., Flynn G. J., Clemett S., Wirick S., and Jacobsen C. (2004) The nature of molecular cloud material in interplanetary dust. *Geochim. Cosmochim. Acta, 68,* 2577–2589.

Kimura M., Hiyagon H., Palme H., Spettel B., Wolf D., Clayton R. N., Mayeda T. K., Sato T., Suzuki T., and Kojima H. (2002) Anomalous H-chondrites, Y792947, Y93408 and Y82038, with abundant refractory inclusions and very low metamorphic grade. *Meteoritics & Planet. Sci., 37,* 1417–1434.

Kita N. T., Nagahara H., Togashi S., and Morishita T. (2000) A short duration of chondrule formation in the solar nebula: Evidence from ^{26}Al in Semarkona ferromagnesian chondrules. *Geochim. Cosmochim. Acta, 64,* 3913–3922.

Kita N. T., Ikeda Y., Togashi S., Liu Y., Morishita Y., and Weisberg M. K. (2004) Origin of ureilites inferred from a SIMS oxygen isotopic and trace element study of clasts in the Dar al Gani 319 polymict ureilite. *Geochim. Cosmochim. Acta, 68,* 4213–4235.

Kleine T., Munker C., Mezger K., and Palme H. (2002) Rapid accretion and early core formation on asteroids and the terrestrial planets from Hf-W chronometry. *Nature, 418,* 952–955.

Krot A. N., Scott E. R. D., and Zolensky M. E. (1995) Mineralogical and chemical modification of components in CV3 chondrites: Nebular or asteroidal processing? *Meteoritics, 30,* 748–775.

Krot A. N., Meibom A., Weisberg M. K., and Keil K. (2002a) The CR chondrite clan: Implications for early solar system processes. *Meteoritics & Planet Sci., 37,* 1451–1490.

Krot A. N., McKeegan K. D., Leshin L. A., MacPherson G. J., and Scott E. R. D. (2002b) Existence of a ^{16}O-rich gaseous reservoir in the solar nebula. *Science, 295,* 1051–1054.

Kuebler K. E., McSween H. Y., Carlson W. D., and Hirsch D. (1999) Sizes and masses of chondrules and metal-troilite grains in ordinary chondrites: Possible implications for nebular sorting. *Icarus, 141,* 96–106.

Lauretta D. S., Lodders K., and Fegley B. (1997) Experimental simulations of sulfide-formation in the solar nebula. *Science, 277,* 358–360.

Lee D.-C. and Halliday A. N. (1997) Hf-W isotopic evidence for rapid accretion and differentiation in the early solar system. *Science, 274,* 1876–1879.

Lewis, R. S., Amari S., and Anders E. (1990) Meteoritic silicon carbide: Pristine material from carbon stars. *Nature, 348,* 293–298.

Lodders K. (2003) Solar system abundances and condensation temperatures of the elements. *Astrophys. J., 591(2),* 1220–1247.

Lugmair G. W. and Shukolyukov A. (2001) Early solar system events and timescales. *Meteoritics & Planet. Sci., 36,* 1017–1026.

MacPherson G. J. (2003) Calcium-aluminum-rich inclusions in chondritic meteorites. In *Treatise on Geochemistry, Vol. 1: Meteorites, Comets, and Planets* (A. M. Davis, ed.), pp. 201–246. Elsevier, Oxford.

MacPherson G. J., Davis A. M., and Zinner E. K. (1995) The distribution of Al-26 in the early solar system — a reappraisal. *Meteoritics, 30,* 365–386.

MacPherson G. J., Huss G. R., and Davis A. M. (2003) Extinct ^{10}Be in type A calcium-aluminum-rich inclusions from CV chondrites. *Geochim. Cosmochim. Acta, 67,* 3165–3179.

Marhas K. K., Goswami J. N., and Davis A. M. (2002) Short-lived nuclides in hibonite grains from Murchison: Evidence for solar system evolution. *Science, 298,* 2182–2184.

McCoy T. J., Keil K., Clayton R. N., Mayeda T. K., Bogard D. D., Garrison D. H., Huss G. R., Hutcheon I. D., and Wieler R. (1996) A petrologic, chemical, and isotopic study of Monument Draw and comparison with other acapulcoites: Evidence for formation by incipient partial melting. *Geochim. Cosmochim. Acta, 60,* 2681–2708.

McCoy T. J., Keil K., Clayton R. N., Mayeda T. K., Bogard D. D., Garrison D. H., and Wieler R. (1997) A petrologic study of lodranites: Evidence for early formation as partial melt residues from heterogeneous precursors. *Geochim. Cosmochim. Acta, 61,* 623–637.

McKay D. S., Gibson E. K., Thomas-Keprta K. L., Vali H., Romanek C. S., Clemett S. J., Chelier X. D. F., Maechling C. R., and Zare R. N. (1996) Search for past life on Mars: Possible relic biogenic activity in Martian meteorite ALH84001. *Science, 273,* 924–930.

McKeegan K. D. and Davis A. (2004) Early solar system chronology. In *Treatise on Geochemistry, Vol. 1: Meteorites, Comets, and Planets* (A. M. Davis, ed.), pp. 431–460. Elsevier, Oxford.

McKeegan K. D., Chaussidon M., and Robert F. (2000) Incorporation of short-lived ^{10}Be in a calcium-aluminum-rich inclusion from the Allende meteorite. *Science, 289,* 1334–1337.

McSween H. Y. and Labotka T. C. (1993) Oxidation during metamorphism of the ordinary chondrites. *Geochim. Cosmochim. Acta, 57,* 1105–1114.

McSween H. Y., Ghosh A., Grimm R. E., Wilson L., and Young E. D. (2002) Thermal evolution models of asteroids. In *Asteroids III* (W. F. Bottke Jr. et al., eds.), pp. 559–571. Univ. of Arizona, Tucson.

Meibom A. and Clark B. E. (1999) Evidence for the insignificance of ordinary chondritic material in the asteroid belt. *Meteoritics & Planet. Sci., 34,* 7–24.

Meibom A., Pataev M. I., Krot A. N., Wood J. A., and Keil K. (1999) Primitive FeNi metal grains in CH carbonaceous chondrites formed by condensation from a gas of solar composition. *J. Geophys. Res., 104,* 22053–22059.

Mendybaev R. A., Beckett J. R., Grossman L., Stolper E., Cooper R. F., and Bradley J. P. (2002) Volatilization kinetics of silicon carbide in reducing gases: An experimental study with applications to the survival of presolar grains in the solar nebula. *Geochim. Cosmochim. Acta, 66,* 661–682.

Messenger S. (2000) Identification of molecular-cloud material in interplanetary dust particles. *Nature, 404,* 968–971.

Messenger S., Keller L. P., Stademann F. J., Walker R. M., and Zinner E. (2003) Samples of stars beyond the solar system: Silicate grains in interplanetary dust. *Science, 300,* 105–108.

Metzler K., Bishoff A., and Stoffler D. (1992) Accretionary dust mantles in CM chondrites: Evidence for solar nebula processes. *Geochim. Cosmochim. Acta, 56,* 2873–2897.

Mittlefehldt D. W. (1994) ALH84001, a cumulate orthopyroxenite member of the martian meteorite clan. *Meteoritics, 29,* 214–221.

Mittlefehldt D. W. (2002) Geochemistry of the ungrouped carbonaceous chondrite Tagish Lake, the anomalous CM chondrite Bells, and comparison with CI and CM chondrites. *Meteoritics & Planet. Sci., 37,* 703–712.

Mittlefehldt D. W. and Lindstrom M. M. (1990) Geochemistry and genesis of the angrites. *Geochim. Cosmochim. Acta, 54,* 3209–3218.

Mittlehehldt D. W., McCoy T. J., Goodrich C. A., and Kracher A. (1998) Non-chondritic meteorites from asteroidal bodies. In *Planetary Materials* (J. J. Papike, ed.), pp. 4-1 to 4-195. Reviews in Mineralogy, Vol. 36, Mineralogical Society of America.

Moustefaoui S., Kita N. T., Togashi S., Tachibana S., Nagahara H., and Morishita Y. (2002) The relative formation ages of ferromagnesian chondrules inferred from their initial $^{26}Al/^{27}Al$. *Meteoritics & Planet. Sci., 37,* 421–438.

Nakamura K., Zolensky M. E., Tomita S., Nakashima S., and Tomeoka K. (2002) Hollow organic globules in the Tagish Lake meteorite as possible products of primitive organic reactions. *Intern. J. Astrobiol., 1,* 179–189.

Nehru C. E., Prinz M., Weisberg M. K. Ebihara M. E., Clayton R. N., and Mayeda T. K. (1992) Brachintes: A new primitive achondrite group (abstract). *Meteoritics, 27,* 267.

Nguyen A. and Zinner E. (2004) Discovery of ancient silicate stardust in a meteorite. *Science, 303,* 1496–1499.

Nier A. O. and Schlutter D. J. (1993) The thermal history of interplanetary dust particles collected in the Earth's stratosphere. *Meteoritics, 28,* 675–681.

Nittler L. R. (2003) Presolar stardust in meteorites: Recent advances and scientific frontiers. *Earth Planet. Sci. Lett., 209,* 259–273.

Nittler L. R., Alexander C. M. O'D., Gao X., Walker R. M., and Zinner E. (1997) Stellar sapphires: The properties and origins of presolar Al_2O_3 in meteorites. *Astrophys. J., 483,* 475–495.

Nyquist L. E., Bogard D. D., Shih C.-Y., Greshake A., Stoffler D., and Eugster O. (2001) Ages and geologic histories of martian meteorites. *Space Sci. Rev., 96,* 105–164.

Ott U. (1993) Interstellar grains in meteorites. *Nature, 364,* 25–33.

Ott U. and Begemann F. (2000) Spallation recoil and age of presolar grains in meteorites. *Meteoritics & Planet. Sci., 35,* 53–63.

Palme H. (2001) Chemical and isotopic heterogeneity in protosolar matter. *Philos. Trans. R. Soc., 359,* 2061–2075.

Palme H. and Jones A. (2003) Solar system abundances of the elements. In *Treatise on Geochemistry, Vol. 1: Meteorites, Comets, and Planets* (A. M. Davis, ed.), pp. 41–61. Elsevier, Oxford.

Pearson V. K., Sephton M. A., Kearsley A. T., Bland P. A., Franchi I. A., and Gilmour I. (2002) Clay mineral — organic matter relationships in the early solar system. *Meteoritics & Planet. Sci., 37,* 1829–1833.

Pizzarello S., Huang Y., Becker L., Roreda R. J., Nieman R. A., Cooper G., and Williams M. (2001) The organic content of the Tagish Lake meteorite. *Science, 293,* 2236–2239.

Richter K. and Drake M. J. (1997) A magma ocean on Vesta: Core formation and petrogenesis of eucrites and diogenites. *Meteoritics & Planet. Sci., 32,* 929–944.

Rietmeijer F. J. M. (1998) Interplanetary dust particles. In *Planetary Materials* (J. J. Papike, ed.), pp. 2-1 to 2-95. Reviews in Mineralogy, Vol. 36, Mineralogical Society of America.

Rubin A. E. and Kallemeyn G. W. (1994) Pecora Escarpment 91002: A member of the new Rumuruti (R) chondrite group. *Meteoritics, 29,* 255–264.

Rubin A. E. and Krot A. N. (1996) Multiple heating of chondrules. In *Chondrules and the Protoplanetary Disk* (R. H. Hewins et al., eds.), pp. 173–180. Cambridge Univ., New York.

Rubin A. E., Keil K., and Scott E. R. D. (1997) Shock metamorphism of enstatite chondrites. *Geochim. Cosmochim. Acta, 61,* 847–858.

Rubin A. E., Kallemeyn G. W., Wasson J. T., Clayton R. N., Mayeda T. K., Grady M., Verchovsky A. B., Eugster O., and Lorenzetti S. (2003) Formation of metal and silicate globules in Gujba: A new Bencubbin-like meteorite fall. *Geochim. Cosmochim. Acta, 67,* 3283–3298.

Russell S. S., Srinivasan G., Huss G. R., Wasserburg G. J., and MacPherson G. J. (1996) Evidence for widespread ^{26}Al in the solar nebula and constraints for nebula time scales. *Science, 273,* 757–762.

Russell S. S., Ott U., Alexander C. M., Zinner E. K., Arden J. W., and Pillinger C. T. (1997) Presolar silicon carbide from the Indarch (EH4) meteorite: Comparison with silicon carbide populations from other meteorite classes. *Meteoritics & Planet. Sci., 32,* 719–732.

Ruzicka A., Snyder G. A., and Taylor L. A. (1997) Vesta as the howardite, eucrite and diogenite parent body: Implications for the size of a core and for large-scale differentiation. *Meteoritics & Planet. Sci., 32,* 825–840.

Schoenberg R., Kamber B. S., Collerson K. D., and Eugster O. (2002) New W-isotope evidence for rapid terrestrial accretion and very early core formation. *Geochim. Cosmochim. Acta, 66,* 3151–3160.

Schulte M. and Shock E. (2004) Coupled organic synthesis and mineral alteration on meteorite parent bodies. *Meteoritics & Planet. Sci., 39,* 1577–1590.

Schulze H., Bischoff A., Palme H., Spettel B., Dreibus G., and Otto J. (1994) Mineralogy and chemistry of Rumuruti: The first meteorite fall of the new Rumuruti group. *Meteoritics, 29,* 275–286.

Scott E. R. D. (1988) A new kind of primitive achondrite, Allan Hills 85085. *Earth Planet. Sci. Lett., 91,* 1–18.

Scott E. R. D., Keil K., and Stoffler E. (1992) Shock metamorphism of carbonaceous chondrites. *Geochim. Cosmochim. Acta, 56,* 4281–4293.

Sephton M. A., Pillinger C. T., and Gilmour I. (1998) δ^{13}C of free and macromolecular polyaromatic structures in the Murchison meteorite. *Geochim. Cosmochim. Acta, 62,* 1821–1828.

Sephton M. A., Pillinger C. T., and Gilmour I. (1999) Small-scale hydrous pyrolysis of macromolecular material in meteorites. *Planet. Space Sci., 47,* 181–187.

Sephton M. A., Pillinger C. T., and Gilmour I. (2000) Aromatic moieties in meteoritic macromolecular materials: Analysis by hydrous pyrolysis and δ^{13}C of individual compounds. *Geochim. Cosmochim. Acta, 64,* 321–328.

Sephton M. A., Verchovsky A. B., Bland P. A., Gilmour I., Grady M. M., and Wright I. P. (2003) Investigating variations in carbon and nitrogen isotopes in carbonaceous chondrites. *Geochim. Cosmochim. Acta, 67,* 2093–2108.

Shen J. J., Papanastassiou D. A., and Wasserburg G. J. (1996) Precise Re-Os determinations and systematics of iron meteorites. *Geochim. Cosmochim. Acta, 60,* 2887–2900.

Shu F. H., Shang H., and Lee T. (1996) Toward an astrophysical model of chondrites. *Science, 271,* 1545–1552.

Shukolyukov A. and Lugmair G. W. (1993) Live iron-60 in the early solar system. *Science, 259,* 1138–1142.

Simon S. B. and Grossman L. (2003) Petrography and mineral chemistry of the anhydrous component of the Tagish Lake carbonaceous chondrite. *Meteoritics & Planet. Sci., 38,* 813–825.

Singletary S. J. and Grove T. L. (2003) Early petrologic processes on the ureilite parent body. *Meteoritics & Planet. Sci., 38,* 95–108.

Smoliar M. I., Walker R. J., and Morgan J. W. (1996) Re-Os ages of group IIA, IIIA, IVA, and IVB iron meteorites. *Science, 271,* 1099–1102.

Srinivasan G., Ulyanov A. A., and Goswami J. N. (1994) ^{41}Ca in the early solar system. *Astrophys. J. Lett., 431,* L67–L70.

Srinivasan G., Goswami J. N., and Bhandari N. (1999) ^{26}Al in eucrite Piplia Kalan: Plausible heat source and formation chronology. *Science, 284,* 1348–1350.

Steele A., Goddard D. T., Stapleton D., Toporski J. K. W., Peters V., Bassinger V., Sharples G., Wynn-Williams D. D., and McKay D. S. (2000) Investigations into an unknown organism on the martian meteorite Allan Hills 84001. *Meteoritics & Planet. Sci., 35,* 237–241.

Stöffler D., Keil K., and Scott E. R. D. (1991) Shock metamorphism of ordinary chondrites. *Geochim. Cosmochim. Acta, 55,* 3845–3867.

Swindle T. D., Davis A. M., Hohenberg C. M., MacPherson G. J., and Nyquist L. E. (1996) Formation times of chondrules and Ca-Al-rich inclusions: Constraints from short-lived radionuclides. In *Chondrules and the Protoplanetary Disk* (R. H. Hewins et al., eds.), pp. 77–86. Cambridge Univ., Cambridge.

Tachibana S. and Huss G. R. (2003) The initial abundance of ^{60}Fe in the solar system. *Astrophys. J. Lett., 588,* L41–L44.

Tang M. and Anders E. (1988) Interstellar silicon carbide: How much older than the solar system? *Astrophys. J. Lett., 335,* L31–L34.

Taylor G. J. (1992) Core formation in asteroids. *J. Geophys. Res., 97,* 14717–14726.

Theimens M. H. (1996) Mass-independent isotopic effects in chondrites: The role of chemical processes. In *Chondrules and the Protoplanetary Disk* (R. H. Hewins et al., eds.), pp. 107–118. Cambridge Univ., Cambridge.

Trieloff M., Jessberger E. K., Herrwerth I., Hopp J., Fleni C., Ghells M. Bourot-Denise M., and Pellas P. (2003) Structure and thermal history of the H-chondrite parent asteroid revealed by thermochronometry. *Nature, 422,* 502–506.

Walker D. and Grove T. (1993) Ureilite smelting. *Meteoritics, 28,* 629–636.

Walker R. J., Horan M. F., Morgan J. W., Becker H., Grossman

J. N., and Rubin A. E. (2002) Comparative ^{187}Re-^{187}Os systematics of chondrites: Implications regarding early solar system processes. *Geochim. Cosmochim. Acta, 66,* 4187–4201.

Weisberg M. K., Prinz M., Clayton R. N., and Mayeda T. K. (1993) The CR (Rennazo-type) carbonaceous chondrite group and its implications. *Geochim. Cosmochim. Acta, 57,* 1567–1586.

Weisberg M. K., Prinz M., Clayton R. N., and Mayeda T. K. (2001) A new metal-rich chondrite grouplet. *Meteoritics & Planet. Sci., 36,* 401–418.

Weyer S., Munker C., and Mezger K. (2003) Nb/Ta, Zr/Hf and REE in the depleted mantle: Implications for the differentiation history of the crust-mantle system. *Earth Planet. Sci. Lett., 205,* 309–324.

Whitby J., Burgess R., Turner G., Gilmour J., and Bridges J. (2000) Extinct ^{129}I in halite from a primitive meteorite: Evidence for evaporite formation in the early solar system. *Science, 288,* 1819–1821.

Wieler R. and Graf T. (2001) Cosmic ray exposure history of meteorites. In *Accretion of Extraterrestrial Material on Earth over Time* (B. Peucker-Ehrenbrink. and B. Schmitz, eds.), pp. 221–240. Kluwer Academic/Plenum, New York.

Wieler R., Pedroni A., and Leya I. (2000) Cosmogenic neon in mineral separates from Kapoeta: No evidence for an irradiation of its parent body regolith by an early active Sun. *Meteoritics & Planet. Sci., 35,* 251–257.

Wilson L. and Keil K. (1991) Consequences of explosive eruptions on small bodies: The case of the missing basalts on the aubrite parent body. *Earth Planet. Sci. Lett., 104,* 505–512.

Woolum D. S. and Hohenberg C. (1993) Energetic particle environment in the early solar system — extremely long precompaction meteoritic ages or an enhanced particle flux. In *Protostars and Planets III* (E. H. Levy and J. I. Lunine, eds.), pp. 903–919. Univ. of Arizona, Tucson.

Yamaguchi A., Clayton R. N., Mayeda T. K., Ebihara M., Oura Y., Miura Y. N., Haramura H., Misawa K., Kojima H., and Nagao K. (2002) A new source of basaltic meteorites inferred from Northwest Africa 011. *Science, 296,* 334–336.

Yang C.-W., Williams D. B., and Goldstein J. I. (1996) A revision of the Fe-Ni phase diagram at low temperatures (<400°C). *J. Phase Equilibria, 17,* 522–531.

Yang C.-W., Williams D. B. and Goldstein J. I. (1997) Low-temperature phase decomposition in metal from iron, stony-iron and stony meteorites. *Geochim. Cosmochim. Acta, 61,* 2943–2956.

Yin Q., Jacobsen S. B., Yamashita K., Blichert-Toft J., Telouk P., and Albarede F. (2002) A short timescale for terrestrial planet formation from Hf-W chronometry of meteorites. *Nature, 418,* 949–952.

Young E. D. (2000) Assessing the implications of K isotope cosmochemistry for evaporation in the preplanetary solar nebula. *Earth Planet. Sci. Lett., 183,* 321–333.

Young E. D. and Russell S. S. (1998) Oxygen reservoirs in the early solar nebula inferred from Allende CAI. *Science, 282,* 452–455.

Yu Y. and Hewins R. H. (1998) Transient heating and chondrule formation: Evidence from sodium loss in flash heating simulation experiments. *Geochim. Cosmochim. Acta, 62,* 159–172.

Zanda B., Bourot-Denise M., Perron C., and Hewins R. H. (1994) Origin and metamorphic redistribution of silicon, chromium, and phosphorus in the metal of chondrites. *Science, 265,* 1846–1849.

Zinner E. (1998) Stellar nucleosynthesis and the isotopic composition of presolar grains from primitive meteorites. *Annu. Rev. Earth Planet. Sci., 26,* 147–188.

Zolensky M. E., Bodnar R. J., Gibson E. K., Nyquist L. E., Reese Y., Shih C.-Y., and Wiesmann H. (1999) Asteroidal water within fluid-inclusion-bearing halite in an H5 chondrite, Monahans. *Science, 285,* 1377–1379.

Zolensky M. E., Nakamura K., Gounelle M., Mikouchi T., Kasama T., Tachikawa O., and Tonui E. (2002) Mineralogy of Tagish Lake: An ungrouped type 2 carbonaceous chondrite. *Meteoritics & Planet. Sci., 27,* 737–761.

Part II:

The Presolar Epoch: Meteoritic Constraints on Astronomical Processes

Nucleosynthesis

Bradley S. Meyer
Clemson University

Ernst Zinner
Washington University

Primitive meteorites contain nano- to micrometer-sized dust grains with large isotopic anomalies relative to the average solar system composition. These large anomalies convincingly demonstrate that the dust grains condensed in the outflow from their parent stars and survived destruction in the interstellar medium and solar-system processing to retain memory of the astrophysical setting of their formation. In addition to the dust grains, certain primitive refractory inclusions in meteorites also show isotopic anomalies, which suggests that the inclusions formed in the solar system from large collections of anomalous dust. Finally, the inferred presence of short-lived radioisotopes in the early solar nebula yields important clues about the circumstances of the Sun's birth. We review these topics in cosmochemistry and their implications for our ideas about stellar nucleosynthesis.

1. INTRODUCTION AND HISTORY

Today it is well established that all elements from C on up were made by stellar nucleosynthesis. After the Big Bang the universe consisted only of H and He and trace amounts of Li, Be, and B, and it was not until the formation of the first stars that the heavier elements came into existence. There exists some evidence now (e.g., *Umeda and Nomoto,* 2003) that the earliest stars, which started out with essentially only H and He, were very heavy. Such stars run through their evolution relatively rapidly (in millions of years) and produce nearly all the heavier elements by various nuclear processes in their hot interiors. They end their lives in gigantic explosions as supernovae that expel the newly synthesized elements in their ejecta into the interstellar medium (ISM), where they are mixed with interstellar gas. The next generation of stars incorporated these elements during their formation and the material in our galaxy underwent several such cycles before our solar system (SS) formed. In contrast to massive stars that become supernovae (SNe), stars with a mass less than ~8 $M_\odot$ experience a different history. They take hundreds of millions to billions of years to burn the H and He in their interior into C and O and in their late stages lose their outer envelopes in the form of stellar winds and planetary nebulae, thereby leaving behind a C-O white dwarf (WD) star that does not undergo further nuclear processes. The stellar wind and planetary nebula material are enriched in certain nuclei that were produced in hot shells on top of the WD.

When the SS formed it incorporated material from many different stellar sources: supernovae, late-type stars, and novae. This material was mixed extremely well so that the SS as a whole has a very uniform elemental and isotopic composition. Differences in elemental abundances among different chondrite groups are considered to reflect variations in the environment where their parent asteroids formed (*Palme,* 2000). However, major elemental differentiation occurred through planet formation and geological processes taking place on larger planetary bodies. Thus, terrestrial rocks have elemental compositions that are quite different from the overall mix from which the whole SS was made. In contrast, primitive meteorites to a large extent preserve the original composition of the SS. They originate from small asteroids that did not experience any planetary differentiation. Their elemental abundances agree very well with that of the Sun except for the volatile elements H, C, N, and O, the noble gases, and Li (*Anders and Grevesse,* 1989; *Grevesse et al.,* 1996). Since geological processes do not affect isotopic ratios very much (they introduce only some mass-dependent fractionation), the isotopic compositions of solar materials from different planetary bodies (Earth, Moon, Mars, and the parent bodies of different types of meteorites) are essentially identical. An exception are isotopic anomalies in certain elements that are due to the decay of radioactive isotopes. Some of these radioisotopes were short-lived, i.e., they existed only early in the SS and evidence for their decay products are found in early-formed SS solids.

Although stellar nucleosynthesis had been proposed before (*Hoyle,* 1946), in the early 1950s it was far from clear that most elements are synthesized in stars. The first observational evidence for stellar nucleosynthesis came from the discovery of the unstable element Tc in the spectra of S-stars (*Merrill,* 1952). A few years later in their classical papers, *Burbidge et al.* (1957) and *Cameron* (1957) established the theoretical framework for stellar nucleosynthesis. They proposed a scheme of eight nucleosynthetic processes taking place in different stellar sources under different con-

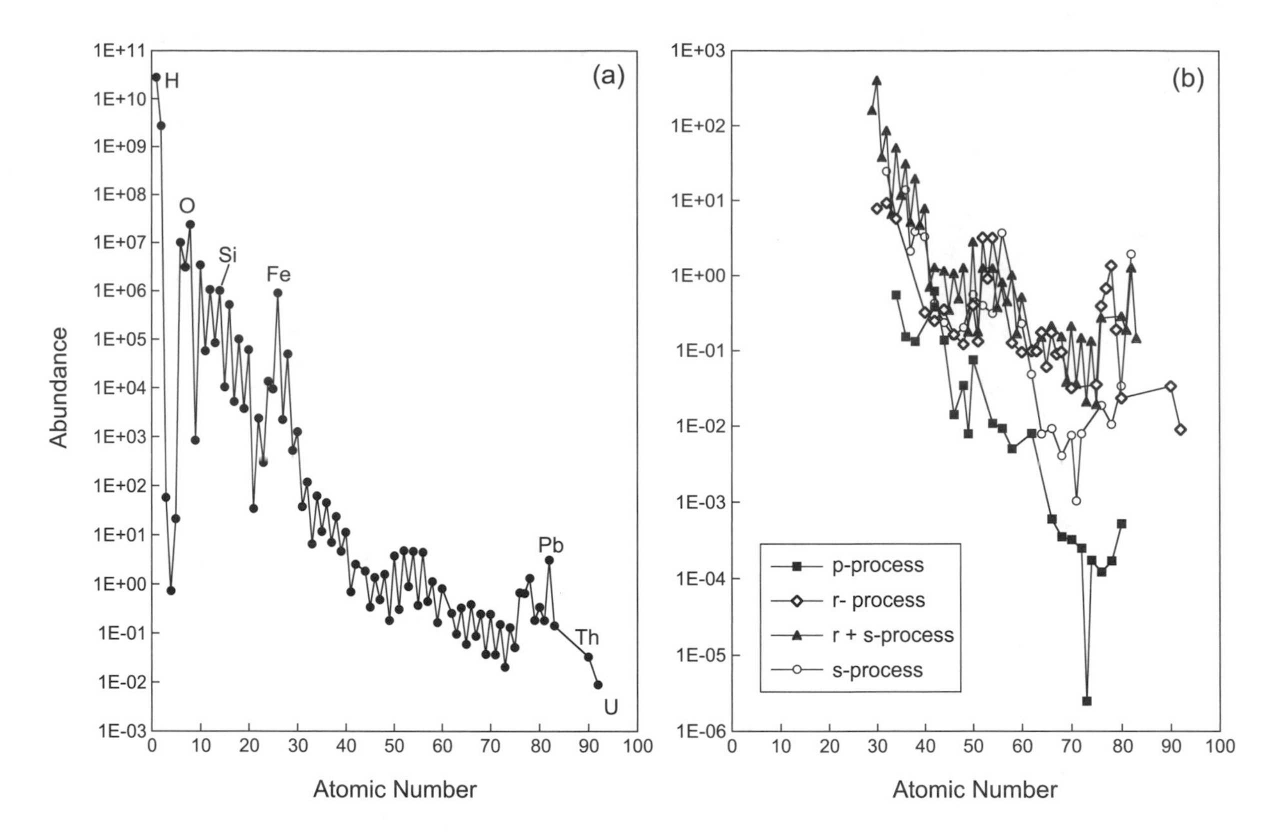

Fig. 1. **(a)** Abundances of the elements in the solar system (*Anders and Grevesse,* 1989). **(b)** Abundances of the nuclides of a given element produced by the p-, r-, and s-process, respectively.

ditions. One important motivation for this work was to explain regularities in the abundance of the nuclides in the SS (Fig. 1) as obtained by the study of meteorites (*Suess and Urey,* 1956). (For a table of the solar system abundances of the nuclides and a brief description of their principal nucleosynthesis production processes, the reader is invited to visit the Web site *http://nucleo.ces.clemson.edu/pages/solar_abundances/.*) While the SS represents only a grand average of many distinct stellar sources, the SS (sometimes also termed cosmic) abundances of the elements and isotopes (*Cameron,* 1973; *Anders and Grevesse,* 1989; *Grevesse et al.,* 1996; *Lodders,* 2003) became an important reference for astronomers. They remained the touchstone for testing models of stellar nucleosynthesis as two generations of nuclear astrophysicists tried to reproduce them (e.g., *Käppeler et al.,* 1989; *Timmes et al.,* 1995). The complete homogenization of all presolar material in a hot solar nebula (*Cameron,* 1962) seemed to be confirmed by the observation that the isotopic compositions of SS materials, including primitive meteorites, are very uniform. The prevalent view was that all solids had been vaporized and thus had lost all memory of their stellar sources, and that the solids we find now in primitive meteorites were produced by condensation from a well-mixed gas (*Grossman,* 1972).

The first hints that not all presolar material had been homogenized and some components found in primitive meteorites preserved a record of their presolar history came in the form of isotopic anomalies, i.e., deviations from the SS ratios, in the element H (*Boato,* 1954) and the noble gases Xe (*Reynolds,* 1960a) and Ne (*Black and Pepin,* 1969). However, these early signs were largely ignored, and it was not until the discovery of an ^{16}O excess (*Clayton et al.,* 1973), an isotopic anomaly in a major rock-forming element, in refractory inclusions (Ca-Al-rich inclusions, or CAIs) that the idea of the survival of presolar isotopic signatures was widely accepted. This discovery was followed by the discovery of anomalies in elements such as Ca, Ti, Cr, Fe, Zn, Sr, Ba, Nd, and Sm (for an overview see *Lee,* 1979, 1988; *Wasserburg,* 1987; *Clayton et al.,* 1988). Important was the realization that certain isotopic patterns represented distinct nucleosynthetic components. For example, Ba, Nd, and Sm in the FUN inclusion EK1-4-1 from the Allende meteorite (*McCulloch and Wasserburg,* 1978a,b) show an r-process signature and thus provided direct evidence for the distinct origin of the r- and s-process nuclides (see also *Begemann,* 1993). Another example is the correlated excesses and depletions of the neutron-rich (n-rich) isotopes ^{48}Ca and ^{50}Ti in the FUN inclusions EK1-4-1 and C-1 and in hibonite grains (*Lee et al.,* 1978; *Niederer et al.,* 1980; *Fahey et al.,* 1985; *Zinner et al.,* 1986) that indicate a separate nucleosynthetic process for the production of the n-rich isotopes of the Fe-peak elements. While some of these isotopic anomalies can be fairly large — ^{50}Ti excesses in hibonite grains range up to almost 30% (*Ireland,* 1990) — it has to be emphasized that the solids exhibiting these anomalies are of SS origin and only incorporated certain presolar components. This is not only evidenced by the size of the effects that are generally much smaller than those expected for bona fide stellar material (*Begemann,* 1993), but also by the fact that other elements (e.g., O, Mg, Si) in CAIs carrying isotopic anomalies in certain elements have essentially normal isotopic compositions.

In addition to isotopic anomalies of nucleosynthetic origin, solids from primitive meteorites show isotopic effects due to the decay of short-lived, now extinct radioisotopes. The first evidence for the presence of a short-lived isotope at the time of SS formation came from excess ^{129}Xe (*Reynolds,* 1960b) and its correlation with ^{127}I (*Jeffery and Reynolds,* 1961), indicating the decay of ^{129}I ($T_{1/2}$ = 16 m.y.). Another relatively long-lived radionuclide, ^{244}Pu ($T_{1/2}$ = 82 m.y.), was identified from excesses in the heavy Xe isotopes resulting from fission (*Rowe and Kuroda,* 1965) and subsequently from fission tracks (*Wasserburg et al.,* 1969). While the observed abundances of these two radioisotopes could be attributed to their steady-state production in the galaxy, the later discovery of evidence for isotopes with much shorter half-lives indicated that the time between their nucleosynthetic production and the formation of the first SS objects could not have been much longer than 1 m.y. This made it necessary to invoke a stellar event (a supernova or an asymptotic giant branch, or AGB, star) shortly preceding SS formation and possibly even triggering it (*Cameron and Truran,* 1977). These short-lived isotopes include ^{26}Al ($T_{1/2}$ = 0.73 m.y.) (*Lee et al.,* 1976, 1977), ^{53}Mn ($T_{1/2}$ = 3.7 m.y.) (*Birck and Allègre,* 1985), ^{60}Fe ($T_{1/2}$ = 1.5 m.y.) (*Shukolyukov and Lugmair,* 1993), and ^{41}Ca ($T_{1/2}$ = 0.1 m.y.) (*Srinivasan et al.,* 1994). Short-lived isotopes contain several types of important information. They not only can serve as chronometers for early solar system (ESS) history (see *Chaussidon and Gounelle,* 2006), but their abundances at the time of SS formation provide information about their nucleosynthetic production, throughout the galaxy as well as in a late stellar event, and possibly even about the mechanism of SS formation.

While the previously discussed types of isotopic anomalies (anomalies of nuclear origin and from the decay of short-lived radionuclides) are carried by solids that themselves formed in the SS, primitive meteorites contain also grains that are of presolar origin. These grains condensed in the outflows of late-type stars and supernova and nova ejecta and thus represent true stardust. Their stellar origin is indicated by their isotopic compositions, which are completely different from those of the SS and, for some elements, cover extremely wide ranges. They can be located in and extracted from their parent meteorites and studied in detail in the laboratory. The finding of an anomalous Ne component in meteorites (*Black and Pepin,* 1969) prompted *Black* (1972) to propose presolar solid carriers, but it took another 15 years until presolar grains were isolated and identified by E. Anders and colleagues at the University of Chicago. These scientists were led by the presence of anomalous ("exotic") noble gas components in Ne and Xe to subject primitive meteorites to increasingly harsher chemical dissolution and physical separation procedures in order to track the gases' solid carriers (*Amari et al.,* 1994). This approach, termed "burning down the haystack to find the needle," resulted in the discovery of presolar diamond, the carrier of Xe-HL (*Lewis et al.,* 1987); silicon carbide (SiC), the carrier of Ne-E(H) and Xe-S (*Bernatowicz et al.,* 1987; *Tang and Anders,* 1988b); and graphite, the carrier of Ne-E(L) (*Amari et al.,* 1990). Subsequently, other presolar grain types, not tagged by exotic noble gases, were identified by isotopic analysis of individual grains in the ion microprobe. These types include silicon nitride (Si_3N_4); oxides such as corundum, spinel, and hibonite; and silicates. With the exception of diamonds, which are too small to be analyzed as single grains, the presolar grains were found to be anomalous in *all* their isotopic ratios and it is this feature that identifies them as samples of stellar material. Their isotopic compositions thus provide direct information on nucleosynthetic processes in their stellar sources.

Besides isotopic anomalies of nucleosynthetic origin, meteorites and especially interplanetary dust particles contain isotopic signatures that are apparently caused by chemical processes. These include D and ^{15}N excesses that indicate an origin in dense molecular clouds (but possibly also in the solar nebula) where ion-molecule reactions at low temperature can lead to large isotopic fractionation effects (*Messenger and Walker,* 1997; *Messenger,* 2000). It is ironic that the ^{16}O enrichments in CAIs, which led to the acceptance of the existence of presolar material in meteorites and for which a nucleosynthetic origin has been invoked (*Clayton,* 1978), might actually be of chemical origin (*Thiemens and Heidenreich,* 1983; *Clayton,* 2002). These effects are covered in *Robert* (2006) and *Nuth et al.* (2006). In addition to these anomalies, there are also possible isotopic effects of irradiation in the SS such as the production of short-lived ^{10}Be (*McKeegan et al.,* 2000). These are discussed in *Robert* (2006). There are also all kinds of isotopic variations in meteorites such as isotopic anomalies of N in iron meteorites (*Murty and Marti,* 1994), for which an unambiguous explanation has not been found. We will not treat them here.

In this chapter we will provide a general introduction to nucleosynthesis and stellar evolution, followed by the discussion of nuclear anomalies in CAIs and other materials, short-lived isotopes in the ESS, and presolar grains. Since the first edition of *Meteorites and the Early Solar System* (*Kerridge and Matthews,* 1988), these last two topics have seen significant advances, and the study of presolar grains in particular is essentially a completely new field. They will therefore receive most of our attention.

2. REVIEW OF NUCLEOSYNTHESIS AND STELLAR EVOLUTION

Stars live most of their lives in hydrostatic equilibrium, the stable configuration in which the stellar pressure supports the star against its self-gravity. These stable states, however, belie the constant struggle stars endure to maintain this equilibrium: The light that we see from their surfaces carries away energy and lessens the internal pressure, and the star must replenish this lost energy. By the turn of the twentieth century, it was becoming clear that no then-known energy source could account for the age and luminosity of the Sun. Work by Helmholtz, Lord Kelvin, and others showed that energy radiated by gravitational collapse could only sustain the Sun's luminosity for some tens of millions of years. This clearly conflicted with geological data, which indicated an age of billions of years for Earth. Rutherford's work on nuclear transmutations suggested a resolution of this problem, namely, that the centers of stars were intense furnaces in which nuclear reactions converted rest mass energy of nuclei into heat to replenish the energy lost from the star by radiation and maintain the star's balance between gravity and pressure. By 1939, Bethe and collaborators (*Bethe and Critchfield,* 1938; *Bethe,* 1939) had worked out the reactions responsible for energy generation by H burning in stars in general and in the Sun in particular, and the connection between stars and nuclear reactions had been firmly established.

In 1946, Hoyle turned this early attention on the role of nuclear reactions in stellar energy generation to their consequences for genesis of the chemical elements (*Hoyle,* 1946). This work led to the classic papers of the 1950s that laid out the essential framework of stellar nucleosynthesis (*Hoyle,* 1954; *Burbidge et al.,* 1957; *Cameron,* 1957). In particular, *Burbidge et al.* (1957), in their seminal paper known as B^2FH, classified and named the principal processes by which nuclei form in stars, and because their terminology has proven so useful, we largely continue to use it today. In what follows, we review the basics of stellar evolution and nucleosynthesis as we currently understand them. The goal is to set the stage for our discussion of extinct radioisotopes and isotopic effects in presolar grains and primitive SS minerals. For the sake of brevity, our review must necessarily only touch on the key points. The interested reader may turn to any number of recent reviews for further information (e.g., *Wallerstein et al.,* 1997; *Busso et al.,* 1999; *Woosley et al.,* 2002).

2.1. The Main Stages of Hydrostatic Stellar Evolution and Nucleosynthesis

In their hydrostatic phases, stars consume nuclear fuel to replenish energy lost by radiation. The evolution proceeds through a sequence of burning stages as the ashes from the previous stage serve as the fuel for the subsequent phase. Because the temperature is usually highest in the stellar core, a stellar burning phase first occurs there. Once the fuel is exhausted, the star contracts until burning of that same fuel commences in a shell surrounding the core. Continued contraction ignites the next burning phase in the core. This process repeats itself and each progressive nuclear burning cycle occurs at higher temperature and on a shorter timescale than the previous one. Finally, the core becomes sufficiently dense that pressure from the degenerate electrons can support the star or nuclear reactions can no longer release energy. We now summarize the principal hydrostatic burning stages.

2.1.1. Hydrogen burning. The initial nuclear fuel available to stars is H, and we know that *H burning*, in which four H nuclei are converted into He, is the principal energy source in main sequence stars. In the Sun, this proceeds via the PP chains (see Table 1) in which the p + p reaction produces a deuteron via a weak decay. Deuterium subsequently captures a proton to make ^{3}He, which, in the principal PP chain (PPI), captures another ^{3}He to make ^{4}He and two protons. In stars of somewhat higher mass, core temperatures are greater and the burning proceeds by the CNO bicycle in which the burning is catalyzed by capture of protons by isotopes of C, N, and O. In these cycles, the ^{14}N(p,γ)^{15}O reaction is typically the slowest of these reactions at H-burning temperatures, thus CNO-burning material is characteristically enriched in ^{14}N. It is also worth noting for the presolar grains that, at the typical temperatures of CNO burning (some tens of millions of degrees Kelvin), the nuclear flows tend to enhance the ^{13}C/^{12}C and ^{17}O/^{16}O abundance ratios over their starting values. Material processed by core H burning is convectively mixed to the stellar surface in the in so-called dredge-up processes during later burning stages.

Our basic understanding of the key reaction sequences in H burning has remained largely unchanged since B^2FH, although we now have a better understanding of the related side reactions in H burning that generate the neutrinos seen in Earth-based detectors and that make isotopes such as ^{17}O and ^{26}Al. The most significant advances since 1957, however, have been in experimental nuclear physics, which have led to improvements in our knowledge of the rates for many of the nuclear reactions involved and their consequences for predicted neutrino fluxes from the Sun and other stars. Interestingly, the reactions occurring in the center of the Sun take place at lower energies than are easily measured in laboratories on Earth. This is due to low count rates and to atomic interactions in the targets that mask the effects of the reaction; thus, in most cases, experimental data must be extrapolated down to energies appropriate for the centers of stars. In order to counter some of these difficulties, experimentalists are moving some of their experiments underground to reduce background due to cosmic rays (*Bonetti et al.,* 1999).

2.1.2. Helium burning. After the star has consumed the H in its core, it contracts and heats. It will begin burning H in a shell. Once the core gets hot enough, it will begin *He burning,* which produces ^{12}C from ^{4}He in the triple-α process and ^{16}O via ^{12}C(α,γ)^{16}O. Core He burning in stars typically occurs at about (1–3) × 10^8 K. B^2FH envisioned He burning also producing ^{20}Ne and possibly ^{24}Mg, but we now know that C burning is largely responsible for ^{20}Ne and C and Ne burning for ^{24}Mg. Interestingly, the ^{12}C(α,γ)^{16}O reaction remains rather uncertain, and this uncertainty strongly affects stellar evolution models because of the reaction's importance in determining the central ^{12}C and ^{16}O abundances, which, in turn, strongly affect the subsequent evolution of

TABLE 1. Principal burning stages in hydrostatic stellar evolution.

Stage	Key Reactions	Important Products	Notes
Hydrogen	PPI: $^{1}H + ^{1}H \rightarrow ^{2}H + \gamma$ $^{2}H + ^{1}H \rightarrow ^{3}He + \gamma$ $^{3}He + ^{3}He \rightarrow ^{4}He + 2^{1}H$	^{4}He	In the solar interior, PP chain percentages are PPI (85%), PPII (15%), PPIII (0.02%). The percentages are expected to be somewhat different in other stars.
	PPII: $^{3}He + ^{4}He \rightarrow ^{7}Be + \gamma$ $^{7}Be + e^{-} \rightarrow ^{7}Li + \nu$ $^{7}Li + ^{1}H \rightarrow ^{4}He + ^{4}He$	^{4}He	
	PPIII: $^{3}He + ^{4}He \rightarrow ^{7}Be + \gamma$ $^{7}Be + ^{1}H \rightarrow ^{8}B + \gamma$ $^{8}B \rightarrow 2^{4}He + e^{+} + \nu$	^{4}He	
	CN cycle: $^{12}C + ^{1}H \rightarrow ^{13}N + \gamma$ $^{13}N \rightarrow ^{13}C + e^{+} + \nu$ $^{13}C + ^{1}H \rightarrow ^{14}N + \gamma$ $^{14}N + ^{1}H \rightarrow ^{15}O + \gamma$ $^{15}O \rightarrow ^{15}N + e^{+} + \nu$ $^{15}N + ^{1}H \rightarrow ^{12}C + ^{4}He$	^{4}He, ^{14}N, ^{26}Al	Together, the CN and NO cycles comprise the CNO bicycle. During shell burning of H in the CN and NO cycles temperatures are high enough to produce ^{26}Al via $^{25}Mg + ^{1}H \rightarrow ^{26}Al + \gamma$
	NO cycle: $^{15}N + ^{1}H \rightarrow ^{16}O + \gamma$ $^{16}O + ^{1}H \rightarrow ^{17}F + \gamma$ $^{17}F \rightarrow ^{17}O + e^{+} + \nu$ $^{17}O + ^{1}H \rightarrow ^{15}N + ^{4}He$	^{4}He, ^{14}N, ^{17}O, ^{26}Al	The NO cycle operates at higher temperature than the CN cycle and provides a leakage out of the CN cycle.
Helium	$^{4}He + ^{4}He + ^{4}He \rightarrow ^{12}C + \gamma$ $^{12}C + ^{4}He \rightarrow ^{16}O + ^{4}He$ $^{14}N + ^{4}He \rightarrow ^{18}F + \gamma$ $^{18}F \rightarrow ^{18}O + e^{+} + \nu$ $^{18}O + ^{4}He \rightarrow ^{22}Ne + \gamma$ $^{22}Ne + ^{4}He \rightarrow ^{25}Mg + n$	^{12}C, ^{16}O, ^{18}O, ^{22}Ne ^{25}Mg, s-process isotopes	End burning stage of stars under 8 $M_{\odot}$.
Carbon	$^{12}C + ^{12}C \rightarrow ^{24}Mg^{*}$ $^{24}Mg^{*} \rightarrow ^{20}Ne + ^{4}He$ $\rightarrow ^{23}Na + ^{1}H$ $\rightarrow ^{24}Mg + \gamma$	^{20}Ne, ^{23}Na, ^{24}Mg	End burning stage of stars in mass range 8–10 $M_{\odot}$.
Neon	$^{20}Ne + \gamma \rightarrow ^{16}O + ^{4}He$ $^{20}Ne + ^{4}He \rightarrow ^{24}Mg + \gamma$	^{16}O, ^{24}Mg	
Oxygen	$^{16}O + ^{16}O \rightarrow ^{32}S^{*}$ $^{32}S^{*} \rightarrow ^{28}Si + ^{4}He$ $\rightarrow ^{31}P + ^{1}H$ $\rightarrow ^{32}S + \gamma$	^{28}Si, ^{31}P, ^{32}S	QSE clusters develop
Silicon	$^{28}Si + ^{28}Si \rightarrow ^{54}Fe + 2^{1}H$ $^{28}Si + ^{28}Si \rightarrow ^{56}Ni$ $^{56}Ni \rightarrow ^{56}Co + e^{+} + \nu$ $^{56}Co \rightarrow ^{56}Fe + e^{+} + \nu$	Fe-peak isotopes ^{56}Fe, ^{54}Fe	$^{28}Si + ^{28}Si$ are effective reactions. The true character of silicon burning is a QSE shift from ^{28}Si to iron isotopes.

the star through advanced burning stages. Determining the $^{12}C(\alpha,\gamma)^{16}O$ reaction rate is a subject of intense experimental study (*Kunz et al.,* 2002).

After He is exhausted in the core, the star will begin He burning in a shell outside the He-exhausted core. In stars of mass greater than about 8 $M_{\odot}$, this proceeds in a relatively quiescent fashion. For lower-mass stars, however, which undergo He-shell burning during their ascent of the asymptotic giant branch (AGB) in the Hertzsprung-Russell diagram, the densities will be high enough that electron degeneracy provides a significant component of the overall pressure. This means that the He shells will undergo periodic shell flashes, which drive convection and mix the material in the shell with the overlying stellar envelope.

Helium burning is the end stage of burning for stars of mass below 8 $M_\odot$. The strong pressure from degenerate electrons means that the star can support itself against gravity without having to continue nuclear fusion reactions in its core. As shell burning continues on top of the degenerate C/O core, strong stellar winds also occur and drive off mass in the planetary nebula phase. Once the envelope of the star is completely lost, a degenerate C/O WD is left behind to cool and fade away. The star may brighten again if accretion from a companion star drives explosive H burning on the surface, which we see as a *nova.* If enough mass has accreted, the C and O may burn explosively and completely disrupt the star. We believe such explosions are the outbursts we see as *Type Ia supernovae.*

From the SS abundances, we know that two main neutron-capture processes are responsible for most of the elements heavier than Fe. These are the s- and r-processes, each of which contributed about one-half the SS's supply of the heavy nuclei (see Fig. 1b). It is during He burning that neutron-liberating reactions can occur and drive the s-process of nucleosynthesis. The s in s-process denotes the fact that neutron densities are sufficiently low to make neutron captures *slow* compared to β decays. This means that the nuclear flow is always near stable nuclei, in contrast to the r-process, in which the neutron density is high and neutron captures are typically much more *rapid* than β decays (see below).

The s-process has two principal components, the weak and main components. The weak component is the dominant producer of s-process isotopes up to mass number A = 90. It occurs primarily during core He burning in massive stars. Here, ^{14}N left over from H burning reacts with two α particles to form ^{22}Ne. The reaction $^{22}Ne(\alpha,n)^{25}Mg$ liberates neutrons that then may be captured by preexisting seed nuclei.

The main component of the s-process occurs during H- and He-shell burning in the AGB phase of a low-mass star. In this phase, the star has developed an inert C/O core from previous He burning. Most of the time H burning proceeds relatively quiescently in a shell at the base of the H envelope. As the He ashes from this burning build up, however, the temperature and density in the He intershell between the H envelope and C/O core rise and the He ignites in a thermal pulse. This drives a convective mixing of the He intershell and temporarily extinguishes the H burning above the He intershell. After the brief pulse has occurred, the C and O ashes settle onto the core, thereby increasing its mass, and the H-burning zone reignites. Dredge-up transports matter from inner to outer stellar layers after the pulse and carries the products of shell burning into the envelope where they may be observed. This process repeats itself roughly every 1000 to 10,000 yr. In this way, the star progressively grows its C/O core and enriches its envelope with heavy elements. The reader is referred to Fig. 5 in the review by *Busso et al.* (1999) and the attendant discussion for more details of this rich phase of a low-mass star's life.

It is in the He intershell that the main component of the s-process most probably occurs. The essential idea is that protons leak down into the shell from the H-rich overlying zones by a process that is still not well understood since an entropy gradient exists between the He and H layers that should prevent such mixing. A "^{13}C pocket" forms as these protons are captured by abundant ^{12}C. Once the shell heats up sufficiently, the reaction $^{13}C(\alpha,n)^{16}O$ occurs. The neutrons thus liberated are captured by preexisting nuclei. When the thermal pulse occurs, the $^{22}Ne(\alpha,n)^{25}Mg$ reaction marginally ignites, which gives a high-temperature but short-duration burst of neutrons. Careful recent work has shown that the He shell is radiative (i.e., the heat transfer to the star's outer layers occurs through radiation; this is in contrast to convective transfer) in the phase in which the ^{13}C is burning but convective during the burning of the ^{22}Ne (*Straniero et al.,* 1995; *Gallino et al.,* 1998; *Busso et al.,* 1999). The isotopic ratios of presolar SiC grains confirm this double-neutron-exposure picture (see sections 5.4.1 and 5.4.2).

2.1.3. Carbon burning. Upon exhaustion of He, a star will next begin to burn C at a temperature near (0.8–1) × 10^9 K by the reaction $^{12}C + ^{12}C \rightarrow ^{24}Mg^*$, where $^{24}Mg^*$ is an excited nucleus of ^{24}Mg. The $^{24}Mg^*$ mostly decays into ^{20}Ne and an α particle; thus, C burning is the major producer of ^{20}Ne in the universe. Light particles released during C burning can also be captured by heavier species, thereby altering their abundances (e.g., *The et al.,* 2000).

Once C is exhausted in the core, shell C burning develops. The amount of ^{12}C left over from He burning strongly affects the nature of the shell C burning (e.g., *Imbriani et al.,* 2001; *El Eid et al.,* 2004) which also affects the later stellar burning stages. For stars in the roughly 8–10 $M_\odot$ range, degeneracy pressure dominates thermal pressure after C burning, and, after the star loses its envelope, an O/Ne/Mg WD is left behind.

2.1.4. Neon burning. The abundant ^{20}Ne remaining after C burning is the next nuclear fuel in the star's life. As *B²FH* recognized, ^{20}Ne burns in the following sequence. First, a ^{20}Ne nucleus disintegrates via $^{20}Ne(\gamma,\alpha)^{16}O$. This is an endothermic reaction; however, another ^{20}Ne nucleus may capture the α particle thus liberated in an exothermic reaction to produce ^{24}Mg: $^{20}Ne(\alpha,\gamma)^{24}Mg$. The net exothermic reaction in *Ne burning* is thus $^{20}Ne + ^{20}Ne \rightarrow ^{16}O + ^{24}Mg$, and it is the major producer of ^{24}Mg. Neon burning typically occurs at a temperature near 1.5 × 10^9 K.

2.1.5. Oxygen burning. As a star contracts and heats to temperatures in excess of about 1.8 × 10^9 K, O burns by the reaction $^{16}O + ^{16}O \rightarrow ^{32}S^*$, where $^{32}S^*$ is an excited nucleus of ^{32}S. The $^{32}S^*$ predominantly decays into ^{28}Si and an α particle, though deexcitation to ^{32}S also occurs; thus, ^{28}Si and ^{32}S are the principal products of *O burning*.

Upon exhaustion of O in the core, the star begins burning O in a shell. At this point, the structure of the massive star is quite complex: The O-burning shell is surrounded by Ne-,

C-, He-, and H-burning shells, and all the shells are typically convective. Oxygen-shell burning is the first of these phases in which the nuclear burning timescales become comparable to the convective timescales, and the standard treatment of convective burning breaks down. Detailed models show that the burning is rather inhomogeneous as convective blobs burn before they have a chance to mix turbulently with their surroundings (*Bazan and Arnett,* 1998). Realistic treatment of this multidimensional burning still lies at the frontiers of computational astrophysics, and we must await further advances to see all the implications for nucleosynthesis.

2.1.6. Silicon burning. At the end of O burning, the matter is dominated by ^{28}Si and ^{32}S. It is also slightly neutron rich due to weak interactions that occurred during O burning. As the star contracts further, *Si burning* is the next stage in the star's life. The mechanism for Si burning is more complicated than that for the previous stages because the high charges of the Si and S isotopes prevent them from interacting directly. In Si burning, disintegration reactions break some of the nuclei down to lighter nuclei and neutrons, protons, and α particles. The remaining nuclei capture these light particles to produce heavier isotopes (up to Fe and Ni). At the conditions present in Si burning, the nuclear reactions are sufficiently fast that nuclei begin to form equilibrium clusters, i.e., subsets of isotopes in equilibrium under exchange of light particles. As the temperature and density rise, the clusters merge into larger clusters until a full quasi-statistical equilibrium (QSE) develops. A full QSE is the condition in which all nuclear species heavier than C are in equilibrium under exchange of light particles (*Bodansky et al.,* 1968; *Meyer et al.,* 1998a). At still higher temperatures, even the three-body reactions that assemble heavy nuclei from light ones, namely, $^{4}He + ^{4}He + ^{4}He \rightarrow ^{12}C$ and $^{4}He + ^{4}He + n \rightarrow ^{9}Be$ followed by $^{9}Be + ^{4}He \rightarrow ^{12}C + n$, become fast enough to establish a full nuclear statistical equilibrium (NSE), the condition in which all nuclear species are in equilibrium under exchange of light particles. In such equilibria, nuclei with strong nuclear binding win out in the struggle for abundance. These are typically the Fe and Ni isotopes. Further nuclear reactions will not release binding energy; hence, when the star has developed an Fe core, it has reached the end of its hydrostatic equilibrium.

Table 1 summarizes the main hydrostatic stages of stellar evolution and some of the key products. We now turn our attention to explosive burning.

2.2. Explosive Nuclear Burning

Explosive burning occurs at higher temperatures and on shorter timescales than the corresponding hydrostatic burning phases. For our purposes, the explosive events we consider are either (1) thermonuclear, in which the nuclear reactions themselves deposit energy too quickly for the system to readjust itself hydrodynamically and allow the burning to take place in a steady fashion; or (2) gravitational, in which part of the star falls down a gravitational well and transfers energy to the remaining layers, which may be ejected. Novae and Type Ia supernovae are of the first type; core-collapse (Type II, Ib, Ic) supernovae are of the second type.

2.2.1. Novae. When WD stars accrete material from a companion star, which is usually a main-sequence star, that accreted matter can ignite under degenerate conditions and burn explosively. The resulting outburst is visible as a *nova.* The accreted matter is mostly a mix of H and He, so the burning is typically explosive H burning, which means that the important products, such as ^{15}N, ^{22}Na, and ^{26}Al, are from proton captures on CNO nuclei. The burning is usually sufficiently energetic to drive some matter out of the gravitational well of the WD and into interstellar space. Moreover, the burning can drive some mixing, which may draw underlying WD matter into the ejecta. The composition of the WD (i.e., whether it is a C/O or O/Ne/Mg WD) may be reflected in the outflow from the star.

2.2.2. Type Ia supernovae. The classification Type I for supernovae is observational and indicates that the ejecta show little H in their spectra. We are now confident that Type Ia supernovae are thermonuclear disruptions of WD stars. If a C/O WD has accreted enough mass, the pressure and temperature can rise high enough to cause the C and O making up the star to fuse. This happens under highly degenerate conditions, so the pressure is insensitive to changes in temperature. Energy deposition increases the temperature further, leading to an increased rate for the nuclear reactions and energy generation; thus, a thermonuclear runaway occurs. This eventually leads to the disruption of the entire star. Because the temperatures become high enough in the explosion to drive the C and O into Fe and Ni isotopes, Type Ia supernovae are prodigious producers of ^{56}Fe and other Fe-peak isotopes. Furthermore, because they are fairly uniform in their light curve characteristics, they have become preferred standard candles in astronomy (*Phillips,* 2003). Observation of these standard candles has led to the astonishing conclusion that the expansion of our universe is in fact accelerating (*Riess et al.,* 1998; *Perlmutter et al.,* 1999).

The literature frequently discusses Type Ia as either deflagrations or detonations. The difference between these two types of events is that in deflagrations the burning front is subsonic while in detonations it is supersonic. Because the flame is subsonic in a deflagration, the matter ahead of the front has a chance to adjust itself before the flame arrives; therefore deflagrations tend to have lower temperatures than detonations. This allows deflagrations to burn less of their matter into Fe, which leaves more Si and S in the ejecta. Detonations, on the other hand, burn more matter into Fe. Spectroscopy on Type Ia supernovae thus yields insights into the nature of the burning front in the explosion. Such spectroscopy has suggested that a plausible scenario for Type Ia events is that of a deflagration turning into a detonation. Recent work has demonstrated the complicated manner in which such a transition may occur (*Plewa et al.,* 2004).

It is also possible that massive WD stars undergo accretion-induced collapse (AIC) directly to a neutron star and explode in a core-collapse supernova (see the next section). Because of the lack of H in their spectra, these would be Type I (perhaps Type Ic) supernovae. Some nucleosynthetic constraints exist on this scenario (e.g., *Woosley and Baron,* 1992), but more work clearly needs to be done. There is interesting speculation that such explosions could be the site of r-process nucleosynthesis (*Qian and Wasserburg,* 2003).

2.2.3. Core-collapse supernovae. Massive stars die in the second type of explosion, the gravitational explosion mentioned above. The literature often refers to such supernovae as Type II, Ib, or Ic, which are spectral classifications based on the light curve (*Filippenko,* 1997). Since the core of the star has already evolved to Fe, no further useful fuel remains. A combination of disintegrations and electron captures initiate the sudden collapse of the core. The inner part of the core collapses subsonically, thus when it reaches nuclear matter density, the pressure rises dramatically due to nucleon-nucleon interactions and causes the collapse to halt and the core to "bounce." The outer part of the collapsing core, on the other hand, falls in supersonically; thus, it does not receive the signal from the inner core to cease collapsing and it crashes onto the inner core. This highly nonadiabatic process generates a shock, which works its way out through the remainder of the star, expelling these layers into the interstellar medium. Current SN models show that the energy lost by the shock in traversing the outer part of the Fe core causes it to stall. The outward motion must be regenerated.

Left behind the shock is an extremely dense proto-neutron star at temperatures of tens of billions of degrees. This star cools over a timescale of seconds by emission of neutrinos, which are the only particles that can escape the dense core. The current thinking is that these neutrinos deposit energy behind the stalled shock and thereby provide the push to revive its outward motion (e.g., *Colgate and White,* 1966; *Bethe and Wilson,* 1985; *Burrows,* 2000).

As the shock traverses the outer layers of the star, it compresses and heats them. This initiates explosive burning in these stellar shells, thereby modifying the composition established by the previous hydrostatic burning. The different presupernova and explosive burning phases give rise to widely varying compositions in the ejecta. The reader may find it useful to keep in mind the "onion-skin" picture, such as that in Fig. 1 of *Meyer et al.* (1995), although the true nature of the ejecta is probably more complicated.

We now discuss some of the particular explosive burning phases in the shells. For detailed output from modern stellar models, we recommend that the interested reader visit *http://www.nucleosynthesis.org* and *http://www.ces.clemson.edu/physics/nucleo/pages/cugce.*

2.2.4. Explosive silicon burning. Shock heating of the Si-rich layer drives explosive Si burning. In the innermost regions, the heating is sufficient for the matter to achieve NSE. The dominant products are Fe and Ni isotopes; however, the expansion and cooling is sufficiently rapid to leave a large abundance of free α particles. This expansion is thus known as the *α-rich freezeout* from NSE. Some reassembly of the free α does occur in the late stages of the freezeout, which makes abundant ^{44}Ti, among other interesting isotopes.

In the outer layers of the Si shell, the matter is less heated from shock passage. The nuclear burning does not achieve full NSE, though it can reach QSE. This is a principal site for production of the short-lived radioisotope ^{53}Mn.

2.2.5. Explosive oxygen burning. Explosive burning in the O-rich layer occurs at even lower temperatures than explosive Si burning. Nevertheless, in the inner regions QSE clusters develop and significant production of isotopes such as ^{40}Ca, ^{44}Ti, and ^{48}Ti occurs. It is also in these regions that the γ process occurs, which is thought to be responsible for many of the p-process isotopes (see below).

2.2.6. Explosive helium burning. Explosive burning in the C-rich zone is not a significant modifier of nuclear abundances except in the innermost layers. The reason is that the outer layers do not achieve high enough temperatures to do much burning by the main C-burning channels, and the preexisting abundances of light particles are too low for other reaction channels to play much of a role. The situation changes in the inner zones of the He-burning shell, however, where there are abundant ^{4}He nuclei as well as a large supply of ^{22}Ne. The sudden heating of this zone causes the ^{22}Ne(α,n)^{25}Mg reaction to release a burst of neutrons, which can be captured by preexisting nuclei and drive an "n-process" (*Blake and Schramm,* 1976). In this way, the n-process is intermediate between the s-process, in which the β-decay rates typically dominate the neutron-capture rates, and the standard r-process, in which neutron captures are more rapid than β decays and certainly fast enough to establish an (n,γ)–(γ,n) equilibrium.

Some researchers in the 1970s and 1980s studied the possibility that the r-process nuclei were in fact produced in an n-process occurring in the He-burning shell (*Hillebrandt et al.,* 1976; *Truran et al.,* 1978; *Thielemann et al.,* 1979). Subsequent work showed that the ^{13}C and ^{22}Ne abundances required to explain the r-process yields were too large (*Blake et al.,* 1981), and thinking on the site of the r-process has returned to matter ejected from the SN core or from neutron star-neutron star collisions (see section 2.3). Nevertheless, the work on the He-shell n-process showed that interesting quantities of the isotopes ^{26}Al and ^{60}Fe could realistically be produced, a point that identified this site as important for cosmochemistry. This was bolstered by independent and contemporaneous efforts to understand the Xe-isotopic patterns in meteoritic nanodiamonds in terms of a "neutron burst" (*Heymann and Dziczkaniec,* 1980; *Heymann,* 1983).

Interest in this nucleosynthetic process is reviving, motivated by the measurements of ^{95,97}Mo excesses in presolar SiC grains of type X (see section 5.6.1) (*Pellin et al.,* 1999, 2000b), which show the signature of a neutron burst (*Meyer et al.,* 2000). Modern stellar models with full nuclear reaction networks confirm that narrow zones with neutron burst signatures are produced in Type II supernovae (e.g., *Rauscher et al.,* 2002). Of additional interest is the fact that sufficiently large quantities of the short-lived radioisotopes ^{60}Fe and ^{182}Hf are produced in the neutron bursts in He shells to explain their abundances in the ESS (*Meyer et al.,* 2003, 2004). The contribution of these neutron bursts to the bulk galactic supply of the heavy elements is small, so traditional astronomical observations or decompositions of SS abundances would not detect this nucleosynthetic component. To date, its signal is only apparent in isotopic anomalies in presolar grains and primitive solar system minerals!

2.2.7. The r-process. *B²FH* recognized that a high-temperature, high-neutron-density explosive process was responsible for many of the heavy elements (Fig. 1). They termed this the r-process since the neutron captures are more *rapid* than the β decays. Unlike the s-process, which is a *secondary* nucleosynthesis process since seed nuclei must already exist, the r-process is *primary* since it assembles its own seed nuclei. The seed nuclei capture neutrons rapidly and establish an (n,γ)–(γ,n) equilibrium for each isotope. Beta decay allows nuclei to increase their charge and thus continue to capture neutrons. This process persists until the supply of neutrons is exhausted.

Since the early 1960s, attention has focused on ejection of neutron-rich matter from SN cores as the likely site for the r-process, but difficulties in ejecting the right amount of such matter plagued the models. In the early 1990s, researchers began to recognize the importance of neutrino-heated ejecta for formation of the r-process isotopes (*Woosley and Hoffman,* 1992; *Meyer et al.,* 1992). The neutrino-energized "hot bubble" that develops near the SN mass cut provides a high-entropy setting in which inefficient assembly of heavy seed nuclei lessens the neutron richness required for formation of the heavy r-process nuclei. *Takahashi et al.* (1994) and *Woosley et al.* (1994) performed detailed SN calculations and found good production of the r-process nuclei but only when they artificially enhanced the entropy of the models by a factor of roughly 3. Subsequently, *Qian and Woosley* (1996) followed neutrino driven wind calculations and found that conditions necessary for the r-process were not obtained. Also posing a problem for the r-process in this setting was the finding that strong neutrino-nucleus interactions during the r-process would severely limit production of heavy nuclei (*Meyer,* 1995; *Fuller and Meyer,* 1995; *Meyer et al.,* 1998b). The problem of too low an entropy (or too low a neutron richness) continues to plague the idea of the r-process in neutrino-driven ejecta unless the proto-neutron star is quite massive (e.g., *Thompson et al.,* 2001) or, perhaps, highly magnetic or rapidly rotating (e.g., *Thompson,* 2003).

Despite the lack of success of the neutrino-driven wind models in explaining the r-process abundances, those models led to several advances in our understanding of the r-process. In particular, we better understand the role of neutrinos in the r-process (e.g., *Meyer,* 1995; *Meyer et al.,* 1998b), the nature and role of statistical equilibria that arise in the r-process (e.g., *Meyer et al.,* 1998b), and the requirements for a successful r-process (*Hoffman et al.,* 1996a; *Meyer and Brown,* 1997). The unsuccessful neutrino-driven models also motivated further work on the r-process in other settings such as directly ejected core matter (*Wheeler et al.,* 1998; *Wanajo et al.,* 2003) or neutron-star collisions (*Freiburghaus et al.,* 1999). While galactic chemical evolution arguments favor supernovae over neutron-star collisions (*Argast et al.,* 2004), a proper understanding of the site of the r-process remains elusive. Nevertheless, the 1990s and early 2000s have seen advances in r-process theory.

The other significant recent advance in understanding the r-process came from observations of heavy-element abundances in metal-poor stars. These observations show that these very old stars have nearly pure and highly enriched r-process heavy-element abundances (*Sneden et al.,* 1996) that match the solar r-process abundance distribution. The principal conclusions from this work are that (1) the r-process dominated the s-process early in the galaxy's history and (2) the r-process, at least for nuclei with mass number greater than 140, is a unique event in that it always gives the same abundance pattern. This second conclusion does not seem to hold for lower-mass r-process elements, which show real star-to-star abundance variations in metal-poor stars (*Sneden et al.,* 2003). Such observations support the notion of diverse r-process events, as inferred from the extinct radioisotopes ^{129}I and ^{182}Hf (*Wasserburg et al.,* 1996; *Qian et al.,* 1998). As discussed further in section 4, the idea is that one type of r-process event is responsible for ^{129}I and a second type of r-process, which occurs about 10 times more frequently than the first, produces ^{182}Hf and the actinides. The challenge to our astrophysical model is to understand how the easier-to-produce low-mass isotopes are made less often than the harder-to-produce heavier isotopes.

2.2.8. Neutron-rich quasi-statistical equilibrium. The neutron-rich Fe-peak isotopes ^{48}Ca, ^{50}Ti, ^{54}Cr, ^{58}Fe, ^{62}Ni, and ^{66}Zn do not seem to have been made in mainline stellar evolution since the main reaction pathways largely bypass them. *Hartmann et al.* (1985) explained the synthesis of the neutron-rich Fe-peak isotopes as a freezeout from a neutron-rich NSE, and cosmochemists embraced that model as an explanation for the isotopic anomalies in ^{48}Ca, ^{50}Ti, ^{54}Cr, ^{58}Fe, and ^{66}Zn in the FUN inclusions (see section 3). The difficulty with this idea, however, is that in the models ^{66}Zn (produced as ^{66}Ni in the NSE) is always more overabundant than ^{48}Ca. This made it difficult to understand how nature in fact could have produced ^{48}Ca.

Subsequently, *Meyer et al.* (1996) showed that expanding and cooling neutron-rich matter would evolve from NSE to a quasi-statistical equilibrium (QSE). The reason is that the three-body reactions that assemble heavier nuclei from α particles, namely, $^{4}He + ^{4}He + ^{4}He \rightarrow ^{12}C$ and the reaction sequence $^{4}He + ^{4}He + n \rightarrow ^{9}Be$ followed by $^{9}Be + ^{4}He \rightarrow ^{12}C + n$, become slower than the expansion at temperatures of roughly 6×10^9 K. This means that the number of heavy nuclei deviates from that required by NSE. Other reactions continue to be fast, and a new equilibrium develops that has one additional constraint, namely, a fixed number of heavy nuclei. These authors also showed that a low-entropy QSE tends to have too many heavy nuclei compared to NSE while a high-entropy QSE tends to have too few. In the former case, ^{48}Ca, ^{50}Ti, and ^{54}Cr are favored over the heavier neutron-rich Fe-peak isotopes, and their synthesis may be understood.

The fact that low-entropy expansions are favored for ^{48}Ca production pointed to deflagrations of near Chandrasekhar-mass WD stars as the likely nucleosynthetic site. The Chandrasekhar mass is the maximum mass a hydrodynamically stable WD star may have, and is typically about 1.4 $M_{\odot}$. Such WD stars have high central densities so that, when they explode, nuclei capture electrons, and the matter becomes sufficiently neutron rich to make ^{48}Ca and the other neutron-rich Fe-peak isotopes. Detailed models of such explosions by *Woosley* (1997) show robust production of ^{48}Ca and the other neutron-rich Fe-peak isotopes. These events likely comprise only about 2% of all Type Ia events; thus, ^{48}Ca is produced only rarely in the galaxy but, when it is, in huge quantities.

2.2.9. The p-processs. A number of proton-rich heavy isotopes, collectively known as the p-process nuclei, are completely bypassed by neutron-capture pathways. *B²FH* envisioned them to be the result of proton captures in a H-rich environment. It now seems clear that this idea was wrong and that they were more likely produced in the "γ process" (*Woosley and Howard,* 1978). In this process, preexisting r- and s-process isotopes are suddenly heated. A sequence of disintegration reactions then begins to strip off first neutrons then protons and α particles. If this processing quenches before the nuclei "melt" all the way down to Fe, an abundance of p-process isotopes results. Such processing occurs in the shock-heated O-rich layers of a core-collapse supernova (*Rayet et al.,* 1995; *Rauscher et al.,* 2002), although it may also occur on the outer layers of a deflagrating C/O WD, a Type Ia supernova (*Howard et al.,* 1991). The important radioisotope ^{146}Sm is made in the γ process.

Two puzzles still surround production of the p-process isotopes. First, the overall yields from the models of explosions of massive stars are a factor of 2 or 3 too low to explain the SS abundances (Fig. 1b). Perhaps production in Type Ia supernovae contributes the missing amount. Second is the fact that all models tend to drastically underproduce the light p-process isotopes $^{92,94}Mo$ and $^{96,98}Ru$. Unlike in heavier elements, the p-process isotopes of Mo and Ru are nearly as abundant as their s- and r-process counterparts. This suggests that some other site is responsible for the bulk of their production. A Type Ia supernova initially seemed a plausible site, but nucleosynthesis calculations done in the context of realistic astrophysical models showed the same underabundances as before (*Howard and Meyer,* 1993). Another possible source is neutrino-heated ejecta in core collapse supernovae that undergo neutron-rich, α-rich freezeouts (*Fuller and Meyer,* 1995; *Hoffman et al.,* 1996b). Those models showed robust production of the light p-process isotopes, but other isotopes were made in greater quantities. New calculations of these α-rich freezeouts that include interactions of the copious neutrinos with nuclei during the nucleosynthesis show that $^{92,94}Mo$ can be the most produced isotopes (*Meyer,* 2003), but it remains to be seen whether the SN models can actually achieve the high neutrino fluxes re-quired for this model. In any event, these neutrino-heated ejecta models produce abundant ^{92}Nb, a radioisotope of great interest for cosmochemistry.

3. SURVIVAL OF NUCLEOSYNTHETIC COMPONENTS IN METEORITES

Although it is clear from nuclear systematics that three different nucleosynthetic processes (p, r, and s) must have contributed to the synthesis of the heavy elements (Fig. 1b), these components as well as the products of other nuclear burning stages and processes were apparently thoroughly mixed during the formation of the SS and, until 30 years ago, it was not possible to detect their separate signatures. The p-process, and certain r-process, nuclides are removed from the s-process path, thus their abundances in the SS are well determined. So are those of s-only nuclides, which are shielded by isobars from r-process contributions. However, there are isotopes that are produced by both the s- and the r-process. The classical approach to determine the respective contributions of these two processes is to calculate their s-process abundances from those of s-only isotopes of the same elements and the requirement of a steady s-process flow, i.e., that the s-process abundances are inversely proportional to the neutron capture cross sections of the s-process isotopes of a given element (*Käppeler et al.,* 1989; *Woolum,* 1988). The r-process abundances (the "residuals") of the isotopes that have both s- and r-process contributions are then obtained by subtracting the calculated s-process from the observed SS abundances. Increasingly detailed astrophysical models now allow r-process residuals to be computed from realistic s-process calculations rather than from the requirement of steady s-process flow (*Straniero et al.,* 1995; *Gallino et al.,* 1998; *Arlandini et al.,* 1995).

The preservation of different nucleosynthetic components had been proposed by Clayton (e.g., *Clayton,* 1975a, 1978; *Clayton and Ward,* 1978), who argued that, because of their distinct stellar sources, these components were transported into the SS by chemically distinct carriers that might have remained separated during SS formation ("cosmic chemical memory" model). It thus created great excitement when in 1978 a CAI from the carbonaceous chondrite Allende, EK 1-4-1, was discovered that exhibited excesses in the r-process isotopes of the elements Ba, Nd, and Sm (*McCulloch and Wasserburg,* 1978a; *McCulloch and Wasserburg,* 1978b) and a depletion in p-only ^{84}Sr (*Papanastassiou and Wasserburg,* 1978). The same year saw the discovery of a Xe component in a chemical separate that exhibited an almost pure s-process pattern (*Srinivasan and Anders,* 1978). The CAI EK 1-4-1 is one of a series of refractory inclusions that exhibit mass-dependent fractionation (F) in the elements O, Mg, and Si together with (at that time UNknown) nuclear anomalies in a large number of elements, and were therefore named "FUN" inclusions. The nuclear anomalies in these inclusions consist of p-, r- and s-process effects in the heavy elements and in excesses and/or depletion in the n-rich isotopes of the Fe-peak elements.

3.1. FUN Inclusions and the Heavy Elements

Two other FUN inclusions, C1 from Allende, and 1623-5 from the CV3 chondrite Vigarano (*Loss et al.,* 1994), show essentially identical anomalies in Sr, Ba, Nd, and Sm [and in many other elements (see *Loss et al.,* 1994)] but the isotopic abundance patterns in these elements are very different from those exhibited by EK 1-4-1. These two different types of patterns have been discussed in detail by *Lee* (1988) in terms of the p-, r-, and s-processes for the production of the heavy elements and we do not want to repeat this discussion here. It is instructive, however, to compare the r-process patterns found in EK 1-4-1 with the s-process patterns found in presolar SiC (see section 5). It was found that SiC is the carrier of the s-process Xe component discovered by *Srinivasan and Anders* (1978) and that SiC is the carrier of s-process Ba, Nd, and Sm as well. Figure 2, adopted from *Begemann* (1993), shows the excesses in the r-process isotopes in EK 1-4-1 and the large depletions of the same isotopes in presolar SiC. These almost perfect complementary patterns give clear evidence of the separate nature of the r- and s-process.

One dramatic difference between the CAI and SiC is the size of the anomalies, which are more than 100-fold larger in SiC than in EK 1-4-1. The reason is that the SiC is true stardust and condensed in the expanding atmosphere of stars whereas the CAIs formed in the SS. Their isotopic anomalies are a memory of presolar material that was not completely homogenized during SS formation and was incorporated into the inclusions. Evidence for a SS system origin of CAIs, even the FUN CAIs, is provided not only by the size of the anomalies, which are much smaller than the expected pure nucleosynthetic components, but also by the fact that, except for mass-dependent fractionation, the major elements such as O, Mg, and Si are isotopically normal. These fractionation effects must be the result of evaporation during the formation of the inclusions, and while part of the small nonlinear deviations in Mg and Si (*Clayton and Mayeda,* 1977; *Wasserburg et al.,* 1977; *Molini-Velsko et al.,* 1986) probably derived from the fact that evaporation does not produce strictly linear mass fractionation, some deviations cannot be explained in this way. It is still a puzzle why in FUN inclusions nuclear anomalies are associated with large mass fractionation effects but, as will be seen in section 3.2, this association of F and UN effects is not the rule.

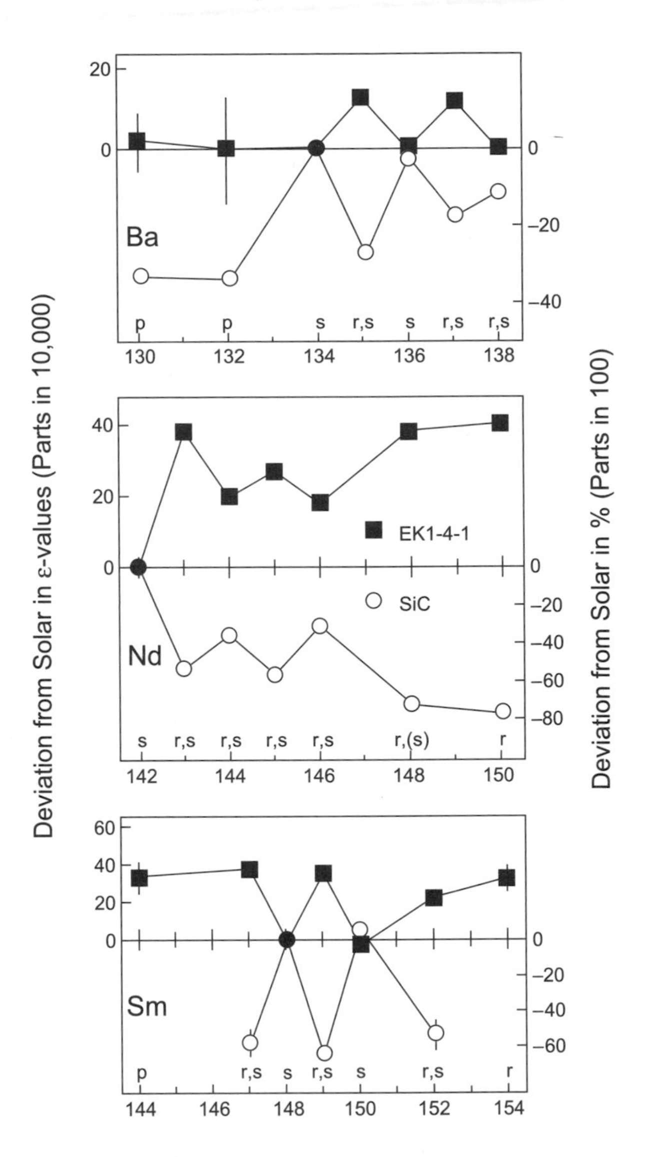

Fig. 2. The isotopic patterns of Ba, Nd, and Sm measured in the FUN inclusion EK 1-4-1 are compared with those measured in presolar SiC. The patterns were normalized to the s-process isotopes ^{134}Ba, ^{142}Nd, and ^{148}Sm. For the EK 1-4-1 data a second normalization was made, requiring that there be no anomalies in ^{138}Ba and ^{150}Ba and equal excesses in ^{143}Nd and ^{148}Nd (*McCulloch and Wasserburg,* 1978a,b). The EK 1-4-1 patterns show excesses in the r-process isotopes; in the case of Sm, also in the p-only isotope ^{144}Sm. The SiC patterns are complementary and are characteristic of the s-process. Adapted from *Begemann* (1993).

3.2. FUN and Hibonite Inclusions and the Iron-Peak Elements

The FUN inclusion EK 1-4-1 has large excesses in the n-rich isotopes ^{48}Ca (*Lee et al.,* 1978), ^{50}Ti (*Niederer et al.,* 1980), ^{54}Cr (*Papanastassiou,* 1986), ^{58}Fe (*Völkening and Papanastassiou,* 1989), and ^{66}Zn (*Völkening and Papanastassiou,* 1990). On the other hand, the Allende inclusions C1 and BG82HB1 and the Vigarano inclusion 1623-5 (all FUN) have large deficiencies in ^{48}Ca (*Lee et al.,* 1978; *Papanastassiou and Brigham,* 1989; *Loss et al.,* 1994), ^{50}Ti (*Niederer et al.,* 1980; *Papanastassiou and Brigham,* 1989; *Loss et al.,* 1994), and ^{54}Cr (*Papanastassiou,* 1986; *Papanastassiou and Brigham,* 1989; *Loss et al.,* 1994). However, this is where the complementarity ends; C1 and BG82HB1 have no clear deficits in ^{58}Fe (*Völkening and Papanastassiou,* 1989) and C1 has normal ^{66}Zn (*Völkening and Papanastassiou,* 1990). There are also minor anomalies in other isotopes such as ^{42}Ca, ^{46}Ca, ^{47}Ti, ^{49}Ti, ^{53}Cr, and ^{70}Zn, and

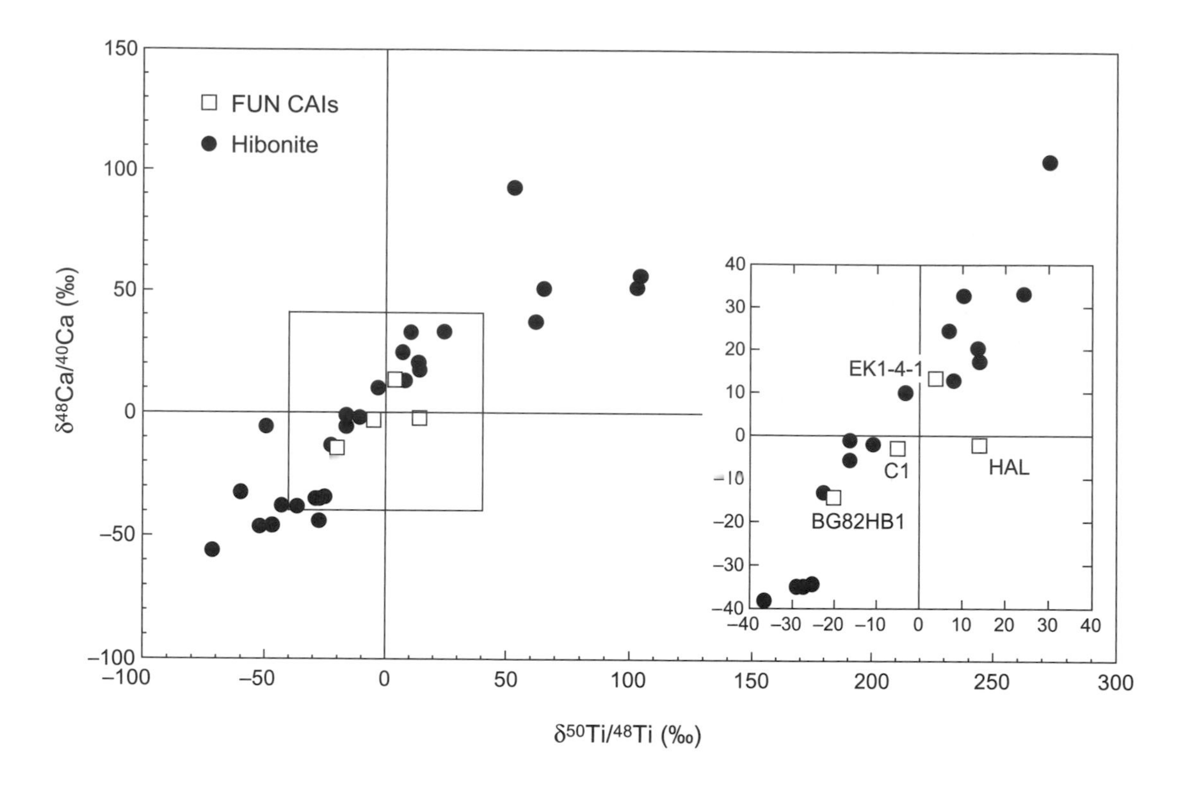

Fig. 3. Isotopic anomalies in the neutron-rich isotopes ^{48}Ca and ^{50}Ti measured in FUN and hibonite-bearing inclusions. The isotopic ratios are expressed as δ values; deviations from the normal ratios are in permil (‰). Isotopic anomalies are qualitatively correlated in that most inclusions with ^{48}Ca excesses have ^{50}Ti excesses and vice versa. Data from *Lee et al.* (1978), *Niederer et al.* (1980), *Fahey et al.* (1987), *Papanastassiou and Brigham* (1989), *Ireland* (1990), and *Sahijpal et al.* (2000).

the exact size of these anomalies depends on the choice of the isotope pair used for internal calibration [see *Lee* (1988) for a detailed discussion]. However, the anomalies in the above-listed n-rich isotopes are so dominant that, even if their exact magnitude is uncertain, their nature cannot be in doubt. Another source of uncertainty is the chemical behavior of the elements involved. For example, Ca and Ti are much more refractory than Zn and partial isotopic equilibration of the latter cannot be excluded (*Völkening and Papanastassiou,* 1990).

Even larger isotopic anomalies than in the FUN inclusions are observed for the elements Ca and Ti in single hibonite grains. While the FUN inclusions have sizes on the order of 1 cm, these hibonite grains are 100 μm or smaller. However, they should be distinguished from presolar hibonite grains discussed in section 5.5. While presolar hibonites have highly anomalous O-isotopic compositions, the hibonites described here have O-isotopic ratios similar to those of other CAIs and are undoubtedly of SS origin. In general, they do not have any pronounced fractionation effects (*Ireland et al.,* 1992), thus they can be termed UN inclusions. For completeness it should be mentioned that there exist also mostly F inclusions such as Allende TE with large fractionations in O, Mg, and Si (*Clayton et al.,* 1984) but only small anomalies in Ca and Ti. Figure 3 shows the anomalies in ^{48}Ca and ^{50}Ti in FUN inclusions and hibonite grains (*Fahey et al.,* 1987; *Papanastassiou and Brigham,* 1989; *Ireland,* 1990; *Sahijpal et al.,* 2000). While there is no perfect linear correlation, with very few exceptions the inclusions and hibonite grains have either excesses or depletions in these two isotopes. Because chemical fractionation between Ca and Ti is possible between the nucleosynthetic source and incorporation into CAIs, a strictly linear correlation is not to be expected.

The general correlation is clear evidence that nucleosynthetic components with large preferential enhancements of the n-rich Fe-peak isotopes and lacking these isotopes were incompletely mixed in the early SS. In the standard Type II SN models (*Woosley and Weaver,* 1995; *Rauscher et al.,* 2002) both ^{48}Ca and ^{50}Ti are underproduced. As discussed below, certain rare Type Ia supernovae are likely responsible for the production of the bulk of the n-rich Fe-group isotopes; thus, the CAIs with depletions in these two isotopes probably formed from collections of SS dust with a lower than average complement of the n-rich component. While components with excesses or depletions in the n-rich isotopes dominate, more detailed analysis of the isotopic data revealed that there are more than only two nucleosynthetic components for Ca and Ti. For example, since Ti has five stable isotopes and two of them (^{46}Ti and ^{48}Ti) are used for making corrections for isotopic mass fractionation, either intrinsic or due to the analysis procedure, we are left with three independent ratios ($^{47}Ti/^{48}Ti$, $^{49}Ti/^{48}Ti$, and $^{50}Ti/^{48}Ti$). If only two components were present, the data would lie on a straight line connecting these components in the three-dimensional space spanned by the three independent ratios. In case of three components, the data would lie on a plane. This kind of analysis showed that for Ti at least four and for Ca at least three components are required to account for the variations in the isotopic ratios (*Jungck et al.,* 1984; *Ireland,* 1990).

Possible nuclear processes that would produce the component with large excesses in the n-rich isotopes were discussed in detail by *Lee* (1988), and the reader is referred to that chapter for the status of the field at that time. Subsequent discovery of ^{58}Fe and ^{66}Zn excesses in EK 1-4-1 seemed to confirm the n-rich nuclear statistical equilibrium (NSE) process (*Hartmann et al.,* 1985) and to exclude n-rich Si burning (*Cameron,* 1979). Neutron-rich NSE burning, even in a multizone mixing model, predicts much larger ^{66}Zn excesses relative to ^{48}Ca than observed (*Völkening and Papanastassiou,* 1990). Chemical fractionation may be responsible for this discrepancy, but it is more likely that the appropriate nucleosynthesis model for the n-rich Fe-peak isotopes is that of a QSE with too many heavy nuclei. As discussed in section 2, matter expanding and cooling from high temperature and density in very low-entropy, neutron rich explosive environments establishes QSEs with too many heavy nuclei relative to NSE, which, in turn, leads to enhanced synthesis of ^{48}Ca, ^{50}Ti, and ^{54}Cr relative to ^{66}Zn (*Meyer et al.,* 1996). Ejection of very-low-entropy, neutron-rich matter arises in deflagrations of near-Chandrasekhar-mass WD stars, and models confirm strong production of ^{48}Ca in these settings (*Woosley,* 1997). Interestingly, such supernovae probably comprise only ~2% of all Type Ia events; therefore, the n-rich Fe-group isotopes are apparently produced infrequently but in prodigious quantities. This is likely to lead to strong inhomogeneities in their abundances distributions in interstellar dust, which, in turn, likely plays an important role in the origin of their anomalies in UN inclusions.

3.3. Other Nuclear Anomalies

In addition to the large and striking anomalies exhibited by FUN and hibonite-bearing inclusions, endemic effects in "normal" CAIs have been found for Ti (*Niederer et al.,* 1980, 1981, 1985; *Niemeyer and Lugmair,* 1981, 1984), Ca (*Jungck et al.,* 1984; *Niederer and Papanastassiou,* 1984), Cr (*Birck and Allègre,* 1984, 1985; *Birck and Lugmair,* 1988) and Ni (*Birck and Lugmair,* 1988). These effects are much smaller than those found in FUN and hibonite inclusions. They are mostly excesses in the n-rich isotopes of these elements and again indicate incomplete mixing of an n-rich component in the ESS.

Improvement in laboratory and instrumental capabilities of mass spectrometry [thermal ionization mass spectrometry (TIMS) and inductively-coupled plasma mass spectrometry (ICP-MS)] has led to further discovery of isotopic anomalies in a series of elements. The isotopic measurements were made either on bulk samples of various meteorites or on different chemical separates (leachates and residues). *Rotaru et al.* (1992) and *Podosek et al.* (1997) have found deficits and large excesses of ^{54}Cr in certain leach fractions from carbonaceous chondrites, the largest effects (excesses of more than 20%) being exhibited by the CI chondrite Orgueil. Although this indicates the presence of the n-rich component of FUN inclusions, no effects in other Fe-peak elements have been detected. Also, no carriers of the ^{54}Cr have yet been detected. Another example is anomalies in the Mo isotopes (*Dauphas et al.,* 2002a,b; *Yin et al.,* 2002a). The dominating pattern found in these measurements is excesses in the r-process isotopes ^{95}Mo, ^{97}Mo, and ^{100}Mo, but also in the p-process isotopes ^{92}Mo and ^{94}Mo. This pattern is the inverse of the s-process pattern present in mainstream SiC grains (see section 5). However, an s-process pattern was also observed in a leachate from Orgueil (*Dauphas et al.,* 2002a). There might be still questions of unresolved interferences but there is no doubt that endemic isotopic anomalies exist not only in CAIs and primitive carbonaceous chondrites but also in planetary differentiates (*Chen et al.,* 2004; *Yin and Jacobsen,* 2004) and the finding of uniform Mo-isotopic compositions (*Becker and Walker,* 2003) seems unfounded. Endemic anomalies are also found for Ru (*Papanastassiou et al.,* 2004). The pattern of these anomalies depends on the normalization scheme used and at the moment it is not clear which nucleosynthetic components can be distinguished. No evidence for the presence of short-lived ^{97}Tc ($T_{1/2} = 2.6 \times 10^6$ yr) or ^{99}Tc ($T_{1/2} = 2.1 \times 10^5$ yr) has been found. Several experimental groups have measured Zr-isotopic ratios in a large variety of meteorites and have found mostly excesses but also some deficits in ^{96}Zr (*Harper,* 1993; *Sanloup et al.,* 2000; *Yin et al.,* 2001; *Schönbächler et al.,* 2003). These patterns are similar to what is found in presolar graphite and SiC grains (section 5) and indicate the presence (or absence) of an r-process component. The isotope ^{92}Zr shows also anomalies but these could be due to the decay of short-lived ^{92}Nb ($T_{1/2} = 3.5 \times 10^7$ yr). Evidence for its presence in the ESS will be discussed in section 4. The Ba-isotopic ratios are also potentially affected by contribution from a radioisotope, ^{135}Cs ($T_{1/2} = 2.3 \times 10^6$ yr) (see section 4). However, Ba-isotopic analyses of leaching fractions from different meteorites (*Hidaka et al.,* 2001, 2003) show excesses as well as deficits in both ^{135}Ba and ^{137}Ba. The patterns with excesses are similar to the r-process pattern exhibited by the FUN inclusion EK 1-4-1, while the pattern with depletions is reminiscent of the s-process pattern found in presolar SiC (see section 5). This again provides evidence that contributions from both nucleosynthetic processes were not perfectly homogenized in the ESS.

There have been many more attempts to detect isotopic anomalies in meteorites (e.g., *Podosek et al.,* 1999) but data need to be on a firmer basis to be discussed here. In some cases, as for example for Cu (*Luck et al.,* 2003; *Russell et al.,* 2003), which has only two stable isotopes, it is not possible to assign any nucleosynthetic pattern to an isotopic anomaly.

TABLE 2. Short-lived radioisotopes for which evidence has been found in meteorites.

Radioisotope*	Half-life (m.y.)	Daughter Isotope	Reference Isotope	Initial Abundance	Reference
^{41}Ca	0.10	^{41}K	^{40}Ca	1.5×10^{-8}	[1]
^{26}Al	0.74	^{26}Mg	^{27}Al	5×10^{-5}	[2]
^{10}Be	1.5	^{10}B	^{9}Be	$\sim 5 \times 10^{-4}$	[3]
^{60}Fe	1.5	^{60}Ni	^{56}Fe	$\sim 1.5 \times 10^{-6}$	[4]
^{53}Mn	3.7	^{53}Cr	^{55}Mn	$\sim 2 \times 10^{-5}$	[5]
^{107}Pd	6.5	^{107}Ag	^{108}Pd	4.5×10^{-5}	[6]
^{182}Hf	9	^{182}W	^{180}Hf	2×10^{-4}	[7]
^{129}I	16	^{129}Xe	^{127}I	10^{-4}	[8]
^{244}Pu	81	Fission Xe	^{238}U	$(4–7) \times 10^{-3}$	[9]
^{146}Sm	103	^{142}Nd	^{144}Sm	8×10^{-3}	[10]
^{7}Be	53 d	^{7}Li	^{9}Be	5×10^{-3}	[11]
^{36}Cl	0.3	^{36}Ar, ^{36}S	^{35}Cl	$(0.14–1) \times 10^{-5}$	[12]
^{135}Cs	2.3	^{135}Ba	^{133}Cs	$(1–5) \times 10^{-4}$	[13]
^{205}Pb	15	^{205}Tl	^{204}Pb	10^{-4}	[14]
^{92}Nb	35	^{92}Zr	^{93}Nb	$\sim 1.5 \times 10^{-5}$	[15]

*For the five isotopes in the lower panel, only hints exist that need confirmation.

References: [1] *Srinivasan et al.* (1994, 1996); [2] *Lee et al.* (1976, 1977); [3] *McKeegan et al.* (2000); [4] *Shukolyukov and Lugmair* (1993), *Tachibana and Huss* (2003), *Mostefaoui et al.* (2005); [5] *Birck and Allègre* (1985), *Lugmair and Shukolyukov* (1998); [6] *Kelly and Wasserburg* (1978); [7] *Lee and Halliday* (1995), *Harper and Jacobsen* (1996), *Kleine et al.* (2002), *Yin et al.* (2002); [8] *Jeffery and Reynolds* (1961); [9] *Rowe and Kuroda* (1965); [10] *Lugmair and Marti* (1977), *Lugmair and Galer* (1992); [11] *Chaussidon et al.* (2004); [12] *Murty et al.* (1997), *Lin et al.* (2005); [13] *McCulloch and Wasserburg* (1978), *Hidaka et al.* (2001); [14] *Nielsen et al.* (2004); [15] *Harper* (1996), *Schönbächler et al.* (2002).

4. SHORT-LIVED NUCLEI

Short-lived radioisotopes have lifetimes (up to 10^8 yr) that are too short for them to have survived for the age of the SS. However, there is evidence that some interesting isotopes were "live" during the early stages of the SS. This evidence comes from excesses in the daughter isotopes above their normal SS abundance. The initial presence of short-lived isotopes (SLI) in ESS materials is a dramatic demonstration of stellar nucleosynthesis occurring as recently as 1 m.y. before SS formation. They thus connect the birth of the SS to the presence of nearby stars.

Short-lived isotopes in the ESS have many important implications. One of them is that the decay of nuclei such as ^{26}Al and ^{60}Fe could have provided the heat for the melting of small planetary bodies (*Urey,* 1955; *Fish et al.,* 1960; *Schramm et al.,* 1970). Another is that SLI can serve as fine-scale chronometers for ESS events (e.g., *Lugmair and Shukolyukov,* 2001; *Gilmour,* 2002). These questions will be treated in other chapters (*Halliday and Kleine,* 2006; *Nichols,* 2006; *Russell et al.,* 2006; *Wadhwa et al.,* 2006) and will not be discussed here. Rather we concentrate on the question of what information SLI can provide about stellar nucleosynthesis. A detailed review of SLI in the ESS greatly exceeds the scope of the present chapter. The interested reader is therefore directed to some recent reviews (*MacPherson et al.,* 1995; *Podosek and Nichols,* 1997; *Goswami and Vanhala,* 2000; *Busso et al.,* 2003; *Nittler and Dauphas,* 2006). Our emphasis will be on recent developments.

Table 2 gives a list of the SLI for which there is clear evidence that they were present and live in the ESS. Hints exist for several other SLI, but these still need confirmation. Positive evidence for the one-time presence of a SLI requires demonstration that the excess of the daughter isotope is linearly correlated with the parent element abundance. In the case of ^{129}I it has been shown that excesses in the decay product ^{129}Xe are correlated with the stable isotope of I, ^{127}I (*Jeffery and Reynolds,* 1961). After the discovery of ^{26}Al in CAIs, a linear correlation was demonstrated between excesses in ^{26}Mg and the stable ^{27}Al in several minerals with different Al/Mg ratios (*Lee et al.,* 1976, 1977). Arguments for a "fossil" origin of ^{26}Mg excesses (*Clayton,* 1986), i.e., ^{26}Al decay long before SS formation and introduction of "fossil" ^{26}Mg into refractory inclusions in the ESS, have not been accepted [see discussion and references in *MacPherson et al.* (1995)]. Presently, there is little doubt that the SLI listed in Table 2 were present in the ESS.

In Table 2 we give the half-lives of the SLI, their daughter isotopes, and the ratios relative to a stable isotope observed in ESS objects. Plutonium has no isotope that still exists today and ^{238}U is used instead. Because these ratios change with time as the different radioisotopes decay at different rates, ratios should be given at the same time. In Table 2 the reference time is the formation time of CAIs that had a $^{26}Al/^{27}Al$ ratio of 5×10^{-5} [the "canonical ratio" (see *MacPherson et al.,* 1995). Recent high-precision inductively-coupled plasma mass spectrometry and high-precision SIMS measurements of low-Al/Mg phases in CAIs (*Galy et al.,* 2004; *Liu et al.,* 2005; *Young et al.,* 2005) have indicated an initial $^{26}Al/^{27}Al$ ratio in CAIs as high as 7×10^{-5}. However, for the time being we use 5×10^{-5} as the canonical ratio.]. For such CAIs there exists an absolute age of 4567.2 ± 0.6 m.y. based on the U-Pb system ("Pb/Pb" age) (*Amelin et al.,* 2002). Another advantage of using CAIs as a reference is that they contain evidence and reliable initial ratios also for ^{41}Ca and ^{10}Be (*Srinivasan et al.,* 1994, 1996; *McKeegan et al.,* 2000; see also *Goswami and Vanhala,* 2000). Hints for the initial presence of ^{53}Mn, ^{60}Fe, and ^{129}I have also been reported for CAIs, but it is not clear whether the inferred parent/daughter ratios refer to the same time as those for the above three isotopes (e.g., *Goswami et al.,* 2001).

A fundamental problem is that not all SLI can be measured in the same ESS objects. The reason is that the determination of excesses in the daughter isotopes requires a reasonably (sometimes very) large ratio of the parent to daughter element. While this is the case for ^{41}Ca, ^{26}Al, and ^{10}Be in CAIs, there are simply no samples in which this is the case for all SLI. If absolute ages could be obtained for objects in which SLI ratios have been measured, these ratios can be extrapolated to the reference time, but even this is not possible in all cases. It should therefore be realized that uncertainties are involved in the ratios for the SLI ^{60}Fe, ^{53}Mn, and ^{107}Pd given in Table 2. In the case of ^{60}Fe, *Birck and Lugmair* (1988) inferred an $^{60}Fe/^{56}Fe$ ratio of 1.6×10^{-6} from their Fe-Ni measurements in CAIs, but because of possible nuclear anomalies in the Ni isotopes, no firm conclusion on the initial presence of ^{60}Fe in the ESS was reached. Subsequently, evidence for ^{60}Fe was found in eucrites, which are differentiated meteorites, and from bulk samples in the eucrite Chervony Kut a $^{60}Fe/^{56}Fe$ ratio of 4×10^{-9} was derived (*Shukolyukov and Lugmair,* 1993). If we adopt the age based on ^{53}Mn for Chervony Kut (*Lugmair and Shukolyukov,* 1998) and the above-cited Pb/Pb age for CAIs, then this ratio extrapolates to only 2×10^{-8} at the time of CAI formation. Recently, much higher values have been obtained from the analysis of sulfides from unequilibrated ordinary chondrites and enstatite chondrites (*Tachibana and Huss,* 2003a; *Mostefaoui et al.,* 2003, 2004a, 2005; *Guan et al.,* 2003). *Mostefaoui et al.* (2005) reported a $^{60}Fe/^{56}Fe$ ratio of 9.2×10^{-7} for troilite from the unequilibrated ordinary chondrite Semarkona, and *Guan et al.* (2003) obtained ratios from various sulfides in enstatite chondrites ranging all the way up to 8.5×10^{-7}. If we assume that these sulfides have ages comparable to those of chondrules and thus are about 2 m.y. younger than CAIs, then these ratios are consistent with the CAI value obtained by *Birck and Lugmair* (1988).

Similarly, considerable uncertainties exist about the $^{53}Mn/^{52}Mn$ ratio. *Birck and Allègre* (1985) inferred a ratio of 4.4×10^{-5} from the analysis of an Allende CAI. However, the Mn/Cr ratio in CAIs is not high and CAIs contain nucleosynthetic anomalies in several elements, including the Cr isotopes, making the $^{53}Mn/^{52}Mn$ ratio in CAIs suspect (*Lugmair and Shukolyukov,* 1998, 2001). Another approach is to extrapolate the ratio measured in the angrite Lewis Cliff 85010 (*Nyquist et al.,* 1994; *Lugmair and Shukolyukov,* 1998) by using the difference in the Pb/Pb ages obtained for the angrites Lewis Cliff 85010 and Angra dos Reis (*Lugmair and Galer,* 1992) and for CAIs (*Amelin et al.,* 2002). In fact, the angrite age has been used as an anchor point for a Mn-Cr chronology (*Lugmair and Shukolyukov,* 2001; *Nyquist et al.,* 2001). However, extrapolation of the Lewis Cliff 85010 ratio to the CAI Pb/Pb formation time gives a $^{53}Mn/^{52}Mn$ ratio of only $\sim 8 \times 10^{-6}$ and higher ratios have been obtained for Kaidun carbonates (*Hutcheon et al.,* 1999), chondrules from the ordinary chondrites Bishunpur and Chainpur (*Nyquist et al.,* 2001), and in samples from CI and CM carbonaceous chondrites (*Rotaru et al.,* 1992; *Birck et al.,* 1999), that range up to 2×10^{-5}. In Table 2 we adopt this as a compromise value.

A $^{107}Pd/^{108}Pd$ ratio of 2×10^{-5} has been obtained in iron meteorites (*Kelly and Wasserburg,* 1978; *Kaiser and Wasserburg,* 1983) and extrapolation to the time of CAI formation gives a value of $\sim 4.5 \times 10^{-5}$ (*Goswami and Vanhala,* 2000). Recently, a consistent value of 1.1×10^{-4} has been obtained for the $^{182}Hf/^{180}Hf$ ratios from samples of the H4 chondrites Ste. Marguerite and Forest Vale (*Kleine et al.,* 2002) and from whole-rock samples of several carbonaceous and ordinary chondrites and an Allende CAI (*Yin et al.,* 2002b). For discussions and references of the ratios for the longer-lived SLI we refer to the literature (*Goswami and Vanhala,* 2000; *Busso et al.,* 2003).

Hints for the presence of several other short-lived isotopes in the ESS exist but they need confirmation. Chlorine-36 can β^--decay into ^{36}Ar (98.1%) or turn into ^{36}S (1.9%) by β^+-decay or electron capture. *Murty et al.* (1997) inferred a $^{36}Cl/^{35}Cl$ ratio of 1.4×10^{-6} from ^{36}Ar excesses, whereas *Lin et al.* (2005) obtained ratios as high as 10^{-5} from ^{36}S excesses that are correlated with Cl/S ratios in Cl-rich sodalite and nepheline in a CAI. Because these phases are secondary and were most likely formed later than the CAI, the $^{36}Cl/^{35}Cl$ ratio at the time of CAI formation could be as high as 4×10^{-4}. Uncertainty also exists about the initial abundance of ^{135}Cs (*Hidaka et al.,* 2001, 2003; *Nichols et al.,* 2002) and inferred $^{135}Cs/^{133}Cs$ ratios range from $\sim 10^{-4}$ to 5×10^{-4}. One complication in these measurements is that ^{135}Ba excesses can also have an r-process origin, which must be corrected for; another is the volatile nature of Cs so that no samples with high Cs/Ba ratios are found. Niobium-92 represents an interesting case because, in contrast to the other short-lived isotopes, it is produced by the p-process. Inferred $^{92}Nb/^{93}Nb$ ratios in the ESS range from $\sim 1.5 \times 10^{-5}$ (*Harper,* 1996; *Schönbächler et al.,* 2002) to $\sim 10^{-3}$ (*Sanloup et al.,* 2000; *Münker et al.,* 2000; *Yin et al.,* 2000). A possible problem is the presence of nucleosynthetic effects in ^{92}Zr, and only the study by *Schönbächler et al.* (2002) obtained an internal isochron. The high ratio would imply a very high production ratio, which is not achieved by the standard explosive processing in shells in Type II SN models (*Rauscher et al.,* 2002). As discussed in section 2, ^{92}Nb is likely made in the same, but as yet mysterious, environment as the light p-process isotopes ^{92,94}Mo and ^{96,98}Ru. Neutrino-driven winds from proto-neutron stars can produce these isotopes (e.g., *Hoffman et al.,* 1996b), perhaps with a necessary boost from the copious neutrinos that drive the wind (*Meyer,* 2003). Such wind scenarios robustly produce ^{92}Nb, at a level ^{92}Nb/^{93}Nb $\approx$ 1.0. *Nielsen et al.* (2004) reported evidence for the initial presence of ^{205}Pb in the ESS from ^{205}Tl excesses in two group IA iron meteorites. Lead-205 is an s-process isotope and decays by electron capture. The inferred ^{205}Pb/^{204}Pb ratio is 10^{-4}, but the possibility that the ^{205}Tl excesses are due to mass-dependent isotopic fraction of Tl, which has only two stable isotopes, has yet to be excluded. Finally, ^{7}Li excesses in an Allende CAI have been interpreted as evidence for the existence of ^{7}Be (*Chaussidon et al.,* 2004, 2005). What makes this finding so interesting is that ^{7}Be can be produced only by energetic particle irradiation and decays by electron capture with a half-life of only 53 d. This sets a severe constraint on the formation time of CAIs and requires an extremely high particle flux from the early Sun.

4.1. Galactic Production of Short-lived Isotopes

As had been proposed early for ^{129}I (*Wasserburg et al.,* 1960), the abundances of ^{244}Pu, ^{146}Sm, ^{182}Hf, and ^{129}I (and possibly ^{53}Mn) seen in the ESS can be explained by the steady-state concentrations of these isotopes in the galaxy at the time of SS formation (*Cameron et al.,* 1995; *Wasserburg et al.,* 1996; *Meyer and Clayton,* 2000). Plutonium-244, ^{182}Hf, and ^{129}I are produced by the r-process and thus have a SN origin. Hafnium-182 can also be produced by the neutron burst in the He shell of an exploding massive star (*Meyer et al.,* 2003, 2004). Samarium-146 is produced by the p-process, also in supernovae, whereas ^{53}Mn has also such an origin but is the result of explosive Si burning. For all these SLI, a steady state is established between SN production throughout galactic history and radioactive decay, and the SS apparently inherited them with the steady-state concentrations they had at the time of SS formation. For all the above SLI except ^{129}I, the abundances in the ESS are consistent with one another if SN production essentially lasted until SS formation; however, in such a case ^{129}I would be overproduced by a large factor (*Wasserburg et al.,* 1996). In fact, a decay interval of ~10^{8} yr had been inferred earlier (*Wasserburg et al.,* 1960). This discrepancy prompted *Qian et al.* (1998) to propose different r-process sources for the heavy nuclei including ^{182}Hf and for nuclei around mass 130 including ^{129}I. According to this proposal the second process would occur much less frequently and would thus explain the observed relative abundances of ^{182}Hf and ^{129}I in the ESS (see also *Qian and Wasserburg,* 2000, 2003). A more detailed treatment of the galactic production of SLI can be found in *Nittler and Dauphas* (2006).

4.2. Production of the Short-lived Isotopes Calcium-41, Aluminum-27, Beryllium-10, Iron-60, and Palladium-107

Galactic steady-state production cannot account for the ESS abundances of ^{41}Ca, ^{27}Al, ^{10}Be, ^{60}Fe, and ^{107}Pd. For ^{26}Al this has been clear from its present-day abundance in the galaxy as inferred from the 1.8 MeV γ-ray line (*Mahoney et al.,* 1984). Gamma-ray data, mostly from COMPTEL (*Diehl et al.,* 1995), show that the abundance of ^{26}Al in the galaxy is at least a factor of 10 lower than that in the ESS (*Clayton and Leising,* 1987; *Knödlseder,* 1999). The presence of ^{26}Al and the other SLI in the above list thus require either a stellar source closely preceding SS formation or production by energetic particles from the early Sun.

4.2.1. Production by solar energetic particles. The original proposal of ^{26}Al production by charged particle irradiation from an early active sun by *Fowler et al.* (1962) has received its most recent detailed expression in the X-wind model put forward by Frank Shu and coworkers (*Shu et al.,* 1996, 2001; *Gounelle et al.,* 2001). These authors postulated not only that ^{26}Al, ^{41}Ca, and ^{53}Mn are produced by local irradiation in the X-wind region of the early Sun but also that CAIs and chondrules are produced in this region but at different distances from the Sun. This would explain that initial ^{26}Al/^{27}Al ratios in chondrules are systematically lower than in CAIs. The X-wind model of SLI production received a boost by the discovery in CAIs of initial ^{10}Be (*McKeegan et al.,* 2000), because this radionuclide is only produced by energetic particle irradiation and not by stellar nucleosynthesis. In their model, *Gounelle et al.* (2001) and *Leya et al.* (2003) could produce the ESS abundances of ^{10}Be, ^{26}Al, ^{41}Ca, and ^{53}Mn. However, these predictions required very special assumptions about target composition and structure and rather extreme assumptions about the energy and composition (extremely high ^{3}He abundance) of the solar cosmic rays. In addition, the X-wind model cannot account for ^{107}Pd, ^{60}Fe (see below), and the high ^{36}Cl/^{35}Cl obtained by *Lin et al.* (2004).

Additional evidence against the X-wind model has been provided by recent measurements. *Marhas et al.* (2002) found evidence for ^{10}Be in hibonite grains devoid of ^{41}Ca and ^{26}Al that indicated that the production of ^{10}Be is decoupled from that of these two SLI. Lack of a correlation between ^{10}Be and ^{26}Al was also seen in other measurements (*MacPherson et al.,* 2003). While ^{10}Be, which is not produced by stellar nucleosynthesis, could have been produced by energetic particles in the ESS, a stellar source seems to be needed for the other SLI. Manganese-53 can be made by proton bombardment of solids in the ESS (*Wasserburg and Arnould,* 1987), but its presence in samples from differentiated parent bodies (*Lugmair and Shukolyukov,* 1998) argues against the X-wind model. It was known all along that ^{60}Fe cannot be produced by energetic particle irradiation and the high ^{60}Fe/^{56}Fe inferred from recent studies (*Guan et al.,* 2003; *Mostefaoui et al.,* 2003, 2004a, 2005; *Tachibana and Huss,* 2003a,b) indicates a recent stellar source. Further evidence against the contention of the X-wind model that the short-lived isotopes, specifically ^{26}Al, had much higher abundances in CAIs than other materials has been provided by two recent studies that found consistency between the Al-Mg and U-Pb clock (*Gilmour,* 2002) for CAIs and chondrules (*Amelin et al.,* 2002) and for CAIs and feldspar from H4 chondritic meteorites (*Zinner and Göpel,* 2002).

In summary, an origin by energetic particle irradiation in the ESS appears to be valid only for ^{10}Be. Even for this SLI other production mechanisms have been suggested. *Cameron* (2002) has proposed production in so-called r-process jets associated with core-collapse supernovae, but the lack of ^{26}Al and ^{41}Ca, which would also be produced by this mechanism, in samples with ^{10}Be seems to rule it out. A more promising alternative has been proposed by *Desch et al.* (2004), who considered ^{10}Be production by spallation from galactic cosmic rays in the ISM and trapping of the cosmic-ray ^{10}Be in the molecular cloud from which the SS formed. The authors estimate that a major portion of the initial ^{10}Be found in CAIs could originate from this source.

4.2.2. Production in a nearby stellar source. Ever since the proposal that a SN explosion triggered the formation of the SS (*Cameron and Truran,* 1977), various stellar sources have been investigated that could have produced and injected the SLI into the protosolar cloud just before its collapse (*Goswami and Vanhala,* 2000). In addition to a supernova (*Cameron et al.,* 1995), an AGB star (*Cameron,* 1993; *Wasserburg et al.,* 1994; *Busso et al.,* 1999, 2003) and a Wolf-Rayet (WR) star (*Arnould et al.,* 1997b) have been proposed as a source for the SLI in the ESS. The question is whether any of these sources can produce the SLI in sufficient amounts and in the proper ratios so that after a certain free decay time between production and formation of CAIs, the abundances of the SLI in the ESS are reproduced. While the free decay time is an adjustable parameter, because of the short half-life of ^{41}Ca, it cannot be much larger than 10^6 yr. While each of these sources can produce some of the SLI, they all have problems with accounting for all of the SLI or accounting for their correct ratios.

The problem with core-collapse supernovae is not that they cannot produce the SLI but that they produce too much of some of them. *Wasserburg et al.* (1998) and *Busso et al.* (2003) have investigated the production of the SLI theoretically predicted from a variety of SN models (*Woosley and Weaver,* 1995; *Rauscher et al.,* 2002). While the models can account for the right proportions of ^{26}Al, ^{41}Ca, and ^{60}Fe, they produce too much ^{53}Mn and far too much ^{107}Pd (*Wasserburg et al.,* 1998; *Meyer et al.,* 2004). A possible solution to this problem has been proposed by Meyer and coworkers (*Meyer and Clayton,* 2000; *Meyer et al.,* 2003). These authors assumed that only the outer layers of the supernova (the layers above the so-called mass cut) held responsible for the SLI were injected into the protosolar cloud and not the inner layers where ^{53}Mn (and also ^{60}Fe to a certain extent) is produced. Various theoretical studies have been devoted to triggered star formation and the injection of SLI into the presolar cloud (*Boss and Foster,* 1997, 1998; *Foster and Boss,* 1997; *Vanhala and Boss,* 2002), and they show that it appears possible to introduce and mix the SLI from a stellar source. There is also astronomical evidence for triggered star formation, most likely from the shock wave associated with SN explosions (*Preibisch,* 1999), and SN remnants are observed close to dense molecular clouds. Injection of only the outer layers of a supernova may require fine-tuning of the mass cut. No detailed quantitative investigations of different SN models have been made to see whether the correct ESS abundances of the SLI are obtained this way. Conclusions on this issue must await detailed models of the evolution of the ejecta from an exploding massive star and their interaction with the surrounding medium.

Wasserburg et al. (1994) and *Busso et al.* (1999, 2003) considered production of the SLI in an AGB star and mixing of its stellar wind or the ejected planetary nebula with the presolar cloud. They have found that the ESS abundances of ^{26}Al, ^{41}Ca, and ^{107}Pd can be produced by a low-mass (1.5 $M_\odot$) star of solar metallicity. This assumes that most of the ^{26}Al is produced during cool bottom processing (*Nollett et al.,* 2003) because production in the H-burning shell is insufficient. Asymptotic giant branch stars do not make ^{53}Mn but, as pointed out above, steady-state galactic production can account for this SLI. The major problem with an AGB source is ^{60}Fe. This SLI can be produced by neutron capture on ^{58}Fe and the unstable ^{59}Fe ($T_{1/2}$ = 44.5 d), but this requires fairly high neutron densities. This excludes low-mass AGB stars if the recent high values of $\sim 10^{-6}$ obtained for the ^{60}Fe/^{56}Fe are correct. Models of AGB stars of 5 $M_\odot$ and solar metallicity or 3 $M_\odot$ but much lower metallicity can produce sufficient amounts of ^{60}Fe but run into problems matching the abundance ratios of the other SLI (*Busso et al.,* 2003). Non-exploding WR stars (*Arnould et al.,* 1997b) can produce the right abundances of ^{26}Al, ^{41}Ca, and ^{107}Pd but fail completely when it comes to the production of ^{60}Fe.

If we exclude a WR star as the source for the SLI, we are still left with the question of whether a supernova or an AGB star produced them. An issue that is often discussed in connection with this question is the probability of a close stellar source, be it a supernova or an AGB star, immediately before SS formation (see, e.g., *Busso et al.,* 2003). We do not deem such discussions to be very fruitful. The SS did form endowed with SLI and there is overwhelming evidence that a stellar source provided these SLI. As with any singular event (the presence of intelligent life on Earth is another example), it is of little use to consider the *a priori* probability of this event.

What is most important is to find out whether a supernova or an AGB did produce the SLI from their abundances in the ESS. Any stellar source that injected the SLI into the presolar cloud would also have injected other, stable isotopes (*Wasserburg et al.,* 1998; *Nichols et al.,* 1999). While such contributions would not be detectable if the injected and cloud material has been thoroughly mixed, a heterogeneous distribution of the injected stellar material would show up in the form of isotopic anomalies and these anomalies would be very distinct and thus diagnostic for a SN or AGB source. But even in the absence of such fingerprints we can only hope that more experimental measurements (e.g., Fe-Ni studies), discovery of other SLI in ESS materials, experimental determination of important cross sections (e.g., only theoretical values exist for the neutron-capture cross sections of ^{59}Fe and ^{60}Fe), and improved models of stellar nucleosynthesis and of the dynamics of stellar ejecta and their interaction with cloud material will provide an answer.

5. PRESOLAR GRAINS

5.1. Introduction

Although the study of presolar grains is a relatively new field of astrophysics (the first presolar grains were isolated from meteorites and identified in 1987), it has grown to such an extent that not all aspects of presolar grains can be treated here. These grains provide information on stellar nucleosynthesis and evolution, mixing in supernovae, galactic chemical and isotopic evolution, physical and chemical conditions in stellar atmospheres and supernova and nova ejecta, and conditions in the ESS and in the parent bodies of the meteorites in which the grains are found. The study of these grains allows us to obtain information about individual stars, complementing astronomical observations of elemental and isotopic abundances in stars (e.g., *Lambert,* 1991), by extending measurements to elements that cannot be measured astronomically. Rather than giving a detailed review we will concentrate on elemental and isotopic features of the grains that provide information and constraints on nucleosynthesis and stellar evolution. For more information the interested reader is referred to some reviews (*Anders and Zinner,* 1993; *Ott,* 1993; *Zinner,* 1998a,b, 2004; *Hoppe and Zinner,* 2000; *Nittler,* 2003; *Clayton and Nittler,* 2004) and to the compilation of papers found in *Bernatowicz and Zinner* (1997).

5.2. Types of Presolar Grains

Table 3 shows a list of the presolar grain types identified to date and lists the sizes, approximate abundances, and stellar sources. The carbonaceous phases diamond, SiC, and graphite were discovered because they carry exotic noble gas components that led to their isolation from meteorites. Diamond carries Xe-HL, Xe enriched in the light and heavy isotopes (*Lewis et al.,* 1987; *Anders and Zinner,* 1993; *Huss and Lewis,* 1994a,b); SiC carries Xe-S, Xe with an s-process isotopic signature (*Srinivasan and Anders,* 1978; *Clayton and Ward,* 1978), and Ne-E(H), almost pure ^{22}Ne released at high temperature (*Tang and Anders,* 1988b; *Lewis et al.,* 1994); graphite carries Ne-E(L), released at low temperatures (*Amari et al.,* 1990, 1995b).

Because noble gases are trace elements, their isotopic compositions have mostly been measured in so-called "bulk samples," collections of large numbers of grains. Bulk measurements have also been made for many other elements (C, N, Mg, Si, Ca, Ti, Cr, Fe, Sr, Ba, Nd, Sm, and Dy) (*Ott and Begemann,* 1990; *Zinner et al.,* 1991; *Prombo et al.,* 1993; *Richter et al.,* 1993, 1994; *Russell et al.,* 1996, 1997; *Amari et al.,* 2000; *Podosek et al.,* 2004). However, because presolar grains come from different stellar sources, information on individual stars is obtained by the study of single

TABLE 3. Presolar grain types.

Grain Type	Abundance* (ppm)	Size (μm)	Stellar Sources	Nucleosynthetic Processes† Exhibited by Grains
Silicates in IDPs	900	≤1	RG, AGB and SN	Core H, He
Silicates in meteorites	150	≤0.5	RG and AGB	Core H, CBP
Spinel‡	1.5	0.15–2	RG, AGB, SNe	Core H, CBP, Shell H
Corundum‡	100	0.15–3	RG, AGB, SNe	Core H, CBP, HBB, Shell H, He, s
Nanodiamonds	1400	0.002	SNe	r, p
Mainstream SiC	14	0.3–20	AGB	Core H, Shell H, Shell He, s
SiC type A + B	0.2	0.5–5	J stars?	Shell He and H
SiC type X	0.06	0.3–5	SNe	H, He, O, e, s, n-burst
Graphite	1	1–20	SNe, AGB	H, He, O, e, s, n-burst; Core H, Shell He
Nova grains	0.001	~1	Novae	Ex H
Si nitride	0.002	≤1	SNe	He, O
TiC	~0.001	0.01-0.5	SNe, AGB	He, O, e

*Abundances vary with meteorite type. Shown here are maximum values.

†*In low-to-intermediate-mass stars:* Core H: Core H burning followed by first (and second) dredge-up. Shell H: Shell H burning during the RG and AGB phase. Shell He: He burning during thermal pulses of AGB phase followed by third dredge-up. CBP: Cool bottom processing. HBB: Hot bottom burning. s: s-process, neutron capture at low neutron density, followed by third dredge-up. *In supernovae:* H, He, O: H, He, and O burning in different stellar zones in the massive star before explosion. s: s-process taking place in several zones. e: equilibrium process, leading to the Fe-Ni core. n-burst: Neutron capture at intermediate neutron density. r: Neutron capture at high neutron density. p: p-process, photo disintegration and proton capture. *In novae:* Ex H: Explosive H burning.

‡Here and in the text we refer to grains with elemental compositions close to $MgAl_2O_4$ and Al_2O_3 as spinel and corundum, respectively, as has been done in many publications. However, it should be emphasized that it has been shown that presolar Al_2O_3 occurs both as the mineral corundum and in amorphous form (*Stroud et al.,* 2004). The same is probably the case for $MgAl_2O_4$.

TABLE 4. Isotopic measurements on presolar grains.

Grain Type	Analysis Type	Element	Grain Size	Analysis Technique
SiC	Bulk	Noble gases, C, N, Mg, Ca, Fe, Sr, Ba, Nd, Sm, Dy		Gas MS, TIMS, ion probe
	Single grain	C, Si, N, Mg, K, Ca, Ti, Fe, Ba	≥0.25 μm	Ion probe
		Fe, Sr, Zr, Mo, Ru, Ba	>1 μm	Laser RIMS
		He, Ne	>3 μm	Laser evap. + gas MS
Graphite	Bulk	Noble gases		Gas MS
	Single grain	C, N, O, Mg, Si, K, Ca, Ti	≥1 μm	Ion probe
		Zr, Mo	>1 μm	Laser RIMS
		He, Ne	>3 μm	Laser evap. + Gas MS
Nanodiamonds	Bulk	Noble gases, N, Sr, Te, Ba		Gas MS, TIMS
Oxides	Single grain	O, Mg, K, Ca, Ti, Cr	≥0.1 μm	Ion probe
Si_3N_4	Single grain	N, Si, C	≥0.5 μm	Ion probe
TiC	Single grain	O, Ti	~0.1 μm	Ion probe

grains. For isotopic analysis secondary ion mass spectrometry (SIMS) with the ion microprobe has become the method of choice and by far most single grain measurements have been made with this instrument. Resonance ionization mass spectrometry (RIMS) has distinct advantages of high sensitivity and elimination of isobaric interferences for analysis of trace elements, albeit at some cost in precision and data acquisition rate compared to the ion probe. While most SIMS measurements to date have been on grains 1 μm in size or larger, a new type of ion probe, the NanoSIMS, allows measurements of grains an order of magnitude smaller (e.g., *Zinner et al.,* 2003b, 2005b). Ion probe analysis of individual grains has led to the discovery of new types of presolar grains such as corundum, spinel, hibonite, and titanium oxide (*Hutcheon et al.,* 1994; *Nittler et al.,* 1994, 2005; *Nittler and Alexander,* 1999; *Choi et al.,* 1998, 1999; *Zinner et al.,* 2003b), Si_3N_4 (*Nittler et al.,* 1995), and silicates (*Messenger et al.,* 2003; *Nguyen and Zinner,* 2004). Finally, Ti-, Zr-, and Mo-rich carbides, cohenite (Fe-Ni)C), kamacite (Fe-Ni), and elemental Fe were found as tiny subgrains in TEM studies of graphite spheres (*Bernatowicz et al.,* 1991, 1996; *Croat et al.,* 2003). While TiC inside a SiC grain (*Bernatowicz et al.,* 1992) could have formed by exsolution, there can be little doubt that interior grains in graphite must have formed prior to the condensation of the spherules.

5.3. Elemental and Isotopic Compositions of Presolar Grains

With the exception of Si_3N_4, kamacite, and elemental Fe (which constitute only a tiny fraction of presolar grains discovered to date), all presolar grains identified so far are either C, carbides, or oxides, demonstrating the importance of the elements C and O in the formation of high-temperature phases from stellar atmospheres. These elements also dominate the chemistry of the envelopes of low-to-intermediate mass stars and the outer layers of presupernova stars. Because of the tight bond of these elements in the CO molecule, it is the excess of one over the other that determines whether carbonaceous (C > O) or O-rich (O > C) phases will be formed. *Clayton et al.* (1999) and *Deneault et al.* (2003) suggested condensation of carbonaceous phases in Type II SN ejecta even while C < O because of the limitation of CO buildup in the high-radiation environment of the ejecta (see section 5.6.1).

Table 4 gives a list of the elements for which isotopic ratios have been measured in presolar grains, on bulk samples with gas, thermal ionization, or secondary ion mass spectrometry (GMS, TIMS, or SIMS) and on individual grains by laser GMS, with the ion microprobe (SIMS), or by resonance ionization mass spectrometry (RIMS). Most valuable are isotopic analyses of as many elements as possible on a single grain because the results provide the most information on the parent star of this grain and set the tightest constraints on the nuclear processes that determined the composition of this stellar source.

One fundamental problem in interpreting information obtained from a given grain is that its stellar source is unknown and has to be inferred from the elemental and isotopic composition of the grain. Fortunately, this is possible for most of the grains. For example, there is little doubt that the majority of SiC grains come from low-mass AGB stars that had become carbon stars. Another example are SiC grains of type X (see below), Si_3N_4, and low-density graphite grains whose isotopic compositions, especially the initial presence of short-lived ^{44}Ti, provide clear evidence that these grains have a SN origin. There are also grains whose stellar origin cannot (yet) be determined with certainty. Silicon carbide grains of type A + B have low $^{12}C/^{13}C$ ratios in the range observed in type J and a few other types of carbon stars (*Amari et al.,* 2001c). However, the nucleosynthesis and mixing processes that would lead to the observed surface compositions are not well understood, and the A + B grains cannot be unambiguously assigned to such stars.

5.4. Silicon Carbide and Graphite Grains from Asymptotic Giant Branch Stars

Although most SiC grains are less than 0.5 μm in diameter, some grains are as large as 20 μm. The availability of a large number of grains in the size range of a few micrometers and the fact that SiC is relatively rich in trace elements

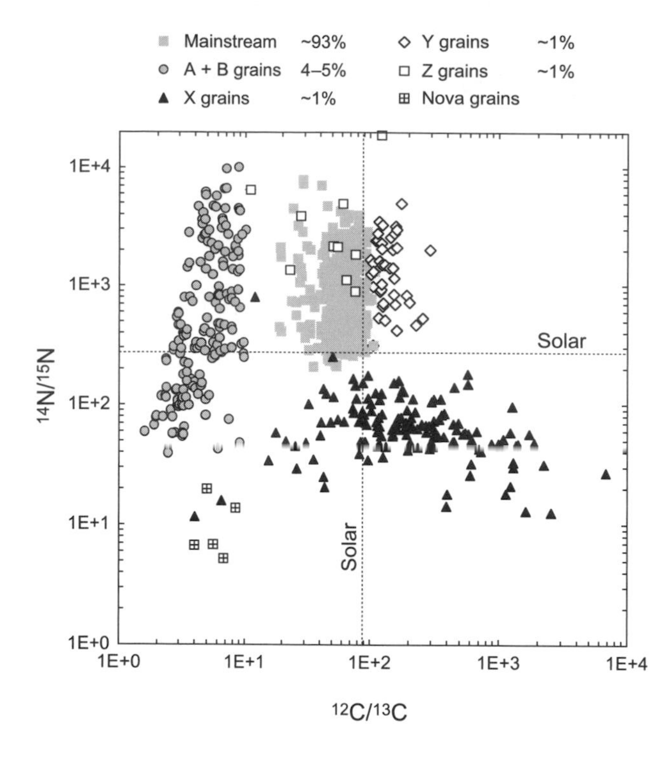

Fig. 4. Nitrogen and C-isotopic ratios of individual presolar SiC grains. Different grain types can be distinguished on the basis of their C, N, and Si isotopic ratios. Because rare grain types were located by automatic ion imaging, the numbers of grains of different types do not correspond to their relative abundances in the meteorites; these abundances are given in the legend. Data are from *Alexander* (1993), *Hoppe et al.* (1994, 1996a, 1997, 2000), *Nittler et al.* (1995), *Huss et al.* (1997), *Amari et al.* (2001a–c), and *Nittler and Hoppe* (2004a).

(*Amari et al.,* 1995a; *Kashiv et al.,* 2002) is the reason that SiC is the best studied presolar grain type (see Table 4). Ion microprobe isotopic analysis has revealed huge variations in many elements. This has led to the classification into different populations (Figs. 4–6) on the basis of C-, N-, and Si-isotopic ratios and inferred $^{26}Al/^{27}Al$ ratios: mainstream grains (~93% of the total), and the minor subtypes A, B, X, Y, Z, and nova grains (*Hoppe and Ott,* 1997; *Amari et al.,* 2001a). It should be noted that the numbers of data points for different grain populations plotted in Figs. 4–6 do not correspond to the abundance of these grains in primitive meteorites because most grains of rare types have been identified through specific searches by ion imaging (*Nittler et al.,* 1995; *Hoppe et al.,* 1996b, 2000; *Lin et al.,* 2002; *Besmehn and Hoppe,* 2003).

Mainstream grains constitute ~93% of all presolar SiC. They have $^{12}C/^{13}C$ ratios between 10 and 100, and mostly ^{14}N, ^{29}Si, and ^{30}Si excesses relative to solar (*Zinner et al.,* 1989, 2003a; *Stone et al.,* 1991; *Virag et al.,* 1992; *Alexander,* 1993; *Hoppe et al.,* 1994, 1996a; *Nittler et al.,* 1995; *Huss et al.,* 1997; *Amari et al.,* 2002; *Nittler and Alexander,* 2003). On a δ-value Si three-isotope plot the data fall along a line with slope of 1.35, which is shifted slightly to the right of the SS composition (Fig. 5 inset). Type Y grains have $^{12}C/^{13}C > 100$ and Si-isotopic compositions that lie to the right of the mainstream correlation line (*Hoppe et al.,* 1994; *Amari et al.,* 2001b). Type Z grains have even larger ^{30}Si excesses relative to ^{29}Si and, on average, lower $\delta^{29}Si$ values than Y grains but have $^{12}C/^{13}C < 100$ (*Alexander,* 1993; *Hoppe et al.,* 1997).

There is abundant evidence that mainstream grains, and very likely also Y and Z grains, originated from carbon stars of low mass (1–3 $M_\odot$). Dust from such stars has been proposed already before identification of presolar SiC to be a minor constituent of primitive meteorites (*Clayton and Ward,* 1978; *Srinivasan and Anders,* 1978; *Clayton,* 1983). Mainstream grains have $^{12}C/^{13}C$ ratios similar to those found in carbon stars (*Lambert et al.,* 1986), which are considered to be the most prolific injectors of carbonaceous dust grains into the ISM (*Tielens,* 1990). Many carbon stars show the 11.3-μm emission feature typical of SiC (*Treffers and Cohen,* 1974; *Speck et al.,* 1997). The presence of Ne-E(H) and the s-process isotopic patterns of the heavy elements exhibited by mainstream SiC provide the most convincing argument for their origin in carbon stars.

5.4.1. Light and intermediate mass elements. Of the elements whose isotopic compositions have been measured in SiC grains, nucleosynthesis occurring in the core during the main sequence of the parent stars affects only C and N. Their isotopic ratios in mainstream, Y, and Z grains are the result of core H burning in the CN cycle followed by the first (and second) dredge-up, and those of C also that of shell He burning and the third dredge-up (TDU) during the thermally pulsing AGB (TP-AGB) phase (*Busso et al.,* 1999). Whereas the first process enriches the star's surface in ^{13}C and ^{14}N, the TDU adds ^{12}C to the envelope, increases the $^{12}C/^{13}C$ ratio from the low values resulting from the first dredge-up, and, by making C > O, causes the star to become a carbon star.

Envelope $^{12}C/^{13}C$ ratios are predicted by canonical stellar evolution models to range from ~20 after the first dredge-up in the RG phase to ~300 in the late TP-AGB phases (*El Eid,* 1994; *Gallino et al.,* 1994; *Amari et al.,* 2001b) and agree with those observed in the above three types of SiC grains (see Fig. 4). The predicted $^{14}N/^{15}N$ ratios of 600–1600 (*Becker and Iben,* 1979; *El Eid,* 1994) fall short of the range observed in the grains. However, the assumption of deep mixing ("cool bottom processing," or CBP) of envelope material to deep hot regions in $M < 2.5\ M_\odot$ stars during their RG and AGB phases (*Charbonnel,* 1995; *Wasserburg et al.,* 1995; *Langer et al.,* 1999; *Nollett et al.,* 2003) results in partial H burning, producing higher $^{14}N/^{15}N$ and lower $^{12}C/^{13}C$ ratios in the envelope than canonical models (see also *Huss et al.,* 1997). Such a process has originally been proposed to explain the low (as low as 3) $^{12}C/^{13}C$ ratios in K giants and early AGB stars (*Gilroy,* 1989; *Gilroy and Brown,* 1991) and the low $^{18}O/^{16}O$ ratios in certain presolar oxide grains (see below). The high $^{14}N/^{15}N$ ratios observed in many SiC grains provide a confirmation that this process occurs in low-mass stars.

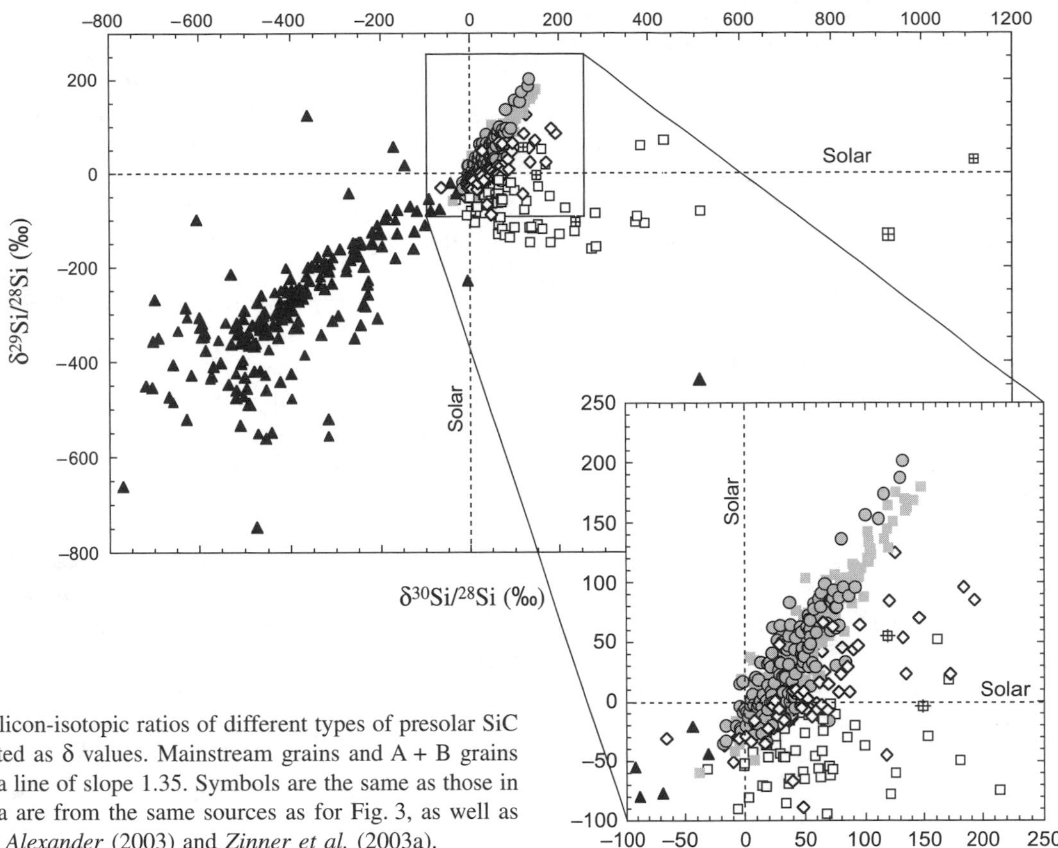

Fig. 5. Silicon-isotopic ratios of different types of presolar SiC grains plotted as δ values. Mainstream grains and A + B grains plot along a line of slope 1.35. Symbols are the same as those in Fig. 4. Data are from the same sources as for Fig. 3, as well as *Nittler and Alexander* (2003) and *Zinner et al.* (2003a).

Two other isotopes that are a signature of AGB stars are ^{26}Al and ^{22}Ne. Figure 6 shows inferred $^{26}Al/^{27}Al$ ratios in different types of SiC grains. Aluminum-26 is produced in the H shell by proton capture on ^{25}Mg and mixed to the surface by the third dredge-up (*Forestini et al.,* 1991; *Karakas and Lattanzio,* 2003). It can also be produced during "hot bottom burning" (HBB) when the convective envelope extends into the H-burning shell (*Lattanzio et al.,* 1997; *Karakas and Lattanzio,* 2003), but this process is believed to prevent carbon-star formation (*Frost and Lattanzio,* 1996). Neon-22, the main component in Ne-E, is produced in the He shell by $^{14}N + 2\alpha$. The Ne-isotopic ratios measured in SiC bulk samples (*Lewis et al.,* 1990, 1994) are very close to those expected for He-shell material (*Gallino et al.,* 1990) without much dilution with envelope material, indicating a special implantation mechanism by an ionized wind during the planetary nebula phase of the parent star (*Verchovsky et al.,* 2004). Evidence that the Ne-E(H) component originated from the He shell of AGB stars and not from ^{22}Na decay (*Clayton,* 1975b) is provided by the fact that in individual grains, of which only ~5% carry ^{22}Ne, it is always accompanied by ^{4}He (*Nichols et al.,* 1995). Excesses in ^{21}Ne in SiC relative to the predicted He-shell composition have been interpreted as being due to spallation by galactic cosmic rays (*Tang and Anders,* 1988a; *Lewis et al.,* 1990, 1994), which allows the determination of grain lifetimes in the ISM. Inferred exposure ages depend on grain size and range from 10 to 130 m.y. (*Lewis et al.,* 1994). However, this interpretation has been challenged (*Ott and Begemann,* 2000), and the question of IS ages of SiC is not settled.

In contrast to the light elements C, N, Ne, and Al and the heavy elements (see below), the Si-isotopic ratios of mainstream grains cannot be explained by nuclear processes

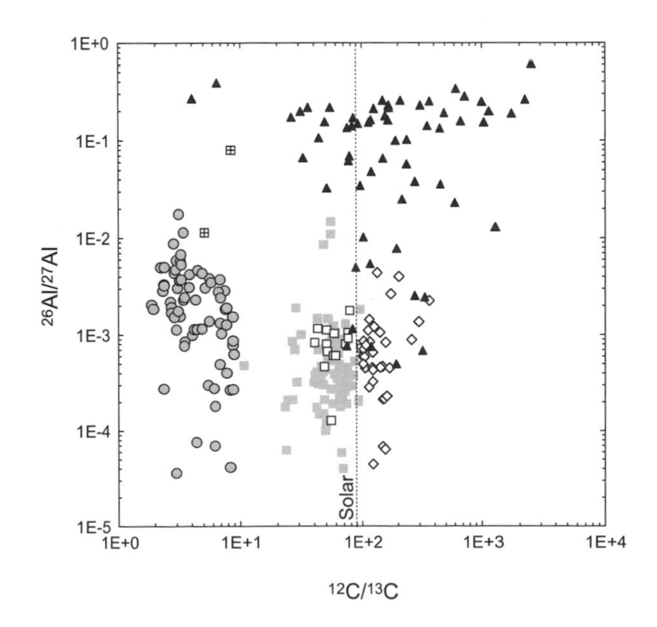

Fig. 6. Aluminum and C-isotopic ratios of individual presolar SiC grains. Symbols and sources for the data are the same as those for Figs. 4 and 5 and *Zinner et al.* (2005a).

taking place within their parent stars. In AGB stars the Si isotopes are affected by neutron capture in the He shell, leading to excesses in ^{29}Si and ^{30}Si along a line with slope 0.1–0.8 (depending on model parameters such as mass, metallicity, mass loss rate, and n-capture cross sections chosen) in a δ-value Si three-isotope plot (*Gallino et al.,* 1990, 1994; *Brown and Clayton,* 1992; *Lugaro et al.,* 1999; *Amari et al.,* 2001b; *Guber et al.,* 2003). Depending on the adopted cross sections (*Bao et al.,* 2000; *Guber et al.,* 2003), predicted excesses are only on the order of 20% for ^{29}Si and 40% for ^{30}Si in low-mass AGB stars of close-to-solar metallicity. This led to the proposal that many stars with varying initial Si-isotopic compositions contributed SiC grains to the SS (*Clayton et al.,* 1991; *Alexander,* 1993) and that neutron-capture nucleosynthesis in these stars only plays a secondary role in modifying these compositions. One explanation for the initial Si ratios in the parent stars of SiC grains is the evolution of the Si-isotopic ratios through galactic history as different generations of supernovae produced Si with increasing ratios of the secondary isotopes ^{29}Si and ^{30}Si to the primary ^{28}Si (*Gallino et al.,* 1994; *Timmes and Clayton,* 1996; *Clayton and Timmes,* 1997a,b). Another is based on local heterogeneities in the galaxy caused by the stochastic nature of the admixture of the ejecta from supernovae of varying type and mass (*Lugaro et al.,* 1999). *Clayton* (1997) has addressed the problem that most SiC grains have higher-than-solar $^{29}Si/^{28}Si$ and $^{30}Si/^{28}Si$ ratios by considering the possibility that the mainstream grains originated from stars that were born in central, more metal-rich regions of the galaxy and moved to the molecular cloud from which our Sun formed. *Alexander and Nittler* (1999), on the other hand, suggested that the Sun has an atypical Si-isotopic composition. Recently, *Clayton* (2003) has invoked a merger of our galaxy (high metallicity) with a satellite galaxy (low metallicity) some time before SS formation to explain the Si-isotopic ratios of mainstream grains. A more detailed treatment of the role of GCE on Si-isotopic ratios in SiC grains from AGB stars is found in *Nittler and Dauphas* (2006).

The Si-isotopic compositions of Y and Z grains have been explained by lower-than-solar metallicities for the parent stars of these grains, approximately half the solar metallicity ($Z = 0.01$ instead of $Z_{\odot} = 0.02$) for Y grains (*Amari et al.,* 2001b) and about a third the solar metallicity ($Z = 0.006$) for Z grains (*Hoppe et al.,* 1997). [This value for the solar metallicity was used for the models in the cited papers. Recently, the solar metallicity has been revised downward to $Z_{\odot} = 0.0122$, based on the newly determined solar abundances of C and O by *Asplund et al.* (2005).] Such stars are predicted to have larger isotopic shifts in the Si isotopes in agreement with the deviations of the Si-isotopic ratios of the Y and Z grains to the right of the mainstream correlation line (Fig. 5). If one projects the Si ratios measured in these grains back onto the galactic evolution line $\delta^{29}Si = \delta^{30}Si$ along lines predicted for neutron capture, the average $^{29}Si/^{28}Si$ (or $\delta^{29}Si$) value of all Y grains is lower than the average of the mainstream grains and the Z grain average is even lower, in agreement with a continuous increase of the Si-isotopic ratios with metallicity during galactic evolution (*Zinner et al.,* 2001; *Nittler and Alexander,* 2003). The larger $^{12}C/^{13}C$ ratios of Y grains agree with an origin from low-metallicity parent stars, which during their TP-AGB phase dredge up more ^{12}C relative to that in the envelope, whose composition is depleted in C, and elements that experienced neutron capture, from the He shell (see also *Lugaro et al.,* 1999). However, for Z grains, which have $^{12}C/^{13}C$ ratios in the range of mainstream grains, one has to invoke cool bottom processing (*Wasserburg et al.,* 1995; *Nittler and Alexander,* 2003; *Nollett et al.,* 2003) during the red giant and AGB phase of their parent stars.

Titanium isotopic ratios in single mainstream and Y grains (*Ireland et al.,* 1991; *Hoppe et al.,* 1994; *Alexander and Nittler,* 1999; *Amari et al.,* 2001b) and in bulk samples (*Amari et al.,* 2000) usually show excesses in all isotopes relative to ^{48}Ti, a result expected of neutron capture in AGB stars. Titanium ratios are correlated with those of Si, and, as for Si, theoretical models (*Lugaro et al.,* 1999) cannot explain the range of Ti ratios in single grains, indicating that the Ti-isotopic compositions are also dominated by galactic evolution effects (*Alexander and Nittler,* 1999; *Nittler and Dauphas,* 2006; *Nittler,* 2005). This interpretation is corroborated by Ti-isotopic data recently obtained for Z grains (*Zinner et al.,* 2005a). These grains show depletions in all isotopes relative to ^{48}Ti, analogous to ^{29}Si depletions in Z grains, except for ^{50}Ti where relative excesses are roughly correlated with ^{30}Si excesses. Apparently, for Ti the secondary isotopes ^{46}Ti, ^{47}Ti, ^{49}Ti, and ^{50}Ti increase in abundance relative to the α-isotope ^{48}Ti with galactic time and excesses in ^{50}Ti because of n-capture in AGB stars are larger in low-metallicity stars.

Calcium-isotopic data for $^{40,42,43,44}Ca$ have been obtained only in bulk samples (*Amari et al.,* 2000), but $^{44}Ca/^{40}Ca$ ratios have been measured in many individual SiC grains (see section 5.6.1). Excesses of ^{42}Ca and ^{43}Ca relative to ^{40}Ca agree with predictions for neutron capture. Large ^{44}Ca excesses are apparently due to the presence of type X grains (see section 5.6.1). Measurements of Fe-isotopic ratios have been reported by *Marhas et al.* (2004), who did not find any significant anomalies.

5.4.2. Heavy elements and the s-process. Heavy elements measured in bulk SiC samples, clearly dominated by mainstream grains, include the noble gases Kr and Xe (*Lewis et al.,* 1990; 1994) and the elements Sr (*Podosek et al.,* 2004), Ba (*Ott and Begemann,* 1990; *Zinner et al.,* 1991; *Prombo et al.,* 1993), Nd and Sm (*Zinner et al.,* 1991; *Richter et al.,* 1993), and Dy (*Richter et al.,* 1994). Resonance ionization mass spectrometry has made it possible to measure the isotopic composition of heavy elements in individual grains, and analyses of Sr (*Nicolussi et al.,* 1998b), Zr (*Nicolussi et al.,* 1997), Mo (*Nicolussi et al.,* 1998a), Ru (*Savina et al.,* 2004a), and Ba (*Savina et al.,* 2003a) have been reported (see also *Lugaro et al.,* 2003; *Savina et al.,* 2003b). Secondary ion mass spectrometry isotopic meas-

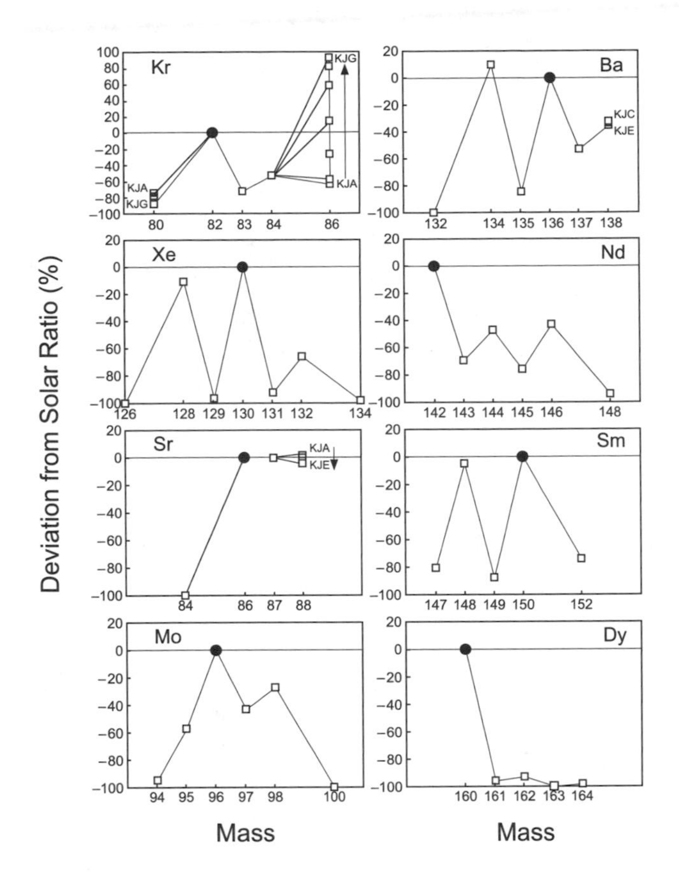

Fig. 7. Isotopic patterns of heavy elements derived from measurements of bulk samples of SiC extracted from the Murchison meteorite. What is plotted are the extrapolated s-process components [the so-called G-components (see *Hoppe and Ott,* 1997)]. Isotopic ratios are relative to the reference isotope plotted as a solid circle and are normalized to the solar isotopic ratios. Data are from *Lewis et al.* (1994) (Kr and Xe), *Podosek et al.* (2004) (Sr), *Prombo et al.* (1993) (Ba), *Richter et al.* (1993) (Nd and Sm), *Richter et al.* (1994) (Dy), and *Lugaro et al.* (2003) (Mo). For ^{80}Kr, ^{86}Kr, ^{138}Ba, and ^{88}Sr the isotopic ratios vary with grain size and the names of different grain size fractions of SiC from the Murchison meteorite (*Amari et al.,* 1994) are indicated in the figure. Grain sizes increase from KJA (average diameter 0.38 μm) to KJG (3.0 μm).

urements of Ba in single grains have also been reported (*Marhas et al.,* 2003). All these elements show the signature of the s-process.

The s-process isotopic patterns of Xe, Kr, Sr, Mo, Ba, Nd, Sm, and Dy are shown in Fig. 7. The isotopic compositions of SiC grains from AGB stars are a mixture of a component with close-to-solar ratios (the so-called N-component) that apparently represents the original composition of the parent star(s) and a pure s-process component that results from nucleosynthesis in the He intershell and TDU. The s-process patterns in Fig. 7 are obtained by extrapolating the measured ratios on three-isotope plots to zero contributions of either p-only or r-only isotopes. It should be mentioned that these patterns are averages obtained from the analysis of many grains. Individual grain analysis shows large variations among grains. One reason is the fact that in a given star SiC grains condense while more and more s-process nuclei are dredged up, enriching the envelope in these and thereby changing the isotopic ratios of the heavy elements. However, individual grains have a wider range in their isotopic ratios than would be expected from a single combination of stellar mass, metallicity, and ^{13}C pocket, confirming multiple stellar sources for presolar SiC.

For all elements listed above (except for Dy) there is good agreement with theoretical models of the s-process in low-mass (mostly <3 $M_{\odot}$) AGB stars (*Gallino et al.,* 1993, 1997; *Lugaro et al.,* 2003; *Savina et al.,* 2004a). Discrepancies with earlier model calculations were caused by incorrect nuclear cross sections and could be resolved by improved experimental determinations (e.g., *Guber et al.,* 1997; *Wisshak et al.,* 1997; *Koehler et al.,* 1998). The s-process isotopic patterns observed from grains allow the determination of different parameters such as neutron exposure, temperature, and neutron density (*Hoppe and Ott,* 1997). These parameters depend in turn on stellar mass and metallicity, as well as on the neutron source operating in AGB stars. Comparisons of isotopic patterns in grains and composition predictions from stellar models thus allow information to be obtained about the parent stars of the grains. For example, the Ba-isotopic ratios indicate a neutron exposure that is only half that required to explain the bulk s-process element abundances in the SS, and also indicate that most grains come from stars of less than 3 $M_{\odot}$ (*Ott and Begemann,* 1990; *Gallino et al.,* 1993). Another example is provided by the abundance of ^{96}Zr in single grains, which is sensitive to neutron density because of the relatively short half-life of ^{95}Zr (~64 d). While the ^{13}C(α,n) source with its low neutron density does not significantly produce ^{96}Zr, production by activation of the ^{22}Ne(α,n) source during later thermal pulses in AGB stars depends sensitively on stellar mass. Some grains have very low ^{96}Zr/^{94}Zr ratios, indicating that the ^{22}Ne(α,n) source was weak in their parent stars and points to low-mass AGB stars as the source of mainstream grains (*Lugaro et al.,* 2003). An exciting new result is evidence for the initial presence of extinct ^{99}Tc in presolar SiC grains (*Savina et al.,* 2004a). The Ru isotopes measured by RIMS display an s-process isotopic pattern and show good agreement with theoretical prediction for low-mass AGB stars except for an excess in ^{99}Ru. This excess is well explained by the decay of ^{99}Tc ($T_{1/2}$ = 0.21 m.y.). It is fitting that signatures of this element, whose presence in stars (*Merrill,* 1952) was the first direct astronomical evidence for stellar nucleosynthesis, are now found in stardust analyzed in the laboratory.

5.4.3. High-density graphite grains. Graphite grains of four different density fractions have been isolated from the Murchison meteorite (*Amari et al.,* 1995b,d). Low-density (1.6–2.05 g cm^{-3}) graphite grains have isotopic signatures indicating a SN origin (see section 5.6). Noble gas measurements have revealed two different s-process components of

Kr in the four graphite fractions (*Amari et al.,* 1995b). The three lower-density fractions have Kr-isotopic compositions consistent with a massive-star origin. In contrast, Kr in the highest density (2.15–2.20 g cm^{-3}) fraction, KFC1, indicates an AGB origin. Such an origin is also indicated by the high Zr, Mo, and Ru contents of these grains (*Bernatowicz et al.,* 1996); these s-process elements are expected to be enhanced in the envelope of carbon stars (*Lodders and Fegley,* 1997). *Nicolussi et al.* (1998c) have reported RIMS measurements of Zr- and Mo-isotopic ratios in KFC1 grains in which no other isotopic ratios had been measured. Several grains show s-process patterns for Zr and Mo, similar to those exhibited by mainstream SiC grains, although two grains with a distinct s-process pattern for Zr have normal Mo. Two grains have extreme ^{96}Zr excesses, indicating an origin in SNe or in low-metallicity AGB stars (*Davis et al.,* 2003), but the Mo isotopes in one are almost normal. Molybdenum, like N, might have suffered isotopic equilibration in graphite. High-density graphite grains apparently come from AGB stars as well as from supernovae. They have a wide range of $^{12}C/^{13}C$ ratios with peaks around 10 and around 300–400 in the distribution (*Hoppe et al.,* 1995). If the grains with isotopically light C have a carbon-star origin, they must originated from low-metallicity or intermediate-mass AGB stars, which are expected to have higher $^{12}C/^{13}C$ ratios than the low-mass solar-metallicity AGB stars that produced the mainstream grains (*Amari et al.,* 2001b).

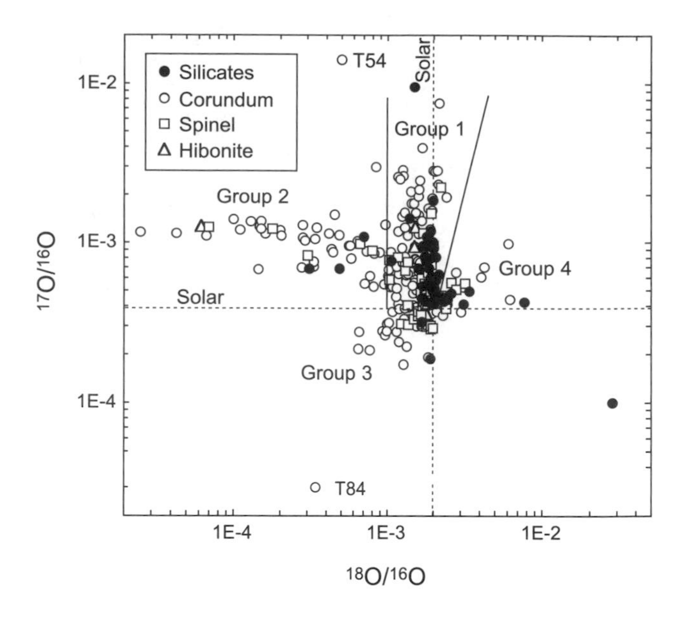

Fig. 8. Oxygen-isotopic ratios in individual oxide grains. Data are from *Nittler et al.* (1997, 1998), *Choi et al.* (1998, 1999), *Krestina et al.* (2002), *Zinner et al.* (2003b, 2004), *Messenger et al.* (2003), *Nguyen and Zinner* (2004), *Floss and Stadermann* (2004), and unpublished data from N. Krestina (2003), L. Nittler (2005), and R. Strebel (2003).

5.5. Oxide Grains from Red Giant and Asymptotic Giant Branch Stars

Presolar oxide grains identified to date include Al oxide (Al_2O_3) (*Huss et al.,* 1994; *Hutcheon et al.,* 1994; *Nittler et al.,* 1994, 1997, 1998; *Nittler and Alexander,* 1999; *Strebel et al.,* 1996; *Choi et al.,* 1998, 1999; *Krestina et al.,* 2002), spinel (*Nittler et al.,* 1997; *Choi et al.,* 1998; *Nguyen et al.,* 2003; *Zinner et al.,* 2003b), hibonite (*Choi et al.,* 1999; *Krestina et al.,* 2002; *Nittler et al.,* 2005), and silicates (*Messenger et al.,* 2003; *Floss and Stadermann,* 2004; *Nguyen and Zinner,* 2004; *Mostefaoui et al.,* 2004b; *Mostefaoui and Hoppe,* 2004; *Nagashima et al.,* 2004). In contrast to SiC grains in meteorites, all of which are of presolar origin, presolar oxides constitute only a small fraction of all oxides and have been identified by O-isotopic measurements in the ion microprobe.

The O-isotopic ratios of presolar oxide and silicate grains are shown in Fig. 8. Because different search techniques were used to locate these grains, the numbers of grains of different mineral phases does not reflect their abundances in meteorites (see also Table 3). Nor does the distribution of isotopic ratios accurately describe the actual distribution among presolar oxide grains. The reason is that searches by direct isotopic imaging (*Nittler et al.,* 1997) discriminate against grains with close-to-normal $^{18}O/^{16}O$ ratios, while raster imaging of tightly packed grains (*Nguyen et al.,* 2003) tends to dilute anomalous isotopic ratios.

Nittler et al. (1997) have classified presolar oxide grains into four different groups according to their O-isotopic ratios (see Fig. 8). The isotopic composition of grains is the result of the original compositions of the parent stars and several different episodes of H burning during the evolution of these stars. Group 1 grains have O-isotopic ratios similar to those observed in red giant and AGB stars (*Harris and Lambert,* 1984; *Harris et al.,* 1987; *Smith and Lambert,* 1990), indicating such an origin also for the grains. These compositions can be explained by H burning in the core of low-to-intermediate mass stars followed by mixing of core material into the envelope during the first dredge-up (also second dredge-up in low-metallicity stars with $M > 3 M_{\odot}$) (*Boothroyd et al.,* 1994; *Boothroyd and Sackmann,* 1999). Variations in $^{17}O/^{16}O$ ratios mainly correspond to differences in stellar mass, while those in $^{18}O/^{16}O$ correspond to differences in the metallicity of the parent stars. According to galactic chemical evolution models, $^{17}O/^{16}O$ and $^{18}O/^{16}O$ ratios are expected to increase as a function of stellar metallicity (*Timmes et al.,* 1995). Group 3 grains could thus come from low-mass stars (producing only small ^{17}O enrichments) with lower-than-solar metallicity (originally having lower-than solar $^{17}O/^{16}O$ and $^{18}O/^{16}O$ ratios). Group 2 grains have ^{17}O excesses and large ^{18}O depletions ($^{18}O/^{16}O < 0.001$). Such depletions cannot be produced by the first and second dredge-up but have been successfully explained by extra mixing (cool bottom processing) of low-mass ($M < 1.65 M_{\odot}$) stars during the AGB phase that circulates material from the envelope through regions close to the H-burning shell (*Wasserburg et al.,* 1995; *Denissenkov and Weiss,* 1996; *Nollett et al.,* 2003). Group 4 grains have both ^{17}O and ^{18}O excesses. If they originated from AGB stars they could either come from low-mass stars, in which ^{18}O produced by He burning of ^{14}N during early pulses was mixed into the envelope by third dredge-up (*Boothroyd and Sackmann,* 1988), or from stars with high metallicity. The O-isotopic ratios of the unusual grain T54 (*Nittler et al.,* 1997) has been interpreted to have an origin in a star with >5 $M_{\odot}$ that experienced hot bottom burning, a condition during which the convective envelope extends into the H-burning shell (*Boothroyd et al.,* 1995; *Lattanzio et al.,* 1997).

The Mg isotopes and Al have been measured in corundum (*Nittler et al.,* 1997; *Choi et al.,* 1998; 1999; *Krestina et al.,* 2002), spinel (*Nittler et al.,* 2003; *Zinner et al.,* 2005b), hibonite (*Nittler et al.,* 2005), and a silicate grain (*Nguyen and Zinner,* 2004). Because Mg in corundum is very low, with the exception of one grain that has a ^{25}Mg excess of ~25% (*Choi et al.,* 1998), only ^{26}Mg excesses due to ^{26}Al decay have been seen. Some but not all grains in the four groups show evidence for initial ^{26}Al. Because ^{26}Al is produced in the H-burning shell (*Forestini et al.,* 1991; *Mowlavi and Meynet,* 2000; *Karakas and Lattanzio,* 2003), dredge-up of material during the TP-AGB phase is required, and grains without ^{26}Al must have formed before their parent stars reached this evolutionary stage. Initial $^{26}Al/^{27}Al$ ratios are highest in Group 2 grains (*Nittler et al.,* 1997; *Choi et al.,* 1998), which is explained if these grains formed in the later stages of the AGB phase when more ^{18}O had been destroyed by cool bottom processing and more ^{26}Al dredged up (*Choi et al.,* 1998). The high Mg content of spinel and hibonite grains also provides the opportunity to obtain more precise $^{25}Mg/^{24}Mg$ ratios (*Nittler et al.,* 2003, 2005; *Zinner et al.,* 2005b). Oxygen and Mg isotopes in grains from RG and AGB stars are affected by nucleosynthesis occurring at different stages of the parent stars' evolution. Oxygen-isotopic ratios are mostly determined by core H burning during the main-sequence phase followed by the first (and second) dredge-up, but can also be affected by CBP and HBB during the RG and AGB phases. In contrast, Mg isotopes and ^{26}Al are changed during the AGB phase as well as by CBP and HBB (*Karakas and Lattanzio,* 2003). In addition, both O and Mg carry the original isotopic signatures of the parent stars. The O- and Mg-isotopic ratios of the grains appear to show the effects of all these processes. In particular, one spinel with large ^{25}Mg and ^{26}Mg excesses and a large ^{18}O depletion must come from an intermediate-mass AGB star that experienced HBB (*Nittler et al.,* 2003). One presolar pyroxene grain with ^{18}O depletion has a ^{26}Mg excess, most likely indicating CBP (*Nguyen and Zinner,* 2004).

Mostefaoui and Hoppe (2004) have measured Si-isotopic ratios in eight presolar silicate grains. The data points lie close to a line that is parallel to the mainstream correlation line of presolar SiC grains but shifted to the left on a δ-value three-isotope plot. The authors interpreted this to indicate that the parent stars of the silicate grains had not experienced TDU of s-processed material that shifted the Si-isotopic compositions of SiC grains to the right.

Titanium-isotopic ratios have been determined in a few presolar corundum grains (*Choi et al.,* 1998; *Hoppe et al.,* 2003). The observed ^{50}Ti excesses agree with those predicted to result from neutron capture in AGB stars. The depletions in all Ti isotopes relative to ^{48}Ti found in two grains indicate that, just as for SiC grains, galactic evolution affects the isotopic compositions of the parent stars of oxide grains. Several hibonite grains were found to have large ^{41}K excesses (*Choi et al.,* 1998; *Nittler et al.,* 2005), corresponding to a $^{41}Ca/^{40}Ca$ ratio of 2×10^{-4}, within the range of values predicted for the envelope of AGB stars (*Wasserburg et al.,* 1994).

5.6. Grains from Supernovae

5.6.1. Silicon carbide, silicon nitride, and graphite. According to their isotopic compositions SiC grains of type X, Si_3N_4, and low-density graphite grains are believed to have a SN origin. The smoking gun is the initial presence of short-lived ^{44}Ti ($T_{1/2}$ = 60 yr) inferred from large ^{44}Ca excesses in many grains (*Amari et al.,* 1992, 1995c; *Hoppe et al.,* 1996b, 2000; *Nittler et al.,* 1996; *Besmehn and Hoppe,* 2003). This isotope is only produced in supernovae (*Timmes et al.,* 1996). Another smoking gun is the presence of even shorter-lived ^{49}V ($T_{1/2}$ = 330 d) inferred from large ^{49}Ti excesses (*Hoppe and Besmehn,* 2002).

Type X grains are characterized by mostly ^{12}C and ^{15}N excesses relative to solar (Fig. 4), excesses in ^{28}Si (Fig. 5), and very large $^{26}Al/^{27}Al$ ratios, ranging up to 0.6 (Fig. 6) (*Amari et al.,* 1995c; *Hoppe et al.,* 1996b, 2000; *Nittler et al.,* 1995; *Lin et al.,* 2002). In ~15% of the grains evidence for ^{44}Ti has been found. Other isotopic signatures also point to a SN origin. In Type II supernovae ^{44}Ti is produced in the Ni zone that experienced an α-rich freezeout and the Si/S zone that experienced explosive O or incomplete Si burning. Silicon in the latter consists of almost pure ^{28}Si and substantial contributions must come from this zone to explain the large ^{28}Si excesses in the X grains. High $^{12}C/^{13}C$ and low $^{14}N/^{15}N$ ratios are the signatures of He burning and are predicted for the He/C zone; and high $^{26}Al/^{27}Al$ ratios can be reached in the He/N zone by H burning. Some SiC X grains also show large excesses in ^{49}Ti (*Amari et al.,* 1992; *Nittler et al.,* 1996; *Hoppe and Besmehn,* 2002). The correlation of these excesses with the V/Ti ratio (*Hoppe and Besmehn,* 2002) indicates that they come from the decay of short-lived ^{49}V ($T_{1/2}$ = 330 d) and that the grains must have formed within a few months of the explosion. Like ^{44}Ti, ^{49}V is produced in the Si/S zone. Resonance ionization mass spectrometry isotopic measurements of Fe, Sr, Zr, Mo, and Ba have been made on X grains (*Pellin et al.,* 1999, 2000a; *Davis et al.,* 2002). The most complete are the Mo measurements, which reveal large excesses in ^{95}Mo and ^{97}Mo

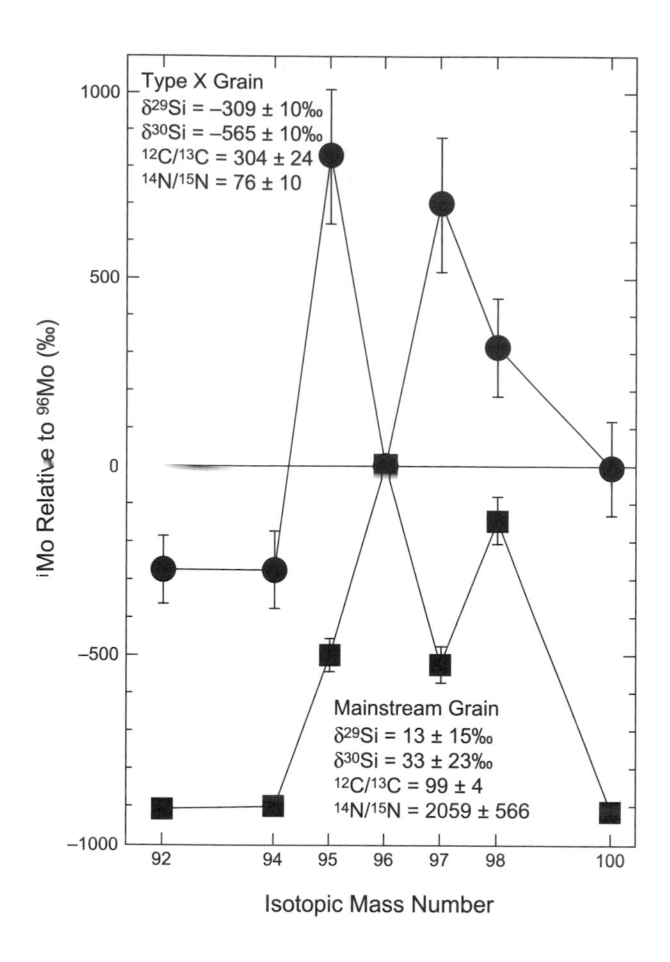

Fig. 9. Molybdenum-isotopic patterns measured by RIMS in a type X and mainstream SiC grain. Figure is from *Pellin et al.* (1999), courtesy of A. Davis.

(Fig. 9). Whereas the mainstream grain shown in Fig. 9 has a typical s-process pattern, in agreement with bulk measurements of other heavy elements such as Xe, Ba, and Nd (Fig. 7), the Mo pattern of the X grain is completely different and indicates neutron capture at much higher neutron densities. It is successfully explained by a neutron-burst model (*Meyer et al.,* 2000). In the Type II SN models by *Rauscher et al.* (2002), an intense neutron burst is predicted to occur when the SN shock heats the He-burning shell and liberates neutrons by the $^{22}Ne(\alpha,n)^{25}Mg$ reaction (see section 2.2.6). This neutron burst can account for the Mo-isotopic patterns observed in X grains.

Presolar Si_3N_4 grains are extremely rare (in Murchison SiC-rich separates ~5% of SiC of type X) but automatic ion imaging has been successfully used to detect those with large ^{28}Si excesses (*Nittler et al.,* 1998; *Besmehn and Hoppe,* 2001; *Lin et al.,* 2002; *Nittler and Alexander,* 2003). The C-, N-, Al-, and Si-isotopic signatures of these grains are the same as those of SiC grains of type X (see Figs. 4–6). Although no resolvable ^{44}Ca excesses have been detected so far (*Besmehn and Hoppe,* 2001), the similarity with X grains implies a SN origin for these grains.

Low-density (LD) graphite grains have in general higher trace-element concentrations than those with higher densities and for this reason have been studied in detail for their isotopic compositions (*Travaglio et al.,* 1999). Many LD grains have ^{15}N and large ^{18}O excesses (*Amari et al.,* 1995e) and high $^{26}Al/^{27}Al$ ratios. Excesses in ^{18}O are correlated with $^{12}C/^{13}C$ ratios. Many grains for which Si-isotopic ratios could be determined with sufficient precision show ^{28}Si excesses, although large ^{29}Si and ^{30}Si excesses are also seen. The similarities of the isotopic signatures with those of SiC X point to a SN origin of LD graphite grains. The ^{18}O excesses are compatible with such an origin. Helium burning produces ^{18}O from ^{14}N, which dominates the CNO isotopes in material that had undergone H burning via the CNO cycle. As a consequence, the He/C zone in pre-SNII massive stars, which experienced partial He burning, has a high ^{18}O abundance (*Woosley and Weaver,* 1995). Wolf-Rayet stars during the WN-WC transitions are predicted to also show ^{12}C, ^{15}N, and ^{18}O excesses and high $^{26}Al/^{27}Al$ ratios (*Arnould et al.,* 1997a) but also large excesses in ^{29}Si and ^{30}Si and are therefore excluded as parent stars of LD graphite grains with ^{28}Si excesses.

Additional features that indicate a SN origin of LD graphite grains are evidence for ^{44}Ti (*Nittler et al.,* 1996) and large excesses of ^{41}K, which must be due to the decay of the radioisotope ^{41}Ca ($T_{1/2} = 1.05 \times 10^5$ yr) (*Amari et al.,* 1996). Inferred $^{41}Ca/^{40}Ca$ ratios are much higher (0.001 to 0.01) than those predicted for the envelopes of AGB stars (*Wasserburg et al.,* 1994) but are predicted to be the result of neutron capture in the C- and O-rich zones of Type II supernovae (*Woosley and Weaver,* 1995). Measurements of Ca-isotopic ratios in grains without evidence for ^{44}Ti show excesses in ^{42}Ca, ^{43}Ca, and ^{44}Ca, with ^{43}Ca having the largest excess (*Amari et al.,* 1996; *Travaglio et al.,* 1999). This pattern is best explained by neutron capture in the He/C and O/C zones of Type II supernovae. Titanium-isotopic ratios show large excesses in ^{49}Ti and smaller ones in ^{50}Ti (*Amari et al.,* 1996; *Nittler et al.,* 1996; *Travaglio et al.,* 1999). This pattern also indicates neutron capture and is well matched by predictions for the He/C zone (*Amari et al.,* 1996). However, large ^{49}Ti excesses in grains with relatively low (10–100) $^{12}C/^{13}C$ ratios can only be explained by the decay of ^{49}V (*Travaglio et al.,* 1999).

The isotopic compositions of SN grains pose the most serious challenge to nuclear astrophysicists. The reason is that the isotopic signatures found in the grains occur in massive stars in very different stellar zones, which experienced different stages of nuclear burning before the SN explosion (e.g., *Woosley and Weaver,* 1995; *Rauscher et al.,* 2002). The isotopic signatures of the SN grains thus suggest deep and inhomogeneous mixing of matter from these different zones in the SN ejecta. While the Ti- and Si-isotopic signature of the grains requires contributions from the Ni, O/Si, and Si/S zones, which experienced Si-, Ne-, and O-burning, significant contributions must also come from the He/N and He/C zones that experienced H and incomplete He burning in order to achieve C > O, the requirement for SiC or graphite condensation from a gas under equilibrium conditions (*Larimer and Bartholomay,* 1979; *Lodders and Fegley,* 1997). Also the C-, N-, O-, and Al-isotopic ratios indicate contributions from these zones. Furthermore, addition of material from the intermediate O-rich layers must be severely limited. Astronomical observations indicate extensive mixing of SN ejecta (e.g., *Ebisuzaki and Shibazaki,* 1988; *Hughes et al.,* 2000) and hydrodynamic models of SN explosions predict mixing in the ejecta initiated by the formation of Rayleigh-Taylor instabilities (e.g., *Herant et al.,* 1994). In order to better constrain theoretical models of SN nucleosynthesis, *Travaglio et al.* (1999), *Hoppe et al.* (2000), and *Yoshida and Hashimoto* (2004) tried to match the isotopic compositions of SN grains by performing multizone mixing calculations. While these calculations qualitatively account for the grains' isotopic signatures, they cannot quantitatively reproduce all the measured ratios. One example is that the grains show excesses of ^{29}Si over ^{30}Si compared to solar. This is a long-standing problem: SN models cannot account for the solar $^{29}Si/^{30}Si$ ratio (*Timmes and Clayton,* 1996). Studies of SiC X grains isolated from the Qingzhen enstatite chondrite (*Lin et al.,* 2002) suggest that there are two population of X grains with different trends in the Si-isotopic ratios, the minor population having lower-than-solar $^{29}Si/^{30}Si$ ratios. The lowest measured $^{14}N/^{15}N$ and highest $^{26}Al/^{27}Al$ and $^{44}Ti/^{48}Ti$ ratios can only be explained by including substantial contributions from the Ni shell (*Yoshida and Hashimoto,* 2004).

However, it still has to be seen whether mixing can occur on a microscopic scale and whether instabilities allow mixing of matter from non-neighboring zones while excluding large contributions from the intermediate O-rich zones. *Clayton et al.* (1999) and *Deneault et al.* (2003) took an alternative approach by suggesting condensation of carbonaceous phases in Type II SN ejecta even while C < O because of the destruction of CO in the high-radiation environment of the ejecta. This relaxes the chemical constraint on mixing. While it might work for graphite, there are doubts whether SiC and Si_3N_4 can condense from a gas with C < O (*Ebel and Grossman,* 2001). Recently, *Clayton et al.* (2002) and *Deneault et al.* (2003) have tried to account for isotopic signatures from different SN zones by considering implantation into newly condensed grains. When the SN shock interacts with density drops at composition boundaries in the star or with mass lost from the pre-supernova star, reverse shocks are generated and work their way back into the ejecta. The gas is slowed but the newly condensed grains, because of their greater inertia, rush ahead of the gas from which they condensed. As the grains overtake the overlying layers, atoms from those layers may be implanted into the grains.

5.6.2. Diamond. Although diamond is the most abundant presolar grain species in meteorites (see Table 3), it remains the least understood. The only presolar isotopic signatures indicating a SN origin are those of Xe-HL (*Lewis et al.,* 1987) and Te (*Richter et al.,* 1998), and to a marginal extent also those of Sr and Ba (*Lewis et al.,* 1991). Nitrogen shows a ^{15}N depletion of 343‰, but isotopically light N is produced by the CN cycle in all stars and is therefore not very diagnostic. Furthermore, while the ^{15}N depletion (corresponding to $^{14}N/^{15}N = 414$) is relative to the terrestrial ratio ($^{14}N/^{15}N = 272$), the diamond ratio is close to the currently preferred solar system ratio of 435 ± 55 (*Owen et al.,* 2001). The C-isotopic composition of bulk diamond is essentially the same as that of the SS (*Russell et al.,* 1991, 1996) and diamonds are too small (the average size is ~2.6 nm, hence the term nanodiamonds) to be analyzed as single grains. Clayton and coworkers (*Clayton,* 1989; *Clayton et al.,* 1995) have tried to also attribute the diamonds and their C- and N-isotopic compositions to a Type II supernova. This requires mixing of contributions from different SN zones. However, because only one diamond grain in a million contains a Xe atom, a solar origin of a large fraction of the nanodiamonds remains a distinct possibility (*Dai et al.,* 2002). The light- and heavy-isotope enrichments in Xe-HL have been interpreted as being due to the p-process (the Xe-L) and a neutron burst (the Xe-H), signaling a SN origin for the diamonds (*Heymann and Dziczkaniec,* 1979, 1980; *Clayton,* 1989). The Xe-H isotopic pattern does not match that expected from the r-process; however, *Ott* (1996) proposed that if r-process Xe is separated from iodine and tellurium precursors on a timescale of a few hours after their production, a Xe-H pattern matching the diamonds results. *Richter et al.* (1997) found that this timescale model also provides a better match to their measured Te-isotopic pattern than the neutron burst model. Still, at present, we do not have an unambiguous identification of the origin of the Xe-HL and Te, and of the diamonds (in case they have a different origin).

5.6.3. Oxide grains. Of the more than 500 presolar oxide grains that have been analyzed to date, only one has the typical isotopic signature expected for SN condensates, namely a large ^{16}O excess (Al-rich oxide grain T84 in Fig. 8) (*Nittler et al.,* 1998). All SN zones dominated by O (O/C, O/Ne, O/Si) are dominated by ^{16}O (*Woosley and Weaver,* 1995; *Thielemann et al.,* 1996; *Rauscher et al.,* 2002) as a result of the $^{12}C(\alpha,\gamma)^{16}O$ reaction during the latter stages of He burning in the pre-supernova evolution. The paucity of such grains, whose abundance is expected to dominate that of carbonaceous phases with a SN origin, remains a puzzle. It has been suggested that oxide grains from supernovae are smaller than those from red giant stars but recent measurements of submicrometer grains have not uncovered any additional oxides with large ^{16}O excesses (*Zinner et al.,* 2003b; *Nguyen et al.,* 2003). A SN origin has been proposed by *Choi et al.* (1998) for the oxide grains with the largest ^{18}O excesses. None of the O-rich SN layers have large ^{18}O excesses, but O in the He/C zone is dominated by ^{18}O, thus material from this zone has to be admixed to material from O-rich zones in order to reproduce the compositions of these grains. Recently, *Messenger and Keller* (2004) reported a silicate grain from a cluster IDP whose $^{18}O/^{16}O$ ratio is much higher than has been observed in any other O-rich grains (see Fig. 8) and is comparable to the ratios found in some LD graphite grains of SN origin. Furthermore, the grain has a large depletion in ^{17}O. There is little doubt that it originated from a SN and its O-isotopic ratios reflect mixing between zones dominated by ^{16}O (O/C, O/Ne) and ^{18}O (He/C zone).

5.7. Grains from Novae

A few SiC and graphite grains have isotopic ratios that are best explained by a nova origin (*Amari et al.,* 2001a; *Nittler and Hoppe,* 2004a,b). These grains have low $^{12}C/^{13}C$ and $^{14}N/^{15}N$ (the SiC grains) ratios (Fig. 4), large ^{30}Si excesses (Fig. 5), and high $^{26}Al/^{27}Al$ ratios (Fig. 6). All these features are predicted to be produced by explosive H burning taking place in classical novae (e.g., *Kovetz and Prialnik,* 1997; *Starrfield et al.,* 1998; *José et al.,* 1999), but the predicted anomalies are much larger than those found in the grains, and the nova ejecta have to be mixed with material of close-to-solar isotopic compositions. A comparison of the data with the models implicates ONe novae with a WD mass of at least 1.25 $M_{\odot}$ as the most likely sources (*Amari et al.,* 2001a; *José et al.,* 2004). A nova origin of some graphite grains (with densities 2.1–2.2 g cm^{-3}) is also indicated by their Ne-isotopic ratios. Laser extraction gas mass spectrometry of single grains show that, like SiC grains, only a small fraction of them contains evidence for Ne-E. Two of these grains have $^{20}Ne/^{22}Ne$ ratios that are lower than ratios predicted to result from He burning in any known stellar sources, implying decay of ^{22}Na (*Nichols et al.,* 2004). Furthermore, their ^{22}Ne is not accompanied by ^{4}He, which is expected if Ne was implanted. The $^{12}C/^{13}C$ ratios of these two grains are 4 and 10.

5.8. Silicon Carbide Grains of Type A + B

Grains of type A + B have $^{12}C/^{13}C < 10$ but their Si-isotopic ratios plot along the mainstream line (Figs. 4 and 5). In contrast to mainstream grains, many A + B grains have lower than solar $^{14}N/^{15}N$ ratios (*Hoppe et al.,* 1995, 1996a; *Huss et al.,* 1997; *Amari et al.,* 2001c). While the isotopic ratios of mainstream, Y, and Z grains find an explanation in nucleosynthetic models of AGB stars, a satisfactory explanation of the isotopic ratios of A + B grains has not been achieved. The low $^{12}C/^{13}C$ ratios of these grains combined with the requirement for a C-rich environment during their formation indicate He burning followed by limited H burning in their stellar sources. Astrophysical sites for this process are not well known. Candidates for the parent stars of grains with no s-process enhancements (*Amari et al.,* 1995a; *Pellin et al.,* 2000b; *Savina et al.,* 2003c) are J-type carbon stars, which have low $^{12}C/^{13}C$ ratios (*Lambert et al.,* 1986). Unfortunately, J stars are not well understood and there are essentially no astronomical observations of N-isotopic ratios in such stars. The only exception, the observation of $^{14}N/^{15}N > 70$ (*Wannier et al.,* 1991), does not provide much of a constraint. Furthermore, the low $^{14}N/^{15}N$ ratios observed in some of the grains as well as the C-rich nature of their parent stars appear to be incompatible with the consequences of H burning in the CNO cycle, which seems to be responsible for the low $^{12}C/^{13}C$ ratios of J stars and the grains. A + B grains with s-process enhancements might come from post-AGB stars that undergo a very late thermal pulse. An example of such a star is Sakurai's object (e.g., *Asplund et al.,* 1999; *Herwig,* 2001). However, grains with low $^{14}N/^{15}N$ ratios pose a problem. *Huss et al.* (1997) proposed that the currently used $^{18}O(p,\alpha)^{15}N$ reaction rate is too low by a factor of 1000. This would result in low $^{12}C/^{13}C$ and $^{14}N/^{15}N$ ratios if an appropriate level of CBP is considered. Savina and coworkers (*Savina et al.,* 2003c, 2004b) have analyzed Mo and Ru isotopes in several A + B SiC grains by RIMS. Most grains have close-to-solar isotopic ratios, indicating no neutron exposure. However, one grain shows a Mo pattern similar to those shown by X grains (see Fig. 9), indicating a neutron burst, and another grain exhibits a p-process signature in the form of ^{92}Mo, ^{94}Mo, ^{96}Ru, and ^{98}Ru excesses. While these findings seems to indicate a SN origin, it is difficult to reconcile the C- and Si-isotopic ratios with such an origin. The authors suggested the transfer of material in a binary system.

6. CONCLUSION AND OUTLOOK

Astronomy, as an observational science, requires us to glean information about stars from any relevant source nature provides. Traditionally, this source has been electromagnetic radiation, but, as we have seen in this review, novel techniques in chemistry and advances in mass spectrometry have initiated an entirely new type of astronomy, the astronomy of presolar meteoritic grains, or stardust. The detailed isotopic abundances available from stardust are heretofore undreamt-of probes of stellar nucleosynthesis, which, as apparent from this review, already challenge our understanding of stellar astrophysics. It is certain that new isotopic measurements and discovery of new stardust grains will be major forces in driving stellar models to the next level of sophistication.

As our understanding of the production and survival of stardust grows, so too will our understanding of the formation of the SS. The nuclear anomalies in primitive solar system minerals are no doubt inherited in a complicated way from their anomalous precursor dust. Unraveling the complex history of these samples from their initial building blocks to their incorporation into meteorite parent bodies will certainly help us decipher the principal aggregation processes in the ESS. Further isotopic measurements and characterization of new samples will be key in this effort. Equally important will be further insights into the short-lived radioisotopes. As the initial abundances of those isotopes in the ESS become better known, the constraints they provide will increasingly challenge our scenarios for the Sun's birth and early history.

Auguste Comte, in the nineteenth century, opined that we would never know much about the stars given their great distance from us. The capabilities of astronomy from atomic spectroscopy, beginning in the late nineteenth century, certainly demolished that view. Today, as we directly study isotopic anomalies in meteoritic samples in our laboratories on Earth, we begin to understand our connection to the stars in a new, more tangible way. And it is inspiring to know that our appreciation of this connection will grow in the coming decades!

Acknowledgments. Detailed reviews by A. Davis, P. Hoppe, and L. Nittler greatly helped in improving this chapter and are deeply appreciated. We thank L. Nittler for providing unpublished data.

REFERENCES

Alexander C. M. O'D. (1993) Presolar SiC in chondrites: How variable and how many sources? *Geochim. Cosmochim. Acta, 57,* 2869–2888.

Alexander C. M. O'D. and Nittler L. R. (1999) The galactic evolution of Si, Ti and O isotopic ratios. *Astrophys. J., 519,* 222–235.

Amari S., Anders E., Virag A., and Zinner E. (1990) Interstellar graphite in meteorites. *Nature 345,* 238–240.

Amari S., Hoppe P., Zinner E., and Lewis R. S. (1992) Interstellar SiC with unusual isotopic compositions: Grains from a supernova? *Astrophys. J. Lett., 394,* L43–L46.

Amari S., Lewis R. S., and Anders E. (1994) Interstellar grains in meteorites: I. Isolation of SiC, graphite, and diamond; size distributions of SiC and graphite. *Geochim. Cosmochim. Acta, 58,* 459–470.

Amari S., Hoppe P., Zinner E., and Lewis R. S. (1995a) Trace-element concentrations in single circumstellar silicon carbide grains from the Murchison meteorite. *Meteoritics, 30,* 679–693.

Amari S., Lewis R. S., and Anders E. (1995b) Interstellar grains in meteorites: III. Graphite and its noble gases. *Geochim. Cosmochim. Acta, 59,* 1411–1426.

Amari S., Zinner E., and Lewis R. S. (1995c) ^{41}Ca in circumstellar graphite from supernovae. *Meteoritics, 30,* 480.

Amari S., Zinner E., and Lewis R. S. (1995d) Interstellar graphite from the Murchison meteorite. In *Nuclei in the Cosmos III* (M. Busso et al., eds.), pp. 581-584. American Institute of Physics, New York.

Amari S., Zinner E., and Lewis R. S. (1995e) Large ^{18}O excesses in circumstellar graphite grains from the Murchison meteorite: Indication of a massive star-origin. *Astrophys. J. Lett., 447,* L147–L150.

Amari S., Zinner E., and Lewis R. S. (1996) ^{41}Ca in presolar graphite of supernova origin. *Astrophys. J. Lett., 470,* L101–L104.

Amari S., Zinner E., and Lewis R. S. (2000) Isotopic compositions of different presolar silicon carbide size fractions from the Murchison meteorite. *Meteoritics & Planet. Sci., 35,* 997–1014.

Amari S., Gao X., Nittler L. R., Zinner E., José J., Hernanz M., and Lewis R. S. (2001a) Presolar grains from novae. *Astrophys. J., 551,* 1065–1072.

Amari S., Nittler L. R., Zinner E., Gallino R., Lugaro M., and Lewis R. S. (2001b) Presolar SiC grains of type Y: Origin from low-metallicity AGB stars. *Astrophys. J., 546,* 248–266.

Amari S., Nittler L. R., Zinner E., Lodders K., and Lewis R. S. (2001c) Presolar SiC grains of type A and B: Their isotopic compositions and stellar origins. *Astrophys. J., 559,* 463–483.

Amari S., Jennings C., Nguyen A., Stadermann F. J., Zinner E., and Lewis R. S. (2002) NanoSIMS isotopic analysis of small presolar SiC grains from the Murchison and Indarch meteorites (abstract). In *Lunar and Planetary Science XXXIII,* Abstract #1205. Lunar and Planetary Institute, Houston (CD-ROM).

Amelin Y., Krot A. N., Hutcheon I. D., and Ulyanov A. A. (2002) Lead isotopic ages of chondrules and calcium-aluminum-rich inclusions. *Science, 297,* 1678–1683.

Anders E. and Grevesse N. (1989) Abundances of the elements: Meteoritic and solar. *Geochim. Cosmochim. Acta, 53,* 197–214.

Anders E. and Zinner E. (1993) Interstellar grains in primitive meteorites: Diamond, silicon carbide, and graphite. *Meteoritics, 28,* 490–514.

Argast D., Samland M., Thielemann F.-K., and Qian Y.-Z. (2004) Neutron star mergers versus core-collapse supernovae as dominant r-process sites in the early Galaxy. *Astron. Astrophys., 416,* 997–1011.

Arlandini C., Gallino R., Busso M., and Straniero O. (1995) New calculations of evolution, dredge-up and nucleosynthesis for low mass stars in the TP-AGB phases. In *Stellar Evolution: What Should Be Done* (A. Noels et al., eds.), pp. 447–452. 32nd Liège Intl. Astrophys. Colloquium, Université de Liège, Liège, Belgium.

Arnould M., Meynet G., and Paulus G. (1997a) Wolf-Rayet stars and their nucleosynthetic signatures in meteorites. In *Astrophysical Implications of the Laboratory Study of Presolar Materials* (T. J. Bernatowicz and E. Zinner, eds.), pp. 179–202. American Institute of Physics, New York.

Arnould M., Paulus G., and Meynet G. (1997b) Short-lived radionuclide production by non-exploding Wolf-Rayet stars. *Astron. Astrophys., 321,* 452–464.

Asplund M., Lambert D. L., Kipper T., Pollacco D., and Shetrone M. D. (1999) The rapid evolution of the born-again giant Sakurai's object. *Astron. Astrophys., 343,* 507–518.

Asplund M., Grevesse N., and Sauval A. J. (2005) The solar chemical composition. In *Cosmic Abundances as Records of Stellar Evolution and Nucleosynthesis* (T. G. Barnes and F. N. Bash, eds.), pp. 25–38. ASP Conference Series, Vol. 336, Astronomical Society of the Pacific, San Francisco.

Bao Z. Y., Beer H., Käppeler F., Voss F., Wisshak K., and Rauscher T. (2000) *Atom. Data Nucl. Data Tables, 75,* 1.

Bazan G. and Arnett D. (1998) Two-dimensional hydrodynamics of pre-core collapse: Oxygen shell burning. *Astrophys. J., 496,* 316–332.

Becker H. and Walker R. J. (2003) Efficient mixing of the solar nebula from uniform Mo isotopic composition of meteorites. *Nature, 425,* 152–155.

Becker S. A. and Iben I. Jr. (1979) The asymptotic giant branch evolution of intermediate-mass stars as a function of mass and composition. I. through the second dredge-up phase. *Astrophys. J., 232,* 831–853.

Begemann F. (1993) Isotope abundance anomalies and the early solar system: MuSiC vs. FUN. In *Origin and Evolution of the Elements* (N. Prantzos et al., eds.), pp. 517–526. Cambridge Univ., Cambridge.

Bernatowicz T. J. and Zinner E., eds. (1997) *Astrophysical Implications of the Laboratory Study of Presolar Materials.* American Institute of Physics, New York. 750 pp.

Bernatowicz T., Fraundorf G., Tang M., Anders E., Wopenka B., Zinner E., and Fraundorf P. (1987) Evidence for interstellar SiC in the Murray carbonaceous meteorite. *Nature, 330,* 728–730.

Bernatowicz T. J., Amari S., Zinner E. K., and Lewis R. S. (1991) Interstellar grains within interstellar grains. *Astrophys. J. Lett., 373,* L73–L76.

Bernatowicz T. J., Amari S., and Lewis R. S. (1992) TEM studies of a circumstellar rock (abstract). In *Lunar and Planetary Science XXIII,* pp. 91–92. Lunar and Planetary Institute, Houston.

Bernatowicz T. J., Cowsik R., Gibbons P. C., Lodders K., Fegley B. Jr., Amari S., and Lewis R. S. (1996) Constraints on stellar grain formation from presolar graphite in the Murchison meteorite. *Astrophys. J., 472,* 760–782.

Besmehn A. and Hoppe P. (2001) Silicon- and calcium-isotopic compositions of presolar silicon nitride grains from the Indarch enstatite chondrite (abstract). In *Lunar and Planetary Science XXXII,* Abstract #1188. Lunar and Planetary Institute, Houston (CD-ROM).

Besmehn A. and Hoppe P. (2003) A NanoSIMS study of Si- and Ca-Ti-isotopic compositions of presolar silicon carbide grains from supernovae. *Geochim. Cosmochim. Acta, 67,* 4693–4703.

Bethe H. A. (1939) Energy production in stars. *Phys. Rev., 55,* 103.

Bethe H. A. and Critchfield C. L. (1938) The formation of deuterons by proton capture. *Phys. Rev., 54,* 248–254.

Bethe H. A. and Wilson J. R. (1985) Revival of a stalled supernova shock by neutrino heating. *Astrophys. J., 295,* 14–23.

Birck J.-L. and Allègre C. J. (1984) Chromium isotopic anomalies in Allende refractory inclusions. *Geophys. Res. Lett., 11,* 943–946.

Birck J.-L. and Allègre C. J. (1985) Evidence for the presence of ^{53}Mn in the early solar system. *Geophys. Res. Lett., 12,* 745–748.

Birck J.-L. and Lugmair G. W. (1988) Nickel and chromium isotopes in Allende inclusions. *Earth Planet. Sci. Lett., 90,* 131–143.

Birck J. L., Rotaru M., and Allègre C. J. (1999) ^{53}Mn-^{53}Cr evolution of the early solar system. *Geochim. Cosmochim. Acta, 63,* 4111–4117.

Black D. C. (1972) On the origins of trapped helium, neon and argon isotopic variations in meteorites II. Carbonaceous meteorites. *Geochim. Cosmochim. Acta, 36,* 377–394.

Black D. C. and Pepin R. O. (1969) Trapped neon in meteorites. II. *Earth Planet. Sci. Lett., 6,* 395–405.

Blake J. B. and Schramm D. N. (1976) A possible alternative to the r-process. *Astrophys. J., 209,* 846–849.

Blake J. B., Woosley S. E., and Weaver T. A. (1981) Nucleosynthesis of neutron-rich heavy nuclei during explosive helium burning in massive stars. *Astrophys. J., 248,* 315–320.

Boato G. (1954) The isotopic composition of hydrogen and carbon in the carbonaceous chondrites. *Geochim. Cosmochim. Acta, 6,* 209–220.

Bodansky D., Clayton D. D., and Fowler W. A. (1968) Nuclear quasi-equilibrium during silicon burning. *Astrophys. J., 16,* 299–371.

Bonetti R., Broggini C., Campajola L., Corvisiero P., D'Alessandro A., Dessalvi M., D'Onofrio A., Fubini A., Gervino G., Gialanella L., Grieife U., Guglielmetti A., Gustavino C., Imbriani G., Junker M., Prati P., Roca V., Rolfs C., Romano M., Schuemann F., Strieder F., Terrasi F., Trautvetter H. P., and Zavatarelli S. (1999) First measurement of the ^{3}He (^{3}He, 2p) ^{4}He cross section down to the lower edge of the solar Gamow peak. *Phys. Rev. Lett., 82,* 5205–5208.

Boothroyd A. I. and Sackmann I.-J. (1988) Low-mass stars. III. Low-mass stars with steady mass loss: Up to the asymptotic giant branch and through the final thermal pulses. *Astrophys. J., 328,* 653–670.

Boothroyd A. I. and Sackmann I.-J. (1999) The CNO isotopes: Deep circulation in red giants and first and second dredge-up. *Astrophys. J., 510,* 232–250.

Boothroyd A. I., Sackmann I.-J., and Wasserburg G. J. (1994) Predictions of oxygen isotope ratios in stars and of oxygen-rich interstellar grains in meteorites. *Astrophys. J. Lett., 430,* L77–L80.

Boothroyd A. I., Sackmann I.-J., and Wasserburg G. J. (1995) Hot bottom burning in asymptotic giant branch stars and its effect on oxygen isotopic abundances. *Astrophys. J. Lett., 442,* L21–L24.

Boss A. and Foster P. (1997) Triggering presolar cloud collapse and injecting matrial into the presolar nebula. In *Astrophysical Implications of the Laboratory Study of Presolar Materials* (T. Bernatowicz and E. Zinner, eds.), pp. 649–664. American Institute of Physics, New York.

Boss A. P. and Foster P. N. (1998) Injection of short-lived isotopes into the presolar cloud. *Astrophys. J. Lett., 494,* L103–L106.

Brown L. E. and Clayton D. D. (1992) SiC particles from asymptotic giant branch stars: Mg burning and the s-process. *Astrophys. J. Lett., 392,* L79–L82.

Burbidge E. M., Burbidge G. R., Fowler W. A., and Hoyle F. (1957) Synthesis of the elements in stars. *Rev. Mod. Phys., 29,* 547–650.

Burrows A. (2000) Supernova explosions in the universe. *Nature, 403,* 727–733.

Busso M., Gallino R., and Wasserburg G. J. (1999) Nucleosynthesis in asymptotic giant branch stars: Relevance for galactic enrichment and solar system formation. *Annu. Rev. Astron. Astrophys., 37,* 239–309.

Busso M., Gallino R., and Wasserburg G. J. (2003) Short-lived nuclei in the early solar system: A low mass stellar source? *Publ. Astron. Soc. Australia, 20,* 356–370.

Cameron A. G. W. (1957) Stellar evolution, nuclear astrophysics and nucleogenesis. *Publ. Astron. Soc. Pacific, 69,* 201–222.

Cameron A. G. W. (1962) The formation of the sun and planets. *Icarus, 1,* 13–69.

Cameron A. G. W. (1973) Abundance of the elements in the solar system. *Space Sci. Rev., 15,* 121–146.

Cameron A. G. W. (1979) The neutron-rich silicon-burning and equilibrium processes of nucleosynthesis. *Astrophys. J. Lett., 230,* L53–L57.

Cameron A. G. W. (1993) Nucleosynthesis and star formation. In *Protostars and Planets III* (E. H. Levy and J. Lunine, eds.), pp. 47–73. Univ. of Arizona, Tucson.

Cameron A. G. W. (2002) Meteoritic isotopic abundance effects from r-process jets (abstract). In *Lunar and Planetary Science XXXIII,* Abstract #1112. Lunar and Planetary Institute, Houston (CD-ROM).

Cameron A. G. W. and Truran J. W. (1977) The supernovae trigger for formation of the solar system. *Icarus, 30,* 447–461.

Cameron A. G. W., Höflich P., Myers P. C., and Clayton D. D. (1995) Massive supernovae, orion gamma rays, and the formation of the solar system. *Astrophys. J. Lett., 447,* L53–L57.

Charbonnel C. (1995) A consistent explanation for ^{12}C/^{13}C, ^{7}Li, and ^{3}He anomalies in red giant stars. *Astrophys. J. Lett., 453,* L41–L44.

Chaussidon M. and Gounelle M. (2006) Irradiation processes in the early solar system. In *Meteorites and the Early Solar System II* (D. S. Lauretta and H. Y. McSween Jr., eds.), this volume. Univ. of Arizona, Tucson.

Chaussidon M., Robert F., and McKeegan K. D. (2004) Li and B isotopic variations in Allende type B1 CAI 3529-41: Trace of incorporation of short-lived ^{7}Be and ^{10}Be. In *Lunar and Planetary Science XXXV,* Abstract #1568. Lunar and Planetary Institute, Houston (CD-ROM).

Chaussidon M., Robert F., and McKeegan K. D. (2005) Li and B isotopic variation in an Allende CAI: Evidence for in situ decay of ^{10}Be and for the possible presence of the short-lived nuclide ^{7}Be in the early solar system. *Geochim. Cosmochim. Acta,* in press.

Chen J. H., Papanastassiou D. A., Wasserburg G. J., and Ngo H. H. (2004) Endemic Mo isotopic anomalies in iron and carbonaceous meteorites (abstract). In *Lunar and Planetary Science XXXV,* Abstract #1431. Lunar and Planetary Institute, Houston (CD-ROM).

Choi B.-G., Huss G. R., Wasserburg G. J., and Gallino R. (1998) Presolar corundum and spinel in ordinary chondrites: Origins from AGB stars and a supernova. *Science, 282,* 1284–1289.

Choi B.-G., Wasserburg G. J., and Huss G. R. (1999) Circumstellar hibonite and corundum and nucleosynthesis in asymptotic giant branch stars. *Astrophys. J. Lett., 522,* L133–L136.

Clayton D. D. (1975a) Extinct radioactivities: Trapped residuals of presolar grains. *Astrophys. J., 199,* 765–769.

Clayton D. D. (1975b) Na-22, Ne-E, extinct radioactive anomalies and unsupported Ar-40. *Nature, 257,* 36–37.

Clayton D. D. (1983) Discovery of s-process Nd in Allende residue. *Astrophys. J. Lett., 271,* L107–L109.

Clayton D. D. (1986) Interstellar fossil ^{26}Mg and its possible relationship to excess meteoritic ^{26}Mg. *Astrophys. J., 310,*490–498.

Clayton D. D. (1989) Origin of heavy xenon in meteoritic diamonds. *Astrophys. J., 340,* 613–619.

Clayton D. D. (1997) Placing the sun and mainstream SiC particles in galactic chemodynamic evolution. *Astrophys. J. Lett., 484,* L67–L70.

Clayton D. D. (2003) Presolar galactic merger spawned the SiC grain mainstream. *Astrophys. J., 598,* 313–324.

Clayton D. D. and Leising M. D. (1987) ^{26}Al in the interstellar medium. *Phys. Rept., 144,* 1–50.

Clayton D. D. and Nittler L. R. (2004) Astrophysics with presolar stardust. *Annu. Rev. Astron. Astrophys., 42,* 39–78.

Clayton D. D. and Timmes F. X. (1997a) Implications of presolar grains for galactic chemical evolution. In *Astrophysical Implications of the Laboratory Study of Presolar Materials* (T. J. Bernatowicz and E. Zinner, eds.), pp. 237–264. American Institute of Physics, New York.

Clayton D. D. and Timmes F. X. (1997b) Placing the Sun in galactic chemical evolution: Mainstream SiC particles. *Astrophys. J., 483,* 220–227.

Clayton D. D. and Ward R. A. (1978) s-Process studies: Xenon and krypton isotopic abundances. *Astrophys. J., 224,* 1000–1006.

Clayton D. D., Obradovic M., Guha S., and Brown L. E. (1991) Silicon and titanium isotopes in SiC from AGB stars (abstract). In *Lunar and Planetary Science XXII,* pp. 221–222. Lunar and Planetary Institute, Houston.

Clayton D. D., Meyer B. S., Sanderson C. I., Russell S. S., and Pillinger C. T. (1995) Carbon and nitrogen isotopes in type II supernova diamonds. *Astrophys. J., 447,* 894–905.

Clayton D. D., Liu W., and Dalgarno A. (1999) Condensation of carbon in radioactive supernova gas. *Science, 283,* 1290–1292.

Clayton D. D., Meyer B. S., The L.-S., and El Eid M. F. (2002) Iron implantation in presolar supernova grains. *Astrophys. J. Lett., 578,* L83–L86.

Clayton R. N. (1978) Isotopic anomalies in the early solar system. *Annu. Rev. Nucl. Part. Sci., 28,* 501–522.

Clayton R. N. (2002) Self-shielding in the solar nebula. *Nature, 415,* 860–861.

Clayton R. N. and Mayeda T. K. (1977) Correlated oxygen and magnesium isotope anomalies in Allende inclusions, I: Oxygen. *Geophys. Res. Lett., 4,* 295–298.

Clayton R. N., Grossman L., and Mayeda T. K. (1973) A component of primitive nuclear composition in carbonaceous meteorites. *Science, 182,* 485–488.

Clayton R. N., MacPherson G. J., Hutcheon I. D., Davis A. M., Grossman L., Mayeda T. K., Molini-Velsko C., Allen J. M., and El Goresy A. (1984) Two forsterite-bearing FUN inclusions in the Allende meteorite. *Geochim. Cosmochim. Acta, 48,* 535–548.

Clayton R. N., Hinton R. W., and Davis A. M. (1988) Isotopic variations in the rock-forming elements in meteorites. *Phil. Trans. R. Soc. Lond., A325,* 483–501.

Colgate S. A. and White R. H. (1966) The hydrodynamic behavior of supernovae explosions. *Astrophys. J., 143,* 626–681.

Croat T. K., Bernatowicz T., Amari S., Messenger S., and Stadermann F. J. (2003) Structural, chemical, and isotopic microanalytical investigations of graphite from supernovae. *Geochim. Cosmochim. Acta, 67,* 4705–4725.

Dai Z. R., Bradley J. P., Joswiak D. J., Brownlee D. E., Hill H. G. M., and Genge M. J. (2002) Possible *in situ* formation of meteoritic nanodiamonds in the early solar system. *Nature, 418,* 157–159.

Dauphas N., Marty B., and Reisberg L. (2002a) Molybdenum evidence for inherited planetary scale isotope heterogeneity of the protosolar nebula. *Astrophys. J., 565,* 640–644.

Dauphas N., Marty B., and Reisberg L. (2002b) Molybdenum nucleosynthetic dichotomy revealed in primitive meteorites. *Astrophys. J. Lett., 569,* L139–L142.

Davis A. M., Gallino R., Lugaro M., Tripa C. E., Savina M. R., Pellin M. J., and Lewis R. S. (2002) Presolar grains and the nucleosynthesis of iron isotopes (abstract). In *Lunar and Planetary Science XXXIII,* Abstract #2018. Lunar and Planetary Institute, Houston (CD-ROM).

Davis A. M., Gallino R., Straniero O., Dominguez I., and Lugaro M. (2003) Heavy element nucleosynthesis in low metallicity, low mass AGB stars (abstract). In *Lunar and Planetary Science XXXIV,* Abstract #2034. Lunar and Planetary Institute, Houston (CD-ROM).

Deneault E. A.-N., Clayton D. D., and Heger A. (2003) Supernova reverse shocks and SiC growth. *Astrophys. J., 594,* 312–325.

Denissenkov P. A. and Weiss A. (1996) Deep diffusive mixing in globular-cluster red giants. *Astron. Astrophys., 308,* 773–784.

Desch S. J., Connolly H. C. Jr., and Srinivasan G. (2004) An interstellar origin for the beryllium 10 in calcium-rich, aluminum-rich inclusions. *Astrophys. J., 602,* 528–542.

Diehl R., Dupraz C., Bennett K., Bloemen H., Hermsen W., Knödlseder J., Lichti G., Morris D., Ryan J., Schönfelder V., Steinle H., Strong A., Swanenburg B., Varendorff M., and Winkler C. (1995) COMPTEL observations of galactic ^{26}Al emission. *Astron. Astrophys., 298,* 445–460.

Ebel D. S. and Grossman L. (2001) Condensation from supernova gas made of free atoms. *Geochim. Cosmochim. Acta, 65,* 469–477.

Ebisuzaki T. and Shibazaki N. (1988) The effects of mixing of the ejecta on the hard X-ray emissions from SN 1987A. *Astrophys. J. Lett., 327,* L5–L8.

El Eid M. (1994) CNO isotopes in red giants: Theory versus observations. *Astron. Astrophys., 285,* 915–928.

El Eid M. F., Meyer B. S., and The L.-S. (2004) Evolution of massive stars up to the end of central oxygen burning. *Astrophys. J., 611,* 452–465.

Fahey A., Goswami J. N., McKeegan K. D., and Zinner E. (1985) Evidence for extreme ^{50}Ti enrichments in primitive meteorites. *Astrophys. J. Lett., 296,* L17–L20.

Fahey A. J., Goswami J. N., McKeegan K. D., and Zinner E. (1987) ^{16}O excesses in Murchison and Murray hibonites: A case against a late supernova injection origin of isotopic anomalies in O, Mg, Ca, and Ti. *Astrophys. J. Lett., 323,* L91–L95.

Filippenko A. V. (1997) Optical spectra of supernovae. *Annu. Rev. Astron. Astrophys., 35,* 309–355.

Fish R. A., Goles G. G., and Anders E. (1960) The record in the meteorites. III. On the development of meteorites in asteroidal bodies. *Astrophys. J., 132,* 243–258.

Floss C. and Stadermann F. J. (2004) Isotopically primitive interplanetary dust particles of cometary origin: Evidence from nitrogen isotopic compositions (abstract). In *Lunar and Planetary Science XXXV,* Abstract #1281. Lunar and Planetary Institute, Houston (CD-ROM).

Forestini M., Paulus G., and Arnould M. (1991) On the production of ^{26}Al in AGB stars. *Astron. Astrophys., 252,* 597–604.

Foster P. N. and Boss A. P. (1997) Injection of radioactive nuclides from the stellar source that triggered the collapse of the presolar nebula. *Astrophys. J., 489,* 346–357.

Fowler W. A., Greenstein J. L., and Hoyle F. (1962) Nucleosynthesis during the early history of the solar system. *Geophys. J., 6,* 148–220.

Freiburghaus C., Rosswog S., and Thielemann F.-K. (1999) R-process in neutron star mergers. *Astrophys. J. Lett., 525,* L121–L124.

Frost C. A. and Lattanzio J. C. (1996) AGB stars: What should be done? In *Stellar Evolution: What Should Be Done* (A. Noel et al., eds.), pp. 307–325. 32nd Liège Intl. Astrophys. Colloquium, Université de Liège, Liège, Belgium.

Fuller G. M. and Meyer B. S. (1995) Neutrino capture and supernova nucleosynthesis. *Astrophys. J., 453,* 792–809.

Gallino R., Busso M., Picchio G., and Raiteri C. M. (1990) On the astrophysical interpretation of isotope anomalies in meteoritic SiC grains. *Nature, 348,* 298–302.

Gallino R., Raiteri C. M., and Busso M. (1993) Carbon stars and isotopic Ba anomalies in meteoritic SiC grains. *Astrophys. J., 410,* 400–411.

Gallino R., Raiteri C. M., Busso M., and Matteucci F. (1994) The puzzle of silicon, titanium and magnesium anomalies in meteoritic silicon carbide grains. *Astrophys. J., 430,* 858–869.

Gallino R., Busso M., and Lugaro M. (1997) Neutron capture nucleosynthesis in AGB stars. In *Astrophysical Implications of the Laboratory Study of Presolar Materials* (T. J. Bernatowicz and E. Zinner, eds.), pp. 115-153. American Institute of Physics, New York.

Gallino R., Arlandini C., Busso M., Lugaro M., Travaglio C., Straniero O., Chieffi A., and Limongi M. (1998) Evolution and nucleosynthesis in low-mass asymptotic giant branch stars. II. Neutron capture and the s-process. *Astrophys. J., 497,* 388–403.

Galy A., Hutcheon I. D., and Grossman L. (2004) (^{26}Al/^{27}Al)$_0$ of the solar nebula inferred from Al-Mg systematics in bulk CAIs from CV3 chondrites (abstract). In *Lunar and Planetary Science XXXV,* Abstract #1790. Lunar and Planetary Institute, Houston (CD-ROM).

Gilmour J. (2002) The solar system's first clocks. *Science, 297,* 1658–1659.

Gilroy K. K. (1989) Carbon isotope ratios and lithium abundances in open cluster giants. *Astrophys. J., 347,* 835–848.

Gilroy K. K. and Brown J. A. (1991) Carbon isotope ratios along the giant branch of M67. *Astrophys. J., 371,* 578–583.

Goswami J. N. and Vanhala H. A. T. (2000) Extinct radionuclides and the origin of the solar system. In *Protostars and Planets IV* (V. Mannings et al., eds.), p. 963. Univ. of Arizona, Tucson.

Goswami J. N., Marhas K. K., and Sahijpal S. (2001) Did solar energetic particles produce the short-lived nuclides present in the early solar system? *Astrophys. J., 549,* 1151–1159.

Gounelle M., Shu F. H., Shang H., Glassgold A. E., Rehm K. E., and Lee T. (2001) Extinct radioactivities and protosolar cosmic rays: Self-shielding and light elements. *Astrophys. J., 548,* 1051–1070.

Grevesse N., Noels A., and Sauval A. J. (1996) Standard abundances. In *Cosmic Abundances* (S. S. Holt and G. Sonneborn, eds.), pp. 117–126. BookCrafters, Inc., San Francisco.

Grossman L. (1972) Condensation in the primitive solar nebula. *Geochim. Cosmochim. Acta, 36,* 597–619.

Guan Y., Huss G. R., Leshin L. A., and MacPherson G. J. (2003) Ni isotope anormalies and ^{60}Fe in sulfides from unequilibrated enstatite chondrites (abstract). *Meteoritics & Planet. Sci., 38,* A138.

Guber K. H., Spencer R. R., Koehler P. E., and Winters R. R. (1997) New ^{142,144}Nd (n,γ) cross sections and the s-process origin of the Nd anomalies in presolar meteoritic silicon carbide grains. *Phys. Rev. Lett., 78,* 2704–2707.

Guber K. H., Koehler P. E., Derrien H., Valentine T. E., Leal L. C., Sayer R. O., and Rauscher T. (2003) Neutron capture reaction rates for silicon and their impact on the origin of presolar mainstream SiC grains. *Phys. Rev. C, 67,* 062802-1 to 062802-4.

Halliday A. N. and Kleine T. (2006) Meteorites and the timing, mechanisms, and conditions of terrestrial planet accretion and early differentiation. In *Meteorites and the Early Solar System II* (D. S. Lauretta and H. Y. McSween Jr., eds.), this volume. Univ. of Arizona, Tucson.

Harper C. L. Jr. (1993) Isotopic astronomy from anomalies in meteorites: Recent advances and new frontiers. *J. Phys. G, 19,* S81–S94.

Harper C. L. Jr. (1996) Evidence for ^{92g}Nb in the early solar system and evaluation of a new p-process cosmochronometer from ^{92g}Nb/^{92}Mo. *Astrophys. J., 466,* 437–456.

Harper C. L. Jr. and Jacobsen S. B. (1996) Evidence for ^{182}Hf in the early solar system and constraints on the timescale of terrestrial accretion and core formation. *Geochim. Cosmochim. Acta, 60,* 1131–1153.

Harris M. J. and Lambert D. L. (1984) Oxygen isotopic abundances in the atmospheres of seven red giant stars. *Astrophys. J., 285,* 674–682.

Harris M. J., Lambert D. L., Hinkle K. H., Gustafsson B., and Eriksson K. (1987) Oxygen isotopic abundances in evolved stars. III. 26 carbon stars. *Astrophys. J., 316,* 294–304.

Hartmann D., Woosley S. E., and El Eid M. F. (1985) Nucleosynthesis in neutron-rich supernova ejecta. *Astrophys. J., 297,* 837–845.

Herant M., Benz W., Hix W. R., Fryer C. L., and Colgate S. A. (1994) Inside the supernova: A powerful convective engine. *Astrophys. J., 435,* 339–361.

Herwig F. (2001) The evolutionary timescale of Sakurai's Object: A test of convection theory? *Astrophys. J. Lett., 554,* L71–L74.

Heymann D. (1983) Barium from a mini r-process in supernovae. *Astrophys. J., 267,* 747–756.

Heymann D. and Dziczkaniec M. (1979) Xenon from intermediate zones of supernovae. *Proc. Lunar Planet. Sci. Conf. 10th,* pp. 1943–1959.

Heymann D. and Dziczkaniec M. (1980) A first roadmap for kryptology. *Proc. Lunar Planet. Sci. Conf. 11th,* pp. 1179–1213.

Hidaka H., Ohta Y., Yoneda S., and DeLaeter J. R. (2001) Isotopic search for live ^{135}Cs in the early solar system and possibility of ^{135}Cs-^{135}Ba chronometer. *Earth Planet. Sci. Lett., 193,* 459–466.

Hidaka H., Ohta Y., and Yoneda S. (2003) Nucleosynthetic components of the early solar system inferred from Ba isotopic compositions in carbonaceous chondrites. *Earth Planet. Sci. Lett., 214,* 455–466.

Hillebrandt W., Kodama T., and Takahashi K. (1976) R-process nucleosynthesis — A dynamical model. *Astron. Astrophys., 52,* 63–68.

Hoffman R. D., Woosley S. E., Fuller G. M., and Meyer B. S. (1996a) Nucleosynthesis in neutrino-driven winds. I. The physical conditions. *Astrophys. J., 460,* 478–488.

Hoffman R. D., Woosley S. E., Fuller G. M., and Meyer B. S. (1996b) Production of the light p-process nuclei in neutrino-driven winds. *Astrophys. J., 460,* 478–488.

Hoppe P. and Besmehn A. (2002) Evidence for extinct vanadium-49 in presolar silicon carbide grains from supernovae. *Astrophys. J. Lett., 576,* L69–L72.

Hoppe P. and Ott U. (1997) Mainstream silicon carbide grains from meteorites. In *Astrophysical Implications of the Laboratory Study of Presolar Materials* (T. J. Bernatowicz and E. Zinner, eds.), pp. 27–58. American Institute of Physics, New York.

Hoppe P. and Zinner E. (2000) Presolar dust grains from meteorites and their stellar sources. *J. Geophys. Res., 105,* 10371–10385.

Hoppe P., Amari S., Zinner E., Ireland T., and Lewis R. S. (1994) Carbon, nitrogen, magnesium, silicon and titanium isotopic compositions of single interstellar silicon carbide grains from the Murchison carbonaceous chondrite. *Astrophys. J., 430,* 870–890.

Hoppe P., Amari S., Zinner E., and Lewis R. S. (1995) Isotopic compositions of C, N, O, Mg, and Si, trace element abundances, and morphologies of single circumstellar graphite grains in four density fractions from the Murchison meteorite. *Geochim. Cosmochim. Acta, 59,* 4029–4056.

Hoppe P., Strebel R., Eberhardt P., Amari S., and Lewis R. S. (1996a) Small SiC grains and a nitride grain of circumstellar origin from the Murchison meteorite: Implications for stellar evolution and nucleosynthesis. *Geochim. Cosmochim. Acta, 60,* 883–907.

Hoppe P., Strebel R., Eberhardt P., Amari S., and Lewis R. S. (1996b) Type II supernova matter in a silicon carbide grain from the Murchison meteorite. *Science, 272,* 1314–1316.

Hoppe P., Annen P., Strebel R., Eberhardt P., Gallino R., Lugaro M., Amari S., and Lewis R. S. (1997) Meteoritic silicon carbide grains with unusual Si-isotopic compositions: Evidence for an origin in low-mass metallicity asymptotic giant branch stars. *Astrophys. J. Lett., 487,* L101–L104.

Hoppe P., Strebel R., Eberhardt P., Amari S., and Lewis R. S. (2000) Isotopic properties of silicon carbide X grains from the Murchison meteorite in the size range 0.5–1.5 μm. *Meteoritics & Planet. Sci., 35,* 1157–1176.

Hoppe P., Nittler L. R., Mostefaoui S., Alexander C. M. O'D., and Marhas K. K. (2003) A NanoSIMS study of titanium-isotopic compositions of presolar corundum grains (abstract). In *Lunar and Planetary Science XXXIV,* Abstract #1570. Lunar and Planetary Institute, Houston (CD-ROM).

Howard W. M. and Meyer B. S. (1993) The p-process and Type Ia supernovae. In *Nuclei in the Cosmos 2* (F. Käppeler and K. Wisshak, eds.), pp. 575–580. Institute of Physics, Bristol and Philadelphia.

Howard W. M., Meyer B. S., and Woosley S. . (1991) A new site for the astrophysical gamma-process. *Astrophys. J. Lett., 373,* L5–L8.

Hoyle F. (1946) The synthesis of the elements from hydrogen. *Mon. Not. R. Astron. Soc., 106,* 343–383.

Hoyle F. (1954) On nuclear reactions occuring in very hot stars. I. The synthesis of elements from carbon to nickel. *Astrophys. J. Suppl., 1,* 121–146.

Hughes J. P., Rakowski C. E., Burrows D. N., and Slane P. O. (2000) Nucleosynthesis and mixing in Cassiopeia A. *Astrophys. J. Lett., 528,* L109–L113.

Huss G. R. and Lewis R. S. (1994a) Noble gases in presolar diamonds I: Three distinct components and their implications for diamond origins. *Meteoritics, 29,* 791–810.

Huss G. R. and Lewis R. S. (1994b) Noble gases in presolar diamonds II: Component abundances reflect thermal processing. *Meteoritics, 29,* 811–829.

Huss G. R., Fahey A. J., Gallino R., and Wasserburg G. J. (1994) Oxygen isotopes in circumstellar Al_2O_3 grains from meteorites and stellar nucleosynthesis. *Astrophys. J. Lett., 430,* L81–L84.

Huss G. R., Hutcheon I. D., and Wasserburg G. J. (1997) Isotopic systematics of presolar silicon carbide from the Orgueil (CI) carbonaceous chondrite: Implications for solar system formation and stellar nucleosynthesis. *Geochim. Cosmochim. Acta, 61,* 5117–5148.

Hutcheon I. D., Huss G. R., Fahey A. J., and Wasserburg G. J. (1994) Extreme ^{26}Mg and ^{17}O enrichments in an Orgueil corundum: Identification of a presolar oxide grain. *Astrophys. J. Lett., 425,* L97–L100.

Hutcheon I. D., Weisberg M. K., Phinney D. L., Zolensky M. E., Prinz M., and Ivanov A. V. (1999) Radiogenic ^{53}Cr in Kaidun carbonates: Evidence for very early aqueous activity (abstract). In *Lunar and Planetary Science XXX,* Abstract #1722. Lunar and Planetary Institute, Houston (CD-ROM).

Imbriani G., Limongi M., Gialanella L., Terrasi F., Straniero O., and Chieffi A. (2001) The $^{12}C(\alpha,\gamma)^{16}O$ reaction rate and the evolution of stars in the mass range $0.8 \leq M/M_\odot \leq 25$. *Astrophys. J., 558,* 903–915.

Ireland T. R. (1990) Presolar isotopic and chemical signatures in hibonite-bearing refractory inclusions from the Murchison carbonaceous chondrite. *Geochim. Cosmochim. Acta, 54,* 3219–3237.

Ireland T. R., Zinner E. K., and Amari S. (1991) Isotopically anomalous Ti in presolar SiC from the Murchison meteorite. *Astrophys. J. Lett., 376,* L53–L56.

Ireland T. R., Fahey A. J., Zinner E., and Esat T. M. (1992) Evidence for distillation in the formation of HAL and related hibonite inclusions. *Geochim. Cosmochim. Acta, 56,* 2503–2520.

Jeffery P. M. and Reynolds J. H. (1961) Origin of excess Xe^{129} in stone meteorites. *J. Geophys. Res., 66,* 3582–3583.

José J., Coc A., and Hernanz M. (1999) Nuclear uncertainties in the NeNa-MgAl cycles and production of ^{22}Na and ^{26}Al during nova outbursts. *Astrophys. J., 520,* 347–360.

José J., Hernanz M., Amari S., Lodders K., and Zinner E. (2004) The imprint of nova nucleosynthesis in presolar grains. *Astro-*

phys. J., 612, 414–428.

Jungck M. H. A., Shimamura T., and Lugmair G. W. (1984) Ca isotope variations in Allende. *Geochim. Cosmochim. Acta, 48,* 2651–2658.

Kaiser T. and Wasserburg G. J. (1983) The isotopic composition and concentration of Ag in iron meteorites and the origin of exotic silver. *Geochim. Cosmochim. Acta, 47,* 43–58.

Käppeler F., Beer H., and Wisshak K. (1989) s-Process nucleosynthesis — nuclear physics and the classic model. *Rept. Prog. Phys., 52,* 945–1013.

Karakas A. I. and Lattanzio J. C. (2003) Production of aluminium and the heavy magnesium isotopes in asymptotic giant branch stars. *Publ. Astron. Soc. Australia, 20,* 279–293.

Kashiv Y., Cai Z., Lai B., Sutton S. R., Lewis R. S., Davis A. M., Clayton R. N., and Pellin M. J. (2002) Condensation of trace elements into presolar SiC stardust grains (abstract). In *Lunar and Planetary Science XXXIII,* Abstract #2056. Lunar and Planetary Institute, Houston (CD-ROM).

Kelly W. R. and Wasserburg G. J. (1978) Evidence for the existence of ^{107}Pd in the early solar system. *Geophys. Res. Lett., 5,* 1079–1082.

Kerridge J. F. and Matthews M. S., eds. (1988) *Meteorites and the Early Solar System.* Univ. of Arizona, Tucson. 1269 pp.

Kleine T., Münker C., Mezger K., and Palme H. (2002) Rapid accretion and early core formation on asteroids and the terrestrial planets from Hf-W chronometry. *Nature, 418,* 952–955.

Knödlseder J. (1999) Implications of 1.8 MeV gamma-ray observations for the origin of ^{26}Al. *Astrophys. J., 510,* 915–929.

Koehler P. E., Spencer R. R., Guber K. H., Winters R. R., Raman S., Harvey J. A., Hill N. W., Blackmon J. C., Bardayan D. W., Larson D. C., Lewis T. A., Pierce D. E., and Smith M. S. (1998) High resolution neutron capture and transmission measurement on ^{137}Ba and their impact on the interpretation of meteoritic barium anomalies. *Phys. Rev. C, 57,* R1558–R1561.

Kovetz A. and Prialnik D. (1997) The composition of nova ejecta from multicycle evolution models. *Astrophys. J., 477,* 356–367.

Krestina N., Hsu W., and Wasserburg G. J. (2002) Circumstellar oxide grains in ordinary chondrites and their origin (abstract). In *Lunar and Planetary Science XXXIII,* Abstract #1425. Lunar and Planetary Institute, Houston (CD-ROM).

Kunz R., Fey M., Jaeger M., Mayer A., Hammer J. W., Staudt G., Harissopulos S., and Paradellis T. (2002) Astrophysical reaction rate of ^{12}C(a, γ)^{16}O. *Astrophys. J., 2002,* 643–650.

Lambert D. L. (1991) The abundance connection — the view from the trenches. In *Evolution of Stars: The Photospheric Abundance Connection* (G. Michaud and A. Tutukov, eds.), pp. 451–460. Kluwer, Dordrecht.

Lambert D. L., Gustafsson B., Eriksson K., and Hinkle K. H. (1986) The chemical composition of carbon stars. I. Carbon, nitrogen, and oxygen in 30 cool carbon stars in the galactic disk. *Astrophys. J. Suppl., 62,* 373–425.

Langer N., Heger A., Wellstein S., and Herwig F. (1999) Mixing and nucleosynthesis in rotating TP-AGB stars. *Astron. Astrophys., 346,* L37–L40.

Larimer J. W. and Bartholomay M. (1979) The role of carbon and oxygen in cosmic gases: Some applications to the chemistry and mineralogy of enstatite chondrites. *Geochim. Cosmochim. Acta, 43,* 1455–1466.

Lattanzio J. C., Frost C. A., Cannon R. C., and Wood P. R. (1997) Hot bottom burning nucleosynthesis in 6 $M_\odot$ stellar models. *Nucl. Phys., A621,* 435c–438c.

Lee D.-C. and Halliday A. N. (1995) Hafnium-tungsten chronometry and the timing of terrestrial core formation. *Nature, 378,* 771–774.

Lee T. (1979) New isotopic clues to solar system formation. *Rev. Geophys. Space Phys., 17,* 1591–1611.

Lee T. (1988) Implications of isotopic anomalies for nucleosynthesis. In *Meteorites and the Early Solar System* (J. F. Kerridge and M. S. Matthews), pp. 1063–1089. Univ. of Arizona, Tucson.

Lee T., Panastassiou D. A., and Wasserburg G. J. (1976) Demonstration of ^{26}Mg excess in Allende and evidence for ^{26}Al. *Geophys. Res. Lett., 3,* 109–112.

Lee T., Papanastassiou D. A., and Wasserburg G. J. (1977) Aluminum-26 in the early solar system: Fossil or fuel? *Astrophys. J. Lett., 211,* L107–L110.

Lee T., Papanastassiou D. A., and Wasserburg G. J. (1978) Calcium isotopic anomalies in the Allende Meteorite. *Astrophys. J. Lett., 220,* L21–L25.

Lewis R. S., Tang M., Wacker J. F., Anders E., and Steel E. (1987) Interstellar diamonds in meteorites. *Nature, 326,* 160–162.

Lewis R. S., Amari S., and Anders E. (1990) Meteoritic silicon carbide: Pristine material from carbon stars. *Nature, 348,* 293–298.

Lewis R. S., Huss G. R., and Lugmair G. (1991) Finally, Ba & Sr accompanying Xe-HI in diamonds from Allende (abstract). In *Lunar and Planetary Science XXII,* pp. 807–808. Lunar and Planetary Institute, Houston.

Lewis R. S., Amari S., and Anders E. (1994) Interstellar grains in meteorites: II. SiC and its noble gases. *Geochim. Cosmochim. Acta, 58,* 471–494.

Leya I., Halliday A. N., and Wieler R. (2003) The predictable collateral consequences of nucleosynthesis by spallation reactions in the early solar system. *Astrophys. J., 594,* 605–616.

Lin Y., Amari S., and Pravdivtseva O. (2002) Presolar grains from the Qingzhen (EH3) meteorite. *Astrophys. J., 575,* 257–263.

Lin Y., Guan Y., Leshin L. A., Ouyang Z., and Wang D. (2004) Evidence for live ^{36}Cl in Ca-Al-rich inclusions from the Ningqiang carbonaceous chondrite (abstract). In *Lunar and Planetary Science XXXV,* Abstract #2084. Lunar and Planetary Institute, Houston (CD-ROM).

Lin Y., Guan Y., Leshin L. A., Ouyang Z., and Wang D. (2005) Short-lived chlorine-36 in Ca- and Al-rich inclusion from the Ningqiang carbonaceous chondrite. *Proc. Natl. Acad. Sci., 102,* 1306–1311.

Liu M.-C., Iizuka Y., McKeegan K. D., Tonui E. K., and Young E. D. (2005) Supracanonical ^{26}Al/^{27}Al ratios in an unaltered Allende CAI (abstract). In *Lunar and Planetary Science XXXVI,* Abstract #2079. Lunar and Planetary Institute, Houston (CD-ROM).

Lodders K. (2003) Solar system abundances and condensation temperatures of the elements. *Astrophys. J., 591,* 1220–1247.

Lodders K. and Fegley B. Jr. (1997) Condensation chemistry of carbon stars. In *Astrophysical Implications of the Laboratory Study of Presolar Materials* (T. J. Bernatowicz and E. Zinner, eds.), pp. 391–423. American Institute of Physics, New York.

Loss R. D., Lugmair G. W., Davis A. M., and MacPherson G. J. (1994) Isotopically distinct reservoirs in the solar nebula: Isotope anomalies in vigarano meteorite inclusions. *Astrophys. J. Lett., 436,* L193–L196.

Luck J. M., Ben Othman D., Barrat J. A., and Albarède F. (2003) Coupled ^{63}Cu and ^{16}O excesses in chondrites. *Geochim. Cosmochim. Acta, 67,* 143–151.

Lugaro M., Zinner E., Gallino R., and Amari S. (1999) Si isotopic ratios in mainstream presolar SiC grains revisited. *Astrophys. J., 527,* 369–394.

Lugaro M., Davis A. M., Gallino R., Pellin M. J., Straniero O., and Käppeler F. (2003) Isotopic compositions of strontium, zirconium, molybdenum, and barium in single presolar SiC grains and asymptotic giant branch stars. *Astrophys. J., 593,* 486–508.

Lugmair G. W. and Galer S. J. G. (1992) Age and isotopic relationships among the angrites Lewis Cliff 86010 and Angra dos Reis. *Geochim. Cosmochim. Acta, 56,* 1673–1694.

Lugmair G. W. and Marti K. (1977) Sm-Nd-Pu timepieces in the Angra dos Reis meteorite. *Earth Planet. Sci. Lett., 35,* 273–284.

Lugmair G. W. and Shukolyukov A. (1998) Early solar system timescales according to ^{53}Mn-^{53}Cr systematics. *Geochim. Cosmochim. Acta, 62,* 2863–2886.

Lugmair G. W. and Shukolyukov A. (2001) Early solar system events and timescales. *Meteoritics & Planet. Sci., 36,* 1017–1026.

MacPherson G. J., Davis A. M., and Zinner E. K. (1995) The distribution of aluminum-26 in the early solar system — A reappraisal. *Meteoritics, 30,* 365–386.

MacPherson G. J., Huss G. R., and Davis A. M. (2003) Extinct ^{10}Be in type A calcium-aluminum-rich inclusions from CV chondrites. *Geochim. Cosmochim. Acta, 67,* 3165–3179.

Mahoney W. A., Ling J. C., Wheaton W. A., and Jacobsen A. S. (1984) HEAO 3 discovery of ^{26}Al in the interstellar medium. *Astrophys. J., 286,* 578–585.

Marhas K. K., Goswami J. N., and Davis A. M. (2002) Short-lived nuclides in hibonite grains from Murchison: Evidence for solar system evolution. *Science, 298,* 2182–2185.

Marhas K. K., Hoppe P., and Ott U. (2003) A NanoSIMS study of C-, Si-, and Ba-isotopic compositions of presolar silicon carbide grains from the Murchison meteorite (abstract). *Meteoritics & Planet. Sci., 38,* A58.

Marhas K. K., Hoppe P., and Besmehn A. (2004) A NanoSIMS study of iron-isotopic compositions in presolar silicon carbide grains (abstract). In *Lunar and Planetary Science XXXV,* Abstract #1834. Lunar and Planetary Institute, Houston (CD-ROM).

McCulloch M. T. and Wasserburg G. J. (1978a) Barium and neodymium isotopic anomalies in the Allende Meteorite. *Astrophys. J. Lett., 220,* L15–L19.

McCulloch M. T. and Wasserburg G. J. (1978b) More anomalies from the Allende meteorite: Samarium. *Geophys. Res. Lett., 5,* 599–602.

McKeegan K. D., Chaussidon M., and Robert F. (2000) Incorporation of short-lived ^{10}Be in a calcium-aluminum-rich inclusion from the Allende meteorite. *Science, 289,* 1334–1337.

Merrill P. W. (1952) Spectroscopic observations of stars of class S. *Astrophys. J., 116,* 21–26.

Messenger S. (2000) Identification of molecular-cloud material in interplanetary dust particles. *Nature, 404,* 968–971.

Messenger S. and Keller L. P. (2004) A supernova silicate from a cluster IDP (abstract). *Meteoritics & Planet. Sci., 39,* A68.

Messenger S. and Walker R. (1997) Evidence for molecular cloud material in meteorites and interplanetary dust. In *Astrophysical Implications of the Laboratory Study of Presolar Materials* (T. Bernatowicz and E. Zinner, eds.), pp. 545–566. American Institute of Physics, New York.

Messenger S., Keller L. P., Stadermann F. J., Walker R. M., and Zinner E. (2003) Samples of stars beyond the solar system: Silicate grains in interplanetary dust. *Science, 300,* 105–108.

Meyer B. S. (1995) Neutrino reactions on ^{4}He and the r-process. *Astrophys. J. Lett., 449,* L55–L58.

Meyer B. S. (2003) Neutrinos, supernovae, molybdenum, and extinct ^{92}Nb. *Nucl. Phys. A, 719,* C13–C20.

Meyer B. S. and Brown J. S. (1997) Survey of r-process models. *Astrophys. J., 112,* 199–220.

Meyer B. S. and Clayton D. D. (2000) Short-lived radioactivities and the birth of the Sun. *Space Sci. Rev., 92,* 133–152.

Meyer B. S., Mathews G. J., Howard W. M., Woosley S. E., and Hoffman R. (1992) The r-process in supernovae. *Astrophys. J., 399,* 656–664.

Meyer B. S., Weaver T. A., and Woosley S. E. (1995) Isotope source table for a 25 $M_{\odot}$ supernova. *Meteoritics, 30,* 325–334.

Meyer B. S., Krishnan T. D., and Clayton D. D. (1996) ^{48}Ca production in matter expanding from high temperature and density. *Astrophys. J., 462,* 825–838.

Meyer B. S., Krishnan T. D., and Clayton D. D. (1998a) Theory of quasi-equilibrium nucleosynthesis and applications to matter expanding from high temperature and density. *Astrophys. J., 498,* 808–830.

Meyer B. S., McLaughlin G. C., and Fuller G. M. (1998b) Neutrino-capture and r-process nucleosynthesis. *Phys. Rev. C, 58,* 3696–3710.

Meyer B. S., Clayton D. D., and The L.-S. (2000) Molybdenum and zirconium isotopes from a supernova neutron burst. *Astrophys. J. Lett., 540,* L49–L52.

Meyer B. S., Clayton D. D., The L.-S., and El Eid M. F. (2003) Injection of ^{182}Hf into the early solar nebula (abstract). In *Lunar and Planetary Science XXXIV,* Abstract #2074. Lunar and Planetary Institute, Houston (CD-ROM).

Meyer B. S., The L.-S., and Clayton D. D. (2004) Helium-shell nucleosynthesis and extinct radioactivities (abstract). In *Lunar and Planetary Science XXXV,* Abstract #1908. Lunar and Planetary Institute, Houston (CD-ROM).

Molini-Velsko C., Mayeda T. K., and Clayton R. N. (1986) Isotopic composition of silicon in meteorites. *Geochim. Cosmochim. Acta, 50,* 2719–2726.

Mostefaoui S. and Hoppe P. (2004) Discovery of abundant in situ silicate and spinel grains from red giant stars in a primitive meteorite. *Astrophys. J. Lett., 613,* L149–L152.

Mostefaoui S., Lugmair G. W., Hoppe P., and El Goresy A. (2003) Evidence for live iron-60 in Semarkona and Chervony Kut: A NanoSIMS study (abstract). In *Lunar and Planetary Science XXXI,V* Abstract #1585. Lunar and Planetary Institute, Houston (CD-ROM).

Mostefaoui S., Lugmair G. W., and Hoppe P. (2004a) In-situ evidence for live iron-60 in the solar system: A potential heat source for planetary differentiation from a nearby supernova explosion (abstract). In *Lunar and Planetary Science XXXV,* Abstract #1271. Lunar and Planetary Institute, Houston (CD-ROM).

Mostefaoui S., Marhas K. K., and Hoppe P. (2004b) Discovery of an in-situ presolar silicate grain with GEMS-like composition in the Bishunpur matrix (abstract). In *Lunar and Planetary Science XXXV,* Abstract #1593. Lunar and Planetary Institute, Houston (CD-ROM).

Mostefaoui S., Lugmair G. W., and Hoppe P. (2005) Iron-60: A heat source for planetary differentiation from a nearby supernova explosion. *Astrophys. J., 625,* 271–277.

Mowlavi N. and Meynet G. (2000) Aluminum 26 production in asymptotic giant branch stars. *Astron. Astrophys., 361,* 959–976.

Münker C., Weyer S., Mezger K., Rehkämper M., Wombacher F., and Bischoff A. (2000) ^{92}Nb-^{92}Zr and the early differentiation history of planetary bodies. *Science, 289,* 1538–1542.

Murty S. V. S. and Marti K. (1994) Nitrogen isotopic signatures in Cape York: Implications for formation of Group IIIA irons. *Geochim. Cosmochim. Acta, 58,* 1841–1848.

Murty S. V. S., Goswami J. N., and Shukolyukov Y. A. (1997) Excess ^{36}Ar in the Efremovka meteorite: A strong hint for the presence of ^{36}Cl in the early solar system. *Astrophys. J. Lett., 475,* L65–L68.

Nagashima K., Krot A. N., and Yurimoto H. (2004) Stardust silicates from primitive meteorites. *Nature, 428,* 921–924.

Nguyen A. N. and Zinner E. (2004) Discovery of ancient silicate stardust in a meteorite. *Science, 303,* 1496–1499.

Nguyen A., Zinner E., and Lewis R. S. (2003) Identification of small presolar spinel and corundum grains by isotopic raster imaging. *Publ. Astron. Soc. Australia, 20,* 382–388.

Nichols R. H. Jr. (2006) Chronological constraints on planetesimal accretion. In *Meteorites and the Early Solar System II* (D. S. Lauretta and H. Y. McSween Jr., eds.), this volume. Univ. of Arizona, Tucson.

Nichols R. H. Jr., Kehm K., and Hohenberg C. M. (1995) Microanalytic laser extraction of noble gases: Techniques and applications. In *Advances in Analytic Geochemistry, Vol. 2* (M. Hyman and M. Rowe, eds.), pp. 119–140. JAI, Greenwich, Connecticut.

Nichols R. H. Jr., Podosek F. A., Meyer B. S., and Jennings C. L. (1999) Collateral consequences of the inhomogeneous distribution of short-lived radionuclides in the solar nebula. *Meteoritics & Planet. Sci., 34,* 869–884.

Nichols R. H. Jr., Brannon J. C., and Podosek F. A. (2002) Excess 135-barium from live 135-cesium in Orgueil chemical separates (abstract). In *Lunar and Planetary Science XXXIII,* Abstract #1929. Lunar and Planetary Institute, Houston (CD-ROM).

Nicolussi G. K., Davis A. M., Pellin M. J., Lewis R. S., Clayton R. N., and Amari S. (1997) s-process zirconium in presolar silicon carbide grains. *Science, 277,* 1281–1283.

Nicolussi G. K., Pellin M. J., Lewis R. S., Davis A. M., Amari S., and Clayton R. N. (1998a) Molybdenum isotopic composition of individual presolar silicon carbide grains from the Murchison meteorite. *Geochim. Cosmochim. Acta, 62,* 1093–1104.

Nicolussi G. K., Pellin M. J., Lewis R. S., Davis A. M., Clayton R. N., and Amari S. (1998b) Strontium isotopic composition in individual circumstellar silicon carbide grains: A record of s-process nucleosynthesis. *Phys. Rev. Lett., 81,* 3583–3586.

Nicolussi G. K., Pellin M. J., Lewis R. S., Davis A. M., Clayton R. N., and Amari S. (1998c) Zirconium and molybdenum in individual circumstellar graphite grains: New isotopic data on the nucleosynthesis of heavy elements. *Astrophys. J., 504,* 492–499.

Niederer F. R. and Papanastassiou D. A. (1984) Ca isotopes in refractory inclusions. *Geochim. Cosmochim. Acta, 48,* 1279–1293.

Niederer F. R., Papanastassiou D. A., and Wasserburg G. J. (1980) Endemic isotopic anomalies in titanium. *Astrophys. J. Lett., 240,* L73–L77.

Niederer F. R., Papanastassiou D. A., and Wasserburg G. J. (1981) The isotopic composition of Ti in the Allende and Leoville meteorites. *Geochim. Cosmochim. Acta, 45,* 1017–1031.

Niederer F. R., Papanastassiou D. A., and Wasserburg G. J. (1985) Absolute isotopic abundances of Ti in meteorites. *Geochim. Cosmochim. Acta., 49,* 835–851.

Nielsen S. G., Rehkämper M., and Halliday A. N. (2004) First evidence of live ^{205}Pb in the early solar system. *Geochim. Cosmochim. Acta, 68,* A727.

Niemeyer S. and Lugmair G. W. (1981) Ubiquitous isotopic anomalies in Ti from normal Allende inclusions. *Earth Planet. Sci. Lett., 53,* 211–225.

Niemeyer S. and Lugmair G. W. (1984) Titanium isotopic anomalies in meteorites. *Geochim. Cosmochim. Acta, 48,* 1401–1416.

Nittler L. R. (2003) Presolar stardust in meteorites: Recent advances and scientific frontiers. *Earth Planet. Sci. Lett., 209,* 259–273.

Nittler L. R. (2005) Constraints on heterogeneous galactic chemical evolution from meteoritic stardust. *Astrophys. J., 618,* 281–296.

Nittler L. R. and Alexander C. M. O'D. (1999) Automatic identification of presolar Al- and Ti-rich oxide grains from ordinary chondrites (abstract). In *Lunar and Planetary Science XXX,* Abstract #2041. Lunar and Planetary Institute, Houston (CD-ROM).

Nittler L. R. and Alexander C. M. O'D. (2003) Automated isotopic measurements of micron-sized dust: Application to meteoritic presolar silicon carbide. *Geochim. Cosmochim. Acta, 67,* 4961–4980.

Nittler L. R. and Dauphas N. (2006) Meteorites and the chemical evolution of the Milky Way. In *Meteorites and the Early Solar System II* (D. S. Lauretta and H. Y. McSween Jr., eds.), this volume. Univ. of Arizona, Tucson.

Nittler L. R. and Hoppe P. (2004a) High initial ^{26}Al/^{27}Al ratios in presolar SiC grains from novae (abstract). *Meteoritics & Planet. Sci., 39,* A78.

Nittler L. R. and Hoppe P. (2004b) New presolar silicon carbide grains with nova isotope signatures (abstract). In *Lunar and Planetary Science XXXV,* Abstract #1598. Lunar and Planetary Institute, Houston (CD-ROM).

Nittler L. R., Alexander C. M. O'D., Gao X., Walker R. M., and Zinner E. K. (1994) Interstellar oxide grains from the Tieschitz ordinary chondrite. *Nature, 370,* 443–446.

Nittler L. R., Hoppe P., Alexander C. M. O'D., Amari S., Eberhardt P., Gao X., Lewis R. S., Strebel R., Walker R. M., and Zinner E. (1995) Silicon nitride from supernovae. *Astrophys. J. Lett., 453,* L25–L28.

Nittler L. R., Amari S., Zinner E., Woosley S. E., and Lewis R. S. (1996) Extinct ^{44}Ti in presolar graphite and SiC: Proof of a supernova origin. *Astrophys. J. Lett., 462,* L31–L34.

Nittler L. R., Alexander C. M. O'D., Gao X., Walker R. M., and Zinner E. (1997) Stellar sapphires: The properties and origins of presolar Al_2O_3 in meteorites. *Astrophys. J., 483,* 475–495.

Nittler L. R., Alexander C. M. O'D., Wang J., and Gao X. (1998) Meteoritic oxide grain from supernova found. *Nature, 393,* 222.

Nittler L. R., Hoppe P., Alexander C. M. O'D., Busso M., Gallino R., Marhas K. K., and Nollett K. (2003) Magnesium isotopes in presolar spinel (abstract). In *Lunar and Planetary Science XXXIV,* Abstract #1703. Lunar and Planetary Institute, Houston (CD-ROM).

Nittler L. R., Alexander C. M. O'D., Stadermann F. J., and Zinner E. K. (2005) Presolar Al-, Ca-, and Ti-rich oxide grains in the Krymka meteorite (abstract). In *Lunar and Planetary Science XXXVI,* Abstract #2200. Lunar and Planetary Institute, Houston (CD-ROM).

Nollett K. M., Busso M., and Wasserburg G. J. (2003) Cool bottom processes on the thermally pulsing asymptotic giant branch and the isotopic composition of circumstellar dust grains.

Astrophys. J., 582, 1036–1058.

Nuth J. A. III, Charnley S. B., and Johnson N. M. (2006) Chemical processes in the interstellar medium: Source of the gas and dust in the primitive solar nebula. In *Meteorites and the Early Solar System II* (D. S. Lauretta and H. Y. McSween Jr., eds.), this volume. Univ. of Arizona, Tucson.

Nyquist L. E., Bansal B., Wiesmann H., and Shih C.-Y. (1994) Neodymium, strontium and chromium isotopic studies of the LEW86010 and Angra dos Reis meteorites and the chronology of the angrite parent body. *Meteoritics, 29,* 872–885.

Nyquist L., Lindstrom D., Mittlefehldt D., Shih C.-Y., Wiesmann H., Wentworth S., and Martinez R. (2001) Manganese-chromium formation intervals for chondrules from the Bishunpur and Chainpur meteorites. *Meteoritics & Planet. Sci., 36,* 911–938.

Ott U. (1993) Interstellar grains in meteorites. *Nature, 364,* 25–33.

Ott U. (1996) Interstellar diamond xenon and timescales of supernova ejecta. *Astrophys. J., 463,* 344–348.

Ott U. and Begemann F. (1990) Discovery of s-process barium in the Murchison meteorite. *Astrophys. J. Lett., 353,* L57–L60.

Ott U. and Begemann F. (2000) Spallation recoil and age of presolar grains in meteorites. *Meteoritics & Planet. Sci., 35,* 53–63.

Owen T., Mahaffy P. R., Niemann H. B., Atreya S., and Wong M. (2001) Protosolar nitrogen. *Astrophys. J. Lett., 553,* L77–L79.

Palme H. (2000) Are there chemical gradients in the inner solar system? In *From Dust to Terrestrial Planets 9* (W. Benz et al., eds.), pp. 237–262. Kluwer, Dordrecht.

Papanastassiou D. A. (1986) Cr isotopic anomalies in the Allende meteorite. *Astrophys. J. Lett., 308,* L27–L30.

Papanastassiou D. A. and Brigham C. A. (1989) The identification of meteorite inclusions with isotope anomalies. *Astrophys. J. Lett., 338,* L37–L40.

Papanastassiou D. A. and Wasserburg G. J. (1978) Strontium isotopic anomalies in the Allenda meteorite. *Geophys. Res. Lett., 5,* 595–598.

Papanastassiou D. A., Chen J. H., and Wasserburg G. J. (2004) More on Ru endemic isotope anomalies in meteorites (abstract). In *Lunar and Planetary Science XXXV,* Abstract #1828. Lunar and Planetary Institute, Houston (CD-ROM).

Pellin M. J., Davis A. M., Lewis R. S., Amari S., and Clayton R. N. (1999) Molybdenum isotopic composition of single silicon carbide grains from supernovae (abstract). In *Lunar and Planetary Science XXX*, Abstract #1969. Lunar and Planetary Institute, Houston (CD-ROM).

Pellin M. J., Calaway W. F., Davis A. M., Lewis R. S., Amari S., and Clayton R. N. (2000a) Toward complete isotopic analysis of individual presolar silicon carbide grains: C, N, Si, Sr, Zr, Mo, and Ba in single grains of type X (abstract). In *Lunar and Planetary Science XXXI,* Abstract #1917. Lunar and Planetary Institute, Houston (CD-ROM).

Pellin M. J., Davis A. M., Calaway W. F., Lewis R. S., Clayton R. N., and Amari S. (2000b) Zr and Mo isotopic constraints on the origin of unusual types of presolar SiC grains (abstract). In *Lunar and Planetary Science XXXI,* Abstract #1934. Lunar and Planetary Institute, Houston (CD-ROM).

Perlmutter S., Aldering G., Goldhaber G., Knop R. A., Nugent P., Castro P. G., Deustua S., Fabbro S., Goobar A., Groom D. E., Hook I. M., Kim A. G., Kim M. Y., Lee J. C., Nunes N. J., Pain R., Pennypacker C. R., Quimby R., Lidman C., Ellis R. S., Irwin M., McMahon R. G., Ruiz-Lapuente P., Walton N., Schaefer B., Boyle B. J., Filippenko A. V., Matheson T., Fruchter A. S., Panagia N., Newbert H. J. M., and Couch W. J. (1999) Measurements of Ω and Λ from 42 high-redshift supernovae. *Astrophys. J., 517,* 565–586.

Phillips M. (2003) Stellar candles for the extragalactic distance scale. In *Lecture Notes in Physics, Vol. 635* (D. Alloin and W. Gieren, eds.), pp. 175–185. Springer-Verlag, Berlin.

Plewa T., Calder A. C., and Lamb D. Q. (2004) Type Ia supernova explosion: Gravitationally confined detonation. *Astrophys. J. Lett., 612,* L37–L40.

Podosek F. and Nichols R. (1997) Short-lived radionuclides in the solar nebula. In *Astrophysical Implications of the Laboratory Study of Presolar Materials* (T. Bernatowicz and E. Zinner, eds.), pp. 617–648. American Institute of Physics, New York.

Podosek F. A., Ott U., Brannon J. C., Neal C. R., Bernatowicz T. J., Swan P., and Mahan S. E. (1997) Thoroughly anomalous chromium in Orgueil. *Meteoritics & Planet. Sci., 32,* 617–627.

Podosek F. A., Nichols R. H. Jr., Brannon J. C., Meyer B. S., Ott U., Jennings C. L., and Luo N. (1999) Potassium, stardust, and the last supernova. *Geochim. Cosmochim. Acta, 63,* 2351–2362.

Podosek F. A., Prombo C. A., Amari S., and Lewis R. S. (2004) s-process Sr isotopic compositions in presolar SiC from the Murchsion meteorite. *Astrophys. J., 605,* 960–965.

Preibisch T. (1999) The history of low-mass star formation in the upper Scorpius OB association. *Astronom. J., 117,* 2381–2397.

Prombo C. A., Podosek F. A., Amari S., and Lewis R. S. (1993) s-process Ba isotopic compositions in presolar SiC from the Murchison meteorite. *Astrophys. J., 410,* 393–399.

Qian Y.-Z. and Wasserburg G. J. (2000) Stellar abundances in the early galaxy and two r-process components. *Phys. Rept., 333-334,* 77–108.

Qian Y.-Z. and Wasserburg G. J. (2003) Stellar sources for heavy r-process nuclei. *Astrophys. J., 588,* 1099–1109.

Qian Y.-Z. and Woosley S. E. (1996) Nucleosynthesis in neutrino-driven winds from protoneutron stars II. The r-process. *Astrophys. J., 471,* 331–351.

Qian Y.-Z., Vogel P., and Wasserburg G. J. (1998) Diverse supernova sources for the r-process. *Astrophys. J., 494,* 285–296.

Rauscher T., Heger A., Hoffman R. D., and Woosley S. E. (2002) Nucleosynthesis in massive stars with improved nuclear and stellar physics. *Astrophys. J., 576,* 323–348.

Rayet M., Arnould M., Hashimoto M., Prantzos N., and Nomoto K. (1995) The p-process in Type II supernovae. *Astron. Astrophys., 298,* 517–527.

Reynolds J. H. (1960a) Determination of the age of the elements. *Phys. Rev. Lett., 4,* 8–10.

Reynolds J. H. (1960b) Isotopic composition of primordial xenon. *Phys. Rev. Lett., 4,* 351–354.

Richter S., Ott U., and Begemann F. (1993) s-process isotope abundance anomalies in meteoritic silicon carbide: New data. In *Nuclei in the Cosmos 2* (F. Käppeler and K. Wisshak, eds.), pp. 127–132. Institute of Physics, Bristol and Philadelphia.

Richter S., Ott U., and Begemann F. (1994) s-process isotope abundance anomalies in meteoritic silicon carbide: Data for Dy. In *Proceedings of the European Workshop on Heavy Element Nucleosynthesis* (E. Somorjai and Z. Fülöp, eds.), pp. 44–46. Hungarian Academy of Sciences, Debrecen.

Richter S., Ott U., and Begemann F. (1997) Tellurium-H in interstellar diamonds (abstract). In *Lunar and Planetary Science XXVIII,* pp. 1163–1164. Lunar and Planetary Institute, Houston.

Richter S., Ott U., and Begemann F. (1998) Tellurium in pre-solar diamonds as an indicator for rapid separation of supernova ejecta. *Nature, 391,* 261–263.

Riess A. G., Flippenko A. V., Challis P., Clocchiatti A., Diercks A., Garnavich P. M., Gilliland R. L., Hogan C. J., Jha S., Kirshner R. P., Leibundgut B., Phillips M. M., Reiss D., Schmidt B. P., Schommer R. A., Smith R. C., Spyromilio J., Stubbs C., Suntzeff N. B., and Tonry J. (1998) Observational evidence from supernovae for an accelerating universe and a cosmological constant. *Astron. J., 116,* 1009–1038.

Robert F. (2006) The solar system deuterium/hydrogen ratio. In *Meteorites and the Early Solar System II* (D. S. Lauretta and H. Y. McSween Jr., eds.), this volume. Univ. of Arizona, Tucson.

Rotaru M., Birck J. L., and Allègre C. J. (1992) Clues to early solar system history from chromium isotopes in carbonaceous chondrites. *Nature, 358,* 465–470.

Rowe M. W. and Kuroda P. K. (1965) Fissiogenic xenon from the Pasamonte meteorite. *J. Geophys. Res., 70,* 709–714.

Russell S. S., Arden J. W., and Pillinger C. T. (1991) Evidence for multiple sources of diamond from primitive chondrites. *Science, 254,* 1188–1191.

Russell S. S., Arden J. W., and Pillinger C. T. (1996) A carbon and nitrogen isotope study of diamond from primitive chondrites. *Meteoritics & Planet. Sci., 31,* 343–355.

Russell S. S., Ott U., Alexander C. M. O'D., Zinner E. K., Arden J. W., and Pillinger C. T. (1997) Presolar silicon carbide from the Indarch (EH4) meteorite: Comparison with silicon carbide populations from other meteorite classes. *Meteoritics & Planet. Sci., 32,* 719–732.

Russell S. S., Zhu X., Guo Y., Belshaw N., Gounelle M., Mullane E., and Coles B. (2003) Copper isotope systematics in CR, CH-like, and CB meteorites: A preliminary study (abstract). *Meteoritics & Planet. Sci., 38,* A124.

Russell S. S., Hartmann L., Cuzzi J., Krot A. N., Gounelle M., and Weidenschilling S. (2006) Timescales of the solar protoplanetary disk. In *Meteorites and the Early Solar System II* (D. S. Lauretta and H. Y. McSween Jr., eds.), this volume. Univ. of Arizona, Tucson.

Sahijpal S., Goswami J. N., and Davis A. M. (2000) K, Mg, Ti, and Ca isotopic compositions and refractory trace element abundances in hibonites from CM and CV meteorites: Implications for early solar system processes. *Geochim. Cosmochim. Acta., 64,* 1989–2005.

Sanloup C., Blichert-Toft J., Télouk P., Gillet P., and Albarède F. (2000) Zr isotope anomalies in chondrites and the presence of ^{92}Nb in the early solar system. *Earth Planet. Sci. Lett., 184,* 75–81.

Savina M. R., Davis A. M., Tripa C. E., Pellin M. J., Clayton R. N., Lewis R. S., Amari S., Gallino R., and Lugaro M. (2003a) Barium isotopes in individual presolar silicon carbide grains from the Murchison meteorite. *Geochim. Cosmochim. Acta, 67,* 3201–3214.

Savina M. R., Pellin M. J., Tripa C. E., Veryovkin I. V., Calaway W. F., and Davis A. M. (2003b) Analyzing individual presolar grains with CHARISMA. *Geochim. Cosmochim. Acta, 67,* 3215–3225.

Savina M. R., Tripa C. E., Pellin M. J., Davis A. M., Clayton R. N., Lewis R. S., and Amari S. (2003c) Isotopic composition of molybdenum and barium in single presolar silicon carbide grains of type A + B (abstract). In *Lunar and Planetary Science XXXIV,* Abstract #2079. Lunar and Planetary Institute, Houston (CD-ROM).

Savina M. R., Davis A. M., Tripa C. E., Pellin M. J., Gallino R., Lewis R. S., and Amari S. (2004a) Extinct technetium in presolar silicon carbide grains. *Science, 303,* 649–652.

Savina M. R., Pellin M. J., Tripa C. E., Davis A. M., Lewis R. S., and Amari S. (2004b) Excess p-process molybdenum and ruthenium in a presolar SiC grain. *Nuclei in the Cosmos VIII,* Abstract #C160. TRIUMF, Vancouver.

Schönbächler M., Rehkämper M., Halliday A. N., Lee D.-C., Bourot-Denise M., Zanda B., Hattendorf B., and Günther D. (2002) Niobium-zirconium chronometry and early solar system development. *Science, 295,* 1705–1708.

Schönbächler M., Lee D.-C., Rehkämper M., Halliday A. N., Fehr M. A., Hattendorf B., and Günther D. (2003) Zirconium isotope evidence for incomplete admixing of r-process components in the solar nebula. *Earth Planet. Sci. Lett., 216,* 467–481.

Schramm D. N., Tera F., and Wasserburg G. J. (1970) The isotopic abundance of ^{26}Mg and limits on ^{26}Al in the early solar system. *Earth Planet. Sci. Lett., 10,* 44–59.

Shu F. H., Shang H., and Lee T. (1996) Toward an astrophysical theory of chondrites. *Science, 271,* 1545–1552.

Shu F. H., Shang H., Gounelle M., Glassgold A. E., and Lee T. (2001) The origin of chondrules and refractory inclusions in chondritic meteorites. *Astrophys. J., 548,* 1029–1050.

Shukolyukov A. and Lugmair G. W. (1993) Live iron-60 in the early solar system. *Science, 259,* 1138–1142.

Smith V. V. and Lambert D. L. (1990) The chemical composition of red giants. III. Further CNO isotopic and s-process abundances in thermally pulsing asymptotic giant branch stars. *Astrophys. J. Suppl. Ser., 72,* 387–416.

Sneden C., McWilliam A., Preston G. W., Cowan J. J., Burris D. L., and Armosky B. J. (1996) The ultra — metal-poor, neutron-capture — rich giant star CS 22892-052. *Astrophys. J., 467,* 819–840.

Sneden C., Cowan J. J., Lawler J. E., Ivans I. I., Burles S., Beers T. C., Primas F., Hill V., Truran J. W., Fuller G. M., Pfeiffer B., and Kratz K. (2003) The extremely metal-poor, neutron capture-rich star CS 22892-052: A comprehensive abundance analysis. *Astrophys. J., 591,* 936–953.

Speck A. K., Barlow M. J., and Skinner C. J. (1997) The nature of silicon carbide in carbon star outflows. *Mon. Not. R. Astron. Soc., 234,* 79–84.

Srinivasan B. and Anders E. (1978) Noble gases in the Murchison meteorite: Possible relics of s-process nucleosynthesis. *Science, 201,* 51–56.

Srinivasan G., Ulyanov A. A., and Goswami J. N. (1994) ^{41}Ca in the early solar system. *Astrophys. J. Lett., 431,* L67–L70.

Srinivasan G., Sahijpal S., Ulyanov A. A., and Goswami J. N. (1996) Ion microprobe studies of Efremovka CAIs: II. Potassium isotope composition and ^{41}Ca in the early solar system. *Geochim. Cosmochim. Acta, 60,* 1823–1835.

Starrfield S., Truran J. W., Wiescher M. C., and Sparks W. M. (1998) Evolutionary sequences for Nova V1974 Cygni using new nuclear reaction rates and opacities. *Mon. Not. R. Astron. Soc., 296,* 502–522.

Stone J., Hutcheon I. D., Epstein S., and Wasserburg G. J. (1991) Correlated Si isotope anomalies and large ^{13}C enrichments in a family of exotic SiC grains. *Earth Planet. Sci. Lett., 107,* 570–581.

Straniero O., Gallino R., Busso M., Chiefei A., Raiteri C. M., Limongi M., and Salaris M. (1995) Radiative ^{13}C burning in asymptotic giant branch stars and s-processing. *Astrophys. J. Lett., 440,* L85–L87.

Strebel R., Hoppe P., and Eberhardt P. (1996) A circumstellar Al-

and Mg-rich oxide grain from the Orgueil meteorite (abstract). *Meteoritics, 31,* A136.

Stroud R. M., Nittler L. R., and Alexander C. M. O'D. (2004) Polymorphism in presolar Al_2O_3 grains from asymptotic giant branch stars. *Science, 305,* 1455–1457.

Suess H. E. and Urey H. C. (1956) Abundances of the elements. *Rev. Mod. Phys., 28,* 53–74.

Tachibana S. and Huss G. R. (2003a) The initial abundance of ^{60}Fe in the solar system. *Astrophys. J. Lett., 588,* L41–L44.

Tachibana S. and Huss G. R. (2003b) Iron-60 in troilites from an unequilibrated ordinary chondrite and the initial $^{60}Fe/^{56}Fe$ in the early solar system (abstract). In *Lunar and Planetary Science XXXIV,* Abstract #1737. Lunar and Planetary Institute, Houston (CD-ROM).

Takahashi K., Witti J., and Janka H.-T. (1994) Nucleosynthesis in neutrino-driven winds from protoneutron stars II. The r-process. *Astron. Astrophys., 286,* 857–869.

Tang M. and Anders E. (1988a) Interstellar silicon carbide: How much older than the solar system? *Astrophys. J. Lett., 335,* L31–L34.

Tang M. and Anders E. (1988b) Isotopic anomalies of Ne, Xe, and C in meteorites. II. Interstellar diamond and SiC: Carriers of exotic noble gases. *Geochim. Cosmochim. Acta, 52,* 1235–1244.

The L.-S., El Eid M. F., and Meyer B. S. (2000) A new study of s-process nucleosynthesis in massive stars. *Astrophys. J., 533,* 998–1015.

Thielemann F.-K., Arnould M., and Hillebrandt W. (1979) Meteoritic anomalies and explosive neutron processing of helium-burning shells. *Astron. Astrophys., 74,* 175–185.

Thielemann F.-K., Nomoto K., and Hashimoto M.-A. (1996) Core-collapse supernovae and their ejecta. *Astrophys. J., 460,* 408–436.

Thiemens M. H. and Heidenreich J. E. I. (1983) The mass independent fractionation of oxygen: A novel isotope effect and its possible cosmochemical implications. *Science, 219,* 1073–1075.

Thompson T. A. (2003) Magnetic protoneutron star winds and r-process nucleosynthesis. *Astrophys. J. Lett., 585,* L33–L36.

Thompson T. A., Burrows A., and Meyer B. S. (2001) The physics of proto-neutron star winds: Implications for r-process nucleosynthesis. *Astrophys. J., 562,* 887–908.

Tielens A. G. G. M. (1990) Carbon stardust: From soot to diamonds. In *Carbon in the Galaxy: Studies from Earth and Space* (J. C. Tarter et al., eds.), pp. 59–111. NASA Conference Publication 3061, U.S. Government Printing Office, Washington DC.

Timmes F. X. and Clayton D. D. (1996) Galactic evolution of silicon isotopes: Application to presolar SiC grains from meteorites. *Astrophys. J., 472,* 723–741.

Timmes F. X., Woosley S. E., and Weaver T. A. (1995) Galactic chemical evolution: Hydrogen through zinc. *Astrophys. J. Suppl. Ser., 98,* 617–658.

Timmes F. X., Woosley S. E., Hartmann D. H., and Hoffman R. D. (1996) The production of ^{44}Ti and ^{60}Co in supernovae. *Astrophys. J., 464,* 332–341.

Travaglio C., Gallino R., Amari S., Zinner E., Woosley S., and Lewis R. S. (1999) Low-density graphite grains and mixing in type II supernovae. *Astrophys. J., 510,* 325–354.

Treffers R. and Cohen M. (1974) High-resolution spectra of cold stars in the 10- and 20-micron regions. *Astrophys. J., 188,* 545–552.

Truran J. W., Cowan J. H., and Cameron A. G. W. (1978) The He-driven r-process in supernovae. *Astrophys. J. Lett., 222,* L63–L67.

Umeda H. and Nomoto K. (2003) First generation black-hole forming supernovae and the metal abundance pattern of a very iron-poor star. *Nature, 422,* 871–873.

Urey H. C. (1955) The cosmic abundances of potassium, uranium and thorium and the heat balances of the Earth, the Moon, and Mars. *Proc. Natl. Acad. Sci., 41,* 127–144.

Vanhala H. A. T. and Boss A. P. (2002) Injection of radioactivities into the forming solar system. *Astrophys. J., 575,* 1144–1150.

Verchovsky A. B., Wright I. P., and Pillinger C. T. (2004) Astrophysical significance of asymptotic giant branch stellar wind energies recorded in meteoritic SiC grains. *Astrophys. J., 607,* 611–619.

Virag A., Wopenka B., Amari S., Zinner E., Anders E., and Lewis R. S. (1992) Isotopic, optical, and trace element properties of large single SiC grains from the Murchison meteorite. *Geochim. Cosmochim. Acta, 56,* 1715–1733.

Völkening J. and Papanastassiou D. A. (1989) Iron isotope anomalies. *Astrophys. J. Lett., 347,* L43–L46.

Völkening J. and Papanastassiou D. A. (1990) Zinc isotope anomalies. *Astrophys. J. Lett., 358,* L29–L32.

Wadhwa M., Srinivasan G., and Carlson R. W. (2006) Timescales of planetesimal differentiation in the early solar system. In *Meteorites and the Early Solar System II* (D. S. Lauretta and H. Y. McSween Jr., eds.), this volume. Univ. of Arizona, Tucson.

Wallerstein G., Iben I. Jr., Parker P., Boesgaard A. M., Hale G. M., Champagne A. E., Barnes C. A., Käppeler F., Smith V. V., Hoffman R. D., Timmes F. X., Sneden C., Boyd R. N., Meyer B. S., and Lambert D. L. (1997) Synthesis of the elements in stars: Forty years of progress. *Rev. Mod. Phys., 69,* 995–1084.

Wanajo S., Tamamura M., Itoh N., Nomoto K., Ishimaru Y., Beers T. C., and Nozawa S. (2003) The r-process in supernova explosions from the collapse of O-Ne-Mg cores. *Astrophys. J., 593,* 968–979.

Wannier P. G., Andersson B. G., Olofsson H., Ukita N., and Young K. (1991) Abundances in red giant stars: Nitrogen isotopes in carbon-rich molecular envelopes. *Astrophys. J., 380,* 593–605.

Wasserburg G. J. (1987) Isotopic abundances: Inferences on solar system and planetary evolution. *Earth Planet. Sci. Lett., 86,* 129–173.

Wasserburg G. J. and Arnould M. (1987) A possible relationship between extinct ^{26}Al and ^{53} Mn in meteorites and early solar activity. In *Nuclear Astrophysics, Vol. 287* (W. Hillebrandt et al., eds.), pp. 262–276. Springer-Verlag, Berlin.

Wasserburg G. J., Fowler W. A., and Hoyle F. (1960) Duration of nucleosynthesis. *Phys. Rev. Lett., 4,* 112–114.

Wasserburg G. J., Huneke J. C., and Burnett D. S. (1969) Correlation between fission tracks and fission type xenon in meteoritic whitlockite. *J. Geophys. Res., 74,* 4221–4232.

Wasserburg G. J., Lee T., and Papanastassiou D. A. (1977) Correlated oxygen and magnesium isotopic anomalies in Allende inclusions: II. Magnesium. *Geophys. Res. Lett., 4,* 299–302.

Wasserburg G. J., Busso M., Gallino R., and Raiteri C. M. (1994) Asymptotic giant branch stars as a source of short-lived radioactive nuclei in the solar nebula. *Astrophys. J., 424,* 412–428.

Wasserburg G. J., Boothroyd A. I., and Sackmann I.-J. (1995) Deep circulation in red giant stars: A solution to the carbon and oxygen isotope puzzles? *Astrophys. J. Lett., 447,* L37–L40.

Wasserburg G. J., Busso M., and Gallino R. (1996) Abundances of actinides and short-lived nonactinides in the interstellar medium: Diverse supernova sources for the r-processes. *Astrophys. J. Lett., 466,* L109–L113.

Wasserburg G. J., Gallino R., and Busso M. (1998) A test of the supernova trigger hypothesis with ^{60}Fe and ^{26}Al. *Astrophys. J. Lett., 500,* L189–L193.

Wheeler J. C., Cowan J. J., and Hillebrandt W. (1998) The r-process in collapsing O/Ne/Mg cores. *Astrophys. J. Lett., 493,* L101–L104.

Wisshak K., Voss F., Käppeler F., and Kazakov L. (1997) Neutron capture in neodymium isotopes: Implications for the s-process. *Nucl. Phys., A621,* 270c–273c.

Woolum D. S. (1988) Solar-system abundances and processes of nucleosynthesis. In *Meteorites and the Early Solar System* (J. F. Kerridge and M. S. Matthews. eds.), pp. 995–1020. Univ. of Arizona, Tucson.

Woosley S. E. (1997) Neutron-rich nucleosynthesis in carbon deflagration supernovae. *Astrophys. J., 476,* 801–810.

Woosley S. E. and Baron E. (1992) The collapse of white dwarfs to neutron stars. *Astrophys. J., 391,* 228–235.

Woosley S. E. and Hoffman R. D. (1992) The alpha-process and the r-process. *Astrophys. J., 395,* 202–239.

Woosley S. E. and Howard W. M. (1978) The p-process in supernovae. *Astrophys. J. Suppl. Ser., 36,* 285–304.

Woosley S. E. and Weaver T. A. (1995) The evolution and explosion of massive stars, II. Explosive hydrodynamics and nucleosynthesis. *Astrophys. J. Suppl. Ser., 101,* 181–235.

Woosley S. E., Wilson J. R., Mathews G. J., Hoffman R. D., and Meyer B. S. (1994) The r-process and neutrino-heated supernova ejecta. *Astrophys. J., 433,* 229–246.

Woosley S. E., Heger A., and Weaver T. A. (2002) The evolution and explosion of massive stars. *Rev. Mod. Phys., 74,* 1015–1071.

Yin Q.-Z. and Jacobsen S. B. (2004) On the issue of molybdenum isotopic anomalies in meteorites: Is it still fun? (abstract). In *Lunar and Planetary Science XXXV,* Abstract #1942. Lunar and Planetary Institute, Houston (CD-ROM).

Yin Q. Z., Jacobsen S. B., McDonough W. F., Horn I., Petaev M. I., and Zipfel J. (2000) Supernova sources and the ^{92}Nb-^{92}Zr p-process chronometer. *Astrophys. J. Lett., 536,* L49–L53.

Yin Q., Jacobsen S. B., Blichert-Toft J., Télouk P., and Albarède F. (2001) Nb-Zr and Hf-W isotope systematics: Applications to early solar system chronology and planetary differentiation (abstract). In *Lunar and Planetary Science XXXII,* Abstract #2128. Lunar and Planetary Institute, Houston (CD-ROM).

Yin Q., Jacobsen S. B., and Yamashita K. (2002a) Diverse supernova sources of pre-solar material inferred from molybdenum isotopes in meteorites. *Nature, 415,* 881–883.

Yin Q., Jacobsen S. B., Yamashita K., Blichert-Toft J., Télouk P., and Albarède F. (2002b) A short timescale for terrestrial planet formation from Hf-W chronometry of meteorites. *Nature, 418,* 949–952.

Yoshida T. and Hashimoto M. (2004) Numerical analyses of isotopic ratios of presolar grains from supernovae. *Astrophys. J., 606,* 592–604.

Young E. D., Simon J. I., Galy A., Russell S. S., Tonui E., and Lovera O. (2005) Supracanonical $^{26}Al/^{27}Al$ and the residence time of CAIs in the solar protoplanetary disk. *Science, 308,* 223–227.

Zinner E. (1998a) Stellar nucleosynthesis and the isotopic composition of presolar grains from primitive meteorites. *Annu. Rev. Earth Planet. Sci., 26,* 147–188.

Zinner E. (1998b) Trends in the study of presolar dust grains from primitive meteorites. *Meteoritics & Planet. Sci., 33,* 549–564.

Zinner E. (2004) Presolar grains. In *Treatise on Geochemistry, Vol. 1: Meteorites, Comets, and Planets* (A. M. Davis, ed.), pp. 17–39. Elsevier, Oxford.

Zinner E. and Göpel C. (2002) Aluminum-26 in H4 chondrites: Implications for its production and its usefulness as a fine-scale chronometer for early-solar system events. *Meteoritics & Planet. Sci., 37,* 1001–1013.

Zinner E. K., Fahey A. J., Goswami J. N., Ireland T. R., and McKeegan K. D. (1986) Large ^{48}Ca anomalies are associated with ^{50}Ti anomalies in Murchison and Murray hibonites. *Astrophys. J. Lett., 311,* L103–L107.

Zinner E., Tang M., and Anders E. (1989) Interstellar SiC in the Murchison and Murray meteorites: Isotopic composition of Ne, Xe, Si, C, and N. *Geochim. Cosmochim. Acta, 53,* 3273–3290.

Zinner E., Amari S., and Lewis R. S. (1991) s-Process Ba, Nd, and Sm in presolar SiC from the Murchison meteorite. *Astrophys. J. Lett., 382,* L47–L50.

Zinner E., Amari S., Gallino R., and Lugaro M. (2001) Evidence for a range of metallicities in the parent stars of presolar SiC grains. *Nucl. Phys., A688,* 102–105.

Zinner E., Amari S., Guinness R., and Jennings C. (2003a) Si isotopic measurements of small SiC and Si_3N_4 grains from the Indarch (EH4) meteorite (abstract). *Meteoritics & Planet. Sci., 38,* A60.

Zinner E., Amari S., Guinness R., Nguyen A., Stadermann F., Walker R. M., and Lewis R. S. (2003b) Presolar spinel grains from the Murray and Murchison carbonaceous chondrites. *Geochim. Cosmochim. Acta, 67,* 5083–5095.

Zinner E., Amari S., Jennings C., Mertz A. F., Nguyen A. N., Nittler L. R., Hoppe P., Gallino R., and Lugaro M. (2005a) Al and Ti isotopic ratios of presolar SiC grains of type Z (abstract). In *Lunar and Planetary Science XXXVI,* Abstract #1691. Lunar and Planetary Institute, Houston (CD-ROM).

Zinner E., Nittler L. R., Hoppe P., Gallino R., Straniero O., and Alexander C. M. O'D. (2005b) Oxygen, magnesium and chromium isotopic ratios of presolar spinel grains. *Geochim. Cosmochim. Acta, 69,* 4149–4165.

Origin and Evolution of Carbonaceous Presolar Grains in Stellar Environments

Thomas J. Bernatowicz
Washington University

Thomas K. Croat
Washington University

Tyrone L. Daulton
Naval Research Laboratory

Laboratory microanalyses of presolar grains provide direct information on the physical and chemical properties of solid condensates that form in the mass outflows from stars. This information can be used, in conjunction with kinetic models and equilibrium thermodynamics, to draw inferences about condensation sequences, formation intervals, pressures, and temperatures in circumstellar envelopes and in supernova ejecta. We review the results of detailed microanalytical studies of the presolar graphite, presolar silicon carbide, and nanodiamonds found in primitive meteorites. We illustrate how these investigations, together with astronomical observation and theoretical models, provide detailed information on grain formation and growth that could not be obtained by astronomical observation alone.

1. INTRODUCTION

In recent years the laboratory study of presolar grains has emerged as a rich source of astronomical information about stardust, as well as about the physical and chemical conditions of dust formation in circumstellar mass outflows and in supernova (SN) ejecta. Presolar grains isolated from primitive meteorites are studied by a variety of microanalytical techniques, leading to a detailed knowledge of their isotopic and chemical compositions, as well as their physical properties. The isotopic compositions of individual grains, obtained by ion microprobe analyses and interpreted with the aid of nucleosynthesis models and astrophysical data, help to identify the specific types of stellar sources in whose mass outflows the grains condensed. They also yield insights into stellar nucleosynthesis at an unprecedented level of detail (see *Bernatowicz and Walker,* 1997; *Bernatowicz and Zinner,* 1997; *Zinner,* 1998; *Nittler,* 2003; *Clayton and Nittler,* 2004; *Meyer and Zinner,* 2006). Mineralogical, microchemical, and microstructural studies by scanning electron microscopy (SEM) and by transmission electron microscopy (TEM) of presolar grains provide unambiguous information about the kinds of astrophysical solids that actually form. With the help of kinetic and thermochemical models, SEM and TEM observations may also be used to infer the conditions of grain condensation in circumstellar dust shells. In addition, because the grains must have traversed the interstellar medium (ISM) prior to their incorporation into the solar nebula, they serve as monitors of physical and chemical processing of grains in the ISM (*Bernatowicz et al.,* 2003).

In this review we focus on carbonaceous presolar grains. Even though presolar silicates (*Messenger et al.,* 2003; *Nguyen and Zinner,* 2004; *Mostefaoui and Hoppe,* 2004; *Nagashima et al.,* 2004) and oxide grains such as spinel, corundum, and hibonite (see *Meyer and Zinner,* 2006, and references therein) are important and abundant presolar minerals, few detailed laboratory data have been obtained thus far on their mineralogy and structure. An exception is the correlated TEM-ion microprobe study of presolar corundum grains by *Stroud et al.* (2004).

The situation is different for presolar silicon carbide (SiC) and graphite. Detailed structural and chemical microanalytical studies have been performed on circumstellar SiC (*Daulton et al.,* 2002, 2003) and graphite (*Bernatowicz et al.,* 1996; *Croat et al.,* 2005b) from asymptotic giant branch (AGB) stars, and on graphite formed in SN outflows (*Croat et al.,* 2003). In the present work we illustrate how microanalytical data on presolar SiC and graphite grains, interpreted with the aid of kinetic and equilibrium thermodynamics models, can be combined with astronomical observations to yield new insights into the grain formation conditions in circumstellar envelopes and SN ejecta. We also discuss the results of microanalytical studies of nanodiamonds from

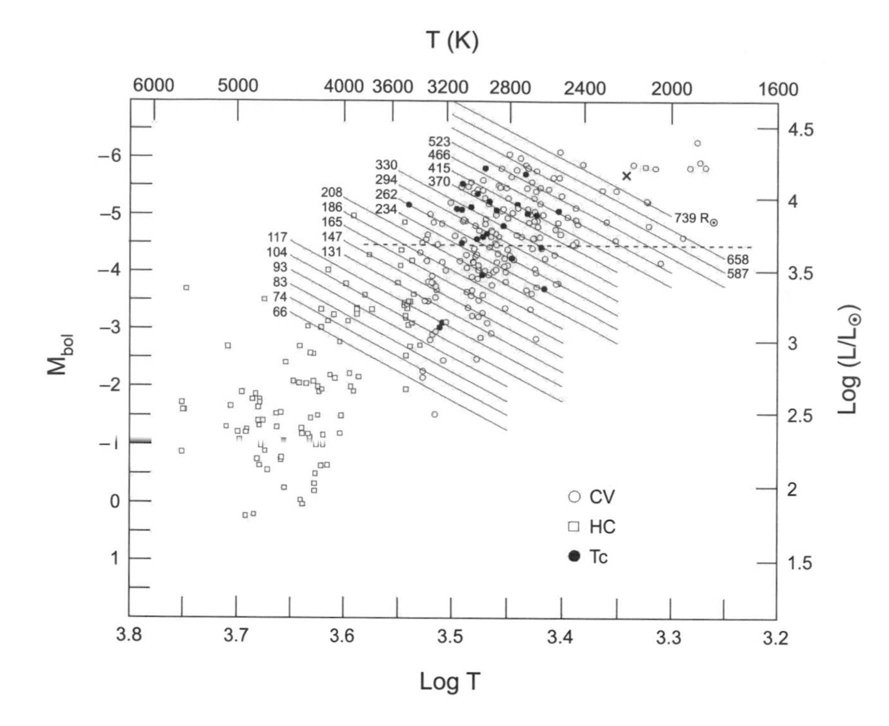

Fig. 1. Bolometric magnitude (M_{bol}) vs. effective temperature (T) for ~300 carbon stars (HR diagram), adapted from *Bergeat et al.* (2002a). CV = variable carbon giants, HC = hot carbon stars, Tc = variable carbon giants with detectable Tc. Loci of constant stellar radius (in terms of the solar radius $R_\odot$) are indicated. The dashed horizontal line at 4700 $L_\odot$ corresponds to carbon stars with M = 1.1 $M_\odot$, the minimum mass needed for a star to have evolved sufficiently rapidly to have contributed grains to the solar nebula. The closest and best studied variable carbon star IRC +10216 is shown as the x-symbol.

meteorites and interplanetary dust particles (IDPs), some components of which probably originate in supernovae.

2. ASYMPTOTIC GIANT BRANCH CARBON STARS: ASTRONOMICAL CONTEXT OF PRESOLAR GRAPHITE AND SILICON CARBIDE

Isotopic studies of presolar SiC grains from meteorites indicate that the vast majority of them originate in the mass outflows of low- to intermediate-mass carbon AGB stars, as discussed by *Meyer and Zinner* (2006). A small fraction (~1%) of SiC grains originate in SN ejecta, but lacking TEM data on such grains, we will set them aside in the present review. For presolar graphite the distribution of grains among astronomical sources is less clear. Low-density graphite spherules from the KE3 separate of the Murchison meteorite [1.65–1.72 g/cm³ (*Amari et al.,* 1994)] tend to be large (mean diameter ~6 μm) and generally show evidence of a SN origin. High-density graphite spherules from the KFC1 separate [2.15–2.20 g/cm³ (*Amari et al.,* 1994)] are more numerous but are typically smaller [mean diameter ~2 μm (cf. *Bernatowicz et al.,* 1996)], and seem to originate predominately in carbon AGB stars (*Croat et al.,* 2005a,b). The specification of *carbon* AGB stars comes from the simple thermochemical consideration that in order for these carbonaceous grains to form it is required that C/O > 1 in the gas phase, because at lower values all C is tied up in the very stable CO molecule. It is only when the C number density exceeds that of O that sufficient C is available to form graphite and carbides (*Lodders and Fegley,* 1995; *Sharp and Wasserburg,* 1995). Silicon carbide is indeed observed astronomically around carbon stars (e.g., *Speck et al.,* 1999), supporting the identification of carbon stars as a source of presolar SiC.

In the present work we do not give a treatment of stellar evolution and nucleosynthesis, so the reader may wish to consult the review by *Lattanzio and Boothroyd* (1997) on the evolution of stars from the main sequence to the mass-losing AGB phase. We also do not treat the rather complex nomenclature of carbon stars. Instead, we refer the reader to the discussion of their classification and properties presented by *Lodders and Fegley* (1997).

The physical characteristics of present-day carbon stars are represented in Fig. 1, a small portion of the Hertzsprung-Russell (HR) diagram that we have adapted from Fig. 9 of *Bergeat et al.* (2002a), depicting a large number (~300) of carbon stars studied by HIPPARCHOS (*ESA,* 1997). Loci of constant stellar radius (in multiples of the solar radius $R_\odot$) are indicated as diagonal lines. For reference, the Sun resides on the main sequence (off the diagram at lower left), at an effective temperature of 5770 K and bolometric magnitude M_{bol} = 4.75 (log $L/L_\odot$ = 0). The carbon stars in Fig. 1 are broadly classified as HC ("hot carbon") stars, CV ("carbon variable") stars, and Tc ("technetium") stars. The HC group consists largely of stars that have nonvariable light, but also includes some irregular and small-amplitude variables, carbon cepheids, and RCrB variables. The R stars are mostly included in this group. Carbon variable stars are cool, long-period variable stars (Miras) whose luminosities, effective temperatures, and radii vary periodically with time. They include the N stars and the late-stage R stars that can be explained by the third dredge-up, and are close to the tip of the thermally pulsing AGB region of the HR diagram. The Tc stars are CV giants that have detectable technetium in their atmospheres, indicating a recent dredge-up of s-process material. The general trends for CV stars are that with increasing stellar luminosity and decreasing effective temperature, the stellar masses, C/O ratios, and atmospheric opacities increase.

Not all the types of carbon stars shown in Fig. 1 are possible contributors of dust to the solar nebula, as discussed by *Bernatowicz et al.* (2005). In this regard, the most fundamental distinction that can be made is on the basis of stellar mass, because it is only stars that have sufficient mass that would have evolved rapidly enough on the main sequence to have contributed dust to the solar nebula. The lifetime t_* of a star on the main sequence varies in proportion to its mass M_*, and varies inversely as the power radiated by the star (its luminosity L_*). The luminosity is in turn proportional to M_*^K, where K = 3.5 classically. Data from HIPPARCOS, however, suggest that the exponent is better determined as K = 3.7 for $M_* \approx 0.5–5\ M_\odot$ (*Lampens et al.,* 1997; *Martin et al.,* 1998). The stellar main-sequence lifetime t_* is then given in terms of the solar main sequence lifetime $t_\odot$ as

$$t_* = t_\odot(M_*/M_\odot)^{-2.7} \qquad (1)$$

with $t_\odot \approx$ 10–11 G.y. (*Iben,* 1967; *Sackmann et al.,* 1993). Since significant mass loss occurs only at the end of their lifetimes, presolar stars must have evolved beyond the main sequence to have contributed dust to the solar nebula. The maximum lifetime t_* of stars (and therefore their minimum mass) that contributed matter to the solar system is then constrained to be approximately the difference between the age of the galaxy t_{galaxy} and the age t_{ss} of the solar system, that is, $t_* \le t_{galaxy} - t_{ss}$. The latter quantity is well determined as t_{ss} = 4.6 G.y., while the former quantity may be estimated from the recent WMAP age of the universe [13.7 ± 0.2 G.y. (*Bennett et al.,* 2003)]. Using these numbers we would find $t_* \le 9.1$ G.y. (and from equation (1), $M_* \gtrsim 1.04\ M_\odot$).

This t_* is, however, really an extreme upper limit on the lifetime of stars that could have contributed dust to the solar mixture for two reasons. First, t_{galaxy} is estimated to be less than the age of the universe by ~1 G.y., based on recent constraints on the age of globular clusters (*Krauss and Chaboyer,* 2003). Second, the limit on t_* implicitly includes the residence lifetime in the ISM of the matter expelled from stars prior to its incorporation in the solar nebula. There are currently no reliable data on the ISM residence lifetimes of presolar grains, but an estimate may be made using theoretical lifetimes of ≤0.5 G.y. against grain destruction in the ISM (see *Jones et al.,* 1997). With these considerations, we estimate $t_* \lesssim$ 7.5–8 G.y. As equation (1) shows, M_* is relatively insensitive to the uncertainty in the stellar lifetime, varying approximately as the inverse cube root of t_*. For t_* in the estimated range, we find $M_* \gtrsim 1.1\ M_\odot$. We therefore take this as the minimum mass for stars *of any kind* that could have contributed grains to the solar nebula.

It is important to note here that the mass range in which *carbon* stars are capable of forming is more restricted on theoretical grounds, both at low and high mass limits. *Gallino et al.* (1997) argue that below ~1.3 $M_\odot$, mass loss may cause such stars to undergo too few of the third dredge-up events necessary to evolve into carbon stars, prior to their evolving into white dwarfs. On the other hand, stars with masses much greater than ~4–5 $M_\odot$ may be subject to hot-bottom burning, which also drastically limits C and s-process element enrichment. These limits are subject to uncertainties that are difficult to evaluate, but are broadly consistent with observation.

The mass-luminosity relation for AGB carbon stars derived by *Bergeat et al.* (2002b), for which the stellar luminosity varies approximately as the square root of the stellar mass, implies a minimum luminosity $L_* \gtrsim 4700\ L_\odot$ (M_{bol} = –4.46) for carbon stars of $M_* \gtrsim 1.1\ M_\odot$ that could have contributed SiC and graphite grains to the solar nebula (*Bernatowicz et al.,* 2005). This luminosity threshold is shown as a dashed horizontal line in Fig. 1. It can be seen that this threshold excludes all but two of the HC stars, effectively eliminating stars of this group as potential contributors of presolar grains. Consequently, it is the long-period variable carbon (CV) stars of mass $M_* \gtrsim 1.1\ M_\odot$ that are implicated as the probable sources of meteoritic presolar graphite and SiC grains that carry enhancements of s-process elements.

The lower limit on the luminosity of carbon stars that were capable of contributing grains to the solar nebula constrains their range of effective temperatures from T ~ 1900–3400 K, their luminosities from L ~ 4700–19,500 $L_\odot$, and their radii from R ~ 200–1300 $R_\odot$ (Fig. 1). A comparatively small number of stars, namely very cool, highly luminous ones (including the nearest and well-studied carbon star IRC +10216), populate one extreme edge of this range. All the stars above the luminosity threshold of 4700 $L_\odot$ (for $M_* \gtrsim 1.1\ M_\odot$) are giants, as can be appreciated from noting that even the smallest have radii of about 1 AU (215 $R_\odot$). The most densely populated portions of this region of the HR diagram correspond to stars with temperatures from T = 2240–3350 K and luminosities from L = 4700–15,500 $L_\odot$. This luminosity range translates into masses from M = 1.1–8 $M_\odot$, using the mass-luminosity relation for CV stars. The CV stars in this range of temperature and luminosity have mean fundamental pulsation periods of about 1 ± 0.4 yr (*Bergeat et al.,* 2002b).

To relate the origin of presolar SiC and graphite more securely to carbon AGB stars, it is also necessary to consider the isotopic compositions of the grains. Nucleosynthesis models suggest that low-mass (~1–3 $M_\odot$) AGB carbon stars are responsible for the majority (>95%) of presolar SiC grains (see *Daulton et al.,* 2003, and references therein). The $^{12}C/^{13}C$ distribution for presolar SiC lies mainly within $^{12}C/^{13}C$ = 30–100, with a maximum in the distribution at 50–60 (*Hoppe and Ott,* 1997) (cf. solar $^{12}C/^{13}C$ = 89), similar to the distribution for N-type carbon star atmospheres (*Lambert et al.,* 1986). The similarity of these isotopic distributions is compelling evidence for the AGB origin of most presolar SiC. A small fraction of SiC (~5%) grains have $^{12}C/^{13}C$ ratios < 10 and possibly originate from J-type carbon stars. An even smaller fraction (~1%) originates in SN outflows, based mainly on observed excesses of ^{28}Si [which can only be produced in the deep interiors of massive stars (see *Zinner,* 1998; *Nittler,* 2003; *Meyer and Zinner,* 2006)]. The majority of these SiCs of SN origin have $^{12}C/^{13}C > 100$, up to ratios of several thousand.

The $^{12}C/^{13}C$ ratios of presolar graphites from AGB stars span a much larger range than those of SiCs from AGB stars. In the Murchison meteorite KFC1 high-density graphite separate [mean diameter ~2 μm; 2.15–2.20 g/cm³ (*Amari et al.,* 1994)] about 10% of the graphites have $^{12}C/^{13}C < 20$, but about two-thirds of them have isotopically light C [$100 \leq {}^{12}C/^{13}C \leq 5000$ (cf. *Bernatowicz et al.,* 1996)]. Noble-gas-isotopic compositions of these graphites suggest an AGB origin (*Amari et al.,* 1995b). However, there is a prominent gap in the distribution of $^{12}C/^{13}C$ ratios for graphite that corresponds roughly to the peak in the distribution of AGB SiC. This suggests that graphite may be produced at a different stage in the evolution of low-mass carbon stars, or else by carbon stars of different mass and/or metallicity than those that produce SiC. Two lines of evidence suggest the latter possibility. First, AGB graphites often contain internal crystals of carbides enriched in s-process transition metal elements [e.g., Zr, Mo, and Ru (*Bernatowicz et al.,* 1996; *Croat et al.,* 2004; 2005a,b)]. This is a clear indication of formation around late-stage AGB stars, where third dredge-up events transport ^{12}C from 3α He-burning as well as s-process elements to the stellar surface. Nucleosynthesis models (*Amari et al.,* 2001) suggest that low-metallicity AGB stars can produce the observed light C as well as the s-process enrichments in Zr, Mo, and Ru needed to condense carbides of these elements. Second, NanoSIMS determinations of Ti-isotopic compositions in TiC within KFC1 graphites (*Amari et al.,* 2004) often show excesses in ^{46}Ti and ^{49}Ti relative to ^{48}Ti (with $^{49}Ti/^{48}Ti$ as high as 5× the solar ratio), consistent with neutron capture in the He intershell during the third dredge-up in thermally pulsing AGB stars. The light-C-isotopic compositions of most KFC1 graphites, as well as their Ti-isotopic compositions, are consistent with an origin around intermediate-mass carbon stars with $M \geq 3 M_{\odot}$.

3. SILICON CARBIDE FROM CARBON STARS

Silicon carbide was first observed in the dusty envelopes of carbon stars from a relatively broad 11.3-μm infrared (IR) feature attributed to emission between the transverse and longitudinal optical phonon frequencies (*Treffers and Cohen,* 1974; *Forrest et al.,* 1975). On the basis of laboratory studies of SiC synthesis, it is expected that the physical characteristics of SiC formed around carbon stars, such as grain size and microstructure, will depend on physical conditions in the circumstellar outflows, and thus could provide information on these conditions. Astronomical data are of limited use in this regard. For example, SiC is known to form on the order of a hundred different polytypes in the laboratory (the unique cubic 3C polytype, also called β-SiC, and hexagonal and rhombohedral polytypes, collectively termed α-SiC; Fig. 2). However, astronomical data have only been shown capable of distinguishing between the 3C and 6H polytypes, and whether this distinction can even be made on the basis of astronomical IR-spectra has been the subject of controversy (see *Speck et al.,* 1997, 1999). On the other hand, if the presolar SiC extracted from meteorites can be taken to represent a fair sample of the types of SiC produced by various stellar sources (with ≥95% from carbon stars), then laboratory studies of these grains provide direct information about the microstructure of the grains formed in these astronomical environments, and can help to infer the physical conditions of their formation.

With this idea in mind, *Daulton et al.* (2002, 2003) determined the polytype distribution of astronomical SiC by

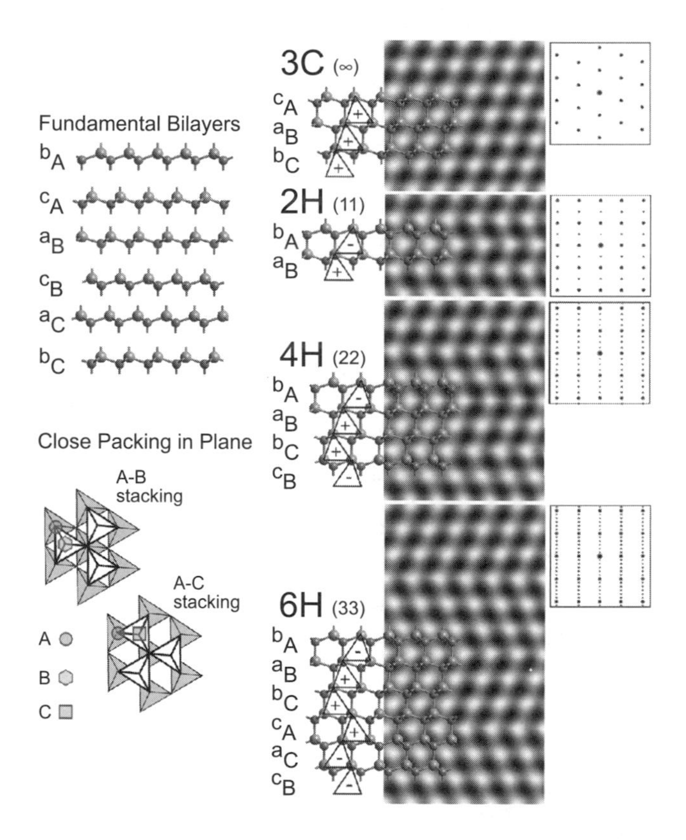

Fig. 2. Silicon carbide polytype structures. Polytypes of SiC are formed by those periodic stacking sequences of bilayers that produce tetrahedral sheets. Atomic models of the six unique (fundamental) bilayers (bA, cA, aB, cB, aC, and bC) of SiC (top left) based on three principle close-packed planes (A, B, and C) (lower left) are shown. Smaller atoms represent C and larger atoms represent Si. The two basic stacking arrangements, A-B and A-C, that form planes of vertex-sharing parallel and antiparallel tetrahedra, respectively, are shown (lower left). Atomic models of the four simplest (3C, 2H, 4H, and 6H) polytypes are shown superimposed on calculated high-resolution TEM lattice images produced using defocus conditions that reproduce the symmetry of the projected lattice (center column). Schematic illustrations of electron diffraction patterns (including forbidden reflections in some cases) are also shown (right column). From *Daulton et al.* (2003).

analytical and high-resolution transmission electron microscopy (HR-TEM) of many hundreds of presolar SiC grains from the Murchison CM2 carbonaceous meteorite. They demonstrated that, aside from a small (1% by number) population of one-dimensionally disordered SiC grains, only two polytypes are present — the cubic 3C (β-SiC) polytype (~79%) and the hexagonal 2H (α-SiC) polytype (~3%) — as well as intergrowths (Fig. 3) of these two polytypes (17%). It is interesting to note that, prior to the *Daulton et al.* (2002) study of SiC from Murchison, 2H SiC had never been observed to occur in nature.

Broadly speaking, studies of SiC synthesis show that the 2H polytype has the lowest formation temperature. As temperature is increased, the next SiC polytype to form is 3C. If temperature is increased further, 2H SiC formation ceases. Cubic 3C SiC has a wide range of growth temperatures and the general trend is that higher-order SiC polytypes begin to grow along with 3C SiC, until temperatures exceed the 3C formation range. Thus, 2H and 3C can be considered the lowest-temperature SiC polytypes. Temperatures at which 2H SiC are known to grow and remain stable [~1470–1720 K (see *Daulton et al.,* 2003, and references therein)] largely fall within the temperature range predicted by equilibrium thermodynamics for SiC formation in circumstellar outflows (see section 4).

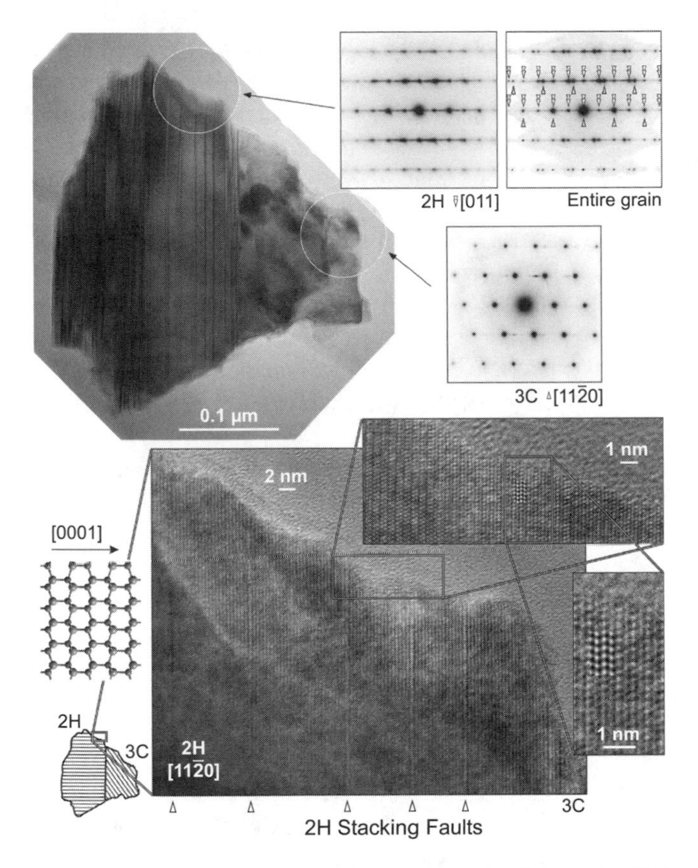

Fig. 3. Transmission electron microscopy study of a presolar SiC grain from the Murchison meteorite, an intergrowth of 2H and 3C polytypes. The selected area electron diffraction pattern for the entire grain is a composite of diffraction patterns from the 2H [11$\bar{2}$0] and 3C [011] oriented SiC domains. A bright field image of the grain is shown along with progressively magnified high-resolution lattice images showing cross-lattice planes. Inset in the high-resolution image is a simulated lattice image calculated under defocus/thickness conditions that match the lattice fringes in the micrograph. Long-range 2H order is evident in the lattice images of the 2H SiC domain. From *Daulton et al.* (2003).

Because pressures in these outflows are low, SiC condensation cannot take place until temperatures drop well below 2000 K, less than the experimental formation temperatures of most SiC polytypes. Indeed, comparison of the equilibrium thermodynamics condensation temperatures with experimental data on the formation of SiC polytypes leads to the conclusion that *only* 2H and 3C polytypes are likely to form in circumstellar atmospheres, in agreement with observation.

Daulton et al. (2002, 2003) proposed a simple hypothesis to account for the observed polytypes and their relative abundances, namely that 3C SiC first condensed at small radii (higher temperatures) and 2H SiC condensed later at larger radii (lower temperatures) in AGB atmospheres. At intermediate radii in the SiC growth region, intergrowth grains might form directly or by 2H heteroepitaxial growth on preexisting 3C SiC grains that were transported to cooler regions in AGB mass outflows by stellar radiation pressure. The observed relative abundances 3C and 2H SiC are consistent with this hypothesis. Since the rate of grain nucleation from the gas varies as the square of the number density of gas species contributing to grain growth (*McDonald,* 1963), polytypes nucleating at higher temperatures and higher gas number densities should be more abundant than polytypes nucleating later at lower temperatures and lower number densities in the expanding outflow. The nucleation rate of lower temperature forms of SiC should also be reduced because of the prior removal of Si into the SiC already formed at higher temperatures. In accordance with these expectations, single crystals of 3C SiC, which formed at higher temperatures than 2H SiC and therefore at higher gas phase Si number densities, are far more numerous than single crystals of 2H SiC.

Reliable constraints on the conditions in AGB star outflows also come from the application of equilibrium thermodynamics to the chemical composition of presolar grains (see section 4). For example, *Lodders and Fegley* (1997) noted that the pattern of trace-element enrichment seen in "mainstream" presolar SiC is exactly mirrored in the elemental depletion patterns observed astronomically in the atmospheres of N-type carbon stars. They showed that these depletion patterns are consistent with the trace-element partitioning into condensing SiC predicted by equilibrium thermodynamics.

Information obtained by studying the external morphology of presolar SiC grains complements that gained by studying its internal structure. *Bernatowicz et al.* (2003) developed a technique involving gentle disaggregation of primitive meteorites, X-ray mapping by SEM, followed by high-resolution imaging in the field emission SEM, to locate presolar SiC grains and study their surfaces. These "pristine" SiC grain surfaces are in a state that reflects their natural history — their formation, journey through the ISM, incorporation into the solar nebula and meteorite parent body — rather than laboratory artifacts induced by the chemical etching that is generally used to isolate SiC for isotopic studies.

About 90% of pristine SiC grains are bounded by one or more crystal faces. Polygonal depressions (generally <100 nm deep) are observed in more than half of these crystal faces, and commonly have symmetries consistent with the structure of the 3C polytype (Figs. 4a–c). Comparison of these features with the surface features present on heavily etched presolar SiC grains from Murchison separate KJG (*Amari et al.,* 1994) indicate that the polygonal depressions on pristine grains are primary features, resulting from incomplete convergence of surface growth fronts during grain formation. The chemically etched presolar SiCs have high densities of surface pits (probably etched linear defects) in addition to the polygonal depressions seen on pristine grains. The inferred defect densities are quite high (as much as 10^8–10^9/cm^2), about 10^3–10^4× higher than in synthetic SiCs that are engineered to minimize defects. Taken together, the polygonal depressions and the high defect densities indicate rapid initial growth of presolar SiC that was kinetically quenched when the gas phase became too rarefied.

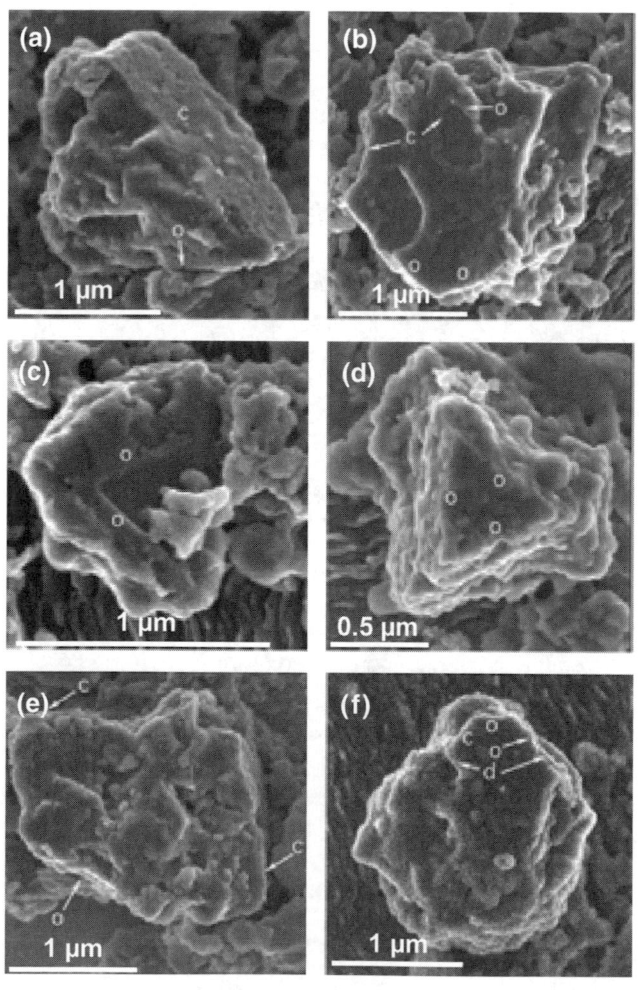

Fig. 4. Field emission SEM images of pristine presolar SiC grains from Murchison exhibiting primary growth crystal faces [**(a)**–**(f)**] and polygonal depressions [**(a)**–**(c)**]. Letter labels within panels indicate forms for cubic crystals: c = cube, o = octahedron; and d = dodecahedron. Panels **(a)**–**(d)** show "euhedral" crystals with well-developed faces. Panels **(e)** and **(f)** show subhedral crystals with few and/or imperfectly developed faces. From *Bernatowicz et al.* (2003).

In their study of pristine SiC, *Bernatowicz et al.* (2003) confirmed an observation of *Alexander et al.* (1990), based on *in situ* characterization of SiC in meteorite matrices, that no other primary condensates are intergrown with or overgrown on presolar SiC. The absence of other primary condensates indicates that further growth or back-reaction with the gas phase became kinetically inhibited as the gas densities in the expanding AGB stellar atmospheres became too low. This has the important implication that primary oxide or silicate minerals sheathing SiC grains cannot have been responsible for their survival in the solar nebula. Silicon carbide volatilization experiments by *Mendybaev et al.* (2002) have shown that the lifetimes of presolar SiC grains exposed to a hot (T ≥ 900°C) solar nebula are quite short (less than several thousand years) compared to nebular cooling timescales. The survival of the SiC grains that are found in meteorites, as well as the often exquisite state of preservation of their surfaces, thus implies that some SiC entered the solar nebula late and/or in its outer, cooler parts. However, there is evidence that some of the surviving presolar SiCs were surficially oxidized. *Stroud and Bernatowicz* (2005) studied a focused ion beam section of a pristine SiC grain by TEM that had a 10–30-nm rim of silicon oxide.

4. GRAPHITE FROM CARBON STARS

As noted in section 2, the isotopic compositions of graphites from the KFC1 separate of the Murchison meteorite (*Amari et al.,* 1994) indicate an origin in AGB stars for most of them. Ultramicrotome sections of many hundreds of these graphites have been the subject of detailed TEM studies (*Bernatowicz et al.,* 1996; *Croat et al.,* 2004, 2005a,b). Two basic spherule types are present. These are designated "onion-type" and "cauliflower-type" based on their external morphologies in SEM images. Transmission electron microscropy images of ultramicrotome slices of the graphite spherules show that structural differences, in the regularity of the stacking and in the long-range continuity of graphene sheets, lead to their two distinct external morphologies (Fig. 5).

The exteriors of onion graphites are well-crystallized graphitic layers that form concentric shells. Selected area electron diffraction (SAED) patterns from the onion layers (Fig. 6) show strong (002) rings formed by planes of atoms viewed edge-on (with their c-axes perpendicular to the electron beam). Low-magnification dark-field images show that the graphene planes are coarsely aligned over hundreds of nanometers, gradually curving to form the concentric layers of the onion. About two-thirds of the onion-type graph-

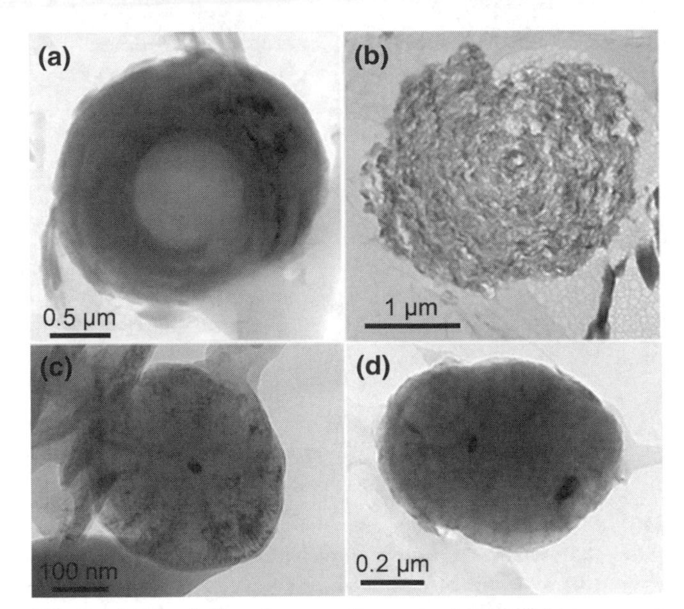

Fig. 5. Transmission electron microscopy bright-field images of 70-nm-thick ultramicrotome sections of presolar graphite grains from Murchison graphite separate KFC1 (*Amari et al.,* 1994): **(a)** Onion-type graphite with a nanocrystalline carbon core; **(b)** cauliflower-type graphite; **(c)** onion-type graphite with 26-nm central carbide crystal (metal composition $Ti_{80}V_7Fe_6Mo_3Zr_3Ru_1$); **(d)** onion-type graphite with Zr-Mo-Ti noncentral carbides (70 nm and 30 nm). These carbides have metal compositions $Zr_{83}Mo_{13}Ti_2Fe_2$ and $Zr_{86}Mo_{13}Ti_1$, respectively.

ites have cores of nanocrystalline C (Figs. 5a and 6a), consisting of small, randomly oriented graphene sheets with a mean diameter of ~3–4 nm. There is no evidence of (002) layer stacking, as in normal graphite, indicating that the graphene sheets in the cores are probably curled due to insertion of pentagonal configurations of C atoms into the overall hexagonal structure of the sheets (*Fraundorf and Wackenhut,* 2002). The shape of diffraction peak intensity profiles also indicates that as much as one-quarter of the mass of the cores may be in the form of polycyclic aromatic hydrocarbons (PAHs) or related structures (*Bernatowicz et al.,* 1996). Indeed, the presence of PAHs in the onion-type graphites has been confirmed directly by two-step laser desorption–laser ionization mass spectrometry (*Messenger et al.,* 1998).

Considering various possibilities for the formation of the onion-type graphites with nanocrystalline C cores, *Bernatowicz et al.* (1996) speculated that they formed in environments where few refractory minerals (see below) were available to serve as sites for heterogeneous nucleation, so that C condensation was delayed until homogeneous nucleation could occur at high supersaturation levels. In this scenario, the ensuing rapid condensation resulted in the formation of isolated aromatic networks (PAHs and graphene sheets) that coalesced to form the nanocrystalline cores, which subsequently were overgrown by graphite at reduced C partial pressures. Rare aggregates of onion-type graphites are also observed. These are composite objects that apparently were created when graphites that originally formed and grew separately became attached and were cemented together during subsequent C condensation.

Cauliflower-type graphites (Fig. 5b) show turbostratic layering, which consists of graphene sheets that are wavy and contorted. Some do show a roughly concentric structure with layers that diffract coherently over several hundred nanometers. Unlike the onion-type layering, however, these coherently scattering domains are of limited thickness (<50 nm). The lack of orderly stacking in the c-axis direction leads to loose-packed structures, with visible gaps in cross-section, and to an apparently lower density than for

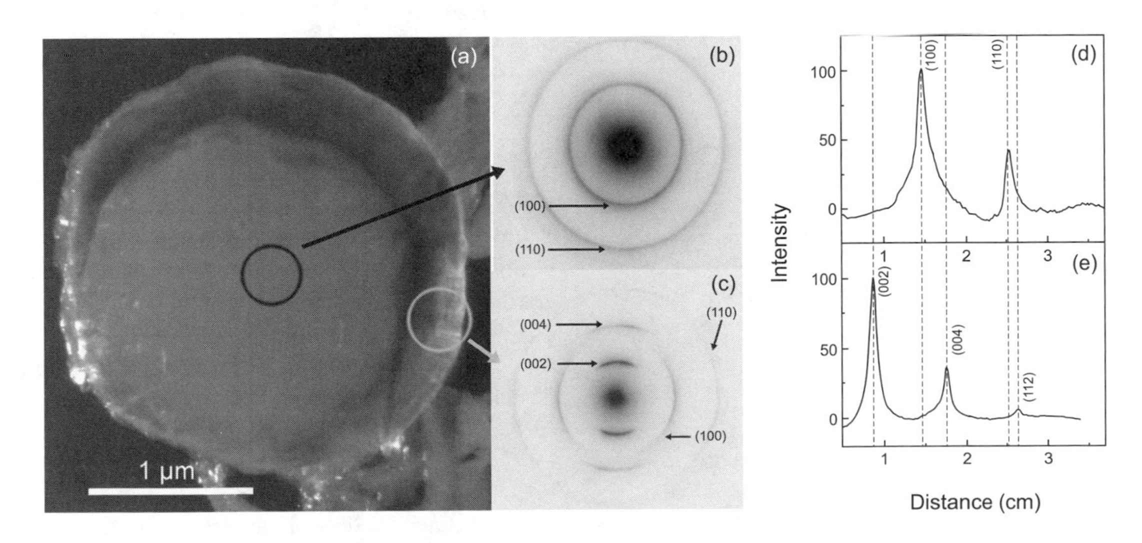

Fig. 6. **(a)** Dark-field image of an ultramicrotome (~70-nm-thick) section of a rim-core onion-type graphite from Murchison separate KFC1 (*Amari et al.,* 1994). **(b),(c)** Selected area electron diffraction (SAED) patterns from separate ≈300-nm regions of the section (with positions indicated by arrows). Note the powder-like diffraction pattern and absence of (002) and other stacking layer reflections from the core region. Line profiles of the diffracted intensity [**(d),(e)**] were taken from digitized SAED patterns (along the dotted line) and major diffraction peaks are indexed with the family of planes (hkl) from the known graphite structure.

onion-type graphites. Other cauliflowers are even less ordered, being devoid of concentric layers and consisting of coherent scattering domains with a maximum diameter of 20–30 nm. Whether onion-type and cauliflower-type graphites are formed in the same astrophysical environments is not known. Despite the fact that their C-isotopic compositions in KFC1 span the same range and are similarly distributed, and that both types contain internal crystals of refractory carbides often enriched in s-process elements like Mo, Zr, and Ru (see below), it seems probable that their different structures reflect (unknown) differences in the conditions of their formation.

Internal crystals of refractory carbides are commonly found inside graphite from AGB stars, in both the onion (Figs. 5c,d) and cauliflower types (e.g., *Bernatowicz et al.,* 1991). These high-temperature condensates were evidently ubiquitous in the gas at the time that graphite began to form. Because only one ~70-nm ultramicrotome section of any given AGB graphite is typically examined, it is clear that the 16% frequency of occurrence of internal carbides is a strict lower limit on the fraction of graphite spherules actually containing internal grains. In some cases the carbides are nearly pure TiC (Fig. 5c). Many of the carbides, however, are solid solutions that show substantial enrichments above the solar ratios in Zr, Mo, and Ru, which are elements predominantly produced by the s-process. Carbide compositions can be dominated by these s-process elements, as in Zr-Mo-carbides with Ti present at only a few atomic percent (Fig. 5d). Carbides from onion-type graphites have an average geometrical mean size of 24 nm, and range from ~7 nm to a maximum of ~90 nm (*Croat et al.,* 2005b). Textural evidence as well as compositional variation among carbides within a single graphite indicate that carbides form first and are then incorporated into the developing graphite, rather than forming later by exsolution (*Bernatowicz et al.,* 1996). About 40% of carbide-containing graphites have an internal grain located at the spherule center (as in Fig. 5c), a clear indication of heterogeneous nucleation of graphite on preexisting carbides.

Because the AGB graphite spherules with carbides consist of at least two entirely different mineral phases, equilibrium thermodynamics may be used to constrain temperatures, pressures, and gas compositions in the grain formation environment, and these predictions can be checked against astronomical observation. Figure 7 displays condensation temperatures for graphite and various carbides as a function of the gas pressure and C/O ratio, under conditions of thermodynamic equilibrium. Only ratios C/O > 1 are shown, because it is only when the C number density exceeds that of O that sufficient C is available to form graphite and carbides (see *Lodders and Fegley,* 1995; *Sharp and Wasserburg,* 1995). The condensation temperatures for the carbides in Fig. 7 reflect solar abundances of these elements. It is clear that ZrC will condense before TiC, and MoC after TiC, over all pressures in the range shown. With increasing degrees of s-process enrichment of Mo and Zr, however, condensation temperatures of carbides formed from these elements increase (*Lodders and Fegley,* 1995). For s-process

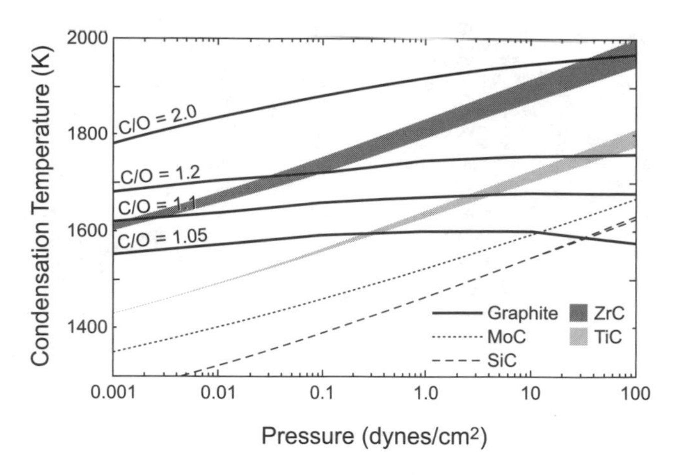

Fig. 7. Equilibrium condensation temperatures for various carbides and graphite as a function of total pressure and C/O ratio (from 1.05 to 2) for solar abundances of all elements but C. Condensation temperatures for ZrC, TiC, and SiC vary slightly with C/O and are displayed as envelopes. For these compounds, the lower boundary of the envelope is for C/O = 1.05, and the upper boundary is for C/O ≥ 2.0. For MoC, the condensation temperature does not vary with C/O in the displayed pressure range. The condensation temperature of Mo_2C (not shown) is only a few degrees higher than for MoC at any given pressure. From *Bernatowicz et al.* (1996) using data from *Lodders and Fegley* (1995).

elemental abundances typical of carbon star outflows (~10–100× solar), MoC condensation temperatures become comparable to those of TiC, and ZrC condensation temperatures increase to values ~7–10% greater than those shown.

Inspection of Fig. 7 shows that the graphite condensation temperature is relatively insensitive to pressure, but strongly dependent on C/O, while carbide condensation temperatures depend on pressure but are insensitive to C/O. Since the condensation curves of the carbides cross those of graphite, whether a given carbide will start forming before or after graphite depends on the ambient pressure. The higher the C/O ratio, the greater is the pressure required for carbides to condense before graphite. It is noteworthy that the condensation temperature for SiC is well below that of graphite at even the lowest C/O ratios, for all but the highest pressures shown. This is consistent with the fact that SiC has been observed in only one KFC1 graphite out of several hundred examined.

Presolar graphites with internal crystals of TiC can be used, for example, to place firm constraints on the C/O ratios of the AGB atmospheres in which they formed. Titanium carbide will condense before graphite at P ≥ 0.3 dynes/cm^2 for C/O = 1.05, at P ≥ 3 dynes/cm^2 for C/O = 1.1, and at P ≥ 30 dynes/cm^2 for C/O = 1.2. However, the C/O ratio cannot be increased much beyond this if TiC is to condense before graphite and possibly serve as a nucleation center. As C/O approaches ~1.5, the required pressure reaches photospheric values for low- to intermediate-mass AGB stars [~700–1400 dynes/cm^2 (*Soker and Harpaz,* 1999)], but at the photosphere the temperature is far too high (~2200 K–3400 K) for any grains to condense. Grains must instead condense in regions above the photosphere where both temperatures and pressures are much lower. The presolar graphites with nuclei of TiC thus indicate ratios of $1 \leq C/O \leq 1.2$ in the circumstellar atmospheres where they formed, as originally argued by *Sharp and Wasserburg* (1995).

This prediction is robust, and moreover is consistent with astronomical observations of AGB stars. Using C/O ratios for carbon stars determined by *Lambert et al.* (1986), *Bergeat et al.* (2001) calculated the mean C/O for variable carbon (CV) stars, and arrived at a mean C/O = 1.15 ± 0.16. These astronomical C/O ratios compare very favorably with the constraints on C/O derived from the interpretation of presolar graphite compositions and microstructures based on thermochemical calculations. The plausibility of the C/O ratios inferred from equilibrium thermodynamics means that reasonable confidence can be placed in the correlated physical inferences about temperatures and pressures in the circumstellar environments where the graphite and carbide grains formed.

Circumstellar grains are important not only because they provide solid matter to the ISM, but also because they play an important role in the dynamical evolution of AGB stars. The general understanding is that shocks in the stellar atmosphere above the photosphere enhance the gas density, promoting the condensation of dust. Copious mass loss from the star occurs largely because the dust is coupled to the stellar radiation field, which accelerates grains by radiation pressure. Momentum is in turn transferred to gas molecules by collisions with grains. The dust/gas mixture is effectively a two-component fluid whose motion and atmospheric structure are dynamically coupled: The radiation pressure on the grains determines the velocity field of the outflow and thus the density distribution, while the density distribution itself determines the conditions of radiative transfer within the outflow and thus the effective radiation pressure (see *Bowen,* 1988; *Netzer and Elitzur,* 1993; *Habing et al.,* 1994; *Ivezic et al.,* 1998; *Winters et al.,* 2000).

For the special case of spherically symmetric mass outflows, the mass-loss rate M is given by the continuity relation

$$\dot{M} = 4\pi R^2 \rho(R) v(R) \qquad (2)$$

where R is the radial distance from the center of the star, and ρ(R) and v(R) are the density and outflow speed at R, respectively. If one applies the density constraints provided by equilibrium thermodynamics discussed above to equation (2), the predicted mass-loss rates are far in excess of the maximum observed mass-loss rates for carbon stars (a few $10^{-4}\ M_\odot$/yr). As noted above, C/O ~ 1.1 for CV stars requires that the pressure $P \geq 3$ dynes/cm^2 in the stellar atmosphere for TiC to condense before graphite. For outflow speeds in grain formation regions (typically at radii R ~ 3 AU) that are as little as 5% of terminal outflow speeds (~10–20 km/s) of AGB stars, equation (2) yields mass-loss rates $\dot{M}$ in excess of $10^{-3}\ M_\odot$/yr.

As the above calculation shows, if we use the gas pressures demanded by equilibrium thermodynamics, the inferred mass-loss rate for *symmetrically* distributed circumstellar matter is simply too large. Alternatively, if we use gas pressures that would be consistent with the observed mass loss rates in (assumed) spherically symmetric mass outflows, it is impossible to produce grains nearly as large as observed for presolar condensates. Indeed, were it not for the fact that circumstellar grains of micrometer-sized graphite are available for study in the laboratory, one would probably infer that only submicrometer-sized graphite grains could be produced in spherically symmetric AGB outflows, and we would certainly not have predicted the occurrence of refractory carbides inside of them. Yet, such grains *do* exist, and we must account for them. The necessary resolution to this problem is that the mass outflows of the carbon stars responsible for presolar graphites that contain carbides must depart radically from spherical symmetry, perhaps in the form of clumps or jets with enhanced densities and pressures (*Bernatowicz et al.,* 1996, 2005; *Chigai et al.,* 2002). This inference is consistent with astronomical observations of carbon stars with the highest mass-loss rates, such as the nearest (and best studied) carbon star IRC +10216. *Weigelt et al.* (1998), using speckle-masking interferometry, showed that the mass outflow from this star is neither spherically symmetric nor homogeneous, but is highly fragmented. They concluded that IRC +10216 appears to be in a very advanced stage of its evolution, probably in a phase immediately preceding its departure from the AGB.

The outflow velocity fields in the AGB circumstellar environment that are derived from detailed dynamical models (e.g., *Ivezic et al.,* 1998), considered in conjunction with formation temperatures implied by equilibrium thermodynamics and radiative equilibrium in the outflow, imply that the time intervals available for the growth of graphite are relatively short, on the order of a few years (*Bernatowicz et al.,* 2005). These short time intervals cannot produce circumstellar graphites of the sizes observed as presolar grains (~1 μm diameter in KFC1) in spherically symmetric outflows, even under the most ideal grow conditions (perfect sticking efficiency, no evaporation, no depletion of gas species contributing to grain growth). This again points to the origin of circumstellar graphite in highly evolved CV stars that have inhomogeneous and clumpy mass outflows with enhanced densities and gas pressures.

5. GRAPHITE FROM SUPERNOVAE

The physical and chemical properties of graphite spherules formed in SN ejecta are quite different from those of spherules originating around AGB stars. The distinctions seem, at least in part, to reflect the disparities in hydrodynamics, in pressures, in gas compositions, and in grain-formation timescales in the two environments. Although dust condensation in SN ejecta has been unambiguously observed (*Meikle et al.,* 1993; *Roche et al.,* 1993; *Wooden et al.,* 1993; *McCray,* 1993), only limited inferences can be drawn from the astronomical observations regarding the composition and physical characteristics of the dust. Silicate formation has been inferred in various spectra from SN remnants (e.g., *Arendt et al.,* 1999). Unfortunately, SN 1987A, which provided the best evidence for dust condensation in core-collapse SN, had a featureless spectrum with no evidence of silicate or SiC formation. This spectrum is consistent with the condensation of graphite and/or Fe grains (*Wooden,* 1997), but the identification is tenuous because of the featureless nature of the IR spectra of these minerals. In general, IR astronomy can determine when and where dust has formed around SN, but is limited in its ability to characterize SN dust in terms of its mineralogy, chemical composition, or size. Given these limitations, the isotopic, chemical, and mineralogical studies of SN dust grains in the laboratory presently provide an alternative and exquisitely detailed source of information on grain formation in SN ejecta.

Low-density graphite spherules from density separate KE3 (1.65–1.72 g cm^{-3}) of the Murchison meteorite often have large excesses in ^{18}O and ^{28}Si, indicative of a supernova origin (*Amari et al.,* 1995a). Transmission electron microscope examination of ultramicrotome sections of these SN graphites (*Bernatowicz et al.,* 1998, 1999; *Croat et al.,* 2003) reveal concentric turbostratic layers with a degree of graphitization intermediate between the onion and cauliflower types of graphite from AGB stars. The KE3 SN graphites are generally larger (4–12 μm) than the KFC1 graphites from carbon stars and have high abundances (25–2400 ppm) of internal TiC crystals (Figs. 8a,b), with a single graphite in some cases containing hundreds of them. The TiC size distributions among SN graphites are quite variable, with the geometrical mean TiC size ranging from 30 to 230 nm. Some, despite having similar total TiC abundances, have very different grain size distributions, and there is no obvious scaling relationship between the size of the graphite and the mean size of its internal grains.

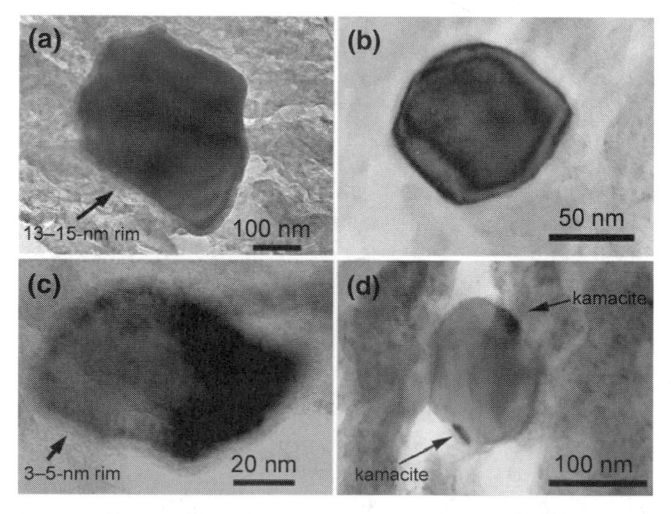

Fig. 8. Transmission electron microscopy bright-field images of internal grains in supernova graphites from Murchison graphite separate KE3 (*Amari et al.*, 1995b): **(a)** TiC with an amorphous rim from spherule KE3e10; **(b)** euhedral TiC crystal from KE3e6 (the light and dark bands are thickness fringes); **(c)** kamacite grain with an amorphous rim from spherule KE3d8 (*Bernatowicz et al.*, 1999); **(d)** subhedral TiC from KE3e6 with epitaxial kamacites on opposite (111) TiC faces. From *Croat et al.* (2003).

Composite TiCs with iron-nickel subgrains and solitary kamacite grains are also found in KE3 graphites (Figures 8c,d). In the composite grains, the Fe phases are kamacite (0–24 atom% Ni) and taenite (up to 60 atom% Ni) epitaxially grown onto one or more TiC faces. This is direct, unambiguous evidence of Fe condensation in SN ejecta. The chemical variations among internal TiC grains (see below), as well as the presence of epitaxial Fe phases on some TiCs, clearly indicate that the phase condensation sequence was TiC, followed by the Fe phases (only found in some KE3 graphites) and finally graphite. Since graphite typically condenses at a higher temperature than Fe at low pressures ($<10^{-3}$ bars) in a gas with C > O and otherwise solar composition, the observed condensation sequence implies a relative Fe enrichment in the gas or greater supersaturation of graphite relative to iron.

Examination of multiple ultramicrotome sections of each large spherule permits a determination of three-dimensional radial distances of internal crystals from the spherule center (based on the location of the crystal within the section and the section diameter). Since these radial distances are a rough measure of the time at which the internal grain was incorporated into the graphite, temporal trends in various chemical and physical properties of grains in the ejecta can sometimes be discerned. Figure 9 shows the V/Ti atomic ratio vs. radial distance from the graphite center in spherule KE3e3, for roughly 150 TiCs. Clear increases in the V/Ti ratio vs. radial distance are evident. This trend, also seen in other KE3 graphites, is probably due to chemical evolution of the TiCs in the ambient gas with time. Vanadium carbide, although soluble in TiC, is less refractory and condenses at a substantially lower temperature (cf. *Lodders and Fegley,* 1995). Thus, TiCs found toward the graphite exterior, which were probably exposed to the gas for longer times and at lower temperatures prior to capture by the graphite, incorporated more V. The observed trend in V/Ti suggests that this ratio was continually evolving in the TiCs before they were incorporated into the graphite, but ceased evolving once the TiC was fully encased.

There is typically a decrease in the number density of internal grains as a function of increasing radial distance (therefore later graphite condensation times) from the SN graphite centers. This is shown in Fig. 9, in which the volume of graphite spherule KE3e3 has been subdivided into 10 spherical shells of equal volume, with the number of TiCs counted in each shell. The number of internal TiCs is clearly higher in the central shells than in the outermost ones. If the TiCs already present in the gas were accreted onto growing graphite spherules, the mechanism of their capture is essentially the same as that of the C-bearing molecules contributing to graphite growth. That the relative numbers of molecules and TiCs captured in a given time interval is not constant requires a change either in their relative abundance or in their relative sticking probabilities. One possibility is a progressive increase in the effective

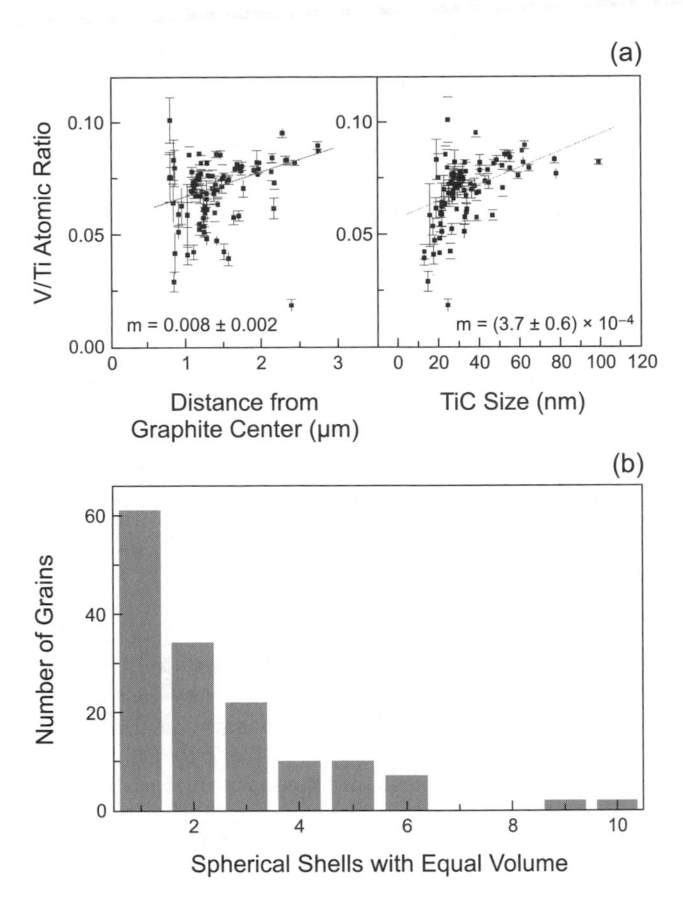

Fig. 9. Trends of internal TiC crystals in supernova graphite spherule KE3e3 from Murchison. **(a)** Vanadium/titanium atomic ratios as a function of three-dimensional radial distance from the spherule center, and as a function of TiC grain size. Slopes obtained from linear least-squares fits to the data are indicated. **(b)** Titanium carbide number density in 10 successive spherical shells of equal volume, starting at the graphite center. From *Croat et al.* (2003).

sticking probability of C-bearing molecules onto graphite, relative to TiC, over the time of graphite growth. This is most simply explained if we assume that the sticking probability of the molecules contributing to graphite growth increases with decreasing ambient temperature, resulting in a smaller number density of TiCs in graphite at greater spherule radii. Alternatively, the relative abundance of TiCs in the gas could have decreased over time if the TiCs had a substantially higher sticking probability than C-bearing molecules, and were more rapidly depleted from the gas.

Titanium carbide morphologies reflect variable degrees of "weathering" in the SN outflow. Most TiCs (about 90%) are euhedral to subhedral, with primary growth faces somewhat corroded but still clearly defined (Fig. 8b). Some KE3 TiCs are surrounded by an amorphous or nearly amorphous rim layer 3–15 nm thick, partially or completely enveloping the grain (Fig. 8a). Rims have also been observed on metal grains in KE3 graphite (Fig. 8c). Rimmed grains are ubiquitous, and constitute up to nearly half of the TiC population in some graphites. The partially amorphous rims (3–15 nm) seen on both TiC and kamacite grains are comparably thick and related in composition to their host grains (*Bernatowicz et al.,* 1999). Given these observations, at least two grain alteration mechanisms may be conceived: particle irradiation and chemical corrosion. The rims resemble the solar-wind damage features seen on lunar soil grains, which have ~50-nm-thick amorphous rims caused by irradiation with ~1 keV/nucleon (~400 km/s) H and He ions (cf. *Keller and McKay,* 1997). Titanium carbides and kamacite grains forming in SN ejecta may also have been exposed to particle irradiation, a kind of "reverse solar wind," due to collisions between the grains and slower-moving parcels of gas. Fingers of gas are ejected at high velocities (~10^4 km/s) in SN due to Rayleigh-Taylor instabilities (*Wooden,* 1997). In such an environment, a drift velocity Δv ~ 100 km/s between the grains and slower-moving parcels of gas is plausible, being at most only a few percent of the mass outflow speed. A drift velocity of this magnitude is a factor of several less than the solar wind speed, and so damaged regions less thick (~10 nm) than those observed on lunar grains are reasonable.

Alternatively, damaged rims could result from chemical corrosion, provided that gas conditions were to change with time to a state in which the TiCs and kamacites were no longer stable. However, this seems less likely given that comparably thick rims are found on compositionally dissimilar phases. Moreover, in the case of KE3e3, where ~6% of the TiCs have rims, the TiC size vs. radial distance trends indicate that most of the TiCs were still growing at the time they were incorporated into their host graphite, so the gas environment could not be simultaneously corroding and growing TiCs. In general, the diversity in TiC properties suggests that TiCs formed first and had substantially diverse histories prior to incorporation into the graphite, implying some degree of turbulent mixing in the SN outflows.

The TEM observations allow inferences to be made about the typical physical conditions in the SN ejecta where the grains condensed. Given the TiC sizes and abundances, the gas was evidently quite dusty. From the observed range in TiC sizes of ~20 nm to ~500 nm, the *minimum* Ti number densities in the gas (assuming ~1 yr growth time and T ~ 1800 K) are inferred to be ~7×10^4 to ~2×10^6 atoms/cm^3, respectively. Although the gas composition is clearly *not* solar, for scale, these number densities would correspond to a range in total pressure from ~0.2 dynes/cm^2 to ~5.0 dynes/cm^2 in a gas of solar composition. They also correspond to minimum TiC grain number densities of ~3×10^{-4} to ~0.2 grains/cm^3, assuming complete condensation of Ti in TiC. The TiC number densities imply a maximum average ratio of TiC grain separation distance in the gas to grain diameter of ~3×10^5 to ~1×10^6 (*Croat et al.,* 2003).

In contrast to the Murchison KFC1 graphites, Zr- Ru- Mo-carbides are *not* observed in SN graphites. This is independent evidence that the weak s-process, from $^{22}Ne(\alpha,n)^{25}Mg$ in the He-burning cores of stars massive enough to for SN, does not give rise to large enough neutron exposures to elevate the abundances of these elements to anything close the levels seen in the carbides from AGB stars (see *Meyer and Zinner,* 2006).

6. METEORITIC NANODIAMONDS

The most abundant refractory carbonaceous mineral in chondrites is nanometer-sized diamond. Nanodiamonds were first isolated from chondrites by destructive chemical dissolution of meteorite matrices (*Lewis et al.,* 1987). The bulk C-isotopic compositions of nanodiamond separates measured by stepped combustion of acid dissolution residues of various carbonaceous, ordinary, and enstatite chondrites are all close to the solar mean, with $\delta^{13}C$ = –32.5 to –38.8‰ (*Russell et al.,* 1996). Although the mean C-isotopic composition of nanodiamonds is essentially solar, there is sufficient isotopic evidence implying that at least *some* meteoritic nanodiamonds are of presolar origin. Nanodiamond separates from acid dissolution residues of chondrites are linked to various isotopic anomalies in H (*Virag et al.,* 1989), N (*Arden et al.,* 1989; *Russell et al.,* 1991, 1996), Sr (*Lewis et al.,* 1991), Pd (*Mass et al.,* 2001), Te (*Richter et al.,* 1997; *Mass et al.,* 2001), Xe (*Lewis et al.,* 1987), and Ba (*Lewis et al.,* 1991). However, only the anomalous Xe and Te associated with SN provide tenable evidence for a presolar origin for at least a subpopulation of the diamonds (*Zinner,* 2004; *Daulton,* 2005). Definitive isotopic evidence of a presolar origin for all or part of the population remains elusive because of the extremely small size of the diamonds. Meteoritic nanodiamonds have effective diameters, defined as the square root of their TEM projected cross-sectional areas, that range between 0.1 nm and 10 nm with a median of the distribution of 2.58 nm and 2.84 nm for diamonds isolated from the Murchison and Allende carbonaceous chondrites, respectively (*Daulton et al.,* 1996). This is consistent with earlier, indirect and lower spatial-resolution size measurements by TEM (*Lewis et al.,* 1989; *Fraundorf et al.,* 1989).

The smallest mineral grains that can be analyzed isotopically by the most advanced generation of secondary ion mass spectrometry (SIMS) instruments, the NanoSIMS, are ~0.1 μm. This is several orders of magnitude larger than meteoritic nanodiamonds. Presently, there are no instruments capable of measuring the isotopic compositions of an individual nanocrystal. In fact, even if there were, the only element in an individual nanodiamond whose isotopic composition could be feasibly measured is that of the primary element C. For example, an average meteoritic nanodiamond contains a mere several thousand C atoms (between 1.0×10^3–7.5×10^3) and only tens of N atoms (<100), where N is the second most abundant trapped element [1800–13,000 ppm by mass (*Russell et al.,* 1996)] following surface bound H [10–40 atom% (*Virag et al.,* 1989)]. Since there is only one trapped noble gas atom per tens of average-sized meteoritic nanodiamonds, measurement of isotopic compositions of trapped noble gases in individual meteoritic nanodiamonds is impossible. The situation is more extreme for isotopically anomalous Xe and Te, for which there is only one trapped Xe or Te atom per millions of mean-sized meteoritic nanodiamonds (see *Zinner,* 2004; *Daulton,* 2005). Consequently, all isotopic measurements of meteoritic nanodiamonds are of elements/gases extracted from billions of individual diamonds. In sharp contrast, the other known presolar minerals are submicrometer to micrometer in size and large enough that isotopic measurements can be performed on individual grains. The origin of meteoritic nanodiamond thus remains largely enigmatic because of its grain size.

Several hypotheses have been put forth for the origin of meteoritic nanodiamonds. Supernovae have been suggested as a source based on the inseparable presence of isotopically anomalous Xe [termed Xe-HL (*Lewis et al.,* 1987)] of probable SN origin, indicating some meteoritic nanodiamonds formed in, or at least were associated with, SN. However, the high abundance of nanodiamonds in chondrites [matrix normalized ~1400 ppm in Murchison (*Huss,* 1997)] suggests a prolific dust source. Although SN are the major contributors of gaseous matter in the ISM, they are not the dominant source of condensed matter. This has led to the suggestion that AGB stars or even the solar nebula are dominant sources. The latter is consistent with normal bulk C-isotopic composition of meteoritic nanodiamonds. In fact, *Dai et al.* (2002) interpreted the presence of nanodiamonds in IDPs *conjectured* to originate from asteroids, and their absence in IDPs *conjectured* to originate from comets, as a possible argument for nanodiamond formation in the inner solar system. Alternatively, the ensemble average of myriad nanodiamonds from a wide variety of stellar sources may simply mirror the solar mean.

A number of mechanisms have been proposed for nanodiamond formation in SN; these include formation by low-pressure condensation, similar to chemical vapor deposition (CVD), in expanding gas ejecta (*Clayton et al.,* 1995); shock metamorphism of graphite or amorphous-C grains in the ISM driven by high-velocity (grain impact grain) collisions in SN shock waves (*Tielens et al.,* 1987); annealing of graphite particles by intense ultraviolet radiation (*Nuth and Allen,* 1992); and irradiation-induced transformation of carbonaceous grains by energetic ions (*Ozima and Mochizuki,* 1993). Condensation by CVD mechanisms in circumstellar atmospheres of carbon stars has been proposed (*Lewis et al.,* 1987). Nanodiamond formation in the solar nebula (*Dai et al.,* 2002), presumably by CVD, has also been proposed.

Microstructures of materials are heavily dependent on growth mechanisms, physical conditions during formation, and postformation alteration mechanisms. Therefore, to evaluate several of the formation theories for meteoritic nanodiamonds, *Daulton et al.* (1996) used HR-TEM to compare nanodiamonds synthesized by shock metamorphism (*Greiner et al.,* 1988) and CVD (*Frenklach et al.,* 1991) to those isolated from the Allende and Murchison carbonaceous chondrites. Although changes in experimental conditions can result in a range of growth features and affect the proportion of nanodiamonds synthesized by any given process, microstructural features should exist that are

TABLE 1. Nanodiamond microstructures.

	Shock	Meteoritic	CVD
Twins/Single-Crystals	2.48	1.28	1.25
Linear Twins/Nonlinear Twins	2.72	0.87	0.36
Star Twins/All Twins	0.04	0.09	0.23
Dislocations	yes	no*	no*

*Not observed.

uniquely characteristic of specific formation mechanisms exclusive to either condensation or shock metamorphism. In fact, such features were identified in the synthesized nanodiamonds, and when compared to the microstructures of nanodiamonds from Allende and Murchison, they indicated that the predominant mechanism for meteoritic nanodiamond formation is vapor condensation. The results of the *Daulton et al.* (1996) study represent the only detailed microstructural data on meteoritic nanodiamonds to date, and are summarized in Table 1.

In cubic diamond, twinning along {111} planes is common and results when the stacking sequence of {111} planes is abruptly reversed, e.g., {AaBbCcBbAa}. In coincident site lattice notation, this twin structure is described as a first-order $\Sigma = 3$ {111} twin. The interface at a $\Sigma = 3$ twin boundary is one of the lowest-energy lattice defects, so cubic nanocrystals can form $\Sigma = 3$ twin structures relatively easily to accommodate growth constraints. Thus, $\Sigma = 3$ twin microstructures can provide a diagnostic indicator for different nanodiamond formation mechanisms.

Multiple $\Sigma = 3$ twins are relatively common microstructures in nanodiamonds and occur in two configurations. The first type (linear) exhibits parallel $\Sigma = 3$ {111} twin boundaries that terminate at the crystal surface (Fig. 10). The second type (nonlinear) is characterized by oblique $\Sigma = 3$ {111} twin boundaries that terminate either at crystal surfaces, twin boundary intersections, or both (Fig. 10). Important differences become apparent when the relative abundances of twin microstructures are compared in the synthesized diamonds (Fig. 11). First, the ratio of twinned crystals to single crystals in shock-synthesized diamonds (2.48; Table 1) is a factor of 2 higher than for those synthesized by CVD (1.25). Since reentrant corners of twinned crystals are associated with increased growth rates over single crystals (*Angus et al.,* 1992), this suggests that the mean growth rate for shock-synthesized diamonds is greater than the mean growth rate of CVD-synthesized diamonds. Nanodiamonds synthesized by detonation must have experienced rapid thermal quenching to escape graphitization after the passage of the shock front. Those nanodiamonds that survived must have experienced high growth rates. The ratio of twinned crystals to single crystals in the meteoritic nanodiamonds (1.28; Table 1) suggests growth rates similar to those of the CVD-synthesized diamonds.

Second, linear twins dominate over nonlinear twins in shock-synthesized nanodiamonds. During detonation syn-

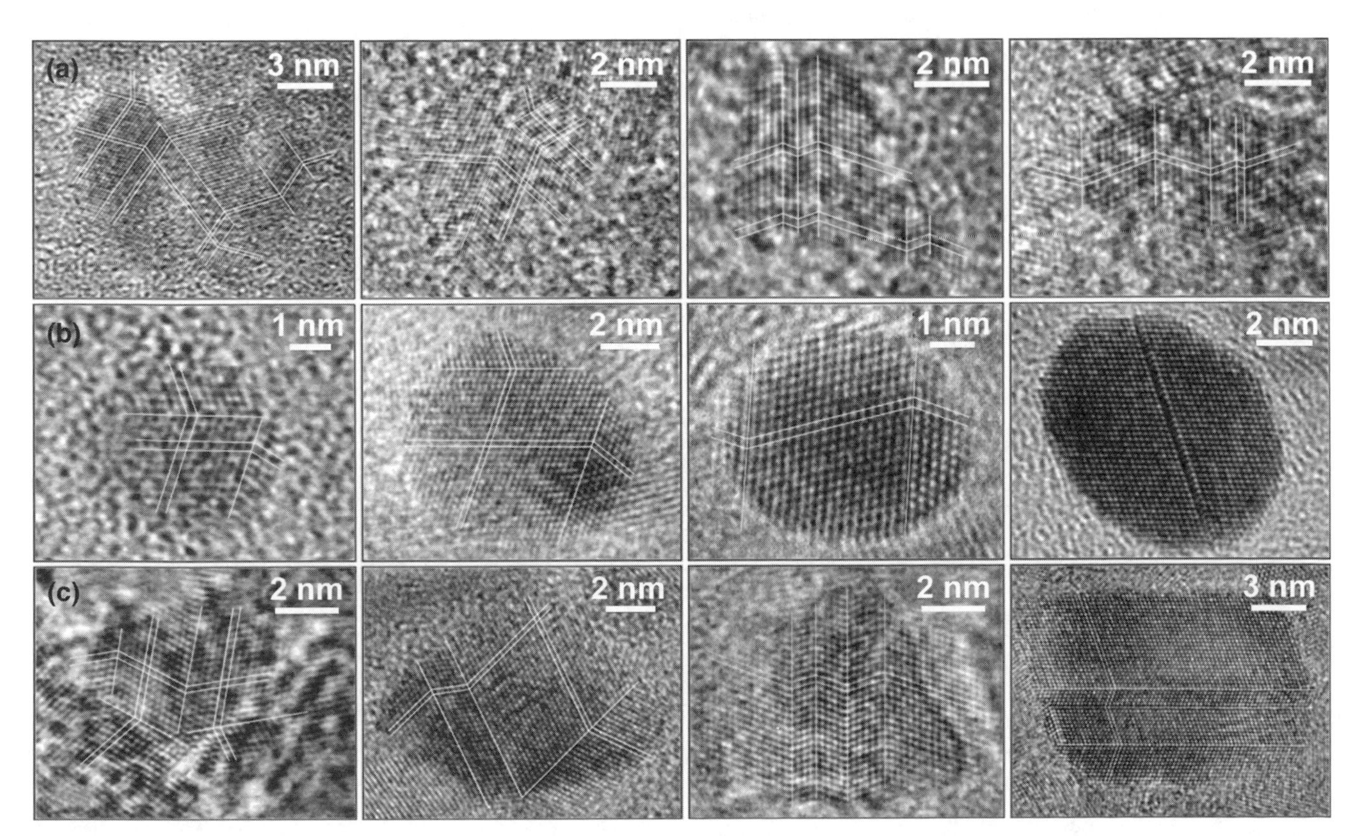

Fig. 10. Multiply twinned microstructures in **(a)** meteoritic, **(b)** CVD, and **(c)** detonation nanodiamonds. The two columns to the left are nonlinear multiple twins and the two columns to the right are linear multiple twins.

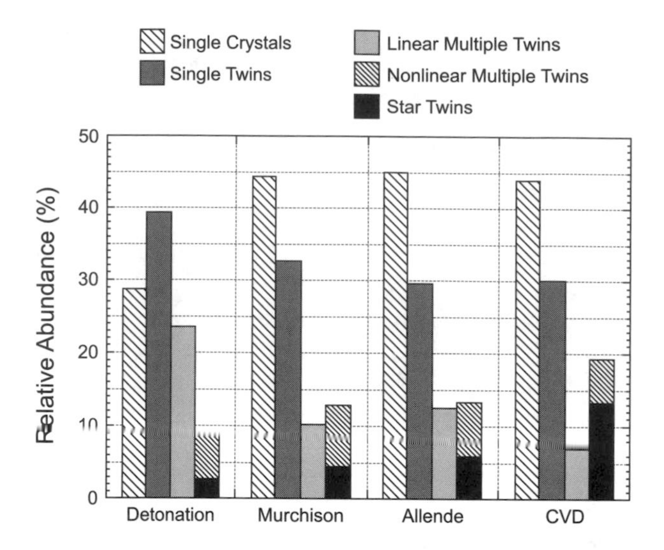

Fig. 11. Normalized distribution of $\Sigma = 3$ twin microstructures for nanodiamonds synthesized by detonation shock (*Greiner et al.,* 1988) and CVD (*Frenklach et al.,* 1991) as well as those isolated from the chondrites Murchison and Allende. Unobstructed and isolated nanodiamonds exhibiting clear cross-lattice fringes were classified by their twin type. Statistics are based on 209 (detonation), 372 (Murchison), 257 (Allende), and 130 (CVD) individual nanodiamonds (a total of 968 nanodiamonds).

thesis of nanodiamonds, large, highly anisotropic shock pressure gradients would momentarily exist. Following a shock-induced carbonaceous grain-on-grain collision, partially molten material would rapidly solidify behind planar shock fronts. Furthermore, any potential nanodiamond condensation at high-pressure shock fronts would occur within highly anisotropic conditions. In both cases, crystallization should occur along planar growth fronts producing microstructures dominated by parallel twin boundaries. Consistent with this interpretation, the shock-synthesized nanodiamonds display a distribution of multiple twins dominated by parallel twin boundaries. The direction of growth is presumably related to the geometry of the shock front and direction of the pressure gradients. In sharp contrast, the CVD- synthesized nanodiamonds are dominated by nonlinear multiple twins indicative of isotropic growth. The multiply twinned structures of meteoritic nanodiamonds more closely resemble CVD nanodiamonds.

The most striking nonlinear, multiply twinned configurations correspond to symmetric multiply twinned particles (MTPs). Nanodiamonds synthesized by direct nucleation and homoepitaxial growth from the vapor phase have a large abundance of decahedra MTPs that have pseudo-five-fold or star morphology (Fig. 12). Similar to CVD nanodiamonds, star twins are relatively common growth features in the meteoritic nanodiamonds. In contrast, star twins are relatively rare in shock synthesized nanodiamonds with an abundance at least a factor of 2 less than meteoritic nanodiamonds (Fig. 11). The coherent twin boundaries present in the star twins are indicative of radial (isotropic) growth, as would be possible from the direct nucleation and homoepitaxial

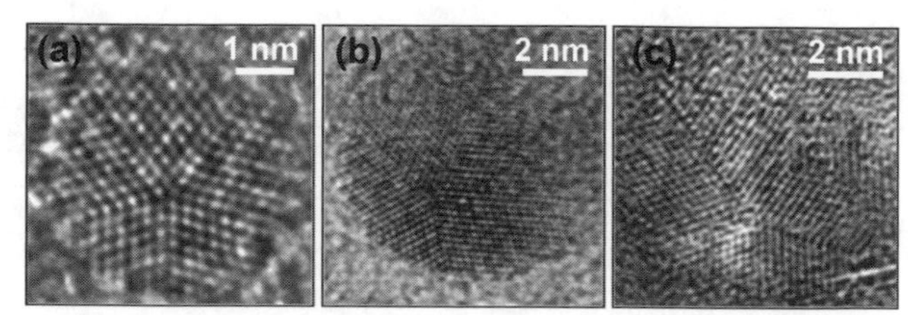

Fig. 12. Star twin microstructures in **(a)** meteoritic nanodiamond from Allende, **(b)** CVD-synthesized nanodiamond, and **(c)** shock-synthesized nanodiamond.

growth from a locally uniform supersaturated gas such as in a CVD type process. Star-twin microstructures would not be expected from the highly anisotropic shock-induced metamorphism of carbonaceous grains. The nonlinear multiple-twin crystals (including MTPs) observed in detonation soot residues might have formed by a less-efficient mechanism occurring within the rarefaction wave of the expanding shock front. Vaporization of a fraction of the precursor carbonaceous material in the shock heating event could supersaturate the partially ionized gas in C. Given favorable conditions, vapor condensation might occur after the passage of the shock front leading to nanodiamond nucleation and growth.

In contrast to twin and stacking fault structures, dislocations represent relatively high-energy defects because their cores contain disrupted nearest-neighbor bonds and significant bond distortion. Whereas the formation of twins and stacking faults is influenced by low-energy processes, the formation of dislocations requires relatively high-energy processes. This is especially the case for diamond, which has strong sp^3 C bonds. For example, epitaxial dislocations are common in CVD diamond films and result from strain induced at the substrate interface and at interfaces where two growing crystals impinge at an arbitrary angle. High dislocation densities also develop in natural diamond to accommodate lattice distortions around mineral inclusions and plastic deformation caused by shear stresses in Earth's upper mantle. In contrast, the situation is disparate for nanocrystals, where perhaps only a few mechanisms are available to form high-energy defects. Martensitic-type mech-

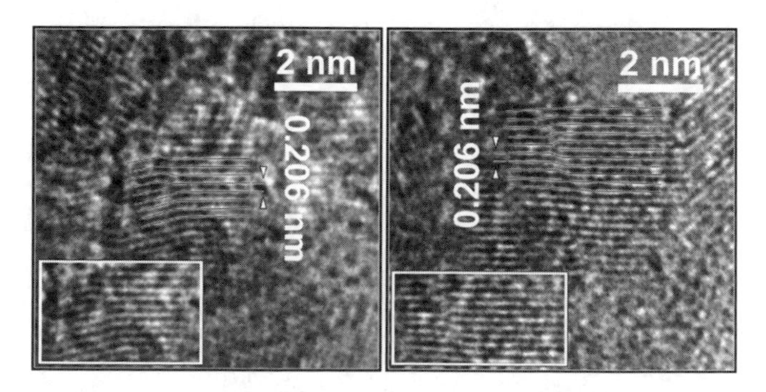

Fig. 13. Shock-synthesized nanodiamonds containing dislocations. The superimposed lines illustrate the missing <111> half planes. Insets display nonannotated dislocation images.

anisms occurring behind shock fronts can leave a high density of residual dislocation defects in the transformation product (*Daulton et al.,* 1996). Shock-synthesized nanodiamonds exhibit a small population of grains that contain dislocations with disorder (Fig. 13). In contrast, dislocations are not expected to form in nanometer grains that nucleated directly from the vapor phase, and indeed, no such defect structures were observed in the CVD-synthesized nanodiamonds.

No analogous dislocations were observed in the Allende and Murchison nanodiamonds. It is unlikely that the acid dissolution used to isolate meteoritic nanodiamonds destroyed those with dislocations. Detonation soot residues were subjected to acid treatments to purify diamond products (*Greiner et al.,* 1988), although not as extensive as the treatments used to isolate nanodiamonds from chondrites. Nevertheless, the detonation soot residues still retain nanodiamonds with dislocations. Furthermore, thermal processing of the Murchison and Allende parent bodies was probably insufficient to remove dislocations by annealing or destroying dislocation-containing diamonds. Noble gas abundances in nanodiamonds suggest mild thermal processing in Murchison with only minimal loss of nanodiamonds (*Huss and Lewis,* 1995). Although Allende experienced greater thermal processing than Murchison, a significant amount of surface-bound H with D excesses is retained by its nanodiamonds (*Carey et al.,* 1987; *Virag et al.,* 1989). Temperatures required to anneal dislocations would be more than sufficient to desorb surface H, resulting in a significant depletion of any D excess. Lack of dislocation microstructures in meteoritic nanodiamonds is probably a primary feature of formation, and therefore suggests that the majority of meteoritic nanodiamonds did not form by shock metamorphism.

Microstructures of nanodiamonds isolated from Allende and Murchison, for the most part, are similar to those of CVD-synthesized nanodiamonds and dissimilar from those of the shock-synthesized nanodiamonds (Table 1). Twin microstructures of meteoritic nanodiamonds suggest isotropic growth conditions, and the lack of dislocations is inconsistent with shock processes. These observations suggest that the predominant mechanism responsible for the formation of meteoritic nanodiamonds is low-pressure condensation, similar to CVD. This conclusion is based on direct microstructural evidence, independent of any models for astrophysical processes or environments (e.g., AGB stars, SN, or the solar nebula). If shock metamorphism (most probably occurring in SN) contributed to the population of nanodiamonds in meteorites, then certainly shock-formed nanodiamonds represent a minor subpopulation, since the majority of the meteoritic nanodiamonds do not exhibit microstructures indicative of anisotropic shock.

Acknowledgments. This chapter is based upon work partially supported by NASA under contracts NNG04GG13G (T.J.B., T.K.C.) and W-10246 (T.L.D.), issued through the Office of Space Science and NRL through the Office of Naval Research (T.L.D.).

REFERENCES

Alexander C. M. O'D., Swan P., and Walker R. M. (1990) In situ measurement of interstellar silicon carbide in two CM chondrite meteorites. *Nature, 348,* 715–717.

Amari S., Lewis R. S., and Anders E. (1994) Interstellar grains in meteorites: I. Isolation of SiC, graphite, and diamond; size distributions of SiC and graphite. *Geochim. Cosmochim. Acta, 58,* 459–470.

Amari S., Zinner E., and Lewis R. S. (1995a) Large ^{18}O excesses in circumstellar graphite grains from the Murchison meteorite: Indication of a massive-star origin. *Astrophys. J. Lett., 447,* L147–L150.

Amari S., Lewis R. S., and Anders E. (1995b) Interstellar grains in meteorites: III. Graphite and its noble gases. *Geochim. Cosmochim. Acta, 59,* 1411–1426.

Amari S., Nittler L. R., Zinner E., Gallino R., Lugaro M., and Lewis R. (2001) Presolar SiC grains of type Y: Origin from low metallicity asymptotic giant branch stars. *Astrophys. J., 546,* 248–266.

Amari S., Zinner E., and Lewis R. S. (2004) Isotopic study of presolar graphite in the KFC1 separate from the Murchison meteorite (abstract). *Meteoritics & Planet. Sci., 39,* A13.

Angus J. C., Sunkara M., Sahaida S. R., and Glass J. T. (1992) Twinning and faceting in early stages of diamond growth by chemical vapor deposition. *J. Mater. Res., 7,* 3001–3009.

Arden J. W., Ash R. D., Grady M. M., Wright I. P., and Pillinger C. T. (1989) Further studies on the isotopic composition of interstellar grains in Allende: 1. Diamonds (abstract). In *Lunar and Planetary Science XX,* pp. 21–22. Lunar and Planetary Institute, Houston.

Arendt R. G., Dwek E., and Moseley S. H. (1999) Newly synthesized elements and pristine dust in the Cassiopeia A supernova remnant. *Astrophys. J., 521,* 234–245.

Bennett C. L., Halpern M., Hinshaw G., Jarosik N., Kogut A., Limon M., Meyer S. S., Page L., Spergel D. N., Tucker G. S., Wollack E., Wright E. L., Barnes C., Greason M. R., Hill R. S., Komatsu E., Nolta M. R., Odegard N., Peiris H. V., and Weiland J. L. (2003) First year Wilkinson Microwave Anisotropy Probe (WMAP) observations: Preliminary maps and basic results. *Astrophys. J. Suppl. Ser., 148,* 1–27.

Bergeat J., Knapik A., and Rutily B. (2001) The effective temperatures of carbon-rich stars. *Astron. Astrophys., 369,* 178–209.

Bergeat J., Knapik A., and Rutily B. (2002a) Carbon-rich giants in the HR diagram and their luminosity function. *Astron. Astrophys., 390,* 967–986.

Bergeat J., Knapik A., and Rutily B. (2002b) The pulsation modes and masses of carbon-rich long period variables. *Astron. Astrophys., 390,* 987–999.

Bernatowicz T. J. and Walker R. M. (1997) Ancient stardust in the laboratory. *Phys. Today, 50(12),* 26–32.

Bernatowicz T. J. and Zinner E., eds. (1997) *Astrophysical Implications of the Laboratory Study of Presolar Materials.* American Institute of Physics, New York. 750 pp.

Bernatowicz T. J., Amari S., Zinner E. K., and Lewis R. S. (1991) Interstellar grains within interstellar grains. *Astrophys. J. Lett., 373,* L73–L76.

Bernatowicz T. J., Cowsik R., Gibbons P., Lodders K., Fegley B., Amari S., and Lewis R. (1996) Constraints on stellar grain formation from presolar graphite in the Murchison meteorite. *Astrophys. J., 472,* 760–782.

Bernatowicz T. J., Amari S., Messenger S., and Lewis R. (1998)

Internal structure and composition of presolar graphites from supernovae (abstract). In *Lunar and Planetary Science XXIX,* Abstract #1393. Lunar and Planetary Institute, Houston (CD-ROM).

Bernatowicz T. J., Bradley J., Amari S., Messenger S., and Lewis R. (1999) New kinds of massive star condensates in a presolar graphite from Murchison (abstract). In *Lunar and Planetary Science XXX,* Abstract #1392. Lunar and Planetary Institute, Houston (CD-ROM).

Bernatowicz T. J., Messenger S., Pravdivtseva O., Swan P., and Walker R. M. (2003) Pristine presolar silicon carbide. *Geochim. Cosmochim. Acta, 67,* 4679–4691.

Bernatowicz T. J., Akande O. W., Croat T. K., and Cowsik R. (2005) Constraints on grain formation around carbon stars from laboratory studies of presolar graphite (abstract). In *Lunar and Planetary Science XXXVI,* Abstract #1509. Lunar and Planetary Institute, Houston (CD-ROM).

Bowen G. H. (1988) Dynamical modeling of long-period variable star atmospheres. *Astrophys. J., 329,* 299–317.

Carey W., Zinner E., Fraundorf P., and Lewis R. S. (1987) Ion probe and TEM studies of a diamond bearing Allende residue. *Meteoritics, 22,* 349–350.

Chigai T., Yamamoto T., and Kozasa T. (2002) Heterogeneous condensation of presolar titanium carbide core-graphite mantle spherules. *Meteoritics & Planet. Sci., 37,* 1937–1951.

Clayton D. D. and Nittler L. R. (2004) Astrophysics with presolar stardust. *Annu. Rev. Astron. Astrophys., 42,* 39–78.

Clayton D. D., Meyer B. S., Sanderson C. I., Russell S. S., and Pillinger C. T. (1995) Carbon and nitrogen isotopes in type II supernova diamonds. *Astrophys. J., 447,* 894–905.

Croat T. K., Bernatowicz T., Amari S., Messenger S., and Stadermann F. J. (2003) Structural, chemical and isotopic microanalytical investigations of graphite from supernovae. *Geochim. Cosmochim. Acta, 67,* 4705–4725.

Croat T. K., Stadermann F. J., Zinner E., and Bernatowicz T. J. (2004) Coordinated isotopic and TEM studies of presolar graphites from Murchison (abstract). In *Lunar and Planetary Science XXXV,* Abstract #1353. Lunar and Planetary Institute, Houston (CD-ROM).

Croat T. K., Stadermann F. J., and Bernatowicz T. J. (2005a) Internal grains within KFC graphites: Implications for their stellar source (abstract). In *Lunar and Planetary Science XXXVI,* Abstract #1507. Lunar and Planetary Institute, Houston (CD-ROM).

Croat T. K., Stadermann F. J., and Bernatowicz T. J. (2005b) Presolar graphite from AGB stars: Microstructure and s-process enrichment. *Astrophys. J.,* in press.

Dai Z. R., Bradley J. P., Joswiak D. J., Brownlee D. E., Hill H. G. M., and Genge M. J. (2002) Possible *in situ* formation of meteoritic nanodiamonds in the early solar system. *Nature, 418,* 157–159.

Daulton T. L. (2005) Nanodiamonds in the cosmos. Microstructural and trapped element isotopic data. In *Syntheses, Properties, and Applications of Ultrananocrystalline Diamond* (D. M. Gruen et al., eds.), pp. 49–62. Springer-Verlag, Berlin.

Daulton T. L., Eisenhour D. D., Bernatowicz T. J., Lewis R. S., and Buseck P. R. (1996) Genesis of presolar diamonds: Comparative high-resolution transmission electron microscopy study of meteoritic and terrestrial nano-diamonds. *Geochim. Cosmochim. Acta, 60,* 4853–4872.

Daulton T. L., Bernatowicz T. J., Lewis R. S., Messenger S., Stadermann F. J., and Amari S. (2002) Polytype distribution in circumstellar silicon carbide. *Science, 296,* 1852–1855.

Daulton T. L., Bernatowicz T. J., Lewis R. S., Messenger S., Stadermann F. J., and Amari S. (2003) Polytype distribution in circumstellar silicon carbide: Microstructural characterization by transmission electron microscopy. *Geochim. Cosmochim. Acta, 67,* 4743–4767.

ESA (1997) *The HIPPARCHOS Catalog.* ESA SP-1200, European Space Agency, The Netherlands.

Forrest W. J., Gillett F. C., and Stein W. A. (1975) Circumstellar grains and the intrinsic polarization of starlight. *Astrophys. J., 195,* 423–440.

Fraundorf P. and Wackenhut M. (2002) The core structure of presolar graphite onions. *Astrophys. J. Lett., 578,* L153–L156.

Fraundorf P., Fraundorf G., Bernatowicz T., Lewis R., and Tang M. (1989) Stardust in the TEM. *Ultramicroscopy, 27,* 401–412.

Frenklach M., Howard W., Huang D., Yuan J., Spear K. E., and Koba R. (1991) Induced nucleation of diamond powder. *Appl. Phys. Lett., 59,* 546–548.

Gallino R., Busso M., and Lugaro M. (1997) Neutron capture nucleosynthesis in AGB stars. In *Astrophysical Implications of the Laboratory Study of Presolar Materials* (T. J. Bernatowicz and E. Zinner, eds.), pp. 115–153. American Institute of Physics, New York.

Greiner N. R., Phillips D. S., Johnson J. D., and Volk F. (1988) Diamonds in detonation soot. *Nature, 333,* 440–442.

Habing H. J., Tignon J., and Tielens A. G. G. M. (1994) Calculations of the outflow velocity of envelopes of cool giants. *Astron. Astrophys., 286,* 523–534.

Hoppe P. and Ott U. (1997) Mainstream silicon carbide grains from meteorites. In *Astrophysical Implications of the Laboratory Study of Presolar Materials* (T. J. Bernatowicz and E. Zinner, eds.), pp. 27–58. American Institute of Physics, New York.

Huss G. R. (1997) The survival of presolar grains in solar system bodies. In *Astrophysical Implications of the Laboratory Study of Presolar Materials* (T. J. Bernatowicz and E. Zinner, eds.), pp. 721–748. American Institute of Physics, New York.

Huss G. R. and Lewis R. S. (1995) Presolar diamond, SiC, and graphite in primitive chondrites: Abundances as a function of meteorite class and petrologic type. *Geochim. Cosmochim. Acta, 59,* 115–160.

Iben I. (1967) Stellar evolution. VI. Evolution from the main sequence to the Red-Giant branch for stars of mass $1 M_{\odot}$, $1.25 M_{\odot}$, and $1.5 M_{\odot}$. *Astrophys. J., 147,* 624–649.

Ivezic Z., Knapp G. R., and Elitzur M. (1998) Stellar outflows driven by radiation pressure. In *Proceedings of the Sixth Annual Conference of the CFD Society of Canada,* IV–13. Computational Fluid Dynamics Society of Canada, Ottawa.

Jones A. P., Tielens A. G. G. M., Hollenbach D. J., and McKee C. F. (1997) The propagation and survival of interstellar grains. In *Astrophysical Implications of the Laboratory Study of Presolar Materials* (T. J. Bernatowicz and E. Zinner, eds.), pp. 595–613. American Institute of Physics, New York.

Keller L. P. and McKay D. S. (1997) The nature and origin of rims on lunar soil grains. *Geochim. Cosmochim. Acta, 61,* 2331–2341.

Krauss L. M. and Chaboyer B. (2003) Age estimates of globular clusters in the Milky Way: Constraints on cosmology. *Science, 299,* 65–69.

Lambert D. L., Gustafsson B., Eriksson K., and Hinkle K. H. (1986) The chemical compositions of carbon stars. I. Carbon, nitrogen, and oxygen in 30 cool carbon stars in the galactic disk. *Astrophys. J. Suppl. Ser., 62,* 373–425.

Lampens P., Kovalevsky J., Froeschle M., and Ruymaekers G. (1997) On the mass-luminosity relation. In *Proceedings of the ESA Symposium "Hipparchos — Venice '97,"* pp. 421–424. ESA SP-402, European Space Agency, The Netherlands.

Lattanzio J. C. and Boothroyd A. I. (1997) Nucleosynthesis of elements in low to intermediate mass stars through the AGB phase. In *Astrophysical Implications of the Laboratory Study of Presolar Materials* (T. J. Bernatowicz and E. Zinner, eds.), pp. 85–114. American Institute of Physics, New York.

Lewis R. S., Tang M., Wacker J. F., Anders E., and Steel E. (1987) Interstellar diamonds in meteorites. *Nature, 326,* 160–162.

Lewis R. S., Anders E., and Draine B. T. (1989) Properties, detectability and origin of interstellar diamonds in meteorites. *Nature, 339,* 117–121.

Lewis R. S., Huss G. R., and Lugmair G. (1991) Finally, Ba & Sr accompanying Xe-HL in diamonds from Allende (abstract). In *Lunar and Planetary Science XXII,* pp. 807–808. Lunar and Planetary Institute, Houston.

Lodders K. and Fegley B. Jr. (1995) The origin of circumstellar silicon carbide grains found in meteorites. *Meteoritics, 30,* 661–678.

Lodders K. and Fegley B. Jr. (1997) Condensation chemistry of carbon stars. In *Astrophysical Implications of the Laboratory Study of Presolar Materials* (T. J. Bernatowicz and E. Zinner, eds.), pp. 391–423. American Institute of Physics, New York.

Maas R., Loss R. D, Rosman K. J. R., De Laeter J. R., Lewis R. S., Huss G. R., and Lugmair G. W. (2001) Isotope anomalies in tellurium and palladium from Allende nanodiamonds. *Meteoritics & Planet. Sci., 36,* 849–858.

Martin C., Mignard F., Hartkopf W. I., and McAlister H. A. (1998) Mass determination of astrometric binaries with Hipparcos. *Astron. Astrophys. Suppl. Ser., 133,* 149–162.

McCray R. (1993) Supernova 1987A revisited. *Annu. Rev. Astron. Astrophys., 31,* 175–216.

McDonald J. E. (1963) Homogeneous nucleation of vapor condensation II: Kinetic aspects. *Am. J. Phys., 31,* 31–41.

Meikle W. P. S, Spyromilio J., Allen D. A., Varani G.-F., and Cummings R. J. (1993) Spectroscopy of supernova 1987A at 1–4 μm — II. Days 377 to 1114. *Mon. Not. R. Astron. Soc., 261,* 535–572.

Mendybaev R. A., Beckett J. R., Grossman L., Stolper E., Cooper R. F., and Bradley J. P. (2002) Volatilization kinetics of silicon carbide in reducing gases: An experimental study with applications to the survival of presolar grains in the solar nebula. *Geochim. Cosmochim. Acta, 66,* 661–682.

Messenger S., Amari S., Gao X., Walker R. M., Clemett S. J., Chillier X. D. F., Zare R. N., and Lewis R. S. (1998) Indigenous polycyclic aromatic hydrocarbons in circumstellar graphite grains from primitive meteorites. *Astrophys. J., 502,* 284–295.

Messenger S., Keller L. P., Stadermann F. J., Walker R. M., and Zinner E. (2003) Samples of stars beyond the solar system: Silicate grains in interplanetary dust. *Science, 300,* 105–108.

Meyer B. S. and Zinner E. (2006) Nucleosynthesis. In *Meteorites and the Early Solar System II* (D. S. Lauretta and H. Y. McSween Jr., eds.), this volume. Univ. of Arizona, Tucson.

Mostefaoui S. and Hoppe P. (2004) Discovery of abundant in situ silicate and spinel grains from red giant stars in a primitive meteorite. *Astrophys. J. Lett., 613,* L149–L152.

Nagashima K., Krot A. N., and Yurimoto H. (2004) Stardust silicates from primitive meteorites. *Nature, 428,* 921–924.

Netzer N. and Elitzur M. (1993) The dynamics of stellar outflows dominated by interaction of dust and radiation. *Astrophys. J., 410,* 701–713.

Nguyen A. N. and Zinner E. (2004) Discovery of ancient silicate stardust in a meteorite. *Science, 303,* 1496–1499.

Nittler L. R. (2003) Presolar stardust in meteorites: Recent advances and scientific frontiers. *Earth Planet. Sci. Lett., 209,* 259–273.

Nuth J. A. III and Allen J. E. Jr. (1992) Supernovae as sources of interstellar diamonds. *Astrophys. Space Sci., 196,* 117–123.

Ozima M. and Mochizuki K. (1993) Origin of nanodiamonds in primitive chondrites: (1) Theory. *Meteoritics, 28,* 416–417.

Richter S., Ott U., and Begemann F. (1997) Tellurium-H in interstellar diamonds (abstract). In *Lunar and Planetary Science XXVIII,* p. 1185. Lunar and Planetary Institute, Houston.

Roche P. F., Aitken D. K., and Smith C. H. (1993) The evolution of the 8–13 micron spectrum of supernova 1987A. *Mon. Not. R. Astron. Soc., 261,* 522–534.

Russell S. S., Arden J. W., and Pillinger C. T. (1991) Evidence for multiple sources of diamond from primitive chondrites. *Science, 254,* 1188–1191.

Russell S. S., Arden J. W., and Pillinger C. T. (1996) A carbon and nitrogen isotope study of diamond from primitive chondrites. *Meteoritics & Planet. Sci., 31,* 343–355.

Sackmann I. J., Boothroyd A. I., and Kramer K. E. (1993) Our Sun. III. Present and future. *Astrophys. J., 418,* 457–468.

Sharp C. M. and Wasserburg G. J. (1995) Molecular equilibria and condensation temperatures in carbon-rich gases. *Geochim. Cosmochim. Acta, 59,* 1633–1652.

Soker N. and Harpaz A. (1999) Stellar structure and mass loss on the upper asymptotic giant branch. *Mon. Not. R. Astron. Soc., 310,* 1158–1164.

Speck A. K., Barlow M. J., and Skinner C. J. (1997) The nature of the silicon carbide in carbon star outflows. *Mon. Not. R. Astron. Soc., 288,* 431–456.

Speck A. K., Hofmeister A. M., and Barlow M. J. (1999) The SiC problem: Astronomical and meteoritic evidence. *Astrophys. J. Lett., 513,* L87–L90.

Stroud R. M. and Bernatowicz T. J. (2005) Surface and internal structure of pristine presolar silicon carbide (abstract). In *Lunar and Planetary Science XXXVI,* Abstract #2010. Lunar and Planetary Institute, Houston (CD-ROM).

Stroud R. M., Nittler L. R., and Alexander C. M. O'D. (2004) Polymorphism in presolar Al_2O_3: Grains from asymptotic giant branch stars. *Science, 305,* 1455–1457.

Tielens A. G. G. M., Seab C. G., Hollenbach D. J., and McKee C. F. (1987) Shock processing of interstellar dust: Diamonds in the sky. *Astrophys. J. Lett., 319,* L109–L113.

Treffers R. and Cohen M. (1974) High-resolution spectra of cool stars in the 10- and 20-micron regions. *Astrophys. J., 188,* 545–552.

Virag A., Zinner E., Lewis R. S., and Tang M. (1989) Isotopic compositions of H, C, and N in Cδ diamonds from the Allende and Murray carbonaceous chondrites (abstract). In *Lunar and Planetary Science XX,* pp. 1158–1159. Lunar and Planetary Institute, Houston.

Weigelt G., Balega Y., Blöcker T., Fleischer A. J., Osterbart R., and Winters J. M. (1998) 76 mas speckle-masking interferometry of IRC +10216 with the SAO 6 m telescope: Evidence for a clumpy shell structure. *Astron. Astrophys., 333,* L51–L54.

Winters J. M., Le Bertre T., Jeong K. S., Helling Ch., and Sedlmayr E. (2000) A systematic investigation of the mass loss mechanism in dust forming long-period variable stars. *Astron.*

Astrophys., 361, 641–659.
Wooden D. H. (1997) Observational evidence for mixing and dust condensation in core-collapse supernovae. In *Astrophysical Implications of the Laboratory Study of Presolar Materials* (T. J. Bernatowicz and E. Zinner, eds.), pp. 317–376. American Institute of Physics, New York.
Wooden D. H., Rank D. M., Bregman J. D., Witteborn F. C., Tielens A. G. G. M., Cohen M., Pinto P. A., and Axelrod T. S. (1993) Airborne spectrophotometry of SN 1987A from 1.7 to 12.6 microns — Time history of the dust continuum and line emission. *Astrophys. J. Suppl., 88,* 477–507.
Zinner E. (1998) Stellar nucleosynthesis and the isotopic composition of presolar grains from primitive meteorites. *Annu. Rev. Earth Planet. Sci., 26,* 147–188.
Zinner E. K. (2004) Presolar grains. In *Treatise on Geochemistry, Vol. 1: Meteorites, Comets, and Planets* (A. M. Davis, ed.), pp. 17–39. Elsevier, Oxford.

Meteorites and the Chemical Evolution of the Milky Way

Larry R. Nittler
Carnegie Institution of Washington

Nicolas Dauphas
The University of Chicago

The theory of galactic chemical evolution (GCE) describes how the chemical and isotopic composition of galaxies changes with time as succeeding generations of stars live out their lives and enrich the interstellar medium with the products of nucleosynthesis. We review the basic astronomical observations that bear on GCE and the basic concepts of GCE theory. In addition to providing a standard set of abundances with which to compare GCE predictions, meteorites also provide information about how the galaxy has evolved through the study of preserved presolar grains and radioactive isotopes.

1. INTRODUCTION

The solar system is situated in the disk of the Milky Way galaxy, some 8.5 kiloparsecs from the galactic center. It formed 4.567 G.y. ago (*Amelin et al.,* 2002) and its composition represents a snapshot of the composition of the Milky Way in the solar neighborhood at that time. The bulk composition of the solar system has often been referred to as the "cosmic composition" (*Anders and Grevesse,* 1989), with the underlying assumption that it represents the average composition of the galaxy. However, stellar nucleosynthesis is a phenomenon that is discrete in both time and space. The concept of cosmic composition therefore breaks down when the chemical and isotopic composition of the galaxy is examined at a fine scale and astronomy abounds with examples of objects and environments with nonsolar compositions. Galactic chemical evolution (GCE) is the name given to the theory of how the chemical composition of a galaxy varies with time and space as succeeding generations of stars live out their lives and enrich the interstellar medium (ISM) with the products of nucleosynthesis. Note that the word "chemical" is somewhat misleading in this context, since GCE refers only to the abundances of nuclei in the galaxy, not to the chemical state in which they might appear.

In this chapter, we discuss some of the ways in which meteorites can help unravel the presolar nucleosynthetic history of the Milky Way. A key constraint for models of GCE has long been the solar isotopic and elemental abundance pattern, largely determined by measurements of CI chondrites. However, meteorites also preserve a record of GCE in the form of preserved presolar dust grains and extinct radioactivities. Here, we will review the basic concepts of GCE theory and astronomical constraints before considering the role of GCE in determining the isotopic compositions of presolar grains and the abundances of radioactivities in the early solar system. Although the subject of GCE has a rich history, we concentrate on recent developments of the field. The reader is referred to the excellent books by *Pagel* (1997) and *Matteucci* (2003) for comprehensive reviews of the rich astronomical literature on the topic.

2. BASIC CONCEPTS AND ASTRONOMICAL CONSTRAINTS

Excluding dark matter, the Milky Way consists of several components, including a thin disk (to which the solar system belongs), a thick disk, a bright inner bulge, and a large spherical diffuse stellar halo. Just how these components formed is not known, but there are several competing models (*Matteucci,* 2003). Most likely, both primordial collapse of a protogalactic gas cloud and subsequent accretion and merger of smaller systems have played a role (*Gibson et al.,* 2003). However, many aspects of GCE modeling do not depend strongly on the overall model of galactic formation. Because the solar system belongs to the thin disk, we will primarily concentrate on observations and GCE models of this component.

A crucial quantity involved in any discussion of GCE is *metallicity*, which is the abundance of elements heavier than He ("metals" for astronomers). The letter Z is usually used to indicate the total metallicity. However, the normalized Fe abundance ($[Fe/H] = \log(Fe/H) - \log(Fe/H)_{\odot}$) is often used as a proxy for total metallicity, as Fe is relatively easy to measure in a large number of astronomical environments. The metallicity of the Sun ($Z_{\odot}$) has long been thought to be ~2%, but a recent downward revision in the solar O abundance by *Allende Prieto et al.* (2001) now indicates that it is closer to 1.4%.

Descriptions and models of GCE require a number of key ingredients:

1. *Boundary conditions:* The initial composition (often taken to be that generated by the Big Bang) and whether the system is closed or open must be defined.

2. *Stellar yields:* The abundances of isotopes produced by nucleosynthesis in stars of various types are required. These are determined by stellar evolutionary calculations coupled to nuclear reaction networks (e.g., *Meyer and Zinner,* 2006). In general, the nucleosynthesis abundance patterns ejected by stars depend critically on the stellar mass and metallicity. Moreover, there are still large uncertainties in the predicted yields due to uncertainties in the stellar evolution physics and nuclear reaction cross sections. A key concept related to stellar yields is the definition of *primary* and *secondary* species. A primary specie is one that can be synthesized in a zero-metallicity star, consisting initially of pure H and He. Examples of primary species are ^{16}O and ^{12}C, both made by stellar He burning. In contrast, nucleosynthesis of a secondary specie requires some preexisting metals to be present in the star. Some secondary species include the heavy isotopes of O, ^{14}N, and the *s*-process elements (*Meyer and Zinner,* 2006).

3. The *star formation rate* (SFR) is usually parameterized in GCE models. A very common parameterization is to assume that the SFR is proportional to the disk gas surface density (σ_g) to some power: $\Psi \propto \sigma^n$, where n = 1–2 (*Schmidt,* 1959). However, there are many examples of more complicated expressions for the SFR (e.g., *Dopita and Ryder,* 1994; *Wyse and Silk,* 1989).

4. The *initial mass function* [ϕ(m)] describes the number distribution of stars that form in a given mass interval at a given time. It is usually parameterized as a single- or multicomponent power law; for example, the common *Salpeter* (1955) IMF is: $\phi(M) \propto M^{-2.35}$. Other parameterizations have different power-law indices for different mass ranges (*Scalo,* 1986). In most models it is assumed that the IMF is constant in space and time; this is consistent with observational evidence (*Kroupa,* 2002).

5. Except for the simplest models (see next section), *infall* of gas from the halo onto the disk and *outflows* from galactic winds are often included in GCE models. The infall rate is usually assumed to vary with galactocentric radius and decrease with time. There is evidence for at least two independent episodes of infall leading to the formation of the halo and the thin disk respectively (*Chiappini et al.,* 1997).

Galactic chemical evolution models are constrained by a number of disparate observational data. These include the present-day values of the star formation and supernova rates, the present-day values of the surface mass density of stars and gas (Fig. 1), stellar mass function, and gas infall rate, as well as chemical abundances measured in a wide variety of stars and interstellar gas. The abundance data can be divided into a number of constraints:

1. *Solar abundances*: The composition of the Sun (e.g., *Anders and Grevesse,* 1989), determined both by spectroscopy of the Sun and analysis of CI chondrites, represents a sample of the ISM 4.6 G.y. ago and GCE models of the Milky Way disk should reproduce it (*Timmes et al.,* 1995). For many elements and isotopes, the solar system abundances are the only extant data with which to compare GCE models.

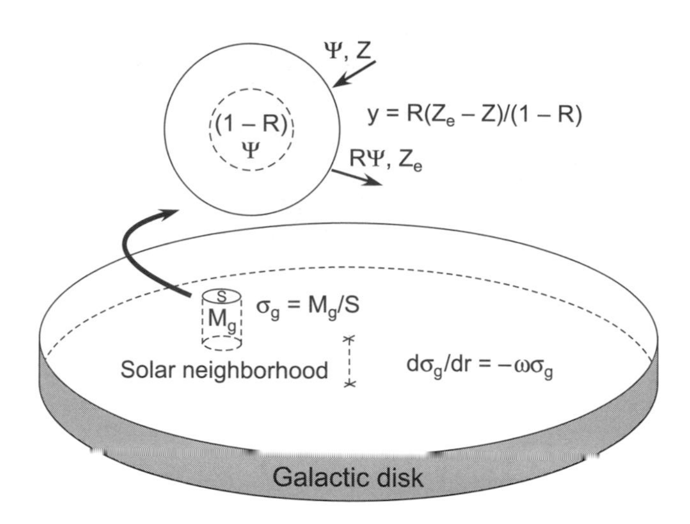

Fig. 1. Closed-box linear model with instantaneous recycling approximation. The gas surface density is the mass of gas contained in a cylinder divided by the area of the top surface of this cylinder. The rate of star formation is assumed to scale linearly with the gas density (linear model) The galactic disk is assumed to have been isolated for most of its history (closed-box model).

2. *Age-metallicity relationship (AMR)*: The measured metallicity of stars decreases on average with stellar age (*Twarog,* 1980; *Edvardsson et al.,* 1993), as would be expected from the basic idea of GCE. However, there is a large observed scatter in metallicity, greater than a factor of 2, for stars of a given age in the solar neighborhood (*Edvardsson et al.,* 1993; *Rocha-Pinto et al.,* 2000). This scatter is still not well-explained and makes the AMR a rather weak constraint for GCE models.

3. *Abundance ratio evolution*: Because different elements are formed by different nucleosynthetic processes, they can evolve at different rates. Thus, studies of element abundance ratios as a function of metallicity can provide important information about GCE (*McWilliam,* 1997). For example, low-metallicity stars have higher-than-solar ratios of so-called α elements (e.g., ^{16}O, ^{24}Mg) to Fe, but these ratios decrease to solar as solar [Fe/H] is reached. This reflects the fact that the α elements are made primarily in Type II supernovae, which evolve rapidly ($\lesssim 10^7$ yr), whereas a major fraction of Fe in the galaxy is made by Type Ia supernovae, which evolve on much longer timescales (~10^9 yr). Thus, high α/Fe ratios at low metallicities indicate that Type Ia supernovae had not yet had time to evolve and enrich the ISM with their ejecta. The exact shape of abundance trends are determined by the relative fractions and timescales of Type Ia and II supernovae as well as other details of GCE.

4. *G-dwarf metallicity distribution*: G dwarfs are low-mass stars (~0.9–1.1 $M_\odot$) that have lifetimes greater than or equal to the age of the disk. These stars are not active sites of nucleosynthesis and the compositions of their envelopes reflect the compositions of the interstellar gas from which they formed. Thus, their metallicity distribution (Fig. 2) represents a history of the star formation rate since the Milky

Way formed (*van den Bergh,* 1962; *Schmidt,* 1963; *Jørgensen,* 2000). As discussed below, the simplest closed-box GCE models overpredict the number of low-metallicity G dwarfs. G dwarfs are actually massive enough that some of them have begun to evolve away from the main sequence, which requires that a correction be applied to the metallicity distribution. Note, however, that stars with lower masses, such as K or M dwarfs, show the same discrepancy between the observed and the predicted abundance of metal deficient stars (*Kotoneva et al.,* 2002). Various approaches have been adopted for solving this problem, including preenrichment of the gas, varying initial mass function, or gas infall. Among these, infall of low-metallicity gas on the galactic plane is the most likely culprit. At present, so-called high-velocity clouds are seen falling on the galactic disk. Some of these clouds have the required low metallicity (down to ~0.1 $Z_\odot$) to solve the G-dwarf problem (*Wakker et al.,* 1999).

5. *Abundance gradients*: Observations of abundances in molecular clouds, stars, and planetary nebulae at a range of distances from the galactic center indicate the presence of metallicity gradients, where the inner galaxy is more metal-rich than the outer galaxy (*Matteucci,* 2003). This result indicates that the Milky Way disk formed in an inside-out fashion, with the inner disk forming on a shorter timescale than the outer disk. The precise values for metallicity gradients expected for the disk depend strongly on the balance between the radial dependences of the SFR and the infall rate.

3. GALACTIC CHEMICAL EVOLUTION MODELS

With the ingredients and observational constraints described in the previous section, models of GCE can be constructed. We first consider *homogeneous* models. These (semi-)analytic models make simplifying assumptions that allow the calculation of the mean properties and elemental abundance evolutionary trends of galactic systems. Most homogeneous GCE models of the galactic disk assume cylindrical symmetry and neglect both the finite thickness of the disk and any possible radial flows of matter. Thus, the only relevant spatial variable is the galactocentric radius and what is calculated is the mean abundance evolution within annuli about the galactic center.

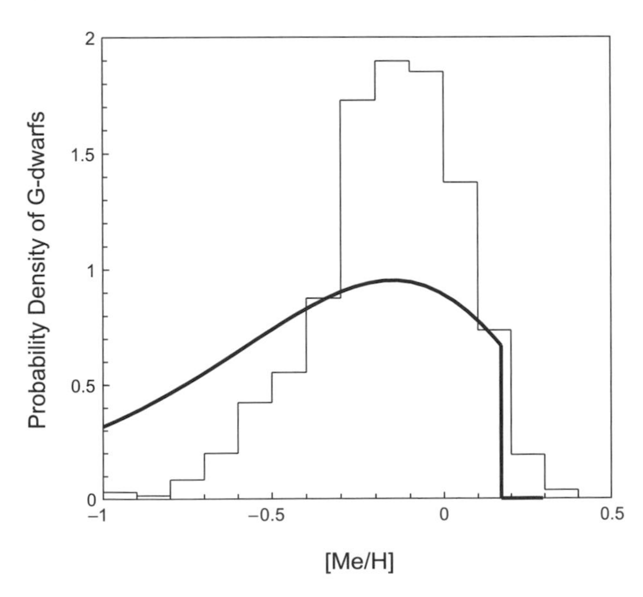

Fig. 2. G-dwarf metallicity distribution in the solar neighborhood. Metallicity is expressed as [Me/H] = log(Me/H)–log(Me/H)$_\odot$, where "Me" is a measurable element (usually O or Fe) Histogram: observations (*Nordström et al.,* 2004). Solid line: closed-box model (*Pagel,* 1997).

We first consider an oversimplified, but illustrative, case: the so-called "simple" model (e.g., *Schmidt,* 1963; *Pagel,* 1997; *Matteucci,* 2003) (Fig. 1). Let us denote Z_i as the mass fraction of nuclide i in the ISM. The galaxy formed from gas of low metallicity ($Z_{i0} = 0$). No mass loss or gain occurs during the galactic history (closed-box model). If E(t) is the rate of gas ejection [per unit surface area of the disk (Fig. 1)] from late-evolving stars and $\Psi(t)$ is the rate of gas accretion on nascent stars, then the rate of change of the gas surface density can be written as

$$d\sigma_g/dt = E - \Psi \quad (1)$$

The rate of change of the abundance of any nuclide i is

$$d(Z_i\sigma_g)/dt = Z_{i,e}E - Z_i\Psi - Z_i\sigma_g/\tau \quad (2)$$

where $Z_{i,e}$ is the mass fraction of i in the ejecta of all stars and τ is the mean-life (if the nuclide is radioactive). There is no delay between accretion of gas in stars and return of the nucleosynthetically enriched gas to the ISM (instantaneous recycling approximation). This assumption allows us to write $E(t) = R\Psi(t)$, where R is the so-called "return fraction," the rate at which mass is returned to the ISM. That is, as soon as a generation of stars is formed, a fraction R instantaneously comes back out, and a fraction (1 – R) remains locked up in stars (white dwarfs, etc.), which do not return mass to the ISM.

After some rearrangement and combining equations (1) and (2), it is straightforward to show that

$$dZ_i/dt = R\Psi(Z_{i,e} - Z_i)/\sigma_g - Z_i/\tau \quad (3)$$

Defining the yield y as the quantity of newly synthesized matter per unit mass of stellar remnants

$$y = R(Z_{i,e} - Z_i)/(1 - R) \quad (4)$$

and assuming a metallicity-independent yield ($Z_{i,e} - Z_i$ is constant) and a linear star formation rate

$$\Psi = \omega\sigma_g/(1 - R) \quad (5)$$

it follows that

$$dZ_i/dt = y\omega - Z_i/\tau \quad (6)$$

This equation can be integrated by the method of varying constant

$$Z_i = y\omega\tau(1 - e^{-t/\tau}) \quad (7)$$

In the limit of a stable nuclide ($\tau \rightarrow \infty$), the previous equation assumes the form

$$Z_i(\tau \rightarrow \infty) = y\omega t \quad (8)$$

An immediately obvious result is that the simple model predicts that the abundance of a stable primary isotope (whose yield y is independent of time and metallicity) increases linearly with time. Moreover, and of great importance to the interpretation of presolar grain data (section 6), the simple model also predicts that the ratio of a secondary isotope to a primary one increases linearly with total metallicity (*Pagel,* 1997). Note that a great deal of physics is hidden by the formulation above. For example, actual calculation of y requires integrating nucleosynthetic yields of stars of different mass over the initial mass function. We will revisit the linear closed-box model and its predictions for radioactive nuclei in section 4.

The main virtue of this model is its simplicity. However, the simple model fails to explain the G-dwarf metallicity distribution (Fig. 2) in that it predicts far more metal deficient stars than what is observed. This is known as the G-dwarf problem. As discussed in the previous section, it now appears most likely that the discrepancy is due to the closed-box assumption; gas flows, especially infall of low-metallicity gas on the disk, must be included in GCE models. Beyond the G-dwarf problem, the simple model also breaks down when trying to describe the evolution of elements produced by long-lived stars, for which the instantaneous recycling approximation is not valid. Moreover, the observed age-metallicity relationship for stars in the solar neighborhood is not well described by equation (8). Thus, more realistic models of GCE are needed. Including infall in an analytic GCE model requires parameterization of the infall rate (e.g., *Larson,* 1974; *Lynden-Bell,* 1975). *Clayton* (1985) provided a very flexible family of GCE models, which he called the "standard model." These models are exactly soluble and allow great freedom in parameterizing both the star formation rate and the infall rate. Clayton's standard model is very useful for understanding the physical behavior of galactic gas without resorting to numerical calculations. *Pagel* (1989) modified Clayton's model to include a fixed time delay for elements produced by long-lived stars (e.g., Fe, *s*-process elements).

Although sophisticated analytical GCE formalisms, like those of *Clayton* (1985) and *Pagel* (1989), are very useful for providing gross physical understanding of GCE, more realistic models require complete relaxation of the instantaneous recycling approximation and closed-box assumptions. Such models do not usually have analytical solutions and require numerical calculation. Typical formalisms (e.g., *Matteucci and Greggio,* 1986; *Timmes et al.,* 1995) require solving coupled integro-differential equations with separate terms describing ISM enrichment by stellar ejecta, star formation, infall, outflow and radioactive decay, respectively

$$\begin{aligned}\frac{d}{dt}\sigma_i(r,t) = \int_{M_L}^{M_U} & \psi(r,t-\tau_m)X_{mi}(t-\tau_m)\phi(m)dm \\ & -Z_i(r,t)\psi(r,t) \\ & +\frac{d}{dt}\sigma_i(r,t)_{infall} - \frac{d}{dt}\sigma_i(r,t)_{outflow} \\ & -\sigma_i(r,t)/\tau_i\end{aligned} \quad (9)$$

where σ_i (r,t) is the surface mass density of gas in the form of isotope i at galactocentric radius r and time t; M_L and M_U are the lower and upper mass limits, respectively, of stars that enrich the ISM at a given time; ψ is the star formation rate; X_{mi} is the mass fraction of i ejected by a star of mass m; ϕ is the initial mass function; τ_m is the lifetime of a star of mass m; $Z_i(t)$ is the mass fraction of i in the ISM at time t; and τ_i is the mean lifetime for isotope i. In fact, the first term is often divided into separate integrals for the ejecta from single stars and binary stars, since a fraction of the latter will result in supernovae of Type Ia and/or novae (*Matteucci and Greggio,* 1986; *Roman and Matteucci,* 2003).

A large number of numerical homogeneous GCE calculations including infall and neglecting the instantaneous recycling approximation have been reported in recent years. Some general conclusions can be drawn from many of these models: (1) The solar abundances of most isotopes up to Zn can be reproduced to within a factor of 2 (*Timmes et al.,* 1995); in most cases discrepancies are due to uncertainties in the nucleosynthetic processes and yields responsible for the specific isotope. (2) The G-dwarf metallicity distribution can be quite well approximated if the local disk formed by infall of extragalactic gas over a period of several gigayears. (3) The observed abundance ratio trends (e.g., O/Fe vs. Fe/H) are well explained by the time delay between supernovae of Type Ia and Type II. (4) To explain abundance gradients, the disk must have formed inside-out with a strong dependence of the star formation rate on galactocentric radius. Despite the success of modern GCE models in reproducing a large number of observational constraints, it should be remembered that there are still many crucial uncertainties. Of particular concern are remaining uncertainties in the nucleosynthetic yields of many isotopes, especially those of the Fe-peak elements, as well as the precise form and possible spatial or temporal variability of the initial mass function.

The homogeneous models described above explain well many of the average properties of the galaxy, for example, the average element/Fe ratios measured in stars of a given metallicity. However, because of the discrete and stochastic nature of star formation and evolution, local chemical heterogeneities about mean trends are to be expected. Observationally, the scatter in elemental abundances in stars increases with decreasing metallicity. This is especially true for the low-metallicity stars of the galactic halo. Although there is a very large scatter in metallicity for disk stars of a given age in the solar neighborhood (*Edvardsson et al.,* 1993; *Reddy et al.,* 2003), metal abundance ratios (e.g., Mg/Fe) in the disk do not show resolvable scatter around the mean trends with metallicity. A number of *heterogeneous* GCE models have been published that attempt to explain abundance scatter (and its decrease with increasing metallicity) in the halo (*Argast et al.,* 2000; *Oey,* 2000; *Travaglio et al.,* 2001) and the large scatter of metallicity in the disk (e.g., *Copi,* 1997; *van den Hoek and de Jong,* 1997). We will consider the issue of heterogeneous GCE and its effects on presolar grain isotopic compositions in section 6.1.

In recent years, an additional class of GCE models has been explored, *galactic chemodynamical* (GCD) models, in which the chemical evolution of the galaxy is explicitly tied to its dynamical evolution. Galactic chemodynamical models range from relatively simple models exploring the effects of radial diffusion of stellar orbits coupled to abundance gradients (*Grenon,* 1987; *François and Matteucci,* 1993; *Clayton,* 1997) to quite sophisticated three-dimensional chemodynamic codes that attempt to self-consistently treat the dynamics of galactic gas, dust, and dark matter along with abundance evolution (*Raiteri et al.,* 1996; *Brook et al.,* 2003).

4. NUCLEAR COSMOCHRONOLOGY AND EXTINCT RADIOACTIVITIES

Nuclear cosmochronology, also sometimes termed nucleocosmochronology, is the study of radioactive nuclei with the goal of constraining the timescales of nucleosynthesis and galaxy formation. This field takes its root in a paper published in 1929, where E. Rutherford first used uranium to estimate the age of Earth and erroneously concluded that "the processes of production of elements like uranium were certainly taking place in the sun 4×10^9 years ago and probably still continue today" (*Rutherford,* 1929). Actually, the U in the solar system was produced, together with other actinides, by the *r*-process of nucleosynthesis in very energetic events, before the birth of the solar system (*Meyer and Zinner,* 2006). The primary aspiration of nuclear cosmochronology historically has been to retrieve the age of the Milky Way and the duration of nucleosynthesis from the abundances of unstable nuclei measured in meteorites.

The abundances of radioactive nuclides in the interstellar medium (ISM) represent a balance between production in stars and decay in the ISM (*Tinsley,* 1977, 1980; *Yokoi et al.,* 1983; *Clayton,* 1985, 1988a; *Pagel,* 1997). These abundances depend very strongly on the dynamical evolution of the galaxy. For instance, the abundances of radioactivities in the ISM would be different if all nuclides had been synthesized in a stellar burst shortly after the formation of the galaxy or if they had been synthesized throughout the galactic history. Radioactivities therefore provide invaluable tools for probing the nucleosynthetic history of matter. In order to investigate the formation of the solar system, one can either proceed forward or backward in time. The abundances of radioactive nuclides in the ISM at solar system birth can be theoretically predicted from models of GCE and stellar nucleosynthesis. The abundances of radioactive nuclides can also be determined from laboratory measurements of extraterrestrial materials, for example, by detection of decay products of now-extinct nuclides. Comparisons between the predicted and the observed abundances provide unequaled pieces of information on galactic nucleosynthesis history and the origin of short-lived nuclides in the early solar system.

4.1. Modeling the Remainder Ratio

When investigating radionuclides in the ISM, it is useful to introduce the remainder ratio ($\mathfrak{R}$), which is the ratio of the abundance of an unstable nuclide to its abundance if it had been stable (*Clayton,* 1988a; *Dauphas et al.,* 2003)

$$\mathfrak{R} = N(\tau)/N(\tau \to \infty) \tag{10}$$

If R is the ratio of the unstable nuclide to another stable nuclide cosynthesized at the same site and P is the production ratio, then the remainder ratio in the early solar system (ESS) can be calculated as

$$\mathfrak{R}_{ess} = R/P \tag{11}$$

The remainder ratio in the ISM at the time of solar system formation can be calculated in the framework of GCE models. Let us begin with the "simple" GCE model discussed in section 3. (Fig. 1). The remainder ratio in the ISM (equation (10)) can be calculated from equation (7)

$$\mathfrak{R}_{ism} = \frac{\tau}{T}(1 - e^{-T/\tau}) \tag{12}$$

The presolar age of the galaxy is denoted T ($T = T_G - T_\odot$, where T_G is the age of the galaxy and $T_\odot$ is the age of the solar system). Thus, in the closed-box linear model, the remainder ratio of a given nuclide depends on its mean-life τ and on the presolar age of the galactic disk T. For short-lived nuclides ($\tau \ll T$), the remainder ratio becomes

$$\mathfrak{R}_{ism} = \tau/T \tag{13}$$

For very short-lived species, the timescale of ISM mixing is longer than the mean-life of the nuclide, granularity of nucleosynthesis must be taken into account, and the notion of steady-state abundances is inappropriate. *Meyer and Clayton* (2000) estimated that the cut where the concept of steady-state abundances breaks down must be for mean-lives around 5 m.y. As discussed earlier, the simple model fails to explain important astronomical observations, notably the G-dwarf metallicity distribution. *Clayton* (1985, 1988a) estimated the remainder ratio for short-lived nuclides in the context of a parameterized linear infall model. More recently, *Dauphas et al.* (2003) improved over this model using a more realistic nonlinear parameterization of the star formation rate [$d\sigma_g/dt = -\omega\sigma_g^n$, with n close to 1.4 (*Gerritsen and Icke,* 1997; *Kennicutt,* 1998; *Kravtsov,* 2003)] and a parameterized infall rate following *Chang et al.* (1999). As discussed by *Clayton* (1988a) and *Dauphas et al.* (2003), when infall of low-metallicity gas is taken into account, equation (13) for the remainder ratio for short-lived nuclides is not valid and the expression

$$\Re_{ISM} = \kappa\tau/T \quad (14)$$

should be used instead. The numerical GCE model of *Dauphas et al.* (2003) gives $\kappa = 2.7 \pm 0.4$ (unless otherwise indicated, all errors in this chapter are 2σ), which is within the range of values advocated by *Clayton* (1985), $2 < \kappa < 4$. For long-lived radionuclides, *Clayton* (1988a) derived an analytic solution for the remainder ratio. If nonlinearity is taken into account, the remainder ratio must be calculated numerically (equation (4) of *Dauphas et al.,* 2003).

4.2. Age of the Galaxy and the Uranium/Thorium Production Ratio

As illustrated in the previous section, the remainder ratio in the ISM at the time of solar system formation depends on the age of the galactic disk. For short-lived radionuclides, it is possible that significant decay can occur between the last nucleosynthesis event and actual incorporation into the solar system's parent molecular cloud core (see section 4.3). For long-lived radionuclides, however, such a free-decay interval can be neglected and the remainder ratio in the ISM must be equal to that in the ESS. It is thus possible to determine the age of the galaxy if a GCE model is specified, if the abundance of the considered radionuclide in the ESS is known, and if its production ratio normalized to a neighbor nuclide co-synthesized at the same site is known. The first meaningful attempt to calculate the radiometric age of the Milky Way was reported in the seminal paper of *Burbidge et al.* (1957). Using a U/Th production ratio of 0.64 and a model of constant production, they estimated an age of approximately 10 G.y. for the galaxy. In the last 50 years, multiple studies have addressed this question and the reader will find ample details in some of these contributions (e.g., *Tinsley,* 1980; *Yokoi et al.,* 1983; *Clayton,* 1988a; *Cowan et al.,* 1991; *Meyer and Truran,* 2000). *Clayton* (1988a) evaluated potential nuclear cosmochronometers and concluded that the pair ^{238}U ($\tau = 6.446 \times 10^9$ yr)–^{232}Th ($\tau = 2.027 \times 10^{10}$ yr) gives the most stringent constraint on the age of the galactic disk. We shall therefore focus our discussion on the $^{238}U/^{232}Th$ ratio, which will simply be denoted U/Th hereafter. There are two approaches that can be used to estimate the age of the Milky Way based on the U/Th ratio. One relies on the determination of this ratio in the spectra of low-metallicity stars in the halo of the galaxy. The second relies on the U/Th ratio measured in meteorites and makes use of galactic chemical evolution models.

The U/Th ratio measured in meteorites (*Chen et al.,* 1993; *Goreva and Burnett,* 2001) is the result of an interplay between production in stars, enrichment of the gas by stellar ejecta, and decay in the ISM. Because U decays faster than Th, its ratio in the ISM changes with time. If one specifies the history of nucleosynthesis before solar system formation, a relationship can be found between the U/Th production ratio, the ratio measured in the ISM at solar system formation (as recorded in meteorites), and the age of the galaxy. For instance, let us consider that actinides were all synthesized at the time the galaxy formed and that they were not subsequently replenished (initial stellar burst scenario). In this case, we can write a simple free-decay equation

$$R_{\odot}^{U/Th} = P^{U/Th}e^{(\lambda_{Th} - \lambda_U)(T_G - T_{\odot})} \quad (15)$$

where $R_{\odot}^{U/Th}$ is the ratio in the ISM at solar system formation, $P^{U/Th}$ is the production ratio, and T_G is the total age of the galaxy (to present). The ratio in the ISM ($R_{\odot}^{U/Th}$) is measured in meteorites; the production ratio ($P^{U/Th}$) can be derived from the theory of *r*-process nucleosynthesis. The age (T_G) can therefore be calculated. Of course, all actinides were not produced in an initial burst and it is thus necessary to consider more realistic GCE models. Such models, describing enrichment of the ISM in actinides through time, can be constrained by a host of astronomical observations (*Yokoi et al.,* 1983; *Clayton,* 1988a; *Dauphas,* 2005a), but a detailed discussion of how the models are parameterized and constrained is beyond the scope of this chapter. The most important feature of the models is that they incorporate infall of low-metallicity gas on the galactic plane. In a recent paper, *Dauphas* (2005a) showed that the relationship between the production ratio, the meteorite ratio, and the age of the galaxy can be approximated by a simple formula, valid between 10 and 20 G.y.

$$P^{U/Th} = R_{\odot}^{U/Th}/(aT_G + b) \quad (16)$$

where $a = -1.576 \times 10^{-2}$ and $b = 0.9946$ (see *Dauphas,* 2005a, for details). The U/Th ratio in the ISM at solar system formation is 0.438 ± 0.006 (*Chen et al.,* 1993). *Goriely and Arnould* (2001) and *Schatz et al.* (2002) recently quantified the influence of nuclear model uncertainties on the *r*-process nucleosynthesis of actinides. The $^{238}U/^{232}Th$ production ratio is estimated by *Schatz et al.* (2002) to be 0.60 ± 0.14, while *Goriely and Arnould* (2001) propose a more conservative range of $0.435^{+0.329}_{-0.137}$ (error bars represent 68% confidence intervals). The range of production ratios estimated by modern *r*-process calculations encompasses the initial solar composition and the approach based on GCE can therefore only provide an upper limit on the age of the galaxy. If we adopt $P^{U/Th} < 0.7$, we can derive an upper limit for the age of the Milky Way of approximately 20 G.y., which is useless in comparison to the precision with which the age of the universe is known [13.7 G.y. (*Spergel et al.,* 2003)]. This shows that the solar U/Th ratio alone cannot be used to constrain the duration of nucleosynthesis.

The U/Th ratio measured in low-metallicity halo stars can also be used as a potential chronometer to determine the age of the Milky Way (*Cayrel et al.,* 2001; *Hill et al.,* 2002; *Cowan et al.,* 2002). These stars formed very early in the galactic history and they inherited at their formation a U/Th ratio that must have been equal to the production ratio by *r*-process nucleosynthesis. *Hill et al.* (2002) measured the most precise U/Th ratio in the low-metallicity halo star CS 31082-001 of 0.115 ± 0.029. For such stars, a simple free decay equation can be written

$$P^{U/Th} = R^{U/Th}_{LMHS} e^{(\lambda_U - \lambda_{Th})T_G} \quad (17)$$

Again, the ratio $R^{U/Th}_{LMHS}$ in low-metallicity halo stars can be measured, the production ratio $P^{U/Th}$ can be derived from the theory of *r*-process nucleosynthesis, and it is therefore possible to calculate the age of the galaxy T_G. *Goriely and Arnould* (2001) propagated the uncertainty on the production ratio and concluded that the age cannot be constrained to better than 9–18 G.y. As in the case of GCE and the solar U/Th ratio, this range is of limited use when trying to establish the chronology of structure of formation in the universe and other methods give more precise estimate of the age of the galaxy (*Krauss and Chaboyer,* 2003; *Hansen et al.,* 2004).

The main source of uncertainty in calculations of the age of the Milky Way based on low-metallicity halo stars and galactic chemical evolution is the U/Th production ratio. In a recent contribution, *Dauphas* (2005a) argued that because there are two equations (equations (16) and (17)) in two unknowns ($P^{U/Th}$ and T_G), the system could actually be solved (Fig. 3). The values that he derived for the age and the production ratio are $14.5^{+2.8}_{-2.2}$ G.y. and $0.571^{+0.037}_{-0.031}$, respectively. The virtue of this approach is that a probabilistic meaning can be ascribed to the uncertainty interval and the U/Th production ratio can be determined independently of *r*-process calculations. The oldest stars in our galaxy formed shortly after the birth of the universe [13.7 ± 0.2 G.y. (*Spergel et al.,* 2003)].

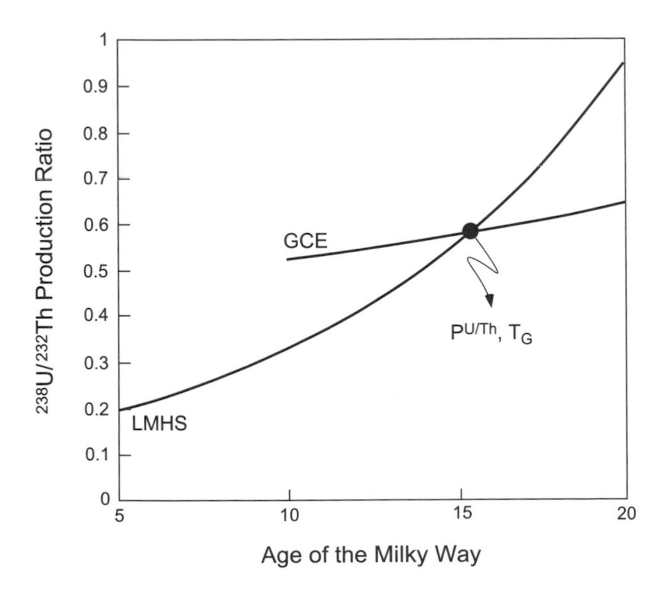

Fig. 3. Calculated $^{238}U/^{232}Th$ production ratio as a function of the age of the Milky Way based on the U/Th ratio measured in low-metallicity halo stars (curve LMHS, equation (17)) and the U/Th ratio measured in solar system material (curve GCE, equation (16)). By combining these two approaches, it is possible to estimate both the age T_G and the production ratio $P^{U/Th}$ (*Dauphas,* 2005a).

4.3. Short-Lived Nuclides in the Early Solar System

The GCE models presented previously all assume that nucleosynthesis is a smooth function of time. Stars are actually discrete in both time and space. Long-lived radionuclides retain a long memory and deterministic models can be applied (*Clayton,* 1988a; *Dauphas et al.,* 2003). In contrast, short-lived radionuclides may be affected by the nucleosynthetic history of the solar neighborhood right before solar system formation. For these nuclides, a stochastic treatment should be applied (e.g., *Meyer and Luo,* 1997). Very short-lived extinct radionuclides, such as ^{26}Al, might have been injected from a nearby giant star [asymptotic giant branch (AGB) or supernova (SN)] that might have triggered the protosolar nebula into collapse. In the present contribution, the discussion is limited to extinct radionuclides that have mean-lives long enough that the composition of the ISM can be estimated in a deterministic way [$\tau > 5$ m.y. (*Meyer and Clayton,* 2000)]. This comprises the nuclides ^{53}Mn (mean-life 5.4 m.y.), ^{92}Nb (50.1 m.y.), ^{107}Pd (9.4 m.y.), ^{129}I (22.6 m.y.), ^{146}Sm (148.6 m.y.), ^{182}Hf (13.0 m.y.), and ^{244}Pu (115.4 m.y.). The abundances of all these nuclides in the ESS are well known. For calculating the remainder ratio, the abundances of these nuclides must be normalized to the abundances of neighbor nuclides produced in the same stellar environment. Hence, the normalizing nuclides used in the present contribution are not always identical to those used in the initial publications reporting extinct nuclide abundances. For evaluating the remainder ratio in the ESS, the basic ingredients are the normalized abundances of the short-lived nuclides in the ESS at the time of calcium-aluminum-rich inclusion (CAI) formation and the associated production ratios.

Manganese-53 is synthesized together with ^{55}Mn in massive stars. As discussed by *Meyer and Clayton* (2000), because solar system ^{53}Cr must have been primarily synthesized as ^{53}Mn, the production ratio $^{53}Mn/^{55}Mn$ can be approximated by the solar system $^{53}Cr/^{55}Mn$ ratio of 0.13 (*Anders and Grevesse,* 1989). The SN II models of *Rauscher et al.* (2002) predict a comparable production ratio, $^{53}Mn/^{55}Mn = 0.15$, when individual yields from SN II of different masses are weighted by a typical initial mass function. Note that Sne II underproduce Fe-peak nuclides relative to ^{16}O by a factor of 2–3 (*Rauscher et al.,* 2002); the rest must be produced in SN Ia. This underproduction feature is reflected in the abundance patterns of low-metallicity halo stars (*Wheeler et al.,* 1989; *McWilliam,* 1997). The initial $^{53}Mn/^{55}Mn$ ratio in CAIs is estimated to be $2.81 \pm 0.31 \times 10^{-5}$ (*Birck and Allègre,* 1985; *Nyquist et al.,* 2001). However, a lower initial value may be required in order to bring the chronologies based on the various extinct and extant radionuclides into agreement (*Lugmair and Shukolyukov,* 1998; *Dauphas et al.,* 2005). We shall adopt here an initial ratio of $1.0 \pm 0.2 \times 10^{-5}$. The remainder ratio is therefore $\mathfrak{R}^{53}_{ess} = 7.7 \pm 1.5 \times 10^{-5}$.

Niobium-92 was most likely synthesized by the *p*-process in SN. This radionuclide cannot be normalized to another isotope of the same element because the only stable isotope of Nb, ^{93}Nb, was synthesized by the *s*-process. It can instead be normalized to ^{92}Mo, which is also a pure *p*-process nuclide. The $^{92}Nb/^{92}Mo$ production ratio during photodisintegration of seed nuclei in SN II is $1.5 \pm 0.6 \times 10^{-3}$ (*Rauscher et al.,* 2002; *Dauphas et al.,* 2003). The initial ratio in meteorites is $2.8 \pm 0.5 \times 10^{-5}$ (*Harper,* 1996; *Schönbächler et al.,* 2002). The remainder ratio is therefore $\mathfrak{R}^{92}_{ess} = 1.9 \pm 0.8 \times 10^{-2}$ (*Dauphas et al.,* 2003). Note that *Münker et al.* (2000) and *Yin et al.* (2000) found higher initial ratios, but the CAI measurements of *Münker et al.* (2000) might have been affected by nucleosynthetic effects, and the zircon measurement of *Yin et al.* (2000) was not replicated by *Hirata* (2001).

The initial $^{107}Pd/^{110}Pd$ in the ESS was $5.6 \pm 1.1 \times 10^{-5}$ (*Kelley and Wasserburg,* 1978; *Chen and Wasserburg,* 1996). This corresponds to $^{107}Pd/^{110}Pd^r = 5.8 \pm 1.2 \times 10^{-5}$, where $^{110}Pd^r$ is the *r*-process contribution to the solar abundance of ^{110}Pd (*Arlandini et al.,* 1999). Palladium-107 is primarily a *r*-process nuclide that could also have received a contribution from the *s*-process. For most *r*-process radionuclides, their production ratios can be reliably estimated by decomposing the abundances of their daughter nuclides into *s*- and *r*-process contributions. For instance, the entire *r*-process abundance of ^{107}Ag must have been channeled through ^{107}Pd. The sites of *r*-nucleosynthesis are not well established but they very likely correspond to the late stages of rapidly evolving stars. The $^{107}Pd/^{110}Pd$ production ratio can therefore be approximated to the solar system $^{107}Ag^r/^{110}Pd^r$, where $^{107}Ag^r$ and $^{110}Pd^r$ are obtained by subtracting the *s*-process contribution to solar abundances (*Arlandini et al.,* 1999). The $^{107}Pd/110Pd$ *r*-process production ratio is therefore 1.36 and the remainder ratio is $\mathfrak{R}^{107}_{ess} = 4.3 \pm 0.9 \times 10^{-5}$.

Iodine-129 is also primarily an *r*-process nuclide. It was historically the first short-lived nuclide to have been found to have been alive in the ESS (*Reynolds,* 1960). When the Pb/Pb age of Efremovka CAIs [4567.2 ± 0.6 Ma (*Amelin et al.,* 2002)] is combined with the observed $^{129}I/^{127}I$-Pb/Pb age correlation in ordinary chondrite phosphates (*Brazzle et al.,* 1999), the early solar system $^{129}I/^{127}I$ is estimated to be $1.19 \pm 0.20 \times 10^{-4}$. This corresponds to a $^{129}I/^{127}I^r$ initial ratio of $1.25 \pm 0.21 \times 10^{-4}$. As discussed in the case of ^{107}Pd, the $^{129}I/^{127}I$ *r*-process production ratio can be estimated from decomposition of ^{129}Xe into *r*- and *s*-processes (*Arlandini et al.,* 1999). Because ^{129}Xe is predominantly synthesized by the *r*-process, little uncertainty affects the production ratio, $^{129}I/^{127}I = 1.45$. The remainder ratio for ^{129}I is $\mathfrak{R}^{129}_{ess} = 8.6 \pm 1.5 \times 10^{-5}$.

Samarium-146 is a pure *p*-process isotope. The ESS $^{146}Sm/^{144}Sm$ ratio is $7.6 \pm 1.3 \times 10^{-3}$ (*Lugmair et al.,* 1983; *Prinzhofer et al.,* 1992). Its production ratio as obtained in the most recent models of the *p*- or γ-process in Sne II is $1.8 \pm 0.6 \times 10^{-1}$ (*Rauscher et al.,* 2002; *Dauphas et al.,* 2003). The remainder ratio is therefore $\mathfrak{R}^{146}_{ess} = 4.2 \pm 1.6 \times 10^{-2}$ (*Dauphas et al.,* 2003).

Hafnium-182 is presumably an *r*-process isotope. From the decomposition of the abundance of its daughter isotope ^{182}W into *r*- and *s*-processes (*Arlandini et al.,* 1999), the $^{182}Hf/^{177}Hf$ production ratio is estimated to be 0.81. The initial $^{182}Hf/^{177}Hf$ ratio in the ESS is $1.89 \pm 0.15 \times 10^{-4}$ (*Yin et al.,* 2002). Note that *Quitté and Birck* (2004) have derived a higher initial ratio from W-isotopic measurements of the Tlacotepec iron meteorite, but this may be affected by cosmogenic effects. The initial ratio of *Yin et al.* (2002) corresponds to $^{182}Hf/^{177}Hf^r = 2.31 \pm 0.18 \times 10^{-4}$. The remainder ratio is therefore $\mathfrak{R}^{182} = 2.86 \pm 0.23 \times 10^{-4}$.

Plutonium-244 is an *r*-process isotope. As with other actinides, its production ratio is uncertain because the closest stable *r*-process nuclide that can anchor the models is ^{209}Bi, 35 amu away. *Goriely and Arnould* (2001) evaluated nuclear model uncertainties on the production of actinides. Among the various models, only those that give a $^{238}U/^{232}Th$ production ratio consistent with meteoritic abundances are retained. The $^{244}Pu/^{238}U$ production ratio is thus estimated to be 0.53 ± 0.36. The initial $^{244}Pu/^{238}U$ in the solar system is estimated to be 0.0068 ± 0.0010 (*Rowe and Kuroda,* 1965; *Hudson et al.,* 1989). The ratio of the remainder ratios $\mathfrak{R}^{244}_{ess}/\mathfrak{R}^{238}_{ess}$ is therefore $1.28 \pm 0.89 \times 10^{-2}$. The remainder ratio of ^{238}U can be estimated in the framework of the open nonlinear GCE model (*Dauphas et al.,* 2003) to be 0.71 for a galactic age of 13.7 G.y. (0.53 in the closed-box model). The remainder ratio of ^{244}Pu is thus $\mathfrak{R}^{244}_{ess} = 9.1 \pm 6.3 \times 10^{-3}$ (*Dauphas,* 2005b). All these ratios are compiled in Table 1.

The remainder ratios in the ESS can be compared with those in the ISM as predicted by GCE models. In order to account for the possible isolation of the solar system parent molecular cloud core from fresh nucleosynthetic inputs, a free-decay interval (Δ) is often introduced [see *Clayton* (1983) for a more complicated treatment]. This corresponds to a time when radioactive species decay without being replenished by stellar sources. The remainder ratio in the ESS is related to that in the ISM through

$$\mathfrak{R}_{ess} = \mathfrak{R}_{ism} \exp\left(-\frac{\Delta}{\tau}\right) \quad (18)$$

The remainder ratio in the ISM is itself a function of the mean-life of the considered nuclide (equation (14)). For radionuclides whose mean-lives are long enough that their abundances in the ESS can be explained by their steady-state abundance in the ISM, combining equations (14) and

TABLE 1. Extinct radionuclides in the ESS (only those with mean-lives $\tau > 5$ m.y. are listed, *Dauphas et al.*, 2003; *Dauphas*, 2005b).

Nuclide	τ (Ma)	Norm.	R	P	$\Re^i_{ess}$	$\Re^i_{ism}$
Manganese-53	5.4	^{55}Mn	$1.0 \pm 0.2 \times 10^{-5}$	$\sim 1.3 \times 10^{-1}$	$7.7 \pm 1.5 \times 10^{-5}$	$1.67 \pm 0.38 \times 10^{-3}$
Niobium-92	50.1	^{92}Mo	$2.8 \pm 0.5 \times 10^{-5}$	$1.5 \pm 0.6 \times 10^{-3}$	$1.9 \pm 0.8 \times 10^{-2}$	$1.55 \pm 0.35 \times 10^{-2}$
Palladium-107	9.4	^{110}Pd^r	$5.8 \pm 1.2 \times 10^{-5}$	~ 1.36	$4.3 \pm 0.9 \times 10^{-5}$	$2.91 \pm 0.66 \times 10^{-3}$
Iodine-129	22.6	^{127}I^r	$1.25 \pm 0.21 \times 10^{-4}$	~ 1.45	$8.6 \pm 1.5 \times 10^{-5}$	$7.03 \pm 1.6 \times 10^{-3}$
Samarium-146	148.6	^{144}Sm	$7.6 \pm 1.3 \times 10^{-3}$	$1.8 \pm 0.6 \times 10^{-1}$	$4.2 \pm 1.6 \times 10^{-2}$	$4.61 \pm 1.05 \times 10^{-2}$
Hafnium-182	13.0	^{177}Hf^r	$2.31 \pm 0.18 \times 10^{-4}$	~ 0.81	$2.86 \pm 0.23 \times 10^{-4}$	$4.03 \pm 0.92 \times 10^{-3}$
Plutonium-244	115.4	^{238}U	$6.8 \pm 1.0 \times 10^{-3}$	$5.3 \pm 3.6 \times 10^{-1}$	$9.1 \pm 6.3 \times 10^{-3}$	$3.58 \pm 0.81 \times 10^{-2}$

The superscript r refers to the *r*-process component of the cosmic abundances [obtained after subtracting the *s*-process contribution (*Arlandini et al.*, 1999)]. R is the ratio observed in the ESS, P is the production ratio, $\Re^i_{ess}$ is the remainder ratio in the ESS calculated as R/P (except for ^{244}Pu, for which the normalizing isotope is unstable and the remainder ratio must be corrected for $\Re^{238} = 0.71$), $\Re^i_{ism}$ is the remainder ratio in the ISM as obtained from an open nonlinear GCE model (*Dauphas et al.*, 2003). The remainder ratio in the ISM is equal to $\Re^i_{ism} = \kappa\tau/T$, with $\kappa = 2.7 \pm 0.4$ and $T = 8.7 \pm 1.5$ G.y. All errors are 2σ. See text for details and references.

(18) gives a relationship between the remainder ratios in the ISM and the ESS (*Dauphas et al.*, 2003; *Dauphas*, 2005b)

$$\Re_{ess} = \Re_{ism} \exp\left(-\frac{\kappa\Delta}{T\Re_{ism}}\right) \quad (19)$$

The remainder ratios of several short-lived radionuclides determined in the ESS are plotted vs. the ISM ratios derived from GCE models in Fig. 4. Also shown are curves corresponding to equation (19) calculated with different values of the free-decay interval Δ. Some implications of the presence of extinct radionuclides in meteorites on stellar nucleosynthesis and solar system formation are discussed in detail in the following sections.

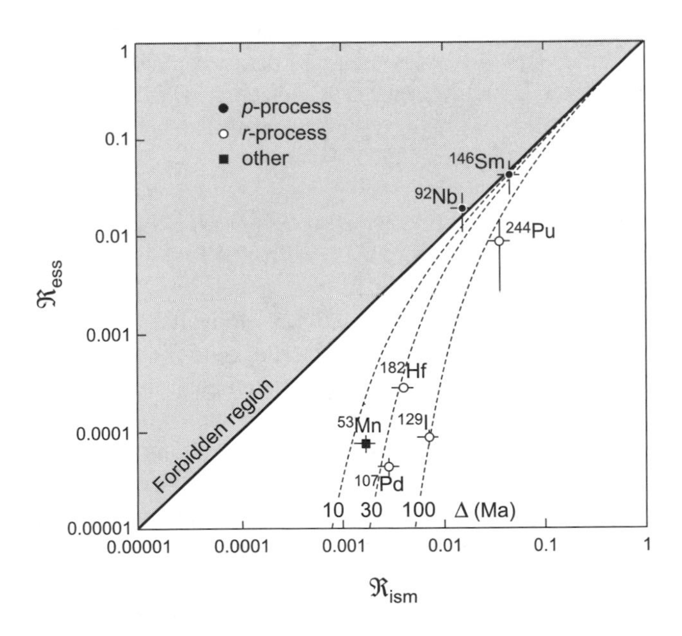

Fig. 4. Remainder ratios of several extinct nuclides in the ESS ($\Re_{ess}$ = R/P) are plotted against the corresponding remainder ratios in the ISM (see text). Theoretical expectations corresponding to different free-decay intervals (Δ) are represented as dashed-curves (equation (19)) (*Dauphas et al.*, 2003; *Dauphas*, 2005b). Extinct radionuclides are grouped according to nucleosynthetic processes. The forbidden region is the place where $\Re_{ess}$ is higher than $\Re_{ism}$ (the extinct radionuclides cannot be inherited from GCE). Error bars are 2σ.

4.3.1. Niobium-92 and the nucleosynthesis of molybdenum-ruthenium p-isotopes. The radionuclides ^{53}Mn and ^{146}Sm define the same free-decay interval within uncertainties ($\Delta \simeq 10$ m.y.). They were probably synthesized in supernovae and were inherited in the ESS from GCE. Niobium-92 is a special case because it lies in a mass region of the nuclide chart where SN models underproduce *p*-process isotopes (^{92}Mo, ^{94}Mo, ^{96}Ru, and ^{98}Ru) by a factor of 10 (*Rauscher et al.*, 2002). This nuclide can therefore be used to test the various hypotheses that have been advanced to remedy the underproduction feature of supernovae in the Mo-Ru mass region (*Yin et al.*, 2000; *Dauphas et al.*, 2003).

4.3.2. The puzzling origins of extinct r-radioactivities. The *r*-process radionuclides ^{129}I and ^{244}Pu were inherited from GCE with a free-decay interval of approximately 100 m.y. (Fig. 4). If the other *r*-process nuclides ^{107}Pd and ^{182}Hf had been inherited from GCE with the same free-decay interval (100 m.y.), then their abundances in the ESS would have been much lower than what is observed in meteorites (they require a shorter free-decay interval, 30 m.y.). The extinct radionuclides ^{107}Pd and ^{182}Hf must therefore have a different origin. Two distinct scenarios have been advocated for explaining the origin of these two short-lived nuclides.

Wasserburg et al. (1996) and *Qian et al.* (1998) questioned the universality of the so-called *r*-process. They suggested that two kinds of *r*-process events are responsible for the nucleosynthesis of neutron-rich nuclei. One of these events would synthesize heavy *r*-nuclei and actinides (^{182}Hf and ^{244}Pu, H events) while the other would synthesize light *r*-nuclei (^{129}I, L events). The H events would occur 10 times more frequently than the L events, which would explain why the free-decay interval inferred from ^{182}Hf is lower than that inferred from ^{129}I. Observations of elemental abundances in low-metallicity stars support the view that all *r*-nuclides are not produced at the same site (*Sneden et al.,* 2000; *Hill et al.,* 2002). Such stars formed early enough in the galactic history that contributions of a limited number of stars can be seen in their spectra. *Sneden et al.* (2000) analyzed the ultra-low-metallicity ([Fe/H] = –3.1) halo star CS 22892-052 and found that low-mass *r*-process elements such as Y and Ag were deficient compared to expectations based on heavier *r*-process elements. More recently, *Hill et al.* (2002) determined the abundance of U, Th, and Eu in another metal-poor ([Fe/H] = –2.9) halo star (CS 31082-001). The age of this star based on the U/Th ratio is $14.5^{+2.8}_{-2.2}$ G.y. (section 4.2), which agrees with independent estimates of galactic ages. In contrast, the Th/Eu ratio corresponds to an age that is younger than the age of the solar system, which is impossible (*Hill et al.,* 2002). This suggests that actinides were produced independently of lighter *r*-process nuclides. Observations of low-metallicity stars therefore point to a multiplicity of *r*-process events, possibly as many as three. *Wasserburg et al.* (1996) and *Qian et al.* (1998) grouped ^{182}Hf with actinides, including ^{244}Pu. However, *Dauphas* (2005b) showed that ^{244}Pu requires a longer free-decay interval ($\Delta \simeq$ 100 m.y.) compared to ^{182}Hf ($\Delta \simeq$ 30 m.y.). This discrepancy may indicate that, in addition to the L and H events, another event must be added to explain actinides (A events). Note that this would be consistent with observations of low-metallicity stars, which require three distinct production sites. According to the multiple *r*-processes model, ^{107}Pd was synthesized by the *s*-process in a low-metallicity, high-mass AGB star that polluted the ESS with a stellar wind (*Wasserburg et al.,* 1994; *Gallino et al.,* 2004). Alternatively, it could have been injected by the explosion of a nearby SN, where it would have been synthesized by the weak *s*-process (*Meyer and Clayton,* 2000).

Another possible scenario is that the nuclides that are overabundant in the ESS compared to GCE expectations (^{107}Pd and ^{182}Hf, as well as ^{26}Al, ^{36}Cl, ^{41}Ca, and ^{60}Fe) were injected in the presolar molecular cloud core by the explosion of a nearby SN that might have triggered the protosolar cloud into collapse (*Cameron and Truran,* 1977; *Meyer and Clayton,* 2000; *Meyer et al.,* 2004). The dynamical feasibility of injecting fresh nucleosynthetic products in the ESS has been studied carefully by *Vanhala and Boss* (2002). In the most recent version of the pollution model, it is assumed that only a fraction of the stellar ejecta is efficiently injected in the nascent solar system [the injection mass cut is the radius in mass coordinates above which the envelope of the SN is injected (*Cameron et al.,* 1995; *Meyer et al.,* 2004)]. For a 25 $M_\odot$ star, an injection mass cut of 5 $M_\odot$, and a time interval of 1 m.y. between the SN explosion and incorporation in the ESS, the abundances of ^{26}Al, ^{41}Ca, ^{60}Fe, and ^{182}Hf are successfully reproduced (*Meyer et al.,* 2004). Chlorine-36 and ^{107}Pd are slightly overproduced but this may reflect uncertainties in ESS abundances and input physics. Because the injection scenario requires only one source for explaining all extinct radionuclides that cannot be produced by GCE or irradiation in the ESS while the multiple *r*-processes scenario requires many, the principle of Ockham's Razor favors the SN pollution model.

4.3.3. Predicted abundance of curium-247 in the early solar system. Among the short-lived nuclides that have mean-lives higher than 5 m.y., only one has eluded detection, ^{247}Cm [τ = 22.5 m.y., decays to ^{235}U (*Chen and Wasserburg,* 1981; *Friedrich et al.,* 2004)]. Modeling of stellar nucleosynthesis and GCE allows the prediction of its abundance in the ESS. Curium-247 is expected to be produced at the same site that synthesized ^{129}I and ^{244}Pu. The same free-decay interval can therefore be applied (Δ = 100 ± 25 m.y.). *Goriely and Arnould* (2001) estimated uncertainties on actinide production ratios. As already discussed in the case of ^{244}Pu, we only consider the models that give a ^{238}U/^{232}Th production ratio consistent with the meteoritic measurement. The inferred ^{247}Cm/^{238}U production ratio is 0.138 ± 0.054. The remainder ratio in the ISM of ^{247}Cm can be calculated using the open nonlinear GCE model of *Dauphas et al.* (2003), $\Re^{247}_{ism} = 7.0 \pm 1.6 \times 10^{-3}$. Using the estimated remainder ratio of ^{238}U (0.71, see discussion on ^{244}Pu), we get the ratio of the remainder ratios of ^{247}Cm and ^{238}U, $\Re^{247}_{ism}/\Re^{238}_{ism} = 9.8 \pm 2.2 \times 10^{-3}$. Allowing for free decay Δ = 100 ± 25 m.y., the ratio in the ESS is $\Re^{247}_{ess}/\Re^{238}_{ess} = 1.2 \pm 0.4 \times 10^{-4}$. This value must be multiplied by the production ratio to get the expected ratio in meteorites. The ratio in the ESS is thus estimated to be ^{247}Cm/^{238}U = 1.6 ± 0.8 × 10^{-5} (^{247}Cm/^{235}U = 5.0 ± 2.6 × 10^{-5} at solar system formation). The present upper limit for the ^{247}Cm/^{235}U ratio obtained from U-isotopic measurements of meteoritic materials is 4 × 10^{-3} (*Chen and Wasserburg,* 1981), which is entirely consistent with the predicted abundance based on modeling of GCE and nucleosynthesis.

5. GALACTIC CHEMICAL EVOLUTION OF STABLE-ISOTOPE RATIOS

As discussed above in section 2, elemental abundance ratios measured in stars are a powerful tool for constraining models of GCE, because different elements are made by different processes in different types of stars with different evolutionary timescales. This statement is of course not limited to elements, but applies equally well to stable-isotope ratios. Many elements are comprised of both primary and secondary isotopes, as defined in section 2, so GCE theory anticipates that many isotopic ratios should evolve in the galaxy. As discussed above, the simple closed-box GCE model predicts that the ratio of a secondary to a primary isotope increases linearly with metallicity in the galaxy. As an example, let us consider the stable isotopes of silicon. Numerical supernova nucleosynthetic calculations (*Woosley and Weaver,* 1995; *Timmes and Clayton,* 1996) indicate that ^{28}Si is a primary isotope, whereas ^{29}Si and ^{30}Si are secondary. Thus, it is not surprising that the numerical GCE model of *Timmes and Clayton* (1996) predicts that the ^{29}Si/^{28}Si and ^{30}Si/^{28}Si ratios increase monotonically with metallicity. The

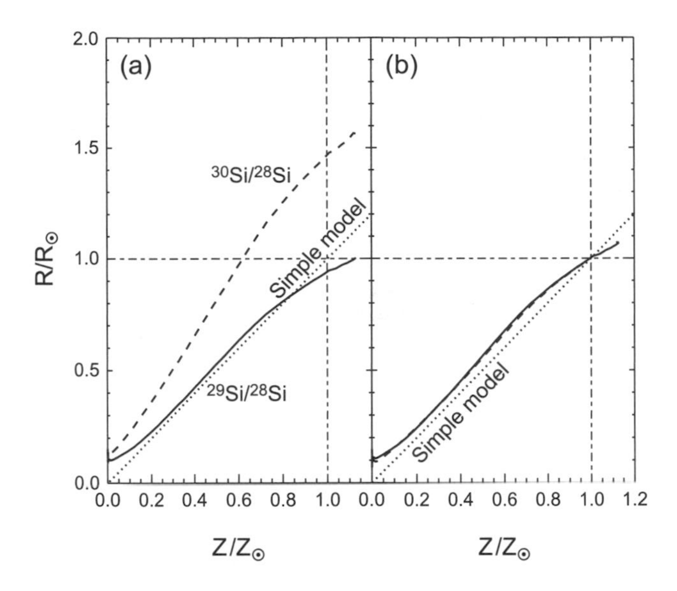

Fig. 5. Silicon-isotopic evolution predicted in the solar neighborhood by *Timmes and Clayton* (1996) and by the "simple" closed-box GCE model (R stands for isotopic ratio). Silicon-isotope ratios increase with metallicity, due to secondary nature of the heavy isotopes ^{29}Si and ^{30}Si. **(a)** Unnormalized calculation: the predicted ratios miss the solar composition due to errors in nucleosynthesis models. **(b)** Isotopic ratios "renormalized" to reproduce the solar values at solar metallicity. The calculated isotope trends are shallower near solar metallicity than is predicted by the simple model of GCE.

Si-isotopic ratios as a function of metallicity predicted by this model are shown in Fig. 5a. The secondary nature of the rare Si isotopes is clear. However, it is also immediately apparent that the model does not exactly reproduce the solar isotopic ratios at solar metallicity; the $^{30}Si/^{28}Si$ ratio is high by ~50% and the $^{29}Si/^{28}Si$ ratio is slightly subsolar. The discrepancy is almost certainly due to errors in the supernova nucleosynthesis calculations that went into the GCE model, and agreement within a factor of 2 is usually considered a success in GCE modeling. However, this is not sufficient for comparison with presolar grain data measured with 1% precision (see next section). As discussed at length by *Timmes and Clayton* (1996), to compare high-precision isotope data with the GCE models in a self-consistent way requires that the models be "renormalized" so that they reproduce the solar abundances. The simplest approach is to rescale the GCE trends so that they pass through the solar composition at solar metallicity. The renormalized Si-isotope trends, scaled in this fashion, are shown in Fig. 5b. With this renormalization, it is clear that the GCE model predicts that the ratios vary in lockstep with one another during galactic evolution. This reflects that ^{29}Si and ^{30}Si are made in similar processes in supernovae. The dotted line indicates the prediction of the simple model for a pure secondary/primary ratio. The renormalized numerical trends clearly show a shallower metallicity dependence near solar metallicity than is predicted by the simple model.

An early application of the idea that isotope ratios evolve in the galaxy was proposed by *Clayton* (1988b) to explain the well-known ^{16}O excess measured in CAIs in meteorites (*Clayton et al.,* 1973; *Clayton,* 1993). In this model, the meteoritic data represents a "chemical memory" of interstellar dust. The interstellar dust is postulated to consist of refractory cores of Al_2O_3 mantled by more volatile O-bearing materials. Because the cores are more robust to destructive processes, they are on average older than the dust mantles. Analogous to Si, $^{17}O/^{16}O$ and $^{18}O/^{16}O$ ratios are expected to increase in the galaxy with time, since ^{16}O is a primary isotope and the others are secondary. Thus, the older refractory dust cores are expected to be rich in ^{16}O, relative to the bulk ISM, which is dominated by more recent stellar ejecta. If the CAIs formed from materials preferentially enriched in the refractory interstellar dust cores, their ^{16}O-richness, compared to typical solar system materials, could be naturally explained. Current belief favors chemical processes in the early solar system to explain the CAI O-isotope data (e.g., *Thiemens and Heidenreich,* 1983; *Clayton,* 2002), but the GCE suggestion of *Clayton* (1988b) has never been disproved.

6. PRESOLAR GRAINS IN METEORITES AND GALACTIC CHEMICAL EVOLUTION

Presolar grains are micrometer-sized and smaller solid samples of stars that can be studied in the laboratory (*Bernatowicz et al.,* 1987; *Lewis et al.,* 1987; *Zinner,* 1998; *Nittler,* 2003; *Clayton and Nittler,* 2004; *Meyer and Zinner,* 2006). They formed in stellar outflows more than 4.6 G.y. ago, became part of the Sun's parent molecular cloud, survived formation of the solar system, and became trapped in asteroids and comets, samples of which now intersect Earth as meteorites and interplanetary dust. They are recognized as presolar grains by their extremely unusual isotopic compositions. These compositions reflect those of the gas from which they condensed and thus provide a great deal of information about the nuclear history of their parent stars. Two types of presolar grains are believed to provide information about GCE (in addition to information about the evolutionary processes of the individual parent stars): SiC and oxides, mostly Al_2O_3 and $MgAl_2O_4$.

As discussed in the previous section, many isotopic ratios are expected to evolve in the galaxy. Because the isotopic ratios of individual presolar grains can in many cases be measured with much higher precision than element ratios can be measured in stars, these grains have the potential to provide additional constraints on GCE. Note, however, that the grains formed in disk stars, probably between ~5 and 10 G.y. ago, and thus span a narrower range of time than can be probed by astronomical observations. Moreover, their parent stars have been dead for eons and thus their properties (stellar type, mass, etc.) have to be inferred from the grain properties, primarily isotopic compositions, themselves.

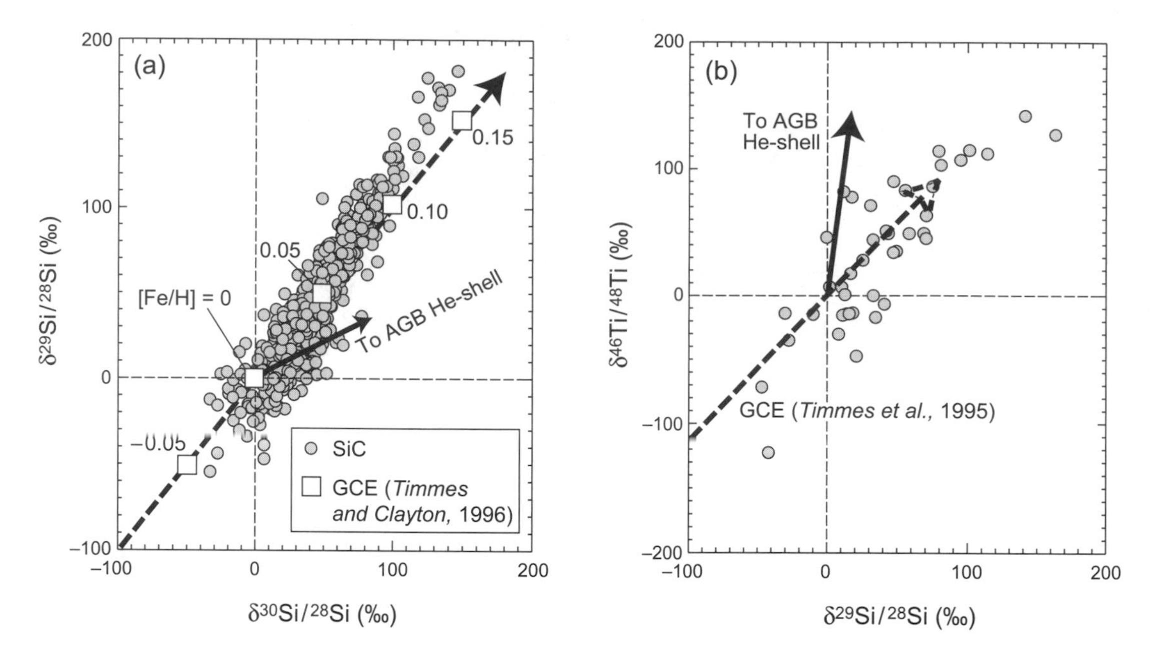

Fig. 6. Silicon- and Ti-isotopic ratios of mainstream presolar SiC grains (*Hoppe et al.,* 1994; *Alexander and Nittler,* 1999). Ratios are expressed as δ values, permil deviations from a terrestrial isotope standard: $\delta R = [R_{meas}/R_{standard} - 1] \times 10^3$. The grain data are correlated in both plots, but the slopes are different from expectations for dredge-up in single AGB stars (*Lugaro et al.,* 1999). Also shown are renormalized GCE calculations (see Fig. 5) of isotopic evolution (*Timmes and Clayton,* 1996; *Timmes et al.,* 1995); metallicity values are indicated in **(a)**.

6.1. Galactic Chemical Evolution and Presolar Silicon Carbide

The best-studied presolar phase in meteorites is silicon carbide (SiC). The vast majority of these grains (the "mainstream" population) is now believed to have originated in C-rich red giant stars during the AGB phase of evolution. This conclusion is supported both by the close similarity of the measured distributions of $^{12}C/^{13}C$ ratios (mainly between 20 and 100) in the grains and the stars, and by stunning agreement with models of AGB stars of the isotopic compositions of heavy elements such as Ba, Zr, and Mo, measured in individual grains by resonance ionization mass spectrometry (e.g., *Lugaro et al.,* 2003). However, the good agreement between the grain compositions and those observed or expected for AGB stars does not extend to the 50% of the grains' atoms that are Si! The Si-isotopic ratios, expressed as δ values, for the mainstream SiC grains are shown in Fig. 6. The grains form a linear array of slope 1.3 on this plot, with isotopic ratios ranging from ~0.95 to 1.2× solar. In AGB stars, heavy element isotopic compositions can be modified by n-capture reactions in the He-burning shell followed by convective mixing with the stellar envelope. As shown in Fig. 6a, this mixing results in Si-isotopic compositions distinct from the observed grain trend. The slope of the grain data is much steeper than predicted Si-isotopic evolution for single AGB stars [~0.3–0.8 (*Lugaro et al.,* 1999)]. Moreover, the range of ratios is larger than that predicted for low-mass, ~solar-metallicity AGB stars (<4% shifts in ratios, compared to the observed 25% range). It is now believed that the mainstream Si array reflects a spread in the initial compositions of a large number of individual stellar sources (*Alexander,* 1993; *Gallino et al.,* 1994; *Timmes and Clayton,* 1996; *Alexander and Nittler,* 1999).

In many presolar SiC grains, Ti is in high enough abundance to determine its isotopic composition. Titanium-isotopic measurements of mainstream SiC grains have indicated a similar behavior to that of Si. Namely, the $^{46,47,49,50}Ti/^{48}Ti$ ratios are correlated with the Si-isotopic ratios, forming arrays on three-isotope plots with slopes distinct from those predicted for the mixing of He-shell material with the envelopes of individual AGB stars (*Ireland et al.,* 1991; *Hoppe et al.,* 1994; *Alexander and Nittler,* 1999; *Lugaro et al.,* 1999). This is illustrated in Fig. 6b, showing the observed correlation between $^{29}Si/^{28}Si$ and $^{46}Ti/^{48}Ti$ [see Fig. 4 of *Lugaro et al.* (1999) for all Ti ratios]. Note that ^{49}Ti and ^{50}Ti are in fact more strongly affected by n-capture in AGB stars than are the Si and other Ti isotopes, so that the $^{49}Ti/^{48}Ti$ and $^{50}Ti/^{48}Ti$ ratios are somewhat less strongly correlated with Si-isotopic ratios in the mainstream grains.

The most obvious explanation for a spread in initial isotopic compositions of individual stars is that it reflects some sort of GCE process. As discussed in the previous section, GCE theory predicts that $^{29}Si/^{28}Si$ and $^{30}Si/^{28}Si$ ratios increase as the metallicity of the ISM does (Fig. 5). Galactic chemical evolution predictions of Ti isotopes are much less secure than those of Si due to lingering uncertainties in the nucleosynthesis processes responsible for some of them (*Timmes et al.,* 1995). Nonetheless, ^{48}Ti is believed to be a primary isotope while the rarer ^{46}Ti, ^{47}Ti, ^{49}Ti, and ^{50}Ti are secondary ones, and the $^{i}Ti/^{48}Ti$ ratios probably increase with metallicity.

The renormalized Si- and Ti-isotope GCE predictions of *Timmes et al.* (1995) and *Timmes and Clayton* (1996) are plotted with the SiC data in Fig. 6. Clearly, GCE predicts a better fit to the grain data than does He-shell mixing in single AGB stars. A GCE interpretation of the Si and Ti data is further supported by the rare SiC subgroups known as Y and Z grains (*Meyer and Zinner,* 2006). Y grains (~2%) have $^{12}C/^{13}C$ ratios higher than mainstream grains and Si-isotopic compositions that plot to the right of the mainstream grains in Fig. 6a. These grains are believed to have originated in AGB stars of ~0.5 $Z_{\odot}$ (*Amari et al.,* 2001). When a component due to dredge-up of He-shell material is subtracted from their Si-isotopic compositions, they are inferred to have lower initial $^{29}Si/^{28}Si$ and $^{30}Si/^{28}Si$ ratios than mainstream grains, as expected for lower metallicity stars. Z grains have similar C isotopes to mainstream grains, but have ^{29}Si depletions and ^{30}Si enrichments, relative to mainstream grains (*Hoppe et al.,* 1997). These grains most likely formed in AGB stars of even lower metallicity than did the Y grains, perhaps as low as $^{1}/_{3}$ $Z_{\odot}$ (*Hoppe et al.,* 1997; *Nittler and Alexander,* 2003), and their inferred initial Si-isotopic ratios are even lower than the Y grains, consistent with GCE expectations. Moreover, Ti-isotopic measurements of several Z grains indicate depletions of ^{46}Ti, ^{47}Ti, and ^{49}Ti, relative to solar (*Amari et al.,* 2003; *Zinner et al.,* 2005a), as expected if Ti-isotopic ratios decrease with decreasing metallicity.

There are significant problems with a simple homogeneous GCE interpretation of the SiC data, however, as illustrated in Fig. 6. First, the slope of the mainstream line is 1.3, steeper than the slope-1 line predicted by GCE theory. Second, most of the grains are enriched in the secondary, neutron-rich isotopes relative to the Sun, but formed in stars born earlier than the Sun. In a homogeneous GCE model that produces the Sun, older stars should have $^{29,30}Si/^{28}Si$ ratios lower than the Sun. Finally, taking the *Timmes and Clayton* (1996) GCE calculation at face value, the mainstream SiC data require a difference of about 5 G.y. to explain the range of Si isotopes from the bottom to the top of the mainstream. From stellar evolutionary considerations, it seems most likely that the SiC grains originated in stars at least as massive as ~1.5–2 $M_{\odot}$ (*Lugaro et al.,* 1999). Such stars have evolutionary timescales much shorter than 5 G.y., so the time difference indicated by the GCE model would indicate an exceedingly large range in interstellar residence ages of the grains themselves. This hardly seems credible, since there are many ISM processes destructive to dust grains (*Jones et al.,* 1997), making a simple temporal interpretation of the Si data implausible. Even if a temporal interpretation is discarded, the grains still appear to have originated in donor AGB stars with higher initial metallicities than the Sun and the problem of the 1.3 slope remains. A number of attempts have been made to resolve these problems, while maintaining a GCE interpretation of the data.

Clayton and Timmes (1997) postulated that the Sun's Si-isotopic composition is strongly peculiar, compared to the mean GCE of the ISM. If so, then the slope-1 line on a Si three-isotope plot predicted by GCE theory could be rotated into a slope-1.3 line as observed for the grains, when normalized to the unusual solar isotope ratios. For this to work quantitatively requires that the Sun must lie far to the right (^{30}Si-rich) side of the initial GCE line in Fig. 6a. However, in this case dredge-up of He-shell material in the parent AGB stars would have to increase the surface $^{30}Si/^{28}Si$ ratio so much that the final (observed mainstream) line falls, as if by a miracle, very near the Sun's abnormal composition. Moreover, such a large increase of $^{30}Si/^{28}Si$ in AGB stars is not regarded as possible, based both on the grains' C-isotopic ratios (*Alexander and Nittler,* 1999) and on AGB nucleosynthesis calculations (*Lugaro et al.,* 1999).

A second approach, proposed by *Clayton* (1997), invokes outward diffusion of stars from the metal-rich inner regions of the galaxy. He suggested that the SiC parent stars formed on circular orbits at smaller galactocentric radii than did the Sun and subsequently scattered from massive molecular clouds into more elongated orbits, eventually ending their lives (during their AGB phases) near the radius of solar birth. Because metallicity gradients exist in the galactic disk, these stars could have higher metallicities than the Sun, despite forming earlier. Stellar orbital diffusion models like this have been advanced to explain the large scatter in metallicity for local disk stars of the same age (*Edvardsson et al.,* 1993; *François and Matteucci,* 1993; *Wielen et al.,* 1996). A semianalytic model by *Nittler and Alexander* (1999), using astronomically derived parameters, indicates that such outward orbital diffusion of stars probably would not result in the observed Si-isotopic distribution of the SiC grains. Moreover, unpublished Monte Carlo calculations (D. D. Clayton, personal communication, 2004) do not indicate large-scale outward scattering of presolar AGB stars.

Alexander and Nittler (1999) took a different approach, attempting to use the grain data themselves to directly infer the relative GCE trends of the isotopic ratios. They took advantage of the fact that the different isotope ratios of Si and Ti are affected to differing degrees by n-capture in the AGB He-shell and performed a χ^2 fit to the mainstream SiC Si- and Ti-isotopic data. The composition of each grain was assumed to be a linear mixture of an initial composition and a He-shell composition, predicted by an AGB nucleosynthesis model (*Gallino et al.,* 1994). Contrary to what was claimed in their paper, this fit cannot uniquely determine the relationship between isotopic ratios and metallicity. However, it can infer the relative isotopic ratio GCE trends, i.e., the slopes of mean GCE trends on three-isotope plots. Based on their fit, *Alexander and Nittler* concluded that the true slope of the Si-isotope GCE trend on the Si δ-value plot is closer to 1.5 than to 1. One possibility to obtain such a slope would be that the initial mass function changes with time in the galaxy, since high-mass supernovae models produce Si with $^{29}Si/^{30}Si >$ solar and low-mass ones produce $^{29}Si/^{30}Si <$ solar (*Woosley and Weaver,* 1995). Alternatively, a faster evolution of ^{29}Si than ^{30}Si near solar metallicity would imply a faster evolution of ^{30}Si at low metallicities.

Thus, if there were a large source of ^{30}Si relative to ^{29}Si at low metallicity, the required steep slope on the Si three-isotope plot might be obtained. There is no hint of an "extra" source of ^{30}Si in low-metallicity supernova calculations, but other possibilities include ONe novae and low-metallicity AGB stars. Recent calculations of each of these stellar types indicate relatively large production factors of ^{30}Si and smaller or no production of ^{29}Si (*Amari et al.,* 2001; *José and Hernanz,* 1998). Galactic chemical evolution calculations taking these sources into account to test this idea are still lacking.

An alternative approach was anticipated by *Timmes and Clayton* (1996), who showed that when the nucleosynthetic yields from supernovae are normalized to the solar-metallicity ISM composition calculated by their GCE model, heterogeneous mixing of individual supernova ejecta into an initially ~solar composition could possibly reproduce a slope 1.3 line on the Si three-isotope plot. *Lugaro et al.* (1999) followed up on this suggestion and explicitly modeled, using a Monte Carlo technique, the effect of inhomogeneous mixing of SN ejecta into localized regions of the ISM. They showed that with a range of model parameters, the model could easily explain the range and scatter of mainstream SiC Si-isotope ratios. They further argued that the observed distribution of SiC Si isotopes probably reflects an interplay of homogeneous GCE gradually increasing the average ISM $^{29,30}Si/^{28}Si$ ratios and heterogeneous GCE leading to local variations about the mean, although the balance between heterogeneous and homogeneous GCE in shaping the mainstream distribution would depend on the range of ages of the parent stars. An attractive feature of this model is that it can naturally account for the isotopic heaviness of the grains with respect to the Sun. However, *Nittler* (2005) extended this model to Ti and O isotopes and showed that it could explain neither the high degree of correlation between Si and Ti isotopes in the grains nor the range of O-isotopic compositions observed in presolar oxide grains (see next section). The observed correlation between $^{29}Si/^{28}Si$ and $^{46}Ti/^{48}Ti$ ratios in the grains allows at most ~40% of the total spread in Si-isotopic composition to be explained by this specific heterogeneous GCE model. It is also highly unlikely that any model of this sort could account for the Si-Ti correlations because the isotopes of these elements are made in different types and/or masses of supernovae and the isotope ratios in the ejecta of supernovae of different masses and types are hence uncorrelated. In fact, the observed Si-Ti isotope correlations indicate that the disk ISM in the vicinity of solar birth 4.6 G.y. ago was remarkably well-mixed with regard to the ejecta of individual supernovae.

Recently, *Clayton* (2003) suggested a radically different explanation for the mainstream SiC Si- (and Ti-) isotopic ratio distribution. In his model, the isotopic correlation lines are two-component mixing lines due to a merger of a low-metallicity dwarf galaxy with the Milky Way disk some 6–7 G.y. ago. At that time, Clayton postulates that the $^{29}Si/^{28}Si$ and $^{30}Si/^{28}Si$ ratios of the local Milky Way ISM were higher than the solar ratios and the merging galaxy had lower-than-solar ratios. The merger induced a period of star formation during which the SiC parent stars were born with a range of initial isotopic compositions due to variable mixing of the interstellar gas from the two components. Since the Sun formed in the same region of the disk, it incorporated its own mix of the two galaxies as well as the local ejecta of stars that occurred between the time of merger and that of solar birth. This model has many attractions, including explanations for the tight correlation between the mainstream SiC Si- and Ti-isotope ratios, the placement of the Sun at the bottom of the SiC mainstream, and the unusual $^{18}O/^{17}O$ ratio of the Sun, compared with the local ISM (*Penzias,* 1981). Moreover, mergers like that postulated are now commonly believed to be how much of the mass of the galaxy was acquired (e.g., *Shetrone et al.,* 2003; *Wyse,* 2003). However, it clearly needs much critical scrutiny by assorted scientific disciplines before it can be accepted. For example, it might be possible to find a fossil record of such a merger in the chemical compositions of nearby stars formed at that time.

We note that the rare SiC Y grains might provide some support for the galactic merger model outlined above (*Clayton,* 2003). The initial Si-isotopic compositions of most Y grains are inferred to lie near the lower end of the mainstream line. Moreover, there is a smooth and rapid increase in the fraction of SiC grains with high $^{12}C/^{13}C$ (a rough proxy for AGB mass) with decreasing $^{29}Si/^{28}Si$ ratios (*Nittler and Alexander,* 2003). Because more-massive stars evolve faster than less-massive ones, these results imply a more recent formation for the parent stars of grains near the bottom of the mainstream compared with those with higher $^{29,30}Si/^{28}Si$ ratios. This runs counter to the expectations for the temporal galactic evolution interpretation of the mainstream line, but is compatible with the physical mixing model, assuming that the mixing fraction of the accreting dwarf galaxy material increases with time during the merger event.

6.2. Galactic Chemical Evolution and Presolar Oxide Grains

Galactic chemical evolution has also been implicated in interpretations of isotope data for presolar oxide grains in meteorites, primarily corundum, Al_2O_3, and spinel, $MgAl_2O_4$ (e.g., *Nittler et al.,* 1997; *Choi et al.,* 1998; *Meyer and Zinner,* 2006). The O-isotopic ratios for several hundred identified presolar oxide grains are plotted in Fig. 7; most of the grains have been assigned to four groups by *Nittler et al.* (1997). We focus here on the Group 1 and 3 grains, since, as discussed below, these provide information about GCE. The initial O-isotopic compositions of the parent stars of the highly ^{18}O-depleted Group 2 grains are believed to have been completely erased by nucleosynthetic processes and the origin of the ^{18}O-rich Group 4 grains is enigmatic (*Meyer and Zinner,* 2006).

The majority of the presolar oxide grains are believed to have formed in low-mass red giant and AGB stars. In par-

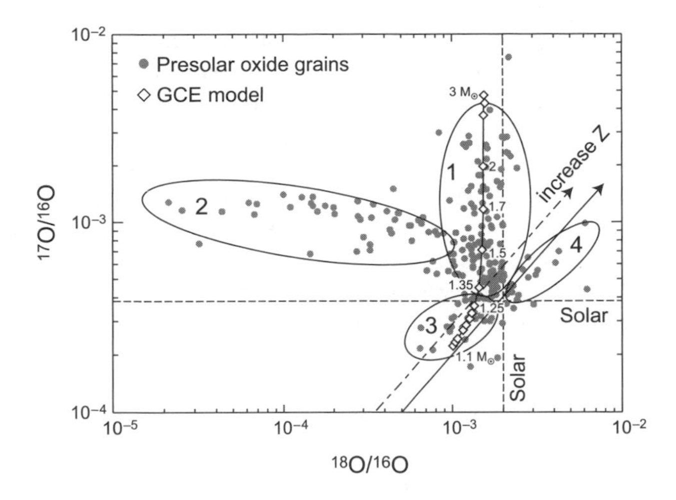

Fig. 7. Oxygen-isotopic ratios measured in presolar oxide minerals (mostly Al_2O_3 and $MgAl_2O_4$) from meteorites. Ellipses indicate group definitions of *Nittler et al.* (1997). Arrows indicate the expected GCE of O isotopes in the ISM (solid line: $^{18}O/^{17}O$ = 5.2; dash-dot arrow: $^{18}O/^{17}O$ = 3.5). Open diamonds indicate predicted average O-isotopic compositions of presolar red giant stars of different masses, taking into account GCE, the first dredge-up, and stellar lifetimes (*Nittler et al.,* 1997); stellar masses are indicated for some points. The good agreement between the GCE model and the Group 1 and 3 oxide grains indicates that these grains formed in red giant stars and that the $^{18}O/^{17}O$ ratio of the Sun is not atypical for its age and location in the galaxy. Data from *Nittler* (1997, and references therein), *Choi et al.* (1998), and *Zinner et al.* (2003).

ticular, the O-isotopic ratios of the dominant Group 1 grains are consistent both with spectroscopic observations of O-rich red giants and AGB stars (*Harris and Lambert,* 1984) and with model calculations of nucleosynthesis and mixing processes in these stars (*Boothroyd and Sackmann,* 1999; *Dearborn,* 1992; *El Eid,* 1994). The $^{17}O/^{16}O$ ratios of these grains are explained by a range of masses of the parent stars, but the range of $^{18}O/^{16}O$ ratios is larger than can be explained by mixing processes in the stars themselves. Analogous to the Si isotopes in SiC grains, a range of initial $^{18}O/^{16}O$ ratios in the parent stars is required. As discussed in section 5, this ratio is expected to increase with time in the galaxy, so GCE is an obvious explanation for the data.

Models of O-isotopic GCE have been reported over the last decade (*Timmes et al.,* 1995; *Prantzos et al.,* 1996; *Romano and Matteucci,* 2003). Unfortunately, although the more recent models can well explain the $^{17}O/^{16}O$ ratio of both the Sun and of molecular clouds throughout the galaxy, a quantitative understanding of the GCE of the $^{18}O/^{16}O$ ratio is still lacking. For example, there are gradients with galactocentric radius in the molecular cloud $^{17}O/^{16}O$ and $^{18}O/^{16}O$ ratios, but no such gradient in the $^{18}O/^{17}O$ ratio (*Penzias,* 1981; *Wilson and Rood,* 1994). This is consistent with the expected primary(secondary) behavior of $^{16}O(^{17,18}O)$, but the molecular cloud $^{18}O/^{17}O$ ratio is ~3.5, much lower than the solar value of 5.2. There is still no good explanation for this discrepancy, although both systematic errors in the molecular cloud observations and a local enrichment by massive supernova ejecta of the solar system's progenitor molecular cloud have been suggested (e.g., *Prantzos et al.,* 1996).

The secondary nature of ^{18}O implies that presolar oxide grains with lower $^{18}O/^{16}O$ ratios originated in stars with lower metallicity than the parent stars of grains with higher ratios. However, the precise relationship between $^{18}O/^{16}O$ and metallicity is unknown. *Nittler et al.* (1997) presented a simple model to explain the distribution of O-isotopic ratios of Group 1 and 3 presolar oxide grains. This model predicted the average $^{17}O/^{16}O$ and $^{18}O/^{16}O$ ratios of red giant stars as a function of mass (from ~1.2 to 3 $M_{\odot}$), such that the stars ended their life 4.6 G.y. ago. Lower-mass stars formed earlier in galactic history (since they have longer lifetimes) than higher-mass stars and hence would be expected to have lower metallicity and lower initial $^{18}O/^{16}O$ ratios. Higher-mass stars also are predicted to have higher $^{17}O/^{16}O$ ratios following mixing of interior material to the stellar surface during the red giant phase of evolution ("first dredge-up"). For lack of a good O-isotopic GCE trend, this model assumed that the $^{17}O/^{16}O$ and $^{18}O/^{16}O$ ratios in the ISM increase linearly with metallicity (*Boothroyd and Sackmann,* 1999) and that at solar metallicity the solar $^{18}O/^{17}O$ (rather than the molecular cloud one) is reproduced. Predictions of the resulting model are shown on Fig. 7; the model curve passes through the center of the distribution of both Group 1 and 3 oxide grains. This result strongly suggests that the Group 3 grains also formed in red giants and AGB stars, albeit ones with initial O-isotopic ratios lower than solar. Moreover, note that the first dredge-up in red giants can only decrease the surface $^{18}O/^{17}O$ ratio. Thus, if the typical presolar ISM had an $^{18}O/^{17}O$ ratio similar to that observed today in molecular clouds, one would expect all Group 1 and 3 presolar oxide grains to have $^{18}O/^{17}O < 3.5$. But in fact, some 30% of the grains have ratios higher than this value. Thus the grain data imply that typical ISM $^{18}O/^{17}O$ ratios were greater in presolar times than is observed today in molecular clouds throughout the galactic disk, perhaps indicating a problem with the molecular cloud observations themselves. Alternatively, the galactic merger model proposed by *Clayton* (2003) to explain mainstream SiC grains (previous section) might also explain the discrepancy between $^{18}O/^{17}O$ in the present-day ISM and in the Sun.

Because low-mass stars have long evolutionary timescales and the parent stars of the presolar grains must have ended their lives prior to the formation of the Sun 4.6 G.y. ago, the grain data also can be used to constrain the age of our galaxy (*Nittler and Cowsik,* 1997). The age of the galaxy must be larger than the longest lifetime of a grain parent star added to that of the Sun. *Nittler and Cowsik* (1997) calculated a lower bound on the age of the disk as 10.5 G.y. and an actual age of 14.4 G.y. The systematic uncertainties affecting this estimate are potentially large (several gigayears), but the age agrees well with those estimated by other means and was determined in a fundamentally new way.

Since the distribution of $^{18}O/^{16}O$ ratios observed in the Group 1 and 3 presolar oxide grains is in good agreement with expectations for GCE, the metallicities of the parent stars can be estimated from grain O-isotope ratios and theoretical models. The grains can then be used to trace evolutionary histories of other isotope systems, if the relevant isotope measurements can be made on the grains. For example, Mg is another element that has both primary (^{24}Mg) and secondary (^{25}Mg, ^{26}Mg) isotopes. Both theoretical models (e.g., *Timmes et al.,* 1995) and spectroscopic observations of main-sequence stars (e.g., *Gay and Lambert,* 2000) indicate that the $^{25,26}Mg/^{24}Mg$ ratios decrease with decreasing metallicity, but the stellar data have large error bars. Presolar grains often have large ^{26}Mg excesses due to *in situ* decay of radioactive ^{26}Al (*Meyer and Zinner,* 2006). Measurements of $^{25}Mg/^{24}Mg$ ratios in mainstream SiC grains (*Huss et al.,* 1997) revealed no variations due to GCE or otherwise, but because Mg contents are very low in the grains, terrestrial contamination cannot be ruled out. Recently, Mg-isotopic data have been reported for a number of presolar $MgAl_2O_4$ grains (*Nittler et al.,* 2003; *Zinner et al.,* 2005b). Although the isotopic systematics of the grains are complex, the $^{25}Mg/^{24}Mg$ ratios suggest that the evolution of this ratio as a function of metallicity is relatively shallow near solar metallicity. This is consistent both with recent observational data (*Gay and Lambert,* 2000) and GCE models taking AGB star nucleosynthesis into account (*Fenner et al.,* 2003).

Titanium-isotopic ratios have been measured in a few presolar Al_2O_3 grains as well (*Choi et al.,* 1998; *Hoppe et al.,* 2003). Because of low Ti concentrations, error bars for most reported measurements are large. However, a few grains have large anomalies and/or small error bars and allow a comparison with SiC Ti data. In general, the observations are consistent with the GCE interpretation of Ti isotopes in SiC (section 6.1) (*Alexander and Nittler,* 1999). For example, there is a general correlation between $^{46}Ti/^{48}Ti$ and $^{18}O/^{16}O$ as expected since both are secondary/primary isotope ratios.

7. CONCLUDING REMARKS

It might seem audacious to attempt to use microscopic constituents of rare and unusual rocks to draw broad inferences about the vast reaches of space in time encompassed by the chemical history of the Milky Way. However, it is clear at this point that galactic evolution has left an isotopic record within meteorites, both in the fossil remnants of radioactive nuclei and in presolar grains of stardust. Our attempts to decode this record are, of course, far from complete. For example, although a signature of GCE is clearly imprinted on some isotopic ratios measured in presolar grains, the exact process or processes involved (e.g., homogeneous GCE, heterogeneous GCE, a galactic merger) have not been unambiguously identified. Full exploitation of the potential of the grains for GCE science will require both additional theoretical and observational work. For example, a presolar galactic merger in the solar neighborhood should have left some record in the chemical compositions of the stars formed at the time. Moreover, GCE models including the relevant isotopic ratios and current nucleosynthetic yields from all types of stars (including, for instance, novae and low-Z AGB stars) are sparse or lacking. There will no doubt be considerable progress in the coming decades as the sophistication of computer models increases as well as new astronomical and meteoritical data are obtained by advances in technology.

As mentioned in section 1, the solar chemical composition was long considered the "cosmic" composition, and a key question in cosmochemistry remains: Just how typical is the solar composition in the context of the galactic evolution? Based on spectroscopic determination of many elements in the atmospheres of dwarf stars in the solar neighborhood, *Edvardsson et al.* (1993) concluded that the Sun's composition is quite typical for its age and location in the galaxy. Moreover, the metallicity distribution for G dwarfs in the solar neighborhood (*Nordström et al.,* 2004) has a peak very close to the solar value ([Me/H] = 0; Fig. 2). A very important observation to be gleaned from section 6 is that the solar composition is also apparently *isotopically* quite typical. Most presolar SiC grains formed in stars with initial Si-isotopic compositions within ~25% of the solar value and the distribution of O isotopes in presolar oxide grains indicate that their parent stars had $^{18}O/^{17}O$ ratios closer to the solar value than to that measured today in molecular clouds. Furthermore, trace heavy elements in individual presolar grains from AGB stars (*Nicolussi et al.,* 1997; *Lugaro et al.,* 2003) have isotopic compositions consistent with mixing of He-shell material with envelope material of essentially solar isotopic composition. Thus, if nothing else, the meteorite data discussed in this chapter help confirm the utility of using the solar composition, about which we know so much, as a standard with which to compare the chemical makeup of the rest of the remote universe, about which we know much less.

Acknowledgments. This work benefited from many fruitful discussions with C. Alexander, D. D. Clayton, A. M. Davis, R. Gallino, T. Rauscher, B. Marty, L. Reisberg, F. Timmes, J. W. Truran, M. Wadhwa, and R. Yokochi. This chapter was improved by constructive reviews by B. Meyer and E. Zinner.

REFERENCES

Alexander C. M. O'D. (1993) Presolar SiC in chondrites: How variable and how many sources? *Geochim. Cosmochim. Acta, 57,* 2869–2888.

Alexander C. M. O'D. and Nittler L. R. (1999) The galactic evolution of Si, Ti, and O isotopic ratios. *Astrophys. J., 519,* 222–235.

Allende Prieto C., Lambert D. L., and Asplund M. (2001) The forbidden abundance of oxygen in the sun. *Astrophys. J. Lett., 556,* L63–L66.

Amari S., Nittler L. R., Zinner E., Gallino R., Lugaro M., and Lewis R. S. (2001) Presolar SiC grains of type Y: Origin from low-metallicity asymptotic giant branch stars. *Astrophys. J., 546,* 248–266.

Amari S., Zinner E., Gallino R., and Lewis R. S. (2003) Silicon and titanium isotopic analysis of silicon carbide grains of type X and Z (abstract). *Meteoritics & Planet. Sci., 38,* A66.

Amelin Y., Krot A. N., Hutcheon I. D., and Ulyanov A. A. (2002) Lead isotopic ages of chondrules and calcium-aluminum-rich inclusions. *Science, 297,* 1678–1683.

Anders E. and Grevesse N. (1989) Abundances of the elements: Meteoritic and solar. *Geochim. Cosmochim. Acta, 53,* 197–214.

Argast D., Samland M., Gerhard O. E., and Thielemann F.-K. (2000) Metal-poor halo stars as tracers of ISM mixing processes during halo formation. *Astron. Astrophys., 356,* 873–887.

Arlandini C., Käppeler F., Wisshak K., Gallino R., Lugaro M., Busso M., and Straniero O. (1999) Neutron capture in low-mass asymptotic giant branch stars: Cross sections and abundance signatures. *Astrophys. J., 525,* 886–900.

Bernatowicz T., Fraundorf G., Tang M., Anders E., Wopenka B., Zinner E., and Fraundorf P. (1987) Evidence for interstellar SiC in the Murray carbonaceous meteorite. *Nature, 330,* 728–730.

Birck J. L. and Allègre C. J. (1985) Evidence for the presence of ^{53}Mn in the early solar system. *Geophys. Res. Lett., 12,* 745–748.

Boothroyd A. I. and Sackmann I.-J. (1999) The CNO-isotopes: Deep circulation in red giants and first and second dredge-up. *Astrophys. J., 510,* 232.

Brazzle R. H., Pravdivtseva O., Meshik A., and Hohenberg C. (1999) Verification and interpretation of the I-Xe chronometer. *Geochim. Cosmochim. Acta, 63,* 739–760.

Brook C. B., Kawata D., and Gibson B. K. (2003) Simulating a white dwarf dominated galactic halo. *Mon. Not. R. Astron. Soc., 343,* 913–923.

Burbidge E. M., Burbidge G. R., Fowler W. A., and Hoyle F. (1957) Synthesis of the elements in stars. *Rev. Mod. Phys., 29,* 547–650.

Cameron A. G. W. and Truran J. W. (1977) The supernova trigger for formation of the solar system. *Icarus, 30,* 447–461.

Cameron A. G. W., Höflich P., Myers P. C., and Clayton D. D. (1995) Massive supernovae, Orion gamma rays, and the formation of the solar system. *Astrophys. J. Lett., 447,* L53–L57.

Cayrel R., Hill V., Beers T. C., Barbuy B., Spite M., Spite F., Plez B., Andersen J., Bonifacio P., François P., Molaro P., Nordström B., and Primas F. (2001) Measurement of stellar age from uranium decay. *Nature, 409,* 691–692.

Chang R. X., Hou J. L., Shu C. G., and Fu C. Q. (1999) Two-component model for the chemical evolution of the galactic disk. *Astron. Astrophys., 350,* 38–48.

Chen J. H. and Wasserburg G. J. (1981) The isotopic composition of uranium and lead in Allende inclusions and meteorite phosphates. *Earth Planet. Sci. Lett., 52,* 1–15.

Chen J. H. and Wasserburg G. J. (1996) Live ^{207}Pd in the early solar system and implications for planetary evolution. In *Earth Processes: Breaking the Isotopic Code* (A. Basu and S. Hart, eds.), pp. 1–20. AGU Geophysical Monograph 95, American Geophysical Union, Washington, DC.

Chen J. H., Wasserburg G. J., and Papanastassiou D. A. (1993) Th and U abundances in chondritic meteorites (abstract). In *Lunar and Planetary Science XXIV,* pp. 277–278. Lunar and Planetary Institute, Houston.

Chiappini C., Matteucci F., and Gratton R. (1997) The chemical evolution of the galaxy: The two-infall model. *Astrophys. J., 477,* 765–780.

Choi B.-G., Huss G. R., Wasserburg G. J., and Gallino R. (1998) Presolar corundum and spinel in ordinary chondrites: Origins from AGB stars and a supernova. *Science, 282,* 1284–1289.

Clayton D. D. (1983) Extinct radioactivities — a three-phase mixing model. *Astrophys. J., 268,* 381–384.

Clayton D. D. (1985) Galactic chemical evolution and nucleocosmochronology: A standard model. In *Nucleosynthesis: Challenges and New Developments* (W. D. Arnett and J. W. Truran, eds.), pp. 65–88. Univ. of Chicago, Chicago.

Clayton D. D. (1988a) Nuclear cosmochronology within analytic models of the chemical evolution of the solar neighbourhood. *Mon. Not. R. Astron. Soc., 234,* 1–36.

Clayton D. D. (1988b) Isotopic anomalies: Chemical memory of galactic evolution. *Astrophys. J., 334,* 191–195.

Clayton D. D. (1997) Placing the sun and mainstream SiC particles in galactic chemodynamical evolution. *Astrophys. J. Lett., 484,* L67–L70.

Clayton D. D. (2003) A presolar galactic merger spawned the SiC-grain mainstream. *Astrophys. J., 598,* 313–324.

Clayton D. D. and Nittler L. R. (2004) Astrophysics with presolar stardust. *Annu. Rev. Astron. Astrophys., 42,* 39–78.

Clayton D. D. and Timmes F. X. (1997) Placing the Sun in galactic chemical evolution: Mainstream SiC particles. *Astrophys. J., 483,* 220–227.

Clayton R. N. (1993) Oxygen isotopes in meteorites. *Annu. Rev. Earth Planet. Sci., 21,* 115–149.

Clayton R. N. (2002) Solar system: Self-shielding in the solar nebula. *Nature, 415,* 860–861.

Clayton R. N., Grossman L., and Mayeda T. K. (1973) A component of primitive nuclear composition in carbonaceous meteorites. *Science, 182,* 485–488.

Copi C. J. (1997) A stochastic approach to chemical evolution. *Astrophys. J., 487,* 704–718.

Cowan J. J., Thielemann F.-K., and Truran J. W. (1991) Radioactive dating of the elements. *Annu. Rev. Astron. Astrophys., 29,* 447–497.

Cowan J. J., Sneden C., Burles S., Ivans I. I., Beers T. C., Truran J. W., Lawler J. E., Primas F., Fuller G. M., Pfeiffer B., and Kratz K.-L. (2002) The chemical composition and age of the metal-poor halo star BD + 17°3248. *Astrophys. J., 572,* 861–879.

Dauphas N. (2005a) The U/Th production ratio and the radiometric age of the Milky Way. *Nature, 439,* 1203–1205.

Dauphas N. (2005b) Multiple sources or late injection of short-lived r-nuclides in the early solar system? *Nucl. Phys. A, 758,* 757–760.

Dauphas N., Rauscher T., Marty B., and Reisberg L. (2003) Short-lived p-nuclides in the early solar system and implications on the nucleosynthetic role of X-ray binaries. *Nucl. Phys. A., 719,* 287c–295c.

Dauphas N., Foley C. N., Wadhwa M., Davis A. M., Janney P. E., Qin L., Göpel C., Birck J.-L. (2005) Protracted core differentiation in asteroids from ^{182}Hf-^{182}W systematics in the Eagle Station pallasite (abstract). In *Lunar and Planetary Science XXXVI,* Abstract #1100. Lunar and Planetary Institute, Houston (CD-ROM).

Dearborn D. S. P. (1992) Diagnostics of stellar evolution: The oxygen isotopes. *Phys. Rept., 210,* 367–382.

Dopita M. A. and Ryder S. D. (1994) On the law of star formation in disk galaxies. *Astrophys. J., 430,* 163–178.

Edvardsson B., Anderson J., Gustaffson B., Lambert D. L., Nissen P. E., and Tomkin J. (1993) The chemical evolution of the galactic disk. *Astron. Astrophys., 275,* 101–152.

El Eid M. (1994) CNO isotopes in red giants: Theory versus observations. *Astron. Astrophys., 285,* 915–928.

Fenner Y., Gibson B. K., Lee H.-c., Karakas A. I., Lattanzio J. C., Chieffi A., Limongi M., and Yong D. (2003) The chemical evolution of magnesium isotopic abundances in the solar neighbourhood. *Publ. Astron. Soc. Australia, 20,* 340–344.

François P. and Matteucci F. (1993) On the abundance spread in solar neighbourhood stars. *Astron. Astrophys., 280,* 136–140.

Friedrich J. M., Ott U., and Lugmair G. W. (2004) Revisiting extraterrestrial U isotope ratios (abstract). In *Lunar and Planetary Science XXXV,* Abstract #1575. Lunar and Planetary Institute, Houston (CD-ROM).

Gallino R., Raiteri C. M., Busso M., and Matteucci F. (1994) The puzzle of silicon, titanium and magnesium anomalies in meteoritic silicon carbide grains. *Astrophys. J., 430,* 858–869.

Gallino R., Busso M., Wasserburg G. J., and Straniero O. (2004) Early solar system radioactivity and AGB stars. *New Astron. Rev., 48,* 133–138.

Gay P. L. and Lambert D. L. (2000) The isotopic abundances of magnesium in stars. *Astrophys. J., 533,* 260–270.

Gerritsen J. P. E. and Icke V. (1997) Star formation in N-body simulations. I. The impact of the stellar ultraviolet radiation on star formation. *Astron. Astrophys., 325,* 972–986.

Gibson B. K., Fenner Y., Renda A., Kawata D., and Lee H. (2003) Galactic chemical evolution. *Publ. Astron. Soc. Australia, 20,* 401–415.

Goreva J. S. and Burnett D. S. (2001) Phosphate control on the Th/U variations in ordinary chondrites: Improving solar system abundances. *Meteoritics & Planet. Sci., 36,* 63–74.

Goriely S. and Arnould M. (2001) Actinides: How well do we know their stellar production? *Astron. Astrophys., 379,* 1113–1122.

Grenon M. (1987) Past and present metal abundance gradient in the galactic disc. *J. Astrophys. Astron., 8,* 123–139.

Hansen B. M. S., Richer H. B., Fahlman G. G., Stetson P. B., Brewer J., Currie T., Gibson B. K., Ibata R., Rich R. M., and Shara M. M. (2004) Hubble Space Telescope observations of the white dwarf cooling sequence of M4. *Astrophys. J. Suppl. Ser., 155,* 551–576.

Harper C. L. (1996) Evidence for ^{92g}Nb in the early solar system and evaluation of a new p-process cosmochronometer from ^{92g}Nb/^{92}Mo. *Astrophys. J., 466,* 437–456.

Harris M. J. and Lambert D. L. (1984) Oxygen isotopic abundances in the atmospheres of seven red giant stars. *Astrophys. J., 285,* 674–682.

Hill V., Plez B., Cayrel R., Beers T. C., Nordström B., Andersen J., Spite M., Spite F., Barbuy B., Bonifacio P., Depagne E., François P., and Primas F. (2002) First stars. I. The extreme r-element rich, iron-poor halo giant CS 31082-001. Implications for the r-process site(s) and radioactive cosmochronology. *Astron. Astrophys., 387,* 560–579.

Hirata T. (2001) Determinations of Zr isotopic composition and U-Pb ages for terrestrial Zr-bearing minerals using laser ablation-inductively coupled plasma mass spectrometry: Implications for Nb-Zr isotopic systematics. *Chem. Geol., 176,* 323–342.

Hoppe P., Amari S., Zinner E., Ireland T., and Lewis R. S. (1994) Carbon, nitrogen, magnesium, silicon and titanium isotopic compositions of single interstellar silicon carbide grains from the Murchison carbonaceous chondrite. *Astrophys. J., 430,* 870–890.

Hoppe P., Annen P., Strebel R., Eberhardt P., Gallino R., Lugaro M., Amari S., and Lewis R. S. (1997) Meteoritic silicon carbide with unusual Si-isotopic compositions: Evidence for an origin in low-mass low-metallicity AGB stars. *Astrophys. J. Lett., 487,* L101–L104.

Hoppe P., Nittler L. R., Mostefaoui S., Alexander C. M. O'D., and Marhas K. K. (2003) A NanoSIMS study of titanium-isotopic compositions of presolar corundum grains (abstract). In *Lunar and Planetary Science XXXIV,* Abstract #1570. Lunar and Planetary Institute, Houston (CD-ROM).

Hudson G. B., Kennedy B. M., Podosek F. A., and Hohenberg C. M. (1989) The early solar system abundance of ^{244}Pu as inferred from the St. Severin chondrite. *Proc. Lunar Planet. Sci. Conf. 19th,* pp. 547–557.

Huss G. R., Hutcheon I. D., and Wasserburg G. J. (1997) Isotopic systematics of presolar silicon carbide from the Orgueil (CI) chondrite: Implications for solar-system formation and stellar nucleosynthesis. *Geochim. Cosmochim. Acta, 61,* 5117–5148.

Ireland T. R., Zinner E. K., and Amari S. (1991) Isotopically anomalous Ti in presolar SiC from the Murchison meteorite. *Astrophys. J. Lett., 376,* L53–L56.

Jones A. P., Tielens A. G. G. M., Hollenbach D. J., and McKee C. F. (1997) The propagation and survival of interstellar grains. In *Astrophysical Implications of the Laboratory Study of Presolar Materials* (T. J. Bernatowicz and E. Zinner, eds.), pp. 595–613. AIP Conference Proceedings 402, American Institute of Physics, New York.

Jørgensen B. R. (2000) The G dwarf problem. Analysis of a new data set. *Astron. Astrophys., 363,* 947–957.

José J. and Hernanz M. (1998) Nucleosynthesis in classical novae: CO versus ONe white dwarfs. *Astrophys. J., 494,* 680–690.

Kelley W. R. and Wasserburg G. J. (1978) Evidence for the existence of Pd-107 in the early solar system. *Geophys. Res. Lett., 5,* 1079–1082.

Kennicutt R. C. (1998) The global Schmidt law in star-forming galaxies. *Astrophys. J., 498,* 541–552.

Kotoneva E., Flynn C., Chiappini C., and Matteucci F. (2002) K dwarfs and the chemical evolution of the solar cylinder. *Mon. Not. R. Astron. Soc., 336,* 879–891.

Krauss L. M. and Chaboyer B. (2003) Age estimates of globular clusters in the Milky Way: Constraints on cosmology. *Science, 299,* 65–69.

Kravtsov A. V. (2003) On the origin of the global Schmidt law of star formation. *Astrophys. J. Lett., 590,* L1–L4.

Kroupa P. (2002) The initial mass function of stars: Evidence for uniformity in variable systems. *Science, 295,* 82–91.

Larson R. B. (1974) Dynamical models for the formation and evolution of spherical galaxies. *Mon. Not. R. Astron. Soc., 166,* 585–616.

Lewis R. S., Tang M., Wacker J. F., Anders E., and Steel E. (1987) Interstellar diamonds in meteorites. *Nature, 326,* 160–162.

Lugaro M., Zinner E., Gallino R., and Amari S. (1999) Si isotopic ratios in mainstream presolar SiC grains revisited. *Astrophys. J., 527,* 369–394.

Lugaro M., Herwig F., Lattanzio J. C., Gallino R., and Straniero O. (2003) s-process nucleosynthesis in asymptotic giant branch stars: A test for stellar evolution. *Astrophys. J., 586,* 1305–1319.

Lugmair G. W. and Shukolyukov A. (1998) Early solar system timescales according to ^{53}Mn-^{53}Cr systematics. *Geochim. Cosmochim. Acta, 62,* 2863–2886.

Lugmair G. W., Shimamura T., Lewis R. S., and Anders E. (1983)

Samarium-146 in the early solar system: Evidence from neodymium in the Allende meteorite. *Science, 222,* 1015–1018.
Lynden-Bell D. (1975) The chemical evolution of galaxies. *Vistas Astron., 19,* 299–316.
Matteucci F. (2003) *The Chemical Evolution of the Galaxy.* Kluwer, Dordrecht. 293 pp.
Matteucci F. and Greggio L. (1986) Relative roles of type I and II supernovae in the chemical enrichment of the interstellar gas. *Astron. Astrophys., 154,* 279–287.
McWilliam A. (1997) Abundance ratios and galactic chemical evolution. *Annu. Rev. Astron. Astrophys., 35,* 503–556.
Meyer B. S. and Clayton D. D. (2000) Short-lived radioactivities and the birth of the Sun. *Space Sci. Rev., 92,* 133–152.
Meyer B. S. and Luo N. (1997) Monte-Carlo modeling of extinct short-lived r-process radioactivities (abstract). In *Lunar and Planetary Science XXXIV,* Abstract #1841. Lunar and Planetary Institute, Houston (CD-ROM).
Meyer B. S. and Truran J. W. (2000) Nucleocosmochronology. *Phys. Rept., 333–334,* 1–11.
Meyer B. S. and Zinner E. (2006) Nucleosynthesis. In *Meteorites and the Early Solar System II* (D. S. Lauretta and H. Y. McSween Jr., eds.), this volume. Univ. of Arizona, Tucson.
Meyer B. S., The L.-S., Clayton D. D., and El Eid M. F. (2004) Helium-shell nucleosynthesis and extinct radioactivities (abstract). In *Lunar and Planetary Science XXXV,* Abstract #1908. Lunar and Planetary Institute, Houston (CD-ROM).
Münker C., Weyer S., Mezger K., Rehkämper M., Wombacher F., and Bischoff A. (2000) ^{92}Nb-^{92}Zr and the early differentiation history of planetary bodies. *Science, 289,* 1538–1542.
Nicolussi G. K., Davis A. M., Pellin M. J., Lewis R. S., Clayton R. N., and Amari S. (1997) s-process zirconium in presolar silicon carbide grains. *Science, 277,* 1281–1283.
Nittler L. R. (1997) Presolar oxide grains in meteorites. In *Astrophysical Implications of the Laboratory Studies of Presolar Materials* (T. J. Bernatowicz and E. Zinner, eds.), pp. 59–82. AIP Conference Proceedings 402, American Institute of Physics, New York.
Nittler L. R. (2003) Presolar stardust in meteorites: Recent advances and scientific frontiers. *Earth Planet. Sci. Lett., 209,* 259–273.
Nittler L. R. (2005) Constraints on heterogeneous galactic chemical evolution from meteoritic stardust. *Astrophys. J., 618,* 281–296.
Nittler L. R. and Alexander C. M. O'D. (1999) Can stellar dynamics explain the metallicity distributions of presolar grains? *Astrophys. J., 526,* 249–256.
Nittler L. R. and Alexander C. M. O'D. (2003) Automated isotopic measurements of micron-sized dust: Application to meteoritic presolar silicon carbide. *Geochim. Cosmochim. Acta, 67,* 4961–4980.
Nittler L. R. and Cowsik R. (1997) Galactic age estimates from O-rich stardust in meteorites. *Phys. Rev. Lett., 78,* 175–178.
Nittler L. R., Alexander C. M. O'D., Gao X., Walker R. M., and Zinner E. (1997) Stellar sapphires: The properties and origins of presolar Al_2O_3 in meteorites. *Astrophys. J., 483,* 475–495.
Nittler L. R., Hoppe P., Alexander C. M. O'D., Busso M., Gallino R., Marhas K. K., and Nollett K. (2003) Magnesium isotopes in presolar spinel (abstract). In *Lunar and Planetary Science XXXIV,* Abstract #1703. Lunar and Planetary Institute, Houston (CD-ROM).
Nordström B., Mayor M., Andersen J., Holmberg J., Pont F., Jørgensen B. R., Olsen E. H., Udry S., and Mowlavi N. (2004) The Geneva-Copenhagen survey of the solar neighbourhood. *Astron. Astrophys., 418,* 989–1019.
Nyquist L., Lindstrom D., Mittlefehldt D., Shih C.-Y., Wiesmann H., Wentworth S., and Martinez R. (2001) Manganese-chromium formation intervals for chondrules from the Bishunpur and Chainpur meteorites. *Meteoritics & Planet. Sci., 36,* 911–938.
Oey M. S. (2000) A new look at simple inhomogeneous chemical evolution. *Astrophys. J. Lett., 542,* L25–L28.
Pagel B. E. J. (1989) An analytical model for the evolution of primary elements in the galaxy. *Rev. Mexicana Astron. Astrofis., 18,* 161–172.
Pagel B. E. J. (1997) *Nucleosynthesis and Chemical Evolution of Galaxies.* Cambridge Univ., Cambridge. 378 pp.
Penzias A. A. (1981) The isotopic abundances of interstellar oxygen. *Astrophys. J., 249,* 518–523.
Prantzos N., Aubert O., and Audouze J. (1996) Evolution of the carbon and oxygen isotopes in the galaxy. *Astron. Astrophys., 309,* 760–774.
Prinzhofer A., Papanastassiou D. A., and Wasserburg G. J. (1992) Samarium-neodymium evolution of meteorites. *Geochim. Cosmochim. Acta, 56,* 797–815.
Qian Y.-Z., Vogel P., and Wasserburg G. J. (1998) Diverse supernova sources for the r-process. *Astrophys. J., 494,* 285–296.
Quitté G. and Birck J. (2004) Tungsten isotopes in eucrites revisited and the initial ^{182}Hf/^{180}Hf of the solar system based on iron meteorite data. *Earth Planet. Sci. Lett., 219,* 201–207.
Raiteri C. M., Villata M., and Navarro J. F. (1996) Simulations of galactic chemical evolution. I. O and Fe abundances in a simple collapse model. *Astron. Astrophys., 315,* 105–115.
Rauscher T., Heger A., Hoffman R. D., and Woosley S. E. (2002) Nucleosynthesis in massive stars with improved nuclear and stellar physics. *Astrophys. J., 576,* 323–348.
Reddy B. E., Tomkin J., Lambert D. L., and Allende Prieto C. (2003) The chemical compositions of galactic disc F and G dwarfs. *Mon. Not. R. Astron. Soc., 340,* 304–340.
Reynolds J. H. (1960) Determination of the age of the elements. *Phys. Rev. Lett., 4,* 8–10.
Rocha-Pinto H. J., Maciel W. J., Scalo J., and Flynn C. (2000) Chemical enrichment and star formation in the Milky Way disk. I. Sample description and chromospheric age-metallicity relation. *Astron. Astrophys., 358,* 850–868.
Romano D. and Matteucci F. (2003) Nova nucleosynthesis and galactic evolution of the CNO isotopes. *Mon. Not. R. Astron. Soc., 342,* 185–198.
Rowe M. W. and Kuroda P. K. (1965) Fissiogenic xenon from the Pasamonte meteorite. *J. Geophys. Res., 70,* 709–714.
Rutherford E. (1929) Origin of actinium and age of the Earth. *Nature, 123,* 313–314.
Salpeter E. E. (1955) The luminosity function and stellar evolution. *Astrophys. J., 121,* 161–167.
Scalo J. M. (1986) The stellar initial mass function. *Fund. Cosmic Phys., 11,* 1–278.
Schatz H., Toenjes R., Pfeiffer B., Beers T. C., Cowan J. J., Hill V., and Kratz K. (2002) Thorium and uranium chronometers applied to CS 31082-001. *Astrophys. J., 579,* 626–638.
Schmidt M. (1959) The rate of star formation. *Astrophys. J., 129,* 243–258.
Schmidt M. (1963) The rate of star formation. II. The rate of formation of stars of different mass. *Astrophys. J., 137,* 758–769.
Schönbächler M., Rehkämper M., Halliday A. N., Lee D.-C., Bourot-Denise M., Zanda B., Hattendorf B., and Gunther D.

(2002) Niobium-zirconium chronometry and early solar system development. *Science, 295,* 1705–1708.

Shetrone M., Venn K. A., Tolstoy E., Primas F., Hill V., and Kaufer A. (2003) VLT/UVES abundances in four nearby dwarf spheroidal galaxies. I. Nucleosynthesis and abundance ratios. *Astron. J., 125,* 684–706.

Sneden C., Cowan J. J., Ivans I. I., Fuller G. M., Burles S., Beers T. C., and Lawler J. E. (2000) Evidence of multiple r-process sites in the early galaxy: New observations of CS 22892-052. *Astrophys. J. Lett., 533,* L139–L142.

Spergel D. N., Verde L., Peiris H. V., Komatsu E., Nolta M. R., Bennett C. L., Halpern M., Hinshaw G., Jarosik N., Kogut A., Limon M., Meyer S. S., Page L., Tucker G. S., Weiland J. L., Wollack E., and Wright E. L. (2003) First-year Wilkinson Microwave Anisotropy Probe (WMAP) observations: Determination of cosmological parameters. *Astrophys. J. Suppl. Ser., 148,* 175–194.

Thiemens M. H. and Heidenreich J. E. (1983) The mas independent fractionation of oxygen: A novel isotope effect and its possible cosmochemical implications. *Science, 219,* 1073–1075.

Timmes F. X. and Clayton D. D. (1996) Galactic evolution of silicon isotopes: Application to presolar SiC grains from meteorites. *Astrophys. J., 472,* 723–741.

Timmes F. X., Woosley S. E., and Weaver T. A. (1995) Galactic chemical evolution: Hydrogen through zinc. *Astrophys. J., 98,* 617–658.

Tinsley B. M. (1977) Chemical evolution in the solar neighborhood. III — Time scales and nucleochronology. *Astrophys. J., 216,* 548–559.

Tinsley B. M. (1980) Evolution of the stars and gas in galaxies. *Fund. Cosmic Phys., 5,* 287–388.

Travaglio C., Galli D., and Burkert A. (2001) Inhomogeneous chemical evolution of the galactic halo: Abundance of r-process elements. *Astrophys. J., 547,* 217–230.

Twarog B. A. (1980) The chemical evolution of the solar neighborhood. II — The age-metallicity relation and the history of star formation in the galactic disk. *Astrophys. J., 242,* 242–259.

van den Bergh S. (1962) The frequency of stars with different metal abundances. *Astron. J., 67,* 486–490.

van den Hoek L. and de Jong T. (1997) Inhomogeneous chemical evolution of the galactic disk: Evidence for sequential stellar enrichment? *Astron. Astrophys., 318,* 231–251.

Vanhala H. A. T. and Boss A. P. (2002) Injection of radioactivities into the forming solar system. *Astrophys. J., 575,* 1144–1150.

Wakker B. P., Howk J. C., Savage B. D., van Woerden H., Tufte S. L., Schwarz U. J., Benjamin R., Reynolds R. J., Peletier R. F., and Kalberla P. M. W. (1999) Accretion of low-metallicity gas by the Milky Way. *Nature, 402,* 388–390.

Wasserburg G. J., Busso M., Gallino R., and Raiteri C. M. (1994) Asymptotic giant branch stars as a source of short-lived radioactive nuclei in the solar nebula. *Astrophys. J., 424,* 412–428.

Wasserburg G. J., Busso M., and Gallino R. (1996) Abundances of actinides and short-lived nonactinides in the interstellar medium: Diverse supernova sources for the r-processes. *Astrophys. J. Lett., 466,* L109–L113.

Wheeler J. C., Sneden C., and Truran J. W. (1989) Abundance ratios as a function of metallicity. *Annu. Rev. Astron. Astrophys., 27,* 279–349.

Wielen R., Fuchs B., and Dettbarn C. (1996) On the birth-place of the Sun and the places of formation of other nearby stars. *Astron. Astrophys., 314,* 438–447.

Wilson T. L. and Rood R. T. (1994) Abundances in the interstellar medium. *Annu. Rev. Astron. Astrophys., 32,* 191–226.

Woosley S. E. and Weaver T. A. (1995) The evolution and explosion of massive stars, II. Explosive hydrodynamics and nucleosynthesis. *Astrophys. J. Suppl. Ser., 101,* 181–235.

Wyse R. (2003) Galactic encounters. *Science, 301,* 1055–1057.

Wyse R. F. G. and Silk J. (1989) Star formation rates and abundance gradients in disk galaxies. *Astrophys. J., 339,* 700–711.

Yin Q., Jacobsen S. B., McDonough W. F., Horn I., Petaev M. I., and Zipfel J. (2000) Supernova sources and the ^{92}Nb-^{92}Zr p-process chronometer. *Astrophys. J. Lett., 536,* L49–L53.

Yin Q., Jacobsen S. B., Yamashita K., Blichert-Toft J., Télouk P., and Albaréde F. (2002) A short timescale for terrestrial planet formation from Hf-W chronometry of meteorites. *Nature, 418,* 949–952.

Yokoi K., Takahashi K., and Arnould M. (1983) The Re-187-Os-187 chronology and chemical evolution of the galaxy. *Astron. Astrophys., 117,* 65–82.

Zinner E. (1998) Stellar nucleosynthesis and the isotopic composition of presolar grains from primitive meteorites. *Annu. Rev. Earth Planet. Sci., 26,* 147–188.

Zinner E., Amari S., Guinness R., Nguyen A., Stadermann F., Walker R. M., and Lewis R. S. (2003) Presolar spinel grains from the Murray and Murchison carbonaceous chondrites. *Geochim. Cosmochim. Acta, 67,* 5083–5095.

Zinner E., Amari S., Jennings C., Mertz A. F., Nguyen A. N., Nittler L. R., Hoppe P., Gallino R., and Lugaro M. (2005a) Al and Ti isotopic ratios of presolar SiC grains of type Z (abstract). In *Lunar and Planetary Science XXXVI,* Abstract #1691. Lunar and Planetary Institute, Houston (CD-ROM).

Zinner E., Nittler L. R., Hoppe P., Gallino R., Straniero O., and Alexander C. M. O'D. (2005b) Oxygen, magnesium and chromium isotopic ratios of presolar spinel grains. *Geochim. Cosmochim. Acta,* in press.

Chemical Processes in the Interstellar Medium: Source of the Gas and Dust in the Primitive Solar Nebula

Joseph A. Nuth III
NASA Goddard Space Flight Center

Steven B. Charnley
NASA Ames Research Center

Natasha M. Johnson
NASA Goddard Space Flight Center

Everything that became part of the primitive solar nebula first passed through the interstellar medium and the dense molecular cloud from which the nebula collapsed. We begin with a brief discussion of the connection between dense molecular clouds and the formation of solar-type stars. We then review processes that occur in these environments in order to constrain the chemical and physical properties of the matter that might still be preserved in primitive meteorites and that was the source material for every asteroid, comet, moon, and planet that exists at the present time in our own or in any other planetary system. Much of this history may be completely overprinted via processes occurring in the nebula itself or in larger planetary bodies, yet some signatures, such as those carried by isotopes of H, O, or N can be much more persistent.

1. INTRODUCTION

Primitive meteorites have a great deal in common with metamorphosed sedimentary rocks. The provenance of their individual components is somewhat uncertain, as is the precise time-temperature-pressure history that resulted in their formation. Yet these initial materials, no matter how different in origin, are found in a single, meter-scale object. We examine the likely input to primitive meteorites from the interstellar medium (ISM), starting from the source regions of the chemical elements and cosmic dust: asymptotic giant branch (AGB) stars, novae, and supernovae, and follow the likely chemical evolution of this material through regions of the interstellar medium and into the nebula. We follow the gas and dust from astrophysical sources separately with the goal of highlighting the signatures of interstellar processes that may be preserved in meteorites. Specifically, we describe the processing that dust grains and large molecules such as polycyclic aromatic hydrocarbons (PAHs) experience in the warm, diffuse ISM due to ultraviolet radiation, cosmic-ray bombardment, and shock chemistry. We also describe the ensuing chemistry when this material is incorporated into cold, molecular clouds where dense prestellar cores form and in which solar-type stars are born.

We examine these interstellar processes to highlight the potential correlations between the isotopic and chemical signatures that might be observable in meteorites or interplanetary dust particles (IDPs). For example, it has often been suggested that circumstellar grains are efficiently destroyed in the ISM. We look into this premise by comparing the likely signatures of solids produced in the ISM to those produced by circumstellar sources. Similarly, ion-molecule reactions in moderate density regions, and reactions in and on ice mantles in dense clouds, are suspected of producing a number of distinctive isotopic signatures in carbonaceous materials. Therefore, we review the gas-phase and solid-phase fractionation pathways for D, ^{15}N, and ^{13}C occurring in interstellar molecules and in the meteoritic data. We focus on recent detections of "superdeuteration" in specific compounds and on the possible relationship of these compounds' isotopic ratios to those found in meteoritic samples and IDPs. Finally, many molecules observed in molecular clouds have been proposed as precursors for chemical pathways that occurred on the parent bodies of carbonaceous chondrites, leading to the rich organic inventory in meteorite specimens.

2. INTERSTELLAR MATTER IN THE SOLAR SYSTEM

This section gives a brief overview of the connection between dense molecular clouds and the formation of solar-type stars. We describe the physical conditions in dense clouds and the processes believed to be involved in star formation, and list some of the measured chemical characteristics of primitive meteorites that can be attributed to interstellar chemistry.

2.1. The Interstellar Medium

The ISM is that region in space between the stars and is composed mainly of H and He gas as well as dust. The ISM ranges from extremely hot (10,000 K) diffuse (1 atom cm^{-3}) regions, to cold (10 K) dense (>10^4 atoms/molecules cm^{-3}) clouds. Gases make up the bulk of the mass of the ISM. The dust is a combination of grains generated in dying stars and modified in the ISM. Although the dust comprises only a small amount of material in the ISM by mass (i.e., 1%), it plays a significantly larger role in interstellar chemistry. While gas is generally transparent to photons from the near-ultraviolet to the far-infrared, dust absorbs and scatters this same energy. This process is referred to as "extinction." Dust can then serve as a "shield" from ultraviolet energy in denser regions so that more complex reactions might occur in the surrounding gas and on the grains themselves in surface-mediated reactions [see *Bakes* (1997) for an overview]. In order to learn more about the ISM, numerous spectral observations have been made of the intervening space using a variety of techniques and instruments (e.g., *Blitz,* 1990). These spectral techniques have revealed a wealth of information and more discoveries are made with each passing year.

The extinction of starlight is caused by the absorbance and scattering of energy by the solid material in the ISM. The degree and wavelength dependence of the interstellar extinction curve is affected by the grain size, composition, and morphology of the dust between the source of the starlight and the observer. As such, the curve has distinct features (see *Zubko et al.,* 2004, and references therein; *Mathis,* 1990). For example, the cause of the 217.5-nm absorption bump/peak has been under debate for many years and is most certainly due to C in some form. This, and many other dust features, is discussed in more detail in section 5, particularly in section 5.2. However, in the next section we focus on the chemistry that occurs within dense molecular clouds, the diffuse ISM, and in the regions surrounding protostellar cores.

2.2. Dense Molecular Clouds and Star Formation

Dissipation of interstellar turbulence in shock waves leads to the formation of dense molecular clouds from the diffuse interstellar medium (see *Mac Low and Klessen,* 2004). The high extinction efficiency of dust grains implies that most of the interstellar radiation impinging a cloud does not penetrate to the innermost dense regions. However, cosmic-ray particles can penetrate throughout the cloud to heat the gas, which can cool through emission of radiation in the rotational lines of the CO molecule. These processes lead to typical gas temperatures of around 10 K (*Goldsmith and Langer,* 1978).

Dense clouds show significant substructure (*Evans,* 1999). Hydrogen nucleon densities of 10^3–10^5 cm^{-3} are found in large-scale [~few parsec (pc)] dense clumps, which are in turn contained in a more diffuse (~10^2–10^3 cm^{-3}) interclump medium. Much higher densities (>10^5 cm^{-3}) are found in the small-scale (~ 0.1 pc) cores that are the future sites of gravitational collapse and low-mass star formation. If molecular clouds are long-lived structures (~10^7 yr), then the low galactic star formation rate implied by observations (*Zuckerman and Evans,* 1974) suggests that their cores must be supported against gravity by magnetic and/or turbulent pressures. There are then three important timescales for the evolution of molecular clouds and star formation. The first is the gravitational free-fall timescale, $\tau_{ff} \sim 4.35 \times 10^7\ n_H^{-1/2}$ yr, where n_H is the H nucleon density. In the case where support is due to the combined effects of magnetic field pressure and the turbulent pressure exerted by magnetohydrodynamic (MHD) waves [the "standard model" of low-mass star formation (see *Shu et al.,* 1987)], support is lost by the decline in magnetic flux due to the relative drift between ions and neutrals. It follows that the second timescale is associated with this ambipolar diffusion process, τ_{AD}. For standard assumptions concerning the cosmic-ray ionization rate and depletion of the elements, $\tau_{AD} \sim 7.3 \times 10^{13}\ x_e\ yr^{-1}$, where the fractional ionization, x_e, is ~10^{-8}–10^{-7} at molecular cloud densities. As $\tau_{AD} \gg \tau_{ff}$, the loss of magnetic flux leads to a slow quasistatic increase in the central density of the core until gravity dominates, then a protostar forms in a rapid, approximately isothermal, free-fall collapse (*Larson,* 2003). The third timescale of importance is associated with the dissipation of turbulence (τ_{dt}) and was found to be comparable to the free-fall time using numerical simulations of MHD turbulence. Clouds and cores cannot be sustained against collapse by turbulent support for much longer than τ_{dt} (*Mac Low and Klessen,* 2004).

This fact and other arguments concerning the observed lack of T Tauri stars in dense clouds recently led to the alternative view that the evolution of molecular clouds and star formation are rapid processes governed by supersonic turbulence (*Elmegreen,* 2000; *Hartmann et al.,* 2001; *Mac Low and Klessen,* 2004; *Larson,* 2003). In this scenario, interstellar clouds spend a significant fraction of their lives in a low-density, mostly atomic state where star formation is inefficient. When diffuse material is compressed to higher densities and molecular H forms, star formation occurs rapidly within only a few free-fall times. In this scenario the molecular clouds themselves are short-lived structures, dissipating soon after star formation has begun (*Elmegreen,* 2000).

To summarize, the protosolar nebula was likely formed by the nearly free-fall gravitational collapse (~3×10^5 yr) of a rotating mass of cold gas and dust. This collapsing material was situated near the center of a dense molecular core that itself either formed slowly (several million years) by gradual contraction due to the gradual loss of magnetic support against gravity (e.g., *Shu et al.,* 1987) or more rapidly (~few free-fall times) by the dissipation of turbulence (e.g., *MacLow and Klessen,* 2004). An alternative view, discussed in *Boss and Goswami* (2006), is that a nearby supernova explosion could have triggered the collapse. Irrespective of the uncertainty over the formation timescale and ages of molecular clouds, protostars form by gravitational collapse of a dense molecular core (*Evans,* 1999). However, it should be pointed out that chemical youth better explains the high CO/O_2 ratios commonly found in molecular clouds (*Goldsmith et al.,* 2000), as well as the high abundances of C-chain molecules detected in them (e.g., *Suzuki et al.,* 1992).

Clearly, regions of star formation cannot be characterized in a few sentences: They are complex, time-dependent extremes of hot and cold. Most of the preexisting matter either falls into the central star or is ejected from the system in bipolar outflows. The majority of the tiny fraction of material left orbiting the central star will be found in the form of planets that retain little record of the gas and grains from which they formed. Only comets and the smaller asteroids preserve intact material that might predate the time of star formation, yet these objects also preserve a significant quantity of material produced via the repeated vaporization and recondensation of nebular solids (*Boynton,* 1985; *Clayton et al.,* 1985).

2.3. Meteoritic Evidence for Interstellar Signatures

All the materials incorporated into protostellar nebulae — elemental nuclei, PAHs, large macromolecular aromatic-carbon structures, and dust particles — were once in the diffuse ISM. Many astronomical sources contribute to the composition of the refractory dust found in the diffuse ISM: the atmospheres and outflows of late-type stars on the AGB, winds from Wolf-Rayet stars, novae, and supernovae. The gas-phase molecular composition of the diffuse ISM, on the other hand, is largely determined by the local physical conditions (*van Dishoeck and Black,* 1988). However, it is the active gas phase chemistry occurring in cold molecular clouds that is pivotal in setting the initial chemical inventory of volatile species available to the protosolar nebula. Here, much of the gas is converted to molecular form (H_2, CO, N_2) and an additional solid phase — the icy grain mantle — is produced through the condensation and reaction of gaseous volatiles. Within the meteoritic record there is clear evidence for chemical processes such as isotopic fractionation, including significant levels of D enrichment that would have been favored in the cold, dense environment of molecular clouds.

The organic inventories of Murchison (a well-studied carbonaceous meteorite), other primitive meteorites, and IDPs display large D enrichments relative to terrestrial values. This constitutes one of the most convincing pieces of evidence that at least some of the original C reservoir was interstellar (e.g., *Kerridge and Chang,* 1985). Meteoritic organic material also contains a large enrichment in ^{15}N (e.g., *Alexander et al.,* 1998) where the anomalous N is generally associated with carbonaceous material. Many IDPs are also highly enriched in D and ^{15}N (e.g., *Messenger and Walker,* 1997; *Messenger,* 2000). In contrast, C isotopes do not show similar anomalies, although there is evidence for variations within specific homologous series (*Gilmour and Pillinger,* 1994). These isotopic anomalies are chemical in origin and distinct from the nucleosynthetic signatures found in circumstellar grains (e.g., *Messenger et al.,* 1998). In cold interstellar clouds, enhanced molecular D/H ratios are observed and can easily be explained by gas phase and grain-surface processes (*Tielens,* 1983; *Millar et al.,* 1989; *Sandford et al.,* 2001). The ^{15}N anomalies could be due to nucleosynthetic processes; however, such processes also tend to produce anomalous effects in the C isotopes that are more pronounced, and are not evident in the meteoritic data (e.g., *Messenger and Walker,* 1997). For this reason, the ^{15}N enrichments are also believed to be due to chemical fractionation in interstellar ion-molecule reactions (e.g., *Terzieva and Herbst,* 2000; *Aléon and Robert,* 2004).

Interstellar chemistry can explain many of the molecules found in meteoritic extracts. Organic compounds identified in carbonaceous C1 and C2 chondrites include amines and amides; alcohols, aldehydes, and ketones; aliphatic and aromatic hydrocarbons; sulfonic and phosphonic acids; amino, hydroxycarboxylic, and carboxylic acids; purines and pyrimidines; and kerogen-type material (*Cronin and Chang,* 1993; *Botta and Bada,* 2002). These molecules show specific enrichments in D, ^{13}C, and ^{15}N (e.g., *Cronin and Chang,* 1993), indicative of their retention of an interstellar heritage. Some of these simple organics may be of true interstellar origin, whereas the more complex molecules may have had simpler interstellar precursors that subsequently reacted in the solar nebula or meteorite parent body. The initial interstellar inventory determined which reaction pathways would be most likely for secondary processing in the protosolar nebula, for example, as the result of aqueous alteration on meteoritic parent bodies.

Carbonaceous meteoritic material represents a mixture of highly processed interstellar matter and, perhaps, some pristine components. Polycyclic aromatic hydrocarbons (PAHs) are ubiquitous in the ISM (*Allamandola et al.,* 1989; *Puget and Léger,* 1989), and are also widespread in meteorites (e.g., *Basile et al.,* 1984), as well as in IDPs (*Allamandola et al.,* 1987; *Clemett et al.,* 1998). Most meteoritic carbon is in the form of a kerogen-like material and this may be associated with interstellar PAHs. The PAHs found in meteorites could have originated in the cool envelopes of carbon stars (*Frenklach and Feigelson,* 1989) although other formation sites are possible, such as in diffuse interstellar gas (*Bettens and Herbst,* 1996) or by energetic processing of specific ices in dense clouds (*Kissel et al.,* 1997), or by reaction of CO, N_2, and H_2 on grain surfaces in the solar nebula. The PAH component of meteorites exhibits a D enrichment indicative of a presolar origin (*Allamandola et al.,* 1987). Since D is consumed in stellar burning, the deuteration of astronomical PAHs can occur in the interstellar me-

TABLE 1. Identified astrophysical molecules.

2	3	4	5	6	7	8	9	10	11	13
H_2	C_3	c-C_3H	C_5	C_5H	C_6H	CH_3C_3N	CH_3C_4H	CH_3C_5N?	HC_9N	$HC_{11}N$
AlF	C_2H	1-C_3H	C_4H	1-H_2C_4	CH_2CHCN	$HCOOCH_3$	CH_3CH_2CN	$(CH_3)_2O$		
AlCl	C_2O	C_3N	C_4Si	C_2H_4	CH_3C_2H	CH_3COOH	$(CH_3)_2O$	$HOCH_2CH_2OH$		
C_2	C_2S	C_3O	1-C_3H_2	CH_3CN	HC_5N	C_7H	CH_3CH_2OH	NH_2CH_2COOH		
CH	CH_2	C_3S	c-C_3H_2	CH_3NC	$HCOCH_3$	H_2C_6	HC_7N			
CH^+	HCN	C_2H_2	CH_2CN	CH_3OH	NH_2CH_3	$HOCH_2CHO$	C_8H			
CN	HCO	CH_2D^+?	CH_4	CH_3SH	c-C_2H_4O					
CO	HCO^+	HCCN	HC_3N	HC_3NH^+	CH_2CHOH					
CO^+	HCS^+	$HCNH^+$	HC_2NC	HC_2CHO						
CP	HOC^+	HNCO	HCOOH	NH_2CHO						
CSi	H_2O	HNCS	H_2CHN	C_5N						
HCl	H_2S	$HOCO^+$	H_2C_2O							
KCl	HNC	H_2CO	H_2NCN							
NH	HNO	H_2CN	HNC_3							
NO	MgCN	H_2CS	SiH_4							
NS	MgNC	H_3O^+	H_2COH^+							
NaCl	N_2H^+	NH_3								
OH	N_2O	SiC_3								
PN	NaCN									
SO	OCS									
SO^+	SO_2									
SiN	c-SiC_2									
SiO	CO_2									
SiS	NH_2									
CS	H_3^+									
HF	SiCN									
SH	AlNC									
FeO										

dium, specifically in the cold environment of dense molecular clouds. In their evolutionary cycle, PAH molecules freeze out onto grains in dense clouds, where they can become fractionated through energetically driven reactions with deuterated water ice. Additionally, several ion-molecule gas phase processes will deuterate PAHs (*Sandford et al.,* 2001) (see section 4.1).

3. INTERSTELLAR CHEMISTRY

From the perspective of providing either the simple molecular precursors or the actual molecules themselves, the pivotal phases of the ISM for meteoritic studies are the dense molecular clouds and the star-forming cores within them. Table 1 presents an inventory of gas-phase astrophysical molecules identified using radio observations. These molecules include those known to be present in the diffuse ISM, dense molecular clouds, and the envelopes of evolved stars. The simpler molecules can be formed by gas-phase chemistry but some of them (e.g., H_2O and CO_2) can be more efficiently formed on the surfaces of dust grains. It appears that many more complex molecules are formed exclusively by grain catalysis (e.g., CH_3OH and C_2H_5OH). The great majority of these species, particularly the more complex ones, have been detected in cold molecular clouds or in the warmer environs of forming protostars. By no means is this list representative of the bulk of the material in the ISM. For example, no polyaromatic hydrocarbons (PAHs) are listed even though they are likely to be more abundant than almost all the molecules listed in Table 1.

In this section, we describe the principle chemical processes that occur in dense molecular clouds and in star-forming environments. Note that most of these processes also apply to other astrochemically active regions such as the diffuse ISM (*Wooden et al.,* 2004), protoplanetary disks (*Markwick and Charnley,* 2004; *Ciesla and Charnley,* 2006), the early universe (*Stancil et al.,* 2002) and in the winds from novae, supernovae, and red giant stars (*Rawlings,* 1988; *Lepp et al.,* 1990; *Glassgold,* 1996). These chemical processes are crucial for understanding what is probably the most direct connection between the ISM and the composition of meteorites: isotopic fractionation.

3.1. Gas Phase Processes

Interstellar chemistry generally takes place in extreme cold ($T \ll 50$ K) and at densities where only bimolecular or unimolecular processes are viable ($<10^{13}$ particles cm^{-3}). In dark, starless regions, the chemistry occurs in gas where the H nucleon density, n_H, and the gas kinetic temperature, T, fall in the ranges of ~10^3–10^5 cm^{-3} and ~10–30 K, respectively (e.g., *Goldsmith and Langer,* 1978). Closer to nascent protostars, in so-called hot molecular cores, the physical conditions are markedly different: $n_H \sim 10^6–10^8$ cm^{-3}, and T ~ 50–300 K (e.g., *van Dishoeck and Blake,* 1998). However, it is important to note that densities and temperatures high enough for the gas-phase synthesis of H_2 are not achieved even here. Instead, H_2 is produced in a three-body reaction where the third body is a dust grain (e.g., *Hollenbach and Salpeter,* 1971). Surface-mediated reactions on dust grains are discussed in section 3.2. In the following section, we describe the gas-phase chemistry that ensues when almost all the available H has been converted to molecular form.

Galactic cosmic rays ionize H_2 molecules to produce electrons, H^+ and H_2^+. The latter rapidly reacts with H_2 to form the H_3^+ ion. Since H_3^+ readily transfers a proton to most species, it plays a central role in dense molecular cloud chemistry. Thus, the principal gas-phase chemical processes occurring in molecular clouds are ion-neutral reactions, ion-electron dissociative reactions, radiative recombination, and neutral-neutral reactions. Since He atoms are also ionized by cosmic rays, dissociative charge transfer reactions involving He^+ play an important role in destroying neutral molecules. External photons from outside the molecular clouds are efficiently extinguished by the dust particles, although a small, weak, residual flux derived from the deexcitation of H_2 following the impact of cosmic-ray-produced electrons may have a significant role (*Prasad and Tarafdar,* 1983) in dark cloud chemistry.

The synthesis of interstellar molecules is initiated by charge transfer between H^+ and O atoms that leads, via a sequence of ion-molecule reactions with H_2, to the production of OH and the onset of neutral chemistry. This leads to CO and N_2 as the most abundant molecules after H_2, with OH, NH_3, and SO also produced in significant abundance. Theoretically, the timescale to attain a chemical steady state for this process is on the order of several million years. In this case almost all the C and O become incorporated into CO and O_2. The fact that high atomic C and low O_2 abundances are observed is evidence either that the lifetimes of molecular clouds are not this long (e.g., *Elmegreen,* 2000), or that this process is not dominant.

These types of chemical reactions fix the degree of ionization, which in turn controls the magnetic evolution of dense clouds through the rate of loss of magnetic flux by ambipolar diffusion. The dominant ions are HCO^+ and N_2H^+ with electron fractions relative to H nuclei of ~10^{-8}–10^{-7} (*Caselli et al.,* 1998). It appears that refractory metals (Fe, Mg, Na, Si; i.e., elements heavier than He) are absent from the dense gas, probably by their almost complete incorporation into/onto dust grains. Unsuccessful searches for various related hydrides and oxides provide strong support for this absence (*Turner,* 1991). If PAH molecules are present in appreciable abundances in cold clouds, they will alter the physics of the plasma, as well as the basic nature of the chemistry (*Lepp and Dalgarno,* 1988). In this case, PAH anions become the dominant carriers of negative charge and the major ion neutralization pathway becomes mutual neutralization.

3.2. Grain-Surface Processes

At gas and dust temperatures near 10 K, efficient adsorption of atoms and molecules from the gas, followed by surface diffusion and reaction of atoms with other surface species, leads to the formation of a "dirty" ice mantle covering the more refractory silicate core (e.g., *Sandford and Allamandola,* 1993). Infrared observations of dark clouds and subsequent interpretations indicate that ice mantles cover interstellar dust grains and suggest that a rich catalytic chemistry is possible (*Ehrenfreund and Charnley,* 2000). The Infrared Space Observatory (ISO) observed many massive star-forming regions and these observations have been used to inventory the most abundant ice species in dense clouds (*Ehrenfreund and Schutte,* 2000; *Gibb et al.,* 2000). The ices are predominately H_2O, with CO, CO_2, and CH_3OH as the next most abundant molecules (comprising from ~5% to 40% of the ice). Several other molecules are only present in trace amounts of a few percent or less: CH_4, NH_3, H_2CO, HCOOH, OCS, and possibly OCN^- (see Fig. 1). In Fig. 1, OCN^- is represented by XCN because until recently it was unclear which element X represented. Recent groundbased observations have also identified high abundances of methanol (CH_3OH) ice in regions of low-mass star formation (*Pontopiddan et al.,* 2003).

Cold grain-surface kinetics occurs in two steps following a Langmuir-Hinchelwood catalytic model: First, atoms and molecules stick to the surface, then subsequently diffuse among surface binding sites either by quantum mechanical tunneling, as is the case for H and D atoms, or by thermal hopping. These processes are depicted in Fig. 2. Diffusion of light atoms is many times faster than that of heavier molecules; therefore the chemistry is dominated by exothermic atom-atom reactions, atom-molecular radical reactions, and by reactions involving activation energy barriers where atoms add to molecules with multiple bonds (e.g., *Allen and Robinson,* 1977; *Tielens and Hagen,* 1982;

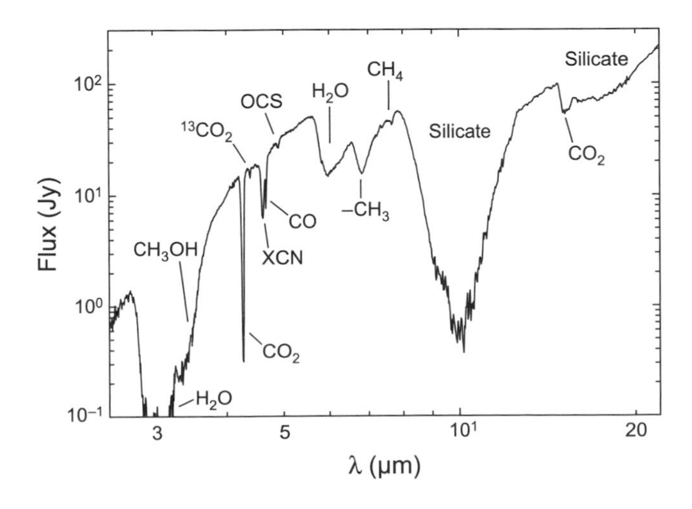

Fig. 1. Molecules detected in interstellar ices surrounding the massive protostar W33A by spacebased IR absorption spectroscopy (*Gibb et al.,* 2000).

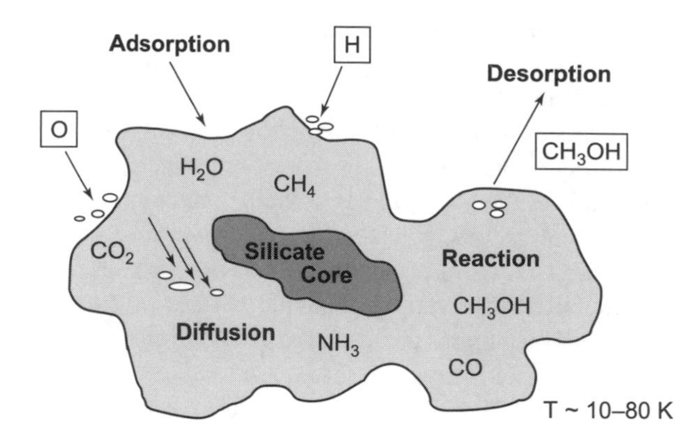

Fig. 2. Schematic depiction of the major physical processes involved in grain-surface catalysis, leading to the growth of interstellar ices (*Ehrenfreund et al.,* 2005).

Charnley, 2001a). On grain surfaces, H_2O, NH_3, H_2S, and CH_4 are formed by successive H atom additions to accreted O, N, S and C. Accreted CO molecules can be reduced, oxidized, and deuterated by H, O, and D atoms in additions that possess activation energy barriers (*Tielens and Hagen,* 1982; *Charnley et al.,* 1997). Extension of the basic kinetic scheme in the CO reduction sequence to additions involving C atoms, with the additional constraint of radical stability, can lead to surface reactions where many known interstellar molecules (e.g., aldehydes, ketones, alcohols, sugars) can be produced (*Charnley,* 1997a, 2001b).

It appears that many of the most abundant interstellar ice molecules can be accounted for either by direct accretion from the gas (e.g., CO) or by one of the above (cold) atom addition processes. Until recently, this assertion was based solely on theory (e.g., *Allen and Robinson,* 1977; *Tielens and Hagen,* 1982), but recent experiments tend to confirm the efficacy of these processes under conditions relevant for interstellar dust. Although the simplest system of H_2 production on various surfaces at temperatures around 10 K has been the primary focus of research efforts (*Pirronello et al.,* 1999; *Manico et al.,* 2001), the hydrogenation of CO to HCHO and CH_3OH at low temperatures has also been studied extensively (*Hiraoka et al.,* 1994, 1998, 2002; *Watanabe and Kouchi,* 2002; *Watanabe et al.,* 2003). The reduction of HC_2H and the production of CO_2 by adding an O atom to CO have also been demonstrated (*Hiraoka et al.,* 2000; *Roser et al.,* 2001).

Once a thick molecular ice mantle has been formed, the bulk ice matrix can itself be subject to further "energetic processing" involving the absorption of UV photons and cosmic-ray impacts. Many experimental studies of these processes have been made (*Allamandola et al.,* 1997; *Hudson and Moore,* 1999), yet it is still unclear what the relative contributions of photolysis and radiolysis are to the observed interstellar inventory. The carrier of the "XCN" band has been identified as OCN^-, and production of this species requires some form of energetic processing in the protostellar environment [see *Ehrenfreund et al.* (2001) for a recent summary].

Thermal heating, prior to complete desorption of the icy mantle, can be viewed as an energetic process that can lead to molecular rearrangement: This is an important aspect of ISM chemistry. *Blake et al.* (1991) presented direct structural evidence of such rearrangements, while *Ehrenfreund et al.* (1998) experimentally showed that the fine structure found by ISO in the 15.2-μm CO_2 band is due to the rearrangement of molecules in the matrix to form Lewis base CO_2-CH_3OH complexes at approximately 70 K.

As ices are formed, there clearly must be retention of the reaction products with a reasonably high (but unknown) efficiency. Nevertheless, some desorption processes do operate and these can be roughly divided into passive and active modes. In passive desorption (e.g., *Willacy and Millar,* 1998), one considers processes directly connected to the surface or bulk chemistry that act to remove molecules. As heavy atom–hydrogen atom additions are generally exothermic, some of the excess energy may be expended in breaking the weak surface bond. Similarly, absorption of UV photons or cosmic rays can lead to direct ejection of mantle molecules. Such energetic processing will also produce a small population of highly reactive species in the bulk ice, and these can undergo a spontaneous "explosion" resulting in the loss of mantle material (e.g., *Schutte and Greenberg,* 1991). Ultraviolet processing has also been shown to produce species found in meteorites such as amino acids (*Bernstein et al.,* 2002), amphiphiles (*Dworkin et al.,* 2001), and aromatic ketones (*Bernstein et al.,* 1999, 2001).

Ices are also returned to the gas in the hot cores surrounding young protostars. Evidence for this lies in the high abundances of many molecules known to have been synthesized on grains: H_2O, CH_4, NH_3, OCS, and CH_3OH (e.g., *Herbst,* 2000). In this case, the external physical environment determines the return of grain mantle material and represents active desorption. Protostellar outflows drive shock waves into the surrounding core and cause ice mantles to be lost by nonthermal sputtering (*Draine et al.,* 1983). Protostars could also affect more distant regions of molecular clouds: The relative streaming of charged and neutral grains in MHD waves may act as an external trigger for the thermal or explosive desorption of ices (*Markwick et al.,* 2000). However, the most elementary mantle loss process is simple thermal evaporation, where the protostellar luminosity heats the infalling icy grains above the evaporation temperatures of the individual mantle molecules (e.g., *Rodgers and Charnley,* 2003).

Many complex organics detected in protostellar gas (e.g., C_2H_5OH, HCO_2CH_3, CH_3CHO, HCO_2H) may have also formed on dust grains. However, other organics such as $(CH_3)_2O$, CH_3CO_2H, and CH_3COCH_5 could have been formed, post-evaporation, by chemical reactions in the hot gas (*Charnley et al.,* 1992). These latter molecules were found by radio astronomers at such low abundances that they would currently be undetectable via infrared ice spectroscopy. Ion-molecule reactions based on cation transfer processes among subliming products of grain-surface chemistry could also form many of the large interstellar molecules (*Charnley et al.,* 1995; *Charnley,* 2001b; *Blagojevic et al.,* 2003). The richness of the organic chemistry found in molecular clouds, particularly in the environments of massive protostars, has long been appreciated (*Cronin and Chang,* 1993) as a source for many of the precursor molecules that would be needed to start various organic syntheses on meteoritic parent bodies (e.g., the Strecker-cyanohydrin production of amino and hydroxy acids). This connection is greatly strengthened by recent molecular line observations that indicate that the environments of low-mass protostars also contain a rich chemical inventory (*Schöier et al.,* 2002; *Cazaux et al.,* 2003; *Kuan et al.,* 2004; *Bottinelli et al.,* 2004).

4. ISOTOPIC FRACTIONATION IN MOLECULAR CLOUDS

The chemistry outlined above leads to significant isotopic fractionation in cold molecular clouds. Although the key issue is that the zero-point energy difference between isotopomers is not negligible in ion-molecule reactions at 10 K, other, subtler, effects play a role in fractionating interstellar molecules. In this section we present an outline of relevant fractionation processes. The challenge is to understand the fractionation found in IDPs, comets, and meteorites, either as a direct result of interstellar chemistry, or as evidence for the action of these processes in protoplanetary disks (*Ciesla and Charnley,* 2006).

4.1. Deuterium

The cosmic D/H ratio is close to 3×10^{-5} (*Vidal-Madjar,* 2002) ($D/H_{Earth} = 1.56 \times 10^{-4}$). Radioastronomical observations of many interstellar molecules and their deuterated counterparts (isotopomers) show that they possess substantial enrichments with D/H enhanced by factors of 10–10,000 over the cosmic value [see *Ceccarelli* (2002) for a recent review]. At low temperatures, such as those found in molecular clouds (~10 K), the ion-molecule reaction

$$H_3^+ + HD \rightarrow H_2D^+ + H_2 \quad (1)$$

breaks D atoms out of HD, allowing D to be redistributed throughout the gas phase (e.g., *Millar et al.,* 2000) and leads to enrichment in the gas phase D/H ratio (~100×). Therefore, a grain-surface chemistry dominated by atom addition reactions (section 3.2) will lead to molecules highly fractionated in D (*Tielens,* 1983). At higher temperatures (>30 K) the reverse reaction efficiently destroys H_2D^+ and deuteration is suppressed. Interstellar chemistry based on these fractionation pathways could account for most of the D/H ratios observed prior to 1998 in both massive hot cores and in cold clouds (e.g., *Charnley et al.,* 1997; *Millar et al.,* 2000).

However, the discovery of enormous fractionation in D_2CO and HDCO (D/H ~ 0.1–0.4) in the circumstellar gas of young low-mass protostars, such as the binary system IRAS 16293-2422 (*Loinard et al.,* 2000, 2001), indicated efficient catalytic formation on grains prior to their deposition. Yet, the discovery of NHD_2 in the cold cloud L134N (*Roueff et al.,* 2000), and subsequently ND_3 in other regions (*van der Tak et al.,* 2002), can best be explained by gas-phase chemistry in an environment where CO and other heavy molecules were depleted from the gas by condensation on grains (*Rodgers and Charnley,* 2001). This depletion mechanism acts to enhance fractionation since reaction with CO is the major destruction pathway for H_2D^+. In such "depletion cores," the abundances of H_2D^+ and D atoms rise dramatically, allowing the former to be easily detected (e.g., *Caselli et al.,* 2003). Thus, it appears that most low-mass stars may form from a "starless" core that has undergone a high depletion phase (*Bergin et al.,* 2002). Depletion has also been proposed to account for the high fractionation seen in D_2CO and HDCO (*Roberts and Millar,* 2000), but is not capable of reproducing the highest ratios observed, or the multiple deuteration in methanol (CHD_2OH and CD_3OH) recently detected by *Parise et al.* (2002, 2004). The problem with attributing a grain surface origin to the D fractionation seen in HCHO and CH_3OH is that the atomic D/H ratios attainable, even in highly depleted gas, are much lower than needed (~0.5). *Roberts et al.* (2003) have shown that extensions of the ion-molecule deuteration scheme to include D_3^+ and D_2H^+ become important at the high densities of prestellar cores (>10^6 cm^{-3}) and can produce the required atomic D/H ratios. D_2H^+ has recently been detected by *Vastel et al.* (2004), who found a D_2H^+/H_2D^+ ratio of unity, as predicted by theory.

4.2. Nitrogen

Apart from D enrichments, the organic matter found in meteorites and IDPs is also known to contain components that exhibit enhancements in $^{15}N/^{14}N$ ratios (e.g., *Messenger,* 2000; *Aleon et al.,* 2003; *Sephton et al.,* 2003; *Aleon and Robert,* 2004) and chemical fractionation in ion-molecule reactions at low temperatures appears to be indicated. *Terzieva and Herbst* (2000) considered several possible processes that could lead to enhanced $^{15}N/^{14}N$ ratios in interstellar molecules and found that ion-molecule reactions such as those shown in reactions (2) and (3) were the most important.

$$^{15}N + {}^{14}N_2H^+ \rightarrow {}^{14}N + {}^{15}N^{14}NH^+ \quad (2)$$

$$^{15}N + {}^{14}N_2H^+ \rightarrow {}^{15}N + {}^{14}N^{15}NH^+ \quad (3)$$

Terzieva and Herbst (2000) calculated the $^{15}N/^{14}N$ ratios for the chemical evolution of a molecular cloud and found enrichment factors to be much smaller (~0.25) than that needed to explain the component "hot spots" in meteoritic/IDP materials (~0.5).

The theoretical prediction (*Charnley,* 1997b) that there could be dense regions where CO would be selectively depleted into CO ice, relative to both H_2 and N_2, has been verified by several recent observations (e.g., *Bergin et al.,* 2002). In this situation, N atoms do not convert to N_2 but instead participate in very restrictive reactions that involve the eventual destruction of the gas-phase molecular nitrogen. *Charnley and Rodgers* (2002) considered such a scenario where all C- and O-bearing species were depleted, except for N_2 and its isotopomers. They showed that reactions (2) and (3) led to more ^{15}N nuclei being incorporated into molecular nitrogen, which, after attack by He^+ ions, then released N^+ and N atoms into the gas with a correspondingly higher $^{15}N/^{14}N$ ratio. This N^+ reacts with H_2 to initiate an ion-molecule sequence leading to ammonia, which is then rapidly accreted as ice with the $^{15}N/^{14}N$ ratio in the ammonia enriched by as much as 80% relative to the starting value.

However, recent experiments by *Geppert et al.* (2004) indicate that there is a destructive channel that dominates the dissociative electron recombination of N_2H^+. In this case, the NH radicals produced in the recombinations can react with N atoms to reform N_2. New calculations show that the peak $^{15}NH_3/^{14}NH_3$ ratios achievable are substantially reduced (*Rodgers and Charnley,* 2004), but high $^{15}NH_3/^{14}NH_3$ ratios can be recovered if either the fractionation chemistry takes place at densities where N_2H^+ recombination occurs primarily on negatively charged macroscopic particles (dust grains or PAHs), or if there is a small barrier to the reaction of N with NH. Recent measurements of $^{15}N/^{14}N$ in anhydrous IDPs by *Floss et al.* (2004) determined enrichment factors of about 1.3, much higher than reported in the theoretical models. However, solid-state $^{15}NH_3/^{14}NH_3$ ratios calculated in these models are the values in the bulk ice mantle; calculations show that, although they are only a fraction of the total number accreted, the late-accreting ammonia molecules exhibit a temporal "spike" in the $^{15}NH_3/^{14}NH_3$ ratio, consistent with the high enrichments found by Floss and coworkers (*Charnley and Rodgers,* 2002; *Rodgers and Charnley,* 2004). This suggests that the $^{15}NH_3/^{14}NH_3$ ratio should vary with depth throughout the mantle, with very high values occurring in the uppermost few monolayers.

Polycyclic aromatic hydrocarbon molecules should also be late-accreting components of these mantles. Due to their relatively slow thermal velocities, PAH molecules will accrete more slowly than lighter atoms and molecules. As the PAH population is depleted, the anion becomes the dominant charge carrier (*Lepp and Dalgarno,* 1988) and, since about 95% of the population of larger "classical" grains are negatively charged, Coulomb repulsion will greatly slow the removal of the remaining PAHs. This effect, coupled with the fact that depleted gas will show very high D/H ratios (*Roberts et al.,* 2003), thus offers a means of spatially colocating, in the uppermost monolayers, carbonaceous molecules enriched in D and ammonia molecules enriched in ^{15}N. Additional energetic processing of these ices could then incorporate this ^{15}N into the carbonaceous matter (*Charnley and Rodgers,* 2002) with the high enrichments found by Floss and co-workers (*Charnley and Rodgers,* 2002; *Rodgers and Charnley,* 2004).

The former result suggests that the cold regions of a protoplanetary disk may be the location of the ^{15}N fractionation. A direct nebular connection is also implied by the fact that this model predicts that enhanced ^{15}N fractionation and depletion of N_2 should be a basic feature of primitive solar system material. The latter is a well-known, but puzzling, aspect of cometary composition (e.g., *Rodgers et al.,* 2004). Recently, *Arpigny et al.* (2003) have shown that cometary $C^{15}N/C^{14}N$ ratios are significantly higher than those of its main photochemical parent, HCN, and are comparable to or greater than those found in IDPs. *Arpigny et al.* (2003) suggested that the secondary source of CN is an unknown organic polymer whose precursor was the isotopically enhanced ammonia predicted by theory. This observation thus strengthens the connection between these fractionation processes and the putative precursors of the organic matter in IDPs.

4.3. Carbon and Oxygen

The insoluble carbonaceous matter that dominates the organic inventory of meteorites exhibits isotopically light $^{13}C/^{12}C$, relative to terrestrial standards. On the other hand, extracts of soluble molecular compounds, such as amino acids and polar hydrocarbons, show enhanced $^{13}C/^{12}C$ ratios, but the enhancements are generally less than those found in D and ^{15}N (*Cronin and Chang,* 1993). The extensive study of ^{13}C and ^{18}O fractionation by *Langer et al.* (1984) considered fairly simple and easily detectable interstellar molecules and was important for showing that the fractionation of interstellar ^{13}C is controlled by the gas temperature, density and metallicity. The CO molecule plays a pivotal role, and at low temperatures (~10 K) and densities, the exchange reaction

$$^{13}C^+ + {}^{12}CO \rightarrow {}^{13}CO + {}^{12}C^+ \quad (4)$$

significantly enhances ^{13}CO. On the other hand, ^{12}C becomes enriched in the other gas-phase species (e.g., C_2H_2 and CH_4, as well as in atomic C), perhaps by large factors leading to $^{12}C/^{13}C$ ratios in these species of ~100–350. Oxygen-isotopic fractionation can also occur in ion-molecule reactions, as shown below in reaction (5)

$$H^{12}C^{16}O^+ + {}^{12}C^{18}O \rightarrow H^{12}C^{18}O^+ + {}^{12}C^{16}O \quad (5)$$

The results of *Langer et al.* (1984) indicate that only modest fractionation in $C^{18}O$ can be expected. The O-isotopic ratios in interstellar molecules would therefore probably largely reflect nuclear processing in the stellar source.

It is unclear whether interstellar chemistry could reproduce the measured O-isotopic ratios in meteoritic matter (see *Clayton,* 2003). Kinetic processes leading to C- and O-isotopic fractionation on grain surfaces have not been considered. It is probable that, as in the case of D and ^{15}N fractionation, the gas-phase-isotope chemistry determines how molecules will be fractionated in ^{13}C (*Charnley et al.,* 2004). For example, if multiply bonded molecules such as CO, C_2H_2, and HC_3N act as seeds for hydrogenation reaction sequences, then the molecules formed from them (including CH_3OH, CH_3CH_3, and C_2H_5CN) should have the same $^{12}C/^{13}C$ ratios as their precursors. Similarly, if C chains grow on 10 K surfaces by single C atom additions, then we may expect these molecules to have $^{12}C/^{13}C$ ratios reflecting that of the atomic C accreting from the gas. One more fractionation process arises from the CO photochemistry. In dense gas, $^{12}C^{16}O$ is sufficiently abundant that it can self-shield against photodissociation from line photons, whereas lower-abundance isotopomers $^{13}C^{16}O$, $^{12}C^{18}O$, and $^{12}C^{17}O$ cannot. External interstellar UV photons in regions of low extinction near cloud surfaces can lead to the isotopically selective photodestruction of $^{13}C^{16}O$ and $^{12}C^{18}O$ with consequent local increases in the number of ^{13}C and ^{18}O atoms available for subsequent chemical reactions. Based on the above processes, regions of varying UV flux and temperature gradients should lead to strongly varying $^{12}C/^{13}C$ fractionation in the ISM. Perhaps the lower-magnitude anomalies in meteoritic ^{13}C reflect this variation in flux and temperature gradients.

Now that we have given a general overview of the gas-phase chemistry that takes place in the dense molecular clouds, we move on to the solid matter in the ISM.

5. STELLAR OUTFLOWS AND INTERSTELLAR SOLIDS

Two separate, major populations of grains arise naturally from the circumstellar outflows of C-rich and O-rich stars. These populations are carbonaceous materials, including graphite, diamond and numerous metal carbides, and oxides consisting primarily of amorphous silicates. Until about 1980 the common assumption had always been that interstellar grain populations had their roots in these stellar outflows (*Mathis et al., 1977*) with only minor modifications arising in special regions of the ISM. Such modifications are attributable to observed phenomena in interstellar environments and include, for example, the growth of a volatile ice mantle or the formation of a carbonaceous residual polymer (*Greenberg,* 1983; *Schutte,* 1996; *Sandford et al.,* 1997). Observations of the shock-induced destruction of interstellar grains (*Seab and Shull,* 1983) have introduced a much more severe modification to this population: the complete destruction of all grains in the ISM on a very rapid timescale (~10^7 yr) compared to the rate at which grains are destroyed by incorporation into new stars (*Jones et al.,* 1996). If this rapid destruction really occurs, then the vast majority of grains found in the ISM must also form there.

The models that predict rapid grain destruction in the ISM seldom mention the problems associated with forming refractory grains from hot diffuse vapors and do not present a viable alternative to forming separate populations of C-rich and O-rich grains in appropriate circumstellar environments. This is a huge flaw, as any reasonable scenario for grain formation in the ISM will yield materials whose spectral properties do not match observations (*Nuth et al.,* 1998). We suggest that the mineralogy of the majority of interstellar silicate grains is established in O-rich circumstellar outflows and is only slightly modified by subsequent interstellar processes. This hypothesis is based on a comparison of the IR spectral properties of amorphous Mg silicates observed in the laboratory at various stages of thermal annealing to the infrared spectra of grains in O-rich circumstellar outflows and in the ISM, as detailed in section 6. We should point out that this hypothesis is also consistent with models of the interstellar C distribution: the formation of interstellar diamond grains in supernovae as well as the formation of carbonaceous and graphitic materials in C-rich stellar outflows (*Tielens,* 1987). However, in the case of C, there is considerable evidence for the growth of grain mantles in cold cloud cores (*Pendleton et al.,* 1990, and references therein; *Zinner,* 1988). Such processes could therefore be responsible for the production of a significant fraction of the amorphous presolar C found in primitive meteorites as discussed in section 3.2.

5.1. Grain Formation in Stellar Outflows

Dying stars are the best natural laboratories for the study of the grain-formation process, especially the formation of oxide grains. Mass outflows from O-rich AGB and red giant stars are roughly steady-state, homogeneous, dynamic atmospheric escape processes (*Kozasa and Hasegawa,* 1987; *Fleischer et al.,* 1991, 1992, 1995; *Jeong et al.,* 2003). The stellar atmosphere initially contains only vapor at temperatures well in excess of 2000 K: Gas expansion away from the stellar surface leads to smooth decreases in both the temperature and pressure of the gas. This cooling eventually leads to the nucleation of refractory grains. These grains then serve as condensation nuclei that rapidly deplete the gas of its remaining, less-refractory vapors. Nucleation and grain growth occur on rapid timescales and produce highly amorphous, chemically heterogeneous grains that become more ordered as they are annealed in the outflow (*Nuth et al.,* 2002; *Nuth,* 1996; *Nuth and Hecht,* 1990). Experiments to characterize the spectral changes brought about by this annealing are discussed in the following section, although it should be noted that much more work along these lines is needed before all the factors controlling the rate and extent of thermal annealing are fully understood.

Grain formation in C-rich AGB stars is more complex than for O-rich stars due to the greater bonding options for C relative to H (as compared to silicon relative to H in O-rich AGB stars) and to the greater thermal stability of species such as metallic carbides and poorly graphitized C. These factors can lead to the formation and growth of grains in the preexpansion atmospheres of C-rich AGB stars. As an example, *Bernatowicz and Cowsik* (1997) have shown that the large, presolar graphite grains isolated from meteorites grew under equilibrium conditions for time periods of at least 1 yr around thermodynamically stable nuclei of titanium carbide prior to their expulsion from the star. Growth of the largest grains may have required periods as long as 100,000 yr (*Michael et al.,* 2003). Even though other carbonaceous grains might nucleate homogeneously in these

C-rich outflows, the presence of preexisting condensation nuclei and large graphite and SiC grains that might act to heterogeneously accrete amorphous C vapor make it more difficult to construct a self-consistent model of grain formation in such environments.

Carbonaceous grains might also form in supernovae (*Clayton et al.,* 1999) by a relatively straightforward chemical reaction mechanism that follows the growth of the C-chain molecules in the ejecta. This mechanism relies on the electron-impact-induced destruction of the extremely stable CO molecule to keep C atoms in the active reaction pathway leading to very large C-chain molecules. More work is needed to predict the isotopic signatures that might appear in such materials in order to distinguish them from more common condensates in C-rich stars. A more thorough discussion of these matters is given in *Bernatowicz et al.* (2006).

5.2. Composition of Interstellar Grains via Spectroscopy

Studies of meteorite matrix and IDPs have shown that, at a minimum, the following materials are present at some level in the interstellar dust: nanodiamonds (*Lewis et al.,* 1989), SiC (*Bernatowicz et al.,* 1987), poorly graphitized C (*Amari et al.,* 1990), amorphous carbonaceous materials that likely formed as grain mantles in cold dark clouds (*Zinner,* 1988), crystalline and amorphous Al_2O_3 and $MgAl_2O_4$ (*Huss et al.,* 1994; *Nittler et al.,* 1997; *Zinner et al.,* 2003; *Stroud et al.,* 2004), and crystalline and amorphous silicates (*Bradley,* 1994; *Messenger et al.,* 2003; *Nguyen and Zinner,* 2004). It is interesting to note that only for amorphous silicates, poorly graphitized C, and amorphous carbonaceous materials is there direct evidence of their existence in the interstellar extinction spectrum though some evidence for the presence of SiC and Al_2O_3 is seen in stellar outflows (*Mutschke et al.,* 2000; *Speck et al.,* 2000; *Hofmeister et al.,* 2000, *Begemann et al.,* 1997) and in meteorites. It may be that, because the relatively large SiC grains found in meteorites are extremely inefficient extinction agents per unit mass, there may be a significant mass fraction of the available silicon (e.g., 10%) tied up in such grains that can never be detected optically. The other option is that SiC grains have a very low abundance.

The situation is further complicated by the fact that a broad, weak SiC feature could easily be hidden under the wider, stronger silicate band (J. Dorschner, personal communication, 1998). In a similar fashion, nanodiamonds in the ISM will retain some spectroscopic resemblance to amorphous carbonaceous materials due to their relatively large surface-to-volume ratio and the necessity to terminate the sp^3 C core of the nanodiamond with sp^2 C, as well as with other species such as H, O, and N. Only in the far-UV (*Saslaw and Gaustad,* 1969) or in the far-IR (*Hill et al.,* 1997) might these tiny diamonds have diagnostic spectroscopic signatures due to the diamond bonds themselves. However, the C-H stretch of H atoms bonded to tertiary sp^3-bonded C has been detected in the ISM (*Allamandola et al.,* 1992, 1993), and this could very well be due to the H atoms terminating the bonds of the C atoms on the surface of nanodiamonds.

The overall mineralogy of grains in the ISM is highly variable, depending on the particular phase under discussion. The mineralogy of cold, dark molecular cloud cores will certainly differ from that of the hot, diffuse ISM, although refractory phases may be common components of both regions. We have therefore tabulated our estimates of the abundances of individual constituents as a fraction of larger groups (Table 2). The abundances of carbonaceous grains are estimated as a fraction of the *total refractory* C and not as a fraction of *total* C, which would include all C-containing compounds, and atomic C as well, in the diffuse ISM. The abundances of oxides are estimated relative to the total refractory oxide grain population while volatiles such as H_2O only enter the solid state in cold clouds. Residues of warmed, irradiated ices are counted as amorphous C grains in this tabulation and we assume that molecular species in ice mantles constitute only a small fraction of the total volume.

The ordering of the constituents in Table 2 reflects the degree of certainty in the existence of the individual components in the ISM with a second level of ordering based on abundance. Each of the first seven constituents has been identified in meteorites. Amorphous C-grain coatings have not been isolated from meteorite matrix directly but are inferred from the presence of highly anomalous isotopic abundances of N, C, and D found in the matrix (*Zinner,* 1988) and in IDPs (e.g., *Zinner et al.,* 1983; *Messenger and Walker,* 1996), indicative of their possible formation via ion-molecule reactions or surface-mediated processes, for example, in cold cloud cores (sections 3.2 and 4.1). The presence of presolar nanodiamond, graphite (better described as poorly graphitized C), and SiC in meteorite matrix is discussed in more detail in *Meyer and Zinner* (2006) and *Bernatowicz et al.* (2005). Note that the primary spectral features listed in Table 2 may not be directly attributable to only one particular component, e.g., the infrared emission bands that have been attributed to PAHs (*Leger et al.,* 1987; *Cohen et al.,* 1986). These PAH-related bands have previously been attributed to various forms of amorphous C such as hydrogenated amorphous C (HAC) (*Duley,* 1985, and references therein) or quenched carbonaceous composites (QCCs) (*Sakata et al.,* 1983), although such materials might simply be aggregated forms of PAHs linked by aliphatic hydrocarbons. These same IR features may also arise from the amorphous, carbonaceous outer layers on nanodiamonds, although measurements of the IR spectra of chemically processed, meteoritic diamonds do not resemble the above-mentioned bands (*Mutschke et al.,* 1995).

The 200-nm region of the interstellar extinction curve contains strong absorption features for a wide range of hydrocarbons, and a very broad "featureless" spectrum has often been attributed to carbonaceous solids containing a range of C bond types (see *Hecht,* 1986; *Sorrell,* 1990, 1991). For almost three decades the increase in extinction

TABLE 2. Mineralogy of interstellar grains.

Component	Fraction of Solid Phase	Primary Spectral Features
Amorphous carbon (sp^2, sp^3), PAHs	>80% C	PAH emission bands, ≈200-nm UV absorption, 3.4-μm C-H stretch
Nanodiamonds (sp^3 carbon)	≈10% C	Tertiary C-H stretch (2.9 μm), far-UV absorption, 22-μm peak
Graphitic carbon (sp^2)	≈5% C	217.5-nm peak
SiC	trace C	Broad 11.2-μm peak
Amorphous silicate $(Fe,Mg)SiO_x$	>95% O	9.7- and 18-μm peaks
Corundum (Al_2O_3)	trace O	12–13-μm peak
Spinel $(Fe,Mg)Al_2O_4$	trace O	12–13-μm peak
Crystalline silicate — rapidly degraded beyond circumstellar environments by cosmic rays	trace O	11.3-μm "olivine" subpeak within the 10-μm silicate feature, far-IR olivine features at 33 and 69 μm
Water ice — polar (plus organics)	>90% V	3-, 6-, 12-, 45-, 60-μm peaks
Nonpolar ices (N_2, O_2)	≈5% V	Indirect effects on dissolved species
Nonpolar ices (CO, CO_2)	≈5% V	4.67- μm (CO), 4.27-μm and 15.2-μm (CO_2) peaks
MgS, CaS, SiS_2 — likely lost by reaction with H_2O in the ISM	Speculation	28-, 40- and 22-μm peaks (respectively) in CS shells

Solid phases: C = solid carbonaceous grains, O = oxide grains, V = volatile ices.

around 217.5 nm has been primarily attributed to graphite (*Stecher,* 1969) yet it is possible that only a very small fraction of the C contributing to this feature need be in this form to ensure that the observed characteristics of this peak are produced (*Hecht,* 1986). Another possibility is that the 217.5-nm peak is due to nanometer-scale particles of hydrogenated amorphous C (*Schnaiter et al.,* 1998) or to carbons on the surfaces of nanodiamonds (*Sandford,* 1996).

Diamond was initially suggested (*Saslaw and Gaustad,* 1969) as a candidate to explain the steep rise in far-UV extinction at wavelengths shorter than about 150 nm. Unfortunately, this effect is not unique and the far-UV rise has also been modeled as due to small MgO particles (*Millar and Duley,* 1978), as well as to small olivine and graphite grains (*Draine and Lee,* 1984). Following the identification of nanodiamonds in meteorites, *Allamandola et al.* (1992, 1993) observed an infrared feature assigned to the tertiary CH stretch in several astronomical sources that they attributed to terminal hydrogens on the outside of nanodiamonds. Again, this identification is not unique, as some degraded forms of fullerenes will also display tertiary CH stretches. Only features intrinsic to diamond itself might serve to unambiguously identify such material in the ISM, and intrinsic diamond absorption features are very weak (*Hill et al.,* 1997) but may be detectable in some circumstances (see *Guillois et al.,* 1999). One such feature may be the 22-μm emission band observed around some C-rich protoplanetary nebulae (*Hill et al.,* 1997; *Kraus et al.,* 1997) previously attributed to SiS_2 (*Goebel,* 1993).

Aluminum oxide grains have occasionally been suggested to explain circumstellar emission peaks near 13 μm around hot stellar sources (*Stencel et al.,* 1990; *Begemann et al.,* 1997; *Posch et al.,* 1999), but these features have never been observed in the ISM. It may be that the population of relatively pure corundum grains in the ISM is quite low due to the fact that aluminum is easily assimilated into silicate grains. Pure corundum grains might only be formed in significant quantity in extremely hot outflows where silicate condensation is inhibited, or in unusual, aluminum-rich environments. Alternatively, Al_2O_3 grains may be one of the metastable eutectics in some Al-rich systems (*Nuth et al.,* 2002). Given the wide range of nucleosynthetic histories of individual corundum grains isolated from meteorites (*Nittler et al.,* 1997), it seems more likely that some very small fraction of relatively pure alumina grains form in all O-rich outflows and that corundum grains in the ISM are only a very small fraction of the oxide grain population. One other such trace Al-bearing constituent is a population of spinel grains, isolated from primitive meteorites, but not seen in the ISM (*Zinner et al.,* 2003). However, the bulk of the cosmic abundance of Al is probably incorporated into amorphous silicate grains.

The chemical compositions of the grains formed in O-rich stellar outflows are predicted to lie at the metastable eutectics in the ternary MgO-FeO-SiO phase diagram (*Rietmeijer et al.,* 1999, 2000; *Rietmeijer and Nuth,* 2000; *Nuth et al.,* 2002). The individual grains should cluster around the serpentine and talc or greenalite and saponite dehydroxalate compositions, and should also include some MgO, SiO_x, and FeO_y grains, depending on the overall composition of the stellar outflow. In condensation experiments, individual condensates never contain both Fe and Mg (*Nuth et al.,* 2000), presumably because FeO and MgO are completely miscible with no stable or metastable eutectics to promote condensation and grain growth. If such materials survive transport through the ISM, as we believe they do (see section 6), then they will constitute the primary refractory oxide input to the solar nebula.

5.3. Infrared Spectra and the Ordering of Amorphous Magnesium Silicates

Amorphous silicates have long been discussed as a significant component of the interstellar dust population (*Ney,* 1977), but only recently have grains potentially representative of this population become available for detailed laboratory studies (*Bradley,* 1994; *Messenger et al.,* 2003; *Keller and Messenger,* 2004). Highly amorphous silicate grains form in the outflows of mass-losing stars, undergo some degree of thermal processing in these shells, and may begin forming more ordered materials. A generalization of this situation was reproduced in the laboratory. Amorphous Mg silicate smokes were prepared in the laboratory by vapor-phase condensation, and then were annealed in vacuum to simulate circumstellar processing. The samples were monitored by IR spectroscopy as a function of annealing time and temperature (1000–1050 K), focusing on the development of the 10- and 20-μm silicate features. The IR spectrum of the initial condensate displayed a broad band at 9.3 μm. As annealing proceeded, the maximum shifted to longer wavelengths and was eventually observed to stabilize at 9.7 μm, the value typically reported for silicates in the interstellar medium and in the circumstellar outflows of O-rich stars (*Hallenbeck et al.,* 1998; *Hallenbeck and Nuth,* 1998). Further thermal evolution of the amorphous Mg silicate smokes led to the appearance of a dual maximum at 9.8 and 11.2 μm, indicative of "crystalline" olivine. The dual maxima 10-μm feature is a natural consequence of the thermal evolution of the amorphous condensate and is not a mixture of distinct populations of amorphous and "crystalline" materials.

In addition, there is a natural pause or "stall" in the spectral evolution of these samples, midway between the initially chaotic condensate and the more-ordered solid. Recent work indicates that this stall phase may represent the nucleation of microcrystals at the outer edge of the amorphous grains (*Kamitsuji et al.,* 2005). Thereafter, individual features sharpen as the sample develops a polycrystalline structure that becomes more ordered with further annealing. For the Mg silicate smoke samples investigated, roughly 11 days of annealing at a grain temperature of 1000 K were required for the peak position of the Si-O stretch to evolve from ~9.3 μm to 9.7 μm. An additional 11 days of thermal annealing (≈22 days total) were required before the grains reached the stall, and it is estimated that nearly 300 additional days at 1000 K would be necessary before the infrared spectrum of the annealing Mg silicate progressed beyond the stall. The stall spectrum is therefore a practical endpoint in the spectral evolution of most Mg silicate condensates in circumstellar outflows, although virtually all grains in the outflow should evolve to exhibit the 9.7-μm silicate stretch. Since grain temperatures range from 300 to 600 K in the outer shells of typical stellar outflows (*Hron et al.,* 1997), any Mg silicate grain sufficiently annealed to exhibit the stall spectrum should remain at this stage of evolution for the remainder of its lifetime barring shock-induced destruction or radiation-induced amorphization in the ISM.

The final infrared spectrum of Mg silicate condensates is heavily dependent on the thermal structure and outflow velocity of the circumstellar shell. Due to the exponential dependence of the annealing rate on grain temperature, the final grain spectrum is extraordinarily sensitive to the peak temperature of the newly formed grains. Based on the laboratory results discussed above, grains formed at temperatures in excess of 1067 K anneal from the initial amorphous condensate to a material that displays a spectrum more evolved than the stall on timescales of minutes. At temperatures of 1000 K these transitions require in excess of 300 days. By "choosing" a temperature at which amorphous silicate grains nucleate in a particular circumstellar outflow, and following the progress of the annealing and growth processes as the smokes move down the temperature gradient, away from the star, it should be possible to construct models of the thermal emission spectrum of any O-rich stellar source of known mass loss rate and outflow velocity (*Hallenbeck et al.,* 2000), accounting for the relative fraction of crystalline materials. These same calculations may also be used to predict the infrared spectrum of the final silicate grain population injected into the ISM from such an outflow.

5.4. Processing of Interstellar Grains

All fresh condensates in these rapidly changing environments start out as amorphous grains. If the temperature of the outflow at the nucleation point is high enough, then some grains might be annealed to crystallinity, while those that condense at lower temperatures (or materials that require much higher temperatures to anneal) will remain amorphous. As an example, Fe and Mg silicates condense out as separate grain populations; iron silicates require much higher temperatures to anneal than do Mg silicates. Therefore, in high-mass-outflow-rate stars, where the condensation temperatures become higher due to higher concentrations of condensable elements, some Mg silicates can anneal to crystallinity while the Fe silicates remain amorphous. That is why some high-mass-outflow-rate stars are observed to contain pure Mg silicate crystals, and yet no Fe-bearing silicate minerals have ever been seen in any of these sources (*Nuth et al.,* 2000, 2002).

Crystalline silicates present an interesting problem and demonstrate a very significant difference between the properties of grains in the ISM vs. the crystalline silicates observed to form in outflows of evolved stars (*Waters et al.,* 1996). Evidence for crystalline forsterite has also been obtained by ISO in the far-IR, where the emission spectrum of a young main-sequence star, HD 100546 (*Waelkens et al.,* 1996), closely matches laboratory spectra of crystalline olivine (*Koike et al.,* 1993). There are a few stellar sources (*Knacke et al.,* 1993) that have been observed to have an emission feature at 11.3 μm that resembles the mid-IR crystalline olivine peak seen in Comet 1P/Halley (*Campins and Ryan,* 1989; *Hanner et al.,* 1994). Given this observational data, there should be some fraction of the interstellar silicate grain population that is crystalline, yet no evidence for crystalline silicates (*Dorschner et al.,* 1995; *Henning et al.,* 1995; *Henning and Mutschke,* 1997) has yet been found in the general ISM. There is the possibility that as much as 5% of interstellar silicates could be crystalline, but these crystalline features would lie below observational detection limits (*Li and Draine,* 2001). Based on the structure of the 10-μm silicate feature measured along a single line of sight toward the galactic center, *Kemper et al.* (2004) calculated an upper limit on the fraction of crystalline silicates along that very long path to be $0.2 \pm 0.2\%$. If interstellar grains are destroyed on rapid timescales in the ISM then the lack of crystalline silicate grains is perfectly understandable. Since these silicates only form in a small fraction of the O-rich stellar outflows, crystalline grains would be relatively rare to begin with. Rapid grain destruction coupled with no feasible mechanism by which crystalline silicates can form or anneal would predict a dearth of crystalline silicates in the ISM and in the presolar grain population. However, given the survival of a host of somewhat delicate carbonaceous materials in the ISM, the new dilemma either becomes explaining the preferential destruction of crystalline interstellar silicates vs. carbonaceous grains or finding an alternative explanation for the observational data. One good alternative is the deterioration of crystalline silicates upon long exposure to cosmic radiation (e.g., *Day,* 1974; *Jäger et al.,* 2003).

Radiation-induced damage to olivine crystals has been shown to produce spectra consistent with observations of interstellar silicates (*Day,* 1974). *Bradley* (1994) has also noted that the amorphous silicates isolated from interplanetary dust grains appear to have been exposed to cosmic rays for a very long time. This damage may in fact be the agent responsible for some aspects of the chemistry of these suspected presolar materials, especially for the chemical gradients of Mg and O observed in the glass with embedded metal and sulfide (GEMS) themselves (*Bradley,* 1994). If crystalline silicates are only formed in very high mass loss rate stellar outflows, and if such outflows constitute only a minor fraction of the mass flux into the ISM, then this minor grain population would be reduced still further by long exposure to cosmic radiation and one would never expect to see signatures of crystalline silicate minerals in the general ISM. However, since some crystalline silicate grains are observed in meteorites and IDPs that are demonstrably presolar (*Messenger et al.,* 2003; *Keller and Messenger,* 2004; *Nguyen and Zinner,* 2004), some small fraction of silicates in the ISM must be crystalline.

6. DUST DESTRUCTION IN THE INTERSTELLAR MEDIUM

Supernovae shock waves destroy refractory grains: This is an observational fact (*Seab and Shull,* 1983). However, establishing the lifetimes of grains in the ISM requires the use of models that contain assumptions ranging from the efficiency of the individual grain destruction mechanisms, such as chemi-sputtering or grain-grain collisions (*Borkowski and Dwek,* 1995; *Dwek et al.,* 1996), to more difficult questions concerning how much time an individual grain might spend in cold cloud cores vs. the warm ISM. Theoretical astrophysicists have done a great deal of work trying to formulate a comprehensive model to address the efficiency of grain destruction as a function of shock velocity. The general conclusion drawn from such efforts is that an average interstellar grain survives on the order of 10^7–10^8 yr before being destroyed (*Jones et al.,* 1994, 1996).

The timescale for grains, formed in the outflows of dying stars, to be incorporated into new protostellar systems is on the order of 10^9 yr (*Dwek and Scalo,* 1980). Given that an average grain injected into the ISM only survives $\sim 10^8$ yr before it is destroyed by a supernova shock wave, the inevitable conclusion is that the vast majority of grains now observed in the ISM must have formed in the ISM. A simple corollary to this conclusion is that only a tiny fraction of grains formed in circumstellar outflows can survive intact to be incorporated into protostellar systems. If we assume that 10^8 yr is the equivalent of a "half-life" for grains in the ISM and that each grain must survive at least 10^9 yr before it becomes part of a new stellar system, then only one grain in a thousand survives the passage from a circumstellar environment to be incorporated into a protostellar environment. If 10^8 yr is assumed to be the e-folding time for grain destruction, then only one grain in 10^5 will survive. If the "half-life" for grain destruction is actually 10^7 yr, then only one grain in 10^{30} can survive transport from a circumstellar outflow to a newly forming protostar, and finding presolar materials would therefore seem to be a very difficult proposition indeed.

The result of taking these models of grain destruction to their natural conclusion — all grains now present in the ISM must have formed in the ISM — is that it becomes impossible to explain the spectral properties inferred from the interstellar extinction curve (section 2.2). The spectral properties of grain analogs formed in the laboratory under conditions appropriate to the ISM are very different from the actual observations. Alternatively, we are forced to postulate that appropriate amorphous silicate and carbonaceous grains simply reform in the ISM by processes that we do not yet understand and have yet to duplicate in the laboratory.

6.1. Meteoritic Data and Dust Destruction

Presolar diamonds, graphite, and SiC have been isolated from the matrices of primitive (undifferentiated) meteorites (*Amari et al.,* 1994; *Anders and Zinner,* 1993; *Zinner,* 1998; *Nittler,* 2003; *Meyer and Zinner,* 2005). In addition, several hundred grains of presolar Al_2O_3 (corundum) and $MgAl_2O_4$ have been recovered from these same meteorites (*Nittler et al.,* 1997; *Zinner et al.,* 2003). Given that a large fraction of the material in the meteorite matrix is of nebular origin (*Boynton,* 1985; *Clayton et al.,* 1985), it is a miracle that any presolar grains survived at all. Considering the generally oxidizing nature of the nebula, it is remarkable that most of the presolar materials isolated to date are carbonaceous grains that are easily destroyed via exposure to even moderately high temperatures in the presence of O. Yet in spite of these caveats, one carbonaceous component — nanodiamonds — accounts for approximately 400 ppm by mass of most primitive meteorites. The total C content of such meteorites is less than 10% by mass. If these diamonds are indeed formed in supernovae and injected into the ISM then one would expect, at most, one grain in a thousand (or as few as one in 10^{30}) to survive transport to a protostellar environment and a much smaller fraction of these grains to survive processing in the oxidizing conditions found in the solar nebula. Yet primitive diamonds represent more than four parts in ten thousand of the total mass (or ≅0.04% of the total C content) of the meteorites. This is barely consistent with the most optimistic projection for grain survival in the ISM if only about half the initial fraction of interstellar diamonds that could have been incorporated into the meteorite parent body were actually destroyed by high-temperature processes in the nebula and *if* the only form of circumstellar carbon was diamond. This large fraction of interstellar diamond in meteorites is inconsistent with high grain destruction rates in the ISM. In addition, a 50% destruction rate is a very low estimate of the fraction of C-bearing grains destroyed via nebular processes (we would estimate the grain destruction efficiency for carbonaceous materials to be well in excess of 90%). The carbonaceous grains that do survive nebular and parent-body processing — such as SiC — show absolutely *no* signs on their surfaces of pitting (e.g., *Bernatowicz et al.,* 2003) due to chemi-sputtering or to chipping and microcratering due to grain-grain collision processes that should be evident if the interstellar medium were as violent as predicted (e.g., *Jones et al.,* 1996).

6.2. Experiments, Predictions, and Timescales

If refractory atoms and molecules slowly accrete onto the surfaces of surviving grains only after they have reached the comparative safety of a molecular cloud (rather than in the diffuse ISM), then such atoms may have the opportunity to react with additional adsorbed atomic and molecular species before being buried beneath the growing grain surface. The most common of these atomic species is H, while the most common molecular components of astrophysical environments are H_2, CO, and H_2O. If these reactions occurred, the SiO, Fe, and Mg atoms generated by the annihilation of grains in supernova shocks might never reform the initial silicate compounds observed in the ISM, but might instead form a complex ice (A. G. G. M. Tielens, personal communication, 1984). Experiments made with mixtures of silane (SiH_4), iron pentacarbonyl ($Fe(CO)_5$), and H_2O ice condensed at very low temperatures (~10 K) have demonstrated that a "silicate" can form from such a mixture, especially if the mixture were irradiated with cosmic-ray energy protons prior to warm-up so as to form chemical radicals in the ice (*Nuth and Moore,* 1989). However, a major problem with the results of this study is that the IR spectrum of the warmed silicate residue does not resemble typical IR spectra observed for silicates in the ISM. Based on previous laboratory work, *Hallenbeck et al.* (1998, 2000) showed that the thermal annealing ($T \geq 1000$ K) of a highly amorphous silicate condensate resulted in more ordered material with an IR spectrum displaying narrower peaks and absorption features that more closely resemble typical ISM silicate spectra. However, no natural process has been identified that would heat grains to temperatures significantly in excess of 1000 K in the ISM for sufficiently long time periods to reproduce the observed silicate spectral profile. Yet this process must occur on a considerably shorter timescale than grain destruction in order to transform all the recondensed refractory material back into the observed amorphous interstellar silicate.

6.3. Evidence from the Interstellar Extinction Curve

If we assume for a moment that grain destruction is as ubiquitous as the models predict, what might be the observable consequences? First, we know from measurements of interstellar depletion that ~99.9% of all refractory materials reside as solids in the ISM (*Savage and Sembach,* 1996; *Sandford,* 1996). Therefore, if grains are destroyed in the ISM, they must reform *in situ,* close to the site of grain destruction. It might be interesting to make measurements of the gas-phase depletion of atomic species at close intervals in the wake of a supernova shock wave to constrain the timescale over which refractory materials are redeposited onto surviving grain surfaces. What spectral properties might characterize such grains?

If a few surviving grains (or grain fragments) serve as substrates onto which refractory atoms are deposited in the low density ISM, then it is relatively easy to predict the spectral properties of the redeposited solids based on measurements of the optical constants of vacuum-deposited refractory materials, models of the size and shape distributions of the grain populations, and the use of Mie theory (*Bohren and Huffman,* 1983). Simple silicates evaporate to form primarily SiO, Mg atoms, Fe atoms, and O_2 vapors. If we assume that a large fraction of the O_2 might be lost, initially via reaction with carbonaceous grains to form CO, or because it simply did not recondense due to its volatile nature, then the recondensed solids will be chemically reduced compared to the initial grain population. Characteristic IR features of these recondensed silicates would be due primarily to solid SiO and possibly to simple MgO and FeO bands, although both Mg and Fe might exist largely as metallic species. A simple grain model of just such a composition was proposed by *Millar and Duley* (1978), but the predictions of this model are inconsistent with observations of the positions or overall shapes of either the 10- or 20-μm IR features that dominate the mid-IR region of the interstellar extinction curve. In this model, any C not converted to CO might be deposited onto the metallic grains — potentially forming carbides or an amorphous C film. To our knowledge, no other models of the interstellar grain population based along these lines have ever been published despite the widespread acceptance of the rapid grain destruction hypothesis.

A second argument against rapid, widespread grain destruction in the ISM has been advanced by Mathis (*O'Donnell and Mathis,* 1997; *Mathis,* 1996). Studies of the interstellar extinction curve along various lines of sight have shown that both the grain size distributions and the overall composition of the grain populations are far from constant. The simplest models that successfully account for such variations in, for example, the shape or relative strength of the 217.5-nm feature vs. the strength of the far-UV rise (characteristic signatures of interstellar spectra), or the depth of the silicate absorption band, all require separate, independently variable populations of carbonaceous and oxide grains. Grain destruction, followed by rapid accretion of refractory (including C) atoms onto grain surfaces, should quickly result in a homogeneous solid state where an average grain contains both SiO and carbonaceous matter. No mechanism has yet been proposed by those advocating the rapid destruction of grains in the ISM that would maintain separate populations of oxide and carbonaceous grains.

Chiar et al. (1998) observed the 3.4-μm C-H stretch in the outflow around the protoplanetary nebula CRL 618, thus proving that this feature does form in the ejecta from evolved C-rich stars. The 3.4-μm feature also appears along several lines of sight in the diffuse interstellar medium (*Butchart et al.,* 1986; *Adamson et al.,* 1990; *Sandford et al.,* 1991; *Pendleton et al.,* 1994). The 3.4-μm feature is not observed in the spectra of dust in dense molecular clouds, although this could simply mean that it becomes obscured by the deposition of an ice mantle. Because it can be formed via UV processing of organic ice mantles (e.g., *Bernstein et al.,* 2002, 2001, 1999; *Dworkin et al.,* 2001; *Greenberg,* 1989; *Greenberg et al.,* 1995; *Hudson and Moore,* 2004), its presence in the ISM is often cited as a measure of the importance of such processes. If subjected to shock, it is easy to imagine that grains could simply dehydrogenate as an effective cooling mechanism. Once the shock passes, grains could react with ambient H to reform C-H bonds on the surfaces of carbonaceous grains. *Sorrell* (1989, 1990) and *Hecht* (1991) both suggested an analogous process that could modulate the relative intensity of the 217.5-nm feature against the much broader, underlying carbonaceous absorption in a manner that was consistent with observations. In their scenario, grains entering the diffuse interstellar medium lost most of their H to form graphitic dust, thus increasing the strength of the 217.5-nm feature. Upon entering denser media, these grains reacted with ambient H to break down many of the "new" aromatic, graphitic centers and decrease the strength of the 217.5-nm peak (*Sorrell,* 1991).

6.4. Summary of Grain Destruction in the Interstellar Medium

As noted above, shock-induced grain destruction is *observed* to occur in the ISM; observations across a shock front show refractory solids in front of the shock and refractory vapor behind. The problem arises in applying this simple observation to the general grain population in order to predict the probability that an individual grain will experience a destructive shock within a specific time. Astrophysicists have used the average number of supernovae observed per year to calculate the frequency of destructive shocks and the mass of grains that such shocks might destroy per year. They divide the mass of dust in our galaxy by this rate of grain destruction to calculate the time it would take to destroy all the grains in the galaxy. This timescale is then defined as the lifetime of the average grain. Alternatively, these shocks may simply serve to break up the grain aggregates favored by *Mathis* (1996) without much erosion of their individual components. These components would then reaggregate in the lee of the shock. A deeper understanding of the dynamics of gas and dust in the ISM is required before any shock-destruction timescale is finalized or even before the relative importance of grain destruction in the ISM can be definitively assessed.

In the preceding sections we have approached the problem of grain destruction from three different perspectives. First, we used the meteoritic record to demonstrate that the unaltered grains from many stars survived passage through the ISM (as well as the destructive processes that occurred in the solar nebula) in vastly greater numbers than would be consistent with a high rate of grain destruction in the ISM. Then we demonstrated that the spectral properties of the grains likely to have formed in either dense clouds or in the diffuse ISM are very different from those actually observed in the ISM. Finally, we argued that the rapid destruction of grains in the ISM should lead to a much more homogenized grain population than is observed. If grains are not efficiently destroyed in the ISM then we must look to the grain populations formed in circumstellar outflows (see *Bernatowicz et al.,* 2006; *Ebel,* 2006) in order to predict the properties of the grains present at the time the solar nebula formed. Indeed, it is quite likely that the mineralogy of interstellar grains is primarily established in the fiery death throes of the stars and only slightly modified in the ISM.

Acknowledgments. The work of S.B.C. was supported by NASA's Origins of Solar Systems program, Exobiology program, and Astrobiology Institute, with funds allocated under Cooperative Agreements with the SETI Institute. J.A.N. acknowledges support from the Cosmochemistry, Origins of Solar Systems, and Laboratory Astrophysics programs. N.M.J. was supported by the NAS/NRC Resident Research Associate program. Thanks to Drs. L. Nittler, S. Sandford, and an anonymous referee for extremely thorough analyses of our original submission and for many constructive comments that greatly improved this manuscript.

REFERENCES

Adamson A. J., Whittet D. C. B., and Duley W. W. (1990) The 3.4-micron interstellar absorption feature in CYG OB2 no. 12. *Mon. Not. R. Astron. Soc., 243,* 400–404.

Aleon J. and Robert F. (2004) Interstellar chemistry recorded by nitrogen isotopes in Solar System organic matter. *Icarus, 167,* 424–430.

Aleon J., Robert F., Chaussidon M., and Marty B. (2003) Nitrogen isotopic composition of macromolecular organic matter in interplanetary dust particles. *Geochim. Cosmochim. Acta, 67,* 3773–3783.

Alexander C. M. O'D., Russell S. S., Arden J. W., Ash R. D., Grady M. M., and Pillinger C. T. (1998) The origin of chondritic macromolecular organic matter: A carbon and nitrogen isotope study. *Meteoritics & Planet Sci., 33,* 603–622.

Allamandola L. J., Sandford S. A., and Wopenka B. (1987) Interstellar polycyclic aromatic hydrocarbons and carbon in interplanetary dust particles and meteorites. *Science, 237,* 56–59.

Allamandola L. J., Tielens A. G. G. M., and Barker J. R. (1989) Interstellar polycyclic aromatic hydrocarbons: The infrared emission bands, the excitation/emission mechanism, and the astrophysical implications. *Astrophys. J. Suppl. Ser., 71,* 733–775.

Allamandola L. J., Sandford S. A., Tielens A. G. G. M., and Herbst T. M. (1992) Infrared-spectroscopy of dense clouds in the C-H stretch region — Methanol and diamonds. *Astrophys. J., 399,* 134–146.

Allamandola L. J., Sandford S. A., Tielens A. G. G. M., and Herbst T. M. (1993) Diamonds in dense molecular clouds — A challenge to the standard interstellar-medium paradigm. *Science, 260,* 64–66.

Allamandola L. J., Bernstein M. P., and Sandford S. A. (1997) Photochemical evolution of interstellar/precometary organic material. In *Astronomical and Biochemical Origins and the Search for Life in the Universe* (C. B. Cosmovici et al., eds.), pp. 23–47. Editrice Compositori, Bologna.

Allen M. and Robinson G. W. (1977) The molecular composition of dense interstellar clouds. *Astrophys. J., 212,* 396–415.

Amari S., Anders E., Virag A., and Zinner E. (1990) Interstellar graphite in meteorites. *Nature, 345,* 238–240.

Amari S., Lewis R. S., and Anders E. (1994) Interstellar grains in meteorites: I. Isolation of SiC, graphite, and diamond; size distributions of SiC and graphite. *Geochim. Cosmochim. Acta, 58,* 459–470.

Anders E. and Zinner E. (1993) Interstellar grains in primitive meteorites: Diamond, silicon, carbide, and graphite. *Meteoritics, 28,* 490–514.

Arpigny C., Jehin E., Manfroid J., Hutesmekers D., Zucconi J.-M., Schulz R., and Stuewe J. A. (2003) Anomalous nitrogen isotope ratio in comets. *Science, 301,* 1522–1524.

Bakes E. L. O. (1997) *The Astrophysical Evolution of the Interstellar Medium.* Twin Press, The Netherlands. 106 pp.

Basile B. P., Middleditch B. S., and Oró J. (1984) Polycyclic aromatic hydrocarbons in the Murchison meteorite (abstract). In *Lunar and Planetary Science XV,* p. 37. Lunar and Planetary Institute, Houston.

Begemann B., Dorschner J., Henning T., Mutschke H., Gurtler J., Kompe C., and Nass R. (1997) Aluminum oxide and the opacity of oxygen-rich circumstellar dust in the 12–17 micron range. *Astrophys. J., 476,* 199–208.

Bergin E. A., Alves J., Huard T., and Lada C. J. (2002) N_2H^+ and $C^{18}O$ depletion in a cold dark cloud. *Astrophys. J. Lett., 570,* L101–L104.

Bernatowicz T. J. and Cowsik R. (1997) Conditions in stellar outflows inferred from laboratory studies of presolar grains. In *Astrophysical Implications of the Laboratory Study of Presolar Materials* (T. J. Bernatowicz and E. K. Zinner, eds.), pp. 452–474. American Institute of Physics, New York.

Bernatowicz T., Fraundorf G., Tang M., Anders E., Wopenka B., Zinner E., and Fraundorf P. (1987) Evidence for interstellar SiC in the Murray carbonaceous meteorite. *Nature, 330,* 728–730.

Bernatowicz T. J., Messenger S., Pravdivtseva O., Swan P., and Walker R. M. (2003) Pristine presolar silicon carbide. *Geochim. Cosmochim. Acta, 67,* 4679–4691.

Bernatowicz T. J., Croat T. K., and Daulton T. L. (2006) Origin and evolution of carbonaceous presolar grains in stellar environments. In *Meteorites and the Early Solar System II* (D. S. Lauretta and H. Y. McSween Jr., eds.), this volume. Univ. of Arizona, Tucson.

Bernstein M. P., Sandford S. A., Allamandola L. J., Gillette J. S., Clemett S .J., and Zare R. N. (1999) UV irradiation of polycyclic aromatic hydrocarbons in ices: Production of alcohols, quinones, and ethers. *Science, 283,* 1135–1138.

Bernstein M. P., Dworkin J. P., Sandford S. A., and Allamandola L. J. (2001) Ultraviolet irradiation of naphthalene in H_2O ice: Implications for meteorites and biogenesis. *Meteoritics & Planet. Sci., 36,* 351–358.

Bernstein M. P., Dworkin J. P., Sandford S. A., Cooper G. W., and Allamandola L. J. (2002) Racemic amino acids from the ultraviolet photolysis of interstellar ice analogues. *Science, 416,* 401–403.

Bettens R. P. A. and Herbst E. (1996) The abundance of very large hydrocarbons and carbon clusters in the diffuse interstellar medium. *Astrophys. J., 468,* 686–693.

Blagojevic V., Petrie S., and Bohme D. K. (2003) Gas-phase syntheses for interstellar carboxylic and amino acids. *Mon. Not. R. Astron. Soc., 339,* L7–L11.

Blake D., Allamandola L., Sandford S., Hudgins D., and Freund F. (1991) Clathrate hydrate formation in amorphous cometary ice analogs in vacuo. *Science, 254,* 548–551.

Blitz L., ed. (1990) *The Evolution of the Interstellar Medium.* ASP Conference Series 12, Astronomical Society of the Pacific, San Francisco. 346 pp.

Bohren C. F. and Huffman D. R. (1983) *Absorption and Scattering of Light by Small Particles.* Wiley, New York.

Borkowski K. J. and Dwek E. (1995) The fragmentation and vaporization of dust in grain-grain collisions. *Astrophys. J., 454,* 254–276.

Boss A. and Goswami J. (2006) Presolar cloud collapse and the formation and early evolution of the solar nebula. In *Meteorites and the Early Solar System II* (D. S. Lauretta and H. Y. McSween Jr., eds.), this volume. Univ. of Arizona, Tucson.

Botta O. and Bada J. (2002) Extraterrestrial organic compounds in meteorites. *Surv. Geophys., 23,* 411–467.

Bottinelli S., Ceccarelli C., Lefloch B., Williams J. P., Castets A., Caux E., Cazaux S., Maret S., Parise B., and Tielens A. G. G. M. (2004) Complex molecules in the hot core of the low-mass protostar NGC 1333 IRAS 4A. *Astrophys. J., 615,* 354–358.

Boynton W. (1985) Meteoritic evidence concerning conditions in the solar nebula. In *Protostars and Planets II* (D. C. Black and M. S. Matthews, eds.), pp. 772–787. Univ. of Arizona, Tucson.

Bradley J. P. (1994) Chemically anomalous, preaccretionally irradiated grains in interplanetary dust from comets. *Science, 265,* 925–929.

Butchart I., McFadzean A. D., Whittet D. C. B., Geballe T. R., and Greenberg J. M. (1986) Three micron spectroscopy of the galactic centre source IRS 7. *Astron. Astrophys., 154,* L5–L7.

Campins H. and Ryan E. V. (1989) The identification of crystalline olivine in cometary silicates. *Astrophys. J., 341,* 1059–1066.

Caselli P., Walmsley C. M., Terzieva R., and Herbst E. (1998) The ionization fraction in dense cloud cores. *Astrophys. J., 499,* 234–249.

Caselli P., van der Tak F. F. S., Ceccarelli C., and Bacmann A. (2003) Abundant H_2D^+ in the pre-stellar core L1544. *Astron. Astrophys., 403,* L37–L41.

Cazaux S., Tielens A. G. G. M., Ceccarelli C., Castets A., Wakelam V. Caux E., Parise B., and Teyssier D. (2003) The hot core around the low-mass protostar IRAS 16293–2422: Scoundrels rule! *Astrophys. J. Lett., 593,* L51–L55.

Ceccarelli C. (2002) Millimeter and infrared observations of deuterated molecules. *Planet. Space Sci., 50,* 1267–1273.

Charnley S. B. (1997a) On the nature of interstellar organic chemistry. In *Astronomical and Biochemical Origins and the Search for Life in the Universe* (C. B. Cosmovici et al., eds.), pp. 89–96. Editrice Compositori, Bologna.

Charnley S. B. (1997b) Chemical models of interstellar gas-grain processes. III — Molecular depletion in NGC 2024. *Mon. Not. R. Astron. Soc., 291,* 455–460.

Charnley S. B. (2001a) Stochastic theory of molecule formation on dust. *Astrophys. J. Lett., 562,* L99–L102.

Charnley S. B. (2001b) Interstellar organic chemistry. In *The Bridge Between the Big Bang and Biology* (F. Giovannelli, ed.), pp. 139–149. Consiglio Nazionale delle Ricerche President Bureau Special Volume, Rome, Italy.

Charnley S. B. and Rodgers S. D. (2002) The end of interstellar chemistry as the origin of nitrogen in comets and meteorites. *Astrophys. J. Lett., 569,* L133–L137.

Charnley S. B., Tielens A. G. G. M., and Millar T. J. (1992) On the molecular complexity of the hot cores in Orion A — Grain surface chemistry as 'The last refuge of the scoundrel.' *Astrophys. J. Lett., 399,* L71–L74.

Charnley S. B., Kress M. E., Tielens A. G. G. M., and Millar T. J. (1995) Interstellar alcohols. *Astrophys. J., 448,* 232–239.

Charnley S. B., Tielens A. G. G. M., and Rodgers S. D. (1997) Deuterated methanol in the Orion compact ridge. *Astrophys. J. Lett., 482,* L203–L206.

Charnley S. B., Ehrenfreund P., Millar T. J., Boogert A. C. A., Markwick A. J., Butner H. M., Ruiterkamp R., and Rodgers S. D. (2004) Observational tests for grain chemistry: Posterior isotopic labeling. *Mon. Not. R. Astron. Soc., 347,* 157–162.

Chiar J. E., Pendleton Y. J., Geballe T. R., and Tielens A. G. G. M. (1998) Near-infrared spectroscopy of the proto-planetary nebula CRL 618 and the origin of the hydrocarbon dust component in the interstellar medium. *Astrophys. J., 507,* 281–286.

Ciesla F. and Charnley S. B. (2006) The physics and chemistry of nebular evolution. In *Meteorites and the Early Solar System II* (D. S. Lauretta and H. Y. McSween Jr., eds.), this volume. Univ of Arizona, Tucson.

Clayton D. D., Liu W., and Dalgarno A. (1999) Condensation of carbon in radioactive supernova gas. *Science, 283,* 1290–1292.

Clayton R. N. (2003) Oxygen isotopes in the Solar System. *Space Sci. Rev., 106,* 19–32.

Clayton R. N., Mayeda T. K., and Molini-Velsko C. A. (1985) Isotopic variations in solar system material: Evaporation and condensation of silicates. In *Protostars and Planets II* (D. C. Black and M. S. Matthews, eds.), pp. 755–771. Univ. of Arizona, Tucson.

Clemett S. J., Chillier X. D. F., Gillette S., Zare R. N., Maurette M., Engrand C., and Kurat G. (1998) Observation of indigenous polycyclic aromatic hydrocarbons in 'giant' carbonaceous Antarctic micrometeorites. *Origins Life Evol. Biosph., 28,* 425–448.

Cohen M., Allamandola L., Tielens A. G. G. M., Bregman J., Simpson J. P., Witteborn F. C., Wooden D., and Rank O. (1986) The infrared emission bands. I. Correlation studies and the dependence on C/O ratio. *Astrophys. J., 302,* 737–749.

Cronin J. R. and Chang S. (1993) Organic matter in meteorites: Molecular and isotopic analyses of the Murchison meteorite. In *The Chemistry of Life's Origins* (J. M. Greenberg et al., eds.), pp. 209–258. Kluwer, Dordrecht.

Day K. L. (1974) A possible identification of the 10-micron "silicate" feature, *Astrophys. J. Lett., 192,* L15–L17.

Dorschner J., Begemann B., Henning Th., Jäger C., and Mutschke H. (1995) Steps toward interstellar silicate mineralogy II. Study of Mg-Fe silicate glasses of variable composition. *Astron. Astrophys., 300,* 503–520.

Draine B. T. and Lee H. K. (1984) Optical properties of interstellar graphite and silicate grains. *Astrophys. J., 285,* 89–108.

Draine B. T., Roberge W. G., and Dalgarno A. (1983) Magnetohydrodynamic shock waves in molecular clouds. *Astrophys. J., 264,* 485–507.

Duley W. W. (1985) Evidence for hydrogenated amorphous carbon in the Red Rectangle. *Mon. Not. R. Astron. Soc., 215,* 259–263.

Dwek E. and Scalo J. M. (1980) The evolution of refractory interstellar grains in the solar neighbourhood. *Astrophys. J., 239,* 193–211.

Dwek E., Foster S. M., and Vancura O. (1996) Cooling, sputtering, and infrared emission from dust grains in fast nonradiative shocks. *Astrophys. J., 457,* 244–252.

Dworkin J. P., Deamer D. W., Sandford S. A., and Allamandola L. J. (2001) Self-assembling amphiphilic molecules: Synthesis in simulated interstellar/precometary ices. *Proc. Natl. Acad. Sci., 98,* 815–819.

Ebel D. S. (2006) Condensation of rocky material in astrophysical environments. In *Meteorites and the Early Solar System II* (D. S. Lauretta and H. Y. McSween Jr., eds.), this volume. Univ. of Arizona, Tucson.

Ehrenfreund P. and Charnley S. B. (2000) Organic molecules in the interstellar medium, comets, and meteorites: A voyage from dark clouds to the early earth. *Annu. Rev. Astron. Astrophys., 38,* 427–483.

Ehrenfreund P. and Schutte W. A. (2000) Infrared observations of interstellar ices. In *Astrochemistry: From Molecular Clouds to Planetary Systems* (Y. C. Minh and E. F. van Dishoeck, eds.), pp. 135–146. IAU Symposium No. 197, Astronomical Society of the Pacific, San Francisco.

Ehrenfreund P., Dartois E., Demyk K., and d'Hendecourt L. (1998) Ice segregation toward massive protostars. *Astron. Astrophys., 339,* L17–L20.

Ehrenfreund P., d'Hendecourt L., Charnley S., and Ruiterkamp R. (2001) Energetic and thermal processing of interstellar ices. *J. Geophys. Res., 106(E12),* 33291–33302.

Ehrenfreund P., Charnley S. B., and Botta O. (2005) A voyage from dark clouds to the early Earth. In *Astrophysics of Life* (M. Livio et al., eds.), pp. 1–20. Space Telescope Institute Symposium Series, Vol. 16, Cambridge Univ., Cambridge.

Elmegreen B. G. (2000) Star formation in a crossing time. *Astrophys. J., 530,* 277–281.

Evans N. (1999) Physical conditions in regions of star formation. *Annu. Rev. Astron. Astrophys., 37,* 311–362.

Fleischer A. J., Gauger A., and Sedlmayr E. (1991) Generation of shocks by radiation pressure on newly formed circumstellar dust. *Astron. Astrophys., 37,* L1–L4.

Fleischer A. J., Gauger A., and Sedlmayr E. (1992) Circumstellar dust shells around long-period variables. I. Dynamical models of C-stars including dust formation, growth and evaporation. *Astron. Astrophys., 266,* 321–339.

Fleischer A. J., Gauger A., and Sedlmayr E. (1995) Circumstellar dust shells around long-period variables III. Instability due to an exterior kappa-mechanism caused by dust formation. *Astron. Astrophys., 297,* 543–555.

Floss C., Stadermann F. J., Bradley J., Dai Z. R., Bajt S. A., and Graham G. (2004) Carbon and nitrogen isotopic anomalies in an anhydrous interplanetary dust particle. *Science, 303,* 1355–1358.

Frenklach M. and Feigelson E. D. (1989) Formation of polycyclic aromatic hydrocarbons in circumstellar envelopes. *Astrophys. J., 341,* 372–384.

Geppert W. D., Thomas R., Semaniak J., Ehlerding A., Millar T. J., Österdahl F., af Ugglas M., Djuric N., Paál A., and Larsson M. (2004) Dissociative recombination of N_2H^+: Evidence for fracture of the NN bond. *Astrophys. J., 609,* 459–464.

Gibb E. and 10 colleagues (2000) An inventory of interstellar ices toward the embedded protostar W33A. *Astrophys. J., 536,* 347–356.

Gilmour I. and Pillinger C. T. (1994) Isotopic compositions of individual polycyclic aromatic hydrocarbons from the Murchison meteorite. *Mon. Not. R. Astron. Soc, 269,* 235–240.

Glassgold A. E. (1996) Circumstellar photochemistry. *Annu. Rev. Astron. Astrophys., 34,* 241–278.

Goebel J. H. (1993) SiS_2 in circumstellar shells. *Astron. Astrophys., 278,* 226–230.

Goldsmith P. F. and Langer W. D. (1978) Molecular cooling and thermal balance of dense interstellar clouds. *Astrophys. J., 222,* 881–895.

Goldsmith P. F. and 19 colleagues (2000) O_2 in interstellar molecular clouds. *Astrophys. J. Lett., 539,* L123–L127.

Greenberg J. M. (1983) The largest molecules in space: Interstellar dust. In *Cosmochemistry and the Origin of Life* (C. Ponnemperuma, ed.), pp. 71–112. Reidel, Dordrecht.

Greenberg J. M. (1989) The core-mantle model of interstellar grains and the cosmic dust connection. In *Interstellar Dust* (L. J. Allamandola and A. G. G. M. Tielens, eds.), pp. 345–355. Reidel, Dordrecht.

Greenberg J. M., Li A., Mendoza-Gomez C. X., Schutte W. A., Gerakines P. A., and De Groot M. (1995) Approaching the interstellar grain organic refractory component. *Astrophys. J. Lett., 455,* L177–L180.

Guillois O., Ledoux G., and Reynaud C. (1999) Diamond infrared emission bands in circumstellar media. *Astrophys. J. Lett., 521,* L133–L136.

Hallenbeck S. L. and Nuth J. A. (1998) Infrared observations of the transition from chaotic to crystalline silicates via thermal annealing in the laboratory. *Astrophys. Space Sci., 255,* 427–433.

Hallenbeck S. L., Nuth J. A., and Daukantas P. L. (1998) Mid-infrared spectral evolution of amorphous magnesium silicate smokes annealed in vacuum: Comparison to cometary spectra. *Icarus, 131,* 198–209.

Hallenbeck S. L., Nuth J. A., and Nelson R. N. (2000) Evolving optical properties of annealing silicate grains: From amorphous condensate to crystalline mineral. *Astrophys. J., 535,* 247–255.

Hanner M. S., Lynch D. K., and Russell R. W. (1994) The 8–13 micron spectra of comets and the composition of silicate grains. *Astrophys. J., 425,* 274–285.

Hartmann L., Ballesteros-Parades J, and Bergin E. A. (2001) Rapid formation of molecular clouds and stars in the solar neighborhood. *Astrophys. J., 562,* 852–868.

Hecht J. H. (1986) A physical model for the 2175Å interstellar extinction feature. *Astrophys. J., 305,* 817–822.

Hecht J. H. (1991) The nature of the dust around R coronae borealis stars: Isolated amorphous carbon or graphite fractals? *Astrophys. J., 367,* 635–640.

Henning Th. and Mutschke H. (1997) Low-temperature infrared properties of cosmic dust analogues. *Astron. Astrophys., 327,* 743–754.

Henning Th., Begemann B., Mutschke H., and Dorschner J. (1995) Optical properties of oxide dust grains. *Astron. Astrophys. Suppl. Ser., 122,* 143–149.

Herbst E. (2000) Models of gas-grain chemistry in star-forming regions. In *Astrochemistry: From Molecular Clouds to Planetary Systems* (Y. C. Minh and E. F. van Dishoeck, eds.), pp. 147–159. IAU Symposium 197, Astronomical Society of the Pacific, San Francisco.

Hill H. G. M., D'Hendecourt L. B., Peron C., and Jones A. P. (1997) Infrared spectroscopy of interstellar nanodiamonds from the Orgueil meteorite. *Meteoritics & Planet. Sci., 32,* 713–718.

Hiraoka K., Ohashi N., Kihara Y., Yamamoto K., Sato T., and Yamashita A. (1994) Formation of formaldehyde and methanol from the reactions of H atoms with solid CO at 10–20K. *Chem. Phys. Lett., 229,* 408–414.

Hiraoka K., Miyagoshi T., Takayama T., Yamamoto K., and Kihara Y. (1998) Gas-grain processes for the formation of CH_4 and H_2O: Reactions of H atoms with C, O, and CO in the solid phase at 12 K. *Astrophys. J., 498,* 710–715.

Hiraoka K., Takayama T., Euchi A., Handa H., and Sato T. (2000) Study of the reactions of H and D atoms with solid C_2H_2, C_2H_4, and C_2H_6 at cryogenic temperatures. *Astrophys. J., 532,* 1029–1037.

Hiraoka K., Sato T., Sato S., Sogoshi, N., Yokoyama T., Takashima H., and Kitagawa S. (2002) Formation of formaldehyde by the tunneling reaction of H with solid CO at 10 K revisited. *Astrophys. J., 577,* 265–270.

Hofmeister A. M., Rosen L. J., and Speck A. K. (2000) IR spectra of nano- and macro-crystals: The overriding importance of optical path. In *Thermal Emission Spectroscopy and Analysis of Dust, Disks and Regoliths* (M. L. Sitko et al., eds.) pp. 291–299. ASP Conference Series 196, Astronomical Society of the Pacific, San Francisco.

Hollenbach D. J. and Salpeter E. E. (1971) Surface recombination of hydrogen molecules. *Astrophys. J., 163,* 155–164.

Hron J., Aringer B., and Kerschbaum F. (1997) Semi regular variables of types SRa and SRb. Silicate dust emission features. *Astron. Astrophys., 322,* 280–290.

Hudson R. L. and Moore M. H. (1999) Laboratory studies of the formation of methanol and other organic molecules by water + carbon monoxide radiolysis: Relevance to comets, icy satellites, and interstellar ices. *Icarus, 140,* 451–461.

Hudson R. L. and Moore M. H. (2004) Reactions of nitriles in ices relevant to Titan, comets, and the interstellar medium: Formation of cyanate ion, ketenimines, and isonitriles. *Icarus, 172,* 466–478.

Huss G. R., Fahey A. J., Gallino R., and Wasserberg G. J. (1994) Oxygen isotopes in circumstellar Al_2O_3 grains from meteorites and stellar nucleosynthesis. *Astrophys. J. Lett., 430,* L81–L84.

Jäger C., Fabian D., Schrempel F., Dorschner J., Henning Th., and

Wesch W. (2003) Structural processing of enstatite by ion bombardment. *Astron. Astrophys., 401,* 57–65.

Jeong K. S., Winters J. M., Le Bertre T., and Sedlmayr E. (2003) Self-consistent modelling of the outflow from the O-rich Mira IRC-20197. *Astron. Astrophys., 407,* 191–206.

Jones A. P., Tielens A. G. G. M., Hollenbach D. J., and McKee C. F. (1994) Grain destruction in shocks in the interstellar medium. *Astrophys. J., 433,* 797–810.

Jones A. P., Tielens A. G. G. M., and Hollenbach D. J. (1996) Grain shattering in shocks: The interstellar grain size distribution. *Astrophys. J., 469,* 740–764.

Kamitsuji K., Sato T., Suzuki H., and Kaito C. (2005) Direct observation of crystallization of amorphous Mg-bearing silicate grains to Mg_2SiO_4 (forsterite). *Astron. Astrophys., 436,* 165–169.

Keller L. P. and Messenger S. (2004) On the origin of GEMS (abstract). In *Lunar and Planetary Science XXXV,* Abstract #1985. Lunar and Planetary Institute, Houston (CD-ROM).

Kemper F., Vriend W. J., and Tielens A. G. G. M. (2004) The absence of crystalline silicates in the diffuse interstellar medium. *Astrophys. J., 609,* 826–837.

Kerridge J. F. and Chang S. (1985) Survival of interstellar matter in meteorites — Evidence from carbonaceous material. In *Protostars and Planets II* (D. C. Black and M. S. Matthews, eds.), pp. 738–754. Univ. of Arizona, Tucson.

Kissel J., Krueger F. R. and Roessler K. (1997) In *Comets and the Origins and Evolution of Life* (P. J. Thomas et al., eds.), pp. 69–110. Springer-Verlag, New York.

Knacke R. F., Fajardo-Acosta S. B., Telesco C. M., Hackwell J. A., Lynch D. K., and Russell R. W. (1993) The silicates in the disk of β Pictoris. *Astrophys. J., 418,* 440–450.

Koike C., Shibai H., and Tuchiyama A. (1993) Extinction of olivine and pyroxene in the mid- and far-infrared. *Mon. Not. R. Astron. Soc., 264,* 654–658.

Kozasa T. and Hasegawa H. (1987) Grain formation through nucleation process in astrophysical environments II. Nucleation and grain growth accompanied by chemical reaction. *Progr. Theor. Phys., 77,* 1402.

Kraus G. F., Nuth J. A., and Nelson R. N. (1997) Is SiS_2 the carrier of the unidentified 21 micron emission feature? *Astron. Astrophys., 328,* 419–425.

Kuan Y.-J. and 10 colleagues (2004) Organic molecules in low-mass protostellar hot cores: Submillimeter imaging of IRAS 16293-2422. *Astrophys. J. Lett., 616,* L27–L30.

Langer W. D., Graedel T. E., Frerking M. A., and Armentrout P. B. (1984) Carbon and oxygen isotope fractionation in dense interstellar clouds. *Astrophys. J., 277,* 581–604.

Larson R. B. (2003) The physics of star formation. *Rep. Progr. Phys., 66,* 1651–1701.

Leger A., d'Hendecourt L., and Boccara N., eds. (1987) *Polycyclic Aromatic Hydrocarbons and Astrophysics.* Reidel, Dordrecht. 402 pp.

Lepp S. and Dalgarno A. (1988) Polycyclic aromatic hydrocarbons in interstellar chemistry. *Astrophys. J., 324,* 553–556.

Lepp S., Dalgarno A., and McCray R. (1990) Molecules in the ejecta of SN 1987A. *Astrophys. J., 358,* 262–265.

Lewis R. S., Anders E., and Draine B. T. (1989) Properties, detectability and origin of interstellar diamonds in meteorites. *Nature, 339,* 117–121.

Li A. and Draine B. T. (2001) Infrared emission from interstellar dust: II. The diffuse interstellar medium. *Astrophys. J., 554,* 778–802.

Loinard L., Castets A., Ceccarelli C., Tielens A. G. G. M., Faur A., Caux E., and Duvert G. (2000) The enormous abundance of D_2CO in IRAS 16293–2422. *Astron. Astrophys., 359,* 1169–1174.

Loinard L., Castets A., Ceccarelli C., Caux E., and Tielens A. G. G. M. (2001) Doubly deuterated molecular species in protostellar environments. *Astrophys. J. Lett., 552,* L163–L166.

Mac Low M.-M. and Klessen R. S. (2004) Control of star formation by supersonic turbulence. *Rev. Mod. Phys., 76,* 125–194.

Manico G., Ragun G., Pirronello V., Roser J. E., and Vidali G. (2001) Laboratory measurements of molecular hydrogen formation on amorphous water ice. *Astrophys. J. Lett., 548,* L253–L256.

Markwick A. J. and Charnley S. B. (2004) Chemistry of protoplanetary disks. In *Astrobiology: Future Perspectives* (P. Ehrenfreund et al., eds.), pp. 33–66. Kluwer, Dordrecht.

Markwick A. J., Millar T. J., and Charnley S. B. (2000) On the abundance gradients of organic molecules along the TMC-1 ridge. *Astrophys. J., 535,* 256–265.

Mathis J. S. (1990) Interstellar dust and extinction. In *The Evolution of the Interstellar Medium* (L. Blitz, ed.), pp. 63–77. ASP Conference Series 12, Astronomical Society of the Pacific, San Francisco.

Mathis J. S. (1996) Dust models with tight abundance constraints. *Astrophys. J., 472,* 643–655.

Mathis J. S., Rumpl W., and Nordsieck K. H. (1977) Size distribution of interstellar grains. *Astrophys. J., 217,* 425–433.

Messenger S. (2000) Identification of molecular-cloud material in interplanetary dust particles. *Nature, 404,* 968–971.

Messenger S. and Walker R. M. (1996) Isotopic anomalies in interplanetary dust particles. In *Physics, Chemistry, and Dynamics of Interplanetary Dust* (B. A. S. Gustafson and M. S. Hanner, eds.), pp. 287–290. ASP Conference Series 104, Astronomical Society of the Pacific, San Francisco.

Messenger S. and Walker R. M. (1997) Evidence for molecular cloud material in meteorites and interplanetary dust. In *Astrophysical Implications of the Laboratory Study of Presolar Materials* (T. J. Bernatowicz and E. Zinner, eds.), pp. 545–564. American Institute of Physics, New York.

Messenger S., Amari S., Gao X., Walker R. M., Clemett S. J., Chillier X. D. F., Zare R. N., and Lewis R. S. (1998) Indigenous polycyclic aromatic hydrocarbons in circumstellar graphite grains from primitive meteorites. *Astrophys. J., 502,* 284–295.

Messenger S., Keller L. P., Stadermann F. J., Walker R. M., and Zinner E. (2003) Samples of stars beyond the solar system: Silicate grains in interplanetary dust. *Science, 300,* 105–108.

Meyer B. S. and Zinner E. (2006) Nucleosynthesis. In *Meteorites and the Early Solar System II* (D. S. Lauretta and H. Y. McSween Jr., eds.), this volume. Univ. of Arizona, Tucson.

Michael B. P., Lilleleht L. U., and Nuth J. A. (2003) Zinc crystal growth in microgravity. *Astrophys. J., 590,* 579–585.

Millar T. J. and Duley W. W. (1978) Diatomic oxide interstellar grains. *Mon. Not. R. Astron. Soc., 183,* 177–185.

Millar T. J., Bennett A., and Herbst E. (1989) Deuterium fractionation in dense interstellar clouds. *Astrophys. J., 340,* 906–920.

Millar T. J., Roberts H., Markwick A. J., and Charnley S. B. (2000) The role of H_2D^+ in the deuteration of interstellar molecules. *Philos. Trans. R. Soc. Lond., A358,* 2535–2547.

Mutschke H., Dorschner J., Henning T., Jäger C., and Ott U. (1995) Facts and artefacts in interstellar diamond spectra. *Astrophys. J. Lett., 454,* L157–L160.

Mutschke H., Henning Th., and Clement D. (2000) Effects of grain morphology and impurities on the infrared spectra of silicon carbide particles. In *Thermal Emission Spectroscopy and Analysis of Dust, Disks and Regoliths* (M. L. Sitko et al., eds.), pp. 273–280. ASP Conference Series 196, Astronomical Society of the Pacific, San Francisco.

Ney E. P. (1977) Star dust. *Science, 195,* 541–546.

Nguyen A. N. and Zinner E. (2004) Discovery of ancient silicate stardust in a meteorite. *Science, 303,* 1496–1499.

Nittler L. R. (2003) Presolar stardust in meteorites: Recent advances and scientific frontiers. *Earth Planet. Sci. Lett., 209,* 259–273.

Nittler L. R., Alexander C. M. O'D., Gao X., Walker R. M., and Zinner E. (1997) Stellar sapphires: The properties and origins of presolar Al_2O_3 in meteorites. *Astrophys. J., 483,* 475–495.

Nuth J. A. (1996) Grain formation and metamorphism. In *The Cosmic Dust Connection* (J. M. Greenberg, ed.), pp. 205–221. Kluwer, Dordrecht.

Nuth J. A. and Hecht J. H. (1990) Signatures of aging silicate dust. *Astrophys. Space Sci., 163,* 79–94.

Nuth J. A. and Moore M. H. (1989) Proton irradiation of SiH_4-$Fe(CO)_5$-H_2O ices: Production of refractory silicates and implications for the solar nebula. *Proc. Lunar Planet. Sci. Conf. 19th,* pp. 565–569.

Nuth J. A., Hallenbeck S. L., and Rietmeijer F. J. M. (1998) Interstellar and interplanetary grains: Recent developments and new opportunities for experimental chemistry. *Earth Moon Planets, 80,* 73–112.

Nuth J. A., Hallenbeck S. L., and Rietmeijer F. J. M. (2000) Laboratory studies of silicate smokes: Analog studies of circumstellar materials. *J. Geophys. Res., 105,* 10387–10396.

Nuth J. A., Rietmeijer F. J. M., and Hill H. G. M. (2002) Condensation processes in astrophysical environments: The composition and structure of cometary grains. *Meteoritics & Planet. Sci., 37,* 1579–1590.

O'Donnell J. Z. and Mathis J. S. (1997) Dust grain size distributions and the abundance of refractory elements in the diffuse interstellar medium. *Astrophys. J., 479,* 806–817.

Parise B., Ceccarelli C., Tielens A. G. G. M., Herbst E., Lefloch B., Caux E., Castets A., Mukhopadhyay I., Pagani L., and Loinard L. (2002) Detection of doubly-deuterated methanol in the solar-type protostar IRAS 16293–2422. *Astron. Astrophys., 393,* L49–L53.

Parise B., Castets A., Herbst E., Caux E., Ceccarelli C., Mukhopadhyay I., and Tielens A. G. G. M. (2004) First detection of triply-deuterated methanol. *Astron. Astrophys., 416,* 159–163.

Pendleton Y. J., Tielens A. G. G. M., and Werner M. W. (1990) Studies of dust grain properties in infrared reflection nebulae. *Astrophys. J., 349,* 107–119.

Pendleton Y. J., Sandford S. A., Allamandola L. J., Tielens A. G. G. M., and Sellgren K. (1994) Near-infrared absorption spectroscopy of interstellar hydrocarbon grains. *Astrophys. J., 437,* 683–696.

Pirronello V., Liu C., Roser J. E., and Vidali G. (1999) Measurements of molecular hydrogen formation on carbonaceous grains. *Astron. Astrophys., 344,* 681–686.

Pontoppidan K. M., Dartois E., van Dishoeck E. F., Thi W. F., and d'Hendecourt L. (2003) Detection of abundant solid methanol toward young low mass stars. *Astron. Astrophys., 404,* L17–L21.

Posch T., Kerschbaum F., Mutschke H., Fabian D., Dorschner J., and Hron J. (1999) On the origin of the 13 μm feature. A study of ISO-SWS spectra of oxygen-rich AGB stars. *Astron. Astrophys., 352,* 609–618.

Prasad S. S. and Tarafdar S. P. (1983) UV radiation field inside dense clouds — Its possible existence and chemical implications. *Astrophys. J., 267,* 603–609.

Puget J. L. and Léger A. (1989) A new component of the interstellar matter — Small grains and large aromatic molecules. *Annu. Rev. Astron. Astrophys., 27,* 161–198.

Rawlings J. M. C. (1988) Chemistry in the ejecta of novae. *Mon. Not. R. Astron. Soc., 232,* 507–524.

Rietmeijer F. J. M. and Nuth J. A. (2000) Metastable eutectic equilibrium brought down to Earth. *Eos Trans. AGU, 81,* 409, 414–415.

Rietmeijer F. J. M, Nuth J. A., and Karner J. M. (1999) Metastable eutectic condensation in a Mg-Fe-SiO-H_2-O_2 vapor: Analogs to circumstellar dust. *Astrophys. J., 527,* 395–404.

Rietmeijer F. J. M., Nuth J. A. III, Jablonska M., and Karner J. M (2000) Metastable eutectic equilibrium in natural environments: Recent developments and research opportunities. In *Res. Trends Geochem., 1,* 30–51.

Roberts H. and Millar T. J. (2000) Gas-phase formation of doubly-deuterated species. *Astron. Astrophys., 64,* 780–784.

Roberts H., Herbst E., and Millar T. J. (2003) Enhanced deuterium fractionation in dense interstellar cores resulting from multiply deuterated H_3^+. *Astrophys. J. Lett., 591,* L41–L44.

Rodgers S. D. and Charnley S. B. (2001) Gas phase production of NHD_2 in L134N. *Astrophys. J., 553,* 613–617.

Rodgers S. D. and Charnley S. B. (2003) Chemical evolution in protostellar envelopes — Cocoon chemistry. *Astrophys. J., 585,* 355–371.

Rodgers S. D. and Charnley S. B. (2004) Interstellar diazenylium recombination and nitrogen isotopic fractionation. *Mon. Not. R. Astron. Soc., 352,* 600–604.

Rodgers S. D., Charnley S. B., Huebner, W. F., and Boice D. C. (2004) Physical processes and chemical reactions in cometary comae. In *Comets II* (M. C. Festou et al., eds.), pp. 505–522. Univ. of Arizona, Tucson.

Roser J. E., Vidali G., Manico G., and Pirronello V. (2001) Formation of carbon dioxide by surface reactions on ices in the interstellar medium. *Astrophys. J. Lett., 555,* L61–L64.

Roueff E., Tine S., Coudert L. H., Pineau des Forets G., Falgarone E., and Gerin M. (2000) Detection of doubly deuterated ammonia in L134N. *Astron. Astrophys., 354,* L63–L66.

Sakata A., Wada S., Okutsu Y., Shintani H., and Nakada Y. (1983) Does a 2,200 Å hump observed in an artificial carbonaceous composite account for UV interstellar extinction? *Nature, 301,* 493–494.

Sandford S. A. (1996) The inventory of interstellar materials available for the formation of the solar system. *Meteoritics & Planet. Sci., 31,* 449–476.

Sandford S. A. and Allamandola L. J. (1993) Condensation and vaporization studies of CH_3OH and NH_3 ices: Major implications for astrochemistry. *Astrophys. J., 417,* 815–825.

Sandford S. A., Allamandola L. J., Tielens A. G. G. M., Sellgren K., Tapia M., and Pendleton Y. J. (1991) The interstellar C-H stretching band near 3.4 microns — Constraints on the composition of organic material in the diffuse interstellar medium. *Astrophys. J., 371,* 607–620.

Sandford S. A., Allamandola L. J., and Bernstein M. P. (1997) The composition and ultraviolet and thermal processing of interstellar ices. In *From Stardust to Planetesimals* (Y. J. Pendleton and A. G. G. M. Tielens, eds.), pp. 201–213. ASP Conference Series 122, Astronomical Society of the Pacific, San Francisco.

Sandford S. A., Bernstein M. P., and Dworkin J. P. (2001) Assessment of the interstellar processes leading to deuterium enrichment in meteoritic organics. *Meteoritics & Planet. Sci., 36,* 1117–1133.

Saslaw W. C. and Gaustad J. E. (1969) Interstellar dust and diamonds. *Nature, 221,* 160–162.

Savage B. D. and Sembach K. R. (1996) Interstellar abundances from absorption-line observations with the Hubble Space Telescope. *Annu. Rev. Astron. Astrophys., 34,* 279–329.

Schnaiter M., Mutschke H., Dorschner J., Henning Th., and Salama F. (1998) Matrix-isolated nano-sized carbon grains as an analog for the 217.5 nanometer feature carrier. *Astrophys. J., 498,* 486–496.

Schöier F. L., Jorgensen J. K., van Dishoeck E. F., and Blake G. A. (2002) Does IRAS 16293–2422 have a hot core? Chemical inventory and abundance changes in its protostellar environment. *Astron. Astrophys., 390,* 1001–1021

Schutte W. A. (1996) Formation and evolution of icy grain mantles. In *The Cosmic Dust Connection* (J. M. Greenberg, ed.), pp. 1–42. Kluwer, Dordrecht.

Schutte W. A. and Greenberg J. M. (1991) Explosive desorption of icy grain mantles in dense clouds. *Astron. Astrophys., 244,* 190–204.

Seab C. G. and Shull J. M. (1983) Shock processing of interstellar grains. *Astrophys. J., 275,* 652–660.

Sephton M. A., Verchovsky A. B., Bland P., Gilmour I., Grady M. M., and Wright I. P. (2003) Investigating the variations in carbon and nitrogen isotopes in carbonaceous chondrites. *Geochim. Cosmochim. Acta, 67,* 2093–2108.

Shu F. H., Adams F. C., and Lizano S. (1987) Star formation in molecular clouds — Observation and theory. *Annu. Rev. Astron. Astrophys., 25,* 23–81.

Sorrell W. H. (1989) Modelling the dust grains around the peculiar Be star HD 45677. *Mon. Not. R. Astron. Soc., 241,* 89–108.

Sorrell W. H. (1990) The 2175-Å feature from irradiated graphitic particles. *Mon. Not. R. Astron. Soc., 243,* 570–587.

Sorrell W. H. (1991) Annealed HAC mantles in diffuse dust clouds. *Mon. Not. R. Astron. Soc., 248,* 439–443.

Speck A. K., Hofmeister A. M., and Barlow M. J. (2000) Silicon carbide: The problem with laboratory spectra. In *Thermal Emission Spectroscopy and Analysis of Dust, Disks and Regoliths* (M. L. Sitko et al., eds.), pp. 281–290. ASP Conference Series 196, Astronomical Society of the Pacific, San Francisco.

Stancil P. C., Lepp S. H., and Dalgarno A. (2002) Atomic and molecular processes in the early universe. *J. Phys., B35,* R57.

Stecher T. P. (1969) Interstellar extinction in the ultraviolet. II. *Astrophys. J. Lett., 157,* L125–L126.

Stencel R. E., Nuth J. A., Little-Marenin I. R., and Little S. J. (1990) The formation and annealing of circumstellar dust as gauged by IRAS low-resolution spectra and the microwave Maser chronology. *Astrophys. J. Lett., 350,* L45–L48.

Stroud R. M., Nittler L. R., and Alexander C. M. O'D. (2004) Polymorphism in presolar Al_2O_3 grains from asymptotic giant branch stars. *Science, 305,* 1455–1457.

Suzuki H., Yamamoto S., Ohishi M., Kaifu N., Ishikawa S-I., Hirahara Y., and Takano S. (1992) A survey of CCS, HC_3N, HC_5N, and NH_3 toward dark cloud cores and their production chemistry. *Astrophys. J., 392,* 551–570.

Terzieva R. and Herbst E. (2000) The possibility of nitrogen isotopic fractionation in interstellar clouds. *Mon. Not. R. Astron. Soc., 317,* 563–568.

Tielens A. G. G. M. (1983) Surface chemistry of deuterated molecules. *Astron. Astrophys., 119,* 177–184.

Tielens A. G. G. M. (1987) Carbon stardust: From soot to diamonds. In *Carbon in the Galaxy: Studies from Earth and Space* (J. C. Tarter et al., eds.), pp. 59–111. NASA Conference Publication 3061, Washington, DC.

Tielens A. G. G. M. and Hagen W. (1982) Model calculations of the molecular composition of interstellar grain mantles. *Astron. Astrophys., 114,* 245–260.

Turner B. E. (1991) Observations and chemistry of interstellar refractory elements. *Astrophys. J., 376,* 573–598.

van Dishoeck E. F. and Black J. H. (1988) Diffuse cloud chemistry. In *Rate Coefficients in Astrochemistry* (T. J. Millar and D. A. Williams, eds.), pp. 209–237. Kluwer, Dordrecht.

van Dishoeck E. F. and Blake G. A. (1998) Chemical evolution of star-forming regions. *Annu. Rev. Astron. Astrophys., 36,* 317–368.

van der Tak F. F. S., Schilke P., Mueller H. S. P., Lis D. C., Phillips T. G., Gerin M., and Roueff E. (2002) Triply deuterated ammonia in NGC 1333. *Astron. Astrophys., 388,* L53–L56.

Vastel C., Phillips T. G., and Yoshida H. (2004) Detection of D_2H^+ in the dense interstellar medium. *Astrophys. J. Lett., 606,* L127–L130.

Vidal-Madjar A. (2002) D/H observations in the interstellar medium. *Planet. Space Sci., 50,* 1161–1168.

Waelkens C. and 20 colleagues (1996) SWS observations of young main-sequence stars with dusty circumstellar disks. *Astron. Astrophys., 315,* L245–L248.

Watanabe N. and Kouchi A. (2002) Efficient formation of formaldehyde and methanol by the addition of hydrogen atoms to CO in H_2O-CO ice at 10 K. *Astrophys. J. Lett., 571,* L173–L176.

Watanabe N., Shiraki T., and Kouchi A. (2003) The dependence of H_2CO and CH_3OH formation on the temperature and thickness of H_2O-CO ice during the successive hydrogenation of CO. *Astrophys. J. Lett., 588,* L121–L124.

Waters L. B. F. M. and 36 colleagues (1996) Mineralogy of oxygen-rich dust shells. *Astron. Astrophys., 315,* L361–L364.

Willacy K. and Millar T. J. (1998) Desorption processes and the deuterium fractionation in molecular clouds. *Mon. Not. R. Astron. Soc., 298,* 562–568.

Wooden D. H., Charnley S. B., and Ehrenfreund P. (2004) Composition and evolution of interstellar clouds. In *Comets II* (M. C. Festou et al., eds.), pp. 33–66. Univ. of Arizona, Tucson.

Zinner E. K. (1988) Interstellar cloud material in meteorites. In *Meteorites and the Early Solar System* (J. F. Kerridge and M. S. Matthews, eds.), pp. 956–983. Univ. of Arizona, Tucson.

Zinner E. K. (1998) Stellar nucleosynthesis and the isotopic composition of presolar grains from primitive meteorites. *Annu. Rev. Earth Planet. Sci., 26,* 147–188.

Zinner E. K., McKeegan K. D., and Walker R. M. (1983) Laboratory measurements of D/H ratios in interplanetary dust. *Nature, 305,* 119–121.

Zinner E., Amari S., Guinness R., Nguyen A., Stadermann F. J., Walker R. M., and Lewis R. S. (2003) Presolar spinel grains from the Murray and Murchison carbonaceous chondrites. *Geochim. Cosmochim. Acta, 67,* 5083–5095.

Zubko V., Dwek E., Arendt R. G. (2004) Interstellar dust models consistent with extinction, emission, and abundance constraints. *Astrophys. J., 152,* 211–249.

Zuckerman B. and Evans N. J. (1974) Models of massive molecular clouds. *Astrophys. J., 192,* L149–L152.

Part III:

Disk Formation Epoch: The Astrophysical Setting and Initial Conditions of the Solar Nebula

Presolar Cloud Collapse and the Formation and Early Evolution of the Solar Nebula

Alan P. Boss
Carnegie Institution of Washington

Jitendra N. Goswami
Physical Research Laboratory

We review our current understanding of the formation and early evolution of the solar nebula, the protoplanetary disk from which the solar system formed. Astronomical understanding of the collapse of dense molecular cloud cores to form protostars is relatively advanced, compared to our understanding of the formation processes of planetary systems. Examples exist of nearly all the phases of protostellar evolution, guiding and validating theoretical models of the star formation process. Astronomical observations of suspected protoplanetary disks are beginning to provide key insights into the likely conditions within the solar nebula, although usually only on much larger scales than the regions where the terrestrial and giant planets formed. The primary mechanism for driving the early evolution of the solar nebula, leading finally to the formation of the planetary system, is still highly uncertain: Magnetic fields, gravitational torques, and baroclinic instabilities are among the competing mechanisms. While the discovery of extrasolar gas giant planets has reinforced the general belief that planetary system formation should be a widespread process, the absence of any undisputed evidence for giant protoplanets in the process of formation makes it hard to decide between competing mechanisms for their formation. The situation regarding extrasolar terrestrial planets is even less constrained observationally, although this situation should improve tremendously in the next decade.

1. INTRODUCTION

The most primitive meteorites are believed to be relatively unaltered samples of the basic building blocks of the inner solar system. As such, the primitive meteorites, as well as comets, are believed to preserve parts of the record of the formational processes that led to the origin of our solar system. Presolar grains contained within them undoubtedly carry much of the history of galactic nucleosynthesis in their isotopic abundances (*Nittler and Dauphas,* 2006; *Meyer and Zinner,* 2006). A major portion of the motivation for observational and laboratory studies of meteorites and comets stems from the desire to use the results to constrain or otherwise illuminate the physical and chemical conditions in the solar nebula, the Sun's protoplanetary disk, in the hopes of learning more about the processes that led to the formation of our planetary system. In addition to meteoritical studies, there are important lessons to be learned about the planet formation process from astrophysical observations of young stellar objects and their accompanying protoplanetary disks. Theoretical models then serve as the ultimate recipient of these constraints and clues, charged with the task of assembling a cohesive explanation for some or all of the data relevant to some particular aspect of the planet formation process. While much progress has been made in the last several decades, we are still far from having an agreed-upon, universal model of planetary system formation and early evolution. An excellent resource for learning more about these and other subjects is the review volume *Protostars and Planets IV* (*Mannings et al.,* 2000). The present chapter is adapted, modified, and updated from two recent review articles about the solar nebula (*Boss,* 2003, 2004c).

2. FORMATION OF THE SOLAR NEBULA

The protosun and solar nebula were formed by the self-gravitational collapse of a dense molecular cloud core, much as we see new stars being formed today in regions of active star formation. The formation of the solar nebula was largely an initial value problem, i.e., given detailed knowledge of the particular dense molecular cloud core that was the presolar cloud, one could in principle calculate the flow of gas and dust subject to the known laws of physics and thus predict the basic outcome. Specific details of the outcome cannot be predicted, however, as there appears to be an inevitable amount of stochastic, chaotic evolution involved, e.g., in the orbital motions of any ensemble of gravitationally interacting particles. Nevertheless, we expect that at least the gross features of the solar nebula should be predictable from theoretical models of cloud collapse, constrained by astronomical observations of young stellar objects. For this reason, the physical structure of likely precollapse clouds is of interest with regard to inferring the formation mechanism of the protosun and the structure of the accompanying solar nebula.

In the following we describe the current state of our knowledge about the nature of the precollapse, dense molecular cloud, and the collapse of such a cloud leading to the formation of a protostar surrounded by a protoplanetary disk, based on the results obtained from analytical modeling and astronomical observations.

2.1. Observations of Precollapse Clouds

Astronomical observations at long wavelengths (e.g., millimeter) are able to probe deep within interstellar clouds of gas and dust, which are opaque at short wavelengths (e.g., visible wavelengths). These clouds are composed primarily of molecular hydrogen gas, helium, and molecules such as carbon monoxide, hence the term molecular clouds. About one percent by mass of these clouds is in the form of submicrometer-sized dust grains, with about another one percent composed of gaseous molecules and atoms of elements heavier than helium. Regions of active star formation are located within molecular clouds and complexes ranging in mass from a few solar masses ($M_\odot$) to over $10^6 M_\odot$. This association of young stars with molecular clouds is the most obvious manifestation of the fact that stars form from these clouds. Many of the densest regions of these clouds were found to contain embedded infrared objects, i.e., newly formed stars whose light is scattered, absorbed, and reemitted at infrared wavelengths in the process of exiting the placental cloud core. Such cores have already succeeded in forming stars. Initial conditions for the collapse of the presolar cloud can be more profitably ascertained from observations of dense cloud cores that do not appear to contain embedded infrared objects, i.e., precollapse cloud cores.

Precollapse cloud cores are composed of cold molecular gas with temperatures in the range of about 7–15 K, and with gas densities of about 1000–100,000 molecules per cubic centimeter. Some clouds may be denser yet, but this is hard to determine because of the limited density ranges for which suitable molecular tracers are abundant (typically isotopes of carbon monoxide and ammonia). Masses of these clouds range from ~1 $M_\odot$ to $10^3 M_\odot$, with the distribution of clump masses fitting a power law such that most of the clumps are of low mass, as is also true of stars in general. In fact, one estimate of the mass distribution of precollapse clouds in Taurus is so similar to the initial mass function of stars that it appears that the stellar mass distribution may be determined primarily by processes occurring prior to the formation of precollapse clouds (*Onishi et al.,* 2002). The cloud properties described below are used to constrain the initial conditions for hydrodynamical models of the collapse of cloud cores. Large radio telescopes have enabled high-spatial-resolution mapping of precollapse clouds and the determination of their interior density structure. While such clouds undoubtedly vary in all three space dimensions, typically the observations are averaged in angle to yield an equivalent, spherically symmetric density profile. These radial density profiles have shown that precollapse clouds typically have flat density profiles near their centers, as is to be expected for a cloud that has not yet collapsed to form a star (*Bacmann et al.,* 2000), surrounded by an envelope with a steeply declining profile that could be fit with a power law. The density profile thus resembles that of a Gaussian distribution, or more precisely, the profile of the Bonnor-Ebert sphere, which is the equilibrium configuration for an isothermal gas cloud (*Alves et al.,* 2001).

While precollapse clouds often have a complicated appearance, attempts have been made to approximate their shapes with simple geometries. Triaxial spheroids seem to be required in general (*Kerton et al.,* 2003), although most lower-mass clouds appear to be more nearly oblate than prolate (*Jones et al.,* 2001). On the larger scale, prolate shapes seem to give a better fit than oblate spheroids. Another study found that the observations could be fit with a distribution of prolate spheroids with axis ratios of 0.54 (*Curry,* 2002), and argued that the prolate shapes derived from the filamentary nature of the parent clouds. We shall see that the precollapse cloud's shape is an important factor for the outcome of the protostellar collapse phase.

Precollapse clouds have significant interior velocity fields that appear to be a mixture of turbulence derived from fast stellar winds and outflows, and magnetohydrodynamic waves associated with the ambient magnetic field. In addition, there may be evidence for a systematic shift in velocities across one axis of the cloud, which can be interpreted as solid-body rotation around that axis. When estimated in this manner, typical rotation rates are found to be below the level needed for cloud support by centrifugal force, yet large enough to result in considerable rotational flattening once cloud collapse begins. Ratios of rotational to gravitational energy in dense cloud cores range from ~0.0001 to ~0.01 (*Goodman et al.,* 1993; *Caselli et al.,* 2002). The presence of a net angular momentum for the cloud is essential for the eventual formation of a centrifugally supported circumstellar disk.

2.2. Onset of Collapse Phase

Dense cloud cores are supported against their own self-gravity by a combination of turbulent motions, magnetic fields, thermal (gas) pressure, and centrifugal force, in roughly decreasing order of importance. Turbulent motions inevitably dissipate over timescales that are comparable to or less than a cloud's freefall time (the time over which an idealized, pressureless sphere of gas of initially uniform density would collapse to form a star), once the source of the turbulence is removed. For a dense cloud core, freefall times are on the order of 0.1 m.y. However, dense clouds do not collapse on this timescale, because once turbulence decays, magnetic fields provide support against self-gravity.

2.2.1. Ambipolar diffusion and loss of magnetic support. Magnetic field strengths in dense clouds are measured by Zeeman splitting of molecular lines, and found to be large enough (about 10–1000 μG) to be capable of supporting dense clouds, provided that both static magnetic fields and magnetohydrodynamic waves are present (*Crutcher,* 1999). Field strengths are found to depend on the density to roughly the 1/2 power, as is predicted to be the case if ambipolar diffusion controls the cloud's dynamics (*Mouschovias,* 1991). Ambipolar diffusion is the process of slippage of the primarily neutral gas molecules past the ions, to which the magnetic field lines are effectively attached. This process occurs over timescales of a few million years or more for dense cloud cores, and inevitably leads to the loss of sufficient magnetic field support such that the slow inward contraction of the cloud turns into a rapid, dynamic collapse phase, when the magnetic field is no longer in control. This is generally believed to be the process through which stars in regions of low-mass (less than ~8 $M_\odot$) star formation begin their life, the "standard model" of star formation (*Shu et al.,* 1987).

2.2.2. Shock-triggered collapse and injection of radioactivity. This standard model has been challenged by evidence for short cloud lifetimes and a highly dynamic star-formation process driven by large-scale outflows (*Hartmann et al.,* 2001). A case for a short lifetime of the protosolar cloud has also been made based on the isotopic records seen in early solar system solids (*Russell et al.,* 2006).

In regions of high-mass (~8 $M_\odot$ or more) star formation, where the great majority of stars are believed to form, quiescent star formation of the type envisioned in the standard model occurs only until the phase when high-mass stars begin to form and evolve. The process of high-mass star formation is less well understood than that of low-mass stars, but observations make it clear that events such as the supernova explosions that terminate the life of massive stars can result in the triggering of star formation in neighboring molecular clouds that are swept up and compressed by the expanding supernova shock front (*Preibisch and Zinnecker,* 1999; *Preibisch et al.,* 2002). Even strong protostellar outflows are capable of triggering the collapse of neighboring dense cloud cores (*Foster and Boss,* 1996; *Yokogawa et al.,* 2003). A supernova shock-triggered origin for the presolar cloud has been advanced as a likely source of the short-lived radioisotopes (e.g., ^{26}Al) that existed in the early solar nebula (*Cameron and Truran,* 1977; *Marhas et al.,* 2002). Detailed models of shock-triggered collapse have shown that injection of shock-front material containing the ^{26}Al into the collapsing protostellar cloud can occur, provided that the shock speed is on the order of 25 km/s (*Boss,* 1995), as is appropriate for a moderately distant supernova or for the wind from an evolved red giant star. High-resolution calculations have shown that this injection occurs through Rayleigh-Taylor instabilities at the shock-cloud boundary, where dense fingers grow downward into the presolar cloud (*Foster and Boss,* 1997; *Vanhala and Boss,* 2000, 2002). However, these results assume an isothermal shock front, and it remains to be seen what happens when a more detailed thermodynamical treatment is employed.

2.3. Outcome of Collapse Phase

Once a cloud begins to collapse as a result of ambipolar diffusion or triggering by a shock wave, supersonic inward motions develop and soon result in the formation of an optically thick first core, with a size on the order of 10 AU. This central core is supported primarily by the thermal pressure of the molecular hydrogen gas, while the remainder of the cloud continues to fall onto the core. For a 1 $M_\odot$ cloud, this core has a mass of about 0.0 $M_\odot$ (*Larson,* 1969). Once the central temperature reaches about 2000 K, thermal energy goes into dissociating the hydrogen molecules, lowering the thermal pressure and leading to a second collapse phase, during which the first core disappears and a second, final core is formed at the center, with a radius a few times that of the Sun. This core then accretes mass from the infalling cloud over a timescale of about 1 m.y. (*Larson,* 1969). In the presence of rotation or magnetic fields, however, the cloud becomes flattened into a pancake, and may then fragment into two or more protostars. At this point, we cannot reliably predict what sort of dense cloud core will form in precisely what sort of star or stellar system, much less what sorts of planetary systems will accompany them, but certain general trends are evident.

2.3.1. Multiple fragmentation leading to ejection of single stars. The standard model pertains only to formation of single stars, whereas most stars are known to be members of binary or multiple star systems. There is growing observational evidence that multiple star formation may be the rule, rather than the exception (*Reipurth,* 2000). If so, then it may be that many (or most) single stars like the Sun are formed in multiple protostar systems, only to be ejected soon thereafter as a result of the decay of the multiple system into an orbitally stable configuration (*Bate et al.,* 2002). In that case, the solar nebula would have been subject to strong tidal forces during the close encounters with other protostars prior to its ejection from the multiple system. This hypothesis has not been investigated in detail (but see *Kobrick and Kaula,* 1979). Detailed models of the collapse of magnetic cloud cores, starting from initial conditions defined by observations of molecular clouds, show that while initially prolate cores tend to fragment into binary protostars, initially oblate clouds form multiple protostar systems that are highly unstable and likely to eject single protostars and their disks (*Boss,* 2002a). Surprisingly, magnetic fields were found to enhance the tendency for a collapsing cloud to fragment by helping to prevent the formation of a single central mass concentration of the type assumed to form in the standard model of star formation.

2.3.2. Collapse leading to single star formation. In the case of nonmagnetic clouds, where thermal pressure and rotation dominate, single protostars can result from the collapse of dense cloud cores that are rotating slowly enough to avoid the formation of a large-scale protostellar disk that could then fragment into a binary system (*Boss and Myhill,* 1995). Alternatively, the collapse of an initially strongly centrally condensed (power-law), nonmagnetic cloud leads to the formation of a single central body (*Yorke and Bodenheimer,* 1999). In the more idealized case of collapse starting from a uniform density cloud core, fragmentation does not occur prior to reaching densities high enough that the cloud starts heating above its initial temperature if the cloud is initially nearly spherical (*Tsuribe and Inutsuka,* 1999a,b). Considering that most cloud cores are believed to be supported to a significant extent by magnetic fields, the applicability of all these results is uncertain. In the case of shock-triggered collapse, calculations have shown that weakly magnetic clouds seem to form single protostars when triggering occurs after the core has already contracted toward high central densities (*Vanhala and Cameron,* 1998).

In the case of the nonmagnetic collapse of a spherical cloud (*Yorke and Bodenheimer,* 1999), the protostar that forms is orbited by a protostellar disk with a similar mass. When angular momentum is transported outward by assumed gravitational torques, and therefore mass is transported inward onto the protostar, the amount of mass remaining in the disk (~0.5 $M_{\odot}$) at the end of the calculation is still so large that most of this matter must eventually be accreted by the protostar. Hence the disk at this phase must still be considered a protostellar disk, not a relatively late-phase, protoplanetary disk where any objects that form have some hope of survival in orbit. Thus even in the relatively simple case of nonmagnetic clouds, it is not yet possible to compute the expected detailed structure of a protoplanetary disk, starting from the initial conditions of a dense cloud core. Calculations starting from less-idealized initial conditions, such as a segment of an infinite sheet, suffer from the same limitations (*Boss and Hartmann,* 2001).

Because of the complications of multiple protostar formation, magnetic field support, possible shock-wave triggering, and angular momentum transport in the disk during the cloud infall phase, among others, a definitive theoretical model for the collapse of the presolar cloud has not yet emerged.

2.4. Observations of Star-forming Regions

Observations of star-forming regions have advanced our understanding of the star-formation process considerably in the last few decades. We now can study examples of nearly all phases of the evolution of a dense molecular cloud core into a nearly fully formed star (i.e., the roughly solar-mass T Tauri stars). As a result of being able to observe nearly all phases of the star-formation process, our basic understanding of star formation is relatively mature, at least compared to the planet-formation process, where there are few observations of the phases intermediate between protoplanetary disks and mature planets. Future progress in our understanding of star formation is expected to center on defining the role played by binary and multiple stars and on refining observations of the known phases of evolution. Most (70% or more) stars form in rich clusters of stars where high-mass stars eventually form and shut down the local star-formation process (*Lada and Lada,* 2003). Examples are the Orion nebula cluster and the Carina nebula (Fig. 1). The remaining stars form in less-crowded regions where only low-mass stars form, such as Taurus and Rho Ophiuchus (Fig. 2). One major question for the origin of the solar system is its birthplace: Was it a region of low- or high-mass star formation? As we shall see, this difference can have profound effects on the protoplanetary disk. Meteoritical evidence may hold the key to deciding between these two different modes of star formation, e.g., with respect to the injection of short-lived radioactivities and photolysis at the disk surfaces driven by ultraviolet radiation from nearby massive stars.

2.4.1. Protostellar phases: Class –I, 0, I, II, III objects. Protostellar evolution can be conveniently subdivided into six phases that form a sequence in time. The usual starting point is the precollapse cloud, which collapses to form the first protostellar core, which is then defined to be a Class –I object. The first core collapses to form the final, second core, or Class 0 object, which has a core mass less than that of the infalling envelope. Class I, II, and III objects (*Lada and Shu,* 1990) are defined in terms of their spectral energy distributions at midinfrared wavelengths, where the emission is diagnostic of the amount of cold, circumstellar dust. Class I objects are optically invisible, infrared protostars with so much dust emission that the circumstellar gas mass is on the order of 0.1 $M_{\odot}$ or more. Class II objects have less dust emission, and a gas mass of about 0.01 $M_{\odot}$. Class II objects are usually optically visible, T Tauri stars, where most of the circumstellar gas resides in a disk rather than in the surrounding envelope. Class III objects are weak-line T Tauri stars, with only trace amounts of circumstellar gas and dust. While these classes imply a progression in time from objects with more to less gas emission, the time for this to occur for any given object is highly variable: Some Class III objects appear to be only 0.1 m.y. old, while some Class II objects have ages of several million years, based on theoretical models of the evolution of stellar luminosities and surface temperatures. Evidence for dust disks has been found around even older stars, such as Beta Pictoris, with an age of about 10 m.y., although its disk mass is much smaller than that of even Class III objects. Stars with such "debris disks" are often classified as Class III objects.

Multiple examples of all these phases of protostellar evolution have been found, with the exception of the short-lived Class –I objects, which have not yet been detected. It is noteworthy that observations of protostars and young stars find a higher frequency (twice as high in some young clusters) of binary and multiple systems than is the case for mature stars, implying the orbital decay of many of these

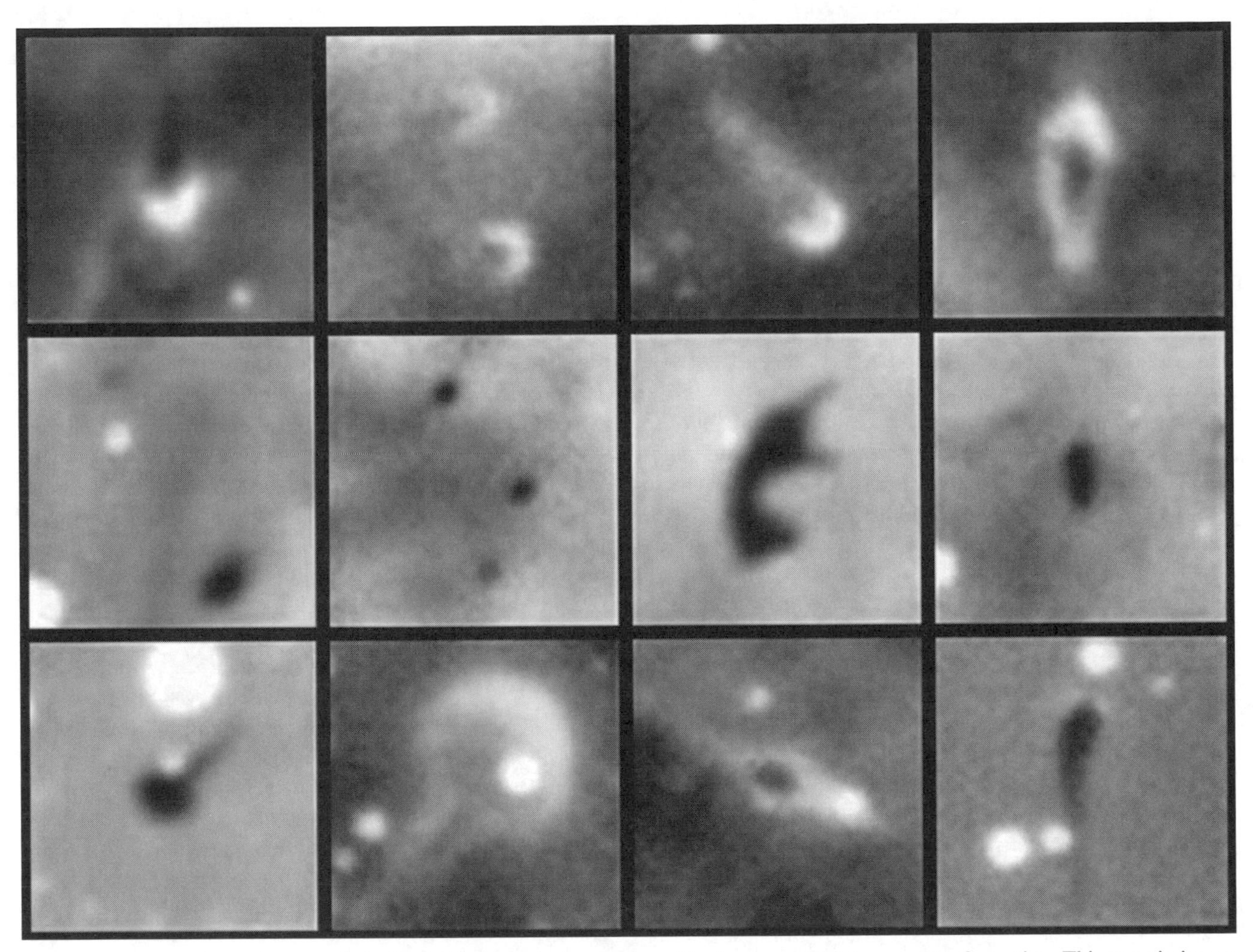

Fig. 1. Suspected protoplanetary disks in the Carina nebula, a region of both low- and high-mass star formation. This mosaic image shows that young stars in the Carina nebula have disks that are similar to the "proplyds" (protoplanetary disks) in Orion, except for being even larger in scale — these disks are on the order of 500 AU in radius. The disks are subjected to an intense flux of ultraviolet radiation from the high-mass stars in Carina, which photoevaporates the outer layers of the disks and blows the debris gas and dust away in a comet-like tail (see also Fig. 3). Image courtesy of University of Colorado/NOAO/AURA/NSF.

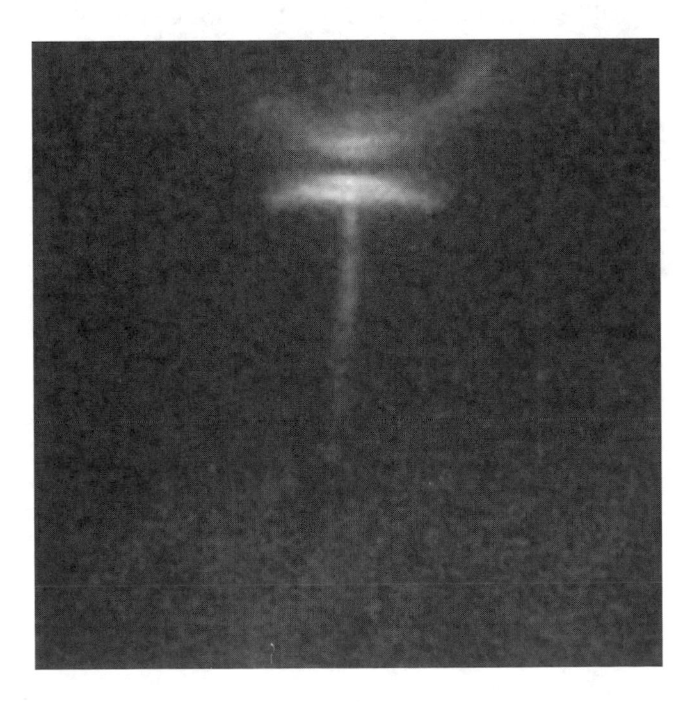

Fig. 2. Hubble Space Telescope image of a typical protoplanetary disk orbiting a solar-type star in a region of low-mass star formation. This Wide Field and Planetary Camera 2 (WFPC2) image shows the edge-on disk orbiting Herbig-Haro 30 (HH 30). The edge-on, flared disk occults the central star, while bipolar outflows from the protostar emerge perpendicular to the disk. Light from the protostar illuminates the bowl-like surfaces of the disk. Region shown is about 400 AU across. Image courtesy of Chris Burrows (STScI), WFPC2 Science Team, and NASA.

young systems (*Reipurth,* 2000; *Smith et al.,* 2000), leading to single stars as ejected members (section 2.3).

2.4.2. Ubiquity of bipolar outflows from earliest phases. A remarkable aspect of young stellar objects is the presence of strong molecular outflows for essentially all young stellar objects, even the Class 0 objects. This means that at the same time that matter is still accreting onto the protostar, it is also losing mass through a vigorous wind directed in a bipolar manner in both directions along the presumed rotation axis of the protostar/disk system (Fig. 2). In fact, the energy needed to drive this wind appears to be derived from mass accretion by the protostar, as observed wind momenta are correlated with protostellar luminosities and with the amount of mass in the infalling envelope (*Andrè,* 1997).

There are two competing mechanisms for driving bipolar outflows, both of which depend on magnetic fields to sling ionized gas outward and to remove angular momentum from the star/disk system. One mechanism is the X-wind model (*Shu et al.,* 2000), where coronal winds from the central star and from the inner edge of the accretion disk join together to form the magnetized X-wind, launched from an orbital radius of a few stellar radii. The other mechanism is a disk wind (*Königl and Pudritz,* 2000), launched from the surface of the disk over a much larger range of distances, from less than 1 AU to as far away as 100 AU or so. In both mechanisms, centrifugal support of the disk gas makes it easier to launch this material outward, and bipolar flows develop in the directions perpendicular to the disk, because the disk forces the outflow into these preferred directions. Toroidal magnetic fields seem to be required in order to achieve the high degree of collimation of the observed outflows, as a purely poloidal magnetic field would launch the wind at a significant angle away from the rotation axis of the protostar/disk system. Because it derives from radii deeper within the star's gravitational potential well, an X-wind is energetically favored over a disk wind. However, observations show that during FU Orionis-type outbursts in young stars, when stellar luminosities increase by factors of 100 in a few years, mass is added onto the central star so rapidly that the X-wind region is probably crushed out of existence, implying that the strong outflows that still occur during these outbursts must be caused by an extended disk wind (*Hartmann and Kenyon,* 1996).

All T Tauri stars are believed to experience FU Orionis outbursts [however, see *Bell et al.* (2000) for a contrary opinion], so disk winds may be the primary driver of bipolar outflows, at least in the early FU Orionis phase of evolution. There is a strong correlation between the amount of mass available for accretion onto the disks and the amount of momentum in the outflow (*Bontemps et al.,* 1996), suggesting that disk mass accretion is directly related to outflow energetics. It is unclear at present what effect an X-wind or a disk wind would have on the planet-formation process, beyond being responsible for the loss of energy (and angular momentum in the latter case), as the winds are thought to be launched either very close to the protostar in the former case, or from the disk's surface in the latter case. However, they can potentially provide the energy needed for the formation of some of the earliest-formed solids in the solar nebula, such as the refractory inclusions (Ca-Al-rich inclusions, or CAIs) and the chondrules that require locally high temperatures in the nebula (*Shu et al.,* 1996; *Connolly et al.,* 2006).

2.4.3. Lifetimes of circumstellar disks. The most robust constraints on the timescale for disk removal come from astronomical observations of young stars. Because of the limited spatial resolution of current interferometric arrays, molecular hydrogen gas or tracer species such as carbon monoxide cannot be mapped at scales of less than about 10 AU in most disks. Instead, the presence of the gaseous portion of a disk is inferred indirectly by the presence of ongoing mass accretion from the disk onto the star. This accretion leads to enhanced emission in the star's Hα line, which is then a diagnostic of the presence of disk gas. Observations of Hα emission in the Orion OB1 association have shown that Hα emission drops toward zero once the stars reach an age of a few 10^6 yr (*Briceño et al.,* 2001). The presence of the dusty portion of disks is signaled by excess infrared emission, derived from dust grains in the innermost region of the disk. Again, observations imply that the dust disk largely disappears by ages of a few 10^6 yr (*Briceño et al.,* 2001). (This does not mean that solids are not left in orbit around these stars: Particles much larger than 1 μm would be undetectable.) *Haisch et al.* (2001) studied a number of young clusters and found that the frequencies of disks around the young cluster stars dropped sharply after an age of about 3×10^6 yr. In the Orion nebula cluster, ages for the outer disks are thought to be even shorter, as a result of photoevaporation caused by UV irradiation by newly formed, massive stars (Figs. 1 and 3), although the inner disks (inside about 10 AU) would be largely unaffected by this process initially. Once the outer disk and infalling envelope is removed, however, there would be no further source of replenishment of the inner disk as it is accreted by the protostar. Hence inner disk lifetimes should be shorter in regions of high-mass star formation (e.g., Orion, Carina) than in regions of low-mass star formation (e.g., Taurus, Ophiuchus). Portions of some disks last for as long as 10^7 yr; Beta Pictoris is about 10^7 yr old and has a remnant dust disk that seems to be replenished by collisions between the members of an unseen population of orbiting bodies.

Constraints on the lifetime of the solar nebula have been derived from isotopic studies of early solar system objects (CAIs and chondrules) that are considered to be products of solar nebula processes. The prevalence of ^{26}Al with nearly uniform initial abundance in CAIs (*MacPherson et al.,* 1995) and the much lower abundance of ^{26}Al in chondrules have been interpreted as implying that CAI formation in the nebula preceded chondrule formation by at least 10^6 yr (*Russell et al.,* 1996; *Kita et al.,* 2000; *Huss et al.,* 2001). This interpretation has been bolstered by the U-Pb dating of CAIs and chondrules that suggests a difference of $\sim 2 \times 10^6$ yr in absolute ages between these two sets of

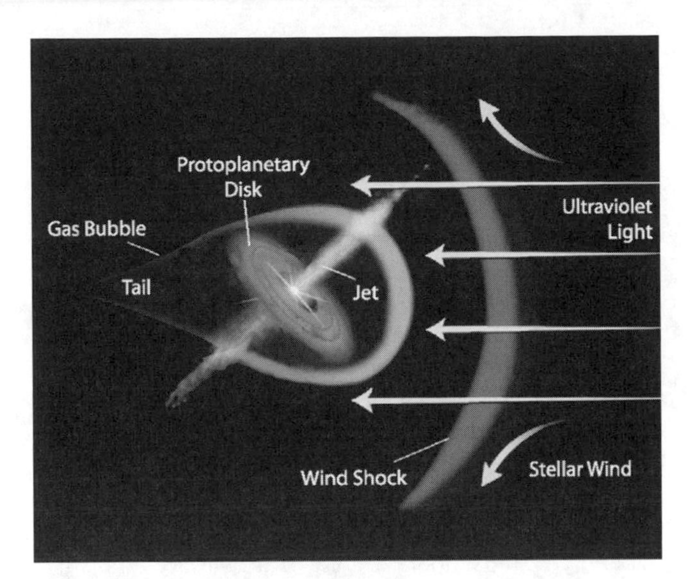

Fig. 3. Schematic diagram of the process of photoevaporation of the protoplanetary disks in regions of high-mass star formation, such as the Orion nebula cluster and the Carina nebula (see Fig. 1). Image courtesy of Space Telescope Science Institute.

objects (*Amelin et al.,* 2002). Furthermore, the presence of a spread in the initial ^{26}Al abundance in chondrules also requires that the nebular processes leading to chondrule formation lasted more than 10^6 yr. Although recent observations (*Bizzarro et al.,* 2004) suggest that some chondrules in the Allende meteorite formed contemporaneously with CAIs, formation of other chondrules in the same meteorite continued for at least 1.5×10^6 yr. Assuming that both the CAIs and chondrules formed within the solar nebula [see *Cameron* (2003) for a contrary view], the isotopic evidence then implies that the solar nebula had a lifetime of at least a few 10^6 yr (*Russell et al.,* 2006), similar to the ages inferred for disks around young cluster stars. The simple picture of the solar nebula being removed by a spherically symmetric T Tauri wind has long since been supplanted by the realization that young stars have directed, bipolar outflows that do not sweep over most of the disk. However, mature stars like the Sun do have approximately isotropic winds, so there must be some transition phase where the bipolar star/disk wind evolves into a more spherically symmetric stellar wind. Presumably this enhanced stellar wind would eventually scour any gas and dust from the system left over from the disk accretion phase, when most of the disk gas and dust is accreted by the growing central protostar.

2.4.4. Star-forming regions. While formation as a single star in an isolated, dense cloud core is usually imagined for the presolar cloud, in reality there are very few examples of isolated star formation. Most stars form in regions of high-mass star formation, similar to Orion, with a smaller fraction forming in smaller clusters of low-mass stars, like Taurus or Ophiuchus. The radiation environment differs considerably between these two extremes, with Taurus being relatively benign, and with Orion being flooded with ultraviolet (UV) radiation once massive stars begin to form (*Hollenbach et al.,* 2000). Even in Taurus, though, individual young stars emit UV and X-ray radiations at levels considerably greater than mature stars (*Feigelson and Montmerle,* 1999). Ultraviolet radiation from the protosun has been suggested as a means of removing the residual gas from the outermost solar nebula (i.e., beyond about 10 AU) through photoevaporation of hydrogen atoms (*Shu et al.,* 1993), a process estimated to require about 10^7 yr.

In regions of high-mass star formation, the low-mass stars must form first, because once the higher-mass stars begin forming, their intense UV radiation heats the remaining molecular gas and drives it away from the newly formed star cluster. This UV radiation also removes the outermost disk gas from any protoplanetary disks that pass close to the high-mass stars, thereby limiting the lifetimes of these disks and potentially the possibilities for planetary system formation. However, UV radiation from massive stars has been invoked in a positive sense, as a means to photoevaporate the gas in the outer solar nebula (beyond about 10 AU) and then to form the ice giant planets, by photoevaporating the gaseous envelopes of the outermost gas giant protoplanets (*Boss et al.,* 2002). The Carina "proplyds" (Fig. 1) imply a considerably different evolution scenario in the outer disk than is the case for isolated disks in regions like Taurus (Fig. 2), where the disks appear to be classic cases of isolated, symmetric, circumstellar disks with perpendicular outflows, similar to that envisioned in simple theoretical models.

3. EARLY EVOLUTION OF THE SOLAR NEBULA

On theoretical grounds, even an initially highly centrally condensed (i.e., power-law density profile) cloud core is likely to collapse to form a protostar surrounded by a fairly massive (~0.5 $M_\odot$) protostellar disk and envelope. Currently available observations of disks around young stars (e.g., *Dutrey et al.,* 2004) imply that at early ages, disk masses are not always a significant fraction (i.e., 10%) of the protostar's mass. However, these observations are unable to probe the innermost regions (i.e., within 50 AU or so) because of limited spatial resolution, so the true amount of disk mass at early phases remains uncertain. Nevertheless, the expectation is that protostellar disks must somehow transport most of their mass inward to be accreted by the protostar, eventually evolving into protoplanetary disks, where planetary bodies should be able to form and survive their subsequent interactions with the disk. This process occurs even as collapse of the presolar cloud onto the growing disk continues, adding significant amounts of mass and angular momentum. Observational evidence is beginning to emerge for decreasing disk masses as protostars become older (*Eisner and Carpenter,* 2003). The transition point from a protostellar disk to a protoplanetary disk is not clear, and the physical mechanisms responsible for disk evolution in either of these two phases remain uncertain, although progress seems to have been made in ruling out several proposed mechanisms.

3.1. Angular Momentum Transport Mechanisms

The basic theory of the evolution of an accretion disk can be derived by assuming that there is some physical mechanism operating that results in an effective viscosity of the gas. Because the intrinsic molecular viscosity of hydrogen gas is far too small to have an appreciable effect on disk evolution in a reasonable amount of time (i.e., within 10^7 yr, given observed disk lifetimes), theorists have sought other sources for an effective viscosity, such as turbulence. In a fully turbulent flow, the effective viscosity can be equal to the molecular viscosity multiplied by a large factor: the ratio of the Reynolds number of the disk (about 10^{10}) to the critical Reynolds number for the onset of turbulence (about 10^3), or a factor of about 10^7. (The Reynolds number is the dimensionless number equal to the product of a mean distance times a mean velocity, divided by the kinematic viscosity of the system.) Under very general conditions, it can be shown (*Lynden-Bell and Pringle,* 1974) that a viscous disk will evolve in such a manner as to transport most of its mass inward, thereby becoming more tightly gravitationally bound, and minimizing the total energy of the system. In order to conserve angular momentum, this means that angular momentum must be transported outward along with a small fraction of the mass, so that the accretion disk expands outside some radius. The loss of significant angular momentum by centrifugally launched winds somewhat relieves this need for the accretion disk to expand; this additional angular momentum sink was not recognized when the theory was first developed (note, however, that in the case of an X-wind, relatively little angular momentum can be lost by the X-wind). While the basic physics of a viscous accretion disk is fairly well developed, the physical mechanism(s) responsible for disk evolution remain contentious.

3.1.1. Hydrodynamic turbulence. Given the high Reynolds number of a protoplanetary disk, one might expect that a turbulent cascade of energy would occur and result in an effective turbulent viscosity that might be sufficient to drive disk evolution. However, because of the strong differential rotation in a Keplerian disk, a high Reynolds number is not a sufficient condition for fully developed turbulence. Instead, the Rayleigh criterion, which applies to rotating fluids but is not strictly applicable to the solar nebula, suggests that Keplerian disks are stable with respect to turbulence. (The Rayleigh criterion deals with the stability of rotating, incompressible, inviscid fluids to turbulence.)

3.1.1.1. Vertical convectively-driven turbulence: While differential rotation may inhibit convective motions in the radial direction in a disk, motions parallel to the rotation axis are relatively unaffected by rotation. In a disk where heat is being generated near the midplane, such as by dissipation of turbulent motions involved in driving mass accretion onto the protostar, and where dust grains are the dominant source of opacity, the disk is likely to be unstable to convective motions in the vertical direction, which carry the heat away from the disk's midplane and deposit it close to the disk's surface, where it can be radiated away. Convective instability was conjectured to lead to sufficiently robust turbulence for the resulting turbulent viscosity to be large enough to drive disk evolution (*Lin and Papaloizou,* 1980), a seemingly attractive, self-consistent scenario that has motivated much of the work on viscous evolution of the solar nebula. However, three-dimensional hydrodynamical models of vertically convectively unstable disks have shown that the convective cells that result are sheared by differential rotation to such an extent that the net transport of angular momentum is very small, and may even be in the wrong direction (see *Stone et al.,* 2000). As a result, convectively-driven disk evolution does not seem to be a major driver. In addition, heating of the surface of the disk by radiation from the central protostar will also act to suppress vertical convection. However, strong evidence of the importance of convection for cooling protoplanetary disks was presented by *Boss* (2004a).

3.1.1.2. Rotational shear-induced turbulence: It has also been suggested that finite amplitude (nonlinear) disturbances to Keplerian flow could result in a self-sustaining shear instability that would produce significant turbulence (*Dubrulle,* 1993). However, when three-dimensional hydrodynamical models were again used to investigate this possibility, it was found that the initially assumed turbulent motions decayed rather than grew (*Stone et al.,* 2000). Evidently purely hydrodynamical turbulence can neither grow spontaneously nor be self-sustained upon being excited by an external perturbation. However, it has been claimed recently that there may exist other possible paths for hydrodynamic shear to produce turbulence (*Chagelishvili et al.,* 2003), such as the bypass concept, where finite-amplitude initial perturbations undergo transient growth until they are large enough to produce positive feedback.

3.1.1.3. Global baroclinic instability-driven turbulence: In spite of these discouraging results for hydrodynamical turbulence, another possibility remains and is under investigation (*Klahr and Bodenheimer,* 2003; *Klahr,* 2004), that of a global baroclinic instability. In this mechanism, turbulence results in essence from steep temperature gradients in the radial direction, which then battle centrifugal effects head-on. Three-dimensional hydrodynamical models imply that this mechanism can drive inward mass transport and outward angular momentum transport, as desired. However, similar models by H. Klahr (personal communication, 2003) with a different numerical code (3D-ZEUS) have reached different conclusions, so the situation regarding this mechanism is unclear.

3.1.1.4. Rossby waves: Rossby waves occur in planetary atmospheres as a result of shearing motions and can produce large-scale vortices such as the Great Red Spot on Jupiter. Rossby waves have been proposed to occur in the solar nebula as a result of Keplerian rotation coupled with a source of vortices. While prograde rotation (cyclonic) vortices are quickly dissipated by the background Keplerian flow, retrograde (anti-cyclonic) vortices are able to survive for longer periods of time (*Godon and Livio,* 1999). Rossby waves have been advanced as a significant source of angular momentum transport in the disk (*Li et al.,* 2001). Rossby vortices could serve as sites for concentrating dust particles, but the difficulty in forming the vortices in the first place, coupled with their eventual decay, makes this otherwise attractive idea somewhat dubious (*Godon and Livio,* 2000). In addition, the restriction of these numerical studies to thin, two-dimensional disk models, where refraction of the waves away from the midplane is not possible, suggests that in a fully three-dimensional calculation, Rossby waves may be less vigorous than in the thin disk calculations (*Stone et al.,* 2000). Recent numerical work (*Davis,* 2002) suggests that vortices have little long-term effect on the disk.

3.1.2. Magneto-rotational-driven turbulence. While a purely hydrodynamical source for turbulence has not yet been demonstrated, the situation is much different when magnetohydrodynamical (MHD) effects are considered in a shearing, Keplerian disk. In this case, the Rayleigh criterion for stability can be shown to be irrelevant: Provided only that the angular velocity of the disk decreases with radius, even an infinitesimal magnetic field will grow at the expense of the shear motions, a fact that had been noted by *Chandrasekhar* (1961) but was largely ignored until it was rediscovered some 30 years later.

3.1.2.1. Balbus-Hawley instability: *Balbus and Hawley* (1991) pointed out that in the presence of rotational shear, even a small magnetic field will grow on a very short timescale. The basic reason is that magnetic field lines can act like rubber bands, linking two parcels of ionized gas. The parcel that is closer to the protosun will orbit faster than the other, increasing its distance from the other parcel. This leads to stretching of the magnetic field lines linking the parcel, and so to a retarding force on the forward motion of the inner parcel. This force transfers angular momentum from the inner parcel to the outer parcel, which means that the inner parcel must fall farther inward toward the protosun, increasing its angular velocity, and therefore leading to even more stretching of the field lines and increased magnetic forces. Because of this positive feedback, extremely rapid growth of an infinitesimal seed field occurs. Consequently, the magnetic field soon grows so large and tangled that its subsequent turbulent evolution must be computed with a fully nonlinear, multidimensional MHD code.

Three-dimensional MHD models of a small region in the solar nebula (*Hawley et al.,* 1995) have shown that, as expected, a tiny seed magnetic field soon grows and results in a turbulent disk where the turbulence is maintained by the magnetic instability. In addition, the magnetic turbulence results in a net outward flow of angular momentum, as desired. The magnetic field grows to a quasi-steady-state value and then oscillates about that mean value, depending on the assumed initial field geometry, which is large enough to result in relatively vigorous angular momentum transport. While promising, these studies of the magneto-rotational instability (MRI) are presently restricted to small regions of the nebula, and the global response of the disk to this instability remains to be determined.

3.1.2.2. Ionization structure and layered accretion: Magneto-rotational instability is a powerful phenomenon, but is limited to affecting nebula regions where there is sufficient ionization for the magnetic field, which is coupled only to the ions, to have an effect on the neutral atoms and molecules. The MRI studies described above all assume ideal MHD, i.e., a fully ionized plasma, where the magnetic field is frozen into the fluid. At the midplane of the solar nebula, however, the fractional ionization is expected to be quite low in the planetary region. Both ambipolar diffusion and resistivity (ohmic dissipation) are effective at limiting magnetic field strengths and suppressing MRI-driven turbulence, but a fractional ionization of only about 1 ion per 10^{12} atoms is sufficient for MRI to proceed in spite of ambipolar diffusion and ohmic dissipation. Close to the protosun, disk temperatures are certainly high enough for thermal ionization to create an ionization fraction greater than this, and thus to maintain full-blown MRI turbulence. Given that a temperature of at least 1400 K is necessary, MRI instability may be limited to the innermost 0.2 AU or so in quiescent phases, or as far out as about 1 AU during rapid mass accretion phases (*Boss,* 1998; *Stone et al.,* 2000).

At greater distances, disk temperatures are too low for thermal ionization to be effective. Cosmic rays were thought to be able to ionize the outer regions of the nebula, but the fact that bipolar outflows are likely to be magnetically driven means that cosmic rays may have a difficult time reaching the disk midplane (*Dolginov and Stepinski,* 1994). However, the coronae of young stars are known to be prolific emitters of hard X-rays, which can penetrate the bipolar outflow and reach the disk surface at distances of about 1 AU or so, where they are attenuated (*Glassgold et al.,* 1997). As a result, the solar nebula is likely to be a layered accretion disk (*Gammie,* 1996), where MRI turbulence results in inward mass transport within thin, lightly ionized surface layers, while the layers below the surface do not participate in MRI-driven transport. Thus the bulk of the disk, from just below the surface to the midplane, is expected to be a magnetically dead zone (*Fleming and Stone,* 2003). Layered accretion is thought to be capable of driving mass inflow at a rate of $\sim 10^{-8}$ $M_\odot$/yr, sufficient to account for observed mass accretion rates in quiescent T Tauri stars (*Calvet et al.,* 2000). Mass accretion rates can vary from less than $\sim 10^{-9}$ $M_\odot$/yr to greater than $\sim 10^{-6}$ $M_\odot$/yr for stars with ages from 10^5 to 10^7 yr, with some evidence for mass accretion rates tapering off for the older stars.

3.1.3. Gravitational torques in a marginally unstable disk. The remaining possibility for large-scale mass transport in the solar nebula is gravitational torques. The likelihood that much of the solar nebula was a magnetically dead zone where MRI transport was ineffective leads to the suggestion that there might be regions where inward MRI mass transport would cease, leading to a local pile-up of mass, which might then cause at least a local gravitational instability of the disk (*Gammie,* 1996). In addition, there is observational and theoretical evidence that protostellar disks tend to start their lives with sufficient mass to be gravitationally unstable in their cooler regions, leading to the formation of nonaxisymmetric structure and hence the action of gravitational torques, and that these torques may be the dominant transport mechanism in early phases of evolution (*Yorke and Bodenheimer,* 1999).

In order for gravitational torques to be effective, a protostellar disk or the solar nebula must be significantly nonaxisymmetric, e.g., threaded by clumps of gas, or by spiral arms, much like a spiral galaxy. In that case, trailing spiral structures, which form inevitably as a result of Keplerian shear, will result in the desired outward transport of angular momentum. This is because in a Keplerian disk, an initial bar-shaped density perturbation will be sheared into a trailing spiral arm configuration. The inner end of the bar rotates faster than the outer end and therefore moves ahead of the outer end. Because of the gravitational attraction between the inner end and the outer end, the inner end will have a component of this gravitational force in the backward direction, while the outer end will feel an equal and opposite force in the forward direction. The inner end will thus lose orbital angular momentum, while the outer end gains this angular momentum. As a result, the inner end falls closer to the protosun, while the outer end moves farther away, with a net outward transport of angular momentum.

3.1.3.1. Rapid mass and angular momentum transport: Models of the growth of nonaxisymmetry during the collapse and formation of protostellar disks show that large-scale bars and spirals can form with the potential to transfer most of the disk angular momentum outward on timescales as short as 10^3–10^5 yr (*Boss,* 1989), sufficiently fast to allow protostellar disks to transport the most of their mass inward onto the protostar and thereby evolve into protoplanetary disks.

Early numerical models of the evolution of a gravitationally unstable disk (e.g., *Cassen et al.,* 1981) suggested that a disk would have to be comparable in mass to the central protostar in order to be unstable, i.e., in order to have the Toomre Q disk parameter be close to unity (*Toomre,* 1964). [Q is proportional to the product of the disk sound speed times the angular velocity, divided by the surface density of the gas; low Q means that the disk is massive enough to overcome the stabilizing effects of thermal gas pressure and shearing motions (Keplerian rotation).] Gravitational instability could then occur in protostellar, but not protoplanetary, disks. Analytical work on the growth of spiral density waves implied that for a 1 $M_\odot$ star, gravitational instability could occur in a disk with a mass as low as 0.19 $M_\odot$ (*Shu et al.,* 1990). Three-dimensional hydrodynamical models have shown that vigorous gravitational instability can occur in a disk with a mass of 0.1 $M_\odot$ or even less, in orbit around a 1 $M_\odot$ star (*Boss,* 2000), because of the expected low midplane temperatures (about 30 K) in the outer disk implied by cometary compositions (*Kawakita et al.,* 2001) and by observations of disks (*D'Alessio et al.,* 2001). Similar models with a complete thermodynamical treatment (*Boss,* 2002b), including convective transport and radiative transfer, show that a marginally gravitationally unstable solar nebula develops a robust pattern of clumps and spiral arms, persisting for many disk rotation periods, and resulting in episodic mass accretion rates onto the central protosun that vary from ~10^{-7} $M_\odot$/yr to as high as ~10^{-3} $M_\odot$/yr. The latter rates are more than high enough to account for FU Orionis outbursts (*Bell et al.,* 2000).

Because angular momentum transport by a strongly gravitationally unstable disk is so rapid, it is unlikely that protostellar or protoplanetary disks are ever strongly gravitationally unstable, because they can probably evolve away from such a strongly unstable state faster than they can be driven into it by, e.g., accretion of more mass from an infalling envelope or radiative cooling. As a result, it is much more likely that a disk will approach gravitational instability from a marginally unstable state (*Cassen et al.,* 1981). Accordingly, models of gravitationally unstable disks have focused almost exclusively on marginally gravitationally unstable disks (e.g., *Boss,* 2000), where primarily the outer disk, beyond about 5 AU, participates in the instability. Inside about 5 AU, disk temperatures appear to be too high for an instability to grow there, although these inner regions may still be subject to shock fronts driven by clumps and spiral arms in the gravitationally unstable, outer region. One-armed spiral density waves can propagate right down to the stellar surface.

3.1.3.2. Global process vs. local viscosity: Gravitational forces are intrinsically global in nature, and their effect on different regions of the nebula can be expected to be highly variable in both space and time. On the other hand, turbulent viscosity is a local process that is usually assumed to operate more or less equally efficiently throughout a disk. As a result, it is unclear if gravitational effects can be faithfully modeled as a single, effective viscosity capable of driving disk evolution in the manner envisioned by *Lynden-Bell and Pringle* (1974). Nevertheless, efforts have been made to try to quantify the expected strength of gravitational torques in this manner (*Lin and Pringle,* 1987). Three-dimensional models of marginally gravitationally unstable disks imply that such an effective viscosity is indeed large and comparable to that in MRI models (*Laughlin and Bodenheimer,* 1994).

3.2. Models of Solar Nebula Evolution

Given an effective source of viscosity, in principle the time evolution of the solar nebula can be calculated in great detail, at least in the context of the viscous accretion disk model. The strength of an effective viscosity is usually quantified by the α parameter. α is often defined in various ways, but typically α is defined to be the constant that, when multiplied by the sound speed and the vertical scale height of the disk (two convenient measures of a typical velocity and length scale), yields the effective viscosity of the disk (*Lynden-Bell and Pringle,* 1974). Three-dimensional MHD models of the MRI imply an typical MRI α of about 0.005–0.5 (*Stone et al.,* 2000). Similarly, three-dimensional models of marginally gravitationally unstable disks imply an α of about 0.03 (*Laughlin and Bodenheimer,* 1994). Steady mass accretion at the low rates found in quiescent T Tauri stars requires an α of about 0.01 (*Calvet et al.,* 2000), in rough agreement with these estimates. Once planets have formed and become massive enough to open gaps in their surrounding disks, their orbital evolution becomes tied to that of the gas. As the gaseous disk is transported inward by viscous accretion, these planets must also migrate inward. The perils of orbital migration for planetary formation and evolution are addressed in *Ward and Hahn* (2000) and *Lin et al.* (2000). Here we limit ourselves to considering the evolution of dust and gas prior to the formation of planetary-sized bodies.

3.2.1. Viscous accretion disk models. The generation of viscous accretion disk models was an active area of research during the period when convective instability was believed to be an effective source of viscosity. *Ruden and Pollack* (1991) constructed models where convective instability was assumed to control the evolution, so that in regions where the disk became optically thin and thus convectively stable, the effective viscosity vanished. Starting with an α of about 0.01, they found that disks evolved for about 10^6 yr before becoming optically thin, often leaving behind a disk with a mass of about 0.1 $M_\odot$. Midplane temperatures at 1 AU dropped precipitously from about 1500 K initially to about 20 K when convection ceased and the disk was optically thin at that radius. Similarly dramatic temperature drops occur throughout the disk in these models, and the outer regions of the models eventually became gravitationally unstable as a result. Given that convective instability is no longer considered to be a possible driver of disk evolution, the *Ruden and Pollack* (1991) models are interesting, but not likely to be applicable to the solar nebula. Unfortunately, little effort has gone into generating detailed viscous accretion models in the interim: The theoretical focus seems to have been more on the question of determining which mechanisms are contenders for disk evolution than on the question of the resulting disk evolution. In particular, the realization that the MRI mechanism is likely to have operated only in the magnetically active surface layers of the disk, and not in the magnetically dead bulk of the disk, presents a formidable technical challenge for viscous accretion disk models, which have usually been based on the assumption that the nebula can be represented by a thin, axisymmetric disk (e.g., *Ruden and Pollack,* 1991), greatly simplifying the numerical solution. The need for consideration of the vertical as well as the radial structure of the disk, and possibly the azimuthal (nonaxisymmetric) structure as well, points toward the requirement of a three-dimensional magnetohydrodynamical calculation of the entire disk. Such a calculation has not been performed, and even the MRI calculations performed to date on small regions of a disk can only be carried forward in time for a small fraction of the expected lifetime of the disk.

Some progress has been made in two-dimensional hydrodynamical models of a thick disk evolving under the action of a globally defined α viscosity, representing the effects of torques in a marginally gravitationally unstable disk (*Yorke and Bodenheimer,* 1999), but in these models the evolution eventually slows down and leaves behind a fairly massive protostellar disk after 10^7 yr, with a radius on the order of 100 AU.

One aspect of particular interest about viscous accretion disk models is the evolution of solid particles, both in terms of their thermal processing and in terms of their transport in the nebula. Interstellar dust grains are small enough (submicrometer-sized) to remain well-coupled to the gas, so they will move along with the gas. During this phase, the gas and dust may undergo trajectories that are outward at first, as the disk accretes matter from the infalling envelope and expands by outward angular momentum transport, followed by inward motion once accretion stops and the disk continues to accrete onto the protostar (*Cassen,* 1996). Once collisional coagulation gets underway and grain growth begins, solid particles begin to move with respect to the gas, suffering gas drag and additional radial migration as a result (*Weidenschilling,* 1988, 2004; *Weidenschilling and Cuzzi,* 2006; *Cuzzi and Weidenschilling,* 2006), with the peak velocity with respect to the gas occurring for roughly meter-sized solids. However, this does not mean that all growing solids will be lost to monotonic inward migration toward the protosun, as the mixing and transport processes in a marginally gravitationally solar nebula are sufficiently robust to mix solids over distances of several AU within 10^3 yr (*Boss,* 2004b). A strong source of generic α turbulent viscosity would contribute to this mixing. Particles that are centimeter-sized or smaller will be effectively tied to the gas during this process. In addition, in a disk with spiral arms, the meter-sized solids are forced by gas drag to move toward the centers of the arms rather than toward the protosun, thereby enhancing their chances for further growth to sizes large enough to become completely decoupled from the gas (*Haghighipour and Boss,* 2003).

3.2.1.1. Volatility patterns and transport in the inner solar system: The bulk compositions of the bodies in the inner solar system show a marked depletion of volatile elements compared to the solar composition. *Cassen* (2001) has shown that these volatile depletions can be explained as a result of the condensation of hot gases and coagulation of the resulting refractory dust grains into solids that are decoupled from the gas through the rapid growth of kilometer-sized planetesimals. The volatile elements remain in gaseous form at these temperatures, and so avoid being incorporated into the planetesimals that will eventually form the terrestrial planets and asteroids. In order for this process to work, significant regions of the nebula must have been hot enough at the midplane to keep volatiles in the gaseous form, a situation that would characterize the nebula when mass accretion rates were on the order of ~10^{-7} $M_\odot$/yr or less. The volatile gases would then be removed from the terrestrial planet region by viscous accretion onto the protosun. The postulated rapid growth from dust grains to kilometer-sized bodies in ~10^5 yr required by this scenario appears to be possible (*Woolum and Cassen,* 1999). *Cuzzi and Weidenschilling* (2006) argue, however, that such rapid growth may not be possible in a nebula with even a low level of generic turbulence.

3.2.2. Clump formation in a marginally gravitationally unstable disk. Given the apparent limitation of MRI-driven accretion to the surfaces of protoplanetary disks, it would appear that gravitational torques may have to be responsible for the evolution in the bulk of the disk. In addition, there are strong theoretical reasons why gravitational torques may be effective, including the difficulty in forming the gas giant planets by the conventional means of core accretion. The standard model for Jupiter formation by core accretion envisions a nebula that has a surface density of solids high enough for a solid core to form within a few 10^6 yr through runaway accretion (*Pollack et al.,* 1996) and then accrete a massive gaseous envelope. However, the gas in such a nebula is likely to be marginally gravitationally unstable, a situation that could result in the rapid formation of gas giant planets in ~10^3 yr by the formation of self-gravitating clumps of gas and dust (*Boss,* 1997, 2000, 2002b). Deciding between these two differing modes of giant planet formation is difficult based on present observations — dating the epoch of giant-planet formation would help to constrain the problem. The ongoing census of extrasolar planetary systems seems to be showing that Jupiter-mass planets are quite common (*Marcy et al.,* 2000), so the giant planet formation mechanism appears to be a robust one. One extrasolar planet survey has found evidence that roughly 15% of nearby stars like the Sun are orbited by Jupiter-mass planets with orbital periods less than 3 yr, while another 25% show evidence for having longer-period, Jupiter-like companions (A. Hatzes, personal communication, 2004). Thus at least 40% of solar-type stars might be orbited by gas giant planets.

In their pioneering study of a marginally unstable disk, *Laughlin and Bodenheimer* (1994) found strong spiral arm formation but no clumps, presumably in large part as a result of the limited spatial resolution that was computationally possible at the time (up to 25,000 particles in a smoothed particle hydrodynamics code). *Boss* (2000) has shown that when 10^6 or more grid points are included in a finite-differences calculation, three-dimensional hydrodynamic models of marginally gravitationally unstable disks demonstrate the persistant formation of self-gravitating clumps, although even these models do not appear to have sufficient spatial resolution to follow the high-density clumps indefinitely in time. Regardless of whether or not such disk instability models can lead to gas-giant-planet formation, the likelihood that the solar nebula was at least episodically marginally gravitationally unstable has important implications for cosmochemistry.

3.2.2.1. Chondrule formation in nebular shock fronts: Perhaps the most well-known, unsolved problem in cosmochemistry is the question of the mechanism whereby dust grain aggregates were thermally processed to form chondrules, which are abundant in all the chondritic meteorites, and the somewhat rare CAIs, present primarily in primitive meteorites. Chondrule compositions and textures in particular require rapid heating and somewhat slower cooling for their explanation; a globally hot nebula is inconsistent with these requirements (*Cassen,* 2001). A wide variety of mechanisms has been proposed and generally discarded, but theoretical work seems to suggest that chondrule formation in nebular shock fronts is a very plausible mechanism (*Desch and Connolly,* 2002; *Connolly et al.,* 2006).

In a marginally gravitationally unstable nebula, clumps and spiral arms at about 8 AU will drive one-armed spiral arms into the inner nebula, which at times results in shock fronts oriented roughly perpendicular to the orbits of bodies in the asteroidal region (*Boss and Durisen,* 2005; Fig. 4). Because of the tendency toward co-rotation in self-gravitating structures, this will lead to solids encountering a shock front at speeds as high as 10 km/s, sufficiently high to result in postshock temperatures of about 3000 K. Detailed one-dimensional models of heating and cooling processes in such a shock front have shown that shock speeds around 7 km/s are optimal for matching chondrule cooling rates and

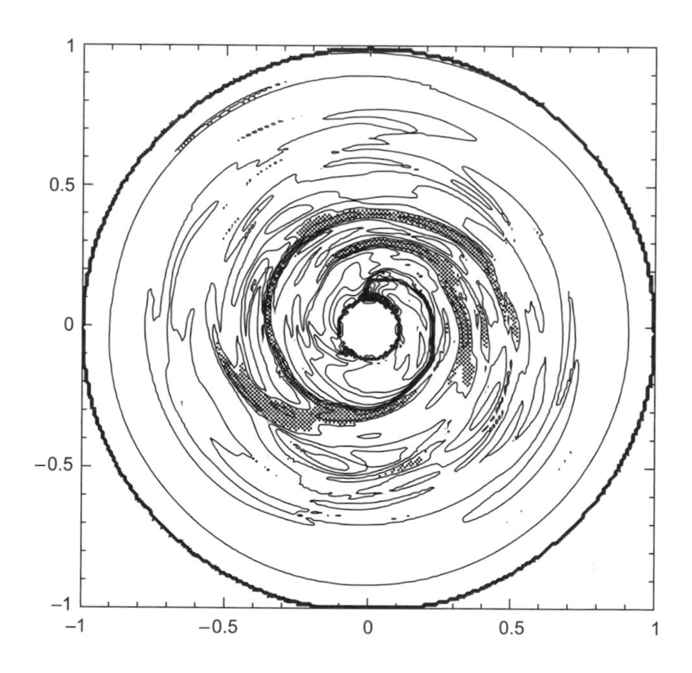

Fig. 4. Density contours in the equatorial plane of a marginally gravitationally unstable protoplanetary disk with a mass of 0.09 $M_\odot$ after 252 yr of evolution (*Boss and Durisen,* 2005). Region shown has an overall radius of 20 AU, while the innermost boundary is at 2 AU. A 1 $M_\odot$ protostar lies at the center of the disk and accretes mass from the disk as it evolves. Cross-hatching defines regions with midplane density above 10^{-10} g cm^{-3}. At 12 o'clock, it can be seen that the spiral structure in the disk has led to the formation of a transient shock front between 2 AU and about 3.5 AU, where solids on Keplerian orbits would strike the gaseous shock front with a relative velocity of ~10 km s^{-1}, sufficient to thermally process solids into chondrules (*Desch and Connolly,* 2002). Image courtesy of A. P. Boss.

therefore textures (*Desch and Connolly,* 2002). Shocks sufficiently strong to melt chondrule precursors are transient in nature and it remains to be seen how often the required conditions are met at 2.5 AU; continuous shock processing is not desired to explain chondrule properties. The chondrules would be transported both inward and outward by the chaotic spiral density waves (*Boss,* 2004b). Once Jupiter forms, it will continue to drive shock fronts in the asteroidal region capable of forming chondrules, for as long as the inner disk gas and dust remains. The meteoritical evidence (*Russell et al.,* 2006) is that chondrule formation and thus the existence of the inner disk persisted for a couple of million years.

3.2.2.2. Mixing processes in gravitationally unstable disks: If disk evolution near the midplane is largely controlled by gravitational torques rather than by a turbulent process such as MRI or convection, then mixing processes might be profoundly different as a result. Gravitational torques could potentially result in matter flowing through the disk without being rapidly homogenized through mixing by turbulence. As a result, spatially heterogeneous regions of the disk might persist for some amount of time, if they were formed in the first place by processes such as the triggered injection of shock-wave material (*Vanhala and Boss,* 2000, 2002) or the spraying and size-sorting of solids processed by an X-wind onto the surface of the nebula (*Shu et al.,* 1996). However, because convective motions appear to play an important part in cooling the disk midplane in recent models of disk instability (*Boss,* 2002b), it is unclear if gravitational torques could act in isolation without interference from convective motions or other sources of turbulence. At any rate, spatially heterogeneous regions might only last for short fraction of the nebular lifetime, requiring rapid coagulation and growth of kilometer-sized bodies if evidence of this phase is to be preserved.

4. CONCLUSIONS

While the meteoritical record provides a rich resource for understanding the early evolution of the solar nebula, eventually one wishes for a more universal theory of protoplanetary disk evolution. Our current understanding of protoplanetary disks in nearby star-forming regions is severely hampered by our inability to resolve processes occurring on the scale of a few AU at distances of 150 pc or more. The advent of the Atacama Large Millimeter Array (ALMA) toward the end of this decade should help alter this situation, as ALMA will have sufficient resolution to begin the probe the innermost regions of protoplanetary disks. ALMA will answer basic questions such as determining the density, temperature, and rotational profiles of the molecular gas in these disks over length scales of a few AU. These studies will be important for determining the astrophysical context for many of the processes inferred to have occurred from the meteoritical record. At the other end of the mass spectrum, we will also be learning in the next decade about the prevalence of Earth-like planets, through the ambitious space telescopes planned by NASA to search for Earth-like planets: the Kepler mission, the Space Interferometry mission, and the Terrestrial Planet Finder. These missions will tell us something about the cosmic efficiency of the collisional accumulation process that produced the planets of the inner solar system, yet failed so spectacularly in the asteroid belt, thereby preserving the window to our own past provided by studies of the most primitive meteorites.

Acknowledgments. This work was partially supported by the NASA Planetary Geology and Geophysics Program under grant NAG 5-10201.

REFERENCES

Alves J. F., Lada C. J., and Lada E. A. (2001) Internal structure of a cold dark molecular cloud inferred from the extinction of background starlight. *Nature, 409,* 159–161.

Amelin Y., Krot A. N., Hutcheon I. D., and Ulyanov A. A. (2002) Lead isotope ages of chondrules and calcium-aluminum-rich inclusions. *Science, 297,* 1678–1683.

Andrè P. (1997) The evolution of flows and protostars. In *Herbig-Haro Flows and the Birth of Low Mass Stars* (B. Reipurth and C. Bertout, eds.), pp. 483–494. Kluwer, Dordrecht.

Bacmann A., Andrè P., Puget J.-L., Abergel A., Bontemps S., and Ward-Thompson D. (2000) An ISOCAM absorption survey of pre-stellar cloud cores. *Astron. Astrophys., 361,* 555–580.

Balbus S. A. and Hawley J. F. (1991) A powerful local shear instability in weakly magnetized disks. I. Linear analysis. *Astrophys. J., 376,* 214–222.

Bate M. R., Bonnell I. A., and Bromm V. (2002) The formation mechanism of brown dwarfs. *Mon. Not. R. Astron. Soc., 332,* L65–L68.

Bell K. R., Cassen P. M., Wasson J. T., and Woolum D. S. (2000) The FU Orionis phenomenon and solar nebula material. In *Protostars and Planets IV* (V. Mannings et al., eds.), pp. 897–926. Univ. of Arizona, Tucson.

Bizzarro M., Baker J. A., and Haack H. (2004) Mg isotope evidence for contemporaneous formation of chondrules and refractory inclusions. *Nature, 431,* 275–277.

Bontemps S., Andrè P., Terebey S., and Cabrit S. (1996) Evolution of outflow activity around low-mass embedded young stellar objects. *Astron. Astrophys., 311,* 858–872.

Boss A. P. (1989) Evolution of the solar nebula I. Nonaxisymmetric structure during nebula formation. *Astrophys. J., 345,* 554–571.

Boss A. P. (1995) Collapse and fragmentation of molecular cloud cores. II. Collapse induced by stellar shock waves. *Astrophys. J., 439,* 224–236.

Boss A. P. (1997) Giant planet formation by gravitational instability. *Science, 276,* 1836–1839.

Boss A. P. (1998) Temperatures in protoplanetary disks. *Annu. Rev. Earth Planet. Sci., 26,* 53–80.

Boss A. P. (2000) Possible rapid gas giant planet formation in the solar nebula and other protoplanetary disks. *Astrophys. J. Lett., 536,* L101–L104.

Boss A. P. (2002a) Collapse and fragmentation of molecular cloud cores. VII. Magnetic fields and multiple protostar formation. *Astrophys. J., 568,* 743–753.

Boss A. P. (2002b) Evolution of the solar nebula. V. Disk instabilities with varied thermodynamics. *Astrophys. J., 576,* 462–472.

Boss A. P. (2003) The solar nebula. In *Treatise on Geochemistry: Volume 1. Meteorites, Planets, and Comets* (A. Davis, ed.), pp. 63–82. Elsevier, Oxford.

Boss A. P. (2004a) Convective cooling of protoplanetary disks and rapid giant planet formation. *Astrophys. J., 610,* 456–463.

Boss A. P. (2004b) Evolution of the solar nebula. VI. Mixing and transport of isotopic heterogeneity. *Astrophys. J., 616,* 1265–1277.

Boss A. P. (2004c) From molecular clouds to circumstellar disks. In *Comets II* (M. C. Festou et al., eds.), pp. 67–80. Univ. of Arizona, Tucson.

Boss A. P. and Durisen R. H. (2005) Chondrule-forming shock fronts in the solar nebula: A possible unified scenario for planet and chondrite formation. *Astrophys. J. Lett., 621,* L137–L140.

Boss A. P. and Hartmann L. (2001) Protostellar collapse in a rotating, self-gravitating sheet. *Astrophys. J., 562,* 842–851.

Boss A. P. and Myhill E. A. (1995) Collapse and fragmentation of molecular cloud cores. III. Initial differential rotation. *Astrophys. J., 451,* 218–224.

Boss A. P., Wetherill G. W., and Haghighipour N. (2002) Rapid formation of ice giant planets. *Icarus, 156,* 291–295.

Briceño C., Vivas A. K., Calvet N., Hartmann L., Pacheco R., Herrera D., Romero L., Berlind P., and Sanchez G. (2001) The CIDA-QUEST large-scale survey of Orion OB1: Evidence for rapid disk dissipation in a dispersed stellar population. *Science, 291,* 93–96.

Calvet N., Hartmann L., and Strom S. E. (2000) Evolution of disk accretion. In *Protostars and Planets IV* (V. Mannings et al., eds.), pp. 377–399. Univ. of Arizona, Tucson.

Cameron A. G. W. (2003) Some nucleosynthesis effects associated with r-process jets. *Astrophys. J., 587,* 327–340.

Cameron A. G. W. and Truran J. W. (1977) The supernova trigger for formation of the solar system. *Icarus, 30,* 447–461.

Caselli P., Benson P. J., Myers P. C., and Tafalla M. (2002) Dense cores in dark clouds. XIV. N2H+(1–0) maps of dense cloud cores. *Astrophys. J., 572,* 238–263.

Cassen P. M., Smith B. F., Miller R., and Reynolds R. T. (1981) Numerical experiments on the stability of preplanetary disks. *Icarus, 48,* 377–392.

Cassen P. (1996) Models for the fractionation of moderately volatile elements in the solar nebula. *Meteoritics & Planet. Sci., 31,* 793–806.

Cassen P. (2001) Nebula thermal evolution and the properties of primitive planetary materials. *Meteoritics & Planet. Sci., 36,* 671–700.

Chagelishvili G. D., Zahn J.-P., Tevzadze A. G., and Lominadze J. G. (2003) On hydrodynamic shear turbulence in Keplerian disks: Via transient growth to bypass transition. *Astron. Astrophys., 402,* 401–407.

Chandrasekhar S. (1961) *Hydrodynamic and Hydromagnetic Stability.* Oxford Univ., Oxford.

Connolly H. C. Jr., Desch S. J., Chiang E., Ash R. D., and Jones R. H. (2006) Transient heating events in the protoplanetary nebula. In *Meteorites and the Early Solar System II* (D. S. Lauretta and H. Y. McSween Jr., eds.), this volume. Univ. of Arizona, Tucson.

Crutcher R. M. (1999) Magnetic fields in molecular clouds: Observations confront theory. *Astrophys. J., 520,* 706–713.

Curry C. L. (2002) Shapes of molecular cloud cores and the filamentary mode of star formation. *Astrophys. J., 576,* 849–859.

Cuzzi J. N. and Weidenschilling S. J. (2006) Particle-gas dynamics and primary accretion. In *Meteorites and the Early Solar System II* (D. S. Lauretta and H. Y. McSween Jr., eds.), this volume. Univ. of Arizona, Tucson.

Davis S. S. (2002) Vorticity-induced wave motion in a compressible protoplanetary disk. *Astrophys. J., 576,* 450–461.

D'Alessio P., Calvet N., and Hartmann L. (2001) Accretion disks around young objects. III. Grain growth. *Astrophys. J., 553,* 321–334.

Desch S. J. and Connolly H. C. (2002) A model of the thermal processing of particles in solar nebula shocks: Application to the cooling rates of chondrules. *Meteoritics & Planet. Sci., 37,* 183–207.

Dolginov A. Z. and Stepinski T. F. (1994) Are cosmic rays effective for ionization of protoplanetary disks? *Astrophys. J., 427,* 377–383.

Dubrulle B. (1993) Differential rotation as a source of angular momentum transfer in the solar nebula. *Icarus, 106,* 59–76.

Dutrey A., Lecavelier des Etangs A., and Augereau J.-C. (2004) The observation of circumstellar disks: Dust and gas components. In *Comets II* (M. C. Festou et al., eds.), pp. 81–95. Univ. of Arizona, Tucson.

Eisner J. A. and Carpenter J. M. (2003) Distribution of circumstellar disk masses in the young cluster NGC 2024. *Astrophys. J., 598,* 1341–1349.

Feigelson E. D. and Montmerle T. (1999) High energy processes in young stellar objects. *Annu. Rev. Astron. Astrophys., 37,* 363–408.

Fleming T. and Stone J. M. (2003) Local magnetohydrodynamic models of layered accretion disks. *Astrophys. J., 585,* 908–920.

Foster P. N. and Boss A. P. (1996) Triggering star formation with stellar ejecta. *Astrophys. J., 468,* 784–796.

Foster P. N. and Boss A. P. (1997) Injection of radioactive nuclides from the stellar source that triggered the collapse of the presolar nebula. *Astrophys. J., 489,* 346–357.

Gammie C. F. (1996) Layered accretion in T Tauri disks. *Astrophys. J., 457,* 355–362.

Glassgold A. E., Najita J., and Igea J. (1997) X-ray ionization of protoplanetary disks. *Astrophys. J., 480,* 344–350.

Godon P. and Livio M. (1999) On the nonlinear hydrodynamic stability of thin Keplerian disks. *Astrophys. J., 521,* 319–327.

Godon P. and Livio M. (2000) The formation and role of vortices in protoplanetary disks. *Astrophys. J., 537,* 396–404.

Goodman A. A., Benson P. J., Fuller G. A., and Myers P. C. (1993) Dense cores in dark clouds. VIII. Velocity gradients. *Astrophys. J., 406,* 528–547.

Haghighipour N. and Boss A. P. (2003) On gas-drag induced rapid migration of solids in an non-uniform solar nebula. *Astrophys. J., 598,* 1301–1311.

Haisch K. E., Lada E. A., and Lada C. J. (2001) Disk frequencies and lifetimes in young clusters. *Astrophys. J. Lett., 553,* L153–L156.

Hartmann L. and Kenyon S. J. (1996) The FU Orionis phenomenon. *Annu. Rev. Astron. Astrophys., 34,* 207–240.

Hartmann L., Ballesteros-Paredes J., and Bergin E. A. (2001) Rapid formation of molecular clouds and stars in the solar neighborhood. *Astrophys. J., 562,* 852–868.

Hawley J. F., Gammie C. F., and Balbus S. A. (1995) Local three-dimensional magnetohydrodynamic simulations of accretion disks. *Astrophys. J., 440,* 742–763.

Hollenbach D. J., Yorke H. W., and Johnstone D. (2000) Disk dispersal around young stars. In *Protostars and Planets IV* (V.

Mannings et al., eds.), pp. 401–428. Univ. of Arizona, Tucson.
Huss G. R., Srinivasan G., MacPherson G. J., Wasserburg G. J., and Russell S. S. (2001) Aluminum-26 in calcium-aluminum-rich inclusions and chondrules from unequilibrated ordinary chondrites. *Meteoritics & Planet. Sci., 36,* 975–997.
Jones C. E., Basu S., and Dubinski J. (2001) Intrinsic shapes of molecular cloud cores. *Astrophys. J., 551,* 387–393.
Kawakita H., Watanabe J., Ando H., Aoki W., Fuse T., Honda S., Izumiura H., Kajino T., Kambe E., Kawanomoto S., Sato B., Takada-Hidai M., and Takeda Y. (2001) The spin temperature of NH_3 in Comet C/1999S4 (LINEAR). *Science, 294,* 1089–1091.
Kerton C. R., Brunt C. M., Jones C. E., and Basu S. (2003) On the intrinsic shape of molecular clouds. *Astron. Astrophys., 411,* 149–156.
Kita N. T., Nagahara H., Togashi S., and Morishita Y. (2000) A short formation period of chondrules in the solar nebula. *Geochim. Cosmochim. Acta, 48,* 693–709.
Klahr H. (2004) The global baroclinic instability in accretion disks. II. Local linear analysis. *Astrophys. J., 606,* 1070–1082.
Klahr H. H. and Bodenheimer P. (2003) Turbulence in accretion disks: Vorticity generation and and angular momentum transport in disks via the global baroclinic instability. *Astrophys. J., 582,* 869–892.
Kobrick M. and Kaula W. M. (1979) A tidal theory for the origin of the solar nebula. *Moon and Planets, 20,* 61–101.
Königl A. and Pudrtiz R. E. (2000) Disk winds and the accretion-outflow connection. In *Protostars and Planets IV* (V. Mannings et al., eds.), pp. 759–787. Univ. of Arizona, Tucson.
Lada C. J. and Shu F. H. (1990) The formation of sunlike stars. *Science, 248,* 564–572.
Lada C. J. and Lada E. A. (2003) Embedded clusters in molecular clouds. *Annu. Rev. Astron. Astrophys., 41,* 57–115.
Larson R. B. (1969) Numerical calculations of the dynamics of a collapsing proto-star. *Mon. Not. R. Astron. Soc., 145,* 271–295.
Laughlin G. and Bodenheimer P. (1994) Nonaxisymmetric evolution in protostellar disks. *Astrophys. J., 436,* 335–354.
Li H., Colgate S. A., Wendroff B., and Liska R. (2001) Rossby wave instability of thin accretion disks. III. Nonlinear simulations. *Astrophys J., 551,* 874–896.
Lin D. N. C. and Papaloizou J. (1980) On the structure and evolution of the primordial solar nebula. *Mon. Not. R. Astron. Soc., 191,* 37–48.
Lin D. N. C. and Pringle J. E. (1987) A viscosity prescription for a self-gravitating accretion disk. *Mon. Not. R. Astron. Soc., 225,* 607–613.
Lin D. N. C., Papaloizou J. C. B., Terquem C., Bryden G., and Ida S. (2000) Orbital evolution and planet-star tidal interaction. In *Protostars and Planets IV* (V. Mannings et al., eds.), pp. 1111–1134. Univ. of Arizona, Tucson.
Lynden-Bell D. and Pringle J. E. (1974) The evolution of viscous disks and the origin of the nebular variables. *Mon. Not. R. Astron. Soc., 168,* 603–637.
MacPherson G. J., Davis A. M., and Zinner E. K. (1995) The distribution of aluminum-26 in the early solar system — A reappraisal. *Meteoritics, 30,* 365–386.
Mannings V., Boss A. P., and Russell S. S., eds. (2000) *Protostars and Planets IV*. Univ. of Arizona, Tucson. 1422 pp.
Marcy G. W, Cochran W. D., and Mayor M. (2000) Extrasolar planets around main-sequence stars. In *Protostars and Planets IV* (V. Mannings et al., eds.), pp. 1285–1312. Univ. of Arizona, Tucson.
Marhas K. K., Goswami J. N., and Davis A. M. (2002) Short-lived nuclides in hibonite grains from Murchison: Evidence for solar system evolution. *Science, 298,* 2182–2185.
Meyer B. S. and Zinner E. (2006) Nucleosynthesis. In *Meteorites and the Early Solar System II* (D. S. Lauretta and H. Y. McSween Jr., eds.), this volume. Univ. of Arizona, Tucson.
Mouschovias T. Ch. (1991) Magnetic braking, ambipolar diffusion, cloud cores, and star formation: Natural length scales and protostellar masses. *Astrophys. J., 373,* 169–186.
Nittler L. R. and Dauphaus N. (2006) Meteorites and the chemical evolution of the Milky Way. In *Meteorites and the Early Solar System II* (D. S. Lauretta and H. Y. McSween Jr., eds.), this volume. Univ. of Arizona, Tucson.
Onishi T., Mizuno A., Kawamura A., Tachihara K., and Fukui Y. (2002) A complete search for dense cloud cores in Taurus. *Astrophys. J., 575,* 950–973.
Pollack J. B., Hubickyj O., Bodenheimer P., Lissauer J. J., Podolak M., and Greenzweig Y. (1996) Formation of the giant planets by concurrent accretion of solids and gas. *Icarus, 124,* 62–85.
Preibisch T. and Zinnecker H. (1999) The history of low-mass star formation in the Upper Scorpius OB association. *Astron. J., 117,* 2381–2397.
Preibisch T., Brown A. G. A., Bridges T., Guenther E., and Zinnecker H. (2002) Exploring the full stellar population of the Upper Scorpius OB association. *Astron. J., 124,* 404–416.
Reipurth B. (2000) Disintegrating multiple systems in early stellar evolution. *Astron. J., 120,* 3177–3191.
Ruden S. P. and Pollack J. B. (1991) The dynamical evolution of the protosolar nebula. *Astrophys. J., 375,* 740–760.
Russell S. S., Srinivasan G., Huss G. R., Wasserburg G. J. and MacPherson G. J. (1996) Evidence for widespread ^{26}Al in the solar nebula and constraints for nebula time scales. *Science, 273,* 757–762.
Russell S. S., Hartmann L., Cuzzi J., Krot A. N., Gounelle M., and Weidenschilling S. (2006) Timescales of the solar protoplanetary disk. In *Meteorites and the Early Solar System II* (D. S. Lauretta and H. Y. McSween Jr., eds.), this volume. Univ. of Arizona, Tucson.
Shu F. H., Adams F. C., and Lizano S. (1987) Star formation in molecular clouds: Observation and theory. *Annu. Rev. Astron. Astrophys., 25,* 23–72.
Shu F. H., Najita J. R., Shang H., and Li Z.-Y. (2000) X-winds: Theory and observation. In *Protostars and Planets IV* (V. Mannings et al., eds.), pp. 780–813. Univ. of Arizona, Tucson.
Shu F. H., Tremaine S., Adams F. C., and Ruden S. P. (1990) Sling amplification and eccentric gravitational instabilities in gaseous disks. *Astrophys. J., 358,* 495–514.
Shu F. H., Johnstone D., and Hollenbach D. (1993) Photoevaporation of the solar nebula and the formation of the giant planets. *Icarus, 106,* 92–101.
Shu F. H., Shang H., and Lee T. (1996) Toward an astrophysical theory of chondrites. *Science, 271,* 1545–1552.
Smith K. W., Bonnell I. A., Emerson J. P., and Jenness T. (2000) NGC 1333/IRAS 4: A multiple star formation laboratory. *Mon. Not. R. Astron. Soc., 319,* 991–1000.
Stone J. M., Gammie C. F., Balbus S. A., and Hawley J. F. (2000) Transport processes in protostellar disks. In *Protostars and Planets IV* (V. Mannings et al., eds.), pp. 589–611. Univ. of Arizona, Tucson.
Toomre A. (1964) On the gravitational stability of a disk of stars. *Astrophys. J., 139,* 1217–1238.
Tsuribe T. and Inutsuka S.-I. (1999a) Criteria for fragmentation

of rotating isothermal clouds revisited. *Astrophys. J. Lett., 523,* L155–L158.

Tsuribe T. and Inutsuka S.-I. (1999b) Criteria for fragmentation of rotating isothermal clouds. I. Semianalytic approach. *Astrophys. J., 526,* 307–313.

Vanhala H. A. T. and Boss A. P. (2000) Injection of radioactivities into the presolar cloud: Convergence testing. *Astrophys. J., 538,* 911–921.

Vanhala H. A. T. and Boss A. P. (2002) Injection of radioactivities into the forming solar system. *Astrophys. J., 575,* 1144–1150.

Vanhala H. A. T. and Cameron A. G. W. (1998) Numerical simulations of triggered star formation. I. Collapse of dense molecular cloud cores. *Astrophys. J., 508,* 291–307.

Ward W. R. and Hahn J. M. (2000) Disk-planet interactions and the formation of planetary systems. In *Protostars and Planets IV* (V. Mannings et al., eds.), pp. 1135–1155. Univ. of Arizona, Tucson.

Weidenschilling S. J. (1988) Formation processes and time scales for meteorite parent bodies. In *Meteorites and the Early Solar System* (J. F. Kerridge and M. S. Matthews, eds.), pp. 348–371. Univ. of Arizona, Tucson.

Weidenschilling S. J. (2004) From icy grains to comets. In *Comets II* (M. C. Festou et al., eds.), pp. 97–104. Univ. of Arizona, Tucson.

Weidenschilling S. J. and Cuzzi J. N. (2006) Accretion dynamics and timescales: Relation to chondrites. In *Meteorites and the Early Solar System II* (D. S. Lauretta and H. Y. McSween Jr., eds.), this volume. Univ. of Arizona, Tucson.

Woolum D. S. and Cassen P. (1999) Astronomical constraints on nebula temperatures: Implications for planetesimal formation. *Meteoritics & Planet. Sci., 34,* 897–907.

Yokogawa S., Kitamura Y., Momose M., and Kawabe R. (2003) High angular resolution, sensitive CS J = 2–1 and J = 3–2 imaging of the protostar L1551 NE: Evidence for outflow-triggered star formation. *Astrophys. J., 595,* 266–278.

Yorke H. W. and Bodenheimer P. (1999) The formation of protostellar disks. III. The influence of gravitationally induced angular momentum transport on disk structure and appearance. *Astrophys. J., 525,* 330–342.

The Population of Starting Materials Available for Solar System Construction

S. Messenger
NASA Johnson Space Center

S. Sandford
NASA Ames Research Center

D. Brownlee
University of Washington

Combined information from observations of interstellar clouds and star-forming regions and studies of primitive solar system materials give a first-order picture of the starting materials for the solar system's construction. At the earliest stages, the presolar dust cloud was comprised of stardust, refractory organic matter, ices, and simple gas phase molecules. The nature of the starting materials changed dramatically together with the evolving solar system. Increasing temperatures and densities in the disk drove molecular evolution to increasingly complex organic matter. High-temperature processes in the inner nebula erased most traces of presolar materials, and some fraction of this material is likely to have been transported to the outermost, quiescent portions of the disk. Interplanetary dust particles thought to be samples of Kuiper belt objects probably contain the least-altered materials, but also contain significant amounts of solar system materials processed at high temperatures. These processed materials may have been transported from the inner, warmer portions of the disk early in the active accretion phase.

1. INTRODUCTION

A principal constraint on the formation of the solar system was the population of starting materials available for its construction. Information on what these starting materials may have been is largely derived from two approaches, namely (1) the examination of other nascent stellar systems and the dense cloud environments in which they form, and (2) the study of minimally altered examples of these starting materials that have survived in ancient solar system materials.

These approaches are complementary to each other and suffer from different limitations. The study of other forming stellar systems is limited by our inability to study the formation process of a single system over the entire formation interval. In addition, since these are distant systems, it is not possible to examine all the processes that are occurring with high spatial resolution. Finally, star formation is observed to occur under a range of environments and one must make educated guesses as to the extent to which observations apply to the specific case of our solar system. Similarly, the search for presolar materials in solar system materials is complicated by overprinting of protostellar nebular and parent-body processes and dilution within a large amount of materials whose original composition has been erased or severely modified.

Despite these limitations, astronomical and meteoritic studies provide insight into the nature of the starting materials from which the solar system formed and we review these constraints here. A central theme of this review is that the nature of the starting materials evolved prior to and together with the nascent solar system and varied with location in the solar system. For instance, terrestrial planets were likely assembled from components that had already undergone substantial cycling through evaporation, condensation, chemical evolution, and fractionation. In contrast, Kuiper belt objects (the sources of short-period comets) are thought to contain abundant interstellar materials because they formed in the relatively cold and quiescent outer (>40 AU) regions of the solar nebula.

In this chapter, we address what we know about the nature of solar system starting materials on the basis of several different approaches. First, we consider constraints derived from observations of dense molecular clouds and star-formation areas within them. We then consider what we can learn about solar system starting materials on the basis of examination of "primitive" materials found within the solar system. In this context, "primitive" refers to those materials that have been largely unaltered, or only slightly altered, from their interstellar forms by nebular or parent-body processing. An emphasis is placed on the properties of interplanetary dust particles (IDPs) that may represent relatively pristine samples of these starting materials. Finally, we consider those fractions of meteoritic materials that clearly have a presolar heritage on the basis of preserved isotopic anomalies. These materials are clearly of presolar origin and provide insights into a variety of stellar nucleosynthetic and interstellar chemical processes that imprinted the materials from which our solar system formed.

2. DENSE MOLECULAR CLOUD MATERIALS

The population of materials from which new stellar and planetary systems form contains materials with wildly disparate histories (*Sandford,* 1996). Materials are injected into the interstellar medium by a diversity of stellar sources (supernovae, novae, AGB stars, etc.) (*Dorschner and Henning,* 1995; *Tielens,* 1995) whose outflows differ significantly in their elemental compositions, thermal environments, densities, and lifetimes. These materials are then processed to varying degrees by interstellar shocks, sputtering, irradiation, etc. [see *Nuth et al.* (2006) for discussion of some of these processes] over timescales of 10^7 yr. Despite their varied histories, all these materials share a similar experience in passing through a dense molecular cloud prior to becoming incorporated into a new stellar system.

Other potentially important sources of material in star-forming regions are the outflows of newly formed stars themselves. It is likely that the majority of stars (70–90%) form in large clusters [see *Lada and Lada* (2003) for a recent review]. Most young protostars likely return an enormous amount of processed matter back into the surrounding dust cloud in bipolar outflows. It is important to appreciate that our own solar system may have been seeded with significant quantities of this "protostellar" dust, and that it ought to have been largely isotopically identical to the solar system. High-mass stars, such as those in OB associations, have dramatic effects on surrounding regions as much as parsecs in size through high-energy radiation that may ablate materials from nearby protostellar cores and disks (e.g., *Smith et al.,* 2003). The most massive stars have such short lifetimes (~10 m.y.) that they may die in a supernova explosion while other stellar systems are still forming (e.g., *Cameron,* 1985). In such clusters, the nature of the materials in a forming protonebular disk will depend, in a stochastic fashion, on the timing of the disk's formation and who its neighbors are. Late-forming stellar systems are thus more likely to be polluted with material originating from young stars and supernovae. The recent demonstration that there was abundant, live ^{60}Fe in the early solar system is best explained by the polluting of the protostellar nebula by the debris of a nearby supernova or AGB star (*Tachibana and Huss,* 2003; *Busso et al.,* 2003). While either source is possible, the timing of an AGB star passing near the Sun would have been fortuitous, whereas there is growing evidence for a supernova source in particular (*Mostefaoui et al.,* 2005).

Dense clouds consist of concentrations of dust, gas, and ice that are sufficiently optically thick that they screen out the majority of the interstellar radiation field. In these relatively protected environments, molecules can form and survive for appreciable times without being photodisrupted. As a result, these clouds are commonly called dense molecular clouds. By far the most abundant molecular species in these clouds is H_2, formed by the reaction of atomic H on dust grains, but a host of other molecular species are present as well (see below).

Dense molecular clouds are not uniform entities, but instead show structure in their density, temperature, turbulence, etc., on a variety of size scales (*Evans,* 1999). On their largest size scales, dense clouds typically show densities of about 10^2–10^3 H atoms/cm^3. Within this general cloud medium are found clumps of various sizes and densities, with typical clumps being a few light years across and having densities of 10^3–10^5 H atoms/cm^3 and temperatures of $10 K < T < 30 K$ (*Goldsmith and Langer,* 1978; *Goldsmith,* 1987). The highest densities occur in small "cores" that are less than a light year across. Such cores typically have densities of 10^6–10^8 H atoms/cm^3 and warmer temperatures ($50 K < T < 300 K$) and are often referred to as "hot molecular cores" (*van Dishoeck and Blake,* 1998). Under conditions in which further gravitation collapse can occur, these cores can become the sites of new star formation (*Mannings et al.,* 2000). Subsequent formation of the protostar is driven by core collapse and the formation and evolution of an accretion disk, during which pressure and temperature conditions may vary over very large ranges depending on time and location (*Cassen and Boss,* 1988; *Boss,* 1998).

Careful examination of dense cloud materials, particularly those in dense cores, provides important insights into the population of materials available for the formation of new stellar and planetary systems (*Sandford,* 1996). However, dense cloud environments span large ranges in densities and temperatures that increase dramatically as star formation proceeds. Consequently, the nature of the starting materials may differ somewhat depending on what one defines as the "start." For example, the temperatures in the general dense cloud medium are quite low ($10 K < T < 30 K$) and under these conditions most molecular species other than H, H_2, He, and Ne are expected to be largely condensed out of the gas phase as ice mantles on refractory grains (*Sandford and Allamandola,* 1993; *Bergin et al.,* 2002; *Walmsley et al.,* 2004). However, warming of these materials during the infall process releases volatiles into the gas phase where they participate in gas phase and gas-grain chemical reactions that are inhibited by the low temperatures of the general cloud medium. Thus, the nature of the population of the "starting materials" evolves as the collapse proceeds.

2.1. Source Materials of Dense Clouds

Dense molecular clouds are initially formed from materials swept together from the diffuse interstellar medium by interstellar shock waves (*Mac Low and Klessen,* 2004), and they are thus mixtures of materials with a variety of origins that have been altered to various degrees while in the diffuse ISM. For a more detailed discussion of many of these sources and processes, see *Nuth et al.* (2006).

As with the rest of the universe, the majority of the matter in dense clouds is in the form of gas phase atomic and molecular H and He. Solid materials, in the form of small (<1 μm diameter) dust grains are also present. Included within the population of solid materials are silicate grains. These are identified by their infrared absorption features seen near 10 and 20 μm due to O-Si-O stretching and bending mode vibrations, respectively (e.g., *Tielens,* 1990; *Jäger et al.,* 1994). These features are seen in absorption along lines of sight through both the diffuse ISM and dense clouds. The lack of spectral structure in the silicate features suggests that the majority of these silicates are amorphous in nature [see *Nuth et al.* (2006) for a detailed discussion on the nature of the silicates].

Refractory carbonaceous grains are also expected to be present in dense clouds. Carbonaceous materials are observed along a number of lines of sight through the diffuse interstellar medium, primarily by the detection of absorption bands due to aromatic and aliphatic C-H stretching bands near 3.3 and 3.4 μm, respectively (*Butchart et al.,* 1986; *Sandford et al.,* 1991; *Pendleton et al.,* 1994). The exact nature of this material is not well understood, but it is unlikely to be due to small, free molecular gas phase molecules as these would be rapidly destroyed in the diffuse ISM. Currently available spectral data is consistent with this material residing in grains that may bear some chemical resemblance to the macromolecular "kerogen" seen in meteorites, i.e., it may consist of aromatic chemical domains that are interlinked by short aliphatic chains (*Pendleton and Allamandola,* 2002). The spatial distributions of the silicate and carbonaceous materials in the diffuse ISM are very similar (*Sandford et al.,* 1995), although it is not clear how significant this is. The two components could show similar distributions because they are in intimate association (for example, carbonaceous coatings on silicate cores), or they may exist as separate particle populations that show similar distributions because of elemental abundance constraints. Strangely, the relatively strong 3.4-μm absorption feature that marks the presence of these organic materials in the diffuse ISM is not clearly seen in dense clouds. The suppression of this feature may be due to evolution of H coverage on the material (*Mennella et al.,* 2002).

One of the most ubiquitous components of the ISM is the class of gas phase molecules known at polycyclic aromatic hydrocarbons (PAHs). These materials consist largely of fused hexagon rings of C with peripheral bonds ending in H atoms, although heteroatoms, like N, in their skeletal structures and chemical functional groups attached to their perimeters are possible in some environments. In most space environments, the PAHs are detected through their characteristic infrared emission features (*Allamandola et al.,* 1989). The emission is driven by the electronic absorption of visible and ultraviolet photons by PAHs, which subsequently convert the absorbed energy into vibrational energy. The molecules then cool through the emission of a cascade of infrared photons whose energies are associated with their vibrational modes (*Allamandola et al.,* 1989). The emission features of PAHs are seen in a vast range of astrophysical environments, including stellar outflows, the diffuse ISM, reflection nebulae, and emission nebula. It is estimated that this class of molecules carries more than 5% of the cosmic C in many of those environments (e.g., *Allamandola et al.,* 1989; *Tielens,* 1995). These molecules should be abundant in dense clouds, but are difficult to detect in these environments because the clouds are opaque to the interstellar radiation field that excites these molecules to emit in the infrared. As a result, PAHs in dense clouds cannot be seen in *emission*, but instead must be detected in *absorption*. This restricts searching for them along lines of sight toward suitable background sources. Absorption features assigned to PAHs have been observed in a limited number of cases (e.g., *Brooke et al.,* 1999; *Chiar et al.,* 2000; *Bregman et al.,* 2000). Laboratory analog experiments suggest that the concentrations of PAHs in these cases fall between 1% and 5% that of H_2O along the same lines of sight (*Sandford et al.,* 2004; *Bernstein et al.,* 2005), which would make them a significant carrier of the C in dense clouds. However, it is not yet clear whether these molecules reside in the gas phase (as they do when seen in emission in other environments), are clustered together in aggregate grains, or are incorporated into icy grain mantles on refractory grains. These possibilities can be distinguished if sufficiently good infrared spectra are obtained (*Sandford et al.,* 2004). Given the very low temperatures found in these clouds, however, it is likely that most of the PAHs in these environments reside in the solid state.

2.2. Probing the Composition of Dense Molecular Clouds

Spectroscopy provides the principal means by which the compositions of dense molecular clouds are studied. At submillimeter and radio wavelength, spectroscopy can be used to detect the rotational transitions of gas phase molecules. Such observations have the advantage that they can rigorously identify molecular species with good rotational dipole moments and can be used to derive the temperature of the molecular population. In addition, they can make measurements in a wide range of dense cloud locations. They suffer, however, from an inability to detect solid-state materials, materials that make up the vast majority of non-H_2 mass of most dense cloud environments. Complementary measurements can be made at infrared wavelengths, where materials can be detected through their interatomic vibrational transitions. Given sufficient spectral resolution, these measurements can also provide identification and temperatures of specific molecular species, although there is a higher degree of band overlap that makes identification more difficult. Infrared spectroscopy is also limited in dense clouds to measuring absorption bands produced by materials along the line of sight to a suitable infrared background source. Infrared spectroscopy enjoys the significant advantage, however, that it can detect solid materials, which constitute the major non-H_2 fraction of the cloud (see, for example, *van Dishoeck,* 2004).

2.3. Chemical Processes in Dense Clouds

Most molecular species are unstable to photodisruption in the radiation field found in the diffuse ISM (PAHs being one of the notable exceptions). Dense clouds effectively screen out interstellar radiation, providing a safe haven in their interiors for gas phase molecules. Furthermore, the greater densities of these clouds provides for increased atomic and molecular interactions that can lead to the formation of new molecules.

Chemistry in dense clouds does not proceed along the same lines we are familiar with in most environments on Earth. At the very low temperatures characteristic of dense clouds ($10 K < T < 50 K$) most neutral molecular species are not reactive because there is insufficient energy available to overcome activation barriers. As a result, much of the gas-phase chemistry in dense clouds occurs via ion-molecule reactions [for more detail on this chemical process, see *Herbst* (1987), *Langer and Graedel* (1989), and *Nuth et al.* (2006)]. Many of the gas phase species so far detected in dense clouds (see Table 1 of *Nuth et al.,* 2006) were probably created in this manner.

At the low temperatures characteristic of dense clouds, most molecular species rapidly condense out onto any dust grains present (*Sandford and Allamandola,* 1993; *Bergin et al.,* 2002; *Walmsley et al.,* 2004). As a result, grains in dense clouds are expected to be coated by an icy mantle (Fig. 1). Studies of scattered light indicate that the average grain size in dense clouds is greater than that in the diffuse ISM (*Pendleton et al.,* 1990). This growth could be due to grain agglomeration and/or the accretion of icy mantles. Modeling of the scattering in conjunction with observations of the profile of the 3.08-μm H_2O ice band in the reflection nebula surrounding OMC-2/IRS1 suggest ice mantle growth is a significant process (*Pendleton et al.,* 1990).

Ice mantles facilitate the production of new kinds of molecules by acting as catalytic surfaces (*Brown and Charnley,* 1990; *Charnley et al.,* 1992; *Ehrenfreund and Charnley,* 2000). These surficial reactions involve simple atom addition to previously condensed species. One of the most important of these gas-grain reactions is the production of molecular H_2 from atomic H (*Hollenbach and Salpeter,* 1971). In environments where atomic H dominates the gas phase ($H/H_2 > 1$), gas-grain atom addition reactions act to hydrogenate species, turning atomic O into H_2O, atomic C into CH_4, molecular CO in CH_3OH, and so on. In environments where molecular H_2 dominates ($H/H_2 < 1$), other atom addition reactions can dominate, resulting in the formation of species like CO_2, N_2, and O_2 (*Tielens and Hagen,* 1982, *Tielens and Allamandola,* 1987). The compositions of interstellar ice mantles thus range from highly polar materials to nonpolar materials, depending on the local environment and history of the material. Both of these basic ice types have been observed by infrared spectroscopy (*Sandford et al.,* 1988; *Tielens et al.,* 1991). The primary components of these ices are simple molecules like H_2O, CO, CO_2,

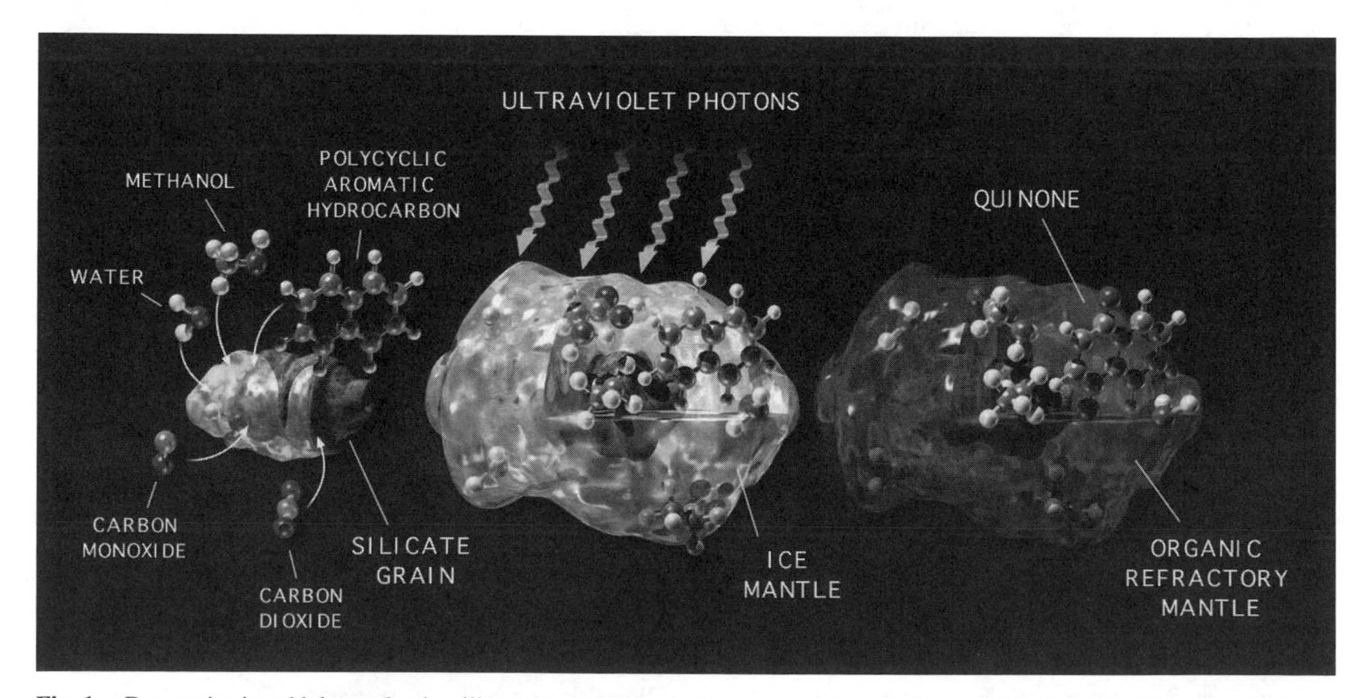

Fig. 1. Dust grains in cold dense clouds will accrete molecules and atoms from the gas phase. Catalytic atom addition reactions can occur on these surfaces and additional chemical reactions may be driven by ionizing radiation in the form of UV photons and cosmic rays. Warming of these grains can drive volatiles from the mantle, resulting in local enrichments of gas phase molecular material and the production of complex refractory organic layers on the grains. Adapted from *Bernstein et al.* (1999a).

HCO, H_2CO, CH_3OH, NH_3, and CH_4 (*Tielens et al.,* 1991; *Lacy et al.,* 1991, 1998; *Allamandola et al.,* 1992; *Boogert et al.,* 1996; *de Graauw et al.,* 1996; *Whittet et al.,* 1996). Molecules like N_2 and O_2 may also be present, but these have no permanent dipole and are difficult to detect spectroscopically.

High-energy irradiation can also drive chemical reactions within the ice mantles. In the general cloud, this ionizing radiation is primarily in the form of galactic cosmic rays, UV photons generated locally by cosmic-ray interactions, and the attenuated diffuse interstellar radiation field (*Norman and Silk,* 1980; *Prasad and Tarafdar,* 1983). In the zones near forming protostars, the radiation density may be greatly augmented by high-energy processes associated with the forming stars. This ionizing radiation breaks chemical bonds in the molecules in the ice and creates chemically reactive ions and radicals. Some of these react immediately with neighboring molecules; others do not react until the mantles are warmed enough to allow some molecular mobility. These processes can produce a wide range of comparatively complex molecules. For example, laboratory studies have shown that the irradiation of interstellar ice analogs results in the production of hundred, if not thousands, of new and more complex and refractory species (e.g., *Allamandola et al.,* 1988; *Bernstein et al.,* 1995, 2003; *Moore et al.,* 1996; *Gerakines et al.,* 2000). These include a range of organics, many of which are identical or similar to compounds identified in primitive meteorites (*Epstein et al.,* 1987; *Engel et al.,* 1990; *Krishnamurthy et al.,* 1992; *Bernstein et al.,* 2001). Many of these species are also of astrobiological interest, for example, aromatic ketones (quinones), amphiphiles, and amino acids (*Bernstein et al.,* 1999b, 2002; *Dworkin et al.,* 2001).

The best evidence that radiation-driven chemistry occurs in dense clouds is the presence of an infrared absorption band near 4.62 μm (2165 cm^{-1}) seen in protostellar spectra (*Lacy et al.,* 1984; *Tegler et al.,* 1993, 1995) that is thought to be due to the OCN^- ion in a molecular complex (*Grim and Greenberg,* 1987; *Park and Woon,* 2004). This feature can be well reproduced in the laboratory whenever simple ices containing H, O, C, and N are irradiated (e.g., *Lacy et al.,* 1984; *Bernstein et al.,* 1995, 2000). Interestingly, this feature appears to be considerably more common and prominent in the spectra of protostars than it is in the spectra of background stars (*Pendleton et al.,* 1999), suggesting that irradiation chemistry may play a more prominent role in the immediate vicinity of star-formation regions.

2.4. Isotopic Fractionation in Dense Cloud Materials

While many of the principal components of dense clouds can be measured remotely, this does not tell us what components of the cloud material actually get incorporated into new stellar systems. In the case of our own solar system, the principal means of identifying surviving presolar materials is through the detection of isotopic anomalies. Some of these anomalies are nucleosynthetic in origin and have signatures set by their original source stars (see below). However, some of the isotopic anomalies seen in meteoritic materials, particularly those of D and ^{15}N, have their origins in interstellar chemistry, and these can be used as tracers of the survival of incoming interstellar materials.

There are four main interstellar chemical processes that can lead to the production of D-enriched species, three of which are confined to environments associated with dense clouds (*Tielens,* 1997; *Sandford et al.,* 2001). The best known of these are the gas phase ion-molecule reactions mentioned earlier. At the very low temperatures typical of dense clouds, differences in the zero point energies of H and D bonds result in the ion-molecule reaction

$$H_3^+ + HD \rightarrow H_2D^+ + H_2$$

preferentially removing D atoms from HD molecules and redistributing them throughout the gas phase during subsequent reactions (e.g., *Millar et al.,* 2000). Thus, H-containing species produced by ion molecule reaction are expected to be D-enriched. As a second process, the difference in zero point energies may also lead to additional D fractionation during catalytic gas-grain reaction on grain surfaces (*Tielens,* 1983). Whether these grain surface reactions produce additional fractionation or not, the ices are expected to be highly D-enriched. This is because the same ion-molecule gas phase reactions that fractionate D also enrich the local gas phase atomic D/H ratio by factors of ~100, and grain surface hydrogenation reactions sequester these enrichments in the ice mantles in the form of D-enriched H_2O, CH_3OH, CH_4, NH_3, etc. Finally, once D-enriched ice mantles form, any radiation processing that occurs will pass these enrichments along to a host of other product molecules (e.g., *Bernstein et al.,* 1995, 2000; *Sandford et al.,* 2000). A fourth process for D enrichment is that of preferential unimolecular photodissociation. This process is restricted to molecular species, like PAHs, that can absorb a high-energy photon without being destroyed, but that will occasionally shed a peripheral H atom if the absorbed energy is large enough. Because of the difference in H and D zero point energies, H atoms should be shed preferentially and repeated loss and capture of peripheral H atoms can lead to D enrichment (*Allamandola et al.,* 1989). All these processes are expected to lead to different distributions of fractionation, both with regard to which molecules the excess D ends up in, and in some cases, where on the molecule it resides (*Sandford et al.,* 2001). Thus, in principle, the placement of D anomalies in meteoritic materials holds clues for assessing the relative roles of the various possible enrichment processes.

Ion-molecule reactions are also expected to yield ^{15}N enrichments in dense clouds (*Langer and Graedel,* 1989; *Terzieva and Herbst,* 2000; *Charnley and Rodgers,* 2002; *Rodgers and Charnley,* 2004). As with D, fractionations established in the gas phase can then be redistributed into the solid phase through subsequent gas-grain reactions and the irradiation of ^{15}N-enriched ice mantles. Since ion-molecule reactions produce both D and ^{15}N enrichments under similar conditions, i.e., in very cold gas, these anomalies would be expected to be correlated, at least qualitatively, in materials made within dense clouds. However, since the extent of the fractionations depend differently on temperature and the specific reaction paths are involved, the relative degrees of fractionation of the two will vary with environmental conditions.

That these various enrichment processes occur for D is amply demonstrated by observation of gas phase D-enriched molecules in dense cloud environments. These include molecules like DCO^+, DCN, HDO, NH_2D, HDCO, CH_3OD, as well as a host of others (*Hollis et al.,* 1976; *Walmsley et al.,* 1987; *Mauersberger et al.,* 1988; *Jacq et al.,* 1990; *Turner,* 2001; *Roberts et al.,* 2002). Interestingly, recent studies have detected unexpectedly large abundances of multiply deuterated species like D_2CO, ND_2H, D_2S, and CD_3OH (*Loinard et al.,* 2000; *Parise et al.,* 2002; *Vastel et al.,* 2003; *Roueff et al.,* 2000). Deuterium enrichments have also been tentatively reported in PAHs (*Peeters et al.,* 2004). There is currently very little observational evidence constraining the level of D-enrichment in interstellar ices, the sole report being the possible detection of HDO-containing ices in two protostellar spectra with HDO/H_2O ratios of 8×10^{-4} and 1×10^{-2}, respectively (*Teixeira et al.,* 1999).

There is currently considerably less evidence for ^{15}N enrichment in interstellar molecules resulting from chemical fractionation. This may be due to the fact that the CN isotopomers are more difficult to resolve spectroscopically and that the predicted isotopic shifts are comparatively much smaller than the case for D/H. Fractionations in ^{13}C and the isotopes of O are also possible in dense clouds. However, in the case of ^{13}C, there are several chemical paths that are expected to fractionate ^{13}C in opposite directions in different species (*Langer et al.,* 1984; *Langer and Graedel,* 1989; *Tielens,* 1997), so it is difficult to assess what net fractionations might be preserved into meteoritic materials. Similar complications occur for O (*Langer et al.,* 1984, *Langer and Graedel,* 1989). As a result, it is somewhat difficult to separate nucleosynthetic signatures from chemical signatures for these two elements in interstellar molecules.

3. TRANSITION TO THE SOLAR NEBULA

As star formation proceeds, the infall of material results in increased densities and temperatures. The rising temperatures lead to the sublimation of volatile components of icy grain mantles. The result is relatively high gas phase densities of compounds that did not previously exist together in the gas phase (*van Dishoeck and Blake,* 1998). These conditions promote a rich gas-phase chemistry that can lead to new, more complex species not efficiently produced by gas-phase chemistry in the general dense cloud medium (e.g., *Charnley et al.,* 1995; *Langer et al.,* 2000; *Rodgers and Charnley,* 2003).

The construction of our planetary system from its natal dust cloud is imagined to have started from smoke-sized (10–1000 nm) "starting materials" that accreted together to form successively larger particles, eventually leading to the formation of large planetesimals and planets. There is some truth to this notion, but the nebula was a messy place and the definition of the starting materials is open to interpretation. Depending on time as well as radial and vertical location in the solar nebula accretion disk, there were actually many types of starting materials. In some cases the building materials that were used to assemble bodies in the solar system were not the original starting materials, but were secondary materials formed within the solar system itself. The nature of the starting materials at each stage of nebular evolution has been obscured by secondary processes, and it will always be a challenge to combine information from astronomical observations and laboratory studies of early solar system materials to see through this veil.

A prime example of a "secondary process" is chondrule formation — a process that, in the region of the nebula where ordinary chondrites accreted, efficiently converted most of the nebular solids in into molten spherules. The dramatic chondrule-formation process overprinted earlier generations of solids. Other important examples of secondary processes include collisions, gas-grain reactions, parent-body heating, and aqueous alteration.

The chondrule-formation process is one manifestation of the dynamic environment in the inner regions of the solar nebula. It is a sobering fact that, over its <10-m.y. lifetime, over a solar mass of material accreted onto the disk and migrated toward its center. The viscous dissipation of the kinetic energy of material spiraling inward is thought to cause disks around very young stars to temporarily outshine the embedded protostar — an enigmatic process known as FU Orionis phenomena (*Hartmann and Kenyon,* 1996). The temperature distribution of protostellar disks may vary significantly with time (as short as 10^2–10^5-yr timescale) and location as the mass accretion rate varies episodically by orders of magnitude. Incredibly, the solar-formation process is thought to have ejected as much as 0.25 $M_\odot$ during the accretion process back into the stellar association from which our Sun formed. This processed, "protostellar material" may have been incorporated into neighboring young stellar systems (and vice versa).

While the inner disk reached temperatures high enough to vaporize silicates, the outer portions of the nebula remained at cryogenic temperatures. It is expected that the grains in the collapsing cloud that formed the nebula were maintained near 10 K during the free-fall phase, assuring that essentially all condensable gas molecules must have condensed on grains. On the basis of astronomical data we would expect this material to be largely composed of submicrometer grains of amorphous silicates, organic material, and condensed ice and other volatiles with an average elemental composition close to solar abundances for condensable elements. A hallmark of the real starting solid materials of the solar system is that they must have contained ices that condensed at such low temperatures. These low temperatures are not found anywhere in the planetary region of the present solar system and it is unlikely that they existed anywhere in the midplane of the solar nebula accretion disk except perhaps in its outermost regions. The ultra-low-temperature condensates present in the starting materials did not fully survive as solids and all currently available primitive solar system materials are at least partially depleted in original volatile compounds.

4. PRIMITIVE SOLAR SYSTEM MATERIALS

Today, the starting materials are probably best preserved in comets. These objects derive from two distinct source regions: the Kuiper belt, a flattened disk ($i < 30°$) of minor planets beyond the orbit of Neptune (30–50 AU); and the Oort cloud, a spherically distributed population of bodies residing very far from the Sun (10^3–10^4 AU). Comets are rich in volatiles and ices because they formed in the coldest regions of the solar nebula (*Boss,* 1998), marking the endpoint of the radial progression from the volatile-depleted terrestrial planets to gas giants and ice-rich outer solar system bodies. Whereas Kuiper belt objects (KBOs) are thought to have formed in place (*Luu and Jewitt,* 2002), the Oort cloud probably originated from the scattering of icy planetessimals that formed in the neighborhood of the giant planets (*Oort,* 1950).

Dynamical arguments point to the Kuiper belt as the source of most short-period comets (*Duncan et al.,* 1988). Long-period comets originate from the Oort cloud, and their original orbits are thus nearly hyperbolic. Close planetary encounters, especially with Jupiter, muddle these distinctions in some cases by strongly altering cometary orbits. For instance, most Halley-type comets, with orbital periods between 20 and 200 years, likely originated from the inner portion of the Oort cloud (*Levison et al.,* 2001). The best-studied comets (e.g., 1P/Halley, C/1995 O1/Hale-Bopp, and C/1996 B2/Hyakutake) probably all originated from the Oort cloud. These comets were well studied for different reasons: Halley was visited by several spacecraft, Hale-Bopp was exceptionally active, and Hyakutake passed close (<0.1 AU) to Earth. In the very near future the Stardust spacecraft will return the first direct samples of cometary solids from the short-period comet Wild-2. Undoubtedly, the analysis of this material will revolutionize our understanding of the nature and origin of these enigmatic bodies.

Comets are rich in volatiles because they formed under cryogenic conditions and have remained in cold storage for billions of years since. Certain cometary volatiles provide constraints on their temperature histories. For instance, the common presence of H_2S and CO ice in cometary nuclei is thought to limit the nebular temperature during comet formation to less than 60 K (*Mumma et al.,* 1993). Still lower temperature limits are inferred from the nuclear spin states (ortho/para ratio) of cometary water. The value derived from comet Halley (2.5 ± 0.1) corresponds to a "spin temperature" of only 29 K. An unresolved issue is whether the spin temperature reflects the original formation temperature or subsequent reequilibration in the cometary nucleus. Short-period comets tend to be more active at great distances from the Sun compared to long-period comets, suggesting that they are richer in low-temperature volatiles (*Whipple,* 2000). This is consistent with the idea that many Oort cloud comets formed in warmer regions of nebula (~100 K, near Jupiter) than Kuiper belt objects (<30 K). As molecular studies of comets have become more sensitive, significant chemical variability among comets has become evident. For instance, the relative production rates of CO, C_2/CN, and simple organic molecules are observed to vary by factors of several (*Mumma et al.,* 2003; *Biver et al.,* 2002). Such compositional variability may be a reflection of the wide heliocentric range (5–50 AU), and the correspondingly diverse environmental conditions, over which comets formed (*Dones et al.,* 2004).

Despite the chemical diversity among comets, the abundances of many simple volatile compounds agree well with those observed in star-forming environments such as hot molecular cores (*Ehrenfreund and Charnley,* 2000). Examples include such species as HCOOH, HCNO, NH_2CHO, and methanol — molecules that are produced by single atom additions on grain surfaces, as described above. Other cometary molecules, including C_2H_2, CH_3CN, HC_3N, and HCN, are more likely to have originated in gas-phase chemistry. These comparisons are tempered by the uncertain role of gas-phase reactions in cometary comae that strongly overprint the nuclear chemical compositions. However, the preponderance of the evidence suggests that the volatile component of comets is a mixture of preserved interstellar ices and differing proportions of material processed in the solar nebula.

The isotopic compositions of cometary volatiles also point toward an interstellar heritage. As reviewed in *Bockelée-Morvan et al.* (2005), several simple molecules have shown anomalous isotopic compositions in one or more comets. The D/H ratio for cometary water is enriched by a factor of 2 relative to the terrestrial value (*Eberhardt et al.,* 1995), and a D enrichment of more than an order of magnitude has been reported for HCN in comet Hale-Bopp (*Meier et al.,* 1998). A significant (~factor of 2) enrichment in ^{15}N has also been observed in CN molecules from two comets (*Arpigny et al.,* 2003). These isotopic measurements represent a small fraction of cometary materials and it is possible that isotopic measurements of additional molecules or on finer spatial scales will reveal a greater isotopic diversity than is apparent so far.

The mineralogy of cometary dust grains provides further insight into their origins. Spectroscopic studies of comets have shown the presence of abundant, fine grained silicates. As with interstellar silicates, the 10-μm feature is generally featureless, indicating the presence of abundant amorphous silicates. However, significant variability in the strength and detailed structure of this band is observed among comets (*Hanner et al.,* 1994). Most importantly, a significant fraction of cometary silicates were found to be crystalline olivine grains, based on a spectral feature at 11.2 μm (*Bregman et al.,* 1987; *Campins and Ryan,* 1989; *Hanner et al.,* 1994). More detailed mineralogical characterization has recently been achieved by far-IR spectroscopy (16–45 μm), where features corresponding to forsterite and enstatite were observed (*Crovisier et al.,* 2000). The observation that comets contain Mg-rich crystalline silicates is intriguing because although a minor fraction of such grains exist around evolved stars are crystalline (10–20%), they are far rarer in the ISM (~0.5%).

The presence of crystalline silicates in comets suggests that they contain a significant, and variable, proportion of solar system materials processed at high temperatures. One possibility is that the crystalline grains were originally amorphous interstellar grains that were annealed by moderate heating. However, it is difficult to envision this process naturally leading to Mg-rich crystalline silicates unless the precursor amorphous grains happened to have had just the right chemical composition. Alternatively, the crystalline grains may be second-generation solids that formed by equilibrium condensation from a high-temperature gas. Either scenario is clearly at odds with the evidence that comets have largely escaped significant thermal processing. A solution to this dilemma is that these materials were processed in the inner portions of the solar nebula, perhaps by shocks (*Harker and Desch,* 2002), and were transported outward to the comet-forming region by turbulence (*Cuzzi et al.,* 2003) or as a consequence of early bipolar outflow of material from the Sun [i.e., the X-wind (*Shu et al.,* 2001)] or its early stellar neighbors.

Thus, even those bodies that formed in the most quiescent portions of the solar nebula appear to have been partially built from materials that were processed in energetic environments. Other processes that may have been important in altering cometary materials include hypervelocity impacts among icy planetessimals, irradiation by galactic cosmic rays of the upper ~1 m, and radiogenic heating (for sufficiently large bodies). Disentangling the imprint of these secondary processes from the original primordial compositions can only be achieved by detailed laboratory analysis of primitive materials.

The most detailed information on the processes, conditions, and timescales of the early history of the solar system has come from the study of meteorites. In addition to preserving the earliest nebular condensates, the least-altered meteorites contain traces of the starting materials, including interstellar dust grains and molecular cloud material (discussed below). However, even the most primitive meteorites are comprised almost entirely of secondary materials formed within the solar nebula or their parent bodies. Two important selection effects account for the rarity of primitive materials among the meteorite collection: Most meteorites are derived from portions of the asteroid belt that lie near orbital resonances with Jupiter, and macroscopic bodies comprised of fragile materials cannot survive the hypervelocity (>12 km/s) passage through the atmosphere.

Interplanetary dust is largely unaffected by the selection effects that limit the diversity of meteorites. First, Earth accretes dust from *all* dust-producing bodies, including comparable proportions of asteroidal and cometary dust. This is due to the fact that all small bodies (<1 cm) in the solar system have unstable orbits owing to their interaction with solar radiation. While the smallest dust grains (~1 μm) are ejected from the solar system by radiation pressure, the orbits of most interplanetary dust grains gradually decay toward the Sun because of Poynting-Robertson light drag and solar wind drag (on a 10^4–10^5-yr timescale). Some of this dust is captured by Earth, amounting to ~4×10^7 kg/yr (*Love and Brownlee,* 1993). The second advantage is that fragile materials easily survive atmospheric entry because their relatively large surface areas cause them to decelerate at very high altitude (~80 km) where the ram pressure is low.

The smallest particles are the least affected by atmospheric entry, although large size (>100 μm), high velocity (>20 km/s), or steep entry angles can lead to strong heating — even melting — among some particles. The best preserved samples of interplanetary dust are collected in the stratosphere (stratospheric IDPs) by high-altitude research aircraft such as the ER-2 and WB-57. Most of these samples are smaller than 20 μm in diameter, although some of the highly porous "cluster particles" probably exceeded 100 μm before they fragmented on the collection surface. Because these particles are collected before they reach the ground, there is minimal potential for chemical weathering. The stratospherically collected IDPs are restricted to small sizes because of the comparative rarity of larger particles.

Unlike primitive meteorites that are widely believed to be samples solely of asteroids, it is nearly certain that stratospheric IDPs include samples of *both* asteroids and comets. Both types of objects are known sources of dust injected into the interplanetary medium (e.g., *Sykes and Greenberg,* 1986; *Sykes et al.,* 1986), and dynamical arguments indicate dust from both sources can be delivered to Earth (e.g., *Sandford,* 1986; *Jackson and Zook,* 1992). Furthermore, observations of the extent of atmospheric heating (*Sandford and Bradley,* 1989; *Nier and Schlutter,* 1992, 1993) and compositional variations (*Bradley et al.,* 1988) suggest both sources are represented in the IDP collections.

Larger particles, accumulated over time, can be collected in certain localities on the ground, particularly from Greenland and Antarctic ice (*Maurette et al.,* 1994; *Duprat et al.,* 2005; *Gattacceca et al.,* 2005; *Noguchi et al.,* 2002; *Taylor et al.,* 2001). By current convention, these particles collected from surface deposits are called micrometeorites. This terminology differs from F. Whipple's original definition of micrometeorites as small particles that were not heated to melting temperatures during atmospheric entry, but it is the current convention and it is consistent with the definition of meteorites as samples that actually impact the surface of Earth. Micrometeorites recovered from Antarctic ice range from 10 μm to millimeter-sized, but most are much larger than collected stratospheric IDPs. Micrometeorites are valuable samples because they are much more massive than typical IDPs, they can be collected in vast numbers, and they include particles near the mass flux peak at ~200 μm size that dominates the bulk of cosmic matter accreted by Earth.

Micrometeorites exhibit a diversity of compositions and structures, with the majority being dominated by fine-grained anhydrous mineralogy. A large fraction of micrometeorites appear to have originated from CM or (less commonly) CI-like chondrite parent bodies — samples that are much rarer among collected meteorites (*Kurat et al.,* 1994). This is likely due to the differing delivery mechanisms between the two types of samples (discussed above). Owing to their larger sizes, the heating effects are more pronounced among micrometeorites with a higher proportion having experienced partial or total melting (*Alexander and Love,* 2001). Some particles up to several hundred micrometers across do survive entry without being heated above the ~1300°C melting points of chondritic composition materials. The larger unmelted micrometeorites probably survived due to a combination of low entry speed (~escape velocity) and low entry angle. However, even moderate atmospheric entry heating has apparently dramatically affected the mineralogy of most micrometeorites, decomposing phyllosilicates into complex fine-grained mixtures of olivine, low-Ca pyroxene, magnetite, and amorphous silicates (*Nakamura et al.,* 2001).

Unlike IDPs, most micrometeorites have been affected by some degree of weathering, for example, particles recovered from solid ice are usually depleted in S, Ni, and Ca relative to chondritic proportions due to terrestrial leaching processes (*Kurat et al.,* 1994). The best-preserved and probably least-biased polar micrometeorites have been recently recovered from Antarctic snow (*Duprat et al.,* 2005). The polar micrometeorites are an increasingly important resource of interplanetary particles (*Rietmeijer,* 2002). These samples are strong compliments to conventional meteorites and IDPs as well as primitive solar system samples to be returned by comet and asteroid missions such as Stardust and Hayabusa.

Of the available extraterrestrial materials, chondritic porous (CP) interplanetary dust particles (Fig. 2) are the most likely samples of Kuiper belt objects and, as such, their properties should most closely resemble the original starting materials. The link between CP IDPs and short-period comets is based on a number of independent lines of evidence. First, CP IDPs are structurally similar to cometary materials in being extremely fine-grained, porous, and fragile (*Bradley and Brownlee,* 1986, *Rietmeijer and McKinnon,* 1987). In fact, CP IDPs are so fragile that these materials are unlikely to survive atmospheric entry as macroscopic bodies, so it is not surprising that similar materials are not represented in the meteorite collections. Second, CP IDPs are highly enriched in C [2–3× CI (*Thomas et al.,* 1993)] and volatile trace elements (*Flynn et al.,* 1993) relative to CI chondrites. Third, as discussed below, detailed chemical, mineralogical, and isotopic studies of these particles show them to have experienced minimal parent-body alteration, and are rich in presolar materials.

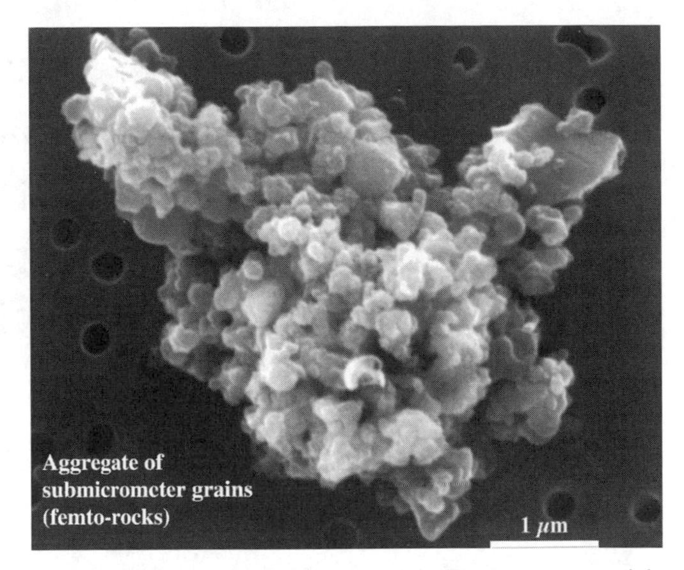

Fig. 2. A CP interplanetary dust particle. The nanogram particle is a porous aggregate of large numbers of submicrometer grains, which individually contain GEMS, glass, crystalline materials, and organic components. The fragile particle has not been subject to compaction since its formation in the solar nebula. The individual components in the particle are likely to be the accretionary units that are representative of the time and location in the solar nebula disk where the particle accreted.

The strongest evidence linking CP IDPs to short-period comets comes from their inferred atmospheric entry velocities. Even after thousands of years in space, cometary and asteroidal dust particles, by and large, retain distinct orbital characteristics (*Sandford,* 1986; *Jackson and Zook,* 1992). Asteroidal particles typically have low eccentricity (e) ~0.1 and low inclination (i) orbits, while cometary particles usually have eccentricities exceeding ~0.4, and a wider range of inclinations. The higher average e and i of cometary dust particles result in higher Earth-encounter velocities. This difference in Earth-encounter velocities should translate into observable differences in the extent of heating experienced during atmospheric entry (*Flynn,* 1989; *Sandford and Bradley,* 1989). Peak temperatures experienced by particles during atmospheric entry were most accurately determined by stepwise He release profiles measured in individual IDPs (*Nier and Schlutter,* 1992, 1993). Based on these measurements, *Joswiak et al.* (2000) showed that the particles with the highest peak temperatures (and therefore the highest inferred entry velocities and orbital eccentricities) were compositionally distinct from those with low peak temperatures. This high-velocity (>18 km/s) subset is dominated by CP IDPs, and it is likely that many of them have cometary origins. Typical hydrated IDPs have lower (<14 km/s) entry speeds consistent with origin from the asteroid belt. It is much less likely for dust from long-period comets to survive atmospheric entry due to the far higher Earth-encoun-

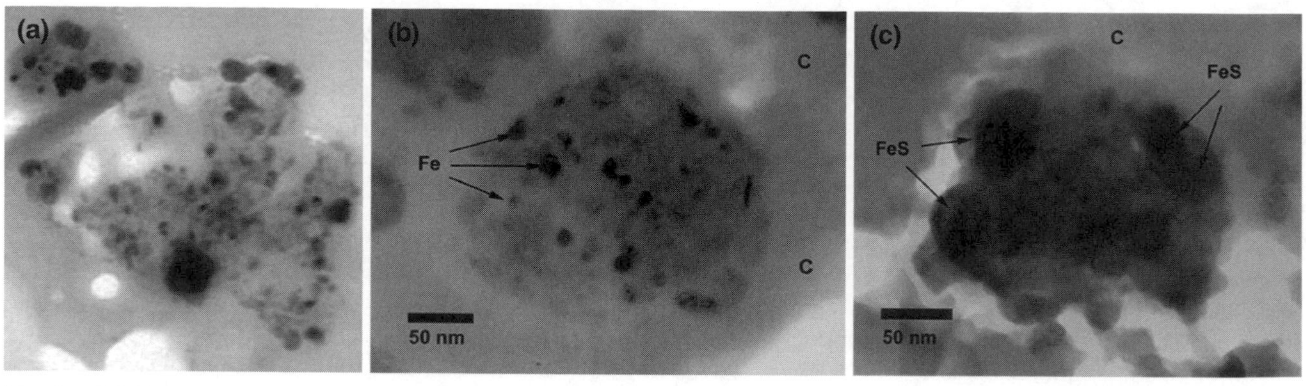

Fig. 3. GEMS exhibit a diversity of compositions and morphologies, ranging from **(a)** irregular grains, which may be a composite object (aggregate); **(b)** glass-rich GEMS grain; **(c)** sulfide-rich GEMS grain.

ter velocities (30–70 km/s). These high-velocity particles are responsible for the most spectacular meteor showers, such as the Leonids.

The CP particles have near-chondritic elemental abundances — an amazing property for samples only a few micrometers across. They are largely aggregates of subgrains <0.5 μm in diameter, with rare grains larger than several micrometers. The subgrains are solid nonporous matter containing a mix of submicrometer glass with embedded metal and sulfides (GEMS) (*Bradley et al.,* 1999), organic materials, olivine, pyroxene, pyrrhotite, less-well-defined materials, and a number of less-abundant phases (*Bradley,* 2004). GEMS grains are submicrometer amorphous Mg-Si-Al-Fe silicate grains that contain numerous 10–50-nm-sized FeNi metal and Fe-Ni sulfides, comprising up to 50 wt% of anhydrous IDPs (Fig. 3).

One of the most fundamental properties of the CP particles is that they are porous aggregates of submicrometer grains. Whether CP particles are made of starting materials or not, their porous aggregate structure of micrometer and submicrometer grains is a predictable property of first-generation primitive materials that formed by accretion of ice-coated submicrometer grains — the size of interstellar dust. The preservation of a highly porous aggregate structure for billions of years inside a parent body probably required the presence of ice, ice that never melted but was lost by sublimation. Sublimation is a gentle process that does not result in compaction. It is possible, but not proven, that the CP particles or at least some subset of them are freeze-dried samples of KBOs. Even if they are comet samples, there is a need of additional information to determine how similar they might be to the original nebular materials. The CP particles have been found to contain presolar grains, but what *fraction* of the subgrains are presolar? Even at Kuiper belt distances, it is possible that nebular shocks modified first-generation grains and it is also possible that there could have been considerable radial transport outward from the inner, warmer regions of the nebular accretion disk.

The CP particles are dominated by anhydrous phases, but they are just one of the types of IDPs collected in the stratosphere. Other IDP types include fine-grained particles composed largely of hydrated silicates and particles dominated by coarse-grained minerals, usually pyrrhotite, forsterite, or enstatite. Both the fine-grained hydrous and anhydrous particles are black and have elemental compositions similar to CI chondrite abundances, often with enhanced volatile contents and usually higher C contents (*Flynn et al.,* 1996). Most of the coarse-grained IDPs have clumps of black fine-grained material with chondritic elemental composition adhering to their surfaces and it is clear that the coarse-grained particles were previously embedded in fine-grained matrix material similar to the fine-grained IDP types.

Hydrated IDPs and micrometeorites are likely to have formed by aqueous alteration of materials inside parent bodies that were heated to temperatures high enough to melt ice. This was a common process in primitive asteroids and it is likely that most of the hydrated IDPs and micrometeorites, like carbonaceous chondrites, are from asteroids and are products of the inner regions of the solar nebula. The CP particles, dominated by anhydrous minerals and glass, do not appear to have ever been exposed to liquid water. Their anhydrous nature is consistent with spectral reflectance data from outer solar system primitive bodies that also lack hydrated phases.

The CP particles are believed to be the most primitive type of IDP and it is likely that they are samples of comets or ice-bearing asteroids. They are distinctive in several respects, including their porosity (lack of compaction and their associated fragility), their content of GEMS, their high He abundances with ^{3}He/^{4}He that are distinct from solar wind, their high volatile contents, their high abundance of presolar grains, and the fact that they do not normally contain hydrated silicates. All the C-rich and volatile-rich meteorite types contain hydrated silicates, presumably formed by secondary processes inside parent bodies. Thus, the CP particles appear to be the best-preserved samples of early solar system fine-grained materials.

It is possible that the CP particles contain or are even dominated by the initial materials that began the accretion process. This first generation of nebular solids has also been referred to as first accretionary particles (FAPs) (*Brownlee and Joswiak,* 2004). The CP particles are largely assemblages of relatively equidimensional rounded components about 0.25 μm in diameter, similar in size to typical interstellar grains. The submicrometer components of the aggregate particles may be FAPs. Figures 4 and 5 show the results of efforts to mechanically separate the individual components of a CP IDP into components that originally accreted to form the aggregates. The subgrains are typically small solid rocks composed of GEMS, crystalline silicates, sulfides, amorphous silicates, and amorphous organic material. They have wildly varying major-element compositions, but groups of a dozen or more of the possible FAPs average close to bulk chondritic compositions. It is interesting that there is a cutoff in size in the potential FAPs particles at about 0.1 μm. Much smaller components are present, but they are all contained inside the potential FAPs, which average about 0.25 μm in diameter. They are solid particles that seem to be the fundamental building blocks of CP IDPs. Some early, preaccretional process assembled these particles into solid femtogram "rocks" that typically range in diameter from 0.1 μm to 0.5 μm.

Although meteorites may be dominated by processed materials such as chondrules, they are also found to contain traces of the original starting materials, including presolar grains. The presolar materials are the only starting materials that can be definitively identified. These presolar components are distinguished by their isotopic compositions, which differ from the well-homogenized solar system materials by degrees that cannot be explained through other local processes such as spallation and isotopic fractionation. Two types of presolar materials are recognized:

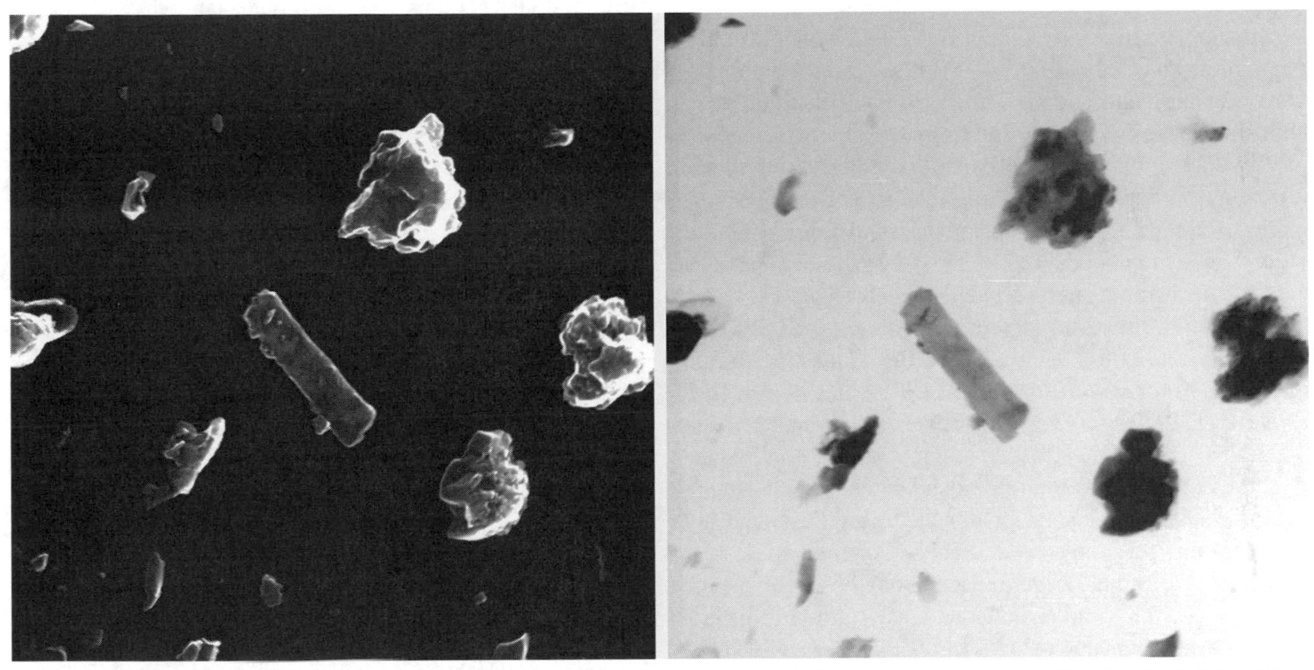

Fig. 4. Individual submicrometer components mechanically separated from a CP interplanetary dust particle. On the left is a SEM image and a transmission image is on the right.

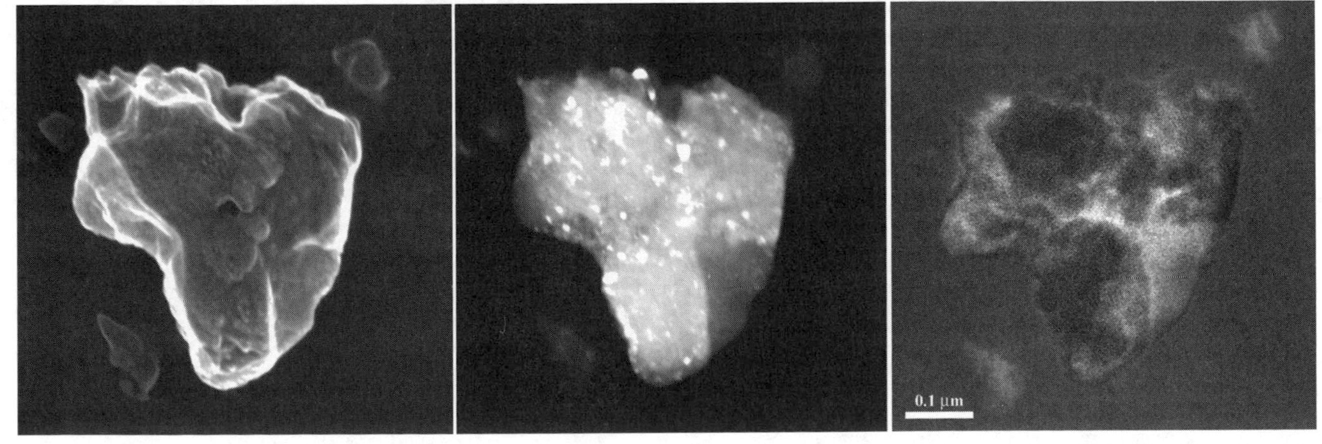

Fig. 5. Images of a single submicrometer grain, a possible first accretionary particle (FAP), shown with an SEM image on the left, a dark field transmission electron microscopy image in the center, and a electron energy loss C map on the right. The femtogram rock is a solid mix of GEMS, glass, mineral grains, and amorphous organic material. Scale bar is 0.1 μm.

presolar grains (stardust) and organic matter that likely originated in a cold molecular cloud environment.

4.1. Interstellar Dust in the Solar System

The vast majority of the various starting materials that ended up in the solar system were obliterated by nebular and parent-body processes. The solar system has been homogenized isotopically on scales ranging from micrometers to astronomical units, but strong chemical fractionation processes have obscured the record of the solar system elemental abundances. Despite being largely comprised of secondary materials, CI chondrite meteorites have bulk abundance patterns (chondritic composition) that closely match that of the Sun and therefore that of the overall solar system (solar composition) for most nonvolatile elements. Determining the elemental abundances of the Sun is challenging because of the difficulty of resolving weak spectral lines, incomplete knowledge of the physical state of the solar photosphere, and insufficiently known atomic transition probabilities. However, only the Sun can yield accurate measures of the relative abundances of volatile elements and all elements heavier than He (metallicity) that are essential parameters for a wide variety of cosmochemical studies. On the other hand, the abundances of most nonvolatile elements can be much more precisely determined from laboratory studies of CI chondrite meteorites (*Anders and Ebihara,* 1982; *Burnett et al.,* 1989). Since the landmark compilations of *Anders and Grevesse* (1989) and *Anders and Ebihara* (1982), the solar abundances of individual elements have been gradually refined. The most recent comprehensive study has shown that the abundances of 31 elements determined from the solar photosphere and CI chondrites agree within 10%, and an additional 5 elements within 15% (*Lodders,* 2003). Further refinements may be possible from analyses of solar wind samples recently returned to Earth by the Genesis spacecraft.

Elemental abundances derived from CI chondrites show smooth functions of mass for odd mass nuclei when broken down into individual isotopic abundances. As first argued by *Suess* (1947), this smooth abundance curve is unlikely to be the result of fractionation processes that occurred during the chondrite-formation process. In fact, the regular variations in solar system elemental and isotopic abundances have retained the imprint of distinct nucleosynthetic processes averaging over many stellar sources. In a seminal paper, *Burbidge et al.* (1957) demonstrated that the average elemental and isotopic abundances of the solar system could be accounted for by a combination of eight nucleosynthetic reactions occurring in stars. The study of the contributions of particular types of stars and nucleosynthetic processes to these abundances is a field in its own right — galactic chemical evolution — which is discussed in *Nittler and Dauphas* (2006).

Because stardust grains sample individual stars and are largely comprised of newly synthesized elements, they are marked by extremely exotic isotopic compositions in both major and trace elements. The isotopic compositions of freshly synthesized elements are functions of the mass, age, and chemical composition of the parent star, leaving an isotopic fingerprint on the stellar ejecta that may differ from solar composition by orders of magnitude. The interpretation of the isotopic compositions of stardust from meteorites in terms of specific stellar sources and nucleosynthesis is reviewed in *Meyer and Zinner* (2006). Here we focus on the nature of these materials.

The types of stardust identified in meteoritic materials to date include nanodiamonds, SiC, graphite, Si_3N_4, TiC, Al_2O_3, TiO_2, hibonite, spinel, forsterite, and amorphous silicates (*Zinner,* 1998) (Table 1). With the exception of nanodiamonds, these grains are large enough (>200 nm) to have their isotopic compositions measured by secondary ion mass spectrometry (Fig. 6). These grains originated from a diversity of stellar sources, including red giant and asymptotic giant branch stars, novae, and supernovae. Their isotopic compositions show that their parent stars had a wide range in mass and chemical composition, requiring contributions from dozens of stars. No single stellar source appears to dominate. Nanodiamonds are by far the most abundant phase (1000 ppm), but their origins remain uncertain because they are much smaller (~2 nm) and their isotopic compositions can only be determined from bulk samples (thousands of nanodiamonds). A minor fraction (10^{-6}) of nanodiamonds is linked to stellar sources by an anomalous Xe component enriched in both heavy and light isotopes (Xe-HL) (*Lewis et al.,* 1987). However, the average C-isotopic composition

TABLE 1. Identified presolar grain types, their observed size range, and abundance in primitive meteorites and IDPs.

Grain Type	Size	Abundance
Nanodiamonds	2 nm	1000–1400 ppm
Amorphous silicates[a]–[d]	0.2–0.5 μm	<20–3600 ppm
Forsterite and enstatite[d]	0.2–0.5 μm	<10–1800 ppm
SiC	0.1–0.5 μm	14–30 ppm
Graphite	1–20 μm	7–13 ppm
Spinel[e] ($MgAl_2O_4$)	0.1–20 μm	1.2 ppm
Corundum (Al_2O_3)[f]	0.2–3 μm	100 ppb
Si_3N_4	1–5 μm	1–20 ppb
TiO_2[g]	n.d.	<10 ppb
Hibonite[h] ($CaAl_{12}O_{19}$)	1–5 μm	20 ppb

The meteorite abundances for SiC, graphite, and nanodiamond are the observed ranges among CI and CM meteorites according to *Huss et al.* (2003), matrix-normalized, and the IDP abundances are bulk values. Notes: [a] Total presolar silicate abundances derived for IDPs vary from 450 to 5500 ppm and for meteorites range from 170 ppm (matrix normalized) to less than 35 ppm. Isotopically anomalous amorphous silicates appear to be twice as abundant as crystalline silicates so far. Further refinements to these numbers are expected. *Messenger et al.* (2003). [b] *Floss and Stadermann* (2004). [c] *Nguyen and Zinner* (2004). [d] *Mostefaoui and Hoppe* (2004), *Nagashima et al.* (2004). [e] *Zinner et al.* (2003). [f] *Nittler et al.* (1995). [g] rough estimate from *Nittler et al.* (2005). [h] *Choi et al.* (1999).

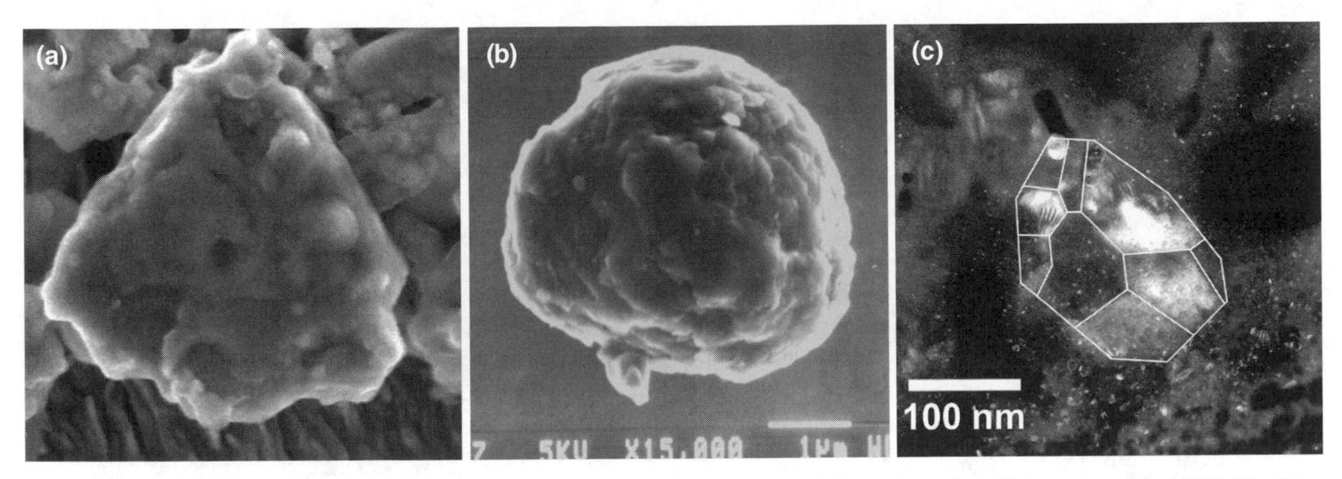

Fig. 6. (**a**) Presolar silicon carbide (*Bernatowicz et al.,* 2003), (**b**) graphite, and (**c**) forsterite grains (*Messenger et al.,* 2005). The lines indicate equilibrium grain boundaries between 50–100 nm forsterites that may have experienced moderate thermal annealing.

of nanodiamonds is identical to that of the solar system, leaving the origins of most nanodiamonds unknown. There is also evidence that some nanodiamonds may have formed in the solar nebula (*Russell et al.,* 1992). In this regard, it is interesting to note that infrared absorption bands charac-ter-istic of nanodiamonds have been detected in the dust surrounding several young stars (*Guillois et al.,* 1999).

Besides the enigmatic nanodiamonds, the most abundant types of stardust are amorphous and crystalline silicates. Despite their high abundance, these grains were only recently identified because of their small size and the fact that they are a minor component of a silicate-rich matrix of solar system origin. As it is not feasible to chemically isolate presolar silicates, they have only been identified through systematic searches in primitive meteorite matrix and IDPs by O-isotopic imaging with the Cameca NanoSIMS and modified IMS-1270 ion microprobes (*Nagashima et al.,* 2004). Presolar silicates appear to be significantly more abundant in IDPs [450–5500 ppm (bulk) (*Floss and Stadermann,* 2004; *Messenger et al.,* 2003)] than in meteorites [<170 ppm matrix normalized (*Nguyen and Zinner,* 2004; *Mostefaoui and Hoppe,* 2004)], consistent with the suggestion that the population of IDPs contains some of the best-preserved samples of early solar system materials.

Because of these difficulties, few presolar silicates have been subjected to detailed mineralogical study by transmission electron microscopy (TEM). Of the six presolar silicates studies by TEM thus far, two are forsterite and four are amorphous silicates including GEMS. The observed proportion of amorphous to crystalline presolar silicates (2 : 1) is at odds with that observed around evolved stars (10 : 1) and in the diffuse ISM (>100 : 1) inferred from fitting the 10-μm spectral feature. This is a major unsolved problem that impacts wide-ranging fields in astrophysics (*Kemper et al.,* 2004). Two possible resolutions to this discrepancy are that (1) the abundances of interstellar amorphous and crystalline silicates have been improperly derived from the ~10-μm feature (*Bowey and Adamson,* 2002), or (2) most of the mass of interstellar silicate grains have been recycled through repetitive sputtering and recondensation in the ISM, rendering them isotopically homogeneous (~solar) (*Bradley and Dai,* 2005). One future test of the interstellar homogenization model will be high-precision isotopic measurements of individual GEMS grains, which would be expected to show mixing trends between solar and evolved-stellar isotopic compositions (^{17}O-rich and ^{18}O-depleted). Whatever their origins, GEMS-like materials (dirty amorphous silicates) were abundant constituents of the protoplanetary disk, and rank among the most important of the starting materials.

The abundances of each type of presolar grain varies considerably among different classes of primitive meteorite, generally following the extent of parent-body metamorphism (*Huss and Lewis,* 1995). In addition, *Huss et al.* (2003) have also argued that nebular heating removed the most volatile and fragile presolar components to varying degrees, generally correlating with the bulk chemical composition (extent of fractionation relative to solar) of the host meteorites. Presolar silicates have also recently been found in Antarctic micrometeorites in abundances (300 ppm) (*Yada et al.,* 2005) intermediate between those observed in meteorites (<170 ppm) (*Mostefaoui and Hoppe,* 2004) and those of IDPs (450–5500 ppm).

The abundance of presolar silicates in primitive extraterrestrial materials is therefore a sensitive probe of the extent of parent-body and nebular alteration. The abundance of presolar silicates in the meteorite Acfer 094 are best known (~170 ppm) because of the large surface area searched. The abundances in Antarctic micrometeorites and IDPs are not as well known in part because of the smaller amount of material studied. These preliminary studies also suggest that there are substantial variations in the presolar grain abundance between different IDPs and micrometeorites. This may reflect the fact that these particles sample a wider range of parent bodies than are represented in the meteorite collection. Again, the highest abundances have so far been observed in some CP IDPs, suggesting that those materials have undergone less-extensive parent-body and/or nebular processing.

4.2. Molecular Cloud Matter in the Solar System

Organic matter in primitive meteorites and IDPs often exhibits large excesses in D and ^{15}N relative to terrestrial values. These anomalies are usually thought to reflect the partial preservation of molecules that formed in a presolar cold molecular cloud by processes that fractionated H and N isotopes as described above (*Geiss and Reeves,* 1981; *Zinner,* 1988; *Messenger and Walker,* 1997; *Cronin and Chang,* 1993). Elucidating the origin and history of this material is complex, as it has since become altered and diluted with local organic matter to an unknown extent. These molecules may have experienced oxidation, polymerization, and isotopic exchange either in the solar nebula or during parent-body hydrothermal alteration. The organic matter in meteorites has been subjected to extensive study, and a detailed discussion of its nature can be found in *Pizzarello et al.* (2006).

Here we are concerned with the question: What was the original state of the organic matter in the solar system? Spectroscopic observations of cold molecular clouds and comets and laboratory simulations of interstellar chemistry provide a useful starting point, but these approaches are of limited value for complex molecules. The value of studying meteorites in this regard is that modern analytical instruments can be brought to bear to reveal the full molecular and isotopic diversity of materials that directly sample remote astrophysical environments. The challenge has been to properly interpret these observations in order to delineate how parent-body alteration may have affected the organic matter present.

The nature of organic matter varies among primitive meteorite and micrometeorite classes and even within individual meteorites (*Cronin et al.,* 1988; *Clemett et al.,* 1993, 1998; *Matrajt et al.,* 2004). However, the majority (>70%) of the organic matter in all cases is comprised of an acid insoluble macromolecular material often likened to terrestrial kerogen (*Cronin et al.,* 1988; *Cody and Alexander,* 2005). The insoluble organic fraction is found to be predominantly comprised of small aromatic domains with heteroatomic substitutions crosslinked by alkyl and ether functional groups. The soluble fraction is a complex assemblage of hundreds of identified compounds including amino acids, amines, carboxylic acids, alcohols, ketones, aliphatic hydrocarbons, and aromatic hydrocarbons. Most molecules are found to exhibit full isomeric diversity and chiral molecules are generally racemic, consistent with their formation through abiogenic processes (*Epstein et al.,* 1987; *Cronin et al.,* 1988).

Considerable variations in the H-, N-, and C-isotopic compositions of the organic matter are observed within and between meteorites. Pyrolysis and combustion experiments have shown that several isotopically distinct components exist within the insoluble organic fraction (*Kerridge et al.,* 1987; *Alexander et al.,* 1998). Several classes of soluble organics have been subjected to compound specific isotopic measurements, including hydroxy, dicarbocylic, and hydroxydicarboxylic acids (*Krishnamurthy et al.,* 1992; *Cronin et al.,* 1988), amino acids (*Pizzarello and Huang,* 2005), and aliphatic, aromatic, and polar hydrocarbons (see *Cronin et al.,* 1988, for a review). Among these, amino acids exhibit the strongest D enrichments, reaching 3600‰ in 2-amino-2,3-dimethylbutyric acid (*Pizzarello and Huang,* 2005).

Comparative studies of organic matter in different meteorites suggest that aqueous processing has had a significant impact on both the soluble and insoluble organic compounds. For instance, amino acids have been hypothesized to have formed during aqueous alteration from preexisting aldehydes and ketones in a so-called Strecker synthesis (*Cronin et al.,* 1988, and references therein). Alternatively, amino acids could have been synthesized directly in the ISM via grain surface catalysis in hot cloud cores (*Kuan et al.,* 2003) or, as discussed above, via radiation processing of interstellar ices (*Bernstein et al.,* 2002). Other evidence for the affect of aqueous alteration is found from the general trend of a sharply decreasing abundance of aliphatic hydrocarbons among meteorites in the order of Tagish Lake < CM2 < CI1 < CR2, inferred to have resulted from low-temperature chemical oxidation (*Cody and Alexander,* 2005).

Future work will benefit from analyzing less-altered materials — IDPs and samples taken directly from comets. For instance, it may be possible to distinguish between the two proposed origins of meteoritic amino acids by analyzing cometary materials that are expected to have escaped aqueous processing. Unfortunately, no direct samples of comets are yet available (although the imminent return of the Stardust spacecraft will change that) and the small size of IDPs (~1 ng) precludes standard chemical analysis techniques. However, isotopic studies of IDPs suggest that some particles contain well-preserved molecular cloud materials.

Because the magnitude of H-isotopic fractionation is so great in the environment of a cold molecular cloud (factors of 100–10,000), D/H measurements provide the most direct means of identifying molecular cloud material. However, recent theoretical studies have suggested that gas-phase chemical reactions in the outer portions of the solar nebula may also have resulted in significant D enrichments in regions that were cold enough and were not opaque to interstellar ionizing radiation (*Aikawa and Herbst,* 2001). In any event, such chemistry would have occurred in conditions that are characteristic of the immediately preceding molecular cloud phase and chemistry in either regime would have isotopically imprinted the starting materials found in the protosolar nebula in a similar manner.

In a simplified view, the most D-rich materials are likely to have experienced the least degree of exchange or dilution. Figure 7 summarizes typical ranges of D/H ratios observed in gas-phase molecules in cold molecular clouds with those of meteorites, IDPs, and comets (where only H_2O and HCN have been measured so far). Deuterium/hydrogen ratios of IDPs reach values (50× terrestrial) that are significantly higher than those observed so far in meteorites (~8× terrestrial) (*Guan et al.,* 1998). However, such extreme values in

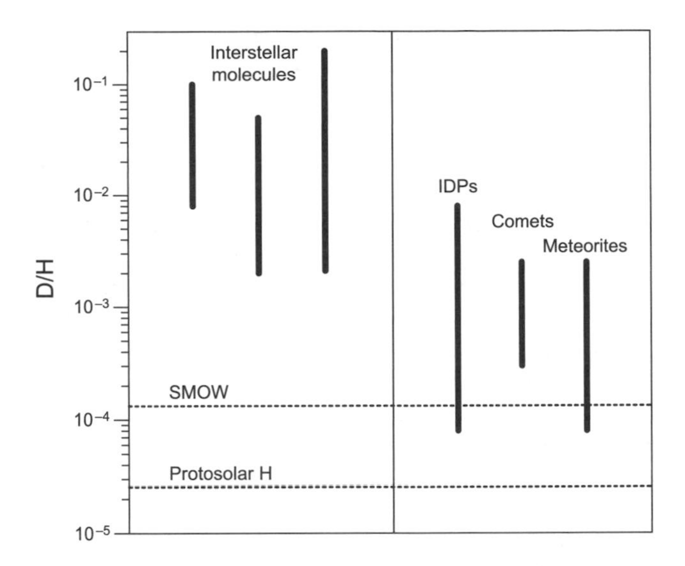

Fig. 7. Comparison of the typical range of D/H ratios observed in gas phase molecules in cold interstellar molecular clouds with those of IDPs, comets (H_2O and HCN), and meteorites. D/H ratios of terrestrial standard mean ocean water (SMOW) and protosolar H_2 are shown for reference. Adapted from *Messenger* (2000).

IDPs are relatively rare, with only ~4% of measured cluster IDP fragments exceeding 10× terrestrial and 8% exceeding 5× terrestrial (*Messenger et al.,* 2002). By comparison, the CR chondrites Renazzo and Al Rais have roughly an order of magnitude lower abundance of these very D-rich materials (*Guan et al.,* 1997, 1998; *Young et al.,* 2004). The average bulk D/H ratio of cluster IDP fragments (δD = 1300‰) (*Messenger et al.,* 2002) is higher than that of most chemically untreated meteorite samples studied so far, but is very similar to some CR chondrites including Renazzo (δD ~ 1000‰) and Al Rais (δD ~ 1280‰) (*Guan et al.,* 1997, 1998; *Young et al.,* 2004; *Robert,* 2003, and references therein). However, the D/H ratios of cluster IDPs exhibit far greater variability in comparison to similar-sized meteorite matrix fragments. Some of this heterogeneity may be due to variable loss of D-rich volatiles during atmospheric entry heating or the loss of soluble D-rich organics during the standard hexane rinse used to clean IDPs of collector (silicone) oil. The similar bulk average D/H ratios of cluster IDPs and CR chondrites suggests that CR chondrites may have accreted organic matter similar to that of CP IDPs that subsequently became isotopically homogenized during parent-body hydrothermal processing. While these considerations serve as a useful guide in identifying the least-altered material, the great majority (>99%) of the organic matter in cold molecular clouds is condensed onto grains whose isotopic compositions have so far defied spectroscopic measurement. It is possible that some *solid-phase* D-rich molecules in meteorites and IDPs are direct samples of interstellar molecules even though their D enrichments (factor of 2–4) are much lower than those observed in *gas-phase* interstellar molecules.

The most D- and ^{15}N-rich materials in meteorites and IDPs have been located by isotopic imaging by ion microprobe. The first D "hotspot" in an IDP was reported by *McKeegan et al.* (1987), where a small (<2 μm) region was found to reach 10× terrestrial, and several other similar cases have since been found (*Messenger,* 2000; *Keller et al.,* 2004). These studies were limited to the 1–2-μm spatial resolution of the IMS-series ion microprobes and were performed on chemically untreated samples. It is possible that significantly more D-rich materials remain to be found at higher spatial resolution or as minor phases that can only be identified by molecular specific isotopic measurements. Recent studies of IDPs using the NanoSIMS ion microprobe have revealed that this is indeed the case for N, finding that some IDPs contain a mixture of moderately (several hundred permil) ^{15}N-rich carbonaceous matter and submicrometer inclusions of very ^{15}N-rich material (>1200‰) (*Floss et al.,* 2004).

Deuterium- and ^{15}N-rich hotspots (Fig. 8) are the best candidates for surviving chunks of molecular cloud material. If this is the case, one would expect any mineral grains entrained within these isotopically anomalous materials to also have presolar origins. Recent studies have indicated that stardust abundances are highest within ^{15}N- and D-rich

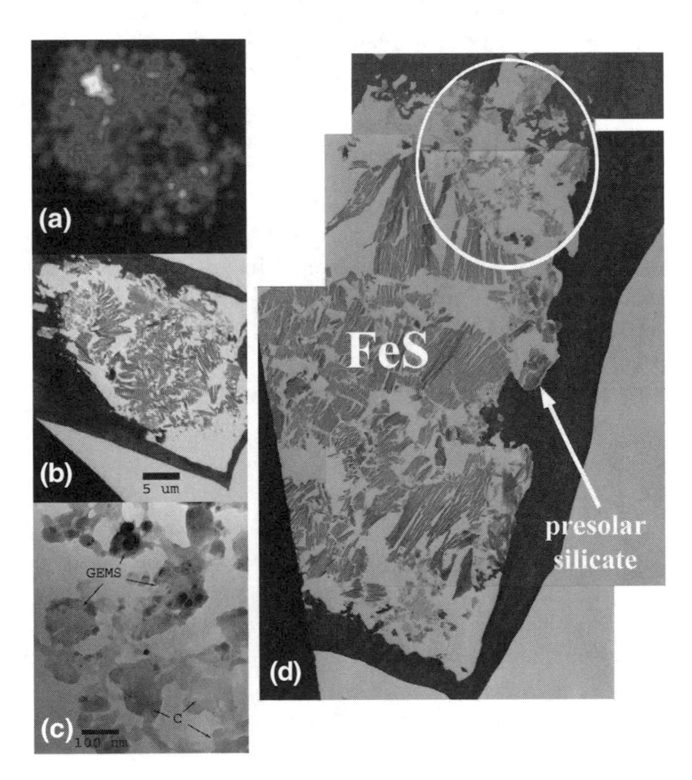

Fig. 8. **(a)** D/H ratio image of IDP L2011 D11. D/H ratios in the image range from 2 to 10× terrestrial. **(b)** TEM brightfield image of the IDP after being extracted from the gold substrate, embedded in elemental S and sliced by diamond ultramicrotomy. **(c)** Representative TEM view of materials found within the D hotspot, including GEMS grains, forsterite, enstatite, FeS, and carbonaceous matter. **(d)** Location of Group 1 presolar silicate grain found by NanoSIMS with location of D hotspot indicated by the ellipse. From *Messenger and Keller* (2005).

ma-terials (*Floss and Stadermann,* 2004; *Messenger and Keller,* 2005). While the first D hotspot investigated so far by Nano-SIMS was found to contain one presolar silicate, many isotopically solar crystalline silicates and GEMS grains were also present. It is likely that in such a case, these materials (presolar organic matter, solar system grains, and stardust) accreted in the solar nebula. In another example of this association (Plate 6), a presolar supernova silicate grain in an IDP was found to be coated with ^{15}N-rich organic matter, perhaps the best candidate so far to resemble the classic model of an interstellar grain coated with a refractory organic mantle (*Li and Greenberg,* 1997) (Fig. 1).

Although organic studies of IDPs are still at an early stage, compositional differences are already apparent between anhydrous IDPs and carbonaceous chondrites (see *Pizzarello et al.,* 2006, for further discussion). So far, the only specific organic molecules identified in IDPs are PAHs, determined by two-step resonant ionization mass spectrometry (*Clemett et al.,* 1993). The particles studied tended to have more high-mass molecules that appeared to be significantly substituted by N-bearing functional groups. A recent FTIR study of IDPs with wide ranging D/H ratios (including the D-rich hotspot in Fig. 8) suggest that the abundance of aliphatic hydrocarbons is higher among D-rich IDPs than D-poor IDPs or meteorites (*Keller et al.,* 2004). This observation is in line with the trend of alteration observed in meteoritic organics and suggests that the organic matter in D-rich anhydrous IDPs most closely resembles the original starting composition of solar system organic matter.

5. OVERVIEW

We have seen that the nature of the starting materials evolved considerably during the prehistory and early history of the solar system. Nevertheless, these materials share several universal traits that extend to other young stellar systems and therefore provide insight into the nature of other planetary systems. Despite the inherent uncertainties and complex overprinting of secondary processes, we can make the following conclusions:

1. The starting materials at all stages were products of long and complex histories beginning with stellar ejection, processing in disparate interstellar environments, and in the protonebular collapse.

2. The starting materials are generally described as aggregates of submicrometer mineral grains dominated by mixtures of amorphous and crystalline silicates, refractory organic matter that contained a significant aromatic component, and condensed volatile ices.

3. The starting materials for planetary construction originated from evolved stars, outflows from young stellar objects, condensation in dense molecular clouds, and variably processed materials in the solar nebula itself.

4. The nature of the starting materials is partially obscured by secondary alteration in multiple environments.

5. Isotopic anomalies are key indicators for the presence of primitive materials and demonstrate that some starting materials had both stellar and interstellar origins.

6. Anhydrous IDPs are the best currently available examples of what the nonvolatile portions of the starting materials probably looked like.

7. An important constituent of the starting materials are volatiles that have not yet been studied in the laboratory, but are likely present in small outer solar system bodies.

The once disparate views of astrophysics and cosmochemistry are beginning to form a unified picture of the complex processes of star formation and the history of the starting materials for their construction. Further advances are anticipated as spatial and spectral resolution of astrophysical observations are improved. Recent advances in cosmochemistry have also been primarily aided by revolutionary advances in analytical instruments. For the future exploration of cosmochemical frontiers, direct samples of comets and asteroids are considered to have overriding importance. The return of cryogenic stored samples from a short-period comet will surely be technically challenging, but such a mission is essential to achieving the clearest view of the starting materials.

REFERENCES

Aikawa Y. and Herbst E. (2001) Two-dimensional distribution and column density of gaseous molecules in protoplanetary discs. *Astron. Astrophys., 371,* 1107–1117.

Alexander C. M. O'D. and Love S. (2001) Atmospheric entry heating of micrometeorites revisited: Higher temperatures and potential biases (abstract). In *Lunar and Planetary Science XXXII,* Abstract #1935. Lunar and Planetary Institute, Houston (CD-ROM).

Alexander C. M. O'D., Russell S. S., Arden J. W., Ash R. D., Grady M. M., and Pillinger C. T. (1998) The origin of chondritic macromolecular organic matter: A carbon and nitrogen isotopic study. *Meteoritics & Planet. Sci., 33,* 603–622.

Allamandola L. J., Sandford S. A., and Valero G. (1988) Photochemical and thermal evolution of interstellar/pre-cometary ice analogs. *Icarus, 76,* 225–252.

Allamandola L. J., Tielens A. G. G. M., and Barker J. R. (1989) Interstellar polycyclic aromatic hydrocarbons: The infrared emission bands, the excitation-emission mechanism and the astrophysical implications. *Astrophys. J. Suppl. Ser., 71,* 733–755.

Allamandola L. J., Sandford S. A., Tielens A. G. G. M., and Herbst T. M. (1992) Infrared spectroscopy of dense clouds in the C-H stretch region: Methanol and "diamonds." *Astrophys. J., 399,* 134–146.

Anders E. and Ebihara M. (1982) Solar system abundances of the elements. *Geochim. Cosmochim. Acta, 46,* 2363–2380.

Anders E. and Grevesse N. (1989) Abundances of the elements: Meteoritic and solar. *Geochim. Cosmochim. Acta, 53,* 197–214.

Arpigny C., Jehin E., Manfroid J., Hutsemékers D., Schulz R., Stüwe J. A., Zucconi J.-M. and Ilyin I. (2003) Anomalous nitrogen isotope ratio in comets. *Science, 301,* 1522–1524.

Bergin E. A., Alves J., Huard T., and Lada C. J. (2002) N_2H^+ and $C^{18}O$ depletion in a cold dark cloud. *Astrophys. J. Lett., 570,* L101–L104.

Bernatowicz T. J., Messenger S., Pravdivtseva O., Swan P., and Walker R. M. (2003) Pristine presolar silicon carbide. *Geochim. Cosmochim. Acta, 67,* 4679–4691.

Bernstein M. P., Sandford S. A., Allamandola L. J., Chang S., and

Scharberg M. A. (1995) Organic compounds produced by photolysis of realistic interstellar and cometary ice analogs containing methanol. *Astrophys. J., 454,* 327–344.

Bernstein M. P., Sandford S. A., and Allamandola L. J. (1999a) Life's far-flung raw materials. *Sci. Am., 281,* 42–49.

Bernstein M. P., Sandford S. A., Allamandola L. J., Gillette J. S., Clemett S. J., and Zare R. N. (1999b) Ultraviolet irradiation of polycyclic aromatic hydrocarbons in ices: Production of alcohols, quinones, and ethers. *Science, 283,* 1135–1138.

Bernstein M. P., Sandford S. A., and Allamandola, L. J. (2000) H, C, N, and O isotopic substitution studies of the 2165 cm^{-1} (4.62 μm) "XCN" feature produced by UV photolysis of mixed molecular ices. *Astrophys. J., 542,* 894–897.

Bernstein M. P., Dworkin J., Sandford S. A., and Allamandola L. J. (2001) Ultraviolet irradiation of naphthalene in H_2O ice: Implications for meteorites and biogenesis. *Meteoritics & Planet. Sci., 36,* 351–358.

Bernstein M. P., Dworkin J. P., Sandford S. A., Cooper G. W., and Allamandola L. J. (2002) The formation of racemic amino acids by ultraviolet photolysis of interstellar ice analogs. *Nature, 416,* 401–403.

Bernstein M. P., Moore M. H., Elsila J. E., Sandford S. A., Allamandola L. J., and Zare R. N. (2003) Side group addition to the PAH coronene by proton irradiation in cosmic ice analogs. *Astrophys. J. Lett., 582,* L25–L29.

Bernstein M. P., Sandford S. A., and Allamandola L. J. (2005) The mid-infrared absorption spectra of neutral PAHs in dense interstellar clouds. *Astrophys. J. Suppl. Ser., 161,* 53–64.

Biver N., Bockelée-Morvan D., Crovisier J., Colom P., Henry F., Moreno R., Paubert G., Despois D., and Lis D. C. (2002) Chemical composition diversity among 24 comets observed at radio wavelengths. *Earth Moon Planets, 90,* 323–333.

Bockelée-Morvan D., Crovisier J., Mumma M. J., and Weaver H. A. (2005) The composition of cometary volatiles. In *Comets II* (M. C. Festou et al., eds.), pp. 391–423. Univ. of Arizona, Tucson.

Boogert A. C. A., Schutte W. A., Tielens A. G. G. M., Whittet D. C. B., Helmich F. P., Ehrenfreund P., Wesselius P. R., de Graauw Th., and Prusti T. (1996) Solid methane toward deeply embedded prototstars. *Astron. Astrophys., 315,* L377–L380.

Boss A. P. (1998) Temperatures in protoplanetary disks. *Annu. Rev. Earth Planet. Sci., 26,* 53–80.

Bowey J. E. and Adamson A. J. (2002) A mineralogy of interstellar silicate dust from 10-mm spectra. *Mon. Not. R. Astron. Soc., 334,* 94–106.

Bradley J. P. (2004) Interplanetary dust particles. In *Treatise on Geochemistry, Vol. 1: Meteorites, Comets, and Planets* (A. M. Davis, ed.), pp. 689–711. Elsevier, Oxford.

Bradley J. P. and Brownlee D. (1986) Cometary particles: Thin sectioning and electron beam analysis. *Science, 231,* 1542–1544.

Bradley J. P. and Dai Z. R. (2005) Mechanism of GEMS formation. *Astrophys. J., 617,* 650–655.

Bradley J. P., Sandford S. A., and Walker R. M. (1988) Interplanetary dust particles. In *Meteorites and the Early Solar System* (J. F. Kerridge and M. S. Matthews, eds.), pp. 861–895. Univ. of Arizona, Tucson.

Bradley J. P., Keller L. P., Snow T. P., Hanner M. S., Flynn G. J., Gezo J. C., Clemett S. J., Brownlee D. E., and Bowey J. E. (1999) An infrared spectral match between GEMS and interstellar grains. *Science, 285,* 1716–1718.

Bregman J. D., Campins H., Witteborn F. C., Wooden D. H., Rank D. M., Allamandola L. J., Cohen M., and Tielens A. G. G. M. (1987) Airborne and ground-based spectrophotometry of comet P/Halley from 5–13 micrometers. *Astron. Astrophys., 187,* 616–620.

Bregman J. D., Hayward T. L., and Sloan G. C. (2000) Discovery of the 11.2 μm PAH band in absorption toward MonR2 IRS3. *Astrophys. J. Lett., 544,* L75–L78.

Breneman H. H. and Stone E. C. (1985) Solar coronal and photospheric abundances from solar energetic particle measurements. *Astrophys. J. Lett., 294,* L57–L62.

Brooke T. Y., Sellgren K., and Geballe T. R. (1999) New 3 micron spectra of young stellar objects with H_2O ice bands. *Astrophys. J., 517,* 883–900.

Brown P. D. and Charnley S. B. (1990) Chemical models of interstellar gas-grain processes — I. Modelling and the effect of accretion on gas abundances and mantle composition in dense clouds. *Mon. Not. R. Astron. Soc., 244,* 432–443.

Brownlee D. E. and Joswiak D. J. (2004) The solar nebula's first accretionary particles (FAPs): Are they preserved in collected interplanetary dust samples? (abstract). In *Lunar and Planetary Science XXXV,* Abstract #1944. Lunar and Planetary Institute, Houston (CD-ROM).

Burbidge E. M., Burbidge G. R., Fowler W. A., and Hoyle F. (1957) Synthesis of the elements in stars. *Rev. Mod. Phys., 29,* 547–650.

Burnett D. S., Woolum D. S., Benjamin T. M., Rogers P. S. Z., Duffy C. J., and Maggiore C. (1989) A test of the smoothness of the elemental abundances of carbonaceous chondrites. *Geochim. Cosmochim. Acta, 53,* 471–481.

Busso M., Gallino R., and Wasserburg G. J. (2003) Short-lived nuclei in the early solar system: A low mass stellar source? *Publ. Astron. Soc. Austral., 20,* 356–370.

Butchart L., McFadzean A. D., Whittet D. C. B, Gavalle T. R., and Greenberg J. M. (1986) Three micron spectroscopy of the galactic centre source IRS7. *Astron. Astrophys., 154,* L5–L7.

Cameron A. G. W. (1985) Formation and evolution of the primitive solar nebula. In *Protostars and Planets II* (D. C. Black and M. S. Matthews, eds.), pp. 1073–1099. Univ. of Arizona, Tucson.

Campins H. and Ryan E. V. (1989) The identification of crystalline olivine in cometary silicates. *Astrophys. J., 341,* 1059–1066.

Cassen P. and Boss A. P (1988) Protostellar collapse, dust grains, and solar system formation. In *Meteorites and the Early Solar System* (J. F. Kerridge and M. S. Matthews, eds.), pp. 304–328. Univ. of Arizona, Tucson.

Charnley S. B. and Rodgers S. D. (2002) The end of interstellar chemistry as the origin of nitrogen in comets and meteorites. *Astrophys. J. Lett., 569,* L133–L137.

Charnley S. B., Tielens A. G. G. M., and Millar T. J. (1992) On the molecular complexity of the hot cores in Orion A: Grain surface chemistry as "the last refuge of a scoundrel." *Astrophys. J. Lett., 399,* L71–L74.

Charnley S. B., Kress M. E., Tielens A. G. G. M., and Millar T. J. (1995) Interstellar alcohols. *Astrophys. J., 448,* 232–239.

Chiar J. E., Tielens A. G. G. M., Whittet D. C. B., Schutte W. A., Boogert A. C. A., Lutz D., van Dishoeck E. F., and Bernstein M. P. (2000) The composition and distribution of dust along the line of sight towards the galactic center. *Astrophys. J., 537,* 749–762.

Choi B.-G., Wasserburg G. J., and Huss G. R. (1999) Circumstellar hibonite and corundum and nucleosynthesis in asymptotic giant branch stars. *Astrophys. J. Lett., 522,* L133–L136.

Clemett S., Maechling C. R., Zare R. N., Swan P., and Walker R. M. (1993) Identification of complex aromatic molecules in individual interplanetary dust particles. *Science, 262,* 721–725.

Clemett S. J., Chillier X. D. F., Gillette S., Zare R. N., Maurette M., Engrand C., and Kurat G. (1998) Observation of indigenous polycyclic aromatic hydrocarbons in 'giant' carbonaceous Antarctic micrometeorites. *Origins Life Evol. Biosph., 28,* 425–448.

Cody G. D. and Alexander C. M. O'D. (2005) NMR studies of chemical structural variation of insoluble organic matter from different carbonaceous chondrite groups. *Geochim. Cosmochim. Acta, 69,* 1085–1097.

Cook W. R., Stone E. C.,and Vogt R. E. (1984) Elemental composition of solar energetic particles. *Astrophys. J., 279,* 827–838.

Cronin J. R. and Chang S. (1993) Organic matter in meteorites: Molecular and isotopic analysis of the Murchison meteorite. In *Chemistry of Life's Origins* (J. M. Greenberg and V. Pirronello, eds.), pp. 209–258. Kluwer, Dordrecht.

Cronin J. R., Pizzarello S., and Cruikshank D. (1988) Organic matter in carbonaceous chondrites, planetary satellites, asteroids, and comets. In *Meteorites and the Early Solar System* (J. F. Kerridge and M. S. Matthews, eds.), pp. 819–857. Univ. of Arizona, Tucson.

Crovisier J. and 14 colleagues (2000) The thermal infrared spectra of comets Hale-Bopp and 103P/Hartley 2 observed with the Infrared Space Observatory. In *Thermal Emission Spectroscopy and Analysis of Dust, Disks, and Regoliths* (M. L. Sitko et al., eds.), pp. 109–117. ASP Conference Series 196, Astronomical Society of the Pacific, San Francisco.

Cuzzi J. N., Davis S. S., and Doubrovolskis A. R. (2003) Blowing in the wind. II. Creation and redistribution of refractory inclusions in a turbulent protoplanetary nebula. *Icarus, 166,* 385 –402.

de Graauw Th. and 31 colleagues (1996) SWS observations of solid CO_2 in molecular clouds. *Astron. Astrophys., 315,* L345–L348.

Dones L., Weissmann P. R., Levison H. F., and Duncan M. J. (2004) Oort cloud formation and dynamics. In *Comets II* (M. C. Festou et al., eds.), pp. 153–174. Univ. of Arizona, Tucson.

Dorschner J. and Henning Th. (1995) Dust metamorphosis in the galaxy. *Astron. Astrophys. Rev., 6,* 271–333.

Duncan M., Quinn T., and Tremaine S. (1988) The origin of short-period comets. *Astrophys. J. Lett., 328,* 69.

Duprat J., Engrand C., Maurette M., Gounelle M., Kurat G., and Leroux H. (2005) Friable micrometeorites from central Antarctica snow (abstract). In *Lunar and Planetary Science XXXVI,* Abstract #1678. Lunar and Planetary Institute, Houston (CD-ROM).

Dworkin J. P., Deamer D. W., Sandford S. A., and Allamandola L. J. (2001) Self-assembling amphiphilic molecules: Synthesis in simulated interstellar/precometary ices. *Proc. Natl. Acad. Sci., 98,* 815–819.

Eberhardt P., Reber M., Krankowsky D., and Hodges R. R. (1995) The D/H and $^{18}O/^{16}O$ ratios in water from comet P/Halley. *Astron. Astrophys., 302,* 301–316.

Ehrenfreund P. and Charnley S. B. (2000) Organic molecules in the interstellar medium, comets, and meteorites: A voyage from dark clouds to the early Earth. *Annu. Rev. Astron. Astrophys., 38,* 427–483.

Engel M. H., Macko S. A., and Silfer J. A. (1990) Carbon isotope composition of the individual amino acids in the Murchison meteorite. *Nature, 348,* 47–49.

Epstein S., Krishnamurthy R. V., Cronin J. R., Pizzarello S., and Yuen G. U. (1987) Unusual stable isotope ratios in amino acid and carboxylic acid extracts from the Murchison meteorite. *Nature, 326,* 477–479.

Evans N. (1999) Physical conditions in regions of star formation. *Annu. Rev. Astron. Astrophys., 37,* 311–362.

Floss C. and Stadermann F. J. (2004) Isotopically primitive interplanetary dust particles of cometary origin: Evidence from nitrogen isotopic compositions (abstract). In *Lunar and Planetary Science XXXV,* Abstract #1281. Lunar and Planetary Institute, Houston (CD-ROM).

Floss C., Stadermann F. J., Bradley J. P., Dai Z. R., Bajt S., and Graham G. (2004) Carbon and nitrogen isotopic anomalies in an anhydrous interplanetary dust particle. *Science, 303,* 1355–1358.

Flynn G. J. (1989) Atmospheric entry heating: A criterion to distinguish between asteroidal and cometary sources of interplanetary dust. *Icarus, 77,* 287–310.

Flynn G. J., Sutton S. R., Bajt S., Klock W., Thomas K. L., and Keller L. P. (1993) The volatile content of anhydrous interplanetary dust. *Meteoritics, 28,* 349–350.

Flynn G. J., Bajt S., Sutton S. R., Zolensky M. E., Thomas K. L., and Keller L. P. (1996) The abundance pattern of elements having low nebular condensation temperatures in interplanetary dust particles: Evidence for a new chemical type of chondritic material. In *Physics, Chemistry, and Dynamics of Interplanetary Dust* (B. Å. S. Gustafson and M. S. Hanner, eds.), p. 291. ASP Conference Series 104, Astronomical Society of the Pacific, San Francisco.

Gattacceca J., Rochette P., Folco L., and Perchiazzi N. (2005) A new micrometeorite collection from Antarctica and its preliminary characterization by microobservation, microanalysis and magnetic methods (abstract). In *Lunar and Planetary Science XXXVI,* Abstract #1315. Lunar and Planetary Institute, Houston (CD-ROM).

Geiss J. and Reeves H. (1981) Deuterium in the solar system. *Astron. Astrophys., 93,* 189–199.

Gerakines P. A., Moore M. H., and Hudson R. L. (2000) Energetic processing of laboratory ice analogs: UV photolysis versus ion bombardment. *J. Geophys. Res., 106,* 3338.

Goldsmith P. F. (1987) Molecular clouds — An overview. In *Interstellar Processes* (D. J. Hollenbach and H. A. Thronson Jr., eds.), pp. 51–70. Reidel, Dordrecht.

Goldsmith P. F. and Langer W. D. (1978) Molecular cooling and thermal balance of dense interstellar clouds. *Astrophys. J., 222,* 881–895.

Grim R. J. A. and Greenberg J. M. (1987) Ions in grain mantles: The 4.62 micron absorption by OCN- in W33A. *Astrophys. J. Lett., 321,* L91–L96.

Guan Y., Messenger S., and Walker R. M. (1997) The spatial distribution of D-enrichments in Renazzo matrix (abstract). In *Lunar and Planetary Science XXVIII,* Abstract #1737. Lunar and Planetary Institute, Houston (CD-ROM).

Guan Y., Hofmeister A., Messenger S., and Walker R. M. (1998) Two types of deuterium-rich carriers in Renazzo matrix (abstract). In *Lunar and Planetary Science XXIX,* Abstract #1760. Lunar and Planetary Institute, Houston (CD-ROM).

Guillois O., Ledoux G., and Reynaud C. (1999) Diamond infrared emission bands in circumstellar media. *Astrophys. J. Lett., 521,* L133–L136.

Hanner M. S., Lynch D. K., and Russell R. W. (1994) The 8–13 μm spectra of comets and the composition of silicate grains. *Astrophys. J., 425,* 274–285.

Harker D. E. and Desch S. J. (2002) Annealing of pre-cometary silicate grains in solar nebula shocks (abstract). In *Lunar and Planetary Science XXXIII,* Abstract #2002. Lunar and Planetary Institute, Houston (CD-ROM).

Hartmann L. and Kenyon S. J. (1996) The FU Orionis phenomenon. *Annu. Rev. Astron. Astrophys., 34,* 207–240.

Herbst E. (1987) Gas phase chemical processes in molecular clouds. In *Interstellar Processes* (D. J. Hollenbach and H. A. Thronson Jr., eds.), pp. 611–629. Reidel, Dordrecht.

Hollenbach D. J. and Salpeter E. E. (1971) Surface recombination of hydrogen molecules. *Astrophys. J., 163,* 155–164.

Hollis J. M., Snyder L. E., Lovas F. J., and Buhl D. (1976) Radio detection of interstellar DCO^+. *Astrophys. J. Lett., 209,* L83–L85.

Huss G. and Lewis R. S. (1995) Presolar diamond, SiC, and graphite in primitive chondrites: Abundances as a function of meteorite class and petrologic type. *Geochim. Cosmochim. Acta, 59,* 115–160.

Huss G., Meshik A. P., Smith J. B., and Hohenberg C. M. (2003) Presolar diamond, silicon carbide, and graphite in carbonaceous chondrites: Implications for thermal processing in the nebula. *Geochim. Cosmochim. Acta, 24,* 4823–4848.

Jackson A. A. and Zook H. A. (1992) Orbital evolution of dust particles from comets and asteroids. *Icarus, 97,* 70–84.

Jacq T., Walmsley C. M., Henkel C., Baudry A., Mauersberger R., and Jewell P. R. (1990) Deuterated water and ammonia in hot cores. *Astron. Astrophys., 228,* 447–470.

Jäger C., Mutschke H., Begemann B., Dorschner J., and Henning Th. (1994) Steps toward interstellar silicate mineralogy I. Laboratory results of a silicate glass of mean cosmic composition. *Astron. Astrophys., 292,* 641–655.

Joswiak D. J., Brownlee D. E., Pepin R. O., and Schlutter D. J. (2000) Characteristics of asteroidal and cometary IDPs obtained from stratospheric collectors: Summary of measured He release temperatures, velocities and descriptive mineralogy (abstract). In *Lunar and Planetary Science XXXI,* Abstract #1500. Lunar and Planetary Institute, Houston (CD-ROM).

Keller L. P. et al. (2004) The nature of molecular cloud material in interplanetary dust. *Geochim. Cosmochim. Acta, 68,* 2577–2589.

Kemper F., Vriend W. J., and Tielens A. G. G. M. (2004) The absence of crystalline silicates in the diffuse interstellar medium. *Astrophys. J., 609,* 826–837.

Kerridge J. F., Chang S., and Shipp R. (1987) Isotopic characterization of kerogen-like material in the Murchison carbonaceous chondrite. *Geochim. Cosmochim. Acta, 51,* 2527–2540.

Krishnamurthy R., Epstein S., Cronin J., Pizzarello S., and Yuen G. (1992) Isotopic and molecular analyses of hydrocarbons and monocarboxylic acids of the Murchison meteorite. *Geochim. Cosmochim. Acta, 56,* 4045–4058.

Kuan Y.-J., Charnley S. B., Huang H.-C., Tseng W.-L., and Kisiel Z. (2003) Interstellar glycine. *Astrophys. J., 593,* 848–867.

Kurat G., Koeberl C., Presper T., Brandstatter F., and Maurette M. (1994) Petrology and geochemistry of Antarctic micrometeorites. *Geochim. Cosmochim. Acta, 58,* 3879–3904.

Lacy J. H., Baas F., Allamandola L. J., Persson S. E., McGregor P. J., Lonsdale C. J., Geballe T. R., and van der Bult C. E. P. (1984) 4.6 micron absorption features due to solid phase CO and cyano group molecules toward compact infrared sources. *Astrophys. J., 276,* 533–543.

Lacy J., Carr J., Evans N., Baas F., Achtermann J., and Arens J. (1991) Discovery of interstellar methane: Observations of gaseous and solid CH_4 absorption toward young stars in molecular clouds. *Astrophys. J., 376,* 556–560.

Lacy J. H., Faraji H., Sandford S. A., and Allamandola L. J. (1998) Unraveling the 10 μm 'silicate' feature of protostars: The detection of frozen interstellar ammonia. *Astrophys. J. Lett., 501,* L105–L109.

Lada C. J. and Lada E. A. (2003) Embedded clusters in molecular clouds. *Annu. Rev. Astron. Astrophys., 41,* 57–115.

Langer W. D. and Graedel T. E. (1989) Ion-molecule chemistry of dense interstellar clouds: Nitrogen-, O-, and carbon-bearing molecule abundances and isotopic ratios. *Astrophys. J. Suppl. Ser., 69,* 241–269.

Langer W. D., Graedel T. E., Frerking M. A., and Armentrout P. B. (1984) Carbon and O isotope fractionation in dense interstellar clouds. *Astrophys. J., 277,* 581–604.

Langer W. D., van Dishoeck E. F.,Bergin E. A., Blake G. A., Tielens A. G. G. M., Velusamy T., and Whittet D. C. B. (2000) Chemical evolution of protostellar matter. In *Protostars and Planets IV* (V. Mannings et al., eds.), pp. 29–57. Univ. of Arizona, Tucson.

Levison H. F., Dones L., and Duncan M. J. (2001) The origin of Halley-type comets: Probing the inner Oort cloud. *Astron. J., 121,* 2253–2267.

Lewis R. S., Tang M., Wacker J. F., Anders E., and Steele E. (1987) Interstellar diamonds in meteorites. *Nature, 326,* 160–162.

Li A. and Greenberg J. M. (1997) A unified model of interstellar dust. *Astron. Astrophys., 323,* 566–584.

Lodders K. (2003) Solar system abundances and condensation temperatures of the elements. *Astrophys. J., 591,* 1220–1247.

Loinard L., Castets A., Ceccarelli C., Tielens A. G. G. M., Faure A., Caux E., and Duvert G. (2000) The enormous abundance of D_2CO in IRAS 16293-2422. *Astron. Astrophys., 359,* 1169–1174.

Love S. G. and Brownlee D. E. (1993) A direct measurement of the terrestrial mass accretion rate of cosmic dust. *Science, 262,* 1993.

Luu J. X. and Jewitt D. C. (2002) Kuiper Belt Objects: Relics from the accretion disk of the Sun. *Annu. Rev. Astron. Astrophys., 40,* 63–101.

Mac Low M.-M. and Klessen R. S. (2004) Control of star formation by supersonic turbulence. *Rev. Mod. Phys., 76,* 125–194.

Mannings V., Boss A. P., and Russell S. S., eds. (2000) *Protostars and Planets IV.* Univ. of Arizona, Tucson. 1422 pp.

Matrajt G., Pizzarello S., Taylor S., and Brownlee D. (2004) Concentration and variability of the AIB amino acid in polar micrometeorites: Implications for the exogenous delivery of amino acids to the primitive Earth. *Meteoritics & Planet. Sci., 39,* 1849–1858.

Mauersberger R., Henkel C., Jacq T., and Walmsley C. M. (1988) Deuterated methanol in Orion. *Astron. Astrophys., 194,* L1–L4.

Maurette M., Immel G., Engrand C., Kurat G., and Pillinger C. T. (1994) The 1994 EUROMET collection of micrometeorites at Cap-Prudhomme, Antarctica. *Meteoritics, 29,* 499.

McKeegan K. D., Swan P. D., Walker R. M., Wopenka B., and Zinner E. (1987) Hydrogen isotopic variations in interplanetary dust particles (abstract). In *Lunar and Planetary Science XVIII,* p. 627. Lunar and Planetary Institute, Houston.

Meier R., Owen T. C., Jewitt D. C., Matthews H. E., Senay M., Biver N., Bockelée-Morvan D., Crovisier J., and Gautier D. (1998) Deuterium in comet C/1995 O1 (Hale-Bopp): Detection of DCN. *Science, 279,* 1707–1710.

Mennella V., Brucato J. R., Colangeli L., and Palumbo P. (2002) C-H bond formation in carbon grains by exposure to atomic hydrogen: The evolution of the carrier of the interstellar 3.4 micron band. *Astrophys. J., 569,* 531–540.

Messenger S. (2000) Identification of molecular cloud material in interplanetary dust particles. *Nature, 404,* 968–971.

Messenger S. and Keller L. P. (2005) Association of presolar grains with molecular cloud material in IDPs (abstract). In *Lunar and Planetary Science XXXVI,* Abstract #1846. Lunar and Planetary Institute, Houston (CD-ROM).

Messenger S. and Walker R. M. (1997) Evidence for molecular cloud material in meteorites and interplanetary dust. In *Astrophysical Implications of the Laboratory Study of Interstellar Materials* (T. J. Bernatowicz and E. K. Zinner, eds.), pp. 545–564. AIP Conference Proceedings 402, American Institute of Physics, New York.

Messenger S., Stadermann F. J., Floss C., Nittler L. R., and Mukhopadhyay S. (2002) Isotopic signatures of presolar materials in interplanetary dust. *Space Sci. Rev., 95,* 1–10.

Messenger S., Keller L. P., Stadermann F. J., Walker R. M., and Zinner E. (2003) Samples of stars beyond the solar system: Silicate grains in interplanetary dust. *Science, 300,* 105–108.

Messenger S., Keller L. P., and Lauretta D. S. (2005) Supernova olivine from cometary dust. *Science, 309,* 737–741.

Meyer B. S. and Zinner E. (2006) Nucleosynthesis. In *Meteorites and the Early Solar System II* (D. S. Lauretta and H. Y. McSween Jr., eds.), this volume. Univ. of Arizona, Tucson.

Millar T. J., Roberts H., Markwick A. J., and Charnley S. B. (2000) The role of H_2D^+ in the deuteration of interstellar molecules. *Philos. Trans. R. Soc. London, A358,* 2535–2547.

Moore M. H., Ferrante R. F., and Nuth J. A. (1996) Infrared spectra of proton irradiated ices containing methanol. *Planet. Space Sci., 44,* 927–935.

Mostefaoui S. and Hoppe P. (2004) Discovery of abundant in situ silicate and spinel grains from red giant stars in a primitive meteorite. *Astrophys. J. Lett., 613,* L149–L152.

Mostefaoui S., Lugmair G. W., and Hoppe P. (2005) ^{60}Fe: A heat source for planetary differentiation from a nearby supernova explosion. *Astrophys. J., 625,* 271–277.

Mumma M. J., Weissman P. R., and Stern A. S. (1993) Comets and the origin of the solar system: Reading the Rosetta stone. In *Protostars and Planets III* (E. H. Levy and J. I. Lunine, eds.), pp. 1177–1252. Univ. of Arizona, Tucson.

Mumma M. J., Disanti M. A., Dello Russo N., Magee-Sauer K., Gibb E., and Novak R. (2003) Remote infrared observations of parent volatiles in comets: A window on the early solar system. *Adv. Space Res., 31,* 2563–2575.

Nagashima K., Krot A. N., and Yurimoto H. (2004) Stardust silicates from primitive meteorites. *Nature, 428,* 921–924.

Nakamura T., Noguchi T., Yada T., Nakamuta Y., and Takaoka N. (2001) Bulk mineralogy of individual micrometeorites determined by X-ray diffraction analysis and transmission electron microscopy. *Geochim. Cosmochim. Acta, 65,* 4385–4397.

Nguyen A. and Zinner E. (2004) Discovery of ancient silicate stardust in a meteorite. *Science, 303,* 1496.

Nier A. O. and Schlutter D. J. (1992) Extraction of helium from individual interplanetary dust particles by step-heating. *Meteoritics, 27,* 166–173.

Nier A. O. and Schlutter D. J. (1993) The thermal history of interplanetary dust particles collected in the Earth's stratosphere. *Meteoritics, 28,* 675–681.

Nittler L. R. and Dauphaus N. (2006) Meteorites and the chemical evolution of the Milky Way. In *Meteorites and the Early Solar System II* (D. S. Lauretta and H. Y. McSween Jr., eds.), this volume. Univ. of Arizona, Tucson.

Nittler L. R., Hoppe P., Alexander C. M. O'D., Amari S., Eberhardt P., Gao X., Lewis R. S., Strebel R., Walker R. M., and Zinner E. (1995) Silicon nitride from supernovae. *Astrophys. J. Lett., 453,* L25–L28.

Nittler L. R., Alexander C. M. O'D., Stadermann F. J., and Zinner E. (2005) Presolar Al-, Ca-, and Ti-rich oxide grains in the Krymka meteorite (abstract). In *Lunar and Planetary Science XXXVI,* Abstract #2200. Lunar and Planetary Institute, Houston (CD-ROM).

Noguchi T., Yano H., Terada K., Imae N., Yada T., Nakamura T., and Kojima H. (2002) Antarctic micrometeorites collected by the Japanese Antarctic Research Expedition teams during 1996–1999. In *Dust in the Solar System and Other Planetary Systems* (S. F. Green et al., eds.), p. 392. IAU Colloquium 181, Pergamon, New York.

Norman C. and Silk J. (1980) Clumpy molecular clouds: A dynamic model self-consistently regulated by T-Tauri star formation. *Astrophys. J., 238,* 158–174.

Nuth J. A. III, Charnley S. B., and Johnson N. M. (2006) Chemical processes in the interstellar medium: Source of the gas and dust in the primitive solar nebula. In *Meteorites and the Early Solar System II* (D. S. Lauretta and H. Y. McSween Jr., eds.), this volume. Univ. of Arizona, Tucson.

Oort J. H. (1950) The structure of the cloud of comets surrounding the solar system and a hypothesis concerning its origin. *Bull. Astron. Inst. Netherlands, 12,* 91–110.

Parise B., Ceccarelli C., Tielens A. G. G. M., Herbst E., Lefloch B., Caux E., Castets A., Mukhopadhyay I., Pagani L., and Loinard L. (2002) Detection of doubly-deuterated methanol in the solar-type protostar IRAS 16293-2422. *Astron. Astrophys., 393,* L49–L53.

Park J.-Y. and Woon D. E. (2004) Computational confirmation of the carrier for the "XCN" interstellar ice band: OCN^- charge transfer complexes. *Astrophys. J. Lett., 601,* L63–L66.

Peeters E. Allamandola L. J., Bauschlicher C. W. Jr., Hudgins D. M., Sandford S. A., and Tielens A. G. G. M. (2004) Deuterated interstellar polycyclic aromatic hydrocarbons. *Astrophys. J., 604,* 252–257.

Pendleton Y. J. and Allamandola L. J. (2002) The organic refractory material in the diffuse interstellar medium: Mid-infrared spectroscopic constraints. *Astrophys. J. Suppl. Ser., 138,* 75–98.

Pendleton Y., Tielens A. G. G. M., and Werner M. W. (1990) Studies of dust grain properties in infrared reflection nebulae *Astrophys. J., 349,* 107–119.

Pendleton Y. J., Sandford S. A., Allamandola L. J., Tielens A. G. G. M., and Sellgren K. (1994) Near-infrared absorption spectroscopy of interstellar hydrocarbon grains. *Astrophys. J., 437,* 683–696.

Pendleton Y. J., Tielens A. G. G. M., Tokunaga A. T., and Bernstein M. P. (1999) The interstellar 4.62 micron band. *Astrophys. J., 513,* 294–304.

Pizzarello S. and Huang Y. (2005) The deuterium enrichment of individual amino acids in carbonaceous meteorites: A case for the presolar distribution of biomolecule precursors. *Geochim. Cosmochim. Acta, 69,* 599–605.

Pizzarello S., Cooper G. W. and Flynn G. J. (2006) The nature and distribution of the organic material in carbonaceous chondrites and interplanetary dust particles. In *Meteorites and the*

Early Solar System II (D. S. Lauretta and H. Y. McSween Jr., eds.), this volume. Univ. of Arizona, Tucson.

Prasad S. S. and Tarafdar S. P. (1983) UV radiation field inside dense clouds: Its possible existence and chemical implications. *Astrophys. J., 267,* 603–609.

Rietmeijer F. J. M. (2002) Collected extraterrestrial materials: Interplanetary dust particles, micrometeorites, meteorites, and meteoric dust. In *Meteors in the Earth's Atmosphere* (E. Murad and I. P. Williams, eds.), p. 215. Cambridge Univ., Cambridge.

Rietmeijer F. J. M. and McKinnon I. D. R. (1987) Cometary evolution: Clues from chondritic interplanetary dust particles. In *Symposium on the Similarity and Diversity of Comets* (E. J. Rolfe and B. Battrick, eds.), pp. 363–367. ESA SP-278, ESTEC, Noordwijk.

Robert F. (2003) The D/H ratio in chondrites. *Space Sci. Rev., 106,* 87–101.

Roberts H., Fuller G. A., Millar T. J., Hatchell J., and Buckle J. V. (2002) A survey of [HDCO]/[H_2CO] and [DCN]/[HCN] ratios towards low-mass protostellar cores. *Astron. Astrophys., 381,* 1026–1038.

Rodgers S. D. and Charnley S. B. (2003) Chemical evolution in protostellar envelopes: Cocoon chemistry. *Astrophys. J., 585,* 355–371.

Rodgers S. D. and Charnley S. B. (2004) Interstellar diazenylium recombination and nitrogen isotopic fractionation. *Mon. Not. R. Astron. Soc., 352,* 600–604.

Roueff E., Tiné S., Coudert L. H., Pineau des Forêts G., Falgarone E., and Gerin M. (2000) Detection of doubly deuterated ammonia in L134N. *Astron. Astrophys., 354,* L63–L66.

Russell S. S., Pillinger C. T., Arden J. W., Lee M. R., and Ott U. (1992) A new type of meteoritic diamond in the enstatite chondrite Abee. *Science, 256,* 206–209.

Sandford S. A. (1986) Solar flare track densities in interplanetary dust particles: The determination of an asteroidal versus cometary source of the zodiacal dust cloud. *Icarus, 68,* 377–394.

Sandford S. A. (1996) The inventory of interstellar materials available for the formation of the solar system. *Meteoritics & Planet. Sci., 31,* 449–476.

Sandford S. A. and Allamandola L. J. (1993) Condensation and vaporization studies of CH_3OH and NH_3 ices: Major implications for astrochemistry. *Astrophys. J., 417,* 815–825.

Sandford S. A. and Bradley J. P. (1989) Interplanetary dust particles collected in the stratosphere: Observations of atmospheric heating and constraints on their interrelationships and sources. *Icarus, 82,* 146–166.

Sandford S. A., Allamandola L. J., Tielens A. G. G. M., and Valero G. J. (1988) Laboratory studies of the infrared spectral properties of CO in astrophysical ices. *Astrophys. J., 329,* 498–510.

Sandford S. A., Allamandola L. J., Tielens A. G. G. M., Sellgren K., Tapia M., and Pendleton Y. (1991) The interstellar C-H stretching band near 3.4 μm: Constraints on the composition of organic material in the diffuse interstellar medium. *Astrophys. J., 371,* 607–620.

Sandford S. A., Pendleton Y. J. and Allamandola L. J. (1995) The galactic distribution of aliphatic hydrocarbons in the diffuse interstellar medium. *Astrohpys. J., 440,* 697–705.

Sandford S. A., Bernstein M. P., Allamandola L. J., Gillette J. S., and Zare R. N. (2000) Deuterium enrichment of PAHs by photochemically induced exchange with deuterium-rich cosmic ices. *Astrophys. J., 538,* 691–697.

Sandford S. A., Bernstein M. P., and Dworkin J. P. (2001) Assessment of the interstellar processes leading to deuterium enrichment in meteoritic organics. *Meteoritics & Planet. Sci., 36,* 1117–1133.

Sandford S. A., Bernstein M. P., and Allamandola L. J. (2004) The mid-infrared laboratory spectra of naphthalene ($C_{10}H_8$) in solid H_2O. *Astrophys. J., 607,* 346–360.

Shu F. H., Shang H., Gounelle M., Glassgold A. E., and Lee T. (2001) The origin of chondrules and refractory inclusions in chondritic meteorites. *Astrophys. J., 548,* 1029–1050.

Smith N., Bally J., and Morse J. A. (2003) Numerous proplyd candidates in the harsh environment of the Carina nebula. *Astrophys. J. Lett., 587,* L105–L108.

Suess H. (1947) Über kosmische Kernhäufigkeiten I. Mitteilung: Einige Häufigkeitsregeln und ihre Anwendung bei der Abschätzung der Häufigskeitwerte für die mittelsschweren und schweren Elemente. II. Mitteilung: Einzelheiten in der Häufigskeitverteilung der mittelschweren und schweren Kerne. *Z. Naturforsch., 2a,* 311–321, 604–608.

Sykes M. V. and Greenberg R. (1986) The formation and origin of the IRAS zodiacal dust bands as a consequence of single collisions between asteroids. *Icarus, 65,* 51–69.

Sykes M. V., Lebofsky L. A., Hunten D. M., and Low F. (1986) The discovery of dust trails in the orbits of periodic comets. *Science, 232,* 1115–1117.

Tachibana S. and Huss G. R. (2003) The initial abundance of ^{60}Fe in the solar system. *Astrophys. J. Lett., 588,* L41.

Taylor S., Lever J. H., and Govoni J. (2001) A second collection of micrometeorites from the South Pole water well (abstract). In *Lunar and Planetary Science XXXII,* Abstract #1914. Lunar and Planetary Institute, Houston (CD-ROM).

Tegler S. C., Weintraub D. A., Allamandola L. J., Sandford S. A., Rettig T. W., and Campins H. (1993) Detection of the 2165 cm^{-1} (4.619 μm) XCN band in the spectrum of L1551 IRS 5. *Astrophys. J., 411,* 260–265.

Tegler S. C., Weintraub D. A., Rettig T. W., Pendleton Y. J., Whittet D. C. B., and Kulesa C. A. (1995) Evidence for chemical processing of precometary icy grains in circumstellar environments of pre-main-sequence stars. *Astrophys. J., 439,* 279–287.

Teixeira T. C., Devlin J. P., Buch V., and Emerson J. P. (1999) Discovery of solid HDO in grain mantles. *Astron. Astrophys., 347,* L19–L22.

Terzieva R. and Herbst E. (2000) The possibility of nitrogen isotopic fractionation in interstellar clouds. *Mon. Not. R. Astron. Soc., 317,* 563–568.

Thomas K. L., Klöck W., Keller L. P., Blanford G. E., and McKay D. S. (1993) Analysis of fragments from cluster particles: Carbon abundance, bulk chemistry, and mineralogy. *Meteoritics, 28,* 448–449.

Tielens A. G. G. M. (1983) Surface chemistry of deuterated molecules. *Astron. Astrophys., 119,* 177–184.

Tielens A. G. G. M. (1990) Towards a circumstellar silicate mineralogy. In *From Miras to PN: Which Path for Stellar Evolution?* (M. O. Mennessier and A. Omont, eds.), pp. 186–200. Editions Frontiere, Gif-sur-Yvette.

Tielens A. G. G. M. (1995) The interstellar medium. In *Airborne Astronomy Symposium on the Galactic Ecosystem: From Gas to Stars to Dust* (M. R. Haas et al., eds.), pp. 3–22. ASP Conference Series 73, Astronomical Society of the Pacific, San Francisco.

Tielens A. G. G. M. (1997) Deuterium and interstellar chemical processes. In *Astrophysical Implications of the Laboratory Study of Presolar Materials* (T. J. Bernatowicz and E. K.

Zinner, eds.), pp. 523–544. AIP Conference Proceedings 402, American Institute of Physics, New York.

Tielens A. G. G. M. and Allamandola L. J. (1987) Evolution of interstellar dust. In *Physical Processes in Interstellar Clouds* (G. E. Morfill and M. Scholer, eds.), pp. 333–376. Reidel, Dordrecht.

Tielens A. G. G. M. and Hagen W. (1982) Model calculations of the molecular composition of interstellar grain mantles. *Astron. Astrophys., 114,* 245–260.

Tielens A. G. G. M., Tokunaga A. T., Geballe T. R., and Baas F. (1991) Interstellar solid CO: Polar and nonpolar interstellar ices. *Astrophys. J., 381,* 181–199.

Turner B. E. (2001) Deuterated molecules in translucent and dark clouds. *Astrophys. J. Suppl. Ser., 136,* 579–629.

van Dishoeck E. (2004) ISO spectroscopy of gas and dust: From molecular clouds to protoplanetary disks. *Annu. Rev. Astron. Astrophys., 42,* 119–167.

van Dishoeck E. F. and Blake G. A. (1998) Chemical evolution of star-forming regions. *Annu. Rev. Astron. Astrophys., 36,* 317–368.

Vastel C., Phillips T. G., Ceccarelli C., and Pearson J. (2003) First detection of doubly deuterated hydrogen sulfide. *Astrophys. J. Lett., 593,* L97–L100.

Walmsley C. M., Hermsen W., Henkel C., Mauersberger R., and Wilson T. L. (1987) Deuterated ammonia in the Orion hot core. *Astron. Astrophys., 172,* 311–315.

Walmsley C. M., Flower D. R., and Pineau des Forêts G. (2004) Complete depletion in prestellar cores. *Astron. Astrophys., 418,* 1035–1043.

Whipple F. L. (2000) Oort-cloud and Kuiper-Belt comets. *Planet. Space Sci., 48,* 1011.

Whittet D. C. B., Schutte W. A., Tielens A. G. G. M., Boogert A. C. A., de Graauw Th., Ehrenfreund P., Gerakines P. A., Helmich F. P., Prusti T., and van Dishoeck E. F. (1996) An ISO SWS view of interstellar ices — First results. *Astron. Astrophys., 315,* L357–L360.

Yada T. and 10 colleagues (2005) Discovery of abundant presolar silicates in subgroups of Antarctic micrometeorites (abstract). In *Lunar and Planetary Science XXXVI,* Abstract #1227. Lunar and Planetary Institute, Houston (CD-ROM).

Young A. F., Nittler L. R., and Alexander C. M. O'D. (2004) Microscale distribution of hydrogen isotopes in two carbonaceous chondrites (abstract). In *Lunar and Planetary Science XXXV,* Abstract #2097. Lunar and Planetary Institute, Houston (CD-ROM).

Zinner E. (1988) Interstellar cloud material in meteorites. In *Meteorites and the Early Solar System* (J. F. Kerridge and M. S. Matthews, eds.), pp. 956–983. Univ. of Arizona, Tucson.

Zinner E. (1998) Stellar nucleosynthesis and the isotopic composition of presolar grains from primitive meteorites. *Annu. Rev. Earth Planet. Sci., 26,* 147–188.

Zinner E., Amari S., Guinness R., Nguyen A., Stadermann F. J., Walker R. M., and Lewis R. S. (2003) Presolar spinel grains from the Murray and Murchison carbonaceous chondrites. *Geochim. Cosmochim. Acta, 67,* 5083–5095.

The Physics and Chemistry of Nebular Evolution

F. J. Ciesla and S. B. Charnley
NASA Ames Research Center

Primitive materials in the solar system, found in both chondritic meteorites and comets, record distinct chemical environments that existed within the solar nebula. These environments would have been shaped by the dynamic evolution of the solar nebula, which would have affected the local pressure, temperature, radiation flux, and available abundances of chemical reactants. In this chapter we summarize the major physical processes that would have affected the chemistry of the solar nebula and how the effects of these processes are observed in the meteoritic record. We also discuss nebular chemistry in the broader context of recent observations and chemical models of astronomical disks. We review recent disk chemistry models and discuss their applicability to the solar nebula, as well as their direct and potential relevance to meteoritical studies.

1. INTRODUCTION

Condensation calculations have been used as the basic guides for understanding solar nebula chemistry over the last few decades. As reviewed by *Ebel* (2006), condensation calculations provide information about the direction of chemical evolution, telling us what the end products would be in a system given a constant chemical inventory, infinite time, and ready access of all chemical species to one another. These studies have been applied to infer the conditions present at the location in the solar nebula where and when a given meteorite, or its components, formed.

While condensation calculations act as a guide for the chemical evolution of the solar nebula, laboratory studies of meteorites and astronomical observations of comets demonstrate that the actual evolution was much more complicated. Protoplanetary disks are known to evolve by transfering mass inward where it is accreted by the central stars, resulting in the disks cooling and becoming less dense over time. Superimposed on this evolution is the transport of chemical species, which will lead to time-varying concentrations of those species at different locations in the disk. In addition, dynamic, transient heating events are known to have occurred within the solar nebula, altering the physical environment over a period of hours to days. All these processes will disrupt or alter the path toward chemical equilibrium. Thus, solar nebula chemistry cannot be studied independently of the physical evolution of the protoplanetary disk.

In this chapter we demonstrate the ways in which the physical processes in the solar nebula affected the chemical reactions that took place within it. This is not a straightforward task, as there is still much uncertainty about what specific processes did occur within the disk as well as the nature of these processes. Despite these uncertainties, we can provide overviews as to what the end result of different facets of disk evolution may have been and discuss what effects this evolution may have had on nebular chemistry. (For example, we will discuss the effect of mass transport through the solar nebula even though the driving mechanism for this transport remains an area of ongoing research.) It should be the goal of future studies and future generations of meteoriticists, astronomers, and astrophysicists to improve upon the overviews provided here and make greater strides toward developing a coherent, unified model for the evolution of protoplanetary disks. In the remainder of this introduction, we describe the record of chemical evolution provided by meteorite studies and astronomical observations. We then describe the processes that would shape the different chemical environments in a protoplanetary disk in section 2. In doing so, we deal mainly with physical and chemical processes intrinsic to the solar nebula and astronomical disks — we do not explicitly consider the chemical effects of the external environment on nebular evolution through, for example, the influence of nearby massive stars (e.g., *Hester et al.,* 2004). In section 3, we discuss how these different processes would play a role in determining the chemical evolution of primitive materials in the nebula, and focus on explaining the meteoritic and astronomical records identified below. In the last section, we summarize the important points from the chapter and discuss future developments that may help us to further our understanding of chemistry in protoplanetary disks.

1.1. The Meteoritic Record

The surviving products of chemical reactions in our solar nebula were eventually incorporated into the asteroids, comets, and planets currently in orbit around our Sun. Once accreted, much of this material was further processed due to heating and differentiation in the bodies they resided in. The unequilibrated chondritic meteorites are the only samples that we have of materials that escaped significant processing. Many of these objects experienced some thermal and aqueous alteration on their parent bodies, although many still preserve the signatures of their formation and processing within the solar nebula. We focus on these primitive meteorites in discussing the chemical evolution of the solar nebula.

Chondritic meteorites have relative elemental abundances that closely mirror those observed in the Sun. The exceptions to this are the most volatile elements (C, N, and the noble gases), which would not be incorporated into solids until very low temperatures were reached. The bulk elemental compositions of these bodies are taken as evidence that they formed directly in the solar nebula. While their bulk compositions reflect this, there are mineralogical and chemical differences that allow these meteorites to be separated into 3 distinct classes (the enstatite, ordinary, and carbonaceous chondrites), and these classes are further divided into 13 chondritic types. The reasons for these differences are unknown.

While differences exist between the chondrite types, the general makeup of the meteorites are fairly similar. A major component of chondritic meteorites are chondrules, which can make up as much as 80% of the volume of a given meteorite. Chondrules are roughly millimeter-sized, igneous, ferromagnesium silicates that are considered to be evidence for transient heating events in the early solar system (*Connolly et al.,* 2006). In addition to chondrules, refractory objects called calcium-aluminum-rich inclusions (CAIs) are found in various (small, <10%) proportions in the chondritic meteorites. Calcium-aluminum-rich inclusions are the oldest objects in the solar system, consisting of the elements that would be the first to condense from a solar composition gas. Interstitial to these different components is matrix material, consisting of chondrule fragments, metal and troilite grains, and fine-grained silicates. It is the sum of all these objects that determine the chemical compositions of the meteorites. While the different components of the meteorites may have unique compositions, the fact that the whole meteorite has a near-solar composition lends support to the idea of chondrule-matrix complementarity, meaning that what is depleted in one meteoritic component is often equally ehanced in another (*Wood,* 1996).

In looking at the bulk properties of chondritic meteorites, they can be divided into groups based on a number of properties. The oxidation history of these meteorites is one example, as the chondritic meteorites record a wide range of nebular oxygen fugacities. This is often looked at in terms of how Fe is distributed in the meteorites, as carbonaceous meteorites often contain Fe in the form of oxides with metal nearly absent, whereas most of the Fe (>75%) in enstatite chondrites is in the form of metal grains. In addition to the oxygen fugacity, each type of chondritic meteorite falls in a diagnostic region on the oxygen three-isotope plot (*Clayton,* 2003). These differences imply that the chondritic meteorites formed in distinct environments. In addition to their chemical makeup, chondritic meteorites also differ in their physical properties: Each group has different mean chondrule sizes and different volume fractions of chondrules. No clear link has been made to explain these physical properties in chemical models, proving that chondritic meteorite study is as much an astrophysical endeavor as it is a cosmochemical one.

Other properties of chondritic meteorites are likely the results of both physical and chemical processes. For example, water was incorporated into some chondrites, but not all, suggesting some chondrites formed in environments where water ice was present while others did not. Also, while chondritic meteorites are dominated by materials that were isotopicaly homogenized (except for O) in the solar nebula, unaltered presolar grains are found mixed into these same objects. These grains were formed prior to the formation of the solar nebula, and understanding how they avoided being processed will constrain the extent of various physical events within the solar nebula. Finally, while chondritic meteorites generally have elemental abundances that closely mirror that of the Sun, the relative abundances of elements that condense in the range ~500–1300 K decrease with condensation temperatures in many carbonaceous and ordinary chondrites. This moderately volatile-element depletion provides clues as to how chemical environments evolved in the solar nebula. It is issues such as these that require investigations of how the physical evolution of the solar nebula affected the chemistry that would take place within it.

1.2. The Astronomical Record

While the main focus of this chapter is on the processes that affected the chemistry of meteorites, astronomical observations help to constrain the chemical and physical natures of the outer solar nebula as well as disks around other stars. Comets are believed to contain the least-altered primitive materials in the solar system. In principle, they can provide us with information about the chemical and physical processes that operated at large heliocentric distances, as well as the relative contribution of pristine interstellar material to their composition. The molecular ice inventory is very similar to that identified in interstellar ices, with an admixture of trace species known to be present in molecular cloud gas (*Ehrenfreund et al.,* 2004). Cometary molecules, as well as some interplanetary dust particles, exhibit enhanced D/H and $^{15}N/^{14}N$ isotopic ratios, indicative of chemistry at low temperatures (*Arpigny et al.,* 2003; *Aléon et al.,* 2003). However, relative to water, the abundances of some molecules (e.g., methanol) is generally much lower than found in interstellar ices. A more striking difference lies in the presence of crystalline silicate dust in comets (e.g., *Wooden et al.,* 2004). This is in marked contrast to interstellar silicates, which are almost entirely amorphous in nature, but similar to the silicates found in circumstellar debris disks (*Wooden,* 2002). This suggests that, for the refractory dust component at least, some cometary material experienced high-temperature conditions, similar to those found in the inner nebula, and hence that outward mixing from this region was important. Thus, a major aim of nebular models is to understand the spatiotemporal chemical evolution in the comet-forming region of the nebula (5–40 AU).

Protoplanetary disks around young stars serve as analogs for our own solar nebula. Understanding their chemical structure will help us to unravel the chemical structure and evolution of the solar nebula. While huge strides are expected in the coming years with the launch of advanced telescopic facilities, we are already making observations that demonstrate that the chemistry in protoplanetary disks deviates from what is predicted under chemical equilibrium. Observations of SVS 13, a young stellar object, show that H_2O is depleted by a factor of 10 relative to CO compared to what chemical equilibrium models predict in the inner disk (*Carr et al.,* 2004). Identifying where other deviations from equilibrium occur in other protoplanetary disks will help us understand how our own solar nebula evolved over time.

2. PHYSICAL EVOLUTION OF THE NEBULA

2.1. Epochs and Timescales

The formation and evolution of the solar nebula can be described as a progression through different stages. While such a description is used by most who study the formation of the solar system, it should be remembered that the boundaries of these stages are not crisp, but instead overlap such that the different stages blur into one another — some of the processes described may have operated during more than just the stages that they are ascribed to below. Here we follow *Cameron* (1995) in defining four main stages of evolution:

Stage 1: Formation of the nebula. This is the period of large-mass infall from the parent molecular cloud from which the solar nebula formed. Due to the angular momentum of the original cloud, most of the material collapses into a disk that orbits the center of mass, with much of it subsequently migrating inward to form the protostar. This phase lasts on the order of 10^5 yr.

Stage 2: High-mass accretion rate through the nebula. During this stage, mass from the molecular cloud continues to rain down onto the nebula, but at a diminished rate. The mass accretion rate through the nebula is large ($\dot{M} > 10^{-7}\ M_\odot$/yr). The accretion rate through the disk may not be steady, and it is during this phase that large episodic increases in the accretion rate may occur, causing FU Orionis (Fuor) or EX Lupi (Exor) events, which are observed to result in increased luminosity for a period of a few decades (*Calvet et al.,* 2000; *Bell et al.,* 2000; *Connolly et al.,* 2006). This stage also lasts on the order of 10^5 yr.

Stage 3: Low-mass accretion rate through the nebula. The mass accretion rate through the disk and onto the central star will decresase over time, falling below $10^{-7}\ M_\odot$/yr, and the protoplanetary disk enters the third stage of evolution. During this stage, the infall of material from the molecular cloud has decreased severely. It is during this stage that the star is thought to be in the T Tauri epoch, which is the phase prior to which it becomes a main sequence star. Observed T Tauri stars are characterized by a median mass accretion rate of $\sim 10^{-8} M_\odot$/yr at an age of $\sim 10^6$ yr, decreasing with time, and a physical extent of up to a few hundred AU (*Hartmann et al.,* 1998; *Mundy et al.,* 2000; *Calvet et al.,* 2000; *Bell et al.,* 2000). This stage likely lasted a few million years.

Stage 4: Quiescent nebula. This is the final stage of evolution for the solar nebula, and the period during which different processes (solar wind interactions, photoevaporation) operate to erode it. Mass transport through the disk has become minor, and the nebula is relatively cool. This phase can last up to ~10 m.y., depending on the conditions of the nebula and its proximity to other stars (*Hollenbach et al.,* 2000; *Matsuyama et al.,* 2003).

The above stages roughly correspond with the different protostellar classifications identifed in the scheme developed by *Lada and Shu* (1990). These classes are identified based on their different radiative properties. In the Class 0 phase, the young protostar and its attendant disk grow by accreting material from the dense core envelope, and typically emit in the microwave regime. The Class I phase is where the natal envelope is being cleared away, most of the protostellar mass has been accreted, and a lower mass disk is present; the central object grows more massive, the circumstellar environment becomes less dusty, and the radiation is dominated by infrared photons. In the Class II phase, only a disk is present around the protostar and it is here, in the protoplanetary disk, that planets can begin to form, and the star can be observed with optical telescopes. Weak-line T Tauri stars with very tenuous disks represent the final Class III phase, and their observed radiation can be fit by a simple blackbody curve, as there is no significant extra mass to add to the emissions. Most of the material accreted during the Class 0–I phases went into forming the protostar, whereas that surviving the Class II–III evolutionary phases is now present in primitive solar system matter.

As the solar nebula evolved through these stages, a wide variety of chemical environments were produced due to the different processes associated with evolutionary stages outlined above. In Table 1, we list the various physical processes that are associated with each stage and the effects that they would have on primitive materials. We now describe these processes in more detail, so that we can understand the specific effects that they would have on solar nebula chemistry as described in section 3.

2.2. Formation of the Solar Nebula

Low-mass stars like the Sun are formed, either singly or as binaries, through the collapse of the dense molecular cores widely observed in molecular clouds such as Taurus, Ophiuchus, and Perseus (*Evans,* 1999). Thus, the formation of the protosolar nebula leads naturally to interstellar chemistry, particularly that of dense "prestellar cores" (e.g., *Caselli et al.,* 2003), as being responsible for setting the initial chemical composition of protosolar matter (e.g., *Sandford,*

TABLE 1. Evolutionary processes associated with each stage of protoplanetary disk evolution.

Physical Process	Effect on Primitive Materials/Environments
Stage 1:	
Infall (rapid) from molecular cloud	Processing (vaporization) of interstellar grains
Formation of the solar nebula	Define initial pressure, temperature structure
	Accretion of most material onto Sun
Stage 2:	
Diminished infall	Processing (vaporization) of ISM grains
Variable $\dot{M}$ ($>10^{-7}$ $M_{\odot}$/yr)	Rapid P, T evolution
Protostellar outflows	Large-scale radial mixing
Stage 3:	
Decreasing $\dot{M}$ ($<10^{-7}$ $M_{\odot}$/yr)	Slow P, T evolution (decreasing values)
Turbulence(?)	Extensive radial mixing
Transient heating events	Chondrule formation
Grain growth/planetesimal formation	Meteorite parent body formation
Giant planet formation	
Stage 4:	
Negligible $\dot{M}$	Parent-body alteration
Removal of nebular gas	
Accretion of protoplanets and planetary embryos	

1996). This chemistry, as described by *Nuth et al.* (2006), proceeds at around 10 K and is comprised of a gaseous phase dominated by H_2 and CO with admixtures of other heavy molecules (e.g., HCN, NH_3, OH, HC_3N, SO, CS). Refractory silicate and carbonaceous dust grains are also present. These become layered with ice mantles predominately of water but also containing various polar and nonpolar molecules (e.g., CO, CO_2, and CH_3OH).

The initial chemical inventory available to the protosolar cloud is sensitive to the details of its dynamical evolution prior to the final collapse phase when the protosun and nebula formed. If magentic fields dominated this evolution, then the final rapid gravitation collapse cannot occur until the sustaining magnetic support against gravity has been lost through ion-neutral drift [i.e., ambipolar diffusion (*Shu et al.,* 1987)]. For typical molecular cloud ionization fractions (~10^{-8}–10^{-7}) this ambipolar diffusion timescale is several million years. Chemically, this timescale is comparable to that needed for a cosmic-ray-driven chemistry to attain a steady state, but much longer than the timescale (~10^5 yr) to freeze out atoms and molecules containing heavy elements through sticking collisions with cold dust grains (*Nuth et al.,* 2006).

Alternatively, molecular clouds may form, collapse to form protostars, and dissipate on much shorter timescales, on the order of the gravitational free-fall time (~10^5–10^6 yr). This scenario (e.g., *Elmegreen,* 2000; *Hartmann et al.,* 2001; *MacLow and Klessen,* 2004) appears to be more in accord with the apparent chemical youth of molecular clouds, as well as with the estimated dynamical lifetimes of protostellar cores and molecular clouds. A third, intermediate possibility is that the dynamical evolution of the protosolar cloud was influenced by either the ejecta of a nearby supernova, or by the wind of a late-type AGB star (*Cameron and Truran,* 1977; *Vanhala and Cameron,* 1998). In either case the SNe or AGB wind could initiate collapse of the natal protosolar core at some arbitrary point in its evolution, irrespective of the timescale on which its chemistry would otherwise evolve. This scenario also allows for the injection of fresh nucleosynthetic products into the protosolar nebula, an appealing aspect from the point of view of explaining the presence of live radionuclides in meteorites (e.g., *Boss and Foster,* 1998; *Vanhala and Boss,* 2002).

2.3. Disk Structure and Evolution

2.3.1. Density and temperature structure. Neglecting the effect of radiation from the forming protostar, to a first approximation the gross physical evolution of the nebula/disk determines its chemical structure. In the gravitational collapse of a rotating core, parcels of infalling gas and dust are spatially redistributed according to their angular momentum. Material with low angular momentum falls toward the center, whereas that with higher angular momentum settles into larger orbits, forming the protostellar disk (e.g., *Cassen and Moosman,* 1981). A molecular disk in differential rotation, where the inner regions rotate more rapidly than the outer ones, forms around the young protostar. The protostar grows both by direct accretion of infalling interstellar matter, and also by the episodic accretion of material from the surrounding disk (e.g., *Hartmann,* 2000).

Differential rotation of the gaseous disk around the star induces shearing motions that provide frictional forces that, in turn, act to drive radial motion. The source of these viscous, dissipative forces could be associated with either gravitational instabilities, magnetic torques, or large-scale convection (*Balbus and Hawley,* 1991; *Ruden and Pollack,* 1991; *Adams and Lin,* 1993). These different mechanisms require different physical conditions in the disk to operate (i.e., mass concentrations, ionization fractions, thermal gradients) and they may operate at different locations or times throughout the lifetime of a disk (*Stone et al.,* 2000). The net result is an evolving disk density structure where most material in the disk loses angular momentum and there is a net radial movement of mass toward the protosun; the remaining disk angular momentum, possessed by a small fraction of the material, is transported outward.

Viscous dissipation will also heat the gas and the thermal structure of the disk is set by how quickly thermal energy is produced by this dissipation and the rate at which the disk can radiate it away (e.g., *Boss,* 1998). The greatest amount of heat is generated at the interior of the disk due to the greater amount of material present and shearing that takes place. These locations also have the largest optical depths (due to the higher density of material present), making it more difficult for radiation to escape from the disk. This creates a thermal structure with high temperatures close to the star and low ones at larger heliocentric distances. As the mass of the disk decreases, the amount of viscous dissipation that occurs and the optical depth of the disk also decrease, allowing the disk to cool over time.

The density and temperature of a given location of the disk will determine what chemical reactions can take place. There will generally be some location of the disk inside of which a chemical species will not be allowed to form, but outside of which the temperature is low enough that the species is thermodynamically stable. The location where this transition takes place is generally called the *condensation front* for that species, as outside that location the species will begin to condense (if it is a solid). An example of such a condensation front is the *snowline,* which represents the radial location outside of which water was found as solid ice. Because the disk cools over time, the condensation front for different species will move inward over time (*Cassen,* 1994). (Note that this is a simplification of the real situation, as the location where a chemical species will condense may depend on pressure in addition to temperature.) This implies that the allowed chemistry at a given location would evolve with time.

2.3.2. Spatial mixing and transport. Over the course of the lifetime of the solar nebula, different processes would have led to radial movement of the gas and solids within it. These transport processes would bring chemical species into regions of the nebula where they would undergo alteration of some sort or remain intact although they would not have formed in the new environment in which they are found. Here we briefly summarize the different transport processes and refer the reader to *Cuzzi and Weidenschilling* (2006) for more details.

2.3.2.1. Large-scale flows: Regardless of the driving mechanism(s) for protoplanetary disk evolution, the net result is that a large fraction of the gas flows inward and is eventually accreted onto the star, while some is carried outward to transport angular momentum. These flows can carry gaseous material and small dust grains with them, leading to the radial transport of different species. The particulars of the flow likely vary with the driving mechanism. In the case of turbulent viscosity, *Takeuchi and Lin* (2002) and *Keller and Gail* (2004) have demonstrated that the direction of the flow may vary with height above the disk midplane: The upper layers of the disk flow inward, while the midplane material moves outward.

2.3.2.2. Diffusion: Diffusion transports things along concentration gradients, from areas of high concentrations to areas of low ones. In the solar nebula, gas molecules would be subject to such transport as would solids, although the mobility of solids due to diffusion would decrease as their masses increased. Thus, dust particles would be those solids most affected by this process. The rate of diffusion would depend on the structure and viscosity (turbulence) of the nebula, and would evolve with time.

2.3.2.3. Gas-drag migration: As the nebular gas rotated around the Sun, it was supported radially by a pressure gradient (hot, dense gas near the Sun and cool, sparse gas further away). Solids in orbit around the Sun did not feel this pressure, and thus the central force experienced by solids (gravity) and gas (gravity minus pressure support) differed, causing the gas to revolve at a slower rate. This difference in orbital velocities led to solids experiencing a headwind in their orbits, causing them to migrate inward with time. The rate of this migration depended on the size of the solids being considered, with meter-sized particles experiencing the most rapid migration (~0.01 AU/yr) (*Weidenschilling,* 1977). In the case of a nonuniform nebula, localized pressure maxima may develop, leading to the outward migration of solids (*Haghighipour and Boss,* 2003). Bodies larger than a kilometer in size, which likely includes all meteorite parent bodies, were essentially immobile due to their large inertia. Therefore, once bodies that large formed, they likely did not migrate due to this effect.

2.3.2.4. Outflows: A feature commonly observed around young stellar objects, and therefore presumably present at some point during the evolution of the solar nebula, are energetic outflows originating from the inner parts of the disk. These outflows are believed to be linked to the accretion of material from the disk onto the star. As material was ejected from near the Sun by outflows, it has been proposed that a fraction of the material may not have reached escape velocity or would travel on trajectories that intersected the nebula. If that was the case, that material would have fallen back and reentered the solar nebula at a radial distance depending on the velocity it was launched at, its trajectory, and the shape of the nebula. Such a process has been proposed as a way of thermally processing chondrules (*Shu et al.,* 1996, 1997; *Liffman and Toscano,* 2000).

2.4. Irradiation

While the rates of gas phase bimolecular and termolecular reactions, and gas-grain processes, depend strongly on disk physics (the local temperature and density), many sources of ionization were also present throughout the neb-

TABLE 2. Molecules detected in protoplanetary disks (adapted from *Markwick and Charnley,* 2004).

Object	Detections	Upper Limits	References
DM Tau	^{12}CO, ^{13}CO, H_2D^+, $C^{18}O$ HCN, HNC, CN, CS $C^{34}S$, H_2CO, HCO^+, C_2H	$H^{13}CO^+$, N_2H^+, SiO SiS, H_2S, C_3H_2, HCO HC_3N, CH_3OH, CO^+ SO, SO_2, SiC_2, HNCS $HCOOCH_3$	1, 2, 3
GG Tau	^{12}CO, ^{13}CO, $C^{18}O$, HCN CN, CS, H_2CO, HCO^+	HNC, $C^{34}S$, $H^{13}CO^+$ C_2H, N_2H^+	3
L1157	CH_3OH		4
LkCa15	CO, ^{13}CO, H_2CO, HCN, CN HCO^+, N_2H^+, CS	DCO^+, H_2D^+, $H^{13}CO^+$, OCS HNC, CH_3OH, HDO, DCN	5, 6, 10
TW Hya	CO, CN, HCN, DCO^+, H_2D^+ HCO^+, $H^{13}CO^+$	HNC, DCN, $H^{13}CN$, CH_3OH SO, N_2H^+	6, 7, 8, 9

References: [1] *Dutrey et al.* (1997); [2] *Dutrey et al.* (2000); [3] *Ceccarelli et al.* (2004); [4] *Goldsmith et al.* (1999); [5] *Qi* (2001); [6] *Thi et al.* (2004); [7] *Kastner et al.* (1997); [8] *van Dishoeck et al.* (2003); [9] *van Zadelhoff et al.* (2001); [10] *Aikawa et al.* (2003).

ula. These processes acted to alter the chemical composition by driving nonequilibrium chemistries, especially close to the protosun, at the disk surface, and deeper in the outer disk. Although the focus here is primarily on chemistry, it should be noted that sources of disk ionization can also strongly influence the global magnetohydrodynamical evolution of a disk (*Hayashi,* 1981; *Balbus and Hawley,* 1991; *Gammie,* 1996) and hence the physics of mass accretion and angular momentum transport.

2.4.1. Ultraviolet radiation. There are two possible sources of UV radiation: interstellar and solar. Ultraviolet photons from the external interstellar radiation field can photodissociate and photoionize a thin surface layer of disk material. The disk is radially and vertically optically thick for stellar UV and optical photons. However, in a flared disk structure (*Chiang and Goldreich,* 1997) stellar UV photons can irradiate the disk surface and produce a hot surface layer, out to large radii (~10^2 AU). In this case the stellar UV contribution at the surface can be ~10^4 times that of the interstellar field (*Willacy and Langer,* 2000). It is now understood that most of of the radicals and ions detected thus far in protoplanetary disks (see Table 2) reside in these surface layers (e.g., *Aikawa and Herbst,* 1999b; *Aikawa et al.,* 2002). It should be noted that, if the protosolar nebula formed in the vicinity of massive O and B stars (*Hester et al.,* 2004), these objects may have provided a much higher flux of UV radiation to the surfaces of the disk. Published nebular chemistry models have not explored this possibility in detail. An additional, low-level source of UV photons may also exist deeper in the disk, including the midplane, where cosmic-ray particles can penetrate. Electrons produced by cosmic-ray ionization of molecular hydrogen can excite other H_2 molecules and UV photons are produced through their subsequent radiative deexcitation (*Prasad and Tarafdar,* 1983).

2.4.2. X-rays. Observational surveys of young Sun-like stars show them to be copious producers of X-ray emission early in their evolution (*Montmerle et al.,* 1983). Studies of regions with ongoing star formation [e.g., with ROSAT (*Feigelson et al.,* 1993)] suggest that X-ray activity continues from the early protostellar phase until around ~10^8 yr, with an almost temporally constant hard X-ray component underlying that associated with flaring. Observations of stellar clusters show a range of X-ray brightnesses with luminosities in the range L ~ 10^{28}–10^{30} erg s^{-1}, and that the constant component probably has a luminosity in excess of ~10^{29} erg s^{-1}. The origin of these X-rays lies in stellar coronal activity, probably associated with the heating from magnetic reconnection and stellar flaring, and this leads to a hot sphere of X-ray-emitting plasma surrounding the central protostar (e.g., *Glassgold et al.,* 1997). Soft X-rays ($E_X < 1$ keV) can be strongly attenuated by the several magnitudes of extinction supplied by the wind. Harder X-rays ($E_X \sim 2$–5 keV) can penetrate into the disk and are the main source of ionization within about 10 AU, especially if the (magnetohydrodynamic) wind from the protostar also shields the inner disk from galactic cosmic rays.

X-rays irradiate the disk obliquely and are efficiently attenuated by the high material column densities of the disk. This leads to the upper layers of the disk being ionized and heated most effectively by X-rays (*Glassgold et al.,* 1997). Dust grain settling lowers the attenuation and allows the higher-energy X-rays to penetrate further into the material of the upper disk layers and to be the dominant source of ionization at disk radii of ~0.1–10 AU. This provides a much higher X-ray ionization rate (ζ_X) in the upper layers of the disk than in the midplane, where a combination of thermal ionization, cosmic rays, and extinct radioactivities provide most of the ionization.

For example, simple disk chemistry models show that between about 200–400 AU above the disk midplane, $\zeta_X \sim 10^{-16}$–10^{-15} s^{-1} for L ~ 10^{31} erg s^{-1}, and $\zeta_X \sim 3 \times 10^{-17}$ s^{-1} at 400 AU if L ~ 10^{29} erg s^{-1} (*Aikawa and Herbst,* 1999b). Apart from ionizing disk material, X-rays can also be an efficient heat source. Heating of the upper layers by X-rays may dominate that from mechanical heating in some cases and lead to extremely hot atomic surface layers (~5000 K) (*Glassgold et al.,* 2004).

2.4.3. Cosmic rays. Unless a powerful protostellar wind is present, protoplanetary disks will experience the same cosmic-ray particle flux as their natal molecular cloud cores. However, the much higher disk surface density leads to their strong attenuation, at the surface of the inner disk region, by ionization and elastic and inelastic scattering (*Umebayashi and Nakano,* 1981). Cosmic rays may be prevented from penetrating deep into the disk by scattering off fluctuations in a disk magnetic field (*Dolginov and Stepinski,* 1994), or from a magnetohydrodynamic wind, if present. If any magnetic attenuation is neglected, then the lower surface density of the outer disk can allow greater penetration, and so cosmic rays could be a significant source of ionization at the midplane. Cosmic-ray ionization of young disks rich in H_2 can in principle initiate a chemistry in the same manner as in the interstellar medium, i.e., by production of H_3^+ and He^+ (*Nuth et al.,* 2006). This has served as the motivation behind recent models of protoplanetary disk chemistry based on interstellar processes.

2.5. Extinct Radionuclides

The decay of extinct radionuclides over the age of the solar system (i.e., with half-lives ~10^5–10^8 yr) also provided a source of ionization. In this case, abundant nuclei with the shortest half-lives, and that produce high-energy particles or γ-rays, can ionize material throughout the disk. The most effective of these are ^{26}Al, ^{60}Fe, and ^{40}K (e.g., *Finocchi and Gail,* 1997). The ionization rate due to both cosmic rays and radioactive decay can thus be parameterized (e.g., *Umebayashi and Nakano,* 1988)

$$\zeta_{CRP}(r) = \zeta_{ISM} \exp\left[-\frac{\sigma(r)}{196 \text{ g cm}^{-2}}\right] + \zeta_{RA}$$

where σ(r) is the mass surface density (g cm^{-2}). The interstellar cosmic-ray ionization rate, ζ_{ISM}, is in the range ≈(1–5) × 10^{-17} s^{-1} and ζ_{RA}, the ionization rate associated with the decay of radioactive nucleotides, can be expressed as $8 \times 10^{-23} f_d$ s^{-1}, where an elemental depletion factor, f_d, has to be accounted for since Al and Fe nuclei are expected to be largely incorporated into refractory dust particles. The relative contribution of these two processes to the midplane disk ionization is such that, early in nebular evolution, ^{26}Al and ^{60}Fe are the major sources of ionization within about 10 AU, outside of which cosmic ray ionization dominates. Later, decay products from the longer-lived ^{40}K decay provide a much weaker (~ few × 10^3 times less) ionization rate that only dominates the cosmic-ray contribution within ≈2–3 AU.

2.6. Transient Heating

As described by *Connolly et al.* (2006), transient heating events are known to have occurred in the solar nebula. Observations of protoplanetary disks have found that Fuor and Exor events, which are episodic increases in the mass accretion rate onto the central star, are common, resulting in luminosity fluctuations on timescales that are short compared to the evolutionary time of protoplanetary disks. While these events significantly increase the luminosity of the protoplanetary system, it is doubtful that they affect material embedded in the disk due to the large optical depth to the disk midplane (*Bell et al.,* 2000).

Primitive meteorites also contain evidence of localized transient heating events in the form of igneous (melted) CAIs and chondrules. The location, timing, and manner of these events are uncertain, and a wide variety of mechanisms have been proposed for generating them. Examples of proposed heating mechanisms include shocks from dynamical instabilities in the nebula, bow shocks around supersonic planetesimals, lightning discharges, migration of objects into outflows, and energetic collisions between planetesimals. Any of these events may have occurred in the nebula, although they may not all have been responsible for the formation of chondrules. The effects that they would have on other primitive materials must still be determined.

3. MODELS OF NEBULAR CHEMISTRY

While the various processes described above operated in the solar nebula, various chemical environments would have formed defined by their pressures, temperatures, and elemental abundances. In this section, we describe what kinds of chemical environments may have been created by the different modes of protoplanetary disk evolution and discuss how the meteoritic and astronomical records preserve signatures of these environments.

3.1. Chemistry of Nebular Formation

The high-energy, dynamical environment encountered by cold interstellar gas and ice-mantled dust grains as they collapsed into the nebular disk presented the first opportunities for their chemical modification. Interstellar matter first experienced direct radiative heating by the central protosun (e.g., *Simonelli et al.,* 1997; *Chick and Cassen,* 1997), as well as by gas-grain drag heating (*Lunine et al.,* 1991). The gas and dust were also dynamically compressed and heated at the accretion shock and subjected to additional heating, photoprocessing, and ionization by UV and X-rays from the shock itself (*Neufeld and Hollenbach,* 1994; *Ruzmaikina and Ip,* 1994). *Chick and Cassen* (1997) calculated the survivability of various materials, based on vaporization temperatures, in model nebulae. They found that, generally, the most refractory compounds could survive entry into the nebula up to almost within 1 AU. However, they found that subsequent survival depended upon the disk accretion rate and optical depth. For low disk accretion rates, and consequent lower disk luminosities, silicates and refractory organic material could survive up to within the terrestrial planet region but water ices would be vaporized within about 5 AU. At higher disk accretion rates, silicates could be destroyed out to a few AU, whereas more volatile organics (such as methanol and formaldeyde) and water ices could only survive outward of 20 AU and 30 AU, respectively.

Subtle chemical kinetic effects, as opposed to thermochemistry arguments based on vaporization temperatures, are probably important for determining change in the gaseous molecules. Except at large heliocentric distances, by the time a parcel of gas and dust has reached the nebular surface, the icy grain mantles will have been deposited into the gas — either by thermal desorption as the grains are heated, or by sputtering at the shock. This gaseous mixture of volatile mantle material is heated and compressed and a postshock chemistry occurs as it cools and selectively recondenses onto dust grains (except for H_2 molecules and He atoms).

At the high number densities ($>10^{12}$ cm^{-3}) and temperatures of the accretion shock (~2000–4000 K), collisional dissociation of H_2 can occur and the key parameter for the chemistry is the H/H_2 ratio in the postshock gas. This can increase by many orders of magnitude from its value in quiescent dense gas and the resulting population of highly reactive hydrogen atoms acts to destroy many other molecules (*Hollenbach and McKee,* 1979; *Neufeld and Hollenbach,* 1994).

For a given mass accretion rate, the highest shock speeds occur closer to the protosun and here the accretion shock can be fully dissociative, i.e., molecules are atomized. The slow (nondissociative) shock speeds and low preshock densities in the outer nebula would favor the survival of interstellar molecules. For example, *Aikawa and Herbst* (1999a) estimated that at 100 AU interstellar material would be accreted in pristine form with its D/H ratios intact. At intermediate disk radii, the shock is only partially-dissociative and the chemistry is sensitive to the actual H/H_2 ratio in the postshock gas. Under these conditions, CO and H_2O can survive the accretion shock and in fact increase from their preshock abundances: Chemical erosion of H_2O to OH and O by H atom abstraction reactions is overwhelmed by hydrogenation reactions with H_2. The abundances of some mantle molecules, like CH_4 and CO_2 are partially reduced. However, other molecules, such as CH_3OH, OCS, and H_2CO, are completely destroyed since they do not have formation pathways with H_2 in the hot gas (S. B. Charnley and S. D. Rodgers, personal communication, 2005). Thus, once the disk was formed, an interstellar chemical signature could have been retained over much of the protosolar disk, apart from within about 1 AU. Subsequent processing of this material was then responsible for spatial and temporal chemical alteration of it throughout the nebula.

3.2. Chemical Models of Nebular Chemistry

Once the disk has formed, chemical reactions will proceed within it, altering the solids that survived the infall from the molecular cloud and creating new ones out of the vapor of those that were destroyed. Not only will the chemical reactions that take place in the disk depend on the chemical environment that is present at a given location in the disk, but the modes of the chemical reactions will as well. This forces us to develop different models for determining what chemistry will take place under the variety of environments that will exist.

3.2.1. Equilibrium chemistry. A basic starting point for looking at the chemical evolution of the nebula is to look at what chemical species are thermodynamically stable under those temperatures, pressures, and elemental abundances that are expected in the nebula, as is done in condensation calculations. Standard condensation sequences present the chemical species expected to be present in a gas of solar composition at constant pressure at different (decreasing) temperatures. This sequence is determined by calculating the distribution of elements among different species using experimentally determined equilibrium coefficients while simultaneously solving the equations of mass balance. The details of calculating a whole condensation sequence are beyond the scope of this chapter, but are discussed and used by a number of authors (*Lewis* 1972; *Grossman,* 1972; *Saxena and Eriksson,* 1986; *Wood and Hashimoto,* 1993; *Ebel and Grossman,* 2000; *Lodders,* 2003; *Ebel,* 2006). Here we review the general results of such calculations, particularly following the recent results of *Lodders* (2003).

Imagine a parcel of solar composition gas at a pressure of 10^{-4} bar, and a temperature high enough such that the only species present are in the gaseous state (while the molecular composition of the gas will change during the following discussion, we focus on the solids that form as they are more relevant to the formation of meteorites). This initial condition may have been appropriate for the inner solar nebula, as it has been speculated that material, out at least to where the current asteroid belt is located, was initially in the vapor phase during the earliest stages of nebular evolution (*Cassen,* 1996). Such a state may be needed to explain the moderately volatile element depletions described below.

As the gas cools (maintaining the constant pressure), the first solids to form are Ca-, Al-, and Ti-oxides such as corundum, hibonite, grossite, gehlenite, perovskite, and titanates. These refractory minerals begin to form at temperatures near 1700 K, with some not appearing until temperatures reach 1500 K. These minerals are commonly found in primitive meteorites as CAIs, and thus support the conclusion that these inclusions are composed of the first condensates in the solar nebula.

As the gas continues to cool, Mg-bearing silicates begin to condense, with the most abundant being forsterite and enstatite. These minerals form over a small temperature range, between 1310 and 1360 K, with metallic Fe appearing at the same time. Magnesium, Si, and Fe are the most abundant rock-forming elements, and make up the bulk of the material found within primitive meteorites, including both chondrules and the surrounding matrix.

After the major rock-forming elements condense at a temperature of ~1300 K, less-abundant elements will continue to be incorporated into solids, adding little to the bulk mass of condensed species. Those elements that condense between this temperature and the condensation temperature of sulfur (~650 K) are called "moderately" volatile elements. As the system cools below the sulfur condensation temperature, leftover Fe is predicted to react with O in gaseous water molecules to form iron oxides, such as magnetite. Significant mass is not added to the solid component until temperatures are low enough for water ice to condense at a temperature of ~160 K. It should be pointed out that the condensation processes may differ from element to element. In the case of Fe, an Fe atom in the gas may condense to become part of an Fe grain

$$\mathrm{Fe(gas)} \rightarrow \mathrm{Fe(solid)}$$

However, under canonical conditions, S condenses when hydrogen sulfide gas reacts with existing Fe grains to form troilite (FeS)

$$\mathrm{Fe(solid)} + \mathrm{H_2S(gas)} \rightarrow \mathrm{FeS(solid)} + \mathrm{H_2(gas)}$$

This reaction could only take place if there was solid Fe available to react with the nebular gas. Thus, these condensation calculations assume that the condensed solids remain in contact with the gas and can react with it to form whichever species are thermodynamically stable. In reality, this may not be the case, as accretion of solids would lead to some material being shielded from contact with the gas (*Petaev and Wood,* 1998). This illustrates the importance of considering various timescales in considering the chemical evolution of the solar nebula.

3.2.2. Kinetic models. As described above, a drawback to calculations such as these is that they only predict what species would exist under the idealized case of thermodynamic equilibrium. However, for a variety of reasons, equilibrium may not have been reached in the solar nebula. As described above, if solids were removed from contact with the nebular gas, predicted chemical reactions may not be able to take place. While diffusion of atoms inside of solids may allow these reactions to occur by transferring reactants from the gas to the solid inteiors, if the solids are large or the temperature is low, then diffusion may not be rapid enough for the equilibrium products to form. This has led to the consideration of chemical kinetics along with chemical models of the solar nebula.

As described previously, the solar nebula was in place around the Sun for a finite time (10^{6-7} yr). In chemical reactions, the formation of a product takes an amount of time dependent on the concentration of the reactants as well as their temperature or mobility — it is not instantaneous. If the amount of time needed for a given reaction to take place was long compared to the lifetime of the solar nebula, then, despite being thermodynamically stable, that reaction product would not form in the nebula.

This idea was extensively developed by *Fegley* (1988) and *Fegley and Prinn* (1989) in looking at the kinetics of gas-solid reactions in the solar nebula in trying to understand the origin of different meteoritic components. To do so, these authors developed what is known as the simple collision theory (SCT), which can be used to calculate the collision rate of gas molecules with the solids that they are to react with. Using the kinetic theory of gases, the fraction of collisions that take place with an energy greater than the activation energy of the reaction can also be calculated. When a collision takes place with an energy greater than the activation energy, the reaction of interest is assumed to take place. Using this idea, the time needed for enough collisions to occur for a reaction to proceed to completion can be calculated. In general, those reactions that require higher activation energies, occur on large grains (small surface area to volume ratios), or occur in low-pressure environments take longer to reach completion. As an example of how this work was applied, it was shown that while predicted to be present in the solar nebula by equilibrium condensation calculations, serpentine (as well as other phyllosilicates) could not form via gas-solid reactions in the canonical solar nebula. In addition to the limitations due to gas-solid collision rates and energetics described above, solid-state diffusion as well as gas-gas collisions may not occur rapidly enough in the solar nebula to reflect the equilibrium predictions of condensation calculations (*Lewis and Prinn,* 1980; *Fegley and Prinn,* 1989).

Kinetic inihibition explains why some equilibrium products are not observed in meteorites. In addition, it also demonstrates that there are other considerations that must be made when looking at the chemical evolution of the solar nebula. For example, all the Fe in some carbonaceous chondrites (CM, CI) is locked up as FeO. As described, as silicates condense, they are expected to be Mg-rich with Fe condensing to form metallic grains. At lower temperatures (<550 K), Fe is predicted to react with water in the gas to form FeO, which will then diffuse into the existing forsterite grains, replacing MgO, to form the observed mineralogy. However, as summarized by *Ebel and Grossman* (2000), this requires equilibrium to be achieved, which is unlikely at these low temperatures due to the slow diffusion rate through olivine. Despite the expected kinetic inhibition of these species, they are still observed in carbonaceous chon-

TABLE 3. Representative gas phase chemical reactions in the nebula.

Reaction Type	Generic	Example	Rate Coefficient at 10 K*
1. Radiative association of neutrals	$A + B \rightarrow AB + \nu$	$CO + S \rightarrow OCS + \nu$	$\sim 10^{-10}$–10^{-17} $cm^3 s^{-1}$
2. Ion-neutral radiative association	$A^+ + B \rightarrow AB^+ + \nu$	$C^+ + H_2 \rightarrow CH_2^+ + \nu$	$\sim 10^{-10}$–10^{-17} $cm^3 s^{-1}$
3. Neutral-neutral	$A + B \rightarrow C + D$	$CN + C_2H_2 \rightarrow HC_3N + H$	$\leq 10^{-10}$ $cm^3 s^{-1}$
4. Ion-neutral	$A^+ + B \rightarrow C^+ + D$	$H_3^+ + CO \rightarrow HCO^+ + H_2$	$\sim 10^{-7}$–10^{-9} $cm^3 s^{-1}$
5. Charge transfer	$A^+ + B \rightarrow B^+ + A$	$S^+ + Mg \rightarrow Mg^+ + S$	$\sim 10^{-9}$ $cm^3 s^{-1}$
6. Radiative recombination	$A^+ + e^- \rightarrow A + \nu$	$H^+ + e^- \rightarrow H + \nu$	$\sim 10^{-12}$ $cm^3 s^{-1}$
7. Dissociative recombination	$AB^+ + e^- \rightarrow A + B$	$H_3O^+ + e^- \rightarrow OH + 2H$	$\sim 10^{-6}$–10^{-7} $cm^3 s^{-1}$
8. Photoionization†	$A + \nu \rightarrow A^+ + e^-$	$S + \nu \rightarrow S^+ + e^-$	$\sim(10^{-9}$–$10^{-10})e^{-2\tau}$ s^{-1}
9. Photodissociation of neutrals†	$AB + \nu \rightarrow A + B$	$CO_2 + \nu \rightarrow CO + O$	$\sim(10^{-9}$–$10^{-10})e^{-2\tau}$ s^{-1}
10. Photodissociation of ions‡	$AB^+ + \nu \rightarrow A^+ + B$	$CH_2^+ + \nu \rightarrow CH^+ + H$	$\sim(10^{-9}$–$10^{-10})e^{-2\tau}$ s^{-1}

*Many reactions, particularly neutral neutral processes, are either endothermic or possess significant activation-energy barriers. These reactions become important at higher temperatures and their temperature-dependent rate coefficients can generally be expressed in the form $\gamma T^{\alpha} \exp(-\beta/T)$ $cm^3 s^{-1}$, where γ, α, and β are constants fitted from experiment.

†Photorates typically contain an exponential factor involving the UV optical depth, τ, to account for the attenuation of dissociating or ionizing photons in the relevant frequency range.

drites. As discussed below, if the products of a reaction thought to be kinetically inhibited are observed in meteorites or protoplanetary disks, they likely are due to parent-body processes or formation in noncanonical nebular conditions.

3.2.3. Low-temperature/density models. Over much of the protosolar nebula, chemical equilibrium would not be attained, thus placing a severe limitation on the "classical" chemical models such as those outlined above. Kinetics must be considered as temperatures and gas densities decrease, which continues to large heliocentric distances or high altitudes above the midplane. The chemistry in the outer regions of the disk midplane should be most reminiscent of that in cold molecular clouds (*Nuth et al.,* 2006), whereas the surface layers and the inner disk will experience chemistries similar to that of the diffuse interstellar medium or in regions of massive star formation (*Tielens and Hollenbach,* 1985; *Maloney et al.,* 1996).

Here we describe how interstellar chemical processes have been incorporated into disk models. Almost all these models include inward radial transport and the chemistry is calculated time-dependently (e.g., *Aikawa et al.,* 1999a). Table 3 shows some of the chemical reactions included in nonequilibrium disk chemistry models [see *Le Teuff et al.* (2000) for more information on astrochemical kinetics]. Isotopic fractionation patterns are crucial discriminants of nebular processes and hence of the origin of presolar material. Recent results on the fractionation chemistry of protoplanetary disks are outlined and discussed primarily in the context of isotopes of H, C, N, and O. Models based on nonequilibrium chemical processes can be regarded as complementary to chemical equilibrium models of the inner nebula (e.g., *Fegley,* 1999, 2000). The efficacy of nebular mixing processes is critical for the relevance of both treatments to meteoritic studies; results from recent nonequilibrium studies are also described here.

3.2.3.1. Ionization and gas-grain interaction: The basic ingredients of modern disk chemistry models can be traced back to early models of the ionization structure of the solar nebula (*Hayashi,* 1981; *Umebayashi and Nakano,* 1988). These steady-state models assumed that cosmic rays would not be shielded and would play an important role in ionizing the outer regions of the nebular midplane. This leads to proton transfer from H_3^+ and He^+ attack on neutral molecules driving ion-neutral and neutral-neutral reactions in the disk midplane. The ionization at these densities is controlled by the gas-grain interaction and so recombination and charge transfer reactions with neutral and charged dust grains also have to be included. However, an unrealistic simplification of these models was that, even at 10 K, this interaction was limited solely to ion-grain, electron-grain, and grain-grain charging processes (e.g., *Umebayashi and Nakano,* 1981, 1990). In fact, at very low temperatures, neutral molecules collide and stick efficiently to dust grains, as is observed for CO (*Dutrey et al.,* 1994). A simple steady-state model of a Hayashi disk in which realistic molecule depletion and (temperature-controlled) evaporation from dust were also included showed that, while the disk ionization structure was broadly similar to the case when these processes were neglected, the chemical structure beyond about 50 AU was completely different (*Charnley,* 1995).

The formation of ice mantles on dust particles, and their exchange with the gas phase, are now seen as fundamental aspects of modern disk models. *Aikawa et al.* (1996, 1997) first constructed models that could account for the observed depletion of CO in disks, and calculated the radial distribution of both the gas- and solid-phase molecular abundances. *Willacy et al.* (1998) showed explicitly that, because of the high removal efficiency of positive ions by neutralization on grains, inclusion of the gas-grain interaction led to the gas-phase chemistry being dominated by neutral-neutral reactions. This result is contrary to the conclusion that ion-molecule reactions dominate the chemistry of the outer regions, as obtained from models that ignored gas-grain charge exchange (*Bauer et al.,* 1997; *Finocchi and Gail,* 1997). *Aikawa et al.* (1997) showed that the precise nature of the gas-grain interaction is also important for the chemical structure, specifically whether molecular ions are simply neutralized or are dissociated upon collision with a negatively charged grain. *Aikawa et al.* (1999b) subsequently showed that for HCO^+-grain collisions, the products were most likely to be CO and H, rather than HCO.

A key issue for the cooler outer disk is the mechanism whereby accreted material can be returned to the gas. In the case where the return of molecules to the gas is by thermal desorption, the rate is most sensitive to the binding energy for physisorption to the grain (e.g., *Tielens and Allamandola,* 1987), and to the dust temperature. The former dependence means that very volatile species, such as CO and N_2, will desorb completely above about 25 K, whereas "stickier" species that can form hydrogen bonds (e.g., water, methanol, ammonia) can be retained at much higher temperatures, up to around 100 K in the case of water. The latter dependence means that the chemical structure of the outer disk depends sensitively on the temperature distribution, and hence on the physics of the disk model adopted (see Fig. 1). In regions where the temperature falls to 10 K, thermal desorption rates become vanishingly small and further nonthermal desorption mechanisms have to be considered. Possible mechanisms previously proposed for the interstellar medium (*Léger et al.,* 1985) are spot or whole-grain heating by cosmic rays, photodesorption, or explosion of irradiated mantles (by UV photons or X-rays) triggered by grain-grain collisions (*Aikawa et al.,* 1996). *Willacy et al.* (1998) found that the midplane chemistry was very sensitive to whether pure thermal desorption or mantle explosions were the sole mechanism in operation. The uncertainties in these models and the detection of radicals and ions in protoplanetary disks (Table 2) has led to greater focus on the chemical effects of irradiation from the central protostar.

3.2.3.2. Photochemistry: Protoplanetary disks like the protosolar nebula are irradiated by UV photons from their central star. Due to the high optical depths of the midplane material, the less-opaque upper layers of the disk will be most affected. The surfaces and underlying regions of protoplanetary disks also experience the interstellar radiation field. This means that the physics and chemistry in both the radial and vertical directions of the disk has to be considered.

Aikawa and Herbst (1999b), in the first of these two-dimensional models, solved for the chemistry in a *Hayashi* (1981) disk, where the vertical structure is isothermal at each radius. They found that they could produce column densities in the outer disk that were broadly similar to those observed by *Dutrey et al.* (1997). However, to overcome the problem that all the heavy gas species are rapidly frozen out as ices, they had to adopt an extremely low sticking efficiency (0.03) for molecule-grain collisions. *Willacy and Langer* (2000) calculated the two-dimensional chemical structure of a flared accretion disk in which stellar photons create a warm (T ~ 120–35 K) surface layer in the outer disk (50–100 AU), in contrast to the cooler midplane region (T ~ 28–10 K). They found that a significant UV flux could penetrate to intermediate depths, and that this could be sufficient to nonthermally detach molecules from the grains by photodesorption, for even unit sticking efficiency. Subsequent disk models included a full two-dimensional radiative transfer solution (*Nomura,* 2002; *Aikawa et al.,* 2002; *van Zadelhoff et al.,* 2003; *Millar et al.,* 2003) and showed the existence of warm molecular layers (T ~ 30–60 K) between the hotter surface and midplane. The gas-grain interaction in these intermediate layers is thus thermally controlled, and so can retain a gaseous phase without the need for low sticking efficiencies or nonthermal desorption processes. The two-dimensional chemical structures calculated in these two-dimensional models gives good agreement with the molecular emission and column densities observed; it is now generally accepted that most of the chemical species detected thus far reside either in the surface layers or in the warm intermediate layers (e.g., Table 2).

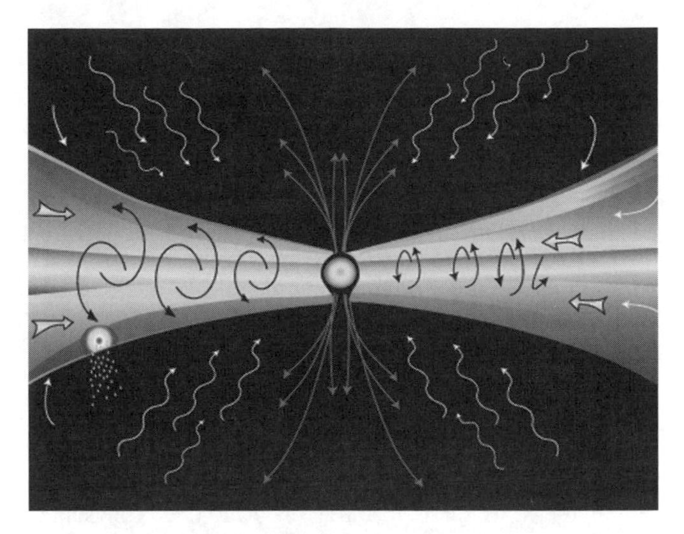

Fig. 1. Schematic representation of the physical structure in a protoplanetary disk. The principal processes that influence the chemical structure are also indicated. Cold material from the *infalling envelope* becomes incorporated into the protostar and its accretion disk. This mixture of gas and dust passes through the *accretion shock* where it is decelerated. The highest postshock temperatures and compressions occur closest to the protostar, where refractory dust, organics, and ice are readily vaporized. Moving farther out in the disk, the increasingly weaker accretion shocks encountered can lead to the survival of even the most volatile interstellar matter. *Interstellar UV photons and cosmic rays* can be important for irradiating the upper disk layers at large radii. The central accreting protostar provides a strong source of *stellar UV photons and X-rays.* These stellar UV photons contribute to heating the uppermost vertical layers of the disk, leading to a hot *UV-illuminated surface* in the outermost regions of a flared disk. In the inner disk, X-rays dominate the heating from the surface to the *disk midplane.* Beyond about a few AU, the midplane region becomes very cold and this allows most of the heavy gas phase molecules to condense onto the (evolving) dust grain population. Between the hot surface and the disk midplane lie the *warm upper layers* where most gaseous molecular emission is currently detected. The physics of disk evolution (see text) allows for the transport and mixing of chemically processed materials from each of these distinct regions. *Inward and outward radial transport* can occur within the disk midplane, whereas vertical diffusion can transfer material between this region and the surface layers.

3.2.3.3. X-ray chemistry: Apart from the very innermost regions of the disk midplane, close to the protostar, material in the upper layers experience the most intense illumination by X-rays (*Glassgold et al.,* 1997; *Igea and Glassgold,* 1999). The ionization and heating of disks by X-rays has been modeled in detail by *Gorti and Hollenbach* (2004) and by *Glassgold et al.* (2004). These studies contain comprehensive summaries of the relevant physics and chemistry but are focused more on making predictions of the atomic and molecular emission expected from protoplanetary accretion disks. *Bergin et al.* (2001) argued that if the disk has a flared structure then X-rays can also affect the upper layers in the outer disk, and that this could provide an additional nonthermal grain desorption mechanism there. *Markwick et al.* (2002) treated the detailed two-dimensional chemistry within 10 AU of the protostar and showed that X-rays have a profound effect on the ionization structure, even within the first 5 AU of the midplane. At present it is unclear precisely what meteoritic signatures would be expected from the effects of disk X-rays on gaseous *molecules*. By contrast, an underlying source of hard X-rays, the flux of protostellar "cosmic rays" produced by flares on protostars, has been proposed as a means of producing extinct radionuclides in solid *rocky* materials similar to those found in meteorites (*Lee et al.,* 1998).

3.2.3.4. Solid-state chemistry: The expectation from the observed CO depletions and gas-grain models is that dust grains in the cold regions of disks should be covered with molecular ice mantles. This has now been directly confirmed by the detection of molecular ices in the disk of CRBR 2422.8–3423 (*Pontoppidan et al.,* 2005). Thus, reactions on the surface of grains is another important means of chemical processing of nebular material. Three chemistries are of interest in the formation and modification of ices in disks.

3.2.3.5. Grain-surface reactions: Ice mantles are laid down in the coolest parts of the disk. As in molecular clouds, below about 15 K highly reactive atoms, primarily H, D, and O, can stick and react on dust grains. Most recent disk chemistry models include an active surface chemistry involving atom addition reactions to other atoms, radicals, and molecules (e.g., *Willacy et al.,* 1998). These schemes tend to lead to mantles showing a mixture of reduced and oxidized states, i.e., H_2O, NH_3, CH_3OH, and CO_2, that depends largely on the local atomic O/H ratio in the gas. It should be emphasized that the mechanisms, reaction rates, and pathways in many of the proposed interstellar schemes are highly uncertain (*Herbst,* 2000; *Charnley,* 2001; *Nuth et al.,* 2006). Hence, detailed conclusions derived from disk chemistry models that are strongly dependent on surface chemistry should be regarded with caution.

3.2.3.6. Heavy particle bombardment of ices: Cosmic-ray impacts can occur in the outer midplane of the disk, and can lead to significant chemical changes in the ice structure and composition. Radiation damage of cosmic ice analogs has been extensively studied in the laboratory and has been important over the lifetime of the solar system (e.g., *Hudson and Moore,* 2001). However, the effects of this process have not yet been evaluated in chemical models.

3.2.3.7. Ultraviolet photolysis of ices: Ultraviolet photolysis also has a long history of study by laboratory astrophysicists (e.g., *Allamandola et al.,* 1988) and was originally proposed as an important means of chemical modification in the cold interstellar medium. Experiments show that, starting from simple ice compositions, consistent with that observed in the ISM, the gross chemical composition of the ices can be altered, and many complex molecules produced, after photolysis and heating (e.g., *Bernstein et al.,* 1995; *Allamandola et al.,* 1997). To date, ice photolysis has also not been considered in disk models. However, this process may in fact be most important for the chemistry in UV-irradiated disks as opposed to interstellar clouds, where ices containing H_2O and other more refractory molecules (e.g., methanol and ammonia) will persist in the UV-illuminated warm molecular layers found to exist above the midplane (*Aikawa et al.,* 2002).

3.2.3.8. Isotopic fractionation: From the meteoritic perspective, perhaps the most interesting aspect of the inclusion of "interstellar" chemical processes in disk models is that this admits many more possibilities for efficient isotopic fractionation throughout the nebula. Isotopic fractionation (in D, ^{13}C, ^{15}N, ^{18}O, and ^{17}O) of the major molecular volatiles (CO, H_2O, NH_3, N_2, O_2, CH_4, C_2H_2), and trace species such as polycyclic aromatic hydrocarbons (PAHs) can occur through ion-molecule reactions in cold gas and on grain surfaces (*Nuth et al.,* 2006).

Deuterium fractionation in the disks has been modeled by *Aikawa and Herbst* (1999a) and in a two-dimensional model by *Aikawa and Herbst* (2001). It was found that the D/H ratios in the disk midplane were higher because of the lower temperature. In two-dimensional models containing warm molecular layers above the midplane, computed column density ratios for the LkCa15 and TW Hya disks were found to agree reasonably well with the observed DCN/HCN and DCO^+/HCO^+ values (*Aikawa et al.,* 2002; *Thi et al.,* 2004). Thermochemical models of a two-dimensional nebula have also been explicitly constructed to explain the apparent reduction of solar system D/H ratios in water and carbonaceous material from the high value seen in the interstellar medium (*Drouart et al.,* 1999; *Robert,* 2003). However, the importance of ion-molecule chemistry is confirmed by the detection of high abundances of H_2D^+ in the disks of DM Tau and TW Hya (*Ceccarelli et al.,* 2004). This molecular ion appears to be present in the CO-depleted gas of the disk midplane, and when interstellar deuterium chemistry (*Roberts et al.,* 2003) is included in disk models (*Ceccarelli and Dominik,* 2005), even higher abundances of D_2H^+ and D_3^+ are predicted in these sources.

Sandford et al. (2001) have reviewed many other interstellar processes that could be responsible for setting the enhanced D/H ratios found in organic meteoritic material. For example, ultraviolet photolysis can also affect the D/H ratios of PAHs. This involves isotope-selective dehydrogenation due to the difference in the C-H and C-D bond energies; simple estimates indicate that about half the peripheral H atoms on a small PAH could be replaced by D (*Allamandola et al.,* 1989). However, apart from ion-molecule reactions and surface reactions (e.g., *Tielens,* 1983), none of these processes (e.g., UV-photolysis of D-rich ices) have thus far been considered in disk models.

Oxygen-isotopic ratios in some meteoritc materials/CAIs exhibit mass-independent fractionation patterns ($\delta^{17}O$ vs. $\delta^{18}O$) that are highly anomalous (e.g., *Clayton,* 2003). One possibility is that mass-independent effects occurred in the primitive solar nebula through specific gas-phase neutral-neutral reactions (e.g., *Thiemens,* 1999). This could occur through the bimolecular process

$$CO + OH \rightarrow CO_2 + H$$

or in isotopic exchange reactions involving excited-state oxygen atoms

$$^{17,18}O(^1D) + CO_2 \rightleftharpoons {}^{17,18}CO_2 + O(^3P)$$

Alternatively, another interstellar fractionation process — isotope-selective photodestruction (ISP) — may be generally important for nebular/disk chemistry (*Clayton,* 2002a,b, 2003). For molecules that are photodissociated by line photons, such as CO, the main isotopomer is sufficiently abundant that a sufficiently large column of molecules can self-shield against the UV radiation field. Lower-abundance isotopomers that cannot self-shield, $^{12}C^{18}O$ and $^{13}C^{16}O$, are thus preferentially destroyed. Since CO is the major molecular repository of carbon and oxygen in disks, this process could have dramatically altered the fractionation chemistry of the major nebular volatiles as atoms of ^{18}O and ^{17}O are selectively released, leading to a ^{16}O-poor gas.

Clayton (2002a) proposed that ISP would probably operate in the inner edge of the disk, close to where the X-wind is formed. Model calculations close to the X-point (~0.1 AU), however, showed that the high temperatures there led to the released oxygen atoms being reincorporated into CO too rapidly for a significant fractionation to occur (*Lyons and Young,* 2003). Using a two-dimensional disk chemistry model, *Young and Lyons* (2003) showed that the lower temperatures of the UV-irradiated diffuse upper layers of the outer disk *could* produce gas enriched in ^{17}O and ^{18}O. Observational support for this picture comes from observations of CO isotopomers in the outer disk of DM Tau. *Dartois et al.* (2003) found that the extent of the outer disk, as traced by $^{12}C^{18}O$ and $^{13}C^{16}O$, is more compact than that inferrred from the $^{12}C^{16}O$ data, a result consistent with ongoing isotopic-selective photodissociation. Vertical diffusion is then necessary to allow material from the disk midplane to access the photochemical region; calculations indicate that a significant mass-independent fractionation can occur in water at R = 30 AU if this mixing is very efficient (*Lyons and Young,* 2004). *Clayton* (2002c) also proposed that isotopic-selective photodissociation could also occur for the $^{15}N^{14}N/^{14}N_2$ ratio, selectively increasing the $^{15}N/^{14}N$ ratio of the gas. It is unclear whether conditions at the X-point would also limit this process there. As with CO, it may be more favored in the outer nebula, but calculations show that here there are competitive gas-grain processes that could provide high midplane $^{15}N/^{14}N$ ratios, without the need for vertical mixing (*Charnley and Rodgers,* 2002).

In conclusion, several "interstellar" fractionation processes will occur in different regions of the nebula, with the results depending mainly upon the local temperature and irradiation flux. Mixing of material between different regions will therefore serve to complicate the problem of determining the origin, in space and time, of particular chemical characteristics of meteorites. In the following section we discuss chemical models that address the mixing question.

3.2.4. Formation of macromolecules. The carbon in carbonaceous meteorites, such as CI and CM chondrites, is mostly present in complex macromolecular networks, with smaller amounts distributed among relatively volatile aromatic and aliphatic hydrocarbons, as well various small (by the standards of terrestrial biochemistry) organic molecules (e.g., *Cronin and Chang,* 1993; *Botta and Bada,* 2002; *Sephton,* 2002; *Pizzarello,* 2004). The enhanced values of δD and $\delta^{15}N$ found in meteoritic materials (including the amino acids) support the view that they, or their precursor molecules, were formed in a low-temperature environment (*Cronin and Chang,* 1993), such as found in the outer nebula or in the natal molecular cloud core.

Apart from the condensation of refractory elements to form dust, the inner nebula was a region where chemical reactions could also have formed large carbonaceous structures and various hydrocarbons (*Zolotov and Shock,* 2001). In the atmospheres of late-type carbon stars, PAH molecules are believed to form by the polymerization of acetylene, in a chemistry similar to that of sooting flames (*Frenklach and Feigelson,* 1989; *Cherchneff et al.,* 1992). It is possible that the PAHs found in meteorites contain a contribution from these regions, although they must have experienced some degree of chemical modification in their passage through the ISM and into the nebula. At the lower pressures of the protosolar nebula, *Morgan et al.* (1991) found that efficient PAH formation could occur within a well-defined pressure-temperature window, such that the optimal sites for PAH growth in the inner nebula would be above the midplane. If this occurred, and these PAHs found their way into meteorites, then any deuterium enrichments would have to be provided within the nebula and this would first require their transport to the cooler outer regions where the known fractionation mechanisms are more favored (e.g., *Sandford et al.,* 2001).

3.3. Chemistry of Disk Evolution

Over the course of the lifetime of the solar nebula, the pressure and temperature structure evolved, causing the allowed chemical reactions throughout the nebula to evolve as well. Thus, understanding how these changes occurred is important to understanding the chemical evolution of the nebula.

Cassen (1994) demonstrated how the thinning and cooling of the disk could have affected the chemical environments within the solar nebula. In this work, it was shown that the evolution of the nebula as described above would cause the condensation fronts of rock and water ice to move inward over the lifetime of the nebula. The rates at which these two fronts would move depend strongly on nebular parameters. The movement of these fronts is important to consider in looking at those meteorites that were exposed to water and those that formed in anhydrous conditions — as the snowline moved into a region where rock had already condensed, water ice would then be available to accrete into larger bodies as well. As this ice was incorporated into large bodies, it would melt and allow the liquid water to react with rock to form hydrated minerals. Anhydrous primitive material would be preserved in those regions of the disk that were not overrun by the snowline or were overrun in the late stages of disk evolution after most of the water was locked away in large planetesimals (*Cuzzi and Zahnle,* 2004).

The change in gas density as the nebula cooled may have also played an important role in determing the bulk compositions of the chondrites. While the elements found in chondrites are near solar in their elemental inventories, a decrease in relative abundance is noticed for the moderately volatile elements that correlates with decreasing condensation temperature (those elements that condense at high temperatures are relatively more abundant than those that condense at lower temperatures). This is thought to have occurred by one of two mechanisms: (1) mixing together of material processed at high temperatures that lost its volatile elements (chondrules or refractory inclusions) with volatile rich material (*Anders,* 1975, 1977; *Hutchison,* 2002) or (2) removal of gas from a given location as the system cooled such that the resevoir of material available to condense at high temperatures would be greater than that available to condense at lower temperatures (*Wasson and Chou,* 1974; *Wai and Wasson,* 1977; *Wasson,* 1977). General consensus, as summarized by *Palme and Boynton* (1993) and *Palme* (2000), has been that the former mechanism does not work due to the presence of volatiles in chondrules and the inability to produce the observed trends by the vaporization of chondritic material. *Alexander* (2005) has recently argued that if evaporation and recycling of chondrule material occurred, then the two-component mixing model may need to be reinvestigated.

The case of the volatile depletion being caused by gas removal from a cooling nebula was modeled in *Cassen* (1996, 2001), where it was quantitatively shown that the depletion trend of moderately volatile elements can be reproduced by the dynamical evolution of the solar nebula. The only elements that were depleted were those that condensed at temperatures below the initial temperature at a given location of the nebula. Thus, if an area of the nebula was initially at a temperature of 800 K, then only those elements that condensed at temperatures below 800 K would show any depletion trend. Therefore, in order for all moderately volatile elements to be part of the depletion trend, temperatures must have originally been above the condensation point at which the depletion begins, which with Si is with a condensation temperature of ~1350 K. This would therefore require all material in the inner nebula to have been completely vaporized to produce the depletion. The level of depletion for the rest of the elements would depend on how quickly the nebula cooled and was depleted of gas, which is determined by the transport of mass and angular momentum as well as the decrease in opacity by the coagulation of dust (*Cassen,* 1994).

3.4. Chemical Effects of Dynamical Mixing

Along with changing the physical structure of the nebula, dynamical mixing between different regions would take place. Chondritic meteorites themselves demonstrate this perfectly as they are agglomerates of materials that formed under a variety of conditions: both high and low temperatures as well as reducing and oxidizing environments. Mixing may have occurred from regions at different radial locations or heights above the midplane. Here we discuss how some of the different mixing processes may have affected what products would form in the solar nebula.

An example of how diffusion might operate in the nebula was studied by *Stevenson and Lunine* (1988) and later by *Cyr et al.* (1998) by looking at water vapor. In the inner nebula, temperatures are expected to have been above the condensation temperature of water, and therefore all water would be present as a gas. The outer nebula was cooler, and therefore the water would have been present as water ice. There would therefore be a concentration gradient that would cause the water vapor to diffuse outward. If no sources of water are present to replenish the vapor, all the water could be removed from the inner nebula on very short timescales (~10^5 yr). Because of the importance of water for determing the oxygen fugacity of the nebula, this process would affect the chemistry of the nebula inside the snowline (*Cyr et al.,* 1999). For example, removal of water vapor is thought to be needed to explain the mineralogy of the enstatite chondrites, such as larger ratios of enstatite to forsterite than expected under solar conditions as well as the presence of Si-bearing metal (*Hutson and Ruzicka,* 2000). The effect described here can be generalized to conclude that, in the absence of a replenishing source, diffusion may also deplete any given species from the part of the nebula that is inside its condensation front.

The inward migration of meter-sized bodies by gas drag could serve as a way of replenishing the inner nebula with material that diffused outward. In the example outlined above, bodies carrying water ice could migrate inward from the snowline and vaporize, replenishing the inner nebula with water vapor. *Cyr et al.* (1998) showed that the migration of those bodies that formed by the condensation of diffusionally redistributed water vapor could slow the dehydration of the inner nebula. This study did not consider the inward migration of material from beyond the snowline, however. *Cuzzi and Zahnle* (2004) showed that if the inward mass flux was high enough, the inward migration and subsequent vaporization of rapidly migrating bodies could lead to enhanced water vapor pressures inside the snowline.

As planetesimals formed in the nebula, increasing the average size of the solids and decreasing the inward mass flux, the vapor would then be diffusionally redistributed. Again, if the water vapor increase in the inner nebula was large enough, it could have significantly affected the chemistry that occurred there. As an example, significantly enhanced oxygen fugacities (at least 3 orders of magnitude) are needed to explain why the FeO contents in chondrites are higher than expected for condensation of a gas of solar composition (*Ebel and Grossman,* 2000). Combining the migration and vaporization of icy bodies along with the diffusion of water vapor may explain the evidence for the orders-of-magnitude range of oxidizing conditions inferred for the nebula by chondritic meteorites (*Krot et al.,* 2000) or how heavy oxygen isotopes were brought inward from the outer nebula (*Krot et al.,* 2005). Kinetics also become important in migration, because material that is chemically altered on long timescales can be delivered to a region of the nebula where it would not normally have formed. *Ciesla and Lauretta* (2005) provided an example of this in looking at the delivery of hydrated minerals from the outer asteroid belt region of the solar nebula to the planetesimals orbiting the Sun where Earth formed.

While radial mixing appears to have been unavoidable in the solar nebula, the chondritic meteorites demonstrate that there were limits to how extensive this mixing was. The observed chondrule-matrix complementarity in meteorites implies that chondrules and matrix are related in some way, perhaps both being processed by the transient heating events that formed chondrules shortly (compared to mixing timescales) before being incorporated meteorite parent bodies. Such a constraint poses problems for the formation of chondrules in protostellar outflows as chondrules and matrix grains would have different aerodynamic properties and would be reaccreted by the solar nebula at different heliocentric distances. There appears to be no reason to expect any chemical relation between chondrules and matrix in such a formation scenario. If chondrules and matrix were chemically processed locally, and then rapidly incorporated into parent bodies, then the chemical relation between the two components can be explained. Unfortunately, the source of these heating events and how meteorite parent bodies are formed are the subjects of ongoing research, and can only provide conceptual constraints rather than detailed quantitative ones.

We conclude this section with a brief review of those recent models that have specifically addressed the chemical effects of radial mixing from distinct regions of a disk. Most of the recent models based on interstellar gas-grain chemistry have included inward radial mixing to some extent. These range from simple implementation of a limiting timescale for solving the time-dependent chemistry at each radius (e.g., *Willacy et al.,* 1998; *Markwick et al.,* 2002) to explicit solution of the advection-reaction-diffusion for gas species (*Aikawa et al.,* 1999a) and dust (*Gail,* 2001; *Wehrstedt and Gail,* 2002). Each approach has its advantages and limitations; with the former, for example, allowing for a broader range of chemical reactions to be more easily studied.

Matter at the midplane edge should initially be mostly pristine ISM material and disk gas-grain chemistry will act to modify it as it is transported inward. There should be significant gas-phase depletion at the outer edge with most of the molecules incorporated as ices. As inflowing material encounters higher temperatures, the most volatile ice species are evaporated. These will be CO and N_2 and chemical reactions in the outer disk can act to convert a significant fraction of this carbon and nitrogen to to less-volatile species: CO_2, CH_4, NH_3, and HCN (*Aikawa et al.,* 1999a). These models can therefore explain the coexistence of oxidized and reduced ices in comets, although pristine interstellar material may provide at least as good an explanation (*Ehrenfreund and Charnley,* 2000). High interstellar D/H ratios should be accreted at the outer disk and *Aikawa and Herbst* (1999a, 2001) have modeled the resulting chemistry as these molecules move inward. However, here it is also difficult to distinguish clear fractionation signatures of the mixing chemistry from that of the pristine ISM. Nevertheless, it does appear that, at least for water and HCN, mixing into the warmer region of giant planet formation tends to reduce molecular D/H ratios to values more in line with those found in comets and meteorites (*Drouart et al.,* 1999; *Hersant et al.,* 2001; *Robert,* 2003).

At the radius where all the volatile (water) ices are desorbed, the gas-grain interaction only sets the charge distribution in the dusty plasma. Inward radial transport to the inner nebula leads to the gas and dust being thermally and chemically modified by the high temperatures there. The nebular chemistry transitions to one close to chemical equilibrium and ultimately the transported dust is destroyed and the release of elements leads to a rich gas-phase chemistry (*Duschl et al.,* 1996; *Bauer et al.,* 1997). The composition of the nebular dust thus formed will depend on the overall elemental composition (i.e., metallicity) of the starting materials (*Wehrstedt and Gail,* 2003), as interstellar dust and gas are converted wholesale to nebular materials. For example, interstellar amorphous Mg-Fe-silicate grains will be converted to crystalline Mg-silicate and Fe particles (*Gail,* 2004). The key issue is how much of the chemistry occurring in this hot inner region can be transported radially outward, either by diffusive turbulent transport (*Morfill and Völk,* 1984) or by large-scale radial flow (*Keller and Gail,* 2004). Many detailed studies have previously considered the potential global impact of the distinct chemistries occurring in the inner nebula and the giant planet subnebulae (*Fegley and Prinn,* 1989; *Fegley,* 1999, 2000; *Kress and Tielens,* 2001; *Mousis et al.,* 2002). A more recent motivation is the fact that crystalline silicates are found in comets (e.g., *Wooden,* 2002) and this strongly suggests efficient radial mixing out to the 5–30 AU region (*Nuth et al.,* 2000; *Hill et al.,* 2001). For example, *Boss* (2004) has considered the transport and mixing of irradiated solid particles containing isotopic anomalies in three-dimensional models of a gravitationally unstable nebula. The grains were presumed to be deposited in an annulus on the disk surface (at 9 AU), either from the X-wind or a nearby supernova; Boss showed that this initial spatial heterogeneity could be maintained for about 1000 years and this material may have eventually been incorporated into chondrites. Recent modeling by Gail and co-workers have considered the chemical effects of the combustion and oxidation of carbon dust, as well as the annealing of amorphous silicates and the mixing of the products to large heliocentric distances. Combustion of carbonaceous dust occurs above about 1000 K and leads to high abundances of gaseous methane (*Gail,* 2001; *Wehrstedt and Gail,* 2002). Amorphous silicates are annealed at about 800 K and are vaporized above 1500 K, leading to the release of metals (Mg, Fe) and oxygen into the hot gas. Oxidation of carbon dust can proceed mainly through attack by reactive hydroxl radicals and leads to its conversion to CO and hydrocarbons. Mixing models show that chemical compositions broadly compatible with that determined for primitive solar system bodies and materials can be accounted for (*Gail,* 2001, 2004; *Bockelée-Morvan et al.,* 2002).

The products of UV or X-ray mediated chemistries in the warm layers and surfaces of the disk require efficient vertical mixing to the midplane to be of relevance to meteoritics. Higher temperatures in the layers above the midplane may also allow specific chemistries to be more viable there, e.g., PAH formation. Vertical chemical mixing has been limited to only a few recent studies. *Ilgner et al.* (2004) studied the effect of mass transport in the inner disk, within a heliocentric distance of 10 AU, by radial advection and vertical diffusion in a 1 + 1-dimensional α-disk model. They showed that the global chemistry of simple sulfur-bearing molecules (e.g., CS, SO, SO_2) was strongly influenced by the transport of material between environments with differing temperature and irradiation flux. However, the focus of this work was to make specific predictions for future radio observations of astronomical disks rather than make a connection with nebular chemistry. *Lyons and Young* (2004) have constructed a simple time-dependent model of the volatile chemistry in the outer nebula (at a heliocentric distance of 30 AU) resulting from the exchange of material between the upper disk layers and the midplane. They showed that vertical diffusion can mix surface material that has experienced the selective photodissociation of the CO isotopomers down into the midplane. Subject to how vigorous the mixing is and the FUV flux at the surface, a mass-independent fractionation in gaseous water molecules can be obtained, consistent with that measured in carbonaceous chondrites.

In summary, the dynamical transport of material could lead to material enhancements or depletions at a given location of the nebula. In addition, if alteration reactions are slow, material can be created in a given location of the nebula and then carried to a region of the nebula where it would not have formed, creating a factory-belt effect for products within the nebula. Because spatial mixing may have been efficient, the diverse chemical products and isotopic abundances seen in chondritic meteorites may reflect temporal changes rather than spatial.

3.5. Chemistry of Transient Phenomena

Transient heating events are known to have operated in the nebula, at least affecting the physical nature of chondrules and igneous CAIs. The effects that such events would have on the chemistry of these objects and/or the matrix of chondrites remain to be determined and are the focus of ongoing study. In the case of chondrules, it is still debated whether they behaved as open or closed systems (reacting with the nebular gas or not) during the transient heating events (*Sears et al.,* 1996; *Wood,* 1996). This issue must be addressed because chondrules can occupy up to 80% of the volume of a given chondrite and understanding the chemical evolution of these components is critical to understanding the chemical evolution of the bodies in which they are located. In the case of matrix material, it is unknown how much of the material was present, if any, during the passage of these events. The presence of presolar grains in these meteorites proves that at least some material escaped significant processing, whereas the chondrule-matrix complementarity of primitive meteorites suggests that a bulk of the material in chondrites was processed in a similar way.

While there has been uncertainty in the chemical behavior of chondrules while they were thermally processed, there is evidence that at least some parts of chondrules did react with the nebular gas during the transient heating event. *Lauretta et al.* (2001) observed layers of fayalite and troilite on metal grains thought to have been expelled from the chondrule melts in the Bishunpur meteorite. These layers likely formed simultaneously, and using equilibrium and kinetic chemical modeling, *Lauretta et al.* (2001) was able to calculate that the corrosion reactions must have taken place at a total pressure of 10^{-5}–10^{-4} bar, between the temperatures of 1261 and 1173 K, and in an environment where dust was enhanced by a factor of ~300 over the canonical solar value, demonstrating the occurrence of chemical reactions that would not be possible under canonical conditions.

Besides determining chondrule compositions, other chemical effects may have occurred in the nebula, depending on the nature of the transient heating event. If lightning discharges occurred, the increase in short-wavelength photons may have triggered some chemical reactions. If collisions among planetesimals were important, then this could have led to mixing of material depending on the source regions of these objects. Once the details of how these various transient heating events are understood, then the chemistry that would take place as a result can be investigated.

As another example, shock waves have been studied in detail as ways of forming chondrules (*Connolly et al.,* 2006). As gas enters a shock wave, its temperature and density increase, leading to pressure enhancements of factors of 10–100. By creating an environment that would not exist under canonical nebular conditions, shocks may have allowed chemical reactions to take place that would not have otherwise occurred. This is demonstrated by looking at the reaction times given by the simple collision theory. The time for a reaction to take place decreases with increasing partial pressure of the reacting species because collisions are more common, and if temperatures are increased, more energetic. Increases in the pressure due to shock waves may allow reactions that are kinetically inhibited under canonical conditions to take place in the nebula. *Ciesla et al.* (2003) explored this possibility by calculating the partial pressure of water that would be produced by a shock wave in an icy region of the nebula. It was found that if ice was locally concentrated in the nebula by factors of a few hundred, similar to the dust enhancements required by *Lauretta et al.* (2001), the time for 10–100-nm phyllosilicates to form via gas-solid reactions would be on the order of hours or days as opposed to billions of years under canonical conditions. This could explain the observed textures of fine-grained rims in CM chondrites that have been interpreted to be due to the accretion of both hydrous and anhydrous material by chondrules (*Metzler et al.,* 1992; *Lauretta et al.,* 2000). The enhanced pressures behind a shock wave may also have helped suppress volatile vaporization or isotopic fractionation from chondrules (*Galy et al.,* 2000). Thus, in addition to being able to match the thermal histories of chondrules, shock waves may also be able to explain aspects of chondrite chemistry.

4. SUMMARY AND OUTLOOK

In considering the chemical evolution of the solar nebula, one must also consider its physical evolution. This physical evolution was responsible for determining the conditions within the nebula that directly affected what chemistry could take place. The key quantities to bear in mind in looking at this evolution are the timescales for the various phenomena: the timescale for the chemical reaction to go to completion (t_{chem}), the timescale for the reactants to move out of the region of the nebula where the reaction could take place (t_{move}), and the timescale for the nebular environment to change such that the reaction is no longer favored (t_{evolve}). If either of the latter two timescales are shorter than the former, then the reaction is not going to occur in the area of interest. Thus, those reactions that occurred in the nebula were those that were kinetically fast and could take place over a range of nebular conditions.

In general, the inner nebula had much more energy (higher temperatures) available to facilitate chemical reactions. This energy made chemical reactants more mobile and decreased the amount of time for a chemical reaction to take place. Thus, equilibrium predictions are more closely matched by objects formed in the inner nebula than in the outer nebula. Despite this, local and transient processes occurred that disrupted the path to equilibrium, and evidence for such processes are also observed.

Our best clues of the processes that operated in the inner solar nebula come in the form of the chondritic meteorites. Chondritic meteorites directly sample the chemical products of the solar nebula. The different physical processes that affected the components of these objects and their accretion into the meteorite parent bodies also determined their chemical evolution. The fact that there are differences in chemistry among the chondrites implies their formation took place in different chemical environments. These environments may have varied with location and/or time. Unraveling the exact cause of these differences remains one of the goals of chemical studies of the solar nebula.

Our understanding of solar nebula chemistry, at finer spatial scales and different evolutionary stages, will benefit greatly in the near future from dynamical-chemical models and observations of astronomical protostellar and protoplanetary disks. Hydrodynamics in three dimensions with a full chemistry is a daunting task, as is even two-dimensional modeling of a realistic, reduced chemistry in a magnetohydrodynamical disk. However, recent approximate methods show that much can be done using simple two-dimension or 1 + 1-dimensional models (*Bockelée-Morvan et al.,* 2002; *Ilgner et al.,* 2004). Particularly important for meteoritics will be studies of specific fractionation chemistries and the redistribution of enhancements and anomalies, by dynamical mixing, throughout the nebula. Additional gas-grain processes will be included and here input from laboratory astrophysics will be important (e.g., *Gardner et al.,* 2003). Observational diagnostics are beginning to be included into disk models (e.g., *Kamp et al.,* 2003; *Gorti and Hollenbach,* 2004; *Semenov et al.,* 2005), and this approach will allow molecular line emission characteristics (line profiles, etc.) and dust spectral energy distributions to be compared directly with observations.

Spacebased and groundbased telescopes will greatly improve our understanding of disk chemical evolution. Observations with the Spitzer Space Telescope (SST) will now permit "snapshots" to be taken of the evolutionary sequence from molecular cores to protoplanetary disks (*Evans et al.,* 2003). Spectroscopic SST observations of dusty astronomical disks, at different stages of their evolution, will provide important data on their chemical and physical evolution. The Herschel satellite (*Pilbratt,* 2005) will make FIR and submillimeter observations of disks. Here, for example, observations of isotopomers of water, CO, and other molecules will be made that could test currently proposed disk fractionation mechanisms.

Groundbased instruments, operating at millimeter and submillimeter wavelengths, will permit protoplanetary disk chemistry to be studied at high spatial resolution. Observations with the Submillimeter Array (SMA) show that this instrument can be an important tool for studying the chemistry of young disks and their surroundings (*Kuan et al.,* 2004; *Qi et al.,* 2004) At present, apart from CO bandhead from emission from the hot inner disk, molecular emission is only detected from the surfaces and warm intermediate layers of the outer disk (~30–60 AU at 150 pc). The Atacama Large Millimeter Array (ALMA) will allow the chemistry of protoplanetary disks to be studied at spatial resolution of a few AU; ALMA will allow the radial and vertical chemical and physical structures of disks to be measured at scales on the order of a few AU (*Wootten,* 2001). In particular, the chemistry of the disk, within the planet-forming region (~10–30 AU at 150 pc) will also be accesible to ALMA. ALMA will allow many trace species, including isotopomers, to be detected. Astronomical observations will therefore constrain the chemical models and hopefully lead to a comprehensive and quantitative understanding of the chemistry in protoplanetary disk analogs of the protosolar nebula.

Acknowledgments. We are grateful to the detailed reviews and suggestions provided by H.-P. Gail and P. Cassen that led to improvements in various parts of the chapter. This work was supported by the NASA Goddard Center for Astrobiology and by NASA's Exobiology and Origins of Solar Systems programs. F.J.C. was supported as a National Research Council Associate.

REFERENCES

Adams F. C. and Lin D. N. C. (1993) Transport processes and the evolution of disks. In *Protostars and Planets III* (E. H. Levy and J. I. Lunine, eds.), pp. 721–748. Univ. of Arizona, Tucson.

Aikawa Y. and Herbst E. (1999a) Deuterium fractionation in protoplanetary disks. *Astrophys. J., 526,* 314–326.

Aikawa Y. and Herbst E. (1999b) Molecular evolution in protoplanetary disks. Two-dimensional distributions and column densities of gaseous molecules. *Astron. Astrophys., 351,* 233–246.

Aikawa Y. and Herbst E. (2001) Two-dimensional distributions and column densities of gaseous molecules in protoplanetary disks. II Deuterated species and UV shielding by ambient clouds. *Astron. Astrophys., 371,* 1107–1117.

Aikawa Y., Miyama S. M., Nakano T., and Umebayashi T. (1996) Evolution of molecular abundance in gaseous disks around young stars: Depletion of CO molecules. *Astrophys. J., 467,* 684–697.

Aikawa Y., Umebayashi T., Nakano T., and Miyama S. M. (1997) Evolution of molecular abundance in protoplanetary disks. *Astrophys. J. Lett., 486,* L51–L54.

Aikawa Y., Umebayashi T., Nakano T., and Miyama S. (1999a) Evolution of molecular abundances in protoplanetary disks with accretion flow. *Astrophys. J., 519,* 705–725.

Aikawa Y., Herbst E., and Dzegilenko F. N. (1999b) Grain surface recombination of HCO^+. *Astrophys. J., 527,* 262–265.

Aikawa Y., van Zadelhoff G.-H., van Dishoeck E. F., and Herbst E. (2002) Warm molecular layers in protoplanetary disks. *Astron. Astrophys., 386,* 622–632.

Aikawa Y., Momose M., Thi W.-F., van Zadelhoff G.-H., Qi C., Blake G. A., and van Dishoeck E. F. (2003) Interferometric observations of formaldehyde in the protoplanetary disk around LkCa15. *Publ. Astron. Soc. Japan, 55,* 11–15.

Aléon J., Robert F., Chaussidon M., and Marty B. (2003) Nitrogen isotopic composition of macromolecular organic matter in interplanetary dust particles. *Geochim. Cosmochim. Acta, 67,* 3773–3783.

Alexander C. M. O'D. (2005) Re-examining the role of chondrules in producing elemental fractionations in chondrites (abstract). In *Lunar and Planetary Science XXXVI,* Abstract #1348. Lunar and Planetary Institute, Houston (CD-ROM).

Allamandola L. J., Sandford S. A., and Valero G. J. (1988) Photochemical and thermal evolution of interstellar/precometary ice analogs. *Icarus, 76,* 225–252.

Allamandola L. J., Tielens G. G. M., and Barker J. R. (1989) Interstellar polycyclic aromatic hydrocarbons — The infrared emission bands, the excitation/emission mechanism, and the astrophysical implications. *Astrophys. J. Suppl. Ser., 71,* 733–775.

Allamandola L. J., Bernstein M. P., and Sandford S. A. (1997) Photochemical evolution of interstellar/precometary organic material. In *Astronomical and Biochemical Origins and the Search for Life in the Universe* (C. B. Cosmovici et al., eds.), pp. 23–47. Editrice Compositori, Bologna.

Arpigny, C., Jehin, E., Manfroid, J., Hutesmekers, D., Zucconi, J.-M., Schulz, R., and Stuewe J. A. (2003) Anomalous nitrogen isotope ratio in comets. *Science, 301,* 1522–1524.

Anders E. (1975) On the depletion of moderately volatile elements in ordinary chondrites. *Meteoritics, 10,* 283–286.

Anders E. (1977) Critique of 'Nebular condensation of moderately volatile elements and their abundances in ordinary chondrites' by Chien M. Wai and John T. Wasson. *Earth Planet. Sci. Lett., 36,* 14–20.

Balbus S. A. and Hawley J. F. (1991) A powerful local shear instability in weakly magnetized disks. I — Linear analysis. II — Nonlinear evolution. *Astrophys. J., 376,* 214–233.

Bauer I., Finocchi F., Duschl W. J., Gail H.-P., and Schloeder J. P. (1997) Simulation of chemical reactions and dust destruction in protoplanetary accretion disks. *Astron. Astrophys., 317,* 273–289.

Bell K. R., Cassen P. M., Wasson J. T., and Woolum D. S. (2000) The FU Orionis phenomenon and solar nebula material. In *Protostars and Planets IV* (V. Mannings et al., eds.), pp. 897–926. Univ. of Arizona, Tucson.

Bergin E. A., Najita J., and Ullom J. N. (2001) X-ray desorption of molecules from grains in protoplanetary disks. *Astrophys. J., 561,* 880–889.

Bernstein M., Sandford S. A., Allamandola L. J., Chang S., and Scharberg M. A. (1995) Organic compounds produced by photolysis of realistic interstellar and cometary ice analogs containing methanol. *Astrophys. J., 454,* 327–344.

Bockelée-Morvan D., Gautier D., Hersant F., Hure J.-M., and Robert F. (2002) Turbulent radial mixing in the solar nebula as the source of crystalline silicates in comets. *Astron. Astrophys., 384,* 1107–1118.

Boss A. P. (1998) Temperatures in protoplanetary disks. *Annu. Rev. Earth Planet. Sci., 26,* 53–80.

Boss A. P. (2004) Evolution of the solar nebula. VI. Mixing and transport of isotopic heterogeneity. *Astrophys. J., 616,* 1265–1277.

Boss A. P. and Foster P. N. (1998) Injection of short-lived isotopes into the presolar cloud. *Astrophys. J. Lett., 494,* L103–106.

Botta O. and Bada J. (2002) Extraterrestrial organic compounds in meteorites. *Surv. Geophys., 23,* 411–467.

Calvet N., Hartmann L., and Strom S. E. (2000) Evolution of disk accretion. In *Protostars and Planets IV* (V. Mannings et al., eds.), pp. 377–399. Univ. of Arizona, Tucson.

Cameron A. G. W. (1995) The first ten million years in the solar nebula. *Meteoritics, 30,* 133–161.

Cameron A. G. W. and Truran J. W. (1977) The supernova trigger for formation of the solar system. *Icarus, 30,* 447–461.

Carr J. S., Tokunaga A. T., and Najita J. (2004) Hot H_2O emission and evidence for turbulence in the disk of a young star. *Astrophys. J., 603,* 213–220.

Caselli P., van der Tak F. F. S., Ceccarelli C., and Bacmann A. (2003) Abundant H_2D^+ in the pre-stellar core L1544. *Astron. Astrophys., 403,* L37–L41.

Cassen P. (1994) Utilitarian models of the solar nebula. *Icarus, 112,* 405–429.

Cassen P. (1996) Models for the fractionation of moderately volatile elements in the solar nebula. *Meteoritics & Planet. Sci., 31,* 793–806.

Cassen P. (2001) Nebular thermal evolution and the properties of primitive planetary materials. *Meteoritics & Planet. Sci., 36,* 671–700.

Cassen P. and Moosman A. (1981) On the formation of protostellar disks. *Icarus, 48,* 353–376.

Ceccarelli C. and Dominik C. (2005) Deuterated H_3^+ in proto-planetary disks. *Astron. Astrophys., 440,* 583–593.

Ceccarelli C., Dominik C., Lefloch B., Caselli P., and Caux E. (2004) Detection of H_2D^+: Measuring the midplane degree of ionization in the disks of DM Tauri and TW Hydrae. *Astrophys. J. Lett., 607,* L51–L54.

Charnley S. B. (1995) The interstellar chemistry of protostellar disks. *Astrophys. Space Sci., 224,* 441–442.

Charnley S. B. (2001) Interstellar organic chemistry. In *The Bridge Between the Big Bang and Biology* (F. Giovannelli, ed.), pp. 139–149. Consiglio Nazionale delle Ricerche President Bureau, Rome.

Charnley S. B. and Rodgers S. D. (2002) The end of interstellar chemistry as the origin of nitrogen in comets and meteorites. *Astrophys. J. Lett., 569,* L133–L137.

Cherchneff I., Tielens A. G. G. M., and Barker J. (1992) Polycyclic aromatic hydrocarbon formation in carbon-rich stellar envelopes. *Astrophys. J., 401,* 269–287.

Chiang E.I. and Goldreich P. (1997) Spectral energy distributions of T Tauri stars with passive circumstellar disks. *Astrophys. J., 490,* 368–376.

Chick K. M. and Cassen P. (1997) Thermal processing of interstellar dust grains in the primitive solar environment. *Astrophys. J., 477,* 398–409.

Ciesla F. J. and Lauretta D. S. (2005) Radial migration of phyllosilicates in the solar nebula. *Earth. Planet. Sci. Lett., 231,* 1–8.

Ciesla F. J., Lauretta D. S., Cohen B. A., and Hood L. L. (2003) A nebular origin for chondritic fine-grained phyllosilicates. *Science, 299,* 549–552.

Clayton R. N. (2002a) Self-shielding in the solar nebula. *Nature, 415,* 860–861.

Clayton R. N. (2002b) Photochemical self-shielding in the solar nebula (abstract). In *Lunar and Planetary Science XXXIII,* Abstract #1326. Lunar and Planetary Institute, Houston (CD-ROM).

Clayton R. N. (2002c) Nitrogen isotopic fractionation by photochemical self-shielding (abstract). *Meteoritics & Planet. Sci., 37,* A35.

Clayton R. N. (2003) Oxygen isotopes in the solar system. *Space Sci. Rev., 106,* 19–32.

Connolly H. C. Jr., Desch S. J., Ash R. D., and Jones R. H. (2006) Transient heating events in the protoplanetary nebula. In *Meteorites and the Early Solar System II* (D. S. Lauretta and H. Y. McSween Jr., eds.), this volume. Univ. of Arizona, Tucson.

Cronin J. R. and Chang S. (1993) Organic matter in meteorites: Molecular and isotopic analyses of the Murchison meteorite. In *The Chemistry of Life's Origins* (J. M. Greenberg et al., eds.), pp. 209–258. Kluwer, Dordrecht.

Cuzzi J. N. and Weidenschilling S. J. (2006) Particle-gas dynamics and primary accretion. In *Meteorites and the Early Solar System II* (D. S. Lauretta and H. Y. McSween Jr., eds.), this volume. Univ. of Arizona, Tucson.

Cuzzi J. N. and Zahnle K. J. (2004) Material enhancement in protoplanetary nebulae by particle drift through evaporation fronts. *Astrophys. J., 614,* 490–496.

Cyr K. E., Sears W. D., and Lunine J. I. (1998) Distribution and evolution of water ice in the solar nebula: Implications for solar system body formation. *Icarus, 135,* 537–548.

Cyr K. E., Sharp C. M., and Lunine J. I. (1999) Effects of the redistribution of water in the solar nebula on nebular chemistry. *J. Geophys. Res., 104,* 19003–19014.

Dartois E., Dutrey A., and Guilloteau S. (2003) Structure of the DM Tau outer disk: Probing the vertical kinetic temperature gradient. *Astron. Astrophys., 399,* 773–787.

Dolginov A. Z. and Stepinski T. F. (1994) Are cosmic rays effective for ionization of protoplanetary disks? *Astrophys. J., 427,* 377–383.

Drouart A., Dubrulle B., Gautier D., and Robert F. (1999) Structure and transport in the solar nebula from constraints on deuterium enrichment and giant planets formation. *Icarus, 40,* 129–155.

Duschl W. J., Gail H.-P., and Tscharnuter W. M. (1996) Destruction processes for dust in protoplanetary accretion disks. *Astron. Astrophys., 312,* 624–642.

Dutrey A., Guilloteau S., and Simon M. (1994) Images of the GG Tauri rotating ring. *Astron. Astrophys., 286,* 149–159.

Dutrey A., Guilloteau S., and Guélin M. (1997) Chemistry of protosolar-like nebulae: The molecular content of the DM Tau and GG Tau disks. *Astron. Astrophys., 317,* L55–L58.

Dutrey A., Guilloteau S., and Guélin M. (2000) Observations of the chemistry in circumstellar disks. In *Astrochemistry: From Molecular Clouds to Planetary Systems* (Y. C. Minh et al., eds.), pp. 415–423. Astronomical Society of the Pacific, Sogwipo.

Ebel D. S. (2006) Condensation of rocky material in astrophysical environments. In *Meteorites and the Early Solar System II* (D. S. Lauretta and H. Y. McSween Jr., eds.), this volume. Univ. of Arizona, Tucson.

Ebel D. S. and Grossman L. (2000) Condensation in dust-enriched systems. *Geochim. Cosmochim. Acta, 64,* 339–366.

Ehrenfreund P. and Charnley S. B. (2000) Organic molecules in the interstellar medium, comets, and meteorites: A voyage from dark clouds to the early Earth. *Annu. Rev. Astron. Astrophys., 38,* 427–483.

Ehrenfreund P., Charnley S. B., and Wooden D. H. (2004) From ISM material to comet particles and molecules. In *Comets II* (M. C. Festou et al., eds.), pp. 115–133. Univ. of Arizona, Tucson.

Elmegreen B. G. (2000) Star formation in a crossing time. *Astrophys. J., 530,* 277–281.

Evans N. (1999) Physical conditions in regions of star formation. *Annu. Rev. Astron. Astrophys., 37,* 311–362.

Evans N. J. and 17 colleagues (2003) From molecular cores to planet-forming disks: A SIRTF legacy. *Publ. Astron. Soc. Pac., 115,* 965–980.

Fegley B. J. (1988) Cosmochemical trends of volatile elements in the solar system. In *Workshop on the Origins of Solar Systems* (J. A. Nuth and P. Sylvester, eds.), pp. 51–59. Lunar and Planetary Institute, Houston.

Fegley B. (1999) Chemical and physical processing of presolar materials in the solar nebula and the implications for preservation of presolar materials in comets. *Space Sci. Rev., 90,* 239–252.

Fegley B. (2000) Kinetics of gas-grain reactions in the solar nebula. *Space Sci. Rev., 92,* 177–200.

Fegley B. J. and Prinn R. G. (1989). Solar nebula chemistry — Implications for volatiles in the solar system. In *The Formation and Evolution of Planetary Systems* (H. A. Weaver et al., eds.), pp. 171–205. Univ. of Arizona, Tucson.

Feigelson E. D., Casanova S., Montmerle T., and Guibert J. (1993) ROSAT X-ray study of the Chamaeleon I dark cloud. I. The stellar population. *Astrophys. J., 416,* 623–646.

Finocchi F. and Gail H.-P. (1997) Chemical reactions in protoplanetary accretion disks. III. The role of ionisation processes. *Astron. Astrophys., 327,* 825–844.

Frenklach M. and Feigelson E. D. (1989) Formation of polycyclic aromatic hydrocarbons in circumstellar envelopes. *Astrophys. J., 341,* 372–384.

Gail H.-P. (2001) Radial mixing in protoplanetary accretion disks. I. Stationary disc models with annealing and carbon combustion. *Astron. Astrophys., 378,* 192–213.

Gail H.-P. (2004) Radial mixing in protoplanetary accretion disks. IV. Metamorphosis of the silicate dust complex. *Astron. Astrophys., 413,* 571–591.

Galy A., Young E. D., Ash R. D., and O'Nions R. K. (2000). The formation of chondrules at high gas pressures in the solar nebula. *Science, 290,* 1751–1754.

Gammie C. F. (1996) Layered accretion in T Tauri disks. *Astrophys. J., 457,* 355–362.

Gardner K., Li J., Dworkin J., Cody G. D., Johnson N., and Nuth J. A. III (2003) A first attempt to simulate the natural formation of meteoritic organics (abstract). In *Lunar and Planetary Science XXXIV,* Abstract #1613. Lunar and Planetary Institute, Houston (CD-ROM).

Glassgold A. E., Najita J., and Igea J. (1997) X-ray ionization of protoplanetary disks. *Astrophys. J., 480,* 344–350.

Glassgold A. E., Najita J., and Igea J. (2004) Heating protoplanetary disk atmospheres. *Astrophys. J., 615,* 972–990.

Goldsmith P. F., Langer W. D., and Velusamy T. (1999) Detection of methanol in a Class 0 Protostellar disk. *Astrophys. J. Lett., 519,* L173–L176.

Gorti U. and Hollenbach D. (2004) Models of chemistry, thermal balance, and infrared spectra from intermediate-aged disks around G and K stars. *Astrophys. J., 613,* 424–447.

Grossman L. (1972) Condensation in the primitive solar nebula. *Geochim. Cosmochim. Acta, 36,* 597–619.

Haghighipour N. and Boss A. (2003) On gas drag-induced rapid migration of solids in a nonuniform solar nebula. *Astrophys. J., 598,* 1301–1311.

Hartmann L. (2000) Observational constraints on transport (and mixing) in pre-main sequence disks. *Space Sci. Rev., 92,* 55–68.

Hartmann L., Calvet N., Gullbring E., and D'Alessio P. (1998) Accretion and the evolution of T Tauri disks. *Astrophys. J., 495,* 385–400.

Hartmann L., Ballesteros-Paredes J., and Bergin E. A. (2001) Rapid formation of molecular clouds and stars in the solar neighborhood. *Astrophys. J., 562,* 852–868.

Hayashi C. (1981) Structure of the solar nebula, growth and decay of magnetic fields and effects of magnetic and turbulent viscosities on the nebula. *Prog. Theor. Phys. Suppl., 70,* 35–53.

Herbst E. (2000) Models of gas-grain chemistry in star-forming regions. In *Astrochemistry: From Molecular Clouds to Planetary Systems* (Y. C. Minh and E. F. van Dishoeck, eds.), pp. 147–159. Astronomical Society of the Pacific, Sogwipo.

Hersant F., Gautier D., and Huré J. (2001). A two-dimensional model for the primordial nebula constrained by D/H measurements in the solar system: Implications for the formation of giant planets. *Astrophys. J., 554,* 391–407.

Hester J. J., Desch S. J., Healy K. R., and Leshin L. A. (2004) The cradle of the solar system. *Science, 304,* 1116–1117.

Hill H. G. M., Grady C. A., Nuth J. A. III, Hallenbeck S. L., and Sitko M. L. (2001) Constraints on nebular dynamics and chemistry based on observations of annealed magnesium silicate grains in comets and in disks surrounding Herbig Ae/Be stars. *Proc. Natl. Acad. Sci., 98,* 2182–2187.

Hollenbach D. and McKee C. F. (1979) Molecule formation and infrared emission in fast interstellar shocks. I Physical processes. *Astrophys. J. Suppl. Ser., 41,* 555–592.

Hollenbach D. J., Yorke H. W., and Johnstone D. (2000) Disk dispersal around young stars. In *Protostars and Planets IV* (V. Mannings et al., eds.), pp. 401–428. Univ. of Arizona, Tucson.

Hudson R. L. and Moore M. H. (2001) Radiation chemical alterations in solar system ices: An overview. *J. Geophys. Res., 106,* 33275–33284.

Hutchison R. (2002) Major element fractionation in chondrites by distillation in the accretion disk of a T Tauri Sun? *Meteoritics & Planet. Sci., 37,* 113–124.

Hutson M. and Ruzicka A. (2000) A multi-step model for the origin of E3 (enstatite) chondrites. *Meteoritics & Planet. Sci., 35,* 601–608.

Igea J. and Glassgold A. E. (1999) X-ray ionization of the disks of young stellar objects. *Astrophys. J., 518,* 848–858.

Ilgner M., Henning Th., Markwick A. J., and Millar T. J. (2004) The influence of transport processes on the chemical evolution in steady accretion disk flows. *Astron. Astrophys., 415,* 643–659.

Kamp I., van Zadelhoff G. J., van Dishoeck E. F., and Stark R. (2003) Line emission from circumstellar disks around A stars. *Astron. Astrophys., 397,* 1129–1141.

Kastner J. H., Zuckerman B., Weintraub D. A., and Forveille T. (1997) X-ray and molecular emission from the nearest region of recent star formation. *Science, 277,* 67–71.

Keller Ch. and Gail H.-P. (2004) Radial mixing in protoplanetary accretion disks. VI. Mixing by large-scale radial flows. *Astron. Astrophys., 415,* 1177–1185.

Kress M. E. and Tielens A. G. G. M. (2001) The role of Fischer-Tropsch catalysis in solar nebula chemistry. *Meteoritics & Planet. Sci., 36,* 75–92.

Krot A. N., Fegley B., Lodders K., and Palme H. (2000) Meteoritical and astrophysical constraints on the oxidation state of the solar nebula. In *Protostars and Planets IV* (V. Mannings et al., eds.), pp. 1019–1054. Univ. of Arizona, Tucson.

Krot A. N., Hutcheon I. D., Yurimoto H., Cuzzi J. N., McKeegan K. D., Scott E. R. D., Libourel G., Chaussidon M., Aléon J., and Petaev M. I. (2005) Evolution of oxygen isotopic composition in the inner solar nebula. *Astrophys. J., 622,* 1333–1342.

Kuan Yi-J. and 10 colleagues (2004) Organic molecules in low-mass protostellar hot cores: Submillimeter imaging of IRAS 16293-2422. *Astrophys. J. Lett., 616,* L27–L30.

Lada C. J. and Shu F. H. (1990) The formation of sunlike stars. *Science, 248,* 564–572.

Lauretta D. S., Hua X., and Buseck P. R. (2000) Mineralogy of fine-grained rims in the ALH 81002 CM chondrite. *Geochim. Cosmochim. Acta, 64,* 3263–3273.

Lauretta D. S., Buseck P. R., and Zega T. J. (2001) Opaque minerals in the matrix of the Bishunpur (LL3.1) chondrite: Constraints on the chondrule formation environment. *Geochim. Cosmochim. Acta, 65,* 1337–1353.

Lee T., Shu F. H., Shang H., Glassgold A., and Rehm K. E. (1998) Protostellar cosmic rays and extinct radioactivities in meteorites. *Astrophys. J., 506,* 898–912.

Léger A., Jura M., and Omont A. (1985) Desorption from interstellar grains. *Astron. Astrophys., 144,* 147–160.

Le Teuff Y.-H., Millar T. J., and Markwick A. J. (2000) The UMIST database for astrochemistry 1999. *Astron. Astrophys. Suppl. Ser., 146,* 157–168.

Lewis J. S. (1972) Metal/silicate fractionation in the solar system. *Earth Planet. Sci. Lett., 15,* 286–290.

Lewis J. S. and Prinn R. G. (1980) Kinetic inhibition of CO and N_2 reduction in the solar nebula. *Astrophys. J., 238,* 357–364.

Liffman K. and Toscano M. (2000) Chondrule fine-grained mantle formation by hypervelocity impact of chondrules with a dusty gas. *Icarus, 143,* 106–125.

Lodders K. (2003) Solar system abundances and condensation temperatures of the elements. *Astrophys. J., 591,* 1220–1247.

Lunine J. I., Engel S., Rizk B., and Horanyi M. (1991) Sublimation and reformation of icy grains in the primitive solar nebula. *Icarus, 94,* 333–344.

Lyons J. R. and Young E. D. (2003) Towards an evaluation of self-shielding at the X-point as the origin of the oxygen isotope anomaly in CAIs (abstract). In *Lunar and Planetary Science XXXIV,* Abstract #1981. Lunar and Planetary Institute, Houston (CD-ROM).

Lyons J. R. and Young E. D. (2004) Evolution of oxygen isotopes in the solar nebula (abstract). In *Lunar and Planetary Science XXXV,* Abstract #1970. Lunar and Planetary Institute, Houston (CD-ROM).

Mac Low M.-M. and Klessen R. S. (2004) Control of star formation by supersonic turbulence. *Rev. Mod. Phys., 76,* 125–194.

Maloney P. R., Hollenbach D. J., and Tielens A. G. G. M. (1996) X-Rayirradiated molecular gas. I. Physical processes and general results. *Astrophys. J., 466,* 561–584.

Markwick A. J. and Charnley S. B. (2004) Physics and chemistry of protoplanetary disks: Relation to primitive solar system material. In *Astrobiology: Future Perspectives* (P. Ehrenfreund et al., eds.), pp. 32–65. Kluwer, Dordrecht.

Markwick A. J., Ilgner M., Millar T. J., and Henning Th. (2002) Molecular distributions in the inner regions of protostellar disks. *Astron. Astrophys., 385,* 632–646.

Matsuyama I., Johnstone D., and Hartmann L. (2003). Viscous diffusion and photoevaporation of stellar disks. *Astrophys. J., 582,* 893–904.

Metzler K., Bischoff A., and Stoeffler D. (1992). Accretionary dust mantles in CM chondrites — Evidence for solar nebula processes. *Geochim. Cosmochim. Acta, 56,* 2873–2897.

Millar T. J., Nomura H., and Markwick A. J. (2003) Two-dimensional models of protoplanetary disk chemistry. *Astrophys. Space Sci., 285,* 761–768.

Montmerle T., Koch-Miramond L., Falgarone E., and Grindlay J. E. (1983) Einstein observations of the Rho Ophiuchi dark cloud — an X-ray Christmas tree. *Astrophys. J., 269,* 182–201.

Morfill G. E. and Völk H. J. (1984) Transport of dust and vapor and chemical fractionation in the early protosolar cloud. *Astrophys. J., 287,* 371–395.

Morgan W. A. Jr., Feigelson E. D., Wang H., and Frenklach M. (1991) A new mechanism for the formation of meteoritic kerogen-like material. *Science, 252,* 109–112.

Mousis O., Gautier D., and Bockelée-Morvan D. (2002) An evolutionary turbulent model of Saturn's subnebula: Implications for the origin of the atmosphere of Titan. *Icarus, 156,* 162–175.

Mundy L. G., Looney L. W., and Welch W. J. (2000). The structure and evolution of envelopes and disks in young stellar systems. In *Protostars and Planets IV* (V. Mannings et al., eds.), pp. 355–376. Univ. of Arizona, Tucson.

Neufeld D. and Hollenbach D. J. (1994) Dense molecular shocks and accretion onto protostellar disks. *Astrophys. J., 428,* 170–185.

Nomura H. (2002) Structure and instabilities of an irradiated viscous protoplanetary disk. *Astrophys. J., 567,* 587–595.

Nuth J. A., Hill H. G. M., and Kletetschka G. (2000) Determining the ages of comets from the fraction of crystalline dust. *Nature, 406,* 275–276.

Nuth J. A. III, Charnley S. B., and Johnson N. M. (2006) Chemical processes in the interstellar medium: Source of the gas and dust in the primitive solar nebula. In *Meteorites and the Early Solar System II* (D. S. Lauretta and H. Y. McSween Jr., eds.) this volume, Univ. of Arizona, Tucson.

Palme H. (2000) Are there chemical gradients in the inner solar system? *Space Sci. Rev., 92,* 237–262.

Palme H. and Boynton W. V. (1993) Meteoritic constraints on conditions in the solar nebula. In *Protostars and Planets III* (E. H. Levy and J. I. Lunine, eds.), pp. 979–1004. Univ. of Arizona, Tucson.

Petaev M. I. and Wood J. A. (1998) The condensation with partial isolation model of condensation in the solar nebula. *Meteoritics & Planet. Sci., 33,* 1123–1137.

Pilbratt G. L. (2005) *The Herschel Mission: Overview and Observing Opportunities.* ESA Special Publication SP-577, in press.

Pizzarello S. (2004) Chemical evolution and meteorites: An update. *Origins Life Evol. Biosph., 34,* 25–34.

Pontoppidan K. M., Dullemond C. P., van Dishoeck E. F., Blake G. A., Boogert A. C. A., Evans N. J. II, Kessler-Silacci J. E., and Lahuis F. (2005) Ices in the edge-on disk CRBR 2422.8-3423: Spitzer spectroscopy and Monte Carlo radiative transfer modeling. *Astrophys. J., 622,* 463–481.

Prasad S. S. and Tarafdar S. P. (1983) UV radiation field inside dense clouds: Its possible existence and chemical implications. *Astrophys. J., 267,* 603–609.

Qi C. (2001) Aperture synthesis studies of the chemical composition of protoplanetary disks and comets. Ph.D. thesis, California Institute of Science and Technology, Pasadena. 141 pp.

Qi C. and 12 colleagues (2004) Imaging the disk around TW Hydrae with the submillimeter array. *Astrophys. J. Lett., 616,* L11–L14.

Robert F. (2003) The D/H ratio in chondrites. *Space Sci. Rev., 106,* 87–101.

Roberts H., Herbst E., Millar T. J. (2003) Enhanced deuterium fractionation in dense interstellar cores resulting from multiply deuterated H_3^+. *Astrophys. J. Lett., 591,* L41–L44.

Ruden S. P. and Pollack J. B. (1991). The dynamical evolution of the protosolar nebula. *Astrophys. J., 375,* 740–760.

Ruzmaikina T. V. and Ip W. H. (1994) Chondrule formation in radiative shock. *Icarus, 112,* 430–447.

Sandford S. A. (1996) The inventory of interstellar materials available for the formation of the solar system. *Meteoritics, 31,* 449–476.

Sandford S. A., Bernstein M. P., and Dworkin J. P. (2001) Assessment of the interstellar processes leading to deuterium enrichment in meteoritic organics. *Meteoritics & Planet. Sci., 36,* 1117–1133.

Saxena S. K. and Eriksson G. (1986) Chemistry of the formation of the terrestrial planets. In *Chemistry and Physics of Terrestrial Planets* (S. K. Saxena, ed.), pp. 30–105. Springer, New York.

Sears D. W. G., Huang S., and Benoit P. H. (1996) Open-system behaviour during chondrule formation. In *Chondrules and the*

Protoplanetary Disk (R. H. Hewins et al., eds.), pp. 221–232. Cambridge Univ., Cambridge.

Semenov D., Pavlyuchenkov Y., Schreyer K., Henning Th., Dullemond C., and Bacmann A. (2005) Millimeter observations and modeling of the AB Aurigae system. *Astrophys. J., 621,* 853–874.

Sephton M. A. (2002) Organic compounds in carbonaceous meteorites. *Natl. Prod. Rep., 19,* 292–311.

Shu F. H., Adams F. C., and Lizano S. (1987) Star formation in molecular clouds: Observation and theory. *Annu. Rev. Astron. Astrophys., 25,* 23–81.

Shu F. H., Shang H., and Lee T. (1996) Toward an astrophysical theory of chondrites. *Science, 271,* 1545–1552.

Shu F. H., Shang H., Glassgold A. E., and Lee T. (1997) X-rays and fluctuating X-winds from protostars. *Science, 277,* 1475–1479.

Simonelli D. P., Pollack J. B., and McKay C. P. (1997) Radiative heating of interstellar grains falling toward the solar nebula: 1-D diffusion calculations. *Icarus, 125,* 261–280.

Stevenson D. J. and Lunine J. I. (1988) Rapid formation of Jupiter by diffuse redistribution of water vapor in the solar nebula. *Icarus, 75,* 146–155.

Stone J. M., Gammie C. F., Balbus S. A., and Hawley J. F. (2000). Transport processes in protostellar disks. In *Protostars and Planets IV* (V. Mannings et al., eds.), pp. 589–612. Univ. of Arizona, Tucson.

Takeuchi T. and Lin D. N. C. (2002) Radial flow of dust particles in accretion disks. *Astrophys. J., 581,* 1344–1355.

Thi W.-F., van Zadelhoff G.-J., and van Dishoeck E. F. (2004) Organic molecules in protoplanetary disks around T Tauri and Herbig Ae stars. *Astron. Astrophys., 425,* 955–972.

Thiemens M. H. (1999) Mass-independent isotope effects in planetary atmospheres and the early solar system. *Science, 283,* 341–345.

Tielens A. G. G. M. (1983) Surface chemistry of deuterated molecules. *Astron. Astrophys., 119,* 177–184.

Tielens A. G. G. M. and Allamandola L. J. (1987) Composition, structure, and chemistry of interstellar dust. In *Interstellar Processes* (D. J. Hollenbach, eds.), pp. 397–469. Reidel, Dordrecht.

Tielens A. G. G. M. and Hollenbach D. J. (1985) Photodissociation regions. I. Basic model. II. A model for the Orion photodissociation region. *Astrophys. J., 291,* 722–754.

Umebayashi T. and Nakano T. (1981) Fluxes of energetic particles and the ionization rate in very dense interstellar clouds. *Publ. Astron. Soc. Japan, 33,* 617–636.

Umebayashi T. and Nakano T. (1988) Ionization state and magnetic fields in the solar nebula. *Prog. Theor. Phys. Suppl., 96,* 151–160.

Umebayashi T. and Nakano T. (1990) Magnetic flux loss from interstellar clouds. *Mon. Not. R. Astron. Soc., 243,* 103–113.

van Dishoeck E. F., Thi W.-F., and van Zadelhoff G.-H. (2003) Detection of DCO^+ in a circumstellar disk. *Astron. Astrophys., 400,* L1–L4.

van Zadelhoff G.-J., van Dishoeck E. F., Thi W.-F., and Blake G. A. (2001) Submillimeter lines from circumstellar disks around pre-main sequence stars, *Astron. Astrophys., 377,* 566–580.

van Zadelhoff G.-J., Aikawa Y., Hogerheijde M. R., and van Dishoeck E. F. (2003) Axi-symmetric models of ultraviolet radiative transfer with applications to circumstellar disk chemistry. *Astron. Astrophys., 397,* 789–802.

Vanhala H. A. T. and Boss A. P. (2002) Injection of radioactivities into the forming solar system. *Astrophys J., 575,* 1144–1150.

Vanhala H. A. T. and Cameron A. G. W. (1998) Numerical simulations of triggered star formation. I. Collapse of dense molecular cloud cores. *Astrophys. J., 508,* 291–307.

Wai C. M. and Wasson J. T. (1977). Nebular condensation of moderately volatile elements and their abundances in ordinary chondrites. *Earth Planet. Sci. Lett., 36,* 1–13.

Wasson J. T. (1977) Reply to Edward Anders: A discussion of alternative models for explaining the distribution of moderately volatile elements in ordinary chondrites. *Earth Planet. Sci. Lett., 36,* 21–28.

Wasson J. T. and Chou C. (1974) Fractionation of moderately volatile elements in ordinary chondrites. *Meteoritics, 9,* 69–84.

Wehrstedt M. and Gail H.-P. (2002) Radial mixing in protoplanetary disks. II. Time dependent disk models with annealing and carbon combustion. *Astron. Astrophys., 385,* 181–204.

Wehrstedt M. and Gail H.-P. (2003) Radial mixing in protoplanetary accretion disks. V. Models with different element mixtures. *Astron. Astrophys., 410,* 917–935.

Weidenschilling S. J. (1977) Aerodynamics of solid bodies in the solar nebula. *Mon. Not. R. Astron. Soc., 180,* 57–70.

Willacy K. and Langer W. D. (2000) The importance of photoprocessing in protoplanetary disks. *Astrophys. J., 544,* 903–920.

Willacy K., Klahr H. H., Millar T. J., and Henning Th. (1998) Gas and grain chemistry in a protoplanetary disk. *Astron. Astrophys., 338,* 995–1005.

Wood J. (1996) Major unresolved issues in the formation of chondrules and chondrites. In *Chondrules and the Protoplanetary Disk* (R. H. Hewins et al., eds.), pp. 55–70. Cambridge Univ., Cambridge.

Wood J. A. and Hashimoto A. (1993) Mineral equilibrium in fractionated nebular systems. *Geochim. Cosmochim. Acta, 57,* 2377–2388.

Wooden D. H. (2002) Comet grains: Their IR emission and their relation to ISM grains. *Earth Moon Planets, 89,* 247–287.

Wooden D. H., Woodward C. E., and Harker D. E. (2004) Discovery of crystalline silicates in Comet C/2001 Q4 (NEAT). *Astrophys. J. Lett., 612,* L77–L80.

Wootten A., ed. (2001) *Science with the Atacama Large Millimeter Array.* ASP Conference Proceedings 235, Astronomical Society of the Pacific, San Francisco.

Young E. D. and Lyons J. R (2003) CO self shielding in the outer solar nebula: An astronomical explanation for the oxygen isotope slope-1 line (abstract). In *Lunar and Planetary Science XXXIV,* Abstract #1923. Lunar and Planetary Institute, Houston (CD-ROM).

Zolotov M. Y. and Shock E. L. (2001) Stability of condensed hydrocarbons in the solar nebula. *Icarus, 150,* 323–337.

Part IV:

The First Nebular Epoch: Genesis of the First Solar System Materials

Timescales of the Solar Protoplanetary Disk

Sara S. Russell
The Natural History Museum

Lee Hartmann
Harvard-Smithsonian Center for Astrophysics

Jeff Cuzzi
NASA Ames Research Center

Alexander N. Krot
Hawai'i Institute of Geophysics and Planetology

Matthieu Gounelle
CSNSM–Université Paris XI

Stu Weidenschilling
Planetary Science Institute

We summarize geochemical, astronomical, and theoretical constraints on the lifetime of the protoplanetary disk. Absolute Pb-Pb-isotopic dating of CAIs in CV chondrites (4567.2 ± 0.7 m.y.) and chondrules in CV (4566.7 ± 1 m.y.), CR (4564.7 ± 0.6 m.y.), and CB (4562.7 ± 0.5 m.y.) chondrites, and relative Al-Mg chronology of CAIs and chondrules in primitive chondrites, suggest that high-temperature nebular processes, such as CAI and chondrule formation, lasted for about 3–5 m.y. Astronomical observations of the disks of low-mass, pre-main-sequence stars suggest that disk lifetimes are about 3–5 m.y.; there are only few young stellar objects that survive with strong dust emission and gas accretion to ages of 10 m.y. These constraints are generally consistent with dynamical modeling of solid particles in the protoplanetary disk, if rapid accretion of solids into bodies large enough to resist orbital decay and turbulent diffusion are taken into account.

1. INTRODUCTION

Both geochemical and astronomical techniques can be applied to constrain the age of the solar system and the chronology of early solar system processes (e.g., *Podosek and Cassen,* 1994). We can study solid material that is believed to date from that time, and that is now preserved in minor bodies in the solar system (asteroids and comets); we can observe circumstellar disks around young solar-mass stars that are similar to the young Sun by measuring the emission of dust around them at appropriate wavelengths (infrared and millimeter), and we can make theoretical models of how disk material is likely to behave. In this chapter, we will address and review the current state of knowledge of these three strands of evidence.

A necessary first stage is to define what we mean by the accretion disk lifetime. We assume that the process of planet formation in the solar system was multistage. First, solids evolved in the protoplanetary disk. These solids began as interstellar and circumstellar grains, mostly *amorphous* material less than 1 μm in size. The constituents of chondritic meteorites — chondrules, refractory inclusions [Ca,Al-rich inclusions (CAIs) and amoeboid olivine aggregates (AOAs)], Fe,Ni-metal grains, and fine-grained matrices — are largely *crystalline*, and were formed from thermal processing of these grains and from condensation of solids from gas. Calcium-aluminum-rich inclusions are tens of micrometers to centimeter-sized irregularly shaped or rounded objects composed mostly of oxides and silicates of Ca, Al, Ti, and Mg, such as corundum (Al_2O_3), hibonite ($CaAl_{12}O_{19}$), grossite ($CaAl_4O_7$), perovskite ($CaTiO_3$), spinel ($MgAl_2O_4$), Al,Ti-pyroxene (solid solution of $CaTi^{4+}Al_2O_6$, $CaTi^{3+}AlSiO_6$, $CaAl_2SiO_6$, and $CaMgSi_2O_6$), melilite (solid solution of $Ca_2MgSi_2O_7$ and $Ca_2Al_2SiO_7$), and anorthite ($CaAl_2Si_2O_8$). Ameboid olivine aggregates are physical aggregates of individual condensate particles: forsterite (Mg_2SiO_4), Fe,Ni-metal, and CAIs composed of spinel, anorthite, and Al,Ti-pyroxene. Evaporation and condensation appear to have been the dominant processes during formation of refractory inclusions. Subsequently, some CAIs experienced melting to various degrees and crystallization over timescales of days under highly reducing (solar nebula) conditions. Chondrules are igneous, rounded objects, 0.01–10 mm in size, composed largely of ferromagnesian olivine ($Mg_{2-x}Fe_xSiO_4$) and pyroxene ($Mg_{1-x}Fe_xSiO_3$, where $1 < x < 0$), Fe,Ni-metal, and glassy or microcrystalline mesostasis. Multiple episodes of melting of preexisting solids accompanied by evaporation-recondensation are believed to have been the dominant processes during chondrule formation. Chondrules appear to have formed in more oxidizing environments than CAIs, and cooled more quickly.

It is widely believed that from these building blocks, asteroids were ultimately formed, and subsequently experienced parent-body processes, such as aqueous alteration and thermal metamorphism. Larger bodies capable of retaining heat, and/or bodies rich in radioactive nuclides, may have, in addition, experienced melting and differentiation. Shock processing, by impact within the asteroid belt, may also have occurred throughout asteroidal history. These processes are often assumed to have occurred sequentially, although this is almost certainly an oversimplification. For example, ^{53}Mn and ^{129}I evidence suggest that aqueous alteration of chondrites lasted over >20 m.y. (*Endreß et al.,* 1996; *Krot et al.,* 2006), longer than the time taken for some small asteroids to differentiate (*Wadhwa and Lugmair,* 1996; *Lugmair and Shukolyukov,* 1998). However, *Hoppe et al.* (2004) suggested that aqueous activity on the Orgueil chondrite parent asteroid lasted <10 m.y. Iodine-129 data further suggest that chondrule formation overlapped with the formation of the first melted planetesimals (*Gilmour et al.,* 2000). From a cosmochemical point of view (sections 2–4), we can assume that the disk "begins" at the time that the first solids are processed, and "ends" at the time at which asteroidal accretion stops. The time of asteroidal accretion can be constrained by dating early processes that must have occurred in an asteroidal environment. From an observational point of view, we assume that the disk lifetime dates from the point at which the star first forms, to the point at which the dust around it disappears, presumably because it has accreted into planetesimals or planets. Theoretical modeling can assist in determining how long dust grains can realistically remain within a solar accretion disk environment.

2. ABSOLUTE AGES OF PROTOPLANETARY DISK MATERIALS DEFINED BY LONG-LIVED RADIONUCLIDES

2.1. Introduction

Many radioactive isotopes can be used as geologic clocks. Each isotope decays at a fixed rate. Once this decay rate is known, the length of time over which decay has been occurring can be estimated by measuring the amount of remaining radioactive parent compared to the amount of stable daughter isotopes. Older objects will have built up a higher fraction of daughter isotopes than younger ones with the same parent/daughter element ratios. For all radioactive decay schemes, the clock "starts" at the time at which the object cools below a temperature (the "closure temperature") at which the parent and daughter isotopes become essentially immobile; this temperature depends on the decay scheme that is used, and on the minerals in which they are trapped. The age is then given by

$$t = 1\frac{1}{\lambda}\ln\left(1 + \frac{D}{P}\right)$$

where t = age, D = abundance of the daughter isotope, and P = abundance of the parent isotope. The ratio D/P can most easily be measured by constructing an isochron. An example is for the Rb-Sr system, where Rb decays to Sr with a half-life of 4.88×10^{10} yr. From a single rock, coexisting minerals with varying Rb/Sr ratios are measured both for Sr isotopes and $^{87}Rb/^{86}Sr$. A plot of $^{87}Rb/^{86}Sr$ as a function of $^{87}Sr/^{86}Sr$ will form a correlation (the isochron) that has a slope defining $^{87}Sr/^{87}Rb$ that can be plugged into the equation above to yield age information. The intercept of the isochron indicates whether the system has experienced late-stage disturbance; this will reequilibrate the Sr isotopes and cause the intercept to fall at higher $^{87}Sr/^{86}Sr$.

Radioactive isotopes with a half-life on the order of the age of the solar system are the most accurate ones to be used to gain absolute age dates of early solar system material. The most common long-lived dating systems used in meteoritics are Pb-Pb, Rb-Sr, and K-Ar [and its sister technique Ar-Ar (e.g., *Turner,* 1970)]. The practical aspects of the use of these systems are described in detail by *Tilton* (1988a). In this chapter, we will focus on results from the Pb-Pb system, which has given the most precise age-dating information for the very oldest meteoritic objects, especially for those that are refractory and thus have a high initial ratio of U to volatile Pb.

The major assumptions made in calculating the ages are that (1) the initial isotopic composition of the radiogenic element is known, or can be accurately calculated from the acquired data, or it was initially so rare in the mineral being measured that it can be ignored; (2) no resetting of the isotopic clock was made, by gain or loss of the parent or daughter isotopes since "isotopic closure," i.e., since the initial cooling of the rock; (3) the half-lives of the isotopes are accurately known; and (4) the initial isotopic composition of the parent element is known to sufficient accuracy. The most accurate results can be obtained for samples that have a high natural initial ratio of parent/daughter element, since in these samples the isotopic ratio of the daughter element is most affected by the addition of a radiogenic component.

The Pb-Pb system utilizes two different decay schemes: the decay of ^{235}U to ^{207}Pb, and ^{238}U to ^{206}Pb, both via several intermediary decay products. Measurements of ^{206}Pb and ^{207}Pb, normalized to nonradiogenic ^{204}Pb, are correlated (therefore producing an "isochron"), which yields age information (Fig. 1a). Modern studies acquire more accurate results by normalizing to ^{206}Pb (Fig. 1b). This dating system

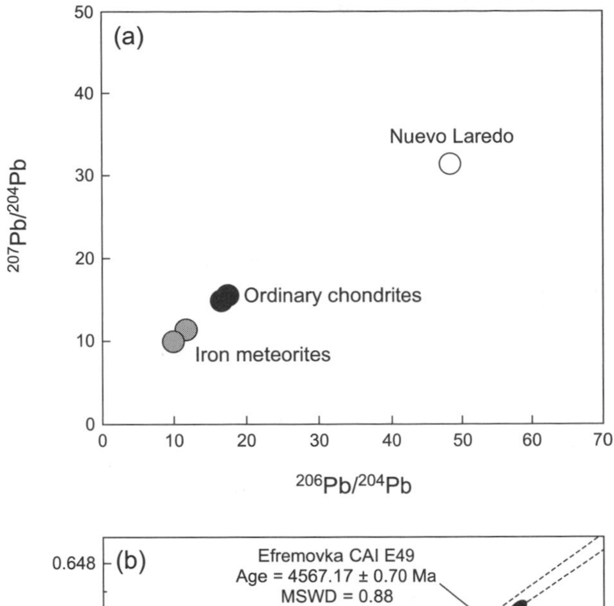

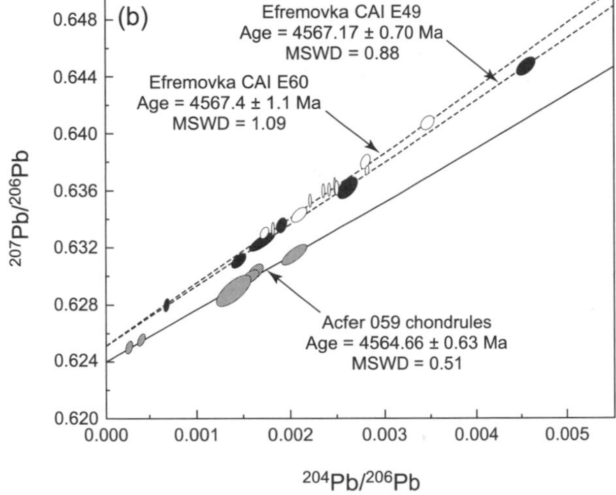

Fig. 1. **(a)** The first Pb-Pb isochron for meteorites. The isochron is composed from data from the ordinary chondrites Forest City (H5) and Modoc (L5), the iron meteorites Henbury and Canyon Diablo, and the eucrite Nuevo Laredo. Using modern assumptions about decay constants and initial Pb-isotopic composition, the slope of the line indicates an age of 4.50 ± 0.07 G.y. (see text). Data from *Patterson* (1956). **(b)** Pb-Pb isochrons for the six most radiogenic Pb-isotopic analyses of acid-washed chondrules from the CR chondrite Acfer 059 (solid line), and for acid-washed fractions from the Efremovka CAIs (dashed lines). $^{207}Pb/^{206}Pb$ ratios are not corrected for initial common Pb. Error ellipses are 2σ. Isochron age errors are 95% confidence intervals. From *Amelin et al.* (2002).

requires only measurement of the isotopic ratios of one element: Pb. No element/element ratios are required, which can be a major source of error in other dating systems, thus the data are potentially very accurate. However, Pb is mobile under many geological conditions, and so the possibility of post-crystallization redistribution must be carefully considered. For more details of the Pb-Pb dating technique, see, e.g., *Faure* (1986), *Henderson* (1982), and *Tilton* (1988a).

2.2. History of Attempts to Learn the Age of the Solar System

Early attempts to uncover the age of Earth utilized several ingenious, although inaccurate, techniques (e.g., *Darwin,* 1859; *Thomson,* 1897; *Holmes,* 1946). Estimates obtained using these techniques were in gross disagreement with each other.

A major breakthrough was provided by the discovery of radioactivity by *Becquerel* (1896a,b). Less than 10 years later, *Rutherford and Boltwood* (1905) realized the potential of radioactive decay as a tool to measure the age of Earth. The first accurate age dates of meteorites, and hence Earth, were published by Clair C. Patterson (*Patterson et al.,* 1955; *Patterson,* 1956). *Patterson et al.* (1955) recorded a Pb-Pb age of 4.57 ± 0.07 G.y. using an isochron composed of data for the eucrite Nuevo Laredo, the ordinary chondrites Forest City (H5) and Modoc (L5), and the iron meteorites Canyon Diablo (IIIAB) and Henbury (IIIAB) (Fig. 1a). Using more accurate values for the initial Pb composition and decay constants, the age was reduced to 4.50 ± 0.07 G.y. (*Tilton,* 1988b). The known age of the solar system has essentially not changed, within error, since this time.

2.3. Petrological Considerations

Calcium-aluminum-rich inclusions, composed of refractory minerals, have long been considered to represent the first solid condensates in the solar system (e.g., *Grossman,* 1972). The occasional presence of relict CAIs within Al-rich and in a few ferromagnesian chondrules (Figs. 2a–c) suggest that most CAIs probably formed earlier than most chondrules (*Krot and Keil,* 2002; *Krot et al.,* 2002, 2004). A single recent observation of a pyroxene-rich chondrule fragment apparently embedded in a CAI suggests that the formation of chondrules and CAIs may have overlapped in time and space (*Itoh and Yurimoto,* 2003). However, to date there is no quantitative chronological information available for this unique object. Recently, *Krot et al.* (2005a) described two additional igneous, anorthite-rich (Type C) CAIs containing relict chondrule fragments in the Allende meteorite. These authors concluded, however, that the chondrule-bearing Type C CAIs experienced late-stage remelting (~2 m.y. after formation of the CAIs with a canonical $^{26}Al/^{27}Al$ ratio of 5×10^{-5}) with addition of chondrule material in the chondrule-forming region. This interpretation is consistent with an age difference between the formation of CAIs and chondrules.

2.4. Absolute Ages of Calcium-Aluminum-rich Inclusions

Calcium-aluminum-rich inclusions yield the oldest reliably measured age dates of any solar system solid. Calcium-aluminum-rich inclusions typically have a high ratio of refractory U to volatile Pb, and thus are good targets for age

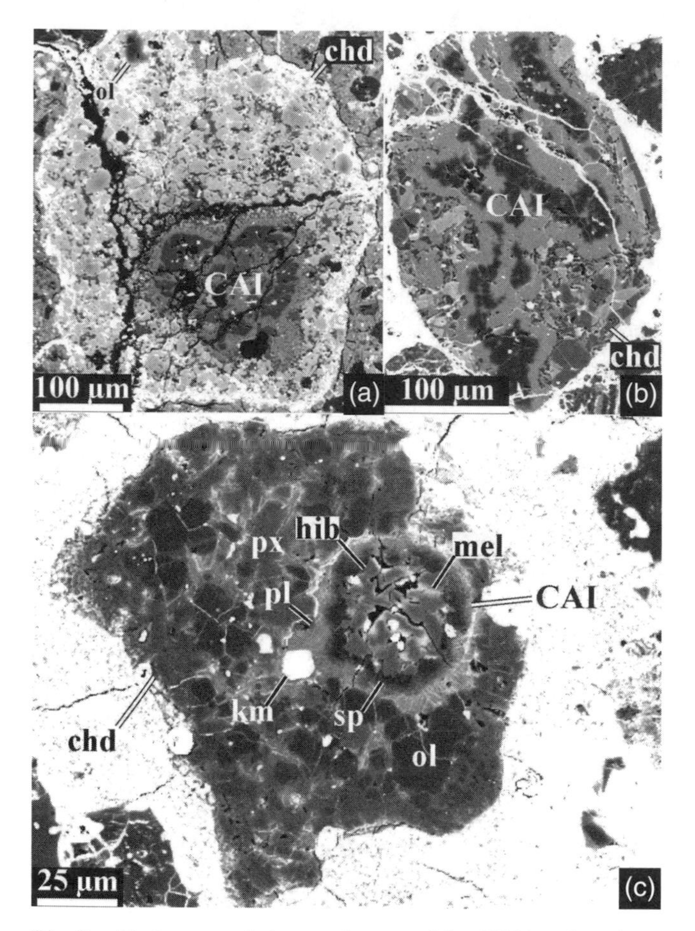

Fig. 2. Backscattered electron images of the CAI-bearing chondrules [**(a)**–**(c)**] and a chondrule-bearing CAI **(d)** in the ungrouped carbonaceous chondrites Acfer 094 **(a)** and Adelaide **(c)**, CH carbonaceous chondrite Acfer 182 **(b)**. **(a)** Relict CAI composed of hibonite and spinel is enclosed by a ferrous olivine chondrule. **(b)** Relict CAI consisting of spinel and anorthite is surrounded by an Al-rich chondrule composed of low-Ca pyroxene, high-Ca pyroxene, and fine-grained mesostasis. **(c)** Relict CAI composed of hibonite, perovskite, melilite, and spinel is enclosed by a magnesian chondrule consisting of olivine, low-Ca pyroxene, glassy mesostasis, and metal. aug = augite; en = enstatite; hib = hibonite; km = kamacite; mel = melilite; ol = forsteritic olivine; pg = pigeonite; sp = spinel; pv = perovskite; tr = troilite. From *Krot et al.* (2004).

dating using the Pb-Pb system. Early studies of chondrite ages typically included CAIs, chondrules, and matrix. *Tatsumoto et al.* (1976) studied several chondrules and refractory inclusions from the Allende meteorite. They reported that both chondrules and CAIs fall on a single isochron line, yielding an age of 4554 ± 4 m.y. A study by *Chen and Tilton* (1976) reported a significantly older age for Allende constituents of 4565 ± 4 m.y. Again, the ages of chondrules and CAIs were found to be indistinguishable from each other within analytical error. Matrix analyses reported during the same study (*Chen and Tilton,* 1976) yielded age dates several million years younger than the chondrules and CAIs, probably reflecting more prolonged and extensive asteroidal alteration of the matrix. Later studies focused on measuring mineral separates from CAIs alone. *Chen and Wasserburg* (1981) made the first measurement of this type, with samples of several Type B CAIs giving an age of 4559 ± 4 m.y.

Recognizing that post-formational redistribution of Pb from matrix to CAIs was a potential problem for measuring Pb-Pb ages of Allende inclusions, *Manhès et al.* (1988) [summarized in English by *Allègre et al.* (1995)] attempted a new approach. They analyzed several CAIs using "differential dissolution." This is a procedure that dissolves the CAIs in stages, using progressively more aggressive reagents. The model ages obtained for the CAIs from the progressive dissolution fractions increased, suggesting that Pb from the matrix was contaminating the early dissolution steps. The ages they obtained range between 4565 ± 1 m.y. and 4568 ± 3 m.y., and they interpreted the latter value as representing the true age of Allende CAIs; no significant age difference was found between different CAIs.

One way around the problem of Pb redistribution after asteroidal accretion is to study meteorites that have experienced much less alteration than Allende. Precise Pb-Pb ages were reported for two CAIs from the reduced CV3 chondrite Efremovka (*Amelin et al.,* 2002). A Type B inclusion yielded an age of 4567.4 ± 1.1 m.y., and a compact Type A inclusion yielded an age of 4567.2 ± 0.7 m.y. These values are within error of the age of Allende inclusions determined by *Manhès et al.* (1988). However, these measurements are model ages, and make the assumption that a single age can be obtained for CAIs, i.e., that they experienced no processing after this formation event. Only by analyzing several separates from one igneous object can the effects of later disturbance be accounted for. Because CAIs have high U/Pb ratios, the Pb isotopes are particularly susceptible to disturbance due to postformation heating events. Clearly, additional analyses, preferably of careful mineral separates of single large CAIs, need to be made.

Perovskite grains in CAIs are present as phenocrysts and are clearly among the first crystals to form within the CAI, thus there is a possibility that they may be relict. Lead-isotopic measurements of individual perovskite grains in the Allende (CV3) and Murchison (CM2) CAIs yielded ages of 4565 ± 34 m.y. and 4569 ± 26 m.y. respectively (*Ireland et al.,* 1990). Thus, there is no evidence at the moment to suggest that any grains within a CAI are relicts that significantly predated the rest of the inclusion.

Calcium-aluminum-rich inclusions have been age dated using the Rb-Sr system (e.g., *Podosek et al.,* 1991). Rubidium-87 has a half-life of 4.88 × 10^{10} yr and decays to ^{87}Sr. These measurements have yielded age dates in line with those for Pb-Pb dating (~4.56 G.y.). Exact age dates are often difficult to obtain because typical CAI measurements indicate a period of Sr remobilization at younger ages. The very old nature of CAIs is also indicated by the observation that the isochron intercept for the Rb-Sr system falls at very low $^{87}Sr/^{86}Sr$ values. The intercept value for Allende CAIs, coined "ALL" by *Gray et al.* (1973), provides the lowest $^{87}Sr/^{86}Sr$ of any solar system material.

2.5. Absolute Ages of Chondrules

The absolute age of chondrules has proved difficult to measure because they typically have lower U/Pb ratios than CAIs, and are typically smaller than the largest CAIs (at least in CV3 chondrites). In one of the first studies of the Allende CV3 meteorite, *Fireman* (1970) reported a K/Ar age of ~4.44 G.y. for a single large chondrule. Other scientists working in the 1970s (*Chen and Tilton,* 1976; *Tatsumoto et al.,* 1976) (see section 2.2 above) found that Allende chondrules are indistinguishable in Pb-Pb age to Allende CAIs within analytical errors (typically ~4 m.y.). *Tera et al.* (1996) pointed out that many Pb-Pb measurements of chondrules are contaminated by terrestrial Pb. More precise age dates have only recently been reported by Amelin and coworkers (*Amelin et al.,* 2002, 2004), but so far only whole chondrules, not mineral separates, have been possible to measure; thus isochrons from individual objects have not yet been obtained. An isochron for a population of Allende chondrules yields an age of 4566.7 ± 1.0 m.y., indistinguishable from the ages of Efremovka CAIs (*Amelin et al.,* 2004). An isochron plotting four whole chondrules from the CR chondrite Acfer 059 yielded an age of 4564.7 ± 0.6 m.y., 2.5 ± 1.2 m.y. younger than the mineral separates from CAIs measured in the Efremovka CV meteorite by the same technique (*Amelin et al.,* 2002). These authors noted that the low error on the isochron regression line suggests that the chondrules analyzed had a similar age to each other, probably all forming within ~1.2 m.y. *Amelin and Krot* (2005) have recently reported absolute ages of chondrule-like silicate clasts from the CB meteorites Gujba (4562.7 ± 0.5 m.y.) and Hammadah al Hamra 237 (4562.8 ± 0.9 m.y.). We would like to emphasize that the origin of the chondrule-like clasts in CB chondrites is controversial, with both nebular (*Weisberg et al.,* 2001; *Campbell et al.,* 2002) and asteroidal (*Rubin et al.,* 2003) models proposed. It is not, therefore, clear whether they can be considered chondrules in the usually understood sense of the word. For example, *Amelin and Krot* (2005) and *Krot et al.* (2005b) concluded that chondrules in CB chondrites formed during a giant impact between Moon-sized planetary embryos after the protoplanetary disk largely (but not completely) dissipated. As a result, the significance of the CB "chondrule" absolute ages for constraining the lifetime of the solar nebula remains unclear.

In summary, Pb-Pb ages currently provide the most precise ages of early solar system solids. The span of ages represented by Pb-Pb dates of chondritic objects is several million years, with CV CAIs and chondrules apparently the oldest, and objects from CB meteorites around 4–5 m.y. younger. If we assume that objects formed in a nebular setting prior to the final accretion of the asteroidal parent bodies and dissipation of the disk, then this constrains a minimum age for the solar accretion disk.

3. RELATIVE AGES OF PROTOPLANETARY DISK MATERIALS DEFINED BY SHORT-LIVED RADIONUCLIDES

3.1. Introduction

Isotopes with half-lives that are very much shorter than the age of the solar system (<100 m.y.) may once have been present in solar system material, but are now considered extinct. However, their initial presence in some early solar system materials can be detected by an excess of their daughter isotope. Evidence for the existence of several short-lived isotopes has now been found in meteorites; these are summarized in Table 1. In principle, short-lived isotopes can provide relative ages of old components with very high precision.

The initial abundance of the short-lived isotope can be quantified if there are regions of variable parent/daughter elemental ratios within a single object that formed during one event (e.g., different minerals within an igneous object, or between several objects that are assumed to be the same age). Figure 3a shows an example of evidence for the extinct short-lived isotope ^{26}Al within an igneous (Type B) CAI. A correlation ("isochron") is observed between $^{27}Al/^{24}Mg$ and $^{26}Mg/^{24}Mg$. Minerals with lower Al/Mg ratios, such as spinels, contain a lower $^{26}Mg/^{24}Mg$ ratio than minerals with high Al/Mg ratios such as anorthite. This is because the effect of ^{26}Al decay on the Mg-isotopic composition is most significant for the minerals with high Al/Mg ratios. Since the excess in ^{26}Mg is due to the decay of ^{26}Al, the slope of the isochron is equal to the initial $^{26}Al/^{27}Al$ ratio. If several objects formed at different times from an isotopically similar reservoir, then the older objects will yield a

TABLE 1. Short-lived isotopes initially present in meteorites.

Radioactive Isotope (R)	T (m.y.)	Daughter Isotope	Stable Isotope (S)	Initial Abundance (R/S)
^{7}Be	52 d	^{7}Li	^{9}Be	6×10^{-3}
^{41}Ca	0.1	^{41}K	^{40}Ca	1.5×10^{-8}
^{36}Cl	0.3	^{36}S	^{35}Cl	$> 1.1 \times 10^{-5}$
^{26}Al	0.74	^{26}Mg	^{27}Al	5×10^{-5}
^{10}Be	1.5	^{10}B	^{9}Be	$4–14 \times 10^{-3}$
^{60}Fe	1.5	^{60}Ni	^{56}Fe	$0.1–1.6 \times 10^{-6}$
^{53}Mn	3.7	^{53}Cr	^{55}Mn	$1–12 \times 10^{-5}$
^{107}Pd	6.5	^{107}Ag	^{108}Pd	$>4.5 \times 10^{-5}$
^{182}Hf	9	^{182}W	^{180}Hf	$>1.0 \times 10^{-4}$
^{129}I	16	^{129}Xe	^{127}I	1.0×10^{-4}
^{92}Nb	36	^{92}Zr	^{93}Nb	$10^{-5} – 10^{-3}$
^{244}Pu	81	Fission products	^{238}U	$4–7 \times 10^{-3}$
^{146}Sm	103	^{142}Nd	^{144}Sm	7×10^{-3}

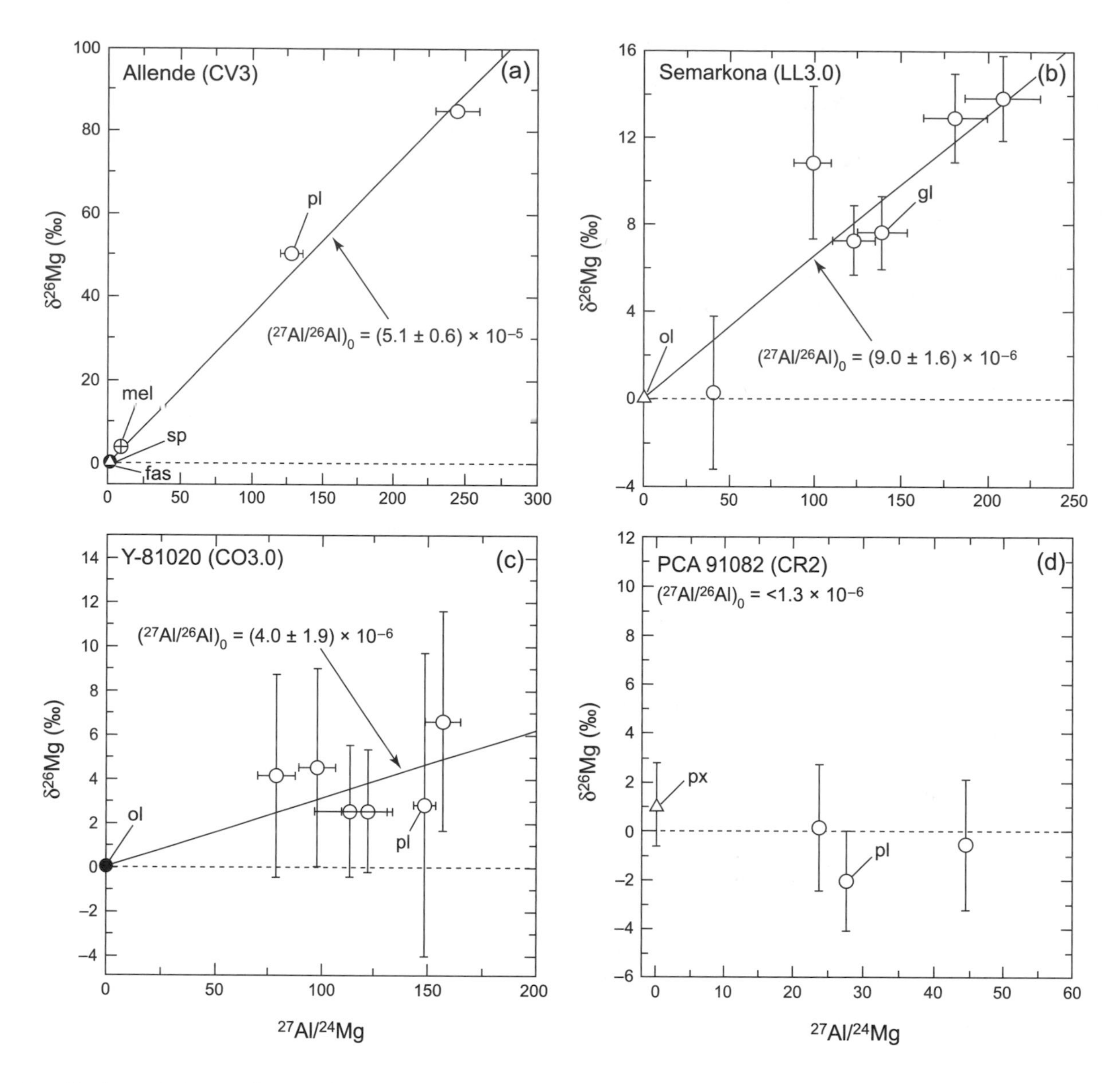

Fig. 3. Al-Mg evolutionary diagram for **(a)** an igneous (Type B) CAI from the CV3 chondrite Allende (from *Lee et al.,* 1976), **(b)** a ferromagnesian chondrule from the LL3.0 chondrite Semarkona (from *Kita et al.,* 2000), **(c)** a ferromagnesian chondrule from the CO3.0 chondrite Y-81020 (from *Kunihiro et al.,* 2004), and **(d)** an Al-rich chondrule from the CR2 chondrite PCA 91082 (from *Hutcheon et al.,* 2004). The observed range in the initial $^{26}Al/^{27}Al$ ratios, if interpreted chronologically, suggests that chondrule formation lasted for at least 2–3 m.y. after CAI formation. fas = fassaite; gl = glass; ol = olivine; pl = plagioclase; px = pyroxene; sp = spinel.

higher initial $^{26}Al/^{27}Al$ ratio than the younger ones (i.e. produce an isochron with a steeper slope), thus age information can be extracted, as the half-life of ^{26}Al is known. The intercept $^{26}Mg/^{24}Mg$ indicates the initial $^{26}Mg/^{24}Mg$ of the rock, and can indicate whether resetting of the isotope clock has taken place. If the system is equilibrated with respect to Mg isotopes after some component of ^{26}Al has decayed, then this caused the bulk $^{26}Mg/^{24}Mg$ of the system to be raised, i.e., the intercept to increase (e.g., *Podosek et al.,* 1991). The initial abundance of other short-lived isotopes is calculated in a similar way. For reviews of the use of short-lived isotopes as chronometers, see, for example, *Swindle et al.* (1996), *Gilmour and Saxton* (2001), and *McKeegan and Davis* (2003).

The use of short-lived isotopes as chronometers involves making several assumptions. The main assumptions are that (1) all the components used to make an isochron formed from a reservoir that was initially isotopically homogeneous with respect to the radioactive parent of the short-lived isotope, (2) the components measured are coeval and initially had the same daughter isotopic composition, (3) no post-formational redistribution occurred, and (4) the half-lives are accurately known. Post-formational redistribution of isotopes, and the relationship between analyzed phases, can to some extent be monitored by geochemical and mineralogical studies. For example, isotope redistribution may have occurred in objects for which there is mineralogical evidence for secondary mobilization by metamorphism or aqueous processing. The assumption of homogeneity is considered in detail in the next section.

3.2. Initial Abundances and Distribution

The initial abundances, as far as we know them, of the known short-lived isotopes in the early solar system are summarized in Table 1. The initial values for ^{10}Be, ^{26}Al, ^{41}Ca, ^{53}Mn, and ^{129}I are taken from values measured in CAIs, which are known to be extremely ancient solids (see section 2 above). Values for the other known short-lived isotopes are taken from other chondritic components, achondrites, and iron meteorites. The parent elements of these isotopes are not refractory enough to have initially been present at a sufficiently high level in CAIs. In this review, we will focus on the isotopes ^{26}Al and ^{53}Mn. This is because these isotopes have been measured in a wide range of components (chondrules and CAIs) assumed to have formed prior to asteroidal accretion, and thus they can potentially provide the most information about early solar system chronology. The short-lived isotope ^{129}I has also been measured in a wide range of samples, including individual chondritic components (e.g., *Whitby et al.,* 2004). Some of these measurements are of primary age dates (e.g., *Gilmour et al.,* 2000; *Whitby et al.,* 2002), but the use of I-Xe as a chronometer of primary mineral crystallization is hampered by the extreme mobility of both I and radiogenic Xe under many geological conditions.

Both the formation mechanism of these nuclides and their distribution mechanism are currently debated. The short-lived isotopes must have been formed soon before the formation of the first solar system solids, because they were still "live" — i.e., present in measurable quantities — when these solids formed. Mechanisms for the formation of these isotopes are that they all formed in a star, or in the local solar environment. A stellar source for short-lived isotopes could have been a supernova, a Wolf-Rayet star, or, less likely, because they are not associated with star-formation regions, a thermally pulsing asymptotic giant branch (AGB) star, and they were then injected into the solar system (e.g., *Arnould et al.,* 1997; *Cameron et al.,* 1995; *Goswami and Vanhala,* 2000; *Boss and Vanhala,* 2001; *Busso et al.,* 2003). If an external stellar source for the short-lived isotopes is assumed, then the existence of ^{10}Be is at first glance problematic, as this isotope is not formed in stellar environments (*McKeegan et al.,* 2000). However, *Desch et al.* (2004) presented a model in which 80% of the ^{10}Be inferred to have been present in CAIs is a trapped galactic cosmic ray (GCR) component, and the remainder formed from spallation reactions by GCRs.

Alternatively, all, or a component of, the isotopes ^{10}Be, ^{26}Al, ^{41}Ca, and ^{53}Mn may have formed by irradiation, i.e., spallation mainly due to energetic protons and ^{3}He nuclei, in the early solar system itself, close to the young Sun (*Lee et al.,* 1998; *Gounelle et al.,* 2001; *Goswami et al.,* 2001; *Leya et al.,* 2003; *Chaussidon and Gounelle,* 2006), and could then have been distributed to asteroidal distances in the protoplanetary disk via an outflow, for example, in CAIs and chondrules within an X-wind (*Shu et al.,* 1996, 2001).

The irradiation model fails to produce the extinct isotope ^{60}Fe. An upper limit for the initial presence of ^{60}Fe in CAIs is at levels of $(^{60}\text{Fe}/^{56}\text{Fe})_0 = 1.6 \times 10^{-6}$ (*Birck and Lugmair,* 1988). Recent measurements of its decay product, ^{60}Ni, have been reported in chondritic sulphides from Bishunpur (LL3.1), demonstrating it was initially present in these minerals at levels of $^{60}\text{Fe}/^{56}\text{Fe} = \sim 1 \times 10^{-7}$ (*Tachibana and Huss,* 2003; *Guan et al.,* 2003). Silicate chondrules appararently formed in an environment in which the $^{60}\text{Fe}/^{56}\text{Fe}$ was $2–5 \times 10^{-7}$ (*Tachibana et al.,* 2005). Sulfides in Semarkona (LL3.0) contained ^{60}Fe at levels that are even higher — $(^{60}\text{Fe}/^{56}\text{Fe})_0 = (1.00 \pm 0.15) \times 10^{-6}$ — although the Fe-Ni system may have experienced remobilization in these samples (*Moustefoui et al.,* 2003, 2004). This is the expected level for late-formed constituents of a long-lasting solar nebula initially seeded by supernova material (*Tachibana et al.,* 2003), and above expected normal galactic background levels. While the proportions of the isotopes formed by each mechanism is unknown, it is probable that the roster of short-lived isotopes as a whole formed by more than one mechanism (e.g., *Russell et al.,* 2001; *Huss et al.,* 2001). The formation of short-lived isotopes is discussed in more detail in *Chaussidon and Gounelle* (2006).

The formation and distribution mechanism of the short-lived isotopes will define the likelihood of whether the initial abundances of short-lived isotopes recorded in CAIs represent values that were once homogeneously present in the solar system as a whole. It is highly improbable that the isotopes that formed by irradiation close to the Sun would be initially homogeneously distributed in all chondritic components, as this would require all components to have been subjected to the same initial irradiation conditions onto the same starting composition and subsequent homogenization in the solar nebula. If these isotopes formed by production in a star, then they may or may not have been homogeneously mixed into the solar accretion disk (*Vanhala,* 2001; *Boss and Vanhala,* 2001; *Vanhala and Boss,* 2002; *Boss,* 2004).

Thus the use of short-lived isotopes as age-dating tools for early solar system objects is limited by the current lack of understanding of their initial distribution. Most authors make an assumption of homogeneity in order to tentatively draw chronological conclusions; this assumption was recently strengthened, at least on an asteroidal scale, by precise measurements of Mg isotopes in bulk chondrite groups (*Bizzarro et al.,* 2004). Below, we summarize the data acquired so far and discuss its implications and limitations for understanding solar system chronology.

3.3. Short-lived Isotope Abundance in Chondrules and Calcium-Aluminum-rich Inclusions

Quantitative information about the age difference between chondritic objects can be obtained from isotope measurements using short-lived isotopes.

3.3.1. Aluminum-26. From thousands of individual measurements by ion microprobe, it has emerged that isochrons determined from high Al/Mg minerals from most unaltered, typical CAIs are characterized by a canonical initial ratio $^{26}Al/^{27}Al = 4–5 \times 10^{-5}$ (*MacPherson et al.,* 1995; *Russell et al.,* 1996, 1998). High-precision ICP-MS analyses of unaltered CAIs suggest that the initial $^{26}Al/^{27}Al$ content of many CAIs may have been as high as 7×10^{-5} (e.g., *Galy et al.,* 2004; *Young et al.,* 2005). The study of *Young et al.* (2005) showed that a single CAI may contain some minerals with an initial $^{26}Al/^{27}Al$ ratio of $\sim 7 \times 10^{-5}$ igneously associated with minerals with initial $^{26}Al/^{27}Al$ of $\sim 5 \times 10^{-5}$, suggesting an extended high-temperature nebular history. We note, however, that the "supracanonical" $^{26}Al/^{27}Al$ ratio in the CV CAIs reported by *Young et al.* (2005) is not without controversy; e.g., some bulk Mg-isotopic measurements of Allende CAIs by ICP-MS revealed a lower model isochron of $(5.25 \pm 0.1) \times 10^{-5}$ (*Bizzarro et al.,* 2004).

A significant proportion of CAIs contain much less, or ummeasurably low, levels of initial ^{26}Al. These include the so-called FUN inclusions (*MacPherson et al.,* 1988), some Type C CAIs (*Wark,* 1987; *Hutcheon et al.,* 2004; *Krot et al.,* 2005a), grossite- and hibonite-rich CAIs in CH chondrites (*MacPherson et al.,* 1989; *Weber et al.,* 1995), and R chondrites (*Bischoff and Srinivasan,* 2003) and CAIs that have experienced extensive secondary elemental redistribution (*MacPherson et al.,* 1995). There are very little data for the related refractory inclusion type, AOAs, but the data that exists point to these having an initial $^{26}Al/^{27}Al$ ratio of around 3×10^{-5} (*Itoh et al.,* 2002).

In situ measurements of chondrules yield initial $^{26}Al/^{27}Al$ values clearly lower than the canonical CAI value, and in general, the values recorded are more variable than for the majority of CAIs. Chondrules from unmetamorphosed ordinary chondrites typically formed with $^{26}Al/^{27}Al \sim 1 \times 10^{-5}$ (e.g., *Russell et al.,* 1996; *Kita et al.,* 2000, *McKeegan et al.,* 2001; *Mostefaoui et al.,* 2002). Other chondrules, from higher metamorphic subgrades (>3.1) of ordinary chondrites, have lower initial values, which could be ascribed to metamorphic resetting (*Huss et al.,* 2001). In carbonaceous chondrites, there is evidence from unaltered CR and CV meteorites for a range in initial $^{26}Al/^{27}Al$ values in chondrules, ranging from 3×10^{-6} to 1×10^{-5} in samples for which evidence for initial ^{26}Al could be found (*Tachibana et al.,* 2003; *Hutcheon et al.,* 2004). *Kunihiro et al.* (2004) also suggested that the initial ^{26}Al abundance varied between chondrules from carbonaceous (initial $^{26}Al/^{27}Al \sim 4 \times 10^{-6}$) and ordinary chondrite groups (initial $^{26}Al/^{27}Al \sim 7 \times 10^{-6}$), suggesting differences in asteroidal accretion times between chondritic parent bodies. The difference in initial ^{26}Al abundance between different chondrules, and the much higher values observed in CAIs, could be interpreted as indicating that the period of formation of nebular solids, from CAIs to the youngest chondrules, extended over at least ~4 m.y. Alternatively, chondrules and CAIs could have formed in regions different to each other in Al-isotopic composition.

Recent bulk Mg-isotopic measurements of Allende chondrules yield a range of $(^{26}Al/^{27}Al)_I$ from $(5.66 \pm 0.80) \times 10^{-5}$ to $(1.36 \pm 0.52) \times 10^{-5}$ (*Bizzarro et al.,* 2004). However, because these measurements involve much lower amounts of $^{26}Mg^*$ compared to CAIs, the $(^{26}Al/^{27}Al)_I$ in chondrules is much more sensitive to fractionation laws and assumed initial Mg-isotopic composition, and so further work is required to explore these results and their implications.

More clear-cut chronological information can be obtained from single objects that have experienced more than one stage in their history. *Hsu et al.* (2000) reported a CAI with an inner zone once containing $^{26}Al/^{27}Al$ at the canonical level of $\sim 5 \times 10^{-5}$ and outer zone depleted in initial ^{26}Al; the authors interpreted this to indicate that the whole CAI initially contained $^{26}Al/^{27}Al$ at $\sim 5 \times 10^{-5}$ and had experienced nebular remelting after the decay of ^{26}Al, indicating a several-million-year preasteroidal accretionary history for this object. Even more compelling is the rare presence of relic CAIs with disturbed Al-Mg systematics found within chondrules (*Krot et al.,* 2004). These inclusions are mineralogically similar to CAIs that formed with canonical $^{26}Al/^{27}Al$, and the disturbance of the Al-Mg systematics in these objects implies that the ^{26}Al in the CAI decayed ≥2 m.y. prior to their incorporation in the chondrule-forming event.

Overall, the data for ^{26}Al suggest that the formation of CAIs and chondrules may have occurred over several million years, assuming it was homogeneously distributed. There is firm evidence for an extended, >2-m.y. nebular history for some individual objects.

3.3.2. Manganese-53. *Birck and Allègre* (1988) reported the first Mn-Cr-isotopic measurements of two Allende CAIs. They concluded that CAIs had formed with an average initial $^{53}Mn/^{55}Mn$ ratio of $(4.37 \pm 1.07) \times 10^{-5}$. *Nyquist et al.* (2001) inferred an initial $^{53}Mn/^{55}Mn$ of $(2.81 \pm 0.31) \times 10^{-5}$ for the Efremovka CAIs, whereas *Papanastassiou et al.* (2002) reported initial values ranging from 1.0×10^{-5} to 12.5×10^{-5} for the Allende CAIs. *Papanastassiou et al.* (2002) suggested that the measured span in the initial $^{53}Mn/^{55}Mn$ ratio in CAIs results from metamorphic redistribution of Mn. Accurate determination of Mn-Cr systematics in CAIs is complicated by the fact that Mn is too volatile to have been initially present in high abundance in CAIs. This problem is discussed by *Lugmair and Shukolyukov* (1998), who suggest using an initial $^{53}Mn/^{55}Mn$ value for CAIs of $\sim 1.4 \times 10^{-5}$, a value chosen to fit well with other isotope systems.

Data for chondrules are limited, but appear more consistent than for CAIs. *Nyquist et al.* (1994, 2001) reported Mn-Cr data on unequilibrated ordinary chondrite chondrules. They found that Chainpur and Bishunpur chondrules formed when $^{53}Mn/^{55}Mn$ was equal to $(0.94 \pm 0.17) \times 10^{-5}$ and $(0.95 \pm 0.13) \times 10^{-5}$ respectively.

These data, if interpreted chronologically, would imply a formation time difference of ~6 m.y. between Allende CAIs and ordinary chondrite chondrules if an average value for the initial CAI value is used, or ~12 m.y. if the highest (i.e., apparently the oldest) value for CAI initial $^{53}Mn/^{55}Mn$ is used. This timescale, using the highest initial $^{53}Mn/^{55}Mn$, is clearly at odds with the timescale inferred from ^{26}Al data.

However, if the initial $^{53}Mn/^{55}Mn$ estimate of ~1.4 × 10^{-5} is used (*Lugmair and Shukolyukov,* 1998), the time difference between the initial value and the chondrules is only ~2 m.y., similar to what is inferred from ^{26}Al.

3.4. Short-lived Isotopes in Achondrites

Timescales of achondrite solidification may give some clues as to planetesimal formation timescales and therefore the end of the disk lifetime (although the disk may not have terminally dissipated at the time of the first planetesimal formation; see section 6). A few occurrences of short-lived ^{26}Al within achondrites have been recorded, including within eucrites (e.g., *Srinivasan et al.,* 1999; *Nyquist et al.,* 2003a), angrites (e.g., *Nyquist et al.,* 2003b), and ureilites (*Kita et al.,* 2003); for details, see *Wadhwa et al.* (2006). For the eucrite Asuka 881394, which is thought to have been a very early-formed rock on the HED parent body [which is possibly the asteroid 4 Vesta (e.g., *Consolmagno and Drake,* 1977)], Mn-Cr-isotopic analyses determined initial $^{53}Mn/^{55}Mn$ = (4.6 ± 1.7) × 10^{-6}. This value has been recently more precisely measured at (4.02 ± 0.26) × 10^{-6} (*Wadhwa et al.,* 2006). This initial ^{53}Mn abundance corresponds to a formation interval Δt_{LEW} = −6 ± 2 m.y. relative to the LEW 86010 angrite if ^{53}Mn was homogeneous, implying an "absolute" age of 4564 ± 2 m.y. The initial $^{26}Al/^{27}Al$ ratio of Asuka 881394 is (1.18 ± 0.14) × 10^{-6}, suggesting a formation time of ~4 m.y. after CAIs, if ^{26}Al was homogeneous. The angrites LEW 86010, D'Orbigny, and Sahara 99555 all contain evidence for live ^{26}Al and ^{53}Mn when they formed. Plagioclase-bearing clasts in the Dar al Gani 319 polymict ureilite yielded an initial $^{26}Al/^{27}Al$ ratio of 4 × 10^{-7} (*Kita et al.,* 2003). This corresponds to formation ~5 m.y. after CAIs, if an initially homogeneous distribution of ^{26}Al is assumed. The significance of a few-million-year delay between CAI formation and achondrite crystallization is discussed in more detail in section 6.2. Absolute ages also suggest a significant delay between CAI formation and achondrite crystallization. The Pb-Pb mineral isochron ages of cumulate eucrites range from 4.40 G.y. to 4.48 G.y. (*Tera et al.,* 1996). Angrites have an absolute age of 4557.8 ± 0.5 m.y. (*Lugmair and Galer,* 1992), i.e., approximately 10 m.y. younger than CV CAIs. These data imply that the formation of some differentiated asteroids occurred several millions of years after the formation of the first chondritic components.

Recently, *Gounelle and Russell* (2005) have proposed a chronological model based on the assumption that ^{26}Al and ^{53}Mn were heterogeneously distributed in the early solar system. They assume that CAIs and chondrules have intrinsic different initial contents of ^{26}Al and ^{53}Mn, and that differentiated parent bodies are the result of the agglomeration of CAIs and chondrules in chondritic proportions. With such assumptions, it is possible to calculate a model relative age that dates the time difference between the agglomeration of CAIs and chondrules of a given parent body until the isotopic closure after differentiation of that same parent body. Note that the meaning of age in such a model is radically different from that of models assuming a homogeneous distribution of ^{26}Al and ^{53}Mn (see above). In such a model, Asuka 881394 crystallized 2.8 ± 1 m.y. after the agglomeration of its precursors, while D'Orbigny crystallized 4.3 ± 1 m.y. after the crystallization of its precursors. Tying these relative numbers to absolute Pb-Pb ages provides absolute ages for the agglomeration of precursors of Asuka 881394 and D'Orbigny (*Gounelle and Russell,* 2005).

4. SUMMARY OF COSMOCHEMICAL EVIDENCE FOR DISK LIFETIMES

The evidence from several long- and short-lived isotope systems and mineralogical observations suggests that CAIs are the oldest known solids in the solar system. The latest Pb-Pb and petrographic data suggest that the first chondrules may have formed at around the same time as CAI formation (see *Amelin et al.,* 2004; *Bizzarro et al.,* 2004). However, chondrule formation appears to have been a protracted process lasting several millions of years. Since both these components are generally believed to have formed while the solar system had a protoplanetary disk, the most straightforward conclusion that can be drawn is that high-temperature processing in the disk may have taken several millions of years. However, results from the different isotope systems often contradict each other in precise detail. For example, the Pb-Pb system suggests that the maximum age difference between CV3 chondrules and CAIs is 2 m.y. or less, whereas the ^{26}Al data suggest an age gap between CAI formation and formation of the first chondrules of ~>2 m.y. The (albeit limited) recent precise Pb-Pb data points to each of the components of a particular meteorite having a similar age to each other, but differences of 4–5 m.y. are seen in bulk age between meteorites from different groups. In contrast, ^{26}Al-^{26}Mg dating suggests that CAIs from most groups have a similar age, distinguishable from chondrules in the same meteorite groups. While Al-Mg systems suggest a 2–3 m.y. age gap between different components from the meteorite groups, Mn-Cr systematics point to a possibly longer age gap of up to 12 m.y. (again between different meteorite groups). Petrologic evidence suggests that the formation of CAIs and chondrules may have even overlapped, albeit from very rare observations.

Clearly, we are far from having a complete understanding of the chronology of the early solar system, and resolution of these inconsistencies requires a detailed consideration of the weaknesses of each isotopic system as a chronometer and an assessment of their initial distribution. The Mn-Cr system can probably be disregarded as an effective chronometer of the CAI-chondrule age gap, since the primary Mn content of CAIs is so low as to make initial abundance of ^{53}Mn in CAIs problematic. In addition, there is evidence that ^{53}Mn was initially heterogeneous in the solar system (*Shukolyukov and Lugmair,* 2000). Both the Al-Mg and Pb-Pb systems may have been affected by later elemental mobilization, either in the nebula or parent body. This may have caused homogenization of Pb isotopes in the parent body, or preferential resetting of Al-Mg systematics in chondrules rather than in CAIs. A final reason for the discrepancy is that the $^{26}Al/^{27}Al$ of the early solar system may have been heterogeneous, diminishing the use of Al-Mg-isotopic systematics as a chronometer in the context of determining the relative age of CAIs and chondrules.

Two working interpretations of these observations will be considered in section 6 below. First, the isotopic data may point to an extended nebular history for the components of each chondrite class. In this case, a viable storage mechanism for the early-formed components must be considered. A second possibility is that all the components from a single class formed at around the same time, but with individual components having very different ^{26}Al contents (e.g., if the short-lived isotopes formed by irradiation close to a young Sun). In this case, the apparent several-million-year age difference between the chondrite groups (*Amelin et al.,* 2002, 2004) is due to the formation of chondrites, from isotopically heterogeneous components, lasting several millions of years.

While the isotope systems do not yield an entirely consistent picture in detail, all the data point to an extended period of formation of nebular solids that lasted on the order of a few million years. In section 6 we discuss several ways to explain these observations. However, precise quantification of the time periods involved will require further analyses of well-characterized chondrite components on unaltered meteorite samples, using several different isotope techniques (see section 7).

5. OBSERVATIONS OF DISK LIFETIMES AROUND YOUNG STELLAR OBJECTS

Our understanding of the lifetimes of the disks of low-mass, pre-main-sequence stars based on astronomical constraints has changed only modestly since the review by *Podosek and Cassen* (1994). Those authors concluded that the observational evidence was strong for disk lifetimes longer than 1 m.y. but typically less than 10 m.y.; recent research has added a number of details, but has not changed the overall picture.

To understand the limitations of astronomical "disk lifetimes," it is necessary to consider what is actually being observed. To begin with, our ability to detect disk gas is poor, especially molecular hydrogen, the dominant constituent, and so little is known about gas-disk lifetimes. A few estimates of disk gas masses have been made using warm molecular hydrogen emission (e.g., *Thi et al.,* 2001), although at least one detection has been called into question (*Lecavelier des Etangs et al.,* 2001). In any case, these detections do not constrain the amount of material at temperatures much lower than about 100 K; in addition, if the disk is optically thick at mid-infrared wavelengths, as is often the case for T Tauri stars, the observed emission may come only from disk surface layers.

Detections of emission from molecules other than H_2 are much more common. Both hot (1000 K) and cold (10–30 K) CO emission has been detected from disks around T Tauri stars (e.g., *Aikawa et al.,* 2002; *Najita et al.,* 2003), and low-temperature molecular emission has also been studied (e.g., *Dutrey et al.,* 1996). The problem with the detections of molecular emission, apart from abundance uncertainties, is that they generally arise from warm(er) upper layers of T Tauri star disks, due to optical depth effects and requirements for excitation, and thus directly trace only a small fraction of the total gas mass present. The situation is different for the so-called "debris disk" systems such as β Pic, which are relatively optically thin and have small gas masses, but these objects probably represent the very end stages, when collisions between planetesimals probably produce most of the observed emitting dust.

The one set of gas diagnostics that appear to reflect the bulk of local gaseous material are the emission lines produced as disk gas accretes through the stellar magnetosphere and continuum emission produced as the accreting gas crashes onto the stellar surface (Fig. 4). Although the amount of gas in these inner regions is very small, the rate of mass flow onto the star is significant in an evolutionary sense. Estimates indicate that typical mass flow rates are $\sim 10^{-8}$ $M_\odot$ per year, so that over the T Tauri lifetime of a few million years the typical star accretes a few times the minimum mass solar nebula (*Hartmann et al.,* 1998). While this accreted mass must be a typical minimum gas mass of T Tauri disks, the measurement of accreting gas obviously does not directly constrain the instantaneous disk mass.

The main astronomical indicator of disk lifetimes is dust emission. Here one must make additional distinctions, depending upon the wavelength of observation. The most sensitive observations so far have been made at near-infrared wavelengths ($\lambda \sim 2$–3 μm) using groundbased telescopes. However, such data only trace dust in the innermost disk (Fig. 4), typically at a distance of about 0.1 AU for T Tauri stars. Dust appears to sublimate at temperatures of about

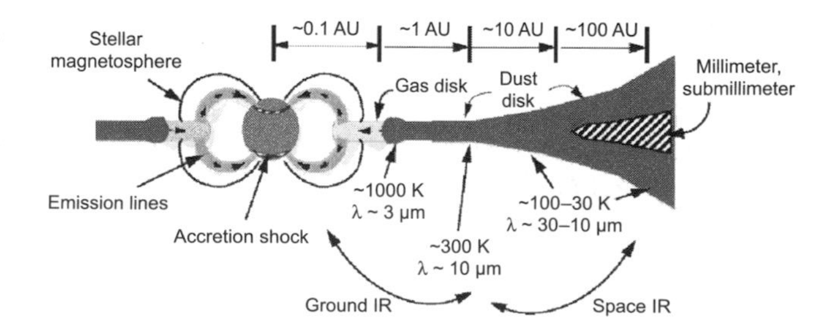

Fig. 4. Schematic diagram of a typical T Tauri star with accretion disk. Differing diagnostics of disks, along with approximate typical distance scales and wavelengths of observation, are indicated. Most infrared dust emission tends to arise from surface layers, due to optical depth effects; submillimeter and millimeter-wave emission is more likely to trace midplane regions where most of the mass resides (see text).

1500 K in pre-main-sequence stellar accretion disks (*Natta et al.,* 2001; *Muzerolle et al.,* 2003), setting the effective inner edge of disk dust. Observations at λ ~ 3 μm generally provide the most sensitive indication of inner dusty disks, especially for T Tauri stars. Groundbased observations on large telescopes can also be used to detect dusty disk emission at λ ~ 10 μm with reasonable sensitivity, probing material at distances of about 1 AU in low-mass systems. Spacebased observations can detect colder dust emission, from regions extending out to a hundred astronomical units or more (Fig. 4). Current datasets from the Infrared Astronomical Satellite (IRAS) and Infrared Space Observatory (ISO) missions provide modest sensitivity to disk emission; forthcoming results from the Spitzer Space Telescope (previously known as the Space Infrared Telescope Facility, or SIRTF) should provide substantial advances in this area. Finally, submillimeter- and millimeter-wave observations can detect the coldest disk dust residing at large distances from the central star. Current sensitivity in this region is limited; the Atacama Large Millimeter Array (ALMA) promises major advances in a few years. Overall, it becomes more difficult to detect dust emission as the wavelength of observation increases, thus our constraints on inner disk dust are much stronger than on outer disk dust.

Dust emission is a poor indicator of disk mass. The shortest-wavelength observations are sensitive to extremely tiny amounts of small dust, but rapidly become insensitive to dust mass for particles much larger than a few micrometers in size. In addition, T Tauri disks are usually optically thick at wavelengths <100 μm; thus, the level of infrared disk emission does not directly reflect the amount of mass, even if the size distribution were known. Submillimeter- and millimeter-wave observations detect more optically thin material, and thus can be used with more certainty to estimate disk masses (e.g., *Beckwith et al.,* 1990), but are subject to substantial uncertainties in the particle size distribution (*D'Alessio et al.,* 2001), and generally do not have sufficient sensitivity to trace disk evolution to smaller masses.

Despite the above caveats, detections of dust emission have played an important role in understanding disk evolution. *Strom et al.* (1989) were the first to constrain the evolution of dusty disk emission, finding that most disks "disappeared" over timescales <10 m.y. Their data indicated a strong decrease in emission from the hottest, innermost disk regions even at ages as young as 3 m.y. *Skrutskie et al.* (1990) performed more sensitive groundbased measurements of mid-infrared emission (λ = 10 μm; Fig. 4), and concluded that about half of all stars exhibited dusty disk emission, while less than 10% of their sample did so at ages >10 m.y. (see review by *Strom et al.,* 1993).

Haisch et al. (2001) studied several young clusters using more sensitive observations in the L-band (3.5-μm wavelength), which essentially probes the inner disk edge (Fig. 4). As indicated by hot disk emission, Haisch et al. inferred that the initial fraction of young stars with dusty disks is high, ≥80% (see also *Hillenbrand et al.,* 1998). At ages of <~3 m.y., the disk fraction was about half, with most disk emission disappearing by about 6 m.y., confirming the results of *Skrutskie et al.* (1990) and *Strom et al.* (1999). No clear evidence has been developed for any difference in disk lifetimes between clustered and dispersed star-forming environments (see also *Brice et al.,* 2001).

There is a generally a close correspondence between the disappearance of inner disk dust emission and the ending of accretion onto the central star (*Hartigan et al.,* 1990; *Muzerolle et al.,* 2000). A notable exception is the older (10 m.y.) T Tauri star TW Hya. As shown in Fig. 5, TW Hya exhibits strong disk emission at long wavelengths, but little

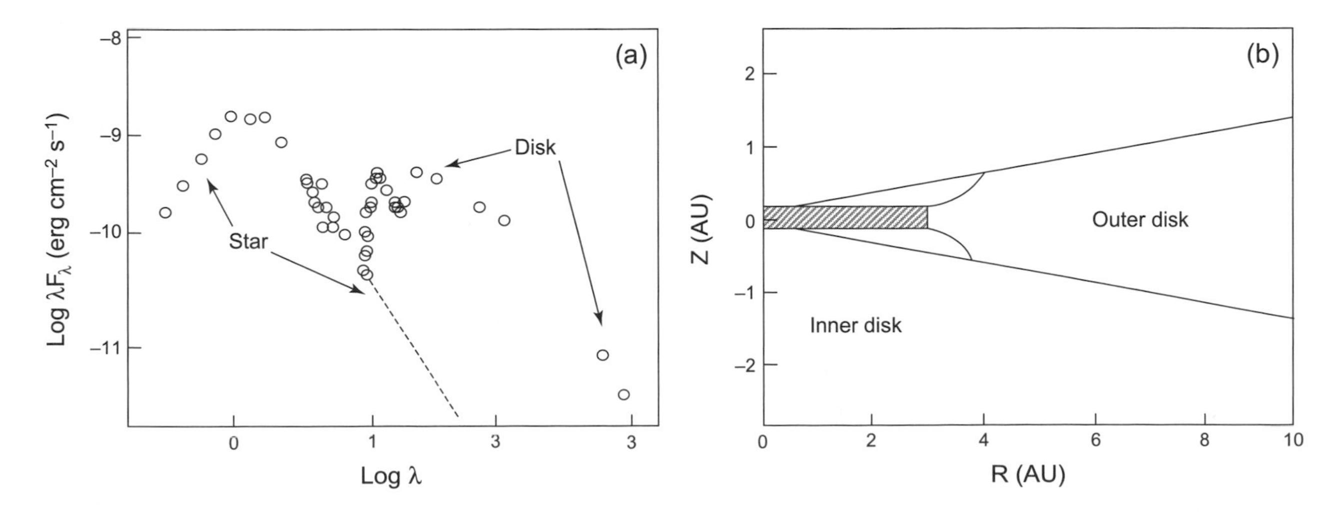

Fig. 5. **(a)** Spectral energy distribution of the 10-m.y.-old star TW Hya. Emission from the dusty disk is strong at wavelengths ≥10 μm, but is nearly absent at shorter wavelengths. **(b)** Models of the spectral energy distribution imply that the disk must be relatively free of small dust particles at radii <4 AU, even though the star is actively accreting, suggesting gas may be present within this "hole" in the dust distribution (see text).

or no hot dust emission. Models of the spectral energy distribution suggest large dust masses beyond 4 AU but very minor amounts of small dust particles within this radius (*Calvet et al.,* 2002). At the same time, TW Hya is accreting gas at highly time-variable rates (*Muzerolle et al.,* 2000; *Alencar and Batalha,* 2002), which suggests that substantial amounts of gas are passing through this region even if small dust is not.

The least-model-dependent way of characterizing the observations of disk emission is to refer to the "disk lifetimes" as the timescale at the wavelength in question when the disk becomes optically thin (e.g., *Strom et al.,* 1989). Objects like TW Hya seem to be rare, suggesting that the timescale for the transition between optically thick and optically thin disk dust emission is very short (*Skrutskie et al.,* 1990). The most straightforward interpretation of the observations is that the disk lifetimes estimated from dust emission basically all correspond to times at which the small particles are incorporated into larger bodies, rather than accretion onto the central star or dispersal. Long-wavelength observations of TW Hya suggest particle growth to sizes of 1 cm or so at radii ~100 AU (*Calvet et al.,* 2002); one would suspect even larger particle growth at smaller distances, where timescales of evolution should be shorter.

The correlation of the decline in infrared emission with the cessation of accretion onto central stars suggests that gas as well as dust is being accumulated. The case of TW Hya is extremely interesting as a transition object in which small dust and gas flow has not completely disappeared.

In summary, between 50% and 80% of low-mass stars initially have circumstellar disks, as indicated by dust emission and stellar accretion. The typical timescale for the disappearance of substantial dust emission and disk is roughly 3–5 m.y.; however, considerable variation in this timescale is seen among individual objects. Very few objects survive with strong dust emission and gas stellar accretion to ages of 10 m.y. These conclusions mostly correspond to disk properties in the inner few astronomical units; more sensitive observations with the Solar Space Telescope (SST) and ALMA should provide better constraints on disk evolution at large radii in a few years. The timescales for evolution on much larger distance scales, and the evolution and ultimate fate of the majority of disk gas (see, e.g., *Hollenbach et al.,* 2000) are poorly understood at present.

6. DYNAMICAL CONSIDERATIONS

The Pb-Pb age data suggest that chondrule formation lasted a few million years after the formation of most CAIs, based on chondrule ages in several different groups (section 2.5). Regardless of the models adopted for the formation of CAIs and chondrules, or the source of the short-lived radioisotope ^{26}Al, there are several known examples of individual objects having a several-million-year nebula history (section 3). As discussed below in section 6.1, there has been concern as to whether individual particles can survive in the nebula for this long. In addition, the formation of planetesimals *too soon* after CAIs may lead to their including so much heat-producing ^{26}Al and ^{60}Fe that far more melted asteroids and meteorites should be observed than actually are. Different scenarios for asteroidal accretion that might account for these observations are discussed in sections 6.2 and 6.3.

6.1. The Role of Nebular Gas

Solid particles in the disk are controlled by their interactions with the gas. Any plausible circumstellar disk is partially supported by a radial pressure gradient, so that it rotates at slightly less than the Kepler velocity. Gas drag acting on particles causes them to drift inward. As described by *Adachi et al.* (1976) and *Weidenschilling* (1977), the peak radial velocity of several tens of meters per second is reached by approximately meter-sized bodies, whose orbits decay at ~1 AU/century. Radial velocities decrease on both sides of this peak, and are approximately proportional (inversely) to size for smaller (larger) bodies. In the absence of other processes, both kilometer-sized planetesimals and millimeter-sized particles would migrate inward 1 AU in ~10^5 yr, while centimeter-sized bodies would travel this distance in ~10^4 yr. These timescales are short compared with the formation intervals and age differences for CAIs and chondrules inferred from ^{26}Al data, causing a concern about their survival. Assuming these age differences are real, two processes that reconcile the theory and observations are discussed in the next two sections.

6.2. Rapid Asteroidal Accretion in Nonturbulent Nebulae

Particle collisions occur due to a variety of processes that operate in different size ranges, including thermal motion, differential settling and radial drift, and turbulence, if present. Assuming that collisions result in coagulation, the timescales for growth can be estimated analytically (*Weidenschilling,* 1988). Because the first aggregates are likely to be porous, crushable particles that can dissipate collisional energy, growth to meter size is likely to be easy regardless of the presence or absence of turbulence (e.g., *Weidenschilling,* 1997). If coagulation is inefficient, growth timescales scale inversely with sticking coefficient (although any sticking coefficient would not be a simple constant, but a function of particle sizes and relative velocities). Various forms of collective effects have been offered that circumvent uncertainties in sticking mechanisms; none of these have been quantified from the standpoint of asteroidal accretion timescales (see *Cuzzi and Weidenschilling,* 2006).

At any heliocentric distance an initial population of micrometer-sized grains can produce centimeter-sized aggregates in a few hundred orbital periods; the principal growth mechanism is sweeping of smaller aggregates by larger ones, which settle more rapidly. If the nebula is nonturbulent, such aggregates settle into a thin layer in the central plane during growth to centimeter sizes.

Size-dependent drift rates cause collisions in any ensemble of particles with a range of sizes. In a nonturbulent nebula, centimeter- to meter-sized particles settle into a fairly dense midplane layer where relative velocities are low and growth is rapid (*Cuzzi et al.,* 1993; *Dobrovolskis et al.,* 1999; *Cuzzi and Weidenschilling,* 2006). For perfect sticking, this process can form kilometer-scale planetesimals in a few thousand orbital periods, i.e., $\sim 10^4$ yr in the asteroid region. More detailed numerical modeling (*Weidenschilling,* 1980, 1997) is in good agreement with these estimates, even when realistic collisional properties are assumed rather than perfect sticking. Such growth times are short enough to prevent much loss of solids into the Sun.

A short formation time for planetesimals in the asteroid belt is consistent with the asteroid region initially containing its full complement of solids relative to H/He, with mass exceeding that of Earth, or $>10^3\times$ the present mass of the belt. Once kilometer-sized or larger planetesimals formed, this high surface density would result in the rapid gravitational accretion of protoplanetary embryos with masses as large as ~0.01-0.1 $M_\oplus$ on timescales $\sim 10^6$ yr, while leaving a significant fraction of the total mass in small bodies. These embryos, and most of the smaller bodies, would be removed on timescales $\sim 10^7$–10^8 yr by scattering into unstable resonances with Jupiter and Saturn (*Chambers and Wetherill,* 2001).

However, such rapid formation of planetesimals encounters two potential problems. One is the thermal consequences of rapid accretion of bodies containing ^{26}Al. If planetesimals formed shortly after CAI formation and if ^{26}Al was uniformly distributed in the solar nebula with an initial ^{26}Al/^{27}Al ratio of 5×10^{-5}, still-alive ^{26}Al would heat their interiors enough to reset isotopic ages in bodies as small as ~30 km diameter (*LaTourrette and Wasserburg,* 1998), and would melt larger bodies. The heating effects of ^{60}Fe would be even more significant over longer timescales. It is not clear whether such thermal histories are compatible with the evidence from asteroids and meteorites. Most meteorites show evidence of thermal metamorphism, and a significant fraction of CAIs do appear to have been altered (whether this alteration occurred in the nebula or within parent bodies is unclear) (*Brearley,* 2006). Iron meteorites provide evidence for melting within multiple parent bodies, but only a few studies of formation age have been made (*Suguira and Hoshino,* 2003; *Scherstén et al.,* 2005), and these suggest accretion before ~2 m.y. after CAIs. However, there are very few asteroids that, from spectral interpretations, appear to be differentiated; 4 Vesta is the only surviving large asteroid of these. We believe we have meteoritic samples of 4 Vesta, in the form of HED meteorites, some of which show clear evidence for having initially contained ^{26}Al (*Srinivasan et al.,* 1999; *Nyquist et al.,* 2003a). Hafnium-tungsten isotopes imply that Vesta differentiated ~3 m.y. after CAI formation (*Yin et al.,* 2002), consistent with the amount of active ^{26}Al incorporated in its melts. The stochastic nature of asteroidal accretion allows the growth of Vesta-sized bodies on such a timescale, but the absence of other large igneous asteroids is puzzling, if the accretion timescale for large planetesimals is indeed $<10^5$ yr. Other achondrite groups (ureilites and angrites) contained ^{26}Al at even lower levels than HEDs when they crystallized (*Kita et al.,* 2003; *Nyquist et al.,* 2003b), although these are more highly evolved rocks that presumably were relatively late to form on their parent bodies. In fact, melted asteroids of any size (and their achondrite meteorites) are in the minority. In general, thermal modeling implies that the initial ^{26}Al/^{27}Al ratio in most parent bodies was much lower than in CAIs, consistent with either a large-scale inhomogeneity in short-lived isotopes across the solar accretion disk, or a delay in the onset of asteroidal accretion (*Bennett and McSween,* 1996; *Ghosh et al.,* 2006).

Another problem with rapid formation of planetesimals is that the apparent CAI-chondrule age difference, the spread in apparent formation ages for different chondrules, and an extended nebular history for some individual objects seem to indicate formation over an interval greater than 1 m.y. If these age differences are real, a simple model in which small solid particles form in the solar nebula and accrete directly into parent bodies of chondritic meteorites is not tenable. Either planetesimal formation had to be delayed, or material that accreted quickly had to be recycled. Preservation of primitive components and isotopic anomalies after such recycling requires that most meteorite parent bodies (i.e., asteroids) accreted from debris of smaller first-generation bodies *that never became hot enough for their isotope systematics to be reset* (i.e., <30 km in diameter). This is discussed in more detail in *Cuzzi and Weidenschilling* (2006). We revisit this concept below, after discussing the role of turbulence.

6.3. Delayed Asteroidal Accretion in Turbulent, Diffusive Nebulae

When, where, and even whether the nebular gas was turbulent continues to be debated (*Stone et al.,* 2000; *Cuzzi et al.,* 2001). The observation that million-year-old disks retain a considerable fraction of material in small dust particles has been used to argue that widespread turbulence must be present at least that long, regardless of our current ignorance about how this is accomplished (*Dullemond and Dominik,* 2005). Turbulence potentially plays a fundamental role in the accretion process, however, as highlighted below. Particle-gas interactions in turbulence are reviewed by *Cuzzi and Weidenschilling* (2006); here, we touch only on some highlights and implications for meteoritic timescales.

6.3.1. Vertical diffusion and growth timescales. Global turbulence diffuses particles into a thick, low-density layer in which particle growth is much slower than in globally nonturbulent nebulae (*Cuzzi et al.,* 1993; *Dubrulle et al.,* 1995; *Dobrovolskis et al.,* 1999; *Cuzzi et al.,* 1996; *Cuzzi and Weidenschilling,* 2006), and meter-sized particles drift into the Sun before growing significantly (*Weidenschilling,* 1988, cf. equation (15)). Naturally, the growth of planetesimals and planetary embryos is thus also frustrated. *Weidenschilling* (2004) has found that the ability to retain enough material in the Kuiper belt region practically requires turbulence to be absent there. However, the situation may be different in the inner solar system where, as noted above, asteroidal and meteoritic evidence suggests that growth needs to be slowed, not enhanced.

6.3.2. Radial diffusion and redistribution of material; application to calcium-aluminum-rich inclusions. Because small particles are quite well trapped to the nebular gas, their random velocities in turbulence are practically the same as the fluctuating gas velocities (10^3–10^4 cm/s), and far exceed their small radial drift velocities (1–10 cm/s). Random motions of this sort, in the presence of a radial gradient in abundance, generate a diffusive mass flux that can easily swamp the mass flux due to inward radial drift. If CAIs are only produced in the inner nebula, an outward gradient will certainly exist, and outward radial diffusion can populate the entire nebula with small particles from the inner nebula; thus "loss into the Sun" is not the problem it once appeared (*Cuzzi et al.,* 2003, 2005). *Bockelée-Morvan et al.* (1995) reached similar conclusions in the context of crystalline silicate grains in comets. Calcium-aluminum-rich inclusions can be formed in the inner solar system and persist for 1–3 m.y. until, or during, the chondrule-formation era. Larger CAIs are not retained as long in this model, which thus predicts that CV chondrites were accumulated earlier than most other meteorite types. There is a hint from the recent work of *Amelin et al.* (2004) that this may indeed be the case. It has also been suggested that CAIs were ejected from the inner solar system on orbital timescales by stellar winds or jets (*Shu et al.,* 1996). Even then, without turbulence, they still drift inward and are lost from the asteroid-belt region on a 10^5 yr or less timescale, thus the transporting wind must have been active for several million years, and accretion must have been fast after emplacement into the asteroid belt. Observation and interpretation of the mineralogy of CAIs might be able to distinguish between these hypotheses (see *Cuzzi and Weidenschilling,* 2006).

In either case, persistence of CAIs as independent nebular constituents requires that they *not* accrete into sizeable planetesimals for the approximately several-million-year chondritic-component-forming period, and suggests that asteroidal accretion might have been *difficult* in the inner solar system for that amount of time. This frustration of accretion is consistent with the apparent delay required in accreting planetesimals large enough to melt under the influence of their ^{26}Al. If CAIs formed early, they might have been absorbed into meter-sized particles, but these either got disrupted and dispersed, evaporated in the inner solar system and got recycled as condensates, or were lost into the Sun, with only the nonaccreted ones surviving.

6.3.3. Recycled planetesimals. An alternative to "storage" of CAIs by outward turbulent diffusion or stellar wind ejection is the rapid formation of planetesimals large enough to be unaffected by the gas drag (at least a few kilometers), so that their orbits do not decay significantly during an interval of ~>1 m.y. Such a scenario would imply the following sequence of events: CAIs form early, perhaps in the hot inner region of the nebula, and are preserved (with less refractory material) within first-generation planetesimals, which are too small to melt. These bodies are later disrupted by collisions, returning CAIs to the nebula and producing large amounts of dust and small fragments. [*Wood* (2004) has noted some difficulties this possibility faces, given the lack of reset Al-Mg ages and the lack of shock features in CAIs.] Some heating mechanism (shock waves?) melts this material, producing chondrules and remelting or altering some CAIs. Chondrules, CAIs, and dust reaccrete to form chondrite parent bodies before the nebula dissipates. In this model, chondrites are secondary products of a complex and violent evolution rather than the first planetesimals to form. The most plausible cause of such events so long after the formation of the solar nebula is the late formation of Jupiter; the model and its implications are explored by *Weidenschilling and Cuzzi* (2006).

The two scenarios of diffusion and recycling are not necessarily mutually exclusive. A period of turbulence could have been followed by asteroidal accretion, with an episode of disruption, processing, and reaccretion after Jupiter formed.

7. SUMMARY AND OPEN QUESTIONS

Cosmochemical measurements and observations of young stellar objects both indicate a disk lifetime of several millions of years. Absolute Pb-Pb measurements of components believed to have formed within the disk point to a minimum duration of 4–5 m.y. Aluminum-26 measurements, less reliable because of questions about its initial distribution, also point to a >5-m.y. age of the solar accretionary disk. This is compatible with known achondrite ages. A disk lifetime of this order seems plausible by comparisons to young stellar objects. The typical disk lifetime (observed from the duration of dust emission) is 3–5 m.y., albeit with a large range of ages between different stars. Preserving nebular components in a disk for this period of time is possible provided that there is a mechanism for preventing the early-formed objects from drifting into the Sun; this could be achieved by turbulent diffusion, or by storing CAIs and early-formed chondrules in planetesimals that were too small to melt.

A number of important open questions on this subject are discussed in *Cuzzi and Weidenschilling* (2006), and *Weidenschilling and Cuzzi* (2006). Some of the outstanding areas are highlighted below.

More measurements of ages of individual components (CAIs and chondrules) within the same meteorite may shed light on asteroidal accretion timescales. It is important to establish if the hints we have seen so far for differences in ages between chondrites is real (*Kunihiro et al.,* 2004; *Amelin et al.,* 2004), and this could potentially lessen the CAI storage problem. One important question is whether the formation of CAIs and chondrules overlapped in time, or if there was a true gap between them. Overlap would suggest that they formed by a single process, perhaps with energy density and temperatures decreasing with time and/or heliocentric distance. An age gap of ~1 m.y. would imply either more than one separate process or the same process occurring multiple times, separated in time by a period of relative inactivity. The distribution of relative ages inferred from ^{26}Al show little overlap between unaltered CAIs and chondrules (*Wadhwa and Russell,* 2000). This distribution appears to favor a gap, but may be due at least in part to selection of types of objects with measurable ^{26}Al ages. Further measurements are required, especially on objects intermediate between CAIs and chondrules. Only three Al-Mg ages exist in the literature for AOAs (*Itoh et al.,* 2002), which are intermediate to CAIs and chondrules in properties and composition. These preliminary measurements on AOAs suggest that they may have formed around 1 half-life (0.7 m.y.) after CAIs, indicating that these objects may "fill the gap" between chondrules and CAIs. Measurements are also required to compare different isotope systems on the same object, and especially to combine absolute age dates by Pb-Pb and Al-Mg and Mn-Cr measurements, in order to establish the distribution of these isotopes in the early solar system.

Continuing studies of asteroid thermal evolution can help to constrain planetesimal accretion timescales. Such studies need to include more realistic physics, such as time-variable growth rates and collisional disruption occurring concurrently with accretion. Associated isotopic analyses of objects as diverse as unequilibrated chondrites to the many different iron cores of differentiated bodies of which we have samples can also help to constrain formation ages.

The distribution and lifetime of cold dusty material around young stars is not currently well constrained. More extensive measurements, particularly of submillimeter- to millimeter-wavelength observations, will allow us to better observe dust at smaller stellocentric distances.

It is critical that we try to understand the prevalence and intensity of turbulence, since it plays a key role in setting the timescales and even radial distribution of planetesimal formation. Theoretical studies of nebula fluid dynamics need to continue, and it may turn out that inferences from various aspects of primitive bodies themselves might help us to understand this issue (*Cuzzi and Weidenschilling,* 2006).

REFERENCES

Adachi I., Hayashi C., and Nakazawa K. (1976) The gas drag effect on the elliptical motion of a solid body in the primordial solar nebula. *Prog. Theor. Phys., 56,* 1756–1771.

Aikawa Y., van Zadelhoff G. J., van Dishoeck E. F., and Herbst E. (2002) Warm molecular layers in protoplanetary disks. *Astron. Astrophys., 386,* 622–632.

Alencar S. H. P. and Batalha C. (2002) Variability of southern T Tauri stars. II. The spectral variability of the classical T Tauri star TW Hydrae. *Astrophys. J., 571,* 378–393.

Allègre C. J., Manhes G., and Gopel C. (1995) The age of the Earth. *Geochim. Cosmochim. Acta, 59,* 1445–1456.

Amelin Y. and Krot A. N. (2005) Young Pb-isotopic ages of chondrules in CB carbonaceous chondrites (abstract). In *Lunar and Planetary Science XXXVI,* Abstract #1247. Lunar and Planetary Institute, Houston (CD-ROM).

Amelin Y., Krot A. N., Hutcheon I. D., and Ulyanov A. A. (2002) Lead isotopic ages of chondrules and calcium-aluminium-rich inclusions. *Science, 297,* 1678–1683.

Amelin Y., Krot A. N., and Twelker E. (2004) Duration of the chondrule formation interval: A Pb isotope study (abstract). *Geochim. Cosmochim. Acta, 68,* A759.

Arnould M., Paulus G., and Meynet G. (1997) Short-lived radionuclide production by non-exploding Wolf-Rayet stars. *Astron. Astrophys., 321,* 452–464.

Beckwith S. V. W., Sargent A. I., Chini R. S., and Guesten R. (1990) A survey for circumstellar disks around young stellar objects. *Astron. J., 99,* 924–945.

Becquerel H. (1896a) On the rays emitted by phosphorescence [translated title]. *Compt. Rendus, 122,* 420–421.

Becquerel H. (1896b) On the invisible rays emitted by phosphorescent bodies [translated title]. *Compt. Rendus, 122,* 501–503.

Bennett M. E. III and McSween H. Y. Jr. (1996) Revised model calculations for the thermal histories of ordinary chondrite parent bodies. *Meteoritics & Planet. Sci., 31,* 783–792.

Birck J.-L. and Allègre C. J. (1988) Manganese-chromium isotope systematics and the development of the early solar system. *Nature, 331,* 579–584.

Bischoff A. and Srinivasan G. (2003) ^{26}Mg excess in hibonites of the Rumuruti chondrite Hughes 030. *Meteoritics & Planet. Sci., 38,* 5–12.

Bizzarro M., Baker J. A., and Haack H. (2004) Mg isotope evidence for contemporaneous formation of chondrules and refractory inclusions. *Nature, 431,* 275.

Bockelée-Morvan D., Brooke T. Y., and Crovisier J. (1995) On the origin of the 3.2 to 3.6 μm emission feature in comets. *Icarus, 133,* 147–162.

Boss A. P. (2004) Evolution of the solar nebula. VI. Mixing and transport of isotopic heterogeneity. *Astrophys. J., 616,* 1265–1277.

Boss A. P. and Vanhala H. A. T. (2001) Origin and early evolution of solid matter in the solar system. *Philos. Trans. R. Soc. Lond., A359,* 2005–2017.

Brearley A. J. (2006) The action of water. In *Meteorites and the Early Solar System II* (D. S. Lauretta and H. Y. McSween Jr., eds.), this volume. Univ. of Arizona, Tucson.

Brice C., Vivas A. K., Calvet N., Hartmann L., Pacheco R., Herrera D., Romero L., Berlind P., Sanchez G., Snyder J. A., and Andrews P. (2001) The CIDA-QUEST large-scale survey of Orion OB1: Evidence for rapid disk dissipation in a dispersed stellar population. *Science, 291,* 93–96.

Busso M., Gallino R., and Wasserburg G. J. (2003) Short-lived nuclei in the early solar system: A low mass stellar source? *Publ. Astron. Soc. Austral., 20,* 356–370.

Calvet N., D'Alessio P., Hartmann L., Wilner D., Walsh A., and Sitko M. (2002) Evidence for a developing gap in a 10 m.y. old protoplanetary disk. *Astrophys. J., 568,* 1008–1016.

Cameron A. G. W., Hoflich P., Myers P. C., and Clayton D. D. (1995) Massive supernovae, Orion gamma rays, and the formation of solar system. *Astrophys. J. Lett., 447,* L53–L57.

Campbell A. J., Humayun M., and Weisberg M. K. (2002) Siderophile element constraints on the formation of metal in the metal-rich chondrites Bencubbin, Weatherford, and Gujba. *Geochim. Cosmochim. Acta, 66,* 647–660.

Chambers J. and Wetherill G. (2001) Planets in the asteroid belt.

Meteoritics & Planet. Sci., 36, 381–399.

Chaussidon M. and Gounelle M. (2006) Irradiation processes in the early solar system. In *Meteorites and the Early Solar System II* (D. S. Lauretta and H. Y. McSween Jr., eds.), this volume. Univ. of Arizona, Tucson.

Chen J. H. and Tilton G. R. (1976) Isotopic lead investigations on the Allende carbonaceous chondrite. *Geochim. Cosmochim. Acta, 40,* 617–634.

Chen J. H. and Wasserburg G. J. (1981) The isotopic composition of U and Pb in Allende inclusions and meteoritic phosphates. *Earth Planet. Sci. Lett., 52,* 1–15.

Consolmagno G. and Drake M. (1977) Composition and evolution of the eucrite parent body — Evidence from rare earth elements. *Geochim. Cosmochim. Acta, 41,* 1271–1282.

Cuzzi J. N. and Weidenschilling S. J. (2006) Particle-gas dynamics and primary accretion. In *Meteorites and the Early Solar System II* (D. S. Lauretta and H. Y. McSween Jr., eds.), this volume. Univ. of Arizona, Tucson.

Cuzzi J. N., Dobrovolskis A. R., and Champney J. M. (1993) Particle gas dynamics in the midplane of a protoplanetary nebula. *Icarus, 106,* 102–134.

Cuzzi J. N., Dobrovolskis A. R., and Hogan R. C. (1996) Turbulence, chondrules and planetesimals. In *Chondrules and the Protoplanetary Disk* (R. H. Hewins et al., eds.,) pp. 35–43. Cambridge Univ., Cambridge.

Cuzzi J. N., Hogan R. C., Paque J. M., and Dobrovolskis A. R. (2001) Size-selective concentration of chondrules and other small particles in protoplanetary nebula turbulence. *Astrophys. J., 546,* 496–508.

Cuzzi J. N., Davis S. S., and Dobrovolskis A. R. (2003) Blowing in the wind. II. Creation and redistribution of refractory inclusions in a turbulent protoplanetary nebula. *Icarus, 166,* 385–402.

Cuzzi J. N., Petaev M., Ciesla F., Krot A. N., and Scott E. R. D. (2005) Nebula models of non-equilibrium mineralogy: Wark-Lovering rims (abstract). In *Lunar and Planetary Science XXXVI,* Abstract #2095. Lunar and Planetary Institute, Houston (CD-ROM).

D'Alessio P., Calvet N., and Hartmann L. (2001) Accretion disks around young objects. III. Grain growth. *Astrophys. J., 553,* 321–334.

Darwin C. (1859) *On the Origin of the Species.* Republished in 1988 by New York Univ., New York. 478 pp.

Desch S. J., Srinivasan G., and Connolly H. C. Jr. (2004) An interstellar origin for the beryllium-10 in CAIs. *Astrophys. J., 602,* 528–542.

Dobrovolskis A. R., Dacles-Mariani J. S., and Cuzzi J. N. (1999) Production and damping of turbulence by particles in the solar nebula. *J. Geophys. Res., 104,* 30805.

Dubrulle B., Morfill G., and Sterzik M. (1995) The dust subdisk in the protoplanetary nebula. *Icarus, 114,* 237–246.

Dullemond C. P. and Dominik C. (2005) Dust coagulation in protoplanetary disks: A rapid depletion of small grains. *Astron. Astrophys., 434,* 971–986.

Dutrey A., Guilloteau S., Duvert G., Prato L., Simon M., Schuster K., and Menard F. (1996) Dust and gas distribution around T Tauri stars in Taurus-Auriga. I. Interferometric 2.7 mm continuum and 13 CO J = 1–0 observations. *Astron. Astrophys., 309,* 493–504.

Endreß M., Zinner E., and Bischoff A. (1996) Early aqueous activity on primitive meteorite parent bodies. *Nature, 379,* 701–703.

Faure G. (1986) *Principles of Isotope Geology.* Wiley, New York. 608 pp.

Fireman E. L., DeFelice J., and Norton E. (1970) Ages of the Allende meteorite. *Geochim. Cosmochim. Acta, 34,* 873–881.

Galy A., Hutcheon I., and Grossman L. (2004) $(^{26}Al/^{27}Al)_0$ of the nebula inferred from Al-Mg systematics in bulk CAIs from CV3 chondrites (abstract). In *Lunar and Planetary Science XXXV,* Abstract #1790. Lunar and Planetary Institute, Houston (CD-ROM).

Ghosh A., Weidenschilling S. J., McSween H. Y. Jr., and Rubin A. (2006) Asteroidal heating and thermal stratification of the asteroid belt. In *Meteorites and the Early Solar System II* (D. S. Lauretta and H. Y. McSween Jr., eds.), this volume. Univ. of Arizona, Tucson.

Gilmour J. D. and Saxton J. M. (2001) A time-scale of formation of the first solids. *Philos. Trans. R. Soc. London, 359,* 2037–2048.

Gilmour J. D., Whitby J. A., Turner G., Bridges J. C., and Hutchison R. (2000) The iodine-xenon system in clasts and chondrules from ordinary chondrites: Implications for early solar system chronology. *Meteoritics & Planet. Sci., 35,* 445–456.

Goswami J. N. and Vanhala H. A. T. (2000) Short-lived nuclides in the early solar system: Meteoritic evidence and plausible sources. In *Protostars and Planets IV* (V. M. Mannings et al., eds.), pp. 963–995. Univ. of Arizona, Tucson.

Goswami J. N., Marhas K. K., and Sahijpal S. (2001) Did solar energetic particles produce the short-lived radionuclides present in the early solar system? *Astrophys. J., 549,* 1151–1159.

Gounelle M. and Russell S. S. (2005) On early solar system chronology: Implications of an heterogeneous spatial distribution of ^{26}Al and ^{53}Mn. *Geochim. Cosmochim. Acta, 69,* 3129–3144.

Gounelle M., Shu F. H., Shang H., Glassgold A. E., Rehm E. K., and Lee T. (2001) Extinct radioactivities and protosolar cosmic-rays: Self-shielding and light elements. *Astrophys. J., 548,* 1051–1070.

Gray C. M., Papanastassiou D., and Wasserburg G. J. (1973) The identification of early condensates from the solar nebula. *Icarus, 20,* 213–239.

Grossman L. (1972) Condensation in the primitive solar nebula. *Geochim. Cosmochim. Acta, 38,* 47–64.

Guan Y., Huss G. R., Leshin L. A., and MacPherson G. J. (2003) Ni isotope anomalies and ^{60}Fe in sulfides from unequilibrated enstatite chondrites (abstract). In *Meteoritics & Planet. Sci., 38,* Abstract #5268.

Haisch K. E., Lada E. A., and Lada C. J. (2001) Disk frequencies and lifetimes in young clusters. *Astrophys. J. Lett., 553,* L153–L156.

Hartigan P., Hartmann L., Kenyon S. J., Strom S. E., and Skrutskie M. F. (1990) Correlations of optical and infrared excesses in T Tauri stars. *Astrophys. J. Lett., 354,* L25–L28.

Hartmann L., Calvet N., Gullbring E., and D'Alessio P. (1998) Accretion and the evolution of T Tauri disks. *Astrophys. J., 495,* 385.

Henderson P. (1982) *Inorganic Geochemistry.* Pergamon, Oxford. 353 pp.

Hillenbrand L. A., Strom S. E., Calvet N., Merrill K. M., Gatley I., Makidon R. B., Meyer M. R., and Skrutskie M. F. (1998) *Astrophys. J., 116,* 1816.

Holmes A. (1946) An estimate of the age of the Earth. *Nature, 57,* 680–684.

Hollenbach D. J., Yorke H. W., and Johnstone D. (2000) Disk dispersal around young stars. In *Protostars and Planets IV* (V. Mannings et al., eds.), pp. 401–428. Univ. of Arizona, Tucson.

Hoppe P., Macdougall J. D., and Lugmair G. W. (2004) High spatial resolution ion microprobe measurements refine chronology of Orgueil carbonate formation (abstract). In *Lunar and Planetary Science XXXV,* Abstract #1313. Lunar and Planetary Institute, Houston (CD-ROM).

Hsu W., Wasserburg G. J., and Huss G. R. (2000) High time resolution by use of the ^{26}Al chronometer in the multistage formation of a CAI. *Earth Planet. Sci. Lett., 182,* 15–29

Huss G. R., MacPherson G. J., Wasserburg G. J., Russell S. S., and Srinivasan G. (2001) Aluminium-26 in calcium-rich inclusions and chondrules from unequilibrated ordinary chondrites. *Meteoritics & Planet. Sci., 36,* 975–997.

Hutcheon I. D., Krot A. N., Marhas K., and Goswami J. (2004) Magnesium isotopic compositions of igneous CAIs in the CR carbonaceous chondrites: Evidence for an early and late-stage melting of CAIs. In *Lunar and Planetary Science XXXV,* Abstract #2124. Lunar and Planetary Institute, Houston (CD-ROM).

Ireland T. R., Compston W., Williams I. S., and Wendt I. (1990) U-Th-Pb systematics of individual perovskite grains from the Allende and Murchison carbonaceous chondrites. *Earth Planet. Sci. Lett., 101,* 379–387.

Itoh S. and Yurimoto H. (2003) Contemporaneous formation of chondrules and refractory inclusions in the early solar system. *Nature, 423,* 728–731.

Itoh S., Rubin A. E., Kojima H., Wasdson J. T., and Yurimoto H. (2002) Amoeboid olivine aggregates and AOA-bearing chondrule from Y-81020 CO 3.0 chondrite: Distribution of oxygen and magnesium isotopes (abstract). In *Lunar and Planetary Science XXXIII,* Abstract #1490. Lunar and Planetary Institute, Houston (CD-ROM).

Kita N. T., Nagahara H., Togashi S., and Morishita Y. (2000) A short duration of chondrule formation in the solar nebula: Evidence from ^{26}Al in Semarkona ferromagnesian chondrules. *Geochim. Cosmochim. Acta, 64,* 3913–3922.

Kita N. T., Ikeda Y., Shimoda H., Morishita Y., and Togashi S. (2003) Timing of basaltic volcanism in ureilite parent body inferred from the ^{26}Al ages of plagioclase-bearing clasts in DaG-319 polymict ureilite (abstract). In *Lunar and Planetary Science XXXII,* Abstract #1557. Lunar and Planetary Institute, Houston (CD-ROM).

Krot A. N. and Keil K. (2002) Anorthite-rich chondrules in CR and CH carbonaceous chondrites: Genetic link between Ca, Al-rich inclusions and ferromagnesian chondrules. *Meteoritics & Planet. Sci., 37,* 91–111.

Krot A. N., Hutcheon I. D., and Keil K. (2002) Anorthite-rich chondrules in the reduced CV chondrites: Evidence for complex formation history and genetic links between CAIs and ferromagnesian chondrules. *Meteoritics & Planet. Sci., 37,* 155–182.

Krot A. N., McKeegan K. D., Huss G. R., Liffman K., Sahijpal S., Hutcheon I. D., Srinivasan G., Bischoff A., and Keil K. (2004) Aluminum-magnesium and oxygen isotope study of relict Ca-Al-rich inclusions in chondrules. *Astrophys. J.,* in press.

Krot A. N., Yurimoto H., Hutcheon I. D., and MacPherson G. J. (2005a) Relative chronology of CAI and chondrule formation: Evidence from chondrule-bearing igneous CAIs. *Nature, 434,* 998–1001.

Krot A. N., Amelin Y., Cassen P., and Meibom A. (2005b) Young chondrules in CB chondrites from a giant impact in the early solar system. *Nature, 436,* 989–992.

Krot A. N., Hutcheon I. D., Brearley A. J., Pravdivtseva O. V., Petaev M. I., and Hohenberg C. M. (2006) Timescales and settings for alteration of chondritic meteorites. In *Meteorites and the Early Solar System II* (D. S. Lauretta and H. Y. McSween Jr., eds.), this volume. Univ. of Arizona, Tucson.

Kunihiro T., Rubin A. E., McKeegan K. D., and Wasson J. T. (2004) Initial ^{26}Al/^{27}Al in carbonaceous-chondrite chondrules: Too little ^{26}Al to melt asteroids. *Geochim. Cosmochim. Acta, 68,* 2947–2957.

LaTourrette T. and Wasserburg G. J. (1998) Mg diffusion in anorthite: Implications for the formation of early solar system planetesimals. *Earth Planet. Sci. Lett., 158,* 91–108.

Lecavelier des Etangs A. and 13 colleagues (2001) Deficiency of molecular hydrogen in the disk of β Pictoris. *Nature, 412,* 706–708.

Lee T., Papanastassiou D. A., and Wasserburg G. J. (1976) Demonstration of ^{26}Mg excess in Allende and evidence for ^{26}Al. *Geophys. Res. Lett., 3,* 109–112.

Lee T., Shu F. H., Shang H., Glassgold A. E., and Rehm K. E. (1998) Protostellar cosmic rays and extinct radioactivities in meteorites. *Astrophys. J., 506,* 892–912.

Leya I., Halliday A. N., and Wieler R. (2003) The predictable collateral consequences of nucleosynthesis by spallation reactions in the early solar system. *Astrophys. J., 594,* 605–616.

Lugmair G. W. and Gailer S. J. G. (1992) Age and isotopic relationships among the angrites Lewis Cliff 86010 and Angra dos Reis. *Geochim. Cosmochim. Acta, 56,* 1673–1694.

Lugmair G. W. and Shukolyukov A. (1998) Early solar system timescales according to ^{53}Mn-^{53}Mn systematics. *Geochim. Cosmochim. Acta, 62,* 2863–2886.

MacPherson G. J., Wark D. A., and Armstrong J. T. (1988) Primitive material surviving in chondirtes: Refractory inclusions. In *Meteorites and the Early Solar System* (J. F. Kerridge and M. S. Matthews, eds), pp. 746–807. Univ. of Arizona, Tucson.

MacPherson G. J., Davis A. M., and Grossman J. N. (1989) Refractory inclusions in the unique chondrite ALH85085 (abstract). In *Meteoritics, 24,* 297.

MacPherson G. J., Davis A. M., and Zinner E. K. (1995) The distribution of aluminum-26 in the early solar system: A reappraisal. *Meteoritics, 30,* 365–386.

Manhès G., Göpel C., and Allègre C. (1988) Systematique U-Pb dans les inclusions refractaires d'Allende: Le plus vieux materiau solaire. *Compt. Rend. de l'ATP Planetol.,* 323–327.

McKeegan K. D. and Davis A. M. (2003) Early solar system chronology. In *Treatise on Geochemistry, Vol. 1: Meteorites, Comets, and Planets* (A. M. Davis, ed.), pp. 431–461. Elsevier, Oxford.

McKeegan K. D., Chaussidon M., and Robert F. (2000) Incorporation of short-lived ^{10}Be in a calcium-aluminum-rich inclusion from the Allende meteorite. *Science, 289,* 1334–1337.

McKeegan K. D., Greenwood J. P., Leshin L. A., and Cosarinsky M. (2001) Abundance of ^{26}Al in ferromagnesian chondrules of unequilibrated ordinary chondrites (abstract). In *Lunar and Planetary Science XXXII,* Abstract #2009. Lunar and Planetary Institute, Houston (CD-ROM).

Mostefaoui S., Kita N. T., Togashi S., Tachinaba S., Nagahara H., and Morishita Y. (2002) The relative formation ages of ferromagnesian chondrules inferred from their initial aluminium-26/aluminium-27 ratios. *Meteoritics & Planet. Sci., 37,* 421–438.

Mostefaoui S., Lugmair G. W., Hoppe P. and El Goresy A. (2003) Evidence for live iron-60 in Semarkona and Chervony Kut: A Nanosims study (abstract). In *Lunar and Planetary Science XXXIV,* Abstract #1585. Lunar and Planetary Institute, Houston (CD-ROM).

Mostefaoui S., Lugmair G. W., and Hoppe P. (2004) In-situ evidence for live ^{60}Fe in the solar system: A potential heat source for planetary differentiation from a nearby supernova explosion (abstract). In *Lunar and Planetary Science XXXV,* Abstract #1271. Lunar and Planetary Institute, Houston (CD-ROM).

Muzerolle J., Calvet N., Brice C., Hartmann L., and Hillenbrand L. (2000) Disk accretion in the 10 m.y. old T Tauri stars TW Hydrae and Hen 3-600A. *Astrophys. J. Lett., 535,* L47–L50.

Muzerolle J., Calvet N., Hartmann L., and D'Alessio P. (2003) Unveiling the inner disk structure of T Tauri stars. *Astrophys. J. Lett., 597,* L149–L152.

Najita J., Carr J. S., and Mathieu R. D. (2003) Gas in the terrestrial planet region of disks: CO fundamental emission from T Tauri stars. *Astrophys. J., 589,* 931–952.

Natta A., Prusti T., Neri R., Wooden D., Grinin V. P., and Mannings V. (2001) A reconsideration of disk properties in Herbig Ae stars. *Astron. Astrophys., 371,* 186–197.

Nyquist L., Lindstrom D., Wiesman H., Martinez R., Bansal B., Mittlefehldt D., Shih C-Y., and Wentworth S. (1994) Mn-Cr systematics of individual Chainpur chondrules. *Meteoritics, 29,* 512.

Nyquist L., Lindstrom D., Mittlefehldt D., Shih C.-Y., Wiesmann H., Wentworth S., and Martinez R. (2001) Manganese-chromium formation intervals for chondrules from the Bishunpur and Chainpur meteorites. *Meteoritics & Planet. Sci., 36,* 911–938.

Nyquist L. E., Reese Y., Wiesmann H., Shih C.-Y., and Takeda H. (2003a) Fossil ^{26}Al and ^{53}Mn in the Asuka 881394 eucrite: Evidence of the earliest crust on asteroid 4 Vesta. *Earth Planet. Sci. Lett., 214,* 11–25.

Nyquist L., Shih C. Y., Wiesmann H., and Mikouchi T. (2003b) Fossil ^{26}Al and ^{53}Mn in Orbigny and Sahara 99555 and the timescale for angrite magmatism (abstract). In *Lunar and Planetary Science XXXII,* Abstract #1388. Lunar and Planetary Institute, Houston (CD-ROM).

Papanastassiou D. A., Bogdanovski O., and Wasserburg G. J. (2002) ^{53}Mn-^{53}Cr systematics in Allende refractory inclusions (abstract). In *Meteoritics & Planet. Sci., 36,* A114.

Patterson C. C. (1956) Age of meteorites and the Earth. *Geochim. Cosmochim. Acta, 10,* 230–237.

Patterson C. C., Tilton G. R., and Inghram M. G. (1955) Age of the Earth. *Science, 121,* 69–75.

Podosek F. A., Zinner E. K., MacPherson G. J., Lundberg L. L., Brannon J. C., and Fahey A. J. (1991) Correlated study of initial Sr-87/Sr-86 and Al-Mg isotopic systematics and petrologic properties in a suite of refractory inclusions from the Allende meteorite. *Geochim. Cosmochim. Acta., 55,* 1083–1110.

Podosek F. A. and Cassen P. (1994) Theoretical, observational, and isotopic estimates of the lifetime of the solar nebula. *Meteoritics, 29,* 6–25.

Rubin A. E., Kallemeyn G. W., Wasson J. T., Clayton R. N., Mayeda T. K., Grady M., Verchovsky A. B., Eugster O., and Lorenzetti S. (2003) Formation of metal and silicate globules in Gujba: A new Bencubbin-like meteorite fall. *Geochim. Cosmochim. Acta, 67,* 3283–3298.

Russell S. S., Srinivasan G., Huss G. R., Wasserburg G. J., and MacPherson G. J. (1996) Evidence for widespread ^{26}Al in the solar nebula and constraints for nebula time scales. *Science, 273,* 757–762.

Russell S. S., Huss G. R., Fahey A. J., Greenwood R. C., Hutchison R., and Wasserburg G. J. (1998) An isotopic and petrologic study of calcium-aluminum-rich inclusions from CO3 meteorites. *Geochim. Cosmochim. Acta, 62,* 689–714.

Russell S. S., Gounelle M., and Hutchison R. (2001) Origin of short-lived radionuclides. *Philos. Trans. R. Soc. Lond., A359,* 1991–2004.

Rutherford E. (1929) Origin of actinium and the age of the Earth. *Nature, 123,* 313–314.

Rutherford E. and Boltwood B. (1905) The relative proportion of radium and uranium in radio-active minerals. *Am. J. Sci., 20,* 55–56.

Scherstén A., Elliott T., Hawkesworth C. J., Russell S. S., and Masarik J. (2005) Rapid differentiation of small planets, high precision evidence from Hf-W chronometry. *European Geophysical Union Abstracts, 7,* EGU05-A-05462. European Geosciences Union, Vienna.

Srinivasan G., Goswami J. N., and Bhandari N. (1999) ^{26}Al in eucrite Piplia Kalan: Plausible heat source and formation chronology. *Science, 284,* 1348–1350.

Shu F. H., Shang H., and Lee T. (1996) Toward an astrophysical theory of chondrites. *Science, 271,* 1545–1552.

Shu F. H., Shang S., Gounelle M., Glassgold A. E., and Lee T. (2001) The origin of chondrules and refractory inclusions in chondritic meteorites. *Astrophys. J., 548,* 1029–1050.

Shukolyukov A. and Lugmair G. W. (2000) On the ^{53}Mn heterogeneity in the early solar system. *Space Sci. Rev., 92,* 225–236.

Skrutskie M. F., Dutkevitch D., Strom S. E., Edwards S., Strom K. M., and Shure M. A. (1990) A sensitive 10-micron search for emission arising from circumstellar dust associated with solar-type pre-main-sequence stars. *Astron. J., 99,* 1187–1195.

Stone J. M., Gammie C. F., Balbus S. A., and Hawley J. F. (2000) Transport processes in protostellar disks. In *Protostars and Planets IV* (V. Mannings et al., eds.), pp. 589–611. Univ. of Arizona, Tucson.

Strom K. M., Strom S. E., Edwards S., Cabrit S., and Skrutskie M. F. (1989) Circumstellar material associated with solar-type pre-main-sequence stars — A possible constraint on the timescale for planet building. *Astron. J., 97,* 1451–1470.

Strom S. E., Edwards S., and Skrutskie M. F. (1993) Evolutionary time scales for circumstellar disks associated with intermediate- and solar-type stars. In *Protostars and Planets III* (E. H. Levy and J. I. Lunine, eds.), pp. 837–866. Univ. of Arizona, Tucson.

Sugiura N. and Hoshino H. (2003) Mn-Cr chronology of five IIIAB iron meteorites. *Meteoritics & Planet. Sci., 38,* 117–144.

Swindle T. D., Davis A. M., Hohenberg C. M., MacPherson G. J., and Nyquist L. E. (1996) Formation times of chondrules and Ca-Al-rich inclusions: Constraints from short-lived nuclides. In *Chondrules and the Protoplanetary Disk* (R. H. Hewins et al., eds.), pp. 77–87. Cambridge Univ., Cambridge.

Tachibana S. and Huss G. R. (2003) The initial abundance of ^{60}Fe in the solar system. *Astrophys. J. Lett., 588,* L41–L44.

Tachibana S., Nagahara H., Mostefaoui S., and Kita N. T. (2003) Correlation between relative ages inferred from ^{26}Al and bulk compositions of ferromagnesian chondrules in least equilibrated ordinary chondrites. *Meteoritics & Planet. Sci., 38,* 939–963.

Tachibana S., Huss G. R., Kita N. T., Shimoda H., and Morishita Y. (2005) The abundance of iron-60 in pyroxene chondrules from unequilibrated ordinary chondrites (abstract). In *Lunar and Planetary Science XXXVI,* Abstract #1529. Lunar and Planetary Institute, Houston (CD-ROM).

Tatsumoto M., Unruh D. M., and Desborough G. A.. (1976) U-Th-Pb and Rb-Sr systematics of Allende and U-Th-Pb systematics of Orgueil. *Geochim. Cosmochim. Acta, 40,* 617–634.

Tera F., Carlson R. W., and Boctor N. Z. (1996) Radiometric ages of basaltic achondrites and their relation to the early history of the solar system. *Geochim. Cosmochim. Acta, 61,* 1713–1731.

Thi W. F., Blake G. A., van Dishoeck E. F., van Zadelhoff G. J., Horn J. M. M., Becklin E. E., Mannings V., Sargent A. I., van den Ancker M. E., and Natta A. (2001) Substantial reservoirs of molecular hydrogen in the debris disks around young stars. *Nature, 409,* 60–63.

Thomson W. (Lord Kelvin) (1897) The age of the Earth as an abode for life. *Philos. Mag., 47,* 66–90.

Tilton G. R. (1988a) Radiometric dating. In *Meteorites and the Early Solar System* (J. F. Kerridge and M. S. Matthews, eds), pp. 249–256. Univ. of Arizona, Tucson.

Tilton G. R. (1988b) Age of the solar system. In *Meteorites and the Early Solar System* (J. F. Kerridge and M. S. Matthews, eds.), pp. 257–275. Univ. of Arizona, Tucson.

Turner G. (1970) $^{40}Ar/^{39}Ar$ dating of lunar rock samples. *Science, 167,* 3918.

Vanhala H. A. T. (2001) Injection of radioactivities into the forming solar system: High resolution simulations (abstract). In *Lunar and Planetary Science XXXII,* Abstract #1170. Lunar and Planetary Institute, Houston (CD-ROM).

Vanhala H. A. T. and Boss A. (2002) Injection of radioactivities into the forming solar system. *Astrophys. J., 575,* 1144–1150.

Yin Q., Jacobsen S. B., Yamashita K., Blichert-Toft J., Télouk P., and Albarède F. (2002) A short timescale for terrestrial planet formation from Hf-W chronometry of meteorites. *Nature, 418,* 949–952.

Wadhwa M. and Lugmair G. W. (1996) Age of the eucrite 'Caldera' from convergence of long-lived and short-lived chronometers. *Geochim. Cosmochim. Acta, 60,* 4889–4893.

Wadhwa M. and Russell S. S. (2000) Timescales of accretion and differentiation in the early solar system: The meteoritic evidence. In *Protostars and Planets IV* (V. M. Mannings et al., eds.), pp. 1017–1029. Univ. of Arizona, Tucson.

Wadhwa M., Amelin A., Bogdanovski O., Shukolyukov A., and Lugmair G. (2005) High precision relative and absolute ages for Asuka 881394, a unique and ancient basalt (abstract). In *Lunar and Planetary Science XXXVI,* Abstract #2126. Lunar and Planetary Institute, Houston (CD-ROM).

Wadhwa M., Srinivasan S., and Carlson R. W. (2006) Timescales of planetesimal differentiation in the early solar system. In *Meteorites and the Early Solar System II* (D. S. Lauretta and H. Y. McSween Jr., eds.), this volume. Univ. of Arizona, Tucson.

Wark D. A. (1987) Plagioclase-rich inclusions in carbonaceous chondrite meteorites: Liquid condensates? *Geochim. Cosmochim. Acta, 51,* 221–242.

Weber D., Zinner E. K., and Bischoff A. (1995) Trace element abundances and magnesium, calcium, and titanium isotopic compositions of grossite-containing inclusions from the carbonaceous chondrite Acfer 182. *Geochim. Cosmochim. Acta, 59,* 803–823.

Weidenschilling S. J. (1977) Aerodynamic of solid bodies in the solar nebula. *Mon. Not. R. Astron. Soc., 180,* 57–70.

Weidenschilling S. J. (1980) Dust to planetesimals — Settling and coagulation in the solar nebula. *Icarus, 44,* 172–189.

Weidenschilling S. J. (1988) Formation processes and time scales for meteorite parent bodies. In *Meteorites and the Early Solar System* (J. F. Kerridge and M. S. Matthews, eds.), pp. 348–371. Univ. of Arizona, Tucson.

Weidenschilling S. J. (1997) When the dust settles — Fractal aggregates and planetesimal formation (abstract). In *Lunar and Planetary Science XXVIII,* p. 517. Lunar and Planetary Institute, Houston.

Weidenschilling S. J. (2004) From icy grains to comets. In *Comets II* (M. C. Festou et al., eds.), pp. 97–104. Univ. of Arizona, Tucson.

Weidenschilling S. J. and Cuzzi J. N. (2006) Accretion dynamics and timescales: Relation to chondrites. In *Meteorites and the Early Solar System II* (D. S. Lauretta and H. Y. McSween Jr., eds.,) this volume. Univ. of Arizona, Tucson.

Weisberg M. K., Prinz M., Clayton R. N., Mayeda T. K., Sugiura N., Zashu S., and Ebihara M. (2001) A new metal-rich chondrite grouplet. *Meteoritics & Planet. Sci., 36,* 401–418.

Whitby J., Gilmour J. D., Turner G., Prinz M., and Ash R. D. (2002) Iodine-xenon dating of chondrules from the Qingzhen and Kota Kota enstatite chondrites. *Geochim. Cosmochim. Acta, 66,* 347–359.

Whitby J., Russell S. S., Turner G., and Gilmour J. D. (2004) I-Xe measurements of CAIs and chondrules from the CV3 chondrites Mokoia and Vigarano. *Meteoritics & Planet. Sci., 39,* 1387–1403.

Wood J. A. (2004) Formation of chondritic refractory inclusions: The astrophysical setting. *Geochim. Cosmochim. Acta, 68,* 4007–4021.

Young E. D., Simon J. I., Galy A., Russell S. S., Tonui E., and Lovera O. (2005) Supra-canonical $^{26}Al/^{27}Al$ and the thermal evolution residence time of CAIs in the solar protoplanetary disk nebula. *Science, 308,* 223–227.

Condensation of Rocky Material in Astrophysical Environments

Denton S. Ebel
American Museum of Natural History

Volatility-dependent fractionation of the rock-forming elements at high temperatures is an early, widespread process during formation of the earliest solids in protoplanetary disks. Equilibrium condensation calculations allow prediction of the identities and compositions of mineral and liquid phases coexisting with gas under presumed bulk chemical, pressure, and temperature conditions. A graphical survey of such results is presented for systems of solar and nonsolar bulk composition. Chemical equilibrium was approached to varying degrees in the local regions where meteoritic chondrules, Ca-Al-rich inclusions, matrix, and other components formed. Early, repeated vapor-solid cycling and homogenization, followed by hierarchical accretion in dust-rich regions, is hypothesized for meteoritic inclusions. Disequilibrium chemical effects appear to have been common at all temperatures, but increasingly so in less-refractory meteoritic components. Work is needed to better model high-temperature solid solutions, indicators of these processes.

1. INTRODUCTION

It may be that much of the early solar nebula consisted of completely vaporized primordial material (>1800 K) that cooled monotonically (*Cameron,* 1963; *Kurat,* 1988). This scenario explains the astonishing isotopic homogeneity (except O) of solar system materials from interplanetary dust grains (IDPs), to chondrites, to planets (e.g., *Zhu et al.,* 2001), but it appears to violate current theories about the thermal structure of the nebula and protoplanetary disk (*Cameron,* 1995; *Gail,* 1998; *Woolum and Cassen,* 1999). Yet equilibrium condensation sequences capture many first-order observations of the volatility-dependent fractionations of the elements, the identities of their host phases in meteorites, and even the chemical structure of the solar system (*Humayun and Cassen,* 2000). In the first edition of this volume, *MacPherson et al.* (1988) concluded that "... good chemical and sparse textural evidence indicate that vapor-solid condensation played a major role in the genesis of many refractory inclusions in the early solar nebula. The condensation probably reflected neither perfect equilibrium nor perfect fractionation." Volatility-related fractionations among rocky elements were established at high temperatures, and are recorded in CAIs and chondrules (*Grossman,* 1996). The focus here is the vapor-liquid-solid equilibria relevant to these fractionation processes.

As investigations focus on disk structure, descriptions of chemical equilibrium between gas + silicate liquid + solid phases constrain the likelihood and extent of local processes that might occur in dense, cool disk regions subject to gravitational instabilities and shock heating (e.g., *Wood,* 1996; *Iida et al.,* 2001; *Desch and Connolly,* 2002; *Ciesla and Hood,* 2002); in hotter, partially ionized regions subject to magnetorotational turbulence and current sheet heating (e.g., *Joung et al.,* 2004); or in streams of material subject to solar flares and winds [the x-wind of *Shu et al.* (2001)]. Parameterizations of chemical equilibrium calculations can be integrated into large-scale nebula models (e.g., *Cassen,* 2001). Condensation calculations are also applicable to other astrophysical environments, such as the stellar atmospheres that produce refractory interstellar dust (*Lattimer et al.,* 1978; *Lodders and Fegley,* 1997; *Ebel,* 2000). With new analytical techniques, meteoriticists have sharpened the search for pristine solar nebula condensates and their isotopic and trace-element signatures, but also recognized that factors other than their thermal stability in a cooling gas have affected the chemistry of the most primitive objects found in meteorites.

The cosmochemical classification of the elements is based on their relative volatilities in a system of solar bulk composition. In a cooling (or heating) solar gas, elements are calculated to condense (or evaporate) in groups dependent upon their volatility: the abundant elements Ca, Al, and Ti are highly refractory; Si, Mg, and Fe are less refractory; K and Na are moderately volatile; S and C are more volatile. The rare earth (REE), platinum group (PGE), and other trace elements (e.g., highly volatile Hg, Tl) can be similarly classified (e.g., *Kerridge and Matthews,* 1988, Part 7; *Ireland and Fegley,* 2000). These groups may be correlated, or combined, with *Goldschmidt*'s (1954) geochemical classification of lithophile, siderophile, chalcophile, and atmophile elements (*Larimer,* 1988; *Hutchison,* 2004). The chemical compositions and associations of the most ancient minerals in meteorites strongly suggest that their formation was controlled in large part by the relative volatilities of the oxides of their constituent elements. Observed fractionations of the elements between meteoritic (asteroidal), cometary, and planetary reservoirs are also correlated to volatility (*Lewis,*

1972a,b). We model high-temperature equilibrium and disequilibrium processes to understand the formation of meteorite components, and the fractionation of the elements, one from another, in our solar system and in other stellar environments.

1.1. Chemical Equilibrium Calculations

"Condensation calculation" is a generic term for models describing the equilibrium distribution of the elements between coexisting phases (solids, liquid, vapor) in a closed chemical system with vapor always present. Predictions of chemical equilibrium made using the equations of state of the phases involved, for example, at fixed total pressure (P^{tot}) and temperature (T), are independent of whether or not the system is cooling. They do not predict the sizes of resultant crystals. They can, however, be used to characterize the degree to which a chemical system is supersaturated in a potential condensate, which is a key parameter in understanding nucleation of condensates. Condensation calculations have been performed to describe chemistry in astrophysical environments since before digital computers became commonplace (*Latimer,* 1950; *Urey,* 1952; *Lord,*1965; cf. historical review by *Lodders and Fegley,* 1997). *Lodders* (2003) has most recently computed the condensation temperatures of all the elements into pure, stoichiometric solid minerals from a cooling vapor of solar composition at $P^{tot} = 10^{-4}$ bar. Others have recently explored the stability of condensed liquid oxide solutions as well as solid solutions, all coexisting with vapor, in systems enriched in previously vaporized chondritic dust (e.g., *Yoneda and Grossman,* 1995; *Ebel and Grossman,* 2000). Chemical equilibrium calculations describe the identity and chemical composition of each chemical phase in the assemblage toward which a given closed system of elements will evolve, to minimize the chemical potential energy of the entire system. This knowledge is a prerequisite to any discussion of whether, where, how, and why equilibrium was or was not achieved in any particular scenario predicted by astrophysical models, or in relation to the origin of any particular meteoritic object.

Equilibrium calculations strictly apply only to systems that can be assumed to have reached or closely approached chemical equilibrium at a given point in their history. Disequilibrium effects can be explored by modification of equilibrium calculations, for example, fractionation or diffusion, by removing high-T condensates from further reaction at lower T (*Larimer,* 1967; *Larimer and Anders,* 1967; *Petaev and Wood,* 1998a,b, 2000; *Meibom et al.,* 2001); retarded nucleation, by constraining the onset of condensation for particular phases (*Blander et al.,* 2005; but cf. *Stolper and Paque,* 1986); evaporation, by computing saturation vapor pressures over evaporating phases, for input into kinetic models (e.g., *Grossman et al.,* 2000; *Nagahara and Ozawa,* 2000; *Alexander,* 2001; *Richter et al.,* 2002); or photodissociation of gaseous molecules, by adjusting their thermal stabilities (*Ebel and Grossman,* 2001).

Reaction network calculations are a very different kind of simulation, in which the rates of individual reactions are addressed explicitly through rate constants. Parameters (e.g., reaction rates, sticking coefficients, surface energies of small clusters, and heterogeneous catalytic effects of dust grains) for full gas-dynamical treatments (e.g., *Draine,* 1979; *Chigai et al.,* 2002) of heterogeneous reactions among rock-forming elements at high temperature are very poorly constrained. Network calculations appear to be most effective for low-temperature phase relations, and gas-phase reactions in the H-C-N-O-S system (e.g., *Hasegawa and Herbst,* 1993; *Bettens and Herbst,* 1995; *Gail,* 1998; *Aikawa et al.,* 2003; *Semenov et al.,* 2004). Partial ionization of gas would certainly affect condensation, and photodissociation would also occur in regions subject to ionizing radiation. Thermal ionization is not significant at the conditions discussed here, and calculations to the present do not address the condensation of rocky material in astrophysical environments with significant flux of ionizing radiation.

The formalism of classical thermodynamics is rooted in the phenomenology of materials, calibrated against more than a century of laboratory investigation (e.g., *Bowen,* 1928; *Nurse and Stutterheim,* 1950; *Newton,* 1987; *Ganguly,* 2001). Powers of interpolation, extrapolation, and prediction stem from this formalism (cf., *DeHeer,* 1986; *Anderson and Crerar,* 1993). This tool, and detailed observations of meteoritic inclusion and matrix chemistries and textures, constrain the conditions and processes these rocky materials might have experienced. Hypotheses for their formation must predict chemical, thermal, and mechanical conditions that are in accord with those constraints. Pressures, temperatures, oxygen fugacities, and chemical environments are deduced from analyses of meteoritic materials, and constrain astrophysical models for the protoplanetary disk in which those materials formed. The resulting understanding of the universal process of disk formation is central to learning how and where habitable planets form, and how unique our own Earth really is.

1.2. Earlier Reviews and Present Focus

Several excellent reviews cover condensation of crystalline solid phases from a gas of solar composition (e.g., *Larimer,* 1988; *Lewis,* 1997; *Fegley,* 1999), particularly *Grossman and Larimer* (1974). *Davis* (2003) has compared kinetically controlled processes with equilibrium processes. The present chapter briefly describes data and algorithms for modeling the thermal stability of condensates relative to vapors of various compositions, presenting results in the form of phase diagrams with a discussion of their use. I also briefly discuss the meteoritic evidence that may indicate direct condensation, or the establishment of chemical equilibrium between gas, solid, and liquid phases in the protoplanetary disk before accretion of the meteorite parent bodies.

Dust-gas chemistry has received substantial attention from astrophysicists interested in interstellar and cometary material (e.g., *Millar and Williams,* 1993; *Altwegg et al.,* 1999). The role of interstellar processes in establishing the meteorite record is increasingly recognized (e.g., *Wood,*

1998; *Irvine et al.,* 2000). Some grains found in meteorites were formed at high temperatures around stars with C/O ≥ 1, or in supernovae (cf. *Zinner,* 1998; *Nittler and Dauphas,* 2006). Astrophysical observations (e.g., *Chen et al.,* 2003) are yielding new data about stars of nonsolar metallicity (i.e., the abundance of H relative to all the heavier elements; in astrophysics, *metals* — here, *metal* will refer to Fe, Ni, Co, etc., and their alloys). Measurement of the atmospheric compositions of stellar grain sources is an active field, as is modeling of nucleosynthesis in AGB stars and novae. Condensation calculations using these results as inputs are vital to understanding the origins of interstellar and presolar grains (e.g., *Amari et al.,* 2001; *Ebel and Grossman,* 2001; cf. review by *Lodders and Fegley,* 1997). Grain formation in stellar outflows is also central to understanding the volatility-dependent depletions of rock-forming elements observed in the interstellar medium (*Field,* 1974; *Irvine et al.,* 1987; *Ebel,* 2000; *Kemper,* 2002). Condensation calculations have been pursued down to absolute zero (*Lewis,* 1972a); however, the equilibrium model is of very limited utility at such low temperatures. High-temperature disequilibrium effects, such as temperature differences between gas and solids (e.g., *Arrhenius and Alfvén,* 1971), can be addressed by modification of equilibrium calculations. This chapter is about the fates of condensable atoms in the nebula and disk, at temperatures above 1000 K, and total pressures between 10^{-8} and 1.0 bar (number densities of molecules $>3 \times 10^{16}$ cm^{-1}). These are conditions, plausible in some but not in all astrophysical models of portions of the protoplanetary disk, where ensembles of atoms may be assumed to approach chemical equilibrium on timescales that make such calculations relevant and useful.

2. TECHNIQUES

2.1. Elemental Abundances

2.1.1. Solar and carbon-enriched systems. The calculations here consider closed chemical systems containing the 22 elements having abundance greater than Zn, plus F: H, He, C, N, O, F, Ne, Na, Mg, Al, Si, P, S, Cl, Ar, K, Ca, Ti, Cr, Mn, Fe, Co, Ni. Equilibrium among these elements fixes their partial pressures in the vapors, from which thermal stabilities of trace-element-bearing compounds can be calculated. The relative atomic fractions of the elements in a vapor of solar composition are from *Anders and Grevesse* (1989), which give a C/O ratio of 0.43. Recent work by *Allende Prieto et al.* (2002) resets solar C/O to 0.50, and *Lodders* (2003) has most recently reevaluated solar abundances of all the elements (cf. *Grevesse and Sauval,* 2000). These revisions have very minor effects on the relative thermal stabilities of rocky condensates. Because C and O combine to form the highly stable CO gaseous molecule, the excess O makes a gas of solar composition oxidizing. If C/O > ~1.0 by addition of C to, or removal of H_2O from, a gas of otherwise solar composition, the system becomes highly reducing, and very different solid assemblages become stable. Carbon-enriched gas compositions are relevant to the formation of presolar graphite, SiC, TiC, and other grains in C-rich stellar environments (*Sharp and Wasserburg,* 1995; *Lodders and Fegley,* 1997), and possibly also to the formation of enstatite chondrites (*Ebel and Alexander,* 2005).

2.1.2. Dust-enriched systems. Innumerable scenarios can be explored for adjusting elemental abundances in a gas of solar composition, by the addition or subtraction of plausibly fractionated components. *Wood* (1963, 1967) proposed enrichment of vapor of solar composition by a fractionated dust of previously condensed rocky material, perhaps in the midplane of a protoplanetary disk to as far as 10 AU (*Millar et al.,* 2003). In this scenario, dust consisting of oxides and sulfides of rock-forming elements is concentrated relative to H_2 and other volatile species. The resulting bulk composition is assumed to be heated to vaporization, and the condensation history of the cooling vapor is predicted. *Wood and Hashimoto* (1993) explored the condensation of bulk compositions enriched or depleted in refractory, organic, and icy components and in H. *Yoneda and Grossman* (1995) explored condensation in solar bulk compositions enriched by a dust of ordinary chondrite composition. Their objective was to find the conditions where CaO-MgO-Al_2O_3-SiO_2 (CMAS) liquids are stable. *Ebel and Grossman* (2000) extended this work to a vapor of solar composition, enriched in a dust of carbonaceous chondrite (Orgueil, CI) composition, and they calculated the thermal stability of CMAS-TiO_2-FeO-Na_2O-K_2O liquids and FeO-bearing silicates. The portion of a vapor of solar composition that *Ebel and Grossman* (2000) considered to be condensable as chondritic (CI) dust is 5.66×10^{-4} (atomic), or 6.69×10^{-3} (by mass), of a canonical solar vapor (*Anders and Grevesse,* 1989). Their dust enrichment factor of 1000× CI corresponds to a dust/gas mass ratio of ~6.7, which is not unjustifiable on astrophysical grounds. Chondritic material is highly depleted in C and H relative to solar. The addition of such "dust" to a gas of solar composition increases the partial pressure of O in particular, but also condensable elements present in the dust.

2.2. Thermodynamic Data

If everyone used the same data and algorithms, there would be a unique canonical "condensation sequence." Such is not the case, and in recent years differences have been exaggerated by the employment of ever more complex models for mineral solid solutions: minerals where multiple combinations of elements can occupy sites in the same crystalline lattice. There is no single "*most* correct" database, or set of solid-solution models; however, there are carefully selected, internally consistent sets of data, and haphazardly selected sets collected from disparate sources. The former should be *more* correct, in application, than the latter. The power and limitations of a particular calculation are most clear when algorithms, but most importantly the core data, are described as completely as editors will allow. For example, the mixing of endmember olivine species fayalite (Fe_2SiO_4) and forsterite (Mg_2SiO_4) can be treated as a mechanical mixture of separately condensing endmembers, or as a real solid solution with accounting (based on experimental data) for the energetic consequences of ionic mixing on the olivine crystalline lattice. The particulars of data, and how it is implemented in calculation, are crucial to evaluating the results of such calculations.

There is reasonable agreement as to the first-order phenomena predicted during equilibrium cooling of a gas of solar composition at fixed P^{tot} (*Larimer,* 1967; *Grossman,* 1972; *Grossman and Larimer,* 1974; *Wood and Hashimoto,* 1993; *Yoneda and Grossman,* 1995; *Ebel and Grossman,* 2000; *Lodders,* 2003) (see section 3.1.2). Calculations of silicate liquid stability as a function of P^{tot} and enrichment in chondritic dust also agree where models overlap (*Yoneda and Grossman,* 1995; *Ebel and Grossman,* 2000; *Ebel,* 2005).

An important consideration is the fundamental thermodynamic data for the elements, depending upon what forms of equations of state are used for solids and liquids. The equations of state for compounds used here combine their enthalpy and entropy of formation from the elements in their standard states at 298 K and 1 bar with the heat capacity of the compound at all relevant temperatures. From these data, suitable equations of state (e.g., apparent Gibbs energy of formation from the elements at constant P and T) are readily calculated from free-energy (Giaque) functions for the constituent elements (cf. *Ghiorso and Kelemen,* 1987; *Berman,* 1988; *Anderson and Crerar,* 1993). Direct use of tabulated Gibbs energies (e.g., *Chase,* 1998) can be hazardous (cf. *Lodders,* 2004).

2.2.1. Major elements. Equilibria between gas, solid, and liquid phases are most conveniently calculated by referring all formation reaction energies to the monatomic gaseous species of the elements. The elements can be divided into major, minor, and trace categories based upon their absolute abundances. The abundant elements are atmophile (gaseous) H, C, N, O; lithophile ("rock-loving") Si, Mg, Al, Ca, Ni, Na, K, Ti; and siderophile ("iron-loving") Fe, Ni, Co, Cr, and S. Just the oxides of Si and Mg, with metallic Fe, constitute >85% of matter condensable above 400 K from a vapor of solar composition. Most of the other elements have minor effects on the identity and compositions of solids and liquids into which they substitute in equilibrium with vapor under most assumed astrophysical conditions.

2.2.2. Minor and trace elements. These elements usually occur as trace substituents in minerals, so they are frequently considered separately from the major, or abundant, rock-forming elements that make up those minerals. The rare earth elements (REE), U, Th, Li, and others are lithophile; siderophile trace elements include the platinum group elements (PGE) (Ru, Rh, Pd, Re, Os, Ir, Pt). The moderately volatile elements (e.g., Mn, Cu, Na, Se, Zn, Cd) were defined by *Larimer* (1988) as "never depleted by more than a factor of 5 in any chondrite, relative to CI," and displaying "no detectable variations within a chondrite group" (cf. *Xiao and Lipschutz,* 1992). *Hutchison* (2004) delineates the 50% condensation T of Cr as separating refractory from moderately volatile (Au, Mn, alkalies) elements, and the onset of FeS condensation as marking the T below which "highly volatile" elements (Pb, Tl, Bi, Cd, Hg, In) condense. In most calculations, these trace elements are allowed to condense as pure elemental, oxide, or sulfide solids, which are then assumed to dissolve into major-element phases present at corresponding conditions. For example, most Cd condenses as CdS, at temperatures where FeS is stable (e.g., *Lodders,* 2003), hence CdS is assumed to dissolve into FeS. Equations of state for the gaseous and solid oxides and other compounds of many minor elements are poorly known or not readily available (*Davis and Grossman,* 1979). The relative solubilities of these elements between gas, solid, and liquid phases are in most cases also not well understood, so analysis of meteoritic minerals usually guides the choice of host phase. Thus in "oxidized" systems (e.g., solar), the REE condense into perovskite or hibonite (*Lodders,* 2003), and in reduced systems into oldhamite [*Lodders and Fegley* (1993) and formulae in Table 1], based on measurements of those phases in ordinary and enstatite chondrites (e.g., *Crozaz and Lundberg,* 1995), and on the condensation temperatures of pure REE oxides and sulfides. The refractory siderophile elements including PGE are expected to condense into refractory metal nuggets at high temperatures (*El Goresy et al.,* 1979; *Fegley and Palme,* 1985; *Lodders,* 2003; *Campbell et al.,* 2001), and into Fe-Ni alloy when Fe condenses. The behavior of trace elements becomes less predictable as their volatility increases, because the assumption of equilibrium becomes less tenable at lower T (*Kornacki and Fegley,* 1986).

2.2.3. Gaseous species. The primary reservoir of the nebula and disk is a vapor containing hundreds of gaseous molecular species (e.g., H, C, O, H_2, CO, CO_2, H_2O). At and below P^{tot} = 1 bar the gas can be considered as an ideal mixture of ideal gases. Because most gas species data is derived from vibrational spectroscopic measurements (*Chase,* 1998), it is assumed to be internally consistent. That is, an equilibrium calculation using data for reacting species accurately reproduces equilibria involving those species as observed in the laboratory. Algorithms to minimize the chemical potential energy in a large, speciated gaseous system stem from the work of *White et al.* (1958) (cf. *Smith and Missen,* 1982).

2.2.4. Pure solid species. Of the 275 mineral species identified in meteorites (*Rubin,* 1997), only a small subset is abundant, stable at high T, and has well-known thermodynamic properties. Much of the high-quality data for solid phases of geologic interest was determined many years ago, in labor-intensive systematic investigations by dedicated experimentalists. Mineral phases common in terrestrial geology are well studied; however, data are absent for some high-temperature phases found in chondrites [e.g., rhönite (*Fuchs,* 1978; see *Beckett et al.,* 2006)]. Data used in calculations must account for phase transitions that occur in many solids (e.g., α-β quartz–cristobalite for SiO_2), as their most stable crystalline lattice structure changes with temperature. Most magnetic transitions are below the temperatures at which equilibrium calculations might be assumed to apply, and they are not considered here. Multiple data compilations and laboratory studies contain equations of state for mineral phases of interest (cf. *Berman,* 1988; *Kuzmenko et al.,* 1997). For example, for six published values for the enthalpy of formation of grossite from the elements at 298 K, the average is –4012.261 kJ, with a standard deviation of 11.413 kJ (*Berman,* 1983; *Geiger et al.,* 1988, and references therein). A similar selection exists for hibonite (*Kumar and Kay,* 1985) and perovskite (formulae in Table 1). These differences affect the stabilities of Ca-aluminates relative to competing oxides, and account for many differences in published condensation sequences (see section 3.1.2), as reviewed in detail by *Ebel and Grossman* (2000). In nature, Mg and Ti dissolve into hibonite (*Hinton et al.,* 1988; *Beckett and Stolper,* 1994; *Simon et al.,* 1997; *Beckett et al.,* 2006), and this behavior is not captured in any calculations to date. Treating these phases as solid solutions is a goal of future work. Various permutations of all available data allow a variety of predictions for the relative stabilities of endmember (pure) corundum, hibonite, Ca-aluminate, grossite, spinel, and perovskite (Table 1) without appreciably altering the relative proportions of Ca, Al, and Ti oxides that are condensed at a particular T and P^{tot}. This is because the differences in the free energies for the solid phases are small, compared to the differences in free energies of the solids relative to the vapor.

2.2.5. Solid solutions. Many minerals found in meteorites, and predicted to be thermodynamically stable at high temperatures in a gas of solar composition, are solutions of two or more endmember components with the same crystal structure. Because these solid-solution minerals incorporate variable amounts of several elements depending upon the conditions in which they form, they are potentially powerful indicator minerals for constraining the condensation, evaporation, and crystallization histories of meteoritic inclusions in which they occur (e.g., *Fraser and Rammensee,* 1982). Examples include

melilite: $Ca_2Al_2SiO_7 - Ca_2MgSi_2O_7$
gehlenite – åkermanite
substitution: Al_2 for MgSi

olivine: $Mg_2SiO_4 - CaMgSiO_4 - Fe_2SiO_4$
forsterite – monticellite – fayalite
substitution: Fe and Ca for Mg

metal alloy: Fe,Cr,Co,Si,Ni
(note: C, P, S not considered here)

Melilite is a binary solid solution, exhibiting complete solid solution between the endmembers gehlenite and åkermanite. It is a charge-coupled substitution, since two Al^{3+} are replaced by the Mg^{2+}–Si^{4+} pair. Magnesium-olivine has limited substitution of Ca, but complete substitution of Fe. Efforts to account for such crystal-chemical effects in equations of state comprise a very large geological, metallurgical, and materials science literature, both experimental and theoretical (*Geiger,* 2001). For example, existing models are not capable of accurately predicting the crystallization of melilite from silicate liquid (*L. Grossman et al.,* 2002, their Fig. 1), therefore they are also unlikely to model gas-melilite equilibria correctly. Yet there is no guarantee that tweaking the model for melilite thermodynamics to work with a particular liquid model will improve the situation. Consistency of calibration across all the models used in a calculation is a goal of future work.

The most successful efforts at modeling groups of pure endmember minerals (e.g., *Berman,* 1988), mineral solid solutions (e.g., *Sack and Ghiorso,* 1989), or complex geochemical systems such as crystallizing magmas (e.g., *Berman,* 1983; *Ghiorso and Sack,* 1995), are based on simultaneous optimization of equations of state that describe multiple experimentally determined equilibrium phase relations. Results of these efforts to describe terrestrial rocks are called "internally consistent," because each piece of data is evaluated and weighted in the context of the entire dataset to optimize agreement with measurements. Less data is available to describe less-common substitutional elements (e.g., Cr, Mn, Ni in olivines and pyroxenes) that may be more important in extraterrestrial rocks.

The most complex solid solution important to the condensation of rocky material at high T is Ca-rich pyroxene. A comprehensive formulation of a thermodynamic model of pyroxenes with the general formula

$$(Ca, Mg, Fe^{2+}, Ti^{4+}, Fe^{3+}, Al)_2 (Si, Al, Fe^{3+})_2 O_6$$

is by *Sack and Ghiorso* (1989, 1994), who concentrated their calibration range on the pyroxene compositions common to basaltic igneous rocks. In particular, their Ti-,Al-pyroxene properties were constrained mainly by studies of basaltic FeO-bearing systems (*Sack and Carmichael,* 1984). Their model reproduces many of the chemical characteristics of experimental and natural pyroxene-bearing mineral assemblages, but not the Ti^{3+}-bearing pyroxenes (formerly known as fassaite) found in CAIs. *Beckett* (1986) identified four endmembers critical to describing pyroxenes in CAIs — diopside = $CaMgSi_2O_6$, CaTs = $CaAl_2SiO_6$–, T_3P = $CaTi^{3+}AlSiO_6$, and T_4P = $CaTi^{4+}Al_2O_6$ — estimated Gibbs free energies for the Ti species; and provided a robust method for calculating these component molecular fractions from oxide weight percent data. Although it includes the Ti^{3+} substitution, the model used by *Yoneda and Grossman* (1995) ignores Fe, and is not calibrated against models of other solid solutions, or against liquids.

A thermodynamic activity model for solid metal alloy was developed by *Grossman et al.* (1979) and revised by *Ebel and Grossman* (2000). Although treatments of more complex systems are reported in the metallurgical literature (e.g., *Fernández-Guillermet,* 1988), they have not yet been incorporated into published condensation calculations. Implementations used in industry are proprietary (e.g., *Eriksson,* 1975; *Eriksson and Hack,* 1990; *Bale et al.,* 2002; see also *http://www.esm-software.com/chemsage/*). Other workers have treated specific metal subsystems (e.g., *Wai and Wasson,* 1979; *Fegley and Palme,* 1985). *Kelly and Larimer* (1977) used thermodynamic theory to trace the history of siderophiles from condensation through planetary differentiation; however, the effects of C, S, and P were not considered. *Petaev et al.* (2003) have coupled a model for diffusion in metal (*Wood,* 1967) with condensation chemistry to address the formation of zoned metal grains (*Meibom et al.,* 2001). The results described here do not include any assumptions about grain sizes or residence times, nor is diffusion data considered strong enough to warrant modeling diffusion in large-scale dust-gas systems.

In reduced chemical systems with high C/O ratios, carbide, nitride, and sulfide minerals are calculated to be stable, and also intermetallic compounds (e.g., FeSi). We do not yet know how to model many reduced solid solutions. For example, a wide range of niningerite (Fe, Mg, Mn)S compositions is found in meteorites (*Ehlers and El Goresy,* 1988), yet this solid solution is not accounted for in existing equilibrium stability calculations for highly reducing systems (e.g., *Lodders and Fegley,* 1997). This is a fertile area for future work.

2.2.6. Liquid solutions. Two well-tested models exist for describing the thermodynamic activities of oxide components in silicate liquids, suitable for modeling crystal-liquid equilibria in magmatic systems. Igneous CAIs and chondrules are subsets of the universe of such systems. Both models use a classical nonideal thermodynamic formalism (cf. *Anderson and Crerar,* 1993) to describe the chemical potential energy of silicate liquids, and are calibrated against internally consistent datasets describing coexisting mineral phases (but not metal alloy). *Berman*'s (1983) model is restricted to $CaO-MgO-Al_2O_3-SiO_2$ (CMAS) liquids. *Ghiorso and Sack*'s (1995) "MELTS" model includes additional components containing the oxides TiO_2, FeO, Cr_2O_3, P_2O_5, Na_2O, and K_2O, but because of their choice of stoichiometric endmember components, it requires that liquids have molar $SiO_2 > [½ (MgO-Cr_2O_3) + ½ FeO + (CaO-3 P_2O_5) + Na_2O + ½ K_2O]$. In the absence of a tested, universal liquid model, one must use either one model or the other, depending upon the composition region of interest (cf. *Ebel and Grossman,* 2000; *Ebel,* 2005). *Berman*'s (1983) model can be applied to CAIs and FeO-poor chondrules, while the MELTS model applies to most chondrule compositions. The limit on MELTS liquid compositions is particularly restrictive in considering FeO evaporation (*Ebel,* 2005).

To calibrate the thermodynamic parameters of a model liquid requires multiple statements of crystal-liquid equilibria at known P and T derived from laboratory data, combined with accurate thermodynamic models of the crystalline phases involved (*Ghiorso,* 1985, 1994; *Berman and Brown,* 1987). The *Berman* (1983) CMAS model is calibrated using an optimized endmember mineral thermodynamic dataset (*Berman,* 1983), but all solid solutions are treated as mechanical mixtures of endmembers (e.g., *L. Grossman et al.,* 2002, their Fig. 1). The MELTS model (*Ghiorso and Sack,* 1995) is calibrated using an internally consistent endmember dataset for solid mineral endmembers (*Berman,* 1988), and complex, nonideal solid-solution models for olivine (*Sack and Ghiorso,* 1989), pyroxene (*Sack and Ghiorso,* 1994), feldspar (*Elkins and Grove,* 1990), spinel (*Sack and Ghiorso,* 1991a,b), etc. These models include energetic effects of atomic order/disorder in crystalline lattices, but not magnetic effects that may contribute below ~800 K. In calculations involving solid solutions and liquids, there are many choices to be made; none is perfect.

2.3. Algorithms

2.3.1. Gas-solid-liquid equilibria. In all the calculations considered here, the gas is considered as the primary reservoir of material. Two approaches may be taken. First, the entire system can be solved as a whole, which amounts to treating each pure solid species as if it were a gaseous molecule, but with a very different equation of state. *Yoneda and Grossman* (1995) were the first to incorporate true solid-solution behavior into such an approach. Alternatively, *Ebel and Grossman* (2000) solved speciation in the gas separately (cf. *White et al.,* 1958; *Grossman and Larimer,* 1974; *Lattimer et al.,* 1978; *Smith and Missen,* 1982; *Lewis,* 1997), treating the gas as a separate solution phase similar to pyroxene or liquid. These approaches yield identical results for the same inputs. *Ebel and Grossman* (2000) adapted algorithms of *Ghiorso* (1985, 1994) to facilitate use of the solid- and liquid-solution models embodied in the MELTS code (*Ghiorso and Sack,* 1995). They substituted gas for liquid as the material reservoir in the MELTS algorithm, and treated liquids analogously to solid-solution phases.

2.3.2. Solid stability and optimization. This is the most laborious part of condensation calculations, particularly when a rigorous treatment of solid solutions is included. The goal is to find the optimal distribution of elements between gas and all possible condensates, with no *a priori* knowledge of what the stable condensates are. At any step in a particular calculation, the thermodynamic stability of all potential condensates can be calculated from their equations of state, given the partial pressures of their constituent elements in the gas. Each stable condensate must be added to the system in a small mass increment, with complementary subtraction of mass from the gas. Minimization of the chemical potential energy (e.g., Gibbs energy) of the gas + condensate system follows, by redistribution of matter between all the coexisting phases. Thermodynamic stability of potential condensates is then tested again. Condensates whose mass decreases below a threshold are returned to the gas reservoir (cf. *Ebel et al.,* 2000), where their component elements are removed into more stable condensates in subsequent steps.

3. RESULTS

Results are presented here in terms of total pressure (P^{tot}, bar) and temperature (T, in Kelvin) (e.g., Plate 7). Abbreviations and chemical formulae used in Plate 7 and elsewhere are listed in Table 1. Discussion logically follows the behavior of cooling systems of particular bulk composition

TABLE 1. Names, abbreviations, and chemical formulae of mineral phases.

Mineral	Abbreviation	Chemical Formula
corundum	cor	Al_2O_3
hibonite	hib	$CaAl_{12}O_{19}$
grossite	grs	$CaAl_4O_7$
Ca-monoaluminate	CA1	$CaAl_2O_4$
perovskite	prv	$CaTiO_3$
melilite	mel	$Ca_2(Al_2, MgSi)SiO_7$
Al-spinel	Al-spn	Al-rich $(Fe,Mg,Cr,Al,Ti)_3O_4$
Cr-spinel	Cr-spn	Cr-rich $(Fe,Mg,Cr,Al,Ti)_3O_4$
olivine	olv	$(Mg_2, Fe_2, MgCa)SiO_4$
metal alloy	met	Fe,Ni,Co,Cr,Si
feldspar	fsp	$(CaAl, NaSi, KSi)AlSi_2O_8$
Ca-pyroxene	Ca-px	$Ca(Mg,Fe,Ti^{4+},Al,Si)_3O_6$
orthopyroxene	opx	$MgSiO_3$-$FeSiO_3$
rhombohedral oxide	rhm oxide	$(Mg,Fe^{2+},Ti^{4+},Mn)_2O_3$
pyrophanite	—	$MnTiO_3$
whitlockite	wt	$Ca_3(PO_4)_2$
troilite	—	FeS
oldhamite	—	CaS
osbornite	—	TiN
graphite	—	C
cohenite	—	Fe_3C
Suppressed:		
cordierite	—	$Mg_2Al_4Si_5O_{18}$
sapphirine	—	$Mg_4Al_8Si_2O_{20}$

Formulae are restricted to solid solution ranges present in calculations (cf. *Ebel and Grossman,* 2000). Ferric iron (Fe^{3+}), although present in some solution models (spinel, Ca-px, rhm-ox), is insignificant in all calculated results shown here.

at fixed P^{tot}, following descending vertical paths. It must be emphasized that temperature is only *one* state variable affecting stable phases; pressure is just as important, as is the bulk vapor composition. Boundaries mark the conditions at which particular condensed phases either appear (become stable), or disappear along such cooling paths. Stable mineral assemblages, in *phase fields*, are listed from most- to least-refractory phase where space permits (e.g., Plate 7). In all cases, vapor is present, and it should be remembered that at most temperatures shown, major elements are continually condensing from gas into solid and/or liquid phases with cooling: The condensate assemblage, alone, does *not* represent a closed system at any point in any diagram.

Although models for solid solutions (Table 1) contain many potential compositions, less-refractory endmembers are not commonly stable in the systems investigated here. Feldspar is nearly always pure anorthite $CaAl_2Si_2O_8$, olivine pure forsterite Mg_2SiO_4 with minor Ca. Melilite and Ca-pyroxene show the greatest variation in composition at high temperatures (cf. *Ebel and Grossman,* 2000).

3.1. Equilibria in a Gas of Solar Composition

3.1.1. Effect of pressure. Plate 7 illustrates equilibrium stability relations of vapor, solid, and liquid phases calculated by this author, with $10^{-8} < P^{tot} < 1$ bar, and $1100 < T < 2000$ K, in a closed system of solar composition (*Anders and Grevesse,* 1989). The method and data are those reported in *Ebel and Grossman* (2000), who found that the refractory minerals cordierite and sapphirine should become stable by reaction with other solids at T and P^{tot} below the temperature at which virtually all Mg, Al, Si, and Ca are condensed. Cordierite and sapphirine are extremely rare in meteorites (*Sheng et al.,* 1991; *Rubin,* 1997), they are calculated to be stable only at relatively low temperatures, and assemblages bearing them are only stable, relative to those reported here, by order $<10^{-2}$ J (e.g., at $P^{tot} = 10^{-4}$, 8×10^{-4} J at 1300 K, 3×10^{-3} at 1200 K) relative to other solids (i.e., well within error); they have been omitted from all the calculations reported here. Although Na-,Fe-free cordierite is calculated to be near the minimum free-energy configuration, the cordierite found by *Fuchs* (1969) has 3–6 wt% Na_2O. The absence of cordierite in meteorites suggests the importance of mineral reconstructive kinetics even at temperatures near the stability field of feldspar.

Plate 8 shows a subportion of Plate 7, illustrating phase relations involving feldspar at low P^{tot}. Phase boundaries are marked "in"/"out" to signify phase appearance/disappearance upon cooling at constant P^{tot}. A common occurrence in this and other results (e.g., liquid field in Plate 7) is the stability of a particular phase over two different T ranges at a fixed P^{tot}. This results from the continued condensation of major elements (e.g., Si) upon cooling (cf., *Yoneda and Grossman,* 1995, liquid field in their Fig. 5). The con-

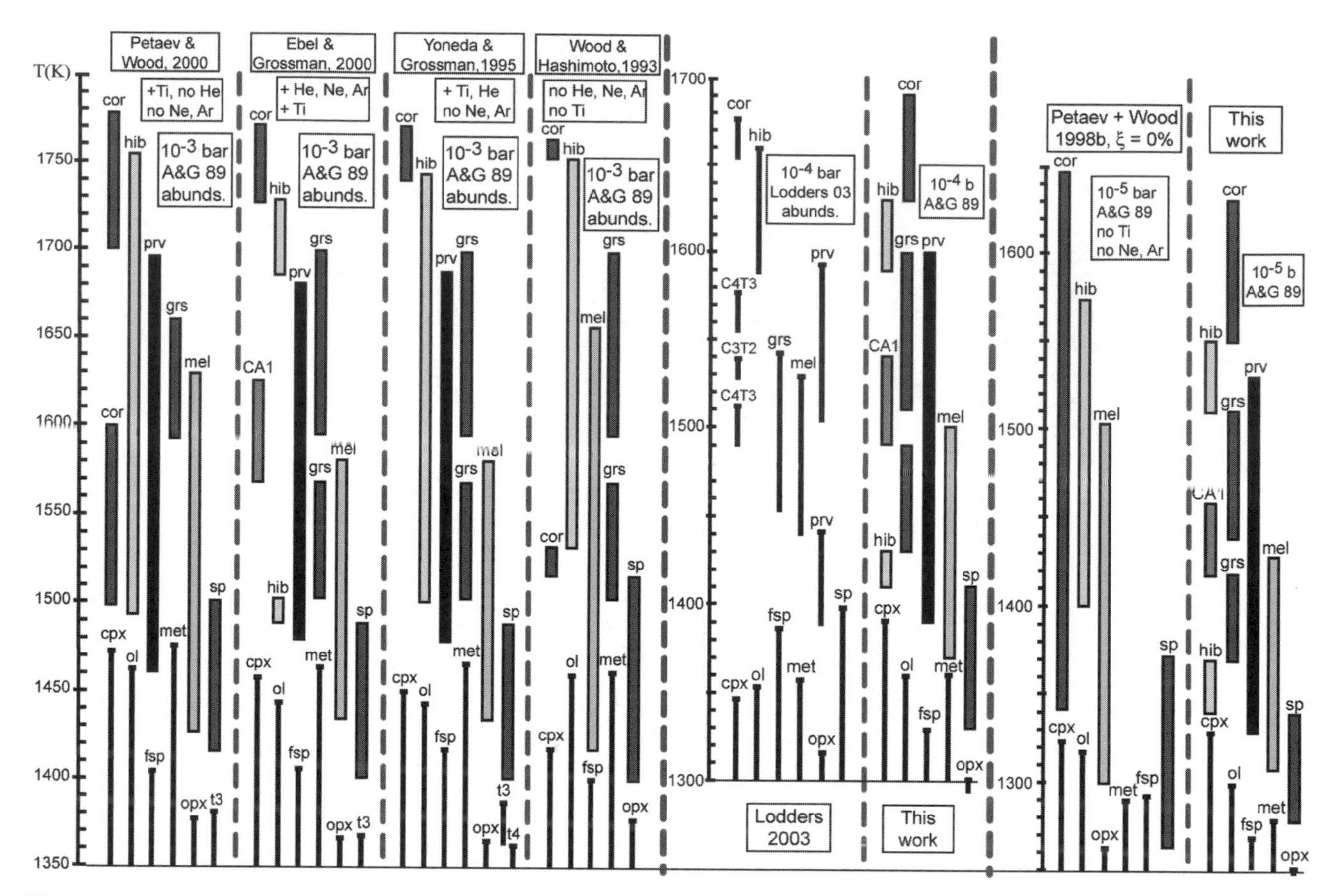

Fig. 1. Comparison of published results for vapor of solar composition (see section 3.1.2). Mineral abbreviations are t3 = Ti_3O_5, t4 = Ti_4O_7, C4T3 = $Ca_4Ti_3O_{10}$, C3T2 = $Ca_3Ti_2O_7$, sp = Al-spinel, cpx = Ca-pyroxene, and as listed in Table 1.

densate assemblages continually adjust so that the chemical potential energy of the system is minimized. Feldspar is nearly pure anorthite. Textures in CAIs can be interpreted to show that anorthite was thermodynamically stable at higher temperatures than olivine and other phases, inconsistent with calculated phase diagrams. *MacPherson et al.* (2004) discussed this phenomenon in the context of calculations showing anorthite condenses at a higher T than forsterite below $P^{tot} \sim 10^{-4}$ bar (their Fig. 2), with dramatic effect (their Fig. 3). Calcium-pyroxene is stable in all these assemblages. For the calculation by *MacPherson et al.* (2004) to predict the correct phase relations, the endmember pyroxene energies of *Sack and Ghiorso* (1994) must be innaccurate, but must be fortuitously corrected by *MacPherson et al.* (2004) ignoring the mixing properties included in the pyroxene model of *Sack and Ghiorso* (1994) (see section 3.1.2). Results of a calculation using the MELTS implementation of that model (Plate 8) illustrate that a narrow high-T anorthite stability field does exist at very low $P^{tot} < \sim 10^{-6.8}$. If feldspar (anorthite) were to form in this T-P range, then it might survive metastably, relative to other phases, in a cooling system. Other arguments for anorthite stability have been discussed by *Weisberg et al.* (2005).

3.1.2. Comparison with other results. Several recent calculations of equilibrium condensation are compared in Fig. 1. Temperatures of mineral phase appearance and disappearance are shown for three specific P^{tot} (bar) published as tables by *Wood and Hashimoto* (1993) and *Petaev and Wood* (1998b), who did not include Ti or noble gases; *Yoneda and Grossman* (1995), who included Ti and He but not Ne or Ar; *Petaev and Wood* (2000), with Ti but no noble gases; *Lodders* (2003) who included the entire periodic table, but did not publish disappearance temperatures of phases; and *Ebel and Grossman* (2000), whose methods were the same as this work. Elemental abundances are from *Anders and Grevesse* (1989) in all cases except for *Lodders* (2003). Inclusion or absence of noble (inert) gases slightly changes the effective P^{tot} in the calculation. The absence of Ti strongly affects phase stability, particularly by preventing perovskite formation. There are many subtle comparisons that could be made using these results; however, it is their overall similarity, particularly for the major phases found in chondrites, that is striking.

All three groups used different data for the chemical potential energy (Gibbs free energy) of endmember mineral phases, with varying degrees of documentation. Differences between *Yoneda and Grossman* (1995) and earlier work are detailed in that paper, and differences between that work and *Ebel and Grossman* (2000), and by extension the present work, are explained in detail in the 2000 paper. In particular, the subtle equilibria among the Ca-aluminates (CA1, hib, grs) among the papers by Grossman's group have all been documented to result from choices of basic data made for internal consistency with liquid solution models (*Ebel and Grossman,* 2000). The other major difference between *Ebel and Grossman* (2000), this work, and that of *Petaev and Wood* (2000) appears to result from the formers' use of the complete MELTS (*Sack and Ghiorso,* 1994) calculation engine for Ca-pyroxene. *Petaev and Wood* (1998a,b, 2000) use endmember data from, e.g., *Sack and Ghiorso* (1989, 1994), but they used ideal mineral-solution models for the mixing properties of mineral solid solutions between those endmembers (*Petaev and Wood,* 2000). *Lodders* (2003) did not consider solid solutions, so, for example, her "melilite" is pure gehlenite. For purposes of parameterizing condensation for astrophysical calculations, all these results are adequate, and that of *Lodders* (2003) is preferred for its complete coverage of all the (astrophysical) "metals" (see section 5.1). The work of *Ebel and Grossman* (2000), and the present calculations, are most consistent with models of liquid solutions at conditions where the latter are thermodynamically stable, and include full implementation of the internally consistent MELTS models for crystalline solutions that coexist with mafic (Mg-Fe-rich multicomponent silicate) liquids.

3.2. Equilibria in Carbon-enriched Systems

Reduced phases are common constituents of enstatite chondrites (e.g., CaS, MgS, etc.) (*Keil,* 1968, 1989; *Brearley and Jones,* 1998). Graphite, SiC, and TiC are common presolar grains (*Meyer and Zinner,* 2006). Both hydrous and anhydrous interplanetary dust particles (IDPs) are suffused by C-rich intergranular material (*Flynn et al.,* 2003), and CI chondrites contain up to 5 wt% C (*Lodders and Osborne,* 1999). A diversity of carbonaceous material was abundantly present in the protoplanetary disk (*Kerridge,* 1999), so exploration of mineral equilibria in hot, C-enriched systems of otherwise solar composition has been an active area of research (*Larimer,* 1975; *Larimer and Bartholomay,* 1979; *Wood and Hashimoto,* 1993; *Sharp and Wasserburg,* 1995; *Lodders and Fegley,* 1993, 1998).

Plate 9a illustrates stability relations among Ti-bearing phases, with other selected phases, shown in a narrow range of C/O where equilibrium chemistry changes dramatically (cf. *Lodders and Fegley,* 1997, their Fig. 3). At $C/O < 0.85$, equilibria are grossly similar to $C/O \sim 0.43$ (solar). Only Ti-bearing phases are listed in most fields, while phase boundaries of Ti-free phases x upon cooling at constant C/O are labeled "+x." The calculations were done using the data and methods of *Ebel and Grossman* (2000). As the C/O ratio approaches unity, reduced phases become stable (e.g., CaS, TiN, TiC). At $C/O > 1$, carbides and graphite are stable at the highest temperatures, and graphite condenses at progressively higher T as C/O increases (Plate 9b). Oxides (e.g., Mg-olivine, spinel) are also stable at high C/O, because condensation of graphite liberates O from gaseous CO. Metal becomes much more Ni-, Si-, and Co-rich than under oxidizing conditions. Intermetallic Fe-Si alloys are not treated rigorously in the calculation, but information on these compounds (e.g., FeSi, $FeSi_2$) does exist (e.g., *Meco and Napolitano,* 2005). Unfortunately, while some data exist on binary solutions, data is sparse regarding the thermochemical behavior of more complex reduced solid solutions [e.g., (Ti, Si, Zr)C, (Ti, Al, W)N, (Ca,Fe,Mg,Mn)S].

3.3. Equilibria in Dust-enriched Systems

Some mixture of dust grains is expected to accumulate at the midplane of the protoplanetary disk, and this dust may be the precursor material for chondrule and/or CAI formation. *Ebel and Grossman* (2000) investigated enrichment of the solar composition in a dust approximating the composition of the meteorite Orgueil (*Anders and Grevesse,* 1989). Chondritic dust is a logical assumption, given the chondritic abundances observed in bulk protoplanetary material. Various other kinds of dust can be supposed, for example, a C-rich dust similar in composition to unequilibrated, anhydrous, interstellar organic- and presolar-silicate-bearing cluster IDPs (*Ebel and Alexander,* 2005). Local vaporization of micrometer-sized dust during rapid heating events (e.g., *Desch and Connolly,* 2002; *Joung et al.,* 2004) might affect chondrule composition by suppressing evaporation (*Galy et al.,* 2000; *Alexander et al.,* 2000). Vaporized dust would promote preaccretionary metasomatic alteration, particularly affecting chondrule mesostasis (*J. Grossman et al.,* 2002), or forming refractory alkali-bearing minerals in CAIs (e.g., sodalite) (*Allen et al.,* 1978; *McGuire and Hashimoto,* 1989; *Krot et al.,* 1995; *Nagahara,* 1997).

Plate 10 illustrates the assemblages of solid and liquid phases that are calculated to be thermodynamically stable relative to vapor, at fixed total pressure, $P^{tot} = 10^{-3}$ bar, as functions of temperature (K) and enrichment in chondritic dust (cf. *Ebel and Grossman,* 2000). The oxides become thermodynamically stable at progressively higher temperatures with increasing dust enrichment, leading to the stability of liquids, and substantial FeO dissolved in silicates (*Ebel and Grossman,* 2000, their Fig. 8). Detailed results such as calculated oxygen fugacity, FeO content of silicates, composition of metal alloy, and melilite composition are presented in *Ebel and Grossman* (2000) for dust enrichment increments of 100×. Cordierite and sapphirine are calculated to be stable below 1330 K, at dust enrichments <20×, but they are entirely omitted from this calculation for reasons detailed above. Calculations to even higher dust enrichments, essentially for a pure chondritic dust, were presented by *Ebel* (2001).

4. METEORITIC EVIDENCE FOR CONDENSATION

4.1. Volatility-related Differences Between Chondrites

Bulk compositions of chondrites are reviewed in detail by *Hutchison* (2004), who offers a somewhat strained dichotomy of causal models for differences between them (his Table 7.1). He concludes that "the chemical compositions

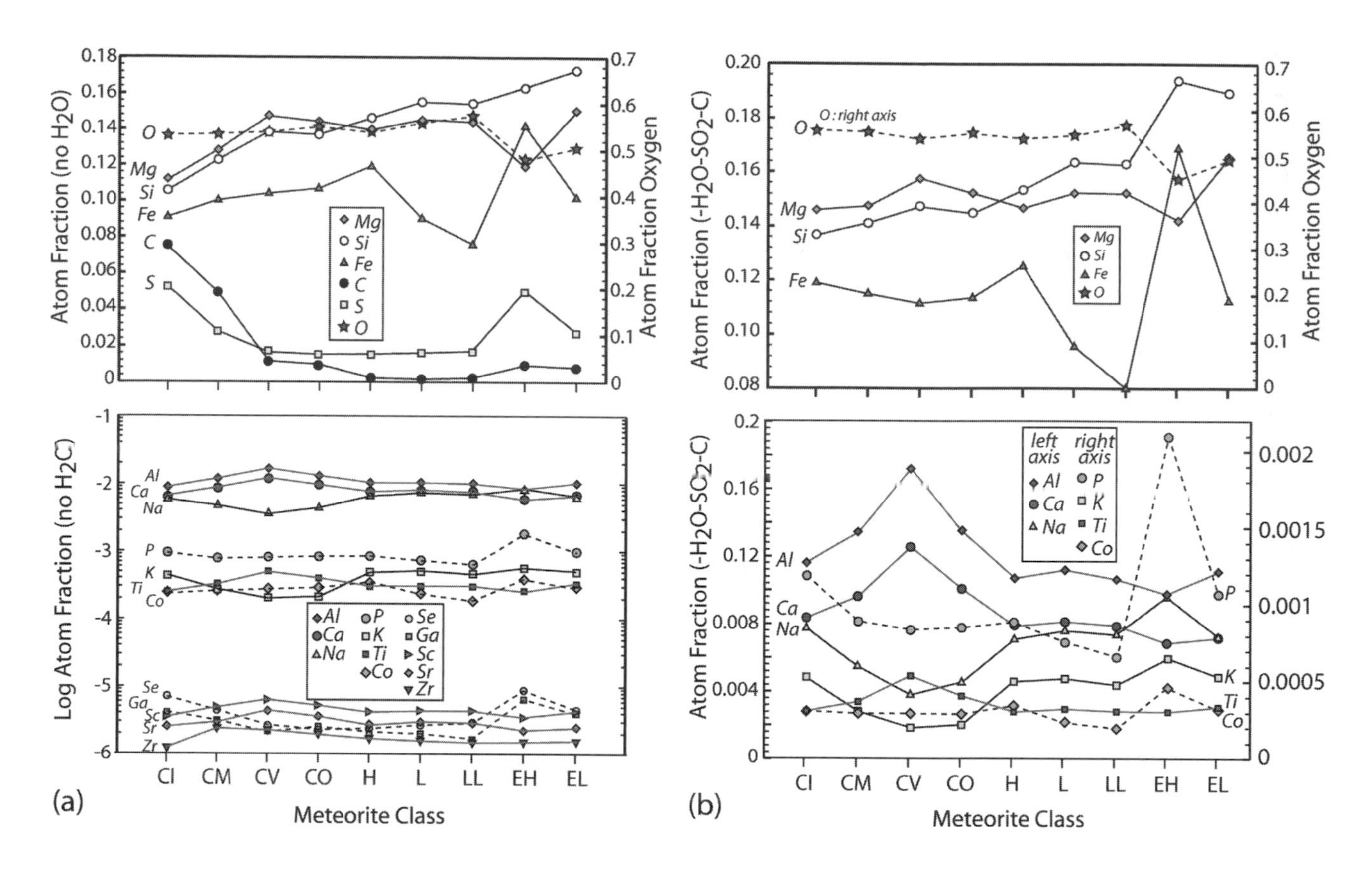

Fig. 2. Volatility-related elemental fractionation among chondrite groups, for selected elements with different cosmochemical affinities. **(a)** All H and O = 0.5*H subtracted. **(b)** All C and S (as SO_2) subtracted.

of the classes and groups of chondrites were established before chondrule formation. Elemental fractionation correlates with volatility . . . ," when CI chondrites are considered to be the most primitive, basic composition, due to their closest equivalence to the composition observed in the solar photosphere (*Hutchison,* 2004, p. 188). Regarding the constituents of chondrites, *Chiang* (2002) characterizes chondrule (and CAI) formation as "a zeroth order" unsolved problem. *Hutchison* (2004, p. 192) considers the problem "even more intractable and emotive than identifying the mechanisms for fractionating CI chondrite material to make the chondrite groups." Chondrite compositions are not simply additive combinations of Mg-Si chondrules, metal, and Ca-,Al-rich inclusions. Instead, some complementarity must be invoked in which chondritic material was fractionated, based on relative volatilities of elements, and then chondrules, etc., formed from these fractionated batches (*Hutchison,* 2004, p. 193; *Palme,* 2001).

One approach to deciphering the elemental differences among the chondrites, and comparing them to planetary bodies, is to assume that volatile elements were added as discrete components. An obvious first choice is water ice (H_2O). Figure 2a illustrates the abundances of selected elements [atomic fraction of total; calculated from data of *Wasson and Kallemeyn* (1988); cf. *Hutchison* (2004)], with all H and O = 0.5 × H removed. The relative abundances of C and S are still very high in CI and CM chondrites, relative to the other classes (cf. *Weisberg et al.,* 2006), with complementary depletion in Si, Mg, and Fe. Relatively volatile alkalies, and Se and Ga, are enriched in the same classes. Large variations in Fe are seen in the enstatite chondrites, particularly EH, and these are reflected in siderophile Co, Ga, and P, and chalcophile Se. These data suggest that Fe alloy fractionation, perhaps even prior to chondrule formation, contributed to the differences between these classes.

Figure 2b shows selected element abundances, recalculated after subtraction of all C as pure C (consistent with studies of interplanetary dust), and all S as SO_2, which may be less valid because it is not clear how important FeS was as a chondrite precursor. This recalculation removes most bias in illustrating the enrichment of CV and CO chondrites in highly refractory Al, Ca, Ti, Sc, and Zr. This is the major volatility-related fractionation seen in carbonaceous chondrites, and extends to other highly-refractory elements not shown here. The variation in alkalies (Na, K) is also likely to be related to their volatility. Variation in the siderophile elements (Fe, Co, P) is most pronounced among the ordinary (H, L, LL) and enstatite (EH, EL) chondrites. The cause of abundance variations of the siderophile elements among chondrites is not clear to this author. Loss (or gain) of metal grains, carrying siderophile elements in chondritic relative abundances, has been hypothesized, but no cause is clearly established. More volatile chalcophile elements ("S-loving," e.g., Se, Sb) would go along with these metal grains, if they were condensed with them. Alternatively, some fraction of the chalcophile component, and also of similarly volatile elements up to the temperatures of stability of solids containing alkalies, was added to the chondritic components (chondrules, CAIs, metal grains, matrix) before (*Allen et al.,* 1978) or during the components' accretion into meteorite parent bodies.

4.2. Meteorite Components

A wide variety of inclusions and matrix grains is found in meteorites, comprehensively reviewed by *Brearley and Jones* (1998). Many have been hypothesized to be condensates, or to have equilibrated with vapor at high T in the protoplanetary disk or nebula. These include, but are not limited to, (1) "fluffy" type A (FTA) inclusions (*MacPherson et al.,* 1984; *Beckett,* 1986; but cf. *MacPherson,* 2003), spinel-pyroxene aggregates (*Kornacki and Fegley,* 1984; *MacPherson et al.,* 1983), hibonite-bearing objects (*MacPherson et al.,* 1983; *Hinton et al.,* 1988; *Ireland et al.,* 1991; *Russell et al.,* 1998; *Simon et al.,* 1998, 2002; *MacPherson,* 2003), and other highly refractory inclusions (*Bischoff et al.,* 1985); (2) CAIs and chondrules (*Wood,* 1963; *Grossman and Clark,* 1973; *Wänke et al.,* 1974; *Beckett and Grossman,* 1988; *Boss,* 1996; *Lin and Kimura,* 2000; *Scott and Krot,* 2001; *Kurat et al.,* 2002; *Lin et al.,* 2003), particularly those with unfractionated REE abundance patterns (*Russell et al.,* 1998); (3) minerals deposited in voids in type B CAIs (*Allen et al.,* 1978); (4) amoeboid olivine aggregates (*Grossman and Steele,* 1975; *Komatsu et al.,* 2001; *Aléon et al.,* 2002; *Krot et al.,* 2004a; *Weisberg et al.,* 2005); (5) metal grains in CB and CH chondrites (*Meibom et al.,* 2001; *Petaev et al.,* 2001, 2003; *Campbell et al.,* 2001), and metal in CR chondrules (*Connolly et al.,* 2001); (7) PGE nuggets (*El Goresy et al.,* 1979; *Palme and Wlotzka,* 1976) or opaque assemblages [a.k.a.-Fremdlinge (*Sylvester et al.,* 1990)] in CAIs; (8) relict hibonite (*El Goresy et al.,* 1984) and fassaite (*Kuehner et al.,* 1989; *Lin and Kimura,* 2000) in CAIs and relict olivine (with included pockets of silicate melt) in chondrules and matrix (*Olsen and Grossman,* 1978; *Steele,* 1986, 1995; *Weinbruch et al.,* 2000; *Pack et al.,* 2004, 2005); (9) fine-grained components in matrix and inclusion rims (*Wood,* 1963; *Wark and Lovering,* 1977; *Huss et al.,* 1981; *Kornacki and Wood,* 1984; *Brearley,* 1993; *Greshake,* 1997; *Ruzicka,* 1997; *Bischoff,* 1998; *Wark and Boynton,* 2001; *Zolensky et al.,* 2003); (10) xenoliths in Allende (*Kurat et al.,* 1989); and (11) FeO-rich matrix olivine (*Weisberg et al.,* 1997; *Weisberg and Prinz,* 1998). Some fraction of IDPs, probably sourced in comets, may be composed almost entirely of condensates (*Rietmeijer,* 1998).

There is intense debate about what the extinct radioisotopic systematics (*MacPherson et al.,* 1995; *Russell et al.,* 1998, 2006; *McKeegan et al.,* 2000; *Guan et al.,* 2000), O-isotopic signatures (*McKeegan et al.,* 1998; *Young and Russell,* 1998; *Krot et al.,* 2002; *Young et al.,* 2002; *Clayton,* 2004), and refractory mineralogies (*Grossman,* 1975; *Simon et al.,* 1997, 1998, 2002; *Russell et al.,* 1998) of CAIs imply about their formation histories, as reviewed by *MacPherson* (2003) and in this volume. Fractional condensation, condensation followed by gas-phase metasomatism, and gas-solid equilibrium interrupted by melting events all seem to have affected the most refractory inclusions (e.g., *Stolper and Paque,* 1986). For example, condensation and aggregation of solids to make fluffy type A CAIs, followed by reaction with vapor in the Ca-pyroxene stability field, followed by melting, may produce type B CAIs [e.g., *Beckett* (1986); perhaps at ② in Plate 7]. Evaporation of many type B CAIs is evident in their heavy isotope enrichment (*Clayton et al.,* 1988), and the effect of evaporation on these objects is receiving much attention (e.g., *Nagahara and Ozawa,* 2000; *Galy et al.,* 2000; *L. Grossman et al.,* 2002; *Richter et al.,* 2002; *Davis,* 2003; *Alexander,* 2004). The precursor dust aggregates of many melted CAIs and chondrules are highly likely to correspond in bulk composition to some assemblages predicted in Plates 7 or 10. Predictions of other inclusion compositions (e.g., radiating pyroxene chondrules) require very extraordinary conditions (*Ebel et al.,* 2003).

Chondrules and igneous CAIs can be thought of as part of a continuum including Al-rich objects (*Ebel and Grossman,* 2000, their Figs. 16 and 17; *MacPherson,* 2003; *MacPherson and Huss,* 2005). Principal component analyses (*Grossman and Wasson,* 1983) and other evidence (*Bischoff and Keil,* 1984; *Weisberg and Prinz,* 1996) indicate these inclusions formed by rapid melting of mixtures of precursor components. The precursors included refractory, Ca-,Al-rich material; less-refractory Mg-,Si-rich material; metal alloy and/or metal sulfide; and a more-volatile, Na- and K bearing component (*Grossman,* 1996). Some CAIs may bear trace-element signatures that were established during stellar condensation of their precursor grains (*Ireland,* 1990). The sizes and relative proportions of inferred precursors were highly variable, and some may be preserved as relict grains, for example, olivine (e.g., *Steele,* 1986; *Jones,* 1996; *Pack et al.,* 2005) and spinel (e.g., *Misawa and Fujita,* 1994; *Maruyama and Yurimoto,* 2003) in chondrules (cf. *Misawa and Nakamura,* 1996); perovskite in CAIs (e.g., *Stolper,* 1982; *Beckett,* 1986). From a cursory inspection of Plates 7 or 10, these precursor components are consistent with progressive, volatility-controlled fractionation. The very existence of CAIs, chondrules, and more volatile-rich matrix demonstrates that such a fractionation process isolated these objects from thermochemical interaction in some locations prior to their accretion into chondrites.

Disequilibrium processes are implicated in the formation of all the meteoritic material listed above. Chemical zoning is frequently observed in the crystals formed from melt in igneous objects. Some chondrules contain heterogeneities in alkaline element contents and silicate mineral zoning that may have resulted from gas-melt interactions (continued condensation) during cooling (*Matsunami et al.,* 1993; *Tissandier et al.,* 2002; *Krot et al.,* 2004a,b; *Alexander and Grossman,* 2005). Matrix assemblages of amorphous + crystalline silicates have been attributed both to disequilibrium condensation (amorphous) followed by crystallization during annealing (e.g., *Brearley,* 1993), or to equilibrium condensation (crystalline) followed by amorphitization due to radiation damage (*Zolensky et al.,* 2003). Fractional evaporation may result in PGE-rich nuggets (*Cameron and Fegley,* 1982), and is certainly important in forming some CAIs (*Davis,* 2003). Partial recondensation has been invoked in several contexts, for example, to explain compositions of rims on CAIs (*Wark and Boynton,* 2001).

4.3. Refractory Aluminum-Calcium-Silicon-Magnesium Phases

4.3.1. Calcium-aluminates. In most cosmochemically plausible protoplanetary chemical systems with C/O < 1, where liquid is not thermodynamically stable, corundum (Al_2O_3) is the highest-T solid to condense (Plates 7, 9a, and 10; Fig. 1). Nearly pure corundum, enclosed in hibonite, with perovskite or grossite, has been reported in five inclusions (*Bar-Matthews et al.,* 1982; *Fahey,* 1988; *Krot et al.,* 2001). At least one (*Simon et al.,* 2002) is convincingly not melted, nor formed by reactions among less-refractory components, nor formed by evaporation (*Floss et al.,* 1998). The grossite, perovskite, and spinel ($MgAl_2O_4$) found with hibonite in the most refractory CAIs are nearly pure phases (Table 1) (e.g., *Weber and Bischoff,* 1994; *Simon et al.,* 1994; *Brearley and Jones,* 1998; *Aléon et al.,* 2002). Calcium-monoaluminate has been reported in exactly one occurrence, with grossite, perovskite, and melilite (*Ivanova et al.,* 2002). These assemblages are consistent with predictions in Plate 7 for fractional condensation at high T and low P^{tot}, and at small dust enrichments (Plate 10).

Hibonite is calculated to condense after corundum in the absence of liquid in systems with C/O < 1. Hibonites in CAIs can contain up to 6 wt% $TiO_2 + Ti_2O_3$ (*MacPherson and Grossman,* 1984; *Brearley and Jones,* 1998; *Aléon et al.,* 2002). The Ti^{+3}/Ti^{+4} ratio in hibonite is a sensitive indicator of oxygen fugacity (*Ihinger and Stolper,* 1986; *Beckett et al.,* 1988). Spherules with nearly pure hibonite, surrounded by spinel, with or without perovskite, are rare components of several meteorite types that have been vigorously investigated (*MacPherson et al.,* 1983; *Hinton et al.,* 1988; *Ireland et al.,* 1991; *Floss et al.,* 1998; *Simon et al.,* 2002). Textures of some inclusions [e.g., SH-6 of *MacPherson et al.* (1984)] provide compelling evidence of vapor-solid condensation of euhedral hibonite plates, and subsequent reaction with vapor to directly form spinel. *Beckett and Stolper* (1994) argued, based on experimental results, that hibonites in the core regions of melted Allende type A CAIs are surviving precursor grains.

4.3.2. Calcium-,aluminum-silicates. Melilite and Ca-rich pyroxene are, with spinel, the most abundant constituent minerals of CAIs that are found in all chondrite classes except CI (*Brearley and Jones,* 1998; *Fagan et al.,* 2000; *MacPherson,* 2003; *Beckett et al.,* 2006). They sometimes contain measurable Sc and Zr (*El Goresy et al.,* 2002; *Simon et al.,* 1991, 1996), and are highly depleted in more volatile elements (Na, Fe). Zoning in melilite crystals can be diagnostic of their origin, whether they crystallized from a closed-system melt (igneous origin), a cooling melt in chemical equilibrium with cooling vapor (metasomatic igneous), or directly from cooling vapor either fractionally or in equilibrium (condensation *senso stricto*). Anorthite appears to have formed by a vapor-solid reaction in many CAIs.

The least-melted CAIs record complex processes in the textural relationships among the mineral phases. These textures are disequilibrium phenomena that record sequential, incomplete, and partial equilibration of minerals with surrounding vapor. *MacPherson* (2003) and *Beckett et al.* (2006) have reviewed Wark-Lovering rim mineralogy and reverse zoning of melilite in CAIs. An unresolved problem is the enclosure of spinel in melilite in most type A CAIs, because spinel is calculated to form at lower T than melilite in most conditions (Plates 7–10; Fig. 1). *Beckett and Stolper* (1994) argued that spinel growth on a hibonite substrate would be kinetically favored (over melilite growth) based on their crystallochemical similarities. Alternatively, *Beckett* (1986) suggested that removal of hibonite-bearing objects rich in Al and the most refractory REE would both suppress melilite condensation relative to spinel, and also explain Group II REE patterns.

In the condensation calculations shown here, the Al-,Ti-rich pyroxenes formerly known as "fassaite" (now, subsilicic titanoan aluminian pyroxene) become stable at higher T than diopside ($CaMgSi_2O_6$), consistent with the experiments by *Stolper* (1982). The calculated compositions of those fassaites cannot be accurate, however, given the lack of Ti^{+3} in the model pyroxene. Although Ti^{+4} does not readily enter tetrahedral sites in silicates, the behavior of the reduced species Ti^{+3} is relatively unknown, and could have significant effects on fassaite stability relative to other phases. *Yoneda and Grossman* (1995) modified the estimates of pyroxene thermodynamic properties by *Beckett* (1986), and assumed ideal solution, to simulate condensation of fassaite from solar and dust-enriched vapors at various total pressures (P^{tot}) below 1 bar. They condensed fassaite at slightly lower temperature than melilite, but their predicted compositions (their Fig. 9) correspond poorly to observed CAI fassaites (e.g., *Beckett,* 1986, Table 6; *Simon et al.,* 1998). Scandium may also stabilize pyroxenes at high T. The thermodynamic properties of refractory Ca-,Al-,Ti-, Sc-pyroxenes remain mysterious. Sequential, incomplete equilibration of CAI pyroxenes with surrounding vapor is indicated by their chemical zoning (*Brearley and Jones,* 1998), particularly in CAI rims. Texturally late FeO-enrichment probably results from aqueous alteration of rim pyroxenes on chondrite parent asteroids.

4.3.3. Rare earth elements. Hibonite is the mineral stable at the highest temperatures that incorporates significant REE (*MacPherson and Davis,* 1994; *Ireland and Fegley,* 2000; *Simon et al.,* 2002; *Lodders,* 2003). Rare earth element enrichments from 5 to 100×, and 0.6 to 5×, CI chondritic abundances are seen, respectively, in CAIs (*Russell et al.,* 1998) and in rapidly quenched chondrules (*Engler et al.,* 2003) with unfractionated REE patterns. These objects may retain the full complement of refractory lithophile REEs condensed directly into their precursors. Other CAIs and chondrules have patterns depleted in the more refractory REE (Group II pattern) (cf. *Mason and Taylor,* 1982). These depletions have been interpreted to result from the earlier condensation and removal of an ultrarefractory component that primarily retained the more-refractory REEs (*Boynton,* 1975, 1978; *Simon et al.,* 1994; *Ireland and Fegley,* 2000). Their volatility-related REE abundances, other isotopic constraints, and high melting temperatures have been seen as evidence for CAI and chondrule origins as condensates near the Sun or in the Sun's photosphere, and subsequent transport outward by stellar winds (*Shu et al.,* 2001; *Ireland and Fegley,* 2000). This idea resonates with the reasoning, before any knowledge of REEs, of *Sorby* (1877), who argued that "some at least of the constituent particles of meteorites were originally detached glassy globules, like drops of fiery rain."

4.4. Metal Alloy and Ferromagnesian Silicates

4.4.1. Metal and siderophile elements. Metallic Fe-Ni alloy is found as subgrains in presolar rocks (*Bernatowicz et al.,* 1999) and is common in chondritic meteorites. The chemical state of Fe (metal, sulfide, oxide) in chondrites, and the relative abundance of Fe among H, L, and LL ordinary chondrites, may result from fractionation of metals and S caused by their relative volatilities or susceptibility to oxidation and reduction in different regions of the protoplanetary disk. Efforts to trace these processes are focussed on the behavior of trace elements (siderophile) that dissolve in Fe (e.g., *Blum et al.,* 1989). Some of the platinum-group elements (PGE) are extremely refractory (Mo, Ru, Os, W, Re, Ir, and Pt), others less so (Fe, Ni, Pd, Co, Rh, Au, Cu) (*Palme and Wlotzka,* 1976; *Humayun et al.,* 2002). The redox behavior of the siderophiles varies as well. Assemblages of PGE-rich grains in CAIs have chondritic (solar) abundances of metals, although individual nuggets are highly heterogeneous (*MacPherson,* 2003).

Zoned metal grains in some CB and CH chondrites have core-to-rim gradients in composition that are correlated to the relative volatilities of the metals. *Campbell et al.* (2001) could not reconcile zoning patterns with equilibrium fractional condensation. Yet these zoned grains have been interpreted by others to form by sequential equilibration with vapor, in calculations that account for diffusive processes (*Meibom et al.,* 2001; *Petaev et al.,* 2003). These models are capable of quantitatively matching many features of the metal grains, but they require a different set of parameters for each grain (*Petaev et al.,* 2001). Exactly how these zoned metal grains formed is uncertain, pending better diffusion data on PGEs in metal.

Metal grains in chondrules have also been investigated for evidence of interaction with vapor. Recondensation is proposed to explain metal grains rimming CR chondrules, and enriched in more volatile siderophile elements: Initially FeS- and FeO-bearing chondrules are heated and partially evaporate; crystals coarsen; metal is reduced and forms alloy grains, while volatile-enriched metal vapor recondenses as rims (*Kong and Palme,* 1999; *Zanda et al.,* 2002). Alternatively, reduction occurs by equilibration with either a reducing gas, or C present with the precursors (*Huang et al.,* 1996; *Hewins et al.,* 1997). Only a few weight percent C are necessary to reduce most of the fayalite in typical chondrules (*Connolly et al.,* 1994). *Connolly et al.* (2001) concluded from experimental work (cited therein) that vapor is a less likely reductant than included C (cf. *Hanon et al.,* 1998). *Lauretta et al.* (2001) came to a similar conclusion, and estimated 4.6 wt% Si in chondrule metal formed by this process in ordinary chondrites, far higher than would condense in an oxidizing (solar or dust-enriched) environment (*Grossman et al.,* 1979; *Ebel and Grossman,* 2000). They went further, and assumed volatile (O, S, and Na) addition after metal formation, through interaction with surrounding vapor at lower T, and used commercially available code [HSC (*Eriksson and Hack,* 1990)] to predict the corrosion of Si-, P-, Ni-bearing metal to silica (SiO_2), troilite (FeS), and other fine-grained phases they observed in SEM images. One difficulty with the *in situ* C-reduction hypothesis is the absence of C from metal in carbonaceous chondrites (*Mostefaoui et al.,* 2000).

4.4.2. Magnesium-(iron)-silicates. The elements Fe, Mg, and Si are ~10× more abundant than the more refractory Al and Ca. Olivine (Mg-orthosilicate, Table 1) has been called "astrophysical silicate," because the olivine species forsterite (Mg_2SiO_4) is the most refractory silicate to condense from a solar gas in quantity. For example, at $P^{tot} = 10^{-3}$ bar, ~0.0027% of all atoms are condensed at 1444 K when forsterite appears, and ~0.0150% are condensed as olivine, out of ~0.0165% total, at 1370 K. This small T (or, alternatively, P^{tot}) interval coincides with the transition from refractory inclusion (CAI) assemblages to chondrule mineral assemblages. It is unlikely that most meteoritic inclusions or their precursors passed through this interval in a smooth, monotonic way; they probably experienced fluctuations in T and P^{tot}.

Olivine is the most abundant silicate in most chondrites; however, there is only controversial evidence that any olivine is a primary condensate, or reached equilibrium with vapor (*McSween,* 1977; *Olsen and Grossman,* 1978; *Palme and Fegley,* 1990; *Steele,* 1986, 1995; *Weisberg et al.,* 1997; *Weisberg and Prinz,* 1996, 1998; *Weinbruch et al.,* 2000; *Pack et al.,* 2004). The FeO and Na_2O content of many meteoritic silicates has been altered during secondary processing on parent bodies, obscuring the signatures of preaccretionary processes (*Bischoff,* 1998; *Sears and Akridge,* 1998; *Krot et al.,* 2000; *Brearley,* 2003). If all Fe in a gas of solar composition (Mg/Si ~1.07) existed as metal, orthopyroxene (enstatite, $MgSiO_3$) would dominate the silicate assemblage, as it does in enstatite chondrites. In condensation from solar gas, FeO-bearing silicates only form by reaction of metal and metal sulfide with Mg-silicates and vapor at very low temperatures, where diffusion is slow. Alternatively, FeO-bearing chondrules may have formed from flash-melted FeO-bearing precursors that crystallized at low temperatures or in an oxidizing environment, without losing FeO by evaporation during melting. Negligible heavy Fe-isotopic enrichment is observed in chondrule silicates (*Zhu et al.,* 2001; *Alexander and Wang,* 2001; *Mullane et al.,* 2003). Flash melting would then have had to occur in a vapor enriched in Fe (*Galy et al.,* 2000; *Alexander,* 2001; *Ebel,* 2005). Similar issues are involved with the Na content of chondrules and CAIs (e.g., *Sears et al.,* 1996).

One way to increase the FeO content of silicates when they first become chemically stable is for the surrounding vapor to be enriched in O. *Ebel and Grossman* (2000, their Fig. 8) showed that above 1200 K, systems enriched in chondritic vapor yield ferromagnesian silicates with molar FeO/(FeO + MgO) ratios of 0.1–0.4, and also stabilize chondrule liquids against evaporation (Plate 10). Alternatively, enrichment of vapor in H_2O ice can stabilize FeO in silicates at high temperature (*Krot et al.,* 2000, their Fig. 6). Mechanisms capable of melting chondrule precursors [shock waves, *Desch and Connolly* (2002); current sheets, *Joung et al.* (2004)] would simultaneously vaporize small silicate dust grains, providing a local dust-enriched region during chondrule cooling.

4.5. Carbides, Sulfides, and Phosphides

Presolar grains have survived ejection from AGB stars or novae, interstellar processing (*Bernatowicz et al.,* 2006; *Nuth et al.,* 2006), presolar cloud collapse (*Boss and Goswami,* 2006) heating and irradiation in the early solar nebula (*Connolly et al.,* 2006; *Chaussidon and Gounelle,* 2006; *Chick and Cassen,* 1997), accretion into chondrite parent bodies (*Cuzzi and Weidenschilling,* 2006; *Weidenschilling and Cuzzi,* 2006), followed by aqueous alteration (*Brearley,* 2006), impacts (*Bischoff et al.,* 2006), lithification, and thermal metamorphism on the parent bodies (*Huss and Lewis,* 1995; *Zinner,* 1998; *Mendybaev et al.,* 2002; *Krot et al.,* 2006; *Huss et al.,* 2006). Graphite, SiC, TiC, Fe-Ni alloy, spinel, corundum, and other presolar phases are all refractory minerals predicted to condense from vapors with variable C/O ratio but solar proportions of other elements (Plate 9b) (*Ebel,* 2000). Isotopic constraints on grains formed in supernovae require mixing between different chemical zones of supernovae, if the grains condensed in systems approaching equilibrium (*Travaglio et al.,* 1999; *Ebel and Grossman,* 2001; *Meyer and Zinner,* 2006). Quantitatively addressing the condensation of these rocky materials would appear to require better models of zone mixing in supernova explosions, to reduce the universe of possible mixtures capable of producing the observed condensates.

Enstatite chondrites contain mixed phases in the FeS-MnS-MgS system (niningerite, alabandite), which may record primary equilibrium with vapor (*El Goresy et al.,* 1988; *Crozaz and Lundberg,* 1995; reviewed by *Brearley and Jones,* 1998). Metal (Plate 9b) is Si-rich, and the Si content of metal increases steadily as coexisting vapor becomes more highly reducing, until the stability fields of Fe-Si intermetallic compounds are reached at high C/O ratios of the bulk system. These compounds are not observed in meteorites. Other common phases include oldhamite, nearly pure CaS; troilite, FeS; daubreelite, (Fe, Mn, Zn)Cr_2S_4; schreibersite, $FeNi_3P$; and perryite, ~$(Ni,Fe)_2(Si,P)$. Other sulfides have been discovered in enstatite chondrites, but their cosmochemical significance is unknown (e.g., *Nagahara,* 1991). The role of vapor-solid equilibria in the origin of components of enstatite chondrites remains poorly constrained.

5. DISCUSSION AND CONCLUSIONS

5.1. The 50% Condensation Temperature

Wasson (1985, his Table G1) has compiled a table of "50% condensation temperatures" for the elements from previous calculations (e.g., *Wai and Wasson,* 1979; *Lodders,* 2003). This single metric has been useful in categorizing the elements by their volatility in a gas of solar composition at fixed P^{tot} (e.g., *Hobbs et al.,* 1993), but can be misleading because elements condense over different T ranges (*Ebel,* 2000). *Lodders* (2003) treated the entire periodic table in a self-consistent manner, to calculate 50% condensation temperatures at $P^{tot} = 10^{-4}$ bar, using her assessment of the most recent elemental abundance data for a vapor of solar composition (see section 3.1.2).

5.2. Contexts of Condensation and Gas-Vapor Equilibrium

Incontrovertible evidence for direct condensation of rocky meteoritic material from vapor to solid or liquid phases remains elusive, apart from micrometer- to submicrometer-sized presolar grains formed around other stars, probably in single outflow episodes. These grain sizes cannot be taken as exemplary of solar nebular processes. Except for some isotopic anomalies that may have resulted from local processes [e.g., O self-shielding (*Navon and Wasserburg,* 1985; *Clayton,* 2004)], the initial gas and dust of the protoplanetary disk appear to have been thoroughly mixed (*Palme,* 2001). Most grains that may represent direct vapor-solid condensates, even in unmelted IDPs, are isotopically solar at the spatial resolution of current instruments. It is not clear how this homogenization occurred, if the solar system was hot only inside ~3 AU, unless mixing in the accretion shock was highly efficient. Perhaps the interstellar medium, the ultimate source of precursor solids, is also mostly well-mixed, most grains are submicrometer-sized, and a "scoop" of them is essentially solar (*Brearley,* 1993). It is also possible that thin planar heating events, such as current sheets and shock waves, driven by magnetorotational and gravitational instabilities, were commonplace in different regions of the protoplanetary disk well beyond 3 AU. These forces would have driven repeated cycling of micrometer-sized solids through the vapor phase, building up a supply of ever-larger condensate grains, and depleting regions of the disk in the same volatility-controlled patterns observed in dense molecular clouds (*Ebel,* 2000). Resulting larger dust and dust aggregates would be isotopically close to solar if evaporating materials were unmelted. Both shocks and current sheets would produce local T excursions decreasing with radial distance from the central star. This is one context for direct condensation in the early solar system. Future astronomical observations of disks at high spatial resolution, by coronography and/or interferometry, should reveal whether such fractionation exists and at what radii from central stars. The precursors to chondrules and CAIs were probably similar to IDP components, isotopically solar in most respects, and fractionated into Ca-,Al-, Ti-rich oxides; Mg-,Si-rich silicates; metal; and weakly bonded components rich in volatile elements.

Rietmeijer (1998) has developed a theory of "hierarchical accretion" in explaining the textures and compositions of IDPs. Meteoritic inclusions experienced many episodes of heating to high temperatures, so their textures would not reveal hierarchical accretion as clearly as do IDPs. But the paradigm may apply most evidently to more refractory CAIs, and to AOAs, where sequential layers of minerals, mechanically accreted, were also subsequently changed by sequential metasomatic reactions with vapor. Many objects preserve layers of minerals of different volatility, as if they passed through regions rich in, and accreted rims of, dust that had been condensed at a particular temperature, P^{tot}, and vapor composition (*MacPherson et al.,* 1985). As the objects themselves occasionally encountered more than one such heating event, dust would progressively sinter and acquire an igneous texture, becoming part of the chondrule or CAI, and also reacting with gas, for example, a gas with higher P_{SiO} than encountered previously. Hierarchical accretion may be caught in the act, in concentrically zoned CAIs (e.g., *Simon et al.,* 1994), or in chondrules with thick layers of dusty rim material, and with concentric internal structure. Partial or complete remelting of such objects would sinter the rim or homogenize the entire object, even as it accreted still more dusty material. This scenario requires a high dust-to-gas ratio, with heating events evaporating nearby fine dust particles so that vapor pressures of O and condensable elements are locally increased around melted chondrules. This would suppress evaporative enrichment in heavy isotopes, make liquids stable against evaporation, and increase the stable FeO content of silicates. This kind of accretion, resulting in chondrules and CAIs, would occur in disk regions closer to the Sun, with higher densities of dusty material.

Vapor-solid-liquid equilibria appear to have been approached inefficiently, due to the rapid timescales of most heating events, resulting in the abundant evidence of disequilibrium in the heterogeneous and zoned mineral grains of so many meteoritic inclusions. It is also likely that kinetic and surface effects are important in stabilizing particular minerals relative to others, in direct gas-solid reactions (*Beckett,* 1986). Differences in thermochemical stability between these phases are small, so surface energy effects become important at small grain sizes. Surface forces may affect macroscopic grain compositions, as they do in sector-zoned crystals. Displacive reactions of preexisting, more refractory phases (e.g., melilite) with vapor during cooling might also be energetically favored, relative to reconstructive reactions (i.e., adjusting a few vs. breaking many bonds). This may explain textures in unmelted objects, in which anorthite appears to have formed at a higher T than olivine (e.g., *Krot et al.,* 2002; *Weisberg et al.,* 2005).

6. OUTLOOK

6.1. Observations of Meteorites

The equilibrium chemical behaviors of the abundant elements (Si, Fe, Ni, Mg, Al, Ca, Ti) are well understood except in highly reduced systems. Attention has shifted to the less-abundant elements. We can expect increasing application of new, high-resolution analytical techniques to elemental and isotopic abundances in particular components of meteorites, and correlation of these results with other evidence (e.g., *Mullane et al.,* 2001; *Young et al.,* 2002; *Kehm et al.,* 2003; *Friedrich et al.,* 2003; *Dauphas et al.,* 2004; *Johnston et al.,* 2004). New techniques of laser ablation ion-coupled plasma mass spectrometry (LA-ICPMS) enable investigation of ever-smaller spatial regions of metal grains (e.g., 30 μm × 15 μm deep) at very high mass resolution (*Campbell and Humayun,* 1999). The moderately volatile elements (e.g., Mn, Cu, Na, Se, Zn, Cd) are increasingly accessible to chemical and isotopic analysis in meteoritic materials. High compositional accuracy and fine spatial resolution of analyses of these and other trace elements will allow new constraints to be placed on the vapor-solid-liquid equilibria operative in the formation of chondritic components.

The search for extrasolar planets is driving astrophysical observations of *disks* around nearby young stars that will yield images of increasingly higher spatial and spectral resolution. It will be possible to ascribe radial gradients in elemental abundances of disk gas to grain formation and/or destruction. Better constraints on temperatures in evolving disks should also result from these observations. It is hoped that these results will have direct relevance to the origin of the rocky material in chondritic meteorites.

6.2. Theory

Condensation calculations relevant to meteorites are grounded in over a century of systematic experimental petrology, mineralogy, and basic research (e.g., *Bowen,* 1914; *Ferguson and Buddington,* 1920; *Robie et al.,* 1978; *Yoder,* 1979; *Chase,* 1998). Results of this difficult work, with modern computational tools, provide the means to calibrate mathematical models for the thermodynamic behavior of minerals and melts (*Charmichael and Eugster,* 1987; *Geiger,* 2001). These fields are perhaps not as actively supported as they once were. The public availability of high-quality primary experimental data by metallurgists and materials scientists appears to have declined since the 1970s. At the same time, the computer has enabled collation and evaluation of subsets of the mountain of existing data, resulting in open databases (e.g., *Berman,* 1988; *Kuzmenko et al.,* 1997) usually optimized for particular subsets of the cosmochemical universe [e.g., THERMOCALC for metamorphic petrology (*Holland and Powell,* 1998)]. Proprietary databases [e.g., FactSage (*Bale et al.,* 2002)] also exist, but their reliability is difficult to assess.

Strong backgrounds in thermodynamics and experimental petrology are necessary for critical assessment of existing data; for the building of optimized, internally consistent databases, both for equilibrium and kinetic approaches; for improvement of solid- and liquid-solution models applicable to extraterrestrial rocky materials; and for informed use of computational tools. Non-open-source, "black-box" software is increasingly available to perform equilibrium thermodynamic calculations with some degree of presumed rigor. These must be used with caution. Progress is most likely to result from the combination of kinetic constraints with thermodynamic models, in open-sourced codes. Early and rigorous training of students in thermodynamics, isotope chemistry, and experimental petrology must be the foundation for future progress in reading the solar system's earliest rocks.

6.3. Experiments

New techniques are being developed to directly probe condensation phenomena (e.g., *Toppani and Libourel,* 2003; *Toppani et al.,* 2005). The Knudsen cell has, belatedly, begun to see more application in petrologically relevant experiments (e.g., *Dohmen et al.,* 1998). The combination of laser-heating devices and quadrupole mass spectrometers to measure gas speciation promises to allow direct investigation of both condensation and evaporation. The parameters necessary to model evaporative mass fractionation are being discovered in the laboratory (*Davis et al.,* 1990; *Richter et al.,* 2002; *Dauphas et al.,* 2004). Amorphous grains 20–30 nm in size condense via homogeneous nucleation from hot gases quenched in short times (*Nuth et al.,* 2002). Their annealing to crystalline phases as a function of time and temperature has been quantified with application to observed AGB outflows, and to make inferences about protoplanetary disk grain cycling (*Brucato et al.,* 1999; *Nuth et al.,* 2002). Other recent work at slightly longer timescales has resulted in partially or fully crystalline olivine condensed from vapor (*Tsukamoto et al.,* 2001). Even longer timescales are being explored, in heterogeneously nucleated systems (*Toppani et al.,* 2004). Gas-grain reactions at lower temperatures (<600 K) are also being explored (*Llorca and Casanova,* 1998). These kinds of experiments become most relevant to the processes recorded by objects in meteorites, when combined with kinetic models grounded in an understanding of equilibrium phase relations. We can expect substantial progress both in direct laboratory explorations, and their understanding using better models, by the next edition of this volume.

Acknowledgments. This research has made use of NASA's Astrophysics Data System Bibliographic Services. This work was partially supported by the NASA Cosmochemistry program under grant NAG5-12855.

REFERENCES

Aikawa Y., Ohashi N., and Herbst E. (2003) Molecular evolution in collapsing prestellar cores. II. The effect of grain-surface reactions. *Astrophys. J., 593,* 906–924.

Aléon J., Krot A. N., and McKeegan K. D. (2002) Calcium-aluminum-rich inclusions and amoeboid olivine aggregates from the CR carbonaceous chondrites. *Meteoritics & Planet. Sci., 37,* 1729–1755.

Alexander C. M. O'D. (2001) Exploration of quantitative kinetic models for the evaporation of silicate melts in vacuum and hydrogen. *Meteoritics & Planet. Sci., 36,* 255–283.

Alexander C. M. O'D. (2004) Chemical equilibrium and kinetic constraints for chondrule and CAI formation conditions. *Geochim. Cosmochim. Acta, 68,* 3943–3969.

Alexander C. M. O'D. and Grossman J. N. (2005) Alkali elemental and potassium isotopic compositions of Semarkona chondrules. *Meteoritics & Planet. Sci., 40,* 541–556.

Alexander C. M. O'D. and Wang J. (2001) Iron isotopes in chondrules: Implications for the role of evaporation during chondrule formation. *Meteoritics & Planet. Sci., 36,* 419–428.

Alexander C. M. O'D., Grossman J. N., Wang J., Zanda B., Bourot-Denise M., and Hewins R. R. H. (2000) The lack of potassium-isotopic fractionation in Bishunpur chondrules. *Meteoritics & Planet. Sci., 35,* 859–868.

Allen J. M., Grossman L., Davis A. M., and Hutcheon I. D. (1978) Mineralogy, textures and mode of formation of a hibonite-bearing Allende inclusion. *Proc. Lunar Planet. Sci. Conf. 9th,* pp. 1209–1233.

Allende Prieto C., Lambert D. L., and Asplund M. (2002) A reappraisal of the solar photospheric C/O ratio. *Astrophys. J. Lett., 573,* L137–L140.

Altwegg G., Ehrenfreund P., Geiss J., and Huebner W. F., eds. (1999) *Composition and Origin of Cometary Materials.* Space Science Reviews, Vol. 90, Kluwer, Dordrecht. 412 pp.

Amari S., Nittler L. R., Zinner E., Lodders K., and Lewis R. S. (2001) Presolar SiC grains of type A and B: Their isotopic compositions and stellar origins. *Astrophys. J., 559,* 463–483.

Anders E. and Grevesse N. (1989) Abundances of the elements: Meteoritic and solar. *Geochim. Cosmochim. Acta, 53,* 197–214.

Anderson G. M. and Crerar D. (1993) *Thermodynamics in Geochemistry.* Oxford Univ., New York. 588 pp.

Arrhenius G. and Alfvén H. (1971) Fractionation and condensation in space. *Earth Planet. Sci. Lett., 10,* 253–267.

Bale C. W., Chartrand P., Degterov S. A., Eriksson G., Hack K., Mahfoud R. Ben, Melançon J., Pelton A. D., and Petersen S. (2002) FactSage thermochemical software and databases. *Calphad, 26,* 189–228.

Bar-Matthews M., Hutcheon I. D., MacPherson G. J., and Grossman L. (1982) A corundum-rich inclusion in the Murchison carbonaceous chondrite. *Geochim. Cosmochim. Acta, 46,* 31–41.

Beckett J. R. (1986) The origin of calcium-, aluminum-rich inclusions from carbonaceous chondrites: An experimental study. Ph.D. dissertation, University of Chicago, Chicago, Illinois.

Beckett J. R. and Grossman L. (1988) The origin of type C inclusions from carbonaceous chondrites. *Earth Planet. Sci. Lett., 89,* 1–14.

Beckett J. R. and Stolper E. (1994) The stability of hibonite, melilite and other aluminous phases in silicate melts: Implications for the origin of hibonite-bearing inclusions from carbonaceous chondrites. *Meteoritics, 29,* 41–65.

Beckett J. R., Live D., Tsay F-D., Grossman L., and Stolper E. (1988) Ti^{3+} in meteoritic and synthetic hibonite. *Geochim. Cosmochim. Acta, 52,* 1479–1495.

Beckett J. R., Connolly H. C., and Ebel D. S. (2006) Chemical processes in igneous calcium-aluminum-rich inclusions: A mostly CMAS view of melting and crystallization. In *Meteorites and the Early Solar System II* (D. S. Lauretta and H. Y. McSween Jr., eds.), this volume. Univ. of Arizona, Tucson.

Berman R. G. (1983) A thermodynamic model for multicomponent melts, with application to the system CaO-MgO-Al_2O_3-SiO_2. Ph.D. thesis, University of British Columbia, Vancouver. 178 pp.

Berman R. G. (1988) Internally consistent thermodynamic data for minerals in the system Na_2O-K_2O-CaO-MgO-FeO-Fe_2O3-Al_2O_3-SiO_2-TiO_2-H_2O-CO_2. *J. Petrol., 29,* 445–522.

Berman R. G. and Brown T. (1987) Development of models for multicomponent melts: Analysis of synthetic systems. In *Thermodynamic Modeling of Geological Materials* (I. S. E. Carmichael and H. P. Eugster, eds.), pp. 405–442. Reviews in Mineralogy, Vol. 17, Mineralogical Society of America.

Bernatowicz T., Bradley J., Amari S., Messenger S., and Lewis R. (1999) New kinds of massive star condensates in a presolar graphite from Murchison (abstract). In *Lunar and Planetary Science XXX,* Abstract #1392. Lunar and Planetary Institute, Houston (CD-ROM).

Bernatowicz T. J., Croat T. K., and Daulton T. L. (2006) Origin and evolution of carbonaceous presolar grains in stellar environments. In *Meteorites and the Early Solar System II* (D. S. Lauretta and H. Y. McSween Jr., eds.), this volume. Univ. of Arizona, Tucson.

Bettens R. P. A. and Herbst E. (1995) The interstellar gas phase production of highly complex hydrocarbons: Construction of a model. *Intl. J. Mass Spectrom. Ion Processes, 149/150,* 321–343.

Bischoff A. (1998) Aqueous alteration of carbonaceous chondrites: Evidence for preaccretionary alteration — A review. *Meteoritics & Planet. Sci., 33,* 1113–1122.

Bischoff A. and Keil K. (1984) Al-rich objects in ordinary chondrites: Related origin of carbonaceous and ordinary chondrites and their constituents. *Geochim. Cosmochim. Acta, 48,* 693–709.

Bischoff A., Keil K., and Stöffler D. (1985) Perovskite-hibonite-spinel-bearing inclusions and Al-rich chondrules and fragments in enstatite chondrites. *Chem. Erde, 44,* 97–106.

Bischoff A., Scott E. R. D., Metzler K., and Goodrich C. A. (2006) Nature and origins of meteoritic breccias. In *Meteorites and the Early Solar System II* (D. S. Lauretta and H. Y. McSween Jr., eds.), this volume. Univ. of Arizona, Tucson.

Blander M., Pelton A. D., Jung I-H., and Weber R. (2005) Non-equilibrium concepts lead to a unified explanation of the formation of chondrules and chondrites. *Meteoritics & Planet. Sci., 39,* 1897–1910.

Blum J. D., Wasserburg G. J., Hutcheon I. D., Beckett J. R., and Stolper E. M. (1989) Origin of opaque assemblages in C3V meteorites: Implications for nebular and planetary processes. *Geochim. Cosmochim. Acta, 53,* 543–556.

Boss A. P. (1996) A concise guide to chondrule formation models. In *Chondrules and the Protoplanetary Disk* (R. H. Hewins et al, eds.), pp. 257–264. Cambridge Univ., Cambridge.

Boss A. P. and Goswami J. N. (2006) Presolar cloud collapse and the formation and early evolution of the solar nebula. In *Meteorites and the Early Solar System II* (D. S. Lauretta and H. Y. McSween Jr., eds.), this volume. Univ. of Arizona, Tucson.

Bowen N. L. (1914) The ternary system: Diopside-forsterite-silica. *Am. J. Sci., 38,* 207–264.

Bowen N. L. (1928) *The Evolution of the Igneous Rocks.* Princeton Univ., Princeton. 332 pp. (Reissued by Dover Publications, 1956.)

Boynton W. V. (1975) Fractionation in the solar nebula: Condensation of yttrium and the rare earth elements. *Geochim. Cosmochim. Acta, 37,* 1119–1140.

Boynton W. V. (1978) The chaotic solar nebula: Evidence for episodic condensation in several distinct zones. In *Protostars and Planets* (T. Gehrels, ed.), pp. 427–438. Univ. of Arizona, Tucson.

Brearley A. (1993) Matrix and fine-grained rims in the unequilibrated CO3 chondrite, ALHA77307: Origins and evidene for diverse, primitive nebular dust components. *Geochim. Cosmochim. Acta, 57,* 1521–1550.

Brearley A. J. (2003) Nebular versus parent body processing. In *Treatise on Geochemistry, Vol. 1: Meteorites, Comets and Planets* (A. M. Davis, ed.), pp. 247–268. Elsevier, Oxford.

Brearley A. J. (2006) The action of water. In *Meteorites and the Early Solar System II* (D. S. Lauretta and H. Y. McSween Jr., eds.), this volume. Univ. of Arizona, Tucson.

Brearley A. and Jones R. (1998) Chondritic meteorites. In *Planetary Materials* (J. J. Papike, ed.), pp. 3-1 to 3-95. Reviews in Mineralogy, Vol. 36, Mineralogical Society of America.

Brucato J. R., Colangeli L., Mennella V., Palumbo P., and Bussoletti E. (1999) Mid-infrared spectral evolution of thermally annealed amorphous pyroxene. *Astron. Astrophys., 348,* 1012–1019.

Cameron A. G. W. (1963) Formation of the solar nebula. *Icarus, 1,* 339–342.

Cameron A. G. W. (1995) The first ten million years in the solar nebula. *Meteoritics, 30,* 133–161.

Cameron A. G. W. and Fegley B. Jr. (1982) Nucleation and condensation in the primitive solar nebula. *Icarus, 52,* 1–13.

Campbell A. J. and Humayun M. (1999) Trace element microanalysis in iron meteorites by laser ablation ICPMS. *Anal. Chem., 71,* 939–946.

Campbell A. J., Humayun M., Meibom A., Krot A. N., and Keil K. (2001) Origin of zoned metal grains in the QUE94411 chondrite. *Geochim. Cosmochim. Acta, 65,* 163–180.

Cassen P. (2001) Nebular thermal evolution and the properties of primitive planetary materials. *Meteoritics & Planet. Sci., 36,* 671–700.

Charmichael I. S. E. and Eugster H. P., eds. (1987) *Thermodynamic Modeling of Geological Materials: Minerals, Fluids and Melts.* Reviews in Mineralogy, Vol. 17, Mineralogical Society of America. 499 pp.

Chase M. W. Jr. (1998) *NIST-JANAF Thermochemical Tables, 4th edition.* Journal of Physical and Chemical Reference Data, Monograph No. 9, published for the National Institute of Standards and Technology by the American Chemical Society and the American Institute of Physics.

Chaussidon M. and Gounelle M. (2006) Irradiation processes in the early solar system. In *Meteorites and the Early Solar System II* (D. S. Lauretta and H. Y. McSween Jr., eds.), this volume. Univ. of Arizona, Tucson.

Chen Y. Q., Zhao G., Nissen P. E., Bai G. S., and Qiu H. M. (2003) Chemical abundances of old metal-rich stars in the solar neighborhood. *Astrophys. J., 591,* 925–935.

Chiang E. I. (2002) Chondrules and nebular shocks. *Meteoritics & Planet. Sci., 37,* 151–153.

Chick K. M. and Cassen P. (1997) Thermal processing of interstellar dust grains in the primitive solar environment. *Astrophys. J., 477,* 398–409.

Chigai T., Yamamoto T., and Kozasa T. (2002) Heterogeneous condensation of presolar titanium carbide core-graphite mantle spherules. *Meteoritics & Planet. Sci., 37,* 1937–1951.

Ciesla F. J. and Hood L. L. (2002) The nebular shock wave model for chondrule formation: Shock processing in a particle-gas suspension. *Icarus, 158,* 281–293.

Clayton R. N. (2004) The origin of oxygen isotope variations in the early solar system (abstract). In *Lunar and Planetary Science XXXV,* Abstract #1682. Lunar and Planetary Institute, Houston (CD-ROM).

Clayton R. N., Hinton R. W., and Davis A. M. (1988) Isotopic variations in the rock-forming elements. *Phil. Trans. R. Soc. London, A325,* 483–501.

Connolly H. C. Jr., Hewins R. H., Ash R. D., Zanda B., Lofgren G. E., and Bourot-Denise J. (1994) Carbon and the formation of reduced chondrules. *Nature, 371,* 136–139.

Connolly H. C. Jr., Huss G. R., and Wasserburg G. J. (2001) On the formation of Fe-Ni metal in Renazzo-like carbonaceous chondrites. *Geochim. Cosmochim. Acta, 65,* 4567–4588.

Connolly H. C. Jr., Desch S. J., Ash R. D., and Jones R. H. (2006) Transient heating events in the protoplanetary nebula. In *Meteorites and the Early Solar System II* (D. S. Lauretta and H. Y. McSween Jr., eds.), this volume. Univ. of Arizona, Tucson.

Crozaz G. and Lundberg L. L. (1995) The origin of oldhamite in unequilibrated enstatite chondrites. *Geochim. Cosmochim. Acta, 59,* 3817–3831.

Cuzzi J. N. and Weidenschilling S. J. (2006) Particle-gas dynamics and primary accretion. In *Meteorites and the Early Solar System II* (D. S. Lauretta and H. Y. McSween Jr., eds.), this volume. Univ. of Arizona, Tucson.

Dauphas N., Janney P. E., Mendybaev R. A., Wadhwa M., Richter F. M., Davis A. M., van Zuilen M., Hines R., Foley C. N. (2004) Chromatographic separation and multicollection-ICPMS analysis of iron. Investigating mass-dependent and -independent isotope effects. *Anal Chem., 76,* 5855–5863.

Davis A. M. (2003) Condensation and evaporation of solar system materials. In *Treatise on Geochemistry, Vol. 1: Meteorites, Comets and Planets* (A. M. Davis, ed.), pp. 407–430. Elsevier, Oxford.

Davis A. M. and Grossman L. (1979) Condensation and fractionation of rare earths in the solar nebula. *Geochim. Cosmochim. Acta, 43,* 1611–1632.

Davis A. M., Hashimoto A., Clayton R. N., and Mayeda T. K. (1990) Isotope mass fractionation during evaporation of Mg_2SiO_4. *Nature, 347,* 655–658.

DeHeer J. (1986) *Phenomenological Thermodynamics, with Application to Chemistry.* Prentice-Hall, Englewood Cliffs, New Jersey. 406 pp.

Desch S. J. and Connolly H. C. Jr. (2002) A model of the thermal processing of particles in solar nebula shocks: Application to the cooling rates of chondrules. *Meteoritics & Planet. Sci., 37,* 183–207.

Dohmen R., Chakraborty S., Palme H., and Rammensee W. (1998) Solid-solid reactions mediated by a gas phase: An experimental study of reaction progress and the role of surfaces in the system olivine+iron metal. *Am. Mineral., 83,* 970–984.

Draine B. T. (1979) Time-dependent nucleation theory and the formation of interstellar grains. *Astrophys. Space Sci., 65,* 313–335.

Ebel D. S. (2000) Variations on solar condensation: Sources of interstellar dust nuclei. *J. Geophys. Res., 105,* 10363–10370.

Ebel D. S. (2001) Vapor/liquid/solid equilibria when chondrites collide (abstract). *Meteoritics & Planet. Sci., 36,* A52–A53.

Ebel D. S. (2005) Model evaporation of FeO-bearing liquids: Application to chondrules. *Geochim. Cosmochim. Acta, 69,* 3183–3193.

Ebel D. S. and Alexander C. M. O'D. (2005) Condensation from cluster-IDP enriched vapor inside the snow line: Implications for Mercury, asteroids, and enstatite chondrites (abstract). In *Lunar and Planetary Science XXXVI,* Abstract #1797. Lunar and Planetary Institute, Houston (CD-ROM).

Ebel D. S and Grossman L. (2000) Condensation in dust-enriched systems. *Geochim. Cosmochim. Acta, 64,* 339–366.

Ebel D. S. and Grossman L. (2001) Condensation from supernova gas made of free atoms. *Geochim. Cosmochim. Acta, 65,* 469–477.

Ebel D. S, Ghiorso M. S., Sack R. O., and Grossman L. (2000) Gibbs energy minimization in gas + liquid + solid systems. *J. Computational Chem., 21,* 247–256.

Ebel D. S., Engler A., and Kurat G. (2003) Pyroxene chondrules from olivine-depleted, dust-enriched systems (abstract). In *Lunar and Planetary Science XXXIV,* Abstract #2059. Lunar and Planetary Institute, Houston (CD-ROM).

Ehlers K. and El Goresy A. (1988) Normal and reverse zoning in niningerite: A novel key parameter to the thermal histories of EH-chondrites. *Geochim. Cosmochim. Acta, 53,* 877–887.

Elkins L. T. and Grove T. L. (1990) Ternary feldspar experiments and thermodynamic models. *Am. Mineral., 75,* 544–559.

El Goresy A., Nagel K., and Ramdohr P. (1979) Spinel framboids and fremdlinge in Allende inclusions: Possible sequential markers in the early history of the solar system. *Proc. Lunar Planet. Sci. Conf. 10th,* pp. 833–850.

El Goresy A., Palme H., Yabuki H., Nagel K., Herrwerth I., and Ramdohr P. (1984) A calcium-aluminum-rich inclusion from the Essebi (CM2) chondrite: Evidence for captured spinel-hibonite spherules and for an ultra-refractory rimming sequence. *Geochim. Cosmochim. Acta, 48,* 2283–2298.

El Goresy A., Yabuki H., Ehlers K., Woolum D., and Pernicka E. (1988) Qingzhen and Yamato-691: A tentative alphabet for the EH chondrites. *Proc. NIPR Symp. Antarct. Meteorites, 1,* 65–101.

El Goresy A., Zinner E., Matsunami S., Palme H., Spettel B., Lin Y., and Nazarov M. (2002) Efremovka 101.1: A CAI with ultrarefractory REE patterns and enormous enrichments of Sc, Zr, and Y in fassaite and perovskite. *Geochim. Cosmochim. Acta, 66,* 1459–1491.

Engler A., Kurat G., and Sylvester P. J. (2003) Trace element abundances in micro-objects from the Tieschitz (H3.6), Krymka (LL3.1), Bishumpur (LL3.1) and Mezö-Madaras (L3.7): Implications for chondrule formation (abstract). In *Lunar and Planetary Science XXXIV*, Abstract #1689. Lunar and Planetary Institute, Houston (CD-ROM).

Eriksson G. (1975) Thermodynamic studies of high temperature equilibria XII. SOLGASMIX, a computer program for calculation of equilibrium compositions in multiphase systems. *Chem. Scripta 8,* 100–103.

Eriksson G. and Hack K. (1990) Chemsage — A computer-program for the calculation of complex chemical-equilibria. *Metall. Trans., 21B,* 1013–1023.

Fagan T. J., Krot A. N., and Keil K. (2000) Calcium-aluminum-rich inclusions in enstatite chondrites: I. Mineralogy and textures. *Meteoritics & Planet. Sci., 35,* 771–781.

Fahey A. J. (1988) Ion microprobe measurements of Mg, Ca, Ti and Fe isotopic ratios and trace element abundances in hibonite-bearing inclusions from primitive meteorites. Ph.D. thesis, Washington University, St. Louis, Missouri. 232 pp.

Fegley B. Jr. (1999) Chemical and physical processing of presolar materials in the solar nebula and the implications for preservation of presolar materials in comets. *Space Sci. Rev., 90,* 239–252.

Fegley B. Jr. and Palme H. (1985) Evidence for oxidizing conditions in the solar nebula from Mo and W depletions in refractory inclusions in carbonaceous chondrites. *Earth Planet. Sci. Lett., 72,* 311–326.

Ferguson J. B. and Buddington A. F. (1920) The binary system Åkermanite-gehlenite. *Am. J. Sci., L(296),* 131–140.

Fernández-Guillermet A. (1988) Thermodynamic properties of the Fe-Co-Ni-C system. *Z. Metallkunde, 70,* 524–536.

Field G. B. (1974) Interstellar abundances: Gas and dust, *Astrophys. J., 187,* 453–459.

Floss C., El Goresy A., Zinner E., Palme H., Weckwerth G., and Rammensee W. (1998) Corundum-bearing residues produced through the evaporation of natural and synthetic hibonite. *Meteoritics & Planet. Sci., 33,* 191–206.

Flynn G. J., Keller L. P., Feser M., Wirick S., and Jacobsen C. (2003) The origin of organic matter in the solar system: Evidence from the interplanetary dust particles. *Geochim. Cosmochim. Acta, 67,* 4791–4806.

Fraser D. G. and Rammensee W. (1982) Activity measurements by Knudsen cell mass spectrometry — the system Fe-Co-Ni and implications for condensation processes in the solar nebula. *Geochim. Cosmochim. Acta, 46,* 549–556.

Friedrich J. M., Wang M.-S., and Lipschutz M. E. (2003) Chemical studies of L chondrites. V: Compositional patterns for forty-nine trace elements in fourteen L4–6 and seven LL4–6 falls. *Geochim. Cosmochim. Acta, 67,* 2467–2479.

Fuchs L. H. (1969) Occurrence of cordierite and aluminous orthoenstatite in the Allende meteorite. *Am. Mineral., 54,* 1645–1653.

Fuchs L. H. (1978) Mineralogy of a rhonite-bearing calcium aluminum rich inclusion in the Allende meteorite. *Meteoritics, 13,* 73–88.

Galy A., Young E. D., Ash R. D., and O'Nions R. K. (2000) The formation of chondrules at high gas pressures in the solar nebula. *Science, 290,* 1751–1754.

Gail H-P. (1998) Chemical reactions in protoplanetary accretion disks IV. Multicomponent dust mixture. *Astron. Astrophys., 332,* 1099–1122.

Ganguly J. (2001) Thermodynamic modelling of solid solutions. In *Solid Solutions in Silicate and Oxide Systems* (C. A. Geiger, ed.), pp. 37–70. EMU Notes in Mineralogy, Vol. 3, European Mineralogical Union.

Geiger C. A., ed. (2001) *Solid Solutions in Silicate and Oxide Systems.* EMU Notes in Mineralogy, Vol. 3, European Mineralogical Union. 465 pp.

Geiger C. A., Kleppa O. J., Mysen B. O., Lattimer J. M., and Grossman L. (1988) Enthalpies of formation of $CaAl_4O_7$ and $CaAl_{12}O_{19}$ (hibonite) by high temperature alkali borate solution calorimetry. *Geochim. Cosmochim. Acta, 52,* 1729–1736.

Ghiorso M. S. (1985) Chemical mass transfer in magmatic processes I. Thermodynamic relations and numerical algorithms. *Contrib. Mineral. Petrol., 90,* 107–120.

Ghiorso M. S. (1994) Algorithms for the estimation of phase stability in heterogenous thermodynamic systems. *Geochim. Cosmochim. Acta, 58,* 5489–5501.

Ghiorso M. S. and Kelemen P. B. (1987) Evaluating reaction stoichiometry in magmatic systems evolving under generalized thermodynamic constraints: Examples comparing isothermal and isenthalpic assimilation. In *Magmatic Processes: Physicochemical Principles* (B. O. Mysen, ed.), pp. 319–336. Geochemical Society Special Publication #1.

Ghiorso M. S. and Sack R. O. (1995) Chemical mass transfer in magmatic processes IV. A revised and internally consistent thermodynamic model for the interpolation and extrapolation of liquid-solid equilibria in magmatic systems at elevated temperatures and pressures. *Contrib. Mineral. Petrol., 119,* 197–212.

Goldschmidt V. M. (1954) *Geochemistry.* Oxford University, Oxford. 730 pp.

Greshake A. (1997) The primitive matrix components of the unique carbonaceous chondrite Acfer 094: A TEM study. *Geochim. Cosmochim. Acta, 61,* 437–452.

Grevesse N. and Sauval A. J. (2000) Abundances of the elements in the sun. In *Origin of Elements in the Solar System: Implications of Post-1957 Observations* (O. Manuel, ed.), pp. 261–277. Kluwer, Dordrecht.

Grossman J. N. (1996) Chemical fractionations of chondrites: Signatures of events before chondrule formation. In *Chondrules and the Protoplanetary Disk* (R. H. Hewins et al., eds.), pp. 243–253. Cambridge Univ., Cambridge.

Grossman J. N. and Wasson J. T. (1983) Refractory precursor components of Semarkona chondrules and the fractionation of refractory elements among chondrites. *Geochim. Cosmochim. Acta, 47,* 759–771.

Grossman J. N, Alexander C. M. O'D. , Wang J., and Brearley A. J. (2002) Zoned chondrules in Semarkona: Evidence for high-and low-temperature processing. *Meteoritics & Planet. Sci., 37,* 49–74.

Grossman L. (1972) Condensation in the primitive solar nebula. *Geochim. Cosmochim. Acta, 36,* 597–619.

Grossman L. (1975) Petrology and mineral chemistry of Ca-rich inclusions in the Allende meteorite. *Geochim. Cosmochim. Acta, 39,* 433–454.

Grossman L. and Clark S. P. Jr. (1973) High-temperature condensates in chondrites and the environment in which they formed. *Geochim. Cosmochim. Acta, 37,* 635–649.

Grossman L. and Larimer J. W. (1974) Early chemical history of the solar system. *Rev. Geophys. Space Phys., 12,* 71–101.

Grossman L. and Steele I. M. (1975) Amoeboid olivine aggregates in the Allende meteorite. *Geochim. Cosmochim. Acta, 40,* 149–155.

Grossman L., Olsen E., and Lattimer J. M. (1979) Silicon in carbonaceous chondrite metal: Relic of high-temperature condensation. *Science, 206,* 449–451.

Grossman L., Ebel D. S., Simon S. B., Davis A. M., Richter F. M., and Parsad N. M. (2000) Major element chemical and isotopic compositions of refractory inclusions in C3 chondrites: The separate roles of condensation and evaporation. *Geochim. Cosmochim. Acta, 64,* 2879–2894.

Grossman L., Ebel D. S., and Simon S. B. (2002) Formation of refractory inclusions by evaporation of condensate precursors. *Geochim. Cosmochim. Acta, 66,* 145–161.

Guan Y., McKeegan K., and MacPherson G. J. (2000) Oxygen isotopes in calcium-aluminum-rich inclusions from enstatite chondrites: New evidence for a single CAI source in the solar nebula. *Earth Planet. Sci. Lett., 181,* 271–277.

Hanon P., Robert F., and Chaussidon M. (1998) High carbon concentrations in meteoritic chondrules. *Geochim. Cosmochim. Acta, 62,* 903–913.

Hasegawa T. I. and Herbst E. (1993) New gas-grain chemical models of quiescent dense interstellar clouds: The effects of H_2 tunnelling reactions and cosmic ray induced desorption. *Mon. Not. R. Astron. Soc., 261,* 83–102.

Hewins R. H., Yu Y., Zanda B., and Bourot-Denise M. (1997) Do nebular fractionations, evaporative losses, or both, influence chondrule compositions? *Proc. NIPR Symp. Antarct. Meteorites, 10,* 275–298.

Hinton R. W., Davis A. M., Scatena-Wachel D. E., Grossman L., and Draus R. J. (1988) A chemical and isotopic study of hibonite-rich refractory inlclusions in primitive meteorites. *Geochim. Cosmochim. Acta, 52,* 2573–2598.

Hobbs L. M, Welty D. E., Morton D. C., Spitzer L., and York D. G. (1993) The interstellar abundances of tin and four other heavy elements. *Astrophys. J., 411,* 750–755.

Holland T. J. G. and Powell R. (1998) An internally consistent thermodynamic data set for phases of petrological interest. *J. Metamorph. Geol., 16,* 309–343.

Huang S. X., Lu J., Prinz M., Weisberg M. K., Benoit P. H., and Sears D. W. G. (1996) Chondrules: Their diversity and the role of open-system processes during their formation. *Icarus, 122,* 316–346.

Humayun M. and Cassen P. (2000) Processes determining the volatile abundances of the meteorites and terrestrial planets. In *Origin of the Earth and Moon* (R. M. Canup and K. Righter, eds.), pp. 3–23. Univ. of Arizona, Tucson.

Humayun M., Campbell A. J., Zanda B., and Bourot-Denise M. (2002) Formation of Renazzo chondrule metal inferred from siderophile elements (abstract). In *Lunar and Planetary Science XXXIII,* Abstract #1965. Lunar and Planetary Institute, Houston (CD-ROM).

Huss G. R. and Lewis R. S. (1995) Presolar diamond, SiC, and graphite in primitive chondrites: Abundances as a function of meteorite class and petrologic type. *Geochim. Cosmochim. Acta, 59,* 115–160.

Huss G. R., Keil K., and Taylor G. J. (1981) The matrices of unequilibrated ordinary chondrites: Implications for the origin and history of chondrites. *Geochim. Cosmochim. Acta, 45,* 33–51.

Huss G. R., Rubin A. E., and Grossman J. N. (2006) Thermal metamorphism in chondrites. In *Meteorites and the Early Solar System II* (D. S. Lauretta and H. Y. McSween Jr., eds.), this volume. Univ. of Arizona, Tucson.

Hutchison R. (2004) *Meteorites: A Petrologic, Chemical and Isotopic Synthesis.* Cambridge Univ., Cambridge. 506 pp.

Ihinger P. D. and Stolper E. (1986) The color of meteoritic hibonite: An indicator of oxygen fugacity. *Earth Planet. Sci. Lett., 78,* 67–79.

Iida A., Nakamoto T., and Susa H. (2001) A shock heating model for chondrule formation in a protoplanetary disk. *Icarus, 153,* 430–450.

Ireland T. R. (1990) Presolar isotopic and chemical signatures in hibonite-bearing refractory inclusions from the Murchison carbonaceous chondrite. *Geochim. Cosmochim. Acta, 54,* 3219–3237.

Ireland T. R. and Fegley B. Jr. (2000) The solar system's earliest chemistry: Systematics of refractory inclusions. *Intl. Geol. Rev., 42,* 865–894.

Ireland T. R., Fegley B. Jr., and Zinner E. (1991) Hibonite-bearing microspherules: A new type of refractory inclusions with large isotopic anomalies. *Geochim. Cosmochim. Acta, 55,* 367–379.

Irvine W. M., Goldsmith P. F., and Hjalmarson Å. (1987) Chemical abundances in molecular clouds. In *Interstellar Processes* (D. J. Hollenbach and H. A. Thronson Jr., eds.), pp. 561–610. Reidel, Dordrecht.

Irvine W. M., Schloerb F. P., Crovisier J., Fegley B. Jr., and Mumma M. J. (2000) Comets: A link between interstellar and nebular chemistry. In *Protostars and Planets IV* (V. Mannings et al., eds.), pp. 1159–2000. Univ. of Arizona, Tucson.

Ivanova M. A., Petaev M. I., MacPherson G. J., Nazarov M. A., Taylor L. A., and Wood J. A. (2002) The first known natural occurrence of calcium monoaluminate, in a calcium-aluminum-rich inclusion from the CH chondrite Northwest Africa 470. *Meteoritics & Planet. Sci., 37,* 1337–1344.

Johnson C. M., Beard B. L., and Albarède F., eds. (2004). *Geochemistry of Non-Traditional Stable Isotopes.* Reviews in Mineralogy, Vol. 55, Mineralogical Society of America. 454 pp.

Jones R. H. (1996) Relict grains in chondrules: Evidence for chondrule recycling. In *Chondrules and the Protoplanetary Disk* (R. H. Hewins et al., eds.), pp. 163–180. Cambridge Univ., Cambridge.

Jones R. H. and Scott E. R. D. (1989) Petrology and thermal history of type IA chondrules in the Semarkona (LL3.0) chondrite. *Proc. Lunar Planet. Sci. Conf. 19th,* pp. 523–536.

Joung M. K. R., Mac Low M-M., and Ebel D. S. (2004) Chondrule formation and protoplanetary disk heating by current sheets in non-ideal magnetohydrodynamic turbulence. *Astrophys. J., 606,* 532–541.

Kehm K., Hauri E. H., Alexander C. M. O'D., and Carlson R. W. (2003) High precision iron isotope measurements of meteoritic material by cold plasma ICP-MS. *Geochim. Cosmochim. Acta, 67,* 2879–2891.

Keil K. (1968) Mineralogical and chemical relationships among enstatite chondrites. *J. Geophys. Res., 73,* 6945–6976.

Keil K. (1989) Enstatite meteorites and their parent bodies. *Meteoritics, 24,* 195–208.

Kelly W. R. and Larimer J. W. (1977) Chemical fractionations in meteorites VIII. Iron meteorites and the cosmochemical history of the metal phase. *Geochim. Cosmochim. Acta, 41,* 93–111.

Kemper F. (2002) Carbonates in dust shells around evolved stars (abstract). In *Lunar and Planetary Science XXXIII,* Abstract #1193. Lunar and Planetary Institute, Houston (CD-ROM).

Kerridge J. F. (1999) Formation and processing of organics in the early solar system. *Space Sci. Rev., 90,* 275–288.

Kerridge J. F. and Matthews M. S., eds. (1988) *Meteorites and the Early Solar System.* Univ. of Arizona, Tucson. 1269 pp.

Komatsu M., Krot A. N., Petaev M. I., Ulyanov A. A., Keil K., and Miyamoto M. (2001) Mineralogy and petrography of amoeboid olivine aggregates from the reduced CV3 chondrites Efremovka, Leoville and Vigarano: Products of nebular condensation, accretion and annealing. *Meteoritics & Planet. Sci., 36,* 629–641.

Kong P. and Palme H. (1999) Compositional and genetic relationship between chondrules, chondrule rims, metal, and matrix in the Renazzo chondrites. *Geochim. Cosmochim. Acta, 63,* 3673–3682.

Kornacki A. S. and Fegley B. Jr. (1984) Origin of spinel-rich chondrules and inclusions in carbonaceous and ordinary chondrites. *Proc. Lunar Planet. Sci. Conf. 14th,* in *J. Geophys. Res., 89,* B588–B596.

Kornacki A. S. and Fegley B. Jr. (1986) The abundance and relative volatility of refractory trace elements in Allende Ca, Al-rich inclusions: Implications for chemical and physical processes in the solar nebula. *Earth Planet. Sci. Lett., 79,* 217–234.

Kornacki A. S. and Wood J. A. (1984) The mineral chemistry and origin of inclusion matrix and meteorite matrix in the Allende CV3 chondrite. *Geochim. Cosmochim. Acta, 48,* 1663–1676.

Krot A. N., Scott E. R. D., and Zolensky M. E. (1995) Mineralogical and chemical modification of components in CV3 chondrites: Nebular or asteroidal processing? *Meteoritics & Planet. Sci., 30,* 748–775.

Krot A. N., Fegley B. Jr., Lodders K., and Palme H. (2000) Meteoritical and astrophysical constraints on the oxidation state of the solar nebula. In *Protostars and Planets IV* (V. Mannings et al., eds.) pp. 1019–1054. Univ. of Arizona, Tucson.

Krot A. N., Huss G. R., and Hutcheon I. D. (2001) Corundum-hibonite refractory inclusions from Adelaide: Condensation or crystallization from melt? *Meteoritics & Planet. Sci., 36,* A105.

Krot A. N., McKeegan K. D., Leshin L. A., MacPherson G. J., and Scott E. R. D. (2002) Existence of an ^{16}O-rich gaseous reservoir in the solar nebula. *Science, 295,* 1051–1054.

Krot A. N., Petaev M. I., Russell S. S., Itoh S., Fagan T. J., Yurimoto H., Chizmadia L., Weisberg M. K., Komatsu M., Olyanov A. A., and Keil K. (2004a) Amoeboid olivine aggregates and related objects in carbonaceous chondrites: Records of nebular and asteroid processes. *Chem. Erde, 64,* 185–239.

Krot A. N., Libourel G., Goodrich C. A., and Petaev M. I. (2004b) Silica-rich igneous rims around magnesian chondrules in CR carbonaceous chondrites: Evidence for condensation origin from fractionated nebular gas. *Meteoritics & Planet. Sci., 39,* 1931–1955.

Krot A. N., Hutcheon I. D., Brearley A. J., Pravdivtseva O. V., Petaev M. I., and Hohenberg C. M. (2006) Timescales and settings for alteration of chondritic meteorites. In *Meteorites and the Early Solar System II* (D. S. Lauretta and H. Y. McSween Jr., eds.), this volume. Univ. of Arizona, Tucson.

Kuehner S. M., Davis A. M., and Grossman L. (1989) Identification of relict phases in a once-molten Allende inclusion. *Geophys. Res. Lett., 16,* 775–778.

Kumar R. V. and Kay D. A. R. (1985) The utilization of galvanic cells using Caβ"-alumina solid electrolytes in thermodynamic investigations of the $CaO-Al_2O_3$ system. *Metall. Trans. B, 16B,* 107–112.

Kurat G. (1988) Primitive meteorites: An attempt towards unification. *Philos. Trans. R. Soc. London, A325,* 459–482.

Kurat G., Palme H., Brandstätter F., and Huth J. (1989) Allende xenolith AF: Undisturbed record of condensation and aggregation of matter in the solar nebula. *Z. Naturforsch., 44a,* 988–1004.

Kurat G., Zinner E., and Brandstätter F. (2002) A plagioclase-olivine-spinel-magnetite inclusion from Maralinga (CK): Evidence for sequential condensation and solid-gas exchange. *Geochim. Cosmochim. Acta, 66,* 2959–2979.

Kuzmenko V. V., Uspenskaya I. A., and Rudnyi E. B. (1997) Simultaneous assessment of thermodynamic functions of calcium aluminates. *Bull. Soc. Chim. Belg., 106,* 235–243.

Larimer J. W. (1967) Chemical fractionation in meteorites, I. Condensation of the elements. *Geochim. Cosmochim. Acta, 31,* 1215–1238.

Larimer J. W. (1975) The effect of C/O ratio on the condensation of planetary material. *Geochim. Cosmochim. Acta, 39,* 389–392.

Larimer J. W. (1988) The cosmochemical classification of the elements. In *Meteorites and the Early Solar System* (J. F. Kerridge and M. S. Matthews eds.), pp. 375–389. Univ. of Arizona, Tucson.

Larimer J. W. and Anders E. (1967) Chemical fractionation in meteorites: II. Abundance patterns and their interpretation. *Geochim. Cosmochim. Acta, 31,* 1239–1270.

Larimer J. W. and Bartholomay H. A. (1979) The role of carbon and oxygen in cosmic gases: Some applications to the chemistry and mineralogy of enstatite chondrites. *Geochim. Cosmochim. Acta, 43,* 1455–1466.

Latimer W. M. (1950) Astrochemical problems in the formation of the earth. *Science, 112,* 101–104.

Lattimer J. M., Schramm D. N. and Grossman L. (1978) Condensation in supernova ejecta and isotopic anomalies in meteorites. *Astrophys. J., 219,* 230–249.

Lauretta D. S., Buseck P. R., and Zega T. J. (2001) Opaque minerals in the matrix of the Bishunpur (LL3.1) chondrite: Constraints on the chondrule formation environment. *Geochim. Cosmochim. Acta, 65,* 1337–1353.

Lewis J. S. (1972a) Low temperature condensation from the solar nebula. *Icarus, 16,* 241–252.

Lewis J. S. (1972b) Metal/silicate fractionation in the solar system. *Earth Planet. Sci. Lett., 15,* 286–290.

Lewis J. S. (1997) *Chemistry and Physics of the Solar System, revised edition.* Academic, San Diego. 591 pp.

Lin Y. and Kimura M. (2000) Two unusual Type B refractory inclusions in the Ningqiang carbonaceous chondrite: Evidence for relicts, zenoliths and multi-heating. *Geochim. Cosmochim. Acta, 64,* 4031–4047.

Lin Y., Kimura M., and Wang D. (2003) Fassaites in compact type A Ca-al-rich inclusions in the Ningqiang carbonaceous chondrite: Evidence for partial melting in the nebula. *Meteoritics & Planet. Sci., 38,* 407–417.

Llorca J. and Casanova I. (1998) Formation of carbides and hydrocarbons in chondritic interplanetary dust particles: A laboratory study. *Meteoritics & Planet. Sci., 33,* 243–251.

Lodders K. (2003) Solar system abundances and condensation temperatures of the elements. *Astrophys. J., 591,* 1220–1247.

Lodders K. (2004) Revised and updated thermochemical properties of the gases mercapto (HS), disulfur monoxide (S_2O), thiazyl (NS), and thioxophosphino (PS). *J. Phys. Chem. Ref. Data, 33,* 357–367.

Lodders K. and Fegley B. Jr. (1993) Lanthanide and actinide chemistry at high C/O ratios in the solar nebula. *Earth Planet. Sci. Lett., 117,* 125–145.

Lodders K. and Fegley B. Jr. (1997) Condensation chemistry of carbon stars. In *Astrophysical Implications of the Laboratory Study of Presolar Materials* (T. J. Bernatowicz and E. Zinner, eds.), pp. 391–424. AIP Conference Proceedings 402, American Institute of Physics, New York.

Lodders K. and Fegley B. Jr. (1998) Presolar silicon carbide grains and their parent stars. *Meteoritics & Planet. Sci., 33,* 871–880.

Lodders K. and Osborne R. (1999) Perspectives on the comet-asteroid-meteorite link. *Space Sci. Rev., 90,* 289–297.

Lord H. C. III (1965) Molecular equilibria and condensation in a solar nebula and cool stellar atmospheres. *Icarus, 4,* 279–288.

MacPherson G. J. (2003) Calcium-aluminum-rich inclusions in chondritic meteorites. In *Treatise on Geochemistry, Vol. 1: Meteorites, Comets and Planets* (A. M. Davis, ed.), pp. 201–246. Elsevier, Oxford.

MacPherson G. J. and Davis A. M. (1994) Refractory inclusions in the prototypical CM chondrite, Mighei. *Geochim. Cosmochim. Acta, 58,* 5599–5625.

MacPherson G. J. and Grossman L. (1984) "Fluffy" Type A Ca-, Al-rich inclusions in the Allende meteorite. *Geochim. Cosmochim. Acta, 48,* 29–46.

MacPherson G. J. and Huss G. R. (2005) Petrogenesis of Al-rich chondrules: Evidence from bulk compositions and phase equilibria. *Geochim. Cosmochim. Acta, 69,* 3099–3127.

MacPherson G. J., Bar-Matthews M., Tanaka T., Olsen E., and Grossman L. (1983) Refractory inclusions in the Murchison meteorite. *Geochim. Cosmochim. Acta, 47,* 823–839.

MacPherson G. J., Grossman L., Hashimoto A., Bar-Matthews M., and Tanaka T. (1984) Petrographic studies of refractory inclusions from the Murchison meteorite. *Proc. Lunar Planet. Sci. Conf. 15th,* in *J. Geophys. Res., 89,* C299–C312.

MacPherson G. J., Hashimoto A., and Grossman L. (1985) Accretionary rims on inclusions in the Allende meteorite. *Geochim. Cosmochim. Acta, 49,* 2267–2279.

MacPherson G. J., Wark D. A., and Armstrong J. T. (1988) Primitive material surviving in chondrites: Refractory inclusions. In *Meteorites and the Early Solar System* (J. F. Kerridge and M. S. Matthews), pp. 746–807. Univ. of Arizona, Tucson.

MacPherson G. J., Davis A. M., and Zinner E. K. (1995) The distribution of aluminum-26 in the early solar system — A reappraisal. *Meteoritics, 30,* 365–386.

MacPherson G. J., Petaev M., and Krot A. N. (2004) Bulk compositions of CAIs and Al-rich chondrules: Implications of the reversal of the anorthite/forsterite condensation sequence at low nebular pressures (abstract). In *Lunar and Planetary Science XXXV,* Abstract #1838. Lunar and Planetary Institute, Houston (CD-ROM).

Maruyama S. and Yurimoto H. (2003) Relationship among O, Mg isotopes and the petrography of two spinel-bearing compound chondrules. *Geochim. Cosmochim. Acta, 67,* 3943–3957.

Mason B. and Taylor S. R. (1982) *Inclusions in the Allende Meteorite.* Smithsonian Contributions to the Earth Sciences No. 25, Smithsonian Institution, Washington, DC.

Matsunami S., Ninagawa K., Nishimura S., Kubono M., Yamamoto I., Kohata M., Wada T., Yamashita Y., Lu J., Sears D. W. G., and Nishimura H. (1993) Thermoluminescence and compositional zoning in the mesostasis of a Semarkona group A1 chondrule and new insights into the chondrule-forming process. *Geochim. Cosmochim. Acta, 57,* 2101–2110.

McGuire A. V. and Hashimoto A. (1989) Origin of fine-grained inclusions in the Allende meteorite. *Geochim. Cosmochim. Acta, 53,* 1123–1133.

McKeegan K. D., Leshin L. A., Russell S. S., and MacPherson G. J. (1998) Oxygen isotopic abundances in calcium-aluminum-rich inclusions from ordinary chondrites: Implications for nebular heterogeneity. *Science, 280,* 414–418.

McKeegan K. D, Chaussidon M., and Robert F. (2000) Incorporation of short-lived ^{10}Be in a calcium-aluminum-rich inclusion from the Allende meteorite. *Science, 289,* 1334–1337.

McSween H. Y. Jr. (1977) On the origin of isolated olivine grains in type 2 carbonaceous chondrites. *Earth Planet. Sci. Lett., 41,* 111–127.

Meco H. and Napolitano R. E. (2005) Liquidus and solidus boundaries in the vicinity of order-disorder transitions in the Fe-Si system. *Scripta Materialia, 52,* 221–226.

Meibom A., Petaev M. I., Krot A. N., Keil K., and Wood J. A. (2001) Growth mechanism and additional constraints on FeNi metal condensation in the solar nebula. *J. Geophys. Res., 106,* 32797–32801.

Mendybaev R. A., Beckett J. R., Grossman L., Stolper E., Cooper R. F., and Bradley J. P. (2002) Volatilization kinetics of silicon carbide in reducing gases: An experimental study with applications to the survival of presolar grains in the solar nebula. *Geochim. Cosmochim. Acta, 66,* 661–682.

Meyer B. S. and Zinner E. (2006) Nucleosynthesis. In *Meteorites and the Early Solar System II* (D. S. Lauretta and H. Y. McSween Jr., eds.), this volume. Univ. of Arizona, Tucson.

Millar T. J. and Williams D. A., eds. (1993) *Dust and Chemistry in Astronomy.* Institute of Physics, London. 335 pp.

Millar T. J., Nomura H., and Markwick A. J. (2003) Two-dimensional models of protoplanetary disk chemistry. *Astrophys. Space Sci. 285,* 761–768.

Misawa K. and Fujita T. (1994) A relict refractory inclusion in a ferromagnesian chondrule from the Allende meteorite. *Nature, 368,* 723–726.

Misawa K. and Nakamura N. (1996) Origin of refractory precursor components of chondrules from carbonaceous chondrites. In *Chondrules and the Protoplanetary Disk* (R. H. Hewins et al., eds.), pp. 221–232. Cambridge Univ., Cambridge.

Mostefaoui S., Perron C., Zinner E., and Sagon G. (2000) Metal-associated carbon in primitive chondrites: Structure, isotopic composition, and origin. *Geochim. Cosmochim. Acta, 65,* 1945–1964.

Mullane E., Mason T. D. F., Russell S. S., Weiss D., Horstwood M., and Parrish R. R. (2001) Copper and zinc stable isotope compositions of chondrules and CAIs: Method development and possible implications for early solar system processes (abstract). In *Lunar and Planetary Science XXXII,* Abstract #1545. Lunar and Planetary Institute, Houston (CD-ROM).

Mullane E., Russell S. S., Gounelle M., and Mason T. D. F. (2003) Iron isotope composition of Allende and Chainpur chondrules: Effects of equilibration and thermal history (abstract). In *Lunar and Planetary Science XXXIV,* Abstract #1027. Lunar and Planetary Institute, Houston (CD-ROM).

Nagahara H. (1991) Petrology of Yamato-75261 meteorite: An enstatite (EH) chondrite breccia. *Proc. 4th NIPR Symp. Antarct. Meteorites,* 144–162.

Nagahara H. (1997) Importance of solid-liquid-gas reactions for chondrules and matrices. *Meteoritics & Planet. Sci. 32,* 739–741.

Nagahara H. and Ozawa K. (2000) Isotopic fractionation as a probe of heating processes in the solar nebula. *Chem. Geol., 169,* 45–68.

Navon O. and Wasserburg G. J. (1985) Self-shielding in O_2 — a possible explanation for oxygen isotopic anomalies in meteorites? *Earth Planet. Sci. Lett., 73,* 1–16.

Newton R. C. (1987) Thermodynamic analysis of phase equilibria in simple mineral systems. In *Thermodynamic Modeling of Geological Materials: Minerals, Fluids and Melts* (I. S. E.

Carmichael and H. P. Eugster, eds.), pp. 1–34. Reviews in Mineralogy, Vol. 17, Mineralogical Society of America.

Nittler L. R. and Dauphas N. (2006) Meteorites and the chemical evolution of the Milky Way. In *Meteorites and the Early Solar System II* (D. S. Lauretta and H. Y. McSween Jr., eds.), this volume. Univ. of Arizona, Tucson.

Nurse R. W. and Stutterheim N. (1950) The system gehlenite-spinel: Relation to the quaternary system CaO-MgO-Al_2O_3-SiO_2. *J. Iron Steel Inst., 165,* 137–138.

Nuth J. A. III, Rietmeijer F. J. M., and Hill H. G. M. (2002) Condensation processes in astrophysical environments: The composition and structure of cometary grains. *Meteoritics & Planet. Sci., 31,* 807–815.

Nuth J. A. III, Charnley S. B., and Johnson N. M. (2006) Chemical processes in the interstellar medium: Sources of the gas and dust in the primitive solar nebula. In *Meteorites and the Early Solar System II* (D. S. Lauretta and H. Y. McSween Jr., eds.), this volume. Univ. of Arizona, Tucson.

Olsen E. and Grossman L. (1978) On the origin of isolated olivine grains in type 2 carbonaceous chondrites. *Earth Planet. Sci. Let., 41,* 111–127.

Pack A., Yurimoto H., and Palme H. (2004) Petrographic and oxygen-isotopic study of refractory forsterites from R-chondrite Dar al Gani 013 (R3.5–6), unequilibrated ordinary and carbonaceous chondrites. *Geochim. Cosmochim. Acta, 68,* 1135–1157.

Pack A., Palme H., and Shelley J. M. G. (2005) Origin of chondritic forsterite grains. *Geochim. Cosmochim. Acta, 69,* 3159–3182.

Palme H. (2001) Chemical and isotopic heterogeneity in protosolar matter. *Philos. Trans. R. Soc. London, A259,* 2061–2075.

Palme H. and Fegley B. Jr. (1990) High-temperature condensation of iron-rich olivine in the solar nebula. *Earth Planet. Sci. Lett., 101,* 180–195.

Palme H. and Wlotzka F. (1976) A metal particle from a Ca, Al-rich inclusion from the meteorite Allende, and the condensation of refractory siderophile elements. *Earth Planet. Sci. Lett., 33,* 45–60.

Petaev M. I. and Wood J. A. (1998a) The condensation with partial isolation (CWPI) model of condensation in the solar nebula. *Meteoritics & Planet. Sci., 33,* 1123–1137.

Petaev M. I. and Wood J. A. (1998b) The CWPI model of nebular condensation: Effects of pressure on the condensation sequence. *Meteoritics & Planet. Sci., 33,* A122.

Petaev M. I. and Wood J. A. (2000) The condensation origin of zoned metal grains in bencubbin/CH-like chondrites: Thermodynamic model (abstract). In *Lunar and Planetary Science XXXI,* Abstract #1608. Lunar and Planetary Institute, Houston (CD-ROM).

Petaev M. I., Meibom A., Krot A. N., Wood J. A., and Keil K. (2001) The condensation origin of zoned metal grains in Queen Alexandria Range 94411: Implications for the formation of the Bencubbin-like chondrites. *Meteoritics & Planet. Sci., 36,* 93–106.

Petaev M. I., Wood J. A., Meibom A., Krot A. N., and Keil K. (2003) The ZONMET thermodynamic and kinetic model of metal condensation. *Geochim. Cosmochim. Acta, 67,* 1737–1751.

Richter F. M., Davis A. M., Ebel D. S., and Hashimoto A. (2002) Elemental and isotopic fractionation of Type B CAIs: Experiments, theoretical considerations, and constraints on their thermal evolution. *Geochim. Cosmochim. Acta, 66,* 521–540.

Rietmeijer F. J. M. (1998) Interplanetary dust particles. In *Planetary Materials* (J. J. Papike, ed.), pp. 2-1 to 2-95. Reviews in Mineralogy, Vol. 36, Mineraleralogical Society of America.

Robie R. A., Hemingway B. S., and Fisher J. R. (1978) *Thermodynamic Properties of Minerals and Related Substances at 298.15 K and 1 Bar (10^5 Pascals) Pressure and at Higher Temperatures.* U.S. Geological Survey Bulletin 1452.

Rubin A. E. (1997) Mineralogy of meteorite groups. *Meteoritics & Planet. Sci., 32,* 231–247.

Russell S. S., Huss G. R., Fahey A. J., Greenwood R. C., Hutchison R., and Wasserburg G. J. (1998) An isotopic and petrologic study of calcium-aluminum-rich inclusions from CO3 meteorites. *Geochim. Cosmochim. Acta, 62,* 689–714.

Russell S. S., Hartmann L., Cuzzi J., Krot A. N., Gounelle M., and Weidenschilling S. (2006) Timescales of the solar protoplanetary disk. In *Meteorites and the Early Solar System II* (D. S. Lauretta and H. Y. McSween Jr., eds.), this volume. Univ. of Arizona, Tucson.

Ruzicka A. (1997) Mineral layers around coarse-grained, Ca-Al-rich inclusions in CV3 carbonaceous chondrites: Formation by high-temperature metasomatism. *J. Geophys. Res., 102,* 13387–13402.

Sack R. O. and Carmichael I. S. E. (1984) Fe^2=Mg^2 and $TiAl_2$=$MgSi_2$ exchange reactions between clinopyroxenes and silicate melts. *Contrib. Mineral. Petrol., 85,* 103–115.

Sack R. O. and Ghiorso M. S. (1989) Importance of considerations of mixing properties in establishing an internally consistent thermodynamic database: Thermochemistry of minerals in the system Mg_2SiO_4-Fe_2SiO_4-SiO_2. *Contrib. Mineral. Petrol., 102,* 41–68.

Sack R. O. and Ghiorso M. S. (1991a) An internally consistent model for the thermodynamic properties of Fe-Mg-titanomagnetite-aluminate spinels. *Contrib. Mineral. Petrol., 106,* 474–505.

Sack R. O. and Ghiorso M. S. (1991b) Chromian spinels as petrogenetic indicators: Thermodynamic and petrological applications. *Am. Mineral., 76,* 827–847.

Sack R. O. and Ghiorso M. S. (1994) Thermodynamics of multicomponent pyroxenes III: Calibration of $Fe^{2+}(Mg)_{-1}$, $TiAl(MgSi)_{-1}$, $TiFe^{3+}(MgSi)_{-1}$, $AlFe^{3+}(MgSi)_{-1}$, $NaAl(CaMg)_{-1}$, $Al_2(MgSi)_{-1}$ and $Ca(Mg)_{-1}$ exchange reactions between pyroxenes and silicate melts. *Contrib. Mineral. Petrol., 118,* 271–296.

Scott E. R. D. and Krot A. N. (2001) Oxygen isotopic compositions and origins of calcium-aluminum-rich inclusions and chondrules. *Meteoritics & Planet. Sci., 36,* 1307–1319.

Sears D. W. G. and Akridge D. G. (1998) Nebular or parent body alteration of chondritic material: Neither or both? *Meteoritics & Planet. Sci., 33,* 1157–1167.

Sears D. W. G., Huang S., and Benoit P. H. (1996) Open-system behaviour during chondrule formation. In *Chondrules and the Protoplanetary Disk* (R. H. Hewins et al., eds.), pp. 221–232. Cambridge Univ., Cambridge.

Semenov D., Pavlyuchenkov Ya., Henning Th., Herbst E., and van Dishoeck E. (2004) On the feasibility of chemical modeling of a protoplanetary disk. *Baltic Astron., 13,* 454–458.

Sharp C. M. and Wasserburg G. J. (1995) Molecular equilibria and condensation temperatures in carbon-rich gases. *Geochim. Cosmochim. Acta, 59,* 1633–1652.

Sheng Y. J., Hutcheon, I. D., and Wasserburg G. J. (1991) Origin of plagioclase-olivine inclusions in carbonaceous chondrites. *Geochim. Cosmochim. Acta, 55,* 581–599.

Shu F. H., Shang H., Gounelle M., Glassgold A. E., and Lee T.

(2001) The origin of chondrules and refractory inclusions in chondritic meteorites. *Astrophys. J., 548,* 1029–1050.

Simon S. B., Grossman L., and Davis A. M. (1991) Fassaite composition trends during crystallization of Allende Type B refractory inclusion melts. *Geochim. Cosmochim. Acta, 55,* 2635–3655.

Simon S. B., Yoneda S., Grossman L., and Davis A. M. (1994) A $CaAl_4O_7$-bearing refractory spherule from Murchison: Evidence for very high-temperature melting in the solar nebula. *Geochim. Cosmochim. Acta, 58,* 1937–1949.

Simon S. B., Davis A. M., and Grossman L. (1996) A unique ultrarefractory inclusion from the Murchison meteorite. *Meteoritics, 31,* 106–115.

Simon S. B., Grossman L., and Davis A. M. (1997) Multiple generations of hibonite in spinel-hibonite inclusions from Murchison. *Meteoritics & Planet. Sci., 32,* 259–269.

Simon S. B., Davis A. M., Grossman L., and Zinner E. K. (1998) Origin of hibonite-pyroxene spherules found in carbonaceous chondrites. *Meteoritics & Planet. Sci., 33,* 411–424.

Simon S. B., Davis A. M., Grossman L., and McKeegan K. D. (2002) A hibonite-corundum inclusion from Murchison: A first-generation condensate from the solar nebula. *Meteoritics & Planet. Sci., 37,* 533–548.

Smith W. R. and Missen R. W. (1982) *Chemical Reaction Equilibrium Analysis: Theory and Algorithms.* Wiley, New York. 360 pp.

Sorby H. C. (1877) On the structure and origin of meteorites. *Nature, 15,* 495–498.

Steele I. M. (1986) Compositions and textures of relic forsterite in carbonaceous and unequilibrated ordinary chondrites. *Geochim. Cosmochim. Acta, 50,* 1379–1395.

Steele I. M. (1995) Oscillatory zoning in meteoritic forsterites. *Am. Mineral., 80,* 823–832.

Stolper E. (1982) Crystallization sequences of Ca-Al-rich inclusions from Allende: An experimental study. *Geochim. Cosmochim. Acta, 46,* 2159–2180.

Stolper E. and Paque J. M. (1986) Crystallization sequences of Ca-Al-rich inclusions from Allende: The effects of cooling rate and maximum temperature. *Geochim. Cosmochim. Acta, 50,* 1785–1806.

Sylvester P. J., Ward B. J., Grossman L., and Hutcheon I. D. (1990) Chemical compositions of siderophile element-rich opaque assemblages in an Allende inclusion. *Geochim. Cosmochim. Acta, 54,* 3491–3508.

Tissandier L., Libourel G., and Robert F. (2002) Gas-melt interactions and their bearing on chondrule formation. *Meteoritics & Planet. Sci., 37,* 1377–1389.

Toppani A. and Libourel G. (2003) Factors controlling compositions of cosmic spinels: Application to atmospheric entry conditions of meteoritic materials. *Geochim. Cosmochim. Acta, 67,* 4261–4638.

Toppani A., Libourel G., Robert F., Ghanbaja J., and Zimmermann L. (2004) Synthesis of refractory minerals by high-temperature condensation of a gas of solar composition (abstract). In *Lunar and Planetary Science XXXII,* Abstract #1846. Lunar and Planetary Institute, Houston (CD-ROM).

Toppani A., Libourel G., Robert F., and Ghanbaja J. (2005) Laboratory condensation of refractory dust in protosolar and circumstellar conditions. *Geochim. Cosmochim. Acta, 69,* in press.

Travaglio C., Gallino R., Amari S., Zinner E., Woosley S., and Lewis R. S. (1999) Low-density graphite grains and mixing in type II supernovae, *Astrophys. J., 510,* 325–354.

Tsukamoto K., Kobatake H., Nagashima K., Satoh H., and Yurimoto H. (2001) Crystallization of cosmic materials in microgravity (abstract). In *Lunar and Planetary Science XXXII,* Abstract #1846. Lunar and Planetary Institute, Houston (CD-ROM).

Urey H. (1952) *The Planets: Their Origin and Development.* Yale Univ., New Haven. 236 pp.

Wai C. M. and Wasson J. T. (1979) Nebular condensation of Ga, Ge and Sb and the chemical classification of iron meteorites. *Nature, 282,* 790–793.

Wänke H., Baddenhausen H., Palme H., and Spettel B. (1974) On the chemistry of the Allende inclusions and their origin as high temperature condensates. *Earth Planet. Sci. Lett., 23,* 1–7.

Wark D. A. and Boynton W. V. (2001) The formation of rims on calcium-aluminum rich inclusions: Step I — Flash heating *Meteoritics & Planet. Sci., 36,* 1135–1166.

Wark D. A. and Lovering J. F. (1977) Marker events in the early evolution of the solar system: Evidence from rims on Ca-Al-rich inclusions in carbonaceous chondrites. *Proc. Lunar Planet. Sci. Conf. 8th,* pp. 95–112.

Wasson J. T. (1985) *Meteorites: Their Record of Early Solar System History.* Freeman, New York. 267 pp.

Wasson J. T. and Kallemeyn G. W. (1988) Composition of chondrites. *Philos. Trans. R. Soc. London, A325,* 535–544.

Weber D. and Bischoff A. (1994) The occurrence of grossite ($CaAl_4O_7$) in chondrites. *Geochim. Cosmochim. Acta, 58,* 3855–3877.

Weidenschilling S. J. and Cuzzi J. N. (2006) Accretion dynamics and timescales: Relation to chondrites. In *Meteorites and the Early Solar System II* (D. S. Lauretta and H. Y. McSween Jr., eds.), this volume. Univ. of Arizona, Tucson.

Weinbruch S., Palme H., and Spettel B. (2000) Refractory forsterite in primitive meteorites: Condensates from the solar nebula? *Meteoritics & Planet. Sci., 35,* 161–171.

Weisberg M. K. and Prinz M. (1996) Agglomeratic chondrules, chondrule precursors, and incomplete melting. In *Chondrules and the Protoplanetary Disk* (R. H. Hewins et al., eds.), pp. 119–128. Cambridge Univ., Cambridge.

Weisberg M. K. and Prinz M. (1998) Fayalitic olivine in CV3 chondrite matrix and dark inclusions: A nebular origin. *Meteoritics & Planet. Sci., 33,* 1087–1099.

Weisberg M. K., Zolensky M. W., and Prinz M. (1997) Fayalitic olivine in matrix of the Krymka LL3.1 chondrite: Vapor-solid growth in the solar nebula. *Meteoritics & Planet. Sci., 32,* 791–801.

Weisberg M. K., Connolly H. C., and Ebel D. S. (2005) Petrology and origin of amoeboid olivine aggregates in CR chondrites. *Meteoritics & Planet. Sci., 39,* 1741–1753.

Weisberg M. K., McCoy T. J., and Krot A. N. (2006) Systematics and evaluation of meteorite classification. In *Meteorites and the Early Solar System II* (D. S. Lauretta and H. Y. McSween Jr., eds.), this volume. Univ. of Arizona, Tucson.

White W. B., Johnson S. M., and Dantzig G. B. (1958) Chemical equilibrium in complex mixtures. *J. Chem. Phys., 28,* 751–755.

Wood J. A. (1963) On the origin of chondrules and chondrites. *Icarus, 2,* 152–180.

Wood J. A. (1967) Chondrites: Their metallic minerals, thermal histories, and parent planets. *Icarus, 6,* 1–49.

Wood J. A. (1996) Processing of chondritic and planetary material in spiral density waves in the nebula. *Meteoritics, 31,* 641–645.

Wood J. A. (1998) Meteoritic evidence for the infall of large interstellar dust aggregates during the formation of the solar system. *Astrophys. J. Lett., 503,* L101–L104.

Wood J. A. and Hashimoto A. (1993) Mineral equilibrium in fractionated nebular systems. *Geochim. Cosmochim. Acta, 57,* 2377–2388.

Woolum D. S. and Cassen P. (1999) Astronomical constraints on nebular temperatures: Implications for planetesimal formation. *Meteoritics & Planet. Sci., 34,* 397–907.

Xiao X. and Lipschutz M. E. (1992) Labile trace elements in carbonaceous chondrites: A survey. *J. Geophys. Res., 97,* 10199–10211.

Yoneda S. and Grossman L. (1995) Condensation of CaO-MgO-Al_2O_3-SiO_2 liquids from cosmic gases *Geochim. Cosmochim. Acta, 59,* 3413–3444.

Young E. D. and Russell S. S. (1998) Oxygen reservoirs in the early solar nebula inferred from an Allende CAI. *Science, 282,* 452–455.

Young E. D., Ash R. D., Galy A., and Belshaw N. S. (2002) Mg isotope heterogeneity in the Allende meteorite measured by UV laser ablation-MC-ICPMS and comparisons with O isotopes. *Geochim. Cosmochim. Acta, 66,* 683–698.

Yoder H. S., ed. (1979) *The Evolution of the Igneous Rocks: Fiftieth Anniversary Perspectives.* Princeton Univ., Princeton. 588 pp.

Zanda B., Bourot-Denise M., Hewins R. H., Cohen B. A., Delaney J. S., Humayun M., and Campbell A. J. (2002) Accretion textures, iron evaporation and re-condensation in Renazzo chondrules (abstract). In *Lunar and Planetary Science XXXIII,* Abstract #1852. Lunar and Planetary Institute, Houston (CD-ROM).

Zhu X. K., Guo Y., O'Nions R. K., Young E. D., and Ash R. D. (2001) Isotopic homogeneity of iron in the early solar nebula. *Nature, 412,* 311–312.

Zinner E. (1998) Stellar nucleosynthesis and the isotopic composition of presolar grains from primitive meteorites. *Annu. Rev. Earth Planet. Sci., 26,* 147–188.

Zolensky M., Nakamura K., Weisberg M. K., Prinz M., Nakamura T., Ohsumi K., Saitow A., Mukai M., and Gounelle M. (2003) A primitive dark inclusion with radiation-damaged silicates in the Ningqiang carbonaceous chondrite. *Meteoritics & Planet. Sci., 38,* 305–322.

The Fayalite Content of Chondritic Olivine: Obstacle to Understanding the Condensation of Rocky Material

A. V. Fedkin
University of Chicago

L. Grossman
University of Chicago

Solar gas is too reducing for the equilibrium X_{Fa} in condensate olivine to reach the minimum X_{Fa} of the precursors of chondrules in unequilibrated ordinary chondrites (UOCs), 0.145, at temperatures above those where Fe-Mg interdiffusion in olivine stops. Vaporization of a region enriched in dust relative to gas compared to solar composition yields higher f_{O_2}, and condensate grains with higher equilibrium X_{Fa}, than in a solar gas at the same temperature. Only dust enrichment factors near the maximum produced in coagulation and settling models, together with C1 chondrite dust whose O content has been enhanced by admixture of water ice, can yield olivine condensate grains with radii ≥1 μm whose mean X_{Fa} exceeds the minimum X_{Fa} of the precursors of UOC chondrules over the entire range of nebular midplane cooling rates. This unlikely set of circumstances cannot be considered a robust solution to the problem of the relatively high fayalite content of UOC olivine.

1. INTRODUCTION

1.1. Minimum Fayalite Content of Chondrule Precursors in Unequilibrated Ordinary Chondrites

Olivine in primitive solar system matter contains significant amounts of fayalite. The mean mole fraction of fayalite, X_{Fa}, in olivine is 0.05–0.10 in CI and CM chondrites, ~0.07 in chondrules of CV chondrites, 0.38 in R chondrites, ~0.50 in the matrices of CO and CV chondrites (*Brearley and Jones,* 1998), and ~0.10 in interplanetary dust particles (*Rietmeijer,* 1998). *Huss et al.* (1981) showed that the mean X_{Fa} of submicrometer olivine grains in the opaque matrix of a given type 3 unequilibrated ordinary chondrite (UOC) is systematically higher than that of the larger olivine grains inside chondrules in the same meteorite. *Alexander et al.* (1989) suggested that the matrix olivine grains may have formed during parent-body metamorphic processes under relatively oxidizing conditions. Within the type 3 UOCs, *McCoy et al.* (1991) noted a progressive increase in mean X_{Fa} of olivine grains inside chondrules with increasing metamorphic grade, and attributed this to reactions between phases in chondrules and matrices during metamorphism. Consequently, in UOCs, the largest olivine grains in the least-metamorphosed types 3.0 and 3.1 are the ones whose compositions most likely reflect premetamorphic conditions. *Dodd et al.* (1967) obtained ~100 electron microprobe analyses of randomly selected, relatively large olivine grains, probably mostly in chondrules, in each of 31 ordinary chondrites. For the purpose of the present work, the mean X_{Fa} of olivine in each of Bishunpur, Krymka, and Semarkona from that study was averaged together with the mean X_{Fa} of chondrule olivine for the same meteorites from the study of *Huss et al.* (1981). The average X_{Fa} so obtained, 0.145, is only a lower limit to the mean X_{Fa} of olivine in chondrule precursors, as such materials are known to have undergone reduction to varying and generally unknown degrees during chondrule formation (*Rambaldi,* 1981). Regardless of the precise value of the characteristic X_{Fa} of the prechondrule, nebular material of these meteorites, it will be clear from what follows that its lower limit is much higher than can be produced during condensation of a solar gas.

1.2. Dust-enriched Systems

While thermodynamic treatments of condensation from a gas of solar composition have been enormously successful at explaining the mineralogy of such chondritic assemblages as Ca-, Al-rich inclusions (CAIs) (*Grossman,* 1980), it is also clear that these models are unable to account for all the mineralogical features of chondrites, including even these high-temperature objects (*MacPherson et al.,* 1983). Nowhere is this failure more evident than in the case of the fayalite content of olivine, the most abundant phase in ordinary chondrites. Equilibrium thermodynamic calculations predict that metallic Fe is at or near stability when olivine first condenses from a gas of solar composition at high temperature. As a result, X_{Fa} is initially nearly zero, and rises with falling temperature as metallic Fe is gradually oxidized (*Grossman,* 1972). A solar gas is so reducing, however, that

X_{Fa} is predicted to reach the levels found in UOCs only below 600 K, where Fe-Mg interdiffusion rates in olivine are negligible, preventing X_{Fa} from reaching these levels and leaving the problem of how chondrites obtained their FeO contents unsolved. This situation was reviewed thoroughly by *Palme and Fegley* (1990). They concluded that chondritic olivine with non-negligible X_{Fa} formed at higher temperature, and thus from a nebular region more oxidized than a solar gas, presumably due to enrichment in dust prior to its vaporization. Indeed, *Dohmen et al.* (1998) observed rapid uptake of FeO by forsterite in contact with metallic Fe under relatively oxidizing conditions at 1573 K.

Wood (1967) was the first to recognize that enrichment of dust in parts of the nebula, followed by its vaporization, could lead to regions having higher f_{O_2} than a gas of solar composition. The dust could either be presolar, having condensed in circumstellar envelopes prior to formation of the solar nebula, or could have condensed in cooler regions of the solar nebula and been transported to the location where ordinary chondrites were about to condense. Because the final product of condensation of such a dust-enriched region must still be chondritic in composition, however, the dust cannot have random composition but must be constrained to have chondritic proportions of condensable elements. If all the rock-forming elements were present as dust, a much greater proportion of the total O compared to the total C or H would be present in the dust, with the exact fractions depending on the precise composition of the dust. If, for example, the dust were of ordinary chondrite composition, it would contain ~27% of the total O, ~0.4% of the C, and ~3 ppm of the H. If it were of C1 chondrite composition, it would contain ~55% of the O, ~11% of the C, and ~190 ppm of the H. If, before nebular temperatures reached their maxima, such dust concentrated in certain regions relative to the gas compared to solar composition, then total vaporization of those regions would have produced a gas richer in O relative to H and C, and thus of higher f_{O_2}, than a solar gas. In an effort to avoid completely the problem of diffusing Fe^{2+} into previously formed forsterite crystals, *Ebel and Grossman* (2000) investigated the extreme conditions under which significant amounts of fayalite are stable in olivine at the very high temperatures at which the olivine first begins to condense. Under such conditions, Fe^{2+} is incorporated into the structure of the olivine crystals while they grow from the vapor. Because the dust enrichment factors required to produce those conditions are so unrealistically large, a different question is asked in the present paper. At the temperatures where fayalite is stabilized at the largest dust enrichment factors predicted by solar nebular coagulation and settling models, what is the maximum grain size that would allow diffusion to increase the X_{Fa} of previously condensed forsterite crystals to the minimum level of the nebular precursors of UOC chondrules?

Over the years, understanding the relatively high fayalite content of chondritic olivine has become even more difficult due to additional data on solar abundances of C and O (*Allende Prieto et al.,* 2001, 2002), which make the nebula more reducing; Fe-Mg interdiffusion rates in olivine (*Chakraborty,* 1997), which make diffusion slower; and non-ideality in olivine solid solutions, lowering the solubility of Fa in it (*Sack and Ghiorso,* 1989). In this paper, full condensation calculations are used to compute the equilibrium distribution of Fe as a function of temperature in a system of solar composition and in more oxidizing systems. These data are then combined with diffusion coefficients to estimate the mean X_{Fa} of condensate olivine grains of various sizes at the temperature at which diffusion stops, with a view toward understanding the formation conditions of the precursors of chondrules in the least-equilibrated ordinary chondrites. Preliminary versions of this work appeared in *Grossman and Fedkin* (2003) and *Fedkin and Grossman* (2004).

2. TECHNIQUE

2.1. Condensation Calculations

The computer program used by *Ebel and Grossman* (2000) employs the most up-to-date and mutually compatible thermodynamic dataset ever used in solar nebular condensation calculations. The numerical procedure used in that program fails at the relatively low temperatures of interest in the present work, however, due to the fact that gas-phase concentrations of refractory elements become vanishingly small. As a result, the immediate predecessor of that program, the one used by *Yoneda and Grossman* (1995), was used herein. It has the additional advantage of being able to compute the complete condensation sequence at 1° temperature intervals from 2000 to 500 K in minutes on a Pentium 4-based PC, whereas the *Ebel and Grossman* (2000) program requires many days on an SGI Origin 2000 with an R10000 processor. For high-temperature condensation of a gas of solar composition, results of both programs are compared and discussed by *Ebel and Grossman* (2000).

In this work, the same thermodynamic data were used as in *Yoneda and Grossman* (1995), with the following exceptions. Errors in the *Chase et al.* (1985) data for gaseous C_2N_2, C_2H_2, CN, and HS were corrected as in *Ebel and Grossman* (2000), and an error was corrected in the *Chase et al.* (1985) tables for another species, $S_2O_{(g)}$. Data for troilite, FeS, were taken from *Gronvold and Stolen* (1992), instead of from *Robie et al.* (1978). Data for olivine and orthopyroxene end members were taken from *Berman* (1988), instead of from *Robinson et al.* (1982). While *Yoneda and Grossman* (1995) treated olivine and orthopyroxene as ideal solutions, they were modeled as nonideal, regular solutions in the present work, as in *Sack and Ghiorso* (1989). For olivine, the solution parameter for the excess free energy of mixing was taken directly from that work, and, for orthopyroxene, we used from that work the equivalent solution parameter in the absence of ordering at its reference pressure. For calibration of the solution models for olivine and orthopyroxene, *Sack and Ghiorso* (1989) used distribution coefficients for the Fe-Mg exchange reaction between them. These were taken from natural and synthetic assemblages, most of which equilibrated in the temperature range 873 to 1273 K, thus bracketing the temperature range of interest in this work.

As in *Yoneda and Grossman* (1995), the *Anders and Grevesse* (1989) estimates of the relative atomic abundances of the elements in the solar system were used, except for that of S, which was taken from the compilation of *Lodders* (2003), and for those of O and C, which were taken from new solar photospheric determinations by *Allende Prieto et al.* (2001) and (2002), respectively. Using *Lodders*' (2003) average S concentration in C1 chondrites of 5.41 wt%, instead of the 6.25 wt% used by *Anders and Grevesse* (1989), reduces the solar system abundance of S by 13.4%, to 4.46×10^5 atoms per 10^6 Si atoms. Compared to the data of *Anders and Grevesse* (1989), the new O and C abundances of 1.37×10^7 and 6.85×10^6 relative to Si = 1×10^6 are 42% and 32% lower, respectively, and yield a slightly higher C/O ratio, 0.50 compared to 0.42.

2.2. Diffusion Calculations

The dependence of the Fe-Mg interdiffusion rate in olivine on temperature, f_{O_2}, olivine composition, and crystallographic direction has been the subject of several experimental studies. *Miyamoto et al.* (2002) corrected the bulk interdiffusion coefficients of *Nakamura and Schmalzried* (1984) as well as the c-axis data from the work of *Buening and Buseck* (1973), *Misener* (1974), and *Chakraborty* (1997) to a constant X_{Fa}(= 0.14) and f_{O_2}(= 10^{-12}), and plotted the logarithms of these corrected interdiffusion coefficients, D, vs. 1/T. On this plot, the data from these four studies lie along four subparallel straight lines offset from one another by a total range of more than 2 orders of magnitude parallel to the log D axis. Strictly speaking, because *Nakamura and Schmalzried* (1984) measured bulk interdiffusion coefficients, their data must be resolved into interdiffusion coefficients parallel to each crystallographic axis in order to compare them with data from the other studies. Although the degree of anisotropy differs, both *Buening and Buseck* (1973) and *Misener* (1974) found that the interdiffusion rates parallel to the c-axis are higher than those parallel to the a-axis, which are greater than those parallel to the b-axis. Using the anisotropies of *Misener* (1974), a geometric correction was made to the bulk interdiffusion coefficients of *Nakamura and Schmalzried* (1984) in order to estimate the interdiffusion coefficient parallel to the c-axis consistent with the latter data. This calculation results in a D parallel to the c-axis, which is approximately twice the bulk D of *Nakamura and Schmalzried* (1984).

Comparing only c-axis data, the largest values of D come from the work of *Buening and Buseck* (1973) and the lowest from *Chakraborty* (1997). At 1323 K, the D of *Misener* (1974) is a factor of ~14 higher, and that of *Nakamura and Schmalzried* (1984) a factor of ~10 higher, than the value of *Chakraborty* (1997). As discussed by *Chakraborty* (1997), poor f_{O_2} control may have affected the *Misener* (1974) data, and use of polycrystalline materials may have compromised the work of *Buening and Buseck* (1973) and *Nakamura and Schmalzried* (1984). Accordingly, the *Chakraborty* (1997) data were employed in this study, by using all his data points with <20% error in a regression of their logarithms against 1/T and X_{Fa}. Because all of *Chakraborty*'s (1997) measurements were at one f_{O_2}, the same dependence of log D on log f_{O_2} was adopted as in *Buening and Buseck* (1973) and *Nakamura and Schmalzried* (1984). The resulting expression for diffusion along the c-axis is

$$\log D = -6.511 + (1/6)\log f_{O_2} - 12092.0/T + 3.351X_{Fa} \quad (1)$$

According to *Tscharnuter and Boss* (1993), the midplane of a dust-enshrouded solar nebular disk experiencing a declining mass accretion rate cools from 1500 K to very low temperature in ~10^5–10^6 yr. If this global cooling occurs at a rate proportional to temperature, the temperature falls exponentially with time at decay constants of 4.32×10^{-5} and 4.32×10^{-6} yr^{-1} for these respective cooling times, hereinafter referred to as "fast" and "slow" cooling respectively. At these rates, the time required to cool from 1200 to 800 K, for example, is ~9×10^3 yr in the fast cooling case and ~9×10^4 yr for slow cooling. For each decay constant, the cooling rate of the solar nebula was calculated at 10^{-6} K temperature intervals and, from these, the cooling time was computed for all intervals.

A finite-difference technique was used to solve the diffusion equation for spherical geometry (*Crank,* 1975) with both a diffusion coefficient and surface X_{Fa} that vary with time, assuming that the surface of the olivine condensate crystal equilibrates with the nebular gas instantaneously. For calculation of the fayalite concentration profile, the grain was assumed to consist of 145 shells of equal thickness from core to rim. The mole fraction fayalite was assumed to be zero in each shell as an initial condition. In all cases investigated in this work, an initial temperature below the condensation temperature of olivine was selected such that, upon cooling, diffusion caused the fayalite concentration throughout the interior of the grain to rise to the level of the surface concentration, producing a uniform profile at some lower temperature. For the first temperature step, the c-axis interdiffusion coefficient was calculated from equation (1), using the f_{O_2} corresponding to the initial temperature taken from a condensation calculation for the appropriate system composition. The surface X_{Fa} and f_{O_2} were calculated by the condensation program at 1 K intervals. Each temperature step was further subdivided into 10^6 equal steps, at which the f_{O_2} and equilibrium X_{Fa} were interpolated. The interdiffusion coefficient in each shell of the next temperature step was calculated from the interpolated temperature and f_{O_2}, and the radial variation of X_{Fa} in the previous step. Because the time taken for the nebula to cool through each 10^{-6} K temperature interval is known, the interdiffusion coefficient and equilibrium X_{Fa} are also known as a function of time for each of the slow and fast cooling cases.

3. RESULTS AND DISCUSSION

3.1. Solar Gas

Although it is commonly assumed that the P_{H_2O}/P_{H_2} ratio is constant for a system of solar composition, it actually varies continuously with temperature, as does the oxygen fugacity, f_{O_2}. Shown in Fig. 1 is the variation with temperature of $\log f_{O_2}$ of the gas in equilibrium with condensates in a system of solar composition at a total pressure, P^{tot}, of 10^{-3} bar over the range of temperatures relevant to high-temperature condensation. Shown for comparison is the curve for equilibrium between iron and wüstite, $Fe_{0.947}O$. Above the latter curve, pure metallic Fe is unstable; below it, pure FeO is unstable. While $\log f_{O_2}$ of a solar gas varies from –15.3 at 1800 K to –19.6 at 1400 K, the $\log f_{O_2}$ for iron-wüstite (IW) equilibrium also falls over this temperature interval, and by almost the same amount. Thus, over this temperature interval, the f_{O_2} of a solar gas remains 6.6 to 6.7 log units more reducing than that necessary to equilibrate iron and wüstite, and its $\log f_{O_2}$ is abbreviated as IW-6.6.

Also shown in Fig. 1 is the curve for $\log f_{O_2}$ of a gas that is solar in composition except that its C, O, and S abundances are those of *Anders and Grevesse* (1989). Note that it is ~0.7 log units more oxidizing than the current solar gas curve over this temperature interval. This is due to the well-known effect of $CO_{(g)}$ (*Larimer,* 1975). This molecule is so stable in cosmic gases that it consumes the entirety of whichever of C and O is the lower in abundance, leaving the excess of the more-abundant element to form other molecules and condensates. The lower C/O ratio of *Anders and Grevesse* (1989) leaves a greater fraction of the O available to form the next most abundant O-containing species, $H_2O_{(g)}$. Coupled with the greater O/H ratio of the *Anders and Grevesse* abundance table, this leads to a higher P_{H_2O}/P_{H_2} ratio and thus a higher f_{O_2} than given by the abundances adopted herein.

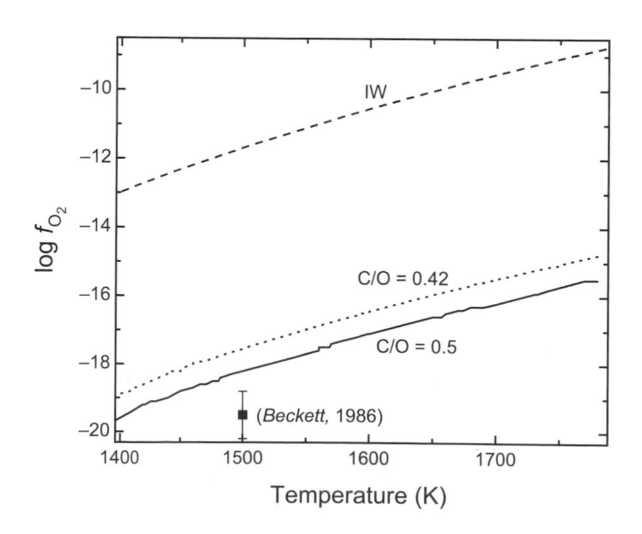

Fig. 1. Variation of the logarithm of the oxygen fugacity with temperature in equilibrium with condensates in a system of solar composition at a total pressure of 10^{-3} bar for C/O ratios of 0.42 (*Anders and Grevesse,* 1989) and the value used in this work, 0.50 (*Allende Prieto et al.,* 2001, 2002), compared to that of the iron-wüstite (IW) buffer. Also shown is the solar nebular oxygen fugacity at 1500 K, estimated from Allende refractory inclusions by *Beckett* (1986).

An independent estimate of the f_{O_2} of solar gas comes from the mineralogy of type B refractory inclusions in C3 chondrites. These objects are widely regarded as having condensed, melted, and recrystallized in a gas of solar composition (*Grossman et al.,* 2000, 2002). A major phase in all such inclusions is fassaite (*Dowty and Clark,* 1973), a Ca-,Al-rich clinopyroxene containing several weight percent of Ti, one-third to two-thirds of which is Ti^{3+} and the remainder Ti^{4+}. This may be the only pyroxene known to contain trivalent Ti and, if so, formed under lower f_{O_2} than all other Ti-bearing pyroxenes in terrestrial and lunar rocks. *Beckett* (1986) crystallized pyroxenes of these compositions from melts having the compositions of type B inclusions at controlled temperature and f_{O_2}. From these experiments, he was able to calculate equilibrium constants at 1500 K for redox reactions involving Ti^{3+}-bearing and Ti^{4+}-bearing pyroxene components, $O_{2(g)}$, and coexisting spinel, anorthite, and melilite. Because fassaite crystallizes from these melts at the solidus, 1500 K, he was able to use these equilibrium constants together with analyses of coexisting phases in actual type B inclusions to calculate the f_{O_2} at which the fassaite formed. The average $\log f_{O_2}$ so calculated, –19.5 ± 0.7, is the sole data point plotted on Fig. 1. Although it lies closer to the f_{O_2} curve for the current C/O ratio than to the one for the *Anders and Grevesse* (1989) ratio, the upper end of its error bar still lies 0.6 log units below the current solar gas curve.

Regardless of whether the actual solar nebular f_{O_2} was that of a system whose C/O ratio was 0.42, 0.50, or even slightly higher than this, Fig. 1 shows why the degree of oxidation of Fe in primitive solar system materials is so difficult to understand. In the high-temperature interval where reaction kinetics are most favorable, the f_{O_2} of a gas of solar composition lies so far below IW that only vanishingly small concentrations of FeO would be expected in silicates that equilibrate with the major Fe-containing high-temperature condensate, metallic Ni-Fe.

In the temperature range considered in Fig. 1, virtually all the C in a system of solar composition is present as $CO_{(g)}$. Upon cooling at equilibrium, a temperature is reached below which a significant fraction of the $CO_{(g)}$ begins to form $CH_{4\,(g)}$ via the reaction,

$$CO_{(g)} + 3H_{2(g)} \leftrightarrow CH_{4(g)} + H_2O_{(g)} \qquad (2)$$

which also increases the P_{H_2O}/P_{H_2} ratio, making the system more oxidizing. In reality, however, reaction (2) is very

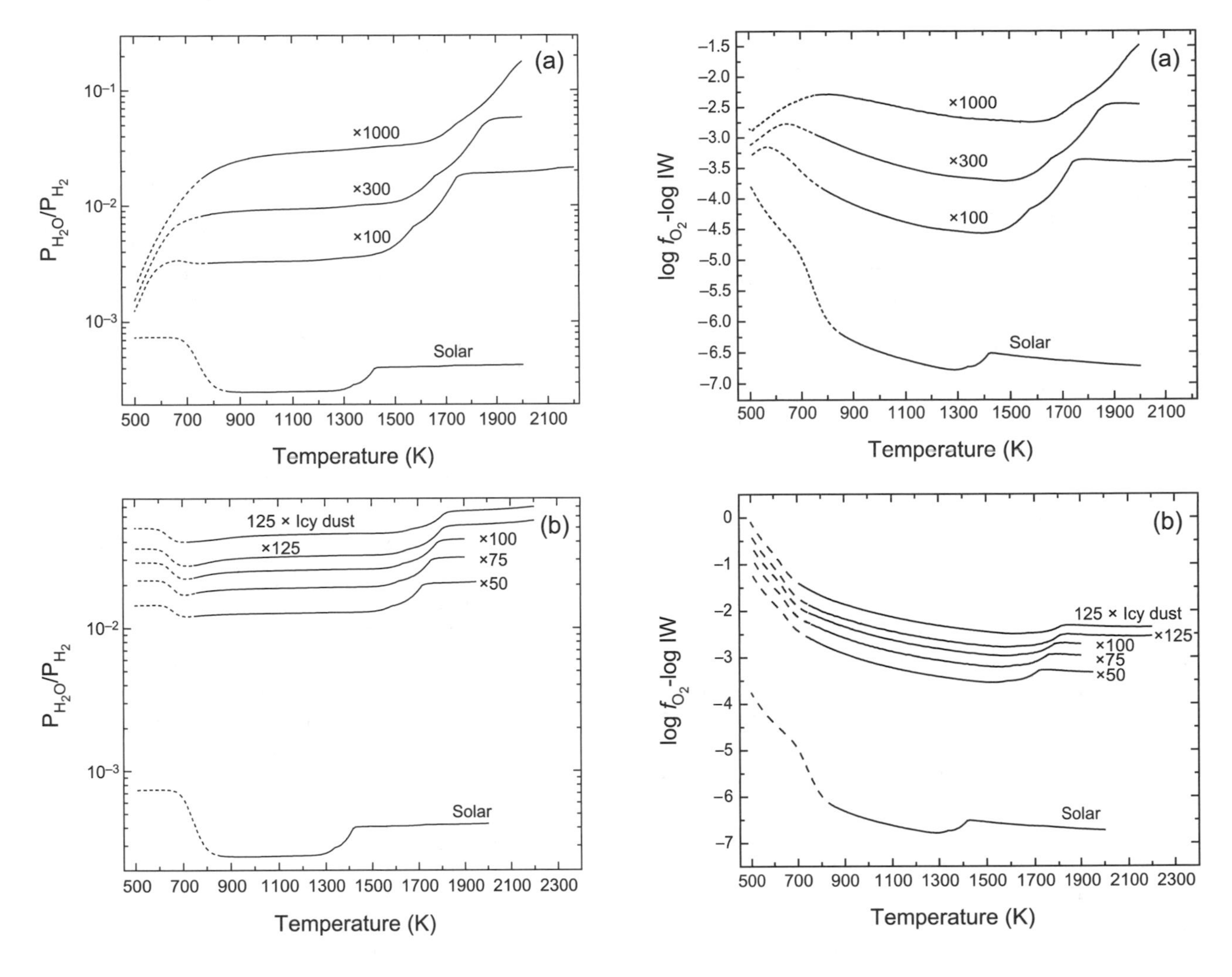

Fig. 2. Variation of the equilibrium P_{H_2O}/P_{H_2} ratio with temperature at $P^{tot} = 10^{-3}$ bar in systems enriched in **(a)** OC dust by factors of 100, 300, and 10^3 and **(b)** C1 dust by factors of 50, 75, 100, and 125 or in icy dust (see text) by a factor of 125 relative to solar composition, in all cases compared to that of a system of solar composition. Each curve is dashed below the temperature at which the P_{CH_4}/P_{CO} ratio becomes ≥0.01, where the P_{H_2O}/P_{H_2} ratio becomes significantly affected by reaction (2).

Fig. 3. Variation with temperature of the equilibrium $\log f_{O_2}$ relative to that of the iron-wüstite (IW) buffer at $P^{tot} = 10^{-3}$ bar in systems enriched in **(a)** OC dust by factors of 100, 300, and 10^3 and **(b)** C1 dust by factors of 50, 75, 100, and 125 or in icy dust (see text) by a factor of 125 relative to solar composition, in all cases compared to that of a system of solar composition. Each curve is dashed below the temperature at which the P_{CH_4}/P_{CO} ratio becomes ≥0.01, where the P_{H_2O}/P_{H_2} ratio becomes significantly affected by reaction (2).

slow. If the reaction were to occur homogeneously in the gas phase at ~800 K (*Lewis and Prinn,* 1980) or by catalysis on metallic Fe surfaces at ~600 K (*Prinn and Fegley,* 1989), for example, the $CO_{(g)}$ destruction time would be comparable to the solar nebular cooling time. Nevertheless, full equilibrium calculations were extended to 500 K in order to determine what the equilibrium state of the system actually is, for comparison with dust-enriched systems where reaction (2) becomes important only below the temperatures of interest in this work. The P_{H_2O}/P_{H_2} ratio is plotted in Fig. 2a, and the difference between the $\log f_{O_2}$ of solar gas and that of IW in Fig. 3a for the entire temperature range investigated here. The P_{H_2O}/P_{H_2} ratio is nearly constant at ~4.1×10^{-4} between 1800 K and 1400 K, falls to ~2.5×10^{-4} by 1300 K and remains constant until ~850 K. The decline that begins at ~1400 K is due to consumption of gaseous O by the major O-containing condensate, forsterite. Assuming equilibrium gas phase speciation, the P_{CH_4}/P_{CO} ratio is <0.01 above 850 K but begins to increase significantly below this temperature due to a shift to the right of reaction (2). Because of this, the P_{H_2O}/P_{H_2} ratio begins to rise below 850 K, reaching ~7.4×10^{-4} at 625 K where it remains until 500 K. Because reaction (2) is kinetically hindered, the solar gas curve in Fig. 2a is dashed below 850 K.

The difference between $\log f_{O_2}$ of solar gas and that of IW decreases slightly with falling temperature from 2000 to 1400 K, then increases slightly where the P_{H_2O}/P_{H_2} ratio drops due to olivine condensation. The difference begins to decrease gradually below 1300 K and then more steeply below 850 K. Because the steep rise is due to the kinetically hindered reaction (2), the solar gas curve is again dashed below 850 K in Fig. 3a. It should be noted that, since wüstite is unstable below 843 K, the $\log f_{O_2}$ of solar gas is being compared to the metastable extension of the IW curve below this temperature. In summary, a gas of solar composition becomes progressively more capable of converting metallic Fe into oxidized Fe with decreasing temperature.

The effect of the temperature variation of $\log f_{O_2}$ on the equilibrium X_{Fa} in condensate olivine is shown in Fig. 4. In a solar gas, X_{Fa} is insignificant, and the olivine is virtually pure forsterite above 850 K. Below this temperature, where reaction (2) begins to contribute significantly to the P_{H_2O}/P_{H_2} ratio at equilibrium, the curve is dashed, indicating the kinetic inhibition of this reaction. Even if gas phase equilibrium were to persist below this temperature, allowing f_{O_2} to rise sharply, X_{Fa} would only reach 0.11 by 500 K. The increase in fayalite content is not due simply to an exchange of Fe^{2+} for Mg^{2+} in the existing olivine; rather, the total amount of olivine increases. This is because the source of Fe^{2+} is progressive oxidation of metallic Fe with falling temperature, creating a progressively greater $(Mg + Fe^{2+})/Si$ ratio in the condensate. A source of Si is needed to consume the additional cations and, since the bulk of the Si not contained in olivine is predicted to be present as orthopyroxene in this temperature range, the latter must be converted into additional olivine. The reaction is written

$$2MgSiO_{3(c)} + 2Fe_{(c)} + 2H_2O_{(g)} \leftrightarrow Fe_2SiO_{4(c)} + Mg_2SiO_{4(c)} + 2H_{2(g)} \quad (3)$$

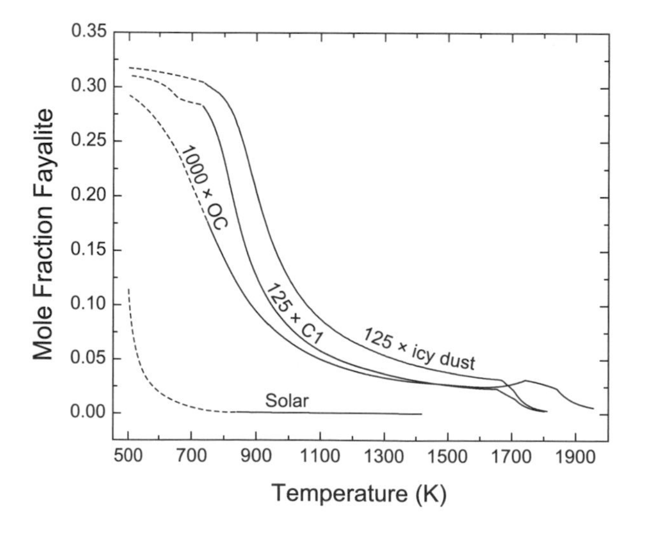

Fig. 4. Equilibrium fayalite content of condensate olivine as a function of temperature at a total pressure of 10^{-3} bar in a gas of solar composition and in cosmic gases enriched by a factor of 10^3 in OC dust, by a factor of 125 in dust of C1 chondrite composition, and by a factor of 125 in icy dust (see text). Each curve is dashed below the temperature at which the P_{CH_4}/P_{CO} ratio becomes ≥ 0.01, where the P_{H_2O}/P_{H_2} ratio becomes significantly affected by reaction (2).

At equilibrium, all olivine must have the same X_{Fa} at a given temperature, both the preexisting olivine and the newly created olivine. At 721 K, residual metallic Fe begins to react with $H_2S_{(g)}$ to form troilite, which consumes virtually all the sulfur and 48.6% of the Fe by 550 K. Because 37.1% of the total Fe remains in metallic form and is thus available for oxidation at 500 K, X_{Fa} is rising steeply when the calculation terminates at this temperature.

Reaction (3) is a solid-state reaction requiring the prior existence of enstatite, and controlled by the f_{O_2} of a coexisting fluid phase. Although reaction rate studies were used by *Imae et al.* (1993) to show that very little enstatite would be expected to form in the solar nebula, this phase has nevertheless been produced by *Toppani et al.* (2004) in 1-hr condensation experiments in high-temperature, low-pressure, multicomponent gases, and has been found by *Nguyen and Zinner* (2004) among presolar grains believed to have formed in stellar atmospheres in less than a few years. Equilibrium fayalite formation depends on a number of other factors, including equilibrium fluid phase speciation, intimate contact between reacting crystalline phases, high collision rates of gas species with the surfaces of these phases, and rapid solid-state diffusion (*Palme and Fegley,* 1990; *Krot et al.,* 2000). In this work, attention is focused on the degrees to which the sluggishness of gas phase reaction (2), the slow collision rate of Fe-bearing gaseous species with olivine crystal surfaces, and the slow rate of Fe-Mg interdiffusion in olivine at these temperatures affect the mean X_{Fa} of condensate olivine crystals.

Using equation (1), Fe-Mg interdiffusion coefficients were calculated from the f_{O_2}-temperature combinations along the solar gas curve in Fig. 3a. Using these, fayalite concentration profiles were calculated at successively lower temperatures until either a temperature was reached at which diffusion became so slow that further evolution of the profile was insignificant, or 850 K, where significant production of $H_2O_{(g)}$ via the kinetically hindered reaction (2) is required to occur at equilibrium, whichever was highest. Resulting concentration profiles for olivine crystals with radii of 3 μm and 1 μm immersed in a gas of solar composition during slow cooling are shown in Figs. 5a and 5b, respectively, and during fast cooling in Figs. 5c and 5d, respectively. In a solar gas at $P^{tot} = 10^{-3}$ bar, olivine first condenses at 1417 K with $X_{Fa} = 1.6 \times 10^{-4}$. The diffusion calculation for slow cooling begins at 1280 K, where the equilibrium X_{Fa} at the surface has risen to 2.8×10^{-4}. At this temperature, diffusion is fast enough that, for a 3-μm grain, the initial profile evolves into a uniform profile at $X_{Fa} = 3.0 \times 10^{-4}$ by 1250 K. Below 1100 K, however, the diffusion rate slows relative to the rate of increase of the surface concentration, and the profiles become steeply curved near the surface, as seen in Fig. 5a. By 850 K, the central X_{Fa} has reached 3.9×10^{-4} and the surface $X_{Fa} = 1.0 \times 10^{-3}$. At this point, the average X_{Fa} of the grain is 5.6×10^{-4}. When a 1-μm grain undergoes slow cooling, the concentration profile begins to curve below ~1050 K. By 850 K, the central $X_{Fa} = 4.6 \times 10^{-4}$, the surface $X_{Fa} = 1.0 \times$

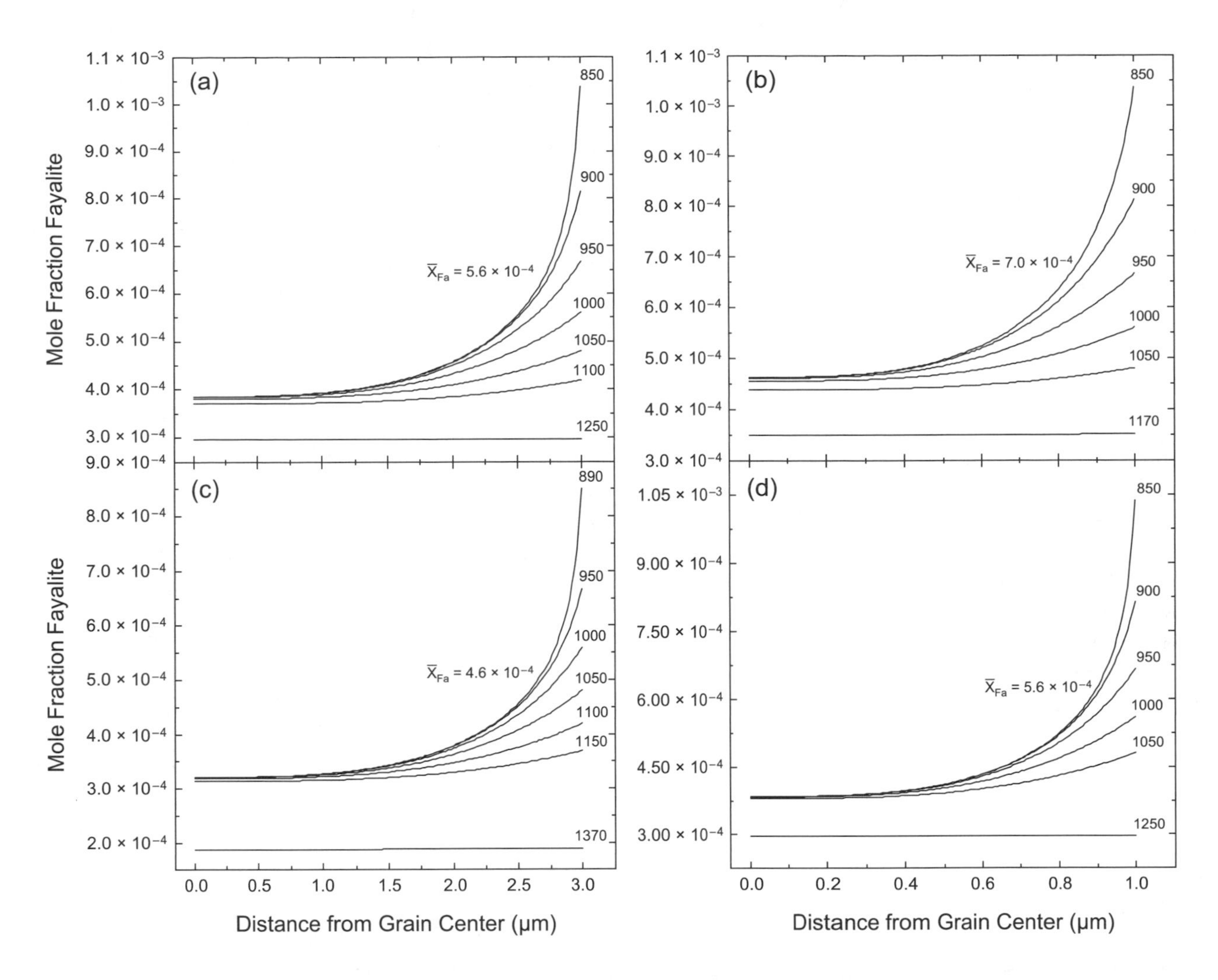

Fig. 5. Calculated variation of fayalite concentration with distance from the centers of condensate olivine crystals with radii of **(a,c)** 3 μm and **(b,d)** 1 μm whose surfaces are at equilibrium with the gas in a system of solar composition at $P^{tot} = 10^{-3}$ bar. For each grain size, profiles are shown at several temperatures during exponential cooling along the f_{O_2}-T path of a system of this composition, with decay constants of **(a,b)** 4.32×10^{-6} yr^{-1} and **(c,d)** 4.32×10^{-5} yr^{-1}. In each case, the numerical label associated with each curve is the temperature in Kelvins at which the profile was calculated, and the curve labeled with the mean X_{Fa} of the grain is the profile for the temperature below which diffusion becomes negligible.

10^{-3} and the mean $X_{Fa} = 7.0 \times 10^{-4}$ (Fig. 5b). In the case of fast cooling, the central X_{Fa} of a 3-μm grain has reached 3.2×10^{-4} at 890 K, and the surface $X_{Fa} = 8.5 \times 10^{-4}$ (Fig. 5c). At this point, the shapes of the profiles change very little with decreasing temperature except for the surface concentration, and the average X_{Fa} of the grain becomes virtually invariant with temperature, at a value of 4.6×10^{-4}. During fast cooling, the central $X_{Fa} = 4.0 \times 10^{-4}$, the surface $X_{Fa} = 1.0 \times 10^{-3}$, and the mean $X_{Fa} = 5.6 \times 10^{-4}$ for a 1-μm grain at 850 K (Fig. 5d).

The ability of olivine grains to reach their equilibrium X_{Fa} also depends on the rate of collision of gaseous Fe-containing species with olivine grain surfaces. In all cases considered in this work, the most abundant Fe-containing species is $Fe_{(g)}$ at high temperature and $Fe(OH)_{2(g)}$ at low temperature. At each temperature step of the condensation program, the partial pressure of the most abundant Fe-containing species was used to calculate the collision rate of gaseous Fe atoms with olivine grain surfaces. This was compared to the rate of change of the number of Fe atoms that must enter each olivine grain to produce the above diffusion profiles, calculated from the change in mean X_{Fa} for the same temperature step and olivine grain size. The collision rate is much higher than the entry rate at high temperature but falls very sharply with decreasing temperature, while the entry rate is relatively constant. The collision rate is assumed to be insufficient to support the entry rate at the temperature where the collision rate falls below 10× the entry rate. This temperature is well below the temperature at which the concentration profile becomes invariant, not only for each of the grain sizes and cooling rates considered above in a system of solar composition, but also for all other cases considered in this work. Thus, the collision rate is not the limiting factor in establishing the mean X_{Fa} in all cases considered herein.

Having established that the mean X_{Fa} that can be reached by olivine grains in solar gas is so much lower than the minimum X_{Fa} of precursors of UOC chondrules, we now repeat the above calculations in various dust-enriched systems.

3.2. Enrichment in Ordinary Chondrite Dust

Yoneda and Grossman (1995) computed a dust composition, hereinafter abbreviated OC, closely approximating that of ordinary chondrites but with a solar S/Si ratio. Full equilibrium condensation calculations were performed over the temperature range 2000–500 K at $P^{tot} = 10^{-3}$ bar in a system enriched in OC dust by factors of 100, 300, and 10^3. The P_{H_2O}/P_{H_2} ratios in these systems are compared to that of a solar gas in Fig. 2a. In these systems, the equilibrium P_{CH_4}/P_{CO} ratio becomes >0.01 below 790, 770, and 750 K respectively. This indicates that the P_{H_2O}/P_{H_2} ratios are not significantly affected by reaction (2) above these temperatures. Because this reaction is kinetically hindered, the curves in Fig. 2a are dashed below these temperatures. As expected, the equilibrium P_{H_2O}/P_{H_2} ratios are much higher than in a gas of solar composition over almost all of the investigated temperature range, but they converge toward one another below 900 K and actually meet at 430 K, below the temperature range of this figure. This behavior is due to the fact that it is the dust whose composition is in equilibrium with the gas in a system of solar composition at 430 K, which is added to the gas in various proportions to yield the OC dust-enriched systems in this figure. All this dust must condense back out of each of the dust-enriched systems by 430 K, leaving the coexisting gas with the original P_{H_2O}/P_{H_2} ratio at this temperature. In other words, a gas of solar composition is saturated in OC dust at 430 K and, no matter how much of this dust is dissolved in the gas at higher temperature, all of it reaches its solubility limit when the system is cooled to that temperature, leaving the same gas composition behind in each case.

In the system enriched in OC dust by a factor of 10^3, the P_{H_2O}/P_{H_2} ratio is very high, ~0.18, at 2000 K, declines sharply with falling temperature to ~0.035 by ~1600 K due to condensation of large amounts of O in such major silicate phases as forsterite, remains approximately constant until ~1000 K, and then declines sharply again, this time due to oxidation of Fe according to reaction (3). The difference between $\log f_{O_2}$ in each of these systems and that of the IW buffer is plotted as a function of temperature in Fig. 3a, where it is compared to that of a solar gas. Below the temperature at which reaction (2) begins to contribute significantly to the P_{H_2O}/P_{H_2} ratio at equilibrium, these curves are dashed, indicating the kinetic inhibition of this reaction. The shapes of these curves largely parallel those of the P_{H_2O}/P_{H_2} ratios, but the important point is that all these systems are much more oxidizing than a gas of solar composition over almost all of the investigated temperature range. The f_{O_2} of a system enriched by a factor of 10^3 in OC dust relative to solar composition is 5 log units closer to IW and thus 5 log units more oxidizing at 2000 K, and more than 3 log units more oxidizing from 1700 K to 800 K, than a solar gas. In Fig. 4, the temperature variation of the equilibrium X_{Fa} of condensate olivine in this system is compared to that in a solar gas. At such large dust enrichments, condensation occurs at temperatures so high that silicate melt is stable and olivine forms above 1950 K. At equilibrium, its X_{Fa} is higher than in a solar gas over the entire temperature range, reaching 0.05 at ~1100 K, 0.25 at ~640 K, and 0.29 at 500 K. At 750 K, where reaction (2) begins to affect the P_{H_2O}/P_{H_2} ratio significantly, X_{Fa} has reached 0.18.

Fayalite zoning profiles in condensate olivine grains are shown in Fig. 6 for a system enriched in OC dust by a factor of 10^3. The curves in this figure are calculated as in Fig. 5, except that the Fe-Mg interdiffusion coefficients are computed from the f_{O_2}-temperature combinations along the curve labeled "×1000" in Fig. 3a instead of along the solar gas curve. The higher f_{O_2} at each temperature that is characteristic of this dust-enriched system leads to higher X_{Fa} at a given temperature, in accord with Fig. 4. Figure 6a shows that, during slow cooling in a system enriched in OC dust by a factor of 10^3 at $P^{tot} = 10^{-3}$ bar, the mean X_{Fa} of a 3-μm olivine grain becomes virtually invariant at 0.078 at 815 K, where its central $X_{Fa} = 0.054$ and its surface $X_{Fa} = 0.13$. The profile of a 1-μm grain becomes invariant at 780 K, where the mean $X_{Fa} = 0.093$ (Fig. 6b). During fast cooling, the 3-μm grain achieves a mean $X_{Fa} = 0.065$ before diffusion stops, and the 1-μm grain reaches $X_{Fa} = 0.078$. For each case considered, diffusion stops at a higher temperature than that where reaction (2) would begin to contribute significantly to the P_{H_2O}/P_{H_2} ratio at equilibrium, so these results are unaffected by the failure of this reaction to reach equilibrium. Because the predicted values for the mean X_{Fa} are below the minimum X_{Fa} of precursors of UOC chondrules, neither the fast- nor the slow-cooling condition is capable of producing condensate olivine grains with the desired properties in a system enriched in OC dust by a factor of 10^3, unless the grains are significantly smaller than 1 μm in radius.

On the other hand, the discrepancy between predicted and observed X_{Fa} is relatively small, suggesting that grains of the desired compositions could be produced at OC dust enrichments only slightly greater than 10^3. As can be seen in Fig. 3a, any gas composition enriched in OC dust by less than that amount will be less oxidizing, will yield lower X_{Fa} at each temperature, and will fall short of matching the minimum X_{Fa} of precursors of UOC chondrules by an amount that increases with decreasing dust enrichment factor. The problem with using enrichment in dust of this composition as an explanation of the minimum X_{Fa} of chondrule precursors of UOCs is that the necessary degree of dust enrichment, $>10^3$, is far beyond what seems possible from models of coagulation and settling in the solar nebula. *Cas-*

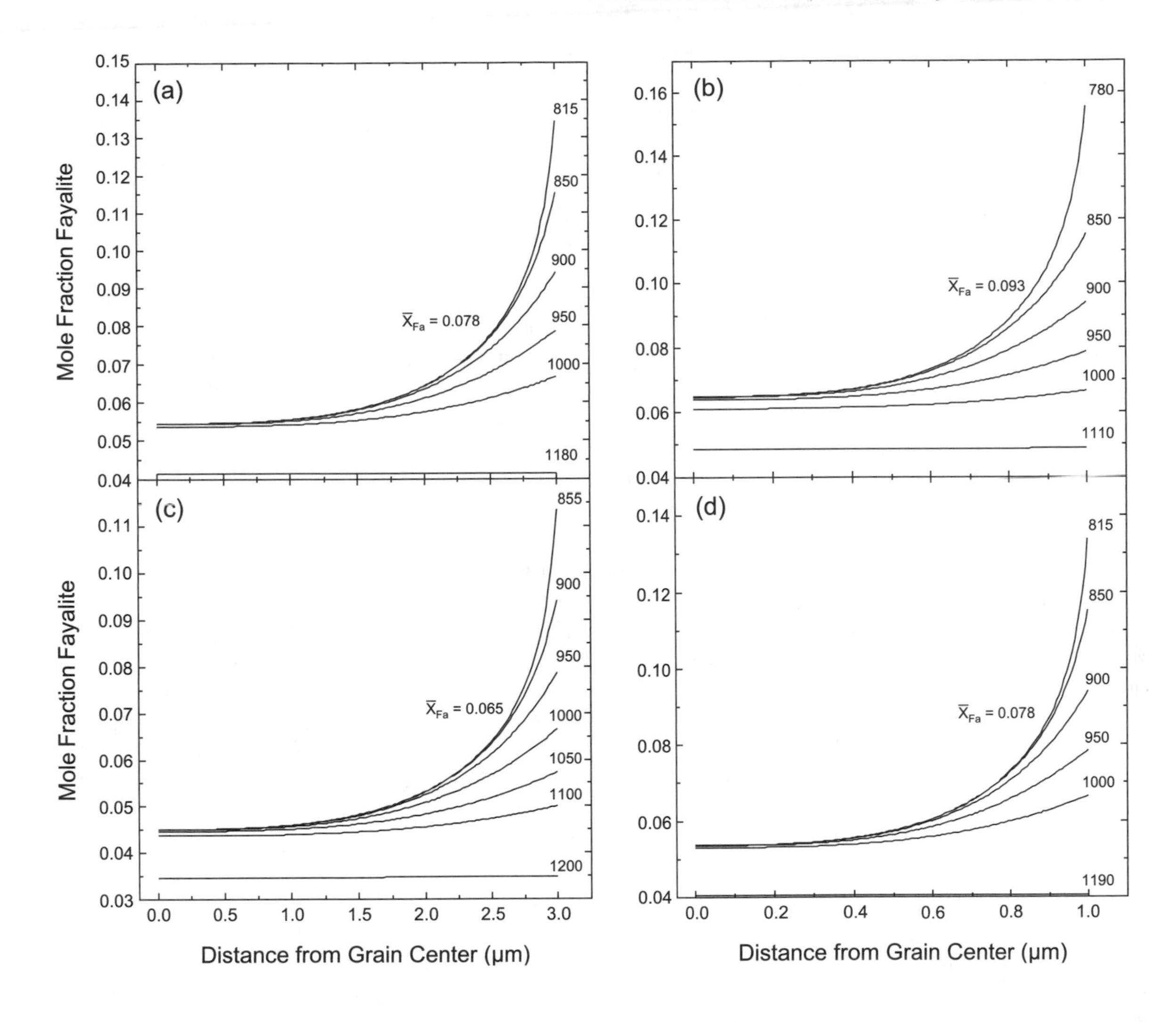

Fig. 6. Calculated variation of fayalite concentration with distance from the centers of condensate olivine crystals with radii of **(a,c)** 3 μm and **(b,d)** 1 μm whose surfaces are at equilibrium with the gas in a system enriched by a factor of 10^3 in OC dust at $P^{tot} = 10^{-3}$ bar. For each grain size, profiles are shown at several temperatures during exponential cooling along the f_{O_2}-T path of a system of this composition, with decay constants of **(a,b)** 4.32×10^{-6} yr^{-1} and **(c,d)** 4.32×10^{-5} yr^{-1}. In each case, the numerical label associated with each curve is the temperature in Kelvins at which the profile was calculated, and the curve labeled with the mean X_{Fa} of the grain is the profile for the temperature below which diffusion becomes negligible.

sen (2001) showed that, while dust-enrichment factors of up to ~130 may be produced under some circumstances, much lower enrichments are more frequently encountered in such models. Thus, if the mean X_{Fa} of precursors of UOC chondrules was established by condensation in dust-enriched regions of the nebula, dust more O-rich than the OC composition must have been responsible in order for the necessary f_{O_2} to be produced at lower dust enrichments.

3.3. Enrichment in C1 Chondrite Dust

C1 chondrites have much higher H_2O, and therefore O, concentrations than ordinary chondrites. Full equilibrium condensation calculations were performed over the temperature range 1900–500 K at $P^{tot} = 10^{-3}$ bar in systems enriched in C1 dust by factors of 50, 75, 100, and 125 relative to solar composition. The P_{H_2O}/P_{H_2} ratios in these systems are plotted as a function of temperature and compared to the ratios in a solar gas in Fig. 2b. In these systems, the equilibrium P_{CH_4}/P_{CO} ratio would become >0.01, and therefore reaction (2) would begin to affect the P_{H_2O}/P_{H_2} ratios significantly, below 760, 750, 740, and 740 K respectively. Because this reaction is kinetically hindered, the curves in Fig. 2b are dashed below these temperatures. The shapes of the curves for the C1 dust-enriched cases are very similar to that of solar gas but are displaced upward by 1–2 orders of magnitude relative to the latter. The P_{H_2O}/P_{H_2} ratios in the system enriched in C1 dust by a factor of 125 are quite comparable to those of the system enriched in OC dust by a factor of 10^3 over the temperature interval from 1600 to 900 K (Fig. 2a). Unlike the latter case, however, the equilibrium P_{H_2O}/P_{H_2} ratios in the system enriched in

C1 dust by a factor of 125 remain 50–120× higher than in a solar gas from 900 to 500 K. The differences between $\log f_{O_2}$ of the C1 dust-enriched systems and that of IW are plotted as a function of temperature in Fig. 3b, where they are compared to the same function for solar gas. In order to investigate the effect of P^{tot} on the f_{O_2} of a system enriched in C1 dust by a factor of 125, an identical condensation calculation was run in this system at $P^{tot} = 10^{-6}$ bar. Although not shown on Fig. 3b, the resulting curve for this system at $P^{tot} = 10^{-6}$ bar is coincident with that for the same dust enrichment at 10^{-3} bar below 1340 K. The shapes of the curves for the C1 dust-enriched cases are very similar to that of solar gas but are displaced upward by more than 3 orders of magnitude relative to the latter. Over the temperature range 1700 to 1000 K, $\log f_{O_2}$ of the system enriched in C1 dust by a factor of 125, ~IW-2.5, is very similar to that of the system enriched in OC dust by a factor of 10^3 (Fig. 3a). Below 1000 K, however, where the equilibrium $\log f_{O_2}$ of the system enriched in OC dust plunges with falling temperature to IW-2.8 by 500 K, that of the system enriched in C1 dust rises to ~IW-0.4 over the same temperature interval. The effect on X_{Fa} is shown in Fig. 4. From 1800 to 1000 K, the temperature interval where $\log f_{O_2}$ of the system enriched in C1 dust is similar to that of the system enriched in OC dust, X_{Fa} is also very similar in the two systems. Below 1000 K, however, the enhanced f_{O_2} of the system enriched in C1 dust causes X_{Fa} to rise more steeply with falling temperature than in the system enriched in OC dust, reaching 0.28 at ~740 K. Below this temperature, reaction (2) would begin to contribute significantly to the P_{H_2O}/P_{H_2} ratio if it were not kinetically inhibited. Even if equilibrium were maintained, however, the rate of increase of X_{Fa} with falling temperature would diminish, and X_{Fa} would level off at 0.31 at 500 K, as shown by the dashed extension of the curve labeled "125 × C1" in Fig. 4. For this enrichment, the curve of X_{Fa} vs. temperature for $P^{tot} = 10^{-6}$ bar is identical to the one shown at 10^{-3} bar below 1340 K.

In Fig. 7, fayalite zoning profiles in condensate olivine grains are shown at $P^{tot} = 10^{-3}$ bar in a system enriched in C1 dust by a factor of 125, a dust enrichment very close to the maximum found in *Cassen*'s (2001) coagulation and settling models. The curves in this figure are calculated as in Figs. 5 and 6, except that the Fe-Mg interdiffusion coefficients are computed from the f_{O_2}-temperature combinations along the curve labeled "×125" in Fig. 3b. The higher f_{O_2} of this system below 1000 K compared to the system enriched in OC dust by a factor of 10^3 results in higher values of the equilibrium X_{Fa} at each temperature, leading to higher average X_{Fa} in condensate olivine crystals. Figure 7a shows that, during slow cooling, the mean X_{Fa} of an olivine grain 3 μm in radius becomes virtually invariant at 0.11 at 790 K, where its central $X_{Fa} = 0.064$ and its surface $X_{Fa} = 0.24$. The profile of a grain whose radius is 1 μm becomes invariant at 770 K, where the mean $X_{Fa} = 0.14$. During fast cooling, the 3-μm grain achieves a mean $X_{Fa} = 0.08$ before diffusion stops, and the 1-μm grain reaches $X_{Fa} = 0.11$. For each case considered, diffusion stops at a temperature higher than 740 K, where reaction (2) begins to contribute significantly to the P_{H_2O}/P_{H_2} ratio at equilibrium, so these results are unaffected by the failure of reaction (2) to reach equilibrium. Because the temperature variations of both $\log f_{O_2}$ and X_{Fa} are identical at 10^{-3} bar and 10^{-6} bar in this temperature range, zoning profiles identical to those in Fig. 7 are obtained at a P^{tot} of 10^{-6} bar.

Thus, in a slowly cooled system enriched in C1 dust by a factor of 125 relative to solar composition, 3-μm olivine grains are produced whose average X_{Fa} is less than the minimum X_{Fa} of precursors of UOC chondrules. If faster cooling or larger grains are considered, the discrepancy between the calculated average X_{Fa} and the minimum value for UOC chondrule precursors becomes even greater. For grains with radii of 1 μm, the average X_{Fa} is very close to this value for slow cooling but not for fast cooling. Thus, in C1 dust-enriched systems, tiny olivine grains (radii ≤1 μm) can achieve an average X_{Fa} above the minimum X_{Fa} of precursors of UOC chondrules, but only at the highest dust enrichments and slowest cooling rates considered here. Only very tiny grains, with radii <~0.5 μm, can do so over the entire range of cooling rates considered here combined with the highest dust enrichments found in *Cassen*'s (2001) study. At much lower, but much more likely, dust enrichments, the average olivine condensate grain would have to be much smaller than 1 μm in order to reach the minimum X_{Fa}, 0.145. This cannot be considered a robust solution to the problem of how the Fe in the precursors of the chondrules of ordinary chondrites achieved its oxidation state.

Even if the precursor material of the UOCs did form at the highest dust enrichments and lowest cooling rates, the radius of the average olivine condensate grain would have to have been ≤1 μm in order to achieve the minimum X_{Fa} of this material. This small size does not seem reasonable. Isotopically anomalous single crystals of olivine and clinopyroxene, whose origin as condensates in the envelopes of AGB stars is indisputable, have radii up to 0.25 μm (*Nguyen and Zinner,* 2004). Because of the very high speed with which such grains escape AGB stars in stellar winds (*Bowen,* 1988), the observed grains must have grown in less than a few years (*Sharp and Wasserburg,* 1995). Combining the temperature interval over which $Mg_{(g)}$ is converted to forsterite in the systems considered in the present work with the solar nebular cooling times, the forsterite crystals into which the Fe must later diffuse grew over a period of at least several thousand years. While gas densities and supersaturation conditions must certainly differ between condensation in AGB stars and solar nebular condensation, it is unlikely that the typical size of a solar nebular condensate grain is less than that of a supernova condensate grain, and this seems to be borne out by observations in meteorites. It has been argued, for example, that fayalitic olivine laths in the matrix of Allende (*Wark,* 1979), hibonite crystals in the Murchison inclusion SH-6 (*MacPherson et al.,* 1984), "refractory" forsterite crystals in CM and CV chondrites (*Steele,* 1986; *Weinbruch et al.,* 2000), and radially zoned metal grains in CB chondrites (*Campbell et al.,* 2001) are primi-

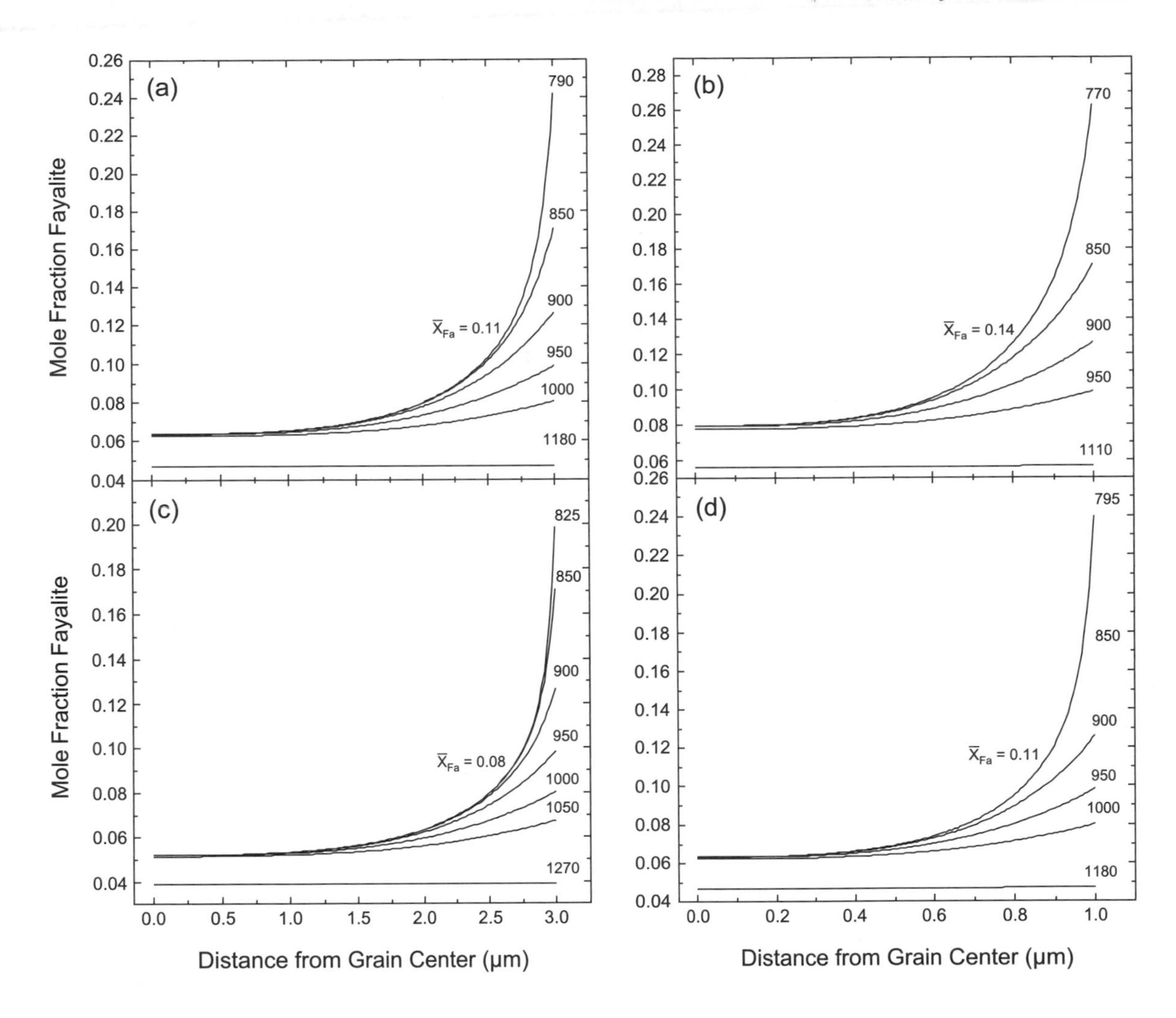

Fig. 7. Calculated variation of fayalite concentration with distance from the centers of condensate olivine crystals with radii of **(a,c)** 3 μm and **(b,d)** 1 μm whose surfaces are at equilibrium with the gas in a system enriched by a factor of 125 in dust of C1 chondrite composition at $P^{tot} = 10^{-3}$ bar. For each grain size, profiles are shown at several temperatures during exponential cooling along the f_{O_2}-T path of a system of this composition, with decay constants of **(a,b)** 4.32×10^{-6} yr^{-1} and **(c,d)** 4.32×10^{-5} yr^{-1}. In each case, the numerical label associated with each curve is the temperature in Kelvins at which the profile was calculated, and the curve labeled with the mean X_{Fa} of the grain is the profile for the temperature below which diffusion becomes negligible.

tive condensate grains from the solar nebula. While the hibonite crystals and fayalitic laths range from several micrometers up to 10 μm in length, the refractory forsterite and radially zoned metal grains can be up to several hundred micrometers in radius. Thus, the problem of the minimum X_{Fa} of olivine in precursors of UOC chondrules could be considered solved if a way could be found to produce olivine grains of at least several micrometers in radius whose mean $X_{Fa} \geq 0.145$.

3.4. Icy Dust with Higher Water Content than C1 Chondrites

A possible solution may be enrichment in dust that is even more O-rich than C1 chondrites. Suppose a type of nebular dust existed that contained solar proportions of condensable elements but was impregnated or coated with water ice such that the H_2O content exceeded that of C1 chondrites. Such icy dust is imagined to have been isolated from its complementary gas at a slightly lower temperature than C1 chondrites, such that more H_2O condensed into it, but no additional C. Total vaporization of a nebular region enriched in such dust would produce a system of higher f_{O_2} and stabilize higher fayalite contents at a given temperature than a system enriched by the same amount in C1 dust. For the purpose of investigating such a system, the primitive dust was assumed to have the composition of a mixture consisting of 1 part H_2O to 10 parts Orgueil by weight. Using the composition of Orgueil from *Anders and Grevesse* (1989), such dust would contain 25.5 wt% H_2O compared to 18.1% for Orgueil, and would have atomic O/Si and C/O ratios of 9.1 and 0.083, compared to 7.6 and 0.099 re-

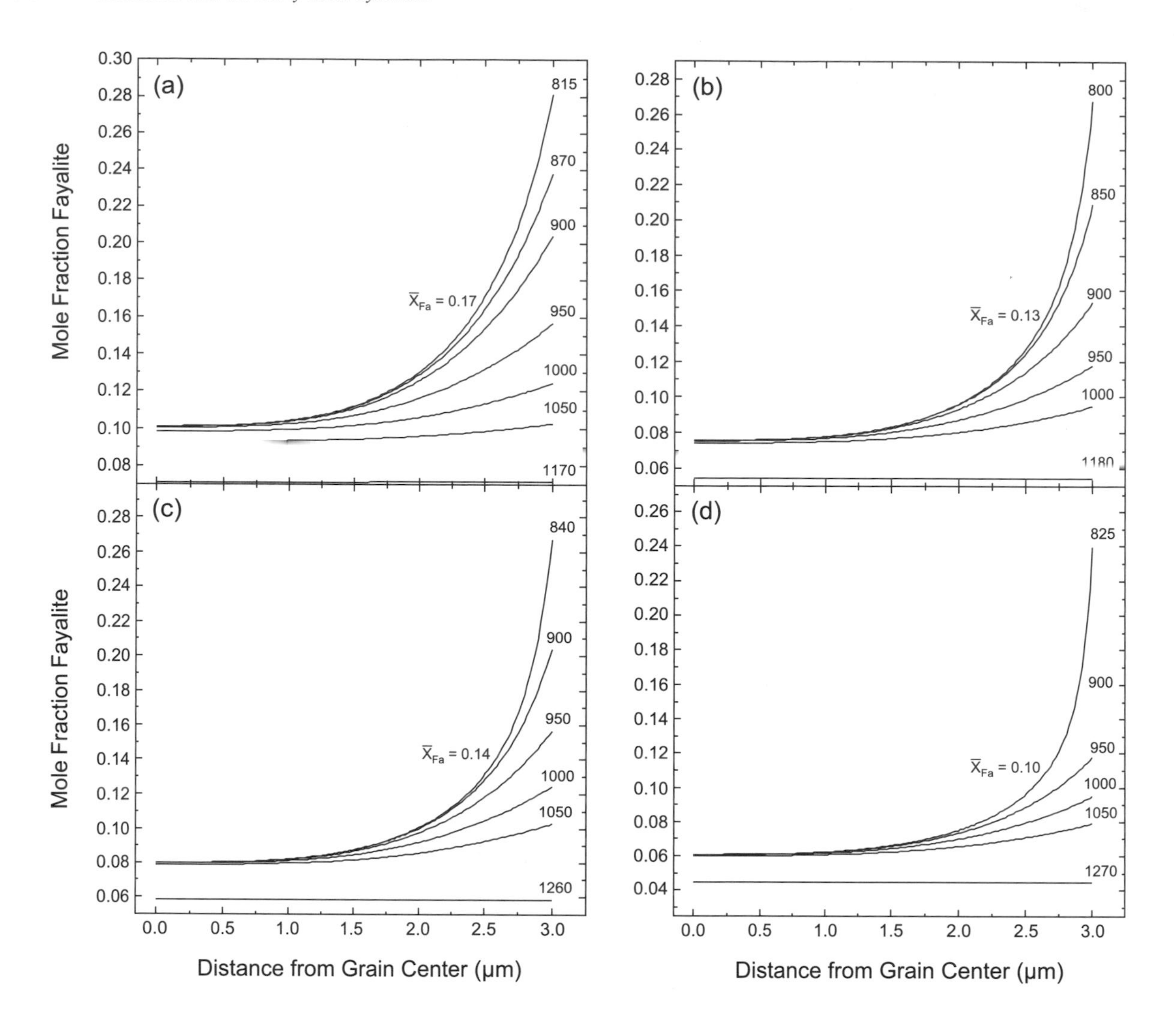

Fig. 8. Calculated variation of fayalite concentration with distance from the centers of condensate olivine crystals with radii of 3 μm whose surfaces are at equilibrium with the gas in a system enriched by a factor of **(a,c)** 125 and **(b,d)** 100 in icy dust (see text) at $P^{tot} = 10^{-3}$ bar. For each system composition, profiles are shown at several temperatures during exponential cooling along the appropriate f_{O_2}-T path, with decay constants of **(a,b)** 4.32×10^{-6} yr^{-1} and **(c,d)** 4.32×10^{-5} yr^{-1}. In each case, the numerical label associated with each curve is the temperature in Kelvins at which the profile was calculated, and the curve labeled with the mean X_{Fa} of the grain is the profile for the temperature below which diffusion becomes negligible.

spectively for Orgueil. Total condensation of such a region would produce material having chondritic relative proportions of nonvolatile elements.

The P_{H_2O}/P_{H_2} ratio of a system enriched in such icy dust by a factor of 125 relative to solar composition is plotted as a function of temperature, and compared to the ratios for systems enriched in C1 dust in Fig. 2b. In this system, reaction (2) would begin to contribute significantly to the P_{H_2O}/P_{H_2} ratio at 730 K if it were not kinetically inhibited. The shape of the curve for the icy dust-enriched system is very similar to that of the system enriched in C1 dust by a factor of 125 but is displaced upward by about 40% relative to the latter, below 1000 K. The difference between $\log f_{O_2}$ of the icy dust-enriched system and that of IW is plotted as a function of temperature in Fig. 3b, where it is compared to the same function for C1 dust-enriched systems. The shape of the curve for the icy dust-enriched system is very similar to that of the system enriched in C1 dust by a factor of 125 but is displaced upward by 0.3 log units relative to the latter, below 1000 K. The effect on X_{Fa} is shown in Fig. 4. The enhanced f_{O_2} of the system enriched in icy dust yields a higher X_{Fa} at each temperature than in the system enriched in C1 dust by a factor of 125. Critically important is the fact that X_{Fa} reaches 0.145 at 965 K in the icy dust-enriched system, compared to 875 K in the system enriched in C1 dust by a factor of 125.

In Figs. 8a and 8c, fayalite zoning profiles in condensate olivine grains are shown at $P^{tot} = 10^{-3}$ bar in a system enriched in icy dust by a factor of 125. The curves in this figure are calculated as in Figs. 5, 6, and 7, except that the Fe-Mg interdiffusion coefficients are computed from the f_{O_2}-temperature combinations along the curve labeled "125 ×

Icy dust" in Fig. 3b. The higher f_{O_2} of this system below 1000 K compared to the system enriched in C1 dust by a factor of 125 results in higher average X_{Fa} in condensate olivine crystals. Figure 8a shows that, during slow cooling in a system enriched in icy dust by a factor of 125, the mean X_{Fa} of an olivine grain 3 μm in radius becomes virtually invariant at 0.17 at 815 K, where its central $X_{Fa} = 0.10$ and its surface $X_{Fa} = 0.28$. During fast cooling, the grain achieves a mean $X_{Fa} = 0.14$ before diffusion stops. In a system enriched in icy dust by a factor of 100, f_{O_2} is ~0.1 log units higher than that for the system enriched in C1 dust by a factor of 125 at the same temperature. In such an icy dust-enriched system, the profile of a 3-μm grain becomes invariant at 800 K, where the mean $X_{Fa} = 0.13$, during slow cooling (Fig. 8b) and at 825 K, where the mean $X_{Fa} = 0.10$, during fast cooling (Fig. 8d). For each case considered, diffusion stops at a temperature higher than 730 K, where reaction (2) would begin to contribute significantly to the P_{H_2O}/P_{H_2} ratio at equilibrium, so these results are unaffected by the failure of reaction (2) to reach equilibrium.

Thus, in a system enriched in icy dust by a factor of 125 relative to solar composition, olivine grains with radii up to 3 μm will attain a mean X_{Fa} equal to or greater than the minimum X_{Fa} of precursors of UOC chondrules over the entire range of cooling rates investigated here. Even when the enrichment in icy dust is only a factor of 100 relative to solar composition, lower than the maximum found in coagulation and settling models, the mean X_{Fa} of 3-μm grains almost reaches the minimum needed for chondrule precursors at slow cooling but not at fast cooling. Thus, condensation of slowly cooled, icy dust-enriched regions could be considered a possible, minimal solution to the problem of how the mean X_{Fa} in chondrule precursors of primitive ordinary chondrites reached values of at least 0.145.

The opaque matrices of the least-equilibrated ordinary chondrites contain micrometer- to submicrometer-sized olivine grains whose average X_{Fa} ranges as high as 0.4 to 0.5, considerably greater than the mean X_{Fa} of the olivine in the chondrules in these meteorites, which is the focus of this study. While *Alexander et al.* (1989) argued that the fayalitic olivine in these matrices formed by parent-body metamorphism under relatively oxidizing conditions, *Huss et al.* (1981) suggested that this olivine could be a primitive condensate. If so, it is conceivable that such fayalite-rich material was the precursor of chondrules, and that olivine in the latter formed after reduction of this relatively FeO-rich material during chondrule formation. It should be emphasized, however, that no mechanism has been found in this study to produce, by equilibration with nebular gases, olivine of any grain size with such high fayalite contents. It was shown above that, at a given temperature, the equilibrium X_{Fa} of condensate olivine increases in systems with progressively higher f_{O_2}, and that, in each system, X_{Fa} reaches a maximum with falling temperature before leveling off. As seen in Fig. 4, however, these maxima on the equilibrium curves are only reached below the temperatures at which reaction (2) begins to affect the P_{H_2O}/P_{H_2} ratios significantly and, because this reaction does not proceed at equilibrium, these maximum values of X_{Fa} may not be reached. Furthermore, even if equilibrium were maintained, there is virtually no difference in the maximum X_{Fa} between the system enriched by a factor of 125 in icy dust and the one enriched by the same factor in C1 dust, despite the difference in f_{O_2} between the two systems. The reason for this behavior is shown in Figs. 9a and 9b, in which the equilibrium distribution of Fe between crystalline phases and vapor is compared in the two systems. In both systems, metallic Ni-Fe condenses at ~1710 K and begins to react with $H_2S_{(g)}$ to form troilite at ~1250 K. As the temperature falls in both systems, metallic Fe is gradually consumed to form fayalite and ferrosilite, but the higher f_{O_2} of the icy dust-enriched system causes the proportion of the total Fe in silicates to that in metal to increase more rapidly with falling temperature. This is the reason why the curve of X_{Fa} in Fig. 4 rises more steeply with falling temperature for the icy dust-enriched system than for the C1 dust-enriched system. Because the total S ultimately condenses as troi-

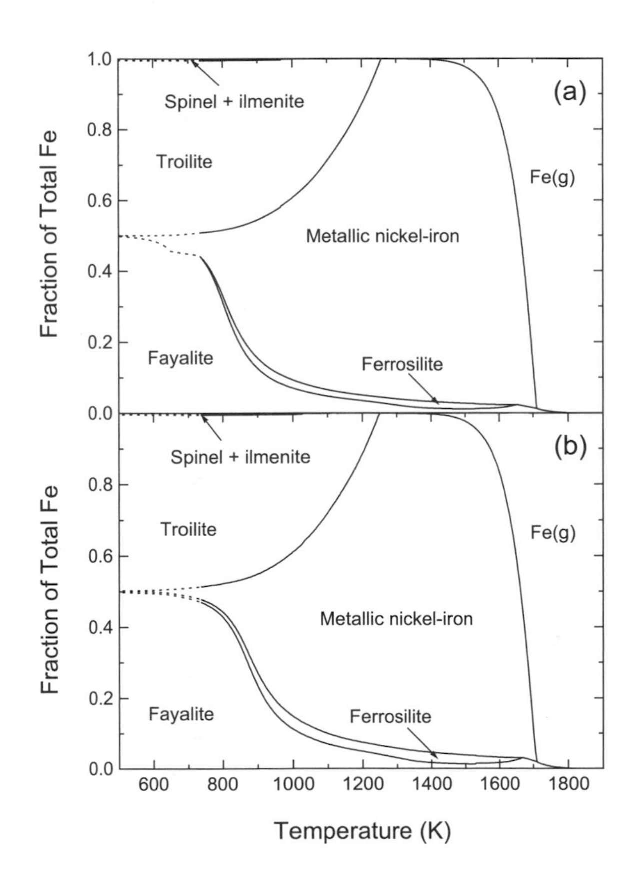

Fig. 9. Equilibrium distribution of Fe between crystalline phases and vapor as a function of temperature at $P^{tot} = 10^{-3}$ bar in a system enriched by a factor of 125 in **(a)** dust of C1 chondrite composition and **(b)** icy dust. Each curve is dashed below the temperature at which the P_{CH_4}/P_{CO} ratio becomes ≥0.01, where the P_{H_2O}/P_{H_2} ratio becomes significantly affected by reaction (2). Despite the higher f_{O_2} in **(b)**, the maximum fraction of the total Fe in silicates is the same.

lite, thereby consuming 49.6% of the total Fe in both systems, however, the fraction of the total Fe that forms fayalite can never exceed 50.4%, corresponding to X_{Fa} = 0.32, regardless of the difference between the two systems in f_{O_2} or in the rate of increase of the proportion of Fe in silicates with temperature. This value is much lower than the range of X_{Fa} seen in the matrices of UOCs.

The minimum dust enrichment required to produce condensate olivine grains with radii ≥1 μm and mean X_{Fa} above the desired value over the entire range of nebular midplane cooling rates lies within the range of dust enrichments encountered in astrophysical models of coagulation and settling only if the dust is more O-rich than C1 chondrites. Such conditions by themselves, however, are insufficient to yield at least one other fundamental mineralogical characteristic of ordinary chondrites: the distribution of Fe among coexisting metallic, sulfide, and silicate phases. In the above models, a maximum temperature is found at which the mean X_{Fa} of zoned condensate olivine grains is frozen in at the minimum X_{Fa} of the precursors of UOC chondrules. In the system enriched in icy dust by a factor of 125, this temperature ranges from 815 to 840 K for 3-μm grains, depending on cooling rate. In this temperature range, ~94% of the sulfur will have condensed, causing ~47% of the Fe to be in troilite, ~41% in silicates, and only ~12% in metallic Ni-Fe, as seen in Fig. 9b. All ordinary chondrites are depleted in S relative to Si by ~75% compared to solar abundances, presumably because metallic Ni-Fe grains were isolated from further reaction with the gas at a temperature above that necessary for complete condensation of S, due either to their being armored or buried during accretion. As a result, only 12.6%, 16.7%, and 18.7% of the total Fe in H, L, and LL chondrites, respectively, occurs as troilite (*Jarosewich,* 1990), much lower than would be expected had equilibrium conversion of metallic Fe to troilite continued to the same temperature as that where the minimum X_{Fa} of the precursors of chondrules in these meteorites was established by oxidation of the same metallic Fe grains. For condensate grains with mean X_{Fa} greater than the minimum value, the discrepancy between the cessation temperature of oxidation of the metal grains and that of sulfidation of the same population of grains would be even greater. This is an important drawback of using any dust containing solar proportions of condensable elements to enhance the f_{O_2} of the nebular source of the ordinary chondrites. A possible solution is to enrich a nebular region in dust that is even more ice-rich than that used above. This would lower the magnitude of the dust enrichment required to produce a mean X_{Fa} = 0.145 at 850 K. Although the lower dust enrichment would, in turn, lower the condensation temperature of troilite, the effect would probably be insufficient to prevent 75% of the S from condensing by 850 K.

3.5. Limitations

As seen above, computation of the mean X_{Fa} of an olivine grain of a given size is based on a diffusion calculation. In all the experimental determinations of interdiffusion rates in olivine reviewed above, the Fe-Mg interdiffusion coefficients, Ds, have similar temperature- and f_{O_2}-dependences as those in equation (1), but the lowest temperature at which any of those measurements was made is 1173 K, and the lowest f_{O_2} ~10^{-13}. In this work, extrapolation of these data to ~800 K and f_{O_2} ~10^{-30} assumes no change in diffusion mechanism in the intervening intervals of temperature and f_{O_2} that would alter these dependences. Although D is known to be anisotropic in olivine, and olivine crystals exhibit a variety of aspect ratios, the calculations in this work were simplified by assuming spherical geometry and isotropic diffusion, and by employing Ds parallel to the c-axis. The error introduced by doing so is limited, however, as Ds parallel to the different crystallographic axes vary by only a factor of ~3. This is relatively small compared to the difference between the Ds of *Chakraborty* (1997), which are employed in this study, and those of *Nakamura and Schmalzried* (1984), which are a factor of 10 higher at 1323 K. When the latter data are used in the case of enrichment in icy dust by a factor of 125, the 3-μm grains reach X_{Fa} = 0.19 and 0.16 during slow and fast cooling, respectively, instead of 0.17 and 0.14, which were estimated above. The differences in mean X_{Fa} produced by substituting the *Nakamura and Schmalzried* (1984) data for the *Chakraborty* (1997) data are thus so small that our conclusions about icy dust-enriched systems would be little affected. Correcting for diffusion anisotropy is expected to produce even smaller differences.

The diffusion calculations in this work require the results of condensation calculations as input parameters. Although the condensation calculations assume complete equilibrium, requiring that all grains of solid solution minerals be uniform and homogeneous in composition, the diffusion calculation shows that condensate olivine grains will, in reality, be zoned in composition. Thus, a possible limitation of the results presented here is that the effect of olivine zonation has not been fed back into the condensation calculation. This effect is expected to be minor in the dust-enriched systems, as it results in more excess metallic Fe and only slightly higher P_{H_2O}/P_{H_2} ratios than in the equilibrium cases calculated here. Another possible limitation is that negligible olivine grain growth is assumed to occur as its concentration changes.

4. CONCLUSIONS

A gas of solar composition is too reducing to allow the equilibrium X_{Fa} in condensate olivine to reach the minimum X_{Fa} of the precursors of UOC chondrules, 0.145, at temperatures above that where gas phase equilibrium breaks down or where Fe-Mg interdiffusion in olivine stops. When a region enriched in dust relative to gas compared to solar composition is vaporized, the resulting vapor has higher f_{O_2} than a gas of solar composition. In such dust-enriched systems, the equilibrium X_{Fa} of olivine at almost any temperature is higher than in a system of solar composition, and the amount by which it is increased increases with the degree of dust enrichment. For OC dust, even enrichments of

10^3 relative to solar composition are insufficient to produce a mean X_{Fa} in condensate olivine crystals with radii of 1 μm and 3 μm that is above the desired level at temperatures where diffusion occurs. Only dust-enrichment factors near the maximum allowed in coagulation and settling models, together with C1 chondrite dust whose O content has been enhanced by admixture of water ice, can yield olivine condensate grains with radii ≥1 μm whose mean X_{Fa} exceeds the minimum X_{Fa} of the precursors of UOC chondrules over the entire range of nebular midplane cooling rates. This unlikely set of circumstances cannot be considered a robust solution to the problem of the relatively high fayalite content of UOC olivine, which remains unsolved.

Acknowledgments. We are grateful for valuable advice from A. J. Campbell and J. Ganguly on diffusion calculations, and from M. Ghiorso on treatment of non-ideality in olivine and orthopyroxene. We thank T. Bernatowicz, J. N. Grossman, G. R. Huss, R. Jones, V. Kress, S. B. Simon, and S. Yoneda for helpful discussions. The manuscript benefited from the in-depth comments of three anonymous reviewers. This research was supported by funds from the National Aeronautics and Space Administration through grant NAG5-11588.

REFERENCES

Alexander C. M. O., Hutchison R., and Barber D. J. (1989) Origin of chondrule rims and interchondrule matrices in unequilibrated ordinary chondrites. *Earth Planet. Sci. Lett., 95,* 187–207.

Allende Prieto C., Lambert D. L., and Asplund M. (2001) The *forbidden* abundance of oxygen in the sun. *Astrophys. J. Lett., 556,* L63–L66.

Allende Prieto C., Lambert D. L., and Asplund M. (2002) A reappraisal of the solar photospheric C/O ratio. *Astrophys. J. Lett., 573,* L137–L140.

Anders E. and Grevesse N. (1989) Abundances of the elements: Meteoritic and solar. *Geochim. Cosmochim. Acta, 53,* 197–214.

Beckett J. R. (1986) The origin of calcium-, aluminum-rich inclusions from carbonaceous chondrites: An experimental study. Ph.D. thesis, University of Chicago.

Berman R. G. (1988) Internally-consistent thermodynamic data for minerals in the system Na_2O-K_2O-CaO-MgO-FeO-Fe_2O_3-Al_2O_3-SiO_2-TiO_2-H_2O-CO_2. *J. Petrol., 29,* 445–522.

Bowen G. H. (1988) Dynamical modeling of long-period variable star atmospheres. *Astrophys. J., 329,* 299–317.

Brearley A. J. and Jones R. H. (1998) Chondritic meteorites. In *Planetary Materials* (J. J. Papike, ed.), pp. 3-1 to 3-398. Reviews in Mineralogy, Vol. 36, Mineralogical Society of America.

Buening D. K. and Buseck P. R. (1973) Fe-Mg lattice diffusion in olivine. *J. Geophys. Res., 78,* 6852–6862.

Campbell A. J., Humayun M., Meibom A., Krot A. N., and Keil K. (2001) Origin of metal grains in the QUE94411 chondrite. *Geochim. Cosmochim. Acta, 65,* 163–180.

Cassen P. (2001) Nebular thermal evolution and the properties of primitive planetary materials. *Meteoritics & Planet. Sci., 36,* 671–700.

Chakraborty S. (1997) Rates and mechanisms of Fe-Mg interdiffusion in olivine at 980°–1300°C. *J. Geophys. Res., 102,* 12317–12331.

Chase M. W. Jr., Davies C. A., Downey J. R., Frurip D. J., McDonald R. A., and Syverud A. N. (1985) *JANAF Thermochemical Tables, 3rd edition. J. Phys. Chem. Ref. Data, 14, Suppl. 1.* Dow Chemical Company, Midland, Michigan.

Crank J. (1975) *The Mathematics of Diffusion, 2nd edition.* Oxford Univ., Oxford.

Dodd R. T. Jr., Van Schmus W. R., and Koffman D. M. (1967) A survey of the unequilibrated ordinary chondrites. *Geochim. Cosmochim. Acta, 31,* 921–951.

Dohmen R., Chakraborty S., Palme H., and Ramensee W. (1998) Solid-solid reactions mediated by a gas phase: An experimental study of reaction progress and the role of surfaces in the system olivine + iron metal. *Am. Mineral., 83,* 970–984.

Dowty E. and Clark J. R. (1973) Crystal structure refinement and optical properties of a Ti^{3+} fassaite from the Allende meteorite. *Am. Mineral., 58,* 230–242.

Ebel D. S. and Grossman L. (2000) Condensation in dust-enriched systems. *Geochim. Cosmochim. Acta, 64,* 339–366.

Fedkin A. V. and Grossman L. (2004) Nebular formation of fayalitic olivine: Ineffectiveness of dust enrichment (abstract). In *Lunar and Planetary Science XXXV,* Abstract #1823. Lunar and Planetary Institute, Houston (CD-ROM).

Gronvold F. and Stolen S. (1992) Thermodynamics of iron sulfides II. Heat capacity and thermodynamic properties of FeS and of $Fe_{0.875}S$ at temperatures from 298.15K to 1000K, of $Fe_{0.98}S$ from 298.15K to 800K, and of $Fe_{0.89}S$ from 298.15K to about 650K. Thermodynamics of formation. *J. Chem. Thermodyn., 24,* 913–936.

Grossman L. (1972) Condensation in the primitive solar nebula. *Geochim. Cosmochim. Acta, 36,* 597–619.

Grossman L. (1980) Refractory inclusions in the Allende meteorite. *Annu. Rev. Earth Planet. Sci., 8,* 559–608.

Grossman L., Ebel D. S., Simon S. B., Davis A. M., Richter F. M., and Parsad N. M. (2000) Major element chemical and isotopic compositions of refractory inclusions in C3 chondrites: The separate roles of condensation and evaporation. *Geochim. Cosmochim. Acta, 64,* 2879–2894.

Grossman L., Ebel D. S., and Simon S. B. (2002) Formation of refractory inclusions by evaporation of condensate precursors. *Geochim. Cosmochim. Acta, 66,* 145–161.

Grossman L. and Fedkin A. V. (2003) Nebular oxidation of iron (abstract). *Geochim. Cosmochim. Acta, 67, Suppl.,* A125.

Huss G. R., Keil K., and Taylor G. J. (1981) The matrices of unequilibrated ordinary chondrites: Implications for the origin and history of chondrites. *Geochim. Cosmochim. Acta, 45,* 33–51.

Imae N., Tsuchiyama A., and Kitamura M. (1993) An experimental study of enstatite formation reaction between forsterite and Si-rich gas. *Earth Planet. Sci. Lett., 118,* 21–30.

Jarosewich E. (1990) Chemical analyses of meteorites: A compilation of stony and iron meteorite analyses. *Meteoritics, 25,* 323–337.

Krot A. N., Fegley B. Jr., Lodders K., and Palme H. (2000) Meteoritical and astrophysical constraints on the oxidation state of the solar nebula. In *Protostars and Planets IV* (V. Mannings et al., eds.), pp. 1019–1054. Univ. of Arizona, Tucson.

Larimer J. W. (1975) The effect of C/O ratio on the condensation of planetary material. *Geochim. Cosmochim. Acta, 39,* 389–392.

Lewis J. S. and Prinn R. G. (1980) Kinetic inhibition of CO and N_2 reduction in the solar nebula. *Astrophys. J., 238,* 357–364.

Lodders K. (2003) Solar system abundances and condensation temperatures of the elements. *Astrophys. J., 591,* 1220–1247.

MacPherson G. J., Bar-Matthews M., Tanaka T., Olsen E., and Grossman L. (1983) Refractory inclusions in the Murchison meteorite. *Geochim. Cosmochim. Acta, 47,* 823–839.

MacPherson G. J., Grossman L., Hashimoto A., Bar-Matthews M., and Tanaka T. (1984) Petrographic studies of refractory inclusions from the Murchison meteorite. In *Proc. Lunar Planet. Sci. Conf. 15th,* in *J. Geophys. Res., 89,* C299–C312.

McCoy T. J., Scott E. R. D., Jones R. H., Keil K., and Taylor G. J. (1991) Composition of chondrule silicates in LL3–5 chondrites and implications for their nebular history and parent body metamorphism. *Geochim. Cosmochim. Acta, 55,* 601–619.

Misener D. J. (1974) Cationic diffusion in olivine to 1400°C and 35 kbar. In *Geochemical Transport and Kinetics* (A. W. Hofmann et al., eds.), pp. 117–129. Carnegie Institution of Washington Publication 634, Washington, DC.

Miyamoto M., Mikouchi T., and Arai T. (2002) Comparison of Fe-Mg interdiffusion coefficients in olivine. *Antarct. Meteorite Res., 15,* 143–151. National Institute of Polar Research, Tokyo.

Nakamura A. and Schmalzried H. (1984) On the Fe^{2+}-Mg^{2+} interdiffusion in olivine (II). *Ber. Bunsenges. Phys. Chem., 88,* 140–145.

Nguyen A. N. and Zinner E. (2004) Discovery of ancient silicate stardust in a meteorite. *Science, 303,* 1496–1499.

Palme H. and Fegley B. Jr. (1990) High-temperature condensation of iron-rich olivine in the solar nebula. *Earth Planet. Sci. Lett., 101,* 180–195.

Prinn R. G. and Fegley B. Jr. (1989) Solar nebula chemistry: Origin of planetary, satellite, and cometary volatiles. In *Origin and Evolution of Planetary and Satellite Atmospheres* (S. K. Atreya et al., eds.), pp. 78–136. Univ. of Arizona, Tucson.

Rambaldi E. R. (1981) Relict grains in chondrules. *Nature, 293,* 558–561.

Rietmeijer F. J. M. (1998) Interplanetary dust particles. In *Planetary Materials* (J. J. Papike, ed.), pp. 2-1 to 2-95. Reviews in Mineralogy, Vol. 36, Mineralogical Society of America.

Robie R. A., Hemingway B. S., and Fisher J. R. (1978) *Thermodynamic Properties of Minerals and Related Substances at 298.15K and 1 Bar (10^3 Pascals) Pressure and at Higher Temperatures.* U.S. Geological Survey Bulletin No. 1452.

Robinson G. R. Jr., Haas J. L. Jr., Schafer C. M., and Haselton H. T. Jr. (1982) *Thermodynamic and Thermophysical Properties of Selected Phases in the MgO-SiO_2-H_2O-CO_2, CaO-Al_2O_3-SiO_2-H_2O-CO_2, and Fe-FeO-Fe_2O_3-SiO_2 Chemical Systems, with Special Emphasis on the Properties of Basalts and Their Mineral Components.* U.S. Geological Survey Open-File Report No. 83-79.

Sack R. O. and Ghiorso M. S. (1989) Importance of considerations of mixing properties in establishing an internally consistent thermodynamic database: Thermochemistry of minerals in the system Mg_2SiO_4-Fe_2SiO_4-SiO_2. *Contrib. Mineral. Petrol., 102,* 41–68.

Sharp C. M. and Wasserburg G. J. (1995) Molecular equilibria and condensation temperatures in carbon-rich gases. *Geochim. Cosmochim. Acta, 59,* 1633–1652.

Steele I. M. (1986) Compositions and textures of relic forsterite in carbonaceous and unequilibrated ordinary chondrites. *Geochim. Cosmochim. Acta, 50,* 1379–1395.

Toppani A., Libourel G., Robert F., Ghanbaja J., and Zimmermann L. (2004) Synthesis of refractory minerals by high-temperature condensation of a gas of solar composition (abstract). In *Lunar and Planetary Science XXXV,* Abstract #1726. Lunar and Planetary Institute, Houston (CD-ROM).

Tscharnuter W. M. and Boss A. P. (1993) Formation of the protosolar nebula. In *Protostars and Planets III* (E. H. Levy and J. I. Lunine, eds.), pp. 921–938. Univ. of Arizona, Tucson.

Wark D. A. (1979) Birth of the presolar nebula: The sequence of condensation revealed in the Allende meteorite. *Astrophys. Space Sci., 65,* 275–295.

Weinbruch S., Palme H., and Spettel B. (2000) Refractory forsterite in primitive meteorites: Condensates from the solar nebula? *Meteoritics & Planet. Sci., 35,* 161–171.

Wood J. A. (1967) Olivine and pyroxene compositions in Type II carbonaceous chondrites. *Geochim. Cosmochim. Acta, 31,* 2095–2108.

Yoneda S. and Grossman L. (1995) Condensation of CaO-MgO-Al_2O_3-SiO_2 liquids from cosmic gases. *Geochim. Cosmochim. Acta, 59,* 3413–3444.

Volatile Evolution and Loss

Andrew M. Davis
University of Chicago

Relative to bulk solar system composition, most meteorites and the terrestrial planets are depleted in volatile elements. This abundance pattern likely arose in the solar nebula rather than during planet formation. The solar system record of volatile element abundances is reviewed and constraints on its origin from isotopic compositions and abundances of presolar grains in primitive meteorites are explored. Major models for the origin of volatility fractionations are described.

1. INTRODUCTION

The CI chondrites are generally regarded as being representative of the bulk elemental composition of the solar system, with the exception of the highly volatile elements hydrogen, carbon, nitrogen, oxygen, and the noble gases (*Anders and Grevesse,* 1989; *Lodders,* 2003; *Palme and Jones,* 2003). Relative to CI chondrites, all other groups of meteorites, as well as bulk terrestrial planets, are depleted in volatile elements. These depletions differ from group to group, but for the most part have the common characteristic of being reasonably smooth functions of volatility (expressed as 50% condensation temperatures). The volatile element patterns of the terrestrial planets and achondrite parent bodies are less smooth, because igneous differentiation has modified trace-element abundances. However, some incompatible elements are not significantly decoupled by igneous differentiation and ratios of volatile to refractory incompatible elements can be used to gauge the level of volatile elements in the bulk planets and parent bodies. The depletion patterns of meteorites have fascinated cosmochemists for decades, yet remain incompletely understood. In this chapter, the volatility patterns of meteorites will be reviewed and their implications for early solar system processes explored.

The subject of volatile elements in meteorites was reviewed in the first *Meteorites and the Early Solar System* volume (*MESS I*) (*Palme et al.,* 1988). Since that time, there has been considerable progress in the field. Although little has changed concerning the basic elemental abundance patterns that are to be explained, the new tools of isotopic mass fractionation and presolar grain abundances allow new constraints to be placed on the mechanism of volatile-element fractionation.

2. THE RECORD

2.1. Elemental Abundances

The CI chondrites are believed to be representative of the bulk elemental composition of the solar system, because abundances of all but the most volatile elements in them match those in the solar photosphere (*Anders and Grevesse,* 1989; *Lodders,* 2003; *Palme and Jones,* 2003). Thus, it is customary to normalize elemental abundances in other solar system objects to those in CI chondrites. In comparing elemental abundances in meteorites and planets with CI chondrites, it is useful to divide the elements into different groups based on their physical and chemical properties (*Larimer,* 1988). Since chondrites and planets have varying quantities of water and organic matter, it is also common to normalize to the abundant moderately volatile element silicon. Magnesium is sometimes used in place of silicon, particularly in sets of analyses collected by neutron activation analysis (silicon cannot be determined by this method).

The primary cosmochemical classification of the elements is based on volatility, with the elements divided into those that are refractory (having equilibrium condensation temperatures higher than the most abundant rock-forming elements, magnesium, silicon, and iron), moderately volatile (having condensation temperatures lower than those of the refractory elements, but higher than that of FeS), and highly volatile (having condensation temperatures below that of FeS). For a gas of solar composition at a total pressure of 10^{-4} atm, refractory elements have 50% condensation temperatures above 1335 K, moderately volatiles between 1335 and 665 K, and volatiles below 665 K [condensation temperatures from *Lodders* (2003)]. All these elements have the same relative abundances as the solar photosphere, within uncertainties (*Anders and Grevesse,* 1989; *Lodders,* 2003; *Palme and Jones,* 2003). A fourth group, the atmophile elements (hydrogen, carbon, nitrogen, oxygen, and the noble gases, having condensation temperatures of 180 K or below), are clearly depleted in CI chondrites relative to the Sun. The other important classification of elements is based on what type of phases elements concentrate in: siderophile elements concentrate into iron-nickel metal; chalcophile elements into troilite, the most abundant sulfide in meteorites; and lithophile elements into oxides and silicates. A cosmochemical periodic table of the elements is given in Fig. 1, in which the 50% condensation temperatures for a gas of solar composition at 10^{-4} atm total pressure (*Lodders,* 2003) are given.

The highly volatile elements can be lost during parent-body processes such as shock heating and redistributed dur-

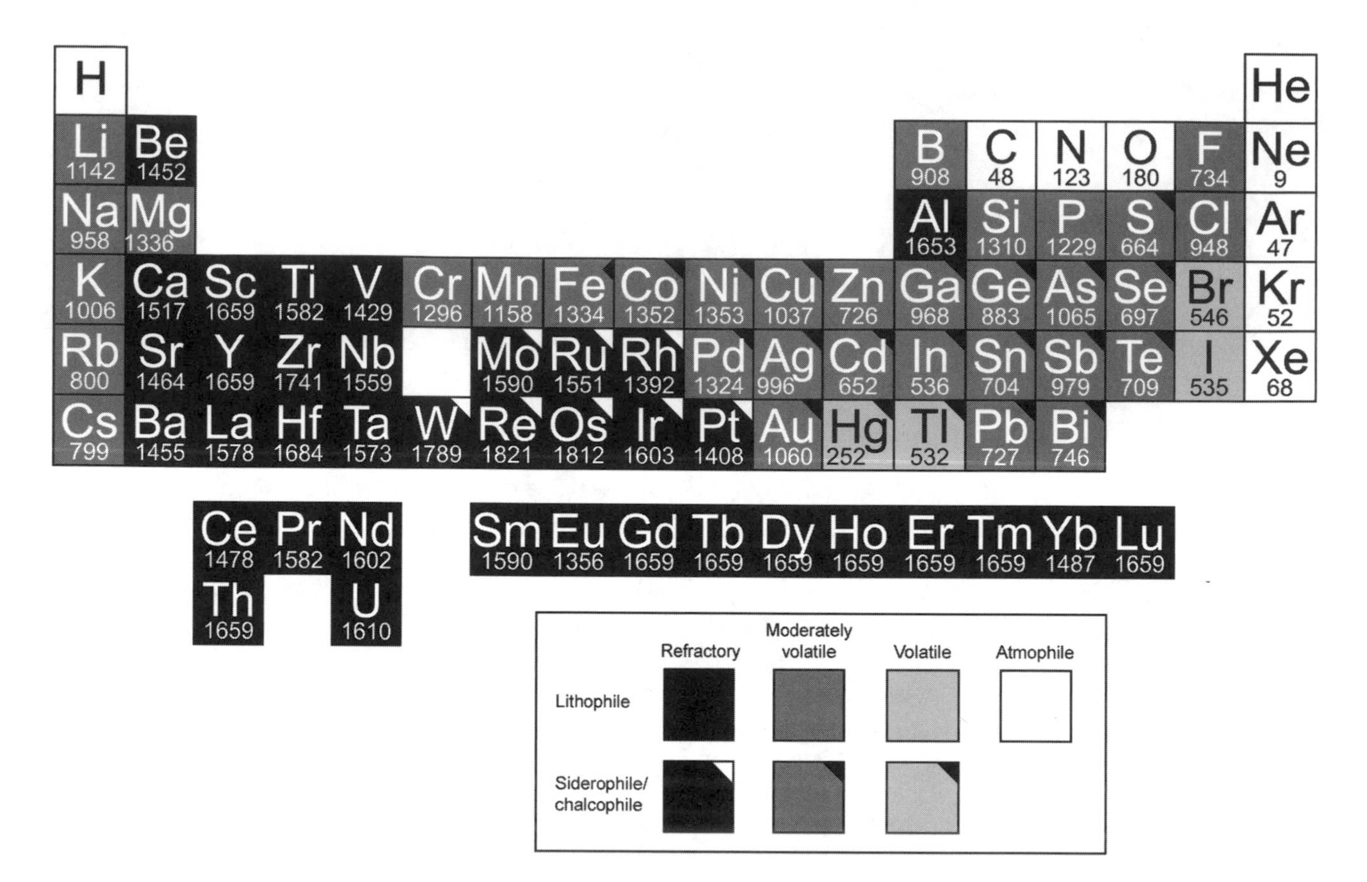

Fig. 1. A cosmochemical periodic table of the elements. Only stable or very long-lived unstable elements (Th and U) are shown. Under each element is given the equilibrium 50% condensation temperature calculated by *Lodders* (2003) for a gas of solar composition at 10^{-4} atm total pressure. The elements are shaded by cosmochemical classification, with siderophile/chalcophile elements indicated by a small triangle at the upper right. Iron is given a half-triangle, because it has lithophile, siderophile, and chalcophile tendencies.

ing parent-body thermal metamorphism (e.g., *Lipschutz and Woolum,* 1988; *Friedrich et al.,* 2003, 2004). It is important not to confuse these effects with nebular fractionations, so data from shocked and/or equilibrated chondrites must be treated with caution.

The primary source of data on elemental abundances in carbonaceous and ordinary chondrites is a series of papers by G. W. Kallemeyn, J. T. Wasson, and coworkers (*Kallemeyn and Wasson,* 1981, 1982, 1985; *Kallemeyn et al.,* 1989, 1991, 1994, 1996). These workers analyzed relatively large samples, generally two 300-mg replicates, by instrumental neutron activation analysis (sometimes supplemented by radiochemical neutron activation analysis) and reported concentrations of 25–30 elements in each meteorite. More recently, the technique of inductively coupled plasma mass spectrometry has been applied to primitive carbonaceous chondrites, allowing determination of over 50 elements (*Friedrich et al.,* 2002). Elemental abundance patterns in bulk chondritic meteorites are shown in Fig. 2, plotted as a function of volatility [where the 50% condensation temperatures at 10^{-4} atm in a gas of solar composition (*Lodders,* 2003) are used as a measure of volatility]. Data are normalized to CI chondrites and to magnesium to remove the effects of dilution by water and organic matter in CI and CM chondrites. It can be seen that among the carbonaceous chondrite groups, abundances are a smooth function of volatility, with the more volatile elements being more depleted. The carbonaceous chondrite groups, with the exception of CR chondrites, also show some enrichment in refractory elements, whereas ordinary and enstatite chondrites are not enriched in refractory elements (EL chondrites are slightly depleted in refractories). The abundance patterns in carbonaceous chondrites are smooth functions of volatility, regardless of geochemical (lithophile or siderophile) character. In contrast, the patterns in ordinary and enstatite chondrites show effects of metal-silicate fractionation. Among the ordinary chondrites, the volatile lithophile elements lie along straight lines on Fig. 2 that pass through the magnesium point at 1327 K. In contrast, the siderophile elements have enrichment factors of ~1 down to ~800 K, indicating that volatile-element fractionation of lithophile and siderophile elements occurred under different conditions. The EH chondrites are also unusual in that they are enriched in moderately volatile elements and only slightly depleted in highly volatile elements.

Elemental abundances in individual samples of planets and differentiated meteorites are modified by separation of melt and crystalline material during differentiation. In order

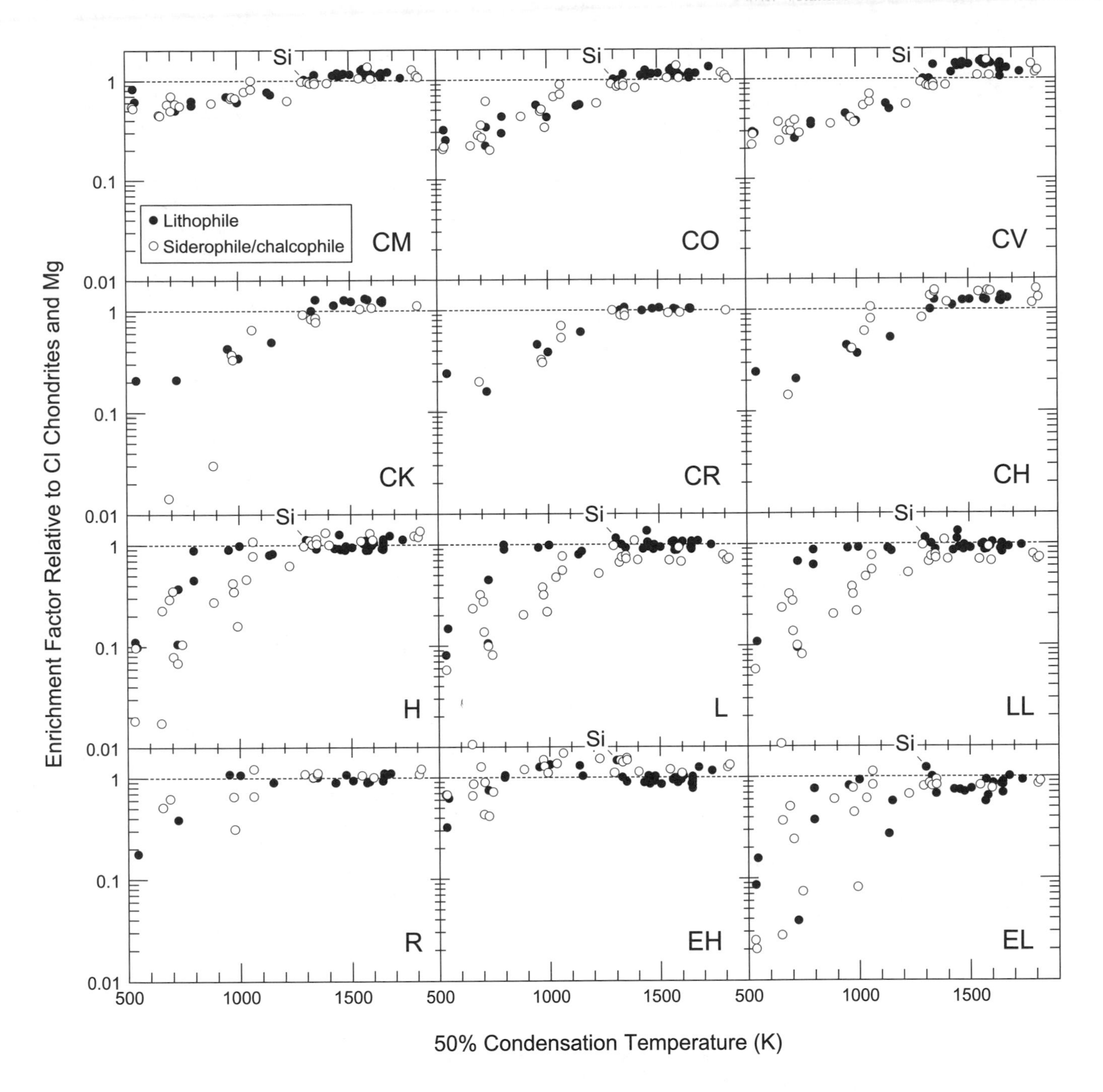

Fig. 2. When plotted vs. 50% condensation temperatures (*Lodders*, 2003), the CI-normalized elemental enrichment factors of the volatile elements lie along linear arrays on semilog plots. The data plotted are largely from the compilation of *Wasson and Kallemeyn* (1988), supplemented with more recent data on the groups CM (*Friedrich et al.*, 2002), CK (*Kallemeyn et al.*, 1991), CR (*Kallemeyn et al.*, 1994), CH (*Wasson and Kallemeyn*, 1990; *Bischoff et al.*, 1993), and R (*Schulze et al.*, 1994).

to see through these effects, it is common practice to plot ratios of incompatible elements that are not strongly fractionated from one another during differentiation. Shown in Fig. 3 are CI-normalized Rb/Sr ratios vs. K/U ratios in planets and bulk meteorites. Each of these ratios is of a volatile to a refractory element. It can be seen that in planets and differentiated meteorites, the extent of volatility fractionation greatly exceeds that seen in chondritic meteorites, with the most extremely fractionated body, the angrite parent body, depleted by 3 orders of magnitude in Rb/Sr relative to CI chondrites. All the planets and differentiated meteorite parent bodies are more fractionated than the most extremely fractionated chondrites, the CVs.

Large volatility fractionations are also apparent in iron meteorites. Shown in Fig. 4 are CI-normalized Ge/Ni vs. Ga/Ni ratios in iron and metal from stony-iron meteorites. Gallium and germanium are moderately volatile elements and nickel is slightly more refractory than iron. The partition

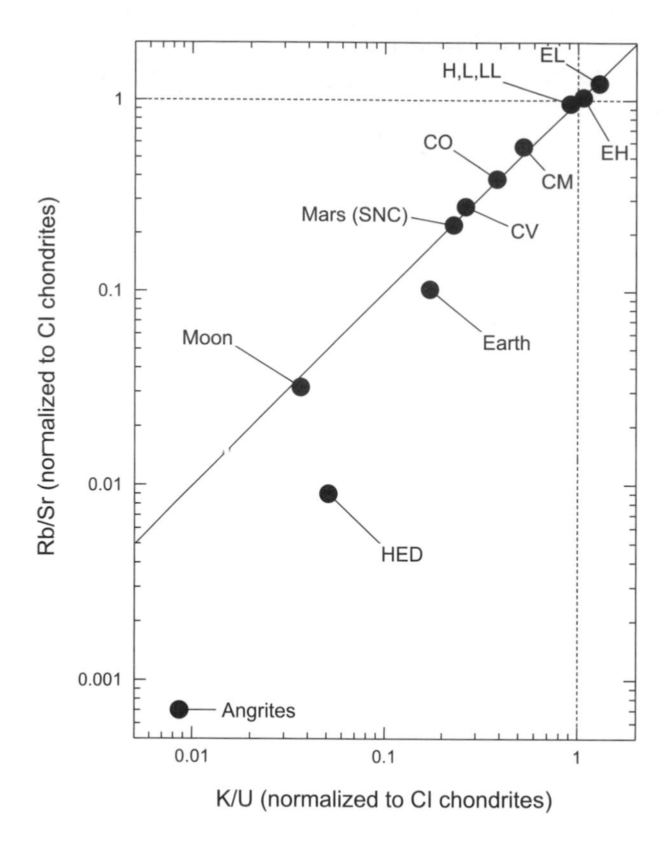

Fig. 3. CI-normalized Rb/Sr vs. K/U ratios in bulk meteorites and planets. Data for chondrites are the same as in Fig. 2; data for HED are from *Kitts and Lodders* (1998); data for Mars are from the SNC meteorite compilation of *Lodders* (1998); data for the Moon are from *Anders* (1977a); data for the bulk silicate Earth (BSE) are from *McDonough and Sun* (1995); data for angrites are from *Lugmair and Galer* (1992) and *Mittlefehldt* (2003). The line has a slope of 1 and shows that Earth, the HED meteorites, and angrites have increasingly fractionated Rb/K ratios compared to the bulk solar system.

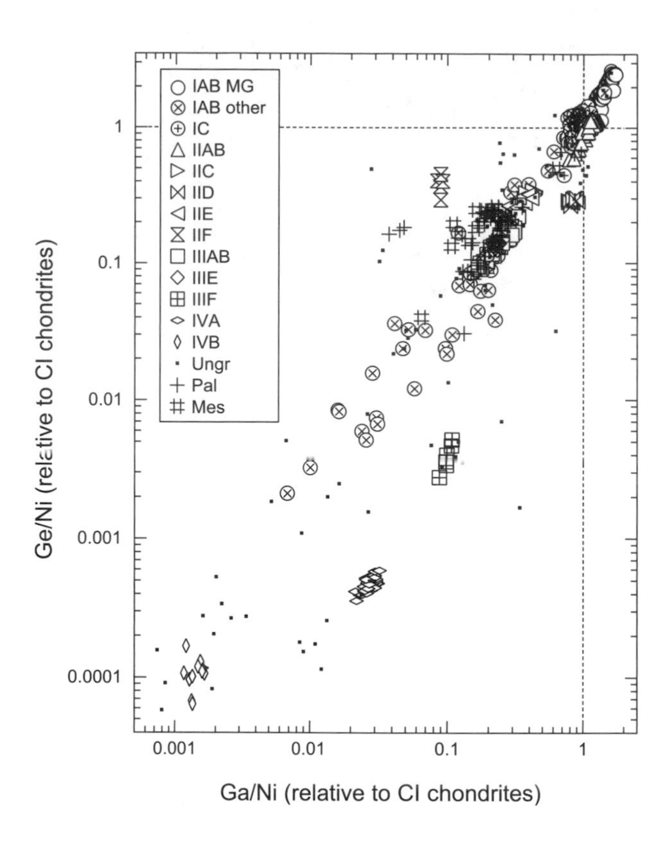

Fig. 4. CI-normalized Ge/Ni vs. Ga/Ni ratios in iron meteorites and metal of stony iron meteorites. These ratios are mostly controlled by volatility: Fractional crystallization of metal does not significantly fractionate these ratios. Data of *Davis* (1977), *Kracher et al.* (1980), *Malvin et al.* (1984), *Rasmussen et al.* (1984), *Schaudy et al.* (1972), *Scott and Wasson* (1976), *Scott et al.* (1973), *Wasson* (1967, 1969, 1970), *Wasson and Choi* (2003), *Wasson and Kallemeyn* (2002), *Wasson and Kimberlin* (1967), *Wasson and Richardson* (2001), *Wasson and Schaudy* (1971), and *Wasson et al.* (1989, 1998).

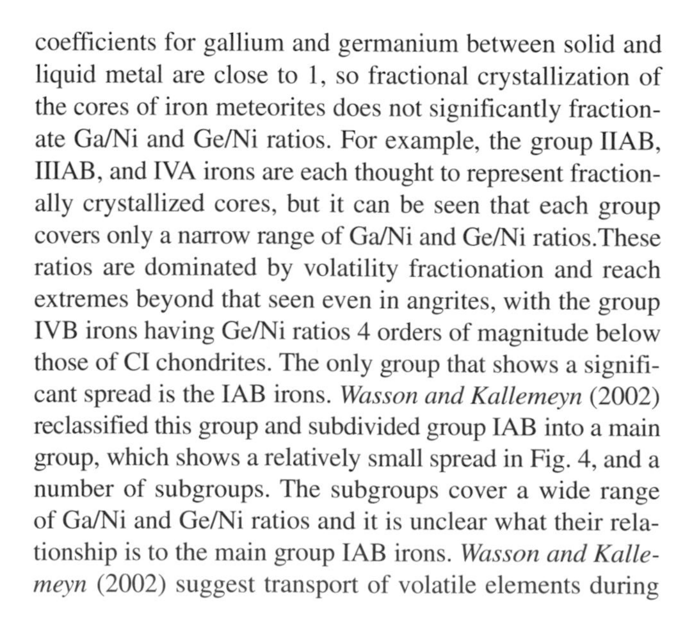

coefficients for gallium and germanium between solid and liquid metal are close to 1, so fractional crystallization of the cores of iron meteorites does not significantly fractionate Ga/Ni and Ge/Ni ratios. For example, the group IIAB, IIIAB, and IVA irons are each thought to represent fractionally crystallized cores, but it can be seen that each group covers only a narrow range of Ga/Ni and Ge/Ni ratios.These ratios are dominated by volatility fractionation and reach extremes beyond that seen even in angrites, with the group IVB irons having Ge/Ni ratios 4 orders of magnitude below those of CI chondrites. The only group that shows a significant spread is the IAB irons. *Wasson and Kallemeyn* (2002) reclassified this group and subdivided group IAB into a main group, which shows a relatively small spread in Fig. 4, and a number of subgroups. The subgroups cover a wide range of Ga/Ni and Ge/Ni ratios and it is unclear what their relationship is to the main group IAB irons. *Wasson and Kallemeyn* (2002) suggest transport of volatile elements during impacts on the parent body, as fractional crystallization and partial melting models do not produce such a spread in Ga/Ni and Ge/Ni.

2.2. Isotopic Fractionation

It is clear from experiments done in vacuum and in hydrogen at plausible nebular pressures (*Davis et al.,* 1990; *Wang et al.,* 1999, 2001; *Grossman et al.,* 2000; *Richter et al.,* 2002), as well as from theoretical considerations (*Grossman et al.,* 2000; *Richter et al.,* 2002; *Richter,* 2004), that high-temperature evaporation can lead to substantial isotopic mass fractionation effects, due to the kinetic isotope effect. It has been clear since detailed isotopic studies began on CAIs in the 1970s that while CAIs can be fractionated in the heavy isotopes of magnesium and silicon by a few permil per amu (*Clayton et al.,* 1988), bulk meteorites of all kinds do not show significant fractionation in these elements.

The first targeted search for isotopic fractionation effects in a volatile element was the potassium isotope survey of meteorites and planets of *Humayun and Clayton* (1995). They studied a wide variety of solar system materials, including chondrites, achondrites, Earth, the Moon, and Mars (SNC meteorites), objects with depletions ranging from undepleted CI chondrites to the highly potassium depleted Moon and eucrites. They found no measurable isotopic fractionation effects within ±0.2–0.5‰/amu and concluded that condensation, not evaporation, must have been responsible for variable depletion in potassium. *Alexander et al.* (2000) used an ion microprobe to measure potassium isotopic compositions within chondrules, which can have potassium depletions even larger than those in some achondrites, and found no isotopic fractionation within ±1.5‰/amu. Since *Yu et al.* (2003) showed that evaporative loss of potassium in vacuum and low-pressure hydrogen fractionates potassium isotopes, *Alexander et al.* (2000) suggested gas-solid exchange during evaporation of chondrules.

Sulfur is the most volatile element for which searches for isotopic fractionation effects have been searched. *Gao and Thiemens* (1993a,b) reported that bulk C, O, and E chondrites have mass fractionation effects of <0.5‰/amu. *Tachibana and Huss* (2005) report that troilite in chondrules has sulfur mass fractionation effects of <1‰/amu. Although evaporation experiments have not proven that evaporation of sulfur produces isotopic mass fractionation, it is hard to believe that it would not. Thus, sulfur isotopic data in chondrites and chondrules also points to gas-solid exchange during evaporative loss of sulfur.

Searches have been made for isotopic fractionation effects in iron, magnesium, and silicon, but effects are limited to less than 1‰/amu in chondrules. Bulk meteorites and planets are perhaps less likely than chondrules to show mass fractionation effects and no effects have been reported. These elements are not as volatile as potassium, but significant losses of all these elements are expected during high-temperature evaporation. Improvements in analytical techniques, particularly the development of multicollector inductively coupled mass spectrometry, have allowed searches for mass fractionation effects at the 0.1‰/amu precision level in recent years. *Poitrasson et al.* (2004) reported that Mars and Vesta are lighter than Earth by 0.03‰/amu and the Moon is 0.07‰/amu heavier than Earth in iron isotopes. They interpreted the difference between Earth and the Moon as being caused by isotopic fractionation due to 1% iron loss during Moon formation.

2.3. Presolar Grain Abundances

Presolar grains were discovered in carbonaceous chondrites in 1987 (*Lewis et al.,* 1987), barely missing publication in *MESS I*. The abundances of the major types of presolar grains — diamond, silicon carbide, and graphite — in meteorites can be inferred from a fairly simple chemical separation followed by step-heating vacuum pyrolysis for noble gas abundances and isotopic compositions (*Huss and Lewis,* 1994a,b, 1995; *Huss et al.,* 1996). These abundances are quite diagnostic of nebular and parent-body thermal events, since the different kinds of noble gas carriers (presolar grains) are destroyed at different temperatures. A recent extensive survey of carbonaceous chondrites (*Huss et al.,* 2003; *Huss,* 2004) has shown that presolar grain abundances and bulk meteorite elemental abundance patterns are related: After correction for parent-body metamorphic effects, abundance variations among carbonaceous chondrites remain and appear to be related to volatile element depletion patterns.

For the purpose of studying thermal processing in the solar system, it is enough to know that there are several noble gas components with a range of sensitivity to thermal effects. However, it is useful to discuss what these components are and the identity of their carriers. Ne-E is a component highly enriched in ^{22}Ne relative to ^{20}Ne and ^{21}Ne. It comes in two flavors, Ne-E(L), which is nearly pure ^{22}Ne, and Ne-E(H), which is not quite as strongly enriched in ^{22}Ne. Ne-E(L) is thought to be the product of decay of ^{22}Na ($T_{1/2}$ = 2.6 yr), which is produced in supernova explosions. Ne-E(H) is thought to be material from the helium shell of an asymptotic giant branch star, implanted into grains after the star lost its hydrogen-rich envelope. Xenon has a number of isotopic components that are released at various temperatures from various host phases. Xenon has nine stable isotopes that are produced by a variety of nucleosynthesis mechanisms, which explains in part why so many components have been identified. Perhaps the most famous is Xe-HL, in which the isotope enrichment pattern relative to the solar system is V-shaped, with enrichments in both light and heavy xenon isotopes. This component is associated with presolar diamonds and is released at high temperature (<1000°C), which led to their discovery (*Lewis et al.,* 1987). There are several components containing approximately normal "planetary" isotopic compositions of noble gases (*Huss and Lewis,* 1994a). Xe-P1 is sited in an unknown carbon-rich carrier "Q" that releases noble gas under fairly mild heating conditions. Xe-P3 and Xe-P6 are released from diamond at fairly low (~500°C) and high (>1000°C) temperatures, respectively. Noble gas components in meteorites are reviewed by *Podosek* (2003).

Ordered from most susceptible to most resistant to thermal metamorphism, the noble gas components are (1) Ne-E(H) in excess of that in silicon carbide in an unknown carrier; (2) graphite, inferred from Ne-E(L); (3) Xe-P3 in diamond; (4) silicon carbide, inferred from Ne-E(H) and *s*-process Xe; (5) Xe-P1, an easily oxidized, but unknown major carrier of planetary noble gases; (6) diamond, inferred from Xe-HL; and (7) Xe-P6 in diamond. The patterns for these components in several kinds of unequilibrated chondrites are shown in Fig. 5. Relative to CI chondrites and CM chondrite matrices, CR chondrites seem to have experienced simple thermal processing and CM, CO, CV, and CH chondrites contain both highly processed and unprocessed ma-

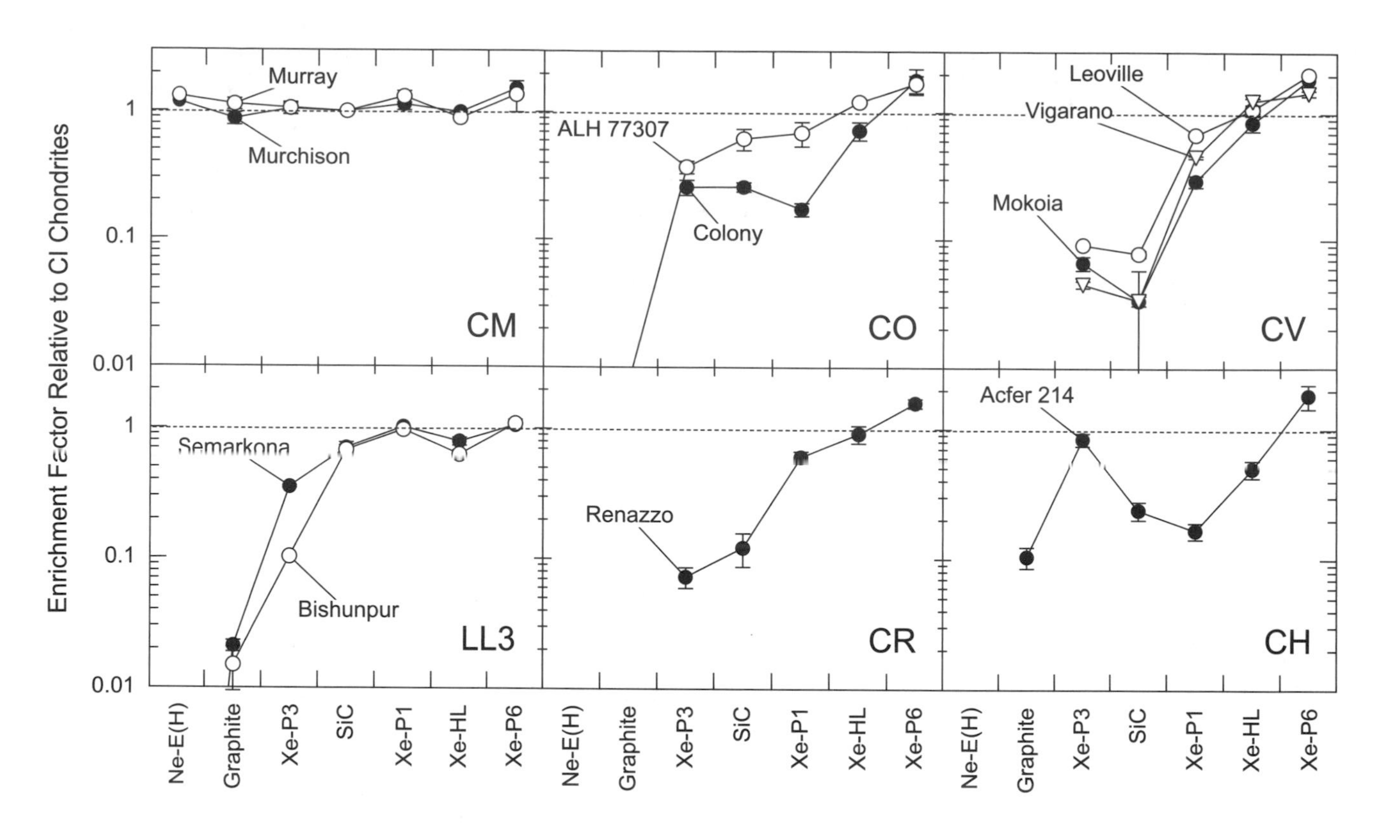

Fig. 5. Abundances of presolar noble gas carriers in chondritic meteorites. Data of *Huss et al.* (2003) and *Huss* (2004).

terials. *Huss et al.* (2003) and *Huss* (2004) suggest that the variations in presolar grain abundances and their correlations with volatile element depletions represent varying degrees of processing of presolar molecular cloud material in the solar nebula.

3. ELEMENTAL AND ISOTOPIC FRACTIONATION DURING EVAPORATION AND CONDENSATION

Isotopic mass fractionation during evaporation and condensation and its relationship to elemental fractionation provides powerful clues about formation conditions. Condensation and evaporation are terms widely used to describe volatility fractionation processes, but they are incomplete. It is important to distinguish between equilibrium and kinetically controlled processes. Condensation is usually used as shorthand for equilibrium fractionation with falling temperature, and evaporation is often taken to mean a kinetically controlled fractionation with increasing temperature, but kinetically controlled condensation and near-equilibrium evaporation can also occur. At this point it is useful to briefly review evaporation and condensation theory. For more extensive treatment of these subjects, see *Wood and Hashimoto* (1993), *Yoneda and Grossman* (1995), *Petaev and Wood* (1998), *Ebel and Grossman* (2000), *Grossman et al.* (2000), *Nagahara and Ozawa* (2000), *Ozawa and Nagahara* (2001), *Richter et al.* (2002), *Davis and Richter* (2003), *Richter* (2004), and *Davis et al.* (2005).

3.1. Evaporation Theory

The evaporative flux from a solid or molten surface into a surrounding gas is given by the Hertz-Knudsen equation

$$J_{i,net} = \sum_{j=1}^{n} \frac{n_{ij}\left(\gamma_{ij}^{evap}P_{ij}^{sat} - \gamma_{ij}^{cond}P_{ij}\right)}{\sqrt{2\pi m_{ij}RT}} \qquad (1)$$

where $J_{i,net}$ is the net flux of element or isotope i (moles per unit area per unit time), n_{ij} is the number of atoms of i in species j, γ_{ij}^{evap} and γ_{ij}^{cond} are the evaporation and condensation coefficients for the jth gas species containing i, P_{ij}^{sat} and P_{ij} are the saturation vapor pressure and the pressure at the evaporating surface of the jth gas species containing the element or isotope i, m_{ij} is the molecular weight of j, R is the gas constant, and T is the absolute temperature. The summation is over all gas species j containing i. It is usually assumed that condensation coefficients are the same as the evaporation coefficients for free evaporation, which must be true at least in the limit $P_{ij} \rightarrow P_{ij}^{sat}$ when equilibrium is approached. Except for the evaporation and condensation coefficients, the various quantities on the righthand side of equation (1) can be calculated given a suitable thermodynamic model for the condensed phase and the surrounding gas. Since there is no formalism for calculating the evaporation coefficients, these must be determined by laboratory experiments.

The gas in equilibrium with liquids of chondritic composition is dominated by species containing single atoms of silicon, magnesium, iron, and potassium (i.e., n = 1, n_{ij} = 1 in equation (1)), which allows us to write a simpler version of equation (1) in the form

$$J_{i,net} = \left[\frac{\gamma_i P_i^{sat}}{\sqrt{2\pi m_i RT}}\right]\left(1 - \frac{P_i}{P_i^{sat}}\right) \qquad (2)$$

where $\gamma_i = \gamma_i^{evap} = \gamma_i^{cond}$. The quantity in square brackets is the free evaporation flux and the net evaporation rate is simply the free evaporation rate reduced by a fraction P_i/P_i^{sat}. Equation (2) shows why vacuum evaporation experiments (limit $P_{ij}/P_{ij}^{sat} \rightarrow 0$) are used to determine evaporation coefficients in that the duration of evaporation and the change in composition of the condensed phase gives a measure of J_i, while P_i^{sat} and m_i can be calculated from the thermodynamics of the system, allowing determination of the only remaining unknown, γ_i^{evap}. Equation (2) remains valid for $P_i/P_i^{sat} > 1$, in which case the condensation flux will be greater than the evaporation flux and the net effect will be condensation rather than evaporation.

Equation (2) can also be used to calculate the evaporation rate of isotopes of a given element, and thus the isotopic fractionation of evaporation residues as a function of the amount of the element evaporated. The relative rate of loss by evaporation of two isotopes of i (denoted by subscripts i,1 and i,2) as

$$\frac{J_{i,2}}{J_{i,1}} = \frac{\gamma_{i,2} P_{i,2}^{sat}}{\gamma_{i,1} P_{i,1}^{sat}}\sqrt{\frac{m_{i,1}}{m_{i,2}}} = R_{2,1}\frac{\gamma_{i,2}}{\gamma_{i,1}}\sqrt{\frac{m_{i,1}}{m_{i,2}}} \qquad (3)$$

where $R_{2,1}$ is the atom ratio of isotope 2 to isotope 1 at the evaporating surface. In writing equation (3), it is assumed that isotopes mix ideally and that at the high temperatures of interest here equilibrium isotope fractionations between the condensed phase and the gas are negligible. Given these assumptions, the ratio of the saturation vapor pressures of the isotopically distinct gas species is the same as the isotope ratio in the condensed phase. The standard assumption until recently has been that the evaporation coefficients of isotopes of a given element are the same and therefore that the righthand side of equation (3) can be further simplified to $R_{2,1}\sqrt{m_{i,1}/m_{i,2}}$, but increasingly precise measurements of the isotopic composition of evaporation residues have shown that in all cases involving evaporation from crystalline and molten silicates this simplification is not warranted.

Equation (3) shows that the isotopic composition of the evaporation flux differs from that of the substrate by the factor $(\gamma_{i,2}/\gamma_{i,1})\sqrt{m_{i,1}/m_{i,2}}$, usually called α. If the transport processes in the evaporating material are able to keep it homogeneous and α is independent of the evolving composition of the evaporating residue, then the isotopic composition of the residue will evolve by Rayleigh fractionation given by

$$R_{2,1}^{res} = R_{2,1}^{0} f_1^{(\alpha - 1)} \qquad (4)$$

where $R_{2,1}^0$ is the initial isotopic composition of the condensed phase and f_1 is the fraction of isotope 1 remaining in the residue. A good way of testing the validity of the assumption made in writing equation (4) is to plot the experimental data as $\ln(R_{2,1}^{res}/R_{2,1}^0)$ vs. $\ln(f_1)$, which according to equation (4) should fall along a straight line with slope α − 1. Recent experiments with high-precision isotopic analyses are very well behaved in this sense (*Richter,* 2004; *Richter et al.,* 2005).

The relatively simple picture outlined above regarding the isotopic fractionation of evaporation residues becomes a bit more complicated as some of the simplifying assumptions are relaxed. If there is a finite pressure of the evaporating species at the surface of the condensed phase (i.e., $P_i/P_i^{sat} \neq 0$) then the isotopic fractionation factor $\alpha = (\gamma_{i,2}/\gamma_{i,1})\sqrt{m_{i,1}/m_{i,2}}$ in equation (4) should for most practical applications be replaced by $\alpha' = 1 + (\alpha - 1)(1 - P_i/P_i^{sat})$ (*Richter et al.,* 2002). The isotopic fractionation of a residue for a given amount of evaporation of the parent element will be reduced when the system becomes diffusion-limited to the extent that diffusion is not sufficiently fast to maintain the homogeneity of the condensed phase. Diffusion limitations are insignificant in the case of molten CAIs or chondrules (*Richter,* 2004), but evaporation of solids such as forsterite will be severely diffusion-limited with isotopic fractionations limited to a very thin boundary layer at the surface. *Wang et al.* (1999) studied the effect of diffusion on the distribution of isotopic fractionation effects within forsterite. Theoretical studies of diffusion effects as a function of forsterite grain size have been done in considerable detail (*Tsuchiyama et al.,* 1999; *Nagahara and Ozawa,* 2000). Diffusion limitation of bulk isotopic fractionation can be mitigated to some extent by isotopic mass fractionation during diffusion (*Richter,* 2004; *Richter et al.,* 1999, 2003).

3.2. Condensation Theory

"Condensation calculations," in which the thermodynamic equilibrium condensate assemblage is calculated as a function of temperature in a gas of fixed pressure and composition (usually solar), provide an extremely useful framework for interpreting bulk compositions of CAIs, chondrules, meteorites, and planets. The first modern calculations for the solar system were done by *Grossman* (1972); a comparison of the condensation sequence at 10^{-3} bar in a gas of solar composition from various workers is given in *Petaev and Wood* (2005). More recent calculations have explored a wider range of parameters. *Wood and Hashimoto* (1993) explored the effects of varying the proportions of four volatility components of solar material: refractory dust, carbonaceous matter, ice, and hydrogen gas. They found that dust enrichment (1) increases the condensation temperatures of all minerals at fixed pressure, (2) increases the iron content of mafic minerals, and (3) permits melts to be in equilibrium with gas. *Yoneda and Grossman* (1995) explored condensation of melts in the CaO-MgO-Al_2O_3-SiO_2 (CMAS) system using a reasonably accurate thermodynamic model for melts for the first time and found that melts can coexist with gas at pressures as low as 10^{-2} atm. *Ebel and Grossman* (2000) extended such calculations to chondritic compositions. Dust-enriched systems have two important differences from solar system composition: condensation temperatures are higher, so melts can be stabilized; and conditions can become oxidizing enough at high dust enrichments to allow condensation of a significant fraction of total iron into ferromagnesian minerals rather than metal (*Wood and Hashimoto,* 1993; *Ebel and Grossman,* 2000). *Fedkin and Grossman* (2006) explored the conditions needed to explain the high fayalite content of chondrule precursors and pointed out the difficulty in achieving sufficiently oxidizing conditions in solar nebular models.

Equilibrium condensation calculations have been applied to trace elements as well. Whereas high-quality thermodynamic data are available for major phases in condensation calculations described above, most trace elements condense in solid solution in one of the major phases. However, for most trace elements, thermodynamic data are only available for pure elements and simple compounds. Sometimes good solid solution models exist for minor and trace elements, but in many cases assumptions must be made about host phases and solution behavior, guided by microbeam trace-element analyses of individual phases in meteorites indicating which minerals concentrate different trace elements. The most up-to-date table of 50% condensation temperatures of all the elements is that of *Lodders* (2003), which has been adopted for use in Figs. 1 and 2.

Rapid, kinetically controlled high-temperature condensation can lead to isotopic fractionation (*Richter,* 2004). In this case, $P_i/P_i^{sat} > 1$ in equation (2) and isotopic fractionations can be predicted as outlined above. The only case for which isotopic fractionation during condensation can be made is for fine-grained group II calcium-, aluminum-rich inclusions (CAIs), which have REE patterns clearly indicating condensation (*Boynton,* 1975; *Davis and Grossman,* 1979) and which contain isotopically light magnesium and silicon (*Clayton et al.,* 1988). Equilibrium isotopic fractionation is minimal at high temperature, and condensation in most plausible nebular environments is expected to produce elemental fractionation without isotopic fractionation.

4. VOLATILITY FRACTIONATION MODELS

Palme et al. (1988) divided models to explain moderately volatile element abundance patterns into two categories, evaporation and condensation models, and discussed their merits in detail. A third class of model, in which volatile element abundances are inherited from the molecular cloud parental to the solar system, has been advocated more recently.

Anders (1964) and *Larimer and Anders* (1967) proposed that volatile element fractionation occurred by evaporation during chondrule formation and that the variations in volatile element abundances in bulk meteorite parent bodies and planets were simply due to variations in the mixing ratio of volatile-depleted chondrules to undepleted matrix. There was a vigorous debate (*Wasson and Chou,* 1974; *Anders,* 1975; *Wai and Wasson,* 1977; *Anders,* 1977b; *Wasson,* 1977) between E. Anders and J. T. Wasson over the issue of whether moderately volatile element abundances were a continuous function of volatility or "a terraced landscape where gently sloping plateaus alternate with steep declines" (*Anders,* 1977b). The latter description was what was predicted from volatile element loss from chondrules during melting, and Anders argued that Wasson's continuous curve was based in part on poorly known condensation temperatures or poorly determined elemental abundances. Current data (Fig. 2) with a quarter century of improvements in analytical techniques and thermodynamic equilibrium calculations favor a continuous function, especially for carbonaceous chondrites, although ordinary chondrites continue to show some scatter. *Shu et al.* (1996) proposed a new location for chondrule formation, close to the Sun, at the X point, where infalling material divides into matter that falls into the Sun and that which is thrown out, either out of the solar system or back out to the solar nebula. This region provides a qualitative explanation for short-lived thermal events needed to explain chondrule formation and would permit mixing of volatile-depleted chondrules with unprocessed material further out in the solar nebula. *Connolly and Love* (1998) and *Desch and Connolly* (2002) have proposed that chondrules are heated by the passage of nebular shock waves, and that the complementary volatile element patterns of chondrules and matrix arise from this process. *Alexander et al.* (2001) also favor a two-component explanation of volatile element abundances, with kinetically controlled volatility fractionation accompanying chondrule formation (shock-wave heating is one way to do this). They postulated some equilibration with nebular gas on cooling to explain the lack of isotopic mass fractionation effects in potassium and iron in chondrules and partial loss of nebular gas to explain the relatively smooth relationship between CI-normalized abundances and condensation temperature. In summary, the lack of isotopic fractionation effects in the moderately volatile elements iron and potassium in chondrules argues against significant volatile loss during melting, at least in the simple picture of Anders, so current models of this sort appeal to back-reaction to reduce mass fractionation effects, chondrule formation in restricted regions of the solar nebula in order not to lose presolar grains from material now in carbonaceous chondrite matrix, and partial loss of nebular gas in chondrule-forming regions to produce smooth fractionation patterns (*Alexander et al.,* 2001; *Alexander,* 2004).

Wasson and Chou (1974) and *Wai and Wasson* (1977) proposed that the abundances of moderately volatile elements in ordinary chondrites arose by equilibrium condensation with continuous loss of nebular gas. They argued that the abundance pattern was a smooth function of condensation temperature and that complete volatile loss during chondrule formation was unlikely because of the high cooling rates inferred for chondrules. Although they calculated condensation temperatures, they did not present a quantitative model of exactly how the process worked. *Cassen* (1996, 2001) presented such a model and found that the patterns in CO and CV chondrites could be reproduced by simple disk models, but had difficulty matching CM chondrites. These models envision a hot solar nebula, with significant evaporation out to 3 AU and a high mass accretion rate, $>10^{-7}\ M_\odot\ yr^{-1}$, with chondrites achieving their volatility fractionation patterns in roughly their final formation locations. *Wulf et al.* (1995) evaporated Allende under reducing conditions in a gas-mixing furnace and showed that volatile element patterns in residues did not match those in bulk CAIs. They concluded that condensation, not evaporation, must be the controlling process in the solar nebula. The major problem with models envisioning a hot solar nebula are the presence of presolar grains in a wide variety of chondritic meteorites with minimal parent-body metamorphism and the presence of isotopic differences in titanium (*Niemeyer and Lugmair,* 1984; *Niederer et al.,* 1985), chromium (*Lugmair and Shukolyukov,* 1998), and molybdenum (*Dauphas et al.,* 2002; *Yin et al.,* 2002) among different bulk meteorites.

Huss et al. (2003) and *Huss* (2004) noted relationships between volatile-element abundances and presolar grain abundances and suggested that initially cold interstellar cloud material was heated in the solar nebula to varying degrees. They pointed out that volatile element enrichment patterns (e.g., Fig. 2) are generally flat at high temperatures, and that elements below a certain temperature are progressively depleted relative to CI chondrites. The temperature at which this break occurs differs for different meteorite groups and appears to be correlated with the abundance patterns of presolar grains. The sequence is from CI to LL to CO to CR to CV. Since presolar grains are in the matrix and the elemental abundance patterns are controlled by high-temperature components (chondrules, metal, and CAIs), Huss argues that these patterns must have been established prior to chondrule-forming events and probably represent a process common to all chondritic parent material. The presence of presolar grains in a wide variety of chondritic meteorites makes it clear that some fraction of material in the solar nebula escaped vaporization and subsequent condensation, because such conditions would have destroyed presolar grains. Huss proposes that a fraction of material from the solar system's parental molecular cloud was partially evaporated and the evolved gas lost, producing high-temperature materials and volatile element depletion patterns in bulk meteorites. Some of the dust was thermally processed, but did not go through chondrule formation, producing the variety of presolar grain abundance patterns. Huss points out unsolved problems, such as the fact that the temperatures of thermal processing of matrix dust ($<\sim 200$°C for CI and LL chondrites, <700°C for CO, ~ 700°C for CR, and >700°C for CV) are much lower than the temperatures needed to produce volatile element enrichment patterns. It is also not at all clear why the patterns in low-temperature and high-temperature components should be correlated.

Spectroscopic studies of absorption lines show that the gas phase of the interstellar medium (ISM) is depleted in refractory elements, with the more refractory elements more depleted (*Palme and Jones,* 2003). Dust in the ISM presumably has a complementary pattern and is enriched in refractory elements. Since this pattern is qualitatively similar to moderately volatile element depletion patterns in chondrites, *Palme* (2001) and *Palme and Jones* (2003) raised the question of whether the pattern in chondrites was inherited from the ISM. They dismiss this possibility on two grounds. (1) Interstellar medium dust must represent many individual stars with a wide variety of isotopic compositions (as is seen in presolar grains), yet the solar system is remarkably isotopically uniform. However, this may not be a strong argument. Presolar grains are circumstellar condensate grains that retain the isotopic signature (generally very different from solar system composition) characteristic of their parent star. But it may be that most dust is reprocessed in the ISM, where shocks are expected to efficiently sputter away circumstellar grains. Modelers of the ISM have great difficulty understanding how the presolar grains found in meteorites survived. Furthermore, the recent discovery of "abundant" presolar silicate in meteorites (*Nguyen and Zinner,* 2004; *Nagashima et al.,* 2004) only corresponds to ~25 ppm of primitive meteorites. (2) In the ISM, most of the strontium is in the dust and most of the rubidium is in the gas. *Palme* (2001) argues that over hundreds of millions of years this should produce a significant variety in $^{87}Sr/^{86}Sr$ ratios, yet the initial solar system strontium ratio is relatively uniform.

For the past several years, *Yin* (2002, 2004) has advocated the idea that the solar nebula inherited its volatile-element abundance pattern from the ISM. In a recent paper, *Yin* (2005) goes into more depth, pointing out the complementarity of carbonaceous chondrite and ISM gas elemental abundance patterns. He points out that in dense molecular cloud material, refractory elements are in dusty cores and volatiles (apart from hydrogen and helium) are in icy mantles. He proposes that in the solar nebula, evaporation of icy mantles and loss of gases allowed solar system materials to inherit the elemental abundance pattern of the dust. He counters Palme's strontium-isotope argument by proposing that the recycling in the long-lived diffuse ISM homogenizes strontium isotopes and that the molecular cloud lived for only a relatively short time, so large heterogeneities could not develop. The ISM inheritance model surely needs more development, as many problems are left unsolved. No explanation is offered for metal-silicate fractionation, for which meteoritic evidence is quite clear. It is also not clear whether ISM fractionations are sufficiently large to explain some meteoritic patterns. For example, Ge/Ni ratios among iron meteorites vary by more than 4 orders of magnitude, but the ISM Ge/Ni ratio only varies by a factor of 100.

In summary, it is clear from the presence of presolar grains in meteorites and isotopic variations in titanium, chromium, and molybdenum among bulk meteorites that the inner solar system cannot have been completely vaporized and homogenized. It is also clear that chondrules and CAIs were exposed to high temperatures, at least for brief periods of time. What is still not clear is whether volatile element depletions in solar system objects arose from transient high-temperature events somewhere in the solar system or were inherited from the interstellar medium.

5. CONCLUSIONS

Moderately volatile elements are depleted in meteorite parent bodies relative to bulk solar system composition. The depletions are smooth functions of condensation temperature when plotted on semilog plots. The lack of isotopic mass-fractionation effects in planets and bulk meteorites indicates that the elemental fractionation process, be it condensation or evaporation, whether it occurred in the solar nebula or somewhere else, occurred under near-equilibrium conditions, but further tests are needed. The survival of presolar grains in primitive meteorites indicates that some material was never vaporized and recondensed in the solar system.

Acknowledgments. This work was supported by the National Aeronautics and Space Administration. I thank reviewers S. Tachibana, Q.-Z. Yin, and J. M. Friedrich for their thoughtful and constructive comments, and G. R. Huss for providing the data used to plot Fig. 5.

REFERENCES

Alexander C. M. O'D. (2004) Chemical equilibrium and kinetic constraints for chondrule and CAI formation conditions. *Geochim. Cosmochim. Acta, 68,* 3943–3969.

Alexander C. M. O'D., Grossman J. N., Wang J., Zanda B., Bourot-Denise M., and Hewins R. H. (2000) The lack of potassium-isotopic fractionation in Bishunpur chondrules. *Meteoritics & Planet. Sci., 35,* 859–868.

Alexander C. M. O'D., Boss A. P., and Carlson R. W. (2001) The early evolution of the inner solar system: A meteoritic perspective. *Science, 293,* 64–68.

Anders E. (1964) Origin, age, and composition of meteorites. *Space Sci. Rev., 3,* 583–714.

Anders E. (1975) On the depletion of moderately volatile elements in ordinary chondrites. *Meteoritics, 10,* 283–286.

Anders E. (1977a) Chemical compositions of the moon, earth, and eucrite parent body. *Philos. Trans. R. Soc. Lond., A285,* 23–40.

Anders E. (1977b) Critique of "Nebular condensation of moderately volatile elements and their abundances in ordinary chondrites" by Chien M. Wai and John T. Wasson. *Earth Planet. Sci. Lett., 36,* 14–20.

Anders E. and Grevesse N. (1989) Abundances of the elements: Meteoritic and solar. *Geochim. Cosmochim. Acta, 53,* 197–214.

Bischoff A., Palme H., Schultz L., Weber D., Weber H., and Spettel B. (1993) Acfer 182 and paired samples, an iron-rich carbonaceous chondrite: Similarities with ALH85085 and relationship to CR chondrites. *Geochim. Cosmochim. Acta, 57,* 2631–2648.

Boynton W. V. (1975) Fractionation in the solar nebula: Condensation of yttrium and the rare earth elements. *Geochim. Cosmochim. Acta, 39,* 569–584.

Cassen P. (1996) Models for the fractionation of the volatile elements in the solar nebula. *Meteoritics & Planet. Sci., 31,* 793–806.

Cassen P. (2001) Nebular thermal evolution and the properties of primitive planetary materials. *Meteoritics & Planet. Sci., 36,* 671–700.

Clayton R. N., Hinton R. W., and Davis A. M. (1988) Isotopic variations in the rock-forming elements in meteorites. *Philos. Trans. R. Soc. Lond., A325,* 483–501.

Connolly H. C. Jr. and Love S. G. (1998) The formation of chondrules: Petrologic tests of the shock wave model. *Science, 280,* 62–67.

Dauphas N., Marty B., and Reisberg L. (2002) Molybdenum evidence for inherited planetary scale isotope heterogeneity of the protosolar nebula. *Astrophys. J., 565,* 640–644.

Davis A. M. (1977) The cosmochemical history of the pallasites. Ph.D. thesis, Yale University.

Davis A. M. and Grossman L. (1979) Condensation and fractionation of rare earths in the solar nebula *Geochim. Cosmochim. Acta, 43,* 1611–1632.

Davis A. M. and Richter F. M. (2003) Condensation and evaporation of solar system materials. In *Treatise on Geochemistry, Vol. 1: Meteorites, Planets, and Comets* (A. M. Davis, ed.), pp. 407–430. Elsevier, Oxford.

Davis A. M., Hashimoto A., Clayton R. N., and Mayeda T. K. (1990) Isotopic mass fractionation during evaporation of forsterite (Mg_2SiO_4). *Nature, 347,* 655–658.

Davis A. M., Alexander C. M. O'D., Nagahara H., and Richter F. M. (2005) Evaporation and condensation during CAI and chondrule formation. In *Chondrites and the Protoplanetary Disk* (A. N. Krot et al., eds.), pp. 432–455. ASP Conference Series 341, Astronomical Society of the Pacific, San Francisco.

Desch S. J. and Connolly H. C. Jr. (2002) A model of the thermal processing of particles in solar nebula shocks: Application to the cooling rates of chondrules. *Meteoritics & Planet. Sci., 37,* 183–207.

Ebel D. S. and Grossman L. (2000) Condensation in dust-enriched systems. *Geochim. Cosmochim. Acta, 64,* 339–366.

Fedkin A. V. and Grossman L. (2006) The fayalite content of chondritic olivine: Obstacle to understanding the condensation of rocky material. In *Meteorites and the Early Solar System II* (D. S. Lauretta and H. Y. McSween Jr., eds.), this volume. Univ. of Arizona, Tucson.

Friedrich J. M., Wang M.-S., and Lipschutz M. E. (2002) Comparison of the trace element composition of Tagish Lake with other primitive carbonaceous chondrites. *Meteoritics & Planet. Sci., 37,* 677–686.

Friedrich J. M., Wang M.-S., and Lipschutz M. E. (2003) Chemical studies of L chondrites. V: Compositional patterns for forty-nine trace elements in fourteen L4–6 and seven LL4–6 falls. *Geochim. Cosmochim. Acta, 67,* 2467–2479.

Friedrich J. M., Wang M.-S., and Lipschutz M. E. (2004) Chemical studies of L chondrites. VI: Variations with petrographic type and shock-loading among equilibrated falls. *Geochim. Cosmochim. Acta, 68,* 2889–2904.

Gao X. and Thiemens M. H. (1993a) Isotopic composition and concentration of sulfur in carbonaceous chondrites. *Geochim. Cosmochim. Acta, 57,* 3159–3169.

Gao X. and Thiemens M. H. (1993b) Isotopic composition and concentration of sulfur in enstatite and ordinary chondrites. *Geochim. Cosmochim. Acta, 57,* 3171–3176.

Grossman L. (1972) Condensation in the primitive solar nebula. *Geochim. Cosmochim. Acta, 36,* 597–619.

Grossman L., Ebel D. S., Simon S. B., Davis A. M., Richter F. M., and Parsad N. M. (2000) Major element chemical and isotopic compositions of refractory inclusions in C3 chondrites: The separate roles of condensation and evaporation. *Geochim. Cosmochim. Acta, 64,* 2879–2894.

Humayun M. and Clayton R. N. (1995) Potassium isotope geochemistry: Genetic implications of volatile element depletion. *Geochim. Cosmochim. Acta, 59,* 2131–2148.

Huss G. R. (2004) Implications of isotopic anomalies and presolar grains for the formation of the solar system. *Antarct. Meteorite Res., 17,* 133–153.

Huss G. R. and Lewis R. S. (1994a) Noble gases in presolar diamonds I: Three distinct components and their implications for diamond origins. *Meteoritics, 29,* 791–810.

Huss G. R. and Lewis R. S. (1994b) Noble gases in presolar diamonds II: Component abundances reflect thermal processing. *Meteoritics, 29,* 811–829.

Huss G. R. and Lewis R. S. (1995) Presolar diamond, SiC, and graphite in primitive chondrites: Abundances as a function of meteorite class and petrologic type. *Geochim. Cosmochim. Acta, 59,* 115–160.

Huss G. R., Lewis R. S., and Hemkin S. (1996) The "normal planetary" noble gas component in primitive chondrites: Compositions, carrier, and metamorphic history. *Geochim. Cosmochim. Acta, 60,* 3311–3340.

Huss G. R., Meshik A. P., Smith J. B., and Hohenberg C. M. (2003) Presolar diamond, silicon carbide, and graphite in carbonaceous chondrites: Implications for thermal processing in the solar nebula. *Geochim. Cosmochim. Acta, 67,* 4823–4848.

Kallemeyn G. W. and Wasson J. T. (1981) The compositional classification of chondrites — I. The carbonaceous chondrite groups. *Geochim. Cosmochim. Acta, 45,* 1217–1230.

Kallemeyn G. W. and Wasson J. T. (1982) The compositional classification of chondrites — III. Ungrouped carbonaceous chondrites. *Geochim. Cosmochim. Acta, 46,* 2217–2228.

Kallemeyn G. W. and Wasson J. T. (1985) The compositional classification of chondrites — IV. Ungrouped chondritic meteorites and clasts. *Geochim. Cosmochim. Acta, 49,* 261–270.

Kallemeyn G. W., Rubin A. E., Wang D., and Wasson J. T. (1989) Ordinary chondrites: Bulk compositions, classification, lithophile-element fractionations, and composition-petrographic type relationships. *Geochim. Cosmochim. Acta, 53,* 2747–2767.

Kallemeyn G. W., Rubin A. E., and Wasson J. T. (1991) The compositional classification of chondrites — V. The Karoonda (CK) group of carbonaceous chondrites. *Geochim. Cosmochim. Acta, 55,* 881–892.

Kallemeyn G. W., Rubin A. E., and Wasson J. T. (1994) The compositional classification of chondrites — VI. The CR carbonaceous chondrite group. *Geochim. Cosmochim. Acta, 58,* 2873–2888.

Kallemeyn G. W., Rubin A. E., and Wasson J. T. (1996) The compositional classification of chondrites — VII. The chondrite group. *Geochim. Cosmochim. Acta, 60,* 2243–2256.

Kitts K. and Lodders K. (1998) Survey and evaluation of eucrite bulk compositions. *Meteoritics & Planet. Sci., 33,* A197–A213.

Kracher A., Willis J., and Wasson J. T. (1980) Chemical classification of iron meteorites. IX. A new group (IIF), revision of IAB and IIICD, and data on 57 additional irons. *Geochim. Cosmochim. Acta, 44,* 773–787.

Larimer J. W. (1988) The cosmochemical classification of the elements. In *Meteorites and the Early Solar System* (J. F. Kerridge and M. S. Matthews, eds.), pp. 375–389. Univ. of Arizona, Tucson.

Larimer J. W. and Anders E. (1967) Chemical fractionations in meteorites. II. Abundance patterns and their interpretation. *Geochim. Cosmochim. Acta, 31,* 1239–1270.

Lewis R. S., Tang M., Wacker J. F., Anders E., and Steel E. (1987) Interstellar diamond in meteorites. *Nature, 326,* 160–162.

Lipschutz M. E. and Woolum D. S. (1988) Highly labile elements. In *Meteorites and the Early Solar System* (J. F. Kerridge and M. S. Matthews, eds.), pp. 462–487. Univ. of Arizona, Tucson.

Lodders K. (1998) A survey of shergottite, nakhlite and chassigny meteorites whole-rock compositions. *Meteoritics & Planet. Sci., 33,* A183–A190.

Lodders K. (2003) Solar system abundances and condensation temperatures of the elements. *Astrophys. J., 591,* 1220–1247.

Lugmair G. W. and Galer S. J. G. (1992) Age and isotopic relationships among the angrites Lewis Cliffs-86010 and Angra dos Reis. *Geochim. Cosmochim. Acta, 56,* 1673–1694.

Lugmair G. W. and Shukolyukov A. (1998) Early solar system timescales according to ^{53}Mn-^{53}Cr systematics. *Geochim. Cosmochim. Acta, 62,* 2863–2886.

Malvin D. J., Wang D., and Wasson J. T. (1984) Chemical classification of iron meteorites. X. Multielement studies of 43 irons, resolution of group IIIE from IIIAB, and evaluation of copper as a taxonomic parameter. *Geochim. Cosmochim. Acta, 48,* 785–804.

McDonough W. F. and Sun S.-S. (1995) The composition of the Earth. *Chem. Geol., 120,* 223–253.

Mittlefehldt D. W. (2003) Achondrites. In *Treatise on Geochemistry, Vol. 1: Meteorites, Planets, and Comets* (A. M. Davis, ed.), pp. 291–324. Elsevier, Oxford.

Nagahara H. and Ozawa K. (2000) Isotopic fractionation as a probe of heating processes in the solar nebula. *Chem. Geol., 169,* 45–68.

Nagashima K., Krot A. N., and Yurimoto H. (2004) Stardust silicates from primitive meteorites. *Nature, 428,* 921–924.

Nguyen A. and Zinner E. (2004) Discovery of ancient silicate stardust in a meteorite. *Science, 303,* 1496–1499.

Niederer F. R., Papanastassiou D. A., and Wasserburg G. J. (1985) Absolute isotopic abundances of Ti in meteorites. *Geochim. Cosmochim. Acta, 49,* 835–851.

Niemeyer S. and Lugmair G. W. (1984) Titanium isotopic anomalies in meteorites. *Geochim. Cosmochim. Acta, 48,* 1401–1416.

Ozawa K. and Nagahara H. (2001) Chemical and isotopic fractionation by evaporation and their cosmochemical implications. *Geochim. Cosmochim. Acta, 65,* 2171–2199.

Palme H. (2001) Chemical and isotopic heterogeneity in protosolar matter. *Philos. Trans. R. Soc. Lond., A359,* 2061–2075.

Palme H. and Jones A. (2003) Solar system abundances of the elements. In *Treatise on Geochemistry, Vol. 1: Meteorites, Comets, and Planets* (A. M. Davis, ed.), pp. 41–61. Elsevier, Oxford.

Palme H., Larimer J. W., and Lipschutz M. E. (1988) Moderately volatile elements. In *Meteorites and the Early Solar System* (J. F. Kerridge and M. S. Matthews, eds.), pp. 436–460. Univ. of Arizona, Tucson.

Petaev M. I. and Wood J. A. (1998) The condensation with parial isolation (CWPI) model of condensation in the solar nebula. *Meteoritics & Planet. Sci., 33,* 1123–1137.

Petaev M. I. and Wood J. A. (2005) Meteoritic constraints on temperatures, pressures, cooling rates, chemical compositions, and modes of condensation in the solar nebula. In *Chondrites and the Protoplanetary Disk* (A. N. Krot et al., eds.), pp. 373–406. ASP Conference Series 341, Astronomical Society of the Pacific, San Francisco.

Podosek F. A. (2003) Noble gases. In *Treatise on Geochemistry, Vol. 1: Meteorites, Planets, and Comets* (A. M. Davis, ed.), pp. 381–405. Elsevier, Oxford.

Poitrasson F., Halliday A. N., Lee D.-C., Levasseur S., and Teutsch

N. (2004) Iron isotope differences between Earth, Moon, Mars and Vesta as possible records of contrasted accretion mechanisms. *Earth Planet. Sci. Lett., 223,* 253–266.

Rasmussen K. L., Malvin D. J., Buchwald V. F., and Wasson J. T. (1984) Compositional trends and cooling rates of group IVB iron meteorites. *Geochim. Cosmochim. Acta, 48,* 805–813.

Richter F. M. (2004) Timescales determining the degree of kinetic isotope fractionation by evaporation and condensation. *Geochim. Cosmochim. Acta, 68,* 4971–4992.

Richter F. M., Liang Y., and Davis A. M. (1999) Isotopic fractionation by diffusion in molten oxides. *Geochim. Cosmochim. Acta, 63,* 2853–2861.

Richter F. M., Davis A. M., Ebel D. S., and Hashimoto A. (2002) Elemental and isotopic fractionation of Type B calcium-, aluminum-rich inclusions: Experiments, theoretical considerations and constraints on their thermal evolution. *Geochim. Cosmochim. Acta, 66,* 521–540.

Richter F. M., Davis A. M., DePaolo D. J., and Watson E. B. (2003) Isotopic fractionation by chemical diffusion between molten basalt and rhyolite. *Geochim. Cosmochim. Acta, 67,* 3905–3923.

Richter F. M., Janney P. E., Mendybaev R. A., Davis A. M., and Wadhwa M. (2005) On the temperature dependence of the kinetic isotope fractionation of Type B CAI-like melts during evaporation (abstract). In *Lunar and Planetary Science XXXVI,* Abstract #2124. Lunar and Planetary Institute, Houston (CD-ROM).

Schaudy R., Wasson J. T., and Buchwald V. F. (1972) Chemical classification of iron meteorites. VI. Reinvestigation of irons with germanium concentrations lower than 1 ppm. *Icarus, 17,* 174–192.

Schultze H., Bischoff A., Palme H., Spettel B., Dreibus G., and Otto J. (1994) Mineralogy and chemistry of Rumuruti: The first meteorite fall of the new R chondrite group. *Meteoritics & Planet. Sci., 29,* 275–286.

Scott E. R. D. and Wasson J. T. (1976) Chemical classification of iron meteorites. VIII. Groups IC, IIE, IIIF, and 97 other irons. *Geochim. Cosmochim. Acta, 40,* 103–115.

Scott E. R. D., Wasson J. T., and Buchwald V. F. (1973) Chemical classification of iron meteorites. VII. Reinvestigation of irons with germanium concentrations between 25 and 80 ppm. *Geochim. Cosmochim. Acta, 37,* 1957–1983.

Shu F. H., Shang H., and Lee T. (1996) Toward an astrophysical theory of chondrites. *Science, 271,* 1545–1552.

Tachibana S. and Huss G. R. (2005) Sulfur isotope composition of putative primary troilite in chondrules from Bishunpur and Semarkona. *Geochim. Cosmochim. Acta, 69,* 3075–3097.

Tsuchiyama A., Tachibana S., and Takahashi T. (1999) Evaporation of forsterite in the primordial solar nebula; rates and accompanied isotopic fractionation. *Geochim. Cosmochim. Acta, 63,* 2451–2466.

Wai C. M. and Wasson J. T. (1977) Nebular condensation of moderately volatile elements and their abundances in ordinary chondrites. *Earth Planet. Sci. Lett., 36,* 1–13.

Wang J., Davis A. M., Clayton R. N., and Hashimoto A. (1999) Evaporation of single crystal forsterite: Evaporation kinetics, magnesium isotopic fractionation and implications of mass-dependent isotopic fractionation of a diffusion-controlled reservoir. *Geochim. Cosmochim. Acta, 63,* 953–966.

Wang J., Davis A. M., Clayton R. N., Mayeda T. K., and Hashimoto A. (2001) Chemical and isotopic fractionation during the evaporation of the $FeO-MgO-SiO_2-CaO-Al_2O_3-TiO_2$-REE melt system. *Geochim. Cosmochim. Acta, 65,* 479–494.

Wasson J. T. (1967) The chemical classification of iron meteorites. I. Iron meteorites with low concentrations of gallium and germanium. *Geochim. Cosmochim. Acta, 31,* 161–180.

Wasson J. T. (1969) Chemical classification of iron meteorites. III. Hexahedrites and other irons with germanium concentrations between 80 and 200 ppm. *Geochim. Cosmochim. Acta, 33,* 859–876.

Wasson J. T. (1970) Chemical classification of iron meteorites. IV. Irons with germanium concentrations greater than 190 ppm and other meteorites associated with group I. *Icarus, 12,* 407–423.

Wasson J. T. (1977) Reply to Edward Anders: A discussion of alternative models for explaining the distribution of moderately volatile elements in ordinary chondrites. *Earth Planet. Sci. Lett., 36,* 21–28.

Wasson J. T. and Choi B.-G. (2003) Main-group pallasites: Chemical composition, relationship to IIIAB irons, and origin. *Geochim. Cosmochim. Acta, 67,* 3079–3096.

Wasson J. T. and Chou C.-L. (1974) Fractionation of moderately volatile elements in ordinary chondrites. *Meteoritics, 9,* 69–84.

Wasson J. T. and Kallemeyn G. W. (1988) Compositions of chondrites. *Philos. Trans. R. Soc. Lond., A325,* 535–544.

Wasson J. T. and Kallemeyn G. W. (1990) Allan Hills 85085: A subchondritic meteorite of mixed nebular and regolithic heritage. *Earth Planet. Sci. Lett., 101,* 148–161.

Wasson J. T. and Kallemeyn G. W. (2002) The IAB iron-meteorite complex: A group, five subgroups, numerous grouplets, closely related, mainly formed by crystal segregation in rapidly cooling melts. *Geochim. Cosmochim. Acta, 66,* 2445–2473.

Wasson J. T. and Kimberlin J. (1967) Chemical classification of iron meteorites. II. Irons and pallasites with germanium concentrations between 8 and 100 ppm. *Geochim. Cosmochim. Acta 31,* 2065–2093.

Wasson J. T. and Richardson J. W. (2001) Fractionation trends among IVA iron meteorites: Contrasts with IIIAB trends. *Geochim. Cosmochim. Acta, 65,* 951–970.

Wasson J. T. and Schaudy R. (1971) The chemical classification of iron meteorites. V. Groups IIIC and IIID and other irons with germanium concentrations between 1 and 25 ppm. *Icarus, 14,* 59–70.

Wasson J. T., Choi B.-G., Jerde E., and Ulff-Møller F. (1998) Chemical classification of iron meteorites. XII. New members of the magmatic groups. *Geochim. Cosmochim. Acta, 62,* 715–724.

Wasson J. T., Ouyang X., Wang J., and Jerde E. (1989) Chemical classification of iron meteorites. XI. Multi-element studies of 38 new irons and the high abundance of ungrouped irons from Antarctica. *Geochim. Cosmochim. Acta, 53,* 735–744.

Wood J. A. and Hashimoto A. (1993) Mineral equilibrium in fractionated nebular systems. *Geochim. Cosmochim. Acta, 57,* 2377–2388.

Wulf A. V., Palme H., and Jochum K. P. (1995) Fractionation of volatile elements in the early solar system: Evidence from heating experiments on primitive meteorites. *Planet. Space Sci., 43,* 451–468.

Yoneda S. and Grossman L. (1995) Condensation of $CaO-MgO-Al_2O_3-SiO_2$ liquids from cosmic gases. *Geochim. Cosmochim. Acta, 59,* 3413–3444.

Yin Q.-Z. (2002) Chemical signatures of interstellar dusts preserved in primitive chondrites and inner planets of the solar system (abstract). In *Lunar and Planetary Science XXX,* Abstract #1436. Lunar and Planetary Institute, Houston (CD-ROM).

Yin Q.-Z. (2004) From dust to planets: The tale told by moderately volatile element depletion (MOVED). In *Chondrites and the Protoplanetary Disk,* Abstract #9066. Lunar and Planetary Institute, Houston (CD-ROM).

Yin Q.-Z. (2005) From dust to planets: The tale told by moderately volatile elements. In *Chondrites and the Protoplanetary Disk* (A. N. Krot et al., eds.), pp. 632–644. ASP Conference Series 341, Astronomical Society of the Pacific, San Francisco.

Yin Q.-Z., Jacobsen S. B, and Yamashita K. (2002) Diverse supernova sources of pre-solar material inferred from molybdenum isotopes in meteorites. *Nature, 415,* 881–883.

Yu Y., Hewins R. H., Alexander C. M. O'D., and Wang J. (2003) Experimental study of evaporation and isotopic mass fractionation of potassium in silicate melts. *Geochim. Cosmochim. Acta, 67,* 773–786.

Origin of Water Ice in the Solar System

Jonathan I. Lunine
Lunar and Planetary Laboratory

The origin and early distribution of water ice and more volatile compounds in the outer solar system is considered. The origin of water ice during planetary formation is at least twofold: It condenses beyond a certain distance from the proto-Sun — no more than 5 AU but perhaps as close as 2 AU — and it falls in from the surrounding molecular cloud. Because some of the infalling water ice is not sublimated in the ambient disk, complete mixing between these two sources was not achieved, and at least two populations of icy planetesimals may have been present in the protoplanetary disk. Added to this is a third reservoir of water ice planetesimals representing material chemically processed and then condensed in satellite-forming disks around giant planets. Water of hydration in silicates inward of the condensation front might be a separate source, if the hydration occurred directly from the nebular disk and not later in the parent bodies. The differences among these reservoirs of icy planetesimals ought to be reflected in diverse composition and abundance of trapped or condensed species more volatile than the water ice matrix, although radial mixing may have erased most of the differences. Possible sources of water for Earth are diverse, and include Mars-sized hydrated bodies in the asteroid belt, smaller "asteroidal" bodies, water adsorbed into dry silicate grains in the nebula, and comets. These different sources may be distinguished by their deuterium-to-hydrogen ratio, and by predictions on the relative amounts of water (and isotopic compositional differences) between Earth and Mars.

1. INTRODUCTION

Water is, by number, the most important condensable in a cosmic composition soup of material. By mass, it rivals that of rock — depending upon the extent to which oxygen is also tied up in carbon monoxide and carbon dioxide (*Prinn,* 1993). And yet many workers fail to consider water ice as a "planet-building material" in the same way as silicates and metals are — in large part because we do not have samples of cometary material. Water ice is simply not stable on small bodies in the region of the solar system inhabited by Earth and Mars, inhabiting instead the polar and high-altitude regions of Earth and the poles and subcrustal reservoirs on Mars. Therefore, traditional meteorite studies ignore water ice in favor of the rocky and metallic phases.

Beyond the asteroid belt, water ice is abundant. It is a minor component of Jupiter's moon Europa, but constitutes almost half the mass of Jupiter's moons Ganymede and Callisto and Saturn's moon Titan. It is the dominant, or at least key, constituent of the intermediate-sized moons of Saturn (e.g., Enceladus), the moons of Uranus, Neptune's moon Triton, and the Kuiper belt object Pluto and its moon Chiron. It was almost certainly an important core-building material of the giant planets. Likewise, water ice is an important component of comets, ranging from being the major solid in fresh comets to a minor component of old comets in asteroid-like orbits. Thus, understanding the formation of planets and smaller bodies in our solar system requires consideration not only of the meteorite record but of icy bodies as well.

Unlike meteorites, icy bodies have been studied only by remote sensing (even the sample collection by the *Stardust* mission captures only, or primarily, atoms from the dust component of Comet 83P/Wild 2). Future opportunities to study icy bodies directly will come from the *Rosetta* mission, already launched, the *Deep Impact* mission to impact the surface of a comet, and a proposed Europa lander or penetrator. Ice in the high latitudes of Mars will be directly sampled and studied by the Project Phoenix Mars lander in 2008. Beyond these four opportunities, icy bodies in the outer solar system will continue to be studied primarily by remote sensing in the near future.

This brief chapter sketches the possible origins of condensed phases of water. Sources of water via direct condensation and infall from the surrounding molecular cloud are considered first, followed by consideration of how giant planet formation may have led to a chemically distinct class of water ice. The implications of the origin and distribution of water ice in the solar system for the source of Earth's water is considered, and the chapter closes with a brief consideration of the density of outer solar system icy bodies for the existence of different reservoirs of condensed water.

2. THE NEBULAR SNOWLINE

The protoplanetary disk out of which the solar system formed (which is referred to in the meteoritical literature as the solar nebula, a term we will use as well here) is a natural outgrowth of the interaction between gravitationally driven collapse of a dense clump of molecular cloud gas, and the conservation of angular momentum contained in the material. The disk is the physical medium of gas and solids through which much of the material out of which the Sun itself formed traveled, and dissipation in the disk transported mass inward to the center and angular momentum outward (*Nelson et al.,* 1998). The dissipation is accomplished through net gravitational forces (torques) that one portion of the rotating disk will exert on another, generating waves of various types (*Lin et al.,* 2000), through torques associated with a possible magnetic field embedded in the disk, or through the shear associated with the radial variation of Keplerian rotation and motion perpendicular to the disk ("vertical") caused by convection of heat away from the disk midplane to the colder upper regions (*Stevenson,* 1990). [An alternative birth site to a clump in a dense molecular cloud, offered recently on the basis of the apparent existence of live ^{60}Fe in early solar system materials (*Hester et al.,* 2004), in the midst of an assemblage of short-lived and massive high-mass stars, has potential implications for typical solar nebula disk models that have yet to be evaluated.]

In a disk with sufficient gas density to be optically thick (that is, with an optical depth — the product of scale length, material density, and absorption coefficient — in excess of unity), the temperature drop along the midplane is determined by the nature of the dissipative processes, and will decline with distance r from the center along the midplane as 1/r. For an optically thin (optical depth less than 1) disk, the temperature drop is determined by absorption of the Sun's radiation by midplane material and drops as $1/r^{1/2}$. In either case, this drop in density ensures that some sort of radial gradation in solid-forming material will occur within the disk. Refractory silicates such as corundum will be stable as solids closer to the proto-Sun than will the more-abundant magnesium silicates, and water ice will appear at even greater distances. The radial distance along the midplane at which water ice may first stably appear is referred to in the planetological literature as the *snowline*.

The simplest calculation of the snowline requires comparing the saturation vapor pressure — a function of the ambient temperature — to the partial pressure of water vapor at the midplane, as a function of distance outward from the center. The partial pressure of water vapor is the total nebular pressure times the mole (i.e., number) fraction of water, and the latter inward of the condensation front is simply the mole fraction based on cosmic abundance of the elements and the distribution of oxygen among several molecular species. Where the partial pressure exceeds the saturation vapor pressure it is thermodynamically possible for condensation to occur, and hence water ice (perhaps with the consistency of snow) will form there. A typical nebular temperature for the water condensation is between 160 and 170 K, with a radial distance that depends on nebular models and the stage of nebular evolution under consideration. The snowline radius, r_{snow}, has a rather large potential range between 1 AU and 5 AU, the former being an extreme value for older, dusty, cold disks (*Sasselov and Lecar,* 2000). Figure 1 shows the snowline position for several different models of the temperature of the solar nebula. The warmest is purely schematic, in which the temperature at 5 AU is set to 160 K and that at Saturn's orbit to 100 K, consistent with the volatiles seen in Saturn's moon Phoebe, likely captured from solar orbit in the vicinity of Saturn. The other profiles are displaced downward and come from *Sasselov and Lecar* (2000), but have the same slope as the schematic model. Thus there is no clear preference based on nebular temperature profiles for a snowline at 1 vs. 5 AU, but both may obtain at different times in the history of the nebula as accretion ceases and temperatures decline.

The above calculation of the snowline radius assumes that the gas and the grain temperatures are identical, which is not necessarily the case. Sizes of grains falling into the nebula range from 0.1 to 10 μm or larger, although the larger grains will be fluffy aggregates of, and hence behave thermally as, the smaller particles (*Weidenschilling and Ruzmaikina,* 1994). Radiative properties of the grains depend on their composition but especially on their size, since the wavelength range over which the peak emission occurs is comparable to or larger than the particle size. Put simply, the smaller particles are poorer radiators and absorbers of thermal energy than are the larger ones (*Lunine et al.,* 1991). The cross section for interaction between a grain of radius a and radiation of wavelength λ can be written as the prod-

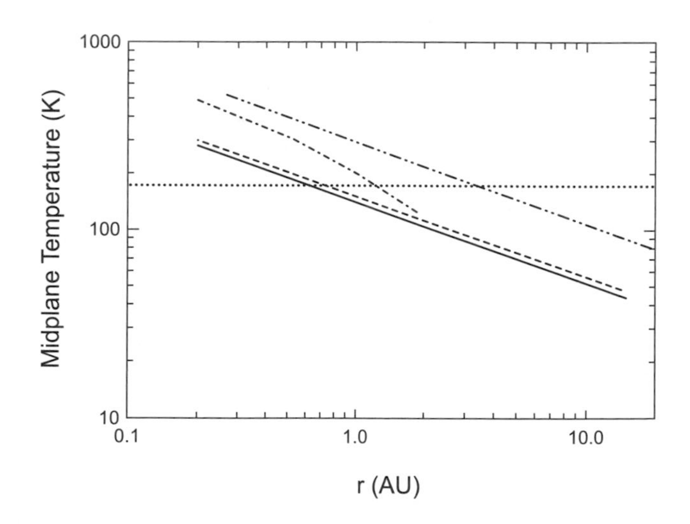

Fig. 1. The nebular snowline in the formal thermodynamic definition for several possible profiles of the midplane temperature in the solar system's protoplanetary disk. The snowline temperature is the dotted horizontal line, and the snowline distance (condensation front) occurs where the diagonal lines cross it. The upper line (dash-dot-dot) is a profile from the author and represents a relatively warm nebula, one in which temperatures in the formation region of Saturn are 80 K — consistent with the constraints from saturnian satellites and objects further out. The lower two models (solid and dashed) are cold models with no accretion, while the dash-dotted model includes some active accretion — all three from *Sasselov and Lecar* (2000), upon which the figure was based.

uct of the geometric area of the grain and an efficiency factor Q; the latter is near unity if the grain radius (actually, the circumference 2π a) is comparable to or larger than λ, and decreases rapidly as 2π a/λ drops progressively below unity corresponding to shrinking grain size (*Podolak and Zucker,* 2004).

The primary consequence of this analysis is that heating and cooling fluxes must be computed in a time-dependent fashion to determine the particle temperatures as a function of the gas temperature and the particle size and composition (icy vs. rocky). In practice this leads to little change in the position of the snowline at the nebular midplane in terms of temperature: Depending on the assumptions regarding heating and cooling rates, the snowline is between 150 K and 170 K, and the temperatures of the gas and icy grains are nearly equal to each other there. However, grain temperatures may be well below the nebular gas temperature inward of the snowline, where only silicates are stable. This is a consequence of the nebular midplane being optically thick, because then the gas is the only source of heating for the grains and the small size and low emissivity of the silicate grains makes the process inefficient (*Podolak and Zucker,* 2004). The situation is much more complicated at the optical surface, or photosphere, of the nebula, where the gas becomes (by definition) optically thin, and grains lofted from the midplane evaporate due both to the lower gas pressure and the direct radiation from the Sun. Although evaporation rates at 150–170 K are very small, dirty ice grains (those with darkening agents) may evaporate within the lifetime of a 10-m.y. disk. Clean ice grains have longer lifetimes (Fig. 2), and so the grain longevity may depend on what other materials are trapped in the water ice during grain formation, and how radiation from the proto-Sun will alter them. For example, methane and methanol may darken rather quickly and increase the grain evaporation rates, and these might be incorporated into icy grains at nebular distances as small as 5 AU (*Hersant et al.,* 2004), or be transported inward by radial gas drag after formation (*Cyr et al.,* 1998) — a complication to which we next turn.

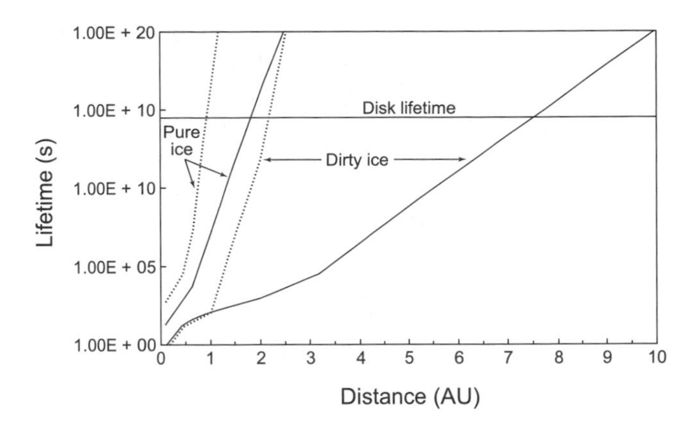

Fig. 2. Lifetime of grains against evaporation for 0.1- (dashed) and 10-μm (solid) grain sizes; the smaller grains are poorer absorbers and radiators of photons and hence evaporate more slowly. The horizontal line corresponds to a disk (solar nebula) lifetime of 10 m.y. From *Podolak and Zucker* (2004).

Inward transport of ice particles by gas drag constitutes an important modification to the simple vapor-pressure-driven picture. Because the gaseous phase of the solar nebula is supported to a small extent by the pressure force, the gas molecules orbit slightly more slowly than the Keplerian speed associated with their radial distance from the disk center. Solid particles thus experience a wind as they move at the Keplerian orbital speed through the gas, which is a function of grain size (*Weidenschilling,* 1977). Very small grains — those less than the mean free path in the gas — act like gas molecules, embedded in the flow and affected only by the collisions with surrounding molecules; particles larger than the mean free path experience frictional drag and are slowed down, causing them to spiral inward toward the center of the disk. Because in the drag regime the ratio of surface area to mass (volume) of the particle determines the efficiency of the drag force, the largest particles will experience essentially no drag. Hence there is a maximum in the drag force, or equivalently the inward radial drift, corresponding to an intermediate particle size. This intermediate particle size depends on the gas density and the fluffiness of the particles themselves, and at 1 AU, for example, can range from subcentimeter to tens of meters in radius for various plausible choices of the density of the solar nebula and particle density (*Weidenschilling,* 1977). The peak drift velocity depends primarily on the nebular gas density and Keplerian rate (hence disk position; it can range from ten to a hundred meters per second) (*Cuzzi and Zahnle,* 2004).

As water ice condenses at the snowline the initial particle size is small enough that the particles remain embedded in the gas; as they grow, they begin to move inward from the boundary and experience nebular temperatures that force evaporation to occur. The primary effect of the evaporation is to deliver water vapor back to its source in the inner nebula from the sink of ice particles at the snowline. Absent an inward drift the water vapor abundance inward of the snowline should exhibit a profile approaching that of a simple "cold-finger" solution to the diffusion equation in a cylindrical disk — this abundance could eventually reach arbitrarily small values (*Stevenson and Lunine,* 1988) and impose significant changes on the oxidation state of the nebular gas inward of the snowline (*Cyr et al.,* 1999). With inward drift of icy planetesimals, however, this decline of the water vapor inward of the snowline is modified in a complicated way that depends on the details of the particle growth, particle properties, and nebular temperature profile. Indeed, inward drift associated with gas drag is not the only mechanism that could carry particles inward (or outward) in a turbulent nebula; advective or convective flows could exist as well (*Prinn,* 1990; *Supulver and Lin,* 2000). Furthermore, water ice grains falling directly into the inner part of the disk from the molecular cloud will contribute water vapor

(*Chick and Cassen,* 1997). For the purely diffusive solution, the region of "enhanced" (over the depleted background) water vapor is about 1 AU (*Cuzzi and Zahnle,* 2004), and the nondiffusive effects cited above will tend to broaden this somewhat.

The snowline is undoubtedly more complicated than has been outlined here, and will not be a unique phenomenon in a nebula where other major species (silicates, sulfur-bearing compounds, ammonia, carbon dioxide) may condense out directly, although many molecular species such as methane and the noble gases are more likely to become trapped in the water ice as an adsorbate or clathrate hydrate guest molecule (*Gautier et al.,* 2001). However, the basic idea that there is a particular distance at which water ice grains become stable and abundant, and which consequently alters the water vapor abundance inward of that point, is a robust one at any given time in the evolution of the nebula.

The snowline has been interpreted traditionally as setting the dividing line between icy (cometary) and rocky (asteroidal) small bodies, with water-rich (C-type) asteroids receiving their water inventory from hydration reactions occurring at temperatures too high for ice itself to be stable. Thus, a somewhat more poorly defined "hydration line" could be defined as extending inward some 1 or 2 AU from the snowline itself, based on the thermodynamic stability of hydrated silicates at temperatures below between 225 K and 250 K (*Fegley,* 2000), and hence defining the parent bodies of the carbonaceous chondrites. However, the kinetics of the hydration reaction between the water-laden gas and preexisting silicate grains are such that the time for conversion may greatly exceed the disk lifetime (*Fegley,* 2000). Others have argued that the process may be much faster on the basis of laboratory studies of dehydration and a model for the relationship between the forward and back reactions (*Ganguly and Bose,* 1995), or because of shock effects near the snowline itself (*Ciesla et al.,* 2003).

Laboratory studies of carbonaceous chondrites suggest that the hydration reactions occurred inside meteorite parent bodies upon exposure to liquid water, rather than in the nebula itself (*Clayton,* 2003). This would implicate water ice within the asteroidal parent bodies, acquired perhaps as the snowline moved inward prior to the accretion of these objects, as the source of the water of hydration. The primary reservoir of water within the hydrated asteroids (or their parent bodies) would then be that of the snowline, with the consequent implication that icy bodies formed outside the snowline ought to have an isotopic composition, for the water, the same as that found in carbonaceous chondrites — in particular, a deuterium-to-hydrogen ratio with a mean value of standard mean ocean water ("SMOW"). If, instead, the silicates were hydrated from the nebula itself, inward of the snowline and hence in the absence of the formation of water ice itself, then the isotopic composition of the hydrated and icy material may not be related. This is an important issue that pertains to the origin of Earth's water, which we will discuss in section 5.

3. INFALLING WATER-ICE GRAINS

Water ice is formed directly in interstellar clouds, mostly from direct adhesion of atoms of preexisting silicate grains at very low temperatures yielding amorphous solid-water phases (*Irvine and Knacke,* 1989), but these may be modified from the sublimation and recondensation of such amorphous structures as they cycle in and out of hot cores in the clouds. Grains grow as they fall into a protoplanetary disk (*Weidenschilling and Ruzmaikina,* 1994), but sublimation of the water ice will occur both in gas dynamical heating associated with shocks (*Lunine et al.,* 1991) and in the ambient disk within a certain radial distance from the Sun, due simply to the radial-temperature profile of the disk. This latter effect is the same as that which determines the nebular snowline distance, but here preexisting, infalling ice particles are considered rather than the formation of the ice particles from the nebula vapor phase.

As one might expect, the survival zone for infalling ice particles could be as close as the nebular snowline, when dynamical heating due to infall is assumed minimal, or could be much farther out if shocks play an important role as a heating mechanism (*Chick and Cassen,* 1997). The extent to which shocks heat the grains depends on the type of shock formed as accreting material accelerates toward the disk and then is decelerated by interaction with the disk itself. *Chick and Cassen* (1997) examined a range of nebular models and found water-ice grain survival could occur as close as 5 AU for an optically thick disk, moving inward to as close as 2 AU if an optically thin cavity forms. The ice stability line could have been as far as 30 AU from the proto-Sun for high accretion rates and warm disk temperatures. That the ice survival line should range from 2 to 30 AU simply reflects the wide range of possible conditions during the lifetime of the solar nebula disk, as well as uncertainties in the model parameters and the environment surrounding the disk. Most of the water vapor in the protoplanetary disk must originally have had its source in sublimated amorphous ice grains in the standard picture of a cold molecular cloud phase as a precursor of collapse of clumps to form stars and planets. However, birth in a more supernova-rich environment with strong UV erosion of a disk (*Hester et al.,* 2004) could have implied a different history of the water in vapor and amorphous phases and led to very different patterns of ice survival, but these possibilities have yet to be quantified.

Ice that has been sublimated (or vaporized, if a surface layer of transient liquid forms first on the grain, although this has no measurable effect on the outcome) will recondense if the ambient gas temperature into which the vapor is mixed is low enough, i.e., if the final "resting place" of the vapor is beyond the snowline. And vapor inward of the snowline will eventually diffuse or be advected or convected into the snowline region, leading to the picture of a water-vapor-depleted inner zone and a water-rich outer zone, the boundary being the snowline. The practical effects of the sublimation and recondensation are two-fold: (1) to convert amorphous ice into crystalline ice and (2) to redistribute substances more volatile than water ice from trapping sites in interstellar grains into the nebular gas, and hence into other phases in the reformed ice and silicate grains.

Effect (1) requires that amorphous ice formed at very low temperatures in molecular clouds be sublimated (at any temperature) and then recondensed at a temperature in excess of 130–140 K (*Kouchi et al.,* 1994). If the nebular snowline is at a temperature of 150–160 K, driven by the solar abundance of available oxygen in the form of water vapor, then this requirement is satisfied. We therefore expect the solar nebula to have contained a mixture of amorphous and crystalline ice: the crystalline ice a result of the water vapor sublimated during nebular infall and then recondensed at the snowline, the amorphous representing the grain component that survived infall to the disk. Some protostellar regions show evidence for crystalline ice (*Malfait et al.,* 1999), and it is possible that disk reprocessing is the source, although direct conversion from amorphous ice in the surrounding hot core should also be examined. Likewise, fresh comets should be a mixture of crystalline and amorphous ice, unless other heat sources (such as decay of the short-lived isotope ^{26}Al) trigger conversion early in the history of the solar system. The low spin temperature seen in the hydrogen in some comets (*Kawakita et al.,* 2001; *Irvine et al.,* 2000) argues against wholesale reheating and in favor of the preservation of delicate amorphous grains.

Effect (2) is potentially much more complex. Interstellar grains are cold enough that virtually all elements and molecules may adsorb at some point on the grains; laboratory experiments down to 20–30 K have been conducted to confirm that the resulting abundances of (for example) noble gases are essentially proportional directly to the gas phase abundances, and it has been argued that this pattern is seen, for example, in elemental abundances in Jupiter (*Owen et al.,* 1999). However, somewhere within 30 AU of the proto-Sun, water ice will be partially or completed sublimated during nebular infall, and the more volatile species released into the gas phase. These species in turn will be retrapped as icy grains reform, but at different pressures and temperatures than obtained in the original molecular cloud (*Lunine et al.,* 1991). Indeed, formation of crystalline ice or clathrate hydrate — in which the volatiles are trapped in a regular crystal structure of water ice with a well-defined entropy and thermodynamic free energy — might have taken place in the region where Jupiter now exists (*Gautier et al.,* 2001) or even further out (*Hersant et al.,* 2004). The pattern of volatile enrichment in the ice depends, in this case, on the details of the reformation and cooling history of the ice.

One potential way of distinguishing between the two alternatives requires a definitive measurement of the oxygen abundance within Jupiter. Adsorption of volatiles on small ice grains at 20 K should be extremely efficient, leading to an enrichment of elemental oxygen (relative to the solar abundance) comparable to that of the noble gases. If, instead, the trapping occurred at a later stage in the nebula, in crystalline (possible clathrate hydrate) water ice, it would have been much less efficient, and the oxygen abundance correspondingly much higher. Specifically, the "primitive adsorption" model implies an oxygen abundance in Jupiter three times solar, while the sublimation and retrapping model implies an oxygen abundance at least nine times solar (*Hersant et al.,* 2004).

4. WATER ICE IN THE REGION OF GIANT PLANET FORMATION

The environments around the giant planets during their formation were vastly different from that in the ambient solar nebula. Unfortunately, quantifying these conditions is extremely difficult, because the mechanisms by which Jupiter and Saturn (as well as extrasolar giant planets) formed remain controversial (*Lunine et al.,* 2004). The traditional model relies on the so-called "nucleated instability," in which the growth of a rock and ice core triggers rapid accretion, if not collapse, of a large gaseous envelope to make a giant planet (*Pollack et al.,* 1996; *Wuchterl et al.,* 2000). Direct collapse, a process dismissed many years ago based on preliminary mathematical models of protoplanetary disks, has been found to produce bodies of jovian mass given the right, marginally unstable, background disk conditions of low temperature and relative high density (*Boss,* 2001). The timescales of the two processes are very different — nucleated collapse requiring on the order of a million years, and direct collapse taking as little as hundreds of years (*Mayer et al.,* 2002). Thus, the temperature and density perturbations of the background nebula must clearly have been different in the two cases. The nucleated instability formation of giant planets is associated with short-lived global perturbations on the entire protoplanetary disk, and so it is not useful to try to distinguish a special set of conditions near the collapsing giant planet that might produce a distinct reservoir of icy planetesimals. Indeed, the disk collapse into giant planets, and subsequent gravitational interactions among these planets leading to a subset being ejected, would likely have erased any radial gradations in icy grain properties through direct dynamical effects and changes in the radial temperature profile. These have yet to be evaluated quantitatively in the context of a detailed model of an unstable protoplanetary disk.

The longer timescales and more nearly quasistatic conditions during the nucleated-instability growth of the giant planets renders practical the modeling of nebular conditions in the vicinity of the growing giant planets. Optically thick disk models with temperature profiles determined by an adiabat are a possible end member for Jupiter and Saturn (*Lunine and Stevenson,* 1982), but an oversimplified one because such a disk does not take into account the transport of material through the disk into the giant planet itself. More recent models include disk transport, and find that an optically thick circumjovian disk is too hot to per-

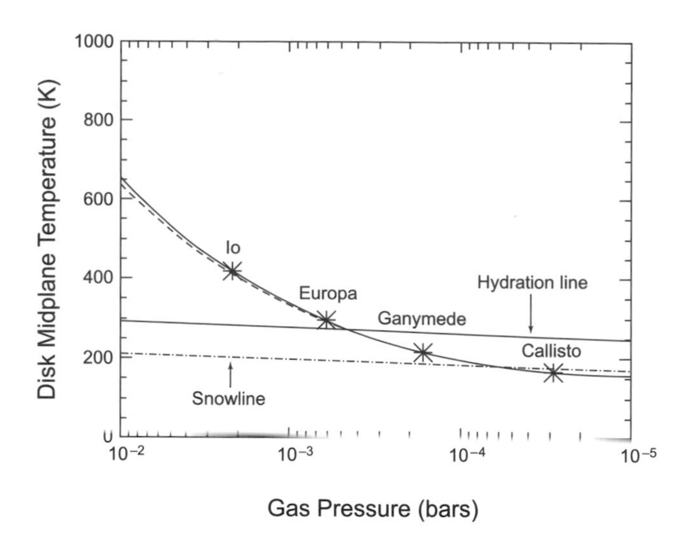

Fig. 3. Example of a temperature-pressure profile in the midplane of a disk surrounding Jupiter as the giant planet formed. This particular disk has an inflow rate that is slow enough to allow temperatures near or below the water condensation line ("snowline") in the region of Ganymede and Callisto, but fast enough that Jupiter itself forms within 5 m.y. (*Canup and Ward,* 2002). It also has a temperature profile that is appropriate to the hydration of phyllosilicates ("hydration line") between the orbits of Ganymede and Callisto. This disk is approximately optically thin so that the photospheric temperature (dashed line) is nearly that of the midplane temperature. Adapted from *Canup and Ward* (2002).

mit a snowline at the orbits of Ganymede or Callisto — or anywhere within a reasonable migration distance. Disks may be cooler, less massive, and with a slower influx of material into proto-Jupiter (*Canup and Ward,* 2002) — limited by the 5–10-m.y. timescale over which Jupiter forms (*Lunine et al.,* 2004). Such disks exhibit snowlines located so as to permit ice to exist where Ganymede and Callisto formed (with a small amount of inward migration), and water of hydration to occur at the orbit of Europa (Fig. 3). Disk pressures are elevated above that in the surrounding solar nebula, so that the snowline and hydration fronts are at correspondingly higher temperatures. The 200 K temperature appropriate for the snowline in the circumjovian disk almost certainly means that crystalline water ice condenses in preference to amorphous ice, and the trapping of more volatiles species depends on clathrate formation rather than adsorption in amorphous ice layers. Thus, the planetesimal population in the circumplanetary disks will differ in their physical properties and volatile compositions from those in the solar nebula, and hence constitute a distinct, third, reservoir of water ice.

5. DELIVERY OF WATER TO THE TERRESTRIAL PLANETS

Although the outer solar system contains many tens of Earth masses of water, located within the giant planets, their moons, and objects in the Kuiper belt and Oort cloud, it is the water in the terrestrial-planet region that has garnered the most persistent interest from the astrobiological point of view [despite the possibility of a subsurface ocean on Europa (*Chyba et al.,* 1998)]. The roughly 0.001 $M_\oplus$ of water in the Earth's crust and mantle is the critical factor in the existence of life — together with a surface temperature salubrious for liquid water. Evidence continues to accumulate that Mars once possessed a surface water budget equivalent to somewhere between 0.01 and 1 times that of Earth (*Baker,* 2001). The highly enriched deuterium abundance in the atmosphere of Venus suggests that this nearly Earth-sized planet once contained amounts of water comparable to Earth (*Hunten et al.,* 1989). Thus, although the amounts of water are small relative to those in the outer solar system, the implications for the question of planetary habitability are large, and hence the question of the origin of water on the terrestrial planets deserves close attention.

It is possible, based on the temperature profiles shown in Fig. 1, that water was directly available in the form of ice or water of hydration in the nebula at or near 1 AU when grains were still small. However, temperature profiles low enough for pure ice to condense are extreme ones, and likely existed (if at all) in the very late stages of disk accretion, when silicate bodies were already large (kilometer-sized or larger). The constraints on the temperature profile are less severe for water of hydration, but as discussed in section 2, doubts have been raised as to whether hydration of silicates could have occurred in the nebula. Were hydration to be widespread, we might expect to find a chondritic composition for planetesimals in the 1 AU region, and hence for Earth itself. However, Earth itself cannot contain more than a few percent, perhaps only 1%, of chondritic material (*Drake and Righter,* 2002). An alternative is that small bodies migrated inward from a zone of hydration at 2–3 AU via gas drag in the disk and retained sufficient water to supply Earth's inventory (*Ciesla et al.,* 2004). Adsorption of water vapor onto dry silicate grains that have the isotopic and elemental composition of Earth is also possible, but the grains must have a highly fractal nature in order to adsorb sufficient water (*Stimpfl et al.,* 2004). Both the adsorption and the inward migration model create constraints on the properties of planetesimals at the time of the growth of Earth (size distribution, porosity, ambient gas density) that have yet to be explored in detail.

Bringing water to Earth from more distant regions of the protoplanetary disk could also have been accomplished through gravitational scattering of icy and hydrated bodies onto highly eccentric orbits that extend from their colder regions of origin to the orbit of Earth. At first sight, this would seem to be a very slow and inefficient process, because planetesimals are initially on circular orbits (determined by the effects of gas drag), and grow rapidly in the presence or absence of gas to an "isolation mass" at which the mean separation between bodies is vastly larger than their cross sections for gravitational interaction (*Goldreich et al.,* 2004; *Raymond et al.,* 2004). That mass is somewhere between the mass of the Moon and Mars, and dramatically slows further accretion to timescales much longer than the tens of millions of years for Earth constrained by radioisotopic data (*Cameron,* 2002). However, the growth of Jupiter provided a large perturbing mass that rapidly increased the eccentricity of these lunar- to Mars-sized "embryos," particularly in what is now the outer asteroid belt, sending them on trajectories to the inner solar system and, in some cases, on Earth-crossing trajectories. Since the formation of Jupiter had to occur when gas was present (*Wuchterl et al.,* 2000), this must have occurred within the several-million-year period of the existence of the nebular gas (*Hartmann,* 2000), and various models of giant planet formation yield formation times comparable to this (*Ida and Lin,* 2004) or much shorter (*Boss,* 2003; *Mayer et al.,* 2002).

Numerical simulations of this process (*Morbidelli et al.,* 2000; *Chambers and Cassen,* 2002; *Raymond et al.,* 2004) produce Earth-sized planets in the region around 1 AU on timescales consistent with the geochemical constraints. All the simulations assume that the region of the asteroid belt contained at least as much rocky material, prior to the formation of Jupiter, as did the equivalent area of the disk in the terrestrial planet region — a reasonable assumption given the current architecture of the asteroid belt and no compelling mechanism for creating an early "pre-Jupiter" gap corresponding to the present-day dearth of material between 2 and 4 AU (*Petit et al.,* 2001). In most of the outcomes of the planet-forming simulations from lunar- to Mars-sized embryos, the majority of material from which these bodies are accreted is local, and determines the geochemical composition of the final planet. But a fraction comes from what is now the asteroid belt, and a portion of that from the region beyond 2 or 2.5 AU where the relatively water-rich chondrites are believed to have originated. The amount of water delivered to Earth, or to any of the terrestrial planets, is highly variable from one simulation to another [ranging from less than an Earth ocean to well over 100 Earth oceans (*Raymond et al.,* 2004)] because the number of embryos is small — on the order of 10^2–10^3. Thus, small changes in the starting conditions can lead to very different results in terms of planet position, mass, and water abundance.

Even assuming a significant loss of water during impact of the embryos with the growing Earth (approximately 50% loss when the mass of the growing Earth is close to the final value, solely an "educated guess" absent detailed tracking of the water during giant impact simulations), the dynamical simulations often yield 1-$M_\oplus$ bodies near 1 AU with an amount of water that overlaps estimates for the surface and mantle — about 3 to 5 times the mass of the surface ocean of Earth today (*Abe et al.,* 2001). The simulations will also produce very water-rich Earths in some cases, up to the wet primordial mantles envisioned by some workers (*Dreibus and Wanke,* 1987).

A potential problem with this scenario arises from the amount of chondritic material added by Earth — assumed to be the prototype for the water-rich embryos beyond 2–3 AU from the Sun — which must be less than 1–3% (depending on the timing of the addition relative to core formation) to satisfy (1) the abundance of siderophile elements in the mantle and (2) the oxygen-isotopic similarity between Earth and the Moon (*Drake and Righter,* 2002). Except for the lower end of the estimates of total water added to Earth, the dynamical model modestly violates (by a factor of 2 or 3) the two geochemical constraints. However, constraint (1) can be removed if the carbonaceous embryo that delivered the water was differentiated, or partially differentiated. In this case, its core, containing most of the siderophile elements, would not mix with Earth's mantle, and thus a much larger fraction of carbonaceous material could have been delivered to Earth without exceeding the amount of siderophile in Earth's mantle. Constraint (2) can be removed if there is a way to homogenize the oxygen-isotopic composition of Earth and the Moon soon after the giant impact that formed the Moon. More work must be done to evaluate these possibilities.

The model for adding water to Earth from the primordial asteroid belt has the virtue that the deuterium-to-hydrogen ratio (D/H) measured in Earth's oceans, the so-called "SMOW," is consistent with the most probable value of D/H computed from the broad range seen in carbonaceous chondrites (*Robert,* 2001; see also *Robert,* 2006). A cometary contribution of water would have a higher D/H: Three comets that fall into the "long-period" comet class, thought to be derived from the Oort cloud, all have D/H approximately twice that of SMOW (*Meier et al.,* 1998a,b). If typical, this should then limit severely the amount of Earth's water that could have been contributed by comets, because no plausible mechanism has been identified for reducing the D/H value after accretion (contact with a nebular composition atmosphere without loss of the water to space seems implausible). Indeed, dynamical calculations limit the amount of water contributed by comets (essentially, from icy bodies resident at and beyond the orbit of Jupiter) to 10% of the total brought in from sources in the primordial asteroid belt (*Morbidelli et al.,* 2000).

Is it possible that the D/H measured in comets represents an alteration of an original value via phase changes in the cometary nucleus? The three comets analyzed have different amounts of exposure to sunlight, Halley having been around the Sun many times, with Hyakutake and Hale-Bopp appearing to be relatively "fresh" comets. This question can be addressed experimentally (R. H. Brown and D. S. Lauretta, personal communication, 2005) or by measurement of additional long- and short-period comets. *Podolak et al.* (2002) have proposed a mechanism for altering the measured coma value of D/H relative to that in the nucleus of a comet, although concerns have been raised about their model (*Krasnopolsky,* 2004).

Mars is an important test of the origin of water in the terrestrial planets because of its greater proximity to the snowline and its small mass, which suggests that it could be a planetary embryo left behind from the accretion process, either by chance (*Lunine et al.,* 2003), or through the orbital damping effect of residual nebular gas (*Kominami and Ida,* 2002). In either case, the source of water can no longer be large embryos, but is a mixture of smaller bodies, some of which are comets. The proportionate amount of water contributed by comets depends on the importance of various local and distal sources of water, i.e., adsorbed water on grains, hydrated or icy bodies brought in by gas drag, or small primordial asteroids whose orbits were gravitationally perturbed by Jupiter. Although the martian data are not yet precise enough to choose among these possibilities, in principle better constraints on the initial amount of water on Mars (currently two orders of magnitude uncertain) will allow such a test to be made. The contribution of high D/H water to the inner solar system is perhaps hinted at in the analysis of hydrated minerals in martian meteorites, with values tending toward twice SMOW (*Leshin,* 2000), although it appears to be lower in some samples (*Boctor et al.,* 2003; *Gillet et al.,* 2002). The complex situation associated with D/H on Mars is discussed in more detail in *Robert* (2006).

6. THE OUTER PLANET SATELLITES: CONSTRAINTS ON RESERVOIRS OF WATER ICE

The satellites of Jupiter, Saturn, and Uranus are a diverse group of objects that illustrate well the idiosyncracies of the formation of secondary bodies around giant planets (*Lunine et al.,* 2004). Satellite formation in the Jupiter environment encouraged, for whatever reason (*Mosqueira and Estrada,* 2003), the formation of four large moons with roughly comparable amounts of silicate — perhaps reflecting a limit on the silicate abundance (essentially, surface density of refractories) in the disk. The total mass of the satellites was then determined by the amount of ice added, and this was in turn determined by the strong temperature gradient in the circumjovian disk (*Canup and Ward,* 2002). Conditions at Io were too warm even for hydrated silicates — or tidal heating result in dehydration of the phyllosilicates and loss of water. Europa began as a hydrated-silicate body, with subsequent internal tidal and radiogenic heating leading to the generation of a thin water mantle atop the Io-mass silicate body. Ganymede and Callisto formed with a full complement of water in the disk, but the accretional energy released per unit mass toward the end of accretion was comparable to the latent heat of vaporization of water, and so accretion tailed off as water vaporization and loss occurred (*Stevenson et al.,* 1986). Thus none of the jovian satellites give us the rock-to-ice ratio of the primordial circumjovian disk.

The saturnian system, on the other hand, consists of (excepting Ganymede-like Titan) intermediate-sized satellites with sizes and densities that are not systematic. The formation of this system remains enigmatic, but evidently less material was available than at Jupiter, and the process was not determined by a radial temperature gradient in the circumsaturnian disk. The mean density of the intermediate-sized saturnian satellites, mass weighted and excluding Titan, is determined from Voyager data to be 1.3 g/cm^3 (*Jacobson,* 2004), well below the 1.9 g/cm^3 of Ganymede, Callisto, and Titan, and indicative of a circumsatellite nebular chemistry distinct from the solar nebula. If the dominant form of carbon in the solar nebula was CO and not CH_4 — an assertion consistent with models of the gas chemistry of the disk — then the amount of oxygen available to make water ice implies a density for ice bodies somewhere around 2.3 g/cm^3 — a number increased over previously published values by the elemental oxygen abundance in the Sun, which has recently been redetermined and substantially lowered relative to earlier studies (*Asplund et al.,* 2004).

The one intermediate-sized satellite that is demonstrably a captured object (in a loose retrograde orbit), Phoebe, has a Cassini-determined density around 1.63 g/cm^3, contains water ice (*Clark et al.,* 2005), and is irregularly shaped. It is significantly denser than the icier saturnian satellites, yet is small enough to be a porous body. For a reasonable porosity of 10–20%, Phoebe's material density would be essentially identical to the 2 g/cm^3 measured for Pluto and Triton (*Stern et al.,* 1997), and hence consistent with formation in solar orbit somewhere in the outer solar system. The inferred rock-to-ice ratios for Pluto and Triton imply a solar nebula whose carbon budget is largely (but not exclusively) in the form of carbon monoxide (*Johnson and Lunine,* 2005). The densities of the other saturnian satellites demand a circumsaturnian disk chemistry much richer in water and hence a carbon budget in which carbon monoxide is a minor or absent component, and methane (or other nonoxidized carbon species) dominate. This more reduced carbon budget is also chemically consistent with the circumstantial evidence for circumsaturnian ammonia (NH_3) as the original source of Titan's atmosphere (*Owen,* 1982), suggested by the absence of nonradiogenic argon in measurements of Titan by the Cassini Ion and Neutral Mass Spectrometer (*Waite et al.,* 2005). If ammonia were absent from the warmer jovian protoplanetary disk, it would explain why Ganymede and Callisto do not have Titan-like dense atmospheres.

The major uranian satellites have water ice on their surfaces (*Brown and Clark,* 1984) but a mean density determined by *Voyager 2* higher than that for the saturnian satellites (*Johnson et al.,* 1987), implying higher rock-to-ice ratios in their interiors. Either the circumuranian disk was not dense or hot enough for its composition to be altered from the solar nebula value, or the circumstances of satellite formation were affected by the impact that altered the Uranus obliquity (*Korycansky et al.,* 1990). Finally, while Pluto and its moon Charon show water ice on their surfaces (*Cruikshank et al.,* 1997), Neptune's moon Triton does not (*Brown et al.,* 1995). Since Triton and Pluto have similar densities and are thought to have similar origins (*McKinnon et al.,* 1995), it is assumed that the water ice crust of Triton is buried beneath other ices.

7. SUMMARY

Water ice, and water of hydration, were major planet-building materials in the protoplanetary disk. Multiple sources of water ice are suggested by modeling of the evolution of the planet-forming disk, but little evidence of these sources can be gleaned from the isotopic and chemical evidence currently available. On the other hand, the densities of the icy bodies of the outer solar system — giant planet moons and Pluto — do suggest that regions around the forming giant planets existed that were chemically distinct from the surrounding solar nebula. Further progress in elucidating the origin of water ice during the formation of the planets, its abundance distribution, and content of more volatile gases depend on measurements of the water abundance and isotopic composition of water in martian materials, more accurate satellite densities, and isotopic measurements of water in comets, among others. These are daunting goals, but important ones if we are to understand how water finds its way into both icy bodies and habitable worlds in planetary systems.

Acknowledgments. Helpful discussions with A. Morbidelli form a part of the discussion on the limits to the carbonaceous contributions to Earth's water budget. Reviews by F. Robert, T. Owen, and A. Morbidelli are also gratefully acknowledged. The preparation of this chapter was supported by NASA's Planetary Atmospheres Research and Analysis Program.

REFERENCES

Abe Y., Ohtani E., Okuchi T., Righter K., and Drake M. (2001) Water in the early Earth. In *Origin of the Earth and the Moon* (R. Canup and K. Righter, eds.), pp. 413–433. Univ. of Arizona, Tucson.

Asplund M., Grevesse N., Sauval A. J., Prieto Allende C., and Kiselman D. (2004) Line formation in solar granulation: IV. [O I], O I, and OH lines and the photospheric O abundance. *Astron. Astrophys., 417,* 751–768.

Baker V. R. (2001) Water and the martian landscape. *Nature, 412,* 228–236.

Boctor N. Z., Alexander C. M. O., Wang J., and Hauri E. (2003) The sources of water in Martian meteorites: Clues from hydrogen isotopes. *Geochim. Cosmochim. Acta., 67,* 3971–3989.

Boss A. P. (2001) Gas giant protoplanet formation: Disk instability models with thermodynamics and radiative transfer. *Astrophys J., 563,* 367–373.

Boss A. P. (2003) Formation of gas and ice giant planets. *Earth Planet. Sci. Lett., 202,* 513–523.

Brown R. H. and Clark R. N. (1984) Surface of Miranda. Identification of water ice. *Icarus, 58,* 288–292.

Brown R. H., Cruikshank D. P., Veverka J., Helfenstein P., and Eluszkiewicz J. (1995) Surface composition and photometric properties of Triton. In *Neptune and Triton* (D. P. Cruikshank, ed.), pp. 991–1030. Univ. of Arizona, Tucson.

Cameron A. G. W. (2002) Birth of a solar system. *Nature, 418,* 924–925.

Canup R. M. and Ward W. R. (2002) Formation of the Galilean satellites: Conditions of accretion. *Astrophys. J., 124,* 3404–3423.

Chambers J. E. and Cassen P. (2002) Planetary accretion in the inner solar system: Dependence on nebula surface density profile and giant planet eccentricities (abstract). In *Lunar and Planetary Science XXXIII,* Abstract #1049. Lunar and Planetary Institute, Houston (CD-ROM).

Chick K. M. and Cassen P. (1997) Thermal processing of interstellar dust grains in the primitive solar environment. *Astrophys. J., 477,* 398–409.

Chyba C. F., Ostro S. J., and Edwards B. C. (1998) Radar detectability of a subsurface ocean on Europa. *Icarus, 134,* 292–302.

Ciesla F. J., Lauretta D. S., Cohen B. A., and Hood L. L. (2003) A nebular origin for chondritic fine-grained phyllosilicates. *Science, 299,* 549–552.

Ciesla F., Lauretta D. S. and Hood L. L. (2004) Radial migration of phyllosilicates in the solar nebula (abstract). In *Lunar and Planetary Science XXXV,* Abstract #1219. Lunar and Planetary Institute, Houston (CD-ROM).

Clark R. N., Brown R. H., Jaumann R., Cruikshank D. P., Nelson R. M., Buratti B. J., McCord T. B., Lunine J., Hoefen T. M., Curchin J. M., Hansen G., Hibbits J., Matz K-D., Baines K. H., Bellucci G., Bibring J.-P., Capaccioni F., Cerroni P., Coradini A., Formisano V., Langevin Y., Matson D. L., Mennella V., Nicholson P. D., Sicardy B., and Sotin C. (2005) The surface composition of Saturn's moon Phoebe as seen by the Cassini Visual and Infrared Mapping Spectrometer. *Nature, 435,* 66–69.

Clayton R. N. (2003) Oxygen isotopes in the solar system. *Space Sci. Rev., 106,* 19–32.

Cruikshank D. P., Roush T. L., Moore J. M., Sykes M. V., Owen T. C., Bartholomew M. J., Brown R. H., and Tryka K. A. (1997) The surfaces of Pluto and Charon. In *Pluto and Charon* (S. A. Stern and D. J. Tholen, eds.), pp. 221–267. Univ. of Arizona, Tucson.

Cuzzi J. N. and Zahnle K. (2004) Material enhancement in protoplanetary nebulae by particle drift through evaporation fronts. *Astrophys. J., 614,* 490–496.

Cyr K. E., Sears W. D., and Lunine J. I. (1998) Distribution and evolution of water ice in the solar nebula: Implications for solar system body formation. *Icarus, 135,* 537–548.

Cyr K. E., Lunine J. I., and Sharp C. (1999) Effects of the redistribution of water in the solar nebula on nebular chemistry. *J. Geophys. Res., 104,* 19003–19014.

Drake M. J. and Righter K. (2002) What is the Earth made of? *Nature, 416,* 39–44.

Dreibus G. and Wanke H. (1987) Volatiles on Earth and Mars: A comparison. *Icarus, 71,* 225–240.

Fegley B. (2000) Kinetics of gas-grain reactions in the solar nebula. *Space Sci. Rev., 92,* 200.

Ganguly J. and Bose K. (1995) Kinetics of formation of hydrous phyllosilicates in the solar nebula (abstract). In *Lunar and Planetary Science XXVI,* pp. 441–442. Lunar and Planetary Institute, Houston.

Gautier D., Hersant F., Mousis O., and Lunine J. I. (2001) Enrichments in volatiles in Jupiter: A new interpretation of the Galileo measurements. *Astrophys. J. Lett., 550,* L227–L230 (erratum, *Astrophys. J. Lett., 559,* L183).

Gillet Ph., Barrat J. A., Deloule E., Wadhwa M., Jambon A., Sautter V., Devouard B., Neuville D., Benzerara K., and Lesourd M. (2002) Aqueous alteration in the Northwest Africa 817 (NWA 817) Martian meteorite. *Earth Planet. Sci. Lett., 203,* 431–444.

Goldreich P., Lithwick Y., and Sari R. (2004) Final stages of planet formation. *Astrophys. J., 614,* 497–507.

Hartmann L. (2000) Observational constraints on transport (and mixing) in pre-main sequence disks. *Space Sci. Rev., 92,* 55–68.

Hersant F., Gautier D., and Lunine J. I. (2004) Enrichment of volatiles in the giant planets of the solar system. *Planet. Space Sci., 52,* 623–641.

Hester J. J., Desch S. J., Healy K. R., and Leshin L. A. (2004) The cradle of the solar system. *Science, 304,* 1116–1117.

Hunten D. M., Donahue T. M., Walker J. C. G., and Kasting J. F. (1989) Escape of atmospheres and loss of water. In *Origin and Evolution of Planetary and Satellite Atmospheres* (S. K. Atreya et al., eds.), pp. 386–422. Univ. of Arizona, Tucson.

Ida S. and Lin D. N. C. (2004) Toward a deterministic model of planetary formation. I. A desert in the mass and semimajor axis distributions of extrasolar planets. *Astrophys. J., 604,* 388–413.

Irvine W. M. and Knacke R. F. (1989) The chemistry of interstellar gas and grains. In *Origin and Evolution of Planetary and Satellite Atmospheres* (S. K. Atreya et al., eds.), pp. 3–34. Univ. of Arizona, Tucson.

Irvine W. M., Crovisier J., Fegley B., and Mumma M. J. (2000) Comets: A link between interstellar and nebular chemistry. In *Protostars and Planets IV* (V. Mannings et al., eds.), pp. 1159–1200. Univ. of Arizona, Tucson.

Jacobson R. A. (2004) The orbits of the major Saturnian satellites and the gravity fields of Saturn from spacecraft and Earth-based observations. *Astrophys. J., 128,* 492–501.

Johnson T.V. J. and Lunine J. I. (2005) Saturn's moon Phoebe as a captured body from the outer solar system. *Nature, 435,* 69–71.

Johnson T. V., Brown R. H., and Pollack J. B. (1987) Uranus satellites — Densities and composition. *J. Geophys. Res., 92,* 14884–14894.

Kawakita H. and 18 colleagues (2001) The spin temperature of NH_3 in Comet C/1999 S4 (LINEAR). *Science, 294,* 1089–1091.

Kominami J. and Ida S. (2002) The effect of tidal interaction with a gas disk on formation of terrestrial planets. *Icarus, 157,* 43–56.

Korycansky D. G., Bodenheimer P., Cassen P., and Pollack J. B. (1990) One-dimensional calculations of a large impact on Uranus. *Icarus, 84,* 528–541.

Kouchi A., Yamamoto Y., Kozasa T., Kuroda T., and Greenberg J. M. (1994) Conditions for condensation and preservation of amorphous ice and crystallinity of astrophysical ices. *Astron. Astrophys., 290,* 1009–1018.

Krasnopolsky V. (2004) Comment on "Is the D/H ratio in the comet coma equal to the D/H ratio in the comet nucleus?" by M. Podolak, Y. Mekler, and D. Prialnik. *Icarus, 168,* 220.

Leshin L. A. (2000) Insights into martian water reservoirs from analyses of martian meteorite QUE 94201. *Geophys. Res. Lett., 27,* 2017–2020.

Lin D. N. C., Papaloizou J. C. B., Terquem C., Bryden G., and Ida S. (2000) Orbital evolution and planet-star tidal interaction. In *Protostars and Planets IV* (V. Mannings et al., eds.), pp. 1111–1134. Univ. of Arizona, Tucson.

Lunine J. I. and Stevenson D. J. (1982) Formation of the Galilean satellites in a gaseous nebula. *Icarus, 52,* 14–39.

Lunine J. I., Engel S., Rizk B., and Horanyi M. (1991) Sublimation and reformation of icy grains in the primitive solar nebula. *Icarus, 94,* 333–343.

Lunine J., Chambers J., Morbidelli A., and Leshin L. A. (2003) The origin of water on Mars. *Icarus, 165,* 1–8.

Lunine J. I., Coradini A., Gautier D., Owen T., and Wuchterl G. (2004) The origin of Jupiter. In *Jupiter* (F. Bagenal, ed.), pp. 19–34. Cambridge Univ., Cambridge.

Malfait K., Waelkens C., Bouwman J., de Kouter A., and Waters L. B. F. M. (1999) The ISO spectrum of the young star HD 142527. *Astron. Astrophys., 345,* 181–186.

Mayer L., Quinn T., Wadsley J. and Stadel J. (2002) Formation of giant planets by fragmentation of protoplanetary disks. *Science, 298,* 1756–1759.

McKinnon W. B., Lunine J. I. and Bandfield D. (1995) Origin and evolution of Triton. In *Neptune and Triton* (D. P. Cruikshank, ed.), pp. 807–877. Univ. of Arizona, Tucson.

Meier R., Owen T. C., Jewitt D. C., Matthews H. E., Senay M., Biver N., Bockelée-Morvan D., Crovisier J., and Gautier D. (1998a) Deuterium in comet C/1995 O1 (Hale-Bopp): Detection of DCN. *Science, 279,* 1707–1710.

Meier R., Owen T. C., Matthews H. E., Jewitt D. C., Bockelée-Morvan D., Biver N., Crovisier J., and Gautier D. (1998b) A determination of the HDO/H_2O ratio in comet C/1995 O1 (Hale-Bopp). *Science, 279,* 842–844.

Morbidelli A., Chambers J., Lunine J. I., Petit J. M., Robert F., Valsecchi G. B., and Cyr K. E. (2000) Source regions and timescales for the delivery of water on Earth. *Meteoritics & Planet. Sci., 35,* 1309–1320.

Mosqueira I. and Estrada P. R. (2003) Formation of the regular satellites of giant planets in an extended gaseous nebula I: Subnebula model and accretion of satellites. *Icarus, 163,* 198–231.

Nelson A. F., Benz W., Adams F. C., and Arnett D. (1998) Dynamics of circumstellar disks. *Astrophys. J., 502,* 342.

Owen T. C. (1982) The composition and origin of Titan's atmosphere. *Planet. Space Sci., 30,* 833–838.

Owen T., Mahaffy P., Niemann H. B., Atreya S. K., Donahue T. M., Bar-Nun A., and de Pater I. (1999) A new constraint on the formation of giant planets. *Nature, 402,* 269–270.

Petit J. M., Morbidelli A., and Chambers J. (2001) The primordial excitation and clearing of the asteroid belt. *Icarus, 153,* 338–347.

Podolak M. and Zucker S. (2004) A note on the snow line in protostellar accretion disks. *Meteoritics & Planet. Sci., 39,* 1859–1868.

Podolak M., Meckler Y., and Prialnik D. (2002) Is the D/H ratio in the comet coma equal to the D/H ratio in the comet nucleus? *Icarus, 160,* 208–211 (rebuttal to critique, *Icarus, 168,* 221–222).

Pollack J. B., Hubickyi O., Bodenheimer P., Lissauer J. J., Podolak M., and Greenzweig Y. (1996) Formation of the giant planets by concurrent accretion of solids and gas. *Icarus, 124,* 62–85.

Prinn R. G. (1990) On neglect of non-linear terms in solar nebula accretion disk models. *Astrophys. J., 348,* 725–729.

Prinn R. G. (1993) Chemistry and evolution of gaseous circumstellar disks. In *Protostars and Planets III* (E. H. Levy and J. I. Lunine, eds.), pp. 1005–1028. Univ. of Arizona, Tucson.

Raymond S., Quinn T., and Lunine J. I. (2004) Making other Earths: Dynamical simulations of terrestrial planet formation and water delivery. *Icarus, 168,* 1–17.

Robert F. (2001) The origin of water on Earth. *Science, 293,* 1056–1058.

Robert F. (2006) Solar system deuterium/hydrogen ratio. In *Meteorites and the Early Solar System II* (D. S. Lauretta and H. Y. McSween Jr., eds.), this volume. Univ. of Arizona, Tucson.

Sasselov D. D. and Lecar M. (2000) On the snow line in dusty protoplanetary disks. *Astrophys. J., 528,* 995–998.

Stern S. A., McKinnon W. B., and Lunine J. I. (1997) On the origin of Pluto, Charon, and the Pluto-Charon binary. In *Pluto and Charon* (S. A. Stern and D. J. Tholen, eds.), pp. 605–663. Univ. of Arizona, Tucson.

Stevenson D. J. (1990) Chemical heterogeneity and imperfect mixing in the solar nebula. *Astrophys. J., 348,* 730–737.

Stevenson D. J. and Lunine J. I. (1988) Rapid formation of Jupiter by diffusive redistribution of water vapor in the solar nebula. *Icarus, 75,* 146–155.

Stevenson D. J., Harris A. W., and Lunine J. I. (1986) Origins of satellites. In *Satellites* (J. A. Burns and M. S. Matthews, eds.), pp. 39–88. Univ. of Arizona, Tucson.

Stimpfl M., Lauretta D. S., and Drake M. J. (2004) Adsorption as a mechanism to deliver water to the Earth (abstract). *Meteoritics & Planet. Sci., 39,* A99.

Supulver K. D. and Lin D. N. C. (2000) Formation of icy planetesimals in a turbulent solar nebula. *Icarus, 146,* 525–540.

Waite J. H., Niemann H. Yelle R. V., Kasprzak W., Cravens T., Luhmann J., McNutt R., Ip W.-H., and the Data Analysis and Operations Teams (2005) Ion-neutral mass spectrometer measurements from Titan (abstract). In *Lunar and Planetary Science XXXVI,* Abstract #2057. Lunar and Planetary Institute, Houston (CD-ROM).

Weidenschilling S. J. (1977) Aerodynamics of solid bodies in the solar nebula. *Mon. Not. R. Astron. Soc., 180,* 57–70.

Weidenschilling S. J. and Ruzmaikina T. V. (1994) Coagulation of grains in static and collapsing protostellar clouds. *Astrophys. J., 430,* 713–726.

Wuchterl G., Guillot T., and Lissauer J. J. (2000) Giant planet formation. In *Protostars and Planets IV* (V. Mannings et al., eds.), pp. 1081–1109. Univ. of Arizona, Tucson.

Part V:

The Second Nebular Epoch: Materials Processing in the Nebula

Irradiation Processes in the Early Solar System

Marc Chaussidon
Centre de Recherches Pétrographiques et Géochimiques

Matthieu Gounelle
Centre de Spectrométrie Nucléaire et de Spectrométrie de Masse–Université Paris XI

During recent years, a growing body of evidence, both astrophysical observations of young stellar objects and cosmochemical observations made on meteorites, has strengthened the case that a fraction of the dust and gas of the solar accretion disk has been irradiated by cosmic rays emitted by the young forming active Sun. We review this meteoritic evidence, specifically the most convincing data obtained on Ca-Al-rich inclusions (CAIs) of primitive chondrite meteorites. Boron-isotopic variations in CAIs can be explained by decay products of short-lived ^{10}Be ($T_{1/2}$ = 1.5 m.y.) while a combination of extinct ^{7}Be ($T_{1/2}$ = 53 d) and of spallogenic Li may explain ^{7}Li/^{6}Li variations. We then discuss irradiation models that attempt to account for the origin of short-lived radionuclides and evaluate the predictions of these models in light of experimental data. Several anomalies, such as ^{21}Ne and ^{36}Ar excesses, suggest an exposure of CAIs and of chondrules (or their precusors) to cosmic rays emitted by the young Sun. The possible effects of early solar irradiation processes on the mineralogy, chemistry, and light-element isotopic compositions (e.g., O, H, N) of early solar system solids are also considered and evaluated.

1. INTRODUCTION

It has long been recognized that, because light elements (D, Li, Be, and B) could not be synthesized by thermonuclear reactions taking place at temperatures higher than $\approx 10^6$ K in stellar interiors, other nucleosynthetic processes had to be invoked to explain their presence in the galaxy and the solar system (*Burbidge et al.,* 1957). This is especially true for ^{6}Li, B, and Be since ^{7}Li can also be produced by Big Bang nucleosynthesis and during stellar evolution either in the red giant phase or in the H-explosive-burning phase of nova (e.g., *Audouze and Vauclair,* 1980). Among the different mechanisms studied for the production of Li-Be-B elements, it was suggested that endothermic spallation reactions that occur at high energies (typically higher than a few MeV) can produce these elements in significant amounts.

From the observation that concentrations of Li-Be-B elements were higher by a factor of $\approx 10^6$ in galactic cosmic rays (GCRs) relative to solar system abundances, *Reeves et al.* (1970) found that these elements were synthesized by spallation reactions between high-energy GCR particles (mostly protons and α particles) and the heavy nuclei ^{12}C and ^{16}O. However, before this understanding of the nucleosynthetic origin of Li-Be-B elements was reached, it was originally proposed that these elements were produced in the early solar system by irradiation of meter-sized planetesimals with high-energy solar particles (*Fowler et al.,* 1962; *Burnett et al.,* 1965). Other scenarios were envisaged, such as the production of Li-Be-B elements by irradiation of interstellar graphite grains, which would be incorporated in the presolar nebula (*Dwek,* 1978a).

It is important to point out here that, historically, the situation with short-lived ^{26}Al is quite similar to that of Li-Be-B. In fact, two possible sources were proposed for ^{26}Al when it was discovered in CAIs: either an external seeding of the solar system by fresh supernova products or a local production within the solar system by irradiation processes around the young active Sun (*Lee et al.,* 1976). The idea of an internal source by irradiation was nearly abandoned for many years, but the original calculations of the fluence required to produce the amount of ^{26}Al observed in CAIs yielded values ($\approx 10^{20}$ protons cm^{-2}) of the same order of magnitude as those calculated in the latest irradiation models.

This old idea that a fraction of solar system material was irradiated by solar cosmic rays (SCRs) around the young active Sun was recently renewed from two different lines of evidence. First, observations of solar-type stars during their pre-main-sequence phases (protostars, "classical" T Tauri stars with accretion disks, and "weak-lined" T Tauri stars without accretion disks) show powerful X-ray flares that are very likely accompanied by intense fluxes of accelerated particles (*Feigelson and Montmerle,* 1999; *Gudel,* 2004). The X-ray flares are seen in virtually all young stars, while radio gyrosynchrotron emission from flare MeV electrons are detected in some cases. Recent studies of the pre-main-sequence population in the Orion nebula have shown that analogs of the young Sun exhibit X-ray flare enhancements by factors of 10^4 and inferred proton fluence enhancements by factors of 10^5 (*Feigelson et al.,* 2002; *Wolk et al.,* 2005). X-ray absorption and Fe fluorescent emission line measurements give direct evidence that X-rays efficiently irradiate the protoplanetary disks (*Tsujimoto et al.,* 2005; *Kastner et al.,* 2005). Second, the discovery of short-lived ^{10}Be in CAIs from primitive chondrites (*McKeegan et al.,* 2000; *Sugiura et al.,* 2001; *MacPherson et al.,* 2003) was taken as a strong argument in favor of the presence of irradiation products among the precursors of CAIs. Although this view was recently challenged from a calculation of the amount of ^{10}Be trapped during stopping of GCRs in the molecular cloud core, parent of the solar system (*Desch et al.,* 2004), the presence of radioactive ^7Be in one Allende CAI (*Chaussidon et al.,* 2004, 2005) would be a decisive argument for an irradiation of a fraction of the solar nebula. However, the evidence for ^7Be needs to be secured from other CAIs in the future.

In the following sections, we review the few traces of irradiation processes that have been reported so far in meteorites. This includes either interactions with accelerated protons or α particles that result in spallation reactions or interactions with X-rays or UV radiation. The strongest evidence that we discuss in more detail is the variation in the Li- and B-isotopic compositions in CAIs and the inferred presence in the early solar system of short-lived ^7Be and ^{10}Be. In addition, we discuss more briefly (1) the case of ^{21}Ne and ^{36}Ar excesses reported in gas-rich meteorites and in chondrules, (2) the recent claims for UV irradiation as a source of non-mass-dependent O-isotopic anomalies and of ^{15}N enrichment in the early solar system, and (3) the chemical and mineralogical traces of electron and ion irradiation in irradiated silicates. We also mention the possible effects of X-rays as a source of ionization that can promote ion-molecule reactions responsible for D- and N-isotopic variations. The existing models of irradiation are then described and their contrasted predictions evaluated concerning the production of the different short-lived radioactivities observed in CAIs (^7Be, ^{10}Be, ^{26}Al, ^{41}Ca, and ^{53}Mn).

2. TRACES OF EARLY IRRADIATION PROCESSES IN METEORITES

2.1. Variations of Lithium-Beryllium-Boron Isotopic Compositions in Calcium-Aluminum-rich Inclusions and Chondrules

2.1.1. The lithium-beryllium-boron system. Small variations of Li- and B-isotopic compositions are generally reported in δ units, i.e., in ‰ relative to a standard: $\delta^{11}B$ are given relative to the NBS 951 borate, which has $^{11}B/^{10}B$ = 4.04558, and δ^7Li are given relative to the NIST LSVEC Li carbonate, which has $^7Li/^6Li$ = 12.0192. In bulk, chondrites δ^7Li and $\delta^{11}B$ values show significant but relatively limited variations. $\delta^{11}B$ values range from –7.5 ± 0.1‰ to +19.2 ± 0.3‰ in chondrites, CM and CI chondrites having $\delta^{11}B$ of –6.4 ± 1.1‰ (*Zhai et al.,* 1996). Reported δ^7Li values range from –3.8 + 0.6‰ to +3.9 ± 0.6‰ for bulk carbonaceous chondrites (*James and Palmer,* 2000; *McDonough et al.,* 2003). However, the isotopic compositions of Li-Be-B elements in components of chondrites (CAIs, refractory grains, chondrules, etc.) are anticipated to be much more variable for several reasons. These variations, if present, might be a very rich source of information concerning irradiation processes in the presolar or protosolar nebula.

First, Li-Be-B elements are efficiently produced by spallation reactions during collisions between accelerated protons and C and O nuclei (e.g., *Reeves,* 1994). Because the solar system abundance of Li-Be-B elements (*Anders and Grevesse,* 1989) is lower by 5 to 7 orders of magnitude relative to that of the target elements C and O (which have a different nucleosynthetic origin than Li-Be-B), even a small addition of a spallation component to a solar system Li-Be-B abundance might be detectable. Second, the spallation Li and B components have *a priori* isotopic compositions that will be very different from bulk chondrites. This comes from the fact that the cross sections for the production of the different Li and B isotopes vary as a function of the energy: *Ramaty et al.* (1996) calculated $^7Li/^6Li$ production ratios ranging from ≈1.5 to ≈8 and $^{11}B/^{10}B$ ratios ranging from ≈2.5 to ≈8 for spallation reactions at energies between a few and 100 MeV/nucleon using variable compositions for the cosmic ray and the target at rest (spallation $^7Li/^6Li$ and $^{11}B/^{10}B$ ratios are the lowest at high energies >100 MeV). The difference between the chondritic and the spallation isotopic ratio is large for Li because solar system Li also contains ^7Li produced during Big Bang nucleosynthesis. Thus, in any case the spallogenic $^7Li/^6Li$ will be significantly lower than the chondritic $^7Li/^6Li$ of ≈12. Third, two radioactive isotopes of Be having very different half-lives exist: ^7Be (half-life of 53 d) decays by electron capture to ^7Li and ^{10}Be (half-life of 1.5 m.y.) decays by β decay to ^{10}B. These two isotopes are produced efficiently by spallation reactions: the cross sections for the production by collisions between accelerated protons and O nuclei are, for instance, at 650 MeV of ≈11 mb, ≈3 mb, and ≈1.5 mb for ^7Be, ^9Be, and ^{10}Be, respectively (*Read and Viola,* 1984; *Michel et al.,* 1997; *Sisterson et al.,* 1997).

2.1.2. Evidence for the incorporation of short-lived beryllium-10 in calcium-aluminum-rich inclusions. Since the discovery of the incorporation of short-lived ^{10}Be in one type B CAI from the Allende CV3 chondrite (*McKeegan et al.,* 2000), ^{10}Be has been found in several other CAIs, either type A or type B, in CV3 chondrites Allende, Efremovka, and Leoville (*Sugiura et al.,* 2001; *Chaussidon et al.,* 2001; *Srinivasan,* 2002; *MacPherson et al.,* 2003) and in single refractory hibonites from the Murchison CM chondrite and the Allende CV3 chondrite (*Marhas et al.,* 2002; *Marhas and Goswami,* 2003). Statistics show that, for CAIs having mineral phases with high Be/B ratios, ^{10}B excesses due to the *in situ* decay of ^{10}Be are always observed. The lack of high Be/B ratios in some CAIs might be due to late addition of B to the CAI (parent-body alteration or even contamination during sample preparation); since Be is more refractory than B, CAIs should have high Be/B ratios in bulk. In fact, Be is predicted from thermodynamic equilibrium calculations to have a 50% condensation temperature of 1500 K and to condense either as gugaïte ($Ca_2BeSi_2O_7$) in solid solution with melilite or into spinel, while B is less refractory with a predicted 50% condensation temperature of 964 K for danburite ($NaBSi_3O_8$) in solid solution with anorthite (*Lauretta and Lodders,* 1997). Melilites are the phases in which the highest $^9Be/^{11}B$ ratios have been observed and variations of this ratio are anticipated to be produced during crystal fractionation of melilite. An experimental study of the partitioning of Be between crystal and melt for melts of CAI composition showed that Be was incompatible in gehlenitic melilites with $D_{melilite/melt} = 0.5$ for $X_{Åk} = 0.3$ and compatible in akermanitic melilites with $D_{melilite/melt} = 1.9$ for $X_{Åk} = 0.75$ (*Beckett et al.,* 1990).

One example of the $^{10}B/^{11}B$ variations due to the *in situ* decay of ^{10}Be is shown in Fig. 1 for one type B CAI (3529-41) from Allende. This CAI has been studied in most details for Li and B isotopic variations: It shows among the highest (close to $\approx 10^{-3}$) $^{10}Be/^9Be$ ratios observed (*Chaussidon et al.,* 2004, 2005). Note in Fig. 1 that the different major minerals (melilite, anorthite, and fassaite) define a single correlation line between $^{10}B/^{11}B$ and $^9Be/^{11}B$ ratios but that a few spots plot significantly off this line. It has been shown for the different CAIs studied that this type of correlation line can only be interpreted as an isochron due to the *in situ* decay of radioactive ^{10}Be present in the CAI

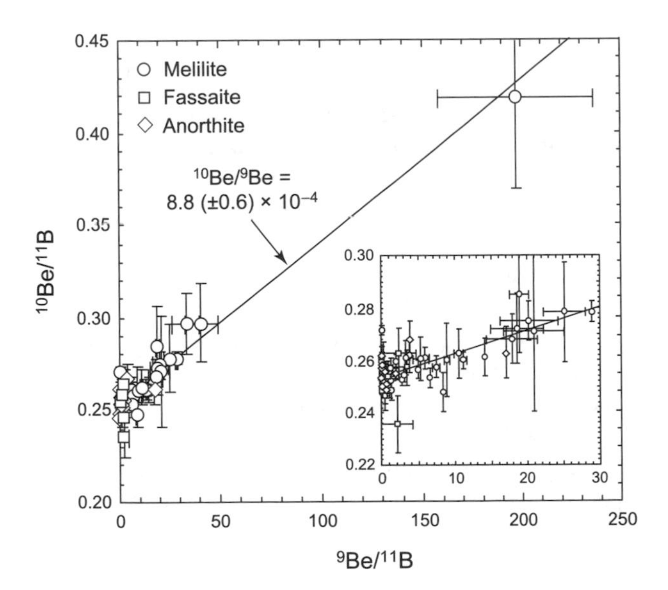

Fig. 1. Variations of $^{10}Be/^9Be$ ratios vs. $^9Be/^{11}B$ ratios in Allende type B CAI 3529-41 [data from *Chaussidon et al.* (2004)]. A blowup of the diagram is shown for low $^9Be/^{11}B$ ratios from 0 to 30. The positive correlation indicates the *in situ* decay of radioactive ^{10}Be incorporated (with a $^{10}Be/^9Be$ ratio $\approx 0.9 \times 10^{-3}$) in the CAI or in its precursors at the time of its formation. Such a ^{10}Be isochron-type relationship has been found for approximately 20 type A and B CAIs and 5 hibonites (see text for references).

melt. This comes mostly from the fact that it cannot be due to a mixing between two components having different $^{10}B/^{11}B$ and $^9Be/^{11}B$ ratios because (1) no correlation is observed between the $^{10}B/^{11}B$ ratios and the B concentrations and (2) B-isotopic variations, if present within the CAI precursors, would have been erased during the melting process (*McKeegan et al.,* 2000; *Sugiura et al.,* 2001; *MacPherson et al.,* 2003; *Chaussidon et al.,* 2004, 2005). In addition, perturbations of the Be/B system, such as those evidenced in Fig. 1 by the few spots plotting off the ^{10}Be isochron, were noted in several CAIs (some show more significant perturbations) and investigated by comparison with those observed for the Al/Mg system (*Sugiura et al.,* 2001; *McKeegan et al.,* 2001; *MacPherson et al.,* 2003; *Chaussidon et al.,* 2003). On average the Be/B system seems to be less perturbed in CAIs than the Al/Mg system. This might seem at first glance rather astonishing because Be and B are two trace elements present below the ppm level in CAIs and thus *a priori* much more sensitive to chemical redistribution during the postmagmatic history of the CAI than the major elements Al and Mg. However, two simple reasons were invoked to explain the observation of a "robustness" of the Be/B system relative to the Al/Mg system: First, the diffusion rate of B in minerals (data exist for anorthite only) seems lower by a factor of ≈ 100 than that of Mg (*Sugiura et al.,* 2001), and second, the half-life of ^{10}Be is twice that of ^{26}Al, implying that perturbations that occur in the first ≈ 1 m.y. following the crystallization of the CAI will significantly affect the Al/Mg system while they will be more or less transparent for the Be/B system.

The $^{10}Be/^9Be$ and the initial $^{10}B/^{11}B$ ratios inferred for the various CAIs and hibonites studied show significant variations. For CAIs the $^{10}Be/^9Be$ ratios range from $\approx 1.8 \pm 0.5 \times 10^{-3}$ to $0.4 \pm 0.1 \times 10^{-3}$ and the $^{10}B/^{11}B$ ratios range from 0.2372 to 0.2576 (or from −40‰ to +42‰ in the $\delta^{11}B$ notation), while hibonites show slightly more extreme ratios, but with lower precision because of the smaller sizes of the grains studied and the limited numbers. It is important to stress here that even if it is tempting to define a "canonical" $^{10}Be/^9Be$ ratio of $\approx 10^{-3}$ by analogy with the $^{26}Al/^{27}Al$ ratio of 5×10^{-5} defined for CAIs, there is at present no reason and/or compelling evidence to do so. Contrary to the $^{26}Al/^{27}Al$ ratio for which ratios higher or lower than 5×10^{-5} were demonstrated in several cases to reflect secondary perturbations (see review by *MacPherson et al.,* 1995), no sign of perturbation was evidenced for $^{10}Be/^9Be$ ratios lower than 10^{-3}. There is thus up to now no reason not to believe that the observed variations in $^{10}Be/^9Be$ and initial $^{10}B/^{11}B$ ratios do reflect isotopic variations present in the early solar nebula.

2.1.3. Evidence for the incorporation of short-lived beryllium-7 in calcium-aluminum-rich inclusions. The evidence for the incorporation in CAIs of short-lived ^7Be is less clear than for the incorporation of ^{10}Be, but because of the significant implications of the presence of ^7Be in the early solar system, it deserves consideration. The very short half-life (53 d) makes ^7Be a "smoking gun" for irradiation processes within the early solar system. The search for ^{7}Be and other very short-lived radionuclides will certainly be an important subject of research in the next few years. Beryllium-7 decay products in CAIs were originally claimed to explain the ^{7}Li/^{6}Li isotopic variations observed in one CAI (USNM 3515) from Allende (*Chaussidon et al.,* 2002). No isochron-type relationship was observed between the ^{7}Li/^{6}Li and the ^{9}Be/^{6}Li ratios for this CAI as a whole. However, different isochrons were tentatively recognized in this CAI: Partial relaxation by solid-state diffusion of the Li-isotopic anomalies due to the *in situ* decay of ^{7}Be were proposed to explain the large variations of the ^{7}Li/^{6}Li ratios observed at the scale of ≈50 μm in USNM 3515 (*Chaussidon et al.,* 2002). In such a model the most undisturbed isochron yielded a ^{7}Be/^{9}Be ratio of 0.076 ± 0.038 and an initial ^{7}Li/^{6}Li ratio of 10.4 ± 0.7. No argument to support or deny this interpretation could be found either from the distribution of ^{10}Be (because USNM 3515 had too high B concentrations for ^{10}Be to be detectable) or from the detailed petrography of USNM 3515 (*Paque et al.,* 2003). Another claim for the incorporation of ^{7}Be in a CAI was made from the analysis of Efremovka CAI E65 (*Srinivasan,* 2002) in which ^{7}Li excesses indicated a ^{7}Be/^{9}Be of 0.059 ± 0.024 and an initial ^{7}Li/^{6}Li of 11.938 ± 0.044.

Recently, large Li-isotopic variations were found in an Allende type B CAI (3529-41), which was previously studied in detail for its petrography and for its ^{26}Al and ^{10}Be systematics (*Podosek et al.,* 1991; *McKeegan et al.,* 2000). In Allende CAI 3529-41, relationships between the Li and Be concentrations and the Akermanite content of melilite allows the identification of zones in the CAI where the original Li and Be distribution inherited from the fractional crystallization of the CAI melt has been preserved (*Chaussidon et al.,* 2004, 2005). Similarly, non-perturbed magmatic Li and Be distributions can be observed in fassaite and anorthite. In 3529-41, all the analytical data from zones in which the Li and Be distributions resulting from the fractional crystallization of the CAI melt were preserved, show a positive correlation between ^{7}Li/^{6}Li and ^{9}Be/^{6}Li ratios (Fig. 2). This positive correlation can be interpreted as reflecting the *in situ* decay of radioactive ^{7}Be (*Chaussidon et al.,* 2004, 2005) implying a ^{7}Be/^{9}Be ratio at the time of crystallization of the CAI of 6.1 ± 1.3 × 10^{-3} and an initial ^{7}Li/^{6}Li ratio of the CAI melt of 11.49 ± 0.13 (Fig. 2).

The case for 3529-41 is at present the only strong indication for the incorporation of short-lived ^{7}Be in CAI, and other evidence must be found for the presence of ^{7}Be in the early solar system to be definitively ascertained. Hints exist in a few other CAIs from Allende and Efremovka, but several CAIs do not show any evidence for incorporation of ^{7}Be (*Chaussidon et al.,* 2001, 2003). This is at variance with the observation of omnipresent ^{10}Be in CAIs and might be due to the very short half-life of ^{7}Be of 53 d. In fact, for ^{7}Be to be detectable in a CAI, this CAI (1) must have formed very shortly after the nucleosynthesis of ^{7}Be and (2) must not have been totally remelted or subjected to strong perturbations of the trace Li and Be distributions after its formation.

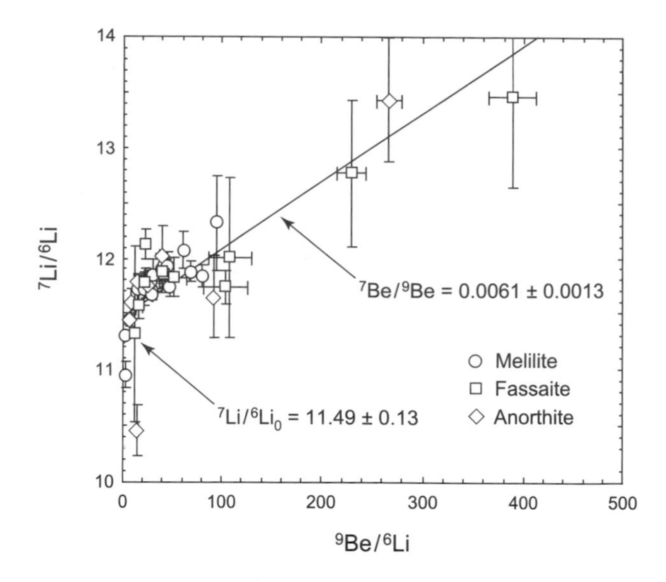

Fig. 2. Variations of ^{7}Li/^{6}Li ratios vs. ^{9}Be/^{6}Li ratios in Allende type B CAI 3529-41 [data from *Chaussidon et al.* (2004)]. The positive correlation is observed in 3529-41 for all the zones in the CAI where the Li and Be distributions do not show obvious signs of perturbation since the crystallization of the CAI (i.e., where the Li and Be concentrations follow experimentally determined partition coefficients between major phases and melt). The slope indicates that when 3529-41 crystallized it (or its precursors) incorporated radioactive ^{7}Be with a ^{7}Be/^{9}Be ≈ 6 × 10^{-3}. Other hints exist in a few CAIs for the *in situ* decay of ^{7}Be (see text), but 3529-41 can be considered as the only demonstrative case reported so far.

2.1.4. Relationship between beryllium-10 and other short-lived radionuclides. The variation of short-lived radionuclides abundances in the different components of primitive chondrites carry fundamental information about the chronology and nature of the different processes that occurred in the early solar system (*McKeegan and Davis,* 2004; *Gounelle and Russell,* 2005). Among the working hypotheses that are required to develop such a chronology is the assumption of a time zero at which an homogeneity of the different radionuclide abundances (e.g., homogeneity of ^{26}Al/^{27}Al ratios and/or of ^{10}Be/^{9}Be ratios) was established in the solar nebula. One obvious way to test these hypotheses is to investigate whether there is a synchronism between different short-lived chronometers. Because of their general presence in varying abundances in CAIs, ^{26}Al and ^{10}Be are two such potential chronometers that need to be compared. Figure 3 shows available data for ^{26}Al and ^{10}Be abundances in CAIs and in hibonites together with two curves showing the covariation with time of the ^{26}Al/^{27}Al and ^{10}Be/^{9}Be ratios calculated for two couples of arbitrary isotopic compositions (either 5 × 10^{-5} and 0.8 × 10^{-3}, or 5 × 10^{-5} and 2 × 10^{-3}, for the ^{26}Al/^{27}Al and ^{10}Be/^{9}Be ratios, respectively). The error bars are large for ^{26}Al/^{27}Al and

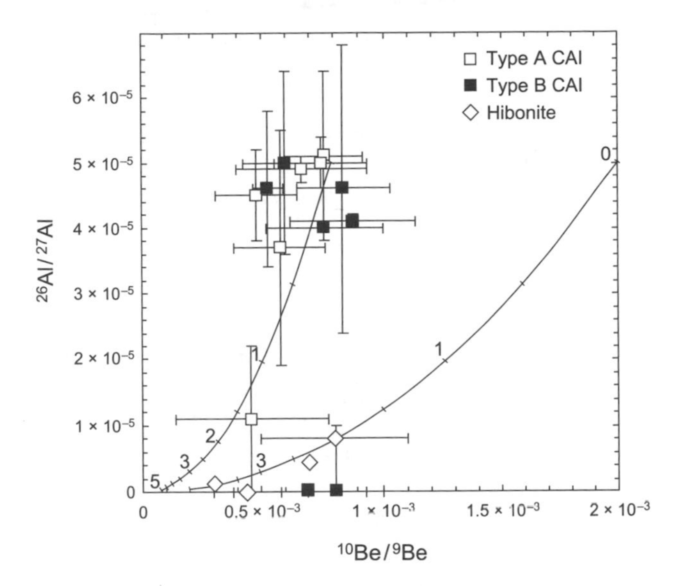

Fig. 3. Variations of initial $^{26}Al/^{27}Al$ ratios vs. $^{10}Be/^{9}Be$ ratios in type A and type B CAIs and in refractory hibonites [data from *McKeegan et al.* (2000, 2001), *Sugiura et al.* (2001), *MacPherson et al.* (2003), *Marhas et al.* (2002), *Marhas and Goswami* (2003), and *Chaussidon et al.* (2003)]. In a few cases the $^{26}Al/^{27}Al$ ratio is the upper limit inferred for the sample (see references). A few CAIs having very obviously disturbed $^{26}Al/Mg$ systems are not shown. The error bars are 1σ; for the sake of clarity they were omitted for a few hibonites where they are generally higher than in CAIs. Two curves showing the field for synchronicity between the $^{26}Al/Mg$ and $^{10}Be/B$ systems are shown for arbitrarily chosen initial $^{26}Al/^{27}Al$ and $^{10}Be/^{9}Be$ ratios of 5×10^{-5} and 0.8×10^{-3} and 5×10^{-5} and 2×10^{-3} respectively. Note that, despite the fact that the error bars are large, there is no possible synchronicity between ^{26}Al and ^{10}Be for hibonites relative to CAIs. This is less clear for CAIs because secondary perturbations might more significantly modify the $^{26}Al/^{27}Al$ ratios than the $^{10}Be/^{9}Be$ ratios (see text).

$^{10}Be/^{9}Be$ ratios in CAIs and hibonites, and prevent firm conclusions on a lack of synchronicity between ^{26}Al and ^{10}Be. However, it is obvious that hibonites and a few type B CAIs have very low $^{26}Al/^{27}Al$ and high $^{10}Be/^{9}Be$ ratios. This lack of synchronicity between ^{26}Al and ^{10}Be points either to different nucleosynthetic sources for ^{26}Al and ^{10}Be (*Marhas and Goswami,* 2003), or to a production mechanism that is able to decouple the two radioactive isotopes (*Gounelle et al.,* 2004).

2.1.5. Lithium and boron isotopic variations in calcium-aluminum-rich inclusions and chondrules. Another important signature of irradiation is carried by the Li- and B-isotopic variations, which are not due to the radioactive decay of ^{10}Be and of ^{7}Be. Although some of these variations may be attributed in the case of Li to magmatic or postmagmatic processes [up to 10–15‰ variations likely due to diffusion processes have recently been evidenced for $^{7}Li/^{6}Li$ ratios in lunar and martian meteorites (*Beck et al.,* 2004; *Barrat et al.,* 2005)], variations of higher amplitude (several tens of permil) reflect the presence in the solar system of Li and B having different nuclear sources. Lithium- and B-isotopic variations have been initially reported for chondrules (*Chaussidon and Robert,* 1995, 1998) with the $^{7}Li/^{6}Li$ ratio ranging from 11.86 to 12.44 (i.e., $\delta^{7}Li$ values ranging from −13‰ to +35‰, 85% of the chondrule data being at +10 ± 10‰) and $^{10}B/^{11}B$ ratios ranging from 0.2364 to 0.2591 (i.e., $\delta^{11}B$ values ranging from −42‰ to +44‰, 87% of the chondrule data being clustered between ≈−40‰ and +10‰). The range of B-isotopic variations in chondrules has been questioned from another ion probe study that was not able to resolve significant variations at the 2σ level (*Hoppe et al.,* 2001): This may be due to (1) the low precision of B-isotopic measurements in B-poor chondrules, or (2) the scarcity of chondrules with large B-isotopic variations. More recently a reappraisal of Li- and B-isotopic variations in chondrules was done using a high-sensitivity ims 1270 ion probe and using chondrules from the Mokoia and Semarkona chondrites, chondrules that were not polished and were not embedded in epoxy to minimize the risks of B contamination (*Robert and Chaussidon,* 2003). This latter study found Li- and B-isotopic variations broadly consistent with the range observed in the earlier studies.

Since the discovery of ^{10}Be, another evidence exists now for B-isotopic variations in the early solar system. The initial $^{10}B/^{11}B$ ratio of CAIs and hibonites do show significant variations from 0.2372 to 0.2576 (*McKeegan et al.,* 2000; *Sugiura et al.,* 2001; *MacPherson et al.,* 2003; *Marhas et al.,* 2002; *Marhas and Goswami,* 2003; *Goswami,* 2003), i.e., $\delta^{11}B$ values ranging from −40‰ to +42‰, which is the range previously found in chondrules. The variations of initial $^{10}B/^{11}B$ ratios vs. $^{10}Be/^{9}Be$ ratios in CAIs and hibonites are shown in Fig. 4a. Although errors bars are large, most of the variations could be explained by the presence in CAI precursors of a B component affected by spallation B together with radiogenic ^{10}B. Using a spallation production ratio of $^{10}B^*/^{9}Be^* \approx 6.2$, $^{10}Be^*/^{9}Be^* \approx 0.1$, and $^{11}B^*/^{10}B^* \approx 3$ (*Ramaty et al.,* 1996) we calculate that irradiation of chondritic matter with $^{10}B/^{11}B = 0.2472$ (*Zhai et al.,* 1996) and B/Be = 28.8 (*Anders and Grevesse,* 1989) would shift the $^{10}B/^{11}B$ ratio to a value of 0.2479 for $^{10}Be/^{9}Be$ of $\approx 10^{-3}$. For comparison, irradiation of a CAI-type composition (B/Be ≈ 0.4) would result in $^{10}B/^{11}B = 0.2811$, for $^{10}Be/^{9}Be = 10^{-3}$. In order to evaluate mixing scenarios, mixing lines are drawn between these two calculated compositions and the chondritic B composition in Fig. 4a. Many of the type B and type A CAI data are consistent with the scenario that initial $^{10}B/^{11}B$ ratios are altered by incorporation of a fraction of spallation B produced together with ^{10}Be. However, the situation seems more complicated for hibonites, even if spallation is *a priori* the only source known for B with $^{10}B/^{11}B$ ratios higher than chondritic.

Not only initial $^{10}B/^{11}B$ ratios, but also $^{7}Li/^{6}Li$ ratios, are found to be variable in CAIs (*Chaussidon et al.,* 2001, 2004; *MacPherson et al.,* 2003). These ratios are generally

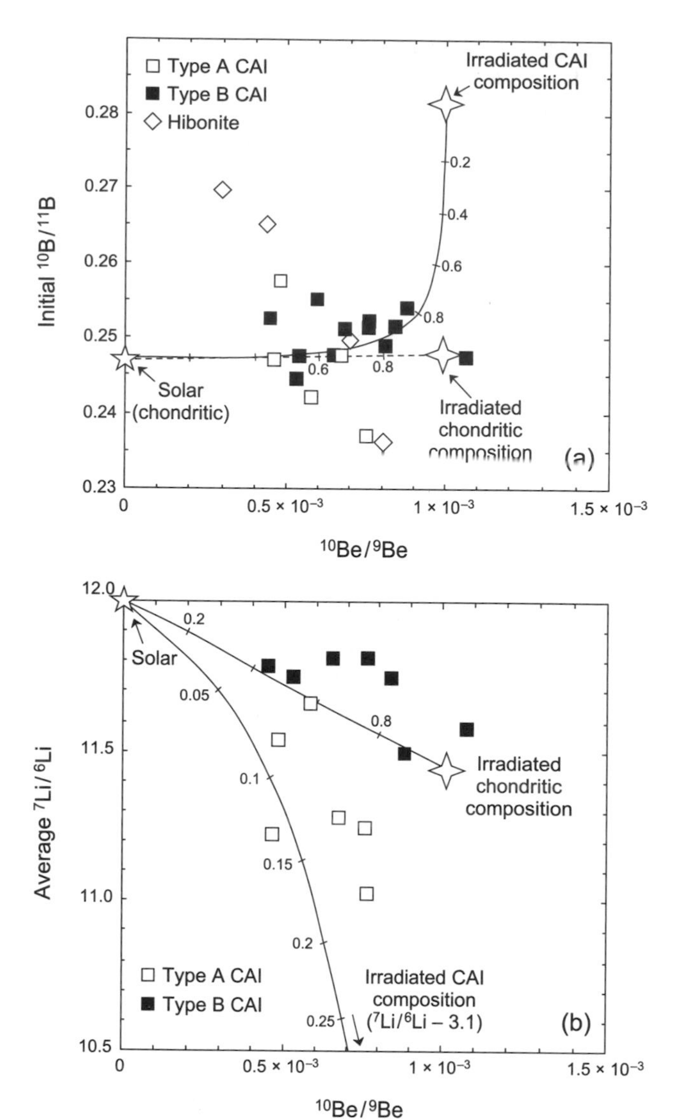

Fig. 4. Variations of **(a)** initial $^{10}B/^{11}B$ ratios vs. $^{10}Be/^{9}Be$ ratios in CAIs and hibonites and **(b)** average $^{7}Li/^{6}Li$ ratios vs. $^{10}Be/^{9}Be$ ratios in CAIs [data from *McKeegan et al.* (2000, 2001), *Sugiura et al.* (2001), *MacPherson et al.* (2003), *Marhas et al.* (2002), *Marhas and Goswami* (2003), and *Chaussidon et al.* (2003)]. For the sake of clarity of the diagrams, errors are not shown. The two curves are mixing curves calculated between a solar system (or chondritic) composition and two compositions corresponding to a chondritic or CAI composition in which spallation Li, Be, and B have been produced in order to reach a $^{10}Be/^{9}Be$ ratio of 10^{-3} (see text for details). These latter two compositions simply model two possible end members containing irradiation products. Note that CAIs have average $^{7}Li/^{6}Li$ that differ strongly from bulk chondrites, perhaps more marked in type A. The mixing curves show that most of the type A and type B CAIs have initial $^{10}B/^{11}B$ variations and average $^{7}Li/^{6}Li$, which can be understood at first order as reflecting the presence of various amounts of spallation B and Li produced together with ^{10}Be. The situation seems perhaps different with hibonites, but at present this cannot be ascertained because the precision on the initial $^{10}B/^{11}B$ of the hibonites is rather poor, as the number of analyses for each hibonite is very limited (individual grains are small).

lower than chondritic $^{7}Li/^{6}Li \approx 12$ (*James and Palmer,* 2000; *McDonough et al.,* 2003), and the two lowest ones are of 9.20 ± 0.22 in Allende CAI 3529-41 (*Chaussidon et al.,* 2004) and of 9.9 ± 1.1 in Axtell CAI 2771 (*MacPherson et al.,* 2003). In addition, the initial $^{7}Li/^{6}Li$ ratio inferred from the ^{7}Be isochron in Allende CAI 3529-41 is of 11.49 ± 0.13. Although the low $^{7}Li/^{6}Li$ ratios observed in extremely Li-poor phases of CAIs (<1 ppb Li) may be due in part to a recent exposure to GCRs (*Chaussidon et al.,* 2005), most of the low $^{7}Li/^{6}Li$ ratio likely reflect spallogenic Li components incorporated in CAI precursors. Note that Li-isotopic fractionation by evaporation would result in a $^{7}Li/^{6}Li$ ratio higher than chondritic. It seems reasonable to also explain the low $^{7}Li/^{6}Li$ ratios of CAIs by shifts due to Li spallation components that could have the same origin as ^{10}Be and ^{7}Be (Fig. 4b). The spallation ratio of B ($^{10}B/^{11}B \approx 0.3333$) differs less from the chondritic ratio ($^{10}B/^{11}B = 0.2472$) than is the case for Li (spallation $^{7}Li/^{6}Li$ ratio ranging from ≈1.5 to 5, compared to a chondritic ratio $^{7}Li/^{6}Li$ of ≈12). Therefore, the $^{7}Li/^{6}Li$ ratio of CAIs may differ more from chondritic than is the case for the $^{10}B/^{11}B$ ratio. Finally, it is important to stress here that the invoked spallogenic component for CAIs (responsible for their low $^{7}Li/^{6}Li$ ratio and for ^{7}Be and ^{10}Be) seems not to be present in bulk chondrites. A calculation of the type used for B isotopes in the previous paragraph (using a spallation production ratio of $^{7}Li^{*}/^{6}Li^{*} \approx 2.5$) yields for a $^{10}Be/^{9}Be$ ratio of $\approx 10^{-3}$ (Fig. 4b) a $^{7}Li/^{6}Li$ ratio ≈11.45 and of ≈3.1 for irradiation of a chondritic and CAI composition, respectively. Such ratios are much lower than that of bulk chondrules (*McDonough et al.,* 2003) or of bulk chondrites (*James and Palmer,* 2000).

2.1.6. Summary of the constraints brought by lithium-beryllium-boron isotopic data on irradiation processes. The major conclusions from the Li-Be-B isotopic variations found in CAIs, hibonites, chondrules, and chondrites are: (1) ^{10}Be ($T_{1/2}$ = 1.5 m.y.) and most probably ^{7}Be ($T_{1/2}$ = 53 d) were present in the solar system when CAIs and hibonites formed with $^{10}Be/^{9}Be$ ratio ranging from $\approx 1.5 \times 10^{-3}$ to 0.4×10^{-3}, and with $^{7}Be/^{9}Be \approx 6 \times 10^{-3}$; (2) there is *a priori* no synchronicity that can be established between $^{10}Be/^{9}Be$ and $^{26}Al/^{27}Al$ ratios of CAIs and hibonites; and (3) significant Li- and B-isotopic variations occurred in CAI precursors, to a lesser extent in chondrule precursors, and none are observed for Li in bulk chondrites. In addition, it seems most likely that the variations observed for Li- and B-isotopic ratios in CAIs reflect the presence of a fraction of spallogenic Li and B of similar origin as ^{10}Be and ^{7}Be. Thus, acceptable models must be able to explain not only the abundance of ^{10}Be observed in CAIs, but also why ^{10}Be seems decoupled from ^{26}Al and why associated spallation Li seems restricted to CAIs (and chondrules to a lesser extent) and of course how ^{7}Be can be incorporated in CAIs. This latest constraint relies at the moment on observations from only one Allende CAI and, in the case of the documentation of ^{7}Be in other CAIs, would provide extremely restricting conditions because of the very short half-life of ^{7}Be (see section 3).

2.2. Neon-21 and Argon-36 Excesses in Meteorites

A vast literature exists on the exposure history of meteorites from studies of noble gas compositions (see reviews by *Caffee et al.,* 1988; *Woolum and Hohenberg,* 1993). It has been shown that meteorites can have a rather complex exposure history. While all have been exposed recently to GCRs and to SCRs during their travel to Earth after ejection from their parent body (see *Eugster et al.,* 2006), some may have been exposed to energetic particles much earlier in their history.

Neon-21 is a good indicator of irradiation since it is low in abundance and easily produced by spallation reactions between energetic particles and common rock-forming elements such as Mg, Al, Si, and Fe. Large ^{21}Ne excesses relative to the spallogenic ^{21}Ne produced during recent exposure to GCRs have been found in meteoritic minerals belonging to CM chondrites and howardites (*Caffee et al.,* 1987). Two extreme views were proposed to explain the preexposition of meteoritic mineral grains to energetic particles: an extremely long exposure to GCRs in an asteroidal regolith or an exposition to an enhanced flux of early SCRs (e.g., *Caffee et al.,* 1987; *Hohenberg et al.,* 1990). Because the ^{21}Ne excesses correlate with very heavy ion nuclear track densities and because the long inferred preexposure ages (on the order of a few hundred million years) are incompatible with what is known of asteroidal regoliths (*Housen and Wilkening,* 1982), *Caffee et al.* (1987) advocated in favour of an active early Sun for the ^{21}Ne excesses in CM chondrites. For the Fayetteville chondrite, *Wieler et al.* (1989) claimed, on the basis of noble gas data and a complex regolith model, that ^{21}Ne excesses can be accounted for by exposure to GCRs in an asteroidal parent body during ≈20 m.y. Since the last review on the subject (*Woolum and Hohenberg,* 1993), work and debate have focused on the gas-rich howardite, Kapoeta. Rao and coworkers (*Rao et al.,* 1997), combining new noble gas data on separated feldspars with literature data, support the view of an early active Sun (10× the flux of the contemporary Sun, which could have lasted 500 m.y. or longer). *Pun et al.* (1998) contend that their noble gas data for a diversity of Kapoeta clasts can be accounted for by a regolith exposure of 1.5–6.5 m.y. The *Wieler et al.* (2000) model calculations indicate that the observed excess ^{21}Ne can be produced by GCRs, and that there is no need for a SCR contribution. These authors, however, acknowledge that the situation may be different for the CM chondrites studied by *Caffee et al.* (1987) and *Hohenberg et al.* (1990). Cosmogenic radionuclides (^{10}Be, ^{26}Al, ^{36}Cl, and ^{53}Mn) analyzed in separated phases from the Kapoeta howardite show that all these activities can be explained by a single 4π exposure stage of Kapoeta lasting ≈3 m.y., and that the cosmogenic ^{21}Ne excesses likely result from a 2π exposure at least 10 m.y. before the recent exposure to GCRs (*Caffee and Nishiizumi,* 2001).

Finally, recent *in situ* laser ablation rare gas analyses of chondrules in the Yamato 791790 enstatite chondrite reveal the presence of a subsolar (i.e., Ar-rich) component evidenced by high quantities of trapped ^{36}Ar (*Okazaki et al.,* 2001). One proposed scenario for the presence of this trapped ^{36}Ar is that it was implanted in chondrules (or in their precursors) by irradiation with SCRs during the T Tauri phase of the Sun (*Okazaki et al.,* 2001). At variance, the Ne- and Ar-isotopic compositions recently analyzed in several CAIs from the Allende chondrite can be interpreted as reflecting the presence of cosmogenic Ne and Ar only; they do not require the presence in CAIs of a trapped component (*Vogel et al.,* 2004).

2.3. Protons, Alpha Particles, X-Rays, and Ultraviolet Irradiation as a Source of Oxygen-, Hydrogen-, and Nitrogen-Isotopic Variations in the Solar Nebula

Young stellar objects (YSOs) are characterized by intense and highly variable emission of thermal X-rays from gas heated to temperatures higher than 10^{7} K, most likely resulting from magnetohydrodynamical reconnection events similar to solar flares (*Feigelson and Montmerle, 1999*). Pre-main-sequence analogs of the young Sun exhibit time-averaged emission around 10^{30} erg/s, about 10^{3}× higher than the contemporary Sun, and individual flares with peaks of 10^{31}–10^{32} erg/s occur on timescales of weeks (*Wolk et al.,* 2005). A mild secular decline in luminosity during the 0.1–10-m.y. pre-main-sequence phase is seen, followed by a steeper decline during the later main-sequence phase (*Preibisch and Feigelson, 2005*). Considerable star-to-star variation in flaring levels is seen. X-rays are accompanied by an intense far-UV radiation. Although there is not yet any definite proof for traces in meteorites of X-ray and/or UV irradiation around the early Sun, such an irradiation is called upon more and more to explain part of the large isotopic anomalies observed in the solar system for several light elements.

This is the case first for the non-mass-dependent O-isotopic anomaly (see reviews by *Clayton,* 1993; *McKeegan and Leshin,* 2001). By analogy with recent observations of non-mass-dependent O-isotopic variations in molecular clouds (*Sheffer et al.,* 2002), it has been recently re-proposed [this idea was first proposed by *Thiemens and Heidenreich* (1983) but was abandoned over the years] that self-shielding of CO from UV photodissociation could significantly enhance the trace species $C^{17}O$ and $C^{18}O$ relative to the dominant species $C^{16}O$ in the interior of the nebula, where the photodissociation lines of trace species are less shielded than that of $C^{16}O$ (*Clayton,* 2002a). In such a view, the ^{16}O-rich composition of CAIs from primitive chondrites ($\delta^{17}O = \delta^{18}O \approx -50‰$) would be that of the protosolar gas before irradiation. At the time of this article, the *Genesis* samples have been returned to Earth with their solar wind collectors, but O-isotopic analyses have not yet been performed, so the $\delta^{17}O$ and $\delta^{18}O$ values of the Sun and by inference the protosolar gas are still unknown. However, recent ion microprobe analyses by depth profiling in metallic grains from the lunar regolith (*Hashizume and Chaussidon,* 2005) indicate that solar energetic particles are indeed enriched in ^{16}O by ≈20‰ relative to bulk meteorites. This would imply $\delta^{17}O$ and $\delta^{18}O$ values for the protosolar nebula of –40 ± 8‰ or lower along the slope 1 line in the classical three-oxygen-isotope diagram (*Hashizume and Chaussidon,* 2005), in agreement with the idea that CAIs have an unmodified protosolar O-isotopic composition. Calculations developed to model isotope-selective photodissociation of CO by UV light show that very few wavelengths may be available for CO self-shielding (*Chakraborty and Thiemens,* 2005), and that in any case conditions must be found in which the isotopic anomaly produced can be transferred to a solid fast enough to avoid reequilibration with the ambient gas. Different astrophysical settings have been proposed: the presolar molecular cloud (*Yurimoto and Kuramoto,* 2004), the surface region of the protosolar nebula (*Lyons and Young,* 2005), or the inner protosolar nebula irradiated by the young Sun (*Clayton,* 2002a).

A similar self-shielding of N_2 gas has also been proposed (*Clayton,* 2002b) to explain the systematic enrichment in ^{15}N of meteoritic components relative to the gas of the protosolar nebula, which was ^{15}N-poor as indicated by the isotopic composition of solar wind N implanted at the surface of lunar soil grains [$\delta^{15}N$ < –240‰ (*Hashizume et al.,* 2000)] and of ammonia in the atmosphere of Jupiter [$\delta^{15}N$ = –380 ± 80‰ (*Owen et al.,* 2001)]. Interestingly, the protosolar nebula was also probably ^{13}C-poor [$\delta^{13}C$ of solar wind implanted at the surface of lunar soils grains <–120‰ (*Hashizume et al.,* 2004)] compared to meteorites and self-shielding of CO, already proposed to cause ^{17}O and ^{18}O enrichments, might cause similarly ^{13}C enrichments.

Another possible way for irradiation processes to produce large isotopic anomalies in the early solar system could be by promoting ion-molecule reactions. This has recently been proposed to explain the correlated D enrichments (relative to the protosolar D/H ratio of 21 ± 5 × 10^{-6}) observed in water and organic matter contained in meteorites, interplanetary dust particles (IDPs), and comets (*Robert,* 2002). In such a model, X-rays emitted from the T Tauri Sun will penetrate deeply into the gaseous disk, increase its temperature and ionization, and promote ion-molecule reactions in which isotopic exchange will take place between, for instance, H_3^+ and HD or between CH_3^+ and HD (*Glassgold et al.,* 2000; *Robert,* 2002; *Montmerle,* 2002; *Semenov et al.,* 2004). Similar reactions can also fractionate N isotopes and explain the relationship observed between the H- and N-isotopic compositions of organic matter in IDPs, meteorites, and comets (*Aléon and Robert,* 2004). Finally, it must also be noted here that theoretically D might be produced during spallation reactions on, for instance, O by high-energy protons accelerated around the T Tauri Sun (*Robert et al.,* 1979).

Finally, it must be mentioned that chondrites may have preserved in their matrix grains condensed from the irradiated protosolar gas. Recently, silica-rich grains with large $^{18}O/^{16}O$ and $^{17}O/^{16}O$ ratios up to 1.2 × 10^{-1} and up to 7.7 × 10^{-2}, respectively, were discovered embedded in the acid-insoluble organic matter extracted from the Murchison CM2 carbonaceous chondrite (*Aléon et al.,* 2005a). Although such extremely high isotope ratios were observed once in a post-asymptotic-giant-branch star (*Cami and Yamamura,* 2001), they are not predicted by evolved star nucleosynthesis. Rather, the observed O-isotopic ratios are predicted in case of selective trapping of O produced by irradiation of a gas of solar composition with particles akin to ^3He-rich impulsive solar flares. Because of the high abundance (almost 1 ppm) of these grains in the host meteorite and the absence of large Si- and Mg-isotopic anomalies (*Aléon et al.,* 2005a,b), it has been suggested that these grains could be condensates of an irradiated gas around the young Sun.

Thus, it seems quite feasible that irradiation processes by protons, α particles, X-rays, and UV around the T Tauri Sun might have produced large isotope variations of the light elements (H, C, N, and O) in the protosolar nebula. However, at the present stage it is extremely difficult to quantify these effects at the scale of the nebula because this will critically depend on parameters such as the optical thickness, the ionization rate, or the temperature of the disk.

2.4. Chemical and Structural Modifications of Silicates Induced by Irradiation with Protons, Alpha Particles, and Electrons

Since the discovery of the presence of thin amorphous rims in the Apollo lunar soil grains (*Dran et al.,* 1970; *Bibring et al.,* 1972), several studies were conducted in order to assess how the implantation of solar wind ions (few KeV/nucleon) can produce amorphous rims within the first 50 nm below the surface of silicate grains (e.g., *Borg et al.,* 1980). Other processes such as vapor deposition also operate in the formation of the rims of the lunar soil grains (*Keller and McKay,* 1997). Apart from the lunar soils, the irradiation by protons and α particles of silicates was studied experimentally in order to understand the peculiar chemistry of irradiated grains present in IDPs (*Bradley,* 1994). These experiments showed that irradiation can induce a preferential sputtering of cations with the lowest binding energy and result, for instance, in Mg depletion and O excesses (*Bradley et al.,* 1996). Following these first results, a number of experimental studies were conducted in order to determine the structural and chemical changes induced in silicate grains during their possible irradiation in the interstellar space and in a circumstellar environment (*Demyk et al.,* 2001; *Carrez et al.,* 2002a,b, and references therein). These studies documented different effects induced by He^+ irradiation of olivine such as amorphization, preferential sputtering leading to a decrease of O/Si and Mg/Si ratios, and production of metallic iron by reduction. In addition, it was also shown that nucleation and growth of MgO and forsterite crystals was promoted by electron irradiation of

$MgSiO_3$ glasses previously irradiated by He^+ ions. These processes might be relevant for the evolution of grains around the forming Sun in the accretion disk as suggested by the recent finding in the Ningqiang carbonaceous chondrite of an inclusion containing olivines and pyroxenes partially annealed and surrounded by amorphous rims (*Zolensky et al.,* 2003). These rims were suggested to be formed during an energetic solar event such as a bipolar outflow or a FU-Orionis flare. Interestingly, these amorphous rims were shown to contain high concentrations of heavy primordial rare gases (^{36}Ar, ^{84}Kr, and ^{132}Xe) whose concentration ratios are compatible with an origin by implantation of solar gases fractionated by incomplete ionization under plasma conditions at 8000 K electron temperature (*Nakamura et al.,* 2003).

3. IRRADIATION MODELS FOR THE ORIGIN OF SHORT-LIVED RADIONUCLIDES

3.1. Introduction

Extinct short-lived radionuclides (SRs) are radioactive isotopes that were alive when the first solar system solids formed. They are now extinct, because their half-life is short compared to the age of the solar system. Some SRs were present within the most primitive (oldest) components of meteorites (i.e., CAIs and chondrules) at a level higher than what is expected by a continuous galactic nucleosynthesis (e.g., *Meyer and Clayton,* 2000). This is especially the case of the ones with the shorter half-lives: ^{7}Be ($T_{1/2}$ = 53 d), ^{10}Be ($T_{1/2}$ = 1.5 m.y.), ^{26}Al ($T_{1/2}$ = 0.74 m.y.), ^{36}Cl ($T_{1/2}$ = 0.30 m.y.), ^{41}Ca ($T_{1/2}$ = 0.10 m.y.), ^{53}Mn ($T_{1/2}$ = 3.7 m.y.), and ^{60}Fe ($T_{1/2}$ = 1.5 m.y.). The presence in the solar accretion disk of these radionuclides is usually attributed to a last-minute origin (see recent reviews by *Goswami and Vanhala,* 2000; *Russell et al.,* 2001). A favored model is a stellar nucleosynthesis followed by an immediate injection in the nascent solar system (e.g., *Busso et al.,* 2003). An alternative theory is production by irradiation of dust and/or gas by accelerated particles (e.g., *Goswami and Vanhala,* 2000). A special origin [trapping of GCRs in the molecular cloud core; see also *Clayton and Jin* (1995)] has been recently proposed for ^{10}Be (*Desch et al.,* 2004). Here, we will discuss only the irradiation models for the production of SRs, and refer to *Meyer and Zinner* (2006) for the stellar production of SRs. The major aim of these various models is to establish if a fraction of the SRs can be produced in the progenitor molecular cloud core or in the protosolar nebula. This excludes ^{60}Fe since this radionuclide is neutron-rich and therefore it is not possible to produce it at appreciable levels by irradiation (e.g., *Lee et al.,* 1998). Because ^{60}Fe can be brought within the solar system by a supernova without coproducing other SRs (e.g., *Busso et al.,* 2003), the presence of ^{60}Fe within meteorites does not rule out an irradiation origin for other SRs such as ^{26}Al.

Irradiation models for the origin of SRs (and other elements) have been introduced very early by astrophysicists (*Fowler et al.,* 1962). Their principle is extremely simple: Accelerated particles with MeV or higher energies react with ambient matter to produce new isotopes via nuclear reactions. For a nuclear reaction between a cosmic ray CR and a target T producing the radionuclide R, the production ratio of radionuclide isotopes (N_R) relative to a stable isotope (N_S) can be written as

$$\frac{N_R}{N_S} = F_0 \Delta t \sum_i y_{CR}^i \sum_j \frac{x_j^T}{x_S} \int \sigma(E) N(E) dE \qquad (1)$$

where F_0 is the proton flux, Δt the irradiation time, y_{CR}^i the abundance of cosmic rays (^{3}He, ^{4}He) relative to protons, x^T the target abundance, x_S the stable-isotope abundance number, $\sigma(E)$ the nuclear cross section, N(E) the energy spectrum of protons, and E the energy of protons. Because, for a given radionuclide, there are a diversity of production reactions, the yield is summed over all the target possibilities (j) and all the impinging particle possibilities (i).

What mainly distinguishes the different irradiation models are (1) the astrophysical context of irradiation (e.g., molecular clouds vs. early solar system), (2) the physical nature of the targets (solid or gaseous), (3) the chemistry of the target, (4) the source of the cosmic rays, and (5) the location of the irradiated targets relative to the source of the cosmic rays. Models are not easy to compare, as the roster of observed SRs having a possible irradiation origin has dramatically increased over the years; they include from ^{26}Al alone to ^{7}Be, ^{10}Be, ^{26}Al, ^{36}Cl, ^{41}Ca, and ^{53}Mn. For example, extinct ^{36}Cl, for which only hints of its existence were reported eight years ago (*Murty et al.,* 1997), is now a well-documented SR (*Lin et al.,* 2005). In addition, it is not always clear for a given SR if a last-minute origin is needed, either because its initial abundance is disputed, or because models of continuous galactic production are uncertain. The best example of this case is ^{53}Mn, whose initial abundance in CAIs varies within an order of magnitude (*Birck and Allègre,* 1985; *Lugmair and Shukolyukov,* 1998; *Papanastassiou et al.,* 2002), and for which the modeling of steady-state abundance in the galaxy is complicated by the diversity of its stellar sources (e.g., *Meyer and Clayton,* 2000). Finally, it has been recently demonstrated that the "canonical" ^{26}Al/^{27}Al ratio, taken as the early solar system initial value and used in all models aiming at calculating the production of SRs, is the result of thermal resetting (*Young et al.,* 2005). We shall present each model independently, without trying to cross-compare them.

3.2. Irradiation Models

Early attempts to produce SRs via irradiation have been developed soon after the discovery of ^{26}Al (*Lee et al.,* 1976, 1977). They were aiming at reproducing the abundance of ^{26}Al only, since other SRs possibly produced by irradiation had not yet been discovered. All workers considered irradiation of a gas within the solar system by protons (*Heymann and Dziczkaniec,* 1976; *Clayton et al.,* 1977; *Dwek,*

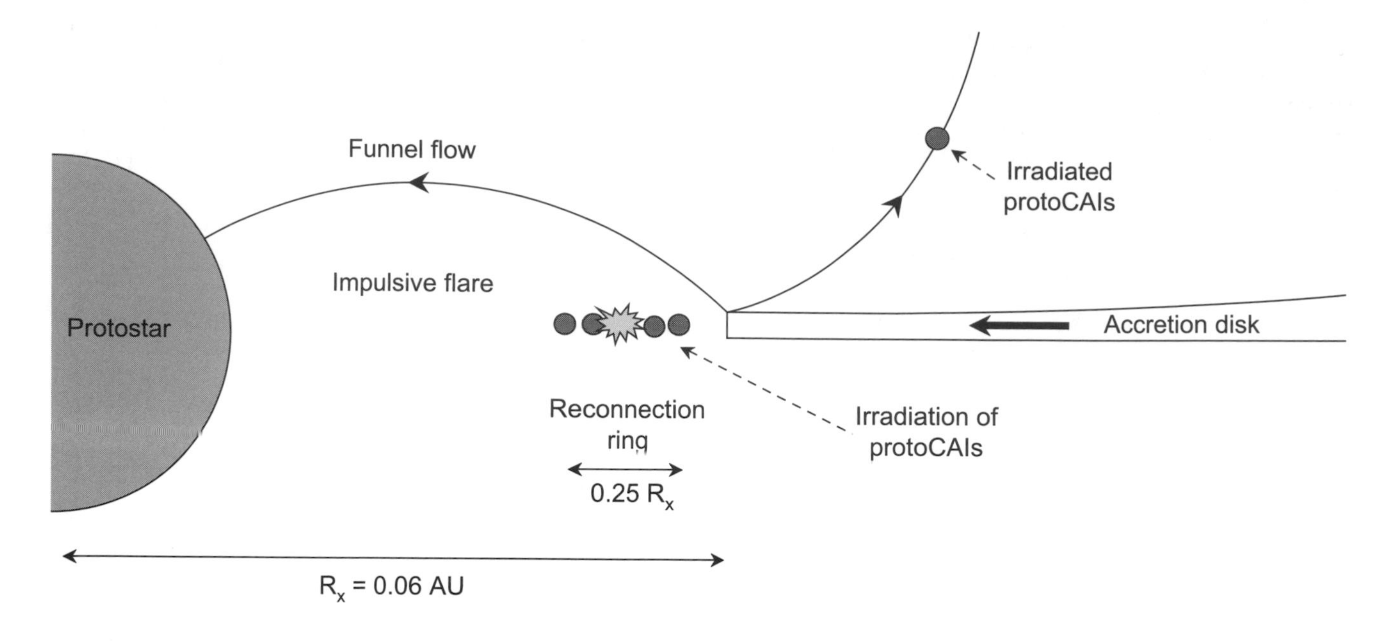

Fig. 5. Simplified cartoon illustrating the irradiation scheme in the context of the x-wind model (e.g., *Shu et al.,* 1994, 1996, 1997; *Lee et al.,* 1998; *Gounelle et al.,* 2001, 2004). This is also the broad picture used by *Leya et al.* (2003). Proto-CAIs are irradiated close to the proto-Sun at a distance of 0.06 AU and transported at asteroidal distance. Irradiating proto-CAIs close to the proto-Sun and subsequently transporting them to asteroidal distances has the distinct advantage of solving the problem of nebular gas shielding (*Lee et al.,* 1998).

1978b; *Heymann et al.,* 1978; *Lee,* 1978). While *Lee*'s (1978) model was specifically aiming at reproducing the ^{16}O anomalies discovered a few years before by *Clayton et al.* (1973), *Dwek* (1978b) predicted a collateral anomaly of ^{41}Ca that was to be discovered some years later (*Srinisavan and Goswami,* 1994).

The discovery of γ-ray lines attributed to the interaction of accelerated particles (cosmic rays) with ambient matter in the Orion molecular cloud (*Bloemen et al.,* 1994) prompted studies examining the possibility of producing light elements (*Cassé et al.,* 1995) and SRs in the context of the molecular cloud core progenitor of our solar system. *Clayton* (1994) first proposed that ^{26}Al could be produced at the meteoritic level by nuclear reaction of low-energy heavy cosmic rays with ambient cloud core matter. In addition to nuclear reactions, *Clayton and Jin* (1995) added the possibility of direct stopping of ^{26}Al low-energy (~100 MeV/A) cosmic rays within the cloud core. *Ramaty et al.* (1996) have suggested that ^{41}Ca could have been produced by nuclear reaction of low-energy heavy cosmic-rays with ambient cloud core matter, but that production of ^{26}Al at a level comparable to that of meteoritic matter would have overproduced ^{6}Li and ^{9}Be. All these studies were largely based on the observations of γ-ray lines in Orion that have been subsequently retracted (*Bloemen et al.,* 1999). It must, however, be mentioned that the retraction of the observations does not preclude the existence of this process in molecular clouds, as long as low-energy cosmic-ray fluxes are high enough.

Goswami et al. (2001) have introduced α reactions for producing SRs via irradiation, and looked at the proton flux enhancement and irradiation time needed to reproduce the meteoritic abundances of ^{26}Al, ^{36}Cl, ^{41}Ca, and ^{53}Mn. They assumed that protons and α particles had free access to nebular regions where meteorites are supposed to have been irradiated. Since nebular gas shielding is neglected in this model, only minimum flux enhancements and irradiation times are calculated. Solid particles having sizes between 10 μm and 1 cm and distributed according to a power law were considered. The flux of solar energetic particles (SEP) was represented as a power law in kinetic energy ($dN \sim E^{-\gamma}dE$) or an exponential law in rigidity ($dN \sim \exp(-R/R_0)$). They concluded that a flux enhancement of 10^5 and an irradiation time of 10^6 yr were needed to produce ^{26}Al in amounts necessary to match the meteorite data, and that no combination of flux enhancement factor and irradiation time can lead to the co-production of ^{26}Al and other SRs (*Goswami et al.,* 2001). This work was complemented by a study by *Marhas et al.* (2002) aimed at understanding the decoupling between ^{26}Al and ^{41}Ca (*Sahijpal et al.,* 1998) and ^{10}Be (*Marhas et al.,* 2002) in some rare hibonite-bearing CAIs. Using the same irradiation model, and including ^{10}Be in the calculations, they show that it was possible to synthetize ^{10}Be without co-producing ^{26}Al and ^{41}Ca as long as the SEP power law index is low (γ ~ 2). The same group concluded that, if ^{10}Be was produced by a SEP having a flat energy spectrum (γ ~ 2), ^{53}Mn was underproduced relative to the meteorite value (*Marhas and Goswami,* 2003).

The x-wind model for the formation of chondrites (*Shu et al.,* 1994, 1996, 1997, 2001) provides a natural solution to a stringent problem of irradiation models: the shielding due to the nebular gas. In the x-wind model, CAIs and chondrules are formed close to the Sun and transported at asteroidal distances by the x-wind (see Fig. 5). Because CAIs and chondrules are formed in a gas-free region, they can be irradiated as bare rocks, minimizing energy losses of the accelerated particles in the gas. Another conceptual advantage of this scheme is that CAIs and chondrules are irradiated close to where proton and He nuclei are accelerated, making irradiation more efficient. *Lee et al.* (1998) calculated yields for ^{26}Al, ^{41}Ca, and ^{53}Mn considering irradiated solids of chondritic composition and distinguishing, by analogy with the contemporary Sun, between impulsive events (^{3}He-rich and steep SEP spectrum) and gradual events (^{3}He-poor and flat SEP spectrum) (e.g., *Reames,* 1995). Introducing ^{3}He reactions leads to the possibility of producing ^{26}Al at high levels, via the reaction $^{24}Mg(^{3}He,p)^{26}Al$. In the case of impulsive flares, *Lee et al.* (1998) could reproduce the meteoritic abundance of ^{26}Al and ^{53}Mn, but overproduced ^{41}Ca by 2 orders of magnitude via the reaction $^{40}Ca(^{3}He,pn)^{41}Ca$. Calcium-41 overproduction was solved by postulating targets with a layered structure (*Gounelle et al.,* 2001; *Shu et al.,* 2001). This layered structure has the advantage of self-shielding ^{40}Ca, the main target for ^{41}Ca production, while keeping ^{24}Mg (the main target for ^{26}Al production) unshielded. It is also possible that the $^{41}Ca/^{40}Ca$ ratio recorded in CAIs is, similarly to ^{26}Al, the result of a protracted sojourn of the CAIs in the solar nebula, and that the initial $^{41}Ca/^{40}Ca$ was higher than 1.5×10^{-8}. In these calculations, the proton luminosity (4.5×10^{29} erg s^{-1}) is deduced from the X-ray observations of protostars (*Montmerle,* 2002; *Feigelson et al.,* 2002), scaled to the contemporary Sun. A $^{7}Be/^{9}Be$ ratio of 0.003 was calculated (*Gounelle et al.,* 2003), close to the experimental value later found by *Chaussidon et al.* (2004). The issue of the decoupling of ^{10}Be with ^{26}Al and ^{41}Ca in rare hibonites was addressed by *Gounelle et al.* (2004), who proposed that impulsive flares events can coproduce $^{7,10}Be$, ^{26}Al, ^{41}Ca, and ^{53}Mn while rare gradual flare events can produce ^{10}Be without producing other SRs.

Leya et al. (2003) calculated the production via irradiation of some SRs whose presence had been reported in the early solar system (^{7}Be, ^{10}Be, ^{26}Al, ^{41}Ca, ^{53}Mn, ^{60}Fe, and ^{92}Nb) and of some SRs yet to be discovered or confirmed (^{14}C, ^{22}Na, ^{36}Cl, ^{44}Ti, ^{54}Mn, ^{63}Ni, and ^{91}Nb). Following the x-wind model (*Shu et al.,* 1996, 1997), irradiation is achieved close to the Sun, and irradiated CAIs and chondrules are subsequently transported to asteroidal distances by the x-wind. They explore gas targets having a CI and a CAI composition, and consider the ratio between impulsive and gradual events as a free parameter. Shielding from the gas is ignored. Calculations are presented relative to the measured meteorite value of $^{10}Be/^{9}Be$. They show that pure gradual events can coproduce ^{10}Be, ^{26}Al, ^{41}Ca, ^{53}Mn, and ^{92}Nb within a factor of 10 in the case of a solar chemistry and provides useful prediction for other SRs. *Leya et al.* (2003) show that it is not possible to co-produce all the SRs of interest in the case of CAI chemistry. They can produce ^{7}Be at the high level first announced by *Chaussidon et al.* (2002), 1 order of magnitude larger than the recently reported ratio (*Chaussidon et al.,* 2004). The production of ^{10}Be, ^{26}Al, ^{41}Ca, ^{53}Mn, and ^{92}Nb requires irradiation times of ~ 1 m.y., if one assumes a 10^{5} enhancement factor of the proton flux in the early Sun (*Feigelson et al.,* 2002). Because of its very short half-life, ^{7}Be is produced on timescales of ~1 yr, and would need proton fluxes 5 orders of magnitude larger than the ones inferred for protostars (*Leya et al.,* 2003). Although they appear different at first glance, *Leya et al.*'s (2003) conclusions are in line with that of *Lee et al.* (1998): The calculated $^{41}Ca/^{26}Al$ ratio is roughly 2 orders of magnitude higher than the measured ratio. Because *Leya et al.* (2003) estimate their uncertainties to be within a factor of 10, they accept that ^{26}Al could have been underproduced by a factor of 10, and ^{41}Ca overproduced by a factor of 10 (Fig. 6). The $^{36}Cl/^{35}Cl$ ratio calculated by *Leya et al.* (2003) (~1.3×10^{-4}) is 1 order of magnitude higher than the recently measured $^{36}Cl/^{35}Cl$ ratio in alteration phases of CAIs [~up to $1.1 \pm 0.2 \times 10^{-5}$ (*Lin et al.,* 2005)]. One should, however, keep in mind that the initial ^{36}Cl content of CAIs was probably higher than the one measured in CAI alteration phases (*Lin et al.,* 2005).

3.3. Irradiation Model Uncertainties

From equation (1), it is clear that the yield of a given ST depends on several poorly known parameters. Among the major uncertainties are the nuclear cross sections. For many reactions, there are no experimental data, and calculated cross sections provided by nuclear codes are used. These calculated cross sections are usually estimated to be accurate within a factor of 2 (*Lee et al.,* 1998; *Leya et al.,* 2003), although discrepancies might be larger. A test of such predictions was recently obtained (*Fitoussi et al.,* 2004) from the experimental determination of the cross sections for the reaction $^{24}Mg(^{3}He,p)^{26}Al$, crucial in the production of ^{26}Al in the context of the x-wind model (*Lee et al.,* 1998). The measured cross sections are shown in Fig. 7 to illustrate the accuracy of the nuclear codes used, and to give an idea of the nuclear cross sections values at the energies considered. The experimental determinations (*Fitoussi et al.,* 2004) are within a factor of 2–3 of the calculated cross section (*Lee et al.,* 1998).

The nature and intensity of accelerated particles in the proto-Sun are another source of uncertainty. Undoubtedly, the high and variable X-ray activity of protostars suggests that there might have been associated accelerated protons and He nuclei at MeV energies (*Feigelson et al.,* 2002). How the X-ray luminosity observed within protostars can be translated into a proton luminosity depends on the nature of the flares, on their spatial position relative to the stars (*Montmerle et al.,* 2000; *Montmerle,* 2002), and on the flaring physics. It is also still unknown exactly when the enhanced

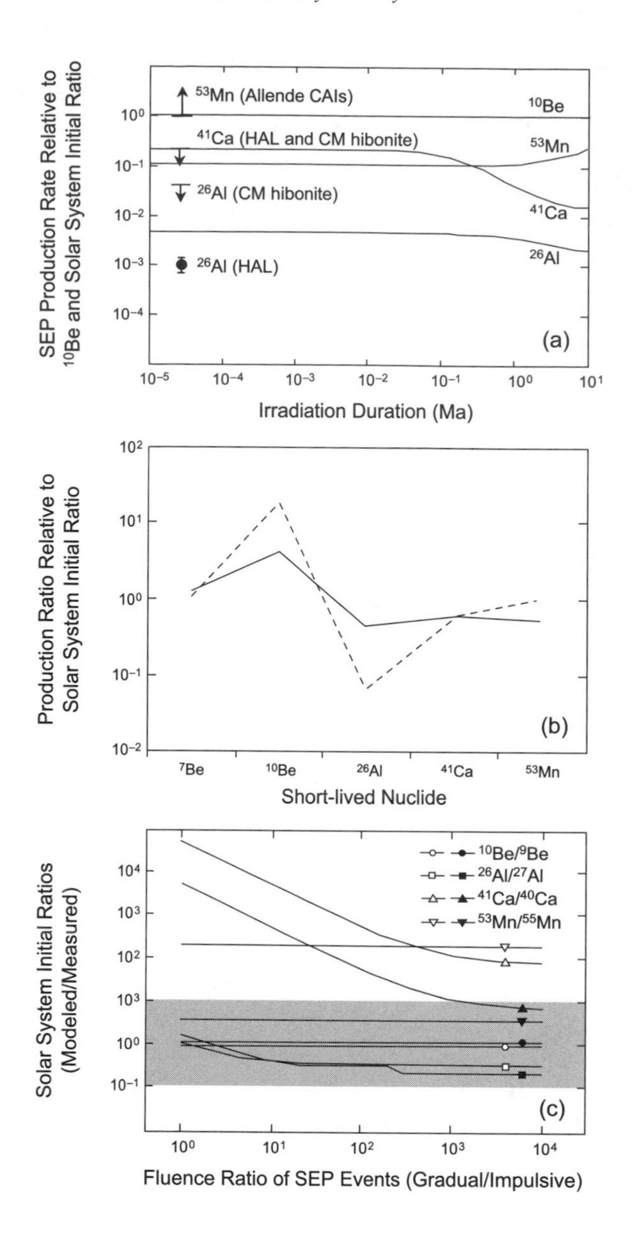

Fig. 6. Results of three different irradiation models for the production of SRs, illustrating the diversity of approaches. Details of the models are given in the text. **(a)** Model of *Marhas et al.* (2002). All calculations are normalized to the meteoritic value of $^{10}Be/^{9}Be$. Yields are calculated for ^{26}Al, ^{53}Mn, and ^{41}Ca as a function of the irradiation time for a SEP spectral index $\gamma = 2$ (with $dN/dE = K E^{-\gamma}$, see text). For such a spectral index, no SR is coproduced with ^{10}Be. **(b)** Model of *Gounelle et al.* (2001, 2004). Absolute yields of ^{7}Be, ^{10}Be, ^{26}Al, ^{41}Ca, and ^{53}Mn normalized to their respective meteoritic values. Impulsive flares ($\gamma = 4$, thick line) and gradual flares ($\gamma = 2.7$, dotted line) are shown. For gradual flares, all SRs are underproduced relative to ^{10}Be as in the previous model, while in impulsive flares all SRs are produced at the meteoritic level within a factor of a few. **(c)** Model of *Leya et al.* (2003). Yields of ^{26}Al, ^{41}Ca, and ^{53}Mn are normalized to their meteoritic value, and relative to the meteoritic abundance of ^{10}Be. Yields for solar chemistry and CAI chemistry are shown as a function of the fluence ratio of SEP events (gradual/impulsive). For pure gradual events, in the case of solar chemistry, *Leya et al.* (2003) have been able to reproduce the early solar system abundance of SRs under scrutiny (gray field) within an order of magnitude.

X-ray activity of protostars begins or ends. Ignorance of the nature of accelerated particles gives uncertainties about F_0, y^i_{CR}, and N(E). At the moment, the best approach is to infer X-ray data from protostars and the expected flux levels of accelerated protons from solar physics. Better constraining these parameters is an important task for the coming years since some models (*Gounelle et al.,* 2001) need abundances of ^{3}He nuclei compared to protons that belong to the high range of what is observed in the contemporary Sun (*Torsti et al.,* 2002).

The irradiation time interval Δt is unknown. Most modelers (*Goswami et al.,* 2001; *Marhas et al.,* 2002; *Marhas and Goswami,* 2003; *Leya et al.,* 2003) use it as a free parameter that vary between short (1 yr) and longer timescales (1 m.y.). In the x-wind model, it is constrained by the fluctuations of the x-point on timescales of 1–10 yr (*Gounelle et al.,* 2001). If irradiation in the early solar system is linked to the protostar and T Tauri stage of our star, it constrains Δt to be inferior to a few million years (see *Russell et al.,* 2006). More important constraints on Δt may be obtained from radionuclides with very short half-lives, such as ^{7}Be.

The last major source of uncertainty is the chemical composition of the irradiated solid or gas (x^T, x_S). This is especially true in the x-wind model (*Gounelle et al.,* 2001), where a layered structure of proto-CAIs has to be used to produce ^{41}Ca at the correct abundance. However, one should note that all modelers use either chondritic chemistry, or CAI chemistry, or a combination of both as a reasonable assumption for the composition of CAI precursor material. This chemistry should not be far away from the real chemistry of CAIs precursors. Note also that irradiation of phases like olivines, which are observed in the interstellar medium and are very likely to be a major component of the dust in

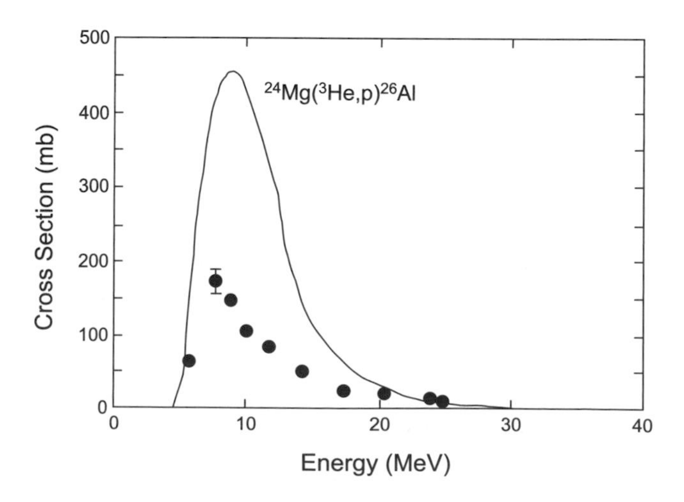

Fig. 7. Predicted and measured nuclear cross sections of the reaction $^{24}Mg(^{3}He,p)^{26}Al$. The predicted excitation function has been calculated using Hauser-Feshbach and fragmentation codes (*Lee et al.,* 1998). The measured cross section (circles) recently determined by *Fitoussi et al.* (2004) are less than 4× smaller and within the range of predicted values (*Lee et al.,* 1998).

the solar accretion disk, can produce the ^{10}Be, ^{26}Al, and ^{41}Ca abundances observed in CAIs (*McKeegan et al.,* 2000) because of their Al- and Ca-poor chemistry.

4. CONCLUSIONS

A growing suite of observations indicate that a fraction of the gas and dust present in the protosolar nebula around the forming Sun was irradiated with energetic photons and particles. Astronomical observations show the ubiquitous presence of powerful X-ray flares in pre-main-sequence stellar systems (e.g., *Montmerle,* 2002; *Feigelson et al.,* 2002; *Preibisch and Feigelson,* 2005; *Wolk et al.,* 2005), plus new evidence that the X-rays penetrate the disk (*Tsujimoto et al.,* 2005; *Kastner et al.,* 2005). Radio gyrosynchrotron emission from MeV electrons is seen in some systems (*Feigelson and Montmerle,* 1999). Together, these astronomical studies provide a strong empirical basis for models of energetic irradiation of early solar system materials. Traces of these irradiation processes (e.g., ^{21}Ne, ^{10}Be, etc.) are most likely present in meteorites. If new evidence of the *in situ* decay of ^{7}Be is found in the near future in other CAIs, the presence of ^{7}Be in the protosolar nebular when CAIs formed will provide strong support for irradiation having taken place in the early solar system.

The origin of SRs is an especially controversial aspect of meteoritics. A whole-stellar origin has undoubtedly been the predominant theory for the last decades (for a review, see *Goswami and Vanhala,* 2000). The discovery of ^{10}Be (*McKeegan et al.,* 2000) has partly changed the situation since Be isotopes are not made in stars (*Reeves,* 1994). Although it has been proposed that ^{10}Be could have had a GCR origin (*Desch et al.,* 2004), most modelers contend that it has been produced by irradiation within the early solar system *(McKeegan et al.,* 2000*; Gounelle et al.,* 2001; *Leya et al.,* 2003; *Marhas et al.,* 2002*)*. Thus, the presence of ^{10}Be (and possibly of ^{7}Be) justifies reevaluating the possibility that some SRs are created by irradiation around the young Sun. Models of irradiation produced remarkably different results, some predicting quite well the observed abundances of the major SRs found in CAIs (^{7}Be, ^{10}Be, ^{26}Al, ^{41}Ca, and ^{53}Mn). Nevertheless, there is at present no model that can account for the initial abundance of these major SRs without requiring special conditions (e.g., high proton fluxes) or a special nature or composition of the irradiated targets (e.g., unobserved structure for proto-CAIs). However, the uncertainties in nuclear cross sections leave much room for model improvements. New observations of SRs [e.g., ^{36}Cl in CAI alteration products (*Lin et al.,* 2005)], as well as a refined knowledge of the initial solar system content of ^{26}Al and ^{41}Ca, will prompt such improvements.

Finally it must be stressed that irradiation processes around the early Sun are not only relevant for the production of SRs in the early solar system, but are also considered to explain (1) some of the large isotopic variations observed for light elements (H, C, N, O) in meteorites through UV-self shielding or X-ray enhanced ion-molecule reactions and (2) chemical and mineralogical modifications of grains in the nebula.

Acknowledgments. We are grateful to E. Feigelson for his help and suggestions concerning astrophysical observations of young stars, to J. Aléon for making available his latest results concerning the discovery of silica grains with large O-isotopic anomalies, and to S. Russell and K. Marti for reviews and numerous comments. This work was supported by Région Lorraine and PNP/INSU grants. This is CRPG-CNRS contribution #1776.

REFERENCES

Aléon J. and Robert F. (2004) Interstellar chemistry recorded by nitrogen isotopes in solar system organic matter. *Icarus, 167,* 424–430.

Aléon J., Robert F., Duprat J., and Derenne S. (2005a) Extreme oxygen isotope ratios in the early solar system. *Nature, 437,* 385–388.

Aléon J., Hutcheon I. D., Weber P. K., and Duprat J. (2005b) Magnesium isotopic mapping of silica-rich grains with extreme oxygen isotopic anomalies(abstract). In *Lunar and Planetary Science XXXVI,* Abstract #1901. Lunar and Planetary Institute, Houston (CD-ROM).

Anders E. and Grevesse N. (1989) Abundance of the elements: Meteoritic and solar. *Geochim. Cosmochim. Acta, 53,* 197–214.

Barrat J. A., Chaussidon M., Bohn M., Gillet Ph., Göpel C., and Lesourd M. (2005) Lithium behavior during cooling of a dry basalt: An ion microprobe study of the lunar meteorite Northwest Africa 479 (NWA 479). *Geochim. Cosmochim. Acta,* in press.

Beck P., Barrat J. A., Chaussidon M., Gillet Ph., and Bohn M. (2004) Li isotopic variations in single pyroxenes from the Northwest Africa 480 shergottite (NWA 480): A record of degasing of Martian magmas? *Geochim. Cosmochim. Acta, 68,* 2925–2933.

Bibring J.-P., Duraud J. P., Durrieu L., Jouret C., Maurette M., and Meunier R. (1972) Ultrathin amorphous coatings on lunar dust grains. *Science, 264,* 1779–1780.

Birck J.-L. and Allègre C.-J. (1985) Evidence for the presence of ^{53}Mn in the early solar system. *Geophys. Res. Lett., 12,* 745–748.

Bloemen H. and 12 colleagues (1994) COMPTEL observations of the Orion complex: Evidence for cosmic-ray induced gamma-ray lines. *Astron. Astrophys., 281,* L5–L8.

Bloemen H. and 13 colleagues (1999) The revised COMPTEL Orion results. *Astrophys. J. Lett., 521,* L137–L140.

Borg J., Chaumont J., Jouret C., Langevin Y., and Maurette M. (1980) Solar wind radiation damage in lunar dust grains and the characteristics of the ancient solar wind. In *Proceedings of the Conference on the Ancient Sun,* pp. 431–461. Pergamon, New York.

Bradley J. P. (1994) Chemically anomalous, preacrretionally irradiated grains in interplanetary dust from comets. *Science, 265,* 925–929.

Bradley J. P., Dukes C., Baragiola R., McFadden L., Johnson R. E., and Brownlee D. E. (1996) Radiation processing and the origins of interplanetary dust particles (abstract). In *Lunar and Planetary Science XXVII,* pp. 149–150. Lunar and Planetary Institute, Houston.

Burbidge E. M., Burbidge G. R., Fowler W. A., and Hoyle F.

(1957) Synthesis of the elements in stars. *Rev. Mod. Phys., 29,* 547–650.

Burnett D. S., Fowler W. A., and Hoyle F. (1965) Nucleosynthesis in the early solar system. *Geochim. Cosmochim. Acta, 29,* 1209–1241.

Busso M., Gallino R., and Wasserburg G. J. (2003) Short-lived nuclei in the early solar system: A low mass stellar source? *Publ. Astron. Soc. Australia, 20,* 356–370.

Caffee M. W. and Nishiizumi K. (2001) Exposure history of separated phases from the Kapoeta howardite. *Meteoritics & Planet. Sci., 36,* 429–437.

Caffee M. W., Hohenberg C. M., Swindle T. D., and Goswami J. N. (1987) Evidence in meteorites for an active early sun. *Astrophys. J. Lett., 313,* L31–L35.

Caffee M. W., Goswami J. N., Hohenberg C. M., Marti K., and Reedy R. C. (1988) Irradiation records in meteorites. In *Meteorites and the Early Solar System* (J. F. Kerridge and M. S. Matthews, eds.), pp. 205–245. Univ. of Arizona, Tucson.

Cami J. and Yamamura I. (2001) Discovery of anomalous oxygen isotopic ratios in HR 4049. *Astron. Astrophys., 367,* L1–L4.

Carrez P., Demyk K., Cordier P., Gengembre L., Grimblot J., d'Hendecourt L., Jones A. P., and Leroux H. (2002a) Low-energy helium iom irradiation-induced amorphization and chemical changes in olivine: Insights for silicate dust evolution in the interplanetary medium. *Meteoritics & Planet. Sci., 37,* 1599–614.

Carrez P., Demyk K., Leroux H., Cordier P., Jones A. P., and d'Hendecourt L. (2002b) Low-temperature crystallization of $MgSiO_3$ glasses under electron irradiation: Possible implications for silicate dust evolution in circumstellar environments. *Meteoritics & Planet. Sci., 37,* 1615–1622.

Cassé M., Lehoucq R., and Vangioni-Flam E. (1995) Production and evolution of light elements in active star forming regions. *Nature, 373,* 318–319.

Chakraborty S. and Thiemens M. H. (2005) Evaluation of CO self-shielding as a possible mechanism for anomalous oxygen isotopic composition of early solar system materials (abstract). In *Lunar and Planetary Science XXXVI,* Abstract #1113. Lunar and Planetary Institute, Houston (CD-ROM).

Chaussidon M. and Robert F. (1995) Nucleosynthesis of ^{11}B-rich boron in the pre-solar cloud recorded in meteoritic chondrules. *Nature, 374,* 337–339.

Chaussidon M. and Robert F. (1998) ^{7}Li/^{6}Li and ^{11}B/^{10}B variations in chondrules from the Semarkona unequilibrated chondrite. *Earth Planet Sci. Lett., 164,* 577–589.

Chaussidon M., Robert F., and McKeegan K. D. (2001) Lithium and boron isotopic compositions of refractory inclusions from primitive chondrites: A record of irradiation in the early solar system (abstract). In *Lunar and Planetary Science XXXII,* Abstract #1862. Lunar and Planetary Institute, Houston (CD-ROM).

Chaussidon M., Robert F., and McKeegan K. D. (2002) Incorporation of short-lived ^{7}Be in one CAI from the Allende meteorite (abstract). In *Lunar and Planetary Science XXXIII,* Abstract #1563. Lunar and Planetary Institute, Houston (CD-ROM).

Chaussidon M., Robert F., Russell S. S., Gounelle M., and Ash R. D. (2003) Variations of apparent ^{10}Be/^{9}Be ratios in Leoville MRS-06 type B1 CAI: Constraints on the origin of ^{10}Be and ^{26}Al (abstract). In *Lunar and Planetary Science XXXIV,* Abstract #1347. Lunar and Planetary Institute, Houston (CD-ROM).

Chaussidon M., Robert F., and McKeegan K. D. (2004) Li and B isotopic variations in Allende type B1 CAI 3529-41: Traces of incorporation of short-lived ^{7}Be and ^{10}Be (abstract). In *Lunar and Planetary Science XXXV,* Abstract #1568. Lunar and Planetary Institute, Houston (CD-ROM).

Chaussidon M., Robert F., and McKeegan K. D. (2005) Li and B isotopic variations in an Allende CAI: Evidence for the in situ decay of short-lived ^{10}Be and for the possible presence of the short-lived nuclide ^{7}Be in the early solar system. *Geochim. Cosmochim. Acta,* in press.

Clayton D. D. (1994) Production of ^{26}Al and other extinct radionuclides by low-energy heavy cosmic rays in molecular clouds. *Nature, 368,* 222–224.

Clayton D. D. and Jin L. (1995) Gamma-rays, cosmic rays, and extinct radioactivities in molecular clouds. *Astrophys. J., 451,* 681–699.

Clayton D. D., Dwek E., and Woosley S. E. (1977) Isotopic anomalies and proton irradiation in the early solar system. *Astrophys. J., 214,* 300–315.

Clayton R. N. (1993) Oxygen isotopes in meteorites. *Annu. Rev. Earth Planet. Sci., 21,* 115–149.

Clayton R. N. (2002a) Self-shielding in the solar nebula. *Nature, 415,* 860–861.

Clayton R. N. (2002b) Nitrogen isotopic fractionation by photochemical self-shielding (abstract). *Meteoritics & Planet. Sci., 37,* A35.

Clayton R. N., Grossman L., and Mayeda T. K. (1973) A component of primitive nuclear composition in carbonaceous meteorites. *Science, 182,* 485–488.

Demyk K., Carrez P., Leroux H., Cordier P., Jones A. P., Borg J., Quirico E., Raynal P. I., and d'Hendecourt L. (2001) Structural and chemical alteration of crystalline olivine under low energy He^+ irradiation. *Astron. Astrophys., 368,* L38–L41.

Desch S. J., Connolly H. C. Jr., and Srinivasan G. (2004) An interstellar origin for the beryllium 10 in calcium-rich, aluminum-rich inclusions. *Astrophys. J., 602,* 528–542.

Dran J. C., Durrieu L., Jouret C., and Maurette M. (1970) Habit and texture studies of lunar and meteoritic material with the 1 MeV electron microscope. *Earth Planet. Sci. Lett., 9,* 391–400.

Dwek E. (1978a) Proton-associated alpha-irradiation in the early solar system: A possible ^{41}K anomaly. *Astrophys. J., 221,* 1026–1031.

Dwek E. (1978b) The beryllium and boron abundance in the meteorites: A synthesis in interstellar grains. *Astrophys. J. Lett., 225,* L149–L152.

Eugster O., Herzog G. F., Marti K., and Caffee M. W. (2006) Irradiation records, cosmic-ray exposure ages, and transfer times of meteorites. In *Meteorites and the Early Solar System II* (D. S. Lauretta and H. Y. McSween Jr., eds.), this volume. Univ. of Arizona, Tucson.

Feigelson E. D. and Montmerle T. (1999) High energy processes in young stellar objects. *Annu. Rev. Astron. Astrophys., 37,* 363–408.

Feigelson E. D., Garmire G. P., and Pravdo S. H. (2002) Magnetic flaring in the pre-main-sequence Sun and implications for the early solar system. *Astrophys. J., 572,* 335–349.

Fitoussi C. and 17 colleagues (2004) AMS measurement of ^{24}Mg(^{3}He,p)^{26}Al cross section: Implications for the ^{26}Al production in the early solar system (abstract). In *Lunar and Planetary Science XXXV,* Abstract #1586. Lunar and Planetary Institute, Houston (CD-ROM).

Fowler W. A., Greenstein J. L., and Hoyle F. (1962) Nucleosynthesis during the early history of the solar System. *Geophys. J. R. Astron. Soc., 6,* 148–220.

Glassgold A. E., Feigelson E. D., and Montmerle T. (2000) Effects of energetic radiation in young stellar objects. In *Protostars and Planets IV* (V. Mannings et al., eds.), pp. 429–456. Univ. of Arizona, Tucson.

Goswami J. N. (2003) Solar system initial boron isotopic composition: Signature of an anomalous cosmogenic component. *Meteoritics & Planet. Sci., 38,* A48.

Goswami J. N. and Vanhala H. A. T. (2000) Extinct radionuclides and the origin of the solar system. In *Protostars and Planets IV* (V. Mannings et al., eds.), pp. 963–994. Univ. of Arizona, Tucson.

Goswami J. N., Marhas K. K., and Sahijpal S. (2001) Did solar energetic particles produce the short-lived radionuclides present in the early solar system? *Astrophys. J., 549,* 1151–1159.

Gounelle M. and Russell S. S. (2005) On early solar system chronology: Implications of an heterogeneous distribution of extinct short-lived radionuclides. *Geochim. Cosmochim. Acta, 69,* 3129–3144.

Gounelle M., Shu F. H., Shang H., Glassgold A. E., Rehm E. K., and Lee T. (2001) Extinct radioactivities and protosolar cosmic-rays: Self-shielding and light elements. *Astrophys. J., 548,* 1051–1070.

Gounelle M., Shang H., Glassgold A. E., Shu F. H., Rehm E. K., and Lee T. (2003) Early solar system irradiation and beryllium-7 synthesis (abstract). In *Lunar and Planetary Science XXXIV,* Abstract #1833. Lunar and Planetary Institute, Houston (CD-ROM).

Gounelle M., Shu F. H., Shang H., Glassgold A. E., Rehm E. K., and Lee T. (2004) The origin of short-lived radionuclides and early solar system irradiation (abstract). In *Lunar and Planetary Science XXXV,* Abstract #1829. Lunar and Planetary Institute, Houston (CD-ROM).

Gudel M. (2004) X-ray astronomy of stellar coronae. *Rev. Astron. Astrophys., 12,* 71–237.

Hashizume K., Chaussidon M, Marty B., and Kentaro T. (2004) Protosolar carbon isotopic composition: Implications for the origin of meteoritics organics. *Astrophys. J. Lett., 600,* L480–L484.

Hashizume K. and Chaussidon M. (2005) A non-terrestrial ^{16}O-rich isotopic composition for the protosolar nebula. *Nature, 434,* 619–622.

Hashizume K., Chaussidon M., Marty B., and Robert F. (2000) Solar wind record on the Moon: Deciphering presolar from planetary nitrogen. *Science, 290,* 1142–1145.

Heymann D. and Dziczkaniec M. (1976) Early irradiation of matter in the solar system: Magnesium (proton, neutron) scheme. *Science, 191,* 79–81.

Heymann D., Dziczkaniec M., Walker A., Huss G., and Morgan J. A. (1978) Effects of proton irradiation on a gas phase in which condensation takes place. I. Negative ^{26}Mg anomalies and ^{26}Al. *Astrophys. J., 225,* 1030–1044.

Hohenberg C. M., Nichols R. H. J., Olinger C. T., and Goswami J. N. (1990) Cosmogenic neon from individual grains of CM meteorites: Extremely long pre-compaction histories or an enhanced early particle flux. *Geochim. Cosmochim. Acta, 54,* 2133–2140.

Hoppe P., Goswami J. N., Krähenbühl U., and Marti K. (2001) Boron in chondrules. *Meteoritics & Planet. Sci., 36,* 1321–1330.

Housen K. R. and Wilkening L. L. (1982) Regoliths in small bodies in the solar system. *Annu. Rev. Earth Planet. Sci., 10,* 355–376.

James R. H. and Palmer M. R. (2000) The lithium isotope composition of international rock standards. *Chem. Geol., 166,* 319–326.

Kastner J. H., Franz G., Grosso N., Bally J., McCaughrean M., Getman K., Schulz N. S., and Feigelson E. D. (2005) X-ray emission from young stars with protoplanetary disks and jets in the Orion Nebula. *Astrophys. J. Suppl. Ser., 160,* 511–529.

Keller L. P. and McKay D. S. (1997) The nature and origin of rims on lunar soil grains. *Geochim. Cosmochim. Acta, 61,* 2331–2341.

Lauretta D. S. and Lodders K. (1997) The cosmochemical behavior of beryllium and boron. *Earth Planet. Sci. Lett., 146,* 315–327.

Lee T. (1978) A local irradiation model for isotopic anomalies in the solar system. *Astrophys. J., 224,* 217–226.

Lee T., Papanatassiou D. A., and Wasserburg G. J. (1976) Demonstration of ^{26}Mg excess in Allende and evidence for ^{26}Al. *Geoph. Res. Lett., 3,* 109–112.

Lee T., Papanastassiou D. A., and Wasserburg G. J. (1977) Aluminium-26 in the early solar system: Fossil or fuel. *Astrophys. J. Lett., 211,* L107–L110.

Lee T., Shu F. H., Shang H., Glassgold A. E., and Rhem K. E. (1998) Protostellar cosmic rays and extinct radioactivities in meteorites. *Astrophys. J., 506,* 898–912.

Leya I., Halliday A. N., and Wieler R. (2003) The predictable collateral consequences of nucleosynthesis by spallation reactions in the early solar system. *Astrophys. J., 594,* 605–616.

Lin Y., Guan Y., Leshin L. A., Ouyand Z., and Wang D. (2005) Short-lived chlorine-36 in a Ca-and Al-rich inclusion from the Ningqiang carbonaceous chondrite. *Proc. Natl. Acad. Sci., 102,* 1306–1311.

Lugmair G. W. and Shukolyukov A. (1998) Early solar system timescales according to ^{53}Mn-^{53}Cr systematics. *Geochim. Cosmochim. Acta, 62,* 2863–2886.

Lyons J. R. and Young E. D. (2005) CO self-shielding as the origin of oxygen isotope anomalies in the early solar nebula. *Nature, 435,* 317–320.

MacPherson G. J., Davis A. M., and Zinner E. K. (1995) The distribution of aluminum-26 in the early solar system: A reappraisal. *Meteoritics, 30,* 365–386.

MacPherson G. J., Huss G. R., and Davis A. M. (2003) Extinct ^{10}Be in type A calcium-aluminum-rich inclusions from CV chondrites. *Geochim. Cosmochim. Acta, 67,* 3165–3179.

Marhas K. K. and Goswami J. N. (2003) Be-B isotope systematics in CV and CM hibonites: Implications for solar energetic particle production of short-lived nuclides in early solar system (abstract). In *Lunar and Planetary Science XXXIV,* Abstract #1303. Lunar and Planetary Institute, Houston (CD-ROM).

Marhas K. K., Goswami J. N., and Davis A. M. (2002) Short-lived nuclides in hibonite grains from Murchison: Evidence for solar system evolution. *Science, 298,* 2182–2185.

McDonough W. F., Teng F. Z., Tomascak P. B., Ash R. D., Grossman J. N., and Rudnick R. L. (2003) Lithium isotopic composition of chondritic meteorites (abstract). In *Lunar and Planetary Science XXXIV,* Abstract #1931. Lunar and Planetary Institute, Houston (CD-ROM).

McKeegan K. D. and Davis A. M. (2004) Early solar system chronology. In *Treatise on Geochemistry, Vol. 1: Meteorites, Comets, and Planets* (A. M. Davis, ed.), pp. 431–460. Elsevier, Oxford.

McKeegan K. D. and Leshin L. A. (2001) Stable isotope variations in extraterrestrial materials *Rev. Mineral., 43,* 279–318.

McKeegan K. D., Chaussidon M., and Robert F. (2000) Incorporation of short-lived ^{10}Be in a calcium-aluminium-rich inclusion from the Allende meteorite. *Science, 289,* 1334–1337.

McKeegan K. D., Chaussidon M., Krot A. N., Robert F., Goswami J. N., and Hutcheon I. D. (2001) Extinct radionuclide abundances in Ca, Al-rich inclusions from the CV chondrites Allende and Efremovka: A search for synchronicity (abstract). In *Lunar and Planetary Science XXXII,* Abstract #2175. Lunar and Planetary Institute, Houston (CD-ROM).

Meyer B. S. and Clayton D. D. (2000) Short-lived radioactivities and the birth of the sun. *Space Sci. Rev., 92,* 133–152.

Meyer B. S. and Zinner E. (2006) Nucleosynthesis. In *Meteorites and the Early Solar System II* (D. S. Lauretta and H. Y. McSween Jr., eds.), this volume. Univ. of Arizona, Tucson.

Michel R., Bodemann R., Busemann H. et al. (1997) Cross sections for the production of residual nuclides by low and medium energy protons from the target elements C, N, O, Mg, Al, Si, Ca, Ti, V, Mn, Fe, Co, Ni, Cu, Sr, Y, Zr, Nb. *Nucl. Inst. Phys. Res. Meth. B, 129,* 153–193.

Montmerle T. (2002) Irradiation phenomena in young solar-type stars and the early solar system: X-ray observations and γ-ray constraints. *New Astron. Rev., 46,* 573–583.

Montmerle T., Grosso N., Tsuboi Y., and Koyama K. (2000) Rotation and X-ray emission from protostars. *Astrophys. J., 532,* 1089–1102.

Murty S. V. S., Goswami J. N., and Shukolyukov Y. A. (1997) Exces ^{36}Ar in the Efremovka meteorite: A strong hint for the presence of ^{36}Cl in the early solar system. *Astrophys. J. Lett., 475,* L65–L68.

Nakamura T., Zolensky M., Sekiya M., Okazaki R., and Nagao K. (2003) Acid-susceptive material as a host phase of argon-rich noble gas in the carbonaceous chondrite Ningqiang. *Meteoritics & Planet. Sci., 38,* 243–250.

Okasaki R., Takaoka N., Nagao K., Sekiya M., and Nakamura T. (2001) Noble-gas-rich chondrules in an enstatite meteorite. *Nature, 412,* 795–798.

Papanastassiou D. A., Bogdanovski O., and Wasserburg G. J. (2002) ^{53}Mn-^{53}Cr systematics in Allende refractory inclusions (abstract). *Meteoritics & Planet. Sci., 36,* A114.

Paque J. M., Burnett D. S., and Chaussidon M. (2003) UNSM 3515: An Allende CAI with Li isotopic variations (abstract). In *Lunar and Planetary Science XXXIV,* Abstract #1401. Lunar and Planetary Institute, Houston (CD-ROM).

Podosek F. A., Zinner E. K., MacPherson G. J., Lundberg L. L., Brannon J. C., and Fahey A. J. (1991) Correlated study of initial ^{87}Sr/^{86}Sr and Al-Mg systematics and petrologic properties in a suite of refractory inclusions from the Allende meteorite. *Geochim. Cosmochim. Acta, 55,* 1083–1110.

Preibisch T. and Feigelson E. D. (2005) The evolution of X-ray emission in young stars. *Astrophys. J. Suppl. Ser., 160,* 390–400.

Pun A., Keil K., Taylor G. J., and Wieler R. (1998) The Kapoeta howardite: Implications for the regolith evolution of the howardite-eucrite-diogenite parent body. *Meteoritics & Planet. Sci., 35,* 835–851.

Ramaty R., Kozlovsky B., and Lingenfelter R. E. (1996) Light isotopes, extinct radioisotopes, and gamma-ray lines from low-energy cosmic-ray interactions. *Astrophys. J., 456,* 525–540.

Rao M. N., Garrisson D. H., Palma R. L., and Bogard D. D. (1997) Energetic proton irradiation history of the howardite parent-body regolith and implications for ancient solar activity. *Meteoritics & Planet. Sci., 32,* 531–543.

Read S. M. and Viola V. E. (1984) Excitation function for A>6 fragments formed in ^{1}H and ^{4}He-induced reactions on light nuclei. *Atomic Data and Nuclear Tables, 31,* 359–385.

Reames D. V. (1995) Solar energetic particles: A paradigm shift. *Rev. Geophys. Suppl., 33,* 585–589.

Reeves H. (1994) On the origin of the light elements (Z<6). *Rev. Mod. Phys., 66,* 193–216.

Reeves H., Fowler W., and Hoyle F. (1970) Galactic cosmic rays origin of Li, Be and B in stars. *Nature, 226,* 727–729.

Robert F. (2002) Water and organic matter D/H ratios in the solar system: A record of an early irradiation of the nebula. *Planet. Space Sci., 50,* 1227–1334.

Robert F. and Chaussidon M. (2003) Boron and lithium isotopic composition in chondrules from the Mokoia meteorite (abstract). In *Lunar and Planetary Science XXXIV,* Abstract #1344. Lunar and Planetary Institute, Houston (CD-ROM).

Robert F., Merlivat L., and Javoy M. (1979) Deuterium concentration in the early solar system: Hydrogen and oxygen isotope study. *Nature, 282,* 785–789.

Russell S. S., Gounelle M., and Hutchison R. (2001) Origin of short-lived radionuclides. *Philos. Trans. R. Soc. Lond., A359,* 1991–2004.

Russell S. S., Hartmann L., Cuzzi J., Krot A. N., Gounelle M., and Weidenschilling S. (2006) Timescales of the solar protoplanetary disk. In *Meteorites and the Early Solar System II* (D. S. Lauretta and H. Y. McSween Jr., eds.), this volume. Univ. of Arizona, Tucson.

Sahijpal S., Goswami J. N., Davis A. M., Grossman L., and Lewis R. S. (1998) A stellar origin for the short-lived nuclides in the early solar system. *Nature, 391,* 559–561.

Semenov D., Wiebe D., and Henning T. (2004) Reduction of chemical networks. II. Analysis of the fractional ionisation in protoplanetary discs. *Astron. Astrophys., 417,* 93–106.

Sheffer Y., Lambert D. L., and Federman S. R. (2002) Ultraviolet detection of interstellar ^{12}C^{17}O and the CO isotopomeric ratios towards X Persei. *Astrophys. J. Lett., 574,* L171–L174.

Shu F. H., Najita J., Ostriker E., Wilkin F., Rude S., and Lizano S. (1994) Magnetocentrifugally friven flows from young stars and disks. I: A generalized model. *Astrophys. J., 429,* 781–796.

Shu F. H., Shang H., and Lee T. (1996) Towards an astrophysical theory of chondrites. *Science, 271,* 1545–1552.

Shu F. H., Shang H., Glassgold A. E., and Lee T. (1997) X-rays and fluctuating X-winds from protostars. *Science, 277,* 1475–1479.

Shu F. H., Shang S., Gounelle M., Glassgold A. E., and Lee T. (2001) The origin of chondrules and refractory inclusions in chondritic meteorites. *Astrophys. J., 548,* 1029–1050.

Sisterson J. M. and 10 colleagues (1997) Measurement of proton production cross sections of ^{10}Be and ^{26}Al from elements found in lunar rocks. *Nucl. Inst. Meth. Phys. Res. B, 123,* 324–329.

Srinivasan G. (2002) ^{10}Be-^{10}B and ^{7}Be-^{7}Li isotope systematics in CAI E65 from Efremovka CV3 chondrite (abstract). *Meteoritics & Planet. Sci., 37,* 135.

Srinisavan G. and Goswami J. N. (1994) ^{41}Ca in the early solar system. *Astrophys. J. Lett., 431,* L67–L70.

Sugiura N., Shuzou Y., and Ulyanov A. A. (2001) Beryllium-boron and aluminum-magnesium chronology of calcium-aluminum-rich inclusions in CV chondrites. *Meteoritics & Planet. Sci., 36,* 1397–1408.

Thiemens M. and Heidenreich J. E. III (1983) Mass independent fractionation of oxygen: A novel isotope effect and its possible cosmochemical implications. *Science, 219,* 1073–1075.

Torsti J., Kocharov L., Laivola J., Lehtinen N., Kaiser M. L., and Reiner M. J. (2002) Solar particle event with exceptionally high ^{3}He enhancement in the energy range up to 50 MeV nucleon^{-1}. *Astrophys. J. Lett., 573,* L59–L63.

Tsujimoto M., Feigelson E. D., Grosso N., Micela G., Tsuboi Y. Favata F., Shang H., and Kastner J. (2005) Iron fluorescent line emission from young stellar objects in the Orion Nebula. *Astrophys. J. Suppl. Ser., 160,* 503–510.

Vogel N., Baur H., Bischoff A., Leya I., and Wieler R. (2004) Noble gas studies in CAIs from CV3 chondrites: No evidence for primordial noble gases. *Meteoritics & Planet. Sci., 39,* 767–778.

Wieler R., Baur H., Pedroni A., Signer P., and Pellas P. (1989) Exposure history of the regolithic chondrite Fayetteville: I. Solar-gas-rich matrix. *Geochim. Cosmochim. Acta, 53,* 1441–1448.

Wieler R., Pedoni A., and Leya I. (2000) Cosmogenic neon in mineral separates from Kapoeta: No evidence for an irradiation of its parent body regolith by an active early sun. *Meteoritics & Planet. Sci., 35,* 251–257.

Wolk S., Harnden F. R., Flaccomio E., Micela G., Favata F., Shang H., and Feigelson E. D. (2005). Stellar activity on the young suns of Orion: COUP observations of K5-7 pre-main sequence stars. *Astrophys. J. Suppl. Ser., 160,* 423–449.

Woolum D. S. and Hohenberg C. M. (1993) Energetic particle environment in the early solar system: Extremely long precompaction meteoritic ages or an enhanced early particle flux. In *Protostars and Planets III* (E. H. Levy and J. I. Lunine, eds.), pp. 903–919. Univ. of Arizona, Tucson.

Young E. D., Simon J. I., Galy A., Russell S. S., Tonui E., and Lovera O. (2005) Supra-canonical $^{26}Al/^{27}Al$ and the residence time of CAIs in the solar protoplanetary disk. *Science, 308,* 223–227.

Yurimoto H. and Kuramoto K. (2004) Molecular cloud origin for the oxygen isotope heterogeneity in the solar system. *Science, 305,* 1763–1766.

Zhai M., Nakamura E., Shaw D. M., and Nakano T. (1996) Boron isotope ratios in meteorites and lunar rocks. *Geochim. Cosmochim. Acta, 60,* 4877–4881.

Zolensky M., Nakamura K., Weisberg M. K., Prinz M., Nakamura T., Ohsumi K., Saitow A., Mukai M., and Gounelle M. (2003) A primitive dark inclusion with radiation-damaged silicates in the Ningqiang carbonaceous chondrite. *Meteoritics & Planet. Sci., 38,* 305–322.

Solar System Deuterium/Hydrogen Ratio

François Robert
Muséum National d'Histoire Naturelle

The D/H variations in chondritic meteorites are reviewed and compared with astronomical observations in the interstellar medium. The observed D enrichment in organic molecules — and to a lesser extent in water — is a general rule in the universe. The D is preferentially transferred from the universal reservoir (H_2) by low-temperature reactions (10–100 K) from neutral molecules (H_2 or H) to hot radicals or ions. The signature of this process is preserved in meteorites and comets. Although this D-enrichment procedure is well understood, in this review a special emphasis is placed on the unsolved issues raised by solar system D/H data. Specifically, it is shown that (1) the usual interstellar interpretation for the origin of the D-rich compounds in chondrites is not quantitatively substantiated and (2) the isotopic heterogeneity observed in LL chondrites is difficult to reconcile with any known natural reservoir or process.

1. INTRODUCTION

The origin of organic compounds and water in the solar system is an important problem that has increasingly attracted the interest of astronomers and cosmochemists following the recognition that H-bearing molecules in comets, planets, and chondritic meteorites show a systematic D enrichment relative to the molecular hydrogen of the solar nebula (*Robert et al.,* 1979, 1987; *Kolodny et al.,* 1980; *Robert and Epstein,* 1982; *McNaughton et al.,* 1981, 1982; *Kerridge,* 1985, 1987; *Deloule and Robert,* 1995, 1998). Because there is no nuclear source for D in the universe (*Epstein et al.,* 1976; *Galli,* 1995), the observed isotopic enrichment must have its origin in chemical reactions having faster reaction rates for D than for H (deuterium and hydrogen, respectively). Such processes lead to "kinetic isotopic fractionation." The D/H ratio in H_2 (or H) in the interstellar medium (ISM) represents the universal isotopic abundance ratio, the D-rich molecular species representing only a minute fraction of the total H present in the ISM. Therefore, deciphering the origin of D/H variations in organic compounds and in water in the solar system hinges on a correct identification of these reactions and their respective magnitude of isotopic fractionation.

The isotopic *exchange* between water and molecular hydrogen is a classical example that can be used for the definition of the isotopic fractionation parameters (*Geiss and Reeves,* 1972; *Lécluse and Robert,* 1996)

$$HD + H_2O \underset{k_r}{\overset{k_f}{\rightleftarrows}} HDO + H_2 \qquad (1)$$

The D enrichment factor f in equation (1) is defined as

$$f_{(H_2-H_2O)} = \{1/2[HDO]/[H_2O]\}/\{1/2[HD]/[H_2]\} = (D/H)_{H_2O}/(D/H)_{H_2} \qquad (2)$$

The factor 1/2 is explicitly written in equation (2) to emphasize that the molecular isotopic abundance in H_2O and H_2 is twice the elemental D/H ratio. Equation (1) does not imply that the thermodynamical equilibrium is reached between the reactants. The notation $f_{(H_2-H_2O)}$ indicates that the isotopic exchange takes place between H_2O and H_2 (this notation will be used hereafter for all types of reactions). In this reaction, water is systematically enhanced in D relative to molecular hydrogen; i.e., $f_{(H_2-H_2O)} \geq 1$. Under equilibrium conditions, the enrichment factor, f_{equi}, stands for the forward to reverse reaction rate ratio; i.e., $f_{equi} = k_f/k_r$. f_{equi} depends exclusively on the temperature (*Richet et al.,* 1977). In the geochemical literature f_{equi} is often noted as $\alpha(T)$ and is referred to as "the isotopic fractionation factor." Reaction (1) can be extended to all kinds of H-bearing molecules (organic molecules, ions, etc.) and (almost) always yields an enrichment in D in H-bearing molecules relative to H_2; i.e., $f_{(H_2-XH)} \geq 1$.

In space, isotopic exchange reactions can take place primarily in three environments: (1) In the solar nebula via a thermal isotopic exchange between molecular hydrogen and H-bearing compounds (*Geiss and Reeves,* 1972; *Lécluse and Robert,* 1994). (2) In the dense ISM at T < 50 K via isotopic exchanges between ionized species and molecules (*Brown and Millar,* 1989; *Brown and Rice,* 1981; *Watson,* 1976; *Willacy and Millar,* 1998; *Yung et al.,* 1988). (3) In denser interstellar clouds (the so-called hot cores), at intermediate temperatures (T < 200K), via isotopic exchanges between radicals (H or D) and neutral molecules (*Rodgers and Millar,* 1996). Calculations and/or experimental determinations of the f values for different molecules and for these three types of environments are available in the literature. The present paper compares these theoretical f values with the galactic and planetary data in order to investigate the possible relations between solar system and interstellar molecules.

Unequilibrated meteorites contain clay minerals (≈90% of their bulk H content) and insoluble organic macromol-

TABLE 1. Selected D/H ratios in the galaxy and solar system.

D/H (× 10^6)	Species	Location	f	References
Galactic[6]				
20–50	H	Big Bang (theoretical)		*Schramm* (1998)
16 ± 1	H	Local interstellar medium		*Linsky* (1993)
110 (+150/–65)	H_2O	Hot cores	7	*Gensheimer et al.* (1996)
900–43[8]	H_2O	ISM (20–80 K, respectively)	36–1.7	*Brown and Millar* (1989)
1 (+3/–0.2) × 10^3	OM[1]	Interstellar clouds (hot cores)	63	Personal compilation[3]
2 (+0.7/–1.8) × 10^4	OM[1]	Cold interstellar clouds	(<20 K) 1250	Personal compilation[3]
Proto-Sun				
25 ± 5	H_2	Protosolar nebula 4.5 × 10^9 yr ago	=1	*Geiss and Gloecker* (1998)
Gaseous Planets[7]				
21 ± 8	H_2	Jupiter (spectroscopic)	=1	
26 ± 7	H_2	Jupiter (MS *in situ*)	=1	*Mahaffy et al.* (1998)
15–35	H_2	Saturn (spectroscopic)	=1	*Griffin et al.* (1996)
65 (+2.5/–1.5)	H_2	Neptune (spectroscopic)	2.6	*Feuchtgruber et al.* (1998)
55 (+35/–15)	H_2	Uranus (spectroscopic)	2.2	*Feuchtgruber et al.* (1998)
Comets[7]				
310 ± 30	H_2O	Comet P/Halley	12.4	*Eberhardt et al.* (1995)
290 ± 100	H_2O	Comet Hyakutake	11.6	*Bockelée-Morvan et al.* (1998)
330 ± 80	H_2O	Comet Hale-Bopp	13.2	*Meier et al.* (1998a)
2300 ± 400	OM[1]	HCN in Comet Hale-Bopp	92	*Meier et al.* (1998b)
Interplanetary Particles[7] (IDPs and Antarctic Micrometeorites)				
90–4000		Bulk	3.6–160	Personal compilation[3]
1500–4000	IOM		60–160	*Messenger* (2000)
150–300	H_2O	(four determinations)	6 to 12	*Aléon et al.* (2001)
Telluric Planets[7]				
149 ± 3	H_2O	Bulk Earth	6	*Lecuyer et al.* (1998)
16 000 ± 200	H_2O	Venus (atmosphere *in situ*)	640	*Donahue et al.* (1982)
780 ± 80	H_2O	Mars (atmosphere *in situ*)	31	*Owen et al.* (1988)
100–780	–OH	SNC (Mars mantle?)	4–31	*Leshin et al.* (1994); *Gillet et al.* (2002)
LL3.0 and LL3.1 Meteorites[7]				
730 ± 120	–OH	in clays and in chondrules ($MgSiO_3$)	29	*Deloule and Robert* (1995)
88 ± 11	–OH	in chondrules (glass)	3.5	*Deloule et al.* (1998)
800–1100	IOM[1]	IOM	32–44	Personal compilation[3]
Carbonaceous Meteorites[7]				
140 ± 10	–OH	Mean statistical value[2]	5.6	Personal compilation[3]
380–620	IOM[1]	(CM,CV,CR)[4]	15–25	Personal compilation[3]
370 ± 6	IOM[1]	(Orgueil CI)[5]	14.8	*Halbout et al.* (1990)
170 ± 10	–OH	(Orgueil CI)[5]	6.8	Personal compilation[3]
315–545	SOM[1]	Amino-acids from CM	12.5–22.0	*Pizzarello et al.* (1991)
185–310	SOM[1]	Hydrocarbons, carboxylic acids from CM	7.5–12.5	*Epstein et al.* (1987)

The f value is defined as $(D/H)_{sample}/(D/H)_{reference}$. The "reference" D/H ratio (f = 1) stands for the molecular hydrogen reservoir from which the H-bearing compounds are formed (see notes [6] and [7]).
Notes: [1] IOM = insoluble organic matter; SOM = soluble organic matter; [2] Statistical distribution: 130 < D/H < 170 (66%); [3] F. Robert, personal compilation of published data; [4] CM, CV, CR: chondrite types; [5] reference meteorite sample for the protosolar chemical composition; [6] $(D/H)_{reference} = 16 \times 10^{-6}$; [7] $(D/H)_{reference} = 25 \times 10^{-6}$; [8] theoretical values.

ecules (hereafter referred to as IOM; ≈10% of H) with IOM systematically enriched in D by a factor of 2 to 3 relative to the minerals. The soluble organic molecules are present at the 100-ppm level and their isotopic compositions are presented in another chapter of this book. Because only a few known natural processes are able to cause such large isotopic heterogeneity, a knowledge of the internal distribution of the D/H ratio in chondrites might become an efficient tool in reconstructing the conditions of synthesis of water and organic molecules. However, little progress has been made in this direction because of the complexity of the organic materials and because of a lack of experimental efforts to determine precisely the distribution of the D/H ratio at a molecular or mineralogical level.

2. NATURAL VARIATIONS OF THE DEUTERIUM/HYDROGEN RATIO

In this section, the natural variations of the D/H ratio in stars, ISM, planets, meteorites, and comets are reviewed. Selected data are reported in Table 1.

2.1. The Universal Deuterium/Hydrogen Ratio

Deuterium formed during the "Big Bang" and has subsequently been destroyed in stars. The formation of D in supernova shocks, if possible, does not seem relevant in the galactic context (*Epstein et al.,* 1976). Therefore the D/H ratio of the ISM should have decreased with time as D-free H has been injected into space by supernovae or stellar winds (*Galli et al.,* 1995). However, exact estimates of the overall decrease in D abundance in the galaxies with time remain controversial. Measurements with the Hubble Space Telescope have given an accurate determination of the local ISM: *Linsky et al.* (1993) reports D/H ratios of $16.5 \pm 1.8 \times 10^{-6}$ and $14.0 \pm 1 \times 10^{-6}$ within 1 kpc of the Sun and *McCullough* (1991) reported 14 ratios averaging at 15×10^{-6}. The remarkable agreement between these two sets of data suggests that there is a single D/H ratio for the average ISM (cf. review by *Linsky,* 2003). Note, however, that *Gry et al.* (1983) or *Vidal-Madjar et al.* (1983) have argued that the D/H ratio may vary according to the line of sight and that the concept of a single ratio for the ISM might be meaningless. In this text, $(D/H)_{H_2} = 16 \pm 1 \times 10^{-6}$ is used as a reference ratio to calculate the different f values for the present-day interstellar molecules (see equation (2) and Table 1).

The standard Big Bang model of nucleosynthesis predicts a universal abundance of D that, in practice, depends on the baryon-to-photon ratio (*Schramm,* 1998). The initial D/H ratio of the universe could be around 50×10^{-6} (*Geiss and Gloeckler,* 1998) (see Table 1).

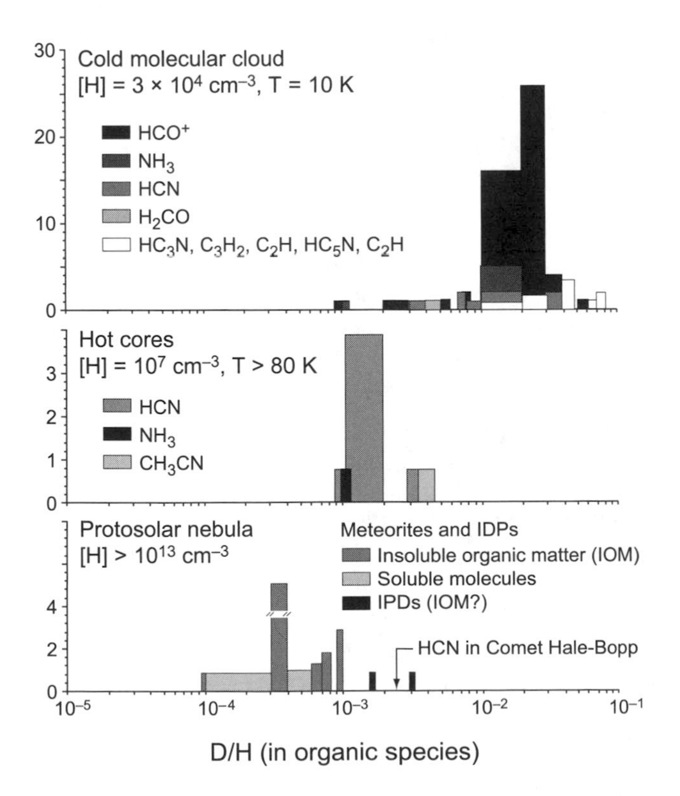

Fig. 1. Distributions of the D/H ratio in organic species for three different environments: ISM (cold molecular clouds), hot cores, and solar system (protosolar nebula). The origin of the difference in the D/H ratios between ISM and solar system organic molecules is still an open issue (see discussion in text).

2.2. The Present-Day Interstellar Deuterium/Hydrogen Ratio

A large enrichment in D is observed in dense molecular clouds in all detectable C-bearing molecules (so-called organic molecules). In these clouds $T \approx 10$ K and $H < 10^3$ cm^{-3}. No reliable data exist on H_2O because the low density of HDO in the gas phase prevents its detection. L. D'Hendecourt and F. Robert (unpublished results) estimated from laboratory spectra of deuterated ice the detection limit of HDO in the infrared spectra of the ISM recorded by the Infrared Space Observatory (ISO). According to this laboratory calibration, the D/H ratio of interstellar water vapor should be $\leq 9 \times 10^{-4}$. A compilation of the isotopic composition of various interstellar organic molecules is reported in Fig. 1 in the form of histograms; weighted means are reported in Table 1. The measured large D enhancement ($f > 1000$) does not seem explainable by grain-surface chemistry or by gas-phase reactions involving neutrals, but might result from ion-molecule reactions (*Watson,* 1976; *Guélin et al.,* 1982; *Brown and Millar,* 1989; *Duley and Williams,* 1984; cf. review by *Millar,* 2003). Ions are formed in the gas by UV irradiation from nearby stars;UV photons can penetrate deep inside molecular clouds. At a first approximation, the enrichment in D reflects the H_2D^+/H_3^+ and CH_2D^+/CH_3^+ ratios that result from

$$H_3^+ + HD \leftrightarrow H_2D^+ + H_2 \quad (3)$$

$$CH_3^+ + HD \leftrightarrow CH_2D^+ + H_2 \quad (4)$$

For these two reactions, $f_{equi\ (H_3^+-H_2)} = \exp(227/T)$ and $f_{equi\ (CH_3^+-H_2)} = \exp(370/T)$, respectively (*Dishoeck,* 1999). In the form of H_2D^+ and CH_2D^+, D is transferred to the organic molecules via reactions such as

$$H_2D^+ + CO \rightarrow DCO^+ + H_2 \quad (5)$$

$$CH_2D^+ + N \rightarrow DCNH^+ + H_2 \quad (6)$$

leading, in these examples, to $f_{(HCO^+-H_2)}$ and $f_{(HCN-H_2)} \geq 1000$ for $T < 50$ K (*Guélin et al.,* 1982; *Duley and Williams,*1984). The steady state is reached in $10^{6\pm1}$ yr, consistent with the estimated lifetime of the cold molecular clouds. *Brown and Millar* (1989) and *Millar et al.* (1989) have shown that, besides reactions (3) and (4), many other reactions are involved in the transfer of D from HD to ions (for example, $C_2H_2^+$ + HD). They have calculated that the measured distribution of D/H ratios in most observed organic molecules corresponds to reaction durations close to 3×10^5 yr. In recent and more detailed models, numerous additional reactions are considered (*Willacy and Millar,* 1998), but these efforts have not changed the main findings described above: Deuterium chemistry in cold interstellar clouds seems well understood.

Determinations of the D/H ratios in organic molecules and in water have been reported in molecular "hot cores." Hot cores are warm (>100 K) and dense ($H > 10^7$ cm^3) and result — as in the case of Orion-KL — from the formation of young massive stars that heat the surrounding ISM. Spatial resolution of the D/H ratio in HCN molecules reveals the occurrence of a transition between the cold molecular cloud and the hot core region (*Hatchell et al.,* 1998; *Schilke et al.,* 1992): the HCN D/H ratio is systematically lower in hot cores than their cold interstellar counterpart (see Fig. 1). It has been proposed that, in hot cores, water results from the evaporation of icy mantles that were condensed at much lower temperature (≈10 K) in the surrounding ISM (*Rodgers and Millar,* 1996). In hot cores the composition of the gas is mainly neutral (*Schilke et al.,* 1992) and the D exchange proceeds via reactions such as

$$H + DCN \leftrightarrow HNC + D \tag{7}$$

During this reaction $f_{(HCN-H)}$ decreases with time because D-rich HCN molecules reequilibrate with H at ~150 K and becomes isotopically lighter than the surrounding cold ISM (≈10 K). A similar situation seems to exist for H_2O: Measured D/H ratios in H_2O in hot cores are systematically lower than those calculated for H_2O in cold molecular clouds. Therefore it has been proposed that, after its evaporation from the grains, water exchanges its D with atomic H, producing a decrease in the D/H ratio as the temperature increases.

2.3. The Early Sun and the Giant Gaseous Planets

The D/H ratio in the solar nebula ($25 \pm 5 \times 10^{-6}$) is derived from from two independent lines of evidence: (1) the jovian and saturnian D/H ratios and (2) the present-day solar $^3He/^4He$ and $^4He/H$ ratios.

1. Numerous spectroscopic determinations of the D/H ratio in the upper atmospheres of the giant planets have been attempted (*Beer and Taylor,* 1973, 1978; *Bezard et al.,* 1986; *de Bergh et al.,* 1986, 1990; *Feuchtgruber et al.,* 1997, 1998; *Griffin et al.,* 1996; *Lellouch et al.,* 1996; *Mahaffy et al.,* 1998; *Niemann et al.,* 1996; *Smith et al.,* 1989a,b; cf. review by *Owen and Encrenaz,* 2003). According to *Gautier and Owen* (1983) the D/H ratio of the two planets should reflect the value of the solar nebula. In these planets, D is essentially in the form of HD whose abundance has been measured by the Infrared Space Observatory. These observations are in good agreement with the HD/H_2 ratio measured *in situ* by the Galileo probe mass spectrometer (D/H = $26 \pm 7 \times 10^{-6}$; see Table 1) (*Niemann et al.,* 1996).

2. The D in the early Sun has been converted in ^{3}He by the thermonuclear reaction $D + H \rightarrow {}^3He$. Therefore the D/H ratio of the early Sun can be derived from the present-day $^3He/^4He$ measured in the solar wind ($^3He/^4He_{solar} \approx 3.8 \pm 0.5 \times 10^{-4}$) (*Geiss and Gloecker,* 1998), provided the initial $^3He/^4He$ ratio, $^3He/^4He_{sn}$, of the Sun is known

$$(D/H)_{early\ Sun} = ({}^3He/{}^4He_{solar} - {}^3He/{}^4He_{sn}) \times ({}^4He/H)_{sn} \tag{8}$$

The $(^3He/^4He)_{sn}$ estimated from chondrites or from Jupiter data is the most inaccurate parameter in equation (8). Using the recent jovian estimates, a $(^3He/^4He)_{sn} \approx 1.5 \times 10^{-4}$ (*Mahaffy et al.,* 1998) corresponds to a $(D/H)_{early\ Sun} = 21 \pm 5 \times 10^{-6}$ (*Geiss and Gloecker,* 1998) [with $(^4He/H)_{sn} \approx 10^{-1}$].

Combining the two determinations we adopt the usual canonical value for the solar nebula D/H ratio (25×10^{-6}). This ratio is therefore used here to calculate the f values reported in Table 1 for all solar system data; i.e., $(D/H)_{H_2} = 25 \times 10^{-6}$ in equation (2).

As noted previously, there is a slight difference between this solar ratio and the present-day interstellar ratio (D/H = 16×10^{-6}). This difference is caused by the destruction of ISM D in stars since the solar system formed, 4.5 G.y. ago. The latter ratio is consistent with the galactic rate of destruction with time of D in stars (*Geiss and Gloecker,* 1998). Therefore for an identical D/H ratio in the ISM and in the solar system, the calculated enrichment factors f are different.

2.4. The Water-rich Giant Planets

Uranus and Neptune have large icy cores. They exhibit D/H ratios significantly higher than in Jupiter and Saturn (*Feuchtgruber et al.,* 1997). An interpretation of this enrichment is based on formation models of these two planets where the cores of the planets grew up by accretion of icy planetesimals with high (D/H) ratio. Assuming that in the interior, water from the planetesimals and molecular hydrogen from the gaseous envelope was isotopically equilibrated at high temperature [$f_{equi\ (H_2-H_2O)} = 1$] at least once during the lifetime of the planet, the initial (D/H) ratio of the water must have been somewhat higher than the present-day (D/H) ratio in the H_2,but lower than the cometary (D/H) ratio. In this model, comets stand as examples of the icy planetesimals at the origin of these two planets. However, the isotopic equilibrium assumption may not apply if the planets are not fully convective.

2.5. The Terrestrial Planets

In the case of Mars and Venus, the photodissociation of water in their upper atmospheres produces H and D that is subsequently lost to space. Since H is lost faster than D, the two atmospheres have been enriched in D over the last

4.5 G.y. by a Rayleigh distillation process (*Donahue et al.,* 1982; *Owen et al.,* 1988). However, the theory is too imprecise to accurately derive the primordial D/H ratio of these planets. In SNC meteorites, hydroxylated minerals exhibit D/H ratios similar to the martian atmospheric value (*Leshin et al.,* 1994, 1996). However, *Gillet et al.* (2002) have shown that other mafic minerals in SNCs have D/H ratios even lower than the terrestrial mantle. Thus the exact value of the pristine martian D/H ratio is still considered to be an open question.

The D/H ratio on Earth has been recently estimated accurately [D/H = 149 ± 3 × 10^{-6} (*Lécuyer et al.,* 1998)]. Escape from the top of the atmosphere should not have decreased the D/H ratio of the oceans significantly (<1%) since the end of the planetary accretion. This has been shown theoretically [by modeling the flux of H escaping to space (*Hunten and McElroy,* 1974)] and observationally [on Earth, kerogen and mineral D/H ratios are invariable within ±1% through geological times (e.g., *Robert,* 1989)].

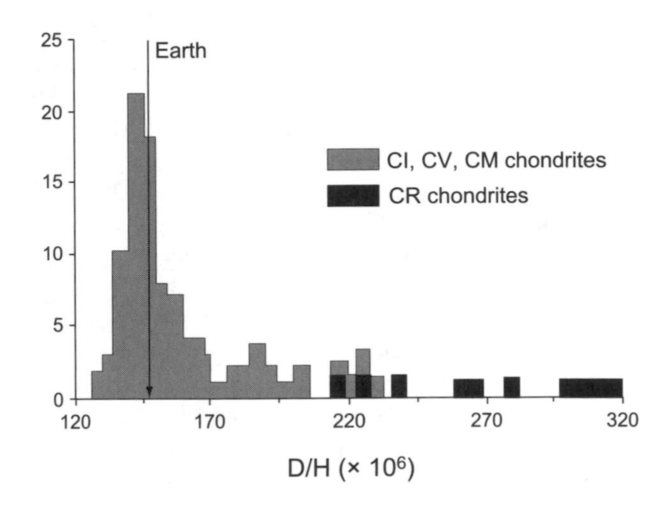

Fig. 2. Distribution of the D/H ratio in whole rock carbonaceous meteorites (N = 106). The large range in D/H likely results from variations in the alteration water D/H ratio. This water may have recorded different degree of isotope exchange during parent body hydrothermalism, with D-rich organic compounds. The tail toward high D/H ratios is likely related to a heliocentric distribution, with the highest D/H ratios located in the coldest and remote regions of the solar system (see IDPs in Fig. 3).

2.6. Meteorites and Some Remarks on Interplanetary Dust Particles

2.6.1. Variations at the whole-rock scale. Little attention has been paid to the origin of variations in the whole-rock H-isotopic composition of carbonaceous chondrites [see the review of the available literature data by *Robert* (2003)]. These meteorites share several common isotopic characteristics: (1) At a bulk scale they are all enriched in D with f values between 5 and 12. (2) They contain IOM systematically enriched in D by a factor 2 to 3 relative to the minerals (*Robert and Epstein,* 1982; *Kerridge et al.,* 1987; *Epstein et al.,* 1987; *Robert et al.,* 1987; *Pizzarello et al.,* 1991).

In Fig. 2, the D/H ratio of the carbonaceous chondrites is reported as a histogram. In meteorites H is present as hydroxyls in phyllosilicates (and to a lesser extend bound to C in organic molecules) but it is collected in the form of water in most experiments when the sample is outgassed by step-heating. All data in Fig. 2 were obtained by outgassing bulk samples under vacuum or in presence of He or molecular oxygen and by analyzing the mass spectrum of the collected molecular hydrogen after the reduction of water.

What fraction of water extracted from an extraterrestrial sample is terrestrial contamination? Surprisingly, detailed experimental studies indicate that this fraction is small, i.e., that the relative contribution of the terrestrial contamination is <15% (*Robert and Deloule,* 2002). These studies have been performed by comparing the D/H ratio of a sample in contact with liquid D-rich water to the same sample in contact with liquid water having a terrestrial D/H ratio. For example, after outgassing the Orgueil meteorite (≈10 wt% H_2O) overnight at 50°C under vacuum, water extracted above 50°C shows almost no (additional?) contamination [≤0.04 wt% H_2O (*Engrand et al.,* 1999)]. Similarly, altered olivine in the Semarkona meteorite shows a relative contamination fraction ≤15%, although the water concentration in this mineral does not exceed 0.2 wt% H_2O (*Robert and Deloule,* 2002). These studies have addressed the specific problem of the terrestrial contamination by isotopic exchange. However, it should be kept in mind that alteration on Earth of meteoritic minerals — by contact with liquid water — yields the addition of terrestrial H in the form of hydroxyl groups. This addition is irreversible at room temperature and the resulting –OH cannot be separated from the indigenous –OH by a simple stepwise heating pyrolysis.

Most samples reported in Fig. 2 are "falls" and thus have never been in contact with liquid water on Earth. Therefore the range in D/H ratio reported in Fig. 2 cannot result from different degrees of terrestrial contamination.

In several CI and CM meteorites (Cold-Bokkeveld, Murray, Murchison, Orgueil) the IOM has been chemically extracted and purified and its D/H ratio has been determined by stepwise temperature pyrolysis (*Robert and Epstein,* 1982; *Kerridge,* 1987; *Yang and Epstein,* 1983). Although there are large uncertainties about these determinations because of the contamination linked to the acid-extraction procedures, the IOM D/H ratios lie between 310 and 370 × 10^{-6} for these meteorites. *Halbout et al.* (1990) have argued that all these extracts have a similar indigeneous D/H ratio of 370 × 10^{-6}, with lower values due to acid contamination. In the case of the Renazzo meteorite, the IOM D/H ratio is markedly different from CI-CM with a ratio of 550 × 10^{-6}. As observed in Fig. 2 the Renazzo-type meteorites (reported as CR meteorites) are systematically enriched in D relative to CI-CM. Mass-balance calculation using the C concentration in the bulk samples and the C/H ratio in the OM indicates that the clay minerals have a D/H ratio close to 300 ×

10^{-6} in Renazzo. Thus, the difference between CI-CM and CR cannot be attributed to the presence of D-rich IOM but must be due to an enhancement in D abundance at the scale of the bulk sample. The most puzzling aspect of the variations reported in Fig. 2 is that they greatly exceed the variations observed on Earth for clay minerals formed at temperatures between 0° and 100°C.

An isotopic fractionation could occur during liquid water circulation in a warm parent body (*Bunch and Chang,* 1980; *Sears et al.,* 1995). As a general rule, clay minerals are depleted in D relative to water from which they formed. As a rule of thumb, for temperatures <100°C one can add $\approx 10 \times 10^{-6}$ to the measured D/H ratio of the minerals in order to obtain the D/H ratio of the water in isotopic equilibrium with that mineral. This correction is small compared to the total variations reported in Fig. 2. Therefore these variations cannot be attributed to different degrees of isotopic exchange with a water reservoir having a unique D/H ratio.

Part of the observed variations in bulk samples may be caused by different mixing ratios between clays and OM. The maximum value for the relative contribution of the organic hydrogen is ≈10% (i.e., 90% of H is hosted by clays), corresponding to a shift in the D/H ratio of $\approx 20 \times 10^{-6}$, i.e., 1 order of magnitude smaller than the total D/H distribution reported in Fig. 2. Several authors have attempted to relate the C/H ratio to the D/H ratio of the carbonaceous chondrites in order to trace the relative contribution of the organic matter to the bulk D/H ratio. Although positive correlations have been found between these two ratios (*Kerridge,* 1985), these correlations have been met with several counter examples. Thus, the variations in Fig. 2 cannot simply be caused by the differences in the mixing ratio between water and organic matter.

Two small peaks are present in the tail of the distribution in Fig. 2, at D/H equal to 190 and 220×10^{-6}. In the distribution reported here, all analyses have been used and thus the same meteorite can be counted more than once. However, reporting an average value for each meteorite — i.e., having a unique D/H ratio per meteorite — does not alter the peaks at D/H equal to 190 and 220×10^{-6}. This observation reinforces the statistical value of the distribution. As a firm conclusion of this discussion, the asymmetry of the distribution reported in Fig. 2 results from a wide range of D/H ratios both in IOM and in clays. *Eiler and Kitchen* (2004) have shown that the least-altered CM samples (e.g., Murchison) have D/H ratios in their matrix and whole rock close to 150×10^{-6}. Conversely, both whole-rock and matrix samples of altered CM chondrites and their chondrules (e.g., Cold Bokkeveld) have D/H close to 125×10^{-6}. The degree of alteration in these samples is traced by several geochemical criteria that correlate with the D/H ratio. Following this interpretation, the lowest D/H ratio of the distribution reported in Fig. 2 stands for the isotopic composition of water that was circulating in the parent bodies and that was at the origin of the mineralogical alteration features; i.e., water D/H $\approx 120 \times 10^{-6}$. In this interpretation, each whole-rock sample is an open system and a large fraction of the alteration water has been lost, perhaps to space.

Finally, it should be noted that the observed distribution of the bulk sample D/H ratio (Fig. 2) could be biased by the dynamics of meteorite delivery to Earth. Fragments of disrupted asteroids have a greater chance to intercept the Earth's orbit if the parent bodies were initially located at distances <2.5 AU (*Vokrouhlický and Farinella,* 2000). At greater distances from the Sun, the fragments have a high probability of being expelled in the outer regions of the solar system. It is thus unlikely that the asymmetry in Fig. 2 reflects the relative proportion of parent bodies in the meteorite source [i.e., the asteroid belt (cf. *Meibom and Clark,* 1999)], which instead might simply reflect the probability distribution for collecting samples on Earth as a function of the distance of their sources. The D/H distribution of IDPs supports this interpretation (Fig. 3). Although the sources of IDPs extend far beyond the asteroid belt, their distribution mimics that of the carbonaceous chondrites. Therefore, parent bodies located in the far regions of the solar system have higher D/H ratios than those in the inner asteroid belts.

2.6.2. Molecular and mineralogical hosts of the deuterium/hydrogen ratio. There are two types of primitive meteorites in which high D/H ratios have been found: LL3 and

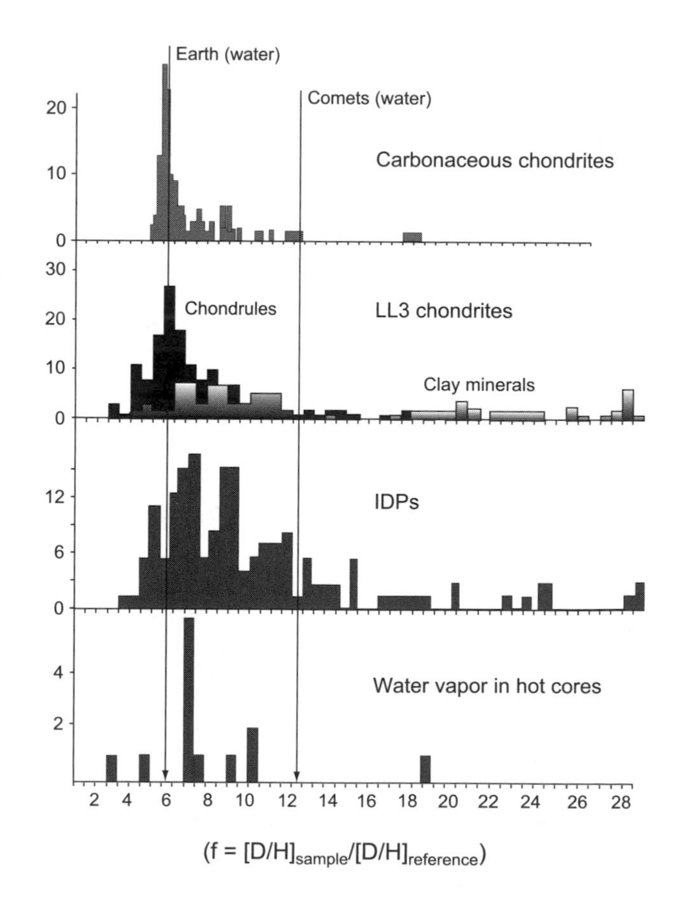

Fig. 3. Distributions of the D/H ratio in (1) whole-rock samples of carbonaceous chondrites, (2) LL3 chondrites (chondrules and clay minerals in the matrix), (3) IDPs, and (4) water vapor from hot cores (*Jacq et al.,* 1990; *Gensheimer et al.,* 1996; personal compilation of published data). The f unit is used to normalized the enrichment in D relative to universal H: D/H = 25×10^{-6} and 16×10^{-6} for meteorites-comets-IDPs and hot cores, respectively.

carbonaceous chondrites (referred hereafter to as CCs). The IOM has been extracted chemically from these two types of rocks and has been found to be systematically enriched in D: $f_{(H_2-IOM)}$ = 15–25 in CCs (with most values around 15; i.e., D/H = 375 to 625 × 10^{-6}) and $f_{(H_2-IOM)}$ up to 44 in LL3 (D/H ≈ 1100 × 10^{-6}). The possible contamination of IOM during the chemical procedure has been carefully evaluated and is negligible (<7%) (*Halbout et al.,* 1990).

The usual interpretation of this isotopic enrichment is "interstellar chemistry." However, as shown in Fig. 1, there is a marked gap in D/H ratios between interstellar and solar system organic matter. Such a difference is impossible to reconcile with a simple temperature effect; i.e., the higher the temperature of ion-molecule reactions, the lower the D/H ratio. The corresponding calculated temperature (≈120 K using reaction (4)) is never reached in the ISM (except around massive hot stars). In addition, if such a high calculated temperature corresponded to a protosolar environment, the corresponding gas density would prevent H from being ionized. Therefore the "interstellar interpretation" should be considered more as an analog with the ISM, rather than as a quantitative model supported by facts.

Interstellar medium organic molecules are simple. Chondritic IOM is highly polymerized. It does not seem plausible that the polymerization of such simple D-rich molecules would be associated with a decrease in their D/H ratios because (1) polymerization does not yield an addition of H (or D) and (2) the isotopic exchange between organic H and molecular H_2 is prohibitively slow at temperatures <250 K. A solution to this problem was proposed (see section 3): In the protosolar nebula, the organic polymer D/H ratio may have decreased by an isotopic exchange with molecular H_2 at temperatures as high as 650 K. However, this issue should nevertheless be considered open because (1) no D/H determination has been reported for interstellar polycyclic aromatic hydrocarbon (PAH) and (2) although several types of ISM synthesis have been proposed to account for the chondritic IOM [UV photolysis of D-rich ices or gas-grain reactions (*Sandford et al.,* 2001; *Sandford,* 2002)], a quantitative estimate of their corresponding D/H ratios has not been reached.

In the case of the Murchison carbonaceous chondrite, the soluble organic fraction (amino-acids, fatty acids, hydrocarbons; see Table 1) has been separated from the bulk sample by organic solvent procedures and yields $f_{(H_2-SOM)}$ = 12.5–22 [D/H = 310 to 550 × 10^{-6} (*Epstein et al.,* 1987; *Pizzarello et al.,* 1991)]. Such D/H ratios are comparable with those measured in the IOM. Their possible "interstellar" origins face similar issues as those previously mentioned for the IOM.

After the formation of CM and CI chondrites a late circulation of water occurring at (or near) the surface of their parent bodies could have formed the clay minerals (*Bunch and Chang,* 1980). An often-quoted terrestrial analogy for this mechanism is "hydrothermal alteration." Consequently, the D/H ratio of these minerals can be used to determine the water D/H ratio because the isotopic fractionation between clay minerals and liquid water is negligibly small (≤6.5%) at the scale of the isotopic variations in the solar system. The D/H ratios of these clay minerals has been determined by two methods: (1) by mass balance in CCs [i.e., by subtracting from whole-rock analysis the measured IOM D/H ratio (cf. *Robert and Epstein,* 1982)] or (2) by *in situ* measurements with the ion microprobe (*Deloule and Robert,* 1995; *Deloule et al.,* 1998). As general rule, whole-rock D/H ratios in CCs can be taken at face value for the D/H ratio of water that has circulated in these bodies (within ±10%). This rule clearly does not apply to the LL3 chondrites.

The LL3 chondrites contain D-rich (720 × 10^{-6}) and D-depleted (hydroxylated?) minerals (75 × 10^{-6}) mixed in various proportions with D-rich IOM (≤1000 × 10^{-6}) (*McNaughton et al.,* 1981; *Yang and Epstein,* 1983; *Deloule and Robert,* 1995). In LL3 meteorites, pyroxene in chondrules is enriched in D [$f_{(H_2-H_2O)}$ = 16; D/H = 400 × 10^{-6}] relative to other minerals in the same chondrule or relative to the clay minerals of the matrix [$f_{(H_2-H_2O)}$ = 3.5; D/H = 88 × 10^{-6}]. In the matrix surrounding the chondrules, some clay minerals also exhibit high D/H ratios [$f_{(H_2-H_2O)}$ = 29; D/H = 725 × 10^{-6}] that seem to have escaped — along with the pyroxenes in chondrules — the isotopic reequilibration during the late hydrothermal alteration. The mechanism for the incorporation and preservation of these high D/H ratios in chondrules is not understood, nor is the origin of such an isotopic heterogeneity among the matrix clay minerals that are considered to be the products of alteration by water that should be — by definition — isotopically homogeneous! Thus, contrary to CCs, no coherent classification can be proposed to account for the molecular distribution of the D/H ratio in LL3 meteorites.

Interplanetary dust particles have been collected in space and in Antarctic ice. They show extremely large variations in their D/H ratios [$f_{(H_2-H_2O)}$ = 3.6 to 161; D/H = 90 to 4000 × 10^{-6}] (*McKeegan et al.,* 1985; *Zinner et al.,* 1983; *Messenger,* 2000; *Engrand et al.,* 1999). The organic nature of the D-rich carriers have been unambiguously identified through the correlated variations of the elemental C/H and isotopic D/H ratios (*Aléon et al.,* 2001; cf. review by *Messenger et al.,* 2003). Water in IDPs has D/H ratios either close to the chondritic value or close to the cometary value for water. In the histogram in Fig. 3, it can be seen that bulk IDPs have a distribution of f values close to CCs, suggesting that cometary fragments are rare in the IDP population.

Important observations have been made through the analysis of IDPs at a 1-μm scale with the ion microprobe (*Aléon et al.,* 2001). In one IDP, three intimately mixed components have been identified: (1) water with D/H ≈ 150 × 10^{-6}, (2) organic molecules with C/H ratios close to those of the CCs and a D/H ≈ 500 × 10^{-6}, and (3) organic molecules with C/H ≈ 1 (an aliphatic molecule?) and D/H ≈ 1500 × 10^{-6} (f = 60). These observations suggest that, under the chemical and physical conditions that prevailed at the formation of IDPs, a wide range of D/H ratios were established in the primitive components of the solar system. It can be thus proposed that the precursors of these different species correspond to different ΔE values for ion-molecule reactions (ΔE is defined as the exothermicity in reactions such as reaction (3) or reaction (4)). The isotopic exchange reactions corresponding to these different ΔE values have not been yet identified but, as suggested by *Sandford et al.* (2001), may reflect the different isotopic fractionation pathways taking place during homogeneous (gas-gas) or heterogeneous (gas-grain) reactions. In other words, grains and gas may have carried a wide range of different D/H ratios.

2.7. Comets

The D/H ratios in water have been reported for three different comets (see Table 1) (*Meier et al.,* 1998a,b; *Eberhardt et al.,* 1995; *Bockelée-Morvan et al.,* 1988) with a mean $f_{(H_2-H_2O)}$ value = 12 (D/H = 300 × 10^{-6}). Such a value is a factor of 2 higher than the mean chondritic value but a factor of 2.5 lower than clay minerals in some LL3 chondrites. In HCN molecules (?), $f_{(H_2-HCN)}$ = 92; such a high f value does not have a chondritic counterpart [the maximum $f_{(H_2-IOM)}$ = 44 in LL3].

It can be seen (cf. Fig. 3) that the weighted mean composition of CCs corresponds precisely to that of Earth [$f_{(H_2-H_2O)}$ = 6]. If CCs are taken as the carrier of water on Earth, a minimum f value can be assigned for the primitive Earth: f ≥ 4 (i.e., the minimum f of the CCs distribution). This implies that (1) no major isotopic fractionation has occurred during the formation of the oceans and (2) no important (>10%) late addition of D-rich water has taken place during the subsequent evolution of the planet.

3. ORIGIN OF WATER IN THE SOLAR SYSTEM

The D/H variations in Fig. 3 correspond to D enrichments by factors of 5–12 relative to universal H_2. Three interpretations for this enrichment have been discussed in detail in the literature. They can be summarized by the following simple reactions:

1. Equilibrium isotopic exchange in the protosolar nebula between molecular hydrogen and water:

$$HD + H_2O \leftrightarrow HDO + H \quad (9)$$

2. Kinetic isotopic exchange in the protosolar nebula between D-rich water and molecular hydrogen:

$$H_2 + HDO \rightarrow HD + H_2O \quad (10)$$

In this case, the water is enriched in D prior to the formation of the solar nebula, presumably through interstellar ion-molecule reactions at low temperature.

3. Low temperature ion-molecule reactions in the ISM:

$$HD + H_3^+ \rightarrow H_2D + H_2 \quad (11)$$

The third reaction represents the first step of the synthesis of water whose D/H ratio reflects that acquired by H_3^+ through reaction (11).

The first two interpretations were addressed by *Drouart et al.* (1999) (see also *Dubrulle,* 1992; *Mousis et al.,* 2000; *Hersant et al.,* 2001; *Robert et al.,* 2000). Reaction (9) has been shown to be impossible because of kinetic limitations (*Lecluse and Robert,* 1994). Reaction (10) has been studied in detail in a model of a turbulent protosolar nebula where the temperature and pressure vary as a function of time and heliocentric distance. It has been shown that the water D/H ratio can decrease in such a nebula by a factor of 10 through the exchange reaction (10), from an initial value of 750 × 10^{-6} to a final value of 75 × 10^{-6}. Water condensed in the outer regions of the nebula should exhibit values ≈300 × 10^{-6} (i.e., the cometary value), while water condensed at a close distance from the Sun (1 AU) can reach a value of ≈75 × 10^{-6}. According to this model, the terrestrial and chondritic D/H value is reached in the protosolar nebula at 3–5 AU in ≤10^6 yr.

Reaction (11) was investigated empirically by *Robert* (2002), who suggested that ion-molecule reactions could have taken place at the surface of the protosolar nebula, where high ion concentration can be reached as a result of UV and/or X-ray irradiation from the early (T Tauri stage) Sun (*Glassgold et al.,* 2000). T. Millar (personal communication, 2003) has performed an unpublished detailed calculation of the D enrichment that can be reached in water through ion-molecule reactions. According to these calculations, the solar system water D/H ratio is also in agreement with that obtained by reaction (11) and would correspond to a formation temperature between 40 and 50 K at densities on the order of 10 H cm^{-3}. However, the corresponding D/H ratio in organic molecules — which can be calculated to first order through reaction (4) — exceed by far those measured in chondritic meteorites. Therefore water and organic matter D/H ratios seem to require different temperatures and densities. In this respect, the different interpretations are not consistent.

Following the interpretation according to which hot atom chemistry took place in the early solar system, *Aléon and Robert* (2004) have attempted to calibrate the temperature of ion-molecule reactions by correlating the D/H and $^{15}N/^{14}N$ ratios in solar system organic molecules from several classes of chondrites, cometary data, and IDPs (*Aléon et al.,* 2003); this gives 70 ≤ T ≤ 90 K. If correct, this implies that — contrary to models derived from reaction (10) — water and organic matter formed at the surface of the disk never thermally reequilibrated with H_2 in the protosolar nebula. This interpretation has not yet been numerically modeled in a turbulent protosolar nebula.

To conclude this section on chondritic data, we emphasize that no quantitative model has successfully accounted for the D/H ratio of coexisting water and organic molecules in comets and carbonaceous meteorites.

4. ORIGIN OF WATER ON EARTH

As extensively discussed in the literature (*Balsiger et al.,* 1995; *Bockelée-Morvan et al.,* 1998; *Meier et al.,* 1998a; *Eberhardt et al.,* 1995; *Robert,* 2001), detailed mass-balance calculations between the D/H ratio in CCs and the D/H ratio in comets show that cometary water cannot represent more than 10% [and perhaps as little as 0% (*Dauphas et al.*, 2000)] of the total budget of water on Earth (*Deloule et al.*, 1998). Two scenarios (*Drake and Righter*, 2002; *Morbidelli et al.*, 2000) have been proposed to import water on Earth during its accretion. The two scenarios are in conflict, but each one has interesting implications.

Morbidelli et al. (2000) proposed on the basis of the dynamical evolution of planetary embryos (1000-km-sized planetesimals) during the first 50 m.y. of the solar system that the carriers of water on Earth originated from ≈2.5 AU. Few impacts would have been sufficient to import the oceanic mass in the final stages of Earth accretion. This scenario is in agreement with the fact that the terrestrial water D/H ratio [$149 \pm 3 \times 10^{-6}$ (*Lecuyer et al.*, 1998)] and the mean D/H of the CCs (cf. Fig. 2 and 3) are indistinguishable.

Drake and Righter (2002) suggested that the water-rich planetesimals at the origin of the terrestrial oceans were formed at the same heliocentric distance as the silicate material that formed Earth. According to these authors, the chemical and isotopic compositions of many elements in these planetesimals have to be different from any known type of CCs in order to account for the present-day bulk composition of Earth.

The interpretation of Drake and Righter implies that the D/H ratio of water condensed at a terrestrial orbit had a D/H ratio similar to the mean value defined by the CCs. The Drouart et al. model shows that this may be indeed be the case (F. Hersan, unpublished data, 2002): In a turbulent nebula the D/H ratio in water should not vary significantly for heliocentric distances <2 AU.

5. FUTURE RESEARCH DIRECTIONS

A number of key experimental determinations could shed some light on the problems discussed above. A determination of the D/H ratio of the icy satellites of Jupiter and Saturn would allow a (de)verification that the D/H ratio is indeed correlated with heliocentric distance in the protosolar nebula. A determination of the D/H ratio of comets from the Kuipper belt region may yield new constraints to the problem of water delivery to the early Earth. Similarly, a determination of the D/H ratio of enstatite chondrites might bring us closer to understanding the isotopic composition of indigenous H on Earth, because these meteorites seem to have formed in conditions similar to those under which proto-Earth material formed. Finally, a determination of the distribution of the D/H ratio at a molecular level among the organic moieties of the IOM remains the best possible way to detect interactions between water and organic molecules in space or in meteoritic parent bodies.

REFERENCES

Aléon J. and Robert F. (2004) Interstellar chemistry recorded by nitrogen isotopes in solar system organic matter. *Icarus, 167,* 424–430.

Aléon J., Engrand C., Robert F., and Chaussidon M. (2001) Clues to the origin of interplanetary dust particles from the isotopic study of their hydrogen-bearing phases. *Geochim. Cosmochim. Acta., 65,* 4399–4412.

Aléon J., Robert F., Chaussidon M., and Marty B. (2003) Nitrogen isotopic composition of macromolecular organic matter in interplanetary dust particles. *Geochim. Cosmochim. Acta, 67,* 3773–3783.

Balsiger H., Altwegg K., and Geiss J. (1995) D/H and $^{18}O/^{16}O$ ratio in the hydronium ion and in neutral water from in situ ion measurements in comet P/Halley. *J. Geophys. Res., 100,* 5827–5834.

Beer R. and Taylor F. W. (1973) The abundance of CH_3D and the D/H ratio in Jupiter. *Astrophys. J., 179,* 309–327.

Beer R. and Taylor F. W. (1978) The D/H and C/H ratios in Jupiter from CH_3D phase. *Astrophys. J., 219,* 763–767.

Bezard B., Gautier D., and Marten A. (1986) Detectability of HD and non equilibrium species in the upper atmospheres of giant planets from their submillimeter spectrum. *Astron. Astrophys., 161,* 387–402.

Bockelée-Morvan D., Gautier D., Lis D. C., Young K., Keene J., Phillips T., Owen T., Crovisier J., Goldsmith P. F., Bergin E. A., Despois D., and Wooten A. (1998) Deuterated water in comet C/1996 B2 (Hyakutake) and its implications for the origin of comets. *Icarus, 193,* 147–162.

Brown P. D. and Millar T. J. (1989) Models of the gas-grain interaction-deuterium chemistry. *Mon. Not. R. Astron. Soc., 237,* 661–671.

Brown R. D. and Rice E. (1981) Interstellar deuterium chemistry. *Philos. Trans. R. Soc. London, A303,* 523–533.

Bunch T. E. and Chang S. (1980) Carbonaceous chondrites: II. Carbonaceous chondrites phyllosilicates and light element geochemistry as indicators of parent body processes and surface conditions. *Geochim. Cosmochim. Acta, 44,* 1543–1577.

Dauphas N., Robert F., and Marty B. (2000) The late asteroidal and cometary bombardment of Earth as recorded in water deuterium to protium ratio. *Icarus, 148,* 508–512.

de Bergh C., Lutz B. L., Owen T., Brault J., and Chauville J. (1986) Monodeuterated methane in the outer solar system. II. Its detection on Uranus at 1.6 μm. *Astrophys. J., 311,* 501–510.

de Bergh C., Lutz B., Owen T., and Maillard J.-P. (1990) Monodeuterated methane in the outer solar system. IV. Its detection and abundance on Neptune. *Astrophys. J., 355,* 661–666.

Deloule E. and Robert F. (1995) Interstellar water in meteorites? *Geochim. Cosmochim. Acta, 59,* 4695–4706.

Deloule E., Doukhan J.-C., and Robert F. (1998) Interstellar hydroxyle in meteorite chondrules: Implications for the origin of water in the inner solar system. *Geochim. Cosmochim. Acta, 62,* 3367–3378.

Dishoeck Van E. F. (1999) Models and observations of gas-grain interactions in star-forming regions. In *Formation and Evolution of Solids in Space* (J. M. Greenberg and A. Li, eds.), pp. 91–121. Kluwer, Dordrecht.

Donahue T. M., Hoffman J. H., Hodges R. R. Jr., and Watson A. J. (1982) Venus was wet: A measurement of the ratio of deuterium to hydrogen. *Science, 216,* 630–633.

Drake M. J. and Righter K. (2002) Determining the composition of the Earth. *Nature, 416,* 39–44.

Drouart A., Dubrulle B., Gautier D., and Robert F. (1999) Structure and transport in the solar nebula from constraints on deuterium enrichment and giant planet formation. *Icarus, 140,* 129–155.

Dubrulle B. (1992) A turbulent closure model for thin accretion disks. *Astron. Astrophys., 266,* 592–604.

Duley W. W. and Williams D. A. (1984) *Interstellar Chemistry.* Harcourt, Brace, and Jovanovich, New York.

Eberhardt P., Reber M., Krankowsky D., and Hodges R. R. (1995)

The D/H and $^{18}O/^{16}O$ ratios in water from comet P/Halley. *Astron. Astrophys., 302,* 301–316.

Engrand C., Deloule E., Robert F., Maurette M., and Kurat G. (1999) Extraterrestrial water in micrometeorites and cosmic spherules from Antarctica: An ion microprobe study. *Meteoritics & Planet. Sci., 34,* 773–786.

Epstein R. I., Lattimer J. M., and Schramm D. N. (1976) The origin of deuterium. *Science, 263,* 198–202.

Epstein S., Krishnamuty R. V., Cronin J. R., Pizzarello S., and Yuen G. U. (1987) Unusual stable isotope ratios in amino acids and carboxylic acid extracts from the Murchison meteorite. *Nature, 326,* 477–479.

Eiler J. M. and Kitchen N. (2004) Hydrogen isotope evidence for the origin and evolution of the carbonaceous chondrites. *Geochim. Cosmochim. Acta, 68,* 1395–1411.

Feuchtgruber H., Lellouch E., de Graauw T., Encrenaz Th., and Griffin M. (1997) Detection of HD on Neptune and determinations of D/H ratio from ISO/SWS observations. *Bull. Am. Astron. Soc., 29,* 995.

Feuchtgruber H., Lellouch E., Encrenaz Th., Bezard B., de Graauw T., and Davies G. R. (1998) Detection of HD in the atmospheres of Uranus and Neptune: A new determination of the D/H ratio. *Astron. Astrophys., 341,* L17–L21.

Galli D., Palla F., Ferrini F., and Penco U. (1995) Galactic evolution of D and ^{3}He. *Astrophys. J., 443,* 536–550.

Gautier D. and Owen T. (1983) Cosmological implication of helium and deuterium abundances of Jupiter and Saturn. *Nature, 302,* 215–218.

Geiss J. and Gloecker G. (1998) Abundances of deuterium and helium in the protosolar cloud. *Space Science Rev., 84,* 239–250.

Geiss J. and Reeves H. (1972) Cosmic and solar system abundances of deuterium and helium-3. *Astron. Astrophys., 18,* 126–132.

Gensheimer P. D., Mauersberger R., and Wilson T. L. (1996) Water in galactic hot cores. *Astron. Astrophys., 314,* 281–294.

Gillet Ph., Barrat J. A., Deloule E., Wadhwa M., Jambon A., Sautter V., Devouard B., Neuville D., Benzerara K., and Lesroud M. (2002) Aqueous alteration in the Northwest Africa (NWA 817) Martian meteorite. *Earth Planet. Sci. Lett., 203,* 431–444.

Glassgold A., Feigelson E. D., and Montmerle T. (2000) Effects on energetic radiation in young stellar objects. In *Protostars and Planets IV* (V. Mannings et al., eds.), pp. 429–455. Univ. of Arizona, Tucson.

Griffin M. J., Naylor D. A., Davis G. R., Ade P. A. R., Oldman P. G., Swinyard B. M., Gautier D., Lellouch E., Orton G. S., Encrenaz Th., de Graauw T., Furniss I., Smith I., Armand C., Burgdorf M., Del Giorgio A., Ewart D., Gry C., King K. J., Lim T., Molinari S., Price M., Sidher S., Smith A., Texier D. N., Trams S. J., Unger S. J., and Salama A. (1996) First detection of the 56 mm rotational line of HD in Saturn's atmosphere. *Astron. Astrophys., 315,* L389–L392.

Gry C., Laurent C., and Vidal-Madjar A. (1983) Evidence of hourly variations in the deuterium Lyman line profiles toward Epsilon Persei. *Astron. Astrophys., 124,* 99–104.

Guélin M., Langer W. D., and Wilson R. W. (1982) The state of ionization in dense molecular clouds. *Astron. Astrophys., 107,* 107–127.

Halbout J., Robert F., and Javoy M. (1990) Hydrogen and oxygen isotope compositions in kerogens from the Orgueil meteorite: Clues to solar origin. *Geochim. Cosmochim. Acta, 54,* 1453–1462.

Hatchell J., Millar T. J., and Rodgers S. D. (1998) The DCN/HCN abundance ratio in hot molecular clouds. *Astron. Astrophys., 332,* 695–703.

Hersant F., Gautier D., and Huré J-M (2001) A two dimensional model for the primordial nebula constrained by D/H measurements in the solar system: Implications for the formation of giant planets. *Astrophys. J., 554,* 391–407.

Hunten D. M. and McElroy M. B. (1974) Production and escape of terrestrial hydrogen. *J. Atmos. Sci., 31,* 305–317.

Jacq T., Walmsley C. M., Henkel C., Baudry A., Mauersberger R., and Jewell P. R. (1990) Deuterated water and ammonia in hot cores. *Astron. Astrophys., 228,* 447–470.

Kerridge J. F. (1985) Carbon, hydrogen and nitrogen in carbonaceous meteorites: Abundances and isotopic compositions in bulk samples. *Geochim. Cosmochim. Acta, 49,* 1707–1714.

Kerridge J. F., Chang S., and Shipp R. (1987) Isotopic characterisation of kerogen-like material in Murchison carbonaceous chondrite. *Geochim. Cosmochim. Acta, 51,* 2527–2540.

Kolodny Y., Kerridge J. F., and Kaplan I. R. (1980) Deuterium in carbonaceous chondrites. *Earth Planet. Sci. Lett., 46,* 149–158.

Lécluse C. and Robert F. (1994) Hydrogen isotope exchange rates: Origin of water in the inner solar system. *Geochim. Cosmochim. Acta, 58,* 2297–2939.

Lécluse C., Robert F., Gautier D., and Guiraud M. (1996) Deuterium enrichment in giant planet. *Planet. Space Sci., 44,* 1579–1592.

Lécuyer C., Gillet Ph., and Robert F. (1998) The hydrogen isotope composition of sea water and the global water cycle. *Chem. Geol., 145,* 249–261.

Lellouch E., Encrenaz Th., Graauw Th., Scheid S., Fëuchtgruber H., Benteima D. A., Bézard B., Drossart P., Griffin M., Heras A., Kesselr M., Leech K., Morris A., Roelfserna P. R., Roos-Serote M., Salama A., Vandenbussche B., Valentijn E. A., Davies G. R., and Naylor D. A. (1996) Determinations of D/H ratio on Jupiter from ISO/SWS observations. *Bull. Am. Astron. Soc., 28,* 1148.

Leshin L., Hutcheon I. D., Epstein S., and Stolper E. (1994) Water on Mars: Clues from deuterium/hydrogen and water contents of hydrous phases in SNC meteorites. *Science, 265,* 86–90.

Leshin L., Epstein S., and Stolper E. (1996) Hydrogen isotope geochemistry of SNC meteorites. *Geochim. Cosmochim. Acta, 60,* 2635–2650.

Linsky J. L. (2003) Atomic deuterium/hydrogen in the galaxy. In *Solar System History from Isotopic Signatures of Volatile Elements* (R. Kallenbach et al., eds.), pp. 49–60. ISSI Space Science Series, Vol. 16.

Linsky J. L., Brown A., Gayley K., Diplas A., Savage B. D., Landsman T. R., Shore S. N., and Heap S. R. (1993) High-resolution spectrograph observations of the local interstellar medium and the deuterium/hydrogen ratio along the line of sight toward Capella. *Astrophys. J., 402,* 694–709.

Mahaffy P. R., Donahue T. M., Atreya S. K., Owen T. C., and Niemann H. B. (1998) Galileo probe measurements of D/H and $^3He/^4He$ in Jupiter's atmosphere. *Space Science Rev., 84,* 251–263.

McCullough P. R. (1991) The interstellar deuterium to hydrogen ratio. A reevaluation of Lyman adsorption line measurements. *Astrophys. J., 390,* 213–225.

McKeegan K. D., Walker R. M., and Zinner E. (1985) Ion microprobe isotopic measurements of individual interplanetary dust particles. *Geochim. Cosmochim. Acta, 49,* 1971–1987.

McNaughton N. J., Borthwicks S., Fallick A. K., and Pillinger C. T. (1981) D/H ratio in unequilibrated ordinary chondrites.

Nature, 294, 639–641.

McNaughton N. J., Fallick A. E., and Pillinger C. T. (1982) Deuterium enrichments in type 3 ordinary chondrites. *Proc. Lunar Planet. Sci. Conf. 13th,* in *J. Geophys. Res., 87,* A294–A302.

Meibom A. and Clark B. E. (1999) Invited review: Evidence for the insignificance of ordinary chondritic material in the asteroid belt. *Meteoritics & Planet. Sci., 34,* 7–24.

Meier R., Owen T. C., Jewitt D. C., Matthews H. M., Senay M., Biver N., Bockelée-Morvan D., Crovisier J., and Gautier D. (1998a) Deuterium in Comet C/1995 O1 (Hale-Bopp). Detection of DCN. *Science, 279,* 1707–1710.

Meier R., Owen T. C., Matthews H. E., Jewitt D. C., Bockelée-Morvan D., Biver N., Crovisier J., and Gautier D. (1998b) A determination of the HDO/H_2O ratio in Comet C/1995 O1 (Hale-Bopp). *Science, 279,* 842–898.

Messenger S. (2000) Identification of molecular cloud material in interplanetary dust. *Nature, 404,* 968–971.

Messenger S., Stadermann F. J., Floss C., Nittler L. R., and Mukhopadhyay S. (2003) Isotopic signature of presolar materials in interplanetary dust. *Space Sci. Rev., 106,* 155–172.

Millar T. J. (2003) Deuterium fractionation in interstellar clouds. *Space Sci. Rev., 106,* 73–86.

Millar T. J., Bennett A., and Herbst E. (1989) Deuterium fractionation in dense interstellar clouds. *Astrophys. J., 340,* 906–920.

Morbidelli A., Chambers J., Lunine J., Petit J. M., Robert F., and Valsecchi G. B. (2000) Source regions and timescales for the delivery of water on Earth. *Meteoritics & Planet. Sci., 35,* 1309–1320.

Mousis O., Gautier D., Bockelée-Morvan D., Robert F., Dubrulle B., and Drouart A. (2000) Constraints on the formation of comets from D/H ratios measured in H_2O and HCN. *Icarus, 148,* 513–525.

Niemann H. B., Atreya S. K., Carignan G. R., Donahue T. M., Haberman J. A., Harpold D. N., Hartle R. E., Hunten D. M., Kasprzak W. T., Mahaffy P. R., Owen T. C., Spencer N. W., and Way S. H. (1996) The Galileo probe mass spectrometer composition of Jupiter's atmosphere. *Science, 272,* 846–849.

Owen T. and Encrenaz T. (2003) Element abundances and isotope ratios in the giant planets and Titan. *Space Sci. Rev., 106,* 121–138.

Owen T., Maillard J. P., Debergh C., and Lutz B. (1988) Deuterium on Mars: The abondance of HDO and the value of D/H. *Science, 240,* 1767–1770.

Pizzarello S., Krisnamurthy R. V., Epstein S., and Cronin J. R. (1991) Isotopic analyses of amino acids from the Murchison meteorite. *Geochim. Cosmochim. Acta, 55,* 905–910.

Richet P., Bottinga Y., and Javoy M. (1977) A review of hydrogen, carbon, nitrogen oxygen, sulphur, and chlorine stable isotope fractionation among gaseous molecules. *Annu. Rev. Earth Planet. Sci., 5,* 65–110.

Robert F. (1989) Hydrogen isotope composition of insoluble organic matter from cherts. *Geochim. Cosmochim. Acta, 53,* 453–460.

Robert F. (2001) The origin of water on Earth. *Science, 293,* 1056–1058.

Robert F. (2002) Water and organic matter D/H ratios in the solar system: A record of an early irradiation of the nebula? *Planet. Space Sci., 50,* 1227–1234.

Robert F. (2003) The D/H ratio in chondrites. *Space Sci. Rev., 106,* 87–101.

Robert F. and Deloule E. (2002) Using the D/H ratio to estimate the terrestrial water contamination in chondrites (abstract). In *Lunar and Planetary Science XXXIII,* Abstract #1299. Lunar and Planetary Institute, Houston (CD-ROM).

Robert F. and Epstein S. (1982) The concentration of isotopic compositions of hydrogen carbon and nitrogen in carbonaceous chondrites. *Geochim. Cosmochim. Acta, 16,* 81–95.

Robert F., Merlivat L., and Javoy M. (1979) Deuterium concentration in the early solar system: An hydrogen and oxygen isotopic study. *Nature, 282,* 785–789.

Robert F. M., Javoy J., Halbout B., Dimon B., and Merlivat L. (1987) Hydrogen isotopes abondances in the solar system, Part 1. Unequilibrated chondrites. *Geochim. Cosmochim. Acta, 51,* 1787–1806.

Robert F., Gautier D., and Dubrulle B. (2000) The solar system D/H ratio: Observations and theories. *Space Sci. Rev., 92,* 201–224.

Rodgers S. D. and Millar T. J. (1996) The chemistry of deuterium in hot molecular cores. *Mon. Not. R. Astron. Soc., 280,* 1046–1054.

Sandford S. A. (2002) Interstellar processes leading to molecular deuterium enrichment and their detection. *Planet. Space Sci., 50,* 1145–1154.

Sandford S. A., Bernstein M. P., and Dworkin J. P. (2001) Assessment of the interstellar processes leading to deuterium enrichment in meteoritic organics. *Meteoritics & Planet. Sci., 36,* 1117–1133.

Schilke P., Walmsley C. M., Pineau des Forêts G., Roueff E., Flower D. R., and Guilloteau S. (1992) A study of HCN, HNC and their isotopomers in OMC-1. I. Abundances and chemistry. *Astron. Astrophys., 256,* 595–612.

Schramm D. D. (1998) Big-Bang nucleosynthesis and the density of baryons in the universe. *Space Sci. Rev., 84,* 3–14.

Sears D. W. G., Morse A. D., Hutchison R., Guimon K. R., Kie L., Alexander C. M. O'D., Benoit P. H., Wright I., Pilinger C., Kie T., and Lipschutz M. E. (1995) Metamorphism and aqueous alteration in low petrographic type ordinary chondrites. *Meteoritics, 30,* 169–181.

Smith M. D., Schempp W. V., Simon J., and Baines K. H. (1989a) The D/H ratio for Uranus and Neptune. *Astrophys. J., 336,* 962–966.

Smith M. D., Schempp W. V., and Baines K. H. (1989b) The D/H ratio for Jupiter. *Astrophys. J., 336,* 967–970.

Vidal-Madjar A., Laurent C., Gry C., Bruston P., Ferlet R., and York D. G. (1983) The ratio of deuterium to hydrogen in interstellar space. V. The line of sight to Epsilon Persei. *Astron. Astrophys., 120,* 58–65.

Vokrouhlický D. and Farinella P. (2000) Efficient delivery of meteorites to the Earth from a wide range of asteroid parent bodies. *Nature, 407,* 606–608.

Watson W. D. (1976) Interstellar molecule reactions. *Rev. Mod. Phys., 48,* 513–552.

Willacy K. and Millar T. J. (1998) Adsorption processes and the deuterium fractionation in molecular clouds. *Mon. Not. R. Astron. Soc., 228,* 562–568.

Yang J. and Epstein S. (1983) Interstellar organic matter in meteorites. *Geochim. Cosmochim. Acta, 47,* 2199–2216.

Yung Y., Friedl R., Pinto J. P., Bayes K. D., and Wen J. S. (1988) Kinetic isotopic fractionation and the origin of HDO and CH_3D in the solar system. *Icarus, 74,* 121–132.

Zinner E., McKeegan K. D., and Walker R. M. (1983) Laboratory measurements of D/H ratios in interplanetary dust. *Nature, 305,* 119–121.

Particle-Gas Dynamics and Primary Accretion

Jeffrey N. Cuzzi
NASA Ames Research Center

Stuart J. Weidenschilling
Planetary Science Institute

We review the basic physics of particle-gas interactions, and describe the various nebula epochs and regimes where these interactions are important. The potential role of turbulence is of special interest in a number of ways. Processes discussed include growth by sticking and incremental accretion, enhancement of abundance due to radial drift across evaporation boundaries, outward transport of small particles by diffusion and stellar winds, various midplane instabilities, and size-selective aerodynamic concentration of chondrule-sized particles. We provide examples of the structure and/or composition of primitive meteorites where these processes might have played a defining role, or where their signatures might be diagnostic.

1. OVERVIEW

It is likely that aerodynamic effects dominate the evolution of the main meteorite constituents (chondrules, refractory inclusions, fine-grained accretion rims, and matrix grains) into the first sizeable objects. We will refer to the first growth from free-floating nebula constituents to objects the size of meteorite parent bodies as primary accretion. In section 2, we review the underlying principles — the relevant properties of the nebula gas and its flow regimes, the physics of gas drag and particle stopping times, and the derivation of particle velocities in the presence or absence of turbulence. In section 3, we review the most well-developed incremental growth models for primary accretion, noting the role of sticking and how the growth regime changes as particles grow from fine dust to meters or larger. The key role of turbulence in determining growth timescales is first encountered here. In section 4, we describe a less-traditional path to primary accretion, in which turbulence selectively concentrates chondrule-sized constituents into dense zones that might subsequently evolve directly into objects with at least the mass of a many-meter-sized particle. In section 5, we note how solids can decouple from the gas and become widely redistributed, interacting with evaporation fronts in ways that can profoundly affect the regional mass distribution and chemistry of planet-forming materials. In section 6, we sketch an evolutionary scenario that might provide some helpful context for interpretation of the meteorite record. We summarize in section 7. Table 1 provides a list of symbols used frequently throughout this chapter.

2. UNDERLYING PRINCIPLES

2.1. The Nebula Gas

The properties of protoplanetary nebulae are as uncertain and controversial as those of meteorites. These are summarized in *Boss and Goswami* (2006) and *Russell et al.* (2006); see also the review by *Calvet et al.* (2000). The typical nebula mass M (98% hydrogen and helium) is a small fraction (0.02–0.2) of the stellar mass M_*, and cannot yet be measured directly. Current estimates of the surface gas mass density $\sigma_g(R)$ (g cm^{-2}), at distance R from the star, are based on assumptions about emission from solid particles, and will underestimate the disk mass once particles have grown past millimeter size. The concept of the minimum mass nebula (MMN) (e.g., *Hayashi,* 1981; *Hayashi et al.,* 1985) is a handy benchmark, but has no solid physical basis and (by construction) is an underestimate to the degree that solids were lost either into the Sun along with hydrogen, or by whatever subsequent processes depleted the asteroid region and ejected comets into the Oort cloud.

The nebula gas flows into the star at a rate $\dot{M}$ and velocity $V_n = \dot{M}/2\pi R\sigma_g$. The inflow rate $\dot{M}$, which decreases by orders of magnitude over the likely disk lifetime $M/\dot{M}$, also seems to vary by over an order of magnitude even for systems with the same apparent age (*Calvet et al.,* 2000) and thus the lifetime is both uncertain and probably system-dependent. Outward transport of angular momentum must occur to create this mass inflow. However, the mechanism for producing this transport is not well understood and remains a matter of debate. Turbulent viscosity ν_T, once the mechanism of choice, has been questioned by many theorists in regions (such as the asteroid belt region) where the nebula gas is too dense and too cool to sustain a magnetorotational instability; this is primarily because numerical models of the differential rotation law obeyed by Keplerian disks have not demonstrated turbulent instability, and linear analysis does not predict it (*Stone et al.,* 2000). However, the situation regarding production of turbulence by differentially rotating neutral gas is in flux (e.g., *Richard and Zahn,* 1999; *Fleming and Stone,* 2003; *Richard,* 2003; *Mosqueira et al.,* 2003; *Klahr and Bodenheimer,* 2003a; *Arldt and Urpin,* 2004; *Umurhan and Regev,* 2004; *Garaud and Ogilvie,*

TABLE 1. List of frequently used symbols.

Parameter	Definition
c	gas molecule thermal speed, or sound speed
C, C_o	concentration or mass fraction of a tracer, and its cosmic value (section 5.4)
G	gravitational constant
h_p	particle midplane layer vertical scale height (section 3.3.1)
H	nebula gas vertical scale height
ℓ, L	generic, and largest, eddy scales in turbulence (section 2.2)
$M_\odot$	solar mass
M_*	stellar mass
$\dot{M}$	nebula mass accretion rate, $M_\odot$/yr
r	particle radius
R	distance from the star
R_{ev}	distance of an evaporation front from the star (section 5.4)
Re, Re*	turbulent Reynolds number (section 2); critical Re for midplane turbulence (section 3.3.1)
Ro	turbulent Rossby number of the near-midplane nebula gas (section 3.3.1)
St, St_L, St_η	particle Stokes number = $t_s\Omega$, $t_s\Omega_L$, $t_s\Omega(\eta)$ (section 2.2)
t_s	particle stopping time due to gas drag (section 2.2)
V_K	Keplerian velocity
V_L	"large" (typically, meter-sized) particle drift velocity ~ βV_K (sections 2.4.1 and 5.4)
V_n	nebula gas radial accretion velocity (sections 2.1 and 5.4)
V_p	particle fluctuating velocity in turbulence (section 2.4)
V_{pg}	relative velocity between particle and gas in turbulence (section 2.4)
V_{pp}	relative velocity between identical particles in turbulence (section 2.4)
α	turbulent intensity parameter (section 2.1)
β	pressure gradient parameter (section 2.4.1, η of *Nakagawa et al.,* 1986)
$\mathcal{D}$, ν_T	nebula turbulent diffusivity and viscosity (section 2.1)
ΔV	maximum velocity difference or headwind between particles and gas (section 2.4.1) = βV_K
η	Kolmogorov scale; smallest eddy scale in turbulence (section 2.2)
ν_m	nebula gas molecular viscosity
Ω	angular velocity
Ω_K	Keplerian (orbital) angular velocity
Ω_L	largest eddy angular velocity
Ω_x	angular velocity at x-point near accreting star (section 5.1)
ρ_s	individual particle internal density
ρ_p	particle phase volume mass density
σ_g, ρ_g	nebula gas surface and volume mass density
σ_L	large particle surface mass density

2005). Of special emerging interest is the possibility that even very small changes in some of the governing parameters, when extrapolated over the very large range of Reynolds numbers separating the nebula from current simulations, might lead to qualitatively different results (*Sreenivasan and Stolovitsky,* 1995a, and references therein; *Afshordi et al.,* 2005; *Mukhopadyay et al.,* 2005; also *Busse,* 2004).

Whatever the physics that drives nebular evolution, nebulae do evolve, and the ongoing inflow means that gravitational energy is constantly being released; this energy alone can maintain turbulent fluid motions. Some evidence for ongoing turbulence through the million-year timeframe is that disks of these ages, and older, commonly show evidence for small grains that are well distributed vertically (*Dullemond and Dominik,* 2005). Based only on the amount of energy being released by the evolving nebula, values of $\alpha \sim 10^{-5}$–10^{-3} are not hard to justify (*Cuzzi et al.,* 2001). These fluid motions create a diffusivity $\mathcal{D}$ even if their particular correlations do not lead to a viscosity ν_T (cf. *Prinn,* 1990). We will finesse the distinction by taking their ratio, the Prandtl number Pr, equal to unity, and parameterize both $\mathcal{D}$ and ν_T by the dimensionless parameter α: $\mathcal{D} = \nu_T = \alpha cH$, where c is the sound speed and H the nebula vertical scale height. Then $V_n \approx \nu_T/R$. Stellar winds and jets are observed in many, but not all, systems, and their strengths generally decrease along with $\dot{M}$ (*André et al.,* 2000).

The nebula gas is heated by a combination of stellar radiation and local dissipation of the released gravitational energy $GM_*\dot{M}/R$. The relative strength of these energy sources depends on the actual viscous dissipation and the steepness of flaring of the vertical scale height H(R). Infrared observations of disk thermal emission can be related to gas temperatures at various vertical and radial locations by making assumptions about the opacity of the solids in the nebula, which is a function of their (time-variable, and mostly unknown) particle size. The temperature is higher near the star, and (if viscous heating is important) closer to the midplane, but just where and when the temperature reaches interesting values such as the condensation points of refractory oxides, silicates, or water ice is model dependent and might have varied from one system to another. For any given $\dot{M}$, nebula temperatures are higher for smaller α (the disk must be more massive) and run higher in recent models and analyses (*Papaloizou and Terquem,* 1999; *Bell et al.,* 1997; *Woolum and Cassen,* 1999) than those reviewed in the original volume of this series (*Wood and Morfill,* 1988).

Nebula models are usually described by their gas density $\rho_g = \sigma_g/2H$; however, meteoriticists usually prefer pressure. The conversion is simply done using the ideal gas law $P = \rho_g R_g T/\mu = \rho_g c^2/3$, where $R_g = 8.3 \times 10^7$ erg/mole-°K = 82.1 cm^3 atm/mole-°K is the universal gas constant, $\mu = 2$ g/mole for molecular hydrogen, local T is degrees Kelvin, and local ρ_g is g/cm^3.

For the purpose of this chapter we will adopt a canonical nebula, keeping in mind that key parameters such as the mass M, turbulent intensity parameter α, and lifetime $M/\dot{M}$ of our own nebula are uncertain by at least an order of magnitude. We will subsequently discuss how different values of M and/or $\dot{M}$ might relate to different nebula epochs, radial regimes, and mineral products. For a specific example, take a MMN accreting at $\dot{M} = 10^{-7}$ $M_\odot$/yr (*Bell et al.,*1997). At 3 AU, with a temperature of 500 K and $\rho_g = 10^{-10}$ g/cm^3, $P = 2 \times 10^{-6}$ bar. At 0.3 AU with a temperature of 1500 K and $\rho_g = 3 \times 10^{-8}$g/cm^3, $P = 1.8 \times 10^{-3}$ bar (see also *Wood,* 2000). Densities and pressures might have been higher in the early days of the nebula ($<10^6$ yr) when formation of refractory inclusions occurred, and lower by the classical T Tauri stage when $\dot{M}$ probably dropped to 10^{-8} $M_\odot$/yr (*Woolum and Cassen,* 1999).

2.2. Particle Stopping Times and Turbulent Eddy Times

The particle stopping time t_s is the time in which the particle equilibrates with a gas moving at some relative velocity. Small, light particles have short stopping times and adjust much more quickly to gas fluctuations than large, dense particles. Particle-gas coupling is by a drag force, which depends on the gas density ρ_g, the particle radius r, and its internal density ρ_s (which is lower than the "solid" density for porous aggregates such as fluffy CAIs). If the particle radius is larger than the molecular mean free path $\lambda \approx 50$ $(10^{-10}$ g cm$^{-3}/\rho_g)$ cm, t_s can depend on its relative velocity as well. Full descriptions of the various regimes of interest are given by *Weidenschilling* (1977), *Nakagawa et al.* (1986), and *Cuzzi et al.* (1993). Since the gas mean free path even at 1 AU is several centimeters, all typical constituents of chondrites satisfy $r < \lambda$, so we will generally use the so-called Epstein drag stopping time expression

$$t_s = \frac{r\rho_s}{c\rho_g} \tag{1}$$

For particles with $r > \lambda$ (the Stokes drag regime), up to nearly 10 m radius under nebula conditions, t_s remains nearly velocity-independent and may be approximated by multiplying the expression above by the factor $r/2\lambda$ (*Cuzzi et al.,* 1993).

In the case of a turbulent gas, particles experience constantly fluctuating velocity perturbations from eddies with a range of length scales ℓ and associated fluctuation times $t_e(\ell)$. Turbulence is an essentially lossless cascade of energy from large, slowly rotating eddies with lengthscale L and velocity V_g, which are forced by nebula-scale processes, through smaller and smaller scales, with correspondingly faster eddy timescales, to some minimum lengthscale η, called the Kolmogorov scale, where molecular viscosity ν_m can dissipate the macroscopic gas motions and turbulence ceases. Instead of presuming that large eddy scales L are on the order of the scale height H, which leads to $V_g = c\alpha$, we believe it more realistic to presume that large eddy frequencies Ω_L are no slower than the orbit frequency Ω_K due to coriolis effects; this logic leads to $V_g = c\alpha^{1/2}$ and $L = H\alpha^{1/2}$ (for details see *Shakura et al.,* 1978; *Cuzzi et al.,* 2001). This relationship for $V_g(\alpha)$ differs from that presented by *Shakura and Sunyaev* (1973), and subsequently followed by a number of authors, including *Weidenschilling and Cuzzi* (1993); thus turbulent velocities are larger than previously thought for any given α, with implications for settling and diffusion of particles from chondrule size to meter size (sections 3 and 5). The turbulent Reynolds number is $Re = (L/\eta)^{4/3}$ (*Tennekes and Lumley,* 1972), where for canonical nebula parameters at 3 AU $Re = \alpha cH/\nu_m \sim 3 \times 10^7$ for $\alpha = 10^{-4}$. In this case, using $L = H\alpha^{1/2}$, $\eta \sim LRe^{-3/4} \sim$ 1 km. In most known cases of real turbulence, the Kolmogorov energy spectrum is a good approximation. Then, for a wide range of lengthscales $\eta < \ell < L$, the turbulent kinetic energy density $E(\ell)$ is given by the inertial range expression $E(\ell) = (V_g^2/2L)(\ell/L)^{-1/3}$, where the factor $V_g^2/2L$ is obtained from the normalization $\int_\eta^L E(\ell)d\ell = V_g^2/2$. The energy characterizing a typical lengthscale ℓ is $E(\ell, d\ell = \ell) = \ell E(\ell) = \ell(V_g^2/2L)(\ell/L)^{-1/3} = 0.5\, V_g^2(\ell/L)^{2/3}$. Then the eddy frequencies for arbitrary ℓ scale as $\Omega(\ell) = 1/t_e(\ell) = v(\ell)/\ell = (2\ell E(\ell))^{1/2}/\ell = \Omega_L(\ell/L)^{-2/3}$ (e.g., cf. *Tennekes and Lumley,* 1972; *Cuzzi et al.,* 1996, 2001).

In calculating the response of a particle to fluctuating eddy motions of this type, the dimensionless Stokes number $St = t_s\Omega(\ell)$ is used. Particles with $St \ll 1$ with respect to some eddy scale ℓ have stopping times much less than the overturn time $\Omega(\ell)^{-1}$ and are strongly coupled to those eddies. Particles with t_s much less than even the shortest (Kolmogorov eddy) overturn time $t_e(\eta)$ are coupled to the gas and, effectively, move as gas molecules. In most cases we will define $St = St_L = t_s\Omega_L = t_s\Omega_K$; however, in section 4 we describe interesting and relevant effects associated with particles having stopping time t_s equal to $t_e(\eta) = \Omega_K^{-1}(\eta/L)^{2/3} = \Omega_K^{-1}(Re^{-3/4})^{2/3} = \Omega_K^{-1}Re^{-1/2}$.

2.3. Meteorite Observations Implicating Aerodynamics

Dodd (1976) first discovered evidence for aerodynamic sorting in unequilibrated ordinary chondrites (UOCs). He compared the sizes of metal grains, and silicate grains of different metal content, and found that the sizes and densities were consistent with $r\rho_s$ = constant (cf. equation (1)). Shortly thereafter, *Hughes* (1978) determined, from thin section and disaggregation analysis, that complete, rounded chondrules and obvious fragments of larger chondrules obeyed the same size distribution. This was reemphasized by *Leenhouts and Skinner* (1991) and *Skinner and Leenhouts* (1991), who also noted the presence of pre-accretionary fine-grained dust rims on chondrule fragments. They concluded that aerodynamics operating on solid particles, rather than something about the formation process, was deterministic to the characteristic narrow chondrule size distributions. *Scott and Haack* (1993) obtained similar results for Lancè (CO3). *Nelson and Rubin* (2002) commented that chondrule fragments in Bishunpur (LL3) were notably smaller than whole chondrules in Bishunpur; however, their Table 2 reveals that chondrules and fragments generally cover the same size range, with aerodynamic differences possibly attributable to shape. As did *Dodd* (1976) for UOCs, *Skinner and Leenhouts* (1993) showed that metal and silicate particles from Acfer 059 (CR2) had the same $r\rho_s$. *Hughes* (1980) found that the density and diameter of disaggregated silicate chondrules were inversely correlated; similarly, *Cuzzi et al.* (1999) found that, for disaggregated chondrules for which both r and ρ_s could be measured, distribution histograms narrowed if the correct $r\rho_s$ product were used for each chondrule. *Kuebler et al.* (1999) introduced tomographic techniques and found that metal grains (although irregularly shaped) and silicate chondrules in three UOCs have nearly the same $r\rho_s$. Consistent with their more irregular shapes, the metal grains of *Kuebler et al.* (1999) were less well sorted than the more equidimensional chondrules. *Haack and Scott* (1993) have suggested that aerodynamical sorting of high-FeO and low-FeO chondrules might have been responsible for the differences between the UOC types (more-dense, low-FeO chondrules sorted primarily into the H chondrites). In sections 4 and 6.3, we describe newer, possibly diagnostic results regarding turbulent concentration of aerodynamically selected particles, and acquisition of fine-grained dust rims, in nebula turbulence.

Not all meteorites and components are so simply interpreted, however. Calcium-aluminum-rich inclusion sizes are poorly known (*May et al.,* 1999) but CAIs appear to be less narrowly sorted (and also somewhat larger) than chondrules and chondrule fragments in the same chondrites. Perhaps this is due to fluffiness, irregular shapes, or lower material densities of the CAIs. However, the chondrite types having the smallest (largest) chondrules also tend to have the smallest (largest) CAIs, supportive of aerodynamic sorting (*Scott et al.,* 1996). It is also true that in CH and CB chondrites, the metal particles are as large, or larger, than the silicate particles (A. Meibom, unpublished data, 2000; personal communication, 2003; see, e.g., *Greshake et al.,* 2002, Fig. 1). This latter fact, if indeed true, seems to be incompatible with aerodynamic sorting. It is emerging that CH and CB chondrites are the youngest of all chondrites (*Amelin et al.,* 2002, 2004; *Krot et al.,* 2005a; *Bizarro et al.,* 2004), and they may have formed in very different environments than normal chondrules and chondrites (*Connolly et al.,* 2006; *Weisberg et al.,* 2006). For more discussion of meteoritic constraints on the environment of these small particles, see *Cuzzi et al.* (2005a).

2.4. Particle Velocities

As one simple example of how t_s works, under a nonfluctuating acceleration such as that due to the vertical component of solar gravity $g_z(z) = -\Omega_K^2 z$, particles reach a terminal velocity $V_z = g_z t_s$, which is size dependent through $t_s(r)$ and altitude dependent through both g_z and $t_s(\rho_g(z))$. Below, we expand on how particles respond to the forces that act upon them; first we discuss the case without turbulence, and then show how turbulence (fluctuating acceleration) affects things.

2.4.1. The headwind and radial drift. *Whipple* (1972) first pointed out that the nebula gas can rotate either more slowly or more rapidly than the Keplerian velocity of a solid particle, depending on the local pressure gradient. The general decrease of gas density and temperature with increasing distance from the Sun, and the corresponding outward pressure gradient, provide a slight outward acceleration on the gas that opposes the dominant inward acceleration of solar gravity. Solid particles, responding only to solar gravity, experience a headwind from the more slowly rotating gas, which saps their angular momentum, causing them to orbit more slowly than the local Keplerian velocity and to drift inward. Subsequent work by *Adachi et al.* (1976), *Weidenschilling* (1977), and *Nakagawa et al.* (1986) describes this physics very clearly and it is only paraphrased here.

The ratio of the differential pressure gradient acceleration Δg to the dominant central gravity acceleration g is

$$\beta \equiv \frac{\Delta g}{2g} = \frac{dP/dR}{2\Omega_K^2 R\rho_g} \sim 10^{-3} - 10^{-2} \qquad (2)$$

in the asteroid belt region, where the range in β is due to the uncertain range of radial density and temperature gradients in the nebula (see *Cuzzi et al.,* 1993, Table 1). A useful approximation is that $\beta \approx (c/V_K)^2 \approx (H/R)^2$.

A parcel of gas experiencing an outward pressure gradient characterized by β will orbit at a velocity slower than Keplerian by an amount $\Delta V = \beta V_K$; at 2.5 AU, V_K = 18 km/s, so $\Delta V \sim$ 36–144 m/s. Large bodies ($\gg$1 m) are not significantly influenced by the gas and orbit at Keplerian velocity, thus incurring a headwind of this magnitude. The very smallest particles are forced to orbit at the gas velocity, but do not feel the outward gas pressure acceleration, so have an imbalance of gravitational and centrifugal accelerations $\Delta g = 2\beta R\Omega_K^2 = 2\beta\Omega_K V_K$. These particles drift slowly inward under this constant acceleration at terminal radial velocity $V_R = \Delta g t_s = 2St\beta V_K = 2St\Delta V$. Radial drift rates increase with size through increasing t_s, and this expression is actually quite good over a wide range of particle sizes <30 cm or so. Particles of roughly meter size, with $t_s \approx t_e(L) = 1/\Omega_K$, expe-

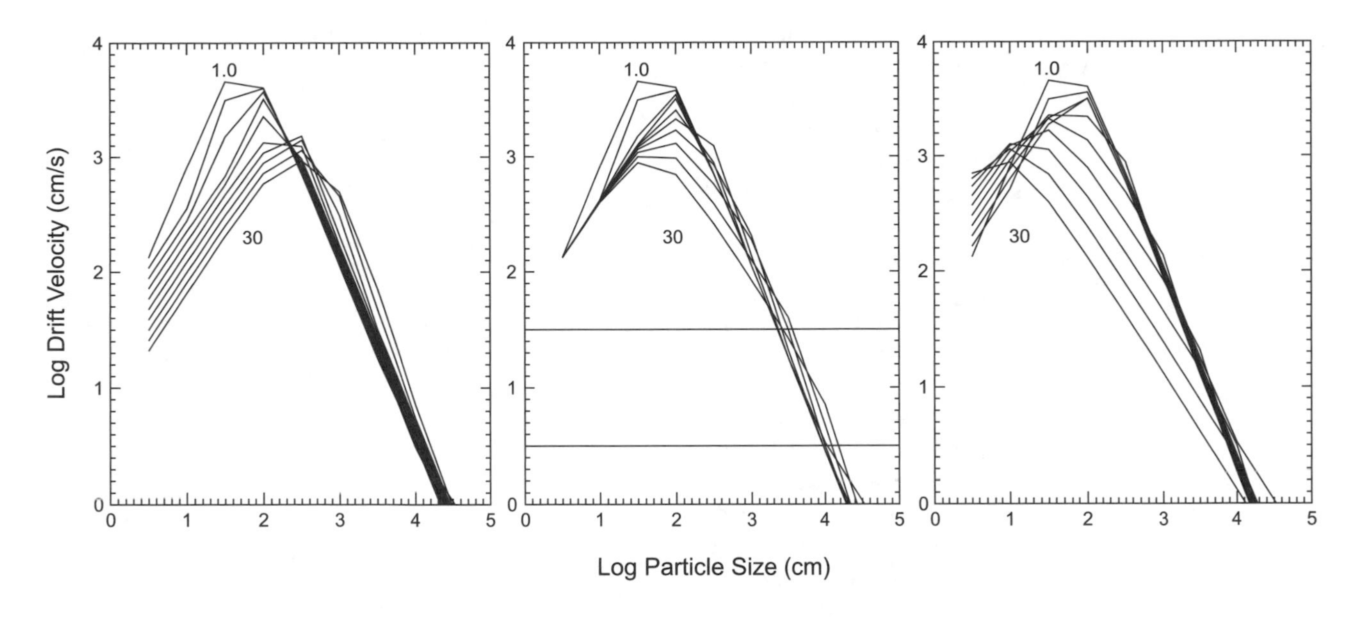

Fig. 1. Radial (inward) drift velocity for unit density particles at different locations in a nominal nebula model ($\sigma_g = 1700(1\ AU/R)^{-p}$ g cm^{-2}), as functions of particle radius (cm) and distance from the Sun (AU). Left: p = 1/2; center: p = 1; right: p = 3/2. For comparison, the horizontal band in the p = 1 plot indicates the range of nebula gas advection, or inward drift, velocities V_n due to angular momentum transport at 5 AU from the Sun for a typical model with $\alpha = 10^{-3}$ (sections 2.1 and 2.4.1). Curves are labeled at radii of 1 and 30 AU, and at equal factors of 1.4 between.

rience nearly the full headwind yet are not massive enough to avoid drifting; they achieve the maximum radial drift rate $V_R \approx \Delta V$ (*Weidenschilling,* 1977). An example of radial drift velocities V_R for a range of nebula models is shown in Fig. 1. Over this range, gas densities decrease radially outward, leading to an outwardly increasing t_s that partially offsets the outwardly decreasing V_K; thus radial dependence of V_R is not particularly strong (especially for $\sigma_g \propto R^{-1}$).

As the local particle mass density ρ_p grows, such as near the midplane, particles with $\rho_p \geq \rho_g$ can drive the entrained gas toward Keplerian velocity, which decreases their headwind and drift rates V_θ and V_R (see below). Analytical expressions for all particle headwind and drift velocities relative to the gas have been derived by *Nakagawa et al.* (1986) for arbitrary ratios of local particle mass density to gas mass density. These can conveniently be used in the limit $\rho_p/\rho_g \ll 1$ to obtain headwinds and drift velocities for isolated particles of any size and density.

We note that drift velocities are systematic, and depend on particle size. Identical particles would have the same velocity components and no relative velocity; they would experience no collisions due to drift. If turbulence is present, it can produce relative velocities and collisions between identical particles (as well as those of different sizes). Unless the nebula is perfectly laminar, particle velocities will be due to both sources, with the actual values dependent on nebular parameters, turbulence properties, and particle sizes.

Local pressure fluctuations and particle concentrations: Even though the nebula has an overall outward pressure gradient, strong local effects might arise in which pressure gradients, and particle drift, could go both ways. For instance, particles could quickly drift *into* local radial pressure maxima in the nebula gas (*Haghighipour and Boss,* 2003a,b; *Fromang and Nelson,* 2005; *Johansen et al.,* 2005). If the nebula gas is globally gravitationally unstable, large-scale spiral density waves will provide such localized nearly radial pressure maxima. Transient local enhancement in spiral density waves of the most rapidly drifting (meter-sized) particles by factors of 10–100 has been seen (*Rice et al.,* 2004). The end result is similar to that seen associated with large vortices (section 3.3.3) in that some potential for particle mass density enhancements exists. As discussed in section 3.3.3, increased collision rates in these regions may or may not lead to faster particle growth, depending on the relative collisional velocities, which will be connected to the strength of turbulence associated with these large-scale fluid dynamical structures (*Boley et al.,* 2005).

2.4.2. Particle velocities in turbulence. While particle velocities in turbulence have been studied extensively in the fluid dynamics literature (see *Cuzzi and Hogan,* 2003), the first and main contribution in the astrophysics literature was made by *Völk et al.* (1980, hereafter *VJMR*). It was updated by *Markiewicz et al.* (1991, hereafter *MMV*), who included the importance of the cutoff in turbulent forcing at the Kolmogorov scale η as suggested by *Weidenschilling* (1984), and also studied by *Clarke and Pringle* (1988).

We will refer to particle velocities with respect to inertial space as V_p, those with respect to the gas (combined radial and angular components) as V_{pg}, and those with respect to each other as V_{pp}. In general, V_p is used to determine the diffusive nature of particle motions, and plays a role in determining the thickness of the midplane particle layer and thus the planetesimal growth rate (section 3.3), as well as diffusing small particles such as CAIs and chondrules radially throughout the nebula (section 6.2). V_{pg} determines how fast the particle sweeps through the gas (and smaller particles more firmly tied to the gas), and plays a role in constraining how chondrules and CAIs acquire fine-grained accretion rims (section 6.3). V_{pp} determines how often like particles encounter each other and with what collisional speed (section 3.3.2). *Cuzzi and Hogan* (2003) expand on the *VJMR* prescription and derive simple, fully analytical expressions for V_p, V_{pg}, and V_{pp} for arbitrary (identical) particles in turbulence of arbitrary intensity. An alternate derivation, but restricted to V_p, was presented by *Cuzzi et al.* (1993).

The response of a particle to turbulence, which has a spectrum of eddy scales and frequencies, is much like the classical problem of the response of an oscillator to periodic forces of different frequency. The oscillator responds well to forces that vary more slowly than its natural response time, and poorly to those that vary more rapidly. For a particle, the natural response time is just t_s. Particles will simply be entrained in eddies that have overturn time $>t_s$, and will share their velocities and spatial excursions. However, nearby, identical size particles in the same eddy will obtain almost no *relative* velocity from these eddies. Particles do experience random perturbations, in different directions, from smaller eddies with smaller timescales than t_s, and even nearby particles can incur random relative velocities this way. However, the absolute velocities of small, high-frequency eddies are smaller than for large, low-frequency ones (section 2.2). So small particles have slow relative velocities, and those with t_s smaller than the smallest eddy overturn time have only very small relative velocities (section 3.2).

Figure 2 shows V_p, V_{pg}, and V_{pp} for (identical) particles of a wide range of sizes, in turbulence of three different Re (or α). All are normalized by the turbulent gas velocity $V_g = c\alpha^{1/2}$. The particle size is presented on the lower axis in full generality as the Stokes number $St_L = t_s/t_L$ where t_L is the overturn time of the largest eddy — presumed to be the local orbit period. Very roughly, centimeter-radius, unit density particles at 2.5 AU from the Sun in a standard MMN have $t_s \sim 10^5$ s and thus $St_L \sim 3 \times 10^{-3}$. It is of special note that meter-radius particles have $t_s \sim$ orbit period and thus $St \sim 1$.

Several points of interest are shown in this figure. There is a single curve for V_p, because only the largest eddies contribute to V_p and so the value of Re, which essentially determines the size of the Kolmogorov scale η and its eddy properties, is unimportant. Particles of all sizes up to nearly a meter share a significant fraction of the gas turbulent velocity. The particle diffusion coefficient $\mathcal{D}_p$, which we will use in section 3.3 to calculate the density of midplane particle layers, and in section 6.2 to explain the preservation of CAIs for millions of years after their formation, can be written as $\mathcal{D}_p = \mathcal{D}V_p^2/V_g^2 = \mathcal{D}/(1 + St)$ (*VJMR*; *Cuzzi et al.,* 1993; *Cuzzi and Hogan,* 2003). Recent numerical studies are generally in agreement with these predictions (*Schräpler and Henning,* 2004; *Carballido et al.,* 2005). Second, for large particles, V_{pp} decreases with size as $St_L^{-1/2}$; for particles smaller than $St_L \approx 1$, V_{pp} increases with size as $St_L^{1/2}$; however, V_{pp} drops sharply to zero for particles with $St_L < Re^{-1/2}$, which translates into particles having $t_s < t_e(\eta)$. We will use V_{pp} in section 3.3.2 to calculate the collisional destruction time of meter-sized particles. Third, there is a steepening in slope of V_{pg} for particle sizes with $t_s < t_e(\eta)$. We will use V_{pg} to model the thickness of fine-grained accretion rims on chondrules and CAIs in section 6.3.

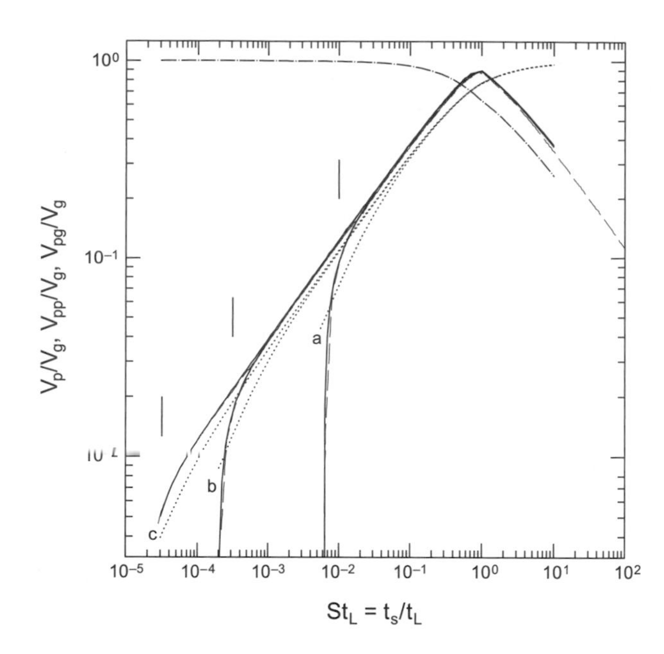

Fig. 2. V_p (dot-dash), V_{pg} (dotted), and V_{pp} (solid) for (identical) particles of a wide range of sizes, as normalized by the gas turbulent velocity V_g, and in turbulence of three different Re (or α). (a) Re = 10^4; (b) Re = 10^7; (c) Re = 10^9. Dashed curve = exact numerical solutions from *MMV*; vertical hash marks = location where $St_\eta = 1$ (discussed in section 4). Figure from *Cuzzi and Hogan* (2003).

3. INCREMENTAL GROWTH

3.1. Sticking

The preceding discussion of particle velocities makes it clear that their relative motions lead to collisions. Even in a nonturbulent nebula, differential drift velocities are typically larger than the escape velocity for bodies smaller than about a kilometer, so growth must involve nongravitational sticking. This requires an attractive force between particles, such as van der Waals or other forms of surface energy, electrostatic or magnetic forces, and some mechanism of dissipating energy during a collision. Outcomes of collisions depend on the impact velocities and the absolute and relative sizes of the particles, along with their physical properties, such as density, impact strength, surface energy, elastic modulus, etc. The result may be sticking, rebound, net gain or erosional loss of mass, or disruption (*Blum,* 2004). For a thorough review of recent progress, see *Dominik et al.* (2006).

Theoretical analyses of collisional mechanics are limited to idealized spherical particles and aggregates thereof. *Chokshi et al.* (1993) considered head-on collisions of spherical particles, with dissipation of energy by elastic vibrations, and derived a critical velocity for the transition between sticking and rebound. The analysis was extended by *Dominik and Tielens* (1997) to include energy dissipation by sliding, twisting, and rolling of contact points between particles; they derived criteria for growth, restructuring, and disruption of aggregates. The theory of *Dominik and Tielens* (1997) predicts that two particles of radius r have a critical velocity for sticking

$$V_{stick} \sim \frac{2(\mathcal{E}/r)^{5/6}}{E^{1/3}\rho_s^{1/2}} \quad (3)$$

where $\mathcal{E}$ is the surface energy and E is Young's modulus. For properties of quartz, this expression gives $V_{stick} \sim 2 \times 10^{-3}$ (r/1 cm)$^{-5/6}$ cm s^{-1}. A micrometer-sized grain would stick to another grain (or the surface of a much larger body such as a chondrule) at an impact velocity <5 cm s^{-1}. For chondrule-sized particles (r = 0.05 cm), $V_{stick} \sim 0.025$ cm s^{-1}. *Blum* (2000) performed laboratory experiments on dust aggregation, and found sticking velocities about an order of magnitude larger than predicted by this theory for the measured values of elastic constants and surface energy. This difference might be due to electrostatic binding, nonsphericity of grains, or other factors. The transition between sticking and fragmentation occurs at around 1 m/s, with some nonzero sticking probability for velocities exceeding 10 m/s. Irregular micrometer-sized grains stick with efficiency of about 50% even at relative velocities greater than 10 m/s (*Blum,* 2004). Particles in the solar nebula probably were softer than silica, with lower elastic moduli. Thus, micrometer-sized grains in the solar nebula should coagulate readily, and chondrules can acquire a coating of grains due to van der Waals bonding alone (section 6.3). Additional sticking forces, such as electrostatic or magnetic interactions, may have been present, but were not necessary.

Collision velocities between chondrule-sized particles depend on the turbulent intensity α and the gas density ρ_g (Fig. 2). Even if these velocities preclude sticking of individual, solid, chondrule-sized particles by van der Waals forces, electrostatic forces might play a role. *Marshall and Cuzzi* (2001) and *Marshall et al.* (2005) presented microgravity studies of chondrule-sized silicate particles that show no tendency at all to stick under terrestrial conditions — much like the familiar "dry grains of sand on a beach" analogy suggests. Under microgravity conditions, tribocharging generates dipole moments on these particles that make them stick quite readily to each other, sometimes forming large dust bunnies having net dipole moments that actively attract nearby particles, and are stable to being bumped up against their containers at speeds of tens of centimeters per second or more. The nebula conditions under which triboelectric (collisional) charging balances charge leakage to the nebula are not at all understood, but a high particle concentration may be needed (see *Desch and Cuzzi,* 2000). Further studies of the sticking of aggregates under microgravity conditions would be valuable.

3.2. Early Growth of Small Aggregates

The following discussion assumes nominal nebular parameters at 3 AU, with temperature 500 K, density 10^{-10} g cm^{-3}, and turbulence characterized by $\alpha = 10^{-4}$ (Re = 3×10^7). The largest eddies have velocity scale $V_g \sim c\alpha^{1/2} \sim 2 \times 10^3$ cm s^{-1}. From the relations in section 2.2, the smallest eddies have scales of size, velocity, and time on the order of 1 km, 25 cm/s, and 1 h, respectively. Consider an initial population of grains with radius 0.5 μm (mass m ~ 10^{-12} g), having abundance 3×10^{-3} times that of the gas. The response time to the gas (equation (1)) is $t_s \sim 5$ s, much less than the large eddy timescale, so $St_L \sim 10^{-7}$. From Fig. 2, it is apparent that relative velocities due to turbulence would be negligible for identical particles. However, collisions at this scale are driven by thermal motions. Particles have thermal velocities $V_{th} = (8\,kT/\pi m)^{1/2} \sim 0.4$ cm/s, from which one can show that the mean collision time is about 10 yr. The collision velocity is less than the threshold for perfect sticking, so grains would coagulate to form fractal structures of low density (*Kempf et al.,* 1999). For simplicity we assume the fractal dimension is 2, which leads to the density of an aggregate being inversely proportional to its radius (mass proportional to the square of the radius). The thermal velocity then varies as r^{-1}, and the number of particles per unit volume as r^{-2}, so the thermal coagulation rate decreases as r^{-1}. For fractal dimension of 2, t_s is independent of the size ($r\rho_s$ is constant), so it seems that particles would coagulate into ever-more-diaphanous cobwebs, and never become compact enough to decouple from the turbulence. However, in the regime of small Stokes numbers, other subtle effects contribute to particle collisions. The gas velocity within the smallest eddies changes on the turnover timescale, subjecting it to accelerations of magnitude $\dot{V}_\eta = V(\eta)\Omega(\eta) \sim V(\eta)\Omega_K Re^{1/2} \sim 10^{-2}$ cm s^{-2} (*Weidenschilling,* 1984). Particles have velocities relative to the gas of magnitude $\dot{V}_\eta t_s$. Because coagulation is stochastic, individual aggregates have a dispersion of fractal dimensions about the mean of about ±10% (*Kempf et al.,* 1999), with corresponding dispersion of t_s. This implies relative velocities of order 0.1 $\dot{V}_\eta t_s \sim 10^{-2}$ cm/s; relative velocities due to turbulence exceed thermal velocities for aggregates larger than tens of micrometers [an additional growth mechanism for tiny grains was described by *Saffman and Turner* (1956)]. The characteristic growth time is a few hundred years. Fractal aggregates will become compacted once collisional energies become large enough to rearrange bonds between grains. *Dominik and Tielens* (1997) predicted that the threshold energy for rolling of micrometer-sized quartz spheres is ~10^{-10} erg. Collisional energies would exceed this value for millimeter-sized aggregates with masses ~10^{-6} g containing 10^6 grains. If the energy threshold is higher by two orders of magnitude (section 3.1), compaction would begin for roughly centimeter-sized aggregates (*Blum and Wurm,*

2000). Thus, there is no problem with the initial growth of aggregates from grains in a turbulent nebula. Indeed, turbulence fosters growth in this size range. In a laminar nebula, growth and compaction would occur on longer timescales, driven by differential settling and radial drift at lower velocities. Once compaction occurs, particles begin to decouple from the turbulence ($St_L \geq 10^{-3}$), and relative velocities increase (Fig. 2), along with collision rates. The particles will continue to grow, unless specific energies of collisions exceed their impact strength.

The impact strength of a primitive aggregate in the solar nebula is uncertain; we have no pristine samples of such material. Theoretical analysis by *Dominik and Tielens* (1997; cf. their Fig. 18) suggests that an aggregate of 100 silicate grains, each 10^{-4} cm in size, would be disrupted by an impact of energy ~10^{-6} erg, implying ~10^{-8} erg per bond between grains. For grain mass 10^{-12} g, this implies a bonding energy ~10^4 erg/g for a chain-like aggregate. A compacted aggregate would have more bonds per grain; also, for purely geometrical reasons, disruption of such an aggregate would involve making and breaking multiple contacts between grains as the impact energy propagated through the aggregate. The experimental results of *Blum* (2000) suggest that the break-up energy per monomer is actually about two orders of magnitude higher than predicted by theory, so impact strengths of ~10^6 erg/g seem reasonable. The actual values will depend on grain sizes and compositions; any additional bonding besides van der Waals forces (e.g., sintering, chemical reactions, or electrostatic forces) would increase the strength. For example, *Sirono and Greenberg* (2000) used the results described above to estimate the strengths of aggregates. In their model, strengths have complex dependence on grain size and porosity; for our purposes it suffices to note that for monomer sizes of a few tenths of a micrometer, and porosities less than 70%, tensile and compressive strengths are in the ranges ~10^5–10^7 and 10^6–10^8 dyn cm^{-2}, respectively. These values may be compared with the (smaller) measured tensile strengths of one powder by *Blum and Schräpler* (2004) and the (larger) experimentally determined strengths of small, but cohesive, silicate objects (*Love and Ahrens,* 1996, their Fig. 7).

3.3. Larger Particles: Settling Toward the Midplane

As particles grow, they acquire larger differential velocities whether or not the nebula is turbulent (section 3.2). Growth accelerates and the packing density of accreted material increases (*Dominik and Tielens,* 1997). Larger particles increasingly decouple from the gas and settle toward the midplane, where their volume density can become much larger than the gas. In this section we describe how the vertical thickness of midplane particle layers is determined, and how this in turn affects accretion.

3.3.1. Thickness of "the midplane": Role of turbulence. The intensity of nebula turbulence determines the degree to which particles settle toward the midplane. Even while particles continue to settle, turbulent diffusion produces a mass flux upward, away from the dense midplane. A steady state occurs when these two mass fluxes balance (*Cuzzi et al.,* 1993, equation (52))

$$\rho_p V_z = \mathcal{D}_p \rho_g \frac{d}{dz}\left(\frac{\rho_p}{\rho_g}\right) \approx \mathcal{D}_p \frac{d\rho_p}{dz} = \frac{\mathcal{D}}{1+St}\frac{d\rho_p}{dz} \tag{4}$$

where $\mathcal{D}_p$ is the particle diffusion coefficient (section 2.4.2) and the two last expressions assume a constant gas density ρ_g, satisfactory for z < H. The situation can be greatly simplified by assuming that the local vertical particle velocity V_z is merely the local terminal velocity $g_z t_s = \Omega_K^2 z t_s$ (*Dubrulle et al.,* 1995) (see also section 2.4). This is not always valid (*Cuzzi et al.,* 1993), but it is a fairly good approximation for particles smaller than a meter or so in size in the 2–3 AU region. *Dubrulle et al.* (1995) derived analytical solutions for the thickness of a layer of particles with different radii and densities, and several nebula α values.

The essence of the results can be understood using dimensional scaling of the derivatives in equation (4): Taking ρ_{po} to be some average density near the midplane and h_p the vertical half-thickness of the particle layer

$$\rho_p V_z \approx \rho_{po}\Omega_K^2 h_p t_s = \frac{\mathcal{D}}{1+St}\frac{\rho_{po}}{h_p} \tag{5}$$

Giving

$$h_p^2 = \frac{\mathcal{D}}{\Omega_K St(1+St)} \tag{6}$$

For global turbulence, $\mathcal{D} = \alpha cH = \alpha H^2 \Omega_K$, thus

$$\left(\frac{h_p}{H}\right)^2 = \frac{\alpha}{St(1+St)} \tag{7}$$

Equation (7) shows that the results of *Dubrulle et al.* (1995), which include other terms extending their validity to the St → 0 limit, can be generalized in terms of a single nondimensional parameter $\mathcal{S}$ = St/α in the limit St < 1 (*Cuzzi et al.,* 1996; their Fig. 1). Settling occurs for $\mathcal{S}$ > 1; as $\mathcal{S}$ increases, either due to larger particles or weaker turbulence, particles settle into thinner layers.

There is a limit as to how thin a layer can be even if global nebula turbulence is zero, which is set by the influence of the layer itself on the gas. As particle mass density ρ_p reaches and exceeds the gas density, the orbiting particles force the entrained gas to rotate at more nearly Keplerian velocity. Because the gas above this dense layer has no such forcing, it orbits at its slower, pressure-supported velocity (section 2.4.1). The associated vertical gradient in gas velocity can be unstable to turbulence, which diffuses the particle layer vertically until it reaches a steady state under its own self-generated turbulence (*Weidenschilling,* 1980). Extensive numerical calculations of this effect have been done by *Cuzzi et al.* (1993), *Champney et al.* (1995) (who studied multiple particle sizes), and *Dobrovolskis et al.* (1999). *Dobrovolskis et al.* (1999) improved on the original work of *Cuzzi et al.* (1993) in several important ways, revising a key parameter, and including a new detailed model for damping of local turbulence *by* the entrained particles; the end results of these changes were particle layer thicknesses fairly similar to those found by *Cuzzi et al.* (1993). One can express the effective gas diffusivity within the near-midplane, self-generated turbulence (cf. *Dobrovolskis et al.,* 1999, equation (12); *Cuzzi et al.,* 1993, equation (21)) as

$$\mathcal{D} \approx \frac{(\beta V_K)^2}{\Omega_K Ro^2} \quad (8)$$

where Ro is the Rossby number of the turbulent layer, defined such that the largest eddies in the self-generated turbulence have frequency $\Omega_e = Ro\Omega_K$. To order unity, Ro = Re∗, where Re∗ is the so-called critical Reynolds number that characterizes the relationship of velocity shear to turbulent intensity for a given flow. The constant Re∗ is derived from measured properties of flows in different regimes. Going back to the basic fluid dynamics literature (e.g., *Wilcox,* 1998, Chapter 5 and references therein), *Dobrovolskis et al.* (1999) found that for flows with geometry most similar to that of the nebula midplane, Re∗ ≈ 20–30 rather than the higher values used previously by others. The range of constants seen in various flows introduces perhaps a factor of 2 uncertainty into Re∗.

Using equation (8) for $\mathcal{D}$, and recalling that β can also be written as c^2/V_K^2, we obtain

$$\left(\frac{h_p}{H}\right)^2 = \frac{(\beta/Ro)}{StRo(1 + StRo)} \quad (9)$$

For particles small enough that StRo < 1, comparing equations (7) and (9) shows that the presence of the layer alone can be identified with a local turbulence equivalent to a global value of $\alpha = \beta/Ro^2 \approx \beta/Re*^2 \approx 10^{-6}$. That is to say, global turbulence must be less than this level for the globally laminar solutions (*Cuzzi et al.,* 1993; *Dobrovolskis et al.,* 1999), which are the regime assumed by all midplane particle layer instabilities, to be valid. For larger particles (StRo > 1), the assumptions going into this simple estimate become invalid. In such particle-laden layers, relative velocities between particles all diminish by a factor on the order of ρ_p/ρ_g (*Dominik et al.,* 2006), making incremental growth very plausible.

As pointed out by *Sekiya* (1998) and *Youdin and Shu* (2002), for sufficiently small particles (the one-phase-fluid regime), a different physics takes over. The steady-state layer described by the physics of equation (4) becomes sufficiently extended that its combination of vertical velocity and density gradients stabilizes it against the Kelvin-Helmholz shear instability, which is responsible for generating gas turbulence, so turbulence ceases (section 3.3.3). In this regime, the thickness of this layer is independent of particle size and is given by $h_p/H = Ri_c^{1/2}H/r$, where Ri_c = 1/4 is the critical Richardson number (*Sekiya,* 1998). In this limit, $h_p > L_E$ and $Ro = \Delta V/(\Omega \max(h_p,L_E)) \sim 1$. Decreasing St further will not lead to thicker layers. Setting the small-St limit of equation (9) equal to the critical-Ri thickness above defines the critical Stokes number for which equation (9) loses validity: $St < 10^{-2}$. Thus the regime studied by *Sekiya* (1998) and *Youdin and Shu* (2002), where for millimeter-radius particles at 2 AU $St \sim 10^{-3}$, falls in the Ri-dominated regime. In the outer solar nebula where gas densities are lower, only correspondingly smaller particles will behave this way.

The simple global turbulence model of equation (7) retains its validity for all St. Taking the entire MMN solid density $\sim 10^{-2}\sigma_g(R)$ in particles of a single size, we can estimate the maximum level of turbulence that would allow unit density particles of a given size to settle into a layer where $\rho_p > \rho_g$ and collective effects become important: $h_p/H < 10^{-2}$. For meter-radius particles, with St ~ 1, $\alpha < 2(h_p/H)^2 = 2 \times 10^{-4}$. For millimeter-radius particles, with $St \sim 10^{-3}$, $\alpha < 10^{-3}(h_p/H)^2 \sim 10^{-7}$. We return to arguments of this sort in section 3.3.3.

3.3.2. Timescales for growth and loss by incremental accretion. In this section we make crude estimates of particle growth and removal rates. We assume the mass growth rate of a particle $\dot{m}$ is dominated by sweep-up of smaller particles, which indeed dominates for all but the very smallest particles. This is a good approximation if velocities are dominated by systematic drift, which is the case if turbulence is low enough to allow settling into a midplane layer. Although larger bodies may dominate the mass distribution, their relative velocities are lower, reducing the collision rate (see *Weidenschilling,* 1988a, 1997, and section 3.2). Then

$$\dot{m} = \pi r^2 \int_0^r V_{rel}(r,r')\rho_p(r')dr' \sim \pi r^2 \overline{V_{rel}(r)}\rho_p(<r) \quad (10)$$

where $V_{rel}(r,r')$ is the relative velocity between particles of radii r and r', $\overline{V_{rel}(r)}$ is its weighted average over r', and $\rho_p(<r)$ is the particle density in sizes smaller than r. Then the e-folding growth time for mass m is crudely

$$\frac{m}{\dot{m}} = \frac{4\rho_s r}{3\overline{V_{rel}(r)}\rho_p(<r)} \quad (11)$$

Since both $\overline{V_{rel}(r)}$ and, to a lesser degree, $\rho_p(<r)$ increase with r, the growth time is only weakly dependent on r, plausibly leading to a particle size distribution that evolves in a self-similar way with equal mass per radius bin as seen in

numerical models (*Weidenschilling,* 1997, 2000). Oversimplifying the above expression further by ignoring fractions of order unity, we estimate the growth times in a nebula with average solid particle density ρ_p as

$$\frac{m}{\dot{m}} \sim \frac{\rho_s r}{\overline{V_{rel}(r)}\rho_p} \sim \frac{(100\ \text{cm})\rho_s(r/1\text{m})}{\rho_p \beta V_K(r/1\text{m})} \sim \frac{(10^4\ \text{cm})\rho_s}{\rho_g \beta V_K \xi}\left(\frac{\alpha}{\text{St}}\right)^{1/2} \sim \frac{10^3}{\xi}\left(\frac{\alpha}{\text{St}}\right)^{1/2} \text{yr} \quad (12)$$

where ξ is a sticking coefficient. We took $\overline{V_{rel}(r)} \sim \beta V_K$ for meter-sized particles (section 2.4.1) and used equation (7) to get $\rho_p \sim 10^{-2}\rho_g(H/h_p) \sim 10^{-2}\rho_g(\text{St}/\alpha)^{1/2}$. If nebula turbulence is lower, denser midplane layers can form in which particles grow much faster (see below).

One can imagine two removal mechanisms for meter-sized particles: They can drift radially inward at the typical rate of 1 AU/century (but will largely be replaced by other particles drifting from further out), or they can collide with comparable-sized particles and be destroyed. We will return to the fate of drifting particles in section 5. Collisional disruption is primarily a worry in turbulent regimes. In turbulence, meter-sized particles, with St ~ 1, couple to the largest, highest velocity turbulent eddies and acquire random velocities relative to each other on the order of $V_{pp} \sim V_g \sim c\alpha^{1/2} \sim 10$ m/s for $\alpha = 10^{-4}$. As shown in Fig. 2, collisional velocities V_{pp} for both smaller and larger particles are slower. Collision times for meter-sized particles with number density n_L and fraction f_L of total solid mass are estimated as

$$t_{coll} \sim \frac{1}{n_L \pi r^2 V_{rel}} \sim \frac{100\rho_s r}{f_L c \rho_g} \sim \frac{30}{f_L} \text{yr} \quad (13)$$

where we used equation (7), $n_L = f_L\rho_p/4\rho_s r^3$, and $V_{rel} \sim V_g \sim c\alpha^{1/2}$. Note that t_{coll} is independent of α.

Recreating a similar calculation by *Weidenschilling* (1988a), we estimate the radius of the largest particle that can survive such collisions as follows: We require $V_{pp}^2 = \text{St}V_g^2 = t_s\Omega_K V_g^2 < E_{crit}/\rho_s$, where $E_{crit} \sim 10^6$ erg cm^{-3} is a plausible strength for a somewhat compacted, meter-sized rubble particle with unit density (section 3.2). The drag regime transitions from Epstein to Stokes for particle sizes in the range of interest. For Epstein drag, we find

$$r < \frac{E_{crit}\rho_g}{\alpha c \rho_s^2 \Omega} \sim \frac{1}{10\alpha} \text{cm} \quad (14)$$

under nominal asteroid belt conditions ($\rho_g = 10^{-10}$ g cm^{-3}, c =10^5 cm/s, and $\Omega_K = 5 \times 10^{-8}$ s^{-1}). Thus particles of this plausible strength grow to meter radius for $\alpha < 10^{-3}$. However, for this size, the Stokes expression is more relevant, which leads to (see section 2.2)

$$r < \left(\frac{2E_{crit}\rho_g\lambda}{\alpha c \rho_s^2 \Omega}\right)^{1/2} \sim \left(\frac{2}{\alpha}\right)^{1/2} \text{cm} \quad (15)$$

where we took $\lambda = 50$ cm as the gas mean free path. Here, meter-radius particles survive $\alpha \sim 2 \times 10^{-4}$. Useful expressions that bridge the transition between Epstein and Stokes drag are given by *Supulver and Lin* (2000) and *Haghighipour and Boss* (2003a).

For $\alpha \sim 10^{-4}$, the growth and collisional disruption times are comparable for meter-sized particles, and the drift distance is less than a few AU. Clearly the local population of meter-sized particles is some mix of material originating at a range of locations. Since particle growth times decrease weakly with r (equation (12)), and removal times are more strongly dependent on r, with drift removal and collision velocity approaching maxima at about 1 m radius, one suspects that growth proceeds readily to meter-sized particles, but further growth is frustrated.

This logic suggests that a balance between turbulence and impact strength might allow bodies to approach meter size, but not to exceed that threshold. As velocities induced by turbulence decrease for larger bodies (St > 1), it might seem that once this threshold is passed there is no further obstacle to growth. However, at low levels of turbulence, or even in a laminar nebula, growth may be limited by erosion due to impacts of small particles upon large bodies. The relative velocity between a small particle coupled to the gas and large body moving at Keplerian velocity is $\beta V_K \approx$ tens of meters per second. Cratering experiments with regolith targets (*Hartmann,* 1985; *Weidenschilling,* 1988b) indicate that a transition from net accretion (i.e., ejecta mass less than the impactor mass) occurs at velocities ≤10 m s^{-1}, with net erosion at higher velocities. *Wurm et al.* (2001) suggested that gas drag would cause ejecta to reimpact the target, but *Sekiya and Takeda* (2003) showed that they would be swept away by the flow. A target body may experience a range of impacts with gains and losses, with the net effect on its mass dependent on the size distribution of the impactors.

When relative velocities are due to differential drift, impact velocities of small particles onto large bodies approach a constant value of βV_K, regardless of size. If impacts at this speed are erosive, there may be no threshold size beyond which further growth is assured, as the large bodies would be subject to continued "sandblasting" by small impactors. For accretion to overcome erosion, it may be necessary for impact velocities to decrease. As β is proportional to the gas temperature, collisional growth may be delayed as the nebula cools. Another possibility is that the effective value of β in the midplane may be decreased by particle mass loading. If particles can settle into a layer with $\rho_p > \rho_g$, then the particles drag the gas within the layer, and its rotation becomes more nearly Keplerian. The effect of such mass loading was analyzed by *Nakagawa et al.* (1986) and included in the coagulation model of *Weidenschilling* (1997), who showed that it could decrease collision velocities to a maximum of ~10 m s^{-1}. Attaining a midplane density sufficient to affect the layer's rotation depends on the particle size distribution (cf. equation (9ff)), but probably requires $\alpha < 10^{-6}$ to ensure easy growth of meter-sized particles without disruption. Thus, it is plausible that a gradually decreasing level of turbulence may lead to a rather abrupt transition from small submeter-sized bodies to large planetesimals.

3.3.3. Midplane instabilities to breach the meter-sized barrier? A number of midplane collective effects and instabilities have been pointed out over the years, with the goal of helping primary accretion proceed to and beyond the perceived meter-sized barrier. All these require particle densities well in excess of the gas density, and sometimes far in excess of it, which places rather strict limits on the level of global turbulence. Below we discuss these ideas.

Particle layer gravitational instabilities: There are two distinct regimes in which midplane gravitational instabilities have been studied. The first is two-phase models in which the particles are treated as moderately decoupled from the gas, while responding to gas drag, and the gas as having constant density (*Safronov,* 1969; *Goldreich and Ward,* 1973, hereafter *GW73*; *Weidenschilling,* 1980, 1984, 1995; *Cuzzi et al.,* 1993, 1994; *Dobrovolskis et al.,* 1999; *Ward,* 2000). The second is one-phase models in which the particles are so small and firmly coupled to the gas that together they comprise a single fluid with vertically varying density (*Sekiya,* 1998; *Sekiya and Ishitsu,* 2000, 2001; *Youdin and Shu,* 2002; *Ishitsu and Sekiya,* 2002, 2003; *Youdin and Chiang,* 2004; *Yamoto and Sekiya,* 2004; *Garaud and Lin,* 2004). In both cases conditions are sought under which the particle density can become so large that the layer undergoes a gravitational instability.

The original two-phase formulation (*GW73*) was attractive because it suggested that planetesimal-sized objects could be formed directly from centimeter-sized particles in a very thin particle layer, surrounded by a turbulent boundary layer. As discussed in section 3.3.1 (see equation (7)), $\alpha \ll 10^{-6}$ for this scenario to be considered at all. Moreover, *Weidenschilling* (1980) showed that centimeter-sized particles were prevented from settling into a sufficiently thin layer by this very self-generated turbulence, and *Cuzzi et al.* (1993) and *Dobrovolskis et al.* (1999) showed that this conclusion remained true even up to meter-sized particles. Moreover, for particles large enough and velocities small enough to approach the particle density threshold normally cited for instability, only transient gravitational clumps result [like the "wakes" in Saturn's rings (*Salo,* 1992)], and not direct collapse to planetesimals as usually envisioned (*Cuzzi et al.,* 1994). *Tanga et al.* (2004) follow large particles (St ≫ 1) in regimes of nominal (marginal) gravitational instability and reproduce just this expected behavior. They then apply an *ad hoc* damping of relative velocities, treated as gas drag acting on much smaller particles with St = 1; this damping allows bound clumps to form that merge with each other over time. However, this study assumed single-sized particles, and *Weidenschilling* (1995) showed that even differential drift velocities in a realistic size distribution would stir the layer too much for it to become unstable. Essentially, by the time particles have formed that are sufficiently large to reach interestingly high midplane densities, the premises of the original instability scenario are no longer valid [see also *Ward* (2000) for a similar conclusion, with caveats about model parameters]. This objection would also apply in the outer solar system where *Möhlmann* (1996) has noted that gravitational instability is more favored. Conceivably, a two-phase gravitational instability might be allowed if the surface mass density of 1–10-cm particles could somehow greatly exceed the nominal solar abundance. Some suggestions have been made along these lines, as discussed below and in section 5. Further studies along the lines of *Tanga et al.* (2004) would be of interest, in which particles of different sizes, with the appropriate St values, were treated self-consistently. Recall that these studies, as with all other midplane instability models, assume a nonturbulent nebula.

The one-phase approach mentioned above (*Sekiya,* 1998, et seq.) assumes particles so small that they cannot separate from the fluid on dynamical timescales. They can, however, settle very slowly into a moderately dense layer *if the global turbulence allows it* (see section 3.3.1 and below). In this regime, the entire (gas plus particle) medium has a strong vertical density gradient that can stabilize the medium against turbulence (the so-called Richardson number criterion). That is, a fluid parcel attempting to become turbulent and rise must work against its natural lack of buoyancy. It turns out that under nominal nebula solid/gas ratios of about 10^{-2}, the layer cannot become gravitationally unstable in this way. However, it was noted by *Sekiya* (1998), and more emphatically by *Youdin and Shu* (2002), that if the solid/gas ratio were significantly enhanced (factor of 10 or so), a thin layer very near the midplane could become unstable. This is essentially because in such dense layers, the mixed-fluid vertical velocity profile is nearly flat for a considerable vertical range, independent of density — it is saturated, as it were, so unstable shear does not develop. *Gomez and Ostriker* (2005) have found that a more careful treatment of vertical shear leads to a layer that is more prone to turbulence — thicker and less dense by about an order of magnitude — than found by *Sekiya* (1998) and *Youdin and Shu* (2002).

Whether true gravitational instability can can occur via this scenario is problematic. First, particles small enough to satisfy the one-phase approach must have $t_s < (dv/dz)^{-1}$ in the marginally stable layer. The layer must have $h_p/H < 10^{-3}$ to approach gravitational instability ($\rho_p > 10^2 \rho_g$), even if the solid/gas ratio were *enhanced* by an order of magnitude over solar. The shear in this layer, which determines the frequency of nascent eddies, is approximately $\Omega_e \sim \Delta V/h_p \sim 10^3 \Delta V/H \sim 10^3 \beta V_K/H \sim 2\Omega_K(R/H) \sim 40\ \Omega_K$. For the one-phase nature, and stabilization by stratification, to be preserved at the onset of instability, $t_s\Omega_e \ll 1$ or St < 10^{-2} is thus required. From equation (7) we see this requires the global $\alpha < 10^{-8}$ for the enhanced solids scenario (or 10^{-10}

for a nominal solid/gas ratio). Second, for such weak nebula turbulence it is hard to imagine particles settling all the way to the midplane without growing at all [see section 3 and *Dullemond and Dominik* (2005)]. Any significant growth would produce self-generated turbulence that would stir the layer (e.g., *Cuzzi et al.,* 1993; *Dobrovolskis et al.,* 1999). Third, mechanisms for enhancing the particle/gas abundance for precisely those particles that are most firmly coupled to the gas are not easily found. *Weidenschilling* (2003a) and *Youdin and Chiang* (2004) get different results regarding the role radial drift might play in enhancing solids (see also below). Achieving this goal by removing gas (by, e.g., photoionization or a disk wind) increases t_s, making it harder to satisfy the requirement that $t_s < (dv/dz)^{-1}$ for density stabilization to apply.

Perhaps most difficult of all for the one-phase gravitational instability scenario is a basic physical obstacle that was pointed out by *Sekiya* (1983) but has apparently been overlooked by subsequent workers except for *Safronov* (1991). In a system where the particles are firmly trapped to the gas ($t_s \ll$ the dynamical collapse time ~ $(G\rho_p)^{-1/2}$), gas pressure precludes gravitational instability, in the traditional sense of inexorable collapse on a dynamical timescale, until the solid/gas mass ratio exceeds 10^7 (at 2.5 AU)! This is 3 orders of magnitude larger than the typically cited thresholds for gravitational instability, and a rather unlikely enhancement. Recent numerical simulations and analytical models support this result (K. Shariff, personal communication and in preparation, 2005). Prior work on midplane instabilities dating back to *Goldreich and Ward* (1973) has treated the "pressure support" term in the dispersion relation as if due to particle random velocities, as decoupled from the gas (as in *Toomre,* 1964), whereas in a one-phase fluid the pressure term involves the far larger gas sound speed, divided by a mass-loading factor. Only for huge loading factors does the gas pressure support term become as small as the traditional pressure support term and allow instabilities to grow (*Sekiya,* 1983). Physically, what happens is that incipient collapse by the mass-dominant particles compresses the entrained gas, leading to an outward pressure gradient that forestalls further collapse of the gas (and trapped particles).

Sekiya (1983) suggests that an *incompressible* three-dimensional mode can emerge for solid/gas mass-loading ratios comparable to the so-called "Roche density" (ρ_R ~ $15(3\ M_\odot/4\pi r^3)$ ~ $10^3\ \rho_g$). The nature of these modes has never been followed, and the fate of solid material within them would seem to be less in the nature of an instability than merely further settling at, effectively, slow terminal velocity. Of course, achieving even the Roche density in a midplane environment still faces the obstacles mentioned above (see section 4, however).

Can enhancement of particle surface density by global radial drift foster gravitational instability? In section 5 we describe how radial drift might lead to significant changes in the local density of solids relative to nominal solar abundance. One naturally asks whether this effect might play a role in fostering instabilities. *Stepinski and Valageas* (1997) and *Kornet et al.* (2001) noted large enhancements in the surface mass density of solids in a range of sizes, due to this effect. Of course, the nebula must still be globally nonturbulent, and one still has to deal with the distinction between transient wavelike instabilities and direct collapse to planetesimals (*Cuzzi et al.,* 1994). Application of radial drift enhancement to the one-phase instability (*Youdin and Shu,* 2002; *Youdin and Chiang,* 2004), besides facing the serious obstacle of gas pressure mentioned above, requires millimeter-sized particles to drift many AU, over 10^5–10^6 yr, in a dense midplane layer with low relative velocities, without growing at all. This process depletes the outer nebula of solids while enhancing the inner region; numerical models (*Youdin and Shu,* 2002; *Youdin and Chiang,* 2004) achieve the critical density inside a few AU, but effectively denude the nebula of solids at larger distances. This migration would have to occur before the formation of the outer planets, which would otherwise act as barriers to drift. In that case, no material would be left from which to make their cores, or comets (*Weidenschilling,* 2003a). Also, the extent of drift would mean that much of the material that formed the terrestrial planets and asteroids originated at much larger heliocentric distances; the implications for the chemistry of this matter has not been addressed.

Can enhancement of large particles inside nebula vortices or spiral density waves foster gravitational instability? It was pointed out by *Barge and Sommeria* (1995), and studied further by *Tanga et al.* (1996), *Bracco et al.* (1998), and *Klahr and Bodenheimer* (2003b), that large, two-dimensional, circulating vortices (*not* true turbulent, randomly fluctuating eddies) had the ability to concentrate meter-sized boulders near their centers. This is a specific form of fostering gravitational instability by increasing surface mass density. Comparable effects related to spiral density waves (in a globally gravitationally unstable nebula) were mentioned in section 2.4.1.

Long-lived vortices rotate clockwise, in the sense of the local radial gradient. Meter-sized boulders are most affected because their t_s is comparable to the rotation time of these vortices. In the enhanced headwind of these vortices (~$\beta^{1/2}V_K$ rather than βV_K), meter-sized boulders drift radially by about the extent of a scale-height-size eddy in one orbital period. When in the outer part of the vortex, meter-sized particles suffer a stronger than normal headwind and their semimajor axis shrinks. When on the inner part, they incur a comparably strong tailwind and their semimajor axis expands. In the local co-rotating frame, they appear to spiral in to the center of the vortex. While some studies indicate vortices of this sort are long-lived (*Godon and Livio,* 2000), other studies indicate they are not (*Davis et al.,* 2000), so it is not clear what the net effect of this concentration might be. Recently *Barranco and Marcus* (2004) have obtained a solution for the full three-dimensional flow in such a large vortex; surprisingly, the rotational flow vanishes near the midplane where meter-sized particles would lie. Furthermore, one must wonder if all the motion in these giant vortices — with speeds approaching the sound speed — can exist without producing some degree of internal turbulence that would prevent even meter-sized particles from settling to the midplane, so any instability mechanisms would remain precluded. As particles collided and broke into smaller particles, the rubble would escape the vortex. Overall, for several reasons, we think this effect probably does not play a significant role in primary accretion.

In spiral density waves, the compressional effect is systematic and is not thought to add to the random velocity (W. Rice, personal communication, 2005), but preexisting three-dimensional dispersions of material are retained and the wave itself might increase nebula turbulence (*Boley et al.,* 2005). Whether or not the increased abundance (and thus collision rate) leads to increased accretion in either waves or vortices depends on the relative collision velocities. Similar concerns apply to concentrations of boulders found in less systematic radial bands of locally high gas pressure (*Fromang et al.,* 2005; *Johansen et al.,* 2005).

Secular instability? A third class of midplane instability model is the drift instability (*Goodman and Pindor,* 2000), which also presumes a globally nonturbulent nebula with quite low α ($\ll 10^{-6}$; section 3.3.1) in order for particles to drive gas velocities. The midplane layer is thought of as a unit drifting inward at a single speed that depends on its surface mass density. As shown by *Nakagawa et al.* (1986), *Cuzzi et al.* (1993), and *Dobrovolskis et al.* (1999), dense particle layers are self-shielded from the gas — they drive the entrained gas closer to the Keplerian velocity, and suffer a much weaker headwind than isolated particles — thus drifting inward more slowly. The premise is that, if some patch of the layer became slightly more dense than its radial neighbors, its drift rate would slow down and material from outside could catch up with it, further increasing its density. This suggestion is interesting, but under perhaps more realistic conditions, where particles in the layer have a range of different sizes and drift rates, as well as particles at different vertical levels in the layer having different drift rates, it would probably be washed out. *Weidenschilling* (2003a) points out that there is a feedback between a local density enhancement and turbulent stress. If a section of the particle layer becomes more dense, then its orbital speed increases, becoming more nearly Keplerian. This increases the shear between the layer and the surrounding gas, and acts to counter the slower drift rate that would result from the density enhancement. This effect depends on the degree of coupling of the particles and the gas, i.e., on their size. It is not taken into account by *Goodman and Pindor* (2000), and its significance is still a matter of controversy. *Safronov* (1991) and *Ward* (1976, 2000) also mention several longer-term, *secular* or *dissipative* instabilities that thin layers are subject to, again if global turbulence is vanishingly small. These are worthy of more study, but the dispersion relations on which these are based are only valid for nonturbulent nebulae, and turbulence will add diffusion terms that will stabilize short lengthscales.

Summarizing: Midplane instabilities of various kinds have been suggested as ways to allow growth to continue into and beyond the problematic meter-sized "barrier." However, as described above in section 3, recent incremental growth models find that the growth beyond meter size is not a problem in nonturbulent regimes, and the instability approaches are invalid in even weakly turbulent regimes where the meter-sized barrier may be a problem. Furthermore, long-neglected gas pressure effects make gravitational instability much more difficult for small particles than widely believed. It may be that local, large-scale, gas dynamical structure in the nebula might be able to increase the density of particles, and thus their collision frequency. However, the net result for accretion depends on the specifics of random collision velocities, which have not been modeled in these scenarios.

4. TURBULENT CONCENTRATION: FINGERPRINTS IN THE ROCKS

In this section we describe an aerodynamic process that is new since the original volume in this series (*Kerridge and Matthews,* 1988) and has the potential for breaching the meter-sized barrier and producing objects that look like actual meteorite parent bodies. The process is preferential concentration in turbulence, or turbulent concentration.

4.1. The Basic Process

One's intuition is that turbulence is a dispersive, homogenizing, and mixing process, and indeed this is true from the standpoint of the global scale trajectories of particles of *all* sizes (i.e., the V_p of section 2.4.2). However, numerical simulations (*Squires and Eaton,* 1990, 1991; *Wang and Maxey,* 1993; *Cuzzi et al.,* 1996, 2001; *Hogan et al.,* 1999) and laboratory experiments (*Eaton and Fessler,* 1994) show that particles having a narrowly selected aerodynamic stopping time t_s actually concentrate in dense zones in turbulent, incompressible fluids. Particles avoid zones of higher vorticity in the gas, so the more densely populated zones are those of lower vorticity. The dense zones move in inertial space, following the fluid properties that foster them. This effect is entirely different in its basic physics, and applies to particles with entirely different properties, than vortex-center concentration of meter-sized boulders as discussed in section 3.3.3. Turbulent concentration is a way to aerodynamically select particles of a very specific t_s for significant local concentration — by orders of magnitude. In the next two subsections we address (1) the size-selection criteria and (2) the concentration factor *C*.

4.2. Size Selection and Size Distribution

Preferentially concentrated particles have stopping time t_s equal to the overturn time of the Kolmogorov scale eddy, $t_e(\eta)$ (*Wang and Maxey,* 1993). We will sometimes refer to these particles as having Stokes number $St_\eta = t_s\Omega(\eta) = t_s/$

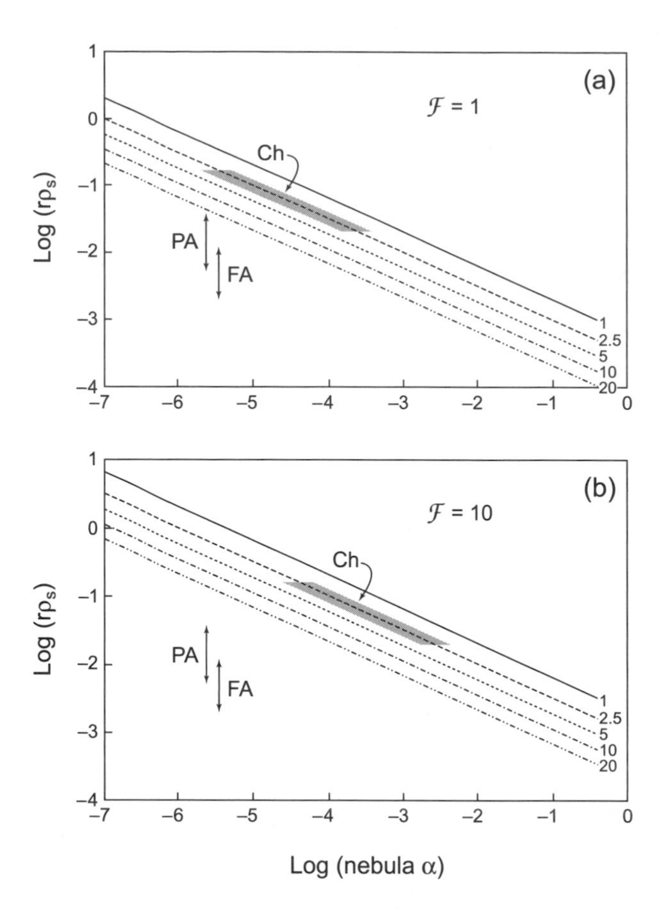

Fig. 3. The radius-density product for preferentially concentrated particles in nebula turbulence of different intensity levels α (equation (16)), depicted at several distances (AU from the Sun) in a typical protoplanetary nebula model (*Cuzzi et al.,* 2001) with **(a)** no enhancement ($\mathcal{F}$ = 1) and **(b)** a factor of 10 enhancement ($\mathcal{F}$ = 10) over the total surface density of a MMN. Denser nebulae concentrate the same $r\rho_s$ at higher α. Observed chondrules fall in the vertical range indicated by the shaded area on the line at 2.5 AU. Porous aggregates (PA) and fluffy aggregates (FA) would have lower $r\rho_s$ products, and would be similarly concentrated at lower gas densities.

$t_e(\eta) = 1$. *Cuzzi et al.* (1994, 1996, 2001) showed that this constraint pointed directly to chondrule-sized silicate particles when nominal nebula conditions were assumed. The stopping time $t_s = r\rho_s/c\rho_g$ is easily related to nominal nebula and chondrule properties. We recall from section 2.2 the eddy time $t_e(\eta) = \Omega_K^{-1}Re^{-1/2}$, and $Re = \alpha cH/\nu_m$. These relationships can be easily combined into a very general expression for the preferentially concentrated particle, having radius r and density ρ_s

$$r\rho_s = 6.3 \times 10^{-4}\left(\frac{\mathcal{F}}{\alpha}\right)^{1/2}\left(\frac{R}{1\ \text{AU}}\right)^{-3/4}\ \text{g cm}^{-2} \qquad (16)$$

In the above expression, a nebula has been assumed with surface mass density $\sigma(R) = 1700\ \mathcal{F}(R/1\text{AU})^{-3/2}$ g cm^{-2}, where $\mathcal{F}$ is the enhancement over a MMN (*Cuzzi et al.,* 2001). Predictions are shown in Fig. 3. Note that a number of combinations of α and nebula density (here represented by distance from the Sun) can concentrate the same particle radius-density product $r\rho_s$. The actual value of nebula α remains very uncertain (section 2.1). Based on nebula lifetimes, and assuming that nebula viscosity is basically the mechanism responsible for disk evolution, values in the 10^{-5}–10^{-2} range are usually inferred. Note also that smaller particles, and/or particles of much lower density — porous or even "fluffy" aggregates — are also concentrated by this process, but in regions of much lower gas density (or higher α) than for solid chondrules. This might be highly relevant for outer solar system accretion, in regions where, perhaps, nearly solid "chondrule"-like particles do not form; however, it might be hard to find evidence for such an effect.

Not only the typical size, but also the characteristic size distribution of meteorite constituents seems to be in good agreement with the predictions of turbulent concentration. *Paque and Cuzzi* (1997) and *Cuzzi et al.* (2001) compared the typical size and density of chondrules separated from five different carbonaceous and ordinary chondrites with model predictions of relative abundance as a function of particle stopping time (as expressed in terms of St_η) and aligning the peaks of the theoretical and observed plots to compare the shapes of the curves. The agreement is very good, especially given that the theoretical values have no free or adjustable parameters (*Hogan and Cuzzi,* 2001).

Aerodynamic size sorting subsequent to chondrule formation might not be the whole story, of course; the process(es) by which chondrules, CAIs, and metal grains formed in the first place certainly did not produce an infinitely broad size distribution. Some heating mechanisms, notably shock waves in the nebular gas (see *Connolly et al.,* 2006), might have preferentially melted millimeter-sized objects; tiny chondrule precursors might have evaporated, and very large ones would not be melted thoroughly, if at all. Alternatively, it has been suggested that surface tension in silicate melts played a role in helping determine chondrule size distributions (e.g., *Liffman and Brown,* 1995, and others).

However, as discussed in section 2.2.1, extensive studies of the size distribution of chondrule-sized particles in meteorites, using thin section and disaggregation techniques (*Hughes,* 1978, 1980), have determined that both true, rounded chondrules and broken "clastic fragments" of once-larger chondrules follow the same size distribution. Apparently, some chondrules were formed with sizes much larger than those in the meteorite, but only after being fragmented by some other process at a later time could their fragments be size-sorted into the parent body along with whole chondrules of similar size. Also, *Krot et al.* (1997) report tiny "microchondrules" in accretionary rims around many chondrules in type 3 chondrites; they are especially abundant in some cases. Since the chondrule-formation process seems to have resulted in size distributions broader than those that are typically found in the chondrites, the narrow observed size distribution probably is not best explained by a process acting only on the liquid state, such as surface tension, or perhaps by the formation process at all. Furthermore, *Cuzzi et al.* (1999) note that the fit of Fig. 4 is considerably improved if individual chondrule radius-density products are used, rather than assuming some average density; this implicates aerodynamics at a stage between formation of chondrules and similar objects, and their accretion. Other approaches to aerodynamic sorting (*Huang et al.,* 1996; *Ackridge and Sears,* 1998) favor a postaccretionary, parent-body process using effluent vapor to create a fluidized bed in which denser particles settle relative to less-dense ones.

An open question that would be valuable to resolve is the detailed size distribution of the irregularly shaped and often porous CAIs, as well as the irregularly shaped metal grains, found in different meteorite types. While the fact that chondrites with large chondrules (CVs) also have large CAIs (type B) and chondrites with small chondrules have small CAIs, it also appears that most type B CAIs are larger than most CV chondrules. Can this be reconciled by density/porosity differences? It is hard to say, as quantitative measurements of CAI size distributions are minimal at best (*May et al.,* 1999).

The agreement of both the characteristic size, and size distribution, of chondrules with the simplest predictions of turbulent concentration theory are what we call fingerprints of the process. A third fingerprint might be present in the relationship of fine-grained accretion rims to their underlying chondrules (section 6.3).

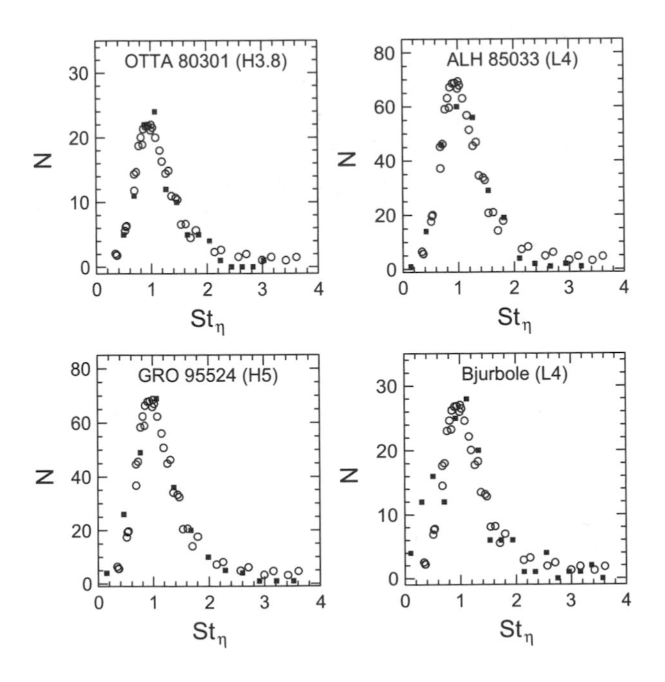

Fig. 4. Comparison of observed chondrule size-density distributions $N(r\rho_s)$ from four different meteorites (solid symbols) with theoretical predictions (open symbols) (*Cuzzi et al.,* 2001). The observed distributions are merely assumed to peak at $St_\eta = t_s/t_\eta = 1$ and are aligned with the predictions (which have no free parameters and are the same in all panels).

4.3. How Much Concentration?

Particle concentration shares a number of properties with other properties of turbulence that are termed *intermittent.* For instance, it is widely known that dissipation of turbulent kinetic energy ε occurs on the Kolmogorov scale; it is less widely known that the spatial distribution of this quantity is highly nonuniform. While locally and temporally unpredictable and fluctuating widely (i.e., intermittent), ε has well-determined statistical properties — its probability distribution function (PDF) is well determined on any lengthscale. Furthermore, it has been shown that the PDFs of both ε and particle concentration share a fractal-like descriptor called the singularity spectrum that is independent of Reynolds number. Determining this fundamental rule of the process (*Hogan et al.,* 1999, and references therein) allows one to predict the PDFs of both ε and particle concentration at any Re (*Cuzzi et al.,* 2001), as long as the process remains independent of lengthscale. This behavior is closely related to the idea that turbulence is a scale-independent cascade process. That is, transport of kinetic energy, vorticity, and dissipation from their sources at large scales to their sinks at small scales obeys rules that are independent of scale. Scale-independence and Re-independence are connected because Re determines the depth of the inertial range: $Re^{3/4} = L/\eta$ (section 2.2). That is, larger Re means a larger number of eddy bifurcations between L and η, and stronger fluctuations in intermittent properties (see, e.g., *Meneveaux and Sreenivasan,* 1987).

The behavior of both ε and *C* can be explained nicely using a cascade model, where upon bifurcation of a unit volume parent eddy into two equal-volume daughter eddies, some quantity x is partitioned unequally, following some multiplier p. Then the density of x becomes p/0.5 in one subeddy and (1 – p)/0.5 in the other. If $p \neq 0.5$, fluctuations in the density of x emerge that grow with successive levels in the cascade. Cascades can be crudely modeled with constant p; more realistically, p is chosen at each level from a PDF that itself is independent of level in the cascade (*Sreenivasan and Stolovitsky,* 1995b). The PDF of multipliers is the rule that is level, or Re, independent. Excellent agreement has been obtained for high-Re turbulence with models of this sort.

However, changing physics can change the rule. Specifically, increasing local particle concentration and the associated mass loading $\rho_p/\rho_g \gg 1$ can locally kill off the turbulence that leads to intermittency. In new work that is as yet incomplete, cascades are being modeled in which the multiplier PDF can be made conditional on local particle density, to account for the effects of mass loading and turbulence damping. Full three-dimensional simulations are being used to determine how multiplier PDFs depend on ρ_p/ρ_g. Preliminary results (R. C. Hogan and J. N. Cuzzi, in preparation, 2004) indicate that concentration continues to grow well after local particle density significantly exceeds the local gas density: regions of $\rho_p \sim 100\ \rho_g$ are seen with low but interesting probability even at fairly low Re, and with increasing probability at higher nebula Re. The goals of ongoing work include determining the lengthscales on which, and the probability with which, zones of different density occur.

4.4. Getting to Planetesimals

In this section we speculate on how, or whether, the fingerprints of turbulent concentration might become directly manifested in actual meteorite parent bodies. Even a particle density in the densest zones is still far from solid density; other physics must come into play to transform dense, particle-rich parcels of the nebula, with mass density less than that of cotton candy, into a meteorite parent body or its immediate precursor. This process remains unstudied and is fraught with obstacles. The scenarios sketched below by which this evolution may proceed are not in any way proven, but intended to provoke further thought and discussion.

Dense, particle-rich zones experience solar gravity as a unit; at lower density than gravitational instability can set in, they begin to settle toward the midplane and move azimuthally through the surrounding gas. The ensuing gas ram pressure and shear flow of displaced gas around the sides of the clump, and perhaps buffeting by turbulent fluctuations it must traverse as it moves, can destroy a strengthless clump in the absense of a counteracting force. However, counteracting forces might exist. Analogies can be sought in the behavior of fluid drops settling in lower density fluids. Pressure stresses quickly destroy fully miscible drops, but surface tension allows strengthless drops to survive and settle as units (*Thompson and Newhall,* 1885; *Clift et al.,* 1978; *Frohn and Roth,* 2000). In the nebula context, the self-gravity of the dense clumps might play a role analogous to surface tension. To the extent that turbulent concentration can produce sufficiently dense regions before viscous disruption occurs, self-gravity might be able to sustain these entities as they settle toward the midplane, or coagulate with similar dense clumps. Once densities and collision rates became high enough, collective electrostatic (dipole-dipole) forces, due to tribocharging in dense clusters (section 3.1), might also increase the binding energy of clumps that were still far from solid.

Candidate dense zones must be considerably larger than the Kolmogorov scale η. The dynamical time of a clump with the Roche density 10^3–$10^4\ \rho_g$ (section 3.3.3) is $t_{coll} \sim (G\rho_p)^{-1/2}$, roughly a few weeks. In order for this material to survive being torn apart by eddy motions, it must be larger than eddies with this timescale. Since eddy times scale as $\ell^{2/3}$ (section 2.2), and the Kolmogorov scale eddy $\eta \sim 1$ km has an eddy time of several thousand seconds, stably bound zones with $\rho_p \sim 10^3$–$10^4\ \rho_g$ must also be of a size scale $\geq 100\ \eta$. The abundance of very dense zones, at fairly large scales, remains to be determined (section 4.3). If spiral density waves repeatedly pass through the nebula, condensing gas and embedded particles by an order of magnitude or more as they pass (e.g., *Rice et al.,* 2004), achieving critical density for gravitational instability might become easier. Clumps with these properties have masses greatly exceeding the "meter-sized barrier," and might be the direct precursors of large planetesimals composed entirely of similarly-sized objects ("chondrules" and their like).

Certainly, these scenarios for the terminal stage of primary accretion remain only suggestive and are untested. We return to the issue of primary accretion in section 6.4, after discussing one more process with important implications for planetesimal formation and meteorite properties — radial decoupling of planetesimal-forming materials from the nebula gas.

5. RADIAL MIXING: DECOUPLING OF PLANETARY MATERIALS FROM THE GAS

Aerodynamic effects can lead to significant radial redistribution of material. These motions are of several possible types. As mentioned in section 2.4.1, large particles generally incur a headwind and drift rapidly inward. In the presence of turbulence, particles in the chondrule-CAI size range and smaller are coupled to gas motions quite well, and their random velocities V_p induce radial and vertical diffusion much like that of gas molecules. While hydrogen molecules are all the same, small solid particles retain a memory of the chemistry, mineralogy, petrology, isotopic content, and crystallinity of their formation region. Diffusion in the presence of radial gradients in any of these properties can lead to significant, and potentially observable, radial evolution of small particles — both inward and outward. Also, direct outward radial transport by stellar or disk winds well above the disk can also be significant for small particles, primarily in the early stages of disk evolution when they are most vigorous.

In this section, we discuss outward ejection of small particles by winds, and inward drifts due to gas drag (especially in connection with evaporation boundary transitions). We discuss diffusion in section 6.

5.1. Outward Ejection by Stellar and Disk Winds

Some theories (see *Connolly et al.,* 2006; *Chaussidon and Gounelle,* 2006) posit formation of CAIs and/or chondrules very close to the protosun (several protostellar radii; about 15 $R_\odot$, or less than 0.1 AU), not far from where nebula gas funnels into the star along magnetic field lines. In these theories, outward transport to parent-body formation regions near 2–3 AU is attributed to entrainment in the stellar wind (*Skinner,* 1990a,b; *Liffman and Brown,* 1995;

Shu et al., 1996, 1997). *Shu et al.* (1996) define a parameter that we simplify slightly and call ζ

$$\zeta = \frac{3\dot{M}_w}{16\pi R_x^2 \Omega_x \rho_s r} \quad (17)$$

where $\dot{M}_w$ is the total mass flux in the stellar wind, and R_x and Ω_x are the radius and rotation frequency at the x-point near 15 $R_\odot$, from where the stellar wind originates in this theory. For simplicity we have used Ω_x rather than the comparable rotation frequency of the star itself, and dropped a constant of order unity. The density of the emerging wind at the launch site is

$$\rho_x \sim \frac{\dot{M}_w}{4\pi R_x^2 V_w} \sim \frac{\dot{M}_w}{4\pi R_x^3 \Omega_x} \quad (18)$$

and so the coupling parameter ζ can be seen to be

$$\zeta = \frac{3}{4} \frac{\dot{M}_w}{4\pi R_x^3 \Omega_x} \frac{R_x}{r\rho_s} \sim \frac{c\rho_x}{r\rho_s} \frac{R_x}{c} = \frac{R_x/c}{t_s} = \frac{R_x/H}{\Omega_x t_s} \quad (19)$$

For present purposes we can ignore the difference between the wind speed and the sound speed c, so ζ can be regarded as the ratio of the time for the wind to travel out of the region to a particle stopping time. For the very hot plasma at the x-point, $R_x/H \sim 0.3$, so $\zeta \sim 0.3/St_x$ where $St_x = t_s\Omega_x$, a Stokes number with Ω_x serving as the normalizing frequency.

A similar derivation for the ratio of the gravitational force on a particle to the gas drag force on it ($\pi r^2 \rho_x V_x c$) reveals them to be equal when $St_x = 1$. Particles experiencing a drag force comparable to gravity, even if it begins in the tangential direction, have their orbits severely perturbed and are slung outward, as well as being lifted out of the plane by the wind. *Shu et al.* (1996) find that, to reach the several-AU region, particles with $St_x \sim 1$ are required. A similar derivation by *Liffman and Brown* (1995; their equation 3.9) gets a similar result.

Shu et al. (1996) found that the throw distance is fairly sensitive to ζ, and suggested that the wind could thereby play a role in size-sorting of chondrules and/or CAIs. However, a more recent version of the x-wind model (*Shu et al.,* 1997) proposes radial fluctuations in the launching point, on decadal timescales, caused by magnetic flux variations in the x-region (typically <0.1 AU). There are several-orders-of-magnitude changes in gas density across the x-point. The wind density fluctuations experienced by particles at different distances, as the launch point sweeps back and forth across them, would cause wide variations in the launch aerodynamic parameter ζ, calling into question whether this mechanism can, by itself, be responsible for the very narrow chondrule size distributions.

It should also be noted that other theories for the observed stellar winds have them originating over a much wider range of disk radii — up to several AU (so-called "disk winds") (e.g., *Königl and Pudritz,* 2000). Recall from equation (18) above that the entrainment gas density, and thus the drag force, is derived by assuming the entire observed wind mass flux $\dot{M}_w$ arises from a very small region of scale $R_x < 0.1$ AU. Disk winds have a much more extended source and thus a far lower entrainment gas density, and therefore may not be capable of ejecting chondrule- or CAI-sized particles. Some recent papers have provided support for disk winds, at least as causing the lower-velocity winds where most of the mass flow $\dot{M}_w$ appears (*Anderson et al.,* 2003; *Hartmann and Calvet,* 1995). It would be useful to resolve this issue.

5.2. Inward Radial Drift of Large Particles and Mass Redistribution

Global scale radial redistribution of large solid particles relative to the gas, under gas drag, was first discussed in detail by *Stepinski and Valageas* (1996, 1997). Since that time it has been studied further by *Kornet et al.* (2001) and *Weidenschilling* (2003a). *Morfill and Völk* (1984) had addressed the problem earlier in general terms, but with a specific application to particles that were too small to show much of a drift-related effect. *Stepinski and Valageas* (1996, 1997) (and *Kornet et al.,* 2001) simplified complex processes into a semi-analytical model to obtain insights into the global parameter space within which solids evolve. Among their simplifications was a single local particle size and collisions driven primarily by turbulence; however, *Weidenschilling* (1997, 2003a) argued that collisional growth is driven primarily by size differences and corresponding systematic velocity differences rather than turbulence (section 3). Nevertheless, some interesting qualitative conclusions emerge. In all these models, outcomes are determined by a race between growth and drift. In general, solids grow and drift fast enough to deplete the outer disk of solids while enhancing the solid/gas ratio further inward. This redistribution in surface mass density is essentially a radial convergence effect associated with the cylindrical geometry and the dependence of drift velocity on location and particle size (cf. section 3.3.3). However, migration stops when (if) coagulation produces bodies large enough (kilometer-sized or larger) for the drift rate to decrease to low values (cf. Fig. 1), and does not necessarily deplete the outer nebula of solids or produce a large enhancement in the inner nebula. Fairly abrupt outer edges are seen in the final radial distribution of solid bodies. *Stepinski and Valgeas* (1997) noted that this might explain the abrupt outer boundary of the Kuiper belt. This edge effect is enhanced in the models of *Weidenschilling* (2003b), which include a distribution of particle sizes. At some distance that depends on the nebular parameters, bodies grow large enough to stop drifting inward; they then effectively capture smaller particles coming from further out, causing mass to pile up at a distance that is typically a few tens of AU.

5.3. Enhancement of Solids Outside Condensation/Evaporation Boundaries

Both radially inward drift of solids, and radially outward diffusion of vapor, can lead to enhancements in surface mass density of solids relative to solar abundance. *Morfill and Völk* (1984) and *Stevenson and Lunine* (1988) addressed enhancement of solid surface mass density just outside a condensation/evaporation boundary at R_{ev}. Stevenson and Lunine assumed the presence of a *cold finger* — a complete sink of material — just outside the boundary. Then, vapor would be steadily diffused outward towards this sink and the inner solar system dried out. Stevenson and Lunine emphasized water and its possible role in augmenting the surface mass density just outside $R_{ev}(H_2O)$, speeding the formation of Jupiter's core; moreover, this process would work in a similar way for any volatile. *Morfill and Völk* (1984) combined outward diffusion with inward particle drift, but ended up treating only particles that drift only slightly faster than the nebula gas and never developed an efficient sink at R_{ev}, so the net effects were not striking by astrophysical standards. *Cuzzi and Zahnle* (2004) show how the buildup of solid material outside evaporation fronts may be dominated by radial inward drift of meter-sized rubble, rather than outward diffusive transport from the inner nebula, and find that order(s) of magnitude effects are possible.

5.4. Vapor Enhancement and/or Depletion Inside Evaporation Fronts

In addition to the enhancement of surface mass density in solids just outside of condensation/evaporation boundaries, there is the potential for significant enhancement of vapor abundances of volatile material at radii further inward than solids can exist. *Cyr et al.* (1998, 1999) noted how the complete removal of water from the inner solar system by the *Stevenson and Lunine* (1988) cold finger might lead to problems explaining the oxidation states of many meteorites, and explored how a leaky cold finger might allow, say, meter-sized particles to carry water vapor back into the inner solar system. Unfortunately, these calculations are inconsistent with very similar calculations by *Supulver and Lin* (2000), which show that even meter-sized particles do not survive far inside the evaporation radius. Independent estimates of evaporation and drift rates, by J.N.C. and by F. Ciesla (personal communication, 2003) support the results of *Supulver and Lin* (2000); however, Supulver and Lin do not address vapor abundances inside R_{ev}. *Morfill and Völk* (1984) found that the vapor abundance inside R_{ev} was always solar; however, this was partly because their sink at R_{ev} remained quite leaky, and partly because of what seems to be an overly restrictive outer boundary condition (their equation B7).

The variation of vapor abundance inside evaporation fronts has recently been reinvestigated by *Cuzzi et al.* (2003) and *Cuzzi and Zahnle* (2004), who emphasize the role of rapidly drifting large boulders in transferring mass. They find that large enhancements can occur in the vapor phase. Below we sketch the approach as slightly simplified from *Cuzzi and Zahnle* (2004) for some volatile species with local concentration C(R), nominal solar abundance C_o, and evaporation boundary at R_{ev}. The process is illustrated in Fig. 5.

The equation for the evolution of C in a one-dimensional cartesian form, with no distributed sources or sinks, and all properties held constant except C(R,t), is

$$\frac{\partial}{\partial t}(\sigma_g C(R,t)) + \frac{\partial}{\partial R}(\Phi(R,t)) = 0 \qquad (20)$$

where the radial mass flux $\Phi(R,t)$ is the sum of nebula advection, diffusion, and midplane mass drift respectively

$$\Phi(R) = -C\sigma_g V_n - \mathcal{D}\sigma_g \frac{dC}{dR} - \sigma_L V_L \frac{C}{C_o} \qquad (21)$$

In the last term of $\Phi(R,t)$, σ_L is the surface mass density of solids in the meter-sized range away from the boundary; these particles drift rapidly at velocity V_L (section 2.4.1). Their abundance is proportional to the local abundance C close to the boundary. A dimensionless sink term $\mathcal{L}$ (defined below) is introduced at R_{ev}, due to accretion onto a band of planetesimals having optical depth τ_{PL}, in a region of width H just outside R_{ev}. These can only provide a true sink if they are too large to drift inward past R_{ev} (have radii r_{PL}>1 km or so).

Inside the evaporation front ($R < R_{ev}$) the steady-state solution to this simple system is an enhancement factor over cosmic abundance, which can be approximated by

$$\frac{C}{C_o} = \frac{E_o}{1 + \mathcal{L}} \qquad (22)$$

where

$$E_o = 1 + \frac{\sigma_L V_L}{C_o \sigma_g V_n} = 1 + \frac{f_L V_L}{V_n} \qquad (23)$$

and

$$\mathcal{L} \equiv \tau_{PL} \frac{f_L V_L}{V_n \alpha^{1/2}} \qquad (24)$$

In the equations above, $f_L = \sigma_L / C_o \sigma_g$ and for simplicity here is assumed to be $\ll 1$. The planetesimal accretion rate is determined by the large particles lying in a layer of thickness given by equation (17), and perfect sticking is assumed. E_o is easily determined from its component parts

$$E_o \approx \frac{f_L V_L}{V_n} \sim \frac{f_L \beta V_K}{(\alpha c H / R_{ev})} \sim \frac{f_L}{\alpha} \qquad (25)$$

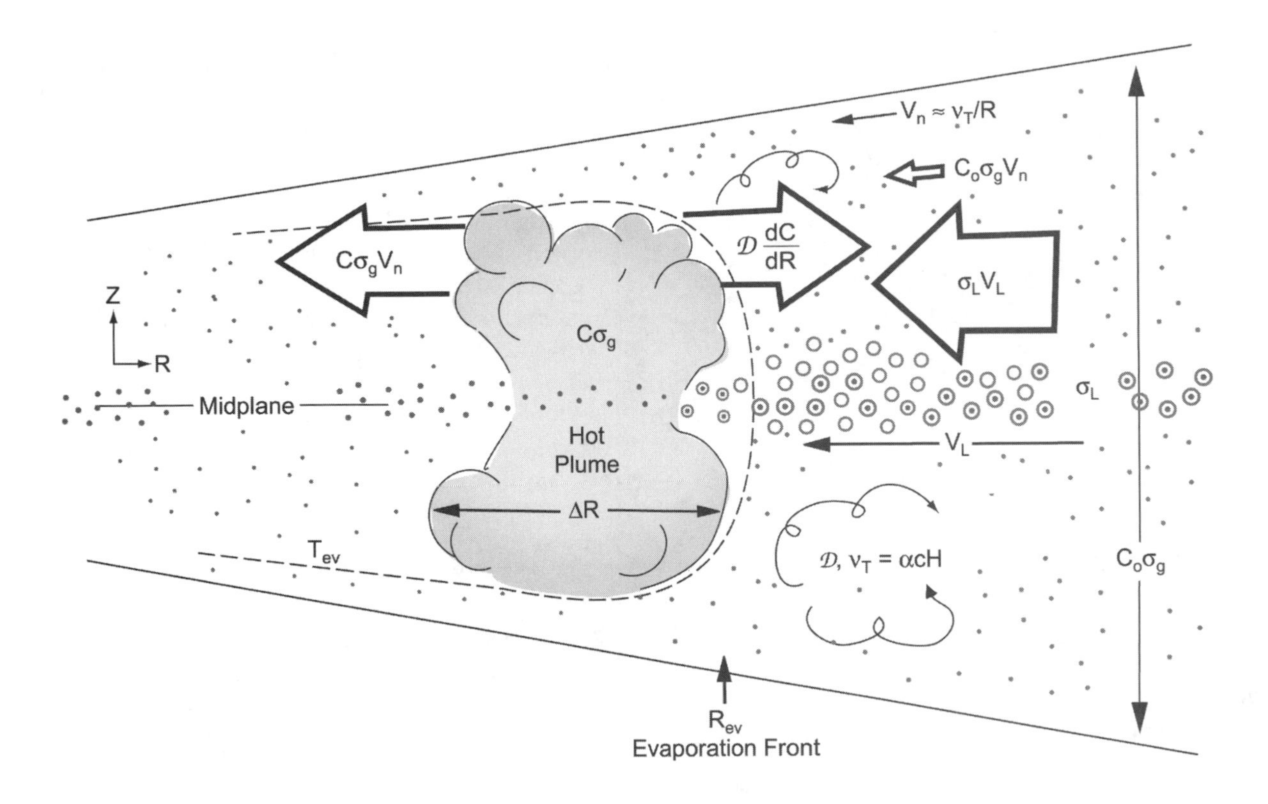

Fig. 5. Sketch illustrating inwardly drifting volatile material crossing its evaporation front R_{ev}, with midplane temperature T_{ev}. The large inward drift flux of midplane solids $\sigma_L V_L$ cannot be offset by vapor removal processes $C\sigma_g V_n + \mathcal{D}\sigma_g dC/dR$ until the concentration of the vapor C is much greater than nominal solar C_o. More refractory material, shown here as a minor constituent, simply goes on drifting and growing.

where $V_L \sim \beta V_K$ and $\beta \sim (H/R)^2$ (section 2.4.1), V_K is the Keplerian velocity, and $c = H\Omega$. As discussed in section 3.3.2 (cf. also *Cuzzi et al.,* 2003), size distributions quickly form by incremental growth containing equal mass per decade from micrometers to meters or so, so $f_L \approx 0.1$ is not unreasonable. Then for plausible values of α, $E_o \geq 10$–100. *Morfill and Völk* (1984) had a parameter analogous to E_o (their $\lambda = V_L/V_n$), but set it to unity, underestimating the power of mass supply by radial drift of large particles. In the limit of very small α, collective effects cause V_L to decrease, so E_o cannot increase without limit (*Cuzzi et al.,* 1993) (see also section 3.3.2). Furthermore, in realistic situations, the supply of drifting solids is finite, and this limits the extent to which E_o can grow. More detailed numerical modeling of this process as applied to water (*Ciesla and Cuzzi,* 2005a,b) indicates that enhancements are smaller — typically a factor of 3–30 — for model parameters thus far explored, because of the finite outer nebula source for the enhancement.

There is an initial, transient, regime of interest (regime 1) in which enhanced, evaporated material is found only within a radial band of width determined by the evaporation time of drifting rubble, which is species-dependent. For the peak enhancement to propagate throughout the inner nebula and approach steady state takes a time $t_{ss} \approx R_{ev}/V_n \approx R_{ev}^2/\alpha cH \approx 40/\alpha$ orbit periods. Considering water as the volatile of interest, $R_{ev} \sim 5$ AU and $\alpha \sim 10^{-3}$–10^{-4}, $t_{ss} \sim 0.5$–5 m.y.; this is long enough to be interesting for chemical variations in the inner nebula due to varying H_2O abundance and associated oxygen fugacity. In steady state, two regimes of interest for C(R) can be distinguished. If $\mathcal{L} \ll 1$ (no sink; regime 2), the entire region inward of R_{ev} is enhanced by E_o over solar. If significant planetesimal growth occurs, leading to a sink and $\mathcal{L} \gg 1$ (regime 3), the inner nebula becomes depleted in the volatile species, with $C/C_o = E_o/(1 + \mathcal{L})$ (the Stevenson and Lunine regime). The regimes are illustrated in Fig. 6. Depending on the rate at which $\mathcal{L}$ grows, the nebula might evolve between these regimes in different ways. Clearly, these are very idealized models, and the process is worthy of future numerical study incorporating accretion. In the next section we highlight other meteoritics applications.

In summary, the essence of this process is that rapid radial drift of meter-sized particles brings mass into the vicinity of an evaporation boundary much faster than it can be removed once it evaporates and becomes coupled to the gas. This is why we refer to it as an *evaporation front*. Its concentration in the gas thus builds significantly. This increased abundance slowly propagates inwards, both by diffusion and by advection with the nebula gas. Formation of a sink at R_{ev} can cause the situation to reverse, with outward diffusion then drying out the inner regions.

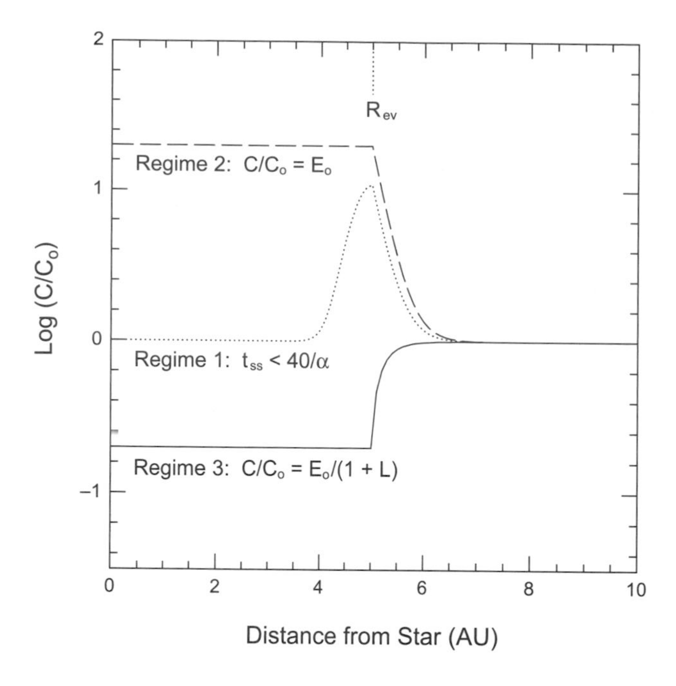

Fig. 6. Schematic of the radial (and temporal) variation of enhancement C/C_o for "water" with an evaporation boundary at R_{ev} = 5 AU, taking for illustration a modest E_o = 20. In regimes 1 and 2, there is no sink at R_{ev} ($\mathcal{L}$ = 0); regime 1 (dotted; schematic only) represents the transient situation, where the inner nebula retains C/C_o = 1 for typically $200/\pi\alpha$ orbit periods. Regime 2 (dashed) is the steady-state solution for $\mathcal{L}$ = 0. As time proceeds and planetesimals grow in the enhanced solid density outside R_{ev}, $\mathcal{L}$ increases; regime 3 (solid) illustrates the steady-state solution for E_o = 20 and $\mathcal{L}$ = 100.

6. SCENARIOS FOR METEORITE PARENT BODIES

Below we present some mental cartoons that illustrate how the physical principles discussed in this chapter might be relevant to parent-body properties. These are intended to be thought-provoking rather than definitive. We expand on these thoughts in *Cuzzi et al.* (2005a).

We start with some of the fundamental challenges of the meteorite data. From studies of thermal histories of ordinary chondrites with different grades of thermal metamorphism, an onion-shell model has arisen (reviewed by *McSween et al.,* 2002; and most recently by *Trieloff et al.,* 2003). Most, if not all, H chondrites of all metamorphic grades are thought to come from a single object that is in the 100-km-radius range, originally composed almost entirely of chondrules of very similar and well-characterized properties; the more deeply buried portions were thermally altered after accretion. While evidence of this sort is somewhat less convincing for other ordinary chondrite types, there is no particular reason to believe their parents, chondrite parents in general, or even the original achondrite parents, should be qualitatively different in this regard. Thus we seek an accretionary process that can construct 100-km bodies out of chondrule-sized particles that are very similar in size and composition within a given parent body, but vary noticeably among parent bodies. The growing evidence for a 1–3-m.y. age gap between CAIs and chondrules [depending on chondrite type; see *Russell et al.* (2006)] is consistent with both the rarity of melted asteroids and the fact that even the oldest achondrites are not measurably older than the oldest chondrules. If chondrules formed with the canonical CAI ^{26}Al abundance — that is, much earlier than a million years after CAI formation — any object as large as 10 km radius would have melted, at least in its central regions (*Lee et al.,* 1976; *Woolum and Cassen,* 1999). Objects larger than 15 km radius, forming even 0.5 m.y. after CAIs, would have reset the Al-Mg systematics in more than half their volume (*LaTourrette and Wasserburg,* 1998). Larger, observable asteroids in the 100-km-radius range must have waited more than 2 m.y. after CAI formation to avoid large-scale melting (*McSween et al.,* 2002, their Fig. 5; also references therein). All the above constraints neglect the additional heating due to ^{60}Fe, which now seems more important than previously thought (compare *Bennett and McSween,* 1996, and *Russell et al.,* 2006). Thus, the process by which these parent bodies formed must have waited 10^6 years or so after CAIs to really get rolling, consistent with observations of the persistence of abundant dust in million-year-old protoplanetary nebulae (*Dullemond and Dominik,* 2005).

If the primary accretion process were too efficient once it started, accretion would complete too quickly for the apparent duration of chondrule and chondrite formation. The diversity of mineralogies seen would rely on spatial, rather than temporal, variations. Because different chondrite types (and apparently different regions of the asteroid belt) have distinct chemical, isotopic, and physical properties, a temporally extended accretionary process must accumulate a given parent body quickly in some sense — that is, before the local gas and solid properties change substantially, and perhaps before chondrules produced in neighboring regions under different conditions can be intermingled (see *Cuzzi et al.,* 2005a, for more discussion).

Chondrites contain mixtures of materials formed in different environments, which argues for significant spatial and/or temporal mixing. Nonequilibrium mineralogies are common in many of these meteorite constituents, and a way must be found to prevent some particles from reaching equilibrium with the gas in which they formed. Fine-grained rims, probably accretionary, envelop pretty much everything in the most primitive, unbrecciated samples. Moreover, 99% of the material in chondrites is isotopically homogeneous in nearly all elements (with the important exception of oxygen). The rare isotopically anomalous grains are ordinarily assumed to be presolar, thus most chondritic silicates seem to have been nebula condensates at some point. These properties are discussed in more detail below.

The first accretion process was incremental accretion from tiny, interstellar grains into a broad, plausibly power-law distribution extending up to a meter or so in size (sections 3.2 and 3.3.2). The level of turbulence plays a critical role in setting the pace of further particle growth; in even quite low levels of turbulence, growth past the meter-sized range might be greatly slowed, or even precluded (section 3.3.2). Few meter-sized aggregates survive as entities, but probably get collisionally disrupted and reaccreted many times over many years, all the time drifting inward.

6.1. Chemical, Mineralogical, and Isotopic Variations Driven by Global Redistribution of Material

If meter-sized particles form quickly, extensive radial redistribution of solid material must have occurred in the early solar system due to their extensive inward radial drift. In combination with this strong mass flow, evaporation fronts, with varying locations, can play a role in enhancing both solid and vapor components of different species relative to each other. More needs to be done to explore the implications of this new complication for meteoritics (sections 5.2–5.4).

The two major volatiles for meteoritics are ferromagnesian silicates ($T_C \approx 1450$ K) and water ($T_C \approx 160$ K). The higher temperature was probably only attained in the innermost solar nebula, early in nebula evolution, when the accretion rate was as high as $10^{-7}\ M_\odot$/yr or more. *Cuzzi et al.* (2003) have described in some detail a scenario in which primitive, carbon-rich, drifting silicate rubble greatly enhanced the vapor abundance of silicate materials inside R_{ev} (silicates), while the combustion of abundant associated carbon maintained the nebula oxygen fugacity at sufficiently low values to explain the (reduced) mineralogy of CAIs that formed in the region. Estimates of the silicate mass flux that passed through R_{ev} as midplane boulders and evaporated can reach tens of Earth masses over the 10^5-yr duration of the high-temperature inner nebula phase. The subsequent fate of this very large amount of material has not been modeled. Surely some — perhaps much — was lost into the Sun, but a considerable amount was probably recycled, by outward diffusion or stellar/disk winds, back into cooler regions where it recondensed. Perhaps this abundant reservoir of evaporated and recondensed material can provide the isotopically homogeneous contents of the meteoritic condensates that accreted more than 10^6 yr later.

The other condensible of interest is water. Solutions presented by *Cuzzi and Zahnle* (2004) (section 5.4) indicate that the warm plume of water vapor created at the water evaporation boundary, or waterline, propagates into the inner solar system on a timescale $\leq 10^6$ yr, bringing with it an enhanced water/hydrogen ratio in the vapor phase. While the analytical solutions of Cuzzi and Zahnle allow enhancements by one or two orders of magnitude, recent numerical modeling by *Ciesla and Cuzzi* (2005a,b) indicates that finite-source and finite-duration factors limit this enhancement to a factor of 3–30 for nebula parameters studied so far. At some point later in time, planetesimal growth just outside the waterline creates a sink that can dry out the inner solar system, leading to water vapor abundances far lower than nominal cosmic. Strongly time- or location-dependent water vapor abundance might help account for the varying FeO abundances in different chondrite components (*Palme and Fegley,* 1990; *Fedkin and Grossman,* 2004). Forming FeO-rich silicates requires nebula oxygen fugacity higher than nominal; on the other hand, forming enstatite chondrites requires oxygen fugacity lower than nominal. Another possible application is the aqueously altered fine-grained material that rims chondrules and other objects in CM chondrites (*Metzler et al.,* 1992). Many believe that these grains can only have been altered on the parent body by liquid water; however, *Ciesla et al.* (2003) argue that nebula shocks can aqueously alter isolated fine-grained silicate particles in the nebula, if the water abundance is enhanced over solar in the nebula gas by a factor of about 100.

This effect might also help us understand mass-independent oxygen-isotopic fractionations. If the ^{17}O and ^{18}O isotopes are preferentially exchanged from CO to water molecules by photodissociation in the parent molecular cloud, or in the rarified upper reaches of the nebula, the early inner nebula where CAIs formed would have been elevated in ^{16}O until sufficient ^{17}O and ^{18}O-rich icy material drifted inward to change the relative abundance (*Yurimoto and Kuramoto,* 2004; *Krot et al.,* 2005b) (cf. sections 5.2–5.4). In this case, correlations between oxygen fugacity and oxygen-isotopic content would be expected. *Yurimoto and Kuramoto* (2004) argue that only a factor of several enhancement in water is needed.

Future work might profitably address how other interesting volatiles such as FeS, or metallic Fe, might behave. Another interesting angle to pursue in this regard might be the volatility-dependent abundance variations seen in moderately volatile elements in all chondrites (*Cassen,* 2001; *Alexander et al.,* 2001).

The strong redistribution of volatiles resulting from radial drift of solids past evaporation boundaries might be observable by remote astronomical observations; enhancements of CO have been found in actively accreting disks out to several AU from the star (*Najita et al.,* 2003) — perhaps without the expected amount of water (*Carr et al.,* 2004).

6.2. Radial Mixing of Small Particles After Formation

Inward drift of large particles is not the only way solids can be redistributed globally. The mixture in primitive meteorites of high-temperature condensates manifesting solar (or lower) oxygen fugacity (CAIs), and lower-temperature condensates manifesting much higher oxygen fugacity (some chondrules, most matrix) — not to mention their age and isotopic differences — has always been a puzzle. Some of the appeal of the stellar wind models for CAI and chondrule formation comes from their potential ability to transport particles directly from regions where the temperature is known to be high enough to melt rocks, to regions where they may accrete. Temporal mixing is another possibility, in principle; CAIs might form at an earlier, higher-temperature epoch, but in the same radial region as chondrules later form and accrete.

It has been a widespread and persistent concern that CAI-sized particles cannot survive in the nebula for the apparent time difference between them and most chondrules, because of their radial drift due to gas drag (section 2.4, Fig. 1). This has even led some to question the reality of the age difference. This concern is traceable to early work by *Weidenschilling* (1977), who quoted drift rates for small particles in a nonturbulent nebula (see section 2.4.1); however, Weidenschilling had included caveats about turbulence, and *Morfill* (1983) argued qualitatively for the vigor of turbulent diffusion of particles.

Recently, *Cuzzi et al.* (2003) showed quantitatively how turbulent diffusion can transport CAIs outward from an inner solar system source to asteroidal distances, and allow them to survive in appropriate abundances for 1–3 m.y. This happens because outwardly radial diffusive mass flux can be as large as inwardly radial gas drag drift flux, in the presence of an outward concentration gradient such as would exist if CAIs were only created in the inner solar system. Tiny grains have the slowest inward drift and are the most easily transported outward by turbulent diffusion. Simple diffusion is capable of explaining outward radial mixing of 1–10-μm-sized crystalline silicate grains in cometary IDPs (*Bockelée-Morvan et al.,* 2002); tiny refractory grains might be mixed outward from the innermost nebula diffusively into comet formation regions, as well as by stellar wind ejection. *Harker and Desch* (2002) suggest that crystalline silicates are formed in the outer nebula by annealing in spiral density waves, and *Boss* (2004) has shown that such waves can themselves produce significant radial mixing even in the absense of diffusive turbulence.

Because larger particles in this size range have faster inward radial drift rates than smaller ones (section 2.4.1), it is somewhat more difficult for diffusion to retain the larger particles; *Cuzzi et al.* (2003) showed that centimeter-sized CAIs could be diffused to the asteroid belt region, but that they disappeared more quickly than the smaller, more ubiquitous millimeter-sized CAIs (which also have different mineralogical and chemical properties). They therefore predicted that the meteorites containing the largest CAIs (CV chondrites) would thus need to have accreted earlier than other meteorite types. It is consistent with this argument that CVs contain not only the largest CAIs, but also the largest AOAs, which are nearly as old but clearly formed in somewhat cooler regions. Some preliminary data is consistent with this prediction (*Amelin et al.,* 2004; *Bizzarro et al.,* 2004; see *Russell et al.,* 2006). Age-dating techniques should be able to test this prediction in the near future.

Some minerals in chondritic particles (all CAIs and AOAs, for instance) are not in their equilibrium state — as if they had not had time to equilibrate with their cooling parent gas before being transported into cooler regions (e.g., *Wood,* 2004). Future studies of minerals found in non-equilibrium states might be able to distinguish between different hypotheses for outward transport of CAIs from hot, inner creation regions to the asteroid belt. Minerals undergo a condensation-reaction sequence as they traverse a cooling gas. If transport is by stellar wind ejection, it occurs almost instantaneously (on an orbital timescale) and at very low density. If outward radial transport is by diffusion, it occurs on a considerably longer timescale. Minerals react slowly with the gas phase, and the kinetics of these reactions are not well understood. Solid-state diffusion coefficients are only known for a limited range of cases and are strongly temperature dependent (e.g., *LaTourrette and Wasserburg,* 1998); however, reaction times could well be in the 10^3–10^4-yr range, depending on the thermal profile experienced by the evolving particle. This subject is covered in more detail by *Cuzzi et al.* (2005a,b).

6.3. Fine-grained Rims: Accretion and/or Erosion

As small, solid particles such as chondrules and CAIs move through the nebula gas, they constantly encounter finer-grained material, both as monomers and as fractal aggregates of monomers, which are more firmly coupled to the gas. Depending on the relative velocity of encounter, these grains might stick to, compress, or erode the granular surfaces of the chondrules and CAIs. *Cuzzi* (2004) has modeled accretion rims on chondrules and CAIs. Using quantitative modeling of relative velocities of particles in gas turbulence, and using sticking outcomes based on the theory of *Dominik and Tielens* (1997), chondrule-sized particles are found to accrete fairly compact rims of fine-grained material in times of 10^3–10^4 yr depending on the relative abundances of chondrules and dust. This model explains how chondrule rims can be fairly compact, with thicknesses nearly proportional to the size of the underlying chondrule (as found by *Metzler et al.,* 1992; *Paque and Cuzzi,* 1997) under plausible conditions of weak nebula turbulence, as long as all chondrules in the same parent body share approximately the same rim accretion time. Some cautionary comments are presented by *Wasson et al.* (2005). One easy way to understand how this can happen is if the chondrules in a region sweep up all the dust in the region, as suggested by *Morfill et al.* (1998). *Cuzzi* (2004) shows that the detailed form of the dependence of rim thickness on core particle size (nearly linear) might be a third indicator that chondrule-sized particles are of the right size for turbulent concentration to act on them (cf. section 4). More measurements of rim thicknesses in different chondrite types would be useful.

Fine-grained accretionary rims on CAIs are prevalent, and many CAI accretion rims show clear affiliations with the nebula regime in which the CAIs themselves formed (*Krot et al.,* 2002) lying beneath rims from the environment in which chondrules formed (*MacPherson et al.,* 1985).

Calcium-aluminum-rich inclusion accretion rims studied to date are much thinner, relative to their core particle size, than fine-grained rims on chondrules in the same chondrites. *Cuzzi* (2004) suggests that this is because the CAIs with the most obvious fine-grained rims are larger than chondrules, and would be expected to have correspondingly larger velocities relative to the gas and coupled fine grains, which may erode, rather than accrete onto, their surfaces. Especially in an erosive regime, protected hollows in complex surfaces will preferentially accumulate material, such as is often seen in CAIs (*MacPherson et al.,* 1985; *Krot et al.,* 2002). This could be studied more using CAIs with sizes closer to those of chondrules. These differences would have to be reconciled with sorting together in the same meteorite of these CAIs and chondrules, which also depends on their stopping time t_s.

This scenario would seem to be consistent with the general observation that chondrule fine-grained rim material is similar in composition to (if somewhat smaller in size than) matrix grains in the same chondrites (*Scott et al.,* 1989; *Ashworth,* 1977) and complementary to the chemistry of the chondrule cores themselves in the sense that iron and other volatiles lost from chondrules are complemented in their rims and matrix (*Wood,* 1985). More studies of rim-matrix relationships would be valuable.

By contrast, the concept that chondrules and CAIs acquire their rims in high-speed stellar winds (*Liffman and Toscano,* 2000) seems rather unlikely. The relative velocities involved are estimated as the terminal velocity of the grain, under solar gravity, in the very low density wind: $V_T \sim gt_s$. Recall from section 5.1 that $t_s \sim \Omega_x^{-1} \sim 10^5$ s; thus anywhere inside 2 AU, $V_T \sim 1$ km/s. This velocity is hard to reconcile with sticking of grains and much more likely to result in net erosion of rims, if not pitting and erosion of the igneous object itself.

6.4. Primary Accretion

By primary accretion, we refer to the process by which primitive planetesimals having the size and content of meteorite parent bodies were put together from the individual components discused above. These are objects large enough to escape gas-drag-driven drift loss "into the Sun," and to have identifiable macroscopic-to-megalithic properties that are connected to their nebula precursors, but small enough that they have escaped melting, preserving the primary characteristics of their nebular constituents. Inferring the true "primary" characteristics from the current crop of meteorites requires us to look backward through a number of planetary processes — collisions, grinding, abrasion, thermal and aqueous alteration, etc. (cf. *Weidenschilling,* 2006). Here we speculate on only the very earliest stages of primary accretion. Some of these speculations are pursued further in *Cuzzi et al.* (2005a).

As mentioned above, several arguments support a 1–3-m.y. hiatus between the earliest days of the solar system (the formation of CAIs) and the primary accretion of the parent bodies of most of our meteorites (*Russell et al.,* 2006). It seems to us that maintaining a small amount of turbulence ($10^{-5} < \alpha < 10^{-2}$) in the inner solar system throughout most of this time is the most obvious way to account for this (section 3.3.2). The situation in the outer nebula might have been different; the need to preserve the limited amount of solid material there by making accretion easier easier and faster might lead one to believe that turbulence there was much weaker (*Weidenschilling,* 1997). However, current estimates suggest that the magnetorotational instability is easily capable of making the entire outer nebula turbulent (*Sano et al.,* 2000). If the outer nebula were turbulent, turbulent concentration might have played a role in fostering planetesimal growth (cf. section 4, Fig. 3). Evidence for this might be hard to find.

In 1–3 m.y., the nebula evolves dramatically, losing most of its initial mass, dropping to a much lower accretion rate and a much lower photospheric and magnetospheric activity level overall, and cooling at all radial locations. Evaporation fronts move, possibly by several AU, as a result of this evolution. The minimum mass nebula with which most planetary accumulation models start, often identified with the revealed stage of T Tauri star evolution, is only a mere shadow of its former self. Once primary accretion starts, one has the (poorly quantified) impression that it happens quickly. Why was primary accretion delayed for over a million years, until after chondrules formed, and yet proceeded with alacrity thereafter?

Given the prevailing expectation that Jupiter itself formed on the 1–3-m.y. timescale, one must at least consider what influence the formation of Jupiter (by this we mean the rapid accretion of its massive gaseous envelope) might have had on the asteroid-formation region. For example, nebula shock waves driven into the inner solar system by the fully formed Jupiter might both have melted the chondrules and influenced the local level of turbulence (more likely to have increased it, than to have decreased it).

A possibility favored by S.J.W. is that decreasing gas density (due to nebula evolution into the Sun) and decreasing opacity (due to accretion) might have led to a substantial cutoff in turbulence at some stage. In section 3.3.1 we showed how sensitive the accretion rate is to α decreasing below a certain (very low) threshold value ($<10^{-6}$–10^{7}) at which point the particle layer becomes self-shielded from the nebula headwind. In this scenario, formation of Jupiter perturbs some planetesimals into eccentric orbits crossing large radial regions, causing bow shocks that generate the chondrules we now see, which, because turbulence is low, accrete quickly. Under this scenario one would need to understand how the distinct parent bodies of the different chondrite classes could be formed, with such well-determined chemical properties and size-density sorting characteristics. Furthermore, the turbulence threshold required for this to occur is extremely low. While it is widely believed that differential rotation alone cannot maintain continuous turbulence, this question is far from settled at the extremely low levels of α that permit rapid midplane accretion (section 2.1).

A different possibility, favored by J.N.C., is that weak turbulence was ongoing and ubiquitous, allowing growth only to meter size, until some confluence of events (perhaps also involving Jupiter's formation) led to chondrule formation and turbulent concentration. The decreasing gas density and varying turbulent intensity might have only then reached combinations (Fig. 3) at which chondrules and their like could incur significant concentration to leapfrog the meter-sized particle barrier that frustrates incremental accretion in turbulence. For an extended time, these processes sporadically produced dense, if not fully packed, protoplanetesimals that continued to compact under collisional and gravitational processes (section 4). Turbulent concentration does present the advantage of allowing small variations in local gas density, or α, or both to determine the properties of the mineral objects that are subjected to strong concentration (sections 4 and 6.3). This scenario predicts widespread radial mixing of chondrite constituents prior to their accretion, which must be reconciled with the distinct properties of the meteorite classes (see *Cuzzi et al.,* 2005a).

The challenge remains to end up with planetesimals made out of the rocks we have in our collections — plausibly, 100-km-radius objects, each composed almost entirely of millimeter-sized components with well-defined properties, but varying noticeably from one to another. It has been a widespread assumption that the chondrite and asteroid groups represent spatial variation at some nominal point in time; however, the alternate must be considered that chondrite groups represent temporal (but still local) grab samples of an evolving nebula mix, or perhaps at least a combination of both spatial and temporal variations. One must also keep in mind that the current crop of meteorite types remains a biased and incomplete sample of what exists out in the asteroid belt. As only one example, a set of Antarctic meteorites has recently been identified as H chondrites, but they incorporate abundant CAIs (rare in normal H chondrites) and possess a matrix like that seen in CO chondrites (*Kimura et al.,* 2002)! Clearly there is a lot to learn, and, while progress is being made in understanding different components of the process, we remain a long way from any coherent scenario of meteorite parent-body formation.

Acknowledgments. We are grateful to many colleagues for helping educate us over the years to the nuances of the meteorite record, against which any theory must be judged. In the context of this specific chapter, we have benefitted from conversations with J. Beckett, L. Grossman, J. Paque Heather, M. Humayan, A. Krot, A. Meibom, S. Russell, and E. Scott. In the theoretical areas we have benefitted from useful conversations with P. Cassen, L. Hartmann, and K. Zahnle, and helpful internal reviews by K. Zahnle, I. Mosqueira, and P. Estrada. We thank our three reviewers for their careful reading of the chapter, and the editors for their flexibility regarding length. Contributions of J.N.C. were supported by grants from NASA's Planetary Geology and Geophysics and Origins of Solar Systems programs. Research of S.J.W. was supported by NASA grant NAG5-13156. This chapter has made ample use of the invaluable ADS Astronomy/Planetary Abstract Service.

REFERENCES

Ackridge G. and Sears D. W. G. (1998) Chondrule and metal size-sorting in asteroidal regoliths: Experimental results with implications for chondrites (abstract). In *Lunar and Planetary Science XXIX,* pp. 1198–1199. Lunar and Planetary Institute, Houston.

Adachi I., Hayashi C., and Nakazawa K. (1976) The gas drag effect on the elliptical motion of a solid body in the primordial solar nebula. *Prog. Theor. Phys., 56,* 1756–1771.

Afshordi N., Mukhopadhyay B., and Narayan R. (2005) Bypass to turbulence in hydrodynamic accretion: Lagrangian analysis of energy growth. *Astrophys. J., 629,* 373–382.

Alexander C. M. O'D., Boss A. P., and Carlson R. W. (2001) Early evolution of the inner solar system: A meteoriticist's perspective. *Science, 293,* 64–68.

Amelin Y., Krot A. N., Hutcheon I. D., and Ulyanov A. A. (2002) Lead isotopic ages of chondrules and calcium-aluminum-rich inclusions. *Science, 297,* 1678–1683.

Amelin Y., Krot A. N., Hutcheon I. D., and Ulyanov A. A. (2004) Duration of the chondrule formation interval; a Pb isotope study (abstract). In *Planetary Timescales: From Stardust to Continents,* February 2004, Canberra, Australia.

Anderson J. M., Li Z.-Y., Krasnopolsky R., and Blandford R. D. (2003) Locating the launching region of T Tauri winds: The case of DG Tauri. *Astrophys. J. Lett., 590,* L107–L110.

André P., Ward-Thompson D., and Barsony M. (2000) From prestellar cores to protostars: The initial conditions of star formation. In *Protostars and Planets* (V. Mannings et al., eds.), pp. 589–612. Univ. of Arizona, Tucson.

Arldt R. and Urpin V. (2004) Simulations of vertical shear instability in accretion disks. *Astron. Astrophys., 426,* 755–765.

Ashworth J. R. (1977) Matrix textures in unequilibrated ordinary chondrites. *Earth Planet. Sci. Lett., 35,* 25–34.

Barranco J. and Marcus P. (2004) Three-dimensional vortices in protoplanetary disks. *Astrophys. J., 623,* 1157–1170.

Barge P. and Sommeria J. (1995) Did planet formation begin inside persistent gaseous vortices? *Astron. Astrophys., 1,* L1–L4.

Bell K. R., Cassen P. M., Klahr H. H., and Henning Th. (1997) The structure and apeparance of protostellar accretion disks: Limits on disk flaring. *Astrophys. J., 486,* 372–387.

Bennett M. E. and McSween H. Y. Jr. (1996) Revised model calculations for the thermal histories of ordinary chondrite parent bodies. *Meteoritics & Planet. Sci., 31,* 783–792.

Bizzarro M., Baker J. A., and Haack H. (2004) Mg isotopic evidence for contemporaneous formation of chondrules and refractory inclusions. *Nature, 431,* 275–278.

Blum J. (2000) Laboratory experiments on preplanetary dust aggregation. *Space Sci. Rev., 192,* 265–278.

Blum J. (2004) Grain growth and coagulation. In *Astrophysics of Dust* (A. N. Witt et al., eds.), pp. 369–391. ASP Conference Series 309, Astronomical Society of the Pacific, San Francisco.

Blum J. and Schräpler R. (2004) Structure and mechanical properties of high-porosity macroscopic aggregates formed by random ballistic deposition. *Phys. Rev. Lett., 93,* 115503–1 to 115503–4.

Blum J. and Wurm G. (2000) Experiments on sticking, restructuring, and fragmentation of preplanetary dust aggregates. *Icarus, 143,* 138–146.

Bockelée-Morvan D., Gautier D., Hersant F., Huré J.-M., and Robert F. (2002) Turbulent radial mixing in the solar nebula as the source of crystalline silicates in comets. *Astron. Astrophys., 384,* 1107–1118.

Boley A. C., Durisen R. H., and Pickett M. K. (2005) The three-dimensionality of spiral shocks: Did chondrules catch a breaking wave? In *Chondrites and the Protoplanetary Disk* (A. N. Krot et al., eds.), pp. 839–848. ASP Conference Series 341, Astronomical Society of the Pacific, San Francisco.

Boss A. P. (2004) Evolution of the solar nebula. VI. Mixing and transport of isotopic heterogeneity. *Astrophys. J., 616,* 1265–1277.

Boss A. P. and Goswami J. N. (2006) Presolar cloud collapse and the formation and early evolution of the solar nebula. In *Meteorites and the Early Solar System II* (D. S. Lauretta and H. Y. McSween Jr., eds.), this volume. Univ. of Arizona, Tucson.

Bracco A., Provenzale A., Spiegel E. A., and Yecko P. (1998) Spotted disks. In *Theory of Black Hole Accretion Disks* (A. Marek et al., eds.), p. 254. Cambridge Univ., Cambridge.

Busse F. (2004) Visualizing the dynamics of the onset of turbulence. *Science, 305,* 1574–1575.

Calvet N., Hartman L., and Strom S. E. (2000) Evolution of disk accretion. In *Protostars and Planets* (V. Mannings et al., eds.), pp. 377–400. Univ. of Arizona, Tucson.

Carballido A., Stone J. M., and Pringle J. E. (2005) Diffusion coefficient of a passive contaminant in a local MHD model of a turbulent accretion disk. *Mon. Not. R. Astron. Soc., 358,* 1055–1060.

Carr J. S., Tokunaga A. T., and Najita J. (2004) Hot H_2O emission and evidence for turbulence in the disk of a young star. *Astrophys. J., 603,* 213–220.

Cassen P. M. (2001) Nebular thermal evolution and the properties of primitive planetary materials. *Meteoritics & Planet. Sci., 36,* 671–700.

Champney J. M., Dobrovolskis A. R., and Cuzzi J. N. (1995) A numerical turbulence model for multiphase flow in a non-equilibrium protoplanetary nebula. *Phys. Fluids, 7,* 1703–1711.

Chaussidon M. and Gounelle M. (2006) Irradiation processes in the early solar system. In *Meteorites and the Early Solar System II* (D. S. Lauretta and H. Y. McSween Jr., eds.), this volume. Univ. of Arizona, Tucson.

Chokshi A., Tielens A. G. G. M., and Hollenbach D. (1993) Dust coagulation. *Astrophys. J., 407,* 806–819.

Ciesla F. J. and Cuzzi J. N (2005a) The distribution of water in a viscous protoplanetary disk (abstract). In *Lunar and Planetary Science XXXVI,* Abstract #1479. Lunar and Planetary Institute, Houston (CD-ROM).

Ciesla F. J. and Cuzzi J. N. (2005b) The evolution of the water distribution in a viscous protoplanetary disk. *Icarus,* in press.

Ciesla F., Lauretta D. S., Cohen B. A., and Hood L. L. (2003) A nebular origin for chondritic fine-grained phyllosilicates. *Science, 299,* 549–552.

Clarke C. J. and Pringle J. E. (1988) The diffusion of contaminant through an accretion disk. *Mon. Not. R. Astron. Soc., 235,* 365–373.

Clift R., Grace J. R., and Weber M. E. (1978) *Bubbles, Drops, and Particles.* Academic, New York. 380 pp.

Connolly H. C. Jr., Desch S. J., Chiang E., Ash R. D., and Jones R. H. (2006) Transient heating events in the protoplanetary nebula. In *Meteorites and the Early Solar System II* (D. S. Lauretta and H. Y. McSween Jr., eds.), this volume. Univ. of Arizona, Tucson.

Cuzzi J. N. (2004) Blowing in the wind: III. Accretion of dust rims by chondrule-sized particles in a turbulent protoplanetary nebula. *Icarus, 168,* 484–497.

Cuzzi J. N., Petaev M., Ciesla F. J., Krot A. N., Scott E. R. D., and Weidenschilling S. J. (2005a) History of thermally processed solids: Reconciling models and meteoritics. In *Chondrites and the Protoplanetary Disk* (A. N. Krot et al., eds.). ASP Conference Series, in press.

Cuzzi J. N., Petaev M., Ciesla F. J., Krot A. N., and Scott E. R. D. (2005b) Nebula models of non-equilibrium mineralogy: Wark-Lovering rims (abstract). In *Lunar and Planetary Science XXXVI,* Abstract #2095. Lunar and Planctary Institute, Houston (CD-ROM).

Cuzzi J. N., Davis S. S., and Dobrovolskis A. R. (2003) Blowing in the wind: II. Creation and redistribution of refractory inclusions in a turbulent protoplanetary nebula. *Icarus, 166,* 385–402.

Cuzzi J. N., Dobrovolskis A. R., and Champney J. M. (1993) Particle-gas dynamics near the midplane of a protoplanetary nebula. *Icarus, 106,* 102–134.

Cuzzi J. N., Dobrovolskis A. R., and Hogan R. C. (1994) What initiated planetesimal formation? (abstract). In *Lunar and Planetary Science XXV,* pp. 307–308. Lunar and Planetary Institute, Houston.

Cuzzi J. N., Dobrovolskis A. R., and Hogan R. C. (1996) Turbulence, chondrules, and planetesimals. In *Chondrules and the Protoplanetary Disk* (R. Hewins et al., eds.), pp. 35–44. Cambridge Univ., Cambridge.

Cuzzi J. N. and Hogan R. C. (2003) Blowing in the wind: I. Velocities of chondrule-sized particles in a turbulent protoplanetary nebula. *Icarus, 164,* 127–138.

Cuzzi J. N., Hogan R. C., and Paque J. M. (1999) Chondrule size density distributions: Predictions of turbulent concentration and comparison with chondrules disaggregated from L4 ALH85033 (abstract). In *Lunar and Planetary Science XXX,* Abstract #1274. Lunar and Planetary Institute, Houston (CD-ROM).

Cuzzi J. N., Hogan R. C., Paque J. M., and Dobrovolskis A. R. (2001) Size-selective concentration of chondrules and other small particles in protoplanetary nebula turbulence. *Astrophys. J., 546,* 496–508.

Cuzzi J. N. and Zahnle K. R. (2004) Material enhancement in protoplanetary nebulae by particle drift through evaporation fronts. *Astrophys. J., 614,* 490–496.

Cyr K., Sears W. D., and Lunine J. I. (1998) Distribution and evolution of water ice in the solar nebula: Implications for solar system body formation. *Icarus, 135,* 537–548.

Cyr K., Sharp C. M., and Lunine J. I. (1999) Effects of the redistribution of water in the solar nebula on nebular chemistry. *J. Geophys. Res., 104,* 19003–19014.

Davis S. S., Sheehan D. P., and Cuzzi J. N. (2000) On the persistence of small regions of vorticity in the protoplanetary nebula. *Astrophys. J., 545,* 494–503.

Desch S. and Cuzzi J. N. (2000) Formation of chondrules by lightning in the protoplanetary nebula. *Icarus, 143,* 87–105.

Dobrovolskis A. R., Dacles-Mariani J. M., and Cuzzi J. N. (1999) Production and damping of turbulence in the solar nebula. *J. Geophys. Res.–Planets, 104,* 30805–30815

Dodd R. T. (1976) Accretion of the ordinary chondrites. *Earth Planet. Sci. Lett., 30,* 281–291.

Dominik C. and Tielens A. G. G. M. (1997) The physics of dust coagulation and the structure of dust aggregates in space. *Astrophys. J., 480,* 647–673a.

Dominik C. P., Blum J., Cuzzi J. N., and Wurm G. (2006) Growth of dust as initial step towards planet formation. In *Protostars and Planets V* (B. Reipurth et al., eds.). Univ. of Arizona, Tucson, in press.

Dubrulle B., Morfill G. E., and Sterzik M. (1995) The dust subdisk in the protoplanetary nebula. *Icarus, 114,* 237–246.

Dullemond C. P. and Dominik C. (2005) Dust coagulation in protoplanetary disks: A rapid depletion of small grains. *Astron. Astrophys., 434,* 971–986.

Eaton J. K. and Fessler J. R. (1994) Preferential concentration of particles by turbulence. *Intl. J. Multiphase Flow, Suppl., 20,* 169–209.

Fedkin A. V. and Grossman L. (2004) Nebular formation of fayalitic olivine: Ineffectiveness of dust enrichment (abstract). In *Lunar and Planetary Science XXXV,* Abstract #1823. Lunar and Planetary Institute, Houston (CD-ROM).

Fleming T. and Stone J. M. (2003) Local magnetohydrodynamic models of layered accretion disks. *Astrophys. J., 585,* 908–920.

Frohn A. and Roth N. (2000) *Dynamics of Droplets.* Springer, Berlin. 292 pp.

Fromang S. and Nelson R. P. (2005) On the accumulation of solid bodies in global turbulent protoplanetary disk models. *Mon. Not. R. Astron. Soc.,* in press.

Garaud P. and Lin D. N. C. (2004) On the evolution and stability of a protoplanetary disk dust layer. *Astrophys. J., 608,* 1050–1075.

Garaud P. and Ogilvie G. I. (2005) A model for the nonlinear dynamics of turbulent shear flows. *J. Fluid Mech., 530,* 145–176.

Godon P. and Livio M. (2000) The formation and role of vortices in protoplanetary disks. *Astrophys. J., 537,* 396–404.

Goldreich P. and Ward W. R. (1973) The formation of planetesimals. *Astrophys. J., 183,* 1051–1061.

Gomez G. C. and Ostriker E. C. (2005) The effect of the Coriolis force on Kelvin-Helmholz-driven mixing in protoplanetary disks. *Astrophys. J.,* in press.

Goodman J and Pindor B. (2000) Secular instability and planetesimal formation in the dust layer. *Icarus, 148,* 537–549.

Greshake A., Krot A. N., Meibom A., Weisberg M. K., Zolensky M. E., and Keil K. (2002) Heavily-hydrated lithic clasts in CH chondrites and the related, metal-rich chondrites QUE94411 and Hammadah al Hamra 237. *Meteoritics & Planet. Sci., 37,* 281–293.

Haack H. and Scott E. R. D. (1993) Nebula formation of the H, L, and LL parent bodies from a single batch of chondritic materials. *Meteoritics, 28,* 358.

Haghighipour N. and Boss A. P. (2003a) On pressure gradients and rapid migration of solids in a non-uniform solar nebula. *Astrophys. J.., 583,* 996–1003.

Haghighipour N. and Boss A. P. (2003b) On gas drag-induced rapid migration of solids in a nonuniform solar nebula. *Astrophys. J., 598,* 1301–1311.

Harker D. E. and Desch S. J. (2002) Annealing of silicate dust by nebular shocks at 10 AU. *Astrophys. J. Lett., 565,* L109–L112.

Hartmann L. and Calvet N. (1995) Observational constraints on FU Ori winds. *Astrophys. J., 109,* 1846–1855.

Hartmann W. K. (1985) Impact experiments 1. Ejecta velocity distributions and related results from regolith targets. *Icarus, 63,* 69–98.

Hayashi C. (1981) Structure of the solar nebula, growth and decay of magnetic fields and effects of magnetic and turbulent viscosities on the nebula. *Prog. Theor. Phys. Suppl., 70,* 35–53.

Hayashi C., Nakazawa K., and Nakagawa Y. (1985) Formation of the solar system. In *Protostars and Planets II* (D. C. Black and M. S. Matthews, eds.), pp. 1100–1153. Univ. of Arizona, Tucson.

Hogan R. C. and Cuzzi J. N. (2001) Stokes and Reynolds number dependence of preferential particle concentration in simulated 3D turbulence. *Phys. Fluids, 13,* 2938–2945.

Hogan R. C. and Cuzzi J. N. (2005) A cascade model for particle concentration and enstrophy in fully developed turbulence with mass loading feedback. *Phys. Rev. E.,* in press.

Hogan R. C., Cuzzi J. N., and Dobrovolskis A. R. (1999) Scaling properties of particle density fields formed in turbulent flows. *Phys. Rev. E, 60,* 1674–1680.

Huang S., Akridge G., and Sears D. W. G. (1996) Metal-silicate fractionation in the surface layers of accreting planetesimals: Implications for the formation of ordinary chondrites and the nature of asteroid surfaces. *J. Geophys. Res., 101,* 29373–29385.

Hughes D. W. (1978) A disaggregation and thin section analysis of the size and mass distribution of the chondrules in the Bjurbole and Chainpur meteorites. *Earth Planet. Sci. Lett., 38,* 391–400.

Hughes D. W. (1980) The dependence of chondrule density on chondrule size. *Earth Planet. Sci. Lett., 51,* 26–28.

Ishitsu N. and Sekiya M. (2002) Shear instabilities in the dust layer of the solar nebula III. Effects of the Coriolis force. *Earth Planets Space, 54,* 917–926.

Ishitsu N. and Sekiya M. (2003) The effects of the tidal force on shear instabilities in the dust layer of the solar nebula. *Icarus, 165,* 181–194.

Johansen A., Klahr H., and Henning Th. (2005) Gravoturbulent formation of planetesimals. *Astrophys. J.,* in press.

Kempf S., Pfalzner S., and Henning T. (1999) N-particle simulations of dust growth 1. Growth driven by brownian motion. *Icarus, 141,* 388–398.

Kerridge J. F. and Matthews M. S., eds. (1988) *Meteorites and the Early Solar System.* Univ. of Arizona, Tucson. 1269 pp.

Kimura M., Hiyagon H., Palme H., Spettel B., Wolf D., Clayton R. N., Mayeda T. K., Sato T., Suzuki A., and Kojima H. (2002) Yamato 792947, 793408, and 82038: The most primitive H chondrites, with abundant refractory inclusions. *Meteoritics & Planet. Sci., 37,* 1417–1434.

Klahr H. H. and Bodenheimer P. (2003a) Turbulence in accretion disks: Vorticity generation and angular momentum transport via the global baroclinic instability. *Astrophys. J., 582,* 869–892.

Klahr H. H. and Bodenheimer P. (2003b) The formation of a planet in the eye of a hurricane — A three phase model for planet formation (abstract). In *Abstracts for the 35th AAS/DPS Meeting,* Abstract #25.01.

Königl A. and Pudritz R. (2000) Disk winds and the accretion-outflow connection. In *Protostars and Planets IV* (V. Mannings et al., eds.), pp. 759–788. Univ. of Arizona, Tucson.

Kornet K., Stepinski T. F., and Rozyczka M. (2001) Diversity of planetary systems from evolution of solids in protoplanetary disks. *Astron. Astrophys., 378,* 180–191.

Krot A. N., Rubin A. E., Keil K., and Wasson J. (1997) Microchondrules in ordinary chondrites: Implications for chondrule formation. *Geochim. Cosmochim. Acta, 61,* 463–473.

Krot A. N., McKeegan K. D., Leshin L. A., MacPherson G. J., and Scott E. R. D. (2002) Existence of an ^{16}O-rich gaseous

reservoir in the solar nebula. *Science, 295,* 1051–105.
Krot A. N., Amelin Y., Cassen P., and Meibom A. (2005a) Young chondrules in CB chondrites from a giant impact in the early solar system. *Nature, 436,* 989–992.
Krot A. N., Hutcheon I. D., Yurimoto H., Cuzzi J. N., McKeegan K. D., Scott E. R. D., Libourel G., Chaussidon M., Aleon J., and Petaev M. I. (2005b) Evolution of oxygen isotopic composition in the inner solar nebula. *Astrophys. J., 622,* 1333–1342.
Kuebler K. E., McSween H. Y. Jr., Carlson W. D., and Hirsch D. (1999) Sizes and masses of chondrules and metal-troilite grains in ordinary chondrites: Possible implications for nebular sorting. *Icarus, 141,* 96–106.
LaTourrette T. and Wasserburg G. J. (1998) Mg diffusion in anorthite: Implications for the formation of early solar system planetesimals. *Earth Planet. Sci. Lett., 158,* 91–108.
Lee T., Papanastassiou D. A., and Wasserburg G. J. (1976) Demonstration of ^{26}Mg excess in Allende and evidence for ^{26}Al. *Geophys. Res. Lett., 3,* 109–112.
Leenhouts J. M. and Skinner W. R. (1991) Shape-differentiated size distributions of chondrules in type 3 ordinary chondrites. *Meteoritics, 26,* 363.
Liffman K. and Brown M. (1995) The motion and size sorting of particles ejected from a protostellar accretion disk. *Icarus, 116,* 275–290.
Liffman K. and Toscano M. (2000) Chondrule fine-grained mantle formation by hypervelocity impact of chondrules with a dusty gas. *Icarus, 143,* 106–125.
Love S. G. and Ahrens T. (1996) Catastrophic impacts on gravity dominated asteroids. *Icarus, 124,* 141–155.
MacPherson G. J., Hashimoto A., and Grossman L. (1985) Accretionary rims on Allende inclusions: Clues to the accretion of the Allende parent body. *Geochim. Cosmochim. Acta, 49,* 2267–2279.
Markiewicz W. J., Mizuno H., and Völk H. J. (1991) Turbulence-induced relative velocity between two grains. *Astron. Astrophys., 242,* 286–289.
Marshall J. and Cuzzi J. N. (2001) Electrostatic enhancement of coagulation in protoplanetary nebulae (abstract). In *Lunar and Planetary Science XXXII,* Abstract #1262. Lunar and Planetary Institute, Houston (CD-ROM).
Marshall J., Sauke T. B., and Cuzzi J. N. (2005) Microgravity studies of aggregation in particulate clouds. *Geophys. Res. Lett., 32,* L11202.
May C., Russell S. S., and Grady M. M. (1999) Analysis of chondrule and CAI size and abundance in CO3 and CV3 chondrites: A preliminary study (abstract). In *Lunar and Planetary Science XXX,* Abstract #1688. Lunar and Planetary Institute, Houston (CD-ROM).
McSween H. Y., Ghosh A., Grimm R. E., Wilson L., and Young E. D. (2002) Thermal evolution models of asteroids. In *Asteroids III* (W. F. Bottke Jr. et al., eds.), pp. 559–571. Univ. of Arizona, Tucson.
Meneveaux C. and Sreenivasan K. R. (1987) Simple multifractal cascade model for fully developed turbulence. *Phys. Rev. Lett., 59,* 1424–1427.
Metzler K., Bischoff A., and Stöffler D. (1992) Accretionary dust mantles in CM chondrites: Evidence for solar nebula processes. *Geochim. Cosmochim. Acta, 56,* 2873–2897.
Möhlmann D. (1996) Origin of comets in the extended preplanetary disk. *Planet. Space Sci., 44,* 731–734.
Morfill G. E. (1983) Some cosmochemical consequences of a turbulent protoplanetary cloud. *Icarus, 53,* 41–54.
Morfill G. E., Durisen R. H., and Turner G. W. (1998) Note: An accretion rim constraint on chondrule formation theories. *Icarus, 134,* 180–184.
Morfill G. E. and Völk H. J. (1984) Transport of dust and vapor and chemical fractionation in the early protosolar cloud. *Astrophys. J., 287,* 371–395.
Mosqueira I., Kassinos S., Shariff K., and Cuzzi J. N. (2003) Hydrodynamical shear instability in accretion disks? (abstract). In *Abstracts for the 35th AAS/DPS Meeting,* Abstract #25.05.
Mukhopadyay B., Afshordi N., and Narayan R. (2005) Bypass to turbulence in hydrodynamic accretion: An eigenvalue approach. *Astrophys. J., 629,* 383–396.
Najita J., Carr J. S., and Mathieu R. D. (2003) Gas in the terrestrial planet region of disks: CO fundamental emission from T Tauri stars. *Astrophys. J., 589,* 931–952.
Nakagawa Y., Sekiya M., and Hayashi C. (1986) Settling and growth of dust particles in a laminar phase of a low-mass solar nebula. *Icarus, 67,* 375–390.
Nelson V. E. and Rubin A. E. (2002) Size-frequency distributions of chondrules and chondrule fragments in LL3 chondrites: Implications for parent-body fragmentation of chondrules. *Meteoritics & Planet. Sci., 37,* 1361–1376.
Palme H. and Fegley B. (1990) High temperature condensation of iron-rich olivine in the solar nebula. *Earth Planet. Sci. Lett., 101,* 180–195.
Papaloizou J. and Terquem C. (1999) Critical protoplanetary core masses in protoplanetary disks and the formation of short-period giant planets. *Astrophys. J., 521,* 825–838.
Paque J. and Cuzzi J. N. (1997) Physical characteristics of chondrules and rims, and aerodynamic sorting in the solar nebula (abstract). In *Lunar and Planetary Science XXVIII,* pp. 1071–1072. Lunar and Planetary Institute, Houston.
Prinn R. G. (1990) On neglect of angular momentum terms in solar nebula accretion disk models. *Astrophys. J., 348,* 725–729.
Rice W. K. M., Lodato G., Pringle J. E., Armitage P. J., and Bonnell I. A. (2004) Accelerated planetesimal growth in self-gravitating protoplanetary disks. *Mon. Not. R. Astron. Soc., 355,* 543–552.
Richard D. and Zahn J.-P. (1999) Turbulence in differentially rotating flows. What can be learned from the Couette-Taylor experiment. *Astron. Astrophys., 347,* 734–738.
Richard D. (2003) On non-linear hydrodynamic instability and enhanced transport in differentially rotating flows. *Astron. Astrophys., 408,* 409–414.
Russell S. S., Hartmann L., Cuzzi J., Krot A. N., Gounelle M., and Weidenschilling S. (2006) Timescales of the solar protoplanetary disk. In *Meteorites and the Early Solar System II* (D. S. Lauretta and H. Y. McSween Jr., eds.), this volume. Univ. of Arizona, Tucson.
Saffman P. G. and Turner J. S. (1956) On the collision of drops in turbulent clouds. *J. Fluids Mech., I,* 16; *Corrigendum, J. Fluids Mech., 196,* 599 (1988).
Safronov V. S. (1969) *Evolution of the Protoplanetary Cloud and the Formation of the Earth and Planets.* NASA TTF-677, U.S. Government Printing Office, Washington, DC.
Safronov V. S. (1991) Kuiper Prize Lecture — Some problems in the formation of the planets. *Icarus, 94,* 260–271.
Salo H. (1992) Gravitational wakes in Saturn's rings. *Nature, 359,* 619–621.
Sano T., Miyama S. M., Umebayashi T., and Nakano T. (2000) Magnetorotational instability in protoplanetary disks, II. Ionization state and unstable regions. *Astrophys. J., 543,* 486–501.

Schräpler R. and Henning Th. (2004) Dust diffusion, sedimentation, and gravitational instabilities in protoplanetary disks. *Astrophys. J., 614,* 960–978.

Scott E. R. D. and Haack H. (1993) Chemical fractionation in chondrites by aerodynamic sorting of chondritic material. *Meteoritics, 28,* 434.

Scott E. R. D., Barber D. J., Alexander C. M., Hutchison R., and Peck J. A. (1989) Primitive material surviving in chondrules: Matrix. In *Meteorites and the Early Solar System* (J. F. Kerridge and M. S. Matthews, eds.), pp. 718–745. Univ. of Arizona, Tucson.

Scott E. R. D., Love S. G., and Krot A. N. (1996) Formation of chondrules and chondrites in the protoplanetary nebula. In *Chondrules and the Protoplanetary Disk* (R. Hewins et al., eds.), pp. 87–96. Cambridge Univ., Cambridge.

Sekiya M. (1983) Gravitational instabilities in a dust-gas layer and formation of planetesimals in the solar nebula. *Prog. Theor. Phys., 69,* 1116–1130.

Sekiya M. (1998) Quasi-equilibrium density distributions of small dust aggregations in the solar nebula. *Icarus, 133,* 298–309.

Sekiya M. and Ishitsu N. (2000) Shear instabilities in the dust layer of the solar nebula I. The linear analysis of a non-gravitating one-fluid model without the Coriolis and the solar tidal forces. *Earth Planets Space, 52,* 517–526.

Sekiya M. and Ishitsu N. (2001) Shear instabilities in the dust layer of the solar nebula II. Different unperturbed states. *Earth Planets Space, 53,* 761–765.

Sekiya M. and Takeda H. (2003) Were planetesimals formed by dust accretion in the solar nebula? *Earth Planets Space, 55,* 263–269.

Shakura N. and Sunyaev R. A. (1973) Black holes in binary systems. Observational appearance. *Astron. Astrophys., 24,* 337–355.

Shakura N. I., Sunyaev R. A., and Zilitinkevich S. S. (1978) On the turbulent energy transport in accretion disks. *Astron. Astrophys., 62,* 179–187.

Shu F. H., Shang H., and Lee T. (1996) Toward an astrophysical theory of chondrites. *Science, 271,* 1545–1552.

Shu F. H., Shang H., Glassgold A. E., and Lee T. (1997) X-rays and fluctuating X-winds from protostars. *Science, 277,* 1475–1479.

Sirono S. and Greenberg J. M. (2000) Do cometesimal collisions lead to bound rubble piles or to aggregates held together by gravity? *Icarus, 145,* 230–238.

Skinner W. R. (1990a) Bipolar outflows and a new model of the early solar system. Part I: Overview and implications of the model (abstract). In *Lunar and Planetary Science XXI,* p. 1166. Lunar and Planetary Institute, Houston.

Skinner W. R. (1990b) Bipolar outflows and a new model of the early solar system. Part II: The origins of chondrules, isotopic anomalies, and fractionations (abstract). In *Lunar and Planetary Science XXI,* p. 1168. Lunar and Planetary Institute, Houston.

Skinner W. R. and Leenhouts J. M. (1991) Implications of chondrule sorting and low matrix contents of type 3 ordinary chondrites. *Meteoritics, 26,* 396.

Skinner W. R. and Leenhouts J. M. (1993) Size distributions and aerodynamic equivalence of metal chondrules and silicate chondrules in Acfer 059 (abstract). In *Lunar and Planetary Science XXIV,* pp. 1315–1316. Lunar and Planetary Institute, Houston.

Squires K. D. and Eaton J. A. (1990) particle response and turbulence modification in isotropic turbulence. *Phys. Fluids, A2,* 1191–1203.

Squires K. D. and Eaton J. A. (1991) Preferential concentration of particles by turbulence. *Phys. Fluids, A3,* 1169–1178.

Sreenivasan K. R. and Stolovitsky K. (1995a) Intermittency, the second-order structure function, and the turbulent energy dissipation rate. *Phys. Rev. E, 52,* 3242–3244.

Sreenivasan K. R. and Stolovitsky K. (1995b) Turbulent cascades. *J. Stat. Phys., 78,* 311–333.

Stepinski T. F. and Valageas P. (1996) Global evolution of solid matter in turbulent protoplanetary disks. I. Aerodynamics of solid particles. *Astron. Astrophys., 309,* 301–312.

Stepinski T. F. and Valageas P. (1997) Global evolution of solid matter in turbulent protoplanetary disks. II. Development of icy planetesimals. *Astron. Astrophys., 319,* 1007–1019.

Stevenson D. J. and Lunine J. I. (1988) Rapid formation of Jupiter by diffuse redistribution of water vapor in the solar nebula. *Icarus, 75,* 146–155.

Stone J. M., Gammie C. F., Balbus S. A., and Hawley J. F. (2000) Transport processes in protostellar disks. In *Protostars and Planets IV* (V. Mannings et al., eds.), pp. 589–599. Univ. of Arizona, Tucson.

Supulver K. and Lin D. N. C. (2000) Formation of icy planetesimals in a turbulent solar nebula. *Icarus, 146,* 525–540.

Tanga B. A., Dubrulle B., and Provenzale A. (1996) Forming planetesimals in vortices. *Icarus, 121,* 158–170.

Tanga P., Weidenschilling S. J., Michel P., and Richardson D. C. (2004) Gravitational instability and clustering in a disk of planetesimals. *Astron. Astrophys., 427,* 1105–1115.

Tennekes H. and Lumley J. L. (1972) *A First Course in Turbulence.* Massachusetts Institute of Technology, Cambridge. 300 pp.

Thompson J. J. and Newhall H. F. (1885) On the formation of vortex rings by drops falling into liquids, and some allied phenomena. *Proc. R. Soc., 39,* 417.

Toomre A. (1964) On the gravitational stability of a disk of stars. *Astrophys. J., 139,* 1217–1238.

Trieloff M., Jessberger E. K., Herrwerth I., Hopp J., Fiéni C., Ghéllis M., Bourot-Denis M., and Pellas P. (2003) Surface and thermal history of the H-chondrite parent asteroid revealed by thermochronometry. *Nature, 422,* 502–506.

Umurhan O. M. and Regev O. (2004) Hydrodynamic stability of rotationally supported flows: Linear and nonlinear 2D shearing box results. *Astron. Astrophys., 427,* 855–872.

Völk H. J., Jones F. C., Morfill G. E., and Röser S. (1980) Collisions between grains in a turbulent gas. *Astron. Astrophys., 85,* 316–325.

Wang L. and Maxey M. R. (1993) Settling velocity and concentration distribution of heavy particles in homogeneous isotropic turbulence. *J. Fluids Mech., 256,* 27–68.

Ward W. R. (2000) On planetesimal formation: The role of collective particle behavior. In *Origin of the Earth and Moon* (R. M. Canup and K. Righter, eds.), pp. 75–84. Univ. of Arizona, Tucson.

Wasson J. T., Trigo-Rodriguez J. M., and Rubin A. E. (2005) Dark mantles around CM chondrules are not accretionary rims (abstract). In *Lunar and Planetary Science XXXVI,* Abstract #2314. Lunar and Planetary Institute, Houston (CD-ROM).

Weidenschilling S. J. (1977) Aerodynamics of solid bodies in the solar nebula. *Mon. Not. R. Astron. Soc., 180,* 57–70.

Weidenschilling S. J. (1980) Dust to planetesimals: Settling and coagulation in the solar nebula. *Icarus, 44,* 172–189.

Weidenschilling S. J. (1984) Evolution of grains in a turbulent solar nebula. *Icarus, 60,* 553–567.

Weidenschilling S. J. (1988a) Formation processes and timescales for meteorite parent bodies. In *Meteorites and the Early Solar*

System (J. F. Kerridge and M. S. Matthews, eds.), pp. 348–371. Univ. of Arizona, Tucson.

Weidenschilling S. J. (1988b) Dust to dust: Low-velocity impacts of fragile projectiles (abstract). In *Lunar and Planetary Science XIX,* pp. 1253–1254. Lunar and Planetary Institute, Houston.

Weidenschilling S. J. (1995) Can gravitational instability form planetesimals? *Icarus, 116,* 433–435.

Weidenschilling S. J. (1997) The origin of comets in the solar nebula: A unified model. *Icarus, 127,* 290–306.

Weidenschilling S. J. (2000) Formation of planetesimals and accretion of the terrestrial planets. *Space Sci. Rev., 92,* 295–310.

Weidenschilling S. J. (2003a) Radial drift of particles in the solar nebula: Implications for planetesimal formation. *Icarus, 165,* 438–442.

Weidenschilling S. J. (2003b) Planetesimal formation in two dimensions: Putting an edge on the solar system (abstract). In *Lunar and Planetary Science XXXIV,* Abstract #1707. Lunar and Planetary Institute, Houston (CD-ROM).

Weidenschilling S. J. (2006) Accretion dynamics and timescales: Relation to chondrites. In *Meteorites and the Early Solar System II* (D. S. Lauretta and H. Y. McSween Jr., eds.), this volume. Univ. of Arizona, Tucson.

Weidenschilling S. and Cuzzi J. N. (1993) Formation of planetesimals in the solar nebula. In *Protostars and Planets III* (E. Levy and J. Lunine, eds.), pp. 1031–1060. Univ. of Arizona, Tucson.

Weisberg M. K., McCoy T. J., and Krot A. N. (2006) Systematics and evaluation of meteorite classification. In *Meteorites and the Early Solar System II* (D. S. Lauretta and H. Y. McSween Jr., eds.), this volume. Univ. of Arizona, Tucson.

Whipple F. (1972) On certain aerodynamic processes for asteroids and comets. In *From Plasma to Planet* (A. Elvius, ed.), pp. 211–232. Wiley, New York.

Wilcox D. (1998) *Turbulence Modeling for CFD, 2nd edition,* Chapter 3. DCW Industries, Inc., La Canada, California.

Wood J. A. (1985) Meteoritic constraints on processes in the solar nebula. In *Protostars and Planets II* (D. C. Black and M. S. Matthews, eds.), pp. 687–702. Univ. of Arizona, Tucson.

Wood J. A. (2000) Pressure and temperature profiles in the solar nebula. *Space Sci. Rev., 92,* 87–93.

Wood J. A. (2004) Formation of chondritic refractory inclusions: The astrophysical setting. *Geochim. Cosmochim. Acta, 68,* 4007–4021.

Wood J. A. and Morfill G. E. (1988) A review of solar nebula models. In *Meteorites and the Early Solar System* (J. F. Kerridge and M. S. Matthews, eds.), pp. 329-347. Univ. of Arizona, Tucson.

Woolum D. and Cassen P. (1999) Astronomical constraints on nebular temperatures: Implications for planetesimal formation. *Meteoritics & Planet. Sci., 34,* 897–907.

Wurm G., Blum J., and Colwell J. (2001) A new mechanism relevant to the formation of planetesimals in the solar nebula. *Icarus, 151,* 318–321.

Yamoto F. and Sekiya M. (2004) The gravitational instability in the dust layer of a protoplanetary disk: Axisymmetric solutions for nonuniform dust density distributions in the direction vertical to the midplane. *Icarus, 170,* 180–192.

Youdin A. and Shu F. (2002) Planetesimal formation by gravitational instability. *Astrophys. J., 580,* 494–505.

Youdin A. N. and Chiang E. I. (2004) Particle pileups and planetesimal formation. *Astrophys. J., 601,* 1109–1119.

Yurimoto H. and Kuramoto N. (2004) Molecular cloud origin for the oxygen isotope heterogeneity in the solar system. *Science, 305,* 1763–1766.

Transient Heating Events in the Protoplanetary Nebula

Harold C. Connolly Jr.
Kingsborough College of the City University of New York, and American Museum of Natural History

S. J. Desch
Arizona State University

Richard D. Ash
University of Maryland

Rhian H. Jones
University of New Mexico

1. INTRODUCTION

The most fundamental reason that transient heating events were important in the formation of the solar system is that they provided the energy to produce chondrules and refractory inclusions (*Beckett et al.,* 2006; *Lauretta et al.,* 2006). From standard astrophysical models for the formation of the solar system, these objects are not predicted to exist and are certainly not predicted to be part of the terrestrial-planet-forming process as has been argued by meteoriticists for decades. They do, however, comprise up to approximately 80% of the mass of ordinary chondrites and approximately 50% of the mass of most carbonaceous chondrites (*Brearley and Jones,* 1998). Thus, the ubiquity of these objects within chondrites and the abundance of chondrites suggest that transient heating must have been common. Furthermore, *Levy* (1988) estimates that the mass of chondrules melted in the inner solar system was at least 10^{24} g, which, although not a planetary mass, is a significant amount, especially considering that it is a minimum quantity.

Today, the solar system is chemically differentiated at many scales. This differentiation is reflected in the composition of planetary materials from the millimeter-scale (chondrules and refractory inclusions) to meter-scale (chondrites), to planetesimals (asteroids) and finally the planetary scale. The compositional differences include both elemental and, in some cases, isotopic abundances, and arose from physical and chemical processes in the nebula and on parent bodies. The major issue is, did transient heating events that produced chondrules and refractory inclusions contribute, at least in part, to the chemical differentiation observed in planetary materials within the solar system? If indeed the bulk of these planetary materials are the same type of materials from which the terrestrial planets were produced, understanding the mechanism that melted chondrules and refractory inclusions is an important component to constraining solar system formation.

Numerous reviews of the petrographic and geochemical characteristics of chondrules and refractory inclusions have been published in the last decade (*Beckett et al.,* 2006; *Brearley and Jones,* 1998; *Connolly and Desch,* 2004; *Jones et al.,* 2000, 2005; *Hewins,* 1997; *Krot et al.,* 2004a,b; *Lauretta et al.,* 2006; *MacPherson,* 2004; *MacPherson et al.,* 2005; *Rubin,* 2000; *Zanda,* 2004). Many assumptions exist in the literature when discussing these objects that we will attempt to clarify as part of our goal to communicate to a wide audience within this chapter. First, when discussing chondrules, most researchers are referring to those objects dominated by FeO- and MgO-rich silicate minerals such as olivine $(Mg,Fe)_2SiO_4$ and enstatite $(Mg,Fe)SiO_3$, with varying abundances of glass, Fe,Ni-rich metal, iron sulfide, and other minor phases. These are referred to as ferromagnesian chondrules.

Refractory inclusions are divided into calcium-aluminum-rich inclusions (CAIs) and amoeboid olivine aggregates (AOA). Typically, the centimeter-sized igneous CAIs known as type B are what comes to mind for most researchers when discussing CAIs, largely due to their striking appearance, resulting in their size, color, and coarse grain-size and consequent popularity in the earlier days of CAI research on the carbonaceous chondrite Allende. It is important to note that all but the fluffy type A (FTA) CAIs experienced some degree of melting and are arguably, in some cases undisputedly, igneous. Because of their highly diverse chemistries and textures we refer the reader to reviews by *Brearley and Jones* (1998), *MacPherson et al.* (1988), *MacPherson* (2004), and *Krot et al.* (2004a) for detailed discussion of these objects.

Throughout this text we will mostly focus discussion on how ferromagnesian chondrules and those refractory inclusions that have experienced melting can constrain the nature and history of transient heating events. Igneous objects such as chondrules and type B CAIs found within chondrites have textures reflective of their thermal histories. The differences in igneous textures of chondrules and CAIs combined with the elemental fractionations within crystals provide powerful constraints on their melting temperatures, duration of melting, and rate of heat loss or cooling rate that they experienced. These same data tell us that many chondrules and type B CAIs have been heated and cooled multiple times, and in specific cases evidence suggests that such reheating or melting occurred after a period of alteration occurred, attesting to their complex petrogenesis (*Beckett et al.,* 2000, 2006; *Connolly and Desch,* 2004; *Connolly et al.,* 2003; *Davis and MacPherson,* 1996; *Jones et al.,* 2000; *Lauretta et al.,* 2006; *Rubin,* 2000; *Zanda,* 2004).

The importance of transient heating events is summed up as (1) they produced chondrules and refractory inclusions; (2) the chemical differences in chondrules, refractory inclusions, chondrites, asteroids, and planets may have, at least in part, been produced through transient heating; (3) they may have processed the majority of rocky planetary materials that were the precursors to terrestrial planets; and (4) the accretion of planets does not, from astronomical theories, have to include a phase of high-temperature processing of chondritic materials, yet meteoritics argues that it must have. Hence, assessing and investigating potential relationships between transient heating events and planet formation will enable us to put constraints on viable planet-formation mechanisms and to further understand meteorite evolution. Large energetic heating events implied by meteorite data place significant constraints on protoplanetary disk evolution not currently accommodated by astrophysical models, especially if most planetary materials went through a high-temperature processing phase before terrestrial planet formation occurred.

In this chapter, we (1) explore the evidence that transient heating events occurred (section 2), (2) discuss time and temperature constraints on transient heating (section 3), (3) investigate ideas and models of mechanisms that may have produced the heating (section 4), and (4) offer a blueprint for issues that must be addressed through further research (section 5).

2. EVIDENCE FOR TRANSIENT HEATING EVENTS

In this section, we review the evidence from the meteorite record and observational astronomy that transient heating events occurred early in solar system history. We make the reasonable assumption that events observed to occur in association with young stellar objects (YSO) with predicted masses similar to that of the Sun also occurred in the early stages of solar system formation. We divide the discussion into evidence that is well established and that which can be interpreted or is implied to support transient heating.

1.1. The Rock Record

2.1.1. Well-established evidence. The most compelling evidence for transient heating within the lifetime of the protoplanetary nebula is the presence of chondrules and refractory inclusions in primitive meteorites. These objects formed on timescales of several hours to a few days (e.g., *Beckett et al.,* 2006; *Connolly and Desch,* 2004; *Hewins et al.,* 1996; *Lauretta et al.,* 2006). Although there is currently some discussion within meteoritics as to the extent of melting many chondrules experienced, in order to at least partially melt they must have experienced temperatures at or above their solidus (conservatively ~1300–1500 K). Many chondrules (as we discuss in more detail below) experienced temperatures near or slightly above their liquidus temperatures [~1700–2000 K (*Lofgren and Lanier,* 1990; *Radomsky and Hewins,* 1990)]. Similarly, refractory inclusons, in particular type B CAIs, experienced melting at ~1700 K (*Stolper and Paque,* 1986), which is within the range of temperatures experienced by chondrules.

2.1.2. Supporting evidence. The existence of other evidence that can be interpreted as having been, at least in part, the result of a transient heating event (or events) is often the subject of intense debate. Because such data may be important to our discussion, we must consider them.

The bulk compositions of chondrites are diverse (*Brearley and Jones,* 1998; *Krot et al.,* 2004a; *Weisberg et al.,* 2006), as are those of chondrules and refractory inclusions. Chondrite compositions deviate from primitive solar CI values, with varying amounts of depletions of moderately volatile and volatile elements, depending on the chondrite type (*Wasson,* 1985). How these differences arose is a major question. One potential explanation relevant to this chapter is that, at least in part, chondritic components and/or their precursors (including those of chondrules and refractory inclusions) may have undergone evaporative processing during transient heating followed by partial or total recondensation of evaporated species (*Alexander,* 2004; *Cohen et al.,* 2000; *Haung et al.,* 1996; *Hewins,* 1997; *Hewins et al.,* 2005; *MacPherson et al.,* 2005). It is clear that at least some CAIs experienced mass-dependent isotopic fractionation (*Ash et al.,* 2002; *Clayton et al.,* 1988; *Galy et al.,* 2000; *Richter et al.,* 2002); however, the extent to which chondrules experienced evaporation while molten is unknown (*Hewins,* 1997; *Connolly et al.,* 2001; *Cohen et al.,* 2000; *Cohen and Hewins,* 2004). There is no isotopic evidence for Rayleigh fractionation resulting from evaporation during chondrule formation, suggesting that free evaporation under canonical conditions (i.e., an ambient 10^{-3} bar gas of solar composition) did not occur as might be expected (*Humayan and Clayton,* 1995; *Galy et al.,* 2000). *Palme et al.* (1993, and references therein) argued that a chemical relationship exists between chondrules and chondrite matrix, matrix being more enriched in volatiles than chondrules, and that the chondrules lost volatiles that recondensed onto or in the form of matrix. Such a relationship may be due to the processing of chondritic materials during transient heating, but it is by no means the only explanation. Although the role that transient heating played in contributing to the bulk compositions of chondrites is a topic of debate, the potential for it to have been an important process cannot be ignored.

The existence of short-lived radionuclides (^{7}Be, ^{10}Be, ^{26}Al, ^{41}Ca, and ^{53}Mn) in refractory inclusions and potentially chondrules has been suggested to be the product of irradiation of precursor materials by the early active Sun (*Gounelle et al.,* 2001). If, as has been suggested by *Gounelle et al.* (2001), the irradiation process is also related to the mechanism that melted refractory inclusions (and perhaps chondrules), then the evidence for the existence of these isotopes is indirect evidence for transient heating, even if it does appear to be circular reasoning.

2.2. Observational Astronomy

2.2.1. Well-established evidence. From observational astronomy, clear evidence exists for transient heating events occurring during the lifetime of a protoplanetary disk. These observations, however, correspond to timescales of months to 10^3–10^5 yr, which is far longer than the kinds of transient heating that chondrules and refractory inclusions experienced. Nevertheless, from an astronomical view they are transient and their relationship to the production mechanisms of shorter transient heating events that may be the heat source for producing igneous melt spheres found within chondrites is unknown. It is therefore important that we review the observations.

FU Orionis outbursts (e.g., *Hartmann and Kenyon,* 1996; *Bell et al.,* 2000) are increases in the optical luminosity of accreting protostars by about a factor of 100. The rise in luminosity occurs on a timescale of a few years, and is followed by decay in the luminosity that takes place over several decades (*Bell et al.,* 2000; *Calvet et al.,* 2000). The increased optical luminosity is likely due to increased mass accretion of material through the disk onto the YSO. Statistics suggest that each protostar experiences about 10 FU Orionis outbursts during its evolution (*Bell et al.,* 2000; *Calvet et al.,* 2000). FU Orionis outbursts are believed to occur only during the initial stages (Class 0, I) of protostellar evolution, or the first 10^5 yr (*Bell et al.,* 2000; *Konigl and Pudritz,* 2000; *Shu et al.,* 2000). Based on this estimate and stellar statistics, the interval between outbursts is estimated at about 10^4 yr. However, from the spacing between the knots within bipolar outflows, the time between FU Orionis outbursts is estimated to be about 10^3 yr (*Bally et al.,* 1995). It would appear that every thousand years or so, protoplanetary disks spend about a century in a stage of enhanced accretion onto the protostar; this sporadic accretion continues for perhaps 10^5 yr.

Similar phenomena to the FU Orionis outbursts are Exor outbursts. They involve an increase in luminosity by a factor of approximately 30, and the timescales for the rise and fall in luminosity are months to years (*Herbig,* 1977, 1989). Exor outbursts are not restricted to the first 10^5 yr of disk evolution and appear to occur even in more classical T Tauri disks [Class 0 through II (*Konigl and Pudritz,* 2000)].

Bell et al. (2000) argued that the midplanes of protoplanetary disks are too deeply embedded to respond to the temperature increases during the outburst phases, but hypothesized that the surfaces of disks can experience significant temperature increases. These temperature increases would, however, necessarily last as long as the outburst, i.e., for years. While transient on disk evolution timescales (several million years), this is much longer than the timescales for chondrule and refractory inclusion formation (*Beckett et al.,* 2006; *Connolly and Desch,* 2004; *Hewins et al.,* 2005; *Jones et al.,* 2000; *Lauretta et al.,* 2006).

Another transient heating mechanism known to occur in protoplanetary disks is X-ray flares. X-ray emission from protostars is observed and is known to be highly variable, changing by an order of magnitude over the course of hours (*Feigelson et al.,* 2003). X-ray flares persist throughout the T Tauri stage of protostellar evolution (Class I through III). However, the actual X-ray energy output from protostars is typically no more than 0.1% of the star's luminosity. The X-rays, and possibly energetic particles (keV to MeV energies) accelerated in the flares, can interact with the surface layers of the disk and ionize local gas very effectively (*Glassgold et al.,* 2000). This in turn may trigger transient energy dissipation mechanisms that rely on coupling between the magnetic field and the gas in the disk, such as magnetic flares in the disk (*Mannings et al.,* 1996). For the X-ray flares to directly heat meteoritic material requires that that material reside close to the star. Rocky material residing in the magnetosphere of the protostar could interact and be heated with energetic particles accelerated in flares (*Shu et al.,* 2001).

In summary, there are several events taking place in protoplanetary disks that imply transient heating. FU Orionis events take place over decades, but Exor events occur with smaller amplitude and faster variation. It is tempting to suggest that even smaller heating events related to these different outbursts might occur in disks on even shorter timescales. Observations of DG Tau show that it exhibits behavior over timescales of only months that may suggest some type of transient heating (*Wooden et al.,* 2000). X-ray flares are observed around protostars and could heat material, although it is unclear how efficient such energetic events would be for producing chondrules and refractory inclusions.

2.2.2. Supporting evidence. Bipolar outflows alone cannot be considered the transient heating events that processed materials within the disk. They are far too hot (*Konigl and Pudritz,* 2000) to have been the direct source of the melted materials found in meteorites. However, winds are hypothesized to occur during the outflow stages and there may be interactions between outflows and disks (*Konigl and Pudritz,* 2000; *Shu et al.,* 2000). Although the winds themselves do not directly heat or melt solid materials, heating may occur when particles are entrained within the winds and irradiated in the optical and X-ray wavelengths by the YSO (*Shu et al.,* 1996, 1997, 2000; *Gounelle et al.,* 2001).

A lesser energetic component of the outflow is not directed along the protostar's poles and could potentially interact with the disk, although it is not yet observed. This process may produce shocks as the wind collides with the disk gas and thus produces heating of gas and any rocky materials or dust grains (*Nakamoto et al.,* 2004). Such heating would be restricted to a very thin surface layer of the protoplanetary disk (≪1 g cm^{-2}) and has yet to be defined numerically in detail.

The exact timing of the start of planet formation is only constrained to within a few million years (*Nichols,* 2006). If it began very early in the disk's lifetime, then the process of accretion would itself have produced transient heating events. The energy associated with accretion, largely through collisional events, could have processed primitive planetary materials at high temperatures. As we discuss below, however, a collisional mechanism for the production of chondrules and refractory inclusions is problematic for several reasons.

3. CONSTRAINTS ON TRANSIENT HEATING EVENTS: TEMPERATURE/TIME VARIABLE

Of critical importance to understanding the energetic nature of transient heating events is gaining some understanding of the temperatures and durations of the events. It is difficult to quantify with confidence the exact peak temperature and duration of any given transient heating event(s). We do not have direct measurements for all aspects of astronomically observed events and in meteorites only an indirect record of these events. Both petrographic and experimental investigations of chondrules and refractory inclusions provide constraints on the temperature and heating duration these objects experienced. As a framework to explore the temperature/time variable of transient heating, we discuss two different aspects of timescale: the first is the hours to days during which individual chondrules and refractory inclusions formed, and the second is the 10^5–10^6 yr over which the mechanism or mechanisms must have operated.

Most constraints on the temperatures and times of transient heating events within the protoplanetary nebula are derived from comparing natural materials with the results of experimental investigations of chondrule and refractory inclusion formation (e.g., *Beckett et al.,* 2006; *Connolly and Desch,* 2004; *Hewins,* 1997; *Hewins et al.,* 2005; *Lofgren,* 1996). Since the chondrules and refractory inclusions we discuss are igneous, we can use their chemical and physical characteristics to constrain the thermal conditions that characterize their melting and crystallization, and this, in turn, can constrain the heating regime of these objects. The thermal histories of chondrules and igneous refractory inclusions (compact type A, type B, type C, and some AOAs) comprise four stages: (1) pre-melting, (2) melting, (3) cooling and crystallization, and (4) post-crystallization. Numerous factors affect each one of these stages and we discuss some of the more important issues below or refer the reader to an appropriate reference.

As was briefly mentioned above, it is important to understand the types of objects we are discussing. Chondrules are diverse, both in their textural types and bulk chemistry (*Grossman et al.,* 1988; *Hewins et al.,* 1996; *Brearley and Jones,* 1998; *Connolly and Desch,* 2004; *Lauretta et al.,* 2006). To date, the only chondrules that have been investigated through experimental petrology in enough detail to provide clear constraints on their thermal history are the Fe,Mg-rich chondrules (*Connolly and Desch,* 2004; *Grossman et al.,* 1988; *Hewins et al.,* 2005; *Lauretta et al.,* 2006). The formation of Al-rich or any other chondrule type is essentially experimentally unconstrained; thus our discussion is limited to the origin of Fe,Mg-rich chondrules (*Connolly and Desch,* 2004). The majority of constraints on the thermal histories of chondrules have not fully integrated free evaporation (elemental fractionation) from the molten chondrules or back-reactions between molten spheres and gas. We are therefore forced to limit our discussion to data on the thermal histories of chondrules that experienced negligible evaporation while molten.

As with chondrules, refractory inclusions are diverse in both their textures and bulk chemistry. To date, only an average type B1 CAI bulk composition has been investigated experimentally in detail (*Beckett et al.,* 2006; *Stolper,* 1982; *Stolper and Paque,* 1986; *Connolly and Burnett,* 2003). The role of free evaporation from molten type B CAIs is not yet well constrained, although evidence is compelling that it did occur (*Richter et al.,* 2002).

3.1. Thermal Histories

1. The pre-melting thermal conditions of chondrules and, to an even greater extent, igneous refractory inclusions are poorly constrained. The primary constraint on pre-melting temperatures that chondrules experienced is derived from the abundance of moderately volatile elements such as S and Na. Chondrules contain varying abundances of volatile and moderately volatile elements. Sulfur is observed mainly in the form of troilite, which is argued to be primary in a small percentage of chondrules (*Rubin et al.,* 1999; *Jones et al.,* 2000; *Tachibana and Huss,* 2005; *Zanda,* 2004). The presence of primary S within chondrules suggests that these chondrules did not experience prolonged heating (more than a few hours) between ~650 and 1200 K before, or more than several minutes at, their peak temperatures (*Hewins et al.,* 1996; *Connolly and Love,* 1998; *Jones et al.,* 2000; *Lauretta et al.,* 2001; *Tachibana and Huss,* 2005). However, it should be noted that the quoted temperature range is dependent on the total pressure and dust/gas ratio of the system during heating and these calculations make certain assumptions about the diffusion and kinetics of reactions of S in a silicate melt at the tested pressures and dust/gas ratios (*Lauretta et al.,* 2001; *Tachibana and Huss,* 2005). Thus, some caution must be taken when using the temperature/time constraints as further experiments might refine these data.

The pre-melting thermal conditions experienced by type B CAIs are essentially unconstrained. These objects contain a very minor amount of S-rich phases argued to be troilite, but it is rather unlikely that such a phase is primary in these objects. Because these objects are composed of refractory phases compared with Fe,Mg-rich chondrules, it is possible that they or their precursors experienced premelting thermal conditions that were hotter than Fe,Mg-rich chondrules for longer periods than the chondrules. However, this issue is essentially quantitatively unconstrained.

2. Constraints on the melting history of chondrules and type B CAIs have been determined from experimental petrology studies using analog compositions. These experiments are generally performed at total pressures of 1 atm and an f_{O_2} controlled to be comparable to those experienced by natural chondrules.

An essential concept for understanding the melting of chondrules is that the degree of melting and the survival of potential nucleation centers depends, to a great extent, on a temperature/time function, which, in part, constrains the type of texture that can be produced. For example, a shorter heating time at a higher temperature may leave many nucleation sites as a longer heating time at somewhat lower temperatures. Hence, there is rarely a unique thermal regime solution for any observed texture (*Jones et al.,* 2000). However, some constraints can be placed from experimental petrology. Peak melting temperatures experienced by the majority of chondrules were between 1770 and 2120 K and these temperatures were maintained for several seconds to minutes (*Connolly and Desch,* 2004; *Connolly and Love,* 1998; *Hewins,* 1997; *Hewins and Connolly,* 1996; *Hewins et al.,* 2005; *Jones et al.,* 2000; *Lofgren,* 1996; *Lofgren and Lanier,* 1990; *Lauretta et al.,* 2006; *Radomsky and Hewins,* 1990). For barred olivine chondrules (BO) it is possible that with very short heating times the absolute maximum temperature was somewhat higher [2200 K (*Connolly et al.,* 1997)]. The number of crystal nuclei that must survive the melting event determines this peak temperature range. Nucleation may also be produced by collisions of molten chondrules with dust particles or other chondrules (*Connolly and Hewins,* 1995), in which case, because survival of nuclei is not required to produce textures, the *peak temperature* might be higher (at least 2200 K).

The duration of heating must be on the order of several seconds to a few minutes. This constraint was determined mainly from the abundance of volatile and moderately volatile elements such as Na and S (*Yu et al.,* 1996; *Yu and Hewins,* 1998). The only alternative to rapid, or flash, melting of chondrule precursors and chondrules is if the ambient nebular gas was enriched in volatile or moderately volatile elements. If this was the case, molten chondrules could have exchanged with the gas during melting and cooling to replenish these elements (*Lewis et al.,* 1993). It is important to note that parent-body processing (hydrous alteration and metamorphism) can also redistribute moderately volatile elements, thereby complicating evidence from volatile-element abundances (*Grossman and Brearley,* 2005). The implication of this parent-body redistribution process is that great caution is needed when assuming that observed moderately volatile-element abundances in chondrules represent only a signature of melting and cooling history.

The time a chondrule spends at peak temperature affects the amount of relict material, be it crystals or nuclei that remain in a melt after heating (*Connolly and Desch,* 2004; *Hewins et al.,* 2005; *Lofgren,* 1996). The number of nuclei remaining in a melted chondrule constrain the type of texture that is produced if nucleation is not affected by external sources such as collisions with dust grains (*Connolly and Hewins,* 1995). Another independent constraint on chondrule heating times is derived from the dissolution rates of silicates in silicate melt (*Greenwood and Hess,* 1996; *Jones et al.,* 2000). Approximately 15% of chondrules in ordinary chondrites contain grains that are interpreted to be relict (*Jones,* 1996), meaning they survived the melting of the chondrule in which they are observed. Since these grains did not melt or completely dissolve, their existence constrains the temperature/time variable of heating to tens of seconds to several minutes.

The presence of relict grains also demostrates that chondrules experienced multiple melting events. Texture and chemistry of these grains imply that they were derived from a previous generation of chondrules. Those chondrules where such grains were originally formed were broken up and their contents found their way into the precursors of another generation of chondrules. Thus, chondrule formation occurred many times and is not restricted to a single heating and cooling episode.

Based on the studies of the crystallization of melilite, the major phase in type B CAIs, these objects experienced peak melting temperatures of ~1700 K (*Beckett et al.,* 2006; *Stolper,* 1982; *Stolper and Paque,* 1986). However, it is possible that heating for shorter times to higher peak temperatures could also have produced similar textures, but detailed experimental data for such a thermal history are not yet available. *Connolly and Burnett* (2003) place limits on the maximum duration of type B CAI heating to less than a few tens of hours based on the inhomogeneous concentrations of V, Ti, and Cr within spinel grains. Like chondrules, type B CAIs contain evidence that they experienced multiple melting events (*Beckett et al.,* 2000, 2006; *Davis and MacPherson,* 1996; *Connolly and Burnett,* 1999; *Connolly et al.,* 2003): In some cases this reheating or remelting occurred after an epoch of alteration (that occurred while the objects were free-floating wanders) was experienced by the object (*Beckett et al.,* 2000, 2006; *Simon et al.,* 2005).

3. The cooling and crystallization of chondrules and type B CAIs is to some level well understood. *Desch and Connolly* (2002) reviewed in detail the experimental data on cooling rates and discussed the limitations of these data. To summarize, the cooling rates of chondrules are constrained from the development of chondrule textures (the arrangement and shape of the crystals) and the major- and to a lesser extent the minor-element zoning behavior within individual crystals. Overall the cooling rates experienced by chondrules based on their texture and chemistry are constrained to a range of 10–3000 K h^{-1}, with the majority of chondrules (i.e., porphyritic textures) cooling at 100 K h^{-1} or less through the temperature range between their liquidus and solidus. Cooling from above the liquidus temperature was very rapid, at least 5000 K h^{-1}. Although in reality chon-

drules probably did not cool linearly, most of their crystallization likely occurred in a temperature range that was approximately linear. Recent research has questioned these cooling rates as being far too slow (*Wasson and Rubin,* 2003), but the faster rates proposed, which are up to 1000°/s, have not been widely accepted and are difficult to reconcile with other observations of natural materials and experimental analogs.

Type B CAIs have been constrained to cool between 0.5 and 50 K h^{-1} with perhaps the most reliable cooling rates appearing to be ~2 K h^{-1} (*Stolper and Paque,* 1986; *MacPherson et al.,* 1984; *Simon et al.,* 1996).

4. The post-crystallization, pre-parent-body-accretion thermal regimes for chondrules and igneous CAIs are largely unconstrained experimentally. For petrological studies of natural samples, the major challenge has been to determine chemical overprints from parent-body processes on these objects. It is not, however, known if chondrules and igneous CAIs could have experienced prolonged heating from subsolidus temperatures (e.g., <1300 K) before accretion. The preservation of igneous zoning profiles (for chondrules and refractory inclusions) and abundant glass (in chondrules) is evidence that any post-formation regime of elevated temperature was limited. The issue has not been quantified in any detail and requires additional research in the future.

3.2. Long-Term Timescales: The Epoch of Chondrule and Igneous Calcium-Aluminum-rich Inclusion Formation

By making certain reasonable assumptions, temporal constraints can be made for the epoch of processing chondritic rock-forming materials into chondrules and refractory inclusions within the nebula. As discussed elsewhere within the volume, the inferred initial $^{26}Al/^{27}Al$ in igneous CAIs of 5×10^{-5} is taken as the solar system initial value (*Bizzarro et al.,* 2004; *Galy et al.,* 2000). Lower values observed in chondrules have been used to argue that chondrules (both Fe,Mg-rich and Al-rich) may have begun to form some 0.7 m.y. later (*Tachibana et al.,* 2003; *Russell et al.,* 1996). The data also suggest that chondrule production lasted for at least 2.4 m.y. after formation of most CAIs (*Tachibana et al.,* 2003; *Huss et al.,* 2001) and potentially up to 5 m.y. later (*Russell et al.,* 1996). Timescales for chondrule forma-tion of several million years are also suggested by Pb-Pb ages (*Amelin et al.,* 2002). There is currently some uncertainty about whether there is a hiatus between CAI and chondrule formation (*Amelin et al.,* 2002). It is important to point out that not all authors of this chapter agree on the interpretation of the relative abundance of ^{26}Al within chondritic material. The apparent age gap between CAI and chondrule formation may or may not be real, depending on whether there was a uniform distribution of ^{26}Al throughout the solar system, which is an assumption that is being actively questioned. With more analyses of minerals with lower Al/Mg and high-precision Pb-Pb dating, the apparent time difference between CAIs and chondrules is diminishing. Interestingly, two amoeboid olivine aggregates (AOAs) have $^{26}Al/^{27}Al$ ratios that correspond to inferred ages of 0.3 and 0.5 m.y. after CAI formation, thus their ages appear to fall between CAI and chondrule formation (*Krot et al.,* 2004b). One of these AOAs experienced very low degrees of melting, suggesting that some processing of planetary materials was occurring after CAI formation but before chondrule formation. A critical point to remember is that the data is widely interpreted as demonstrating that CAI formation was apparently short-lived, whereas chondrule formation lasted for potentially millions of years. Future data may dramatically change our current thinking on the time frame of chondrule and refractory inclusion formation, but for now we are con-strained to work with available data.

Several important implications arise from the discussion of the ages of refractory inclusions and chondrules. First, such data essentially provide no constraint on the mechanism that melted chondrules and CAIs, but on the duration of the process. The zero-order constraint is that transient heating within the disk (CAIs) occurred very close to t = 0 and then definitively at 0.7 m.y. later for a period of at least 2.5 m.y. later. We cannot state with any confidence if the mechanism that produced the heating and melting lasted continuously for at least 2.5 m.y. or if it was episodic, turning off and on. It is not possible at this time to determine if more than one mechanism was responsible for CAI and chondrule production, or even if all chondrules were formed by the same process. Age data on AOAs, intermediate in composition between CAIs and chondrules, suggests that the processing of chondritic material may have been continuous from t = 0 and that the composition of the material being processed changed or evolved from more refractory to less so with time. Simply stated, there is no *a priori* reason that these objects cannot have been melted by the same mechanism or type of mechanism. Unfortunately, this is not a unique solution because no compelling evidence exists to eliminate the potential that refractory inclusions and chondrules were melted by different mechanisms. We stress that this does not mean that the environments of formation were the same, and the reader is cautioned not to automatically equate the environment of formation with a mechanism because they may not be intimately linked.

4. THE PRODUCTION OF CHONDRULES AND REFRACTORY INCLUSIONS

We categorize the numerous proposed transient heating events into three major paradigms that represent various environments associated with protoplanetary disks (*Connolly and Desch,* 2004): (1) collision or impact events between asteroid- to planetary-sized bodies in the earliest stages of planet formation, (2) linking the formation of chondrules and refractory inclusions to the YSO, and (3) what we consider to be mechanisms that are purely hypothetical. The first two paradigms have some undisputed bases in linking the formation of chondrules and refractory inclusions to events or processes that are known to occur in protoplanetary disks or did occur in the solar system. In the following section we explore these models and their merits and weaknesses for producing heating events in the contexts of our criteria established above.

4.1. Chondrules as Objects Formed by Collisional Interactions of Planetary Bodies

The idea of chondrules being the product of some kind of planetary setting rather than free-floating wanderers is not new, appearing as far back as *Tschermak* (1895). A big advantage to invoking this kind of setting for chondrule and refractory inclusion production is that it is well established that, to build planetesimals and planets, collisions and impacts occurred. Many different types of planetary settings for the formation of chondrules have been proposed in the last 100 years (*Connolly and Desch,* 2004, and references therein). Such ideas include magmatic processes, presumably in planetary bodies; ejection from planetary bodies by such forces as volcanos (*Merrill,* 1920; *Lugmair and Shukolyukov,* 2001); and collision events between two bodies, either solid, partially molten, or fully molten in the interior (*Urey and Craig,* 1953; *Urey,* 1967; *Sanders,* 1996). Many of the impact or collision ideas typically involve small bodies or planetesimals, the size of asteroids, colliding with one another or impacting into Moon-sized bodies. It may have been the case that Moon-sized bodies were also colliding into each other in the early nebula. Astrophysical models of hierarchical accretion in the protoplanetary disk provide a plausible basis for hypotheses involving collisions. On the other hand, astrophysical models of these accretion stages are incomplete, and currently there are no models that constrain the magnitude, frequency, or effects of the collisions. The data discussed in previous sections do not refute the planetary collision paradigm, but neither can they be used to support it. However, we know that these events occurred and we know that, in some cases, they produced melts. These melts appear as shock-melt veins in chondrites (*Dodd and Jarosewich,* 1979; *McCoy et al.,* 1995), as meteorites interpreted as crystallized impact melts, and, in some cases, as melt spheres, e.g., in diogenites (*Mittlefelt et al.,* 1998), in howardites, on the Moon, and on Earth [e.g., Lonar Crater (*Fredriksson et al.,* 1973)]. However, the geological and astronomical conditions under which these were formed are quite different from those expected on a chondrite parent body and there are key differences in all these compared with chondrules. But without more sophisticated numerical models of impact melting and chondrule production from a chondritic parent body, this remains a potential mechanism for chondrule and possibly refractory inclusion formation.

The key problem of the planetary setting paradigm is that it is based on untested ideas. Many objections to forming chondrules in some kind of planetary setting and collision process have been discussed by *Taylor et al.* (1983), *Grossman et al.* (1988), and *Hewins et al.* (1996); these objections include the need for multiple heating events, the production of rims around chondrules, etc. Although these authors raise interesting and important points against such models, the main difficulty in assessing any kind of planetary setting for the production of chondrules is that no quantitative model has yet been developed and ideas have not been developed beyond the cartoon and idea stage. The thermal histories of chondrules — their peak temperatures, durations of heating, and cooling rates — have simply never been computed in the context of a planetary setting. This is also precisely the main reason not to reject or quickly dismiss such ideas. At this point in the investigation of chondrule and refractory inclusion formation it is simply unknown if any of these ideas have merit in a rigorous quantitative treatment.

Apart from the scientific issues discussed above, another major challenge for any model that ties chondrule formation to a planetary setting is the implicit assumption almost universally made in the field of meteoritics that chondrules (and potentially refractory inclusions) predate planetesimals. This is an understandable deduction, since the majority of primitive chondrite materials are composed almost entirely of chondrules. However, the age of chondrules, refractory inclusions, and magmatic processing on asteroids could be interpreted to suggest that this is not the case (*Nichols,* 2006), and continued research into this important issue is needed.

4.2. Linking the Formation of Chondrules with the Early Active Sun

Relating the Sun or the byproduct processes of an early active star to the formation mechanism of chondrules was first proposed by *Sorby* (1877) with his suggestion "... that some at least of the constituent particles of meteorites were originally detached glassy globules, like drops of fiery rain ... when conditions now met with only near the surface of the sun extended much further out from the centre of the solar system." A major advantage to linking chondrule formation with the early active Sun is that many of the processes that the early active Sun is thought to have experienced are observed in other YSOs. Linking the Sun's early activity to chondrule and refractory inclusion formation has significant merit in that it can potentially lead to testable, quantitative predictions.

Several ideas that link chondrule production with the early active Sun have been proposed (*Grossman,* 1988), including FU Orionis outbursts (*Boss,* 1996) and magnetic flares (*Levy and Araki,* 1989; *Morfill et al.,* 1993). In the last few years, however, one idea has received considerable attention: chondrules as byproducts of the bipolar outflow stage of the YSO. The idea can first be attributed to *Skinner* (1990), although his model was purely qualitative. *Liffman and Brown* (1996) produced a quantitatively detailed model for the formation of chondrules by ablation off larger bodies in outflows. Their model attempted to quantitatively reproduce chondrule thermal histories but did not successfully reproduce the maximum temperatures experienced by Fe,Mg-rich chondrules or the duration of melting. Furthermore, in their model chondrules are produced as ablation spheres from larger rocks that find their way into the outflows and are pushed along the edge regions of the flow. There is no *a priori* reason to assume that chondrules were or were not formed by ablation from larger bodies. This does add an extra step to chondrule formation: the formation of meter- to kilometer-sized bodies before chondrule production. These models attempt to reproduce chondrule cooling rates quantitatively, but it is not clear whether the required chondrule cooling rates of 10–100 K h^{-1} can be achieved and sustained throughout the chondrule crystallization temperature range. The model tends to predict chondrule cooling rates on the order of a few 1000 K h^{-1}, far too fast to reproduce porphyritic textures. Other problems with this model are that it is not clear how remelting or reheating of chondrules would occur or how CAIs and chondrules would be related to each other. Several other unaddressed issues exist, as outlined by *Jones et al.* (2000). These include total pressure or partial pressure of different elements such as O during chondrule formation.

Of chondrule formation models that rely on the Sun's early activity, the one that has stimulated the most interest is the X-wind model of *Shu et al.* (1996, 1997, 2001). This model hypothesizes that chondrule and CAI precursors were transported radially through the protoplanetary disk to the X-wind region, where they were irradiated by energetic flares and/or optical photons from the Sun, then carried by a magnetocentrifugal outflow and flung outward to land back into the disk in the 2–3 AU region. Even though the X-wind model has received significant attention in the last few years, no one has yet calculated the thermal histories of chondrules and CAIs in the context of the X-wind. Chondrule melting is attributed to X-ray flares, which heat the chondrules and CAIs directly, or which may heat the surrounding disk material. Rough estimates of the peak temperatures and cooling rates have been made (*Jones et al.,* 2000) under the somewhat arbitrary assumptions that X-ray flares triple the temperature of the disk gas, and that the flare decays in about an hour. While data on protostellar X-ray flares make these assumptions plausible, no effort has been made to tie chondrule heating and cooling to the substantial literature on the statistics of protostellar X-ray flares (e.g., their peak luminosities and decay timescales). Important details such as the response of the disk to a flare, its thermal inertia, and its vertical structure, have not been computed in detail. The X-wind model provides a framework in which predictions of chondrule peak temperatures, heating durations, and cooling rates could be made, and the details of pre- and post-melting thermal histories examined. For the most part this potential has remained unfulfilled.

In addition to the absence of calculated thermal histories, the X-wind model is also in apparent conflict with other constraints discussed by *Connolly and Desch* (2004), in particular the fact that the solar bulk composition of chondrites is due to the presence of chondrules and matrix. Each of these components alone does not constitute a solar bulk composition. It has thus been suggested that chondrules and matrix formed and/or experienced thermal processing in very close proximity, if not together (*Palme et al.,* 1993). While calculations of particle trajectories in the X-wind have not been presented in great detail in the refereed literature, it is clear from *Shu et al.* (1996) that micrometer-sized dust grains launched by the X-wind will almost certainly escape the solar system. They will never be able to accompany chondrules flung by the X-wind to the asteroid belt. This begs the question of how chondrules entering the disk at 2–3 AU could be complementary in composition to the matrix grains located and likely formed at 2–3 AU when the chondrules themselves formed at 0.1 AU. Dust and chondrules will not be transported together in the X-wind model. Their size and density are considerably different and thus they are not coupled together when entrained in the wind. This apparent conflict might be resolvable within the context of a detailed model. In its current form, however, the X-wind model remains an association of interesting concepts that makes few specific predictions about chondrule or refractory inclusion properties that can be tested, especially regarding the thermal histories of chondrules.

However, within the last few years the community has stimulated discussion on the formation of CAIs by the X-wind model (or a similar type of mechanism) and chondrule production by another mechanism, potentially by shock waves (the current favorite contender). The concept of X-wind production of CAIs that are subsequently flung out to pepper the protoplanetary nebula as exotic materials does not face the same chemical objections as does the genesis of chondrules in this manner. Calcium-aluminum-rich inclusions do not have any complimentarity with their host rocks; indeed their primary chemistry and isotope systematic bears more resemblance to other CAIs than they do to anything else. Hence, a strong argument may be made that they were formed in one place and then scattered among the chondrite-forming region. An X-wind-like model may shed considerable light onto this issue if future detailed numerical modeling is performed.

4.3. Transient Heating Mechanism Hypothesized to have Occurred Within the Disk

Numerous hypothetical transient heating mechanisms that essentially have no bases in observations have been proposed over the last 100 years (*Boss,* 1996; *Connolly and Desch,* 2004; *Grossman,* 1988). Such mechanisms include processes like a hot inner nebula (*Boss,* 1983; *Cameron and Fegley,* 1982; *Morfill,* 1983), nebular lightning (*Desch and Cuzzi,* 2000; *Eisenhour et al.,* 1994; *Eisenhour and Buseck,* 1995; *Horanyi et al.,* 1995, *Love et al.,* 1995), magnetic sheets (*Joung et al.,* 2004), and shock waves (*Boss and Graham,* 1993; *Ciesla and Hood,* 2002; *Connolly and Love,* 1998; *Desch and Connolly,* 2002; *Hood and Horanyi,* 1991, 1993; *Hood and Kring,* 1996; *Iida et al.,* 2001; *Miura and Nakamoto,* 2005; *Miura et al.,* 2002; *Ruzmaikina and Ip,* 1994; *Weidenschilling et al.,* 1998; *Wood,* 1963, 1984). As of the writing of this chapter, the most popular of all these models is the nebular shock wave model. The main reason for the popularity of this model is two-fold: (1) it is testable in that what is predicted is observed in the rock record (*Connolly and Love,* 1998), and (2) it is to date the most quantitative of all models for chondrule formation and it reproduces the thermal histories of the objects in detail.

In summary, the nebular shock wave model hypothesizes that a shock wave of compressed, hot gas moves through a dust-rich region of the nebula at supersonic speeds. It assumes that the dust component is individual dust grains that make up chondrule precursors and chondrules. Numerical modeling of the shock encountering chondrule precursors, chondrules, and their associated gas predict that the gas is compressed and the grains encounter gas drag for a short duration (seconds) after contact with the shock front. This is the first pulse of heating experienced by materials as they enter a shock wave. During this drag-heating phase, mineral grains and chondrules are rapidly or flash heated to liquidus to superliquidus temperatures (2000–2300 K). After the grains and chondrules achieve the same speed as the gas (also known as their stopping distance), the drag-heating phase ends (and no more work is performed on the particles due to frictional heating between atoms and mineral grains or chondrules) and infrared radiation of hot particles to other particles and of hot, compressed gas to solid particles becomes the dominant heating mechanism. Radiation from the shock region heats particles pre-shock passage and particles that absorbed infrared radiation post-shock are heated for considerable time (hours), but this energy is slowly lost over time to produce cooling of chondrules. Numerical models (*Ciesla and Hood,* 2002; *Desch and Connolly,* 2002) show that the shock wave model predicts cooling rates of particles (chondrules) in the range of 10–100°/h. The rate at which chondrules cool in these models through the range of temperatures where crystallization occurs is set by the overall time it takes for the gas and chondrules to move approximately 5 optical depths from the shock front. A key parameter of the nebular shock wave model is that both particles (chondrules) and gas must be modeled. The more particles that are encountered or added to the shocked region, the faster the chondrules will cool due to interactions between gas and chondrules. The nebular shock wave model robustly predicts that most chondrules would cool at rates that experiments have determined produce porphyritic textures, which are the most abundant of chondrule textural types. The model has the advantage that it naturally predicts an increase in total pressure of almost 2 orders of magnitude, which is critical in order to stabilize chondrule liquids because they are not stable at canonical pressures of 10^{-6}–10^{-5} atm (*Ebel,* 2006). Considerable research is being devoted to continuing to develop the nebular shock wave model (*Desch et al.,* 2005). The addition of evaporation and condensation of silicates from molten chondrules as well as the role H_2O may play (*Ciesla et al.,* 2003) in the calculations and hence the partial pressure of oxygen (f_{O_2}) are being modeled. It is also a goal of future models to fully incorporate the production of refractory inclusions, to which the model has only loosely been applied so far (*Desch and Connolly,* 2002).

4.4. Some Additional Discussion on Transient Heating Mechanisms

It is a priority of meteorite research to explain the existence of objects that were processed by transient heating and link their formation to an astrophysically sound process and mechanism. We summarized the various proposed mechanisms that may have produced chondrules and refractory inclusions. It would be very helpful in understanding the applicability of these models and ideas if we could clearly define the framework of hierarchical constraints on the formation of chondrules and refractory inclusions. This, however, cannot be done in an unbiased fashion. *Jones et al.* (2000) discussed in some detail the cooking recipe for making chondrules and refractory inclusions. In an ideal book chapter we would present a table with all the constraints known for chondrule and refractory inclusion formation, listed in order of importance, from zero order or the absolute ground conditions that any model must satisfy to those that are very clearly able to be tweaked, meaning they are of second or third order of importance to modeling. The only issue that all the authors of this chapter can essentially totally agree upon is that the thermal history of chondrules and refractory inclusions must be reproduced by models. If a model cannot or does not quantitatively satisfy this zero-order constraint then we adopt the conclusion that the model is not viable and we do not discuss it further. This does not mean that the model is wrong or that the authors do not agree with the model, it simply means that we cannot make an informed judgment based on the data available.

5. ADDITIONAL ISSUES AND A BLUEPRINT FOR THE BIG PICTURE

We have highlighted some important observations and models that have gained wide attention since the publication of *Chondrules and the Protoplanetary Disk* (*Hewins et al.,* 1996). Many reviews exist (including within this volume) that detail the petrography, chemistry, and constraints on the formation of both chondrules and refractory inclusions as discussed above. These provide more details on what we have learned about these issues since 1996. From this review several critical questions are raised. J. Wood (*Kerr,* 2001) has pointed out the need to see the big picture — putting the details of the petrographic and chemical analysis of these rocks into a framework for their origin and that origin's relationship to solar system formation. Below we list several issues that future research needs to address for determining additional constraints on the formation of chondrule and refractory inclusions, in order to provide a better blueprint for relating all the issues discussed herein to an astrophysical setting of the formation of the solar system. Some additional issues have been reviewed by *Wood* (1996), *Jones et al.* (2000), *Connolly and Desch* (2004), and *Connolly* (2005).

1. What could produce shock waves in the nebula? Although clearly the nebular shock wave model is the most quantitative model to date for predicting the thermal histories of chondrules, how these shocks were produced is unknown. We cannot currently observe processes within disks, thus we cannot witness shocks or chondrule formation within the disks. Currently, no hypothesis for the production of nebula shocks has gained wide acceptance (*Boss and Durisen*, 2005). The nebular shock wave model has also not been applied in detail to the formation of refractory inclusions, although *Desch and Connolly* (2002) predict that it could be. These two major issues concerning the reality of nebular shock waves as the mechanism that produced chondrules and refractory inclusions must be solved. It is a challenge to astrophysics and astronomy to test the prediction of the nebular shock wave model — to predict a mechanism that produced shocks that could have made chondrules and igneous CAIs. Until a convincing method for producing shocks is observed or hypothesized, the nebular shock wave model remains the only quantitative and most favorable model for producing chondrules even though we do not yet know how they could have been produced.

2. The X-wind model needs detailed predictions on thermal histories. Much attention has been given to the X-wind model, but unfortunately it does not yet quantitatively predict the thermal histories of chondrules and refractory inclusions. It is the obligation of science to generate quantitative predictions and, in this case, match those predictions to the rock record. Thus we challenge interested parties to perform and present quantitative predictions for the thermal histories of chondrules and refractory inclusions in the context of the X-wind or some modified version of the model. Clearly the ideas within the X-wind model have considerable merit and must be researched further. Currently, however, it is in an infancy stage and cannot be applied rigorously to chondrule and refractory inclusion formation, but it also cannot and must not be dismissed.

3. A detailed quantitative model for chondrule and igneous CAI formation by interactions between planetary bodies is required. Any idea or model that associates the formation of chondrules and refractory inclusions with planetary bodies, regardless of their size or hypothesized means of producing melted rock, cannot be dismissed as *potentially unviable*. It may not be palatable to many, but it cannot be ignored. It is perfectly reasonable to associate the formation of these igneous meteorite inclusions with planetary bodies that were colliding within the earliest stages of planet formation. Collisions occurred; that is a fact. Some of those collisions were energetic enough to produce melted rock — as evinced by the formation of melt veins within chondrites. However, we need to have numerical models of the thermal evolution of such melt material in the nebula if we are to have informed discussion of the viability of such models for chondrule formation. At present these remain cartoon models.

4. What is the relationship between the mechanism of transient melting and the environment of formation? The nebular shock wave model has very few requirements as to where it could have occurred. It is assumed to have occurred at the 2–3 AU region, or at least that is where the melting of mineral aggregates occurred to produce chondrules and possibly refractory inclusions. The X-wind model *requires* that potentially chondrules and certainly refractory inclusions formed above the disk (although later versions of the model do form chondrules within the disk). It is unknown if the region where these two kinds of objects formed were similar or different. We do not known if the total or partial pressure of H_2 and other elements that comprised the nebular gas during chondrule formation was similar to that experienced by refractory inclusions. It appears that, at least at the writing of this chapter, a strong argument can be made that the environments where refractory inclusions and chondrules were formed had different properties. This does not mean that the mechanism had to be different. Thus, the relationship between the environments where chondrules and refractory inclusions formed and a transient heating mechanism need to be defined — if any relationship even exists. The bottom line is that we really do not understand precisely where chondrules or CAIs formed within the lifetime of the protoplanetary disk.

5. Were chondrules and refractory inclusions produced by the same or different mechanisms? The end of section 4 above presents an important issue that remains unanswered: We do not know if chondrules and refractory inclusions were formed by the same mechanisms or different ones. In the last few years it appears that a movement is afoot that suggests chondrules and refractory inclusions were produced by different transient heating mechanisms. It is, however, not clear what *data* support such an idea or hypothesis. Many in the field appear to associate refractory inclusion formation with the X-wind model and chondrule formation with the nebular shock wave model. It is not required that this be the case. It is also not required that chondrules and refractory inclusions experienced different melting histories. The cooling rates experienced by these objects overlap — no large gap in the rate exists. Is there any reason why the mechanism that produced these objects cannot have been the same? Is it possible that the mechanism was the same, but the environments in which they were produced were different in physical characteristics such as pressure or dust abundance? This issue must be resolved before any relationship between the production of chondrules and igneous CAIs, transient heating events, environments of formation, dynamical processes in the disk, and terrestrial planet formation can be accurately constrained.

6. When were chondrules and refractory inclusions produced? Earlier we discussed the current data and hypothesis for the temporal relationship between chondrules and

refractory inclusions. As we also discussed, however, the verdict is still out as to the exact nature of this relationship. It is critical that the exact nature of this relationship be illuminated. We also need to understand why, apparently, CAI formation was limited to far less than 1 m.y. whereas chondrule formation lasted for at least 1.5 up to 4 m.y. after t = 0. In addition to this issue, we do not yet understand the temporal relationship between igneous CAI and chondrule formation to the collapse of the molecular cloud and formation of the disk and the location within the disk where these objects were produced. As discussed elsewhere in this volume, the times of differentiation of planetesimals and the temporal relationship to chondrule and refractory inclusion formation needs to be clearly resolved. Defining these relationships with higher confidence will provide additional, much-needed constraints on their relationship to transient heating mechanism(s) and potentially terrestrial planet formation.

7. Will we be able to observe the processing of chondritic materials? With many new generations of observing tools coming on line in the near future, observations of disks at finer and finer resolution in the next decade or two may provide new insights into the processing of chondritic materials, either mineral grains that compose these rocks or the actual melting of mineral aggregates that are potentially chondrules and refractory inclusions. It is important to obtain an accurate survey of disks in various stages of evolution in a search for chondrule and refractory inclusio production.

8. Are current observations of dust on the edges of disks related to transient heating events? Determining the composition of dust around protoplanetary disks is a rapidly emerging field of study in observational astronomy (*Forrest et al.*, 2004; *Sargent et al.*, 2004; *Watson et al.*, 2004). In addition to determination of the composition of dust is the goal of understanding the structure of the dust — amorphous or crystalline. Such studies are indicating that amorphous dust is somehow processed into crystalline material within (or at least at the surface of) disks. An obvious question related to our discussions is whether this process could be some type of transient heating event, and whether such an event could be related to the production of chondrules and refractory inclusions? Future research with the Spitzer Space Telescope and other missions may provide powerful constraints on this issue.

9. Continued interdisciplinary approaches to solving the problem of transient heating are needed. The greatest strength in solving the problem of transient heating mechanisms, chondrule and CAI formation, and indeed the fundamental issue of planet formation within the early solar system is to continue to exploit an interdisciplinary approach that blends meteoritics, astronomy, and astrophysics.

Acknowledgments. We are grateful to C. M. O'D. Alexander, S. S. Russell, and F. Ciesla for their helpful reviews of this manuscript, as well as the comments of A. Rubin and D. S. Lauretta. We thank E. Chiang for his early participation in this book chapter and are grateful to him for numerous helpful discussions. The senior author wishes to additionally thank the editors for inviting him to serve on the editorial board and to write this chapter; it was a powerful learning experience. The senior author also thanks V.K.R. for so many things. This work was in part supported by grants from PSC-CUNY (H.C.C.) and NASA NAG5-9463 (R.H.J.).

REFERENCES

Alexander C. M. O'D. (2004) Chemical equilibrium and kinetic constraints for chondrule and CAI formation conditions. *Geochim. Cosmochim. Acta, 68,* 3943–3969.

Amelin Y., Krot A. N., Hutcheon I., and Ulyanov A. (2002) Lead isotopic ages of chondrules and calcium-aluminum-rich inclusions. *Science, 297,* 1678–1683.

Ash R. D., Russell S. S., Belshaw N. C., Young E. D., and Gounelle M (2002) Mg isotopes in melilite, fassaite and spinels in CAIs: Evidence for evaporation, equilibration and late stage alteration (abstract). In *Lunar and Planetary Science XXXIII,* Abstract #2063. Lunar and Planetary Institute, Houston (CD-ROM).

Bally J., Devine D., Fesen R. A., and Lane A. R. (1995) Twin Herbig-Haro jets and molecular outflows in L1228. *Astrophys. J., 454,* 345–360.

Beckett J. R., Simon S. B., and Stolper E. (2000) The partitioning of Na between melilite and liquid. Part II: Applications to type B inclusions from carbonaceous chondrites. *Geochim. Cosmochim. Acta, 64,* 2519–2534.

Beckett J. R., Connolly H. C. Jr., and Ebel D. S. (2006) Chemical processes in calcium-aluminum-rich inclusions: A mostly CMAS view of melting and crystallization. In *Meteorites and the Early Solar System II* (D. S. Lauretta and H. Y. McSween Jr., eds.), this volume. Univ. of Arizona, Tucson.

Bell K. R., Cassen P. M., Wasson J. T., and Woolum D. S. (2000) The FU Orionis phenomenon and solar nebula material. In *Protostars and Planets IV* (V. Mannings et al., eds.), pp. 897–926. Univ. of Arizona, Tucson.

Bizarro M., Baker J. A., and Haack H. (2004) Mg isotope evidence for contemporaneous formation of chondrules and refractory inclusions. *Nature, 431,* 275–278.

Boss A. P. (1983) Evolution of the solar nebula. II. Thermal structure during nebula formation. *Astrophys. J., 417,* 351–367.

Boss A. P. (1996) A concise guide to chondrule formation models. In *Chondrules and the Protoplanetary Disk* (R. H. Hewins et al., eds.), pp. 257–264. Cambridge Univ., Cambridge.

Boss A. P. and Durisen R. H. (2005) Chondrule-forming shock fronts in the solar nebular: A possible unified scenario for planet and chondrite formation. *Astrophys. J. Lett., 62,* L137–L140.

Boss A. P. and Graham J. A. (1993) Clumpy disk accretion and chondrule formation. *Icarus, 106,* 168–176.

Brearley A. J. and Jones R. H. (1998) Chondritic meteorites. In *Planetary Materials* (J. J. Papike, ed.), pp. 3-1 to 3-398. Reviews in Mineralogy, Vol. 36, Mineralogical Society of America.

Calvet N., Hartmann L., and Strom S. E. (2000) Evolution of disk accretion. In *Protostars and Planets IV* (V. Mannings et al., eds.), pp. 377–395. Univ. of Arizona, Tucson.

Cameron A. G. W. and Fegley M. B. (1982) Nucleation and condensation in the primitive solar nebula. *Icarus, 52,* 1–13.

Cassen P. (2001) Nebular thermal evolution and the properties of primitive planetary materials. *Meteoritics & Planet. Sci., 36,* 671–700.

Ciesla F. J. and Hood L. L. (2002) The nebular shock wave model for chondrule formation: Shock processing in a particle-gas suspension. *Icarus, 158,* 281–293.

Ciesla F. J., Lauretta D. S., Cohen B. A., and Hood L. L. (2003) A nebular origin for chondritic fine-grained phyllosilicates. *Science, 299,* 549–552.

Clayton R. N., Hinton R. W., and Davis A. M. (1988) Isotopic variations in the rock-forming elements in meteorites. *Phil. Trans. R. Soc. Lond., A325,* 483–501.

Cohen B. A. and Hewins R. H. (2004) An experimental study of the formation of metallic iron in chondrules. *Geochim. Cosmochim. Acta, 68,* 1677–1689.

Cohen B. A., Hewins R. H., and Yu Y. (2000) Evaporation in the young solar nebula as the origins of "just-right" melting of chondrules. *Nature, 406,* 600–602.

Connolly H. C. Jr. (2005) Refractory inclusions and chondrules: Insights into a protoplanetary disk and planet formation. In *Chondrites and the Protoplanetary Disk* (A. N. Krot et al., eds.), pp. 215–224. ASP Conference Series 341, Astronomical Society of the Pacific, San Francisco.

Connolly H. C. Jr. and Burnett D. S. (1999) A study of the minor element concentrations of spinels from two type B CAIs: An investigation into potential formation conditions of CAIs. *Meteoritics & Planet. Sci., 34,* 829–848.

Connolly H. C. Jr. and Burnett D. S. (2003) On type B CAI formation: Experimental constraints on f_{O_2} variations in spinel, minor element partitioning and reequilibration effects. *Geochim. Cosmochim. Acta, 67,* 4429–4434.

Connolly H. C. Jr. and Desch S. J. (2004) On the origin of the "kleine Kugelchen" called chondrules. *Chem. Erde, 64,* 95–125.

Connolly H. C. Jr. and Hewins R. H. (1995) Chondrules as products of dust collisions with totally molten droplets within dust-rich nebular environment: An experimental investigation. *Geochim. Cosmochim. Acta, 59,* 3231–3246.

Connolly H. C. Jr. and Love S. (1998) The formation of chondrules: Petrologic tests of the shock wave model. *Science, 280,* 62–67.

Connolly H. C. Jr., Jones B. D., and Hewins R. H. (1997) The flash melting of chondrules: An experimental investigation into the melting history and physical nature of chondrule precursors. *Geochim. Cosmochim. Acta, 62,* 2725–2735.

Connolly H. C. Jr., Huss G. R., and Wasserburg G. J. (2001) On the formation of Fe-Ni metal in Renazzo-like carbonaceous chondrites. *Geochim. Cosmochim. Acta, 65,* 4567–4588.

Connolly H. C. Jr., Burnett D. S., and McKeegan K. D. (2003) The petrogenesis of type B1 Ca-Al-rich inclusions: The spinel connection. *Meteoritics & Planet. Sci., 34,* 829–848.

Davis A. M. and MacPherson G. J. (1996) Thermal processing in the solar nebula: Constraints from refractory inclusions. In *Chondrules and the Protoplanetary Disk* (R. H. Hewins et al., eds.), pp. 71–76. Cambridge Univ., Cambridge.

Desch S. J. and Connolly H. C. Jr. (2002) A model of the thermal processing of particles in solar nebula shocks: Application to the cooling rates of chondrules. *Meteoritics & Planet. Sci., 37,* 183–207.

Desch S. J. and Cuzzi J. N. (2000) The generation of lightning in the solar nebula. *Icarus, 143,* 87–105.

Desch S. J., Ciesla F. J., Hood L. L., and Nakamoto T. (2005) Heating of chondritic materials in solar nebula shocks. In *Chondrites and the Protoplanetary Disk* (A. N. Krot et al., eds.), pp. 849–872. ASP Conference Series 341, Astronomical Society of the Pacific, San Francisco.

Dodd R. T. and Jarosewich E. (1979) Incipient melting in and shock classification of L-group chondrites. *Earth Planet. Sci. Lett., 44,* 335–340.

Ebel D. S. (2006) Condensation of rocky material in astrophysical environments. In *Meteorites and the Early Solar System II* (D. S. Lauretta and H. Y. McSween Jr., eds.), this volume. Univ. of Arizona, Tucson.

Eisenhour D. D. and Buseck P. R. (1995) Chondrule formation by radiative heating: A numerical model. *Icarus, 117,* 197–211.

Eisenhour D. D., Daulton T. L., and Buseck P. R. (1994) Electromagnetic heating in the early solar nebula and the formation of chondrules. *Science, 265,* 1067–1070.

Feigelson E. D., Gaffney J. A. III, Garmire G., Hillenbrand L. A., and Townsley L. (2003) X-rays in the Orion nebula cluster: Constraints on the origins of magnetic activity in pre-main-sequence stars. *Astrophys. J., 584,* 911–930.

Forrest W. J., Sargent B., Furlan E., Chen C. H., Kemper F., Calvet N., Hartmann L., Uchida K. I., Watson D. M., Green J. D., Keller L. D., Sloan G. C., Herter T. L., Brandl B. R., Houck J. R., Barry D. J., Hall P., Morris P. W., Najita J., Myers P. C., D'Alessio P., and Jura M. (2004) Mid-infrared spectroscopy of disks around classical T Tauri stars. *American Astrophysical Society Meeting, 204,* Abstract #41.10.

Fredriksson K., Noonan A., and Nelen J. (1973) Meteoritic, lunar and Lonar impact chondrules. *Moon, 7,* 475–482.

Galy A., Young E. D., Ash R. D., and O'Nions R. K. (2000) The formation of chondrules at high gas pressure in the solar nebula. *Science, 290,* 1751–1753.

Glassgold A. E., Feigelson E. D., and Montmerle T. (2000) Effects of energetic radiation in young stellar objects. In *Protostars and Planets IV* (V. Mannings et al., eds.), pp. 429–455. Univ. of Arizona, Tucson.

Greenwood J. P. and Hess P. C. (1996) Congruent melting kinetics: Constraints on chondrule formation. In *Chondrules and the Protoplanetary Disk* (R. H. Hewins et al., eds.), pp. 205–212. Cambridge Univ., Cambridge.

Grossman J. N. (1988) Formation of chondrules. In *Meteorites and the Early Solar System* (J. F. Kerridge and M. S. Matthews, eds.), pp. 680–696. Univ. of Arizona, Tuscon.

Grossman J. N. and Brearley A. J. (2005) The onset of metamorphism in ordinary and carbonaceous chondrites. *Meteoritics & Planet. Sci., 40,* 87–122.

Grossman J. N., Rubin A. E., Nagahara H., and King E. A. (1988) Properties of chondrules. In *Meteorites and the Early Solar System* (J. F. Kerridge and M. S. Matthews, eds.), pp. 619–659. Univ. of Arizona, Tuscon.

Gounelle M., Shu F. H., Shang H., Glassold A. E., Rhem K. E., and Lee T. (2001) Extinct radioactivities and protosolar cosmic rays: Self-shielding and light elements. *Astrophys. J., 548,* 1051–1070.

Hartmann L. and Kenyon S. J. (1996) The FU Orionis phenomenon. *Annu. Rev. Astron. Astrophys., 34,* 207–240.

Herbig G. H. (1977) Eruptive phenomena in early stellar evolution. *Astrophys. J., 217,* 693–715.

Herbig G. H. (1989) FU Orionis eruptions. In *Proceedings of the ESO Workshop on Low-Mass Star Formation and Pre-Main Sequence Objects* (B. Reipurth, ed.), p. 233. European Southern Observatory, Garching.

Hewins R. H. (1991) Retention of sodium during chondrule melting. *Geochim. Cosmochim. Acta, 55,* 935–942.

Hewins R. H. (1997) Chondrules. *Annu. Rev. Earth Planet. Sci., 25,* 61–83.

Hewins R. H. and Connolly H. C. Jr. (1996) Peak temperatures of flash-melted chondrules. In *Chondrules and the Protoplanetary Disk* (R. H. Hewins et al., eds.), pp. 197–204. Cambridge Univ., Cambridge.

Hewins R. H., Jones R. H., and Scott E. R. D., eds. (1996) *Chondrules and the Protoplanetary Disk.* Cambridge Univ., Cambridge. 343 pp.

Hewins R. H., Connolly H. C. Jr., Lofgren G. E., and Libourel G. (2005) Experimental constraints on chondrule formation. In *Chondrites and the Protoplanetary Disk* (A. N. Krot et al., eds.), pp. 286–316. ASP Conference Series 341, Astronomical Society of the Pacific, San Francisco.

Hood L. L. and Horanyi M. (1991) Gas dynamic heating of chondrule precursor grains in the solar nebula. *Icarus, 93,* 259–269.

Hood L. L. and Horanyi M. (1993) The nebular shock wave model for chondrule formation — One-dimensional calculations. *Icarus, 106,* 179–188.

Hood L. L. and Kring D. A. (1996) Models for multiple heating mechanisms. Peak temperatures of flash-melted chondrules. In *Chondrules and the Protoplanetary Disk* (R. H. Hewins et al., eds.), pp. 265–278. Cambridge Univ., Cambridge.

Horanyi M., Morfill G., Goertz C. K., and Levy E. H. (1995) Chondrule formation in lightning discharges. *Icarus, 114,* 174–185.

Huang S., Lu J., Prinz M., Weisberg M. K., Benoit P. H., and Sears D. W. G. (1996) Chondrules: Their diversity and the role of open-system processes during their formation. *Icarus, 122,* 316–346.

Humayun M. and Clayton R. N. (1995) Potassium isotope cosmochemistry: Genetic implications of volatile element depletion. *Geochim. Cosmochim. Acta, 59,* 2131–2148.

Huss G. R., MacPherson G. J., Wasserburg G. J., Russell S. S., and Srinivasan G. (2001) ^{26}Al in CAIs and chondrules from unequilibrated ordinary chondrites. *Meteoritics & Planet. Sci., 36,* 975–997.

Iida A., Nakamoto T., Susa H., and Nakagawa Y. (2001) A shock heating model for chondrule formation in a protoplanetary disk. *Icarus, 153,* 430–450.

Jones R. H. (1996) Relict grains in chondrules: Evidence for chondrule recycling. In *Chondrules and the Protoplanetary Disk* (R. H. Hewins et al., eds.), pp. 163–172. Cambridge Univ., Cambridge.

Jones R. H., Lee T., Connolly H. C. Jr., Love S. G., and Shang H. (2000) Formation of chondrules and CAIs: Theory vs. observations. In *Protostars and Planets IV* (V. Mannings et al., eds.), pp. 927–962. Univ. of Arizona, Tucson.

Jones R. H., Grossman J. N., and Rubin A. E. (2005) Chemical, mineralogical and isotopic properties of chondrules: Clues to their origin. In *Chondrites and the Protoplanetary Disk* (A. N. Krot et al., eds.), pp. 251–285. ASP Conference Series 341, Astronomical Society of the Pacific, San Francisco.

Joung M. K. R., MacLow M-M., and Ebel D. S. (2004) Chondrule formation and protoplanetary disk heating by current sheets in nonideal magnetohydrodynamic turbulence. *Astrophys. J., 606,* 523–534.

Kerr R. A. (2001) A meteoriticist speaks out, his rocks remain mute. *Science, 293,* 1581–1584.

Konigl A. and Pudritz R. E. (2000) Disk winds and the accretion-outflow connection. In *Protostars and Planets IV* (V. Mannings et al., eds.), pp. 759–787. Univ. of Arizona, Tucson.

Krot A. N., Keil K., Goodrich C. A., Scott E. R. D., and Weisberg M. K. (2004a) Classification of meteorites. In *Treatise on Geochemistry, Vol. 1: Meteorites, Comets, and Planets* (A. M. Davis, ed.), pp. 83–128. Elsevier, Oxford.

Krot A. N., Petaev M. I., Russell S. S., Itoh S., Fagan T. J., Yurimoto H., Chzmadia L., Weisberg M. K., Komatsu M., Ulyanov A. A., and Keil K. (2004b) Amoeboid olivine aggregates in carbonaceous chondrites: Records of nebular and asteroidal processes. *Chem. Erde, 64,* 185–239.

Lauretta D. S., Buseck P. R., and Zega T. J. (2001) Opaque minerals in the matrix of the Bishunpur (LL3.1) chondrite: Constraints on the chondrule formation environment. *Geochim. Cosmochim. Acta, 65,* 1337–1353.

Lauretta D. S., Nagahara H., and Alexander C. M. O'D. (2006) Petrology and origin of ferromagnesian silicate chondrules. In *Meteorites and the Early Solar System II* (D. S. Lauretta and H. Y. McSween Jr., eds.), this volume. Univ. of Arizona, Tucson.

Levy E. H. (1988) Energetics of chondrule formation. In *Meteorites and the Early Solar System* (J. F. Kerridge and M. S. Matthews, eds.), pp. 697–714. Univ. of Arizona, Tucson.

Levy E. H. and Araki S. (1989) Magnetic reconnection flares in the proto-planetary nebula and the possible origin of meteoritic chondrules. *Icarus, 81,* 622–628.

Lewis R. D., Lofgren G. E., Franzen H. F., and Windom K. E. (1993) The effect of Na vapor on the Na content of chondrules. *Meteoritics, 28,* 622–628.

Liffman K. and Brown M. (1996) The protostellar jet model of chondrule formation. In *Chondrules and the Protoplanetary Disk* (R. H. Hewins et al., eds.), pp. 197–204. Cambridge Univ., Cambridge.

Lofgren G. E. (1996) A dynamic crystallization model for chondrule melts. In *Chondrules and the Protoplanetary Disk* (R. H. Hewins et al., eds.), pp. 187–196. Cambridge Univ., Cambridge.

Lofgren G. E. and Lanier A. B. (1990) Dynamic crystallization study of barred olivine chondrules. *Geochim. Cosmochim. Acta, 54,* 3537–3551.

Love S. G., Keil K., and Scott E. R. D. (1995) Electrical discharge heating of chondrules in the solar nebula. *Icarus, 115,* 97–108.

Lugmair G. W. and Shukolyukov A. (2001) Early solar system events and timescales. *Meteoritics & Planet. Sci., 36,* 1017–1026.

Mannings V., Boss A. P., and Russell S. S., eds. (1996) *Protostars and Planets IV.* Univ. of Arizona, Tucson. 1422 pp.

MacPherson G. J. (2004) Calcium-aluminum-rich inclusions in chondritic meteorites. In *Treatise on Geochemistry, Vol. 1: Meteorites, Comets, and Planets* (A. M. Davis), pp. 201–241. Elsevier, Oxford.

MacPherson G. J., Paque J. M., Stolper E., and Grossman L. (1984) The origin and significance of reverse zoning in melilite from Allende type B inclusions. *J. Geol., 92,* 289–305.

MacPherson G. J., Wark D. A., and Armstrong J. T. (1988) Primitive materials surviving in chondrites: Refractory inclusions. In *Meteorites and the Early Solar System* (J. F. Kerridge and M. S. Matthews, eds.), pp. 746–807. Univ. of Arizona, Tucson.

MacPherson G. J., Simon S. B., Davis A. M., Grossman L., and Krot A. N. (2005) Calcium-aluminum-rich inclusions: Major unanswered questions. In *Chondrites and the Protoplanetary Disk* (A. N. Krot et al., eds.), pp. 225–250. ASP Conference Series 341, Astronomical Society of the Pacific, San Francisco.

McCoy T. J., Keil K., Bogard D. H., Casanova I., Lindstrom M., Brearley A. J., Kehm K., Nichols R. H., and Hohenberg C. M.

(1995) Origin and history of impact-melt rocks of enstatite chondritic parentage. *Geochim. Cosmochim. Acta, 59,* 161–175.

Mittlefehldt D. W., McCoy T. J., Goodrich C. A., and Kracher A. (1998) Non-chondritic meteorites from asteroidal bodies. In *Planetary Materials* (J. J. Papike, ed.), pp. 3-1 to 3-398. Reviews in Mineralogy, Vol. 36, Mineralogical Society of America.

Merrill G. P. (1920) On chondrules and chondritic structure in meteorites. *Proc. Natl. Acad. Sci., 8,* 449–472.

Mirua H. and Nakamoto T. (2005) A shock-wave model for chondrule formation II. Minimum size of chondrule precursors. *Icarus, 175,* 289–304.

Miura H., Nakamoto T., and Susa H. (2002) A shock-wave heating model for chondrule formation: Effects of evaporation and gas flows. *Icarus, 160,* 258–270.

Morfill G. E. (1983) Some cosmochemical consequences of a turbulent protoplanetary cloud. *Icarus, 53,* 47–50.

Nakamoto R., Kita N. T., Tachibana S., and Hayashi M. R. (2004) X-ray flare induced shock waves and chondrule formation in upper solar nebula (abstract). In *Lunar and Planetary Science XXXV,* Abstract #1821. Lunar and Planetary Institute, Houston (CD-ROM).

Nichols R. H. Jr. (2006) Chronological constraints on planetesimal accretion. In *Meteorites and the Early Solar System II* (D. S. Lauretta and H. Y. McSween Jr., eds.), this volume. Univ. of Arizona, Tucson.

Palme H., Spettel B., and Ikeda Y. (1993) Origin of chondrules and matrix in carbonaceous chondrites. *Meteoritics, 28,* 417.

Radomsky P. M. and Hewins R. H. (1990) Formation conditions of pyroxene-olivine and magnesium olivine chondrules. *Geochim. Cosmochim. Acta, 54,* 3475–3490.

Richter F. M., Davis A. M., Ebel D. S., and Hashimoto A. (2002) Elemental and isotopic fractionation of type B calcium-, aluminum-rich inclusions: Experiments, theoretical considerations, and constraints on their thermal evolution. *Geochim. Cosmochim. Acta, 66,* 521–540.

Rubin A. E. (2000) Petrologic, geochemical and experimental constraints on models of chondrule formation. *Earth Sci. Rev., 50,* 3–27.

Rubin A. E., Sailer A., and Wasson J. T. (1999) Troilite in the chondrules of unequilibrated ordinary chondites: Implications for chondrule formation. *Geochim. Cosmochim. Acta, 63,* 2281–2298.

Russell S. S., Srinivasan G., Huss G. R., Wasserburg G. J., and MacPherson G. J. (1996) Evidence for widespread ^{26}Al in the solar nebula and constraints for nebula time scales. *Science, 273,* 757–762.

Ruzmaikina T. V. and Ip W. H. (1994) Chondrule formation in radiative shock. *Icarus, 112,* 430–447.

Sanders I. S. (1996) A chondrule-forming scenario involving molten planetesimals. In *Chondrules and the Protoplanetary Disk* (R. H. Hewins et al., eds.), pp. 327–334. Cambridge Univ., Cambridge.

Sargent B., Forrest W. J., D'Alessio P., Calvet N., Furlan E., Hartmann L., Uchida K. I., Sloan G. C., Chen C. H., Kemper F., Watson D. M., Green J. D., Keller L. D., Herter T. L., Brandl B. R., Houck J. R., Barry D. J., Hall P., Morris P. W., Najita J., and Myers P. C. (2004) Grain processing in YSO disks. *American Astronomical Society Meeting, 205,* Abstract #136.07.

Shu F. H., Shang H., and Lee T. (1996) Towards an astrophysical theory of chondrites. *Science, 271,* 1545–1552.

Shu F. H., Shang H., Glassgold A. E., and Lee T. (1997) X-rays and fluctuating X-winds from protostars. *Science, 277,* 1475–1479.

Shu F. H., Najita J. R., Shang H., and Li Z. Y. (2000) X-winds: Theory and observations. In *Protostars and Planets IV* (V. Mannings et al., eds.), pp. 789–813. Univ. of Arizona, Tucson.

Shu F. H., Shang H., Gounelle M., Glassgold M., and Lee T. (2001) The origin of chondrules and refractory inclusions in chondritic meteorites. *Astrophys. J., 548,* 1029–1050.

Simon S. B., Davis A. M., Richter F. M., and Grossman L. (1996) Experimental investigation of the effect of cooling rate on melilite/liquid distribution coefficients for Sr, Ba, and Ti in type B refractory inclusion melts (abstract). In *Lunar and Planetary Science XXVII,* pp. 1201–1202. Lunar and Planetary Institute, Houston.

Simon S. B., Grossman L., and Davis A. M. (2005) A unique type B inclusion from Allende with evidence from multiple stages of melting. *Meteoritics & Planet. Sci., 40,* 461–506.

Skinner W. R. (1990) Bipolar outflow and a new model of the early solar system. Part II: The origins of chondrules (abstract). In *Lunar and Planetary Science XXI,* pp. 1545–1552. Lunar and Planetary Institute, Houston.

Sorby H. (1877) On the structure and origin of meteorites. *Nature, 15,* 495–498.

Stolper E. (1982) Crystallization sequence of Ca-Al-rich inclusions from Allende: An experimental study. *Geochim. Cosmochim. Acta, 46,* 2159–2180.

Stolper E. and Paque J. (1986) Crystallization sequences of Ca-Al-rich inclusions from Allende: The effects of cooling rate and maximum temperature. *Geochim. Cosmochim. Acta, 50,* 1785–1806.

Tachibana S. and Huss G. R. (2005) Sulfur isotope composition of putative primary troilite in chondrules from Bishunpur and Semarkona. *Geochim. Cosmochim. Acta, 69,* 3075–3097.

Tachibana S., Nagahara H., Mostefaoui S., and Kita N. T. (2003) Correlation between relative ages inferred from ^{26}Al and bulk compositions of ferromagnesian chondrules in least equilibrated ordinary chondrites. *Meteoritics & Planet. Sci., 38,* 939–962.

Taylor G. J., Scott E. R. D., and Keil K. (1983) Cosmic setting for chondrule formation. In *Chondrules and their Origins* (E. A. King, eds.), pp. 262–268. Lunar and Planetary Institute, Houston.

Tschermak G. (1885) The microscopic properties of meteorites. Translated by Wood J. A. and Wood E. M. in *Smithsonian Contrib. Astrophys., 4(6),* Smithsonian Institution Press.

Urey H. C. (1967) Parent bodies of the meteorites and the origins of chondrules. *Icarus, 7,* 350–359.

Urey H. C. and Craig H. (1953) The composition of the stone meteorites and the origin of the meteorites. *Geochim. Cosmochim. Acta, 4,* 36–82.

Wasson J. T. (1985) *Meteorites: Their Record of Early Solar-System History.* Freeman, New York. 267 pp.

Wasson J. T. and Rubin A. E. (2003) Ubiquitous low-FeO relict grains in type II chondrules and limited overgrowths on phenocrysts following the final melting event. *Geochim. Cosmochim. Acta., 67,* 2239–2250.

Watson D. M., Kemper F., Calvet N., Keller L. D., Furlan E., Hartmann L., Forrest W. J., Chen C. H., Uchida K. I., Green J. D., Sargent B., Sloan G. C., Herter T. L., Brandl B. R., Houck J. R., Najita J., D'Alessio P., Myers P. C., Barry D. J., Hall P., and Morris P. W. (2004) Min-infrared spectra of class I protostars in Taurus. *Astrophy. J. Suppl. Ser., 154,* 391–395.

Weidenschilling S. J., Marzari F., and Hood L. L. (1998) The origin of chondrules as jovian resonance. *Science, 279,* 681.

Weisberg M. K., McCoy T. J., and Krot A. N. (2006) Systematics and evaluation of meteorite classification. In *Meteorites and the Early Solar System II* (D. S. Lauretta and H. Y. McSween Jr., eds.), this volume. Univ. of Arizona, Tucson.

Wood J. A. (1963) Chondrites and chondrules. *Sci. Am., 65,* 65–83.

Wood J. A. (1984) On the formation of meteoritic chondrules by aerodynamic drag heating in the solar nebula. *Earth Planet. Sci. Lett., 70,* 11–23.

Wood J. A. (1996) Unresolved issues in the formation of chondrules and chondrites. In *Chondrules and the Protoplanetary Disk* (R. H. Hewins et al., eds.), pp. 55–70. Cambridge Univ., Cambridge.

Wooden D. H., Bell K. R., Harker D. E., and Woodward C. E. (2000) Yearly variations in DG Tau's 10 micron silicate feature: Evidence for dust processing in a disk atmosphere. *Bull. Am. Astron. Soc., 32,* 1482.

Yu Y. and Hewins R. H. (1998) Transient heating and chondrule formation: Evidence from Na loss in flash heating simulation experiments. *Geochim. Cosmochim. Acta, 62,* 159–172.

Yu Y., Hewins R. H., and Zanda B. (1996) Sodium and sulfur in chondrules: Heating time and cooling curves. In *Chondrules and the Protoplanetary Disk* (R. H. Hewins et al., eds.), pp. 213–220. Cambridge Univ., Cambridge.

Zanda B. (2004) Chondrules. *Earth Planet. Sci. Lett., 224,* 1–17.

Chemical Processes in Igneous Calcium-Aluminum-rich Inclusions: A Mostly CMAS View of Melting and Crystallization

John R. Beckett
California Institute of Technology

Harold C. Connolly
Kingsborough College, City University of New York and American Museum of Natural History

Denton S. Ebel
American Museum of Natural History

1. INTRODUCTION

Calcium-,aluminum-rich inclusions (CAIs) and Al-rich chondrules reflect multiple processes in many different environments. In this chapter, we consider constraints on high-temperature processes from the perspective of phase equilibria and dynamic crystallization experiments. With chondrules, whose evolution is discussed in *Lauretta et al.* (2006), it is almost axiomatic that one or more igneous events were involved in processing, so the focus of research becomes one of constraining the nature of the melting event(s). CAIs are not so simple. While many CAIs are thought to have crystallized from partially or completely molten droplets, others did not, and one of the first tasks in constraining processes involved in the evolution of an object is to decide whether or not an igneous event was involved.

Our approach is to first introduce some of the objects thought to have passed through an igneous stage at some point, without getting overly concerned with nomenclature. We then explore the application of phase equilibria to CAIs through classical phase diagrams, which are useful for predicting the order of crystallization of a cooling droplet where the system is sufficiently simple, and via computational models, which draw from both phase equilibria and thermodynamic data and can, in principle, be applied to compositionally complex liquids. The phase diagram analysis proceeds mostly from the perspective of the system CaO-MgO-Al_2O_3-SiO_2 (CMAS). Other elements (e.g., Na, Ti, Fe) are considered primarily to the extent that they affect expectations based on the phase equilibria of CMAS or can help to constrain the liquid composition, but we attempt no formal characterization of the phase relations in these more complex systems. This leads us to ignore ferromagnesian chondrules, which generally have considerable amounts of FeO, and the Fe-rich opaque assemblages, or Fremdlinge, and associated V-rich pyroxenes of CAIs (*El Goresy et al.*, 1978; *Blum et al.*, 1989) even though they may have quite a bit to say about melting events. Since the observed bulk composition may be very different from the one(s) pertinent to an ancient igneous event, we consider the issue of secondary alteration, albeit briefly, in section 2.

The partitioning of minor and trace elements between crystalline phases and a coexisting liquid has a negligible effect on the overall phase equilibria relative to CMAS, but the process leaves clues to the history of a CAI, and in section 3.4, we discuss experimental constraints on trace-element partitioning between crystals and CAI-like and Al-rich chondrule-like liquids. The processes of volatilization and condensation are considered in *Davis* (2006) and *Ebel* (2006) (see also *Davis and Richter,* 2004). We do consider dynamic crystallization experiments in section 3.5.

We assume the reader to have a basic working knowledge of phase equilibria and the use of phase diagrams; we introduce nomenclature for CAIs and related objects as needed and note end-member compositions of mineral solid solutions when they are first encountered. *Bowen* (1928) and *Morse* (1980), among many others, provide basic background in the interpretation of phase diagrams and their use, particularly with respect to the determination of crystallization paths. Readers are referred to the review by *Brearley and Jones* (1998) for an extensive overview of the chemistry of meteoritic minerals, including those discussed here; to *MacPherson and Huss* (2005) for background on Al-rich chondrules; and to *Grossman* (1980), *MacPherson et al.* (1988), *Grossman et al.* (1988), *Kimura et al.* (1993), *Brearley and Jones* (1998), and *MacPherson* (2004) for basic background in CAIs. *MacPherson* (2004) in particular gives a nice introduction to the essential features of CAIs, and a reading of that work will provide more than adequate preparation for this chapter.

2. CALCIUM-ALUMINUM-RICH INCLUSIONS AND ALUMINUM-RICH CHONDRULES

In this section, we first introduce a few of the targets, meteoritic materials thought to have been partially or completely molten at some point and whose bulk compositions can plausibly be treated in terms of CMAS. This is done with motivational and not comprehensive intent since our objective is to introduce tools for analyzing the possible role of liquids in the evolution of an arbitrary meteoritic inclusion rather than summarizing assertions made about specific inclusions. For our purposes, it is the bulk composition of an object during a melting event that is the single most important discriminator of how it will respond to a particular thermal history and environment; the mineralogy, elemental fractionations in individual crystals or zoning, trace-element abundances and distributions, isotopic compositions, and the presence or absence of relict crystals all flow from this fundamental property.

Although the bulk compositions of most CAIs and many Al-rich chondrules are well described by the CMAS system, they range within the system from objects with virtually no Mg or Si to objects with only a few wt% CaO + Al_2O_3. There are large corresponding variations in mineralogy. The most aluminous of CAIs contain substantial amounts of aluminates, most commonly one or more of spinel ($MgAl_2O_4$), hibonite (nominally $CaAl_{12}O_{19}$ but frequently containing significant amounts of Ti and Mg), and/or grossite ($CaAl_4O_7$); rarely, corundum (Al_2O_3) or calcium monoaluminate ($CaAl_2O_4$) may be present. Plate 11a shows an example of a corundum-bearing inclusion, which contains substantial hibonite. There is no settled nomenclature for grossite-bearing or hibonite-rich inclusions and we will not try to impose one here. As the bulk composition becomes more Si-rich, melilite, which can usually be described quite well in CAIs as a binary solid solution between gehlenite ($Ca_2Al_2SiO_7$) and åkermanite ($Ca_2MgSi_2O_7$), becomes important (e.g., Plate 11b shows four oxide phases set in melilite). *Grossman* (1975) established a basic classification structure for relatively melilite-rich inclusions. Where the inclusion is predominately melilite with minor included spinel and perovskite (± hibonite), it is referred to as a type A. Most type As have a porous structure and are called "fluffy" or FTA. These are generally not thought to have crystallized from a melt, but some type As have a compact form, and generally rounded shapes. These are referred to as "compact" type As, or CTAs, an example of which is shown in Plate 11c. With additional Mg and Si in the bulk composition, fassaite, describable in CAIs as Ca(Mg,Al,Ti^{3+},Ti^{4+})$(Al,Si)_2O_6$ and anorthite (nominally $CaAl_2Si_2O_8$) become important. Plate 11d is an example of a classic "type B1" CAI, which is characterized by a coarse-grained, melilite-rich mantle surrounding a core composed of melilite, spinel, fassaite, and anorthite. Type B2 inclusions (Plate 11e,f) have the same mineralogy but lack the melilite-rich mantle. A few type B3 inclusions, which contain significant amounts of forsterite in addition to melilite, have also been described. These objects carry an importance far beyond their sparse numbers because they frequently display a potpourri of isotope anomalies (e.g., *Davis et al.,* 1991). Type C inclusions are similar to type B2s mineralogically, but anorthite dominates over melilite. Types B and C inclusions and CTAs are generally regarded as igneous objects.

Aluminum-rich chondrules (e.g., Plate 11g) are a broad class of objects with compositions intermediate between those of CAIs and ferromagnesian chondrules. Olivine, orthopyroxene, and plagioclase in various ratios usually, but not always, dominate the phase assemblage. Pigeonite and clinopyroxene may be present, but melilite is absent. The bulk compositions of Al-rich chondrules are generally Mg-, Si-rich and Ca-, Al-poor relative to most CAIs but, for an operational distinction between CAIs and Al-rich chondrules, we follow *MacPherson and Huss* (2005), who noted that residual liquids produced during crystallization of bulk compositions of Si-bearing CAIs generally end up producing melilite while those of Al-rich chondrules evolve away from the melilite field and toward the stability field of olivine. *MacPherson and Huss* (2005) distinguished three types of Al-rich chondrules: (1) olivine phyric, Al-rich [Oliv] chondrules and (2) plagioclase phyric Al-rich [Plag] chondrules based on their interpretation of whether plagioclase or olivine crystallized first, and (3) Al-rich [Glass] chondrules for glass-rich objects. The term phyric is a general textural term referring to the presence of large crystals in an igneous rock set in a finer-grained groundmass or glass quenched from a liquid.

All the inclusions shown in Plate 11a–f were interpreted in terms of crystallization from a melt based on both textural and chemical arguments, but any previous existence of a liquid was inferred from compositions of the crystalline phases. There are also a few examples for which cooling rates were sufficiently rapid that the liquid quenched to glass. There is some in the Al-rich chondrule shown in Plate 11g. Plate 11h shows an example of a CAI containing glass. Glass-bearing objects described in the literature encompass most of the range of bulk compositions found for completely crystalline CAIs and Al-rich chondrules.

While we are careful to distinguish between what we call a type C inclusion and an Al-rich chondrule, the demarcation between Al-rich chondrules and ferromagnesian chondrules is left fuzzy. The original definition of *Bischoff and Keil* (1983, 1984) restricted Al-rich chondrules to bulk Al_2O_3 exceeding 10 wt% but, as *MacPherson and Huss* (2005) pointed out, this is an inherently arbitrary boundary given the continuum nature of the bulk compositions. Since we restrict attention in this chapter to objects whose igneous history can plausibly be interpreted in terms of CMAS, the lack of specific pigeonholes for Al-rich and ferromagnesian chondrules will not be problematic.

The classification of CAIs and Al-rich chondrules proceeds primarily from petrography and, particularly, the phase assemblages. This is, however, largely an expression of the bulk compositions. Figure 1 is a ternary projection designed by *MacPherson and Huss* (2005) to show, among other things, the bulk compositions of many different kinds of meteoritic inclusions. We will introduce projections below but for now this figure can simply be considered a representation of bulk composition that eliminates certain ambiguities that tend to occur in binary oxide-oxide plots. The projection of Fig. 1 works well as long as the bulk composition is not too aluminous (e.g., very rich in hibonite or spinel). Grossite-bearing inclusions, CTAs, and type B and C inclusions all plot in distinct regions in Fig. 1 and can

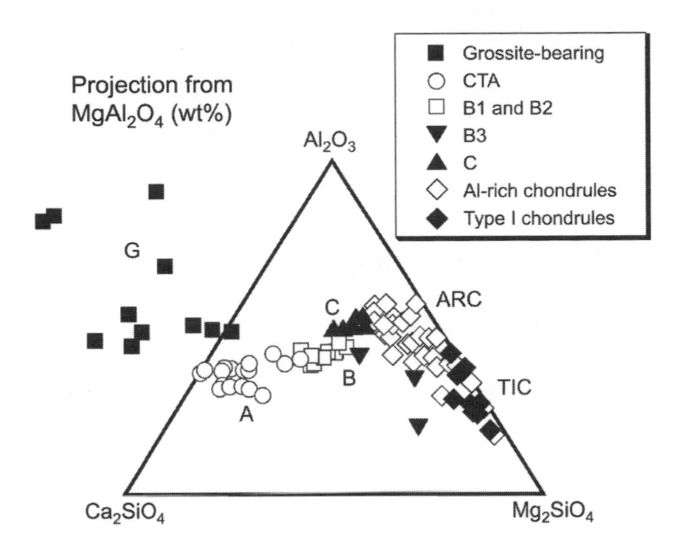

Fig. 1. Projection of bulk compositions of meteoritic inclusions from $MgAl_2O_4$ onto the plane defined by Mg_2SiO_4, Al_2O_3, and Ca_2SiO_4. Regions occupied by different types of inclusions are indicated. A: compact type A; B: type B; C: type C; G: grossite-bearing CAIs; ARC: Al-rich chondrules; TIC: type I chondrules with less than 3 wt% each of Na_2O and FeO. Data are from *Bischoff and Keil* (1983, 1984), *Bischoff et al.* (1985, 1993), *Davis et al.* (1991), *Grossman et al.* (2000), *Jones and Scott* (1989), *Jones* (1994), *Kimura et al.* (1993) *Krot et al.* (2001), *Lin and Kimura,* (1998, 2000, 2003), *MacPherson and Huss* (2005), *Sheng et al.* (1991b), *Srinivasan et al.* (2000), *Wark* (1987), *Wark et al.* (1987), and *Weber and Bischoff* (1994, 1997).

usually be distinguished from one another based solely on the bulk composition. Aluminum-rich chondrules plot separately from the CAIs but there is substantial overlap with ferromagnesian chondrules, a reflection of the lack of a clear distinction between these two types. It should be noted that the distribution of bulk compositions, at least in this projection, is more or less continuous and yet there are sharp distinctions between different inclusion types. Investigators classified the objects based primarily on phase assemblage. This limits the range in bulk composition, but the lack of overlap is largely a consequence of the phase relations during crystallization.

Before considering the application of CMAS-based phase diagrams to meteoritic inclusions, it is important to first divide components into two camps, those present during a postulated melting event and those introduced during some later alteration event. Some meteoritic inclusions contain a lot of secondary Na_2O and/or FeO that postdates melting while in other cases, often in the same inclusion, these same oxides were present during melting (*Grossman,* 1980; *MacPherson et al.,* 1988; *Krot et al.,* 1995; *MacPherson and Huss,* 2005). In section 3.2.5, we note that phase relations based on CMAS are not badly distorted with the addition of percent level Na_2O or FeO and the usual approach at this level is to simply ignore the additional oxides and renormalize the bulk composition to CMAS. This "ignorance is bliss" approach can work well but may backfire in two ways. First, some or all of the non-CMAS oxide may have been present in the liquid during melting. For example, even in CAIs, Na contents of igneous plagioclase are usually nonzero (e.g., *Hutcheon et al.,* 1978; *Steele et al.,* 1997), which implies that Na was present in the coexisting liquid, and they can rise to ~3 wt% Na_2O in the plagioclase from some Al-rich chondrules (*MacPherson and Huss,* 2005). Similar arguments hold for Fe and Ti oxides. Since all three oxides tend to be incompatible in the phases that crystallize at the highest temperatures, they accumulate in the residual liquid, potentially rising upon cooling to concentrations high enough to confound expectations based solely on CMAS phase relations. One should therefore monitor the concentration levels of non-CMAS components in a postulated residual liquid. Second, it is important to remember that, in general, an alteration reaction is not simply a matter of adding some new oxide to an inclusion but usually involves gains or losses of the CMAS oxides themselves. So, normalizing the bulk composition of an altered inclusion to CMAS can lead to a poor estimate of the original bulk composition. At a more sophisticated level, one can attempt to explicitly define the nature of the alteration reactions and reconstitute the primary bulk composition. For example, *MacPherson and Huss* (2005) noted that secondary nepheline preferentially replaced igneous plagioclase in some Al-rich chondrules and they estimated the prealteration bulk composition by assuming that any modal nepheline was originally plagioclase of the average composition for the inclusion. This implies a net gain in Na in the inclusion but also a net loss of Ca relative to Al and Si. *Wark* (1981), in one of his many contributions to our understanding of CAIs, noted that melilite in Allende type A inclusions was altered much more heavily than other phases. He reconstituted primary bulk compositions of mostly type A inclusions from altered ones through mass balance between melilite and the altering medium under the assumption that Al was immobile and that perovskite, hibonite, and spinel were unaffected. He found that Si generally entered the inclusions during alteration while Ca and Mg generally left. The bottom line is that if an inclusion is heavily altered, some effort must be expended to lift that veil before enlisting high-temperature phase relations in an attempt to understand the igneous history of the inclusion.

3. APPLICATION OF PHASE DIAGRAMS IN THE SYSTEM $CaO-MgO-Al_2O_3-SiO_2$

For the most part, bulk compositions of CAIs are sufficiently complex (see, e.g., *Ireland et al.,* 1991; *Weber and Bischoff,* 1994, 1997; *Lin and Kimura,* 1998; *Grossman et al.,* 2000) that the crystallization behavior cannot be described in terms of simple binary or ternary systems. There are, however, some important exceptions. We begin this section with a description of some CAIs for which simple systems have been used to deduce useful information regarding crystallization behavior or thermal history. We then consider relevant portions of the CMAS system in which all four oxide components are required for a good description of the phase relations.

3.1. Application of Binary and Ternary Phase Diagrams

3.1.1. CaO-Al_2O_3. The corundum-hibonite inclusion GR-1 is described by *MacPherson et al.* (1984a) and shown in Plate 11a. It is dominated by corundum (Al_2O_3) and low-Ti hibonite (nominally $CaAl_{12}O_{19}$), so the bulk composition is very close to the CaO-Al_2O_3 binary. Based on the phase diagram of *Nurse et al.* (1965), *MacPherson et al.* (1984a) noted that crystallization of hibonite and corundum from a liquid in GR-1 required temperatures in excess of 1830°C. They rationalized the observed textures in terms of melting of an original hibonite with partial distillation of Ca followed by crystallization of corundum and a second generation of hibonite. A later assessment of the CaO-Al_2O_3 binary by *Hallstedt* (1990) puts the hibonite-corundum peritectic at 1883°C, somewhat higher than given by *Nurse et al.* (1965), but this does not affect the fundamental point made by *MacPherson et al.* (1984a), that very high temperatures were required to produce GR-1 and, based on their observations, an environment in the solar nebula capable of generating such high temperatures must have existed.

3.1.2. MgO-Al_2O_3. Although no CAIs with bulk compositions on this join have been described, aluminous spinels with compositions along the join $MgAl_2O_4$-Al_2O_3 are encountered occasionally in which the molar Al/Mg ratio of the spinel exceeds the 2.0 expected for $MgAl_2O_4$ (*El Goresy et al.,* 1984; *Simon et al.,* 1994b; *Yurimoto et al.,* 2001). The stability limits of phases of known composition can sometimes provide useful constraints on thermal history. The solubility of excess Al in Mg-Al spinels is negligible below 1000°C but increases with temperature to a maximum molar Al/Mg of 20 for spinel in equilibrium with corundum and liquid at 1994°C (*Hallstedt,* 1992). The highly aluminous spinels described by *Yurimoto et al.* (2001), with Al/Mg as high as 7, require temperatures of at least ~1800°C if they formed at equilibrium. Similarly, the 10 mol% $Al_{8/3}O_4$ (a molecule of alumina on a 4-oxygen basis to be consistent with spinel stoichiometry) reported by *El Goresy et al.* (1984) for spinel in the rim of an Essebi inclusion would require a *minimum* temperature of formation of 1590°C according to the assessment of *Hallstedt* (1992), a condition that would have resulted in partial melting had the entire object been equilibrated at such a high temperature. Since *El Goresy et al.* (1984) interpreted the inclusion in terms of condensation/evaporation, this implies that the spinel was either exotic or that nonequilibrium processes were involved.

3.1.3. CaO-Al_2O_3-SiO_2. *Ivanova et al.* (2002) described a fragment of an originally spheroidal inclusion containing calcium monoaluminate ($CaAl_2O_4$), an as-yet-unnamed mineral, from the CH chondrite Northwest Africa (NWA) 470. The inclusion, labeled E1-005, is modally zoned outward from a core containing ragged grossite ($CaAl_4O_7$) with interstitial melilite, very nearly pure gehlenite in composition, and perovskite, to a region dominated by $CaAl_2O_4$ enclosing melilite and perovskite with minor grossite, to an area composed essentially of grossite and late perovskite and, finally, a mantle of essentially monomineralic melilite. *Ivanova et al.* (2002) compared texturally inferred crystallization sequences with those predicted for crystallization of an initially molten droplet from the outside inward using the phase diagram for the CaO-Al_2O_3-SiO_2 (CAS) system. They concluded that the outer melilite rim was inconsistent with crystallization from a melt corresponding to the bulk inclusion. If, however, the melilite mantle was removed from the analysis, then bulk silica was cut in half and a crystallization sequence of grossite → $CaAl_2O_4$ (with grossite in a reaction relationship) → melilite made crystallization from a liquid more plausible. At first glance, the analysis appears flawed because, although the inclusion has virtually no Mg, there is considerable Ti. *Ivanova et al.* (2002) give the bulk TiO_2 as ~6 wt% and because grossite, $CaAl_2O_4$, and melilite are virtually Ti-free, residual liquids produced as these phases crystallized would have become progressively more titaniferous until perovskite finally appeared. In other words, unless most of the perovskite is relict, seemingly unlikely given its petrographic context, the appropriate system for discussing the phase equilibria of E1-005 is CaO-Al_2O_3-SiO_2-TiO_x and not CAS. To the extent Ti affects the topology of the phase diagram and/or positions of boundary curves relative to CAS, ignoring Ti could lead to erroneous predictions of crystallization sequences and therefore incorrect conclusions regarding whether or not the observed phases crystallized from a melt. Certainly, the crystallization sequence of an otherwise CAS liquid with 6 wt% MgO would generally bear little resemblance to the crystallization sequence expected for a melt in the ternary CAS. Unfortunately, there are no experimental data for the phase equilibria of a bulk composition corresponding to that of E1-005, and therefore no direct test of the influence of Ti on the phase relations. Data are available, however, for the ternaries Al_2O_3-SiO_2-TiO_2 (*Kirschen et al.,* 1999), CaO-SiO_2-TiO2 (*DeVries et al.,* 1955), CaO-Al_2O_3-TiO_2 (*Imlach and Glasser,* 1968), and three silica-rich joins, Anorthite ($CaAl_2Si_2O_8$)-Titanite ($CaSiTiO_5$)-Silica, Anorthite-Titanite-Wollastonite ($CaSiO_3$), and Anorthite-Titanite-Silica with 30 wt% Wollastonite, whose compositions also lie within the CaO-Al_2O_3-SiO_2-TiO_2 system (*Agamawi and White,* 1953, 1954a,b). For these data, there are two consistent patterns as TiO_2 is added to a base composition in CAS. First, the liquidus temperature decreases, for 6 wt% TiO_2 by anywhere from 10° to 115°C. Second, and most important, boundary curves involving two Ti-free crystalline phases with compositions in CAS tend, as TiO_2 is added to the liquid, to point toward TiO_2. This means that if the composition of a multiply saturated liquid is projected back to CAS from TiO_2 (i.e., TiO_2 is ignored), it has essentially the same position it would have had in the Ti-free system. An exception occurs for the calcic aluminates $Ca_3Al_2O_6$ and $Ca_{12}Al_{14}O_{33}$, the projected equilibria being more calcic than expected based on CaO-Al_2O_3, but these phases are not observed in CAIs. Thus, the addition of Ti affects the properties of the liquid, leading to some suppression of the liquidus temperature but to first order, it has no effect on relative stabilities of the

Ti-free crystalline phases. It should be remembered that this assertion has not been tested for low-silica compositions in the $CaO-Al_2O_3-SiO_2-TiO_2$ system relevant to grossite- and $CaAl_2O_4$-bearing CAIs but, in the absence of such a study, the approach used by *Ivanova et al.* (2002) of ignoring the effect of Ti on the order of crystallization seems reasonable provided a Ti-rich phase other than rutile is not on the liquidus. The caveat arises from the fact that fractionation of a Ti-rich phase containing significant amounts of another oxide (e.g., CaO in perovskite) depletes the liquid in that oxide relative to the bulk composition, thereby potentially affecting predictions concerning the identity of a second crystallizing phase. In E1-005, however, perovskite is interstitial (*Ivanova et al.,* 2002) and there is no other phase with significant Ti. A final caveat concerning the use of CAS is that if the melt contains significant Mg as well as Ti, this allows the coupled substitution $Mg^{2+} + Ti^{4+} = 2Al^{3+}$ in hibonite and stabilization of hibonite relative to grossite (e.g., *Beckett and Stolper,* 1994). CAS should probably not be used if hibonite is present and bulk MgO exceeds a few tenths of a wt%. *Weber and Bischoff* (1994) used the phase relations for $CaO-Al_2O_3-SiO_2$ taken from *Gentile and Foster* (1963) to predict crystallization sequences for grossite-bearing inclusions with 1–2 wt% MgO but properly cautioned that considerable uncertainty accrued through the presence of minor elements. We discuss these inclusions in section 3.2.

3.1.4. $CaO-Al_2O_3-MgO$. *MacPherson et al.* (1983) described BB-2, a spinel-hibonite spherule from Murchison, and *Simon et al.* (1994b) described a grossite-bearing spherule with a very similar bulk composition. Both inclusions have ~3 wt% TiO_2 tied up in perovskite, but virtually no Si. *Lin and Kimura* (2000) described a spinel-hibonite-grossite spherule included in a type B inclusion and *Yurimoto et al.* (2001) gave a preliminary description of an ultrarefractory inclusion with hibonite, spinel, perovskite, and an alumina phase. In all four studies, the system $CaO-Al_2O_3-MgO$ (CAM) was used as a basis for discussing the phase relations and the potential for the inclusions having crystallized from a liquid. *MacPherson et al.* (1983) used a compilation of data from *Rankin and Merwin* (1916), *Gentile and Foster* (1963), *Nurse et al.* (1965), and *Rao* (1968). *Simon et al.* (1994b) and *Lin and Kimura* (2000) used the thermodynamic model of *Berman* (1983) to construct a CAM phase diagram. This model is discussed in section 3.3.1 but for now it can effectively be considered an alternative phase diagram for CAM that draws on a somewhat larger set of data sources than used by *MacPherson et al.* (1983). A more recent model for the system (*Hallstedt,* 1995) generates a reasonably similar phase diagram but the discovery of two new phases, $Ca_2Mg_2Al_{28}O_{46}$ and $CaMg_2Al_{16}O_{27}$, with compositions between spinel and hibonite (*Göbbels et al.,* 1995) postdates *Hallstedt*'s (1995) assessment and suggests that the phase diagram may need revision. The two new phases have not been observed in CAIs, possibly speaking to very limited stability as Si and/or Ti is added to the system. According to *de Aza et al.* (2000), the phases have adjacent stable primary phase fields bounded by those of hibonite, grossite, and spinel with a configuration such that liquids coexisting simultaneously with hibonite, grossite, and spinel are unstable. Experimental confirmation for the new phases is desirable and also for the de Aza et al. corundum and hibonite primary phase fields, which extend to considerably more magnesian compositions than obtained by previous workers. The de Aza et al. phase diagram also should not be used without correction of its numerous violations of Schreinemakers' rule (*Zen,* 1966). Finally, it should be noted that information on the effect of Ti on the phase relations relative to pure CAM is limited to the ternaries $CaO-MgO-TiO_2$ (*Kirschen and de Capitani,* 1999) and $CaO-Al_2O_3-TiO_2$ (*Imlach and Glasser,* 1968). For liquids in $CaO-MgO-TiO_2$ with less than ~10 wt% TiO_2, the position of the periclase (MgO)-lime-liquid boundary curve projected from TiO_2 onto CaO-MgO is within 1 wt% of its position on the Ti-free binary CaO-MgO although the temperature decreases by as much as ~150°C. This is consistent with the theme presented in the preceding section that additions of Ti tend to reduce liquidus temperatures but affect Ti-free crystalline phases similarly. Since hibonite can accommodate considerable amounts of Ti if Mg is present, it is likely that hibonite is stabilized relative to grossite, as indeed was observed by *Beckett and Stolper* (1994) in the more complex system CMAS + TiO_2 (CMAST). The solubility of Ti in the new phases described by *Göbbels et al.* (1995) is unknown and it is uncertain whether or not they are stable for bulk compositions relevant to CAIs.

3.2. Projections of Quaternary Phase Relations

In a projection, a composition presented in terms of one basis set of components (e.g., end-member oxides) is recast in terms of an alternative basis set. The reason for doing this is to reduce the number of dimensions for graphical depiction of a suite of compositions without too much loss of information. Projections are a basic tool used to depict compositions in multicomponent systems (e.g., *McMurdie and Insley,* 1936; *Greenwood,* 1975; *Spear,* 1993) and they have found widespread application in studies of CAIs (e.g., *Stolper,* 1982; *Stolper and Paque,* 1986; *Davis et al.,* 1991; *MacPherson and Davis,* 1993; *Beckett and Stolper,* 1994; *Lin and Kimura,* 2000; *Huss et al.,* 2001; *Krot et al.,* 2001) and Al-rich chondrules (*Sheng et al.,* 1991a; *Sheng,* 1992; *Srinivasan et al.,* 2000; *MacPherson and Huss,* 2005). There are three basic types of applications. First, projections of liquid compositions saturated with respect to one or more crystalline phases can be used to construct phase diagrams from which the primary phase field of a given bulk composition can be determined. Second, provided a phase diagram is appropriate for a given bulk composition, the liquid line of descent for equilibrium or, more usually, fractional crystallization of a liquid and hence the order of appearance of phases can be determined. A third use is to depict variations in composition for one or more groups of CAIs or glasses, usually with the idea of comparing compositions of meteoritic materials with model predictions for condensation/volatilization trajectories, identifying possible end members for mixtures producing the observed suite of compositions, or showing differences in composition among groups of meteoritic inclusions (*Stolper,* 1982; *Beckett and Grossman,* 1988; *Grossman et al.,* 2000; *MacPherson and Huss,* 2005). Figure 1 is an example of the latter.

3.2.1. Stolper's projection. *Stolper* (1982) was the first to popularize the use of projections to describe phase equilibria relevant to CAIs. The diagram was designed primarily for type B inclusions, for which it remains very useful (e.g., *Lin and Kimura,* 2000; *Simon and Grossman,* 2003). Later projections (e.g., *Beckett and Grossman,* 1988; *Beckett and Stolper,* 1994; *MacPherson and Huss,* 2005; this work) were tailored to other types of CAIs or allowed the depiction of a broader range of compositions than is possible with *Stolper*'s (1982) projection.

Figure 2a shows the tetrahedron defined by the compositions of end member gehlenite ($Ca_2Al_2SiO_7$), anorthite ($CaAl_2Si_2O_8$), forsterite (Mg_2SiO_4), and spinel ($MgAl_2O_4$) (GAFS) within the tetrahedron defined by the oxides CaO, MgO, Al_2O_3, and SiO_2 (CMAS). Any bulk composition lying inside or on the CMAS tetrahedron can be described in terms of positive amounts of CMAS oxides. Similarly, any composition lying inside or on the GAFS tetrahedron can be described in terms of positive amounts of GAFS components and, since GAFS lies within CMAS, such a composition could also be described in terms of positive amounts of CMAS oxides. In Fig. 2a, however, much of the CMAS tetrahedron lies outside the GAFS tetrahedron. All such compositions are physically realizable but if they are defined in terms of GAFS, one or more of the GAFS components will be negative. There is nothing wrong with using negative amounts of components to describe a composition.

The physical significance of the projection of a CMAS composition that lies outside the GAFS tetrahedron from $MgAl_2O_4$ spinel (SP) onto the plane defined by end-member compositions of gehlenite (GE), anorthite (AN), and forsterite (FO) can be seen in Fig. 2a. Consider the join Ca-Tschermaks (CaTs; $CaAl_2SiO_6$)-diopside (Di; $CaMgSi_2O_6$), which is an important binary for CAI fassaites. The compositions of both CaTs and Di lie outside the GAFS tetrahedron. To obtain a projection of Di from SP onto the plane GE-AN-FO, a line defined by the two points Di and SP is drawn and the point where this line intersects the plane defined by GE-AN-FO defines the projection point for Di. As shown in Fig. 2a, this point lies within the GE-AN-FO triangle (see also Fig. 4a of *Stolper,* 1982). A line drawn through spinel (SP) and CaTs also intersects the GE-AN-FO plane but the projection point lies outside the triangle. This illustrates the important concept that there is nothing sacred about projection points having to plot within the composition triangle defining the projection plane.

In the above analysis, we accepted the set of components used by Stolper without pausing to consider what constituted a valid set of components or why one should prefer one particular set of candidate end members to another. These are two distinctly different issues to consider. (1) In general, any set of four end-member compositions within CMAS that defines a tetrahedron will generate a valid set of components and, in fact, end-member compositions lying outside CMAS are equally valid, although such components have not been used in meteoritic applications. The statement that a set of proposed compositions form a tetrahedron within or even outside CMAS is equivalent to saying that the proposed set of components must be linearly independent. Thus, the set of components $MgAl_2O_4$, Al_2O_3, Ca_2SiO_4, and Mg_2SiO_4 selected by *MacPherson and Huss* (2005) provides a valid representation of CMAS compositions, but åkermanite ($Ca_2MgSi_2O_7$), gehlenite ($Ca_2Al_2SiO_7$), Ca-Tschermaks ($CaAl_2SiO_6$), and diopside ($CaMgSi_2O_6$) would be invalid because the compositions lie within the same plane, and therefore a tetrahedron cannot be constructed from them. In the latter case, the proposed components are related through the relationship åkermanite + Ca-Tschermaks = diopside + gehlenite (i.e., they are not independent). For CAI applications, some components are invariably chosen to correspond to major end-member compositions of phases that occur either in CAIs or as stable phases within CMAS, but this is not required. Any set of four independent compositions could, in principle, be selected. In a system containing five or more components, the easily visualized geometric construct usable in a quaternary system fails, but the basic constraint that all compositions of all proposed end members must be independent still holds. (2) The second issue regarding the selection of components for constructing projections depends critically on the application. In Stolper's projection, the operational objectives were to show bulk compositions of type B and C CAIs, multiply saturated liquidus phase fields for such bulk compositions in which spinel and liquid coexist with one or more additional crystalline phases, the crystallization paths of such liquids, and a comparison of bulk CAI compositions with calculated condensates. Since the intent was a consideration of phase assemblages saturated with respect to spinel that is very close in composition to the end member $MgAl_2O_4$ for the CAIs of interest and in his experiments, *Stolper* (1982) chose $MgAl_2O_4$ spinel as one component. For the remaining three components, the choice was mostly one of convenience in presentation. Type B and C inclusions contain spinel and various amounts of melilite, fassaite, anorthite, and occasionally olivine, so likely candidates include the end-member compositions åkermanite, gehlenite, Ca-Tschermaks, diopside, anorthite, and forsterite. The only combination of three end members from this list that led to all bulk compositions of type B and C inclusions available in 1982 being projected from spinel to points within or even close to the base ternary was the one selected by Stolper (i.e., gehlenite, anorthite, and forsterite).

In CMAS, the locus of liquid compositions in equilibrium with two crystalline phases over a range of temperatures will describe a surface, and where two such surfaces intersect, they produce a space curve, which describes the compositions of liquids saturated with respect to three crys-

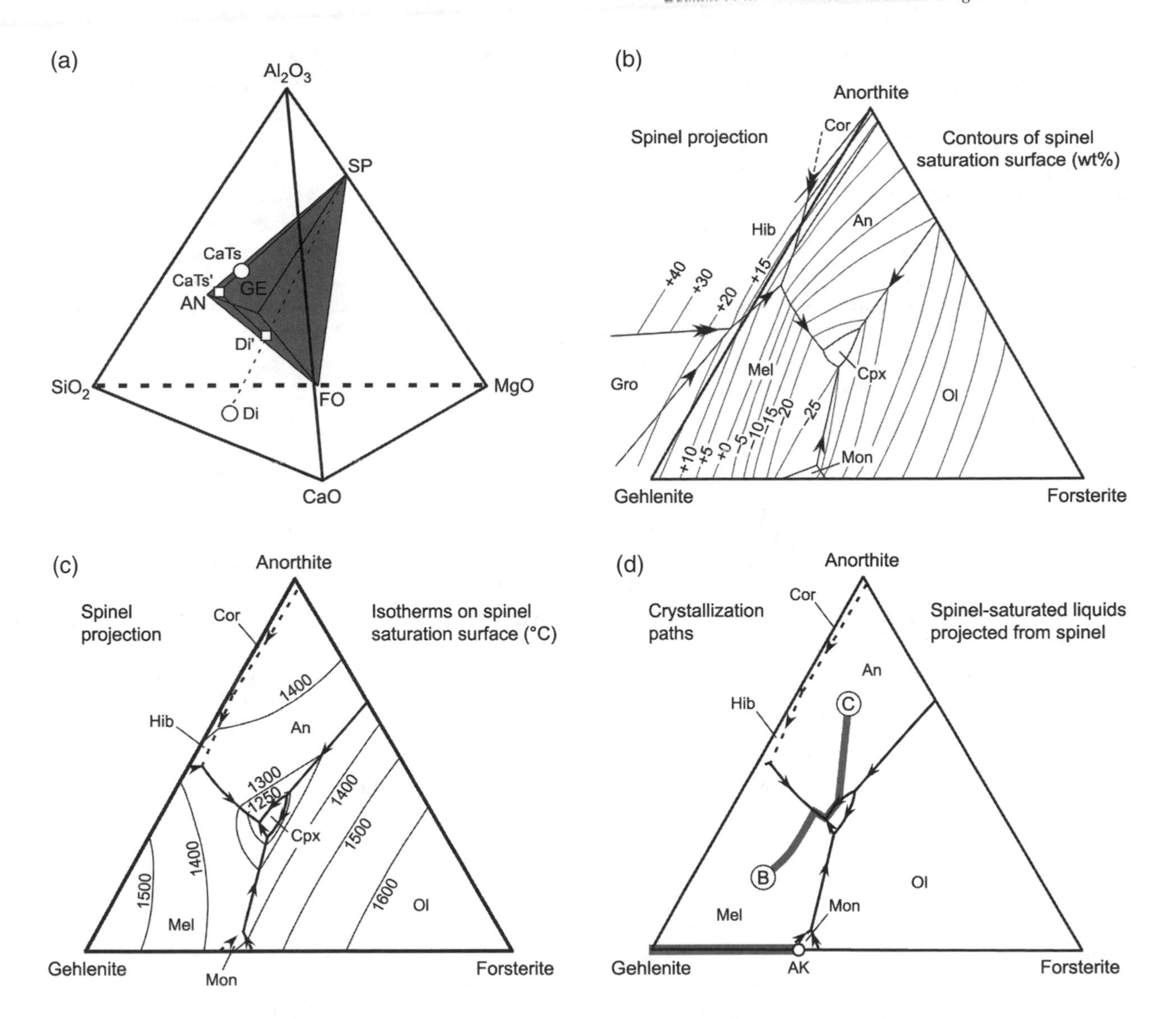

Fig. 2. *Stolper*'s (1982) projection from spinel onto the plane defined by the compositions of gehlenite, anorthite, and forsterite. **(a)** Location of the gehlenite-anorthite-forsterite-spinel (GAFS) tetrahedron within the CMAS tetrahedron. GAFS end members gehlenite (GE), anorthite (AN), forsterite (FO), and spinel (SP) are indicated. The positions of end member diopside (Di) and Ca-Tschermaks (CaTs) and their projected positions (Di', CaTs') are also shown (squares). **(b)** Projection of surfaces, defined by liquid compositions in which spinel is in equilibrium with the liquid and an additional crystalline phase, from $MgAl_2O_4$ spinel onto the GAF plane. Each field in **(b)**, **(c)**, and **(d)** is labeled only by the identity of the additional phase as all surfaces refer to compositions of liquids that are saturated with respect to spinel. The surfaces are contoured in wt% $MgAl_2O_4$ above the gehlenite-anorthite-forsterite plane. Spinel-saturated liquidus phase fields are indicated for melilite (Mel), clinopyroxene (Cpx), olivine (Ol), monticellite (Mon), grossite (Gro), hibonite (Hib), corundum (Cor), and anorthite (An). The figure, after *Beckett and Stolper* (1994), is a slight modification of the projection given by *Stolper* (1982). **(c)** Projection of spinel-saturated liquidus surfaces, contoured in terms of temperature (°C). After *Stolper* (1982). **(d)** Fractional crystallization paths for compositions typical of type B (point B) and type C (point C) inclusions. Phase relations after *Stolper* (1982). The projected composition of åkermanite (AK) is indicated and the melilite (åkermanite-gehlenite) join shaded.

talline phases. In *Stolper*'s (1982) projection, a set of surfaces saturated with respect to spinel were projected from $MgAl_2O_4$ spinel onto the plane defined by the compositions of gehlenite, anorthite, and forsterite. Compositions above the surface have spinel on the liquidus, the temperature above which no crystalline phases are stable; compositions on the surface have spinel and a second crystalline phase on the liquidus. If the liquid composition lies just below the surface, then the second phase will be on the liquidus. There are big advantages to projections in describing crystallization sequences as discussed below but there are also disadvantages. Information is always lost in a projection and it is very important to first make sure that a particular projection is valid for the compositions of interest. Three-dimensional saturation surfaces disappear in projection and the three-dimensional boundary curves marking the intersections of saturation surfaces map as two-dimensional curves. The first step in deciding whether or not a particular projection is suitable for describing the crystallization behavior of a particular liquid is to determine whether or not the composition of that liquid plots within the primary liquidus phase volume of the saturating phase or on a bounding multiply saturated liquidus surface. For Stolper's projection, this is equivalent to asking if a liquid composition is on or above the saturation surface, that is, between spinel and the saturation surface. A simple way to address this is to contour the projected saturation surface in terms of the amount of the projected component, essentially a topographic map of the projected surface, and then ask whether the liquid composition of interest lies above or below the surface. Figure 2b shows the spinel saturation surface for GAFS contoured in terms of wt% of the spinel component. It is worth noting in Fig. 2b that many of the contours have negative amounts of spinel component in the liquid. It is, however, the relative amounts of spinel component in a liquid and on the saturation surface that is important in determining the liquidus phase, not whether the numbers are positive or negative. Thus, a liquid whose composition has –10 wt% spinel component that plots in projection on a –20 wt% isopleth in Fig. 2b will have spinel on the liquidus because the composition lies above the spinel liquidus saturation surface even though it has a negative amount of spinel component. One with +20 wt% spinel component that plots on the +40 wt% spinel isopleth will have some other phase on the liquidus because it plots below the spinel saturation surface even though it has a positive amount of spinel component. In general, if a liquid composition lies above the saturation surface based on the GAFS components calculated according to the method given in *Stolper* (1982), then the projection provides a valid representation of the crystallization behavior of the liquid as long as the saturating phase, spinel in this case, is present. If it plots below the spinel saturation surface, then some phase other than spinel is on the liquidus and Stolper's projection becomes valid only when spinel joins the instantaneously crystallizing phase assemblage. This is a vital point for the use of any projection to infer crystallization sequences in CAIs. If a liquid composition is not above the saturation surface being projected, then the projected phase is not on the liquidus and use of the projection to infer crystallization sequences is invalid.

The saturation surface identifies the phase on the liquidus for compositions lying immediately beneath the saturation surface but it says nothing about how deep that liquidus phase volume might be and, therefore, unless the composition of interest is essentially kissing the surface, the liquidus phase and crystallization behavior prior to the appearance of spinel should be determined before proceeding. For some compositions, the liquid will never actually reach the spinel saturation surface even if it projects comfortably within the GAF triangle of Fig. 2b. *Stolper* (1982) showed for nearly all type B1, B2, and C inclusions that the bulk compositions were on or above the spinel saturation surface and, therefore, his projection scheme was valid for determining crystallization paths of those inclusions. For some type B3 inclusions (e.g., *Davis et al.,* 1991), olivine is the liquidus phase and early crystallization behavior for such inclusions should be determined using a suitable projection from forsterite. For type A inclusions, melilite is generally the liquidus phase.

Temperature is an important intensive variable for constraining astrophysical processes, and *Stolper*'s (1982) projection can be used to place some constraints on what the temperatures may have been during melting and crystallization of type B and C inclusions. This derives from the fact that two crystalline phases, such as melilite and spinel, are in equilibrium with the same CMAS liquid at only one temperature (for a specific total pressure), which means it is possible to contour the spinel-saturated liquidus surface in terms of temperature (Fig. 2c); by similar reasoning, one can also contour in terms of the composition of a solid phase (e.g., *Beckett et al.,* 1999). For bulk compositions of CAIs available in 1982, minimum temperatures above which melilite is unstable, and therefore minimum peak temperatures for the inclusions if melilite crystallized from a melt, ranged from ~1240°C for compositions plotting near the fassaite + spinel field to ~1450°C for melilite-rich type B inclusions. For the average type B composition *Stolper* (1982) investigated experimentally, the corresponding temperature is ~1400°C and this, together with the results of dynamic crystallization experiments discussed in section 3.5, led to a widely held view that maximum temperatures experienced by type B inclusions were roughly 1400°C.

One of the most important properties of projections for the study of meteoritic inclusions is that the determination of crystallization paths and sequences in a quaternary system projected to a ternary is, in many instances, no more complicated than for a true ternary system. Consider points B and C in Fig. 2d. Point B would be typical of a type B inclusion and point C of a type C inclusion. Let's assume, as is observed for the meteoritic inclusions, that both compositions plot above the spinel saturation surface. Fractional crystallization of spinel from bulk composition B drives the liquid composition directly away from spinel until it encounters the spinel saturation surface. Since the projection in Fig. 2d is from the composition of the phase that is crystallizing (i.e., from $MgAl_2O_4$ spinel), bulk composition B and a residual liquid derived from B by crystallization of spinel project to the same point. Once on the spinel saturation surface, however, melilite (plus spinel) crystallize and the projected position of the residual liquid is driven away from the plane åkermanite-gehlenite (i.e. melilite)-spinel, which projects to the base of the triangle in Fig. 2d. The instantaneously crystallizing melilite becomes more åkermanitic with progressive crystallization and this leads to the curved path shown in Fig. 2d. Further crystallization drives the liquid to the melilite + anorthite + spinel + liquid boundary curve, at which point anorthite joins the instantaneously crystallizing phase assemblage. The residual liquid composition then moves down the boundary curve, in the direction away from the composition plane of the instantaneously crystallizing phase assemblage, until clinopyroxene crystallizes and the remaining liquid is consumed or the liquid leaves the spinel saturation surface. Fractional crystalliza-

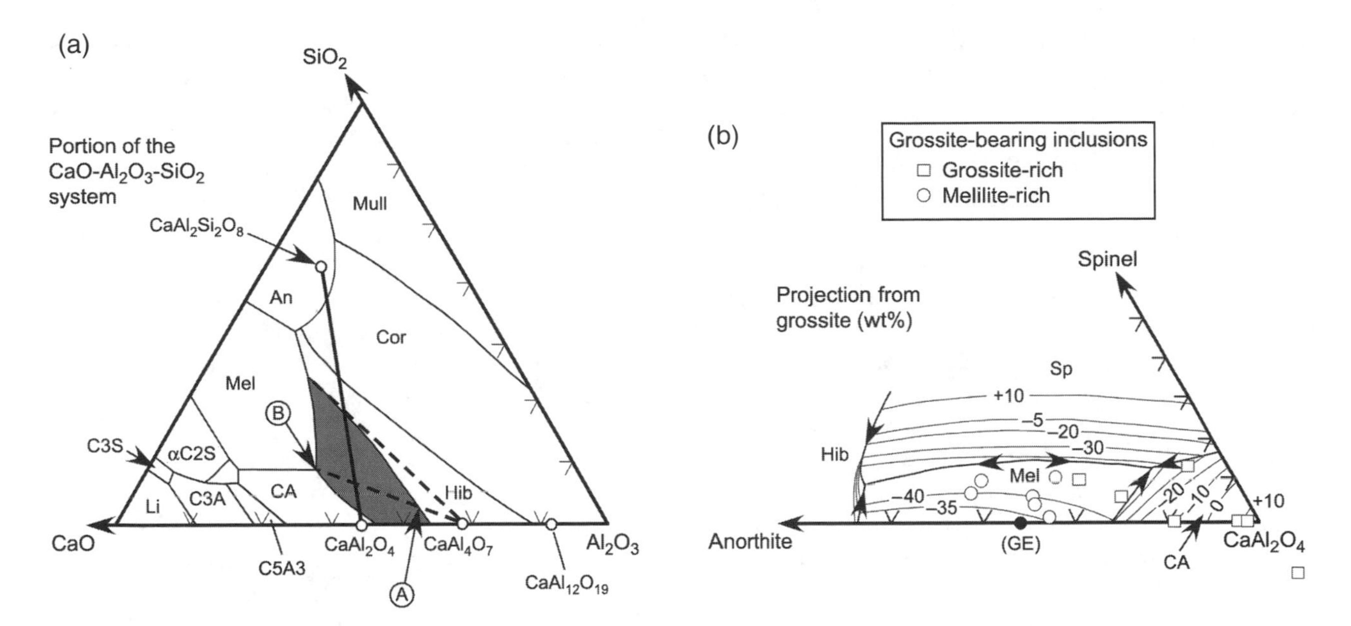

Fig. 3. Projections of liquidus-saturated phase relations in portions of CMAS. **(a)** Liquidus phase fields for a portion of the system CaO-Al_2O_3-SiO_2 (after *Gentile and Foster,* 1963). Two projection lines from grossite onto the join anorthite ($CaAl_2Si_2O_8$)-calcium monoaluminate ($CaAl_2O_4$) are indicated. Points A and B are stability limits for grossite plus liquid along one of these lines. Primary phase fields for anorthite (An), $CaAl_2O_4$ (CA), $Ca_5Al_6O_{14}$ (C5A3), $Ca_3Al_2O_6$ (C3A), Ca_3SiO_5 (C3S), α-Ca_2SiO_4 (αC2S), corundum (Cor), grossite, hibonite (Hib), lime (Li), melilite (Mel), and mullite (Mull) are shown, with the grossite field shaded. For clarity, the β-Ca_2SiO_4 and grossite fields are not labeled. Tic marks are shown at 10 wt% intervals. **(b)** Projection of grossite-saturated liquidus phase fields in CMAS from $CaAl_4O_7$ (grossite) onto the plane defined by $CaAl_2O_4$ (calcium monoaluminate), $MgAl_2O_4$ (spinel), and $CaAl_2Si_2O_8$ (anorthite). Grossite is in a reaction relationship with the liquid along the CA-spinel-grossite boundary curve. Tic marks are shown at 10% intervals along $CaAl_2O_4$-$CaAl_2Si_2O_8$ and 5% intervals along $CaAl_2O_4$-$MgAl_2O_4$. The saturation surface is contoured in terms of wt% $CaAl_4O_7$. Bulk compositions of grossite-bearing inclusions from *Weber and Bischoff* (1994) are shown as open circles (<10 wt% $CaAl_4O_7$ component) and open squares (>70 wt% $CaAl_4O_7$ component) and the projected composition of end member gehlenite (GE) is also indicated (closed circle). Melilite in grossite-bearing inclusions is highly aluminous (e.g., *Weber and Bischoff,* 1994) and compositions plot very close to GE. Coordinates for plotting points in this diagram may be determined from oxide mole fractions as $CaAl_2O_4$ (wt%) = $79.02(4CaO + 2MgO + 2Al_2O_3 - SiO_2)$, $MgAl_2O_4 = 142.65MgO$, $CaAl_2Si_2O_8 = 139.105SiO_2$, and $CaAl_4O_7 = 260.001(Al_2O_3\text{-}CaO\text{-}MgO)$ followed by normalization to $CaAl_2O_4$-$MgAl_2O_4$-$CaAl_2Si_2O_8$. See text for data sources used in constructing the figure.

tion of C can be followed in a similar manner. Spinel crystallizes until the liquid reaches the spinel saturation surface, at which point anorthite joins the crystallizing phase assemblage and the residual liquid is driven directly away in projection from the anorthite vertex. When the residual liquid reaches the anorthite + clinopyroxene + spinel + liquid boundary curve, clinopyroxene crystallizes and the liquid evolves along the boundary curve until it reaches the melilite-anorthite-clinopyroxene-spinel invariant point. Note that the inferred crystallization paths for points B and C look just like they would have been (aside from spinel being present) had Fig. 2d been a true ternary.

3.2.2. Projections for highly aluminous systems. The Stolper projection is well posed for a discussion of phase equilibria relevant to most type B and C inclusions, but is much less useful for the often spinel-undersaturated, highly aluminous CAIs, whose compositions tend to project outside the GAF triangle to the forsterite-poor side of the anorthite-gehlenite join in Fig. 2b or of Al-rich chondrules, whose compositions generally project outside the GAF triangle to the gehlenite-poor side of the anorthite-forsterite join. In this and the following section, we discuss projections that can be applied to some aluminous CAIs and Al-rich chondrules to infer liquidus phases and crystallization paths.

Stolper (1982) projected compositions from $MgAl_2O_4$ because spinel is present in type B1 and B2 CAIs on or near the liquidus and the composition of the phase is consistently close to $MgAl_2O_4$ in these inclusions. Similar considerations apply to projections for which spinel is not the liquidus phase, but if the liquidus phase melts incongruently, as hibonite, enstatite, and grossite do, some complications not apparent in Stolper's projection become important. We illustrate the principles for a two-dimensional case in Fig. 3a, which shows aluminous portions of the CAS system after *Gentile and Foster* (1963). Suppose that we are interested in projecting compositions of liquids saturated with respect to grossite and an additional crystalline phase from $CaAl_4O_7$ (grossite) onto the join $CaAl_2O_4$ (CA)-$CaAl_2Si_2O_8$ (An), which is really a description of CAS compositions in terms of grossite ($CaAl_4O_7$), calcium monoaluminate ($CaAl_2O_4$), and anorthite ($CaAl_2Si_2O_8$). Two examples are shown in

Fig. 3a via dashed line segments drawn from $CaAl_4O_7$ through the CA-An join. The first is drawn grazing the extreme Si-rich end of the grossite primary phase field. It intersects CA-An at the projection point for this liquid. Note that there is only one grossite-saturated liquid that projects to this point and that there are no grossite-saturated liquids whose compositions plot at more An-rich compositions along CA-An. This lack of stable liquids saturated with respect to the projecting phase is a natural consequence of projecting from the composition of an incongruently melting compound. The second dashed line in Fig. 3a is drawn through the central portion of the grossite primary phase field. It intersects two multiply saturated liquid compositions, both of which project to the same point on CA-An. One of these, point A, is hibonite- and grossite-saturated and the other, less $CaAl_4O_7$-rich, point B, is an invariant point, saturated with respect to the three crystalline phases grossite, $CaAl_4O_7$, and melilite. It happens that one liquid (point A) is $CaAl_4O_7$-rich relative to CA-An and the other, point B, is $CaAl_4O_7$-poor (i.e., one above and one below the CA-An line, respectively). If contoured in terms of wt% grossite above or below the CA-An line, one would have positive amounts of grossite component A and one would have negative amounts of grossite component B, but this is an artifact of the selected projection line. If the projection line were CaO-SiO_2, both liquids would be above the projection line (both with positive amounts of grossite component), and if An-$CaAl_{12}O_{19}$ were used, both liquid compositions would be below the projection line (both with negative amounts of grossite component). The existence of an upper and lower multiple saturation curve in this example from CAS and upper and lower surfaces for comparable examples in CMAS is an endemic feature of projections taken from compositions of incongruently melting phases. While it is possible to construct projections from the compositions of congruently melting phases for which more than one liquid composition projects to the same point, these are not important for CAIs.

Grossite-bearing inclusions are relatively common in CH chondrites and rather rare elsewhere. Figure 3b shows the compositions of grossite-saturated liquidus phase fields projected from $CaAl_4O_7$ (grossite) onto the plane defined by the end-member compositions $CaAl_2O_4$ (calcium monoaluminate), $MgAl_2O_4$ (spinel), and $CaAl_2Si_2O_8$ (anorthite), contoured in wt% $CaAl_4O_7$. The depicted fields describe a lower surface of stability for grossite based on data from *Osborn and Gee* (1969), *Gutt and Russell* (1977), *Beckett and Stolper* (1994) and references cited therein, and *de Aza et al.* (1999). There is also an upper stability limit, but this is poorly constrained except in the system CaO-Al_2O_3-SiO_2 (Fig. 3a), which plots along the base of the triangle in Fig. 3b. Here, the maximum wt% of the $CaAl_4O_7$ component that a bulk composition can have and still have grossite on the liquidus ranges from +70 wt% at the $CaAl_2O_4$ vertex (lower surface at ~+12 wt% $CaAl_4O_7$ component) to ~+60 at the calcium monoaluminate-melilite-grossite – liquid invariant point (lower surface at ~–30 wt%) to ~+40 at the 60 wt% $CaAl_2O_4$ tic (lower surface at ~–30 wt% $CaAl_4O_7$ component), after which it drops rapidly to meet the lower surface shown in Fig. 3b. To take a specific example, consider compositions along the $CaAl_4O_7$-$CaAl_2O_4$ join, which plot at the $CaAl_2O_4$ vertex in Fig. 3b and along the base of the triangle in Fig. 3a. Such compositions will have grossite on the liquidus only for wt% of the $CaAl_4O_7$ component between ~+12 and +70 wt%. If the bulk composition has less than +12% $CaAl_4O_7$ component, $CaAl_2O_4$ or some other phase will be on the liquidus (see, e.g., Fig. 3a) and if there is more than +70 wt% $CaAl_4O_7$ component, hibonite or corundum will be on the liquidus. The importance of this can be seen from the bulk compositions of grossite-bearing CAIs described by *Weber and Bischoff* (1994), which come in two distinct populations. Bulk compositions of the melilite-rich inclusions (open circles) have –28 to +9 wt% $CaAl_4O_7$ component; they plot above the lower grossite saturation surface and almost certainly below the upper surface; these inclusions will have grossite on the liquidus, and Fig. 3b can be used to infer the identity of the second crystallizing phase (melilite) and the crystallization sequence provided Ti-contents are sufficiently low. The predicted fractional crystallization sequence for such bulk inclusions would when projected from TiO_2, be grossite → melilite → spinel → either calcium monoaluminate or hibonite, depending on which side of the spinel-melilite thermal divide the residual liquid strikes. $CaAl_2O_4$ is not observed in these inclusions even though the phase relations suggest that most of them would eventually crystallize $CaAl_2O_4$. This may indicate that perovskite is generally relict; the bulk compositions would, if projected from perovskite instead of TiO_2, all plot on the anorthite-rich side of the thermal divide (i.e., some of the Ca in the bulk composition is tied up in perovskite, making the relevant residual liquid compositions less calcic), which in projection, shifts compositions away from the $CaAl_2O_4$ vertex. The second population of grossite-bearing CAIs (open squares) is characterized by amounts of $CaAl_4O_7$ component exceeding +70 wt%. Grossite is not the liquidus phase for these bulk compositions (even pure $CaAl_4O_7$ does not have grossite on the liquidus because the phase melts incongruently), and Fig. 3b cannot be used to infer early portions of the crystallization sequences.

Hibonite is an important phase in many aluminous CAIs, and *Beckett and Stolper* (1994) projected compositions of hibonite-saturated liquid compositions from $CaAl_{12}O_{19}$, the nominal end-member composition of hibonite, onto the plane defined by the end-member compositions of gehlenite, anorthite, and spinel (Fig. 4). This tetrahedron shares one face (anorthite-gehlenite-spinel) with the one used by Stolper (Fig. 2b), but the fourth vertex (hibonite) lies on the CaO-Al_2O_3 binary rather than MgO-SiO_2 (forsterite). Bulk compositions of some hibonite-rich inclusions and hibonite-glass spherules lie above or close to the hibonite-saturation surface (Fig. 4a), implying that hibonite was a liquidus or near-liquidus phase, but others plot far below the saturation surface. All the glass compositions from hibonite-glass spherules considered by *Beckett and Stolper* (1994) are be-

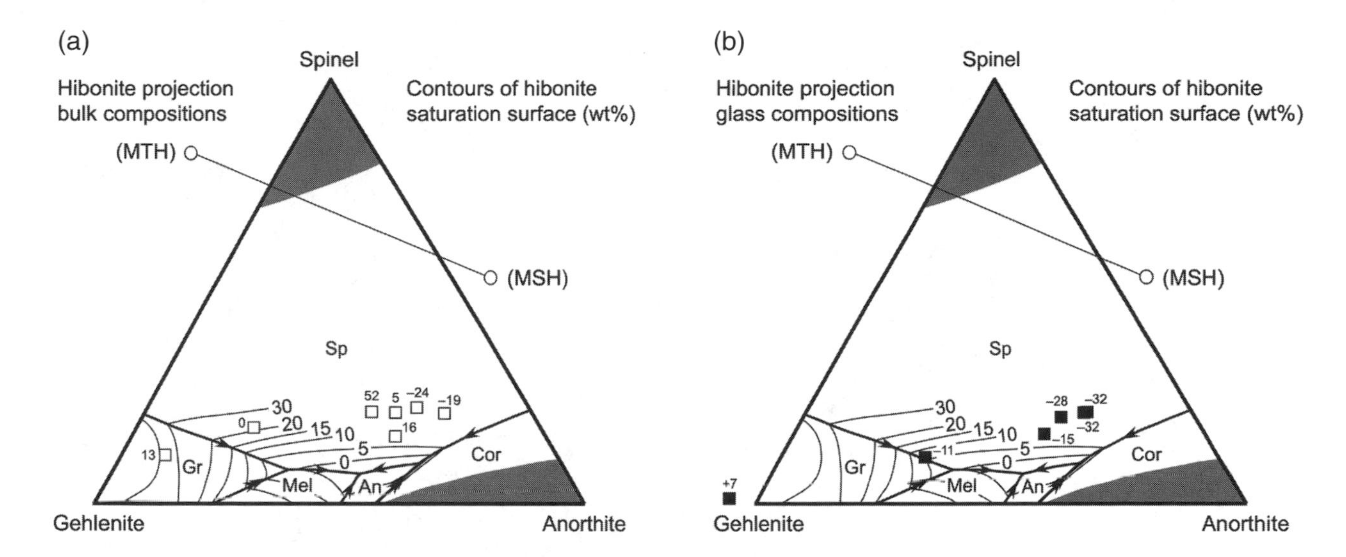

Fig. 4. Projection of hibonite-saturated liquidus surfaces from $CaAl_{12}O_{19}$ (hibonite) onto the plane defined by the compositions of gehlenite ($Ca_2Al_2SiO_7$), spinel ($MgAl_2O_4$), and anorthite ($CaAl_2Si_2O_8$) after *Beckett and Stolper* (1994). The identity of the second saturated crystalline phase is indicated in each field. Shading indicates regions in which there is no stable hibonite-saturated liquid. Projected positions of $CaAl_{10}MgTiO_{19}$ (MTH) and $CaAl_{10}MgTiO_{19}$ (MSH) are also shown. Gr: grossite. Other abbreviations are as in Fig. 2. **(a)** Bulk compositions of hibonite-rich inclusions. Data from G. J. MacPherson (personal communication, 1991) and *Ireland et al.* (1991). **(b)** Glass compositions in hibonite-bearing inclusions. Data from *Kurat* (1975), *Ireland et al.* (1991), and G. J. MacPherson (personal communication, 1991).

low the saturation surface (Fig. 4b). They took this to imply that hibonite was often relict and, where glass was present, that cooling rates were sufficiently rapid that crystallization of stable phases did not take place. In some cases (*Ireland et al.,* 1991), hibonite and glass are not in isotopic equilibrium, also consistent with the idea that some of the hibonite is relict with respect to the last melting event. A similar theme occurs in a consideration of hibonite crystals in type A inclusions (not shown). Here, bulk compositions plot well below the hibonite saturation surface even though hibonite is texturally early, an apparent inconsistency with the phase relations, which was used as one line of evidence by *MacPherson and Grossman* (1984) that fluffy type As did not crystallize from a melt. The important point is that phase diagrams can be used to help establish inconsistencies with observed phase assemblages and textures and this can lead to the identification of possible relict phases and the effects of processes other than simple equilibrium or fractional crystallization.

3.2.3. Projections for Mg-, Si-rich objects. The often high Fe and Na contents inferred for the melted precursors of ferromagnesian chondrules make CMAS an inappropriate analog system but there are objects whose bulk compositions are intermediate between those of CAIs and ferromagnesian chondrules that are generally referred to as Al-rich chondrules (*Bischoff and Keil,* 1983, 1984; *Sheng et al.,* 1991b; *MacPherson and Huss,* 2005). For some of these, CMAS phase equilibria provide useful constraints, and *Sheng et al.* (1991a) introduced a set of three projections for such inclusions: projection from spinel onto forsterite + silica + anorthite (Fig. 5a); projection from forsterite onto the plane formed by end-member compositions of spinel, silica, and diopside (Fig. 5b); projection from anorthite onto spinel + silica + diopside (Fig. 5c). The last two diagrams are given in a larger format by *Srinivasan et al.* (2000), taken from *Sheng*'s (1992) thesis, and also shown by *MacPherson and Huss* (2005). By selecting the appropriate phase diagram (i.e., the one for which the bulk composition plots above the saturation surface and within the contoured portion of the diagram), *Sheng et al.* (1991a), *Sheng* (1992), *Srinivasan et al.* (2000), and *MacPherson and Huss* (2005) were able to determine crystallization paths for the inclusions they studied. For the most part, once the residual liquid intersected a liquidus phase saturation surface, these authors were able to treat the phase relations shown in the projections in the same way they would have had the projections been true ternary systems. As noted above, this reduction in complexity is one of the great advantages of projections. The apparent simplicity is not always real, however, and there are traps lurking for the unwary within these diagrams. These snares come in two broad classes, those reflecting the properties of projections in CMAS, and those reflecting the consequences of using CMAS for objects containing additional components. Here, we consider projection effects for purely CMAS compositions. In section 3.2.5, we consider problems associated with the addition of non-CMAS components.

One potential problem with projections arises when the phase whose composition is being projected from has a reaction relationship with the liquid. More often than not, this information is left off the figure. Consider Fig. 5a, a projection from $MgAl_2O_4$ (spinel) onto the plane formed

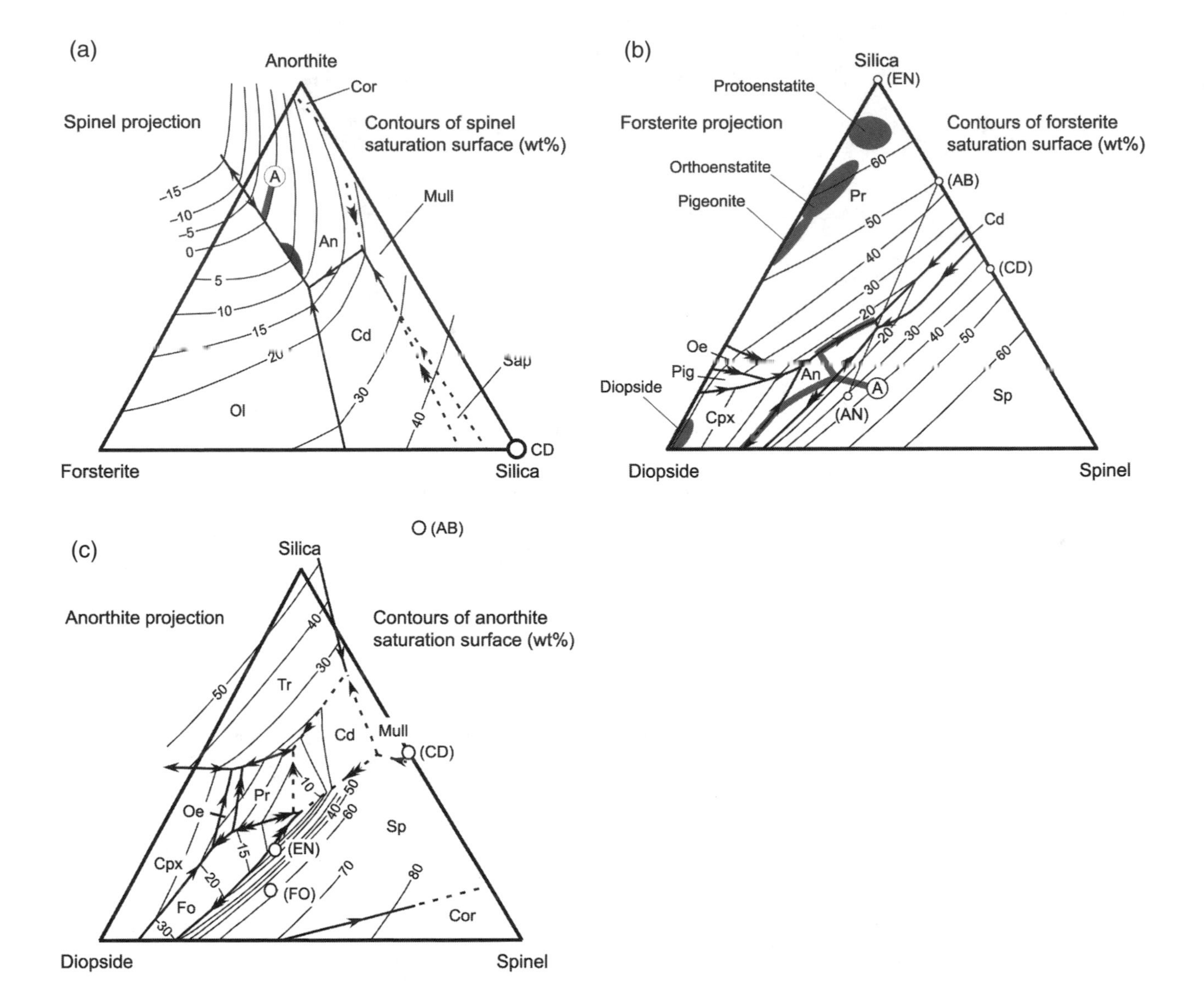

Fig. 5. Projections relevant to the study of Al-rich chondrules after *Sheng et al.* (1991a) and *Sheng* (1992). **(a)** Projection of spinel-saturated liquidus phase fields from the composition of spinel onto the plane defined by the compositions of anorthite, forsterite, and silica. All phase fields are spinel-saturated; the identity of the second crystalline phase is indicated. Contours are in wt% spinel component. The fractional crystallization path for a composition that projects to point A is also shown. The shaded half-ellipse indicates a region on the anorthite + spinel saturation surface where spinel is in a reaction relationship with the liquid. Sap: sapphirine; Cd: cordierite. Abbreviations for other spinel-saturated liquidus phase fields are as in previous figures. CD and the associated open circle refer to the projected position of end member cordierite. **(b)** Projection of olivine-saturated liquidus phase fields from the composition of forsterite onto the plane defined by spinel, silica, and diopside. Note that the panels in this figure appear to have been produced independently and therefore are not consistent with each other in detail. No attempt is made here to map liquid compositions from one panel to the next in order to induce an internal consistency in the phase relations, but the cordierite-spinel-forsterite boundary curve was adjusted to be a reaction curve as it approached the silica-spinel-forsterite join, consistent with behavior expected based on **(a)**. The topology of boundary curves between the anorthite and cordierite fields was also changed from the original diagram to be consistent with expectations based on data of *Longhi* (1987) and *Sheng* (1992) that there is no stable boundary curve in CMAS for forsterite + spinel + protoenstatite + liquid. Oe: orthoenstatite, Pig: pigeonite, and Pr: protoenstatite for spinel-saturated phase fields, and EN (enstatite), AB (albite), and CD (cordierite) for projected positions of end members, which are indicated by open circles. Gray ellipses outline compositions of solids obtained by *Longhi* (1987). Fractional crystallization paths are shown for two bulk compositions, differing only in the presence or absence of Na_2O. The projection is taken from Na_2O so that both bulk compositions plot in the same place, A, in projection. **(c)** Projection of anorthite-saturated liquidus phase fields projected from anorthite onto the plane defined by end-member compositions of spinel, silica, and diopside. Contours are in wt% anorthite component above this plane. Abbreviations include FO for the end member forsterite and Tr to indicate the tridymite-spinel-liquid field, with others as previously defined.

by the compositions of end member forsterite, silica, and anorthite. It shows the compositions of spinel-saturated liquids contoured in wt% spinel component after *Sheng et al.* (1991a) and *Sheng* (1992). Let's ignore the fact that Fig. 5a is a projection for a moment (i.e., pretend that spinel is not present and assume that this is a true ternary with "forsterite," "anorthite," and "cordierite" plotting at the vertices). Ignoring any minor solid-solution effects, we would conclude that "anorthite" was the liquidus phase for point A and that fractional crystallization would drive the composition of the residual liquid directly away from the "anorthite" vertex until it reached the "forsterite" + "anorthite" boundary curve. We could determine the nature of the boundary curve by extending a tangent from the curve to the join formed by the instantaneously crystallizing phase assemblage (i.e. the "anorthite"-"forsterite" join). Since the tangent intersects the join *between* "anorthite" and "forsterite," the composition of the instantaneously crystallizing phase assemblage is described by positive amounts of both "anorthite" and "forsterite"; in our imaginary ternary, this means the boundary curve is a subtraction curve and that "anorthite" and "forsterite" co-crystallize along it. The composition of the residual liquid moves with decreasing temperature, or equivalently, with increasing degree of crystallization along the curve away from the "anorthite"-"forsterite" join until it reaches the "cordierite"-"anorthite"-"forsterite" invariant point. Since the invariant point lies inside the composition triangle defined by "anorthite" + "forsterite" + "cordierite," the liquid composition can be described in terms of positive amounts of these three phases and the liquid would therefore crystallize "anorthite," "forsterite," and "cordierite" in a constant ratio until it was exhausted. Sounds simple. Now, let's consider fractional crystallization for a liquid plotting at point A when Fig. 5a is treated as the spinel projection it actually is. To be specific, let A have 3 wt% spinel component, which would be typical of some Al-rich chondrules studied by Sheng and coworkers (see also *Wark,* 1987). Since point A plots on top of the 0 wt% spinel contour, spinel is the liquidus phase and Fig. 5a is a suitable projection for determining crystallization sequences. If this liquid is cooled from above its liquidus, Fig. 5a predicts that spinel crystallizes first and the liquid composition is driven directly away from spinel until it strikes the saturation surface. While crystallizing spinel alone, the residual liquid continues to plot in projection at point A because it is projected from the composition of the instantaneously crystallizing phase assemblage (i.e., $MgAl_2O_4$ spinel). Once the liquid encounters the spinel-saturated liquidus surface, anorthite becomes a stable phase. At this point, we can ask what happens if anorthite is allowed to crystallize by using the same approach as in the imaginary ternary exercise but taking into account that the projection describes a surface: Take a tangent to the saturation surface in the direction of the join describing compositions of the two solids spinel and anorthite. Since all compositions are projected from spinel, the composition of this instantaneously crystallizing phase assemblage projects to the anorthite vertex. From the contours, we can see qualitatively that the tangent's intersection with this join will occur *between* anorthite, which plots at 0 wt% spinel component, and spinel, which has 100% spinel component. So, anorthite and spinel co-crystallize and the projected liquid composition moves along the saturation surface away (in projection) from anorthite until it strikes the anorthite-forsterite-spinel boundary curve. This is just the sequence expected based on a simple ternary approach ignoring spinel. Now, Fig. 5a shows a single arrow on the boundary curve, which is associated in true ternaries with a subtraction curve. It might therefore be tempting to simply assume that anorthite, olivine, and spinel co-crystallize and that the liquid composition evolves along this curve until it reaches the cordierite-anorthite-forsterite-spinel invariant point, just as expected from a pure ternary approach. We might then conclude that the phase diagram was predicting the presence of cordierite and that no pyroxene of any sort should be expected. Since the inclusions in question generally have little or no cordierite but lots of low-Ca pyroxene (*Sheng et al.,* 1991b), a serious discrepancy would be "discovered" and a speculative discussion of kinetics or relict phases might ensue based on what turns out to be a bad assumption. In general, projections in use for Al-rich objects from meteorites do not indicate where there are reaction relationships involving the projected phase and this includes Fig. 5a. The single arrow on the forsterite-anorthite-spinel boundary curve is saying that neither anorthite nor forsterite is in a reaction relationship with the liquid but it says nothing about whether or not spinel is. The true three-dimensional nature of this boundary curve with respect to spinel can be determined by taking a tangent to the curve, remembering that this is now a space curve, back to the composition plane formed by the three phases anorthite, forsterite, and spinel. This point will plot in projection along the anorthite-forsterite join, but a look at the contours in Fig. 5a shows that the boundary curve slopes down enough to make the tangent intersect this plane outside the triangle formed by the crystallizing phase assemblage, to the spinel-poor side of anorthite-forsterite. This means that the instantaneously crystallizing phase assemblage requires negative amounts of spinel (i.e., spinel is in a reaction relationship with the liquid). For perfect fractional crystallization in which all solids are immediately removed from contact with the liquid, a liquid evolving from point A will leave the spinel-saturation surface once it strikes the anorthite-forsterite-spinel boundary curve, whereupon Fig. 5a is no longer pertinent. Based on Fig. 5a, Al-rich chondrules whose compositions lead them to the anorthite-forsterite-spinel boundary curve will have spinel in a reaction relationship. For such Al-rich chondrules, one should therefore be looking for evidence of resorption or armoring of spinel where present, as is observed in these Al-rich chondrules. Another phase diagram (e.g., Fig. 5b) is needed to help follow later stages of fractional crystallization. The prediction, based on a pure ternary approach to Fig. 5a, that cordierite crystallization should follow olivine but that pyroxene crystallization should not was a mirage.

Finally, we note as an aside that the nature of a boundary curve can usually be determined quantitatively by taking a tangent to the boundary curve, if an equation is given for it, or, much more likely, by writing four mass balance equations relating compositions of the three crystalline phases plus a liquid composition along the boundary curve to the composition of a nearby down-temperature liquid. For CaO, this can be represented by

$$n_{CaO}^{Xl\#1} n_{Xl\#1} + n_{CaO}^{Xl\#2} n_{Xl\#2} +$$
$$n_{CaO}^{Xl\#3} n_{Xl\#3} + n_{CaO}^{Liq\#1} n_{Liq\#1} = n_{CaO}^{Liq\#2}$$

where n_{CaO}^{j} is the number of moles of CaO in phase j, n_j is the number of moles of phase j, and xl#i refers to crystalline phase i. Similar equations can be written for each of the oxides, MgO, Al_2O_3, and SiO_2, and the resulting set of four linear equations in four unknowns solved. One can also use the components of the figure on a molar basis rather than converting to end-member oxides. In either case, the mass balance defines relative amounts of phases produced or consumed. If positive amounts of all the crystalline phases are required to produce the down-temperature liquid, the boundary curve is a subtraction curve. Negative amounts indicate a reaction curve and the signs point the way to which phase(s) has a reaction relationship with the liquid.

3.2.4. Going for a more global view. Projections described in the previous two subsections were designed primarily to portray the phase relations for composition volumes relevant to one or a few groups of CAIs or related objects. If one attempts to plot compositions of CAIs from other groups or objects that might be related to CAIs on these diagrams, they often project to great distances from the composition triangles used to define the projection planes. This makes it difficult to see possible relationships among the different groups. *MacPherson and Huss* (2005) dealt with this problem by projecting spinel-saturated liquidus phase fields from $MgAl_2O_4$ onto the plane defined by Mg_2SiO_4, Al_2O_3, and Ca_2SiO_4 (Fig. 6a). The ternary used in Stolper's projection, in which compositions are also projected from spinel, is outlined within Fig. 6a by joining the points GE (gehlenite), AN (anorthite), and Mg_2SiO_4 (FO; forsterite). Figure 5a (AN-FO-SIL) can also be seen as can a face of the hibonite-anorthite-gehlenite-spinel tetrahedron used by *Beckett and Stolper* (1994).

Figure 6a is after the phase diagram presented by *MacPherson and Huss* (2005), differing only in handling of phase fields plotting toward the base of the triangle, which are not important for the interpretation of phase equilibria relevant to CAIs or Al-rich chondrules. The merwinite (MER; $Ca_3MgSi_2O_8$) - melilite (+ spinel) and merwinite - periclase (+ spinel) boundary curves are reasonably well-determined (*Prince,* 1951; *Osborn et al.,* 1954; *Gutt,* 1964, 1968; *Osborn and Gee,* 1969). Compositions of bredigite

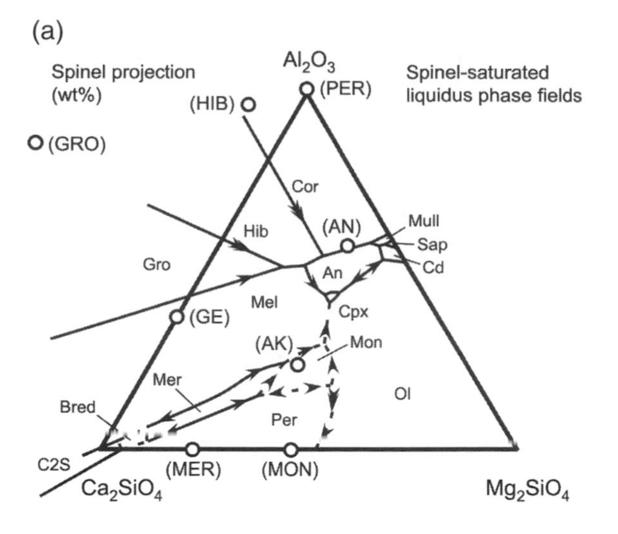

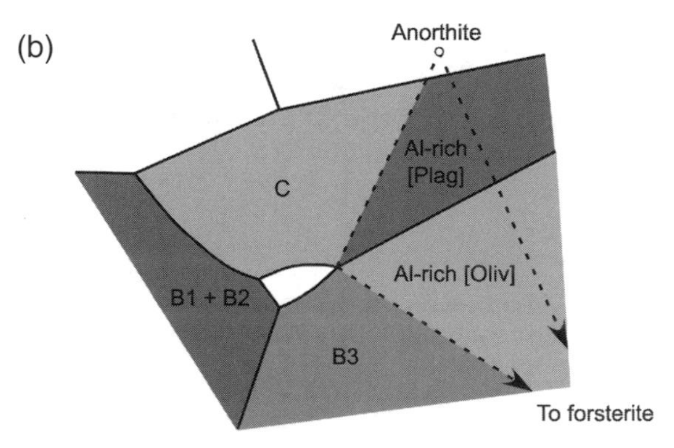

Fig. 6. **(a)** Projection of spinel-saturated liquidus phase fields from $MgAl_2O_4$ onto the plane defined by Mg_2SiO_4, Al_2O_3, and Ca_2SiO_4, modified after *MacPherson and Huss* (2005). In each field, a liquid coexists with spinel and the crystalline phase indicated in or near the field. Projected positions of end-member compositions of crystalline phases are indicated in caps enclosed in parentheses. Bred: bredigite ($Ca_5MgSi_3O_{12}$), C2S: Ca_2SiO_4, and Per: periclase for phases; MER: merwinite ($Ca_3MgSi_2O_8$) and PER: periclase (MgO) for end members. Other abbreviations are as defined in previous figures. See text for data sources. **(b)** A schematic portion of **(a)** centered on the clinopyroxene + spinel + liquid field. Solid curves are boundary curves. Dashed lines are drawn between the end-member compositions of anorthite and forsterite or between an end-member composition and an invariant point.

($Ca_5MgSi_3O_{12}$)-saturated (+ spinel) liquids are not available but a field is shown truncating the merwinite field in Fig. 6a to be consistent with experimental work of *Biggar* (1971); we neglect a likely stable spinel-saturated liquidus field for the compound $Ca_6MgAl_8O_{21}$ (*Biggar,* 1971). Phase relations involving monticellite + spinel are not well constrained, and the positioning of the olivine - melilite (+ spinel) boundary curve is also surprisingly uncertain. Placement of the latter curve in *Stolper* (1982) was based on the intersection of three dashed lines in a figure from *Schairer and Yoder*

(1969). Taking all available experimental data at face value, the periclase-olivine (+ spinel) boundary curve extends past the periclase + monticellite + olivine (+ spinel) invariant point well into the olivine field shown in Fig. 6a but this is inconsistent with data for olivine-, spinel-saturated liquids that would end up on the wrong side of the boundary curve (e.g., *DeVries and Osborn,* 1957; *Gutt,* 1964; *Osborn and Gee,* 1969). We assumed in constructing Fig. 6a that the stable periclase + olivine (+ spinel) boundary curve is truncated by the monticellite field and that extensions of the boundary curve into the olivine field are metastable, perhaps reflecting the metastable persistence of periclase in experimental run products. An alternative extension of the monticellite field toward Mg_2SiO_4-Ca_2SiO_4 seems unlikely given the large collection of experimentally produced spinel + olivine- (no periclase)-saturated liquids in this region and the well-defined position of the periclase + olivine + liquid field in the MAS system, which plots in projection on an extension of Al_2O_3-$MgAl_2O_4$, well below the triangle shown in Fig. 6a. The precise position and extent of the monticellite + spinel + liquid field is highly uncertain, but since igneous monticellite and periclase are not found in CAIs or related objects, this is not a significant problem for the present purposes. Clarification of the phase relations in this portion of the phase diagram would probably require additional experimentation.

Figure 6b, modified after *MacPherson and Huss* (2005), shows a portion of the spinel-saturated phase equilibria displayed in Fig. 6a (see also Fig. 2b) with the regions where bulk compositions of specific types of objects plot indicated by shaded fields. Consider bulk compositions that project to either the Al-rich [Plag] or Al-rich [Oliv] shaded regions; upon fractional crystallization, residual liquids evolve to the olivine-anorthite + spinel boundary curve. Since spinel is in a reaction relationship along this curve, the liquids leave the spinel saturation surface and melilite does not appear during crystallization. These objects will become Al-rich inclusions and not CAIs. This fundamental distinction in phase assemblage wrought by the phase relations is the basis for the *MacPherson and Huss* (2005) classification scheme. The distinction between Al-rich [Plag] and Al-rich [Oliv] is also derived from the phase relations. Bulk compositions of Al-rich [Plag] chondrules project to the spinel + anorthite + liquid saturation surface. Anorthite will precede olivine in the crystallization sequence for these objects. Bulk compositions of Al-rich [Oliv] chondrules project to the spinel + olivine + liquid saturation surface. Olivine precedes anorthite in the crystallization sequence. These behaviors contrast fundamentally with those expected for the shaded regions shown for type B1/B2, B3, and C inclusions. In each, fractional crystallization leads residual liquid compositions to the melilite stability field so that melilite is generally expected to be part of the phase assemblage for these objects. The first silicate, or in some cases, first crystallizing phase will be melilite in the B1 + B2 field, olivine for B3, and anorthite for C. Thus, the classification of many CAIs and Al-rich chondrules is a direct reflection of the phase relations, both in terms of the presence or absence of phases and in terms of order of crystallization.

A major advantage of large-scale phase diagrams like Fig. 6a is that suites of objects ranging from the most aluminous CAIs to the most magnesian of amoeboid olivine aggregates (AOAs), Al-rich chondrules, and even ferromagnesian chondrules can be plotted on the same diagram (*MacPherson and Huss,* 2000, 2005; *Krot et al.,* 2001; *MacPherson et al.,* 2004) (see also Fig. 1). The principle disadvantages of the MacPherson/Huss diagram are consequences of the same features that make the diagram so useful for seeing how broad variations in composition play out for multiple types of objects. Although it is possible to plot virtually any bulk CAI composition on Fig. 6a, the phase relations refer only to spinel-saturated liquids. While spinel is the liquidus phase for many CAIs, melilite is on the liquidus for type As, grossite or hibonite for most highly aluminous CAIs, olivine for some B3s, and olivine or plagioclase for many if not most Al-rich chondrules. For most inclusions, it should be possible to use a version of the *MacPherson and Huss* (2005) phase diagram, in which the spinel saturation surface has been contoured in terms of wt% $MgAl_2O_4$ component, to determine if the phase relations are applicable, but smaller-scale diagrams (e.g., sections 3.2.1–3.2.3) may be necessary.

3.2.5. When to get nervous: Nonquaternary components. All CAIs and all Al-rich chondrules contain non-CMAS components and, if too much of one or more is present during crystallization, the phase relations depicted in a CMAS projection may be so poorly connected to the actual system defined by the object as to be useless or, even worse, misleading. The bounds beyond which the use of CMAS becomes problematic have not been studied rigorously and we provide only general guidance here. The problems associated with extra components fall into two basic categories:

1. CMAS-based phase relations are adequate, at least for determining crystallization sequences, but incorporation of non-CMAS components into a solid solution affects how one interprets the diagram.

One problem with non-CMAS components is that, if they are soluble in a crystalline phase, thermal divides involving that phase will shift even if the boundary curves and saturation surfaces hardly budge, and this can lead to fundamental changes in crystallization sequences. *MacPherson and Huss* (2005) give an example of this for an Al-rich chondrule, 4128-3-2 from Semarkona (Plate 11g), and point A in Fig. 5b shows a similar composition. Let's presume first that point A is in CMAS and plots above the olivine saturation surface. Fractional crystallization of the bulk will drive the composition of residual liquids away from the composition of forsterite until the forsterite-spinel saturation surface is intersected, the compositions of the residual liquids all projecting to point A, neglecting small amounts of Ca_2SiO_4 in the olivine. Spinel now joins the crystallizing phase assemblage and the residual liquid moves away in projection from the spinel vertex until the boundary curve is encountered. Spinel is in a reaction relationship with the liquid along this boundary curve, so the residual liquid strikes out across the anorthite-forsterite saturation surface directly away from the composition of end member anorthite until it hits the forsterite + anorthite + protoenstatite boundary curve, which it follows to the forsterite + anorthite + protoenstatite + orthoenstatite invariant point where protoenstatite is in a reaction relationship and the residual liquid leaves the saturation surface. So, the CMAS phase diagram makes a clear prediction that the phase assemblage produced from composition A in Fig. 5b will have protoenstatite but little or no calcic pyroxene. The problem for *MacPherson and Huss* (2005) was that Al-rich chondrule 4128-3-2 had lots of calcic pyroxene and no low-Ca pyroxene. Now, MacPherson and Huss could have danced around with *ad hoc* errors in the bulk composition, suggested that something odd happened to the chondrule, or, perhaps, argued that Na, which was present during melting based on compositions of igneous plagioclase, dramatically stabilized diopside relative to protoenstatite. *MacPherson and Huss* (2005), however, pointed to a simple alternative. They noted that plagioclase in their chondrule is somewhat sodic (~An_{80}). Such a feldspar composition will not project to the composition of anorthite in Fig. 5b, instead plotting along the anorthite-albite join. How does this affect the predicted crystallization sequence? If enough Na is present, then the CMAS phase diagram is simply inapplicable, but let's assume that the amount of Na is insufficient to substantially affect the positions of boundary curves as given by CMAS phase equilibria. Residual liquids produced by fractional crystallization of A will follow the same path in projection as they would for CMAS down to the anorthite + forsterite + spinel boundary curve provided they are projected from Na_2O. Once plagioclase begins to crystallize, however, the residual liquid is driven away from the composition of the Na-bearing feldspar, whose composition lies further up the anorthite-albite join than does end member anorthite. The residual liquid is still driven across the plagioclase + forsterite saturation surface but the direction is altered relative to expectations based on CMAS and it is the diopside + plagioclase + forsterite boundary curve that is intersected, not protoenstatite + plagioclase + forsterite (Fig. 5b). Moreover, the thermal divide along this curve, which is the intersection of the plane defined by the compositions of coexisting forsterite, plagioclase, and diopside with the boundary curve, migrates toward albite as the plagioclase incorporates Na. Having reached this boundary curve, further crystallization drives residual liquids away from the low-Ca pyroxene fields, consistent with the observed phase assemblage in the chondrule.

Unexpected projection effects can also arise when projecting liquid compositions from the composition of a crystalline phase that contains significant amounts of a non-CMAS component because ignoring an oxide in a bulk composition is equivalent to projecting from that oxide. Suppose, for example, in Fig. 5b, that liquid A is in CMAS and above the olivine saturation surface. Ignoring minor Ca_2SiO_4, the bulk composition and residual liquid compositions after fractional or equilibrium crystallization of olivine plot in the same place (i.e., at A) until spinel joins the crystallization sequence. If the initial liquid composition contains significant amounts of FeO, however, olivine will incorporate some of this as fayalite (Fe_2SiO_4) during crystallization. Since fayalite plots at the silica vertex when projected from Mg_2SiO_4 (forsterite) and FeO, crystallization of the Fe-bearing olivine drives the residual liquid composition in projection directly away from the silica vertex. How far depends on how much olivine crystallization is involved and how Fe-rich the olivine, but it is not difficult to conceive of circumstances under which predicted crystallization sequences would be affected. Similar effects can arise during crystallization of Ti-bearing hibonite (Fig. 4), residual liquid compositions moving away from MTH, and of Na-bearing plagioclase in Fig. 5c, residual liquids moving away from the composition of plagioclase, which plots along the join AN-AB.

2. If too much of a non-CMAS oxide is added to a composition, finessing the phase relations as described in the preceding paragraphs fail, either because the relative stabilities of solid phases change, leading to shifts in saturation surfaces and boundary curves, or the topology of the phase diagram changes because a new CMAS or non-CMAS phase is stabilized or an old one disappears. Even when a projection looks much the same after the addition of some non-CMAS oxide, temperatures on the saturation surfaces may change significantly. Here, we supply a few semi-quantitative constraints with respect to the oxides of Na, Ti, and Fe. We ignore the effects of other oxides, sulfide-rich liquids, and the metallic phase, the first because they are generally present in very low concentrations, and the latter two because they are immiscible with negligible solubility in the coexisting silicate liquids under the reducing conditions of interest here.

Small amounts (<1 wt%) of Na_2O have little effect on melilite + spinel equilibria according to *Beckett and Stolper* (2000), and *Sheng* (1992) showed that 4 wt% Na_2O had a only modest effect on the anorthite-, spinel-, forsterite-saturated phase boundary shown in Fig. 5b. On the other hand, *Kushiro* (1975) noted that the stability field of forsterite + liquid in the MgO-SiO_2 system expands relative to that of enstatite + liquid as Na is added. Also, *Longhi and Pan* (1988), *Pan and Longhi* (1990), and *Sheng* (1992) found that increasing Na generally decreased the liquidus temperature relative to expectations based on CMAS, suppressed plagioclase crystallization, expanded the spinel + forsterite field relative to forsterite + cordierite, and reduced the spinel + forsterite field relative to those of diopside + forsterite and melilite + forsterite, though usually several mol% Na_2O was required to make the changes obvious.

Changes in the relative stabilities of crystalline phases in liquids can be quantified for liquids relevant to CAIs and Al-rich chondrules through a consideration of relative affinities (e.g., *Prigogine and Defay,* 1954). When a liquid is in equilibrium with a solid phase, the difference in chemical potential relative to a common standard state, or affinity, is zero. For example, for olivine coexisting with liquid, we can define the affinity as

$$A_{fo-liq} = \mu^{ol}_{Mg_2SiO_4} - 2\mu^{liq}_{MgO} - \mu^{liq}_{SiO_2}$$

where μ^j_i is the chemical potential of component i in phase j. Similar expressions can be written for other crystalline phases. Now, suppose in some experiment that two crystalline phases are in equilibrium with the same Na-bearing liquid. The affinity for each crystalline phase relative to the liquid is zero since it is in equilibrium with the Na-bearing liquid. If the chemical potentials of components in the liquid are calculated based on CMAS, the affinities will in general be different from zero and, more to the point, different from each other. These differences are a measure of the relative stabilization or destabilization of the crystalline phases as Na is added to the system. Figure 7 shows the affinity for forsterite relative to other phases calculated for Na-bearing liquids using the thermodynamic model of *Berman* (1983), which is described in section 3.3.1. Each point was derived either from a glass analysis taken from an experimental run product or through digitization of a phase diagram. A positive value indicates that forsterite is stabilized relative to the second crystalline phase; a negative value implies relative destabilization of forsterite. Based on reconnaissance calculations using *Berman*'s (1983) model, an affinity difference exceeding several kJ/mole generates shifts in saturation surfaces severe enough to render suspect a graphical determination of crystallization paths based on CMAS figures. From Fig. 7, forsterite is stabilized relative to spinel and protoenstatite and destabilized relative to diopside. At constant $X^{liq}_{Al_2O_3}/X^{liq}_{SiO_2}$ in the liquid, the effect of additional Na on the relative stabilities is muted over some range in Na but, with further additions of Na, the stability of one phase is strongly enhanced over the other. In general, values of $X^{liq}_{Na_2O}$, the mole fraction of Na_2O in the liquid, beyond which large changes in relative affinity occur, increase with increasing Al/Si in the melt. The relevant ranges in $X^{liq}_{Al_2O_3}/X^{liq}_{SiO_2}$ are ~0.25–0.40 for type B and C CAIs and ~0.15–0.30 for Fe-poor, Al-rich chondrules, leading to strong differences in affinity for $X^{liq}_{Na_2O}$ above ~0.05 mol% Na_2O except for spinel + forsterite in low Al/Si liquids relevant to some Al-rich chondrules, for which limiting the use of CMAS projections to liquids with $X^{liq}_{Na_2O} < 0.03$ is a safer bet. Similar diagrams for diopside-anorthite, forsterite-anorthite, and spinel-anorthite based on data of *Biggar* (1984), *Bowen* (1915), *Libourel* (1999), *Pan and Longhi* (1990), *Schairer and Morimoto* (1959), *Schairer and Yoder* (1960, 1961), *Schairer et al.* (1967), and *Soulard et al.* (1992) are consistent with the idea that forsterite and diopside are stabilized relative to plagioclase as Na is added to CMAS liquids independent of Al/Si, both relative affinities being approximately proportional to $(X^{liq}_{Na_2O})^2$ at least for $X^{liq}_{Na_2O} < \sim 0.08$.

We noted above that the addition of several wt% TiO_2 to liquids within some of the subsystems of CMAS led to variably reduced appearance temperatures for crystalline phases but that crystallization sequences were the same as in the

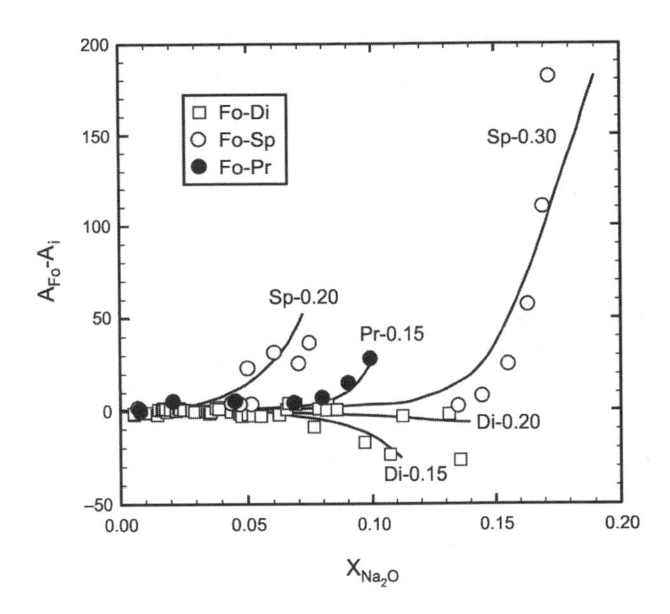

Fig. 7. The effect of Na_2O addition to CMAS liquids on the relative stabilities of liquidus phases. Each point represents a CMAS + Na_2O liquid in equilibrium with two or more crystalline phases but with affinities calculated ignoring Na (i.e., liquid compositions normalized to CMAS). Affinities of diopside, protoenstatite, and spinel relative to the affinity of forsterite are shown. Values were calculated using the model of *Berman* (1983) for CMAS liquids and end-member crystalline phases. The curves are labeled by phase (Di: diopside; Pr: protoenstatite, Sp: spinel) and molar Al_2O_3/SiO_2 in the liquid. The curves are visual aids only but are consistent with the data in that liquids with higher Al_2O_3/SiO_2 than the curves (±0.01) plot at higher values of Na_2O. Data are from *Biggar* (1984), *Bowen* (1915), *Libourel* (1999), *Onuma and Yagi* (1967), *Pan and Longhi* (1989, 1990), *Schairer* (1957), *Schairer and Morimoto* (1959), *Schairer and Yoder* (1960, 1961), *Schairer et al.* (1967), *Soulard et al.* (1992), and *Yagi and Onuma* (1969).

Ti-free system to a good approximation provided perovskite or some other Ti-rich phase was not stabilized. *Stolper* (1982) found for compositions relevant to type B inclusions, and *Beckett and Stolper* (1994) for hibonite-rich inclusions, that the addition of a few wt% TiO_2 had no significant effect on predicted crystallization sequences. *Longhi and Pan* (1988) were also unable to discern a TiO_2 effect in compositionally more complex melts with <5 wt% TiO_2 but they found significant differences for liquids with more than ~5 wt% TiO_2. Thus, the danger point for using CMAS phase relations to interpret basic crystallization sequences for CMAST liquids appears to be several wt% TiO_2 for most liquids and, given the low bulk concentrations for this oxide in most CAIs and Al-rich chondrules, TiO_2 generally becomes problematic only for late-stage crystallization.

In CAIs, FeO contents during melting were negligible but the same is not true for many Al-rich chondrules, and here the issue of how much to trust a CMAS analysis must be confronted. While the topology of CMAS phase diagrams relevant to Al-rich chondrules appear to be largely unaffected except at fairly high concentrations of FeO (*Longhi and Pan,* 1988; *Shi and Libourel,* 1991), there are substantial shifts in the positions of phase boundaries even with modest additions. *Sheng* (1992) (see also *MacPherson and Huss,* 2005) showed that the addition of 6 wt% FeO shifted the forsterite + anorthite + spinel and forsterite + anorthite + protoenstatite boundary curves shown in Fig. 5b substantially toward the silica vertex, largely a consequence of destabilization of protoenstatite. This led *MacPherson and Huss* (2005) to make use of a set of diagrams produced by *Longhi and Pan* (1988) and calculations using the program MELTS (*Ghiorso and Sack,* 1995) to constrain crystallization paths in Al-rich inclusions whenever FeO contents rose above ~3 wt%. *Longhi and Pan* (1988) documented shifts in the spinel + forsterite field and for olivine saturated liquidus fields with protopyroxene, orthopyroxene, pigeonite, and clinopyroxene with changing FeO contents in the liquid or, equivalently, in the olivine. At some point a judgment call needs to be made, but it seems unlikely that usage of CMAS phase diagrams is justified for determining crystallization sequences of FeO-bearing liquid compositions when there is more than 2–3 wt% FeO in the liquid.

3.3. Calculation of Phase Relations in the System CaO-MgO-Al_2O_3-SiO_2 Using Thermodynamic Models

In the previous section, we considered phase relations pertinent to CAIs and related objects based on graphical representations of experimentally determined phase equilibria. This approach is very useful if bulk compositions can be described in terms of quaternary or lower variance systems and if there is enough data in the composition region of interest to allow confident construction of an appropriate phase diagram. If the database is weak, locations of phase boundaries may become little more than intuitive guesses. If more than four components are required, projections carefully tailored to minimize distortions caused by additional components, multiple projections, or empirical fits to boundary curves are usually needed to capture the essence of the phase equilibria (*Walker et al.,* 1979; *Longhi and Pan,* 1988; *Bartels and Grove,* 1991; *Shi,* 1992). An alternative is to determine the thermodynamic properties of all phases in a system and calculate stable phase assemblages through the thermodynamic requirement that a stable phase assemblage have the minimum free energy. In principle, such an approach can be applied to systems with any number of components. Suppose, for example, that we know the composition of a CMAS liquid in equilibrium with anorthite at some temperature and that the thermodynamic model for the liquid uses end-member solid oxides as components. From the reaction $CaO_{(liq)} + Al_2O_{3(liq)} + 2SiO_{2(liq)} = CaAl_2Si_2O_8$(plagioclase), the free energy of reaction can be expressed as

$$G_{\rightarrow} = 0 = G^{\circ}_{\rightarrow} + RT\ln\left(\frac{a^{plagioclase}_{CaAl_2Si_2O_8}}{a^{liq}_{CaO}a^{liq}_{Al_2O_3}\left(a^{liq}_{SiO_2}\right)^2}\right) \qquad (1)$$

where $G^{\circ}_{\rightarrow}$ refers to the standard state free energy of reaction (all components in their standard state), a^j_i to the activity of component i in phase j, R to the gas constant, and T to temperature in degrees Kelvin. If a different set of end-member components is used to describe the liquid composition, a different reaction will result but the number of constraints arising from a given crystal-liquid pair remains the same. If a crystalline phase is a solid solution, additional equations analogous to equation (1) can be written for each component in the solid and if a liquid is multiply saturated, each phase generates one or more equations similar to equation (1).

Making use of equation (1) requires an expression for $G^{\circ}_{\rightarrow}$ and each of the activities (*Anderson and Crerar,* 1993). The usual approach is to define standard-state free energies of formation and activity-composition relationships for the solid phases primarily from solid state thermodynamic data, while phase equilibria and constraints of the type represented by equation (1) are used to determine the thermodynamic mixing properties of the liquid (*Berman,* 1983; *Ghiorso and Sack,* 1995). The basic idea is to then fit expressions for the activity of each component in the liquid as functions of liquid composition and temperature through differentiation of an equation for the excess free energy, a collection of equations analogous to equation (1), and information concerning the thermodynamic properties of the solids, using multiple regression or linear programming (e.g., *Berman and Brown,* 1984) techniques. Once an equation of state for the liquid is defined in this manner, equilibrium phase assemblages can be obtained through free energy minimization for a bulk composition (e.g., *Ghiorso,* 1994) and, therefore, a phase diagram can be constructed even in regions of composition space for which there are no experimental data. Moreover, values for activities of components in the liquid can be specified for any liquid composition at any temperature.

An internally consistent model for silicate liquids will always give an answer to the question of what the activity of a component in the liquid is according to the model, but there are a number of reasons beyond quality of input data for why such predictions may be inaccurate. For example, statements of equilibrium within a given liquidus phase field constrain only relative values of activities for certain components in the liquid. For the anorthite example given above, there is one constraint from the presence of anorthite (equation (1)) and one from the Gibbs-Duhem equation but, with four components, specific values are not determined for any component unless $CaAl_2Si_2O_8$ happens to be selected as one of them. In general, large systematic errors can be accommodated while obeying all available phase equilibria constraints. *Chamberlin et al.* (1994), for example, found experimentally determined MgO and Al_2O_3 activities in CAI-like liquids, most of them spinel-saturated, to be systematically lower than predicted by *Berman*'s (1983) model and silica activities systematically higher. A second difficulty with using solution models to predict the thermodynamic properties of silicate liquids is that most of the constraints on liquid properties are based on liquidus data, which is not very good for separating out contributions to the free energy of reaction due to differences in temperature from those due to differences in liquid composition. Thus, the quality of activity predictions can be expected to decline away from the liquidus, perhaps precipitously. Condensation and/or volatilization calculations, which involve equilibrium between a vapor and a liquid, are often performed for conditions in which no crystalline phases are present (i.e., above the liquidus). While it is hoped that a combination of data for multiple phases in various portions of the system used to calibrate the model adequately constrains activities in all part of the system at all temperatures of interest, there is no guarantee that this is a sound assumption. In the following two subsections, we consider some specific models for the thermodynamic properties of CMAS liquids.

3.3.1. Berman's model. The most frequently used model for CAIs was developed by *Berman* (1983). It has been used to investigate the stability of silicate liquids in nebular environments (*Yoneda and Grossman,* 1995; *Ebel and Grossman,* 2000); establish activities of major-element oxides in trace-element partitioning reactions (*Beckett et al.,* 1990, 2000; *Beckett and Stolper,* 1994, 2000); constrain the thermodynamic properties of UNK, an unnamed Ti-bearing phase (*Paque et al.,* 1994); and predict the thermodynamic contributions to diffusive properties and volatilization behavior of CAI-like liquids (*Richter et al.,* 2002). The principle advantages of Berman's model are that it was calibrated for the entire CMAS system using a considerable amount of phase-equilibrium data for quaternary liquids, and it is generally available. Disadvantages, in addition to general problems discussed in the previous paragraph, include the fact that solid solutions are not explicitly modeled; the solution model for the liquid is based on end-member oxides, which means that the entropy of mixing is very poorly represented; and, as noted by *Hashimoto* (1991), the thermodynamic properties of the solids were allowed to vary in a way that almost certainly led to portions of the free energy function for the liquid being distributed among the various solids. Also, according to *Barron* (1985), *Berman and Brown*'s (1984) model for CAS liquids is inconsistent with liquid immiscibility constraints (i.e., the model for the liquid is flawed), and this criticism can also be leveled at *Berman*'s (1983) model for CMAS. In spite of its faults, Berman's model remains a useful tool.

3.3.2. Other models. Although *Berman*'s (1983) model for CMAS liquids has dominated usage in the CAI literature, perhaps largely through inertia, alternative formulations are becoming available. The IRSID model of *Gaye and Welfringer* (1984) uses a cell model to describe interactions within the liquid. They calculated phase diagrams for CAS and the 30 wt% Al_2O_3 plane of CMAS, and showed good agreement with experimentally determined silica activities in CAS. *Ban-ya* (1993) used a regular solution model with oxide end members for multicomponent liquids in describing alloy-slag equilibria. In principle, the model can be used for CMAS liquids, but agreement between predicted and experimentally determined (*Rein and Chipman,* 1963, 1965; *Chamberlin et al.,* 1994; *Liang et al.,* 1997) activities in CAS and CMAS is generally poor. *Björkvall et al.* (2001) used an ionic model based on binary interactions but, as with *Ban-ya*'s (1993) model, agreement with experimentally determined activities of oxide components is not very good.

Most efforts to model the thermodynamic properties of CMAS liquids involve the selection of a set of end members for the liquid. *Hallstedt et al.* (1994) chose an ionic model and were apparently able to reproduce phase equilibria in CAS but at the cost of using a large number of arbitrary ternary parameters, suggesting that simple ionic models are not viable over large portions of CMAS. The proprietary FactSage package (*Bale et al.,* 2002), developed by Pelton and coworkers, uses a quasichemical model for the liquid (e.g., *Pelton and Blander,* 1984; *Pelton et al.,* 2000). While a full optimization is available for CAS (also CAS + FeO_x), other ternaries in CMAS have either not been evaluated at all or are only partially explored. We regard FactSage as promising but not yet applicable to most CAIs or Al-rich chondrules. *Hashimoto* (1991) also used a quasichemical model in a preliminary report on modeling activities in CMAS and achieved fairly good agreement with silica and CaO activities determined by *Rein and Chipman* (1965). MTDATA (*Davies et al.,* 2002) is a program with associated databases developed at the National Physical Laboratory in the United Kingdom for calculating phase equilibria in a wide variety of systems. There have been some forays into using MTDATA for modeling CMAS liquids using various molecules as end members (*Davies et al.,* 1994; *Blundy et al.,* 1996), but the quality and range of applicability is difficult to gauge based on published results.

3.4. Trace-Element Partitioning

Trace-element systematics are useful recorders of igneous and metamorphic processes. In CAIs and Al-rich inclusions, trace-element distributions within a crystal may reflect simple closed-system fractional crystallization from an initially homogeneous droplet, but the system may also be characterized by significant exchange/volatilization. Complications that need to be considered in applying partition coefficients to CAI data include the possible influence of rapid crystal growth (*Kennedy et al.,* 1997), sector zoning (*Hutcheon et al.,* 1978; *Steele et al.,* 1997), crystallization into a liquid of heterogeneous composition due to volatilization (e.g., *Richter et al.,* 2002) or incomplete dissolution (*Kennedy et al.,* 1997), subsolidus reequilibration with another primary or secondary phase and/or diffusive homogenization (*Meeker et al.,* 1983; *Connolly et al.,* 2003), and multiple heating and secondary alteration events (*MacPherson and Davis,* 1993; *Beckett et al.,* 2000). This may seem a depressing litany of potential complications, but the complications also contain information concerning multiple processes. The study of trace-element distributions within CAIs continues to produce new insights into CAI evolution based on our understanding of the partitioning behavior of trace elements between crystals and CAI-like liquids in controlled conditions.

In this section, we consider crystal/liquid partitioning mostly for trace elements and we emphasize experimental data obtained for essentially CMAS or CMAST liquids containing less than 1 wt% Na_2O. We also ignore data produced through optical beta autoradiography (*Hoover,* 1978) as there are substantial problems with the technique (*Tingle,* 1987; *Beattie,* 1993). General compilations/reviews of experimentally determined partition coefficients include *Irving* (1978), *Green* (1994), *Jones* (1995), and *Wood and Blundy* (2004).

Equilibrium partitioning of trace elements between crystals and liquid can be described in terms of exchange equilibria. If, for example, a trace amount of La substitutes into the M1 site of melilite (mel), then at equilibrium, a reaction $LaO_{3/2(liq)} + Ca_2Al_2SiO_{7(mel)} + MgO_{(liq)} = CaO_{(liq)} + \frac{1}{2}Al_2O_{3(liq)} + CaLaMgAlSiO_{7(mel)}$ can be written using oxide components for the liquid and gehlenite, a usually major component in melilite. This has a free energy of reaction

$$G_{\rightarrow} = 0 = G^{\circ}_{\rightarrow} + RT\ln\left(\frac{a^{liq}_{CaO}\left(a^{liq}_{Al_2O_3}\right)^{1/2} a^{mel}_{CaLaMgAlSiO_7}}{a^{liq}_{MgO} a^{liq}_{LaO_{3/2}} a^{mel}_{Ca_2Al_2SiO_7}}\right)$$

which can be rearranged to isolate the partition coefficient

$$D^{mel/liq}_{La} = \frac{W^{mel}_{La}}{W^{liq}_{La}} =$$

$$\left[-\exp\left(\frac{G^{\circ}_{\rightarrow}}{RT}\right)\right]\left[\left(\frac{a^{liq}_{MgO}}{a^{liq}_{CaO}\left(a^{liq}_{Al_2O_3}\right)^{1/2}}\right)\left(\gamma^{liq}_{LaO_{3/2}}\right)\right] \quad (2)$$

$$\left[\left(a^{mel}_{CaAl_2SiO_7}\right)\left(\frac{1}{\gamma^{mel}_{CaLaMgAlSiO_7}}\right)\right]\left(\frac{1}{\Phi_{La}}\right)$$

where W^j_i and X^j_i are the wt% and mole fraction of i in phase j, $\gamma^j_i = a^j_i/X^j_i$ is the activity coefficient of component i in phase j. Φ_{La} is a conversion factor to transform the mole ratio to a weight ratio. Equation (2) illustrates the variables that influence the value of a partition coefficient. The right-hand side of equation (2) contains three bracketed terms. The first one has variables whose values depend on the selected standard state and temperature but not composition. Standard state free energies of formation and oxide fusion data required for $G^{\circ}_{\rightarrow}$ are usually poorly constrained, so this term tends to be folded in with other multiplicative constants in equation (2) and indirectly determined through experimentation. The second bracketed term contains activities of major components in the liquid (CaO, MgO, Al_2O_3) and an activity coefficient associated with trace La in the liquid. All these variables depend on the temperature and major-element composition of the melt. In principle, $\gamma^{liq}_{LaO_{3/2}}$ also depends on the concentration of La in the liquid but if, as is commonly assumed, trace elements in the liquid obey Henry's law, then $\gamma^{liq}_{LaO_{3/2}}$ is constant for any specific temperature and major-element chemistry of the melt. The fact that trace elements often strongly prefer one of two immiscible liquids (*Watson,* 1976; *Ryerson and Hess,* 1978) implies a significant liquid composition effect if the range in composition is sufficiently large. Recent work emphasizing the role of liquid composition on partitioning includes *Kohn and Schofield* (1994), *Ertel et al.* (1997), and *O'Neill and Eggins* (2002). The third bracketed term contains activities for gehlenite and a trace La component in melilite. Values for these depend on the major-element chemistry of the melilite and the specific local environment for La in the melilite. The conversion factor Φ_{La} is a weak function of the major-element composition of both melilite and the liquid.

Equation (2) summarizes the basic problem facing all applications of trace-element partitioning. There may in principle be significant contributions from the compositions of both phases and from temperature. While it can be argued (e.g., *O'Neill and Eggins,* 2002) that the effect of temperature on equilibrium trace-element partitioning is a secondary factor for most magmatic systems, crystallization can occur over ranges of hundreds of degrees in CAIs and Al-rich chondrules, making a complete dismissal of temperature effects rather dangerous. Also, the relative effects of solid and liquid compositions upon partition coefficients remain controversial. Fortunately, many of the variables in equation (2) cancel out for ratios of partition coefficients for elements of the same valence that substitute into the same site (e.g., for La in melilite, this would mean other trivalent cations substituting into M1), as noted by *Drake and Weill* (1975).

The crystal-oriented perspective for computing trace-element partition coefficients dates to observations of *Brixner* (1967) and *Onuma et al.* (1968) that the partition coefficients of similar trace elements substituting into a given site are roughly parabolic functions of ionic radius. *Brice* (1975) found that a simple isotropic lattice-strain accompanying the substitution of an element into a site that it does not quite fit could account for the differences in partition coefficients observed by Brixner. *Nash and Crecraft* (1985) introduced Brice's theory to the geochemical literature, and Blundy and coworkers in a series of papers (*Blundy and Wood,* 1991, 1994, 2003; *Hill et al.,* 2000; *Wood and Blundy,* 2001, 2004) have extended and greatly popularized it.

Table 1 presents a compilation of ranges for experimentally determined values of trace- and minor-element partition coefficients for phases pertinent to CAIs and Al-rich chondrules in near-CMAS or CMAST liquids. Major-element partition coefficients are given, where supplied by the authors, but no attempt was made to calculate values from compositions of coexisting phases or to extract them from phase diagrams. The table is intended to give the reader a gross sense for what information is available, not what should be used in a particular application. In particular, it should be kept in mind that many of these elements occur in more than one valence state, leading to sensitivity of the partition coefficients to redox conditions. Specific values for selected elements for different crystalline phases are discussed below.

3.4.1. Oxides. To our knowledge, experimentally determined trace-element partition coefficients are not available for $CaAl_2O_4$ or grossite while for corundum, there is but one datum, a preliminary value of 0.12 for Sm in a CAS

TABLE 1. Ranges of experimentally determined partition coefficients.

Element	Anorthite	Clinopyroxene	Corundum	Hibonite	Melilite	Olivine	Perovskite	Protoenstatite	Spinel
Al	1.5–2.0	0.11–0.39		2.2		0.003–0.02	0.07–0.11		
Ba	0.06–0.26	0.003–0.02		0.03	0.025–0.050		0.03		
Be	0.90				0.5–1.8				
Ca	0.7–1.4	0.9–1.2		0.3		0.02–0.05	1.2	0.03	
Co	0.02	0.4–1.4							
Cr	0.016	2–34				0.3–0.8	0.63	1–4.5	
Cs									
Cu						0.27–0.47			
Fe	0.09	0.8–2.3							
Ga	0.8–1.0	0.19–0.24				0.012–0.018			
Ge	0.47–0.56	1.0–4.9		0.74–0.81		0.06–0.8	2.1		0.11
Hf		0.4–6.3		0.67–0.78			5		
In		1.3–2.8							
Li	0.23	0.03–0.21							
Mg	0.001–0.11	1.1		0.05		1.8–6.3	0.01–0.03	2.15	
Mn		0.68–0.80				0.5–0.7			
Mo		0.001–0.005							
Na	0.17–0.57	0.05–3			0.4–2.0				
Nb		0.006–0.07		0.18–0.35			2.7		
Ni		1.9–3.1				4–11		2.1	
Pd									<0.02
Rh		0.23–0.27							78–90
Ru		1.0–4.3							22–29
Sb		0.08–1.7							
Sc	0.012	0.35–29		0.31–0.50	0.01–0.07		0.16		0.05
Si	0.85			0.003–0.06			0.01		0.0006–0.002
Sn		2–29							
Sr	0.9–1.2	0.05–0.08		0.52–0.64	0.65–1.1		0.7–1.1		
Ta		0.02–0.41							
Th		0.01		0.7–1.2			16		
Ti	0.03–0.04	0.1–1.4		0.8–2.1	0.02		2.4		0.04–2.0
U	0.01	0.003–0.009		0.07–0.09			4.7		
V		0.01–0.15					1		~0–2.2
W		0.0003–0.005							
Y	0.003–0.03	0.6–0.9			0.015–0.09		1.1–1.9		
Zr	0.0003–0.0005	0.2–2.4		0.33–0.36	0.001–0.002		0.16–0.58		
La	0.02–0.17	0.07–0.16		4.9–7.2	0.04–0.48		2.6–15		0.01
Ce	0.07–0.10	0.08–0.27		3.9–5.2	0.06–0.21		11.6		
Pr	0.09	0.26–0.41		3.3–4.4			12		
Nd	0.08–0.10	0.18–0.51		2.8–3.5			13.4		
Sm	0.007–0.08	0.05–0.92	0.12	1.6–3.1	0.032–0.75		2.7–17		0.006
Eu	0.05–0.10	0.2–1.0		1.2–1.6	0.6–1.3		2.0–7.6		0.006
Gd	0.05	0.2–1.1		0.9–1.6	1.2		2.6–4.4		
Tb		1.04		0.52–0.76	0.49		1.6–3.6		0.008
Dy	0.02–0.03	0.23–1.07		0.35–0.41			3.6		
Ho		0.77–0.94		0.25–0.27			2.5		
Er	0.01–0.04	0.22–0.76		0.17–0.26			1.6		
Tm	0.03	0.7–1.1		0.10–0.13	0.03–0.14		1.2		
Yb	0.002	0.2–1.0		0.03–0.1	0.009–0.25		0.5–17		0.008
Lu	0.02–0.42	0.26–0.83		0.02–0.09	0.24		0.4–0.7		0.021

In addition to references cited in *Irving* (1978), *Green* (1994), *Jones* (1995), and *Wood and Blundy* (2004), data were drawn from *Beckett and Stolper* (1994, 2000), *Connolly and Burnett* (2003), *Davis et al.* (1996), *Drake* (1981), *Hanson and Jones* (1998), *Kennedy et al.* (1994), *Libourel* (1999), *Miller et al.* (2003), *Pack and Palme* (2003), *Peters et al.* (1995), *Schreiber* (1979), *Simon et al.* (1994a, 1996b), *Soulard et al.* (1992), *Steele et al.* (1997), *Terakado and Masuda* (1979), *Watson* (1977), and *Woolum et al.* (1988).

liquid (*Drake,* 1981). Trace-element partitioning data for spinel are less sparse but not very systematic. In CMAS liquids, *Malvin and Drake* (1987) reported $D_{Ge}^{sp/liq} = 0.1$ while *Capobianco and Drake* (1990) found, in air, that Pd is highly incompatible ($D_{Pd}^{sp/liq} < 0.02$) and Rh and Ru highly compatible ($D_i^{sp/liq} > 20$). *Malvin and Drake* (1987) obtained $D_{Ga}^{sp/liq} = 4–5$ for spinels with ~1 wt% Fe_2O_3. For the transition elements, data are quite limited. *Connolly and Burnett* (2003) obtained $D_{Ti}^{sp/liq} = 0.04$ for CAI-like liquids in air but 0.20 at the C-CO buffer (CCO), which they attributed to the stabilization of Ti^{3+} under reducing conditions. Their data for $D_V^{sp/liq}$ also show a strong f_{O_2} effect, ranging from ~0 in air to 2.2 at IW to 1.4 at CCO. Under oxidizing conditions, Cr^{3+} is so soluble in spinel that pichrochromites (nominally $MgCr_2O_4$) coexist with otherwise CMAS liquids containing 0.1–0.2 wt% Cr as Cr_2O_3 (*Hanson and Jones,* 1998). The solubility of Cr in spinel under reducing conditions, where Cr^{2+} is likely dominant in the liquid (*Schreiber et al.,* 1978), does not appear to have been studied. *Nagasawa et al.* (1980) studied Sc and REE partitioning using INAA analysis of separates for an MAS liquid. All were reported to be incompatible but the $D_{REE}^{sp/liq}$ form an odd concave-up pattern as a function of ionic radius. Contamination of the spinel separate by glass cannot account for this effect. On the other hand, significant contamination by forsterite, which has partition coefficients for the heavy REE, $D_{HREE}^{forst/liq}$, much higher than those for the light REE, $D_{LREE}^{forst/liq}$ (e.g., *Jones,* 1995), could generate the observed pattern as ~40% of the solid phase assemblage in *Nagasawa et al.*'s (1980) experiment which would, according to the MAS phase diagram, have been olivine. If olivine contamination was in fact a problem with no significant contributions from glass, then $D_{LREE}^{sp/liq}$ given by *Nagasawa et al.* (1980) are a little low while those for Sc and the HREE are probably much higher than the correct values. An alternative is that the reported partition coefficients capture the true values for spinel. If so, perhaps the REE substitute into two different sites in spinel, with LREE preferring one and HREE preferring the other, or they substitute into the same site but with local charge balance and/or structural distortions being accommodated differently.

Hibonite has received more thorough study than spinel or corundum. *Drake and Boynton* (1988) determined the partitioning behavior of Sr and various REE between hibonite and CMAS liquids; samples were doped with several wt% of the "trace" elements and analyzed by EPMA. *Kennedy et al.* (1994) determined hibonite/liquid partition coefficients for an extensive suite of trace elements under reducing conditions. Doping levels were tens to hundreds of ppmw, with analysis by SIMS. $D_i^{hib/liq}$ for the two studies are broadly similar and, given the high doping concentrations and absence of Ti in the experiments of *Drake and Boynton* (1988), in good agreement. According to *Kennedy et al.* (1994), $D_{Th}^{hib/liq} \gg D_U^{hib/liq}$, which implies that slowly crystallized igneous hibonite should have strongly fractionated Th/U relative to the liquid. They also noted that, when plotted as a function of ionic radius, $D_{Th}^{hib/liq}$ and $D_U^{hib/liq}$ are near the REE and suggested that these elements substitute into the Ca site. Interestingly, $D_i^{hib/liq}$ for U, Th, and the trivalent REE at a given ionic radius are higher than those of divalent cations substituting into the Ca site, which is the opposite of expectations based solely on charge (*Wood and Blundy,* 2001) and observed in clinopyroxene (e.g., *Hill et al.,* 2000). *Kennedy et al.* (1994) attributed this to short-range order of compensating cations in nearby sites. They also found that $D_i^{hib/liq} > 1$ for the LREE, while middle and HREE are incompatible, with $D_{La}^{hib/liq}/D_{Lu}^{hib/liq} \sim 200$. Thus, igneous hibonite should be LREE-enriched relative to the liquid from which it crystallized and HREE-depleted unless crystallization was so rapid that all the $D_i^{hib/liq}$ approached 1. Small, highly charged cations (Si, Ge, Hf, Zr, Ti) and the trivalent transition elements (e.g., Sc) except for Ti are moderately to highly incompatible (*Drake and Boynton,* 1988; *Beckett and Stolper,* 1994; *Kennedy et al.,* 1994).

The overall pattern of trace-element partitioning between perovskite and liquid (*Kennedy et al.,* 1994; *Corgne and Wood,* 2002) as a function of ionic radius is qualitatively similar to that of hibonite, in that trivalent and quadrivalent cations substituting into the Ca site are more compatible than divalent cations of the same ionic radius and the phase prefers light over heavy REE. Based on data of *Kennedy et al.* (1994), however, $D_{LREE}^{pv/liq}/D_{HREE}^{pv/liq} \sim 10$ (vs. 200 for hibonite), $D_{Th}^{pv/liq}/D_U^{pv/liq} \sim 3$ (vs. 8–17 for hibonite), and generally $D_i^{pv/liq}/D_i^{hib/liq} > 1$ for high field strength elements. Literature determinations of $D_{REE}^{pv/liq}$ (*Ringwood,* 1975; *Nagasawa et al.,* 1980; *Kennedy et al.,* 1994; *Simon et al.,* 1994a) agree only within factors of 6 (La, Sm) to 35 (Yb). *Simon et al.* (1994a) suggest that these differences reflect differing Al contents of the liquid and oxygen fugacities. The former, or some crystal composition parameter strongly correlated with Al in the liquid, appears to be the dominant factor. Fig-ure 8a shows values of $RT\ln D_{La}^{pv/liq}$ (R in terms of kJ) as a function of $X_{Al_2O_3}^{liq}$, the mole fraction of Al_2O_3 in the liquid, with all Ti as Ti^{4+}. All the CMAST data, regardless of whether experiments were conducted under oxidizing or reducing conditions, are consistent with a single linear trend in spite of major differences in liquid composition. This implies that $D_{La}^{pv/liq}$ can be treated as a simple function of Al_2O_3 concentration in the liquid; since *Nagasawa et al.*'s (1980) value for a CST liquid lies well below the regression, the simple functionality must break down at low Al_2O_3. Analogous plots for Sm and Yb show similar systematics. For CAIs, which are invariably aluminous, suitable values for $D_{REE}^{pv/liq}$ of Y and other trivalent REE can be inferred from $D_{La}^{pv/liq}$ through an unweighted fit of *Kennedy et al.*'s (1994) data to *Blundy and Wood*'s (1994) expression converted to log form

$$RT\ln\left(D_i^{pv/liq}/D_{La}^{pv/liq}\right) = -1614\left\{\left(r_{La}^{pv/liq}\right)^2 - \left(r_i^{pv/liq}\right)^2\right\} - 944.6\left\{\left(r_i^{pv/liq}\right)^3 - \left(r_{La}^{pv/liq}\right)^3\right\} \quad (3)$$

where following *Corgne and Wood* (2002) we use ionic radii for 8-coordinated sites. The Ca site in perovskite is actually 12-coordinated, but appropriate radii are not available for all the REE. The possible influence of liquid composition on partitioning for other trace elements is unknown. The experimental database for $D_i^{pv/liq}$ is nevertheless good enough to provide a strong test of igneous partitioning for perovskites in CAIs (e.g., *Davis and Grossman,* 1979; *Simon et al.,* 1994a, 1996a, 1999).

3.4.2. Silicates. Experimentally determined partitioning data are available for three phases, anorthite, olivine, and melilite, that crystallized from melts in at least some CAIs. No experimental data are available for fassaite but there is a substantial database for aluminous clinopyroxenes in equilibrium with CMAS liquids, and *Simon et al.* (1991)

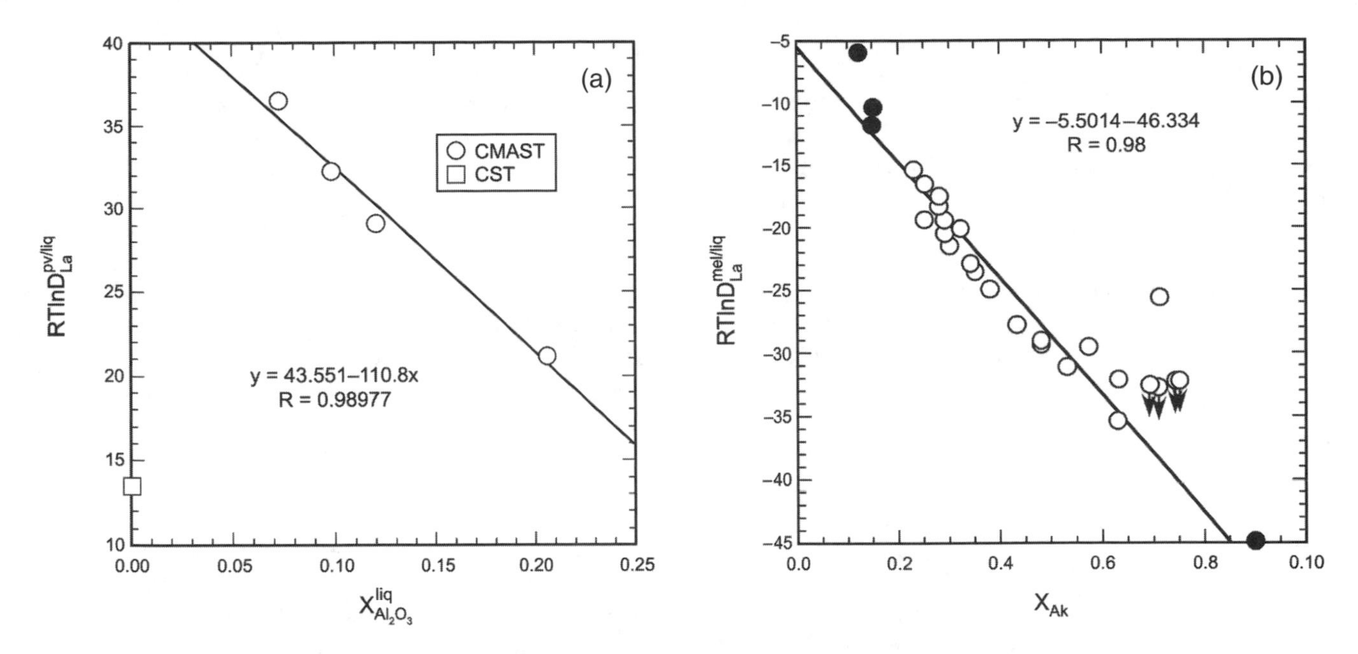

Fig. 8. Partitioning of La between crystals and liquid. **(a)** $RTlnD^{pv/liq}_{La}$ (R in terms of kJ) as a function of $X^{liq}_{Al_2O_3}$. CMAST data are from *Ringwood* (1975), *Kennedy et al.* (1994), and *Simon et al.* (1994a) and CST data from *Nagasawa et al.* (1980). The line is an unweighted linear regression through the CMAST data. **(b)** $RTlnD^{mel/liq}_{La}$ (R in terms of kJ) as a function of X_{Ak}. Data are from *Ringwood* (1975), *Nagasawa et al.* (1980), *Kuehner et al.* (1989), *Beckett et al.* (1990), and *Davis et al.* (1996). Closed and open circles refer to compositions on and off the åkermanite-gehlenite join, respectively. The line is an unweighted linear regression to all data excluding that of *Ringwood* (1975), which lies well off the trend, and the upper limits, shown with arrows, from *Beckett et al.* (1990).

used observed abundances in fassaite from type B inclusions to infer partition coefficients.

Plagioclase has been the subject of several partitioning studies in CMAS liquids. Using a CAI-like liquid as a base composition, *Simon et al.* (1994a) doped samples with several wt% total of Sr, Zr, La, Sm, and Yb oxides and determined $D^{an/liq}_i$ using EPMA and SIMS. Except for Sr ($D^{an/liq}_{Sr}$ = 1.0), all these elements, along with Mg and Ti^{4+} ($D^{an/liq}_{Ti^{4+}}$ = 0.03), are highly incompatible in plagioclase. *Peters et al.* (1995) obtained similar results for Mg and Ti^{4+}, inferred $D^{an/liq}_{Ti^{3+}}$ ~ 0.04, and concluded that both Ti^{3+} and Ti^{4+} are probably in tetrahedral coordination. Barium (*Miller et al.,* 2003), Ga, Ge (*Malvin and Drake,* 1987), and Na (*Soulard et al.,* 1992; *Beckett and Stolper,* 2000) are moderately incompatible. Early work suggesting a Henry's law effect for trace-element partitioning at very low concentrations such that $D^{an/liq}_{REE}$ increased dramatically at low concentrations (e.g., *Hoover,* 1978) are apparently due to analytical artifacts (*Beattie,* 1993), but *Bindeman and Davis* (2000) observed similar effects in some experiments involving sodic plagioclase (An_{45-75}) based on SIMS analyses, with increases of up to a factor of 2 between low (~1 ppm) and high (100 to several hundred ppm) concentrations in plagioclase. They observed no Henry's law effect for Sr-doped experiments and no effect for elements other than the REE.

Experimentally determined $D^{an/liq}_{Mg}$ are rather variable. Some of *Malvin and Drake*'s (1987) data imply $D^{an/liq}_{Mg}$ ~ 0.001, although this is probably an analytical artifact given the low concentrations in plagioclase; most experimentally determined $D^{an/liq}_{Mg}$ are in the range 0.01–0.11 (*Malvin and Drake,* 1987; *Peters et al.,* 1995; *Steele et al.,* 1997; *Beckett and Stolper,* 2000; *Miller et al.,* 2003). Much of this range probably reflects differences in liquid composition (*Peters et al.,* 1995; *Simon et al.,* 1994a; *Steele et al.,* 1997) and some may reflect activation volumes for plagioclase analyses extending through the often thin, tabular, feldspar crystals. A complicating factor for the interpretation of both experimental and meteoritic data is that the anorthite is frequently sector-zoned (*Hutcheon et al.,* 1978; *Steele et al.,* 1997), with concentrations of Mg and Na differing by up to a factor of 2 across sector boundaries. Possible sector-zoning effects for elements other than Mg, Na, and Ti do not appear to have been examined. Finally, it should be mentioned that fits for plagioclase-liquid partitioning of trace elements in more sodic and magmatic systems than considered here (*Blundy and Wood,* 1991; *Bindeman et al.,* 1998; *Bindeman and Davis,* 2000) can be extrapolated to anorthite. Agreement with the above-mentioned CMAS and CMAST studies ranges from fairly good (Na, Mg, Ti) to very bad (REE). Since these expressions include many elements (e.g., Be, Co, Fe, Li, U) for which there are currently no experimental data involving CAI-like liquids, they are nevertheless still worth using if a general idea of compatibility is desired.

Olivine is a major phase in terrestrial rocks and its partitioning behavior in magmatic systems has received considerable attention (e.g., *Irving,* 1978; *Green,* 1994; *Jones,* 1995; *Wood and Blundy,* 2004). With, however, the notable exception of Ca (*Libourel,* 1999), Ga and Ge (*Capobianco and Watson,* 1982; *Malvin and Drake,* 1987), and some of the transition elements (*Watson,* 1977; *Hart and Davis,* 1978; *Schreiber,* 1979; *Hanson and Jones,* 1998; *Dudnikova et al.,* 2001), most data refer to Fe-bearing systems, although extrapolation to forsterite is often possible (e.g., *Jones,* 1995). Perhaps the most immediately relevant data for CAIs is the correlation between $D_{Ca}^{ol/liq}$ and CaO content of the liquid (*Libourel,* 1999). The relationship holds well for CMAS, CMAST, and CMAS + Cr liquids (*Libourel,* 1999), and *Pack and Palme* (2003) found no discernible effect for high cooling rates. The addition of Na to the liquid leads to higher values of $D_{Ca}^{ol/liq}$ than expected based on *Libourel*'s (1999) equation, but this is negligible for <2 wt% Na_2O, the range of interest for CAIs. Although it has not been used for this purpose, the Ca content of olivine from type B3 (olivine-bearing) CAIs may be a powerful monitor of liquid composition.

Experimentally determined $D_i^{mel/liq}$ for melilite are available for minor or trace amounts of Ba, Be, Na, Sc, Sr, Ti, Y, Zr, and the REE (*Ringwood,* 1975; *Nagasawa et al.,* 1980; *Kuehner et al.,* 1989; *Beckett et al.,* 1990; *Soulard et al.,* 1992; *Davis et al.,* 1996; *Simon et al.,* 1996b; *Beckett and Stolper,* 2000). Unlike olivine and anorthite in CAIs, whose variations in major-element chemistry are small and whose compositions therefore have little impact on partitioning, melilite is essentially a binary solid solution in CAIs and, where charge-compensating substitutions are required, there is a strong dependence of $D_i^{mel/liq}$ on the major-element composition, or X_{Ak}. An example is shown for La, which substitutes into the Ca site, in Fig. 8b. Data for Y and the other REE show similar relationships between $D_i^{mel/liq}$ and X_{Ak}. $D_{La}^{mel/liq}$ varies by nearly a factor of 20 but, with the exception of one datum from *Ringwood* (1975), all the data are consistent with the line shown in Fig. 8b. It is possible the aberrant $D_{La}^{mel/liq}$ reflects a liquid composition effect as the bulk composition, which was based on a mixture of melilite and a CAI pyroxene composition from *Mason and Martin* (1974), is significantly different from those of other studies (pyroxene-rich vs. compositions along the join åkermanite-gehlenite, and spinel-poor relative to CAI-like compositions) or even a typo (if $D_{La}^{mel/liq}$ were 0.05 instead of 0.15, the point would have plotted near the line). In any case, suitable $D_{La}^{mel/liq}$ for CAIs can be obtained from Fig. 8b without resolving the relevance of Ringwood's data. Values for other trivalent REE can be obtained from an unweighted regression of Y and trivalent REE data from *Davis et al.* (1996) for X_{Ak} = 0.23 melilite, giving

$$RT\ln\left(D_i^{mel/liq}/D_{La}^{mel/liq}\right) = -1155\left\{\left(r_{La}^{mel/liq}\right)^2 - \left(r_i^{mel/liq}\right)^2\right\} - 692.8\left\{\left(r_i^{mel/liq}\right)^3 - \left(r_{La}^{mel/liq}\right)^3\right\} \quad (4)$$

(see discussion of *Wood and Blundy,* 2004). All the trivalent REE are predicted to be incompatible in melilite with $D_{REE}^{mel/liq}$ generally decreasing with increasing X_{Ak}. Beryllium has the opposite behavior with $D_{Be}^{mel/liq}$ increasing with increasing X_{Ak} such that Be changes from incompatible to compatible with increasing degree of crystallization in a CAI-like bulk composition (*Beckett et al.,* 1990). Based on Sc data of *Nagasawa et al.* (1980) and *Beckett et al.* (1990), $RT\ln D_{Sc}^{mel/liq}$ = –32.04–65.4 X_{Ak} for X_{Ak} < 0.36 and –58 + 6.71 X_{Ak} for X_{Ak} > 0.36 (R being the gas constant in kJ). Scandium probably has a different dominant substitution mechanism in aluminous vs. magnesian melilites. Barium and Sr substitute into the Ca site. $D_{Ba}^{mel/liq}$ is 0.02–0.04 (*Beckett et al.,* 1990; *Simon et al.,* 1994a) and $D_{Sr}^{mel/liq}$ is probably ~0.7 (*Nagasawa et al.,* 1980; *Woolum et al.,* 1988; *Kuehner et al.,* 1989, *Simon et al.,* 1996b), although values as high as 1.1 have been reported (*Ringwood,* 1975; *Kuehner et al.,* 1989). $D_i^{mel/liq}$ for Ba and Sr are essentially independent of X_{Ak} (*Beckett et al.,* 1990; *Simon et al.,* 1994a). *Simon et al.* (1996b) obtained $D_{Ti^{4+}}^{mel/liq}$ = 0.020. Finally, Na deserves special mention. As a trace constituent, Na behaves similarly to Be (*Beckett and Stolper,* 2000), increasing with increasing X_{Ak} by a factor of nearly 4 between X_{Ak} = 0.22 ($D_{Na}^{mel/liq}$ = 0.4) and X_{Ak} = 0.65 ($D_{Na}^{mel/liq}$ = 1.5) for CAI-like liquids. As Na_2O in the liquid is increased to more than a few tenths of a wt%, the influence of liquid composition becomes more apparent, and *Beckett and Stolper* (2000) appealed to both crystal-chemical and liquid composition effects in modeling available Na partitioning data (*Pan and Longhi,* 1989, 1990; *Soulard et al.,* 1992; *Krigman et al.,* 1995; *Beckett and Stolper,* 2000).

While a fairly extensive collection of data exists on trace-element partitioning for clinopyroxenes with compositions close to the join diopside-CaTschermaks, references for which can be found in the reviews of *Irving* (1978), *Green* (1994), and *Wood and Blundy* (2004), there are no experimentally determined $D_i^{fass/liq}$ for the compositionally complex fassaites observed in CAIs. Probably the best option at present is to use apparent $D_i^{fass/liq}$ derived by *Simon et al.* (1991) from three type B1 CAIs. Their model values for Y, REE, Sc, Zr, Hf, and Ta are within the range observed experimentally for Di-CaTs pyroxenes; their model $D_i^{fass/liq}$ for U and Th are lower than obtained by *Benjamin et al.* (1980) and *LaTourette and Burnett* (1992), and the model $D_{Nb}^{fass/liq}$ is higher than values obtained by *Lundstrom et al.* (1998) and *Hill et al.* (2000). In addition to significant differences in the pyroxene compositions, there may also be different valence states in CAI liquids and pyroxenes relative to the experiments. According to the model of *Simon et al.* (1991), Sc, Zr, and Hf are compatible; Nb, Ta, Y, and the REE are moderately incompatible; and U and Th are highly incompatible.

3.5. Dynamic Crystallization Experiments

In previous sections, the primary focus was on tools that can be used for the exploration of phase equilibria of CAI-like liquids and equilibrium constraints on the crystallization of these objects if they were partially or completely melted. This is important, but crystallization of a CAI is an inherently dynamic process. The effect of kinetic processes on the observed phase assemblages and their compositions is constrained through equilibrium phase diagrams and thermodynamic calculations in the sense that boundary conditions at crystal-liquid surfaces and perfect fractional crystallization paths can be determined, metastable extensions of stability fields can be used to help establish liquid lines of descent if a stable phase fails to nucleate, liquid compositions can be constrained from the crystalline phase compositions, and the conditions under which a crystal dissolves and whether or not that dissolution is likely to be congruent (the phase dissolves directly into the melt) or incongruent (one or more crystalline phases form between the original dissolving phase and the liquid) can be predicted (*Cooper,* 1970; *Zhang and Lee,* 2000).

A formal consideration of transport properties in the liquid and/or vapor is required to flesh out time and chemical evolution (see, e.g., *Grossman et al.,* 2000; *Richter et al.,* 2002; *Davis and Richter,* 2004). Some useful constraints on process can also be obtained through dynamic crystallization and partial melting experiments without understanding much about the processes producing the phases, but great care is required to avoid categorical statements based on them because the results are almost always permissive, especially where textural observations are involved (i.e., different sets of conditions not explored in the experiments could lead to a similar result).

The experimental database for chondrules is heavily weighted by dynamic crystallization experiments of one sort or another and the results have strongly influenced thinking about these objects (e.g., *Connolly and Hewins,* 1996; *Hewins and Connolly,* 1996; *Connolly and Desch,* 2004). In the CAI literature, there is only one really influential classical dynamic crystallization study, that of *Stolper and Paque* (1986). They conducted a series of single-stage, dynamic crystallization experiments on one bulk composition typical of type B CAIs, in which the sample was held at a maximum temperature for a specified period of time, cooled at a constant rate, and then quenched at a specific temperature. They found that fast cooling rates (>~100°C/h) and maximum temperatures (T_{max}) significantly above the appearance of melilite (1390°C for their bulk composition) led to dendritic melilite and significant Mg-Tschermaks component in pyroxene, neither of which is observed in type Bs (but see *Simon et al.,* 1998), leading to the conclusion that type B CAIs probably cooled more slowly. *MacPherson et al.* (1984b) had already shown for the same bulk composition that cooling rates faster than 50°C/h would lead exclusively to normally zoned melilites in which compositions become progressively more magnesian from core-to-rim. Reversals back toward more aluminous melilites were observed by *MacPherson et al.* (1984b) in melilites grown at cooling rates less than 50°C/h and since reversely zoned melilites are commonly observed in type B1 CAIs, MacPherson et al. concluded that cooling rates for these CAIs were slower than ~50°C/h. Later single- and multiple-stage dynamic crystallization experiments designed to reproduce textures (*Maharaj and Hewins,* 1994; *Paque,* 1995) and trace-element partitioning (*Davis et al.,* 1996; *Simon et al.,* 1996b) of CAIs have not changed the basic constraints established in these two papers, although the influence of "flash heating" (*Maharaj and Hewins,* 1995), precursor materials (*Paque,* 1995; *Maharaj and Hewins,* 1997), remelting (*Paque et al.,* 1998, 2000), and nonlinear cooling paths have not been fully explored. Production of the mineral "UNK" in dynamic crystallization experiments on various CAI-like compositions if cooling rates were less than a few tens of degrees per hour but not at faster rates (*Paque et al.,* 1994) is also consistent with a maximum cooling rate in the range of several tens of degrees per hour. In an interesting set of experiments, *Mendybaev et al.* (2003) found that a continuous melilite mantle, the basic defining characteristic of type B1 inclusions, could be produced by cooling a CAI-like composition in a flowing H_2 gas at 5°C/h but that no mantle was produced in a similar sample cooled at 10°C/h. They suggested that the difference between type B1 and B2 inclusions was essentially the difference between relatively faster (B2) and more slowly (B1) cooled droplets. Overall, data from all these studies are consistent with cooling rates of a few to a few tens of degrees per hour for type B CAIs at ~1300°–1400°C. *Stolper and Paque* (1986) suggested a lower limit of a few tenths of a degree per hour near the appearance temperature of plagioclase based on the premise that slower rates would *always* lead to the appearance of plagioclase before fassaite, whereas the opposite is frequently inferred for type Bs (*MacPherson et al.,* 1984b). There is growing evidence that many type B CAIs were partially melted more than once (*MacPherson and Davis,* 1993; *Beckett et al.,* 2000; *Connolly and Burnett,* 2003) and some experimental evidence (*Paque et al.,* 2000) that multiple melting events enhance the probability of producing coarse-grained plagioclase before pyroxene, but this order of crystallization constraint should probably be viewed as qualitative at present.

Acknowledgments. We are grateful to J. Paque and S. B. Simon for their reviews of the manuscript, as well as the comments of S. Krot and D. S. Lauretta. This work was supported in part by grants from PSC CUNY (H.C.C., PI) and NASA.

REFERENCES

Agamawi Y. M. and White J. (1953) The quaternary system CaO-Al_2O_3-TiO_2-SiO_2 Part I. — The ternary system anorthite-sphene-silica. *Trans. British Cer. Soc., 52,* 271–310.

Agamawi Y. M. and White J. (1954a) The quaternary system CaO-Al_2O_3-TiO_2-SiO_2 Part II. — The ternary system calcium metasilicate-sphene-silica. *Trans. British Cer. Soc., 53,* 1–22.

Agamawi Y. M. and White J. (1954b) The quaternary system CaO-Al_2O_3-TiO_2-SiO_2 Part III. — The quaternary system calcium metasilicate-anorthite-sphene-silica. *Trans. British Cer. Soc., 53,* 23–38.

Anderson G. M. and Crerar D. A. (1993) *Thermodynamics in Geochemistry: The Equilibrium Model.* Oxford Univ., Oxford. 588 pp.

Bale C. W., Chartrand P., Degterov S. A., Eriksson G., Hack K., Ben Mahfoud R., Melancon J., Pelton A. D., and Petersen S. (2002) FactSage thermochemical software and databases. *Calphad, 26,* 189–228.

Ban-ya S. (1993) Mathematical expression of slag-metal reactions in steelmaking process by quadratic formalism based on the regular solution model. *ISIJ Intl., 33,* 2–11.

Barron L. M. (1985) Comment on "A thermodynamic model for multicomponent melts, with application to the system CaO-Al_2O_3-SiO_2" by Berman and Brown. *Geochim. Cosmochim. Acta, 49,* 611–612.

Bartels K. S. and Grove T. L. (1991) High-pressure experiments on magnesian eucrite compositions: Constraints on magmatic processes in the eucrite parent body. *Proc. Lunar Planet. Sci. Conf., Vol. 21,* pp. 351–365.

Beattie P. (1993) On the occurrence of apparent non-Henry's Law behaviour in experimental partitioning studies. *Geochim. Cosmochim. Acta, 57,* 47–55.

Beckett J. R. and Grossman L. (1988) The origin of type C inclusions from carbonaceous chondrites. *Earth Planet. Sci. Lett., 89,* 1–14.

Beckett J. R. and Stolper E. (1994) The stability of hibonite, melilite and other aluminous phases in silicate melts: Implications for the origin of hibonite-bearing inclusions from carbonaceous chondrites. *Meteoritics, 29,* 41–65.

Beckett J. R. and Stolper E. (2000) The partitioning of Na between melilite and liquid: Part I. The role of crystal chemistry and liquid composition. *Geochim. Cosmochim. Acta, 64,* 2509–2517.

Beckett J. R., Spivack A. J., Hutcheon I. D., Wasserburg G. J., and Stolper E. M. (1990) Crystal chemical effects on the partitioning of trace elements between mineral and melt: An experimental study of melilite with applications to refractory inclusions from carbonaceous chondrites. *Geochim. Cosmochim. Acta, 54,* 1755–1774.

Beckett J. R., Paque J. M., and Stolper E. M. (1999) The use of melilite compositions to constrain the thermal history and liquid line of descent of Type B CAIs (abstract). In *Lunar and Planetary Science XXX,* Abstract # 1920. Lunar and Planetary Institute, Houston (CD-ROM).

Beckett J. R., Simon S. B., and Stolper E. (2000) The partitioning of Na between melilite and liquid: Part II. Applications to Type B inclusions from carbonaceous chondrites. *Geochim. Cosmochim. Acta, 64,* 2519–2534.

Benjamin T., Heuser W. R., Burnett D. S., and Seitz M. G. (1980) Actinide crystal-liquid partitioning for clinopyroxene and $Ca_3(PO_4)_2$. *Geochim. Cosmochim. Acta, 44,* 1251–1264.

Berman R. G. (1983) A thermodynamic model for multicomponent melts, with application to the system CaO-MgO-Al_2O_3-SiO_2. Ph.D. dissertation, Univ. of British Columbia.

Berman R. G. and Brown T. H. (1984) A thermodynamic model for multicomponent melts, with application to the system CaO-Al_2O_3-SiO_2. *Geochim. Cosmochim. Acta, 48,* 661–678.

Biggar G. M. (1971) Phase relationships of bredigite ($Ca_5Mg Si_3O_{12}$) and of the quaternary compound ($Ca_6MgAl_8SiO_{21}$) in the system CaO-MgO-Al_2O_3-SiO_2. *Cement Concrete Res., 1,* 493–513.

Biggar G. M. (1984) The composition of diopside solid solutions, and of liquids, in equilibrium with forsterite, plagioclase, and liquid in the system Na_2O-CaO-MgO-Al_2O_3-SiO_2 and in remelted rocks from 1 bar to 12 kbar. *Mineral. Mag., 48,* 481–494.

Bindeman I. N. and Davis A. M. (2000) Trace element partitioning between plagioclase and melt: Investigation of dopant influence on partition behavior. *Geochim. Cosmochim. Acta, 64,* 2863–2878.

Bindeman I. N., Davis A. M., and Drake M. J. (1998) Ion microprobe study of plagioclase-basalt partition experiments at natural concentration levels of trace elements. *Geochim. Cosmochim. Acta, 62,* 1175–1193.

Bischoff A. and Keil K. (1983) Ca-Al-rich chondrules and inclusions in ordinary chondrites. *Nature, 303,* 588–592.

Bischoff A. and Keil K. (1984) Al-rich objects in ordinary chondrites: Related origin of carbonaceous and ordinary chondrites and their constituents. *Geochim. Cosmochim. Acta, 48,* 693–709.

Bischoff A., Keil K., and Stöffler D. (1985) Perovskite hibonite-spinel-bearing inclusions and Al-rich chondrules and fragments in enstatite chondrites. *Chem. Erde, 44,* 97–106.

Bischoff A., Palme H., Schultz L., Weber D., Weber H. W., and Spettel B. (1993) Acfer 182 and paired samples, an iron-rich carbonaceous chondrite: Similarities with ALH85085 and relationship to CR chondrites. *Geochim. Cosmochim. Acta, 57,* 2631–2648.

Björkvall J., Du S., and Seetharaman S. (2001) Thermodynamic model calculations in multicomponent liquid silicate systems. *Ironmaking Steelmaking, 28,* 250–257.

Blum J. D., Wasserburg G. J., Hutcheon I. D., Beckett J. R., and Stolper E. M. (1989) Origin of opaque assemblages in C3V meteorites: Implications for nebular and planetary processes. *Geochim. Cosmochim. Acta, 53,* 543–556.

Blundy J. D. and Wood B. J. (1991) Crystal-chemical controls on the partitioning of Sr and Ba between plagioclase feldspar, silicate melts, and hydrothermal solutions. *Geochim. Cosmochim. Acta, 55,* 193–209.

Blundy J. D. and Wood B. J. (1994) Prediction of crystal-melt partition coefficients from elastic moduli. *Nature, 372,* 452–454.

Blundy J. D. and Wood B. J. (2003) Partitioning of trace elements between crystals and melts. *Earth Planet. Sci. Lett., 210,* 383–397.

Blundy J. D., Wood B. J., and Davies A. (1996) Thermodynamics of rare earth element partitioning between clinopyroxene and melt in the system CaO-MgO-Al_2O_3-SiO_2. *Geochim. Cosmochim. Acta, 60,* 359–364.

Bowen N. L. (1915) The crystallization of haplobasaltic, haplodioritic and related magmas. *Am. J. Sci., 190,* 161–185.

Bowen N. L. (1928) *The Evolution of the Igneous Rocks.* Princeton Univ., Princeton. 332 pp.

Brearley A. J. and Jones R. H. (1998) Chondritic meteorites. In *Planetary Materials* (J. J. Papike, ed.), pp. 3-1 to 3-398. Reviews in Mineralogy, Vol. 36, Mineralogical Society of America.

Brice J. C. (1975) Some thermodynamic aspects of the growth of strained crystals. *J. Crystal Growth, 28,* 249–253.

Brixner L. H. (1967) Segregation coefficients of the rare earth niobates in $CaMoO_4$. *J. Electrochem. Soc., 114,* 108–110.

Capobianco C. J. and Drake M. J. (1990) Partitioning of ruthenium, rhodium, and palladium between spinel and silicate melt and implications for platinum group element fractionation trends. *Geochim. Cosmochim. Acta, 54,* 869–874.

Capobianco C. J. and Watson E. B. (1982) Olivine/silicate melt partitioning of germanium: An example of a nearly constant partition coefficient. *Geochim. Cosmochim. Acta, 46,* 235–240.

Chamberlin L., Beckett J. R., and Stolper E. (1994) Pd-oxide equilibration: A new experimental method for the direct determination of oxide activities in melts and minerals. *Contrib. Mineral. Petrol., 116,* 169–181.

Connolly H. C. and Burnett D. S. (2003) On type B CAI formation: Experimental constraints on f_{O_2} variations in spinel minor element partitioning and reequilibration effects. *Geochim. Cosmochim. Acta, 67,* 4429–4434.

Connolly H. C. and Desch S. J. (2004) On the origin of the "kleine Kugelchen" called chondrules. *Chem. Erde, 64,* 95–125.

Connolly H. C. and Hewins R. H. (1996) Constraints on chondrule precursors from experimental data. In *Chondrules and the Protoplanetary Disk* (R. H. Hewins et al., eds.), pp. 129–135. Cambridge Univ., Cambridge.

Connolly H. C., Burnett D. S., and McKeegan K. D. (2003) The petrogenesis of type B1 Ca-Al-rich inclusions: The spinel perspective. *Meteoritics & Planet. Sci., 38,* 197–224.

Cooper A. R. (1970) The use of phase diagrams in dissolution studies. In *Phase Diagrams Materials Science and Technology, Vol. III* (A. M. Alper, ed.), pp. 237–251. Academic, New York.

Corgne A. and Wood B. J. (2002) $CaSiO_3$ and $CaTiO_3$ perovskite-melt partitioning of trace elements: Implications for gross mantle differentiation. *Geophys. Res. Lett., 29(19),* 1933,doi: 10.1029/2001GL014398,2002.

Davies A., Wood B., Barry T., Dinsdale A., and Gisby J. (1994) The thermodynamics of mineral-melt equilibria in the system CaO-MgO-Al_2O_3-SiO_2 (abstract). *Mineral. Mag., 58A,* 213–214.

Davies R. H., Dinsdale A. T., Gisby J. A., Robinson J. A. J., and Martin S. M. (2002) MTDATA — Thermodynamic and phase equilibrium software from the National Physical Laboratory. *Calphad, 26,* 229–271.

Davis A. M. (2006) Volatile evolution and loss. In *Meteorites and the Early Solar System II* (D. S. Lauretta and H. Y. McSween Jr., eds.), this volume. Univ. of Arizona, Tucson.

Davis A. M. and Grossman L. (1979) Condensation and fractionation of rare earths in the solar nebula. *Geochim. Cosmochim. Acta, 43,* 1611–1632.

Davis A. M. and Richter F. M. (2004) Condensation and evaporation of solar system materials. In *Treatise on Geochemistry, Vol. 1: Meteorites, Comets, and Planets* (A. M. Davis, ed.), pp. 407–430. Elsevier, Oxford.

Davis A. M., MacPherson G. J., Clayton R. N., Mayeda T. K., Sylvester P. J., Grossman L., Hinton R. W., and Laughlin J. R. (1991) Melt solidification and late-stage evaporation in the evolution of a FUN inclusion from the Vigarano C3V chondrite. *Geochim. Cosmochim. Acta, 55,* 621–637.

Davis A. M., Richter F. M., Simon S. B., and Grossman L. (1996) The effect of cooling rate on melilite/liquid partition coefficients for Y and REE in Type B CAI melts (abstract). In *Lunar and Planetary Science XXVII,* pp. 291–292. Lunar and Planetary Institute, Houston.

de Aza A. H., Pena P., and de Aza S. (1999) Ternary system Al_2O_3-MgO-CaO: I, Primary phase field of crystallization of spinel in the subsystem $MgAl_2O_4$-$CaAl_4O_7$-CaO-MgO. *J. Am. Ceramic Soc., 82,* 2193–2203.

de Aza A. H., Iglesias J. E., Pena P., and de Aza S. (2000) Ternary system Al_2O_3-MgO-CaO: II, Phase relationships in the subsystem Al_2O_3-$MgAl_2O_4$-$CaAl_4O_7$. *J. Am. Ceramic Soc., 83,* 919–927.

DeVries R. C. and Osborn E. F. (1957) Phase equilibria in high-alumina part of the system CaO-MgO-Al_2O_3-SiO_2. *J. Am. Ceramic Soc., 40,* 6–15.

DeVries R. C., Roy R., and Osborn E. F. (1955) Phase equilibria in the system CaO-TiO_2-SiO_2. *J. Am. Ceramic Soc., 38,* 158–171.

Drake M. J. (1981) Partitioning of Sm between hibonite and corundum and silicate melt (abstract). In *Lunar and Planetary Science XII,* pp. 235–236. Lunar and Planetary Institute, Houston.

Drake M. J. and Boynton W. V. (1988) Partitioning of rare earth elements between hibonite and melt and implications for nebular condensation of the rare earth elements. *Meteoritics, 23,* 75–80.

Drake M. J. and Weill D. F. (1975) Partition of Sr, Ba, Ca, Y, Eu^{2+}, Eu^{3+}, and other REE between plagioclase feldspar and magmatic liquid: An experimental study. *Geochim. Cosmochim. Acta, 39,* 689–712.

Dudnikova V. B., Zharikov E. V., Eremin N. N., Zharkova E. V., Lebedev V. F., and Urusov V. S. (2001) Vanadium distribution between forsterite and its melt: The structural and oxidation state of vanadium. *Geochem. Intl., 39(7),* 667–675.

Ebel D. S. (2006) Condensation of rocky material in astrophysical environments. In *Meteorites and the Early Solar System II* (D. S. Lauretta and H. Y. McSween Jr., eds.), this volume. Univ. of Arizona, Tucson.

Ebel D. S. and Grossman L. (2000) Condensation in dust-enriched systems. *Geochim. Cosmochim. Acta, 64,* 339–366.

El Goresy A., Nagel K., and Ramdohr P. (1978) Fremdlinge and their noble relatives. *Proc. Lunar Planet. Sci. Conf. 9th,* pp. 1279–1303.

El Goresy A., Palme H., Yabuki H., Nagel K., Herrwerth I., and Ramdohr P. (1984) A calcium-aluminum-rich inclusion from the Essebi (CM2) chondrite: Evidence for captured spinel-hibonite spherules and for an ultra-refractory rimming sequence. *Geochim. Cosmochim. Acta, 48,* 2283–2298.

Ertel W., Dingwell D. B., and O'Neill H. S. C. (1997) Compositional dependence of the activity of nickel in silicate melts. *Geochim. Cosmochim. Acta, 61,* 4707–4721.

Gaye H. and Welfringer J. (1984) Modelling of the thermodynamic properties of complex metallurgical slags. In *Second International Symposium on Metallurgical Slags and Fluxes* (H. A. Fine and D. R. Gaskell, eds.), pp. 357–375. Metallurgical Society AIME.

Gentile A. L. and Foster W. R. (1963) Calcium hexaluminate and its stability relations in the system CaO-Al_2O_3-SiO_2. *J. Am. Ceramic Soc., 46,* 74–76.

Ghiorso M. S. (1994) Algorithms for the estimation of phase stability in heterogeneous thermodynamic systems. *Geochim. Cosmochim. Acta, 58,* 5489–5501.

Ghiorso M. S. and Sack R. O. (1995) Chemical mass transfer in magmatic processes IV. A revised and internally consistent thermodynamic model for the interpolation and extrapolation of liquid-solid equilibria in magmatic systems at elevated temperatures and pressures. *Contrib. Mineral. Petrol., 119,* 197–212.

Göbbels M., Woermann E., and Jung J. (1995) The Al-rich part of the system CaO-Al_2O_3-MgO Part I. Phase relationships. *J. Solid State Chem., 120,* 358–363.

Green T. H. (1994) Experimental studies of trace-element partioning applicable to igneous petrogenesis — Sedona 16 years later. *Chem. Geol., 117,* 1–36.

Greenwood H. J. (1975) Thermodynamically valid projections of extensive phase relationships. *Am. Mineral., 60,* 1–8.

Grossman J. N., Rubin A. E., and MacPherson G. J. (1988) ALH85085: A unique volatile-poor carbonaceous chondrite with possible implications for nebular fractionation processes. *Earth Planet. Sci. Lett., 91,* 33–54.

Grossman L. (1975) Petrography and mineral chemistry of Ca-rich inclusions in the Allende meteorite. *Geochim. Cosmochim. Acta, 39,* 433–454.

Grossman L. (1980) Refractory inclusions in the Allende meteorite. *Annu. Rev. Earth Planet. Sci., 8,* 559–608.

Grossman L., Ebel D. S., Simon S. B., Davis A. M., Richter F. M., and Parsad N. M. (2000) Major element chemical and isotopic compositions of refractory inclusions in C3 chondrites: The separate roles of condensation and evaporation. *Geochim. Cosmochim. Acta, 64,* 2879–2894.

Gutt W. (1964) High-temperature phase equilibria for the partial system $3CaO.MgO.2SiO_2$-$MgO.Al_2O_3$-$2CaO.Al_2O_3.SiO_2$ in the quaternary system CaO-SiO_2-Al_2O_3-MgO. *J. Iron Steel. Inst., 202,* 770–774.

Gutt W. (1968) Crystallization of merwinite from melilite compositions. *J. Iron Steel. Inst., 206,* 840–841.

Gutt W. and Russell A. D. (1977) Studies of the system CaO-SiO_2-Al_2O_3-MgO in relation to the stability of blastfurnace slag. *J. Mater. Sci., 12,* 1869–1878.

Hallstedt B. (1990) Assessment of the CaO-Al_2O_3 system. *J. Am. Ceramic Soc., 73,* 15–23.

Hallstedt B. (1992) Thermodynamic assessment of the system MgO-Al_2O_3. *J. Am. Ceramic Soc., 75,* 1497–1507.

Hallstedt B. (1995) Thermodynamic assessment of the CaO-MgO-Al_2O_3 system. *J. Am. Ceramic Soc., 78,* 193–198.

Hallstedt B., Hillert M., Selleby M., and Sundman B. (1994) Modelling of acid and basic slags. *Calphad, 18,* 31–37.

Hanson B. and Jones J. H. (1998) The systematics of Cr^{3+} and Cr^{2+} partitioning between olivine and liquid in the presence of spinel. *Am. Mineral., 83,* 669–684.

Hart S. R. and Davis K. E. (1978) Nickel partitioning between olivine and silicate melt. *Earth Planet. Sci. Lett., 40,* 203–219.

Hashimoto A. (1991) Prediction of activities of oxide components in the multicomponent liquid system FeO-MgO-CaO-Na_2O-$AlO_{1.5}$-SiO_2 (abstract). In *Lunar and Planetary Science XXII,* pp. 533–534. Lunar and Planetary Institute, Houston.

Hewins R. H. and Connolly H. C. Jr. (1996) Peak temperatures of flash-melted chondrules. In *Chondrules and the Protoplanetary Disk* (R. H. Hewins et al., eds.), pp. 197–204. Cambridge Univ., Cambridge.

Hill E., Wood B. J., and Blundy J. D. (2000) The effect of Ca-Tschermaks component on trace element partitioning between clinopyroxene and silicate melt. *Lithos, 53,* 203–215.

Hoover J. D. (1978) The distribution of samarium and thulium between plagioclase and liquid in the systems An-Di and Ab-An-Di at 1300°C. *Carnegie Inst. Wash. Yearb., 77,* 703–706.

Huss G. R., MacPherson G. J., Wasserburg G. J., Russell S. S., and Srinivasan G. (2001) Aluminum-26 in calcium-aluminum-rich inclusions and chondrules from unequilibrated ordinary chondrites. *Meteoritics & Planet. Sci., 36,* 975–997.

Hutcheon I. D., Steele I. M., Smith J. V., and Clayton R. N. (1978) Ion microprobe, electron microprobe and cathodoluminescence data for Allende inclusions with emphasis on plagioclase chemistry. *Proc. Lunar Planet. Sci. Conf. 9th,* pp. 1345–1368.

Imlach J. A. and Glasser F. P. (1968) Phase equilibria in the system CaO-Al_2O_3-TiO_2. *Trans. British Cer. Soc., 67,* 581–609.

Ireland T. R., Fahey A. J., and Zinner E. K. (1991) Hibonite-bearing microspherules: A new type of refractory inclusions with large isotopic anomalies. *Geochim. Cosmochim. Acta, 55,* 367–379.

Irving A. J. (1978) A review of experimental studies of crystal/liquid trace element partitioning. *Geochim. Cosmochim. Acta, 42,* 743–770.

Ivanova M. A., Petaev M. I., MacPherson G. J., Nazarov M. A., Taylor L. A., and Wood J. A. (2002) The first known occurrence of calcium monoaluminate, in a calcium-aluminum-rich inclusion from the CH chondrite Northwest Africa 470. *Meteoritics & Planet. Sci., 37,* 1337–1344.

Jones J. H. (1995) Experimental trace element partitioning. In *Rock Physics and Phase Relations: A Handbook of Physical Constants, Vol. 3,* pp. 73–104. American Geophysical Union, Washington, DC.

Jones R. H. (1994) Petrology of FeO-poor, porphyritic pyroxene chondrules in the Semarkona chondrite. *Geochim. Cosmochim. Acta, 58,* 5325–5340.

Jones R. H. and Scott E. R. D. (1989) Petrology and thermal history of Type IA chondrules in the Semarkona (LL3.0) chondrite. *Proc. Lunar Planet. Sci. Conf. 19th,* pp. 523–536.

Kennedy A. K., Lofgren G. E., and Wasserburg G. J. (1994) Trace-element partition coefficients for perovskite and hibonite in meteorite compositions. *Chem. Geol., 117,* 379–390.

Kennedy A. K., Beckett J. R., Edwards D. A., and Hutcheon I. D. (1997) Trace element disequilibria and magnesium isotope heterogeneity in 3655A: Evidence for a complex multi-stage evolution of a typical Allende Type B1 CAI. *Geochim. Cosmochim. Acta, 61,* 1541–1561.

Kimura M., El Goresy A., Palme H., and Zinner E. (1993) Ca-, Al-rich inclusions in the unique chondrite ALH85085: Petrology, chemistry, and isotopic compositions. *Geochim. Cosmochim. Acta, 57,* 2329–2359.

Kirschen M. and De Capitani C. (1999) Experimental determination and computation of the liquid miscibility gap in the system CaO-MgO-SiO_2-TiO_2. *J. Phase Equil., 20,* 593–611.

Kirschen M., De Capitani C., Millot F., Rifflet J.-C., and Coutures J.-P. (1999) Immiscible silicate liquids in the system SiO_2-TiO_2-Al_2O_3. *Eur. J. Mineral., 11,* 427–440.

Kohn S. C. and Schofield P. F. (1994) The importance of melt composition in controlling trace-element behaviour: An experimental study of Mn and Zn partitioning between forsterite and silicate melts. *Chem. Geol., 117,* 73–87.

Krigman L. D., Kogarko L. N., and Veksler I. V. (1995) Melilite-melt equilibrium and the role of melilite in the evolution of ultraalkaline magmas. *Geochem. Intl., 32(8),* 91–101.

Krot A. N., Scott E. R. D., and Zolensky M. E. (1995) Mineralogical and chemical modifications of components in CV3 chondrites: Nebular or asteroidal processing? *Meteoritics, 30,* 748–775.

Krot A. N., McKeegan K. D., Russell S. S., Meibom A., Weisberg M. K., Zipfel J., Krot T. V., Fagan T. J., and Keil K. (2001) Refractory calcium-aluminum-rich inclusions and aluminum-diopside-rich chondrules in the metal-rich chondrites Hammadah al Hamra 237 and Queen Alexandra Range 94411. *Meteoritics & Planet. Sci., 36,* 1189–1216.

Kuehner S. M., Laughlin J. R., Grossman L., Johnson M. L., and Burnett D. S. (1989) Determination of trace element mineral/liquid partition coefficients in melilite and diopside by ion and electron microprobe techniques. *Geochim. Cosmochim. Acta, 53,* 3115–3130.

Kurat G. (1975) Der kohlige Chondrit Lancé: Eine petrologische Analyse der komplexen Genese eines Chondriten. *Tschermaks Mineral. Petrogr., 22,* 38–78.

Kushiro I. (1975) On the nature of silicate melt and its significance in magma genesis: Regularities in the shift of the liquidus boundaries involving olivine, pyroxene, and silica minerals. *Am. J. Sci., 275,* 411–431.

LaTourrette T. Z. and Burnett D. S. (1992) Experimental determination of U and Th partitioning between clinopyroxene and natural and synthetic basaltic liquid. *Earth Planet. Sci. Lett., 110,* 227–244.

Lauretta D. S., Nagahara H., and Alexander C. M. O'D. (2006) Petrology and origin of ferromagnesian silicate chondrules. In *Meteorites and the Early Solar System II* (D. S. Lauretta and H. Y. McSween Jr., eds.), this volume. Univ. of Arizona, Tucson.

Liang Y., Richter F. M., and Chamberlin L. (1997) Diffusion in silicate melts: III. Empirical models for multicomponent diffusion. *Geochim. Cosmochim. Acta, 61,* 5295–5312.

Libourel G. (1999) Systematics of calcium partitioning between olivine and silicate melt: Implications for melt structure and calcium content of magmatic olivines. *Contrib. Mineral. Petrol., 136,* 63–80.

Lin Y. and Kimura M. (1998) Anorthite-spinel-rich inclusions in the Ningqiang carbonaceous chondrite: Genetic links with Type A and C inclusions. *Meteoritics & Planet. Sci., 33,* 435–446.

Lin Y. and Kimura M. (2000) Two unusual Type B refractory inclusions in the Ningqiang carbonaceous chondrite: Evidence for relicts, xenoliths and multi-heating. *Geochim. Cosmochim. Acta, 64,* 4031–4047.

Lin Y. and Kimura M. (2003) Ca-Al-rich inclusions from the Ningqiang meteorite: Continuous assemblages of nebular condensates and genetic link to Type B inclusions. *Geochim. Cosmochim. Acta, 67,* 2251–2267.

Longhi J. (1987) Liquidus equilibria and solid solution in the system $Ca_2Al_2Si_2O_8$-Mg_2SiO_4-$CaSiO_3$-SiO_2 at low pressure. *Am. J. Sci., 287,* 265–331.

Longhi J. and Pan V. (1988) A reconnaisance study of phase boundaries in low-alkali basaltic liquids. *J. Petrol., 29,* 115–147.

Lundstrom C. C., Shaw H. F., Ryerson F. J., Williams Q., and Gill J. (1998) Crystal chemical control of clinopyroxene-melt partitioning in the Di-Ab-An system: Implications for elemental fractionations in the depleted mantle. *Geochim. Cosmochim. Acta, 62,* 2849–2862.

MacPherson G. J. (2004) Calcium-aluminum-rich inclusions in chondritic meteorites. In *Treatise on Geochemistry, Vol. 1: Meteorites, Comets, and Planets* (A. M. Davis, ed.), pp. 201–246. Elsevier, Oxford.

MacPherson G. J. and Davis A. M. (1993) A petrologic and ion microprobe study of a Vigarano Type B refractory inclusion: Evolution by multiple stages of alteration and melting. *Geochim. Cosmochim. Acta, 57,* 231–243.

MacPherson G. J. and Grossman L. (1984) "Fluffy" type A Ca-, Al-rich inclusions in the Allende meteorite. *Geochim. Cosmochim. Acta, 48,* 29–46.

MacPherson G. J. and Huss G. R. (2000) Convergent evolution of CAIs and chondrules: Evidence from bulk compositions and a cosmochemical phase diagram (abstract). In *Lunar and Planetary Science XXXI,* Abstract #1796. Lunar and Planetary Institute, Houston (CD-ROM).

MacPherson G. J. and Huss G. R. (2005) Petrogenesis of Al-rich chondrules: Evidence from bulk compositions and phase equilibria. *Geochim. Cosmochim. Acta, 69,* 3099–3127.

MacPherson G. J., Bar-Matthews M., Tanaka T., Olsen E., and Grossman L. (1983) Refractory inclusions in the Murchison meteorite. *Geochim. Cosmochim. Acta, 47,* 823–839.

MacPherson G. J., Grossman L., Hashimoto A., Bar-Matthews M., and Tanaka T. (1984a) Petrographic studies of refractory inclusions from the Murchison meteorite. *Proc. Lunar Planet. Sci. Conf. 15th,* in *J. Geophys. Res., 89,* C299–C312.

MacPherson G. J., Paque J. M., Stolper E., and Grossman L. (1984b) The origin and significance of reverse zoning in melilite from Allende Type B inclusions. *J. Geol., 92,* 289–305.

MacPherson G. J., Wark D. A., and Armstrong J. T. (1988) Primitive material surviving in chondrites: Refractory inclusions. In *Meteorites and the Early Solar System* (J. F. Kerridge and M. S. Matthews), pp. 746–807. Univ. of Arizona, Tucson.

MacPherson G. J., Petaev M., and Krot A. N. (2004) Bulk compositions of CAIs and Al-rich chondrules: Implications of the reversal of the anorthite/forsterite condensation sequence at low nebular pressures (abstract). In *Lunar and Planetary Science XXXV,* Abstract #1838. Lunar and Planetary Institute, Houston (CD-ROM).

Maharaj S. V. and Hewins R. H. (1994) Alternative thermal histories for type B Ca-Al-rich inclusions (abstract). In *Lunar and Planetary Science XXV,* pp. 825–826. Lunar and Planetary Institute, Houston.

Maharaj S. V. and Hewins R. H. (1995) Flash heating a Type B CAI composition (abstract). In *Lunar and Planetary Science XXVI,* pp. 883–884. Lunar and Planetary Institute, Houston.

Maharaj S. V. and Hewins R. H. (1997) Effect of melt time and precursor phases on type B CAI textures (abstract). In *Lunar and Planetary Science XXVIII,* Abstract #1624. Lunar and Planetary Institute, Houston.

Malvin D. J. and Drake M. J. (1987) Experimental determination of crystal/melt partitioning of Ga and Ge in the system forsterite-anorthite-diopside. *Geochim. Cosmochim. Acta, 51,* 2117–2128.

Mason B. and Martin P. M. (1974) Minor and trace element distribution in melilite and pyroxene from the Allende meteorite. *Earth Planet. Sci. Lett., 22,* 141–144.

McMurdie H. F. and Insley H. (1936) Studies of the quaternary system CaO-MgO-$2CaO{\cdot}SiO_2$-$5CaO{\cdot}3Al_2O_3$. *J. Res. Nat. Bur. Standards, 16,* 467–474.

Meeker G. P., Wasserburg G. J., and Armstrong J. T. (1983) Replacement textures in CAI and implications regarding planetary metamorphism. *Geochim. Cosmochim. Acta, 47,* 707–721.

Mendybaev R. A., Richter F. M., and Davis A. M. (2003) Formation of the melilite mantle of the type B1 CAIs: Experimental simulations (abstract). In *Lunar and Planetary Science XXXIV,* Abstract #2062. Lunar and Planetary Institute, Houston (CD-ROM).

Miller S. A., Burnett D. S., and Asimow P. D. (2003) Experimental divalent element partitioning between anorthite and CAI melt (abstract). In *Lunar and Planetary Science XXXIV,* Abstract #1446. Lunar and Planetary Institute, Houston (CD-ROM).

Morse S. A. (1980) *Basalts and Phase Diagrams.* Springer-Verlag, Berlin. 493 pp.

Nagasawa H., Schreiber H. D., and Morris R. V. (1980) Experimental mineral/liquid partition coefficients of the rare earth elements (REE), Sc and Sr for perovskite, spinel and melilite. *Earth Planet. Sci. Lett., 46,* 431–437.

Nash W. P. and Crecraft H. R. (1985) Partition coefficients for trace elements in silicic magmas. *Geochim. Cosmochim. Acta, 49,* 2309–2322.

Nurse R. W., Welch J. H., and Majumdar A. J. (1965) The CaO-Al_2O_3 system in a moisture-free atmosphere. *Trans. British Cer. Soc., 64,* 409–418.

O'Neill H. S. C. and Eggins S. M. (2002) The effect of melt composition on trace element partitioning: An experimental investigation of the activity coefficients of FeO, NiO, CoO, MoO_2 and MoO_3 in silicate melts. *Chem. Geol., 186,* 151–181.

Onuma K. and Yagi K. (1967) The system diopside-akermanite-nepheline. *Am. Mineral., 52,* 227–243.

Onuma N., Higuchi H., Wakita H., and Nagasawa H. (1968) Trace element partitioning between two pyroxenes and the host lava. *Earth Planet. Sci. Lett., 5,* 47–51.

Osborn E. F. and Gee K. H. (1969) Part IV. Phase equilibria at liquidus temperatures for a part of the system CaO-MgO-Al_2O_3-TiO_2-SiO_2 and their bearing on the effect of titania on the properties of blast furnace slag. *Bull. Earth Mineral. Sci. Expt. Station, 85,* 57–80.

Osborn E. F., DeVries R. C., Gee K. G., and Kraner H. M. (1954) Optimum composition of blast furnace slag as deduced from liquidus data for the quaternary system CaO-MgO-Al_2O_3-SiO_2. *J. Metals (January),* 3–15.

Pack A. and Palme H. (2003) Partitioning of Ca and Al between forsterite and silicate melt in dynamic systems with implications for the origin of Ca, Al-rich forsterites in primitive meteorites. *Meteoritics & Planet. Sci., 38,* 1263–1281.

Pan V. and Longhi J. (1989) Low pressure liquidus relations in the system Mg_2SiO_4-Ca_2SiO_4-$NaAlSiO_4$-SiO_2. *Am. J. Sci., 289,* 1–16.

Pan V. and Longhi J. (1990) The system Mg_2SiO_4-Ca_2SiO_4-$CaAl_2O_4$-$NaAlSiO_4$-SiO_2: One atmosphere liquidus equilibria of analogs of alkaline mafic lavas. *Contrib. Mineral. Petrol., 105,* 569–584.

Paque J. M. (1995) Effect of residence time at maximum temperature on the texture and phase compositions of a Type B Ca-Al-rich inclusion analog (abstract). In *Lunar and Planetary Science XXVI,* pp. 1099–1100. Lunar and Planetary Institute, Houston.

Paque J. M., Beckett J. R., Barber D. J., and Stolper E. M. (1994) A new titanium-bearing calcium aluminosilicate phase: I. Meteoritic occurrences and formation in synthetic systems. *Meteoritics, 29,* 673–682.

Paque J. M., Le L., and Lofgren G. E. (1998) Experimentally produced spinel rims on Ca-Al-rich inclusions bulk compositions (abstract). In *Lunar and Planetary Science XXIX,* Abstract #1221. Lunar and Planetary Institute, Houston.

Paque J. M., Lofgren G. E., and Le L. (2000) Crystallization of calcium-aluminum-rich inclusions: Experimental studies on the effects of repeated heating events. *Meteoritics & Planet. Sci., 35,* 363–371.

Pelton A. D. and Blander M. (1984) Computer-assisted analysis of the thermodynamic properties and phase diagrams of slags. In *Second International Symposium on Metallurgical Slags and Fluxes* (H. A. Fine and D. R. Gaskell), pp. 281–294. Metallurgical Society AIME.

Pelton A. D., Degterov S. A., Eriksson G., Robelin C., and Dessureault Y. (2000) The modified quasichemical model I — Binary solutions. *Metall. Mater. Trans., 31B,* 651–659.

Peters M. T., Shaffer E. E., Burnett D. S., and Kim S. S. (1995) Magnesium and titanium partitioning between anorthite and Type B CAI liquid: Dependence on oxygen fugacity and liquid composition. *Geochim. Cosmochim. Acta, 59,* 2785–2796.

Prigogine I. and Defay R. (1954) *Chemical Thermodynamics.* Longmans, Essex.

Prince A. T. (1951) Phase equilibrium relationships in a portion of the system MgO-Al_2O_3-2CaO·SiO_2. *J. Am. Ceramic Soc., 34,* 44–51.

Rankin G. A. and Merwin H. E. (1916) The ternary system CaO-Al_2O_3-MgO. *J. Am. Chem. Soc., 38,* 568–588.

Rao M. R. (1968) Liquidus relations in the quaternary subsystem $CaAl_2O_4$-$CaAl_4O_7$-$Ca_2Al_2SiO_7$-$MgAl_2O_4$. *J. Am. Ceramic Soc., 51,* 50–54.

Rein R. H. and Chipman J. (1963) The distribution of silicon between Fe-Si-C alloys and SiO_2-CaO-MgO-Al_2O_3 slags. *Trans. Metall. Soc. AIME, 227,* 1193–1203.

Rein R. H. and Chipman J. (1965) Activities in the liquid solution SiO_2-CaO-MgO-Al_2O_3 at 1600°C. *Trans. Metall. Soc. AIME, 233,* 415–425.

Richter F. M., Davis A. M., Ebel D. S., and Hashimoto A. (2002) Elemental and isotopic fractionation of Type B calcium-, aluminum-rich inclusions: Experiments, theoretical considerations, and constraints on their thermal evolution. *Geochim. Cosmochim. Acta, 66,* 521–540.

Ringwood A. E. (1975) Some aspects of the minor element chemistry of lunar mare basalts. *Moon, 12,* 127–157.

Ryerson F. J. and Hess P. C. (1978) Implications of liquid-liquid distribution coefficients to mineral-liquid partitioning. *Geochim. Cosmochim. Acta, 42,* 921–932.

Schairer J. F. (1957) Melting relations of the common rock-forming oxides. *J. Am. Ceramic Soc., 40,* 215–235.

Schairer J. F. and Morimoto N. (1959) The system forsterite-diopside-silica-albite. *Carnegie Inst. Wash. Yearb., 58,* 113–118.

Schairer J. F. and Yoder H. S. (1960) The nature of residual liquids from crystallization, with data on the system nepheline-diopside-silica. *Am. J. Sci., 258A,* 273–283.

Schairer J. F. and Yoder H. S. (1961) Crystallization in the system nepheline-forsterite-silica at one atmosphere pressure. *Carnegie Inst. Wash. Yearb., 60,* 141–144.

Schairer J. F. and Yoder H. S. (1969) Critical planes and flow sheet for a portion of the system CaO-MgO-Al_2O_3-SiO_2 having petrological applications. *Carnegie Inst. Wash. Yearb., 68,* 202–214.

Schairer J. F., Tilley C. E., and Brown G. M. (1967) The join nepheline-diopside-anorthite and its relation to alkali basalt fractionation. *Carnegie Inst. Wash. Yearb., 66,* 467–471.

Schreiber H. D. (1979) Experimental studies of nickel and chromium partitioning into olivine from synthetic basaltic melts. *Proc. Lunar Planet. Sci. Conf. 10th,* pp. 509–516.

Schreiber H. D., Thanyasiri T., Lach J. J., and Legere R. A. (1978) Redox equilibria of Ti, Cr, and Eu in silicate melts: Reduction potentials and mutual interactions. *Phys. Chem. Glasses, 19,* 126–139.

Sheng Y. J. (1992) The origin of plagioclase-olivine inclusions. Ph.D. dissertation, California Institute of Technology, Pasadena.

Sheng Y. J., Beckett J. R., Hutcheon I. D., and Wasserburg G. J. (1991a) Experimental constraints on the origin of plagioclase-olivine inclusions and CA chondrules (abstract). In *Lunar and Planetary Science XXII,* pp. 1231–1232. Lunar and Planetary Institute, Houston.

Sheng Y. J., Hutcheon I. D., and Wasserburg G. J. (1991b) Origin of plagioclase-olivine inclusions in carbonaceous chondrites. *Geochim. Cosmochim. Acta, 55,* 581–599.

Shi P. (1992) Basalt evolution at low pressure: Implications from an experimental study in the system CaO-FeO-MgO-Al_2O_3-SiO_2. *Contrib. Mineral. Petrol., 110,* 139–153.

Shi P. and Libourel G. (1991) The effects of FeO on the system CMAS at low pressure and implications for basalt crystallization processes. *Contrib. Mineral. Petrol., 108,* 129–145.

Simon S. B. and Grossman L. (2003) Insights into the formation of type B2 refractory inclusions (abstract). In *Lunar and Planetary Science XXXIV,* Abstract #1796. Lunar and Planetary Institute, Houston (CD-ROM).

Simon S. B., Grossman L., and Davis A. M. (1991) Fassaite composition trends during crystallization of Allende Type B refractory inclusion melts. *Geochim. Cosmochim. Acta, 55,* 2635–2655.

Simon S. B., Kuehner S. M., Davis A. M., Grossman L., Johnson M. L., and Burnett D. S. (1994a) Experimental studies of trace element partitioning in Ca, Al-rich compositions: Anorthite and perovskite. *Geochim. Cosmochim. Acta, 58,* 1507–1523.

Simon S. B., Yoneda S., Grossman L., and Davis A. M. (1994b) A $CaAl_4O_7$-bearing refractory spherule from Murchison: Evidence for very high-temperature melting in the solar nebula. *Geochim. Cosmochim. Acta, 58,* 1937–1949.

Simon S. B., Davis A. M., and Grossman L. (1996a) A unique ultrarefractory inclusion from the Murchison meteorite. *Meteoritics & Planet. Sci., 31,* 106–115.

Simon S. B., Davis A. M., Richter F.-M., and Grossman L. (1996b) Experimental investigation of the effect of cooling rate on melilite/liquid distribution coefficients for Sr, Ba, and Ti in Type B refractory inclusion melts (abstract). In *Lunar and Planetary Science XXVII,* pp. 1201–1202. Lunar and Planetary Institute, Houston.

Simon S. B., Davis A. M., and Grossman L. (1998) Formation of an unusual compact Type A refractory inclusion from Allende. *Meteoritics & Planet. Sci., 33,* 115–126.

Simon S. B., Davis A. M., and Grossman L. (1999) Origin of compact type A refractory inclusions from CV3 carbonaceous chondrites. *Geochim. Cosmochim. Acta, 63,* 1233–1248.

Soulard H., Provost A., and Boivin P. (1992) CaO-MgO-Al_2O_3-SiO_2-Na_2O (CMASN) at 1 bar from low to high Na_2O contents: Topology of an analogue for alkaline basic rocks. *Chem. Geol., 96,* 459–477.

Spear F. S. (1993) *Metamorphic Phase Equilibria and Pressure-Temperature-Time Paths.* Mineralogical Society of America, Washington, DC.

Srinivasan G., Huss G. R., and Wasserburg G. J. (2000) A petrographic, chemical, and isotopic study of calcium-aluminum-rich inclusions and aluminum-rich chondrules from the Axtell (CV3) chondrite. *Meteoritics & Planet. Sci., 35,* 1333–1354.

Steele I. M., Peters M. T., Shaffer E. E., and Burnett D. S. (1997) Minor element partitioning and sector zoning in synthetic and meteoritic anorthite. *Geochim. Cosmochim. Acta, 61,* 415–423.

Stolper E. (1982) Crystallization sequences of Ca-Al-rich inclusions from Allende: An experimental study. *Geochim. Cosmochim. Acta, 46,* 2159–2180.

Stolper E. and Paque J. M. (1986) Crystallization sequences of Ca-Al-rich inclusions from Allende: The effects of cooling rate and maximum temperature. *Geochim. Cosmochim. Acta, 50,* 1785–1806.

Terakado Y. and Masuda A. (1979) Experimental study of REE partitioning between diopside and melt under atmospheric pressure. *Geochem. J., 13,* 121–129.

Tingle T. N. (1987) An evaluation of the carbon-14 beta track technique: Implications for solubilities and partition coefficients determined by beta track mapping. *Geochim. Cosmochim. Acta, 51,* 2479–2487.

Walker D., Shibata T., and DeLong S. E. (1979) Abyssal tholeiites from the Oceanographer Fracture Zone II. Phase equilibria and mixing. *Contrib. Mineral. Petrol., 70,* 111–125.

Wark D. A. (1981) The pre-alteration compositions of Allende Ca-Al-rich condensates (abstract). In *Lunar and Planetary Science XII,* pp. 1148–1150. Lunar and Planetary Institute, Houston.

Wark D. A. (1987) Plagioclase-rich inclusions in carbonaceous chondrite meteorites: Liquid condensates? *Geochim. Cosmochim. Acta, 51,* 221–242.

Wark D. A., Boynton W. V., Keays R. R., and Palme H. (1987) Trace element and petrologic clues to the formation of forsterite-bearing Ca-Al-rich inclusions in the Allende meteorite. *Geochim. Cosmochim. Acta, 51,* 607–1622.

Watson E. B. (1976) Two-liquid partition coefficients: Experimental data and geochemical implications. *Contrib. Mineral. Petrol., 56,* 119–134.

Watson E. B. (1977) Partitioning of manganese between forsterite and silicate liquid. *Geochim. Cosmochim. Acta, 41,* 1362–1374.

Weber D. and Bischoff A. (1994) The occurrence of grossite ($CaAl_4O_7$) in chondrites. *Geochim. Cosmochim. Acta, 58,* 3855–3877.

Weber D. and Bischoff A. (1997) Refractory inclusions in the CR chondrite Acfer 059 — El Djouf 001: Petrology, chemical composition, and relationship to inclusion populations in other types of carbonaceous chondrites. *Chem. Erde, 57,* 1–24.

Wood B. J. and Blundy J. D. (2001) The effect of cation charge on crystal-melt partitioning of trace elements. *Earth Planet. Sci. Lett., 188,* 59–71.

Wood B. J. and Blundy J. D. (2004) Trace element partitioning under crustal and uppermost mantle conditions: The influences of ionic radius, cation charge, pressure, and temperature. In *Treatise on Geochemistry, Vol. 2: The Mantle and Core* (R. W. Carlson, ed.), pp. 395–424. Elsevier, Oxford.

Woolum D. S., Johnson M. L., Burnett D. S., and Sutton S. R. (1988) Refractory lithophile partitioning in Type B CAI materials (abstract). In *Lunar and Planetary Science XIX,* pp. 1294–1295. Lunar and Planetary Institute, Houston.

Yagi K. and Onuma K. (1969) An experimental study on the role of titanium in alkalic basalts in light of the system diopside-akermanite-nepheline-$CaTiAl_2O_6$. *Am. J. Sci., 267-A,* 509–549.

Yoneda S. and Grossman L. (1995) Condensation of CaO-MgO-Al_2O_3-SiO_2 liquids from cosmic gases. *Geochim. Cosmochim. Acta, 59,* 3413–3444.

Yurimoto H., Rubin A. E., Itoh S., and Wasson J. T. (2001) Non-stoichiometric Al-rich spinel in an ultrarefractory inclusion from CO chondrite (abstract). In *Lunar and Planetary Science XXXII,* Abstract #1557. Lunar and Planetary Institute, Houston (CD-ROM).

Zen E-an (1966) *Construction of Pressure-Temperature Diagrams for Multicomponent Systems after the Method of Shreinemakers — A Geometric Approach.* U.S. Geol. Survey Bulletin 1225. 56 pp.

Zhang S. and Lee W. E. (2000) Use of phase diagrams in studies of refractories corrosion. *Intl. Mater. Rev., 45,* 41–58.

Petrology and Origin of Ferromagnesian Silicate Chondrules

Dante S. Lauretta
University of Arizona

Hiroko Nagahara
University of Tokyo

Conel M. O'D. Alexander
Carnegie Institution of Washington

Ferromagnesian silicate chondrules are major components of most primitive meteorites. The shapes, textures, and mineral compositions of these chondrules are consistent with crystallization of a molten droplet that was floating freely in space in the presence of a gas. The texture and mineralogy of a chondrule reflects the nature and composition of its precursor material as well as its thermal history. There is an enduring debate about the degree to which chondrules interacted with the ambient gas during formation. In particular, it is uncertain whether or not chondrules experienced evaporation during heating and recondensation during cooling. The extent to which these processes took place in chondrule melts varied as a function of the duration of heating as well as the environmental conditions such as pressure, temperature, composition, and size. Thus, locked in chondrule bulk compositions, mineralogy, and textures are clues to the ambient conditions of the solar nebula and the nature of material processing in the inner solar system at the early stages of planet formation. Here we survey what is known about the properties of ferromagnesian chondrules in primitive meteorites and use this information to place constraints on these important parameters. The topics covered in this chapter are listed in Table 1.

TABLE 1. Contents of this chapter.

Section	Topic
1	Introduction
2	Properties of Ferromagnesian Chondrules
3	Precursor Chemistry and Mineralogy
4	Thermal Histories and Crystallization of Chondrules
5	Metallogenesis and Redox Reactions in the Melt
6	Elemental Evaporation and Recondensation
7	Volatile Recondensation and Formation of Chondrule Rims
8	Constraints on Models of Chondrule Formation

1. INTRODUCTION

The most primitive meteorites, chondrites, have textures that suggest formation by accumulation of material that was freely floating in the solar nebula (Fig. 1). In a sense, they can be viewed as sedimentary rocks that swept up material in specific regions of the early solar system. While part of their parent asteroids, chondrites experienced varying degrees of aqueous alteration, thermal metamorphism, and impact shock. Mineral compositions within and among the least-altered chondrites are highly variable. This wide chemical variation indicates that the different components have not been in chemical communication long enough to homog-

Fig. 1. Plane-polarized transmitted light image of the Semarkona (LL3.0) primitive unequilibrated ordinary chondrite (from *Lauretta and Killgore,* 2005). Chondrules make up ~70 vol% of this meteorite. The remainder is composed of dark, opaque, fine-grained matrix. FOV = 1.5 cm.

TABLE 2. Representative bulk compositions for major chondrule types (in wt%).

	IA*	IAB†	IB†	IIA‡	IIAB§	IIB§
SiO_2	44.8	56.4	57.9	45.1	53.8	58.7
TiO_2	0.19	0.11	0.11	0.1	0.16	0.18
Al_2O_3	3.9	2.4	2.4	2.68	4.8	4.6
Cr_2O_3	0.44	0.58	0.59	0.51	0.5	0.59
FeO	1.18	3.4	2.8	14.9	8.8	7.8
MnO	0.12	0.36	0.28	0.39	0.6	0.54
MgO	40.7	33.9	33.4	31.3	25.1	21.4
CaO	3.5	1.7	1.7	1.89	2.7	3.3
Na_2O	0.52	0.56	0.46	1.65	2.1	1.9
K_2O	0.07	0.06	0.05	0.17	0.28	0.29
P_2O_5	—	—	—	0.34	—	—
FeS	0.2	0.3	0.09	0.94	0.28	0.13
FeNi	3.76	0.32	0.28	0.51	—	0.08

**Jones and Scott* (1989).
†*Jones* (1994).
‡*Jones* (1990).
§*Jones* (1996).

enize. These meteorites, referred to as unequilibrated chondrites, preserve a number of chemical, mineralogical, and structural signatures inherited from the protoplanetary disk.

The major components (50–80 vol%) of most chondrites are ~1-mm-diameter igneous spheres, called chondrules. Representative bulk compositions of natural chondrules are summarized in Table 2, and illustrate the range in their bulk chemistry. In this chapter, we focus on the most common types of chondrules, which are primarily composed of the ferromagnesian minerals olivine [$(Mg,Fe)_2SiO_4$], low-Ca pyroxene [$(Mg,Fe)SiO_3$], and high-Ca pyroxene [$Ca(Mg,Fe)Si_2O_6$]. These objects also contain a silicate-glass mesostasis rich in Na, Al, Si, Ca, and K, relative to chondrite bulk composition. Iron-based alloys, sulfides, and oxides are present in varying abundances.

In unequilibrated chondrites, chondrules are often surrounded by rims of fine-grained material. Chondrules and rims are themselves cemented together by a fine-grained material mineral matrix. Like the chondrules, the rims and matrix are dominated by silicates, metals, and sulfides. However, they are compositionally and mineralogically distinct from chondrules.

2. PROPERTIES OF FERROMAGNESIAN CHONDRULES

2.1. Chondrule Classification

Gooding and Keil (1981) performed a comprehensive petrographic survey of chondrules in unequilibrated ordinary chondrites. They classified chondrules based on their textures (see Fig. 2 and Table 3). Porphyritic chondrules contain abundant relatively large (up to one-half the chondrule diameter), fairly uniformly sized crystals set in a fine-grained or glassy mesostasis. Porphyritic chondrules that are dominated by olivine (>10/1 volume ratio) are referred to as porphyritic olivine (PO) chondrules, while those that contain abundant pyroxene (>10/1 volume ratio) are porphyritic pyroxene chondrules (PP). Chondrules with silicate mineral abundances between these two extremes are called porphyritic olivine-pyroxene (POP) chondrules. Granular (G) chondrules contain many small grains of uniform size (<10 μm) with poorly defined crystal outlines. Granular chondrules are also subdivided into GO, GOP, and GP groups. Barred-olivine chondrules (BO) contain olivine crystals that have one dimension that is markedly longer than the other two when viewed in thin section. Olivine grains in barred chondrules either all have the same crystallographic orientation or contain several sets of such oriented grains. Radial-pyroxene chondrules (RP) consist of fan-like arrays of low-Ca pyroxene, which radiate from one or more points near their surface. Cryptocrystalline chondrules are objects that do not exhibit any recognizable crystal structure under an optical microscope. They have a high abundance of glassy material and may contain submicrometer dendrites. Metallic chondrules consist of Fe,Ni metal, some troilite (FeS), and minor amounts of accessory phases such as schreibersite and metallic Cu (*La Blue and Lauretta,* 2004). Neither chondrule size nor shape is strongly correlated with textural type. Since the study of *Gooding and Keil* (1981), two other chondrule classification schemes have been developed. The most commonly used scheme divides chondrules into two broad categories: type-I and type-II, based on bulk FeO contents (*McSween,* 1977; *Scott and Taylor,* 1983). Type-I chondrules are characterized by the presence of FeO-poor olivine and pyroxene (Fo and En > 90). Type-II chondrules contain FeO-rich olivine and pyroxene (Fo and En < 90). The textural characteristics of both the type-I and type-II series are gradational. These two main categories are further subdivided into subtypes A and B based on their abundances of olivine and pyroxene. Type-IA and IIA chondrules contain abundant olivine (>80 vol%), type-IB and IIB chondrules contain abundant pyroxene (>80 vol%), and chondrules with intermediate abundances of olivine and pyroxene are classified as type-IAB or IIAB. This approach along with the *Gooding and Keil* (1981) classification scheme are often used in conjunction with one another.

In addition to the textural and chemical classification schemes presented above, another, independent chondrule classification system has been proposed by *Sears et al.* (1992) and *DeHart et al.* (1992). This system does not rely on textural properties or modal mineralogy. Instead, chondrules are classified based on their cathodoluminescence properties and mineral compositions. Cathodoluminescence refers to the emission of visible light from a solid that is excited by interaction with an electron beam. Cathodoluminescence is highly dependent on trace-element abundances and crystal lattice defects. Chondrules that luminesce brightly are classified as group-A chondrules, while those that show little or no luminescence are group-B chondrules.

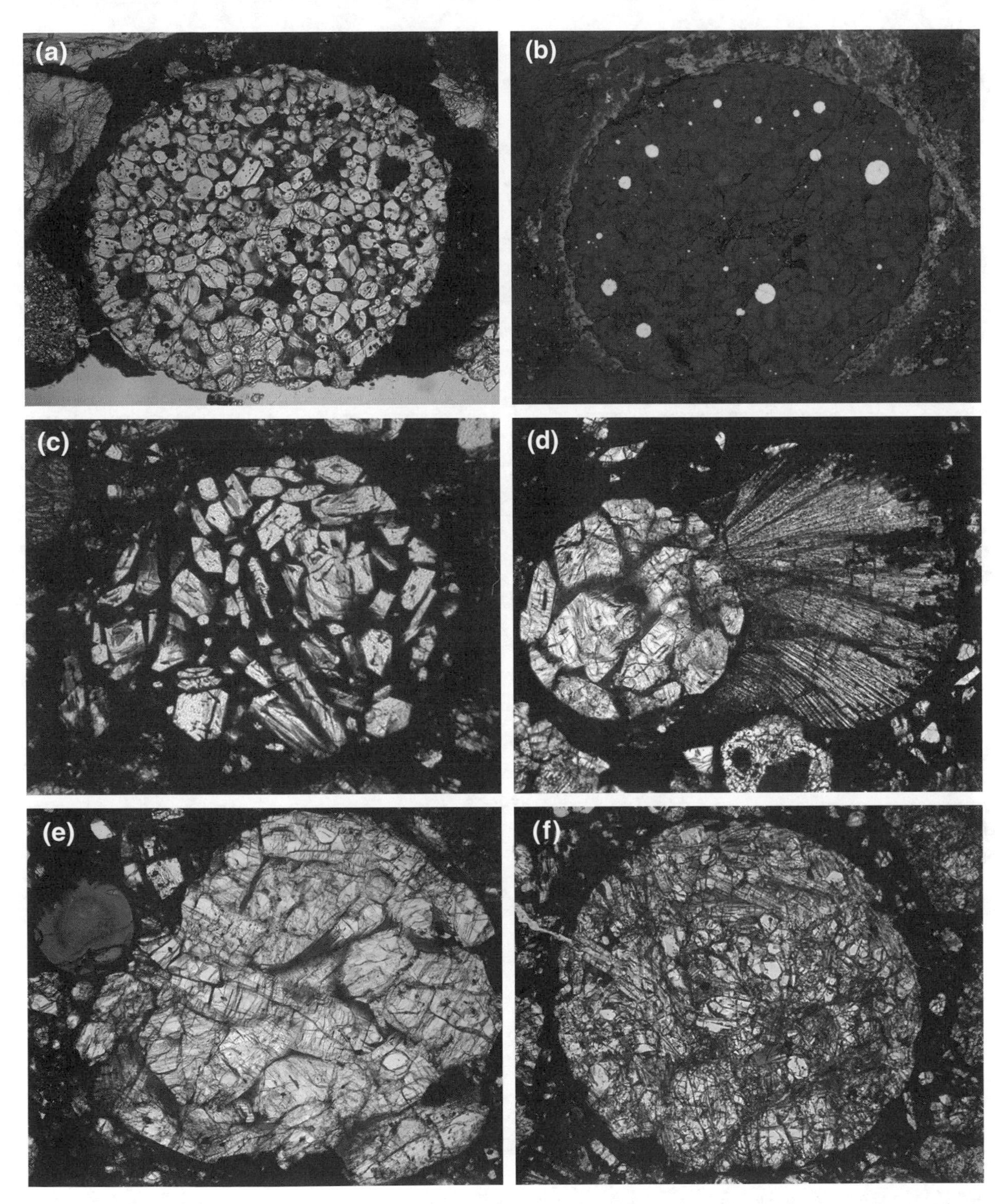

Fig. 2. Optical microscopy images illustrating the variety of chondrule textures. FOV = 1.35 mm in all cases except **(h)** (FOV = 2.7 mm). **(a)** PO chondrule from QUE 97008 (L). **(b)** Reflected image of the same chondrule shown in **(a)**. Note the rounded droplets of metal concentrated near the chondrule boundary. **(c)** Another PO chondrule from Clovis (H) illustrating a larger grain size distribution compared to the chondrule in **(a)**. **(d)** A PO-RP chondrule pair from EET 90066 (L). **(e)** A PP chondrule from ALH 78119 (large chondrule). Note also the small cryptocrystalline chondrule in the upper left. **(f)** A POP chondrule from Bishunpur (LL3.1). Note the olivine crystals concentrated near the center and the pyroxene toward the outer edge.

These two groups are further subdivided based on the mean FeO and CaO concentration in olivine and the composition of the chondrule mesostasis. *Sears et al.* (1992) identified four groups of chondrules as primitive and suggested that the other groups represent alteration of primitive chondrules on meteorite parent asteroids. Cathodoluminescence is a less widely available technique, and as a result the Sears et al. classification is not commonly used.

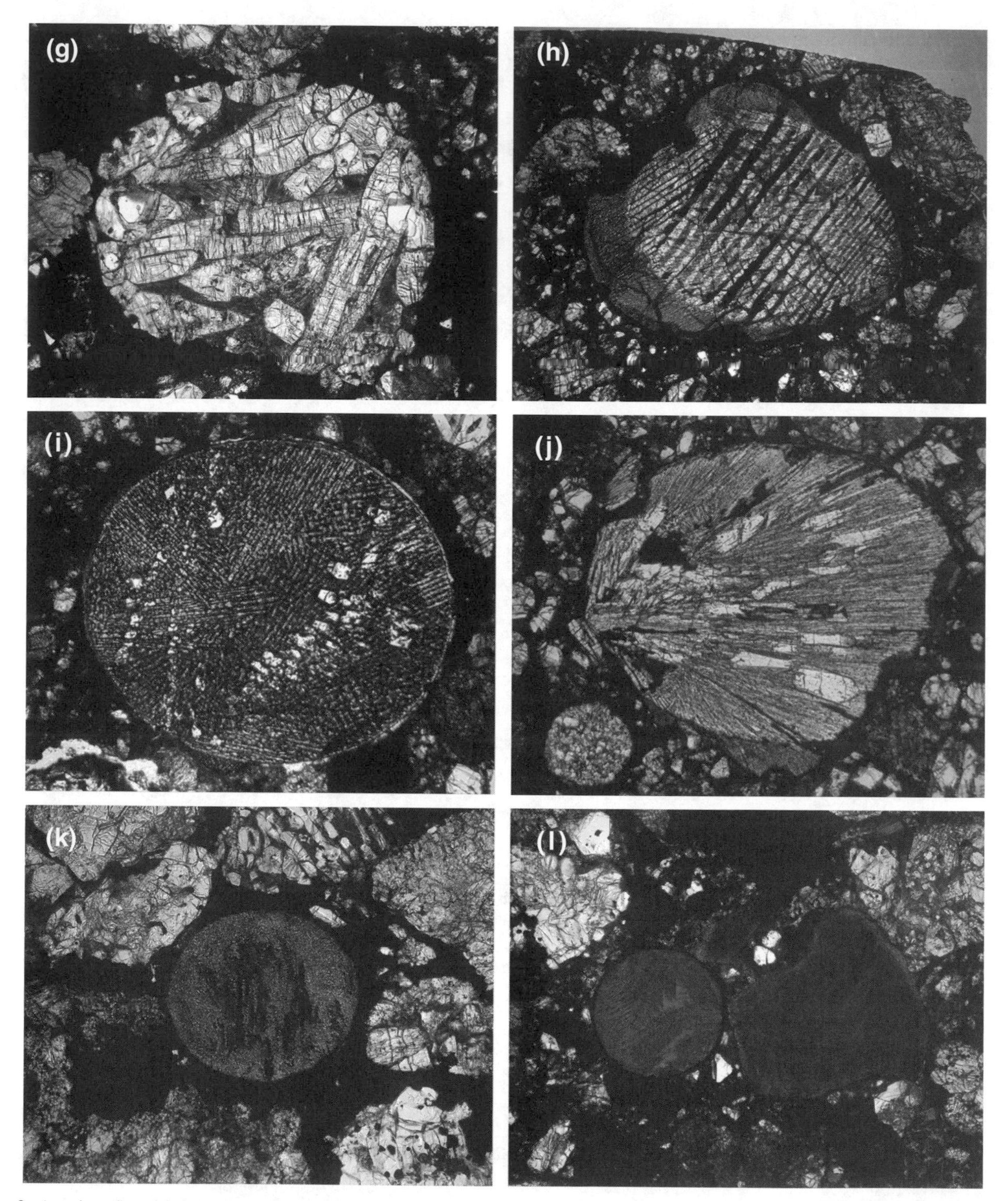

Fig. 2. (continued). **(g)** A coarse-grained porphyritic pyroxene chondrule from EET 90066 (L). **(h)** A barred-olivine chondrule from Bishunpur (LL3.1). **(i)** A barred-olivine chondrule from Saratov (L/LL4) showing multiple barred units. Note the presence of some porphyritic olivine crystals. **(j)** A radiating pyroxene chondrule from ALH 78119 (L). **(k)** A granular-olivine chondrule from QUE 97008 (L). **(l)** A pair of crypto-crystalline chondrules from ALH 78119 (L).

2.2. Chondrule Petrology

The most detailed studies of chondrule petrology have been conducted on Semarkona (LL3.0), one of the least-metamorphosed ordinary chondrites. The results of this work are summarized below. However, it is important to remember that the sizes and relative abundances of chondrules vary significantly among chondrite classes.

Type-IA porphyritic chondrules from the Semarkona (LL3.0) chondrite have been described in detail (*Jones and Scott,* 1989). These objects contain metallic Fe,Ni and abundant small olivine crystals in a glassy to microcrystalline

TABLE 3. Chondrule classification systems.

Textural Classification		
Abbreviation	Abundance in OCs (%)	Texture
POP	47–52	Porphyritic, both olivine and pyroxene
PO	15–27	Porphyritic, dominated by olivine (>10/1 by volume)
PP	9–11	Porphyritic, abundant pyroxene (>10/1 by volume)
RP	7–9	Radial pyroxene
BO	3–4	Barred olivine
CC	3–5	Cryptocrystalline
GOP	2–5	Granular olivine-pyroxene
M	<1	Metallic

Compositional Classification			
Type	Silicate Composition	Subtype	Silicate Abundances
I	FeO-poor (Fo, En > 90)	A	Abundant olivine (>80 vol%)
		AB	Intermediate abundances of olivine and pyroxene
II	FeO-rich (Fo, En < 90)	B	Abundant pyroxene (>80 vol%)

mesostasis. Olivine grain sizes are typically 15–40 μm. This small grain size is a diagnostic property of type-IA porphyritic chondrules. The olivine grains are typically euhedral with well-formed crystal faces in contact with the mesostasis. Some olivines contain inclusions of mesostasis. It is common for the low-Ca pyroxene in these chondrules to be concentrated near the chondrule margins and poikilitically enclose olivine grains. Rims of Ca-rich pyroxene occur on the low-Ca pyroxene grains. Rounded grains of Fe,Ni metal and associated minerals are evenly distributed throughout the chondrules or concentrated near their margins.

In type-IAB porphyritic chondrules in Semarkona, euhedral equant olivine crystals are up to 150 μm in diameter (*Jones,* 1994). Olivine (averaging 25 μm across) also occurs in large poikilitic low-Ca pyroxene crystals. Crystals of low-Ca pyroxene are twinned, generally tabular, and contain lamellar, compositional zoning. Calcium-rich pyroxene occurs as thin (<20 μm), discontinuous overgrowth rims on low-Ca pyroxene. Contacts between low-Ca and high-Ca pyroxene are sharp. Chemical concentration gradients are common in these Ca-rich pyroxene rims from the inner edge in contact with the low-Ca pyroxene to the outer edge next to the mesostasis. Grains of Fe,Ni metal are also common. Mesostasis is glassy and, in some cases, contains needle-like crystallites of Ca-rich pyroxene.

Type-IB porphyritic chondrules in Semarkona are FeO-poor. They are distinguished from type-IAB chondrules by their high proportion of pyroxene and low abundance (<20%) of olivine (*Jones,* 1994). However, olivine is present in even the most pyroxene-rich type-IB chondrules and commonly occurs as relict grains (see section 3.2.2 below).

Type-IIA porphyritic chondrules in Semarkona have an overall rounded morphology but are not, in general, perfectly spherical (*Jones,* 1990). They contain euhedral olivine (10–250 μm in diameter averaging 50 μm). Olivine grains have average compositions of Fo_{85}. They are chemically zoned with FeO increasing toward grain edges. Troilite occurs as small (<10 μm) rounded blebs distributed throughout the phenocrysts and mesostasis. Small (<1 μm) euhedral grains of chromite ($FeCr_2O_4$) also occur in the mesostasis. Low-calcium pyroxene and Fe,Ni metal are rare.

In Semarkona, type-IIB chondrules pyroxene morphologies vary from euhedral crystals to sets of subparallel bars (*Jones,* 1990). The pyroxene cores consist of enstatite or clinoenstatite and are overgrown with narrow rims of Ca-rich pyroxene, typically successive overgrowth of augite on pigeonite. Augite occurs as prisms on the ends of clinoenstatite laths. Mesostasis consists of glass with abundant microcrystallites of augite. Small rounded grains of metal and sulfide are present.

3. PRECURSOR CHEMISTRY AND MINERALOGY

In order to understand the chemical evolution of chondrule melts, it is necessary to place some constraints on the composition and mineralogy of their precursor material. Based on our current models of the early solar system, there are several possible sources of this precursor material, including presolar material, nebular condensates, fragments of earlier generations of chondrules, and fragments from planetesimals.

Presolar material is primordial dust inherited from the protosolar molecular cloud, some of which still survives in the chondrule rims and matrices of chondrites, including circumstellar grains and interstellar organic matter (*Huss et al.,* 1981). A significant proportion of material in the inner solar system may have been vaporized in the high-temperature environment near the proto-Sun. As the nebula cooled, the most refractory material condensed as solid grains. These nebular condensates may have then been reheated during the event (or events) responsible for chondrule formation. Some models suggest that the vapor pressure of the major rock-forming elements was high enough in the inner solar system for chondrule melts to condense directly from the gas phase. If there were multiple heating events, then chondrules could have been fragmented and remelted many times, becoming their own precursors. Finally, it is possible that chondrules formed after larger planetesimals underwent chemical differentiation. In this case, chondrule precursors could be material that was ejected by collisions between planetesimals or during volcanic eruptions on asteroid surfaces. In order to constrain the various models for chondrule precursors, we look to their mineralogy, bulk chemistry, and isotopic compositions, as well as to the information preserved in the relict grains that occur in many chondrules.

3.1. Constraints on Chondrule Precursors from Bulk Compositions

Studies of chondrule bulk compositions have been used to infer the nature of chondrule precursors. In a series of detailed investigations, the bulk compositions, accompanied by petrographic control, were determined for ordinary (*Grossman and Wasson,* 1982, 1983a; *Swindle et al.,* 1991a,b), enstatite (*Grossman et al.,* 1985), and carbonaceous chondrite (*Rubin and Wasson,* 1987, 1988) chondrules. Detailed reviews of these studies are given in *Grossman and Wasson* (1983b), *Grossman et al.* (1988), and *Grossman* (1996).

Elemental correlations are present in chondrules that appear to be related to their volatility and chemical affinity, although there are some variations between chondrite groups. The three most universal correlations occur within the refractory lithophiles (Al, Ca, Ti, Sc, and REEs), common siderophiles and some chalcophiles (Fe, Co, Ni, Se), and the alkalis (Na and K). The behavior of elements like Mg, Mn, and Cr are more complex, and this is reflected in how the compositions of olivine and pyroxene correlate with the different elemental groups. For instance, the refractory lithophiles correlate with the presence of FeO-poor olivine in the ordinary and enstatite chondrites, with FeO-poor pyroxene in the CM and CO chondrites, and with neither in the CV chondrites. As would be expected, the siderophile and chalcophile elements are generally associated with metal and troilite, although the behavior of Au, As, and Ga varies between the chondrite groups and the refractory siderophiles Ir, Os, and Ru appear to have affinities with the common siderophiles and refractory lithophiles.

These elemental correlations are thought to reflect variable abundances of different nebular condensates in their precursors. For example, the refractory component in ordinary chondrite chondrules has been interpreted as being olivine-rich condensates formed above the condensation temperature of enstatite, while the nonrefractory component (FeO-rich pyroxene, moderately volatile lithophiles, etc.) may have formed from fine-grained materials that were able to equilibrate at lower nebular temperatures (*Grossman and Wasson,* 1983b). Type-IIA chondrules from Semarkona (LL3.0) have (Na + K)/Al atomic ratios that approach but do not exceed unity, which has been taken as evidence that the principle alkali-bearing precursor was albite (*Hewins,* 1991). The fact that chondrules from different chondrite groups exhibit different element-mineral correlations would require that the condensate precursors for each chondrite group formed under somewhat different conditions.

There are two potential problems with this interpretation. By the time olivine condenses there are several predicted condensate minerals that would be the hosts for the refractory elements: corundum (Al_2O_3), hibonite ($CaAl_{12}O_{19}$), spinel ($MgAl_2O_4$), perovskite ($CaTiO_3$), melilite [$Ca_2(Mg,Al)(Si,Al)_2O_7$], anorthite ($CaAl_2Si_2O_8$), and diopside ($CaMgSi_2O_6$). The correlation between the refractory lithophiles would require little fractionation between these minerals, despite the fact that some fractionation between condensates is required to produce the less-refractory lithophile component(s). Perhaps more importantly, none of these minerals have been found as relict grains in ordinary chondrite chondrules (see section 3.2.2). Nor have they been found in significant amounts in the rims or matrix of the least-altered chondrites. Yet rims and matrix do preserve primitive materials, including presolar materials.

An alternative explanation for the interelement correlations is that they were produced by some combination of evaporation, metal/sulfide loss, FeO reduction, and/or random sampling of recycled chondrule material (*Sears et al.,* 1996; *Alexander,* 1994, 1996). Starting with a homogenous population of precursors, variable degrees of evaporation, FeO reduction, and metal/sulfide loss would automatically induce a correlation between refractory lithophile elements by enriching them in the residual chondrules. As discussed in section 3.2.2, recycled chondrule material was an important component of chondrule precursors. All the refractory lithophiles behave incompatibly in chondrule melts, so they are concentrated in the glass. Thus, a correlation between the refractory lithophiles could be produced by variable amounts of glass from an earlier generation of chondrules in the precursors. Metal/silicate fractionation during chondrule formation would create siderophile-element correlations. The upper limit of 1 for (Na + K)/Al ratios in type-II Semarkona chondrules may simply reflect the fact that for the alkalis to be easily accommodated in a silicate melt or glass requires the coupled substitution $(Na,K)AlO_2$ — SiO_2 (*Alexander,* 1994).

Ultimately, the relative importance of precursors and chondrule formation conditions depends on the degree to which individual chondrules were open or closed chemical systems during formation (see section 6). However, bulk-chondrule refractory-element fractionations are unlikely to be produced during evaporation and recondensation associated with ferromagnesian chondrule formation. Therefore, if they are present they must reflect variations in precursor compositions.

The rare earth elements (REEs) and other refractory lithophiles are very useful probes of geochemical and cosmochemical processes. The REEs can be strongly fractionated from one another during planetary differentiation, during high-temperature condensation, and during extreme evaporation. Most chondrules from ordinary (*Grossman and Wasson,* 1983a; *Kurat et al.,* 1984; *Alexander,* 1994), enstatite (*Grossman and Wasson,* 1985), and CH (*Krot et al.,* 2001a) chondrites have nearly flat CI-normalized REE patterns, although some have small anomalies in Eu and possibly Yb. There is no evidence in the REE patterns of these chondrules that their precursors had experienced planetary differentiation, evaporation, or condensation.

Some chondrules from CV and CO carbonaceous chondrites exhibit REE patterns that resemble patterns in some CAIs (*Misawa and Nakamura,* 1988a,b, 1996). The abundance of CAIs in CVs and COs makes the inclusion of CAI material in some chondrule precursors very likely, unless CAIs and chondrules were mixed together after chondrule formation. The preservation of a few CAIs inside chondrules (*Itoh and Yurimoto,* 2003) suggests that they were mixed together before chondrule formation. The reason that more CAIs are not found inside chondrules is probably because their liquidus temperatures are lower than those of most chondrules.

There are very few CAIs in ordinary or enstatite chondrites (*Guan et al.,* 2000; *Fagan et al.,* 2001). Thus, it is perhaps not surprising that most chondrules in these meteorites have unfractionated REE patterns. However, *Pack et al.* (2004) have found two chondrules in two ordinary chondrites with large negative anomalies in Sm, Eu, and Yb. The pattern of anomalies are consistent with predictions of condensation under very reducing conditions, but have not been observed even in CAIs from the highly reduced enstatite chondrites. The two chondrules themselves are not very reduced, but reduced condensates could have been minor components of their precursors. If this is the case, it is surprising that similar patterns have not been found in enstatite chondrite chondrules.

3.2. Mineralogical Constraints on Chondrule Precursors

3.2.1. Primitive chondrules. Chondrules exhibit a wide variety of textures and grain sizes. Since many chondrules were heated enough to melt, some chondrules must have experienced temperatures between ambient and melting. The finest-grained chondrules are considered to be the least altered and therefore best preserve precursor mineralogy and bulk compositions. Foremost among such objects are agglomeratic olivine chondrules, also referred to in the literature as "dark zoned chondrules" (*Dodd,* 1971; *Hewins,* 1997) and "fine-grained lumps" (*Rubin,* 1984). A comprehensive description of these objects is given by *Weisberg and Prinz* (1996). They are comparable in size to true droplet chondrules. They are typically composed of a coarser-grained core surrounded by material primarily comprised of fine-grained (<5 μm) olivine, which constitutes 57–94 vol%. Larger grains of olivine, up to 400 μm, can also be present. They also contain lesser amounts of pyroxene and feldspathic glass (<20 vol%). Minor components include chromite, metal, and sulfide (<5 vol%). Some agglomeratic olivine chondrules contain fragments of prior generations of chondrules and refractory inclusions.

Olivine compositions in a single agglomeratic chondrule span a wide range of FeO contents, indicating formation in diverse environments with little subsequent equilibration. Many of these objects are layered and exhibit distinct textural, mineralogical, compositional, and O-isotopic differences between core and rim. Their textures are consistent with a thermal history in which they were briefly (minutes to hours) heated up to temperatures high enough to result in sintering, but not enough to produce significant melting. Based on the characteristics of agglomeratic chondrules, we can infer that chondrule precursor materials were dominated by olivine crystals 2–5 μm in size with varying amounts of pyroxene, metal, troilite, chromite, and alkali feldspar or feldspathic glass.

Metal is a common constituent of type-I and agglomeratic chondrules and was likely a component of chondrule precursors. Additional metal may have formed as a result of *in situ* reduction either by carbonaceous material in the chondrule precursors or reaction with the nebula gas. Carbonaceous material occurs in chondrule interiors as graphite inclusions in metal (*Mostefaoui and Perron,* 1994), as fine-grained carbonaceous material associated with metal and sulfide (*Lauretta and Buseck,* 2003), and as poorly graphitized C included in chondrule silicates (*Fries and Steele,* 2005).

In many chondrules sulfide occurs as droplets and as eutectic mixes when associated with metal. These textures are consistent with crystallization of a sulfide melt. In the finest-grained, least-heated chondrules, a troilite-pentlandite assemblage occurs associated with minor amounts of kamacite. The co-occurrence of these two sulfide phases is consistent with experimental studies of kamacite sulfurization under solar nebula conditions (*Lauretta et al.,* 1997, 1998) and suggests that sulfides were an important component of some chondrule precursors (*Yu et al.,* 1996).

Some researchers have suggested that chondrules could have been produced by volcanism on planetesimals (*Hutchison and Graham,* 1975; *Kennedy et al.,* 1992) or by impacts between partially molten planetesimals (*Hutchison et al.,* 1988). Chondrules formed in this way would contain distinct geochemical or isotopic signatures associated with the partial melting and differentiation of their planetary bodies. Igneous fragments occur in primitive ordinary chondrites that contain just such signatures (*Graham et al.,* 1976;

Hutcheon and Hutchison, 1989; *Hutchison and Graham,* 1975; *Hutchison et al.,* 1988; *Kennedy et al.,* 1992). However, such igneous fragments are relatively rare and do not appear to have been the primary source of chondrule precursors, nor do chondrules show the major- or trace-element fractionations that are characteristic of partial melting and differentiation.

3.2.2. Relict grains in chondrules. Some olivine and orthopyroxene grains have compositions, particularly FeO contents, that are distinctly different from those of the other grains in their host chondrules (*Nagahara,* 1981; *Rambaldi,* 1981). These so-called relict grains often display compositional zoning that is inconsistent with *in situ* crystallization from their host chondrule melt (*Rambaldi,* 1981). They appear to be precursor material that survived chondrule formation without melting or dissolving. Therefore, they provide a record of the compositions and mineralogies of chondrule precursors, although this record is biased toward those grains with higher melting temperatures (and therefore lower dissolution rates) than the peak temperatures experienced by the chondrules. Thus, it is likely that there were many more phases in the precursors that did not survive.

Rambaldi (1981) estimated that approximately half of all porphyritic chondrules in unequilibrated ordinary chondrites contain relict grains, while *Jones* (1996) estimated that 15% of all chondrules contain relict grains. *Jones* (1996) found forsteritic relict grains in 3 out of 11 type-II Semarkona (LL3.0) chondrules and in 5 out of 6 such chondrules in ALHA 77307 (CO3.0). However, it is important to keep in mind that there is no *a priori* reason to assume that all relict grains will have anomalous compositions. Such grains are simply the easiest to identify. It is likely that there are many more relict grains that are not readily identified because their compositions are not disparate.

Dusty relict olivines, which contain numerous submicrometer to 10-μm inclusions of Fe-metal, are probably also relict and are observed in ~10% of all chondrules in ordinary chondrites (*Nagahara,* 1983; *Grossman et al.,* 1988). Iron-metal grains in dusty olivines are severely depleted in Ni, relative to CI abundances and likely formed when the chondrule melted in the presence of a reducing agent, such as H_2 in the surrounding gas or solid carbonaceous material (*Rambaldi,* 1981). Dusty olivine grains are less common in CM and CV chondrites (*Jones,* 1996). Olivine crystals that are completely enclosed in pyroxene grains in porphyritic chondrules may also be relict grains (*Nagahara,* 1983). However, *Lofgren and Russell* (1986) showed that similar textures can be produced during chondrule crystallization. Nearly pure forsterite exhibiting blue cathodoluminescence occurs inside chondrules of carbonaceous and ordinary chondrites and is thought to represent another population of relict grains (*Steele,* 1986).

Misawa and Fujita (1994) report the occurrence of an intact CAI fragment in a ferromagnesian chondrule from Allende, which confirms that some CAIs contributed to chondrule precursor material. Furthermore, *Itoh and Yurimoto* (2003) discovered a ferromagnesian chondrule fragment embedded in a CAI from the Yamato 81020 (CO3.0) chondrite, providing evidence that either chondrules contributed some material to CAI precursors, or that remelting of CAIs took place during or after chondrule formation. Initial $^{26}Al/^{27}Al$ ratios in Allende chondrules and CAIs also support the idea that chondrule formation began contemporaneously with the formation of CAIs (*Bizzaro et al.,* 2004).

Based on major- and minor-element compositions, most relict grains formed in earlier generations of chondrules (*Jones,* 1996; *Nagahara,* 1983); relict forsterite and enstatite grains found in FeO-rich chondrules were derived from FeO-poor chondrules, and dusty olivine grains found in FeO-poor chondrules formed in FeO-rich chondrules (*Jones,* 1996). Oxygen-isotopic data also support the view that the low-FeO relict grains formed in a previous generation of low-FeO porphyritic chondrules that were subsequently fragmented (*Kunihiro et al.,* 2004). Relict grains, chondrule fragments, and isolated olivine and pyroxene grains in matrix all show that fragmentation and recycling of chondrule material was an important process in the chondrule-forming region (*Alexander et al.,* 1989a; *Jones,* 1992). Furthermore, based on the abundances of nonspherical chondrules (arbitrarily defined as having aspect ratios >1.20) in CO3.0 chondrites, *Rubin and Wasson* (2005) suggest that chondrule recycling was the rule in the CO chondrule-formation region and that most melting events produced only low degrees of melting. They also suggest that the rarity of significantly nonspherical, multilobate chondrules in ordinary chondrites may reflect more-intense heating of chondrule precursors in the region of the solar nebula in which they formed.

Weinbruch et al. (2000) suggest that some refractory, relict forsterite grains cannot have formed by crystallization from chondrule melts. They favor an origin by condensation from an oxidized nebular gas. *Rambaldi et al.* (1983) noted that igneous olivine within Qingzhen (EH3) chondrules always contain detectable amounts of CaO, while relict olivines are essentially CaO-free. They also suggest that the relict olivines did not have an igneous origin and might represent condensates from the solar nebula. Regardless of their origin, the sporadic occurrence of large relict grains in such chondrules suggests a mix of occasional large grains among finer material in chondrule precursors (*Hewins,* 1997).

3.3. Experimental Constraints on Chondrule Precursors

There have only been a few experiments that constrain the nature of chondrule precursors. In a series of experiments, it was demonstrated that the grain size of chondrule precursors has a significant effect on the resulting chondrule textures (*Radomsky and Hewins,* 1990; *Connolly and Hewins,* 1995, 1996; *Connolly et al.,* 1998; *Hewins and Fox,* 2004). These experiments suggest that granular chondrule textures (grain size <10 μm) result from the incomplete melting of fine-grained starting materials, less than 20 μm in diameter. Furthermore, *Hewins and Fox* (2004) show that it is very difficult to generate porphyritic chondrules by melting such fine-grained precursors. These textures require precursor grain sizes >40 μm.

Experimental simulations of chondrule-like heating events using hydrous silicates as precursors always yield high abundances of vesicles. The fact that vesicles are exceedingly rare in chondrules is given as evidence that chondrules were generated from anhydrous precursor minerals (*Maharaj and Hewins,* 1994, 1998). However, if chondrule melts were heated to within 50°–100° of their liquidus temperature, vesicles could have migrated out of the spheres. *Deloule et al.* (1998) report high concentrations of water in chondrule silicates from the ordinary chondrites Semarkona and Bishunpur, which they interpret as being due to the presence of hydrous silicates in their precursors. On the other hand, both chondrites experienced parent-body aqueous alteration (*Alexander et al.,* 1989a), and *Grossman et al.* (2002) concluded that the zoning of water and halogens in the Semarkona chondrules was due to this alteration.

3.4. Constraints on Chondrule Precursors from Isotopic Studies

Chondrules in some carbonaceous chondrites show evidence of ^{16}O-rich components. This signature is often interpreted as evidence for the survival or limited processing of a presolar component in the chondrule precursors (*Wood,* 1981; *Clayton et al.,* 1991; *Yu et al.,* 1995). However, the absence of correlated anomalies in any other element, except perhaps Cu (*Luck et al.,* 2003), and the rarity of ^{16}O-rich presolar grains (*Zinner,* 1998) strongly argue against the presolar precursor explanation. Heterogeneity of O-isotopic ratios within these chondrules may be the result of incorporation of relict grains from objects such as amoeboid olivine aggregates, followed by solid-state chemical diffusion without concomitant O equilibration (*Jones et al.,* 2004). More recently, it was argued that the mass-independent fractionations observed in chondrule O isotopes result from chemical processes in the gas phase that accompanied their formation (*Thiemens,* 1996; *Clayton,* 2004; *Lyons and Young,* 2003; *Jones et al,* 2004). Even if the diversity of O-isotopic compositions are due to non-mass-dependent isotopic fractionation, the relationships between the O-isotopic compositions and different chondrite components are still unknown.

Isotopic anomalies of Ti are present in chondrules from Allende (CV3) (*Niemeyer,* 1988). There is no known way of generating such anomalies during or after chondrule formation, requiring that the Ti-isotopic diversity in chondrules is a feature that they inherited from their precursors. The chondrule with the largest measured Ti-isotopic anomaly is also Al-rich, suggesting that these anomalies may reflect the presence of CAIs in their precursors.

The isotopic composition of Sr was obtained for eight olivine chondrules from Allende (CV3) (*Patchett,* 1980). The Allende chondrules are mass fractionated by up to 1.4‰/amu in the lighter Sr isotopes. *Patchett* (1980) suggests that these chondrules derived their fractionated Sr from a region of the nebula made isotopically light by partial kinetic mass separation of elements in the vapor phase. Later, the solid objects may have moved to a more isotopically normal region, where the Allende matrix accreted. However, these signatures could also be due to condensation or metamorphism. Allende chondrule mesostasis is almost always recrystallized, chemically zoned (as are the olivines), and can contain halogens, sodalite, etc.

3.5. Direct Condensation from a Vapor Phase

Several researchers have suggested that chondrule melts condensed directly from the vapor phase (e.g., *Blander and Katz,* 1967; *Blander et al.,* 1976, 2001, 2004). *Nelson et al.* (1972) suggested that chondrule textures can result from rapid crystallization of relatively slowly cooling molten droplets and suggested that this history is consistent with condensation from a nebula of solar composition and solidification in an ambient medium at high temperature.

Ebel and Grossman (2000) performed detailed equilibrium calculations of the sequence of condensation of the elements from gases with various enrichments in a dust of CI-chondrite composition. They determined that the FeO contents typical of type-IIA chondrules can only form at high total pressures (>10^{-3} bar) and high dust enrichments (100–1000× solar) at temperatures high enough for dust-gas equilibration to occur (~900°C). Even under these conditions, the condensates contain much more CaO and Al_2O_3 (relative to MgO and SiO_2) than most ordinary-chondrite chondrules. Furthermore, most chondrule compositions are too Na-rich compared to predictions of condensate bulk compositions. However, the higher chondrule Na contents could reflect fractionation prior to chondrule formation or alkali reintroduction at lower temperatures.

A condensation origin has been proposed for chondrules in the Hammadah al Hamra 237 and QUE 94411 bencubbinite chondrites, as well as the recently recognized CH group of chondrites (*Krot et al.,* 2001a,b). Chondrules in these meteorites generally have barred-olivine and cryptocrystalline textures and are essentially metal-free. Zoned metallic chondrules are also abundant in these meteorites. The presence of inclusions with chondrule-like textures in the zoned metal grains suggests that the two components formed in the same nebular region, with silicate chondrules forming at temperatures above those of metal condensation (1773°C).

Two formation scenarios for these chondrules have been considered. Either these chondrules formed by direct gas-liquid condensation or by prolonged heating of chondrule precursors above their liquidus temperatures, which destroyed all solid nuclei. The complementary behavior of refractory lithophile elements (Ca, Al, Ti, REEs) in barred-olivine and cryptocrystalline chondrules suggests that these chondrules formed in a chemically closed system (*Krot et al.,* 2001a). Furthermore, high Cr_2O_3 (0.4–0.7 wt%) and FeO (1–4 wt%) contents in the chondrules and complementary depletions in the FeNi-metal grains suggests formation in a closed system under relatively oxidizing conditions. These compositional trends are consistent with formation in a dust-enhanced (10–50× solar) region of the solar nebula (*Krot et al.,* 2001b; *Petaev et al.,* 2001). The inferred enhanced initial dust/gas ratios supports the theory that the chondrules in these meteorites formed by direct gas-liquid condensation (*Ebel and Grossman,* 2000).

One of the primary arguments against direct condensation is the extreme conditions required to produce the FeO contents of type-II chondrules. However, *Blander et al.* (2001, 2004) suggest that the condensation of metallic Fe was suppressed because of nucleation constraints. The inhibition of Fe condensation results in supersaturation of Fe in the gas phase. As a result, higher levels of FeO entered silicate melts by condensing directly from the vapor. This model also explains the general absence of metal in type-II chondrules.

The strongest evidence that at least some chondrules formed through melting of preexisting materials and did not condense from a gas is the presence of large, compositionally distinct relict grains (see section 3.2.2 above). These grains clearly survived the chondrule formation process without melting. In addition, porphyritic chondrules, the most abundant type of chondrule, require the preservation of numerous crystal nuclei (see section 4.1), implying that their precursors were also at least partially crystalline.

4. THERMAL HISTORIES AND CRYSTALLIZATION OF CHONDRULES

It has long been established that many of the features seen in chondrules are the result of crystallization from a melt. In the previous section, we describe the constraints that can be placed on the precursor material. Here we identify the constraints that can be placed on the peak temperatures and duration of heating from experimental simulations of chondrule formation, and considerations of melting kinetics of relict grains. We also describe how chondrule textures and disequilibrium features, such as compositional zoning in minerals, can be used to constrain the cooling histories of chondrules.

4.1. Peak Temperatures and Preservation of Nucleation Sites

A prerequisite for the crystallization of a melt is the presence of crystal nuclei, and nuclei abundance is a major factor in determining chondrule textures (*Lofgren,* 1996). On the timescales of chondrule formation, spontaneous formation of nuclei in the melt is a relatively slow process. As a result, nuclei-free chondrule melts can cool well below their liquidus temperatures and become supercooled. When a nucleus forms there is a burst of rapid crystal growth, producing textures similar to those of barred-olivine and radial-pyroxene chondrules (*Lofgren and Lanier,* 1990; *Radomsky and Hewins,* 1990).

Porphyritic and granular textures form when abundant nuclei are present during cooling. The simplest interpretation of this result is that chondrule peak temperatures were slightly below their liquidus temperatures (*Lofgren and Russell,* 1986), which enabled the survival of nuclei. These unmelted relicts provided abundant nucleation sites that generated the characteristic porphyritic textures on cooling (*Lofgren,* 1989; *Hewins and Radomsky,* 1990). Chondrule liquidus temperatures therefore provide important constraints on the peak temperatures experienced by chondrule melt droplets. Droplets that were completely molten (e.g., those that crystallized as barred, excentroradial, and cryptocrystalline chondrules) have liquidus temperatures that range from 1200°C to over 1900°C and are highly dependent on bulk composition (*Herzberg,* 1979; *Hewins and Radomsky,* 1990). Given this wide range of temperatures, it is clear that bulk composition is at least as important as initial temperature in controlling chondrule textures (*Connolly and Hewins,* 1996). Based on the abundances, compositions, and textures of incompletely melted vs. completely melted chondrules in carbonaceous and ordinary chondrites, it has been estimated that few chondrules with liquidus temperatures over 1750°C were completely melted, and few with liquidus temperatures under 1400°C were incompletely melted. Therefore, it was concluded that most chondrules reached peak temperatures of 1500°–1550°C (*Radomsky and Hewins,* 1990).

Further constraints of chondrule thermal histories have been obtained by experimental simulations that accurately reproduce chondrule grain sizes and textures (*Hewins and Radomsky,* 1990; *Lofgren,* 1996; *Connolly et al.,* 1998; *Tsuchiyama et al.,* 2004). These studies show that it is important to understand fully not only the peak temperatures that chondrules experienced, but also the duration of the heating event. This dual consideration is important because, for kinetically driven processes, the effect of heating to a high peak temperature for a short time can be equivalent to that of heating to a lower temperature for longer times (*Jones et al.,* 2000). The total amount of energy, as opposed to the peak temperatures, is therefore constrained to be below that required both to raise the precursor material to the liquidus and to supply the necessary latent heat of fusion to melt all solid precursor material. *Connolly et al.* (1998) showed that a short heat pulse, with a peak temperature well above the liquidus, can also produce porphyritic chondrule textures. Nuclei survive because the timescales are shorter than the melting and dissolution kinetics. Thus, the maximum temperature achieved during chondrule formation could have been approximately 2100°C, 200°C higher than the liquidus temperatures of even the most refractory chondrules. The distinctive olivine rims on barred-olivine chondrules also provide information about peak temperatures. The formation of compact olivine rims on these chondrules requires a heating temperature about 100°C or more above the liquidus to promote heterogeneous nucleation of olivine on the surface of the melt droplet (*Tsuchiyama et al.,* 2004).

It is also possible that chondrule formation occurred in a dusty environment, which provided abundant nucleation sites. Experimental simulations that seeded molten droplets with small grains of dust have reproduced a wide variety of chondrule textures (*Connolly and Hewins,* 1995). Textures vary as a function of the temperature at which dust is introduced to the melt, the number density of dust grains, and their mineralogy.

4.2. Kinetics of Melting and Survival of Relict Grains

Another independent constraint on the duration of chondrule melting is provided by the presence of relict grains that were not in contact with melt for a long enough time to dissolve. Early studies used thermal diffusion calculations to determine the relationship between the duration of the chondrule heating event and the maximum attainable temperature inside chondrules (*Fujii and Miyamoto,* 1983). These researchers determined that the duration of each heating event had to be less than 0.01 s for chondrules with radii of 0.3 mm for the survival of relict grains.

More recently, *Greenwood and Hess* (1996) used a model for the kinetics of congruent melting to place constraints on the duration and maximum temperature experienced by the interiors of relict-bearing chondrules. They argue that relict forsteritic olivine grains in chondrules were not heated above 1900°C, approximately the melting point of forsterite. Their temperature limit for chondrules with Mg-rich pyroxene relicts is 1557°C, the incongruent melting point of enstatite. In both cases, the input of energy could not have lasted for more than a few seconds. At temperatures below their melting points, relict grains will dissolve if they are not in equilibrium with the melt. Dissolution is slower and quantitatively less well understood than melting, but ultimately relict grains may be used to better constrain the thermal histories of a wide range of chondrules.

4.3. Chondrule Cooling Rates

4.3.1. Chondrule textures. Chondrule textures are controlled by nucleation density, degree of undercooling, and cooling rate. The density of nucleation sites is a function of the duration and peak temperature of the heating event as well as the abundance of dust particles in the chondrule formation region (section 4.2). The abundance of nucleation sites in the melt also controls the degree of undercooling that can occur. These factors together with the subsequent cooling rate have direct effects on the development of both crystal morphology and the elemental distributions within the resulting chondrule grains.

Dynamic crystallization experiments reveal that forsterite crystallizes with one of four distinct morphologies: euhedral (well-defined crystal faces), tabular (preferential development in two directions), hopper (hourglass shape with hexagonal outline), and dendritic (intricate branching structures) (*Faure et al.,* 2003). The morphology of forsterite crystals varies as a function of the cooling rate, degree of undercooling, and abundance of crystallization nuclei. As the degree of undercooling increases, forsterite morphology evolves from tabular to hopper. At the most rapid cooling rates, dendritic crystals form.

Euhedral olivine crystals develop in silicate melts with abundant nuclei. No supercooling occurs and crystallization begins as soon as the temperature drops below the liquidus. The resulting textures are uniformly sized olivine phenocrysts in a glassy mesostasis (*Lofgren,* 1989; *Radomsky and Hewins,* 1990). The cooling rates required to produce porphyritic textures in experimental chondule melts are less than 1000°C/h with those on the order of 100°C/h best reproducing the textures in porphyritic chondrules (*Lofgren,* 1989; *Radomsky and Hewins,* 1990).

Barred-olivine chondrules contain olivine crystals with single-plate dendritic texture. These textures are the result of crystallization of nuclei-free melt droplets. A significant amount of undercooling may have occurred in these melt droplets. Comparison of the olivine morphology in barred chondrules with analog textures produced in experiments indicates that these chondrules cooled at rates of 500°–2300°C/h (*Lofgren and Lanier,* 1990; *Tsuchiyama et al.,* 2004).

Pyroxene crystallization textures also vary as a function of melt composition and cooling history. Porphyritic pyroxene textures are produced only in experiments that do not exceed the liquidus temperature or exceed it only for a brief period of time (*Lofgren and Russell,* 1986). These textures form when cooling rates are well in excess of 100°C/h. However, at cooling rates greater than 2000°C/h, the crystal sizes are significantly below those seen in most chondrules of this type.

Crystals in radial pyroxene chondrules are generally highly acicular, radial, and often in the form of spherulites. These morphologies are indicative of growth from a melt that experienced very high degrees of undercooling. Nucleation sites, which are initially absent in the melt, form from crystal embryos that exceed some critical radius. Once nucleated, growth of the pyroxene crystals is very rapid. The large degree of supercooling stabilizes the highly non-equilibrium crystal shapes. The radial or spherulitic textures can develop even at quite slow cooling rates (5°C/h) in any melt where olivine is not a liquidus or near-liquidus phase (*Lofgren and Russell,* 1986).

4.3.2. Zoning profiles. Experimental studies suggest that chondrule textures largely reflect the abundance of nucleation sites in the melt, which directly influences the degree of undercooling that a droplet can experience. Even chondrules that clearly crystallized rapidly, e.g., radial pyroxene chondrules, can form at relatively slow cooling rates. Thus, chondrule texture appears to be controlled predominantly by the intensity and duration of the heating event. However, olivine crystals in many chondrules are heterogeneous and are chemically zoned from core to rim. These zoning profiles indicate that the chondrules cooled too quickly for the olivines to maintain equilibrium with the remaining melts and for diffusion to homogenize the olivines. Thus, detailed characterization of the chemical variation in chondrule olivines can provide additional information about the cooling rates that these objects experienced.

Olivine zonation is particularly prominent in type-IIA porphyritic olivine chondrules in the least-equilibrated ordinary chondrites, such as Semarkona (*Jones,* 1990). Zonation occurs in FeO (9–27 wt%), Cr_2O_3 (0.2–0.6 wt%), MnO (0.2–0.7 wt%), CaO (0.1–0.4 wt%), and P_2O_5 (0–0.5 wt%), with abundances increasing from cores to rims. Both the olivine textures and the normal zoning profiles are consistent with an igneous origin. The zonation likely results from fractional crystallization of the chondrule melts.

The zoning profiles in porphyritic olivine chondrules have been compared to those in experimentally produced analogs (*Jones and Lofgren,* 1993; *Weinbruch et al.,* 1998). In Semarkona chondrules, the shapes of the zoning profiles, as well as the absolute values of FeO and minor-element contents, of the natural and synthetic olivine crystals are similar at cooling rates as low as 5°C/h. Olivine zoning is much more restricted at low cooling rates in bulk compositions, which permit pyroxene nucleation and growth (*Radomsky and Hewins,* 1990).

4.3.3. Exsolution microstructures. Another constraint on cooling rate is exsolution microstructure (*Carpenter,* 1979; *Kitamura et al.,* 1983; *Muller et al.,* 1995; *Watanabe et al.,* 1985; *Weinbruch and Muller,* 1995; *Weinbruch et al.,* 2001). Lamellar exsolution has been observed in some chondrule pyroxenes. In particular, pigeonite/diopside exsolution lamellae occurs in granular-olivine-pyroxene chondrules from the Allende meteorite (*Weinbruch and Muller,* 1995). It is suggested that these textures are produced when cooling rates are on the order of 5°–25°C/h in the temperature range of 1200°–1350°C. These exsolution textures have also been produced in synthetic analogs of chondrules (*Weinbruch et al.,* 2001) under cooling rates of 10°–50°C/h between 1000° and 1455°C. In these experiments, higher cooling rates (50°–450°C/h) yield only modulated structures that are unlike those seen in chondrule pyroxenes.

4.4. Mineral Crystallization Sequence

The sequence in which minerals crystallize in chondrule melts is highly dependent on the bulk composition as well as the thermal history. In many chondrules, olivine is the earliest phase nucleating from the melt, forming between 1600°C and 1400°C (*Ferraris et al.,* 2002). This temperature range is consistent with temperatures derived from chondrule chromite-olivine pairs, which have Fe-Mg partitioning characteristic of phases that crystallized from silicate melts at temperatures >1400°C (*Johnson and Prinz,* 1991).

Liquid Fe-based alloys containing less than 5.2 atom% Ni will preciptitate metal crystals with a body-centered cubic structure (δ-Fe) that have slightly lower Ni contents (<4.1 atom%) between 1536° and 1513°C. Nickel-rich metals (>5.2 atom% Ni) will crystallize with the face-centered cubic structure. Crystallization temperatures decrease with increasing Ni content, reaching a minimum of 1425°C between 60 and 70 atom% Ni.

In FeO-rich porphyritic pyroxene chondrules in Semarkona it appears that the order of crystallization of olivine and pyroxene is variable (*Jones,* 1996). Intergrown pyroxene and olivine phenocrysts indicate that both minerals were crystallizing simultaneously as the chondrule cooled. In several chondrules olivine is dominant and appears to have preceded growth of pyroxene. In the coarse-barred chondrules, olivine is sometimes present as small rounded inclusions, indicating that it crystallized before pyroxene.

Textural relationships between the various pyroxene phases in Semarkona chondrules indicate that the crystallization sequence was one of progressive enrichment in wollastonite (Wo) content. In these objects, low-Ca pyroxenes are overgrown with pigeonite followed by augite, which only occurs as rims on low-Ca pyroxene phenocrysts (*Jones,* 1996). Pigeonitic rims form at temperatures above 1000°C. By subsolidus exsolution, high-pigeonite inverts to twinned low-pigeonite between 935° and 845°C. Augite crystallizes as euhedral grains hosted within the glassy matrix, or as skeletal crystals, most probably just after the pigeonitic rim growth between 1150° and 1000°C. Pigeonitic and augitic rims form augitic and enstatitic sigmoids, respectively, by exsolution between 1000° and 640°C. Based upon glass composition, the overall solidification temperature of the mesostasis occurs close to 990°C, the minimum melt temperature in the quartz-albite-orthoclase system (*Hamilton and Mackenzie,* 1965).

5. METALLOGENESIS AND REDOX REACTIONS IN THE MELT

5.1. Metal in Chondrule Interiors

Metal-bearing type-I chondrules have highly reduced mineral assemblages, such as silicates with low FeO contents (*Jones and Scott,* 1989) and low-Ni metal grains that contain variable amounts of P, Cr, and Si (*Lauretta and Buseck,* 2003; *Rambaldi and Wasson,* 1981; *Zanda et al.,* 1994). Despite much interest, the origin of the metal in these chondrules is still the subject of debate. Some researchers propose that metal grains are derived from metallic nebular condensates that avoided oxidation even during chondrule formation (*Grossman and Wasson,* 1985; *Rambaldi and Wasson,* 1981). Other studies suggest that the metal formed by *in situ* reduction of preexisting sulfides, oxides, and FeO-rich silicates in the melt (*Connolly et al.,* 1994; *Lauretta et al.,* 2001; *Lauretta and Buseck,* 2003). In this section, we examine the evidence for the origin of metal in silicate chondrules.

5.2. Formation of Metal by Condensation

Several researchers have suggested that metal grains in chondrule interiors formed by melting of metal and sulfide nebular condensates. It is suggested that these grains maintained their chemical identity during the melting process, presumably by surviving as immiscible droplets that did not exchange with the surrounding silicate melt. Evidence for this idea comes from the abundances of siderophile and chalcophile elements in Semarkona (LL3.0) chondrules (*Grossman and Wasson,* 1985), which are presumably hosted by the metal and sulfide. The abundances of these elements in chondrules are highly variable. The nonrefractory siderophile abundance patterns in Ni-rich chondrules are smooth functions of volatility, which have been interpreted as nebular signatures. However, the volatile siderophile fractionations may also reflect evaporation during chondrule melting.

Metal from chondrule interiors in Renazzo, Al Rais, and the related chondrite, MacAlpine Hills 87320 (*Lee et al.,* 1992), shows generally high Ni and Co, flat Ni and Co profiles, and moderately large grain-to-grain compositional variations (even within chondrules). Metal from chondrule margins adjacent to matrix has convex (inverted U-shaped) Ni and Co zoning profiles with the highest Ni and Co concentrations at grain centers. These profiles are similar to those observed in the CH chondrites, which have been interpreted as resulting from condensation in the solar nebula (*Meibom et al.,* 1999) or a postimpact vapor plume (*Campbell et al.,* 2002). However, *Lee et al.* (1992) interpreted these patterns to reflect dilution by Fe produced by FeO reduction, which occurred during a brief period of parent-body metamorphism.

5.3. Metallogenesis by Reduction in Chondrule Melts

Fine (~2 μm), Ni-poor (~1 wt%) metal grains are common inclusions in the olivine in porphyritic chondrules in unequilibrated ordinary chondrites (*Rambaldi and Wasson,* 1981, 1982). The most common occurrence of this "dusty" metal is in the core of olivine grains having clear Fe-poor rims. The original compositions of dusty grains, which have been determined by assuming that all Fe metal was originally present as FeO (*Jones and Danielson,* 1997), closely match those of olivines in chondrules from unequilibrated chondrites. These assemblages likely reflect high temperature *in situ* reduction of FeO from olivine (*Rambaldi and Wasson,* 1981; *Jones and Danielson,* 1997).

The evidence that dusty olivine grains formed by *in situ* reduction suggests that similar processes may have affected the compositions of other chondrule metal phases. Chondrule kamacite in Krymka and Chainpur has Ni concentrations (3–4 wt%) that are significantly below that expected for nebular condensates (~5 wt%). This kamacite may have been diluted by Fe reduced from the silicates during chondrule formation by either H_2 gas or carbonaceous precursor matter (*Rambaldi and Wasson,* 1984).

The elements Cr, Si, and P are common in metal grains in the least-metamorphosed ordinary and carbonaceous chondrites and provide additional information on redox reactions in type-I chondrules (*Zanda et al.,* 1994). The concentrations of these elements are inversely correlated with the FeO concentrations of their host chondrule olivines. This relation suggests that Cr, Si, and P concentrations could not have been established by nebular condensation, but are the result of metal-silicate equilibration within chondrules.

A link between chondrule-derived metal and opaque assemblages in the matrix of primitive chondrites has also been established, based on the chemistry of Cr, Si, and P (*Lauretta et al.,* 2001; *Lauretta and Buseck,* 2003). It is likely that the metal melted in and equilibrated with a silicate liquid under high-temperature, reducing conditions. In this environment the molten alloys incorporated varied amounts of Si, Ni, P, Cr, and Co, depending on the oxygen fugacity and temperature of the melt. As the system cooled, some metal oxidation occurred in the chondrule interior, producing metal-associated phosphate, chromite, and silica. Metal that migrated to chondrule boundaries experienced extensive corrosion as a result of exposure to an external, oxidizing atmosphere.

Trace-element data also support a chondrule-melt origin for metal grains in ordinary chondrites (*Kong and Ebihara,* 1996, 1997). Characteristics retained in the bulk metals of unequilibrated H, L, and LL chondrites (abundant C; high contents of Cr, V, Mn and low contents of W, Mo, and Ga compared to metals of equilibrated chondrites; less enrichment of W than Mo; and fractionation of Co from Ni) demonstrate that chondrite metals are not nebular condensates. All these characteristics are consistent with melting. Furthermore, these data suggest that chondrite metals are not melting remnants of previously condensed metals. Rather, they have been produced by reduction of CI- or CM-like material during the melting process.

Volatile-siderophile-element abundance patterns of metallic and nonmagnetic fractions of CR chondrites are complementary and volatility-dependent (*Kong et al.,* 1999). In Renazzo, fine-grained metal is located inside chondrules or in chondrule rims. Matrix contains mostly coarse metal grains. Both types of metal, fine and coarse, exhibit similar chemical signatures, comparatively high in Cr and low in Ni, suggesting a genetic relationship (*Kong and Palme,* 1999) in which Renazzo chondrules and metal were formed by reduction of oxidized precursors compositionally similar to CI chondrites. The differences in siderophile-element pattern between chondrules, chondrule metal, and matrix indicate evaporation and recondensation of volatile elements during chondrule formation.

Connolly et al. (2001) performed a detailed study of metal in CR chondrites from chondrule interiors, chondrule rims, and in the matrix. They found that the refractory platinum group elements Os, Ir, and Pt behave coherently in metal. They are all uniformly enriched, depleted, or unfractionated with respect to Fe. Metal with solar ratios of Os/Fe, Ir/Fe, and Pt/Fe occur primarily in chondrule interiors. All metal grains have essentially CI ratios of Ni/Fe and Co/Fe. Almost all metal grains have lower-than-CI ratios of the volatile siderophile elements, Au and P. Furthermore, the bulk compositions of chondrules are generally unfractionated relative to CI chondrites for elements more refractory than Cr, but are depleted in the more volatile elements. The abundances of siderophile elements correlate strongly with the metal abundance in the chondrules, which implies that the siderophile depletions are due to expulsion of metal from the chondrule melts. These data are interpreted to suggest that most metal formed during melting via reduction of oxides by C that was part of the chondrule precursor.

Nickel has a lower volatility than Fe, and *Humayun et al.* (2002) found that Ni/Fe ratios in metal in CR chondrule interiors tend to be higher than those in the rims and matrix. In addition, Pd has nearly the same volatility as Fe, and Pd/Fe ratios are nearly the same in metal in chondrules, rims and matrix. Copper is more volatile than Fe, and Cu/Fe ratios are much lower in chondrule metal relative to that in rims and matrix. These observations can be explained by evaporation of volatile siderophile elements, including Fe, and subsequent recondensation mostly onto chondrule rims. In very FeO-poor ordinary chondrite chondrules, there is a correlated decrease in Fe, Mn, and Cr (*Alexander,* 1994). Both Fe and Cr are siderophile under reducing conditions but Mn is not, suggesting that these correlated depletions are due to evaporation. Rims with large correlated enrichments of FeO, Si, Mn, and Cu have also been reported in ordinary chondrites, and may result from recondensation of evaporated material (*Alexander,* 1995).

5.4. Experimental Studies of Chondrule Melt Redox Reactions

Experimental studies have added to our understanding of the role of *in situ* reduction in producing type-I chondrule textures and mineral compositions. Experiments in which C, in the form of graphite or diamond, is flash-heated together with magnesian silicates reproduce many mineralogical features characteristic of type-I chondrules, including Mg-rich and dusty olivines, Cr and Si dissolved in metal, and silica inclusions inside metal (*Connolly et al.,* 1994). These results provide strong evidence that C in chondrule melts is an effective reducing agent and that chondrule melts were internally buffered. Furthermore, they suggest that C played a key role in chondrule formation by briefly creating a reducing environment within chondrule melts that was decoupled from the nebula gas.

Further experiments have provided more detail on the nature of the reduction reaction. *Cohen and Hewins* (2004) performed an experimental study of metallic Fe formation in chondrule-like melts. They showed that troilite and Fe-metal are extremely volatile under canonical chondrule-formation conditions. Conversely, at higher total pressures (P H_2 = 1 atm) Fe metal is less volatile and remains in the melt for longer times. These experimental results suggest that metal-bearing chondrules formed in environments with elevated pressures relative to the canonical nebula. Alternatively, unlike in the experiments, the chondrule melt and the vapor phase may have reached equilibrium in the solar nebula, in which case the metal may have stabilized.

Additional studies have investigated the transformations of olivine single crystals in contact with graphite powder. At 1150°C precipitates of Fe,Ni and a thin amorphous layer of silica formed on the surface of the olivine crystals (*Lemelle et al.,* 2001). A reaction layer composed of Fe-Ni precipitates and FeO-depleted olivine develops between the surface and an inner reaction front at 1350°C (*Lemelle et al.,* 2000). The reaction layer contains two reaction fronts (inner and outer) corresponding to the beginning of metal precipitation and to the total depletion of Fe^{2+} from olivine. The propagation of the two reaction fronts follows a parabolic law and is limited by the interdiffusion of Fe/Ni and Mg in the olivine lattice. In both natural and experimental samples, micrometer-sized metal blebs observed in the dusty region often show preferential alignments along crystallographic directions of the olivine grains, have low Ni contents (typically <2 wt%), and are frequently surrounded by a silica-rich glass layer (*Leroux et al.,* 2003) These features suggest that dusty olivines are formed by a subsolidus reduction of initially fayalitic olivines. Comparison with experimentally produced dusty olivines suggests that timescales on the order of minutes usually inferred for chondrule formation are also adequate for the formation of dusty olivines. These observations are consistent with the hypothesis that at least part of the metal phase in chondrites originated from reduction during chondrule formation.

There have also been recent experiments on the nature of chemical reduction in silicate melts in a reducing atmosphere (*Everman and Cooper,* 2003). Melt samples were heated to 1300°–1400°C under a flowing gas mixture of 1-atm CO/CO_2 (f_{O_2} = QIF – 2). The melt compositions are enriched in Al_2O_3 and depleted in MgO and CaO relative to ferromagnesian chondrules. The melt experienced an internal reaction in which a dispersion of nanometer-scale Fe-metal precipitates formed at an internal interface. The distance between this interface and the melt surface increased throughout the experiment. Oxygen was lost from the melt at the free surface via conversion of CO to CO_2 resulting in an increase in the cation-to-oxygen ratio and net reduction of the melt.

Thus, while it seems likely that nebular condensate metal was incorporated into chondrule precursors, it is also likely that this metal experienced significant interaction with the silicate melt. In chondrule melts that incorporated a significant amount of carbonaceous material or that interacted with a highly reduced vapor phase, additional metal was created by reduction of Fe and associated elements from silicate and other oxidized phases.

Subsequent petrologic work supports the model that some chondrules incorporated C-rich material in their precursors (e.g., *Lauretta et al.,* 2001; *Lauretta and Buseck,* 2003; *Fries and Steele,* 2005). Small (<1–20 μm) C inclusions are common in the metal of type-3 ordinary chondrites (*Lauretta et al.,* 2001; *Mostefaoui et al.,* 2000) and fine-grained carbonaceous inclusions occur in some chondrule interiors (*Lauretta and Buseck,* 2003; *Scott et al.,* 1984). It is likely that these inclusions are the remains of carbonaceous matter that escaped oxidation during chondrule formation. There are claims of high C contents (several weight percent) in chondrule olivines from the primitive chondrites Allende (CV3), Semarkona (LL3.0), and Bishunpur (LL3.1) (*Hanon et al.,* 1998). These researchers report that mean

C concentrations are 3× higher in type-I chondrules than in type-II chondrules. However, nuclear reaction analyses show that chondrule olivines have C contents of less than 120 ppm (*Varela and Metrich,* 2000).

6. ELEMENTAL EVAPORATION AND RECONDENSATION

6.1. Chemical and Isotopic Fractionations Resulting from Evaporation

The origin of the wide range of compositions of ferromagnesian chondrules remains controversial. Figure 3 summarizes the variation and the average compositions of Semarkona (*Jones and Scott,* 1989; *Jones,* 1990, 1994, 1996; *Huang et al.,* 1996; *Tachibana et al.,* 2003) and Renazzo (*Kong and Palme,* 1999) chondrules. Type-I chondrules in particular exhibit significant chemical fractionations, with depletions of elements like Cr, Mn, Na, K, and Fe, and enrichments of elements such as Al, Ca, and Ti, relative to Mg.

These compositional variations suggest that elemental volatility is the primary factor controlling chondrule bulk compositions. The relative volatility of elements can be estimated in two ways: (1) from the calculated equilibrium temperature at which 50% of an element has condensed from a gas of solar composition typically at $P_{tot} = 10^{-4}$ bar (e.g., *Lodders,* 2003), and (2) from evaporation experiments of chondritic melts (e.g., *Hashimoto,* 1983; *Floss et al.,* 1996; *Wang et al.,* 2001). Both approaches suggest that the order of decreasing volatility is roughly S, Na-K, Fe-Mn-Cr, Si, Mg, and Ca-Ti-Al. This order is qualitatively consistent with the variations in chondrule compositions. However, the degree to which these variations were inherited from chondrule precursors or generated during chondrule formation is unknown (e.g., *Gooding et al.,* 1980; *Grossman and Wasson,* 1983b; *Jones,* 1990; *Connolly et al.,* 2001; *Huang et al.,* 1996; *Sears et al.,* 1996; *Alexander,* 1994, 1996, *Hewins et al.,* 1997). The bulk chemistry of chondrules alone is insufficient to resolve this debate. Two other approaches have been used: petrologic evidence for volatile loss and recondensation in chondrule rims and matrix (see section 7) and isotopic signatures of evaporation and recondensation.

Both evaporation and condensation are generally accompanied by mass-dependent isotopic fractionation. Evaporation can result in enrichment in the heavy isotopes in the residue. Condensation enhances the abundances of light isotopes in the condensates, relative to the gas. The largest isotopic fractionations occur during Rayleigh fractionation and the isotopic fractionations correlate with the extent of volatile loss or gain. For Rayleigh fractionation to occur, the evaporating residue or condensing gas must remain isotopically uniform and the evaporation or condensation must be unidirectional (i.e., there is no back reaction). At the other extreme, if diffusion or convection occurs in the evaporating material on timescales that are much slower than evaporation, mass loss and elemental fractionation can occur with little or no isotopic effects. Similarly, high ambient gas pressures suppress isotopic fractionation and absolute evaporation rates (e.g., *Richter et al.,* 2002), as will any back reaction with the evaporated gas.

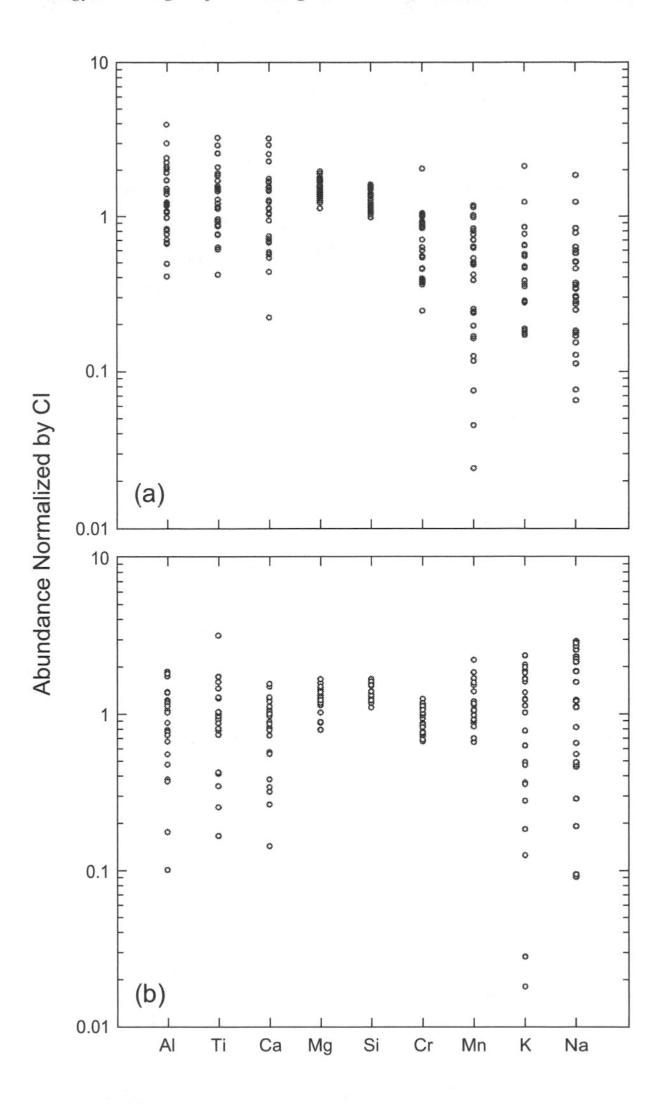

Fig. 3. Major-element abundances, normalized to CI chondrites for **(a)** type-I and **(b)** type-II chondrules.

Whether or not volatile-element fractionation is accompanied by isotopic fractionation depends on the conditions under which the volatile loss took place. Under Rayleigh conditions, large isotopic fractionations will correlate with elemental fractionation. Such correlated elemental and isotopic fractionations are seen in low-pressure evaporation experiments (section 6.3) and in cosmic spherules that are severely heated during atmospheric entry heating (*Alexander et al.,* 2002). If volatile loss occurs far from the Rayleigh condition, there is little associated isotopic fractionation. In this case, it is not possible to tell whether volatile loss was inherited from chondrule precursors or produced by chondrule formation.

Since there is this potential ambiguity in the elemental and isotopic data, considerable effort has been put into developing a theoretical understanding of the kinetics of evaporation and condensation associated with chondrule formation. The ultimate aim of these efforts is to be able to predict the volatile loss, and accompanied isotopic fractionations, that would be expected under a given set of chondrule formation conditions. By comparing these predictions to observations of chondrule compositions, it is hoped the conditions of chondrule formation can be further constrained. Here we review the elemental, isotopic, and petrologic studies of chondrules and the experimental and theoretical progress that has been made in understanding evaporation and condensation, in an attempt to place limits on the conditions of chondrule formation.

6.2. Evidence for Evaporation and Recondensation

6.2.1. Sulfur. Sulfur is the most volatile major element in chondrites. Consequently, it is expected to rapidly evaporate from chondrules and recondense on metal in rims and matrix. Despite it being a potentially good indicator for evaporation, our knowledge of the petrology and cosmochemistry of S in chondrules is fairly limited. Troilite is uncommon in FeO-poor chondrules in Semarkona. *Hewins et al.* (1997) showed that the abundance of S in Semarkona FeO-poor (type-I) chondrules decreases with increasing size of olivine phenocrysts (Fig. 4). They argue that increasing grain size indicates more severe heating and, therefore, fewer crystal nuclei surviving to develop into phenocrysts on cooling. *Tachibana and Huss* (2005) measured S isotopes of small troilite grains included in Fe-Ni metal spherules in six Bishunpur FeO-poor chondrules and four large troilite grains in the matrix. All the troilite grains have the same S-isotopic compositions within analytical errors. Since evaporation of Fe-S melt under Rayleigh conditions does result in isotopic mass fractionation (*McEwing et al.*, 1980), the large elemental fractionation without accompanying isotopic mass fractionation in chondrules indicates either volatilization of S under non-Rayleigh conditions or subsequent isotopic reequilibration.

6.2.2. Sodium and potassium. As with S, Na and K abundances in type-I chondrules decrease with increasing grain size of olivine phenocrysts (Fig. 4), suggesting progressive alkali loss with the increasing severity of heating.

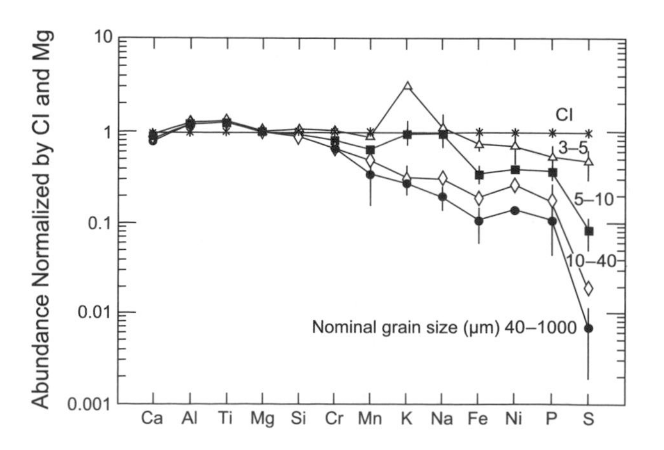

Fig. 4. Graph illustrating the variation in volatile-element abundances as a function of nominal grain size in chondrules.

Kong and Palme (1999) showed that chondrules and matrix in Renazzo are complementary in the elemental abundances; chondrules are depleted in refractory and volatile elements. There are also observations that suggest that Na and K may have recondensed into chondrules after their formation. *Ikeda* (1982) was the first to report an increase in Na, K, and Fe near the edge of a Ca-Al-rich chondrule in ALH 77003. *Matsunami et al.* (1993) described a Semarkona chondrule, in which Na, Mn, and Si increase outward, that they suggest resulted from recondensation after crystallization of forsterite. *Libourel et al.* (2003) reported a positive correlation between SiO_2 and Na_2O in mesostasis and enrichment of Si in the outer portion of some FeO-poor chondrules in CR2 chondrites and Semarkona. They interpret these observations as resulting from recondensation of alkalis and SiO_2. It is, however, not clear whether the zoning pattern is an indication of condensation or later metasomatic exchange of Na and Ca between chondrules and matrix (*Grossman et al.,* 2002).

Sodium is monoisotopic but K has two stable isotopes. The isotopic composition of K has been measured in whole Allende (CV3) chondrules (*Humayun and Clayton,* 1995) and chondrule mesostasis and melt inclusions in Semarkona (LL3.0) and Bishunpur (LL3.1) (*Alexander et al.,* 2000; *Alexander and Grossman,* 2005). Most K-isotopic analyses are normal within error with no evidence of systematic isotopic fractionation with K abundance. These results rule out volatile loss of K under Rayleigh conditions.

6.2.3. Iron, nickel, manganese, and chromium. The reduced state of Fe is the defining criterion for type-I chondrules, and its abundance variation among chondrules is the most significant of all major elements (Fig. 3). The complex behavior of Fe as metal, oxide, or sulfide has made identifying the origin(s) of the depletions difficult. Nevertheless, evidence exists for evaporative loss of some siderophile elements in CR2 chondrules (see section 5.3).

Alexander and Wang (2001) found that olivines ($Fo_{99.7}$ to Fo_{78}) in nine Chainpur (LL3.4) chondrules are Fe-isotopically unfractionated within analytical errors ($2\sigma < 1–2$‰/amu). Bulk Tieschitz (H3.4) chondrules show very small fractionations of <–0.25‰/amu (*Kehm et al.,* 2003). The small light-isotope enrichments may be due either to condensation or parent-body processes. Four Allende chondrules show mass-dependent isotopic fractionation of up to 1.1‰/amu (*Mullane et al.,* 2003). However, Allende chondrules are chemically and isotopically zoned, including in Fe, and this zonation is likely a metamorphic overprint (*Zhu et al.,* 2001). The isotopic mass fractionation of other siderophile elements and Cr in chondrules have not been measured.

The textures of barred-olivine chondrules may also be an indication of evaporation during chondrule formation (*Tsuchiyama et al.,* 2004). Partial evaporation (mainly FeO) enhances the selective nucleation by cooling the surface, relative to the interior. This process produces the compact, coarse-grained rims that are characterisitic of these chondrules. The occurrence of such rims on barred olivine chondrules suggests chondrule formation in an open system.

6.2.4. Silicon and magnesium. If type-I chondrules formed from type-II- or CI-like precursors, there must be loss of both FeO and SiO_2 during chondrule formation (*Jones,* 1990; *Sears et al.,* 1996, *Cohen et al.,* 2004). This scenario is consistent with the FeO- and SiO_2-rich rims that are often present on type-IA chondrules in ordinary chondrites. There is petrologic evidence for recondensation of SiO_2 back into chondrules (*Matsunami et al.,* 1993; *Libourel et al.,* 2003). In the most highly reducing melts, it is possible that a significant amount of Si was incorporated into the Fe-based metallic melt component. Expulsion of some of this material is another mechanism for simultaneous Fe and Si loss from bulk chondrule compositions (*Lauretta and Buseck,* 2003).

The Si-isotopic compositions of 34 individual chondrules from type-3 ordinary chondrites analyzed by *Clayton et al.* (1991) exhibit a range of ±0.25‰/amu about the bulk chondrite value. If type-I chondrules were part of the analyzed suite of chondrules, the observed range of Si-isotopic compositions is less than would be expected if Si was lost via Rayleigh evaporation.

Evaporative loss should be faster for smaller objects because of the large surface area/volume ratio. *Clayton et al.* (1991) analyzed chondrules of varying size from type-3 ordinary chondrites. For Dhajala (H3.8), smaller chondrules are isotopically heavier. However, the total change in Si-isotopic fractionation is only ~0.5‰/amu. For Chainpur (LL3.4) there was no systematic isotopic variation with chondrule size.

Magnesium should not experience much evaporative loss from ferromagnesian chondrules, although significant Mg loss may have occured during formation of Al-rich chondrules. *Galy et al.* (2000) measured Mg-isotopic fractionations of up to ~1.5‰/amu in eight Allende chondrules. They also found that the degree of fractionation is inversely correlated with the bulk Mg/Al ratios of the chondrules. However, the isotopic fractionations are much smaller than expected for Mg evaporation under Rayleigh conditions.

Young et al. (2002) measured Mg-isotopic compositions in four Allende chondrules. One barred olivine chondrule is isotopically zoned; the center has δ^{25}Mg = 1.9–1.7‰ (relative to a standard SRM980), whereas that of the margin is 0.3‰. An interesting result is that the Mg-isotopic variation is correlated with δ^{17}O. The authors suggest that mixing of prechondrule materials with distinct Mg- and O-isotopic compositions and later alteration may be responsible for this pattern.

6.3. Experimental Evaporation of Chondrule Melts

Kinetic parameters for gas-condensed phase reactions along with thermodynamic data can be applied to models of reaction processes relevant to chondrule formation. To date, the range of isotopic fractionations in S, K, Fe, Si, and Mg measured in chondrules are much smaller than expected for elemental fractionation resulting from evaporation under Rayleigh conditions. A number of suggestions have been made about how evaporation from chondrules could be achieved without corresponding isotopic fractionations. Subsolidus evaporation of S- and K-bearing chondrule precursors (*Young,* 2000; *Tachibana and Huss,* 2005) is one possibility. High partial pressures of H could lead to non-Rayleigh evaporation of melts (e.g., *Nagahara and Ozawa,* 1996c; *Galy et al.,* 2000). Diffusion of evaporating species through the ambient gas can also suppress isotopic fractionation (*Richter et al.,* 2002). Alternatively, chondrules may have exchanged with the ambient gas during chondrule formation (*Alexander et al.,* 2000; *Alexander,* 2004).

6.3.1. Types of evaporation experiments. Experimental studies of silicate-melt evaporation are summarized in Table 4. In earlier experiments, continuous or stepwise heating experiments were employed to see how chondritic materials suffered from "evaporation metamorphism" (*Hashimoto et al.,* 1979) or to see how alkali elements were fractionated during evaporation from lunar materials (*Gibson and Hubbard,* 1972; *Gibson et al.,* 1973). These experiments focused on single condensed phases. *Hashimoto* (1983) performed systematic isothermal experiments on multicomponent silicate melts, which provide an excellent dataset for modeling chemical fractionation (e.g., *Alexander,* 2001, 2002). Since then, a large number of experiments have been performed to understand the kinetics of gas-silicate melt reactions aimed at understanding the chemical and isotopic fractionations during chondrule formation.

Free evaporation experiments have been performed at pressures of less than ~10^{-4} bar for various starting compositions, such as chondrites (Murchison and Allende, etc.) (*Floss et al.,* 1996; *Hashimoto et al.,* 1979), a synthetic CI composition (*Hashimoto,* 1983; *Wang et al.,* 1994, 2001), the CMAS system with a type-B CAI composition (*Richter et al.,* 2002; *Richter et al.,* 1999) or near-chondritic composition (*Nagahara and Ozawa,* 2003), and a simple Mg_2SiO_4-SiO_2 system (*Nagahara,* 1996; *Nagahara and Ozawa,* 1996a,b). Experiments that provide unequivocal information on evaporation must be isothermal and in the absence of crystalline phases that complicate evaporation processes by inhibiting direct evaporation from the silicate melt (*Nagahara,* 1996; *Ozawa and Nagahara,* 1997). Free evaporation experiments provide evaporation coefficients, the extent of inhibition of evaporation reactions, and unequivocal proof that chemical and isotopic fractionations are generally coupled.

Experiments performed in the presence of ambient gas are grouped into four categories depending on the type of the ambient gas: (1) air (or residual gas) at pressures less than 1 atm (*Notsu et al.,* 1978; *Shimaoka and Nakamura,* 1991), (2) 1-atm experiments with specified oxygen fugacity controlled by flowing either H_2/CO_2 or CO/CO_2 mixture (*Richter and Davis,* 1997; *Richter et al.,* 1999, 2002; *Shirai et al.,* 2000; *Tsuchiyama et al.,* 1981), (3) H_2 or He gas representing solar-nebula conditions that may enhance evaporation (*Richter et al.,* 2002; *Richter et al.,* 1999; *Wang et al.,* 1999; *Yu et al.,* 2003), and (4) gas with species consisting of evaporating condensed phases that may inhibit evapora-

TABLE 4. Experimental studies of gas-liquid reactions related to ferromagnesian chondrule formation.

Starting Composition	Temperature (°C)	Pressure (Pa)	References
Evaporation in vacuum			
Lunar basalts, breccias, and soil	950–1400	$\sim 2.5 \times 10^{-4}$	*Gibson and Hubbard* (1972), *Gibson et al.* (1973)
Murchison (CM) meteorite	1300, 1900	$<10^{-3}$	*Hashimoto et al.* (1979)
Synthetic chondritic Si-Al-Fe-Mg-Ca	1700–2000	$<10^{-5}$	*Hashimoto* (1983)
Etter (L5) chondrite	1150–1450	$\sim 10^{-3}$	*Shimaoka and Nakamura* (1989)
Allende (CV) meterorite	peak: 1500–2100	$\sim 10^{-4}$	*Floss et al.* (1996)
$Fe_{0.947}O$	1450–1750	$\sim 10^{-4}$	*Wang et al.* (1994)
Oxide mixture with CI composition Fe-Mg-Si-Ca-Al-Ti-REE	peak: 1500–2000	$\sim 10^{-4}$	*Wang et al.* (1994, 2001)
Mg_2SiO_4–SiO_2	600, 1700	$<10^{-4}$	*Nagahara and Ozawa* (1996a,b), *Nagahara* (1996)
Synthetic Type B CAI-like Ca-Al-Si-Mg	1700, 1800	$<10^{-4}$	*Richter et al.* (1999, 2002)
Synthetic chondrule-like CMAS	1500, 1600	$<10^{-4}$	*Nagahara and Ozawa* (2003)
Evaporation in gas			
Allende (CV), JB-1 basalt	~2000 or 1300–2000	10^5 or ~1	*Notsu et al.* (1978)
Etter (L5) chondrite	1300	10–10^{-3}	*Shimaoka and Nakamura* (1991)
Synthetic Type IIAB chondrule	1450, 1485	1–140	*Yu et al.* (2003), *Wang et al.* (1999)
Mixture of silicates from H5 chondrite	1450–1600	10^5	*Tsuchiyama et al.* (1981)
Knippa basalt	1300, 1325, 1350	10^5	*Lewis et al.* (1993)
Synthetic Type IIAB and Type I and IA chondrules + sulfide	peak: 1470–1750	10^5	*Yu et al.* (1996)
Mineral mix with chondrule-like composition	peak: 1470–1750	10^5	*Yu and Hewins* (1998)
Synthetic Type B CAI-like Ca-Al-Si-Mg-Ti	1400	10^5	*Richter and Davis* (1997), *Richter et al.* (1999, 2002)
Synthetic Type IIAB chondrule	1450	10^5	*Yu et al.* (2003), *Wang et al.* (1999)
SiO_2-Na_2O system	1300–1400	10^5	*Shirai et al.* (2000)
Synthetic Type B CAI-like Ca-Al-Si-Mg-Ti	1500	~20	*Richter et al.* (1999, 2002)
Synthetic Type IIAB chondrule	1450, 1485	7–10^3	*Yu et al.* (2003), *Wang et al.* (1999)
Mineral mixture with CI composition	1580	13	*Cohen et al.* (2000)
Jilin (H5) meteorite	1300, 1400	$P_{Na} = 0 \sim 1.0$, $P_K = 0 \sim 0.07$	*Nagahara and Ozawa* (2000), *Nagahara* (2000)
Synthetic chondrule-like CMAS	1500, 1600	NA	*Nagahara and Ozawa* (2003)
Transpiration (condensation?)			
Knippa basalt	1330	10^5	*Lewis et al.* (1993)
Allende (CV), Ornans (CO) (±quartz), Eagle Station pallasite	1400–1500	10^5	*Yu et al.* (1995)
Allende (CV), Ornans (CO) chondrites + quartz	1500–1000	10^5	*Yu et al.* (1995)
SiO_2 + C source, CMAS, or Murchison (CM) melt target	target: 1330–1480	8–30	*Tissandier et al.* (2000; 2002)
K_2CO_3 + C source, CMAS ± K target	source: 1000, target: 1400	10^5	*Georges et al.* (1999)
Dust melt collision			
Synthetic Type IIA, Type IIAB chondrules	melt: 1564–1600, dust: 1100–1560	10^5	*Connolly and Hewins* (1995)

tion (*Nagahara,* 2000; *Nagahara and Ozawa,* 2000, 2003). Evaporation is suppressed at high partial pressures of melt-component species or high total pressures due to significant back reaction. Evaporation is enhanced at low oxygen fugacity and high H pressures.

6.3.2. Major-element fractionations during evaporation. Compositional changes during free evaporation of CI- and type-B CAI-like compositions have also been studied (*Hashimoto,* 1983; *Richter et al.,* 2002; *Wang et al.,* 2001). In both systems, Fe is the most volatile, followed by Si, Mg, Ti, Ca, and Al. Silicon and Mg have fairly similar evaporation behavior; however, Si evaporates more rapidly at earlier stages and Mg at later stages of evaporation. Aluminum, Ca, and Ti do not evaporate until the final stages of mass loss.

6.3.3. Trace-element fractionations during evaporation. Rare earth elements experience little evaporation from silicate melts in vacuum. An exception is Ce, which evaporates at a rate similar to Fe (*Floss et al.,* 1996; *Wang et al.,* 2001). This behavior is thought to result from oxidation of Ce_2O_3 to the more volatile form CeO_2 in the residual melt. The REE abundance, relative to CI, along with the degree of Ce depletion can thus be used as an indicator for evaporation of CI-composition precursors under oxidizing conditions.

6.3.4. Isotopic fractionation due to evaporation. Experimental evaporation of chondritic materials causes mass-dependent isotopic fractionation, which results in heavy isotope enrichment in Mg, Si, Ca, Ti, K, and O. The heavy isotope enrichments follow the Rayleigh law provided that diffusion within the experimental melts is not a rate-limiting process. The degree of mass fractionation is smaller than theoretical expectations based on thermodynamic considerations (*Floss et al.,* 1996; *Wang et al.,* 2001; *Richter et al.,* 2002). However, the relative degree of isotopic fractionation decreases with increasing atomic number (*Wang et al.,* 2001). Kinetic processes at the surface of an evaporating melt may suppress isotopic fractionation.

6.4. Theoretical Modeling of Chondrule Melt Evaporation

6.4.1. Stability of silicate melt in equilibrium conditions. Although it is well known that chondrule formation involved kinetic processes, it is important to know the environmental conditions required to stabilize silicate melts. *Wood and Hashimoto* (1993) suggested that dust enrichment greater than 1000× solar is required to stabilize silicate melt in the solar nebula, assuming ideal mixing of the multicomponent silicate melt. *Nagahara and Ozawa* (1995) and *Ozawa and Nagahara* (1997) examined melt stability over a wide range of dust-enrichment factors, and delineated the stability field as a function of total pressure vs. dust enrichment at a given temperature, and determined that the total pressure of 10^{-6} bar is the lower limit of melt stability at 1627°C. *Yoneda and Grossman* (1995) incorporated nonideality in CMAS melts using the model of *Berman* (1983) and examined the stability of this melt over a wide range of temperatures, pressures, and dust enrichments. They found that CMAS melts are stable at a much lower dust enrichments than suggested by *Wood and Hashimoto* (1993). *Ebel and Grossman* (2000) examined the stability of ferromagnesian melts in the solar nebula and suggested that dust enrichments of 20× solar stabilizes ferromagnesian melts at 10^{-3} bar and 1127°–1527°C.

6.4.2. Kinetic models. Equilibrium may have been achieved between gas and condensed phases at high temperatures in the early solar nebula. However, petrographic observations show that chondrule formation involved kinetic processes, such as rapid heating and cooling. Kinetic aspects that must be considered in modeling chondrule gas-melt reactions are (1) surface reaction kinetics, which is the major barrier for evaporation and condensation; (2) elemental transport in the melt through diffusion or convection; (3) the incongruent nature of evaporation and condensation; and (4) transportation of gas species in the ambient gas.

Surface reaction kinetics have been treated by applying the Hertz-Knudsen equation

$$J_{max} = \frac{P_{eq} - P}{\sqrt{2\pi mkT}} \quad (1)$$

to multicomponent systems, which is, in the strictest sense, good only for metallic phases in the presence of monatomic gas molecules. The maximum possible evaporation rate (J_{max}) is proportional to the difference between the equilibrium partial pressure (P_{eq}) and the actual pressure (P) and the average velocity of gas molecules. Kinetic barriers are incorporated into the above equation by multiplying P and P_{eq} by evaporation and condensation coefficients (α_{evap} and α_{cond}), respectively, to give the actual evaporation rate J as

$$J = \frac{\alpha_{evap}P_{eq} \cdot \alpha_{cond}P}{\sqrt{2\pi mkT}} \quad (2)$$

The reason why this simple equation cannot be extended to more complex oxide systems without ambiguity is that the presence of other components involved in the reaction, such as O or H, cannot be properly treated by the equation. In spite of this fact, the ease of use of this approximation and the absence of well-established alternatives means that most researchers use it in their modeling. One approach combines thermodynamic calculations for a given gas composition to obtain the driving force and the evaporation coefficient, α_{evap}, is estimated from the results of free evaporation experiments (*Alexander,* 2002; *Ebel,* 2002; *Grossman et al.,* 2002; *Richter et al.,* 1999). Another method expresses the term for a specific evaporation (or condensation) reaction as a function of the partial pressure of related gas species in the ambient gas, the activity of the component in the melt multiplied by evaporation coefficient, and the equilibrium constant for the reaction (*Alexander,* 2001, 2002; *Shirai et al.,* 2000; *Tsuchiyama and Tachibana,* 1999). The equilibrium constant is assumed to be extendable to kinetic domains according to the principle of detailed balancing (*Lasaga,* 1998). A third approach uses equation (2) and combines kinetic effects and nonideality of silicate melts into the evaporation coefficient (*Ozawa and Nagahara,* 2001). This approach is applicable to the situation where O is much more abundant than metallic gas species.

6.5. Isotopic Fractionation or Lack Thereof

The fundamental discrepancy between chemical and isotopic fractionations in chondrules is that they are depleted in volatile elements but not isotopically fractionated. This problem was first addressed by *Humayun and Clayton* (1995) and discussed intensively by many later workers. This decoupling between chemical and isotopic fractionations is a major constraint on the origin of chondrules. There are several possible explanations. Highly volatile elements can be fractionated with negligible fractionation if the evaporation Peclet number (P_e) is very large (*Ozawa and Nagahara,* 1998; *Galy et al.,* 2000). The Peclet number reflects the relative rates of diffusion and evaporation and is defined as

$$P_e = \frac{J \cdot r}{D} \tag{3}$$

where J is the elemental evaporation rate, r is the radius of the condensed grain, and D is the elemental diffusion rate. Alternatively, volatile elements can be fractionated with negligible isotopic fractionation if there is extensive back reaction during evaporation (*Ozawa and Nagahara,* 1998). Finally, the presence of a gaseous atmosphere inhibits the escape of evaporated gas species and decreases the isotopic fractionation factor to a value close to unity, thus suppressing isotopic fractionation while preserving notable chemical fractionation (*Richter et al.,* 2002).

Diffusion is not expected to play a limiting role in chondrule melt droplets less than a few millimeters in size. Back reaction with the gas is essential if equilibrium between gas and silicate melt is approached. However, it is questionable if total equilibrium was achieved in the solar nebula. Even if chondrule melts could not attain equilibrium with the surrounding vapors, isotopic fractionation of volatile to moderately volatile elements would have been suppressed while still achieving elemental fractionations if the chondrule melts experienced a limited range of cooling rates (*Ozawa and Nagahara,* 1998). The slow escape rate of evaporating gas species due to the presence of H is only effective if H pressure is higher than 10^{-3} bar (*Richter et al.,* 2002).

Combining the evidence for evaporation and recondensation recorded in natural chondrules, experiments, and modeling, the most important conclusions are summarized as follows: (1) Chondrule precursors were heated to lose moderately volatile components by evaporation. The degree of evaporative loss varied depending on the melt composition, grain size, rate of heating, total heating time, and the physical conditions such as pressure and temperature. (2) The evaporated gas was not lost from the system. Rather, significant recondensation took place during cooling. Lithophile elements reentered crystallizing chondrules to various degrees, but siderophile and chalcophile elements condensed as new phases, which are now in chondrule rims and/or matrix. (3) Isotopic fractionation should have been significant at the early stages of evaporation; however, isotopic reequilibration ocurred during subsequent cooling, which resulted in little isotopic fractionation preserved in chondrules.

7. VOLATILE RECONDENSATION AND FORMATION OF CHONDRULE RIMS

Rims around chondrules are an additional source of information about chondrule formation conditions, as well as their precursors. In addition, rims and interchondrule matrix are important complementary components to chondrules that are required to make up the various chondrite group bulk compositions, and all three components may have been derived from a common reservoir (*Klerner and Palme,* 1999). It should be noted that in many studies no distinction is made between chondrule rims and interchondrule matrix, and they are collectively referred to as matrix. However, rims partially or completely surrounding chondrules are texturally and often chemically distinct from the interchondrule matrix (*Ashworth,* 1977; *Allen et al.,* 1980; *Scott et al.,* 1988).

Chondrule rims can be broadly divided into either accretionary or igneous rims. As the name implies, accretionary rims formed by the accretion of fine-grained dust onto the chondrules. Accretionary rims exhibit a range of compositions and textures from fine-grained, FeO-rich rims with few discernable grains even in secondary electron images, to clastic rims that chemically, texturally, and mineralogically resemble the interchondrule matrix (*Ashworth,* 1977; *Alexander et al.,* 1989b). Igneous rims, on the other hand, have relatively coarse-grained igneous textures indicative of partial melting (*Rubin,* 1984; *Rubin and Wasson,* 1987; *Krot and Wasson,* 1995).

Perhaps the best-developed accretionary rims occur in CM chondrites (*Metzler et al.,* 1992; *Lauretta et al.,* 2000; *Zega and Buseck,* 2003). In the unbrecciated areas of CM chondrites (so-called primary rock), rims surround almost all objects, including chondrules, chondrule fragments, and mineral fragments. The most abundant phase in rims around chondrules in CM chondrites is Mg-rich serpentine. Other minerals embedded within the rims include cronstedtite, tochilinite, chlorite, pentlandite, gypsum, olivine, kamacite, taenite, and chromite. In these rims, regions containing hydrated and anhydrous minerals are in direct contact with each other, which has been interpreted as rim material accreting directly from multiple nebular reservoirs. It has also been suggested that aqueous alteration and metamorphism on the CM chondrite parent body has significantly modified the primary mineralogy of the accretionary rims (*Brearley and Geiger,* 1993; *Chizmadia and Brearley,* 2003). There is a clear relationship between the radius of the core object and the thickness of its rim. A similar relationship has been reported for the CV chondrites (*Paque and Cuzzi,* 1997) and may or may not be present in the ordinary chondrites (*Matsunami,* 1984; *Huang et al.,* 1993).

In a few meteorites, such as Bishunpur (LL3.1) and ALH 77307 (CO3.0), there is an abundance of amorphous material in the rims and interchondrule matrix (*Alexander et al.,* 1989a,b; *Brearley,* 1993). Amorphous material is very susceptible to alteration and recrystallization, and its preservation suggests that some of the primary material has survived. *Brearley and Geiger* (1993) and *Chizmadia and Brearley* (2003) argue from the lack of phyllosilicate pseudomorphs that CM rims were also largely amorphous prior to alteration.

There are likely to have been several sources of amorphous material. In Bishunpur, chondrule mineral fragments are abundant in the interchondrule matrix and rims. These must be accompanied by amorphous, Al-, Ca-, and Si-rich chondrule mesostasis. Indeed, the finest-grained rim and matrix material tends to be enriched in these elements. However, in ALH 77307 clastic mineral fragments are much less abundant, leading *Brearley* (1993) to suggest that much of the amorphous material is a condensate. Numerous authors have suggested that rim and interchondrule matrix material is composed of nebular condensates (e.g., *Larimer and Anders,* 1967; *Huss et al.,* 1981; *Nagahara,* 1984).

An alternative source of condensates is material evaporated during chondrule formation. The shorter timescale of chondrule formation would also be more conducive to formation of amorphous material. There is some direct evidence of recondensation during chondrule formation. Iron, Si, and Mn are among the most volatile of the major and minor elements, and they are depleted in type-I chondrules. *Alexander* (1995) described compositionally and texturally zoned rims around some type-I Bishunpur chondrules, with large FeO, Si, and Mn enrichments at the rim-chondrule interface that often grade into matrix-like compositions and textures at the rims' outer edges. Transmission electron microscopy of rims with similar compositions shows this Fe-,Si-rich material to be amorphous. There is also evidence of partial to complete reequilibration of clastic olivine and pyroxene grains with the FeO-rich material, as well as reaction with protuberances from the chondrules.

Petrologic evidence shows that accretion of rims began before expulsion of metal and sulfide from chondrules was complete, and that the metal/sulfide that was trapped in the rims reacted with the nebular gases as the system cooled (*Allen et al.,* 1980; *Alexander et al.,* 1989b; *Lauretta and Buseck,* 2003). In the CR chondrites, at least, there is evidence of both evaporation of metal from chondrules and its recondensation onto rims (*Kong and Palme,* 1999; *Connolly et al.,* 2001; *Humayun et al.,* 2002; *Zanda et al.,*2002). Finally, rims and matrix-like material are sometimes trapped between plastically deformed or compound chondrules, indicating that the chondrules were still hot when they collided (e.g., *Hutchison and Bevan,* 1983; *Scott et al.,* 1984; *Holmen and Wood,* 1986; *Lauretta and Buseck,* 2003). All these features suggest that the rims began accreting while the chondrules were still hot.

While Fe and Si are relatively volatile, the alkalis are much more so. If there was some evaporation of Fe and Si during chondrule formation, there should have been considerable loss of the alkalis from chondrules, and enrichment of alkalis in rims and interchondrule matrix. Evidence for recondensation of alkalis in rims is hampered by their mobility during even very mild aqueous alteration or metamorphism. Rims and interchondrule matrix are generally enriched in alkalis relative to chondrules in the least altered/ metamorphosed chondrites. Of the OCs, alkali enrichments are greatest in Semarkona (LL3.0), but even in this meteorite there is ample evidence of reentry of alkalis into chondrules (*Grossman et al.,* 2002).

While there is evidence that some rims accreted while their host chondrules were still hot, this cannot be true for some or, perhaps, even most rims. The rims and matrix of chondrites are the host of presolar material: circumstellar grains and interstellar organic matter. Indeed, the abundances of presolar material in rim/matrix in the least-metamorphosed members of the chondrite groups are very similar to that of CI chondrites (*Huss and Lewis,* 1995; *Alexander et al.,* 1998). The CI chondrites are thought to have contained few if any chondrules or CAIs. They most likely escaped high-temperature processes prior to accretion and their presolar abundances should most closely resemble those of the primordial dust. The implication of the presolar abundances, particularly since rims make up the bulk of the fine-grained material in many chondrites, is that the major component of chondrule rims and matrix resembled CI chondrites in composition and did not experience the high temperatures associated with nebula-wide condensation or with chondrule formation. The CI chondrites are almost completely aqueously altered, but perhaps the most likely analogs for them prior to alteration are the anhydrous interplanetary dust particles. They are very fine-grained and can contain abundant amorphous material.

Intuitively, one might expect accretionary rims that formed in a dusty nebula to resemble loosely packed dust balls. Yet, accretionary rims have surprisingly low porosity (<20%). They were also robust enough to survive accretion onto the chondrite parent bodies, as well as any regolith processing that occurred prior to final burial and compaction. Recondensation of vaporized material from chondrules would help cement a rim, as well as fill in porosity. Models and experiments (e.g., *Dominik and Tielens,* 1997; *Blum and Wurm,* 2000) find that sticking of micrometer-sized dust is most efficient when collisions occur at less than 10–20 m/s. At speeds of up to several meters per second, the collisions actually tend to compact fluffy aggregates. Electrostatic attraction between grains may enhance sticking efficiencies, as well as increase the range of relative speeds over which sticking and compaction occurs (*Marshall and Cuzzi,* 2001). The most detailed modeling of rim formation has been done in the context of the turbulent concentration mechanism (*Cuzzi,* 2004). In a weakly turbulent nebula, the timescales for making rims by this mechanism are likely to be 10–1000 yr. Such timescales are possible for rims that accreted onto cold chondrules, but are clearly inconsistent with rims that show evidence of accretion while the chondrules were still hot.

Igneous rims appear to have formed by partial melting of preexisting fine-grained rims (*Rubin,* 1984; *Rubin and Wasson,* 1987; *Krot and Wasson,* 1995). They probably form a continuum with dark-zoned or microporphyritic chondrules, and with so-called enveloping compound chondrules (*Wasson,* 1993). Their fine grain size and their limited interaction with their host chondrules suggest that igneous rims formed at lower peak temperatures and from material with lower solidus temperatures than their host chondrules. More intense heating would presumably lead to the mutual assimilation of rim and host, producing a new chondrule. Thus, it is likely that fine-grained rim and matrix material was one component of the chondrule precursors.

In summary, the presence of fine-grained rims around chondrules suggests that they formed in relatively dusty environments, consistent with chondrule cooling rates. In at least some instances, rims began accreting while the chondrules were still hot and acted as sites for recondensation of material evaporated from the chondrules. The dust may also have provided nuclei for completely melted chondrules (see section 4.1). Whether this early accreted material survived chondrule formation or was mixed into the chondrule-forming region soon after it began to cool is unclear. While recondensation may have aided rim accretion at this stage, it is likely that by the time rims began to accrete the relative velocities of chondrules and micrometer-sized dust were <10–20 m/s, placing some constraints on models that involve gas drag for heating and/or sorting of chondrules. Later, mixing of cold chondrules and unheated material must have occurred. The CI-like abundances of presolar material in rim/matrix requires that most rim formation occurred after chondrules had cooled. Finally, the presence of igneous rims on chondrules point to multiple episodes of chondrule and rim formation, requiring a relatively dynamic environment, and suggests that fine-grained rim/matrix material was one component of the chondrule precursors.

8. CONSTRAINTS ON MODELS OF CHONDRULE FORMATION

The primary purpose of this chapter is to summarize the petrologic history of those chondrules dominated by ferromagnesian silicates. It is not the intent of this chapter to discuss various models for chondrule formation, which place this history in the larger context of environments and processes in the early solar system (see *Connolly et al.,* 2006). However, the information presented here is essential for constraining models of chondrule formation. Any such model must account for all the features present in chondrules, summarized below.

Chondrule precursors were composed of anhydrous minerals, predominantly fine-grained olivine (<5 μm). However, larger grains of olivine, up to 400 μm, were also incorporated. Lesser amounts of pyroxene and feldspathic glass, chromite, metal, and sulfide were likely also present. The presence of sulfide places some constraints on the minimum temperatures of chondrule precursors. In a canonical solar-composition system, this temperature is ~427°C, where sulfide condensation begins (*Lauretta et al.,* 1996). However, in systems with enhanced S/H ratios, relative to solar, this temperature could be as high as 1000°C for total gas pressures $<10^{-4}$ bar (*Lauretta et al.,* 2001). Carbonaceous material was included in some chondrule melts and likely played an important role in controlling the oxidation state of the resulting mineralogy.

At least some chondrules formed through melting of preexisting materials, indicated by the presence of large, compositionally distinct grains. These relict grains were part of the precursor material that survived chondrule formation without melting or dissolving. Many of these grains appear to be fragments of both chondrules and refractory inclusions. These features imply that the process or processes of chondrule formation were repeated many times.

Chondrules in some carbonaceous chondrites show evidence of ^{16}O-rich components. These signatures may be evidence of the survival or limited processing of a presolar component in the chondrule precursors. Alternatively, they may result from chemical processes in the gas phase that occurred prior to or during chondrule formation

Chondrule textures result from a complex interplay between the peak temperature, duration of heating, and preservation of nucleation sites, either due to incomplete melting of chondrule precursors or the presence of fine-grained dust in the chondrule-formation region. Experimental studies place a firm limit of 2100°C on the maximum temperature of chondrule formation. Few chondrules with liquidus temperatures over 1750°C were completely melted, and few with liquidus temperatures under 1400°C were incompletely melted. Most chondrules reached peaked temperatures of 1500°–1550°C.

Simulation experiments and chondrule textures have also placed constraints on chondrule cooling rates. The cooling rates required to produce porphyritic olivine textures are less than 1000°C/h with rates on the order of 100°C/h providing the best experimental analogs. Porphyritic pyroxene textures develop at slow cooling rates, on the order of 2°–50°C/h, regardless of the degree of undercooling. Cooling rates as low as 5°C/h are also consistent with the zoning observed in Semarkona chondrules. Barred-olivine chondrules cooled at rates of 500° to 2300°C/h. Radial pyroxene chondrules grew from a melt that experienced very high degrees of undercooling at quite slow cooling rates (5°C/hr). Pigeonite/diopside exsolution lamellae suggest that these textures are produced when cooling rates are on the order of 5°–25°C/h in the temperature range of 1200°–1350°C.

An important constraint on models of chondrule formation is the variation in the oxidation states of their components. Type-I chondrules experienced *in situ* reduction either by incorporating reducing material in their precursors or interacting with a reducing gas phase. Type-I chondrules also exhibit significant chemical fractionations, with depletions of elements like Cr, Mn, Na, K, and Fe, and enrichments of elements, such as Al, Ca, and Ti, relative to Mg. The order of these depletions and enrichments suggest volatility has played a role.

Chondrule precursors were heated and lost moderately volatile components by evaporation. However, significant recondensation took place during cooling. Lithophile elements recondensed into crystallizing chondrules to various degrees during cooling, but siderophile and chalcophile elements condensed as new phases or recondensed into metal expelled from chondrules that did not completely evaporate, which are now in chondrule rims and matrix. Isotopic fractionation should have been significant at the early stage of evaporation; however, isotopic reequilibration during subsequent cooling must have been almost complete, resulting in little or no isotopic fractionation in chondrules.

Petrologic evidence shows that accretion of rims began while chondrules were still molten, before expulsion of metal and sulfide from chondrules was complete. Metal and sulfide was trapped in the rims and reacted with nebular gases as the system cooled. Finally, rims and matrix-like material are sometimes trapped between plastically deformed or compound chondrules, indicating that the chondrules were still hot when they collided. However, since the rims and matrix of chondrites are the host of presolar material, much of this material must have accreted after chondrules had cooled back down to ambient nebular conditions. Igneous rims appear to have formed by partial melting of preexisting fine-grained rims, suggesting that reheating events occurred well after the volatile-rich accretionary rims were in place.

REFERENCES

Alexander C. M. O'D. (1994) Trace-element distributions within ordinary chondrite chondrules — Implications for chondrule formation conditions and precursors. *Geochim. Cosmochim. Acta, 58,* 3451–3467.

Alexander C. M. O'D. (1995) Trace element contents of chondrule rims and interchondrule matrix in ordinary chondrites. *Geochim. Cosmochim. Acta, 59,* 3247–3266.

Alexander C. M. O'D. (1996) Recycling and volatile loss in chondrule formation. In *Chondrules and the Protoplanetary Disk* (R. H. Hewins et al., eds.), pp. 233–241. Cambridge Univ., New York.

Alexander C. M. O'D. (2001) Exploration of quantitative kinetic models for the evaporation of silicate melts in vacuum and in hydrogen. *Meteoritics & Planet. Sci., 36,* 255–283.

Alexander C. M. O'D. (2002) Application of MELTS to kinetic evaporation models of FeO-bearing silicate melts. *Meteoritics & Planet. Sci., 37,* 245–256.

Alexander C. M. O'D. (2004) Chemical equilibrium and kinetic constraints for chondrule and CAI formation conditions. *Geochim. Cosmochim. Acta, 68,* 3943–3969.

Alexander C. M. O'D. and Grossman J. N. (2005) Alkali elemental and potassium isotopic compositions of Semarkona chondrules. *Meteoritics & Planet. Sci., 40,* 541–556.

Alexander C. M. O'D. and Wang J. (2001) Iron isotopes in chondrules: Implications for the role of evaporation during chondrule formation. *Meteoritics & Planet. Sci., 36,* 419–428.

Alexander C. M., O'D., Barber D. J., and Hutchison R. (1989a) The microstructure of Semarkona and Bishunpur. *Geochim. Cosmochim. Acta, 53,* 3045–3057.

Alexander C. M. O'D., Hutchison R., and Barber D. J. (1989b) Origin of chondrule rims and interchondrule matrices in unequilibrated ordinary chondrites. *Earth Planet. Sci. Lett., 95,* 187–207.

Alexander C. M. O'D., Russell S. S., Arden J. W., Ash R. D., Grady M. M., and Pillinger C. T. (1998) The origin of chondritic macromolecular organic matter: A carbon and nitrogen isotope study. *Meteoritics & Planet. Sci., 33,* 603–622.

Alexander C. M. O'D., Grossman J. N., Wang J., Zanda B., Bourot-Denise M., and Hewins R. H. (2000) The lack of potassium-isotopic fractionation in Bishunpur chondrules. *Meteoritics & Planet. Sci., 35,* 859–868.

Alexander C. M. O., Taylor S., Delaney J. S., Ma P. X., and Herzog G. F. (2002) Mass-dependent fractionation of Mg, Si, and Fe isotopes in five stony cosmic spherules. *Geochim. Cosmochim. Acta, 66,* 173–183.

Allen J. S., Nozette S., and Wilkening L. L. (1980) A study of chondrule rims and chondrule irradiation records in unequilibrated ordinary chondrites. *Geochim. Cosmochim. Acta, 44,* 1161–1175.

Ashworth J. R. (1977) Matrix textures in unequilibrated ordinary chondrites. *Earth Planet. Sci. Lett., 35,* 25–34.

Berman R. G. (1983) A thermodynamic model for silicate melts with application to the system $CaO-MgO-Al_2O_3-SiO_2$. Ph.D. thesis, University of British Columbia. 178 pp.

Bizzarro M., Baker J. A., and Haack H. (2004) Mg isotope evidence for contemporaneous formation of chondrules and refractory inclusions. *Nature, 431,* 275–278.

Blander M. and Katz J. L. (1967) Condensation of primordial dust. *Geochim. Cosmochim. Acta, 31,* 1025–1034.

Blander M., Planner H. N., Keil K., Nelson L. S., and Richardson N. L. (1976) Origin of chondrules — Experimental investigation of metastable liquids in system $Mg_2SiO_4-SiO_2$. *Geochim. Cosmochim. Acta, 40,* 889–896.

Blander M., Unger L., Pelton A., and Ericksson G. (2001) Nucleation constraints lead to molten chondrule precursors in the early solar system. *J. Phys. Chem., B105,* 11823–11827.

Blander M., Pelton A. D., Jung I. H., and Weber R. (2004) Nonequilibrium concepts lead to a unified explanation of the formation of chondrules and chondrites. *Meteoritics & Planet. Sci., 39,* 1897–1910.

Blum J. and Wurm G. (2000) Experiments on sticking, restructuring, and fragmentation of preplanetary dust aggregates. *Icarus, 143,* 138–146.

Brearley A. J. (1993) Matrix and fine-grained rims in the unequilibrated CO3 chondrite, ALHA77307: Origins and evidence for diverse, primitive nebular components. *Geochim. Cosmochim. Acta, 57,* 1521–1550.

Brearley J. and Geiger T. (1993) Fine-grained chondrule rims in the Murchison CM2 chondrite: Compositional and mineralogical systematics (abstract). *Meteoritics, 28,* 328.

Campbell A. J., Humayun M., and Zanda B. (2002) Partial condensation of volatile elements in Renazzo chondrules (abstract). *Geochim. Cosmochim. Acta, 66,* A117.

Carpenter M. A. (1979) Experimental coarsening of antiphase domains in a silicate mineral. *Science, 206,* 681–683.

Chizmadia L. J. and Brearley A. J. (2003) Mineralogy and textural characteristics of fine-grained rims in the Yamato 791198 CM2 carbonaceous chondrite: Constraints on the location of aqueous alteration (abstract). In *Lunar and Planetary Science XXXIV,* Abstract #1419. Lunar and Planetary Institute, Houston (CD-ROM).

Clayton R. N. (2004) The origin of oxygen isotope variations in the early solar system (abstract). In *Lunar and Planetary Science XXXV,* Abstract #1682. Lunar and Planetary Institute, Houston (CD-ROM).

Clayton R. N., Mayeda T. K., Goswami J. N., and Olsen E. J. (1991) Oxygen isotope studies of ordinary chondrites. *Geochim. Cosmochim. Acta, 55,* 2317–2337.

Cohen B. A. and Hewins R. H. (2004) An experimental study of the formation of metallic iron in chondrules. *Geochim. Cosmochim. Acta, 68,* 1677–1689.

Cohen B., Hewins R. H. and Alexander C. M. O'D. (2004) The formation of chondrules by open-system melting of nebular condensates. *Geochim. Cosmochim. Acta, 68,* 1661–1675.

Connolly H. C. and Hewins R. H. (1995) Chondrules as products of dust collisions with totally molten droplets within a dust-rich nebular environment — An experimental investigation. *Geochim. Cosmochim. Acta, 59,* 3231–3246.

Connolly H. C. and Hewins R. H. (1996) Constraints on chondrule precursors from experimental data. In *Chondrules and the Protoplanetary Disk* (R. H. Hewins et al., eds.), pp. 129–136. Cambridge Univ., New York.

Connolly H. C., Hewins R. H., Ash R. D., Zanda B., Lofgren G. E., and Bourot-Denise M. (1994) Carbon and the formation of reduced chondrules. *Nature, 371,* 136–139.

Connolly H. C., Jones B. D., and Hewins R. H. (1998) The flash melting of chondrules: An experimental investigation into the melting history and physical nature of chondrule precursors. *Geochim. Cosmochim. Acta, 62,* 2725–2735.

Connolly H. C., Huss G. R., and Wasserburg G. J. (2001) On the formation of Fe-Ni metal in Renazzo-like carbonaceous chondrites. *Geochim. Cosmochim. Acta, 65,* 4567–4588.

Connolly H. C. Jr., Desch S. J., Ash R. D., and Jones R. H. (2006) Transient heating events in the protoplanetary nebula. In *Meteorites and the Early Solar System II* (D. S. Lauretta and H. Y. McSween Jr., eds.), this volume. Univ. of Arizona, Tucson.

Cuzzi J. N. (2004) Blowing in the wind: III. Accretion of dust rims by chondrule-sized particles in a turbulent protoplanetary nebula. *Icarus, 168,* 484–497.

Dehart J. M., Lofgren G. E., Jie L., Benoit P. H., and Sears D. W. G. (1992) Chemical and physical studies of chondrites 10: Cathodoluminescence and phase-composition studies of metamorphism and nebular processes in chondrules of type-3 ordinary chondrites. *Geochim. Cosmochim. Acta, 56,* 3791–3807.

Deloule E., Robert F., and Doukhan J. C. (1998) Interstellar hydroxyl in meteoritic chondrules: Implications for the origin of water in the inner solar system. *Geochim. Cosmochim. Acta, 62,* 3367–3378.

Dodd R. T. (1971) Petrology of chondrules in Sharps meteorite. *Contrib. Mineral. Petrol., 31,* 201.

Dominik C. and Tielens A. G. G. M. (1997) The physics of dust coagulation and the structure of dust aggregates in space. *Astrophys. J., 480,* 647–673.

Ebel D. S. (2002) Model evaporation of FeO-bearing liquids. *Meteoritics & Planet. Sci., 37,* A43.

Ebel D. S. and Grossman L. (2000) Condensation in dust-enriched systems. *Geochim. Cosmochim. Acta, 64,* 339–366.

Everman R. L. A. and Cooper R. F. (2003) Internal reduction of an iron-doped magnesium aluminosilicate melt. *J. Am. Ceram. Soc., 86,* 487–494.

Fagan T., McKeegan K. D., Krot A. N., and Keil K. (2001) Calcium-aluminum-rich inclusions in enstatite chondrites (II): Oxygen isotopes. *Meteoritics & Planet. Sci., 36,* 223–230.

Faure F., Trolliard G., Nicollet C., and Montel J. M. (2003) A developmental model of olivine morphology as a function of the cooling rate and the degree of undercooling. *Contrib. Mineral. Petrol., 145,* 251–263.

Ferraris C., Folco L., and Mellini M. (2002) Chondrule thermal history from unequilibrated H chondrites: A transmission and analytical electron microscopy study. *Meteoritics & Planet. Sci., 37,* 1299–1321.

Floss C. F., El Goresy A. E., Zinner E., Kransel G., Rammensee W., and Palme H. (1996) Elemental and isotopic fractionations produced through evaporation of the Allende CV chondrite: Implications for the origin of HAL-type hibonite inclusions. *Geochim. Cosmochim. Acta, 60,* 1975–1997.

Fries M. and Steele A. (2005) Carbonaceous inclusions in meteorite chondrules as characterized using confocal Raman imaging. *Astrobiology, 5,* 221.

Fujii N. and Miyamoto M. (1983) Constraints on the heating and cooling processes of chondrule formation. In *Chondrules and Their Origins* (E. A. King, ed.), pp. 53–60. Lunar and Planetary Institute, Houston.

Galy A., Young E. D., Ash R. D., and O'Nions R. K. (2000) The formation of chondrules at high gas pressures in the solar nebula. *Science, 290,* 1751–1753.

Georges P., Libourel G., and Dloule E. (1999) Potassium condensation experiments and their bearing on chondrule formation (abstract). In *Lunar and Planetary Science XXX,* Abstract #1744. Lunar and Planetary Institute, Houston (CD-ROM).

Gibson E. K. Jr. and Hubbard N. J. (1972) Thermal volatilization studies on lunar samples. *Proc. Lunar Sci. Conf. 3rd,* pp. 2003–2014.

Gibson E. K. Jr., Hubbard N. J., Wiesmann H., Bansal B. M., and Moore G. W. (1973) How to lose Rb, K, and change the K/Rb ratio: An experimental study. *Proc. Lunar Sci. Conf. 4th,* pp. 1263–1273.

Gooding J. L. and Keil K. (1981) Relative abundances of chondrule primary textural types in ordinary chondrites and their bearing on conditions of chondrule formation. *Meteoritics, 16,* 17–43.

Gooding J. L., Keil K., Fukuoka T., and Schmitt R. A. (1980) Elemental abundances in chondrules from unequilibrated chondrites — Evidence for chondrule origin by melting of pre-existing materials. *Earth Planet. Sci. Lett., 50,* 171–180.

Graham A. L., Easton A. J., Hutchison R., and Jerome D. Y. (1976) Bovedy meteorite — Mineral chemistry and origin of its Ca-rich glass inclusions. *Geochim. Cosmochim. Acta, 40,* 529.

Greenwood J. P. and Hess P. C. (1996) Congruent melting kinetics: Constraints on chondrule formation. In *Chondrules and the Protoplanetary Disk* (R. H. Hewins et al., eds.), pp. 205–212. Cambridge Univ., New York.

Grossman J. N. (1996) Chemical fractionations of chondrites: Signatures of events before chondrule formation. In *Chondrules and the Protoplanetary Disk* (R. H. Hewins et al., eds.), pp. 243–256. Cambridge Univ., New York.

Grossman J. N. and Wasson J. T. (1982) Evidence for primitive nebular components in chondrules from the Chainpur chondrite. *Geochim. Cosmochim. Acta, 46,* 1081–1099.

Grossman J. N. and Wasson J. T. (1983a) Refractory precursor components of Semarkona chondrules and the fractionation of refractory elements among chondrites. *Geochim. Cosmochim. Acta, 47,* 759–771.

Grossman J. N. and Wasson J. T. (1983b) The composition of chondrules in unequilibrated chondrites: An evaluation of models for the formation of chondrules and their precursor materials. In *Chondrules and Their Origins* (E. A. King, ed.), pp. 88–121. Lunar and Planetary Institute, Houston.

Grossman J. N. and Wasson J. T. (1985) The origin and history of the metal and sulfide components of chondrules. *Geochim. Cosmochim. Acta, 49,* 925–939.

Grossman J. N., Rubin A. E., Rambaldi E. R., Rajan R. S., and Wasson J. T. (1985) Chondrules in the Qingzhen type-3 enstatite chondrite — Possible precursor components and comparison to ordinary chondrite chondrules. *Geochim. Cosmochim. Acta, 49,* 1781–1795.

Grossman J. N., Rubin A. E., Nagahara H., and King E. A. (1988) Properties of chondrules. In *Meteorites and the Early Solar System* (J. F. Kerridge and M. S. Matthews), pp. 619–659. Univ. of Arizona, Tucson.

Grossman J. N., Alexander C. M. O'D., Wang J., and Brearley A. J. (2002) Zoned chondrules in Semarkona: Evidence for high- and low-temperature processing. *Meteoritics & Planet. Sci., 37,* 49–74.

Guan Y., McKeegan, K. D., MacPherson G. J. (2000) Oxygen

isotopes in calcium-aluminum-rich inclusions from enstatite chondrites: New evidence for a single CAI source in the solar nebula. *Earth Planet. Sci. Lett., 181,* 271–277.

Hamilton D. I. and MacKenzie W. S. (1965) Phase equilibrium studies in the system $NaAlSiO_2$-$KAlSiO_4$-SiO_2-H_2O. *Mineral. Mag., 34,* 214–231.

Hanon P., Robert F., and Chaussidon M. (1998) High carbon concentrations in meteoritic chondrules: A record of metal-silicate differentiation. *Geochim. Cosmochim. Acta, 62,* 903–913.

Hashimoto A. (1983) Evaporation metamorphism in the early solar nebula — evaporation experiments on the melt FeO-MgO-SiO_2-CaO-Al_2O_3. *Geochem. J., 17,* 111–145.

Hashimoto A., Kumazawa M., and Onuma N. (1979) Evaporation metamorphism of primitive dust material in the early solar nebula. *Earth Planet. Sci. Lett., 43,* 13–21.

Herzberg C. T. (1979) Solubility of olivine in basaltic liquids — Ionic model. *Geochim. Cosmochim. Acta, 43,* 1241–1251.

Hewins R. H. (1991) Retention of sodium during chondrule melting. *Geochim. Cosmochim. Acta, 55,* 935–942.

Hewins R. H. (1997) Chondrules. *Annu. Rev. Earth Planet. Sci., 25,* 61–83.

Hewins R. H. and Fox G. E. (2004) Chondrule textures and precursor grain size: An experimental study. *Geochim. Cosmochim. Acta, 68,* 917–926.

Hewins R. H. and Radomsky P. M. (1990) Temperature conditions for chondrule formation. *Meteoritics, 25,* 309–318.

Hewins R. H., Yu Y., Zanda B., and Bourot-Denis M. (1997) Do nebular fractionations, evaporative losses, or both, influence chondrule compositions? *Antarct. Meteorite Res., 10,* 275–298.

Holmen B. A. and Wood J. A. (1986) Chondrules that indent one another: Evidence for hot accretion? (abstract). *Meteoritics, 21,* 399.

Huang S., Benoit P. H., and Sears D. W. G. (1993) Metal and sulfide in Semarkona chondrules and rims: Evidence for reduction, evaporation, and recondensation during chondrule formation. *Meteoritics, 28,* 367–368.

Huang S. X., Lu J., Prinz M., Weisberg M. K., Benoit P. H., and Sears D. W. G. (1996) Chondrules: Their diversity and the role of open-system processes during their formation. *Icarus, 122,* 316–346.

Humayun M. and Clayton R. N. (1995) Potassium isotope cosmochemistry — Genetic-implications of volatile element depletion. *Geochim. Cosmochim. Acta, 59,* 2131–2148.

Humayun M., Campbell A. J., Zanda B., and Bourot-Denise M. (2002) Formation of Renazzo chondrule metal inferred from siderophile elements (abstract). In *Lunar and Planetary Science XXXIII,* Abstract #1965. Lunar and Planetary Institute, Houston (CD-ROM).

Huss G. R. and Lewis R. S. (1995) Presolar diamond, SiC, and graphite in primitive chondrites: Abundances as a function of meteorite class and petrologic type. *Geochim. Cosmochim. Acta, 59,* 115–160.

Huss G. R., Keil K., and Taylor G. J. (1981) The matrices of unequilibrated ordinary chondrites: Implications for the origin and history of chondrites. *Geochim. Cosmochim. Acta, 45,* 33–51.

Hutcheon I. D. and Hutchison R. (1989) Evidence from the Semarkona ordinary chondrite for Al-26 heating of small planets. *Nature, 337,* 238–241.

Hutchison R. and Bevan A. W. R. (1983) Conditions and time of chondrule accretion. In *Chondrules and Their Origins* (E. A. King), pp. 162–179. Lunar and Planetary Institute, Houston.

Hutchison R. and Graham A. L. (1975) Significance of calcium-rich differentiates in chondritic meteorites. *Nature, 255,* 471.

Hutchison R., Alexander C. M. O., and Barber D. J. (1988) Chondrules — Chemical, mineralogical and isotopic constraints on theories of their origin. *Philos. Trans. R. Soc. Lond., A325,* 445–458.

Ikeda Y. (1982) Petrology of the ALH-77003 chondrite (C3). *Proc. 7th Symp. Antarct. Meteorites,* 34–65.

Itoh S. and Yurimoto H. (2003) Contemporaneous formation of chondrules and refractory inclusions in the early solar system. *Nature, 423,* 728–731.

Johnson C. A. and Prinz M. (1991) Chromite and olivine in type-II chondrules in carbonaceous and ordinary chondrites — Implications for thermal histories and group-differences. *Geochim. Cosmochim. Acta, 55,* 893–904.

Jones R. H. (1990) Petrology and mineralogy of type-II, FeO-rich chondrules in Semarkona (LL3.0) — Origin by closed-system fractional crystallization, with evidence for supercooling. *Geochim. Cosmochim. Acta, 54,* 1785–1802.

Jones R. H. (1992) On the relationship between isolated and chondrule olivine grains in the carbonaceous chondrite ALHA 77307. *Geochim. Cosmochim. Acta, 56,* 467–482.

Jones R. H. (1994) Petrology of FeO-poor, porphyritic pyroxene chondrules in the Semarkona chondrite. *Geochim. Cosmochim. Acta, 58,* 5325–5340.

Jones R. H. (1996) FeO-rich, porphyritic pyroxene chondrules in unequilibrated ordinary chondrites. *Geochim. Cosmochim. Acta, 60,* 3115–3138.

Jones R. H. and Danielson L. R. (1997) A chondrule origin for dusty relict olivine in unequilibrated chondrites. *Meteoritics & Planet. Sci., 32,* 753–760.

Jones R. H. and Lofgren G. E. (1993) A comparison of FeO-rich, porphyritic olivine chondrules in unequilibrated chondrites and experimental analogs. *Meteoritics, 28,* 213–221.

Jones R. H. and Scott E. R. D. (1989) Petrology and thermal history of type-IA chondrules in the Semarkona (LL3.0) chondrite. *Proc. Lunar Planet. Sci. Conf. 19th,* pp. 523–536.

Jones R. H., Lee T., Connolly H. C., Love S. G., and Shang H. (2000) Formation of chondrules and CAIs: Theory vs. observation. In *Protostars and Planets IV* (V. Mannings et al., eds.), pp. 927–962. Univ. of Arizona, Tucson.

Jones R. H., Leshin L. A., Guan Y. B., Sharp Z. D., Durakiewicz T., and Schilk A. J. (2004) Oxygen isotope heterogeneity in chondrules from the Mokoia CV3 carbonaceous chondrite. *Geochim. Cosmochim. Acta, 68,* 3423–3438.

Kehm K., Hauri E. H., Alexander C. M. O., and Carlson R. W. (2003) High precision iron isotope measurements of meteoritic material by cold plasma ICP-MS. *Geochim. Cosmochim. Acta, 67,* 2879–2891.

Kennedy A. K., Hutchison R., Hutcheon I. D., and Agrell S. O. (1992) A unique high Mn/Fe microgabbro in the Parnallee (LL3) ordinary chondrite — Nebular mixture or planetary differentiate from a previously unrecognized planetary body. *Earth Planet. Sci. Lett., 113,* 191–205.

Kitamura M., Yasuda M., Watanabe S., and Morimoto N. (1983) Cooling history of pyroxene chondrules in the Yamato-74191 chondrite (L3) — an electron-microscopic study. *Earth Planet. Sci. Lett., 63,* 189–201.

Klerner S. and Palme H. (1999) Origin of chondrules and matrix in carbonaceous chondrites (abstract). In *Lunar and Planetary Science XXX,* Abstract #1272. Lunar and Planetary Institute, Houston (CD-ROM).

Kong P. and Ebihara M. (1996) Metal phases of L chondrites: Their formation and evolution in the nebula and in the parent body. *Geochim. Cosmochim. Acta, 60,* 2667–2680.

Kong P. and Ebihara M. (1997) The origin and nebular history of the metal phase of ordinary chondrites. *Geochim. Cosmochim. Acta, 61,* 2317–2329.

Kong P. and Palme H. (1999) Compositional and genetic relationship between chondrules, chondrule rims, metal, and matrix in the Renazzo chondrite. *Geochim. Cosmochim. Acta, 63,* 3673–3682.

Kong P., Ebihara M., and Palme H. (1999) Distribution of siderophile elements in CR chondrites: Evidence for evaporation and recondensation during chondrule formation. *Geochim. Cosmochim. Acta, 63,* 2637–2652.

Krot A. N. and Wasson J. T. (1995) Igneous rims on low-FeO chondrules in ordinary chondrites. *Geochim. Cosmochim. Acta, 59,* 4951–4966.

Krot A. N., McKeegan K. D., Russell S. S., Meibom A., Weisberg M. K., Zipfel J., Krot T. V., Fagan T. J., and Keil K. (2001a) Refractory calcium-aluminum-rich inclusions and aluminum-diopside-rich chondrules in the metal-rich chondrites Hammadah al Hamra 237 and Queen Alexandra Range 94411. *Meteoritics & Planet. Sci., 36,* 1189–1216.

Krot A. N., Meibom A., Russell S. S., Alexander C. M. O., Jeffries T. E., and Keil K. (2001b) A new astrophysical setting for chondrule formation. *Science, 291,* 1776–1779.

Kunihiro T., Rubin A. E., McKeegan K. D., and Wasson J. T. (2004) Oxygen-isotopic compositions of relict and host grains in chondrules in the Yamato 81020 CO3.0 chondrite. *Geochim. Cosmochim. Acta, 68,* 3599–3606.

Kurat G., Pernicka E., and Herrwerth I. (1984) Chondrules from Chainpur (LL-3): Reduced parent rocks and vapor fractionation. *Earth Planet. Sci. Lett., 68,* 43–56.

La Blue A. R. and Lauretta D. S. (2004) Metallic chondrules in NWA 1390 (H3-6): Clues to their history from metallic Cu (abstract). In *Lunar and Planetary Science XXXV,* Abstract #1939. Lunar and Planetary Institute, Houston (CD-ROM).

Larimer J. W. and Anders E. (1967) Chemical fractionation in meteorites: II. Abundance patterns and their interpretation. *Geochim. Cosmochim. Acta, 31,* 1239–1270.

Lasaga T. C. (1998) *Kinetic Theory in the Earth Sciences.* Princeton Univ., Princeton. 811 pp.

Lauretta D. S. and Buseck P. R. (2003) Opaque minerals in chondrules and fine-grained chondrule rims in the Bishunpur (LL3.1) chondrite. *Meteoritics & Planet. Sci., 38,* 59–79.

Lauretta D. S. and Killgore M. (2005) *A Color Atlas of Meteorites in Thin Section.* Golden Retriever Publications/Southwest Meteorite Press.

Lauretta D. S., Kremser D. T., and Fegley B. (1996) The rate of iron sulfide formation in the solar nebula. *Icarus, 122,* 288–315.

Lauretta D. S., Lodders K., and Fegley B. (1997) Experimental simulations of sulfide formation in the solar nebula. *Science, 277,* 358–360.

Lauretta D. S., Lodders K., and Fegley B. (1998) Kamacite sulfurization in the solar nebula. *Meteoritics & Planet. Sci., 33,* 821–833.

Lauretta D. S., Hua X., and Buseck P. R. (2000) Mineralogy of fine-grained rims in the ALH81002 CM chondrite. *Geochim. Cosmochim. Acta, 64,* 3263–3273.

Lauretta D. S., Buseck P. R., and Zega T. J. (2001) Opaque minerals in the matrix of the Bishunpur (LL3.1) chondrite: Constraints on the chondrule formation environment. *Geochim. Cosmochim. Acta, 65,* 1337–1353.

Lee M. S., Rubin A. E., and Wasson J. T. (1992) Origin of metallic Fe-Ni in Renazzo and related chondrites. *Geochim. Cosmochim. Acta, 56,* 2521–2533.

Lemelle L., Guyot F., Fialin M., and Pargamin J. (2000) Experimental study of chemical coupling between reduction and volatilization in olivine single crystals. *Geochim. Cosmochim. Acta, 64,* 3237–3249.

Lemelle L., Guyot F., Leroux H., and Libourel G. (2001) An experimental study of the external reduction of olivine single crystals. *Am. Mineral., 86,* 47–54.

Leroux H., Libourel G., Lemelle L., and Guyot F. (2003) Experimental study and TEM characterization of dusty olivines in chondrites: Evidence for formation by in situ reduction. *Meteoritics & Planet. Sci., 38,* 81–94.

Lewis R. D., Lofgren G. E., Franzen H. F., and Windom K. E. (1993) The effect of Na vapor on the Na content of chondrules. *Meteoritics, 28,* 622–628.

Libourel G., Krot A. N., and Tissandier L. (2003) Evidence for high temperature condensation of moderately-volatile elements during chondrule formation (abstract). In *Lunar and Planetary Science XXXIV,* Abstract #1558. Lunar and Planetary Institute, Houston (CD-ROM).

Lodders K. (2003) Solar system abundances and condensation temperatures of the elements. *Astrophys. J., 591,* 1220–1247.

Lofgren G. (1989) Dynamic crystallization of chondrule melts of porphyritic olivine composition — Textures experimental and natural. *Geochim. Cosmochim. Acta, 53,* 461–470.

Lofgren G. E. (1996) A dynamic crystallization model for chondrule melts. In *Chondrules and the Protoplanetary Disk* (R. H. Hewins et al., eds.), pp. 187–196. Cambridge Univ., New York.

Lofgren G. and Lanier A. B. (1990) Dynamic crystallization study of barred olivine chondrules. *Geochim. Cosmochim. Acta, 54,* 3537–3551.

Lofgren G. and Russell W. J. (1986) Dynamic crystallization of chondrule melts of porphyritic and radial pyroxene composition. *Geochim. Cosmochim. Acta, 50,* 1715–1726.

Luck J. M., Ben Othman D., Barrat J. A., and Albarede F. (2003) Coupled Cu-63 and O-16 excesses in chondrites. *Geochim. Cosmochim. Acta, 67,* 143–151.

Lyons J. R. and Young E. D. (2003) Towards an evaluation of self-shielding at the x-point as the origin of the oxygen isotope anomaly in CAIs (abstract). In *Lunar and Planetary Science XXXIV,* Abstract #1981. Lunar and Planetary Institute, Houston (CD-ROM).

Maharaj S. V. and Hewins R. H. (1994) Clues to chondrule precursors — An investigation of vesicle formation in experimental chondrules. *Geochim. Cosmochim. Acta, 58,* 1335–1342.

Maharaj S. V. and Hewins R. H. (1998) Chondrule precursor minerals as anhydrous phases. *Meteoritics & Planet. Sci., 33,* 881–887.

Marshall J. and Cuzzi J. N. (2001) Electrostatic enhancement of coagulation in protoplanetary nebulae (abstract). In *Lunar and Planetary Science XXXII,* Abstract #1262. Lunar and Planetary Institute, Houston (CD-ROM).

Matsunami S. (1984) Chemical compositions and textures of matrices and chondrule rims in eight unequilibrated ordinary chondrites: A preliminary report. *Mem. Natl. Inst. Polar Res., Spec. Issue 24,* 126–148.

Matsunami S., Ninagawa K., Nishimura S., Kubono N., Yamamoto I., Kohata M., Wada T., Yamashita Y., Lu J., Sears D. W. G., and Nishimura H. (1993) Thermoluminescence and compositional zoning in the mesostasis of a Semarkona group A1 chondrule and new insights into the chondrule-forming process. *Geochim. Cosmochim. Acta, 57,* 2101–2110.

McEwing C. E., Thode H. G., and Rees C. E. (1980) Sulphur isotope effects in the dissociation and evaporation of troilite:

A possible mechanism for ^{34}S enrichment in lunar soils. *Geochim. Cosmochim. Acta, 44,* 565–571.

McSween H. Y. (1977) Chemical and petrographic constraints on origin of chondrules and inclusions in carbonaceous chondrites. *Geochim. Cosmochim. Acta, 41,* 1843–1860.

Meibom A., Petaev M. I., Krot A. N., Wood J. A., and Keil K. (1999) Primitive FeNi metal grains in CH carbonaceous chondrites formed by condensation from a gas of solar composition. *J. Geophys. Res.–Planets, 104,* 22053–22059.

Metzler K., Bischoff A., and Stoffler D. (1992) Accretionary dust mantles in CM chondrites: Evidence for solar nebula processes. *Geochim. Cosmochim. Acta, 56,* 1873–1898.

Misawa K. and Fujita T. (1994) A relict refractory inclusion in a ferromagnesian chondrule from the Allende meteorite. *Nature, 368,* 723–726.

Misawa K. and Nakamura N. (1988a) Demonstration of REE fractionation among individual chondrules from the Allende (CV3) chondrite. *Geochim. Cosmochim. Acta, 52,* 1699–1710.

Misawa K. and Nakamura N. (1988b) Highly fractionated rare-earth elements in ferromagnesian chondrules from the Felix (CO3) meteorite. *Nature, 334,* 47–50.

Misawa K. and Nakamura N. (1996) Origin of refractory precursor components of chondrules from carbonaceous chondrites. In *Chondrules and the Protoplanetary Disk* (R. H. Hewins et al., eds.), pp. 99–106. Cambridge Univ., New York.

Mostefaoui S. and Perron C. (1994) Redox processes in chondrules recorded in metal. *Meteoritics, 29,* 506–506.

Mostefaoui S., Perron C., Zinner E., and Sagon G. (2000) Metal-associated carbon in primitive chondrites: Structure, isotopic composition, and origin. *Geochim. Cosmochim. Acta, 64,* 1945–1964.

Mullane E., Russell S. S., Gounelle M., and Mason T. F. D. (2003) Iron isotope composition of Allende and Chainpur chondrules: Effects of equilibration and thermal history (abstract). In *Lunar and Planetary Science XXXIV,* Abstract #1027. Lunar and Planetary Institute, Houston (CD-ROM).

Muller W. F., Weinbruch S., Walter R., and Mullerbeneke G. (1995) Transmission electron-microscopy of chondrule minerals in the Allende meteorite — Constraints on the thermal and deformational history of granular olivine-pyroxene chondrules. *Planet. Space Sci., 43,* 469–483.

Nagahara H. (1981) Evidence for secondary origin of chondrules. *Nature, 292,* 135–136.

Nagahara H. (1983) Chondrules formed through incomplete melting of the pre-existing minerals and the origin of chondrules. In *Chondrules and Their Origins* (E. A. King, ed.), pp. 211–222. Lunar and Planetary Institute, Houston.

Nagahara H. (1984) Matrices of type-3 ordinary chondrites — Primitive nebular records. *Geochim. Cosmochim. Acta, 48,* 2581–2595.

Nagahara H. (1996) Evaporation induced isothermal crystallization by evaporation. In *Lunar and Planetary Science XXVIII,* pp. 927–928. Lunar and Planetary Institute, Houston.

Nagahara H. (2000) Chemical fractionation in the presence of ambient gas. *Meteoritics & Planet. Sci., 35,* A116.

Nagahara H. and Ozawa K. (1995) Stability of silicate melt in the solar nebula. *Symp. Antarct. Meteorites, 20,* 172–173.

Nagahara H. and Ozawa K. (1996a) Evaporation kinetics of silicate melt (abstract). *Meteoritics & Planet. Sci., 31,* A95–A96.

Nagahara H. and Ozawa K. (1996b) Evaporation of silicate melt in the sytem Mg_2SiO_4-SiO_2. *Symp. Antarct. Meteorites, 21,* 125–27.

Nagahara H. and Ozawa K. (1996c) Evaporation of forsterite in H_2 gas. *Geochim. Cosmochim. Acta, 60,* 1445–1459.

Nagahara H. and Ozawa K. (2000) Isotopic fractionation as a probe of heating processes in the solar nebula. *Chem. Geol., 169,* 45–68.

Nagahara H. and Ozawa K. (2003) Was the diversity of chondrule compositions achieved by evaporation and condensation processes? (abstract). *Meteoritics & Planet. Sci., 38,* A125.

Nelson L. S., Skaggs S. R., Keil K., and Blander M. (1972) Use of a CO_2-laser to prepare chondrule-like spherules from supercooled molten oxide and silicate droplets. *Earth Planet. Sci. Lett., 14,* 338.

Niemeyer S. (1988) Titanium isotopic anomalies in chondrules from carbonaceous chondrites. *Geochim. Cosmochim. Acta, 52,* 309–318.

Notsu K., Onuma N., and Nishida N. (1978) High temperature heating of the Allende meteorite. *Geochim. Cosmochim. Acta, 42,* 903–907.

Ozawa K. and Nagahara H. (1997) Stability of chondrule melt in the solar nebula (abstract). In *Lunar and Planetary Science XXVIII,* pp. 1055–1056. Lunar and Planetary Institute, Houston.

Ozawa K. and Nagahara H. (1998) Chemical and isotopic fractionation during evaporation of a multi-component system: (2) General model and application to Mg-Fe olivine. *Antarct. Meteorites XXIII,* 126–127.

Ozawa K. and Nagahara H. (2001) Chemical and isotopic fractionations by evaporation and their cosmochemical implications. *Geochim. Cosmochim. Acta, 65,* 2171–2199.

Pack A., Shelley J. M. G., and Palme H. (2004) Chondrules with peculiar REE patterns: Implications for solar nebular condensation at high C/O. *Science, 303,* 997–1000.

Paque J. M. and Cuzzi J. N. (1997) Physical characteristics of chondrules and rims and aerodynamic sorting in the solar nebula (abstract). In *Lunar and Planetary Science XXVIII,* pp. 1071–1072. Lunar and Planetary Institute, Houston.

Patchett P. J. (1980) Sr isotopic fractionation in Allende chondrules — A reflection of solar nebular processes. *Earth Planet. Sci. Lett., 50,* 181–188.

Petaev M. I., Meibom A., Krot A. N., Wood J. A., and Keil K. (2001) The condensation origin of zoned metal grains in QUE 94411: Implications for the formation of the Bencubbin-like chondrites. *Meteoritics & Planet. Sci., 36,* 93–107.

Radomsky P. M. and Hewins R. H. (1990) Formation conditions of pyroxene-olivine and magnesian olivine chondrules. *Geochim. Cosmochim. Acta, 54,* 3475–3490.

Rambaldi E. R. (1981) Relict grains in chondrules. *Nature, 293,* 558–561.

Rambaldi E. R. and Wasson J. T. (1981) Metal and associated phases in Bishunpur, a highly unequilibrated ordinary chondrite. *Geochim. Cosmochim. Acta, 45,* 1001–1015.

Rambaldi E. R. and Wasson J. T. (1982) Fine, nickel-poor Fe-Ni grains in the olivine of unequilibrated ordinary chondrites. *Geochim. Cosmochim. Acta, 46,* 929–939.

Rambaldi E. R. and Wasson J. T. (1984) Metal and associated phases in Krymka and Chainpur: Nebular formational processes. *Geochim. Cosmochim. Acta, 48,* 1885–1897.

Rambaldi E. R., Rajan R. S., Wang D., and Housley R. M. (1983) Evidence for relict grains in chondrules of Qingzhen, an E3-type enstatite chondrite. *Earth Planet. Sci. Lett., 66,* 11–24.

Richter F. M. and Davis A. M. (1997) Mass loss kinetics and isotope fractionation of Type B CAI melts at 1400°C and solar f_{O_2} (abstract). In *Lunar and Planetary Science XXVIII,* pp. 1161–1162. Lunar and Planetary Institute, Houston.

Richter F. M., Parsad N., Davis A. M., and Hasegawa H. (1999)

CAI cosmobarometry (abstract). In *Lunar and Planetary Science XXX,* Abstract #1989. Lunar and Planetary Institute, Houston (CD-ROM).

Richter F. M., Davis A. M., Ebel D. S., and Hashimoto A. (2002) Elemental and isotopic fractionation of Type B calcium-aluminum-rich inclusions: Experiments, theoretical considerations, and constraints on their thermal evolution. *Geochim. Cosmochim. Acta, 66,* 521–540.

Rubin A. E. (1984) Coarse-grained chondrule rims in type-3 chondrites. *Geochim. Cosmochim. Acta, 48,* 1779–1789.

Rubin A. E. and Wasson J. T. (1987) Chondrules, matrix and coarse-grained chondrule rims in the Allende meteorite — Origin, interrelationships and possible precursor components. *Geochim. Cosmochim. Acta, 51,* 1923–1937.

Rubin A. E. and Wasson J. T. (1988) Chondrules and matrix in the Ornans CO3 meteorite — Possible precursor components. *Geochim. Cosmochim. Acta, 52,* 425–432.

Rubin A. E. and Wasson J. T. (2005) Non-spherical lobate chondrules in CO3.0 Y-81020: General implications for the formation of low-FeO porphyritic chondrules in CO chondrites. *Geochim. Cosmochim. Acta, 69,* 211–220.

Scott E. R. D. and Taylor G. J. (1983) Chondrules and other components in C-chondrite, O-chondrite, and E-chondrite — Similarities in their properties and origins. *Proc. Lunar Planet. Sci. Conf. 14th,* in *J. Geophys. Res., 88,* B275–B286.

Scott E. R. D., Rubin A. E., Taylor G. J., and Keil K. (1984) Matrix material in type-3 chondrites — Occurrence, heterogeneity and relationship with chondrules. *Geochim. Cosmochim. Acta, 48,* 1741–1757.

Scott E. R. D., Barber D. J., Alexander C. M. O'D., Hutchison R., and Peck J. A. (1988) Primitive material surviving in chondrites: Matrix. In *Meteorites and the Early Solar System* (J. F. Kerridge and M. S. Matthews, eds.), pp. 718–745. Univ. of Arizona, Tucson.

Sears D. W. G., Lu J., Benoit P. H., Dehart J. M., and Lofgren G. E. (1992) A compositional classification scheme for meteoritic chondrules. *Nature, 357,* 207–210.

Sears D. W. G., Huang S., and Benoit P. H. (1996) Open-system behavior during chondrule formation. In *Chondrules and the Protoplanetary Disk* (R. H. Hewins et al., eds.), pp. 221–232. Cambridge Univ., New York.

Shimaoka T. and Nakamura N. (1989) Vaporization of sodium from a partially molten chondritic material. *Proc. NIPR Symp. Antarct. Meteorites, 2,* 252–267.

Shimaoka T. and Nakamura N. (1991) The effect of total pressure on vaporization of alkalis from partially molten chondritic material. In *Origin and Evolution of Interplanetary Dust* (A. C. Levasseur-Regourd and H. Hasegawa, eds.), pp. 79–82. Kluwer, Dordrecht.

Shirai T., Tachibana S., and Tsuchiyama A. (2000) Evaporation rates of Na from Na_2O-SiO_2 melt at 1 atm (abstract). In *Lunar and Planetary Science XXXI,* Abstract #1610. Lunar and Planetary Institute, Houston (CD-ROM).

Steele I. M. (1986) Compositions and textures of relic forsterite in carbonaceous and unequilibrated ordinary chondrites. *Geochim. Cosmochim. Acta, 50,* 1379–1395.

Swindle T. D., Caffee M. W., Hohenberg C. M., Lindstrom M. M., and Taylor G. J. (1991a) Iodine-xenon studies of petrographically and chemically characterized Chainpur chondrules. *Geochim. Cosmochim. Acta, 55,* 861–880.

Swindle T. D., Grossman J. N., Olinger C. T., and Garrison D. H. (1991b) Iodine-xenon, chemical, and petrographic studies of Semarkona chondrules — Evidence for the timing of aqueous alteration. *Geochim. Cosmochim. Acta, 55,* 3723–3734.

Tachibana S. and Huss G. R. (2005) Sulfur isotope composition of putative primary troilite in chondrules from Bishunpur and Semarkona. *Geochim. Cosmochim. Acta, 69,* 3075–3097.

Tachibana S., Nagahara H., Mostefaoui S., and Kita N. T. (2003) Correlation between relative ages inferred from ^{26}Al and bulk compositions of ferromagnesian chondrules in least equilibrated ordinary chondrites. *Meteoritics & Planet. Sci., 38,* 939–962.

Thiemens M. H. (1996) Mass-independent isotopic effects in chondrites: The role of chemical processes. In *Chondrules and the Protoplanetary Disk* (R. H. Hewins et al., eds.), pp. 107–118. Cambridge Univ., New York.

Tissandier L., Libourel G., Robert F., and Chaussidon M. (2000) SiO_2 condensation experiments and implications for protosolar materials (abstract). In *Lunar and Planetary Science XXXI,* Abstract #1553. Lunar and Planetary Institute, Houston (CD-ROM).

Tissandier L., Libourel G., and Robert F. (2002) Gas-melt interactions and their bearing on chondrule formation. *Meteoritics & Planet. Sci., 37,* 1377–1389.

Tsuchiyama A. and Tachibana S. (1999) Evaporation rates of elements, such as sodium, from melts under solar nebula conditions (abstract). In *Lunar and Planetary Science XXX,* Abstract #1535. Lunar and Planetary Institute, Houston (CD-ROM).

Tsuchiyama A., Nagahara H., and Kushiro I. (1981) Volatilization of sodium from silicate melt spheres and its application to the formation of chondrules. *Geochim. Cosmochim. Acta, 45,* 1357–1367.

Tsuchiyama A., Osada Y., Nakano T., and Uesugi K. (2004) Experimental reproduction of classic barred olivine chondrules: Open-system behavior of chondrule formation. *Geochim. Cosmochim. Acta, 68,* 653–672.

Varela M. E. and Metrich N. (2000) Carbon in olivines of chondritic meteorites. *Geochim. Cosmochim. Acta, 64,* 3433–3438.

Wang J., Davis A. M., and Clayton R. N. (1994) Modeling of isotopic fractionation during the phase transition of a diffusion controlled reservoir. In *Eighth International Conference on Geochronology, Cosmochronology, and Isotope Geology.* U.S. Geological Survey Circular 1107.

Wang J., Yu Y., Alexander C. M. O.'D., and Hewins R. H. (1999) The influence of oxygen and hydrogen on the evaporation of K (abstract). In *Lunar and Planetary Science XXX,* Abstract #1778. Lunar and Planetary Institute, Houston (CD-ROM).

Wang J., Davis A. M., Clayton R. N., Mayeda T. K., and Hashimoto A. (2001) Chemical and isotopic fractionation during the evaporation of the FeO-MgO-SiO_2-CaO-Al_2O_3-TiO_2 rare earth element melt system. *Geochim. Cosmochim. Acta, 65,* 479–494.

Wasson J. T. (1993) Constraints on chondrule formation. *Meteoritics, 28,* 13–28.

Watanabe S., Kitamura M., and Morimoto N. (1985) A transmission electron-microscope study of pyroxene chondrules in equilibrated L-group chondrites. *Earth Planet. Sci. Lett., 72,* 87–98.

Weinbruch S. and Muller W. F. (1995) Constraints on the cooling rates of chondrules from the microstructure of clinopyroxene and plagioclase. *Geochim. Cosmochim. Acta, 59,* 3221–3230.

Weinbruch S., Buttner H., Holzheid A., Rosenhauer M., and Hewins R. H. (1998) On the lower limit of chondrule cooling rates: The significance of iron loss in dynamic crystallization experiments. *Meteoritics & Planet. Sci., 33,* 65–74.

Weinbruch S., Palme H., and Spettel B. (2000) Refractory forsterite in primitive meteorites: Condensates from the solar nebula? *Meteoritics & Planet. Sci., 35,* 161–171.

Weinbruch S., Muller W. F., and Hewins R. H. (2001) A transmission electron microscope study of exsolution and coarsening in iron-bearing clinopyroxene from synthetic analogues of chondrules. *Meteoritics & Planet. Sci., 36,* 1237–1248.

Weisberg M. K. and Prinz M. (1996) Agglomeratic chondrules, chondrule precursors, and incomplete melting. In *Chondrules and the Protoplanetary Disk* (R. H. Hewins et al., eds.), pp. 119–128. Cambridge Univ., New York.

Wood J. A. (1981) The interstellar dust as a precursor of Ca,Al-rich inclusions in carbonaceous chondrites. *Earth Planet. Sci. Lett., 56,* 32–44.

Wood J. A. and Hashimoto A. (1993) Mineral equilibrium in fractionated nebular systems. *Geochim. Cosmochim. Acta, 57,* 2377–2388.

Yoneda S. and Grossman L. (1995) Condensation of CaO-MgO-Al_2O_3-SiO_2 liquids from cosmic gases. *Geochim. Cosmochim. Acta, 59,* 3413–3444.

Young E. D. (2000) Assessing the implications of K isotope cosmochemistry for evaporation in the preplanetary solar nebula. *Earth Planet. Sci. Lett., 183,* 321–333.

Young E. D., Ash R. D., Galy A., and Belshaw N. S. (2002) Mg isotope heterogeneity in the Allende meteorite measured by UV laser ablation-MC-ICPMS and comparisons with O isotopes. *Geochim. Cosmochim. Acta, 66,* 683–698.

Yu Y. and Hewins R. H. (1998) Transient heating and chondrule formation: Evidence from sodium loss in flash heating simulation experiments. *Geochim. Cosmochim. Acta, 62,* 159–172.

Yu Y., Hewins R. H., Clayton R. N., and Mayeda T. K. (1995) Experimental study of high-temperature oxygen-isotope exchange during chondrule formation. *Geochim. Cosmochim. Acta, 59,* 2095–2104.

Yu Y., Hewins R. H., and Zanda B. (1996) Sodium and sulfur in chondrules: Heating time and cooling curves. In *Chondrules and the Protoplanetary Disk* (R. H. Hewins et al., eds.), pp. 213–220. Cambridge Univ., New York.

Yu Y., Hewins R. H., Alexander C. M. O.'D., and Wang J. (2003) Experimental study of evaporation and isotopic mass fractionation of potassium in silicate melts. *Geochim. Cosmochim. Acta, 67,* 773–786.

Zanda B., Bourot-Denise M., Perron C., and Hewins R. H. (1994) Origin and metamorphic redistribution of silicon, chromium, and phosphorus in the metal of chondrites. *Science, 265,* 1846–1849.

Zanda B., Bourot-Denise M., Hewins R. H., Cohen B. A., and Delaney J. (2002) Accretion textures, iron evaporation and recondensation in Renazzo chondrules (abstract). In *Lunar and Planetary Science XXXIII,* Abstract #1852. Lunar and Planetary Institute, Houston (CD-ROM).

Zega T. J. and Buseck P. R. (2003) Fine-grained-rim mineralogy of the Cold Bokkeveld CM chondrite. *Geochim. Cosmochim. Acta, 67,* 1711–1721.

Zhu X. K., Guo Y., O'Nions R. K., Young E. D., and Ash R. D. (2001) Isotopic homogeneity of iron in the early solar nebula. *Nature, 412,* 311–313.

Zinner E. (1998) Stellar nucleosynthesis and the isotopic composition of presolar grains from primitive meteorites. *Annu. Rev. Earth Planet. Sci., 26,* 147–188.

Part VI:

The Accretion Epoch: Formation of Planetesimals

Chronological Constraints on Planetesimal Accretion

Robert H. Nichols Jr.
Washington University

This chapter provides a brief overview of the timescales associated with the accretion (assemblage) of planetesimals (asteroidal-sized objects on the order of 10–100 km in diameter) in the early solar system based primarily on meteoritic studies, but also on dynamical and astronomical constraints. In the first edition of this book (*Kerridge and Matthews,* 1988) results from each these fields (meteoritic, theoretical, and astronomical) were consistent with a short, dominant planetesimal accretion period, ΔT_{acc} ~ 10 m.y. within the formation of the first solids in the early solar system, and with a precision in each field on the order of a few million years. The more precise limits available now follow the same general trend as before, ΔT_{acc}(dynamic) ≤ ΔT_{acc}(meteoritic) ≤ ΔT_{acc}(astronomical), with this order being dictated primarily by the available precision in each field and by the specifics of the objects being measured and/or the events that are modeled. Meteoritic constraints suggest that initial planetesimal formation (e.g., 4 Vesta, the presumed HED parent body) easily occurred within ~5 m.y. of the formation of the first solar system solids, and more likely within 1–2 m.y. Recent astronomical measurements have also shortened the likely accretion period, and more specifically point to the nebular clearing of dust to 6 m.y. Dynamical considerations, in particular within the context of extrasolar planet modeling, can potentially push planetesimal accretion to within a few 10–100 k.y. Such rapid "primary" formation timescales from each discipline are contrasted with meteoritic evidence for continued "secondary" accretionary processing for periods of 10–100 m.y., as suggested by measurements of precompaction regolith exposure ages, compaction ages, and impact-related events.

1. INTRODUCTION

The epochs in the formation of the early solar system as laid out by the organizational section headings in this book are not necessarily chronologically disjoint spans of time, i.e., they may overlap, often to uncertain extents depending on one's interest, definition, perspective, and available precision. Such is indeed the case for the planetesimal accretion epoch. In a broad sense the onset of planetesimal accretion could begin with the *onset* of the accretion of the nebular gas into the protoplanetary Sun, roughly coinciding with the formation of the first solids in the solar system. Similarly, again in the broadest sense, this epoch is still ongoing given the presence of planetesimal-sized objects in the present-day solar system that are often mutually interacting — e.g., asteroids, comets, Triton, and arguably Pluto — and that are potentially accreting material still, albeit at extremely low rates. The present population of asteroids has indeed evolved through relatively recent impacts and gravitational reaccretion, some within the last 6 m.y. (e.g., *Nesvorný et al.,* 2002). While such a working definition may be of interest to some, it is obviously of little use for our particular interest in the early solar system, other than providing the first, and rather loose, constraint on the duration of accretion of planetesimals, an upper limit of ~4.5 G.y. Herein the emphasis is rather on the timescales (onset, duration, and termination) for the formation of the *first* asteroidal-sized objects (tens of meters to >100 km) in the early solar system primarily in the context of meteoritic records but also with regard to dynamical models of the solar nebula and astronomical observations of circumstellar disk lifetimes.

Most, if not all, variations on the standard paradigm of planetesimal accretion follow a similar framework, although the factors influencing the rate of accretion may vary widely from study to study. From a dynamical perspective primary planetesimal formation may generally be divided into three initial stages coinciding with (1) gravitational collapse of gas and dust; (2) the coagulation of submicrometer- and micrometer-sized grains into centimeter-sized aggregates as gas and dust settle onto the midplane of the nebula (cf. *Russell et al.,* 2006), followed by the gas-drag-induced collisional growth of centimeter- to meter-sized objects (cf. *Cuzzi and Weidenschilling,* 2006); and (3) the gravitational accretion of objects larger than ~10–100 m (cf. *Weidenschilling,* 2000; *Weidenschilling and Cuzzi,* 2006). During stage 3 planetesimal growth may continue to a size less than or equal to that of a planetary precursor or protoplanet, i.e., a body large enough, on the order of 10–100 km in diameter, to undergo melting and metal-silicate differentiation, taken herein to be equivalent to core formation. While the primary accretion process for asteroids presently residing at 3 AU may have ceased before a planet could be formed in that region, it is possible that planet-sized bodies did form in the asteroid belt, but were subsequently removed dynamically (e.g., *Wetherill,* 1992; *Petit et al.,* 2001). The post-

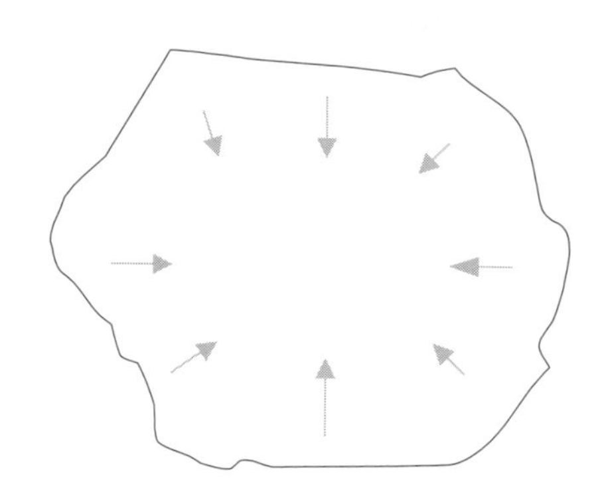

(1) Gravitational collapse of gas and dust.

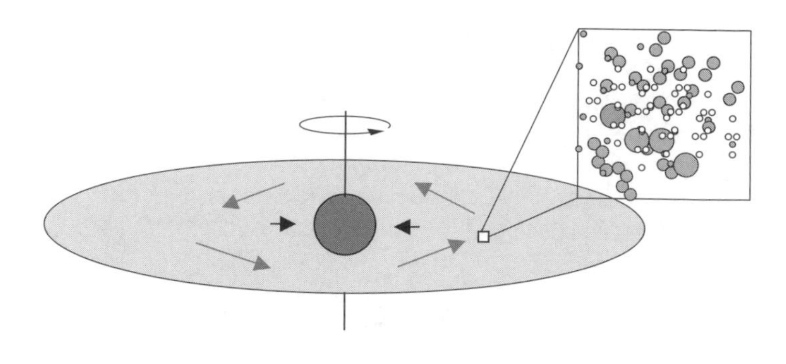

(2) Formation of rotating nebular disk: gas and dust falling onto the midplane and proto-Sun. Formation of initial millimeter-to centimeter-sized objects from gas/dust aggregates.

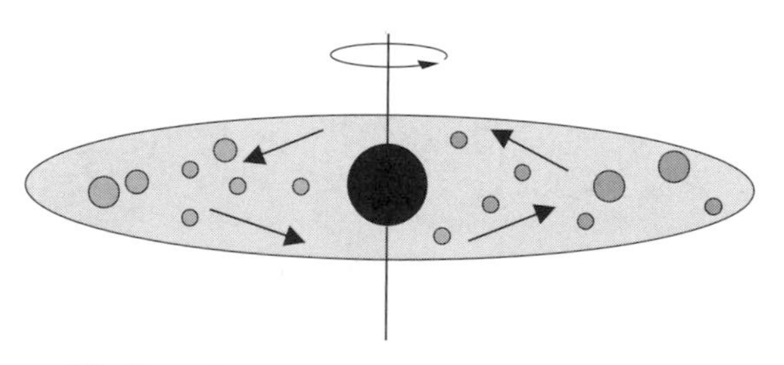

(3) Protoplanetary disk: Gravitational accretion of 10–100-m objects upto 10–100 km.

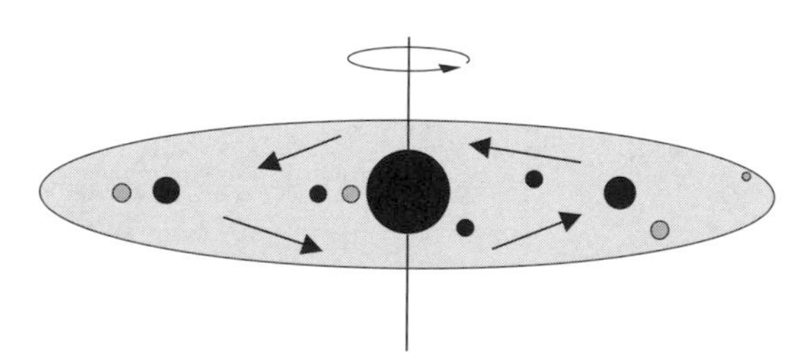

(4) Evolved disk: Nebular gas cleared, collisions and secondary impacts and planetary/post-planetary formation.

Fig. 1. Depiction of four (1–4) dynamical stages of nebular evolution (see text): (1) gravitational collapse of gas and dust; (2) the coagulation of submicrometer- and micrometer-sized grains into centimeter-sized aggregates as gas and dust settle onto the midplane of the nebula, followed by growth of centimeter- to meter-sized objects; (3) the gravitational accretion of objects larger than ~10–100 m, up to planetesimal size, or 10–100 km; and (4) planetary and post-planetary processing.

primary accretion stage (4) coincides with subsequent collisions and impacts within the asteroid belt over the course of the solar system's history, clearly altering to some degree the first generation of planetesimals (see, e.g., *Turner,* 1988; *Bischoff et al.,* 2006).

A graphical depiction for each of these stages (1–4) is illustrated in Fig. 1. From a dynamical perspective, the onset of each stage could formally be associated with a specific physical parameter and perhaps with a specific point in time. In the present context, however, the adjacent stages are assumed to overlap to some extent based on available precision (ranging from ~100–1000 k.y., see below), with the first three stages occurring in the presence of a nebular gas and dust, and the last stage (planetary formation and secondary planetary processing) occurring after the nebular had cleared. The topic herein, the timing of the onset of primary planetesimal accretion (stage 3), thus occurs in the presence of a nebula prior to the formation of planetary-sized bodies (stage 4). (The duration of the solar nebula is a topic that is beyond this discussion; the reader is referred to the chapters in Parts III–V of this volume.)

1.2. Constraining Planetesimal Accretion: The Problem

Each of these stages of planetesimal accretion potentially lends itself to the formation or processing of macroscopic objects that may be found in meteorites and that may still be sufficiently pristine to shed light on the timeline for primary accretion. In particular, the objects of interest, either isolated from meteorites or whole meteorites themselves, are those whose formation effectively coincides with the rapid isotopic closure or resetting of some radiometric system, thereby permitting object formation to be dated by either absolute or relative isotope dating techniques. These objects include primary nebular condensates that formed near or at the beginning of stages 1 or 2 of planetesimal accretion described above such as individual mineral grains, calcium-aluminum-rich inclusions (CAIs), and chondrules (cf. *Russell et al.,* 2006). Other common objects that are dated are associated with secondary processes such as aqueous alteration and thermal and impact metamorphism (cf. *Krot et al.,* 2006) and differentiation (cf. *Wadhwa et al.,*

2006), all of which potentially postdate stage 3 and the cessation of primary accretion.

The subtle point, and hence the problem, is that chronometric techniques thus applied only set upper limits for the duration of the primary accretion, which may, as will be shown on the basis of dynamical and other arguments, have likely proceeded on a much more rapid timescale. This point may also be rephrased in that our sampling of meteoritic material is likely biased and may have not yet provided us with information directly related to the earliest accretionary events (cf. *Wetherill and Chapman,* 1988). Unlike the events and processes that are dated above (such as the formation of isotopic reservoirs associated with differentiation or isotopic closure associated with mineral formation), the accretionary process itself offers few similarly well-defined, rapid closure events clearly marking either its onset or its termination for any given planetesimal associated with a specific meteorite type. Processes directly associated with early accretion, e.g., impacts, regolith mixing, and breccia formation, are probably insufficiently energetic to set or reset isotopic clocks on the scale of a parent body, and are likely to be subsequently masked by parent-body-wide processes such as the formation of isotopic reservoirs due to differentiation. Accretionary processes, if recorded, tend to reflect episodic secondary processing of the larger bodies that postdates the primary accretion phase and may effectively wipe away any record of the primary accretion process. Thus, while the dating of such secondary events may provide an estimate for the duration of all (primary and secondary) planetesimal accretion in the early solar system, such chronometry in general offers few clues to the timescales associated with initial planetesimal accretion.

In this chapter boundary conditions for the duration of primary planetesimal accretion for objects up to ~100 km in size will be reviewed based on a variety of radiometric, dynamical, and astronomical techniques. While many of these topics are discussed in more detail elsewhere within this book, they are not specifically discussed as a group or in regard to planetesimal accretion. A summary of the primary results is provided in Table 1. Given the paucity of available radiometric limits directly related to primary planetesimal accretion, appeal will also be made to two lesser-studied techniques, "precompaction" exposure to energetic particles and compaction ages, to attempt to set tighter constraints on the duration of planetesimal accretion, in particular for the carbonaceous chondrites.

2. PRIMARY PLANETESIMAL ACCRETION

2.1. The First Solids

The inferred presence of several short-lived nuclides extant within solids at their formation in the early solar system, in particular ^{41}Ca ($T_{1/2}$ ~ 0.1 m.y.) (*Srinivasan et al.,* 1994, 1996), suggests that the formation of these first solids must have occurred relatively quickly (within ~1 m.y.) after the stellar nucleosynthesis of these radioactive nuclei (cf. *Meyer and Clayton,* 2000). While a weaker case may be made for the "local" production of some these short-lived nuclei in the early solar system via an energetic particle irradiation from a "T-Tauri" Sun (e.g., *Chaussidon and Gounelle,* 2006), such a scenario does not preclude the formation of solids very early on (within 1 m.y. of the onset of nebular collapse). In either case the reference point for the formation of these first solar system condensates is generally taken to be the formation of "normal" calcium-aluminum-rich inclusions (CAIs).

Lead-lead measurements of CAIs from the CV chondrite Allende indicate formation at 4.566 ± 0.002 Ga (*Manhès et al.,* 1988; *Göpel et al.,* 1991, 1994; *Allègre et al.,* 1995). More recent and more precise Pb-Pb measurements of similar CAIs from the CV chondrite Efremovka indicate formation at 4.5672 ± 0.0006 Ga (*Amelin et al.,* 2002). While the impressive ±0.6 Ma precision of this Pb-Pb measurement is the most precise age determined to date for the oldest nebular condensates in the solar system using any absolute or long-lived chronometer (see, e.g., *Chen and Tilton,* 1976; *Tatsumoto et al.,* 1976; *Chen and Wasserburg,* 1981) appeal to shorter-lived nuclides is required to narrow the formation interval for the first nebular condensates.

The interval over which "normal" CAIs formed appears to be quite short (<~0.5 m.y.), primarily based on the narrow range of initial $^{26}Al/^{27}Al$ ratios [$T_{1/2}(^{26}Al)$ = 0.74 m.y.] in these objects (*MacPherson et al.,* 1995). Given the likelihood that these CAIs are nebular products (and not produced by planetesimal/planetary processes as may arguably be the case for chondrules as noted below) and that they are likely the oldest objects available for dating in the solar system, the narrow 500 k.y. interval at an age of 4.5672 ± 0.0006 Ga has traditionally been taken to be coincident with the onset of planetesimal accretion, at least from an experimental standpoint; any real formation interval for CAIs has been *assumed* to be negligible given the precision of the isotopic measurements. While it is possible that CAIs existed in the nebula for up to a few million years before planetesimals formed (if, for example, nebula turbulence prevented the aggregation of millimeter-sized solids into larger bodies for protracted periods of time), there is, however, at the present time, little isotopic data supporting this. For comparison, dynamical models of nebular collapse and grain formation up to ~CAI-sized objects commonly require only on the order of ~1 k.y. or less (cf. *Boss,* 2003; *Weidenschilling,* 2000; *Weidenschilling and Cuzzi,* 2006).

With the advent of higher-precision analytic instruments, the assumption of negligibility may not hold. Work on the ^{26}Al-^{26}Mg system (e.g., *Bizzarro et al.,* 2004) suggests that all CAIs formed within the quite narrow span of ~50 k.y., generally consistent with previous work, thereby potentially providing the most precise experimentally derived temporal anchor for the early solar system to date. *Young et al.* (2005), however, suggest that the "canonical" initial $^{26}Al/^{27}Al$ used in all previous studies is due to a CAI formation

TABLE 1. Chronological constraints on the accretion of 10–100-km-sized planetesimals (and larger objects for comparison) in the early solar system after the formation of CAIs, and estimated formation durations for selected objects.

Meteoritic Constraints			
		Radiometric Technique	References
Fiducial point: CAI formation	4.5672 ± 0.0006 G.y.	Pb-Pb	[1]
CAI formation duration	50 k.y.	Pb-Pb	[2]
	300 ± 100 k.y.	Al-Mg	[3]
	~500 k.y.	Pb-Pb	[1]
Formation of parent bodies (10–100 km)	~5–10 m.y.	^{26}Al, ^{60}Fe heat sources	see text
HED parent body (Vesta, 525 km)	2 m.y.	Mn-Cr	[4]
	4.2 ± 1.3 m.y.	Hf-W	[5]
	~5 m.y.	Hf-W	[5,6]
	5 m.y.	Al-Mg	[7]
H chondrites	3 m.y.	^{244}Pu, ^{40}Ar-^{39}Ar	[8]
Iron meteorite parent body	1–2 m.y.	Al-Mg, Hf-W	[9,10]
Mars (6800 km)	13 ± 2 m.y.	Hf-W	[5]
Earth (12,756 km)	33 ± 2	Hf-W	[5]
Chondrite parent-body formation duration	2.5 m.y.	Pb-Pb	[1,11,12]
Iron meteorite parent-body formation duration	5 m.y.	Re-Os	[13]
Dynamical Constraints			
		Heliocentric Distance	
10–100 km	~1–10 k.y.		
1–10 km to Moon (3476 km)/Mars (6794 km)	~100 k.y.	at 1 AU	[14]
Mercury (4880 km)/Mars (6794 km)	~100 k.y.		
Earth (12,756 km)	~100 m.y.		
Ceres-sized (933 km)	~50 k.y.	at 1 AU	[14]
	~500 k.y.	at 2 AU	
	~5 m.y.	at 2.5 AU	
Gas giants (for comparison)	~1–10 k.y.		[15]
Astronomical Constraints			
		Technique	
Circumstellar Disk Lifetimes	~5–15 m.y.	IR, millimeter, submillimeter emission	[16,17]
	3–6 m.y.	IR, millimeter, submillimeter emission	[18]

References: [1] *Amelin et al.* (2002); [2] *Bizzarro et al.* (2004); [3] *Young et al.* (2005); [4] *Lugmair and Shykolyukov* (1998); [5] *Kleine et al.* (2002); [6] *Kleine et al.* (2004); [7] *Srinivasan et al.* (1999); [8] *Trieloff et al.* (2003); [9] *Bizzarro et al.* (2005); [10] *Kleine et al.* (2005); [11] *Bizzarro et al.* (2004); [12] *Kunihiro et al.* (2004); [13] *Horan et al.* (1998); [14] *Kortenkamp et al.* (2001); [15] *Boss* (2003); [16] *Strom et al.* (1989); [17] *Skrutskie et al.* (1990); [18] *Haisch et al.* (2001).

interval of 300 ± 100 k.y. Until similarly precise measurements are performed on suites of whole-rock objects dating differentiation, the discussion herein remains effectively unchanged.

All other potential nebular condensates, e.g., chondrules, have formation ages that postdate CAI formation by at least ~1 m.y. and likely range up to 2–3 m.y. later (e.g., *Russell et al.*, 1996; *Kita et al.*, 2000; *Amelin et al.*, 2002; *Kunihiro et al.*, 2004; *Bizzarro et al.*, 2004). By definition the formation of chondrules must predate the "final" assemblage of their respective chondritic parent bodies. Formation ages for the oldest chondrules may be contemporaneous at least with the presence of potentially large planetesimals, based on some dynamical models requiring rapid accretion on the order of 1 m.y. (cf. below) and not radiometric constraints. Unfortunately, there is currently no consensus on how planetesimals formed, nor how long the process took (cf. *Chambers,* 2006). If it could be shown, however, that the "chondrule-forming process" requires the presence of a first generation of planetesimals of at least a given size that formed early on, as some "chondrules-forming scenarios" require, then the formation interval between CAIs and chondrules may provide one of the most stringent radiometric limits on how rapidly accretion did occur. Any physical link between the chondrule formation process and the process of planetesimal accretion is, however, only tenuous at this time and the reader is referred to a variety of reviews on this subject (*Grossman,* 1988; *Hewins et al.,* 1996).

2.2. Planetesimal Heating, Differentiation, and Core Formation

Given the presence of millimeter- to centimeter-sized dust in the solar nebula at ~4.5672 Ga (taken herein to be the fiducial point marking the effective onset of planetesimal accretion), a second, later reference point is required to mark the end of the accretion of the first planetesimals. In addition to the rocky planets, several parent bodies were sufficiently large (~10–100 km) to have undergone differentiation, reflected by the fact that many meteorites show evidence that they were partially or totally molten on their parent planetesimals (e.g. *Scott,* 1979). The heating mechanism and its products provide key chronometric constraints for the onset and duration of planetesimal accretion based on (1) the heat source itself, (2) elemental separation occurring between mineral at the onset of melting, and (3) isotopic closure in minerals as the objects cooled, all of which potentially lend themselves to several short- and long-lived chronometric techniques. The subtleties of the different chronometric techniques will not be reviewed here; for this the reader is referred to each of the other chronological chapters within this volume and references therein. As a general rule, however, if it may be assumed that many of the short-lived nuclides were homogeneously distributed in the solar system, then the tightest constraints may be derived from those nuclides that can easily date the formation of isotopic reservoirs, e.g., ^{26}Al, ^{53}Mn, and possibly ^{182}Hf, rather than those radiometric systems that date isotopic closure occurred after an object has cooled, e.g., K-Ar, Rb-Sr, Sm-Nd, I-Xe, and Pb-Pb. This distinction can lead to differences of several millions of years for the formation of a single object.

2.2.1. Planetesimal heat source. A variety of heat sources have been proposed to explain the partial or total melting of the planetesimals including short-lived nuclei radioactive decay (e.g., *Lee et al.,* 1976), electromagnetic induction heating (cf. *Sonnet and Reynolds,* 1979), and accretion of small objects preheated in the nebula (*Larimer and Anders,* 1967). Other potential heat sources include the accretion process itself and associated impacts, as well as heating by the decay of long-lived nuclides. These latter sources, however, appear to be only effective on a terrestrial planet scale whereby mantle melting and mantel metal-silicate differentiation can be driven by impacts that are substantially more energetic than those occurring during initial planetesimal formation.

Of these mechanisms the most plausible heat source appears to be the decay of short-lived nuclides, dominated by that of ^{26}Al (*Lee et al.,* 1976; *Herndon and Herndon,* 1977; *Hutcheon and Hutchison,* 1989; *Zinner and Göpel,* 2002) and possibly to a lesser degree ^{60}Fe ($T_{1/2}$ = 1.5 m.y.) (*Birck and Lugmair,* 1988). *Mostefaoui et al.* (2004) revised the initial abundance of ^{60}Fe upward such that ^{60}Fe may be considered a more important heat source (see also *Moynier et al.,* 2005; *Quitté et al.,* 2005), while the ^{26}Al abundance in some of the most primitive chondrites may, however, have been too low to cause planet-wide heating (*Kunihiro et al.,* 2004). In either case, the accretion of planetesimals would thus need to occur early and rapidly, before these heat sources had decayed, thereby constraining assemblage periods to within a few half-lives of ^{26}Al and ^{60}Fe, translating to within 5–10 m.y. after CAI formation for objects on the order of 10–100 km (e.g., *Wood,* 2000).

2.2.2. The earliest formation ages. The 5–10-m.y. constraint noted above has been well known for some period of time (*Lee et al.,* 1976), and has been confirmed by many model radiometric age measurements of basaltic meteorites over the last 30 years. The majority of these measurements, as evidenced by the compilation of *MacPherson et al.* (1995), has been performed on the ^{26}Al-^{26}Mg system, e.g., *Hutcheon and Hutchison* (1989), *Russell et al.* (1996), *Kita et al.* (2000), *Nyquist et al.* (2003), *Kunihiro et al.* (2004), and *Bizzarro et al.* (2004), and perhaps provides the most robust age information about processes in the early solar system. Other radiometric systems, such as ^{53}Mn-^{53}Cr (e.g., *Lugmair and Shukolyukov,* 1998), I-Xe (cf. *Gilmour,* 2000), and ^{182}Hf-^{182}W (cf. *Horan et al.,* 1998; *Halliday,* 2000; *Halliday et al.,* 2003; *Kleine et al.,* 2002, 2004), yield results that are consistent with the 5–10-m.y. constraint. Herein only a few of the tightest constraints will be noted; for more detailed discussion of these and other isotopic systems the reader is referred to *Wadhwa et al.* (2006) and *Halliday* (2006).

The initial ^{26}Al/^{27}Al ratio measured in the basaltic eucrite Piplia Kalan indicates that its HED (howardite-eucrite-diogenite) parent object, presumably the ~250-km asteroid "4 Vesta" (*McCord et al.,* 1970; *Binzel and Xu,* 1993), melted and differentiated within ~5 m.y. of CAI formation (*Srinivasan et al.,* 1999). Recent measurements of ^{244}Pu fission-track and ^{40}Ar-^{39}Ar thermochronologies of unshocked H chondrites indicate rapid accretion of 100-km-sized objects within 3 m.y. of CAI formation and provides some of the most substantial support for an ^{26}Al heat source (*Trieloff et al.,* 2003), although heat by impact cannot be ruled out (e.g., *Rubin,* 2004). The ^{182}Hf-^{182}W system (*Kleine et al.,* 2002, 2004) indicates that core formation occurred relatively quickly after the formation of CAIs: 4.2 ± 1.3 m.y.

for "4 Vesta," 13 ± 2 m.y. for Mars, and 33 ± 2 m.y. for Earth; on Mars, mantle-crust differentiation and core formation may have been coeval.

An even earlier differentiation (~2 m.y.) for the basaltic achondrite parent body (Vesta) has been inferred based on the ^{53}Mn-^{53}Cr chronometer (*Lugmair and Shukolyukov,* 1998). They obtained a Mn-Cr age of ~4565 Ma from a eucrite whole-rock isochron (which is ~2 m.y. younger than the Pb-Pb age for CAIs). The same authors, however, argue that the solar system might be older than that given by the Pb-Pb age of CAIs. Using their 4571-Ma estimate for the upper age limit of the solar system would give an age for Vesta's mantle of ~6 Ma. Recent ^{26}Al-^{26}Mg and ^{182}Hf-^{182}W measurements on iron meteorites further indicate that the differentiation of their parent asteroids occurred within 1–2 m.y. after CAI formation (e.g., *Bizzarro et al.,* 2005; *Kleine et al.,* 2005).

For many of these objects differentiation does not coincide with any solar-system-wide cessation of planetesimal accretion. For example, the accretion of the chondrite parent bodies has been inferred to have lasted for a period of at least 2.5 m.y. following CAI formation (*Amelin et al.,* 2002; *Bizzarro et al.,* 2004; *Kunihiro et al.,* 2004). For the bulk of the iron meteorites an accretion period of ~5 m.y. has been inferred (*Horan et al.,* 1998). While protracted accretion rates for objects larger than planetesimals based on Hf-W have been suggested by *Halliday* (2000, 2006), the radiometric data alone do not preclude the accretion of a given object having terminated substantially earlier than its model radiometric differentiation age.

2.3. Dynamical Constraints

Many conventional dynamical models of planetesimal/planetary formation suggest an early and rapid period of accretion, commonly referred to as "runaway growth" (e.g., *Weidenschilling and Cuzzi,* 2006). More precisely, the term "runaway growth" usually does not apply to the formation of 1–10-km planetesimals, but rather their subsequent growth when gravitational interactions between the solid bodies become significant. Most, if not all, of these models independently point to the rapid formation of bodies 1–10 km up to Moon/Mars-sized bodies within ~0.1 m.y. at 1 AU. As an example, one recent theoretical model for the growth of planetesimals and planets (*Kortenkamp et al.,* 2001) indicates extremely rapid formation of 10–100-km planetesimals within 10^4 yr, Mercury- and Mars-sized objects within 10^5 yr, and Earth-sized objects within 10^8 yr, with the rates being contingent on heliocentric distance. In the region of the asteroid belt (~3 AU) strong perturbations with massive gas giants may limit the rate of growth such that several Ceres-sized objects (~900 km) could form at 1 AU within 0.05 m.y., at 1.5 AU within 0.5 m.y., at 2 AU within 1 m.y., and 2.5 AU in 5 m.y. Alternatively, the effects of "dynamical friction," on which "runaway growth" is likely attributed, may also be significantly diminished at 3 AU compared to 1 AU due to lower nebular surface densities and orbital periods (e.g., *Weidenschilling,* 2000; *Wetherill and Inaba,* 2000). The potentially rapid accretion and differentiation in asteroids and the terrestrial planets (e.g., *Kleine et al.,* 2002) taken on its own would indicate that the accretionary process has an effective early termination point (cf. *Boss,* 2003, *Boss and Goswami,* 2006; *Weidenschilling,* 2000; *Weidenschilling and Cuzzi,* 2006). Even more rapid formation timescales within 1000 yr have been developed for the outer planets around massive stars (*Boss,* 2003), but such models are generally focused on giant-planet formation and are not directly relevant to the type of planetesimal accretion discussed herein.

2.4. Astronomical Constraints

Measurements of infrared (~micrometer), submillimeter, and millimeter emission in low- to intermediate-mass circumstellar disks have long suggested disk lifetimes on the order of ~5–15 m.y. (e.g., *Strom et al.,* 1989; *Skrutskie et al.,* 1990), which are consistent with the typical ~10-m.y. lifetime of the solar nebular accretion disk inferred from meteoritic analyses (cf. *Podosek and Cassen,* 1994). A recent astronomical survey of circumstellar disks around young clusters of stars (*Haisch et al.,* 2001) indicates that all stars within a given cluster lose their disks within ~6 m.y. of formation and half within 3 m.y. While this is the tightest astronomical constraint to date for nebular clearing and may be coincident with the cessation of planetesimal accretion, it again only places an upper limit on planetesimal accretion timescales. The distinction must be made between the duration of the nebula, inferred from the presence of dust, and the timescales for the formation of the first planetesimals. The presence of micrometer- to millimeter-sized grains at 10 Ma does not preclude the presence of large (1–100 km) objects, or even planets, that formed early on (e.g., *Natta et al.,* 2000). Poynting-Robertson lifetimes of micrometer-sized dust less than ~10^5 yr coupled with the persistence of dust to 5–10 m.y. suggests that micrometer-sized grains must be replenished by some source, presumably collisions of planetesimal-sized objects (e.g., *Beckwith et al.,* 1990; *Strom,* 1995).

3. POST-PRIMARY ACCRETION

Upper limits for the duration of post-primary accretion events, i.e., stage 4 (Fig. 1), associated with parent-body regolith formation, brecciation, and final compaction of the meteorite may be estimated by a several chronometric methods, including galactic-cosmic-ray (GCR) exposure, fission track dating, and ^{40}Ar-^{39}Ar chronometry. The time at which a meteorite was assembled in its present form on its parent body prior to its final ejection is commonly referred to as its "compaction age." Prior to compaction the meteorite constituents (individual grains, clasts, etc.) may have also been at times in sufficiently close proximity (a few meters) to the near surface of the parent body in its regolith, created by the process of accretion and associated impacts, to record exposure to energetic GCRs. The duration of the residence of meteorite constituents at the near-surface is referred to as its "precompaction age." This age is a lower limit to the time period between formation of the meteorite constituents (either in the nebula or on the parent body) and compaction of the meteorite.

The techniques for measuring precompaction exposure ages based on GCR spallation reactions and compaction ages based primarily on fission track dating are thoroughly covered in *Caffee et al.* (1988) and *Caffee and MacDougall* (1988) and will not be discussed here. Furthermore, only a few new measurements have been reported since the first volume of this book (*Kerridge and Matthews,* 1988). Lower limits on precompaction exposure ages for the CI and CM parent bodies still range from a few million years up to possibly 100 m.y., respectively (*Hohenberg et al.,* 1990; *Nichols et al.,* 1992). While these ages do provide loose limits on the extent of regolith residence and mixing, uncertainties in the energy and flux of solar energetic particles (SCRs) in the early solar system relative to GCRs in the early solar system can introduce factors of a few into the errors on these ages (*Caffee et al.,* 1987; see also *Chaussidon and Gounelle,* 2006). Furthermore, the spallogenic reactions associated with precompaction ages are not energetic enough to reset any other standard radiometric clock, thereby making it difficult or impossible to corroborate the ages.

Macdougall and Kothari (1976) utilized the fission-track method on a suite of mineral grains from CM meteorites, yielding compaction ages on the order of 100–200 m.y. following CAI formation. The technique is limited in that Pu/U fractionation and the presence of actinide-rich matrix adjacent to grains can affect the ages by factors of a few (e.g., *Crozaz et al.,* 1989). A more precise measurement for the compaction age of Orgueil (CI), ~10–15 m.y., utilized Rb-Sr dating of carbonates (*Madougall et al.,* 1984; cf. *Caffee and MacDougall,* 1988), consistent with the few-million-year precompaction exposure noted above. Unfortunately, no known meteorites to date exhibit the characteristics of Orgueil (large, Rb-rich aqueous veins) necessary for a Rb-Sr compaction age determination.

Argon-40–argon-39 chronometry has provided a much larger suite of estimates for post-primary accretion events. Various classes of meteorites exhibit "brecciated" structures, or mixtures of rocks with varying lithologies, created from impacts on parent bodies (cf. *Bischoff et al.,* 2006). The details of "brecciation classification" is often more complicated than a simple single-stage accretionary process, as the parent bodies may have been partially to completely destroyed and reassembled, perhaps multiple times. The presence of brecciated and shocked meteorites and reset radiometric clocks [e.g., ^{40}Ar-^{39}Ar (*Bogard and Garrison,* 2003)] clearly indicate that accretion, disassembly, reaccretion, and impact events have occurred at least for some 50–500 m.y. after CAI formation, culminating with a major period of asteroidal disruption commonly referred to as the "late heavy bombardment" (e.g., *Kring and Cohen,* 2002). Some unbrecciated meteorites, e.g., the Shallowater aubrite, also exhibit complex stages of formation, reassembly, and episodic cooling histories (*Keil et al.,* 1989). Such post-planetary processing, however, makes it difficult to delineate the records associated with the duration of initial accretion, especially when overlain with secondary alteration (cf. *Krot et al.,* 2006), complex cooling histories, and uncertain peak metamorphic temperatures (cf. *Huss et al.,* 2006).

4. SUMMARY

The objective herein has been to present a brief overview of the recent chronological constraints of the accretion of 10–100-km-sized planetesimals in the early solar system as determined by meteoritic, dynamical, and astronomical studies. With advances and increased precision in each field since the first volume of this book (*Kerridge and Matthews,* 1988), the initial planetesimal accretion timescale has become somewhat shorter than 10 m.y.: astronomical to ~5 m.y. from the onset of first solid formation, meteoritic to ~1–2 m.y., and dynamical potentially to less than 1 m.y. (Table 1). While the estimates differ, this is primarily due again to the precision available in each field, or within a field due to the specific objects that are being measured, modeled, or observed. Taken together, it is likely that some planetesimals formed rapidly within 1 m.y. and that the dominant phase of planetesimal accretion occurred within a few million years after the formation of the first solids in the solar system.

At this time it is uncertain whether additional advances and increased precision within each of these fields will be able to significantly improve on what we know about this subject. In the astronomical field, perhaps, room is available for the discovery of solar-type systems whose nebulae clear on timescales significantly shorter than the current ~6-m.y. limit. Whether our solar system acted as fast would still remain uncertain, as is the case for dynamical models that can push planetesimal accretion to a few 10–100 k.y. In the meteoritic field, a combination of increased radiometric precision (substantially less than 1 m.y. for suites of chronometers that can be compared directly with one another) and additional measurements may provide more stringent constraints, but this remains to be seen.

Acknowledgments. I thank T. Kleine and J. Chambers for beneficial reviews. I also thank M. Wadhwa for editorial handling. This work was supported by a grant from the National Aeronautics and Space Administration.

REFERENCES

Allègre C. J., Manhès G., and Göpel C. (1995) The age of the Earth. *Geochim. Cosmochim. Acta, 59,* 1445–1456.

Amelin Y., Krot A. N., Hutcheon I. D., and Ulyanov A. A. (2002) Lead isotopic ages of chondrules and calcium-aluminum-rich inclusions. *Science, 297,* 1678–1683.

Beckwith S. V. W., Sargent A. I., Chini R. S., and Guesten R. (1990) A survey for circumstellar disks around young stellar objects. *Astron. J., 99,* 924–945.

Binzel R. P. and Xu S. (1993) Chips off of asteroid 4 Vesta: Evidence for the parent body of basaltic achondrite meteorites. *Science, 260,* 186–191.

Birck J. L. and G. W. Lugmair (1988) Nickel and chromium isotopes in Allende inclusions. *Earth Planet. Sci. Lett., 90,* 131–143.

Bischoff A., Scott E. R. D., Metzler K., and Goodrich C. A. (2006) Nature and origins of meteoritic breccias. In *Meteorites and the Early Solar System II* (D. S. Lauretta and H. Y. McSween Jr., eds.), this volume. Univ. of Arizona, Tucson.

Bizzarro M., Baker J. A., and Haack H. (2004) Mg isotope evidence for contemporaneous formation of chondrules and refractory inclusions. *Nature, 431,* 275–278.

Bizzarro M., Baker J. A., and Haack H. (2005) Timing of crust formation on differentiated asteroids inferred from Al-Mg chronometry (abstract). In *Lunar and Planetary Science XXXVI,* Abstract #1312. Lunar and Planetary Institute, Houston (CD-ROM).

Bogard D. D. and Garrison D. H. (2003) ^{39}Ar-^{40}Ar ages of eucrites and thermal history of asteroid 4 Vesta. *Meteoritics & Planet. Sci., 38,* 669–710.

Boss A. (2003) Rapid formation of outer giant planets by disk instability. *Astrophys. J., 599,* 577–581.

Boss A. P. and Goswami J. N. (2006) Presolar cloud collapse and the formation and early evolution of the solar nebula. In *Meteorites and the Early Solar System II* (D. S. Lauretta and H. Y. McSween Jr., eds.), this volume. Univ. of Arizona, Tucson.

Caffee M. W. and MacDougall J. D. (1988) Compaction ages. In *Meteorites and the Early Solar System* (J. F. Kerridge and M. S. Matthews, eds.), pp. 289–298, Univ. of Arizona, Tucson.

Caffee M. W., Hohenberg C. M., and Swindle T. D. (1987) Evidence in meteorites for an active early sun. *Astrophys. J. Lett., 313,* L31–L35.

Caffee M. W., Goswami J. N., Hohenberg C. M., Marti K., and Reedy R. C. (1988) Irradiation records in meteorites. In *Meteorites and the Early Solar System* (J. F. Kerridge and M. S. Matthews, eds.), pp. 205–245, Univ. of Arizona, Tucson.

Chambers J. (2006) Meteoritic diversity and planetesimal formation. In *Meteorites and the Early Solar System II* (D. S. Lauretta and H. Y. McSween Jr., eds.), this volume. Univ. of Arizona, Tucson.

Chaussidon M. and Gounelle M. (2006) Irradiation processes in the early solar system. In *Meteorites and the Early Solar System II* (D. S. Lauretta and H. Y. McSween Jr., eds.), this volume. Univ. of Arizona, Tucson.

Chen J. H. and Tilton G. R. (1976) Isotopic lead investigations on the Allende carbonaceous chondrite. *Geochim. Cosmochim. Acta, 40,* 635–643.

Chen J. H. and Wasserburg G. J. (1981) The isotopic composition of uranium and lead in Allende inclusions and meteoritic phosphates. *Earth Planet. Sci. Lett., 52,* 1–15.

Crozaz G., Pellas P., Bourot-Denise M., de Chazal S. M., Fiéni C., Lundberg L. L., and Zinner E. (1989) Plutonium, uranium and rare earths in phosphates of ordinary chondrites-the quest for a chronometer. *Earth Planet. Sci. Lett., 93,* 157–169.

Cuzzi J. N. and Weidenschilling S. J. (2006) Particle-gas dynamics and primary accretion. In *Meteorites and the Early Solar System II* (D. S. Lauretta and H. Y. McSween Jr., eds.), this volume. Univ. of Arizona, Tucson.

Gilmour J. D. (2000) The extinct radionuclide timescale of the early solar system. *Space Sci. Rev., 92,* 123–132.

Göpel C., Manhès G., and Allègre C. J. (1991) Constraints on the time of accretion and thermal evolution of chondrite parent bodies by precise U-Pb dating of phosphates. *Meteoritics, 26,* 338.

Göpel C., Manhès G., and Allègre C. J. (1994) U-Pb systematics of phosphates from equilibrated ordinary chondrites. *Earth Planet. Sci. Lett., 121,* 153–171.

Grossman J. N. (1988) Irradiation records in meteorites. In *Meteorites and the Early Solar System* (J. F. Kerridge and M. S. Matthews, eds.), pp. 680–696. Univ. of Arizona, Tucson.

Haisch K. E. Jr., Lada E. A., and Lada C. J. (2001) Disk frequencies and lifetimes in young clusters. *Astrophys. J. Lett., 553,* L153–L156.

Halliday A. (2000) Hf-W chronometry and inner solar system accretion rates. *Space Sci. Rev., 92,* 355–370.

Halliday A. N. (2006) Meteorites and the timing, mechanisms, and conditions of terrestrial planet accretion and early differentiation. In *Meteorites and the Early Solar System II* (D. S. Lauretta and H. Y. McSween Jr., eds.), this volume. Univ. of Arizona, Tucson.

Halliday A., Quitté G., and Lee D.-C. (2003) Tungsten isotopes and the time-scales of planetary accretion. *Meteoritics & Planet. Sci., 38,* 5256.

Herndon J. M. and Herndon M. A. (1977) Aluminum-26 as a planetoid heat source in the early solar system. *Meteoritics, 12,* 459–465.

Hewins R., Jones R., and Scott E., eds. (1996) *Chondrules and the Protoplanetary Disk.* Cambridge Univ., Cambridge. 360 pp.

Hohenberg C. M., Nichols R. H. Jr., Olinger C. T. and Goswami J. N. (1990) Cosmogenic neon from individual grains of CM meteorites — Extremely long pre-compaction exposure histories or an enhanced early particle flux. *Geochim. Cosmochim. Acta, 54,* 2133–2140.

Horan M. F., Smoliar M. I., and Walker R. J. (1998) W-182 and Re(187)-Os-187 systematics of iron meteorites — Chronology for melting, differentiation, and crystallization in asteroids. *Geochim. Cosmochim. Acta, 62,* 545–554.

Huss G. R., Rubin A. E., and Grossman J. N. (2006) Thermal metamorphism in chondrites. In *Meteorites and the Early Solar System II* (D. S. Lauretta and H. Y. McSween Jr., eds.), this volume. Univ. of Arizona, Tucson.

Hutcheon I. D. and Hutchison R. (1989) Evidence from the Semarkona ordinary chondrite for Al-26 heating of small planets. *Science, 337,* 238–241.

Keil K., Ntaflos Th., Taylor G. J., Brearley A. J., Newsom H. E., and Romig A. D. Jr. (1989) The Shallowater aubrite: Evidence for origin by planetesimal impacts. *Geochim. Cosmochim. Acta, 53,* 3291–3307.

Kerridge J. F. and Matthews M. S., eds. (1988) *Meteorites and the Early Solar System.* Univ. of Arizona, Tucson. 1269 pp.

Kita N. T., Nagahara H., Togashi S., and Morishita Y. (2000) A short duration of chondrule formation in the solar nebula: Evidence from ^{26}Al in Semarkona ferromagnesian chondrules. *Geochim. Cosmochim. Acta, 64,* 3913–3922.

Kleine T., Münker C., Mezger K., and Palme H. (2002) Rapid accretion and early core formation on asteroids and the terrestrial planets from Hf-W chronometry. *Nature, 418,* 952–955.

Kleine T., Mezger K., Munker C., Palme H., and Bischoff A. (2004) ^{182}Hf-^{182}W isotope systematics of chondrites, eucrites, and martian meteorites: Chronology of core formation and early mantle differentiation in Vesta and Mars. *Geochim. Cosmochim. Acta, 68,* 2935–2946.

Kleine T., Mezger K., Palme H., and Scherer E. (2005) Tungsten isotopes provide evidence that core formation in some asteroids predates the accretion of chondrite parent bodies (abstract). In *Lunar and Planetary Science XXXVI,* Abstract #1431. Lunar and Planetary Institute, Houston (CD-ROM).

Kortenkamp S. J., Wetherill G. W. and Inaba S. (2001) Runaway growth of planetary embryos facilitated by massive bodies in a protoplanetary disk. *Science, 293,* 1127–1129.

Kring D. A. and Cohen B. A. (2002) Cataclysmic bombardment throughout the inner solar system 3.9–4.0 Ga. *J. Geophys. Res., 107,* 4–1.

Krot A. N., Hutcheon I. D., Brearley A. J., Pravdivtseva O. V., Petaev M. I., and Hohenberg C. M. (2006) Timescales and settings for alteration of chondritic meteorites. In *Meteorites and the Early Solar System II* (D. S. Lauretta and H. Y. McSween Jr., eds.), this volume. Univ. of Arizona, Tucson.

Kunihiro T., Rubin A. E., McKeegan K. D., and Wasson J. T. (2004) Initial ^{26}Al/^{27}Al in carbonaceous-chondrite chondrules: Too little ^{26}Al to melt asteroids. *Geochim. Cosmochim. Acta, 68,* 2947–2957.

Larimer J. W. and Anders E. (1967) Chemical fractionations in meteorites — II. Abundance patterns and their interpretation. *Geochim. Cosmochim. Acta, 31,* 1239–1270.

Lee T., Papanastassiou D. A., and Wasserburg G. J. (1976) Demonstration of ^{26}Mg excess in Allende and evidence for ^{26}Al. *Geophys. Res. Lett., 3,* 109–112.

Lugmair G. W. and Shukolyukov A. (1998) Early solar system timescales according to ^{53}Mn-^{53}Cr systematics. *Geochim. Cosmochim. Acta, 62,* 2863–2886.

Macdougall J. D. and Kothari B. K. (1976) Formation chronology for C2 meteorites. *Earth Planet. Sci. Lett., 33,* 36–44.

Macdougall J. D., Lugmair G. W., and Kerridge J. F. (1984) Early solar system aqueous activity: Sr isotope evidence from Orgueil CI chondrite. *Science, 307,* 249–251.

MacPherson G. J., Davis A. M., and Zinner E. K. (1995) The distribution of aluminum-26 in the early solar system — A reappraisal. *Meteoritics, 30,* 365–386.

Manhès G., Göpel C., and Allègre C. J. (1988) Systématique U-Pb dans les inclusions réfractaires d'Allende: le plus vieux matériau solaire. *Comptes Rendus de l'ATP Planétologie,* 323–327.

McCord T. B., Adams J. B., and Johnson T. V. (1970) Asteroid Vesta: Spectral reflectivity and compositional implications. *Science, 168,* 1445–1447.

Meyer B. S. and Clayton D. D. (2000) Short-lived radioactivities and the birth of the sun. *Space Sci. Rev., 92,* 133–152.

Mostefaoui S., Lugmair G. W., and Hoppe P. (2004) In-situ evidence for live iron-60 in the early solar system: A potential heat source for planetary differentiation from a nearby supernova explosion (abstract). In *Lunar and Planetary Science XXXV,* Abstract #1271. Lunar and Planetary Institute, Houston (CD-ROM).

Moynier F., Blichert-Toft J., Telouk P., and Albarède F. A. (2005) Excesses of ^{60}Ni in chondrites and iron meteorites (abstract). In *Lunar and Planetary Science XXXV,* Abstract #1593. Lunar and Planetary Institute, Houston (CD-ROM).

Natta A., Grinin V. P., and Mannings V. (2000) Properties and evolution of disks around pre-main-sequence stars of intermediate mass. In *Protostars and Planets IV* (V. Mannings et al., eds.), pp. 559–587. Univ. of Arizona, Tucson.

Nesvorný D., Bottke W. F. Jr., Dones L., and Levison H. F. (2002) The recent breakup of an asteroid in the main belt region. *Nature, 417,* 720–771.

Nichols R. H. Jr., Hohenberg C. M., and Goswami J. N. (1992) Spallation-produced ^{21}Ne in individual Orgueil olivines (abstract). In *Lunar and Planetary Science XXIII,* pp. 987–988. Lunar and Planetary Institute, Houston.

Nyquist L. E., Reese Y., Wiesmann H., Shih C.-Y., and Takeda H. (2003) Fossil ^{26}Al and ^{53}Mn in the Asuka 881394 eucrite: Evidence of the earliest crust on asteroid 4 Vesta. *Earth Planet. Sci. Lett., 214,* 11–25.

Petit J.-M., Morbidelli A., and Chambers J. E. (2001) The primordial excitation and clearing of the asteroid belt. *Icarus, 153,* 338–347.

Podosek F. A. and Cassen P. (1994) Theoretical, observational, and isotopic estimates for the lifetime of the solar nebula. *Meteoritics, 29,* 6–25.

Quitté G., Latkoczy C., Halliday A. N., Schönbächler M., and Günther D. (2005) Iron 60 in the eucrite parent body and the initial ^{60}Fe/^{56}Fe of the solar system (abstract). In *Lunar and Planetary Science XXXVI,* Abstract #1827. Lunar and Planetary Institute, Houston (CD-ROM).

Rubin A. E. (2004) Postshock annealing and postannealing shock in equilibrated ordinary chondrites: Implications for the thermal and shock histories of chondritic asteroids. *Geochim. Cosmochim Acta, 68,* 673–689.

Russell S. S., Srinivasan G., Huss G. R., Wasserburg G. J., and MacPherson G. J. (1996) Evidence for widespread ^{26}Al in the solar nebula and constraints for nebula time scales. *Science, 273,* 757–762.

Russell S. S., Hartmann L., Cuzzi J., Krot A. N., Gounelle M., and Weidenschilling S. (2006) Timescales of the solar protoplanetary disk. In *Meteorites and the Early Solar System II* (D. S. Lauretta and H. Y. McSween Jr., eds.), this volume. Univ. of Arizona, Tucson.

Scott E. R. D. (1979) Origin of anomalous iron meteorites. *Mineral. Mag., 43,* 415–421.

Skrutskie M. F., Dutkevitch D., Strom S. E., Edwards S., Strom K. M., and Shure M. A. (1990) A sensitive 10-micron search for emission arising from circumstellar dust associated with solar-type pre-main-sequence stars. *Astron. J., 99,* 1187–1195.

Sonnet C. P. and Reynolds R. T. (1979) Primordial heating of asteroidal parent bodies. In *Asteroids* (T. Gehrels, ed.), pp. 822–848. Univ. of Arizona, Tucson.

Srinivasan G., Ulyanov A. A., and Goswami J. N. (1994) ^{41}Ca in the early solar system. *Astrophys. J. Lett., 431,* L67–L70.

Srinivasan G., Sahijpal S., Ulyanov A. A., and Goswami J. N. (1996) Ion microprobe studies of Efremovka CAIs: II. Potassium isotope composition and ^{41}Ca in the early solar system. *Geochim. Cosmochim. Acta, 60,* 1823–1835.

Srinivasan G., Goswami J. N., and Bhandari N. (1999) ^{26}Al in eucrite Piplia Kalan: Plausible heat source and formation chronology. *Science, 284,* 1348–1350.

Strom K. M., Strom S. E., Edwards S., Cabrit S., and Skrutskie M. F. (1989) Circumstellar material associated with solar-type pre-main-sequence stars — A possible constraint on the timescale for planet building. *Astron. J., 97,* 1451–1470.

Strom S. E. (1995) Initial frequency, lifetime and evolution of YSO disks. *Rev. Mex. AA (Serie de Conferencias), 1,* 317–328.

Tatsumoto M., Unruh D. M., and Desborough G. A. (1976) U-Th-Pb and Rb-Sr systematics of Allende and U-Th-Pb systematics of Orgueil. *Geochim. Cosmochim. Acta, 40,* 617–634.

Trieloff M., Jessberger E. K., Herrwerth I., Hopp J., Fiéni C., Ghélis M., Bourot-Denise M., and Pellas P. (2003) Structure and thermal history of the H-chondrite parent asteroid revealed by thermochronometry. *Nature, 422,* 502–506.

Turner G. (1988) Dating of secondary events. In *Meteorites and the Early Solar System* (J. F. Kerridge and M. S. Matthews, eds.), pp. 276–288, Univ. of Arizona, Tucson.

Wadhwa M., Srinivasan G., and Carlson R. W. (2006) Timescales of planetesimal differentiation in the early solar system. In *Meteorites and the Early Solar System II* (D. S. Lauretta and H. Y. McSween Jr., eds.), this volume. Univ. of Arizona, Tucson.

Weidenschilling S. (2000) Formation of planetesimals and accretion of the terrestrial planets. *Space Sci. Rev., 92,* 295–310.

Weidenschilling S. J. and Cuzzi J. N. (2006) Accretion dynamics and timescales: Relation to chondrites. In *Meteorites and the Early Solar System II* (D. S. Lauretta and H. Y. McSween Jr., eds.), this volume. Univ. of Arizona, Tucson.

Wetherill G. W. (1992) An alternative model for the formation of the asteroids. *Icarus, 100,* 307–325.

Wetherill G. W. and Chapman C. R. (1988) Irradiation records in meteorites. In *Meteorites and the Early Solar System* (J. F. Kerridge and M. S. Matthews, eds.), pp. 35–67. Univ. of Arizona, Tucson.

Wetherill G. W. and Inaba S. (2000) Planetary accumulation with a continuous supply of planetesimals. *Space Sci. Rev., 92,* 311–320.

Wood J. A. (2000) The beginning: Swift and violent. *Space Sci. Rev., 92,* 97–112.

Young E. D., Simon J. I., Galy A., Russell S. S. and Tonui E. (2005) Supra-canonical $^{26}Al/^{27}Al$ and the residence time of CAIs in the solar protoplanetary. *Science, 308,* 223–227.

Zinner E. and Göpel C. (2002) Aluminum-26 in H4 chondrites: Implications for its production and its usefulness as a fine-scale chronometer for early-solar-system events. *Meteoritics & Planet. Sci., 37,* 1001–1013.

Accretion Dynamics and Timescales: Relation to Chondrites

S. J. Weidenschilling
Planetary Science Institute

J. N. Cuzzi
NASA Ames Research Center

In this chapter we consider aspects of the formation of a presumed initial population of approximately kilometer-sized planetesimals and their accretion to form larger bodies in the asteroid belt, including the parent bodies of meteorites. Emphasis is on timescales predicted from dynamical models and comparison with measured ages of meteoritic components. In the simplest models, the timescales for planetesimal formation and accretion of protoplanetary embryos are not consistent with the apparent age difference between CAIs and chondrules, the inferred duration for chondrule formation, or the expected degree of heating and metamorphism due to ^{26}Al. We suggest alternative scenarios that may delay planetesimal formation and/or produce chondrites from recycled debris of first-generation planetesimals that were collisionally disrupted after Jupiter formed. We discuss collisional evolution during and after accretion, and consequences for lithification of meteorites. The region of the solar nebula that corresponded to the asteroid belt originally contained ~10^3× its present mass of solid matter. The present asteroid belt does not represent an unbiased sample of that material. Meteorites are preferentially derived from bodies originally between a few tens and a few hundreds of kilometers in size; these were sufficiently small and numerous to leave a remnant population after depletion of the belt, but large enough to survive the subsequent 4.5 G.y. of collisional evolution.

1. INTRODUCTION

Collisions played a dominant role in the evolution of the parent bodies of meteorites, from their accretion as (more or less) "pristine" planetesimals, through the stirring and depletion of the asteroid belt and its collisional evolution over the subsequent ~4.5 G.y. of solar system history. Elsewhere in this volume (*Cuzzi and Weidenschilling,* 2006), we discuss the earliest stage of "primary accretion," which occurred during the formation of planetesimals from small grains and particles that were controlled or influenced by the gas in the solar nebula. In the present chapter, we explore the later evolution of the swarm of planetesimals that presumably formed in the asteroid region. We begin by summarizing in section 2 the "conventional wisdom" as of the previous volume in this series, *Meteorites and the Early Solar System (MESS I)* (*Kerridge and Matthews,* 1988), and in section 3 changes to that picture that have resulted from more recent work. We then examine in section 4 the current models for planetesimal formation, accretion of protoplanetary embryos, removal of most of the primordial mass from the asteroid region, consolidation and lithification of asteroids, and the later collisional evolution of the remnant population. Simple models for planetesimal formation and embryo growth predict timescales for these processes that are not consistent with measured ages of CAIs and chondrules. In section 5 we explore two possible scenarios for the formation of parent bodies that may (not without difficulties) meet the age constraints. Section 6 is a summary and call for further investigation of outstanding issues.

2. UNDERSTANDING AS OF METEORITES AND THE EARLY SOLAR SYSTEM I

The principal goal of cosmogonists has been to explain the formation of the planets. The asteroid belt is "the exception that proves the rule," as it is necessary to understand why that region of the solar system does not contain any body of planetary mass. Our understanding of the origins of meteorites, and their parent bodies, the asteroids, has changed significantly since the earlier volumes in this series, *MESS I* (*Kerridge and Matthews,* 1988) and *Asteroids II* (*Binzel et al.,* 1989). Chapters in those volumes called attention to several apparent paradoxes. The isotopic compositions of CAIs showed evidence for decay of ^{26}Al to ^{26}Mg, while chondrules had little or no ^{26}Al. The simplest interpretation was that chondrules formed ~1–2 m.y. later than CAIs, after the ^{26}Al had decayed. Other isotopic systems indicated that chondrules formed over an extended period of a few million years (*Grossman et al.,* 1988; see also *Swindle et al.,* 1996). However, models for the coagulation and settling of grains in the solar nebula yielded much shorter timescales for planetesimal formation, only a few thousand orbital periods, or ~10^4 yr in the asteroid region (*Weidenschilling,* 1988). Due to gas drag in the solar nebula, small particles would be lost into the Sun on timescales of

~10^5 yr for millimeter-sized chondrules and ~10^4 yr for centimeter-sized CAIs. Therefore, rapid formation of kilometer-sized or larger planetesimals appeared to be necessary to preserve them.

The lack of evidence for ^{26}Al (i.e., radiogenic ^{26}Mg*) in chondrules could also be explained as due to its inhomogeneous distribution in the solar nebula, rather than by decay before their formation. If ^{26}Al was present in the reservoir that produced CAIs, but was lacking in the material that produced chondrules, then the difference would have no chronological significance. This interpretation would be consistent with rapid accretion, and solves the loss-into-Sun problem. It has generally been assumed, either explicitly or implicitly, that chondrules formed before planetesimals, and accretion occurred monotonically, starting with dust and proceeding to kilometer-sized planetesimals and larger bodies. However, early production of chondrules before the first planetesimals implies that most of the solid silicate matter in the inner solar system, as much as several Earth masses, was converted into chondrules. The heating events that could melt such a large mass would require a large amount of energy in a short time (*Levy,* 1988). As argued by *Wood* (1996), more energy was available for such heating early in the lifetime of the solar nebula, during infall and redistribution of mass and angular momentum in an accretion disk. Thus, a consistent scenario could be constructed in which ^{26}Al was manufactured locally in the hot innermost nebula by an active early Sun and incorporated into CAIs, while chondrules were produced throughout the rest of the terrestrial and asteroidal regions, and after some mixing, rapidly accreted into planetesimals.

By the time of *MESS I*, it was realized that accretion of protoplanetary embryos proceeded by runaway growth, in which the largest body in a localized region of the planetesimal swarm would gain mass more rapidly than the next largest, and come to dominate that region (*Wetherill and Stewart,* 1989). Repetition of this process at different heliocentric distances would produce multiple embryos with masses comparable to the Moon or Mars, or ~10^2–10^3× that of the largest asteroid (~10^{24} g). Such bodies are too large to be destroyed by collisions (*Davis et al.,* 1989), and at the time no mechanism was recognized that could remove them from the asteroid region. Thus, runaway growth could have occurred in the region of terrestrial planets, but some process appeared to have prevented it in the asteroid belt (*Ruzmaikina et al.,* 1989). This might be the result of Jupiter's perturbations that stirred up eccentricities, causing accretion to stop and erosion to begin (*Wetherill,* 1989). Ceres, at ~1.5×10^{24} g, contains about one-third of the total mass in the asteroid belt, and is about three times as massive as the next largest asteroid. As such, it might be identified as an incipient runaway body that did not grow to completion. The growth times of protoplanetary embryos by runaway were estimated to be ~0.1–1 m.y, requiring precise timing for Jupiter's formation to interrupt the runaway. Based on models of the solar nebula (e.g., *Weidenschilling,* 1977), the mass of condensed solids in the asteroid region was originally comparable to or greater than that of Earth; it was necessary to remove >99% of this mass to produce the final belt. This removal had to occur early, before the formation of Vesta's basaltic crust, or its preservation could not be explained (*Davis et al.,* 1989).

3. RECENT DEVELOPMENTS

As mentioned above, as of a decade ago early production of both CAIs and chondrules, and their rapid incorporation into planetesimals, seemed to be dynamically plausible, or even required. However, recent absolute Pb-Pb isotopic dating of chondrules (*Amelin et al.,* 2002; *Russell et al.,* 2006) supports the age difference of a few million years between CAIs and chondrules implied by the Al-Mg data discussed above. The Pb-Pb and Al-Mg results taken together also support earlier inferences from Mn-Cr and I-Xe (as well as Al-Mg) isotopes (*Swindle et al.,* 1996) that chondrules are younger than CAIs. Whatever process produced CAIs seems to have stopped working early, perhaps lasting ≪1 m.y., while chondrules formed 1–2 m.y. later. If chondrules formed before planetesimals, then some process must have inhibited planetesimal formation for more than 1 m.y. after the formation of the solar nebula. During this interval, CAIs — at least not all of them — were not lost into the Sun.

This finding of a hiatus between CAIs and chondrules leads us to consider alternative models for the preservation of CAIs and accretion of chondritic parent bodies. One possibility, suggested by *Shu et al.* (1997, 2001), is that CAIs formed near the Sun, and were ejected well out into the nebula by a bipolar outflow from near the Sun-nebula interface. Another possibility, discussed by *Cuzzi et al.* (2003) and *Cuzzi and Weidenschilling* (2006), is that CAIs were produced in the hot inner region of the solar nebula, not necessarily very close to the Sun, and transported outward by turbulent diffusion, which can preserve them for times longer than their lifetimes against inward drift. We return to these possibilities in section 5.1. It is not clear from the measured ages whether CAIs and chondrules overlapped in time, or whether the apparent age difference represents a distinct gap between their formation times (*Russell et al.,* 2006). If the age gap between CAIs and chondrules is real, it implies that they were produced by separate mechanisms. Such a gap would also imply that after the nebula formed, it was relatively inactive without significant energetic heating events for ~1 m.y., then some heating process occurred that produced chondrules. Even if there was no gap, the chondrule-forming process was prolonged, and lasted longer than the putative timescales either for loss into the Sun by gas drag, or for the formation of planetesimals.

Another argument against early and rapid formation of planetesimals is provided by their thermal histories. The presence of ^{26}Al in CAIs is confirmed by excess ^{26}Mg produced by its decay. The initial abundance of ^{26}Al in CAIs appears to be quite uniform, with ^{26}Al/^{27}Al ~5×10^{-5} (*MacPherson et al.,* 1995; *Russell et al.,* 2006). If this abundance was typical for other Al-containing bodies, then decay of ^{26}Al would have been a significant heat source within planetesimals. *LaTourette and Wasserburg* (1998) measured diffusion rates of Mg in anorthite as a function of temperature, and concluded that isotopes would have been homogenized in CAIs exposed to temperatures in excess of ~750°C. They calculated thermal histories for planetesimals of various sizes, with a range of initial ^{26}Al abundances corresponding to decay from the "canonical" initial value before their formation. The critical temperature would be exceeded within a volume fraction that increased with planetesimal size, but decreased for lower ^{26}Al abundance, i.e., later formation time. From the observation that ~50% of CAIs retain Mg-isotopic anomalies, they concluded that planetesimals did not exceed ~30 km diameter within 1–2 m.y. of CAI formation. LaTourette and Wasserburg estimated that bodies with diameters >100 km that formed within 1 m.y. of CAIs would have melted. *McSween et al.* (2002) and *Grimm and McSween* (1993) concluded that even smaller bodies would have melted. While most meteorites show evidence of thermal metamorphism and some asteroids (Vesta and parent bodies of iron meteorites) did experience melting, the thermal history of the asteroid belt as a whole is consistent with depletion of ^{26}Al before accretion (*Ghosh et al.,* 2006).

It is no longer necessary to assume that large bodies never accreted in the asteroid region, as a plausible mechanism for their removal has been identified. *Wetherill* (1992) suggested an alternative model for the formation of asteroids and depletion of the primordial mass of solids in that region of the solar nebula. If large bodies accreted by runaway growth, their mutual perturbations could scatter them into resonances with Jupiter. Resonant bodies would have their eccentricities raised until they could encounter Jupiter, collide with the terrestrial planets, or even be thrown into the Sun. Smaller asteroid-sized bodies would also be scattered out of the region during this process. This model was examined in more detail by *Chambers and Wetherill* (2001), who showed that more massive bodies were more effective at scattering; accretion of large protoplanetary embryos may have been necessary for clearing mass from the asteroid region. In context of these developments, we revisit some of the questions raised in *MESS I.*

4. HISTORY OF THE ASTEROID BELT: CURRENT UNDERSTANDING

4.1. Planetesimal Formation Timescale

Models of the formation of planetesimals in the solar nebula involve settling of particles to the central plane, their coagulation in collisions, or both. Various scenarios are discussed in more detail by *Cuzzi and Weidenschilling* (2006). Here we note that settling requires a very low level of turbulence in the nebular gas. Some collisional growth can occur in a turbulent medium, but the maximum size may be limited by erosion and/or disruption of larger particles. A well-known result of analytical and numerical modeling (*Weidenschilling,* 1980; *Nakagawa et al.,* 1981) is that settling and coagulation produce kilometer-sized or larger bodies on timescales of a few thousand orbital periods, or $<10^4$ yr in the asteroid region. However, such models have been developed in detail only for a purely laminar nebula. As discussed by *Cuzzi and Weidenschilling* (2006), very low levels of turbulence can prevent settling and delay accretion. Thus, the canonical short timescale for planetesimal formation is only a lower limit.

A perfectly laminar nebula was probably unrealistic at any time, but in any case it took time for turbulence associated with its formation to decay. The nebula was certainly turbulent during infall of gas from the presolar cloud, for at least the free-fall timescale of $\sim 10^5$ yr. The mismatch in angular momentum between infalling gas and the circumstellar disk would have produced turbulence (*Cassen and Moosman,* 1981). Most of the solar mass was processed through the disk, and most of the infall would have occurred before the formation of the Sun itself, as well as any components preserved in meteorites. However, the infall surely did not end abruptly, and the associated turbulence presumably decayed gradually, probably non-monotonically, as the presolar cloud was depleted. There would have been other possible sources of turbulence after infall ceased, such as thermal convection and/or magnetorotational instability (*Stone et al.,* 2000), not necessarily resulting in major redistribution of mass and angular momentum. Coagulation of grains into aggregates probably took place during this stage, producing bodies of at least centimer, perhaps approaching meter, size. These would spiral inward due to gas drag and evaporate, with vapor diffusing outward and recondensing. Calcium-aluminum-rich inclusions might be produced by this process. These could diffuse outward due to turbulence (*Cuzzi et al.,* 2003). While turbulence lasted, it would prevent grains and small aggregates from settling to form a dense layer in the nebular midplane. As described by *Cuzzi and Weidenschilling* (2006), declining turbulence may have allowed planetesimal formation to begin, perhaps abruptly, once a particle layer formed. Another possibility is that turbulence did not inhibit planetesimal formation, but caused it by concentrating particles within eddies until they reached a critical density for gravitational collapse (*Cuzzi et al.,* 2001). While these scenarios can explain delayed formation of planetesimals and their content of CAIs, it is not clear how chondrules fit into this picture (see section 5).

4.2. Accretion of Larger Bodies from Planetesimals

Once planetesimals formed, with sizes large enough (~1–10 km) to be no longer controlled by nebular gas drag, their accretion into larger bodies was the result of collisions, with gravity as the binding force. The rate of collisions was mediated by feedback between relative velocities (i.e., eccentricities and inclinations) of the bodies, stirring by gravitational perturbations, and damping by collisions and gas drag (*Wetherill and Stewart,* 1993). Larger bodies have collisional cross-sections augmented by gravitational focusing, so that the largest body in a given region gains mass more rapidly than the second largest, and becomes dominant. Rapid growth occurs when the random velocities of the median-mass bodies are less than that due to Keplerian shear relative to the large body. At that point, gravitational focusing maximizes the large body's effective collisional cross-section, and its growth rate becomes large. This condition typically requires that the large body must be ~10^4–10^5× the median mass. Starting from a swarm of initially equal-sized bodies, it takes many collisions before stochastic collisions allow such a dominant body to emerge. Numerical simulations (*Wetherill,* 1989; *Weidenschilling et al.,* 1997) suggest runaway growth in the asteroid region would begin on a timescale ~10^5 yr if the initial swarm consisted of kilometer-sized bodies. If the initial size was larger for a given mass of the swarm, the collision rate is lower, and the onset of runaway growth is delayed; for 10-km planetesimals it would require ~10^6 yr. Most of this interval is the time required for a largest body to become dominant in any localized region of the swarm. Rapid growth continues until the smaller bodies near its orbit are depleted, and/or stirred up to high velocities such that gravitational focusing becomes less effective.

A growing embryo does not have unlimited access to small bodies in the surrounding swarm, but can only accrete objects whose orbits approach it within a critical distance. From conservation of the Jacobi parameter in the restricted three-body problem (an effective energy barrier in a frame moving with the massive body), it can be shown that the distance limit for collisions in the idealized three-body case (in which initial orbits are all circular and coplanar) is $2\sqrt{3}\times$ the Hill radius R_H, defined as $R_H = (M/3M_\odot)^{1/3}a$, where M, $M_\odot$ are the masses of the embryo and the Sun, and a the embryo's heliocentric distance. The total mass available within $2\sqrt{3}R_H$ is $8\pi\sqrt{3}aR_H\sigma$, where σ is the surface density of accretable matter. Equating this to M in the expression for R_H, it can be shown that the "isolation mass" is equal to $M_{iso} \sim 166\ M_\odot^{-1/2}a^3\sigma^{3/2} = 0.0021\ a^3\sigma^{3/2}\ M_\oplus$, where a is in AU and σ in g cm^{-2} (*Lissauer and Stewart,* 1993). The isolation mass is a convenient benchmark, but is not an absolute limit, as accretion in a swarm of planetesimals may violate some conditions of the restricted three-body problem: The small bodies may collide and migrate due to gas drag, be perturbed by more than one embryo, etc., and embryos may perturb and collide with each other rather than grow in isolation. Numerical simulations of accretion (*Weidenschilling et al.,* 1997; *Kokubo and Ida,* 1998) typically result in an "oligarchy" of embryos with masses comparable to M_{iso}, with orbital spacings ~10 R_H. If the asteroid zone was not depleted in mass, so that σ was a few g cm^{-2}, $M_{iso} \sim 0.1\ M_\oplus$.

Somewhat counterintuitively, runaway growth of large embryos does not totally deplete the population of small bodies; rather, the process produces a bimodal size distribution, typically with comparable amounts of the total mass in embryos and small bodies. Once this mass distribution develops, the small bodies are stirred by gravitational perturbations of multiple embryos, and the higher encounter velocities decrease the gravitational cross-sections of the embryos, slowing their growth. The higher velocities also may change the collisional evolution of the small bodies from net accretion to fragmentation. Thus, runaway growth actually tends to preserve a substantial fraction of the swarm in the form of small bodies. The swarm's evolution by runaway growth at any helicentric distance can be divided into three stages: slow emergence of a dominant body from the swarm, its rapid growth to ~M_{iso}, and slow sweeping up of the remaining small bodies.

Figure 1 shows results of an accretion simulation, using the code described by *Weidenschilling et al.* (1997). The swarm contains about 4 $M_\oplus$ between 2 and 4 AU, composed of planetesimals with diameter 10 km at the start of the simulation. It is assumed that the entire population of planetesimals was present at the beginning. If planetesimals formed over some interval [e.g., by turbulent concentration (cf. *Cuzzi and Weidenschilling,* 2006)], the results would be somewhat different in detail. However, simulations by *Wetherill and Inaba* (2000) show similar pattern of runaway growth for a full initial population and continuous formation. The swarm evolves by collisional accretion and mutual gravitational stirring; external perturbations, e.g., by Jupiter, are not included. Fragmentation is allowed, with nominal impact strength of 3 × 10^6 erg cm^{-3}; fragments below 0.1 km in size are lost from the simulation. After 2 m.y. of model time, less than 10% of the initial mass has been lost by fragmentation. At this point, typical eccentricities in the swarm are a few times 10^{-2}, resulting in impact velocities of a few tenths to ~1 km s^{-1}.

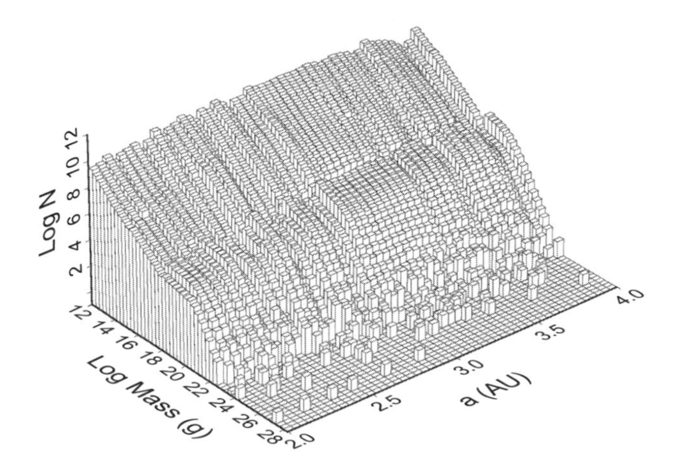

Fig. 1. Size distribution produced by accretional evolution of a swarm of planetesimals with total mass 4 $M_\oplus$, distributed between 2 and 4 AU. At t = 0, all mass was in planetesimals of diameter 10 km (mass 1.4 × 10^{18} g). After 2 m.y., runaway growth has produced 101 bodies larger than Ceres (~2 × 10^{24} g), with the largest ~10^{27} g. The bodies smaller than 10 km are fragments from cratering and/or disruption of larger bodies; less than 10% of the total mass has been lost to sizes smaller than 0.1 km.

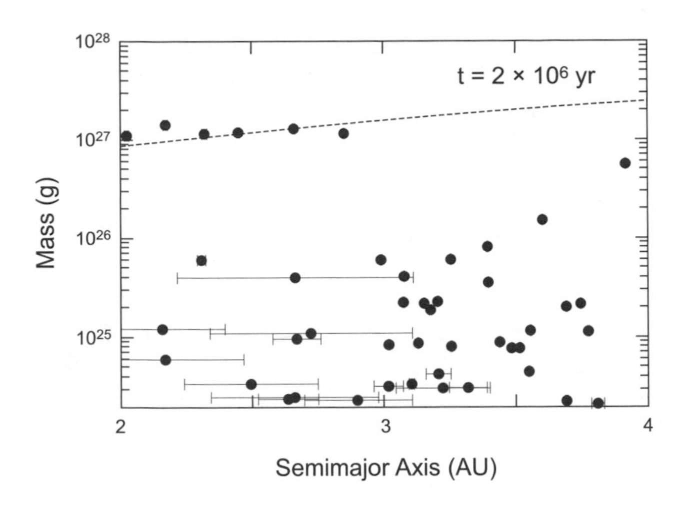

Fig. 2. Masses vs. semimajor axes for the large bodies (>2 × 10^{24} g) of the simulation shown in Fig. 1. Horizontal bars denote the range from perihelion to aphelion. The largest bodies, ~10^{27} g, are concentrated inside 3 AU; their perturbations have stirred the smaller ones to eccentricities to ~0.01–0.1, while most bodies beyond 3 AU have e < 0.01.

Figure 2 shows the masses and semimajor axes of the large (>2 × 10^{24} g) bodies after 2 × 10^6 yr; the largest bodies out to about 3 AU have masses comparable to M_{iso}, ~10^{27} g. Bodies as large as Ceres (~10^{24} g) were produced beginning at about 10^5 yr, and ~0.1 $M_\oplus$ by 10^6 yr. If accretion was assumed to have stopped before any bodies larger than Ceres accreted, e.g., by introducing jovian perturbations, this would have required rapid formation of Jupiter and precise timing. Also, the growth timescale is proportional to the Kepler period and inversely proportional to the surface density. This variation means that growth is most rapid at the inner edge of the asteroid region and propagates outward as a wave; for the assumed surface density profile, the growth timescale is proportional to $a^{5/2}$. Stopping accretion by introducing some external influence at a given time would leave the largest bodies at the inner edge of the belt. It is impossible to halt runaway growth in such a manner as to leave Ceres as the single largest body in the center of the belt (at 2.77 AU); thus, the suggestion of *Weidenschilling* (1988) that Ceres represents a "frozen" runaway does not seem tenable.

If Jupiter formed first, its perturbations could have kept velocities high enough to prevent runaway from starting, as suggested by *Wetherill* (1989). In such "orderly growth," the size distribution would evolve into a power law with a shallow slope, such that most of the mass was in the largest bodies. There would be multiple bodies of comparable mass at the upper end of the distribution, without any prominent gap between Ceres and the rest of the population. It does not seem possible to explain the dominant place of Ceres in the present mass distribution by such a scenario.

Another way to slow growth, and perhaps prevent runaway, is to assume that solids were depleted in the asteroid region before (or during) planetesimal formation. But then could anything grow even to asteroid size? A simulation comparable to that of Fig. 1, but starting with 1% of the mass (0.04 $M_\oplus$), still about an order of magnitude more massive than the present belt, has M_{iso} about equal to the mass of Ceres. However, growth is much slower than in the previous case; only one body as large as 500 km diameter grows in 5 × 10^6 yr, at the inner edge of the belt. A compromise scenario that assumes a lesser degree of depletion before the beginning of accretion could be invoked, but no mechanism for such depletion at that stage of nebular evolution has been identified. It seems likely that the asteroids accreted before their region of the solar nebula was substantially depleted of condensed solids (*Wetherill,* 1989).

4.3. Removal of Large (and Small) Bodies from the Asteroid Region

If large bodies ever formed in the asteroid zone, they were removed early in the solar system's history. The process was studied by *Chambers and Wetherill* (2001), who integrated orbits of systems of massive bodies subjected to perturbations by Jupiter (and Saturn in most cases). They started with systems of 16 to 204 bodies comprising a swarm with total mass 0.6–5 $M_\oplus$. In each case the bodies initially had identical masses, ranging from 0.017 to 0.33 $M_\oplus$, and collisions were assumed to result in accretion. Detailed results varied with initial conditions and the stochastic nature of encounters, but typically gravitational scattering among the bodies would place some of them into various resonances with the giant planets. They developed large eccentricities, resulting in close encounters with Jupiter, collisions with the terrestrial planets, or impacts onto the Sun. In a majority of the cases, no massive bodies were left in the asteroid region beyond 2 AU. The clearing process took from a few million to a few hundred million years, with the timescale weakly dependent on the initial masses, and strongly dependent on the semimajor axes and orbital eccentricities of the giant planets, which were varied in their simulations.

Chambers and Wetherill (2001) examined only the fates of massive bodies much larger than the asteroids, but any initial population of smaller bodies in that region would have been subject to similar gravitational scattering, and would have met a similar fate. While there is a reasonable probability that all the large embryos would be removed, the smaller bodies are more likely to leave some survivors in stable orbits, by reason of their greater initial number (*Petit et al.,* 2001). We illustrate this in Fig. 3, where we take the size distribution of Fig. 1 and multiply the number of bodies in each mass bin by a factor of 0.01; where less than one body remains, the population is set to one body if the fractional value exceeds a random number uniformly distributed between 0 and 1. Before depletion, there were 101 bodies with masses >2 × 10^{23} g. After depletion, there are seven bodies remaining with masses ~10^{23}–10^{24} g, corresponding to diameters of about 410–740 km. The

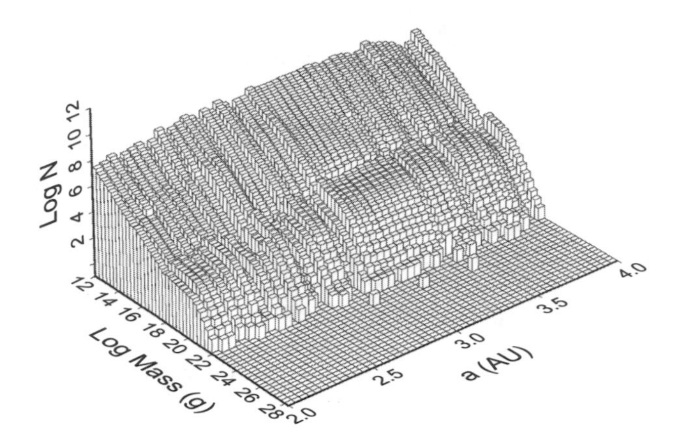

Fig. 3. The mass distribution produced by depleting the population from Fig. 1 to 1% of its value by number, as described in the text. The largest remaining body has mass ~5×10^{23} g, diameter ~740 km. The total mass of the remnant swarm is 0.023 $M_{\oplus}$.

depletion is by number, not mass, so proportionally more mass is removed with the larger bodies. The final mass of the swarm is ~0.023 $M_{\oplus}$, about 50× that of the present belt. A similar depletion to 0.001× the initial number leaves the largest bodies of mass 5×10^{22} g, diameter ~325 km. These examples show that accretional growth followed by removal of ~99–99.9% of the embryos and planetesimals produces a reasonable analog of the asteroid belt.

4.4. Consolidation and Lithification of Meteorites

Most meteorites are fairly strong rocks (no doubt with selection effect due to survival of atmospheric entry). Measured porosities of chondrites are typically less than 10%, with little correlation of porosity with metamorphic grade (*Consolmagno et al.,* 1998). This result suggests that lithification was not simply due to heating within parent bodies, but involved pressure as well. However, lithostatic pressures within asteroid-sized parent bodies are not sufficient for compression and lithification. The mean internal pressure in a body of density ρ_s and radius r due to self-gravity (half the maximum value) is $\pi G \rho^2 r^2/3$, or ~1 MPa (10 bars) within a body with radius 100 km. This is comparable to pressures a few tens of meters below Earth's surface; much higher pressures are involved in lithification of sediments.

The first kilometer-sized planetesimals were presumably porous aggregates. If they were assembled by collapse of self-gravitating condensations (*Cuzzi and Weidenschilling,* 2006), they would have been assembled very gently and have high porosities, at least a few tens of percent. If instead they formed by collisional accretion that was driven by differential aerodynamic drag, relative velocities would cause impacts at speeds of tens of meters per second. The pressures associated with such impacts would produce pressures ~density × V^2, ~10^7 dyn cm^{-2} (1 MPa). Higher pressures (~GPa) can be developed by impacts at velocities approximately a few tenths of kilometers per second, consistent with stirring by embryos produced by runaway growth. Simulations typically show eccentricities ~$1–2 \times 10^{-2}$ for the smaller bodies stirred by embryos and damped by nebular gas drag, without any outside influence by Jupiter. With V_K ~ 19 km s^{-1} at 2.5 AU, impact velocities are ~0.2–0.5 km s^{-1} (several times the escape velocity from a 100-km-diameter asteroid, or ~0.1× the mean impact velocity at present time). The corresponding impact-produced pressures are ~0.1 GPa, comparable to the central pressure in Ceres, but still lower than needed to produce impact melting (*Bischoff et al.,* 2006). It seems plausible that collisional compression was a factor, and possibly dominant, in lithification of meteorites. *Consolmagno et al.* (1998) find that measured porosities of meteorites decrease with increasing shock state, consistent with lithification by impacts. The densities of asteroids for which estimates are available are typically less than that of meteorites, suggesting that they have macroprosity, i.e., extensive cracks, voids, and/or "rubble pile" structure on larger scales. This structure is presumably due to more energetic impacts that occurred after velocities in the asteroid belt were pumped up to their present values, giving mean impact velocities of several kilometers per second.

4.5. Late Collisional Evolution and Depletion of Small Asteroids

The asteroid belt continued to evolve, albeit more slowly, after removal of large bodies and decimation of the smaller ones. Models of collisional evolution of the asteroid belt over the ~4.5 G.y. after velocities were stirred up (*Davis et al.,* 1989) suggest that the "initial" mass after the present high-velocity regime was established was not more than about 5× the present value; this constraint is required for preservation of Vesta's basaltic crust, and is consistent with the observed number of Hirayama families. The collisional loss was size-dependent; the evolutionary models of *Davis et al.* (1989) indicate that primordial asteroids with sizes ≤30 km diameter would have been ground down to dust over the age of the solar system. The present population of small asteroids consists of fragments of formerly larger bodies. Thus, all meteorites that are now being delivered to Earth were once part of bodies that attained sizes at least a few tens of kilometers in diameter. This result appears to explain why nearly all meteorites show evidence of thermal metamorphism (and why aqueous alteration is common). Below a certain size, planetesimals would not have been subject to significant heating by ^{26}Al; their compositions would presumably have been primitive and volatile-rich. This "pristine" material was selectively removed by collisions that depleted the small end of the size distribution, after scattering removed the large protoplanetary embryos. A few large asteroids with diameters >200 km have been disrupted by collisions, producing the prominent Hirayama families. Most of our samples of large asteroids may be from these disrupted parent bodies, rather than from those that are still intact. The fairly high escape velocities from the few largest intact asteroids probably limit their contribution to the meteorite flux. The exception is Vesta, which experienced a very large and statistically unlikely basin-forming impact, producing a family of small fragments that supply the HED meteorites. Ceres is unlikely to be a source of meteorites delivered to Earth. Thus, most meteorites come from bodies that attained a restricted range of sizes, neither too large nor too small.

5. WHERE DO CALCIUM-ALUMINUM-RICH INCLUSIONS AND CHONDRULES FIT IN?

In this section we return to the earliest stage of planetesimal formation, looking more closely at the implications of the most basic properties of the primitive meteorites. Specifically, we focus on the apparent existence of a 1–3-m.y. interval between the formation of (most) CAIs and the production of chondrules and the accretion of chondrite parent bodies. This hiatus is supported by the consistency of radioisotope dating of meteorite constituents (*Russell et al.,* 2006), and constraints posed by models of thermal evolution and melting of planetesimals (*Ghosh et al.,* 2006). The combination of this hiatus with the need to explain how millimeter- to centimeter-sized particles can be stored for millions of years is a fundamental constraint on models of accretion.

In a laminar nebula, planetesimals would form quickly by coagulation and settling (*Weidenschilling,* 1980; *Weidenschilling and Cuzzi,* 1993). If the laminar stage began in the CAI epoch, these planetesimals would have contained short-lived radioisotopes in abundance, with observational consequences due to internal heating. These early planetesimals might have been collisionally recycled later, due to Jupiter's influence. Or, turbulence could have delayed their formation by preventing coagulation until turbulent velocities of particles in the gas fell below a critical threshold, after which rapid accretion proceeded in a laminar environment. Alternatively, turbulence could have persisted throughout the accretion stage, but only after 1–3 m.y. did conditions become suitable for concentrating chondrule-sized particles within eddies to densities sufficient to initiate collapse into actual planetesimals (*Cuzzi et al.,* 2001; *Cuzzi and Weidenschilling,* 2006). Formation in a laminar environment, whether early or delayed, probably implies that planetesimals were produced in a brief episode; formation in ongoing turbulence may imply a more extended process in which planetesimals were produced over an interval of 1 m.y. or longer. These scenarios have different implications for the preservation of CAIs and chondrules, and their incorporation within meteorites.

5.1. Delayed Planetesimal Formation in Ongoing Turbulence

Cuzzi et al. (2003) describe how outward transport by turbulent diffusion along a concentration gradient can solve the "problem" of retaining early high-temperature condensates (CAIs and the like) against gas-drag drift into the Sun for several million years. A planetesimal formation scenario sketched by *Cuzzi and Weidenschilling* (2006) envisions accretion beyond a meter or so in size being thwarted by this ongoing turbulence, at least in the inner solar system, for 1–3 m.y., until conditions became right for chondrule formation to begin (i.e., melting of free-floating nebula particles by some unspecified mechanism). Due to a combination of the changing aerodynamic properties of solid particles as a result of their melting, and the evolving nebula gas density and turbulent intensity, turbulent concentration may have allowed accretion to leapfrog the meter-size barrier and produce sizeable planetesimals for the first time (at least in the inner solar system). The specifics of this process remain only dimly perceived, and obstacles remain (discussed by *Cuzzi and Weidenschilling,* 2006).

If CAIs were produced by condensation and subsequent melting of refractory minerals in the hot inner part of the nebula, then less-refractory particles should have formed contemporaneously at slightly greater heliocentric distances, where the nebula was cooler. In a turbulent nebula, some of this material would also diffuse outward. One would expect to find in meteorites other, less-refractory objects roughly the size of CAIs, with similar ages. It is implausible that they all were lost, if CAIs were preserved. Amoeboid olivine aggregates (AOAs) are intermediate in chemistry between CAIs and chondrules. Few ages have been measured for AOAs; they appear to be older than chondrules (based on ^{26}Al), but as yet none have been found to be as old as CAIs (*Krot et al.,* 2004). However, if CAIs and AOAs formed at slightly different temperatures in adjacent regions of a turbulent nebula and diffused outward, one would expect their ages to have the same distribution. As one indication that they may be nearly contemporary, one might consider the similarity of their O isotopes to those of CAIs (*Hiyagon and Hashimoto,* 1999). If no other objects are found to be as old as CAIs, and CAIs were preserved by outward transport alone rather than stored in first-generation parent bodies, one possibility is that their outward transport, at least its initial stage, involved some mechanism other than turbulent diffusion through the disk, e.g., ejection to large heliocentric distances by a stellar wind (*Shu et al.,* 1997, 2001). Direct outward transport by such a wind into a nonturbulent nebula, however, leaves ejected particles vulnerable to short inward drift timescales, and would have to be repeated many times to preserve CAIs for several million years. Calcium-aluminum-rich inclusions may have been melted by exposure to high ambient temperatures in the inner nebula, as the duration of their exposure to high temperatures is not tightly constrained. Chondrules, of course, do not fit into this sequence, as their mineralogy and compositions indicate that they were melted by brief heating events (*Hewins and Radomsky,* 1990).

If turbulence delayed planetesimal formation, the ages of CAIs and chondrules suggest that it had to be sustained for >1 m.y. If diffusion was effective in transporting CAIs outward for distances of at least 2–3 AU, one might expect them to be well mixed during that process, except perhaps for some sorting by size (larger ones would have higher drift rates superposed on the diffusion, and be transported outward less effectively). Rapid coagulation of planetesimals after turbulence died out would tend to produce bodies with an indiscriminate mixture of components of all ages. In that case, it is hard to account for the variety of meteorite types, or the tendency of chondrules and matrix to have complementary compositions (*Wood,* 1985).

The diversity of characteristic chondrite types, with well-defined properties, might be more easily explained by continual formation of planetesimals over an extended interval, rather than a single episode after chondrule formation was completed. If different types of CAIs and/or chondrules were produced at different times, rather than at different locations in the nebula, and planetesimal formation occurred continually, e.g., by turbulent concentration, over an interval of >1 m.y., then planetesimals that formed at different times would have varied compositions/textures. It seems necessary to assume that planetesimals did not begin to form until after CAI production was complete. If there was a gap between the production of CAIs and chondrules, but planetesimal formation was a continuous process, then one might ask why we do not find meteorites that contain CAIs but lack chondrules. In the context of turbulent concentration, their absence might be explained by the lower abundance of highly refractory material; condensations might not attain a critical density for collapse until chondrules were added to the mix. However, in a turbulent environment one would expect that any chondrite produced in this way would contain a mixture of all types of CAIs and chondrules that were produced in the nebula up to the time that it accreted; the youngest component would define the age of the planetesimal. The exception would be preferential loss of larger objects (i.e., type-B CAIs, which are only found in CV chondrites) by inward drift overcoming diffusive mixing (*Cuzzi et al.,* 2003). And indeed, with the exception of type-B CAIs, the mix of CAI types from one chondrite class to another appears to differ primarily in chemistry that may reflect alteration in the nebula and/or in parent bodies under somewhat different conditions (*Brearley and Jones,* 2000).

The model of planetesimal formation by turbulent diffusion and concentration is not linked to a specific heating mechanism that melted chondrules. However, if chondrules formed before planetesimals (which may have formed as a consequence), then their abundance in meteorites would be indicative of their abundance at least throughout the source region in the asteroid belt, and probably throughout the inner solar system. The total mass of silicates that had to be melted and converted to chondrules would be of the order of a few Earth masses. This requirement places severe demands on the heating mechanism and energy source for melting the chondrule precursors. The total energy required to melt 1 $M_\oplus$ of silicate material is ~3×10^{38} erg (*Levy,* 1988); since this energy could not be applied efficiently only to the silicates [some chondrules show evidence for multiple heating events (*Rubin and Krot,* 1996), and gas was presumably heated as well], the energy source would have to be correspondingly greater. A simple estimate of the energy content of a nebular shock such as now being actively modeled in the context of melting chondrules (*Desch and Connolly,* 2002), with velocity V_s ~ 8× sound speed, volume $H^2 \times a$ (H = scale height), and postshock density 10^{-9} g cm^{-3} is ~10^{40} erg.

Intrinsic gravitational instabilities in the nebula might be responsible for such shocks, but that mechanism would presumably be more effective during the early evolution of the nebula. In that case, at least some chondrules should be as old as CAIs. Gravitational instability and resulting shocks might be triggered by cooling of the nebula, if such cooling could be delayed for ~1 m.y. Another possibility — which has the dual advantage of meeting both time and energy constraints — is shocks produced by density waves in the solar nebula due to Jupiter's perturbations after the giant planet reached its present mass. It remains to be demonstrated that such shocks would be intense enough to produce the requisite heating, yet sufficiently brief to be consistent with chondrule cooling rates, and that the nebula could sustain such perturbations for approximately million-year timescales without driving gas out of the asteroid region. If the onset of chondrule formation was delayed by ~1 m.y. after CAIs (i.e., after the formation of the nebula), this would imply that Jupiter formed by core accretion. That mechanism requires planetesimals to form and accrete into a massive (~10 $M_\oplus$) core that can then accrete gas (*Pollack et al.,* 1996). Such a scenario would imply that planetesimals could form easily beyond the nebular "snow line," but were inhibited in the inner nebula. This could be explained by some combination of radial variation of turbulent intensity and/or enhancement of solids outside the snow line (*Stevenson and Lunine,* 1988; *Cuzzi and Zahnle,* 2004). However, as mentioned above (section 4.2), the sequence Jupiter–chondrules–asteroids has problems accounting for the present size distribution in the main belt.

5.2. Early Planetesimal Formation with Recycling

The apparent disagreement of the short timescale for planetesimal formation and loss into the Sun by gas drag with the measured ages of CAIs and chondrules led *Weidenschilling et al.* (1998) to suggest that chondrites are derived from a second generation of planetesimals. They proposed that kilometer-sized (or larger) planetesimals formed early, before chondrules were produced. These bodies, which were large enough to avoid orbital decay due to drag, incorporated CAIs, and so prevented their loss into the Sun. Larger bodies (embryos) accreted in the asteroid zone, as well as in the terrestrial planet region. After ~1–2 m.y. a "cataclysm" occurred, in which first-generation bodies were broken up by collisions, and dust and CAIs were released into the solar nebula. Shock waves melted the dust (or other debris in the approximately millimeter size range), producing chondrules, which then accreted onto the surfaces of existing planetesimals or formed new ones. These events were assumed to be triggered by the formation of Jupiter, specifically by rapid accretion of gas onto a massive (~10 $M_\oplus$) core that formed beyond the nebula's "snow line." As mentioned above, Jupiter's formation by this mechanism seems to be the only event that could have both this timing and such significant consequences within the solar nebula.

Weidenschilling et al. (1998) and *Marzari and Weidenschilling* (2002) showed that rather small planetesimals, with diameters only a few tens of kilometers, could be trapped in jovian resonances (particularly the 2:1 commensurability), and have eccentricities pumped up to ~0.5, despite the damping effect of nebular gas drag. These resonant bodies would move supersonically through the gas at speeds up to 6–8 km s^{-1}, producing bow shocks with Mach numbers ~5 (depending on the local gas temperature and orbital true anomaly). Such shocks would be adequate to melt small particles, producing chondrules (and resetting ages of those CAIs that passed through the shocks). Dust, CAIs, and chondrules were assumed to reaccrete onto small (asteroid-sized) planetesimals that were not in resonances, and so were damped by drag to low eccentricities. Eventually the nebula dissipated, and large embryos were removed by the Wetherill-Chambers scattering. Most of the smaller bodies were also removed by this process, as described in section 4.3.

As originally envisioned by *Weidenschilling et al.* (1998), the shock waves would be produced by rather small bodies, a few tens to hundreds of kilometers in diameter. They assumed these were brought to the resonance from larger heliocentric distances by orbital decay due to gas drag, which favors transport of smaller bodies (although larger bodies, once trapped, would remain in the resonance for longer times). One advantage of this small-body resonant trapping is that the resonant bodies tend to have low inclinations, keeping them near the nebular midplane where the density of chondrule precursors would be highest. However, it was shown by *Ciesla et al.* (2004) that shock waves from asteroid-sized bodies would give cooling times for chondrules that are too short to be consistent with petrographic evidence. Although the cooling times depend on the amount of fine dust present [contrary to some previous assumptions, less dust results in slower cooling (*Desch and Connolly,* 2002)], it appears that larger shocked regions would be needed, as would be produced by bodies >1000 km in size. As discussed above (section 4.2), runaway growth produces protoplanetary embryos of the required size. The orbital integrations of *Chambers and Wetherill* (2001) show that such bodies can be scattered into resonances, and attain high eccentricities (as large as 0.8) on timescales less than 1 m.y. (from the start of their integrations, which may be taken to be the formation of Jupiter). If the solar nebula persisted in the asteroid region for a few million years after Jupiter's formation, the scattered embryos would produce bow shocks of the appropriate size and strength to melt chondrules. In most of the simulations by Chambers and Wetherill, the timescale for clearing the embryos exceeds 10 m.y.; the duration of chondrule formation would reflect the lifetime of the nebula, rather than that of the embryos.

If such large bodies were the source of shocks that melted chondrules, the fate of the chondrules and other debris is not clear. To produce chondrites, the small particles would have to accrete into asteroid-sized bodies, or be accreted by preexisting planetesimals. Small particles would settle toward the central plane of the nebula, where they could be swept up by nonresonant planetesimals, if these were sufficiently damped to low eccentricities by gas drag. As the chondrules and dust would be coupled to the gas, this sweeping up would be dominated by smaller bodies that comprise most of the surface area, rather than by the gravity of the larger embryos. The collection efficiency would be influenced by the gas flow around the accreting bodies, allowing the possibility of aerodynamic fractionation of chondrules from dust, as originally suggested by *Whipple* (1972). As a result, chondrites would represent a (possibly quite thick) veneer on the surfaces of preexisting objects having more primordial constituents in their centers; that material would likely have been heated strongly, or perhaps melted. As the large embryos would not have been formed from or converted to chondrules, the total mass of chondrules that would have to be produced by this mechanism is much less than the total mass of solids originally present in that region of the solar nebula; perhaps a few percent of 1 $M_\oplus$.

Some outstanding issues with this model involve the dynamics of asteroid-sized planetesimals in such an environment. There must be enough disruptive collisions to liberate stored CAIs from the first-generation bodies and provide the raw material for chondrules, yet the debris must be accreted efficiently. It is far from certain that the smaller bodies would have orbits with sufficiently low eccentricities for such accretion, when subject to scattering by massive embryos moving on eccentric orbits. Heating and melting of the debris must have been quite efficient in order to produce chondrules, but a substantial fraction of CAIs must escape remelting or heating to temperatures that would erase their $^{26}Mg^*$ anomalies. This is necessary for any chondrule-formation scenario, but high efficiency may be hard to achieve if the heating is done by only a few large, highly eccentric bodies. Detailed quantitative modeling is needed to determine if another heating mechanism, e.g., nebula-scale shocks, is necessary.

Another potential problem is the inferred cooling times for chondrules. While large embryos could produce shocks with the requisite size scale, they might be accompanied by smaller bodies; if they also had bow shocks capable of producing chondrules, these would have unacceptably short cooling times. Trapping in the 2:1 resonance is fairly efficient for bodies with low eccentricities that are brought to the resonance by drag (*Marzari and Weidenschilling,* 2002), but may be less likely when they are subjected to perturbations by other, more massive bodies. Velocity impulses due to gravitational scattering by embryos may move them out of resonances before they attain eccentricities sufficient to produce strong shocks. However, scattering by embryos would also decrease the ability of smaller bodies to sweep up the chondrules. Another possibility is that collisions among small bodies were too frequent to allow them to attain such high eccentricities while in resonances.

One of the implications of this model for chondrule production is that velocities were not uniform within the asteroid region, nor did they increase monotonically as larger bodies accreted. Rather, the region may have simultaneously contained bodies with high eccentricities in jovian resonances and also near-circular orbits of nonresonant bodies damped by gas drag (*Weidenschilling et al.,* 2001), and/or the numbers and orbits of the large resonant embryos may have varied with time, allowing periods dominated by disruption and melting to alternate with relatively quiescent periods of accretion. This mixture allows the possibility of destructive hypervelocity impacts occurring with accretional growth, and may explain some anomalous textures, compositional mixtures, or clasts within some meteorites. The existence of high impact speeds before or during chondrule formation is implied by the presence of fragments of shattered chondrules within chondrules. Particles of this size coupled well to the gas, making it difficult for turbulence to produce relative velocities sufficient to shatter them. On the other hand, the shocks that melted chondrules may have been the most intense of a distribution, with weaker events inducing collisions at velocities sufficient to fragment them while floating freely in the nebula. Collisional experiments by *Ueda et al.* (2001) show that impact velocities of tens of meters per second are needed to shatter chondrules; relative velocities of this magnitude imply much higher mean velocities. With recycled planetesimals, as is also the case for delayed planetesimal formation, there should have been some bodies produced that contained CAIs but no chondrules. Furthermore, there should have been some first-generation bodies that melted because of, or retained other evidence of, a high level of ^{26}Al, which would have been fragmented and reaccumulated during this stage. It is hard to understand why no such material is recognized among the inventory of meteorite types (unless irons, stony irons, and some achondrites had such an origin).

One potentially serious problem with producing chondrules from recycled planetesimals is the thermal evolution of the assumed first-generation bodies. If CAIs were preserved within them against loss by gas drag, those bodies had to be at least kilometer-sized. However, they must not have grown much larger, or internal heating would have been excessive. As mentioned above (section 3), the initial abundance of ^{26}Al may have been sufficient to melt planetesimals that grew larger than a few tens of kilometers in diameter within the first 1–2 m.y. after CAIs formed, or at least to erase isotopic anomalies in the CAIs. This constraint, from thermal models by *LaTourette and Wasserburg* (1998), needs to be reexamined. They computed temperature profiles for bodies of various sizes, assuming that they formed instantaneously, with ^{26}Al abundances characteristic of the time of their formation. Bodies that accreted over an interval of time would have had more heating in the early-formed central regions, but large accretionary collisions could have stirred and mixed this material. LaTourette and Wasserburg assumed planetesimals to be dense (3.3 g cm^{-3}), with thermal properties of igneous rock. If ice was condensed in the asteroid region, it would tend to lower peak temperatures by its large heat capacity and dilution of ^{26}Al. However, hydration reactions are exothermic; the net effect would depend on whether hydrothermal convection could effectively transport heat, i.e., on the sizes and internal structure of the planetesimals *(McSween et al.,* 2002). *Bischoff* (1998) has reviewed mineralogical evidence for aqueous alteration of carbonaceous chondrites, and concluded that it occurred within small precursor planetesimals, which were later destroyed by collisions, with the debris accreted into meteorite parent bodies.

It seems likely that the first-generation bodies would have been porous "rubble piles." Their lower density would yield a lower concentration of heat sources, but their thermal conductivity would also be lower, presumably yielding higher internal temperatures. Macroporosity could have had a significant effect on their thermal evolution. *Wurm et al.* (2004) computed rates of gas diffusion through porous planetesimals, where the flow is driven by the ram pressure of orbital motion in a nebula that is supported by a radial pressure gradient (typical velocity of a planetesimal relative to the nebula would be ~50 m s^{-1}). One can adapt their analysis to show that microporosity between millimeter-sized grains is not sufficient to allow significant gas flow within kilometer-sized or larger bodies. However, macroporosity with channels of diameter ~$10^{-3}\times$ that of the planetesimal (e.g., meter-sized voids within a kilometer-sized body) would allow flow at rates that could produce significant cooling (i.e., the mass of gas flowing through the channels during one half-life of ^{26}Al would be comparable to that of the planetesimal).

As mentioned above (section 4.2), runaway growth of embryos tends to maintain a population of small bodies. The size distribution shown in Fig. 3, after 2 m.y. of accretion with fragmentation, followed by depletion of 99% of bodies in the asteroid belt, yields median masses for the remaining bodies corresponding to diameters ~65 km in the inner belt to ~32 km in the outer zones. The initial planetesimal size was taken to be 10 km, but the median size of the remnant is similar if the simulation is started with 1-km bodies; runaway growth begins sooner, but the embryos' perturbations cause accretion to slow at about the same size distribution. These sizes for the survivors could allow a significant fraction of their interiors to remain below the temperature for isotopic resetting; this would be aided if the start of accretion was delayed by turbulence for 10^5 yr or more after CAI formation. Also, smaller (and therefore less heated) bodies would be more easily disrupted to provide raw material for chondrule formation and accretion of second-generation bodies. It is not clear whether this scenario is consistent with the amount of melting required to account for iron meteorites and other igneous types. *Ghosh et al.*

(2000) note that preliminary results of a thermal model of Vesta call for starting its accretion from a kilometer-sized seed 2 m.y. after CAI formation. Also, *Sugiura and Hoshino* (2003) modeled the formation of the parent body of the IIIAB iron meteorites with accretion beginning at least 1.7 m.y. after CAI formation. Clearly, more detailed and realistic thermal models are needed, as well as improved understanding of collisional evolution.

6. SUMMARY AND FUTURE WORK

Meteoritic evidence shows that the parent bodies in the asteroid belt experienced complex histories of accretion, alteration, and disruption. The sequence and timing of these events are unclear, and their evolution may have included multiple episodes of each type. Plausible dynamical models now exist for accretion, excitation, and depletion of the asteroid belt, but they still appear at odds with the measurable properties of the meteorites themselves, particularly the age difference between CAIs and chondrules. We have sketched two possible scenarios for this difference, involving delayed formation of the first planetesimals or their recycling into the parent bodies of the present specimens. Both models contain serious gaps, and neither is likely to be correct in detail. Alternatives and compromise scenarios are possible. If accretion was delayed, and chondrules preceded planetesimals, then bow shocks would be excluded as their heating mechanism. With recycling, bow shocks are allowed but not required; nebular shocks due to Jupiter-produced density waves might be a viable alternative.

It cannot be denied that Jupiter has been the dominant influence on the asteroid belt over the history of the solar system. We need better understanding of the mechanism and timing of Jupiter's formation, as well as its orbital history (did its eccentricity and semimajor axis evolve with time?). More modeling is needed of clearing of the asteroid region by scattering of bodies into jovian resonances, with orbital and collisional evolution of small bodies, as well as the large ones that can be tracked directly by orbital integrations, including effects of nebular gas drag.

The most important observational constraint on the formation of chondrites is the ages of CAIs and chondrules. Specifically, their formation mechanisms may be explained, or at least some mechanisms eliminated, if the following questions are answered: Did formation of CAIs and chondrules overlap in time, or is there a real gap between them? What is the range of ages for each; i.e., how long did the process(es) that produced them operate in the solar nebula? Are these ages and/or ranges different in different types of meteorites? Are there other components (e.g., AOAs) that are intermediate in age?

Models of planetesimal formation by collisional coagulation need to be extended to a dynamically evolving turbulent solar nebula, rather than the static laminar environments that have been studied to date. Both coagulation and turbulent concentration would produce planetesimals continuously over some interval; the former after decay of turbulence below some critical value, the latter during the lifetime of turbulence in the nebula, or at least for some interval while the strength of turbulence remained within some range. Most models of planetary accretion assume starting conditions with the full complement of planetesimals present at a well-defined starting time, and growth commencing from that state. The present results from such models are consistent with the start of accretion 1–2 m.y. after CAI formation (*Ghosh et al.,* 2003, 2006), which appears to favor a delay in planetesimal formation. There is a need for additional studies like that of *Wetherill and Inaba* (2000), in which planetesimals are introduced gradually rather than instantaneously. Similarly, thermal modeling of planetesimals due to heating by ^{26}Al needs to be done with bodies growing over time, at rates consistent with dynamical accretion models, including fragmentation in large collisions and "rubble pile" structure of the accreting bodies. Such models are needed to test the constraints on planetesimal size vs. formation time in order to be consistent with preservation and/or resetting of isotopic anomalies in CAIs, as well as thermal metamorphism within meteoritic parent bodies.

Acknowledgments. This work was supported by grants from the NASA Planetary Geology and Geophysics and Origins of Solar Systems programs.

REFERENCES

Amelin Y., Krot A. N., Hutcheon I., and Ulyanov A. A. (2002) Lead isotopic ages of chondrules and calcium-aluminum-rich inclusions. *Science, 297,* 1678–1683.

Binzel R. P., Gehrels T., and Matthews M. S., eds. (1989) *Asteroids II.* Univ. of Arizona, Tucson. 1258 pp.

Bischoff A. (1998) Aqueous alteration of carbonaceous chondrites: Evidence for preaccretionary alteration — A review. *Meteoritics & Planet. Sci., 33,* 1113–1122.

Bischoff A., Scott E. R. D., Metzler K., and Goodrich C. A. (2006) Nature and origins of meteoritic breccias. In *Meteorites and the Early Solar System II* (D. S. Lauretta and H. Y. McSween Jr., eds.), this volume. Univ. of Arizona, Tucson.

Brearly A. J. and Jones R. H. (2000) Chondritic meteorites. In *Planetary Materials* (J. Papike, ed.), pp. 1–189. Reviews in Mineralogy, Vol. 36, Mineralogical Society of America.

Cassen P. and Moosman A. (1981) On the formation of protostellar disks. *Icarus, 48,* 353–375.

Chambers J. E. and Wetherill G. W. (2001) Planets in the asteroid belt. *Meteoritics & Planet. Sci., 36,* 381–399.

Ciesla F. J., Hood L. L., and Weidenschilling S. J. (2004) Evaluating planetesimal bow shocks as sites for chondrule formation. *Meteoritics & Planet. Sci., 39,* 1809–1821.

Consolmagno G. J., Britt D. T., and Stoll C. P. (1998) The porosities of ordinary chondrites: Models and interpretation. *Meteoritics & Planet. Sci., 33,* 1221–1229.

Cuzzi J. N. and Hogan R. C. (2003) Blowing in the wind I. Velocities of chondrule-sized particles in a turbulent protoplanetary nebula. *Icarus, 164,* 127–138.

Cuzzi J. N. and Weidenschilling S. J. (2006) Particle-gas dynamics and primary accretion. In *Meteorites and the Early Solar System II* (D. S. Lauretta and H. Y. McSween Jr., eds.), this volume. Univ. of Arizona, Tucson.

Cuzzi J. N. and Zahnle K. J. (2004) Material enhancement in protoplanetary nebulae by particle drift through evaporation fronts. *Astrophys. J., 614,* 490–496.

Cuzzi J. N., Hogan R. C., Paque J. M., and Dobrovolskis A. R. (2001) Size-selective concentration of chondrules and other small particles in protoplanetary nebula turbulence. *Astrophys. J., 546,* 496–508.

Cuzzi J. N., Davis S. S., and Dobrovolskis A. R. (2003) Blowing in the wind II. Creation and redistribution of refractory inclusions in a turbulent protoplanetary nebula. *Icarus, 166,* 385–402.

Davis D. R., Weidenschilling S. J., Farinella P., Paolicchi P., and Binzel R. P. (1989) Asteroid collisional history: Effects on sizes and spins. In *Asteroids II* (R. Binzel et al., eds.), pp. 805–826. Univ. of Arizona, Tucson.

Desch S. and Connolly H. (2002) A model for the thermal processing of particles in solar nebula shocks: Application to cooling rates of chondrules. *Meteoritics & Planet. Sci., 37,* 183–207.

Ghosh A., Weidenschilling S. J. and McSween H. Y. Jr. (2000) An attempt at constraining early solar system chronology using thermal accretionary models (abstract). In *Lunar and Planetary Science XXXI,* Abstract #1845. Lunar and Planetary Institute, Houston (CD-ROM).

Ghosh A., Weidenschilling S. J., and McSween H. Y. Jr. (2003) Importance of the accretion process in asteroid thermal evolution: 6 Hebe as an example. *Meteoritics & Planet. Sci., 38,* 711–724.

Ghosh A., Weidenschilling S. J., McSween H. Y. Jr., and Rubin A. (2006) Asteroidal heating and thermal stratification of the asteroid belt. In *Meteorites and the Early Solar System II* (D. S. Lauretta and H. Y. McSween Jr., eds.), this volume. Univ. of Arizona, Tucson.

Grimm R. E. and McSween H. Y. Jr. (1993) Heliocentric zoning of the asteroid belt by aluminum-26 heating. *Science, 259,* 653–655.

Grossman J. N., Rubin A. E., Nagahara H., and King E. A. (1988) Properties of chondrules. In *Meteorites and the Early Solar System* (J. Kerridge and M. S. Matthews, eds.), pp. 619–659. Univ. of Arizona, Tucson.

Hewins R. H. and Radomsky P. M. (1990) Temperature conditions for chondrule formation. *Meteoritics, 26,* 309–318.

Hiyagon H. and Hashimoto A. (1999) ^{16}O excesses in olivine inclusions in Yamato-86009 and Murchison chondrites and their relation to CAIs. *Science, 283,* 828–831.

Kerridge J. F and Matthews M. S., eds. (1988) *Meteorites and the Early Solar System.* Univ. of Arizona, Tucson. 1269 pp.

Kokubo E. and Ida S. (1998) Oligarchic growth of protoplanets. *Icarus, 131,* 171–178.

Krot A. N., Petaev M. I., Russell S. S., Itoh S., Fagan T. J., Yurimoto H., Chizmadia L., Weisberg M. K., Komatsu M., Ulyanov A. A., and Keil K. (2004) Amoeboid olivine aggregates and related objects in carbonaceous chondrites: Records of nebular and asteroidal processes. *Chem. Erde, 64,* 185–239.

LaTourette T. and Wasserburg G. J. (1998) Mg diffusion in anorthite: Implications for the formation of early solar system planetesimals. *Earth Planet. Sci. Lett., 158,* 91–108.

Levy E. H. (1988) Energetics of chondrule formation. In *Meteorites and the Early Solar System* (J. Kerridge and M. S. Matthews, eds.), pp. 697–711. Univ. of Arizona, Tucson.

Lissauer J. J. and Stewart G. R. (1993) Growth of planets from planetesimals. In *Protostars and Planets III* (E. H. Levy and J. I. Lunine, eds.), pp. 1061–1088. Univ. of Arizona, Tucson.

MacPherson G. J., Davis A. M., and Grossman J. N. (1995) The distribution of aluminum-26 in the early solar system: A reappraisal. *Meteoritics, 30,* 365–386.

Marzari F. and Weidenschilling S. J. (2002) Mean motion resonances, gas drag, and supersonic planetesimals in the solar nebula. *Cel. Mech. Dyn. Astron., 82,* 225–242.

McSween H. Y., Ghosh A., Grimm R. E., Wilson L., and Young E. D. (2002) Thermal evolution models of asteroids. In *Asteroids III* (W. F. Bottke Jr. et al., eds.), pp. 559–571. Univ. of Arizona, Tucson.

Nakagawa Y., Nakazawa K., and Hayashi C. (1981) Growth and sedimentation of dust grains in the primordial solar nebula. *Icarus, 45,* 517–528.

Petit J.-M., Morbidelli A., and Chambers J. E. (2001) The primordial excitation and clearing of the asteroid belt. *Icarus, 153,* 338–347.

Pollack J. B., Hubickyj O., Bodenheimer P., Lissauer J. J., Podolak M., and Greenzweig Y. (1996) Formation of the giant planets by concurrent accretion of solids and gas. *Icarus, 124,* 62–85.

Rubin A. E. and Krot A. N. (1996) Multiple heating of chondrules. In *Chondrules and the Protoplanetary Disk* (R. H. Hewins et al., eds.), pp. 173–180. Cambridge Univ., Cambridge.

Russell S. S., Hartmann L., Cuzzi J., Krot A. N., Gounelle M., and Weidenschilling S. J. (2006) Timescales of the solar protoplanetary disk. In *Meteorites and the Early Solar System II* (D. S. Lauretta and H. Y. McSween Jr., eds.), this volume. Univ. of Arizona, Tucson.

Ruzmaikina T. V., Safronov V. S., and Weidenschilling S. J. (1989) Radial mixing of material in the asteroidal zone. In *Asteroids II* (R. Binzel et al., eds.), pp. 681–700. Univ. of Arizona, Tucson.

Shu F. H., Shang H., Glassgold A. E., and Lee T. (1997) X-rays and fluctuating X-winds from protostars. *Science, 277,* 1475–1479.

Shu F. H., Shang S., Gounelle M., Glassgold A. E., and Lee T. (2001) The origin of chondrules and refractory inclusions in chondritic meteorites. *Astrophys. J., 548,* 1029–1050.

Stevenson D. J. and Lunine J. I. (1988) Rapid formation of Jupiter by diffusive redistribution of water vapor in the solar nebula. *Icarus, 75,* 146–155.

Stone J. M., Gammie C. F., Balbus S. A., and Hawley J. F. (2000) Transport processes in protostellar disks. In *Protostars and Planets IV* (V. Mannings et al., eds.), pp. 589–599. Univ. of Arizona, Tucson.

Sugiura N. and Hoshino H. (2003) Mn-Cr chronology of five IIIAB iron meteorites. *Meteoritics & Planet. Sci., 38,* 117–143.

Swindle T.. D., Davis A. M., Hohenberg C. M., MacPherson G. J., and Nyquist L. E. (1996) Formation times of chondrules and Ca-Al-rich inclusions: Constraints from short-lived radionuclides. In *Chondrules and the Protoplanetary Disk* (R. H. Hewins et al., eds.), pp. 77–86. Cambridge Univ., Cambridge.

Ueda T., Murakami Y., Ishitsu N., Kawabe H., Inoue R., Nakamura T., Sekiya M., and Takaoka N. (2001) Collisional destruction experiment of chondrules and formation of fragments in the solar system. *Earth Planets Space, 53,* 927–935.

Weidenschilling S. J. (1977) The distribution of mass in the planetary system and the solar nebula. *Astrophys. Space Sci., 51,* 153–158.

Weidenschilling S. J. (1980) Dust to planetesimals: Settling and coagulation in the solar nebula. *Icarus, 44,* 172–189.

Weidenschilling S. J. (1988) Formation processes and time scales for meteorite parent bodies. In *Meteorites and the Early Solar System* (J. Kerridge and M. S. Matthews, eds.), pp. 348–371. Univ. of Arizona, Tucson.

Weidenschilling S. J. and Cuzzi J. N. (1993) Formation of planetesimals in the solar nebula. In *Protostars and Planets III* (E. H. Levy and J. Lunine, eds.), pp. 1031–1060. Univ. of Arizona, Tucson.

Weidenschilling S. J., Spaute D., Davis D. R., Marzari F., and Ohtsuki K. (1997) Accretional evolution of a planetesimal swarm 2. The terrestrial zone. *Icarus, 128,* 429–455.

Weidenschilling S. J., Marzari F., and Hood L. L. (1998) The origin of chondrules at jovian resonances. *Science, 279,* 681–684.

Weidenschilling S. J., Davis D. R., and Marzari F. (2001) Very early collisional evolution of the asteroid belt. *Earth Planets Space, 53,* 1093–1097.

Wetherill G. W. (1989) Origin of the asteroid belt. In *Asteroids II* (R. Binzel et al., eds.), pp. 661–680. Univ. of Arizona, Tucson.

Wetherill G. W. (1992) An alternative model for the formation of the asteroid belt. *Icarus, 100,* 307–325.

Wetherill G. W. and Inaba S. (2000) Planetary accumulation with a continuous supply of planetesimals. *Space Sci. Rev., 92,* 311–320.

Wetherill G. W. and Stewart G. R. (1989) Accumulation of a swarm of small planetesimals. *Icarus, 77,* 350–357.

Wetherill G. W. and Stewart G. R. (1993) Formation of planetary embryos: Effects of fragmentation, low relative velocity, and independent variation of eccentricity and inclination. *Icarus, 106,* 190–209.

Whipple F. L. (1972) On certain aerodynamic processes for asteroids and comets. In *From Plasma to Planet* (A. Elvius, ed.), pp. 211–232. Wiley, New York.

Wood J. A. (1996) Unresolved issues in the formation of chondrules and chondrites. In *Chondrules and the Protoplanetary Disk* (R. H. Hewins et al., eds.), pp. 55–69. Cambridge Univ., Cambridge.

Wood J. A. (1985) Meteoritic constraints on processes in the solar nebula. In *Protostars and Planets II* (D. C. Black and M. S. Matthews, eds.), pp. 687–702. Univ. of Arizona, Tucson.

Wurm G., Paraskov G., and Krauss O. (2004) On the importance of gas flow through porous bodies for the formation of planetesimals. *Astrophys. J., 606,* 983–987.

Meteoritic Diversity and Planetesimal Formation

John Chambers
Carnegie Institution of Washington

Meteorite parent bodies vary widely in composition and structure even though they all formed in the same protoplanetary nebula over a limited range of time and space. Nebula material underwent substantial mixing at an early stage leading to approximate isotopic homogeneity in meteorites. Turbulence and gas drag transported gas, dust, and solid bodies smaller than 1 km over large radial distances. After Jupiter formed, bodies up to 100 km in size moving on eccentric orbits experienced rapid inward drifting due to gas drag. The parent bodies of most primitive meteorites required at least 2–3 m.y. to form. This provided time for diverse meteorite components (various types of chondrule, refractory inclusions, and metal grains) to form and evolve due to nebular heating events and interactions with nebular gas. Parent bodies acquired different mixtures of each component depending on when and where they accreted. Much information about where parent bodies formed has been erased by the unknown process that removed the vast majority of the primordial solid material from the asteroid belt. However, it appears that time of formation was more important than location in determining the composition of parent bodies.

1. INTRODUCTION

Meteorites and their parent bodies are a highly diverse group of objects. Different meteorite classes have a range of elemental compositions and O-isotope abundances. Meteorite parent bodies have undergone varying degrees of thermal and aqueous alteration, and experienced different collisional histories. Some meteorites formed under extremely reducing conditions while others are highly oxidized. Meteorite components range from ones that are dominated by highly refractory minerals to ones that are rich in volatiles and have never experienced high temperatures. Short-lived isotopes were clearly present when some meteorite components formed but not others.

Despite this broad diversity, meteorite parent bodies all formed in the Sun's protoplanetary nebula over a limited range of time and space. Astronomical observations, theoretical studies, and some characteristics of meteorites themselves suggest these parent bodies formed in a nebula that was turbulent and characterized by widespread mixing of material. Gas, fine dust, millimeter-sized particles, boulders, kilometer-sized planetesimals, and even large asteroids were probably transported over radial distances of an astronomical unit (AU) or more. Thus, it is something of a paradox that meteorite parent bodies came to be so different from one another. In this chapter, I will discuss the evidence for widespread mixing due to a number of physical processes, and try to reconcile this with the existence of the diverse group of meteorites we see today.

2. NEBULAR MIXING AND HETEROGENEITY

Many young stars lie at the center of disks containing gas and dust, surrounded by more tenuous envelopes of material (*Beckwith et al.,* 1990). These disks are thought to be close analogs of the Sun's protoplanetary nebula. Circumstellar disks are heated by radiation from the central star, and also by the release of gravitational energy as material moves inward through the disk and accretes onto the star. It is unclear what mechanism causes material to flow through the disk, but it is generally assumed to involve turbulence (density waves and vortices may also play a role). In very young disks, temperatures at the midplane may be high enough to vaporize common silicates as far as 3 AU from the star (*Boss,* 1993; *Woolum and Cassen,* 1999). However, this high-temperature phase is short-lived. Disks older than a few hundred thousand years probably have midplane temperatures of 200–750 K at 1 AU from the star and lower still at 2.5 AU (*Woolum and Cassen,* 1999). Temperatures may even be low enough for water ice to condense throughout this region (*Chiang and Goldreich,* 1999).

If the Sun's protoplanetary nebula was hot and turbulent early in its history, it is plausible that material would have been mixed and homogenized. There is some evidence that this happened. The Mg-isotope ratios observed on Earth, the Moon, Mars, and a group of pallasite meteorites suggest they formed in a nebula where Mg isotopes had been homogenized (*Norman et al.,* 2004). Similarly, the mix of Fe isotopes in several groups of chondritic (primitive) and achondritic (melted) meteorites can be derived from a uniform reservoir by mass-dependent fractionation (*Zhu et al.,* 2001), i.e., purely as a result of physical processes that depend on mass differences between isotopes.

The situation is more complicated for O, an element that was abundant in both solid and gaseous phases in the solar nebula. Oxygen-isotope ratios vary significantly from one meteorite group to another, and these differences cannot be explained by mass-dependent fractionation alone. Some of the variation seen in carbonaceous chondrites can be attributed to aqueous alteration in the parent bodies of these meteorites (*Clayton,* 2004). However, several O-isotopic reservoirs are still required to explain differences between the chondrite groups. Calcium-aluminium-rich inclusions (CAIs) (meteoritic components that probably formed at a very early stage) lie along a single mixing line in O-isotope phase space with a common ^{16}O-rich end member (*Clayton,* 2004). This suggests that the nebula was initially uniformly rich in ^{16}O, with the variations seen in primitive meteorites arising at a later stage.

A small fraction of grains in primitive meteorites contain highly unusual isotopic ratios in many elements (*Zinner,* 1998; *Nittler,* 2003). These isotopic mixtures cannot be explained in terms of mass fractionation. Apparently, these "presolar" grains formed outside the solar nebula and avoided homogenization with the other material in the solar system (*Nittler,* 2003).

Many CAIs once contained short-lived isotopes such as ^{41}Ca, ^{26}Al, ^{10}Be, and ^{53}Mn. The inferred initial amounts of ^{41}Ca and ^{26}Al are often correlated (*Sahijpal et al.,* 1998), suggesting they came from the same source, possibly a nearby supernova. The ^{10}Be was distributed in a different manner (*Marhas et al.,* 2002), and may have been produced in the solar nebula when material lying close to the Sun was irradiated by solar particles (*McKeegan et al.,* 2000). [See *Desch et al.* (2004) for an alternative source for ^{10}Be.] Calcium-aluminum-rich inclusions typically formed with ^{26}Al/^{27}Al ratios close to a "canonical" value of 4.5×10^{-5} (*MacPherson et al.,* 1995; *Huss et al.,* 2001). This suggests that these particles formed in a region of the nebula that was isotopically uniform in Al. A few CAIs (the "FUN" kind), and some inclusions in CH chondrites, contained little or no ^{26}Al when they formed (*Krot et al.,* 2002). FUN inclusions contain anomalous amounts of ^{50}Ti and other supernova-generated isotopes that were not produced via radioactive decay (*Clayton et al.,* 1988). It is possible that FUN inclusions predate other CAIs and formed before the nebula was homogenized.

The decay product of ^{53}Mn appears to be present in solar system bodies in amounts that correlate with distance from the Sun (*Lugmair and Shukolyukov,* 2001). This may indicate that ^{53}Mn was heterogeneously distributed in the early solar nebula. Computer simulations of the injection of supernova-generated material into a protoplanetary nebula predict an uneven distribution of material, at least initially (*Vanhala and Boss,* 2002), although it is strange that ^{26}Al was later homogenized while ^{53}Mn was not. This may indicate that ^{53}Mn, like ^{10}Be, came from a different source than ^{26}Al.

Many primitive meteorites are depleted in moderately volatile elements, compared to the Sun, in a way that correlates with each element's volatility (*Palme et al.,* 1988; *Alexander et al.,* 2001). This correlation may be due to incomplete condensation of the more volatile elements during the early hot phase of the nebula (*Cassen,* 1996). If this interpretation is correct, the various components in meteorites inherited their elemental depletion patterns from precursors that formed at an early stage, and subsequent events did not alter these patterns. In this model, the progressive depletion of volatile elements implies that solids were preferentially retained in the nebula while gas accreted onto the Sun. Recent theoretical models suggest that centimeter- to meter-sized solid objects were actually highly mobile in the nebula (*Stepinski and Valageas,* 1997). It appears that solid particles could have been lost at a *faster* rate than the gas, unless solids rapidly aggregated to form bodies that were on the order of 1 km in size.

Primitive meteorites are mostly made up of chondrules: rounded, millimeter-sized particles that are believed to have formed by flash heating in the nebula. Different types of chondrule exhibit different elemental depletion patterns. Type-I (FeO-poor) chondrules tend to be substantially depleted in moderately volatile elements, while type II (FeO-rich) chondrules are not (*Zanda,* 2004). If elemental abundance patterns are the result of condensation in a hot nebula, we would infer that the precursors of type-I chondrules formed closer to the Sun than precursors of the type-II chondrules. However, this conclusion is hard to reconcile with the traditional viewpoint that ordinary chondrite meteorites, which contain many type-II chondrules, come from the inner asteroid belt, while carbonaceous chondrites, composed largely of type-I chondrules, are thought to come from the outer asteroid belt.

An alternative interpretation is that the elemental depletions seen in chondritic meteorites were caused by partial evaporation of chondrules during the events that melted them. This idea remains controversial because chondrules show few signs of having undergone mass-dependent fractionation in elements such as K (*Humayun and Clayton,* 1995). In general, one would expect to see some mass fractionation during partial evaporation. However, this need not be the case if the chondrule-heating events took place in regions with high partial pressures of the rock-forming elements (*Alexander et al.,* 2000; *Alexander,* 2004). Thus, we can understand the lack of strong mass-dependent fractionation if chondrules formed in regions where the solid-to-gas ratio was higher than the nebula as a whole.

In most chondrites, the refractory lithophile (rock-loving) elements are present in proportions similar to those seen in the Sun (*Krot et al.,* 2004). This suggests these elements were distributed quite evenly in the nebula. However, there is some variation, particularly in the Mg/Si and Al/Si ratios of chondritic meteorites (*Drake and Righter,* 2002). Differences in the Al/Si ratio may be explained by the fact that some chondrites contain more Al-rich CAIs than others, although the correlation is not perfect (*Krot et al.,* 2004). The different Mg/Si ratios are reflected in the different amounts of olivine and pyroxene seen in various chondrules. These differences are hard to explain unless chemical reactions took place between chondrules and nebular gas (*Scott and Krot,* 2004). If we abandon the idea that chondrules were closed systems, then a chondrule's composition could have changed each time it was heated, due to partial evaporation and acquisition of new material (*Alexander,* 1996; *Sears et al.,* 1996; *Scott and Krot,* 2004). The relative volatility of Mg and Si would have fluctuated according to local nebula conditions such as the dust/gas ratio (*Alexander,* 1996, 2004). Thus, differences in the Mg/Si ratio of primitive meteorites may be explained in terms of the different histories of their components, provided that these components remained in the nebula for an extended period of time.

In a turbulent nebula, the timescale required to transport gas throughout the inner solar system would have been quite short. Theoretical models predict that water could have been removed from the entire inner nebula in $\sim 10^5$ yr by "cold trapping": a combination of turbulent diffusion of nebula gas and condensation of water ice near the "snow line" (*Stevenson and Lunine,* 1988). Nebula models suggest the snow line was located somewhere between 4 and 7 AU from the Sun during the first few hundred thousand years, when disk mass accretion rates were high (*Boss,* 1996). Later, when the nebula cooled, the snow line moved closer to the Sun. Many primitive asteroids apparently experienced aqueous alteration as a result of reactions between water ice and dry rock. This implies the asteroid belt was never completely dry. However, the abundances of highly volatile materials could have been distinctly nonsolar for much of the time that meteorite parent bodies were forming, due to cold trapping and other processes. Changes in the composition of the nebula would have led to differences in the compositions of parent bodies that formed at different times.

3. FORMATION OF METEORITE COMPONENTS

Calcium-aluminum-rich inclusions from a variety of meteorite classes have broadly similar primary mineralogies, similar O-isotope abundances, and similar inferred initial $^{26}Al/^{27}Al$ ratios (*MacPherson,* 2004). These properties suggest that many CAIs formed under similar nebula conditions over an interval of no more than a few hundred thousand years. It is currently unclear where CAIs formed. The highly refractory nature of CAIs suggests they formed in a region where the ambient temperature was high, otherwise the volatile elements could have condensed onto them. The former presence of ^{10}Be in CAIs suggests they formed close to the Sun, although this interpretation remains a matter of debate (e.g., *Desch et al.,* 2004).

Some ferromagnesian and Al-rich chondrules in carbonaceous and ordinary chondrites contained ^{26}Al when they formed (*Mostefaoui et al.,* 2002; *Kurahashi et al.,* 2004). However, the inferred $^{26}Al/^{27}Al$ ratios were substantially lower than those of CAIs. If this difference is due to the decay of ^{26}Al from an initially uniform reservoir, it means most chondrules formed about 1–3 m.y. after CAIs. This interpretation is supported by high-precision measurements using the U-Pb isotope system to estimate the age of chondrules in a CR chondrite (*Amelin et al.,* 2002). The textures of chondrules suggest they were heated to peak temperatures of around 1500–1850 K (*Hewins and Connolly,* 1996), and then cooled in a matter of hours. The short cooling timescale strongly suggests the heated regions were much smaller than the nebula itself (*Desch and Connolly,* 2002). Since chondrules are a major component of primitive meteorites, there must have been many heating events. The presence of relict fragments of type-I chondrules in type-II chondrules and vice versa (*Jones,* 1996) also argues for multiple heating episodes.

Many meteorites contain compound chondrules formed by collisions between chondrules while they were still hot. These were unlikely to have formed in large numbers unless the concentration of solid material was enhanced in the chondrule-forming regions. Models for the chemical evolution of chondrules during heating events also infer that chondrules formed in regions with dust-to-gas ratios much higher than the nebula as a whole (*Alexander,* 2004). Compound chondrules often have nonporphyritic (i.e., fully melted) textures, suggesting they achieved very high peak temperatures. This can be understood if chondrules were melted by passage through a shock wave in the nebula (*Desch and Connolly,* 2002). Shock waves are not the only possible mechanism for producing chondrules, but such an origin does appear to explain many of their properties (*Chiang,* 2002). The origin of these shock waves remains unclear, although computer models of disks thought to be similar to the Sun's protoplanetary nebula show that shock waves can occur (*Boss,* 2000).

Chondrules typically have fine-grained rims of volatile-rich dust, similar in composition to the matrix material that fills the gaps between the chondrules. The rim and matrix material contains interstellar organic compounds and circumstellar dust grains that apparently have not been heated above ~700 K since their formation (*Ott,* 1993). This means the chondrules encountered dusty nebular regions that had escaped high-temperature events before the chondrules were incorporated into larger bodies. It has been estimated that chondrules required 100–1000 yr to accrete their rims in a dusty environment (*Cuzzi,* 2004). Iron and Si are distributed in a complementary way in chondrules and matrix of CM chondrites (*Wood,* 1985). Some elements also have complementary abundances in the chondrules and matrix of CV chondrites (*Palme et al.,* 1992). The most obvious way to interpret these correlations is if chondrule precursors and matrix material coexisted in a low-temperature region of the nebula. A subset of this material was heated to form chondrules while the remainder escaped heating and was accreted subsequently.

Amoeboid olivine inclusions (AOAs) are minor components seen in many chondrites, and these have compositions intermediate between CAIs and ferromagnesian chondrules. The initial ^{26}Al abundances of AOAs suggest they formed before most chondrules and roughly 0.1–0.5 m.y. after CAIs (*Itoh et al.,* 2002). Amoeboid olivine inclusions have ^{16}O-rich compositions similar to CAIs (*Itoh et al.,* 2002), suggesting they formed in the same nebular environment, and indeed some AOAs are observed to contain relict CAIs (*MacPherson,* 2004).

From these observations, we can construct a plausible sequence of events for the formation of meteorite components. FUN inclusions formed first, the other CAIs followed soon afterward, AOAs formed a little later, and chondrules formed last of all, about 1–3 m.y. after CAIs. Chondrules, and possibly CAIs, were formed in multiple events, with each generation having a somewhat different composition. All these components underwent some spatial mixing before incorporation into parent bodies.

There were probably some exceptions to this sequence of events. At least one CAI appears to contain a relict chondrule (*Itoh and Yurimoto,* 2003), which suggests this chondrule formed before the CAI. In addition, CH chondrites and some CB chondrites do not fit comfortably into the chronological scheme outlined above. These meteorites contain zoned metal grains that appear to have condensed directly from the hot early nebula, cooled rapidly, and subsequently avoided reheating (*Meibom et al.,* 2000). Most chondrules in CH meteorites are nonporphyritic, highly depleted in volatiles, and lack fine-grained dust rims (*Krot et al.,* 2002), suggesting they also formed in a very hot nebula. Calcium-aluminum-rich inclusions and chondrules in CH chondrites seem to have a close relationship, and they may have shared a common origin (*Krot et al.,* 2002). It seems likely that these meteorites and their components formed at a time and/or place quite different from other primitive meteorites.

4. RADIAL REDISTRIBUTION OF SMALL SOLIDS

In most models for protoplanetary disks, gas pressure decreases with radial distance from the central star. As a result, the gas is partially supported against the star's gravitational pull and revolves around the star more slowly than a solid body moving on a circular orbit. Solid bodies in the Sun's protoplanetary nebula would have experienced "gas drag" that gradually removed angular momentum from their orbit so that they spiralled inward toward the Sun (*Weidenschilling,* 1977). Small particles drifted inward at a rate determined by their terminal velocity, which increased with particle size. Large particles drifted at a rate that depended on the ratio of their surface area to volume, which decreased with particle size. Drift rates were highest for meter-sized particles (*Weidenschilling,* 1977). These particles would have drifted 1 AU every few hundred years in a nonturbulent nebula (see Fig. 1).

If a significant fraction of solid material was present in meter-sized bodies at any one time, gas drag would have been the dominant process causing radial redistribution of material in the nebula (*Stepinski and Valageas,* 1997). Drifting particles may not have fallen all the way into the Sun. Instead, small solids would have evaporated once they reached a region that was hot enough (*Supulver and Lin,* 2000). Ices evaporated near the snow line, while silicate-rich bodies evaporated closer to the Sun. It is plausible that solid material drifted inward faster than turbulent diffusion could act to redistribute the gas, so that the vapor phase of each condensible material became enhanced near its condensation distance (*Cuzzi and Weidenschilling,* 2006). At early times, therefore, volatile materials such as water and

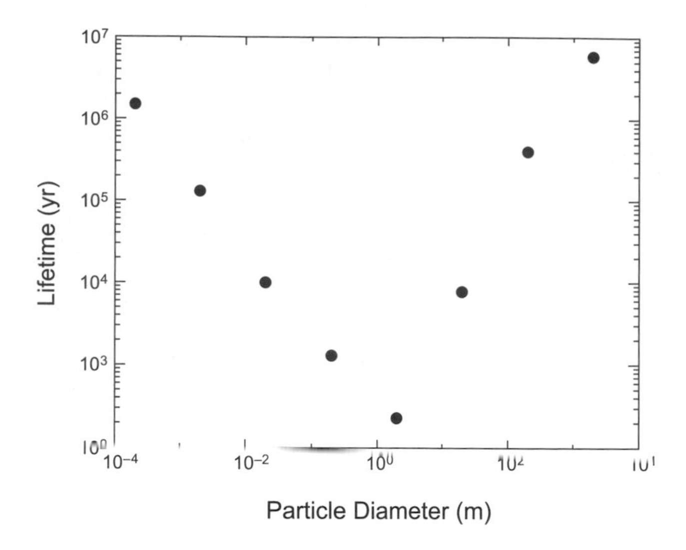

Fig. 1. Lifetime of spherical particles against drift due to gas drag, at 2.5 AU in the solar nebula, as a function of particle size. Gas density is 10^{-9} g/cm^3 at 1 AU, varying as $1/r^2$. Adapted from *Weidenschilling* (1977).

CO were probably enhanced in the inner nebula rather than depleted by cold trapping. Later, once planetesimals formed in large numbers near the snow line, the inner nebula became depleted in volatile materials (*Cuzzi and Weidenschilling,* 2006). Thus, we would expect meteorite parent bodies that formed at different times to have different chemistries and oxidation states.

In the absence of turbulence, a 1-mm-diameter CAI would have drifted from the asteroid belt to a point where it would vaporize in around ~10^5 yr. This means that essentially all CAIs would have been lost by the time the majority of chondrules formed 1–3 m.y. later. The fact that chondrules and CAIs are seen in the same meteorites suggests the nebula was turbulent for much of the first 3 m.y. in order to preserve some CAIs for this length of time. Millimeter-sized particles would have coupled strongly to turbulent eddies in the nebula at the level of turbulence envisioned in most nebular models (*Cuzzi et al.,* 2001). Theoretical models show that although most CAIs would continue to drift inward in a turbulent nebula, some would also diffuse outward, thereby extending their lifetime by several million years (*Cuzzi et al.,* 2003). These models suggest that large type-B CAIs, seen in CV chondrites, could only be retained if they formed in a region close to the asteroid belt, or were deposited there by an x-wind (see below). Smaller type-A CAIs, seen in most chondrite classes, would have diffused into the asteroid belt and survived for several million years even if they initially formed close to the Sun (*Cuzzi et al.,* 2003).

It is conceivable that some CAIs were incorporated in an early generation of planetesimals shortly after the CAIs formed. These planetesimals could have survived for several million years, safe from the effects of gas drag. Later catastrophic collisions could have returned CAIs to the nebula, allowing them to be accreted along with newly formed chondrules to form new planetesimals (*Lugmair and Shukolyukov,* 2001). Currently, there is no evidence to support this hypothesis. We lack examples of primitive meteorites composed primarily of CAIs, while the known achondrites appear to have had chondritic precursors (*Meibom and Clark,* 1999) rather than precursors made of CAI-like material. If CAI-rich planetesimals did form, they are now rare or absent in the asteroid belt. Instead, it appears that most CAIs remained in the nebula for several million years before they and the chondrules coalesced to form planetesimals. Some CAIs show signs of multiple episodes of modification in a nebular setting (*Hsu et al.,* 2000; *MacPherson and Davis,* 1993), which lends weight to this interpretation.

Nebular gas pressure may have varied nonmonotonically with distance from the Sun. In this case, drag effects would have collected small solids at each local pressure maximum (*Haghighipour and Boss,* 2003), and these locations would become preferred sites for planetesimal formation. Even in a nebula with a smooth pressure gradient, drift rates due to gas drag would have varied with distance from the Sun. Recent studies of individual and collective drift rates for millimeter-sized particles find that the number of particles in the inner nebula probably increased over time as material drifted inward from the outer nebula (*Youdin and Shu,* 2002; *Youdin and Chiang,* 2004) [but see comments by *Weidenschilling* (2003)]. This radial concentration may have been a necessary prerequisite for the formation of larger bodies by gravitational instability (see next section).

Solid particles may have been redistributed in another way. Young, actively accreting stars typically have collimated jets of gas that arise near the inner edge of the disk and travel away from the star perpendicular to the disk plane. This "x-wind" may be strong enough to launch millimeter-sized particles on trajectories that carry them outward for several astronomical units before they fall back into the disk (*Shu et al.,* 1996). It is possible that a significant fraction of the material contained in meteorites drifted close to the Sun and was transported back into the asteroid belt via an x-wind, possibly several times, during the early active phase of the nebula. However, the compositions of most chondrules makes it hard to understand how they could have formed in an x-wind (*Hutchison,* 2002). In addition, it is unclear why the CI class of chondrites failed to incorporate chondrules if these particles were distributed throughout the nebula by an x-wind. It seems likely that the x-wind mechanism was not effective during the time when most chondrules formed. Equally, recycling by an x-wind was presumably not the only process that acted to preserve CAIs until chondrules and parent bodies formed.

5. PLANETESIMALS AND PARENT BODIES

Chondrules are a major component in almost all primitive meteorites. Carbonaceous chondrites commonly contain 50% chondrules by volume, while ordinary and enstatite chondrites contain up to 80% (*Scott and Krot,* 2004). Chondrules within each chondrite group typically have a narrow size distribution. The mean size differs from group to group but is generally in the range 0.1–1 mm (*Scott and Krot,* 2004). There are two ways to interpret these observations. Either compact millimeter-sized particles had to be abundant in the nebula in order to trigger the formation of primitive planetesimals, or compact millimeter-sized particles happened to be abundant in the nebula when conditions became right for these planetesimals to form. In the former case, chondrules were the perpetrators, while in the latter case they were merely bystanders caught up in events around them.

There is growing evidence from radioactive isotope dating that chondrule formation extended over several million years (see Fig. 2), and that chondrules in a given meteorite have a wide range of ages (*Amelin et al.,* 2002; *Mostefaoui et al.,* 2002; *Kurahashi et al.,* 2004). Many primitive meteorite parent bodies apparently required 2–3 m.y. to form even though chondrules had existed in the nebula for at least 1 m.y. at that point. Compact CAIs, with mechanical and aerodynamic properties similar to chondrules, had been present for even longer. This suggests that the formation of millimeter-sized particles did not trigger planetesimal formation. Unless CAIs and early generations of chondrules remained uniformly spaced in the nebula, it seems unlikely that planetesimal formation had to wait for the global number density of millimeter-sized particles to reach a certain level. CI chondrites, which are composed almost entirely of fine-grained matrix material, formed at a time and place where chondrules were essentially absent, which also suggests that chondrule production and planetesimal formation were separate processes.

The formation times of achondritic asteroids are difficult to pin down, although the iron IIIAB parent body probably formed in about ~2 m.y. (*Sugiura and Hoshino,* 2003).

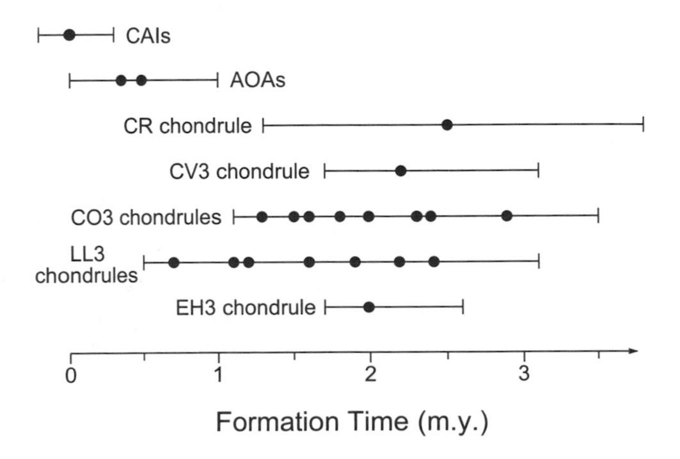

Fig. 2. Formation times of refractory inclusions and chondrules in various primitive meteorites based on the U-Pb (CR chondrule and CAIs) and Al-Mg (all except CR chondrule) isotope systems. Data taken from *Amelin et al.* (2002); *Guan et al.* (2002); *Hutcheon et al.* (2000); *Itoh et al.* (2002); *Mostefaoui et al.* (2002); and *Kurahashi et al.* (2004). Horizontal lines show age error bars. For the CO3 and LL3 chondrites, the errors are larger than the separation of neighboring data points but smaller than the total spread in chondrule ages.

Models for the thermal evolution of Vesta suggest it took roughly 2.9 m.y. to accrete (*Ghosh and McSween,* 1998). Thus, at least some differentiated bodies apparently formed long after chondrules first appeared. This reinforces the notion that planetesimal formation did not coincide with the first appearance of chondrules. It is also possible that many achondritic asteroids formed before chondrules, in which case the two processes were entirely separate.

Several models for the origin of planetesimals are being pursued, although none of these models is fully developed at present. Calculations suggest that pairwise collisions between small particles could have formed meter-sized bodies on timescales of ~10^4 yr in the absence of turbulence, provided that particle-sticking mechanisms were effective (*Weidenschilling,* 1980). Aggregation may have continued in this way until kilometer-sized planetesimals were produced, at which point gravity would aid further accretion. However, as particles grew in size, collisions became more likely to cause disruption rather than accretion, especially if the nebula was turbulent (*Cuzzi and Weidenschilling,* 2006). It is possible that pairwise accretion stalled at sizes much less than 1 km.

If the level of turbulence was very low, solid material would have collected near the nebular midplane, and this region could have become gravitationally unstable, forming planetesimals directly (*Goldreich and Ward,* 1973). However, even with low levels of turbulence, gravitional instability requires substantial enhancement in the local dust-to-gas ratio (*Sekiya,* 1998). This might be achieved either by radial drift and concentration of particles as described in the previous section, or if gas is lost from the nebula at a greater rate than solids (*Youdin and Shu,* 2002).

A third model for planetesimal formation invokes "turbulent concentration" of small solids in stagnant regions in a turbulent nebula. This could have produced very high concentrations of particles with particular aerodynamic properties, up to a factor of 10^5 higher than normal (*Cuzzi et al.,* 2001). High particle concentrations presumably would have aided further growth by one of the mechanisms described above. Encouragingly, models for turbulent concentration predict chondrule size distributions similar to those seen in primitive meteorites (*Cuzzi et al.,* 2001).

These models for planetesimal formation share a common property: In each case the growth of large bodies depends on local conditions in the nebula. In particular, the dust-to-gas ratio and the level of turbulence seem to be key factors. The long delay between the formation of CAIs and the appearance of many parent bodies suggests that a threshhold effect was at work. It is plausible that planetesimal formation was delayed until turbulence fell below a certain level, or until the local solid-to-gas ratio was enhanced to a certain point (*Youdin and Shu,* 2002). Once suitable conditions arose, the formation of planetesimals could have proceeded rapidly.

There was presumably a time lag between the appearance of kilometer-sized planetesimals and the formation of meteorite parent bodies that were large enough to survive until the present day. Numerical accretion models suggest this delay was relatively short, however. Once planetesimals appeared in large numbers, only 10^5–10^6 yr were required to form bodies as large as Ceres (*Wetherill and Stewart,* 1993). In general, planetesimals would have undergone less radial migration than smaller objects, so the differences between meteorite parent bodies probably reflect differences in the populations of planetesimals that formed them. However, there is some evidence for heterogeneity within parent bodies. For example, the abundance of CAIs in different meteorites from the same ordinary chondrite or enstatite chondrite group can vary by an order of magnitude (*Kimura et al.,* 2000, 2002). These differences presumably arose because the planetesimals that aggregated to form a given parent body were not all the same. Similarly, the bimodal olivine abundance seen in L chondrites (*Wood,* 1996) may reflect differences among the parent body's component planetesimals.

The different O-isotopic compositions of chondrite groups have been interpreted to mean that each meteorite parent body formed in a different region of the nebula, and that mixing between these regions was minimal during parent-body formation (e.g., *Wood,* 1996). The decreasing amount of aqueous alteration seen in carbonaceous, ordinary, and enstatite chondrites, respectively, is usually attributed to formation at distances progressively closer to the Sun, where nebula temperatures were higher and water ice was less likely to condense. This notion of narrow, isolated accretion zones receives some support from the preservation of different $^{52}Cr/^{53}Cr$ ratios in bulk samples from different meteorites (*Lugmair and Shukolyukov,* 1998), and also the observed radial zoning of the asteroid belt with respect to asteroid spectral class, although the different spectral classes overlap substantially (*Gradie and Tedesco,* 1982).

However, this picture is almost certainly too simplistic. The preservation of CAIs apparently requires that the nebula remained turbulent for several million years. As a result, different meteorite components should have become thoroughly mixed together. Most types of chondrule and refractory inclusion are seen in most meteorite groups, but the proportions are different in each case. Some of these differences may be attributed to the aerodynamic processes that caused the widespread size sorting of small solids seen in chondrites. This may explain why large, type-B CAIs are seen only in CV chondrites. This group of chondrites also tends to contain large chondrules and large type-A CAIs (*Scott and Krot,* 2004). If all type-B CAIs were large, we might convince ourselves that they would tend to appear only in those meteorites with large components.

More perplexing are the different numbers of CAIs seen in different meteorite groups. Several carbonaceous chondrite groups contain 4–13% refractory inclusions by volume (*Scott et al.,* 1996; *Scott and Krot,* 2004). In ordinary and enstatite chondrites, CAIs are present in only trace amounts, usually ~0.1% (*Scott and Krot,* 2004). This is hard to explain if meteorite parent bodies formed at similar times in a well-mixed nebula, even if they formed at different locations. It seems more likely that the parent bodies formed at *different times* in a nebula that was continually losing CAIs as a result of gas drag. (Chondrules were also lost by gas drag, but these losses were at least partially offset by new generations of chondrules.) This suggests the ordinary and enstatite chondrites formed later than carbonaceous chondrites. The very low abundance of CAIs in ordinary and enstatite chondrites could mean these meteorites sample some of the last planetesimals to form. This would explain why ordinary-chondrite-like parent bodies are much rarer in the asteroid belt than carbonaceous chondrites (*Meibom and Clark,* 1999). In addition, there may have been a substantial decrease in the level of nebula turbulence at around this time, with a corresponding increase in the loss rate of CAIs. For this reason, a small difference in planetesimal formation time could have resulted in a large difference in the number of CAIs these planetesimals contained.

It is interesting to pursue this hypothesis a little further to see if other meteorite characteristics can be explained in terms of differences in formation time rather than location. In the popular core-accretion model for the formation of giant planets (*Pollack et al.,* 1996), Jupiter formed at about 5 AU from the Sun because this is where the snowline was located. However, the snowline could have remained here for no more than a few hundred thousand years, so planetesimal formation at 5 AU must have occurred more rapidly than in the asteroid belt. It seems likely that the inner nebula became progressively depleted in water over time as large numbers of planetesimals formed near 5 AU (*Stevenson and Lunine,* 1988). Early generations of planetesimals forming in the asteroid region would have done so in a water-rich, oxidizing environment, precisely the characteristics needed to explain the chemistry of many carbonaceous chondrites. At later times, planetesimals would have formed in a reducing environment, highly depleted in water, yielding ordinary chondrites, and finally the enstatite chondrites. Conceivably, the enstatite chondrites formed in a region that was nominally beyond the snowline, yet they remained dry because almost all the water had already been incorporated into planetesimals lying further from the Sun.

In this interpretation, the timescale for planetesimal formation *decreases* with distance from the Sun, rather than increasing. This is also true if we adopt the traditional assumption that enstatite, ordinary, and carbonaceous chondrites formed at progressively larger distances from the Sun. The greater degree of thermal metamorphism seen in ordinary chondrites compared to carbonaceous chondrites is often attributed to the fact that ordinary chondrites formed first, and so formed with a higher abundance of ^{26}Al, the primary heat source for asteroid-sized bodies (e.g., *Grimm and McSween,* 1993). However, more recent studies suggest that the degree of thermal metamorphism in carbonaceous chondrites was moderated by hydrothermal convection due to the presence of large amounts of water ice (*McSween et al.,* 2002). In this case, the requirement that ordinary chondrites formed before carbonaceous chondrites no longer applies.

Differences in the time of formation of various meteorite parent bodies may ultimately explain many of their properties, perhaps along the lines described above. However, we are still left with a challenging conundrum: If the chondrules within a single meteorite group truly have ages spanning 1 m.y. or more, yet turbulent mixing was effective during much of the early history of the nebula, why do the various chondrite groups have distinct elemental abundances and O-isotope ratios? Unfortunately, there is no clear answer to this problem at present. A possible explanation is that the O-isotopic composition, Mg/Si ratios, and other chondrule properties were easily altered during chondrule-heating events, while resetting of the Al/Mg isotope system used to age date most chondrules occurred only occasionally. This possibility will necessarily remain speculative until better age data for chondrules become available.

6. DEPLETION OF THE ASTEROID BELT

Collisions were a necessary part of the formation of meteorite parent bodies and must have involved some mixing of material between different planetesimal populations. Impact breccias and foreign clasts are seen in many meteorite classes (*Wilkening,* 1973; *Bunch,* 1988; *Lipschutz et al.,* 1989), so collisions continued to be important after parent bodies formed, causing some mixing and processing of material. In general, foreign clasts make up a small fraction of the total mass of most meteorites. This suggests that collision velocities were large, since relatively little impactor material is incorporated in a target body during a high-speed collision (*Scott,* 2002). This is presumably why most meteorite groups have retained a distinct identity rather than being equal mixtures of pieces from many sources. However, a few meteorites such as Kaidun appear to be composed mainly of assorted pieces of other parent bodies (*Zolensky et al.,* 1996).

Carbonaceous-chondrite clasts are quite common in ordinary chondrites, whereas the opposite is not the case (*Wasson and Wetherill,* 1979; *Lipschutz et al.,* 1989). This can be interpreted to mean that ordinary chondrite parent bodies were rare in the early asteroid belt (*Meibom and Clark,* 1999). Alternatively, this may indicate that carbonaceous chondrites formed in the outer asteroid belt while ordinary chondrites formed closer to the Sun, because collision fragments destined to become clasts would have drifted inward due to gas drag. A third interpretation is simply that carbonaceous chondrite parent bodies accreted at an early stage, so they were unable to incorporate ordinary chondrite fragments since these did not yet exist. Each of these interpretations could be correct since they are not mutually exclusive.

There are signs that many large bodies were collisionally disrupted and reassembled early in the solar system. The ureilite meteorites apparently formed at depth on a body at least 200 km in size that was later partially disrupted (*Goodrich et al.,* 2002). The mesosiderites plausibly formed when an impact mixed the molten core of a 200–400-km asteroid with mantle and crustal material from the target and projectile (*Scott et al.,* 2001). Ordinary chondrite breccias exhibit a wide range of cooling rates, which suggests they come from different depths on a body at least 200 km in diameter (*McSween et al.,* 2002). Today, 200-km asteroids have collisional lifetimes that are longer than the age of the solar system, so the collision rate in the early asteroid belt must have been substantially higher than now. This suggests the asteroid belt was also more massive early in its history.

A high early collision rate would explain why iron meteorites, derived from the cores of differentiated asteroids, are common, while the corresponding mantle and crustal material is rarely seen. The cooling rates deduced for most iron meteorites suggest they came from small asteroids with diameters <100 km (*Haack and McCoy,* 2004) that probably accreted at an early stage while ^{26}Al was still present. The paucity of mantle-derived meteorites suggests that all the iron meteorite parent bodies have been severely eroded or catastrophically disrupted, and the mantle material destroyed. Some core material has survived, in part because it was buried more deeply in the asteroid, but mostly because iron meteoroids have collisional lifetimes 1–2 orders of magnitude longer than stony bodies, according to the cosmic-ray exposure ages seen in meteorites (*Herzog,* 2004).

The existence of a largely intact basaltic crust on Vesta places strong constraints on the collisional evolution of the asteroid belt after this crust formed. Numerical models suggest the mass of the main belt has been within a factor of 5 of its current mass ever since the present regime of high relative velocities was established (*Davis et al.,* 1994). The high relative velocities in the belt today were almost certainly caused directly or indirectly by gravitational perturbations from Jupiter and Saturn. The giant planets are thought to have formed in less than 10 m.y. (*Pollack,* 1996), while thermal models for Vesta suggest its crust formed in about 6.6 m.y. (*Ghosh and McSween,* 1998). It seems likely that both the mass of material in the asteroid belt and the collision rate were approaching their current values only 10 m.y. after the solar system formed. In contrast, the collision rate may have been higher by several orders of magnitude before Vesta's crust formed.

The asteroid belt currently contains about $5 \times 10^{-4}\ M_\oplus$ of material. Accretion models suggest the region once contained at least 100× as much material in order to form bodies such as Vesta within a few million years (*Wetherill,* 1992). A smooth interpolation of the amount of rocky material contained in the terrestrial and giant planets suggests the asteroid belt once contained $\sim 2\ M_\oplus$ of solid material. If true, this means the modern belt is depleted by a factor of roughly 4000. While collisional erosion has certainly played a role in removing mass from the asteroid belt, it seems highly unlikely that collisions alone could have removed the great majority of the primordial material from this region in only a few million years. For this reason, attention has shifted to dynamical processes that would have depleted the belt.

Two dynamical models are being actively pursued at present. Each invokes the effects of dynamically unstable resonances associated with the giant planets. Asteroids located in many of the orbital resonances undergo large changes in their orbits, and are likely to fall into the Sun or become ejected from the solar system on a timescale of $\sim 10^6$ yr (*Gladman et al.,* 1997). When the giant planets formed, asteroids located in resonances would have been perturbed onto highly eccentric orbits. Gas drag is very effective for bodies moving on eccentric orbits. If nebular gas was still present in the asteroid belt at this time, many asteroids would have drifted inward as a result. In addition, secular resonances probably swept across the asteroid region during the dissipation of the nebula (*Nagasawa et al.,* 2000). Numerical simulations suggest that a combination of resonance sweeping and gas drag was sufficient to remove most bodies up to 100 km in diameter from the asteroid region (*Franklin and Lecar,* 2000).

The asteroid belt may once have contained bodies larger than Ceres. Such objects would have been too massive to be affected appreciably by gas drag, but they probably would not have remained in the asteroid region for long. Gravitational interactions between such large bodies would have caused frequent changes in their orbits. Sooner or later, each body would have ended up in an unstable resonance zone, leading to its removal. Numerical simulations suggest the most likely outcome would be the loss of all large objects together with ~99% of objects the size of modern asteroids (*Wetherill,* 1992; *Petit et al.,* 2001). Simulations that take into account changes in the orbits of the giant planets find that the asteroid belt would have approached its current mass on a timescale of ~10 m.y. (*Chambers and Wetherill,* 2001).

It is unclear which of these two mechanisms was responsible for removing most of the original solid material from the asteroid belt. However, the outcome would have been similar in a number of ways. Most surviving asteroids would have undergone substantial radial migration as a result of the clearing process. The degree of radial displacement depends sensitively on the initial orbital parameters of an asteroid in each model (*Franklin and Lecar,* 2000; *Chambers and Wetherill,* 2001). As a result, any primordial compositional gradient in the asteroid region has been partially erased. This is reflected in the degree of overlap seen in the radial distributions of the major asteroid spectral classes (*Gradie and Tedesco,* 1982). The degree of mixing cannot have been too extreme, however, since some radial structure is apparent today.

7. SUMMARY

Understanding the diversity of meteorites is a fundamental problem for planetary science. The solution to this problem should tell us a great deal about the formation of the solar system and planetary systems in general. While we are not there yet, some aspects of the solution are becoming apparent. The high mass accretion rates seen in very young protoplanetary disks and the approximate isotopic homogeneity of many meteorite components suggests the Sun's protoplanetary nebula was hot and turbulent at an early stage, leading to substantial mixing of material. Subsequent events mainly acted to introduce heterogeneity. It now seems likely that interactions between small solids and nebula gas produced variations in the O-isotope ratios and Mg/Si ratios and depletions in the moderately volatile elements seen in primitive meteorites. Mixing of different amounts of these solids led to differences in meteorite bulk compositions. Variations in the amounts of water ice and short-lived radionuclides gave rise to different thermal histories for meteorite parent bodies of similar size. Some differences between meteorites may be attributed to formation in different parts of the nebula. However, it is difficult to draw firm conclusions here because the process responsible for clearing most of the primordial mass from the asteroid belt also caused widespread radial mixing of the surviving asteroids. The several-million-year spread in the ages of components seen in primitive meteorites makes it unlikely that the properties of meteorites can be explained solely in terms of their formation location. Instead, many of the differences seen in meteorites may reflect differences in the time of parent-body formation. Different formation times can explain the varying number of refractory inclusions seen in the chondrite groups, as well as the meteorites' different oxidation states, primordial water abundances, and thermal histories.

Acknowledgments. I would like to thank C. Alexander, A. Boss, L. Chambers, L. Nittler, A. Youdin, and two anonymous reviewers for helpful comments and suggestions during the preparation of this manuscript. This work was supported by NASA's Origins of Solar Systems and Planetary Geology and Geophysics programs.

REFERENCES

Alexander C. M. O'D. (1996) Recycling and volatile loss in chondrule formation. In *Chondrules and the Protoplanetary Disk* (R. H. Hewins et al., eds.). pp. 233–242. Cambridge Univ., Cambridge.

Alexander C. M. O'D. (2004) Chemical equilibrium and kinetic constraints for chondrule and CAI formation conditions. *Geochim. Cosmochim. Acta, 68,* 3943–3969.

Alexander C. M. O'D., Grossman J. N., Wang J., Zanda B., Bourot-Denise M., and Hewins R. H. (2000) The lack of potassium isotopic fractionation in Bishunpur chondrules. *Meteoritics, 35,* 859–868.

Alexander C. M. O'D., Boss A. P., and Carlson R. W. (2001) The early evolution of the inner solar system: A meteoritic perspective. *Science, 293,* 64–69.

Amelin Y., Krot A. N., Hutcheon I. D., and Ulyanov A. A. (2002) Lead isotopic ages of chondrules and calcium-aluminum rich inclusions. *Science, 297,* 1678–1683.

Beckwith S. V. W., Sargent A. I., Chini R. S., and Guesten R. (1990) A survey for circumstellar disks around young stellar objects. *Astron. J., 99,* 924–945.

Boss A. P. (1993) Evolution of the solar nebula II. Thermal structure during nebula evolution. *Astrophys. J., 417,* 351.

Boss A. P. (1996) Evolution of the solar nebula III. Protoplanetary disks undergoing mass accretion. *Astrophys. J., 469,* 906.

Boss A. P. (2000) Possible rapid gas giant planet formation in the solar nebula and other protoplanetary disks. *Astrophys. J. Lett., 536,* L101–L104.

Bunch T. E. and Rajan R. S. (1988) Meteorite regolithic breccias. In *Meteorites and the Early Solar System* (J. F. Kerridge and M. S. Matthews, eds.), pp. 144–164. Univ. of Arizona, Tucson.

Cassen P. (1996) Models for the fractionation of moderately volatile elements in the solar nebula. *Meteoritics, 31,* 793–806.

Chambers J. E. and Wetherill G. W. (2001) Planets in the asteroid belt. *Meteoritics, 36,* 381–400.

Chiang E. I. (2002) Chondrules and nebula shocks. *Meteoritics, 37,* 151–153.

Chiang E. I. and Goldreich P. (1999) Spectral energy distribution of passive T Tauri disks: Inclination. *Astrophys. J., 519,* 279–284.

Clayton R. N. (2004) Oxygen isotopes in meteorites. In *Treatise on Geochemistry, Vol. 1: Meteorites, Comets, and Planets* (A. M. Davis, ed.), pp. 129–142. Elsevier, Oxford.

Clayton R. N., Hinton R. W., and Davis A. M. (1988) Isotopic variations in the rock forming elements in meteorites. *Philos. Trans. R. Soc., 325,* 483–501.

Cuzzi J. N. (2004) Blowing in the wind III: Accretion of dust rims by chondrule sized particles in a turbulent protoplanetary nebula. *Icarus, 168,* 484–497.

Cuzzi J. N. and Weidenschilling S. J. (2006) Particle-gas dynamics and primary accretion. In *Meteorites and the Early Solar System II* (D. S. Lauretta and H. Y. McSween Jr., eds.), this volume. Univ. of Arizona, Tucson.

Cuzzi J. N., Hogan R. C., Paque J. M., and Dobrovolskis A. R. (2001) Size selective concentration of chondrules and other small particles in protoplanetary nebula turbulence. *Astrophys. J., 546,* 496–508.

Cuzzi J. N., Davis S. S., and Dobrovolskis A. R. (2003) Blowing in the wind II. Creation and redistribution of refractory inclusions in a turbulent protoplanetary nebula. *Icarus, 166,* 385–402.

Davis D. R., Ryan E. V., and Farinella P. (1994) Asteroid collisional evolution: Results from current scaling algorithms. *Planet. Space Sci., 42,* 599–610.

Desch S. J. and Connolly H. C. (2002) A model of the thermal processing of particles in solar nebula shocks: Application to the cooling rates of chondrules. *Meteoritics, 37,* 183–207.

Desch S. J., Connolly H. C., and Srinivasan G. (2004) An interstellar origin for the beryllium 10 in calcium-rich, aluminum-rich inclusions. *Astrophys. J., 602,* 528–542.

Drake M. J. and Righter K. (2002) Determining the composition of the Earth. *Nature, 416,* 39–44.

Franklin F. A. and Lecar M. (2000) On the transport of bodies within and from the asteroid belt. *Meteoritics, 35,* 331–340.

Ghosh A. and McSween H. Y. (1998) A thermal model for the differentiation of asteroid 4 Vesta based on radiogenic heating. *Icarus, 134,* 187–206.

Gladman B. J., Migliorini F., Morbidelli A., Zappala V., Michel P., Cellino A., Froeschlé C., Levison H. F., Bailey M., and Duncan M. (1997) Dynamical lifetimes of objects injected into asteroid belt resonances. *Science, 277,* 197–201.

Goldreich P. and Ward W. R. (1973) The formation of planetesimals. *Astrophys. J., 183,* 1051–1062.

Goodrich C. A., Krot A. N., Scott E. R. D., Taylor G. J., Fioretti

A. M., and Keil K. (2002) Formation and evolution of the ureilite parent body and its offspring (abstract). In *Lunar and Planetary Science XXXIII,* Abstract #1379. Lunar and Planetary Institute, Houston (CD-ROM).

Gradie J. and Tedesco E. (1982) Compositional structure of the asteroid belt. *Science, 216,* 1405–1407.

Grimm R. E. and McSween H. Y. (1993) Heliocentric zoning of the asteroid belt by aluminum-26 heating. *Science, 259,* 653–655.

Guan Y., Huss G. R., MacPherson G. J., and Leshin L. A. (2002) Aluminum-magnesium isotopic systematics of aluminum-rich chondrules in unequilibrated enstatite chondrites (abstract). In *Lunar and Planetary Science XXXIII,* Abstract #2034. Lunar and Planetary Institute, Houston (CD-ROM).

Haack H. and McCoy T. J. (2004) Iron and stony iron meteorites. In *Treatise on Geochemistry, Vol. 1: Meteorites, Comets, and Planets* (A. M. Davis, ed.), pp. 325–346. Elsevier, Oxford.

Haghighipour N. and Boss AP. (2003) On gas drag-induced rapid migration of solids in a nonuniform solar nebula. *Astrophys. J., 598,* 1301–1311.

Herzog G. F. (2004) Cosmic-ray exposure ages of meteorites. In *Treatise on Geochemistry, Vol. 1: Meteorites, Comets, and Planets* (A. M. Davis, ed.), pp. 347–380. Elsevier, Oxford.

Hewins R. H. and Connolly H. C. (1996) Peak temperatures of flash-melted chondrules. In *Chondrules and the Protoplanetary Disk* (R. H. Hewins et al., eds.), pp. 197–204. Cambridge Univ., Cambridge.

Hsu W., Wasserburg G. J., and Huss G. R. (2000) High time resolution by use of the ^{26}Al chronometer in the multistage formation of a CAI. *Earth Planet. Sci. Lett., 182,* 15–29.

Humayun M. and Clayton R. N. (1995) Potassium isotope cosmochemistry: Genetic implications of volatile element depletion. *Geochim. Cosmochim. Acta, 59,* 2131–2148.

Huss G. R., MacPherson G. J., Wasserburg G. J., Russell S. S., and Srinivasan G. (2001) ^{26}Al in CAIs and chondrules from unequilibrated ordinary chondrites. *Meteoritics, 36,* 975–997.

Hutcheon I. D., Krot A. N., and Ulyanov A. A. (2000) ^{26}Al in anorthite-rich chondrules in primitive carbonaceous chondrites: Evidence chondrules postdate CAI (abstract). In *Lunar and Planetary Science XXXI,* Abstract #1869. Lunar and Planetary Institute, Houston (CD-ROM).

Hutchison R. (2002) Major element fractionation in chondrites by distillation in the accretion disk of a T Tauri sun? *Meteoritics, 37,* 113–124.

Itoh S., Rubin A. E., Kojima H., Wasson J. T., and Yurimoto H. (2002) Amoeboid olivine aggregates and AOA-bearing chondrule from Y-81020 CO3.0 chondrite: Distribution of oxygen and magnesium isotopes (abstract). In *Lunar and Planetary Science XXXIII,* Abstract #1490. Lunar and Planetary Institute, Houston (CD-ROM).

Itoh S. and Yurimoto H. (2003) Contemporaneous formation of chondrules and refractory inclusions in the early Solar System. *Nature, 423,* 728–731.

Jones R. H. (1996) Relict grains in chondrules: Evidence for chondrule recycling. In *Chondrules and the Protoplanetary Disk* (R. H. Hewins et al., eds.), pp. 163–172. Cambridge Univ., Cambridge.

Kimura M., Lin Y., and Hiyagon H. (2000) Unusually abundant refractory inclusions and iron-oxide-rich silicates in an EH3 chondrite: Sahara 97159 (abstract). *Meteoritics, 35,* A87.

Kimura M., Hiyagon H., Palme H., Spettel B., Wolf D., Clayton R. N., Mayeda T. K., Sato T., Suzuki A., and Kojima H. (2002) Yamato 792947, 793408 and 82038: The most primitive H chondrites, with abundant refractory inclusions. *Meteoritics, 37,* 1417–1434.

Krot A. N., Meibom A., Weisberg M. K., and Keil K. (2002) The CR chondrite clan: Implications for early solar system processes. *Meteoritics, 37,* 1451–1490.

Krot A. N., Keil K., Goodrich C. A., Scott E. R. D., and Weisberg M. K. (2004) Classification of meteorites. In *Treatise on Geochemistry, Vol. 1: Meteorites, Comets, and Planets* (A. M. Davis, ed.), pp. 83–128. Elsevier, Oxford.

Kurahashi E., Kita N. T., Nagahara H., and Morishita Y. (2004) Contemporaneous formation of chondrules in the ^{26}Al-^{26}Mg system for ordinary and CO chondrites (abstract). In *Lunar and Planetary Science XXXV,* Abstract #1476. Lunar and Planetary Institute, Houston (CD-ROM).

Lipschutz M. E., Gaffey M. J., and Pellas P. (1989) Meteoritic parent bodies: Nature, number, size, and relation to present day asteroids. In *Asteroids II* (R. P. Binzel et al., eds.), pp. 740–777. Univ. of Arizona, Tucson.

Lugmair G. W. and Shukolyukov A. (1998) Early solar system timescales according to ^{53}Mn-^{53}Cr systematics. *Geochim. Cosmochim. Acta, 62,* 2863–2886.

Lugmair G. W. and Shukolyukov A. (2001) Early Solar System events and timescales. *Meteoritics, 36,* 1017–1026.

MacPherson G. J. and Davis A. M. (1993) A petrologic and ion microprobe study of a Vigarano type-B refractory inclusion-evolution by multiple stages of alteration and melting. *Geochim. Cosmochim. Acta, 57,* 231–243.

MacPherson G. J. (2004) Calcium-aluminum-rich inclusions in chondritic meteorites. In *Treatise on Geochemistry, Vol. 1: Meteorites, Comets, and Planets* (A. M. Davis, ed.), pp. 201–246. Elsevier, Oxford.

MacPherson G. J., Davis A. M., and Zinner E. K. (1995) The distribution of aluminum-26 in the early solar system — A reappraisal. *Meteoritics, 30,* 365.

Marhas K. K., Goswami J. N., and Davis A. M. (2002) Short-lived nuclides in hibonite grains from Murchison: Evidence for solar system evolution. *Science, 298,* 2182–2185.

McKeegan K. D., Chaussidon M., and Robert F. (2000) Incorporation of short-lived ^{10}Be in a calcium-aluminium-rich inclusion from the Allende meteorite. *Science, 289,* 1334–1337.

McSween H. Y., Ghosh A., Grimm R. E., Wilson L., and Young E. D. (2002) Thermal evolution models of asteroids. In *Asteroids III* (W. F. Bottke Jr. et al., eds.), pp. 559–571. Univ. of Arizona, Tucson.

Meibom A. and Clark B. E. (1999) Evidence for the insignificance of ordinary chondritic material in the asteroid belt. *Meteoritics, 34,* 7–24.

Meibom A., Desch S. J., Krot A. N., Cuzzi J. N., Petaev M., Wilson L., and Keil K. (2000) Large-scale thermal events in the solar nebula: Evidence from Fe,Ni metal grains in primitive meteorites. *Science, 288,* 839–841.

Mostefaoui S., Kita N. T., Togashi S., Tachibana S., Nagahara H., and Morishita Y. (2002) The relative formation ages of ferromagnesian chondrules inferred from their initial aluminum-26/aluminum-27 ratios. *Meteoritics & Planet. Sci., 37,* 421–438.

Nagasawa M., Tanaka H., and Ida S. (2000) Orbital evolution of asteroids during depletion of the solar nebula. *Astron. J., 119,* 1480–1497.

Nittler L. R. (2003) Presolar stardust in meteorites: Recent advances and scientific frontiers. *Earth Planet. Sci. Lett., 209,* 259–273.

Norman M., McCulloch M., O'Neill H., and Brandon A. (2004) Magnesium isotopes in the Earth, Moon, Mars and pallasite parent body: High-precision analysis of olivine by laser-ablation multi-collector ICPMS (abstract). In *Lunar and Planetary Science XXXV,* Abstract #1447. Lunar and Planetary Institute, Houston (CD-ROM).

Ott U. (1993) Interstellar grains in meteorites. *Nature, 364,* 25–33.

Palme H., Larimer J. W., and Lipschutz M. E. (1988) Moderately volatile elements. In *Meteorites and the Early Solar System* (J. F. Kerridge and M. S. Matthews, eds.), pp. 436–461. Univ. of Arizona, Tucson.

Palme H., Spettel B., Kural G., and Zinner E. (1992) Origin of Allende chondrules (abstract). In *Lunar and Planetary Science XXIII,* pp. 1021–1022. Lunar and Planetary Institute, Houston.

Petit J. M., Morbidelli A., and Chambers J. E. (2001) The primordial excitation and clearing of the asteroid belt. *Icarus, 153,* 338–347.

Pollack J., Hubickyj O., Bodenheimer P., Lissauer J., Podolak M., and Greenzweig Y. (1996) Formation of giant planets by concurrent accretion of solids and gas. *Icarus, 124,* 62–85.

Sahijpal S., Goswami J. N., Davis A. M., Lewis R. S., and Grossman L. (1998) A stellar origin for the short lived nuclides in the early solar system. *Nature, 391,* 559.

Scott E. R. D., Love S. G., and Krot A. N. (1996) Formation of chondrules and chondrites in the protoplanetary nebula. In *Chondrules and the Protoplanetary Disk* (R. H. Hewins et al., eds.), pp. 87–98. Cambridge Univ., Cambridge.

Scott E. R. D., Haack H., and Love S. G. (2001) Formation of mesosiderites by fragmentation and reaccretion of a large differentiated asteroid. *Meteoritics, 36,* 869–891.

Scott E. R. D. (2002) Meteorite evidence for the accretion and collisional evolution of asteroids. In *Asteroids III* (W. F. Bottke Jr et al., eds.), pp. 697–709. Univ. of Arizona, Tucson.

Scott E. R. D. and Krot A. N. (2004) Chondrites and their components. In *Treatise on Geochemistry, Vol. 1: Meteorites, Comets, and Planets* (A. M. Davis, ed.), pp. 143–200. Elsevier, Oxford.

Sears D. W. G., Huang A., and Benoit P. H. (1996) Open-system behaviour during chondrule formation. In *Chondrules and the Protoplanetary Disk* (R. H. Hewins et al., eds.), pp. 221–232. Cambridge, Cambridge.

Sekiya M. (1998) Quasi-equilibrium density distributions of small dust aggregations in the solar nebula. *Icarus, 133,* 298–309.

Shu F. H., Shang H., and Lee T. (1996) Toward an astrophysical theory of chondrites. *Science, 271,* 1545–1552.

Stepinski T. F. and Valageas P. (1997) Global evolution of solid matter in turbulent protoplanetary disks II. Development of icy planetesimals. *Astron. Astrophys., 319,* 1007–1019.

Stevenson D. J. and Lunine J. I. (1988) Rapid formation of Jupiter by diffuse redistribution of water vapour in the solar nebula. *Icarus, 75,* 146–155.

Sugiura N. and Hoshino H. (2003) Mn-Cr chronology of type IIIAB iron meteorites. *Meteoritics, 38,* 117–144.

Supulver K. and Lin D. N. C. (2000) Formation of icy planetesimals in a turbulent solar nebula. *Icarus, 146,* 525–540.

Vanhala H. A. T. and Boss A. P. (2002) Injection of radioactivities into the forming solar system. *Astrophys. J., 575,* 1144–1150.

Wasson J. T. and Wetherill G. W. (1979) Dynamical, chemical and isotopic evidence regarding the formation locations of asteroids and meteorites. In *Asteroids* (T. Gehrels, ed.), pp. 926–974. Univ. of Arizona, Tucson.

Weidenschilling S. J. (1977) Aerodynamics of small bodies in the solar nebula. *Mon. Not. R. Astron. Soc., 180,* 57–70.

Weidenschilling S. J. (1980) Dust to planetesimals — Settling and coagulation in the solar nebula. *Icarus, 44,* 172–189.

Weidenschilling S. J. (2003) Radial drift of particles in the solar nebula: Implications for planetesimal formation. *Icarus, 165,* 438–442.

Wetherill G. W. (1992) An alternative model for the formation of the asteroids. *Icarus, 100,* 307–325.

Wetherill G. W. and Stewart G. R. (1993) Formation of planetary embryos — Effects of fragmentation, low relative velocity, and independent variation of eccentricity and inclination. *Icarus, 106,* 190.

Wilkening L. L. (1973) Foreign inclusions in stony meteorites — I. Carbonaceous chondritic xenoliths in the Kapoeta howardite. *Geochim. Cosmochim. Acta, 37,* 1985–1986.

Wood J. A. (1985) Meteoritic constraints on processes in the solar nebula. In *Protostars and Planets II* (D. C. Black and M. S. Matthews, eds.), pp. 687–702. Univ. of Arizona, Tucson.

Wood J. A. (1996) Unresolved issues in the formation of chondrules and chondrites. In *Chondrules and the Protoplanetary Disk* (R. H. Hewins et al., eds.), pp. 55–70. Cambridge Univ., Cambridge.

Woolum D. S. and Cassen P. (1999) Astronomical constraints on nebular temperatures: Implications for planetesimal formation. *Meteoritics, 34,* 897–907.

Youdin A. N. and Shu F. (2002) Planetesimal formation by gravitational instability. *Astrophys. J., 580,* 494–505.

Youdin A. N. and Chiang E. I. (2004) Particle pileups and planetesimal formation. *Astrophys. J., 601,* 1109–1119.

Zanda B. (2004) Chondrules. *Earth Planet. Sci. Lett., 224,* 1–18.

Zhu X. K., Guo Y., O'Nions R. K., Young E. D., and Ash R. D. (2001) Isotopic homogeneity of iron in the early solar nebula. *Nature, 412,* 311–313.

Zinner E. (1998) Stellar nucleosynthesis and the isotopic composition of presolar grains from primitive meteorites. *Annu. Rev. Earth Planet. Sci., 26,* 147–188.

Zolensky M. E., Ivanov A. V., Yang S. V., Mittlefehldt D. W, and Ohsumi K. (1996) The Kaidun meteorite: Mineralogy of an unusual CM1 lithology. *Meteoritics, 31,* 484–493.

Trapping and Modification Processes of Noble Gases and Nitrogen in Meteorites and Their Parent Bodies

Rainer Wieler
Eidgenössische Technische Hochschule (ETH) Zürich

Henner Busemann
University of Bern (now at Carnegie Institution of Washington)

Ian A. Franchi
Open University

We review the inventories of primordial noble gases and nitrogen in meteorites, their carrier phases, how and where they may have been incorporated, as well as processes modifying their abundances on meteorite parent bodies. Some of the many distinct noble gas and nitrogen components have an isotopic composition very different from that in the Sun. These components reside in presolar grains. "Anomalous" noble gas and nitrogen components thus are used to infer parent stars of presolar grains as well as theories of stellar nucleosynthesis. Other noble gas components have an isotopic signature roughly similar to the solar composition. Some of these "normal" components are also carried by presolar grains and probably approximately represent the average isotopic composition of their parent stars. Carriers of other normal components remain ill-defined and their origin unclear. Their isotopic identity was possibly established in the solar nebula, but it appears increasingly likely that this often also happened earlier somewhere in the presolar molecular cloud. Apart from allowing us to study meteorite formation, primordial noble gases and nitrogen also are important tracers to constrain the metamorphic history of meteorite parent bodies.

1. INTRODUCTION

Noble gases are extremely rare in meteorites. A remarkable consequence of this scarcity is the impressive variety of different noble gas "components" that can be distinguished in meteorites, i.e., reservoirs with a relatively well-defined elemental and/or isotopic composition of a more or less well-understood origin. *Podosek* (2003) gives an insightful explanation of this basic feature of noble gas cosmochemistry. Some components were produced *in situ*, e.g., the cosmogenic and the radiogenic noble gases discussed elsewhere in this book (*Eugster et al.,* 2006; *Krot et al.,* 2006). Other components were trapped by the meteoritic materials or their precursors. These trapped noble gases — in particular the primordial noble gases trapped very early in solar system history — are the focus of this chapter.

Among all elements, N is most similar to the noble gases in cosmochemical behavior. Although in a few minerals N belongs to the crystal structure, more often than not it is present as an ultratrace element only. Often N has also been trapped (or produced *in situ*) by similar processes as noble gases, and its variability in isotopic composition in meteorites is among the largest of all elements. Furthermore, N and noble gases are often analyzed in parallel on the same sample. For all these reasons, it is logical to include some aspects of N in the discussion here.

The multitude of trapped gas components allows one to study a wide variety of processes throughout most of the epochs discussed in this book, and some longstanding problems in noble gas cosmochemistry have their roots in the fact that it is not always clear in what particular epoch a certain component received its identity. It is nevertheless appropriate that this chapter is found in Part VI of this volume, because the "accretionary epoch" certainly has left many imprints on the record of trapped noble gases and N.

The multitude of distinct noble gas and N components makes the subject notoriously difficult to follow. In the words of *Ozima and Podosek* (2002), "noble gas geochemistry often seems to non-practitioners to have much the air of the secret society and its dark art." This certainly holds no less for noble gas cosmochemistry. We try to mitigate this problem by focussing on processes that may have shaped the noble gas and N record in meteorites rather than on the details of the various components. Readers are referred to a recent review by *Ott* (2002) for a comprehensive presentation of the component structure of trapped noble gases in meteorites and details of their composition. We will adopt Ott's nomenclature, which is an attempt to structure the noble gas "alphabet." Another comprehensive recent review on noble gases in meteorites is by *Podosek* (2003), while *Grady and Wright* (2003) review the available N data. Several review papers in books (*Porcelli et al.,* 2002; *Kallenbach et al.,* 2003) also discuss various aspects of noble gas and N cosmochemistry.

In the following we first define the relevant terminology before we discuss carriers and inferred trapping processes for various components and constraints from primordial noble gases and N on the accretionary history of meteorites as well as the metamorphic history on parent bodies. In the final section we give an outlook.

2. COMPONENTS

Trapped noble gases in meteorites have traditionally been subdivided into "solar" and "planetary" varieties (*Signer and Suess,* 1963). Signer and Suess recognized that in some meteorites the elemental abundances are similar to those expected for the Sun, whereas in other meteorites the lighter noble gases are heavily depleted relative to Xe and solar abundances, with abundance patterns resembling that of Earth's atmosphere. Later it was realized that the isotopic composition of the solar noble gases in meteorites indeed is very similar to that in the solar wind (*Geiss,* 1973) and that elemental abundances of the "planetary" gases in meteorites also resembled those of Venus and Mars.

Early workers called both the solar and the planetary varieties "primordial." This term implies that the gases had been incorporated very early, either in precursor solids of meteoritic materials or during formation of meteorites and their parent bodies. However, it was soon recognized that the solar noble gases had been trapped from the solar wind in regoliths on asteroids (*Eberhardt et al.,* 1965), similar to the large amounts of solar-wind noble gases implanted into the dust grains on the lunar surface (e.g., *Pepin et al.,* 1970). While at least for some meteorites this trapping may have occurred early in solar system history, the solar component nevertheless lost the status of being "primordial," since it is not thought that the solar noble gases in asteroidal regoliths generally reflect the accretionary or preaccretionary history of meteorites. *Wieler* (1998, 2002) and *Podosek* (2003) provide recent reviews on regolithic solar noble gases. We will use solar gases as a baseline composition, but not discuss them otherwise. However, we will address recent work indicating that some meteorites contain solar-like gases that appear to deserve again the attribute "primordial," as they seemingly were trapped during the formation of the solar system. We will also review attempts to derive the solar N-isotopic composition from lunar dust samples. Note also that the true abundances and isotopic compositions of noble gases and N in the Sun (more precisely, its outer convective zone) might be slightly different from measured values in the solar wind (*Wieler,* 2002). However, for the purposes of this paper these differences are small and will hardly be of concern.

Unlike the solar component, the origin of Signer and Suess' planetary noble gases has resisted a straightforward understanding, and partly remains mysterious despite decades of effort. What has become clear is that they represent a complex mixture of components of diverse origin, residing in different carrier phases, only some of which have been identified. It has also become clear that many parts of this mixture — if not all — were trapped into their carriers prior to accretion, either in the solar nebula or in presolar environments, and it is highly questionable whether these gases are the major source of the noble gases in the terrestrial planets. In order not to perpetuate such a misleading link, many workers thus discontinued to use the term "planetary" for noble gases that actually are found in meteorites only. This is also the position taken here. Note, however, that in the view of others the term is still useful to denote the ensemble of meteoritic components strongly depleted in the light elements relative to solar abundances (e.g., *Podosek,* 2003), in part because no well-accepted alternative term has emerged. We will sometimes use the term "primordial noble gases" instead, which, however, would also encompass primordially trapped noble gases of solar-like composition.

Table 1 (modified from *Ott,* 2002) lists the most important primordial noble gas components. A few additional important terms should be introduced here (cf. *Podosek,* 2003). Some primordial noble gas components have isotopic compositions not too different from solar. These are termed "isotopically normal." Normal components may be of "local" origin, i.e., derived from a solar-like composition after the solar nebula had been isolated from the surrounding molecular cloud, or they may represent a presolar mixture from various nucleosynthetic sources that resembles the solar mix. The counterpart to "normal" is "anomalous." An anomalous component has an isotopic composition sufficiently different from solar in at least one noble gas as to exclude a possible derivation from a reservoir of solar (or approximately solar) composition. Anomalous components thus indicate (possibly diluted) contributions from specific nucleosynthetic processes in stars. Anomalous components are carried by presolar grains (which may also carry normal components). Some of these grains formed around other stars, and their components are also often called "exotic." We will avoid using this term, because such components can at the same time be normal. Note, however, that some authors use "exotic" as a synonym for "anomalous."

Figure 1 shows several isotopic ratios of one anomalous (*HL*) and two normal (*P3* and *Q*) components, normalized to solar composition. Details are given in the figure caption and below in the sections on presolar diamonds and phase *Q*. The figure illustrates that anomalous components often have isotopic compositions very distinct from solar (e.g., the factor of 2 enrichment of both lightest and heaviest isotopes in Xe-*HL*). However, for the light noble gases He and Ne the characterization as normal or anomalous is often not based primarily on isotopic composition but rather on relations with the respective components in the heavier gases. An example is again the *HL* component.

Nitrogen, with only two stable isotopes, is much more ambiguously assigned to such headings as normal or anomalous. The willingness of N to participate in chemical reac-

TABLE 1. Major primordial noble gas components.

Component	Alternative Names*	Carriers	Remarks
Q gases	*P1, OC*-Xe†	phase *Q*	dominates Ar, Kr, Xe $\delta^{15}N$ = –50‰ to –15‰
Ureilite gases	Kenna type†	diamond, graphite	very similar to Q gases $\delta^{15}N$ = ≈–120‰
P3 gases	Ne-*A1*	presolar diamond	isotopically normal
P6 gases		presolar diamond	isotopically normal?
HL gases	Ne-*A2*‡	presolar diamond	nuclear component (r-process?)
N component		SiC and presolar graphite§	isotopically normal
G component	Ne-*E(H)*, Kr-*S*, Xe-*S*	SiC and presolar graphite§	nuclear component (s-process)
R component (Ne only)	Ne-E(L)¶	presolar graphite	radiogenic ^{22}Ne from ^{22}Na
Primordial solar		bulk/silicates**	

* Alternative names often refer to one noble gas only, although the component may be defined for all five noble gases.
† As Q gases are ubiquitous in all primitive meteorite classes, we discourage the use of host-specific names for this component.
‡ Ne-*A2* = Ne-*HL* + Ne-*P6*.
§ SiC and graphite probably carry discrete *N* and *G* components respectively.
¶ Ne-E(L) = Ne-*R* + Ne-*G* (*Amari et al.,* 1995a).
** "Subsolar" gases, mainly found in enstatite chondrites, are a mixture containing a small fraction of solar noble gases (section 3.3.5).

Table adapted from *Ott* (2002), where the isotopic compositions of these and other components are also compiled. Most components occur in all primitive meteorite classes, some in achondrites as well.

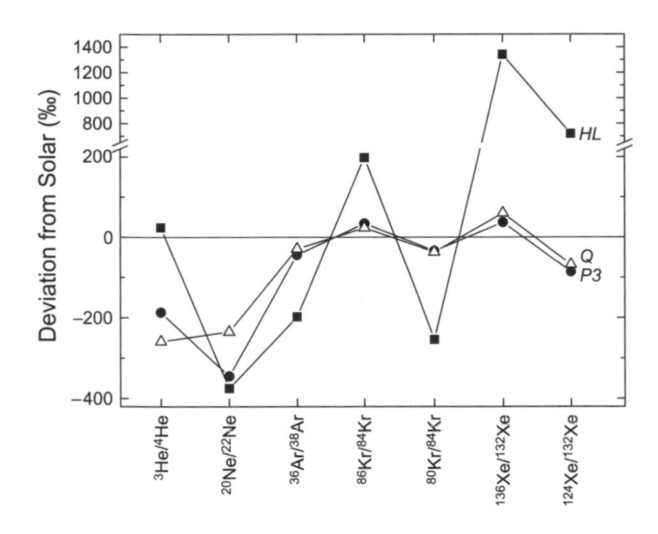

Fig. 1. Selected isotopic ratios of one isotopically "anomalous" and two "normal" noble gas components in meteorites (data from *Ott,* 2002). For Xe compare also with Fig. 4. The anomalous *HL* as well as the normal *P3* component are both carried by presolar diamonds; the *Q* component resides in the ill-defined phase *Q*. Isotopic ratios are normalized to the protosolar composition for He [assumed to be identical to that in Jupiter (*Mahaffy et al.,* 1998)], to best estimates for the bulk solar composition for Ne and Ar (*Wieler,* 2002) and to solar-wind values for Kr and Xe (*Wieler,* 2002). Protosolar He rather than solar wind He is used for normalization because the latter has become enriched in the Sun by D burning. In the three heavy noble gases, the anomalous *HL* component is much less close to solar composition than the normal *Q* and *P3* gases. Most diagnostic is the factor of ~2 enrichment in both the lightest and the heaviest isotopes in Xe-*HL*. However, in He and Ne no such clear distinction is seen. The figure therefore also shows that for the light gases the attributes "normal" and anomalous" are not based on isotopic composition but on affinities with the respective heavy gases, such as similar release characteristics.

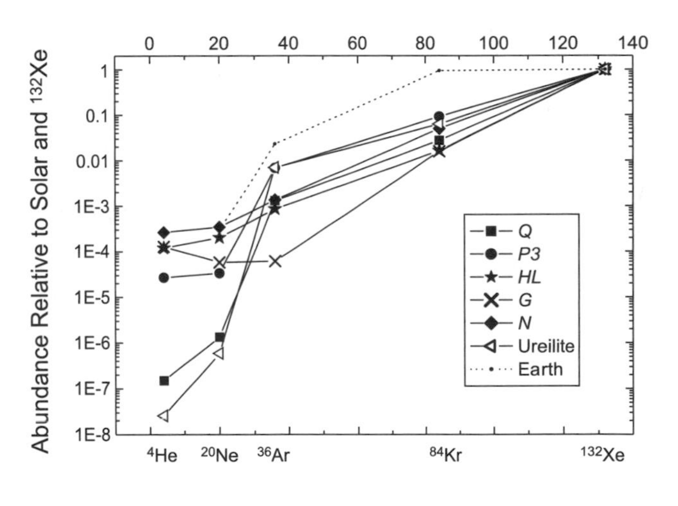

Fig. 2. Elemental abundances of major primordial noble gas components in meteorites (*Ott,* 2002), normalized to abundances estimated for the Sun (*Wieler,* 2002). The abundance pattern of the terrestrial atmosphere is also shown (except for He, which escapes from the atmosphere).

tions means that large isotopic fractionations associated with reactions at low temperature are difficult to resolve unambiguously from the contribution of anomalous components unless clearly associated isotope signatures from other elements are present.

Figure 2 displays the elemental abundances of major components, normalized to solar composition. It should be noted that it is not always possible to quantify or even identify specific N components associated with all of the noble gas components, although some success has been achieved in identifying N-isotopic signatures associated with some of the noble gas components. Details about the elemental and isotopic composition of the various noble gas and N components in meteorites are given by *Ott* (2002) and *Grady and Wright* (2003).

3. ORIGIN OF CARRIERS AND TRAPPING PROCESSES OF PRIMORDIAL NITROGEN AND NOBLE GASES

3.1. Component Separation

Early workers separated various noble gas components by stepwise heating or physical separation of bulk meteorite samples (*Reynolds and Turner,* 1964; *Black and Pepin,* 1969; *Eberhardt,* 1974). Such work provided clear-cut evidence that presolar solids had survived in meteorites. Investigating these, *Lewis et al.* (1975) found that most of the primordial noble gases of the Allende carbonaceous chondrite reside in the residue that remains after dissolving the bulk meteorite in HF and HCl. Since this landmark paper, work on primordial noble gases largely concentrated on acid-resistant residues and much of the subsequent discussion will be focused on such data. Notwithstanding the huge success of this approach, it also has its potential pitfalls; however, since primordial gas components residing in acid-soluble phases might go largely undetected, just as the "burning the haystack to find the needle" approach to isolate presolar grains will fail to recognize less-resistant types of such grains. Indeed, we will discuss also recent *in vacuo* etch studies of bulk meteorite samples that have allowed characterization of primordial noble gas components in a way never before possible.

In the following, we first discuss the noble gas and N components in bona fide presolar grains, before treating the components whose origin or carriers are less clear.

3.2. Presolar Grains and Their Noble Gases and Nitrogen

Arguably, the most remarkable fact about noble gases in meteorites is that preserved presolar grains contain a sizeable fraction of the bulk noble gas budget. Therefore, noble gases are ideal to trace enrichments or depletions of presolar grains in separated phases. Nitrogen shows similarly extreme variations, but in most cases the analyses are performed independently of the noble gases, often benefiting from analyses of individual grains with corresponding isotopic measurements of other elements in the same grains, although hardly ever noble gases. Therefore, it is not always straightforward or even possible to link the N information to that obtained from the noble gases.

All types of presolar grains studied for noble gases (SiC, graphite, and nanodiamonds) contain both anomalous and isotopically normal components. The anomalous components can be linked to specific nucleosynthetic sources, e.g., the s-process [for a discussion of nucleosynthetic processes see *Meyer and Zinner* (2006)], whereas the normal components presumably represent a mixture of various sources representative of the various parent stars of the grain assemblage, just as the solar noble gas composition represents the average of all sources having contributed to the Sun. We now present first an overview of the noble gas and N inventories in the various types of presolar grains and then discuss trapping and modification processes.

3.2.1. Noble gas and nitrogen components in presolar silicon carbide, graphite, and diamonds. 3.2.1.1. Silicon carbide. A normal (*N*) and an anomalous (*G*) noble gas component are present in SiC separates from the Murchison CM2 chondrite (*Lewis et al.,* 1990, 1994; cf. *Ott,* 2002) (To distinguish the *N* component from the elemental symbol N for nitrogen, the *N* component, and hence all other acronyms for noble gas components, are italicized.) Silicon carbide grains show isotopic anomalies in many elements as expected for contributions from s-process nucleosynthesis (*Meyer and Zinner,* 2006), strongly indicating an origin in asymptotic giant branch (AGB) stars for the large majority of the SiC grains. Indeed, both the *N* and the *G* component are remarkably close in composition to values expected for the envelope and the s-process region of AGB stars respectively (*Gallino et al.,* 1990, *Lewis et al.,* 1994) (see Fig. 3).

The *N* component is thus a mixture of the noble gases that made up the bulk of the many AGB stars that contributed SiC grains to the Murchison meteorite. Although isotopically normal, the *N* component is enriched in the two heaviest noble gases Kr and Xe by more than 3 orders of magnitude relative to He and Ne and solar composition.

In contrast, the *G* component ("*G*" stands for "giant") represents products of nucleosynthesis in the s-process region in C-rich red giant stars in their AGB phase, as is discussed next. Neon very highly enriched in ^{22}Ne was among the first anomalous components recognized (*Black and Pepin,* 1969; *Eberhardt,* 1974). Known as Ne-*E*, it was originally thought to derive from the decay of very short-lived ^{22}Na (half-life = 2.6 yr). This remains the accepted interpretation for the major fraction of the ^{22}Ne in presolar graphite (see below). However, the theoretical and experimental work on SiC has shown that the subcomponent known as Ne-*E*(*H*) actually also contains minor amounts of ^{20,21}Ne and was produced by stellar nucleosynthesis in the same zone as the s-process-enriched Kr and Xe in SiC grains (*Lewis et al.,* 1990, 1994; *Gallino et al.,* 1990; *Straniero et al.,* 1997). This has been confirmed by He and Ne analyses on single SiC grains by *Nichols* (1992), *Nichols et al.* (1992a, 1994), and *Heck et al.* (2005). Only SiC grains rich in ^{22}Ne also contain ^{4}He above background, with ^{4}He/^{22}Ne ratios indicating an origin of this component in the region of s-process nucleosynthesis (*Straniero et al.,* 1997) in (low-mass) AGB stars. Ne-*E*(*H*) has thus been rebaptized Ne-*G*.

Kr-*G* and Xe-*G* are both strongly enriched in the isotopes produced only by the s-process (relative to the *N* component or solar composition). Xenon-*G* as well as several other important Xe components discussed later are shown in Fig. 4. Kr-*G* is particularly interesting, because its composition depends on the neutron density and temperature during the s-process (*Ott et al.,* 1988, *Gallino et al.,* 1990). *Lewis et al.*

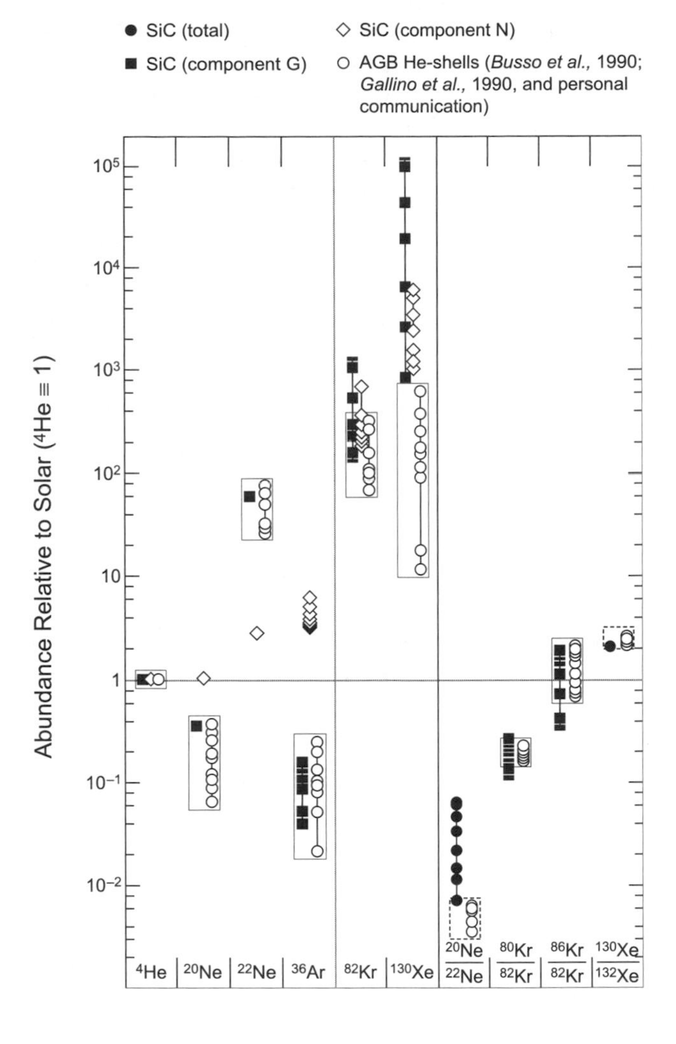

Fig. 3. Comparison of elemental abundances (left) and isotopic ratios (right) of noble gases in SiC from the Murchison meteorite with values calculated for production in s-process region in AGB stars. Isotopic ratios are normalized to solar ratios, elemental abundances to solar abundances and ^{4}He. Boxes encompass range of values calculated for the s-process zone (*Busso et al.,* 1990; *Gallino et al.,* 1990). Measured data and figure are from *Lewis et al.* (1994).

(1990, 1994) showed that the sensitive Kr-isotopic ratios vary with the grain size of the SiC. The Kr-isotopic composition correlates with variations in the elemental abundances, especially the Ne/Xe ratio. *Gallino et al.* (1990) and *Pignatari et al.* (2003) showed that the measured Kr-*G* and Xe-*G* compositions are strikingly similar to values expected for AGB stars of ~1.5–3 $M_{\odot}$ and metallicities slightly less than solar. The Ne and Ar data support this conclusion (*Lewis et al.,* 1994).

Stepped heating or combustion extractions of SiC have revealed a bulk N-isotopic composition enriched in ^{14}N with $\delta^{15}N$ values down to –650‰ (e.g., *Ash et al.,* 1989) and corresponding ^{13}C enrichments ($\delta^{13}C \approx$ 1430‰ for bulk residue) (e.g., *Russell et al.,* 1997). While no evidence for multiple components has been uncovered by stepped combustion, ion microprobe measurements of individual SiC grains have revealed a number of different grain types. About 90% of the grains (the mainstream grains) have isotopic signatures consistent with those identified by the stepped-heating experiments, with elevated $^{14}N/^{15}N$ ratios up to 5000 ($\delta^{15}N \approx$ –950‰) coupled with $^{12}C/^{13}C$ ratios from ≈40 to 100 ($\delta^{13}C \approx$ +1250‰ to –100‰) (for details see *Zinner,* 1998; *Hoppe and Zinner,* 2000; *Meyer and Zinner,* 2006). This indicates an input from H burning via the CNO cycle in red giant carbon stars. The additional presence of s-process noble gases indicates an origin in the thermally pulsing asymptotic giant branch (TP-AGB) phase.

Apart from the mainstream grains, several further groups of SiC grains, labeled A, B, Y, and Z, also contain N with elevated $^{14}N/^{15}N$ ratios. They formed in similar types of nucleosynthetic environments as the mainstream grains. Variations in the isotopic and elemental abundances of these grains reveal much about the metallicity or evolutionary history of the source star (for details see *Hoppe et al.,* 1997; *Amari et al.,* 2001a,b; *Ott,* 2002; *Meyer and Zinner,* 2006). Approximately 1% of the SiC grains are so-called X grains, with $^{14}N/^{15}N$ ratios of 13–180, much less than the solar estimate of 435 (section 5.3) and $^{12}C/^{13}C$ ratios generally higher than solar — indicative of He burning. Most probably they

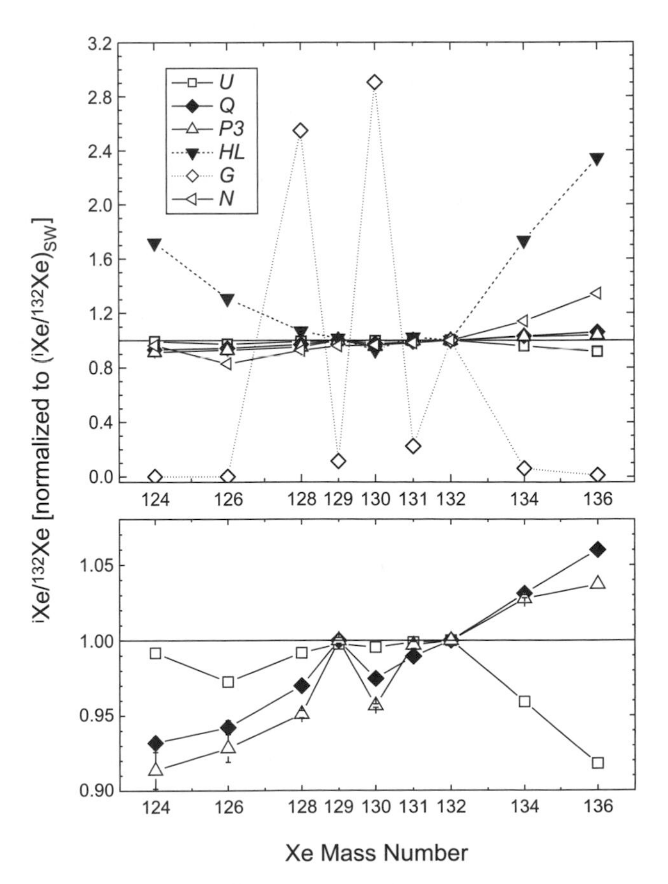

Fig. 4. Isotopic compositions of some major Xe components discussed in the text. Data are from *Ott* (2002), who also lists further Xe components not shown here. Ratios are normalized to ^{132}Xe and solar-wind composition as given by *Wieler* (2002).

originate in Type II supernovae, as indicated by large overabundances of ^{28}Si among other isotopes (*Hoppe and Zinner,* 2000; *Ott,* 2002; *Meyer and Zinner,* 2006).

3.2.1.2. Graphite. Unlike SiC, which is dominated by grains from AGB stars, presolar graphite has sizeable contributions from several sources (e.g., *Zinner,* 1998; *Meyer and Zinner,* 2006). The noble gases are again mixtures of normal and anomalous components. As in SiC, Kr and Xe are both dominated by a *G* component from s-process nucleosynthesis. Hence, noble gases are in line with conclusions based on other elements that AGB stars also supply a large part of the presolar graphite grains. Unlike SiC, the Ne budget in graphite is dominated by Ne-*R*, however (*Amari et al.,* 1995a), with only 5–50% being Ne-*G*. Ne-*R* [also known as Ne-*E*(*L*)] is monoisotopic ^{22}Ne from the decay of ^{22}Na (hence *R* = "radiogenic"). About 30–40% of the single graphite grains studied by *Nichols et al.* (1992a, 1994) contain ^{22}Ne above background levels, but not even the most ^{22}Ne-rich grains contain detectable amounts of ^{4}He or ^{20}Ne. This implies that the bulk of the ^{22}Ne in graphite separates does not come from AGB stars, in agreement with *Amari et al.* (1995a). Strictly speaking, however, only for the most ^{22}Ne-rich grains is an origin from ^{22}Na decay definitely needed, implying explosive nucleosynthesis from novae or supernovae. Additional single grain analyses would be highly desirable to further constrain the actual fraction of graphite grains containing Ne from explosive nucleosynthesis.

Ion microprobe analysis of individual graphite particles reveals that almost 50% contain N and C with isotopic signatures similar to the SiC X grains and indicative of an explosive nucleosynthetic origin (e.g., *Hoppe et al.,* 1995; *Amari et al.,* 1995b; *Zinner et al., 1995*; *Ott,* 2002). Other graphite grains have isotopic signatures indicative of the AGB stars identified as the source of the mainstream and the A and B SiC grains. A final group has solar-like ^{12}C/^{13}C ratios and N-isotopic compositions that concentrate strongly around the terrestrial value, with only a few examples extending to more ^{15}N-rich values. This has been taken to indicate that these grains condensed from the molecular cloud from which the solar system formed (*Hoppe et al.,* 1995; *Zinner et al.,* 1995). However, it should be noted that according to the most recent estimate the solar N-isotopic composition is more ^{14}N-enriched than terrestrial N (see section 5.3). Therefore, the graphite grains with terrestrial-like N may have more in common with the source of the terrestrial atmosphere than the bulk solar system value.

3.2.1.3. Nanodiamonds. This presolar carrier contributes typically some 90% to the primordial He and Ne inventory and still some 10% to the heavier primordial noble gases in bulk primitive meteorites. The isotopic compositions of the various noble gas components carried by diamonds have been compiled elsewhere (e.g., *Ott,* 2002). The most diagnostic gas is the anomalous Xe-*HL*, which is enriched in both the heaviest (*H*) and the lightest (*L*) isotopes (Fig. 1). Xenon-*HL* was first detected by *Reynolds and Turner* (1964) and has been instrumental in separating the diamond carriers. It is accompanied by a distinct component in all other noble gases, which have thus also been assigned the *HL* label, although the isotopic composition of He-*HL* and Ne-*HL* is not obviously anomalous. Also, Ar-*HL* and Kr-*HL* are considerably less different from solar composition than is Xe-*HL*. To complicate matters further, the diamonds appear to contain two additional primordial components of apparently "normal" (but clearly distinct) composition in all five noble gases, called *P3* (Fig. 1) and *P6* (*Huss and Lewis,* 1994) (see section 3.3.4).

Xenon-*HL* is of particular interest because it contains large enrichments of isotopes produced only in the p-process (^{124,126}Xe) *and* large enrichments of isotopes made in the r-process only (^{134,136}Xe). At least those few diamonds carrying Xe-*HL* are thus of likely supernova origin (*Meyer and Zinner,* 2006). It is a major question whether diamonds carrying Xe-*L* differ in some property from those carrying Xe-*H*. Attempts to partially separate the two subcomponents by selective absorption of laser light have remained inconclusive so far (*Meshik et al.,* 2001), although *Verchovsky et al.* (1998) reported variations in Xe-isotopic composition with grain size that they argue indicate different Xe components were implanted into the diamonds at different energies. Note that on average only 1 out of 1,000,000 nanodiamonds contains an atom of Xe-*HL*, and hence in all likelihood is presolar. This is often stressed because the isotopic composition of C and probably also N (see next paragraph) of nanodiamond samples are solar-like, suggesting the possibility that most of the nanodiamonds have a solar system or molecular cloud origin (*Meyer and Zinner,* 2006). However, it should not be forgotten that possibly 1 out of 10 nanodiamonds contains an atom of He-*HL* (the extreme alternative that 1 out of 1,000,000 diamonds contains 100,000 He-*HL* atoms is unrealistic, given that an individual nanodiamond consists of a mere 1000 or so C atoms). Since He-*HL* is closely tied to the exotic Xe-*HL*, this may thus indicate that the presolar fraction of the nanodiamonds is considerably larger than the minimum implied by the Xe-*HL* concentration.

Nitrogen is abundant in the nanodiamonds, with concentrations of up to 1.3 wt%. There is no evidence for any significant variation in the ^{14}N/^{15}N ratio of the diamonds, although there may be a N-rich and N-poor population (*Russell et al.,* 1996). The distinctive δ^{15}N value of about –350‰ (^{14}N/^{15}N ≈ 420) was initially taken as outside normal solar system values and therefore indicated that many, if not all, of the nanodiamonds were presolar (*Lewis et al.,* 1983; *Russell et al.,* 1996). The measured δ^{15}N value in the diamonds is, however, remarkably similar to several recently measured values for ammonia in the atmosphere of Jupiter by the Infrared Space Observatory (ISO) (*Fouchet et al.,* 2000), Galileo (*Atreya et al.,* 2003), and Cassini (*Fouchet et al.,* 2004): -480^{+240}_{-280}‰, –374 ± 82‰, and –395 ± 140‰, respectively. The jovian value is generally interpreted as a good proxy for the solar value. Therefore the similarity with the value for the nanodiamonds does at least call into question whether the N-rich component deserves the label presolar. The near-solar ^{12}C/^{13}C ratio (≈92) of the nanodiamonds (e.g., *Lewis et al.,* 1983) also suggests a local origin for the vast bulk of the nanodiamonds. However, the very large range in $\delta^{15}N$ values displayed by materials generally accepted as forming in the solar nebula (e.g., many meteoritic components, Jupiter, the Sun — see section 5.3) or later means that it is difficult to determine unequivocally a presolar origin for the N-containing nanodiamonds. On the other hand, the inferred $^{12}C/^{13}C$ ratio possibly as high as 120 in the N-rich nanodiamond fraction postulated by *Russell et al.* (1996) does offer some support to the argument for a presolar origin for the N-rich nanodiamonds.

3.2.2. Trapping and modification processes of noble gases in presolar grains. Ion implantation from a stellar wind is the most likely trapping mechanism for most — if not all — noble gas components in the various types of presolar grains. For the *G* component in SiC the main argument for this inference is that the measured elemental abundances agree remarkably well with predictions for He shell production (*Lewis et al.,* 1994), hence a mechanism that does not strongly fractionate elemental abundances is required. Subsequent diffusive losses of the lighter gases apparently did not occur and SiC therefore seems to contain a pristine sample of matter from the s-process zone in AGB-stars. It appears likely that also graphite trapped the *G* component by ion implantation (*Ott,* 2002), but Ne possibly was lost later, as its abundance is lower than expected by 1–2 orders of magnitude (*Amari et al.,* 1995a).

The relatively high Kr and Xe abundances in the *N* component in SiC (Fig. 3) indicate a fractionation process. *Lewis et al.* (1990) proposed a scenario in which SiC grains trapped ions from a stellar wind during condensation in the envelopes of AGB stars. The variable elemental ratios in *N* and *G* components appear to require two wind components, one from a fully ionized region and one from a cooler, partly ionized zone. The latter would contribute most of the Kr and Xe of the *N* component. Support for this scenario comes from model calculations by *Verchovsky et al.* (2004), who suggest that the *N* component was implanted with low energy (wind speed 10–30 km/s) during the main AGB stage, whereas the *G* component was implanted with a few thousand kilometers per second at the very end of the AGB phase.

It is remarkable that the isotopic composition of Kr-*G*, which is established in the hot interior of a star, correlates with the size of the carrier grains, which depends on conditions in the cool atmosphere. *Lewis et al.* (1994) point out that both properties ultimately depend on mass and metallicity of the star and might be coupled this way. *Nuth et al.* (2003, and personal communication) indeed predict large grains (those mainly analyzed so far) to come from low-mass stars (1–3 $M_{\odot}$), whereas larger stars would produce smaller (0.1 μm) grains.

Ion implantation is also the most viable trapping process for both the anomalous and the normal noble gas components in diamonds. Arguments for this are summarized by *Ott* (2002) and include decreasing gas content with grain size (*Verchovsky et al.,* 1998) and similar release patterns from artificially implanted and meteoritic nanodiamonds (*Koscheev et al.,* 2001). Artificial ion implantation experiments into terrestrial nanodiamonds call for caution when interpreting noble gas data from meteoritic diamonds, because their very small size may lead to isotopic fractionations of the implanted portion and also to a bimodal temperature release pattern of a single implanted component (*Koscheev et al.,* 2001). One possible consequence of this is that the true Xe-*HL* composition may actually be free of any s-process contribution. Another consequence of this work is that the isotopic fractionation may compromise our understanding of the relationship between Xe-*HL* on the one hand and its accompanying He-*HL* to Kr-*HL* on the other, because the isotopic compositions of the implanted gases may be different from what we measure (*Huss et al.,* 2000). In fact, *Huss and Lewis* (1994) suggested previously that the *HL* component has a normal and an anomalous part, the anomalous one becoming more prominent in the heavy gases.

The Ne-*R* very likely was incorporated into presolar graphite by condensation of ^{22}Na during grain growth rather than by implantation as ^{22}Ne already, as otherwise substantial amounts of ^{20}Ne would also be expected (*Nichols et al.,* 1992a).

In the case of N there is no systematic variation in the abundance levels with nanodiamond grain size, indicating that ion implantation is not the primary trapping mechanism (*Verchovsky et al.,* 1998). Rather, the N content is thought to be determined by the partial pressure of N or N-bearing species (e.g., HCN) at the time and place where the diamonds formed (*Verchovsky et al.,* 1998), probably by chemical vapor deposition (section 3.3.4). High, and sometimes variable, concentrations of N in the SiC and graphite grains (e.g., *Ash et al.,* 1989; *Smith et al.,* 2004) also show that the N is trapped chemically and that implantation plays a minor role at most.

As described above, the noble gas component structure in nanodiamonds is even more complicated than those of SiC or graphite, which is one reason why the former is less well understood than the latter two. It therefore remains open whether the various components may reside in different subpopulations produced in different circumstellar environments. *Huss and Lewis* (1994) conclude that the *HL* component resides in bulk diamonds rather than trace phases and that the carriers of *HL* and *P6* gases are probably different.

In summary, grains having entered the solar nebula as solids and having remained solid ever since contribute a remarkably large fraction of the primordial noble gases, but not N, in meteorites. Some of these gas components appear to have remained unfractionated, testifying to the excellent state of preservation of some presolar grains. On the other hand, a large fraction of the presolar grains may have lost their primordial noble gases (*Nichols et al.,* 1992a).

3.3. Components of Less-Certain — Presolar or Early Solar System — Origin

In addition to noble gas and N components that can directly be attributed to nucleosynthesis in stars, meteorites contain components whose origin is not unequivocally presolar or whose identity most likely was established in the early solar system. These are the "*Q* gases" and the related N residing in the enigmatic "phase *Q*" in chondrites and their close relatives found in ureilites, as well as the "*P3*" and "*P6*" noble gases residing in nanodiamonds. Furthermore, primordial noble gases of solar composition (not to be confused with solar wind noble gases incorporated more recently in a planetary regolith) are being found in an increasing number of meteorites of various classes.

3.3.1. Noble gases in phase Q. *Lewis et al.* (1975) observed that a large portion of the primordial noble gases in the carbonaceous HF/HCl-resistant residues of the CV3 chondrite Allende were lost upon oxidation. These authors dubbed the oxidizable, almost mass-less gas-rich carrier "phase *Q*" (for "quintessence"). Numerous subsequent analyses yielded relatively uniform Ne-Xe element and normal isotope abundances of the *Q* gases in all primitive meteorite classes, implying that phase *Q* was widespread in the early solar system (e.g., *Srinivasan et al.,* 1977, 1978; *Reynolds et al.,* 1978; *Alaerts et al.,* 1979a,b; *Matsuda et al.,* 1980; *Moniot,* 1980; *Huss et al.,* 1996). Comparing the noble gases in unoxidized and oxidized residues does not allow a good characterization of He and Ne in phase *Q* due to their low abundance relative to the presolar gases present in oxidized residues. A precise determination of the He- and Ne-isotopic compositions in phase *Q* became possible with *in vacuo* etch experiments releasing essentially the pure *Q* gases (*Wieler et al.,* 1991, 1992; *Busemann et al.,* 2000, 2001).

Q gases are heavily fractionated relative to solar composition favoring the heavier elements and isotopes (Figs. 1, 2, and 4) (*Ott,* 2002). Up to 10% of the primordial He and 90% of the primordial Xe in primitive meteorites originate from phase *Q*. Solely Ne and He appear to show isotopic variations, whereas Ar, Kr, and Xe in phase *Q* are isotopically uniform. The *Q* component encompasses "OC-Xe" in ordinary chondrites (*Lavielle and Marti,* 1992). Average carbonaceous chondrite (AVCC) composition represents the sum of all primordial components in carbonaceous chondrites (*Eugster et al.,* 1967). Despite many efforts, the carrier phase *Q* and its origin remain unknown. In the following, we will discuss proposed trapping mechanisms, starting with models that assume a presolar origin of phase *Q*, followed by hypotheses for a "local" origin, in the vicinity of the early Sun or on parent bodies.

Much remains to be determined regarding the origin of phase *Q*. The most critical issues to be accounted for are (1) the fractionated isotope and element abundances of the *Q* gases relative to solar, especially the isotopic composition of He-*Q* ($^3He/^4He = (1.23 \pm 0.02) \times 10^{-4}$) (*Busemann et al.,* 2001) and variations in Ne-*Q* (*Busemann et al.,* 2000); (2) the presence of small but significant amounts of He and Ne (*Busemann et al.,* 2000; *Wieler et al.,* 1991); (3) the high *Q* gas release temperatures of ~1200°–1600°C (*Huss et al.,* 1996), (4) the chemical resistance to HF and HCl combined with the strong susceptibility to oxidizing agents (*Moniot,* 1980; *Ott et al.,* 1981; *Huss et al.,* 1996); and finally (5) the gas concentration in phase *Q*, which is extremely high irrespective of the ill-constrained extent to which the carrier phase is actually destroyed during noble gas release. In addition, physical conditions of the suggested trapping environments (e.g., partial pressure of the noble gases in the nebula) must be considered.

Various workers favor a presolar origin of phase *Q*. The presence of *Q* gases in all primitive meteorite classes implies complete mixing and widespread distribution of phase *Q* in the solar system. This appears easier to achieve in the presolar cloud than in planetary environments. *Huss and Alexander* (1987) pointed out that the noble gases in the Sun's parent molecular cloud, mixed from different stellar sources, were already isotopically normal. They suggested that at the low dust temperatures of 10–30 K Ar-Xe would have been trapped in icy mantles around presolar grains. Exposure to UV photons or energetic ions would form radicals and other organic molecules. Evaporation of the ice would leave carbonaceous mantles that keep preferentially the heavy noble gases. While this model explains the enrichment in the heavy noble gases, the trapping efficiency for He and Ne remains uncertain as well as the general isotopic fractionation of the *Q* gases relative to solar abundances. *Sandford et al.* (1998) simulated trapping in the Sun's parent molecular cloud (10–25 K, ionizing UV irradiation). They succeeded at producing carbonaceous residues that carry heavy noble gases in similar concentrations as acid-resistant residues and with *Q*-like fractionation. However, the major gas release from the residues occurred already by 250°C. Enormous amounts of Xe were also trapped in vapor-deposited silicate condensates by anomalous chemical adsorption. These grains may subsequently have been coated by stable organic mantles under UV irradiation in the cold presolar molecular cloud (*Nichols et al.,* 1992b). The trapped Xe was indeed released at relatively high temperatures and showed the *Q*-typical isotope fractionation. However, Ne and Ar were not significantly captured and the presence of noble gases in coated silicates in meteorites is speculative.

Hohenberg et al. (2002) suggested active capture and anomalous adsorption to explain the large Kr and Xe concentrations of phase *Q*. Anomalous adsorption on growing surfaces involves chemical bonds, in contrast to the Van der Waals forces of normal adsorption. When chemically active species are insufficiently available, Kr and Xe can form transient chemical bonds on fresh surfaces. Subsequent deposition of material enhances capture efficiency and gas retentivity. Active capture experiments reproduced both the large Xe concentrations in phase *Q* as well as the fractionation of the Xe isotopes relative to solar composition. Possibly, an enhanced energetic proton flux from the pre-main-sequence Sun formed radicals on the surface of presolar diamonds, which adsorbed Kr and Xe, although it is unclear whether the process can replicate the isotopic fractionation of Kr-*Q* relative to solar, or is capable of accounting for the lighter *Q* gases.

The correlation of the abundances of both *Q* gas and presolar grains with petrologic type in various meteorite classes implies an intimate mixing of phase *Q* and presolar dust in the Sun's parent molecular cloud and thus a presolar origin of phase *Q* (*Huss,* 1997). The putative formation of phase *Q* in the presolar cloud as carbonaceous grain mantles requires presolar grains as "condensation nuclei." *Busemann et al.* (2000) observed a correlated release of Ne-*R/G* (most likely from presolar graphite) and a subfraction of the *Q* gases upon mild etching of the CM2 chondrite Cold Bokkeveld. A close relationship between phase *Q* and presolar diamonds was suggested by *Amari et al.* (2003), based on the similar *Q*/diamond ratios in various separates of Allende. However, *Q* has higher gas-release temperatures and a better resistance to metamorphism than presolar diamonds, which leads to variable *Q*/diamond ratios (*Huss,* 1997). This argues against a close physical relationship between the two carriers.

"Local" models aim to derive *Q* gases from solar noble gases after formation of the Sun. However, most of these models have not identified adequate regions within the solar system for the required fractionation and the formation of the chemically resistant, carbonaceous carrier. Furthermore, phase *Q* must have been well mixed into the meteorite forming regions together with presolar grains and dust. *Ozima et al.* (1998) suggested that Rayleigh fractionation of initially solar noble gases would explain the isotopic differences between *Q*- and solar gases. An additional fractionation mechanism would be responsible for the elemental composition of the *Q* gases, presumably somewhere in the gaseous nebula. In this scenario, the starting He is isotopically similar to present-day solar wind, i.e., the *Q* gases originate from the Sun after pre-main-sequence D burning (but see also *Pepin and Porcelli,* 2002).

Based on experiments with laboratory diamonds, *Matsuda and Yoshida* (2001) proposed that *Q* gases may have been trapped on the surfaces of presolar diamonds in a plasma (see section 3.3.3 for similar mechanisms suggested for the *Q* gases in ureilitic diamonds). However, as pointed out above, a close physical relationship between *Q* and presolar diamonds appears to be unlikely.

Two models suggested by *Pepin* (1991, 2003) explore the possibility that the *Q* gas pattern was generated by hydrodynamic escape. The first model invokes isotopic fractionation during the blow-off of transient atmospheres of early planetesimals. Primordial solar Ne-Xe accreted in ice-mantles of presolar grains later became concentrated below surface ice layers on these bodies. Impacts caused hydrodynamic blow-off of the noble gases, driven by H_2 or — more likely — CH_4. The residual gas is enriched in the heavier species and reproduces the isotopic patterns observed in the *Q* gases. Subsequent impacts induced the formation of phase *Q*. Among other constraints on the dynamics of the early solar nebula, this model requires that the *Q* gas carriers from various planetesimals must have been thoroughly mixed before having been incorporated into present-day meteorite parent bodies.

The second model involves fractionation during the dissipation of the solar accretion disk. Ultraviolet irradiation from a nearby massive star initiated photo dissociation and strong H outflows from the disk. The consequence is hydrodynamic escape of the noble gases that are assumed to be initially solar in composition (*Pepin,* 2003). However, the noble gases were sufficiently fractionated to resemble *Q* element patterns only at the accretion disk's boundary, while the isotopic compositions observed for *Q* were not reproduced at all. Furthermore, the hot environment would hamper retentive trapping of the gas.

The origin of phase *Q* is closely related to its physical properties and will most likely only be resolved by establishing correlations of *Q* gas release with compositional properties of the releasing phase(s), such as the D/H or $^{15}N/^{14}N$ ratios, C contents, or changes in their carrier structures upon gas release. A number of attempts have been made to identify and characterize the N associated with phase *Q*. The results indicate variable $\delta^{15}N$ values, from ≈–50‰ in CO3 meteorites to –15‰ in ordinary chondrites (*Murty,* 1996; *Hashizume and Nakaoka,* 1998), while a component in ureilites most similar to *Q* (see below) has a $\delta^{15}N$ value of –21‰ (*Rai et al.,* 2003a). However, the significance of these values is unclear. Nitrogen is present in a wide range of phases whereas high-resolution stepped combustion of HF/HCl residues of a suite of enstatite chondrites reveals that *Q* noble gases appear to be highly concentrated in a very minor part of the acid-resistant C inventory (<5%). The N in these residues does not appear to be related to either the main carbon or noble gas releases (*Verchovsky et al.,* 2002).

3.3.2. The nature of phase Q. Substantial work has also focused on revealing the nature of phase *Q*. The earliest suggestion was chromite or an Fe-Cr-Ni-rich sulfide (*Anders et al.,* 1975; *Gros and Anders,* 1977), but both were excluded by *Frick and Chang* (1978). Subsequently, trapping experiments on synthesized carbonaceous matter as well as density separation and combustion of meteoritic residues have shown that phase *Q* is mainly carbonaceous (*Frick and Chang,* 1978; *Frick and Pepin,* 1981; *Ott et al.,* 1981). Later, *Wacker et al.* (1985) and *Wacker* (1989) suggested that the *Q* gases might have been physically adsorbed at low pressures in the solar nebula in micropore labyrinths of amorphous carbon near grain surfaces. Adsorption experiments at low pressures by *Marrocchi et al.* (2005) showed that carbon blacks can trap gas amounts larger than those found in phase *Q*, with relative Ar and Kr (less so Xe) abundances in *Q* being reproduced. Best matches were achieved for trapping temperatures of 75–100 K, which the authors argue might be realistic in a later-stage solar nebula in the meteorite-forming region. It remains to be seen whether it is also possible to reproduce the release temperatures of phase *Q* of ~1000°C and to trap the small amounts of He and Ne. *Verchovsky et al.* (2002) concluded from stepped-combustion experiments that phase *Q* is presolar, volume rather than surface correlated, and a minor component of the total macromolecular C in chondrites (see also *Schelhaas et al.,* 1990). Phase *Q* appears to be more resistant to thermal metamorphism than the bulk macromolecular C.

Fullerenes (C_{60}) can enclose He-Xe atoms, if produced under high gas pressure (*Saunders et al.,* 1996). Becker and coworkers (e.g., *Becker et al.,* 2000) suggested fullerenes as carriers of the *Q* gases, following an earlier proposal by *Heymann* (1986) but withdrawn by *Heymann and Vis* (1998). However, convincing evidence for a correlation of *Q* gas and fullerene content in meteorites is lacking. *Ott and Herrmann*

(2003) showed that the noble gas trapping efficiency of artificial fullerenes decreases as mass increases from He to Xe, inverse to the trend required to explain the *Q* gas elemental patterns (Fig. 2). Moreover, most workers could not confirm the presence of C_{60} in carbonaceous chondrites at all (*Buseck,* 2002). *Heymann and Vis* (1998) and *Vis and Heymann* (1999) suggested that carbon nanotubes could be the carrier of the *Q* gases. However, nanotubes, as well as carbynes, suggested by *Whittaker et al.* (1980) to carry *Q* gas, have not been found in Allende, Leoville, and Vigarano (*Vis et al.,* 2002). Nevertheless, these authors suggested various three-dimensional closed carbon structures, such as nanotubes, carbon onions, or adsorption sites in labyrinths of pores in amorphous carbon, but not fullerenes, as possible *Q* gas carriers. A similar suggestion has been made by *Garvie and Buseck* (2004), who found hollow nanospheres and nanotubes in Tagish Lake and various CM chondrites.

Matsuda et al. (1999) and *Amari et al.* (2003) succeeded in significantly enriching phase *Q* and presolar diamonds in the Allende meteorite by physical separation techniques. The material most enriched in phase *Q* (the carrier itself?) appears to have a low density of (1.65 ± 0.04) g/cm^3, consistent with earlier observations by *Ott et al.* (1981).

There are hints on the existence of subfractions of phase *Q*. *Gros and Anders* (1977) defined phase Q_1 and Q_2 based on the susceptibility to nitric acid. While the identification of metal-sulfide and chromite as gas carriers turned out to be wrong (see above), the presence of two different subphases with different resistance to nitric acid and different noble gas compositions has been confirmed (*Busemann et al.,* 2000). Distinct release systematics of Ar and Xe also suggest different *Q* subcarriers (*Verchovsky et al.,* 2002).

3.3.3. Primordial noble gases and nitrogen in ureilites. Ureilites typically contain concentrations of C and primordial noble gases comparable to those in the most primitive carbonaceous chondrites. It appears difficult to reconcile this with the achondritic, igneous nature of the ureilites (*Goodrich,* 1992). A major carrier is diamond (not identical to the presolar nanodiamonds). The much more abundant graphite appears to be mostly gas-free (*Weber et al.,* 1976; *Göbel et al.,* 1978); gas-rich graphite has only been reported by *Nakamura et al.* (2000). A third carrier, even more gas-rich than diamond (*Göbel et al.,* 1978), was identified as amorphous fine-grained C (*Wacker,* 1986). *Ott et al.* (1985, 1986) and *Rai et al.* (2002) postulated a close affinity of this poorly characterized carrier to phase *Q*, which is supported by similar release characteristics of residues from enstatite chondrites and ureilites (*Verchovsky et al.,* 2002). The overall elemental fractionation pattern and the Ne-Xe-isotopic compositions in ureilites resemble those found in phase *Q* (*Marti,* 1967; *Wilkening and Marti,* 1976; *Ott et al.,* 1985; *Busemann et al.,* 2000).

Two widely different scenarios for the origin of the unique, gas-rich diamonds are discussed, according to which they either formed in the solar nebula or by shock on the ureilite parent body. The shock hypothesis (e.g., *Lipschutz,* 1964; *Vdovykin,* 1970; *Bischoff et al.,* 1999; *Nakamuta and Aoki,* 2000) may account for the igneous origin of the ureilites. It can explain the absence of diamonds in the least-shocked ureilites, the presence of compressed graphite, and also the good retention of noble gases. However, the character of the carbonaceous precursor and the reason for its high abundance remain uncertain. Graphite may be a potential precursor, but it is then difficult to explain why most graphite is gas-free (*Göbel et al.,* 1978). Possibly a more viable candidate is the amorphous fine-grained C (*Wacker,* 1986; *Goodrich et al.,* 1987).

The correlation between diamond concentration and shock stage in ureilites is subject to debate. Furthermore, N-isotopic data may argue against shock conversion of graphite or amorphous C into diamonds (*Rai et al.,* 2002, 2003a; see below). These observations may rather speak in favor of a scenario where ureilitic diamonds and the amorphous C have been formed by chemical vapor deposition (CVD) in the nebula (*Arrhenius and Alfvén,* 1971; *Fukunaga and Matsuda,* 1997; *Fukunaga et al.,* 1987).

Most noble gas trapping mechanisms suggested for the ureilites resemble those described for the *Q* gases (section 3.3.1). The elemental abundances relative to solar composition correlate well with the first ionization potential of the noble gases, which might indicate that the abundances were established at high nebular plasma temperatures (*Weber et al.,* 1971; *Göbel et al.,* 1978; *Rai et al.* 2003b).

Plasma-induced fractionation in the nebula agrees well with a CVD origin of the ureilitic diamonds, because synthetic CVD diamonds can trap significant noble gas amounts with a fractionation pattern similar to that observed in ureilites (*Matsuda et al.,* 1991). However, it is unclear how the required plasma conditions could have been established, especially as the presence of solar nebula diamonds exclusively in ureilites would require explanation. Furthermore, the (much smaller) presolar nanodiamonds (also probably produced by CVD, see next section) would have been destroyed in ureilites but at the same time CVD-produced diamonds from the nebula would have survived processing on the ureilite parent body.

Ozima et al. (1998) suggested mass as the main parameter governing noble gas fractionation in ureilites and in phase *Q* (section 3.1.1). However, the mass of the noble gases correlates less perfectly than the first ionization potential with the element depletion in ureilites (*Weber et al.,* 1971), and the isotope ratios are not fractionated to the extent expected from a mass-dependent element-depletion process (*Rai et al.,* 2003b).

As noted previously, there are a number of different N components in ureilites, the origins of which are not always clear. Some of the N does appear to be associated with the noble gases in the diamond phase but most of it is contained in amorphous or graphitic C (with and without noble gases, respectively) and more importantly associated with metal and silicate phases (without noble gases) (*Yamamoto et al.,* 1998; *Rai et al.,* 2003a). The N with the most extreme isotopic composition resides in the diamond phase (≈−120‰ or less) and in a poorly characterized phase in the polymict ureilites (540‰) (*Grady and Pillinger,* 1988). The N in the diamond phase is quite distinct from that in the graphitic or amorphous C (>+50‰ and ~–20‰ respectively). Therefore *Rai et al.* (2003a) concluded that the diamond could not have been produced *in situ* by shock from one of these phases but rather prefer a nebula origin. In contrast, *Yamamoto et al.* (1998) argue that the heavy N seen previously in the polymict ureilites is also present in some of the monomict ureilites and therefore argue for late-stage injection of several volatile-rich objects into ureilitic silicates.

3.3.4. P3 and P6 noble gases in diamonds. Two isotopically apparently normal components, "*P3*" and the minor "*P6*," have been found in presolar diamonds in addition to the anomalous "*HL*" noble gases (*Huss and Lewis,* 1994) (see also section 3.2.1). Both *P3* and *P6* are elementally and isotopically distinguishable from the *Q* gases (Figs. 1, 2, and 4) (*Ott,* 2002). Despite their "normal" compositions, *P3* and *P6* are considered to be presolar by *Huss and Lewis* (1994). While *P3* is released by pyrolysis below 800°C, which possibly reflects superficial trapping, *P6* is released at higher temperatures than the *HL* gases, implying distinct sites for these components.

The most popular formation scenario for presolar nanodiamonds is chemical vapor deposition in the vicinity of supernovae or carbon stars (e.g., *Saslaw and Gaustad,* 1969; *Jørgensen,* 1988; *Lewis et al.,* 1989; *Daulton et al.,* 1996), but annealing of hydrocarbons (*Nuth and Allen,* 1992) and the transformation of C by shockwaves in the ISM (*Tielens et al.,* 1987) have also been suggested. The *P3* and *P6* noble gases could have been enclosed during the formation of the diamonds or trapped later into existing diamonds, e.g., by ion implantation in stellar environments or interstellar space. Ion implantation is supported by experiments that succeeded in separating diamond grain size fractions (*Verchovsky et al.,* 1998) and trapping noble gases in synthetic diamonds in concentrations even exceeding those in presolar diamonds (*Koscheev et al.,* 2001). Moreover, the gas release pattern of the irradiated artificial diamonds closely resembles that of natural diamonds. This suggests that the components *HL* and *P3* may be located in the same fraction of diamonds, implanted by distinct events. Most of the Ne, Ar, and ^{130}Xe accompanying the anomalous Xe-*HL* may actually be fractionated *P3* (*Huss et al.,* 2000; *Ott,* 2003).

3.3.5. Primordial solar-like noble gases in non-carbonaceous meteoritic matter. The noble gas and N components discussed in the previous sections reside in minor carbonaceous phases, which remain after demineralization of the host meteorites. In chondrites, this carbonaceous matter carries most of the primordial noble gases, whereas primordial gases in major phases have hardly received attention. This appears unjustified, because there is increasing evidence that silicates also carry primordial noble gases and that these are elementally and isotopically considerably closer to solar composition than the gases in carbonaceous matter. Note that we consider here primordial solar-like noble gases, in contrast to solar-wind noble gases trapped later in an asteroidal regolith.

The most prominent example for primordial noble gases in meteoritic silicates is the so-called subsolar or Ar-rich "component" in enstatite chondrites (*Crabb and Anders,* 1981; *Patzer and Schultz,* 2002). Gas concentrations may be substantial. For example, the trapped ^{36}Ar in the silicates of South Oman, which is commonly used as reference sample for the subsolar gases, is higher than typical contributions of *Q* gases in carbonaceous chondrites. Hints of subsolar gases have also been found in other meteorite classes (e.g., *Alaerts et al.,* 1979b; *Matsuda et al.,* 1980; *Schelhaas et al.,* 1990). High-resolution gas release experiments revealed that the subsolar noble gases in the E chondrite St. Mark's represent a mixture of probably primordially trapped solar gases, *Q* gases, and terrestrial contamination (*Busemann et al.,* 2003a,b). Hence, the subsolar component appears to be a mixture and does not represent a well-defined noble gas composition.

High concentrations of solar-like heavy noble gases (Ar-Xe on the order of values in South Oman) have been found in chondrules of the enstatite chondrite Y 791790 (*Okazaki et al.,* 2001). This is surprising in view of the general scarcity of trapped noble gases in high-temperature objects like chondrules (*Vogel et al.,* 2004a,b) (section 4). Okazaki and co-workers (*Okazaki et al.,* 2001) concluded that chondrules carry the "subsolar" noble gas component and proposed that solar gases were implanted into chondrule precursors by energetic particles from the early Sun. The primordial solar He and Ne would subsequently have been lost. This interpretation is not easy to reconcile with the observation by *Busemann et al.* (2003a) that the "subsolar" component in another enstatite chondrite actually contains a fraction of elementally unfractionated solar gases. A survey of noble gas concentrations in chondrules of other enstatite chondrites is thus highly desirable.

Particularly noble-gas-rich, acid-susceptible rims have been observed in a primitive dark inclusion in the C3 chondrite Ningqiang (*Nakamura et al.,* 2003b). Again, the concentrations are comparable to those found in the bulk of South Oman. The element composition resembles that of the "subsolar" mixture in E chondrites and ureilitic diamonds, implying a common origin, e.g., by implantation of nebular noble gases that were incompletely ionized in a plasma (*Nakamura et al.,* 2003b).

There are also indications of primordial noble gases of solar-like composition in differentiated meteorites. *Mathew and Begemann* (1997) and *Busemann et al.* (2004) found solar-like Ne in silicates of the Brenham pallasite and the D'Orbigny angrite, respectively. Busemann and co-workers suggest that solar gases were transported from the interior of the angrite parent body by volcanic CO_2 and trapped in the quenched D'Orbigny glass near the parent-body surface, analogous to the trapping of terrestrial mantle gases in ocean basalts.

Although many details remain unclear, the observations of primordial noble gases of solar composition in some meteorite parent bodies may indicate that meteoritic precursor material trapped solar-wind noble gases during an early irradiation prior to accretion to large parent bodies (*Podosek et al.,* 2000; *Okazaki et al.,* 2001; *Busemann et al.,* 2003b). However, most of the dust accreted to larger objects while being efficiently shielded from solar wind (*Nakamura et al.,*

1999b, 2003a). This may indicate that some early solar system material has been irradiated, e.g., off-disk or at the inner edge of the solar system (*Wetherill,* 1981).

3.4. Primordial Noble Gases in Interplanetary Dust Particles and Micrometeorites

Noble gases in interplanetary dust particles (IDPs) and micrometeorites are mostly dominated by the solar wind trapped while the particles were on their way to Earth (*Olinger et al.,* 1990; *Nier and Schlutter,* 1992, 1993; *Pepin et al.,* 2000, 2001; *Osawa et al.,* 2003). The solar gases hinder the assessment of the primordial noble gas inventory, which is regrettable since many IDPs probably are fragments of parent bodies so far unsampled otherwise. Yet in some cases primordial noble gases have been detected. *Osawa et al.* (2003) report that Kr and Xe in micrometeorites from Antarctic ice are dominated by a primordial component. It has not yet been possible, however, to establish whether the signatures of the heavy primordial gases in micrometeorites differ in any respect from those of "normal" meteorites.

Remarkably high $^{3}He/^{4}He$ ratios of up to 40× the solar-wind value have been reported in some IDPs (*Nier and Schlutter,* 1993; *Pepin et al.,* 2000, 2001). Production of the required amounts of ^{3}He in parent-body regoliths by cosmic-ray interactions would imply exposure times of up to more than 10^9 yr (*Pepin et al.,* 2001). Although this might not be unreasonable for the regoliths of Kuiper belt comets, *Pepin et al.* (2001) also consider the intriguing possibility that the ^{3}He excesses represent a ^{3}He-rich primordial component from an unknown source.

3.5. Nitrogen in Organic Macromolecular Material and Interplanetary Dust Particles

While phase *Q* is most likely associated with some minor C-rich phase, this represents only a tiny part of the total C found in the carbonaceous chondrites. Most of the C in these meteorites exists as organic material and probably contains no distinct noble gas components. It does contain considerable amounts of N, however, contributing a few thousand parts per million N to the most primitive chondritic meteorites (e.g., CI, CM, and CR). The organics consist of an insoluble macromolecular material plus lesser amounts of a wide variety of soluble compounds, which can be extracted with organic solvents.

The N-isotopic composition varies considerably between and even within the different organic components. It is beyond the scope of this chapter to detail the extensive body of research that covers the organic material (see recent reviews by *Kerridge,* 1999; *Botta and Bada,* 2002; *Sephton,* 2002; *Gilmour,* 2003; *Pizzarello et al.,* 2006). The macromolecular material can be subdivided into two operationally defined components, refractory organic matter that is stable under harsh conditions with $\delta^{15}N$ around –25‰ and a more labile component with measured $\delta^{15}N$ values up to 90‰ (*Sephton et al.,* 2003). The free organic compounds include amino acids with $\delta^{15}N$ values of up to 184‰ (e.g., *Engel et al.,* 1990; *Engel and Macko,* 1997) and other N-rich fractions up to 103‰ (see review by *Botta and Bada,* 2002). In many cases, elevated $\delta^{15}N$ values are associated with elevated D/H ratios (up to 3400‰), which has been taken to infer formation of these compounds in the ISM where gas-phase ion-molecule reactions at low temperature are believed capable of producing large isotopic fractionations (e.g., *Adams and Smith,* 1981).

Recent calculations have shown that ammonia and organic compounds can be enriched in ^{15}N by such interstellar chemistry by almost a factor of 2 (e.g., *Terzieva and Herbst,* 2000; *Charnley and Rodgers,* 2002). If the solar $\delta^{15}N$ value of –374‰ (section 5.3) is taken as representative of the local ISM value at the time of solar system formation, then these calculations indicate that condensed forms of N could have $\delta^{15}N$ values as high as 250‰. As reviewed by *Grady and Wright* (2003), and shown in Fig. 5, most meteorites contain N with isotopic compositions between these two values. Mixing of these two components therefore could generate much of the wide range of values observed.

A number of meteorites such as bencubbinites and some ordinary chondrites contain ^{15}N enrichments well beyond the limits calculated for the ISM, with $\delta^{15}N$ values of up to 1500‰ (e.g., *Prombo and Clayton,* 1985; *Franchi et al.,* 1986; *Mostefaoui et al.,* 2000, 2002). The original carrier of such a N component remains unknown, as this ^{15}N-rich N now appears to reside in secondary, or very modified carbonaceous phases (e.g., *Mostefaoui et al.,* 2000, 2002). Therefore, it may be that an additional source of heavy N is required to account for these components. Alternatively, our understanding of the isotopic enrichments that can be generated in the ISM is incomplete.

Further evidence of very ^{15}N-enriched N in association with organic C and elevated D/H ratios is observed in IDPs. These tiny particles contain abundant carbonaceous, largely organic matter. Significant amounts of N are also present, primarily in macromolecular-like material but also as free compounds such as amino acids (e.g., *Brinton et al.,* 1998; *Maurette et al.,* 2000; *Aléon et al.,* 2003). The range of isotopic compositions of this N is almost equal to that seen in meteorites with $\delta^{15}N$ values ranging from –373‰ to +1250‰ (see compilation of data by *Messenger et al.,* 2003; *Aléon et al.,* 2003). Some of the ^{15}N-rich areas ($\delta^{15}N$ > 400‰) are intimately associated with elevated D/H ratios (>5000‰) and have C/H ratios similar to chondritic insoluble macromolecular material but with N contents up to an order of magnitude greater (*Aléon et al.,* 2003). Similarly high N/C ratios have been measured in CHON grains from Comet Halley (e.g., *Jessberger et al.,* 1988). Coupled with observed ^{15}N excesses in CN radicals from two comets, this suggests a cometary origin for organic matter in IDPs (*Arpigny et al.,* 2003). However, HCN in Comet Hale-Bopp has been reported to be depleted in ^{15}N (*Jewitt et al.,* 1997), indicating a very heterogeneous N-isotopic distribution in comets at the molecular level and making it difficult to generalize about the relative abundance of the N isotopes in cometary material.

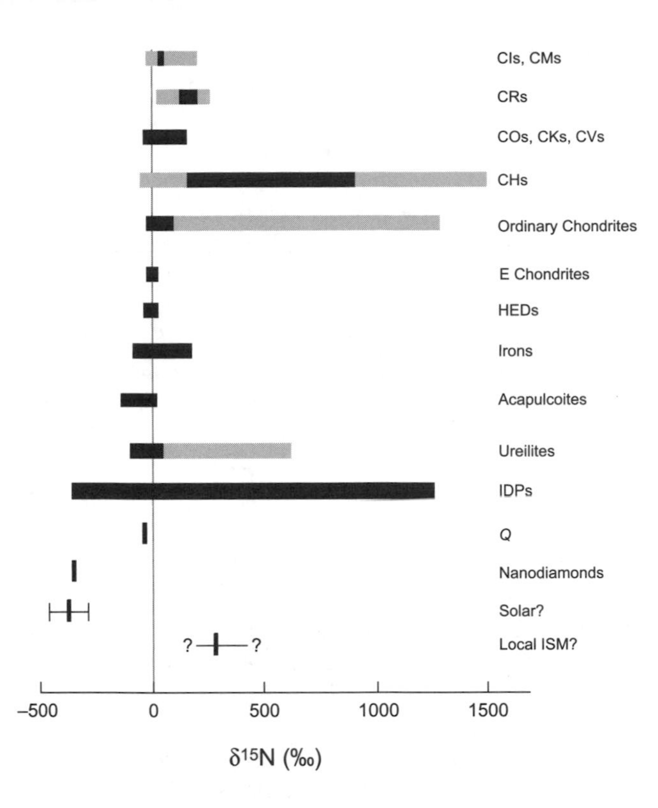

Fig. 5. Nitrogen-isotopic variation of the main meteorite groupings. Solid bars show range of $\delta^{15}N$ values for whole-rock or well-characterized significant components. Hashed bars show the range of $\delta^{15}N$ values for minor components (other than known presolar grains and nanodiamonds) — e.g., organic fractions, peaks in stepped heating extractions. Developed from reviews by *Botta and Bada* (2002), *Grady and Wright* (2003), and *Messenger et al.* (2003), as well as other papers referenced in the text. ISM and solar values inferred — see sections 3.5 and 5.3 and Fig. 7.

CI and CM chondrites have the highest abundances of presolar noble gas carriers, in particular, the most fragile types. Other meteorite classes show considerably different noble gas abundances after accounting for gas loss induced by parent-body thermal metamorphism. Meteorites relatively poor in volatile elements are also depleted in presolar noble gas carriers, and the most fragile carriers are most efficiently lost. An example is shown in Fig. 6 (adapted from *Huss et al.,* 2003). If bulk meteorite compositions and abundances of noble gas carriers indeed reflect the same process, volatile-element and noble gas signatures are the result of different levels of heating of presolar materials rather than variably efficient condensation, because the presolar grains would not have survived in a completely vaporized nebula (*Huss et al.,* 2003).

Primordial noble gases predominantly reside in fine-grained constituents ("matrix"), whereas high-temperature phases such as chondrules and CAIs are gas-poor (*Göbel et al.,* 1982; *Nakamura et al.,* 1999a; *Vogel et al.,* 2004a,b). The one intriguing exception, the high concentrations of trapped heavy noble gases in the chondrules of an enstatite chondrite (*Okazaki et al.,* 2001), has been discussed in section 3.3.5. The most detailed study of noble gases in chondrules is by *Vogel et al.* (2004a). At least about 20% of all studied chondrules contain primordial Ne that cannot be explained by possible contamination with gas-rich matrix. Even most chondrules contain primordial Ar, on levels of up to a few percent of the respective matrix values. Primordial noble gases were thus not quantitatively expelled from all chondrules, although these were once at least partly molten. *Vogel et al.* (2004a) conclude that chondrules retain *HL* gases better than *Q* gases. The latter and their carrier appear to have been removed from chondrules by metal-sulfides.

4. NOBLE GAS CONSTRAINTS ON THE ACCRETIONARY HISTORY OF METEORITE PARENT BODIES

In the previous section we discussed gas trapping from a "microscopic" point of view, e.g., when, where, and how primordial gases were incorporated into their carriers such as presolar grains or phase *Q*. In this section, we summarize inferences about the accretionary history of meteoritic material from studies of trapped noble gases in acid-resistant residues of bulk meteorites but also in "bulk samples" separated from mineralogically or texturally well-defined entities such as chondrules, chondrule rims, or matrix.

Comprehensive studies of the abundances of the various primordial noble gas components in acid-resistant residues from a large number of chondrites of different types have been performed by *Huss and Lewis* (1995) and *Huss et al.* (1996, 2003). These data were used to determine abundances and characteristics of the different carriers in the different meteorite classes, with the goal of studying the metamorphic history of meteorite parent bodies (*Huss and Lewis,* 1995) as well as pre-parent-body modifications of precursor materials of carbonaceous chondrites (*Huss et al.,* 2003). The former topic is discussed in section 5, the latter here.

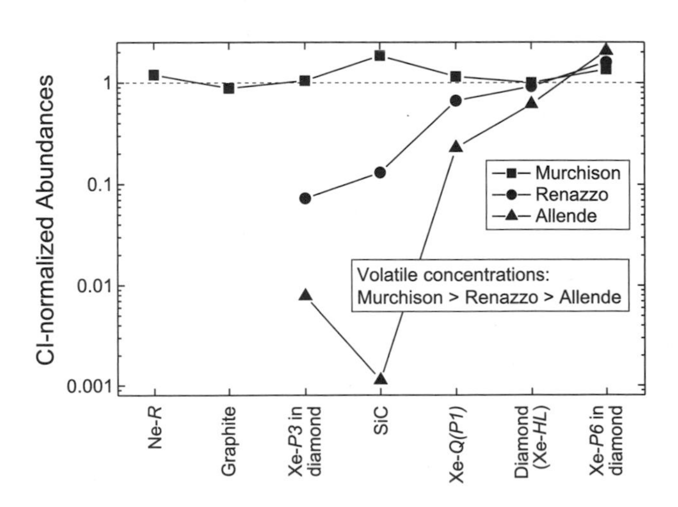

Fig. 6. Abundances of primordial noble gas components or carriers in three different carbonaceous chondrites normalized to abundances in the CI chondrite Orgueil. The components are arranged in order of increasing resistance to chemical and thermal destruction of the carriers from left to right. The figure illustrates that a correlation exists between volatile-element depletion in bulk meteorites and a depletion of fragile noble gas carriers. Figure adapted from *Huss et al.* (2003).

In contrast to chondrules, CAIs do not contain measurable amounts of primordial Ne (*Göbel et al.,* 1982; *Vogel et al.,* 2004b). Occasional reports of the presence of ^{22}Ne-rich Ne-*G* or Ne-*R* should be viewed with caution, because cosmogenic Ne from Na-rich minerals (abundantly present in CAIs) mimicks additions of pure ^{22}Ne. Presumably, CAIs are also devoid of primordial Ar (*Göbel et al.,* 1982; *Vogel et al.,* 2004b). Unless CAI precursors were much poorer in primordial noble gases than chondrule precursors, the noble gas data thus imply a more efficient degassing of the CAI precursors, in agreement with evidence that CAIs once were hotter and cooled more slowly than chondrules (*Jones et al.,* 2000).

A few studies attempted to elucidate the accretionary history of primitive meteorites based on the distribution of noble gases in fine-grained components on a submillimeter scale (*Nakamura et al.,* 1999a; *Vogel et al.,* 2003). In most cases fine-grained rims around chondrules show the highest concentrations of primordial gases. This supports rim accretion around chondrules in the nebula and argues against rim formation by aqueous alteration on parent bodies. The noble gas carriers phase *Q* and presolar diamonds are well mixed on a scale of a few micrograms (*Nakamura et al.,* 1999a). The decrease of noble gas concentrations from rim to matrix appears to reflect a progressive dilution of noble-gas-rich nebular dust with gas-poor material (*Vogel et al.,* 2003). However, in parts of the nebula, an opposite effect seems to have happened, since, e.g., in the LL chondrite Krymka the chondrule rims are less gas-rich than the matrix. Macroscopic assemblages of the presumed noble-gas-rich dust added to the nebula in the Krymka region are present as dark inclusions in this meteorite (*Vogel et al.,* 2003).

5. NOBLE GAS AND NITROGEN CONSTRAINTS ON THE METAMORPHIC HISTORY OF METEORITE PARENT BODIES

In this section we will discuss the evolution of primordial noble gas and N inventories during parent-body differentiation and alteration. Ureilites are discussed in section 3, however, because they show primitive features such as large C and noble gas contents that are incompatible with a more evolved, igneous origin.

5.1. Primordial Noble Gases and Nitrogen in Chondrites

The trapped noble gas content in chondrites is one criterion for their metamorphic classification (*Heymann and Mazor,* 1968; *Sears et al.,* 1980; *Anders and Zadnik,* 1985). This presumes that all members of a certain class had the same original primordial gas inventories and that the major gas carriers react similarly to the various planetary processes. Indeed, within a meteorite class, the concentrations of *Q*, *HL*, *P3*, and *P6* are all strongly inversely correlated with the metamorphic grade (*Huss,* 1997). Phase *Q* is more resistant than presolar grains to thermal and aqueous alteration (*Huss and Lewis,* 1995; *Huss,* 1997; *Nakasyo et al.,* 2000), whereas diamonds are more resistant to shock than phase *Q* (*Nakamura et al.,* 1997). Phase *Q* may consist of two subpopulations that are distinctly susceptible to thermal and aqueous alteration and carry *Q* gases of slightly different elemental composition (*Busemann et al.,* 2000). Hence, subclassifications based on planetary alteration such as those suggested for CV or CM chondrites (*Browning et al.,* 1996; *Guimon et al.,* 1995) are reflected in the elemental composition of the *Q* gases (*Busemann et al.,* 2000).

Only a few studies on the N-isotopic signatures of ordinary chondrites have been reported, possibly because analyses of the low N contents encountered are challenging. The most unequilibrated ordinary chondrites (type 3) contain significant amounts of organic material. *Alexander et al.* (1998) argued that this was initially similar to the ^{15}N-rich macromolecular material present in CR carbonaceous chondrites. This organic material shows decreasing δ^{15}N with increasing C/N ratio as graphitization progresses as the result of thermal metamorphism. This shows that the ^{15}N-rich component is more labile (*Alexander et al.* (1998) and accompanies an overall decrease in N abundance with increasing metamorphic grade (*Hashizume and Sugiura,* 1995). At higher petrologic types (4–6) the measured N abundance reaches a plateau as it approaches the detection limit of the analytical systems (*Hashizume and Sugiura,* 1995). While large isotopic heterogeneity exists (δ^{15}N from –200‰ to +750‰) between different ordinary chondrites (*Sugiura and Hashizume,* 1992), some homogenization of internal isotopic variation would be expected during the metamorphic event, potentially diluting the isotopic extremes measured (*Hashizume and Sugiura,* 1995). Taking this further, numerical simulation of the redistribution of reactive volatile elements during thermal metamorphism of an OC parent body indicates that the observed N concentration in taenite can only be achieved with a parent-body interior with low permeability and that the isotopic heterogeneity observed can be retained on a single parent body provided that the size of isotopically distinct components is large, i.e., >0.1 of the body (*Hashizume and Sugiura,* 1998).

5.2. Primordial Noble Gases and Nitrogen in Differentiated Meteorites

We noted in the previous section that primordial noble gas concentrations in chondrites strongly decrease with increasing metamorphic processing. One might thus expect that differentiated meteorites contain even less primordial gases than highly metamorphosed chondrites, reflecting losses during igneous activity. Indeed, many differentiated meteorites contain negligible amounts of primordial Ne and Ar, but only very few are largely devoid of any primordial Xe (see *Busemann and Eugster,* 2002, for compilation). Nevertheless, the usually small amounts of trapped heavy noble gases are often overprinted by terrestrial contamination, since atmospheric Xe in particular can be adsorbed very tightly (e.g., *Niedermann and Eugster,* 1992). In cases where contamination is small, the primordial Xe-isotopic signature resembles mostly *Q*-Xe (*Busemann and Eugster,* 2002), supporting the idea that the achondrites were formed from chondritic material.

Many acapulcoites and lodranites, primitive achondrites being residues of partial melting of chondritic precursors, show remarkably high concentrations of primordial noble gases. Values are much higher than those of more evolved achondrites and sometimes almost reach those of primitive chondrites (*Busemann and Eugster,* 2002). This indicates that the degree of igneous planetary processing of originally primitive matter may be one parameter that controls the survival of primitive noble gases. Metal- and troilite-rich phases of Lodran show the largest concentrations of heavy *Q* gases, indicating that carbonaceous phase *Q* might have partitioned into the metal during partial melting (*Busemann and Eugster,* 2002). The acapulcoite Y 74063 contains much higher Xe concentrations, approaching those of the most gas-rich ureilites. These gases might reside in gas bubbles formed during partial melting and oxidation of originally carbonaceous carriers (*Takaoka et al.,* 1994). This is in line with the observation that much of the Xe in Acapulco is lost upon crushing (*Kim and Marti,* 1994).

In the case of N, the acapulcoites show minimum δ^{15}N values comparable or even lower than those in ureilites with values of –120‰ to –154‰ (*El Goresy et al.,* 1995; *Kim et al.,* 1995). The light N is present in a number of phases including graphitic particles associated with, but most likely not exsolved from, metal (*El Goresy et al.,* 1995), or in the metal itself (*Kim et al.,* 1995). However, other phases such as the chromites and silicates have much higher δ^{15}N values, around 15‰ (*Kim and Marti,* 1994). This indicates that isotopic homogeneity was not achieved from what was clearly a very heterogeneous body with respect to N, despite temperatures reaching the melting point of the rock. The lodranites, which have experienced hotter or longer heating, contain little N, with a uniform isotopic composition around 10‰ (*Kim and Marti,* 1994), indicating that the more extensive heating resulted in major loss of volatiles and homogenization of what remained.

Observations of trapped noble gas components in iron and stony-iron meteorites are rare, in agreement with the high temperatures probably experienced by these meteorites. Trapped gases were mainly found in silicate inclusions of IAB iron meteorites, often associated with C (e.g., *Bogard et al.,* 1971; *Crabb,* 1983; *Mathew and Begemann,* 1995), and in silicates of the pallasite Brenham. The latter contains solar Ne and possibly Ar (*Mathew and Begemann,* 1997) and solar or possibly U-Xe, a hypothetical component that will be discussed below (*Nagao and Miura,* 1994), implying an early incorporation of the gases, prior to the formation of the pallasite parent body.

The Xe-isotopic composition in iron meteorites appears surprisingly inhomogeneous. Primordial Xe in most silicate inclusions is similar to that in chondrites (*Bogard et al.,* 1971). Nonmagnetic acid residues from the Campo del Cielo ("El Taco") IAB iron meteorite may show a Xe composition similar to terrestrial Xe, but depleted in the light isotopes (*Murty et al.,* 1983). However, these anomalies could not be found in acid-resistant graphite residues from the same meteorite (*Mathew and Begemann,* 1995). Those samples contain Xe that appears distinct from all known Xe components ("El-Taco-Xe"), which could be fractionated U-Xe plus ^{244}Pu fission Xe or fractionated solar Xe plus a fission component of unknown origin. Otherwise, Xe in the silicates is common *Q*-Xe (*Mathew and Begemann,* 1995). Xenon components similar to El-Taco-Xe have also been reported in graphite from the IAB iron meteorite Bohumilitz (*Maruoka et al.,* 2001).

Xenon-*HL* from presolar diamonds has been found in the IAB iron meteorite Mugura (*Maruoka et al.,* 1998) and may also be present in the Campo del Cielo residues of Mathew and Begemann (*Maruoka,* 1999). Similarly, Ne-*Q* and Ne-*HL* may be present in carbonaceous residues of the Canyon Diablo iron meteorite, implying that their carriers were present when the carbonaceous material in the iron meteorites was incorporated (*Namba et al.,* 2000). The observation of terrestrial Ar, Kr, and Xe in iron meteorites is commonly attributed to terrestrial weathering, although an extraterrestrial origin has also been postulated (*Hwaung and Manuel,* 1982). It has been suggested that these inhomogeneities imply a low-temperature formation of iron meteorites directly from the nebula (*Kurat,* 2003). However, abundant presolar grains and phase *Q* would be expected in primitive nebular condensates. Their absence argues against such a hypothesis. In any case, the results discussed above require further confirmation.

The presence of U-Xe or fractionated U-Xe has been reported for some differentiated meteorites (*Michel and Eugster,* 1994; *Nagao and Miura,* 1994; *Weigel and Eugster,* 1994; *Mathew and Begemann,* 1995). U-Xe is a theoretically derived component, supposed to represent the most primitive solar system Xe. It is similar to solar Xe except for lower abundances in the two heaviest isotopes (*Pepin,* 2000) (Fig. 4) and is essential to account for the Xe composition in the present terrestrial atmosphere. However, most of the experimental evidence for U-Xe could not be confirmed. Xenon in the silicates of Brenham can be explained as a mixture of the known Xe components solar, air, fission, and spallogenic (*Mathew and Begemann,* 1997) and Xe in Tatahouine and Lodran appears to be Xe-*Q* plus air (*Busemann and Eugster,* 2002). Thus, the alleged presence of the most primitive U-Xe in differentiated meteorites cannot be used to imply their direct formation in the nebula as proposed by *Kurat* (2003).

The iron meteorite groups show a wide range of δ^{15}N values from –95‰ to +155‰ although quite limited within individual groups (*Franchi et al.,* 1993; *Prombo and Clayton,* 1993). Given that measurable quantities of O are rarely present in these highly reduced meteorites, the N-isotopic signature is a useful tool in determining genetic links between some stony and iron meteorite groups — e.g., IIE irons and H chondrites have δ^{15}N values around 20‰ and IVA irons and L chondrites around 0‰ (e.g., *Prombo and*

Clayton, 1993). Some isotopic fractionation of the N may occur during slow degassing of the molten or solid metal as it cools (*Franchi et al.,* 1993) — this has been taken to extreme levels in the case of the pallasites, which, although strongly related to the IIIAB irons (*Scott,* 1977), contain much less N and $\delta^{15}N$ values up to 100‰ higher than the latter (*Prombo and Clayton,* 1993). Correlation of the N content with $\delta^{15}N$ for the pallasites indicates that extensive loss of molecular N has occurred during their formation (*Franchi et al.,* 1993), the extent of which may offer information on the location of the pallasites within the IIIAB parent body.

The origin of the N in the iron meteorites is unclear. The N content of metal condensing from the solar nebula should be very low — around 0.1 ppm or less for reasonable pressure-temperature (PT) conditions (*Fegley,* 1983), yet the iron meteorites contain several orders of magnitude higher concentrations of up to 85 ppm (e.g., *Prombo and Clayton,* 1993; *Franchi et al.,* 1993). This indicates that the N must have been acquired by the metal at a later stage, most probably by a redistribution of organic N during the heating/melting event on the parent body (*Fegley,* 1983). As outlined above, the $\delta^{15}N$ values of organic matter in meteorites and IDPs are generally much larger than most of the values found in the irons, indicating that an additional source of accretable ^{14}N-rich N may be required.

The distribution of N within the iron meteorites is quite heterogeneous, with γ-Fe containing up to 1 wt% N while the α-Fe was recorded as containing 200 ppm — most of which was deemed to be contamination (*Sugiura,* 1998). This diffusion-controlled redistribution of N during cooling and subsolidus crystallization of the metal phase generates distinctive and variable N-distribution curves within the γ-Fe that appear related to the cooling rate of the meteorites (*Sugiura,* 1998). Calibration of such profiles may offer valuable insight into the thermal history of these parent bodies.

5.3. Solar Nitrogen in the Lunar Regolith

Although this chapter does not review noble gases trapped from the solar wind in dust on the surfaces of asteroids and the Moon (see *Wieler,* 1998, 2002; *Podosek,* 2003), we address here the controversial origin of N in the lunar regolith, which is relevant for the inferred isotopic composition of N in the Sun and the solar nebula. The most recent comprehensive reviews of the topic are by *Becker et al.* (2003) and *Marty et al.* (2003).

At the center of the controversy is the observation that the N-isotopic composition in lunar soils varies dramatically — by up to ~35% — in samples of different antiquity as well as in different extraction steps of individual samples (*Kerridge,* 1975; *Clayton and Thiemens,* 1980). For a long time the prevailing interpretation has been that this implies a secular increase of the $^{15}N/^{14}N$ ratio in the solar wind, with the present-day value being higher than the atmospheric ratio by perhaps ~100‰ (*Kerridge,* 1993). The alternative interpretation is that only a minor part of the N in lunar soils is

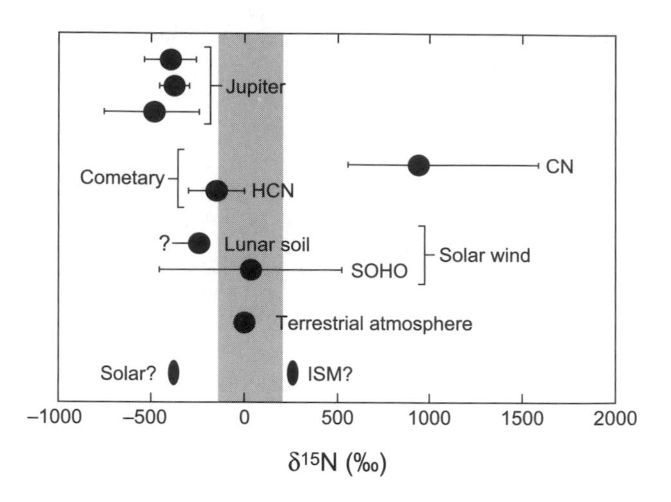

Fig. 7. Nitrogen-isotopic composition of major components in the solar system. Jupiter NH_3 values from *Fouchet et al.* (2000), *Owen et al.* (2001), and *Fouchet et al.* (2004). Cometary data from Comet Hale-Bopp from *Jewitt et al.* (1997) and *Arpigny et al.* (2003). Solar-wind data from *Kallenbach* (2003) and *Hashizume et al.* (2000). The shaded box shows the range of whole-rock meteoritic values with the exception of the CH meteorites, which may be strongly influenced by late-stage additional components. The "Solar?" and "ISM?" values are inferred from the other reservoirs and calculated fractionation effects respectively (see sections 5.3 and 3.5).

of solar wind origin and that the isotopic variability is due to variable admixture of nonsolar sources, presumably meteorites, micrometeorites, or comets (*Wieler et al.,* 1999; *Hashizume et al.,* 2000, 2002; *Marty et al.,* 2003). *Hashizume et al.* (2000) derived an upper limit of the $\delta^{15}N$ value in the solar wind of –240‰ relative to the terrestrial atmosphere, close to the value inferred earlier for ancient solar wind (*Clayton and Thiemens,* 1980). The isotopically light composition of solar-wind N advocated by *Hashizume et al.* (2000) is in agreement with measurements in Jupiter's atmosphere (*Fouchet et al.,* 2000; *Atreya et al.,* 2003). Within its large limits of uncertainty, the reevaluated solar-wind $^{15}N/^{14}N$ ratio measured by the Solar and Heliospheric Observatory (SOHO) mission (*Kallenbach,* 2003) is consistent with either of the values ($\delta^{15}N$ ~+100‰, <–240‰) inferred for the solar wind from lunar samples. Figure 7 shows a summary of these main isotopic compositions. Here we adopt the Jupiter value $^{15}N/^{14}N = (2.3 \pm 0.3) \times 10^{-3}$ or ($\delta^{15}N$ = –374 ± 82)‰ measured by the Galileo probe mass spectrometer (*Atreya et al.,* 2003) as the isotopic composition of protosolar N.

6. OUTLOOK

Much progress has been made since the publication of *Meteorites and the Early Solar System* (*Kerridge and Matthews,* 1988) toward a more detailed characterization of the primordial noble gas and N record in meteorites, as has been summarized in this chapter. Nevertheless, some of the major questions asked in 1988 remain unanswered today. Here we attempt to sketch out top-priority issues that we hope may have come closer to a solution at the time *Meteorites and the Early Solar System III* will go to press.

Presolar grains will remain one of the hot issues in meteoritics in the foreseable future. Parallel to the continuous improvements of microanalytical techniques to study single presolar grains (*Nittler,* 2003), noble gas studies on individual grains as pioneered by Nichols and co-workers (*Nichols et al.,* 1992a, 1994) should be extended to include smaller grains and, if possible, the heavy noble gases. Ideally, noble gas studies should be combined with prior isotope analyses of other elements on the ion probe. In particular, combining the extensive dataset of single-grain N data from secondary ion mass spectrometry (SIMS) analyses with noble gas data and more conventional step-heating data would help in the identification of populations and the ability to track them through secondary processes. For nanodiamonds individual grain analyses are not feasible, but it will be important to continue efforts to separate the *H* and the *L* branches of Xe-*HL* in diamond separates (*Verchovsky et al.,* 1998; *Meshik et al.,* 2001), as this would allow the recognition of different diamond populations.

A better understanding of carriers and trapping mechanisms of the multitude of primordial noble gas components not unambiguously residing in presolar carriers remains essential as well. The nature of the elusive noble gas carrier phase *Q* needs to be studied further, presumably by developing increasingly sophisticated physical separation methods (e.g., *Amari et al.,* 2003) combined with modern tools of C-compound characterization. Trapping and release of noble gases by laboratory simulations also needs further investigation (*Hohenberg et al.,* 2002, *Koscheev et al.,* 2001). The presence of primordial solar noble gases needs confirmation, e.g., by further *in vacuo* etch analyses of bulk samples from meteorites known to contain "subsolar" noble gases. Further miniaturization may allow combination of the "semimicroscopic" and the "microscopic" approach to study the siting of primordial noble gases, by analyzing acid-resistant residues of well-characterized phases instead of bulk meteorites.

A related topic of high priority is the improved determination of the isotopic composition of the solar wind by the Genesis mission. Apart from O, most urgent in this context are N and the heavy noble gases Kr and Xe. In the case of N such a result is very important, given the large range of isotopic ratios observed in meteorites and IDPs. Further measurements of the N-isotopic variation in comets and the ISM are required in order to better constrain the range of values present. Also, more refined modeling or simulation of the possible reactions capable of producing large isotopic fractionations may help close the gap with the measured range in $\delta^{15}N$.

A central open question in noble gas cosmochemistry is the origin and evolution of noble gases in planets. This topic has recently been reviewed by *Pepin and Porcelli* (2002) and *Porcelli and Pepin* (2003). It would have been far beyond the scope of this contribution to address it in detail here, but it is clear that elucidating the role of meteoritic noble gases and N in this context — if there is one — is crucial. The fact that isotopic patterns of noble gases in planetary interiors often are "solar-like" (*Becker et al.,* 2003) led many workers to favor mechanisms to incorporate noble gases in planets directly from a solar nebula source, avoiding the need to invoke large contributions of fractionated meteoritic gases [of the "planetary type" in the terminology of *Signer and Suess* (1963)]. However, recent observations of solar-like primordial noble gases in meteorites (e.g., *Busemann et al.,* 2003a, 2004) and suggestions that solar-wind implantation on planetesimals might be a source of terrestrial noble gases (*Podosek et al.,* 2000) may require a reassessment of the importance of planet building blocks as suppliers of terrestrial noble gases. Since the isotopic composition of noble gases in planetary atmospheres is distinctly different from solar composition, the present atmospheric noble gas compositions must be the result of fractionating processes, either on the planets themselves or their building blocks. This is comprehensively reviewed by *Pepin and Porcelli* (2002) and *Porcelli and Pepin* (2003).

Acknowledgments. This work has been supported by grants to the Swiss National Science Foundation and PPARC. We appreciate comments by N. Vogel; discussion with A. Verchovsky; thorough and constructive reviews by G. Huss, U. Ott, and R. Pepin; and editorial handling by F. Podosek. H.B. thanks the Carnegie Institution for enabling him to finish this manuscript.

REFERENCES

Adams N. G. and Smith D. (1981) $^{14}N/^{15}N$ isotope fractionation in the reaction $N_2H^+ + N_2$: Interstellar significance. *Astrophys. J. Lett., 247,* L123–L125.

Alaerts L., Lewis R. S., and Anders E. (1979a) Isotopic anomalies of noble gases in meteorites and their origins — III. LL-chondrites. *Geochim. Cosmochim. Acta, 43,* 1399–1415.

Alaerts L., Lewis R. S., and Anders E. (1979b) Isotopic anomalies of noble gases in meteorites and their origins — IV. C3 (Ornans) carbonaceous chondrites. *Geochim. Cosmochim. Acta, 43,* 1421–1432.

Aléon J., Robert F., Chaussidon M., and Marty B. (2003) Nitrogen isotopic composition of macromolecular organic matter in interplanetary dust particles. *Geochim. Cosmochim. Acta, 67,* 3773–3783.

Alexander C. M. O'D., Russell S. S., Arden J. W., Ash R. D., Grady M. M., and Pillinger C. T. (1998) The origin of chondritic macromolecular organic matter: A carbon and nitrogen isotope study. *Meteoritics & Planet. Sci., 33,* 603–622.

Amari S., Lewis R. S., and Anders E. (1995a) Interstellar grains in meteorites: III. Graphite and its noble gases. *Geochim. Cosmochim. Acta, 59,* 1411–1426.

Amari S., Zinner E., and Lewis R. S. (1995b) Large ^{18}O excesses in circumstellar graphite grains from the Murchison meteorite: Indication of a massive-star origin. *Astrophys. J. Lett., 447,* L147–L150.

Amari S., Nittler L. R., Zinner E., Gallino R., Lugaro M., and Lewis R. S. (2001a) Presolar SiC grains of type Y: Origin from low-metallicity asymptotic giant branch stars. *Astrophys. J., 546,* 248–266.

Amari S., Nittler L. R., Zinner E., Lodders K., and Lewis R. S. (2001b) Presolar SiC grains of type A and B: Their isotopic compositions and stellar origins. *Astrophys. J., 559,* 463–483.

Amari S., Zaizen S., and Matsuda J. (2003) An attempt to separate Q from the Allende meteorite by physical methods. *Geochim. Cosmochim. Acta, 67,* 4665–4677.

Anders E. and Zadnik M. G. (1985) Unequilibrated ordinary chondrites: A tentative subclassification based on volatile-element content. *Geochim. Cosmochim. Acta, 49,* 1281–1291.

Anders E., Higuchi H., Gros J., Takahashi H., and Morgan J. W. (1975) Extinct superheavy element in the Allende meteorite. *Science, 190,* 1262–1271.

Arpigny C., Jehin E., Manfroid J., Hutsémekers D., Schulz R., Stüwe J. A., Zucconi J.-M., and Ilyin I. (2003) Anomalous nitrogen isotope ratio in comets. *Science, 301,* 1522–1524.

Arrhenius G. and Alfvén H. (1971) Fractionation and condensation in space. *Earth Planet. Sci. Lett., 10,* 253–267.

Ash R. D., Arden J. W., and Pillinger C. T. (1989) Light nitrogen associated with SiC in Cold Bokkeveld. *Meteoritics, 24,* 248–249.

Atreya S. K., Mahaffy P. R., Niemann H. B., Wong M. H., and Owen T. C. (2003) Composition and origin of the atmosphere of Jupiter — an update, and implications for the extrasolar giant planets. *Planet. Space Sci., 51,* 105–112.

Becker L., Poreda R. J., and Bunch T. E. (2000) Fullerenes: An extraterrestrial carbon carrier phase for noble gases. *Proc. Natl. Acad. Sci., 97,* 2979–2983.

Becker R. H., Clayton R. N., Galimov E. M., Lammer H., Marty B., Pepin R. O., and Wieler R. (2003) Isotopic signatures of volatiles in terrestrial planets — Working group report. *Space Sci. Rev., 106,* 377–410.

Bischoff A., Goodrich C. A., and Grund T. (1999) Shock-induced origin of diamonds in ureilites (abstract). In *Lunar and Planetary Science XXX,* Abstract #1100. Lunar and Planetary Institute, Houston (CD-ROM).

Black D. C. and Pepin R. O. (1969) Trapped neon in meteorites — II. *Earth Planet. Sci. Lett., 6,* 395–405.

Bogard D. D., Huneke J. C., Burnett D. S., and Wasserburg G. J. (1971) Xe and Kr analyses of silicate inclusions from iron meteorites. *Geochim. Cosmochim. Acta, 35,* 1231–1254.

Botta O. and Bada J. L. (2002) Extraterrestrial organic compounds in meteorites. *Surv. Geophys., 23,* 411–467.

Brinton K. L. F., Engrand C., Glavin D. P., Bada J. L., and Maurette M. (1998) A search for extraterrestrial amino acids in carbonaceous Antarctic micrometeorites. *Origins Life Evol. Biosphere, 28,* 413–424.

Browning L. B., McSween H. Y. Jr., and Zolensky M. E. (1996) Correlated alteration effects in CM carbonaceous chondrites. *Geochim. Cosmochim. Acta, 60,* 2621–2633.

Buseck P. R. (2002) Geological fullerenes: Review and analysis. *Earth Planet. Sci. Lett., 203,* 781–792.

Busemann H. and Eugster O. (2002) The trapped noble gas component in achondrites. *Meteoritics & Planet. Sci., 37,* 1865–1891.

Busemann H., Baur H., and Wieler R. (2000) Primordial noble gases in "phase Q" in carbonaceous and ordinary chondrites studied by closed-system stepped etching. *Meteoritics & Planet. Sci., 35,* 949–973.

Busemann H., Baur H., and Wieler R. (2001) Helium isotopic ratios in carbonaceous chondrites: Significant for the early solar nebula and circumstellar diamonds? (abstract). In *Lunar and Planetary Science XXXII,* Abstract #1598. Lunar and Planetary Institute, Houston (CD-ROM).

Busemann H., Eugster O., Baur H., and Wieler R. (2003a) The ingredients of the "subsolar" noble gas component (abstract). In *Lunar and Planetary Science XXXIV,* Abstract #1674. Lunar and Planetary Institute, Houston (CD-ROM).

Busemann H., Baur H., and Wieler R. (2003b) Primordial solar noble gases in E-chondrites; a planetary connection? *Geochim. Cosmochim. Acta, 67,* A50.

Busemann H., Lorenzetti S., and Eugster O. (2004) Solar noble gases in the angrite parent body — Evidence from volcanic volatiles trapped in D'Orbigny glass (abstract). In *Lunar and Planetary Science XXXV,* Abstract #1705. Lunar and Planetary Institute, Houston (CD-ROM).

Busso M., Gallino R., Picchio G., and Raiteri C. M. (1990) Dredge-up of thermal pulse nucleosynthesis products. A clue to interpret photospheric abundances in AGB stars and isotopic anomalies in the solar system. In *Nuclei in the Cosmos* (H. Oberhummer and H. Hillebrandt, eds.), pp. 233–237. Max Planck Institut für Physik Astrophysik, Garching.

Charnley S. B. and Rodgers S. D. (2002) The end of interstellar chemistry as the origin of nitrogen in comets and meteorites. *Astrophys. J. Lett., 569,* L133–L137.

Clayton R. N. and Thiemens M. H. (1980) Lunar nitrogen: Evidence for secular change in the solar wind. In *Proceedings of the Conference on the Ancient Sun* (R. O. Pepin et al., eds.), pp. 463–473. Pergamon, New York.

Crabb J. (1983) On the siting of noble gases in silicate inclusions of the El Taco iron meteorite (abstract). In *Lunar and Planetary Science XIV,* pp. 134–135. Lunar and Planetary Institute, Houston.

Crabb J. and Anders E. (1981) Noble gases in E-chondrites. *Geochim. Cosmochim. Acta, 45,* 2443–2464.

Daulton T. L., Eisenhour D. D., Bernatowicz T. J., Lewis R. S., and Buseck P. R. (1996) Genesis of presolar diamonds: Comparative high resolution transmission electron microscopy study of meteoritic and terrestrial nano diamonds. *Geochim. Cosmochim. Acta, 60,* 4853–4872.

Eberhardt P. (1974) A neon-E rich phase in the Orgueil carbonaceous chondrite. *Earth Planet. Sci. Lett., 24,* 182–187.

Eberhardt P., Geiss J., and Grögler N. (1965) Further evidence on the origin of trapped gases in the meteorite Khor Temiki. *J. Geophys. Res., 70,* 4375–4378.

El Goresy A., Zinner E., and Marti K. (1995) Survival of isotopically heterogeneous graphite in a differentiated meteorite. *Nature, 373,* 496–499.

Engel M. H. and Macko S. A. (1997) Isotopic evidence for extraterrestrial non-racemic amino acids in the Murchison meteorite. *Nature, 389,* 265–268.

Engel M. H., Macko S. A., and Silfer J. A. (1990) Carbon isotope composition of individual amino acids in the Murchison meteorite. *Nature, 348,* 47–49.

Eugster O., Eberhardt P., and Geiss J. (1967) Krypton and xenon isotopic composition in three carbonaceous chondrites. *Earth Planet. Sci. Lett., 3,* 249–257.

Eugster O., Herzog G. F., Marti K., and Caffee M. W. (2006) Irradiation records, cosmic-ray exposure ages, and transfer times of meteorites. In *Meteorites and the Early Solar System II* (D. S. Lauretta and H. Y. McSween Jr., eds.), this volume. Univ. of Arizona, Tucson.

Fegley B. Jr. (1983) Primordial retention of nitrogen by terrestrial planets and meteorites. *Proc. Lunar Planet. Sci. Conf. 13th,* in *J. Geophys. Res., 88,* A853–A868.

Fouchet T., Lellouch E., Bézard B., Encrenaz T., Drossart P., Feuchtgruber H., and de Graauw T. (2000) ISO-SWS observations of Jupiter: Measurement of the ammonia tropospheric profile and of the N-15/N-14 isotopic ratio. *Icarus, 143,* 223–243.

Fouchet T., Irwin P. G. J., Parrish P., Calcutt S. B., Taylor F. W., Nixon C. A., and Owen T. (2004) Search for spatial variation in the jovian $^{15}N/^{14}N$ ratio from Cassini/CIRS observations. *Icarus, 172,* 50–58.

Franchi I. A., Wright I. P., and Pillinger C. T. (1986) Heavy nitrogen in Bencubbin — a light-element isotopic anomaly in a stony-iron meteorite. *Nature, 323,* 138–140.

Franchi I. A., Wright I. P., and Pillinger C. T. (1993) Constraints on the formation conditions of iron meteorites based on concentrations and isotopic compositions of nitrogen. *Geochim. Cosmochim. Acta, 57,* 3105–3121.

Frick U. and Chang S. (1978) Elimination of chromite and novel sulfides as important carriers of noble gases in carbonaceous meteorites. *Meteoritics, 13,* 465–470.

Frick U. and Pepin R. O. (1981) On the distribution of noble gases in Allende: A differential oxidation study. *Earth Planet. Sci. Lett., 56,* 45–63.

Fukunaga K. and Matsuda J. (1997) Vapor-growth carbon and the origin of carbonaceous material in ureilites. *Geochem. J., 31,* 263–273.

Fukunaga K., Matsuda J., Nagao K., Miyamoto M., and Ito K. (1987) Noble-gas enrichment in vapour-growth diamonds and the origin of diamonds in ureilites. *Nature, 328,* 141–143.

Gallino R., Busso M., Picchio G., and Raiteri C. M. (1990) On the astrophysical interpretation of isotope anomalies in meteoritic SiC grains. *Nature, 348,* 298–302.

Garvie L. A. J. and Buseck P. R. (2004) Nanosized carbon-rich grains in carbonaceous chondrite meteorites. *Earth Planet. Sci. Lett., 224,* 431–439.

Geiss J. (1973) Solar wind composition and implications about the history of the solar system. In *Proceedings of the 13th International Cosmic Ray Conference* (R. L. Chasson, ed.), pp. 3375–3398. University of Denver, Colorado Associated University Press, Boulder.

Gilmour I. (2003) Structural and isotopic analysis of organic matter in carbonaceous chondrites. In *Treatise on Geochemistry, Vol. 1: Meteorites, Comets and Planets* (A. M. Davis, ed.), pp. 269–290. Elsevier, Oxford.

Göbel R., Ott U., and Begemann F. (1978) On trapped noble gases in ureilites. *J. Geophys. Res., 83,* 855–867.

Göbel R., Begemann F., and Ott U. (1982) On neutron-induced and other noble gases in Allende inclusions. *Geochim. Cosmochim. Acta, 46,* 1777–1792.

Goodrich C. A. (1992) Ureilites: A critical review. *Meteoritics, 27,* 327–352.

Goodrich C. A., Keil K., Berkley J. L., Laul J. C., Smith M. R., Wacker J. F., Clayton R. N., and Mayeda T. K. (1987) Roosevelt County 027: A low-shock ureilite with interstitial silicates and high noble gas concentrations. *Meteoritics, 22,* 191–218.

Grady M. M. and Pillinger C. T. (1988) ^{15}N-enriched nitrogen in polymict ureilites and its bearing on their formation. *Nature, 331,* 321–323.

Grady M. M. and Wright I. P. (2003) Elemental and isotopic abundances of carbon and nitrogen in meteorites. *Space Sci. Rev., 106,* 231–248.

Gros J. and Anders E. (1977) Gas-rich minerals in the Allende meteorite: Attempted chemical characterization. *Earth Planet. Sci. Lett., 33,* 401–406.

Guimon R. K., Symes S. J. K., Sears D. W. G., and Benoit P. H. (1995) Chemical and physical studies of type 3 chondrites XII: The metamorphic history of CV chondrites and their components. *Meteoritics, 30,* 704–714.

Hashizume K. and Nakaoka Y. T. (1998) Q nitrogen in CO3 chondrites (abstract). *Meteoritics & Planet. Sci., 33,* A65–A66.

Hashizume K. and Sugiura N. (1995) Nitrogen isotopes in bulk ordinary chondrites. *Geochim. Cosmochim. Acta, 59,* 4057–4069.

Hashizume K. and Sugiura N. (1998) Transportation of gaseous elements and isotopes in a thermally evolving chondritic planetesimal. *Meteoritics & Planet. Sci., 33,* 1181–1195.

Hashizume K., Chaussidon M., Marty B., and Robert F. (2000) Solar wind record on the moon: Deciphering presolar from planetary nitrogen. *Science, 290,* 1142–1145.

Hashizume K., Marty B., and Wieler R. (2002) Analyses of nitrogen and argon in single lunar grains: Towards a quantification of the asteroidal contribution to planetary surfaces. *Earth Planet. Sci. Lett., 202,* 201–216.

Heck P. R., Marhas K. K., Baur H., Hoppe P., and Wieler R. (2005) Presolar He and Ne in single circumstellar SiC grains extracted from the Murchison and Murray meteorites (abstract). In *Lunar and Planetary Science XXXVI,* Abstract #1938. Lunar and Planetary Institute, Houston (CD-ROM).

Heymann D. (1986) Buckminsterfullerene, its siblings, and soot: Carriers of trapped inert gases in meteorites? *Proc. Lunar Planet. Sci. Conf. 17th,* in *J. Geophys. Res., 91,* E135–E138.

Heymann D. and Mazor E. (1968) Noble gases in unequilibrated ordinary chondrites. *Geochim. Cosmochim. Acta, 32,* 1–19.

Heymann D. and Vis R. D. (1998) A novel idea about the nature of phase Q (abstract). In *Lunar and Planetary Science XXIX,* Abstract #1098. Lunar and Planetary Institute, Houston (CD-ROM).

Hohenberg C. M., Thonnard N., and Meshik A. (2002) Active capture and anomalous adsorption: New mechanisms for the incorporation of heavy noble gases. *Meteoritics & Planet. Sci., 37,* 257–267.

Hoppe P. and Zinner E. (2000) Presolar dust grains from meteorites and their stellar sources. *J. Geophys. Res.–Space Phys., 105,* 10371–10385.

Hoppe P., Amari S., Zinner E., and Lewis R. S. (1995) Isotopic compositions of C, N, O, Mg, and Si, trace element abundances, and morphologies of single circumstellar graphite grains in four density fractions from the Murchison meteorite. *Geochim. Cosmochim. Acta, 59,* 4029–4056.

Hoppe P., Annen P., Strebel R., Eberhardt P., Gallino R., Lugaro M., Amari S., and Lewis R. S. (1997) Meteoritic silicon carbide grains with unusual Si isotopic compositions: Evidence for an origin in low mass, low metallicity asymptotic giant branch stars. *Astrophys. J. Lett., 487,* L101–L104.

Huss G. R. (1997) The survival of presolar grains in solar system bodies. In *Astrophysical Implications of the Laboratory Study of Presolar Materials* (T. J. Bernatowicz and E. Zinner, eds.), pp. 721–748. AIP Conference Proceedings 402, American Institute of Physics, New York.

Huss G. R. and Alexander E. C. (1987) On the presolar origin of the "normal planetary" noble gas component in meteorites. *Proc. Lunar Planet. Sci. Conf. 17th,* in *J. Geophys. Res., 92,* E710–E716.

Huss G. R. and Lewis R. S. (1994) Noble gases in presolar diamonds I: Three distinct components and their implications for diamond origins. *Meteoritics, 29,* 791–810.

Huss G. R. and Lewis R. S. (1995) Presolar diamond, SiC, and graphite in primitive chondrites — Abundances as a function of meteorite class and petrologic type. *Geochim. Cosmochim. Acta, 59,* 115–160.

Huss G. R., Lewis R. S., and Hemkin S. (1996) The "normal planetary" noble gas component in primitive chondrites: Compositions, carrier, and metamorphic history. *Geochim. Cosmochim. Acta, 60,* 3311–3340.

Huss G. R., Ott U., and Koscheev A. P. (2000) Implications of ion-implanatation experiments for understanding noble gases in presolar diamonds (abstract). *Meteoritics & Planet. Sci., 35,* A79–A80.

Huss G. R., Meshik A. P., Smith J. B., and Hohenberg C. M. (2003) Presolar diamond, silicon carbide, and graphite in carbonaceous chondrites: Implications for thermal processing in the solar nebula. *Geochim. Cosmochim. Acta, 67,* 4823–4848.

Hwaung G. and Manuel O. K. (1982) Terrestrial-type xenon in meteoritic troilite. *Nature, 299,* 807–810.

Jessberger E. K., Christoforidis A., and Kissel J. (1988) Aspects of the major element composition of Halley dust. *Nature, 332,* 691–695.

Jewitt D. C., Matthews H. E., Owen T., and Meier R. (1997) Measurements of $^{12}C/^{13}C$, $^{14}N/^{15}N$, and $^{32}S/^{34}S$ ratios in comet Hale Bopp (C/1995 O1). *Science, 278,* 90–93.

Jones R. H., Lee T., Connolly H. C., Love S. G., and Shang H. (2000) Formation of chondrules and CAIs: Theory vs. observation. In *Protostars and Planets IV* (V. Mannings et al., eds.), pp. 927–962. Univ. of Arizona, Tucson.

Jørgensen U. G. (1988) Formation of Xe-HL-enriched diamond grains in stellar environments. *Nature, 332,* 702–705.

Kallenbach R. (2003) Isotopic fractionation by plasma processes. *Space Sci. Rev., 106,* 305–316.

Kallenbach R., Encrenaz T., Geiss J., Mauersberger K., Owen T., and Robert F., eds. (2003) Solar system history from isotopic signatures of volatile elements. *Space Science Series of ISSI, Vol. 16.* Kluwer, Dordrecht. 425 pp.

Kerridge J. F. (1975) Solar nitrogen: Evidence for a secular increase in the ratio of nitrogen-15 to nitrogen-14. *Science, 188,* 162–164.

Kerridge J. F. (1993) Long-term compositional variation in solar corpuscular radiation: Evidence from nitrogen isotopes in the lunar regolith. *Rev. Geophys., 31,* 423–437.

Kerridge J. F. (1999) Formation and processing of organics in the early solar system. *Space Sci. Rev., 90,* 275–288.

Kerridge J. F. and Matthews M. S., eds. (1988) *Meteorites and the Early Solar System.* Univ. of Arizona, Tucson. 1269 pp.

Kim Y. and Marti K. (1994) Isotopic evolution of nitrogen and trapped xenon in the Acapulco parent body. *Meteoritics, 29,* 482–483.

Kim Y., Zipfel J., and Marti K. (1995) Evolutionary trends in acapulcoites and lodranites: Evidence from N and Xe signatures (abstract). In *Lunar and Planetary Science XXVI,* pp. 751–752. Lunar and Planetary Institute, Houston.

Koscheev A. P., Gromov M. D., Mohapatra R. K., and Ott U. (2001) History of trace gases in presolar diamonds inferred from ion-implantation experiments. *Nature, 412,* 615–617.

Krot A. N., Hutcheon I. D., Brearley A. J., Pravdivtseva O. V., Petaev M. I., and Hohenberg C. M. (2006) Timescales and settings for alteration of chondritic meteorites. In *Meteorites and the Early Solar System II* (D. S. Lauretta and H. Y. McSween Jr., eds.), this volume. Univ. of Arizona, Tucson.

Kurat G. (2003) Why iron meteorites cannot be samples of planetesimal smelting. In *Symposium on the Evolution of Solar System Materials,* pp. 65–66. National Institute of Polar Research, Tokyo.

Lavielle B. and Marti K. (1992) Trapped xenon in ordinary chondrites. *J. Geophys. Res.–Planets, 97,* 20875–20881.

Lewis R. S., Srinivasan B., and Anders E. (1975) Host phase of a strange xenon component in Allende. *Science, 190,* 1251–1262.

Lewis R. S., Anders E., Wright I. P., Norris S. J., and Pillinger C. T. (1983) Isotopically anomalous nitrogen in primitive meteorites. *Nature, 305,* 767–771.

Lewis R. S., Anders E., and Draine B. T. (1989) Properties, detectability and origin of interstellar diamonds in meteorites. *Nature, 339,* 117–121.

Lewis R. S., Amari S., and Anders E. (1990) Meteoritic silicon carbide: Pristine material from carbon stars. *Nature, 348,* 293–298.

Lewis R. S., Amari S., and Anders E. (1994) Interstellar grains in meteorites: II. SiC and its noble gases. *Geochim. Cosmochim. Acta, 58,* 471–494.

Lipschutz M. E. (1964) Origin of diamonds in the ureilites. *Science, 143,* 1431–1434.

Mahaffy P. R., Donahue T. M., Atreya S. K., Owen T. C., and Niemann H. B. (1998) Galileo probe measurements of D/H and $^{3}He/^{4}He$ in Jupiter's atmosphere. *Space Sci. Rev., 84,* 251–263.

Marrocchi Y., Razafitianamaharavo A., Michot L. J., and Marty B. (2005) Low pressure adsorption of Ar, Kr and Xe on carbonaceous materials (kerogen and carbon blacks), ferrihydrite and montmorrilonite: Implications for the trapping of noble gases onto meteoritic matter. *Geochim. Cosmochim. Acta, 69,* 2419–2430.

Marti K. (1967) Isotopic composition of the trapped krypton and xenon in chondrites. *Earth Planet. Sci. Lett., 3,* 243–248.

Marty B., Hashizume K., Chaussidon M., and Wieler R. (2003) Nitrogen isotopes on the Moon: Archives of the solar and planetary contributions to the inner solar system. *Space Sci. Rev., 106,* 175–196.

Maruoka T. (1999) Re-definition of "El Taco Xe" based on ^{132}Xe-normalized data: Multiple primordial components in IAB irons. *Geochem. J., 33,* 343–350.

Maruoka T., Matsuda J., and Kurat G. (1998) Multiple primordial components of Xe in the Magura IAB iron. *Antarct. Meteorites XXIII,* 69–71.

Maruoka T., Matsuda J., and Kurat G. (2001) Abundance and isotopic composition of noble gases in metal and graphite of the Bohumilitz IAB iron meteorite. *Meteoritics & Planet. Sci., 36,* 597–609.

Mathew K. J. and Begemann F. (1995) Isotopic composition of xenon and krypton in silicate-graphite inclusions of the El Taco, Campo Del Cielo, IAB iron meteorite. *Geochim. Cosmochim. Acta, 59,* 4729–4746.

Mathew K. J. and Begemann F. (1997) Solar-like trapped noble gases in the Brenham pallasite. *J. Geophys. Res., 102,* 11015–11026.

Matsuda J. and Yoshida T. (2001) The plasma model for the origin of the phase Q: An experimental approach and the comparison with the labyrinth model (abstract). *Meteoritics & Planet. Sci., 36,* A127.

Matsuda J., Lewis R. S., Takahashi H., and Anders E. (1980) Isotopic anomalies of noble gases in meteorites and their origins — VII. C3V carbonaceous chondrites. *Geochim. Cosmochim. Acta, 44,* 1861–1874.

Matsuda J., Fukunaga K., and Ito K. (1991) Noble gas studies in vapor-growth diamonds: Comparison with shock-produced diamonds and the origin of diamonds in ureilites. *Geochim. Cosmochim. Acta, 55,* 2011–2023.

Matsuda J., Amari S., and Nagao K. (1999) Purely physical separation of a small fraction of the Allende meteorite that is highly enriched in noble gases. *Meteoritics & Planet. Sci., 34,* 129–136.

Maurette M., Duprat J., Engrand C., Gounelle M., Kurat G., Matrajt G., and Toppani A. (2000) Accretion of neon, organics, CO_2, nitrogen and water from large interplanetary dust particles on the early Earth. *Planet Space Sci., 48,* 1117–1137.

Meshik A. P., Pravdivtseva O. V., and Hohenberg C. M. (2001) Selective laser extraction of Xe-H from Xe-HL in meteoritic nanodiamonds: Real effect or experimental artifact? (abstract). In *Lunar and Planetary Science XXXII,* Abstract #2158. Lunar and Planetary Institute, Houston (CD-ROM).

Messenger S., Stadermann F. J., Floss C., Nittler L. R., and Mukhopadhyay S. (2003) Isotopic signatures of presolar materials in interplanetary dust. *Space Sci. Rev., 106,* 155–172.

Meyer B. S. and Zinner E. (2006) Nucleosynthesis. In *Meteorites and the Early Solar System II* (D. S. Lauretta and H. Y. McSween Jr., eds.), this volume. Univ. of Arizona, Tucson.

Michel T. and Eugster O. (1994) Primitive xenon in diogenites and plutonium-244-fission xenon ages of a diogenite, a howardite, and eucrites. *Meteoritics, 29,* 593–606.

Moniot R. K. (1980) Noble-gas rich separates from ordinary chondrites. *Geochim. Cosmochim. Acta, 44,* 253–271.

Mostefaoui S., Perron C., Zinner E., and Sagon G. (2000) Metal-associated carbon in primitive chondrites: Structure, isotopic composition, and origin. *Geochim. Cosmochim. Acta, 64,* 1945–1964.

Mostefaoui S., El Goresy A., Hoppe P., Gillet P., and Ott U. (2002) Mode of occurrence, textural settings and nitrogen-isotopic compositions of in situ diamonds and other carbon phases in the Bencubbin meteorite. *Earth Planet. Sci. Lett., 204,* 89–100.

Murty S. V. S. (1996) Isotopic composition of nitrogen in 'phase Q.' *Earth Planet. Sci. Lett., 141,* 307–313.

Murty S. V. S., Goel P. S., Minh D. Vu., and Shukolyukov Yu. A. (1983) Nitrogen and xenon in acid residues of iron meteorites. *Geochim. Cosmochim. Acta, 47,* 1061–1068.

Nagao K. and Miura Y. N. (1994) Trapped Xe component in a silicate phase of the Brenham pallasite. *Meteoritics, 29,* 509.

Nakamura T., Zolensky M. E., Hörz F., Takaoka N., and Nagao K. (1997) Shock effects on phase Q and HL diamonds inferred from experimental shock loading on Allende meteorite (abstract). In *Lunar and Planetary Science XXVIII,* Abstract #1416. Lunar and Planetary Institute, Houston (CD-ROM).

Nakamura T., Nagao K., and Takaoka N. (1999a) Microdistribution of primordial noble gases in CM chondrites determined by in situ laser microprobe analysis: Decipherment of nebular processes. *Geochim. Cosmochim. Acta, 63,* 241–255.

Nakamura T., Nagao K., Metzler K., and Takaoka N. (1999b) Heterogeneous distribution of solar and cosmogenic noble gases in CM chondrites and implications for the formation of CM parent bodies. *Geochim. Cosmochim. Acta, 63,* 257–273.

Nakamura T., Nagao K., and Takaoka N. (2000) Microdistribution of heavy primordial noble gases in ureilites (abstract). *Meteoritics & Planet. Sci., 35,* A117.

Nakamura T., Noguchi T., Zolensky M. E., and Tanaka M. (2003a) Mineralogy and noble-gas signatures of the carbonate-rich lithology of the Tagish Lake carbonaceous chondrite: Evidence for an accretionary breccia. *Earth Planet. Sci. Lett., 207,* 83–101.

Nakamura T., Zolensky M., Sekiya M., Okazaki R., and Nagao K. (2003b) Acid-susceptive material as a host phase of argon-rich noble gas in the carbonaceous chondrite Ningqiang. *Meteoritics & Planet. Sci., 38,* 243–250.

Nakamuta Y. and Aoki Y. (2000) Mineralogical evidence for the origin of diamond in ureilites. *Meteoritics & Planet. Sci., 35,* 487–494.

Nakasyo E., Maruoka T., Matsumoto T., and Matsuda J. (2000) A laboratory experiment on the influence of aqueous alteration on noble gas compositions in the Allende meteorite. *Antarct. Meteorite Res., 13,* 135–144.

Namba M., Maruoka T., Amari S., and Matsuda J. (2000) Neon isotopic composition of carbon residues from the Canyon Diablo iron meteorite. *Antarct. Meteorite Res., 13,* 170–176.

Nichols R. H. Jr. (1992) The origin of Ne-E: Neon-E in single interstellar silicon carbide and graphite grains. Ph.D. thesis, Washington University, St. Louis, Missouri.

Nichols R. H. Jr., Hohenberg C. M., Hoppe P., Amari S., and Lewis R. S. (1992a) ^{22}Ne-E(H) and ^{4}He in single SiC grains and ^{22}Ne-E(L) in single C-α grains of known C-isotopic compositions (abstract). In *Lunar and Planetary Science XXIII,* pp. 989–990. Lunar and Planetary Institute, Houston.

Nichols R. H. Jr., Nuth J. A., Hohenberg C. M., Olinger C. T., and Moore M. H. (1992b) Trapping of noble gases in proton-irradiated silicate smokes. *Meteoritics, 27,* 555–559.

Nichols R. H. Jr., Kehm K., Brazzle R., Amari S., and Hohenberg C. M. (1994) Ne, C, N, O, Mg, and Si isotopes in single interstellar graphite grains: Multiple stellar sources for Neon-E(L) (abstract). *Meteoritics, 29,* 510–511.

Niedermann S. and Eugster O. (1992) Noble gases in lunar anorthositic rocks 60018 and 65315: Acquisition of terrestrial krypton and xenon indicating an irreversible adsorption process. *Geochim. Cosmochim. Acta, 56,* 493–509.

Nier A. O. and Schlutter D. J. (1992) Extraction of helium from individual interplanetary dust particles by step-heating. *Meteoritics, 27,* 166–173.

Nier A. O. and Schlutter D. J. (1993) The thermal history of interplanetary dust particles collected in the Earth's stratosphere. *Meteoritics, 28,* 675–681.

Nittler L. R. (2003) Presolar stardust in meteorites: Recent advances and scientific frontiers. *Earth Planet. Sci. Lett., 209,* 259–273.

Nittler L. R., Hoppe P., Alexander C. M. O'D., Amari S., Eberhardt P., Gao X., Lewis R. S., Strebel R., Walker R. M., and Zinner E. (1995) Silicon nitride from supernovae. *Astrophys. J. Lett., 453,* L25–L28.

Nuth J. A. III and Allen J. E. (1992) Supernovae as sources of interstellar diamonds. *Astrophys. Space Sci., 196,* 117–123.

Nuth J. A. III, Wilkinson G. M., and Johnson N. (2003) Radiation pressure, long-term stability, and growth of large silicon carbide crystals (abstract). *Meteoritics & Planet. Sci., 38,* A126.

Okazaki R., Takaoka N., Nagao K., Sekiya M., and Nakamura T. (2001) Noble-gas-rich chondrules in an enstatite meteorite. *Nature, 412,* 795–798.

Olinger C. T., Maurette M., Walker R. M., and Hohenberg C. M. (1990) Neon measurements of individual Greenland sediment particles: Proof of an extraterrestrial origin and comparison with EDX and morphological analyses. *Earth Planet. Sci. Lett., 100,* 77–93.

Osawa T., Nakamura T., and Nagao K. (2003) Noble gas isotopes and mineral assemblages of Antarctic micrometeorites collected at the meteorite ice field around the Yamato Mountains. *Meteoritics & Planet. Sci., 38,* 1627–1640.

Ott U. (2002) Noble gases in meteorites — Trapped components. *Rev. Mineral. Geochem., 47,* 71–100.

Ott U. (2003) Isotopes of volatiles in pre-solar grains. *Space Sci. Rev., 106,* 33–48.

Ott U. and Herrmann S. (2003) Trapping of noble gases during fullerene synthesis (abstract). *Meteoritics & Planet. Sci., 38,* A91.

Ott U., Mack R., and Chang S. (1981) Noble-gas-rich separates from the Allende meteorite. *Geochim. Cosmochim. Acta, 45,* 1751–1788.

Ott U., Löhr H. P., and Begemann F. (1985) Trapped noble gases in 5 more ureilites and the possible role of Q (abstract). In *Lunar and Planetary Science XVI,* pp. 639–640. Lunar and Planetary Institute, Houston.

Ott U., Löhr H. P., and Begemann F. (1986) Noble gases in ALH 82130: Comparison with ALHA 78019 and diamond-bearing ureilites. *Meteoritics, 21,* 477–478.

Ott U., Begemann F., Yang J., and Epstein S. (1988) S-process krypton of variable isotopic composition in the Murchison meteorite. *Nature, 332,* 700–702.

Owen T., Mahaffy P. R., Niemann H. B., Atreya S. K., and Wong M. (2001) Protosolar nitrogen. *Astrophys. J. Lett., 553,* L77–L79.

Ozima M. and Podosek F. A. (2002) *Noble Gas Geochemistry.* Cambridge Univ., Cambridge. 286 pp.

Ozima M., Wieler R., Marty B., and Podosek F. A. (1998) Comparative studies of solar, Q-gases and terrestrial noble gases, and implications on the evolution of the solar nebula. *Geochim. Cosmochim. Acta, 62,* 301–314.

Patzer A. and Schultz L. (2002) Noble gases in enstatite chondrites II: The trapped component. *Meteoritics & Planet. Sci., 37,* 601–612.

Pepin R. O. (1991) On the origin and early evolution of terrestrial planet atmospheres and meteoritic volatiles. *Icarus, 92,* 2–79.

Pepin R. O. (2000) On the isotopic composition of primordial xenon in terrestrial planet atmospheres. *Space Sci. Rev., 92,* 371–395.

Pepin R. O. (2003) On noble gas processing in the solar accretion disk. *Space Sci. Rev., 106,* 211–230.

Pepin R. O. and Porcelli D. (2002) Origin of noble gases in the terrestrial planets. *Rev. Mineral. Geochem., 47,* 191–246.

Pepin R. O., Nyquist L. E., Phinney D., and Black D. C. (1970) Isotopic composition of rare gases in lunar samples. *Science, 167,* 550–553.

Pepin R. O., Palma R. L., and Schlutter D. J. (2000) Noble gases in interplanetary dust particles, I: The excess helium-3 problem and estimates of the relative fluxes of solar wind and solar energetic particles in interplanetary space. *Meteoritics & Planet. Sci., 35,* 495–504.

Pepin R. O., Palma R. L., and Schlutter D. J. (2001) Noble gases in interplanetary dust particles, II: Excess helium-3 in cluster particles and modeling constraints on interplanetary dust particle exposures to cosmic-ray irradiation. *Meteoritics & Planet. Sci., 36,* 1515–1534.

Pignatari M., Gallino R., Reifarth R., Käppeler F., Amari S., Davis A. M., and Lewis R. S. (2003) S-process xenon in presolar silicon carbide grains and AGB models with new cross sections (abstract). *Meteoritics & Planet. Sci., 38,* A152.

Pizzarello S., Cooper G. W., and Flynn G. J. (2006) The nature and distribution of the organic material in carbonaceous chondrites and interplanetary dust particles. In *Meteorites and the Early Solar System II* (D. S. Lauretta and H. Y. McSween Jr., eds.), this volume. Univ. of Arizona, Tucson.

Podosek F. A. (2003) Noble gases. In *Treatise on Geochemistry, Vol. 1: Meteorites, Comets and Planets* (A. M. Davis, ed.), pp. 381–405. Elsevier, Oxford.

Podosek F. A., Woolum D. S., Cassen P., and Nichols R. H. (2000) Solar gases in the Earth by solar wind irradiation. *J. Conf. Abstr., 5(2),* 804.

Porcelli D. and Pepin R. O. (2003) The origin of noble gases and major volatiles in the terrestrial planets. In *Treatise on Geochemistry, Vol. 4: Atmospheres* (R. F. Keeling, ed.), pp. 319–347. Elsevier, Oxford.

Porcelli D., Ballentine C. J., and Wieler R. (2002) *Noble Gases in Geochemistry and Cosmochemistry.* Reviews in Mineralogy and Geochemistry, Vol. 47, Geochemical Society and Mineralogical Society of America.

Prombo C. A. and Clayton R. N. (1985) A striking nitrogen isotope anomaly in the Bencubbin and Weatherford meteorites. *Science, 230,* 935–937.

Prombo C. A. and Clayton R. N. (1993) Nitrogen isotopic compositions of iron meteorites. *Geochim. Cosmochim. Acta, 57,* 3749–3761.

Rai V. K., Murty S. V. S., and Ott U. (2002) Nitrogen in diamond-free ureilite Allan Hills 78019: Clues to the origin of diamond in ureilites. *Meteoritics & Planet. Sci., 37,* 1045–1055.

Rai V. K., Murty S. V. S., and Ott U. (2003a) Nitrogen components in ureilites. *Geochim. Cosmochim. Acta, 67,* 2213–2237.

Rai V. K., Murty S. V. S., and Ott U. (2003b) Noble gases in ureilites: Cosmogenic, radiogenic, and trapped components. *Geochim. Cosmochim. Acta, 67,* 4435–4456.

Reynolds J. H. and Turner G. (1964) Rare gases in the chondrite Renazzo. *J. Geophys. Res., 69,* 3263–3281.

Reynolds J. H., Frick U., Neil J. M., and Phinney D. L. (1978) Rare-gas-rich separates from carbonaceuos chondrites. *Geochim. Cosmochim. Acta, 42,* 1775–1797.

Russell S. S., Arden J. W., and Pillinger C. T. (1996) A carbon and nitrogen isotope study of diamond from primitive chondrites. *Meteoritics & Planet. Sci., 31,* 343–355.

Russell S. S., Ott U., Alexander C. M. O'D., Zinner E. K., Arden J. W., and Pillinger C. T. (1997) Presolar silicon carbide from the Indarch (EH4) meteorite: Comparison with silicon carbide populations from other meteorite classes. *Meteoritics & Planet. Sci., 32,* 719–732.

Sandford S. A., Bernstein M. P., and Swindle T. D. (1998) The trapping of noble gases by the irradiation and warming of interstellar ice analogs (abstract). *Meteoritics & Planet. Sci., 33,* A135.

Saslaw W. C. and Gaustad J. E. (1969) Interstellar dust and diamonds. *Nature, 221,* 160–162.

Saunders M., Cross R. J., Jimenez-Vazquez H. A., Shimshi R., and Khong A. (1996) Noble gas atoms inside fullerenes. *Science, 271,* 1693–1697.

Schelhaas N., Ott U., and Begemann F. (1990) Trapped noble gases in unequilibrated ordinary chondrites. *Geochim. Cosmochim. Acta, 54,* 2869–2882.

Scott E. R. D. (1977) Pallasites — Metal composition, classification and relationships with iron meteorites. *Geochim. Cosmochim. Acta, 41,* 349–360.

Sears D. W., Grossman J. N., Melcher C. L., Ross L. M., and Mills A. A. (1980) Measuring metamorphic history of unequilibrated ordinary chondrites. *Nature, 287,* 791–795.

Sephton M. A. (2002) Organic compounds in carbonaceous meteorites. *Natural Products Reports, 19,* 292–311.

Sephton M. A., Verchovsky A. B., Bland P. A., Gilmour I., Grady M. M., and Wright I. P. (2003) Investigating the variations in carbon and nitrogen isotopes in carbonaceous chondrites. *Geochim. Cosmochim. Acta, 67,* 2093–2108.

Signer P. and Suess H. E. (1963) Rare gases in the sun, in the atmosphere, and in meteorites. In *Earth Science and Meteorites* (J. Geiss and E. D. Goldberg, eds.), pp. 241–272. North-Holland, Amsterdam.

Smith J. B., Weber P. K., Huss G. R., and Hutcheon I. D. (2004)

Nitrogen and carbon isotopic composition of silicon carbide in the CO3.0 meteorite ALHA77307 (abstract). In *Lunar and Planetary Science XXXV,* Abstract #2006. Lunar and Planetary Institute, Houston (CD-ROM).

Srinivasan B., Gros J., and Anders E. (1977) Noble gases in separated meteoritic minerals: Murchison (C2), Ornans (C3), Karoonda (C5), and Abee (E4). *J. Geophys. Res., 82,* 762–778.

Srinivasan B., Lewis R. S., and Anders E. (1978) Noble gases in the Allende and Abee meteorites and a gas-rich mineral fraction: Investigation by stepwise heating. *Geochim. Cosmochim. Acta, 42,* 183–198.

Straniero O., Chieffi A., Limongi M., Busso M., Gallino R., and Arlandini C. (1997) Evolution and nucleosynthesis in low-mass asymptotic giant branch stars. I. Formation of population I carbon stars. *Astrophys. J., 478,* 332–339.

Sugiura N. (1998) Ion probe measurements of carbon and nitrogen in iron meteorites. *Meteoritics & Planet. Sci., 33,* 393–409.

Sugiura N. and Hashizume K. (1992) Nitrogen isotope anomalies in primitive ordinary chondrites. *Earth Planet. Sci. Lett., 111,* 441–454.

Takaoka N., Motomura Y., Ozaki K., and Nagao K. (1994) Where are noble gases trapped in Yamato-74063 (unique)? *Proc. NIPR Symp. Antarct. Meteorites, 7,* 186–196.

Terzieva R. and Herbst E. (2000) The possibility of nitrogen isotopic fractionation in interstellar clouds. *Mon. Not. R. Astron. Soc., 317,* 563–568.

Tielens A. G. G. M., Seab C. G., Hollenbach D. J., and McKee C. F. (1987) Shock processing of interstellar dust: Diamonds in the sky. *Astrophys. J. Lett., 319,* L109–L113.

Vdovykin G. P. (1970) Ureilites. *Space Sci. Rev., 10,* 483–510.

Verchovsky A. B., Fisenko A. V., Semjonova L. F., Wright I. P., Lee M. R., and Pillinger C. T. (1998) C, N, and noble gas isotopes in grain size separates of presolar diamonds from Efremovka. *Science, 281,* 1165–1168.

Verchovsky A. B., Sephton M. A., Wright I. P., and Pillinger C. T. (2002) Separation of planetary noble gas carrier from bulk carbon in enstatite chondrites during stepped combustion. *Earth Planet. Sci. Lett., 199,* 243–255.

Verchovsky A. B., Wright I. P., and Pillinger C. T. (2004) Astrophysical significance of AGB stellar wind energies recorded in meteoritic SiC grains. *Astrophys. J., 607,* 611–619.

Vis R. D. and Heymann D. (1999) On the Q-phase of carbonaceous chondrites. *Nucl. Instrum. Meth., B158,* 538–543.

Vis R. D., Mrowiec A., Kooyman P. J., Matsubara K., and Heymann D. (2002) Microscopic search for the carrier phase Q of the trapped planetary noble gases in Allende, Leoville and Vigarano. *Meteoritics & Planet. Sci., 37,* 1391–1399.

Vogel N., Wieler R., Bischoff A., and Baur H. (2003) Microdistribution of primordial Ne and Ar in fine-grained rims, matrices, and dark inclusions of unequilibrated chondrites — Clues on nebular processes. *Meteoritics & Planet. Sci., 38,* 1399–1418.

Vogel N., Wieler R., Bischoff A., and Baur H. (2004a) Noble gases in chondrules and associated metal-sulphide-rich samples: Clues on chondrule formation and the behaviour of noble gas carrier phases. *Meteoritics & Planet. Sci., 39,* 117–135.

Vogel N., Baur H., Leya I., and Wieler R. (2004b) Noble gas studies in CAIs from CV3 chondrites: No evidence for primordial noble gases. *Meteoritics & Planet. Sci., 39,* 767–778.

Wacker J. F. (1986) Noble gases in the diamond-free ureilite, ALHA 78019: The roles of shock and nebular processes. *Geochim. Cosmochim. Acta, 50,* 633–642.

Wacker J. F. (1989) Laboratory simulation of meteoritic noble gases. III. Sorption of neon, argon, krypton, and xenon on carbon: Elemental fractionation. *Geochim. Cosmochim. Acta, 53,* 1421–1433.

Wacker J. F., Zadnik M. G., and Anders E. (1985) Laboratory simulation of meteoritic noble gases I. Sorption of xenon on carbon: Trapping experiments. *Geochim. Cosmochim. Acta, 49,* 1035–1048.

Weber H. W., Hintenberger H., and Begemann F. (1971) Noble gases in the Haverö ureilite. *Earth Planet. Sci. Lett., 13,* 205–209.

Weber H. W., Begemann F., and Hintenberger H. (1976) Primordial gases in graphite-diamond-kamacite inclusions from the Haverö ureilite. *Earth Planet. Sci. Lett., 29,* 81–90.

Weigel A. and Eugster O. (1994) Primitive trapped Xe in Lodran minerals and further evidence from EET84302 and Gibson for break-up of the Lodranite parent asteroid 4 Ma ago (abstract). In *Lunar and Planetary Science XXV,* pp. 1479–1480. Lunar and Planetary Institute, Houston.

Wetherill G. W. (1981) Solar wind origin of ^{36}Ar on Venus. *Icarus, 46,* 70–80.

Whittaker A. G., Watts E. J., Lewis R. S., and Anders E. (1980) Carbynes: Carriers of primordial noble gases in meteorites. *Science, 209,* 1512–1514.

Wieler R. (1998) The solar noble gas record in lunar samples and meteorites. *Space Sci. Rev., 85,* 303–314.

Wieler R. (2002) Noble gases in the solar system. *Rev. Mineral. Geochem., 47,* 21–70.

Wieler R., Anders E., Baur H., Lewis R. S., and Signer P. (1991) Noble gases in "phase Q": Closed-system etching of an Allende residue. *Geochim. Cosmochim. Acta, 55,* 1709–1722.

Wieler R., Anders E., Baur H., Lewis R. S., and Signer P. (1992) Characterisation of Q-gases and other noble gas components in the Murchison meteorite. *Geochim. Cosmochim. Acta, 56,* 2907–2921.

Wieler R., Humbert F., and Marty B. (1999) Evidence for a predominantly non-solar origin of nitrogen in the lunar regolith revealed by single grain analyses. *Earth Planet. Sci. Lett., 167,* 47–60.

Wilkening L. L. and Marti K. (1976) Rare gases and fossil particle tracks in the Kenna ureilite. *Geochim. Cosmochim. Acta, 40,* 1465–1473.

Yamamoto T., Hashizume K., Matsuda J., and Kase T. (1998) Multiple nitrogen isotopic components coexisting in ureilites. *Meteoritics & Planet. Sci., 33,* 857–870.

Zinner E. (1998) Stellar nucleosynthesis and the isotopic composition of presolar grains from primitive meteorites. *Annu. Rev. Earth Planet. Sci., 26,* 147–188.

Zinner E., Amari S., Wopenka B., and Lewis R. S. (1995) Interstellar graphite in meteorites: Isotopic compositions and structural properties of single graphite grains from Murchison. *Meteoritics, 30,* 209–226.

Part VII:

The Parent-Body Epoch:
A. Alteration and Metamorphism

Timescales and Settings for Alteration of Chondritic Meteorites

Alexander N. Krot
University of Hawai'i at Manoa

Ian D. Hutcheon
Lawrence Livermore National Laboratory

Adrian J. Brearley
University of New Mexico

Olga V. Pravdivtseva
Washington University

Michael I. Petaev
Harvard University

Charles M. Hohenberg
Washington University

Most groups of chondritic meteorites experienced diverse styles of secondary alteration to various degrees that resulted in formation of hydrous and anhydrous minerals (e.g., phyllosilicates, magnetite, carbonates, ferrous olivine, hedenbergite, wollastonite, grossular, andradite, nepheline, sodalite, Fe,Ni-carbides, pentlandite, pyrrhotite, and Ni-rich metal). Mineralogical, petrographic, and isotopic observations suggest that the alteration occurred in the presence of aqueous solutions under variable conditions (temperature, water/rock ratio, redox conditions, and fluid compositions) in an asteroidal setting, and, in many cases, was multistage. Although some alteration predated agglomeration of the final chondrite asteroidal bodies (i.e., was preaccretionary), it seems highly unlikely that the alteration occurred in the solar nebula, nor in planetesimals of earlier generations. Short-lived isotope chronologies (^{26}Al-^{26}Mg, ^{53}Mn-^{53}Cr, ^{129}I-^{129}Xe) of the secondary minerals indicate that the alteration started within 1–2 m.y. after formation of the Ca,Al-rich inclusions and lasted up to 15 m.y. These observations suggest that chondrite parent bodies must have accreted within the first 1–2 m.y. after collapse of the protosolar molecular cloud and provide strong evidence for an early onset of aqueous activity on these bodies.

1. INTRODUCTION

In this chapter, we review the mineralogy, petrology, and timescales for secondary alteration of type 1–3 carbonaceous (CI, CM, CR, CV, and ungrouped carbonaceous chondrite MAC 88107), enstatite, and ordinary chondrites that resulted in the formation of hydrous and anhydrous minerals (e.g., phyllosilicates, carbonates, magnetite, Ni-bearing sulfides, fayalite, ferrous olivine, andradite, hedenbergite, wollastonite, grossular, nepheline, and sodalite). Although thermal and shock metamorphism are also among the secondary processes that affected most chondritic meteorites and resulted in some mineralogical modifications, the ages of these processes are not discussed here.

Chondrites consist of four major components: chondrules, Fe,Ni-metal grains and/or metal-troilite aggregates, refractory inclusions [Ca,Al-rich inclusions (CAIs) and amoeboid olivine aggregates (AOAs)], and fine-grained matrix material. The only exception is CI chondrites, which lack chondrules, refractory inclusions, and Fe,Ni-metal grains. In addition, some chondrites contain foreign lithic clasts. It is generally believed that the refractory inclusions, chondrules, and Fe,Ni-metal formed in the solar nebula by high-temperature processes that included evaporation and condensation. Many CAIs and most chondrules and Fe,Ni-metal were subsequently melted during multiple brief heating episodes. The refractory inclusions are considered to be the oldest solids formed in the solar nebula 4567.2 ± 0.6 m.y. ago (*Amelin et al.,* 2002). Chondrule formation appears to have started less than 1 m.y. after CAIs and lasted for at least 4 m.y. (*Amelin et al.,* 2004; *Bizzarro et al.,*2004). Chondrules and matrices in a primitive chondrite are chemically complementary (P. Bland, personal communication, 2004), suggesting that most of the matrix materials could have been thermally processed during chondrule formation (*Scott and Krot,* 2005).

Most chondrite groups show evidence for relatively low-temperature alteration that affected all their chondritic components (*Brearley and Jones,* 1998). The nature of this alteration remains controversial and has been attributed to nebular (or preaccretionary) and/or asteroidal processing (e.g., *Brearley,* 2003). Timing of the alteration using short-lived chronology such as ^{26}Al-^{26}Mg, ^{53}Mn-^{53}Cr, and ^{129}I-^{129}Xe can potentially resolve this controversy and constrain ages of chondrule formation and time of accretion of the chondrite parent asteroids. We note that because the lifetime of the solar nebula is poorly constrained (*Podosek and Cassen,* 1994), dating of secondary alteration alone typically cannot distinguish between nebular and asteroidal settings of alteration, which should be based on mineralogical and isotopic (e.g., O) observations, thermodynamic analysis, and petrologic experiments. At the same time, the prolonged duration of alteration and similar ages of alteration to other asteroidal processes, such as thermal metamorphism and igneous differentiation, favor asteroidal settings of alteration. Since the environment of alteration (nebular vs. asteroidal) remains controversial, in each section we briefly summarize the mineralogical, petrologic, and isotopic (O-isotopic compositions) arguments supporting nebular or asteroidal settings for the alteration of a chondrite group (see also *Brearley,* 2003, 2006).

2. SHORT-LIVED ISOTOPE CHRONOLOGY OF SECONDARY ALTERATION OF CHONDRITIC METEORITES

2.1. Aluminum-26–Magnesium-26 Ages

Aluminum-26 is a short-lived radionuclide that β-decays to ^{26}Mg with a half-life of ~0.73 m.y. Excess ^{26}Mg (^{26}Mg*) can be detected by secondary ionization mass spectrometry (SIMS) or by other mass spectrometric techniques in bulk samples or mineral fractions [e.g., thermal ionization mass spectrometry (TIMS) and inductively coupled plasma mass spectrometry (ICP-MS)]. If ^{26}Mg* is derived from *in situ* decay of ^{26}Al, then the data points plotted as δ^{26}Mg [permil (‰) deviation from the terrestrial ^{26}Mg/^{24}Mg ratio of 0.13932] against the ^{27}Al/^{24}Mg ratio will define a straight line (Al-Mg isochron) with the slope proportional to ^{26}Al/^{27}Al at the time of Al-Mg-isotopic system closure. Based on the measured abundances of ^{26}Mg* in numerous CAIs, the solar system initial ^{26}Al/^{27}Al ratio, referred to as "canonical," is estimated to be ~5 × 10^{-5} (e.g., *MacPherson et al.,* 1995; *Bizzarro et al.,* 2004). The difference in the initial ^{26}Al/^{27}Al ratios between the unknown sample and the canonical ^{26}Al/^{27}Al ratio in CAIs corresponds to their relative formation age

$$\Delta t_{\text{sample-CAI}}(\text{Ma}) = 1/\lambda \times \ln[(^{26}\text{Al}/^{27}\text{Al})_{\text{CAI}}/(^{26}\text{Al}/^{27}\text{Al})_{\text{sample}}] \quad (1)$$

where $\lambda = \ln 2/0.73$ is the ^{26}Al decay constant; negative/positive values correspond to older/younger ages than CAIs with a canonical ^{26}Al/^{27}Al ratio.

2.2. Manganese-53–Chromium-53 Ages

Manganese-53 is a short-lived radionuclide that β-decays to ^{53}Cr with a half-life of ~3.7 m.y. (*Lugmair and Shukolyukov,* 1998). This half-life and the fact that Mn and Cr are reasonably abundant elements that experienced extensive fractionation during aqueous alteration make the ^{53}Mn-^{53}Cr chronometer very useful for dating aqueous activity on chondrite parent asteroids (e.g., formation of carbonates and fayalite).

The excess of ^{53}Cr (^{53}Cr*) relative to the terrestrial ^{53}Cr/^{52}Cr ratio of 0.113458 (*Papanastassiou,* 1986) can be detected by SIMS in individual minerals having high (>100) Mn/Cr ratios, which can yield a high concentration of radiogenic ^{53}Cr, with minimal interference from nonradiogenic Cr. If ^{53}Cr* is derived from *in situ* decay of ^{53}Mn, then the data points plotted as δ^{53}Cr (‰ deviation from the terrestrial ^{53}Cr/^{52}Cr ratio) against ^{55}Mn/^{52}Cr ratio will define a straight line (Mn-Cr isochron) with the slope proportional to ^{53}Mn/^{55}Mn ratio at the time of the isotope closure of Mn-Cr system. The relative ages of two samples, 1 and 2, are then calculated from their ^{53}Mn/^{55}Mn ratios

$$\Delta t_{1-2}(\text{Ma}) = 1/\lambda \times \ln[(^{53}\text{Mn}/^{55}\text{Mn})_2/(^{53}\text{Mn}/^{55}\text{Mn})_1] \quad (2)$$

where $\lambda = \ln 2/3.7$ is the ^{53}Mn decay constant. Due to the uncertainty in the solar system initial abundance of ^{53}Mn [estimates range from 0.84 × 10^{-5} (*Lugmair and Shukolyukov,* 1998) to 1.4 × 10^{-5} (*Lugmair and Shukolyukov,* 2001) to (2.8 ± 0.3) × 10^{-5} (*Nyquist et al.,* 2001) to 4.4 × 10^{-5} (*Birck and Allègre,* 1988; *Birck et al,* 1999)], the Mn-Cr ages discussed below are given relative to the $(^{53}\text{Mn}/^{55}\text{Mn})_0$ ratio of (1.25 ± 0.07) × 10^{-6} for the angrite Lewis Cliff (LEW) 86010 (Δt_{LEW}) that has the absolute age determined by Pb/Pb of 4557.8 ± 0.5 Ma (*Lugmair and Shukolyukov,* 1998).

2.3. Iodine-129–Xenon-129 Ages

The incorporation of live ^{129}I into solid matter in the early solar system and subsequent β-decay resulted in production of its stable ^{129}Xe daughter (^{129*}Xe) at I-bearing sites. Isotopic closure, achieved when I and Xe migration ceased, preserved a parent-daughter record that is observable today in whole-rock samples or mineral separates in many meteorites. If no Xe losses have occurred, the ratio of radiogenic ^{129*}Xe to stable ^{127}I equals the value for the initial I (^{129}I/^{127}I) at the time of isotopic closure. Due to the 15.7-m.y. half-life of ^{129}I, ^{129}I/^{127}I evolved rapidly in the early solar system. Differences in this initial I among meteoritic samples form the basis of I-Xe dating (*Reynolds,* 1960). The analytical technique of I-Xe dating involves neutron irradiation in a reactor, which converts a fraction of ^{127}I to ^{128*}Xe [^{127}I (n, γβ) → ^{128*}Xe]. Correlated quantities of two I-derived Xe isotopes (^{129*}Xe and ^{128*}Xe) were released in stepwise pyrolysis and measured by ion counting mass spectrometry (*Hohenberg,* 1980). The simplicity of this technique is enhanced by including in the irradiation a meteorite standard of known age (*Nichols et al.,* 1994) with the relative I-Xe age then given by the relative slopes of the isochrons. Typically, the ratio of ^{129}Xe to some Xe isotope not produced in the irradiation, such as ^{130}Xe or ^{132}Xe, is plotted against the ratio of ^{128}Xe to that same isotope. The choice of ^{130}Xe or ^{132}Xe normalization is usually determined by the relative correction (if any) for spallation or fission effects, respectively, at these isotopes. If the ^{128*}Xe and ^{129*}Xe are both derived from I of uniform isotopic composition, then the data points will define a straight line (I-Xe isochron), with the slope proportional to the ^{129}I/^{127}I ratio at the last time Xe isotopes were in equilibrium (*Swindle and Podosek,* 1988). The I-Xe isochron is thus a two-component mixture of trapped and I-derived Xe. The trapped Xe component is confined to lie at the lower end of this isochron and is typically of "planetary" composition (*Lavielle and Marti,* 1992). Therefore, I-Xe ages are calculated directly from the differences in isochron slopes (^{129*}Xe/^{128*}Xe)$_{sample}$ with that of the standard (^{129*}Xe/^{128*}Xe)$_{standard}$ [Shallowater aubrite or Bjurböle L4 ordinary chondrite; Bjurböle predates Shallowater by 460,000 yr (*Brazzle et al.,* 1999)]

$$\Delta t_{\text{sample-Shallowater}}(\text{Ma}) = 1/\lambda \times \ln[(^{129}\text{I}/^{127}\text{I})_{\text{Shallowater}}/(^{129}\text{I}/^{127}\text{I})_{\text{sample}}] \quad (3)$$

where $\lambda = \ln 2/15.7$ is the ^{129}I decay constant; negative/positive values correspond to older/younger ages than Shallowater. Based on the comparison of I-Xe and Mn-Cr systems with the absolute Pb-Pb chronometer for samples analyzed by mulitple systems, J. Gilmour et al. (personal communication, 2005) concluded that the I-Xe system closed in Shallowater aubrite 5.7 ± 1.1 m.y. earlier than the Mn-Cr system closed in LEW 86010 angrite, at 4563.5 ± 1.0 m.y. before the present. Use of St. Severin as internal standard during the early measurements (e.g., *Zaikowski,* 1980), which was later shown to be inhomogeneous (*Hohenberg et al.,* 1981), makes these measurements difficult to compare with recent results, although the relative ages should be meaningful (*Pravdivtseva et al.,* 2003a).

Because I is a mobile element, the I-Xe chronometry has been shown to be a promising technique for dating secondary alteration processes, i.e., capable of resolving age differences of a few hundred thousand years between closure times of different mineral phases from the same meteorite (e.g., *Swindle,* 1998; *Brazzle et al.,* 1999; *Pravdivtseva and Hohenberg,* 2001; *Pravdivtseva et al.,* 2001, 2003a,b,c; *Hohenberg et al.,* 2004). Special attention, however, must be paid to proper selection and preparation of the samples for I-Xe dating, and to its interpretation (what mineral phases or process are being dated), which should be based on detailed mineralogical study (e.g., *Krot et al.,* 1999). If samples contain more than one I-bearing phase and if the different mineralogical sites degas with different time-temperature profiles, stepwise pyrolysis can simulate mineral separation (*Swindle,* 1998). Whole-rock samples occasionally produce well-defined isochrons, but the results cannot be adequately interpreted if the major I carrier phase is unknown (e.g., *Kennedy et al.,* 1988). On the other hand, some chondritic components such as chondrules, CAIs, and lithic chondritic clasts (often called "dark inclusions"), although representing mixtures of several mineral phases, can often be studied as simple objects if the major I carrier can be identified (e.g., *Kirschbaum,* 1988).

3. TIMESCALES AND SETTINGS FOR SECONDARY ALTERATION OF CHONDRITIC METEORITES

3.1. Timescale of Aqueous Alteration of CI Chondrites

Although CI chondrites are chemically the most primitive meteorites in that they provide the best compositional match to the solar photosphere (*Anders and Grevesse,* 1989; *Palme and Jones,* 2003), their primary mineralogy and petrography were erased by extensive aqueous alteration at ~50°–150°C on their parent body (e.g., *Richardson,* 1978; *McSween,* 1979; *Kerridge et al.,* 1979a,b; *Bunch and Chang,* 1980; *Clayton and Mayeda,* 1984; *Zolensky et al.,* 1989; *Endress and Bischoff,* 1996; *Endress et al.,* 1996; *Leshin et al.,* 1997). Subsequently, some CI chondrites experienced thermal metamorphism (e.g., *Tonui et al.,* 2003). All known CI chondrites are regolith breccias consisting of various types (lithologies) of heavily hydrated lithic fragments composed of a fine-grained phyllosilicate-rich matrix containing magnetite, sulfides, sulfates, and carbonates. The fragments are cemented by networks of secondary Ca- and Mg-sulfate veins that could be of terrestrial origin (*Gounelle and Zolensky,* 2001).

Four chemically distinct types of carbonates are found in CI chondrites: dolomite [$CaMg(CO_3)_2$], breunnerite [$Mg(Fe,Mn)(CO_3)_2$], calcite ($CaCO_3$), and Mg,Ca-bearing siderite ($FeCO_3$), with dolomite being the dominant phase (*Richardson,* 1978; *Fredriksson and Kerridge,* 1988; *Johnson and Prinz,* 1993; *Riciputi et al.,* 1994; *Endress and Bischoff,* 1996). Mineralogical and isotopic (O, C) observations suggest that the carbonates precipitated from aqueous solutions circulating on the CI parent body (e.g., *Clayton and Mayeda,* 1984; *Grady et al.,* 1988; *Endress and Bischoff,* 1996; *Leshin et al.,* 2001). Carbonates are commonly intergrown with magnetite of different textural types (platelet, framboidal, spherulitic), phosphates, and sulfides, suggesting a related paragenesis (*Endress and Bischoff,* 1996). Based on the chemical differences among dolomites within and among CI chondrites and petrographic observations of dissolution textures, composite grains, etc., *Endress et al.* (1996) concluded that several episodes of aqueous alteration occurred on the CI parent body. In spite of such complexity, there have been no attempts yet to combine petrographic observations with isotopic measurements to date the different episodes of aqueous activity of the CI parent body.

3.1.1. Strontium-isotopic dating of CI carbonates. Strontium-isotopic measurements of carbonate separates from the CI chondrite Orgueil reveal that dolomite and breunnerite

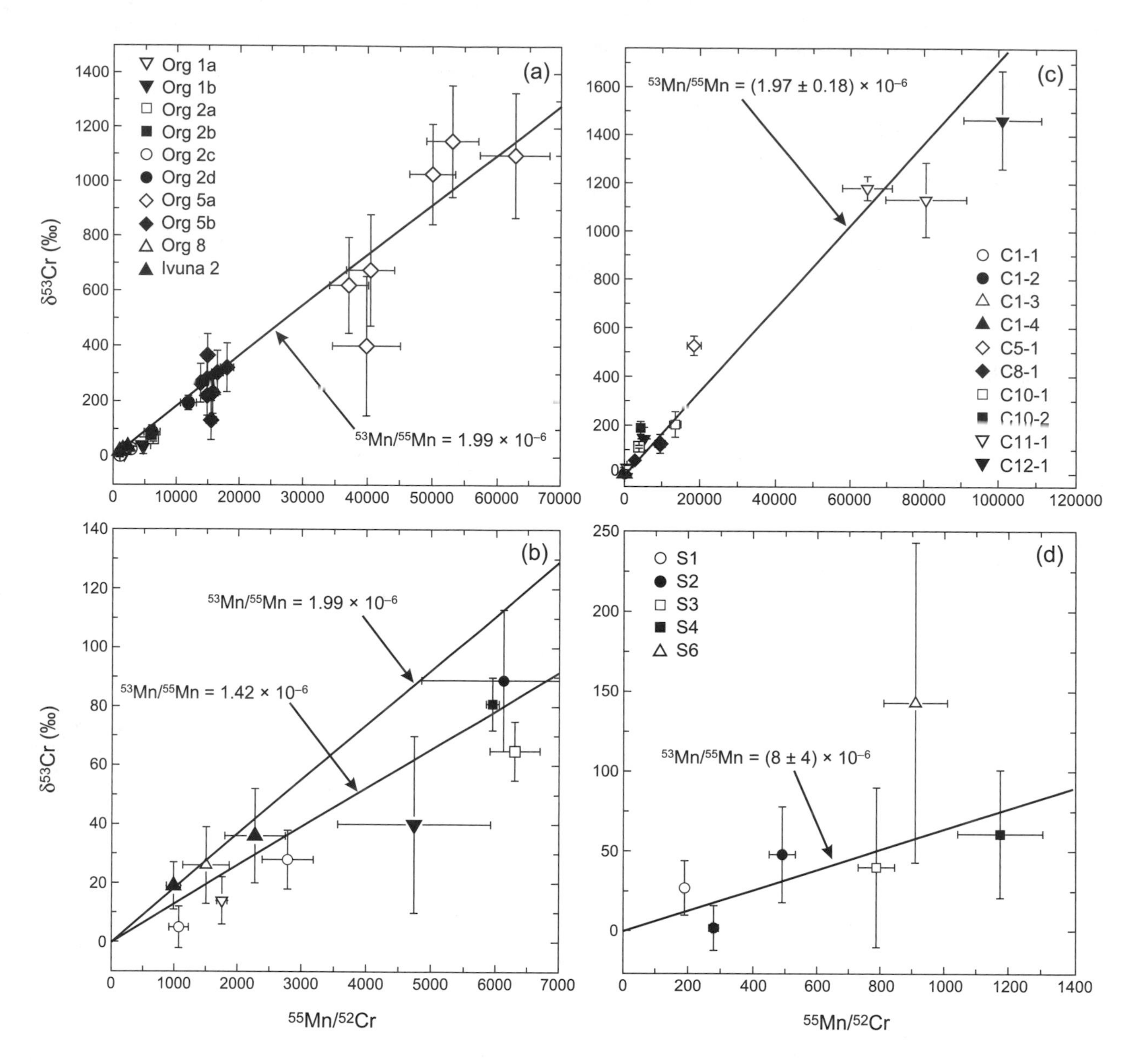

Fig. 1. ^{53}Mn-^{53}Cr evolution diagrams for carbonates in CI chondrites. **(a)** Dolomite fragments from Orgueil and Ivuna; **(b)** an expanded-scale view of the lower lefthand corner of this plot. **(a),(b),(c)** Different spots on a given dolomite fragment. The line of slope 1.99×10^{-6} is a best-fit line through all data points for Orgueil 5 and normal Cr (that is, δ^{53}Cr = 0 at ^{55}Mn/^{52}Cr = 0). The data points for Orgueil 8 and Ivuna 2 fragments are consistent with this line; the data points for the remaining two fragments, Orgueil 1 and 2, fall close to the line of slope 1.42×10^{-6} but, compared to the analytical errors, the deviations are not large enough to clearly establish that different carbonates formed at different times (from *Endress et al.,* 1996). **(c)** Breunnerite and dolomite in Orgueil CI carbonaceous chondrites; **(d)** dolomite in Supuhee CI carbonaceous chondrites. Different symbols represent different grains. The lines of slope $(1.97 \pm 0.18) \times 10^{-6}$ and $(8 \pm 4) \times 10^{-6}$ are the best-fit lines through all data points and normal Cr; error bars are 2σ (from *Hutcheon and Phinney,* 1996; *Hutcheon et al.,* 1997).

formed within 50 m.y. after accretion of its parent body (*Macdougal et al.,* 1984; *Macdougal and Lugmair,* 1989). Relatively large variations of ^{87}Sr/^{86}Sr ratios (0.699–0.702) observed among different carbonates suggest different formation times for different types of CI carbonates (*Macdougal et al.,* 1984).

3.1.2. Chromium-isotopic dating of CI carbonates. Scatena-Wachel et al. (1984) reported ^{53}Cr* corresponding to an upper limit for the ^{53}Mn/^{55}Mn ratio of 3.8×10^{-7} in a breunnerite grain from Orgueil, but did not draw any conclusion about the timescale of aqueous activity. Subsequently *Endress et al.* (1996) measured Cr-isotopic compositions of five dolomite fragments from the CI chondrites Orgueil and Ivuna. These fragments occur between lithic clasts and are not genetically related to lithological units; they may represent debris of former carbonate veins, which were subsequently destroyed and distributed during impact-induced regolith gardening (*Endress and Bischoff,* 1996).

All five fragments show $^{53}Cr^*$ linearly correlated with the $^{55}Mn/^{52}Cr$ ratios, indicative of *in situ* decay of ^{53}Mn (Figs. 1a,b). The data points for two fragments from Orgueil and for an Ivuna fragment plot along a line corresponding to the initial $^{53}Mn/^{55}Mn$ ratio of $(1.99 \pm 0.16) \times 10^{-6}$; the data points for two other fragments from Orgueil define a line with slope $(1.42 \pm 0.16) \times 10^{-6}$. The difference between the lines, if significant, corresponds to a time difference of 1.8 m.y. Alternatively, all five carbonates formed contemporaneously, but some of the Orgueil carbonates experienced partial isotopic equilibration of the Mn-Cr system (*Endress et al.,* 1996).

Subsequently, evidence for live ^{53}Mn in isolated carbonate grains from Orgueil [Fig. 1c; see also Fig. 1 in *Hoppe et al.* (2004)] and a CI-like clast from the Supuhee (H6) chondrite breccia (Fig. 1d) have been reported by *Hutcheon and Phinney* (1996), *Hutcheon et al.* (1997), and *Hoppe et al.* (2004). Chromium-isotopic measurements in Orgueil carbonates imply $^{53}Mn/^{55}Mn$ ratio ranging from $(1.77 \pm 0.15) \times 10^{-6}$ to $(3.88 \pm 0.39) \times 10^{-6}$ at the time of formation of these grains. This difference in slope was interpreted as possibly (1) indicating isotopic disequilibrium among carbonates subjected to several alteration events reflecting either the growth of carbonates from isotopically disparate fluids or partial Cr-isotopic reequilibration, or (2) having chronological significance, corresponding to an interval of ~4 m.y. between episodes of aqueous activity. The latter is consistent with the variation in the times of formation of dolomites and breunnerites suggested from Sr-isotopic studies (*Macdoudal and Lugmair,* 1989). Chromium-isotopic measurements of carbonates in Supuhee define a correlation line with a slope corresponding to $(^{53}Mn/^{55}Mn)_0$ ratio of $(8 \pm 4) \times 10^{-6}$, suggesting that the aqueous activity on the CI-like parent body started earlier than recorded by the Orgueil carbonates (*Hutcheon et al.,* 1997).

The observed range in the initial $^{53}Mn/^{55}Mn$ ratios in CI carbonates [from $(1.42 \pm 0.16) \times 10^{-6}$ to $(8 \pm 4) \times 10^{-6}$] corresponds to an age difference of ~9 Ma and may represent the duration of aqueous activity of the CI parent body that started ~10 ± 3 m.y. prior to differentiation of the angrite parent body.

3.1.3. Iodine-xenon-isotopic dating of CI magnetites. Jeffery and Anders (1970) showed that the trapped Xe resided mostly in phyllosilicates and the radiogenic ^{129}Xe in magnetite of Orgueil. Later, *Herzog et al.* (1973) and *Lewis and Anders* (1975) reported the apparent I-Xe age of the Orgueil magnetites as being ~7 m.y. older than Shallowater and interpreted this age as the condensation time of the solar nebula. These results and interpretation are clearly inconsistent with petrographic evidence for asteroidal formation of magnetite (e.g., *Kerridge et al.,* 1979a) and with relatively young ^{53}Mn-^{53}Cr and $^{87}Sr/^{86}Sr$ ages of carbonates described above. Recent reexamination of the anomalous ^{129}I-^{129}Xe age of Orgueil magnetite (fraction containing >90% magnetite) showed much later closing time of the I-Xe system, 1.9 ± 0.2 m.y. older than Shallowater (Fig. 2) (*Hohenberg et al.,* 2000; *Pravdivtseva et al.,* 2003b). The differences between the two studies are probably due to difficulties with the irradiation monitors in the early studies (*Hohenberg et al.,* 2000).

For a highly magnetic fraction (composed of ~14% magnetite and ~86% hydrated carbonaceous material) separated from Orgueil with a hand magnet, the I-Xe ages are 3.0 ± 0.4 m.y. (*Hohenberg et al.,* 2000) and 1.5 ± 0.3 m.y. (*Pravdivtseva et al.,* 2003b) younger than Shallowater, suggest-

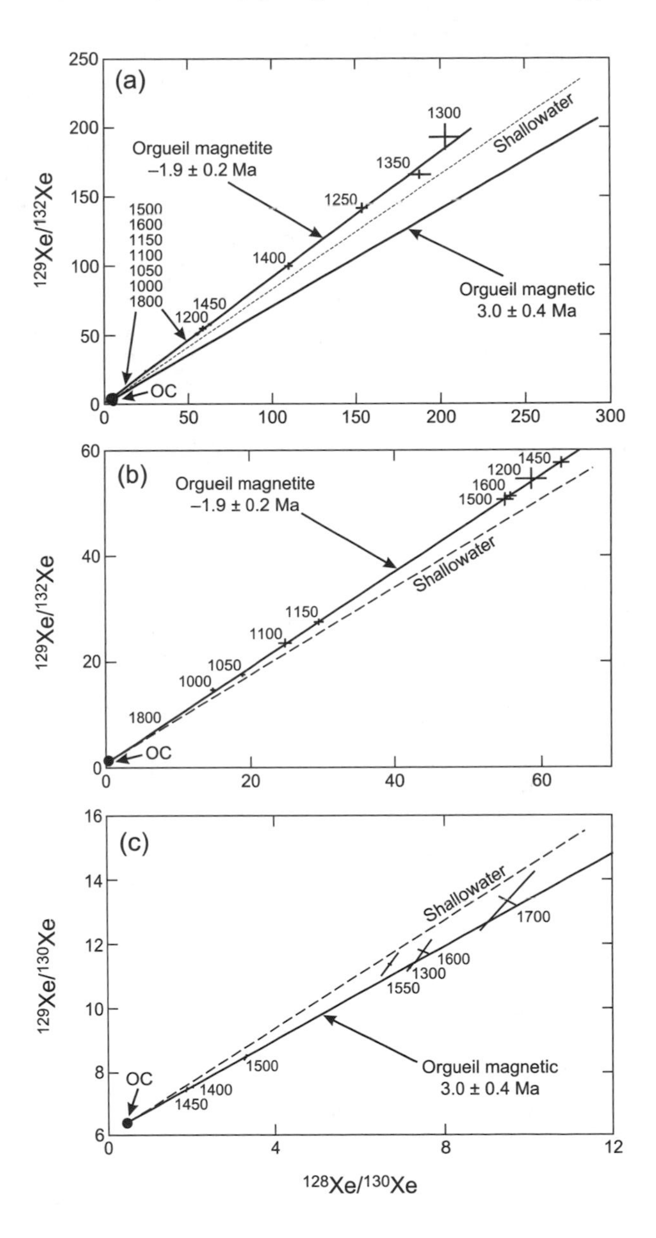

Fig. 2. ^{129}I-^{129}Xe evolution diagrams for the Orgueil magnetic separates and Shallowater aubrite. The ^{129}I-^{129}Xe ages for the nearly pure magnetite fraction containing >90% magnetite predate Shallowater by 1.9 ± 0.2 Ma. For the highly magnetic fraction composed of ~14% magnetite and ~86% hydrated carbonaceous material, the I-Xe age is 3.0 ± 0.4 m.y. younger than Shallowater (4563.5 ± 1 Ma) (J. Gilmour et al., personal communication, 2005), suggesting that the magnetic fraction may contain several I carriers recording different stages of aqueous activity on the CI parent body; error bars are 1σ (data from *Hohenberg et al.,* 2000; *Pravdivtseva et al.,* 2003b).

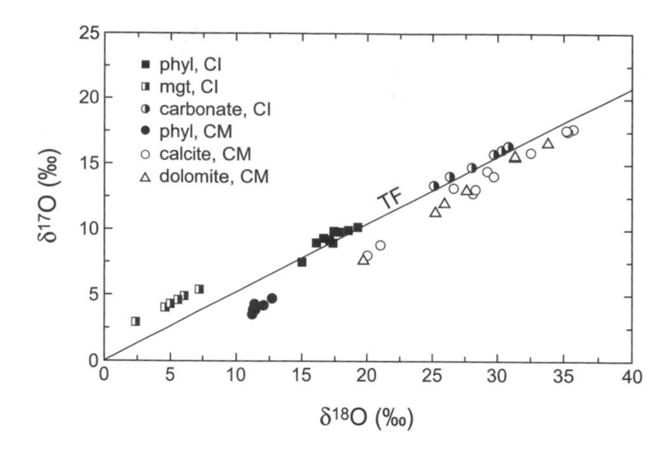

Fig. 3. Oxygen-isotopic compositions of separated components from CI and CM chondrites. Phyllosilicate-rich matrix (phyl) + carbonates and magnetite (mgt) in CI chondrites, as well as phyllosilicate-rich matrix and carbonates in CM chondrites, are out of isotopic equilibrium. Similar $\Delta^{17}O$ values for calcite and dolomite fractions from the same splits of the same CM chondrites indicate that both minerals in each split precipitated from a single fluid reservoir. The terrestrial fractionation (TF) line is shown for reference (data from *Rowe et al.,* 1994; *Leshin et al.,* 2001; *Benedix et al.,* 2003).

ing that the magnetic fraction may contain several I carriers recording different stages of aqueous activity on the CI parent body. This is consistent with O-isotopic compositions of separated components from CI chondrites (Fig. 3). In each CI chondrite, magnetite ($\Delta^{17}O$ = 1.3‰ to 1.8‰) is out of O-isotopic equilibrium with the phyllosilicates ($\Delta^{17}O$ = –0.3‰ to 0.3‰), suggesting that phyllosilicates continued to equilibrate with water as anhydrous silicates are progressively altered to phyllosilicates, whereas isotope exchange between magnetite and fluid was kinetically slow (*Rowe et al.,* 1994). Carbonates ($\Delta^{17}O$ = 0.3‰ to 0.5‰) are in isotopic equilibrium with phyllosilicates (*Leshin et al.,* 2001), suggesting precipitation from a fluid of similar O-isotopic composition.

3.2. Timescale of Aqueous Alteration of the Polymict Chondrite Breccia Kaidun: Evidence from Chromium-Isotopic Compositions of Carbonates

Kaidun is a polymict chondrite breccia containing lithic clasts of the C1, CM-like, CR-like, CV, R, EH, and EL chondrites (*Clayton et al.,* 1994; *Clayton and Mayeda,* 1999; *Zolensky and Ivanov,* 2003). Most clasts have been extensively altered at ~250°–450°C by hydrothermal fluids that resulted in formation of phyllosilicates, and carbonate- and phyllosilicate-filled veins (*Johnson and Prinz,* 1993; *Weisberg et al.,* 1994; *Zolensky et al.,* 1996). All lithic clasts contain carbonates, with calcite being the dominant phase; dolomite is less abundant (*Johnson and Prinz,* 1993; *Weisberg et al.,* 1994). Calcite occurs within altered chondrules, CAIs, and mineral fragments, and as fragments dispersed throughout the matrix and in veins. The veins occur along the boundaries between lithic clasts, suggesting some calcite formed after agglomeration of the Kaidun breccia.

Chromium-isotopic compositions of five calcite and one dolomite grain from three different lithologies (CR-like, CM1, and C1) measured by *Hutcheon et al.* (1999) are plotted in Fig. 4a. The slope of the correlation line on a ^{53}Mn-^{53}Cr evolution diagram corresponds to an initial $^{53}Mn/^{55}Mn$ ratio of $(9.4 \pm 1.6) \times 10^{-6}$, suggesting nearly contemporaneous formation of carbonates in the lithologies studied ($\Delta t_{LEW} \sim 10.8 \pm 1$ m.y.).

3.3. Timescale of Aqueous Alteration of CM Chondrites

The CM carbonaceous chondrites are a diverse group of petrologic type 1–2 meteorites that experienced low-temperature (~0°–25°C) aqueous alteration to various degrees in an asteroidal setting that resulted in formation of a variety of secondary phases, including phyllosilicates, magnetite, Fe,Ni-sulfides, and carbonates (e.g., *Kerridge and Bunch,* 1979; *Zolensky and McSween,* 1988; *Zolensky et al.,* 1993; *Brearley and Jones,* 1998). Some CM chondrites [e.g., Belgica 7904, Yamato (Y) 86720] subsequently experienced thermal metamorphism and partial dehydration (e.g., *Tomeoka et al.,* 1989; *Tomeoka,* 1990; *Ikeda,* 1992; *Clayton and Mayeda,* 1999).

Detailed mineralogical and isotopic studies of CM carbonates revealed their complex formation history, involving periods of dissolution and reprecipitation due to interactions with fluids of different compositions (e.g., *Zolensky et al.,* 1989; *Johnson and Prinz,* 1993; *Riciputi et al.,* 1994; *Brearley et al.,* 1999, 2001; *Brearley and Hutcheon,* 2000, 2002; *Benedix et al.,* 2003). Carbonates occur in fine-grained rims around chondrules and mineral fragments and within altered CAIs (e.g., *Bunch and Chang,* 1980). There is no common association of carbonates with other phases in CM chondrites, although textural observations suggest that carbonates must have coprecipitated with phyllosilicates, magnetite, and tochilinite (e.g., *Kerridge and Bunch,* 1979; *Bunch and Chang,* 1980; *Barber,* 1981; *Mackinnon et al.,* 1984; *Johnson and Prinz,* 1993). However, O-isotopic compositions of carbonates (*Clayton and Mayeda,* 1984; *Brearley et al.,* 1999; *Benedix et al.,* 2003), magnetite (*Rowe et al.,* 1994), and phyllosilicate-rich matrix (*Clayton and Mayeda,* 1984) in the CM chondrite Murchison (Fig. 3) indicate that the phyllosilicates are not in isotope equilibrium with carbonates and magnetite. It is suggested that carbonates and magnetites precipitated from an isotopically evolving water reservoir, prior to formation of phyllosilicates (*Rowe et al.,* 1994; *Brearley et al.,* 1999; *Benedix et al.,* 2003). Timing of aqueous activity on the CM parent body remains poorly constrained and is largely based on a limited number of Cr-isotopic measurements in carbonates (*Brearley and Hutcheon,* 2000; *Brearley and Hutcheon,* 2002).

3.3.1. Chromium-isotopic dating of CM carbonate formation. In the relatively weakly altered and virtually un-

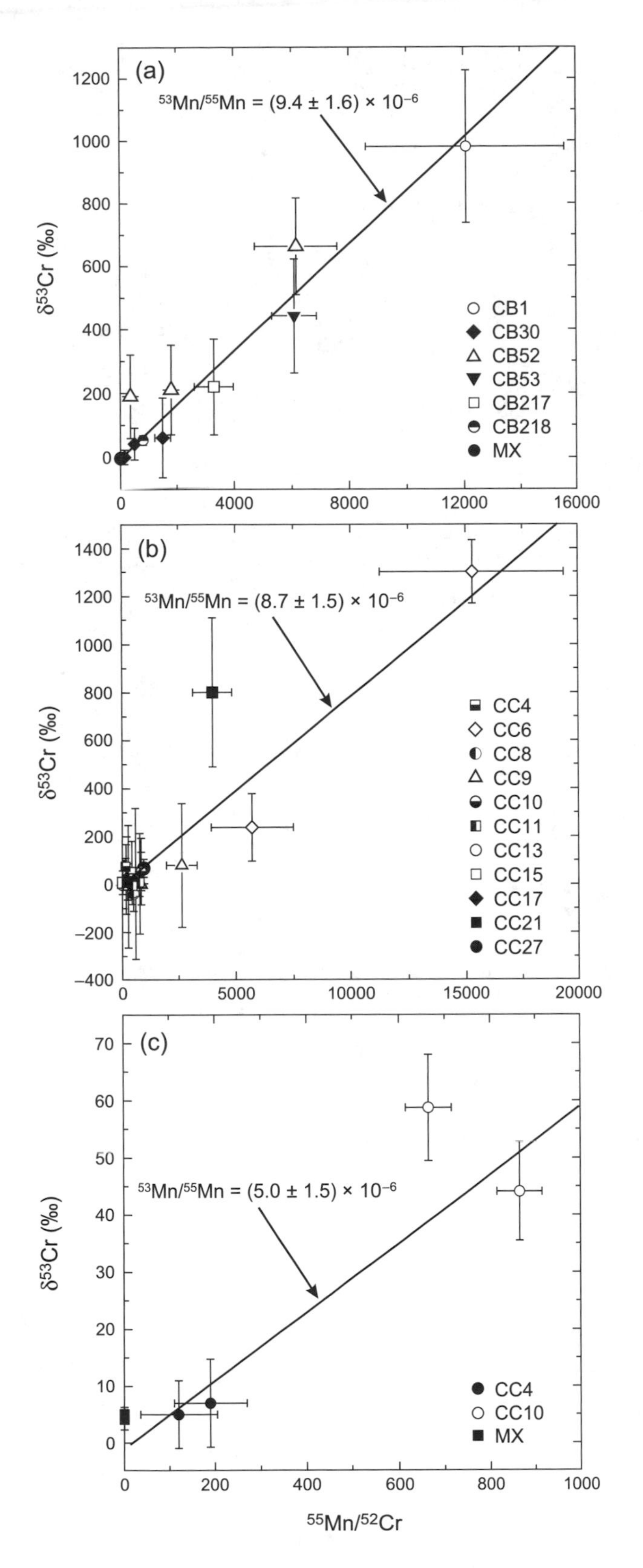

Fig. 4. ^{53}Mn-^{53}Cr evolution diagrams for carbonates from **(a)** Kaidun and CM carbonaceous chondrites **(b)** Y 791198 and **(c)** ALH 84034. Different symbols represent different grains. The lines of slope $(9.4 \pm 1.6) \times 10^{-6}$, $(8.7 \pm 1.5) \times 10^{-6}$, and $(5.0 \pm 1.5) \times 10^{-6}$ are the best-fit lines through all data points and normal Cr; error bars are 2σ (data from *Hutcheon et al.,* 1999; *Brearley et al.,* 2001; *Brearley and Hutcheon,* 2000).

brecciated CM chondrite Y 791198, calcite is the only carbonate present. Calcite grains show complex zoning indicating periods of dissolution and reprecipitation (*Brearley et al.,* 2001). Three out of six calcite grains analyzed for Cr-isotopic compositions showed the presence of $^{53}Cr^*$ correlated with the respective $^{55}Mn/^{52}Cr$ ratios, indicating *in situ* decay of ^{53}Mn. The slope of the correlation line on a ^{53}Mn-^{53}Cr evolution diagram corresponds to an initial $^{53}Mn/^{55}Mn$ ratio of $(8.7 \pm 1.5) \times 10^{-6}$ (Fig. 4b).

In the heavily altered CM1 chondrite Allan Hills (ALH) 84034, both calcite and dolomite are present; calcite is much less abundant (*Brearley and Hutcheon,* 2000). The two phases always occur separately, except within altered CAIs where they can coexist. Dolomites are commonly intergrown with serpentines and pentlandite; calcites are inclusion-free. One of the dolomite grains exhibits resolvable $^{53}Cr^*$ corresponding to an initial $^{53}Mn/^{55}Mn$ ratio of $(5.0 \pm 1.5) \times 10^{-6}$ (Fig. 4c).

The observed range in the initial $^{53}Mn/^{55}Mn$ ratios in CM carbonates [from $(1.3 \pm 0.6) \times 10^{-5}$ to $(5.0 \pm 1.5) \times 10^{-6}$] corresponds to an age difference of ~5 Ma and may represent duration of aqueous activity of the CM parent body that started ~12.5 ± 2.5 m.y. prior to differentiation of the angrite parent body (*Lugmair and Shukolyukov,* 1998).

3.4. Timescale of *In Situ* Aqueous Alteration of the Ungrouped Carbonaceous Chondrite MacAlpine Hills 88107: Evidence from Petrographic Observations and Chromium-Isotopic Compositions of Secondary Fayalite

The ungrouped carbonaceous chondrite MacAlpine Hills (MAC) 88107 has a bulk chemical composition intermediate between CO and CM chondrites, and O-isotopic composition similar to CO-CV-CK chondrites (*Clayton and Mayeda,* 1999). In contrast to CK and most CO chondrites, MAC 88107 shows no evidence for thermal metamorphism; its thermoluminescence properties (*Sears et al.,* 1991) suggest low petrographic type (3.0–3.1). Chondrules and CAIs are surrounded by continuous fine-grained, accretionary rims, indicating that the meteorite largely escaped postaccretional brecciation (*Krot et al.,* 2000a).

The meteorite experienced a small degree of *in situ* alteration that resulted in formation of saponite, serpentine, magnetite, pentlandite, fayalite, and hedenbergite (*Krot et al.,* 2000a). We emphasize that similar secondary phases are observed in the Bali-like oxidized CV chondrites, where their origin remains controversial (see below). Because MAC 88107 may provide a clue for understanding the origin of secondary mineralization in CV chondrites, it is discussed in detail in this chapter. Fayalite (Fa_{90-100}) and hedenbergite ($\sim Fs_{50}Wo_{50}$) occur as veins, which start at the opaque nodules in the chondrule peripheries, crosscut fine-grained rims, and either terminate at the boundaries with the neighboring fine-grained rims or continue as layers between these rims (Fig. 5a,b). Fayalite also overgrows isolated forsteritic (Fa_{1-5}) and fayalitic (Fa_{20-40}) olivine grains

Fig. 5. Backscattered electron images of **(a),(b)** a porphyritic olivine-pyroxene type I chondrule, **(c)** isolated fayalitic olivine grain, and **(d)** magnetite-sulfide nodule in the ungrouped carbonaceous chondrite MAC 88107. **(a),(b)** The chondrule is surrounded by a continuous fine-grained rim (FGR) crosscut by fayalite (fa) – hedenbergite (hed) – magnetite (mgt) veins. The veins start at the opaque nodules composed of Ni-bearing sulfide and magnetite in the peripheral portion of the chondrule. Chondrule mesostasis (lm) is largely leached out, whereas forsteritic olivine (fo) and low-Ca pyroxene (px) phenocrysts appear to be unaltered. **(c)** Fayalitic olivine (fa ol) is overgrown by fayalite. **(d)** Fayalite preferentially replaces magnetite of the sulfide-magnetite nodule. Low-calcium pyroxene grains at the contact with fayalite and magnetite and forsteritic olivine grain overgrown by fayalite appear to be unaltered (after *Krot et al.,* 2000a).

without any evidence for Fe-Mg interdiffusion (Fig. 5c), and replaces magnetite-sulfide grains (Fig. 5d). All textural varieties of fayalite are compositionally similar and characterized by high MnO content (0.4–0.85 wt%) and nearly complete absence of Cr_2O_3.

Based on the petrographic observations and thermodynamic analysis of phase relations in the Si-Fe-Ca-O-H system, *Krot et al.* (2000a) concluded that phyllosilicates, magnetite, pentlandite, fayalite, and hedenbergite in MAC 88107 formed during low-temperature (~150°–200°C) alteration in the presence of aqueous solution capable of transporting Si, Fe, Ca, Mn, and Mg. Because most fayalite grains in MAC 88107 are too small (<10 µm) for Mn-Cr-isotopic study by an ion microprobe, the Cr-isotopic compositions were measured only for a coarse-grained fayalite replacing a magnetite-sulfide nodule (Fig. 5d) and adjacent matrix. Both analyses of the fayalite grain show large $^{53}Cr^*$ correlated with the respective $^{55}Mn/^{52}Cr$ ratios, indicative for the *in situ* decay of ^{53}Mn (Fig. 6a). The slope of the correlation line fitted to the data, passing through the normal Cr-isotopic composition of matrix ($\delta^{53}Cr = 0$) at Mn/Cr ≅ 1, corresponds to the initial $^{53}Mn/^{55}Mn$ ratio of $(1.58 \pm 0.26) \times 10^{-6}$ at the time the fayalites formed ($\Delta t_{LEW} = -1.25 \pm 0.83$ Ma).

3.5. CV Chondrites and Settings of Their Alteration

The CV carbonaceous chondrites are currently subdivided into the reduced (CV_{red}) and two oxidized subgroups, Allende-like (CV_{oxA}) and Bali-like (CV_{oxB}) (*McSween,* 1977; *Weisberg and Prinz,* 1998), which largely reflect their complex alteration history and may represent different

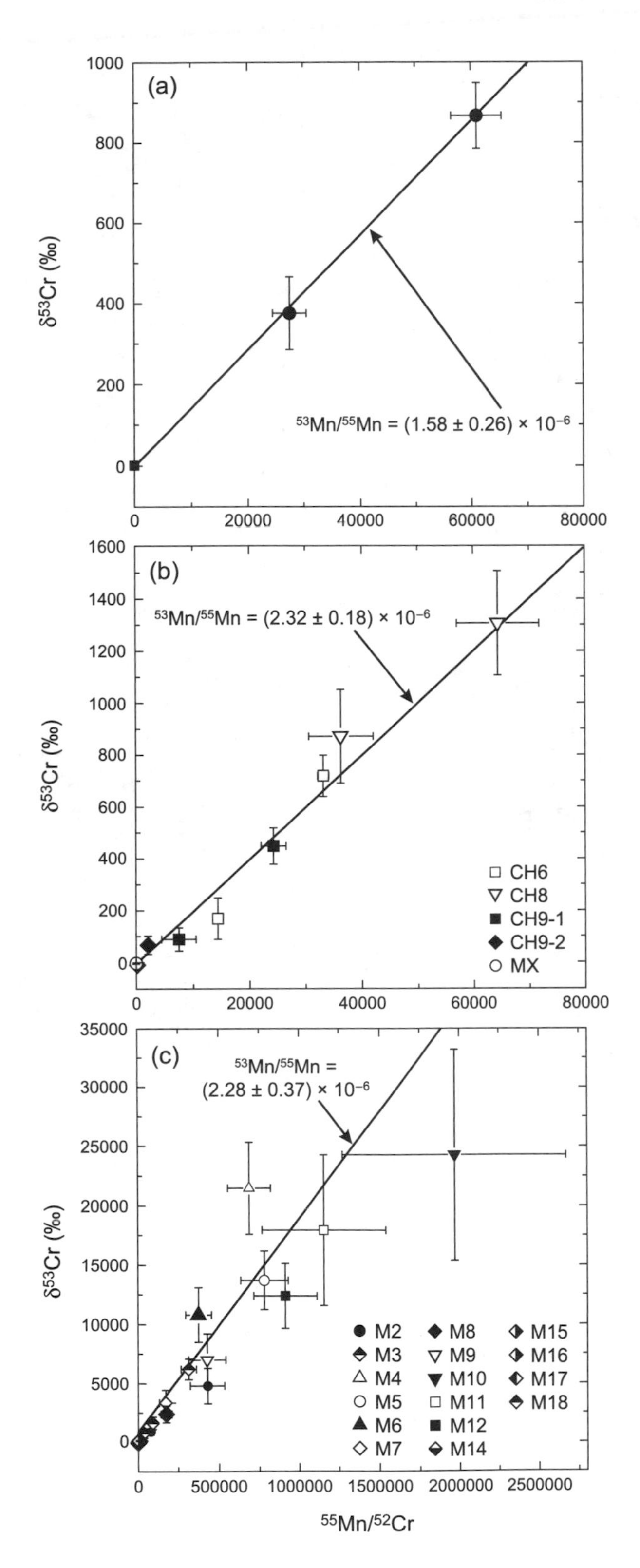

Fig. 6. ^{53}Mn-^{53}Cr evolution diagrams for **(a)** a fayalite grain from the ungrouped carbonaceous chondrite MAC 88107, **(b)** 4 fayalite grains in 3 porphyritic olivine-pyroxene type I chondrules from the CV_{oxB} chondrite Mokoia, and (c) 12 fayalite grains in matrix of the CV_{oxB} chondrite. The lines of slope $(1.58 \pm 0.26) \times 10^{-6}$, $(2.32 \pm 0.18) \times 10^{-6}$, and $(2.28 \pm 0.37) \times 10^{-6}$ are the best-fit lines through the data points and normal Cr; error bars are 2σ (data from *Hutcheon et al.,* 1998; *Krot et al.,* 2000a; *Hua et al.,* 2002, 2005).

lithological varieties of the same asteroidal body (*Krot et al.,* 1998a).

The CV_{oxB} chondrites (e.g., Kaba, Bali) experienced aqueous alteration resulting in replacement of primary minerals in chondrules, CAIs, and AOAs by secondary phyllosilicates, magnetite, Fe,Ni-sulfides, Fe,Ni-carbides, fayalite, salite-hedenbergite pyroxenes ($Fs_{10-50}Wo_{45-50}$), and andradite. Their matrices largely consist of the secondary minerals, including concentrically zoned nodules of Ca,Fe-pyroxene and andradite, coarse (>10 μm) grains of nearly pure fayalite (>Fa_{90}), abundant phyllosilicates, and very fine-grained (<1–2 μm) ferrous olivine (~Fa_{50}) (Fig. 7a).

The CV_{oxA} chondrites (e.g., Allende, ALH 84128) are more extensively altered than the CV_{oxB}, but contain very minor phyllosilicates (this alteration is often referred as Fe-alkali metasomatism). Chondrules and refractory inclusions in the CV_{oxA} chondrites contain secondary nepheline, sodalite, Ca,Fe-pyroxenes, andradite, Fe,Ni-sulfides, magnetite, Ni-rich metal, and ferrous olivine. Their matrices are coarser grained than those in the CV_{oxB} and largely consist of Ca,Fe-pyroxene ± andradite nodules, lath-shaped ferrous olivine (~Fa_{50}), and nepheline (Fig. 7c).

The CV chondrite Mokoia is a complex breccia containing clasts of the CV_{oxA} and CV_{oxB} lithologies and heavily metamorphosed oxidized chondritic clasts (*Krot et al.,* 1998a). The CV_{oxA} clasts experienced aqueous alteration that overprints "anhydrous" Allende-like alteration (*Kimura and Ikeda,* 1998). Some oxidized CVs (e.g., MET 00430) are mineralogically intermediate between the CV_{oxB} and CV_{oxA} chondrites (Fig. 7b) (*Krot et al.,* 2004a).

The CV_{red} chondrites Efremovka and Leoville experienced alteration similar to that of CV_{oxA}, but to a smaller degree. The reduced CV chondrite breccia Vigarano contains clasts of the CV_{oxB} and CV_{oxA} materials (*Krot et al.,* 2000b); the reduced portion experienced aqueous alteration resulting in formation of phyllosilicates and magnetite.

In addition to the oxidized and reduced CV subgroups, CV chondrites contain dark inclusions that are chemically and petrographically similar to their host meteorites (Fig. 7d), but appear to have experienced more extensive alteration (e.g., *Fruland et al.,* 1978; *Kurat et al.,* 1989; *Johnson et al.,* 1990; *Buchanan et al.,* 1997; *Krot et al.,* 1997a, 1998a, 1999, 2001).

3.5.1. Nebular versus asteroidal alteration of CV chondrites. The origin of secondary mineralization in CV chondrites remains controversial; nebular and asteroidal settings have been proposed. According to the nebular models (*Palme and Wark,* 1988; *Weisberg and Prinz,* 1998), chondrules and refractory inclusions in the CV_{oxA} were exposed to a highly oxidized nebular gas resulting in their alteration; matrix minerals directly condensed from this gas. This model is, however, inconsistent with (1) the presence of poorly graphitized carbon (PGC) and pentlandite inclusions in matrix olivine (*Brearley,* 1999); (2) the lack of volatility-controlled rare earth element (REE) patterns in matrix Ca,Fe-pyroxenes and andradite (*Brearley and Shearer,* 2000); (3) the large mass-dependent fractionation of O iso-

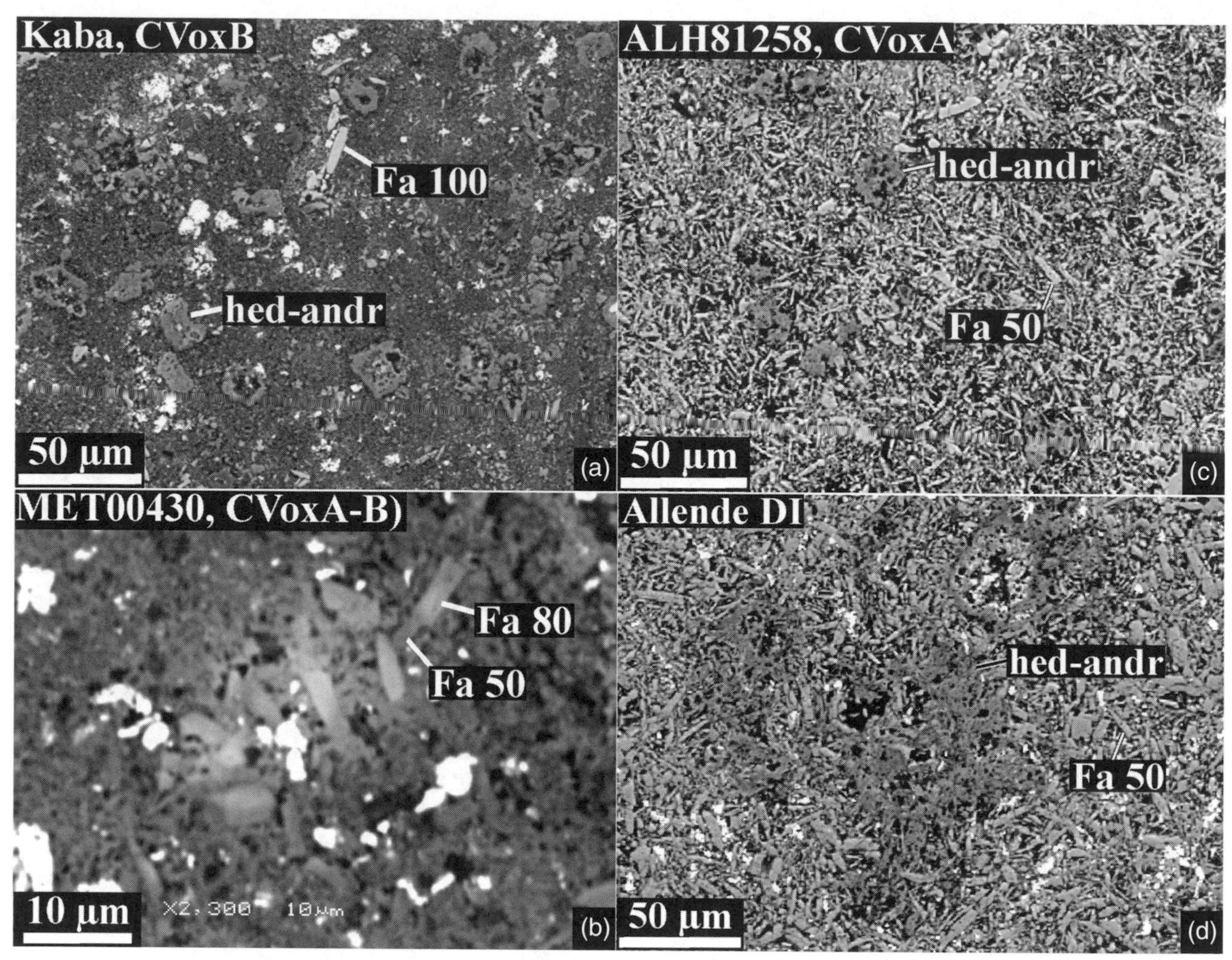

Fig. 7. Backscattered electron images of matrices in the **(a)**,**(b)** oxidized CV chondrites and **(d)** Allende dark inclusions. All matrices contain Ca,Fe-pyroxenes-andradite (hed-andr) nodules. Matrix in the CV_{oxB} Kaba contains nearly pure fayalite (~Fa_{100}) and very fine-grained groundmass largely composed of ferrous olivine (~Fa_{50}) and phyllosilicates. Matrix in the $CV_{oxA\text{-}B}$ MET 00430 contains fayalite grains showing inverse compositional zoning (Fa_{80-50}) and coarser-grained lath-shaped ferrous olivine (~Fa_{50}). Matrices in the CV_{oxA} ALH 81258 and Allende dark inclusion contain relatively coarse-grained, lath-shaped compositionally uniform (~Fa_{50}) ferrous olivine.

topes ($\delta^{18}O$ ~ 20‰) in secondary fayalite, magnetite, Ca,Fe-rich pyroxenes, and andradite (*Krot et al.,* 2000c; *Choi et al.,* 2000; *Cosarinsky et al.,* 2003); and (4) the thermodynamic analysis of condensation of ferrous olivine (*Grossman and Fedkin,* 2003).

According to the asteroidal models (*Krot et al.,* 1995, 1997a, 1998a,b, 2004a; *Kojima and Tomeoka,* 1996), CV chondrites experienced fluid-assisted thermal metamorphism to various degrees, which resulted in mobilization of Ca, Si, Fe, Mg, Mn, Na, and S and replacement of primary phases in chondrules, CAIs, and matrices by secondary minerals. It was originally suggested that secondary ferrous olivine in CV chondrites formed by dehydration of phyllosilicates during thermal metamorphism (*Krot et al.,* 1995, 1997a). This mechanism, however, appears to be inconsistent with the lack of mass-dependent fractionation of O isotopes in bulk CV chondrites (*Clayton and Mayeda,* 1999), which is expected for extensively aqueously altered and dehydrated meteorites (e.g., metamorphosed CI/CM). Recently, *Krot et al.* (2004a) concluded that ferrous olivine in CV chondrites formed by several mechanisms during fluid-assisted metamorphism, including replacement of opaque nodules, magnesian olivine, and pyroxene; direct precipitation from a supersaturated fluid; and, possibly, dehydration of phyllosilicates.

3.5.2. Settings and timescale of secondary alteration of the CV_{oxB} chondrites. There are several lines of evidence suggesting that the secondary minerals in the CV_{oxB} formed during relatively low-temperature aqueous alteration of the CV asteroidal body, rather than by high-temperature gas-solid reactions in the solar nebula. (1) The secondary minerals occur in all CV_{oxB} chondritic components, including chondrules, CAIs, AOAs, and matrices. (2) Fine-grained rims around chondrules are commonly crosscut by fayalite-bearing veins that start at the opaque nodules in the chondrule peripheries (Figs. 8a,b). (3) Fayalite replacing magnetite-sulfide nodules in type I chondrules (Figs. 8c,d) show large mass-dependent fractionation of O isotopes and con-

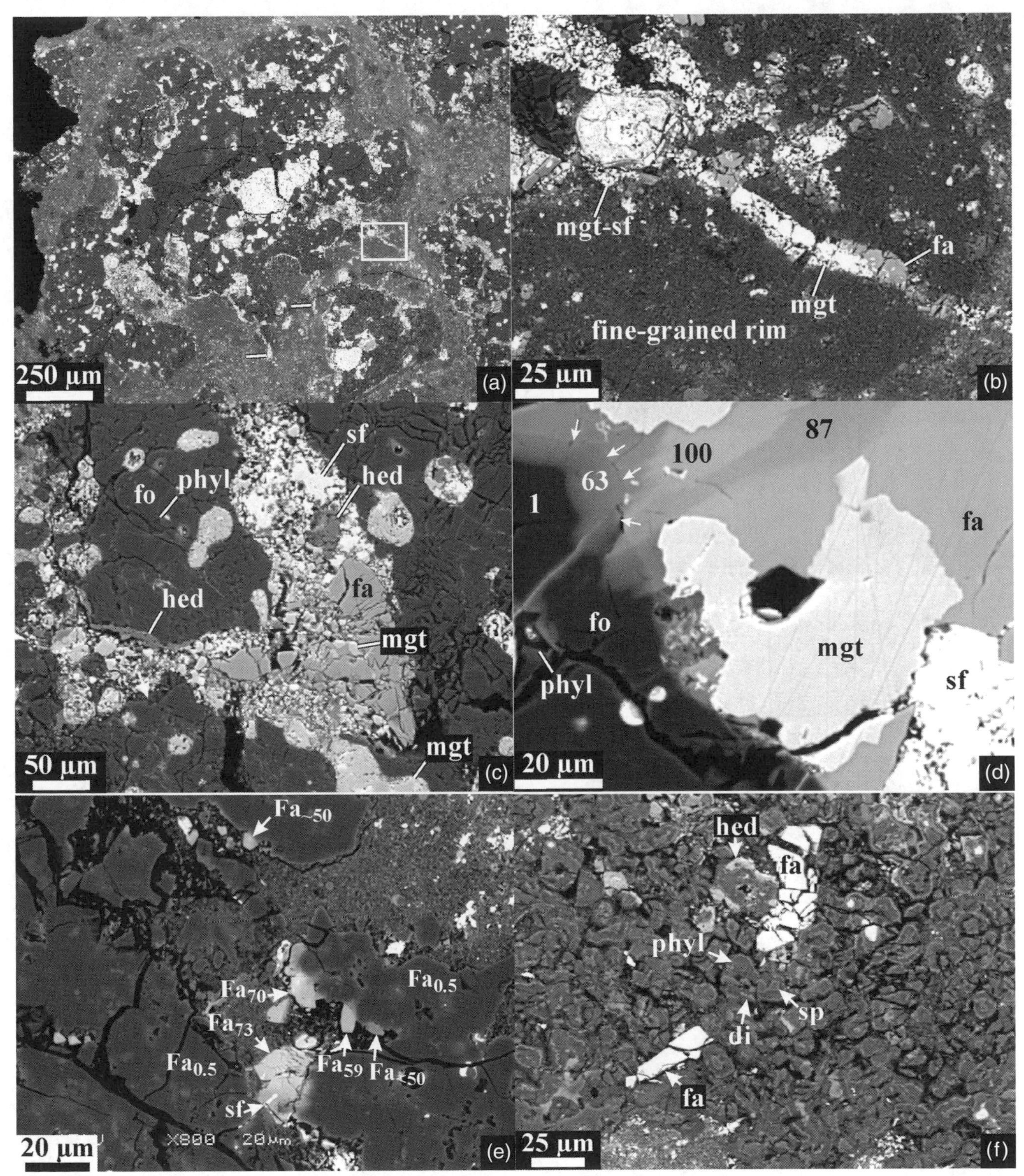

Fig. 8. Backscattered electron images of different textural occurrences of secondary fayalite in the CV_{oxB} chondrites Kaba and Mokoia. **(a)**,**(b)** Porphyritic olivine-pyroxene (POP) type I chondrule surrounded by a continuous fine-grained rim crosscut by fayalite (fa) – magnetite (mgt) veins. The veins start at the opaque nodules composed of Ni-bearing sulfide (sf) and magnetite in the peripheral portion of the chondrule. Region outlined in **(a)** is shown in detail in **(b)**. **(c)** Opaque nodule in type I chondrule replaced by magnetite, Ni-bearing sulfides, fayalite, and salite-hedenbergite pyroxenes (hed). **(d)** Opaque nodule within type I POP chondrule; numbers correspond to fayalite content (in mol%). Magnetite is replaced by pure fayalite (Fa_{100}); forsterite phenocrysts (Fa_1) are partly pseudomorphed by ferrous olivine (Fa_{63}); an outline of one of the grains is indicated by arrows. Fayalite is crosscut by a vein of ferrous olivine (Fa_{87}), suggesting that forsterite is the source of Mg. **(e)** Amoeboid olivine aggregate composed of forsterite, spinel, Al-diopside, and anorthite. Forsterite grains are overgrown by euhedral ferrous olivines ranging in compositions from $Fa_{<50}$ to Fa_{73}; some of the fayalite grains contain inclusions of Fe,Ni-sulfides (sf). **(f)** Fine-grained CAI consisting of concentrically zoned objects composed of spinel (sp) surrounded by phyllosilicates (phyl) and Al-diopside (di); phyllosilicates probably replace primary anorthite or melilite. Euhedral fayalite grains occur between these bodies; Ca,Fe-pyroxenes (hed) overgrow Al-diopside.

tain sulfide inclusions (Fig. 9a), suggesting low-temperature formation. (4) Euhedral fayalite grains of variable compositions overgrow forsterite grains of AOAs without any evidence for Fe-Mg interdiffusion in the neighboring forsterite grains, suggesting precipitation from a low-temperature fluid of variable chemical composition (Fig. 8e). Occasionally, ferrous olivine pseudomorphs chondrule phenocrysts (Fig. 8d), supporting the presence of aqueous solutions during the alteration. (5) Low Al contents in secondary Ca,Fe-pyroxenes, indicating large Ca/Al fractionation during their formation, is inconsistent with their high-temperature condensation origin (both Ca and Al are refractory lithophile elements of similar volatility and are not expected to be fractionated from each other during condensation). These observations and thermodynamic analysis of phase relations in the Si-Fe-Ca-O-H system (*Krot et al.,* 1998a) suggest that secondary minerals in CV_{oxB} chondrites in the presence of aqueous solutions capable of transporting Si, Fe, Ca, Mn, and Mg.

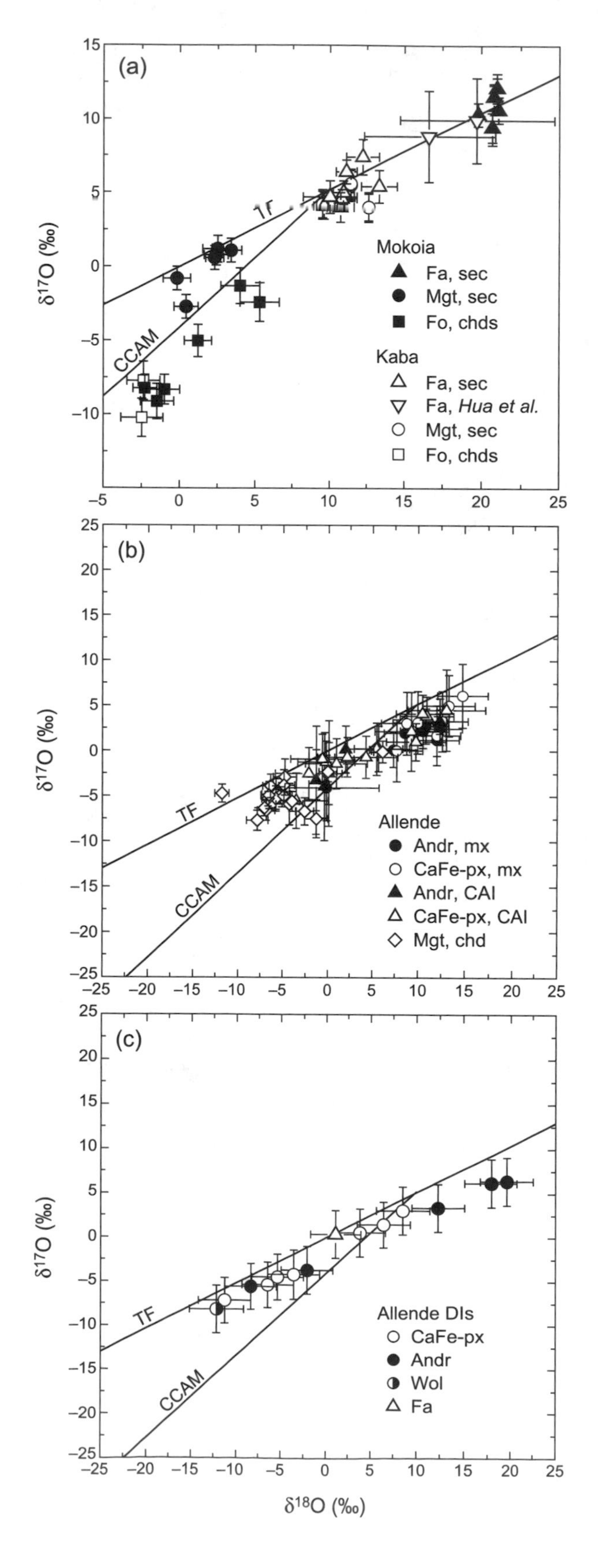

Petrographic observations of type I chondrules (*Hua and Buseck,* 1995; *Krot and Todd,* 1998; *Krot et al.,* 1998a,b) suggest the following sequence of secondary mineral formation. Magnetite and Fe,Ni-sulfides replacing Fe,Ni-nodules formed first. Phyllosilicates replacing chondrule mesostases and phenocrysts formed either subsequently or contemporaneously with magnetite. Fayalite, Ca,Fe-pyroxenes, and andradite replace magnetite and coexist with phyllosilicates, possibly indicating contemporaneous formation of these phases; occasionally, fayalite is corroded by Ca,Fe-pyroxenes.

3.5.2.1. Manganese-chromium-isotopic dating of secondary fayalite in Kaba and Mokoia: High MnO contents (up to 1.5 wt%) in secondary fayalite and nearly complete absence of Cr (Mn/Cr ratios range up to 2×10^6) favor Cr-isotopic measurements of fayalite to constrain its crystalli-

Fig. 9. Oxygen-isotopic compositions of secondary magnetite (Mgt), fayalite (Fa), Ca,Fe-rich pyroxenes (CaFe-px), andradite (Andr), and wollastonite (Wol), and primary forsteritic olivine (Fo) **(a)** in type I chondrules in the CV_{oxB} chondrites Kaba and Mokoia (data from *Choi et al.,* 2000; *Hua et al.,* 2005); **(b)** in chondrules, matrix (mx), and rims around CAIs (data from *Choi et al.,* 2000; *Cosarinsky et al.,* 2003); and **(c)** in and around Allende dark inclusions (data from *Krot et al.,* 2000c); error bars are 2σ. The terrestrial fractionation (TF) line and carbonaceous chondrite anhydrous mineral (CCAM) line are shown for reference. **(a)** In Mokoia, the magnetite and fayalite differ in $\delta^{18}O$ by ~20‰, suggesting formation at low temperatures. In Kaba, the compositions of fayalite and magnetite reported by *Choi et al.* (2000) are nearly identical, and very close to the intersection of the TF and CCAM lines. The compositions of Kaba fayalites reported by *Choi et al.* (2000) are inconsistent with those reported by *Hua et al.* (2005); the latter are similar to those of Mokoia fayalites. We note that compositions of fayalite and magnetite in Kaba reported by *Choi et al.,* 2000) were collected within a three-month interval and might be in error. Compositions of forsteritic olivine phenocrysts plot along the CCAM line and are not in equilibrium with those of the secondary minerals. **(b)** Oxygen-isotopic compositions of Ca,Fe-rich pyroxenes and andradite in matrix (mx) and in rims around CAIs are similar and plot parallel to the TF with a range in $\delta^{18}O$ of ~20‰, suggesting formation at low temperatures. Oxygen-isotopic compositions of magnetite overlap with those of Ca,Fe-pyroxenes and andradite, but plot largely to the left from the CCAM line.

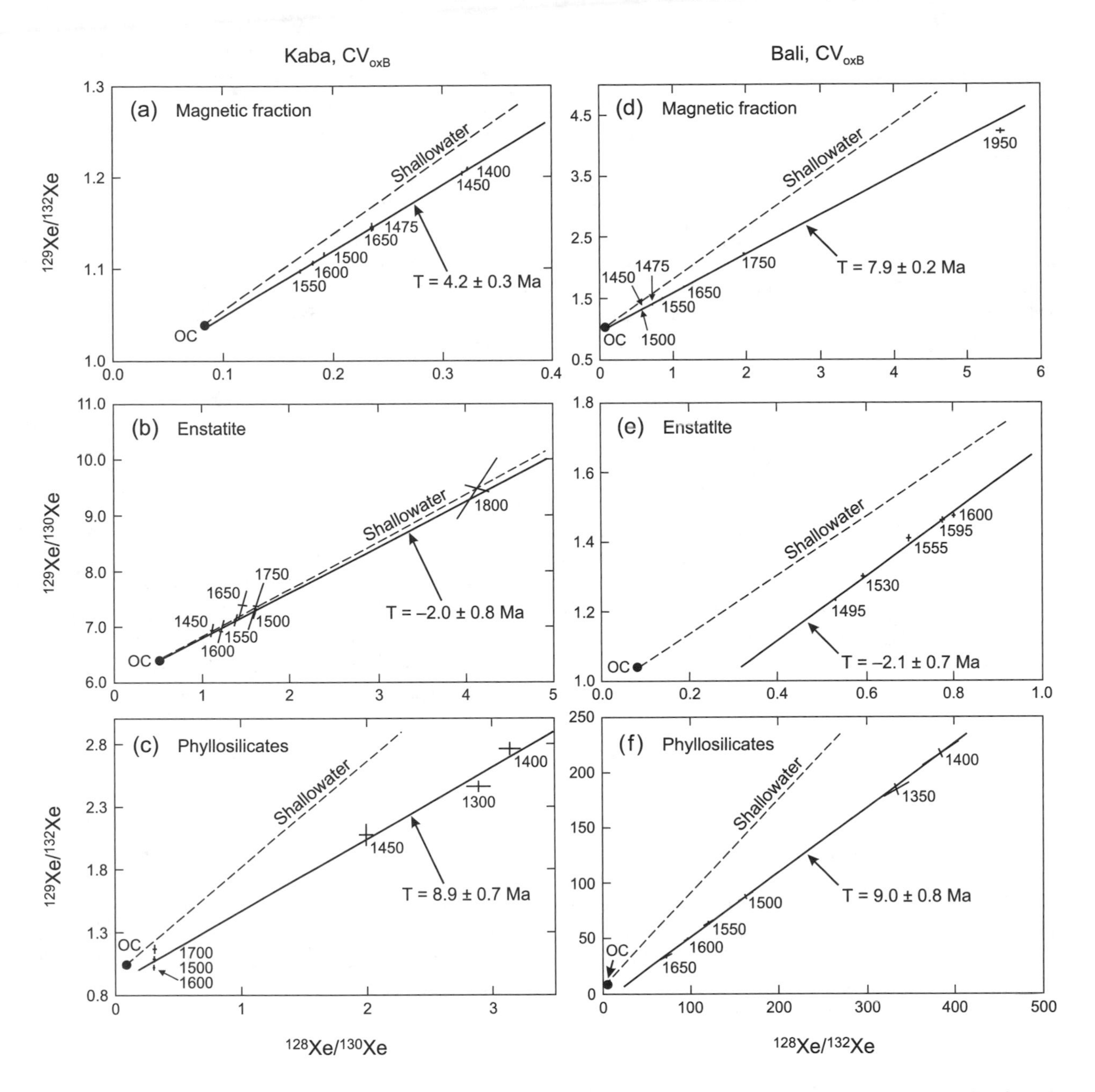

Fig. 10. ^{129}I-^{129}Xe evolution diagrams for mineral fractions separated from the CV_{oxB} chondrites **(a)–(c)** Kaba and **(d)–(f)** Bali; error bars are 1σ. The I-Xe ages shown are relative to the Shallowater internal standard (4563.5 ± 1 Ma) (J. Gilmour et al., personal communication, 2005). Numbers next to points represent extraction W-coil temperatures in °C (the sample is probably 150°–200°C cooler).

zation time (*Hutcheon et al.,* 1998; *Hua et al.,* 2003, 2004). *Hutcheon et al.* (1998) measured Cr-isotopic compositions of six fayalite grains replacing magnetite-sulfide nodules within type I chondrules from Mokoia. *Hua et al.* (2002, 2005) analyzed Cr-isotopic compositions of 12 fayalite grains associated with magnetite and sulfides in Kaba matrix. Fayalite grains in both textural occurrences have large $^{53}Cr^*$ correlated with $^{55}Mn/^{52}Cr$ ratios, indicative for *in situ* decay of ^{53}Mn, which define similar initial $^{53}Mn/^{55}Mn$ ratios of (2.32 ± 0.18) × 10^{-6} and (2.28 ± 0.37) × 10^{-6}, respectively (Δt_{LEW} = ~3.0 ± 0.7 m.y.) (Figs. 6b,c).

3.5.2.2. Iodine-xenon dating of magnetite and phyllosilicates formation in Kaba and Bali: *Pravdivtseva and Hohenberg* (2001) measured Xe-isotopic compositions of magnetic fractions separated with a hand magnet from fine-crushed Kaba, Bali, and Mokoia. The ^{128}Xe and ^{129}Xe release profiles in Kaba and Bali suggest one major I carrier in magnetic separates that yield well-defined isochrons in temperature ranges of 1400°–1750°C and 1450°–1950°C, respectively (Figs. 10a,d). The isochrons correspond to closure time of the I-Xe system in the Kaba and Bali magnetite of 4.2 ± 0.3 Ma and 7.9 ± 0.2 Ma relative to the Shallowater internal reference standard, respectively.

The ^{128}Xe and ^{129}Xe release profiles at 1400°–1900°C of the Mokoia magnetic fraction suggest multiple I carriers; no isochron was obtained. These results are consistent with the complex brecciated nature and multistage alteration history of this meteorite (*Krot et al.,* 1998a; *Kimura and Ikeda,* 1998).

Two nonmagnetic fractions hand picked from coarsely crushed samples of Kaba and Bali and tentatively identified (using EDS) as enstatite and a mixture of plagioclase-rich mesostasis and Al-rich phyllosilicates were measured for Xe-isotopic compositions (*Pravdivtseva et al.,* 2001). Considering the very fine scale of primary and secondary mineral intergrowths in the CV_{oxB} chondrules, it is difficult to expect good mineral separation, and the results should be treated cautiously. The enstatite separates define precise high-temperature isochrons from ~1400°C to ~1800°C with similar I-Xe ages: –2.0 ± 0.8 Ma for Kaba and –2.1 ± 0.7 Ma for Bali, relative to the Shallowater internal standard (Figs. 10b,e). The mixture of plagioclase-rich mesostasis and Al-rich phyllosilicates yield lower-temperature isochrons corresponding to I-Xe ages of 8.9 ± 0.7 Ma and 9.0 ± 0.8 Ma for Kaba and Bali, respectively (Figs. 10c,f). These ages are systematically younger than the corresponding magnetite ages and may suggest that either magnetite formation predates formation of phyllosilicates or that I-Xe-isotopic closure in magnetite occurred prior to that in phyllosilicates. The overall I-Xe data suggest that the aqueous alteration on the CV_{oxB} parent body lasted for at least 10 m.y.

3.5.3. Settings and timescale of alteration of the CV_{oxA} chondrites. There are several lines of evidence suggesting that Fe-alkali metasomatic alteration of the CV_{oxA} chondrites resulted from fluid-assisted thermal metamorphism of the CV asteroidal body (e.g., *Krot et al.,* 1998a), rather than from high-temperature gas-solid reactions in the solar nebula (e.g., *Palme and Wark,* 1988). (1) The secondary minerals occur in all CV_{oxA} chondritic components, including chondrules, CAIs, AOAs, and matrices (e.g., *Hashimoto and Grossman,* 1987; *MacPherson et al.,* 1988; *Krot et al.,* 1995), and show evidence for *in situ* formation (e.g., veins, rims, chondrule pseudomorphs; Fig. 11) (*Krot et al.,* 1997a, 1998a,b, 2001; *MacPherson and Krot,* 2002). (2) Oxygen-isotopic compositions of the secondary Ca,Fe-pyroxenes, andradite, and wollastonite in matrix and rims around CAIs plot parallel to the terrestrial fractionation line at $\Delta^{17}O$ ~ –2.5‰ with a large range in $\delta^{18}O$ (~20‰) (Fig. 9b), comparable to the range reported for the secondary magnetites and fayalites in the CV_{oxB} chondrites (Fig. 9a), suggesting low-temperature formation. This mechanism is also consistent with the presence of sulfide inclusions (Fig. 11b) and lack of volatility-controlled REE patterns in Ca,Fe-pyroxenes and andradite in the Allende matrix (*Brearley and Shearer,* 2000). (3) Secondary ferrous olivine replacing low-Ca pyroxene phenocrysts in type I chondrules coexists with talc and amphibole (*Brearley,* 1997), suggesting that Fe was transported by low-temperature aqueous solutions (*Krot et al.,* 2004a), rather than by a high-temperature gas phase (*Dohmen et al.,* 1998).

Although there are many textural and mineralogical similarities in secondary alteration of the CV_{oxB} and CV_{oxA} chondrites (Figs. 8, 11a–d), the latter are more extensively altered and contain secondary ferrous olivine (Fa_{40-60}), nepheline, and sodalite instead of fayalite (Fa_{90-100}) and phyllosilicates. There are also some difference in $\Delta^{17}O$ values of the secondary phases (fayalite, magnetite, and Ca,Fe-rich silicates) in the CV_{oxB} (~–0.6‰) and CV_{oxA} chondrites (–2.6‰) (Figs. 9a,b). Secondary fayalites in the intermediate CV_{oxA-B} meteorites (e.g., MET 00430) show inverse compositional zoning (Figs. 7b, 11e) and evidence for dissolution of fayalite and precipitation of more forsteritic olivine (Fig. 11f). These observations may indicate that the CV_{oxA} experienced alteration at higher temperatures than the CV_{oxB}.

Petrographic observations on type I chondrules in the CV_{oxA} (Figs. 11a–d) (*Krot et al.,* 1998a,b) suggest the following sequence of secondary mineral formation. Magnetite and Fe,Ni-sulfides replacing Fe,Ni-nodules formed first. Ferrous olivine and Ca,Fe-pyroxenes formed later; they preferentially replace magnetite of the nodules and contain abundant inclusions of Fe,Ni-sulfides (Figs. 11a–c). Ferrous olivine also replaces low-Ca pyroxene phenocrysts and overgrows or replaces forsteritic olivine phenocrysts (Figs. 10c,d) (*Krot et al.,* 1997a). Nepheline and sodalite replace chondrule mesostasis and may have formed prior to, contemporaneously with, or after ferrous olivine (e.g., *Kimura and Ikeda,* 1995).

In addition to Fe- and alkali-rich minerals in the CV_{oxA} chondrites, coarse-grained CAIs in Allende contain secondary grossular, monticellite, wollastonite, and forsterite that typically replace melilite-anorthite assemblages (*Hutcheon et al.,* 1978; *Hutcheon and Newton,* 1981; *Wark,* 1987; *Krot et al.,* 2004b). Based on petrographic observations and thermodynamic analysis, *Hutcheon and Newton* (1981) concluded that grossular and monticellite formed during a prolonged heating in the solar nebula at ~950 K via the closed-system reaction

$$3Ca_2MgSi_2O_7 + Ca_2Al_2SiO_7 + CaAl_2Si_2O_8 = 2Ca_3Al_2Si_3O_{12} + 3CaMgSiO_4 \quad (4)$$

Krot et al. (2004b) concluded instead that other closed-system reactions took place

$$3Ca_2MgS_2O_7 + Ca_2Al_2SiO_7 + 2CaAl_2Si_2O_8 = 3Ca_3Al_2Si_3O_{12} + CaMgSiO_4 \quad (5)$$

and

$$4Ca_2MgSi_2O_7 + Ca_2Al_2SiO_7 + CaAl_2Si_2O_8 = 2Ca_3Al_2Si_3O_{12} + 4CaMgSiO_4 + CaSiO_3 \quad (6)$$

Although under equilibrium conditions these reaction occur below 950 K, the common presence of unaltered melilite-anorthite intergrowths in the Allende Type C CAIs implies the lack of equilibrium (i.e., temperature estimates should be considered with caution).

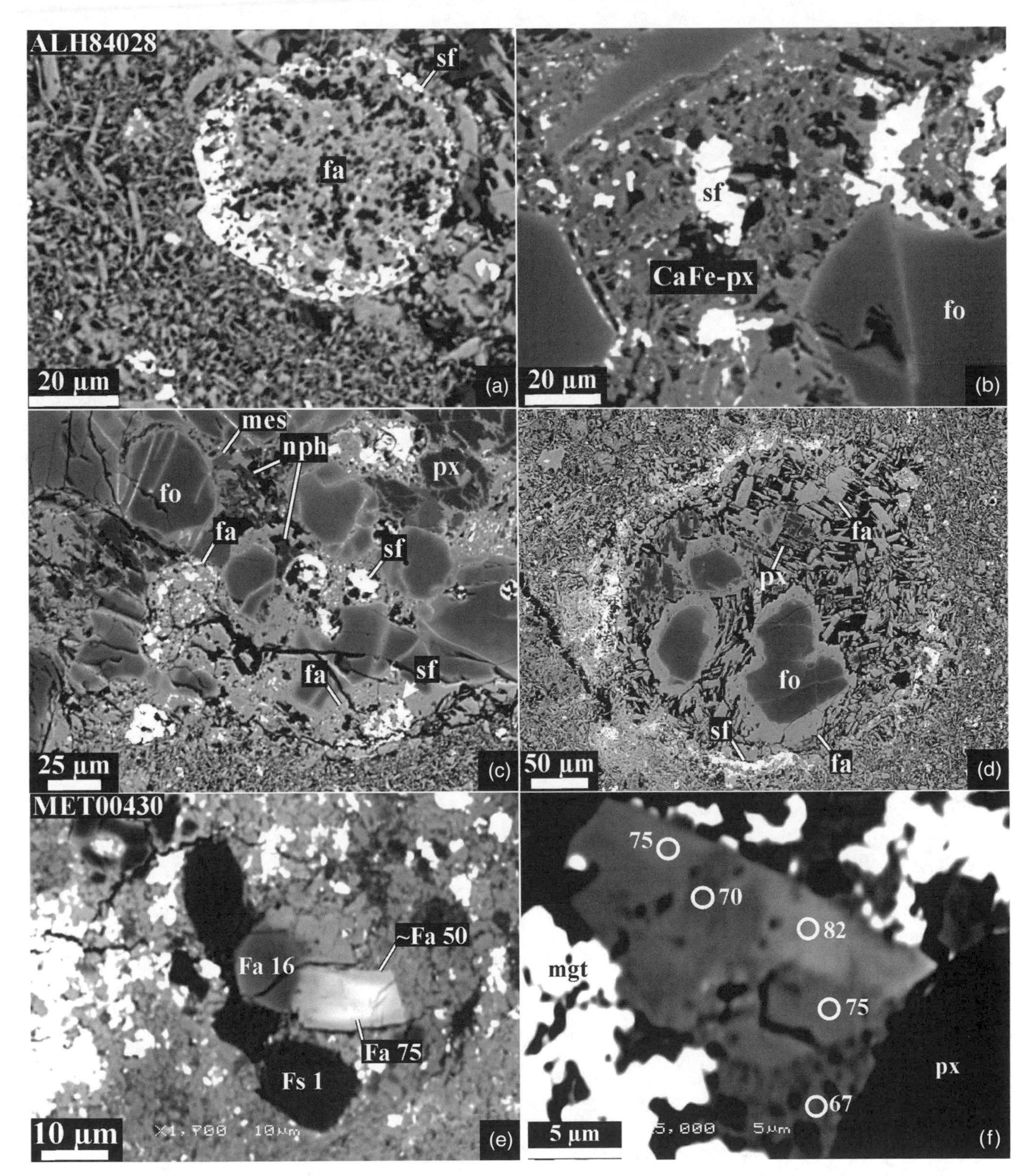

Fig. 11. **(a)–(d)** Backscattered electron images of secondary minerals in POP type I chondrules in the CV_{oxA} chondrite ALH 84028. Magnetite-sulfide nodules are replaced by ferrous olivine (fa) and Ca,Fe-pyroxenes (CaFe-px); both contain abundant inclusions of sulfides (sf). Low-calcium pyroxene phenocryststs (px) are replaced by ferrous olivine. Chondrule mesostasis is replaced by nepheline (nph). Forsteritic olivine (fo) phenocrysts are largely unaltered, but show enrichment in fayalite contents near the edges and along the fractures. **(e),(f)** Backscattered electron images of secondary fayalite in the $CV_{oxA\text{-}B}$ chondrite MET 00430. **(e)** Fayalite overgrowing olivine-pyroxene chondrule fragment shows inverse compositional zoning (Fa_{75-50}). **(f)** Euhedral fayalite grain overgrowing low-Ca pyroxene (px) phenocryst in outer part of a type I chondrule shows complex chemical zoning suggesting dissolution of fayalite and precipitation of more forsteritic olivine from a fluid phase. Numbers correspond to fayalite contents (from *Krot et al.,* 2004a).

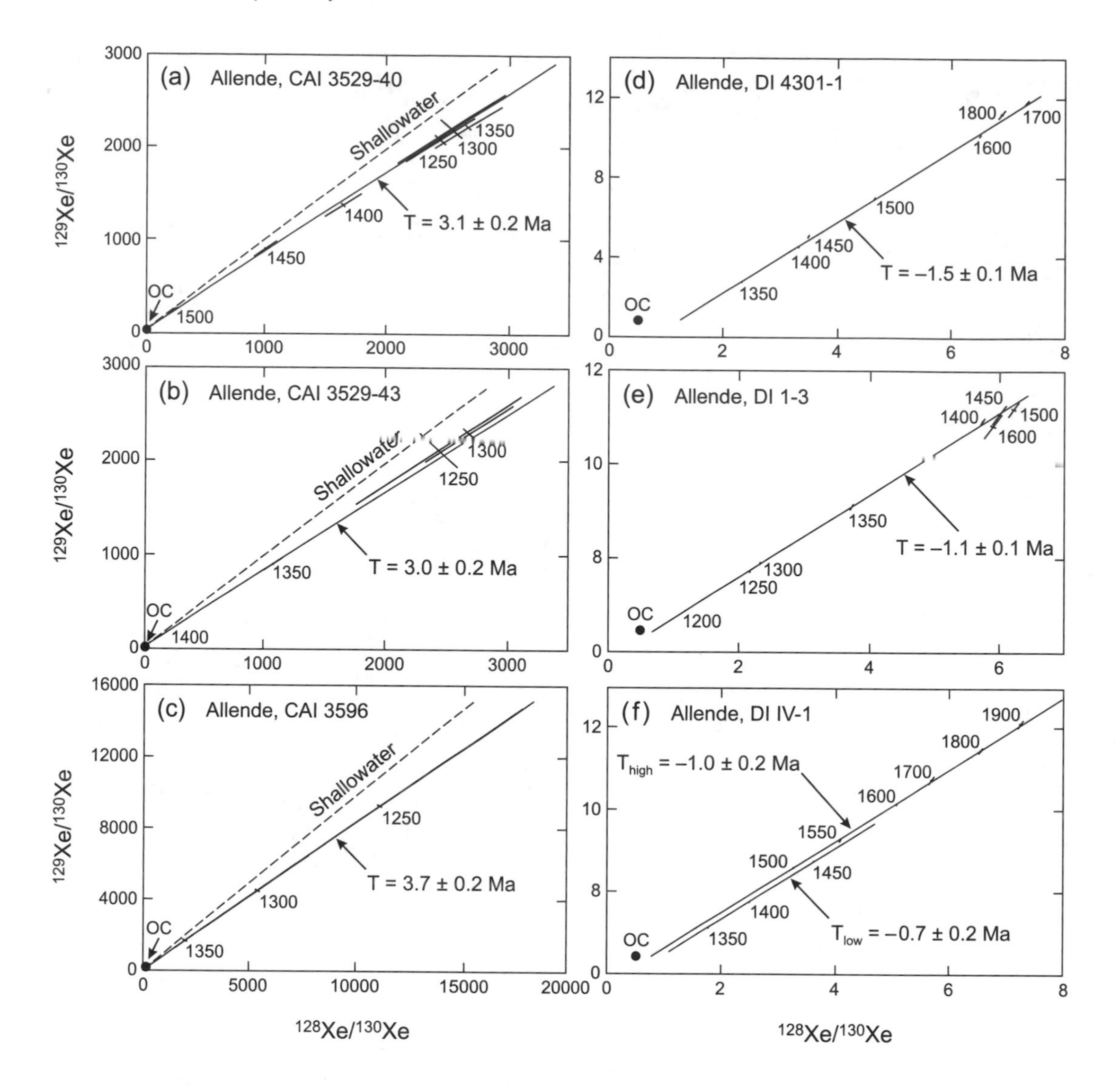

Fig. 12. **(a)–(c)** ^{129}I-^{129}Xe evolution diagrams for fine-grained CAIs in Allende (from *Pravdivtseva et al.,* 2003b). The contribution from the trapped Xe component is within experimental uncertainty consistent with the "planetary" OC-Xe (*Lavielle and Marti,* 1992). **(d)–(f)** ^{129}I-^{129}Xe evolution diagrams for Allende dark inclusions. Two isochrons plotted for the dark inclusion IV-1 correspond to low- and high-temperature Xe released. All isochrons suggest "subplanetary" trapped components (*Hohenberg et al.,* 2004). Error bars are 1σ. Iodine-xenon ages are relative to the Shallowater internal standard (4563.5 ± 1 Ma) (J. Gilmour et al., personal communication, 2005).

3.5.3.1. Aluminum-magnesium-isotopic dating of secondary alteration of the CV calcium-aluminum-rich inclusions: Secondary minerals (nepheline, sodalite, grossular) in the CV CAIs generally show no evidence for ^{26}Mg*, suggesting that the alteration took place at least several half-lives of ^{26}Al after the formation of the primary phases typically having canonical ^{26}Al/^{27}Al ratios of ~5×10^{-5} (e.g., *Hutcheon and Newton,* 1981). The only Allende CAI with excesses of ^{26}Mg in secondary nepheline and sodalite is a fine-grained spinel-rich inclusion analyzed by *Brigham et al.* (1986). The observed ^{26}Mg* corresponds to an initial ^{26}Al/^{27}Al ratio of $(6–7) \times 10^{-5}$. *MacPherson et al.* (1995) interpreted these data as evidence for an early, nebular formation of the secondary minerals that continued over an extended (several million years) period of time. However, taking into account the low ^{27}Al/^{24}Mg ratios in the analyzed minerals, the anomalously high initial ^{26}Al/^{27}Al ratio inferred for this CAI, and the clear evidence for metamorphic redistribution of Mg isotopes in the Allende CAIs (e.g., *Yurimoto et al.,* 2000), it seems more likely that this "isochron" resulted from Mg-isotopic exchange between the primary and secondary minerals of the CAI and does not have a chronological meaning [see also Fig. 2 in *MacPherson et al.* (1995)].

3.5.3.2. Iodine-xenon dating of secondary alteration of calcium-aluminum-rich inclusions and chondrules in Allende: Iodine-xenon-isotopic data for the coarse-grained and fine-grained CAIs in Allende, which experienced iron-alkali metasomatic alteration, encompass a spread of ≥10 m.y., supporting an asteroidal setting of alteration (*Swindle* et al., 1988). The strong correlation of I with Cl in two fine-grained CAIs analyzed by *Kirschbaum* (1988), together with the fact that sodalite is the only significant Cl-bearing mineral in these CAIs, verified sodalite as the major I-carrier phase.

Recent Xe-isotopic measurements (*Pravdivtseva et al.,* 2003b) showed that heavily altered fine-grained CAIs in Allende define isochrons with ages between 3.1 ± 0.2 and 3.7 ± 0.2 m.y. younger than Shallowater (Figs. 12a–c). The CAIs have nearly identical release profiles for radiogenic ^{129*}Xe and ^{128*}Xe, suggesting the same I carrier for both (probably sodalite) (*Kirschbaum,* 1988).

Although Allende chondrules often contain large fractions of radiogenic Xe, and a chronometry suggestive of an I-Xe association, they rarely yield isochrons that are well-defined at the level of precision provided by the isotopic data. Among nine chondrules studied by *Swindle et al.* (1983), eight have a pattern of increasing model age with each incrementally increased temperature step. This was attributed to relatively slow cooling [~10°–20°C/m.y. using the lower release temperatures (600°–1100°C) or of 50°–300°C/m.y. using release temperatures above 1300°C] or the monotonic (with release temperature) relaxation of other conditions during thermal metamorphism or alteration (*Swindle et al.,* 1983; *Nichols et al.,* 1990). One chondrule gave a well-defined isochron with an apparent age 0.53 ± 0.15 m.y. younger than the Bjurböle whole-rock age standard (*Swindle et al.,* 1983), and the I-Xe ages of four different chondrules gave ages ranging from –0.37 ± 0.16 Ma to 1.54 ± 0.07 Ma, relative to Bjurböle (*Nichols et al.,* 1990). Four coarse-grained chondrule rims tended to be slightly older than the interiors, but these rims were separated from a different set of chondrules, and the only chondrule/rim pair combination yielded concordant ages.

3.5.4. Setting and iodine-xenon dating of alteration of the CV dark inclusions. Dark inclusions in Allende experienced similar secondary alteration to their host meteorite, but to a higher degree (Figs. 13a,b). The very heavily altered dark inclusions consist almost entirely of secondary ferrous olivine, Ca,Fe-pyroxenes, andradite, nepheline, sodalite, and Fe,Ni-sulfides, and, if not brecciated, are surrounded by continuous, multilayered Ca,Fe-rich rims (Fig. 14) composed of Ca,Fe-pyroxenes, andradite, ± wollastonite, ± kirschteinite. The outermost layer of the rims is often intergrown with chondrule fragments and matrix olivine of the Allende host [Fig. 13f in *Krot et al.* (1998a); Fig. 10 in *Krot et al.* (2001)]. Some of the dark inclusions are crosscut by multiple Ca,Fe-rich veins (Fig. 14a), which are mineralogically similar to the Ca,Fe-rich rims and often connected to them. The outer portions of the rimmed dark inclusions are depleted in Ca, whereas the Allende matrix just outside the rims contain abundant Ca,Fe-rich silicate inclusions (Figs. 14a,b). Oxygen-isotopic compositions of Ca,Fe-rich silicates within and around dark inclusions in Allende (Fig. 9c) plot parallel to the terrestrial fractionation line at $\Delta^{17}O \sim -2‰$ with a large range in $\delta^{18}O$ (~30‰), suggesting low-temperature formation of these minerals (*Krot et al.,* 2000c).

Based on these observations and thermodynamic analysis of phase relations in the Si-Fe-Ca-O-H system, *Krot et al.* (2001) concluded that the rimmed dark inclusions in Allende experienced at least two stages of alteration in the presence of aqueous solutions. During an early stage of the alteration, which took place in an asteroidal setting, but not in the current location of the dark inclusions, chondrule silicates were replaced by secondary ferrous olivine, nepheline, and sodalite. Calcium lost from the chondrules was redeposited as Ca,Fe-pyroxene-andradite veins and nodules in the dark inclusion matrices. The second stage of alteration took place *in situ*, during the alteration of the Allende host, and resulted in mobilization of Ca from the dark inclusions and its redeposition as Ca,Fe-rich rims around the dark inclusions and as Ca,Fe-rich nodules in the neighboring matrix of Allende.

Xenon-isotopic compositions were measured in bulk samples of 17 Allende dark inclusions (*Pravdivtseva et al.,* 2003b). All dark inclusions yielded similar release profiles with two major peaks, suggesting two major I carriers (sodalite, and possibly Ca,Fe-pyroxenes or nepheline), and well-defined I-Xe isochrons (Figs. 12d–f) with ages ranging from 0.5 ± 0.3 to 2.8 ± 0.3 m.y. older than the Shallowater internal standard (Table 1). In contrast, three heavily altered fine-grained CAIs in Allende yielded well-defined isochrons with ages 3.1 ± 0.2, 3.0 ± 0.2, and 3.7 ± 0.2 m.y. younger than Shallowater (*Pravdivtseva et al.,* 2003b). The I-Xe ages of the dark inclusions were interpreted as the time of their early alteration prior incorporation into Allende. The younger I-Xe ages of the fine-grained spinel-rich CAIs may reflect hydrothermal alteration of the Allende host, which could have occurred contemporaneously with the second stage of alteration of the Allende dark inclusions. The lack of evidence for the disturbance of I-Xe system in the Allende dark inclusions suggests that fluid responsible for the alteration of the Allende CAIs must have been in equilibrium with the I- and Xe-bearing phases of the dark inclusions, so the latter were not affected by the second stage of alteration.

Dark inclusions in the CV_{red} chondrites Efremovka, Leoville, and Vigarano experienced different styles of aqueous alteration to various degrees (Figs. 13c–f) that resulted in formation of ferrous olivine, andradite, magnetite, and phyllosilicates (*Kracher et al.,* 1985; *Tomeoka and Kojima,* 1998; *Brearley,* 1998; *Krot et al.,* 1999). The presence of aqueous solutions during the alteration is supported by the textural observations (e.g., chondrule pseudomorphs), the presence of minor phyllosilicates (*Krot et al.,* 1999), and bulk O-isotopic compositions, which on a three-O-isotopic plot deviate to the right from the CCAM line (*Clayton and Mayeda,* 1999). Xenon-isotopic compositions were measured in bulk samples of six dark inclusions from the reduced CVs (*Swindle et al.,* 1998; *Krot et al.,* 1999; *Pravdivtseva et al.,* 2003c). The I carriers in the dark inclusions have not been identified; phyllosilicates and magnetite are two possible candidates. The I-Xe ages of the dark inclusions range from –4.9 ± 1.8 to 9.5 ± 2.3 Ma relative to Shallowater and

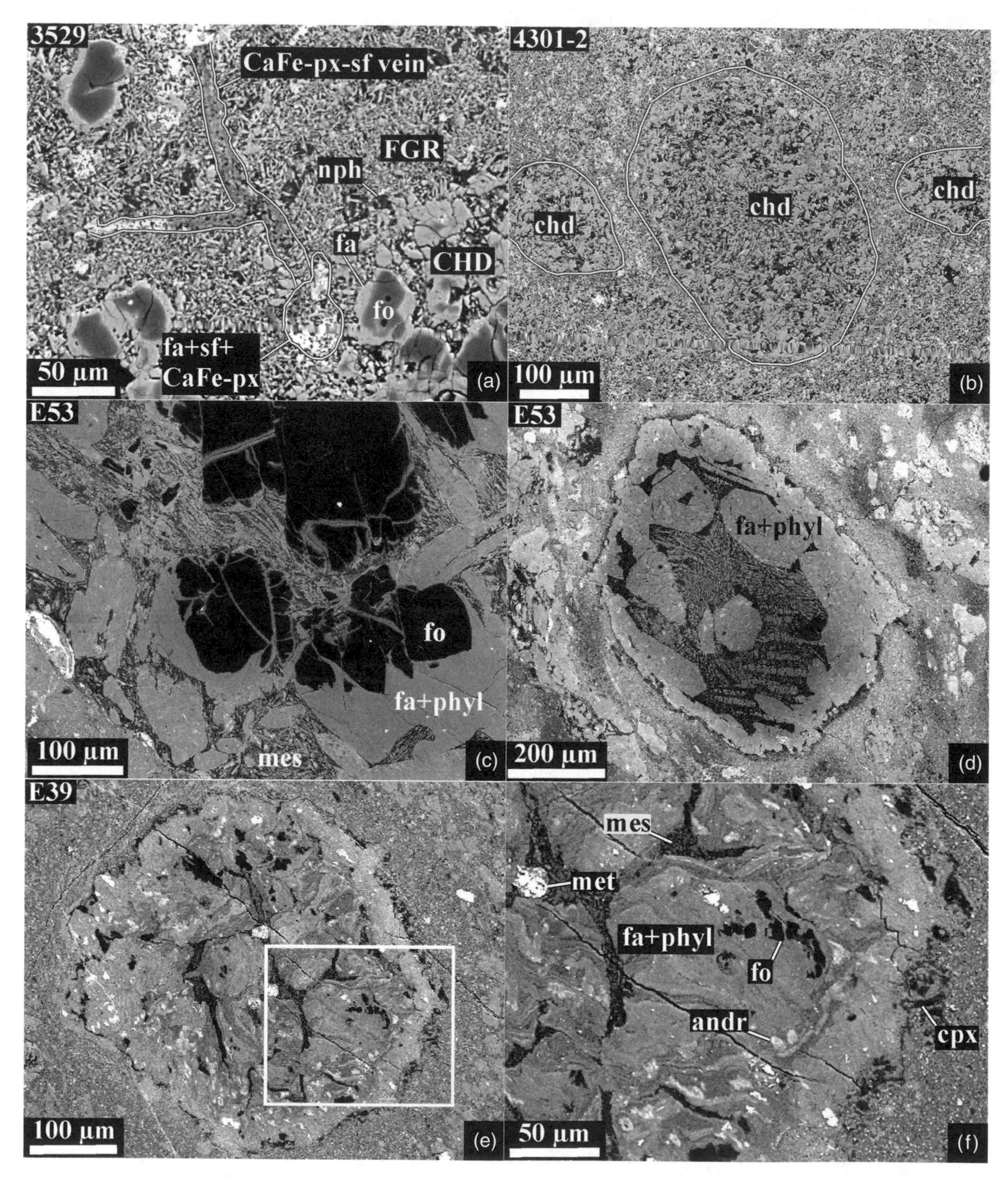

Fig. 13. Backscattered electron images of altered chondrules in the dark inclusions (DI) **(a)** 3529, **(b)** 4301-2, **(c)**,**(d)** E53, and **(e)**,**(f)** E39 in **(a)**,**(b)** the oxidized CV chondrite Allende and **(c)**–**(f)** the reduced CV chondrite Efremovka. **(a)**,**(b)** Chondrules are replaced to various degrees by ferrous olivine, nepheline [nph; black in **(d)**], and Ca,Fe-pyroxenes (Ca,Fe-px). Chondrule shown in **(a)** is surrounded by a fine-grained rim composed of lath-shaped ferrous olivine (fa) and nepheline. The rim is crossut by a vein composed of Ca,Fe-pyroxenes and Fe,Ni-sulfides (white). The vein starts at the opaque nodule (outlined) that is replaced by Ca,Fe-pyroxenes and ferrous olivine; sulfide grains (white) are relict. **(c)**,**(d)** Chondrules in E53 are pseudomorphed to a various degree by a fine-grained mixture of ferrous olivine (fa) and very minor phyllosilicates (phyl). **(e)**,**(f)** Chondrules in E39 are nearly completely replaced by a fine-grained mixture of ferrous olivine, phyllosilicates, and andradite (andr). Forsteritic olivine (fo) and high-Ca pyroxenes (cpx) are relict. met = mesostasis; met = Fe,Ni-metal (from *Krot et al.,* 1998a, 1999).

are generally younger than those of the Allende dark inclusions (Table 1). For the Efremovka dark inclusions, there is a correlation between the degree of alteration and I-Xe closure times: E39 and E80 are more altered than E53 and show an apparent closure time ~4–6 ± 2 m.y. later than E53 (Fig. 15).

In spite of the different degrees and styles of alteration of dark inclusions in the reduced and oxidized CV chondrites, all of them require "subplanetary" trapped Xe, which has been interpreted as a result of shock or thermal metamorphism that occurred after precipitation of the I host, while Xe and some I were still in solution (*Hohenberg et al.,* 2003). This interpretation is generally consistent with the aqueous alteration–dehydration model proposed for the dark inclusions (e.g., *Kojima and Tomeoka,* 1996) and with shock metamorphism features observed in some of the dark inclusions [e.g., lineation of chondrule pseudomorphs; Fig. 1 in *Krot et al.* (1999)].

To summarize, the I-Xe ages of the CV dark inclusions, which probably represent fragments of the CV asteroidal body, span ~14 m.y., suggesting a long period of aqueous alteration on the CV parent body.

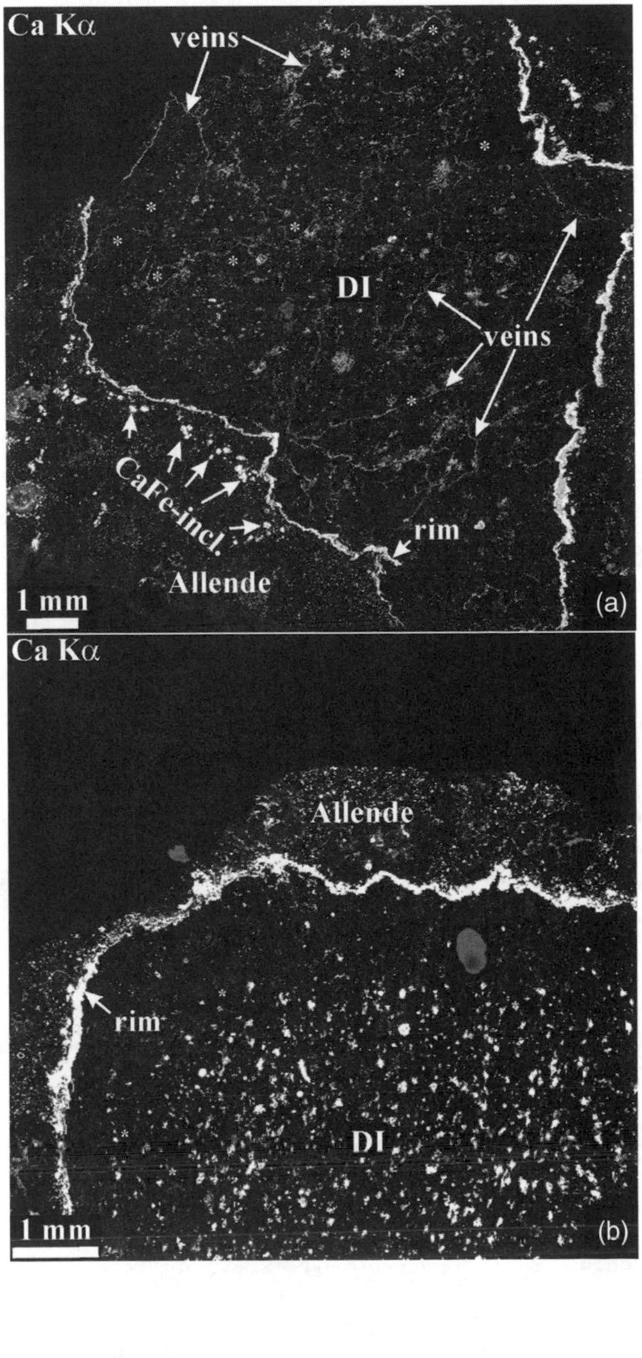

TABLE 1. The I-Xe ages of the dark inclusions from the reduced and oxidized CV chondrites relative to the age of Shallowater (4563.5 ± 1.0 Ma) (J. Gilmour et al., personal communication, 2005) internal standard.

Chondrite/Classification	Sample	I-Xe Age (Ma)	Reference
Allende CV_{oxA}	1a-1	−2.8 ± 0.3	[1]
	12b-1	−2.0 ± 0.3	
	4294-1	−1.9 ± 0.3	
	4a1/b1	−1.9 ± 0.3	
	IV-1	−1.9 ± 0.2	
	14b-1	−1.6 ± 0.2	
	4884-2	−1.5 ± 0.2	
	4301-1	−1.5 ± 0.1	
	4884-1	−1.4 ± 0.8	
	4314-3	−1.1 ± 0.2	
	1-3	−1.1 ± 0.2	
	25sl-tw1	−1.1 ± 0.2	
	4320-1	−1.0 ± 0.3	
	4884-6	−1.0 ± 0.2	
	IV-2	−0.8 ± 0.3	
	4884-5	−0.7 ± 0.2	
	4884-3	−0.5 ± 0.3	
Efremovka CV_{red}	E53	−4.9 ± 1.8	[2,3]
	E39	+0.8 ± 2.0	
	E80	−1.0 ± 0.5	
Leoville CV_{red}	LV1	+3.0 ± 0.1	[4]
	LV2	+9.5 ± 2.3	
Vigarano $CV_{breccia}$	2226	+8.8 ± 0.6	

References: [1] *Pravdivtseva et al.* (2003b); [2] *Krot et al.* (1999); [3] *Swindle et al.* (1998); [4] *Pravdivtseva et al.* (2003c).

Fig. 14. Calcium Kα X-ray elemental maps of the heavily altered dark inclusions **(a)** 4301-2 and **(b)** IV-1 in the oxidized CV chondrite Allende. The dark inclusions (DI) contain chondrule pseudomorphs (indicated by stars) that are depleted in Ca and consist of the secondary ferrous olivine, nepheline, sodalite, and Fe,Ni-sulfides [see **(b)**]. The dark inclusion 4301-2 is crosscut by multiple veins composed of Ca,Fe-pyroxenes and andradite. Both dark inclusions are surrounded by continuous Ca-rich rims composed of Ca,Fe-pyroxenes, andradite, wollastonite, and kirschsteinite. The outer portions of the dark inclusions are depleted in Ca, whereas the neighboring matrix of Allende contains abundant Ca,Fe-rich nodules composed of Ca,Fe-pyroxenes, andradite, and wollastonite, suggesting that Ca lost from the dark inclusions precipitated as rims and nodules around them (from *Krot et al.,* 2001).

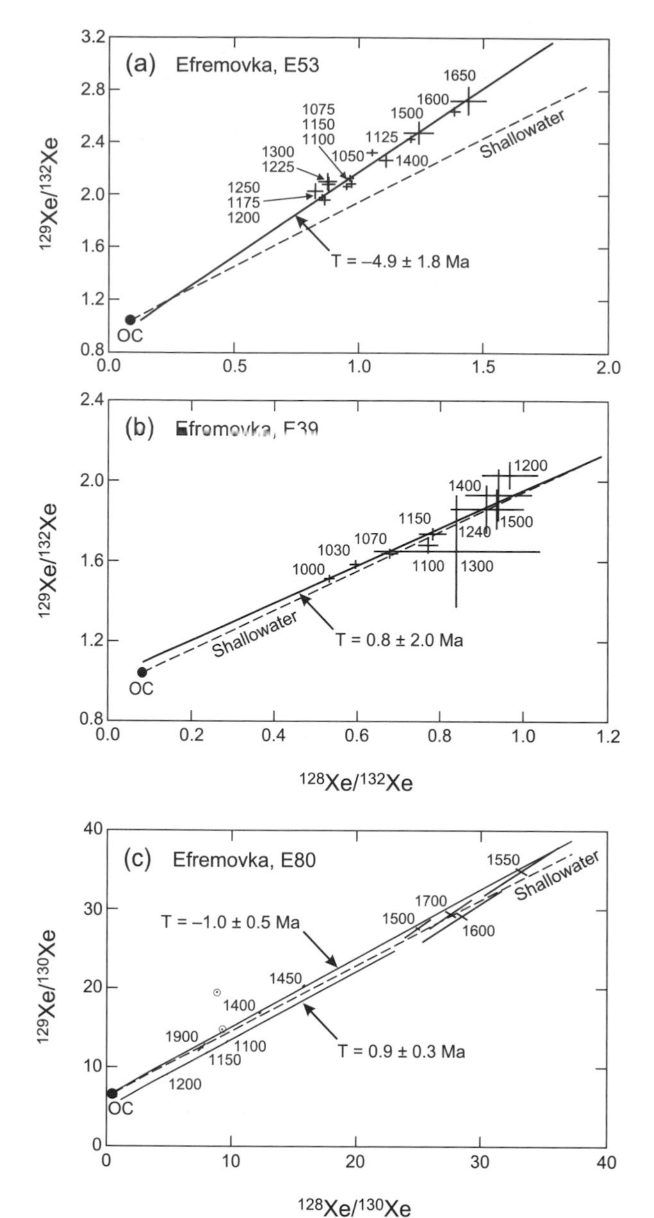

Fig. 15. ^{129}I-^{129}Xe evolution diagrams for the Efremovka dark inclusions E53, E39 (data from *Swindle et al.,* 1998; *Krot et al.,* 1999), and E80; error bars are 1σ (data from *Pravdivtseva et al.,* 2003c). The ages shown are relative to the Shallowater internal standard (4563.5 ± 1 Ma) (J. Gilmour et al., personal communication, 2005). The two apparent isochrons for E80 correspond to different peaks in the release profiles of radiogenic ^{128}Xe and ^{129}Xe, suggesting that E80 contains two different I-carrying mineral phases with the same closure time but different trapped components. The circled temperature points represent intermediate extraction steps between these two release peaks where radiogenic ^{128}Xe and ^{129}Xe do not correlate.

3.6. Timescale of Aqueous Alteration of Ordinary Chondrites

The effects of aqueous alteration are best documented in chondrules and matrices of the type 3 ordinary chondrites Semarkona (LL3.0), Bishunpur (LL3.1), Krymka (LL3.1), Parnallee (LL3.4), Chainpur (LL3.4), and Tieschitz (H/L3.6) (*Hutchison et al.,* 1987, 1998; *Alexander et al.,* 1989a,b; *Bridges et al.,* 1997; *Krot et al.,* 1997b; *Keller,* 1998; *Choi et al.,* 1998; *Grossman et al.,* 2000, 2002). This alteration must have occurred in an asteroidal setting and resulted in formation of secondary phyllosilicates, magnetite, maghemite, Fe,Ni-carbides, calcite, Ni-bearing sulfides, ferrous olivine, and alkali-rich secondary phases. Chondrules in some of the altered ordinary chondrites were dated using I-Xe systematics (*Swindle et al.,* 1991a,b; *Ash et al.,* 1995).

In Semarkona, evidence for aqueous alteration in an asteroidal setting includes (1) the large range in mass-dependent fractionation of O-isotopic compositions of magnetite grains (δ^{18}O ~ 13‰), indicative of Rayleigh fractionation as a result of growth in the presence of a limited water reservoir (*Choi et al.,* 1998); (2) the presence of carbide-magnetite-

TABLE 2. The I-Xe ages of chondrules from type 3 ordinary and enstatite chondrites relative to the age of Shallowater (4563.5 ± 1.0 Ma) (J. Gilmour et al., personal communication, 2005) internal standard.

Chondrite/Classification	chd#	I-Xe Age (Ma)	Reference
Semarkona, LL3.0	CD-159(l)	+4.9 ± 0.5	[1]
	CD-159(h)	–4.9 ± 2.9	
	CD-92	–1.9 ± 1.1	
	CD-95	+4.1 ± 1.2	
	CD-54	–0.4 ± 1.2	
	CD-79	+0.6 ± 1.7	
	CD-60	–2.4 ± 1.7	
	CD-173	–1.8 ± 2.1	
	CD-160	+4.7 ± 1.1	
	CD-8(l)	+0.6 ± 1.1	
	CD-8(h)	–4.5 ± 2.3	
	CD-84(l)	+0.8 ± 0.3	
	CD-84(h)	–2.5 ± 0.9	
	CD-129	+1.3 ± 1.0	
	CD-153	–1.8 ± 2.8	
	CD-169	0 ± 0.3	
	CD-139	–2.6 ± 2.5	
	CD-101	–4.2 ± 0.6	
	CD-174	–1.9 ± 1.7	
Parnallee, LL3.4	CB1	+4.16 ± 0.44	[2]
	CB2	+1.29 ± 0.16	
	P6	+4.54 ± 0.70	
	P9	no ^{129}Xe*	
	Feline	+5.05	
	P32	+1.94 ± 0.26	
	MC1	no ^{129}Xe*	
Qingzhen, EH3	QC1	+1.98	[3]
	QC3	+0.44	
	QC4	–1.08	
	QC5	+1.41	
	QC6	+0.10	
	QC7	+0.64	
	QC8	+1.70	

References: [1] *Swindle et al.* (1991a); [2] *Ash et al.* (1995); [3] *Ash et al.* (1997).

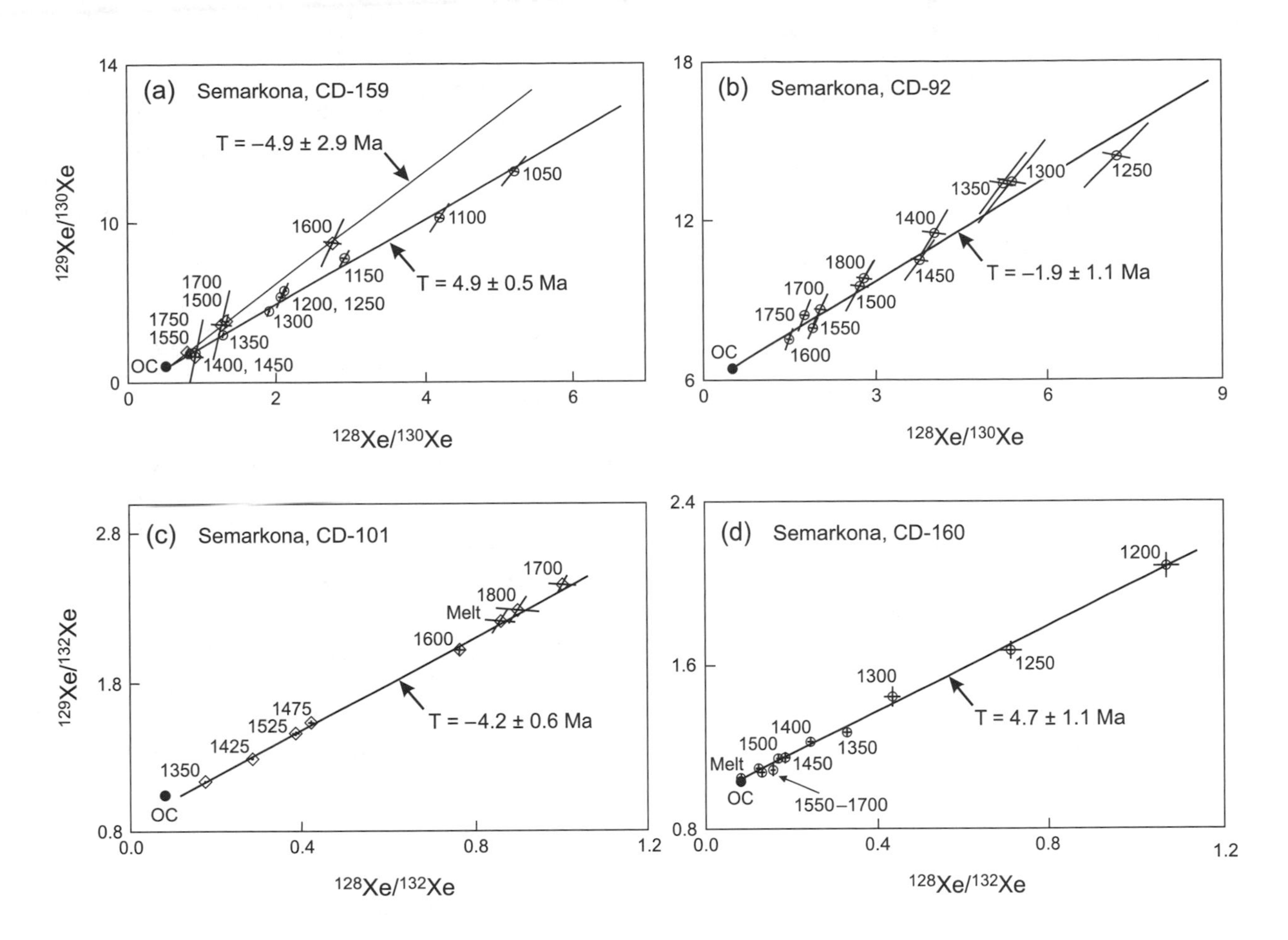

Fig. 16. Representative three-isotope plots for Xe from irradiated Semarkona chondrules. These include one sample with dual isochrons (CD-159), and samples with single isochrons with apparent old (CD-92 and CD-101) and young (CD-160) I-Xe ages. Diamonds denote points included in high-temperature isochrons and circles are those included in low-temperature (or single) isochrons; error bars are 1σ. Numbers next to points represent extraction coil temperatures in °C (the sample is probably 200°–300°C cooler) (from *Swindle et al.,* 1991a).

sulfide veins crosscutting fine-grained rims around chondrules (*Krot et al.,* 1997b); (3) the presence of phyllosilicates in the chondrules and matrix (*Hutchison et al.,* 1987; *Alexander et al.,* 1989a,b); (4) the presence of bleached chondrules and evidence for removal of chondrule mesostasis by dissolution; and (5) the elevated D/H ratios in the bleached chondrules and matrix of Semarkona, suggesting exchange with an isotopically similar reservoir, most likely an aqueous solution (*Grossman et al.,* 2000). *Swindle et al.* (1991a) observed a range of >10 Ma in apparent I-Xe-isotopic ages (from –4.4 ± 2.9 Ma to 5.4 ± 0.5 Ma, relative to Shallowater) for 17 chondrules analyzed in Semarkona (Table 2; Fig. 16). The oldest I-Xe ages were attributed to chondrule formation, whereas the younger ages were attributed to aqueous alteration. Taking into account the petrographic evidence for multistage aqueous alteration of the Semarkona chondrules (e.g., *Grossman et al.,* 2000) and the possible presence of several I carriers (e.g., magnetite, phyllosilicates), we suggest instead that the entire range of I-Xe ages may reflect the duration of aqueous alteration on the LL asteroidal body.

The H/L3.6 chondrite Tieschitz contains secondary nepheline, albite, and unidentified hydrous (?) phases that precipitated from a halogen-bearing aqueous fluid in interchondrule voids and replaced chondrule mesostasis leached out by the fluid (*Hutchison et al.,* 1998). Based on the evidence for partial resetting of the Sm-Nd and K-Ar systems at ~2 Ga (*Turner et al.,* 1978; *Krestina et al.,* 1996), *Hutchison et al.* (1998) speculated that aqueous activity on the Tieschitz parent body occurred ~2 G.y. ago. However, the I-Xe ages of the Tieschitz chondrules (*Nichols et al.,* 1991) do not support this hypothesis [decoupling of the I-Xe chronometer from the Ar-Ar chronometer has also been observed for chondrules from the EH3 chondrite Qingzhen (*Ash et al.,* 1997)]. The best isochrons for three chondrules define closure ages of 1.3, 3.6, and 4.9 m.y. after Bjurböle (Fig. 17). All the chondrules display the regular I-Xe structure; the high-temperature sites have higher values of ^{129}I/^{127}I than the low-temperature sites, suggesting slow cooling or monotonic relaxation of the conditions during metamorphism (*Nichols et al.,* 1991). Using a nondiffusive, activation-energy-dependent model, cooling rates corresponding to a few hun-

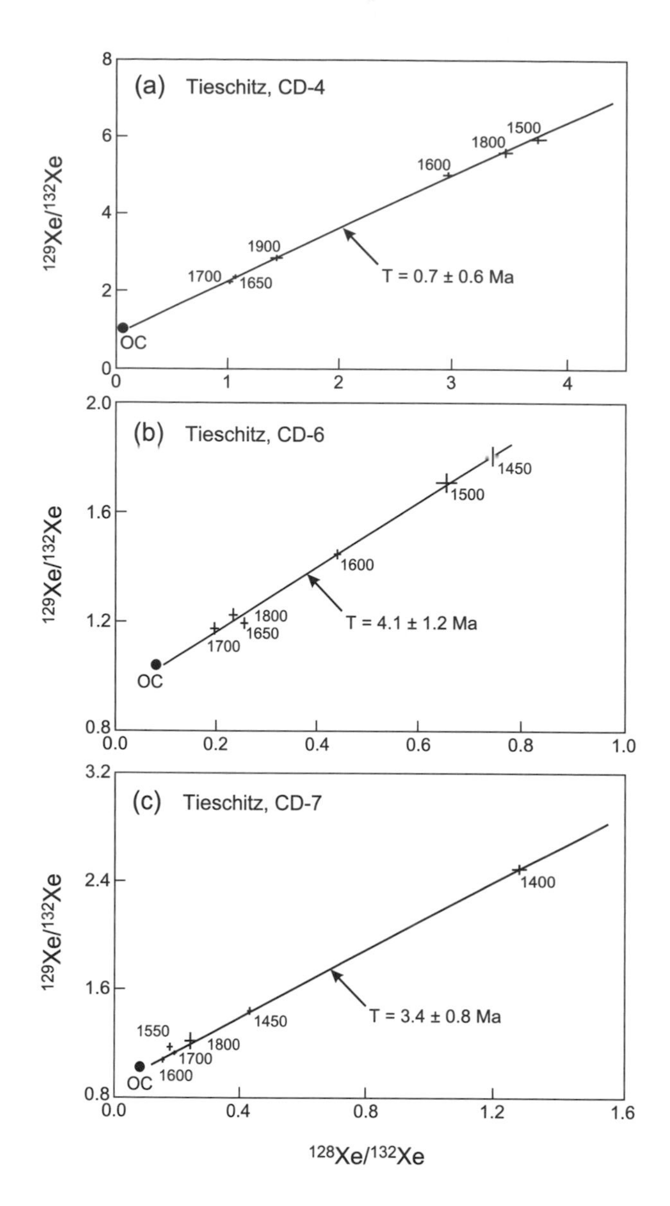

Fig. 17. ^{129}I-^{129}Xe evolution diagrams for Tieschitz chondrules. Numbers next to points represent extraction coil temperatures in °C (the sample is probably 200°–300°C cooler); error bars are 2σ (from *Nichols et al.,* 1991).

dred degrees per million years for the high-temperature sites, down to a few degrees per million years for the low-temperature sites, are estimated (*Nichols et al.,* 1991). This is the same range of values observed for the Allende CAIs and chondrules (*Swindle et al.,* 1983, 1988). These slow "cooling" rates suggest that the postformational processes in the regolith are likely responsible for the I-Xe fine structure.

Bridges et al. (1997) described a number of chondrules separated from Chainpur (LL3.4) and Parnallee (LL3.6) that contain mesostasis enriched in Na and Cl and contain microcrystalline sodalite, nepheline, and scapolite, and attributed these features to a preaccretionary (which could be nebular or asteroidal) metasomatism. The I-Xe ages of the Parnallee chondrules (*Ash et al.,* 1995), which range from 1.75 ± 0.16 m.y. to 5.0 ± 0.70 m.y. after Bjurböle chondrule closure (Table 2), favor an asteroidal setting for the alteration. Two chondrules contain ^{128}Xe* but no ^{129}Xe*, suggesting that they formed after the decay of ^{129}I, possibly by impact (*Ash et al.,* 1995). *Swindle et al.* (1991b) showed that the range of apparent I-Xe ages of Chainpur chondrules is ~50 Ma and that the chondrules evolved in a common reservoir with a chondritic I/Xe ratio. Based on these observations, *Swindle et al.* (1991b) concluded that these ages reflect asteroidal processing in a regolith.

The presence of halite (NaCl) and sylvite (KCl) containing inclusions of aqueous salt solutions in the H-chondrite regolith breccias Monahans (1998) (H5) and Zag (H3–6) indicates that some of the aqueous alteration on the H-chondrite parent body postdated thermal metamorphism (*Zolensky et al.,* 1999). We note, however, that there is no evidence that the halite in Zag and Monahans formed *in situ* (e.g., *Rubin et al.,* 2002). Based on the presence of secondary fluid inclusions in halite of both meteorites, *Zolensky et al.* (1999) concluded that aqueous activity occurred at low temperature (<50°C) and was episodic. A Rb-Sr model age for a halite crystal in Monahans (1998), calculated for an initial ratio of ^{87}Sr/^{86}Sr = 0.69876 ± 0.00040, the average for H-group chondrites, is 4.7 ± 0.2 Ga (*Zolensky et al.,* 1999). Subsequently, *Whitby et al.* (2000) reported essentially pure radiogenic ^{129}Xe in halite from Zag. Correlated release of ^{129}Xe and ^{128}Xe corresponds to an initial (^{129}I/^{127}I) ratio of (1.35 ± 0.05) × 10^{-4} and an apparent formation time for the halite of 4.8 ± 0.9 m.y. before the formation of the Bjurböle reference chondrite, suggesting an early onset of aqueous activity on the Zag parent body. The retention of a high ^{129}Xe*/^{127}I ratio implies that halite has not been subjected to substantial dissolution and recrystallization in the 4.5 G.y. since its formation, suggesting that the processes that led to aqueous activity on the Zag parent body may have ended quickly after evaporation of water into space (*Whitby et al.,* 2000).

3.7. Timescale of Alteration of Enstatite Chondrites

Whitby et al. (1997) and *Ash et al.* (1997) reported apparent I-Xe ages of chondrules from the EH3 enstatite chondrite Qingzhen. Most chondrules give excellent isochrons with errors <1 Ma; only one of the chondrules shows evidence for a slight isotopic disturbance. The observed range in I-Xe ages, from –1.08 to +1.98 relative to Shallowater (Table 2), is comparable to those in most unequilibrated ordinary chondrites. Based on the apparent lack of evidence for secondary alteration in Qingzhen, these ages were interpreted as primary, corresponding to the ages of chondrule formation. The I carrier in enstatite chondrites is unknown, but the presence of sodalite-like mesostases in some type I chondrules in Qingzhen suggests that it could be sodalite. The origin of these mesostases, and the interpretation of I-Xe ages, remain unclear.

4. SUMMARY AND FUTURE WORK

Mineralogical, petrographic, and isotopic observations indicate that most groups of chondritic meteorites experienced asteroidal alteration to various degrees, resulting in formation of secondary minerals such as phyllosilicates, magnetite, carbonates, ferrous olivine (Fa_{40-100}), salite-hedenbergite pyroxenes ($Fs_{10-50}Wo_{45-50}$), wollastonite, andradite, nepheline, pentlandite, pyrrhotite, Fe,Ni-carbides, and Ni-rich metal. The alteration occurred in the presence of aqueous solutions under variable conditions (temperature, water/rock ratio, f_{O_2}, and fluid compositions) and in many cases was multistage. Although some alteration predated agglomeration of the final chondrite asteroidal bodies (e.g., dark inclusions in CV chondrites), there is no compelling evidence that the alteration occurred in the solar nebula nor in planetesimals of earlier generations. The ^{26}Al-^{26}Mg, ^{53}Mn-^{53}Cr, and ^{129}I-^{129}Xe dating of secondary minerals suggests that alteration may have started within 1–2 m.y. after formation of the CV CAIs having an absolute Pb-Pb age of 4567.2 ± 0.6 Ma and lasted up to 15 m.y. (Tables 1–3; Figs. 18, 19). Based on these observations, we infer that the chondrite parent bodies must have accreted within the first 1–2 m.y. after collapse of the protosolar molecular cloud.

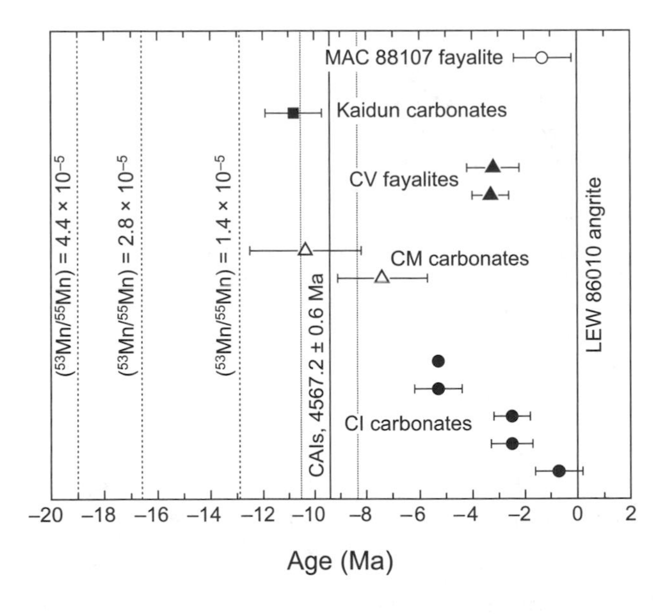

Fig. 18. Manganese-chromium ages of the secondary carbonates and fayalite in carbonaceous chondrites relative to the LEW 86010 angrite; error bars are 2σ (data from *Endress et al.,* 1996; *Hutcheon and Phinney,* 1996; *Hutcheon et al.,* 1997, 1998, 1999; *Brearley and Hutcheon,* 2000; *Brearley et al.,* 2001; *Hua et al.,* 2002, 2005; *Krot et al.,* 2000a). Absolute ages of CAIs from CV chondrites (4567.2 ± 0.6 m.y.) (*Amelin et al.,* 2002) and ages calculated based on the initial $^{53}Mn/^{55}Mn$ ratios of 1.4×10^{-5} (*Lugmair and Shukolyukov,* 2001), $(2.8 \pm 0.3) \times 10^{-5}$ (*Nyquist et al.,* 2001), and 4.4×10^{-5} (*Birck and Allègre,* 1988; *Birck et al.,* 1999) are plotted for reference.

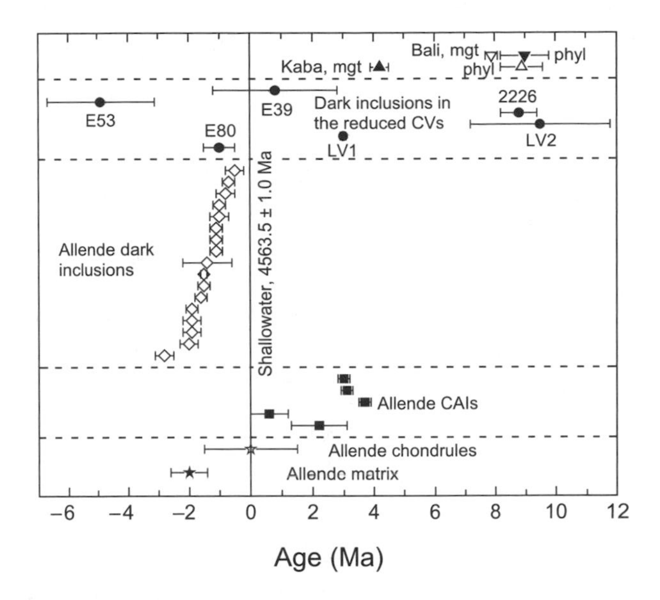

Fig. 19. Iodine-xenon ages of the CV chondritic components (CAIs, chondrules, matrix, dark inclusions) and mineral fractions (magnetite, phyllosilicates) relative to the Shallowater aubrite internal standard; errors are 1σ (data from *Swindle et al.,* 1983, 1988, 1998; *Krot et al.,* 1999; *Hohenberg et al.,* 2001; *Pravdivtseva et al.,* 2003b,c). Based on the comparison of I-Xe and Mn-Cr systems with the absolute Pb-Pb chronometer for samples analyzed by multiple-isotope systems, J. Gilmour et al. (personal communication, 2005) infer that the I-Xe system closed in Shallowater aubrite at 4563.5 ± 1.0 m.y. before the present, i.e., 5.7 ± 1.1 m.y. earlier than the Mn-Cr system closed in LEW 86010 angrite.

There are several carbonaceous chondrite groups not discussed in this chapter with clear evidence for secondary alteration; these include CR, CH, CB, and CO chondrites. The CR chondrites experienced aqueous alteration to various degrees that resulted in formation of phyllosilicates, magnetite, and carbonates (e.g., *Krot et al.,* 2002). The CB and CH chondrites contain heavily aqueously altered clasts composed of phyllosilicates, framboidal and platelet magnetite, and carbonates (*Greshake et al.,* 2002). The CO chondrites experienced alteration similar to that observed in CV chondrites (see section 3.5). The alteration resulted in formation of nepheline, sodalite, ferrous olivine, magnetite, Fe,Ni-carbides, and Ni-bearing sulfides (e.g., *Jones,* 1997a,b; *Rubin,* 1998; *Russell et al.,* 1998; *Chizmadia et al.,* 2002; *Itoh and Tomeoka,* 2003). A degree of alteration correlates with petrologic types of the host meteorites, suggesting that it occurred in an asteroidal setting (e.g., *Itoh and Tomeoka,* 2003). Although the secondary mineralization in the CR, CO, CB, and CH chondrites has been well documented, there have yet been no attempts made to date it.

Future studies of isotopic dating of secondary mineralization of chondritic meteorites should also be focused on understanding the multistage alteration histories using combinations of analytical tools, including SEM, EPMA, CL, TEM, SIMS, and ICP-MS. This approach has already been

TABLE 3. The initial $^{53}Mn/^{55}Mn$ ratios in secondary carbonate and fayalite in carbonaceous chondrites and their Mn-Cr ages relative to angrite LEW 86010 (4557.8 ± 0.5 Ma).

Chondrite	Classification	Mineral Analyzed	$(^{53}Mn/^{55}Cr)_0$	Age Relative to LEW 86010	Reference
Orgueil and Ivuna	CI	dolomite	$(1.99 \pm 0.16) \times 10^{-6}$	–2.5 ± 0.7	[1]
Orgueil	CI	dolomite	$(1.42 \pm 0.16) \times 10^{-6}$	–0.7 ± 0.9	[1]
Orgueil	CI	breunnerite	$(1.97 \pm 0.18) \times 10^{-6}$	–2.5 ± 0.8	[2]
Orgueil	CI	breunnerite	$(3.4 \pm 0.4) \times 10^{-6}$	–5.3 ± 0.9	[2]
Orgueil	CI	Cr-carbonates	3.4×10^{-6}	–5.3	[3]
Supuhee, clast	CI-like	carbonates	$(8 \pm 4) \times 10^{-6}$	–9.9 ± 2.5	[3]
Kaidun*	breccia	calcite, dolomite	$(9.4 \pm 1.6) \times 10^{-6}$	–10.8 ± 1.1	[4]
ALH 84034	CM1	dolomite	$(5.0 \pm 1.5) \times 10^{-6}$	–7.4 ± 1.7	[5]
Y 791198	CM2	calcite	$(8.7 \pm 1.5) \times 10^{-6}$	–10.3 ± 2.2	[6]
Kaba	CV_{oxB}	fayalite	$(2.32 \pm 0.18) \times 10^{-6}$	–3.3 ± 0.7	[7]
Mokoia	CV_{oxB}	fayalite	$(2.28 \pm 0.37) \times 10^{-6}$	–3.2 ± 1.0	[8]
MAC 88107	ungrouped	fayalite	$(1.58 \pm 0.28) \times 10^{-6}$	–1.3 ± 1.2	[9]

*Contains CR-, C1-, and CM-like materials.

References: [1] *Endress et al.* (1996); [2] *Hutcheon and Phinney* (1996); [3] *Hutcheon et al.* (1997); [4] *Hutcheon et al.* (1999); [5] *Brearley and Hutcheon* (2000); [6] *Brearley et al.* (2001); [7] *Hutcheon et al.* (1998); [8] *Hua et al.* (2002, 2005); [9] *Krot et al.* (2000a).

successfully used in dating carbonate formation in CM carbonaceous chondrites (*Brearley et al.,* 1999, 2001; *Brearley and Hutcheon,* 2000, 2002). Small grain sizes of the secondary minerals suitable for *in situ* Mn-Cr-isotopic dating (e.g., carbonates, ferrous olivine) will probably require use of NanoSIMS (e.g., *Hoppe et al.,* 2004).

Finally, we would like to emphasize that progress in the chronology of the early solar system processes requires better understanding of the origin of short-lived radionuclides [external (injection) vs. internal (irradiation)] and their distribution (homogeneous vs. heterogeneous) in the protoplanetary disk (e.g., *Goswami and Vanhala,* 2000; *Goswami et al.,* 2004; *Gounelle et al.,* 2001), and establishing a unified chronology of the early solar system processes using these radionuclides (e.g., *Gilmour and Saxton,* 2001; *Gilmour et al.,* 2004). These issues remain unresolved.

Acknowledgments. This work was supported by NASA Grants NAG5-10610 and NAG6-57543 (A. Krot, P.I.), NAG5-11591 (K. Keil, P.I.), NAG5-10523 (I. Hutcheon, P.I.), and NAG5-11682 (A. Brearley, P.I.). We thank T. D. Swindle, M. E. Zolensky, and L. A. Leshin for constructive reviews. This is Hawai'i Institute of Geophysics and Planetology Publication No. 1418 and School of Ocean and Earth Science and Technology Publication No. 6687.

REFERENCES

Alexander C. M. O'D., Barber D. J., and Hutchison R. (1989a) The microstructure of Semarkona and Bishunpur. *Geochim. Cosmochim. Acta, 53,* 3045–3057.

Alexander C. M. O'D., Hutchison R., and Barber D. J. (1989b) Origin of chondrule rims and interchondrule matrix in unequilibrated ordinary chondrites. *Earth Planet. Sci. Lett., 95,* 187–207.

Amelin Y., Krot A. N., Hutcheon I. D., and Ulyanov A. A. (2002) Lead isotopic ages of chondrules and calcium-aluminum-rich inclusions. *Science, 297,* 1678–1683.

Amelin Y., Krot A. N., and Twelker E. (2004) Pb isotopic age of the CB chondrite Gujba, and the duration of the chondrule formation interval. *Geochim. Cosmochim. Acta, 68,* Abstract #E958.

Anders E. and Grevesse N. (1989) Abundances of the elements: Meteoritic and solar. *Geochim. Cosmochim. Acta, 53,* 197–214.

Ash R. D., Gilmour J. D., Whitby J. A., Turner G., Bridges J. C., and Hutchison R. (1995) The history of the Parnallee meteorite as revealed by iodine-xenon dating (abstract). *Meteoritics, 30,* 483–484.

Ash R. D., Gilmour J. D., Whitby J., Prinz M., and Turner G. (1997) I-Xe dating of chondrules from the Qingzhen unequilibrated enstatite chondrite (abstract). In *Lunar and Planetary Science XXVIII,* pp. 61–62. Lunar and Planetary Institute, Houston.

Barber D. J. (1981) Matrix phyllosilicates and associated minerals in C2M carbonaceous chondrites. *Geochim. Cosmochim. Acta, 45,* 945–970.

Benedix G. K., Leshin L. A., Farquhar J., Jackson T., and Thiemens M. H. (2003) Carbonates in CM2 chondrites: Constraints on alteration conditions from oxygen isotopic compositions and petrographic observations. *Geochim. Cosmochim. Acta, 67,* 1577–1589.

Birck J.-L. and Allègre C. J. (1988) Manganese-chromium isotope systematics and the development of the early solar system. *Nature, 331,* 579–584.

Birck J.-L., Rotaru M., and Allègre C. J. (1999) ^{53}Mn-^{53}Cr evolution of the early solar system. *Geochim. Cosmochim. Acta, 54,* 4111–4117.

Bizzarro M., Baker J. A., and Haack H. (2004) Mg isotope evidence for contemporaneous formation of chondrules and refractory inclusions. *Nature, 431,* 275–278.

Brazzle R. H., Pravdivtseva O. V., Meshik A. M., and Hohenberg C. M. (1999) Verification and interpretation of the I-Xe chronometer. *Geochim. Cosmochim. Acta, 63,* 739–760.

Brearley A. J. (1997) Disordered biopyriboles, amphibole, and talc in the Allende meteorite; products of nebular or parent body aqueous alteration? *Science, 276,* 1103–1105.

Brearley A. J. (1998) Dark inclusions in the Leoville CV3 carbonaceous chondrite (abstract). In *Lunar and Planetary Science XXIX,* Abstract #1245. Lunar and Planetary Institute, Houston (CD-ROM).

Brearley A. J. (1999) Origin of graphitic carbon and pentlandite in matrix olivines in the Allende meteorite. *Science, 285,* 1380–1382.

Brearley A. J. (2003) Nebular *vs.* asteroidal processing. In *Treatise on Geochemistry, Vol. 1: Meteorites, Comets and Planets* (A. M. Davis, ed.), pp. 247–269. Elsevier, Oxford.

Brearley A. J. (2006) The action of water. In *Meteorites and the Early Solar System II* (D. S. Lauretta and H. Y. McSween Jr., eds.), this volume. Univ. of Arizona, Tucson.

Brearley A. J. and Hutcheon I. D. (2000) Carbonates in the CM1 chondrite ALH84034: Mineral chemistry, zoning and Mn-Cr systematics (abstract). In *Lunar and Planetary Science XXXI,* Abstract #1407. Lunar and Planetary Institute, Houston (CD-ROM).

Brearley A. J. and Hutcheon I. D. (2002) Carbonates in the Y791918 CM2 chondrite: Zoning and Mn-Cr systematics (abstract). *Meteoritics & Planet. Sci., 37,* A23.

Brearley A. J. and Jones R. H. (1998) Chondritic meteorites. In *Planetary Materials* (J. J. Papike, ed.), pp. 3-1 to 3-398. Reviews in Mineralogy, Vol. 36, Mineralogical Society of America.

Brearley A. J. and Shearer C. K. (2000) Origin of calcium-iron-rich pyroxenes in Allende matrix: Clues from rare-earth-element abundances (abstract). *Meteoritics & Planet. Sci., 35,* A33.

Brearley A. J., Saxton J. M., Lyon I. C., and Turner G. (1999) Carbonates in the Murchsion CM chondrite: CL characteristics and oxygen isotopic compositions (abstract). In *Lunar and Planetary Science XXX,* Abstract #1301. Lunar and Planetary Institute, Houston (CD-ROM).

Brearley A. J., Hutcheon I. D., and Browning L. (2001) Compositional zoning and Mn-Cr systematics in carbonates from the Y791198 CM2 carbonaceous chondrite (abstract). In *Lunar and Planetary Science XXXII,* Abstract #1458. Lunar and Planetary Institute, Houston (CD-ROM).

Bridges J. C., Alexander C. M. O'D., Hutchison R., Franchi I. A., and Pillinger C. T. (1997) Sodium-, chlorine-rich mesostases in Chainpur (LL3) and Parnallee (LL3) chondrules. *Meteoritics & Planet. Sci., 32,* 555–565.

Brigham C. A., Hutcheon I. D., Papanastassiou D. A., and Wasserburg G. J. (1986) Evidence for ^{26}Al and Mg isotopic heterogeneity in a fine-grained CAI (abstract). In *Lunar and Planetary Science XVII,* pp. 85–86. Lunar and Planetary Institute, Houston.

Buchanan P. C., Zolensky M. E., and Reid A. M. (1997) Petrology of Allende dark inclusions. *Geochim. Cosmochim. Acta, 61,* 1733–1743.

Bunch T. E. and Chang S. (1980) Carbonaceous chondrites — II. Carbonaceous chondrite phyllosilicates and light element geochemistry as indicators of parent body processes and surface conditions. *Geochim. Cosmochim. Acta, 44,* 1543–1577.

Chizmadia L. J., Rubin A. E., and Wasson J. T. (2002) Mineralogy and petrology of amoeboid olivine inclusions in CO3 chondrites; relationship to parent-body aqueous alteration. *Meteoritics & Planet. Sci., 37,* 1781–1796.

Choi B.-G., McKeegan K. D., Krot A. N., and Wasson J. T. (1998) Extreme oxygen-isotope compositions in magnetite from unequilibrated ordinary chondrites. *Nature, 392,* 577–579.

Choi B.-G., Krot A. N., and Wasson J. T. (2000) Oxygen-isotopes in magnetite and fayalite in CV chondrites Kaba and Mokoia. *Meteoritics & Planet. Sci., 35,* 1239–1249.

Clayton R. N. and Mayeda T. K. (1984) The oxygen isotope record in Murchison and other carbonaceous chondrites. *Earth Planet. Sci. Lett., 67,* 151–161.

Clayton R. N. and Mayeda T. K. (1999) Oxygen isotope studies of carbonaceous chondrites. *Geochim. Cosmochim. Acta, 63,* 2089–2104.

Clayton R. N., Mayeda T. K., Ivanov A. V., and MacPherson G. J. (1994) Oxygen isotopes in Kaidun (abstract). In *Lunar and Planetary Science XXV,* pp. 269–270. Lunar and Planetary Institute, Houston.

Cosarinsky M., Leshin L. A., MacPherson G. J., Krot A. N., and Guan Y. (2003) Oxygen isotope composition of Ca-Fe-rich silicates in and around al Allende Ca-Al-rich inclusion (abstract). In *Lunar and Planetary Science XXXIV,* Abstract #1043. Lunar and Planetary Institute, Houston (CD-ROM).

Dohmen R., Chakraborty S., Palme H., and Rammensee W. (1998) Solid-solid reactions mediated by a gas phase; an experimental study of reaction progress and the role of surfaces in the system olivine+iron metal. *Am. Mineral., 83,* 970–984.

Endress M. and Bischoff A. (1996) Carbonates in CI chondrites: Clues to parent body evolution. *Geochim. Cosmochim. Acta, 60,* 489–507.

Endress M., Zinner E., and Bischoff A. (1996) Early aqueous activity on primitive meteorite parent bodies. *Nature, 379,* 701–703.

Fredriksson K. and Kerridge J. F. (1988) Carbonates and sulfates in CI chondrites: Formation by aqueous activity on the parent body. *Meteoritics, 23,* 35–45.

Fruland R. M., King A. E., and McKay D. S. (1978) Allende dark inclusions. In *Proc. Lunar Sci. Conf. 9th,* pp. 1305–1329.

Gilmour J. D. and Saxton J. M. (2001) A time-scale of formation of the first solids. *Philos. Trans. R. Soc. Lond., A359,* 2037–2048.

Gilmour J. D., Pravdivtseva O. V., Busfield A., and Hohenberg C. M. (2004) I-Xe and the chronology of the early solar system (abstract). In *Chondrites and the Protoplanetary Disk,* pp. 39–40. SOEST Publication No. 04-03, University of Hawai'i, Manoa.

Goswami J. N. and Vanhala H. A. T. (2000) Extinct radionuclides and the origin of the solar system. In *Protostars and Planets IV* (V. Mannings et al., eds.), pp. 963–995. Univ. of Arizona, Tucson.

Goswami J., Marhas K., Chaussidon M., Gounelle M., and Meyer B. (2004) Origin of short-lived radionuclides in the solar system (abstract). In *Chondrites and the Protoplanetary Disk, Kauai,* pp. 43–44. SOEST Publication No. 04-03, University of Hawai'i, Manoa.

Gounelle M. and Zolensky M. E. (2001) A terrestrial origin for sulfate veins in CI1 chondrites. *Meteoritics & Planet. Sci., 36,* 1321–1329.

Gounelle M., Shu F. H., Shang H., Glassgold A. E., Rehm K. E., and Lee T. (2001) Extinct radioactivities and protosolar cosmic-rays: Self-shielding and light elements. *Astrophys. J., 548,* 1051–1070.

Grady M M., Wright I. P., Swart P. K., and Pillinger C. T. (1988) The carbon and oxygen isotopic composition of meteoritic carbonates. *Geochim. Cosmochim. Acta, 52,* 2855–2866.

Greshake A., Krot A. N., Meibom A., Weisberg M. K., and Keil K. (2002) Heavily-hydrated matrix lumps in the CH and metal-rich chondrites QUE 94411 and Hammadah al Hamra 237. *Meteoritics & Planet. Sci., 37,* 281–294.

Grossman J. N., Alexander C. M. O'D., Wang J. H., and Brearley A. J. (2000) Bleached chondrules: Evidence for widespread aqueous alteration on the parent asteroids of ordinary chondrites. *Meteoritics & Planet. Sci., 35,* 467–486.

Grossman J. N., Alexander C. M. O'D., Wang J. H., and Brearley A. J. (2002) Zoned chondrules in Semarkona: Evidence for high- and low-temperature processing. *Meteoritics & Planet. Sci., 37,* 49–73.

Grossman L. and Fedkin A. V. (2003) Elemental abundance constraints on condensation of Allende matrix olivine (abstract). In *NIPR Symposium on Evolution of Solar System Materials, a New Perspective from Antarctic Meteorites,* pp. 31–32. National Institute of Polar Research, Tokyo.

Hashimoto A. and Grossman L. (1987) Alteration of Al-rich inclusions inside amoeboid olivine aggregates in the Allende meteorite. *Geochim. Cosmochim. Acta, 51,* 1685–1704.

Herzog G. F., Anders E., Alexander E. C. Jr., Davis P. K., and Lewis R. S. (1973) Iodine-129/xenon-129 age of magnetite from the Orgueil meteorite. *Science, 180,* 489–491.

Hohenberg C. M. (1980) High sensitivity pulse-counting mass-spectrometer system for noble gas analysis. *Rev. Sci. Instrum., 51,* 1075–1082.

Hohenberg C. M., Hudson B., Kennedy B. M., and Podosek F. A. (1981) Noble gas retention chronologies for the St Severin meteorite. *Geochim. Cosmochim. Acta, 45,* 535–546.

Hohenberg C. M., Pravdivtseva O., and Meshik A. (2000) Reexamination of anomalous I-Xe ages: Orgueil and Murchison magnetites and Allegan feldspar. *Geochim. Cosmochim. Acta, 64,* 4257–4262.

Hohenberg C. M., Meshik A. P., Pravdivtseva O. V., and Krot A. N. (2001) I-Xe dating: Dark inclusions from Allende CV3 (abstract). *Meteoritics & Planet. Sci., 36,* A83.

Hohenberg C. M., Pravdivtseva O. V., and Meshik A. P. (2003) Trapped Xe in dark inclusions II: New data from reduced CV3 meteorites (abstract). *Meteoritics & Planet. Sci., 38,* A149.

Hohenberg C. M., Pravdivtseva O. V., and Meshik A. P. (2004) Trapped Xe and I-Xe ages in aqueously altered CV3 meteorites. *Geochim. Cosmochim. Acta, 68,* 4745–4765.

Hoppe P., Macdougal D., and Lugmair G. W. (2004) High spatial resolution ion microprobe measurements refine chronology of Orgueil carbonate formation (abstract). In *Lunar and Planetary Science XXXV,* Abstract #1313. Lunar and Planetary Institute, Houston (CD-ROM).

Hua X. and Buseck P. R. (1995) Fayalite in the Kaba and Mokoia carbonaceous chondrites. *Geochim. Cosmochim. Acta, 59,* 563–579.

Hua X., Huss G. R., Sharp T. G. (2002) ^{53}Mn-^{53}Cr dating of fayalite formation in the Kaba CV3 carbonaceous chondrite (abstract). In *Lunar and Planetary Science XXXIII,* Abstract #1660. Lunar and Planetary Institute, Houston (CD-ROM).

Hua X., Huss G. R., Tachibana S., and Sharp T. G. (2003) Oxygen isotopic compositions of fayalite in the Kaba CV3 carbonaceous chondrite (abstract). In *Lunar and Planetary Science XXXIV,* Abstract #1702. Lunar and Planetary Institute, Houston (CD-ROM).

Hua X., Huss G. R., Tachibana S., and Sharp T. G. (2005) Oxygen, Si, and Mn-Cr isotopes of fayalite in the oxidized Kaba CV3 chondrite: Constraints for its formation history. *Geochim. Cosmochim. Acta, 69,* 1333–1348.

Hutcheon I. D. and Newton R. C. (1981) Mg isotopes, mineralogy, and mode of formation of secondary phases in C3 refractory inclusions (abstract). In *Lunar and Planetary Science XII,* pp. 491–493. Lunar and Planetary Institute, Houston.

Hutcheon I. D. and Phinney D. L. (1996) Radiogenic ^{53}Cr* in Orgueil carbonates: Chronology of aqueous activity on the CI parent body (abstract). In *Lunar and Planetary Science XXVII,* pp. 577–578. Lunar and Planetary Institute, Houston.

Hutcheon I. D., Steele I. M., Smith J. V., and Clayton R. N. (1978) Ion microprobe, electron microprobe and cathodoluminescence data for Allende inclusions with emphasis on plagioclase chemistry. *Proc. Lunar Planet. Sci. Conf. 9th,* pp. 1345–1368.

Hutcheon I. D., Steele I. M., Smith J. V., and Clayton R. N. (1981) Ion microprobe, electron microprobe and cathodoluminescence data for Allende inclusions with emphasis on plagioclase chemistry. In *Proc. Lunar Planet. Sci. Conf. 9th,* pp. 1345–1368.

Hutcheon I. D., Phinney D. L., and Hutchison R. (1997) Radiogenic chromium-53 in CI carbonates: New evidence of early aqueous activity (abstract). *Meteoritics & Planet. Sci., 32,* A63.

Hutcheon I. D., Krot A. N., Keil K., Phinney D. L., and Scott E. R. D. (1998) ^{53}Mn-^{53}Cr dating of fayalite formation in the CV3 chondrite Mokoia: Evidence for asteroidal alteration. *Science, 282,* 1865–1867.

Hutcheon I. D., Weisberg M. K., Phinney D. L., Zolensky M. E., Prinz M., and Ivanov A. V. (1999) Radiogenic ^{53}Cr in Kaidun carbonates; evidence for very early aqueous activity (abstract). In *Lunar and Planetary Science XXX,* Abstract #1722. Lunar and Planetary Institute, Houston (CD-ROM).

Hutchison R., Alexander C. M. O'D., and Barber D. J. (1987) The Semarkona meteorite; first recorded occurrence of smectite in an ordinary chondrite, and its implications. *Geochim. Cosmochim. Acta, 51,* 1875–1882.

Hutchison R., Alexander C. M. O'D., and Bridges J. C. (1998) Elemental redistribution in Tieschitz and the origin of white matrix. *Meteoritics & Planet. Sci., 33,* 1169–1179.

Ikeda Y. (1992) An overview of the research consortium, "Antarctic carbonaceous chondrites with CI affinities, Yamato-86720, Yamato-82162, and Belgica-7904." *Proc. NIPR Symp. Antarct. Meteorites 5th,* pp. 49–73.

Itoh D. and Tomeoka K. (2003) Nepheline formation in chondrules in CO3 chondrites: Relationship to parent-body thermal metamorphism (abstract). In *NIPR Symposium on Evolution of Solar System Materials, a New Perspective from Antarctic Meteorites,* pp. 45–46. National Institute of Polar Research, Tokyo.

Jeffery P. M. and Anders E. (1970) Primordial noble gases in separated meteoritic minerals, I. *Geochim. Cosmochim. Acta, 34,* 1175–1198.

Johnson C. A. and Prinz M. (1993) Carbonate compositions in CM and CI chondrites, and implications for aqueous alteration. *Geochim. Cosmochim. Acta, 57,* 2843–2852.

Johnson C. A., Prinz M., Weisberg M. K., Clayton R. N. and Mayeda T. K. (1990) Dark inclusions in Allende, Leoville, and Vigarano: Evidence for nebular oxidation of CV3 constituents. *Geochim. Cosmochim. Acta, 54,* 819–831.

Jones R. H. (1997a) Alteration of plagioclase-rich chondrules in CO3 chondrites; evidence for late-stage sodium and iron metasomatism in a nebular environment (abstract). In *Workshop on Parent-Body and Nebular Modification of Chondritic Materials* (M. E. Zolensky et al., eds), pp. 30–31. LPI Tech. Rpt. 97-02, Lunar and Planetary Institute, Houston.

Jones R. H. (1997b) Ubiquitous anorthitic plagioclase in type I chondrules in CO3 chondrites; implications for chondrule formation and parent-body metamorphism (abstract). *Meteoritics & Planet. Sci., 32,* pp. 67–68.

Keller L. P. (1998) A transmission electron microscope study of iron-nickel carbides in the matrix of the Semarkona unequilibrated ordinary chondrite. *Meteoritics & Planet. Sci., 33,* 913–919.

Kennedy B. M., Hudson B., Hohenberg C. M., and Podosek F. A. (1988) ^{129}I-^{127}I variations among enstatite chondrites. *Geochim. Cosmochim. Acta, 52,* 101–111.

Kerridge J. F. and Bunch T. E. (1979) Aqueous activity on asteroids: Evidence from carbonaceous meteorites. In *Asteroids* (T. Gehrels, ed.), pp. 745–764. Univ. of Arizona, Tucson.

Kerridge J. F., Mackay A. L., and Boynton W. V. (1979a) Magnetite in CI carbonaceous meteorites: Origin by aqueous activity on a planetesimal surface. *Science, 205,* 395–397.

Kerridge J. F., Macdougall J. D., and Marti K. (1979b) Clues to the origin of sulfide minerals in CI chondrites. *Earth Planet. Sci. Lett., 43,* 359–367.

Kirschbaum C. (1988) Carrier phases for iodine in the Allende meteorite and their associated $^{129}Xe_r/^{129}I$ ratios; a laser microprobe study. *Geochim. Cosmochim. Acta, 52,* 679–699.

Kimura M. and Ikeda Y. (1995) Anhydrous alteration of Allende chondrules in the solar nebula; II, Alkali-Ca exchange reactions and formation of nepheline, sodalite and Ca-rich phases in chondrules. *Proc. NIPR Symp. Antarct. Meteorites 8th,* pp. 123–138.

Kimura M. and Ikeda Y. (1998) Hydrous and anhydrous alterations of chondrules in Kaba and Mokoia CV chondrites. *Meteoritics & Planet. Sci., 33,* 1139–1146.

Kojima T. and Tomeoka K. (1996) Indicators of aqueous alteration and thermal metamorphism on the CV3 parent body: Microtextures of a dark inclusion from Allende. *Geochim. Cosmochim. Acta, 60,* 2651–2666.

Kracher A., Keil K., Kallemeyn G. W., Wasson J. T., Clayton R. N., and Huss G. R. (1985) The Leoville (CV3) accretionary breccia. *Proc. Lunar Planet. Sci. Conf. 16th,* in *J. Geophys. Res., 90,* D123–D135.

Krestina N., Jagoutz E., and Kurat G. (1996) Sm-Nd system in single chondrules from Tieschitz (H3) (abstract). In *Lunar and Planetary Science XXVII,* pp. 701–702. Lunar and Planetary Institute, Houston.

Krot A. N. and Todd C. S. (1998) Metal-carbide-magnetite-fayalite association in a Bali-like clast in the reduced CV3 chondrite breccia Vigarano (abstract). *Meteoritics & Planet. Sci., 34,* A88–A89.

Krot A. N., Scott E. R. D., and Zolensky M. E. (1995) Mineralogic and chemical variations among CV3 chondrites and their components: Nebular and asteroidal processing. *Meteoritics, 30,* 748–775.

Krot A. N., Scott E. R. D., and Zolensky M. E. (1997a) Origin of fayalitic olivine rims and plate-like matrix olivine in the CV3 chondrite Allende and its dark inclusions. *Meteoritics, 32,* 31–49.

Krot A. N., Zolensky M. E., Wasson J. T., Scott E. R. D., Keil K., and Ohsumi K. (1997b) Carbide-magnetite-bearing type 3 ordinary chondrites. *Geochim. Cosmochim. Acta, 61,* 219–237.

Krot A. N., Petaev M. I., Scott E. R. D., Choi B.-G., Zolensky M. E., and Keil K. (1998a) Progressive alteration in CV3 chondrites: More evidence for asteroidal alteration. *Meteoritics & Planet. Sci., 33,* 1065–1085.

Krot A. N., Zolensky M. E., Keil K., Scott E. R. D., and Nakamura K. (1998b) Secondary Ca-Fe-rich minerals in the Bali-like and Allende-like oxidized CV3 chondrites and Allende dark inclusions. *Meteoritics & Planet. Sci., 33,* 623–645.

Krot A. N., Brearley A. J., Ulyanov A. A., Biryukov V. V., Swindle T. D., K. Keil, Mittlefehldt D. W., Scott E. R. D., Clayton R. N., and Mayeda T. K. (1999) Mineralogy, petrography and bulk chemical, I-Xe, and oxygen isotopic compositions of dark inclusions in the reduced CV3 chondrite Efremovka. *Meteoritics & Planet. Sci., 34,* 67–89.

Krot A. N., Brearley A. J., Petaev M. I., Kallemeyn G. W., Sears D. W. G., Benoit P. H., Hutcheon I. D., Zolensky M. E., and Keil K. (2000a) Evidence for *in situ* growth of fayalite and hedenbergite in MacAlpine Hills 88107, ungrouped carbonaceous chondrite related to CM-CO clan. *Meteoritics & Planet. Sci., 35,* 1365–1387.

Krot A. N., Meibom A., and Keil K. (2000b) A clast of Bali-like oxidized CV3 material in the reduced CV3 chondrite breccia Vigarano. *Meteoritics & Planet. Sci., 35,* 817–827.

Krot A. N., Hiyagon H., Petaev M. I., and Meibom A. (2000c) Oxygen isotopic compositions of secondary Ca-Fe-rich silicates from the Allende dark inclusions: Evidence against high-temperature formation (abstract). In *Lunar and Planetary Science XXXI,* Abstract #1463. Lunar and Planetary Institute, Houston (CD-ROM).

Krot A. N., Petaev M. I., Meibom A., and Keil K. (2001) *In situ* growth of Ca-rich rims around Allende dark inclusions. *Geochem. Intl., 36,* 351–368.

Krot A. N., Meibom A., Weisberg M. K., and Keil K. (2002) The CR chondrite clan: Implications for early solar system processes. *Meteoritics & Planet. Sci., 37,* 1451–1490.

Krot A. N., Petaev M. I., and Bland P. A. (2004a) Multiple formation mechanisms of ferrous olivine in CV3 carbonaceous chondrites during fluid-assisted metamorphism. *Antarct. Meteorite Res., 17,* 154–172.

Krot A. N., Yurimoto H., Petaev M. I., Hutcheon I. D., and Wark D. A. (2004b) Type C CAIs: New insights into the early solar system processes (abstract). *Meteoritics & Planet. Sci., 39,* A57.

Kurat G., Palme H., Brandstätter F. and Huth J. (1989) Allende xenolith AF: Undisturbed record of condensation and aggregation of matter in the solar nebula. *Z. Naturforsch., 44a,* 988–1004.

Lavielle B. and Marti K. (1992) Trapped xenon in ordinary chondrites. *J. Geochem. Res., 97,* 875–881.

Leshin L. A., Rubin A. E., and McKeegan K. D. (1997) The oxygen isotopic composition of olivine and pyroxene from CI chondrites. *Geochim. Cosmochim. Acta, 61,* 835–845.

Leshin L. A., Farquhar J., Guan Y., Pizzarello S., Jackson T. L., and Thiemens M. H. (2001) Oxygen isotopic anatomy of Tagish Lake: Relationship to primary and secondary minerals in CI and CM chondrites (abstract). In *Lunar and Planetary Science XXXII,* Abstract #1843. Lunar and Planetary Institute, Houston (CD-ROM).

Lewis R. S. and Anders E. (1975) Condensation time of the solar nebula from the extinct ^{129}I in primitive meteorites. In *Proc. Natl. Acad. Sci. USA, 72,* 268–273.

Lugmair G. W. and Shukolyukov A. (1998) Early solar system timescales according to ^{53}Mn-^{53}Mn systematics. *Geochim. Cosmochim. Acta, 62,* 2863–2886.

Lugmair G. W. and Shukolyukov A. (2001) Early solar system events and timescales. *Meteoritics & Planet. Sci., 36,* 1017–1026.

Macdougall J. D. and Lugmair G. W. (1989) Chronology of chemical change in the Orgueil CI chondrite based on Sr isotope systematics (abstract). *Meteoritics, 24,* 297.

Macdougall J. D., Lugmair G. W., and Kerridge J. F. (1984) Early solar system aqueous activity: Sr isotope evidence from the Orgueil CI meteorite. *Science, 307,* 249–251.

Mackinnon I. D. R. and Zolensky M. E. (1984) Proposed structures for poorly characterized phases in C2M carbonaceous chondrite meteorites. *Nature, 309,* 240–242.

MacPherson G. J. and Krot A. N. (2002) Distribution of Ca-Fe-silicates in CV3 chondrites: Possible controls by parent body compaction (abstract). *Meteoritics & Planet. Sci., 37,* A91.

MacPherson G. J., Wark D. A., and Armstrong J. T. (1988) Primitive material surviving in chondrites: Refractory inclusions. In *Meteorites and the Early Solar System* (J. F. Kerridge and M. S. Matthews, eds.), pp. 746–807. Univ. of Arizona, Tucson.

MacPherson G. J., Davis A. M., and Zinner E. K. (1995) The distribution of aluminum-26 in the early solar system: A reappraisal. *Meteoritics, 30,* 365–386.

McSween H. Y. Jr. (1977) Petrographic variations among carbonaceous chondrites of the Vigarano type. *Geochim. Cosmochim. Acta, 41,* 1777–1790.

McSween H. Y. Jr. (1979) Are carbonaceous chondrites primitive or processed? A review. *Rev. Geophys. Space Phys., 17,* 1059–1078.

Nichols R. H. Jr., Hohenberg C. M. and Olinger C. T. (1990) Allende chondrules and rims: I-Xe systematics (abstract). In *Lunar and Planetary Science XXI,* pp. 879–880. Lunar and Planetary Institute, Houston.

Nichols R. H. Jr., Hagee B. E., and Hohenberg C. M. (1991) Tieschitz chondrules; I-Xe systematics (abstract). In *Lunar and Planetary Science XXII,* pp. 975–976. Lunar and Planetary Institute, Houston.

Nichols R. H. Jr., Hohenberg C. M., Kehm K., Kim Y., and Marti K. (1994) I-Xe studies of the Acapulco meteorite: Absolute I-Xe ages of individual phosphate grains and the Bjurböle standard. *Geochim. Cosmochim. Acta, 58,* 2553–2561.

Nyquist L. E., Reese Y., Wiesman H., Shih C. Y., and Takeda H. (2001) Live ^{53}Mn and ^{26}Al in an unique cumulate eucrite with very calcic feldspar (abstract). *Meteoritics & Planet. Sci., 36,* 151–152.

Palme H. and Jones A. (2003) Solar system abundances of the elements. In *Treatise on Geochemistry, Vol. 1: Meteorites, Comets and Planets* (A. M. Davis, ed.), pp. 41–63. Elsevier, Oxford.

Palme H. and Wark D. A. (1988) CV-chondrites; high temperature gas-solid equilibrium vs. parent body metamorphism (abstract). In *Lunar and Planetary Science XXIX,* pp. 897–898. Lunar and Planetary Institute, Houston.

Papanastassiou D. A. (1986) Chromium isotopic anomalies in the Allende meteorite. *Astrophys. J. Lett., 308,* L27–L30.

Podosek F. A. and Cassen P. (1994) Theoretical, observational, and isotopic estimates of the lifetime of the solar nebula. *Meteoritics, 29,* 6–25.

Pravdivtseva O. V. and Hohenberg C. M. (2001) (abstract). The I-Xe system of magnetic fractions from CV3 meteorites (abstract). In *Lunar and Planetary Science XXXII,* Abstract #2176. Lunar and Planetary Institute, Houston (CD-ROM).

Pravdivtseva O. V., Hohenberg C. M., Meshik A. P., and Krot A. N. (2001) I-Xe ages of different mineral fractions from Bali and Kaba (CV3) (abstract). *Meteoritics & Planet. Sci., 36,* A168.

Pravdivtseva O. V., Krot A. N., Hohenberg C. M., Meshik A. M., Weisberg M. K., and Keil K. (2003a) The I-Xe record of alteration in the Allende CV chondrite. *Geochim. Cosmochim. Acta, 67,* 5011–5026.

Pravdivtseva O. V., Hohenberg C. M., and Meshik A. M. (2003b) The I-Xe age of Orgueil magnetite: New results (abstract). In *Lunar and Planetary Science XXXIV,* Abstract #1863. Lunar and Planetary Institute, Houston (CD-ROM).

Pravdivtseva O. V., Hohenberg C. M., Meshik A. M., Krot A. N., and Brearley A. J. (2003c) I-Xe age of the dark inclusions from the reduced CV3 chondrites Leoville, Efremovka, and Vigarano (abstract). *Meteoritics & Planet. Sci., 38,* A140.

Reynolds J. H. (1960) I-Xe dating of meteorites. *J. Geophys. Res., 65,* 3843–3846.

Richardson S. M. (1978) Vein formation in the C1 carbonaceous chondrites. *Meteoritics, 13,* 141–159.

Riciputi L. R., McSween H. Y. Jr., Johnson C. A., and Prinz M. (1994) Minor and trace element concentrations in carbonates of carbonaceous chondrites, and implications for the compositions of coexisting fluids. *Geochim. Cosmochim. Acta, 58,* 1343–1351.

Rowe M. W., Clayton R. N., and Mayeda T. K. (1994) Oxygen isotopes in separated components of CI and CM meteorites. *Geochim. Cosmochim. Acta, 58,* 5341–5347.

Rubin A. E. (1998) Correlated petrologic and geochemical characteristics of CO3 chondrites. *Meteoritics & Planet. Sci., 33,* 385–391.

Rubin A. E., Zolensky M. E., and Bodnar R. J. (2002) The halite-bearing Monahans (1998) and Zag meteorite breccias: Shock metamorphism, thermal metamorphism and aqueous alteration on the H-chondrite parent body. *Meteoritics & Planet. Sci., 37,* 125–142.

Russell S. S., Huss G. R., Fahey A. J., Greenwood R. C., Hutchison R., and Wasserburg G. J. (1998) An isotopic and petrologic study of calcium-aluminum-rich inclusions from CO3 meteorites. *Geochim. Cosmochim. Acta, 62,* 689–714.

Scatena-Wachel D. E., Hinton R. W., and Davis A. M. (1984) Preliminary ion microprobe study of chromium isotopes in Orgueil (abstract). In *Lunar and Planetary Science XV,* pp. 718–719. Lunar and Planetary Institute, Houston.

Scott E. R. D. and Krot A. N. (2005) Thermal history of silicate dust in the solar nebula: Clues from primitive chondrite matrices. *Astrophys. J., 623,* 571–578.

Sears D. W. G., Hasan F. A., Batchelor J. D., and Lu J. (1991) Chemical and physical studies of type 3 chondrites; XI, Metamorphism, pairing, and brecciation of ordinary chondrites. *Proc. Lunar Planet. Sci. Conf., Vol. 21,* pp. 493–512.

Swindle T. D. (1998) Implications of iodine-xenon studies for the dating and location of secondary alteration. *Meteoritics & Planet. Sci., 33,* 1147–1157.

Swindle T. D. and Podosek F. A. (1988) Iodine-xenon dating. In *Meteorites and the Early Solar System* (J. F. Kerridge and M. S. Matthews, eds.), pp. 1127–1146. Univ. of Arizona, Tucson.

Swindle T. D., Caffee M. W., Hohenberg C. M., and Lindstrom M. M. (1983) I-Xe studies of individual Allende chondrules. *Geochim. Cosmochim. Acta, 47,* 2157–2177.

Swindle T. D., Caffee M. W., and Hohenberg C. M. (1988) Iodine-xenon studies of Allende inclusions: EGGs and the Pink Angel. *Geochim. Cosmochim. Acta, 52,* 2215–2229.

Swindle T. D., Grossman J. N., Olinger C. T., and Garrison D. H. (1991a) Iodine-xenon, chemical, and petrographic studies of Semarkona chondrules: Evidence for the timing of aqueous alteration. *Geochim. Cosmochim. Acta, 55,* 3723–3734.

Swindle T. D., Caffee M. W., Hohenberg C. M., Lindstrom M. M., and Taylor G. J. (1991b) Iodine-xenon studies of petrographically and chemically characterized Chainpur chondrules. *Geochim. Cosmochim. Acta, 55,* 861–880.

Swindle T. D., Cohen B., Li B., Olson E., Krot A. N., Birjukov V. V., and Ulyanov A. A. (1998) Iodine — Xenon studies of separated components of the Efremovka (CV3) meteorite (abstract). In *Lunar and Planetary Science XXIX,* Abstract #1005. Lunar and Planetary Institute, Houston (CD-ROM).

Tomeoka K. (1990) Mineralogy and petrology of Belgica-7904: A new kind of carbonaceous chondrite from Antarctica. *Proc. NIPR Symp. Antarct. Meteorites 3rd,* pp. 40–54.

Tomeoka K. and Kojima T. (1998) Arcuate band texture in a dark inclusion from the Vigarano CV3 chondrite; possible evidence for early sedimentary processes. *Meteoritics & Planet. Sci., 33,* 519–525.

Tomeoka K., Kojima H., and Yanai K. (1989) Yamato-86720: A CM carbonaceous chondrite having experienced extensive aqueous alteration and thermal metamoprhism. *Proc. NIPR Symp. Antarct. Meteorites 2nd,* pp. 55–74.

Tonui E. K., Zolensky M. E., Lipschutz M. L., Wang M.-S., and Nakamura T. (2003) Yamato 86029: Aqueously altered and thermally metamorphosed CI-like chondrite with unusual textures. *Meteoritics & Planet. Sci., 38,* 269–292.

Turner G., Enright M. C., and Cadogan P. H. (1978) The early history of chondrite parent bodies inferred from ^{40}Ar-^{39}Ar ages. *Proc. Lunar Planet. Sci. Conf. 9th,* pp. 989–1025.

Wark D. A. (1987) Plagioclase-rich inclusions in carbonaceous chondrite meteorites — Liquid condensates? *Geochim. Cosmochim. Acta, 51,* 221–242.

Weisberg M. K. and Prinz M. (1998) Fayalitic olivine in CV3 chondrite matrix and dark inclusions: A nebular origin. *Meteoritics & Planet. Sci., 33,* 1087–1111.

Weisberg M. K., Prinz M., Zolensky M. E., and Ivanov A. V. (1994) Carbonates in Kaidun chondrite (abstract). *Meteoritics, 29,* pp. 549–550.

Whitby J. A., Gilmour J. D., Ash R. D., Prinz M., and Turner G. (1997) Iodine-xenon dating of chondrules and matrix from the Qingzhen and Kota-Kota EH3 chondrite (abstract). *Meteoritics & Planet. Sci., 32,* A140.

Whitby J., Burgess R., Turner G., Gilmour J., and Bridges J. (2000) Extinct ^{129}I in halite from a primitive meteorite: Evidence for evaporite formation in the early solar system. *Science, 288,* 1819–1821.

Yurimoto H., Koike O., Nagahara H., Morioka M., and Nagasawa H. (2000) Heterogeneous distribution of Mg isotopes in anorthite single crystal from type B CAI in Allende meteorite (abstract). In *Lunar and Planetary Science XXXI,* Abstract #1593. Lunar and Planetary Institute, Houston (CD-ROM).

Zaikowski A. (1980) I-Xe dating of Allende inclusions: Antiquity and fine structure. *Earth Planet. Sci. Lett., 47,* 211–222.

Zolensky M. and Ivanov A. V. (2003) The Kaidun microbreccia meteorite: A harvest from the inner and outer asteroid belt. *Chem. Erde, 63,* 185–246.

Zolensky M. E. and McSween H. Y. Jr (1988) Aqueous alteration. In *Meteorites and the Early Solar System* (J. F. Kerridge and M. S. Matthews, eds.), pp. 114–143. Univ. of Arizona, Tucson.

Zolensky M. E., Bourcier W. L., and Gooding J. L. (1989) Aqueous alteration on the hydrated asteroids: Results of EQ3/6 computer simulations. *Icarus, 78,* 411–425.

Zolensky M., Barrett R., and Browning L. (1993) Mineralogy and composition of matrix and chondrule rims in carbonaceous chondrites. *Geochim. Cosmochim. Acta, 57,* 3123–3148.

Zolensky M. E., Ivanov A. I., Yang S. V., Mittlefehldt D. W., and Ohsumi K. (1996) The Kaidun meteorite: Mineralogy of an unsual CM1 lithology. *Meteoritics & Planet. Sci., 31,* 484–493.

Zolensky M. E., Bodnar R. J., Gibson E. K. Jr., Nyquist L. E., Reese Y., Shih C.-Y., and Wiesmann H. (1999) Asteroidal water within fluid inclusion-bearing halite in an H5 chondrite, Monahans (1998). *Science, 185,* 1377–1379.

Asteroidal Heating and Thermal Stratification of the Asteroid Belt

A. Ghosh
University of Tennessee and NASA Goddard Space Flight Center

S. J. Weidenschilling
Planetary Science Institute

H. Y. McSween Jr.
University of Tennessee

A. Rubin
University of California, Los Angeles

The asteroid belt is thermally stratified, with melted or metamorphosed asteroids dominating the inner belt and relatively unaltered asteroids dominating the outer belt. The compositional structure of the asteroid belt is a unique signature of the heating agent that caused melting and metamorphism of planetesimals in the early solar nebula, thus it constitutes an important test of plausibility of heating mechanisms. Previous work attributes the stratification to a radial thermal gradient in the solar nebula or heating by ^{26}Al decay or electromagnetic induction. Present thinking attributes the thermal stratification of multiple processes that were active in the early solar system, including ^{26}Al, the dependence to accretion timescale on heliocentric distance, the presence of ice in planetesimals that formed in the outer belt, and loss of primitive material by collisional disruption of small asteroids over the age of the solar system.

1. HEAT SOURCES IN THE EARLY SOLAR SYSTEM

Half a century ago, H. Urey recognized that "it is difficult to believe that heating by K, U and Th is a feasible explanation for the high-temperature stage required to produce the meteorites." He proceeded to perform the first back-of-the-envelope calculation and suggested ^{26}Al to be a heat source for asteroidal heating (*Urey,* 1955), and perhaps inadvertently initiated the subdiscipline of asteroid thermal modeling. Ever since, thermal models have been used as plausibility calculations for chondrite metamorphism (or melting) in an asteroidal body. Since Urey advocated ^{26}Al as a heat source, many heat sources have been suggested at various points of time in the literature, like the radiant energy of a Hayashi-phase Sun (*Wasson,* 1974) and the energy released from exothermic reactions between unstable presolar compounds (*Clayton,* 1980). Of the proposed heat sources, three remain a subject of active discussion today: heating by the decay of ^{26}Al, electromagnetic induction, and impact heating.

1.1. Electromagnetic Induction Heating

The theory of electromagnetic induction heating, first proposed by *Sonnett et al.* (1968), was based on the physics of T Tauri outflow (*Kuhi,* 1964). An electric current is generated in planetesimals that move through an embedded magnetic field generated by the T Tauri phase of the Sun. This electric current that flows through the planetesimals causes them to undergo heating. There are two modes of electromagnetic induction: the transverse magnetic mode, which is caused by the steady motion of the solar wind past the planetesimals, and the transverse electric mode of heating, which is caused by the fluctuations in the magnetic field of the solar wind. The transverse magnetic mode is believed to be effective for asteroidal bodies (*Sonett and Colburn,* 1968; *Sonett et al.,* 1968), while the transverse electric mode is effective for larger bodies like the Moon. At the time of publication of the *Sonnett et al.* (1968) paper, the mass loss rates were believed to be very high: on the order of 50% of the initial mass of the Sun over the entire T Tauri phase. The T Tauri outflow was also believed to have been isotropic.

Modeling electromagnetic induction heating is fairly complex: This is reflected by the fact that remarkably few attempts have been made to model the phenomenon, in comparison to ^{26}Al heating models. It is straightforward to list the factors that will influence the degree of heating; it is simulating the relative importance and interdependence of the factors in the heating process that is difficult, particularly since a few of the variable parameters are unconstrained. The intensity as well as duration of the solar wind during the T Tauri phase determines the amount of current generated in the planetesimal, and therefore the amount of heating. Solar wind velocity, electrical conductivity, surface magnetic field, and rotation rate of the T Tauri Sun are all proportional to the amount of heat generated. The size of a planetesimal is a more complicated parameter: It seems that the transverse magnetic (TM) mode in large bodies creates secondary magnetic fields strong enough to deflect part of the magnetic fields that can deflect part of the solar wind, thereby causing the induced electrical current to decrease. Thus, heating increases with asteroidal size until a critical radius is reached, beyond which an increase in size causes the heating to decrease.

Recent studies of T Tauri stars, however, moderate the conditions that would favor the electromagnetic induction heating hypothesis. First, the T Tauri stellar winds have been found to be anisotropic with a greater wind density at high latitudes, avoiding the plane of the circumstellar disk where planetesimals form (*Edwards et al.,* 1987). This goes against the assumption of isotropic dissipation of the solar wind by *Sonett et al.* (1968) and means that the solar wind flux in the equatorial plane where planets form was probably lower than was originally thought. Second, pre-main-sequence mass loss from a T Tauri star has been revised from ~50% (*Kuhi,* 1964) to a few percent (*DeCampli,* 1981), and the rate of mass loss has been revised downward by 1 to 2 orders of magnitude. Consequently, since the intensity of electromagnetic heating is proportional to the intensity of the solar wind (which in turn is governed by the rate of mass loss from the Sun), the amount of heat generation will also be lower than the initial models of *Sonett et al.* (1968). Third, it was believed that T Tauri outflow continued after planetesimals accreted and the disk dissipated. However, *Edwards et al.* (1987) showed that this phase does not exist; in other words, the T Tauri outflow stops as soon as the process of active accretion stops in the disk.

Recent models (*Herbert,* 1989; *Shimazu and Terasawa,* 1995) produce melting in asteroidal-sized bodies. However, this has not been without criticism; for example, *Wood and Pellas* (1991) analyze the parameters used by *Herbert et al.* (1991) and deduce that the latter do not take into account poleward collimation of the solar wind and assume higher-than-defensible values of solar wind velocities. The problem of electromagnetic induction heating hinges on the choice of a reasonable parameter set where, as noted by *Wood and Pellas* (1991), most of the parameters are unconstrained. In the absence of our knowledge of a "reasonable" parameter set for electromagnetic induction, the question of whether this can be the dominant heat source in the early solar system will remain unresolved. Electromagnetic induction continues to be invoked, particularly in scenarios where short-lived radionuclides do not produce the desired result.

1.2. Impact Heating

Impact energy is the heat generated when the gravitational potential energy of the impactor in the target's gravitational field (plus any kinetic energy associated with relative velocities due to orbital motion) is converted to heat. A portion of this heat is lost, carried off by escaping ejecta and/or radiated into space, and the remaining heat is deposited in the target asteroid. The gravitational potential energy of an impactor varies as the square of the target-body radius. *Melosh* (1990) pointed out that during accretion, when impact velocities are comparable to the target's escape velocity, the energy deposited by impacts on targets smaller than a critical radius is not significant, and that heat deposition by impacts is a significant cause of global metamorphism in planets, but insignificant for most asteroids.

Although impacts could not have caused global metamorphism of asteroids, they may well have caused localized heating that mimics the effects of thermal metamorphism. Impacts have commonly affected meteorites, and many ordinary chondrites have been shocked and subsequently annealed. These rocks possess olivines with sharp optical extinction (indicative of unshocked S1 material), but also include relict shock features such as extensive silicate darkening, shock veins, metallic Cu, and polycrystalline troilite. In the case of LL6 MIL 99301, the annealing event occurred 4.26 G.y. ago (*Dixon et al.,* 2003), ~300 m.y. after accretion. At this late epoch, impacts are the only plausible heat source. Proponents of impact heating believe the following to be consistent with the hypothesis: (1) the positive correlation between petrologic type and shock stage among ordinary chondrites; (2) the observation that, among the three ordinary chondrite groups, L chondrites have the highest (and H chondrites the lowest) proportions of highly annealed members and highly shocked members; and (3) the general trend for chondrite groups with many highly metamorphosed members (H, L, LL, EH, EL, CK) to also have shocked members of stage S4, and for chondrite groups with no highly metamorphosed members (CI, CM, CO, CR) to have no shocked members of stage S4 (*Rubin,* 2004). Proponents of impact heating, in addition, believe that it could have operated on porous rubble piles. In such bodies, collisional kinetic energy is distributed through relatively small volumes of material and efficiently converted into heat. Although impacts have clearly heated chondrites, this may be a localized annealing overprint rather than the thermal event that produced the metamorphic sequence in chondrites.

1.3. Heating by the Decay of Aluminum-26

In a landmark paper, *Lee et al.* (1976) observed a large anomaly in the isotopic composition of Mg in a Ca-Al-rich inclusion (CAI) from the Allende meteorite. *Lee et al.* (1976) hypothesized that the most plausible cause of the anomaly was the *in situ* decay of now-extinct ^{26}Al, and measured an initial ratio of ^{26}Al/^{27}Al equal to 5×10^{-5} at CAI formation. The numerical value of the following four parameters provided just the correct amount of heat to metamorphose asteroidal bodies for a few million years: (1) the proportional abundance of Al in chondrites, (2) the initial ratio, (3) the short half-life of ^{26}Al, and (4) the high decay energy per nucleon (2 MeV).The efficacy of the heat source dwindled rapidly because of the short half-life. Once the heat source dwindled in potency, the asteroidal bodies lost their heat relatively quickly due to the large surface area to volume ratio. Thus, ^{26}Al was able to theoretically produce a short period of intense heating in the early solar system, as was being observed in meteorites.

For the last 30 years, ^{26}Al has largely dominated any discussion of asteroidal heating. The case for ^{26}Al has become increasingly stronger over the last decade. Some perceived problems with the ^{26}Al hypothesis (*Wood and Pellas,* 1991) have been addressed: Magnesium-26 due to the decay of ^{26}Al has been detected in noncumulate eucrites (*Srinivasan et al.,* 1999), and reasons as to why it might not be observed in others have been put forward (*Ghosh and McSween,* 1998; *LaTourrette and Wasserburg,* 1998). An explanation for the difference in ages between noncumulate and cumulate eucrites has been offered (*Ghosh and McSween,* 1998). The thermal stratification in the asteroid belt has been explained (*Grimm and McSween,* 1993; *Ghosh et al.,* 2001). A compilation of ^{26}Mg measurements for various meteorite types shows a conspicuous peak at the canonical value of ^{26}Al/^{27}Al of 5×10^{-5} (*MacPherson et al.,* 1995). However, the existence in certain refractory inclusions (e.g., FUN inclusions) of isotopic anomalies with no radiogenic excess of ^{26}Mg is perceived as a reason to invoke isotopic heterogeneity in the distribution of ^{26}Al (*MacPherson et al.,* 1995). The issue of possible nebular heterogeneity of ^{26}Al clouds its efficacy as a heat source; however, the consistency of ^{26}Al/^{27}Al ratios in CAIs and meteorites of different classes argues for homogeneity (*Huss et al.,* 2001).

1.4. Heating by Decay of Iron-60

Iron-60 was proposed as a heat source by *Shukolyukov and Lugmair* (1993) on the basis of measurements in Juvinas. Using relative chronology, they calculated the initial ratio of ^{60}Fe/^{56}Fe to be ~10^{-6} at the time of CAI formation. Because of the high abundance of Fe in the solar system and the high decay energy of ^{60}Fe (3.04 MeV), it was proposed that ^{60}Fe could be a planetary heat source. As of 1995 (G. Lugmair, personal communication), this ratio was revised downward to ~10^{-8}. *Ghosh* (1997) made a first-order calculation of the heat generated per kilogram due to decay of ^{26}Al and ^{60}Fe with initial ratios (at CAI formation) of 5×10^{-5} and 1×10^{-8}, respectively. The calculations show that the rate of heat generation at CAI formation is 1×10^7 and 1×10^3 J/m.y./kg for ^{26}Al and ^{60}Fe, respectively. In other words, if the initial ratio of ^{60}Fe/^{56}Fe was 1×10^{-8}, the rate of heat generation by ^{60}Fe is 4 orders of magnitude lower than that of ^{26}Al. This shows that the initial ratio of ^{60}Fe needs to be 4 orders of magnitude higher to be able to match heat production by ^{26}Al. A recent paper by *Mostefaoui* (2005) analyzed the mineral phases of troilite and magnetite in Semarkona (LL3) and reported an initial ratio of $0.92 \pm 0.24 \times 10^{-6}$ for ^{60}Fe/^{56}Fe. While this ranks among the higher published initial ratios of ^{60}Fe/^{56}Fe, the rate of heat generation at CAI formation for ^{60}Fe in such a scenario would still be 2 orders of magnitude lower than ^{26}Al. Thus, the state of our present knowledge calls for a limited role, if any, for ^{60}Fe heating; i.e., ^{60}Fe could not, for example, have been responsible for the early solar system heating by itself. Having said that, ^{60}Fe can have a secondary effect on planetary and asteroidal heating in two ways. First, ^{60}Fe may cause considerable heating when Fe is segregated in the core during core formation. *Ghosh* (1997) evaluated such a scenario for Vesta with an initial ratio of 1×10^{-8}, and failed to establish any significant heat generation by ^{60}Fe. It will be interesting to revisit the problem using the updated ^{60}Fe/^{56}Fe ratios reported recently by *Mostefaoui* (2005). Second, while ^{60}Fe cannot surpass rate of heat generation by ^{26}Al for the first few million years, it can moderate cooling rates by staying live for a longer time than ^{26}Al, since the half-life of ^{60}Fe is 1.5 m.y. (more than twice that of ^{26}Al).

The primary motivation of this chapter is to discuss the thermal stratification of the asteroid belt; a detailed review of heat sources is beyond the scope of the paper. For a more detailed discussion of thermal evolution models, see *McSween et al.* (2003); for a somewhat dated but otherwise excellent review of the plausibility of heat sources, see *Wood and Pellas* (1991).

2. THERMAL MODELING BASICS

The thermal evolution of the asteroid can be divided into the following stages: accretion, heating (with or without melting and differentiation, metamorphism, or aqueous alteration), and cooling. The heat transfer equation is the basis of the calculations [for detailed discussion of the equation and parameters, see appendix in *Ghosh and McSween* (1998)]: This is the standard diffusion equation with a diffusion term that equates the rate of temperature change to the heat lost by thermal conduction plus any heat generated by a heat source. The heat source term is different depending on the heat source or the combination of heat sources used. Most models assume asteroidal accretion to be instantaneous and heating to occur in a body of a fixed size. *Ghosh and McSween* (2000) presented a model for incremental accretion, in which the body's radius increases with time, and *Ghosh et al.* (2001) used growth rates from a dynamical accretion calculation (*Weidenschilling et al.,* 1997) in an incremental accretion thermal model. As with most theoretical problems, there are uncertainties in initial conditions (e.g., the temperature of the asteroid at the beginning of the simulation), boundary conditions (e.g., the nebular ambient temperature and the emissivity of the asteroid), and model parameters (e.g., specific heat capacity and thermal diffusivity, and their temperature and composition dependence). Although a methodical study has not been conducted, the parameters are tightly constrained in comparison to models of electromagnetic induction heating. Boundary conditions are implemented in two ways: The Dirichlet boundary condition forces the temperature of the surface of the asteroid to the temperature of the ambient nebula, and the radiation boundary condition calculates a heat flux depending on the difference in temperature between the asteroidal surface and the nebula. Although the radiation boundary condition is numerically unstable, it is believed to be more realistic (*Ghosh and McSween,* 1998). Three methods exist for numerical solution of the heat transfer: the classical series solution (e.g., *Miyamoto,* 1981; *Bennett and McSween,* 1996; *Akridge et al.,* 1998), the finite-difference method (e.g., *Wood,* 1979; *Grimm and McSween,* 1989), and the finite-element method (e.g., *Ghosh and McSween,* 1998). The finite-element method, which uses a basis function to minimize approximation error during numerical integration, has been found to be more accurate than either the finite-difference method or the classical series solution (*Baker and Pepper,* 1991).

3. THERMAL STRUCTURE OF THE ASTEROID BELT

An understanding of the thermal evolution of the asteroid belt is one of the most important problems of planetary science and one of its greatest challenges. An understanding of the evolution of the asteroid belt can help constrain the origin and early history of the solar system, the radial gradient in composition, the difference in the dynamical parameters with heliocentric distance, and the disruption of asteroidal bodies at a later stage.

Gradie and Tedesco (1982) showed that the asteroid belt is stratified with different asteroidal classes, as inferred from their photometric properties, distributed at different heliocentric distances. In particular, *Gradie and Tedesco* (1982) observed that asteroids with spectra indicative of high-temperature silicate minerals (like asteroid types E, R, S, M, F) are concentrated between 1.8 and 3 AU, while dark C-rich types (like asteroid types P, C, D) seem to be located farther away from the Sun.

Bell et al. (1989) attempted the first quantification of asteroid peak temperatures as a function of heliocentric distance; Fig. 1a shows the distribution of the various taxonomic types, and the distributions of different spectral classes are seen to peak at various locations in the asteroid belt. *Bell* (1986) assumed various spectral types to correspond to specific meteorite families. Using Tholen taxonomic classes, *Bell* (1986) grouped asteroids into three broader superclasses based on the inferred degree of heating and metamorphism. The three superclasses are: (1) primitive asteroids with little or no discernable evidence of heating; (2) metamorphosed asteroids that have undergone heating sufficiently to exhibit spectral changes; and (3) igneous asteroids whose mineralogy indicates derivation from igneous processes.

A plot of the abundance of superclasses with heliocentric distance (Fig. 1b) by *Bell* (1986) added to the realization that bodies with distinct thermal histories occur at different characteristic distances from the Sun, thereby restricting asteroid evolution models (e.g., *Zellner,* 1979; *Gradie and Tedesco,* 1982). Igneous asteroids dominate inward of 2.7 AU, metamorphosed asteroids around 3 AU, and ice-bearing planetesimals beyond 3.4 AU (*Gradie and Tedesco,* 1982;

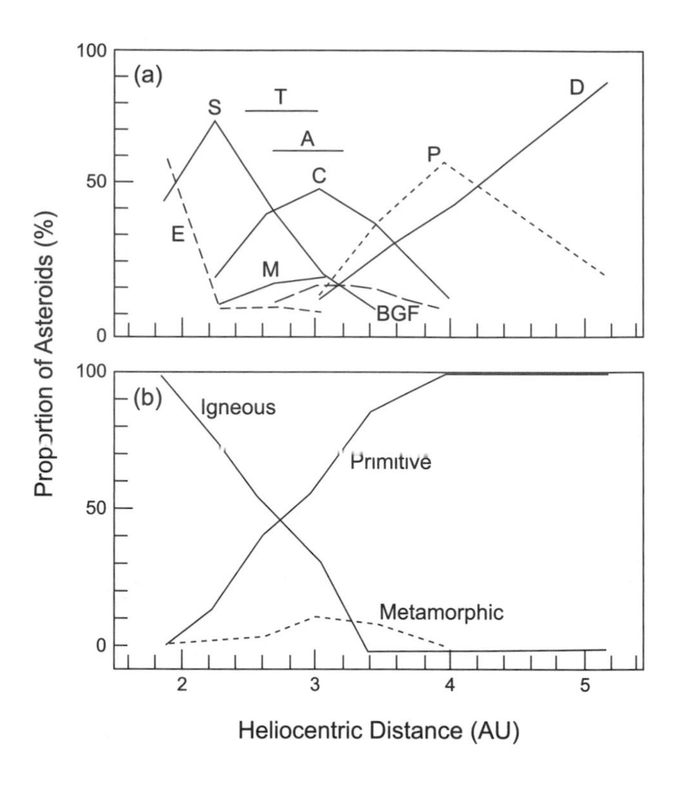

Fig. 1. **(a)** Distribution of various Tholen taxonomic types as a function of heliocentric distance. **(b)** Distribution of superclasses as a function of heliocentric distance. The diagram shows the predominance of melted asteroids in the inner belt and unaltered asteroids in the outer belt.

Bell et al., 1989). The relative abundance of the igneous superclass of *Bell et al.* (1989) decreases from about 100% at ~2 AU to 0% at 3.4 AU, while the relative abundance of the primitive superclass increases from 0% to 100% from 2 AU to about 3.4 AU. The metamorphic class is more broadly distributed with a smaller peak of about 10% around 3 AU. There is some uncertainty attributable to orbital reshuffling, but an overall structure is preserved. Also, the quoted boundaries represent transition zones rather than clean lines of demarcation, since each taxonomic class defines a roughly Gaussian distribution with a full width of distribution at half maximum value (FWHM) of ~1 AU.

There have been a couple of realizations in this area since *Bell et al.* (1986) was published. First, *Bottke et al.* (2000) showed that dynamical models of meteorite delivery indicate that most of the sampled meteorite parent bodies lie near or inside the 5:2 resonance or at 2.82 AU. Thus, according to *Bottke et al.* (2000), the inner belt has been sampled in much greater detail in terrestrial meteorite collections. Second, *Keil* (2000) showed that 80% of the ~135 meteorite types in terrestrial collections underwent either partial or complete melting. The remaining 20% either underwent metamorphism or aqueous alteration.

Hardersen (2003), as part of his doctoral dissertation with M. Gaffey, undertook a more detailed mineralogic mapping of the asteroid belt by taking into account the latest constraints from spectroscopy, dynamical models, and mineralogical studies of asteroids. He started by dividing asteroids into two classes: Class I asteroids, which allow diagnostic determinations of their mineralogy, and Class II asteroids, where the determination is nondiagnostic and open to interpretation. Out of the initial dataset of 152 asteroids, 44 were placed in the Class I group. The Class II asteroids include 32 S-type asteroids from *Gaffey et al.* (1993), 12 asteroids from *Jones et al.* (1990), and 56 C-, D- and T-type asteroids. Based on increasing heating, *Hardersen* (2003) divided the asteroids into four groups, Zones 0, I, II, III, and IV, respectively. *Hardersen* (2003) then tried to account for parent-body radii in evaluating the maximum temperatures possible. The results of *Hardersen* (2003) reinforce the idea of thermal stratification of the asteroid belt proposed by *Gradie and Tedesco* (1982) and *Bell et al.* (1989) except for regional heterogeneity superimposed on this overall trend.

The radial stratification of the asteroid belt would have been disturbed by the dynamical process(es) that removed most of its mass. This depletion may have blurred the primordial variation of compositions with heliocentric distance, but did not erase it. *Petit et al.* (2001) performed numerical integrations of orbits perturbed by Jupiter and massive planetary embryos. They showed that during excitation and depletion of the belt, surviving asteroids would have experienced changes in semimajor axis of several tenths of an AU, consistent with the observed transitions between regions dominated by taxonomic types S, C, and D. After depletion of the belt, the largest surviving bodies were too small to further alter asteroidal orbits by gravitational scattering. Later collisional evolution resulted in only minor dispersion of orbital elements of fragments, as shown by preservation of Hirayama families.

4. HELIOCENTRIC ZONATION IN A MODEL OF ALUMINUM-26 HEATING

Decrease of surface density of planetesimals from the inner to the outer asteroid belt coupled with time-dependent heating by ^{26}Al decay has been advocated to explain the heliocentric zonation (*Grimm and McSween,* 1993). The decrease in surface density with heliocentric distance results in a lower probability of bodies colliding with each other. The decreased likelihood of collisions in the outer belt translates to a slower accretion rate. This results in longer accretion times in the outer belt compared to the inner belt. Variable ^{26}Al heating would have occurred if the initial amount of ^{26}Al was uniform at the time of CAI formation, yet the amount incorporated into planetesimals was controlled by the time at which they accreted. *Grimm and McSween* (1993) assumed accretion to be instantaneous at a given distance (i.e., the asteroid formed at the instant the process of accretion was completed) on a timescale that increased from the inner to the outer belt. As a result, asteroids across the belt incorporated different amounts of live ^{26}Al and therefore underwent variable heating.

According to analytical models of planetary accretion (*Lissauer and Stewart,* 1993), the accretion rate is proportional to the mass in the accretion zone and inversely proportional to the square of the heliocentric distance and the width of the accretion zone (*Grimm and McSween,* 1993, equation (1)). Empirically, it can be stated that accretion time is proportional to the nth power of heliocentric distance a when n = 1.5–3, depending on the assumptions about the radial variations of surface density of the planetesimal swarm (*Grimm and McSween,* 1993), or $\tau = a^n$, for the usual assumption that surface density varied as $a^{-3/2}$, n = 3. Using the derived proportional relationship between accretion time and heliocentric distance, *Grimm and McSween* (1993) parameterized accretion time across the asteroid belt. According to the model, all asteroids at a specific heliocentric distance were assumed to have the same accretion time. The model assumed temperature-dependent properties of ice and rock and included thermodynamic effects of phase transitions. The silicate- and ice-melting isotherms were taken to be 1100°C and 0°C, and were assumed to have occurred at 2.7 AU and 3.4 AU, respectively. Two models were run: one without ice, and another with 30% ice. The results of the first case are illustrated in Fig. 2. The best fit was obtained for n = 3 and a formation time of 3.5 and 3 m.y. for asteroids at 2 AU, for anhydrous and hydrous models, respectively.

By assuming instantaneous accretion (i.e., asteroids with fixed sizes began thermal evolution after a time delay, rather than growing over that interval), however, *Grimm and McSween* (1993) neglected the dynamics of the accretion process as well as heat budgeting during planetesimal growth. As they acknowledged, the inclusion of radioactive heat *during* accretion would result in higher temperatures, thus the accretion times (peak temperatures) calculated bracket

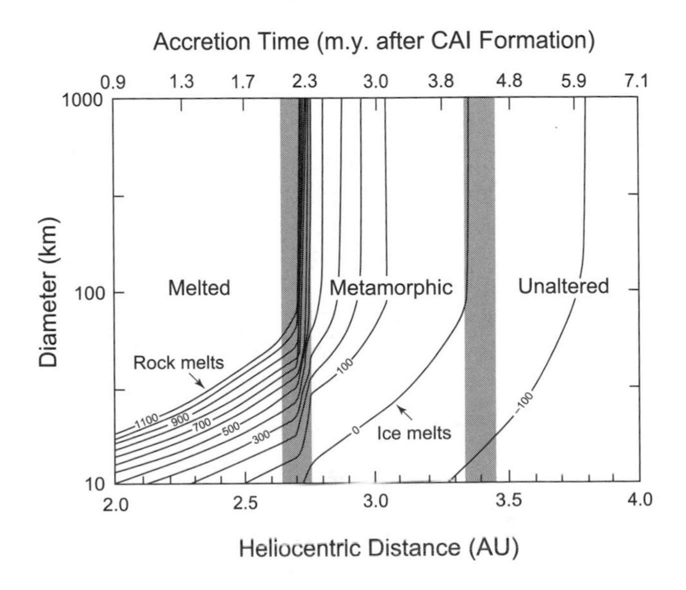

Fig. 2. Peak temperatures as a function of accretion time, heliocentric distance and asteroid radius for bodies that do not contain ice.

the upper bound (lower bound) of possibilities. Also, their results show that at 2 AU bodies of <~50 km radius would have a peak temperature of 870 K or less (the temperature threshold for type 4 metamorphism). Mass distributions from accretion codes (e.g., *Weidenschilling et al.,* 1997) show that bodies <~50 km radius would constitute a significant mass fraction at a particular heliocentric distance. Thus, the peak temperatures calculated for the inner asteroid belt by *Grimm and McSween* (1993) would yield a significant fraction of unaltered asteroids and not melted bodies as is observed. In principle, this problem could be alleviated by assuming a steeper gradient of accretion times with distance and shorter timescales in the main belt. However, the assumption of instantaneous accretion with a time delay is physically not justified.

5. HELIOCENTRIC ZONATION IN A MODEL OF ELECTROMAGNETIC INDUCTION HEATING

The discovery of the heliocentric zonation of the asteroid belt intuitively favored a heating agent with the Sun as an energy source. Thus, intuitively the intensity of solar wind should decrease with heliocentric distance, and this should moderate the heating that takes place with increasing heliocentric distance. Hence, electromagnetic induction heating models were favored initially to explain the thermal stratification of the asteroid belt. However, since electromagnetic induction heating is dependent on multiple parameters, the relationship between heliocentric distance and peak temperature is not that simple. To give an idea of the complexity, let us consider the peak temperatures attained as a function of heliocentric distance and planetesimal size. *Herbert et al.* (1991) made four runs with different parameter sets; in all these runs, bodies of a particular size bracket melted irrespective of heliocentric distance. What this means is, if the size of the body was larger or smaller than the size bracket for melting, irrespective of heliocentric distance, they would never melt. Also, this means that bodies of a particular size range would melt irrespective of heliocentric distance. Thus, it is important to highlight that electromagnetic induction heating proponents do not have an unambiguous model to explain the thermal stratification of the asteroid belt.

6. HELIOCENTRIC ZONATION IN A COMBINED ACCRETION-THERMAL MODEL

6.1. Methodology

As will be shown below, timescales for accretion of planetesimals in the asteroid region are on the same order as the decay timescale for ^{26}Al, thus the thermal evolution of parent bodies of meteorites is coupled to their accretionary history. The characteristic timescale for accretion increases with heliocentric distance, due to the lower surface density of accretable matter and longer orbital periods. These factors lead to differences in thermal histories between the inner and outer belt regions. In this section we evaluate the thermal histories of planetesimals as functions of size and heliocentric distance, using a numerical code for heat transfer and realistic growth rates derived from a numerical simulation of accretion. We use these results to derive a quantitative estimate of the mass fraction of material that experiences various degrees of thermal metamorphism at different locations within the asteroid belt.

6.2. Thermal Code

The thermal code tracks the heat budget of an accreting planetesimal, with production of heat by radioactive decay and transfer of heat by conduction, with a moving boundary condition as the body grows in size (*Ghosh et al.,* 2003). The dominant source of radiogenic heating is ^{26}Al, due to its short half-life and relatively high abundance; contributions by other nuclides are not taken into account since heat generated by them has been shown to be negligible (*Ghosh and McSween,* 1998). The initial abundance of ^{26}Al relative to ^{27}Al is 5×10^{-5} at the time of CAI formation (*Clayton,* 1999; *Huss et al.,* 2001). Accretion is assumed to initiate 2 m.y. after CAI formation (*Amelin and Krot,* 2002; *Russell et al.,* 2006). The rate of heat production decreases with time as the ^{26}Al decays. The initial composition and total Al abundance is taken to be that of H chondrites (*Ghosh and McSween,* 1998). Temperature-dependent specific heat capacity and the latent heat for melting are included (*Ghosh and McSween,* 1998, 1999). Where water ice is assumed to be present, the specific heat, latent heat of melting and vaporization, and heat of hydration (at 298 K) and dehydration (at 648 K) are incorporated from *Grimm and McSween* (1989). The thermal code is used to characterize the mass distribution of material with varying degrees of thermal metamorphism, according to the peak temperatures reached in the centers of planetesimals. Thus, we evaluate radius thresholds for planetesimals to remain unaltered (<870 K), metamorphose (870–1223 K), or undergo Fe-FeS melting (1223–1443 K), silicate melting (1443–2123 K), or complete melting (>2123 K). For bodies containing ice, the following thresholds are used: ice unmelted (<273 K), melted with aqueous alteration (273–648 K), and dehydrated (>648 K). For convenience, the respective bodies are notated as A, B, C, D, E for bodies without ice, and X, Y, Z for bodies with ice, respectively. The nebula is assumed to be optically thick, with temperature decreasing inversely with heliocentric distance. The presence of ice in the outer asteroid belt can be inferred from the spectral signatures of D and P asteroids (*Lebofsky et al.,* 1981), the textures of carbonaceous chondrites that are characteristic of aqueous alteration (*Bunch and Chang,* 1980), and the difficulty of condensing hydrous silicates directly from the nebula (*Prinn and Fegley,* 1987). The temperature is chosen so that water ice condenses (T = 160 K) at 3.4 AU. Thus, we introduce 30% by volume of ice in bodies that accrete beyond this distance (*Grimm and McSween,* 1989). The initial temperature of the kilometer-sized starting bodies and the temperature of the accreted material are assumed to be equal to the temperature of the ambient nebula at any given distance. Accretion of material that condensed in a turbulent nebula would result in a mixture of unequilibrated grains. The low internal pressures within asteroid-sized bodies and the porosity of most meteorites (*Weidenschilling and Cuzzi,* 2006) argue for accretion of unconsolidated material. Thus, we assume the accreting planetesimals to have a thermal diffusivity characteristic of unsintered porous dust, taken to be 10^{-9} m^2 s^{-1} (*Fountain and West,* 1970). Sintering is a complex process occurring at high temperatures (as high at 1050°C for basalt powder) and dependent on static and shock pressures, composition, and the presence or absence of volatiles in voids. As temperature increases, the interior is expected to undergo sintering, but the surficial layers that govern heat loss from the interior would remain unsintered. Sintering of dust into rock could, in principle, increase the diffusivity by 2 orders of magnitude. This process is ignored in the present study since we are interested in peak temperatures that are attained primarily in the first few million years of accretion. Test calculations show that maximum interior temperatures are similar for bodies with the diffusivity of solid rock, provided a regolith of low diffusivity is present, compared to a body composed of unsintered dust.

6.3. Accretion Code

The evolution of the size distribution of the planetesimal swarm is calculated using an accretion code that computes rates of collisions and velocity evolution due to gravitational stirring among bodies in the swarm (*Weidenschilling et al.,* 1997). This calculation gives the size distribution as functions of both heliocentric distance and time. In this simulation, the initial bodies are assumed to be 0.5 km in radius at the start of accretion. The mass is assumed to be distributed with surface density inversely proportional to the heliocentric distance. The surface density of solid matter is chosen to be 8.3 g cm^{-2} at 1 AU, corresponding to a total mass of the swarm of nearly 4 $M_\oplus$ between 2 and 4 AU. These parameters are consistent with a solar nebula about 1.5× the minimum mass. The calculation assumes no fragmentation; i.e., all collisions result in coagulation, which yields a lower limit to the accretion timescale. Growth is fastest in the inner region, where surface density is high and orbital periods are shortest. The formation of large bodies occurs as a wave that propagates outward with time. The size distribution evolves with runaway growth of large bodies in the swarm of smaller bodies. This produces an "oligarchy" of protoplanetary embryos of masses ~10^{26}–10^{27} g (*Kokubo and Ida,* 1993), with comparable masses and roughly uniform orbital spacing. After 1 m.y. of accretion (3 m.y. after CAI formation), the asteroid region contains bodies with radii ranging from the original 0.5 km to ~2500 km (Fig. 3). The resulting mass distribution is bimodal, with comparable fractions of the total mass in the embryos and a tail of smaller bodies with an approximate power-law distribution at radii less than about 100 km. The present asteroid belt contains much less mass than is present in this simulation (by 3 orders of magnitude), and with the exception of Ceres, Pallas, and Vesta, it consists of bodies of radii <~200 km. Most of the original mass, and all the larger embryos, are believed to have migrated to jovian resonances and were eventually removed by collisions with terrestrial planets or encounters with Jupiter (*Chambers and Wetherill,* 2001). Only the small remnant of a formerly much more numerous population is present in the current asteroid belt to act as the source of meteorites.

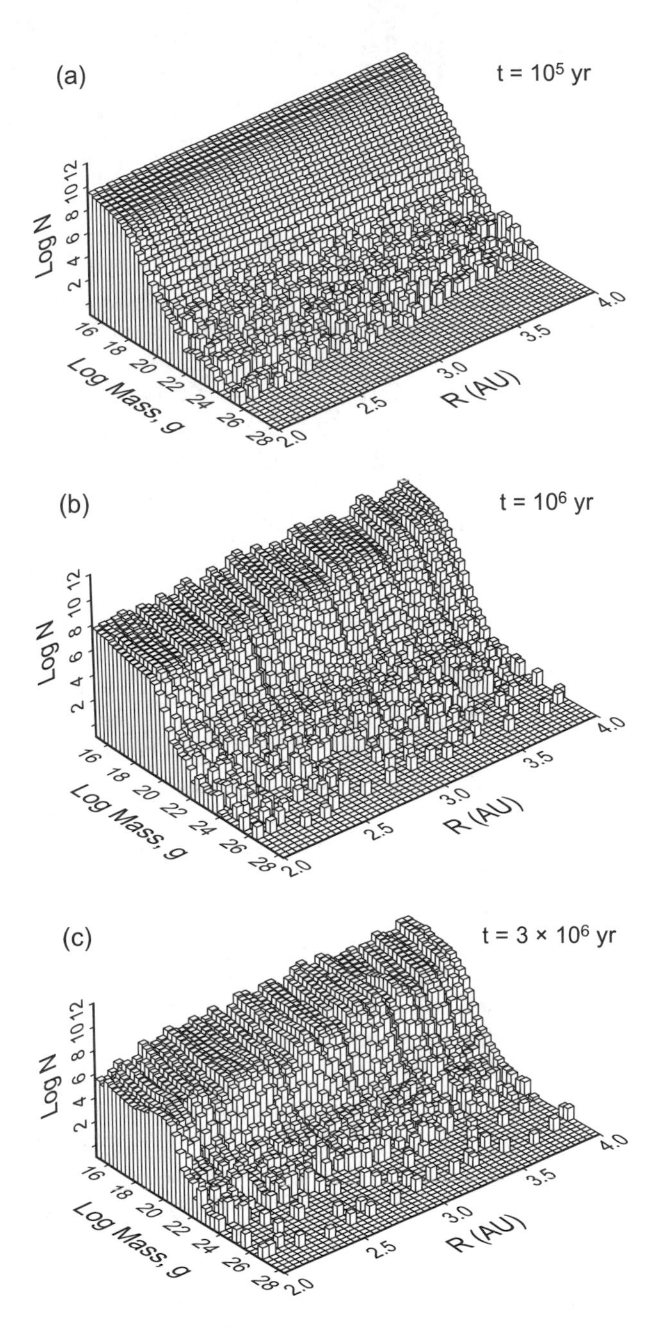

Fig. 3. Size distributions produced by the accretion code at model times of **(a)** 10^5, **(b)** 10^6, and **(c)** 3×10^6 yr from the start of accretion (assumed to be 2×10^6 yr after CAI formation).

6.4. Combining the Results

Growth rates at various heliocentric distances are estimated from the multizone accretion code, and these are used in the thermal evolution code to compute internal temperature profiles. At any given time, the size distribution contains bodies with a range of radii, which have grown at different rates due to the stochastic nature of collisional coagulation. The rate of change of radius of planetesimals (dR/dt) depends on the mass bin; on average, a 200-km body has grown at twice the rate of a 100-km body with the same semimajor axis, with the rate varying with heliocentric distance. For simplicity, the growth of any body's radius is assumed to be linear in time when computing its thermal profile. This assumption ignores the possibility that two bodies of the same size could be produced by different series of collisions, with growth rates that differ in detail. Such stochastic effects, and differences in heliocentric distance, may explain why Vesta experienced melting while the larger Ceres does not appear to be differentiated. In comparing the results of simulations of early evolution of the belt and the currently observable asteroids, it is necessary to consider both the depletion in mass and the effects of ~4.5 G.y. of collisional evolution that occurred after eccentricities and inclinations were stirred up to their current high values. All the large embryos, which would have been melted by ^{26}Al, have been removed. Of the remainder, a significant fraction of mass would have been in the smallest bodies (radii <20 km). Collisional evolution is believed to have preferentially removed these small asteroids. Simulations of model asteroid populations predict that bodies of initial radius less than a specific threshold, ~10–20 km, would be destroyed by collisional fragmentation over the age of the solar system (*Davis et al.,* 1989). The present population of small asteroids in this size range consists of fragments of larger parent bodies, of original radii greater than the radius threshold for destruction by fragmentation. This collisional evolution and destruction of small primordial asteroids accounts for the near-ubiquity of thermal and aqueous alteration in meteorites and the lack of truly pristine material. Thus, *Ghosh et al.* (2001) assume that planetesimals of radius <14.5 km were destroyed by fragmentation, in computing the composition of the asteroid belt as a function of heliocentric distance. Similarly, larger asteroids >240 km are assumed to have migrated to the terrestrial planet region.

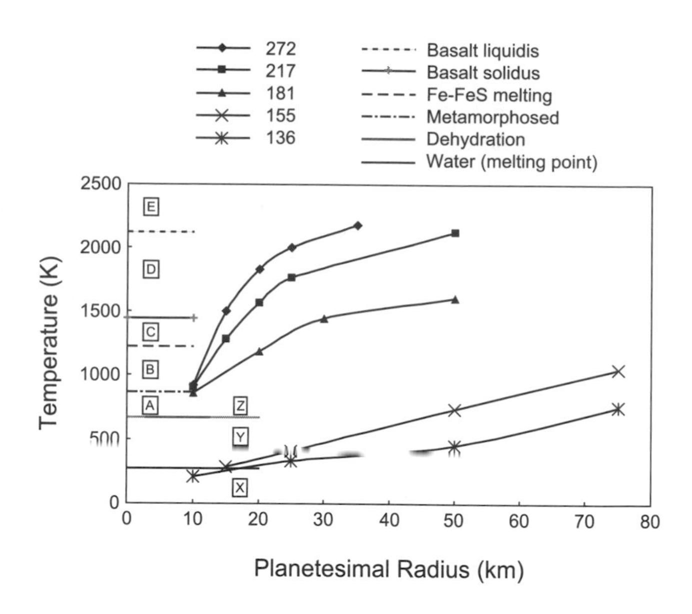

Fig. 4. Plots of maximum temperature realized in the interior of planetesimal against radius of the body. Accretion rates are constrained from *Weidenschilling et al.* (1997). An optically thick nebula with temperature inversely proportional to heliocentric distance is assumed, thus initial temperature of the planetesimals is assumed to decrease from 272 K at 2 AU to 136 AU at 4 AU. Condensation of ice is assumed to take place at approximately 3.4 AU, thus all planetesimals outward of 3.4 AU are assumed to contain 30% ice. The maximum temperature attained at a planetesimal center is calculated as a function of planetesimal radius for each heliocentric distance. These plots are then used to demarcate planetesimals at each heliocentric distance according to thermal regimes, i.e., the radius ranges of planetesimal types A–E and X–Z are evaluated. Planetesimals types A, B, C, D, and E are defined to represent the following thermal regimes: unaltered (<870 K), metamorphosed (870–1223 K), FeS melted (1223–1443 K), silicate melted (1443–2123 K), and totally molten (>2123 K). X, Y, Z indicate a corresponding scale for ice-bearing planetesimals: ice unmelted (<273 K), ice melted with subsequent aqueous alteration (273–648 K), and dehydrated (>648 K), respectively.

6.5. Results

At a specific heliocentric distance, the maximum temperature realized in the asteroid center increases with body size (Fig. 4). This is due to the decrease in the surface area to volume ratio as the size of the asteroid increases. On the other hand, peak temperatures for an asteroid of a given size decrease with heliocentric distance (Fig. 4). Using the data of peak temperature as a function of heliocentric distance and body size from Fig. 4, *Ghosh et al.* (2001) computed radius windows for planetesimal types A, B, C, D, E, X, Y, and Z. Using the radius windows, the mass distributions of the multizone accretion code are categorized according to the thermal history. This result translates into radial zoning of compositions, summarized in Fig. 5. The innermost asteroid belt shows a predominance of type D (or melted) planetesimals. At 3 AU, a predominance of type B (or metamorphosed) planetesimals is observed. Correspondingly, planetesimals with ice in the outer asteroid belt (3.4–4 AU) either undergo aqueous alteration or remain unaltered, if internal temperatures do not rise above the melting point of water. The compositional boundaries are not sharp, due to the distribution of sizes at any distance; these would be further blurred by orbital migration due to gravitational scattering that occurred before the removal of the larger protoplanetary embryos.

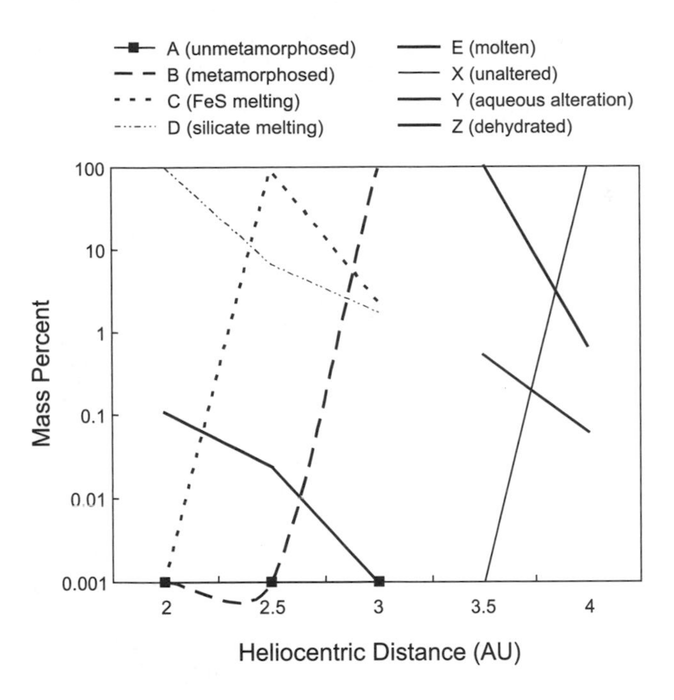

Fig. 5. Mass proportion of various planetesimal types as a function of heliocentric distance. Using the radius ranges for each planetesimal type calculated from Fig. 2, the mass distribution of the multizone accretion code is characterized in terms of planetesimal types. Bodies of radius <14.5 km are assumed to be destroyed by fragmentation. Bodies of radius >231 km are assumed to migrate to the terrestrial planet region. The histogram shows a predominance of melted planetesimals (type D) at 2 AU, FeS melted planetesimals (type C) at 2.5 AU, metamorphosed planetesimals (type B) at 3 AU, ice melted with subsequent aqueous alteration (type Y) at 3.5 AU, and ice unmelted (type X) at 4 AU.

6.6. Discussion

The following discussion examines each of the factors in the model to discern whether they are required to simulate (or, in contrast, serve to merely facilitate) the observed heliocentric zonation in the asteroid belt.

6.6.1. Fragmentation. The nature of accretion in the asteroid belt is such that a significant proportion of the mass remains in smaller size bins after the growth of large embryos. In the thermal model, it is not possible to sustain metamorphism in bodies <~10 km radius, due to the high surface area to volume ratio, which results in efficient removal of heat by radiation. Unless most of these bodies were removed by fragmentation, a substantial amount of the mass in all parts of the belt would consist of primitive, unaltered material. Yet the inner asteroid belt, according to observation, is dominated by melted or metamorphosed bodies. Thus, the only way to reproduce the radial stratification is to remove the smaller bodies by fragmentation.

6.6.2. Effect of nebular temperature. A temperature gradient is expected between the inner and outer asteroid belt, depending upon the assumed total mass of the nebula, its evolutionary stage, and the rate and scale of heat and mass transport by convective processes (*Boss,* 1996; *Cassen,* 1994). A temperature gradient is required to explain the presence of ice in the outer asteroid belt. The presence of ice in the present model is critical in reproducing the observed stratification, as explained in the next paragraph. However, the nature of the decrease in temperature with heliocentric distance is not a critical factor. In the thermal evolution model, we assume the initial bodies and accreted material to have temperatures corresponding to ambient conditions in an optically thick nebula, with T inversely proportional to heliocentric distance and with the ice line (T = 160 K) at 3.4 AU. If instead we assume the nebula to be optically thin, with T varying as $R^{-0.5}$, and keeping the same temperature of 160 K at 3.4 AU, the ambient temperature at 2 AU decreases from 272 K to 208 K. This decrease in temperature causes only a slight shift in peak temperature realized at the asteroid centers (Fig. 4), and a small change in the radius thresholds of various planetesimal types. In other words, the temperature gradient must allow the presence of ice in the outer asteroid belt; however, the exact nature of temperature decrease is not a critical factor in explaining the stratification.

6.6.3. Effect of ice. The incorporation of ice in asteroids produces thermal buffering because of greater specific heats of ice and water relative to silicates, dilution of the mass fraction of ^{26}Al, heats of melting and vaporization, and the endothermic dehydration reaction (*Grimm and McSween,* 1989). Approximately 3 kJ/g of energy is required to melt and vaporize ice, compared to ~1.6 kJ/g required to heat silicates from 200 to 1400 K (*Wood and Pellas,* 1991). In the absence of ice, the peak temperature for a 10-km-radius planetesimal at 4 AU would have exceeded 700 K, greater than the dehydration temperature of serpentine at ~648 K. This is contrary to spectral observations, which show evidence of aqueous alteration in the outer belt and the presence of water of hydration. Thus, inclusion of ice in the thermal model is critical in terms of reproducing the observed stratification.

6.6.4. Effect of sintering. A model run assuming a planetesimal to have the diffusivity of compacted rock calculates the radius threshold for melting and metamorphism at 2 AU of 51 km and 55 km, respectively. Due to the nature of the mass distribution that retains the median mass in smaller size bins, this translates into a predominance of unaltered bodies in the inner asteroid belt. It is not possible to offset this result unless the radius threshold for fragmentation is increased to ~50 km. However, literature on collisional evolution of asteroids (*Davis et al.,* 1989) indicates that bodies with primordial radii larger than 20 km should have survived. Therefore, unsintered bodies, or at least a significant thickness of regolith with low thermal diffusivity, is required.

6.6.5. Effect of the accretion process and the nature of aluminum-26 heating. The characteristic growth rate is proportional to the surface density of accretable matter and the local Kepler frequency (*Lissauer and Stewart,* 1993). Thus, the physics of the accretion process requires faster accretion rates at smaller heliocentric distances; for the nominal model this difference is more than a factor of 5 between the inner and outer edges of the belt. Heating by ^{26}Al with a half-life of 0.7 m.y. acts in conjunction with faster accretion times in the inner asteroid belt to produce higher temperatures within planetesimals in the inner belt (*Grimm and McSween,* 1993). Heat generated during the accretion process decreases by more than an order of magnitude in 3 m.y., thus differences in accretion rates result in greater heat deposition for planetesimals accreting in the inner belt. This in turn results in greater heating in the inner belt in contrast to the outer belt. Had the accretion timescale been much longer or shorter than the half-life of the dominant heat source, the difference in thermal histories would be much smaller.

7. CONCLUSIONS AND FUTURE WORK

Many aspects of the asteroid belt evolution remain unclear. It is possible that the dynamical evolution of the belt, for example, did disrupt the primordial stratification. Thus, it is possible that what we see today does not resemble the initial configuration. However, at a minimum, it is well established that we have melted or metamorphosed asteroids in the inner belt and ice in the outer belt, thus there was at least a rudimentary form of thermal stratification.

Given the uncertainty in parameters of electromagnetic induction heating models (*Wood and Pellas,* 1991), it is probably premature to attribute the stratification to the effect of this mechanism (*Sonett et al.,* 1968; *Herbert,* 1989). Indeed, in certain induction models (*Herbert,* 1989), the relationship between heating and heliocentric distance is so weak that melting is produced throughout the belt, if at all. Also, in a model of electromagnetic induction heating, where planetesimals of a certain size bracket undergo heating (*Herbert,* 1989), smaller planetesimals would remain unmetamorphosed. This will produce the problem of an overabundance of unaltered bodies in the inner belt (although the subsequent collisional depletion of small bodies would also aid this model).

Thermal stratification of the asteroid belt was the result of the interplay of multiple processes. Integrated thermal-accretion models are the first step in trying to realistically simulate how the stratification was produced. Present models (*Ghosh et al.,* 2001) tie the structure of the asteroid belt to a dynamically plausible accretion history, and show that a combination of several factors is required to reproduce its radial compositional stratification. The factors critical to reproducing the stratification are slower accretion in the outer belt combined with decreasing abundance of ^{26}Al with time and low thermal diffusivity of unsintered planetesimals. Temperatures in the asteroid belt should have allowed for condensation of ice in the outer belt; however, the actual temperature gradient is not critical. It is also necessary to allow for the removal of large planetary embryos (as well as most of the original mass at all sizes) by gravitational scattering after the accretionary era, and the selective collisional depletion of small planetesimals of primitive composition over the lifetime of the solar system. A significant improvement in our understanding of the processes that cause the stratification can be attained if nebular models can be anchored to CAI formation. This would enable an integration of nebular models with thermal and accretion models. In addition to providing accurate temperatures of the accreting mass and the ambience, it will constrain the rate at which solids condense in the terrestrial planet region. This is turn will constrain accretion models, which at present assume that the entire mass in the terrestrial planet region existed at the start of the simulation.

Acknowledgments. This work was supported by NASA grants NASG-54387 (H.Y.M.), Center for Innovative Technology Challenge Grant (A.G.), NAGW-4219 (S.J.W.), and NAG5-4766 (A.E.R.). We thank the Joint Institute for Computational Science at the University of Tennessee and the NASA HPCC/ESS Program for the use of computer time.

REFERENCES

Akridge G., Benoit P. H., and Sears D. W. G. (1998) Regolith and megaregolith formation of H-chondrites: Thermal constraints on the parent body. *Icarus, 132,* 185–194.

Amelin Y. and Krot A. N. (2002) Pb isotopic age of chondrules from CR carbonaceous chondrites ACFER 059. *Meteoritics & Planet. Sci., 37,* A12.

Baker A. J. and Pepper D. W. (1991) *Finite Element 1-2-3.* McGraw-Hill, New York.

Bell J. F. (1986) Mineralogical evolution of meteorite parent bodies (abstract). In *Lunar and Planetary Science XVII,* p. 985. Lunar and Planetary Institute, Houston.

Bell J. F., Davis D. R., Hartmann W. K., and Gaffey M. J. (1989) Asteroids: The big picture. In *Asteroids II* (R. P. Binzel et al., eds.), pp. 921–945. Univ. of Arizona, Tucson.

Bennett M. E. III and McSween H. Y. Jr. (1996) Revised model calculations for the thermal histories of ordinary chondrite parent bodies. *Meteoritics, 31,* 783.

Boss A. P. (1996) Evolution of the solar nebula. III. Protoplanetary disks undergoing mass accretion. *Astrophys. J., 469,* 906.

Bottke W. F. Jr., Rubincam D. P., and Burns J. A. (2000) Dynamical evolution of main belt asteroids: Numerical simulations incorporating planetary perturbations and Yarkovsky thermal forces. *Icarus, 145,* 301–331.

Bunch T. E. and Chang S. (1980) Carbonaceous chondrites. II — Carbonaceous chondrite phyllosilicates and light element geochemistry as indicators of parent body processes and surface conditions *Geochim. Cosmochim. Acta, 44,* 1543.

Cassen P. (1994) Utilitarian models of the solar nebula. *Icarus, 112,* 405–429.

Chambers J. E. and Wetherill G. W. (2001) Planets in the asteroid belt. *Meteoritics & Planet. Sci., 36,* 381–399.

Clayton D. D. (1980) Chemical energy in cold-cloud aggregates — The origin of meteoritic chondrules. *Astrophys. J. Lett., 239,* L37–L41.

Clayton R. N. (1999) Oxygen isotope studies of carbonaceous chondrites. *Geochim. Cosmochim. Acta, 63,* 2089.

Davis D. R., Weidenschilling S. J., Farinella P., Paolicchi P., and Binzel R. P. (1989) Asteroid collisional history: Effects on sizes and spin. In *Asteroids II* (R. P. Binzel et al., eds.), pp. 805–826. Univ. of Arizona, Tucson.

DeCampli W. M. (1981) T Tauri winds. *Astrophys. J., 244,* 124–146.

Dixon E. T., Bogard D. D., and Rubin A. E. (2003) ^{39}Ar-^{40}Ar evidence for an ~4.26 Ga impact heating event on the LL parent body (abstract). In *Lunar and Planetary Science XXXIV,* Abstract #1108. Lunar and Planetary Institute, Houston (CD-ROM).

Edwards S., Cabrit D., Strom S. E., Heyer I., Strom K. M., and Anderson E. (1987) Forbidden line and Hα profiles in T Tauri star spectra: A probe of anisotropic mass outflows and circumstellar disks. *Astrophys. J., 321,* 473–495.

Fountain J. A. and West E. A. (1970) Thermal conductivity of particulate basalt as a function of density in simulated lunar and martian environments. *J. Geophys. Res., 75,* 4063–4069.

Gaffey M. J., Burbine T. H., and Binzel R. P. (1993). Asteroid spectroscopy: Progress and perspectives. *Meteoritics, 28,* 161–187.

Ghosh A. (1997) A thermal model for the differentiation of asteroid 4 Vesta based on radiogenic and collisional heating. Ph.D. dissertation, Univ. of Tennessee, Knoxville.

Ghosh A. and McSween H. Y. Jr. (1998) A thermal model for the differentiation of asteroid 4 Vesta, based on radiogenic heating. *Icarus, 134,* 187–206.

Ghosh A. and McSween H. Y. Jr. (1999) Temperature dependence of specific heat capacity and its effect on asteroid thermal models. *Meteoritics & Planet. Sci., 34,* 121–127.

Ghosh A. and McSween H. Y. Jr. (2000) The effect of incremental accretion on the thermal modeling of asteroid 6 Hebe. *Meteoritics & Planet. Sci., 35,* A59.

Ghosh A., Weidenschilling S. J., and McSween H. Y. Jr. (2001) Thermal consequences of the multizone accretion code on the structure of the asteroid belt (abstract). In *Lunar and Planetary Science XXXII,* Abstract #1760. Lunar and Planetary Institute, Houston (CD-ROM).

Ghosh A., Weidenschilling S. J., and McSween H. Y. Jr. (2003) Importance of the accretion process in asteroidal thermal evolution: 6 Hebe as an example. *Meteoritics & Planet. Sci., 38,* 711–724.

Gradie J. C. and Tedesco E. F. (1982) Compositional structure of the asteroid belt. *Science, 216,* 1405–1407.

Grimm R. E. and McSween H. Y. Jr. (1989) Water and the thermal evolution of carbonaceous chondrite parent bodies. *Icarus, 82,* 244–280.

Grimm R. E. and McSween H. Y. Jr. (1993) Heliocentric zoning of the asteroid belt by aluminum-26 heating. *Science, 259,* 653–655.

Hardersen P. S. (2003) Near-IR reflectance spectroscopy of asteroids and study of the thermal history of the main asteroid belt. Ph.D. dissertation, Rensselaer Polytechnic Institute, Troy, New York. 387 pp.

Herbert F. (1989) Primoridal electrical induction heating of asteroids. *Icarus, 78,* 402–410.

Herbert F., Sonett C. P., and Gaffey M. (1991) Protoplanetary thermal metamorphism: The hypothesis of electromagnetic induction in the protosolar wind. In *The Sun in Time* (C. P. Sonett et al., eds.), pp. 710–739. Univ. of Arizona, Tucson.

Huss G. R., MacPherson G. J., Wasserburg G. J., Russell S. S., and Srinivasan G. (2001) Aluminum-26 in calcium-aluminum-rich inclusions and chondrules from unequilibrated ordinary chondrites. *Meteoritics & Planet. Sci., 36,* 975–1025.

Jones T. D., Lebofsky L. A., Lewis J. S., and Marley M. S. (1990). The composition and origin of the C, P, and D asteroids: Water as a tracer of thermal evolution in the outer belt. *Icarus, 88,* 172–192.

Keil K. (2000) Thermal alteration of asteroids: Evidence from meteorites. *Planet. Space Sci., 48,* 887–903.

Kokubo E. and Ida S. (1993) Velocity evolution of disk stars due to gravitational scattering by giant molecular clouds. *Back to the Galaxy, Proceedings of the 3rd October Astrophysics Conference* (S. S. Hold and F. Verter, eds.), p. 380. AIP Conference Proceedings 278, American Institute of Physics, New York.

Kuhi L. V. (1964) Mass loss from T Tauri stars. *Astrophys. J., 140,* 1409.

LaTourrette T. and Wasserburg G. J. (1998) Mg diffusion in anorthite: Implications for the formation of early solar system planetesimals *Earth. Planet. Sci. Lett., 158,* 91–108.

Lebofsky L. A., Feierberg M. A., Tokunaga A. T., Larson H. P., and Johnson J. R. (1981) *Icarus, 48,* 453.

Lee T., Papanastassiou D. A., and Wasserburg G. J. (1976) Demonstration of ^{26}Mg excess in Allende and evidence for ^{26}Al. *Geophys. Res. Lett., 3,* 41–44.

Lissauer J. J. and Stewart G. R. (1993) Planetary accretion in circumstellar disks. In *Planets Around Pulsars*, pp. 217–233. California Institute of Technology, Pasadena.

Macpherson G., Davis A. M., and Zinner E. K. (1995) The distribution of aluminum-26 in the early solar system — A reappraisal. *Meteoritics, 30,* 365–386.

McSween H. Y. Jr., Ghosh A., Grimm R. E., Wilson L., and Young E. D. (2003) Thermal evolution models of asteroids. In *Asteroids III* (W. F. Bottke Jr. et al., eds.), pp. 559–571. Univ. of Arizona, Tucson.

Melosh H. J. (1990) Giant impacts and the thermal state of the early Earth. In *Origin of the Earth* (H. E. Newsom and J. H. Jones, eds.), pp. 69–84. Oxford Univ., New York.

Miyamoto M., Fujii N., and Takeda H. (1981) Ordinary chondrite parent body: An internal heating model. *Proc. Lunar Planet. Sci. 12A,* pp. 1145–1152.

Mostefaoui S., Lugmair G. W., and Hoppe P. (2005) ^{60}Fe: A heat source for planetary differentiation from a nearby supernova explosion. *Astrophys. J., 625,* 271–277.

Petit J.-M., Morbidelli A., and Chambers J. (2001) The primordial excitation and clearing of the asteroid belt. *Icarus, 153,* 338–347.

Prinn R. G. and Fegley B. Jr. (1987) The atmospheres of Venus, Earth, and Mars — A critical comparison. *Annu. Rev. Earth Planet. Sci., 15,* 171.

Rubin A. E. (2004) Postshock annealing and postannealing shock in equiliubrated ordinary chondrites: Implications for the thermal and shock histories of chondritic asteroids. *Geochim. Cosmochim. Acta, 68,* 673–689.

Russell S. S., Hartmann L., Cuzzi J., Krot A. N., Gounelle M., and Weidenschilling S. (2006) Timescales of the solar protoplanetary disk. In *Meteorites and the Early Solar System II* (D. S. Lauretta and H. Y. McSween Jr., eds.), this volume. Univ. of Arizona, Tucson.

Shimazu H. and Terasawa T. (1995) Electromagnetic induction heating of meteorite parent bodies by the primordial solar wind. *J. Geophys. Res., 100,* 16,923–16,930.

Shukolyukov A. and Lugmair G. W. (1993) Live iron-60 in the early solar system. *Science, 259,* 1138–1142.

Sonett C. P. and Colburn D. S. (1968) The principle of solar wind induced planetary dynamos. *Phys. Earth Planet. Inter., 1,* 326–346.

Sonnett C. P., Colburn D. S., and Schwartz K. (1968) Electrical heating of meteorite parent bodies and planets by dynamo induction from a premain sequence T Tauri "solar wind." *Nature, 219,* 924–926.

Srinivasan G., Goswami J. N., and Bhandari N. (1999) ^{26}Al in eucrite Piplia Kalan: Plausible heat source and formation chronology. *Science, 284,* 1348–1350.

Urey H. (1955) The cosmic abundances of potassium, uranium, and thorium and the heat balances of the Earth, the Moon and Mars. *Proc. Natl. Acad. Sci. U.S., 41,* 127–144.

Wasson J. T. (1974) *Meteorites: Classification and Properties.* Springer-Verlag, New York. 327 pp.

Weidenschilling S. J. and Cuzzi J. N. (2006) Accretion dynamics and timescales: Relation to chondrites. In *Meteorites and the Early Solar System II* (D. S. Lauretta and H. Y. McSween Jr., eds.), this volume. Univ. of Arizona, Tucson.

Weidenschilling S. J., Spaute D., Davis D. R., Mazari F., and Ohtsuki K. (1997) Accretional evolution of a planetesimal swarm 2. The terrrestrial zone. *Icarus, 128,* 429–455.

Wood J. A. (1979) Review of the metallographic cooling rates of meteorites and a new model for the planetesimals in which they formed. In *Asteroids* (T. Gehrels, eds.), pp. 740–760. Univ. of Arizona, Tucson.

Wood J. A. and Pellas P. (1991) What heated the meteorite parent planets. In *The Sun in Time* (C. P. Sonett et al., eds.), pp. 740–760. Univ. of Arizona, Tucson.

Zellner B. (1979) Asteroid taxonomy and the distribution of the compositional types. In *Asteroids* (T. Gehrels, eds.), pp. 783–806. Univ. of Arizona, Tucson.

Thermal Metamorphism in Chondrites

Gary R. Huss
Arizona State University

Alan E. Rubin
University of California, Los Angeles

Jeffrey N. Grossman
U.S. Geological Survey

Thermal metamorphism has affected most chondritic meteorites to some extent, and in most ordinary chondrites, some carbonaceous chondrites, and many enstatite chondrites it has significantly modified the primary characteristics of the meteorites. Metamorphic grade, as described by the petrologic type, is one axis of the current two-dimensional system for classifying chondrites. Many changes are produced during thermal metamorphism, including textural integration and recrystallization, mineral equilibration, destruction of primary minerals, and growth of secondary minerals. Understanding these changes is critical if one hopes to infer the conditions under which chondrites originally formed. In addition to summarizing metamorphic changes, we also discuss temperatures, oxidation states, and possible heat sources for metamorphism.

1. INTRODUCTION

Once the various chondrite components had formed, they accreted into meteorite parent bodies. Starting with accretion itself, secondary processes such as shock metamorphism, thermal metamorphism, and aqueous alteration began to modify the primary chondritic components. The nature and pervasiveness of secondary processing must be understood before currently observed properties of chondrites and their components can be used to infer the nature of the solar nebula. In this chapter, we discuss the effects of thermal metamorphism on chondritic meteorites and the conditions under which it occurred. Thermal metamorphism affected most chondrite classes, and its effects play a major role in the classification schemes that have been developed for chondrites (see *Weisberg et al.,* 2006).

Thermal metamorphism is a broad topic without sharp boundaries. For example, heating due to the passage of a shock wave potentially affects components of a meteorite in the same way as heat provided by gravitational compression or radioactive decay. In addition, the style of thermal metamorphism is affected by the presence of water, so it is difficult to set a firm boundary between aqueous alteration and thermal metamorphism. In this chapter, we will mainly discuss the changes induced by heating on the meteorite parent bodies over a long timescale. The pervasive aqueous alteration that affected CI and CM chondrites, and to a lesser extent CR and other carbonaceous chondrites, is covered in the next chapter. However, we will discuss the effects of small amounts of water during the metamorphism of type 3 ordinary chondrites and CO3 chondrites. Although shock metamorphism is not covered in detail, we will mention some of its effects in order to distinguish them as far as possible from the effects of thermal metamorphism.

The current classification system for chondrites is based on the scheme proposed by *Van Schmus and Wood* (1967). These authors divided chondrites into compositional groups and divided the compositional groups into six petrologic types. Petrologic types 3–6 were viewed as reflecting the level of thermal metamorphism, whereas types 1 and 2 were used to designate unmetamorphosed meteorites unrelated to types 3–6. Types 1 and 2 have experienced different degrees of aqueous alteration (e.g., *McSween,* 1979). All well-defined chondrite groups contain little-metamorphosed members. However, many groups of carbonaceous chondrites have no known members of petrologic type ≥4 (e.g., CI, CM, CO, CV, CR, CH).

The petrologic types in the original *Van Schmus and Wood* (1967) classification have been modified in several ways in the nearly 40 years since they were proposed. A few dozen meteorites have been classified as type 7 [e.g., Shaw (*Dodd et al.,* 1975), Happy Canyon (*Olsen et al.,* 1976)]. Many of these meteorites probably are (or contain) impact melts (e.g., *Taylor et al.,* 1979; *McCoy et al.,* 1995) and thus are not part of the continuum of thermal metamorphism. For these meteorites, a type 7 designation is inappropriate, since shock stage is classified independently (*Stöffler et al.,* 1991; *Scott et al.,* 1992). Other chondritic meteorites have been heated to the point of partial melting and are possibly metamorphic rocks of higher grade than type 6. These are known as "primitive achondrites," and are beyond the scope of this chapter. But there are a few highly metamorphosed chondrites, such as Lewis Cliff 88663, that satisfy the *Dodd et al.* (1975) criteria for type 7 without evidence for shock melting or partial melting (*Mittlefehldt and Lindstrom,* 2001). These can conceivably be called type 7 chondrites, although they do fit comfortably within the *Van Schmus and Wood* (1967) definition of type 6. We choose not to make any distinction between these in the following discussion.

A second extension of the *Van Schmus and Wood* (1967) classification involves dividing type 3 into subtypes, 3.0 to 3.9, where type 3.0 designates meteorites that have experienced very little metamorphism and 3.9 designates those that have nearly reached the degree of chemical equilibrium associated with type 4. This scheme was originally proposed by for ordinary chondrites by *Sears et al.* (1980), and has since been extended to several other chondrite groups, including CO (*Scott and Jones,* 1990; *Sears et al.,* 1991a; *Chizmadia et al.,* 2002), CV (e.g., *Guimon et al.,* 1995; *Bonal et al.,* 2004), and R (e.g., *Rubin and Kallemeyn,* 1989; *Bischoff,* 2000).

In this chapter, we discuss the chemical and mineralogical changes in chondrites of all groups, first during the early stages of metamorphism from types 3.0 to 3.9, and then at higher grades represented by types 4–6, as metamorphism proceeds to near the point of partial melting. We conclude with a discussion of the conditions and timescales of metamorphism and the processes responsible for the heating.

2. LOW-DEGREE METAMORPHISM (TYPE 3 CHONDRITES)

2.1. Type 3 Ordinary Chondrites

Type 3 ordinary chondrites, although often considered unmetamorphosed, show a wide range of metamorphic effects. *Wood* (1967) identified a population of primitive chondrites, including several ordinary chondrites, that do not show a coherent relationship between taenite composition and grain size (indicative of slow cooling). He concluded that these were "more lightly touched by metamorphism than the other[s]." *Dodd et al.* (1967) and *Dodd* (1969) described the characteristics of several ordinary chondrites designated as type 3 by *Van Schmus and Wood* (1967) that seemed less equilibrated than others. *Huss et al.* (1978, 1981) attempted to arrange type 3 ordinary chondrites in a metamorphic sequence based on the degree of recrystallization and homogenization of olivine compositions in the fine-grained matrix. Shortly thereafter, *Afiattalab and Wasson* (1980) observed that the Co contents of metal grains also vary systematically with degree of metamorphic heating. *Sears et al.* (1980) discovered that bulk thermoluminescence (TL) sensitivity is an excellent indicator of metamorphic grade in type 3 ordinary chondrites; they used this parameter to define the modern petrologic types, 3.0–3.9 (see also *Sears et al.,* 1982a; *Sears et al.,* 1991b; *Benoit et al.,* 2002). *Grossman and Brearley* (2005) studied the effects of very low degrees of metamorphism on olivine and glass composition, opaque mineralogy, and matrix composition in order to define petrologic types 3.00–3.15. A summary of the modern petrologic classification scheme for ordinary chondrites and the characteristic properties of each type is given in Table 1.

2.1.1. Physical effects (thermoluminescence and cathodoluminescence). Thermoluminescence sensitivity is the amount of TL that is induced in a sample by irradiation under standard conditions. Thermoluminescence sensitivity varies by ~3 orders of magnitude among type 3 ordinary chondrites and is correlated with all metamorphic indicators noted by earlier workers (Table 1), including degree of matrix recrystallization, olivine heterogeneity, matrix and metal compositions, and bulk volatile contents (C and ^{36}Ar) (*Sears et al.,* 1980, 1982a). Differences in TL sensitivity among type 3 ordinary chondrites are due mainly to crystallization of feldspar during metamorphism, primarily via devitrification of chondrule glass (*Sears et al.,* 1980, 1982a), but also via recrystallization of matrix (*Huss et al.,* 1981).

The shape of the TL-sensitivity glow curve (TL sensitivity plotted vs. the temperature of emission) varies as a function of metamorphic grade in type 3 ordinary chondrites (*Sears et al.,* 1980, 1982a). Below type 3.3, the shape of the glow curve is irregular. Between 3.3 and 3.5, a well-defined peak develops. Between types 3.5 and 3.6, this peak broadens and moves to higher temperature, probably due to the order/disorder transition of feldspar that occurs at 500°–600°C (*Guimon et al.,* 1985; *Sears et al.,* 1991b). The irregular shape of the glow curve below type 3.3 may reflect dominance of phosphors other than feldspar in meteorites whose metamorphic temperatures were too low for extensive feldspar growth (e.g., *Guimon et al.,* 1988).

Cathodoluminescence (CL) is visible light emitted by minerals in response to an electron beam. The CL properties of type 3 ordinary chondrites also vary greatly in response to metamorphism [see photomicrographs in *Sears et al.* (1990)]. Low-FeO chondrules containing glass with bright yellow CL are found only in the lowest petrologic types, such as Semarkona (LL3.00), Bishunpur (LL3.15), and Krymka (LL3.2) [petrologic types from *Grossman and Brearley* (2005)]. By type 3.4, the mesostasis in low-FeO chondrules shows only blue CL (*DeHart et al.,* 1992; *Sears et al.,* 1995). With increasing petrologic type, the proportion of chondrules showing blue CL increases, and the matrix also begins to luminesce blue. This behavior parallels that shown by TL and is due to the growth of blue-luminescing albitic plagioclase (*Grossman and Brearley,* 2005). Olivine showing red CL (forsterite) is common in petrologic types ≤3.4, but disappears at higher grades, reflecting the entry of FeO during metamorphism (*Sears et al.,* 1990). Fine-grained matrix forsterite is lost by type ~3.2, while coarser chondrule forsterite is gone by type ~3.6. The matrix of the least-metamorphosed ordinary chondrite,

TABLE 1. Summary of petrologic types of ordinary chondrites.

Category	Parameter	Petrologic types 3.0	3.1	3.2	3.3	3.4	3.5	3.6	3.7	3.8	3.9	4	5	6	Discussed in Section
Bulk composition	Carbon (wt%)	0.3–0.6					0.2–0.5			<0.3					§2.1.5 MetBase*
	H_2O (wt%)	1.0–2.5							<1.0						
	Indium (ppb)	8–100										0.1–8			
	$^{36}Ar_p$ (10^{-8} g^{-1} STP)	45–60		15–50					5–15			0.5–5		0.2–2	
Texture	Chondrule textural equilibration	None						Incipient		Noticeable		Minor	Moderate	Extensive	§3.1.1
Thermoluminescence	Sensitivity (×1000, rel. Dhajala)	<1		1–2.2	2.2–4.6	4.6–10	10–22	22–46	46–100	100–220	220–460	300–600	600–2000		§2.1.1
	Glowcurve shape	Irregular		Sharp peak											
	Peak temperature	170–200		<140			>140								
Cathodoluminescence	Yellow CL in chondrule mesostasis	Common		Rare			Absent								§2.1.1
	Blue CL in chondrule mesostasis	Present		Common			Dominant								
	Red CL in chondrule olivine	Present					Rare to absent					Absent			
	Matrix CL	Red	Red areas	Low CL			Increasingly blue						Matrix absent		
Matrix	Hydrated phases	Abundant	Present		Absent								Matrix absent		§2.1.2
	Presolar graphite	Present			Absent								Matrix absent		
	Presolar diamonds	Decreasing								Absent			Matrix absent		
	F/FM†	>1.4					1.1–1.4			<1.1			Matrix absent		
	Matrix recrystallization (%)		<20	>20								100	Matrix absent		
	Sulfur content (wt%)	>1.0	0.5–1.0		<0.5								Matrix absent		
Feldspathic material	Texture of chondrule mesostasis	Number density of crystallites increases										Recrystallized			§2.1.3 §3.1.2
	Isotropic glass	Common				Less common		Rare				Absent			
	Type I chondrule mesostasis	Modal albite uncommon				Modal albite present						Mesostasis absent			
		Normative anorthite decreases										Mesostasis absent			
	Type II chondrule mesostasis	Albite rare	Modal albite increases			Modal albite present						Mesostasis absent			
	Grain size secondary feld (μm)	Submicrometer										2	2–10	50	
Olivine	PMD‡ Fa (%)		>33	15–3			3			5–15		<5			§2.1.3
	Type I chondrules	FeO < 2 wt% CaO > 0.3 wt% common			FeO increases to equilibrium value CaO decreases to equilibrium value					Near equilibrium Near equilibrium		Uniform Uniform			
	Type II chondrules	FeO converges to equilibrium value								Near equilibrium		Uniform			
		Cr_2O_3 > 0.3 wt%	Cr_2O_3 heter.	Cr_2O_3 > 0.1 wt%											
Pyroxene	PMD* Fs in low-Ca pyroxene (%)		>30	20–3			0					5–20	<5		§2.1.3 §3.1.2 §3.1.3
	Structure of low-Ca pyroxene	Predominantly monoclinic										Mixed	Orthorhombic		
	Grain-size of high-Ca pyroxene (μm)	Small grain in chondrule mesostasis										<1	2–5	5–30	
Nonsilicates	Phosphates	Some Na- and Fe-rich		Ca-rich, increasing abundance and grain size								Large grains Ca-phosphate			§2.1.4 §3.1.2 §3.1.4
	Kamacite	Co heterogeneity high			Co heterogeneity moderate			Co heterogeneity low							
		P, Cr, Si-bearing	P, Cr, Si-free												
	Taenite	Unzoned			Zoned										
	Carbides, magnetite	Common			Present		Rare to absent								

*Compiled from data in MetBase 7.1, a comprehensive meteorite database marketed by *Köblitz* (2005). Primordial ^{36}Ar calculated from ^{36}Ar and ^{38}Ar abundances assuming primordial $^{36}Ar/^{38}Ar = 5.6$ and spallogenic $^{36}Ar/^{38}Ar = 0.65$. Sharps was excluded from ranges due to its anomalously high volatile-element content, probably carried by abundant C- and S-rich xenoliths (J. Grossman, unpublished data).

†FeO/(FeO + MgO) in matrix, normalized to whole rock.

‡PMD = percent mean deviation, defined by *Dodd et al.* (1967) as the mean deviation of the iron concentration (wt%) in individual measurements of olivine or pyroxene from the mean of these measurements divided by the average iron content (wt%) multiplied by 100. Literature data probably dominated by grains in chondrules and chondrule fragments, but may include some in matrix.

Solid vertical lines represent criteria that are most important in distinguishing adjacent subtypes. All increases and decreases that are noted run from low to high petrologic types. The listed ranges of petrologic types associated with each parameter in this table were determined by the present authors based on primary literature data except for TL sensitivity in types 3.2–3.9, which are quoted from *Sears et al.* (1980) and the cutoff in PMD of olivine between types 3.9 and 4, which is taken directly from *Van Schmus and Wood* (1967). See the indicated text section for references.

Semarkona, shows a pervasive smoky-red CL (*DeHart et al.,* 1992). This property is less pronounced in type 3.1 chondrites, and is absent in higher types.

2.1.2. Effects in fine-grained matrix. Evidence of post-accretionary alteration in the matrix of type 3 ordinary chondrites was found in the first detailed studies to utilize the power of the scanning electron microscope (*Christophe Michel-Levy,* 1976; *Ashworth and Barber,* 1976; *Ashworth,* 1977). Later studies identified a series of changes in optical properties, mineralogy, and degree of chemical equilibration that could be confidently attributed to thermal metamorphism (*Huss et al.,* 1981; *Nagahara,* 1984; *Grossman and Brearley,* 2005). Recent work has also given us some idea of what the original, unmetamorphosed matrix material might have looked like.

In type 3.0–3.2 ordinary chondrites, matrix is opaque in standard thin sections. The very mild thermal metamorphism experienced by these meteorites has proceeded in the presence of varying amounts of water. In Semarkona (LL3.00) matrix, most of the primary silicates and sulfides have been converted to phyllosilicates and Ni-rich pyrrhotite (*Hutchison et al.,* 1987; *Alexander et al.,* 1989a). This assemblage gives Semarkona matrix a translucent brown to olive-green color in very thin (~10 μm) sections (*Hutchison et al.,* 1987). In Bishunpur (LL3.15), the matrix contains Na-, Al-, and Si-rich amorphous material instead of phyllosilicate (*Alexander et al.,* 1989a). Such material is highly susceptible to aqueous alteration and in Semarkona was probably converted to phyllosilicates. It survives in Bishunpur because less water was present. Type 3.0–3.2 chondrites contain micrometer-sized matrix olivine grains. In Semarkona, the compositions of these olivine grains range from 2 mol% fayalite [molar FeO/(FeO + MgO)] (Fa_2) to Fa_{40}, whereas in Bishunpur and Krymka (LL3.2), matrix olivines range from Fa_{25} to Fa_{90} (*Huss et al.,* 1981; *Nagahara,* 1984). These olivines were probably produced *in situ* from amorphous material and phyllosilicate (*Alexander et al.,* 1989a,b; *Scott and Krot,* 2003), and the more-restricted range of Semarkona matrix olivines may be a consequence of more-extensive aqueous alteration. Semarkona matrix also exhibits pervasive red CL (*Sears et al.,* 1990) that probably reflects a significant abundance of tiny forsterite grains. These grains are much less abundant in Bishunpur and are apparently absent in Krymka; they may have been converted to Fe-rich olivine by mild thermal metamorphism.

In petrologic types above 3.2, thermal metamorphism seems to have proceeded without the aid of significant amounts of water. Amorphous material is not present, and with increasing petrologic type, increasing proportions of the originally opaque matrix appear translucent due to recrystallization; mineral grains have grown large enough to make up a significant fraction of the thickness of the thin section. Recrystallized matrix is less porous and contains less C than opaque matrix (*Huss et al.,* 1981). Furthermore, as the recrystallized matrix becomes more abundant, the remaining opaque matrix becomes coarser, less friable, and increasingly depleted in minute metallic Fe-Ni and sulfide grains (*Huss et al.,* 1981). For example, matrix olivines in Chainpur (LL3.4) have begun to coalesce to form a mesh-like texture (*Nagahara,* 1984). By type 4, the matrix consists entirely of translucent to transparent crystals, although it can still be identified by grain size, morphology, and location. These changes are illustrated in Fig. 1.

Accompanying the textural changes are major changes in matrix mineralogy, composition, and mineral chemistry. As petrologic type increases, matrix olivine grains show less compositional spread, until by type ~3.8, they have largely equilibrated at the mean Fa content for the bulk meteorite (Fig. 2). With increasing petrologic type, plagioclase, chromite, and chlorapatite nucleate and grow, and minute sulfide and metallic Fe-Ni grains that are abundant in the matrix of type 3.0 ordinary chondrites coalesce into larger metal and sulfide grains (*Dodd et al.,* 1967; *Huss et al.,* 1981; *Grossman and Brearley,* 2005). Low in the metamorphic sequence, the Na, S, and Ni contents of the matrix decrease as alkalis enter chondrules, and sulfides and metal grains recrystallize (*Wood,* 1967; *Grossman and Brearley,* 2005). Above petrologic type ~3.5, matrix begins to show strong blue CL (*Sears et al.,* 1990) and the alkali content again rises, indicating that secondary feldspar grains have begun to grow (*Grossman and Brearley,* 2005).

The matrix of type 3 ordinary chondrites is host to a variety of presolar materials, including diamond, graphite, silicon carbide, oxides, deuterium-rich organic compounds, and crystalline and amorphous silicates (e.g., *Nittler,* 2003; *Mostefaoui et al.,* 2004). The abundances of presolar components correlate strongly with the petrologic type of the host meteorites (Fig. 3). Different presolar components have different susceptibilities to metamorphic breakdown. For example, presolar silicates and graphite have only been found in type 3.0–3.1, whereas diamond persists up to type 3.7–3.8 and the carrier of P1 noble gases (also known as Q gases) survives into type 4 and beyond (*Huss and Lewis,* 1995; *Huss et al.,* 1996; *Mostefaoui et al,* 2004).

2.1.3. Effects on silicate minerals in chondrules.

2.1.3.1. Mesostasis: Chondrule mesostasis in type 3 ordinary chondrites generally consists of feldspathic glass ± very small crystallites (<5 μm) and is highly susceptible to metamorphic alteration. Isotropic glass is common in low-FeO chondrules in type 3.00 Semarkona, but has mostly devitrified in related chondrules in type 3.6 Parnallee (*McCoy et al.,* 1991). *DeHart et al.* (1992) observed that the abundance of crystallites in the mesostasis also increases with metamorphic grade, again consistent with progressive devitrification.

The composition of chondrule mesostasis changes markedly during metamorphism. FeO increases and TiO_2 decreases in mesostasis in FeO-poor type I chondrules between types 3.0 and 3.6, whereas FeO and MnO decrease in mesostasis from FeO-rich type II chondrules along this sequence (*McCoy et al.,* 1991). In type I chondrules, the abundances of Na and K increase as these elements diffuse into the mesostasis and Ca is lost (*Sears et al.,* 1990; *Alexander et al.,* 2000; J. N. Grossman, unpublished data, 2005).

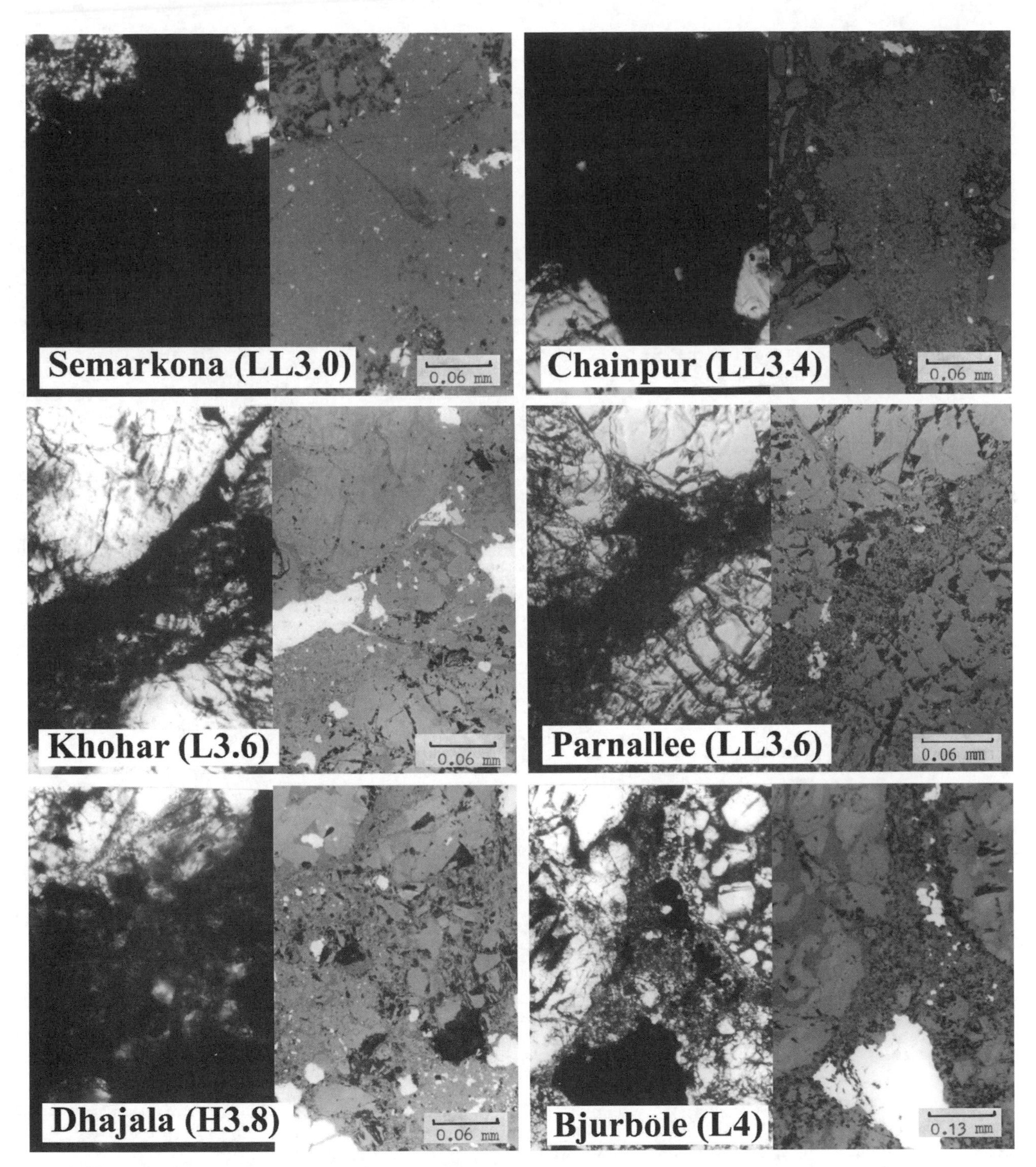

Fig. 1. Photomicrographs of the matrix in six chondrites. For each meteorite, the left image is in transmitted light and the right image is the same area in reflected light. In Semarkona (LL3.0), the matrix is opaque in thin section and consists of very fine-grained silicates, sulfides, and Fe-Ni metal. In Chainpur (LL3.4), the matrix is still largely opaque, but the matrix mineral grains are larger. In Khohar (L3.6) and Parnallee (LL3.6), the matrix is translucent in reflected light because the size of the mineral grains is approaching the thickness of the section. Sulfides and metal have coalesced into larger grains. In Dhajala (H3.8), most matrix minerals are now large enough to see in transmitted light. In Bjurbole (L4), the matrix is completely recrystallized and the grains are larger (note scale), although matrix can still be recognized by its grain size and location between chondrules.

The entry of alkalis is followed by crystallization of albitic feldspar by type ~3.4 (*Grossman and Brearley,* 2005). In the mesostasis of type II chondrules, crystalline albite is present even in type 3.00 Semarkona, but the abundance increases with petrologic type. Development of albite is responsible for the increase in blue CL shown by chondrules in type 3 ordinary chondrites as a function of metamorphic grade (see above) (*Grossman and Brearley,* 2005). As feld-

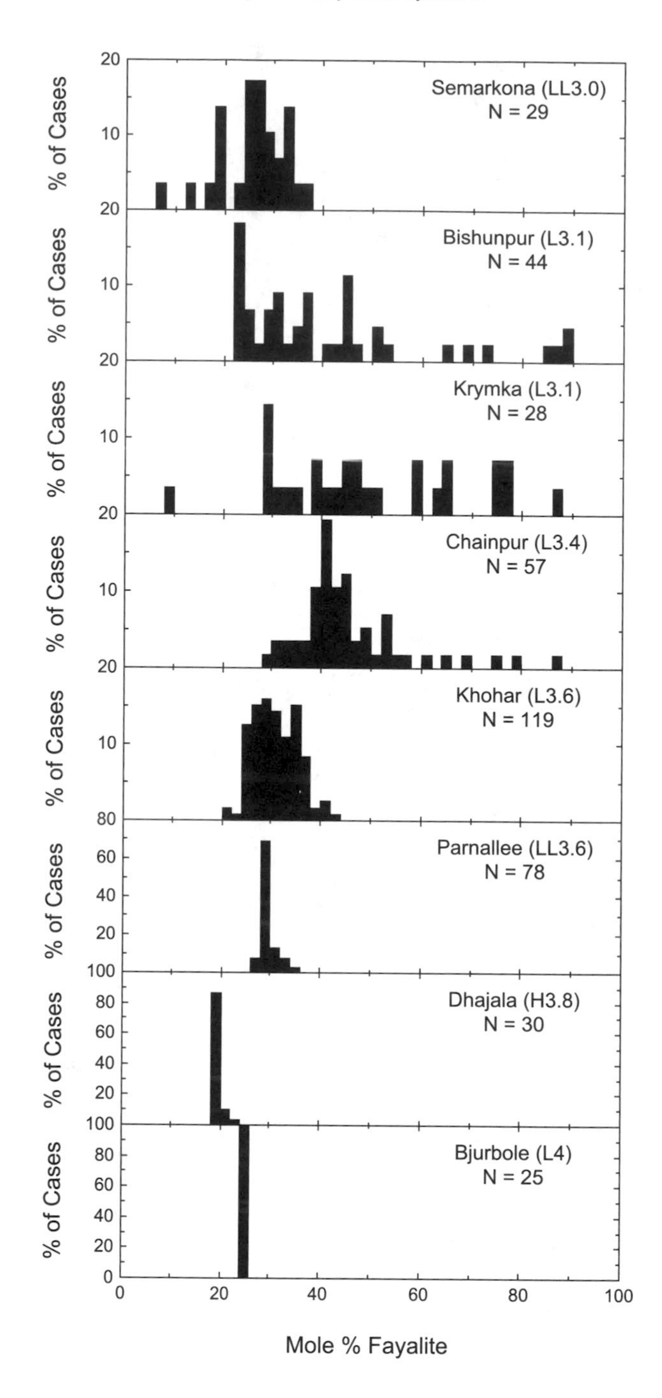

Fig. 2. Histograms showing fayalite contents of matrix olivine as a function of petrologic type. The distribution is widest in type 3.1 chondrites, and becomes systematically narrower with increasing petrologic type. By type 3.6 to 3.8, the matrix olivine has equilibrated at the mean fayalite value for the bulk meteorite. Chondrule olivine equilibrates by type 4. Data from *Huss et al.* (1981).

spar crystallizes from chondrule mesostasis during metamorphism, K/Na in residual glass rises to several times the solar ratio (*Grossman and Brearley,* 2005).

2.1.3.2. Olivine: Olivine compositional variability is a strong indicator of metamorphic grade (*Van Schmus and Wood,* 1967; *Dodd et al.,* 1967). Chondrule olivine from highly unequilibrated ordinary chondrites has a wide range of Fa contents (e.g., Fig. 1 of *Dodd et al.,* 1967). With increasing petrologic type, the Fa content becomes increasingly uniform and converges on the value for equilibrated chondrites of the same group. The traditional measure of olivine heterogeneity (defined by *Dodd et al.,* 1967) is percent mean deviation (PMD, see notes for Table 1). Percent mean deviation remains high between types 3.0 and 3.4 (*Sears et al.,* 1980), but drops systematically as metamorphism proceeds from type 3.4 to type 4 (defined by PMD <5%). During this process, igneous zoning profiles in chondrule olivine are erased and FeO from the matrix diffuses into forsteritic chondrule olivine along grain boundaries and cracks (*McCoy et al.,* 1991). In a similar way, the MnO and P_2O_5 contents of olivine homogenize during metamorphism (*DeHart et al.,* 1992).

Although the PMD of chondrule olivine is not sensitive to metamorphic changes between types 3.0 and 3.4, there are profound changes in olivine chemistry within this range. Olivine in low-FeO chondrules begins to take up FeO by type 3.1, and by type 3.3 many type I chondrules contain significant FeO in olivine (*DeHart et al.,* 1992; *Scott et al.,* 1994). High CaO contents (>0.3 wt%) are common in olivine from types 3.0–3.1 chondrites, but by types 3.3–3.4 these concentrations occur less frequently, and by type 3.8 converge to near the equilibrated value of <0.05 wt%. Similar CaO losses occur in olivine from FeO-rich type II chondrules during metamorphism, although compositions with >0.2 wt% CaO are rare in these chondrules even in type 3.0. Like CaO, both TiO_2 and Al_2O_3 are enriched in some low-

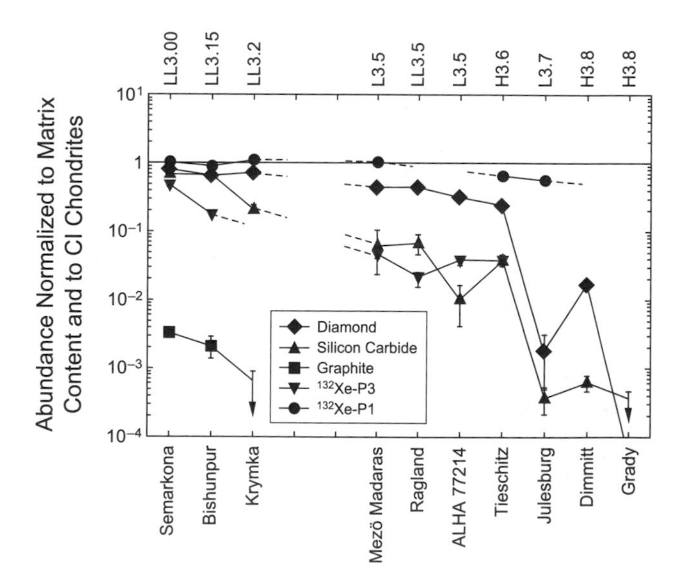

Fig. 3. Abundances of several presolar components found in primitive chondrites as a function of petrologic type of the host meteorites. The abundances have been normalized to vol% matrix (the grains are sited in the matrix) and are plotted as ratios to the abundances in CI chondrites (absolute abundances vary by several orders of magnitude). Although the absolute resistance to metamorphism varies dramatically among the plotted phases, all components decrease with increasing petrologic type. Most are effectively gone by type 3.8, where the matrix is almost completely recrystallized. Data from *Huss and Lewis* (1994, 1995) and *Huss et al.* (1996).

FeO chondrule olivines in type 3.0, but these compositions are lost during metamorphism (*DeHart et al.,* 1992; *Scott et al.,* 1994).

One of the most sensitive indicators of metamorphism in type 3 ordinary chondrites is the Cr_2O_3 content of olivine in type II chondrules. The Cr_2O_3 contents of these olivine grains decrease from ~0.5 wt% in type 3.0 chondrites to 0.15–0.2 wt% in type 3.2–3.3 to below detection limit in type 3.6–3.8 (*McCoy et al.,* 1991; *DeHart et al.,* 1992). *Grossman and Brearley* (2005) showed that in type 3.0–3.2 ordinary chondrites the distribution of Cr_2O_3 within FeO-rich olivine grains changes systematically and may be used to differentiate among the types. Experimental data confirm the extreme sensitivity of Cr_2O_3 in olivine to loss during heating (*Jones and Lofgren,* 1993).

2.1.3.3. Pyroxene: Pyroxene is relatively resistant to metamorphic alteration compared to other silicate minerals in chondrules. Whereas olivine begins equilibrating in the lowest petrologic types, and the PMD of its Fa content begins to decrease by type 3.4, the PMD of the ferrosilite (Fs) content of low-Ca pyroxene remains high until at least type 3.6–3.8 and is still relatively high into type 4 (*Dodd et al.,* 1967). During metamorphism of FeO-poor type I chondrules, FeO and MnO increase and Cr_2O_3 and Al_2O_3 decrease in low-Ca pyroxene (*McCoy et al.,* 1991). *DeHart et al.* (1992) observed that the CaO content of high-Ca pyroxene decreases somewhat with increasing petrologic type.

2.1.4. Effects on nonsilicate minerals. In type 3.0 ordinary chondrites, P, Si, and Cr occur in solid solution in some metallic Fe-Ni grains in type I chondrules (*Zanda et al.,* 1994); phosphides and silicides may also be present as inclusions in metal. By petrologic type 3.1, these elements leave the metal, reacting with atomic species from elsewhere in the meteorite (e.g., Ca, Fe, O, Cl). Even in type 3.00 Semarkona, Cr has been oxidized from metal grains in many type I chondrules. In some metal grains from type 3.15 Bishunpur, Na and Fe phosphates, chromite, and silica are present as inclusions; reduced P-, Si-, and Cr-bearing phases are absent (B. Zanda, personal communication, 2003). At or above type 3.1, Ca-phosphate becomes the dominant P-rich phase associated with chondrule metal, and phosphates form at the edges of metal grains (*Rambaldi and Wasson,* 1981; B. Zanda, personal communication, 2003), forming mantles (*Rubin and Grossman,* 1985). Large phosphates >20 μm in diameter found in the matrix of Parnallee (LL3.6), Bremervörde (H/L3.9), and Dhajala (H3.8) are presumably of metamorphic origin (*Huss et al.,* 1981).

Kamacite in type 3.0 ordinary chondrites is heterogeneous with respect to Ni and Co (e.g., Fig. 163 of *Brearley and Jones,* 1998). With increasing metamorphism, the concentrations of Ni and Co become more uniform, and the mean concentrations tend to increase (*Rubin,* 1990). Nickel in taenite grains is unzoned in type 3.0–3.1 ordinary chondrites, whereas zoned taenite is present at type 3.4 and above (*Wood,* 1967; *Afiattalab and Wasson,* 1980). During metamorphism, Ni diffuses out of kamacite and into taenite. With cooling, diffusion becomes more sluggish. At the kamacite-taenite interface, Ni becomes depleted at the edges of kamacite grains (the so-called Agrell dip) and enriched at the edges of taenite grains, leaving "M-shaped" Ni profiles in the taenite. These frozen-in diffusion profiles have been used to estimate cooling rates of chondrites (e.g., *Wood,* 1967; *Willis and Goldstein,* 1981).

In the most primitive type 3 ordinary chondrites, much of the low-Ni metal has been replaced by a complex mixture of carbides, magnetite, Ni-rich metal, and Ni-rich sulfides (*Huss,* 1979; *Taylor et al.,* 1981). These assemblages occur as large millimeter-sized masses between chondrules in Semarkona (LL3.00) and Ngawi (LL3.0–3.6 breccia), and they probably originated through hydrothermal alteration of metallic iron by a C-O-H fluid on the parent body (*Krot et al.,* 1997). Less complex but similar magnetite-sulfide assemblages surround chondrules, particularly in Semarkona. These assemblages exhibit structures consistent with fluid deposition and grade into the matrix; they appear to have formed in place during fluid-assisted metamorphism (*Huss,* 1979).

2.1.5. Bulk effects. The bulk compositions of ordinary chondrites are largely unaffected by metamorphism (e.g., *Kallemeyn et al.,* 1989). However, the abundances of some volatile and mobile elements decrease with increasing petrologic type (Table 1). Much of this variation occurs among the type 3 ordinary chondrites, although some parameters continue to decrease right through type 6. Bulk contents of C, H_2O, and primordial (=P1 or Q) noble gases decrease with increasing petrologic type in ordinary chondrites (Table 1) (*Van Schmus and Wood,* 1967; *Zähringer,* 1968; *Alaerts et al.,* 1979). Whereas water is truly mobile, much of the C in type 3 ordinary chondrites is in relatively refractory organic compounds, some of which are the noble-gas carriers (e.g., *Huss et al.,* 1996). These compounds are easily oxidized to CO and CO_2, releasing the noble gases and producing the observed correlation between C and noble gases. Among true volatile elements, the Cs, Cd, Bi, Tl, and In contents of type 4–6 chondrites are lower by 1–2 orders of magnitude than those in type 3 [first shown for In by *Tandon and Wasson* (1968); see reviews by *Dodd* (1969) and *Lipschutz and Woolum* (1988)].

2.2. CO3 Chondrites

McSween (1977a) presented the first strong case that CO3 chondrites represent a metamorphic sequence. The CO3 chondrites are now subdivided using a scheme analogous to that of type 3 ordinary chondrites [3.0–3.7, *Scott and Jones* (1990); 3.0–3.8, *Chizmadia et al.* (2002); or 3.0–3.9, *Sears et al.* (1991a)]. In the following discussions, we use the classification of *Chizmadia et al.* (2002) because it seems to correlate better with that of type 3 ordinary chondrites. This system adds 0.1 to the classifications of *Scott and Jones* (1990) for all CO3 chondrites at or above type 3.1.

2.2.1. Matrix. The matrix of CO3 chondrites changes markedly during metamorphism. The matrix of Allan Hills (ALH) A77307 (CO3.0) is highly unequilibrated and consists of an amorphous silicate component, olivine, pyroxene, magnetite, kamacite, pentlandite, pyrrhotite, anhydrite, and disordered mixed-layer phyllosilicates (*Brearley,* 1993). There are two populations of olivine and pyroxene in ALH A77307 matrix. One consists submicrometer-sized grains of nearly pure forsterite and enstatite grains that are probably primary. A more-abundant type of poorly crystalline olivine has compositions ranging from Fa_{18} to Fa_{72} and is probably secondary (*Brearley,* 1993). The matrix of ALH A77307 also contains relatively high abundances of presolar grains, including presolar silicates (*Huss et al.,* 2003; *Kobayashi et al.,* 2005). Matrix in type 3.1–3.8 CO chondrites apparently does not contain tiny forsterite and enstatite grains and is dominated by fine-grained FeO-rich olivine (*Brearley,* 1995). With increasing petrologic type, the range of olivine compositions decreases (*Brearley and Jones,* 1998), and the mean Fa content of matrix Fe-rich olivine decreases systematically from ~Fa_{66-68} in ALH A77307 (CO3.0) to ~Fa_{32} in Warrenton (CO3.7) (*Scott and Jones,* 1990, and references therein). Abundances of presolar grains also decrease systematically with increasing petrologic type (*Huss,* 1990; *Huss et al.,* 2003). Matrix texture changes only slightly with increasing petrologic type, appearing somewhat turbid in Warrenton (CO3.7) and Isna (CO3.8) (*McSween,* 1977a) rather than opaque as in lower types.

Water played a role in the alteration history of some CO3 matrix. Allan Hills A77307 contains phyllosilicates and anhydrite that were probably produced *in situ,* although they might be weathering products from Antarctica (*Brearley,* 1993). The matrix of Lancé (CO3.4), a witnessed fall, also contains phyllosilicates that partially replace matrix olivine, but matrix in Kainsaz (CO3.2) and Warrenton (CO3.7) apparently does not contain phyllosilicates (*Keller and Buseck,* 1990).

2.2.2. Chondrules, calcium-aluminum-rich inclusions, and amoeboid olivine inclusions. Type IA chondrules in CO chondrites show systematic increases in the Fa content of olivine and the Fs content of low-Ca pyroxene and a systematic decrease in the CaO content of olivine from type 3.0 to type 3.8 (*Scott and Jones,* 1990). Type II chondrules show a similar trend for CaO content, but not for mean Fa content because the olivine in type II chondrules started out close to the equilibrium Fa content for CO3 chondrites. These changes are best understood in terms of diffusional exchange between matrix and chondrules, with equilibration approached, but not reached, by type CO3.8 (e.g., *Scott and Jones,* 1990).

Chondrule mesostases show increasing degrees of alteration that correlate with petrologic type. *Itoh and Tomeoka* (2003) found that chondrules in Yamato (Y) 81020 (CO3.0) have unaltered mesostases with little nepheline. In contrast, 60% of the chondrules in CO3.2 chondrites and >90% of the chondrules in CO3.3–3.5 chondrites have minor to moderate degrees of mesostasis alteration and abundant nepheline. The chondrule mesostases in Warrenton (CO3.7) have been nearly completely replaced by nepheline or nepheline-rich aggregates.

Calcium-aluminum-rich inclusions (CAIs) have experienced considerable secondary alteration, both before and after accretion (*Russell et al.,* 1998). Probable metamorphic alteration on the parent body can be identified by correlations between CAI characteristics and petrologic type. For example, melilite-rich inclusions are abundant in CO3.0–3.4 meteorites, rare in CO3.5 meteorites, and absent in CO3.6–3.8. Melilite in CAIs is probably replaced by feldspathoids, pyroxene, and Fe-rich spinel. Spinel in melilite-rich and coarse-grained spinel-pyroxene inclusions becomes more Fe-rich, with development of nearly homogenous hercynitic spinel (~50–60 mol% hercynite) in types >3.5. Perovskite has been converted to ilmenite in types >3.5 (*Russell et al.,* 1998).

Amoeboid olivine inclusions (AOIs) (also called amoeboid olivine aggregates, AOAs) in CO3 chondrites are sensitive to parent-body aqueous alteration and thermal metamorphism (*Rubin,* 1998; *Chizmadia et al.,* 2002). Between types 3.0 and 3.2, forsterite grains in AOIs acquire thin ferroan olivine (Fa_{60-75}) rims. By type 3.4–3.5, the fayalitic rims have thickened, and the olivine compositional distributions of individual AOIs have broadened and become bimodal. At type 3.6–3.7, the compositions of AOI olivine become unimodal at ~Fa_{65} although there is still a low-Fe tail. The high-Fe peaks are rather broad, extending from Fa_{50-55} to Fa_{68-70}. By type 3.8, the forsteritic tails disappear and only ferroan olivine remains in a relatively narrow peak at ~Fa_{65}. In addition to changes in FeO/(FeO + MgO), the abundances and variability of Mn, Cr, and Ca in AOI olivine grains decrease as equilibrium is approached (*Chizmadia,* 2000).

Although matrix and AOIs in CO3.8 meteorites each approach equilibration, they are not equilibrated with each other (*Methot et al.,* 1975; *Scott and Jones,* 1990; *Chizmadia et al.,* 2002). In addition, chondrule silicates are far from equilibrated (*Scott and Jones,* 1990), showing olivine and pyroxene distributions similar to type ≤3.5 ordinary chondrites (cf. *Dodd et al.,* 1967; *Huss et al.,* 1981). Thus, even the highest-grade CO3 meteorites in our collections are not equilibrated chondrites.

2.2.3. Metal. Kamacite grains in CO3.0 chondrites contain high concentrations of Cr, P, and Si, as do metal grains from the least-metamorphosed LL chondrites (see above). Kamacite shows systematic increases in the mean concentrations of Ni and Co from CO3.0 to CO3.8, while Cr content (and presumably P and Si content) in metal decreases (*McSween,* 1977a; *Scott and Jones,* 1990).

2.3. CV3 Chondrites

The CV3 chondrites and their constituents have been heavily affected by secondary processes, including alkali-halogen-Fe metasomatism, aqueous alteration, thermal metamorphism, and shock metamorphism (e.g., *Krot et al.,* 1995). As a result, attempts to divide CV3 chondrites into petrologic subtypes (e.g., *Guimon et al.,* 1995) have not been particularly successful (e.g., *Krot et al.,* 1995). CV3 chondrites have been divided into oxidized and reduced subgroups, and the oxidized subgroup has been subdivided into $CV3_{oxA}$ and $CV3_{oxB}$ largely based on the style of secondary-alteration features (*McSween,* 1977b; *Weisberg et al.,*

1997). In this section, we discuss the secondary features of each subgroup of CV3 chondrites that have been attributed to parent-body metamorphism.

2.3.1. $CV3_{oxB}$ chondrites. Oxidized CV3 chondrites of the Bali type (Bali, Kaba, portions of Mokoia, Grosnaja, and ALH 85006) have experienced significant secondary processing (*Krot et al.,* 1995, 1998a). In Kaba, alteration of chondrule mesostasis (*Keller and Buseck,* 1990) and associated redistribution of Ca, Si, Al, Mg, Fe, Mn, and Na (*Krot et al.,* 1998a) is best understood in terms of aqueous alteration on the parent body. A similar origin is likely for phyllosilicate, tetrataenite, awaruite, and pentlandite in the matrix of Kaba and Mokoia (*Keller and Buseck,* 1990; *Krot et al.,* 1995). Other features in Kaba, including Fe-rich halos around some opaque nodules and veins of fayalitic olivine within forsterite grains, are more likely due to high-temperature processing (e.g., *Hua et al.,* 2005). Similar features in other CV3 chondrites have been linked to processing prior to the final accretion of the parent body (*Peck and Wood,* 1987; *Krot et al.,* 1995). However, the decreasing spread of matrix-olivine Fa compositions going from Fa_0–Fa_{100} in Kaba to Fa_{10}–Fa_{75} in Mokoia to Fa_{35}–Fa_{55} in Grosnaja appears to be a metamorphic sequence (e.g., *Krot et al.,* 1995).

2.3.2. $CV3_{oxA}$ chondrites. Oxidized CV3 chondrites of the Allende type (e.g., Allende, Axtell, Lewis Cliff 86006, and ALH A81258) have experienced considerable alkali-halogen-Fe metasomatism (*Krot et al.,* 1995). Metasomatic effects include replacement of primary minerals in CAIs and their Wark-Lovering rims by secondary nepheline, sodalite, grossular, anorthite, hedenbergite, and other phases, and entry of Fe into spinel. Metasomatism of chondrules in Allende and Axtell has resulted in replacement of mesostasis with secondary minerals (*Kimura and Ikeda,* 1995; *Srinivasan et al.,* 2000), enrichment of olivine and pyroxene in Fe, and replacement of low-Ca pyroxene by fayalitic olivine (*Housley and Cirlin,* 1983; *Krot et al.,* 1995; *Srinivasan et al.,* 2000). Many coarse-grained igneous rims around chondrules contain secondary nepheline and sodalite (*Rubin,* 1984). Matrix has also been extensively affected by alkali-halogen-Fe metasomatism. The site of this metasomatism is still unknown. Earlier workers favored alteration prior to accretion (*Peck and Wood,* 1987; *Palme and Wark,* 1988; *Kimura and Ikeda,* 1995), but in a series of papers, Krot and co-workers have made a case for parent-body alteration (*Krot et al.,* 1995, 1998a, 1998b).

Thermal metamorphism also affected $CV3_{oxA}$ chondrites. In Axtell and Allende, matrix olivine has a restricted range of compositions (~Fa_{40-50}), suggesting a higher metamorphic grade than in other oxidized and reduced CV3s (e.g., *Krot et al.,* 1995). Abundances of presolar grains suggest that Allende is somewhat more metamorphosed than Axtell and that both are significantly more metamorphosed than Mokoia, Leoville, and Vigarano (*Huss et al.,* 2003). This sequence is consistent with that derived from olivine data, but is inconsistent with the one based on TL sensitivity (*Guimon et al.,* 1995).

2.3.3. $CV3_{red}$ chondrites. In many ways, the $CV3_{red}$ chondrites (Leoville, Efremovka, Vigarano, Arch) show the least evidence of secondary processing. Reduced CV3s have primary metal and troilite as their main opaque phases (*Krot et al.,* 1995, 1998a). Their CAIs show little evidence for alkali-halogen-Fe metasomatism (*Krot et al.,* 1995), and secondary phases are rare or absent from their matrix (e.g., *Krot et al.,* 1998a). However, the presence in matrix of abundant fayalitic olivine with a restricted range of composition and absence of amorphous material and forsterites that are common in matrix of the most primitive chondrites suggest that even the least-altered $CV3_{red}$ chondrites have experienced some alteration (*Krot et al.,* 1998a). Abundances of presolar diamond and SiC indicate that Leoville is less metamorphosed than Vigarano (*Huss and Lewis,* 1995), consistent with the order obtained from TL sensitivity (*Guimon et al.,* 1995).

2.3.4. Comparing $CV3_{red}$, $CV3_{oxB}$, and $CV3_{oxA}$ chondrites. The three subclasses of CV3 chondrites have experienced different levels of aqueous alteration, metasomatism, and thermal metamorphism. The $CV3_{oxB}$ meteorites experienced significant aqueous processing, while $CV3_{oxA}$ meteorites primarily show evidence of metasomatism; few meteorites show evidence of both, perhaps implying that metasomatism actually represents aqueous alteration followed quickly by thermal metamorphism. The $CV3_{red}$ meteorites show little evidence of either aqueous alteration or metasomatism (e.g., *Krot et al.,* 1995). Spreads in matrix olivine compositions indicate that Kaba ($CV3_{oxA}$) is among the least metamorphosed, followed by Leoville ($CV3_{red}$) and Mokoia ($CV3_{oxA}$), followed in turn by Vigarano ($CV3_{red}$) and Grosnaja ($CV3_{oxB}$), and then by Axtell and Allende (both $CV3_{oxA}$) (Fig. 8 of *Krot et al.,* 1995; Fig. 10.2.6 of *Scott et al.,* 1988). Abundances of presolar grains imply that Leoville is slightly less metamorphosed than Mokoia, followed in turn by Vigarano, Axtell, and Allende (*Huss et al.,* 2003). Raman spectroscopy of organic material indicates that Leoville is the least altered, followed by Vigarano ≈ Kaba ≈ Efremovka < Grosnaja < Mokoia < Axtell < Allende (*Bonal et al.,* 2004). These studies give a similar order, but this order is not consistent with that inferred from TL sensitivity (*Guimon et al.,* 1995), except perhaps among $CV3_{red}$ meteorites.

2.4. Other Chondrite Groups

2.4.1. Carbonaceous chondrites. Most other carbonaceous chondrite groups consist of nearly unmetamorphosed meteorites. There is little evidence that CH, CB, and CR chondrites have experienced any metamorphic heating (*Weisberg et al.,* 2001; *Krot et al.,* 2002). In most CI and CM chondrites, the effects of severe aqueous alteration obscure any metamorphic history they might have had. However, a small group of CI and CM chondrites do show metamorphic effects (*Zolensky et al.,* 2005). *Tonui et al.* (2001) describe blurring of chondrule boundaries and other effects of significant heating in some of these chondrites, implying metamorphic conditions above type 3. However, highly unequilibrated olivine compositions seem to be more consistent with petrologic types below 3.2 (*Grossman et al.,*

2005). The CK group has metamorphosed and unmetamorphosed members (*Kallemeyn et al.,* 1991), but CK and Kakangari-like chondrites have too few type 3 representatives to permit systematic studies of their metamorphic properties. Further study is clearly required to deduce the thermal histories of these meteorites.

2.4.2. Enstatite chondrites. Type 3 members are well described in both the EH and EL groups. EH3 and EL3 chondrites are texturally similar, having well-delineated chondrules and containing minor olivine (<1% to 5% by volume), but EL chondrites have larger chondrules on average than EH chondrites and their mineral chemistries differ somewhat (*Binns,* 1967; *Prinz et al.,* 1984; *Rubin et al.,* 1997). The matrix in type 3 enstatite chondrites is not well characterized. It is typically not opaque, as it is in type 3 ordinary chondrites, but is fine-grained and recognizable by its grain size, morphology, and location. The matrix of EH3 chondrites (and presumably EL3 chondrites, although no data are yet available) acquired abundances of presolar diamond and SiC similar to other chondrite classes (*Huss and Lewis,* 1995).

In general, we know little about the metamorphic history of type 3 enstatite chondrites. Although they display considerable mineralogical variation, they cannot easily be put into a metamorphic sequence because they have had complex thermal and shock histories compared to ordinary and carbonaceous chondrites (*Zhang et al.,* 1995).

2.4.3. R chondrites. Rumuruti (R) chondrites resemble ordinary chondrites in many ways: They have large chondrules, $\Delta^{17}O > 0$, and similar chondrule types. They show a similar range of petrologic types as well, although most R3 chondrites have experienced metamorphism equivalent to at least type 3.6 in ordinary chondrites (*Rubin and Kallemeyn,* 1989; *Bischoff et al.,* 1994). The only highly unequilibrated material described in the literature consists of clasts of petrologic type <3.5, possibly as low as 3.2, in some R chondrite breccias (*Bischoff,* 2000). Matrix in the R3.8 chondrites shows evidence of recrystallization; it is a low-porosity, translucent material that consists primarily of olivine accompanied by minor 2–3-μm pyroxene grains micrometer-sized sulfide grains, and, in some cases, 10-μm patches of plagioclase (*Kallemeyn et al.,* 1996). Again, little is known about the metamorphic effects in R3 chondrites, although matrix material and chondrule olivine clearly underwent Fe-Mg exchange similar to that in ordinary chondrites.

3. HIGHER-TEMPERATURE EFFECTS (TYPES 4–6)

Metamorphic changes in type 4–6 chondrites have been significant, but many of the criteria used to classify type 3.00–3.9 chondrites do not help to classify these "equilibrated" chondrites. Criteria used to classify type 4–6 ordinary chondrites are summarized in Table 1. Because the most useful ones tend to be somewhat subjective, the boundaries between types 4, 5, and 6 are rather poorly defined. Classification can also be difficult if a meteorite is a breccia containing material of different petrographic types. In such a meteorite, the current assemblage has never been heated above the temperature experienced by the lowest-grade material.

3.1. Ordinary Chondrites

3.1.1. Textural integration. Textural differences between type 3 chondrites and type 4–6 chondrites are significant and well documented (*Dodd et al.,* 1967; *Van Schmus and Wood,* 1967). Textural integration of matrix, chondrules, chondrule fragments, and inclusions proceeds up the metamorphic scale (Fig. 4), and is observable in chondrites of high petrologic type 3 and above. The opaque, fine-grained matrix found in primitive type 3 ordinary chondrites has completely recrystallized by type 4, but is still recognizable as matrix. By type 5, matrix is no longer recognizable, and textural integration between chondrules and matrix has begun. In type 6, small chondrules have lost their identity and only large chondrules are readily recognizable (Fig. 4). The mean modal abundance of recognizable chondrules decreases from 65–75 vol% in H3 chondrites (Table 1 of *Rubin,* 2000) to 22 vol% in H5 and 11 vol% in H6 (Table 2 of *Rubin et al.,* 2001).

3.1.2. Grain growth and appearance of new phases. Coarse-grained plagioclase is essentially absent from type 3 ordinary chondrites, except for anorthitic phenocrysts in rare Al-rich chondrules (e.g., *Russell et al.,* 2000). Small primary plagioclase grains also occur in coarse-grained, igneous rims around chondrules (*Rubin,* 1984). The clear, Na- and Al-rich chondrule mesostasis observed in type 3 chondrules has begun to devitrify in high type 3s, as albitic plagioclase nucleates and begins to grow. By type 4, chondrules have turbid mesostasis. Type 5 ordinary chondrites contain mainly microcrystalline (~2–10 μm) plagioclase grains, whereas type 6s contain abundant plagioclase grains ≥50 μm in size (e.g., *Van Schmus and Wood,* 1967).

Diopside occurs as small, commonly skeletal, crystallites in the mesostases of some chondrules in type 3 ordinary chondrites, but it is rare to absent in the fine-grained silicate matrix. Diopside is typically of submicrometer size in type 4s; it occurs as 2–5-μm grains in type 5s and as coarser grains (up to tens of micrometers) in type 6s.

Phosphate minerals and chromite occur as tiny grains (<5–10 μm) in type II chondrules, and nucleate as submicrometer secondary minerals in type I chondrules in the earliest stages of metamorphism (*Grossman and Brearley,* 2005). By type 5 and 6, grains of merrillite and chlorapatite have grown to tens of micrometers in size (*Van Schmus and Ribbe,* 1969). Chromite grains in unshocked type 4–6 chondrites can be as large as 200 μm (*Rubin,* 2003). Metallic Fe-Ni and troilite coarsen significantly between types 3 and 6. For example, metal and troilite in H3.4 Sharps have mean particle sizes of ~65 μm and ~30 μm, respectively (*Benoit et al.,* 1998), whereas these phases have grown to ~120 μm and ~80 μm, respectively, in H5 and 6 chondrites (*Rubin et al.,* 2001). By type 6, many metal grains are no longer

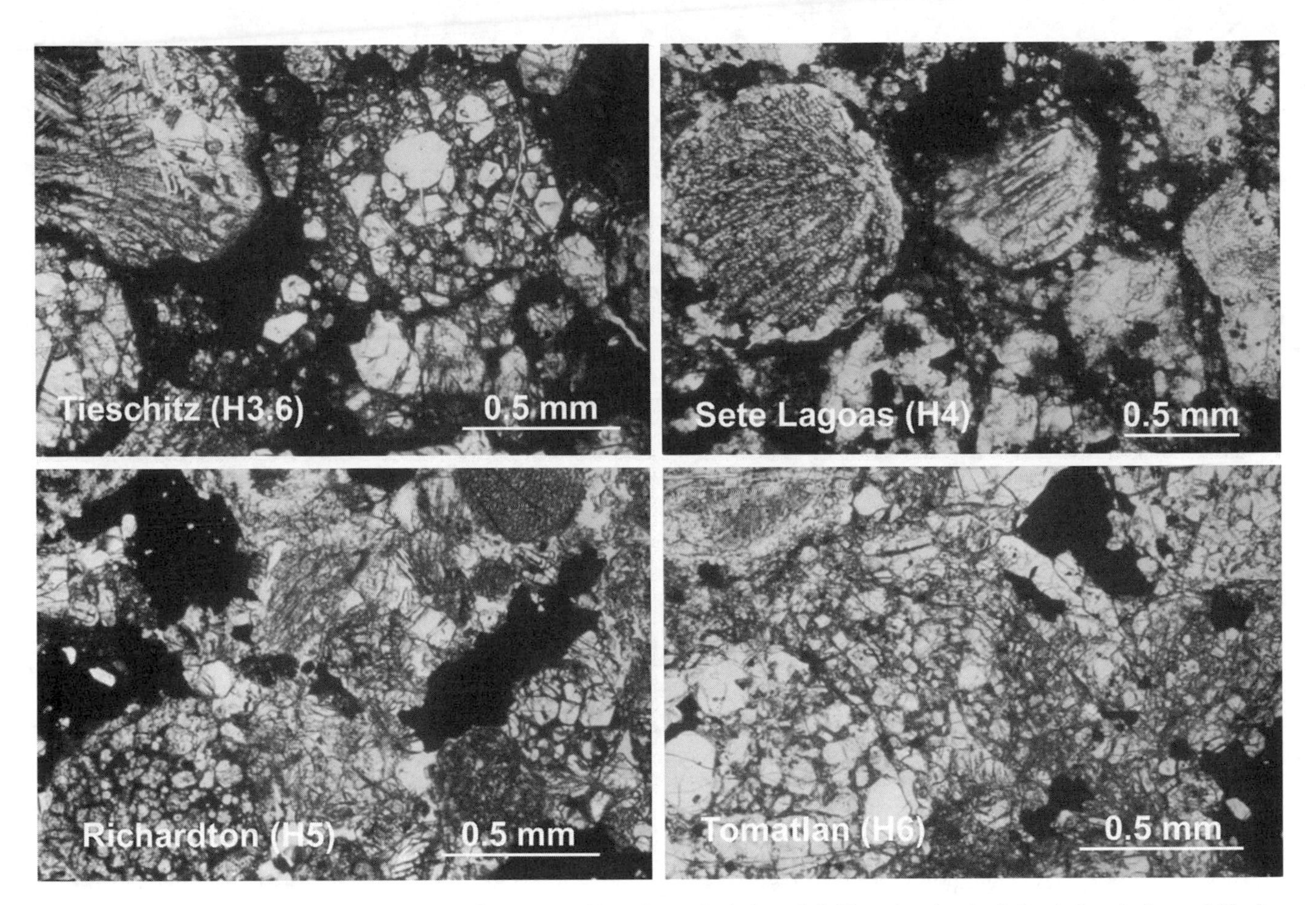

Fig. 4. Transmitted-light photomicrographs of representatives of type 3, 4, 5, and 6. Note the clearly defined chondrules and black, opaque matrix in Tieschitz (H/L3.6), the well-defined chondrules and recrystallized matrix in Sete Lagoas (H4), the integration of the chondrules with surrounding material and large metal grains in Richardton (H5), and the almost complete lack of identifiable chondrules in Tomatlan (H6). The black phases in the H4, H5, and H6 chondrites are metal and sulfide grains.

relatively compact, but have taken on irregular shapes defined by surrounding mafic-silicate grains (*Dodd,* 1969).

3.1.3. Phase transformation. Low-calcium clinopyroxene exhibiting polysynthetic twinning and inclined extinction commonly occurs as phenocrysts in chondrules in type 3 and 4 ordinary chondrites (e.g., Fig. 4.5 of *Dodd,* 1981). Low-calcium clinopyroxene transforms to orthopyroxene at ~630°C (*Boyd and England,* 1965) and thus is largely absent in types 5 and 6 (*Van Schmus and Wood,* 1967). However, low-Ca clinopyroxene can form by shock heating and quenching in chondrites of any petrologic type [stage ≥S3 (*Stöffler et al.,* 1991)].

3.1.4. Changes in mineral compositions. Many of the most striking compositional changes occur in type 3 chondrites as major silicate phases equilibrate. Olivine is equilibrated by type 4, but pyroxene equilibration is not completed until type 5 (*Van Schmus and Wood,* 1967; *Dodd et al.,* 1967). In the most metamorphosed members of type 6, classified by some workers as type 7, the CaO content of low-Ca pyroxene may be somewhat higher (>1 wt%) than in members that have experienced lower peak temperatures (*Dodd et al.,* 1975; *Mittlefehldt and Lindstrom,* 2001).

Metal grains do not equilibrate during metamorphism in ordinary chondrites due to sluggish diffusion. Instead, taenite grains exhibit a characteristic "M-shaped" Ni profile. These frozen-in diffusion profiles provide one means of estimating cooling rates of chondrites (e.g., *Wood,* 1967; *Willis and Goldstein,* 1981).

3.2. Carbonaceous Chondrites

3.2.1. CK chondrites. The CK chondrites are the only carbonaceous chondrites that exhibit the full range of petrologic types, CK3–CK6. *Kallemeyn et al.* (1991) described a number of metamorphic effects in CK chondrites that are analogous to those of ordinary chondrites. Homogeneity of olivine composition increases from type 3 to types 4–6. The mean diameter of plagioclase grains increases from CK3, where plagioclase is absent or microcrystalline, to CK4, where grains are <4 μm, to CK5 (4–50 μm) to CK6 (≥50 μm). Primary glass is found only in CK3 chondrules. Chondrule borders are sharply delineated in CK3, well defined in CK4, readily discernable in CK5, and poorly defined in CK6. The mean grain size of interchondrule material coarsens from CK3 (≤0.1–10 μm) to CK4 (5–50 μm) to CK5 (50–200 μm) to CK6 (50–300 μm). This parameter is the most diagnostic feature of petrologic type in CK chondrites.

3.2.2. Other metamorphosed carbonaceous chondrites. Coolidge and Loongana 001 constitute a distinct grouplet of carbonaceous chondrites (*Kallemeyn and Rubin,* 1995). They have relatively uniform olivine and heterogeneous low-Ca pyroxene compositional distributions, characteristic of petrologic types 3.8 to 4. Several other metamorphosed carbonaceous chondrites have been described [e.g., Mulga (west)] but they are not yet sufficiently well characterized to determine if they are CK chondrites, members of the Coolidge-Loongana-001 subgroup, ungrouped individuals, or members of new grouplets.

3.3. Enstatite Chondrites

Enstatite chondrites comprise two distinct groups, EH and EL, that formed under highly reducing conditions (e.g., *Sears et al.,* 1982b). Both groups include meteorites from petrologic types 3–6. As in ordinary and carbonaceous chondrites, the texture and mineralogy of EH and EL chondrites change during progressive thermal metamorphism. Chondrules are sharply defined in type 3, well-defined in type 4, readily delineated in type 5, and poorly defined, recrystallized, or, in some cases, absent in type 6. Low-Ca pyroxene exhibits significant compositional heterogeneity in type 3 (Fs_0 to Fs_{35}), minor heterogeneity in type 4, and uniformity in types 5 and 6. Monoclinic low-Ca pyroxene is predominant in type 3 enstatite chondrites, abundant in type 4, and essentially absent in types 5 and 6 (which instead contain orthopyroxene). Crystalline plagioclase is absent or fine-grained in type 3, coarser-grained (5–50 μm) in types 4 and 5, and very coarse-grained (≥50 μm) in type 6. Olivine decreases in abundance from 0.2 to 5 vol% in E3 chondrites to 0 to 1 vol% in E4; olivine is absent in E5 and E6. The nickel silicide, perryite [$(Ni,Fe)_5(Si,P)_2$], is relatively abundant in EH3 chondrites (e.g., perryite constitutes 15% of the total opaque phases in Kota-Kota), but it is rare in EH4 and EH5 chondrites (e.g., *Reed,* 1968; *Prinz et al.,* 1984). As perryite disappears during metamorphism, its elemental constituents are partitioned mainly into kamacite.

The abundances of presolar diamond and SiC also correlate with petrologic type in EH chondrites. Unlike in other chondrite groups, these phases persist in EH4 and perhaps even EH5 chondrites, probably because the reduced compositions of the host meteorites allow the presolar grains to survive to higher metamorphic temperatures without undergoing oxidation (*Huss and Lewis,* 1995).

Enstatite chondrites also exhibit properties that, although they correlate with petrologic type, cannot be explained by thermal metamorphism. For example, EH5 and EL6 chondrites contain high abundances of a "subsolar" noble gas component in their silicates that is either absent or present in only very low abundance in EH3 and EH4 chondrites (*Crabb and Anders,* 1981, 1982). These gases have a less-severe elemental fractionation relative to solar-system abundances than do planetary gases, and neither the fractionation nor the abundances can be produced by thermal metamorphism of more primitive material. This implies that these gases were acquired prior to accretion, which presents an interesting problem for models attempting to explain the range of properties of EH or EL chondrites strictly in terms of metamorphism.

3.4. R Chondrites

Many R chondrites are petrologic type 3.8–4 (e.g., Rumuruti, Carlisle Lakes, NWA 2198), and some are regolith breccias containing clasts with a diverse array of petrologic types (e.g., PCA 91002, R3.8–6) (*Kallemeyn et al.,* 1996). The R5 and R6 clasts are moderately to highly recrystallized, and matrix material is not recognizable. Highly recrystallized chondrules occur in R5 clasts, but discernable chondrules are absent in R6 clasts. A few R5 chondrites have been described (including the possibly paired NWA 830, 1585, 1668, 3098, and the unrelated find Sahara 99527). No individual R6 chondrites have been described.

4. TEMPERATURES OF METAMORPHISM

Two types of metamorphic temperatures are discussed in the literature. Peak metamorphic temperature, the highest temperature reached during metamorphism, is typically estimated from phase changes or reactions that occur at specific temperatures. Peak temperatures are particularly relevant for unequilibrated chondrites. Equilibration temperatures are determined from coexisting mineral phases and typically reflect the lowest temperature during the cooling of the system at which minerals remained in diffusive equilibrium. To accurately reflect metamorphism, the temperature must have been high enough to erase partitioning established during chondrule formation or other earlier high-temperature events. Temperatures given for equilibrated chondrites are typically equilibration temperatures, and in some cases they may approach the peak temperature. Typically, pressure is not an issue because metamorphism in chondrites probably occurred at pressures of ≤1 kb (e.g., *Dodd,* 1981; *McSween and Patchen,* 1989).

4.1. Ordinary Chondrites

Dodd (1981) estimated that type 3 ordinary chondrites reached temperatures of 400°–600°C, but these temperatures are clearly too high for the lowest petrologic types. *Alexander et al.* (1989a) set an upper limit of ~260°C for the alteration temperature of Semarkona (LL3.00) based on matrix clay mineralogy. *Rambaldi and Wasson* (1981) used the Ni content of matrix metal to estimate a metamorphic temperature of 300°–350°C for Bishunpur (LL3.15). *Brearley* (1990) used a thermometer based on the degree of ordering of poorly graphitized C (*Reitmeijer and MacKinnon,* 1985) to derive temperatures of 350°–400°C for type 3.4–3.6 chondrites. On the other hand, *Sears et al.* (1991b) inferred temperatures of 500°–600°C for type 3.5 chondrites based on the order/disorder transition of feldspar. *Wlotzka* (1985, 1987) used the Cr-spinel/olivine thermometer to estimate equilibration temperatures of 600°–700°C for Dhajala (H3.8), Study Butte (H3.7), and Clovis (no. 1) (H3.7), whereas *McCoy et al.* (1991) estimated peak temperatures for type 3.7–3.8 ordinary chondrites of 525°–600°C. *Huss and Lewis* (1994) used these data and the amounts of P3 noble gases in presolar diamonds to construct a temperature scale for type 3 ordinary chondrites (and other primitive meteorites) ranging from ~200°C for Semarkona (LL3.00) to ~600°C for Parnallee (LL3.6). This temperature scale is a smooth (but probably not linear) function of petrologic type within a meteorite class.

Type 4–6 ordinary chondrites have reached higher temperatures than type 3s, with an upper limit defined by the onset of melting at ~950°C (*Dodd,* 1981). Some partially melted chondrites (i.e., primitive achondrites) may have reached as much as 1150°C (*McSween and Patchen,* 1989). Equilibration temperatures for type 6 chondrites based on Ca-Mg partitioning between coexisting high-Ca pyroxenes are systematically 50°–150°C higher than those based on similar partitioning in low-Ca pyroxene (*McSween and Patchen,* 1989; *Harvey et al.,* 1993) and ~200°C higher than those based on olivine-spinel equilibrium (*Kessel et al.,* 2002). These temperatures reflect differences in closure temperatures for the three systems (*Kessel et al.,* 2002). High-Ca pyroxene probably gives the best estimate for peak metamorphic temperature in type 6 chondrites: 900°–960°C for LL6, 850°–950°C for L6, and 800°–950°C for H6 (*McSween and Patchen,* 1989; *Harvey et al.,* 1993). Temperatures for LL6 chondrites seem to be the best determined. Several studies have shown evidence for decreasing peak temperature from LL6 to L6 to H6 (*Dodd,* 1981; *Olsen and Bunch,* 1984; *Harvey et al.,* 1993), but equilibration temperatures reflect cooling rate as well as peak temperature. Temperatures for types 4 and 5 chondrites are hard to estimate because some mineral thermometers have not equilibrated (*McSween and Patchen,* 1989).

The widely used temperature scale given by *Dodd* (1981) implies that peak temperatures increase from type 4 to 5 to 6. However, there is some evidence from mineral-equilibration thermometers that type 4 and 5 chondrites may have reached peak temperatures similar to those for type 6 chondrites, but then cooled more quickly (*Harvey et al.,* 1993; *Kessel et al.,* 2002). Additional work to separate peak temperature from cooling profile is clearly needed.

4.2. CO3 Chondrites

The least-metamorphosed CO3 chondrites, ALH A77307 (CO3.0) and Colony (CO3.0), probably experienced metamorphic temperatures of ~200°C, based on the absence of metamorphic effects on a micrometer scale in their silicates and on the abundances and characteristics of presolar grains (*Brearley,* 1993; *Huss et al.,* 2003; *Kobayashi et al.,* 2005). Kainsaz (CO3.2) may have been as hot as ~300°C (*Huss and Lewis,* 1994), similar to ordinary chondrites like Bishunpur (LL3.15). *Sears et al.* (1991a) argued on the basis of TL glow curves that the CO3 chondrites were heated to lower temperatures than type 3 ordinary chondrites, but for longer times. However, this conclusion is based in part on higher temperatures for type 3 ordinary chondrites (400°C for type 3.0–3.1 and 600°–700°C for type 3.6–3.8) than are currently favored (see above).

4.3. CV3 Chondrites

Assigning absolute temperatures to CV3 metamorphism is difficult because of their complex histories before and after accretion (see above). Precursor materials for matrix, the component where most temperature indicators are found, may have been heated significantly prior to accretion (*Huss et al.,* 2003). Despite these complications, *Huss and Lewis* (1994) used gas-release profiles from presolar diamonds and several methods of estimating temperature to infer maximum metamorphic temperatures of ~300°C for Leoville and Vigarano, two of the most primitive $CV3_{red}$ meteorites. Mokoia ($CO3_{ox}$ breccia) may have experienced a similar metamorphic temperature (*Huss et al.,* 2003). *Bonal et al.* (2004) placed the CV3 chondrites into a metamorphic sequence based on Raman spectroscopy of organic material and tied the resulting sequence to that for type 3 ordinary chondrites. Using the *Bonal et al.* (2004) sequence, the characteristics of presolar grains, and temperatures derived by *Huss and Lewis* (1994), the following approximate temperatures can be assigned to the CV3 chondrites: Leoville, ~250°C; Vigarano, Mokoia, Kaba, Efremovka, 300°–400°C; Grosnaja, ~500°C; Axtell, Allende, 550°–600°C.

4.4. Enstatite Chondrites

Enstatite chondrites in general seem to have experienced more-severe thermal and shock metamorphism than most other meteorites. *Huss and Lewis* (1994) used noble gases in presolar diamonds to estimate maximum temperatures for Qingzhen (EH3) and Indarch (EH4) of ~525°C and ~640°C, respectively. Disturbed ^{26}Al-^{26}Mg, ^{53}Mn-^{53}Cr, and ^{60}Fe-^{60}Ni systematics in E3 chondrules and sulfides (*Guan et al.,* 2004, 2005) also indicate that E3 chondrites experienced higher metamorphic temperatures than type 3.0–3.5 ordinary chondrites. Maximum metamorphic temperatures for E5 and E6 chondrites are poorly constrained. Various thermometers suggest that E6 chondrites reached temperatures similar to or somewhat higher than type 6 ordinary chondrites (*Dodd,* 1981). Equilibration temperatures for EH5 and EH6 chondrites were in the range of 800°–1000°C, but EL5 and EL6 chondrites may have equilibrated at 600°–800°C (*Zhang et al.,* 1996). Many enstatite chondrites exhibit evidence of moderate to severe shock, which may have heated them to temperatures equal to or higher than their metamorphic temperatures after which they were rapidly quenched (*Rubin and Scott,* 1997). Separating effects of static and shock metamorphism is particularly difficult for enstatite chondrites.

4.5. Other Classes

Temperatures of metamorphism for other chondrite classes are poorly determined. Renazzo (CR2) apparently was heated to ~150°–300°C on its parent body (*Huss et al.,* 2003, and references therein). CH and CB chondrites show little if any evidence of parent-body metamorphism (*Weisberg et al.,* 2001; *Krot et al.,* 2002), which implies similarly low temperatures. The metamorphic sequence shown by the CK chondrites suggests temperatures similar to those of ordinary chondrites (*Geiger and Bischoff,* 1991).

5. OXIDATION STATE DURING METAMORPHISM

5.1. Ordinary Chondrites

There is disagreement in the literature about the extent to which oxygen fugacity (f_{O_2}) may have changed as a function of metamorphic grade in ordinary chondrites. Observations relevant to this issue are (1) the amount of metal appears to increase from type 3 to 4 [e.g., ~11 vol% in H3 to ~16 vol% in H4–6 (*Lux et al.,* 1980)]; (2) the C abundance decreases from type 3 to 4 (Table 1); (3) the FeO content of olivine and pyroxene increases slightly but systematically from type 4 to 6 (*Scott et al.,* 1986; *Rubin,* 1990); (4) metal abundance decreases slightly from type 4 to 6 (*McSween and Labotka,* 1993); and (5) the Ni and Co content of metal increases slightly from type 4 to 6 (*Rubin,* 1990; *McSween and Labotka,* 1993).

Observations 1 and 2 imply that type 3 material was reduced by reactions with C as it was metamorphosed to type 4 (e.g., *Lux et al.,* 1980). However, the data are not as clear-cut as they first appear. Among H3 chondrite falls, Sharps has an anomalously low metal content (cf. *Jarosewich,* 1990) and is primarily responsible for the trend in H chondrites. Among L-chondrite falls, there is only a hint of a trend, and among LL falls, the type 3s seem to have higher metal contents than types 4–6 (*Jarosewich,* 1990). Thus, while reduction by C is reasonable, the data supporting the idea are not convincing.

Observations 3–5 have been interpreted to indicate oxidation during metamorphism from type 4–6 (*McSween and Labotka,* 1993). These authors also calculated oxygen fugacities from chondrite mineral equilibria and found that f_{O_2} increases slightly between types 4 and 6. Their results are at odds with earlier intrinsic f_{O_2} measurements, which appear to show reduction during metamorphism (*Brett and Sato,* 1984), and with earlier calculated oxygen fugacities (*McSween and Labotka,* 1993, and references therein). More recently, *Kessel et al.* (2004) found no evidence for major shifts in f_{O_2} among type 4–6 H chondrites using olivine-orthopyroxene-metal assemblages, and they suggest that apparent changes in calculated f_{O_2} from type 4–6 might reflect higher peak temperatures for type 4 and 5 chondrites during metamorphism than are currently favored.

5.2. Other Classes

Other chondrite classes were formed and metamorphosed under very different conditions. The R and CK chondrites are highly oxidized. They contain olivine that is enriched in FeO and NiO, they have low modal metal abundances, and the metal that is present tends to be Ni rich (*Rubin and Kallemeyn,* 1989; *Bischoff et al.,* 1994; *Geiger and Bischoff,* 1991). At the other extreme, enstatite chondrites are highly reduced and consist of FeO-free silicates and have high metal contents and a wide variety of exotic sulfides (e.g., *Dodd,* 1981). However, details of the metamorphic environments for these and other meteorite classes are not understood.

6. HEAT SOURCES

Two fundamentally different scenarios have been proposed to explain the sequence of textural and compositional changes among chondritic meteorites. In the autometamorphism model (also known as "hot accretion"), batches of hot (~700°–900°C) material accreted to a parent body and began to cool (*Anders,* 1964; *Larimer and Anders,* 1967; *Hutchison et al.,* 1980; *Hutchison,* 1996). In this model, rapidly cooled material developed into type 3 chondrites, while very slowly cooled material became type 6 chondrites. *Rubin and Brearley* (1996) evaluated the arguments for hot accretion and concluded that this model is untenable. There is now a consensus that chondrites experienced prograde thermal metamorphism. In the prograde-metamorphism model, cold, unequilibrated material was progressively heated on a parent body (*Merrill,* 1921; *Wood,* 1962; *Van Schmus and Koffman,* 1967; *Dodd,* 1969; *Lux et al.,* 1980). The heat sources that have been proposed to drive prograde metamorphism in chondrites include energy from collisions, radioactive decay (^{26}Al and ^{60}Fe), electromagnetic induction, and FU-Orionis-type events. The heat source(s) that metamorphosed chondrites may also have been responsible for melting some asteroids, producing differentiated meteorites (HEDs, mesosiderites, pallasites, angrites, and magmatic irons).

6.1. Accretional or Collisional Heating

The accretion process that built asteroids from smaller planetesimals converted kinetic energy into heat. However, the amount of energy released during the gentle impacts that permitted asteroids to grow rather than disrupt is too small to have resulted in significant heating (*Wood and Pellas,* 1991). Once an asteroid formed, higher-velocity impacts could have deposited significant energy without disrupting the body. There is a positive correlation between petrologic type and shock stage among ordinary chondrites (*Stöffler et al.,* 1991; *Rubin,* 1995, 2003), and there is also a correspondence among chondrite groups between those with many metamorphosed members and those with many shocked members (*Rubin,* 2004). Moreover, many type 5–6 ordinary chondrites of apparently low shock stage (S1) show evidence for annealing after an earlier episode of more intense shock (*Rubin,* 2004). These observations suggest that shock heating has played a significant role in metamorphism of ordinary chondrites.

6.2. Short-lived Radionuclides

Two short-lived radionuclides may have been significant heat sources: ^{26}Al and ^{60}Fe. Other radionuclides were also present, but either they were incorporated into parent bodies with abundances too low to produce significant heating (^{41}Ca, ^{10}Be), or because of long half-lives, heat generation was too slow to overcome heat loss (^{53}Mn).

6.2.1. Aluminum-26. The presence in the early solar system of live ^{26}Al ($t_{1/2}$ = 0.72 m.y.) has been inferred from a wide variety of chondritic materials (e.g., *Lee et al.,* 1976; *MacPherson et al.,* 1995). The highest inferred initial ^{26}Al/^{27}Al ratios [(^{26}Al/^{27}Al)$_0$] are from CAIs, and the canonical initial ratio for the solar system ($\sim 5 \times 10^{-5}$) has been derived from CAI data. Advocates of ^{26}Al heating generally assume that this isotope was homogeneously distributed in the solar system. Chondrules in ordinary chondrites have lower inferred (^{26}Al/^{27}Al)$_0$ ratios ($\sim 1 \times 10^{-5}$), indicating that they formed ~2 m.y. after CAIs (*Kita et al.,* 2000, 2005). Accretion of chondritic asteroids probably postdated chondrule formation. Asteroid thermal evolution models based on these assumptions show that an asteroid of ~100 km in diameter could have reached a maximum temperature of >1000°C at its center and sustained it for more than 10 m.y. (*Miyamoto et al.,* 1981; *Bennett and McSween,* 1996; *McSween et al.,* 2002). This is ample heating to account for metamorphism in ordinary chondrites.

6.2.2. Iron-60. Recent work has shown that ^{60}Fe ($t_{1/2}$ = 1.5 m.y.) was incorporated into chondritic meteorites with inferred (^{60}Fe/^{56}Fe)$_0$ ratios of between $\sim 3 \times 10^{-7}$ and $\sim 1 \times 10^{-6}$ (*Tachibana and Huss,* 2003; *Tachibana et al.,* 2005; *Mostefaoui et al.,* 2003). If incorporated into chondrites at this level, ^{60}Fe would have been able to raise the internal temperatures of asteroids with LL and CV compositions by 140°C and 170°C, respectively (*Kunihiro et al.,* 2004). Thus, ^{60}Fe would have been a significant secondary heat source for metamorphism.

6.3. Electromagnetic Induction

The active pre-main-sequence Sun emitted an intense, high-density, at least partly ionized solar wind at an outflow velocity of several hundred kilometers per second. Such a wind would have carried an embedded magnetic field and might have been capable of driving electrical currents in asteroids that were sufficiently conductive (e.g., *Herbert and Sonett,* 1979; *Herbert,* 1989). Heating would have occurred in the asteroid interior where heat is more efficiently retained. However, T Tauri stars blow off much of their mass at their poles, not at the midplane where asteroids would reside, and induction-heating models typically assume unrealistically high outflow velocities and solar wind fluxes (*Wood and Pellas,* 1991). Most workers do not currently believe that electromagnetic induction was a significant source of asteroidal heating.

6.4. FU-Orionis-type Events

Pre-main-sequence stars apparently experience periodic outbursts during which brightness increases dramatically (up to a factor of 250) over periods as short as 1 year and then decreases to pre-outburst levels over 10–100 years (*Herbig,* 1978; *Bell et al.,* 2000). T Tauri stars may be in an outburst state ~10% of the time. *Wasson* (1992) suggested that these outbursts could have heated the outer ~100 m of planetesimals at a distance of ~3 AU to >1200°C for several hundred years. This type of heating would occur only if the nebula had largely cleared between the Sun and the asteroids. However, the outbursts primarily occur when stars have massive accretion disks (*Bell et al.,* 2000). As a consequence, most workers do not consider this heating mechanism to be viable.

7. TIMESCALE FOR METAMORPHISM

Estimates of metamorphic timescales come primarily from radiometric age dating, supported in some cases by estimates of cooling rates. Essentially all applicable data come from ordinary chondrites. Short-lived nuclides provide constraints on the early stages of the metamorphic epoch, if these nuclides were distributed homogeneously in the early solar system. The inferred (^{26}Al/^{27}Al)$_0$ ratios for chondrules from Semarkona (LL3.00) and Bishunpur (LL3.15) indicate that chondrule formation took place 1–2 m.y. after CAIs (*Kita et al.,* 2000, 2005). Inferred (^{60}Fe/^{56}Fe)$_0$ ratios for chondrules, magnetite, and sulfides in Semarkona, Bishunpur, and Krymka (LL3.2) imply that parent-body alteration of these meteorites took place 0.5–2.0 m.y. after chondrules formed (*Tachibana et al.,* 2005). The (^{26}Al/^{27}Al)$_0$ ratios for a glassy clast from Bovedy (L3.7) and for plagioclase from Ste. Marguerite (H4) imply that these meteorites had cooled below the blocking temperature for the Al-Mg system by ~6 m.y. after CAIs formed, and the upper limit on (^{26}Al/^{27}Al)$_0$ for a basaltic clast from Barwell (L6) implies that Barwell did not cool down until at least 7 m.y. after CAIs formed (*Huss et al.,* 2001). These data imply that, relative to the time of CAI formation, accretion of ordinary chondrites took place ~2 m.y. later, metamorphism of type 3.0–3.2 chondrites took place between 2 and 4 m.y. later, in type 3.7–4 chondrites lasted until ~6 m.y. later, and in type 6 chondrites continued for >7 m.y. after CAIs formed.

Data for ^{244}Pu-fission-track and Pb-Pb dating give a coherent picture of H-chondrite metamorphism and appear to support an onion-shell model for the H-chondrite parent body (*Trieloff et al.,* 2003). The two methods have different closure temperatures and thus date different points in the cooling history of the H-chondrite parent body. The Pb-Pb ages, which reflect cooling below ~750 K, indicate that metamorphism of H4 chondrites was ending 4–6 m.y. after CAIs, consistent with estimates from ^{26}Al. Lead-lead data further suggest that metamorphism ended for H5 and H6 chondrites 10–16 m.y. and 45–63 m.y. after CAIs, respectively. Fission tracks are retained at ~550 K (pyroxene) or ~390 K (merrilite), so ^{244}Pu fission track ages reflect a later stage in the cooling history. These ages indicate that metamorphism ended ~10 m.y. for H4, 50–65 m.y. for H5, and 55–65 m.y. for H6 chondrites after CAIs formed (*Trieloff et al.,* 2003). These cooling times are generally consistent with ^{26}Al (and ^{60}Fe) as the heat source for metamorphism (cf. *Ghosh and McSween,* 1998). However, an extensive set of metallographic cooling rates for H chondrites, many of which were not studied by *Trieloff et al.* (2003), do not show a consistent relationship between cooling rate and petrologic type (*Taylor et al.,* 1987), suggesting that the H-chondrite parent body did not have an onion-shell structure, or if it did, that this structure was disrupted before the parent body cooled below ~800 K.

Acknowledgments. The authors thank R. Jones, M. Miyamoto, and associate editor H. Y. McSween Jr. for their detailed and constructive reviews. This work was supported by NASA grants NAG5-11543 and NNG05GG48G (G.R.H.), NAG5-12967 (A.E.R.), and order numbers W-10,087 and W-19,575 (J.N.G.).

REFERENCES

Afiattalab F. and Wasson J. T. (1980) Composition of the metal phases in ordinary chondrites: Implications regarding classification and metamorphism. *Geochim. Cosmochim. Acta, 44,* 431–446.

Alaerts L., Lewis R. S., and Anders E. (1979) Isotopic anomalies of noble gases in meteorites and their origins — III. LL-chondrites. *Geochim. Cosmochim. Acta, 43,* 1399–1425.

Alexander C. M. O'D., Barber D. J., and Hutchison R. H. (1989a) The microstructure of Semarkona and Bishunpur. *Geochim. Cosmochim. Acta, 53,* 3045–3057.

Alexander C. M. O'D., Hutchison R., and Barber D. J. (1989b) Origin of chondrule rims and interchondrule matrices in unequilibrated ordinary chondrites. *Earth Planet. Sci. Lett., 95,* 187–207.

Alexander C. M. O'D., Grossman J. N., Wang J., Zanda B., Bourot-Denise M., and Hewins R. H. (2000) The lack of potassium-isotopic fractionation in Bishunpur chondrules. *Meteoritics & Planet. Sci., 35,* 859–868.

Anders E. (1964) Origin, age and composition of meteorites. *Space Sci. Rev., 3,* 583–714.

Ashworth J. R. (1977) Matrix textures in unequilibrated ordinary chondrites. *Earth Planet. Sci. Lett., 35,* 25–34.

Ashworth J. R. and Barber D. J. (1976) Lithification of gas-rich meteorites. *Earth Planet. Sci. Lett., 30,* 222–223.

Bell K. R., Cassen P. M., Wasson J. T., and Woolum D. S. (2000) The FU Orionis phenomenon and solar nebula material. In *Protostars and Planets IV* (V. Mannings et al., eds.), pp. 897–926. Univ. of Arizona, Tucson.

Bennett M. E. III and McSween H. Y. Jr. (1996) Revised model calculations for the thermal histories of ordinary chondrite parent bodies. *Meteoritics, 31,* 783–792.

Benoit P. H., Akridge G., and Sears D. W. G. (1998) Size sorting of metal, sulfide, and chondrules in Sharps (H3.4) (abstract). In *Lunar and Planetary Science XXIX,* Abstract #1457. Lunar and Planetary Institute, Houston (CD-ROM).

Benoit P. H., Akridge G. A., Ninagawa K., and Sears D. W. G. (2002) Thermoluminescence sensitivity and thermal history of type 3 ordinary chondrites: Eleven new type 3.0–3.1 chondrites and possible explanations for differences among H, L, and LL chondrites. *Meteoritics & Planet. Sci., 37,* 793–805.

Bischoff A. (2000) Mineralogical characterization of primitive, type-3 lithologies in Rumuruti chondrites. *Meteoritics & Planet. Sci., 35,* 699–706.

Bischoff A., Geiger T., Palme H., Spettel B., Schultz L., Scherer P., Loeken T., Bland P., Clayton R. N., Mayeda T. K., Herpers U., Meltzow B., Michel R. and Dittrich-Hannen B. (1994) Acfer 217 — A new member of the Rumuruti chondrite group (R). *Meteoritics, 29,* 264–274.

Binns R. A. (1967) Olivine in enstatite chondrites. *Am. Mineral., 52,* 1549–1554.

Bonal L., Quirico E., and Bourot-Denise M. (2004) Petrologic type of CV3 chondrites as revealed by Raman spectroscopy of organic matter (abstract). In *Lunar and Planetary Science XXXV,* Abstract #1562. Lunar and Planetary Institute, Houston (CD-ROM).

Boyd F. R. and England J. L. (1965) The rhombic enstatite-clinoenstatite inversion. In *Year Book — Carnegie Institution of Washington,* pp. 117–120. Carnegie Institution of Washington, Washington, DC.

Brearley A. J. (1990) Carbon-rich aggregates in type 3 ordinary chondrites: Characterization, origins, and thermal history. *Geochim. Cosmochim. Acta, 54,* 831–850.

Brearley A. J. (1993) Matrix and fine-grained rims in the unequilibrated CO3 chondrite, ALHA77307: Origins and evidence for diverse, primitive nebular dust components. *Geochim. Cosmochim. Acta, 57,* 1521–1550.

Brearley A. J. (1995) Nature of matrix in unequilibrated chondrites and its possible relationship to chondrules. In *Chondrules and the Protoplanetary Disk* (R. H. Hewins et al., eds.), pp. 137–151. Cambridge Univ., Cambridge.

Brearley A. J. and Jones R. H. (1998) Chondritic meteorites. In *Planetary Materials* (J. J. Papike, ed.,) pp. 3-1 to 3-398. Reviews in Mineralogy, Vol. 36, Mineralogical Society of America.

Brett R. and Sato M. (1984) Intrinsic oxygen fugacity measurements on seven chondrites, a pallasite, and a tektite and the redox state of meteorite parent bodies. *Geochim. Cosmochim. Acta, 48,* 111–120.

Chizmadia L. J. (2000) Amoeboid olivine inclusions in CO3 carbonaceous chondrites: Sensitive indicators of parent body aqueous alteration. Master's thesis, Univ. of California, Los Angeles.

Chizmadia L. J., Rubin A. E., and Wasson J. T. (2002) Mineralogy and petrology of amoeboid olivine inclusions in CO3 chondrites: Relationship to parent-body aqueous alteration. *Meteoritics & Planet. Sci., 38,* 1781–1796.

Christophe Michel-Levy M. (1976) La matrice noire et blanche de la chondrite de Tieschitz (H3). *Earth Planet. Sci. Lett., 30,* 143–150.

Crabb J. and Anders E. (1981) Noble gases in E-chondrites. *Geochim. Cosmochim. Acta, 45,* 2443–2464.

Crabb J. and Anders E. (1982) On the siting of noble gases in E-chondrites. *Geochim. Cosmochim. Acta, 46,* 2351–2361.

DeHart J. M., Lofgren G. E., Lu J., Benoit P. H., and Sears D. W. G. (1992) Chemical and physical studies of chondrules X. Cathodoluminescence and phase composition studies of metamorphism and nebular processes in chondrules of type 3 ordinary chondrites. *Geochim. Cosmochim. Acta, 56,* 3791–3807.

Dodd R. T. (1969) Metamorphism of the ordinary chondrites: A review. *Geochim. Cosmochim. Acta, 33,* 161–203.

Dodd R. T. (1981) *Meteorites — A Petrologic-Chemical Synthesis.* Cambridge, New York. 368 pp.

Dodd R. T., Van Schmus W. R., and Koffman D. M. (1967) A survey of the unequilibrated ordinary chondrites. *Geochim. Cosmochim. Acta, 31,* 921–951.

Dodd R. T., Grover J. E., and Brown G. E. (1975) Pyroxenes in the Shaw (L-7) chondrite. *Geochim. Cosmochim. Acta, 39,* 1585–1594.

Geiger T. and Bischoff A. (1991) The CK chondrites — Conditions of parent body metamorphism. *Meteoritics, 26,* 337.

Ghosh A. and McSween H. Y. Jr. (1998) A thermal model for the differentiation of asteroid 4 Vesta based on radiogenic heating. *Icarus, 134,* 187–206.

Grossman J. N. and Brearley A. J. (2005) The onset of metamorphism in ordinary and carbonaceous chondrites. *Meteoritics & Planet. Sci., 40,* 87–122.

Grossman J. N., Zolensky M. E., and Tonui E. K. (2005) What are the petrologic types of thermally metamorphosed CM chondrites? (abstract). *Meteoritics & Planet. Sci., 40,* A61.

Guan Y., Huss G. R., and Leshin L. A. (2004) Further observations of ^{60}Fe-^{60}Ni and ^{53}Mn-^{53}Cr systems in sulfides from enstatite chondrites (abstract). In *Lunar and Planetary Science XXXV,* Abstract #2003. Lunar and Planetary Institute, Houston (CD-ROM).

Guan Y., Huss G. R., Leshin L. A., MacPherson G. J., and McKeegan K. D. (2005) Oxygen isotopes and ^{26}Al/^{26}Mg systematics of aluminum-rich chondrules from unequilibrated enstatite chondrites. *Meteoritics & Planet. Sci., 40,* in press.

Guimon R. K., Keck D. B., Weeks S. K., Dehart J., and Sears G. (1985) Chemical and physical studies of type 3 chondrites — IV: Annealing studies of a type 3, 4 ordinary chondrite and the metamorphic history of meteorites. *Geochim. Cosmochim. Acta, 49,* 1515–1524.

Guimon R. K., Lofgren G. E., and Sears D. W. G. (1988) Chemical and physical studies of type 3 chondrites. IX: Thermoluminescence and hydrothermal annealing experiments and their relationship to metamorphism and aqueous alteration in type <3.3 ordinary chondrites. *Geochim. Cosmochim. Acta, 52,* 119–127.

Guimon R. K., Symes S. J. K., Sears D. W. G., and Benoit P. H. (1995) Chemical and physical studies of type 3 chondrites XII: The metamorphic history of CV chondrites and their components. *Meteoritics, 30,* 704–714.

Harvey R. P., Bennett M. L., and McSween H. Y. Jr. (1993) Pyroxene equilibration temperatures in metamorphosed ordinary chondrites (abstract). In *Lunar and Planetary Science XXIV,* pp. 615–616. Lunar and Planetary Institute, Houston.

Herbert F. (1989) Primordial electrical induction heating of asteroids. *Icarus, 78,* 402–410.

Herbert F. and Sonett C. P. (1979) Electromagnetic heating of minor planets in the early solar system. *Icarus, 40,* 484–496.

Herbig G. H. (1978) Some aspects of early stellar evolution that may be relevant to the origin of the solar system. In *The Origin of the Solar System* (S. F. Dermott, ed.), pp. 219–235. Wiley, New York.

Housley R. M. and Cirlin E. H. (1983) On the alteration of Allende chondrules and the formation of matrix. In *Chondrules and Their Origins* (E. A. King, ed.), pp. 145–161. Lunar and Planetary Institute, Houston.

Hua X., Huss G. R., Tachibana S., and Sharp T. G. (2005) Oxygen, silicon, and Mn-Cr isotopes of fayalite in the Kaba oxidized CV3 chondrite: Constraints for its formation history. *Geochim. Cosmochim. Acta, 69,* 1333–1348.

Huss G. R. (1979) The matrix of unequilibrated ordinary chondrites: Implications for the origin and subsequent history of chondrites. Master's thesis, Univ. of New Mexico, Albuquerque.

Huss G. R. (1990) Ubiquitous interstellar diamond and SiC in primitive chondrites: Abundances reflect metamorphism. *Nature, 347,* 159–162.

Huss G. R. and Lewis R. S. (1994) Noble gases in presolar diamonds II: Component abundances reflect thermal processing. *Meteoritics, 29,* 811–829.

Huss G. R. and Lewis R. S. (1995) Presolar diamond, SiC, and graphite in primitive chondrites: Abundances as a function of meteorite class and petrologic type. *Geochim. Cosmochim. Acta, 59,* 115–160.

Huss G. R., Keil K., and Taylor G. J. (1978) Composition and recrystallization of the matrix of unequilibrated (type 3) ordinary chondrites (abstract). *Meteoritics, 13,* 495.

Huss G. R., Keil K., and Taylor G. J. (1981) The matrices of unequilibrated ordinary chondrites: Implications for the origin and history of chondrites. *Geochim. Cosmochim. Acta, 45,* 33–51.

Huss G. R., Lewis R. S., and Hemkin S. (1996) The "normal planetary" noble gas component in primitive chondrites: Compositions, carrier, and metamorphic history. *Geochim. Cosmochim. Acta, 60,* 3311–3340.

Huss G. R., MacPherson G. J., Wasserburg G. J., Russell S. S., and Srinivasan G. (2001) Aluminum-26 in calcium-aluminum-rich inclusions and chondrules from unequilibrated ordinary chondrites. *Meteoritics & Planet. Sci., 36,* 975–997.

Huss G. R., Meshik A. P., Smith J. B., and Hohenberg C. M. (2003) Presolar diamond, silicon carbide, and graphite in carbonaceous chondrites: Implications for thermal processing in the solar nebula. *Geochim. Cosmochim. Acta, 67,* 4823–4848.

Hutchison R. (1996) Hot accretion of the ordinary chondrites: The rocks don't lie (abstract). In *Lunar and Planetary Science XXVII,* pp. 579–580. Lunar and Planetary Institute, Houston.

Hutchison R., Bevan A. W. R., Agrell S. O., and Ashworth J. R. (1980) Thermal history of the H-group of chondritic meteorites. *Nature, 287,* 787–790.

Hutchison R., Alexander C. M. O'D., and Barber D. J. (1987) The Semarkona meteorite; first recorded occurrence of smectite in an ordinary chondrite, and its implication. *Geochim. Cosmochim. Acta, 51,* 1875–1882.

Itoh D. and Tomeoka K. (2003) Dark inclusions in CO3 chondrites: New indicators of parent-body processes. *Geochim. Cosmochim. Acta, 67,* 153–169.

Jarosewich E. (1990) Chemical-analyses of meteorites — A compilation of stony and iron meteorite analyses. *Meteoritics, 25,* 323–337.

Jones R. H. and Lofgren G. E. (1993) A comparison of FeO-rich, porphyritic olivine chondrules in unequilibrated chondrites and experimental analogues. *Meteoritics, 28,* 213–221.

Kallemeyn G. W. and Rubin A. E. (1995) Coolidge and Loongana-001 — A new carbonaceous chondrite grouplet. *Meteoritics, 30,* 20–27.

Kallemeyn G. W., Rubin A. E., Wang D., and Wasson J. T. (1989) Ordinary chondrites: Bulk compositions, classification, lithophile-element fractionations, and composition-petrographic type relationships. *Geochim. Cosmochim. Acta, 53,* 2747–2767.

Kallemeyn G. W., Rubin A. E., and Wasson J. T. (1991) The compositional classification of chondrites: V. The Karoonda (CK) group of carbonaceous chondrites. *Geochim. Cosmochim. Acta, 55,* 881–892.

Kallemeyn G. W., Rubin A. E., and Wasson J. T. (1996) The compositional classification of chondrites: VII. The R chondrite group. *Geochim. Cosmochim. Acta, 60,* 2243–2256.

Keller L. P. and Buseck P. R. (1990) Matrix mineralogy of the Lancé CO3 carbonaceous chondrite: A transmission electron microscope study. *Geochim. Cosmochim. Acta, 54,* 1155–1163.

Kessel R., Beckett J. R., and Stolper E. M. (2002) The thermal history of equilibrated ordinary chondrites and the relationship between textural maturity and temperature (abstract). In *Lunar and Planetary Science XXXIII,* Abstract #1420. Lunar and Planetary Institute, Houston (CD-ROM).

Kessel R., Beckett J. R., Huss G. R., and Stolper E. M. (2004) The activity of chromite in multicomponent spinels: Implications for T-f_{O_2} conditions of equilibrated H chondrites. *Meteoritics & Planet. Sci., 39,* 1287–1305.

Kimura M. and Ikeda Y. (1995) Anyhdrous alteration of Allende chondrules in the solar nebula. II: Alkali-Ca exchange reactions and formation of nepheline, sodalite and Ca-rich phases in chondrules. *Proc. NIPR Symp. Antarct. Meteorites, 8,* 123–138.

Kita N. T., Nagahara H., Togashi S., and Morishita Y. (2000) A short duration of chondrule formation in the solar nebula: Evidence from ^{26}Al in Semarkona ferromagnesian chondrules. *Geochim. Cosmochim. Acta, 64,* 3913–3922.

Kita N. T., Huss G. R., Tachibana S., Amelin Y., Nyquist L. E., and Hutcheon I. D. (2005) Constraints on the origin of chondrules and CAIs from short-lived radionuclides. In *Chondrites and the Protoplanetary Disk* (A. N. Krot et al., eds.), pp. 558–587. ASP Conference Series 341, Astronomical Society of the Pacific, San Francisco.

Kobayashi S., Tonotani A., Sakamoto K., Nagashima K, Krot A. N., and Yurimoto H. (2005) Presolar silicate grains from primitive carbonaceous chondrites Y-81025, ALHA 77307, Adelaide and Acfer 094 (abstract). In *Lunar and Planetary Science XXXVI,* Abstract #1931. Lunar and Planetary Institute, Houston (CD-ROM).

Krot A. N., Scott E. R. D., and Zolensky M. E. (1995) Mineralogical and chemical modification of components in CV3 chondrites: Nebular or asteroidal processing? *Meteoritics, 30,* 748–775.

Krot A. N., Zolensky M. E., Wasson J. T., Scott E. R. D., Keil K., and Oshumi K. (1997) Carbide-magnetite assemblages in type 3 ordinary chondrites. *Geochim. Cosmochim. Acta, 61,* 219–237.

Krot A. N., Petaev M. I., Scott E. R. D., Choi B.-G., Zolensky M. E., and Keil K. (1998a) Progressive alteration in CV3 chondrites: More evidence for asteroidal alteration. *Meteoritics & Planet. Sci., 33,* 1065–1085.

Krot A. N., Petaev M. I., Zolensky M. E., Keil K., Scott E. R. D., and Nakamura K. (1998b) Secondary calcium-iron-rich minerals in the Bali-like and Allende-like oxidized CV3 chondrites and Allende dark inclusions. *Meteoritics & Planet. Sci., 33,* 623–645.

Krot A. N., Meibom A., Weisberg M. K., and Keil K. (2002) The CR chondrite clan: Implications for early solar system processes. *Meteoritics & Planet. Sci., 37,* 1451–1490.

Kunihiro T., Rubin A. E., McKeegan K. D., and Wasson J. T. (2004) Initial ^{26}Al/^{27}Al in carbonaceous-chondrite chondrules: Too little ^{26}Al to melt asteroids. *Geochim. Cosmochim. Acta, 68,* 2947–2957.

Larimer J. W. and Anders E. (1967) Chemical fractionation of meteorites, II. Abundance patterns and their interpretation. *Geochim. Cosmochim. Acta, 31,* 1239–1270.

Lee T., Papanastassiou D. A., and Wasserburg G. J. (1976) Demonstration of Mg-26 excess in Allende and evidence for Al-26. *Geophys. Res. Lett., 3,* 41–44.

Lipschutz M. E. and Woolum D. S. (1988) Highly labile elements. In *Meteorites and the Early Solar System* (J. F. Kerridge and M. S. Matthews, eds.), pp. 462–487. Univ. of Arizona, Tucson.

Lux G., Keil K. and Taylor G. J. (1980) Metamorphism of the H-group chondrites: Implications from compositional and textural trends in chondrules. *Geochim. Cosmochim. Acta, 44,* 841–855.

MacPherson G. J., Davis A. M., and Zinner E. K. (1995) The distribution of aluminum-26 in the early solar system — A reappraisal. *Meteoritics, 30,* 365–386.

McCoy T. J., Scott E. R. D., Jones R. H., Keil K., and Taylor G. J. (1991) Composition of chondrule silicates in LL3–5 chondrites and implications for their nebular history and parent body metamorphism. *Geochim. Cosmochim. Acta, 55,* 601–619.

McCoy T. J., Keil K., Bogard D. D., Garrison D. H., Casanova I., Lindstrom M. M., Brearley A. J., Kehm K., Nichols R. H. Jr., and Hohenberg C. M. (1995) Origin and history of impact-melt rocks of enstatite chondrite parentage. *Geochim. Cosmochim. Acta, 59,* 161–175.

McSween H. Y. Jr. (1977a) Carbonaceous chondrites of the Ornans type: A metamorphic sequence. *Geochim. Cosmochim. Acta, 41,* 477–491.

McSween H. Y. Jr. (1977b) Petrographic variations among carbonaceous chondrites of the Vigarano type. *Geochim. Cosmochim. Acta, 41,* 1777–1790.

McSween H. Y. Jr. (1979) Are carbonaceous chondrites primitive or processed? A review. *Rev. Geophys. Space Phys., 17,* 1059–1078.

McSween H. Y. Jr. and Labotka T. C. (1993) Oxidation during metamorphism of the ordinary chondrites. *Geochim. Cosmochim. Acta, 57,* 1105–1114.

McSween H. Y. Jr. and Patchen A. D. (1989) Pyroxene thermobarometry in LL-group chondrites and implications for parent body metamorphism. *Meteoritics, 24,* 219–226.

McSween H. Y. Jr., Ghosh A., Grimm R. E., Wilson L., and Young E. D. (2002) Thermal evolution models of asteroids. In *Asteroids III* (W. F. Bottke Jr. et al., eds.), pp. 559–571. Univ. of Arizona, Tucson.

Merrill G. P. (1921) On metamorphism in meteorites. *Geol. Soc. Am. Bull., 32,* 395–416.

Methot R. L., Noonan A. F., Jarosewich E., Al-Far D. M., and Degasparis A. A. (1975) Mineralogy, petrology and chemistry of the Isna C3 meteorite. *Meteoritics, 10,* 121–130.

Mittlefehldt D. W. and Lindstrom M. M. (2001) Petrology and geochemistry of Patuxent Range 91501, a clast-poor impact melt from the L-chondrite parent body and Lewis Cliff 88663, an L7 chondrite. *Meteoritics & Planet. Sci., 36,* 439–457.

Miyamoto M., Fujii N., and Takeda H. (1981) Ordinary chondrite parent body: An internal heating model. *Proc. Lunar Planet. Sci. 12B,* pp. 1145–1152.

Mostefaoui S., Lugmair G. W., Hoppe P., and El Goresy A. (2003) Evidence for live iron-60 in Semarkona and Chervony Kut: A nanoSIMS study (abstract). In *Lunar and Planetary Science XXXIV,* Abstract #1585. Lunar and Planetary Institute, Houston (CD-ROM).

Mostefaoui S., Marhas K. K., and Hoppe P. (2004) Discovery of an in-situ presolar silicate grain with GEMS-like composition in the Bishunpur matrix (abstract). In *Lunar and Planetary Science XXXV,* Abstract #1593. Lunar and Planetary Institute, Houston (CD-ROM).

Nagahara H. (1984) Matrices of type 3 ordinary chondrites — primitive nebular records. *Geochim. Cosmochim. Acta, 48,* 2581–2595.

Nittler L. R. (2003) Presolar stardust in meteorites: Recent advances and scientific frontiers. *Earth Planet. Sci. Lett., 209,* 259–273.

Olsen E. J. and Bunch T. E. (1984) Equilibration temperatures of the ordinary chondrites: A new evaluation. *Geochim. Cosmochim. Acta, 48,* 1363–1365.

Olsen E., Bunch T., Jarosewich E., and Huss G. I. (1976) Happy Canyon: An E7 enstatite chondrite. *Meteoritics, 11,* 348–349.

Palme H. and Wark D. A. (1988) CV-chondrites: High temperature gas-solid equilibrium vs. parent body metamorphism (abstract). In *Lunar and Planetary Science XIX,* p. 897. Lunar and Planetary Institute, Houston.

Peck J. A. and Wood J. A. (1987) The origin of ferrous zoning in Allende chondrule olivine. *Geochim. Cosmochim. Acta, 51,* 1503–1510.

Prinz M., Nehru C. E., Weisberg M. K., and Delaney J. S. (1984) Type 3 enstatite chondrites: A newly recognized group of unequilibrated enstatite chondrites (UEC's) (abstract). In *Lunar and Planetary Science XIV,* pp. 653–654. Lunar and Planetary Institute, Houston.

Rambaldi E. R. and Wasson J. T. (1981) Metal and associated phases in Bishunpur, a highly unequilibrated ordinary chondrite. *Geochim. Cosmochim. Acta, 45,* 1001–1015.

Reed S. J. B. (1968) Perryite in the Kota-Kota and South Oman enstatite chondrites. *Mineral. Mag., 36,* 850–854.

Reitmeijer F. J. M. and MacKinnon I. D. R. (1985) Poorly graphitized carbon as a new cosmothermometer for primitive extraterrestrial materials. *Nature, 315,* 733–736.

Rubin A. E. (1984) Coarse-grained chondrule rims in type 3 chondrites. *Geochim. Cosmochim. Acta, 48,* 1779–1789.

Rubin A. E. (1990) Kamacite and olivine in ordinary chondrites: Intergroup and intragroup relationships. *Geochim. Cosmochim. Acta, 54,* 1217–1232.

Rubin A. E. (1995) Petrologic evidence for collisional heating of chondritic asteroids. *Icarus, 113,* 156–167.

Rubin A. E. (1998) Correlated petrologic and geochemical characteristics of CO3 chondrites. *Meteoritics & Planet. Sci., 33,* 385–391.

Rubin A. E. (2000) Petrologic, geochemical and experimental constraints on models of chondrule formation. *Earth Sci. Rev., 50,* 3–27.

Rubin A. E. (2003) Chromite-plagioclase assemblages as a new shock indicator; implications for the shock and thermal histories of ordinary chondrites. *Geochim. Cosmochim. Acta, 67,* 2695–2709.

Rubin A. E. (2004) Post-shock annealing and post-annealing shock in equilibrated ordinary chondrites: Implications for the thermal and shock histories of chondritic asteroids. *Geochim. Cosmochim. Acta, 68,* 673–689.

Rubin A. E. and Brearley A. J. (1996) A critical evaluation of the evidence for hot accretion. *Icarus, 124,* 86–96.

Rubin A. E. and Grossman J. N. (1985) Phosphate-sulfide assemblages and Al/Ca ratios in type-3 chondrites. *Meteoritics, 20,* 479–489.

Rubin A. E. and Kallemeyn G. W. (1989) Carlisle Lakes and Allan Hills 85151: Members of a new chondrite grouplet. *Geochim. Cosmochim. Acta, 53,* 3035–3044.

Rubin A. E. and Scott E. R. D. (1997) Abee and related EH chondrite impact-melt breccias. *Geochim. Cosmochim. Acta, 61,* 425–435.

Rubin A. E., Scott E. R. D., and Keil K. (1997) Shock metamorphism of enstatite chondrites. *Geochim. Cosmochim. Acta, 61,* 847–858.

Rubin A. E., Ulff-Møller F., Wasson J. T., and Carlson W. D. (2001) The Portales Valley meteorite breccia: Evidence for impact-induced melting and metamorphism of an ordinary chondrite. *Geochim. Cosmochim. Acta, 65,* 323–342.

Russell S. S., Huss G. R., Fahey A. J., Greenwood R. C., Hutchison R. and Wasserburg G. J. (1998) An isotopic and petrologic study of calcium-aluminum-rich inclusions from CO3 meteorites. *Geochim. Cosmochim. Acta, 62,* 689–714.

Russell S. S., MacPherson G. J., Leshin L. A., and McKeegan K. D. (2000) ^{16}O enrichments in aluminum-rich chondrules from ordinary chondrites. *Earth Planet. Sci. Lett., 184,* 57–74.

Scott E. R. D. and Jones R. H. (1990) Disentangling nebular and asteroidal features of CO3 carbonaceous chondrite meteorites. *Geochim. Cosmochim. Acta, 54,* 2485–2502.

Scott E. R. D. and Krot A. N. (2003) Chondrites and their components. In *Treatise on Geochemistry, Vol. 1: Meteorites, Comets, and Planets* (A. M. Davis, ed.), pp. 143–200. Elsevier, Oxford.

Scott E. R. D., Taylor G. J., and Keil K. (1986) Accretion, metamorphism, and brecciation of ordinary chondrites: Evidence from petrologic studies of meteorites from Roosevelt County, New Mexico. *Proc. Lunar Planet. Sci. Conf. 17th,* in *J. Geophys. Res., 91,* E115–E123.

Scott E. R. D., Barber D. J., Alexander C. M. O'D., Hutchison R., and Peck J. A. (1988) Primitive material surviving in chondrites; matrix. In *Meteorites and the Early Solar System* (J. F. Kerridge and M. S. Matthews, eds.), pp. 718–745. Univ. of Arizona, Tucson.

Scott E. R. D., Keil K., and Stöffler D. (1992) Shock metamorphism of carbonaceous chondrites. *Geochim. Cosmochim. Acta, 56,* 4281–4293.

Scott E. R. D., Jones R. H., and Rubin A. E. (1994) Classification, metamorphic history, and pre-metamorphic composition of chondrules. *Geochim. Cosmochim. Acta, 58,* 1203–1209.

Sears D. W. G., Grossman J. N., Melcher C. L., Ross L. M., and Mills A. A. (1980) Measuring metamorphic history of unequilibrated ordinary chondrites. *Nature, 287,* 791–795.

Sears D. W. G., Grossman J. N., and Melcher C. L. (1982a) Chemical and physical studies of type 3 chondrites — I: Metamorphism related studies of Antarctic and other type 3 ordinary chondrites. *Geochim. Cosmochim. Acta, 46,* 2471–2481.

Sears D. W., Kallemeyn G. W., and Wasson J. T. (1982b) The compositional classification of chondrites: II. The enstatite chondrite groups. *Geochim. Cosmochim. Acta, 46,* 597–608.

Sears D. W. G., DeHart J. M., Hasan F. A., and Lofgren G. E. (1990) Induced thermoluminescence and cathodoluminescence studies of meteorites. In *Spectroscopic Characterization of Minerals and Their Surfaces* (L. M. Coyné et al., eds.), pp. 190–222. American Chemical Society, Washington, DC.

Sears D. W. G., Batchelor J. D., Lu J., and Keck B. D. (1991a) Metamorphism of CO and CO-like chondrites and comparisons with type 3 ordinary chondrites. *Proc. NIPR Symp. Antarct. Meteorites, 4,* 319–343.

Sears D. W. G., Hasan E. A., Batchelor J. D., and Lu J. (1991b) Chemical and physical studies of type 3 chondrites — XI: Metamorphism, pairing, and brecciation of ordinary chondrites. *Proc. Lunar Planet. Sci. Conf., Vol. 21,* pp. 493–512.

Sears D. W. G., Huang S., and Benoit P. H. (1995) Chondrule formation, metamorphism, brecciation, an important new primary chondrule group, and the classification of chondrules. *Earth Planet. Sci. Lett., 131,* 27–39.

Srinivasan G., Huss G. R., and Wasserburg G. J. (2000) A petrographic, chemical, and isotopic study of calcium-aluminum-rich inclusions and aluminum-rich chondrules from the Axtell (CV3) chondrite. *Geochim. Cosmochim. Acta, 35,* 1333–1354.

Stöffler D., Keil K., and Scott E. R D. (1991) Shock metamorphism of ordinary chondrites. *Geochim. Cosmochim. Acta, 55,* 3845–3867.

Tachibana S. and Huss G. R. (2003) The initial abundance of ^{60}Fe in the solar system. *Astrophys. J. Lett., 588,* L41–L44.

Tachibana S., Huss G. R., Kita N. T., Shimoda H., and Morishita Y. (2005) The abundances of iron-60 in pyroxene chondrules from unequilibrated ordinary chondrites (abstract). In *Lunar and Planetary Science XXXVI,* Abstract #1529. Lunar and Planetary Institute, Houston (CD-ROM).

Tandon S. N. and Wasson J. T. (1968) Gallium, germanium, indium and iridium variations in a suite of L-group chondrites. *Geochim. Cosmochim. Acta, 32,* 1087–1110.

Taylor G. J., Keil K., Berkley J. L., and Lance D. E. (1979) The Shaw meteorite: History of a chondrite consisting of impact-melt and metamorphic lithologies. *Geochim. Cosmochim. Acta, 43,* 323–337.

Taylor G. J., Okada A., Scott E. R. D., Rubin A. E., Huss G. R., and Keil K. (1981) The occurrence and implications of carbide-magnetite assemblages in unequilibrated ordinary chondrites (abstract). In *Lunar and Planetary Science XII,* pp. 1076–1078. Lunar and Planetary Institute, Houston.

Taylor G. J., Maggiore P., Scott E. R. D., Rubin A. E., and Keil K. (1987) Original structures, and fragmentation and reassembly histories of asteroids: Evidence from meteorites. *Icarus, 69,* 1–13.

Tonui E., Zolensky M., Lipschutz M., and Okudaira K. (2001) Petrographic and chemical evidence of thermal metamorphism in new carbonaceous chondrites (abstract). *Meteoritics & Planet. Sci., 36,* A207.

Trieloff M., Jessberger E. K., Herrwerth I, Hopp J., Fiéni C., Ghélis M, Bourot-Denise M., and Pellas P. (2003) Structure and thermal history of the H-chondrite parent asteroid revealed by thermochronometry. *Nature, 422,* 502–506.

Van Schmus W. R. and Koffman D. M. (1967) Equilibration temperatures of iron and magnesium in chondritic meteorites. *Science, 155,* 1009–1011.

Van Schmus W. R. and Ribbe P. H. (1969) Composition of phosphate minerals in ordinary chondrites. *Geochim. Cosmochim. Acta, 33,* 637–640.

Van Schmus W. R. and Wood J. A. (1967) A chemical-petrologic classification for the chondritic meteorites. *Geochim. Cosmochim. Acta, 31,* 747–765.

Wasson J. T. (1992) Planetesimal heating by FU-Orionis-Type events (abstract). *Eos Trans. AGU, 73,* 336.

Weisberg M. K., Prinz M., Clayton R. N., and Mayeda T. K. (1997) CV3 chondrites: Three subgroups, not two (abstract). *Meteoritics & Planet. Sci., 32,* A138–A139.

Weisberg M. K., Prinz M., Clayton R. N., Mayeda T. K., Sugiura N., Zashu S., and Ebihara M. (2001) A new metal-rich chondrite grouplet. *Meteoritics & Planet. Sci., 36,* 401–418.

Weisberg M. K., McCoy T. J., and Krot A. N. (2006) Systematics and evaluation of meteorite classification. In *Meteorites and the Early Solar System II* (D. S. Lauretta and H. Y. McSween Jr., eds.), this volume. Univ. of Arizona, Tucson.

Willis J. and Goldstein J. I. (1981) A revision of metallographic cooling rate curves for chondrites. *Proc. Lunar Planet. Sci. 12B,* pp. 1135–1143.

Wlotzka F. (1985) Olivine-spinel and olivine-ilmenite thermometry in chondrites of different petrologic type (abstract). In *Lunar and Planetary Science XVI,* pp. 918–919. Lunar and Planetary Institute, Houston.

Wlotzka F. (1987) Equilibration temperatures and cooling rates of chondrites: A new approach (abstract). *Meteoritics, 22,* 529–531.

Wood J. A. (1962) Metamorphism in chondrites. *Geochim. Cosmochim. Acta, 26,* 739–749.

Wood J. A. (1967) Chondrites: Their metallic minerals, thermal histories, and parent planets. *Icarus, 6,* 1–49.

Wood J. A. and Pellas P. (1991) What heated the meteorite planets? In *The Sun in Time* (C. P. Sonett et al., eds.), pp. 740–760. Univ. of Arizona, Tucson.

Zähringer J. (1968) Rare gases in stony meteorites. *Geochim. Cosmochim. Acta, 32,* 209–237.

Zanda B., Bourot-Denise M., Perron C., and Hewins R. H. (1994) Origin and metamorphic redistribution of silicon, chromium and phosphorus in the metal of chondrites. *Science, 265,* 1846–1849.

Zhang Y., Benoit P. H., and Sears D. W. G. (1995) The classification and complex thermal history of the enstatite chondrites. *J. Geophys. Res., 100,* 9417–9438.

Zhang Y., Huang S., Schneider D., Benoit P. H., DeHart J. M., Lofgren G. E., and Sears D. W. G. (1996) Pyroxene structures, cathodoluminescence and the thermal history of the enstatite chondrites. *Meteoritics & Planet. Sci., 31,* 87–96.

Zolensky M., Abell P., and Tonui E. (2005) Metamorphosed CM and CI carbonaceous chondrites could be from the breakup of the same Earth-crossing asteroid (abstract). In *Lunar and Planetary Science XXXVI,* Abstract #2084. Lunar and Planetary Institute, Houston (CD-ROM).

The Action of Water

Adrian J. Brearley
University of New Mexico

A variety of lines of evidence show that water was widely available during the earliest geological history of the solar system. Water played a significant role in the cosmochemical and geological evolution of a range of small asteroidal bodies, by modifying the primary mineralogical and textural characteristics of pristine nebular materials and providing an agent for mass and heat transport from the interior of these bodies. The mineralogical and stable isotopic characteristics of the products of aqueous alteration, combined with geochemical and thermal modeling, can be used to place important constraints on a number of different aspects of the aqueous alteration processes, including the mechanisms of aqueous alteration reactions, the conditions of alteration, the location and timing of alteration, and the scale of mass transport. Aqueous alteration of nebular materials may have occurred in a variety of locations including the solar nebula, within ephemeral protoplanetary bodies, and within asteroidal parent bodies after final accretion.

1. INTRODUCTION

Among the processes that have affected primitive solar system materials, such as chondritic meteorites and interplanetary dust particles (IDPs), aqueous alteration is perhaps the most widespread. Since the discovery over 150 years ago that the CI carbonaceous chondrite fall Orgueil contained water-bearing clay minerals (*Pisani,* 1864), it has become increasingly well-recognized that water played a key role in the early geological evolution of the solar system. It is now apparent that numerous other chondrite groups, including the CM, CR, CV, CO, and ordinary chondrites, also exhibit evidence of aqueous alteration to varying degrees. Even metamorphosed ordinary chondrites, which have traditionally been regarded as being extremely dry meteorites, show evidence that they have been affected by water early in solar system history (e.g., *Zolensky et al.,* 1999; *Grossman et al.,* 2000; *Rubin et al.,* 2002). In addition to these direct observations, the spectral characteristics of C-, D-, G-, F-, and B-class asteroids (e.g., *Vilas and Gaffey,* 1989; *Vilas et al.,* 1993, 1994), which may be related genetically to carbonaceous chondrites (e.g., *Gaffey et al.,* 1993; *Pieters and McFadden,* 1994; *Hiroi et al.,* 2001; *Kanno et al.,* 2003), indicate that hydrous phyllosilicates are present on their surfaces.

These observations provide unequivocal evidence that water was widely available in the early solar system and played a significant role in the evolution of the mineralogy and cosmochemistry of a number of diverse solar system bodies. An understanding of the origin of the alteration assemblages in chondritic meteorites has the potential to place important constraints on when and where aqueous alteration in chondrites occurred, the timescales for alteration, and the effects of alteration on primary nebular components. Furthermore, it can also help constrain the role of volatiles in the solar nebula, as well as during parent-body accretion and subsequently within asteroidal parent bodies. Finally, a knowledge of the behavior of water during the earliest stages of solar system evolution is essential to developing a full understanding of the behavior of the biogenic elements through geologic time. This chapter reviews the body of evidence that demonstrates that aqueous fluids were of major importance in the earliest geological evolution of our solar system.

2. MINERALOGICAL AND COSMOCHEMICAL EFFECTS OF AQUEOUS ALTERATION

The most obvious effect of aqueous alteration on chondritic meteorites is the formation of a variety of secondary phases, usually hydrous minerals such as serpentines and clays, but also carbonates, sulfates, oxides, sulfides, halides, and oxyhydroxides. In addition, alteration has had important effects on both the stable isotopic compositions of the bulk meteorites and individual components that provide very useful insights into the alteration process. In some cases, these effects are significant as in the case of the CI, CM, CR, and some CV chondrites, whereas in other chondrites such as the CO and unequilibrated ordinary chondrites (UOCs), alteration effects are more subtle. In the case of some CV chondrites, such as Allende, dark inclusions (DIs) in CV and CO chondrites, and equilibrated ordinary chondrites, alteration effects are often cryptic, but can be dramatic as in the case of halite in the Monahans (1998) (H5) and Zag (H5) ordinary chondrite breccias (e.g., *Zolensky et al.,* 1999; *Rubin et al.,* 2002).

Perhaps the most remarkable feature of alteration in chondritic meteorites is the diversity in alteration assemblages and styles of aqueous alteration. Although there are general similarities between aqueous alteration in different chondrite groups, there are also significant differences, sug-

TABLE 1. Alteration phases found in carbonaceous and ordinary chondrites (minor phases shown in italics).

CI Chondrites	CM Chondrites	CR Chondrites	CV Chondrites	CO Chondrites	Ordinary Chondrites
Silicates	**Silicates**	**Silicates**	**Silicates**	**Silicates**	**Silicates**
Fe-Mg serpentines	Fe-Mg serpentines	Fe-Mg serpentines	Fe-Mg serpentine	Fe-Mg serpentine	smectite
Saponite (smectite)	cronstedtite	saponite	saponite	*chlorite*	
			amphibole		**Carbonates**
Carbonates	*chlorite*	**Carbonates**	*biopyribole*		*calcite*
calcite	*saponite*	*calcite*	*talc*		
dolomite	*vermiculite*		*margarite*		**Oxides**
breunnerite		**Oxides**	*clintonite*		*maghemite*
siderite	**Tochilinite**	magnetite	*muscovite*		
	tochilinite		*Na phlogopite*		**Sulfides**
Sulfates		**Sulfides**	*montmorillonite*		pyrrhotite
gypsum	**Carbonates**	pyrrhotite	*chlorite*		pentlandite
epsomite	calcite	pentlandite			
bloedite	dolomite		fayalite		**Halides**
Ni bloedite	*aragonite*		hedenbergite		*halite*
			diopside		
Sulfides	**Sulfates**		wollastonite		
pyrrhotite	*gypsum*		andradite		
pentlandite	*hemihydrate*		*grossular*		
cubanite	*anhydrite*		*monticellite*		
	thenardite		kirschteinite		
Oxides			nepheline		
magnetite	**Oxides**		sodalite		
ferrihydrite	*magnetite*				
			Carbonates		
Native elements	**Hydroxides**		*calcite*		
sulfur	*brucite*				
			Sulfides		
	Sulfides		pyrrhotite		
	pyrrhotite		pentlandite		
	pentlandite				
			Oxides		
	Halides		magnetite		
	halite		ferrihydrite		
	sylvite				

gesting that the conditions of alteration and the availability of fluid was quite variable, both within individual asteroids and between asteroid types. In this section, the alteration assemblages and textural characteristics of alteration in these different meteorites will be summarized and the cosmochemical effects of alteration examined. Detailed discussions of the alteration mineralogy of these meteorites have been presented in extensive literature reviews (e.g., *Barber,* 1981; *Zolensky and McSween,* 1988; *Buseck and Hua,* 1993; *Brearley and Jones,* 1998; *Brearley,* 2003) and will only be discussed briefly here. The alteration mineralogy in the main groups of chondritic meteorites is summarized in Table 1 and mineral formulae for the principle alteration phases are presented in Table 2.

Identification of chondrites that have experienced aqueous alteration is usually based on the presence of hydrous phases such as phyllosilicates. However, over the last decade, it has been recognized that secondary alteration effects are present in some chondrites [e.g., CV and CO chondrites and the dark inclusions (DIs) that they contain] that may have been formed by the action of water but did not result in the widespread formation of hydrous phases (e.g., *Krot et al.,* 1995). In these cases, the effects of aqueous fluids are more cryptic and establishing the involvement of water in the formation of these mineral assemblages is more challenging and certainly controversial (see *Brearley,* 2003). In these situations, a combination of mineralogical studies, thermodynamic modeling, and O-isotopic data are required to demonstrate the involvement of aqueous fluids, and in some cases, the results may be equivocal. In general, aqueous alteration, which leaves a somewhat cryptic signature(s) on the rocks that it has affected, appears to have occurred at temperatures higher than those experienced during aqueous alteration that primarily resulted in the formation of

TABLE 2. Characteristics alteration phases in altered carbonaceous and ordinary chondrites.

Mineral Group	End Member	Chemical Formulae
Phyllosilicates		
Serpentines:		
	chrysotile	$Mg_3Si_2O_5(OH)_4$
	lizardite	$Mg_3Si_2O_5(OH)_4$
	antigorite	$Mg_3Si_2O_5(OH)_4$
	cronstedtite	$Fe_2^{2+}Fe^{3+}(SiFe^{3+})O_5(OH)_4$
Smectites		
	montmorillonite	$(Na,Ca)_{0.3}(Al,Mg)_2Si_4O_{10}(OH)_2 \cdot nH_2O$
	saponite	$(Ca/2)_{0.3}(Mg,Fe^{2+},Mg)_3(Si,Al)_4O_{10}(OH)_2 \cdot nH_2O$
Chlorite		
	clinochlore	$(Mg,Fe^{2+})_5Al(Si_3Al)O_{10}(OH)_8$
	chamosite	$(Mg,Fe^{2+},Fe^{3+})_5Al(Si_3Al)O_{10}(OH)_8$
Talc		$Mg_3Si_4O_{10}(OH)_2$
Vermiculite		$Mg_3(Si,Al)_4O_{10}(OH)_2 \cdot 4.5nH_2O[Mg]_{0.35}$
Micas		
	clintonite	$CaMg_3(Al_2Si_2)(OH)_2$
	margarite	$CaAl_2(Al_2Si_2)(OH)_2$
	muscovite	$KAl_2(AlSi_3O_{10})(OH)_2$
	phlogopite	$(K,Na)Mg_3(AlSi_3O_{10})(OH)_2$
Amphibole		
	anthophyllite	$Mg_7Si_8O_{22}(OH)_2$
	hornblende	$(Ca,Na)_{2-3}(Mg,Fe,Al)_5Si_6(Si,Al)_2O_{22}(OH)_2$
Biopyriboles		
	jimthompsonite	$(Mg,Fe)_{10}Si_{12}O_{32}(OH)_4$
Pyroxenes		
	hedenbergite	$(CaFe)Si_2O_6$
	diopside	$(CaMg)Si_2O_6$
Pyroxenoids		
	wollastonite	$CaSiO_3$
Garnet		
	andradite	$Ca_3Fe^{3+}{}_2Si_3O_{12}$
	grossular	$Ca_3Al_2Si_3O_{12}$
Feldspathoids		
	nepheline	$(Na,K)AlSiO_4$
	sodalite	$Na_8(AlSiO_4)_6Cl_2$
Olivines		
	fayalite	Fe_2SiO_4
	kirschteinite	$CaFeSiO_4$
	monticellite	$CaMgSiO_4$
Tochilinite		
	tochilinite	$2[(Fe,Mg,Cu,Ni)S] \cdot 1.57–1.85[Mg,Fe,Ni,Al,Ca)(OH)_2]$
Sulfates		
	epsomite	$MgSO_4 \cdot 7H_2O$
	gypsum	$CaSO_4 \cdot 2H_2O$
	bloedite	$\mathbf{Na_2Mg(SO_4)_2 \cdot 2H_2O}$
	Ni-bloedite	
Carbonates		
	calcite	$CaCO_3$
	aragonite	$CaCO_3$
	dolomite	$(CaMg)CO_3$
	breunnerite	$(Mg,Fe,Mn,Fe)CO_3$
	siderite	$\mathbf{FeCO_3}$

TABLE 2. (continued).

Mineral Group	End Member	Chemical Formulae
Sulfides		
	pyrrhotite	$Fe_{1-x}S$ (x = 0.0.17)
	pentlandite	$(FeNi_9)S_8$
	cubanite	$CuFe_2S_3$
Oxides		
	magnetite	Fe_3O_4
	maghemite	Fe_2O_3
	ferrihydrite	$5Fe_2O_3 \cdot 9H_2O$
Hydroxides		
	brucite	$Mg(OH)_2$
Halides		
	halite	NaCl
	sylvite	KCl

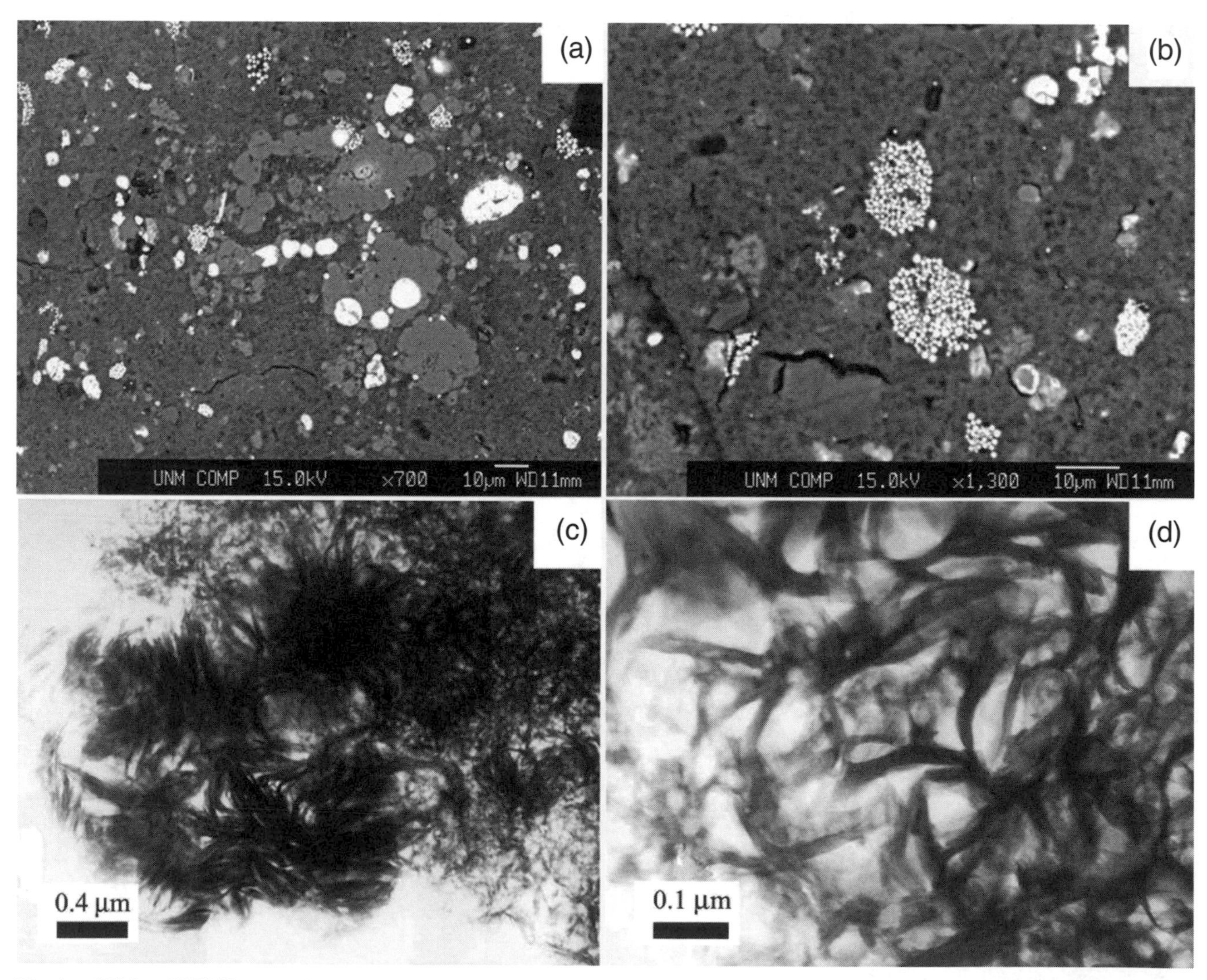

Fig. 1. SEM and TEM images showing the general characteristics of CI chondrites that result from aqueous alteration. **(a)** BSE image of Ivuna matrix showing the presence of magnetite (Mgt) and irregularly shaped carbonate grains (Cb), set in a very fine-grained matrix of phyllosilicates. **(b)** BSE image of magnetite framboids (Mgt) in Ivuna matrix. **(c)** TEM image of coarse- and fine-grained phyllosilicates (serpentine interlayered with saponite) in Ivuna matrix. **(d)** TEM image showing a closeup of the textures of phyllosilicates in the CI chondrite, Ivuna.

phyllosilicates. For example, the widespread development of hydrous phases that have been described in many carbonaceous and ordinary chondrites, as discussed below, is generally estimated to have occurred at temperatures below ~100°C, although some chondrites may have experienced somewhat higher temperatures. In contrast, the formation of largely anhydrous mineral assemblages as a result of the interaction of chondritic protoliths with hydrous fluids probably occurred at temperatures in the range 200°–400°C (*Krot et al.,* 1998a,b; *Brearley,* 1999). To describe alteration processes in this temperature regime, terms such as metasomatism and fluid-assisted metamorphism should be used to distinguish between lower-temperature aqueous alteration, whose primary mineralogical products are hydrous phases such as phyllosilicates.

2.1. CI Carbonaceous Chondrites

CI chondrites are among the rarest type of carbonaceous chondrite (only seven are currently known), but their cosmochemical significance far outweighs their relative scarcity. The CI chondrites have bulk elemental ratios that show a 1 : 1 correlation with those of the solar photosphere for most elements (*Anders and Grevesse,* 1989; *Lodders,* 2003) and are considered to have the most primitive compositions of all known solar system materials. However, despite their unfractionated chemical compositions, the CI chondrites are the most hydrated of all the chondrite groups.

The CI chondrites are complex meteorites (Figs. 1a,b) that consist of a dark, fine-grained matrix comprised largely of phyllosilicates (serpentine interlayered with saponite) with magnetite, sulfides, carbonates, and sulfates embedded within it (*DuFresne and Anders,* 1962; *Böstrom and Fredriksson,* 1966; *Nagy,* 1966; *Tomeoka and Buseck,* 1988). Magnetite development is extensive and occurs with framboidal, plaquette, and platelet morphologies (e.g., *Kerridge et al.,* 1979a) (Fig. 1b). A variety of carbonates (breunnerite, dolomite, and calcite) (*Endress and Bischoff,* 1996) and sulfides (pyrrhotite, pentlandite, and cubanite) occur within the matrix (e.g., *Kerridge et al.,* 1979b). The occurrence of veins of white sulfate (epsomite, $MgSO_4 \cdot 7H_2O$) in the CI chondrite Orgueil has been widely reported in the literature (e.g., *DuFresne and Anders,* 1962; *Mason,* 1962; *Boström and Fredriksson,* 1966; *Richardson,* 1978). This observation has been cited as evidence for aqueous alteration on the CI-chondrite parent asteroid (e.g., *Zolensky and McSween,* 1988). However, recent studies (*Gounelle and Zolensky,* 2001) provide strong evidence that the veins are, in fact, the result of remobilization and reprecipitation of soluble sulfates caused by interaction with the terrestrial atmosphere after the meteorite fell. It should also be noted that no carbonate veins have been observed in CI chondrites, although fragments of what may have been carbonate vein fillings have been described in CI chondrites (e.g., *Richardson,* 1978; *Endress and Bischoff,* 1996).

Although the bulk chemical compositions of CI chondrites are a close match to the solar photosphere abundances, aqueous alteration has caused extensive chemical fractionation between different alteration components. Elements that are generally considered to be immobile during aqueous alteration (e.g., Al and Ti) are unfractionated in CI matrix (Fig. 2) and have solar ratios, but soluble elements such as Ca, Mn, Na, K, and S are strongly fractionated (*McSween and Richardson,* 1977; *Zolensky et al.,* 1993; *Brearley and Prinz,* 1992, *Brearley,* 1997a). These elements were probably mobilized and precipitated as sulfates and carbonates, although whether this process resulted in the formation of veins in the CI chondrites is open to debate, as discussed above.

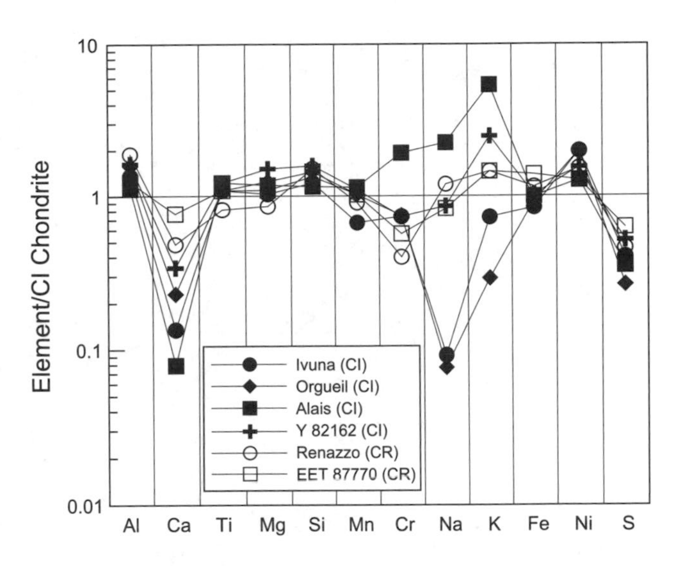

Fig. 2. Normalized element abundance diagram (normalized to bulk CI and Si) showing the compositions of the matrices of several CI- and CR-chondrite matrices. The data are arranged in order of decreasing condensation temperatures. The data show that the fine-grained phyllosilicate component of CI matrices represented by the data is fractionated relative to bulk CI-chondrite values. Soluble elements show variable behavior, but Ca in particular is always depleted in matrices, presumably because it has been fractionated into coarser-grained phases such as calcite and Ca-sulfates. Other soluble elements such as Na and K show variable behavior and may be enriched or depleted. CR chondrites that are less altered than CI chondrites have matrices that show less fractionation in these soluble elements than CI chondrites. Broad-beam electron microprobe data from *Zolensky et al.* (1993).

2.2. CM Carbonaceous Chondrites

The CM chondrites are a diverse, complex group of meteorites that have undergone aqueous alteration and brecciation to different degrees (e.g., *Metzler et al.,* 1992). A significant advance in our understanding of CM chondrites has come from the work of *Metzler et al.* (1992), who recognized that the least-brecciated CM chondrites consist essentially entirely of what they described as a "primary accretionary rock." Such primary accretionary rocks consist of coarse-grained components (chondrules, CAIs, amoeboid olivine aggregates, and mineral fragments) that are all sur-

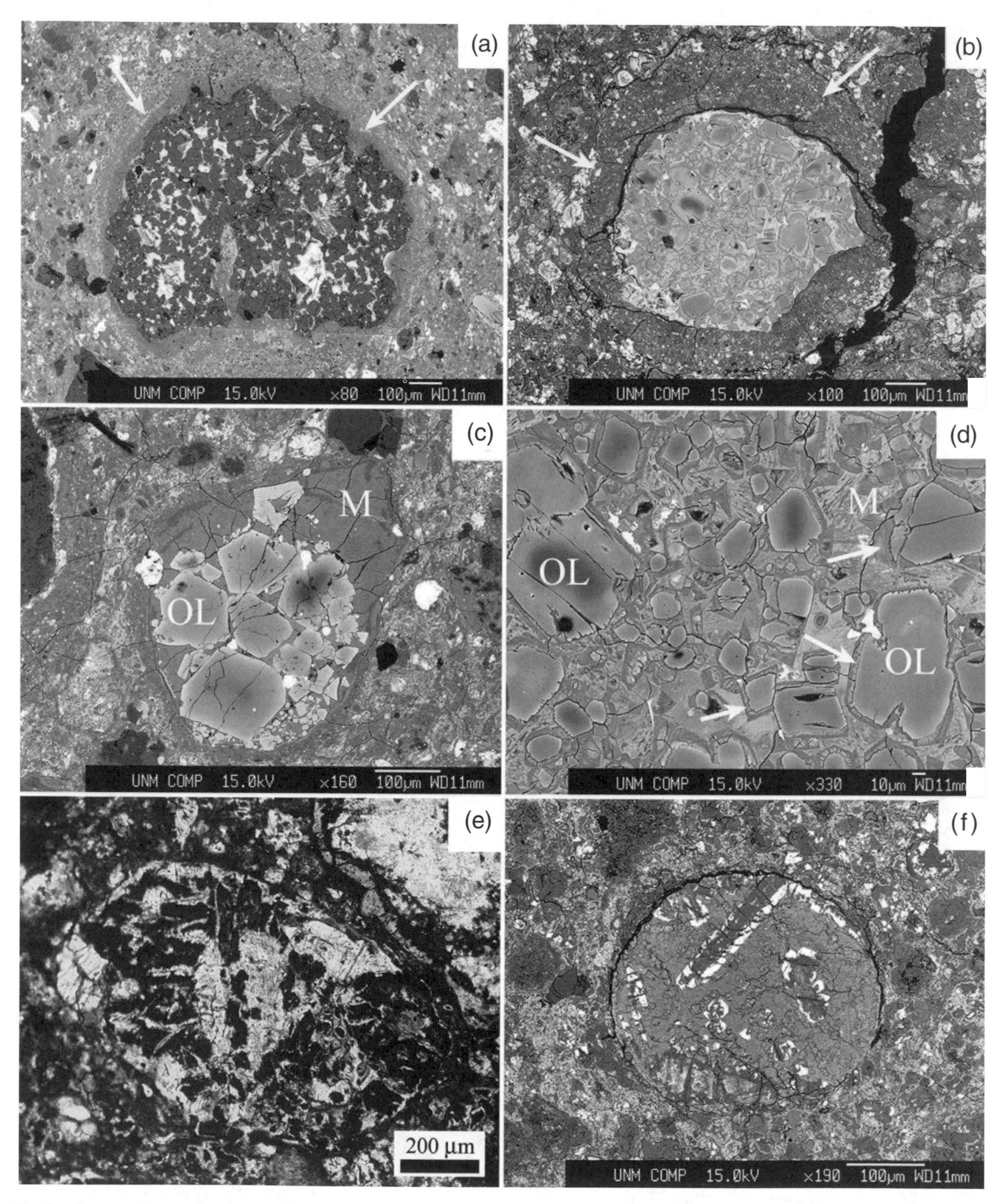

Fig. 3. Images of textural and mineralogical characteristics of the effects of aqueous alteration in CM carbonaceous chondrites. **(a)** BSE image showing a fine-grained rim (arrowed) around a porphyritic olivine-rich type I chondrule in the relatively weakly altered CM2 Murchison; **(b)** BSE image of well-developed fine-grained rim (arrowed) in the unbrecciated, but relatively heavily altered CM2 chondrite ALH 81002. The rim surrounds a heavily altered type IIA chondrule in which of olivine phenocrysts have been partially replaced by serpentine. **(c)** BSE image of a type II FeO-rich chondrule in Y 791198 showing altered mesostasis glass (M), but completely unaltered olivine (OL) phenocrysts. **(d)** BSE closeup showing the details of alteration of olivine (OL) to serpentine in the type IIA chondrule shown in Fig. 3b. Narrow rims of serpentine (arrowed) are present around the peripheries of all olivine grains. **(e)** Optical micrograph showing a completely pseudomorphed porphyritic olivine chondrule from the CM1 chondrite ALH 84034. The chondrule has been replaced by serpentine and Fe-oxides. **(f)** BSE image of a porphyritic chondrule in the CM1 chondrite, ALH 84034, showing perfect pseudomorphism of some of the chondrule phenocrysts.

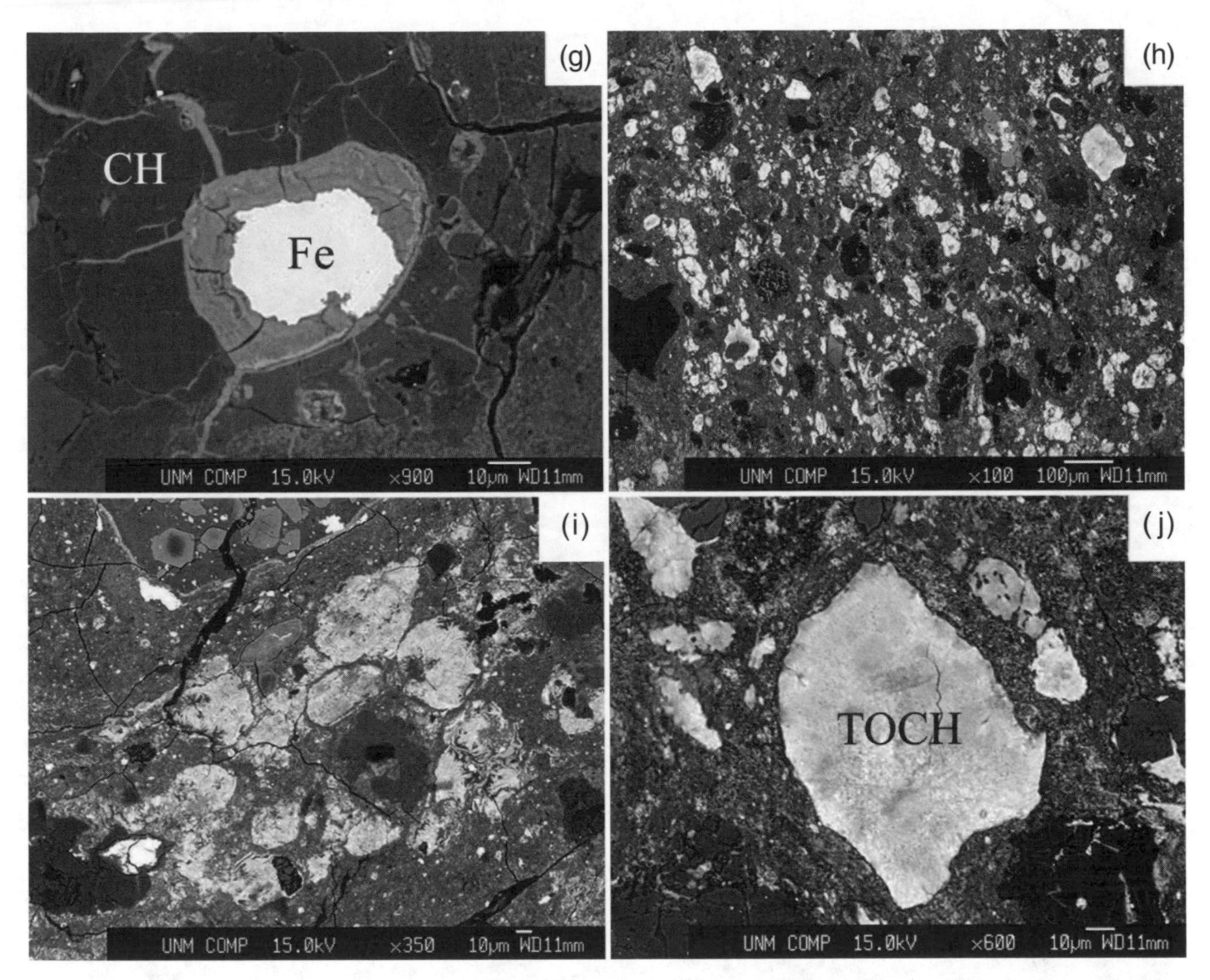

Fig. 3. (continued). **(g)** Partially altered metal grain (Fe) in a type I chondrule (CH) in the weakly altered CM chondrite Y 791198. The metal shows significant corrosion and is surrounded by a rim of tochilinite/cronstedtite replacement products. **(h)** BSE image of a region of matrix in Murchison containing abundant, irregularly shaped masses of tochilinite/cronstedtite intergrowths (formally termed PCP). **(i)** BSE image of tochilinite/cronstedtite intergrowths (high BSE contrast objects) in the matrix of Y 791198, a comparatively weakly altered CM2 carbonaceous chondrite, showing the fibrous characteristics of this material. **(j)** BSE image showing a closeup of a massive tochilinite/cronstedtite intergrowth (TOCH) in Murchison matrix. Compositional zoning is apparent within the aggregate.

rounded by well-developed fine-grained rims (Figs. 3a,b). *Metzler et al.* (1992) argued that the fine-grained rims represent nebular dust that accreted onto chondrules in the solar nebula, although other models for the formation of these rims have also been proposed (e.g., *Sears et al.,* 1993). Brecciation on the CM-chondrite parent body resulted in the progressive fragmentation of this primary texture and produced a clastic texture that is characteristic of many CM chondrites. As a consequence, most CM chondrites are breccias consisting of fragments of primary accretionary rocks and surviving intact chondrules, CAIs, etc., embedded within a matrix of clastic fragments derived from chondrules, fine-grained rims, etc. In extreme cases of brecciation, all evidence of the primary accretionary texture has been destroyed (e.g., in the unusual CM2 chondrite, Bells) (*Brearley,* 1995). The interaction of brecciation with aqueous alteration has resulted in remarkable textural diversity within the CM chondrites and is a complication that makes unraveling the alteration histories of these meteorites particularly challenging.

An important characteristic of CM chondrites is that they display a wide range of degrees of aqueous alteration. Most CM chondrites are classified as petrologic type 2, indicating that they typically have essentially fully hydrated matrices, but within CM2 chondrites, the extent of alteration of chondrules is highly variable from one chondrite to another (e.g., Figs. 3a–f) and is a function of the degree of alteration of the host meteorite and the chondrule type (*Hanowski and Brearley,* 2001). A few CM chondrites exist that show essentially 100% hydration and have been termed CM1 chondrites (*Grady et al.,* 1987; *Zolensky et al.,* 1997). In the CM1s, all chondrules and other coarse-grained objects have been completely pseudomorphed by alteration phases, dominantly serpentine (Figs. 3e,f).

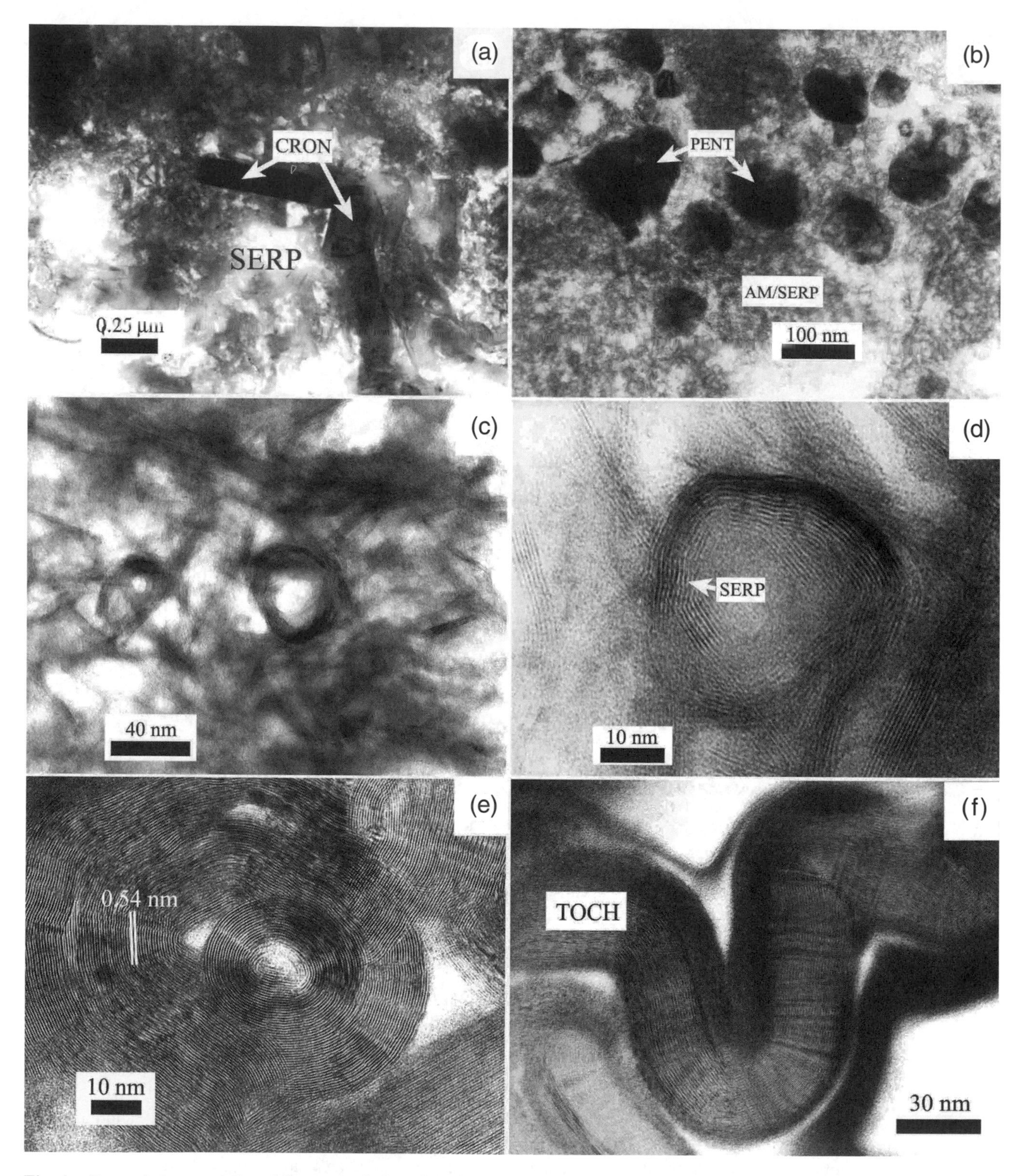

Fig. 4. Textural characteristics of fine-grained alteration phases in CM2 carbonaceous chondrites. **(a)** TEM image of a fine-grained rim in Murchison showing comparatively coarse-grained cronstedtite (CRON) crystals surrounded by a fine-grained matrix of serpentine. **(b)** TEM image of nanophase sulfide (PENT) crystals in a fine-grained rim in Murchison embedded in a groundmass of very fine-grained Mg-serpentine or amorphous material (AM/SERP) with a composition close to that of serpentine. **(c)** TEM image of a fine-grained rim in Murchison showing a region dominated by fine-grained Mg-serpentines with rolled and platy morphologies. **(d)** High-resolution TEM image of a serpentine (SERP) crystal in a fine-grained rim in Murchison showing a characteristic rolled morphology. **(e)** High-resolution TEM image of a cylindrical tochilinite tube from Murchison matrix. Image courtesy of K. Tomeoka. **(f)** High-resolution TEM image showing contorted structures in tochilinite (formerly PCP) from a tochilinite grain in Murray matrix similar to those shown in Murchison in Fig. 3h. Image courtesy of K. Tomeoka after *Tomeoka and Buseck* (1985).

Compared with CI chondrites, the alteration mineralogy of CM2s is quite distinct. Transmission electron microscope (TEM) studies have shown that the matrices and fine-grained rims of these meteorites are dominated by fine-grained serpentines with different morphologies and variable compositions (e.g., *Barber,* 1981; *Zolensky et al.,* 1993; *Lauretta et al.,* 2000; *Zega and Buseck,* 2003; *Zega et al.,* 2003) (Figs. 4a–d). Amorphous/nanocrystalline material is also a common component of fine-grained rims in the least-altered CM chondrites (*Barber,* 1981; *Chizmadia and Brearley,* 2003). Nanometer- to micrometer-sized sulfides (pentlandite and pyrrhotite) commonly occur embedded in this groundmass of phyllosilicates and amorphous material (Fig. 4b). Magnetite is rare except in the unusual CM chondrite Bells (*Rowe et al.,* 1994; *Brearley,* 1995). One of the diagnostic characteristics of CM chondrites is the presence of the unusual phase tochilinite [$6Fe_{0.9}S \cdot 5(Fe,Mg)(OH)_2$], which commonly occurs intergrown on a very fine scale with the Fe^{3+}-rich serpentine, cronstedtite (e.g., *Tomeoka and Buseck,* 1985). These intergrowths have commonly been called "spinach" or "poorly characterized phase (PCP)" in the literature, but both of these terms are now obsolete. These tochilinite/cronstedtite intergrowths occur throughout CM chondrites, often as distinct, irregularly shaped masses, sometimes surrounded by fine-grained rims and as a replacement product of Fe,Ni metal grains in the matrix and in chondrules (Figs. 3g–j and Figs. 4e,f). They are also a common phase within the matrix and fine-grained rims of CM chondrites, occurring in a variety of different morphologies (Figs. 4e,f).

Carbonates are ubiquitous in CM2 chondrites and are also present, with rare exceptions, in all CM1 chondrites (e.g., *Johnson and Prinz,* 1993; *Zolensky et al.,* 1993; *Brearley and Hutcheon,* 2000), occurring as 10–50 μm grains distributed throughout the matrix (Figs. 5a–f). Calcite also occurs commonly in altered CAIs, but is rare in altered chondrules and their fine-grained rims (*Brearley and Hutcheon,* 2002). In CM2s, calcite is often the only type of carbonate present, except in the most heavily altered CM2s such as Nogoya that also contain dolomite (*Johnson and Prinz,* 1993). In some CM1s, both calcite and dolomite occur (*Zolensky et al.,* 1997). Many of these carbonate grains show previously unrecognized, complex minor-element zoning (Fig. 5) that is revealed by cathodoluminescence and electron microprobe studies (*Brearley et al.,* 1999; *Brearley and Hutcheon,* 2002). Compared with CI chondrites, sulfates are much less common, although they do occur in many CMs (*Airieau et al.,* 2001), in some cases as crosscutting veins (*Lee,* 1993). Based on the observations of *Gounelle and Zolensky* (2001) on CI chondrites, it is possible that these sulfate veins may actually be terrestrial in origin, but this remains to be confirmed.

Petrographic studies show that mesostasis, Fe,Ni metal, and troilite in chondrules are highly susceptible to aqueous alteration and, even in the least-altered CM chondrites, are extensively altered (e.g., Figs. 3c,d) (*Chizmadia and Brearley,* 2003). Mesostasis has commonly been replaced by serpentine with variable compositions (*Richardson,* 1981; *Hanowski and Brearley,* 2001). Metal has altered to tochilinite/cronstedtite intergrowths (Fig. 3g) and most troilite has altered to pyrrhotite or pentlandite. More advanced alteration results first in the alteration of more Fe-rich olivines in type IIA chondrules (e.g., Fig. 3d) and low-Ca pyroxene (clinoenstatite). Forsteritic olivine in type I chondrules, augite, and orthopyroxene (when present) are highly resistant to alteration (*Zolensky et al.,* 1993; *Hanowksi and Brearley,* 2001) and only show significant replacement at the most advanced stages of alteration such as those observed in CM1 and very altered CM2 chondrites, such as Nogoya.

Despite their extensive degree of aqueous alteration, CM-chondrite bulk compositions show no evidence of elemental fractionations between water-soluble elements such as the alkalis and alkali-earths and refractory immobile elements, such as Al and Sc. Figure 6 shows Na/Sc, Ca/Al, and Mg/Al ratios for CM chondrites as a function of their mineralogical alteration index (MAI) (*Browning et al.,* 1996). The ratios of water-soluble Ca and Na to refractory elements are essentially constant throughout the range of alteration experienced by the CM2 chondrites. Magnesium/aluminum is also constant regardless of the extent of alteration of the host meteorite, even though Mg was clearly mobile on a localized scale during aqueous alteration as indicated by the replacement of Mg-rich chondrule silicates by Fe-rich phases and the formation of dolomite. The Na/Sc data for three CM1 chondrites are also shown for comparison with CM2 chondrites. The spread of Na/Sc ratios exhibited by these meteorites suggests that Na may have been mobilized to some degree during advanced stages of aqueous alteration. However, all three chondrites are Antarctic finds and so the possibility that the Na depletion is due to terrestrial alteration must also be considered. Calcium/aluminum data are available for only one CM1 chondrite, Yamato (Y) 82042, and suggest that Ca may have been depleted during aqueous alteration.

2.3. CR Carbonaceous Chondrites

CR chondrites are all petrologic type 2 (*Weisberg et al.,* 1993), with the exception of one meteorite, Grosvenor Mountains (GRO) 95577, which appears to be a CR1 (*Weisberg and Prinz,* 2000). Like the CM2 chondrites, the CRs exhibit significant variability of their degree of alteration. Although the general characteristics of alteration in this group of chondrites are known (e.g., *Weisberg et al.,* 1993; *Zolensky et al.,* 1993; *Noguchi,* 1995), the details of alteration are lacking, in comparison with CM or CI chondrites. Most CR chondrites have experienced significantly less alteration than typical CM2 chondrites. This is reflected by the fact that several CR chondrites contain chondrules with unaltered glassy mesostases or mesostases that only show the earliest stages of aqueous alteration on the periphery of chondrules (Fig. 7a) (e.g., *Noguchi,* 1995). In some cases, unaltered chondrule glass can occur in direct contact with hydrous matrix, as described in, for example, Y 8449

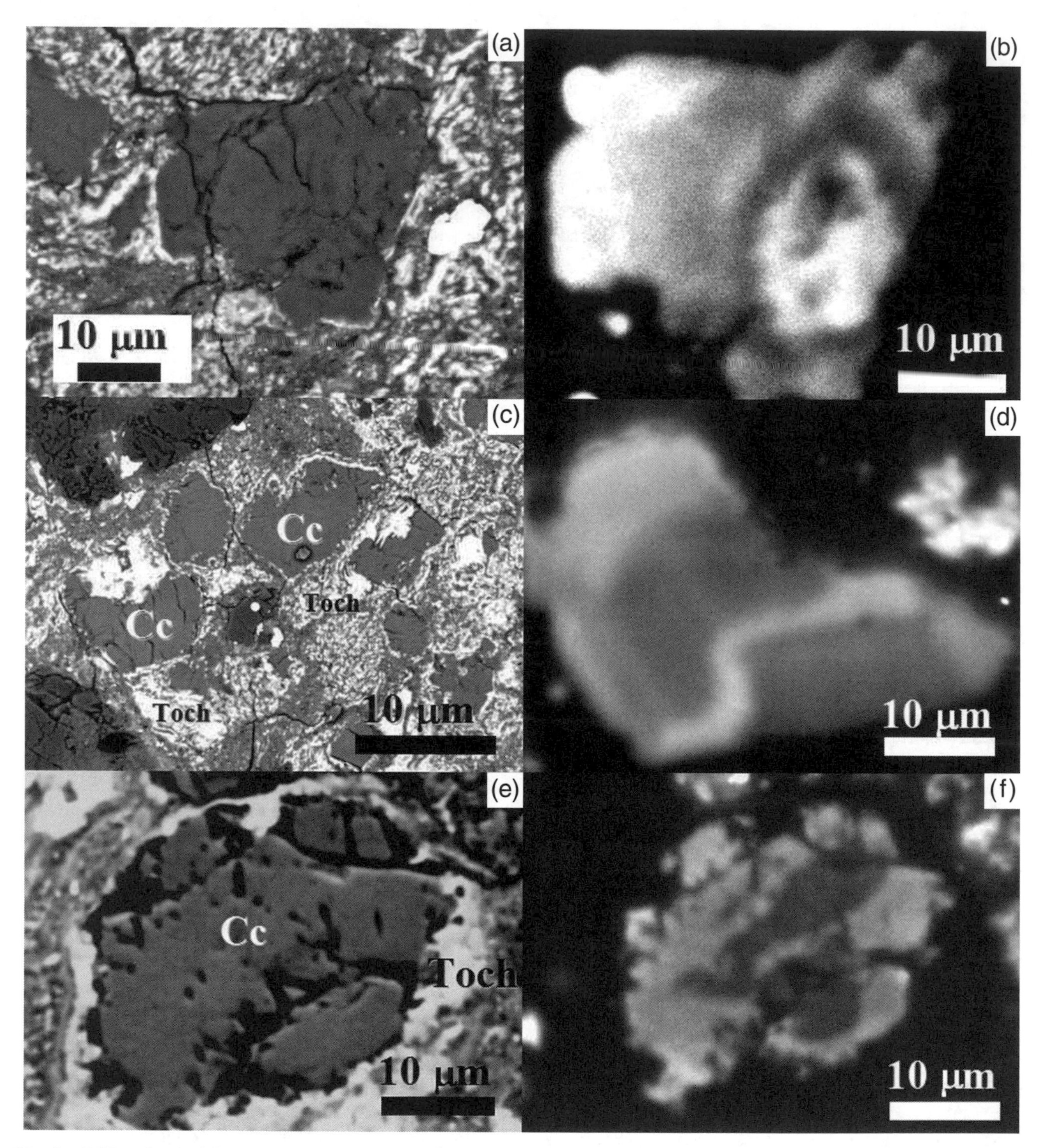

Fig. 5. BSE and cathodoluminescence images of carbonate occurrences in CM2 carbonaceous chondrites. **(a)** Typical occurrence of calcite in Y 791198 — anhedral, isolated grain embedded within fine-grained matrix material. **(b)** CL intensity images of the grain shown in **(a)** illustrating complex zoning characteristics with an irregularly shaped, corroded core with a zoned overgrowth. **(c)** Cluster of anhedral calcite (Cc) grains associated with tochilinite (Toch) in the fine-grained matrix of Y 791198. **(d)** CL intensity image of calcite grain in Y 791198 consisting of two subgrains that show asymmetric, but regular zoning. **(e)** Calcite (Cc) grain in Murchison matrix associated with tochilinite (Toch). **(f)** CL intensity image of grain shown in **(e)** illustrating the grain consists of two subgrains with subrounded cores, possibly indicative of dissolution, overgrown by a homogeneous rim.

(*Ichikawa and Ikeda,* 1995). In addition, abundant unaltered Fe,Ni metal is present in most CR chondrites. However, some CR chondrites, such as Renazzo and Al Rais, show much more advanced alteration, where the glassy mesotases in chondrules have mostly been replaced by phyllosilicates, indicating a degree of alteration that is more comparable with typical CM2 chondrites (*Weisberg et al.,* 1993).

Most evidence of aqueous alteration is present in the matrices of CR chondrites and the style of alteration, although more limited, appears to be most similar to CI chon-

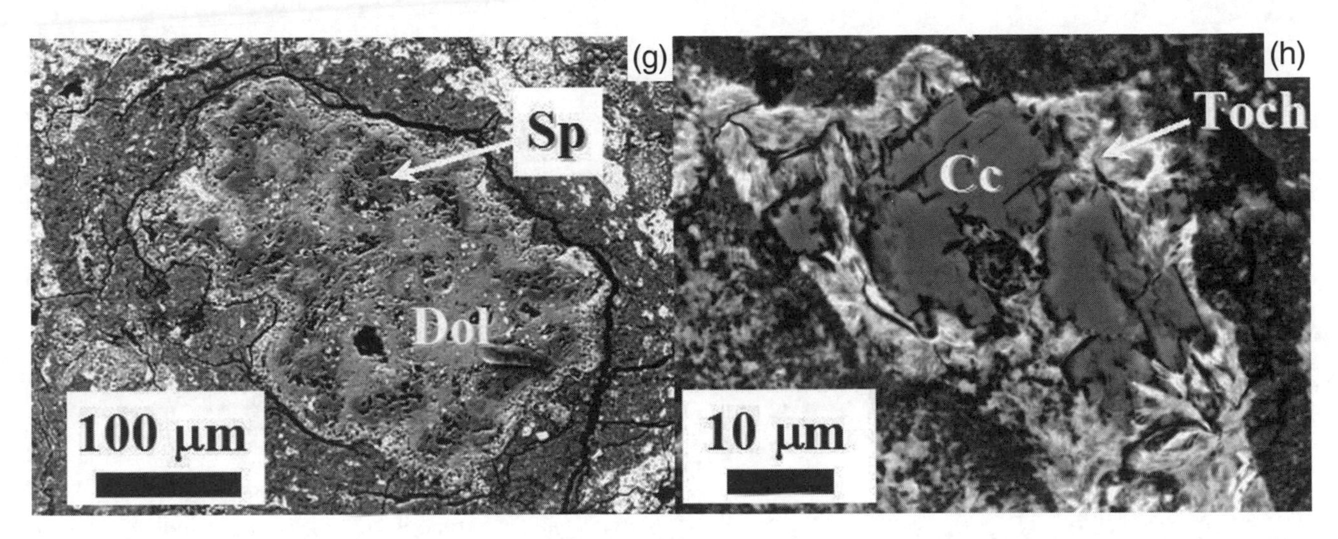

Fig. 5. (continued). **(g)** Compact altered CAI in ALH 81002 that has been extensively replaced by dolomite (Dol). The primary phases in the CAI that are unaltered are spinel (Sp) and small perovskite grains (bright blebs). **(h)** Calcite (cc) in ALH 81002 associated with tochilinite (Toch). The grain has a ragged, corroded appearance indicating that has undergone dissolution and replacement by tochilinite.

drites. Framboidal magnetite is common in the matrices of many CR chondrites (Fig. 7b), and fine-grained, Ca-rich carbonates and sulfides are also present (*Weisberg et al.,* 1993). In the matrices of two CR chondrites, *Zolensky et al.* (1993) reported unaltered olivine in addition to fine-grained phyllosilicates, predominantly serpentine, intergrown with saponite as the main alteration phases. However, in the CR Elephant Moraine (EET) 92042, *Abreu and Brearley* (2004, 2005a) found that the matrix consists almost entirely of fine-grained phyllosilicates that act as a groundmass for ubiquitous, submicrometer Fe,Ni sulfides (Fig. 7c,d).

CR chondrites also contain DIs that occur as lithic fragments within the host chondrites (*Weisberg et al.,* 1993; *Bischoff et al.,* 1993; *Endress et al.,* 1994) and are mineralogically similar to CR chondrite matrices, although they appear to have experienced a higher degree of hydration. These DIs have O-isotopic compositions that clearly relate them to the CR chondrites (*Weisberg et al.,* 1993), suggesting that they are derived from a region of the same parent body that has experienced more extensive aqueous alteration.

The signatures of aqueous alteration are apparent in the composition of matrices in CR chondrites, although the data for this group of chondrites is limited. The matrices of the CR chondrites are much less fractionated than CI matrices (Fig. 2), consistent with their lower degree of aqueous alteration, but nevertheless show characteristic depletions in Ca, which can be attributed to mobilization of this element to form calcite during aqueous alteration.

2.4. CO Carbonaceous Chondrites

The CO3 carbonaceous chondrites as a group have been minimally affected by low-temperature aqueous alteration, although they may have been subjected to fluid-assisted metamorphism (*Rubin,* 1998). Evidence of alteration is variable within the CO group; most CO chondrites (e.g., Kainsaz, Ornans, and Warrenton) show no evidence of alteration and, in those that do, the effects are minimal even at the TEM scale (*Keller and Buseck,* 1990a; *Brearley,* 1994). The effects of alteration are almost exclusively observed within the matrix, although *Ikeda* (1983) and *Itoh and Tomeoka* (2001) have reported phyllosilicates replacing chondrule glass in Allan Hills (ALH) 77307 (3.0). Minor phyllosilicate formation is evident in the matrices of ALH 77307, Lancé, and Ornans (*Brearley,* 1993; *Kerridge,* 1964; *Keller and Buseck,* 1990a) with very fine-grained serpentine and saponite occurring interstitially to matrix olivines and sometimes within veins in the larger matrix olivines. CO chondrites show no evidence of the development of carbonates or sulfates and in this respect are similar to the CV chondrites, but quite distinct from CI, CR, and CM chondrites.

2.5. CV Carbonaceous Chondrites and Dark Inclusions

Evidence of aqueous alteration in the CV carbonaceous chondrites is complex and controversial. In members of the oxidized and reduced CV carbonaceous chondrites, unambiguous evidence of aqueous alteration is indicated by the presence of hydrous phases, such as phyllosilicates. However, members of the oxidized CV3 chondrites also contain a variety of anhydrous phases that are also generally agreed to be of secondary origin. Although these phases have been traditionally regarded as being of nebular origin, it has been proposed more recently that these alteration assemblages are in fact the result of metasomatic processes involving an aqueous fluid on an asteroidal parent body (*Krot et al.,* 1995, 1998a,b, 2000a). The origin of these assemblages has therefore been the subject of intense debate over the last decade (see *Brearley,* 2003, for a recent summary).

Unlike all the other effects of aqueous alteration discussed elsewhere in this chapter, the proposed metasomatism of the oxidized CV chondrites occurred at temperatures

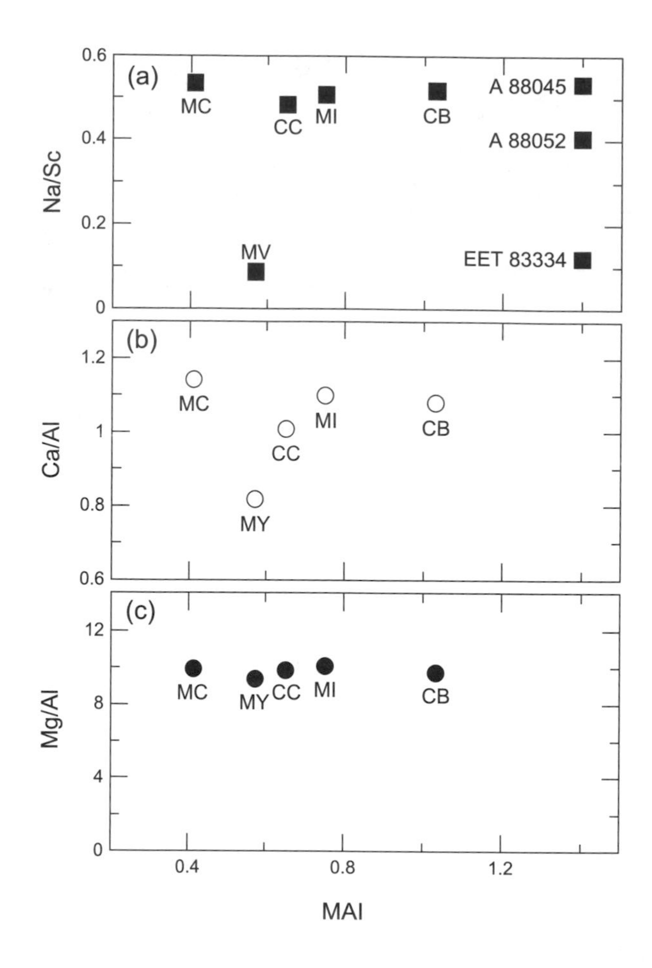

Fig. 6. Bulk-rock element/refractory element ratios for CM chondrites vs. mineralogical alteration index (MAI) as defined by *Browning et al.* (1996). The chondrites range from relatively weakly altered CM2 chondrites to fully hydrated CM1 chondrites. The CM1 chondrites are arbitrarily designated as having MAIs of 1.4. **(a)** Ca/Al ratio showing that despite aqueous alteration, this refractory element ratio remains constant irrespective of degree of aqueous alteration. Ca is clearly highly mobile on the scale of hundreds of micrometers to millimeters in CM chondrites, because of its high solubility in aqueous fluids, but was not depleted in the bulk rock, regardless of degree of aqueous alteration. **(b)** Na/Sc, showing that the moderately volatile and highly soluble element Na is not fractionated from the refractory and immobile element Sc during aqueous alteraton. **(c)** Mg/Al vs. MAI, demonstrating that refractory element ratios remain constant during aqueous alteration. Murray appears to be anomalously low in its Ca and Na abundances, possibly because it has been extensively brecciated. Data from *Kallemeyn and Wasson* (1981); *Grady et al.* (1987); *Wlotzka et al.* (1989); *Browning et al.* (1996); *Zolensky et al.* (1997).

significantly higher than the alteration in most other chondrites (~200°–300°C), based on thermodynamic calculations (e.g., *Krot et al.,* 1998a,b, 2000a). As a result the dominant alteration products are anhydrous rather than hydrous phases such as phyllosilicates. It is the absence of this "smoking gun" that is, in part, responsible for the extensive debate as to the origin of these anhydrous mineral assemblages in the Allende subgroup of the oxidized CV3 chondrites ($CV3_{OxA}$). However, it should be noted that some authors have argued that the rare hydrous phases that occur in CAIs and chondrules in meteorites such as Allende were also formed during this relatively high-temperature aqueous metasomatism (e.g., *Brearley,* 1999). In the case of the Bali subgroup of the oxidized CV3 chondrites ($CV3_{OxB}$), there is less ambiguity because the anhydrous phases coexist with phyllosilicate phases.

Aqueous alteration has affected the reduced CV chondrites to a very limited extent. Vigarano, the only reduced CV chondrite fall, exhibits incipient aqueous alteration within its matrix, indicated by the presence of ferrihydrite and rare, fine-grained saponite, that occur interstitially to and replacing FeO-rich matrix olivines (*Lee et al.,* 1996). Alteration in Vigarano is heterogeneous as indicated by the presence of phyllosilicate-rich fine-grained rims on some chondrules in Vigarano that have experienced more advanced alteration than typical regions of Vigarano matrix (*Tomeoka and Tanimura,* 2000). These variations in the degree of alteration are probably a reflection of the fact that Vigarano is a breccia consisting of both oxidized and reduced types that may have experienced variably alteration histories. Very minor development of carbonate that appears to be extraterrestrial in origin has also been reported in Vigarano associated with chondrules and CAIs (*Mao et al.,* 1990; *Davis et al.,* 1991; *Abreu and Brearley,* 2005b).

In the oxidized CV chondrites, phyllosilicates have been widely reported, and the range of alteration in this group varies from essential zero to regions of some chondrites where alteration is almost 50%. In members of the Bali subgroup, such as Bali, Kaba, Mokoia, and Grosnaja, matrix olivines have commonly been replaced by saponite and a variety of phyllosilicate phases including Fe-bearing saponite, Na phlogopite, Al-rich serpentine, and Na-K mica are found in altered chondrules and CAIs (*Cohen et al.,* 1983; *Keller and Buseck,* 1990b; *Tomeoka and Buseck,* 1990, *Keller et al.,* 1994; *Krot et al.,* 1998a,b). In general, evidence of significant elemental redistribution is limited in the CV_{oxBali} subgroup, with the exception of Bali itself. Bali contains distinct regions of highly altered material characterized by the formation of secondary phosphates, magnetite, and carbonates.

In members of the Bali subgroup, anhydrous phases that may have formed as a result of metasomatic processes involving aqueous fluids coexist with the phyllosilicate minerals described above. These secondary phases include magnet-ite, Ni-rich sulfides, fayalitic olivine, Ca-Fe-rich pyroxenes (Di-Hd), and andradite (Figs. 8a,b). Textural relationships between these phases indicate a complex sequence of replacement reactions that involved extensive mobilization of Ca from Ca-rich phases in chondrules and CAIs. Metal nodules in chondrules have been replaced by magnetite that has itself been replaced to varying degrees by fayalitic olivine, Ca-Fe-rich pyroxenes, and andradite. Thermodynamic analysis constrains formation of these mineral assemblages to temperatures of ~300°C (*Krot et al.,* 1998a,b).

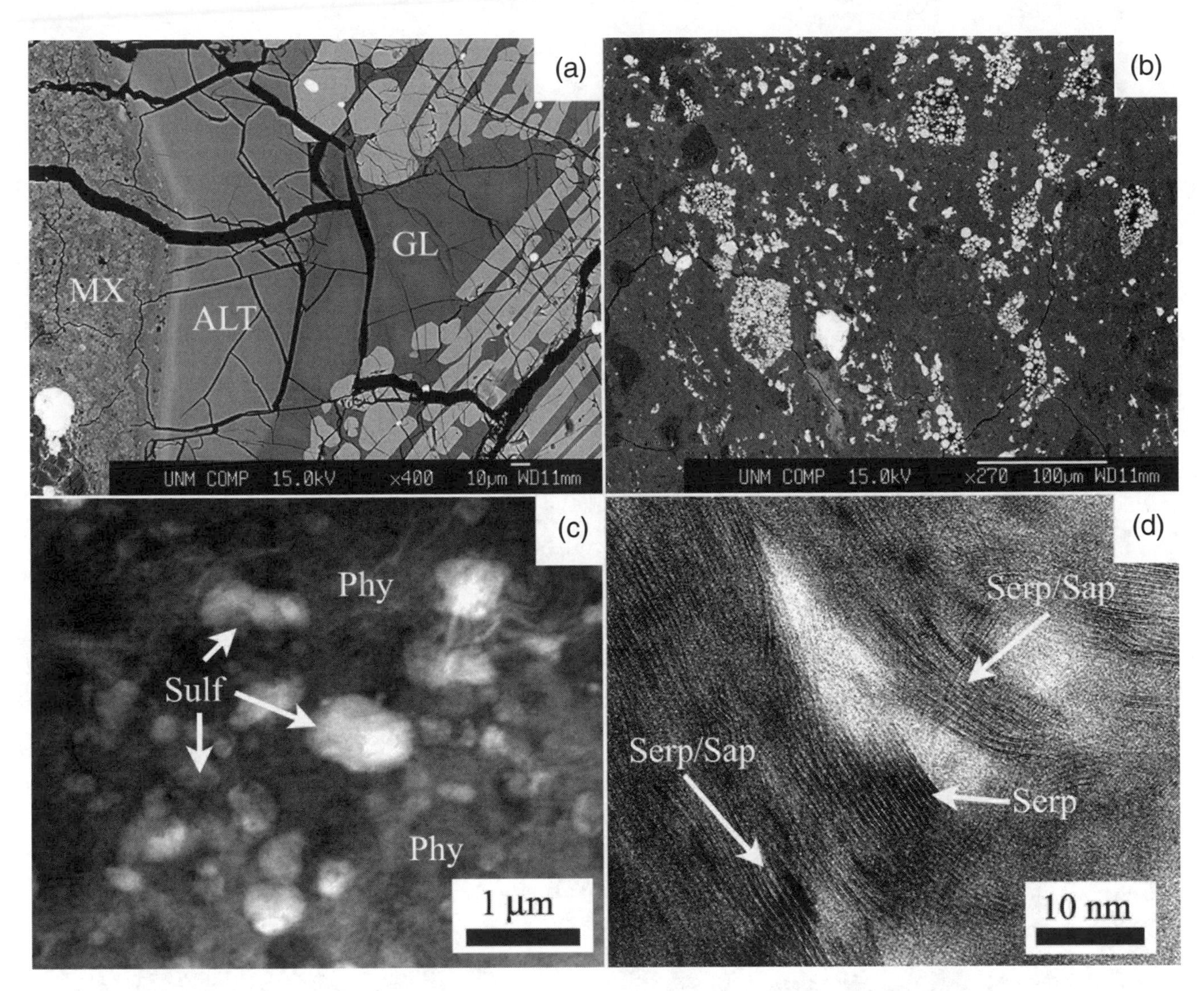

Fig. 7. Textural and mineralogical characteristics of aqueous alteration in CR chondrites. **(a)** BSE image showing a zone of altered chondrule glass (ALT) in contact with fine-grained matrix material (MX) on the edge of a type II chondrule in the weakly altered CR2 EET 87770. In the interior of the chondrule, chondrule mesostasis glass (GL) is unaffected by the alteration process. **(b)** BSE image of typical region of matrix in EET 92024 showing abundant framboids of magnetite, embedded in a very fine-grained matrix. **(c)** Dark field scanning transmission electron microscope Z-contrast image of a region of typical matrix in the EET 92024 CR carbonaceous chondrite. In this image, phases with bright contrast have the highest average atomic number. Abundant submicrometer Fe,Ni sulfides, pentlandite, and pyrrhotite (Sulf) are present embedded in a groundmass of fine-grained phyllosilicates (Phy), principally serpentine and saponite. **(d)** High-resolution TEM image of typical fine-grained phyllosilicates in the matrix of the CR carbonaceous chondrite EET 92024. The matrix phyllosilicates consist of individual serpentine grains (labeled Serp) and grains that are disordered intergrowths of serpentine and the smectite group mineral saponite (labeled Serp/Sap). Image courtesy of A. M. Abreu.

The situation in the Allende-like subgroup of the oxidized CV chondrites is extremely complex, in part because Allende itself is the only chondrite of this subgroup that has been studied in any detail. Allende has an exceptionally low water content (<0.2 wt%) (*Jarosewich et al.,* 1987) and contains no hydrous phases within its fine-grained matrix (*Toriumi,* 1989; *Brearley,* 1999). Nevertheless, several occurrences of rare phyllosilicates have been described in Allende in CAIs and to a lesser extent in chondrules. Among the phases that have been described in Allende CAIs are phlogopite, montmorillonite, clintonite, margarite, saponite, and chlorite (*Wark and Lovering,* 1977; *Dominik et al.,* 1978; *Hashimoto and Grossman,* 1987; *Keller and Buseck,* 1991; *Tomeoka and Buseck,* 1982a,b). In low-Ca pyroxene-bearing chondrules, *Brearley* (1999) reported the widespread occurrence of small amounts of talc, amphibole, and disordered biopyriboles (Fig. 8b). Compared with the hydrous phases observed in the Bali-subgroup, it is notable that with the exception of montmorillonite and saponite, the phases present in Allende all have higher ther-

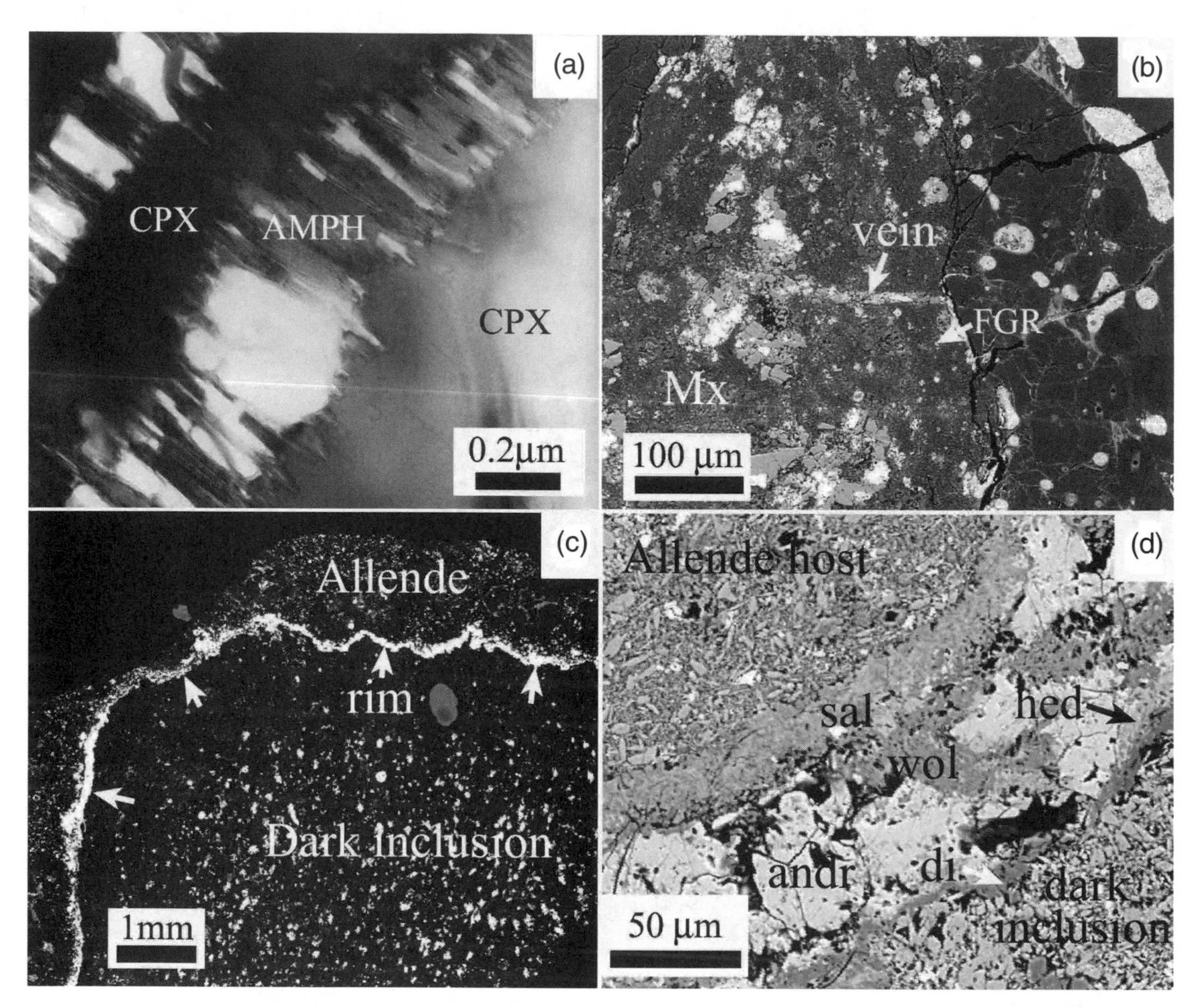

Fig. 8. BSE SEM and TEM images showing a variety of styles of alteration in CV3 carbonaceous chondrites and DIs. **(a)** TEM image showing development of fibrous amphiboles replacing low-Ca pyroxene in a chondrule from Allende. Although such phases are rare, they appear to be widely developed in pyroxene-bearing chondrules in this meteorite. **(b)** BSE image showing development of a vein of alteration that crosscuts a fine-grained rim in the Mokoia CV3 (oxidized) chondrite. The vein consists of the anhydrous phases, fayalite, magnetite and Fe-sulfide. Image courtesy of A. N. Krot (after *Krot et al.,* 1998b). **(c)** Ca K_α X-ray map of part of a DI in the oxidized CV3 chondrite Allende. The DI is surrounded by a narrow rim of Ca-rich phases that is in direct contact with the Allende host matrix. A zone around the periphery of the inclusions shows a depletion in Ca that is interpreted as a zone in which soluble Ca has been leached from the DI by aqueous fluids and has been precipitated at the boundary between the DI and the host Allende matrix. Image courtesy of A. N. Krot (after *Krot et al.,* 2000a). **(d)** BSE image showing the detailed characteristics of the Ca-rich rim between the DI and Allende host shown in **(c)**. The rim consists of a complex layered sequence of andradite garnet, wollastonite, salitic pyroxene, hedenbergite, and diopside. Image courtesy of A. N. Krot (after *Krot et al.,* 1998b).

mal stabilities. The style of aqueous alteration in Allende is therefore quite different from that observed in the Bali-like oxidized CV chondrites.

Members of the Allende subgroup have also experienced secondary alteration that has resulted in the formation of anhydrous, rather than hydrous, mineral assemblages that share some similarities with those observed in the Bali subgroup, but there are notable differences. The effects of this secondary alteration include (1) Fe-alkali-halogen metasomatism of CAIs, chondrules, and matrix; (2) formation of ferrous olivine rims on chondrules, isolated olivine grains, etc.; and (3) oxidation and sulfidization of opaque assemblages in CAIs, chondrules, and matrix. Additional possible effects of alteration noted by *Krot et al.* (1995) also include the formation of platy FeO-rich olivines, Ca-Fe pyroxenes, wollastonite, and andradite in the matrix. Unlike members of the Bali subgroup, phyllosilicates and pure fayalitic olivine are extremely rare. Instead, secondary alkali-rich phases such as nepheline and sodalite are common (*Krot et al.,* 1998a,b).

The bulk compositions of the reduced and oxidized CV carbonaceous chondrites show some compositional differences that are probably a reflection of complex secondary processing. These differences are particularly apparent for elements that are mobile during aqueous alteration and metasomatism, with the oxidized CV chondrites showing enrichments in elements such as Fe and Na, relative to the reduced CV chondrites (*Krot et al.,* 1995, 1998a,b). The origin of these compositional variations appears to be strongly correlated with the fact that the oxidized subgroup has undergone extensive aqueous alteration (CV_{oxBali}) and/or metasomatism (CV_{oxAll}), showing that alteration for this group of chondrites did not occur under closed-system conditions.

CV chondrites commonly contain dark, lithic clasts, termed dark inclusions (DIs), that are considered to have a close genetic relationship to the host CV chondrites (*see Krot et al.,* 1995; *Brearley,* 2003). The same alteration features that are observed within the oxidized CV chondrites are also apparent within DIs, but the range of alteration in DIs is much wider. Consequently, DIs provide additional insights into the alteration processes that affected the CV3 chondrites. Dark inclusions have been classified into two types: type A and type B (e.g., *Johnson et al.,* 1990; *Krot et al.,* 1995). Type A inclusions resemble the host CV chondrites, but have smaller chondrules and CAIs. Type B inclusions contain no chondrules and CAIs, but consist largely of aggregates of fine-grained FeO-rich olivine with chondrule-like shapes (*Kojima et al.,* 1993; *Kojima and Tomeoka,* 1996; *Krot et al.,* 1997a) set in a matrix of finer-grained olivine. In DIs in the oxidized CV chondrites, the olivine aggregates contain platy olivine whereas in the reduced CVs, such as Leoville and Efremovka, chondrules have been perfectly pseudomorphed by dense aggregates of fine-grained FeO-rich olivine (*Kracher et al.,* 1985; *Johnson et al.,* 1990; *Krot et al.,* 1999).

With a few exceptions, the alteration mineralogy of DIs lacks hydrous phases such as phyllosilicates. Hence the mineralogical evidence for aqueous alteration is cryptic and other indicators such as the O-isotopic compositions of the DIs are used to infer that these complex materials interacted with water. Dark inclusions from reduced CV chondrites such as Leoville and Efremovka have bulk O-isotopic compositions that do not lie on the carbonaceous chondrite anhydrous mineral (CCAM) line. Instead, these inclusions show heavy isotope enrichments that are generally interpreted as resulting from interaction with aqueous fluids, probably on an asteroidal parent body (e.g., *Clayton and Mayeda,* 1999). However, not all inclusions show such heavy isotope enrichments, including most inclusions that have been studied from the Allende meteorite. In the case of Allende, the involvement of water in the genesis of these inclusions has been questioned (*Clayton and Mayeda,* 1999), despite the fact that some textural lines of evidence appear to provide strong support for the role of fluids in the alteration processes (*Krot et al.,* 1998a,b, 2000a), such as the presence of Ca-rich rims at the interface of the dark inclusions with the Allende matrix (Fig. 8c,d) and veins of Ca-rich pyroxenes that crosscut the dark inclusions (see below).

2.6. Unique Carbonaceous Chondrites

There are several unique carbonaceous chondrites such as Tagish Lake, MacAlpine Hills (MAC) 88107, Kaidun, and Lewis Cliff (LEW) 85332 that exhibit evidence of aqueous alteration. However, for brevity only Tagish Lake and MAC 88107 are discussed here. Details of aqueous alteration in LEW 85332 and Kaidun can be found in *Brearley* (1997a) and *Zolensky and Ivanov* (2003), respectively. Tagish Lake is probably the most pristine carbonaceous chondrite ever recovered. This meteorite is clearly a type 2 carbonaceous chondrite with affinities to both CI1 and CM2 chondrites, but defies a more definitive classification. *Grady et al.* (2002) have suggested that Tagish Lake is best classified as a CI2 chondrite. Tagish Lake is unusual in having an extremely low density (1.66 g/cc) (*Zolensky et al.* 2002), much lower than any other carbonaceous chondrite measured to date. Tagish Lake appears to be a breccia with two major distinct lithologies (carbonate-poor and carbonate-rich) in addition to clasts that have affinities to CM1 chondrites (*Zolensky et al.,* 2002). Sparse chondrules and CAIs, showing partial alteration, are embedded in a phyllosilicate-rich matrix. The alteration mineralogy of Tagish Lake consists of fine-grained phyllosilicates (saponite and serpentine), magnetite, Fe,Ni sulfides, and carbonates (calcite, dolomite, siderite). The fine-grained textural characteristics of the matrix phyllosilicates are typical of CI chondrites, but differ markedly from CM chondrites.

MacAlpine Hills 88107 is a unique, unequilibrated carbonaceous chondrite that has mineralogical and compositional characteristics that are intermediate between the CM and CO chondrites (*Krot et al.,* 2000b). Aqueous alteration in this chondrite has resulted in the formation of both hydrous and anhydrous phases. Saponite and serpentine occurring as intergrowths are the dominant phyllosilicates in the fine-grained matrix and coarser-grained fayalite, hedenbergite, magnetite, pyrrhotite, and pentlandite are also widely present. This alteration assemblage bears close similarities to alteration in the Bali subgroup of the oxidized CV3 carbonaceous chondrites.

2.7. Ordinary Chondrites

2.7.1. Unequilibrated ordinary chondrites (UOCs). Evidence of aqueous alteration is best developed in the matrices of a few UOCs such as Semarkona (LL3.0), Bishunpur (LL3.1), and Chainpur (LL3.4) (*Hutchison et al.,* 1987; *Alexander et al.,* 1989), but alteration effects have also been recognized recently in chondrules (*Hutchison et al.,* 1998; *Grossman et al.,* 2000, 2002). In Semarkona, alteration in the fine-grained rims and matrix is pervasive and few primary phases have survived the alteration event(s) (*Hutchison et al.,* 1987; *Alexander et al.,* 1989). Extremely fine-grained Fe-rich smectite (Fig. 9a) is the dominant alteration mineral although roughly circular areas of coarser-grained smectite up to few micrometers in size are also present. These regions are sometimes rimmed by fine-grained maghemite and calcite. Rare primary phases that have survived alteration are Mg-rich olivine and twinned monoclinic low-Ca pyroxene, as well as partially altered grains of what appear to be densely packed aggregates of FeO-rich olivine. Sulfides of probable secondary origin, such as laths of pentlandite, also occur in the matrix. Primary sulfides in Semarkona matrix have been partially replaced along well-defined crystallographic planes by Fe-rich phyllosilicates. In the least-altered sulfides, the phyllosilicates have compositions that resemble Fe-rich chlorite or septachlorite. Sulfides that show more advanced alteration have been replaced by

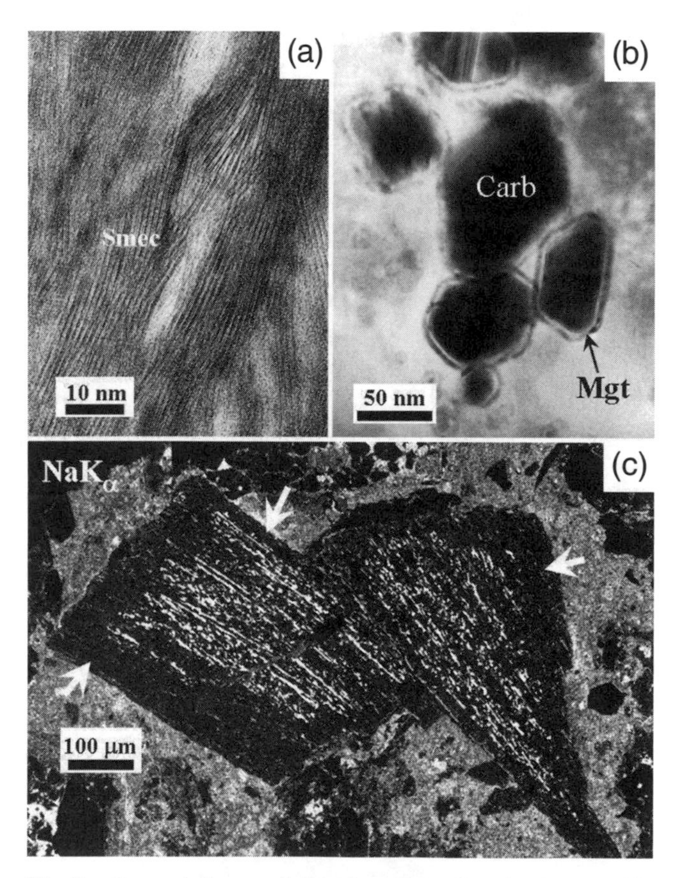

Fig. 9. Textural characteristics of aqueous alteration in unequilibrated ordinary chondrites (UOCs). **(a)** High-resolution TEM image of very fine-grained FeO-rich smectite (Smec) in the matrix of the unequilibrated ordinary chondrite Semarkona. **(b)** TEM image of Fe-carbides (Carb) in the matrix of Semarkona surrounded by thin oxidation rims of magnetite (Mag). **(c)** Na K_α X-ray map of fragments of a bleached radial pyroxene chondrule in Semarkona. The chondrule clearly fragmented presumably as a result of brecciation and the two fragments were offset from one another. The bleached zone (arrowed) is up to 100 µm thick on essentially all the peripheral regions of the fragment, except along the plane where fragmentation occurred. Here the bleached zone is significantly thinner, indicating that the bleaching process occurred both before and after the brecciation event that disrupted the fragment. After *Grossman et al.* (2000).

smectite with compositions that match those of smectite elsewhere in Semarkona matrix. In addition to phyllosilicates, some regions of the matrix contain the fine-grained Fe,Ni carbides (cohenite and haxonite) that are surrounded by thin oxidation rinds of magnetite (Fig. 9b) (*Keller,* 1998). Coarse-grained carbide-magnetite assemblages are widespread in Semarkona, replacing metal-sulfide nodules (*Taylor et al.,* 1981; *Hutchison et al.,* 1987; *Krot et al.,* 1997b). These carbide-magnetite assemblages appear to be the result of the interaction of C-O-H fluids with primary metal and sulfides during mild asteroidal metamorphism (e.g., *Krot et al.,* 1997b) and will not be discussed further here.

Compared with Semarkona, alteration of matrix and fine-grained rims in Bishunpur (*Alexander et al.,* 1989) is much less advanced. The precursor mineralogy of matrix and fine-grained rims consists of fragments of olivine and low-Ca pyroxene embedded in an amorphous material that is rich in normative feldspar. Alteration appears to have affected the amorphous material exclusively, which has been replaced by smectite with lower FeO, but higher CaO contents than smectite in Semarkona. Smectite in Bishunpur is also Cl-bearing with up to 0.3 wt% Cl, indicating that Cl-bearing fluids were certainly involved in the alteration process.

Aqueous alteration of chondrules is somewhat more cryptic and comes in the form of so-called bleached chondrules that have been reported in several UOCs (*Kurat,* 1969; *Christophe Michel-Levy,* 1976; *Grossman et al.,* 2000). In these meteorites, radial pyroxene and cryptocrystalline chondrules have porous outer zones that have a lighter appearance when viewed by optical microscopy, where mesostasis has been destroyed and alkalis and Al removed (Fig. 9c), probably by dissolution (*Grossman et al.,* 2000). Transmission electron microscope studies of chondrules in Semarkona show that partial replacement of chondrule glass by fine-grained smectite has occurred. This smectite is FeO-rich and has a composition that is essentially identical to that of smectite in the adjacent matrix (*Grossman et al.,* 2002). *Hutchison et al.* (1987) noted that calcite sometimes occurs as trains of crystals lying approximately parallel to the margins of chondrules that may represent a sink for Ca that has been mobilized from chondrule glass by aqueous fluids.

2.7.2. Equilibrated ordinary chondrites. Metamorphosed ordinary chondrites have been viewed for decades as being free of any effects of aqueous alteration. This view has required radical revision after the discovery of halite crystals that contain fluid inclusions, up to 15 µm in size, in the Monahans (H5) and Zag (H4–6) ordinary chondrite breccias (*Zolensky et al.,* 1999; *Rubin et al.,* 2002). In both meteorites, halite occurs in the porous clastic matrix, which is interstitial to coarser-grained clasts of metamorphosed, and in some cases, shocked ordinary chondrite material. In Zag, halite occurs as grains up to 100 µm in size whereas in Monahans, euhedral crystals of halite, sometimes with sylvite inclusions, can be up to 5 mm in size.

Additional evidence that metamorphosed ordinary chondrites have experienced aqueous alteration is provided by the presence of bleached chondrules in rare petrologic type 4 and 6 chondrites (*Grossman et al.,* 2000). These chondrules have essentially identical features to those in type 3 ordinary chondrites, but appear to have experienced metamorphism that postdates aqueous alteration.

3. LOCATION OF ALTERATION

A critical question that has become increasingly controversial over the last 15 years concerns the environment where aqueous alteration of chondritic meteorites took place. This controversy has focused largely on the CM chondrites and, to a lesser extent, the CVs and CRs. Alteration of other chondrite groups such as the CIs, COs, and UOCs is much less contentious and there appears to be a general consensus that most evidence is compatible with asteroidal alteration.

At present, four different alteration scenarios have been proposed, three involving hydration prior to accretion of the final asteroidal parent body. These models can be summarized as (1) reaction of anhydrous, high-temperature condensate phases with water vapor as the solar nebula cooled to the condensation temperature of water ice [~160 K at P ~ 10^{-6} bar, e.g., *Cyr et al.* (1998)]; (2) hydration of silicate dust in the solar nebula during the passage of shock waves through regions of elevated ice/dust ratios (*Ciesla et al.,* 2003); (3) alteration within small (tens of meters), ephemeral, water-bearing protoplanetary bodies that were later disrupted (e.g., *Metzler et al.,* 1992; *Bischoff,* 1998) and their altered components dispersed and then accreted with unaltered materials into the final asteroidal parent bodies (preaccretionary alteration); and (4) the more traditional parent-body alteration model (e.g., *Kerridge and Bunch,* 1979; *McSween,* 1979; *Browning et al.,* 1996; *Hanowski and Brearley,* 2001) in which aqueous alteration occurs entirely after asteroidal accretion. These models are not necessarily mutually exclusive: Preaccretionary alteration may have occurred, but was, in some or most cases, followed by asteroidal alteration.

In order to evaluate the evidence for and against these different alteration scenarios, it is useful to review possible criteria that can potentially be used to discriminate between alteration in different environments. These different criteria have been used in a rather *ad hoc* fashion in the literature, leading to confusion. In most cases, none of the criteria listed below provide exclusive proof of alteration in particular environment, but generally support alteration in one environment or the other. Specific criteria that can and have been used are (1) heterogeneity vs. homogeneity of the mineralogical effects of alteration within different primitive components of chondrites (e.g., CAIs, chondrules, matrix, isolated mineral grains, i.e., presence of disequilibrium mineral assemblages) (*Brearley,* 1997b; *Hanowski and Brearley,* 2001); (2) stable isotopic heterogeneity vs. homogeneity of alteration products in altered chondrites as an indicator of alteration in different or similar isotopic reservoirs; (3) correlation of mineralogical alteration effects with stable isotopic data (*Browning et al.,* 1996); (4) evidence of mass-independent fractionation (nebular) vs. mass-dependent fractionation (asteroidal) in the O-isotopic composition of bulk chondrites and individual components (*Weisberg et al.,* 1993; *Clayton and Mayeda,* 1999); (5) evidence (or lack thereof) of bulk compositional variations that might constrain the location of alteration; (6) constraints provided by the thermodynamic stabilities of mineral phases under solar nebular and asteroidal conditions (e.g., *Krot et al.,* 1998a,b); (7) experimental and theoretical constraints on the kinetics of alteration reactions (i.e., gas-solid, liquid-solid etc); and (8) timing of alteration, i.e., evidence of the formation of alteration phases more than several million years after CAI formation is most consistent with asteroidal alteration, assuming nebular lifetimes of 5–10 m.y. (*Podosek and Cassen,* 1994).

A variety of lines of evidence to support both asteroidal and preaccretionary alteration have been presented in the literature. Some of the key lines of evidence are reviewed below; for an exhaustive review of the evidence for preaccretionary alteration, see *Bischoff* (1998).

3.1. CI Chondrites: Evidence for Parent-Body Alteration

Since early work on the CI chondrites (*DuFresne and Anders,* 1962; *Richardson,* 1978; *Fredriksson and Kerridge,* 1988), this group of meteorites has been regarded as representing the definitive example of asteroidal aqueous alteration. As discussed above, one of the key lines of evidence for asteroidal aqueous alteration, i.e., crosscutting sulfate veins, has been brought into serious doubt. *Gounelle and Zolensky* (2001) have reinterpreted the veining as reflecting dissolution, local transport, and reprecipitation of extraterrestrial sulfates by absorbed terrestrial water. Nevertheless, other lines of textural evidence are supportive of an asteroidal scenario for alteration, such as (1) crosscutting phyllosilicate-rich veins (*Tomeoka,* 1990), (2) the presence of carbonates that appear to represent fragments of an earlier generation of carbonate veins or resemble vein fillings (*Richardson,* 1978; *Endress and Bischoff,* 1996), and (3) the ubiquitous presence of framboidal magnetite (*Kerridge et al.,* 1979a). In the latter case, *Kerridge et al.* (1979a) have argued that precipitation of magnetite from a gel-like phase is likely.

Analyses of bulk CI chondrites show that they have unfractionated bulk compositions relative to the solar photosphere (*Anders and Grevesse,* 1989). This observation appears to be most consistent with alteration in an essentially closed system, i.e., within a parent body under conditions where fluid flow was minimal. Refractory and moderately volatile alkalis and alkali earths such as K, Na, Ca, Rb, and Sr, as well as the rare earth elements, have variable solubilities in aqueous fluids (e.g., *Hass et al.,* 1995) and are certainly leached at different rates from carbonaceous chondrites (*Mittlefehldt and Wetherill,* 1979). Alteration in anything other than a closed system on an asteroidal parent body would invariably cause fractionation of these elements from one another, as well as from less-soluble elements such as Ti and Al. Alteration of components to varying degrees in the nebula or within ephemeral protoplanetary bodies would therefore result in a final parent body that showed fractionated elemental abundance ratios, unless all the materials from the same region of the nebula were accreted together at the same time. This is plausible, but is a scenario that is not supported by the O-isotopic composition of different components in CI chondrites. CI chondrites have bulk O-isotopic compositions that lie close to the terrestrial fractionation line on a three-isotope plot (Fig. 10a), with a $\Delta^{17}O$ of +0.38 ± 0.09‰ ($2\sigma_m$) (*Rowe et al.,* 1994). For Orgueil, the bulk rock, carbonates ($\Delta^{17}O$ = ~+0.37‰), phyllosilicates (matrix: $\Delta^{17}O$ = ~+0.24‰), and water released from stepwise heating ($\Delta^{17}O$ = ~+0.6‰) (*Baker et al.,* 2002) all lie on or close to a mass-dependent fractionation line, indicating that they equilibrated with the same fluid. These data are most consistent with asteroidal alteration (*Clayton and Mayeda,* 1999; *Leshin et al.,* 2001; *Baker et al.,* 2002), although there are clearly complexities. For example, magnetites from Alais, Ivuna, and Orgueil have $\Delta^{17}O$ = +1.70 ± 0.05‰ and are clearly not at isotopic equilibrium with the host rock.

Grady et al. (1988) have also argued that the C-isotopic composition of carbonates in CI chondrites is most consistent with formation on a parent body. The CI carbonates have median $\delta^{13}C$ values of +50‰ to +60‰ compared with –15‰ to –30‰ for CI organic carbon. Formation of these isotopically-heavy carbonates has been attributed to the selective alteration of interstellar grains such as SiC that have $\delta^{13}C$ values >1200‰, within the CI parent body. If the carbonates had formed prior to accretion, their isotopic composition would reflect the typical C-isotopic composition of the nebular gas, which is considered to have had a light isotopic composition similar to that of organic carbon (i.e., $\delta^{13}C$ ~ –15‰) (*Kerridge,* 1993). Preservation of these heavy $\delta^{13}C$ values requires a selective alteration process within a closed system such as an asteroidal parent body. Similar arguments can also be advanced for formation of carbonates in the CM and CR chondrites, which also have isotopically-heavy carbonates (*Clayton,* 1963; *Halbout et al.,* 1986; *Grady et al.,* 1988; *Clayton and Mayeda,* 1999).

Manganese-chromium dating of carbonates in CI chondrites also appears to be most consistent with parent-body alteration [see also *Krot et al.* (2006) and later discussion]. Carbonates in Orgueil and Ivuna (*Endress et al.,* 1996; *Hutcheon and Phinney,* 1996) yield formation ages of 16.5–18.3 m.y. after CAI formation, implying that alteration of CI chondrites occurred over periods of time that are significantly longer than the estimated lifetimes for the solar nebula (*Podosek and Cassen,* 1994) and hence are more consistent with asteroidal alteration.

3.2. CM Chondrites

There is a significant body of evidence to support both preaccretionary and asteroidal alteration of these complex meteorites. As a consequence, alteration in this group of

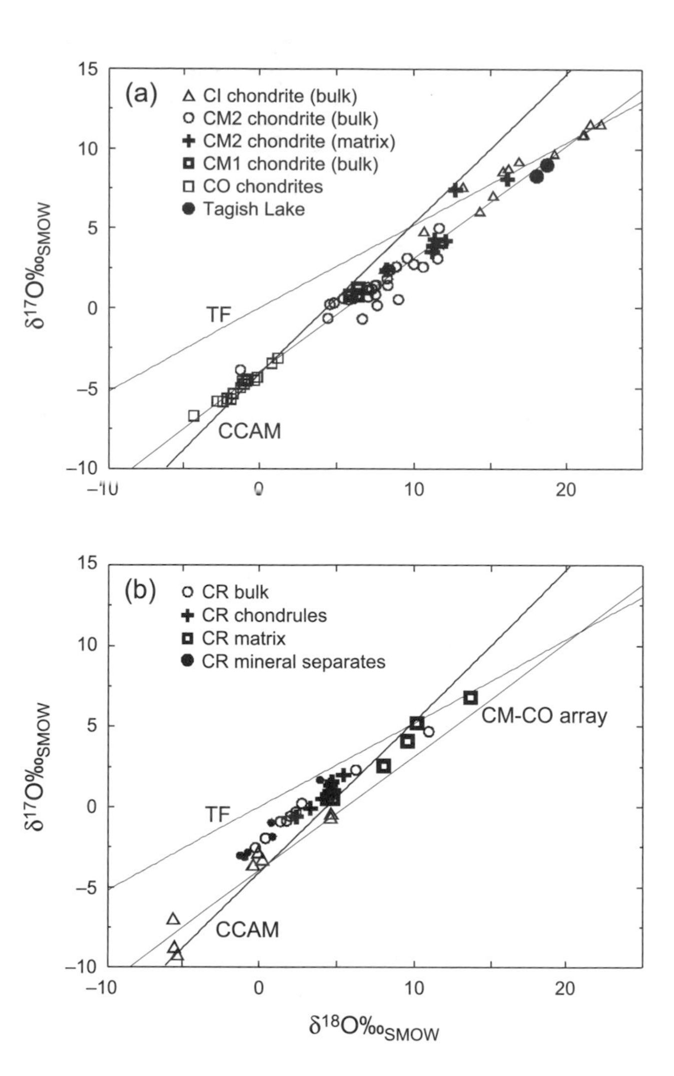

Fig. 10. **(a)** Oxygen three-isotope diagram showing the composition of bulk CI, CO, CM2, and CM1 chondrites, as well as the unique chondrite Tagish Lake. The compositions of matrix in CM2 chondrites are also shown. All the data for the CO and CM chondrites lie along the same mixing line with slope 0.7. According to the model of *Clayton and Mayeda* (1984), the CM chondrites were altered under fluid/rock ratios ranging from 0.3 to 0.6, whereas the CI chondrites and Tagish Lake experienced alteration under much higher water/rock ratios (~1.2). CM1 chondrites, despite their high degree of hydration, were altered under water/rock ratios similar to the least-altered CM2 carbonaceous chondrites. Data from *Clayton and Mayeda* (1999, 2001) and *Zolensky et al.* (1997). **(b)** Oxygen-isotope diagram showing the composition of bulk CR chondrites and mineral separates. The data lie along a line of slope 0.59 that is distinct from that defined by the CM and CO chondrites. Data from *Clayton and Mayeda* (1999).

chondrites is currently an area of significant controversy. A brief summary of the key lines of evidence that support each of the scenarios is discussed below.

3.2.1. Evidence for asteroidal alteration.

3.2.1.1. Veining in CM chondrites: Unlike the CI chondrites, crosscutting veins are extremely rare in CM chondrites. However, examples of veining have been documented, including Ca sulfate veins crosscutting the matrix of Cold Bokkeveld (*Lee,* 1993) and thin veins of Fe oxyhydroxide originating from altered metal in chondrules and extending across fine-grained rims into the clastic matrix in Murchison (*Hanowski and Brearley,* 2000). In both cases, vein formation clearly occurred after final accretion of the CM parent body.

3.2.1.2. Iron-rich aureoles: *Hanowski and Brearley* (2000) documented aureoles of Fe-enrichment surrounding large, altered metal grains in several CM chondrites that incorporate matrix, chondrules, and mineral fragments. The peripheries of the aureoles sometimes crosscut fine-grained rims and interfinger between chondrules and mineral fragments, demonstrating that they formed after asteroidal accretion. Aureole formation appears to be the result of oxidation of metal to soluble Fe^{2+} that was transported in solution through unconsolidated chondritic materials and was finally precipitated as insoluble Fe^{3+} oxyhydroxides.

3.2.1.3. Bulk compositional homogeneity of CM chondrites: Alteration in CM chondrites is quite variable in extent, even within CM2s. However, despite this variability, the bulk compositions of CM chondrites are quite homogeneous, suggesting that alteration was essentially isochemical in character. There is, for example, no evidence of the addition or removal of soluble elements, such as Ca, by fluids (e.g., Fig. 6). Instead, elemental mass transfer occurred on a localized scale and alteration largely involved exchange between fine-grained matrix and chondrules. Highly soluble Ca was leached from chondrules and CAIs during aqueous alteration and appears to have been precipitated in the matrix as Ca carbonate. This evidence is incompatible with the preaccretionary model, in which chondrules and matrix experienced variable degrees of alteration in different protoplanetary bodies. Preserving the homogeneous bulk composition of CM chondrites in the preaccretionary model requires mixing of altered chondrules with depleted Ca contents with an exact proportion of fine-grained dust with enriched Ca contents, in order to retain the solar Ca/Al ratio of the final chondrite.

3.2.1.4. Elemental exchange between chondrules and matrix during progressive alteration: Prior to aqueous alteration, the CM chondrites can, simplistically, be regarded as consisting of a mixture of Fe-rich, fine-grained matrix and Mg-rich chondrules. As proposed by *McSween* (1979, 1987), progressive aqueous alteration of these meteorites should result in the redistribution of these elements between these two components. The model predicts that with progressive aqueous alteration, matrix should become more Mg-rich and chondrules Mg-depleted. Data from CM chondrites show that the Mg content of matrix and fine-grained rims does, in fact, increase with progressive alteration (*McSween,* 1979; *Chizmadia and Brearley,* 2004), although the trend is complicated by brecciation effects. In addition, the Fe-rich serpentine alteration products of chondrule silicates become more Mg-rich as progressive alteration of chondrules occurs (*Hanowski and Brearley,* 1997).

3.2.1.5. Homogeneity of chondrule alteration: Asteroidal aqueous alteration requires that all components within the same meteorite were altered together under the same physicochemical conditions and for the same duration. Provided the effects of primary bulk compositions, mineralogy, grain size, and textures of different objects (e.g., chondrules) can be constrained, it would be expected that for *in situ* alteration, there should be a high degree of consistency in the style and degree of aqueous alteration of different components. Detailed studies of alteration of chondrules in two unbrecciated CM2 chondrites that have been altered to different degrees, Y 791198 (*Chizmadia and Brearley,* 2003) and ALH 81002 (*Hanowski and Brearley,* 2001) show that this is indeed the case. In the comparatively weakly altered Y 791198, only mesostasis glass, Fe,Ni metal, and sulfides are altered in all chondrules, whereas in ALH 81002 individual chondrules show highly variable degrees of alteration with extensive replacement of primary chondrule phenocryst phases such as olivine and low-Ca pyroxene. Pseudomorphic replacement of the chondrules and their phases is essentially perfect such that the morphologies of the precursors are retained exactly. When chondrules of the same type are compared, the style and extent of alteration is consistent, as are the compositions of the serpentine alteration products. These observations are consistent with *in situ* alteration in which all chondrules in an individual, unbrecciated CM chondrite interacted with the same fluid. This evidence does not necessarily preclude preaccretionary alteration. However, the probability of transporting groups of chondrules with the same degrees of alteration through a turbulent nebula as proposed in the preaccretionary model (e.g., *Metzler et al.,* 1992), without mixing them with chondrules with different degrees of alteration, is small.

3.2.1.6. Oxygen-isotopic compositions: The O-isotopic data show systematic relationships between the bulk isotopic composition of CM chondrites and the degree of aqueous alteration (*Browning et al.,* 1996), a relationship that would be extremely difficult to maintain in a preaccretionary environment where disruption of planetessimals occured prior to accretion of the final parent body. Furthermore, the O-isotopic compositions of carbonates in CM chondrites also evolve with increasing degree of alteration in a manner that is most consistent with formation from a fluid that is progressively equilibrating with the host rock (e.g., *Benedix et al.,* 2003). In addition, as noted earlier, the O-isotopic composition of water evolved from phyllosilicates in Murchison also appears to lie on the same mass-dependent fractionation line as carbonates (*Baker et al.,* 2002), supporting the view that these two components equilibrated with the same fluid. However, it should also be noted that not all alteration components in CM2 chondrites were in isotopic equilibrium with the fluid, as indicated by the fact that Murchison matrix has a distinctly lower $\Delta^{17}O$ than either the carbonates or water released from phyllosilicates (e.g., *Baker et al.,* 2002). In addition, sulfates in CM chondrites have $\Delta^{17}O$ values that are more positive than either carbonates or phyllosilicates (*Airieau et al.,* 2001). Clearly the process of isotopic equilibration between a heterogeneous mixture of materials with variable isotopic compositions during aqueous alteration was a complex process, aspects of which are not yet fully understood.

3.2.2. Evidence for preaccretionary alteration. In the CM2 chondrites, most of the evidence for preaccretionary alteration focuses on the textural relationships between fine-grained rims (Fig. 3) and the objects (i.e., chondrules) they surround (*Metzler et al.,* 1992; *Bischoff,* 1998). In addition, the characteristics of the mineral assemblages present in the rims themselves are also important. Implicit in the preaccretionary model is the assumption that the rims formed by accretion of fine-grained dust onto chondrules within the solar nebula, a model that has received widespread support. The presence of enrichments of primordial noble gases in fine-grained rims appears to be most consistent with a nebular model (*Nakamura et al.,* 1999). However, asteroidal formation mechanisms for fine-grained rims have also been proposed (e.g., *Richardson,* 1981; *Sears et al.,* 1993).

3.2.2.1. Occurrence of unaltered chondrule glass in contact with hydrated rim materials: *Metzler et al.* (1992) described the rare occurrence of apparently unaltered chondrule glass juxtaposed against hydrated fine-grained rim materials in some CM chondrites. Chondrule glass is generally considered to be among the most susceptible of chondrule phases to aqueous alteration (see below) and therefore *Metzler et al.* (1992) argued that alteration of fine-grained rims could not have occurred without alteration of chondrule glass. They concluded that alteration of dusty rim materials must have occurred prior to accretion of rims onto the chondrules in the solar nebula, with no subsequent aqueous alteration within the asteroidal parent body. However, it is known that hydration of glasses can occur under certain conditions, but the amorphous structure of the glass is retained (e.g., *Mungall and Martin,* 1994). *Metzler et al.* (1992) did not report compositional data for the unaltered glass and so it is not possible to evaluate alternative explanations for this textural relationship. However, studies of other CM chondrites indicate that preserved chondrule glass is extremely rare. For example, *Chizmadia and Brearley* (2003) found no evidence of preserved chondrule glass in a large population of chondrules in Y 791198, a chondrite that is one of the least altered of the CM2 group.

3.2.2.2. Unaltered chondrule olivine fracture surfaces in contact with hydrated fine-grained rim material: In some CM chondrites, partially altered chondrule olivines with clean, unaltered fracture surfaces occur in direct contact with hydrous, fine-grained rim materials. *Metzler et al.* (1992) and *Bischoff* (1998) interpreted these relationships as evidence for preaccretionary alteration. In their alteration scenario, individual chondrules were altered prior to fracturing, then a fine-grained rim consisting of anhydrous and hydrated dust was accreted onto the chondrule fragments in the solar nebula. Finally, these rimmed objects were accreted into the CM parent body. *Metzler et al.* (1992) argued that if further aqueous alteration had occurred on an asteroidal parent body, the exposed fracture surfaces on the exterior of the chondrule would have undergone hydration to serpentine. However, *Metzler et al.* (1992) did not consider the possibility that the local chemical microenvironment in the fine-grained rim may have inhibited aqueous alteration of the olivine, as appears to have occurred in some chondrules in CM chondrites (e.g., *Hanowski,* 1998). Hanowski reported examples of chondrules that showed evidence of advanced alteration of olivine in their interiors, but no alteration where the olivine is in direct contact with fine-grained rim materials, features that cannot be interpreted as being due to preaccretionary alteration.

3.2.2.3. Disequilibrium mineral assemblages within fine-grained rims: In some CM chondrites, the fine-grained rim materials appear to be a disequilibrium mixture of unaltered and altered mineral grains. In Y 791198, perhaps the least-altered CM2 chondrite currently known, *Metzler et al.* (1992) found that fine-grained rims contain an intimate mixture of primary phases such as Fe,Ni metal, troilite, and olivine and alteration phases such as phyllosilicates, tochilinite, and sulfides. In addition, in ALH 81002 *Lauretta et al.* (2000) argued that regions of unaltered olivine, pyroxene, and metal were present in fine-grained rims. These complex mineral assemblages have been interpreted as evidence that unaltered and altered nebular dust were mixed together prior to asteroidal accretion (*Metzler et al.,* 1992; *Bischoff,* 1998). It is expected that subsequent asteroidal alteration would have caused oxidation of Fe,Ni metal and troilite to secondary alteration products. However, *Chizmadia et al.* (2003) proposed an alternative scenario in which the survival of Fe,Ni metal is the result of hydrolysis reactions during the early stages of aqueous alteration that consume protons and drive the fluid pH to very alkaline conditions under which metal alteration is inhibited.

There is a substantial body of textural evidence to support aqueous alteration in a preaccretionary environment. However, many of these interpretations are based on simple assumptions as to how mineral phases react with aqueous fluids and do not take into consideration the complex and changing geochemical environments that were prevalent during aqueous alteration. In particular, the role of localized geochemical microenvironments and their effects on promoting or inhibiting alteration reactions needs to be examined in much greater detail. Once this understanding is developed, it will be possible to evaluate the evidence for preaccretionary alteration in a much more robust manner. Although preaccretionary alteration may have been important in CM chondrites, its effects have certainly been overprinted by later parent-body alteration. Distinguishing these preaccretionary effects from asteroidal alteration remains a significant challenge in CM chondrites.

3.3. CR Chondrites

3.3.1. CR chondrites: Evidence for asteroidal alteration. There is some ambiguity as to the location of alteration of the CR chondrites that arises in part from the fact that most CR chondrites are breccias. *Weisberg et al.* (1993) noted that chondrules in CR chondrites often show variable degrees of alteration that could be the result of alteration prior to final lithification of these meteorites. However, they also suggested that this alteration could have taken place within the CR parent body and that the components were mixed together by later brecciation, rather than alteration occurring in a preaccretionary environment.

Evidence for parent-body alteration is provided by the alteration of chondrule glass in two chondrules from the CR2 chondrites, EET 87770 and EET 92105 (*Burger and Brearley,* 2004). In both meteorites, type II chondrules show evidence of partial alteration of chondrule glass to phyllosilicates where it is in direct contact with matrix. Calcium and P have been leached from the chondrule glass and the depletion in Ca is mirrored by an enrichment of Ca in the matrix that is due to the presence of fine-grained Ca phosphate. This evidence suggests that elemental exchange between chondrules and matrix has occurred during alteration with reprecipitation of Ca as Ca phosphate in the matrix immediately outside the chondrule.

Oxygen-isotope data for CR chondrites also support a parent-body alteration model. The O-isotopic compositions of bulk CR chondrites and separated components (chondrules, matrix, and mineral fragments) lie along a line of slope 0.59 (Fig. 10b) (*Weisberg et al.,* 1993; *Clayton and Mayeda,* 1999) that is interpreted as a mixing line between at least two O-isotopic reservoirs: a ^{16}O-rich anhydrous component and a ^{16}O-poor hydrous component. However, magnetite and phyllosilicate-rich matrix separates from Renazzo have O-isotopic compositions that lie on the terrestrial fractionation line, indicating that both materials formed from the same fluid (*Clayton and Mayeda,* 1977), a scenario that is most compatible with alteration within an asteroidal environment (*Clayton and Mayeda,* 1999).

3.3.2. CR chondrites: Evidence for preaccretionary alteration. Although *Weisberg et al.* (1993) concluded that alteration of the components in CR chondrites probably occurred within an asteroidal environment, they did not rule out the possibility of some preaccretionary alteration. *Ichikawa and Ikeda* (1995) have described unaltered chondrule glass in contact with hydrous matrix in Y 8449, which they attributed to preaccretionary alteration, i.e., hydrous alteration of matrix materials that were later mixed with unaltered chondrules prior to asteroidal accretion. They also noted marked compositional differences between phyllosilicates in chondrules and those in the adjacent matrix, which they argued were also the result of preaccretionary alteration.

3.4. CV Chondrites

3.4.1. CV Chondrites: Evidence for asteroidal alteration. Among the oxidized CV chondrites, members of the Bali subgroup show the strongest evidence for parent-body alteration. Aqueous alteration has affected all components (i.e., chondrules, CAIs, and matrix) of Bali, Kaba, and Mokoia, indicating that the fluid interaction occurred after all these components were assembled together (e.g., *Cohen et al.,* 1983; *Keller et al.,* 1994; *Kimura and Ikeda,* 1998). This conclusion is supported by evidence that Ca has been lost from chondrule mesostasis during alteration and has been redistributed into the matrix (*Krot et al.,* 1998a,b). The presence of alteration veins that postdate formation of a prominent shock-produced foliation is also consistent with this interpretation. Alteration phases in the matrices of these meteorites have clearly developed interstitially to fine-grained olivines, as well as replacing individual grains, indicating that alteration occurred after accretion.

Krot et al. (1998a,b) described textures in members of the Bali subgroup of the oxidized CV3 chondrites that provide strong evidence for parent-body alteration. As discussed above, in this group of chondrites, secondary phyllosilicates coexist with a variety of secondary anhydrous phases. Of particular significance is the observation that veins of fayalite, magnetite, and sulfide are present that originate in chondrules, crosscut fine-grained rims, and extend into the matrix of the meteorites (Fig. 8a), clearly indicating formation of the veins after accretion. Members of the Bali subgroup have bulk O-isotopic compositions that lie on the CCAM line. However, their O-isotopic compositions are lower in ^{16}O than either the Allende subgroup or the reduced CV chondrites, an effect that may be due to aqueous alteration (*Krot et al.,* 1995; *Clayton and Mayeda,* 1996, 1999). The isotopic effects of aqueous alteration are especially well developed in regions of Bali that show extensive development of phyllosilicates (*Keller et al.,* 1994). Samples of such regions have O-isotopic compositions that show significant heavy isotope depletions consistent with low-temperature aqueous alteration.

In addition, ion microprobe studies of coexisting magnetite and fayalite that have replaced metal nodules in chondrules in Mokoia have O-isotopic compositions that lie close to the terrestrial fractionation line (*Krot et al.,* 1998b). The two phases appear to lie on a mass-dependent fractionation line and have a large $\Delta^{18}O_{fayalite\text{-}magnetite}$ fractionation of ~20‰, indicative of a process that occurred at relatively low temperatures, most consistent with asteroidal alteration involving a fluid.

There are many unresolved questions regarding alteration in the Allende subgroup of the oxidized CV chondrites, the location of alteration being but one of them. Evidence of hydrous alteration in this group of chondrites is minimal and the O-isotopic compositions of bulk Allende and Allende matrix show no evidence of the heavy isotope enrichments that are the characteristic signature of aqueous alteration. However, laser fluorination O-isotopic mass spectrometric analyses of individual chondrules in Allende (*Young and Russell,* 1998) show clear evidence of limited heavy isotope enrichments, consistent with aqueous alteration. Mineralogically, *Brearley* (1997b) found that the same rare, hydrous phases occur in all low-Ca pyroxene-bearing chondrules studied within a single thin section and the style and abundance of the phases is consistent from one chondrule to another. This observation suggests that all the chondrules were altered under similar conditions, for similar lengths of time, consistent with an asteroidal environment. However, this evidence is difficult to reconcile with the evidence that Allende matrix shows no direct mineralogical or isotopic evidence of aqueous alteration. An asteroidal environment for alteration is therefore somewhat equivocal at this point, the data also being suggestive of preaccretionary alteration.

Studies of DIs in CV chondrites have provided quite an extensive body of textural and isotopic evidence for *in situ* parent-body alteration. *Krot et al.* (1998a,b, 2000a) have described DIs from Allende that are crosscut by veins of salitic pyroxenes and Fe,Ni sulfides. In addition, each inclusion is surrounded by continuous, layered rims consisting of Ca-Fe-rich mineral phases (Figs. 8c,d). The inner layer consists of diopside-salitic pyroxene, followed by a central zone of hedenbergite, wollastonite, andradite, and kirschsteinite. The outermost zone, in contact with Allende host matrix, is composed of salitic pyroxene. The thickness of the Ca-Fe-rich rim appears to be quite well correlated to observed depletions in Ca in the outer parts of each DI (Fig. 8c).

Krot et al. (2000a) have argued that these characteristics are consistent with at least two periods of aqueous alteration, both of which postdate the aggregation and lithification of the DIs. The first stage resulted in the formation of secondary mineral phases such as fayalitic olivine, nepheline, sodalite, and Fe,Ni sulfides as well as the crosscutting veins of Ca-Fe pyroxene and andradite. Following this stage of alteration, the DIs were transported from their formation location, probably by impact processes on the CV3 asteroid, and were mixed into Allende host material. The alteration assemblages in the DI were out of equilibrium with fluids in host Allende, causing dissolution of Ca-Fe pyroxenes in the DIs. Reprecipitation of these phases occurred at the interface between the DI and Allende host, a process that appears to have been initiated at temperatures of ~250°C, based on thermodynamic calculations (*Krot et al.,* 2000a). Diffusional exchange between the DIs and Allende host evidently occurred during this event, requiring that this episode of alteration occurred *in situ* within an asteroidal parent body.

Oxygen-isotopic data for DIs (*Clayton and Mayeda,* 1999) provide some important insights into their possible origins. Like Allende matrix, Allende DIs have compositions that are not displaced from the CCAM line to ^{16}O-rich composition (*Krot et al.,* 1998b). However, inclusions from Efremovka, Leoville, and Vigarano are displaced to the right of the CCAM line, forming an array that extends toward the terrestrial fractionation line and shares a number of similarities with the arrays found in the CR and CM chondrites. *Clayton and Mayeda* (1999) have argued that large heavy isotope enrichments in these DIs are likely to be the result of low-temperature aqueous alteration processes, although mineralogically there is little evidence of hydrous phases in these inclusions. For some inclusions (e.g., Efremovka) (*Krot et al.,* 1999), there is a clear correlation between degree of replacement of primary phases with heavy isotope enrichment, providing further support for such a model.

3.4.2. CV chondrites: Evidence for preaccretionary alteration. Most studies of aqueous alteration in the Bali subgroup of the oxidized CV chondrites favor alteration in an asteroidal environment (e.g., *Keller et al.,* 1994; *Krot et al.,* 1998a), whereas *Kimura and Ikeda* (1998) did not rule out the possibility of preaccretionary alteration. In the Allende subgroup, the heterogeneous development of rare phyllosilicate phases in CAIs has been widely attributed to formation of these phases in the solar nebula prior to accretion. For example, *Hashimoto and Grossman* (1987) and *Keller and Buseck* (1991) have proposed that phyllosilicate phases formed by reaction of individual CAIs with a nebular gas that also caused Fe-alkali-halogen metasomatism. In this scenario, heterogeneous development of hydrous phases depends on whether an individual CAI interacted with the gas or not.

3.5. CO Chondrites

3.5.1. CO chondrites: Evidence for asteroidal alteration. Definitive evidence for the location of the low-temperature aqueous alteration in the CO chondrites is limited, but points largely toward an asteroidal environment. In the matrix, development of hydrous phases has occurred interstitially to fine-grained matrix phases, filling pore space and partially replacing FeO-rich matrix olivine. This evidence indicates that alteration probably occurred after all the matrix grains accreted together and is most consistent with parent-body alteration. In the CO3.4, for example, Lancé, matrix olivines that have equilibrated Fa contents (Fa_{45-50}), indicative of thermal metamorphism (*Keller and Buseck,* 1990a), also show evidence of aqueous alteration, indicating that interaction with an aqueous fluid was a late-stage event and postdated parent-body metamorphism.

Rubin (1998) has argued that parent-body metamorphism in the CO chondrites occurred in the presence of an aqueous fluid and that aqueous alteration and metamorphism are closely related (i.e., fluid-assisted metamorphism). This suggestion appears to be supported by the bulk O-isotopic data for CO carbonaceous chondrites, which show a weak correlation with petrologic type. *Clayton and Mayeda* (1999) showed that CO bulk-rock data can be fitted well to a single mixing line with slope 0.70 that is also defined by CM whole-rock and matrix separates (Fig. 10a). *Clayton and Mayeda* (1999) tentatively interpreted these data as indicating that the CO chondrites probably interacted with an external O-bearing (aqueous?) reservoir during parent-body metamorphism. At least some alteration in the CO chondrites probably occurred post-metamorphism as a result of interaction of residual aqueous fluids that were not lost during the metamorphic event.

3.5.2. CO chondrites: Evidence for preaccretionary alteration. Despite the body of evidence for parent-body alteration in the CO chondrites, there is some evidence for preaccretionary alteration in the most primitive member of the group, ALH 77307 (3.0). This chondrite has a bulk O-isotopic composition that lies very close to the average value for anhydrous silicate minerals in Murchison, a value that *Clayton and Mayeda* (1984, 1999) have interpreted as being the primary nebular precursor composition of Murchison minerals, prior to aqueous alteration/thermal metamorphism. However, *Itoh and Tomeoka* (2001) have described phyllosilicates in chondrules in ALH 77307 that are embedded in a very fine-grained matrix that shows essentially no evidence of aqueous alteration and is very primitive in its textural and mineralogical characteristics (*Brearley,* 1993). Further studies are needed, but these data appear to be indicative of aqueous alteration of chondrules prior to parent-body accretion, although *Itoh and Tomeoka* (2001) attributed these phyllosilicates to parent-body alteration.

3.6. Unique Carbonaceous Chondrites

The unique carbonaceous chondrite MAC 88107 contains both textural and isotopic evidence of aqueous alteration within an asteroidal parent body (*Krot et al.,* 2000b). The most notable textural characteristic indicative of *in situ* alteration is the presence of widespread veins of fayalite and hedenbergite up to 30 μm in thickness. These veins originate at opaque nodules within chondrules and crosscut the surrounding fine-grained rims, extending into the matrix. These observations indicate that chondrules and their fine-grained rims must have accreted and been compacted together with matrix before alteration occurred. Additional evidence for asteroidal alteration comes from the Mn-Cr-isotopic systematics of fayalite in MAC 88107 (see *Krot et al.,* 2006). The initial ^{53}Mn of fayalite [(1.58 ± 0.26) × 10^{-6}] indicates a formation age ~9 or 18 m.y. after Allende CAI formation, depending on the choice of the solar system initial abundance of ^{53}Mn used in the age calculation.

3.7. Ordinary Chondrites

Aqueous alteration of the ordinary chondrites appears to have occurred dominantly in an asteroidal environment based on several independent lines of evidence. *Hutchison et al.* (1987) showed that alteration veins in the matrix of Semarkona penetrate into chondrules. They also described trains of calcite crystals that decorate the margins of chondrules and clasts, indicating formation after accretion. This conclusion is also supported by the observations of *Grossman et al.* (2000), who described a radial pyroxene chondrule in Semarkona (Fig. 9c) that underwent an early stage of alteration that produced a bleached zone around its exterior. This chondrule was later fragmented as a result of regolith processes and a second bleached zone developed along the surfaces exposed by the fracture. Smectite, with a composition essentially identical to that in the matrix, occurs replacing mesostasis glass, suggesting that alteration products of matrix and glass equilibrated with the same fluid on an asteroid.

The D/H ratios of chondrules and matrix in Semarkona provide evidence to support *in situ* exchange of aqueous fluids between these two different components. *Deloule and Robert* (1995) reported very elevated D/H ratios in the matrix of Semarkona that are indicative of the presence of interstellar water. A similar range of D/H ratios occurs in bleached chondrules in Semarkona (*Grossman et al.,* 2000), although these ion microprobe measurements are complicated by the fact that D/H ratios vary as a function of depth into the sample. This variability may be due to exchange of terrestrial water with interlayer water in smectite. Nevertheless, despite some ambiguity, the elevated D/H ratios in both matrix and altered chondrules indicate that they both exchanged with an isotopically similar reservoir, implying that alteration occurred within an asteroidal environment.

Although halite in the equilibrated ordinary chondrites Zag and Monahans appears to be unambiguously asteroidal in origin (*Zolensky et al.,* 1999; *Rubin et al.,* 2002), there is debate over whether it precipitated from fluids that were derived internally from within the ordinary chondrite parent body. *Rubin et al.* (2002) proposed a scenario in which extensive parent-body processing (metamorphism, breccia, regolith gardening) occurred before precipitation of halides from aqueous solutions within porous regions of the asteroidal regolith. However, *Zolensky et al.* (1999) and *Bridges et al.* (2004) have suggested that the halite was derived from water-rich materials that were added at a late stage to the ordinary chondrite regolith and hence is not indigenous to the ordinary chondrite parent body.

4. MODELS FOR PARENT-BODY ALTERATION

Several different approaches have been taken to develop models for various aspects of the alteration history of carbonaceous chondrite parent bodies and include both empirical and theoretical treatments. These models have attempted to address (1) the evolution of mineralogic alteration assemblages with progressive alteration using observations and computer modeling (e.g., *Zolensky et al,* 1989, *Rosenberg et al.,* 2001), (2) the stable isotopic evolution of solid and fluid phases during the alteration process (e.g., *Clayton and Mayeda,* 1984, 1999; *Young et al.,* 1999), and (3) the thermal evolution of asteroids undergoing alteration (*Grimm and McSween,* 1989; *Wilson et al.,* 1999; *Cohen and Coker,* 2000). In general, the focus of the bulk of these models has been directed toward understanding alteration of CM and CI chondrites, although the work of *Young et al.* (1999) and *Young* (2001) is more global in character.

4.1. CM Chondrite Alteration Models

McSween (1979, 1987) proposed a model for progressive aqueous alteration of CM chondrites, based on mineralogical and compositional criteria, that stems from the recognition that CM chondrites exhibit variable degrees of alteration. This model has been refined progressively by further studies (*Tomeoka and Buseck,* 1985; *Zolensky et al.,* 1993; *Browning et al.,* 1996; *Hanowski and Brearley,* 2001) and various petrographic and isotopic indices to track the degree of aqueous alteration have been proposed (e.g., *Browning et al.,* 1996).

At the core of the progressive alteration model is the concept that as alteration proceeds and phases with different alteration susceptibilities react, elemental exchange occurs between the FeO-rich, fine-grained matrix and chondrules that are dominated by Mg-rich phenocryst phases such as olivine and low-Ca pyroxene. Hence, the model predicts that with increased degrees of alteration, fine-grained matrix should become more Mg-rich. Compositional data for CM matrices suggest that such a general trend does exist, although it shows considerable complexities, indicating that other factors also need to be considered (*McSween,* 1979). For example, in the original model proposed by *McSween* (1979, 1987), the extent of alteration was closely coupled to brecciation. As chondrules altered they were broken apart by regolith comminution, causing a progressive increase in the modal abundance of matrix as these fragments were mixed with existing matrix materials and underwent alteration. Although this model holds for many CM chondrites, more recent studies show that advanced aqueous alteration can also occur entirely independent of brecciation as indicated by relatively heavily altered CM chondrites, such as ALH 81002 and several CM1 chondrites that are essentially unbrecciated (*Llorca and Brearley,* 1992; *Zolensky et al.,* 1997; *Hanowski and Brearley,* 2001).

Petrographic and mineralogic studies of CM chondrites have established a general sequence of alteration of primary phases that is consistent with this model (e.g., *McSween,* 1979; *Tomeoka and Buseck,* 1985; *Zolensky et al.,* 1993; *Browning et al.,* 1996; *Hanowski and Brearley,* 2001). This sequence is based on the observed relative alteration susceptibilities of the major primary phases in the presence of aqueous fluids. Iron-nickel metal, troilite, and chondrule glass are most susceptible to alteration. In comparison, olivine and low-Ca pyroxene appear to be much more resistant, although there is clearly a compositional dependence on olivine alteration, with FeO-bearing olivine in type IIA chondrules (Fa_{20-45}) altering more rapidly than forsteritic olivine in type IA chondrules and amoeboid olivine aggregates. These observations are consistent with experimental studies of olivine dissolution (*Wogelius and Walther,* 1991, 1992). Only in very heavily altered CM2 chondrites and CM1s are Mg-rich olivines altered to a significant degree.

Following *McSween* (1979, 1987), *Browning et al.* (1996) made an important effort to correlate alteration based on mineralogic criteria with other parameters that may be potential monitors of aqueous alteration in the CM chondrites. Their primary mineralogic criterion is based on the progressive increase in Mg content of matrix serpentine as alteration proceeds. This reaction can be formalized as a progressive change from cronstedtite to Mg-rich serpentine, described by the idealized coupled substitution

$$2(Fe^{3+},Al^{3+})_{cronstedtite} = Si^{4+} + (Mg^{2+}, Fe^{2+})_{serpentine}$$

the so-called mineralogic alteration index (MAI). Additional criteria that were used as indicators of aqueous alteration are the abundance of isolated, anhydrous mineral fragments in the matrix and the relative percentage of chondrule alteration. These parameters were chosen based on the assumption that as alteration proceeds, anhydrous phases that occur as isolated mineral grains and in chondrules should progressively alter to serpentine and hence will decrease in abundance. Despite the errors involved in measuring these different parameters because of the effects of brecciation in CM chondrites, *Browning et al.* (1996) were able to demonstrate that there are reasonable correlations between these three different independent measures of extent of alteration. Most importantly, the relative sequence of alteration for different CM chondrites is reproduced to a large extent in each case, with some minor differences. Good correlations are observed between the MAI and key bulk properties of CM chondrites, including H content and planetary noble gas content (Ar) (Fig. 11). As expected, the H content increases progressively as degree of hydration increases, whereas the ^{36}Ar content shows the reverse relationship. The decrease in ^{36}Ar content is attributed to progressive degassing of Ar as a result of destruction of the noble gas carrier phases by aqueous fluids.

Similar results were also obtained by *Eiler and Kitchen* (2004), who investigated the H-isotopic composition and highly volatile element contents of CM chondrites that straddle the range of alteration found in this group. In this study, δD was found to be highest for the least-altered chondrites and shows a decrease as alteration proceeds that is coupled to decreases in bulk-rock Na content and planetary noble gas content (Ar). These correlations are especially strong when δD for matrix separates is compared with whole-rock data for Na and Ar contents. Although Na shows an apparent decrease with increasing degree of alteration, *Eiler and Kitchen* (2004) suggested that this was not necessarily due to removal of Na during alteration, but could be due to a dilution effect caused by water that is structurally bound in the rock during aqueous alteration. The decrease in δD is also correlated with a decrease in whole-rock N/H and C/H ratios that is probably attributable to loss of C and N as CH_4 and NH_3 that are evolved from organic material during asteroidal hydration reactions.

4.2. Isotopic Constraints on Alteration in CM Chondrites

The O-isotopic compositions of CM chondrites provide some important constraints on the alteration process and can be used to constrain water/rock ratios and temperatures of alteration. Bulk CM chondrites and mineral separates have O-isotopic compositions that lie along a mixing line with slope 0.7, rather than the slope 0.5 CCAM line (*Clayton and Mayeda,* 1984, 1999). The slope of this mixing line is attributed to the fact that CM chondrites consist of a mixture of ^{16}O-rich primitive nebular silicate material in chondrules and CAIs with compositions that lie along the CCAM line and hydrated ^{16}O-poor material, principally phyllosilicates that occur in the matrix (Fig. 10a) (*Clayton and Mayeda,* 1984, 1999). Formation of the hydrated material occurs

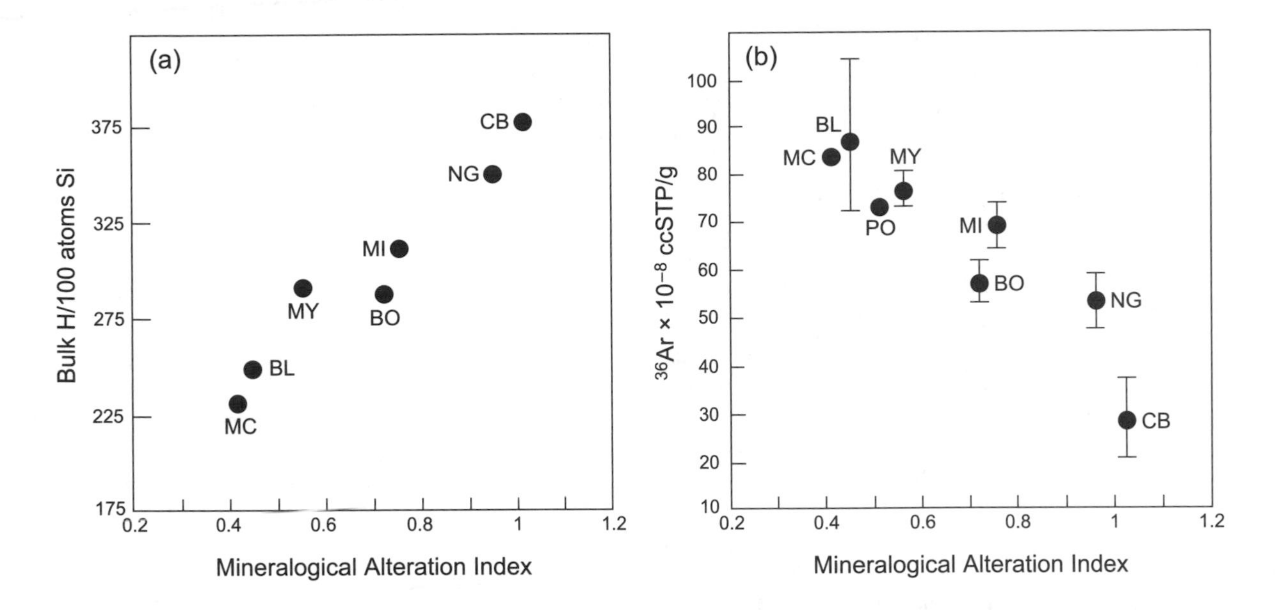

Fig. 11. Correlation of mineralogical alteration index (MAI) of *Browning et al.* (1996) with bulk properties of CM2 carbonaceous chondrite falls. **(a)** MAI vs. bulk H contents (plotted as bulk H/100 atoms Si), showing a progessive increase with increasing degree of alteration. **(b)** MAI vs. bulk ^{36}Ar concentration showing an inverse correlation that is interpreted as the destruction of the carrier(s) of planetary noble gas composition as alteration advances. Meteorite abbreviations: Murchison (MC), Bells (BL), Murray (MY), Boriskino (BO), Mighei (MI), Nogoya, (NG), Cold Bokkeveld (CB), Pollen (PO).

by reaction of anhydrous precursors with a ^{16}O-poor aqueous reservoir with a composition above the terrestrial fractionation line. Based on the fractionation of ^{18}O between phyllosilicates and precursor hydrous phases (17‰), *Clayton and Mayeda* (1996) estimated the water/rock ratio during aqueous alteration to be ~0.35 by oxygen atoms (~the ratio by volume). CM2 chondrites show a range of bulk O-isotopic compositions that appear to be roughly correlated with the degree of alteration (*Browning et al.,* 1996). For the CM2 chondrites, this correlation may be attributable to higher degrees of alteration due to higher water/rock ratios ranging between 0.3 and 0.6, with the most highly altered CM2 having seen the highest water/rock ratios. This model does not appear to hold for the CM1 chondrites that have O-isotopic compositions that are similar to the least-altered CM2 chondrites, despite their more extensive degree of alteration (Fig. 10a). *Zolensky et al.* (1997) have interpreted this observation as indicating that alteration of CM1 chondrites occurred under similar fluid/rock ratios as relatively weakly altered CM2 chondrites. However, alteration of the CM1 chondrites occurred either at a more elevated temperature or for a more extended period of time.

Details of the complex alteration history of the CM chondrites have come from studies of the isotopic compositions of individual components, such as phyllosilicates, carbonates, sulfates, magnetite, and water in these meteorites. *Clayton and Mayeda* (1984) found that calcite and phyllosilicate separates from Murchison lie on a mass-dependent fractionation line with Δ^{17}O of –1.4‰, indicating that they formed at equilibrium from the same aqueous reservoir. The magnitude of the fractionation of ^{18}O (22.0‰) between these two phases requires a very low temperature of formation, probably around 0°C, depending on the initial assumptions made in the model. *Clayton and Mayeda* (1984, 1999) have argued that that this is a process that was most likely to have occurred within an asteroidal parent body. However, more recent measurements of carbonates from Murchison (*Benedix et al.,* 2003) yield Δ^{17}O of –0.72‰, suggesting that the carbonates may not actually be in isotopic equilibrium with the phyllosilicates.

Measurements of the O-isotopic composition of water released by stepwise heating of Murchison (*Baker et al.,* 2002) also provide evidence that isotopic equilibration between various phases in CM chondrites is not complete. *Baker et al.* (2002) found that most water is released in a broad peak at ~350°C, corresponding to the release of structurally bound water in serpentine group minerals. This water has a Δ^{17}O of –0.77‰, very similar to that of carbonates in the same rock, suggesting that these two components lie on a mass-dependent fractionation line and are in isotopic equilibrium. However, the water is not in isotopic equilibrium with the matrix in Murchison, which has a Δ^{17}O of –1.88‰, an effect that *Baker et al.* (2002) suggested may be due to differences in the isotopic composition of O on different structural sites within alteration phases. One possibility is that tetrahedrally coordinated O in precursor silicate phases does not undergo complete isotopic equilibration with O in the altering fluids during the alteration process. Hence, the product phases of alteration, such as serpentine, may retain a proportion of ^{16}O from the precursor silicates, whereas O associated with hydroxyl records the O-isotopic composition of the altering water with higher fidelity.

Hydrogen-isotope studies of bulk CM chondrites and separated chondrules and matrix samples (*Eiler and Kitchen,* 2004) show that the whole-rock δD values have a general negative correlation with $\Delta^{17}O$, decreasing from a maximum value of ~0‰ in the least-altered CMs to ~–200‰ for the most heavily altered examples. This correlation is particular strong when data for matrix separates alone are compared with the difference $\delta^{18}O_{matrix} - \delta^{18}O_{whole\ rock}$, a parameter used by *Clayton and Mayeda* (1999) as a preferred measure of aqueous alteration.

A significant outcome of the *Eiler and Kitchen* (2004) study is that any direct relationship between CI and CM chondrites based on their H-isotopic compositions is highly improbable, although this is a relationship that has been suggested by some authors. Despite their intense degree of aqueous alteration, the CI chondrites have bulk δD values of ~100‰, distinctly higher than the least-altered CM chondrites. The H-isotopic composition of CM chondrites becomes less, rather than more, CI-chondrite-like with increasing aqueous alteration, the reverse of the trend observed for their O-isotopic compositions. These data preclude an origin for the CI chondrites by very advanced aqueous alteration of CM chondrites.

4.3. Thermal Modeling of Asteroidal Aqueous Alteration

Over the last 15 years, there have been important efforts to model aqueous alteration processes in the asteroidal parent bodies of carbonaceous chondrites. These efforts have involved developing viable thermal models for asteroidal-sized parent bodies and provide an important theoretical framework for interpreting petrographic and cosmochemical observations of altered carbonaceous chondrites. Essentially all the models assume that the source of water for asteroidal alteration is water ice that was accreted along with the anhydrous precursor components of carbonaceous chondrites (i.e., chondrules, CAIs, matrix, etc.). A number of assumptions have to be made about the initial physical states of primitive asteroids as well as the important processes that control their development. There are important differences between each of the models in regard to processes such as fluid flow, venting of gases, convection, etc.

Most models assume an asteroid diameter of 100 km and an initial accretion temperature of 170–180 K (*Grimm and McSween,* 1989; *Wilson et al.,* 1999; *Cohen and Coker,* 2000), although *Young* (2001) has explored alteration on asteroids with larger radii. In all cases, melting of accreted water ice occurs as a result of heat released by the decay of ^{26}Al, which is typically assumed to have been accreted into the asteroid at initial $^{26}Al/^{27}Al$ ratios lower than the canonical value of 5×10^{-5} (*MacPherson et al.,* 1995). The models require a knowledge of a large number of physical and chemical parameters, such as porosity, permeability, thermal conductivity, reaction kinetics, etc. Some of these parameters are reasonably well known based on studies of meteorites, but many are poorly constrained and have to be estimated or are varied within what are considered to be a reasonable range of plausible values (e.g., initial ice/silicate ratio, etc.). As with any modeling effort of this type, an assessment of the sensitivity of the models to different parameters has to be made.

Grimm and McSween (1989) investigated aqueous alteration for two endmember models. In both models, the accreted materials consist of a mixture of anhydrous silicate materials and ice (volume fraction <0.4), and the mixing is homogeneous throughout the parent body. In the first model, alteration occurs throughout the interior of an asteroidal parent body and hydrothermal circulation can occur if sufficiently large thermal gradients develop. In the second model, alteration occurs within an accretional regolith and water to drive aqueous alteration is supplied to the regolith by a variety of mechanisms, including melting of ice *in situ*, hydrothermal circulation, venting through fractures, or vapor diffusion through fractures and pores.

In both models, rates of aqueous alteration reactions are considered to be essentially instantaneous, in comparison with the timescales of asteroidal thermal evolution. Among the many parameters that affect thermal evolution, the initial $^{26}Al/^{27}Al$ ratio, the latent heat of fusion of ice, and the exothermic character of dehydration reactions are especially significant. *Wilson et al.* (1999) calculated that for an initial volume fraction of ice of 0.3, almost 6× as much heat is available from the exothermic hydration reactions as is required to melt ice. The consequence of this process is that once aqueous alteration reactions commence, a runaway reaction process occurs.

For an initial $^{26}Al/^{27}Al$ ratio of 1.2×10^{-6}, and an ice fraction of 0.2 (chosen to be consistent with the final hydration level of a CM chondrite, i.e., 1 : 1 hydrous/anhydrous mineral ratio), melting of ice occurs over a period of approximately 3 m.y. and the internal temperature in the asteroid is maintained at 0°C from the core out to within 10 km of the asteroid surface (*Grimm and McSween,* 1989). Once all the ice has melted, however, exothermic alteration reactions cause the interior temperature to rise rapidly to maximum values of 150°C and marked thermal gradients develop within the asteroid (Fig. 12a). All the water in the asteroid is consumed by the hydration reactions. At a higher $^{26}Al/^{27}Al$ ratio (5×10^{-5}), the rise in temperature is so rapid that temperatures high enough to cause dehydration of alteration products (350°C) occurs within 1 m.y. If this occurs, *Grimm and McSween* (1989) argue that internal pressurization of the parent body will occur as the pore volume is filled with near-critical H_2O. With pressures of several kilobars in the interior of a 100-km body, venting of H_2O along fractures will occur, so that H_2O will be lost from the asteroid.

A similar model with a higher starting water volume fraction (0.4) to simulate alteration in CI carbonaceous chondrites (i.e., near complete hydration of anhydrous rock) also predicts a rapid rise in temperature to the dehydration

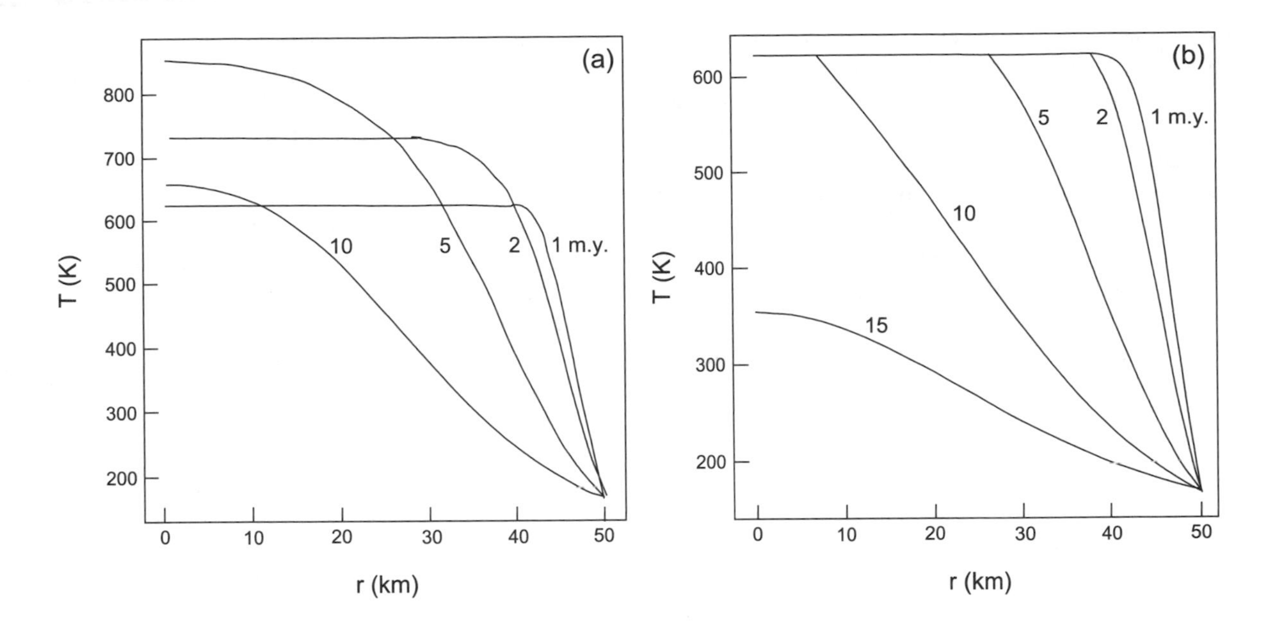

Fig. 12. Examples of calculated thermal histories for 100-km-diameter CM and CI chondrite parent bodies after *Grimm and McSween* (1989). **(a)** Thermal model for CM-chondrite-like alteration with initial ice fraction of 0.2 and $^{26}Al/^{27}Al$ ratio of 1.2×10^{-6}. Following absorption of latent heat of fusion of H_2O after 2 Ma, temperatures show a rapid increase due to heat released by exothermic hydration reactions. **(b)** Calculated model thermal history for CI-chondrite-like parent body with initial ice fraction of 0.4 (required to achieve complete hydration of the precursor anhydrous materials) and an $^{26}Al/^{27}Al$ ratio of 5×10^{-6}.

temperature, but not beyond it (Fig. 12b). The lower maximum temperature (~620 K) reached in this model compared to that for the CM model is due to the fact that at least some of the heat produced from exothermic hydration reactions is consumed in melting the larger fraction of ice.

To extend the work of *Grimm and McSween* (1989), *Cohen and Coker* (2000) have developed a thermal model to clarify the temperature distribution and duration of liquid water on a 100-km-diameter CM chondrite parent body. By solving the one-dimensional radially symmetric heat conduction equation for a range of different initial conditions (i.e., accretion temperature, ice fraction, accretion time, abundance of radionuclides, permeability), they showed that liquid water can exist near the center of the asteroid for up to several million years. If water is not a limiting agent (i.e., is not completely consumed by hydration reactions), high ice fractions (0.4) result in lower temperatures in the asteroid core. This effect is the result of (1) the lower volume fraction of anhydrous material, hence a lower abundance of radionuclides that provide heat; (2) the fact that the higher fraction of ice requires a higher heat budget to melt the higher proportion of ice; and (3) liquid water, where present, buffers the increase in temperature because it has a higher heat capacity than rock.

One important outcome of the model is that liquid water never exists in the regolith of the asteroid over any range of model parameters. Depending on the choice of initial parameter values, the thickness of the outer part of the asteroid that remains frozen varies from as little as 6 km to about 15 km for estimated "canonical" values for CM-chondrite alteration (i.e., accretion 3 m.y. after nebular collapse, 3 AU from the Sun, permeability of 10^{-12} m^2, and ice fraction of 0.3).

Essentially all the thermal models discussed above assume that the asteroid remains as a single body throughout the period of alteration, which may last for several millions of years. However, *Wilson et al.* (1999) have examined the possibility that gas production during aqueous alteration with a subsequent internal pressurization could have resulted in partial to complete disruption of the parent body. Modeling of fluid-rock interactions in carbonaceous chondrites indicates that significant amounts of gas are released, such as H_2 from alteration of silicates, CO_2, and possibly methane due to the breakdown of organic compounds (e.g., *Rosenberg et al.,* 2001).

A possible outcome of rapid gas generation is that the asteroid can become internally pressurized, resulting in the progressive development of fractures, provided the gas pressure exceeds the tensile strength by about a factor of 2. Once a pathway that allows pressurized gas to escape to the surface has developed, runaway fragmentation of the material on either side of the fracture can occur. Expansion of pressurized gas as it escapes to the surface will entrain fragments of material. For small asteroids (3 km to a few tens of kilometers radius), this process could result in disruption of the asteroid. If the fragments produced in this process are small (<a few tens of millimeters in size), complete disruption of the asteroid will occur. On the other hand, for

larger fragments, partial reaccretion may occur on the timescale of hours, resulting in the formation of an asteroidal body in which the original radial distribution of materials has been completely changed. Hence, the outer portion of an asteroid, containing frozen water ice, could be buried in the interior of the asteroid, enabling it to undergo melting and initiating a new wave of aqueous alteration, provided that disruption occurs before ^{26}Al content has decreased too much. Such a disruption process could be responsible for the widespread brecciated characteristics of CI, CM, CV, and CR chondrites, rather than being the result of impact brecciation on asteroidal surfaces.

Young et al. (1999) and *Young* (2001) have developed finite difference codes to model aqueous alteration in carbonaceous chondrite parent bodies with diameters <100 km, which differ from other models in that they include the mineralogical and isotopic effects of aqueous alteration as well as vapor advection. *Young et al.* (1999) have argued that the O-isotopic composition of the components of different carbonaceous chondrites can only be explained by significant fluid flow through asteroidal parent bodies, an important effect that is not considered in detail in other thermal models. The effects of fluid flow on the thermal evolution of asteroidal parent bodies are significant, because fluid-rock reactions are much more limited when fluid flow occurs, compared with stationary fluid. This significantly reduces the heating effects of exothermic hydration reactions that are important in other models.

In comparison with other models, in which pressure buildup in the asteroid interior is released by episodic fracturing and venting, *Young* (2001) argued that the internal pressurization drives fluid flow from the interior of the asteroid through porosity to the asteroid surface, where water can be lost into space. This process is likely to occur on asteroids that have diameters that are too large to disrupt by internal pressurization.

This model provides useful additional insights into the possible evolution of parent bodies that were heated uniformly throughout their interiors. First, it appears that asteroids with radii <50 km maintain high thermal gradients, whereas larger bodies have essentially uniform internal thermal gradients except near their outer surfaces. As with other models, aqueous alteration is predicted to commence within a few 100,000 years after heating commences, assuming an initial $^{26}Al/^{27}Al$ ratio of 1×10^{-5}. As ice melting commences, liquid water migrates toward the exterior of the asteroid driven by pressure gradients and generates a region between the asteroid core and surface where fluid flow occurs for an extended period of time. This is a region where significant aqueous alteration occurs; the O-isotopic compositions of carbonates and hydrous minerals generated in the model calculations are in good agreement with those observed in CI, CM, and CV chondrites, supporting the notion that alteration on the parent bodies of these chondrites occurred under conditions of significant fluid flow. The model is remarkably successful at matching the observed increases in $\Delta^{17}O$ of the solids during the alteration process, assuming an initial $\Delta^{17}O$ of ice of 16.0 and rock of –2.7. The very elevated $\Delta^{17}O$ for ice assumed in the calculation is consistent with values used by *Clayton and Mayeda* (1984) to explain the O-isotopic evolution of CM and CI chondrites.

Second, for parent bodies with radii greater than 50 km in size, the modeling suggests that alteration will occur much later than for small asteroids. If the initial $^{26}Al/^{27}Al$ ratio is quite high (e.g., 5.0×10^{-6}), then the rate of heating for a 1 Ceres-sized body (1025 km diameter) will be fast and water will be lost from the asteroid, before significant alteration can occur. For an initial $^{26}Al/^{27}Al$ ratio of 6.8×10^{-7} heating is sufficiently slow so that maximum temperatures just above the melting temperature of water ice are reached 5–6 m.y. after accretion. In this situation, alteration occurs principally in a region near the asteroid surface, because this is the region of highest fluid flow. These timescales are consistent with the extended durations for aqueous alteration that are indicated by Mn-Cr dating of carbonates in CI and CM chondrites, which suggest that alteration occurred over periods of up 9 m.y. (e.g., *Krot et al.,* 2006).

A significant issue that remains to be addressed for the fluid-flow models concerns the effects that such flow would have on the chemistry of carbonaceous chondrites. If significant fluid movement had occurred through carbonaceous chondrite parent bodies, elemental mass transfer of the most soluble elements, such as the alkalis (e.g., Na, K) and alkali earths (e.g., Ca, Sr), would be expected. The bulk compositions of the most-altered carbonaceous chondrites (CI and CM) show no evidence of elemental fractionations that would indicate either addition or removal of these elements from the protolith material, arguing against fluid flow. In the case of the CI chondrites, these elements are unfractionated relative to the solar photosphere and in the case of the CMs display fractionations that are consistent with their cosmochemical behavior (i.e., refractory Ca is unfractionated relative to other refractory elements; Na is depleted according to its volatility). These data indicate that the elemental abundances in carbonaceous chondrites were established prior to accretion and have been essentially unaffected by later secondary processing, i.e., that alteration occurred isochemically in a closed system. There is little doubt however, that on the scale of millimeters, these elements were clearly highly mobile and were redistributed between components during aqueous alteration.

5. FLUID COMPOSITIONS

Among the most challenging aspects of understanding aqueous alteration in chondritic meteorites is constraining the compositions of the altering fluids in terms of both their chemistry and isotopic composition. This problem is particularly complex, because alteration in chondrites commenced under conditions where the fluid was in chemical and isotopic disequilibrium with the solid phases. The fluid composition therefore changed progressively with time as the altering system evolved toward equilibrium and may also have been variable within different regions of the same parent body (e.g., *Richardson,* 1978; *Fredrikkson and Kerridge,* 1988; *Johnson and Prinz,* 1993, *Riciputi et al.,* 1994). Furthermore, unlike terrestrial rocks where fluid inclusions are quite common, direct samples of altering fluids are extremely rare in chondrites or are very small and difficult to analyze. Despite these problems, progress has been made that puts useful constraints on the elemental and isotopic composition of the fluids, particularly in the CI and CM chondrites.

5.1. Elemental Composition of Altering Fluids: Evidence for Brines

Among the phases that have proved most useful in constraining fluid compositions are highly soluble phases such as carbonates, sulfates, and halides. Assuming appropriate mineral/fluid element partitioning data are available, minor- and trace-element analyses of such phases can be used to constrain element ratios in the altering fluids. *Riciputi et al.* (1994) demonstrated the feasibility of this approach based on ion microprobe analyses of a suite of elements (Fe, Mg, Mn, Sr, Na, B, and Ba) in dolomites and carbonates from CI and CM chondrites. These data indicate that carbonates either precipitated from, or recrystallized in equilibrium with, low-temperature aqueous solutions with element/Ca ratios that are comparable with buried terrestrial brines (defined here as highly saline solutions). Sodium/calcium ratios are notably enriched in both dolomite and calcite in CM and CI chondrites. Dolomite in CI and CM chondrites appears to have precipitated from compositionally similar fluids, but calcite and dolomite in CM chondrites are clearly not in chemical equilibrium. This evidence suggests that these phases grew at different times, possibly during different alteration events, and provides evidence that fluid compositions varied temporally in the CM parent body.

Riciputi et al. (1994) observed significant variability in the element/Ca ratios in the carbonate grains that they analyzed, sometimes up to 2 orders of magnitude. This variation is probably the result, at least in part, of the presence of complex, fine-scale zoning in individual carbonate grains that has only recently been recognized (e.g., *Brearley et al.,* 1999; *Brearley and Hutcheon,* 2002). The presence of such zoning suggests that rather than recording fluid compositions from single events, calcite and dolomite both record complex variations in fluid composition through time.

A unique opportunity to study the composition of asteroidal fluids directly has come from the discovery of fluid inclusions in halite in the Monahans and Zag equilibrated ordinary chondrites (*Zolensky et al.,* 1999; *Rubin et al.,* 2002; *Bridges et al.,* 2004). In Monahans, primary and secondary fluid inclusions up to 15 μm in size are present, whereas in Zag the inclusions are typically <10 μm in size. The primary inclusions were trapped during the initial growth of the halite from asteroidal fluids, whereas the secondary inclusions formed as a result of healing of fluid-filled fractures in existing halite. Fluid inclusion studies show that vapor bubbles are rare in fluid inclusions in Zag halites, suggesting that the trapping of the fluids occurred at temperatures ≤100°C, perhaps in the range 25°–50°C (*Rubin et al.,* 2002). In Monahans, the fluid inclusions appear to contain Fe, Mg, and/or Ca, indicated by the fluid-freezing temperatures (*Zolensky et al.,* 1999).

Halite in these ordinary chondrites appears to have formed by precipitation from concentrated brines. *Rubin et al.* (2002) have speculated that these brines formed in a near-surface environment in which evaporation caused a progressive increase in alkali and chloride ions until the solutions became supersaturated and halite precipitated. As discussed earlier, however, the halite may in fact be derived from water-rich materials that were mixed into the ordinary chondrite during the waning stages of asteroidal metamorphism.

5.2. Oxygen-Isotopic Evolution of Aqueous Fluids

The O-isotopic systematics of altered chondrites have been discussed earlier and in *Clayton and Mayeda* (1984, 1999). Here we focus specifically on data from individual mineral phases that provide information about how the O-isotopic composition of the fluids may have evolved during aqueous alteration. *Benedix et al.* (2003) have reported data for carbonates (dolomite and calcite) from several CM2 chondrites. They found that the carbonate $\Delta^{17}O$ values become progressively lower as a function of the degree of alteration of the host chondrite based on the MAI of *Browning et al.* (1996). *Airieau et al.* (2001) found similar behavior for sulfates in CM2 chondrites, but $\Delta^{17}O$ for sulfates is more positive than for either carbonates or phyllosilicates in the same chondrite. These data indicate that the change in $\Delta^{17}O$ may be due to precipitation of the carbonates and sulfates from a fluid that becomes progressively more equilibrated with the host chondrite as alteration proceeds, assuming that the altering fluid had a higher $\Delta^{17}O$ than anhydrous precursor phases in the chondrite (*Clayton and Mayeda,* 1999). *Airieau et al.* (2001) argued that the differences in $\Delta^{17}O$ between sulfate and carbonate were due to precipitation of sulfate earlier than carbonate, from a fluid that was less isotopically equilibrated with the host rock.

There also appears to be some isotopic variability in carbonates within individual CM chondrites that might represent a record of changing fluid composition with progressive alteration, although alternative explanations are also possible. For example, *Benedix et al.* (2003) found $\Delta^{17}O$ values between −0.54‰ and −0.84‰ for four different splits from Murchison, significantly different from the value measured by *Clayton and Mayeda* (1984) (−1.4‰). This difference may in part reflect the fact that Murchison is a complex breccia and contains clasts with different alteration histories. The data also show a range of $\delta^{18}O$ values (26.6–35.5‰) that is consistent with the range of values measured

TABLE 3. Summary of estimated conditions of parent-body alteration for chondritic meteorites.

Meteorite Group	Temperature (°C)	pH	eH	f_{O_2}	W/R*
CI	~20[1] 50–150[2] 100–150[3]	8–10[1], 7–10[2]	–0.2[1] –0.3 to –0.8[2]	$>10^{-55}$–10^{-70}†	1.1–1.2[4,7]
CM2	0[4], ~20[1], 1–25[2] 80[5]	6–8[1], 7–12[2],	–0.75–0.5[2] 7–13[6]	$<10^{-85}$‡	0.3–0.6[4,7]
CM1	~120[7]	—	—	$<10^{-55}$–10^{-70}†	0.3[7]
CV(All ox)	200–300[9,10] <340[11]	—	—	—	—
CV (Bali ox)	50–150[8]	—	—	$<10^{-55}$–10^{-70}†	0.8–1.1[9]
CO	0–50[8]	—	—	$>10^{-55}$–10^{-70}†	0.1–0.6[8]
CR	50–150[8]	—	—	$>10^{-55}$–10^{-70}†	0.4–1.1[8]
MAC 88107	<200[10]	—	—	—	—
UOC	<260[12]	—	—	—	—
EOC	<100[13]				

*W/R = water/rock ratio — based on proportions of oxygen atoms (~ equal to the relative volumes of water to rock).

†Range of oxygen fugacities based on the calculated position of the sulfide-sulfate boundary in the temperature range 0°–150°C (*Bourcier and Zolensky,* 1992). CI chondrites are the only group of chondrites that contain significant amounts of sulfates.

‡Based on the calculated stability of tochilinite (*Browning and Bourcier,* 1996).

References and methods used for estimations: [1] *DuFresne and Anders* (1962), mineral equilibria; [2] *Zolensky et al.* (1989), thermodynamic modeling of mineral assemblages; [3] *Clayton and Mayeda* (1999), O-isotopic fractionation between carbonate and phyllosilicates; [4] *Clayton and Mayeda* (1999), modeling of bulk-rock O-isotopic composition; [5] *Baker et al.* (2002), O-isotopic fractionation between structural bound water in phyllosilicates and calcite; [6] *Rosenberg et al.* (2001), thermodynamic modeling of mineral assemblages; [7] *Zolensky et al.* (1997), modeling of bulk-rock O-isotopic composition; [8] *Zolensky et al.* (1993), mineral assemblages and thermodynamic modeling; [9] *Krot et al.* (1998a,b), thermodynamic modeling of phase equilibria; [10] *Krot et al.* (2000b), thermodynamic modeling; [11] *Brearley* (1997b), phase equilibria; [12] *Alexander et al.* (1989), thermodynamic calculations; [13] *Rubin et al.* (2002), fluid inclusions in halite.

conventionally by *Grady et al.* (1988) and by ion probe by *Brearley et al.* (1999). These data indicate that carbonate was either forming at constant temperature from an evolving reservoir and that the carbonate grains may show isotopic zoning or have grown at different times, or that temperature was varying during carbonate growth.

6. CONDITIONS OF AQUEOUS ALTERATION

General constraints on the conditions of alteration observed in carbonaceous chondrites come from undepleted volatile trace-element abundances, the survival of organic compounds that are sensitive to thermal processing, highly unequilibrated mineral compositions, and lack of evidence of any metamorphic equilibration of even submicrometer grains. These data indicate collectively that the temperatures of aqueous alteration were certainly less than 300°C, in some cases 200°–300°C lower than this value, with the exception of some of the oxidized CV3 chondrites as discussed above. However, beyond this upper limit, putting exact constraints on the physical conditions (T, f_{O_2}, fluid/rock ratios, pH, etc.) of alteration is one of the most challenging problems of studying alteration in chondritic meteorites. Constraints on the conditions of alteration come from a variety of sources, including stable-isotope data, mineralogical phase equilibria, experimental data, and thermodynamic modeling. In all these approaches, there are significant uncertainties because of assumptions in the models and the fact that mineral assemblages in altered chondrites may not always be in thermodynamic equilibrium. Despite these uncertainties, these different approaches do at least provide a plausible envelope of the alteration conditions experienced by different chondrite types. The possible ranges of alteration conditions estimated for carbonaceous and ordinary chondrites are summarized in Table 3 and are illustrated graphically in Fig. 13 for the carbonaceous chondrites.

6.1. CI Carbonaceous Chondrites

DuFresne and Anders (1962) examined the mineralogical phase relations present in CI chondrites in considerable detail and concluded that aqueous alteration by liquid water occurred under equilibrium conditions at temperatures of ~20°C. They constrained fluid Eh to –0.2 V and pH in the range 8–10. Efforts to constrain the alteration conditions of the CI chondrites by *Zolensky et al.* (1989) using thermo-

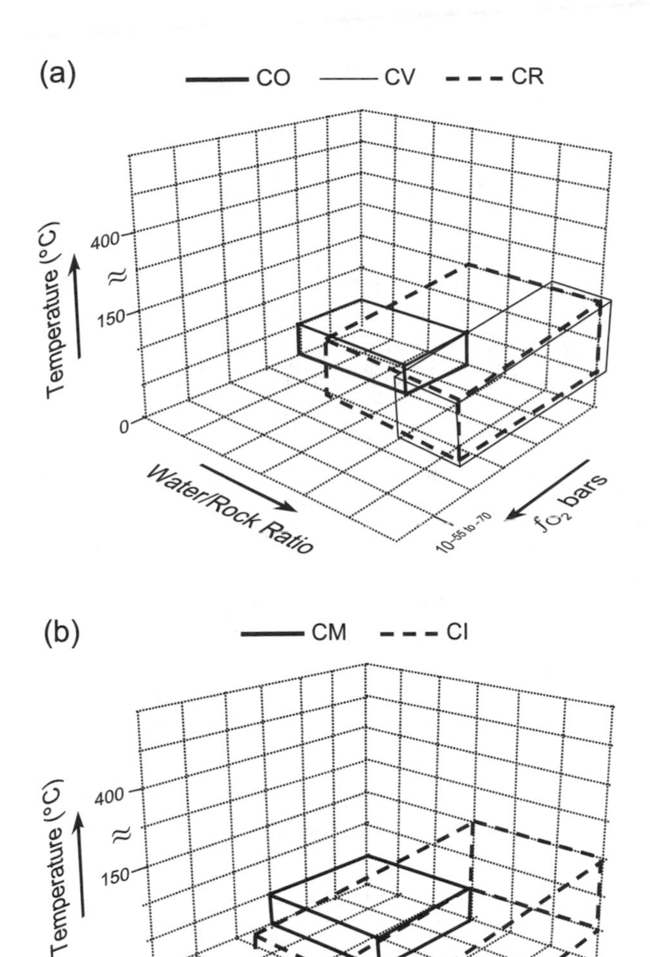

Fig. 13. Estimated temperature-water/rock ratio-oxygen fugacity conditions of aqueous alteration conditions in carbonaceous chondrites after *Zolensky et al.* (1993). **(a)** Alteration conditions for weakly altered CO, CV, CR, and unique carbonaceous chondrites. **(b)** Alteration conditions for more pervasively altered carbonaceous chondrites (CI and CM). The oxygen fugacity indicated (f_{O_2} = 10^{-55}–10^{-70}) represents the range over which the boundary between sulfate and sulfide occurs in the temperature range 0°–150°C, based on the calculations of *Bourcier and Zolensky* (1992). Sulfates are generally absent in all but the CI chondrites. Sulfides are the dominant S-bearing phases in the CO, CV, and CR chondrites. The CM chondrites appear to have formed under more reducing conditions than the other chondrite groups due to the presence of tochilinite that is stable at values of f_{O_2} < ~–85 (*Browning and Bourcier,* 1996).

dynamic modeling yield higher-temperature conditions of alteration ranging from 50 to (at least) 150°C, over a wide range of plausible fluid/rock ratios. In these calculations, solution pH was found to vary from 7 up to between 9 and 10, and Eh from –0.3 to –0.8 V, with the low pH and higher eH values being prevalent during the earliest stages of alteration. Other efforts to constrain the conditions of aqueous alteration of the CI chondrites also suggest higher temperatures of alteration for these meteorites. The relatively small O-isotopic fractionation between carbonate and phyllosilicates in Orgueil of 8‰ suggests that alteration occurred at a relatively high temperature, certainly higher than that of CM2 chondrites, probably in the range 100°–150°C. Based on the bulk O-isotopic composition of CI chondrites, the integrated water/rock ratio (oxygen atoms) during alteration appears to have been high, probably ~1.1–1.2 (*Clayton and Mayeda,* 1999). This appears to be consistent with the presence of saponite, a phase that appears to require relatively high water/rock ratios to be stable (*Bourcier and Zolensky,* 1992). The oxygen fugacity during alteration of CI chondrites appears to be relatively high based on the presence of magnetite rather than tochilinite in this group of meteorites. This observation is supported by the presence of sulfate in CI chondrites. *Bourcier and Zolensky* (1992) determined that in neutral solutions f_{O_2} of the sulfide-sulfate boundary lies between 10^{-70} and 10^{-55} bars in the temperature range 0°–150°C.

6.2. CM Carbonaceous Chondrites

Estimates for the conditions of alteration of the CM chondrites using different approaches are in better agreement than for the CI chondrites. *DuFresne and Anders* (1962) and *Zolensky et al.* (1989) reported temperatures of alteration of ~20°C and 1° to ~25°C, respectively. Using a simple closed-system hydration reaction model, *Clayton and Mayeda* (1999) also found that the O-isotopic systematics of CM chondrites were consistent with alteration near 0°C, with fluid/rock ratios of 0.3–0.6. However, more recent work (*Baker et al.,* 2002) based on the O-isotopic fractionation between calcite and structurally-bound water released from phyllosilicates in Murchison indicates somewhat higher alteration temperatures of ~80°C. The discrepancies between these estimates are a reflection of uncertainties in the extent to which different components are in isotopic equilibrium with one another.

Like the CI chondrites, fluid pH values in CMs appear to have ranged from mildly acidic to alkaline in character. *DuFresne and Anders* (1962) calculated a pH range of 6–8, compared with 7 to slightly above 12 determined by *Zolensky et al.* (1989) for a wide range of rock/fluid ratios.

Mineralogical and thermodynamic constraints show that alteration conditions in the CM2 chondrites were more reducing than the CIs. Mineralogically, this is indicated by the unique presence of tochilinite in CMs as the dominant Fe-bearing phase compared with magnetite in CIs and the rarity of sulfates in CM2 chondrites (e.g., *Lee,* 1993). *Zolensky et al.* (1989) calculated an Eh range from 0.5 to –0.75 V for the development of the mineral assemblages observed in CM chondrites. Very reducing conditions appear to be required to stabilize tochilinite according to phase equilibria calculations based on estimated thermodynamic data (*Browning and Bourcier,* 1996) and observations on the stability of tochilinite in the terrestrial environment (*van de*

Vusse and Powell, 1983). The upper thermal stability of tochilinite appears to be ~120°C. The calculated phase relations are consistent with the mineralogical associations in CM chondrites and indicate that tochilinite is only stable at PO_2 values (log $PO_2 < -80$), conditions under which PH_2 would be high. Tochilinite could plausibly coexist with low Ni,Fe metal in CM chondrites. In contrast, tochilinite is a relatively rare phase in CM1 chondrites, suggesting that the most heavily altered CMs experienced conditions of alteration higher than ~120°C or more oxidizing than the CM2 chondrites (*Zolensky et al.,* 1997).

More detailed geochemical modeling provides support for low-temperature alteration (25°C) under reducing conditions for the CM2 carbonaceous chondrites (*Rosenberg et al.,* 2001). Using two different potential anhydrous precursor CM-chondrite mineral assemblages (*Browning and Bourcier,* 1998), a reduced Fe endmember (Fe present dominantly as Fe^0 in Fe metal) and an oxidized Fe endmember (Fe present as Fe^{2+}), *Rosenberg et al.* (2001) found that good matches to the observed alteration assemblages in CM chondrites could be produced. Chrysotile, greenalite, and tochilinite are formed as alteration products from both precursor assemblages, but cronstedtite only develops from the oxidized precursor. In both cases, the pH of the altering solutions evolve from an initial value of 7 to highly basic values (12–13). Alteration of the reduced precursor results in the generation of large quantities of H_2 due to the oxidation of Fe metal, whereas essentially no H is evolved during alteration of the oxidized precursor assemblage.

6.3. Other Carbonaceous Chondrites

The conditions of alteration in other carbonaceous chondrites remain less well constrained although estimates are available, in part, based on comparisons with other carbonaceous chondrites. *Zolensky et al.* (1993) summarized plausible alteration conditions for CR, CV, and CO chondrites (in addition to CI and CM as discussed above), based principally on mineralogical criteria. The alteration mineralogy of CR chondrites, mainly serpentine/saponite and magnetite, suggests alteration under relatively oxidizing conditions and temperatures, similar to those experienced by the CI chondrites, although under a somewhat more restricted range of f_{O_2}. Water/rock ratios for alteration in both CI and CR chondrites may have been similar as indicated by the presence of saponite. *Krot et al.* (1998a,b) have carried out detailed thermodynamic calculations to constrain the conditions of metasomatic alteration of the oxidized CV chondrites. These studies indicate that the coexistence of phyllosilicates, magnetite, fayalite, Ca-Fe-rich pyroxenes, and andradite in members of the Bali subgroup of CV3 chondrites is consistent with formation at temperatures <300°C during fluid-rock interactions. Using these same calculations, *Krot et al.* (2000b) were able to infer that the coexistence of fayalite, hedenbergite, and magnetite in the unique chondrite MAC 88107 is indicative of alteration at lower temperatures or less-oxidizing conditions than for the CV3 chondrites. Alteration is constrained to have occurred at temperatures between ~120° and 220°C, assuming a confining pressure of 100 bars.

6.4. Unequilibrated Ordinary Chondrites

Alexander et al. (1989) used a variety of mineralogical and thermodynamic constraints to estimate the conditions of aqueous alteration in the UOCs Semarkona and Bishunpur. They concluded, based on the presence of Na-bearing smectite and the behavior of K-rich smectite in the terrestrial environment, that alteration temperatures could not have exceeded 260°C. They also examined the variation in gas phase composition as a function of temperature under two sets of limiting conditions: (1) the minimum partial pressure of H_2O needed to stabilize smectite, and (2) a maximum PH_2O assuming that liquid water is present. The calculations demonstrated that under both these limiting conditions, H_2O and H_2 dominate the gas phase and smectite can be stabilized with or without liquid water in the system.

7. TIMING OF ALTERATION

Constraints on the timing and duration of aqueous alteration are key components in unraveling the alteration history of chondritic meteorites. For example, evidence that alteration occurred for extended periods (5–20 m.y.) after CAI formation favors an asteroidal scenario for alteration vs. time periods <5 m.y. that could also be reconciled with alteration prior to accretion of asteroidal parent bodies. A detailed discussion of the timescales of aqueous alteration based on the Mn-Cr and I-Xe systems are discussed in *Krot et al.* (2006). Only a brief discussion of the implications for aqueous alteration will be discussed here.

Manganese-chromium-isotopic studies have focused on dating the formation of carbonates in CI and CM chondrites and the unusual chondrite Kaidun (*Endress et al,* 1996; *Hutcheon et al.,* 1999; *Brearley et al.,* 2001; *Brearley and Hutcheon,* 2000, 2002). Dating of carbonates in CIs and Kaidun is facilitated by the presence of Mn-bearing breunnerites and dolomites; for the CM chondrites this task is much more challenging because Mn-poor calcite is the dominant carbonate, although Mn-bearing dolomite does occur in the most highly altered CM2s (*Johnson and Prinz,* 1993) and is relatively common in some CM1s (*Zolensky et al.,* 1997).

Using the constraints provided by Al-Mg and Mn-Cr systems, combined with the results of thermal modeling of asteroidal parent bodies, it is possible to put some constraints on the onset and duration of aqueous alteration in asteroidal parent bodies. If it is assumed that the measured difference in initial $^{27}Al/^{26}Al$ ratios between CAIs and chondrules is real, then accretion of the carbonaceous chondrite parent bodies must have occurred at least 3 m.y. after CAI formation. Thermal modeling (see above) indicates that temperatures within these asteroids will have increased to the melting temperature of water ice within 0.03–0.6 m.y., assuming initial $^{26}Al/^{27}Al$ ratios in the range 1.6×10^{-6} to 2×10^{-5} (e.g., *Wilson et al.,* 1999). In addition, aqueous alteration had to have commenced shortly after chondrule formation (i.e., within ~3–4 m.y. of CAI formation), otherwise ^{26}Al would have decayed to values that were too low to heat asteroidal parent bodies to temperatures sufficiently high to melt ice (e.g., *Cohen and Coker,* 2000).

Measured initial $^{53}Mn/^{55}Mn$ ratios in carbonates in carbonaceous chondrites range from $(1.42 \pm 0.6) \times 10^{-6}$ to $(8 \pm 4) \times 10^{-6}$ for CIs to $(1.3 \pm 0.6) \times 10^{-5}$ to $(5.0 \pm 1.5) \times 10^{-6}$ in CMs and $(9.4 \pm 1.6) \times 10^{-6}$ in Kaidun. Assuming a solar system initial $^{53}Mn/^{55}Mn$ ratio of $(4.4 \pm 1.0) \times 10^{-5}$ (*Lugmair and Shukolyukov,* 2001), these differences in initial $^{53}Mn/^{55}Mn$ ratios correspond to a time period of ~9 Ma. Perhaps the key issue that arises from the Mn-Cr system ages is exactly what events are being dated. Aqueous alteration of asteroidal parent bodies was a complex process and probably involved different episodes of carbonate growth. A conservative assumption is that the carbonate ages record the last stage of aqueous alteration in carbonaceous chondrite parent bodies. If this is the case, then the following inferences about the duration of aqueous alteration can be made. Aqueous alteration in CM chondrites clearly lasted for a significantly shorter time than the CI chondrites. If alteration commenced within 4 m.y. of CAI formation, the measured initial $^{53}Mn/^{55}Mn$ ratios for carbonates in CM chondrites suggest aqueous alteration in the CM chondrites could have occurred for periods of ~7.5 ± ~2 m.y. There is some indication that CM2 chondrites may have been altered for a shorter period of time than the more heavily altered CM1 chondrites, but given the significant uncertainties in the initial $^{53}Mn/^{55}Mn$ ratios, this remains rather speculative (*Brearley and Hutcheon,* 2000, 2002). Further studies are necessary to refine the alteration chronology more precisely. The much lower initial $^{53}Mn/^{55}Mn$ in CI chondrites suggests that alteration extended for ~12.5 m.y. after accretion, and possibly longer if carbonates do not record the last period of asteroidal aqueous alteration.

The occurrence of halite in the equilibrated ordinary chondrites Monahans and Zag has provided a unique opportunity to study the timing of formation of this mineral using I-Xe dating (*Whitby et al.,* 2000; *Busfield et al.,* 2004). However, the data from both meteorites show significant scatter in their initial $^{129}I/^{127}Xe$, presumably due to disturbance of the I-Xe system, precluding a simple interpretation of the formation age of these minerals. However, based on these data, *Busfield et al.* (2004) proposed a model involving formation of halite in both meteorites ~5 m.y. after the Shallowater enstatite achondrite, which has an absolute age of 4559 Ma. This would place halite formation at about the same time as carbonate formation in CI chondrites. However, for the case of halite, this does not necessarily imply that water was present continuously for several million years on the ordinary chondrite parent body.

Taken together, these data clearly demonstrate that liquid water was available for extended periods of time (several million years) on asteroidal parent bodies. Based on the thermal modeling discussed above, it appears that these timescales are inconsistent with small asteroidal parent bodies and require larger asteroids with radii >50 km. These bodies would have to accrete with initial $^{26}Al/^{27}Al$ ratios that were high enough to generate enough heat to melt water ice, but not high enough to allow rapid heating and ensuing early loss of water, which would prevent significant aqueous alteration.

8. EPISODIC OR CONTINUOUS AQUEOUS ALTERATION

One issue of importance for aqueous alteration in chondrites is whether alteration was continuous, extending over several million years (as indicated by Mn-Cr and I-Xe dating), or was episodic. Evidence from studies of carbonates in CM chondrites provides some indication that alteration may have been episodic and involved periods of carbonate dissolution followed by reprecipitation (e.g., *Brearley et al.,* 1999; *Brearley and Hutcheon,* 2000, 2002). One possible explanation for the complex zoning in carbonate minerals in some CM chondrites is that carbonate dissolution and/or growth may be related to changes in the fluid composition as a result of perturbations caused by impact-related brecciation or degassing (e.g., *Wilson et al.,* 1999). An additional possibility is that liquid water produced during the main phase of heating may diffuse back from the outer regions of the asteroid into the interior as the asteroid cools (e.g., *Cohen and Coker,* 2000). This process can only occur provided that water is not lost into space either by diffusional loss through the outer icy regolith or by venting along fractures.

9. CONCLUSIONS

Over the last two decades, significant advances have been made in our understanding of the aqueous alteration processes that have affected chondritic meteorites. However, many important questions remain that are currently active areas of debate. The issues of where and over what time periods aqueous alteration took place are first-order questions that require much more extensive investigation. Significant advances to address these questions are likely to come in the future from the application of modern microbeam techniques to understand the microstructural, trace-element, and isotopic characteristics of individual alteration phases. The recognition that phases such as carbonates record compositional zoning at the microscale provides the opportunity to examine the chemical and isotopic evolution of fluids, in addition to constraining the timescale of alteration using short-lived radionuclide chronometers such as the Mn-Cr system. The integration of petrographic observations with improved asteroidal thermal models is likely to be an especially important area of investigation. This work has special significance in integrating observations made at the microscale with macroscopic modeling efforts. Finally, an improved understanding of the role of microchemical environments during all stages of aqueous alteration of complex heterogeneous mineral assemblages that characterize chondritic meteorites is essential in developing robust criteria for evaluating asteroidal vs. preaccretionary environments for aqueous alteration.

Acknowledgments. Supported by NASA Grants NAG5-9798 and NAG5-11862 to A.J.B. (P.I.).

REFERENCES

Abreu N. M. and Brearley A. J. (2004) Characterization of matrix in the EET92042 CR2 carbonaceous chondrite: Insights into textural and mineralogical heterogeneity (abstract). *Meteoritics & Planet. Sci., 39,* A178.

Abreu N. M. and Brearley A. J. (2005a) HRTEM and EFTEM studies of phyllosilicate-organic matter associations in matrix and dark inclusions in the EET92042 CR2 carbonaceous chondrite (abstract). In *Lunar and Planetary Science XXXVI,* Abstract #1744. Lunar and Planetary Institute, Houston (CD-ROM).

Abreu N. M. and Brearley A. J. (2005b) Carbonates in Vigarano: Terrestrial, preterrestrial or both? *Meteoritics & Planet. Sci., 40,* 609–625.

Airieau S. A., Farquhar J., Jackson T. L., Leshin L. A., Thiemens M. H., and Bao H. (2001) Oxygen isotope systematics of CI and CM chondrite sulfate: Implications for evolution and mobility of water in planetesimals (abstract). In *Lunar and Planetary Science XXXII,* Abstract #1744. Lunar and Planetary Institute, Houston (CD-ROM).

Alexander C. M. O., Hutchison R., and Barber D. J. (1989) Origin of chondrule rims and interchondrule matrices in unequilibrated ordinary chondrites. *Earth Planet. Sci. Lett., 95,* 187–207.

Anders E. and Grevesse N. (1989) Abundances of the elements: Meteoritic and solar. *Geochim. Cosmochim. Acta, 53,* 197–214.

Baker L., Franchi I. A., Wright I. P., and Pillinger C. T. (2002) The oxygen isotopic composition of water from Tagish Lake: Its relationship to low-temperature phases and to other carbonaceous chondrites. *Meteoritics & Planet. Sci., 37,* 977–985.

Barber D. J. (1981) Phyllosilicates and other layer-structured minerals in stony meteorites. *Clay Minerals, 20,* 415–454.

Benedix S. A., Leshin L. A., Farquhar J., Jackson T. L., and Thiemens M. H. (2003) Carbonates in CM chondrites: Constraints on alteration conditions from oxygen isotopic compositions and petrographic observations. *Geochim. Cosmochim. Acta, 67,* 1577–1588.

Bischoff A. (1998) Aqueous alteration of carbonaceous chondrites: Evidence for preaccretionary alteration — A review. *Meteoritics & Planet. Sci., 33,* 1113–1122.

Bischoff A., Palme H., Ash R. D., Clayton R. N., Schultz L., Herpers U., Stoffler D., Grady M. M., Pillinger C. T., Spettel B., Weber H., Grund T., Endress M., and Weber D. (1993) Paired Renazzo-type (CR) carbonaceous chondrites from the Sahara. *Geochim. Cosmochim. Acta, 57,* 1587–1603.

Bostrom K. and Fredriksson K. (1966) Surface conditions of the Orgueil meteorite parent body as indicated by mineral associations. *Smithson. Misc. Coll., 151,* 1–39.

Bourcier W. L. and Zolensky M. E. (1992) Computer modeling of aqueous alteration on carbonaceous chondrite parent bodies (abstract). In *Lunar and Planetary Science XXIII,* pp. 143–144. Lunar and Planetary Institute, Houston.

Brearley A. J. (1993) Matrix and fine-grained rims in the unequilibrated CO3 chondrite, ALH A77307: Origins and evidence for diverse, primitive nebular dust components. *Geochim. Cosmochim. Acta, 57,* 1521–1550.

Brearley A. J. (1994) Metamorphic effects in the matrices of CO3 chondrites: Compositional and mineralogical variations (abstract). In *Lunar and Planetary Science XXV,* pp. 165–166. Lunar and Planetary Institute, Houston.

Brearley A. J. (1995) Aqueous alteration and brecciation in Bells, an unusual, saponite-bearing CM carbonaceous chondrites. *Geochim. Cosmochim. Acta, 59,* 2291–2317.

Brearley A. J. (1997a) Phyllosilicates in the matrix of the unique carbonaceous chondrite, LEW 85332 and possible implications for the aqueous alteration of CI chondrites. *Meteoritics & Planet. Sci., 32,* 377–388.

Brearley A. J. (1997b) Disordered biopyriboles, amphibole, and talc in the Allende meteorite: Products of nebular or parent body aqueous alteration? *Science, 276,* 1103–1105.

Brearley A. J. (1999) Origin of graphitic carbon and pentlandite inclusions in matrix olivines in the Allende meteorite. *Science, 285,* 1380–1382.

Brearley A. J. (2003) Nebular vs parent body processing of chondritic meteorites. In *Treatise on Geochemistry, Vol. 1: Meteorites, Comets, and Planets* (A. M. Davis, ed.), pp. 1–22. Elsevier, Oxford.

Brearley A. J. and Jones R. H. (1998) Chondritic meteorites. In *Planetary Materials* (J. J. Papike, ed.), pp. 3-1 to 3-398. Reviews in Mineralogy, Vol. 36, Mineralogical Society of America.

Brearley A. J. and Hutcheon I. D. (2000) Carbonates in the CM1 chondrite ALH84034: Mineral chemistry, zoning and Mn-Cr systematics (abstract). In *Lunar and Planetary Science XXX,* Abstract #1407. Lunar and Planetary Institute, Houston (CD-ROM).

Brearley A. J. and Hutcheon I. D. (2002) Carbonates in the Y791198 CM2 chondrite: Zoning and Mn-Cr systematic (abstract). *Meteoritics & Planet. Sci., 37,* A23.

Brearley A. J. and Prinz M. (1992) CI-like clasts in the Nilpena polymict ureilite: Implications for aqueous alteration processes in CI chondrites. *Geochim. Cosmochim. Acta, 56,* 1373–1386.

Brearley A. J., Saxton J. M., Lyon I. C., and Turner G. (1999) Carbonates in the Murchison CM chondrites: CL characteristics and oxygen isotopic compositions (abstract). In *Lunar and Planetary Science XXX,* Abstract #1301. Lunar and Planetary Institute, Houston (CD-ROM).

Brearley A. J., Hutcheon, I. D., and Browning L. (2001) Compositional zoning and Mn-Cr systematics in carbonates from the Y791198 CM2 carbonaceous chondrite (abstract). In *Lunar and Planetary Science XXXII,* Abstract #1458. Lunar and Planetary Institute, Houston (CD-ROM).

Bridges J. C., Banks D. A., Smith M., and Grady M. M. (2004) Halite and stable chlorine isotopes in the in the Zag H3–6 breccia. *Meteoritics & Planet. Sci., 39,* 657–666.

Browning L. B. and Bourcier W. L. (1996) Tochilinite: A sensitive indicator of alteration conditions on the CM asteroidal parent body (abstract). In *Lunar and Planetary Science XXVII,* pp. 171–172. Lunar and Planetary Institute, Houston.

Browning L. B. and Bourcier W. (1998) Constraints on the anhydrous precursor mineralogy of fine-grained materials in carbonaceous chondrites. *Meteoritics & Planet. Sci., 33,* 1213–1220.

Browning L. B., McSween H. Y. Jr., and Zolensky M. E. (1996) Correlated alteration effects in CM carbonaceous chondrites. *Geochim. Cosmochim. Acta, 60,* 2621–2633.

Burger P. V. and Brearley A. J. (2004) Chondrule glass alteration in Type IIA chondrules in the CR2 chondrites EET 87770 and

EET 92105: Insights into elemental exchange between chondrules and matrices (abstract). In *Lunar and Planetary Science XXXV,* Abstract #1966. Lunar and Planetary Institute, Houston (CD-ROM).
Busfield A., Gilmour J. D., Whitby J. A., and Turner G. (2004) Iodine-xenon analysis of ordinary chondrite halide: Implications for early solar system water. *Geochim. Cosmochim. Acta, 68,* 195–202.
Buseck P. R. and Hua X. (1993) Matrices of carbonaceous chondrite meteorites. *Annu. Rev. Earth Planet. Sci., 21,* 255–305.
Chizmadia L. J. and Brearley A. J. (2003) Mineralogy and textural characteristics of fine-grained rims in the Yamato 791198 CM2 carbonaceous chondrite: Constraints on the location of aqueous alteration (abstract). In *Lunar and Planetary Science XXXIV,* Abstract #1419. Lunar and Planetary Institute, Houston (CD-ROM).
Chizmadia L. J. and Brearley A. J. (2004) Aqueous alteration of carbonaceous chondrites: New insights from comparative studies of two unbrecciated CM2 chondrites, Y-791198 and ALH81002 (abstract). In *Lunar and Planetary Science XXXV,* Abstract #1419. Lunar and Planetary Institute, Houston (CD-ROM).
Chizmadia L. J., Xu Y., Schwappach C., and Brearley A. J. (2003) Insights into Fe,Ni metal survival in the hydrated fine-grained rims in the Y-791198 CM2 carbonaceous chondrite (abstract). *Meteoritics & Planet. Sci., 38,* A137.
Christophe Michel-Lévy M. (1976) La matrice noire et blanche de la chondrite de Tieschitz. *Earth Planet. Sci. Lett., 30,* 143–150.
Ciesla F. J., Lauretta D. S., Cohen B. A., and Hood L. L. (2003) A nebular origin for chondritic fine-grained phyllosilicates. *Science, 299,* 549–552.
Clayton R. N. (1963) Carbon isotopes in meteoritic carbonates. *Science, 140,* 192–193.
Clayton R. N. and Mayeda T. K. (1977) Oxygen isotopic compositions of separated fractions of the Leoville and Renazzo carbonaceous chondrites. *Meteoritics, 12,* A199.
Clayton R. N. and Mayeda T. K. (1984) The oxygen isotope record in Murchison and other carbonaceous chondrites. *Earth Planet. Sci. Lett., 67,* 151–161.
Clayton R. N. and Mayeda T. K. (1996) Oxygen isotope relations among CO, CK, and CM chondrites and carbonaceous chondrite dark inclusions. *Meteoritics & Planet. Sci., 31,* 30.
Clayton R. N. and Mayeda T. K. (1999) Oxygen isotope studies of carbonaceous chondrites. *Geochim. Cosmochim. Acta, 63,* 2089–2104.
Clayton R. N. and Mayeda T. K. (2001) Oxygen isotopic composition of the Tagish Lake carbonaceous chondrite (abstract). In *Lunar and Planetary Science XXXII,* Abstract #1885. Lunar and Planetary Institute, Houston (CD-ROM).
Cohen B. A. and Coker R. A. (2000) Modeling of liquid water on CM meteorite parent bodies and implications for amino acid racemizations. *Icarus, 145,* 369–381.
Cohen R. E., Kornacki A. S., and Wood J. A. (1983) Mineralogy and petrology of chondrules and inclusions in the Mokoia CV3 chondrite. *Geochim. Cosmochim. Acta, 47,* 1739–1757.
Cyr K. E., Sears W. D., and Lunine J. I. (1998) Distribution and evolution of water ice in the solar nebula: Implications for solar system body formation. *Icarus, 135,* 537–548.
Davis A. M., MacPherson G. J., Clayton R. N., Mayeda T. K., Sylvester P. J., Grossman L., Hinton R. W., and Laughlin J. R. (1991) Melt solidification and late-stage evaporation in the evolution of a FUN inclusion from the Vigarano C3V chondrite. *Geochim. Cosmochim. Acta, 55,* 621–637.
Deloule E. and Robert F. (1995) Interstellar water in meteorites? *Geochim. Cosmochim. Acta, 59,* 4695–4706.
Dominik B., Jessberger E. K., Staudacher T., Nagel K., and El Goresy A. (1978) A new type of white inclusion in Allende: Petrography, mineral chemistry $^{40}Ar/^{39}Ar$ ages and genetic implications. *Proc. Lunar Planet. Sci. Conf. 9th,* pp. 1249–1266.
DuFresne E. R. and Anders E. (1962) On the chemical evolution of the carbonaceous chondrites. *Geochim. Cosmochim. Acta, 26,* 1085–1114.
Eiler J. M. and Kitchen N. (2004) Hydrogen isotope evidence for the origin and evolution of carbonaceous chondrites. *Geochim. Cosmochim. Acta, 68,* 1395–1411.
Endress M. and Bischoff A. (1996) Carbonates in CI chondrites: Clues to parent body evolution. *Geochim. Cosmochim. Acta, 60,* 489–507.
Endress M., Keil K., Bischoff A., Spettel B., Clayton R. N., and Mayeda T. K. (1994) Origin of dark clasts in the Acfer 059/El Djouf 001 CR2 chondrite. *Meteoritics, 29,* 26–40.
Endress M., Zinner E., and Bischoff A. (1996) Early aqueous activity on primitive meteorite parent bodies. *Nature, 379,* 701–703.
Fredriksson K. and Kerridge J. F. (1988) Carbonates and sulfates in Cl chondrites: Formation by aqueous activity on the parent body. *Meteoritics, 23,* 35–44.
Gaffey M. J., Burbine T. H., and Binzel R. P. (1993) Asteroid spectroscopy: Progress and perspectives. *Meteoritics, 28,* 161–187.
Gounelle M. and Zolensky M. E. (2001) A terrestrial origin for sulfate veins in CI1 chondrites. *Meteoritics & Planet. Sci., 35,* 1321–1329.
Grady M. M., Graham A. L., Barber D. J., Aylmer D., Kurat G., Ntaflos T., Ott U., Palme H., and Spettel B. (1987) Yamato-82042: An unusual carbonaceous chondrite with CM affinities. *Proc. 11th Symp. Antarct. Meteorites, Mem. Natl. Inst. Polar Res. Spec. Issue, 64,* 162–178.
Grady M. M., Wright I. P., Swart P. K., and Pillinger C. T. (1988) The carbon and oxygen isotopic composition of meteoritic carbonates. *Geochim. Cosmochim. Acta, 52,* 2855–2866.
Grady M. M., Verchovsky A. B., Franchi I. P., Wright I. P., and Pillinger C. T. (2002) Light element geochemistry of the Tagish Lake CI2 chondrite: Comparison with CI and CM chondrites. *Meteoritics & Planet. Sci., 37,* 713–736.
Grimm R. E. and McSween H. Y. Jr. (1989) Water and the thermal evolution of carbonaceous chondrite parent bodies. *Icarus, 82,* 244–280.
Grossman J. N., Alexander, C.M.O'D., Wang J., and Brearley A. J. (2000) Bleached chondrules: Evidence for widespread aqueous processes on the parent asteroids of ordinary chondrites. *Meteoritics & Planet. Sci., 35,* 467–486.
Grossman J. N., Alexander C. M. O., Wang J. H., and Brearley A. J. (2002) Zoned chondrules in Semarkona: Evidence for high- and low-temperature processing. *Meteoritics & Planet. Sci., 37,* 49–73.
Halbout J., Mayeda T. K., and Clayton R. N. (1986) Carbon isotopes and light element abundances in carbonaceous chondrites. *Earth Planet. Sci. Lett., 80,* 1–8.
Hanowski N. P. (1998) The aqueous alteration of CM carbonaceous chondrites — Petrographic and microchemical constraints. Ph.D. thesis, Univ. of New Mexico, Albuquerque. 241 pp.
Hanowski N. P. and Brearley A. J. (1997) Chondrule serpentines as indicators of aqueous alteration in CM carbonaceous chondrites (abstract). In *Lunar and Planetary Science XXVIII,* pp. 501–502. Lunar and Planetary Institute, Houston.
Hanowski N. P. and Brearley A. J. (2000) Iron-rich aureoles in

the CM carbonaceous chondrites, Murray, Murchison and ALH81002: Evidence for in situ alteration. *Meteoritics & Planet. Sci., 35,* 1291–1308.

Hanowski N. P. and Brearley A. J. (2001) Aqueous alteration of chondrules in the CM carbonaceous chondrites, Allan Hills 81002. *Geochim. Cosmochim. Acta, 65,* 495–518.

Hashimoto A. and Grossman L. (1987) Alteration of Al-rich inclusions inside amoeboid olivine aggregates in the Allende meteorite. *Geochim. Cosmochim. Acta, 51,* 1685–1704.

Hass J. R., Shock E. L., and Sassani D. C. (1995) Rare earth elements in hydrothermal solutions: Estimates of standard partial molal thermodynamic properties of aqueous complexes of the rare earth elements at high pressures and temperatures. *Geochim. Cosmochim. Acta, 59,* 4329–4350.

Hiroi T., Zolensky M. E., and Pieters C. M. (2001) The Tagish Lake meteorite: A possible sample of a D-type asteroid. *Science, 293,* 2234–2236.

Hutcheon I. D. and Phinney D. L. (1996) Radiogenic $^{53}Cr^*$ in Orgueil carbonates: Chronology of aqueous activity on the CI parent body (abstract). In *Lunar and Planetary Science XXVII,* pp. 577–578. Lunar and Planetary Institute, Houston.

Hutcheon I. D., Weisberg M. K., Phinney D. L., Zolensky M. E., Prinz M., and Ivanov A. V. (1999) Radiogenic ^{53}Cr in Kaidun carbonates: Evidence for very early aqueous activity (abstract). In *Lunar and Planetary Science, XXX,* Abstract #1722. Lunar and Planetary Institute, Houston (CD-ROM).

Hutchison R., Alexander C. M. O., and Barber D. J. (1987) The Semarkona meteorite: First recorded occurrence of smectite in an ordinary chondrite, and its implications. *Geochim. Cosmochim. Acta, 51,* 1875–1882.

Hutchison R., Alexander C. M. O. D., and Bridges J. C. (1998) Elemental redistribution in Tieschitz and the origin of white matrix. *Meteoritics & Planet. Sci., 33,* 1169–1180.

Ichikawa O. and Ikeda Y. (1995) Petrology of the Yamato-8449 CR chondrite. *Proc. NIPR Symp. Antarct. Meteorites, 8,* 63–78.

Ikeda Y. (1983) Alteration of chondrules and matrices in the four Antarctic carbonaceous chondrites ALH 77307 (C3), Y-790123 (C2), Y-75293(C2) and Y-74662(C2). *Proc. 8th Symp. Antarct. Meteorites, Mem. Natl. Inst. Polar Res., 30,* 93–108.

Itoh S. and Tomeoka K. (2001) Phyllosilicate-bearing chondrules and clasts in the ALHA 77307 CO3 chondrite: Evidence for parent-body processes. *Antarctic Meteorites XXVI, Papers Presented to the 26th Symposium on Antarctic Meteorites,* pp. 47–49. National Institute of Polar Research, Tokyo.

Jarosewich E., Clarke R. S. J., and Barrows J. N. (1987) *The Allende Meteorite Reference Sample.* Smithsonian Contributions to Earth Science, Vol. 27.

Johnson C. A. and Prinz M. (1993) Carbonate compositions in CM and CI chondrites and implications for aqueous alteration. *Geochim. Cosmochim. Acta, 57,* 2843–2852.

Johnson C. A., Prinz M., Weisberg M. K., Clayton R. N., and Mayeda T. K. (1990) Dark inclusions in Allende, Leoville and Vigarano: Evidence for nebular oxidation of CV3 constituents. *Geochim. Cosmochim. Acta, 54,* 819–830.

Kallemeyn G. W. and Wasson J. T. (1981) The compositional classification of chondrites — I. The carbonaceous chondrite groups. *Geochim. Cosmochim. Acta, 45,* 1217–1230.

Kanno A., Hiroi T., Nakamura R., Abe M., Ishiguro M., Hasegawa S., Miyasaka S., Sekiguchi T., Terada H., and Igarashi G. (2003) The first detection of water absorption on a D type asteroid. *Geophys. Res. Lett., 30,* PLA 2-1 (CiteID 1909, DOI 10.1029/2003GL017907).

Keller L. P. (1998) A transmission electron microscope study of iron-nickel carbides in the matrix of the Semakona unequilibrated ordinary chondrite. *Meteoritics & Planet. Sci., 33,* 913.

Keller L. P. and Buseck P. R. (1990a) Matrix mineralogy of the Lancé CO3 carbonaceous chondrite: A transmission electron microscope study. *Geochim. Cosmochim. Acta, 54,* 1155–1163.

Keller L. P. and Buseck P. R. (1990b) Aqueous alteration in the Kaba CV3 carbonaceous chondrite. *Geochim. Cosmochim. Acta, 54,* 2113–2120.

Keller L. P. and Buseck P. R. (1991) Calcic micas in the Allende meteorite: Evidence for hydration reactions in the early solar nebula. *Science, 252,* 946–949.

Keller L. P., Thomas K. L., Clayton R. N., Mayeda T. K., DeHart J. M., and McKay D. S. (1994) Aqueous alteration of the Bali CV3 chondrite: Evidence from mineralogy, mineral chemistry, and oxygen isotopic compositions. *Geochim. Cosmochim. Acta, 58,* 5589–5598.

Kerridge J. F. (1964) Low-temperature minerals from the fine-grained matrix of some carbonaceous meteorites. *Ann. N.Y. Acad. Sci., 119,* 41–53.

Kerridge J. F. (1993) Origins of organic matter in meteorites. *Proc. NIPR Symp. Antarct. Meteorites, 6,* 293–303.

Kerridge J. F. and Bunch T. E. (1979) Aqueous activity on asteroids: Evidence from carbonaceous chondrites. In *Asteroids* (T. Gehrels, ed.), pp. 745–764. Univ. of Arizona, Tucson.

Kerridge J. F., Mackay A. L., and Boynton W. V. (1979a) Magnetite in Cl carbonaceous chondrites: Origin by aqueous activity on a planetesimal surface. *Science, 205,* 395–397.

Kerridge J. F., Macdougall J. D., and Marti K. (1979b) Clues to the origin of sulfide minerals in Cl chondrites. *Earth Planet. Sci. Lett., 43,* 359–367.

Kimura M. and Ikeda Y. (1998) Hydrous and anhydrous alterations of chondrules in Kaba and Mokoia CV chondrites. *Earth Planet. Sci. Lett., 33,* 1139–1146.

Kojima T. and Tomeoka K. (1996) Indicators of aqueous alteration and thermal metamorphism on the CV parent body: Microtextures of a dark inclusion from Allende. *Geochim. Cosmochim. Acta, 60,* 2651–2666.

Kojima T., Tomeoka K., and Takeda H. (1993) Unusual dark clasts in the Vigarano CV3 carbonaceous chondrite: Record of parent body processes. *Meteoritics, 28,* 649–658.

Kracher A., Keil K., Kallemeyn G. W., Wasson J. T., Clayton R. N., and Huss G. I. (1985) The Leoville (CV3) accretionary breccia. *Proc. Lunar Planet Sci. Conf. 16th,* pp. 123–135.

Krot A. N., Scott E. R. D., and Zolensky M. E. (1995) Mineralogical and chemical modification of components in CV3 chondrites: Nebular or asteroidal processing. *Meteoritics, 30,* 748–775.

Krot A. N., Scott E. R. D., and Zolensky M. E. (1997a) Origin of fayalitic olivine rims and lath-shaped matrix olivine in the CV3 chondrite Allende and its dark inclusions. *Meteoritics, 32,* 31–49.

Krot A. N., Zolensky M. E., Wasson J. T., Scott E. R. D., Keil K., and Ohsumi K. (1997b) Carbide-magnetite assemblages in type-3 ordinary chondrites. *Geochim. Cosmochim. Acta, 61,* 219–237.

Krot A. N., Petaev M. I., Zolensky M. E., Keil K., Scott E. R. D., and Nakamura K. (1998a) Seconday calcium-iron-rich minerals in the Bali-like and Allende-like oxidized CV3 chondrites and Allende dark inclusions. *Meteoritics & Planet. Sci., 33,* 623–645.

Krot A. N., Petaev M. I., Scott E. R. D., Choi B.-G., Zolensky M. E., and Keil K. (1998b) Progressive alteration in CV3 chondrites: More evidence for asteroidal alteration. *Meteoritics & Planet. Sci., 33,* 1065–1085.

Krot A. N., Brearley A. J., Ulyanov A. A., Biryukov V. V., Swindle T. D., Keil K., Mittlefehldt D. W., Scott E. R. D., Clayton R. N., and Mayeda T. K. (1999) Mineralogy, petrography and bulk chemical, I-Xe, and oxygen isotopic compositions of dark incusions in the reduced CV3 chondrite Efremovka. *Meteoritics & Planet. Sci., 34,* 67–90.

Krot A. N., Petaev M. I., Meibom, A. and Keil K. (2000a) In situ growth of Ca-rich rims around Allende dark inclusions. *Geochem. Intl., 38,* S351–S368.

Krot A. N., Petaev M. I., Brearley A. J., Kallemeyn G. W., Sears D. W. G., Benoit P. H., Hutcheon I. D. and Keil K. (2000b) MAC88107, ungrouped carbonaceous chondrite with affinities to CM-CO clan: Evidence for in situ growth of fayalite and hedenbergite. *Meteoritics & Planet. Sci., 35,* 1365–1386.

Krot A. N., Hutcheon I. D., Brearley A. J., Pravdivtseva O. V., Petaev M. I., and Hohenberg C. M. (2006) Timescales and settings for alteration of chondritic meteorites. In *Meteorites and the Early Solar System II* (D. S. Lauretta and H. Y. McSween Jr.), this volume. Univ. of Arizona, Tucson.

Kurat G. (1969) The formation of chondrules and chondrites and some observations on chondrules from the Tieschitz meteorite. In *Meteorite Research* (P. M. Millman, ed.), pp. 185–190. Reidel, Dordrecht.

Lauretta D. S., Hua X., and Buseck P. R. (2000) Mineralogy of fine-grained rims in the ALH81002 CM chondrite. *Geochim. Cosmochim. Acta, 64,* 3263–3273.

Lee M. (1993) The petrogaphy, mineralogy and origins of calcium sulphate within the Cold Bokkeveld CM carbonaceous chondrite. *Meteoritics, 28,* 53–62.

Lee M. R., Hutchison R., and Graham A. L. (1996) Aqueous alteration in the matrix of the Vigarano (CV3) carbonaceous chondrite. *Meteoritics & Planet. Sci., 31,* 477–483.

Leshin L. A., Farquhar, J., Guan Y., Pizzarello S., Jackson T. L., and Thiemens M. H. (2001) Oxygen isotopic anatomy of Tagish Lake: Relationship to primary and secondary minerals in CI and CM chondrites (abstract). In *Lunar and Planetary Science XXXII,* Abstract #1843. Lunar and Planetary Institute, Houston (CD-ROM).

Llorca J. and Brearley A. J. (1992) Alteration of chondrules in ALH 84034, an unusual CM2 carbonaceous chondrite (abstract). In *Lunar and Planetary Science XXIII,* pp. 793–794.

Lodders K. (2003) Solar system abundances and condensation temperatures of the elements. *Astrophys. J., 591,* 1220–1247.

Lugmair G. W. and Shukolyukov A. (2001) Early solar system events and timescales. *Meteoritics & Planet. Sci., 36,* 1017–1026.

MacPherson G. J., Davis A. M., and Zinner E. K. (1995) The distribution of aluminum-26 in the early solar system — A reappraisal. *Meteoritics, 30,* 365–386.

Mason B. (1962) The carbonaceous chondrites. *Space Sci. Rev., 1,* 621–646.

Mao X.-Y., Ward B. J., Grossman L., and MacPherson G. J. (1990) Chemical composition of refractory inclusions from the Vigarano and Leoville carbonaceous chondrites. *Geochim. Cosmochim. Acta, 54,* 2121–2132.

McSween H. Y. Jr. (1979) Alteration in CM carbonaceous chondrites inferred from modal and chemical variations in matrix. *Geochim. Cosmochim. Acta, 43,* 1761–1770

McSween H. Y. (1987) Aqueous alteration in carbonaceous chondrites: Mass balance constraints on matrix mineralogy. *Geochim. Cosmochim. Acta, 51,* 2469–2477.

McSween H. Y. and Richardson S. M. (1977) The composition of carbonaceous chondrite matrix. *Geochim. Cosmochim. Acta, 41,* 1145–1161.

Metzler K. Bischoff A., and Stöffler D. (1992) Accretionary dust mantles in CM chondrites: Evidence for solar nebula processes. *Geochim. Cosmochim. Acta, 56,* 2873–2897.

Mittlefehldt D. W. and Wetherill G. W. (1979) Rb-Sr studies of CI and CM chondrites. *Geochim. Cosmochim. Acta, 43,* 201–206.

Mungall J. E. and Martin R. F. (1994) Severe leaching of trachytic glass without devitrification, Terceira, Azores. *Geochim. Cosmochim. Acta, 58,* 75–83.

Nagy B. (1966) Investigations of the Orgueil carbonaceous meteorite. *Geol. Foren. Stockholm Forh., 88,* 235–272.

Nakamura T., Nagao K., and Takaoka N. (1999) Microdistribution of primordial noble gases in CM chondrites determined by in situ laser microprobe analysis: Decipherment of nebular processes. *Geochim. Cosmochim. Acta, 63,* 241–255.

Noguchi T. (1995) Petrology and mineralogy of the PCA 91082 chondrite and its comparison with the Yamato-793495 (CR) chondrite. *Proc. NIPR Symp. Antarct. Meteorites, 8,* 32–62.

Pieters C. M. and McFadden L. A. (1994) Meteorite and asteroid reflectance spectroscopy: Clues to early solar system processes. *Annu. Rev. Earth Planet. Sci., 22,* 457–497.

Pisani F. (1864) Etude chimique et analyse de l'aerolithe d'Orgueil. *Compt. Rend., 59,* 132–135.

Podosek F. A. and Cassen P. (1994) Theoretical, observational, and isotopic estimates of the lifetime of the solar nebula. *Meteoritics, 29,* 6–25.

Richardson S. M. (1978) Vein formation in the C1 carbonaceous chondrites. *Meteoritics, 13,* 141–159.

Richardson S. M. (1981) Alteration of mesostasis in chondrules and aggregates from three C2 carbonaceous chondrites. *Earth Planet. Sci. Lett., 52,* 67–75.

Riciputi L. R., McSween H. Y. J., Johnson C. A., and Prinz M. (1994) Minor and trace element concentrations in carbonates of carbonaceous chondrites, and implications for the compositions of coexisting fluids. *Geochim. Cosmochim. Acta, 58,* 1343–1351.

Rosenberg N. D., Browning L., and Bourcier W. L. (2001) Modeling aqueous alteration of CM carbonaceous chondrites. *Meteoritics & Planet. Sci., 36,* 239–244.

Rowe M. W., Clayton R. N., and Mayeda T. K. (1994) Oxygen isotopes in separated components of CI and CM meteorites. *Geochim. Cosmochim. Acta, 58,* 5341–5348.

Rubin A. E. (1998) Correlated petrologic and geochemical characteristics of CO3 chondrites. *Meteorites & Planet. Sci., 33,* 383–391.

Rubin A. E., Zolensky M. E., and Bodnar R. J. (2002) The halite-bearing Zag and Monahans (1998) meteorite breccias: Shock metamorphism, thermal metamorphism and aqueous alteration on the H-chondrite parent body. *Meteorites & Planet. Sci., 37,* 125–141.

Sears D. W. G., Benoit P. H., and Lu J. (1993) Two chondrule groups each with distinctive rims in Murchison recognized by cathodoluminescence. *Meteoritics, 28,* 669–675.

Taylor G. J., Okada A., Scott E. R. D., Rubin A. E., Huss G. R., and Keil K. (1981) The occurrence and implications of carbide-magnetite assemblages in unequilibrated ordinary chondrites (abstract). In *Lunar and Planetary Science XII,* pp. 1076–1078. Lunar and Planetary Institute, Houston.

Tomeoka K. (1990) Phyllosilicate veins in a CI meteorite: Evidence for aqueous alteration on the parent body. *Nature, 345,* 138–140.

Tomeoka K. and Buseck P. R. (1982a) An unusual layered mineral in chondrules and aggregates of the Allende carbonaceous chondrite. *Nature, 299,* 327–329.

Tomeoka K. and Buseck P. R. (1982b) Intergrown mica and montmorillonite in the Allende carbonaceous chondrite. *Nature, 299,* 326–327.

Tomeoka K. and Buseck P. R. (1985) Indicators of aqueous alteration in CM carbonaceous chondrites: Microtextures of a layered mineral containing Fe, S, O and Ni. *Geochim. Cosmochim. Acta, 49,* 2149–2163.

Tomeoka K. and Buseck P. R. (1988) Matrix mineralogy of the Orgueil CI carbonaceous chondrite. *Geochim. Cosmochim. Acta, 52,* 1627–1640.

Tomeoka K. and Buseck P. R. (1990) Phyllosilicates in the Mokoia CV carbonaceous chondrite: Evidence for aqueous alteration in an oxidizing condition. *Geochim. Cosmochim. Acta, 54,* 1787–1796.

Tomeoka K. and Tanimura I. (2000) Phyllosilicate-rich chondrule rims in the Vigarano CV3 chondrite: Evidence for parent-body processes. *Geochim. Cosmochim. Acta, 64,* 1971–1988.

Toriumi M. (1989) Grain size distribution of the matrix in the Allende chondrite. *Earth Planet. Sci. Lett., 92,* 265–273.

van de Vusse R. and Powell R. (1983) The interpretation of pyrrhotine-pentlandite-tochilinite-magnetite-magnesite textures in serpentinites from Mount Keith, Western Australia. *Mineral. Mag., 47,* 501–505.

Vilas F. and Gaffey M. J. (1989) Phyllosilicate absorption features in main-belt and outer-belt asteroids from reflectance spectroscopy. *Science, 246,* 790–792.

Vilas F., Hatch E. C., Larson S. M., Sawyer S. R., and Gaffey M. J. (1993) Ferric iron in primitive asteroids; a 0.43 mm absorption feature. *Icarus, 102,* 225–231.

Vilas F., Jarvis K. S., and Gaffey M. J. (1994) Iron alteration minerals in the visible and near-infrared spectra of low-albedo asteroids. *Icarus, 109,* 274–283.

Wark D. A. and Lovering J. F. (1977) Marker events in the early evolution of the solar system: Evidence from rims on Ca-Al-rich inclusions from carbonaceous chondrites. *Proc. Lunar Sci. Conf. 8th,* pp. 95–112.

Weisberg M. K. and Prinz M. (2000) The Grosvenor Mountains 95577 CR1 chondrite and hydration of the CR chondrites (abstract). *Meteoritics & Planet. Sci., 35,* A168.

Weisberg M. K., Prinz M., Clayton R. N., and Mayeda T. K. (1993) The CR (Renazzo-type) carbonaceous chondrite group and its implications. *Geochim. Cosmochim Acta, 57,* 1567–1586.

Whitby J. A., Burgess R., Turner G., Gilmour J. D., and Bridges J. C. (2000) Extinct ^{126}I in halite from a primitive meteorite: Evidence for evaporite formation in the early solar system. *Science, 208,* 1819–1821.

Wilson L., Keil K., Browning L. B., Krot A. N., Bourcier W. (1999) Early aqueous alteration, explosive disruption and reprocessing of asteroids. *Meteoritics & Planet. Sci., 34,* 541–557.

Wlotzka F., Spettel B., Palme H., and Schultz L. (1989) New CM chondrites from Antarctica — Different mineralogy, same chemistry. *Meteoritics, 24,* 341–342.

Wogelius R. A. and Walther J. V. (1991) Olivine dissolution at 25°C: Effects of pH, CO_2, and organic acids. *Geochim. Cosmochim. Acta, 55,* 943–954.

Wogelius R. A. and Walther J. V. (1992) Olivine dissolution kinetics at near-surface conditions. *Chem. Geol., 97,* 101–112.

Young E. D. (2001) The hydrology of carbonaceous chondrite parent bodies and the evolution of planet progenitors. *Phil. Trans. R. Soc. London, A359,* 2095–2109.

Young E. and Russell S. S. (1998) Oxygen reservoirs in the early solar nebula inferred from an Allende CAI. *Science, 282,* 452–455.

Young E. D. Ash R., England P., and Rumble D. (1999) Fluid flow in chondritic parent bodies: Deciphering the compositions of planetesimals. *Science, 286,* 1331–1335.

Zega T. J. and Buseck P. R. (2003) Fine-grained-rim mineralogy of the Cold Bokkeveld CM chondrite. *Geochim. Cosmochim. Acta, 67,* 1711–1721.

Zega T. J., Garvie L. A. J., and Buseck P. R. (2003) Nanometer-scale measurements of iron oxidation states of cronstedtite from primitive meteorites. *Am. Mineral., 88,* 1169–1172.

Zolensky M. E. and Ivanov A. (2003) The Kaidun microbreccia meteorite: A harvest from the inner and outer solar asteroid belt. *Chem. Erde, 63,* 185–246.

Zolensky M. E. and McSween H. Y. Jr. (1988) Aqueous alteration. In *Meteorites and the Early Solar System* (J. F. Kerridge and M. Matthews, eds.), pp. 114–143. Univ. of Arizona, Tucson.

Zolensky M. E., Bourcier W. L., and Gooding J. L. (1989) Aqueous alteration on the hydrous asteroids: Results of EQ3/6 computer simulations. *Icarus, 78,* 411–425.

Zolensky M. E., Barrett R., and Browning L. (1993) Mineralogy and composition of matrix and chondrule rims in carbonaceous chondrites. *Geochim. Cosmochim. Acta, 57,* 3123–3148.

Zolensky M. E., Mittlefehldt D. W., Lipschutz M. E., Wang M.-S., Clayton R. N., Mayeda T. K., Grady M. M., Pillinger C., and Barber D. (1997) CM chondrites exhibit the complete petrologic range from type 2 to 1. *Geochim. Cosmochim Acta, 61,* 5099–5115.

Zolensky M. E., Bodnar R. J., Gibson E. K. Jr, Nyquist L. E., Resse Y., Shih Chi-Ti, and Wiesman H. (1999) Asteroidal water within fluid inclusion-bearing halite in an H5 chondrite, Monahans. *Science, 285,* 1377–1379.

Zolensky M. E., Nakamura K., Gounelle M., Mikouchi T., Kasama T., Tachikawa O., and Tonui E. (2002) Mineralogy of Tagish Lake: An ungrouped type 2 carbonaceous chondrite. *Meteoritics & Planet. Sci., 37,* 737–761.

The Nature and Distribution of the Organic Material in Carbonaceous Chondrites and Interplanetary Dust Particles

S. Pizzarello
Arizona State University

G. W. Cooper
NASA Ames Research Center

G. J. Flynn
State University of New York at Plattsburgh

Most of the carbon in carbonaceous chondrites is organic material that displays structures as diverse as kerogen-like macromolecules and simpler soluble compounds ranging from polar amino acids and polyols to nonpolar hydrocarbons. Overall, the large molecular and isotopic diversity of meteorite organics verifies their extraterrestrial origin and points to synthetic pathways in a variety of chemical regimes. These include exothermic reactions in the cold, H-fractionating interstellar gas phase and aqueous reactions in asteroidal parent bodies. Reactions on interstellar grains and during nebular processes were most likely involved but are inferred with less certainty. The unique L-asymmetry of some meteoritic amino acids suggests their possible contribution to terrestrial molecular evolution. Many interplanetary dust particles show a primitive composition that distinguishes them from known meteorites and could offer unique insights into unaltered nebular material. So far, micrometer sizes have prevented their wide-ranging molecular characterization.

1. INTRODUCTION

Organic material has been detected within the solar system in planetary satellites (*Courting et al.,* 1991), the coma of comets (*Kissel and Kruger,* 1987; *Bockelée-Morvan,* 2001; *Mumma et al.,* 2001), interplanetary dust particles (IDPs) (*Clemett et al.,* 1993; *Flynn et al.,* 2000), micrometeorites recovered from polar ice (*Brinton et al.,* 1998; *Matrajt et al.,* 2001), and meteorites. While the first two observations are limited to remote sensing or spacecraft findings, the organics in meteorites and, to a lesser degree, the related smaller IDPs and micrometeorites have received direct scrutiny in laboratory analyses. Studies have been extensive for the carbonaceous chondrites (CC) and several earlier reviews have followed their progress; the most recent ones, *Botta and Bada* (2002), *Sephton* (2002), and *Cronin and Chang* (1993), provide references to earlier reviews as well as to most of this introduction.

The organic content of CC is high in relation to total carbon and is complex in composition; it consists almost exclusively of an insoluble kerogen-like material in the CO, CV, and CK classes of these meteorites, while in the CI, CM, and CR classes it also includes numerous soluble organic compounds (see chapters in Part I of this volume for meteorite classification). The relative abundance of insoluble to soluble organic carbon varies between meteorites; in the more soluble-rich chondrites, e.g., the CM2s Murchison and Murray, the insoluble organic material constitutes ~70% of the total while in others, such as the ungrouped Tagish Lake meteorite, it surpasses 99%.

Variability of composition is one of the common characteristics of the organic material of CC and is shown not only by the relative abundance variations described above but also by compositional differences in both insoluble (*Cody et al.,* 2005) and extracted organic materials between meteorites of the same class. In particular, the complex soluble organic suite of CMs varies significantly in abundance and composition between and even within individual meteorites (*Cronin and Pizzarello,* 1983; *Pizzarello et al.,* 2001, 2003). Several of the organic compounds found in meteorites are also common terrestrial biomolecules; however, in CC they are components of a diverse and random mixture, a fact that is consistent with abiotic processes and that contrasts with the structural specificity seen in their terrestrial counterparts. For example, over 80 isomeric and homologous amino acid species up to C_8 have been identified in the Murchison meteorite in contrast to just 20 that make up terrestrial proteins.

A portion of the soluble and insoluble meteoritic organic material, as well as IDPs, share an enrichment in the heavier stable isotopes of H, C, and N. The findings, in particular the magnitude of the positive δD of some soluble meteoritic compounds, have been related to the large D-enrichments observed in cold interstellar clouds and suggest that meteoritic compounds had interstellar precursors. However, this enrichment is either absent or less pronounced in other meteorite organics, such as much of the insoluble carbon and the aliphatic hydrocarbons.

Given these large compositional and isotopic differences, an inclusive theory for the origin of all the varied organic constituents in CC appears hard to formulate. Moreover, the variability in molecular, chiral, and isotopic distribution shown by discreet soluble organic compounds indicates that their formation was likely a diverse and multistep process. On the basis of the constraints posed by the overall molecular and isotopic data, it is possible to estimate that several processes contributed to meteorite organic syntheses, either singularly or in concurrence, at different points in their histories: presolar syntheses, solar nebular reactions of organic material, and parent-body water chemistry.

Many organic molecules have been observed in interstellar and circumstellar environments where they are, for the most part, detected in the gas phase by infrared and microwave spectroscopy (see The National Radio Astronomy Observatory Web site, www.nrao.edu, for updated lists of interstellar and circumstellar molecules). It is expected that refractory interstellar grains could host the syntheses of an even larger variety of molecules. Many interstellar molecules have shown molecular and isotopic features that are characteristic of the low temperatures and energetic conditions at which they were formed (e.g., *Roueff and Gerin,* 2003). Some of the CC organics appear to have kept, at least in part, some of those features such as D-enrichment and highly branched carbon structures. We have to assume, therefore, that presolar syntheses provided an inventory of simple organics and precursor species to the collapsing solar nebula. A significantly larger D-enrichment found for some of the amino and monocarboxylic acids in the Murchison meteorite further indicates that some of these syntheses must have been more advanced for distinct subgroups of meteoritic compounds and may have produced several of the compounds observed in meteorites or their direct precursors.

Very little is known about the conditions and processes that could have altered, destroyed, or formed organic compounds during the periods of interstellar cloud collapse, stellar origin, and solar nebular formation. Even if poorly understood, however, interactions of gas and solids in the nebula have to be assumed and could include both reactions of infalling interstellar material and the syntheses of new molecular species. Guidelines within the many uncertainties that exist in regard to nebular conditions are provided by spectroscopic observation of young stellar objects (e.g., *Qi et al.,* 2003) and theoretical models of collapsing interstellar cores (e.g., *Aikawa et al.,* 2003), protoplanetary disks (e.g., *Aikawa et al.,* 2002), and planetesimal mass distribution (e.g., *Kortenkamp et al.,* 2001) that could affect the primordial asteroid belt. Also, *Aikawa and Herbst* (2001) estimated large D/H ratios for some molecules in cold protoplanetary disk regions. This D-enrichment could derive from H-fractionating chemistry in the disk similar to that in the interstellar cloud and not just from the incorporation of interstellar material.

Certain scenarios of solar system formation (*Boss et al.,* 2002) allow for a flux of UV radiation in the inner nebula that could lead to photolysis of planetesimal ices and an extensive organic chemistry of the kind observed on Titan (*Boss,* 2002). Classic Fisher Tropsch-type (FTT) reactions, i.e., involving CO, H_2, and the minerals of a condensing nebula, have been invoked for the formation of many meteoritic compounds but failed to be confirmed by the isotopic data. However, catalytic gas-solid reactions could be possible in the nebula (*Kress and Tielens,* 2001) on the surface of condensing grains, small bodies, and planetesimals; catalytic liquid-solid reactions could have occurred during a liquid phase as well. Heating by shock waves, impacts, and gravitational collapse could have contributed to the chemistry of organics at different stages of nebular condensation.

Parent-body water alteration processes are recorded in CC not only by the matrix mineralogy but also by their soluble organic composition. For example, a Strecker-like reaction of aldehyde and ketone precursors with ammonia and HCN in water (Scheme 2) is consistent with the finding in meteorites of comparable suites of α-amino, α-hydroxy, and imino acids. Although similar indirect proofs are not found for other subgroups of the same compounds, in general, the breadth of water-soluble organics found in CC, their coexistence with clay minerals in these meteorites, as well as some of the plausible hypotheses formulated for their synthesis would reasonably suggest that at least the final formation stages of most of these organic compounds were affected by reactions with water, such as hydrolysis of nitriles and saturation of double bonds.

2. THE INSOLUBLE CARBONACEOUS MATERIAL

The larger fraction of the organic carbon in CC, amounting to 70–99% of the total, is present as a complex and insoluble macromolecular material that is obtained from meteorite powders only after repeated extractions with solvents and demineralization by concentrated hydrofluoric and hydrochloric acids (e.g., *Cronin et al.,* 1987) or CsF (*Cody et al.,* 2002). In addition to C, this insoluble organic material (IOM) also contains H, O, N, and S, with elemental composition that varies between classes and subclasses of CC, e.g., similar combustion methods gave the general formulas $C_{100}H_{70}N_3O_{12}S_2$ and $C_{100}H_{46}N_{10}O_{15}S_{4.5}$ for the Murchison (*Hayatsu et al.,* 1980) and Tagish Lake (*Pizzarello et al.,* 2001) meteorites, respectively. The Murchison IOM isotopic composition, as averaged from several studies (*Kerridge et al.,* 1987), is given by the values $\delta D = +727.6‰$, $\delta^{13}C = -14.3‰$, and $\delta^{14}N = +6.6‰$, while two analyses of Tagish Lake IOM gave $\delta D = 930‰$ (*Pizzarello et al.,* 2001) and a summed $\delta^{13}C \approx +24.3‰$ (*Brown et al.,* 2000).

Macromolecular carbonaceous material free of minerals can also be found in untreated CC powders as distinct nanometer-sized entities such as aromatic flakes, solid or hollow spheres, and tubes (*Nagy et al.,* 1962; *Rossignol-Strick and Barghoorn,* 1971; *Pflug,* 1984; *Nakamura et al.,* 2002;

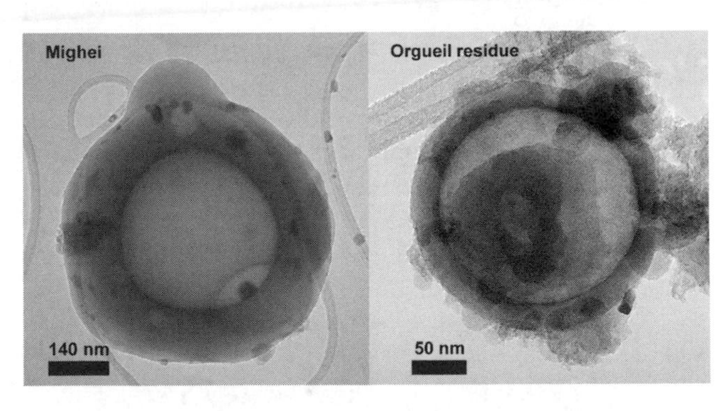

Fig. 1. TEM images of hollow carbonaceous particles from the Mighei (unprocessed powders) and Orgueil (HF/HCl acid residue) carbonaceous chondrites. The Orgueil sphere is surrounded by adhering IOM. From *Garvie and Buseck* (2004b).

Garvie and Busek, 2004b), which are observed in several CI and CM powders as well as in the acid insoluble residues (*Garvie and Buseck,* 2004a) (Fig. 1). They are analyzed by transmission electron microscopy (TEM) and electron energy-loss spectroscopy (EELS). By visual estimate, they appear to have an abundance of up to 10% of total IOM in Orgueil (L. Garvie, personal communication, 2004) and are heterogeneous in composition, size, and distribution in all meteorites analyzed. In Tagish Lake alone, hollow nanospheres range in outside diameter from 166 to 830 nm and in internal volumes from 6.9×10^3 to 3.3×10^5 nm^3 (*Garvie and Buseck,* 2004b). Their carbon is aromatic in nature, varying in composition from close to pure graphitic C (>99%) to containing several percent of O, N, and S; e.g., the ele-mental composition of a Tagish Lake carbonaceous flake was observed to be $C_{100}S_{10.1}$ while that of a globule in the same meteorite was $C_{100}S_{2.7}N_{5.0}O_{10}$ (bonded H is not detected by EELS and is excluded from the above calculations). Functional groups such as C=O and C=N are also observed in some nanostructures. Other well-characterized carbon structures identified in the IOM are sometimes described as exotic C species by their association with isotopically exotic He, Xe, and Ne. These are the nanodiamonds (*Lewis et al.,* 1987; *Blake et al.,* 1988), silicon carbide (*Tang et al.,* 1989), graphite (*Smith and Buseck,* 1981; *Amari et al.,* 1990), and fullerenes (*Becker et al.,* 2000). All have been considered of presolar origin and possibly formed in the outflows of carbon stars.

The remaining larger portion of the insoluble carbon can be seen by TEM to be an unstructured and heterogeneous material that does not show any clear resolution in its continuity (*Derenne et al.,* 2002; *Garvie and Buseck,* 2004a). Given its complex nature and inaccessibility to solvents, so far, the IOM of CC has been mainly analyzed as a whole by either decomposition techniques, such as chemical degradation and pyrolysis, or solid-state nuclear magnetic resonance (NMR) spectroscopy. These approaches, although limited in their capability to provide a complete chemical definition of all IOM macromolecular phases, have been useful to characterize many of their compositional features.

Oxidative degradation carried out by *Hayatsu et al.* (1977, 1980) on Murchison IOM, together with detailed analyses of the composition and abundance of the fragmentation products, gave a general structure for the material that is still plausible today. It would consist of condensed aromatic, hydroaromatic, and heteroaromatic macromolecules that contain alkyl branching, functional groups such as OH and COOH, and are bridged by alkyl chains and ether linkages. Vacuum pyrolysis analyses of Orgueil and Cold Bokkeveld IOM (*Bandurski and Nagy,* 1976) are in general agreement with this model and in addition provided identification of several IOM nitrile species. Hydrous pyrolysis studies conducted by *Sephton et al.* (1998, 2000) on the IOM of several CC have further constrained this macromolecular model. The authors found that these pyrolytic conditions release a set of aromatic compounds such as toluene, naphthalene, methyl naphthalenes, and others that are also found in the solvent extracts of the meteorite, while leaving behind a residue of virtually unaffected material. On the basis of C-isotopic differences between soluble, pyrolysable, and residual material, the hypothesis was put forward (*Sephton et al.,* 1998) that the macromolecular C comprises at least two components: a more labile portion, which during parent-body alteration processes might already have generated some of the aromatic compounds detected in whole meteorite extracts, and a refractory material. This hypothesis of diverse IOM phases agrees with isotopic data (*Kerridge et al.,* 1987) that showed a two-temperature release of carbonaceous components from the IOM (at ~250°–350°C and over 400°C, respectively) where the more labile phase was enriched in D ($\delta D \geq +1165$‰). The $\delta^{15}N$ values measured by step combustion of several CC IOMs (*Alexander et al.,* 1998) showed a two-component distribution for this element also. However, a third component of heavier N ($\delta^{15}N \geq +260$‰) was released above 500°C, a finding that was later expanded by *Sephton et al.* (2003) and would indicate that the makeup of the IOM residue left after hydrous pyrolysis is itself heterogeneous.

The possibility of parent-body thermal alteration of the IOM subsequent to the aqueous processes has also been considered and studied by pyrolysis-GC (*Murae et al.,* 1991; *Kitajima et al.,* 2002). Analyses assessed the relative abundances of lower- and higher-temperature pyrolysates released from several meteorite IOMs, on the assumption that more thermally altered material would be more graphitic and yield less of the lower molecular weight products such as naphthalenes. Pyrolysate results confirmed this assumption and indicated that the possible metamorphic variations observed in the IOMs correlate well with the estimated heating of the corresponding meteorites. *Shimoyama et al.* (1991) reached similar conclusions from their analysis of thermal decomposition products from whole meteorite powders.

Infrared (IR), solid-state ^{13}C and ^{1}H NMR, and electron paramagnetic resonance (EPR) spectroscopic analyses of the IOM in CC have been complimentary to the decomposition analyses described above and offer an overview of the chemical nature of the IOM as a whole, with emphasis on its bonding, functional group composition, and radical abundance properties (e.g., see Fig. 2). The ^{13}C NMR spectra of the Orgueil (*Cronin et al.,* 1987; *Gardinier et al.,* 2000) and Murchison (*Cronin et al.,* 1987; *Gardinier et al.,* 2000; *Cody et al.,* 2002) IOM were collected by magic-angle spinning, employing both cross polarization (CP) and single pulse (SP) techniques [see *Cody et al.* (2002) for a summary of these techniques]. The spectra confirm the complex nature of the IOM revealed upon its degradation and show two main broad features characteristic of sp^2 and sp^3 hybridized C atoms, i.e., olefinic/aromatic and aliphatic features, respectively. The relative abundance of aromatic carbon varies between these meteorites and is largest in Murchison, where it represents 0.61–0.66 of the IOM carbon (*Cody et al.,* 2002). The Tagish Lake meteorite presents an NMR spectrum that is substantially different from those of Orgueil and Murchison, with a predominant aromatic feature that represents over 80% of the IOM carbon (*Pizzarello et al.,* 2001; *Cody et al.,* 2005).

In the meteorites analyzed so far, the aromatic carbon appears to comprise mainly rather small (smaller than C_{24} coronene) and highly substituted (*Gardinier et al.,* 2000; *Cody et al.,* 2005) ring aggregates, where larger unprotonated structures make up, at most, only 10% of total aromatics (*Cody et al.,* 2002). Insoluble organic material aliphatic carbon in Murchison and other meteorites includes methyl, methylene, and methine groups (*Gardinier et al.,* 2000; *Cody et al.,* 2005) and appears to be composed of highly branched short chains that could be related to the aromatic fraction via extensive cross linking. In Murchison, the IOM aliphatic carbon contains the majority (~70%) of total hydrogen (*Cody et al.,* 2002).

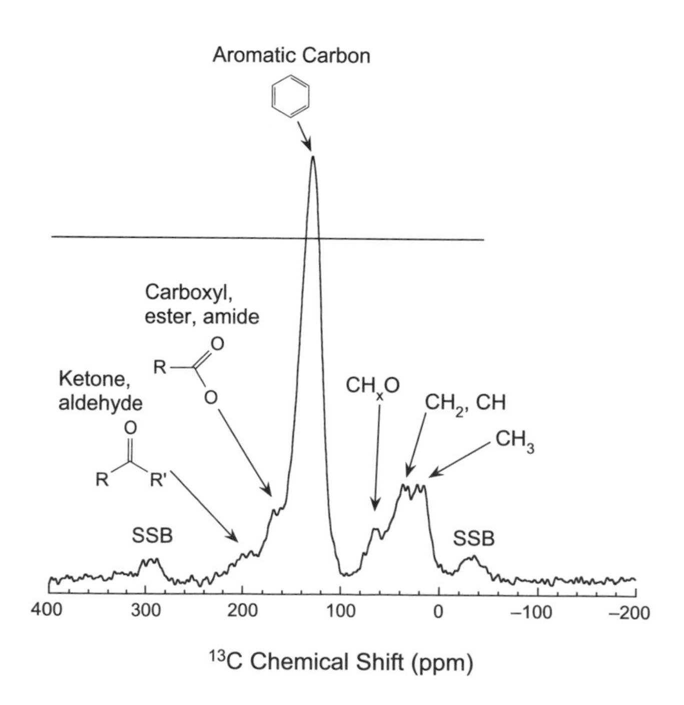

Fig. 2. ^{13}C-NMR spectrum of Murchison IOM showing the range of organic functionalities in the insoluble carbon (by variable amplitude cross polarization magic angle spinning). From *Cody et al.,* 2002.

Nuclear magnetic resonance spectra have also identified several IOM functional groups; most of them are O-containing (e.g., carboxylate and carbonyl) and vary between meteorites in their relative proportion to other functionalities. When their abundance is plotted in relation to aliphatic carbon, which is more susceptible to oxidation than the aromatic material, it shows different levels of IOM oxidation between various CC with the comparative distribution: CR2 < CI1 < CM2 < Tagish Lake (*Cody et al.,* 2005). Insoluble organic matter organic S functionalities, which are not measured by NMR but should be abundant given the material overall elemental composition, were observed by X-ray absorption near-edge structure (XANES) analyses to be of numerous kinds (*Derenne et al.,* 2002). Interestingly, comparative analyses of Orgueil (*Binet et al.,* 2002), Murchison (*Binet et al.,* 2004a), and Tagish Lake (*Binet et al.,* 2004b) IOM by EPR and electron nuclear double resonance studies (ENDOR) have shown that these materials, in spite of dissimilarities in their makeup, all show the same feature of abundant and heterogeneously distributed mono- and di-radicals. The equal presence and similarity of distribution of these aromatic features in meteorites that differ in degrees of mineral alteration and organic content could indicate that at least a portion of the IOM predated and survived parent-body environments.

What conclusions do the analytical studies allow in regard to the formation of the IOM in CC, its origin, and genetic relationship, if any, with the soluble organic material? Carbonaceous material organized in well-defined macromolecular structures, most or all of which do not appear to alter through the harsh demineralization procedures, seems to indicate that the origin of at least a sizable portion of the IOM predated parent-body processes. *Garvie and Buseck* (2004b) point out that some of the nanostructure compositional differences, such as the example of sharp dissimilarities between contiguous formations in the Bells meteorite (a CM2 fall), are not compatible with their asteroidal aggregation during an aqueous phase since such water processes would have likely acted as compositional equalizers, at least at short distances. Moreover, the enrichment in D and ^{15}N found for discrete IOM phases points to a distinct origin related to interstellar syntheses for some of the IOM constituents, in spite of the lack of identification of their carrier moieties.

On the other end, IOM isotopic, pyrolytic, and comparative NMR analyses give convincing evidence that at least a portion of this material can be and probably was modified during parent-body processes, resulting in the graphitization and chemical oxidation we observe and the possible release of soluble organic compounds. However, the latter processes must have been more complex than one directly comparable with laboratory IOM hydrous pyrolysis, which is conducted in the absence of minerals and any control of oxygen fugacity (*Shock and Schulte,* 1998). For example, it has been observed that simpler aromatic species are py-

TABLE 1. Soluble organic compounds in the Tagish Lake and Murchison meteorites.

Class	Tagish Lake[1] Concentration (ppm)	Tagish Lake[1] Compounds Identified[3]	Murchison[2] Concentration (ppm)	Murchison[2] Compounds Identified
Aliphatic hydrocarbons	5	12*	>35	140
Aromatic hydrocarbons	≥1	13	15–28	87
Polar hydrocarbons	n.d.	2*	<120	10*
Carboxylic acids	40.0	7	>300	48[4]*
Amino acids	<0.1	4	60	74*
Hydroxy acids	b.d.	0	15	7
Dicarboxylic acids	17.5	18*	>30	17[5]*
Dicarboximides	5.5	9	>50	2[6]
Pyridine carboxylic acids	7.5	7	>7	7[1]
Sulfonic acids	≥20	1	67	4[7]
Phosphonic acids	n.d.	n.d.	2	4[7]
N-heterocycles	n.d.	n.d.	7	31
Amines	<0.1	3	13	20[8]
Amides	<0.1	1	n.d.	27[6]*
Polyols	n.d.	n.d.	30	19[9]
Imino acids	n.d.	n.d.	n.d.	10[10]

*The compound group comprises significantly more molecular species than those unequivocally identified.

References: [1] *Pizzarello et al.* (2001); [2] *Cronin et al.* (1988) and references therein, unless otherwise indicated; [3] By reference standards or unequivocal matching of compound and library mass spectra; [4] *Huang et al.* (2005); [5] *Pizzarello and Huang* (2002); [6] *Cooper and Cronin* (1995); [7] *Cooper et al.* (1992); [8] *Pizzarello et al.* (1994); [9] *Cooper et al.* (2001); [10] *Lerner and Cooper* (2005).

rolytically released from the IOMs of more altered meteorites such as Orgueil (*Sephton et al.,* 1999); however, if the more open macromolecular structure implied by this release had been the effect of aqueous alteration, the liquid extracts of this meteorite should also show a larger abundance of soluble aromatics than less-altered meteorites, which does not seem to be the case. The Tagish Lake molecular distribution also speaks to a complex relation between IOM and soluble organics. This meteorite displays the largest amount of carbonates among CC, a lack of several soluble compound classes, e.g., most of the amino acids (see section 3.1.1 below), and higher levels of IOM oxidation relative to other meteorites, data that would point to long parent-body aqueous processes and possible oxidative degradation. Yet Tagish Lake solvent extracts display an unusual (compared to Murchison) abundance of nitriles and other cyano-species (*Gilmour et al.,* 2001; *Pizzarello et al.,* 2001), which would have been hydrolyzed by such water exposure. In the case of this meteorite, therefore, the overall data appear to indicate that at least some of the IOM and soluble aromatic hydrocarbons resulted from synthetic and/or accretion histories that were independent from each other.

The less labile of the insoluble carbon remains the most poorly characterized phase in the IOM. It contains "exotic" carbon of dissimilar ^{13}C content, such as nanodiamonds and silicon carbide ($\delta^{13}C \approx -38$ and >+60‰, respectively). Overall, this refractory material appears depleted in the heavy isotopes of carbon but retains δD values of 400‰ (*Kerridge et al.,* 1987), an isotopic content that would be compatible with cold nebular as well as presolar processes. The seemingly constant quantitative distribution of nanodiamonds to IOM, indirectly measured by *Alexander et al.* (1998) in several CC, would seem to be consistent with the latter possibility.

Although the relation between nanostructures and IOM is difficult to ascertain due to the lack of comparable isotopic data, the studies of the structured carbonaceous material may possibly reveal features lost in the overall analysis of the IOM. Their large compositional heterogeneity, when taken together with the isotopic inhomogeneity observed for the IOM, strongly suggest that the various phases of the insoluble carbon in meteorites are the complex end products of formative conditions, chemical regimes, and cosmic environments that varied extensively.

3. THE SOLUBLE METEORITIC ORGANIC COMPOUNDS

Soluble organics in CC comprise a varied and complex suite of compounds that span the range from large species, such as polycyclic aromatic hydrocarbons (PAHs) or long alkyl acids and diacids at the limit of solubility, to small molecules like methane and formic acid. Table 1 lists the organic composition of the Murchison and Tagish Lake meteorites, a CM2 and ungrouped chondrite, respectively, which are two of the most comprehensively studied CC and representative of the compositional variability among these meteorites. The isotopic composition of CC soluble organics varies as well and is given in Tables 3–6 and in Fig. 3. It should be noted that the carbonaceous chondrites studied

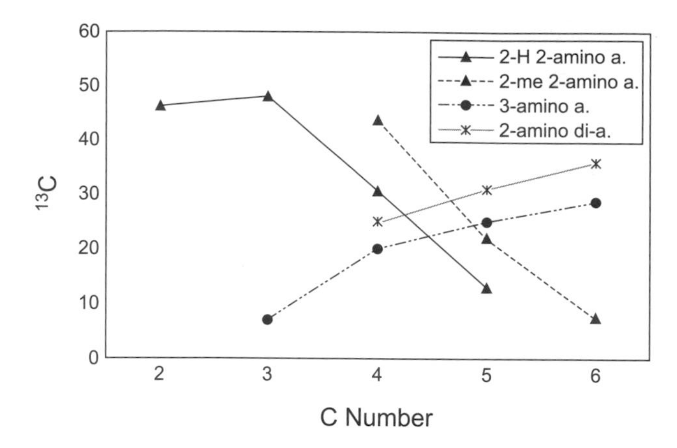

Fig. 3. Plot of $\delta^{13}C$ values vs. C number for 2-H-2-amino acids (solid line), 2-methyl-2-amino acids (dashed line), 3-amino acids (dash-dot line), and 2-amino dicarboxylic acids (gray line). Data points represent averages when more than one is available for given C# (see Table 3).

so far for their content of soluble organic compounds have been relatively few and that future falls, studies of the abundant Antarctica meteorite reservoir, and additional analyses of the known meteorites will likely expand these available data.

3.1. Amino Acids

Amino acids are amphoteric compounds (i.e., contain both the basic amino and acidic carboxyl functions) that are extensively distributed in the biosphere where they are represented primarily by 20 compounds as the monomeric components of proteins. Many other less-common amino acids are found in bacteria and molds or as metabolites in higher organisms. Terrestrial amino acids include both aliphatic and aromatic molecules, are almost exclusively of the L-configuration, and their chiral homogeneity (homochirality) appears to be indispensable to the organization and function of biopolymers. In meteorites, amino acids have been found in the water extracts of CI (*Lawless et al.,* 1972), CM falls (*Kvenvolden et al.,* 1970; *Raia,* 1966; *Cronin and Pizzarello,* 1983; *Komiya et al.,* 2003), Antarctica finds (*Shimoyama et al.,* 1979; *Cronin et al.,* 1979), CR (S. Pizzarello and J. R. Cronin, unpublished data, 1987), and the ungrouped Tagish Lake (*Pizzarello et al.,* 2001) meteorites, although with large compositional variability. When abundant, as in some CM chondrites, they comprise a diverse suite of over 80 aliphatic species of up to eight-carbon (C_8) alkyl chain length that include all possible molecular subgroups [generalized stuctures are shown in Scheme 1, where R is an alkyl group and (R) indicates alkyl or H]. This diverse and random distribution is one of the traits that differentiate meteoritic from terrestrial amino acids, the others being the lack of single chiral configuration (but see section 3.1.1 for discussion of certain anomalous D/L ratios); enrichment in the heavier isotopes of H, C, and N; and declining abundance with increasing chain length. The complex molecular and isotopic composition of meteoritic amino acid has posed many analytical challenges and, in spite of extensive studies of over 30 years, new information is being revealed by novel techniques even for the most analyzed meteorites.

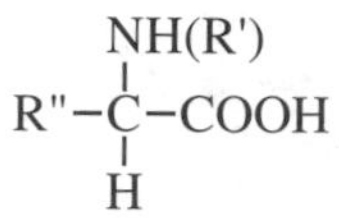

NH(R')
R"–C–COOH
H
(a)

α-H α-amino acids

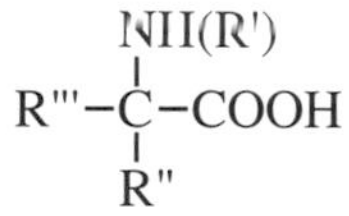

NH(R')
R'''–C–COOH
R"
(b)

α-alkyl α-amino acids

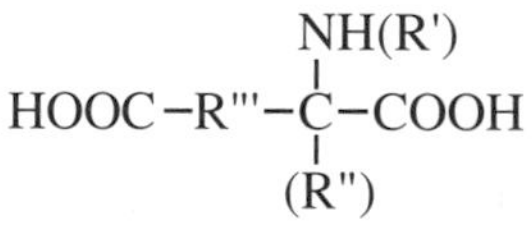

NH(R')
HOOC–R'''–C–COOH
(R")
(c)

α-amino alkanedioic acids

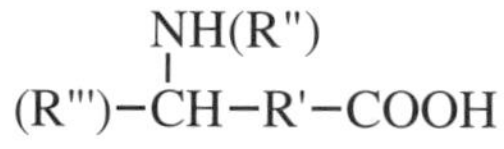

NH(R")
(R''')–CH–R'–COOH
(d)

β-, γ-, etc., amino acids

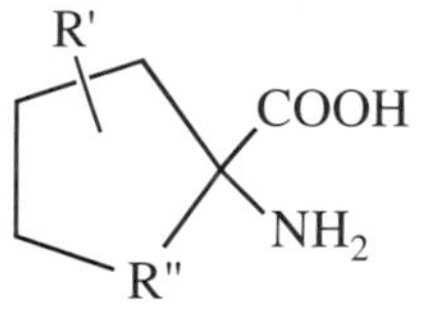

R'
COOH
NH_2
R"
(e)

α-amino cycloalkane acids

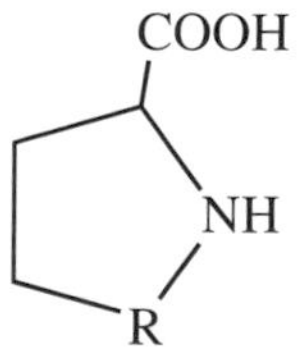

COOH
NH
R
(f)

Proline + homologs and isomers

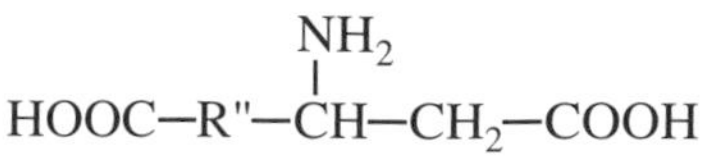

NH_2
HOOC–R"–CH–CH_2–COOH
(g)

β-amino alkendioic acids

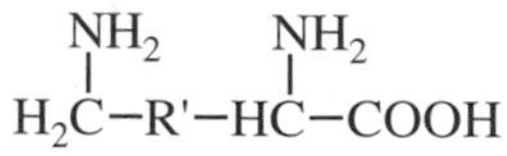

NH_2 NH_2
H_2C–R'–HC–COOH
(h)

diamino acids

Scheme 1

3.1.1. Molecular analyses. Molecular analyses of Murchison meteorite water extracts show that nearly all-possible isomeric and homologous species having the basic formulas (a)–(f) listed in Scheme 1 may be present up to the limit of solubility, approximately C_8, although identifications may vary between subgroups, depending upon the availability of reference standards. Diamino acids have been identified up to C_5 (*Meierhenrich et al.,* 2004) and only one β-aminodicarboxylic acid has been found so far (*Pizzarello et al.,* 2004). For α-amino acids (2-aminoalkanoic acids) that have been extensively characterized with the help of synthetic standards, all compounds up to C_7 have been identified (*Cronin and Pizzarello,* 1983, 1986). For the β-, γ-, and δ-amino acids [3-, 4-, 5-aminoalkanoic acids (*Cronin et al.,* 1985)] or α-amino dicarboxylic acids (*Cronin et al.,* 1993; *Pizzarello et al.,* 2004), complete identification is limited to C_5 and C_6 compounds, respectively. For the remaining subgroups, fewer compounds have been identified: N-methylglycine (sarcosine), N-methylalanine, and N-methyl-α-aminoisobutyric acid; cycloleucine and 1-amino-1-cyclohexane carboxylic acid; proline and pipecolic acid. Besides these positively identified compounds, longer chain amino acid homologs are clearly recognized within all subgroups on the basis of their mass-fragmentation spectra by gas chromatography-mass spectrometry (GC-MS) (e.g., *Cronin et al.,* 1982; *Pizzarello et al.,* 2004). Eight of the proteins amino acids — glycine, alanine, proline, valine, leucine, isoleucine, and aspartic and glutamic acids — are also found in meteorites.

The amino acids of these various isomeric subgroups show differences in their abundance as well as their chemical and/or physical binding in the meteorite. For example, the non-α-amino acids and diamino acids (*Meierhenrich et al.,* 2004) are released in some Murchison extracts only after acid hydrolysis of acid-labile derivatives. Similarly, α-amino α-H species require a higher extraction temperature and hydrolysis of the extract for full release, while the α-alkyl α-amino acids are easily extracted in water at room temperature (*Pizzarello and Cronin,* 2000). Allowing for variability between meteorites, the rough estimate is that the three subgroups of branched α-amino acids, non-α-amino acids, and α-amino dicarboxylic acids are represented approximately in equal amount in Murchison acid hydrolyzed extracts; the linear α-amino acids trail in abundance by about one-third. Cyclic α-amino acids are present in many isomeric forms up to at least C_8 and have consistently higher abundances than branched species of comparable chain length (S. Pizzarello, unpublished data, 2004). The N-substituted amino acids are also present in several isomeric forms and, within approximation, compare in amounts to the linear α-amino acids. Many compounds that release amino acids upon acid hydrolysis have been identified and are discussed in section 3.3.2. Peptides and amino acid dimers have been searched for in two CMs (the Yamato-791198 and Murchison meteorites) and were not found, with the only exception of the glycine dipeptide and diketopiperazine, which were observed in low amounts (*Shimoyama and Ogasawara,* 2002).

The distinction between linear and α-substituted α-amino acids (Scheme 1) is of particular interest and is based on the different molecular, isotopic, and chiral distribution found for the branched compounds in CM chondrites. Although it has been studied in depth for CM α-amino acids, isotopic data seem to indicate that a distinction between compounds based on alkyl chain branching may be justified for other amino acid subgroups as well (see section 3.1.3 below) (*Pizzarello and Huang,* 2005). The α-methyl α-amino acids are rare or unknown in the biosphere but abundant in CM chondrites such as Murchison and Murray where the first homolog of the series, α-amino isobutyric acid, is present at ~10 ppm and is often the most-abundant compound in the amino acid suite. Their abundance declines with chain length, as with other meteoritic compounds; however, their total abundance varies independently from that of α-H amino acids (*Cronin et al.,* 1993; *Pizzarello et al.,* 2003).

Uniquely, the chiral homologs of this amino acid series display L-enantiomeric excesses (ee) (*Cronin and Pizzarello,* 1997; *Pizzarello and Cronin,* 2000), whose magnitude varies within the meteorite and is largest, up to 15%, for isovaline (2-methyl-2-aminobutyric acid). The indigeneity of amino acid ee in CC is supported by compound-specific C- (*Pizzarello et al.,* 2003) (Table 2) and H- (*Pizzarello and Huang,* 2005) isotopic data obtained for Murchison and Murray isovaline enantiomers (Table 3). The origin of these ee is still unknown; the first hypothesis to be proposed was that of an asymmetric photolysis of meteorite organics by UV circularly polarized light (CPL) irradiation during their synthesis (*Rubenstein et al.,* 1983; *Cronin and Pizzarello,* 1997; *Engel and Macko,* 1997). This proposal appeared to set a preplanetary locale for the formation of meteoritic amino acids, or at least the α-methyl α-amino acids, since it is likely that an asymmetry-causing UV radiation would have been shielded in the meteorite parent bodies (*Pizzarello and Cronin,* 2000). Compound specific analyses (*Pizzarello et al.,* 2003), however, have revealed data that were unexpected in the context of this irradiation hypothesis and found that isovaline ee are sometimes larger than would be theoretically allowed (*Balavoine et al.,* 1974) if photolysis by UV CPL had been the only source of this amino-acid asymmetry (Table 2). This work also suggested a possible correla-

TABLE 2. $\delta^{13}C$ (‰ PDB) values of meteoritic L- and D-isovaline.

Sample	L-ival ee%	L-ival $\delta^{13}C$	D-ival $\delta^{13}C$
1. Murchison I*	13.2	+18.0	+16.0
2. Murchison I	6.0	+17.8	+11.5
3. Murchison E†	12.6	+21.9	+19.9
4. Murchison E	0	+17.5	+17.3
5. Murray I	6.0	+20.0	+18.8
6. Standard‡		–28.0	–31.9

*I = interior.

†E = exterior.

‡Combined DL-peak: –30.5‰.

From *Pizzarello et al.* (2003).

tion between the distribution of nonracemic amino acids and the abundance of hydrous silicates; however, the formation of primary asymmetric mineral phases and the possible influence they may have yielded on the molecular asymmetry of meteorite organics is still undefined and will require extensive new studies. It is possible that the α-H-amino acids were also affected by asymmetric processes but their ee were lost due to race-mization, to which substituted species such as α-methyl α-amino acids are resistant. Thus far, no unequivocal sign of this possibility has been found (*Engel and Macko,* 1997, 2001; *Pizzarello and Cronin,* 1998) and isotopic analyses of a Murchison sample have shown that the L-enantiomers of alanine, aspartic, and glutamic acids, each having terrestrial counterparts, all show significantly lower ^{13}C content (*Pizzarello et al.,* 2004), as would be expected by contributions from terrestrial L-amino acid contaminants.

An amino-acid composition with general similarity to that of Murchison has been described for the Murray meteorite (*Cronin and Moore,* 1971; *Pizzarello and Cronin,* 2000) and, by single analyses, for the Yamato-791198 (*Shimoyama et al.,* 1985), Santa Cruz, Mighei, Nogoya, and Cold Bokkeveld (*Cronin and Pizzarello,* 1983) CM meteorites. However, an exponential decay was also observed in the abundance of α-amino isobutyric acid with the terrestrial age of all these meteorites except Nogoya, suggesting that comparisons between data from meteorites with different fall ages may not be accurate (*Cronin and Pizzarello,* 1983). In the newly analyzed Sayama meteorite, whose fall dates to 1986, amino acid abundance was found to be orders of magnitude lower than in Murchison (*Komiya et al.,* 2003). Murray is the only meteorite other than Murchison for which chiral analyses of branched amino acids have been conducted, with results comparable to those of Murchison (*Pizzarello and Cronin,* 2000). The CI chondrites Orgueil and Ivuna show a distinctly different amino-acid composition (*Ehrenfreund et al.,* 2001) that is characterized by less overall abundance and diversity of compounds, where β-alanine is the largest component, and the α-methyl amino-acid series is represented by only the first homolog, α-aminoisobutyric, found in low abundance. In Tagish Lake amino acids are present in far lower amounts than in Murchison (by ~3 orders of magnitude) and are limited to three linear homologs: glycine, alanine, and α-amino-n-butyric acid (*Pizzarello et al.,* 2001).

3.1.2. Synthetic processes. The synthetic processes capable of producing amino acids as chemically diverse as those found in CC were, most likely, diverse as well. For α-amino acids, the current view is that they possibly formed during the parent-body aqueous alteration period by a Strecker-like reaction of precursor aldehydes and ketones, ammonia, and HCN (Scheme 2), as proposed by *Peltzer and Bada* (1978). The supporting evidence for this hypothesis has been the finding in the Murchison meteorite of comparable suites of α-amino and α-hydroxy linear acids (*Peltzer and Bada,* 1978) [although this correspondence is not valid for the α-methyl compounds (*Cronin et al.,* 1993)] and of imino acids (*Pizzarello and Cooper,* 2001; *Lerner and Cooper,* 2005). These are compounds where two carboxyl-containing alkyl chains are bonded at the same amino group and would likely result from a Strecker synthesis, e.g., when an amino-acid product becomes the reactant in place of ammonia.

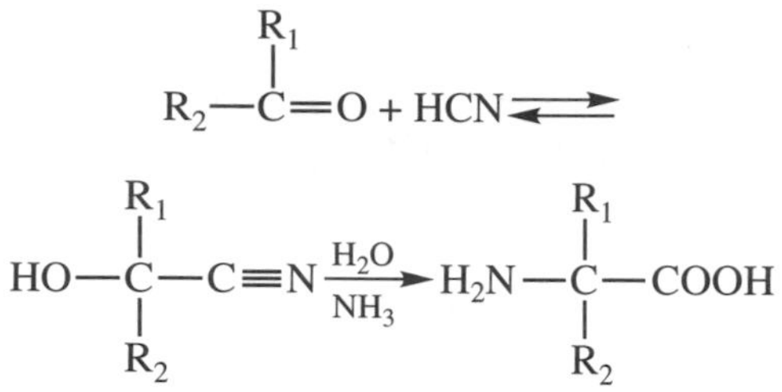

$$R_2{-}C(R_1){=}O + HCN \rightleftharpoons$$

$$HO{-}C(R_1)(R_2){-}C{\equiv}N \xrightarrow[NH_3]{H_2O} H_2N{-}C(R_1)(R_2){-}COOH$$

Scheme 2

This synthesis does not lend itself to the direct formation of asymmetric products, although it could be part of processes that involved asymmetric catalysts (see *Cronin and Reiss,* 2005, for a review). Also, the Strecker route is not allowed for non-α-amino acids (this reaction involves the addition of HCN to the carbonyl carbon, i.e., the α-carbon in the amino and/or hydroxy acid product) and other pathways need to be postulated for their syntheses. For β-amino acids, it has been proposed that they could derive from unsaturated nitriles by the addition of ammonia to the double bond(s), as from acrylonitrile in the case of β-alanine (*Cronin and Chang,* 1993, and references therein).

$$CH_2{=}CH{-}C{\equiv}N \xrightarrow{NH_3}$$

$$NH_2{-}CH_2{-}CH_2{-}C{\equiv}N \xrightarrow{H_2O} H_2N{-}CH_2{-}CH_2{-}COOH$$

Scheme 3

Although the addition of ammonia to a 3–4 double bond of unsaturated nitriles would be less favored (unsaturated aldehydes and ketones with conjugated double bonds are energetically favored), this process could still be invoked for the formation of γ- and δ-amino acids under favorable conditions, e.g., had their precursors evaded the energetic requirements through ion molecule reactions and subsequently avoided the expected isomerization to 2=3 species during an aqueous phase by rapid ammonia addition. A known source of at least some meteoritic γ-amino acids is the hydrolysis of four carbon lactams, such as butanelactam (Scheme 4), which are abundant in Murchison (*Cooper and Cronin,* 1995).

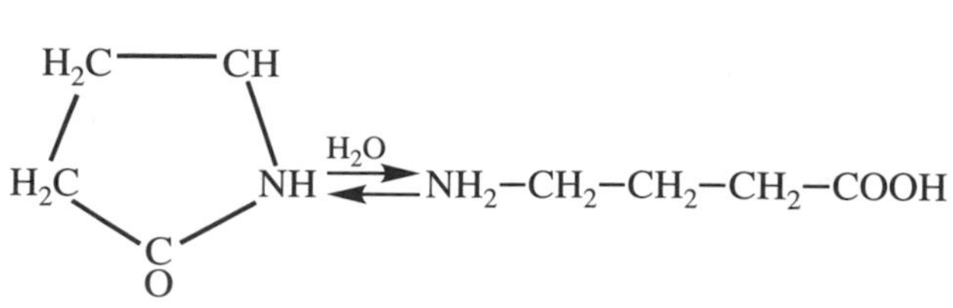

$$\text{(cyclic: } H_2C{-}CH{-}NH{-}C(O){-}H_2C\text{)} \underset{}{\overset{H_2O}{\rightleftharpoons}} NH_2{-}CH_2{-}CH_2{-}CH_2{-}COOH$$

Scheme 4

TABLE 3. Stable-isotope composition of individual amino acids in CM2 chondrites.

Amino Acid	Murray δD‰[1]	Murchison $\delta^{13}C$‰[2]	Murchison $\delta^{15}N$‰[3]
Glycine	399	41, 22[4]	37
D-alanine	614	52, 30[4]	60
L-alanine	510	38, 27[4]	57
D-2-aminobutyric acid	1633	29	
L-2-aminobutyric acid	—	28	
2-aminoisobutyric acid	3097	43, 5[4]	184
DL[1] isovaline	3181	22, 17[4]	66
L-isovaline	3282		
D-norvaline	1505	15	
D-valine	2432		
L-valine	2266		
L-isoleucine	1819		
D-allo isoleucine	2251		
L-allo isoleucine	2465		
DL 2-methylnorvaline	2986	7	
DL 2-amino2,3-methylbutyric a.	3604	22	
β-alanine	1063	5	61
4-aminobutyric a.	763	18	
DL 4-aminovaleric a.	965	29	
5-aminovaleric a.	552	24	
3-amino 2,2-methylpropanoic a.	2590	25	
3-amino-3-methylbutyric a.	1141		
DL 4-amino-2-methylbutyric a. +DL 4-amino-3-methylbutyric a.	1697		
D-glutamic a.	221	29, 33[5]	60
DL 2-aminoadipic a.	619	35	
DL 2-methylglutamic a.	1563	32	
DL threo 3-methylglutamic a.	1343	34	
DL proline	478	4	50
Cyclo leucine	850	24	
Sarcosine	1337	53	129
N-methylalanine	1267		
N-methyl-2-aminoisobutyric a.	3446		

[1] *Pizzarello and Huang* (2005), by GCHT-IRMS and listed by chemical subgroup; [2] *Pizzarello et al.* (2004), unless otherwise stated, by GC-C-IRMS; [3] *Engel and Macko* (1997), by GC-C-IRMS; [4] *Engel et al.* (1990), by GC-C-IRMS; [5] *Pizzarello and Cooper* (2001), by GC-C-IRMS.

Besides these synthetic pathways, the Tagish Lake distribution of low abundance and few linear amino acids may indicate that, in protracted aqueous processes of the parent body, some catalytic syntheses of amino acids could take place, e.g., by magnetite catalysis as observed with model experiments (*Pizzarello,* 2002a).

3.1.3. Isotopic analyses. Isotopic analyses of meteoritic amino acids have significantly sorted out the possible formative processes suggested by their molecular analyses. The first isotopic analyses to target these compounds were conducted by dual-inlet, dual-collector mass spectrometry of combined or partially separated groups of Murchison amino acids (*Epstein et al.,* 1987; *Pizzarello et al.,* 1991). They revealed an enrichment in D, ^{13}C, and ^{15}N with ‰ ranges of $\delta D = 1014–2448$, $\delta^{13}C = 23–44$, and $\delta^{15}N = 90$. The $\delta^{13}C$ values of six amino acids subsequently obtained by gas chromatography-combustion-isotope ratio mass spectrometry (GS-C-IRMS) (*Engel et al.,* 1990; *Pizzarello and Cooper,* 2001) were in general agreement with the previous data ranges while ^{15}N measurements of 12 amino acids expanded the $\delta^{15}N$ values to a range of 37–184‰ (*Engel and Macko,* 1997) (Table 3). In view of the large isotopic enrichment observed for interstellar molecules, these findings were interpreted by *Epstein et al.* (1987) to indicate a relationship between meteoritic amino acids and interstellar chemistry. More specifically, according to an interstellar/parent-body hypothesis (*Cronin and Chang,* 1993), they would result from syntheses of amino acids from highly D-enriched interstellar precursor [e.g., the aldehydes and ketones involved in Strecker reactions (*Lerner et al.,* 1993)] during the asteroidal parent-body water-alteration period.

Comprehensive C- and D-isotopic analyses of several individual amino acids by GC-C-IRMS have provided more details on the isotopic distribution of these compounds and offered additional understanding of their possible synthetic history. Murchison amino acids $\delta^{13}C$ values were found to range from 4.9 (norvaline) to 52.8‰ (sarcosine) (Table 3), with statistically significant differences observed both within and between amino acid subgroups (*Pizzarello et al.,* 2004). Within α-amino acids, the α-methyl compounds have larger $\delta^{13}C$ values than those having an α-H, and the ^{13}C content of homologs in both subgroups declines with increasing chain length (Fig. 3). This isotopic trend is remarkably similar to that observed by *Yuen et al.* (1984) for Murchison carboxylic acids and lower hydrocarbons. Both hypotheses postulated by those authors as to the possible cause of these trends, i.e., the growth of the compound alkyl chains by addition of single moieties and/or ^{13}C intramolecular differences between the acids' alkyl and carboxyl moieties (see section 3.4 below), could apply to meteoritic α-amino acid syntheses as well. For example, a Strecker reaction of interstellar precursors with HCN, water, and ammonia could account for the observed isotopic decline with increasing C-number, had the reactant HCN been heavier in ^{13}C than the precursor alkyl chains. Another possibility is ^{13}C fractionation in cold interstellar clouds; the reaction of ^{12}CO with C ions, such as $^{12}CO + ^{13}C^{+}$ to give $^{13}CO + ^{12}C^{+}$ (*Sandford et al.,* 2001), would result in C species other than CO and its protonated form HCO^{+} (*Charnley et al.,* 2004) to be depleted in ^{13}C. According to this scheme, the interstellar precursors (e.g., aldehydes and ketones) of higher amino acids would be expected to have a ^{13}C enrichment that decreases with increasing chain length.

Non-α-amino acid values do not show a decreasing trend in $\delta^{13}C$ from lower to higher homologs; in fact, those obtained for β-amino acids could show an increasing trend, if not for the possibility that the lower $\delta^{13}C$ value of β-alanine could be due to terrestrial contamination (Fig. 3). The data do not contradict the possibility of nitriles as precursors of non-α-amino acid; nitriles are observed as interstellar species of up to C_{11} (*Bell et al.,* 1992) and their long chains, were they formed through a synthetic pathway that did not involve CO (*Ehrenfreund and Charnley,* 2000) and enriched in ^{13}C, could have sustained little parent-body/nebular C-isotopic alteration and, thereby, account for the high $\delta^{13}C$ values of higher homologs in this subgroup of meteoritic amino acids. The $\delta^{13}C$ values obtained for the α-amino dicarboxylic acids are surprising when viewed in relation to meteoritic α-amino acid. These compounds have been considered likely products of a Strecker synthesis (*Cronin and Chang,* 1993) and are consistently found in meteorite water extracts containing α-amino acids, yet they do not appear to share the same isotopic trend of declining $\delta^{13}C$ with increasing C number. Instead, they resemble isotopically the dicarboxylic acids both in the magnitude of the $\delta^{13}C$ values and their invariance with chain length (see section 3.5).

Overall, the C-isotopic data for Murchison amino acids and the differences observed between the individual isomeric and/or homologous species reflect the molecular complexity of these compounds and support the hypothesis of their diverse syntheses. By contrast, the H-isotopic analyses of individual Murchison and Murray amino acids offer a surprisingly simple distribution (*Pizzarello and Huang,* 2005) (Table 3). In fact, while all the compounds analyzed show δD values that are higher than terrestrial and confirm the general D-enrichment observed by bulk samples, in some of these amino acids the enrichment is much higher and for 2-amino-2,3-methylbutyric acid reaches a δD value of 3604‰, the highest ever recorded for a discrete meteoritic compound. In contrast to their differences in ^{13}C composition, α-amino as well as non-α-amino acids and dicarboxylic amino acids display high D content as long as they have a branching methyl group. That is, the magnitude of amino-acid δD values in this meteorite appears to vary more with the structure of their C-chains than with their number and distribution of functionalities, differences in synthetic histories, or ^{13}C content.

Thus, the ^{13}C data indicate that C-chain elongation for meteoritic amino acid followed at least two synthetic paths, one that produced declining ^{13}C content with increasing carbon number and another where ^{13}C content remained unchanged, if not increased, with increasing chain length. At the same time, H-isotopic analyses show that the formation of amino acids in meteorites involved only two different H-fractionating chemical processes, with each being capable of producing compounds of all homologous series in spite of their ^{13}C differences.

We have to conclude, therefore, that each of these two H-fractionating processes comprised diverse synthetic pathways to amino acids and/or their precursors that encompassed both C-fractionating processes. If we assume that D enrichment indicates a synthetic relation to cold chemical regimes, then we have also to conclude that branched amino acids, or their precursors, were synthesized more abundantly and farther along in C-chain saturation with H and/or D in such regimes. Cold, preaccretionary environments may also have been the source of the chiral and molecular traits of these meteoritic amino acids by providing asymmetric influences toward enantiomeric excesses and an abundance of reactions involving radicals and ions for the syntheses of highly branched species.

For α-methyl α-amino acids, it is plausible that both H-fractionating processes may have occurred to a different extent, thus explaining the large variability of both their amount ratios to α-H amino acids and their ee within and between meteorites. In a less-D-enriching process, a Strecker-type synthesis was likely involved in the production of the α-H-α-amino acids and could have resulted in parent-body D-H exchanges during the aqueous phase as well as racemization, had the precursors of these amino acids been endowed with ee.

3.2. Pyridine Carboxylic Acids

Nicotinic acid and its isomers and homologs are amphoteric compounds that are separated with the amino acids during procedural preparations of meteorite water extracts. As with the amino acids, they have prebiotic significance;

nicotinic acid (niacin) occurs in the biosphere as part of the co-factor (NAD, NADP) of several enzymes that catalyze oxidation-reduction reactions in cell metabolism.

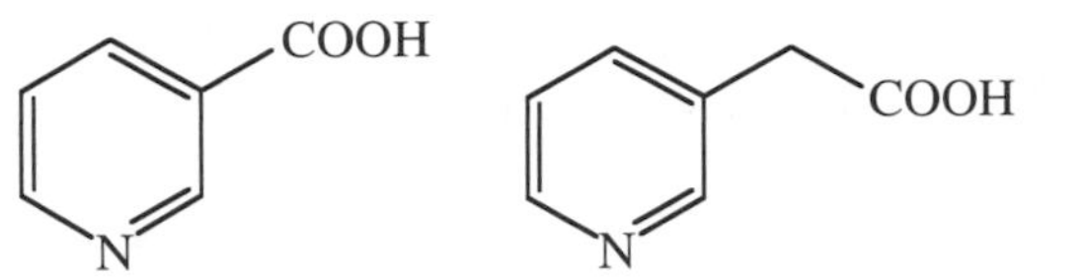

COOH COOH
N N

Nicotinic Acid Pyridine Acetic Acid

Scheme 5

Carboxylated pyridines have been described in the Tagish Lake and Murchison meteorites (*Pizzarello et al.,* 2001) where they occur as very similar suites that contain nicotinic acid, the most abundant species, its two isomers, and at least 12 methyl and dimethyl homologs. Hydrogen- and C-isotopic analyses of nicotinic acid gave the values of δD = 129‰ and $\delta^{13}C$ = 20‰, respectively (*Pizzarello and Huang,* 2005; *Pizzarello et al.,* 2004). Both these isotopic enrichments are lower than those of amino acids with comparable carbon number and, with respect to δD, those of heteroatom-containing aromatic compounds as a group (*Krishnamurthy et al.,* 1992).

The origin of these water-soluble compounds in meteorites is not clear; their approximately equal abundance and distribution in two meteorites that have a largely different content of amino acids would indicate that the synthetic environments of the two suites of compounds were not similar. This finding also hinders the attempt to find a comprehensive relation between the organic distribution and other compositional features of the Tagish Lake meteorite, for example, between this meteorite's protracted water phase, large abundance of carbonates, oxidative alteration of the IOM, and possible decomposition of soluble organics such as amino acids. In fact, it would be difficult to explain how a pyridine carboxylic acid suite exactly matching that of Murchison's could have survived unchanged the oxidative conditions capable of producing those features. The solution of this apparent contradiction, if found, could help in the overall understanding of meteorite organics formation.

3.3. Amines and Aliphatic Amides

3.3.1. Amines. Amines are an abundant group of water-soluble organics found in all meteorites where amino acids are present. In Murchison, they consist of a varied series of alkyl compounds of one- to at least five-carbon chain length (*Jungclaus et al.,* 1976; *Pizzarello et al.,* 1994) whose abundances are comparable to those of the amino acids and decrease with increasing molecular weight. Meteoritic amine preparations show isotopic enrichments in D, ^{13}C, and ^{15}N with bulk values of δD = 1221‰, $\delta^{13}C$ = 22.4‰, and $\delta^{15}N$ = 93.4‰. These values validate the compounds' indigeneity and, as an average, are comparable to those displayed by amino acids in the same meteorite; in particular, the N composition is very similar in both classes of compounds [90 vs. 94‰ (*Pizzarello et al.,* 1994)]. In view of these similarities it has been proposed that at least a fraction of meteoritic amines could have derived from amino acids via decarboxylation, which is a common pathway of thermal decomposition for α-amino acids. The molecular distribution of Murchison amines would be compatible with this suggestion, since 16 of the 20 amines identified in the meteorite could have derived from known Murchison α-amino acids.

Because *sec*-butyl amine could have formed via decarboxylation of isovaline, an amino acid found in Murchison with ee that may exceed 15%, its chiral distribution was analyzed in the Murchison meteorite (*Pizzarello,* 2002b). While one sample showed a fairly large L-ee (~20%), analysis of another stone did not show any significant ee. It is not yet clear if this reflects the same heterogeneity observed for isovaline in Murchison (*Pizzarello et al.,* 2003) or indicates that the first sample had carried terrestrial contamination, possibly by transfer of vapors during curatorial storage (this amine has only minor terrestrial distribution).

3.3.2. Alkyl amides. Alkyl amides are represented in the Murchison meteorite water extracts by an extensive series of linear and cyclic aliphatic compounds (*Cooper and Cronin,* 1995). These include amides that, upon acid hydrolysis, yield mono- and dicarboxylic acids, α- and non-α-amino acids and hydroxy acids, i.e., monocarboxylic acid amides, dicarboxylic acid monoamides, cyclic imides, carboxy lactams, N-acetyl amino acids, amino acid hydantoins, lactams, and hydroxy acid amides. With the possible exception of the hydantoins and N-acetyl amino acids, numerous isomers and homologs through at least C_8 were observed. Carboxy lactams, lactams, hydantoins, and N-acetyl amino acids are converted to amino acids by hydrolysis, thus these compounds qualitatively account for the earlier observation of acid-labile amino acid precursors in the meteorite extracts (*Cronin,* 1976).

One apparent possibility for the formation of at least some of the amides is the partial hydrolysis of known interstellar nitriles, e.g., CH_3CN, CH_3CH_2CN, etc. However, based on laboratory studies, *Cooper and Cronin* (1995) also proposed an unusual formation reaction involving the addition of cyanate (a known interstellar molecule) to amino acids or amino acid derivatives. The reactions of amino acids with cyanate and thiocyanate ions are well known (*Ware,* 1950) and there is at least a suggestive similarity in the distribution of dicarboxylic acids and their corresponding amides in Murchison, where the straight-chained C_4, C_5, and C_6 dicarboxylic monoamides are each more abundant than their branched isomers (*Cooper and Cronin,* 1995). A series of alkyl dicarboximides were also found in both the Tagish Lake and Murchison meteorites and have distributions that reflect those of the corresponding dicarboxylic acids (*Pizzarello and Huang,* 2002; *Cooper and Cronin,* 1995).

Ketene addition to amino acids (or amino acid derivatives) was invoked by *Cooper and Cronin* (1995) as a possible means of producing N-acetyl amino acids in meteorites. Ketene is a molecule detected in the interstellar medium and has been shown to be an effective acetylating agent for amino acids in aqueous solution. Given the well-known reactivity of molecules such as ketene and cyanate in aqueous solution, the authors suggested that amide formation occurred relatively late in the period of aqueous phase chemistry.

3.4. Carboxylic Acids

Carboxlyic acids have been analyzed in the Murchison (*Yuen and Kvenvolden,* 1974; *Lawless and Yuen,* 1979; *Huang et al.,* 2005), Murray (*Yuen and Kvenvolden,* 1974), Yamato (*Shimoyama et al.,* 1989), and Asuka (*Naraoka et al.,* 1996) CM2 chondrites, and in the Tagish Lake meteorite (*Pizzarello et al.,* 2001). In CM2, they are more abundant than the amino acids by factors of 1–2 and most of the possible isomeric species C_1 (*Kimball,* 1988; *Huang et al.,* 2005) to C_{10} (*Yuen and Lawless,* 1979; *Naraoka et al.,* 1996; *Huang et al.,* 2005) and linear species up to C_{11} (*Naraoka et al.,* 1996) have been identified. As with many groups of water-soluble compounds, several more branched acids can be clearly recognized by their mass spectra (*Naraoka et al.,* 1996; S. Pizzarello, unpublished data, 1998). Branched acids appear to be as abundant as linear compounds and all acids show declining abundance with increasing carbon number. By contrast, only the first two homologs of the linear acid series, formic and acetic, were observed in the Tagish Lake meteorite (*Pizzarello et al.,* 2001).

The H- and C-isotopic compositions of combined Murchison carboxylic acids have ranges of δD = 377‰ to 697‰ and $\delta^{13}C$ = –3‰ to +7‰, respectively (*Epstein et al.,* 1987; *Krishnamurthy et al.,* 1992). Carbon-isotopic analyses of individual C_2 to C_5 acids (*Yuen et al.,* 1984) showed that $\delta^{13}C$ values are lower for all acids than for meteorite CO_2, higher for branched than linear C_4 and C_5 acids, and decline with increased C number (Table 4). This isotopic trend would be consistent with syntheses of higher molecular weight compounds from lower homologs either under kinetically controlled conditions (*Yuen et al.,* 1984), such as possible in warmer interstellar or nebular regions, or by ion-molecule processes as described above in regard to meteoritic α-amino acids (*Pizzarello et al.,* 2004). Either process should have been independent of those yielding CO in the meteorite, given that CO displays a far lower $\delta^{13}C$ value (–32‰) than the carboxylic acids and thus was an unlikely source of these compounds (*Yuen et al.,* 1984).

TABLE 4. $\delta^{13}C$ (‰ PDB) of Murchison CO, CO_2, and alkyl monocarboxylic acids.

Compound	$\delta^{13}C$‰
Carbon dioxide	+29.1
Carbon monoxide	–32.0
Acetic acid	+22.7
Propionic acid	+17.4
Isobutyric acid	+16.9
Butyric acid	+11.0
Isovaleric acid	+8.0
Valeric acid	+4.5

From *Yuen et al.* (1984).

Yuen et al. (1984) also proposed that an intramolecular difference in ^{13}C content between the carboxyl and alkyl moieties could be responsible for the declining ^{13}C content of Murchison carboxylic acids. This, in turn, could result from primary synthetic processes or from secondary exchanges, e.g., during the parent-body aqueous phase. The second scenario is suggested by the theoretical work of *Shock* (1988). He proposed that under geologic conditions where high concentrations of aqueous carboxylic acids are in equilibrium with dissolved CO_2, the acids might become unstable toward decarboxylation. On the basis of this possibility, and by extending its implications to include a carboxyl carbon exchange with dissolved CO_2 from inorganic sources, *Diaz et al.* (2002) explained their finding of decreasing $\delta^{13}C$ with increasing chain-length in low molecular weight acids from six oil-producing rocks. Whether a similar exchange between carbonyl carbon and other heavier C, such as found in carbonates, might be invoked for aqueous processes in a meteorite parent body remains to be seen. Also, it cannot be excluded that primary and secondary effects could have both contributed, at unknown extent, to the meteoritic carboxylic acid isotopic composition.

Comprehensive deuterium analyses of several individual Murchison carboxylic acids (*Huang et al.,* 2005) have shown that the δD values of all branched acids in the meteorite are significantly and consistently higher than those of linear species (approximately 1770–2000‰ vs. 600–700‰). These findings are in agreement with those established for amino acids and, similarly, point to a distinct relation of the branched compounds to cold and H-fractionating processes that could have comprised diverse syntheses and C-fractionation pathways.

3.5. Aliphatic Dicarboxylic Acids

Aliphatic dicarboxylic acids are notable among meteoritic compounds in that they have been found in remarkably similar amounts and homolog-distribution patterns in meteorites that differ in their overall soluble organic content, such as Murchison, Tagish Lake, and Orgueil (*Pizzarello and Huang,* 2002), as well as in a Yamato CM2 find (*Shimoyama and Shigematsu,* 1994). In these meteorites, they comprise a homologous series of over 40 saturated and at least 6 unsaturated species of C_3 through C_{10}. Linear saturated acids are predominant; however, variability of the glutaric/methyl succinic ratio has been observed between fragments of a Tagish Lake stone that could indicate a different zonal distribution of linear and branched diacids in the meteorite similar to that observed for amino acids. As for other soluble meteoritic compounds, the diacids show decreasing abundance with increasing chain length and the most abundance, succinic acid, is present at approximately 40 nmol/g of meteorite (~4 ppm). Four chiral diacids, the 2-methyl and 2-ethyl

TABLE 5. Hydrogen- and C-isotopic composition of individual meteoritic alkyl dicarboxylic acids.

	Tagish Lake		Murchison		Orgueil	
Acid	δD‰*	δ^{13}C†	δD‰	δ^{13}C‰	δD‰	δ^{13}C‰
Succinic	+1121	+23	+390	+28		–24
Methylsuccinic	+1109	+15	+1225	+26		–20
Glutaric	+1355	+23	+795	+27		–20
2-methylglutaric	+1364	+19	+1551	+28		–10
3-methylglutaric		+13		+19		
Adipic		+6		+21		

*Relative to VSMOW.
† Relative to PVBD.

From *Pizzarello and Huang* (2002).

isomers of succinic and glutaric acids, were found to be racemic in the Yamato meteorite (*Shimoyama and Shigematsu,* 1994), as was 2-methyl glutaric in the Tagish Lake meteorite (*Pizzarello and Huang,* 2002). These findings, however, may not be relevant toward assessing the distribution of ee in diacids in meteorites, since all the compounds analyzed contain an H-containing carbon adjacent to the carboxyl group and would have the same opportunity to racemize in water, as would, for example, nonracemic α-H amino acids in the Murchison and Murray meteorites.

The C- and H-isotopic composition obtained for the individual dicarboxylic acids of three meteorites (*Pizzarello and Huang,* 2002) (Table 5) shows that all the compounds analyzed in the Murchison and Tagish Lake suites are enriched in the heavier isotopes of both elements while those of Orgueil are depleted in ^{13}C. Given that the dicarboxylic acids display a similar molecular composition in the three meteorites and that these compounds, with the exception of succinic and glutaric acids, are not widely distributed in the biosphere, the data reported for Orgueil should not be attributed solely to terrestrial contamination, even if it is not evident how the observed differences in isotopic composition were produced.

The isotopic data do not show any C- or D-enrichment trends with increasing chain length or the branching of their alkyl chains that would resemble those observed for some amino acids or low molecular weight carboxylic acids. The isotopic values are, however, in the range of those compounds and indicate at least a general relationship to interstellar precursors. Dinitriles and dicyanopolyynes have been proposed for this role (*Pizzarello et al.,* 2001); however, fully unsaturated precursors to alkyl dicarboxylic acids seem unlikely based on the molecular and isotopic data. These precursors would hydrolyze into only linear, odd C compounds and their hydrogenation in a parent body most probably would not have resulted in their high δD values. On the other hand, partially unsaturated nitriles, which like cyanopolyynes have been detected in the interstellar medium, may be considered as presolar precursor candidates to diacids (*Pizzarello and Huang,* 2002). The more direct dinitrile precursors could have derived from nitriles in interstellar and/or nebular and parent-body environments as, for example, by HCN addition (*Peltzer et al.,* 1984).

The singularity of diacid distributions in meteorites of otherwise different composition has been explained with the possibility that precursors to diacids and the majority of other organic compounds in meteorites had different volatilities and only the less volatile survived under parent-body conditions that allowed the more volatile to escape (*Pizzarello et al.,* 2001). Dinitriles, as precursors to dicarboxylic acids, would satisfy the requisite of not being easily sublimed, as could have been the more volatile aldehydes and ketones, precursors to the amino acids (e.g., succino nitrile's boiling point at 20 torr is 158°–160°C while that of butyraldehyde is –12°C).

The synthetic locale of meteoritic dicarboxylic acid is difficult to assess. Their isotopic values point to possible formation in cold interstellar media; however, the preponderance of linear species in the diacid suite, which contrasts with the highly branched distribution of most meteorite soluble organics, suggests the possibility of other preaccretionary processes (*Pizzarello and Huang,* 2002).

3.6. Sulfonic and Phosphonic Acids

These compounds have the general formula R-CH_2SO_3H and R-PO_3H (where R equals methyl-, ethyl-, propyl-, and butyl-groups) and were identified as homologous series in aqueous extracts of the Murchison meteorite (*Cooper et al.,* 1992). The sulfonic acids are roughly 40× more abundant than the phosphonic acids, 380 nmol/g for methanesulfonic acid (MSA) vs. 9 nmol/g for methylphosphonic acid. As with other homologous series of meteoritic organic compounds, these acids show a distribution of declining abundance with increasing carbon number and structural diversity, with the majority of possible structural isomers through C_4 being identified in the meteorite.

Cooper et al. (1997) performed C, H, and S stable-isotopic measurements on the C_1 to C_3 sulfonic acids. The δ^{13}C values ranged from +29.8‰ (MSA) to –0.9‰ (isopropanesulfonic) and δD values were from 483‰ (MSA) to 852‰ (isopropanesulfonic). Sulfur-isotopic measurements showed

MSA to be somewhat apart from the other sulfonic acids, with enrichments in ^{33}S (7.63‰), ^{34}S (11.27‰), and ^{36}S (22.5‰) being clearly above its higher homologs, which have average values 0.28‰ (^{33}S), 1.0‰ (^{34}S), and 1.9‰ (^{36}S). Bulk-isotope measurements of the phosphonic acids gave values of –20‰ for $\delta^{13}C$ and +219‰ for δD; these values were uncorrected for procedural blanks and thus are considered minimums.

Because S has four stable isotopes, one can also construct a three-isotope plot with Murchison sulfonic acid data to detect possible non-mass-dependent isotope effects. Again MSA proved to be unusual relative to the other acids: it displayed an anomalous non-mass-dependent enrichment in ^{33}S of $\Delta^{33}S$ = 2.0‰ ($\Delta^{33}S = \delta^{33}S - 0.5\ \delta^{34}S$ where 0.5 $\delta^{34}S$ is the normal mass-dependent value of $\delta^{33}S$). However, two of the other sulfonic acids did display anomalous depletions in ^{33}S: $\Delta^{33}S = -0.24$ (ethanesulfonic acid) and $\Delta^{33}S = -0.40$ (n-propanesulfonic acid). The authors indicated that extraterrestrial gas-phase irradiation of symmetrical precursors to the meteoritic sulfonic acids, e.g., CS_2 in the case of MSA, SCCS for ESA, etc., may have produced the anomalous $\Delta^{33}S$ values. *Cooper et al.* (1992) had speculated that the precursors to meteoritic sulfonic and phosphonic acids could have been the interstellar molecules CxS and CxP, respectively; the sulfonic acid isotopic analyses give additional indication for an interstellar and/or presolar formation of these compounds.

Cooper et al. (1997) further suggested that CS, ubiquitous in interstellar space, might have played a role in the production of meteoritic macromolecular material due to its known ability to form an insoluble solid phase on surfaces when condensed at very low temperatures (*Moltzen et al.,* 1988; *Klabunde et al.,* 1989). The unusual finding of just one sulfonic acid (MSA) in the Tagish Lake meteorite (*Pizzarello et al.,* 2001) may be consistent with these distinct isotopic properties of MSA relative to the other sulfonic acids. Phosphonic acids were not seen in Tagish Lake possibly due to abundances below the detection limit of GC-MS.

3.7. Alcohols and Carbonyls

Alcohols and carbonyls were identified in Murchison shortly after the meteorite's fall (*Jungclaus et al.,* 1976) but their study has not been revisited. C_4 (R-OH, R-CHO) to C_5 (R-CO-R) molecular species were identified that include all homologs to C_4 and show declining abundance with increasing C number.

3.8. Hydroxy and Keto Acids

Peltzer and Bada (1978) first reported a suite of α-hydroxy acids in the Murchison meteorite and identified seven acids up to C_5, with structures comparable to those of meteoritic amino acids. Enantiomeric analysis on four of these compounds (lactic acid, α-hydroxy-n-butyric acid, α-hydroxy-n-valeric acid, and α-hydroxyisovaleric acid) showed them to be racemic, or nearly so. Based on their abundances and distribution patterns, which are similar to those of Murchison α-H amino acids, the authors suggested that both groups of compounds could have been synthesized in the meteorite by the Strecker-cyanohydrin reaction (Scheme 1). *Cronin et al.* (1993) extended these analyses by the GC-MS characterization of a total of 60 hydroxy monocarboxylic acids up to C_8 (14 of these compounds were identified with reference standards) and of a suite of hydroxy dicarboxylic acids that included several species up to C_9. This study confirmed the correspondence previously observed between α-H-α-amino, and α-H-α-hydroxy acid distributions; however, it also showed that α-methyl- and non-α-hydroxy acids are far less abundant within the meteoritic hydroxy acid suite. Other molecular subgroups have not been searched for.

Murchison bulk-isotopic analyses (*Cronin et al.,* 1993) gave values of $\delta^{13}C$ = 4‰ and δD = 573‰ for hydroxy acids and $\delta^{13}C$ = –6‰ and δD = +357‰ for a combined sample of hydroxydicarboxylic and dicarboxylic acids. The average isotopic composition of hydroxy acids differs from those of amino acids in the same meteorite. The finding in Murchison of large D-enrichment in the individual amino acids in the subset of branched compounds (Table 3) and the low abundance of this subset of hydroxy acids may explain the observed difference in isotopic values. The molecular and isotopic findings for hydroxy acids also support the proposal of different venues of formation for linear and branched amino acids (section 3.1.3) and that the branched amino- and hydroxy acids may have been synthesized less abundantly via the Strecker reaction.

A series of keto acids have been identified in the Murchison and Murray meteorites by *Cooper et al.* (2005); it includes acetoacetic acid and its higher homologs having the keto group adjacent to the terminal C in the alkyl chain. This series shows the common meteoritic features of decreased abundance with increasing C number and extensive isomerism. Hydrolysis of interstellar nitriles, e.g., HCCN, HC_2CN, etc., and their reaction with ketene could be suggested as a possible formation route to these meteoritic compounds.

3.9. Polyols

Several polyhydroxylated organic compounds (polyols) have been discovered in water extracts of the Murchison and Murray meteorites by *Cooper et al.* (2001). These include one sugar and several compounds that are called sugar derivatives on the basis of their molecular structures, i.e., sugar alcohols, sugar mono-acids, sugar di-acids, and deoxy-sugar acids [Scheme 6, R=H or $(-CH_yOH)_x$; for deoxy-sugar acids, the structure shows just one possible placement of the "deoxy" carbon]. In general, these meteoritic sugar derivatives display the abiotic pattern of decreasing abundance with increasing carbon number within a class of compounds, many (if not all) possible isomers at each carbon number, and distributions that are similar in the Murchison and Murray meteorites.

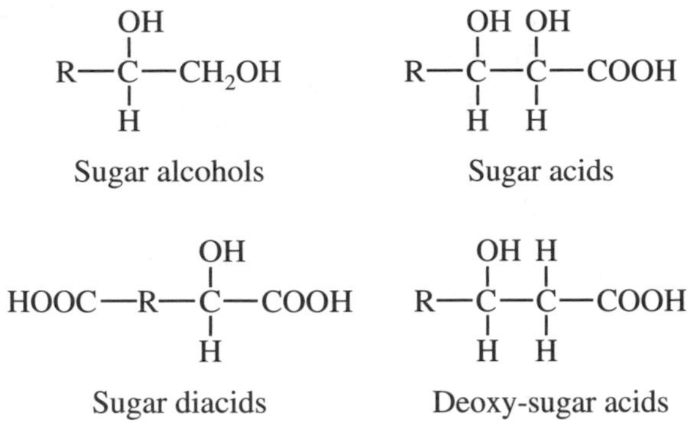

$$R-\underset{H}{\overset{OH}{C}}-CH_2OH$$

Sugar alcohols

$$R-\underset{H}{\overset{OH}{C}}-\underset{H}{\overset{OH}{C}}-COOH$$

Sugar acids

$$HOOC-R-\underset{H}{\overset{OH}{C}}-COOH$$

Sugar diacids

$$R-\underset{H}{\overset{OH}{C}}-\underset{H}{\overset{H}{C}}-COOH$$

Deoxy-sugar acids

Scheme 6

The sugar dihydroxyacetone was found in both Murchison and Murray. A search in Murchison water extracts for the C_2 analog of dihydroxyacetone, glycolaldehyde, was inconclusive and C_4 and C_5 sugars (e.g., erythrose, ribose, etc.) were not found. The sugar alcohol series begins with the C_2 ethylene glycol, and extends through at least the C_6 homologs. In multiple analyses ethylene glycol and glycerol are the most abundant of all identified polyols; glycerol is present in Murchison at ~160 nmol/g, i.e., roughly the same as the highest concentration found for an individual amino acid in the same meteorite. After acid hydrolysis and fractionation by ion exchange chromatography, bulk isotope values (‰) of Murchison sugar alcohols were $\delta^{13}C = -5.89$ and $\delta D = +119$ (*Cooper et al.,* 2001). These values were uncorrected for procedural blanks and are thus considered lower limits. Even so, they lie outside the range of typical terrestrial organic compounds and are consistent with values obtained for other indigenous classes of meteoritic organic compounds.

The sugar mono-acids ("aldonic" acids) comprise compounds from C_3 glyceric acid through at least the C_6 homologs. The abundance of glyceric acid, ~80 nmol/g, is comparable to those of the more-abundant amino acids in CM chondrites. Deoxy-sugar acids ("saccharinic" acids), less common in nature than sugars, were also present: four C_4, one C_5, and two C_6 compounds were identified; other unidentified compounds were possibly branched deoxy acids. Enantiomeric analyses of some of the acids (G. Cooper, unpublished data, 2005) have shown glyceric, 2-methylglyceric, and 2,4-dihydroxybutyric acids to have approximately equal abundances of their respective D- and L-isomers, also suggestive of extraterrestrial origins.

For some of the meteoritic polyols, *Cooper et al.* (2001) suggested low-temperature UV photolysis of interstellar grains. Under simulated interstellar conditions, photolysis has been shown to produce at least the smaller of the meteoritic polyols, e.g., ethylene glycol, glycerol, and glyceric acid, from simple starting compounds found in interstellar environments such as CO, methanol, etc. (*Agarwal et al.,* 1985; *McDonald et al.,* 1996; *Bernstein et al.,* 2002). In the case of the wider range of meteoritic polyols, *Cooper et al.* (2001) suggested the possibility that interstellar formaldehyde chemistry was a contributor to their synthesis. Formaldehyde (CH_2O) condenses to produce a variety of hydroxylated compounds, predominately sugars and sugar alcohols, in neutral to slightly basic aqueous solution, i.e., in a pH range consistently found for meteorite water extracts and also estimated for the meteorite parent body (*DuFresne and Anders,* 1962; *Bunch and Chang,* 1980; *Zolensky et al.,* 1989; *Schulte and Shock,* 2004). From sugars, the formation of sugar-acids is a relatively facile process in alkaline solution; the sugars would undergo various reactions including oxidation to acids of corresponding (or lower) carbon numbers, e.g., ribose to ribonic acid and fragmentation/rearrangement to deoxy sugar acids.

Extraterrestrial origins for these meteoritic polyols, as well as a direct chemical connection between interstellar environments and primitive solar system bodies, are further suggested by the finding of a relatively high abundance of ethylene glycol in Comet Hale-Bopp (*Crovisier et al.,* 2004) and the identification of interstellar ethylene glycol, glycolaldehyde (*Hollis et al.,* 2000, 2002), and diadroxyacetone (*Widicus-Weaver and Blake,* 2005) in Sagittarius B2(N-LMH). All the preceding authors pointed out that the production of these compounds would be consistent with models of interstellar grain chemistry, a possibility supported by the identification on these grains of phases of water, formaldehyde, and methanol (*Schutte et al.,* 1996).

3.10. Hydrocarbons

Hydrocarbons, as a class of compounds, were first identified in meteorites shortly after the first recorded falls in the 1800s (see *Mullie and Reisse,* 1987, for a historical review). At the time, their indigeneity was never in doubt. Later, however, their wide distribution in fossil fuel, soil humic acids, and many other terrestrial environments made the study of meteoritic hydrocarbons, and straight chain aliphatic compounds in particular, difficult and often controversial.

3.10.1. Aliphatic hydrocarbons. Aliphatic hydrocarbons are found in meteorites with a wide range of molecular sizes, from methane and its volatile homologs up to C_{30} nonvolatile molecules that must be extracted with solvents. Obtaining hydrocarbons of the first group is challenging, since most have been lost to evaporation and only a portion that is adsorbed, trapped, or otherwise bound in the matrix can be obtained. For example, *Yuen et al.* (1984) used freeze-thaw disaggregation of meteorite stones and subsequent treatment of the powders with sulfuric acid for the release of volatile hydrocarbons; Table 6 shows the C_1 to C_5 compounds found in this study and their C-isotopic composition.

The analyses of meteoritic nonvolatile hydrocarbons led at first to contradictory results. A Murchison study conducted shortly after the meteorite's fall (*Kvenvolden et al.,* 1970) described an aliphatic suite made up of a large number of isomeric alkanes, alkenes, and bi- and tricyclic saturated species that were poorly separated under the analytical conditions used. *Oró et al.* (1971) confirmed these

TABLE 6. $\delta^{13}C$ (‰ PDB) of Murchison lower molecular weight hydrocarbons.

Compound	$\delta^{13}C$‰
Methane	+9.2
Ethane	+3.7
Propane	+1.2
Isobutane	+4.4
Butane	+2.4
Ethene	−0.1
Benzene	−28.7

From *Yuen et al.* (1984).

results while *Studier et al.* (1972) identified predominantly linear aliphatic hydrocarbons in a Murchison sample, attributing their results to better analytical resolution.

The ensuing debate on the origin of these compounds has been extensively reviewed by others (e.g., *Cronin et al.,* 1988). Briefly, it centered on two models proposed for the formation of the Murchison organics. One hypothesis was that of a FTT catalytic synthesis of organics from simple gases (CO, H_2, NH_3) in a cooling nebula. The second envisioned organic syntheses in a reduced planetary atmosphere by the energetic production, recombination, and ultimate reaction of radicals in water. Both these early theories fell short of matching the molecular and/or isotopic distribution of organics later found in meteorites; particularly decisive was the finding by *Yuen et al.* (1984) that in the Murchison meteorite the CO $\delta^{13}C$ value was lighter than those of the lower alkanes. In fact, the kinetic model of C-fractionation for FTT syntheses of hydrocarbons from CO predicts the opposite, i.e., that residual CO would be isotopically heavier than the hydrocarbon products.

Molecular distribution studies that followed were able to show that the presence of some hydrocarbons was related to increasing contamination in meteorite samples (see section 3.10.4 below) (*Cronin and Pizzarello,* 1990), while recently *Kissin* (2003) argued that some of the alkanes in meteorites could derive from the slow thermal decomposition of meteoritic complex cross-linked polymers.

The long debate on the indigenous nature of the aliphatic hydrocarbon distribution in meteorites has been aided by the prompt analyses of the Tagish Lake meteorite after its fall. This meteorite suite comprises saturated and cyclic and/or unsaturated aliphatic hydrocarbons, where n-alkanes are predominant and C_{23} the most abundant (*Pizzarello et al.,* 2001). Twelve were individually identified, i.e., substantially fewer than were reported in the Murchison meteorite; however, their presence indicates that linear alkanes can be found in meteorites independently of terrestrial contamination. Whether the hydrocarbon suite of this ungrouped meteorite can be validly compared to those of CM2 chondrites, such as Murchison and Murray, or represents oxidative alteration products remains to be seen.

Deuterium- and C-isotopic analyses of combined aliphatic preparations from the Murchison meteorite gave average values of δD = 211‰ and $\delta^{13}C$ = 9‰ (*Krishnamurthy et al.,* 1992). The bulk $\delta^{13}C$ is higher than the weighted average value obtained by compound specific analyses of individual volatile hydrocarbons. *Sephton et al.* (2001) measured the ^{13}C content of individual n-alkanes (C_{12} to C_{26}) from six chondrites, Murchison, Orgueil, Cold Bokkeveld, Vigarano, Ornans, and Bishunpur. They found $\delta^{13}C$ values that fall within the relatively narrow range of approximately −25‰ to −39‰ and concluded that these compounds were likely terrestrial contaminants. However, *Pizzarello et al.* (2001) determined that the $\delta^{13}C$ values for the Tagish Lake n-alkane suite range from −18.8‰ (C_{16}) to −29.0‰ (C_{20}), with those for branched, unsaturated, and cyclic alkanes varying similarly from −18.4‰ to −27.1‰. Although the Tagish Lake data are also in the terrestrial range, several points appear to support their indigeneity. These are the pristine nature of the meteorite sample; the relatively low $\delta^{13}C$ values of other species in Tagish Lake, e.g., phenanthrene at −25‰, that possibly indicate a general depletion of ^{13}C for hydrocarbons; and the data of *Yuen at al.* (1984), showing that $\delta^{13}C$ values were already declining below 2.5‰ at C_4.

Overall, the C-isotopic data of Yuen et al. suggest that the formative processes for aliphatic hydrocarbons in CMs involved chain growth by addition of single moieties and were possibly kinetically controlled. Compound-specific deuterium data are sorely missing and would help to constrain their synthetic locales. If the D-range found by *Krishnamurthy et al.* (1992) for a combined sample were confirmed for individual compounds, their syntheses would have to be framed within cold environments like for other meteoritic compounds, at least at some point in their history. In this regard, it should be noted that the first two members of the n-alkane family, methane and ethane, have been identified in comets (e.g., *Mumma et al.,* 1996; *Kawakita et al.,* 2003), raising the possibility that comets could also contain higher homologs.

3.10.2. Aromatic hydrocarbons. Aromatic hydrocarbons have been analyzed in CC by solvent or supercritical fluid extraction (SFE) combined with GC-MS and by laser-desorption/laser-ionization mass spectrometry (L^2MS). The latter technique, by exploiting the volatility and susceptibility of PAHs to UV ionization without molecular fragmentation, targets aromatic compounds on the surface of meteorite stones, offers the advantage of determining their spatial distribution, and is the most sensitive method for these compounds. As a disadvantage, it lacks the possibility of isomeric differentiation and the more reliable quantification provided by GC-MS.

The most-abundant aromatic hydrocarbons obtained by solvent extraction of the Murchison, as well as other CM2 Antarctic meteorites, are pyrene and fluoranthene, followed by phenanthrene and acenaphthene. Other unsubstituted, alkyl substituted, and partially hydrogenated compounds make up a complex suite that comprises several ring forms and extends to polycyclic $C_{22}H_{12}$ isomers [e.g., benzo(ghi)perylene] (*Sephton,* 2002, and references therein). Extraction with supercritical fluid CO_2 (SFE), which reduces losses of more volatile compounds, demonstrates that lower molecular weight aromatic compounds such as toluene, alkyl ben-

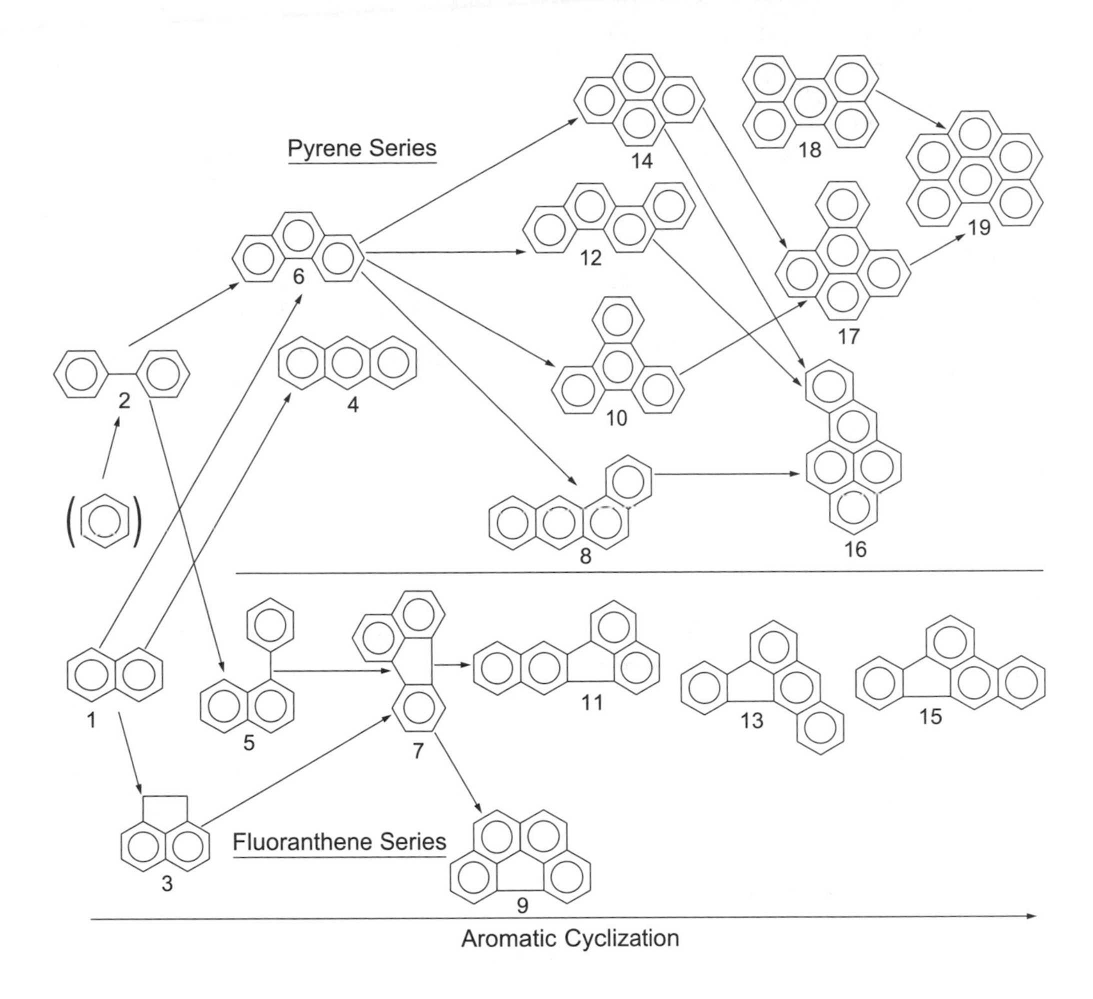

Fig. 4. Possible reaction pathways in the formation of meteoritic PAHs. (1) naphthalene, (2) biphenyl, (3) acenaphthene, (4) anthracene, (5) 1-phenyl naphthalene, (6) phenanthrene, (7) fluoranthene, (8) benz(a)anthracene, (9) benzo(ghi)fluoranthene, (10) triphenylene, (11,13,15) benzofluoranthenes, (12) crysene, (14) pyrene, (16,17) benzopyrenes, (18) perylene, (19) benzo(ghi)perylene. Re-drawn from *Naraoka et al.,* 2000.

zenes, and naphthalene can also be abundant in the Murchison meteorite (*Sephton et al.,* 1998). μL²MS analyses of CM (Murchison, Murray, Mighei, and Harripura) (*Hahn et al.,* 1988) and CV (Allende) (*Hahn et al.,* 1988; *Zenobi et al.,* 1989) meteorites gave results that, although showing differences in the relative distribution of individual polycyclic compounds at m/z higher than 178, were qualitatively similar to the meteorites' solvent extract. By contrast, μL²MS analysis of the Ivuna (CI) meteorite showed only three major masses for naphthalene, phenanthrene/anthracene, and pyrene/fluoranthene. Extraction with supercritical fluid CO_2 extracts of the CI Orgueil meteorite also showed a different distribution of only lower molecular weight species with a predominance of hydrated naphthalenes (*Sephton et al.,* 2001). In the Tagish Lake meteorite many lower molecular weight aromatic compounds, such as alkyl and di-alkyl benzenes, were observed directly in the solvent extracts (*Gilmour et al.,* 2001), a finding likely due to the prompt recovery of this meteorite. Naphthalene, alkyl- and phenylnaphthalenes, phenanthrene, anthracene, pyrene, and fluoranthene were also observed in the extracts (*Pizzarello et al.,* 2001, and unpublished data).

The average isotopic values obtained for preparations of Murchison aromatic hydrocarbons are δD = 353‰ and $\delta^{13}C$ = 5.5‰ (*Krishnamurthy et al.,* 1992). All C-isotopic values established for individual aromatic compounds in the meteorite have been lower than the above average; with the less-volatile larger aromatic clusters giving values from –5.9‰ for fluoranthene to –22.3‰ for benzopyrene (*Gilmour and Pillinger,* 1992, 1994). The only trend apparent in these data was that the addition of a ring to a given cluster structure appeared to decrease the ^{13}C content of the product compound, a finding that would be suggestive of condensation of larger ring clusters from lower homologs. *Naraoka et al.* (2000) confirmed and expanded this dataset by analysis of a CM2 Asuka Antarctic meteorite. The authors evaluated the molecular and isotopic distributions of the meteoritic PAHs to assess any indication of possible thermodynamic equilibration during their formation. They found none and observed instead a significant difference in C-isotopic composition between PAHs having very similar thermodynamic characteristics, such as pyrene and fluoranthene. In view of this, they proposed that the compounds could have formed through two reaction pathways. As shown in Fig. 4, higher ring structures for each PAH series would be produced from the lower homologs under kinetic control and with large isotopic fractionation for each cyclization step.

This C kinetic isotope effect proposed for meteoritic PAHs is not seen in their terrestrial counterparts, which are mostly produced at high temperature, and is suggestive of low-temperature synthetic processes in meteorites. However, the isotopic data available to date do not allow the unambiguous proposal of a single synthetic locale and it is plausible that the aromatic compounds we find in meteorites, both free and condensed in the IOM, have a diverse origin.

The $\delta^{13}C$ values of lower aromatic species have been found to vary from –28.8‰ to –12.6‰ for toluene and naphthalene, respectively (*Sephton et al.,* 1998), a range that is in agreement with the value of –28.7‰ reported for benzene by *Yuen et al.* (1984). The data appear to indicate opposite trends for the lower and higher ring number species; however, *Naraoka et al.* (2000) point to the similarities between these and terrestrial C-isotopic values and warn of a possible mixture of indigenous and contaminant volatile species in meteorites.

3.11. Polar Hydrocarbons

The heteroatom-containing hydrocarbons (i.e., having N, S, and O either in the aromatic ring or attached to it) represent the most abundant hydrocarbon component in the meteorites where it has been analyzed. In the Murchison meteorite polar hydrocarbons are 69–74% of total solvent extract (*Krishnamurthy et al.,* 1992) and include alkyl aryl ketones, e.g., anthracyl/phenanthryl ketones, aromatic ketones, and diketones, e.g., antracenone and anthracenedione, nitrogen-heterocyclic (section 3.10.2), and sulfur-heterocyclic compounds, such as dibenzothiophene. In the Tagish Lake meteorite many polar species have been identified, although some, e.g., the nitrogen heterocycles, could not be specifically searched for. The many identified include both smaller species, such as benzaldehyde, benzonitrile, phenol, di-cyano benzene, phenyl pyridine, isocyano naphthalene (*Gilmour et al.,* 2001), dicyano toluene, and larger compounds such as the ones shown in Scheme 6 (*Pizzarello et al.,* 2001, and unpublished data).

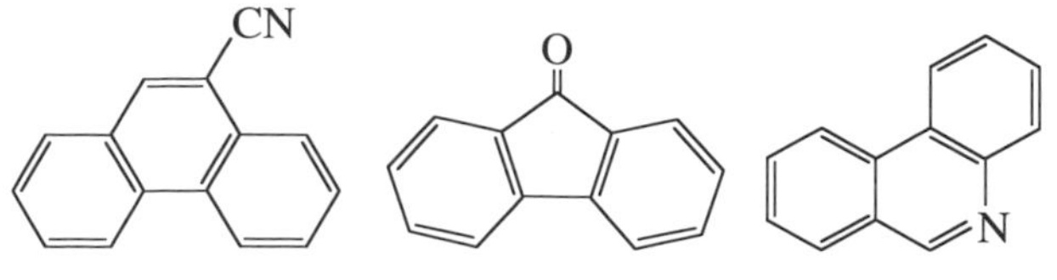

CN O N

Cyanophenanthrene Fluorenone Phenanthridine

Scheme 7

Murchison polar hydrocarbon bulk samples showed larger D- and ^{13}C-isotopic enrichments than the aromatic and aliphatic compounds, with average values of δD = 881‰ and $\delta^{13}C$ = 5.7‰ (*Krishnamurthy et al.,* 1992).

3.11.1. Nitrogen-containing heterocycles. Nitrogen-containing heterocycles are aromatic compounds having one or more N atoms replacing C in their rings. As a group of compounds, they have been systematically searched for and identified only in the Murchison meteorite. They comprise a series of pyridines, quinolines, isoquinolines, including several of their methyl isomers (*Stoks and Schwartz,* 1982), and benzoquinolines (Scheme 7) (*Krishnamurthy et al.,* 1992). Although isotopic analyses have not been performed, this diverse distribution is evidence of their abiotic origin. However, only one pyrimidine (uracil) and four purines (xanthine, hypoxanthine, guanine, and adenine) have been found in low amounts. Because they all have biological distribution, isotopic analyses are needed to prove their indigeneity.

Pyridine carboxylic acids have been described in both Murchison and the Tagish Lake meteorites (section 3.2). Aromatic amides were first described in the early work of *Hayatsu* (1964) and *Hayatsu et al.* (1968), who identified guanine, guanylurea, and melamine, among other compounds, in the Orgueil meteorite. In the Murchison meteorite, *Hayatsu et al.* (1975) found guanine, cyanuric acid, and guanylurea.

3.11.2. Aromatic dicarboximides. Aromatic dicarboximides (Scheme 8) have been identified in the Tagish Lake meteorite, where their indigeneity was corroborated by isotopic analyses (*Pizzarello et al.,* 2001).

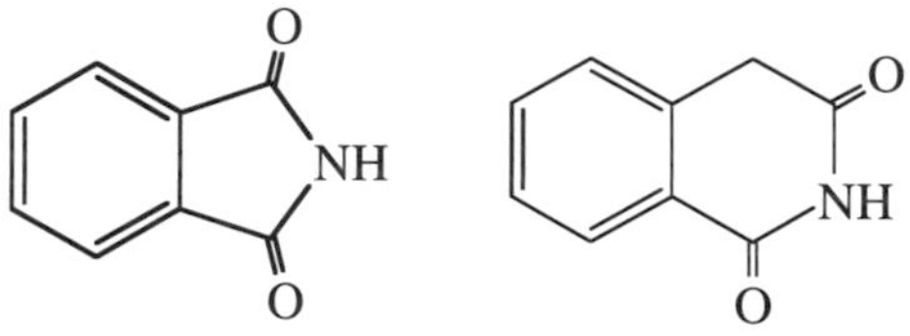

O NH O O NH O

Phthalimide Homophthalimide

Scheme 8

They comprise the imide derivatives of phthalic and homophthalic acids, and some of their methyl and dimethyl homologs; a total of nine have been detected. Phthalic acid, its isomers, and some of their ester derivatives and anhydrides are commonly used as industrial plasticizers and have been viewed before as possible contaminants. Phthalates were, in fact, observed in Murchison samples and attributed to contamination (*Cronin and Pizzarello,* 1990). However, the imide derivatives of these acids were found in a pristine Tagish Lake sample with $\delta^{13}C$ values that indicate a nonterrestrial origin (Table 7). It is plausible that the exclusive presence of these compounds in the Tagish Lake meteorite is related to some synthetic environments that were not shared by Murchison. The abundance of the carbonates in this meteorite (*Zolensky et al.,* 2002), as well as that of magnetite, are good indicators of a particularly long aqueous alteration

TABLE 7. $\delta^{13}C$ (‰ PDB) of aromatic dicarboximides in the Tagish Lake meteorite.

Compound	$\delta^{13}C$‰
Homophthalimide	1.5
Methylphthalimide	7.7
Dimethylphthalimide	5.9
Dimethylphthalimide	10.5
Dimethylphthalimide	17.5

From *Pizzarello et al.* (2001).

process that could be associated with the formation of aromatic dicarboxylic acids, for example, by oxidation of alkyl benzenes or substituted naphthalenes. Alkyl and aromatic diacid anhydrides and imides would have formed from the acids upon loss of water and in the presence of ammonia.

3.11.3. Amphiphilic material. Amphiphilic material, i.e., having both polar and hydrophobic moieties, has been extracted from Murchison powders by dissolution in chloroform. This material contains large, complex molecules and includes a fluorescent yellow pigment of unknown composition that can be observed directly in the matrix of cut meteorite stones (*Deamer,* 1985); infrared spectroscopy suggests it may contain aromatic hydrocarbon polymers as well as phenolic and carboxyl groups. The material assumes different structures in water depending on the pH, apparently from the titration of polar functions, and in alkaline solutions forms hydrated gels that assemble into membranous vesicles (see section 5).

4. INTERPLANETARY DUST PARTICLES

Interplanetary dust particles (IDPs), which are fragments of comets and asteroids ranging in size from ~5 μm to ~50 μm, have been collected from Earth's stratosphere by NASA aircraft since the mid-1970s (*Brownlee,* 1985). Due to their small size, the IDPs may radiate energy efficiently upon atmospheric entry, avoid severe heating during deceleration, and preserve any organic matter either incorporated into the dust at formation or produced on the parent body during later processes, e.g., aqueous activity. Each particle contains a number of "internal thermometers" such as moderately volatile elements (e.g., Zn and S), minerals with low thermal stability (e.g., hydrated minerals and sulfides), solar-wind-implanted noble gases (e.g., He, which is lost at 600° to 1200°C), and solar flare damage tracks. When available, all can be used to select IDPs that have experienced minimal entry heating and subject them to organic analysis (*Flynn,* 1989).

Anhydrous and hydrated IDPs are collected from the stratosphere in roughly equal abundances (*Fraundorf et al.,* 1982) and, for most, their major rock-forming element content is within a factor of 2 of that found for CI meteorites (*Schramm et al.,* 1989). Interplanetary dust particles are enriched in the moderately volatile elements and C by a factor of 2 to 3 over the CI meteorites (*Flynn et al.,* 1996), indicating that temperatures throughout and following their formation were low enough to condense and retain these elements. This chemical composition distinguishes the anhydrous group of IDPs from all known anhydrous meteorites, which are depleted relative to CI in the moderately volatile elements. Thus, the anhydrous IDPs appear to be less altered by thermal and aqueous processing, i.e., they are more primitive than any known meteorite. As a result, these particles have the highest likelihood of preserving the interstellar organic matter that was present in the solar nebula and of providing samples of the first organic matter that formed in our solar system as well.

Few organic analysis techniques have both the required sensitivity and the high spatial resolution necessary to study the organic matter in IDPs. Those that do include L^2MS, laser Raman, IR, and X-ray absorption near-edge structure (XANES) spectroscopy. However, each of these techniques is functional group specific, rather than providing characterization of the entire organic molecule. Thus, the information obtained from the organic matter in IDPs is different from that available for organic matter in carbonaceous meteorites.

The C of the IDPs varies in content between particles and is heterogeneously distributed within most IDPs. *Schramm et al.* (1989) analyzed 30 IDPs, using thin-window energy dispersive X-ray (EDX) fluorescence, and found a mean carbon abundance ~2.5× that of the CI carbonaceous meteorites. The porous IDPs, a group dominated by anhydrous particles, have almost twice the carbon content of the smooth IDPs, which are dominated by hydrated particles. *Thomas et al.* (1994) used the same technique on ~100 IDPs and found that the carbon content was highly variable, with a mean weight value of ~12% (i.e., ~3× CI). *Flynn et al.* (2000) confirmed the high carbon abundance in IDPs by mapping the C spatial distribution in ultramicrotomed thin-sections using a scanning transmission X-ray microscope (STXM).

Wopenka et al. (1988) performed Raman spectroscopy, a technique that is particularly sensitive to the structure of carbon, on 20 IDPs. The spectra were dominated by poorly crystallized carbon exhibiting a wide range of disorder but no specific signature of organic matter; only amorphous elemental carbon was detected. *Raynal et al.* (2001) obtained Raman spectra of six IDPs; two, dominated by Fe-sulfides, showed no Raman features, indicating they contained little or no carbon. The remaining four silicate IDPs contained carbonaceous material having a significantly different structure than that in the acid insoluble residues from carbonaceous meteorites, with the carbonaceous material in the IDPs being slightly more disordered than the meteorite residue but being "marginally similar to . . . immature type II kerogens," i.e., a H-rich (H/C between 1 and 1.3) type of organic matter that is insoluble in nonpolar solvents and nonoxidizing acids.

Clemett et al. (1993) provided the first significant evidence for the presence of indigenous organic matter in some

IDPs using μL²MS. Two of the eight IDPs they analyzed, Aurelian and Florianus, both of which also exhibit large D-enrichments detected by secondary-ion mass spectrometry, gave a series of mass peaks consistent with PAHs, while none of the remaining six IDPs gave detectable signatures of PAHs. The mass spectra of both Aurelian and Florianus show three major mass envelopes near 60, 250, and 370 amu. The lowest mass envelope contains inorganic species including Na, K, and Al as well as hydrocarbon fragments. The two higher envelopes contain signatures of PAHs and their alkylated derivatives separated by 14 amu (i.e., indicative of the successive addition of methylene groups to the PAH skeleton), plus a unique distribution of odd mass peaks in the range of 180–340 amu. This technique was also applied to the analysis of multiple fragments from a single "cluster IDP," L2008#5, which was so weak that it broke apart in hundreds of fragments upon hitting the collector surface (*Thomas et al.,* 1995). Analysis of the multiple fragments indicates this cluster particle contained ~7 wt% carbon, but comparison of μL²MS results from several fragments shows that carbon and organic matter are distributed inhomogeneously and that some of these fragments contain PAHs with masses in the 400–500 amu range. *Thomas et al.* (1995) suggest that these high-mass PAHs may either be organic compounds unique to the IDPs or they may be the product of polymerization of lower molecular weight species, possibly by thermal processing during atmospheric entry.

Swan et al. (1987) detected CH_2 and CH_3 stretching absorption consistent in energy with those of aliphatic hydrocarbons in four anhydrous and four hydrated IDPs using conventional globar-based FTIR. However, the authors could not eliminate the possibility that the weak features observed were due to contamination. *Flynn et al.* (2000, 2003), employing a synchrotron-based FTIR with about 100× the signal/noise of globar-based FTIR, were able to unambiguously characterize the organic matter in 11 anhydrous and 5 hydrated IDPs. They detected the CH_2 and CH_3 stretching absorptions near 2960, 2926, 2880, and 2854 cm^{-1} (Fig. 5) that are characteristic of aliphatic hydrocarbon and distinctly different from the signature of the silicone oil in which the IDPs are collected (*Flynn et al.,* 2000). *Raynal et al.* (2000) obtained similar results on a smaller group of IDPs; they also analyzed a Fe-sulfide IDP, but detected no organic matter (*Raynal et al.,* 2000).

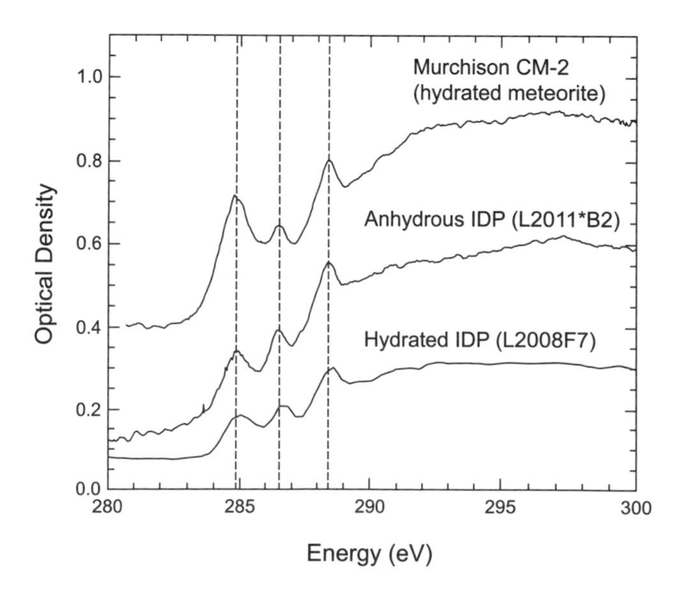

Fig. 5. Comparison of the C XANES spectra of a hydrated IDP, an anhydrous IDP, and the Murchison meteorite. Adapted from *Flynn et al.,* 2003.

The C-H stretching region in the infrared spectra of the anhydrous and the hydrated IDPs differ from the spectrum of Murchison acid insoluble organic residue in two significant ways. First, the aliphatic CH_3 to CH_2 absorption depth ratio is much larger in Murchison than in the IDPs, suggesting that the mean aliphatic chain length in the IDPs is longer than in Murchison (*Flynn et al.,* 2003). Second, the aromatic C-H stretches are easily detected in Murchison, as a broad feature centered near 3050 cm^{-1}, but this absorption is seen in only 1 of the 19 IDPs analyzed by *Flynn et al.* (2003), indicating that most IDPs have a considerably lower aromatic to aliphatic CH content than does the Murchison residue. Thus, we should conclude that in the IDPs the aromatic content is lower and/or the aromatic rings are less hydrogenated than in Murchison acid residues.

In most IDPs the higher-energy features from C- to O- and C- to N-bonding are obscured by the stronger absorption features from the major mineral phases, silicates, oxides, and carbonates. To eliminate the mineral features and reveal underlying organic features *Flynn et al.* (2002) performed FTIR analyses on IDPs that had been acid etched by a technique developed by *Brownlee et al.* (2000) to remove the major minerals. The acid etched IDPs showed several carbonyl (C=O) absorption features; the strongest at ~1705 cm^{-1} is characteristic of the carbonyl of an aliphatic ketone, but the anhydrous IDP also showed a significant absorption near ~1760 cm^{-1}, indicative of an ester carbonyl (*Flynn et al.,* 2002).

Flynn et al. (2003) and *Feser et al.* (2003) also analyzed the IDPs using C-, N-, and O-XANES spectroscopy, a sensitive indicator of functional groups. They found that the dominant type of carbonaceous material in most anhydrous and hydrated IDPs have similar C-XANES spectra (see Fig. 5). The IDP spectra exhibit two strong absorptions, one at ~285 eV attributed to C=C, which is found in both elemental carbon (graphite and amorphous carbon) and organic carbon (aromatic or olefinic); one at ~288.5 eV, which can result from either the C=O or C-H functional group; and a weaker absorption near 286.5 eV. *Flynn et al.* (2001, 2003) performed O-XANES spectroscopy on the same region of one of the IDPs, and detected a 532 eV absorption feature, consistent with C=O, confirming that a significant fraction of the 288.5 eV absorption in the IDPs is from the C=O functional group. The assignment of a specific functional group to the weaker 286.5 eV absorption in the IDPs has not yet been possible because both oxygen bonded to an aromatic ring and the C-N bond of an amide group give rise to a feature near that energy.

The ratio of the area of the C=C absorption feature to the C=O absorption feature is nearly identical in the anhydrous and the hydrated IDPs. This ratio is found to be higher in the Murchison acid insoluble organic residue, indicating that a higher proportion of the C in the IDPs is carbonyl compared to Murchison (*Flynn et al.,* 2003). A comparison of the height of the C-edge to the height of the O-edge provides a measure of the C/O ratio in the organic phase of the IDPs and Murchison. The one anhydrous IDP measured has a C:O ratio of ~2 (*Flynn et al.,* 2001), compared to ~8.3 reported by *Zinner* (1988) for a Murchison acid insoluble organic residue.

Nitrogen is present in IDPs at a much lower concentration than carbon and oxygen and is more difficult to detect and characterize. *Keller et al.* (1995) reported a mean C/N ratio of ~10 in several IDPs measured by electron energy loss spectroscopy (EELS). *Matrajt et al.* (2003) measured three IDPs using a nuclear microprobe that irradiates the sample with an intense beam of D ions and detects the products from $^{12}C(d,p_o)^{13}C$ and $_{14}N(d,p_o)^{15}N$ reactions, and found an average C/N ratio of ~22. Both N-XANES and N-EELS spectra show two broad pre-edge absorption features consistent with literature data for amides or amines (*Feser et al.,* 2003; *Keller et al.,* 1995). As mentioned previously, the two IDPs Aurelian and Florianus display several μL^2MS odd-mass peaks that could be explained by the presence of N-bearing heterocycles (*Clemett et al.,* 1993); by contrast, the spectra of the fragments from L2008#5 were dominated by even mass peaks (*Thomas et al.,* 1995).

Many of the IDPs, particularly the anhydrous cluster IDPs, have regions where the D/H ratio far exceeds that found in terrestrial materials, suggesting that these D-rich IDPs contain relatively well-preserved interstellar material (*Messenger,* 2000). *Aleon et al.* (2001) mapped the spatial distribution of H, D, $^{12,13}C$, $^{16,18}O$, ^{27}Al, and $^{28,29,30}Si$ in five IDPs. They concluded that water in these IDPs had an approximately chondritic D/H ratio, while some of the organic matter was D-rich. *Stephan et al.* (2003), using time of flight-secondary ion mass spectrometry (TOF-SIMS), also concluded that the host of the anomalous deuterium in IDPs is an organic phase. *Messenger* (2000) has mapped the spatial distribution of the deuterium using SIMS, and *Keller et al.* (2004) have prepared ultramicrotomed sections, preserving the spatial relationships, to examine the D-rich spots by C-XANES and FTIR spectroscopy. They found that the organic C, with a C-XANES spectrum showing the same three absorption peaks found in most other IDPs, was spatially associated with the D-rich spot, indicating that the D host is organic matter rather than water. Further, the infrared analysis showed a high aliphatic C-H content in two D-rich IDPs but a low aliphatic C-H content in a D-poor IDP, resulting in *Keller et al.* (2004) suggesting that the D host is aliphatic. However, *Mukhopadhyay and Nittler* (2003) have identified water as the D-host in one IDP.

Based on the similarity of the types and abundance of organic matter in the anhydrous IDPs and the hydrated IDPs, *Flynn et al.* (2003) suggest that the bulk of the prebiotic organic matter in the solar system formed prior to or concurrent with dust condensation in the solar nebula, and was not produced by aqueous processing on an asteroidal parent body.

5. METEORITE ORGANICS AND THE ORIGIN OF LIFE

Organic compounds, as their name originally was meant to indicate, have their largest molecular representation in the terrestrial biosphere. Here, their C is arranged in a multiplicity of chains and rings to which other atoms are attached, mainly H, but also O, N, S, and P. In living systems these compounds attain structures, functions, and selection possibilities that allow and sustain life. Not surprisingly, the study of organic compounds discovered in the solar system and beyond has often been framed in regard to origins of life, i.e., how abiotic organic compounds could possibly be linked to the development of planetary life, either here on Earth or elsewhere. Even before these discoveries, *Haldane* (1928) and *Oparin* (1938) independently proposed that the onset of terrestrial life was preceded by the abiotic formation of simple organic precursors and their subsequent development into increasingly complex molecules that ultimately yielded life. Carbonaceous chondrite meteorites provide an unparalleled record of such organic prebiotic chemistry in the early solar system, as found in a planetary setting and closest to the onset of life. Because the majority of known organic compounds in meteorites are also found in Earth's biosphere and an exogenous delivery of asteroidal material has undoubtedly showered Earth throughout its history (*Anders,* 1989; *Delsemme,* 1992; *Chyba and Sagan,* 1992), it is reasonable to consider the possibility that meteorites and comets could have provided the early Earth with life's precursor molecules.

As noted throughout this chapter, however, the overall abundances and composition of meteorite organics are distinguished by randomness and diversity that contrast with the structural specificity that characterizes terrestrial biomolecules. Also, as *Cronin and Chang* (1993) have pointedly remarked, an inventory of suitable organics ". . . is clearly not the recipe for constructing the progenote . . ." and a basic challenge resides in finding ". . . where diversity that characterizes chemical evolution . . . gives way to the selectivity of life". An intriguing exception to the organic diversity in CC is the presence in some meteorites of amino acids that share, to some degree, the trait of chiral asymmetry with biomolecules. These are the α-methyl-α-amino acids that display enantiomeric excesses that, if not very extensive, have the same sign (L-) as terrestrial amino acids. This finding has raised anew speculations that extraterrestrial prebiotic processes might have provided the early Earth with a "primed" inventory of essential organic molecules holding an advantage in molecular evolution.

The α-methylamino acids, which are generally unimportant in terrestrial biochemistry and that may not be viewed as relevant to the origin of life, are abundant in meteorites and could be well suited for such a role. First, their ee can be as high as 15%. Although the chemical pathways of early bioprecursors may have been largely different from those of contemporary metabolism, homochirality is essential to the structural organization of the biopolymers necessary for life. Therefore, we may assume chiral selection to have been an early evolutionary achievement. Second, α-substituted amino acids do not racemize, as α-H isomers do. This is a definite advantage in water on which, we have also to assume, early life was based. Third, α-methyl amino acids are known to have strong helix inducing and stabilizing effects (*Altman et al.,* 1988; *Formaggio et al.,* 1995) and we know that polymerization accompanied by formation of regular secondary structures can be an effective way to amplify modest initial ee (see *Cronin and Reisse,* 2005, for a review). It is conceivable, therefore, that meteoritic amino acids may have played a significant early role in the chemical evolution of homochirality and perhaps the biochemistry of a pre-RNA world.

The exogenous delivery of prebiotic material may also have contributed to molecular evolution in combination with early Earth endogenous processes. Amino acids are catalysts in reactions that could be pathways to biogenesis; for example, the condensation of simple aldehydes such as formaldehyde and glycolaldehyde to give sugar and amino acids (*Weber,* 2001). It has been shown that ee are catalytically transferred from both isovaline and alanine when tetroses are synthesized from glycolaldehyde in water, that such an effect is larger for isovaline than for alanine, and that it occurs at levels of amino acid ee comparable to those seen in meteorites (*Pizzarello and Weber,* 2004). Therefore, meteoritic amino acids could have provided the symmetry breaking necessary for the evolution of RNA by acting as catalysts or promoters of asymmetric syntheses leading to D-ribose.

Among the possible effects of exogenous delivery of carbonaceous material, the extraction from the Murchison meteorite of vesicle-forming molecules (*Deamer,* 1985) may hold a particular significance. For example, *Mauntner at al.* (1995) demonstrated that hydrothermal conditions release from the meteorite a mixture of amphiphilic components, longer chain carboxylic acids among others, that collaborate efficiently in giving water solutions a sustained increase of surface tension toward the formation of membranes and vesicles. The findings therefore raise the interesting prospect that meteoritic components could also have contributed self-assembly capabilities to the early Earth milieu (*Deamer et al.,* 2003).

In the larger context of the possible distribution of life around other stars, the finding in meteorites of a variety of amino acids with large D-enrichment would show that potential precursors to prebiotic molecules are already abundant and varied in cold interstellar environments. On the assumption that chemical evolution was relevant to the origin of terrestrial life, this abundance of biomolecule precursors in cosmic environments therefore implies that similar life would be facilitated elsewhere as well.

Acknowledgments. We are grateful to G. Cody, L. Garvie, and H. Naraoka for their contribution of figures, discussions, and suggestions. We also thank G. Cody, an anonymous referee, and associate editor J. Nuth for their careful reviews of this chapter.

REFERENCES

Agarwal V. K., Schutte W., Greenberg J. M, Ferris J. P., Briggs R., Connor S., Van De Bult C. P. E. M., and Baas F. (1985) Photochemical reactions in interstellar grains photolysis of CO, NH_3, and H_2O. *Origins of Life, 16,* 21–40.

Aikawa Y. and Herbst E. (2001) Two-dimensional distribution and column density of gaseous molecules in protoplanetary discs. *Astron. Astrophys., 371,* 1107–1117.

Aikawa Y., van Zadelhoff G. J., van Dishoeck E. F., and Herbst E. (2002) Warm molecular layers in protoplanetary discs. *Astron. Astrophys., 386,* 622–623.

Aikawa Y., Ohashi N., and Herbst E. (2003) Molecular evolution in collapsing prestellar cores II. The effect of grain-surface reactions. *Astrophys.. J., 593,* 906–924.

Aleon J., Engrand C., Robert F., and Chaussidon M. (2001) Clues to the origin of interplanetary dust particles from the isotopic study of their hydrogen-bearing phases. *Geochim. Cosmochim. Acta, 65,* 4399–4412.

Alexander C. M. O'D., Russel S. S., Arden J. W., Ash R. D., Grady M. M., and Pillinger C. T. (1998) The origin of chondritic macromolecular organic matter: A carbon and nitrogen isotope study. *Meteoritics & Planet. Sci., 33,* 603–622.

Altman E., Altman K. H., Nebel K., and Mutter M. (1988) Conformational studies on host-guest peptides containing chiral α-methyl-α-amino acids. *Intl. J. Protein Res., 32,* 344.

Amari S., Anders E., Virag A., and Zinner E. (1990) Interstellar graphite in meteorites. *Nature, 345,* 238–240.

Anders E. (1989) Pre-biotic organic matter from comets and asteroids. *Nature, 342,* 255–256.

Balavoine G., Moradpour A., and Kagan H. B. (1974) Preparation of chiral compounds with high optical purity by irradiation with circularly polarized light, a model reaction for the prebiotic generation of optical activity. *J. Am. Chem. Soc., 96,* 5152–5158.

Barduski E. L. and Nagy B. (1976) The polymer-like organic material in the Orgueil meteorite. *Geochim. Cosmochim. Acta, 40,* 1397–1406.

Becker L., Poreda R. J., and Bunch T. E. (2000) Fullerenes: An extraterrestrial carbon carrier phase for noble gases. *Proc. Natl. Acad. Sci., 97(7),* 2979–2983.

Bell M. B., Avery L. W., MacLeod J. M., and Matthews H. E. (1992) The excitation temperature of HC_9N in the circumstellar envelope. *Astrophys. J., 400,* 551–555.

Bernstein M. P., Dworkin J. P., Sandford S. A., Cooper G. W., and Allamandola L. J. (2002) Racemic amino acids from the ultraviolet photolysis of interstellar ice analogues. *Nature, 416,* 401–403.

Binet L., Gourier D., Derenne S., and Robert F. (2002) Heterogeneous distribution of paramagnetic radicals in insoluble organic matter from the Orgueil and Murchison meteorites. *Geochim. Cosmochim. Acta, 66,* 4177–4186.

Binet L., Gourier D., Derenne S., Robert F., and Ciofini I. (2004a) Occurrence of abundant diradicaloid moieties in the insoluble organic matter of the Orgueil and Murchison meteorites: A fingerprint of its extraterrestrial origin? *Geochim. Cosmochim. Acta, 68,* 881–891.

Binet L., Gourier D., Derenne S., Pizzarello S., and Becker L. (2004b) Diradicaloids in the insoluble organic matter from the Tagish Lake meteorite: Comparison with the Orgueil and Murchison meteorites. *Meteoritics & Planet. Sci., 39,* 1649–1654.

Blake D. F., Freund F., Krishnan K. F. M., Echer C. J., Shipp R., Bunch T. E., Tielens A. G., Lipari R. J., Hetherington C. J. D., and Chang S. (1988) The nature and origin of interstellar diamond. *Nature, 332,* 611–613.

Bockelée-Morvan D. and 11 colleagues (2001) Outgassing behavior and composition of comet C/1999 S4 (LINEAR) during its disruption. *Science, 292,* 1339–1343.

Boss A. P. (2002) Astrobiological implications of forming the solar system in an ultraviolet-rich environment. *Astrobiology, 4,* 523–524.

Boss A. P., Wetherill J. W., and Haghighipour N. (2002) Rapid formation of ice giant planets. *Icarus, 156,* 291–295.

Botta O. and Bada J. L. (2002) Extraterrestrial organic compounds in meteorites. *Surv. Geophys., 23,* 411–465.

Brinton K., Engrand C., Glavin D., Bada J., and Maurette M. (1998) A search for extraterrestrial amino acids in carbonaceous Antarctic micrometeorites. *Orig. Life Evol. Biosph., 28,* 413–424.

Brown P. G. and 21 colleagues (2000) The fall, recovery, orbit, and composition of the Tagish Lake meteorite: A new type of carbonaceous chondrite. *Science, 290,* 320–325.

Brownlee D. E. (1985) Cosmic dust — Collection and research. *Annu. Rev. Earth Planet. Sci., 13,* 147–173.

Brownlee D. E., Joswiak D. J., Bradley J. P., Gezo J. C., and Hill H. G. M. (2000) Spatially resolved acid dissolution of IDPs: The state of carbon and the abundance of diamonds in the dust (abstract). In *Lunar and Planetary Science XXXI,* Abstract #1921. Lunar and Planetary Institute, Houston (CD-ROM).

Bunch T. E. and Chang S. (1980) Carbonaceous chondrites — II. Carbonaceous chondrite phyllocilicates and light element geochemistry as indicators of parent body processes and surface conditions. *Geochim. Cosmochim. Acta, 44,* 1543–1577.

Charnley S. B., Ehrenfreund P., Millar T. J., Boogert A. C. A., Markwick A. J., Butner H. M., Ruiterkamp R., and Rodgers S. D. (2004) Observational tests for grain chemistry: Posterior isotopic labeling. *Mon. Not. R. Astron. Soc., 347,* 157–162.

Chyba C. F. and Sagan C. (1992) Endogenous production, exogenous delivery, and impact-shock synthesis of organic molecules: An inventory for the origins of life. *Nature, 355,* 125.

Clemett S. J., Maechling C. R., Zare R. N., Swan P. D., and Walker R. M. (1993) Identification of complex aromatic molecules in individual interplanetary dust particles. *Science, 262,* 721–725.

Cody G., Alexander C. M. O'D., and Tera F. (2002) Solid state (^{1}H and ^{13}C) NMR spectroscopy of the insoluble organic residue in the Murchison meteorite: A self-consistent quantitative analysis. *Geochim. Cosmochim. Acta, 66,* 1851–1865.

Cody G., Alexander C. M. O'D., and Tera F. (2005) NMR studies of chemical structural variation of insoluble organic matter from different carbonaceous chondrite groups. *Geochim. Cosmochim Acta, 69,* 1085–1097.

Cooper G. W. and Cronin J. R. (1995) Linear and cyclic aliphatic carboxamides of the Murchison meteorite: Hydrolyzable derivatives of amino acids and other carboxylic acids. *Geochim. Cosmochim. Acta, 59,* 1003–1015.

Cooper G. W., Onwo W. M., and Cronin J. R. (1992) Alkyl phosphonic acids and sulfonic acids in the Murchison meteorite. *Geochim. Cosmochim. Acta, 56,* 4109–4115.

Cooper G. W., Thiemens M. H., Jackson T., and Chang S. (1997) Sulfur and hydrogen isotope anomalies in meteoritic sulfonic acids. *Science, 277,* 1072–1074.

Cooper G., Kimmich N., Belisle W., Sarinana J., Brabham K., and Garrel L. (2001) Carbonaceous meteorites as a source of sugar-related organic compounds for the early Earth. *Nature, 414,* 879–883.

Cooper G., Dugas A., Byrd A., Chang P. M., and Washington N. (2005) Keto-acids in carbonaceous meteorites (abstract). In *Lunar and Planetary Science XXXVI,* Abstract #2381. Lunar and Planetary Institute, Houston (CD-ROM).

Courtin R., Wagener R., McKay C. P., Caldwell J., Fricke K. H., Raulin F., and Bruston P. (1991) UV spectroscopy of Titan's atmosphere, planetary organic chemistry and prebiological synthesis. II. Interpretation of new IUE observations in the 220–335 nm range. *Icarus, 90,* 43–56.

Cronin J. R. (1976) Acid-labile amino acid precursors in the Murchison meteorite. I. Chromatographic fractionation. *Origins of Life, 7,* 337–342.

Cronin J. R. and Chang S. (1993) Organic matter in meteorites: Molecular and isotopic analyses of the Murchison meteorite. In *The Chemistry of Life's Origins* (J. M. Greenberg *et al.,* eds.), pp. 209–258. Kluwer, Dordrecht.

Cronin J. R. and Moore C. B. (1971) Amino acid analyses of the Murchison, Murray, and Allende carbonaceous chondrites. *Science, 172,* 1327–1329.

Cronin J. R. and Pizzarello S. (1983) Amino acids in meteorites. *Adv. Space Res., 3,* 5–18.

Cronin J. R. and Pizzarello S. (1986) Amino acids of the Murchison meteorite. III. Seven carbon acyclic primary α-amino alkanoic acids. *Geochim. Cosmochim. Acta, 50,* 2419–2427.

Cronin J. R. and Pizzarello S. (1990) Aliphatic hydrocarbons of the Murchison meteorite. *Geochim. Cosmochim. Acta, 54,* 2859–2868.

Cronin J. R. and Pizzarello S. (1997) Enantiomeric excesses in meteoritic amino acids. *Science, 275,* 951–955.

Cronin J. R. and Reisse J. (2005) Chirality and the origin of homochirality. In *Lectures in Astrobiology, Vol. 1* (M. Gargaud, ed.), pp. 473–514. Springer-Verlag, Berlin.

Cronin J. R., Pizzarello S., and Moore C. B. (1979) Amino acids in an Antarctic carbonaceous chondrite. *Science, 206,* 335–337.

Cronin J. R., Yuen G. U., and Pizzarello S. (1982) Gas chromatographic-mass spectral analyses of the five-carbon β-, γ-, and δ-amino alkanoic acids. *Anal. Biochem., 124,* 139–149.

Cronin J. R., Pizzarello S., and Yuen G. U. (1985) Amino acids of the Murchison meteorite: II. Five carbon acyclic primary β-, γ-, and δ-amino alkanoic acids. *Geochim. Cosmochim. Acta, 49,* 2259–2265.

Cronin J. R., Pizzarello S., and Frye J. S. (1987) ^{13}C NMR spectroscopy of insoluble carbon in carbonaceous chondrites. *Geochim. Cosmochim. Acta, 51,* 299–303.

Cronin J. R., Pizzarello S., and Cruikshank D. P. (1988) Organic matter in carbonaceous chondrites, planetary satellites, asteroid, and comets. In *Meteorites and the Early Solar System* (J. F.

Kerridge and M. S. Matthews, eds.), pp. 819–857. Univ. of Arizona, Tucson.

Cronin J. R., Pizzarello S., Epstein S., and Krishnamurthy R. V. (1993) Molecular and isotopic analyses of the hydroxy acids, dicarboxylic acids and hydroxydicarboxylic acids of the Murchison meteorite. *Geochim. Cosmochim. Acta, 57,* 4745–4752.

Crovisier J., Bockelée-Morvan D., Biver N., Colom P., Despois D., and Lis D. C. (2004) Ethylene glycol in comet C/1995 O1 (Hale-Bopp). *Astron. Astrophys., 418,* L35–L38.

Deamer D. (1985) Boundary structures are formed by organic components of the Murchison carbonaceous chondrite. *Nature, 317,* 792–794.

Deamer D., Dworkin J. P., Sandford S. A., Bernstein M. P., and Allamandola L. J. (2003) The first cell membranes. *Astrobiology, 2,* 371–381.

Delsemme A. H. (1992) Cometary origin of carbon, nitrogen, and water on the Earth. *Origins of Life, 21,* 279–298.

Derenne S., Robert F., Binet L., Gourier D., Rouzaud J. N., and Largeau C. (2002) Use of combined spectroscopic and microscopic tools for deciphering the chemical structure and origin of the insoluble organic matter in the Orgueil and Murchison meteorites (abstract). In *Lunar and Planetary Science XXXIII,* Abstract #1182. Lunar and Planetary Institute, Houston (CD-ROM).

Diaz R. F., Freeman K. H., Lewan M. D., and Franks S. G. (2002) $\delta^{13}C$ of low-molecular-weight organic acids generated by hydrous pyrolysis of oil-prone source rocks. *Geochim. Cosmochim. Acta, 66,* 2755–2769.

DuFresne E. R. and Anders E. (1962) On the chemical evolution of the carbonaceous chondrites. *Geochim. Cosmochim. Acta, 26,* 1085–1114.

Ehrenfreund P. and Charnley S. B. (2000) Organic molecules in the interstellar medium, comets, and meteorites: A voyage from dark clouds to the early Earth. *Annu. Rev. Astron. Astrophys., 38,* 427–483.

Engel M. H. and Macko S. A. (1997) Isoptopic evidence for extraterrestrial non racemic amino acids in the Murchison meteorite. *Nature, 389,* 265.

Engel M. H. and Macko S. A. (2001) The stereochemistry of amino acids in the Murchison meteorite. *Precambrian Res., 106,* 35–45.

Engel M. H., Macko S. A., and Silfer J. A. (1990) Carbon isotope composition of individual amino acids in the Murchison meteorite. *Nature, 348,* 47.

Epstein S., Krishnamurthy R. V., Cronin J. R., Pizzarello S., and Yuen G. U. (1987) Unusual stable isotope ratios in amino acid and carboxylic acid extracts from the Murchison meteorite. *Nature, 326,* 477–479.

Ehrenfreund P., Glavin D., Botta O., Cooper G., and Bada J. (2001) Extraterrestrial amino acids in Orgueil and Ivuna: Tracing the parent body of Cl type carbonaceous chondrites. *Proc. Natl. Acad. Sci., 98,* 2138–2141.

Feser M., Wirick S., Flynn G. J., and Keller L. P. (2003) Combined carbon, nitrogen, and oxygen XANES spectroscopy on hydrated and anhydrous interplanetary dust particles (abstract). In *Lunar and Planetary Science XXXIV,* Abstract #1875. Lunar and Planetary Institute, Houston (CD-ROM).

Flynn G. J. (1989) Atmospheric entry heating: A criterion to distinguish between asteroidal and cometary sources of interplanetary dust. *Icarus, 77,* 287–310.

Flynn G. J., Bajt S., Sutton S. R., Zolensky M. E., Thomas K. L., and Keller L. P. (1996) The abundance pattern of elements having low nebular condensation temperatures in interplanetary dust particles: Evidence for a new chemical type of chondritic material. In *Physics, Chemistry, and Dynamics of Interplanetary Dust* (B. A. S. Gustafson and M. S. Hanner, eds.), pp. 291–296. ASP Conference Series 104, Astronomical Society of the Pacific, San Francisco.

Flynn G. J., Keller L. P., Jacobsen C., Wirick S., and Miller M. A. (2000) Organic Carbon in Interplanetary Dust Particles. In *Bioastronomy '99: A New Era in Bioastronomy* (G. A. Lemarchand and K. J. Meech, eds.), pp. 191–194. ASP Conference Series 213, Astronomical Society of the Pacific, San Francisco.

Flynn G. J., Feser M., Keller L. P., Jacobsen C., Wirick S., Avakians S. (2001) Carbon-XANES and oxygen-XANES measurements on interplanetary dust particles: A preliminary measurement of the C to O ratio in the organic matter in a cluster IDP (abstract). In *Lunar and Planetary Science XXXII,* Abstract #1603. Lunar and Planetary Institute, Houston (CD-ROM).

Flynn G. J., Keller L. P., Joswiak D., and Brownlee D. E. (2002) Infrared analysis of organic carbon in anhydrous and hydrated interplanetary dust particles: FTIR identification of carbonyl (C=O) in IDPs (abstract). In *Lunar and Planetary Science XXXIII,* Abstract #1320. Lunar and Planetary Institute, Houston (CD-ROM).

Flynn G. J., Keller L. P., Feser M., Wirick S., and Jacobsen C. (2003) The origin of organic matter in the solar system: Evidence from the interplanetary dust particles. *Geochim. Cosmochim. Acta, 67,* 4791–4806.

Formaggio F., Crisma M., Bonora G. M., Pantano M., Valle G., Toniolo C., Aubry A., Bayeul D., and Kamphuis J. (1995) R-isovaline homopeptides adopt the left-handed 3_{10}-helical structure. *Peptide Res., 8,* 6–14.

Fraundorf P., Walker R. M., and Brownlee D. E. (1982) Laboratory studies of interplanetary dust. In *Comets* (L. Wilkening, ed.), pp. 383–409. Univ. of Arizona, Tucson.

Gardinier A., Derenne S., Robert F., Behar F., Largeau C., and Maquet J. (2000) Solid state CP/MAS ^{13}C NMR of the insoluble organic matter of the Orgueil and Murchison meteorites: Quantitative study. *Earth Planet. Sci. Lett., 184,* 9–21.

Garvie L. A. J. and Buseck P. R. (2004a) Nanosized carbon-rich grains in carbonaceous chondrite meteorites. *Earth Planet. Sci. Lett., 224,* 431–439.

Garvie L. A. J. and Buseck P. R. (2004b) Nanoglobules, macromolecular material, and carbon sulfides in carbonaceous chondrite meteorites (abstract). In *Lunar and Planetary Science XXXV,* Abstract #1795. Lunar and Planetary Institute, Houston (CD-ROM).

Gilmour I. and Pillinger C. T. (1992) Isotopic differences between PAH isomers in Murchison. *Meteoritics, 27,* 224–225.

Gilmour I. and Pillinger C. T. (1994) Isotopic composition of individual polycyclic aromatic hydrocarbons from the Murchison meteorite. *Mon. Not. R. Astron. Soc., 269,* 235–240.

Gilmour I., Sephton M. A., and Pearson V. K. C. T. (2001) The Tagish Lake chondrite and the interstellar parent body hypothesis (abstract). In *Lunar and Planetary Science XXXII,* Abstract #1969. Lunar and Planetary Institute, Houston (CD-ROM).

Hahn J. H., Zenobi R., Bada J. F., and Zare R. N. (1988) Application of two step laser mass spectrometry to cosmogeochemistry: Direct analysis of meteorites. *Science, 239,* 1523–1525.

Haldane J. B. S. (1928) *Possible Worlds.* Hugh and Brothers, New York.

Hayatsu R. (1964) Orgueil meteorite: Organic nitrogen contents. *Science, 146,* 1291–1292.

Hayatsu R., Studier M. H., Oda A., Fuse K., Anders E. (1968)

Origin of organic matter in the early solar system — II. Nitrogen compounds. *Geochim. Cosmochim. Acta, 32,* 175–190.

Hayatsu R., Studier M. H., Moore L. P., and Anders E. (1975) Purines and triazines in the Murchison meteorite. *Geochim. Cosmochim. Acta, 39,* 471–488.

Hayatsu R., Matsuoka S., Scott R. G., Studier M. H., and Anders E. (1977) Origin of organic matter in the early solar system — VII. The organic polymer in carbonaceous chondrites. *Geochim. Cosmochim. Acta, 41,* 1325–1339.

Hayatsu R., Winans R. E., Scott R. G., McBeth R. L., Moore L. P., and Studier M. H. (1980) Phenolic ethers in the organic polymer of the Murchison meteorite. *Science, 207,* 1202–1204.

Hollis J. M., Lovas F. J., and Jewell P. R. (2000) Interstellar glycolaldehyde: The first sugar. *Astrophys. J. Lett., 540,* L107–L110.

Hollis J. M., Lovas F. J., Jewell P. R. and Coudert L. H. (2002) Interstellar antifreeze: Ethylene glycol. *Astrophys. J. Lett., 571,* L59–L62.

Huang Y., Wang Y., Alexandre M. R, Lee T., Rose-Petruck C., Fuller M., and Pizzarello S. (2005) Molecular and compound-specific isotopic characterization of monocarboxylic acids in carbonaceous meteorites. *Geochim. Cosmochim. Acta, 69,* 1073–1084.

Jungclaus G., Cronin J. R., Moore C. B. and Yuen G. U. (1976) Aliphatic amines in the Murchison meteorite. *Nature, 261,* 126–128.

Kawakita H., Watanabe J., Kinoshita D., Ishiguro M., and Nakamura R. (2003) Saturated hydrocarbons in Comet 153P/Ikeya-Zhang: Ethane, methane, and monodeuterio-methane. *Astrophys. J., 590,* 573–578.

Keller L. P., Thomas K. L., Bradley J. P. and McKay D. S. (1995) Nitrogen in interplanetary dust particles. *Meteoritics, 30,* 526–527.

Keller L. P., Messenger S., Flynn G. J., Clemett S, Wirick S., and Jacobsen C. (2004) The nature of molecular cloud material in interplanetary dust. *Geochim. Cosmochim. Acta, 68,* 2577–2589.

Kerridge J. F., Chang S., and Shipp R. (1987) Isotopic characterization of kerogen-like material in the Murchison carbonaceous chondrite. *Geochim. Cosmochim. Acta, 51,* 2527–2540.

Kimball B. A. (1988) Determination of formic acid in chondritic meteorites. M.S. thesis, Arizona State University, Tempe.

Kissel J. and Kruger F. R. (1987) The organic component in dust from comet Halley as measured by the PUMA mass spectrometer onboard Vega 1. *Nature, 326,* 755–760.

Kissin Y. V. (2003) Hydrocarbon components in carbonaceous meteorites. *Geochim. Cosmochim. Acta, 67,* 1723–1735.

Kitajima F., Nakamura T., Takaoka N., and Murae T. (2002) Evaluating the thermal metamorphism of CM chondrites by using the pyrolytic behaviour of carbonaceous macromolecular matter. *Geochim. Cosmochim. Acta, 66,* 163–172.

Klabunde K. J., Moltzen E., and Voska K. (1989) Carbon monosulfide chemistry. Reactivity and polymerization studies. *Phosphorus Sulfur Silica, 43,* 47–61.

Komiya M., Naraoka H., Mita H., and Shimoyama A. (2003) Amino acids and hydrocarbons in the Sayama and Antarctic Asuka carbonaceous chondrites (abstract). *Proc. NIPR Symp. Antarct. Meteorites,* p. 60. National Institute of Polar Research, Tokyo.

Kortenkamp S. J., Wetherill G. W., and Inaba S. (2001) Runaway growth of planetary embryos facilitated by massive bodies in a protoplanetary disk. *Science, 293,* 1127–1129.

Kress M. E. and Tielens A. G. G. M. (2001) The role of Fisher-Tropsch catalysis in solar nebula chemistry. *Meteoritics & Planet. Sci., 36,* 75–91.

Krishnamurthy R. V., Epstein S., Cronin J. R., Pizzarello S., and Yuen G. U. (1992) Isotopic and molecular analyses of hydrocarbons and monocarboxylic acids in the Murchison meteorite. *Geochim. Cosmochim. Acta, 56,* 4045–4058.

Kvenvolden K., Lawless J., Pering K., Peterson E., Flores J., Ponnamperuma C., Kaplan J. R., and Moore C. (1970) Evidence of extraterrestrial amino acids and hydrocarbons in the Murchison meteorite. *Nature, 228,* 923–926.

Jungclaus G., Cronin J. R., Moore C. B., and Yuen G. U. (1976) Aliphatic amines in the Murchison meteorite. *Nature, 261,* 126–128.

Lawless J. G. and Yuen G. (1979) Quantification of monocarboxylic acids in the Murchison carbonaceous meteorite. *Nature, 282,* 396–398.

Lawless J. G., Kvenvolden K. A., Peterson E., Ponnamperuma C., and Jarosewich E. (1972) Evidence for amino acids of extraterrestrial origin the Orgueil meteorite. *Nature, 236,* 66–67.

Lerner N. R. and Cooper G. W. (2005) Iminodicarboxylic acids in the Murchison meteorite: Evidence of Strecker reactions. *Geochim. Cosmochim. Acta, 69,* 2901–2906.

Lerner N. R., Peterson E., and Chang S. (1993) The Strecker synthesis as a source of amino acids in carbonaceous chondrites: Deuterium retention during synthesis. *Geochim. Cosmochim. Acta, 57,* 4713–4723.

Lewis R. S., Tang M., Wacker J. F, Anders E., and Steel E. (1987) Interstellar diamonds in meteorites. *Nature, 326,* 160–162.

Matrajt G., Flynn G. J., Bradley J., and Maurette M. (2001) FTIR and STXM detection of organic carbon in scoriaceous-type Antarctic micrometeorites (abstract). In *Lunar and Planetary Science XXXII,* Abstract #1336. Lunar and Planetary Institute, Houston (CD-ROM).

Matrajt G., Taylor S., Flynn G., Brownlee D., and Joswiak D. (2003) A nuclear microprobe study of the distribution and concentration of carbon and nitrogen in Murchison and Tagish Lake meteorites, Antarctic micrometeorites, and interplanetary dust. *Meteoritics & Planet. Sci., 38,* 1585–1600.

Mautner M. N., Leonard R. L., and Deamer D. W. (1995) Meteorite organics in planetary environments: hydrothermal release, surface activity, and microbial utilization. *Planet. Space Sci., 43,* 139–147.

McDonald G. D., Whited L. J., DeRuiter C., Khare B. N., Patnaik A., and Sagan C. (1996) Production and chemical analysis of cometary ice tholins. *Icarus, 122,* 107–117

Meierhenrich U. J., Muñoz Caro G. M., Bredehöft J. H., Jessberger E. K., and Thiemann W. H.-P. (2004) Identification of diamino acids in the Murchison meteorite. *Proc. Natl. Acad. Sci., 101,* 9182–9186.

Messenger S. (2000) Identification of molecular-cloud material in interplanetary dust. *Nature, 404,* 968–971.

Moltzen E., Klabunde K. J., and Senning A. (1988) Carbon monosulfide: A review. *Chem. Rev., 88,* 391–406.

Mullie F. and Reisse J. (1987) Organic matter in carbonaceous chondrites. In *Topics in Current Chemistry, Vol. 139,* pp. 85–117. Springer-Verlag, Berlin.

Mukhopadhyay S. and Nittler L. R. (2003) D-rich water in interplanetary dust particles (abstract). In *Lunar and Planetary Science XXXIV*, Abstract #1941. Lunar and Planetary Institute, Houston (CD-ROM).

Mumma M. J., DiSanti M. A., DelloRusso N., Fomenkova M., Magee-Sauer K., Kaminski C. D., and Xie D. X. (1996) Detection of abundant ethane and methane, along with carbon monoxide and water, in comet C/1996 B2 Hyakutaki: Evidence for interstellar origin. *Science, 272,* 1310–1314.

Mumma M. J., Dello Russo N., DiSanti M. A., Magee-Sauer K., Novak R. E., Brittain S., Rettig T., McLean I. S., Reuter D. C., and Xu Li-H. (2001) Organic composition of C/1999 S4 (LINEAR); a comet formed near Jupiter? *Science, 292,*1334.

Murae T., Kitajima F., and Masuda A. (1991) Pyrolytic nature of carbonaceous matter in carbonaceous chondrites and secondary metamorphism. *Proc. NIPR Symp. Antarct. Meteorites, 4,* 384–389.

Nagy B., Claus G., and Hennessy D. J. (1962) Organic particles embedded in minerals in the Orgueil and Ivuna carbonaceous chondrites. *Nature, 24,* 1129–1133.

Nakamura T., Zolensky M. E., Tomita S., Nakashima S., and Tomeoka K. (2002) Hollow organic globules in the Tagish Lake meteorite as possible product of primitive organic reactions. *Intl. J. Astrobiol., 1,* 179–189.

Naraoka H., Shimoyama A., and Harada K. (1996) Molecular distribution of monocarboxylic acids in Asuka carbonaceous chondrites from Antarctica. *Orig. Life Evol. Biosph., 29,* 187–201.

Naraoka H., Shimoyama A., and Harada K. (2000) Isotopic evidence from an Antarctica carbonaceous chondrite for two reaction pathways of extraterrestrial PAH formation. *Earth Planet. Sci. Lett., 184,* 1–7.

Oparin A. I. (1938) *The Origin of Life.* Reprinted by Dover, New York, 1952.

Oró J., Lichtenstein H., Wikstrom S., and Flory D. A. (1971) Amino acids, aliphatic, and aromatic hydrocarbons in the Murchison meteorite. *Nature, 230,* 105–106.

Peltzer E. T. and Bada J. L. (1978) α-hydroxycarboxylic acids in the Murchison meteorite. *Nature, 272,* 443–444.

Peltzer E. T., Bada J. L., Schlesinger G., and Miller S. L. (1984) The chemical conditions on the parent body of the Murchison meteorite: Some conclusions based on amino, hydroxy and dicarboxylic acids. *Adv. Space Res., 4,* 69–74.

Pflug H. D. (1984) Microvescicles in meteorites, a model of prebiotic evolution. *Naturwissensch., 71,* 531–533.

Pizzarello S. (2002a) Catalytic syntheses of amino acids: Significance for nebular and planetary chemistry (abstract). In *Lunar and Planetary Science XXXIII,* Abstract # 1236. Lunar and Planetary Institute, Houston (CD-ROM).

Pizzarello S. (2002b) The chiral amines of the Murchison meteorite: A preliminary characterization (abstract). In *Lunar and Planetary Science XXXIII,* Abstract #1233. Lunar and Planetary Institute, Houston (CD-ROM).

Pizzarello S. and Cronin J. R. (1998) On the reported non racemic Alanine from the Murchison meteorite — an analytical perspective. *Nature, 394,* 236.

Pizzarello S. and Cronin J. R. (2000) Non racemic amino acids in the Murchison and Murray meteorites. *Geochim. Cosmochim. Acta, 64,* 329–338.

Pizzarello S. and Cooper G. W. (2001) Molecular and chiral analyses of some protein amino acid derivatives in the Murchison and Murray meteorites. *Meteoritics & Planet. Sci., 36,* 897–909.

Pizzarello S. and Huang Y. (2002) Molecular and isotopic analyses of Tagish Lake alkyl dicarboxylic acids. *Meteoritics & Planet. Sci., 37,* 687–696.

Pizzarello S. and Huang Y. (2005) The deuterium enrichment of individual amino acids in carbonaceous meteorites: A case for the presolar distribution of biomolecules precursors. *Geochim. Cosmochim. Acta, 69,* 599–605.

Pizzarello S. and Weber A. L. (2004) Prebiotic amino acids as asymmetric catalysts. *Science, 303,* 1151.

Pizzarello S., Krishnamurthy R. V., Epstein S. and Cronin J. R. (1991) Isotopic analyses of amino acids from the Murchison meteorite. *Geochim. Cosmochim. Acta, 55,* 905–910.

Pizzarello S., Feng X., Epstein S., and Cronin J. R. (1994) Isotopic analyses of nitrogenous compounds from the Murchison meteorite: Ammonia, amines, amino acids, and polar hydrocarbons. *Geochim. Cosmochim. Acta, 58,* 5579–5587.

Pizzarello S., Huang Y., Becker L., Poreda R. J., Nieman R. A., Cooper G. and Williams M. (2001) The organic content of the Tagish Lake meteorite. *Science, 293,* 2236–2239.

Pizzarello S., Zolensky M., and Turk K. A. (2003) Non racemic isovaline in the Murchison meteorite: Chiral distribution and mineral association. *Geochim. Cosmochim. Acta, 67,* 1589–1595.

Pizzarello S., Huang Y., and Fuller M. (2004) The carbon isotopic distribution of Murchison amino acids. *Geochim. Cosmochim. Acta, 68,* 4963–4969.

Qi C. H, Kessler J. E., Koerner D. W., Sargent A. I., and Blake G. A. (2003) Continuum and CO/HCO^+ emission from the disk around the T Tauri star LkCa 15. *Astrophys. J., 597,* 986–997.

Raia J. C. (1966) An investigation of the carbonaceous material in the Orgueil, Murray, Mokoia, and Lance carbonaceous chondritic meteorites. Ph.D. thesis, University of Texas, Austin.

Raynal P. I., Quirico E., Borg J., Deboffle D., Dumas P., d'Hendecourt L., Bibring J.-P., and Langevin Y. (2000) Synchrotron infrared microscopy of micron-sized extraterrestrial grains. *Planet. Space Sci., 48,* 1329–1339.

Raynal P. I., Quirico E., Borg J., and D'Hendecourt L. (2001) Micro-Raman survey of the carbonaceous matter in stratospheric IDPs and carbonaceous chondrites (abstract). *Meteoritics & Planet. Sci., 36,* A171.

Rossignol-Strick E. S. and Barghoorn P. R. (1971) Extraterrestrial abiogenic organization of organic matter: Hollow spheres of the Orgueil meteorite. *Space Life Sci., 3,* 89–107.

Roueff E. and Gerin M. (2003) Deuterium in molecules of the interstellar medium. *Space Sci. Rev., 106,* 61–72.

Rubenstein E., Bonner W. A., Noyes H. P., and Brown G. S. (1983) Supernovae and life. *Nature, 306,* 118.

Sandford S. A., Bernstein M. P, and Dworkin J. P. (2001) Assessment of the interstellar processes leading to deuterium enrichment in meteoritic organics. *Meteoritics & Planet. Sci., 36,* 1117–1133.

Schramm L. S., Brownlee D. E., and Wheelock M. M. (1989) Major element composition of stratospheric micrometeorites. *Meteoritics, 24,* 99–112.

Schulte M. and Shock E. (2004) Coupled organic synthesis and mineral alteration on meteorite parent bodies. *Meteoritics & Planet. Sci., 39,* 1577–1590.

Schutte W. A., Gerakines P. A., Geballe T. R., van Dishoeck E. F., and Greenberg J. M. (1996) Discovery of solid formaldehyde towards the protostar GL 2136: Observations and laboratory simulation. *Astron. Astrophys., 309,* 633–647.

Sephton M. A. (2002) Organic compounds in carbonaceous meteorites. *Natl. Prod. Rep., 19,* 292–311.

Sephton M. A., Pillinger C. T., and Gilmour I. (1998) $\delta^{13}C$ of free and macromolecular aromatic structures in the Murchison meteorite. *Geochim. Cosmochim. Acta, 62,* 1821–1828.

Sephton M. A., Pillinger C. T., and Gilmour I. (1999) Small-scale hydrous pyrolysis of macromolecular material in meteorites. *Planet. Space Sci., 47,* 181–187.

Sephton M. A., Pillinger C. T., and Gilmour I. (2000) Aromatic moieties in meteoritic macromolecular material: Analyses by hydrous pyrolysis and $^{13}\delta C$ of individual compounds. *Geochim. Cosmochim. Acta, 64,* 321–328.

Sephton M. A., Pillinger C. T., and Gilmour I. (2001) Supercritical fluid extraction of the non-polar organic compounds in meteorites. *Planet. Space Sci., 49,* 101–106.

Sephton M. A., Verchovsky A. B., Bland P. A., Gilmour I, Grady M. M., and Wright I. P. (2003) Investigating the variations in carbon and nitrogen isotopes in carbonaceous chondrites. *Geochim. Cosmochim. Acta, 67,* 2093–2108.

Shimoyama A. and Ogasawara R. (2002) Dipeptides and diketopiperazines in the Yamato-791198 and Murchison carbonaceous chondrites. *Orig. Life Evol. Biosph., 32,* 165–179.

Shimoyama A. and Shigematsu R. (1994) Dicarboxylic-acids in the Murchison and Yamato-791198 carbonaceous chondrites. *Chem. Lett., 3,* 523–526.

Shimoyama A., Ponnamperuma C., and Yanai K. (1979) Amino acids in the Yamato carbonaceous chondrites from Antarctica. *Nature, 282,* 394.

Shimoyama A., Harada K., and Yanai K. (1985) Amino acids in the Yamato-791198 carbonaceous chondrite from Antarctica. *Chem. Lett., 1985,* 1183–1186.

Shimoyama A., Naraoka H., Komiya M., and Harada K. (1989) Analyses of carboxylic-acids and hydrocarbons in Antarctic carbonaceous chondrites, Yamato-74662 and Yamato-793321. *Geochem. J., 23,* 181–193.

Shimoyama A., Komiya M., and Harada K. (1991) Release of organic compounds from some Antarctic CI and CM chondrites by laboratory heating. *Proc. NIPR Symp. Antarct. Meteorites, 4,* 247–260.

Shock E. L. (1988) Organic acid metastability in sedimentary basins. *Geology, 16,* 886–890.

Shock E. L. and Schulte M. D. (1998) Organic synthesis during fluid mixing in hydrothermal systems. *J. Geophys. Res., 103,* 28513–28527.

Smith P. P. K. and Buseck P. R. (1981) Graphytic carbon in the Allende meteorite: A microstructural study. *Science, 212,* 322–324.

Stephan T., Leitner J., Floss C., and Stadermann F. J. (2003) TOF-SIMS analysis of isotopically anomalous phases in interplanetary dust and Renazzo (abstract). In *Lunar and Planetary Science XXXIV,* Abstract #1343. Lunar and Planetary Institute, Houston (CD-ROM).

Stoks P. G. and Schwartz A. W. (1982) Basic nitrogen heterocyclic compounds in the Murchison meteorite. *Geochim. Cosmochim. Acta, 46,* 309–315.

Studier M. H., Hayatsu R., and Anders E. (1972) Origin of organic matter in the early solar system — V. Further studies of meteoritic hydrocarbons and a discussion of their origin. *Geochim. Cosmochim. Acta, 36,* 189–215.

Swan P. D., Walker R. M., Wopenka B., and Freeman J. J. (1987) The 3.4 μm absorption in interplanetary dust particles: Evidence for indigenous hydrocarbons and a further link to Comet Halley. *Meteoritics, 22,* 510–511.

Tang M., Anders E., Hoppe P., and Zinner E. (1989) meteoritic silicon carbide and interstellar sources; implication for galactic chemical evolution. *Nature, 339,* 351–354.

Thomas K. L., Keller L. P., Blanford G. E., and McKay D. S. (1994) Quantitative analyses of carbon in anhydrous and hydrated interplanetary dust particles. In *Analysis of Interplanetary Dust* (M. E. Zolensky et al., eds.), pp. 165–174. AIP Conference Proceedings 310, American Institute of Physics, New York.

Thomas K. L. and 13 colleagues (1995) An asteroidal breccia: The anatomy of a cluster IDP. *Geochim. Cosmochim. Acta, 59,* 2797–2815.

Ware E. (1950) The chemistry of the hydantoins. *Chem. Rev., 46,* 403–470.

Weber A. L. (2001) The sugar model: Catalysis by amines and amino acid products. *Orig. Life Evol. Biosph., 31,* 71–86.

Widicus-Weaver S. L. and Blake G. A. (2005) 1,3-Dihydroxyacetone in Sgr B2(N-LMH): The first interstellar ketose. *Astrophys. J. Lett., 624,* L33–L36.

Wopenka B. (1988) Raman Observations on individual interplanetary dust particles. *Earth Planet. Sci. Lett., 88,* 221–231.

Yuen G. U. and Kvenvolden K. A. (1974) Monocarboxylic acids in Murray and Murchison carbonaceous meteorites. *Nature, 246,* 301–303.

Yuen G. U. and Lawless J. G. (1979) Quantification of monocarboxylic acids in the Murchison carbonaceous meteorite. *Nature, 282,* 396–398.

Yuen G., Blair N., Des Marais D. J., and Chang S. (1984) Carbon isotopic composition of individual, low molecular weight hydrocarbons and monocarboxylic acids from the Murchison meteorite. *Nature, 307,* 252–254.

Zenoby R., Philippoz J. M., Buseck P. R., and Zare R. N. (1989) Spatially resolved organic analysis of the Allende meteorite. *Science, 246,* 1026–1029.

Zinner E. (1988) Interstellar cloud material in meteorites. In *Meteorites and the Early Solar System* (J. F. Kerridge and M. S. Matthews, eds.), pp. 956–983. Univ. of Arizona, Tucson.

Zolensky M. E., Bourcier W. L., and Gooding J. L. (1989) Aqueous alteration on the hydrous asteroids: Results of EQ3/6 computer simulations. *Icarus, 78,* 411–425.

Zolensky M. E., Nakamura K., Gounelle M., Mikouchi T., Kasama T., Tachikawa O., Tonui E., and Lindstrom D. (2002) Mineralogy of Tagish Lake: An ungrouped Type 2 carbonaceous chondrite. *Meteoritics & Planet. Sci., 37,* 737–761.

Shock Effects in Meteorites

Thomas G. Sharp
Arizona State University

Paul S. DeCarli
SRI International

1. INTRODUCTION

Shock waves have played an important role in the history of virtually all meteorites. All models of the early solar system invoke condensation of mineral grains, aggregation of grains to form small bodies, and aggregation of most of the small bodies to form planets. The remaining small bodies, the asteroids, are accepted as the source of most meteorites. Throughout the history of the solar system, these small bodies have repeatedly collided with one another and with the planets. Since collisions produce shock waves in the colliding bodies, an understanding of shock wave effects is important to unraveling the impact history of the solar system as it is revealed in meteorites.

This chapter was originally intended as an update of the chapter by *Stöffler et al.* (1988). Rather than update that chapter, which relies heavily on laboratory shock-recovery experiments in the interpretation of shock effects in meteorites, we have decided to take a different approach. Here we emphasize the use of static high-pressure data on phase equilibria together with shock wave and thermal physics calculations to interpret observed microstructures of shocked meteorites. A number of papers published during the past ten years have shown that this approach can yield new insights on the impact history of meteorites (*Chen et al.,* 1996, 2004b; *Sharp et al.,* 1997; *Langenhorst and Poirier,* 2000a; *Xie and Sharp,* 2000, 2004; *Xie et al.,* 2002, 2005; *DeCarli et al.,* 2004; *Beck et al.,* 2004, 2005; *Ohtani et al.,* 2004). One requirement for use of this approach is that some knowledge of shock wave physics is required. Most general articles on shock wave physics do not present the information in a way that is useful to a reader who has been primarily trained in geology or mineralogy.

One of the objectives of this chapter is to present a useful tutorial on shock waves and shock wave calculations, including shock temperature and postshock temperature calculations. Our emphasis is on simple techniques and useful approximations rather than mathematical rigor. We will even attempt to present simple explanations of complicated phenomena, such as the quasichaotic nature of shock propagation in heterogeneous and/or porous materials. We also present examples to illustrate how the principles of shock wave and thermal physics may be used to interpret the history of naturally shocked materials and how the occurrences and formation mechanisms of high-pressure minerals in meteorites can be used to constrain shock pressures.

2. BACKGROUND

Shock wave effects on materials have been studied for centuries, ever since the invention of the cannon. Military engineers, concerned either with protecting or destroying structures (e.g., castles, forts, ships, armored vehicles), studied semiempirical correlation between impact damage and both target and projectile parameters, e.g., target dimensions and strength, impact velocity, projectile shape, projectile strength, etc. In a parallel effort, physicists were attempting to gain a good theoretical understanding of shock wave propagation in solids. The theoretical approach begins with simplification of the problem, e.g., the assumption that the material is homogeneous and can be treated as a continuum. Additional simplifications, e.g., assuming one-dimensional planar flow and neglecting material strength, lead to a theory of shock wave propagation based on Newton's laws of motion and on classical thermodynamics. The continuum theory and its historical development are covered in *Courant and Friedrichs* (1948). However, those of us who use high-resolution tools to study shock wave effects in materials may find it difficult to reconcile our observations of localized shock effects with continuum theories.

Shock propagation in a heterogeneous material, such as a meteorite, is an extraordinarily complex process when examined on the submicrometer scale of our observational tools. Because the shock properties of meteoritic minerals span a wide range, the initial nanoseconds of shock propagation through the meteorite appear chaotic. The initial peak pressure in the shock front can vary by as much as an order of magnitude from grain to grain or even within a single mineral grain, depending on details of the local environment. One consequence of that chaotic pressure distribution is that the shock temperature distribution is also chaotic. As we will show later, the shock temperature can also vary by as much as an order of magnitude between grains and even within a single grain. Pressure equilibration is achieved rapidly, on the timescale of the elastic wave transit time through the larger mineral grains. For a typical chondrite, having millimeter sized grains, pressure equilibration will be complete within about a microsecond after arrival of the initial shock. In contrast, millimeter-scale temperature heterogeneities equilibrate on a timescale of seconds.

The studies by Kieffer and co-workers (*Kieffer,* 1971; *Kieffer et al.,* 1976) of shocked Coconino sandstone from Meteor Crater, Arizona, detail the complexity of shock interactions on a submicrometer scale. They observed the intermixture of quartz, coesite, and stishovite within small areas and noted that no equilibrium P–T–V conditions exist where all three phases are simultaneously stable. They state: "The coexistence of these phases across regions of less than a thousand Angstroms in diameter indicates great variations of pressure and temperature locally within the shock, or nonequilibrium reaction conditions, or both."

For the most part, shock wave researchers have been able to ignore the complications of initial shock pressure equilibration in heterogeneous materials by using measurement techniques that average over the local fluctuations. This approach of using a continuum model to study shock wave propagation in porous and/or heterogeneous materials is generally satisfactory when one is concerned with centimeter-scale phenomena. Thus, the term "peak pressure," as it is customarily used, refers to the averaged peak pressure rather than to the nanosecond-duration peaks localized in submillimeter-sized regions. In the interest of facilitating interdisciplinary communication, we shall use the term "peak pressure" in its customary sense. In the absence of accepted terminology we use the terms "transient" and "spike" to refer to the nanoseconds-duration initial peak. Following common practice among shock wave researches, we use the term "ring down" to describe the complex series of shock interactions by which an initial high-pressure spike equilibrates to a lower pressure. Similarly, the term "ring up" describes a process beginning at a low pressure and equilibrating at a higher pressure.

As we will show later, shock temperature is sensitive to the detailed loading path experienced by the material during the initial pressure equilibration phase. Consider two feldspar grains in a chondrite. Depending on details such as the geometry and mineralogy of nearby grains, one feldspar grain may experience an initial low pressure that "rings up" to equilibrium and the second grain may experience an initial high-pressure spike that "rings down" to equilibrium. The difference in shock temperatures could be more than a factor of 2. In the case of a short-duration shock, one grain could be solid on release of pressure, whereas the other grain would be molten.

With the evolution of computer methods and capabilities, it has become possible to begin modeling shock wave propagation in real heterogeneous materials. New experimental techniques, including high-spatial-resolution dynamic measurements, permit detailed comparison of experiments with three-dimensional calculations (*Baer and Trott,* 2002, 2004). The goal of modern shock wave researchers is to develop a first-principles understanding of shock wave propagation valid on all scales from the atomic upward (*Gupta,* 2002; *Asay,* 2002).

The term "microscale modeling" is used to describe atomic and molecular scale modeling. Computational modeling on the micrometer scale, the scale of interest to meteorite researchers, must therefore be called "mesoscale modeling." Baer and co-workers have described mesoscale three-dimensional calculations of shock wave propagation in simple heterogeneous materials (*Baer,* 2000; *Baer and Trott,* 2002). Extraordinary computer capability was required to cope with 5-μm cell sizes and nanosecond time steps. One calculation indicated that the process of equilibrating to a "peak pressure" of 3 GPa involved nanosecond-duration transient spikes of over 30 GPa in the model heterogeneous material. One might suggest that these calculations confirm Kieffer's inference of great local variations of pressure and temperature in the initial stages of shock compression of a heterogeneous material. However, it would be more appropriate to state that the calculations are validated by Kieffer's work. Her observations had a spatial resolution at least 2 orders of magnitude higher than Baer's calculations. Kieffer also noted that the effective shock pressure duration in her samples was in the range of tens of milliseconds. Baer's calculations cover only the time required for pressure equilibration, generally less that a microsecond. Although calculations of pressure equilibration in heterogeneous materials require a supercomputer, independent thermal equilibration calculations can be performed on a modest personal computer.

This paper concentrates on the effects of shock pressures *per se*. However, we must mention that much current research is concerned with understanding phenomena, including fracture and plastic deformation, that occur during release of pressure. These phenomena should be of interest to students of meteorites who wonder why many shocked chondrites are so large and so strong. The usual result of shock loading experiments on nonporous rocks is fragmentation into submillimeter-sized pieces. This is the case even when extreme care is taken to use matched impedance momentum traps (*Hörz,* 1968). The wave interactions that result in fracture are associated with rapid release of pressure. One would expect that fragmentation and fracture would be minimized by a relatively slow release of pressure. We know from shock wave theory that the rate of pressure release will vary inversely with distance from a free surface. Thus it appears that appropriate application of dynamic fracture studies could yield further details, such as depth of a meteorite below the surface of a parent body at the time of a major impact. Recent books by *Nesterenko* (2001) and *Antoun et al.* (2003) summarize current knowledge of dynamic deformation and fracture. The book by *Meyers* (1994) is an excellent introduction to the field.

2.1. Elementary Shock Wave Theory and Calculations

Most review articles on shock wave physics take a rigorous approach toward interpretation of the physics of very simple geometries, e.g., planar impact or spherical flow in a homogeneous material. Here we attempt to present a simple, nonrigorous discussion of the complexities of shock wave propagation in a polymineralic and possibly cracked or porous rock. We also present approximate spreadsheet-based techniques for calculation of postshock temperature and shock temperature. For a rigorous analysis and deriva-

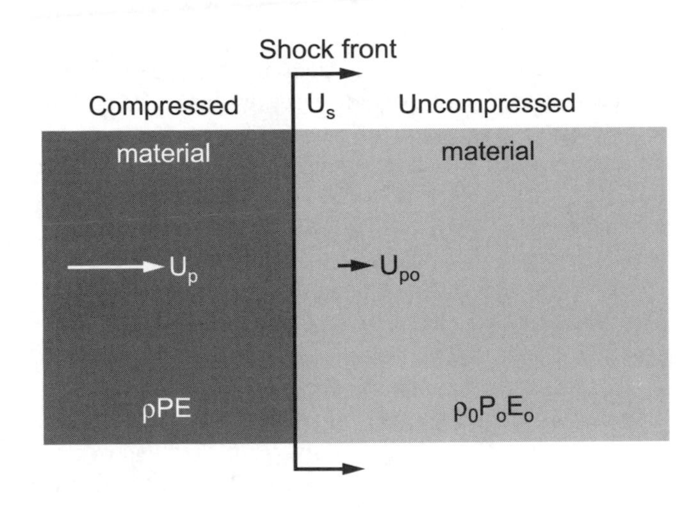

Fig. 1. Planar shock wave in a material.

tion of equations, the reader may consult various standard references (*Courant and Friedrichs,* 1948; *McQueen et al.,* 1970; *Duvall and Fowles,* 1963; *Melosh,* 1989; *Migault,* 1998). The graphical techniques and simplifying assumptions presented in this section are familiar to most shock wave specialists, who still use them to verify that complex computer codes running on supercomputers are giving "reasonable" results.

We begin with Fig. 1, the standard one-dimensional planar shock diagram illustrating the thermodynamic state variables and the Rankine Hugoniot equations. The state variables are pressure P, specific volume V ($V = 1/\rho$), and internal energy E. P_o, V_o, (or ρ_0), and E_o are the values of these variables in the unshocked state. P, V (or ρ), and E are the values in the shocked state. A shock wave is characterized by the kinetic variables U_s, shock velocity, and U_p, particle velocity.

From conservation of mass,

$$\rho(U_s - U_p) = \rho_0 U_s \quad (1)$$

From conservation of momentum,

$$P - P_o = \rho_0 U_s U_p \quad (2)$$

From conservation of energy,

$$PU_p = \rho_0 U_s (E - E_o) + (1/2)\rho_0 U_p^2 U_s \quad (3)$$

These equations may be rewritten as

$$U_p^2 = (P - P_o)(V_o - V) \quad (4)$$

$$U_s^2 = V_o^2 (P - P_o)/(V_o - V) \quad (5)$$

$$E - E_o = (1/2)(P + P_o)(V_o - V) \quad (6)$$

There are three equations in five unknowns, P, V, U_s, U_p, and E. With a fourth relationship between any two of the variables one may completely define the shock state. This fourth relationship is the equation of state of the material. The equation of state may be represented in a variety of equivalent forms: As relationships in the P–V plane, in the P–U_s plane, in the P–U_p plane, or in the U_s–U_p plane. P–U_p representations are useful when calculating interface pressures between different materials. P–V representations are useful when making approximate calculations of shock and postshock temperatures (or internal energy changes).

However, shock wave equations of state (Hugoniots) are most often presented as relationships between U_s, shock velocity, and U_p, particle velocity, in the form

$$U_s = C_o + sU_p \quad (7)$$

McQueen (1964) originally presented Hugoniot data in this form because shock velocity was directly measured and particle velocity was inferred from the measured free surface velocity, Ufs. In the absence of phase transitions, $U_p \sim U_{fs}/2$ (*Walsh and Christian,* 1955). *McQueen* (1964) also noted that the relationship between U_s and U_p, for most materials, is linear over a large range of pressures. Deviations from linearity are often a sign of complexity in material response, e.g., phase transitions. In some cases, the deviations from linearity may be handled by the addition of a quadratic term. Alternatively, one may make a piecewise linear fit to complex U_s–U_p relations. This latter approach has been taken by *Ahrens and Johnson* (1995a,b) in their compilation of Hugoniot data for major rocks and minerals. Data originally presented as P–U_s or P–V Hugoniots were transformed to the U_s–U_p form for these compilations. A further advantage of the U_s–U_p form is that it lends itself well to extrapolations beyond the limits of available data.

Figure 2 is a representation of a Hugoniot in the P–V plane. Strictly speaking, the Hugoniot represents only those states that are reached by a single shock from a given initial state (P_o, V_o, E_o). The Rayleigh line is the loading path, shown as a straight line connecting the initial state with the state on the Hugoniot. The energy increase on shock compression is simply the area of the right triangle ABC: (1/2) P (V_o–V). Here we neglect the P_o (one atmosphere) of equation (6). This energy increase comprises both mechanical and thermal terms. On pressure release, the mechanical portion (equivalent to the area under the release adiabat) is recovered. In general, assuming that no phase transitions have occurred, the release adiabat lies to the right of the Hugoniot because of thermal expansion. However, most rocks and their constituent minerals have low coefficients of thermal expansion. One may therefore approximate the release adiabat with the Hugoniot. Any difference between the two is usually of the same order as the uncertainty in the measurements of the Hugoniot. Indeed, this approximation is equivalent to the approximation that the free surface velocity is twice the particle velocity.

The residual energy, the cross-hatched region of Fig. 2, is often called the waste heat. If one can estimate the waste heat, one can use atmospheric-pressure heat capacity (C_p)

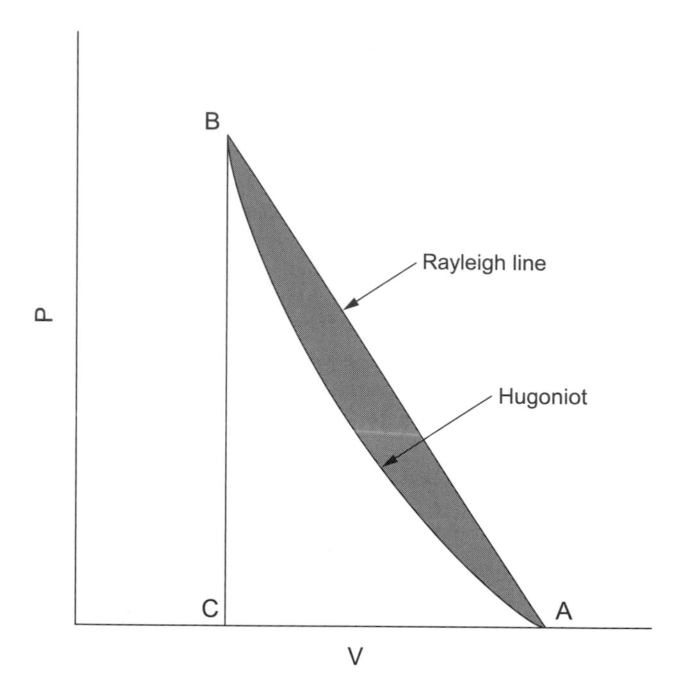

Fig. 2. Pressure-volume Hugoniot.

data on rocks and minerals to determine the postshock temperature, i.e., the temperature at the foot of the release adiabat. With a postshock temperature as a starting point one may estimate the temperature increase on adiabatic compression to the shocked state P, V. We present a simple spreadsheet technique for calculation of waste heat in Appendix A.

2.2. Porous Hugoniots

Figure 3 illustrates the typical response of a porous rock to shock compression and release. With increasing pressure, the porosity is crushed out until the pressure-volume Hugoniot becomes indistinguishable from that of an initially solid rock of the same mineralogy. In general, the porosity is not recovered on release of pressure. In the absence of release measurement data, one may use the Hugoniot of the initially solid rock to approximate the release adiabat of the initially porous rock.

Initial porosity obviously has a very large effect on waste heat. The waste heat for nonporous quartz shocked to 10 GPa is about 60 J/g. Assuming a uniform distribution of the waste heat and an initial temperature of 300 K, the postshock temperature would be ~375 K.

In contrast, the waste heat for sand, having an initial density of 1.4 g/cc (about 40% porosity) and shocked to 10 GPa is about 1740 J/g. Again assuming a uniform distribution of the waste heat and an initial temperature of 300 K, the postshock temperature would be about 1820 K. We know that the initial distribution of waste heat is highly nonuniform, as shown by Baer's calculations and Kieffer's observations. We also know that the pressure history in the sand was chaotic on a nanosecond timescale. Micrometer-sized regions of the sand could have experienced nanosecond-duration spikes of 50–100 GPa. However, the only number we can reliably calculate is the continuum value of 10 GPa to which the chaotic pressure distribution equilibrates within less than a microsecond (for millimeter-sized grains). For millimeter-scale thermal inhomogeneities, the equilibration time would be on the order of seconds. Shock temperature calculations for initially porous materials are valid only for very long-duration shocks.

The waste heat estimate is an approximation of the entropy increase over the cycle of compression and release. Strictly speaking, the waste heat method of estimating postshock temperature applies to the one-dimensional shock propagation geometry of Fig. 1. In most natural impacts, heating due to plastic strain in divergent flow and in tension may add substantially to the postshock temperature. Also note that localized hot spots or bands may form during shock deformation of nonporous homogeneous materials. The phenomenon of adiabatic shear, first observed in metals by *Zener and Hollomon* (1944), may account for some of the melt veins observed in meteorites.

In current models of the early solar system, mineral grains condense as the temperature of the solar nebula decreases. These mineral grains then aggregate, eventually forming larger bodies. One may infer that collisions among porous bodies aid in their compaction and lithification. For 1.4 g/cc sand, 10 GPa corresponds to a particle velocity of ~2 km/s. Two bodies of that porosity would have to collide at a relative velocity of ~4 km/s to achieve a continuum pressure of 10 GPa at the point of impact. We emphasize that the peak pressure is obtained at the point of impact; shocks attenuate rapidly in porous materials because the release wave velocities in compressed material are much higher than the shock velocities in the porous material. *Rubin* (2005) has hypothesized that shock compression of porous material provided a significant heat source for meteorite parent bodies in the early solar system. Rubin suggests that it may even be necessary to invoke shock compression as a heat source. Should this be the case, one could perform a series

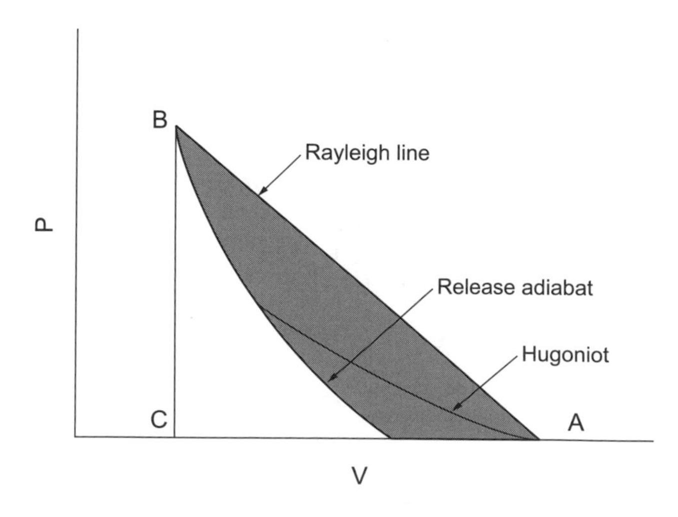

Fig. 3. Pressure-volume response of a porous material.

of hydrocode calculations of impacts to test models of the size distribution and collisional probabilities (including relative collision velocities) of porous parent bodies in the early solar system.

2.3. Phase Transitions on the Hugoniot

In an earlier era, shock wave workers were hard pressed to account for the apparent rapidity of shock-induced phase transitions, e.g., of graphite to diamond *(DeCarli and Jamieson,* 1961). *Alder*'s (1963) arguments were typical of those ascribing "special" characteristics to shock compression. Subsequent work on shock synthesis of diamond indicated that polycrystalline cubic diamond could form via an ordinary nucleation and growth mechanism in hotspots created by shock interactions (jetting) around pores (*DeCarli,* 1979, 1995). Conditions for shock synthesis of diamond were predictable on the basis of static high-pressure experiments (*Bundy and Kasper,* 1967) on the direct (uncatalyzed) synthesis of diamond. Bundy and Kasper observed that conditions for direct diamond synthesis were sensitive to the crystalline perfection of the carbon starting material. Poorly crystalline graphitic carbons required transient (microsecond-duration) temperatures above ~3000 K and pressures above ~15 GPa. The resultant diamond is cubic, polycrystalline, and optically isotropic. If the starting material were a well-ordered natural graphite, transformation occurred at lower transient temperatures in the range of 1300 K to 2000 K. The resulting diamond is a polycrystalline optically anisotropic mixture of cubic and hexagonal (lonsdaleite) diamond. The hexagonal phase has a preferred orientation, (100) parallel to (001) graphite. This latter form of direct-transition diamond has been made in laboratory shock experiments and it is the form commonly found in both meteorites and in impact craters (*DeCarli et al.,* 2002a). To date, we have found no evidence to contradict our working hypothesis that conditions for shock-induced phase transitions can be generally predicted on the basis of static high-pressure data. In other words, the physics of phase transformations during shock are the same as in static experiments. However, one must be aware that the detailed correlations between static high-pressure observations and observations on shocked minerals will necessarily be imperfect because of differences in detailed pressure-temperature-time histories between shock and static loading and unloading paths.

Numerous Hugoniot measurements on rocks and minerals have been interpreted in terms of phase transitions to denser phases. However, the evidence for phase transitions is often circumstantial. For example, Hugoniot data on pyroxenes and olivines indicate that they compress to densities appropriate to high-pressure phases (*Ahrens and Gaffney,* 1971; *Ahrens et al.,* 1971). However, their measured release adiabats are indistinguishable from their Hugoniots, within the limits of measurement error. Furthermore, there is no evidence from shock-recovery experiments, up to 80 GPa, that phase transformations of olivines and pyroxenes have occurred (*Jeanloz,* 1980). There are localized occurrences of high-pressure phases of olivine and pyroxene in many meteorites, as will be discussed in detail later in this paper. The high-pressure phases are invariably found in regions that appear to have been subjected to localized high temperatures, either in or adjacent to the so-called melt veins of meteorites. The shock behavior of olivines and pyroxenes thus appears consistent with static high-pressure studies that show their phase transformations to be very sluggish at modest temperature.

In contrast, there is excellent evidence that quartz and feldspars undergo phase transitions under shock compression. At shock pressures above about 30 GPa, these minerals have volumes appropriate to more densely packed phases. Some workers interpret the Hugoniot data in terms of phase transitions to crystalline phases, e.g., stishovite or hollandite. This interpretation is contradicted by *Panero et al.* (2003), who observe that the high-pressure Hugoniot data on quartz indicate somewhat lower densities at pressure than do static high-pressure measurements on stishovite. Measured Hugoniot release adiabats indicate that the densely packed phases persist on release down to a pressure of about 7 GPa. Upon further release to zero pressure, the volume increases to a final state about 20% greater than the initial volume. These release data support the view that the inferred Hugoniot transitions are primarily to close-packed disordered phases that invert to low-pressure disordered phases (i.e., diaplectic glass) on release of pressure. This view is supported by shock-recovery experiments in which the bulk of the sample is recovered as diaplectic glass (*DeCarli and Jamieson,* 1959; *Milton and DeCarli,* 1963). Again, the shock wave data are consistent with the results of static high-pressure experiments. In room-temperature static experiments at pressures above about 12 GPa, quartz and feldspars transform to dense amorphous phases that invert to low-density amorphous phases on release of pressure (*Hemley et al.,* 1988; *Williams and Jeanloz,* 1989). Hugoniot and release adiabat data on albite (*Ahrens and Gregson,* 1964; *Ahrens and Rosenberg,* 1968) are shown in Fig. 4.

Release paths from pressures above about 35 GPa are very steep, appropriate for complete transformation to a close-packed disordered structure. Release paths from pressures below about 20 GPa are indistinguishable from the Hugoniot and imply no significant transformation. Measured release paths from intermediate pressures, e.g., 28 GPa, are indicative of partial transformation and lie between the extremes illustrated in Fig. 4. Release adiabats and Hugoniots are measured in microsecond-duration experiments. The fact that the Hugoniot transition pressure is higher than observed in static experiments implies that kinetic factors are important. Kinetic factors could account for the observation of complete transformation of feldspars in meteorites that appear to have been subjected to modest pressures (21–25 GPa) over a long duration (0.01–1 s) (e.g., *Xie et al.,* 2005).

As noted earlier, the postshock temperature may be estimated by determination of the waste heat, the difference

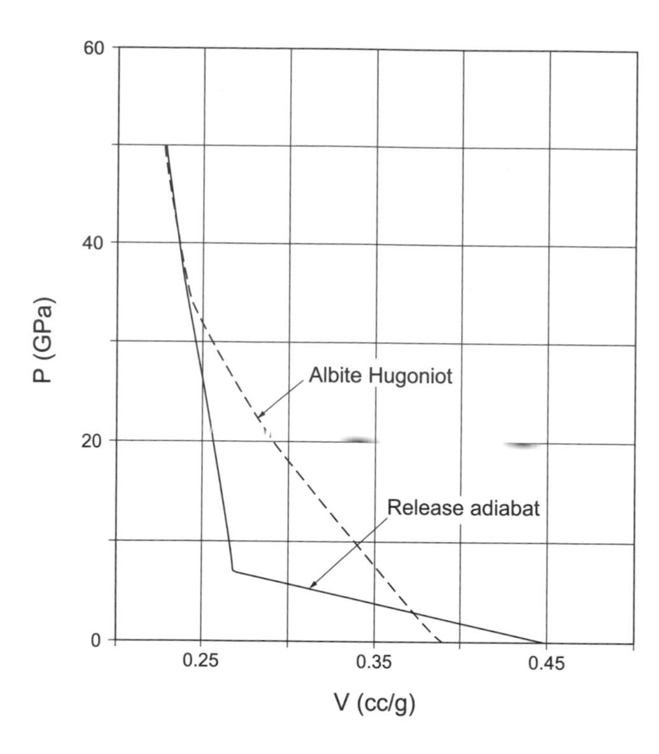

Fig. 4. Hugoniot and release adiabat data for albite (*Ahrens and Rosenberg,* 1968).

between the total energy increase on shock compression, and the mechanical energy recovered on adiabatic release. Phase transitions of the type illustrated in Fig. 4 can have a very large effect on the waste heat. If one uses the Hugoniot to approximate the release adiabat, the calculated waste heat on release for 35 GPa is ~320 J/g, corresponding to a postshock temperature of ~600 K (assuming an initial T of 300 K). If one uses the release adiabat shown in Fig. 3, the calculated waste heat on release is ~1250 J/g, corresponding to a partial melt at 1373 K. Simple postshock temperature calculations thus indicate that diaplectic glass formation, in both quartz and feldspar, is limited to a relatively narrow pressure range.

2.4. Shock Reflections and Loading-Path Effects

Shock interactions such as rarefactions or reflections take place at interfaces between materials of different shock impedance. To illustrate the effects of shock interactions on loading path, we assume the simple one-dimensional geometry illustrated in Fig. 5. Using materials for which we have Hugoniot data, we choose an albite inclusion in an iron meteorite. This example may not be particularly relevant to an actual iron meteorite, but it is similar to the high shock impedance container geometry used in many shock-recovery experiments (*Milton and DeCarli,* 1963).

The inclusion has infinite lateral extent but finite thickness, and the direction of shock propagation is normal to the plane of the inclusion. We assume a flat-topped shock pressure pulse in the iron having an amplitude of 50 GPa. Since albite has a much lower shock impedance than Fe-Ni, the initial shock transmitted into the albite will be less than 50 GPa; pressure equilibration will be achieved by a sequence of shock reflections. The detailed pressure history of the inclusion is calculated through the use of Hugoniots in the P–U_p plane, as shown in Fig. 6.

The albite Hugoniot lies well below the iron Hugoniot. When the shock arrives at the iron-silicate interface, the shock pressure on both sides of the interface must become equal and the particle velocities on both sides must also become equal, in accordance with the Hugoniots and release adiabats of both materials. As shown in Fig. 6, the initial state, 24.4 GPa, is achieved at the intersection of the iron release adiabat originating from 50 GPa with the Hugoniot of the silicate. The result is a transmitted shock into the silicate and a release wave moving backward into the iron. When the shock reaches the downstream silicate-iron interface, pressures and particle velocities must again be matched. The reflected shock state, 38.9 GPa, corresponds to the intersection of the reflected shock Hugoniot of the albite with the iron Hugoniot. A pressure of 46.2 GPa is reached via a second shock reflection. After several more reflections, the pressure in the silicate closely approaches the initial 50 GPa in the iron. The equilibration time is simply the sum of the shock transit times through the inclusion thickness. Albite shock velocities over the pressure regime of interest are 5–7 km/s; a 1-mm-thick inclusion would "ring up" to pressure equilibrium in less than a microsecond.

The thermodynamic final state achieved in the silicate via successive reflections is not the same state that would be achieved in a single shock, as illustrated in Fig. 7. The total energy increase is the sum of the areas under each Rayleigh line. For the reflected shock case illustrated above, the waste heat on release from 50 GPa is ~1040 J/g, equivalent to a postshock temperature of 1220 K (for an initial temperature of 300 K). For albite shocked in one step to 50 GPa, the waste heat on release is ~2060 J/g, equivalent to a liquid at a temperature of 1770 K. Although the same peak pressures may be achieved via single shock and re-

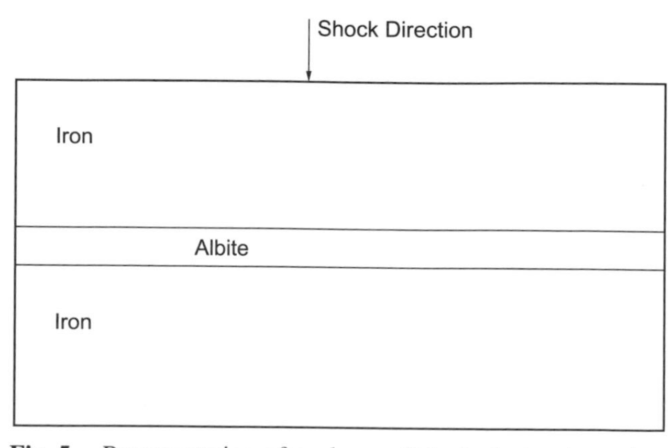

Fig. 5. Representation of a planar albite inclusion in an iron meteorite.

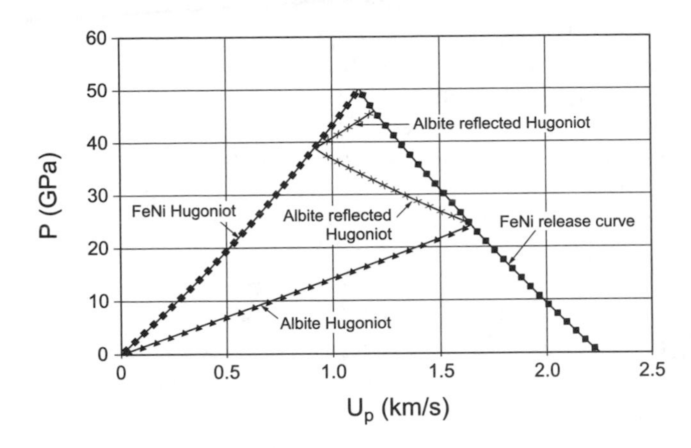

Fig. 6. Pressure equilibration via reflected shocks.

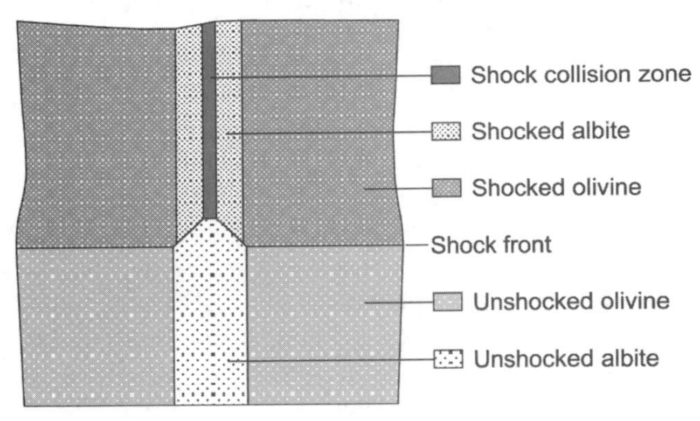

Fig. 8. Shock collision example.

flected shock loading paths, calculated waste heats may differ by a factor of 2 or more.

2.5. Shock Collisions

In the continuum model of plane shock propagation in a polymineralic rock, the Hugoniot represents averaged values of shock velocity, particle velocity, and pressure. On the scale of these averaged values the shock front is planar. On the scale of the grain size (or smaller), the shock front may be very irregular because of shock velocity differences among the different minerals. Analogous with the refraction of light, the shock front may be refracted at interfaces between different minerals. Refracted shock fronts may collide, producing localized (micrometer-scale) pressure spikes (nanoseconds-duration) in the low-impedance mineral that greatly exceed the pressure in the high-impedance minerals. We present a simple example in Fig. 8.

Figure 8 represents a thin plate of albite between two forsterite grains. This is similar in appearance to Fig. 5, except that the shock is traveling parallel to the plane of the low-impedance inclusion. For a pressure of 30 GPa, the shock velocity is 7.86 km/s in the forsterite. The shock velocity in albite at 30 GPa is only 5.84 km/s. As a consequence, the shock fronts are refracted at the forsterite-albite interfaces. The refracted shock fronts collide at the albite midplane, and the collision region (of finite width) is forced to propagate at the 7.86 km/s velocity of the shock in the forsterite. From the Hugoniot of albite, a shock velocity of 7.86 km/s corresponds to a pressure of 70 GPa. This high-pressure, high-temperature albite in the central region will "ring down" to 30 GPa. The calculated waste heat of 3300 J/g corresponds to a postshock temperature of about 2700 K. Assuming a Grüneisen parameter of 1.5, the calculated shock temperature at the continuum pressure of 30 GPa would be ~3280 K, more than sufficient for melting on the Hugoniot. The albite near the interface with the forsterite will "ring up" to 30 GPa. Assuming negligible phase transformation, the calculated shock temperature is ~550 K; assuming complete phase transformation, the calculated shock temperature is ~1300 K.

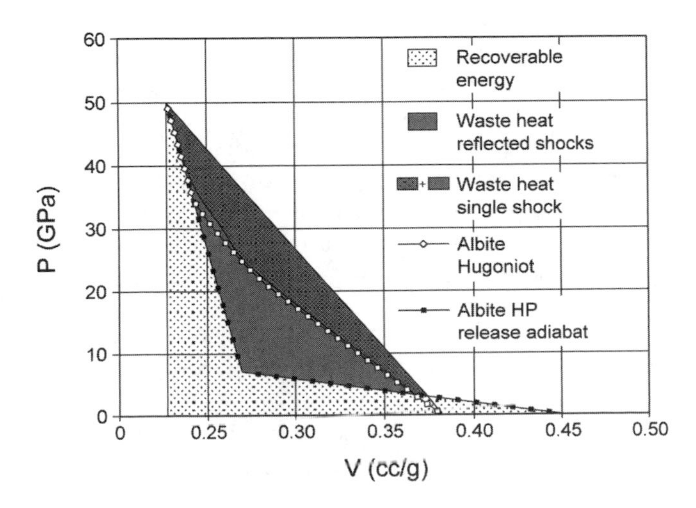

Fig. 7. Comparison of waste heat in albite for single vs. reflected shock compression to 50 GPa.

2.6. Further Comments on Shock Wave Heterogeneity

The simple examples we have presented are intended to provide an introduction to the complexities of shock wave propagation in a real three-dimensional polymineralic rock. However, the greatest source of localized pressure and temperature heterogeneity is the interaction of shock waves with cracks and pores. Kieffer's inferences of shock interaction effects in shocked Coconino sandstone were informed by knowledge of the porosity of the unshocked material (*Kieffer,* 1971; *Kieffer et al.,* 1976). Unfortunately, we do not have direct information on the nature of the porosity in a given meteorite prior to the shock event that produced the effects of interest. We are particularly interested in melt-vein-producing shock events. Shock collisions due to grain-scale heterogeneities as described above could form narrow veins in low-impedance minerals like feldspar and graphite, but cannot account for veins in olivine and pyroxene. Adiabatic shear might account for narrow planar melt veins (*Schmitt and Ahrens,* 1983), but many melt veins and pockets are neither planar nor narrow. Furthermore, adiabatic shear is often correlated with the flow divergence during pressure release (*Nestorenko,* 2001). Shock wave interaction along cracks and pores is one explanation that could account for the melt veins and pockets observed in many highly shocked meteorites. This explanation satisfies simple energy balance estimates. For a nonporous chondrite with the composition of Tenham, we calculate a waste heat of ~80 J/g (about 100° above its preshock temperature) on release from a shock pressure of 25 GPa. If the initial porosity were ~0.07, a typical value for chondrites (*Consolmagno and Britt,* 1998), the waste heat on release from 25 GPa would be ~320 J/g. Assuming the porosity is in the form of cracks, one could predict that the "excess" waste heat of 240 J/g (320–80) would be deposited in the vicinity of the cracks. The energy required to melt forsterite, starting from an initial temperature of 300 K, is about 3000 J/g. Thus 240 J/g would account for a melt vein fraction of about 0.08 in Tenham, consistent with our observations on Tenham (*Xie et al.,* 2005).

Although the melt is the result of a transient and localized high pressure, the melt persists long after the pressure spike has rung down to the continuum pressure. Any observations on the vein should be relevant to the continuum pressure and its decay as the melt solidified and thermally equilibrated with the surrounding cooler material. For example, it has been shown that the effect of shocking a granular sample in a gaseous environment can be to trap gas in the shock compacted aggregate (*Fredriksson and DeCarli,* 1964; *Pepin et al.,* 1964; *Wiens and Pepin,* 1988). In retrospect, one can explain these results as the expected consequence of ordinary diffusion in a thermally heterogeneous high-pressure environment. The high-pressure gas would diffuse into the localized hot (even molten) regions, which are then quenched by conduction. One might have anticipated that melt veins in martian meteorites could contain copious amounts of martian atmosphere (*Bogard and Johnson,* 1983). In contrast, a meteorite that cools from its postshock temperature in a low-pressure environment would be expected to lose some of its radiogenic gas (*Davis,* 1977).

We pointed out earlier that waste heat (a proxy for postshock temperature) is much lower for reflected shock loading (ringing up to continuum pressure) in comparison with single shock loading to the same continuum pressure. Shock-recovery experiments have been performed on preheated samples in an attempt to compensate for the waste heat difference between single shock and reflected shock experiments. However, the preheating does not compensate for the fact that localized pressure and temperature heterogeneities are also substantially reduced in reflected shock loading (*DeCarli et al.,* 2002a). Shock interactions around cracks and pores can occur only during passage of the initial shock because the subsequent shock reflections pass through compressed nonporous material.

Up to this point, we have discussed shock propagation through materials that are heterogeneous on a millimeter scale. One must also consider the possibility of larger-scale heterogeneities in a meteorite parent body. For example, consider a material having open cracks with a spacing of about 10 cm. The initial chaotic period of pressure equilibration to the continuum value could span as much as 100 µs. Approximate analyses indicate that shock collisions within these 10-cm blocks and shear localization now become viable mechanisms for production of melt veins within the blocks. One would also predict that the former open cracks would become large melt veins. In principle, one could refine these approximate analyses with detailed shock propagation calculations for a range of possible crack spacings and crack openings.

It is important to recognize that interpretations of shock effects in meteorites are often influenced by unstated assumptions concerning the overall structure of a parent body and the detailed properties of the region of the parent body from which a meteorite originated. We prefer simple models because they permit quantitative evaluation of shock propagation and heat flow effects. One must admit, however, that quantitative efforts may be irrelevant if the model differs too much from reality.

2.7. Limitations of Hugoniot Measurements and Shock-Recovery Experiments

The major limitation of Hugoniot measurements is that they are made in relatively short-duration (<10 µs) experiments. Many shock wave workers have speculated that kinetic effects on phase transitions might be important in millisecond- or longer-duration shock events, but it is either prohibitively expensive or impossible to perform long-duration shock compression measurements in the appropriate stress range for most mineral phase transitions of interest. Our working hypothesis is that the results of very long-duration shock events will be consistent with the results of static high-pressure experiments. *DeCarli et al.* (2002b) have recently interpreted Meteor Crater shocked Coconino sandstone mineralogy in terms of evidence for kinetic effects, but alternative explanations have not been ruled out.

Shock-recovery experiments have duplicated many, but not all, of the metamorphic features observed in meteorites. However, the range of shock conditions (loading paths, peak pressure, peak pressure duration, unloading paths, etc.) accessible in laboratory shock-recovery experiments is sharply limited in comparison with natural events. As implied by our discussion of the phase transitions in quartz and feldspars, we question the relevance to meteorite studies of the pressure calibration scales developed in shock-recovery experiments. Furthermore, many experimental studies report their results in terms of a single parameter, shock pressure, and do not provide enough detailed information on the experimental design to permit independent assessment of other relevant factors such as loading and unloading paths.

However, one may point to a recent paper by *Tomeoka et al.* (1999) that does provide detailed information on the experiment, including impact velocity, flyer plate material and dimensions, sample container material, sample material and dimensions, and the assumptions of the calculations that were made. They properly interpreted their reflected shock experimental data as equivalent to much-lower-pressure natural events. The Murchison Hugoniot (*Anderson and Ahrens,* 1998) that they used to interpret their results understates the initial porosity of the meteorite. As a consequence, their calculated waste heat is too low, 600 J/g vs. our calculation of 900 J/g, for the sample loaded to 48.8 GPa via reflected shocks. However, their conclusion is unchanged. Reflected shock loading to 48.8 GPa is equivalent, in terms of waste heat, to single shock loading to ~17 GPa. *Tomeoka et al.* (1999) also describe the deformation observed in their recovered samples. In our opinion, this deformation is attributable to the details of pressure release, specific to the conditions of the experiment and not necessarily relevant to a natural event. We will not attempt to estimate the additional heating ascribable to deformation in their experiments.

Experiments reported by *Huffman et al.* (1993) provide a measure of deformational heating. They observed bulk melting in shock-recovery experiments at 5 GPa on samples of Westerly granite preheated to 750°C in high-impedance stainless steel containers. We calculate that a temperature increase of only 20° could be attributed to the waste heat increase on compression and release. However, *Huffman et al.* (1993) also report that some of their samples were reduced in thickness by as much as 50%. One may infer that the observed melting may be attributable to the additional heating from plastic deformation.

The starting point for interpretation of shock effects in meteorites is in detailed observations on the meteorites. To interpret these observations, one uses all the relevant information available, including Hugoniot measurements, results of shock-recovery experiments, and results of static high-pressure studies.

As shown above, it may be useful to reinterpret some experimental results on the basis of simple shock wave calculations. All interpretations should be tested for consistency with existing knowledge of heat flow, diffusion, and phase transformation kinetics, *inter alia*. We have tried to emphasize the importance of details, e.g., presence of pores and cracks and grains of differing shock impedance, that influence shock wave propagation and loading paths. It is also useful to remember, when examining a two-dimensional slice at high resolution, that shock waves propagate in three dimensions. Given one's ignorance of relevant details that are either unknown or out of the plane of observation, one must be willing to admit the possibility of alternate interpretations.

2.8. Shock Metamorphism and Shock Classification in Meteorites

The metamorphic effects seen in shocked meteorites can be described in terms of either deformation or transformation or some combination of the two. Shock effects in meteorites and terrestrial rocks have been discussed at length in other papers (*Chao,* 1967, 1968; *Heymann,* 1967; *Kieffer,* 1971; *Scott et al.,* 1992; *Stöffler,* 1972, 1974; *Stöffler et al.,* 1991; *Ashworth,* 1985; *Bischoff and Stöffler,* 1992; *Leroux,* 2001; and many more) and so only a subset of shock effects will be discussed here.

Deformational effects include fracturing, plastic deformation, twinning, and mosaicism within constituent minerals. Planar shock features, generally referred to as planar deformation features (PDFs) (*Goltrant et al.,* 1991, 1992; *Langenhorst et al.,* 1995), have been attributed to deformational processes, but they have also been shown to contain transformed material, either diaplectic glass or high-pressure phases (*Goltrant et al.,* 1991; *Bowden,* 2002). The term planar deformation feature is therefore misleading because it neglects the transformation component. We will not discuss shock-induced deformation effects in this paper.

Transformational effects seen in shocked meteorites include shock melting, which commonly results in localized melt veins and pockets, transformation of minerals to high-pressure polymorphs, formation of diaplectic glass, and recrystallization of highly deformed material. The high-pressure minerals that occur in shocked meteorites form by either crystallization of silicate liquids in melt veins and pockets, or by solid-state transformation of the constituent minerals in the meteorite (*Chen et al.,* 1996). Solid-state phase transformations can provide important constraints on shock conditions of a meteorite, but transformation pressures are difficult to calibrate accurately because of kinetic effects and the heterogeneous nature of the initial transient shock pressures. Crystallization of chondritic melt provides an alternative means of constraining the crystallization pressure, which can be related to the shock pressure of the sample. In this part of the paper we will discuss high-pressure minerals in shocked meteorites and the constraints that they can provide for the interpretation of shock pressure and duration.

The shock effects in meteorites have been classified and calibrated to shock pressure by many previous workers (*Heymann,* 1967; *Carter et al.,* 1968; *Van Schmus and Ribbe,* 1968; *Taylor and Heymann,* 1969; *Dodd and Jarosewich,* 1979; *Sears and Dodd,* 1988; *Stöffler et al.,* 1988, 1991; *Schmitt,* 2000; *Leroux,* 2001). *Stöffler et al.* (1991) made a significant revision of previous classification systems by concentrating on petrographically observable shock effects in olivine and plagioclase. Incorporating a large database of experimental shock results, *Stöffler et al.* (1991) classified the shock effects in ordinary chondrites as six distinct shock stages, S1–S6. An advantage of the Stöffler et al. system is that the shock features used are those of olivine and plagioclase, which are common in ordinary chondrites and many achondrites. The characteristic shock features are easy to observe with a petrographic microscope and therefore the classification system has become readily used in the shock classification of meteorites. The characteristic shock effects are S1 (unshocked), sharp optical extinction; S2 (very weakly shocked), undulatory extinction in olivine; S3 (weakly shocked), planar fractures in olivine; S4 (moderately shocked), mosaicism in olivine; S5 (strongly shocked), maskelynite formation and planar deformation features in olivine; S6 (very strongly shocked), recrystallization of olivine that can be combined with high-pressure phase transitions such as the formation of ringwoodite. S6 features are described as "restricted to regions adjacent to melted portions of the sample which is otherwise only strongly shocked." Stöffler et al. interpret the S6 features to represent local pressure and temperature "excursions" or spikes that develop as a result of local variations in shock impedance. In this interpretation, the S6 shock effects would have to form under transient pressure spikes before they were rung down to the continuum shock pressure corresponding to S5 conditions. *Spray* (1999) proposed an alternative mechanism of cavitation (bubble implosion) within shock melt to produce local pressure spikes during pressure release that could transform olivine to ringwoodite in and along melt veins. In both models, high-pressure phase transitions are caused by local pressure spikes that exceed the equilibrated peak pressure of the rock. In the *Stöffler et al.* (1991) model, the pressure spikes would be early in the shock event, whereas in the *Spray* (1999) model, the pressure spikes would occur during pressure release. One objection to the Spray model is that the melt veins of many meteorites contain high-pressure phases that must have crystallized from the melt as it cooled via conduction. Spray's model would predict the presence of low-pressure phases in the melt veins. The problem with *Stöffler et al.*'s (1991) interpretation of S6 "pressure-temperature excursions" is that it overemphasizes the role of pressure (*Xie and Sharp,* 2004). As discussed in the background section, the pressure heterogeneities generated by shock wave interactions ring down to the continuum shock pressure within about 1 μs for a material with millimeter-scale impedance heterogeneities. The phase transformations that define S6 shock effects require high-pressure conditions for much longer than 1 μs (*Ohtani et al.,* 2004; *Chen et al.,* 2004b; *Beck et al.,* 2005; *Xie et al.,* 2005).

Although the shock stages of *Stöffler et al.* (1991) correctly reflect the sequence of increasingly more metamorphosed material, there are several reasons why the corresponding pressure calibration provides shock pressure estimates that are too high. The pressure calibration of Stöffler et al., which is based on shock-recovery experiments, defines the S1/S2, S2/S3, S3/S4, S4/S5, and S5/S6 transitions as <5, 5–10, 15–20, 30–35, and 45–55 GPa, respectively, with S6 conditions including pressures up to 90 GPa. *Schmitt* (2000) points out that although ordinary chondrites are porous, the samples used in most shock-recovery experiments, which commonly include single crystals and igneous rock fragments, are not porous. As discussed in the section on porous Hugoniots, the internal energy increase of shock compression is much higher for initially porous materials. Recognizing that shock-recovery experiments on nonporous samples result in relatively low shock temperatures, Schmitt did shock-recovery experiments at elevated starting temperatures (920 K) as well as at low temperature (293 K) to investigate the temperature effect on shock metamorphism pressures for the H chondrite Kernouvé. The importance of kinetics in shock metamorphism is illustrated by the fact that maskelynite formation occurred at 20–25 GPa in the preheated experiments compared to 25–30 GPa in the low-temperature samples and 30–35 GPa in the *Stöffler et al.* (1991) study. These and other high-temperature shock experiments demonstrate that shock metamorphic effects are temperature dependent as well as pressure dependent and that one cannot accurately calibrate the shock pressures without considering kinetic effects. Because kinetic effects are important, one may further question the relevance of microsecond-duration shock-recovery experiments to much-longer-duration natural shock events.

In addition to the porosity problem, shock-recovery experiments done in high-shock-impedance sample containers result in high shock pressures and relatively low shock temperatures (*Bowden,* 2002). As discussed in the section on shock reflections and loading path effects, samples in high impedance containers reach peak (continuum) pressure via a series of shock reflections, with the result that shock and postshock temperatures are substantially lower than for a sample loaded via a single shock to the same peak pressure (*Bowden,* 2002; *DeCarli et al.,* 2002b). For example, *Bowden* (2002), who shocked quartz sand in containers of various shock impedances, produced multiple intersecting planar features at 8 GPa in impedance-matched polyethylene containers and at 19 GPa in high-impedance stainless steel containers. Since most of the shock-recovery data used in the *Stöffler et al.* (1991) calibration were from experiments in high-impedance containers, the calibrated pressures of thermally activated shock effects, such as phase transformations, are likely to be too high.

2.9. High-Pressure Minerals in Meteorites

High-pressure minerals are common in and around melt veins in highly shocked meteorites (*Binns et al.,* 1969; *Binns,* 1970; *Smith and Mason,* 1970; *Putnis and Price,* 1979; *Price et al.,* 1983; *Langenhorst et al.,* 1995; *Chen et al.,* 1996; *Sharp et al.,* 1997; *Tomioka and Fujino,* 1997; *Langenhorst and Poirier,* 2000b; *Gillet et al.,* 2000; and others). The discovery of ringwoodite, the spinel-structured polymorph of olivine, in Tenham (*Binns et al.,* 1969) provided proof that Tenham had experienced very high shock pressures. Ringwoodite had already been observed in the Coorara chondrite by *Mason et al.* (1968), but it was misidentified as garnet based on X-ray diffraction data from the majorite garnet in the sample. *Smith and Mason* (1970) clarified the mistake when they published the discovery of majorite garnet in Coorara. The majorite that they describe occurred in a fine-grained mixture of garnet, Fe-oxide, and Fe, which must have been crystallized chondritic melt such as that described by *Chen et al.* (1996). The fact that these garnets had higher concentrations of Na, Al, and Cr than the orthopyroxenes in the sample confirms that they crystallized from a melt. In the same paper, *Smith and Mason* (1970) describe an isotropic phase with an orthopyroxene composition that they speculated was also majorite. It was clear from this early work that Tenham and Coorara had experienced high pressures. Although the significance may not have been realized at the time, it is clear from this work that ringwoodite and majorite were only observed in close association with melt veins. *Putnis and Price* (1979) used transmission electron microscopy (TEM) to characterize the microstructures of ringwoodite in Tenham and subsequently discovered the β-spinel polymorph of olivine, which they later named wadsleyite. Although the high-pressure minerals were used to interpret shock effects in meteorites (*Price et al.,* 1979, 1983) much of the subsequent work on naturally occurring ringwoodite and wadsleyite in meteorites was focused on defects and transformation mechanisms that might be important in Earth's mantle (*Price et al.,* 1982; *Madon and Poirier,* 1983; *Price,* 1983). More recent work concerning high-pressure minerals in meteorites (*Langenhorst et al.,* 1995; *Chen et al.,* 1996, 2004b; *Sharp et al.,* 1997, 1999; *Tomioka and Fujino,* 1997; *El Goresy et al.,* 2000; *Gillet et al.,* 2000; *Tomioka et al.,* 2000; *Xie et al.,* 2005; and others) have returned the focus back to interpreting shock conditions and durations.

Numerous other high-pressure minerals have been found in shocked meteorites since the time of ringwoodite and majorite discoveries. In nearly all cases, the high-pressure polymorphs occur within shock melt or adjacent to it. Mori appears to have been the first to discover a number of high-pressure phases including magnesiowüstite in melt veins (*Mori and Takeda,* 1985), plagioclase with the hollandite structure, and a glassy $MgSiO_3$-rich material that was inferred to be $MgSiO_3$-perovskite that was vitrified after pressure release (*Mori,* 1994). Magnesiowüstite was rediscovered by *Chen et al.* (1996), who realized that it crystallized along with majorite-pyrope garnet from chondritic melt in Sixiangkou. *Chen et al.* (1996) pointed out that one could use the mineral assemblages that crystallize from chondritic melt to estimate melt-vein crystallization pressure. Akimotoite, the ilmenite-structured polymorph of $MgSiO_3$, was discovered almost simultaneously by *Sharp et al.* (1997) and *Tomioka and Fujino* (1997) in Acfer 040 and Tenham, respectively. In Acfer 040, akimotoite crystallized directly from the melt (*Sharp et al.,* 1997), whereas in Tenham, the akimotoite formed from enstatite by a solid-state phase transition (*Tomioka and Fujino,* 1997). In both of these studies, a form of silicate perovskite was also described. In Tenham, the silicate perovskite, like the akimotoite, formed directly from enstatite by a solid-state mechanism (*Tomioka and Fujino,* 1997). However, in Acfer 040 equant domains of glass surrounded by akimotite and ringwoodite were interpreted as silicate perovskite that had crystallized from the melt and subsequently vitrified after pressure release (*Sharp et al.,* 1997). Similar material had already been found by *Mori* (1994) and has been subsequently found in Zagami (*Langenhorst and Porier,* 2000a) and in Tenham (*Xie et al.,* 2005). Plagioclase with the hollandite structure was rediscovered in Sixiangkou (*Gillet et al.,* 2000) and in Tenham (*Tomioka et al.,* 2000). This material occurs as polycrystalline aggregates with crystals that are tens of nanometers in size. Similar hollandite has been found in Umbarger (*Xie and Sharp,* 2004). Two new poststishovite polymorphs of silica were discovered in the martian meteorite Shergotty (*Sharp et al.,* 1999; *El Goresy et al.,* 2000). These polymorphs have an orthorhombic α-PbO_2 structure and a monoclinic ZrO_2-like structure, respectively, that are both slightly denser than stishovite. *Chen et al.* (2003) also discovered two high-pressure polymorphs of chromite in the Suizhou L6 chondrite.

Maskelynite is a form of amorphous plagioclase feldspar that was first described by *Tschermak* (1872). It was demonstrated to be a diaplectic glass formed by shock metamorphism when it was synthesized in a shock-recovery experiment (*Milton and DeCarli,* 1963). As discussed in the section on shock-induced transformations on the Hugoniot, the transformations of feldspars to maskelynite and crystalline silica to diaplectic quartz appear to take place on compression. Although the high temperatures associated with melt veins are not needed for the formation of maskelynite, higher temperatures do enhance the transformation around shock-melt veins and pockets. The formation of maskelynite is directly linked to the formation of PDFs in plagioclase, which represents partially transformed material. In the melt regions of highly shocked samples, plagioclase has been described as normal glass, which was interpreted to have melted during shock (*Stöffler et al.,* 1991). *Chen and El Goresy* (2000) have also demonstrated that what appears to be maskelynite can actually be melted plagioclase that quenched to glass. However, in very highly shocked samples the glassy plagioclase that occurs within or adjacent to shock melt can be nanocrystalline material with the hollandite structure. This has been documented by Raman spectroscopy in Sixiangkou (*Gillet et al.,* 2000) and by TEM in Tenham (*Tomioka et al.,* 2000; *Xie and Sharp,* 2003; *Xie et al.,* 2005) and by TEM in Umbarger (*Xie and Sharp,* 2004).

2.10. Mechanisms of Solid-State Transformations

The mechanisms by which low-pressure minerals transform into their high-pressure polymorphs during shock are important for understanding the kinetics of shock effects. Here we review the mechanisms of solid-state polymorphic reactions in general and then use the microstructures of high-pressure minerals in chondrites to infer the transformation mechanisms that occur during shock metamorphism.

The transformation of minerals to their high-pressure polymorphs can be described as displacive, reconstructive, or martensitic-like mechanisms. In a displacive phase transition, the structural difference between the polymorphs is small and can generally be described as minor displacements of the atom positions, resulting in a symmetry change that does not require breaking of bonds. These transitions are nonquenchable in that the high-pressure polymorph will spontaneously transform to the low-pressure polymorph upon pressure release. Such a displacive transition occurs between the high-pressure form of clinoenstatite (high-clinoenstatite with space group C2/c), which transforms to low-clinoenstatite ($P2_1/c$) upon decompression at about 10 GPa for $MgSiO_3$ (*Angel et al.,* 1992; *Hugh-Jones et al.,* 1996).

Reconstructive phase transitions involve a much more substantial change in crystal structure that involves the breaking of bonds and the formation of new bonds. These transitions have much higher activation energies, generally require high temperatures, and occur by nucleation and growth. Nucleation can occur homogeneously throughout the crystal of the parent phase or it can occur heterogeneously at defect sites or grain boundaries of the parent phase. In some cases, the new structure has a crystallographic relationship and coherent grain boundary relations with the initial structure. Phase transitions that involve a common interface and generally have only two crystallographic directions in common are called epitaxial, whereas transitions that involve a common sublattice and a three-dimensional crystallographic orientation relationship are known as topotaxial. If the new phase has a chemical composition that is different from that of the initial phase, then growth of the new phase will require diffusion of atoms to and away from the growing crystal. The rate of growth in this case is known as diffusion controlled because the rate is limited by rates of long-range diffusion. An example of nucleation with diffusion-controlled growth is the equilibrium transformation of $(Mg,Fe)_2SiO_4$ olivine to wadsleyite. In this case, the equilibrium composition of the wadsleyite is more Fe_2SiO_4-rich than the coexisting olivine. Growth of the wadsleyite requires long-range Fe-Mg interdiffusion. If the transformation occurs out of equilibrium, by overstepping the equilibrium phase boundary in pressure, the same transformation can occur without a change in composition. In this case, growth of the wadsleyite is controlled by the rate of short-range diffusion across the interface between the two phases and is referred to as interface-controlled growth.

In a martensitic phase transformation, the mechanism does not involve nucleation and growth, but rather the rearrangement of atom positions by shearing. In this case, shearing results from the passage of partial dislocations and the product phase is related to the initial phase by a series of stacking faults. Martensitic transformations occur in metals where the new structure can be created by changing the layer stacking of the metal by shearing (*Porter and Easterling,* 1978). *Madon and Poirier* (1980) proposed that the olivine-ringwoodite transformation could occur by a martensitic-like mechanism, where the glide of partial dislocations on (100) of olivine would change the oxygen sublattice from hexagonal close-packed to cubic close-packed. This is not a true martensitic transition because a change in the oxygen sublattice from hexagonal to cubic close-packed is not enough to make the spinel structure. The shearing of the oxygen sublattice must be accompanied by shifting of the cations to new positions through a process known as "synchro-shear" (*Madon and Poirier,* 1980). *Burnley and Green* (1989) have shown that the transformation of Mg_2GeO_4-olivine to spinel can occur by this mechanism under conditions of relatively high shear stress.

2.10.1. Olivine transformations in meteorites. The ringwoodite that is readily observable by optical petrography in very highly shocked S6 chondrites occurs as polycrystalline aggregates that have the same chemical composition as the olivines in the same samples (*Chen et al.,* 1996; *Langenhorst et al.,* 1995). One may therefore infer that it formed via a solid-state transformation mechanism during shock compression (*Chen et al.,* 1996). Transmission electron microscopy examination of the polycrystalline ringwoodite in chondrites generally shows randomly oriented ringwoodite crystallites that range from about 100 nm (*Putnis and Price,* 1979; *Langenhorst et al.,* 1995) to several micrometers (*Chen et al.,* 1996) (Fig. 9). The random orientations and homogeneous distributions of ringwoodite crystallites suggest homogeneous intracrystalline nucleation throughout the olivine rather than heterogeneous nucleation on grain boundaries, which is the dominant mechanism at pressures closer to the equilibrium phase boundary (*Kerschhofer et al.,* 2000; *Mosenfelder et al.,* 2001). The presence of small amounts of glassy material in ringwoodites from Tenham (*Price et al.,* 1979) have been interpreted as remnants of a prograde high-density olivine glass that was an intermediate phase in the transformation of olivine to ringwoodite. However, such glassy phases have not been reported in more recent studies. In most samples, the ringwoodite composition is constant, implying that there was no Fe-Mg exchange during the transformation, and therefore the crystallites grew by interface-controlled growth rather than diffusion-controlled growth.

High-pressure experiments have been performed on hot-pressed olivine samples to determine reaction mechanisms and kinetics of the olivine-ringwoodite and olivine-wads-

Fig. 9. Bright-field TEM image of a ringwoodite crystal in a large polycrystalline aggregate from Sixiangkou. The image was obtained using g = 220, which highlights the distinctive stacking faults on {110}.

leyite phase transformations (*Brearley et al.,* 1992; *Rubie and Ross,* 1994; *Sharp and Rubie,* 1995; *Kerschhofer et al.,* 1996, 1998, 2000). The dominant transformation mechanism was inferred to be incoherent grain boundary nucleation followed by interface-controlled growth. *Kerschhofer et al.* (1996) discovered a new intracrystalline mechanism that involved the formation of stacking faults in olivine on (100) followed by the nucleation of ringwoodite lamellae on the stacking faults. This produced the crystallographic relationship expected for the martensitic-like transformation mechanism, but clearly involved coherent heterogeneous nucleation followed by interface-controlled growth. *Chen et al.* (2004b) have discovered ringwoodite lamellae on $\{101\}_{ol}$ and $\{100\}_{ol}$ in partially transformed olivines in the Sixiangkou L6 chondrite, which they interpreted to have formed by the same mechanism as that described by *Kerschhofer et al.* (1996, 1998, 2000). However, these ringwoodite lamellae have a slightly higher fayalite content then the surrounding olivine, suggesting a diffusion-controlled growth mechanism rather than the interface-controlled growth mechanism described by *Kerschhofer et al.* (1996, 1998, 2000). The lamellae in Sixiangkou are also much coarser than those observed by *Kerschhoffer et al.* (1996, 1998, 2000). Detailed structural studies of the Sixiangkou ringwoodite lamellae are needed to determine the heterogeneous nucleation mechanism active in Sixiangkou and how it is related to those observed by *Kerschhofer et al.* (1996, 1998, 2000).

2.10.2. Enstatite transformations in meteorites. The transformation of pyroxene to its high-pressure polymorphs in shocked meteorites is less documented and more complicated than the olivine-ringwoodite transformations. In Sixiangkou, the majorite after enstatite (Fig. 10) consists of relatively large crystallites that are randomly oriented but contain subgrain boundaries (*Chen et al.,* 1996). The origin of the subgrain boundaries is not clear, but similar structures are observed in enstatite-pyrope composition garnets synthesized at high pressure and temperature and rapidly quenched (*Heinemann et al.,* 1997). The mechanism appears to be the same as that of ringwoodite: homogenous intracrystalline nucleation followed by interface-controlled growth. In the transformation of enstatite to akimotoite (*Tomioka and Fujino,* 1997), akimotoite occurs in a granular texture as well as in columnar texture where the crystals have a topotaxial relationship with the enstatite. The granular texture consists of randomly oriented crystallites from 100 to 200 nm, which is consistent with homogeneous intracrystalline nucleation and interface-controlled growth similar to that of polycrystalline ringwoodite. *Tomioka and Fujino* (1997) interpret the columnar texture as resulting from a martensitic-like mechanism. However, the TEM data presented by *Tomioka and Fujino* (1997) is also consistent with a coherent nucleation mechanism without martensitic-like shear. Enstatite in Tenham is also partially transformed to a granular intergrowth of 200-nm $MgSiO_3$-perovskite crystallites with the same chemical composition as the precursor enstatite (*Tomioka and Fujino,* 1997). This occurrence is also consistent with homogeneous nucleation and interface-controlled growth.

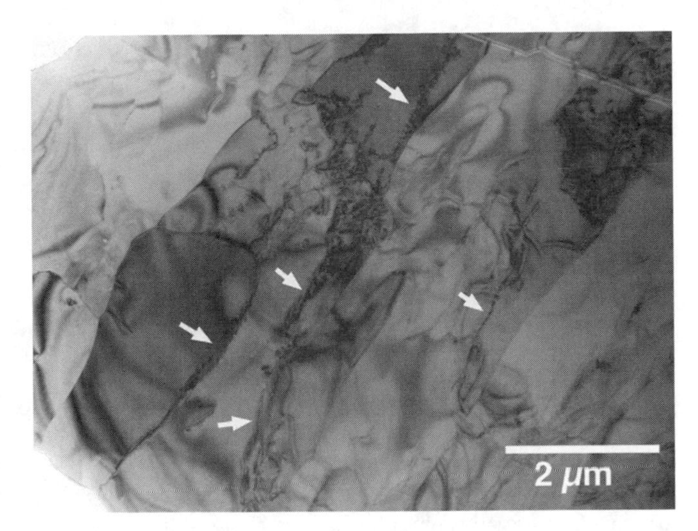

Fig. 10. Bright-field TEM image of a majorite crystal in a polycrystalline aggregate from Sixiangkou. The majorite crystals are >10 μm in size and contain numerous subgrain boundaries.

2.10.3. Plagioclase transformation in meteorites. The transformation of plagioclase to the hollandite structure, like the olivine and pyroxene transformations, occurs in and adjacent to shock melt. The hollandite consists of randomly oriented nanocrystals that range in size from 10 to about 100 nm (Fig. 11). Optically, the hollandite-structured plagioclase is isotropic and looks like maskelynite (diaplectic glass) or melted plagioclase (normal glass). The composition of hollandite-structured feldspars in meteorites ranges from $KAlSi_3O_8$-rich (*Langenhorst and Poirier,* 2000a) to $NaAlSi_3O_8$-rich (*Gillet et al.,* 2000; *Tomioka et al.,* 2000) and intermediate-plagioclase compositions (*Langenhorst and Poirier,* 2000b; *Xie and Sharp,* 2004). The origins of these various hollandite aggregates have been interpreted

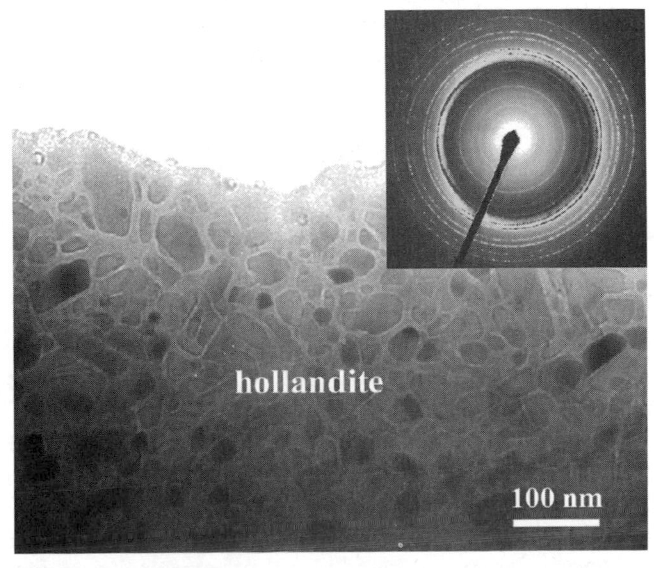

Fig. 11. Bright-field TEM image of nanocrystalline hollandite from Tenham. The individual crystallites range in size from 20 to 100 nm. The selected-area electron diffraction pattern contains diffraction rings that confirm the hollandite structure and indicate random orientations of the grains.

to be solid-state transformation (*Tomioka et al.,* 2000; *Langenhorst and Poirier,* 2000b; *Xie and Sharp,* 2004) and crystallization from melt (*Gillet et al.,* 2000). The nanocrystalline granular texture is consistent with both origins if the melt was a pure feldspar-composition liquid. If formed by a solid-state mechanism, the microstructure suggests homogeneous nucleation and interface-controlled growth, as in the formation of ringwoodite. If the polycrystalline hollandites in melt veins crystallized from feldspar-composition liquids during shock compression, the feldspar-composition liquid did not mix with the surrounding chondritic liquid, as one would expect if the liquids are miscible. This suggests that most of the transformation is via a solid-state mechanism. However, veins of hollandite extending away from polycrystalline aggregates in Sixiangkou indicate that at least some of the plagioclase was molten during shock. The distinction between solid-state transformation and liquid crystallization is an important consideration when using the presence of hollandite to constrain shock pressures.

2.10.4. Graphite transformation to diamond. The graphite-to-diamond transition is a special case. Here we summarize the detailed presentations of *DeCarli* (1995) and *DeCarli et al.* (2002a), which fully reference the relevant literature. As we noted earlier, both static and shock experiments indicate that the direct (uncatalyzed) transformation of disordered graphitic carbon to diamond requires pressures above ~15 GPa and transient high temperatures >3000 K. The resultant diamond is optically isotropic, mechanically strong, and polycrystalline, with randomly oriented crystallites from 20 nm (shock) to 50 nm (static). This reconstructive transformation is similar to the olivine-ringwoodite transition and involves homogeneous nucleation and growth. The diamond must be quenched at pressure to lower temperatures to avoid graphitization on pressure release. For lack of a better term, we will refer to it as HT (high-temperature) shock diamond. We do not preclude the possibility that HT shock diamond could be formed in a meteorite. To the best of our knowledge, however, all descriptions of meteoritic diamond appear to match the characteristics of diamond formed by a low-temperature mechanism in both static and shock experiments. The static flash heating experiments of *Bundy and Kasper* (1967) indicate that well-ordered graphite may be transformed at pressures above ~15 GPa and transient temperatures above ~1300 K. We will refer to the product as LT (low-temperature) shock or static diamond; 1300 K is a very low temperature relative to the melting point of graphite. The LT static diamond is a polycrystalline mixture of cubic and hexagonal (lonsdaleite) structures, optically anisotropic, and has a strong preferred orientation, with londaleite (100) parallel to (001) graphite. Although the preferred orientation and the low transformation temperature imply a martensitic mechanism, Bundy and Kasper noted that simple shear would not suffice to convert hexagonal graphite to either hexagonal or cubic diamond. The requirement for a minimum, albeit low, temperature requirement implies a thermally activated component in the transformation. The results of shock-recovery experiments on formation of LT shock diamond and studies of natural LT shock diamond from impact craters are in general accord with the static data. The common characteristics of the samples we have observed include optical anisotropy and strong preferred orientation. Distinct hexagonal (lonsdaleite) reflections are observed in some, but not all, terrestrial LT impact diamonds (*Koeberl et al.,* 1997). Laboratory shock experiments have also produced nanodiamond, cubic diamond particles having diameters in the range of 2–7 nm. These diamonds are structurally indistinguishable from the diamonds found in the detonation products of oxygen-deficient explosives and resemble the nanodiamonds found in carbonaceous chondrites.

Knowledge of the conditions for shock synthesis of LT diamond and for its subsequent survival can be used to estimate shock pressure in the surrounding rock or metal. The peak shock temperature of the graphite (which must be well-ordered) must be at least 1300 K, based on the static data. However, the temperature of the newly formed shock diamond and the surrounding rock and metal must be less than 2000 K on release of pressure, if graphitization of the diamond is to be avoided. Graphite has a very low shock impedance; shock pressure equilibration with the matrix will be achieved via a sequence of shock reflections. As we pointed out earlier, in the discussion of shock reflections and loading path effects, shock temperature in the graphite will depend on the loading path. Based on the simple geometry illustrated in Fig. 5, we have calculated shock temperature in graphite as a function of continuum pressure in various matrices. For graphite in iron-nickel meteorites, continuum pressures in the iron in the range of 60–120 GPa correspond to shock temperatures in the graphite in the range of 1300–2000 K. The calculated temperature of the iron on release from 120 GPa will be <1700 K. For graphite in granite or quartzite, the continuum pressure range of ~27–38 GPa corresponds to the 1300–2000 K graphite shock-temperature range. This suggests that graphite is more likely to transform to diamond where it is surrounded by minerals with low shock impedance. There are several caveats. The shock-temperature calculations are based on a simple geometric model of the graphite inclusion and shock propagation normal to the plane of the inclusion. In a more realistic geometry the shock temperature in the graphite could be nonuniform. One must also consider the possibility of chemical reactions or diffusion between hot diamond and hot matrix, e.g., diffusion of carbon in iron or reaction of diamond with a molten silicate. These considerations may further narrow the pressure range over which shock synthesis of LT diamond is possible. *El Goresy et al.* (2001) observed in Ries Crater gneiss that impact diamond seemed to be preferentially found in contact with garnet grains, and they infer that the reflected shock from the higher-shock-impedance garnet was responsible for the transformation. However, other interpretations are possible. The garnet may have served as a heat sink that preserved the adjacent diamond from graphitization. Low-temperature shock diamond has been found in heavily shocked iron meteorites and is commonly found in ureilites. The characteristics of LT shock diamond are distinctive and are unlike the characteristics of either terrestrial mantle diamond or low-pressure CVD diamond. The presence of LT shock diamond in some samples of the Canyon Diablo meteorite is correlated with metallographic evidence of high shock pressures in the range above 60 GPa (*Lipschutz,* 1968).

2.11. Transformation Kinetics and Temperature

Nucleation and growth has only been discussed as a viable mechanism for the shock transformation of olivine to ringwoodite by relatively few authors (e.g., *Chen et al.,* 1996, 2004b; *Ohtani et al.,* 2004; *Xie and Sharp,* 2004) because it has been assumed by many that the shock events that produced reconstructive phase transitions in meteorites were too short in duration for such a complex kinetic process. An alternative explanation, based on the presence of interstitial glass, is that the shock event initially produces a high-density prograde glass that subsequently devitrifies to form high-pressure minerals (*Price et al.,* 1979). One problem with this interpretation is that it is difficult to prove that interstitial glass is prograde rather than retrograde. Like the direct transformation, it still requires nucleation and growth of the high-pressure phases during shock compression because solid-state devitrification at low pressure would require temperatures too high for metastable high-pressure minerals to form. The textural evidence for homogeneous nucleation and interface-controlled growth in many polycrystalline high-pressure silicates in melt veins indicate that the shock pulses that caused the transformations in natural samples were of relatively long duration, perhaps up to several seconds (*Ohtani et al.,* 2004; *Chen et al.,* 2004b). The microsecond durations typical of shock experiments, combined with the relatively low shock temperatures produced in shock-recovery experiments, explain why such experiments have not produced high-pressure polymorphs of olivine and pyroxene. The kinetics of such reconstructive phase transformations are simply too slow under conditions of shock-recovery experiments.

Because reconstructive phase transitions in silicates are kinetically sluggish and require high temperatures to overcome the large activation barriers, the transformations that occur in meteorites are strongly dependent on temperature as well as pressure. The shock calibration of *Stöffler et al.* (1991) stresses shock pressure as the primary driver of shock-metamorphic effects and does not discuss temperature effects or reaction kinetics. Static high-pressure kinetic experiments have shown that dry hot-pressed San Carlos olivine transforms to ringwoodite, on an observable timescale, only above 900°C at 18–20 GPa (*Kerschhofer et al.,* 1996, 1998, 2000). The transformation of enstatite to akimotoite is even more sluggish, requiring temperatures in excess of 1550°C for transformation at 22 GPa (*Hogrefe et al.,* 1994). The fact that some olivine and low-Ca pyroxene in meteorites transform to high-pressure polymorphs by nucleation and growth indicates that shock temperatures of the transformed material must have been much higher than the temperatures in the static experiments. This is supported by the observation that solid-state transformations of olivine and pyroxene occur almost exclusively within or in close proximity to shock melt, which represents the hottest part of the sample during shock. Although some solid-state phase transitions may occur in response to localized and transient pressure spikes during the passage of the shock front (*Kieffer,* 1971), the transformations associated with mesoscale shock melting is not limited to the microsecond timescales of transient shock pressures. The ubiquity of transformed silicates within and along shock veins in S6 samples and the spatial scale of transformation indicate that high temperatures associated with melting provide the energy to overcome the kinetic barriers to nucleation and growth of the high-pressure phases. Because the elevated temperatures of the melt-vein regions can last up to about 1 s, the solid-state transformations that occur in and along melt veins are likely to have formed at the equilibrated shock pressure rather than during transient pressure spikes.

2.12. Pressure Constraints from Solid-State Transformations

The use of high-pressure solid-state phase transitions to quantitatively calibrate shock pressures in natural samples is problematic because most of the shock-induced transitions are reconstructive and kinetically sluggish. The fact that olivine and pyroxene do not transform to their high-pressure polymorphs in shock-recovery experiments has been used by *Stöffler et al.* (1991) as evidence that extreme pressure is needed to transform them in meteorites. The key questions are: How important is pressure in driving phase transitions, and what effect does pressure have on reaction kinetics?

The driving energy for nucleation is provided by the volume free energy change ΔG_v of forming the nuclei. Nucleation rates are controlled by the activation energy of nucleation, which is proportional to the inverse square of the free energy change. For the transformation of a low-pressure phase to a high-pressure polymorph, overstepping the phase boundary in pressure rapidly decreases the activation energy for nucleation. All such transformations require some pressure overstepping for nucleation to occur, but there is little nucleation rate data available for phase transitions in silicates. Experimental kinetic data for the olivine-wadsleyite and olivine-ringwoodite transformations are only available for incoherent nucleation on grain boundaries (*Rubie and Ross,* 1994; *Brearley et al.,* 1992; *Kubo et al.,* 1998a,b; *Liu et al.,* 1998) and for coherent intracrystalline heterogeneous nucleation (*Kerschhofer et al.,* 1996, 1998, 2000). There is no experimental data for intracrystalline homogeneous nucleation, which appears to be the mechanism most common in meteorites. For incoherent grain-boundary nucleation, little pressure overstepping is required and nucleation rates are assumed to be very fast (*Mosenfelder et al.,* 2001). Because the activation barrier for intracrystalline homogeneous nucleation should be higher, larger pressure overstepping is expected.

In studies of olivine-ringwoodite transformation kinetics, growth rather than nucleation is the dominant rate-controlling step (*Rubie,* 1993; *Mosenfelder et al.,* 2001). The rate equation for interface-controlled growth is given by

$$\dot{x} = k_0 T\exp[-(\Delta H_a + PV^*)/RT][1-\exp(\Delta G_r/RT)]$$

where $\dot{x}$ is the growth rate, k_0 is a constant, ΔH_a is the activation enthalpy, V^* is the activation volume for growth, ΔG_r is the free energy change of reaction, and R is the gas constant. The first part of the equation is an Arrhenius term that describes the thermally activated diffusion of atoms across the interface. The second part is a thermodynamic term that describes the driving force for growth. Many experimental studies have provided kinetic data for interface-controlled growth in the olivine-spinel transformation (for reviews, see *Rubie,* 1993; *Mosenfelder et al.,* 2001). The temperature dependence of growth is given by the activation enthalpy, which is approximately 350 to 400 kJ mol^{-1} (*Mosenfelder et al.,* 2001). This indicates a strong temperature dependence, which is consistent with the requirement of $T > 900°C$ for observable transformation of dry olivine in quench experiments. The pressure dependence of growth is given by the activation volume V^*. *Kubo et al.* (1998a) have used *in situ* transformation experiments to constrain the value of V^* to between 0 and 4 cm^3 mol^{-1}, whereas *Rubie and Ross* (1994) estimate V^* to range from 12 cm^3 mol^{-1} at 1 bar and 4 cm^3 mol^{-1} at 15 GPa. These positive activation volume estimates indicate that interface-controlled growth rates decrease with increasing pressure. This also suggests that the olivine-ringwoodite transformation that occurs in naturally shocked meteorites should require increasingly higher temperatures with increasing pressure. The fact that olivine and pyroxenes do not transform at very high pressures in diamond anvil experiments without being heated to very high temperatures further supports the idea that high temperatures actually control the distribution of olivine and pyroxene high-pressure polymorphs in chondrites.

Phase equilibrium data from static high-pressure experiments can be used to provide only limited constraints on pressure. The actual phase transitions are commonly metastable, involving polymorphs that are not in stable equilibrium. For example, the post-stishovite phase of SiO_2 known as seifertite (the α-PbO_2 structure) in Shergotty formed directly from either cristobalite or tridymite although there is no stable equilibrium between these phases. The equilibrium phase boundary between the $CaCl_2$ structure and the α-PbO_2 structure, at around 80 GPa, is irrelevant and one must consider the metastable boundaries between either cristobalite and seifertite or tridymite and seifertite. Because these metastable boundaries occur at a lower pressure than the $CaCl_2$-seifertite equilibrium boundary, the minimum transformation pressure is actually much lower than the 80 GPa pressure that one would infer from the minimum stability (*El Goresy et al.,* 2000).

The olivine-ringwoodite, enstatite-majorite, enstatite-akimotoite, and enstatite-$MgSiO_3$-perovskite phase transitions are all metastable. In all these cases, the metastable phase boundaries of interest are lower in pressure than the minimum pressure of stable equilibrium. The phase relations only limit the pressure to be greater than the metastable equilibrium between the low- and high-pressure polymorphs. This issue is especially important for ringwoodite, which is commonly used to indicate S6 shock conditions and pressures in excess of 50 GPa in shocked chondrites (*Stöffler et al.,* 1991). The pressure overstepping required to form ringwoodite is dependent on both the temperature and the time available for the transformation. The fact that ringwoodite has never been synthesized in a shock-recovery experiment is a result of the very short durations of those experiments, combined with the relatively low temperatures that result from using high-impedance containers. However, ringwoodite that occurs in or adjacent to shock melt in S6 chondrites formed at temperatures much higher than those in shock-recovery experiments and over a shock duration that may exceed the experimental shock duration by 5 orders of magnitude. Ringwoodite in shocked chondrites must form at pressures in excess of the metastable phase boundary (~18 GPa), but the amount of pressure overstepping required in nature is not constrained by shock or static high-pressure experiments. An alternative means of constraining shock pressures in meteorites is needed.

2.13. Constraints on Shock Pressure from Melt-Vein Crystallization

Shock-induced melt veins are common in moderately shocked chondrites of shock stage S3–S4 and ubiquitous in highly shocked S5–S6 chondrites. They occur as thin black veins, from 1 μm to several millimeters wide (*Stöffler et al.,* 1991). *Fredriksson et al.* (1963) used shock experiments to demonstrate that these melt veins are indeed the result of shock melting. Shock veins can form by shock-wave collisions, as discussed above, by frictional heating along shear bands, analogous to pseudotachilites, or by the collapse of open fractures or pores during shock compression. Shock veins are of critical importance in the interpretation of shock transformation effects because they are the locations of nearly all crystalline high-pressure minerals that occur in shocked meteorites and the exclusive locations of S6 shock effects.

An alternative to using deformation and solid-state transformation effects to constrain shock pressures is to use the crystallization of high-pressure minerals from the shock-induced melt combined with experimental high-pressure melting relations. This approach was first used by *Chen et al.* (1996), who used TEM to determine melt-vein assemblages in Sixiangkou. They found that the chondritic melt in a large melt vein crystallized to form majoritic garnet and magnesiowüstite (Fig. 12), which, based on the phase diagram of the CV3 chondrite Allende (*Agee et al.,* 1995), is stable at pressures between about 22 and 27 GPa (Fig. 13). *Chen et al.* (1996) inferred a crystallization pressure of about 25 GPa, which is half the value of the low-pressure threshold for S6 shock conditions of *Stöffler et al.* (1991).

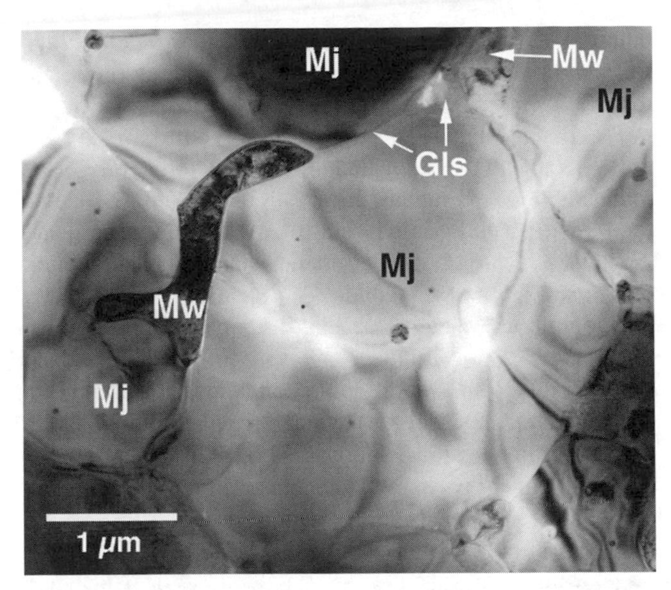

Fig. 12. Bright-field TEM image of majoritic garnets and magnesiowüstite that crystallized in a melt vein in the Sixiangkou L6 S6 chondrite.

Using melt-vein crystallization to estimate shock pressure is controversial in the field of shock metamorphism because it uses phase equilibrium data obtained in static high-pressure experiments. However, there are several reasons why high-pressure melting relations can be applied to the interpretation of melt-vein crystallization. First, the most common melt-vein assemblage seen in S6 chondrites, majorite plus magnesiowüstite, is also produced in static high-pressure melting experiments on both Allende (*Agee et al.,* 1995) and on Kilburn-hole-1 peridotite (*Zhang and Herzberg,* 1994). The textures and crystal sizes in the centers of large chondritic melt veins, such as those in Tenham, Sixiangkou, and RC 106 (*Chen et al.,* 1996; *Aramovich et al.,* 2003; *Xie et al.,* 2005), are very similar to the textures and crystal sizes produced in the static experiments (*Agee et al.,* 1995). Similarly, the chemical compositions of the crystallized majoritic garnets are very similar to the compositions of garnets in the experiments (*Chen et al.,* 1996). Compared to solid-state reconstructive phase transitions, melt-vein crystallization involves much smaller kinetic barriers.

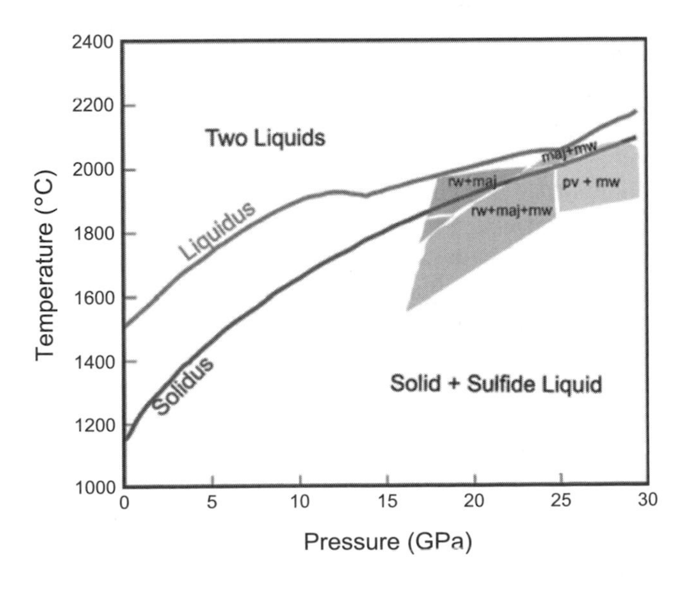

Fig. 13. The crystallization-pressure regions illustrated on a simplified version of the Allende phase diagram (*Agee et al.,* 1995); rw = ringwoodite, maj = majorite, mw = magnesiowüstite, pv = perovskite.

Melt-vein crystallization has a great advantage over solid-state transformation for constraining shock pressure histories. Because the cooling produced by adiabatic pressure release is relatively small, shock melt cools predominantly by conduction to the surrounding, relatively cool, host meteorite. This results in crystallization that starts at the vein margins and moves inward to the melt-vein core as crystallization proceeds. The resulting crystallization sequence provides a record of shock pressure through time. As we will discuss below, this record can be several hundred milliseconds long. If the recorded pressure-temperature-time history exceeds the period of elevated pressure, crystallization assemblages should record the pressure release.

An important aspect of using melt-vein mineralogy to constrain pressure is that melt veins do not crystallize at equilibrium. Silicate liquids must be supercooled relative to the liquidus in order to provide the free energy to overcome the activation energy of nucleation and growth. Unlike the case of solid-state transformations, melt-vein crystallization occurs at very high temperatures, on the order of 2000° to 2400°C at 25 GPa for Allende and KLB-1, respectively (*Agee et al.,* 1995; *Zhang and Herzberg,* 1994). At these temperatures, diffusion and crystal growth rates can be very rapid and therefore the kinetic barriers for crystallizing high-pressure minerals are much lower than those for solid-state transformations. Contrary to the pressure overstepping needed for solid-state transformations, the supercooling of silicate liquid does not require excess pressure to form high-pressure phases, so pressure constraints from melt-vein crystallization are likely to provide lower and more accurate pressure estimates that phase transformations calibrated against shock-recovery experimental results.

A limitation of the melt-vein approach is that the mineral assemblage that crystallizes may not be the equilibrium assemblage inferred from the static high-pressure experiments. Crystallization of metastable phases such as akimotoite, which is a subsolidus phase in the Mg_2SiO_4-$MgSiO_3$ system (*Gasparik,* 1992), occurs along the margins of melt veins in Tenham (Fig. 14) where the melt-vein quench rate was highest and therefore the supercooling was large (*Xie et al.,* 2005). However, the central core of the same melt veins contains the majorite plus magnesiowüstite assemblage that is inferred to be the equilibrium assemblage in the high-pressure experiments. It appears that the melt-vein margins and very thin melt veins, which are most rapidly quenched, are most likely to contain metastable crystallization products. In using melt-vein mineralogy to constrain crystallization and shock pressure, it is not appropriate or

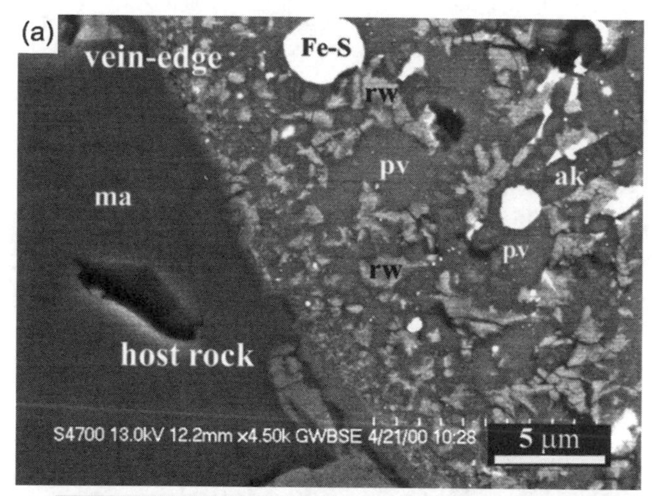

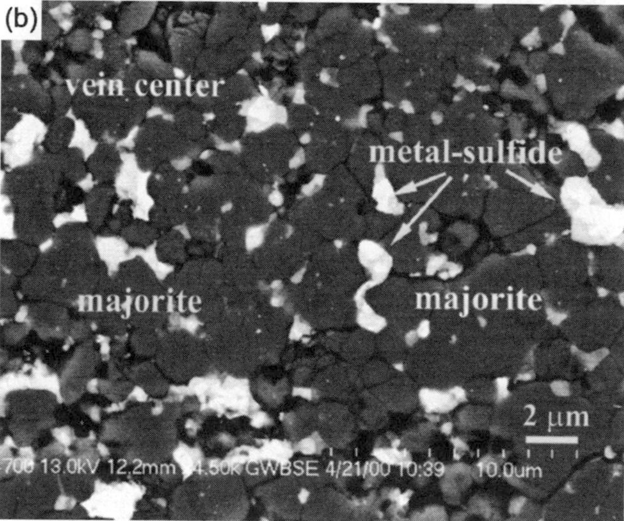

Fig. 14. Field-emission SEM images of a melt vein in Tenham. The melt-vein margin (top) contains ringwoodite (rw), akimotoite (ak), and vitrified silicate perovskite (pv) along with solidified droplets of Fe-sulfide melt. The vein core (bottom) contains the common assemblage of majorititic garnet and magnesiowüstite along with blebs of solidified metal-sulfide melt.

necessary to assume that crystallization was an equilibrium process. Like solid-state transformations, melt-vein assemblages provide constraints on crystallization pressure rather than precise pressure determinations. Because kinetic barriers for crystallization are less than for reconstructive phase transformations, it is likely that crystallization pressure can be more accurately calibrated.

2.14. Crystallization Pressure vs. Shock Pressure

A key issue in the use of melt-vein assemblages for constraining shock pressure is the relationship between crystallization pressure and shock pressure. This relationship is determined by when the melt vein crystallizes relative to shock loading and pressure release. Thermal modeling of melt-vein quench shows that a 1-mm-wide melt vein requires hundreds of milliseconds to crystallize (*Langenhorst and Poirier,* 2000a; *Xie et al.,* 2003, 2005), whereas transient pressure heterogeneities equilibrate in about 1 µs for a sample with a 1-mm grain size. Therefore, nearly all the melt-vein crystallization occurs after pressure equilibration such that the pressure history recorded by melt-vein crystallization does not include transient pressure spikes. Small-scale transformation effects, such as those described by *Kieffer* (1971), record the transient pressure history and are therefore not as useful for determining the equilibrated shock pressure. The timing of melt-vein crystallization relative to pressure release is equally important. One might argue that the relatively low pressure of melt-vein crystallization in S6 chondrites compared to the S6 calibration pressure of *Stöffler et al.* (1991) indicates that crystallization occurs upon pressure release from an equilibrated shock pressure in excess of 50 GPa. Such a hypothesis can be easily tested by looking at the sequence of minerals that crystallize from the melt-vein edge to the melt-vein core. Because the crystallization history of the melt vein is recorded from rim to core, the assemblages across large melt veins should record such a pressure release. This is the case for Zagami (*Langenhorst and Poirier,* 2000b), but clear evidence for crystallization during pressure release is lacking for highly shocked chondrites such as Tenham (*Xie et al.,* 2005) and RC 106 (*Aramovich et al.,* 2003). If crystallization occurred during pressure release for these S6 chondrites, then the pressure release was only ~5 GPa over the relatively long duration (50–250 ms) of cystallization.

The crystallization assemblages in a given melt vein will depend on the time required for melt vein quench vs. the duration of high shock pressures (Fig. 15). If the shock duration were longer than the crystallization time of the melt vein, then we would expect crystallization to have occurred during the period of high shock pressure and therefore record the continuum shock pressure. This appears to be the case for S6 samples such as Tenham (*Langenhorst et al.,* 1995; *Xie et al.,* 2003, 2005), RC 106 (*Aramovich et al.,* 2003), and Sixiangkou (*Chen et al.,* 1996). If the shock duration was the same duration as melt-vein crystallization, then it is likely that melt-vein crystallization would record both the continuum shock pressure and a lower pressure of partial release such that the core of the vein might contain an assemblage that crystallized at a lower pressure than that of the rest of the vein. If the shock pulse were shorter than the melt-vein quench time, one would expect crystallization of low-pressure assemblages in the core of the melt vein. This appears to be the case for S4 samples Kunashak and La Lande, which contain plagioclase-bearing crystallization assemblages. Finally, we note that some veins may form at low pressure during pressure release. The mineralogy of these veins would be unrelated to either the magnitude or duration of peak shock pressure.

2.15. Thermal Modeling of Melt-Vein Quench

The thermal history recorded in melt veins can be extracted by modeling the thermal history of the vein as it quenches and crystallizes (*Langenhorst and Poirier,* 2000a; *Xie et al.,* 2005). Finite-element methods can be used to

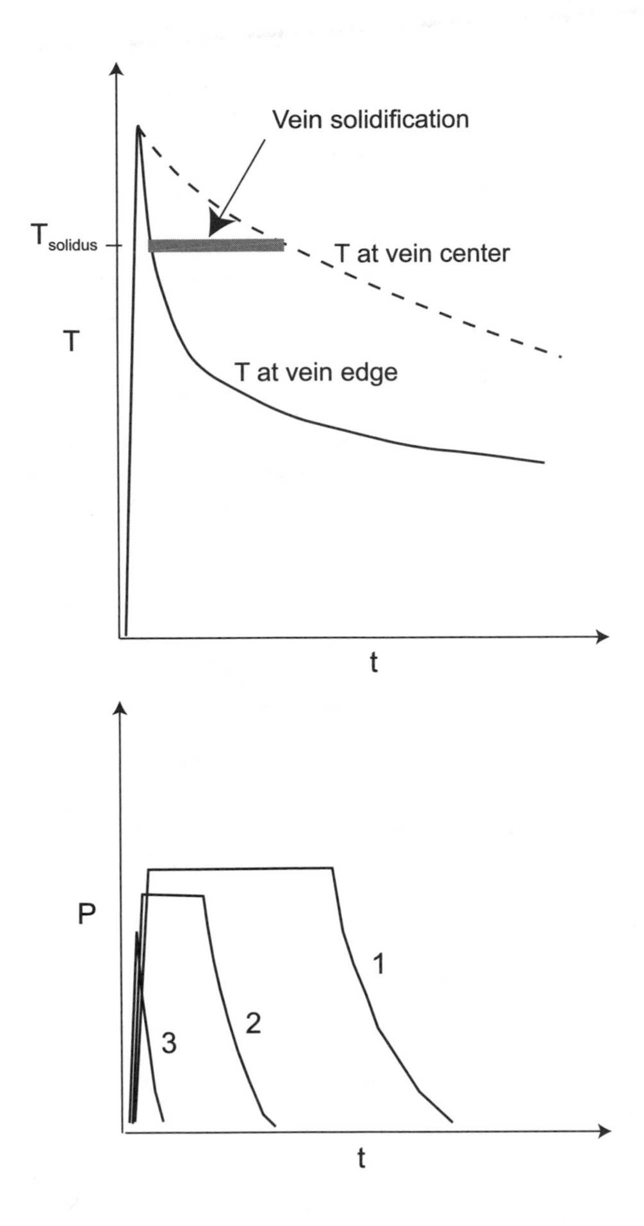

Fig. 15. The temperature profile (top) illustrates the cooling history of the melt-vein edge (solid) and melt-vein center (dashed) as they pass through the solidus temperature. The pressure-time profile (bottom) illustrates three quench scenarios: (1) The pressure pulse exceeds the duration of the total quench time and crystallization occurs at equilibrium shock pressure. (2) The pressure pulse is shorter than the total quench time, resulting in quench through the equilibrium shock pressure and into pressure release. (3) The pressure pulse is much shorter than the total quench time such that most of the quench occurs after pressure release.

model heat transfer to calculate transient heat flow from the melt vein into the chondrite matrix. The temperature difference between the melt vein and the surrounding is estimated by assuming that the melt-vein temperature is above the liquidus at pressure, and calculating the bulk shock temperature of host rock by using the method described above and in Appendix A. The thermal modeling allows one to calculate the temperature of any point within or near the melt vein as a function of time. The total crystallization time for the vein is estimated by calculating the time difference between when the melt-vein center and margin pass through the solidus temperature. This approach has been used to estimate a 50-ms time required to crystallize a 580-μm melt vein in Tenham (*Xie et al.,* 2003, 2005) (Fig. 16).

2.16. Application to Melt Veins in S3–S6 Chondrites

Melt-vein assemblages have been determined for a variety of S6 chondrites including Sixiangkou (*Chen et al.,* 1996), Acfer 040 (*Sharp et al.,* 1997), Tenham (*Xie et al.,* 2005), and RC 106 (*Aramovich et al.,* 2003). Similarly, a series of S3–S5 samples, including Umbarger (*Xie and Sharp,* 2004), Roy, La Landa, and Kunishack have been characterized. The melt-vein assemblages correspond to crystallization over a wide range of pressures from less than 3 GPa (La Landa and Kunishack) to approximately 25 GPa for the S6 samples. These results suggest that the maximum crystallization pressures, and therefore shock pressures, of L chondrites are around 25 GPa. Samples such as Acfer 040 (*Sharp et al.,* 1997) and Tenham (*Xie et al.,* 2005) have melt-vein assemblages that include silicate-perovskite, which is stable above 23 GPa (*Chen et al.,* 2004a). However, the perovskite in both these samples occurs with ringwoodite, suggesting that the pressure was not much greater than 23 GPa. This lack of evidence for melt-vein crystallization above 25 GPa indicates that the calibration of S6 shock effects of *Stöffler et al.* (1991) is about two times too high. A maximum shock pressure of about 25 GPa for chondrites also suggests that either the impact velocities in the early solar system are relatively low, or that samples shocked to significantly higher pressures have not been recognized. Using the synthetic Hugoniot for Tenham and assuming an impact between two Tenham-like bodies, one can calculate that a

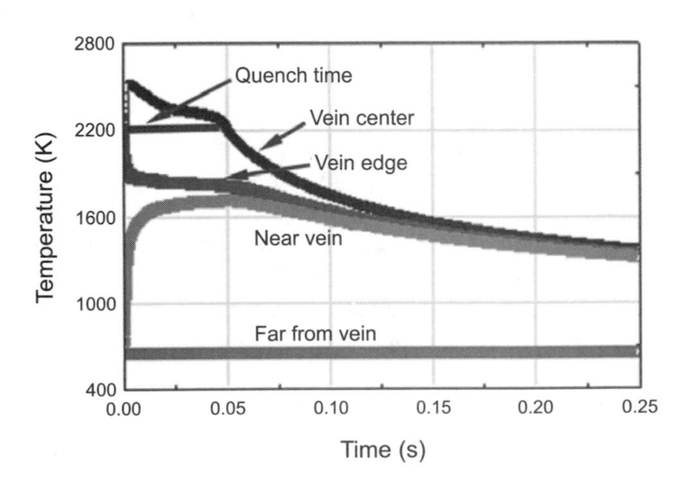

Fig. 16. Temperature vs. time profiles for the vein center, vein edge, and host rock for the 580-μm-wide vein in Tenham. The quench duration is the lag time between crystallization of vein edge and vein center, which is ~50 ms. The distances from vein-host interface to vein center, vein edge inside the vein margin, host rock near but outside the vein, and host rock outside the vein are 0.29 mm, 0.02 mm, 0.04 mm, and 4 mm respectively.

pressure of 25 GPa corresponds to a relative impact velocity of ~2 km/s (*Xie et al.,* 2005). Alternatively, the 25-GPa samples may have sampled a part of the parent body that was relatively far from the impact site and therefore experienced a lower continuum shock pressure than that near the impact site. If this were the case, one would expect to see chondrites that are more highly shocked than the S5–S6 samples that have been studied to date. It is possible that the more highly shocked samples exist, but high postshock temperatures have annealed out shock metamorphic features and transformed high-pressure minerals back to their low-pressure polymorphs. *Rubin* (2004) has provided evidence for nearly complete annealing of shock features in ordinary chondrites that had been previously shocked. Much more work is needed to determine shock pressures and durations that affected meteorites. Only after shock pressures are well known can we constrain the velocities of impact on meteorite parent bodies.

APPENDIX A: SPREADSHEET CALCULATIONS

In this section, we present a specific example of spreadsheet shock wave calculations. All the operations are based on equations (1)–(7) in the theory section. We use the formula notation of Excel in this section.

In column A, labeled U_p, increment particle velocity from 0 in steps of 0.01 km/s (0.00, 0.01, 0.02 . . .) up to 5 km/s. The increment of 0.01 is arbitrary, chosen because it provided adequate resolution; the upper particle velocity limit of 5 km/s covers the pressure range in excess of 100 GPa for meteoritic minerals.

In column B, U_s (shock velocity), enter the formula: = C + s * Ax, where C and s are the equation of state parameters of equation (7), and Ax is the column reference. *Ahrens and Johnson* (1995a) list the following equation of state parameters for Stillwater bronzitite: $U_s = 5.99 + 1.56 * U_p$ (for $U_p <$ 0.483 km/s), $U_s = 6.47 + 0.6 * U_p$ (for 0.483 km/s $< U_p <$ 2.131 km/s), and $U_s = 5.16 + 1.17 * U_p$, (for 2.043 km/s $<$ $U_p < 3.481$). Note that the particle velocity ranges in parentheses refer to the data range over which the straight line fit was made and not to a range of validity. Hugoniot measurements on other pyroxenes indicate no anomalies in the particle velocity range between 3.5 km/s and 5 km/s. It is therefore probably safe to extrapolate the Stillwater bronzitite Hugoniot to a particle velocity of 5 km/s.

For efficiency, one can enter the first equation in column B, the second in column C, and the third in column D. The fill operation is used to calculate all three relationships over the entire range of particle velocities. By inspection, one can determine the smoothest crossover points. The column C equation is entered in column B at the first crossover particle velocity (0.50 km/s, in this case), and the column D equation is entered in column B at the second crossover velocity (2.30 km/s). Columns C and D may now be overwritten.

In column C, P (pressure), enter the formula = ρ_0 * Ax * Bx, (equation (2); neglecting P_o, atmospheric pressure in this case) where ρ_0 is the initial density and Ax and Bx are column references to U_p and U_s. Note: If U_s and U_p have units of km/s and ρ has units of Mg/m^3, P has the dimensions of GPa. Also note that g/cm^3 and Mg/m^3 are numerically equal.

In column D, V (specific volume at pressure P), enter the formula = $1/\rho_0$–Ax^2/Cx (from equation (4), neglecting P_o). At this point, one may plot the Hugoniot data in P–V, P–U_p, P–U_s, or U_s–U_p planes.

One may also calculate the internal energy increase on shock compression, E–E_o in column E by entering the formula = 0.5 * Cx * ($1/\rho_0$–Dx) (equation (6), neglecting P_o). Note that E–E_o has dimensions of kJ/g or the numerically equal MJ/kg.

The area under the Hugoniot, an approximation to the area under the release adiabat, may be calculated by a Simpson's rule integration in column F. The first entry in the column is 0. The second entry is the formula = 0.5 * (Cx + Cx_{-1})*(Dx_{-1}–Dx) + Fx_{-1}, e.g., F10 = 0.5 * (C10 + C9) * (D9–D10) + F9.

The waste heat, column G, is given by the formula = Ex–Fx.

A portion of the spreadsheet is shown in Table A-1.

In column G, one may note that the waste heat on release from a pressure of 50.27 GPa is 223 J/g. It is useful to know that for most common minerals, excluding metals, the average specific heat over the temperature range between 300 K and 1000 K is ~1 J/gK. Assuming a preshock temperature of 300 K, one might estimate the postshock temperature to be ~525 K, ±~60 K. For a more precise estimate of postshock temperature, one would consult a table of H_t–H_{298} for the specific mineral over the temperature range of interest.

The corresponding shock temperature may be estimated from the approximate relationship:

Tshock = T_o * e^ (Γ*(V_o–V)), where To is the waste heat temperature (in K) and Γ is the Grüneisen parameter, a function of thermal expansion, bulk modulus, specific volume and C_v, the heat capacity at constant volume. High-pressure data on Γ are sparse, but for most rocks and minerals over the pressure range up to about 100 GPa, estimated values of Γ are in the range of 0.5 to about 2. Using a Γ of 1 and a postshock temperature of 525 K, the calculated shock temperature would be ~565 K. With a Γ of 2, the calculated shock temperature would be ~615 K. These calculations support the general argument that the reduction of temperature is relatively small on adiabatic expansion from high shock pressures.

Constructing Synthetic Hugoniots

The Hugoniot data compilations of *Ahrens and Johnson* (1995a,b) are relatively complete. Only a few mineral or rock Hugoniots have been published subsequently. One may

TABLE A-1.

A	B	C	D	E	F	G
Stillwater ρ = 3.277 (U_p, km/s)	Bronzitite (U_s, km/s)	P, GPa	Specific Volume (cc/g)	Internal Energy Increase ($E-E_o$, kJ/g)	Recoverable Energy Area under Hugoniot (kJ/g)	Waste Heat (kJ/g)
2	7.67	50.26918	0.225585	2	1.776532311	0.223467689
2.01	7.676	50.56005	0.22525	2.02005	1.793438821	0.226611179
2.02	7.682	50.85131	0.224915	2.0402	1.810416376	0.229783624
2.03	7.688	51.14296	0.224581	2.06045	1.827464866	0.232985134
2.04	7.694	51.43501	0.224247	2.0808	1.84458418	0.23621582
2.05	7.7	51.72745	0.223914	2.10125	1.861774208	0.239475792

note that Hugoniot data are not available for many mineral compositions of interest to meteorite researchers. Hugoniot data are available for only a few olivine, pyroxene, and feldspar compositions. However, available data indicate that Hugoniots are not too sensitive to compositional differences in isomorphous mineral series. The $P-U_p$ Hugoniots of forsterite and fayalite are closely matched, the $P-U_p$ Hugoniots of various feldspars are closely matched and $P-U_p$ Hugoniots of various pyroxenes are also closely matched. In the absence of Hugoniot data on the specific mineral composition of interest, one is forced to pick the closest match available.

However, there are a few minerals for which no Hugoniot data exist. We have noted that most rocks and minerals have small coefficients of thermal expansion to justify our use of the Hugoniot to approximate the release adiabat. By the same argument, one may use isothermal P–V data from static compression experiments to approximate Hugoniots, using equations (4) and (5) to convert P–V data to U_s-U_p and $P-U_p$ planes. If the material of interest is a polymineralic rock, one may construct a P–V relationship for the rock by summing the volumes of the constituent minerals at pressure. (*Ahrens and Johnson,* 1995b). Shock wave workers have been using this technique for over 50 years because it works well in the pressure range above about 10 GPa, where material strength effects may be neglected. The agreement has been excellent between measured Hugoniots and synthetic Hugoniots. In the most extreme example, the measured Hugoniot of brucite is a good match in the pressure regime above 10 GPa to the Hugoniot synthesized by volume addition of MgO and water Hugoniots. One may also synthesize the Hugoniot of a porous material, allowing the initial porosity to crush up over an appropriate range of pressures. Since the Hugoniot range of interest would probably be above 10 GPa, one could simply let the porosity crush up linearly over the range of 0–5 GPa. The details of the crushup would have no effect on the Hugoniot above the crushup range. Note that initial porosity has a much greater effect on a Hugoniot than compositional differences. *Anderson and Ahrens* (1998) reported measurements of the Hugoniot of Murchison. They noted that their measurements disagreed with the Hugoniot calculated by simple addition at pressure of the volumes of the constituent minerals. However, rather than measuring porosity, they used mineral norms and composition data to calculate a porosity of 16%. Measured values of Murchison porosity cluster around 23%. If one uses 23% porosity in calculating the Hugoniot of Murchison, there is excellent agreement between measured and calculated Hugoniots.

Figure A-1 depicts $P-U_p$ Hugoniots of some chondritic minerals and a synthetic Hugoniot for Tenham.

Meteoritic iron, represented by the Hugoniot for Fe-10%Ni, has the highest shock impedance over the entire pressure range considered here. However, relative shock impedances can change with pressure. At 10 GPa, the shock impedances of troilite and albite are closely matched. At about 55 GPa, the shock impedance of troilite becomes higher than the shock impedance of fayalite. The synthetic Hugoniot of Tenham, calculated by volume addition of its components, is very close to the Hugoniot of forsterite. The closeness of the Hugoniots of fayalite and forsterite illustrates the relative insensitivity of $P-U_p$ Hugoniots to the

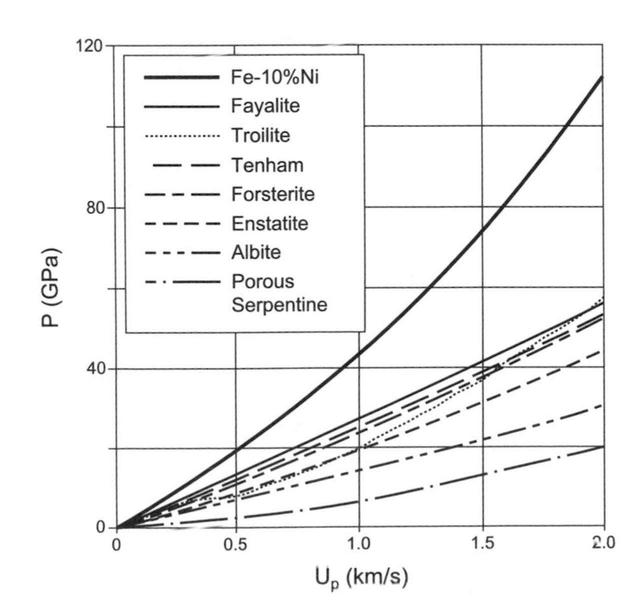

Fig. A-1. $P-U_p$ Hugoniots of some chondritic minerals and a synthetic Hugoniot for Tenham.

substitution of Fe for Mg in isomorphous series. Finally, the Hugoniot for porous serpentine is a proxy for a typical carbonaceous chondrite Hugoniot.

REFERENCES

Agee C. B., Li J., Shannon M. C., and Circone S. (1995) Pressure-temperature phase diagram for the Allende meteorite. *J. Geophys. Res., 100,* 17725–17740.

Ahrens T. J. and Gaffney E. S. (1971) Dynamic compression of enstatite. *J. Geophys. Res., 76,* 5504–5513.

Ahrens T. J. and Gregson V. G. (1964) Shock compression of crustal rocks: Data for quartz, calcite, and plagioclase rocks. *J. Geophys. Res., 69,* 4839–4874.

Ahrens T. J. and Johnson M. L. (1995a) Shock wave data for minerals. In *Mineral Physics and Crystallography: A Handbook of Physical Constants* (T. J. Ahrens, ed.), pp. 143–184. American Geophysical Union, Washington, DC.

Ahrens T. J. and Johnson M. L. (1995b) Shock wave data for rocks. In *Rock Physics and Phase Relations: A Handbook of Physical Constants* (T. J. Ahrens, ed.), pp. 35–44. American Geophysical Union, Washington, DC.

Ahrens T. J. and Rosenberg J. T (1968) Shock metamorphism: Experiments on quartz and plagioclase. In *Shock Metamorphism of Natural Materials* (B. M. French and N. M. Short, eds.), pp. 59–81. Mono, Baltimore.

Ahrens T. J., Lower J. H., and Lagus P. L. (1971) Equation of state of forsterite. *J. Geophys. Res., 76,* 518–528.

Alder B. J. (1963) Physics experiments with strong pressure pulses. In *Solids Under Pressure* (W. Paul and D. W. Warshauer, eds.), pp. 385–420. McGraw-Hill, New York.

Anderson W. W. and Ahrens T. J. (1998) Shock wave equations of state of chondritic meteorites. In *Shock Compression of Condensed Matter–1997* (S. C. Dandekar et al., eds.), pp. 115–118. AIP Conference Proceedings 429, American Institure of Physics, New York.

Angel R. J., Chopelas A., and Ross N. L., (1992) Stability of high-density clinoenstatite at upper-mantle pressure. *Nature, 358,* 322–324.

Antoun T., Seaman L., Curran D. R., Kanel G. I., Razorenov S. V., and Utkin A. V., (2003) *Spall Fracture*. Springer-Verlag, New York. 404 pp.

Aramovich C. J., Sharp T. G., and Wolf G. (2003) The distribution and significance of shock-induced high-pressure minerals in chondrite Skip Wilson (abstract). In *Lunar and Planetary Science XXXIV,* Abstract #1355. Lunar and Planetary Institute, Houston (CD-ROM).

Asay J. R. (2002) Shock wave paradigms and new challenges. In *Shock Compression of Condensed Matter–2001* (M. D. Furnish et al., eds.), pp. 26–35. AIP Conference Proceedings 620, American Institute of Physics, New York.

Ashworth J. R. (1985) Transmission electron microscopy of L-group chondrites, 1. Natural shock effects. *Earth Planet. Sci. Lett., 73,* 17–32.

Baer M. R.(2000) Computational modeling of heterogeneous materials at the mesoscale. In *Shock Compression of Condensed Matter–1999* (M. Furnish et al., eds.), pp. 27–33. AIP Conference Proceedings 505, American Institute of Physics, New York.

Baer M. R. and Trott W. M. (2002) Mesoscale descriptions of shock-loaded heterogeneous porous materials. In *Shock Compression of Condensed Matter–2001* (M. D. Furnish et al., eds.), pp. 517–520. AIP Conference Proceedings 620, American Institute of Physics, New York.

Baer M. R. and Trott W. M. (2004) Mesoscale studies of shock loaded tin sphere lattices In *Shock Compression of Condensed Matter–2003* (M. D. Furnish et al., eds.), pp. 517–520. AIP Conference Proceedings 706, American Institute of Physics, New York.

Beck P., Gillet P., Gautron L., Danielle I., and El Goresy A. (2004) A new natural high-pressure aluminium-silicate in shocked martian meteorites. *Earth Planet. Sci. Lett., 219,* 1–12.

Beck P., Gillet Ph., El Goresy A., and Mostefaoui S. (2005) Timescales of shock processes in chondritic and martian meteorites. *Nature, 435,* 1071–1074.

Binns R. A. (1970) (Mg, Fe)$_2$SiO$_4$ spinel in a meteorite. *Phys. Earth Planet. Inter., 3,* 156–160.

Binns R. A., Davis R. J., and Reed S. B. J. (1969) Ringwoodite, natural (Mg, Fe)$_2$SiO$_4$ spinel in the Tenham meteorite. *Nature, 221,* 943–944.

Bischoff A. and Stöffler D. (1992) Shock metamorphism as a fundamental process in the evolution of planetary bodies: Information from meteorites. *Eur. J. Mineral., 4,* 707–755.

Bowden K. E. (2002) Effects of loading path on the shock metamorphism of porous quartz: An experimental study. Ph.D. thesis, University College London. 228 pp.

Bogard D. D. and Johnson P. (1983) Martian gases in an Antarctic meteorite? *Science, 221,* 651–654.

Brearley A. J., Rubie D. C., and Ito E. (1992) Mechanisms of the transformations between the α-polymorphs, β-polymorphs and γ-polymorphs of Mg$_2$SiO$_4$ at 15 GPa. *Phys. Chem. Minerals, 18,* 343–358.

Bundy F. P. and Kasper J. S. (1967) Hexagonal diamond — a new form of carbon. *J. Chem. Phys., 46,* 3437–3446.

Burnley P. C. and Green H. W. (1989) Stress dependence of the mechanism of the olivine spinel transformation. *Nature, 338,* 753–756.

Carter N. L., Raleigh C. B., and DeCarli P. S. (1968) Deformation of olivine in stony meteorites. *J. Geophys. Res., 73,* 5439–5461.

Chao E. C. T. (1967) Shock effects in certain rock-forming minerals. *Science, 156,* 192–202.

Chao E. C. T. (1968) Pressure and temperature history of impact metamorphosed rocks-based on petrographic observations. In *Shock Metamorphism of Natural Materials* (B. M. French and N. M Short, eds.), pp. 135–158. Mono, Baltimore.

Chen M. and El Goresy A. (2000) The nature of maskelynite in shocked meteorites: Not diaplectic glass but a glass quenched from shock-induced dense melt at high pressures. *Earth Planet. Sci. Lett., 179,* 489–502.

Chen M., Sharp T. G., El Goresy A., Wopenka B., and Xie X. (1996) The majorite-pyrope + magnesiowüstite assemblage. Constraints on the history of shock veins in chondrites. *Science, 271,* 1570–1573.

Chen M., Shu J. F., Xie X. D., and Mao H. K. (2003) Natural CaTi$_2$O$_4$-structured FeCr$_2$O$_4$ polymorph in the Suizhou meteorite and its significance in mantle mineralogy. *Geochim. Cosmochim. Acta, 67,* 3937–3942.

Chen M., El Goresy A., Frost D., and Gillet P. (2004a) Melting experiments of a chondritic meteorite between 16 and 25 GPa: Implication for Na/K fractionation in a primitive chondritic Earth's mantle. *Eur. J. Mineral., 16,* 203–211.

Chen M., El Goresy A, and Gillet P. (2004b) Ringwoodite lamellae in olivine: Clues to olivine-ringwoodite phase transition mechanisms in shocked meteorites and subducting slabs. *Proc. Natl. Acad. Sci. U.S., 101,* 15033–15037.

Consolmagno G. J. and Britt D. T. (1998) The density and porosity of meteorites from the Vatican collection. *Meteoritics & Planet. Sci., 33,* 1231–1241.

Courant R. and Friedrichs K. O. (1948) *Supersonic Flow and Shock Waves*. Interscience, New York.

Davis P. K. (1977) Effects of shock pressure on ^{40}Ar-^{39}Ar radiometric age determinations. *Geochim. Cosmochim. Acta, 41,* 195–205.

DeCarli P. S. (1979) Nucleation and growth of diamond in shock wave experiments. In *High Pressure Science and Technology, Vol. 1* (K. D. Timmerhaus and M. S. Barber, eds.), pp. 940–943. Plenum, New York.

DeCarli P. S. (1995) Shock wave synthesis of diamond and other phases. In *Mechanical Behavior of Diamond and Other Forms of Carbon* (M. D. Drory et al., eds.), pp. 21–31. Materials Research Society Symposium Proceedings 383, Materials Research Society.

DeCarli P. S. and Jamieson J. C. (1959) Formation of an amorphous form of quartz under shock conditions. *J. Chem. Phys., 31,* 1675–1676.

DeCarli P. S. and Jamieson J. C. (1961) Formation of diamond by explosive shock. *Science, 133,* 1821–1822.

DeCarli P. S., Bowden E., Jones A. P., and Price G. D. (2002a) Laboratory impact experiments versus natural impact events. In *Catastrophic Events and Mass Extinctions: Impacts and Beyond* (C. Koeberl and K. G. MacLeod, eds.), pp. 595–605. GSA Special Paper 356, Geological Society of America, Boulder.

DeCarli P. S., Bowden E., Sharp T. G., Jones A. P., and Price G. D. (2002b) Evidence for kinetic effects on shock wave propagation in tectosilicates. In *Shock Compression of Condensed Matter–2001* (M. D. Furnish et al., eds.), pp. 1381–1384. AIP Conference Proceedings 620, American Institute of Physics, New York.

De Carli P. S., Aramovich C. J., Xie Z., and Sharp T. G. (2004) Meteorite studies illuminate phase transition behavior of minerals under shock compression. In *Shock Compression of Condensed Matter–2003* (M. D. Furnish et al., eds.), pp. 1427–1430. AIP Conference Proceedings 706, American Institute of Physics, New York.

Dodd R. T. and Jarosewich E. (1979) Incipient melting in and shock classification of L-group chondrites. *Earth Planet. Sci. Lett. 59,* 335–340.

Duvall G. E. and Fowles G. R. (1963) Shock waves. In *High Pressure Physics and Chemistry, Vol. 2* (R. S. Bradley, ed.), pp. 209–291. Academic, London.

El Goresy A., Dubrovinsky L., Sharp T. G., Saxena S. K., and Chen M. (2000) A monoclinic polymorph of silica in the Shergotty meteorite. *Science, 288,* 37–55.

El Goresy A., Gillet P., Chen M., Künstler F., Graup G., and Stähle V. (2001) In situ discovery of shock-induced graphite-diamond transition in gneisses from the Ries Crater, Germany. *Am. Mineral., 86,* 611–621.

Fredriksson K. and DeCarli P. S. (1964) Shock emplaced argon in a stony meteorite I. *J. Geophys. Res., 69,* 1403–1406.

Fredriksson K., DeCarli P. S., and Aaramäe A (1963) Shock-induced veins in chondrites. In *Space Research III, Proceedings of the Third International Space Science Symposium* (W. Priester, ed.), pp. 974–983. North-Holland, Amsterdam.

Gasparik T. (1992) Melting experiments on the enstatite-pyrope join at 80–152 kbar. *J. Geophys. Res., 97,* 15181–15188.

Gillet P., Chen M., Dubrovinsky L., and El Goresy A. (2000) Natural $NaAlSi_3O_8$-hollandite in the shocked Sixiangkou meteorite. *Science, 287,* 1633–1636.

Goltrant O., Cordier P., and Doukhan J. C. (1991) Planar deformation features in shocked quartz: A transmission electron microscopy investigation. *Earth Planet. Sci. Lett., 106,* 103–155.

Goltrant O., Doukhan J. C., and Cordier P. (1992) Formation mechanisms of planar deformation features in naturally shocked quartz. *Phys. Earth. Planet. Inter., 74,* 219–240.

Gupta Y. M. (2002) The coupling between shock waves and condensed matter: Continuum mechanics to quantum mechanics. In *Shock Compression of Condensed Matter–2001* (M. D. Furnish et al., eds.), pp. 3–10. AIP Conference Proceedings 620, American Institute of Physics, New York.

Heinemann S., Sharp T. G., Seifert F., and Rubie D. C. (1997) The cubic-tetragonal phase transition in the system majorite ($Mg_4Si_4O_{12}$) — pyrope ($Mg_3Al_2Si_3O_{12}$), and garnet symmetry in the Earth's transition zone. *Phys. Chem. Minerals, 24,* 206–221.

Hemley R. J., Jephcoat A. P., Mao H. K., Ming L. C., and Manghani M. H. (1988) Pressure-induced amorphization of crystalline silica. *Nature, 334,* 52–54.

Heymann D. (1967) On the origin of hypersthene chondrites: Ages and shock effects of black chondrites. *Icarus, 6,* 189–221.

Hogrefe A., Rubie D. C., Sharp T. G., and Seifert F. (1994) Metastability of enstatite in deep subducting lithosphere. *Nature, 372,* 351–353.

Hörz F. (1968) Statistical measurements of deformation structures and refractive indices in experimentally shock loaded quartz. In *Shock Metamorphism of Natural Materials* (B. M French and N. M Short, eds.), pp. 243–253. Mono, Baltimore.

Huffman A. R., Brown J. M., Carter N. L., and Reimold W. U. (1993) The microstructural response of quartz and feldspar under shock loading at variable temperature. *J. Geophys. Res., 98,* 22171–22197.

Hugh-Jones T., Sharp T. G., Angel R., and Woodland A. (1996) The transition of orthoferrosilite to high-pressure *C2/c* clinoferrosilite at ambient temperature. *Eur. J. Mineral., 8,* 1337–1345.

Jeanloz R. (1980) Shock effects in olivine and implications for Hugoniot data. *J. Geophys. Res., 85,* 3163–3176.

Kerschhofer L., Sharp T. G., and Rubie D. C. (1996) Intracrystalline transformation of olivine to wadsleyite and ringwoodite under subduction zone conditions. *Science, 274,* 79–81.

Kerschhofer L., Dupas C., Liu M., Sharp T. G., Durham W. B., and Rubie D. C. (1998) Polymorphic transformations between olivine, wadsleyite and ringwoodite: Mechanisms of intracrystalline nucleation and the role of elastic strain. *Mineral. Mag., 62,* 617–638.

Kerschhofer L., Rubie D. C., Sharp T. G., McConnell J. D. C., and Dupas-Bruzek C. (2000) Kinetics of intracrystalline olivine-ringwoodite transformation. *Phys. Earth Planet. Inter., 121,* 59–76.

Kieffer S. W. (1971) Shock metamorphism of the Coconino sandstone at Meteor Crater. Arizona. *J. Geophys. Res., 76,* 5449–5473.

Kieffer S. W., Phakey P. P., and Christie J. M. (1976) Shock processes in porous quartzite: Transmission electron microscope observations and theory. *Contrib. Mineral. Petrol., 59,* 41–93.

Koeberl C., Masaitis V. L., Shafranovsky G. I., Gilmour I., Langenhorst F., and Schrauder M. (1997) Diamonds from the Popigai impact structure. *Geology, 25,* 967–970.

Kubo T., Ohtani E., Kato T., Morishima H., Yamazaki D., Suzuki A., Mibe K., Kikeegawa T., and Shimomura O. (1998a) An in situ X-ray diffraction study of the α-β transformation kinetics of Mg_2SiO_4. *Geophys. Res. Lett., 25,* 695–698.

Kubo T., Ohtani E., Kato T., and Fujino K. (1998b) Experimental investigation of the α-β transformation in San Carlos olivine single crystal. *Phys. Chem. Mineral., 26,* 1–6.

Langenhorst F. and Poirier J. P. (2000a) Anatomy of black veins in Zagami: Clues to the formation of high-pressure phases. *Earth Planet. Sci. Lett., 184,* 37–55.

Langenhorst F. and Poirier J. P. (2000b) 'Eclogitic' minerals in shocked basaltic meteorite. *Earth Planet. Sci. Lett., 176,* 259–265.

Langenhorst F., Joreau P., and Doukhan J. C. (1995) Thermal and shock metamorphism of the Tenham chondrite: A TEM examination. *Geochim. Cosmochim. Acta, 59,* 1835–1845.

Leroux H. (2001) Microstructural shock signatures of major minerals in meteorites. *Eur. J. Mineral., 13,* 253–272.

Lipschutz M. E. (1968) Shock effects in iron meteorites: A review. In *Shock Metamorphism of Natural Materials* (B. M. French and N. M. Short, eds.), pp. 571–599. Mono, Baltimore.

Liu M., Kerschhofer L., Mosenfelder J., and Rubie D. C. (1998) The effect of strain energy on growth rates during the 4 olivine-spinel transformation and implications for olivine metastability in subductoing slabs. *J. Geophys. Res., 103,* 23897–23909.

Madon M. and Poirier J. P. (1980) Dislocations in spinel and garnet high-pressure polymorphs of olivine and pyroxene: Implications for mantle mineralogy. *Science, 207,* 66–68.

Madon M. and Poirier J. P. (1983) Transmission electron microscope observation of α, β, γ $(Mg,Fe)_2SiO_4$ in shocked meteorites: Planar defects and polymorphic transitions. *Phys. Earth Planet. Inter., 33,* 31–44.

Mason B., Nelen J., and White S. J. Jr. (1968) Olivine-garnet transformation in a meteorite. *Science, 160,* 66–67.

McQueen R. G. (1964) Laboratory techniques for very high pressures and the behavior of metals under dynamic loading. In *Metallurgy at High Pressures and High Temperatures* (K. S. Gschneider et al., eds.), pp. 44–132. Gordon and Breach, New York.

McQueen R. G., Marsh S. P., Taylor J. W., Fritz J. N., and Carter W. J. (1970) The equation of state of solids from shock wave studies. In *High-Velocity Impact Phenomena* (R. Kinslow, ed.), pp. 530–568. Academic, New York.

Melosh H. J. (1989) *Impact Cratering: A Geologic Process.* Oxford Univ., New York. 245 pp.

Meyers M. A. (1994) *Dynamic Behavior of Materials.* Wiley and Sons, New York. 688 pp.

Migault A. (1998) Concepts of shock waves. In *Impacts on Earth* (D. Benest and C. Froeshlé, eds.), pp. 79–112. Springer-Verlag, Berlin.

Milton D. J. and DeCarli P. S. (1963) Maskelynite: Formation by explosive shock. *Science, 147,* 144–145.

Mori H. (1994) Shock-induced phase transformations on the Earth and planetary materials. *J. Mineral. Soc. Japan, 23,* 171–178.

Mori H. and Takeda H. (1985) Magnesiowüstite in a shock-produced vein of the Tenham chondrite (abstract). In *Lunar and Planetary Science XVI,* pp. 579–598. Lunar and Planetary Institute, Houston.

Mosenfelder J. L., Marton F. C., Ross C. R. II, Kerschhofer L., and Rubie D. C. (2001) Experimental constraints on the depth of olivine metastability in subducting lithosphere. *Phys. Earth Planet. Inter., 127,* 165–180.

Nesterenko V. F. (2001) *Dynamics of Heterogeneous Materials.* Springer-Verlag, New York. 510 pp.

Ohtani E., Kimura Y., Kimura M., Takata T., Kondo T., and Kubo T. (2004) Formation of high-pressure minerals in shocked L6 chondrite Yamato 791384: Constraints on shock conditions and parent body size. *Earth Planet. Sci. Lett., 227,* 505–515.

Panero W. R., Benedetti L. R., and Jeanloz R. (2003) Equation of state of stishovite and interpretation of SiO_2 shock-compression data. *J. Geophys. Res., 108,* B012015, doi:10.1029/2001JB001663.

Pepin R. O., Reynolds J. H., and Turner G. (1964) Shock emplaced argon in a stony meteorite II. *J. Geophys. Res., 69,* 1406–1411.

Porter D. A. and Easterling K. E. (1978) Dynamic studies of the tensile deformation and fracture of pearlite using high-resolution 200-KV SEM. *Scandinav. J. Metall., 7,* 55–56.

Price G. D. (1983) The nature and significance of stacking-faults in wadsleyite, natural beta-$(Mg,Fe)_2SiO_4$ from the Peace River meteorite. *Phys. Earth Planet. Inter., 33,* 137–147.

Price G. D., Putnis A., and Agrell S. O. (1979) Electron petrography of shock-produced veins in the Tenham chondrite. *Contrib. Mineral. Petrol., 71,* 211–218.

Price G. D., Putnis A., and Smith D. G. W. (1982) A spinel to β-phase transformation mechanism in $(Mg,Fe)_2SiO_4$. *Nature, 296,* 729–731.

Price G. D., Putnis A., Agrell S. O., and Smith D. G. W. (1983) Wadsleyite, natural beta-$(Mg,Fe)_2SiO_4$ from the Peace River meteorite. *Can. Mineral., 21,* 29–35.

Putnis A. and Price G. D. (1979) High-pressure $(Mg, Fe)_2SiO_4$ phases in the Tenham chondritic meteorite. *Nature, 280,* 217–218.

Rubie D. C. (1993) Mechanisms and kinetics of reconstructive phase transformations in the Earth's mantle. In *Short Courses Handbook on Experiments at High Pressure and Applications to the Earth's Mantle, Vol. 21* (R. W. Luth, ed.), pp. 247–303. Mineralogical Assocication of Canada, Edmonton.

Rubie D. C. and Ross C. R. (1994) Kinetics of the olivine-spinel transformation in subducting lithosphere: Experimental constraints and implications for deep slab processes. *Phys. Earth Planet. Inter., 86,* 223–241.

Rubin A. E. (2004) Postshock annealing and postannealing shock in equilibrated ordinary chondrites: Implications for the thermal and shock histories of chondritic asteroids, *Geochim. Cosmochim. Acta, 68,* 673–689.

Rubin A. E. (2005) What heated the asteroids? *Sci. Am., 285(5),* 81–87.

Schmitt D. R. and Ahrens T. J. (1983) Temperatures of shock-induced shear instabilities and their relation to fusion curves. *Geophys. Res. Lett., 10,* 1077–1080.

Schmitt R. T. (2000) Shock experiments with the H6 chondrite Kernouvé: Pressure calibration of microscopic shock effects. *Meteoritics & Planet. Sci., 35,* 545–560.

Scott E. R. D., Keil K., and Stöffler D. (1992) Shock metamorphism of carbonaceous chondrites. *Geochim. Cosmochim. Acta, 56,* 4281–4293.

Sears D. W. and Dodd R. T. (1988) Overview and classification of meteorites. In *Meteorites and the Early Solar System* (J. F. Kerridge and M. S. Matthews, ed.), pp. 3–31. Univ. of Arizona, Tucson.

Sharp T. G. and Rubie D. C. (1995) Catalysis of the olivine to spinel transformation by high clinoenstatite. *Science, 269,* 1095–1098.

Sharp T. G., Lingemann C. M., Dupas C., and Stöffler D. (1997) Natural occurrence of $MgSiO_3$-ilmenite and evidence for $MgSiO_3$-perovskite in a shocked L chondrite. *Science, 277,* 352–355.

Sharp T. G., El Goresy A., Wopenka B., and Chen M. (1999) A post-stishovite SiO_2 polymorph in the meteorite Shergotty: Implications for impact events. *Science, 284,* 1511–1513.

Smith J. V. and Mason B. (1970) Pyroxene-garnet transformation in Coorara meteorite. *Science, 168,* 832.

Spray J. G. (1999) Shocking rocks by cavitation and bubble implosion. *Geology, 27,* 695–698.

Stöffler D. (1972) Deformation and transformation of rock-forming minerals by natural and experimental shock processes: I. Behavior of minerals under shock compression. *Fortsch. Mineral., 49,* 50–113.

Stöffler D. (1974) Deformation and transformation of rock-forming minerals by natural and experimental shock processes: II. Physical properties of shocked minerals. *Fortsch. Mineral., 51,* 256–289.

Stöffler D., Bischoff A., Buchwald V., and Rubin A. E. (1988) Shock effects in meteorites. In *Meteorites and the Early Solar System* (J. F. Kerridge and M. S. Matthews, ed.), pp. 165–202. Univ. of Arizona, Tucson.

Stöffler D., Keil K., and Scott E. R. D. (1991) Shock metamorphism of ordinary chondrites. *Geochim. Cosmochim. Acta, 55,* 3845–3867.

Taylor G. J. and Heymann D. (1969) Shock, reheating, and the gas retention ages of chondrites. *Earth Planet. Sci. Lett., 7,* 151–161.

Tomeoka K., Yamahana Y., and Sekine T. (1999) Experimental shock metamorphism of the Murchison CM carbonaceous chondrite. *Geochim. Cosmochim. Acta, 63,* 3683–3703.

Tomioka N. and Fujino K. (1997) Natural (Mg,Fe)SiO_3-ilmenite and -perovskite in the Tenham meteorite. *Science, 277,* 1084–1086.

Tomioka N., Mori H., and Fujino K. (2000) Shock-induced transition of $NaAlSi_3O_8$ feldspar into a hollandite structure in a L6 chondrite. *Geophys. Res. Lett., 27,* 3997–4000.

Tschermak G. (1872) Die meteoriten von Shergotty und Goalpur. *Sitzber. Akad. Wiss. Wien Math.-naturw., Kl. Abt I, 65,* 122–146.

Van Schmus W. R. and Ribbe P. H. (1968) The composition and structural state of feldspar from chondritic meteorites. *Geochim. Cosmochim. Acta, 32,* 1327–1342.

Walsh J. M. and Christian R. H. (1955) Equation of state of metals from shock wave measurements. *Phys. Rev., 97,* 1544–1556.

Wiens R. C. and Pepin R. O. (1988) Laboratory shock emplacement of noble gases, nitrogen, and carbon dioxide into basalt, and implications for trapped gases in shergottite EETA 79001. *Geochim. Cosmochim. Acta, 52,* 295–307.

Williams Q. and Jeanloz R. (1989) Static amorphization of anorthite at 300 K and comparison with diaplectic glass. *Nature, 338,* 413–415.

Xie Z. (2003) Transmission electron microscopy study of high pressure phases in the shock-induced melt veins of L chondrites: Constraints on the shock pressure and duration. Ph.D. thesis, Arizona State Univ., Tempe. 149 pp.

Xie Z. and Sharp T. G. (2000) Mineralogy of shock-induced melt veins in chondrites as a function of shock grade (abstract). In *Lunar and Planetary Science XXXI,* Abstract #2065. Lunar and Planetary Institute, Houston (CD-ROM).

Xie Z. and Sharp T. G. (2003) TEM observations of amorphized silicate-perovskite, akimotoite, and Ca-rich majorite in a shock-induced melt vein in the Tenham L6 chondrite (abstract). In *Lunar and Planetary Science XXXIV,* Abstract #1469. Lunar and Planetary Institute, Houston (CD-ROM).

Xie Z. and Sharp T. G. (2004) High-pressure phases in shock-induced melt veins of the Umbarger L6 chondrite: Constraints on shock pressure. *Meteoritics & Planet. Sci., 39,* 2043–2054.

Xie Z., Tomioka N., and Sharp T. G. (2002) Natural occurrence of Fe_2SiO_4-spinel in the shocked Umbarger L6 chondrite. *Am. Mineral., 87,* 1257–1260.

Xie Z., Sharp T. G., and DeCarli P. S. (2003) Estimating shock pressures from high-pressure minerals in shock-induced melt veins of the chondrites (abstract). In *Lunar and Planetary Science XXXIV,* Abstract #1280. Lunar and Planetary Institute, Houston (CD-ROM).

Xie Z., Sharp T. G., and DeCarli P. S. (2005) High pressure phases in a shock-induced melt vein of Tenham L6 chondrite: Constraints on shock pressure and duration. *Geochim. Cosmochim. Acta,* in press.

Zener C. and Hollomon J. H. (1944) Effect of strain rate on plastic flow of steel. *J. Appl. Phys., 15,* 22–32.

Zhang J. and Herzberg C. (1994) Melting experiments on anhydrous peridotite KLB-1 from 5.0 to 22.5 GPa. *J. Geophys. Res., 99,* 17729–17742.

Nature and Origins of Meteoritic Breccias

Addi Bischoff
Westfälische Wilhelms-Universität Münster

Edward R. D. Scott
University of Hawai'i

Knut Metzler
Westfälische Wilhelms-Universität Münster

Cyrena A. Goodrich
University of Hawai'i

Meteorite breccias provide information about impact processes on planetary bodies, their collisional evolution, and their structure. Fragmental and regolith breccias are abundant in both differentiated and chondritic meteorite groups and together with rarer impact-melt rocks provide constraints on cratering events and catastrophic impacts on asteroids. These breccias also constrain the stratigraphy of differentiated and chondritic asteroids and the relative abundance of different rock types among projectiles. Accretional chondritic breccias formed at low impact speeds (typically tens or hundreds of meters per second), while other breccias reflect hypervelocity impacts at higher speeds (~5 km/s) after asteroidal orbits were dynamically excited. Iron and stony-iron meteorite breccias only formed, when their parent bodies were partly molten. Polymict fragmental breccias and regolith breccias in some meteorite groups contain unique types of clasts that do not occur as individual meteorites in our collections. For example, ureilite breccias contain feldspathic clasts from the ureilite parent body as well as carbonaceous chondritic projectile material. Such clasts provide new rock types from both unsampled parent bodies and unsampled parts of known parent bodies. We review breccias in all types of asteroidal meteorites and focus on the formation of regolith breccias and the role of catastrophic impacts on asteroids.

1. GENERAL INTRODUCTION

1.1. Meteorite Breccias

Chondrites and differentiated meteorites provide information about collisions during the accretion of asteroids at relatively slow speeds (typically less than a few hundred meters per second), and during and after thermal processing when asteroids collided at speeds of many kilometers per second. Impacts at speeds above ~20 m/s broke rock, while hypervelocity impacts left shock damage and formed breccias from fragments of earlier rocks. The study of meteoritic breccias contributes significantly to our understanding of early solar system processes of accretion, differentiation, and surface (regolith) evolution, and also provides unique information about the primordial, chemical, and mineralogical characteristcs of the accreted components themselves (*Bischoff and Stöffler,* 1992). The latter can best be seen by examining the constituents of primitive accretionary breccias (Lunar and martian breccias are not discussed in this paper.)

In the nineteenth century certain textural features in meteorites were described that we now recognize to result from shock waves during collisional processes. *Partsch* (1843) and *von Reichenbach* (1860) described "polymict breccias" and *Tschermak* (1872) identified "maskelynite." The presence of shock veins and brecciation were used as fundamental criteria of the Rose-Tschermak-Brezina classification scheme of meteorites (*Brezina,* 1904). Numerous early studies on meteoritic breccias are reviewed by *Rajan* (1974), *Bunch* (1975), *Dymek et al.* (1976), *Prinz et al.* (1977), and *Keil* (1982). These papers used comparisons between lunar and meteorite impact breccias to establish a basic understanding of meteorite breccias. However, meteoritic breccias are more difficult to understand because the asteroids are very different in terms of their target size, projectile flux, impact velocities, and physical and chemical properties. In addition, the meteorites come from a set of bodies that experienced very diverse alteration and metamorphic and igneous histories and they preserve evidence for hypervelocity impacts over ~50 m.y. on hot targets and 4.5 G.y. of impacts on cold targets. We have examined only a few asteroids from passing spacecraft and only one has been studied from orbit [prior to a landing on Eros by the *NEAR Shoemaker* spacecraft in 2001 (*Sullivan et al.,* 2002)]. We remain woefully ignorant about the diverse effects of impacts between asteroids spanning many orders of magnitude in size (*Holsapple et al.,* 2002), and we have yet to understand fundamental questions such as the impact rate in the primordial asteroid belt and the excitation process that changed asteroids' orbits, causing hypervelocity impacts (*Petit et al.,* 2002). Despite these problems, considerable progress has been made in characterizing and understanding meteorite breccias in the last 20 years.

1.2. Why Study Meteoritic Breccias?

Meteoritic breccias represent fragmented samples from a variety of parent bodies (*Burbine et al.,* 2002). In many cases, a large fraction of our samples are brecciated. For example, all CI chondrites, mesosiderites, and aubrites and over 80% of all HED meteorites are breccias of various kinds. Only four groups of stony meteorites lack clearly defined breccias: angrites, brachinites, acapulcoites, and lodranites. Thus the vast majority of coherent rocks in their parent bodies cannot be understood without an appreciation for the long history of impacts that have affected them.

The brecciated meteorites provide important information about the history and evolution of the asteroids and impact processes on small bodies (*Keil,* 1982). This includes processes of accretion; the nature of primary parent body lithologies; excavation of those lithologies; and impact-related heating, metamorphism, melting, and mixing, as well as subsequent reaccretion and lithification. Because of the presence of types of clasts in polymict breccias that do not occur as individual meteorites in our collections, it is possible to study samples of new rock types, from both unsampled parent bodies and unsampled parts of known parent bodies. For example, dark inclusions found in ordinary and carbonaceous chondrites are unique, and may be fragments of C-chondrite parent bodies that existed prior to formation of the present host carbonaceous chondrite parent bodies. Alternatively, they may represent fragments of different lithologies from the same parent body. Likewise, feldspathic clasts in polymict ureilites may represent basaltic rocks complementary to the ultramafic monomict ureilites, and are otherwise unrepresented in our collections. In addition, mixtures of various clasts in a breccia (e.g., the presence of clasts of differentiated material in a chondritic breccia, or vice versa) can provide information about relative ages of early solar system processes, as well as the varieties of materials available within one region of the asteroid belt.

Regolith breccias provide us with our best tangible clues to the nature of asteroid surfaces studied by remote sensing techniques. Their solar-wind-implanted noble gases and irradiation records in minerals may provide information about a possible early active phase of the Sun (*Woolum and Hohenberg,* 1993).

An essential step in unraveling this history is to date the formation of the breccias. Compaction ages of regolith breccias can help to understand regolith formation on asteroids and constrain the evolution of the Sun. Dating of impact-melt breccias gives detailed insights into the impact histories of the asteroids over 4.5 G.y. and major impact events.

2. BRECCIA TYPES AND ABUNDANCES

2.1. Nomenclature and Characterization of Breccias

As pointed out by *Bischoff and Stöffler* (1992) various collision scenarios lead to specific combinations of shock metamorphism [and its effects (*Stöffler et al.,* 1988, 1991)] and breccia formation, if the relative sizes and velocities of the colliding bodies and the specific impact energies are considered (see paragraph 3). Breccia formation requires mass transport and therefore the relative movement of rock fragments and their displacement from the primary location in the source material (*Stöffler et al.,* 1988).

Details of the modern classification and nomenclature of breccias and their components are given by *Stöffler et al.* (1979, 1980, 1988), *Keil* (1982), *Scott and Taylor* (1982), *Taylor* (1982), *Bunch and Rajan* (1988), and *Bischoff and Stöffler* (1992). Some characteristics of special types of breccias and their constituents, which are discussed in this chapter, are given in Table 1.

Primitive, accretionary breccias can be formed in a low-velocity regime and mainly occur among carbonaceous and ordinary chondrites (e.g., *Kracher et al.,* 1982; *Scott and Taylor,* 1982). These chondrites have matrices composed almost entirely of primitive components found in type 3 chondrites including chondrules and opaque and recrystallized, fine-grained silicate matrix (*Scott and Taylor,* 1982). Chondrites that contain chondritic clasts that are rimmed by matrix material, lack shock effects, and are comparable in size to chondrules may have accreted together with the other chondritic components — CAIs, chondrules, and matrix — during assembly of the parent body (*Kracher et al.,* 1982). Such chondritic clasts appear to be derived from early-formed planetesimals. Most accretionary breccias apparently lack solar-wind gases (e.g., *Nakamura et al.,* 2003).

The vast majority of meteorite breccias formed during impacts between asteroids at velocities in excess of about 0.5–1 km/s, which shocked and melted minerals (*Stöffler et al.,* 1988). Impact velocities in the main asteroid belt due to mutual collisions currently range from 1 to 12 km/s with a mean of 5.3 km/s (*Bottke et al.,* 1994). (Comets impact at higher speeds, but they are relatively rare and their effects have not been recognized yet.) Impact processes modify the targets and the melts can be incorporated into crater deposits (*Bischoff and Stöffler,* 1992); in asteroidal surface-subsurface units the following types of impact breccias can be found: monomict (for example the brecciated monolithic basement rock) and dimict breccias, polymict breccias (such as regolith breccias), fragmental breccias, impact-melt breccias, and granulitic breccias. Dimict breccias are composed of two distinct lithologies, whereas polymict breccias are consolidated rocks consisting of clasts and/or matrix of different composition and/or origin. If they result from lithification of the upper surface debris and contain grains that were in the top millimeter of the asteroidal surface, they contain solar-wind-implanted noble gases and solar-flare tracks (e.g., *Wänke,* 1965; *Eberhardt et al.,* 1965; *Geiss,*

mental and regolith breccias containing impact melt-rock clasts. Regimes for collision-induced effects in the meteorites' parent bodies as a function of the approach velocity (velocity at infinity) and the size of the colliding bodies are shown in Fig. 1. Figure 1 is modified and redrawn from *Hartmann* (1979) and *Stöffler et al.* (1988) and assumes the collision of similar-sized bodies that have interiors with strength and elasticity of basalt or other igneous rock, but surface layers of loose or weakly bonded fragmental material. The figure describes several distinct regimes of interest (cf. *Hartmann,* 1979): (1) rebound and escape of two bodies; (2) rebound and fallback, producing an unfractured contact binary; (3) fragmentation and reaccretion leading to a brecciated spheroid; and (4) disruption with sufficient energy that most bodies escape entirely leading to total disruption.

Impact-induced breccia formation of porous regolith or C-chondrite materials requires collision velocities of at least 1.3 km/s (Fig. 1) (*Stöffler et al.,* 1988; *Bischoff and Stöffler,* 1992). Regolith breccias or fragmental breccias with melt-rock clasts come from bodies that experienced collisions at velocities of at least 4.5–5 km/s, which might be slightly lower if abundant metals are involved (e.g., mesosiderites). One should point out that all these suggested velocities are well within the range expected for the impact velocities in the main asteroid belt (see above).

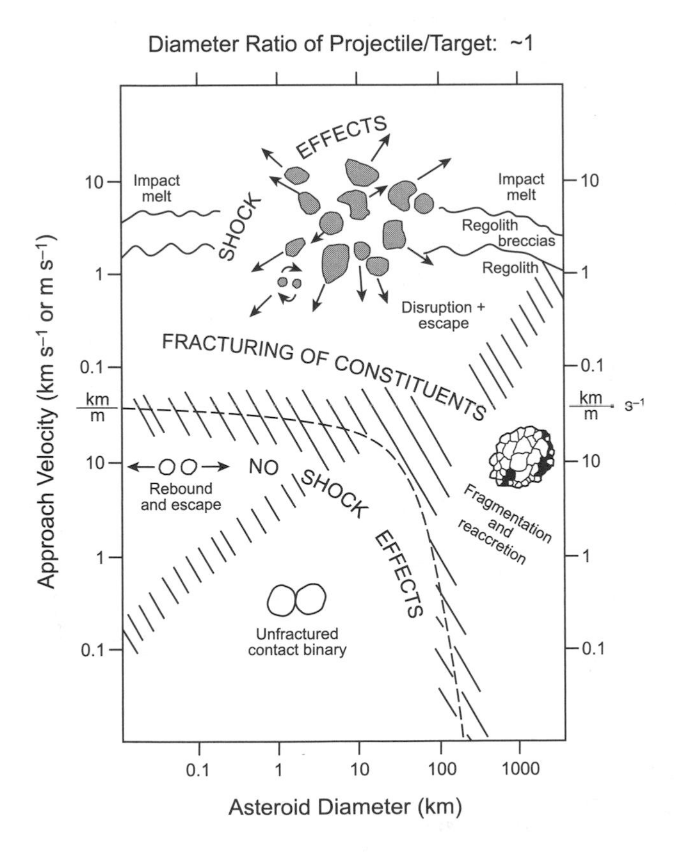

Fig. 1. Regimes of collision-induced effects as a function of the approach velocity (velocity of infinity) and the size of the colliding bodies: Equal-sized bodies are assumed. Minimum velocities for the onset of certain shock effects in meteorites (e.g., formation of impact melts and regolith breccias) are given in wavy lines. The diagram is modified and redrawn from *Hartmann* (1979) and *Stöffler et al.* (1988).

All types of breccia formation require mass transport (ballistic or nonballistic). Relative movement of rock fragments and their displacement from the original location of the source material is involved. The geological scenarios of breccia formation in impact craters are summarized in detail by *Stöffler et al.* (1988) (see their Figs. 3.6.2 and 3.6.3).

4. BRECCIATION IN METEORITE GROUPS

4.1. Ordinary Chondrites

There are an enormous number of observations of lithic clasts in ordinary chondrites (Fig. 2). An outstanding summary of the various types of fragments within H, L, and LL chondritic breccias is given by *Keil* (1982). It is impossible to cite here all the reports listed by *Keil* (1982). The major studies are, e.g., *Wahl* (1952); *Wlotzka* (1963); *Van Schmus* (1967); *Binns* (1968); *Fodor et al.* (1972, 1974, 1976, 1977, 1980); *Keil and Fodor* (1973, 1980); *Fodor and Keil* (1973, 1975, 1976a,b, 1978); *Bunch and Stöffler* (1974); *Dodd* (1974); *Wilkening and Clayton* (1974); *Fredriksson et al.* (1975); *Hoinkes et al.* (1976); *Noonan et al.* (1976); *Wilkening* (1976, 1977, 1978); *Leitch and Grossman* (1977); *Clayton and Mayeda* (1978); *Lange et al.* (1979); *Taylor et al.* (1979); *Wlotzka et al.* (1979); *Grossman et al.* (1980); *Keil et al.* (1980); *Rubin et al.* (1981a,b); *Scott and Rajan* (1981); and *Sears and Wasson* (1981). In this paper the basic findings on brecciated ordinary chondrites of the last 20 years are summarized.

4.1.1. Black and dark inclusions (clasts). The most obvious and easily visible clasts in ordinary chondrite breccias are the so-called black and dark inclusions (or clasts). These terms do not provide any information about the genetic origin and the mineralogy of the fragments. They encom-

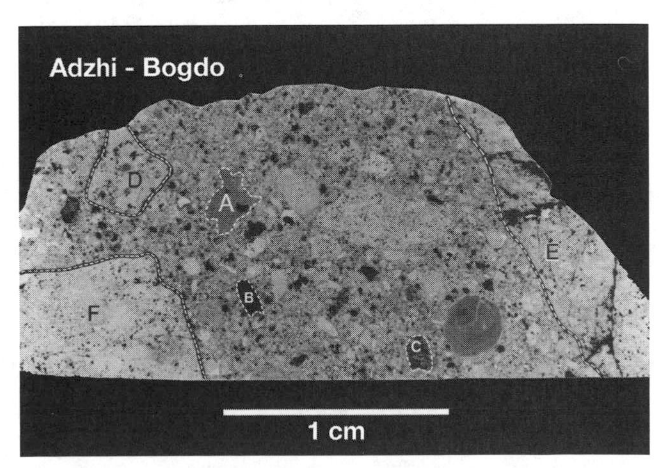

Fig. 2. Transmitted light photomicrograph of a thin section from the Adzhi-Bogdo (LL3–6) chondrite regolith breccia showing various types of fragments: highly recrystallized clasts (D, E, and F) sometimes with internal shock veins (E), melt-rock clast (A), fragmental breccia clast (B), and a shock-darkened fragment (C). After *Bischoff et al.* (1993c).

pass (1) shock-darkened objects; (2) specimens of various types of primitive rocks, mainly of carbonaceous (C) chondrites; (3) fragments of fine-grained breccias (breccia in a breccia); (4) metal-troilite-rich clasts; (5) fine-grained, matrix-like cognate inclusions; and (6) fragments of shock melts with abundant tiny metal/sulfide grains.

4.1.1.1. Shock-darkened clasts: Clasts of this type formed from light-colored lithologies during shock events (shock darkening, shock blackening) that melted and transported a significant fraction of their metallic Fe-Ni and troilite grains into silicates (e.g., *Dodd,* 1981; *Stöffler et al.,* 1991; *Rubin,* 1992; *Rubin et al.,* 2002; *Welzenbach et al.,* 2005). Impact-induced frictional melting is considered as a possible mechanism for the "darkening" in ordinary chondrites (*van der Bogert et al.,* 2003). About 15% of ordinary chondrites are regarded as "black ordinary chondrites" (e.g., *Heymann,* 1967; *Britt and Pieters,* 1991, 1994). A very dark shock-blackened clast has been reported from Nulles (*Williams et al.,* 1985).

4.1.1.2. Clasts of various types of foreign, accretionary rocks, mainly carbonaceous (C) chondrite-like: A black microchondrule- and carbon-bearing L-like chondritic fragment was found in the Mezö-Madaras L3 chondritic breccia, which may represent a new specimen of C-rich ordinary chondrite (*Christophe Michel-Lévy,* 1988). A similar clast with tiny chondrules was found in Krymka (LL3) (*Rubin,* 1989). *Rubin et al.* (1982) also described a fine-grained microchondrule-bearing fragment from Piancaldoli (LL3), which was reclassified by *Krot et al.* (1997b) as being not an individual clast, but part of a chondule rim. One moderately dark fragment in the Krymka ordinary chondrite is clearly a clast of a different type of chondrite, perhaps a carbonaceous chondrite (Fig. 3). It has nearly equilibrated Mg-rich olivine (Fa 8.1 ± 1.6) and somewhat more unequilibrated pyroxene (Fs 6.4 ± 3.3 mol%). Some other dark lithic fragments in Krymka are considered to represent pieces of a primary accretionary precursor rock that has been fragmented and its fragments incorporated into the Krymka host (*Semenenko et al.,* 2001; *Semenenko and Girich,* 2001). Similarly, *Vogel et al.* (2003) report that Krymka dark inclusions must have accreted from regions different from those of their respective rims and matrices and were later incorporated into the host meteorite.

4.1.1.3. Fragments of fine-grained breccias (breccia in a breccia): In many cases, black clasts in the Adzhi-Bogdo (LL3–6) chondritic breccia are fine-grained breccias themselves and also contain abundant tiny opaque phases (Fig. 4) (*Bischoff et al.,* 1993c). Similar dark fragments (portions) occur in Fayetteville (*Xiao and Lipschutz,* 1991).

4.1.1.4. Metal-troilite rich clasts: Metal-troilite-rich clasts, which also could be regarded as some type of dark inclusion or clast, were found in several ordinary chondrites (e.g., Moorabie, Bishunpur, Krymka). Impact melting of metal and sulfide-rich materials close to the eutectic composition of the Fe-S system was suggested by *Scott* (1982) and *Fujita and Kitamura* (1992) for their origin. *Kojima et al.* (2003) argued that troilite-silicate-metal inclusions in Bishunpur were fragmented and dispersed after impact-induced compaction, and then reaccreted onto the parent body. Three varieties of predominantly opaque, shocked metal-troilite-rich clasts were reported from the Northwest Africa (NWA) 428 (L6) chondrite breccia (*Rubin,* 2003). Also, in Krymka (LL3) the occurrence of several sulfide- and metal-enriched fragments has been reported (*Semenenko and Girich,* 2001).

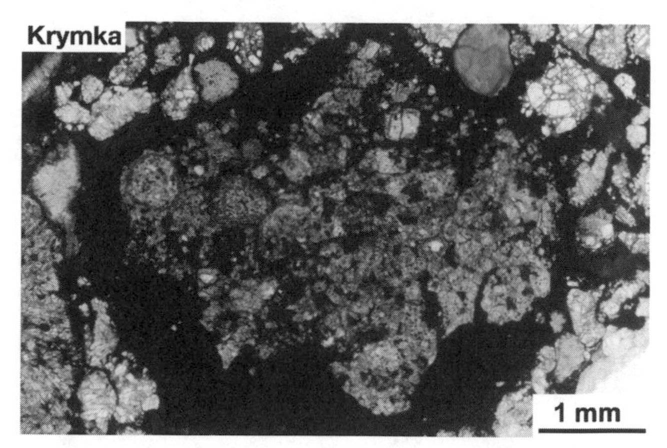

Fig. 3. Fragment of an unusual chondritic fragment in the unequilibrated ordinary chondrite Krymka (LL3). The clast consists chemically of almost equilibrated Mg-rich olivine (Fa: 8.1 ± 1.6) and somewhat more variable pyroxene (Fs: 6.4 ± 3.3 mol%). Due to the ocurrence of abundant opaque grains the fragment has a black appearance at the edge (especially in the lower part). Photomicrograph in transmitted polarized light.

4.1.1.5. Fine-grained, matrix-like cognate inclusions: This type of inclusion appears to consist mostly of typical matrix material, Huss-matrix (*Huss et al.,* 1981). A fine-grained inclusion in Sharps (H3) described by *Zolensky et al.* (1996a) appears to be a matrix "lump" genetically related to the host chondrite. A similar object (BK15) has been identified in Krymka (*Semenenko et al.,* 2001). The same is probably the case for a dark fragment found in Tieschitz (H3) (*Kurat,* 1970).

4.1.1.6. Fragments or areas of shock melts with abundant tiny metal/sulfide grains: A dark inclusion with "augen" in the Manych LL (3.1) ordinary chondrite consisting mainly of Fe-rich olivine, high-Ca pyroxene, and Na-rich feldspathic glass is suggested to represent a shock melt containing some unmelted precursor material (*Kojima et al.,* 2000). This type of inclusion could also be classified as an impact-melt breccia (see below). In all types of ordinary chondrites shock veins exist. Actually, these "veins" are only veins on a two-dimensional scale (for example, in a thin section) — in three dimensions they are irregularly shaped "plates" with variable thickness. In some chondrites (and certainly in thin sections, due to sectioning) these plates can lead to huge optically dark areas. This is especially the case for many of the "very strongly shocked" (S6) (*Stöffler et al.,* 1991), ringwoodite-bearing LL (*Bischoff,* 2002) and L chondrites (e.g., *Binns et al.,* 1969; *Smith and Mason,* 1970; *Stöffler et al.,* 1991). Dark gray impact melt-rock clasts are also known from Nulles (*Williams et al.,* 1985).

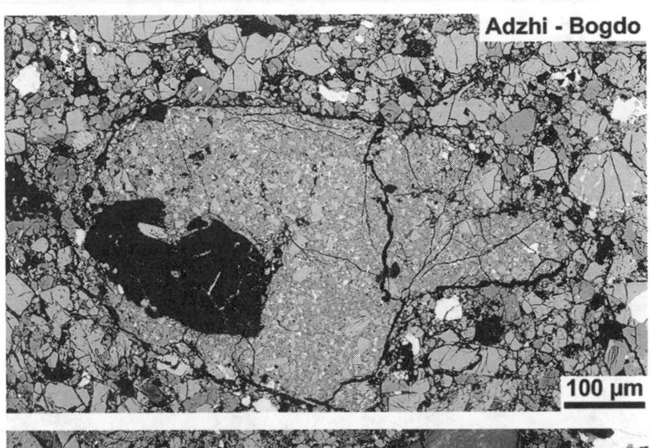

Fig. 4. Backscattered electron image of two fine-grained, optically dark fragments in the Adzhi-Bogdo (LL3–6) ordinary chondrite regolith breccia. Both fragments can be described as individual breccias that occur within the Adzhi-Bogdo host breccia (breccia in a breccia). The large black grain in the upper image is a plagioclase grain (An_{20}).

4.1.2. Impact-melt rocks and breccias. Impact-melt breccias among the ordinary chondrites or clasts of impact-melt breccias in ordinary chondrites offer direct evidence for high-energy collisions. Impact-melt breccias are well-known among the L chondrites, e.g., Chico, Ramsdorf, Shaw, Madrid, Point of Rocks, and Patuxent Range (PAT) 91501 (e.g., *Taylor et al.,* 1979; *Nakamura et al.,* 1990b; *Casanova et al.,* 1990; *Bogard et al.,* 1995; *Yamaguchi et al.,* 1999; *Mittlefehldt and Lindstrom,* 2001; *Norman and Mittlefehldt,* 2002). Dar al Gani (DaG) 896, Orvinio, Spade, and Smyer can be regarded as H-chondrite impact-melt breccias (*Folco et al.,* 2002, 2004; *Rubin,* 2002; *Burbine et al.,* 2003; *Rubin and Jones,* 2003; *Grier et al.,* 2004). The Antarctic LL chondrites Yamato (Y)-790964 and -790143 are regarded as impact-melt rocks that represent nearly total melting of precursor rocks (*Sato et al.,* 1982; *Okano et al.,* 1990; *Yamaguchi et al.,* 1998). Fragments of impact melt or impact-melt breccias occur in many brecciated ordinary chondrites (e.g., *Keil,* 1982, and references therein; *Bischoff et al.,* 1993c; *Welzenbach et al.,* 2005).

The L chondrite Ramsdorf appears to be one of the most heavily shocked ordinary chondrites and has only partly retained some traces of its original texture: Brief shock heating to 1400°–1600°C caused complete melting of metallic Fe,Ni, plagioclase, and partial or complete melting of pyroxenes and olivines (*Yamaguchi et al.,* 1999).

The H6 chondrite Portales Valley rock is an annealed impact-melt breccia with coarse metal interstitial to angular and subrounded chondritic clasts (*Rubin et al.,* 2001). It may be a sample of the brecciated and metal-veined floor of an impact crater that was subsequently buried and cooled at 6.5 K/m.y. (e.g., *Kring et al.,* 1999; *Sepp et al.,* 2001). However, since molten metallic Fe-Ni is not thought to segregate in asteroids from chondritic impact melts (*Keil et al.,* 1997), the source of heat may have been internal.

A C-rich chondritic clast PV1 within the Plainview H-chondrite regolith breccia, originally described by *Scott et al.* (1988), was suggested to be an impact-melted fragment that experienced aqueous alteration and enrichment of C (*Rubin et al.,* 2004).

4.1.3. Igneous textured clasts. In several ordinary chondrites igneous-textured clasts with abundant SiO_2-phases (quartz, cristobalite, tridymite) occur (e.g., *Bischoff,* 1993; *Bischoff et al.,* 1993c; *Bridges et al.,* 1995a; *Ruzicka et al.,* 1995; *Hezel,* 2003). Strong differentiation of chondritic material is required to form silica-oversaturated liquids, leading for example to coarse-grained granitoidal clasts as found in the Adzhi-Bogdo LL3–6 chondrite (Fig. 5) (*Bischoff,* 1993; *Bischoff et al.,* 1993c) or to clasts in Parnallee (LL3) and Farmington (L5) consisting of up to 95 vol% of an SiO_2 phase (*Bridges et al.,* 1995a). A clast with SiO_2-normative mesostasis was found in the Hammadah al Hamra (HH) 180 unique chondrite with affinity to LL-group ordinary chondrites (*Bischoff et al.,* 1997). Other large, igneous-textured clasts without abundant SiO_2-phase(s) were reported from several other chondrites [e.g., Julesburg (L3), Vishnupur (LL4–6)] (*Ruzicka et al.,* 1998; *Bridges and Hutchison,* 1997). *Kennedy et al.* (1992) discuss a microgabbro in the Parnallee chondrite, suggesting that it formed by partial melting in a planetary body after removal of metallic Fe. A feldspar-nepheline achondritic clast in Parnallee has an

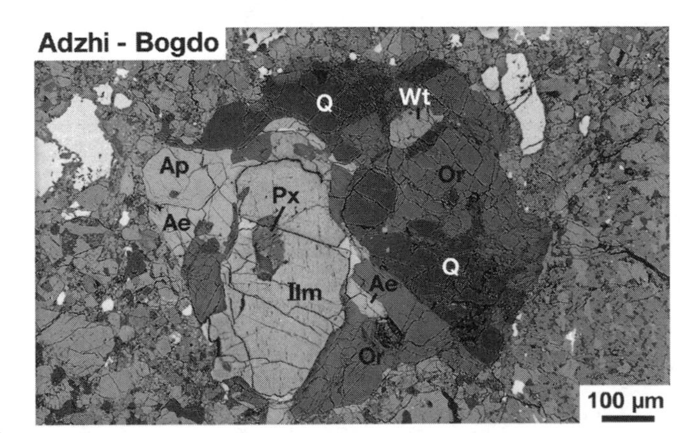

Fig. 5. Alkali-granitoidal clast from Adzhi-Bogdo (LL3–6) ordinary chondrite regolith breccia consisting of apatite (Ap), aenigmatite (Ae), ilmenite (Ilm), whitlockite (Wt), quartz (Q), and K-feldspar (Or). Within quartz tiny grains of zircon are enclosed (white). Photomicrograph in backscattered electrons (cf. *Bischoff et al.,* 1993c; *Bischoff,* 1993).

O-isotopic composition, indicating carbonaceous chondrite affinities (*Bridges et al.,* 1995b).

Other large achondritic clasts include troctolitic and/or dunitic and/or harzburgitic inclusions in Barwell (L6), Y-75097 (L6), Y-794046 (H5), and Y-793241 (e.g., *Prinz et al.,* 1984; *Hutchison et al.,* 1988; *Nagao,* 1994; *Mittlefehldt et al.,* 1995) and pyroxenitic or noritic fragments in Hedjaz (L3.7) (*Nakamura et al.,* 1990a; *Misawa et al.,* 1992). Some of these igneous-textured clasts have sizes and compositions like those of chondrules and CAIs. It is therefore possible that some are related to chondrules and CAIs and are not clasts from an igneously differentiated body.

4.1.4. Foreign clasts. A small number of clasts exist in ordinary chondrite breccias that are unrelated to the host meteorite (e.g., *Dodd,* 1974; *Fodor and Keil,* 1975, 1978; *Keil,* 1982; *Rubin et al.,* 1983b; *Prinz et al.,* 1984; *Wieler et al.,* 1989; *Bischoff et al.,* 1993c; *MacPherson et al.,* 1993). Some more details on these clasts are given in section 6.

4.1.5. Granulitic breccias. Cabezo de Mayo is a recrystallized, metamorphosed L/LL6 chondrite, also described as a pre-metamorphic fragmental breccia (*Casanova et al.,* 1990). It is not clear if this rock can already be grouped with the granulitic breccias.

Fig. 6. Overview of the R chondrite regolith breccia Rumuruti. The photographed hand specimen consists of light- and dark-colored fragments embedded in a clastic matrix. The abundance of large clasts is roughly 50%. See *Schulze et al.* (1994) for details.

4.2. Rumuruti Chondrites

The abundance of regolith breccias among the R chondrites is 50% (Table 2; Fig. 6). The brecciated samples [e.g., Rumuruti, DaG 013, Hughes 030, Pecora Escarpment (PCA) 91002, and Acfer 217] have been studied in great detail (e.g., *Schulze et al.,* 1994, *Bischoff et al.,* 1994a, 1998; *Rubin and Kallemeyn,* 1994; *Jäckel et al.,* 1996; *Kallemeyn et al.,* 1996; *Bischoff,* 2000). These breccias contain cognate, lithic fragments of various type and metamorphic degree. Impact-melt-rock clasts (Fig. 7), dark unequilibrated fragments, and clasts of all petrologic types are embedded in a fine-grained, well-lithified, olivine rich matrix. A detailed characterization of unequilibrated, type 3 lithologies in R chondrites is given by *Bischoff* (2000). So far, fragments of other chondrite classes have not been found within the brecciated R chondrites.

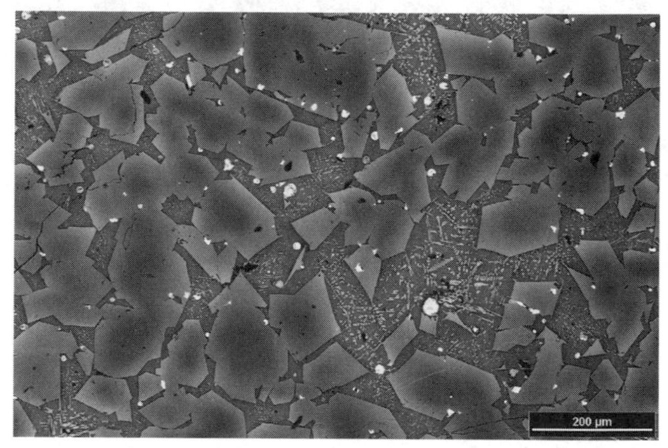

Fig. 7. Typical texture of an impact melt clast within the R chondrite DaG 013. Euhedral to subhedral, zoned olivine grains are embedded in a mesostasis containing tiny, skeletal crystals of pyroxene. Light grains are either chromite or Fe-sulfide. Image in backscattered electrons.

4.3. Enstatite Chondrites

Enstatite chondrites contain a variety of breccias and impact-melted rocks. *Keil* (1989) noted that 7 out of 45 E chondrites appeared to be fragmental breccias; one was a regolith breccia [the EH3 chondrite Allan Hills (ALH) A77156]. Happy Canyon and Ilafegh 009 were both inferred to be clast-free impact melts (*Bischoff et al.,* 1992; *McCoy et al.,* 1995), but this was questioned by *Weisberg et al.* (1997a). Abee and Adhi Kot were characterized as impact melts with ghosts of chondrule-bearing clasts (*Rubin and Scott,* 1997). *Lin and Kimura* (1998) identified two further EH impact-melt rocks, Y-82189 and Y-8404. An EH3 chondrite, Parsa, and five EL6 chondrites including Hvittis contain abundant clasts or large opaque veins that probably formed by impact melting (*Rubin,* 1985; *Rubin et al.,* 1997).

Weisberg et al. (1997a) suggested that the EH chondritic melt rock Queen Alexandra Range (QUE) 94204 is an internally derived melt rock from an EH-like parent body and that the interpretation of other meteorites as impact-melt rocks needs to be reconsidered.

Recent noble gas studies have identified other E chondrites with solar and solar-like gases, but the origin of these gases is controversial (*Ott,* 2002). In their noble gas study of 57 E chondrites, *Patzer and Schultz* (2002) concluded that about 30% of E3 chondrites are solar gas rich, but only one had been described as a fragmental breccia, MacAlpine Hills (MAC) 88138 (*Lin et al.,* 1991). However, solar gases are heterogeneously distributed in ALHA 77156, consistent with the possible presence of unrecognized clasts. The type 4–6

EH and EL chondrites, which appear to lack solar wind gases, contain a so-called subsolar component. *Patzer and Schultz* (2002) argued that the subsolar gases were not simply a reflection of metamorphic heating of samples containing solar-wind gases because ordinary chondrites do show such an effect. They inferred that the subsolar gases were acquired before or during accretion. However, on the basis of their analyses of St. Mark's, *Busemann et al.* (2003a,b) argued that the subsolar gases were actually a mixture of Q (planetary) gases with small amounts of solar gases plus terrestrial contamination, and that the subsolar gases were not a separate component. They further argued that the solar gases were probably trapped *prior* to accretion and were not present on the surface of regolith grains as they were only released after lengthy etching. In addition, neither St. Mark's nor other E chondrites containing subsolar gases appear to be brecciated.

E chondrites also differ from ordinary and carbonaceous chondrites in their breccia properties. Impact melts and well-shocked chondrites are relatively abundant (*Rubin et al.,* 1997), but foreign clasts and mixtures of type 3–6 material appear to be absent. This might reflect the difficulty of identifying clasts in weathered chondrites, the limited amount of material that has been carefully studied, or a significant difference in the formation and evolution of these chondrites. Conceivably, the differences between the enstatite and ordinary chondrites in their trapped noble gases may simply result from more intense impact processing of the enstatite chondrites. Resolution of these issues and the origin of the trapped noble gases will greatly help in understanding the origin of E chondrites.

4.4. Carbonaceous Chondrites

4.4.1. CI chondrites and related chondrites. Although the CI chondrites (Ivuna, Orgueil, Alais, Tonk, and Revelstoke) are regarded as the chemically most primitive rocks in the solar system, all CI chondrites are complex breccias consisting of fragments up to several hundred micrometers in size surrounded by a fine-grained matrix (e.g., *Richardson,* 1978; *Beauchamp and Fredriksson,* 1979; *Endress and Bischoff,* 1993, 1996; *Endress,* 1994; *Endress et al.,* 1994a; *Morlok,* 2002; *Morlok et al.,* 2000; 2001). All analyzed CI chondrites contain solar-wind-implanted noble gases (Table 2) (*Bischoff and Schultz,* 2004), indicating their presence at the upper surface of their parent body(ies). Orgueil is highly brecciated; and the degree of brecciation decreases in the order Orgueil > Ivuna > Alais ≅ Tonk (*Morlok,* 2002). The clasts vary significantly in mineralogy and chemistry (Fig. 8). *Endress* (1994), *Morlok et al.* (2001), and *Morlok* (2002) defined several groups of fragments with similar chemical and mineralogical characteristics ranging from clasts dominated by coarse, Mg-rich phyllosilicates to fragments with high abundance of Fe. Phosphate-rich clasts were also encountered (*Morlok et al.,* 2001; *Morlok,* 2002), as are olivine-bearing clasts (*Endress,* 1994; *Bischoff,* 1998). These various types of clasts represent distinct lithologies found on the CI chondrite parent body(ies) prior to impact brecciation, mixing, and reaccreation.

The slightly metamorphosed sample Y-82162 appears to be related to the CI chondrites. Based on the occurrence of abundant clasts (up to several millimeters in size) it was classified as a chondritic breccia (cf. Figs. 1–5 in *Bischoff and Metzler,* 1991).

Yamato-86029 is a similar CI-like breccia consisting of a variety of clasts. *Tonui et al.* (2003) suggest that olivine aggregates in Y-86029 were mechanically mixed from another environment into the host chondrite and may represent parts of another asteroid (probably of ordinary chondrite material).

The Tagish Lake meteorite shares similarities with CI and CM chondrites (e.g., *Mittlefehldt,* 2002; *Zolensky et al.,* 2002, *Grady et al.,* 2002) and consists of different lithologies: a dominant carbonate-poor and a less-abundant carbonate-rich lithology (*Zolensky et al.,* 2002). A CM1 clast has been studied in detail by *Zolensky et al.* (2002) and *Bullock et al.* (2005).

4.4.2. CM chondrites. CM-like clasts are widespread within impact breccias of other meteorite classes like howardites, polymict eucrites, and ordinary and carbonaceous chondrites, suggesting that their parent asteroids are abundant in the main belt (e.g., *Wilkening,* 1973; *Bunch et al.,* 1979; *Kozul and Hewins,* 1988; *Mittlefehldt and Lindstrom,* 1988; *Hewins,* 1990; *Reid et al.,* 1990; *Zolensky et al.,* 1992b, 1996c; *Buchanan et al.,* 1993; *Pun et al.,* 1998; *Buchanan and Mittlefehldt,* 2003).

CM chondrites themselves are impact breccias, consisting of subangular mineral and lithic clasts set in a fine-grained clastic matrix (e.g., *Fuchs et al.,* 1973; *Dodd,* 1981; *Metzler et al.,* 1992; *Metzler,* 1995). The majority of these lithic clasts belong to a texturally well defined chondritic rock type (primary accretionary rock) (*Metzler et al.,* 1992), which can be described as an agglomerate of chondrules and other

Fig. 8. Part of the severely brecciated CI chondrite Orgueil. Individual fragments are variable in composition as indicated by different gray tones. Image in backscattered electrons; modified after *Morlok* (2002).

fine-grained components, most of which are surrounded by fine-grained rims (*Bunch and Chang,* 1980; *Metzler et al.,* 1992; *Metzler and Bischoff,* 1996).

The lithic clasts display sharp contacts to the surrounding clastic matrix, best visible using scanning electron microscopy (SEM)-techniques (Fig. 9). In the case of Nogoya these features are even visible by the naked eye in the form of light-dark structures (*Heymann and Mazor,* 1967). The clastic matrix essentially shows the same mineralogical and chemical composition as the lithic clasts. Hence, both lithologies seem to originate from the same precursor rock (*Metzler,* 1990; *Metzler et al.,* 1992). CM chondrites differ texturally from each other by variable ratios of clastic matrix to lithic clasts. Based on the degree of brecciation the amount of clastic matrix varies from 100% in Essebi and Bells (*Metzler et al.,* 1992) to almost zero in Y-791198. This Antarctic CM chondrite studied by *Metzler et al.* (1992) is unbrecciated on the centimeter scale (thin-section scale) and may represent a remnant of pristine precursor rock from which brecciated CM chondrites have formed by impact comminution.

Many CM chondrites show evidence of intensive aqueous alteration on the parent body (e.g., *Kerridge and Bunch,* 1979; *Zolensky and McSween,* 1988; *Brearley and Geiger,* 1991; *Browning et al.,* 1995; *Hanowsky and Brearley,* 1997, 2001). In some cases this epoch clearly predates the brecciation events, which led to the formation of the final host breccias. This preaccretionary aqueous alteration may have occurred in small precursor planetesimals (*Metzler et al.,* 1992; *Bischoff,* 1998). Nogoya consists of lithic clasts with different alteration stages (*Metzler,* 1995) mixed together in the same breccia. The observation that all lithic clasts still show their original accretionary texture indicate that a single starting material was affected by liquid water under different alteration conditions, followed by impact brecciation and mixing. In this sense, CM chondrites like Nogoya and Cold Bokkeveld are genomict breccias, consisting of clasts of the same compositional group but of various petrologic types (*Metzler,* 1995; *Zolensky et al.,* 1997).

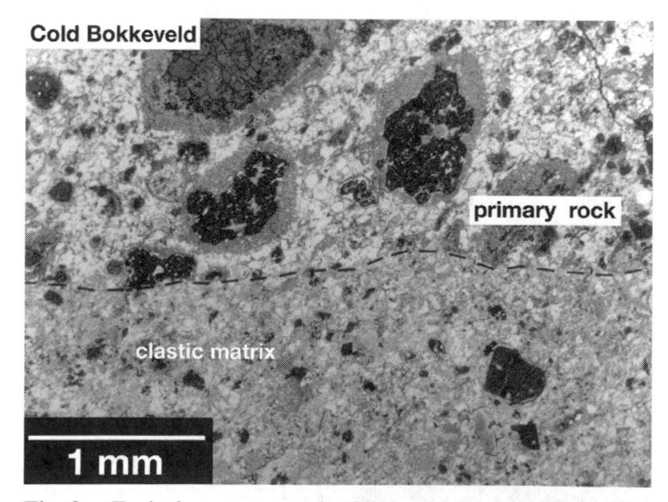

Fig. 9. Typical appearance of a CM breccia as indicated by the boundary between a fragment of primary accretionary rock and the clastic matrix. In the upper part all the coarse-grained components are surrounded by accretionary dust mantles, whereas in the clastic matrix angular fragments occur and all coarse-grained components have lost their dust mantles due to impact-induced fragmentation. Image in backscattered electrons after *Metzler et al.* (1992).

CM chondrites are regolith breccias (Table 2). It has been shown by *Nakamura et al.* (1999a,b) that solar noble gases in brecciated CM chondrites are restricted to the clastic matrix of these meteorites. Lithic clasts are free of solar gases and dominated by planetary noble gas components. The study of track-rich olivines in the CM chondrites Cold Bokkeveld, Mighei, Murchison, and Nogoya revealed that all preirradiated grains occur in the clastic matrix of these breccias as well (*Metzler,* 1993, 1997, 2004). Hence, in close analogy to regolith breccias from ordinary chondrites, the fine-grained clastic matrix of CM chondrites is the host lithology for both solar gases and preirradiated grains.

4.4.3. CV chondrites. Members of the Vigarano-type carbonaceous chondrite group vary considerably. Based on petrology they can be subdivided into the reduced and the Bali-like and Allende-like oxidized subgroups (*Weisberg et al.,* 1997b). Two types of clasts are present in CV3 chondrites: dark inclusions, which are present in many, or even most, CV3 chondrites (e.g., Allende, Efremovka, Leoville, Mokoia, Vigarano, Ningqiang, Y-86751), and other chondritic inclusions, which are generally found in the regolith breccias.

The dark clasts or dark inclusions record a history of fragmentation, mixing, and relithification (e.g., *Fruland et al.,* 1978; *Kracher et al.,* 1985; *Heymann et al.,* 1987; *Bischoff et al.,* 1988; *Kurat et al.,* 1989; *Palme et al.,* 1989; *Johnson et al.,* 1990; *Kojima et al.,* 1993; *Murakami and Ikeda,* 1994; *Kojima and Tomeoka,* 1996, 1997; *Buchanan et al.,* 1997; *Krot et al.,* 1997a, 1998, 1999; *Brearley and Jones,* 1998; *Ohnishi and Tomeoka,* 2002; *Vogel et al.,* 2003; *Zolensky et al.,* 2003). All these dark inclusions are olivine-rich and most probably represent fragments of a parent body that experienced aqueous alteration and subsequent dehydration (e.g., *Kracher et al.,* 1985; *Kojima et al.,* 1993; *Kojima and Tomeoka,* 1996; *Buchanan et al.,* 1997; *Krot et al.,* 1997a, 1998, 1999). Alternatively, some may be fragments of primitive accreted material (e.g., *Palme et al.,* 1989; *Kurat et al.,* 1989; *Zolensky et al.,* 2003). Dark inclusions in Mokoia appear to have experienced a higher degree of thermal metamorphism than the host meteorite (*Ohnishi and Tomeoka,* 2002), whereas one primitive dark inclusion in Ningqiang was added to the host chondrite after any parent-body alteration event (*Zolensky et al.,* 2003). Bulk O-isotopic compositions of some dark inclusions in the reduced CV3 chondrite Efremovka plot in the field of aqueously altered CM chondrites (*Krot et al.,* 1999), indicating significant differences between these inclusions and their host rocks.

In summary, the dark inclusions found in CV3 chondrites are unique and represent an important source of information about early solar system materials not sampled by individual rocks. It is important to note that all dark inclusions have experienced at least the same and in some cases a higher degree of thermal metamorphism than their host meteorites! No phyllosilicate-rich DIs have been reported from CV3 chondrites.

Clasts in regolith breccias are much more diverse. The Vigarano regolith breccia contains lithic clasts of Bali-like oxidized CV materials and abundant reduced materials, while the Mokoia regolith breccia contains Allende-like and Kaba-like oxidized materials (*Krot et al.,* 1998, 2000). These mixtures were attributed to impact mixing and lithification of reduced and oxidized CV material from a single, heterogeneously altered asteroid (*Krot et al.,* 2000). Mixing of oxidized and reduced fragments from different precursor planetesimals can certainly not be ruled out. However, alteration on asteroidal bodies lasted for up to 15 m.y. (*Russell et al.,* 2006), much longer than the accretion timescales for asteroids (~1 m.y.). Mokoia also contains metamorphosed chondritic clasts, which may be CV4/5 material from deep within the CV body (*Krot et al.,* 1998). Camel Donga 040 has been described as a genomict breccia containing unequilibrated material and a metamorphosed lithology (*Zolensky et al.,* 2004). A C2 clast up to several centimeters in size was found in the Leoville regolith breccia (*Keil et al.,* 1969); *Kennedy and Hutcheon* (1992) describe a basaltic plagioclase-olivine inclusion in Allende. We infer that the CV regolith breccias contain valuable clues to the interior of the CV body and the projectiles that modified it.

Metal-sulfide aggregates are common constituents in Y-86751 (CV3) (*Murakami and Ikeda,* 1994); however, it is unclear whether they represent true clasts produced by impact processes on the parent or precursor parent body or parts of a possible impactor.

4.4.4. CO chondrites. Only a very few CO3 chondrites are breccias. Frontier Mountain (FRO) 95002 was classified as a brecciated rock (*Grossman,* 1997), but there are no reports on distinct individual fragments or inclusions within it. Three angular 1–6-mm-sized chondritic clasts were found by *Rubin et al.* (1985) in the Colony breccia. *Scott et al.* (1992) suggested that Felix may be a breccia containing fragments with diverse shock histories. Recently, *Itoh and Tomeoka* (2003) reported the occurrence of dark inclusions in the CO3 chondrites Kainsaz, Ornans, Lance, and Warrenton and suggested that these clasts had undergone aqueous alteration and subsequent dehydration at a location different from the present location in the meteorite. "Basaltic" fragments were also reported from Lance (*Kurat and Kracher,* 1980). Isna was first reported to be a solar-wind-rich regolith breccia (*Scherer and Schultz,* 2000), but this classification has been changed by L. Schultz (personal communication, 2004) (Table 2).

4.4.5. CK chondrites. The CK chondrite group contains no gas-rich regolith breccias (*Scherer and Schultz,* 2000) (Table 2). However, some CK meteorites are described as fragmental breccias. *Geiger* (1991) reports that ALH 82135 (CK4/5) and ALH 84038 (CK4/5) are severely brecciated on a thin-section scale with variable fragment sizes. Similarly,

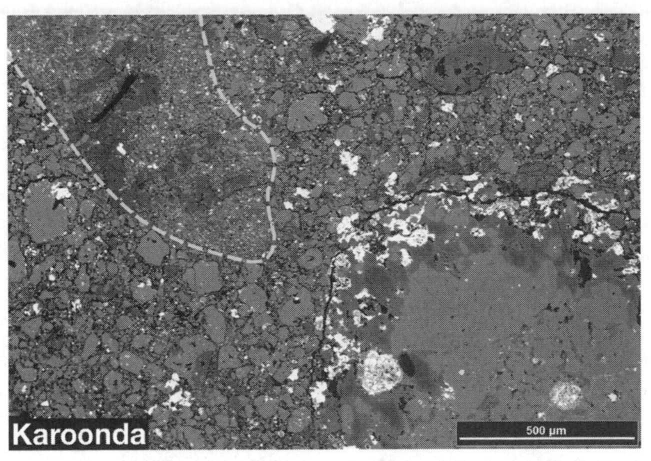

Fig. 10. Brecciated nature of the Karoonda CK chondrite. A fine-grained (upper left) and coarse-grained fragment (lower right) are embedded in a fine-grained clastic matrix. Image in backscattered electrons.

Karoonda (CK4) is heavily brecciated (Fig. 10). Shock darkening and the presence of melt veins in some CK chondrites are also known (e.g., *Rubin,* 1992).

4.4.6. CR clan meteorites (CR, CH, and related chondrites). CR and CH chondrites, the Bencubbin-like (CB; Bencubbin, Gujba, and Weatherford) and CH-like grouplets (HH 237 and QUE 94411), and the unique sample Lewis Cliff (LEW) 85332 are chemically and mineralogically related and form the CR clan (e.g., *Weisberg et al.,* 1990, 1995, 2001; *Bischoff,* 1992; *Bischoff et al.,* 1993a,b; *Krot et al.,* 2002; *Weisberg et al.,* 2006).

Most — if not all — CR chondrites contain dark inclusions (e.g., *Zolensky et al.,* 1992; *Bischoff et al.,* 1993a; *Weisberg et al.,* 1993; *Endress et al.,* 1994b; *Abreu and Brearley,* 2004, 2005), which are the only "xenolithic" lithology known in typical CR chondrites, which are all regolith breccias (Table 2). These dark clasts may represent fragments of different lithologies of the same parent body or accreted as xenoliths to the same time with other components during parent-body formation. *Abreu and Brearley* (2004) found that within the CR chondrite Elephant Moraine (EET) 92042 impact brecciation has formed regions within the matrix that are highly clastic in character.

CH chondrites contain a high proportion of fragmented components (mainly chondrules) (*Bischoff et al.,* 1993b), indicating that the precursor components of the CH constituents were much larger prior to accretion and lithification of the parent body (cf. Fig. 2 in *Bischoff et al.,* 1993b). The most obvious xenolithic components are dark, phyllosilicate-rich inclusions (Fig. 11) (e.g., *Grossman et al.,* 1988; *Scott,* 1988; *Weisberg et al.,* 1988; *Bischoff et al.,* 1993b, 1994b).

The absence of effects of aqueous alteration in the chondrules and metals in CHs indicates that the phyllosilicate-rich dark inclusions experienced aqueous alteration prior to being incorporated into their immediate (CH) parent bodies (*Krot et al.,* 2002).

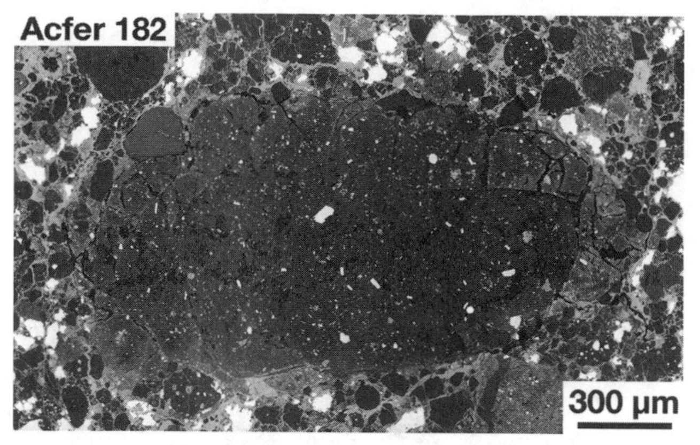

Fig. 11. Dark inclusion in the CH chondrite Acfer 182. The fine-grained inclusion contains some large pyrrhotite grains (white) and abundant pores (dark spots) and fractures. Photomicrograph in backscattered electrons; cf. *Bischoff et al.* (1993b).

The members of the Bencubbin-like grouplet (CB) are interesting breccias consisting of roughly 60 vol% of metal fragments and 40% of silicate-rich fragments (e.g., *Weisberg et al.,* 1990, 2001; *Krot et al.,* 2002). As early as 1932, Simpson and Murray mention the breccia appearance of Bencubbin, which was later described as a shock-welded breccia (*Newsom and Drake,* 1979). The host silicate fragments have texture and mineralogy similar to those of barred olivine chondrules; however, they are much larger and angular rather than submillimeter-sized fluid droplet-shaped objects (*Weisberg et al.,* 1990). Xenolithic ordinary chondrite clasts were reported to occur as components of Bencubbin and Weatherford, as well as dark (carbonaceous) clasts (*Weisberg et al.,* 1990; *Barber and Hutchison,* 1991). These dark clasts are very different from those in CR and CH chondrites (which are phyllosilicate-rich): They contain metals and highly elongated, olivine-rich lenses (augen) set in a fine-grained matrix (*Weisberg et al.,* 1990; *Barber and Hutchison,* 1991). Fragments with R-chondrite characteristics were reported from Weatherford (*Prinz et al.,* 1993).

Members of the CH-like grouplet, HH 237 and QUE 94411, consist of mixtures of metal and silicate chondrules and fragments. These rocks have been classified as CB chondrites by *Weisberg et al.* (2001), although remarkable differences to CB chondrites exist. The rare occurrence of heavily hydrated dark clasts — similar to those found in CR and CH chondrites — has been reported by *Krot et al.* (2001).

Carbonaceous chondrite clasts with affinities to CI and C2 chondrites, troilite-rich clasts, and a schreibersite-bearing fragment were found in the LEW 85332 unique carbonaceous chondrite breccia (*Rubin and Kallemeyn,* 1990).

4.5. Other Types of Chondrites

With respect to Kakangari (K) chondrite grouplet it is known that Kakangari is a gas-rich regolith breccia (*Srinivasan and Anders,* 1977; *Brearley,* 1988; *Weisberg et al.,* 1996).

The unique chondrite Acfer 094 may be a primitive accretionary breccia, although it contains parts having a clastic matrix (*Bischoff and Geiger,* 1994) and fragments. One object contained in it (Fig. 12) is clearly a lithic clast, containing a Ca,Al-rich inclusion. This fragment has been characterized as a CAI within a chondrule by *Krot et al.* (2004).

4.6. Acapulcoites and Lodranites

Acapulco-like achondrites appear to be ultra-metamorphosed chondrites (*Palme et al.,* 1981) and products of parent-body-wide processes including a complex thermal history (*McCoy et al.,* 1996). So far no clearly defined breccias have been reported among the members of this group. All acapulcoites are relatively unshocked (*McCoy et al.,* 1996). However, the sample of LEW 86220 deserves special attention and is linked to the acapulcoites and lodranites, which may be genetically related to each other, through both its O-isotopic and mineral composition (*Clayton and Mayeda,* 1996; *McCoy et al.,* 1997; *Mittlefehldt et al.,* 1998). The two existing lithologies are, however, not regarded as clasts as in meteoritic breccias. The lodranites are unbrecciated.

4.7. Winonaites

Winonaites, which are unshocked or very weakly shocked, show minor evidence for brecciation during metamorphism in Y-75300, Winona, and Mt. Morris (*Benedix et al.,* 1997, 1998). Yamato-75261 is an impact-melt breccia that has been linked to winonaites by virtue of its comparable O-isotopic composition (*Benedix et al.,* 1998), but its mineralogy shows that it is an impact melt from the EH-chondrite parent body (*Nagahara,* 1991). The proposed relationship between the winonaites and IAB irons leads to the suggestion that the partially melted and incompletely differentiated

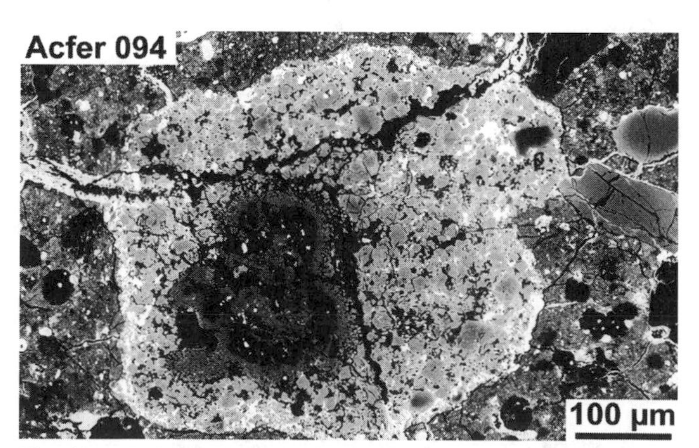

Fig. 12. CAI-bearing lithic fragment in the unique carbonaceous chondrite Acfer 094. This clast contains a Ca,Al-rich inclusion (dark gray; center, lower-left) surrounded by olivine-rich components. Image in backscattered electrons.

IAB iron-winonaite parent body experienced catastrophic breakup and reassembly (*Benedix et al.,* 2000, and discussion and references therein).

4.8. Ureilites

Ureilites comprise the second largest group of achondritic meteorites with over 100 separate meteorites. Although many ureilites are shocked, most are coarse-grained unbrecciated rocks. However, there are about 15% that are breccias, which provide important clues to the geology of the parent asteroid: 14 polymict breccias, 1 dimict breccia, and 1 that appears to be a monomict breccia. Since the earliest studies of polymict ureilites (*Jaques and Fitzgerald,* 1982; *Prinz et al.,* 1986, 1987, 1988), it was recognized that they contain a large variety of clast types, some of which are unlike unbrecciated ureilites. Several recent studies (*Ikeda et al.,* 2000, 2003; *Cohen et al.,* 2004; *Kita et al.,* 2004) have provided comprehensive surveys of these materials. The following survey is based on a review of ureilitic breccias by *Goodrich et al.* (2004).

4.8.1. Monomict ureilites and dimict breccia. All monomict ureilites are coarse-grained, ultramafic (olivine-pyroxene) rocks characterized by high abundances (up to ~5 vol%) of C (graphite and secondary, shock-produced diamond or other high pressure forms) (e.g., *Vdovykin,* 1972; *Bischoff et al.,* 1999; *Grund and Bischoff,* 1999; *El Goresy et al.,* 2004), with metal and sulfide as the only other common accessory phases (see reviews by *Goodrich,* 1992; *Mittlefehldt et al.,* 1998). The majority are olivine-pigeonite or olivine-orthopyroxene assemblages interpreted to be residues of ~25–30% partial melting. A small number are augite-bearing, and appear to be cumulates or paracumulates.

Frontier Mountain 93008 (possibly part of a single meteoroid comprising nine ureilites found in Frontier Mountain, Antarctica) has been recognized as a dimict ureilite (*Fioretti and Goodrich,* 2001; *Smith et al.,* 2000). It consists of two monomict ureilite-like lithologies — an olivine-pigeonite assemblage of Fo 79, and an augite-bearing assemblage of Fo 87 — separated by a brecciated contact containing some exotic materials. The scale of "clasts" in this breccia is much larger than found in polymict ureilites, and it is likely to have formed in a different environment.

4.8.2. Polymict breccias. Eight of the 14 known polymict ureilites [North Haig, Nilpena, DaG 164, DaG 165, DaG 319 (Fig. 13), DaG 665, EET 83309, and EET 87720] are well-studied and consist of lithic and mineral fragments that represent a variety of lithologies and thus can be classified as fragmental breccias. Solar-wind-implanted gases are present in Nilpena, EET 83309, and EET 87720, indicating that they are regolith breccias (*Ott et al.,* 1990, 1993; *Rai et al.,* 2003). The absence of solar gases in other samples does not necessarily imply a grossly different origin; all polymict ureilites are petrographically similar, and most likely formed in the same environment (*Goodrich et al.,* 2004).

The most extensive survey of the types of materials found in polymict ureilites is that of *Ikeda et al.* (2000),

Fig. 13. Overview photograph of the polymict ureilite DaG 319 showing the fragmented character of the sample (1.4 cm in largest dimension).

who developed a petrographic classification scheme consisting of 7 major groups, with 24 types of lithic clasts and 22 types of mineral clasts, for DaG 319. *Cohen et al.* (2004) provide an extensive survey of feldspathic materials in DaG 319, DaG 165, DaG 164, DaG 665, and EET 83309. These two works encompass most of the major types of materials previously observed in North Haig, Nilpena, and EET 83309 (*Jaques and Fitzgerald,* 1982; *Prinz et al.,* 1986, 1987).

More than ~97% of the material in polymict ureilites consists of lithic clasts that are compositionally and texturally similar to monomict ureilites, or mineral clasts that could have been derived from them. Olivine-pigeonite assemblages dominate among the lithic clasts, and pigeonite appears to be the dominant pyroxene among mineral clasts. Only one lithic clast resembling the olivine-orthopyroxene monomict ureilites has been identified (*Goodrich and Keil,* 2002), but isolated orthopyroxene clasts that could have been derived from them are common (*Ikeda et al.,* 2000).

Lithic clasts similar to the augite-bearing monomict ureilites have been identified in DaG 319 (*Ikeda et al.,* 2000; *Ikeda and Prinz,* 2001) and DaG 165 (*Goodrich and Keil,* 2002). *Ikeda and Prinz* (2001) discovered one poikilitic orthopyroxene-olivine-augite clast and two isolated olivine clasts. *Goodrich and Keil* (2002) found a melt-inclusion-bearing olivine clast resembling the augite-bearing ureilite HH 064 (*Weber and Bischoff,* 1998; *Weber et al.,* 2003), and an olivine-augite-orthopyroxene-pigeonite clast with complex poikilitic relationships resembling the augite-bearing ureilite MET A78008 (*Berkley and Goodrich,* 2001).

The remaining 2–3% of material (lithic and mineral clasts) in polymict ureilites is highly diverse, and can be divided into (1) materials that could be indigenous (cognate) to the ureilite parent body, but are not represented among monomict ureilites; and (2) xenolithic materials that were contributed to the regolith by impactors.

4.8.2.1. Indigenous (cognate) clasts: Indigenous clasts can be subdivided into feldspathic (containing plagioclase and/or plagioclase-normative glass), mafic, and metal- or sulfide-rich types. Feldspathic clasts have attracted considerable attention because they may represent "missing" basaltic melts complementary to monomict ureilite residues.

(a) Feldspathic clasts. Feldspathic clasts are described by *Prinz et al.* (1986, 1987, 1988), *Ikeda et al.* (2000), *Guan and Crozaz* (2001), *Goodrich and Keil* (2002), and *Cohen et al.* (2004, 2005). One of the striking features of feldspathic clasts in polymict ureilites is that plagioclase compositions span essentially the entire range from An 0–100. Several distinct feldspathic clast populations have been recognized. Two of these (ferroan anorthitic clasts and chondrule/chondrite fragments) are nonindigenous and are discussed below. Indigenous feldspathic clasts are divided by *Cohen et al.* (2004) and *Goodrich and Keil* (2002) into pristine (retaining primary petrologic characteristics) and nonpristine (shock-melted and/or mixed with other pristine lithologies and/or impactors) clasts:

Pristine feldspathic clasts: An albitic lithology was identified as the most abundant population of feldspathic clasts in all polymict ureilites, while a labradoritic lithology was found as lithic clasts only in DaG 665 and EET 83309 (*Cohen et al.,* 2004). A third pristine feldspathic lithology identified by *Cohen et al.* (2004) comprises clasts in which olivine and augite are the only mafic minerals, and a fourth, rare, type is characterized by very anorthite-rich plagioclase. In addition, *Cohen et al.* (2004) also identified many lithic clasts containing plagioclase with a range of An content similar to the albitic and labradoritic lithologies, and pyroxene and/or olivine with compositions that differ from those of olivine-pigeonite ureilites in being more calcic and more ferroan, but whose relationship to one another is difficult to determine.

Nonpristine feldspathic clasts: A variety of feldspathic clasts that appear to have been shock-melted and possibly mixed with other lithologies have been described. Clasts consisting of glass with sprays of radiating plagioclase microlites (giving them a chondrule-like appearance) occur in North Haig, Nilpena, EET 83309, DaG 319, and DaG 165 (*Prinz et al.,* 1986, 1988; *Goodrich and Keil,* 2002). *Ikeda et al.* (2000) describe pilotaxitic clasts, consisting of masses of irregularly interwoven, small plagioclase laths and minor interstitial pyroxene and silica-rich mesostasis. Another variety of extremely fine-grained clast (*Cohen et al.,* 2004) consists of skeletal to feathery mafic minerals in crystalline plagioclase. *Cohen et al.* (2004) and *Goodrich and Keil* (2002) describe several feldspathic clasts in which pyroxene grains have monomict ureilite like pigeonite cores that are probably relicts, with sharp boundaries to augite. Two clasts described by *Cohen et al.* (2004) consist of abundant euhedral, normally zoned olivine crystals in a glassy groundmass of albitic, non-stoichiometric plagioclase composition with fine crystallites. In addition, some feldspathic clasts are clastic breccias, containing a variety of angular grains of various types in a glassy feldspathic groundmass indicating multiple episodes of brecciation (breccias in a breccia; cf. Fig. 4).

(b) Mafic clasts. Lithic and mineral clasts consisting of Fo-rich olivine (Fo_{90-99}, usually with strong reverse zoning) and/or pyroxene (enstatite) are common in DaG 319 (*Ikeda et al.,* 2000), DaG 165 (C. A. Goodrich, unpublished data, 2004), North Haig and Nilpena (*Prinz et al.,* 1986, 1988), and EET 83309 (*Prinz et al.,* 1987), and are most likely highly shocked and reduced versions of monomict ureilite-like materials. *Goodrich and Keil* (2002) describe one extremely unusual oxidized mafic clast in DaG 165, whose origin is unclear.

(c) Sulfide- and metal-rich clasts. Rare sulfide-rich lithic clasts in DaG 319 are described by *Ikeda et al.* (2000, 2003). They consist of anhedral grains of olivine, sometimes enclosed in massive sulfide (troilite), with a fine-grained, porous silicate matrix containing disseminated sulfide. Fine-grained metal-rich clasts in DaG 319 (*Ikeda et al.,* 2000, 2003) consist mainly of enstatite and metal, with variable amounts of a silica phase, plagioclase, sulfide, and rarely olivine. A few large enstatite grains contain aggregates of submicrometer-sized metal and silica, probably formed by *in situ* reduction.

4.8.2.2. Xenolithic clasts. Several types of xenolithic (nonindigenous) clasts occur in various polymict ureilites (*Goodrich et al.,* 2004):

(a) Ferroan, anorthite-rich plagioclase clasts: Rare, ferroan, anorthitic clasts resembling the angrite meteorites (particularly Angra Dos Reis, or ADOR) were described from Nilpena and North Haig (*Jaques and Fitzgerald,* 1982; *Prinz et al.,* 1986, 1987). One small clast observed in DaG 319 [C4-2 gabbroic type of *Ikeda et al.* (2000)] probably represents the same lithology. Some of the most anorthite-rich plagioclase mineral clasts found in EET 83309, DaG 164/165, DaG 319, and DaG 665 (*Prinz et al.,* 1987; *Ikeda et al.,* 2000; *Goodrich and Keil,* 2002; *Cohen et al.,* 2004) may also be derived from it. *Kita et al.* (2004) showed that the oxygen isotopic composition of one ferroan, anorthite-rich clast in DaG 319 is similar to that of ADOR.

(b) Chondrules and chondrite fragments: *Jaques and Fitzgerald* (1982) described an olivine-clinobronzite clast in Nilpena that appeared to be an unequilibrated H-group chondrule. *Prinz et al.* (1986, 1987) noted that some orthopyroxene mineral fragments in North Haig and Nilpena and rare olivine mineral fragments in EET 83309 are of ordinary chondritic composition, and *Prinz et al.* (1988) recognized barred olivine, radial pyroxene, and cryptocrystalline chondrules. *Ikeda et al.* (2000, 2003) identified barred olivine (type F1-1), porphyritic olivine (type F1-2), porphyritic olivine-pyroxene (type F1-3), and radial pyroxene (type F1-5) chondrules and chondrule fragments, as well as equilibrated chondrite fragments (type F2) in DaG 319. Equilibrated chondrite fragments have homogeneous olivine with lesser amounts of pyroxenes, plagioclase, sulfide, and chromite and are similar in mineralogy and mineral compositions to R-group chondrites.

(c) Dark clasts: Dark clasts resembling carbonaceous chondrite matrix material were first observed as components of polymict ureilities in North Haig and Nilpena (*Prinz et al.,* 1987; *Brearley and Prinz,* 1992; *Brearley and Jones,* 1998). Similar dark clasts, generally a few hundred micrometers to several millimeters in size and angular, are abundant in DaG 319 (*Ikeda et al.,* 2000; 2003). *Ikeda et al.* (2003) divided them into two subtypes: fayalite-free (D1), which are common, and fayalite-bearing (D2), which are rare. *Ikeda et al.* (2003) noted that the occurrence of fayalite in these clasts suggests affinities to oxidized CV chondrites such as Kaba, Bali, and Mokoia. Several dark clasts have been observed in DaG 165 (*Goodrich and Keil,* 2002). They have extremely fine-grained matrices consisting largely of phyllosilicates with bulk compositions similar to those of the dark clasts in Nilpena and North Haig (*Brearley and Prinz,* 1992), and contain abundant grains of Fe,Ni sulfide, framboidal magnetite, and larger magnetite.

4.9. HED Meteorites

Howardites, eucrites, and diogenites are genetically related and form the HED suite of achondrites, which may come from Vesta (*Wahl,* 1952; *Mason,* 1962; *McCarthy et al.,* 1972; *Takeda et al.,* 1983; *Clayton and Mayeda,* 1983). Eucrites and diogenites are magmatic rocks, representing a wide range of chemical compositions and variable crystallization histories (e.g., *Miyamoto and Takeda,* 1977; *Takeda,* 1979; *Takeda et al.,* 1984; *Hewins and Newsom,* 1988; *McSween,* 1989; *Mittlefehldt et al.,* 1998, *McCoy et al.,* 2006). At least 85% of HED meteorites are impact breccias formed in the regolith and megaregolith of their parent body (*Stöffler et al.,* 1988). The megaregolith is the thick layer of fractured and possibly mixed planetary crust beneath the surface regolith, which has been discussed in detail by *Hartmann* (1973, 1980). While many eucrites and diogenites occur as monomict breccias, howardites are mechanical mixtures of diogenites and eucrites and, hence, breccias by definition (e.g., *McCarthy et al.,* 1972; *Duke and Silver,* 1967; *Bunch,* 1975; *Delaney et al.,* 1983; *Buchanan and Reid,* 1996). In the order of increasing amounts of diogenite component, polymict HED breccias are classified as polymict eucrites, howardites, and polymict diogenites (*Delaney et al.,* 1983). According to *Delaney et al.* (1983), eucrites containing up to 10% diogenitic material are called polymict eucrites, whereas diogenites containing up to 10% eucritic material are called polymict diogenites. HED meteorites containing 10–90% eucritic material are defined as howardites. Several members of the HED suite display petrologic features that seem to result from annealing by igneous activity or impact-melt sheets after crystallization and brecciation (e.g., *Labotka and Papike,* 1980; *Takeda et al.,* 1981; *Yamaguchi and Takeda,* 1994, 1995; *Yamaguchi et al.,* 1994, 1997; *Metzler et al.,* 1995; *Buchanan and Reid,* 1996; *Takeda,* 1997; *Mittlefehldt et al.,* 1998; *Saiki and Takeda,* 1999).

4.9.1. Monomict breccias among the eucrites and diogenites. The suite of monomict eucrites and monomict diogenites consist of obviously unbrecciated and monomict brecciated meteorites. Brecciated monomict eucrites and brecciated monomict diogenites are called monomict eucrite breccias and monomict diogenite breccias in the following. Most monomict eucrite breccias and monomict diogenite breccias are characterized by distinct variations of grain size and texture on a millimeter to centimeter scale, mainly due to the existence of large lithic clasts embedded in a fine-grained clastic matrix (*von Engelhardt,* 1963; *Mason,* 1963; *Duke and Silver,* 1967; *Reid and Barnard,* 1979; *Takeda et al.,* 1983; *Palme et al.,* 1988; *Takeda and Yamaguchi,* 1991; *Mittlefehldt,* 1994; *Yamaguchi et al.,* 1994; *Metzler et al.,* 1995). The ratio lithic clasts/clastic matrix varies significantly. Individual meteorites of this kind consist of a single clast type or a limited range of lithic clast types that are of nearly identical bulk chemical composition. Nevertheless, several monomict eucrite breccias are texturally polymict (e.g., *Takeda and Graham,* 1991). This observation indicates that these breccias originate from parent rocks, which were texturally heterogeneous. The initial brecciation events were followed by later stages of parent-body evolution, as documented by impact-induced intrusive melt dikes and shock veins, crosscutting the brecciated texture (e.g., *Duke and Silver,* 1967; *Takeda et al,* 1981; *Bogard et al.,* 1985; *Dickinson et al.,* 1985; *Bobe et al.,* 1989; *Bobe,* 1992). Furthermore, the original textures are overprinted by thermal annealing (e.g., *Labotka and Papike,* 1980; *Fuhrmann and Papike,* 1981; *Takeda et al,* 1981; *Bobe et al.,* 1989; *Yamaguchi and Takeda,* 1992, 1994; *Yamaguchi et al.,* 1994, 1996; *Metzler et al.,* 1995; *Papike et al.,* 2000; *Miyamoto et al.,* 2001) and later periods of impact brecciation (e.g., *Delaney et al.,* 1982; *Metzler et al.,* 1995; *Saiki and Takeda,* 1999). Bilanga is classified as a monomict diogenite breccia. However, *Domanik et al.* (2004) describe exotic fragments that are of a different but possibly related rock type incorporated in the Bilanga breccia.

4.9.2. Polymict breccias (polymict eucrites, howardites, polymict diogenites). There is an enormous literature database concerning the petrology and chemistry of lithic and mineral clasts in polymict HED breccia (e.g., *Mittlefehldt et al.,* 1998, and references therein; *Buchanan and Mittlefehldt,* 2003; *Cohen,* 2004). These breccias are fragmental and regolith breccias consisting of coarse mineral and lithic clasts of eucritic and diogenitic compositions, embedded in a fine-grained clastic matrix (Fig. 14). They formed by impact comminution and local mixing (e.g., *Delaney at al.,* 1983, 1984; *Metzler et al.,* 1995; *Buchanan and Reid,* 1996).

The textural variability of igneous clasts is coupled with large mineralogical and chemical variations of entire clasts and their mineral constituents. Based on the large variety of polymict HED breccias *Saiki and Takeda* (1999) claim that there is a distinct local heterogeneity on the HED parent body and most of these breccias formed locally around the floor of impact craters. As it is observed in monomict eucrite breccias and monomict diogenite breccias, the clastic matrix of polymict HED breccias is either fragmental (e.g., *Reid et al.,* 1990) or recrystallized, the latter due to thermal annealing after brecciation (e.g., *Takeda,* 1991; *Bogard et al.,* 1993; *Metzler et al.,* 1995).

Many howardites contain fragments of impact-melt rocks and fused soils and are enriched in solar gases (Table 2) and

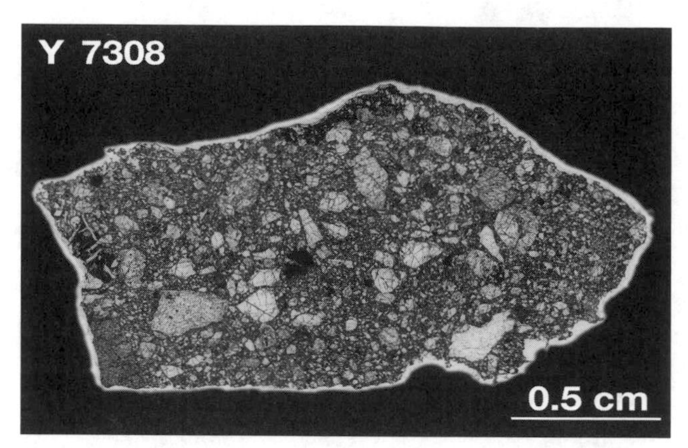

Fig. 14. Overview photograph of the Y-7308 howardite. Different types of fragments are embedded within a fine-grained clastic matrix. Image in transmitted light. See *Metzler* (1985) for details.

track-rich grains (e.g., *Suess et al.,* 1964; *Wilkening et al.,* 1971; *Labotka and Papike,* 1980; *MacDougall et al.,* 1973; *Caffee et al.,* 1988; *Rao et al.,* 1997; *Wieler et al.,* 2000; *Caffee and Nishiizumi,* 2001). Diverse clast types found so far in polymict HED breccias are described below in detail.

4.9.2.1. Eucrite clasts. Polymict eucrites are mainly composed of typical eucrite clasts with a wide range of textures and chemical compositions, which are very similar to the monomict eucrites and to clasts from monomict eucrite breccias (e.g., *Duke and Silver,* 1967; *Delaney et al.,* 1984). These clast types include variolitic and subophitic basalts as well as gabbroic lithologies (e.g., *Bobe,* 1992; *Yamaguchi et al,* 1994; *Metzler et al.,* 1995; *Buchanan et al.,* 2000a,b; *Patzer et al.,* 2003). Clasts of cumulate eucrites and ordinary eucrites (*Takeda,* 1991) occur in the same breccia (*Saiki et al.,* 2001). The same holds for howardites (e.g., *Ikeda and Takeda,* 1985; *Metzler et al.,* 1995), but compared to typical eucrite clasts, these breccias contain fragments of more extreme composition, which are not represented as distinct meteorites (e.g., *Bunch,* 1975). Large eucrite clasts have been found in the polymict diogenite Aioun el Atrouss (*Lomena et al.,* 1976) and small amounts of a eucritic component were observed in the polymict diogenite Garland (*Varteresian and Hewins,* 1983).

4.9.2.2. Diogenite clasts. These fragments are common in polymict eucrites and contribute up to ~50% to the howardites known so far (e.g., *Duke and Silver,* 1967; *Delaney et al.,* 1983, 1984; *Ikeda and Takeda,* 1985; *Warren,* 1985; *Metzler et al.,* 1995; *Mittlefehldt et al.,* 1998). Polymict diogenites are mainly composed of coarse diogenite clasts set in a fine-grained clastic matrix of the same material, containing up to 10% additional eucritic material (*Delaney et al.,* 1983).

4.9.2.3. Granulite clasts. Granulite clasts are common in polymict eucrites and howardites (e.g., *Bobe,* 1992; *Yamaguchi et al.,* 1994; *Metzler et al.,* 1995; *Metzler and Stöffler,* 1995; *Buchanan et al.,* 2000a; *Patzer et al.,* 2003). The common eucrite clasts in Y-791960 are granulitic (*Takeda,* 1991), indicating annealing of the parent rock prior to brecciation. A further good example for granulitic breccias is the meteorite Asuka 881388 (A. Yamaguchi, personal communication, 2004). In the polymict eucrite Pasamonte a granulite clast was found, which shows an enrichment in a chondritic component, indicating contaminations by a chondritic projectile (*Metzler et al.,* 1995). A texturally similar clast, interpreted to be a heavily metamorphosed impact melt breccia, was found in the polymict eucrite Macibini (*Buchanan et al.,* 2000b).

4.9.2.4. Clasts of impact-melt rocks and breccias. Most polymict HED breccias contain different types of impact-melt rock with abundances of up to 15 vol% (e.g., *Labotka and Papike,* 1980; *Metzler and Stöffler,* 1987; *Olsen et al.,* 1987, 1990; *Bogard et al.,* 1993; *Mittlefehldt and Lindstrom,* 1993; *Metzler et al.,* 1995; *Metzler and Stöffler,* 1995; *Pun et al.,* 1998; *Buchanan et al.,* 2000b; *Sisodia et al.,* 2001; *Buchanan and Mittlefehldt,* 2003). In thin sections these lithologies appear as angular to subrounded clasts, often darker than the surroundings due to finely disseminated sulfides. They can be subdivided into glassy, devitrified or crystallized impact melts and impact-melt breccias. In EET 87503 two impact-melt breccia clasts have been detected that are enriched in a chondritic component, indicating intensive mixture of projectile and target melts during impact (*Metzler et al.,* 1995).

The two-component mixing model for HED breccias by *McCarthy et al.* (1972) and *Delaney et al.* (1983) is basically supported by the observation that melt composition of impact-melt rocks from polymict eucrites and howardites follow the mixing line between eucritic and diogenitic lithologies. Up to now not a single clast of pure diogenitic impact-melt rock has been described. This indicates that extended orthopyroxenites were never exposed to such crustal levels, where pure diogenitic whole-rock impact-melt rocks could have formed (*Metzler and Stöffler,* 1995).

The polymict eucrite ALH A81011 represents a vesicular impact melt breccia as a whole (*Metzler et al.,* 1994). In addition, NWA 1240 has been described as an HED parent-body impact-melt rock (*Barrat et al.,* 2003).

4.9.2.5. Breccia clasts (breccia-in-breccia structures). Clasts of clastic matrix (breccia-in-breccia structure) have been found in several howardites (e.g., *Metzler et al.,* 1995; *Pun et al.,* 1998). These inclusions seem to have formed and compacted at different locations near the surface of the parent body and were admixed as lithic clasts to the host breccias by impact.

4.9.2.6. Foreign clasts (xenoliths). Carbonaceous chondritic clasts have been found in some polymict HED breccias, e.g., in LEW 85300 (*Zolensky et al.,* 1992a, 1996c). *Wilkening* (1973) first identified carbonaceous chondrite clasts in the howardite Kapoeta that were mineralogically similar to CM and CV3 chondrites. Later, chondritic clasts were separated by *Bunch et al.* (1979) from the howardite Jodzie and studied mineralogically and chemically. Further reports on such clasts include *Kozul and Hewins* (1988), *Mittlefehldt and Lindstrom* (1988), *Hewins* (1990), *Olsen et al.* (1990), *Reid et al.* (1990), *Buchanan et al.* (1993),

Pun et al. (1998), and *Buchanan and Mittlefehldt* (2003). *Mittlefehldt* (1994) describes a dark gray, fine-grained fragment with a chondritic chemical signature in the diogenite Ellemeet. Unfortunately, a detailed petrographic description is missing.

4.10. Aubrites

Aubrites, which are composed largely of enstatite grains with <1% FeO, resemble howardites in that all meteorites are fragmental or regolith breccias of igneous rocks. But aubrites, unlike howardites, lack closely related unbrecciated meteorites (except possibly for some metal-rich meteorites). Everything that we know about the geology of the parent asteroid of the aubrites has been derived from studies of breccia clasts.

Six of the 20 analyzed aubrites contain solar wind gases: Bustee, EET 90033, Khor Temiki, LEW 87007, Pesyanoe, and Y-793592. Others, including Cumberland Falls, Bishopville, and Mayo Belwa, have Kr-, Sm-, and Gd-isotopic effects indicating neutron capture near the surface of their parent asteroid for periods of up to several hundred million years (*Lorenzetti et al.,* 2003; *Hidaka et al.,* 1999). Aubrites with solar-wind gases are composed of millimeter- to centimeter-sized clasts in a finer-grained matrix (*Poupeau et al.,* 1974). Other aubrites appear to be coarser grained, with enstatite crystal fragments up to 10 cm in size.

The best studied aubrite, Norton County, which lacks solar-wind gases, is largely composed of enstatite crystals derived from orthopyroxenite, plus pyroxenite clasts with igneous textures composed of orthoenstatite, pigeonite, and diopside, and impact-melt-breccia clasts (*Okada et al.,* 1988). In addition, there is a clast composed of diopside, plagioclase, and silica, and olivine grains and feldspathic clasts that are probably derived from separate lithologies. Norton County contains ~1.5 vol% of Fe,Ni metal grains up to a centimeter in size with associated sulfides and schreibersite that probably represent metal that was incompletely separated from silicate during differentiation (*Casanova et al.,* 1993). Taenite compositions suggest most metal grains cooled through 500°C at ~2°C/m.y. but some cooled faster. Thus one or more major impacts must have pulverized the parent asteroid and excavated material from great depths. *Okada et al.* (1988) suggest that the parent body was collisionally disrupted and gravitationally reassembled.

Other clasts derived from the aubrite parent body include a 4-cm-wide enstatite-oldhamite clast with blebby diopside in Bustee (*McCoy,* 1998), an oxide-bearing clast in ALH 84008 (*McCoy et al.,* 1999), basaltic vitrophyre clasts (*Fogel,* 1997), and round inclusions in NWA 2736 (*Love et al.,* 2005). Foreign clasts are present in the three aubrites Cumberland Falls, ALH A78113, and Pesyanoe: They contain abundant FeO-bearing, chondritic inclusions (xenoliths) up to 4 cm in size (*Binns,* 1969; *Lipschutz et al.,* 1988; *Lorenz et al.,* 2005). The O-isotopic composition of some of these chondritic clasts indicates a relationship with ordinary chondrites, but the clasts have been affected by reduction on the aubrite body and their origin is controversial (*Wasson et al.,* 1993).

Three related metal-rich meteorites may be samples from the aubrite parent body, although they are not visibly brecciated. *Watters and Prinz* (1980) suggested that the iron meteorite, Horse Creek, and a stony-iron meteorite, Mt. Egerton, which is composed of coarsely crystalline enstatite with ~20% metallic Fe,Ni, represented the core and core-mantle interface, respectively, of the aubrite parent body. A third meteorite, Itqiy (*Patzer et al.,* 2001), appears to resemble Mt. Egerton. The evidence in the aubritic breccias for impact scrambling of their parent asteroid suggests that metal-rich samples should have been mixed with aubrites.

Shallowater appears to be an unbrecciated, enstatite achondrite that is closely related to the aubritic breccias. However, *Keil et al.* (1989) infer from its mineralogical and thermal properties that it contains chondritic inclusions and formed on a separate asteroid (*Keil,* 1989). Shallowater contains 20 vol% of metal-bearing inclusions, which are inferred to be chondritic material from a projectile or the cool outer layer of a partly molten target that was mixed with enstatitic melt, causing the melt to cool rapidly at >100°C/h. Thus Shallowater can be considered to be an impact-melt breccia in which the melt was not formed by impact but was initially present in the target.

5. BRECCIAS FORMED FROM PARTLY MOLTEN ASTEROIDS

All the breccias that we have described above, apart from the accretionary breccias, have close analogs among the lunar breccias. But there are also meteorite breccias without lunar analogs that appear to have formed <100 m.y. after the asteroids accreted by impact mixing of partly molten asteroids, rather than impact melting of solid bodies. The best examples are the pallasites, which contain angular fragments of olivine embedded in metallic Fe,Ni; the IAB iron meteorites with angular chondritic clasts; and the mesosiderites. The simplest of these are the pallasitic breccias, which probably formed as a result of impact-induced mixing of molten Fe,Ni from the cores of igneously differentiated asteroids with fragmented mantle material located directly above. The group IAB iron meteorites appear to have formed in a partly melted asteroid that was catastrophically broken apart and then reaccreted during a major impact (*Benedix et al.,* 2000).

Mesosiderites are breccias consisting entirely of core and crustal material from a differentiated asteroid, with relatively little of the intervening mantle (*Mittlefehldt et al.,* 1998). The lithic clasts, which may be as large as 2 cm or more, are dominantly from the crust and are broadly similar to eucrites and diogenites. They include basalts, gabbros, and orthopyroxenites, with lesser amounts of dunite, and rare anorthosites (*Rubin and Mittlefehldt,* 1992; *Ikeda et al.,* 1990; *Kimura et al.,* 1991; *Tamaki et al.,* 2004). Other clasts include impact melts and diogenitic monomict breccia clasts. Mineral clasts are mostly coarse-grained orthopyroxene and minor olivine up to 10 cm in size and rarer plagioclase. Mesosiderite clasts differ from HED lithologies in that olivine is rare in HEDs and the REE fractionation patterns of some gabbroic clasts in mesosiderites have extremely high Eu/Sm ratios (*Rubin and Mittlefehldt,* 1992). Other differences are summarized by *Kimura et al.* (1991). Foreign clasts are not observed. Some workers have suggested that the metallic Fe,Ni is derived from the projectile (*Rubin and Mittlefehldt,* 1993), but it seems unlikely that vast amounts of target and projectile material would be mixed by low-speed impacts ~100 m.y. after the asteroids accreted (*Scott et al.,* 2001). The alternative model is that mesosiderites formed during breakup and reaccretion of an asteroid with a molten core (*Scott et al.,* 2001).

6. IMPACT-RELATED MIXING

The abundance of foreign clasts in meteorites gives a good measure of the degree of mixing among asteroids and the relative abundance of different types of material at different times and places in the asteroid belt. Aside from Kaidun, which is described below, and the Cumberland Falls aubrite (*Binns,* 1969), few meteorites contain more than a few volume percent of foreign clasts. The most abundant clasts are CM-like chondritic fragments (*Meibom and Clark,* 1999).

In ordinary chondrites, clasts from different ordinary chondrites groups are relatively rare. In the LL chondrite St. Mesmin, intensely shocked H-group chondrite fragments were found (*Dodd,* 1974). *Rubin et al.* (1983b) describe an LL5 clast in the Dimmitt H-chondrite regolith breccia. An L-group melt-rock fragment was found in the LL chondrite Paragould (*Fodor and Keil,* 1978). Similarly, olivines of fragments in Adzhi-Bogdo (LL3–6) clearly fall in the range of L-group chondrites (*Bischoff et al.,* 1993c, 1996). A troctolitic clast in the Y-794046 (L6) chondrite has an H-chondrite O-isotopic composition (*Prinz et al.,* 1984). In the Fayetteville H-chondrite regolith breccia an L-chondritic inclusion is described by *Wieler et al.* (1989). *Fodor and Keil* (1975) identified a clast of H parentage within the Ngawi LL chondrite.

A small CM-chondrite clast consisting of olivine crystals and an altered barred olivine chondrule embedded in a matrix of phyllosilicates and sulfide was observed in the Magombedze (H3–5) chondrite (*MacPherson et al.,* 1993). Also, a carbonaceous clast was found in the Dimmitt ordinary chondrite breccia (*Rubin et al.,* 1983b). Some further reports on (possibly) carbonaceous clasts in ordinary chondrites are listed in *Keil* (1982).

In some polymict HED breccias (e.g., Kapoeta and LEW 85300) carbonaceous chondrite clasts were found that are mineralogically similar to CM and CV3 chondrites (*Wilkening,* 1973; *Zolensky et al.,* 1992a,b, 1996c). A chondritic clast was separated by *Bunch et al.* (1979) from the howardite Jodzie. Further reports on such clasts include *Kozul and Hewins* (1988), *Mittlefehldt and Lindstrom* (1988), *Hewins* (1990), *Reid et al.* (1990), *Buchanan et al.* (1993), *Mittlefehldt* (1994), *Pun et al.* (1998), and *Buchanan and Mittlefehldt* (2003).

Ordinary chondrite fragments are present in polymict ureilites (e.g., *Jaques and Fitzgerald,* 1982; *Prinz et al.,* 1986, 1987, 1988; *Ikeda et al.,* 2000, 2003). Angrite-like fragments are also known from several samples (e.g., *Jaques and Fitzgerald,* 1982; *Prinz et al.,* 1986, 1987; *Ikeda et al.,* 2000; *Goodrich and Keil,* 2002; *Cohen et al.,* 2004; *Kita et al.,* 2004). Dark clasts that resemble fine-grained carbonaceous chondrite material are known to occur in ureilites (e.g., *Prinz et al.,* 1987; *Brearley and Prinz,* 1992; *Ikeda et al.,* 2000, 2003; *Goodrich and Keil,* 2002).

Enstatite chondrite clasts appear to be especially rare in other meteorite classes. The Galim LL-chondrite breccia appears to have formed after pieces of an EH-chondrite projectile were mixed with LL target material (*Rubin,* 1997b). An EH-chondritic asteroid appears to have formed the Serenitatis Basin on the Moon (*Norman et al.,* 2002), and an EH fragment was found in the Apollo 15 soil sample (*Rubin,* 1997a).

The most spectacular mixing product among the chondrites is probably Kaidun, which consists almost entirely of millimeter- and submillimeter-sized fragments of EH3–5, EL3, CV3, CM1–2, and R chondrites (*Ivanov,* 1989; *Zolensky et al.,* 1996b; *Ivanov et al.,* 2003; *Zolensky and Ivanov,* 2003, and references therein). In addition, it contains C1 and C2 lithologies, alkaline-enriched clasts (similar to the granitoidal clasts found in the Adzhi-Bogdo ordinary chondrite regolith breccia) (*Bischoff et al.,* 1993c), fragments of impact melt products, phosphide-bearing clasts, new enstatite-bearing clasts, fragments of Ca-rich achondrite, and possibly aubritic materials (*Ivanov,* 1989; *Ivanov et al.,* 2003; *Zolensky and Ivanov,* 2003; *Kurat et al.,* 2004). A possible ordinary chondrite clast has been described by

Fig. 15. Kaidun belongs to the most heavily fragmented chondrites showing the huge number of diverse lithologies. Backscattered electron image, 4 cm across, provided by M. Zolensky; see *Zolensky and Ivanov* (2003) for details.

Mikouchi et al. (2005). Clasts that are breccias themselves were also observed (see Fig. 20 in *Zolensky and Ivanov,* 2003) (cf. Fig. 15). According to *Zolensky and Ivanov* (2003), Kaidun may be derived from an especially large asteroid like Ceres, or an unusually located one like Phobos, the larger moon of Mars. Alternatively, Kaidun might be considered as an unusually clast-rich, chondrule-poor chondrite that formed during a lull in chondrule formation or deposition when turbulent accretion concentrated chondritic clasts rather than chondrules (*Scott,* 2002).

7. FORMATION OF BRECCIAS

7.1. Breakup and Reassembly of Asteroids

Asteroids are modified by two kinds of hypervelocity impacts: frequent impacts that crater the surface, and large rare impacts that damage the whole asteroid and create large volumes of rubble. The power-law mass distribution of asteroids is such that the dominant events in the history of meteorites and asteroids are the large infrequent impacts, rather than the cratering events. These large impacts probably create a major fraction of the meteorite breccias (see, e.g., *Scott,* 2002; *Scott and Wilson,* 2005) and reduce many asteroids between 0.1 and 100 km in size to gravitational aggregates of loosely consolidated material (e.g., *Asphaug et al.,* 2002; *Richardson et al.,* 2002); for counterarguments, see *Sullivan et al.* (2002) and *Wilkison et al.* (2002).

Evidence that many asteroids are porous aggregates has accumulated from measurements of asteroid densities (*Britt et al.,* 2002) and numerical models of asteroid impacts to form asteroid families and satellites (*Richardson et al.,* 2002; *Michel et al.,* 2003; *Durda et al.,* 2004). Additional evidence comes from the spin rates of asteroids: asteroids between 150 m and ~10 km in size that approach but do not exceed the upper limit for strengthless asteroids (*Pravec et al.,* 2002). (Asteroids less than 150 m in size spin faster with periods of under 2 h, suggesting they are stronger, more-coherent bodies.)

Simulations using numerical models offer insights into the way that catastrophic events can fragment asteroids (e.g., *Love and Ahrens,* 1996; *Nolan et al.,* 2001). Major impacts on asteroids that form craters comparable in size to the radius of the parent body may cause extensive damage throughout the asteroid: e.g., at the core-mantle boundary (e.g., making pallasites in bodies with molten cores), and at the exterior where material is fractured and briefly lofted (making near-surface monomict breccias like diogenite breccias). Impacts at higher specific energies cause the fragments to reaccrete into a porous rubble pile in which interior fragments have been rotated but not significantly displaced from one another. With increasing specific impact energy, the impact debris forms a cloud of fragments that largely reaccretes within a few hours to days, although some fragments may take up to a year to reaccrete (*Love and Ahrens,* 1996). Eventually, the target will be converted into a family of asteroids with diverse masses, each of which is a gravitationally bound rubble pile (e.g., *Michel et al.,* 2003). Since the specific impact energy for dispersing at least half the target mass is ~100× that for shattering more than half the initial mass, and the number of asteroids is proportional to $r^{-5/2}$, the dispersal lifetime beforehand will be ~50× longer than the shattering lifetime (*Holsapple et al.,* 2002). Thus asteroids will experience numerous shattering impacts before they are dispersed into families of gravitationally bound rubble.

Evidence for catastrophic impacts in meteorites has been derived from meteorite groups with extensive evidence for shock deformation and impact melting, meteorite breccias in which the components are derived from diverse depths of the parent asteroid, and meteorites with anomalous thermal histories (*Keil et al.,* 1994; *Scott,* 2002). The best example of a large impact that formed a body with large volumes of shocked meteorites and several impact-melt breccias is the 500-Ma impact that probably disrupted the L-chondrite parent body, possibly forming the Flora asteroid family (*Haack et al.,* 1996a; *Nesvorný et al.,* 2002). Ureilites, which also have a high abundance of shocked samples, appear to have survived a family-forming impact at 4.5 Ga (*Goodrich et al.,* 2004). Meteorite groups with components derived from diverse depths include fragmental breccias like Mezo-Madaras that contain mixtures of type 3–6 material, and H, L, and LL regolith breccias, which have metal grains with diverse cooling rates (*Grimm,* 1985; *Taylor et al.,* 1987; *Williams et al.,* 2000). In addition, H-, L-, and LL-group chondrites show little correlation between metallographic cooling rates and petrologic types indicating major impact mixing within their parent asteroids (*Taylor et al.,* 1987). Meteorites with anomalous thermal histories indicating quenching in days or less from temperatures above 1000°C include IVA irons with silicate inclusions (*Scott et al.,* 1996; *Haack et al.,* 1996b), mesosiderites (*Scott et al.,* 2001), the Shallowater enstatite achondrite (*Keil et al.,* 1989), and ureilites (*Keil et al.,* 1994; *Goodrich et al.,* 2004). Group IAB irons with silicate inclusions and the closely related winonaites also contain textural evidence for a catastrophic impact that created metal-silicate and silicate breccias (*Benedix et al.,* 2000). All the meteorites with anomalous thermal histories probably formed as a result of major impacts into hot bodies <100 m.y. after the asteroids accreted. In each case, the impact debris reaccreted so that cooling rates at lower temperatures were much slower than those at 1000°C.

The effectiveness of catastrophic impacts for mixing material from diverse depths has not been investigated with high-resolution numerical models. Low-resolution models suggest that effective scrambling requires an impact that at least halves the target's mass (*Scott et al.,* 2001). These models also suggest that impacts on asteroids that are less than a few hundred kilometers in size create relatively little impact melt, and do not mix much projectile material into the target (*Love and Ahrens,* 1996). They simply convert coherent asteroids into fragmental breccias. Impact-melt breccias and projectile clasts are predicted to comprise less than a few volume percent of the residual target material (*Keil et al.,* 1997). Abundances of foreign clasts and impact-melt breccias (excluding those formed when asteroids were partly molten) are much less than the volume of fragmental breccias, consistent with this conclusion.

7.2. Regolith Breccias

Many meteorites rich in solar-wind gases have a prominent brecciated appearance, commonly with light clasts in a dark matrix. The solar-wind gases in these samples were acquired by grains <100 nm from a planetary surface (*Goswami et al.,* 1984; *Caffee et al.,* 1988). A few type 2–3 chondrites with solar-wind gases appear to lack evidence for brecciation (e.g., EH3 and EL3), but this may simply reflect the dark nature of clasts and matrix. Although there are rare components in meteorites that may have been irradiated in space (e.g., *Zolensky et al.,* 2003), comparisons with lunar samples and regolith modeling strongly suggest that fine-grained material was irradiated on the surface of asteroids (*Bunch and Rajan,* 1988; *Caffee et al.,* 1988; *McKay et al.,* 1989). Solar-gas-rich meteorites are therefore called regolith breccias, although *McKay et al.* (1989) cautioned that these meteorites could not be fully understood without better constraints on asteroidal impacts and visits from spacecraft.

Lunar regolith, largely by definition, is formed by relatively small cratering events that distribute soil and rocks locally around craters, rather than large impacts with global ejecta blankets (see *Robinson et al.,* 2002; *Hörz and Cintala,* 1997). Similarly, meteorite regolith breccias are usually interpreted in terms of relatively small impacts that just crater the surface, but larger impacts may play a role, e.g., in dispersing regolith breccias throughout asteroids (*Crabb and Schultz,* 1981). Meteoritic breccias rich in solar wind gases differ from their lunar analogs in having lower abundances of agglutinates and other impact glasses, solar-flare-irradiated grains, and solar-wind gases. They have therefore been interpreted as immature regolith samples that were present in the regolith for relatively brief periods (e.g., *Housen et al.,* 1979a; *Caffee et al.,* 1988). However, this description may be misleading, as the abundance of these features may reflect mixing of diverse materials (see below), as well as the intensity of shock lithification (*Bischoff et al.,* 1983). Typically, carbonaceous chondrites samples are less like lunar regolith samples than howardite and ordinary chondrite samples in the abundances of radiation features. These differences and the high proportion of regolith breccias in many groups (e.g., CM and CI chondrites; Table 2) are difficult to reconcile with detailed models for asteroid regolith development (e.g., *Housen et al.,* 1979a,b). As a result, other sites for the irradiation have been considered: in centimeter- to meter-sized planetesimals prior to accretion of the parent body and in similar-sized components in a megaregolith (see *Caffee et al.,* 1988). However, these models cannot account for the concentration of solar-wind gases in the fine-grained, clastic matrices of meteorites (e.g., *Nakamura et al.,* 1999a,b).

Detailed studies of regolith development on asteroids by *Housen et al.* (1979a,b) assume that regolith forms solely by cratering of an initially coherent planetary surface. For 300-km-diameter asteroids, this model predicts that a 3.5-km regolith would develop in 2.6 G.y. However, predictions that strong 10-km-diameter asteroids would have <1 mm of regolith, and 1–10-km weak asteroids centimeter- to meter-thick regolith appear inconsistent with high-resolution spacecraft images. Four C- and S-class asteroids with mean diameters of 12–50 km have regoliths that are tens of meters thick (*Sullivan et al.,* 2002; *Robinson et al.,* 2002). Regolith models for small asteroids may be inadequate because they do not include the effects of major impacts that cause catastrophic fragmentation or fragmentation and reaccretion and because they overestimate the strength of small bodies (*Asphaug et al.,* 2002).

An important difference between meteoritic and lunar regolith breccias is that meteoritic grains with solar-flare tracks have large excesses of spallation Ne but track-poor grains seldom do (*Caffee et al.,* 1988). In the top few meters of the lunar regolith, many grains lack solar-flare tracks as they have not resided in the upper millimeter, but they all contain spallogenic gases due to exposure to galactic cosmic rays. The excess spallation Ne in track-rich grains in the CM chondrites Murchison and Murray require exposure to galactic cosmic rays for several hundred million years, a period that seems too long to be consistent with regolith models and compaction ages. An alternative explanation is that the track-rich grains were exposed to intense solar cosmic rays from the early Sun (*Woolum and Hohenberg,* 1993). However, *Wieler et al.* (2000) studied grains in the howardite Kapoeta and found no evidence for a high flux of energetic particles from the Sun. They argued instead that the correlated occurrences of solar-flare tracks and cosmogenic Ne excesses could be explained by mixing of mature and immature soils.

Compaction ages of meteorite regolith breccias have been inferred from fission-track techniques that date the time matrix and olivine grains were brought in contact and from radiometric ages of clasts (see *McKay et al.,* 1989). These data suggest that CM chondrite regolith breccias may have formed before 4.3 Ga, whereas ordinary chondrite and achondrite regolith breccias formed much more recently, in one case more recently than 1.3 Ga. However, these data only provide upper limits on compaction ages: New techniques are still needed to date regolith breccias with confidence so that their irradiation effects and records of global geology can be understood better. We do not know, for example, if grains in meteorite regolith breccias were irradiated and lithified when asteroids accreted, during the last billion years in the asteroid belt, or more recently on near-Earth asteroids.

The wide variation in the abundances of regolith breccias in meteorite groups (0–100%; Table 2) can be related to the physical properties and impact histories of the samples. Six classes of meteorites — angrites, brachinites, acapulcoites, winonaites, iron, and stony-iron meteorites — lack both solar-wind-bearing samples and fragmental breccias (excluding metamorphosed and igneously formed breccias). The absence of breccias from these groups may be attributed to poor sampling of parent asteroids, the difficulty of lithifying metal-rich material, and the greater strength of metal-rich rocks. Impact fragments from high-strength asteroids are ejected at high speeds and are liable to escape from small bodies (*Housen,* 1979a,b). In nearly all other groups — ureilites, aubrites, the HED group, and all chondrite groups except CO and CK — regolith breccia and fragmental breccias are both abundant. For the CI-, CM-, CR-, and R-chondrite groups, where 50–100% of the samples are regolith breccias, surface material was remarkably well dispersed throughout the sampled regions of their parent bodies. A plausible explanation is that mature regolith soil was intimately mixed with much larger volumes of deeply buried and poorly consolidated material during an impact that caused breakup and reassembly of the parent asteroid. Thus, if regolith breccias are defined as consolidated debris from a surficial fragmental layer, these meteorites should not be called regolith breccias, but fragmental breccias containing a few percent of grains that were present in the top meter of regolith.

The HED meteorites have a smaller fraction of samples with solar gases (8/124; Table 2) and generally higher concentrations of solar-wind gas than the CI–CM chondrites (*Goswami et al.,* 1984). The lack of olivine-rich mantle material in howardites suggests that mixing was much more limited than in the carbonaceous asteroids. The properties of the solar-gas-rich howardites appear compatible with mixing of mature regolith soil with material that completely lacked irradiation features. Mature soil from the top few meters may have been periodically dispersed throughout a 1-km-thick layer of fragmental material by impact-generated seismic waves that could loft fractured and poorly consolidated material to heights of several kilometers (*Housen et al.,* 1979b; *Asphaug,* 1997).

Because solar-wind-rich meteorite breccias contain only a small fraction of grains that have been in the top meter, they are poor guides to the spectral properties of asteroids. However, they do contain much information about the distribution and diversity of rock types and the impact history of asteroids. New models are clearly needed to understand the formation of asteroidal regolith breccias.

7.3. Simultaneous Accretion of Asteroidal Clasts and Chondrules?

The idea that certain meteorites represent samples of "second-generation" parent bodies (daughter asteroids) formed after collisional destruction of "grandparent" planetary bodies has been discussed earlier (e.g., *Urey,* 1959, 1967; *Zook,* 1980; *Hutchison et al.,* 1988; *Hutchison,* 1996; *Sanders,* 1996; *Bischoff,* 1998; *Bischoff and Schultz,* 2004).

Evidence for the existence of planetesimals before the accretion of the parent asteroids of chondrites have recently been provided by W-isotope data for CAIs, metal-rich chondrites, and iron meteorites (*Kleine et al.,* 2004, 2005a,b). The decay of now extinct ^{182}Hf to ^{182}W (half-life = 9 m.y.) is well suited for dating the formation of metal and refractory phases in the early solar system. Hafnium is lithophile and W is siderophile, such that the separation of metal from silicate (e.g., during core formation) results in a strong Hf-W fractionation. Metals are virtually Hf-free, such that they maintain the W-isotope composition acquired at the time of metal formation. Recent studies have shown that the initial W-isotope composition of magmatic iron meteorites is similar to that of CAIs, indicating that most magmatic iron meteorites formed within less than ~1 m.y. after formation of CAIs (*Kleine et al.,* 2004, 2005a,b). In contrast, the well-preserved U-Pb and ^{26}Al-^{26}Mg age differences between CAIs and chondrules indicate that the formation of chondrules and hence the accretion of the parent asteroids of chondrites persisted for at least 2–3 m.y. and possible even up to ~5 m.y. after the start of the solar system. Based on the combined Hf-W and Al-Mg age constraints, *Kleine et al.* (2004, 2005a,b) argue that certain iron meteorites are remnants of the earliest asteroids and that chondrites derive from relatively late-formed planetesimals that may have formed by reaccretion of debris produced during collision disruption of first-generation planetesimals (the latter represented by the magmatic iron meteorites).

Considering various types of clasts in carbonaceous and ordinary chondrites *Bischoff and Schultz* (2004) suggested that many breccias result from mixing of fragments after total destruction of precursor parent bod(y)ies. They suggest that dark inclusions in CR and CH chondrites (Fig. 11) may be excellent witnesses to document formation of the final parent body by secondary accretion. In many primitive chondrites [Krymka (Fig. 3), Adrar 003, Acfer 094], unusual fragments exist that may represent chondritic fragments of a first-generation parent body. In summary, increasing evidence is found for accretion of planetesimal clasts with chondrules at a time when chondrite parent bodies formed.

Acknowledgments. The authors thank A. Deutsch, L. Schultz, T. Kleine, and I. Weber for fruitful discussions; A. Ruzicka, M. Weisberg, A. Yamaguchi, and A. Krot for their constructive comments and suggestions; and T. Grund, F. Bartschat (Münster), and Matthias Bölke (Lüdinghausen) for technical assistance. We also thank M. Zolensky for supplying the great Kaidun photograph.

REFERENCES

Abreu N. M. and Brearley A. J. (2004) Characterization of matrix in the EET 92042 CR2 carbonaceous chondrite: Insight into textural and mineralogical heterogeneity (abstract). *Meteoritics & Planet. Sci., 39,* A12.

Abreu N. M. and Brearley A. J. (2005) HRTEM and EFTEM studies of phyllosilicate-organic matter associations in matrix and dark inclusions in the EET 92042 CR2 carbonaceous chondrite (abstract). In *Lunar and Planetary Science XXXVI,* Abstract #1826. Lunar and Planetary Institute, Houston (CD-ROM).

Ashworth J. R. and Barber D. J. (1976) Lithification of gas-rich meteorites. *Earth Planet. Sci. Lett., 30,* 222–233.

Asphaug E. (1997) Impact origin of the Vesta family. *Meteoritics & Planet. Sci., 32,* 965–980.

Asphaug E., Ryan E. V., and Zuber M. T. (2002) Asteroid interiors.

In *Asteroids III* (W. F. Bottke Jr. et al., eds.), pp. 463–484. Univ. of Arizona, Tucson.

Barber D. J. and Hutchison R. (1991) The Bencubbin stony-iron meteorite breccia: Electron petrography, shock-history and affinities of a "carbonaceous chondrite" clast. *Meteoritics, 26,* 83–95.

Barrat J. A., Jambon A., Bohn M., Blichert-Toft J., Sautter V., Göpel C., Gillet P., Boudouma O., and Keller F. (2003) Petrology and geochemistry of the unbrecciated achondrite Northwest Africa 1240 (NWA 1240): An HED parent body impact melt. *Geochim. Cosmochim. Acta, 67,* 3959–3870.

Beauchamp R. and Fredriksson K. (1979) Ivuna and Orgueil C-1 chondrites: A new look. *Meteoritics, 14,* 344.

Benedix G. K., McCoy T. J., and Keil K. (1997) Winonaites revisited: New insights into their formation (abstract). In *Lunar and Planetary Science XXVIII,* pp. 91–92. Lunar and Planetary Institute, Houston.

Benedix G. K., McCoy T. J., Keil K., Bogard D. D., and Garrison D. H. (1998) A petrologic and isotopic study of winonaites: Evidence for early partial melting, and metamorphism. *Geochim. Cosmochim. Acta, 62,* 2535–2553.

Benedix G. K., McCoy T. J., Keil K., and Love S. G. (2000) A petrologic study of the IAB iron meteorites: Constraints on the formation of the IAB-Winonaite parent body. *Meteoritics & Planet. Sci., 35,* 1127–1141.

Berkley J. L. and Goodrich C. A. (2001) Evidence for multi-episodic igneous events in ureilite MET 78008 (abstract). *Meteoritics & Planet. Sci., 36,* A18–A19.

Binns R. A. (1967) Structure and evolution of noncarbonaceous chondrites. *Earth Planet. Sci. Lett., 2,* 23–28.

Binns R. A (1968) Cognate xenoliths in chondritic meteorites: Examples in Mezo-Madaras and Ghubara. *Geochim. Cosmochim. Acta, 32,* 299–317.

Binns R. A. (1969) A chondritic inclusion of unique type in the Cumberland Falls meteorite. In *Meteorite Research* (P. M. Millman, ed.), pp. 696–704. Reidel, Dordrecht.

Binns R. A., Davis R. J., and Reed S. J. B. (1969) Ringwoodite, natural $(Mg,Fe)_2SiO_2$ spinel in the Tenham meteorite. *Nature, 221,* 943–944.

Bischoff A. (1992) ALH 85085, Acfer 182, and Renazzo-type chondrites — Similarities and differences. *Meteoritics, 27,* 203–204.

Bischoff A. (1993) Alkali-granitoids as fragments within the ordinary chondrite Adzhi-Bogdo: Evidence for highly fractionated, alkali-granitic liquids on asteroids (abstract). In *Lunar and Planetary Science XXIV,* pp. 113–114. Lunar and Planetary Institute, Houston.

Bischoff A. (1998) Aqueous alteration of carbonaceous chondrites: Evidence for preaccretionary alteration — A review. *Meteoritics & Planet. Sci., 33,* 1113–1122.

Bischoff A. (2000) Mineralogical characterization of primitive, type-3 lithologies in Rumuruti chondrites. *Meteoritics & Planet. Sci., 35,* 699–706.

Bischoff A. (2002) Discovery of purple-blue ringwoodite within shock veins of an LL6 ordinary chondrite from Northwest Africa (abstract). In *Lunar and Planetary Science XXXIII,* Abstract #1264. Lunar and Planetary Institute, Houston (CD-ROM).

Bischoff A. and Geiger T. (1994) The unique carbonaceous chondrite Acfer 094: The first CM3 chondrite (?) (abstract). In *Lunar and Planetary Science XXV,* pp. 115–116. Lunar and Planetary Institute, Houston.

Bischoff A. and Metzler K. (1991) Mineralogy and petrography of the anomalous carbonaceous chondrites Y-86720, Y-82162 and B-7904. *Proc. NIPR Symp. Antarct. Meteorites, 4,* 226–246.

Bischoff A. and Stöffler D. (1992) Shock metamorphism as a fundamental process in the evolution of planetary bodies: Information from meteorites. *Europ. J. Mineral., 4,* 707–755.

Bischoff A. and Schultz L. (2004) Abundance and meaning of regolith breccias among meteorites (abstract). *Meteoritics & Planet. Sci., 39,* A15.

Bischoff A., Rubin A. E., Keil K., and Stöffler D. (1983) Lithification of gas-rich chondrite regolith breccias by grain boundary and localized shock melting. *Earth Planet. Sci. Lett., 66,* 1–10.

Bischoff A., Palme H., Spettel B., Clayton R. N., and Mayeda T. K. (1988) The chemical composition of dark inclusions from the Allende meteorite (abstract). In *Lunar and Planetary Science XIX,* pp. 88–89. Lunar and Planetary Institute, Houston.

Bischoff A., Palme H., Geiger T., and Spettel B. (1992) Mineralogy and chemistry of the EL-chondritic melt rock Ilafegh-009 (abstract). In *Lunar and Planetary Science XXIII,* pp. 105–106. Lunar and Planetary Institute, Houston.

Bischoff A., Palme H., Ash R. D., Clayton R. N., Schultz L., Herpers U., Stöffler D., Grady M. M., Pillinger C. T., Spettel B., Weber H., Grund T., Endreß M., and Weber D. (1993a) Paired Renazzo-type (CR) carbonaceous chondrites from the Sahara. *Geochim. Cosmochim. Acta, 57,* 1587–1603.

Bischoff A., Palme H., Schultz L., Weber D., Weber H. W., and Spettel B. (1993b) Acfer 182 and paired samples, an iron-rich carbonaceous chondrite: Similarities with ALH 85085 and relationship to CR chondrites. *Geochim. Cosmochim. Acta, 57,* 2631–2648.

Bischoff A., Geiger T., Palme H., Spettel B., Schultz L., Scherer P., Schlüter J., and Lkhamsuren J. (1993c) Mineraolgy, chemistry, and noble gas contents of Adzhi-Bogdo — An LL3-6 chondritic breccia with L-chondritic and granitoidal clasts. *Meteoritics, 28,* 570–578.

Bischoff A., Geiger T., Palme H., Spettel B., Schultz L., Scherer P., Loeken T., Bland P., Clayton R. N., Mayeda T. K., Herpers U., Meltzow B., Michel R., and Dittrich-Hannen B. (1994a) Acfer 217 — A new member of the Rumuruti chondrite group (R). *Meteoritics, 29,* 264–274.

Bischoff A., Schirmeyer S., Palme H., Spettel B., and Weber D. (1994b) Mineralogy and chemistry of the carbonaceous chondrite PCA91467 (CH). *Meteoritics, 29,* 444.

Bischoff A., Gerel O., Buchwald V. F., Spettel B., Loeken T., Schultz L., Weber H. W., Schlüter J., Baljinnyam L., Borchuluun D., Byambaa C., and Garamjav D. (1996) Meteorites from Mongolia. *Meteoritics & Planet. Sci., 31,* 152–157.

Bischoff A., Weber D., Spettel B., Clayton R. N., Mayeda T. K., Wolf D., and Palme H. (1997) Hammadah al Hamra 180: A unique unequilibrated chondrite with affinities to LL-group ordinary chondrites (abstract). *Meteoritics & Planet. Sci., 32,* A14.

Bischoff A., Weber D., Bartoschewitz R., Clayton R. N., Mayeda T. K., Spettel B., and Weber H. W (1998) Characterization of the Rumuruti chondrite regolith breccia Hughes 030 (R3–6) and implications for the occurrence of unequilibrated lithologies on the R-chondrite parent body (abstract). *Meteoritics & Planet. Sci., 33,* A15–A16.

Bischoff A., Goodrich C. A., and Grund T. (1999) Shock-induced origin of diamonds in ureilites (abstract). In *Lunar and Plane-*

tary Science XXX, Abstract #1100. Lunar and Planetary Institute, Houston (CD-ROM).

Bobe K. D. (1992) Die monomikten Eukrite und ihre mehrphasige magmatische, impaktmetamorphe und thermische Entwicklungsgeschichte auf dem HED-Mutterkörper. Ph.D. dissertation, Universität Münster. 164 pp.

Bobe K. D., Bischoff A., and Stöffler D. (1989) Impact and thermal metamorphism as fundamental processes in the evolution of the Stannern, Juvinas, Jonzac, Peramiho, and Millbillillie eucrite parent body. *Meteoritics, 24,* 252.

Bogard D. D., Taylor G. J., Keil K., Smith M. R., and Schmitt R. A. (1985) Impact melting of the Cachari eucrite 3.0 Ga ago. *Geochim. Cosmochim. Acta, 49,* 941–946.

Bogard D. D., Nyquist L., Takeda H., Mori H., Aoyoma T., Bansal B., Wiesman H., and Shih C.-Y. (1993) Antarctic polymict eucrite Yamato 792769 and the cratering record on the HED parent body. *Geochim. Cosmochim. Acta, 57,* 2111–2121.

Bogard D. D., Garrison D. H., Norman M., Scott E. R. D., and Keil K. (1995) ^{39}Ar-^{40}Ar age and petrology of Chico: Large scale impact melting on the L chondrite parent body. *Geochim. Cosmochim. Acta, 59,* 1383–1399.

Bottke W. F., Nolan M. C., Greenberg R., and Kolvoord R. A. (1994) Velocity distributions among colliding asteroids. *Icarus, 107,* 255–268.

Brearley A. J. (1988) Nature and origin of matrix in the unique chondrite Kakangari: A TEM investigation (abstract). In *Lunar and Planetary Science XIX,* pp. 130–131. Lunar and Planetary Institute, Houston.

Brearley A. J. and Geiger T. (1991) Mineralogical and chemical studies bearing on the origin of accretionary rims in the Murchison CM2 carbonaceous chondrite (abstract). *Meteoritics, 26,* 323.

Brearley A. J. and Jones R. H. (1998) Chondritic meteorites. In *Planetary Materials* (J. J. Papike., ed.), pp. 3-1 to 3-398. Mineralogical Society of America, Washington, DC.

Brearley A. J. and Prinz M. (1992) CI chondrite-like clasts in the Nilpena polymict ureilite. Implications for aqueous alteration processes in CI chondrites. *Geochim. Cosmochim. Acta, 56,* 1373–1386.

Brezina A. (1904) The arrangement of collections of meteorites. *Trans. Am. Philos. Soc., 43,* 211–247.

Bridges J. C. and Hutchison R. (1997) A survey of clasts and chondrules in ordinary chondrites. *Meteoritics & Planet. Sci., 32,* 389–394.

Bridges J. C., Franchi I. A., Hutchison R., Morse A. D. Long J. V. P., and Pillinger C. T. (1995a) Cristobalite- and tridymite-bearing clasts in Parnallee (LL3) and Farmington (L5). *Meteoritics, 30,* 715–727.

Bridges J. C., Hutchison R., Franchi I. A., Alexander C. M. O'D., and Pillinger C. T. (1995b) A feldspar-nepheline achondrite clast in Parnallee. *Proc. NIPR Symp. Antarct. Meteorites, 8,* 195–203.

Britt D. T. and Pieters C. M (1991) Black ordinary chondrites: An analysis of abundance and fall frequency. *Meteoritics, 26,* 279–285.

Britt D. T. and Pieters C. M (1994) Darkening in black and gas-rich ordinary chondrites: The spectral effects of opaque morphology and distribution. *Geochim. Cosmochim. Acta, 58,* 3905–3919.

Britt D. T., Yeomans D., Housen K., and Consolmagno G. (2002) Asteroid density, porosity and structure. In *Asteroids III* (W. F. Bottke Jr. et al., eds.), pp. 485–500. Univ. of Arizona, Tucson.

Browning L. B., McSween H. Y. Jr., and Zolensky M. E. (1995) Parent body alteration features in CM rims (abstract). In *Lunar and Planetary Science XXVI,* pp. 181–182. Lunar and Planetary Institute, Houston.

Buchanan P. C. and Mittlefehldt D. W. (2003) Lithic components in the paired howardites EET 87503 and EET 87513: Characterization of the regolith of 4 Vesta. *Antarc. Meteorite Res., 16,* 128–151

Buchanan P. C. and Reid A. M. (1996) Petrology of the polymict eucrite Petersburg. *Geochim. Cosmochim. Acta, 60,* 135–146.

Buchanan P. C., Zolensky M. E., and Reid A. M. (1993) Carbonaceous chondrite clasts in the howardites Bholghati and EET87513. *Meteoritics, 28,* 659–669.

Buchanan P. C., Zolensky M. E., and Reid A. M. (1997) Petrology of Allende dark inclusions. *Geochim. Cosmochim. Acta, 61,* 1733–1743.

Buchanan P. C., Mittlefehldt D. W., Hutchison R., Koeberl C., Lindstrom D. J., and Pandit M. K. (2000a) Petrology of the Indian eucrite Piplia Kalan. *Meteoritics & Planet. Sci., 35,* 609–615.

Buchanan P. C., Lindstrom D. J., Mittlefehldt D. W., Koeberl C., and Reimold W. U. (2000b) The South African polymict eucrite Macibini. *Meteoritics & Planet. Sci., 35,* 1321–1331

Bullock E. S., Grady M. M., Russell S. S., and Gounelle M. (2005) Fe-Ni sulfides within a CM1 clast in Tagish Lake (abstract). In *Lunar and Planetary Science 36,* Abstract #1883. Lunar and Planetary Institute, Houston (CD-ROM).

Bunch T. E. (1975) Petrolography and petrology of basaltic polymict breccias (howardites). *Proc. Lunar Sci. Conf. 6th,* pp. 469–492.

Bunch T. E. and Chang S. (1980) Carbonaceous chondrites — II. Carbonaceous chondrite phyllosilicates and light element geochemistry as indicators of parent body processes and surface conditions. *Geochim. Cosmochim. Acta, 44,* 1543–1577.

Bunch T. E. and Rajan R. S. (1988) Meteorite regolith breccias. In *Meteorites and the Early Solar System* (J. F. Kerridge and M. S. Matthews, eds.), pp. 144–164. Univ. of Arizona, Tucson.

Bunch T. E. and Stöffler D. (1974) The Kelly chondrite: A parent body surface metabreccia. *Contrib. Mineral Petrol., 44,* 157–171.

Bunch T. E., Chang S., Frick U, Neil J., and Moreland G. (1979) Carbonaceous chondrites — I. Characterisation and significance of carbonaceous chondrite (CM) xenoliths in the Jodzie howardite. *Geochim. Cosmochim. Acta, 43,* 1727–1742.

Burbine T. H., McCoy T. J., Meibom A., Gladman B., and Keil K. (2002) Meteoritic parent bodies: Their number and identification. In *Asteroids III* (W. F. Bottke Jr. et al., eds.), pp. 653–667. Univ. of Arizona, Tucson.

Burbine T. H., Folco L., Capitani G., Bland P. A., Menvies O. N., and McCoy T. J. (2003) Identification of nanophase iron in a H chondrite impact melt (abstract). *Meteoritics & Planet. Sci., 38,* A111.

Busemann H., Eugster O., Bauer H., and Wieler R. (2003a) The ingredients of the "subsolar" noble gas component (abstract). In *Lunar and Planetary Science XXXIV,* Abstract #1674. Lunar and Planetary Institute, Houston (CD-ROM).

Busemann H., Bauer H., and Wieler R. (2003b) Solar noble gases in enstatite chondrites and implications for the formation of the terrestrial planets (abstract). In *Lunar and Planetary Science XXXIV,* Abstract #1665. Lunar and Planetary Institute, Houston (CD-ROM).

Caffee M. W. and Nishiizumi K. (2001) Exposure history of separated phases from the Kapoeta meteorite. *Meteoritics & Planet. Sci., 36,* 429–437.

Caffee M. W., Goswami J. N., Hohenburg C. M., Marti K., and Reedy R. C. (1988) Irradiation records in meteorites. In *Meteorites and the Early Solar System* (J. F. Kerridge and M. S. Matthews, eds), pp. 205–245. Univ. of Arizona, Tucson.

Casanova I., Keil K., Wieler R., San Miguel A., and King E. A. (1990) Origin and history of chondrite regolith, fragmental and impact-melt breccias from Spain. *Meteoritics, 25,* 127–135.

Casanova I., Keil K., and Newsom H. (1993) Composition of metal in aubrites: Constraints on core formation. *Geochim. Cosmochim Acta, 57,* 675–682.

Christophe Michel-Lévy M. (1988) A new component of the Mezö-Madaras breccia: A microchondrule- and carbon-bearing L-related chondrite. *Meteoritics, 23,* 45–48.

Clayton R. N. and Mayeda T. K. (1978) Multiple parent bodies of polymict-brecciated meteorites. *Geochim. Cosmochim. Acta, 42,* 325–327.

Clayton R. N. and Mayeda T. K. (1983) Oxygen isotopes in eucrites, shergottites, nakhlites, and chassignites. *Earth Planet. Sci. Lett., 67,* 151–161.

Clayton R. N. and Mayeda T. K. (1996) Oxygen-isotope studies of achondrites. *Geochim. Cosmochim. Acta, 60,* 1999–2018.

Cohen B. A. (2004) Survey of clasts in howardite Grosvenor Mountains 95574 (abstract). *Meteoritics & Planet. Sci., 39,* A24.

Cohen B. A., Goodrich C. A., and Keil K. (2004) Feldspathic clast populations in polymict ureilites: Stalking the missing basalts from the ureilite parent body. *Geochim. Cosmochim. Acta, 68,* 4249–4266.

Cohen B. A., Swindle T. D., and Olson E. K. (2005) Geochronology of clasts in polymict ureilite Dar al Gani 665 (abstract). In *Lunar and Planetary Science XXXVI,* Abstract #1704. Lunar and Planetary Institute, Houston (CD-ROM).

Crabb J. and Schultz L. (1981) Cosmic ray exposure ages of the ordinary chondrites and their significance for parent body stratigraphy. *Geochim. Cosmochim. Acta, 45,* 2151–2160.

Delaney J. S., Prinz M., Nehru C. E. and O'Neill C. O. (1982) The polymict eucrites Elephant Moraine A79004 and the regolith history of a basaltic achondrite parent body. *Proc. Lunar Planet. Sci. Conf. 13th,* in *J. Geophys. Res., 87,* A339–A352.

Delaney J. S., Takeda H., Prinz M., Nehru C. E., and Harlow G. E. (1983) The nomenclature of polymict basaltic achondrites. *Meteoritics, 18,* 103–111.

Delaney J. S., Prinz M., and Takeda H. (1984) The polymict eucrites. *Proc. Lunar Planet. Sci. Conf. 15th,* in *J. Geophys. Res., 89,* C251–C288.

Dickinson T., Keil K., Lapaz L., Bogard D. D., Schmitt R. A., Smith M. R., and Rhodes J. M. (1985) Petrology and shock age of the Palo Blanco Creek eucrite. *Chem. Erde, 44,* 245–257.

Dodd R. T. (1974) Petrology of the St. Mesmin chondrite. *Contrib. Mineral. Petrol., 46,* 129–145.

Dodd R. T. (1981) *Meteorites — A Petrologic-Chemical Synthesis.* Cambridge Univ., Cambridge. 368 pp.

Dominak K., Kolar S., Musselwhite D., and Drake M. J. (2004) Accessory silicate mineral assemblages in the Bilanga diogenite: A petrographic study. *Meteoritics & Planet. Sci., 39,* 567–579.

Duke M. B. and Silver L. T. (1967) Petrology of eucrites, howardites, and mesosiderites. *Geochim. Cosmochim. Acta, 31,* 1637–1665.

Durda D. D., Bottke W. F. Jr., Enke B. L., Merline W. J., Asphaug E., Richardson D. C., and Leinhardt Z. M. (2004) The formation of asteroid satellites in large impacts: Results from numerical simulations. *Icarus, 167,* 382–396.

Dymek F. R., Albee A. L., Chodos A. A., and Wasserburg G. J. (1976) Petrology of isotopically-dated clasts in the Kapoeta howardite and petrologic constraints on the evolution of its parent planet. *Geochim. Cosmochim. Acta, 40,* 1115–1130.

Eberhardt P., Geiss J., and Grogler N. (1965) Further evidence on the origin of trapped gases in the meteorite Khor Temiki. *J. Geophys. Res., 70,* 4375–4378.

El Goresy A., Gillet P., Dubrovinsky L., Chen M., and Nakamura T. (2004) A super-hard, transparent carbon form, diamond, and secondary graphite in the Haverö ureilite: A fine-scale micro-raman and synchroton tomography (abstract). *Meteoritics & Planet. Sci., 39,* A36.

Endress M. (1994) Mineralogische und chemische Untersuchungen an CI-Chondriten. Ph.D. thesis, University of Münster, Münster, Germany. 223 pp.

Endress M. and Bischoff A. (1993) Mineralogy, degree of brecciation, and aqueous alteration of the CI chondrites Orgueil, Ivuna, and Alais (abstract). *Meteoritics, 28,* 345–346.

Endress M. and Bischoff A. (1996) Carbonates in CI chondrites: Clues to parent body evolution. *Geochim. Cosmochim. Acta, 60,* 489–507.

Endress M., Spettel B., and Bischoff A. (1994a) Chemistry, petrography, and mineralogy of the Tonk CI chondrite: Preliminary results (abstract). *Meteoritics, 29,* 462–463.

Endress M., Keil K., Bischoff A., Spettel B., Clayton R. N., and Mayeda T. K. (1994b) Origin of dark clasts in the Acfer 059/El Djouf 001 CR2 chondrite. *Meteoritics, 29,* 26–40.

Engelhardt W. von (1963) Der Eukrit von Stannern. *Beitr. Mineral. Petrogr., 9,* 65–94.

Fioretti A. M. and Goodrich C. A. (2001) A contact between an olivine-pigeonite lithology and an olivine-augite-orthopyroxene lithology in ureilites FRO 93008: Dashed hopes? (abstract). *Meteoritics & Planet. Sci., 36,* A58.

Fodor R. V. and Keil K. (1973) Composition and origin of lithic fragments in L- and H-group chondrites (abstract). *Meteoritics, 8,* 366–367.

Fodor R. V. and Keil K. (1975) Implications of poikilitic textures in LL-group chondrites. *Meteoritics, 10,* 325–340.

Fodor R. V. and Keil K. (1976a) Carbonaceous and non-carbonaceous lithic fragments in the Plainview, Texas, chondrite: Origin and history. *Geochim. Cosmochim. Acta, 40,* 177–189.

Fodor R. V. and Keil K. (1976b) A komatiite-like lithic fragment with spinifex texture in the Eva meteorite: Origin from a supercooled impact-melt of chondritic parentage. *Earth Planet. Sci. Lett., 29,* 1–6.

Fodor R. V. and Keil K. (1978) Catalog of lithic fragments in LL-group chondrites. *Institute of Meteoritics Special Publication No. 19,* University of New Mexico, Albuquerque. 38 pp.

Fodor R. V., Keil K., and Jarosewich E. (1972) The Oro Grande, New Mexico, chondrite and its lithic inclusion. *Meteoritics, 7,* 495–507.

Fodor R. V., Prinz M., and Keil K. (1974) Implications of K-rich lithic fragments and chondrules in the Bhola brecciated chondrite (abstract). *Geol. Soc. Am. Abstr. with Progr., 6,* 739–740.

Fodor R.V., Keil K., Wilkening L. L. Bogard D. D., and Gibson E. K. (1976) Origin and history of a meteorite parent-body regolith breccia: Carbonaceous and non-carbonaceous lithic fragments in the Abbott, New Mexico, chondrite. In *Tectonics and Mineral Resources of Southwestern U. S.,* pp. 206–218.

New Mexico Geological Society Special Publication No. 6.

Fodor R. V., Keil K., and Gomes C. B. (1977) Studies of Brazilian meteorites IV. Origin of a dark-collored, unequilibrated lithic fragment in the Rio Negro chondrite. *Revista Brasileira Geociencias, 7,* 45–57.

Fodor R. V., Keil K., Prinz M., Ma M. S., Murali A. V., and Schmitt R. A. (1980) Clast-laden melt-rock fragment in the Adams County, Colorado, H5 chondite. *Meteoritics, 15,* 41–62.

Fogel R. A. (1997) A new aubrite basalt vitrophyre from the LEW 87007 aubrite (abstract). In *Lunar and Planetary Science XXVIII,* pp. 369–370. Lunar and Planetary Institute, Houston.

Folco L., Bland P. A., D'Orazio M., Franchi I. A., and Rocchi S. (2002) Dar al Gani 896: A unique picritic achondrite (abstract). *Meteoritics & Planet. Sci., 39,* A49.

Folco L., Bland P. A., D'Orazio M., Franchi I. A., Kelley S., and Rocchi S. (2004) Extensive impact melting on the H-chondrite parent asteroid during the cataclysmic bombardment of the early solar system: Evidence from the achondritic meteorite Dar al Gani 896. *Geochim. Cosmochim. Acta, 68,* 2379–2397.

Fredriksson K., Noonan A., and Nelen J. (1975) The Bhola stone: A true polymict breccia? (abstract). *Meteoritics, 10,* 87–88.

Fruland R. M., King E. A., and McKay D. S. (1978) Allende dark inclusions. *Proc. Lunar Planet. Sci. Conf. 9th,* pp. 1505–1329.

Fuchs L. H., Olsen E. and Jensen K. J. (1973) Mineralogy, mineral chemistry and composition of Murchison (C2) meteorite. *Smithson. Contrib. Earth Sci., 10.* 39 pp.

Fuhrmann M. and Papike J. J. (1981) Howardites and polymict eucrites: Regolith samples from the eucrite parent body. Petrology of Bholghati, Bununu, Kapoeta, and ALHA 76005. *Proc. Lunar Sci. Conf. 12B,* pp. 1257–1279.

Fujita T. and Kitamura M. (1992) Shock melting origin of a troilite-rich clast in the Moorabie chondrite (L3). *Proc. NIPR Symp. Antarct. Meteorites, 5,* 258–269.

Geiger T. (1991) Metamorphe kohlige Chondrite: Petrologische Eigenschaften und Entwicklung des Mutterkörpers. Ph.D. thesis, University of Münster, Münster, Germany. 147 pp.

Geiss J. (1973) Solar wind composition and implications about the history of the solar system. *13th Intl. Cosmic Ray Conf., 5,* 3375–3398.

Goodrich C. A. (1992) Ureilites: A critical review. *Meteoritics, 27,* 327–352.

Goodrich C. A. and Keil K. (2002) Feldspathic and other unusual clasts in polymict ureilite DaG 165 (abstract). In *Lunar and Planetary Science XXXIII,* Abstract #1777. Lunar and Planetary Institute, Houston (CD-ROM).

Goodrich C. A., Scott E. R. D., and Fioretti A. M. (2004) Ureilitic Breccias: Clues to the petrologic structure and impact disruption of the ureilite parent asteroid. *Chem. Erde, 64,* 283–327.

Goswami J. N., Lal D., and Wilkening L. L. (1984) Gas-rich meteorites: Probes for particle environment and dynamical processes in the inner solar system. *Space Sci. Rev., 37,* 111–159.

Grady M. M., Verchovsky A. B., Franchi I. A., Wright I. P., and Pillinger C. T. (2002) Light element geochemistry of the Tagish Lake CI2 chondrite: Comparison with CI1 and CM2 meteorites. *Meteoritics & Planet. Sci., 37,* 713–735.

Grier J. A., Kring D. A., Swindle T. D., Rivkin A. S., Cohen B. A., and Britt D. T. (2004) Analyses of the chondritic meteorite Orvinio (H6): Insight into the origin and evolution of shocked H chondrite material. *Meteoritics & Planet. Sci., 39,* 1475–1493.

Grimm R. E. (1985) Penecontemporaneous metamorphism, fragmentation, and reassembly of ordinary chondrite parent bodies. *J. Geophys. Res., 90,* 2022–2028.

Grossman L., Allen J. M., and MacPherson G. J. (1980) Electron microprobe study of a "mysterite"-bearing inclusion from the Krymka LL-chondrite. *Geochim. Cosmochim. Acta, 44,* 221–216.

Grossman J. N. (1997) *The Meteoritical Bulletin, No. 81,* 1997 July. In *Meteoritics & Planet. Sci., 32,* A159–A166.

Grossman J. N., Rubin A. E., and MacPherson G. J. (1988) ALH 85085; a unique volatile-poor carbonaceous chondrite with possible implications for nebular fractionation processes. *Earth Planet. Sci. Lett., 91,* 33–54.

Grund T. and Bischoff A. (1999) Cathodoluminescence properties of diamonds in ureilites: Further evidence for a shock-induced origin (abstract). *Meteoritics & Planet. Sci., 34,* A48–A49.

Guan Y. and Grozaz G. (2001) Microdistributions and petrogenetic implications of rare earth elements in polymict ureilites. *Meteoritics & Planet. Sci., 36,* 1039–1056.

Haack H., Farinella P., Scott E. R. D., and Keil K. (1996a) Meteoritic, asteroidal, and theoretical constraints on the 500 Ma disruption of the L chondrite parent body. *Icarus, 119,* 182–191.

Haack H., Scott E. R. D., Love S. G., Brearley A. J., and McCoy T. J. (1996b) Thermal histories of IVA stony-iron and iron meteorites: Evidence for asteroid fragmentation and reaccretion. *Geochim. Cosmochim. Acta, 60,* 3103–3113.

Hanowski N. P. and Brearley A. J. (1997) Parent body alteration of large metal inclusions in the CM carbonaceous chondrite, Murray (abstract). In *Lunar and Planetary Science XXVIII,* pp. 503–504. Lunar and Planetary Institute, Houston.

Hanowski N. P. and Brearley A. J. (2001) Aqueous alteration of chondrules in the CM carbonaceous chondrite, Allan Hills 81002: Implications for parent body alteration. *Geochim. Cosmochim. Acta, 65,* 495–518.

Hartmann W. K. (1973) Ancient lunar mega-regolith and subsurface structure. *Icarus, 18,* 634–636.

Hartmann W. K. (1979) Diverse puzzling asteroids and a possible unified explanation. In *Asteroids* (T. Gehrels, ed.), pp. 446–479. Asteroids, Univ. of Arizona, Tucson.

Hartmann W. K. (1980) Dropping stones in magma oceans: Effects of early lunar cratering. In *Proceedings of the Conference on the Lunar Highlands Crust* (J. J. Papike and R. B. Merrill, eds.), pp. 155–171. Pergamon, New York.

Hartmann W. K. (1983) *Moon and Planets.* Wadsworth, Belmont. 509 pp.

Hewins R. H. (1990) Geologic history of LEW 85300, 85302 and 85303 polymict eucrites (abstract). In *Lunar and Planetary Science XXI,* pp. 509–510. Lunar and Planetary Institute, Houston.

Hewins R. H. and Newsom H. E. (1988) Igneous activity in the early solar system. In *Meteorites and the Early Solar System* (J. F. Kerridge and J. F. Matthews, eds.), pp. 73–101. Univ. of Arizona, Tucson.

Heymann D. (1967) On the origin of hypersthene chondrites: Ages and shock effects of black chondrites. *Icarus, 6,* 189–221.

Heymann D. and Mazor E. (1967) Light-dark structure and rare gas content of the carbonaceous chondrite Nogoya. *J. Geophys. Res., 72,* 2704–2707.

Heymann D., Van der Stap C. C. A. H., Vis R. N., and Verheul H. (1987) Carbon in dark inclusions of the Allende meteorite. *Me-*

teoritics, 22, 3–15.

Hezel D. (2003) Die Bildung SiO_2-reicher Phasen im frühen Sonnensystem. Ph.D. thesis, Universität zu Köln. 131 pp.

Hidaka H., Ebihara M., and Yoneda S. (1999) High fluences of neutrons determined from Sm and Gd isotopic compositions in aubrites. *Earth Planet. Sci. Lett., 173,* 41–51.

Hoinkes G., Kurat G., and Baric L. (1976) Dubrovnik: Ein L3-6 Chondrit. *Ann. Naturhistor. Mus. Wien, 80,* 39–55.

Holsapple K., Giblin I., Housen K., Nakamura A., and Ryan E. (2002) Asteroid impacts: Laboratory experiments and scaling laws. In *Asteroids III* (W. F. Bottke Jr. et al., eds.), pp. 443–462. Univ. of Arizona, Tucson.

Hörz F. and Cintala M. (1997) Impact experiments related to the evolution of planetary regoliths. *Meteoritics & Planet. Sci., 32,* 179–209.

Housen K. R., Wilkening L. L., Chapman C. R., and Greenberg R. (1979a) Asteroidal regoliths. *Icarus, 39,* 7–351.

Housen K. R., Wilkening L. L., Chapman C. R., and Greenberg R. (1979b) Regolith development and evolution on asteroids and the Moon. In *Asteroids* (T. Gehrels and M. S. Matthews, eds.), pp. 601–627. Univ. of Arizona, Tucson.

Huss G. R., Keil K., and Taylor G. J. (1981) The matrices of unequilibrated ordinary chondrites: Implications for the origin and history of chondrites. *Geochim. Cosmochim. Acta, 45,* 33–41.

Hutchison R. (1996) Chondrules and their associates in ordinary chondrites: A planetary connection. In *Chondrules and the Protoplanetary Disk* (R. H. Hewins et al., eds.), pp. 311–318. Cambridge Univ., Cambridge.

Hutchison R., Williams C. T., Din V. K., Clayton R. N., Kirschbaum C., Paul R. L., and Lipschutz M. E. (1988) A planetary, H-group pebble in the Barwell, L6, unshocked chondritic meteorite. *Earth Planet. Sci. Lett., 90,* 105–118.

Itoh D. and Tomeoka K. (2003) Dark inclusions in CO3 chondrites: New indicators of parent-body processes. *Geochim. Cosmochim. Acta, 67,* 153–169.

Ikeda Y. and Prinz M. (2001) Magmatic inclusions and felsic clasts in the Dar al Gani 319 polymict ureilite. *Meteoritics & Planet. Sci., 36,* 481–499.

Ikeda Y. and Takeda H. (1985) A model for the origin of basaltic achondrites based on the Yamato 7308 howardite. *Proc. Lunar Planet. Sci. Conf. 15th,* in *J. Geophys. Res., 90,* C649–C663.

Ikeda Y., Ebihara M., and Prinz M. (1990) Enclaves in the Mt. Padbury and Vaca Muerta mesosiderites: Magmatic and residue (or cumulate) rock types. *Proc. NIPR Symp. Antarct. Meteorites, 3,* 99–131.

Ikeda Y., Prinz M., and Nehru C. E. (2000) Lithic and mineral clasts in the Dar al Gani (DAG) 319 polymict ureilite. *Antarct. Meteorite Res., 13,* 177–221.

Ikeda Y., Kita N. T., Morishita Y., and Weisberg M. K. (2003) Primitive clasts in the Dar al Gani 319 polymict ureilite: Precursors of the ureilites. *Antarct. Meteorite Res., 16,* 105–127.

Ivanov A. V. (1989) The Kaidun meteorite: Composition and history. *Geochem. Intl., 26,* 84–91.

Ivanov A. V., Kononkova N. N., Yang A. V., and Zolensky M. E. (2003) The Kaidun meteorite: Clasts of alkaline-rich fractionated materials. *Meteoritics & Planet. Sci., 38,* 725–737.

Jäckel A. and Bischoff A. (1988) Textural and mineralogical differences between LL-chondritic fragmental and regolith breccias (abstract). *Meteoritics & Planet. Sci., 33,* A77–A78.

Jäckel A., Bischoff A., Clayton R. N., and Mayeda T. K. (1996) Dar al Gani 013 — A new Saharan Rumuruti-chondrite (R3-6) with highly unequilibrated (type 3) fragments (abstract). In *Lunar and Planetary Science XXVII,* pp. 595–596. Lunar and Planetary Institute, Houston.

Jaques A. L. and Fitzgerald M. J. (1982) The Nilpena ureilites, an unusual polymict breccia: Implications for origin. *Geochim. Cosmochim. Acta, 46,* 893–900.

Johnson C. A., Prinz M., Weisberg M. K., Clayton R. N., and Mayeda T. K. (1990) Dark inclusions in Allende, Leoville, and Vigarano: Evidence for nebular oxidation of CV3 constituents. *Geochim. Cosmochim. Acta, 54,* 819–830.

Kallemeyn G. W., Rubin A. E., and Wasson J. T. (1996) The compositional classification of chondrites: VII. The R chondrite group. *Geochim. Cosmochim. Acta, 60,* 2243–2256.

Keil K. (1982) Composition and origin of chondritic breccias. In *Workshop on Lunar Breccias and Soils and Their Meteoritic Analogs* (G. J. Taylor and L. L. Wilkening, eds.), pp. 65–83. LPI Tech. Rpt. 82-02, Lunar and Planetary Institute, Houston.

Keil K. (1989) Enstatite meteorites and their parent bodies. *Meteoritics, 24,* 195–208.

Keil K. and Fodor R. V. (1973) Composition and origin of lithic fragments in LL-group chondrites (abstract). *Meteoritics, 8,* 394–396.

Keil K. and Fodor R. V. (1980) Origin and history of the polymict-brecciated Tysnes Island chondrite and its carbonaceous and non-carbonaceous lithic fragments. *Chem. Erde, 39,* 1–26.

Keil K., Huss G. I., and Wiik H. B. (1969) The Leoville, Kansas, meteorite: A polymict breccia of carbonaceous chondrites and achondrite. In *Meteorite Research* (P. M. Millman, ed.), p. 217. Reidel, Dordrecht.

Keil K., Fodor R. V., Starzyk P. M., Schmitt R. A., Bogard D. D., and Husain L. (1980) A 3.6-b.y.-old impact-melt rock fragment in the Plainview chondrite: Implications for the age of the H-group chondrite parent body regolith formation. *Earth Planet. Sci. Lett., 51,* 235–247.

Keil K., Ntaflos Th., Taylor G. J., Brearley A. J., Newson H. E., and Romig A. D. (1989) The Shallowater aubrite: Evidence for origin by planetesimal impact. *Geochim. Cosmochim Acta, 53,* 3291–3307.

Keil K., Haack H., and Scott E. R. D. (1994) Catastrophic fragmentation of asteroids: Evidence from meteorites. *Planet. Space Sci., 42,* 1109–1122.

Keil K., Stöffler D., Love S. G., and Scott E. R. D. (1997) Constraints on the role of impact heating and melting in asteroids. *Meteoritics & Planet. Sci., 32,* 349–363.

Kennedy A. K. and Hutcheon I. D. (1992) Chemical and isotopic constraints on the formation and crystallization of SA-1, a basaltic Allende plagioclase-olivine inclusion. *Meteoritics, 27,* 539–554.

Kennedy A. K., Hutchison R., Hutcheon I. D., and Agrell S. O. (1992) A unique high Mn/Fe microgabbro in the Parnallee (LL3) ordinary chondrite: Nebular micture or planetary differentiate form a previously unrecognized planetary body? *Earth Planet. Sci. Lett., 113,* 191–205.

Kerridge J. F. and Bunch T. E. (1979) Aqueous activity on asteroids: Evidence from carbonaceous meteorites. In *Asteroids* (T. Gehrels, ed.), pp. 745–764. Univ. of Arizona, Tucson.

Kerridge J. F. and Matthews M. S., eds. (1988) *Meteorites and the Early Solar System.* Univ. of Arizona, Tucson. 1269 pp.

Kieffer S. W. (1975) From regolith to rock by shock. *Moon, 13,* 301–320.

Kimura M., Ikeda Y., Ebihara M., and Prinz M. (1991) New enclaves in the Vaca Muerta mesosiderite: Petrogenesis and comparison with HED meteorites. *Proc. NIPR Symp. Antarct. Meteorites, 4,* 263–306.

Kita N. T., Ikeda Y., Togashi S., Liu Y., Morishita Y., and Weisberg M. K. (2004) Origin of ureilites inferred from a SIMS oxygen isotopic and trace element study of clasts in the Dar al Gani 319 polymict ureilite. *Geochim. Cosmochim. Acta, 68,* 4213–4235.

Kleine T., Mezger K., Palme H., Scherer E., and Münker C. (2004) A new chronology for asteroid formation in the early solar system based on ^{182}W systematics. *EOS Trans. AGU, 85(47),* Fall Meeting Supplement, Abstract P31C–04.

Kleine T., Mezger K., Palme H., and Scherer E. (2005a) Tungsten isotopes provide evidence that core formation in some asteroids predates the accretion of chondrite parent bodies (abstract). In *Lunar and Planetary Science XXXVI,* Abstract #1431. Lunar and Planetary Institute, Houston (CD-ROM).

Kleine T., Mezger K., Palme H., Scherer E., and Münker C. (2005b) A new model for the accretion and early evolution of asteroids in the early solar system: Evidence from ^{182}Hf-^{182}W in CAIs, metal-rich chondrites and iron meteorites. *Geochim. Cosmochim. Acta,* in press.

Kojima T. and Tomeoka K. (1996) Indicators of aqueous alteration and thermal metamorphism on the CV parent body: Microtextures of a dark inclusion from Allende. *Geochim. Cosmochim. Acta, 60,* 2651–2666.

Kojima T. and Tomeoka K. (1997) A fine-grained dark inclusion in the Vigarano CV3 chondrite: Record of accumulation processes on the meteorite parent body. *Antarct. Meteorite Res., 10,* 203–215.

Kojima T., Tomeoka K., and Takeda H. (1993) Unusual dark clasts in the Vigarano CV3 carbonaceous chondrite: Record of parent body process. *Meteoritics, 28,* 649–658.

Kojima T., Yatagai T., and Tomeoka K. (2000) A dark inclusion in the Manych LL (3.1) ordinary chondrite: A product of strong shock metamorphism. *Antarct. Meteorite Res., 13,* 39–54.

Kojima T., Lauretta D. S., and Buseck P. R. (2003) Accretion, dispersal, and reaccumulation of the Bishunpur (LL3.1) brecciated chondrite: Evidence from troilite-silicate-metal inclusions and chondrule rims. *Geochim. Cosmochim. Acta, 67,* 3065–3078.

Kozul J. and Hewins R. H. (1988) LEW 85300,02,03 polymict eucrites consortium — II: Breccia clasts, CM inclusions, glassy matrix and assembly history (abstract). In *Lunar and Planetary Science XIX,* pp. 647–648. Lunar and Planetary Institute, Houston.

Kracher A., Keil K., and Scott E. R. D. (1982) Leoville (CV3) — An "accretionary breccia"? *Meteoritics, 17,* 239.

Kracher A., Keil K., Kallemeyn G. W., Wasson J. T., Clayton R. N., and Huss G. I. (1985) The Leoville (CV3) accretionary breccia. *Proc. Lunar Planet. Sci. Conf. 16th,* in *J. Geophys. Res., 90,* D123–D135.

Kring D. A., Hill D. H., Gleason J. D., Britt D. T., Consolmagno G. J., Farmer M., Wilson S., and Haag R. (1999) Portales Valley: A meteoritic sample of the brecciated and metal-veined floor of an impact crater on an H-chondrite asteroid. *Meteoritics & Planet. Sci., 34,* 663–669.

Krot A. N., Scott E. R. D., and Zolensky M. E. (1997a) Origin of fayalitic olivine rims and lath-shaped matrix olivines in the CV3 chondrite Allende and its dark inclusions. *Meteoritics & Planet. Sci., 32,* 31–49.

Krot A. N., Rubin A. E., Keil K., and Wasson J. T. (1997b) Microchondrules in ordinary chondrites: Implications for chondrule formation. *Geochim. Cosmochim. Acta, 61,* 463–473.

Krot A. N., Petaev M. I., Zolensky M. E., Keil K., Scott E. R. D., and Nakamura K. (1998) Secondary calcium-iron-rich minerals in the Bali-like and Allende-like oxidized CV3 chondrites and Allende dark inclusions. *Meteoritics & Planet. Sci., 33,* 623–645.

Krot A. N., Brearley A. J., Ulyanov A. A., Biryukov V. V., Swindle T. D., Keil K., Mittlefehldt D. W., Scott E. R. D., Clayton R. N., and Mayeda T. K. (1999) Mineralogy, petrography, bulk chemical, iodine-xenon, and oxygen-isotopic compositions of dark inclusions in the reduced CV3 chondrite Efremovka. *Meteoritics & Planet. Sci., 34,* 67–89.

Krot A. N., Meibom A., and Keil K. (2000) A clast of Bali-like oxidized CV material in the reduced CV chondrite breccia Vigarano. *Meteoritics & Planet. Sci., 35,* 817–825.

Krot A. N., McKeegan K. D., Russell S. S., Meibom A., Weisberg M. K., Zipfel J., Krot T. V., Fagan T. J., and Keil K. (2001) Refractory calcium-aluminum-rich inclusions and aluminum-diopside-rich chondrules in the metal-rich chondrites Hammadah al Hamra 237 and Queen Alexandra Range 94411. *Meteoritics & Planet. Sci., 36,* 1189–1216.

Krot A. N., Meibom A., Weisberg M. K., and Keil K. (2002) The CR chondrite clan: Implications for early solar system processes. *Meteoritics & Planet. Sci., 37,* 1451–1490.

Krot A. N., Fagan T. J., Keil K., McKeegan K. D., Sahijpal S., Hutcheon I. D., Petaev M. I., and Yurimoto H. (2004) Ca,Al-rich inclusions, amoeboid olivine aggregates, and Al-rich chondrules from the unique carbonaceous chondrite Acfer 094: I. Mineralogy and petrology. *Geochim. Cosmochim. Acta, 68,* 2167–2184.

Kurat G. (1970) Zur Genese des kohligen Materials im Meteoriten von Tieschitz. *Earth Planet. Sci. Lett., 7,* 317–324.

Kurat G. and Kracher A. (1980) Basalts in the Lance carbonaceous chondrite. *Z. Naturforsch., 35a,* 180–190.

Kurat G., Palme H., Brandstätter F., and Huth J. (1989) Allende Xenolith AF: Undisturbed record of condensation and aggregation of matter in the solar nebula. *Z. Naturforsch., 44a,* 988–1004.

Kurat G., Zinner E., Brandstätter F., and Ivanov A. V. (2004) Enstatite aggregates with niningerite, heideite, and oldhamite from the Kaidun carbonaceous chondrite: Relatives of aubrites and EH chondrites? *Meteoritics & Planet. Sci., 39,* 53–60.

Labotka T. C. and Papike J. J. (1980) Howardites: Samples of the regolith of the eucrite parent-body: Petrology of Frankfort, Pavlovka, Yurtuk, Malvern, and ALHA 77302. *Proc. Lunar Planet. Sci. Conf. 11th,* pp. 1103–1130.

Lange D. E., Keil K., and Gomes C. B. (1979) The Mafra meteorite and its lithic clasts: A genomict L-group chondrite breccia (abstract), *Meteoritics, 14,* 472–473.

Leitch C. A. and Grossman L. (1977) Lithic clasts in the Supuhee chondrite. *Meteoritics, 12,* 125–139.

Lin Y. and Kimura M. (1998) Petrographic and mineralogical study of new EH melt rocks and a new enstatite chondrite grouplet. *Meteoritics & Planet. Sci., 33,* 501–511.

Lin Y. T., Nagel H-J., Lundberg L. L., and El Goresy A. (1991) MAC88136 — The first EL3 chondrite (abstract). In *Lunar and Planetary Science XXII,* pp. 811–812. Lunar and Planetary Institute, Houston.

Lipschutz M. E., Verkouteren R. M., Sears D. W. G., Hasan F., Prinz M., Weisberg M. K., Nehru C. E., Delaney J. S., Grossman L., and Boily M. (1988) Cumberland falls chondritic inclusions: III. Consortium study of relationship to inclusions in Allan Hills 78113 aubrite. *Geochim. Cosmochim. Acta, 52,* 1835–1848.

Lomena I. S. M., Touré F., Gibson E. K. JR, Clanton U. S., and Reid A. M. (1976) Aioun el Atrouss: A new hypersthene achondrite with eucritic inclusions. *Meteoritics, 11,* 51–57.

Lorenz C. A., Ivanova M. A., Kurat G., and Brandstaetter F. (2005) FeO-rich xenoliths in the Staroye Pesyanoe aubrite (abstract). In *Lunar and Planetary Science XXXVI,* Abstract #1612. Lunar and Planetary Institute, Houston (CD-ROM).

Lorenzetti S., Eugster O., Busemann H., Marti, K., Burbine T. H., and McCoy T. (2003) History and origin of aubrites. *Geochim. Cosmochim. Acta, 67,* 557–571.

Love J. J., Hill D. H., Domanik K. J., Lauretta D. S., Drake M. J., and Killgore M. (2005) NWA 2736: An unusual new graphite-bearing aubrite (abstract). In *Lunar and Planetary Science XXXVI,* Abstract #1913. Lunar and Planetary Institute, Houston (CD-ROM).

Love S. G. and Ahrens T. J. (1996) Catastrophic impacts on gravity dominated asteroids. *Icarus, 124,* 141–155.

MacDougall D., Rajan R. S., and Price P. B. (1973) Gas-rich meteorites: Possible evidence for origin on a regolith. *Science, 183,* 73–74.

MacPherson G. J., Jarosewich E., and Lowenstein P. (1993) Magombedze: A new H-chondrite with light-dark structure. *Meteoritics, 28,* 138–142.

Mason B. (1962) *Meteorites.* Wiley, New York. 274 pp.

Mason B. (1963) The hypersthene achondrites. *Am. Mus. Novitates No. 2155.* 13 pp.

McCarthy T. S., Ahrens L. H., and Erlank A. J. (1972) Further evidence in support of the mixing model for howardite origin. *Earth Planet Sci. Lett., 15,* 86–93.

McCoy T. J. (1998) A pyroxene-oldhamite clast in Bustee: Igneous aubritic oldhamite and a mechanism for the Ti enrichment in aubritic troilite. *Antarct. Meteorite Res., 11,* 32–48.

McCoy T. J. and 9 colleagues (1995) Origin and history of impact-melt rocks of enstatite chondrite parentage. *Geochim. Cosmochim. Acta, 59,* 161–175.

McCoy T. J., Keil K., Clayton R. N., Mayeda T. K., Bogard D. D., Garrison D. H., Huss G. H., Hutcheon I. D., and Wieler R. (1996) A petrologic, chemical, and isotopic study of Monument Draw and comparison with other acapulcoites: Evidence for formation by incipient partial melting. *Geochim. Cosmochim. Acta, 60,* 2681–2708.

McCoy T. J., Keil K., Muenow D. W., and Wilson L. (1997) Partial melting and melt migration in the acapulcoite-lodranite parent body. *Geochim. Cosmochim. Acta, 61,* 639–650.

McCoy T. J., Rosenshein E. B., and Dickinson T. I. (1999) A unique oxide-bearing clast in the aubrite Allan Hills 84008: Evidence for oxidation during magmatic processes (abstract). In *Lunar and Planetary Science XXX,* Abstract #1347. Lunar and Planetary Institute, Houston (CD-ROM).

McCoy T. J., Mittlefehldt D. W., and Wilson L. (2006) Asteroid differentiation. In *Meteorites and the Early Solar System II* (D. S. Lauretta and H. Y. McSween Jr., eds.), this volume. Univ. of Arizona, Tucson.

McKay D. S., Swindle T. D., and Greenberg R. (1989) Asteroidal regoliths: What we do not know. In *Asteroids II* (R. P. Binzel et al., eds.), pp. 617–642. Univ. of Arizona, Tucson.

McSween H. Y. Jr. (1989) Achondrites and igneous processes on asteroids. *Annu. Rev. Earth Planet. Sci., 17,* 119–140.

Meibom A. and Clark B. E. (1999) Evidence for the insignificance of ordinary chondrite material in the asteroid belt. *Meteoritics & Planet. Sci., 34,* 7–24.

Metzler K. (1985) Gefüge und Zusammensetzung von Gesteinsfragmenten in polymikten achondritischen Breccien. Diplomarbeit, Univ. Münster, Germany. 167 pp.

Metzler K. (1990) Petrographische und mikrochemische Untersuchungen zur Akkretions- und Entwicklungsgeschichte chondritischer Mutterkörper am Beispiel der CM-Chondrite. Ph.D. thesis, Univ. Münster, Germany. 183 pp.

Metzler K. (1993) In-situ investigation of preirradiated olivines in CM chondrites. *Meteoritics, 28,* 398–399.

Metzler K. (1995) Aqueous alteration of primary rock on the CM parent body (abstract). In *Lunar and Planetary Science XXVI,* pp. 961–962. Lunar and Planetary Institute, Houston.

Metzler K. (1997) Preirradiated olivines in CM chondrites (abstract). *Meteoritics, 32,* A91–A92.

Metzler K. (2004) Formation of accretionary dust mantles in the solar nebula: Evidence from preirradiated olivines in CM chondrites. *Meteoritics & Planet. Sci., 39,* 1307–1319.

Metzler K. and Bischoff A. (1996) Constraints on chondrite agglomeration from fine-grained chondrule rims. In *Chondrules and the Protoplanetary Disk* (R. H. Hewins et al., eds.), pp. 153–161, Cambridge Univ., Cambridge.

Metzler K. and Stöffler D. (1987) Polymict impact breccias on the eucrite parent body: I. Lithic clasts in some eucrites and howardites (abstract). In *Lunar and Planetary Science XVII,* pp. 641–642. Lunar and Planetary Institute, Houston.

Metzler K. and Stöffler D. (1995) Impact melt rocks and granulites from the HED asteroid. *Meteoritics, 30,* 547.

Metzler K., Bischoff A., and Stöffler D. (1992) Accretionary dust mantles in CM chondrites: Evidence for solar nebula processes. *Geochim. Cosmochim. Acta, 56,* 2873–2897.

Metzler K., Bobe K.-D., Kunz J., Palme H., and Spettel B. (1994) ALHA 81011 — A eucritic impact melt breccia formed 350 m.y. ago. *Meteoritics, 29,* 502–503.

Metzler K., Bobe K.-D., Palme H, Spettel B., and Stöffler D. (1995) Thermal and impact metamorphism on the HED parent asteroid. *Planet. Space Sci., 43,* 499–525.

Michel P., Benz W., and Richardson D. C. (2003) Disruption of fragmented parent bodies as the origin of asteroid families. *Nature, 421,* 608–611.

Mikouchi T., Makishima J., Koizumi E., and Zolensky M. E. (2005) Porphyritic olivine-pyroxene clast in Kaidun: First discovery of an ordinary chondrite clast? (abstract). In *Lunar and Planetary Science XXXVI,* Abstract #1956. Lunar and Planetary Institute, Houston (CD-ROM).

Misawa K., Watanabe S., Kitamura M., Nakamura N., Yamamoto K., and Masuda A. (1992) A noritic clast from the Hedjaz chondritic breccia: Implications for melting events in the early solar system. *Geochim. J., 26,* 435–446.

Mittlefehldt D. W. (1994) The genesis of diogenites and HED parent body petrogenesis. *Geochim. Cosmochim. Acta, 58,* 1537–1552.

Mittlefehldt D. W. (2002) Geochemistry of the ungrouped carbonaceous chondrite Tagish Lake, the anomalous CM chondrite Bells, and comparison with CI and CM chondrites. *Meteoritics & Planet. Sci., 37,* 703–712.

Mittlefehldt D. W. and Lindstrom M. M. (1988) Geochemistry of diverse lithologies in antarctic eucrites (abstract). In *Lunar and Planetary Science XIX,* pp. 790–791. Lunar and Planetary Institute, Houston.

Mittlefehldt D. W. and Lindstrom M. M. (1993) Geochemistry

and petrology of a suite of ten Yamato HED meteorites. *Proc. NIPR Antarct. Meteorites, 6,* 268–292.

Mittlefehldt D. W. and Lindstrom M. M. (2001) Petrology and geochemistry of Patuxent Range 91501, a clast-poor impact melt from the L-chondrite parent body and Lewis Cliff 88663, an L7 chondrite. *Meteoritics & Planet. Sci., 36,* 439–457.

Mittlefehldt D. W., Lindstrom M. M., Wang M.-S., and Lipschutz M. E. (1995) Geochemistry and origin of achondritic inclusions in Yamato-75097, -793241 and -794046 chondrites. *Proc. NIPR Symp. Antarct. Meteorites, 8,* 251–271.

Mittlefehldt D. W., McCoy T. J., Goodrich C. A., and Kracher A. (1998) Non-chondritic meteorites from asteroidal bodies. In *Planetary Materials* (J. J. Papike, ed.), pp. 4-1 to 4-195. Mineralogical Society of America, Washington, DC.

Miyamoto M. and Takeda H. (1977) Evaluation of a crust model of eucrites from the width of exsolved pyroxene. *Geochem. J., 11,* 161–169.

Miyamoto M., Mikouchi T., and Kaneda K. (2001) Thermal history of the Ibitira eucrite as inferred from pyroxene exsolution lamella: Evidence for reheating and rapid cooling. *Meteoritics & Planet. Sci., 36,* 231–237.

Morlok A. (2002) Mikrosondenanalytik und Sekundärionenmassenspektrometrie an Fragmenten und Lithologien in CI-Chondriten: Auswirkungen der Brecciierung auf die Verteilung von Elementen in CI-Chondriten. Ph.D. thesis, Univ. of Münster, Germany. 205 pp.

Morlok A., Bischoff A., Henkel T., Rost D., Stephan T., and Jessberger E. K. (2000) The chemical heterogeneity of CI-chondrites on the submillimeter-scale (abstract). *Meteoritics & Planet. Sci., 35,* A113–A114.

Morlok A., Bischoff A., Henkel T., Rost D., Stephan T., and Jessberger E. K. (2001) The chemical heterogeneity of CI chondrites (abstract). In *Lunar and Planetary Science XXXII,* Abstract #1530. Lunar and Planetary Institute, Houston (CD-ROM).

Murakami T. and Ikeda Y. (1994) Petrology and mineralogy of the Yamato-86751 CV3 chondrite. *Meteoritics, 29,* 397–409.

Nagahara H. (1991) Petrology of Yamato-75261 meteorite: An enstatite (EH) chondrite breccia. *Proc. NIPR Symp. Antarct. Meteorites, 4,* 144–162.

Nagao K. (1994) Noble gases in hosts and inclusions from Yamato-75097 (L6), -793241 (L6) and -794046 (H5). *Proc. NIPR Symp. Antarct. Meteorites, 7,* 197–216.

Nakamura N., Misawa K., Kitamura M., Masuda A., Watanabe S., and Yamamoto K. (1990a) Highly fractionated REE in the Hedjaz (L) chondrite: Implications for nebular and planetary processes. *Earth Planet. Sci. Lett., 99,* 290–302.

Nakamura N., Fujiwara T., and Nofda S. (1990b) Young asteroid melting event indicated by Rb-Sr dating of Point of Rocks meteorite. *Nature, 345,* 51–53.

Nakamura T., Nagao K., and Takaoka N. (1999a) Microdistribution of primordial noble gases in CM chondrites determined by in situ laser microprobe analysis: Decipherment of nebular processes. *Geochim. Cosmochim. Acta, 63,* 241–255.

Nakamura T., Nagao K., Metzler K. and Takaoka N. (1999b) Heterogeneous distribution of solar and cosmogenic noble gases in CM chondrites and implications for the formation of CM parent bodies. *Geochim. Cosmochim. Acta, 63,* 257–273.

Nakamura T., Noguchi T., Zolensky M. E., and Tanaka M. (2003) Mineralogy and noble-gas signatures of the Tagish Lake carbonaceous chondrite: Evidence for an accretionary breccia. *Earth Planet. Sci. Lett., 207,* 83–101.

Nesvorný D., Morbidelli A., Vokrouhlický D., Bottke W. F., and Brož M. (2002) The Flora family: A case of the dynamically dispersed swarm? *Icarus, 157,* 155–172.

Newsom H. E. and Drake M. J. (1979) The origin of metal clasts in the Bencubbin meteoritic breccia. *Geochim. Cosmochim. Acta, 43,* 689–707.

Nolan M. C., Asphaug E., Greenberg R., and Melosh H. J. (2001) Impacts on asteroids: Fragmentation, regolith transport, and disruption. *Icarus, 153,* 1–15.

Noonan A. F., Nelen J., and Fredriksson K. (1976) Mineralogy and chemistry of xenoliths in the Weston chondrite — ordinary and carbonaceous (abstract). *Meteoritics, 11,* 344–346.

Norman M. D. and Mittlefehldt D. W. (2002) Impact processing of chondritic planetesimals. Siderophile and volatile element fractionation in the Chico L chondrite. *Meteoritics & Planet. Sci., 37,* 329–344.

Norman M. D., Bennett V. C., and Ryder G. (2002) Targeting the impactors: Siderophile element signatures of lunar impact melts from Serenitatis. *Earth Planet. Sci. Lett., 202,* 217–228.

Ohnishi I. and Tomeoka K. (2002) Dark inclusions in the Mokoia CV3 chondrite: Evidence for aqueous alteration and subsequent thermal and shock metamorphism. *Meteoritics & Planet. Sci., 37,* 1843–1856.

Okada A., Keil K., Taylor G. J., and Newsom H. (1988) Igneous history of the aubrite parent asteroid: Evidence from the Norton County enstatite achondrite. *Meteoritics, 23,* 59–74.

Okano O., Nakamura N., and Nagao K. (1990) Thermal history of the shock-melted Antarctic LL-chondrites from the Yamato-79 collection. *Geochim. Cosmochim. Acta, 54,* 3509–3523.

Olsen E. J., Dod B. D., Schmitt R. A., and Sipiera P. P. (1987) Monticello: A glass-rich howardite. *Meteoritics, 22,* 81–96.

Olsen E. J., Fredriksson K, Rajan S., and Noonan A. (1990) Chondrule-like objects and brown glasses in howardites. *Meteoritics, 25,* 187–194.

Ott U. (2002) Noble gases in meteorites-trapped components. *Rev. Mineral. Geochem., 47,* 71–100.

Ott U., Löhr H. P., and Begemann F. (1990) EET83309: A ureilite with solar noble gases. *Meteoritics, 25,* 396.

Ott U., Löhr H. P., and Begemann F. (1993) Solar noble gases in polymict ureilites and an update on ureilite noble gas data. *Meteoritics, 28,* 415–416.

Palme H., Schultz L., Spettel B., Weber H. W., Wänke H., Christophe Michel-Levy M., and Lorin J. C. (1981) The Acapulco meteorite: Chemistry, mineralogy, and irradiation effect. *Geochim. Cosmochim. Acta, 45,* 727–752.

Palme H., Wlotzka F., Spettel B., Dreibus G., and Weber H. (1988) Camel Donga: A eucrite with high metal content. *Meteoritics, 23,* 49–57.

Palme H., Kurat G., Spettel B., and Burghele A. (1989) Chemical composition of an unusual xenolith of the Allende meteorite. *Z. Naturforsch., 44a,* 1005–1014.

Papike J. J., Shearer C. K., Spilde M. N., and Karner J. M. (2000) Metamorphic diogenite Grosvenor Mountains 95555: Mineral chemistry of orthopyroxene and spinel and comparisons to the diogenite suite. *Meteoritics & Planet. Sci., 35,* 875–879.

Partsch P. (1843) Die Meteoriten oder vom Himmel gefallene Stein- und Eisenmassen. Im K.K. Hofmineralienkabinette zu Wien. Catalog of Vienna Collection, Wien.

Patzer A. and Schultz L. (2002) Noble gases in enstatite chondrites II. The trapped component. *Meteoritics & Planet. Sci., 37,* 601–612.

Patzer A., Hill D. H., and Boynton W. V. (2001) Itqiy: A metal-rich enstatite meteorite with achondritic texture. *Meteoritics & Planet. Sci., 36,* 1495–1505.

Patzer A., Hill D. H., and Boynton W. V. (2003) New eucrite Dar al Gani 872: Petrography, chemical composition, and evolution. *Meteoritics & Planet. Sci., 38,* 783–794.

Petit J-M., Chambers J., Franklin F., and Nagasawa M. (2002) Primordial excitation and depletion of the asteroid belt. In *Asteroids III* (W. F. Bottke Jr. et al., eds.), pp. 711–723. Univ. of Arizona, Tucson.

Poupeau G., Kirsten T., Steinbrunn, and Storzer D. (1974) The records of solar wind and solar flares in aubrites. *Earth Planet. Sci. Lett., 24,* 229–241.

Pravec P., Harris A. W., and Michełowski T. (2002) Asteroid rotations. In *Asteroids III* (W. F. Bottke Jr. et al., eds.), pp. 113–122. Univ. of Arizona, Tucson.

Prinz M., Fodor R. V., and Keil K. (1977) Comparison of lunar rocks and meteorites. In *The Soviet-American Conference on Cosmochemistry of the Moon and Planets* (J. H. Pomeroy and N. J. Hubbard, eds.), pp. 183–199. NASA SP-370, Washington, DC.

Prinz M., Nehru C. E., Weisberg M. K., Delany J. S., Yanai K., and Kojima H. (1984) H-chondritic clasts in a Yamato L6 chondrite: Implications for metamorphism. *Meteoritics, 19,* 292–293.

Prinz M., Weisberg M. K., Nehru C. E., and Delaney J. S. (1986) North Haig and Nilpena: Paired polymict ureilites with Angra dos Reis-related and other clasts (abstract). In *Lunar and Planetary Science XVII,* pp. 681–682. Lunar and Planetary Institute, Houston.

Prinz M., Weisberg M. K., Nehru C. E., and Delaney J. S. (1987) EET83309, a polymict ureilite: Recognition of a new group (abstract). In *Lunar and Planetary Science XVIII,* pp. 802–803. Lunar and Planetary Institute, Houston.

Prinz M., Weisberg M. K., and Nehru C. E. (1988) Feldspathic components in polymict ureilites (abstract). In *Lunar and Planetary Science XIX,* pp. 947–948. Lunar and Planetary Institute, Houston.

Prinz M., Weisberg M. K., Clayton R. N., and Mayeda T. K. (1993) Ordinary and Carlisle Lakes-like chondritic clasts in the Weatherford chondrite breccia (abstract). *Meteoritics, 28,* 419–420.

Pun A., Keil K., Taylor G. J., and Wieler R. (1998) The Kapoeta howardite: Implications for the regolith evolution of the howardite-eucrite-diogenite parent body. *Meteoritics & Planet. Sci., 33,* 835–851.

Rai V. K., Murth A. V. S., and Ott U. (2003) Nobles gases in ureilites: Cosmogenic, radiogenic, and trapped components. *Geochim. Cosmochim. Acta, 67,* 4435–4456.

Rajan R. S. (1974) On the irradiation history and origin of gas-rich meteorites. *Geochim. Cosmochim. Acta, 38,* 777–788.

Rao M. N., Garrison D. H., Palma R. L., and Bogard D. D. (1997) Energetic proton irradiation history of the howardite parent body regolith and implications for ancient solar activity. *Meteoritics & Planet. Sci., 32,* 531–543.

Reichenbach v. (1860) *Poggendorffs Annal. 1860, III,* 353–386.

Reid A. M. and Barnard B. M. (1979) Unequilibrated and equilibrated eucrites (abstract). In *Lunar and Planetary Science X,* pp. 1019–1024. Lunar and Planetary Institute, Houston.

Reid A. M., Buchanan P., Zolensky M. E., and Barrett R. A. (1990) The Bholghati howardite: Petrography and mineral chemistry. *Geochim. Cosmochim. Acta, 54,* 2161–2166.

Richardson D. C., Leinhardt Z. M., Melosh H. J., Bottke W. F. Jr., and Asphaug E. (2002) Gravitational aggregates: Evidence and evolution. In *Asteroids III* (W. F. Bottke Jr. et al., eds.), pp. 501–515. Univ. of Arizona, Tucson.

Richardson S. M. (1978) Vein formation in the C1 carbonaceous chondrites. *Meteoritics, 13,* 141–159.

Robinson M. S., Thomas P. C., Veverka J., Murchie S. L., and Wilcox B. B. (2002) The geology of 433 Eros. *Meteoritics & Planet. Sci., 37,* 1651–1684.

Rubin A. E. (1985) Impact melt products of chondritic material. *Rev. Geophys., 23,* 277–300.

Rubin A. E. (1989) An olivine-microchondrule-bearing clast in the Krymka meteorite. *Meteoritics, 24,* 191–192.

Rubin A. E. (1992) A shock-metamorphic model for silicate darkening and compositionally variable plagioclase in CK and ordinary chondrites. *Geochim. Cosmochim. Acta, 56,* 1705–1714.

Rubin A. E. (1997a) The Hadley Rille enstatite chondrite and its agglutinate-like rim: Impact melting during accretion to the Moon. *Meteoritics & Planet. Sci., 32,* 135–141.

Rubin A. E. (1997b) The Galim LL/EH polymict breccia: Evidence for impact-induced exchange between reduced and oxidized meteoritic material. *Meteoritics & Planet. Sci., 32,* 489–492.

Rubin A. E. (2002) Smyer H-chondrite impact-melt breccia and evidence for sulfur vaporization. *Geochim. Cosmochim. Acta, 66,* 699–711.

Rubin A. E. (2003) Northwest Africa 428: Impact-induced annealing of an L6 chondrite breccia. *Meteoritics & Planet. Sci., 38,* 1499–1506.

Rubin A. E. and Jones R. H. (2003) Spade: An H chondrite impact-melt breccia that experienced post shock annealing. *Meteoritics & Planet. Sci., 38,* 1507–1520.

Rubin A. E. and Kallemeyn G. W. (1990) Lewis Cliff 85332: A unique carbonaceous chondrite. *Meteoritics, 25,* 215–225.

Rubin A. E. and Kallemeyn G. W. (1994) Pecora Escarpment 91002: A member of the new Rumuruti (R) chondrite group. *Meteoritics, 29,* 255–264.

Rubin A. E. and Mittlefehldt D. W. (1992) Classification of mafic clasts from mesosiderites: Implications for endogeneous igneous processes. *Geochim. Cosmochim. Acta, 56,* 827–840.

Rubin A. E. and Mittlefehldt D. W. (1993) Evolutionary history of the mesosiderites asteroid: A chronologic and petrologic synthesis. *Icarus, 101,* 201–212.

Rubin A. E. and Scott E. R. D. (1997) Abee and related EH chondrite impact-melt breccias. *Geochim. Cosmochim. Acta, 61,* 425–435.

Rubin A. E., Scott E. R. D., Taylor G. J., and Keil K. (1981a) The Dimmitt H chondrite regotlith breccia and implications for the structure of the H chondrite parent body (abstract). *Meteoritics, 16,* 382–383.

Rubin A. E., Keil K., Taylor G. J., Ma M. S., Schmitt R. A., and Bogard D. D. (1981b) Derivation of a heterogeneous lithic fragment in the Bovedy L-group chondrite from impact-melted porphyritic chondrules. *Geochim. Cosmochim. Acta, 45,* 2213–2228.

Rubin A. E., Scott E. R. D., and Keil K. (1982) Microchondrule-bearing clast in the Piancaldoli LL3 meteorite: A new kind of type 3 chondrite and its relevance to the history of chondrules. *Geochim. Cosmochim. Acta, 46,* 1763–1776.

Rubin A. E., Rehfeldt A., Peterson E., Keil K., and Jarosewich E. (1983a) Fragmental breccias and the collisional evolution of ordinary chondrite parent bodies. *Meteoritics, 18,* 179–196.

Rubin A. E., Scott E. R. D., Taylor G. J., Keil K., Allen J. S. B.,

Mayeda T. K., Clayton R. N., and Bogard D. D. (1983b) Nature of the H chondrite parent body regolith: Evidence from the Dimmitt breccia. *Proc. Lunar Planet. Sci. Conf. 13th,* in *J. Geophys. Res., 88,* A741–A754.

Rubin A. E., James J. A., Keck B. D., Weeks K. S., Sears D. W. G., and Jarosewich E. (1985) The Colony meteorite and variations in CO3 chondrite properties. *Meteoritics, 20,* 175–196.

Rubin A. E., Scott E. R. D., and Keil K. (1997) Shock metamorphism of enstatite chondrites. *Geochim. Cosmochim. Acta, 61,* 847–858.

Rubin A. E., Ulff-Moller F., Wasson J. T., and Carlson W. D. (2001) The Portales Valley meteorite breccia: Evidence for impact-induced melting and metamorphism of an ordinary chondrite. *Geochim. Cosmochim. Acta, 65,* 323–342.

Rubin A. E., Zolensky M. E., and Bodnar R. J. (2002) The halite-bearing Zag and Monahans (1998) meteorite breccias: Shock metamorphism, thermal metamorphism and aqueous alteration on the H-chondrite parent body. *Meteoritics & Planet. Sci., 37,* 125–141.

Rubin A. E., Trigo-Rodriguez J. M., Kunihiro T., Kallemeyn G. W., and Wasson J. T. (2004) Clues to the formation of PV1, an enigmatic carbon-rich chondritic clast from the Plainview H-chondrite regolith breccia (abstract). In *Lunar and Planetary Science XXXV,* Abstract #1175. Lunar and Planetary Institute, Houston (CD-ROM).

Ruzicka A., Kring D. A., Hill D. H., Boynton W. V., Clayton R. N., and Mayeda T. K. (1995) Silica-rich orthopyroxenite in the Bovedy chondrite. *Meteoritics, 30,* 57–70.

Ruzicka A., Snyder G. A., and Taylor L. A. (1998) Mega-chondrules and large, igneous-textured clasts in Julesberg (L3) and other ordinary chondrites: Vapor-fractionation, shock-melting, and chondrule formation. *Geochim. Cosmochim. Acta, 62,* 1419–1442.

Russell S. S., Hartmann L., Cuzzi J., Krot A. N., Gounelle M., and Weidenschilling S. (2006) Timescales of the solar protoplanetary disk. In *Meteorites and the Early Solar System II* (D. S. Lauretta and H. Y. McSween Jr., eds.), this volume. Univ. of Arizona, Tucson.

Saiki K. and Takeda H. (1999) Origin of polymict breccias on asteroids deduced from their pyroxene fragments. *Meteoritics & Planet. Sci., 34,* 271–283.

Saiki K., Takeda H., and Ishii T. (2001) Mineralogy of Yamato-791192, HED breccia and relationship between cumulate eucrites and ordinary eucrites. *Antarct. Meteorite Res., 14,* 28–46.

Sanders I. S. (1996) A chondrule-forming scenario involving molten planetesimals. In *Chondrules and the Protoplanetary Disk* (R. H. Hewins et al., eds.), pp. 327–334. Cambridge Univ., Cambridge.

Sato G., Takeda H., Yanai K., and Kojima H. (1982) Electron microprobe study of impact-melted regolith breccias (abstract). In *Workshop on Lunar Breccias and Soils and Their Meteoritic Analogs* (G. J. Taylor and L. L. Wilkening, eds.), pp. 120–122. LPI Tech. Rpt. 82-02, Lunar and Planetary Institute, Houston.

Scherer P. and Schultz L. (2000) Noble gas record, collisional history, and pairing of CV, CO, CK, and other carbonaceous chondrites. *Meteoritics & Planet. Sci., 35,* 145–153.

Schulze H., Bischoff A., Palme H., Spettel B., Dreibus G., and Otto J. (1994) Mineralogy and chemistry of Rumuruti: The first meteorite fall of the new R chondrite group. *Meteoritics, 29,* 275–286.

Schultz L. and Kruse H. (1978) Light noble gases in stony meteorites — A compilation. *Nucl. Track Detection, 2,* 65–103.

Schultz L and Kruse H. (1989) Helium, neon, and argon in meteorites — A data compilation. *Meteoritics, 24,* 155–172.

Scott E. R. D. (1982) Origin of rapidly solidified metal-troilite grains in chondrites and iron meteorites. *Geochim. Cosmochim. Acta, 46,* 813–823.

Scott E. R. D. (1988) A new kind of primitive chondrite, Allan Hills 85085. *Earth Planet. Sci. Lett., 91,* 1–18.

Scott E. R. D. (2002) Meteorite evidence for the accretion and collisional evolution of asteroids. In *Asteroids III* (W. F. Bottke Jr. et al., eds.), pp. 697–709. Univ. of Arizona, Tucson.

Scott E. R. D. and Rajan R. S. (1981) Metallic minerals, thermal histories and parent bodies of some xenolithic, ordinary chondrite meteorites. *Geochim. Cosmochim. Acta, 45,* 53–67.

Scott E. R. D. and Taylor G. J. (1982) Primitive breccias among the type 3 ordinary chondrites — origin and relation to regolith breccias. In *Workshop on Lunar Breccias and Soils and Their meteoritic Analogs* (G. J. Taylor and L. L. Wilkening., eds.), pp. 130–134. LPI Tech. Rpt. 82-02, Lunar and Planetary Institute, Houston.

Scott E. R. D. and Wilson L. (2005) Meteoritic and other constraints on the internal structure and impact history of small asteroids. *Icarus, 74,* 46–53.

Scott E. R. D., Brearley A. J., Keil K., Grady M. M., Pillinger C. T., Clayton R. N., Mayeda T. K., Wieler R., and Signer P. (1988) Nature and origin of C-rich ordinary chondrites and chondritic clasts. *Proc. Lunar Planet. Sci. Conf. 18th,* pp. 513–523.

Scott E. R. D., Keil K., and Stöffler D. (1992) Shock metamorphism of carbonaceous chondrites. *Geochim. Cosmochim. Acta, 56,* 4281–4293.

Scott E. R. D., Haack H., and McCoy T. J. (1996) Core crystallization and silicate-metal mixing in the parent body of the IVA iron and stony-iron meteorites. *Geochim. Cosmochim. Acta, 60,* 1615–1631.

Scott E. R. D., Haack H., and Love S. G. (2001) Formation of mesosiderites by fragmentation and reaccretion of a large differentiated asteroid. *Meteoritics & Planet. Sci., 36,* 869–881.

Sears D. W. and Wasson J. T. (1981) Dark inclusions in the Abbott, Cynthiana and Abee chondrites (abstract). In *Lunar and Planetary Science XII,* pp. 958–960. Lunar and Planetary Institute, Houston.

Semenenko V. P. and Girich A. L. (2001) A variety of lithic fragments in the Krymka (LL3.1) chondrite (abstract). *Meteoritics & Planet. Sci., 36,* A187.

Semenenko V. P., Bischoff A., Weber I., Perron C., and Girich A. L. (2001) Mineralogy of fine-grained material in the Krymka (LL3.1) chondrite. *Meteoritics & Planet. Sci., 36,* 1067–1085.

Sepp B., Bischoff A., and Bosbach D. (2001) Low-temperature phase decomposition in iron-nickel metal of the Portales Valley meteorite. *Meteoritics & Planet. Sci., 36,* 587–595.

Simpson E. S. and Murray D. G. (1932) A new siderolite from Bencubbin, Western Australia. *Mineral. Mag., 23,* 33–37.

Sisodia M. S., Shukla A. D., Suthar K. M., Mahajan M. R., Murty S. V. S., Shukla P. N., Bhandari N., and Natarajan R. (2001) The Lohawat howardite: Mineralogy, chemistry and cosmogenic effects. *Meteoritics & Planet Sci., 36,* 1457–1466.

Smith C. L., Wright I. P., Franchi I. A., and Grady M. M. (2000) A statistical analysis of mineralogical data from Frontier Mountain ureilites (abstract). *Meteoritics & Planet. Sci., 35,* A150.

Smith J. V. and Mason B. (1970) Pyroxene-garnet transformation in the Coorara meteorite. *Science, 168,* 832–833.

Srinivasan B. and Anders E. (1977) Noble gases in the unique chondrite Kakangari. *Meteoritics, 12,* 417–424.

Stöffler D., Knöll H.-D., and Maerz U. (1979) Terrestrial and lunar impact breccias and the classification of lunar highland rocks. *Proc. Lunar Planet. Sci Conf. 10th,* pp. 639–675.

Stöffler D., Knöll H.-D., Marvin U. B., Simonds C. H., and Warren P. H. (1980) Recommended classification and nomenclature of lunar highland rocks — a committee report. In *Proceedings of the Conference on the Lunar Highlands Crust* (J. J. Papike and R. B. Merrill, eds.), pp. 51–70. Pergamon, New York.

Stöffler D., Bischoff A., Buchwald V., and Rubin A. E. (1988) Shock effects in meteorites. In *Meteorites and the Early Solar System* (J. F. Kerridge and M. S. Matthews, eds.), pp. 165–202. Univ. of Arizona, Tucson.

Stöffler D., Keil K., and Scott E. R. D. (1991) Shock metamorphism of ordinary chondrites. *Geochim. Cosmochim. Acta, 55,* 3845–3867.

Suess H. E., Waenke H., and Wlotzka F. (1964) On the origin of gas-rich meteorites. *Geochim. Cosmochim. Acta, 28,* 209–233.

Sullivan R. J., Thomas P. C., Murchie S. L., and Robinson M. S. (2002) Asteroid geology from *Galileo* and *NEAR Shoemaker* data. In *Asteroids III* (W. F. Bottke Jr. et al., eds.), pp. 331–350. Univ. of Arizona, Tucson.

Takeda H. (1979) A layered-crust model of a howardite parent body. *Icarus, 40,* 455–470.

Takeda H. (1991) Comparison of Antarctic and non-Antarctic achondrites and possible origin of the differences. *Geochim. Cosmochim. Acta, 55,* 35–47.

Takeda H. (1997) Mineralogical records of early planetary processes on the howardite, eucrite, diogenite parent body with reference to Vesta. *Meteoritics & Planet. Sci., 32,* 841–853.

Takeda H. and Graham A. L. (1991) Degree of equilibration of eucritic pyroxenes and thermal metamorphism of the earliest planetary crust. *Meteoritics, 26,* 129–134.

Takeda H. and Yamaguchi A. (1991) Recrystallization and shock textures of old and new samples of Juvinas in relation to its thermal history. *Meteoritics, 26,* 400.

Takeda H., Mori H., and Yanai K. (1981) Mineralogy of the Yamato diogenites as possible pieces of a single fall. *Mem. Natl. Inst. Polar Res., Spec. Issue 20,* 81–99.

Takeda H., Mori H., Delaney J. S., Prinz M., Harlow G. E., and Ishii T. (1983) Mineralogical comparison of Antarctic and non-Antarctic HED (howardites-eucrites-diogenites) achondrites. *Mem. Natl. Inst. Polar Res., Spec. Issue 30,* 181–205.

Takeda H., Mori H., Ikeda Y., Teruaki I., and Yanai K. (1984) Antarctic howardites and their primitive crust. *Mem. Natl. Inst. Polar Res., Spec. Issue, 31,* 81–101.

Tamaki M., Yamaguchi A., Misawa K., Ebihara M., and Takeda H. (2004) Petrologic study of eucritic clasts in mesosiderites, Mount Pudbury and Vaca Muerta. *Antarct. Meteorite Res., 28,* 85–86.

Taylor G. J. (1982) Petrologic comparison of lunar and meteoritic breccias. In *Workshop on Lunar Breccias and Soils and Their Meteoritic Analogs* (G. J. Taylor and L. L. Wilkening), pp. 153–167. LPI Tech. Rpt. 82-02, Lunar and Planetary Institute, Houston.

Taylor G. J., Keil K., Berkley J. L., Lange D. E., Fodor R. V., and Fruland R. M. (1979) The Shaw meteorite: History of a chondrite consisting of impact-melted and metamorphic lithologies. *Geochim. Cosmochim. Acta, 43,* 323–337.

Taylor G. J., Maggiore P., Scott E. R. D., Rubin A. E., and Keil K. (1987) Original structures, and fragmentation and reassembly histories of asteroids: Evidence from meteorites. *Icarus, 69,* 1–13.

Tonui E. K., Zolensky M. E., Lipschutz M. E., Wang M.-S., and Nakamura T. (2003) Yamato 86029: Aqueously altered and thermally metamorphosed CI-like chondrite with unusual textures. *Meteoritics & Planet. Sci., 38,* 269–292.

Tschermak G. (1872) Die Meteoriten von Shergotty and Gopalpur. Sitzungsber. *Akad. Wiss. Wien, Math.-Naturwiss., Kl 65, Teil 1,* 122–145.

Urey H. C. (1959) Primary and secondary objects. *J. Geophys. Res., 64,* 1721–1737.

Urey H. C. (1967) Parent bodies of the meteorites and origin of chondrules. *Icarus, 7,* 350–359.

Van der Bogert C. H., Schultz P. H., and Spray J. G. (2003) Impact-induced frictional melting in ordinary chondrites: A mechanism for deformation, darkening, and vein formation. *Meteoritics & Planet. Sci., 38,* 1521–1531.

Van Schmus W. R. (1967) Polymict structure of the Mezo-Madaras chondrite. *Geochim. Cosmochim. Acta, 31,* 2072–2042.

Varteresian C. and Hewins R. H. (1983) Magnesian noritic and basaltic clasts in the Garland and Peckelsheim diogenites (abstract). In *Lunar and Planetary Science XIX,* pp. 800–801. Lunar and Planetary Institute, Houston.

Vdovykin G. P. (1972) Forms of carbon in the new Haverö ureilite of Finland. *Meteoritics, 7,* 547.

Vogel N., Wieler R., Bischoff A., and Baur H. (2003) Microdistribution of primordial Ne and Ar in fine-grained rims, matrices, and dark inclusions of unequilibrated chondrites — Clues on nebular processes. *Meteoritics & Planet. Sci., 38,* 1399–1418.

Wänke H. (1965) Der Sonnenwind as Quelle der Uredelgase in Steinmeteoriten. *Z. Naturforsch., B20a,* 946–949.

Wahl W. (1952) The brecciated stony meteorites and meteorites containing foreign fragments. *Geochim. Cosmochim. Acta, 2,* 91–117.

Warren P. H. (1985) Origin of howardites, diogenites and eucrites: A mass balance constraint. *Geochim. Cosmochim. Acta, 49,* 577–586.

Wasson J. T. (1974) *Meteorites: Their Classification and Properties.* Springer-Verlag, Berlin. 316 pp.

Wasson J. T., Rubin A. E., and Kallemeyn G. W. (1993) Reduction during metamorphism of four ordinary chondrites. *Geochim. Cosmochim. Acta, 57,* 1867–1878.

Watters T. and Prinz M. (1980) Mt. Egerton and the aubrite parent body (abstract). In *Lunar and Planetary Science XI,* pp. 1225–1227. Lunar and Planetary Institute, Houston.

Weber I. and Bischoff A. (1998) Mineralogy and chemistry of the ureilites Hammadah al Hamra 064 and Jalanash (abstract). In *Lunar and Planetary Science XIX,* p. 1365. Lunar and Planetary Institute, Houston.

Weber I., Bischoff A., and Weber D. (2003) TEM investigations on the monomict ureilites Jalanash and Hammadah al Hamra 064. *Meteoritics & Planet. Sci., 38,* 145–156.

Welzenbach L. C., McCoy T. J., Grimberg A., and Wieler R. (2005) Petrology and noble gases of the regolith breccia MAC 87302 and implications for the classification of Antarctic meteorites (abstract). In *Lunar and Planetary Science XXXVI,* Abstract #1425. Lunar and Planetary Institute, Houston (CD-ROM).

Weisberg M. K., Prinz M., and Nehru C. E. (1988) Petrology of ALH 85085: A chondrite with unique characteristics. *Earth Planet. Sci. Lett., 91,* 19–32.

Weisberg M. K., Prinz M., and Nehru C. E. (1990) The Bencubbin chondrite breccia and its relationship to CR chondrites and the ALH85085 chondrite. *Meteoritics, 25,* 269–279.

Weisberg M. K., Prinz M., Clayton R. N., and Mayeda T. K. (1993) The CR (Renazzo-type) carbonaceous chondrite group and its implications. *Geochim. Cosmochim. Acta, 57,* 1567–1586.

Weisberg M. K., Prinz M., Clayton R. N., Mayeda T. K., Grady M. M., and Pillinger C. T. (1995) The CR chondrite clan. *Proc. NIPR Symp. Antarct. Meteorites, 8,* 11–32.

Weisberg M. K., Prinz M., Clayton R. N., Mayeda T. K., Grady M. M., Franchi I., Pillinger C. T., and Kallemeyn G. W. (1996) The K (Kakangari) chondrite grouplet. *Geochim. Cosmochim. Acta, 60,* 4253–4263.

Weisberg M. K., Prinz M., and Nehru C. E. (1997a) QUE 94204: An EH-chondritic melt rock (abstract). In *Lunar and Planetary Science XXVIII,* Abstract #1358. Lunar and Planetary Institute, Houston (CD-ROM).

Weisberg M. K., Prinz M., Clayton R. N., and Mayeda T. K. (1997b) CV3 chondrites: Three subgroups, not two (abstract). *Meteoritics & Planet. Sci., 32,* A138–A139.

Weisberg M. K., Prinz M., Clayton R. N., and Mayeda T. K., Sugiura N., Zashu S., Ebihara M. (2001) A new metal-rich chondrite grouplet. *Meteoritics & Planet. Sci., 36,* 401–418.

Weisberg M. K., McCoy T. J., and Krot A. N. (2006) Systematics and evaluation of meteorite classification. In *Meteorites and the Early Solar System II* (D. S. Lauretta and H. Y. McSween Jr., eds.), this volume. Univ. of Arizona, Tucson.

Wieler R., Graf T., Pedroni A., Signer P., Pellas P., Fieni C., Suter M., Vogt S., Clayton R. N., and Laul J. C. (1989) Exposure history of the regolithic chondrite Fayetteville: II. Solar-gas-free light inclusions. *Geochim. Cosmochim. Acta, 53,* 1449–1459.

Wieler R., Pedroni A., and Leya I. (2000) Cosmogenic neon in mineral separates from Kapoeta: No evidence for an irradiation of its parent body by an early active sun. *Meteoritics & Planet. Sci., 35,* 251–257.

Wilkening L. L. (1973) Foreign inclusions in stony meteorites — I. Carbonaceous chondritic xenoliths in the Kapoeta howardite. *Geochim. Cosmochim. Acta, 37,* 1985–1989.

Wilkening L. L. (1976) Carbonaceous chondritic xenoliths and planetary-type noble gases in gas-rich meteorites. *Proc Lunar Sci. Conf. 7th,* pp. 3549–3559.

Wilkening L. L. (1977) Meteorites in meteorites: Evidence for mixing among the asteroids. In *Comets, Asteroids and Meteorites* (A. H. Delsemme), pp. 389–396. Univ. of Toledo, Ohio.

Wilkening L. L. (1978) Tysnes Island: An unusual clast composed of solidified, immiscible, Fe-FeS and silicate melts. *Meteoritics, 13,* 1–9.

Wilkening L. L. and Clayton R. N. (1974) Foreign inclusions in stony meteorites-II. Rare gases and oxygen isotopes in a carbonaceous chondritic xenolith in the Plainview gas-rich chondrite. *Geochim. Cosmochim. Acta, 38,* 937–945.

Wilkening L. L., Lal D., and Reid A. M. (1971) The evolution of the Kapoeta howardite based on fossil track studies. *Earth Planet. Sci. Lett., 10,* 334–340.

Wilkison S. L., Robinson M. S., Thomas P. C., Veverka J., McCoy T. J., Murchie S. L., Prockter L., and Yeomans D. (2002) An estimate of Eros's porosity and implications for internal structure. *Icarus, 155,* 94–103.

Williams C. V., Rubin A. E., Keil K., and San Miguel A. (1985) Petrology of the Cangas de Onis and Nulles regolith breccias: Implications for parent body history. *Meteoritics, 20,* 331–345.

Williams C. V., Keil K., Taylor G. J., and Scott E. R. D. (2000) Cooling rates of equilibrated clasts in ordinary chondrite regolith breccias: Implications for parent body histories. *Chem. Erde, 59,* 287–305.

Wlotzka F. (1963) Über die Hell-Dunkel-Struktur der urgashaltigen Chondrite Breitscheid und Pantar. *Geochim. Cosmochim. Acta, 27,* 419–429.

Wlotzka F., Palme H., Spettel B., Wänke H., Fredriksson K., and Noonan A. F. (1979) Krähenberg and Bhola: LL chondrites with differentiated K-rich inclusions (abstract). *Meteoritics, 14,* 566.

Woolum D. S. and Hohenberg C. (1993) Energetic particle environment in the early solar system — Extremely long pre-compaction ages or enhanced early particle flux. In *Protostars and Planets III* (E. H. Levy and J. I. Lunine, eds.), pp. 903–919. Univ. of Arizona, Tucson.

Xiao X. and Lipschutz M. E. (1991) Chemical studies of H chondrites: III. Regolith evolution of the Fayetteville chondrite parent body. *Geochim. Cosmochim. Acta, 55,* 3407–3415.

Yamaguchi A. and Takeda H. (1992) Mineralogical study of some brecciated antarctic eucrites. *Proc. NIPR Antarct. Meteorites, 5,* 242–257.

Yamaguchi A. and Takeda H. (1994) Granulitic matrices in monomict eucrites (abstract). In *Lunar and Planetary Science XXV,* pp. 1525–1526. Lunar and Planetary Institute, Houston.

Yamaguchi A. and Takeda H. (1995) Mineralogical studies of some antarctic monomict eucrites, including Yamato-74356: A unique rock containing recrystallized clastic matrix. *Proc. NIPR Antarct. Meteorites, 8,* 167–184.

Yamaguchi A., Takeda H., Bogard D. D., and Garrison D. H. (1994) Textural variation and impact history of the Millbillillie eucrite. *Meteoritics, 29,* 237–245.

Yamaguchi A., Taylor G. J., and Keil K. (1996) Global crustal metamorphism of the eucrite parent body. *Icarus, 124,* 97–112.

Yamaguchi A., Taylor G. J. and Keil K. (1997) Shock and thermal history of equilibrated eucrites from Antarctica. *Antarct. Meteorite Res., 10,* 415–436.

Yamaguchi A., Scott E. R. D., and Keil K. (1998) Origin of unusual impact melt rocks, Yamato-790964 and -790143 (LL-chondrites). *Antarct. Meteorite Res., 11,* 18–31.

Yamaguchi A., Scott E. R. D., and Keil K. (1999) Origin of a unique impact-melt rock — the L-chondrite Ramsdorf. *Meteoritics & Planet. Sci., 34,* 49–59.

Zolensky M. and Ivanov A. (2003) The Kaidun microbreccia meteorite: A harvest from the inner and outer asteroid belt. *Chem. Erde, 63,* 185–246.

Zolensky M. E. and McSween H. Y. (1988) Aqueous alteration. In *Meteorites and the Early Solar System* (J. F. Kerridge and M. S. Matthews, eds.), pp. 114–143. Univ. of Arizona, Tucson.

Zolensky M. E., Hewins R. H., Mittlefehldt D. W., Lindstrom M. M., Xiao X., and Lipschutz M. E. (1992a) Mineralogy, petrology and geochemistry of carbonaceous chondritic clasts in the LEW 85300 polymict eucrite. *Meteoritics, 27,* 596–604.

Zolensky M., Weisberg M. K., Buchanan P. C., Prinz M., Reid A., and Barrett R. A. (1992b) Mineralogy of dark clasts in CR chondrites, eucrites and howardites (abstract). In *Lunar and Planetary Science XXIII,* pp. 1587–1588. Lunar and Planetary Institute, Houston.

Zolensky M., Krot A. N., Weisberg M. K., Buchanan P. C., and Prinz M. (1996a) Fine-grained inclusions in type 3 ordinary and carbonaceous chondrites (abstract). In *Lunar and Plane-*

tary Science XXVII, pp. 1507–1508. Lunar and Planetary Institute, Houston.

Zolensky M. E., Ivanov A. V., Yang S. V., Mittlefehldt D. W., and Ohsumi K. (1996b) The Kaidun meteorite: Mineralogy of an unusual CM1 lithology. *Meteoritics & Planet. Sci., 31,* 484–493.

Zolensky M. E., Weisberg M. K., Buchanan P. C., and Mittlefehldt D. W. (1996c) Mineralogy of carbonaceous chondrite clasts in HED achondrites and the moon. *Meteoritics & Planet. Sci., 31,* 518–537.

Zolensky M. E., Mittlefehldt D. W., Lipschutz M. E., Wang M.-S., Clayton R. N., Mayeda T. K., Grady M. M., Pillinger C., and Barber D. (1997) CM chondrites exhibit the complete petrologic range from type 2 to 1. *Geochim. Cosmochim. Acta, 61,* 5099–5115.

Zolensky M., Nakamura K., Gounelle M., Mikouchi T., Kasama T., Tachikawa O., and Tonui E. (2002) Mineralogy of Tagish Lake: An ungrouped type 2 carbonaceous chondrite. *Meteoritics & Planet. Sci., 37,* 737–761.

Zolensky M., Nakamura K., Weisberg M. K., Prinz M., Nakamura T., Ohsumi K., Saitow A., Mukai M., and Gounelle M. (2003) A primitive dark inclusion with radiation-damaged silicates in the Ninqiang carbonaceous chondrite. *Meteoritics & Planet. Sci., 38,* 305–322.

Zolensky M. E., Tonui E. K., Bevan A. W. R., Le L., Clayton R. N., Mayeda T. K., and Norman M. (2004) Camel Donga 040: A CV chondrite genomict breccia with unequilibratedand metamorphosed material. *Antarct. Meteorite Res., 28,* 95–96.

Zook H. A. (1980) A new impact model for the generation of ordinary chondrites. *Meteoritics, 15,* 390–391.

Part VIII:

The Parent-Body Epoch:
B. Melting and Differentiation

Timescales of Planetesimal Differentiation in the Early Solar System

M. Wadhwa
The Field Museum

G. Srinivasan
University of Toronto

R. W. Carlson
Carnegie Institution of Washington

Chronological investigations of a variety of meteoritic materials indicate that planetesimal differentiation occurred rapidly in the early solar system. Some planetesimals only experienced the earliest stage of the onset of melting and were subsequently arrested in that state. On some of these bodies, such incipient melting began ~3–5 m.y. after the formation of calcium-aluminum-rich inclusions (CAI), but possibly extended to ≥10 m.y. on others. Extensive differentiation took place on other planetesimals to form silicate crust-mantle reservoirs and metallic cores. Crust formation began within ~2 m.y. of CAI formation, and likely extended to at least ~10 m.y. Global mantle differentiation, which established the source reservoirs of the crustal materials, occurred contemporaneously for the angrites and eucrites, within ~2–3 m.y. of CAI formation. Metal segregation on different planetesimals occurred within a narrow time interval of ≤5 m.y. In the specific case of the howardite-eucrite-diogenite (HED) parent body, the process of core formation is estimated to have occurred ~1 m.y. before global mantle differentiation. Following metal segregation, core crystallization is likely to have occurred over a time interval of 10 m.y. or less.

1. INTRODUCTION

Within the last decade, significant advances have been made in our understanding of the timescales involved in the differentiation history of planetesimals in the early solar system. In particular, the application of several chronometers based on short-lived radionuclides (i.e., with half-lives less than ~100 m.y.) has allowed unprecedented time resolution (often ≤1 m.y. for events that occurred close to ~4.6 b.y. ago).

The earliest differentiation events in solar system history were those that occurred between nebular gas and the earliest condensates that are represented by components within chondritic meteorites. Timescales involved in such events, which likely occurred during the accretion disk phase of the nebula, are discussed elsewhere in this book (*Russell et al.,* 2006). In this chapter, we will focus on the *timescales* involved in the differentiation events that occurred on planetesimals, defined here as small asteroidal-sized bodies (up to hundreds of kilometers in diameter) that accreted from nebular dust and the earliest condensates. The *processes* likely to be involved in planetesimal differentiation are discussed in more detail in *McCoy et al.* (2006).

The meteoritic record includes a variety of materials resulting from planetesimal differentiation. These materials are primarily composed of two types: (1) primitive achondrites that have near-chondritic bulk compositions (e.g., brachinites, acapulcoites, lodranites, and winonaites) and formed during the earliest stages of melting on planetesimals (which were subsequently arrested at this stage and did not differentiate further); and (2) meteorites that resulted from more extensive differentiation such that their bulk compositions are significantly different from those of chondrites. The latter are in turn comprised of three classes of materials: (1) achondrites that formed within the silicate crust of differentiated planetesimals, such as eucrites, diogenites, and howardites (or HED meteorites), angrites, and HED-like silicate clasts in mesosiderites; (2) metal-silicate meteorites (pallasites) that may have been formed near the core-mantle boundary; and (3) metal-rich meteorites that are the products of metal segregation from essentially chondritic precursors and some of which may represent the cores of extensively differentiated planetesimals.

Age determinations on these meteoritic materials with the appropriate chronometers then make it possible to estimate the timing of planetesimal differentiation events ranging from incipient melting to more extensive processing involving silicate differentiation (including crust formation and fractionation within the mantle), metal segregation, and core crystallization. Thus far, chronometers based on both long- and short-lived radionuclides have been applied extensively for age-dating such events. Long-lived chronom-

TABLE 1. Selected short-lived radioisotopes utilized so far in constraining the timescales of planetesimal differentiation.

Radioisotope (R*)	Half-life (m.y.)	Daughter Isotope (D*)	Reference Isotope (R)	Solar System Initial Ratio $(R^*/R)_0$	Time Anchor* (if any)
^{26}Al	0.72	^{26}Mg	^{27}Al	~5×10^{-5}	CAIs $(R^*/R)_T = 5 \times 10^{-5}$ at 4.567 Ga
^{60}Fe	1.5	^{60}Ni	^{56}Fe	~$3\text{–}10 \times 10^{-7}$	
^{53}Mn	3.7	^{53}Cr	^{55}Mn	~10^{-5}	LEW 86010 Angrite $(R^*/R)_T = 1.25 \times 10^{-6}$ at 4.558 Ga
^{107}Pd	6.5	^{107}Ag	^{108}Pd	~5×10^{-5}	Gibeon (IVA) Iron $(R^*/R)_T = 2.4 \times 10^{-5}$ at ~4.56 Ga
^{182}Hf	9	^{182}W	^{180}Hf	$1.0\text{–}1.6 \times 10^{-4}$	CAIs and Chondrites $(R^*/R)_T = 1 \times 10^{-4}$ at ~4.56 Ga
^{146}Sm	103	^{142}Nd	^{144}Sm	~7×10^{-3}	Angrites $(R^*/R)_T = 7 \times 10^{-3}$ at 4.558 Ga

*References: ^{26}Al: *Lee et al.* (1976), *Amelin et al.* (2002), and references therein; ^{60}Fe: *Tachibana and Huss* (2003), *Mostefaoui et al.* (2004); ^{53}Mn: *Lugmair and Galer* (1992), *Lugmair and Shukolyukov* (1998); ^{107}Pd: *Chen and Wasserburg* (1996); ^{182}Hf: *Kleine et al.* (2002), *Yin et al.* (2002b), *Quitté and Birck* (2004); ^{146}Sm: *Lugmair and Galer* (1992) and references therein. We note that some recent investigations indicate that the initial ^{26}Al/^{27}Al ratio at the time of CAI formation was possibly as high as ~7×10^{-5} (e.g., *Young et al.,* 2005). Since *Amelin et al.* (2002) determined a ^{26}Al/^{27}Al ratio of 4.63 ± 0.44 in the same Efremovka CAI (E60) that had a Pb-Pb age of 4567.4 ± 1.1 Ma, we will assume here that the initial ^{26}Al/^{27}Al ratio at the time of CAI formation was near-canonical (i.e., ~5×10^{-5}).

eters, however, typically do not have the time resolution required to resolve events occurring within the first tens of millions of years of solar system history. In contrast, short-lived chronometers can have high time resolution (i.e., a million years or less) and are thus capable of precisely age dating the earliest solar system events. Nevertheless, the application of chronometers based on short-lived radionuclides is not without its challenges. In particular (and provided there are no other complications such as heterogeneity in the initial distribution of the parent radionuclide and/or later disturbance of the isotope systematics), isochrons based on short-lived radionuclides can provide only relative ages since the slope of such an isochron reflects the abundance of the now-extinct radionuclide at the time of isotopic closure following a parent-daughter fractionation event. Therefore, since the high-resolution chronometers based on short-lived radionuclides can provide only relative time differences between early solar system events, it is essential to have a "time anchor" [i.e., a sample that allows one to determine the value of the parameter $(R^*/R)_T$ at a precisely defined absolute time T, where R* is the abundance of the radioisotope and R is that of a stable reference isotope of the same element] so as to map the relative ages derived from such a chronometer onto an absolute timescale. The only absolute chronometer capable of providing time resolution comparable to that of chronometers based on short-lived radionuclides, and thus providing an appropriate time anchor, is the U-Pb chronometer. A unique attribute of this chronometer is that it is based on the decay of two long-lived radioisotopes, i.e., ^{235}U that decays to ^{207}Pb with a half-life of 703.8 m.y., and ^{238}U that decays to ^{206}Pb with a half-life of 4469 m.y. The extraordinarily high precision afforded by the U-Pb chronometer compared to chronometers based on other long-lived radioisotopes results from the rapid evolution of the radiogenic Pb isotopic composition (i.e., ^{207}Pb/^{206}Pb) due to the relatively short half-life of ^{235}U and the occasionally very high U/Pb ratios in planetesimals and planets affected by volatile loss and/or core formation since Pb is both volatile and chalcophile, but U is neither.

2. INITIAL CONDITIONS

Prior to a discussion of the timescales of planetesimal differentiation, we briefly summarize our current understanding of the initial conditions prevalent in the solar nebula immediately prior to the formation of these planetesimals. More detailed discussions of nebular conditions and processes may be found elsewhere in this book (e.g., the chapters in Part III). We note that an understanding of conditions and processes in the solar nebula (especially the degree to which it was homogenized) is particularly important with regard to the application of the high-resolution chronometers based on short-lived radionuclides, since factors such as the abundance and distribution of the parent radionuclides are critical in determining the feasibility of age-dating events in the early solar system.

Table 1 provides a listing of selected short-lived radionuclides that have so far been applied toward determining the timescales involved in planetesimal differentiation. This table includes current estimates of the solar system initial abundance ratios $(R^*/R)_0$, which are estimates of the initial abundances of these radionuclides in the meteorite-forming region of the solar nebula at the time of formation of the earliest condensates, and the refractory CAIs found in primitive chondrites, the most precise age estimate for which is currently 4567.2 ± 0.6 Ma (*Amelin et al.,* 2002).

In Table 1, the solar system initial abundance ratios $(R^*/R)_0$ of the radionuclides with half-lives ≥10 m.y. (i.e., ^{146}Sm and ^{182}Hf), but perhaps also some shorter-lived radionuclides (such as ^{107}Pd and ^{53}Mn), may be accounted for by continuous galactic nucleosynthesis (e.g., *Meyer and Clayton,* 2000). Radionuclides whose abundances may be accounted for by this means are generally assumed to be homogeneously distributed in the solar nebula.

The initial abundance ratios of several of the radionuclides with half-lives ≤10 m.y. (e.g., ^{26}Al and possibly also ^{53}Mn) are too high to be accounted for by continuous galactic nucleosynthesis alone. Alternatives for the predominant production sites of these shorter-lived radionuclides are that either (1) they were synthesized in a stellar environment and injected into the molecular cloud just prior to its collapse and solar nebula formation (e.g., *Cameron et al.,* 1995; *Cameron,* 2001; *Gallino et al.,* 2004) or (2) they were produced by energetic particle irradiation within the nebula during an early active phase of the Sun (*Shu et al.,* 1996; *Gounelle et al.,* 2001; *Leya et al.,* 2003a). Additionally, in the specific case of ^{10}Be ($t_{1/2}$ ~ 1.5 m.y.), which is produced only by spallation reactions and not by stellar nucleosynthesis, it is suggested that this radionuclide may have been incorporated into the solar nebula by trapping of galactic cosmic rays in the molecular cloud as it collapsed (*Desch et al.,* 2004). Whether the short-lived radioisotopes that may be produced in any of the above scenarios were effectively homogenized prior to the formation of solids in the nebula is presently unresolved. Numerical simulations of processes associated with molecular cloud core collapse triggered by a nearby explosive stellar event indicate that spatial heterogeneities in the distributions of shock-wave-injected short-lived radionuclides may survive on short timescales (*Vanhala and Boss,* 2002). Nevertheless, in the case of an external seeding source for these short-lived radionuclides, differential rotation (and possibly also turbulence) in the nebula is generally anticipated to result in radial and vertical mixing on the timescale of significantly less than a million years subsequent to initiation of collapse (which is assumed to occur immediately prior to formation of the first solids in the nebula). If, however, the short-lived radionuclides were predominantly produced by local irradiation within the early solar system, they are expected to be distributed heterogeneously within the nebular disk at least over the duration of the active phase of the early Sun (e.g., *Gounelle et al.,* 2001).

The application of a chronometer based on a short-lived radionuclide rests critically on the assumption that the initial abundance ratio of the radioactive to the stable isotope, i.e., $(R^*/R)_0$, is uniform in the meteorite-forming region. In the specific case of determining the timing of planetesimal differentiation using such a chronometer, the scale at which the isotopic heterogeneity may be present is also important. Numerous studies have shown that large isotopic heterogeneities were present in the nebula on the small scale, i.e., on the scale of micrometer-sized presolar grains and millimeter- to centimeter-sized inclusions (CAIs) within primitive meteorites (e.g., *Loss et al.,* 1994; *Zinner,* 1996). However, there are only a few instances where isotopic heterogeneities have been documented on the planetesimal scale. Specifically, mass-independent variations in the isotopic compositions of O (*Clayton et al.,* 1973; *Clayton,* 1993) and Cr (*Lugmair and Shukolyukov,* 1998; *Shukolyukov et al.,* 2003) have been documented in bulk samples of primitive and differentiated meteorites. Although disputed by some (*Becker and Walker,* 2003a,b), such variations have also been noted for Mo and Ru (*Dauphas et al.,* 2002, 2004a; *Yin et al.,* 2002a; *Chen et al.,* 2003, 2004; *Papanastassiou et al.,* 2004). By and large, however, most other isotopic systems investigated so far in meteorites indicate homogeneity on the planetesimal scale at the level of precision achievable by current state-of-the-art instrumentation (e.g., *Zhu et al.,* 2001; *Dauphas et al.,* 2004bc).

3. TIMESCALES OF PLANETESIMAL DIFFERENTIATION

3.1. Incipient Melting

Primitive achondrites, such as acapulcoites, lodranites, winonaites, and brachinites, have igneous textures but their bulk compositions, although showing some variations, are relatively close to those of chondrites. Therefore, these meteorites are considered to be the products of the earliest stages of melting and igneous processing on planetesimals (*Mittlefehldt et al.,* 1998, and references therein).

The acapulcoites and lodranites are thought to be the residual products of partial melting of chondritic precursors (*Mittlefehldt et al.,* 1996; *McCoy et al.,* 1997a,b). Chronological constraints obtained so far indicate that these meteorites formed within ~10 m.y. of the formation of the earliest solids in the nebula. Evidence for the presence of live ^{244}Pu ($t_{1/2}$ ~ 82 m.y.) in Acapulco phosphates supports early formation of this meteorite (*Pellas et al.,* 1997). Samarium-147–neodymium-143 systematics determined by *Prinzhofer et al.* (1992) for Acapulco gave a very old age (4.60 ± 0.03 Ga), which the authors interpreted as the time of recrystallization immediately following its formation event. Given this old ^{147}Sm-^{143}Nd age, one would expect a ^{146}Sm/^{144}Sm ratio significantly higher than in angrites at 4558 Ma (i.e., the time anchor for the ^{146}Sm-^{142}Nd system; Table 1). However, the ^{146}Sm/^{144}Sm ratio of 0.0067 ± 0.0019 determined for this sample (*Prinzhofer et al.,* 1992) is within error of that in angrites at their time of formation. Moreover, U-Pb systematics in phosphates from Acapulco give an age of 4557 ± 2 Ma (*Göpel et al.,* 1992, 1994), marginally younger, but much more precise, than the ^{147}Sm-^{143}Nd age. The older ^{147}Sm-^{143}Nd age is therefore questionable and could be due to disturbance during extensive later metamorphism experienced by this meteorite (*McCoy et al.,* 1996). Furthermore, Acapulco has an inferred ^{53}Mn/^{55}Mn ratio of (7.5 ± 1.4) × 10^{-7}; compared to the LEW 86010 angrite (the time anchor for the ^{53}Mn-^{53}Cr system; Table 1); this translates to a ^{53}Mn-^{53}Cr age of 4555.1 ± 1.2 Ma (*Zipfel et al.,* 1996), consistent with its Pb-Pb and ^{146}Sm-^{142}Nd systematics. Iodine-129–xenon-129 ($t_{1/2}$ ~ 17 m.y.) systematics in Acapulco phosphates indicate Xe closure ~9 m.y. after the Bjurbole L4 chondrite, which also appears to be consistent with Pb-Pb systematics (*Nichols et al.,* 1994; *Brazzle et al.,* 1999).

Brachinites are olivine-rich primitive achondrites generally considered to be either partial melt residues (e.g., *Nehru et al.,* 1983, 1992, 1996; *Goodrich,* 1998) or igneous cumulates (e.g., *Warren and Kallemeyn,* 1989). Although precise absolute formation ages for these primitive achondrites are not yet available, there are indications that they formed close to ~4.56 Ga, soon after the beginning of the solar system. These include the presence of fission tracks from the decay of live ^{244}Pu (*Crozaz and Pellas,* 1983) and ^{129}Xe excesses due to the decay of ^{129}I (*Bogard et al.,* 1983; *Ott et al.,* 1987; *Swindle et al.,* 1998). *Wadhwa et al.* (1998) reported a ^{53}Mn-^{53}Cr isochron for Brachina that indicated a ^{53}Mn/^{55}Mn ratio of (3.8 ± 0.4) × 10^{-6} at the time of last equilibration of Cr isotopes (Fig. 1). Comparison of this value with the ^{53}Mn/^{55}Mn ratio of (1.25 ± 0.07) × 10^{-6} in the angrite LEW 86010 at 4558 Ma (Fig. 1; Table 1) implies a Mn-Cr age of 4563.7 ± 0.9 Ma for this meteorite, which is only 3.5 ± 1.1 m.y. after the time of CAI formation (i.e., 4567.2 ± 0.6 Ma) (*Amelin et al.,* 2002).

However, not all primitive achondrites investigated so far have consistent long- and short-lived radioisotope systematics. The Divnoe meteorite is an ultramafic primitive achondrite whose relationship to other primitive achondrite groups is as yet unclear (*Petaev et al.,* 1994; *Weigel et al.,* 1996). This meteorite is inferred to have an extremely high ^{146}Sm/^{144}Sm ratio of 0.0116 ± 0.0016 that in comparison to the

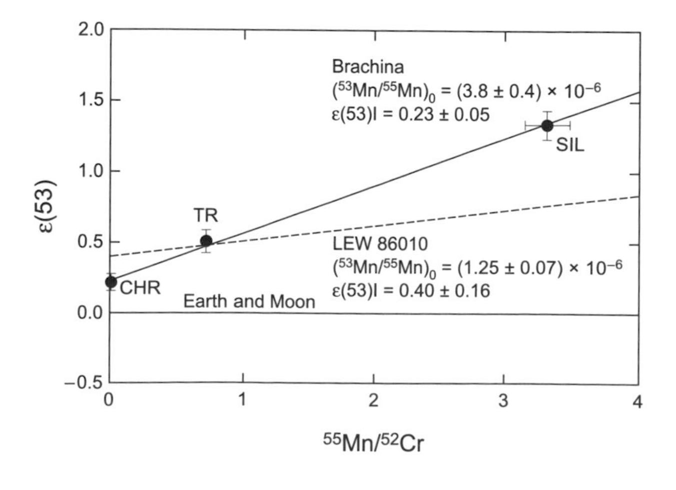

Fig. 1. Excesses in the ^{53}Cr/^{52}Cr ratio in parts per 10^4, or ε(53), relative to a terrestrial standard vs. ^{55}Mn/^{52}Cr ratios in the primitive achondrite Brachina (CHR = chromite; TR = whole rock; SIL = silicates); data from *Wadhwa et al.* (1998). The dashed line is the isochron for the LEW 86010 angrite (*Lugmair and Shukolyukov,* 1998).

^{146}Sm/^{144}Sm ratio of 0.007 in angrite LEW 86010 (Table 1) provides an age of 4.633 ± 0.022 Ga, consistent with its old ^{147}Sm-^{143}Nd age of 4.62 ± 0.07 Ga, although uncertainties are large (*Bogdanovski and Jagoutz,* 1996). These ages are problematic since they are older than the age inferred for CAI formation, which is generally assumed to reflect the time of formation of the first solids in the solar nebula. Manganese-53–chromium-53 systematics in this achondrite, on the other hand, indicate that Cr isotopes were equilibrated at ≤4542 Ma when the ^{53}Mn/^{55}Mn ratio was ≤6 × 10^{-8} (*Bogdanovski et al.,* 1997).

Despite some discrepancies, the above discussion shows that the onset of melting on some planetesimals began as early as ~3–5 m.y. after the formation of the first solids and possibly extended to ≥10 m.y. on others. While the most plausible heat source for early incipient melting on some planetesimals is the decay of short-lived radionuclides such as ^{26}Al and ^{60}Fe, such melting episodes at significantly later times on other planetesimals are likely to be the result of impact heating.

3.2. Crust Formation

Primary crystallization ages, ideally based on internal mineral isochrons obtained from individual members of achondrite groups such as the angrites and noncumulate eucrites that represent basaltic rocks that formed in asteroidal near-surface environments, offer the best means of assessing the timescales of crust formation on planetesimals.

3.2.1. Angrites. Angrites are a small group of mineralogically unique basalts composed mostly of Ca-Al-Ti-rich pyroxenes (fassaite), olivine, and anorthitic plagioclase (*Mittlefehldt et al.,* 1998, and references therein). Early evidence for the presence of fission Xe (from the decay of ^{244}Pu; $t_{1/2}$ ~ 82 m.y.) in Angra dos Reis, the type meteorite of the angrites, established its antiquity (*Hohenberg,* 1970; *Lugmair and Marti,* 1977; *Wasserburg et al.,* 1977). Subsequently, evidence for the former presence of ^{244}Pu has also been reported in two other angrites (*Eugster et al.,* 1991; *Hohenberg et al.,* 1991). Samarium-147–neodymium-143 systematics in Angra dos Reis (ADOR) and LEW 86010 (LEW) are well behaved and give old crystallization ages between 4.53 ± 0.04 and 4.56 ± 0.04 Ga (*Lugmair and Marti,* 1977; *Wasserburg et al.,* 1977; *Jacobsen and Wasserburg,* 1984; *Lugmair and Galer,* 1992; *Nyquist et al.,* 1994). Samarium-147–neodymium-143 systematics have also been determined in the more recently discovered angrite D'Orbigny, and, despite some disturbance evident in the plagioclase, possibly due to late metamorphism and/or terrestrial weathering, are generally consistent with earlier results for ADOR and LEW (*Nyquist et al.,* 2003a; *Tonui et al.,* 2003). Samarium-146–neodymium-142 systematics in ADOR and LEW (^{146}Sm/^{144}Sm ~0.007) are concordant with their ^{147}Sm-^{143}Nd systematics, although preliminary data for D'Orbigny suggest disturbance of the ^{146}Sm-^{142}Nd system (*Tonui et al.,* 2003).

The most precise estimate of the crystallization age ofthe angrites is offered by their Pb-Pb systematics. An internal

Pb-Pb isochron defined by LEW minerals gives an age of 4558.2 ± 3.4 Ma, concordant with the highly precise U-Pb model age of 4557.8 ± 0.5 Ma obtained from the extremely radiogenic Pb compositions in the pyroxenes of ADOR and LEW (*Lugmair and Galer,* 1992). The highly precise Pb-Pb age of the LEW angrite has made it possible to use the formation time of this sample as a precise time anchor for the application of chronometers based on short-lived radionuclides, particularly ^{53}Mn ($t_{1/2}$ ~ 3.7 m.y.) and ^{146}Sm ($t_{1/2}$ = 103 m.y.) (Table 1). Preliminary U-Pb model ages derived from D'Orbigny pyroxenes appeared to be in agreement with those derived from ADOR and LEW pyroxenes (*Jagoutz et al.,* 2003). However, a recent reevaluation of these data by Jagoutz and colleagues has resulted in a somewhat older age of 4563 ± 1 Ma for D'Orbigny (G. W. Lugmair, personal communication, 2005). Finally, *Baker et al.* (2005) have reported an even older Pb-Pb age for the Sahara 99555 angrite of 4566.2 ± 0.1 Ma.

Evidence for the presence of live ^{53}Mn at the time of their formation is found in ADOR and LEW. *Nyquist et al.* (1994) reported a ^{53}Mn/^{55}Mn ratio of (1.44 ± 0.07) × 10^{-6} for the LEW angrite. Subsequently, a study by *Lugmair and Shukolyukov* (1998) gave a ^{53}Mn/^{55}Mn ratio of (1.25 ± 0.07) × 10^{-6} for LEW (with the data point for ADOR olivine being consistent with this value). These values are in reasonable agreement, especially if the Cr-isotopic composition of olivine reported by *Nyquist et al.* (1994) is corrected for contribution from spallogenic Cr. No detectable evidence for the presence of live ^{26}Al has been found in these samples (^{26}Al/^{27}Al < 2 × 10^{-7}) (*Lugmair and Galer,* 1992). However, such a result is not unexpected given that, at the time these angrites formed (4558 Ma), ~12 half-lives of ^{26}Al had elapsed since the time of CAI formation (4567 Ma) (*Amelin et al.,* 2002) when the ^{26}Al/^{27}Al ratio was ~5 × 10^{-5}. The ^{53}Mn/^{55}Mn ratio at the time of formation of D'Orbigny and Sahara 99555 angrites (~3 × 10^{-6}) (*Nyquist et al.,* 2003a; *Glavin et al.,* 2004) is significantly higher than in ADOR and LEW, implying that D'Orbigny and Sahara 99555 formed ~4–5 m.y. prior to ADOR and LEW. This is consistent with the Pb-Pb systematics in D'Orbigny, which indicate that this angrite is 5 ± 1 m.y. older than ADOR and LEW (*Lugmair and Galer,* 1992; G. W. Lugmair, personal communication, 2005). There is also evidence for the presence of live ^{26}Al in the D'Orbigny and Sahara 99555 angrites (^{26}Al/^{27}Al ~5 × 10^{-7}) (*Spivak-Birndorf et al.,* 2005), which is consistent with the higher ^{53}Mn/^{55}Mn ratio and the older Pb-Pb age of D'Orbigny compared to ADOR and LEW. However, the Pb-Pb age of Sahara 99555 (*Baker et al.,* 2005) is not concordant with the ages based on ^{26}Al-^{26}Mg and ^{53}Mn-^{53}Cr internal isochrons (*Spivak-Birndorf et al.,* 2005). Finally, high-precision Mg-isotopic analyses of bulk samples of the Sahara 99555 and NWA 1296 angrites show small but resolvable excesses in ^{26}Mg, which may be indicative of early Al/Mg fractionation in their source resulting from crust formation on their parent body while ^{26}Al was still extant (*Bizzarro et al.,* 2005).

3.2.2. Eucrites. Like the angrites, the noncumulate eucrites are pyroxene-plagioclase rocks. However, there are significantly greater numbers of known noncumulate eucrites than there are angrites. Recent high-precision O-isotopic data of *Wiechert et al.* (2004) demonstrate that most noncumulate eucrites along with the cumulate eucrites, diogenites, and howardites have uniform Δ^{17}O (within ± 0.02‰) and thus lie on a single mass fractionation consistent with their origin on a single parent body. Therefore, these basalts are the most numerous crustal rocks available from any single solar system body other than Earth and the Moon. A handful of the noncumulate eucrites (in particular Ibitira, but possibly also Caldera, Pasamonte, and ALHA 78132) have O-isotopic compositions distinct from the others, implying either that these samples originated on different parent bodies or that isotopic heterogeneity was preserved on the HED parent body (*Wiechert et al.,* 2004).

Unlike angrites (which did not undergo any significant degree of recrystallization or metamorphism), the noncumulate eucrites appear to record a protracted history of extensive thermal processing on their parent body subsequent to their original crystallization. As a result, many of the long- and short-lived chronometers investigated in these samples appear to record secondary thermal events rather than their original crystallization. Nevertheless, there are several lines of evidence that suggest that these basalts crystallized very early in the history of the solar system. Although typically characterized by large uncertainties, the ^{147}Sm-^{143}Nd ages of several of these samples such as Chervony Kut (4580 ± 30 Ma) (*Wadhwa and Lugmair,* 1995), Juvinas (4560 ± 80 Ma) (*Lugmair,* 1974), Pasamonte (4580 ± 120 Ma) (*Unruh et al.,* 1977), Piplia Kalan (4570 ± 23 Ma) (*Kumar et al.,* 1999), and Yamato 792510 (4570 ± 90 Ma) (*Nyquist et al.,* 1997a) are close to ~4.56 Ga. In some cases where ^{147}Sm-^{143}Nd systematics appear to record ages younger than ~4.56 Ga, the ^{146}Sm-^{142}Nd systematics still provide evidence for early crystallization of basaltic eucrites such as Caldera (^{146}Sm/^{144}Sm = 0.0073 ± 0.0011) (*Wadhwa and Lugmair,* 1996) and Ibitira (^{146}Sm/^{144}Sm ~ 0.009 ± 0.001) (*Prinzhofer et al.,* 1992). This may be explained by a model proposed by *Prinzhofer et al.* (1992), according to which the apparent discrepancy between the long-lived ^{147}Sm-^{143}Nd and the short-lived ^{146}Sm-^{142}Nd systems can be interpreted in terms of a short episodic disturbance resulting in partial reequilibration of the rare earth elements (REEs), predominantly between plagioclase (which has very low REE abundances) and phosphates (which are the primary REE carriers). As shown by the modeling results of these authors, such a disturbance could partially reset the ^{147}Sm-^{144}Nd isochron, although the ^{146}Sm-^{142}Nd system would not be resolvably affected.

There is at least one instance (i.e., the basaltic eucrite EET 90020) where the ^{147}Sm-^{143}Nd age (4430 ± 30 Ma) and ^{146}Sm-^{142}Nd systematics (^{146}Sm/^{144}Sm = 0.0048 ± 0.0020) are indeed concordant, although within rather large uncertainties. As will be discussed below, both the ^{147}Sm-^{144}Nd age and initial ^{146}Sm/^{144}Sm ratio for EET 90020 are similar to values obtained for the Moore County and Moama cumulate eucrites (*Jacobsen and Wasserburg,* 1984; *Tera et al.,* 1997). The young ^{147,146}Sm-^{143,142}Nd age for this noncumulate eucrite has been interpreted to possibly record late magmatism on the parent body (*Nyquist et al.*, 1997b). Therefore, EET 90020 could be the crystallization product of a younger melting event that may also be responsible for the formation of the cumulate eucrites. However, the ^{39}Ar-^{40}Ar chronometer, which is more easily reset by shock reheating than the Sm-Nd system, records a slightly older age of ~4.48–4.49 Ga for EET 90020 (*Bogard and Garrison*, 1997). Therefore, it is possible that this basalt underwent a complex petrogenetic history as suggested by *Yamaguchi et al.* (2001), such that its Sm-Nd systematics may not necessarily record its primary igneous formation.

Although the Rb-Sr chronometer tends to be more susceptible to resetting than the Sm-Nd system, it also generally points to an ancient age of formation for the noncumulate eucrites (e.g., *Allègre et al.*, 1975; *Nyquist et al.*, 1986). The most precise of the absolute chronometers, i.e., the U-Pb system, appears to have been affected by postcrystallization events and terrestrial Pb contamination in most noncumulate eucrites, recording mineral isochron ages in the range of 4128 Ma to 4530 Ma (*Tatsumoto et al.*, 1973; *Unruh et al.*, 1977; *Galer and Lugmair*, 1996; *Tera et al.*, 1997). However, Ibitira whole-rock samples with radiogenic Pb-isotopic compositions gave old Pb-Pb model ages of 4556 ± 6 Ma (*Chen and Wasserburg*, 1985) and 4560 ± 3 Ma (*Manhés et al.*, 1987). Furthermore, a recent determination of the Pb-Pb mineral isochron for the Asuka 881394 eucrite yielded a precise and ancient age of 4566.5 ± 0.9 Ma (*Wadhwa et al.*, 2005; Y. Amelin, personal communication, 2005). This is only 0.7 ± 1.1 m.y. younger than the time of CAI formation (4567.2 ± 0.6 Ma) (*Amelin et al.*, 2002) and indicates that crust formation on differentiated planetesimals occurred within ~2 m.y. of the formation of the first solids in the early solar system.

In recent years there has been increasing evidence for the presence of various short-lived radionuclides (particularly ^{26}Al, ^{53}Mn, and ^{60}Fe) in the noncumulate eucrites at their time of crystallization, which further supports the early formation of these crustal basalts. The first evidence for live ^{26}Al in an achondrite was reported by *Srinivasan et al.* (1999) (Fig. 2). These authors demonstrated the presence of excess ^{26}Mg from the decay of ^{26}Al in plagioclase of the Piplia Kalan basaltic eucrite, from which they inferred a $^{26}Al/^{27}Al$ ratio of $(7.5 \pm 0.9) \times 10^{-7}$ in this sample at the time of its formation. However, this value is more likely to be a lower limit on the actual $^{26}Al/^{27}Al$ ratio at the time of crystallization of this eucrite, since the plagioclase appears to have undergone intragrain equilibration (as is evident from the near-constant ^{26}Mg excesses determined in various spot analyses of plagioclase with different Al/Mg ratios; Fig. 2). Compared to the $^{26}Al/^{27}Al$ ratio of $\sim 5 \times 10^{-5}$ in CAIs, the $^{26}Al/^{27}Al$ ratio inferred for Piplia Kalan suggests that crust formation on the parent planetesimal of this eucrite occurred at most ~5 m.y. after the beginning of the solar system. Subsequently, evidence for the presence of live ^{26}Al has been found in additional basaltic eucrites ($^{26}Al/^{27}Al$ ratios in the range of $\sim 6 \times 10^{-7}$ to $\sim 2 \times 10^{-6}$) (*Srinivasan*, 2002; *Nyquist et al.*, 2003b; *Wadhwa et al.*, 2005). Of these, the $^{26}Al/^{27}Al$ ratio is most precisely determined in the Asuka 881394 eucrite, i.e., $(1.34 \pm 0.04) \times 10^{-6}$ (Fig. 3), and translates to an Al-Mg age of 4563.5 ± 0.9 Ma, slightly younger than its ancient Pb-Pb mineral isochron age, possibly indicative of slow subsolidus cooling of this eucrite (*Wadhwa et al.*, 2005; Y. Amelin, personal communication, 2005). Additionally, *Bizzarro et al.* (2005) have recently reported high-precision

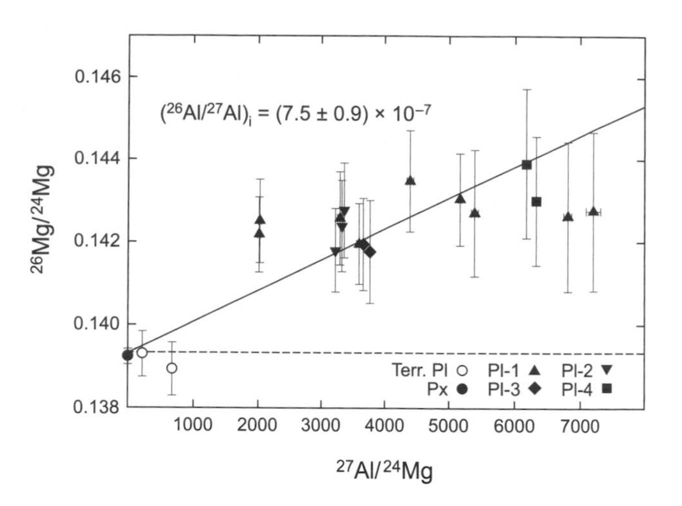

Fig. 2. Excesses in the $^{26}Mg/^{24}Mg$ ratio vs. $^{27}Al/^{24}Mg$ ratios in the noncumulate eucrite Piplia Kalan (Px = pyroxene; Pl = plagioclase). Figure from *Srinivasan et al.* (1999).

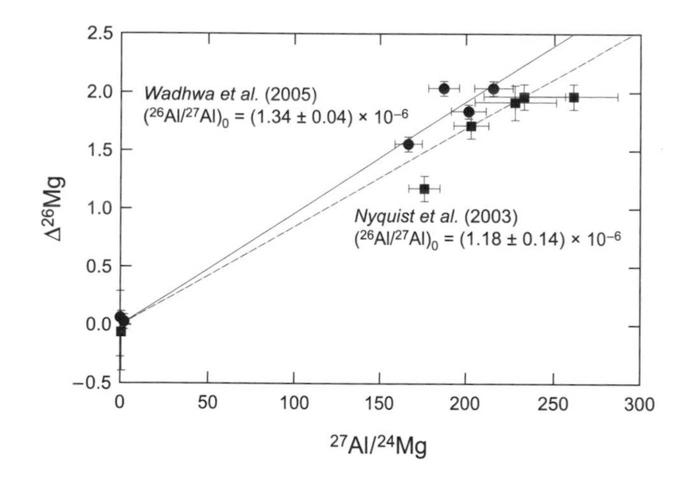

Fig. 3. Excesses in the $^{26}Mg/^{24}Mg$ ratio in per mil, or $\Delta^{26}Mg$, relative to a terrestrial standard plotted vs. $^{27}Al/^{24}Mg$ ratios in the Asuka 881394 eucrite. Solid circles show the data of *Wadhwa et al.* (2005) and the solid line is the isochron defined by these data, the slope of which corresponds to a $^{26}Al/^{27}Al$ ratio of $(1.34 \pm 0.04) \times 10^{-6}$; low Al/Mg data points near the origin are for pyroxene and whole rock, and high Al/Mg data points are for plagioclase. Solid squares are the data of *Nyquist et al.* (2003b) and the dashed line is the isochron defined by these data, the slope of which corresponds to a $^{26}Al/^{27}Al$ ratio of $(1.18 \pm 0.14) \times 10^{-6}$; low Al/Mg data points near the origin are for pyroxene, and high Al/Mg data points are for plagioclase.

Mg-isotopic analyses of bulk samples of several noncumulate eucrites, which show that these samples have small excesses in ^{26}Mg, possibly indicative of early Al/Mg fractionation in their source resulting from basaltic volcanism and crust formation on their parent body while ^{26}Al was extant.

Evidence for the presence of live ^{53}Mn at the time of eucrite formation has been documented in several noncumulate eucrites including Chervony Kut, Juvinas, Ibitira (*Lugmair and Shukolyukov,* 1998), and Asuka 881394 (*Nyquist et al.,* 2003b; *Wadhwa et al.,* 2005). By comparison with the LEW angrite (Table 1), Mn-Cr ages for these eucrites were determined to be 4563.6 ± 0.9 Ma, 4562.5 ± 1.0 Ma, 4557 +2/–4 Ma, and 4563.8 ± 0.7 Ma, respectively (*Lugmair and Shukolyukov,* 1998; *Wadhwa et al.,* 2005). These ages indicate that crust formation on the parent planetesimal of these basalts began possibly within ~3 m.y., and may have continued up to ~10 m.y., after the formation of the first solids (additionally, the Mn-Cr age for the Asuka 881394 eucrite is consistent with its Al-Mg age). Several other basaltic eucrites in which Mn-Cr systematics have been investigated do not show any evidence for live ^{53}Mn (*Lugmair and Shukolyukov,* 1998; *Nyquist et al.,* 1996, 1997a,b). In these cases, it is likely that secondary thermal events have resulted in later resetting of the Mn-Cr system, as appears to be the case for the U-Pb system in most noncumulate eucrites as well. Finally, there is unambiguous evidence for the former presence of live ^{60}Fe in three noncumulate eucrites, Chervony Kut, Juvinas, and Bouvante (*Shukolyukov and Lugmair,* 1993a,b; *Quitté et al.,* 2005), also attesting to early formation of these basalts.

Basaltic noncumulate eucrites thus show clear evidence of having formed close to ~4.56 Ga in the crust of their parent planetesimal. In contrast, cumulate eucrites, which formed in the crust of the same parent planetesimal as the noncumulate eucrites (*Clayton and Mayeda,* 1996; *Wiechert et al.,* 2004), have significantly younger concordant Sm-Nd and Pb-Pb ages, ranging from the oldest of 4456 ± 25 Ma (Sm-Nd) and 4484 ± 19 Ma (Pb-Pb) for Moore County (*Tera et al.,* 1997) to the youngest of 4410 ± 20 Ma (Sm-Nd) (*Lugmair et al.,* 1977) and 4399 ± 35 Ma (Pb-Pb) (*Tera et al.,* 1997) for Serra de Magé. Thus, Sm-Nd and Pb-Pb systematics in the cumulate eucrites indicate that isotopic closure occurred up to ~150 m.y. after the noncumulate eucrites (*Lugmair et al.,* 1977; *Jacobsen and Wasserburg,* 1984; *Lugmair et al.,* 1991; *Tera et al.,* 1997), possibly implying that active magmatism persisted on the eucrite parent body for this extended period (*Tera et al.,* 1997). This may be supported by the 4430 ± 30 Ma ^{147}Sm-^{143}Nd age of the noncumulate eucrite EET 90020 (*Nyquist et al.,* 1997b) and the overlapping, relatively low, initial ^{146}Sm/^{144}Sm ratios of Moore County, Moama, and EET 90020 (*Jacobsen and Wasserburg,* 1984; *Nyquist et al.,* 1997b; *Tera et al.,* 1997). If this is so, it further implies that the process of crust formation on the eucrite parent body possibly extended to ~150 m.y. after solar system formation. Since, as discussed above, the oldest noncumulate eucrites formed within ~3 m.y. of CAI formation, the decay of short-lived radionuclides such as ^{26}Al and ^{60}Fe is the likely heat source for the early and extensive differentiation experienced by their parent planetesimal. Energy sources that can account for later igneous activity (i.e., tens of millions of years after CAI formation) on small planetesimals are not obvious unless the cumulate eucrites are the crystallization products of impact melting on the eucrite parent body. Alternatively (and perhaps more likely), since the cumulate eucrites are slowly cooled rocks that possibly formed deeper within the crust of their parent body than noncumulate eucrites, the long-lived chronometers could be recording the long cooling times required to achieve subsolidus temperatures. This is supported by the modeling results of *Ghosh and McSween* (1998), which show that, assuming reasonable parameters, it is possible to maintain temperatures in excess of the solidus temperature for basalt at a depth of ~100 km for over ~100 m.y. in a Vesta-sized planetesimal.

3.2.3. Other achondrites. Besides the angrites and the eucrites, there are other types of achondrites, such as aubrites and ureilites, whose origins are somewhat enigmatic but that were nevertheless formed in the crusts of extensively differentiated asteroidal bodies. There are very few chronological investigations of the aubrites, which are essentially monomineralic rocks composed of coarse-grained enstatite. *Brazzle et al.* (1999) estimated an I-Xe age of 4566 ± 2 Ma for the Shallowater aubrite, using the I-Xe and Pb-Pb systematics in phosphate of the Acapulco primitive achondrite as an anchor. More recently, however, based on correlations of relative I-Xe ages with absolute Pb-Pb ages of various meteorites and their components, it has been suggested that enstatite in the Shallowater aubrite crystallized at 4563.5 ± 1.0 Ma (*Pravdivtseva et al.,* 2005). Early formation (i.e., within a few million years of CAIs) of the aubrites is further supported by evidence for live ^{53}Mn in Peña Blanca Spring (*Shukolyukov and Lugmair,* 2004).

Uranium-thorium-lead and Sm-Nd isotope systematics in the ureilites generally indicate early formation, although there are apparent complications resulting from later disturbance, i.e., during a metasomatic event on the ureilite parent body and/or by recent terrestrial contamination (*Goodrich and Lugmair,* 1995; *Torigoye-Kita et al.,* 1995a,b). More recent studies have demonstrated evidence for live ^{53}Mn and ^{26}Al at the time of ureilite formation. Specifically, it has been shown that a feldspathic clast from the Dar al Gani 165 ureilite had a ^{53}Mn/^{55}Mn ratio of (2.91 ± 0.16) × 10^{-6} at the time of last equilibration of Cr isotopes (*Goodrich et al.,* 2002). Compared to the LEW angrite (Table 1), a Mn-Cr age of 4562.3 ± 0.6 Ma is determined for this sample. *Kita et al.* (2003) have reported Al-Mg data for a plagioclase-bearing clast from the Dar al Gani 319 ureilite that indicate a ^{26}Al/^{27}Al ratio of (3.95 ± 0.59) × 10^{-7} at the time of its crystallization. Comparison with the canonical value at the time of CAI formation (i.e., 4567.2 ± 0.6 Ma) (*Amelin et al.,* 2002) gives an Al-Mg age of 4562.2 ± 0.6 Ma, which agrees well with the Mn-Cr age discussed above. Finally, a Hf-W-isotopic investigation of bulk samples of several ureilites has yielded subchondritic values for ε^{182}W (i.e., the ^{182}W/^{184}W or ^{182}W/^{183}W ratio relative to the terrestrial standard in parts per 10^4) and Hf/W ratios, indicating that the differentiation process that resulted in the formation of these ureilites occurred while ^{182}Hf was still extant (*Lee et al.,* 2005).

3.3. Global Silicate (Mantle) Differentiation

3.3.1. Whole-rock isochrons. While an internal mineral isochron can provide constraints on the timing of formation of an individual achondrite in the crust of a planetesimal, a whole-rock isochron (based on a long- or a short-lived chronometer) of a particular achondrite group can provide limits on the timing of parent-daughter element fractionation in the source (mantle) reservoir of their parent planetesimal, which in turn is a reflection of the timescale involved in global silicate fractionation (possibly associated with crystallization of a magma ocean). Whole-rock Rb-Sr isochrons for the basaltic eucrites established early on that Rb-Sr fractionation in the mantle source reservoir of these achondrites occurred close to ~4.6 Ga (*Papanastassiou and Wasserburg,* 1969; *Birck and Allègre,* 1978). *Smoliar* (1993) evaluated all available Rb-Sr data for the eucrites and obtained a whole-rock isochron age of 4.55 ± 0.06 Ga for the noncumulate eucrites.

Achondrite whole-rock isochrons based on short-lived chronometers have the potential for more precisely constraining the timing of global silicate differentiation on planetesimals. Such isochrons have recently been reported for the ^{53}Mn-^{53}Cr and the ^{182}Hf-^{182}W systems in the HED group of achondrites (*Lugmair and Shukolyukov,* 1998; *Quitté et al.,* 2000; *Kleine et al.,* 2004). The HED whole-rock ^{53}Mn-^{53}Cr isochron corresponds to a ^{53}Mn/^{55}Mn ratio of (4.7 ± 0.5) × 10^{-6} (Fig. 4) (*Lugmair and Shukolyukov,* 1998). Comparison with the LEW angrite (Table 1) gives a Mn-Cr age of 4564.8 ± 0.9 Ma for global silicate fractionation on the HED parent body, which is 2.4 ± 1.1 m.y. after the formation of CAIs (*Amelin et al.,* 2002). This age is, within error, similar to the Al-Mg and Mn-Cr ages of eucrite Asuka 881394 (*Nyquist et al.,* 2003b; *Wadhwa et al.,* 2005) and the Mn-Cr age of eucrite Chervony Kut (*Lugmair and Shukolyukov,* 1998). The near agreement of the ages provided by the whole-rock isochrons and internal isochrons of these eucrites indicates not only that planetesimal differentiation started soon after solar system formation, but that the duration of the initial HED crust-mantle differentiation may have been as short as ~2 m.y. or less.

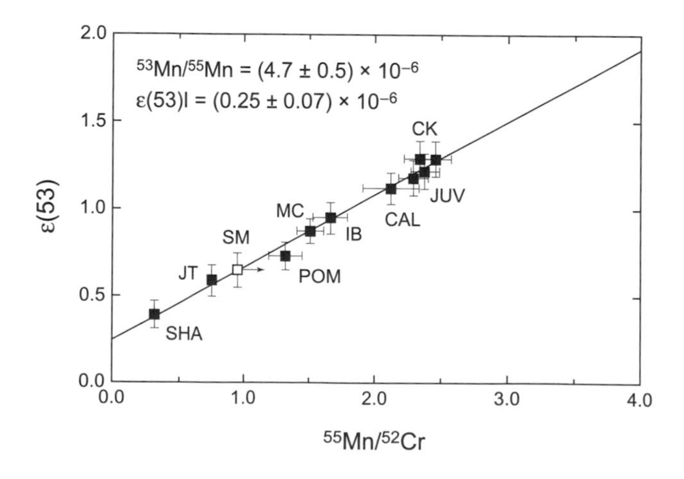

Fig. 4. ^{53}Mn-^{53}Cr systematics in the HED parent body. Data points are whole rocks of noncumulate eucrites (CAL = Caldera; CK = Chervony Kut; IB = Ibitira; JUV = Juvinas; POM = Pomozdino), cumulate eucrites (MC = Moore County; SM = Serra de Magé), and diogenites (JT = Johnstown; SHA = Shalka). ε(53) is as defined in the caption for Fig. 1. The slope of the HED whole-rock isochron corresponds to a ^{53}Mn/^{55}Mn ratio of (4.7 ± 0.5) × 10^{-6} at the time of global silicate differentiation. Figure from *Lugmair and Shukolyukov* (1998).

Trinquier et al. (2005a,b) have recently suggested that HED meteorites have a deficit in the ^{54}Cr/^{52}Cr ratio of 0.72 ± 0.02 ε compared to the terrestrial standard. This is in contrast to previous work on Cr-isotopic compositions of achondrites, including eucrites, which indicates that the ^{54}Cr/^{52}Cr ratio in these samples is the same as the terrestrial value to within ±0.25 ε (*Lugmair and Shukolyukov,* 1998; *Wadhwa et al.,* 2003). A potentially anomalous ^{54}Cr/^{52}Cr ratio in bulk eucrites is relevant here since many previous Cr-isotopic studies use a second-order fractionation correction that assumes a normal (i.e., terrestrial) ^{54}Cr/^{52}Cr ratio. The reason for this apparent discrepancy is not evident at this time. Nevertheless, in the case of eucrites and other achondrites that have undergone complete equilibration of Cr-isotopic composition between their constituent minerals because they crystallized from a melt, a second-order correction such as used by *Lugmair and Shukolyukov* (1998) will not alter the slope of the ^{53}Mn-^{53}Cr isochron or the relative age based on this slope, which is derived from comparison with another magmatic sample, the LEW angrite.

A whole-rock ^{182}Hf-^{182}W isochron for the HEDs was initially reported by *Quitté et al.* (2000). Subsequently, the data of *Quitté et al.* (2000) were combined with additional data by *Kleine et al.* (2004), who reported a ^{182}Hf/^{180}Hf ratio of (7.25 ± 0.50) × 10^{-5} at the time of crust-mantle differentiation on the eucrite parent body (Fig. 5a). Currently there is no consensus on the initial ^{182}Hf/^{180}Hf ratio of the solar system, with proposed values ranging from ~1 × 10^{-4} (*Kleine et al.,* 2002, 2005; *Yin et al.,* 2002b) to ~1.6 × 10^{-4} (*Quitté and Birck,* 2004). An initial ^{182}Hf/^{180}Hf ratio of ~1 × 10^{-4} would imply that global Hf-W fractionation in the HED mantle source occurred ~4 m.y. after the beginning of the solar system, which is consistent with the HED whole-rock ^{53}Mn-^{53}Cr systematics discussed above. However, if an initial ^{182}Hf/^{180}Hf ratio of ~1.6 × 10^{-4} is assumed, then Hf-W fractionation in the mantle of the HED parent body occurred ~9 m.y. after the beginning of the solar system. This is inconsistent not only with the HED whole-rock ^{53}Mn-^{53}Cr systematics but also with the recent evidence for live ^{26}Al in several eucrites (*Srinivasan et al.,* 1999; *Srinivasan,* 2002; *Nyquist et al.,* 2003b; *Wadhwa et al.,* 2005). In this regard, it is worth noting that the initial ^{182}Hf/^{180}Hf ratio of 1.6 × 10^{-4} is based on the reanalysis of the W-isotopic composition of the Tlacotepec (IVB) iron meteorite, which has an

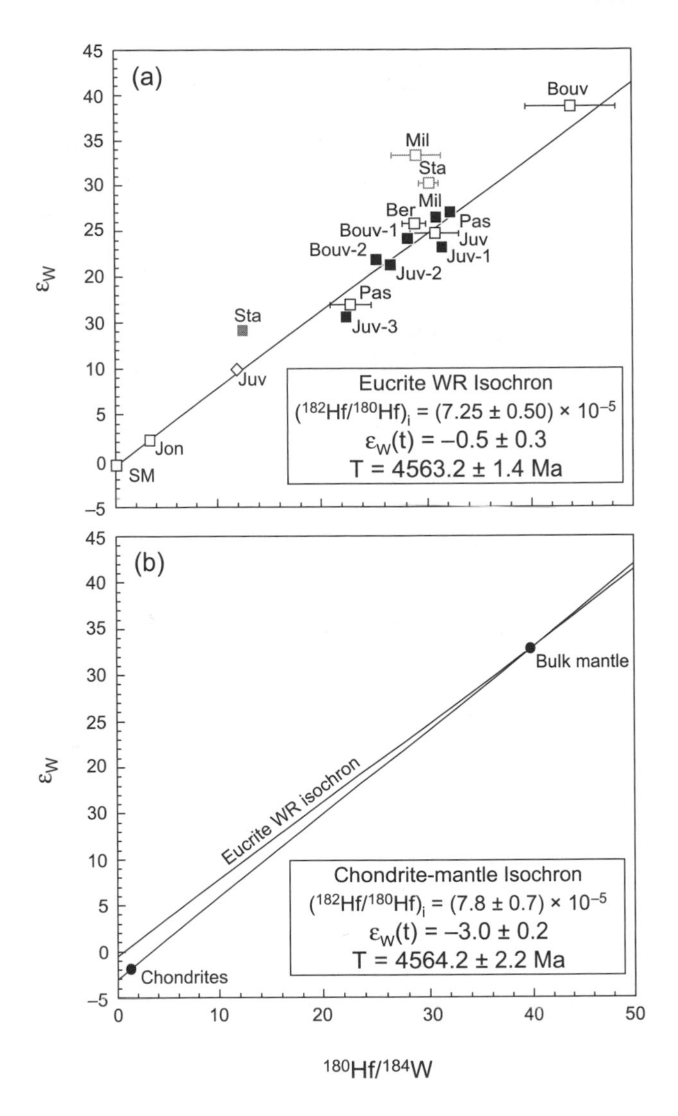

Fig. 5. ^{182}Hf-^{182}W systematics in the HED parent body. ε_W is the ^{182}W/^{184}W relative to a terrestrial standard in parts per 10^4. **(a)** Data points are whole rocks of several noncumulate eucrites (Ber = Bereba; Bouv = Bouvante; Juv = Juvinas; Jon = Jonzac; Mil = Millbillillie; Pas = Pasamonte; Sta = Stannern) and a cumulate eucrite (SM = Serra de Magé). Solid symbols are data from *Kleine et al.* (2005); open symbols: *Quitté et al.* (2000); open diamond: *Yin et al.* (2002b). The slope of the HED whole-rock isochron corresponds to a ^{182}Hf/^{180}Hf ratio of $(7.25 \pm 0.50) \times 10^{-5}$ at the time of global silicate differentiation. **(b)** Comparison of the slope of the eucrite whole-rock isochron from **(a)** with that of the bulk mantle-chondrite model isochron suggests that core formation on the HED parent body occurred 0.9 ± 0.3 m.y. before global silicate differentiation. The W-isotopic composition of the eucrite bulk mantle is estimated from the whole-rock isochron, using a ^{180}Hf/^{184}W ratio of ~40. Figure from *Kleine et al.* (2004).

extremely low ε^{182}W value of –4.5 (*Quitté and Birck,* 2004); similarly low ε^{182}W values (i.e., lower than ~–3.5) for several iron meteorites have recently been called into question (*Yin and Jacobsen,* 2003; *Scherstén et al.,* 2004). However, the proposed initial ^{182}Hf/^{180}Hf ratio of 1×10^{-4} is based on the ^{182}Hf-^{182}W systematics in bulk chondrites that have undergone different histories and on Allende CAIs that have may have undergone some degree of alteration. Therefore, although a value close to 1×10^{-4} appears to be more plausible at this time, the question regarding the initial ^{182}Hf/^{180}Hf ratio for the solar system remains to be resolved.

Recently, *Wadhwa et al.* (2003) reported a whole-rock ^{53}Mn-^{53}Cr isochron for eucritic and diogenitic clasts from the Vaca Muerta mesosiderite that corresponded to a ^{53}Mn/^{55}Mn ratio of $(3.3 \pm 0.6) \times 10^{-6}$. These authors concluded that the lower slope of the isochron defined by these clasts implies that these materials originated on a parent planetesimal distinct from that of the HEDs, and underwent global (mantle) differentiation ~2 m.y. after the HED parent body (i.e., at ~4563 Ma compared to the LEW angrite anchor). Finally, bulk samples of three enstatite achondrites (aubrites) also define an isochron that indicates that the last global Mn/Cr fractionation on their parent body occurred at ~4563 Ma (*Shukolyukov and Lugmair,* 2004).

3.3.2. Initial strontium-isotopic composition. The antiquity of the highly volatile-depleted parent planetesimals of the angrites and the eucrites can also be inferred from their initial ^{87}Sr/^{86}Sr ratios. Assuming that the initial ^{87}Sr/^{86}Sr ratio at the beginning of the solar system is represented by the average initial ^{87}Sr/^{86}Sr ratio measured in Allende CAIs (*Gray et al.,* 1973; *Podosek et al.,* 1991), and thatsubsequent evolution of radiogenic Sr occurred in an environment with solar Rb/Sr ratios, the initial ^{87}Sr/^{86}Sr ratios of the angrites (*Lugmair and Galer,* 1992; *Nyquist et al.,* 1994, 2003a; *Tonui et al.,* 2003) translate to an age difference of ~4 m.y. between CAI formation and the timing of Rb/Sr depletion event that established the angrite source characteristics.

The very low initial ^{87}Sr/^{86}Sr ratios of the eucrites similarly indicate that the volatile depletion characterizing the eucrite parent body may have occurred early in the history of the solar system (*Carlson and Lugmair,* 2000, and references therein). A reevaluation of the Sr-isotopic data for the eucrites by *Smoliar* (1993) shows that whole-rock Rb-Sr isochrons for the noncumulate and the cumulate eucrites define slightly, but resolvably, different initial ^{87}Sr/^{86}Sr ratios [which are both distinctly lower than the eucrite initial, BABI, previously defined by *Papanastassiou and Wasserburg* (1969)]. In fact, the initial ^{87}Sr/^{86}Sr ratio for the cumulate eucrites proposed by *Smoliar* (1993) is, within errors, similar to that for the angrites (e.g., *Lugmair and Galer,* 1992), suggesting that the volatile depletion in their sources was established at similar times (possibly ~4 m.y. after CAI formation; see above). This timescale for the fractionation event that established the low Rb/Sr source characteristics of the cumulate eucrites is essentially consistent with the timing of global mantle differentiation defined by the HED whole-rock ^{53}Mn-^{53}Cr and ^{182}Hf-^{182}W isochrons. However, the slighter higher initial ^{87}Sr/^{86}Sr ratio defined by the noncumulate eucrites is potentially problematic since the simplest interpretation would be that their source evolved with a chondritic Rb/Sr ratio for ~4 m.y. longer (and is thus younger) than that of the cumulate eucrites, further implying that these two types of eucrites originated on distinct parent bodies (*Smoliar,* 1993). This is inconsistent with recent high-precision O-isotopic data (*Wiechert et al.,* 2004) that suggest that the cumulate and noncumulate eucrites (with the possible exception of Ibitira) originated on a common parent planetesimal. As discussed by *Carlson and Lugmair* (2000), a possible explanation could be that the severe volatile depletion on the eucrite parent body did not occur in a single-step process, but rather took place over the course of its accretionary and early differentiation history. Subsequently, the process of magma ocean crystallization may have resulted in a slighter higher Rb/Sr ratio in the source of the noncumulate eucrites compared to that of the cumulate eucrites, thereby resulting in the higher initial $^{87}Sr/^{86}Sr$ ratio indicated by the whole-rock isochron for the noncumulate eucrites.

3.4. Metal Segregation and Core Crystallization

Limits on the timescales involved in metal segregation and core crystallization on planetesimals may be obtained from chronological investigations of iron-rich meteorites, such as magmatic irons (which represent the metallic cores of differentiated asteroidal bodies) and pallasites (considered to have formed near the core-mantle boundary). Depending on the geochemical affinities of the parent and daughter elements, different chronometers applied to such meteorite types can provide constraints on either the process of metal segregation (e.g., the ^{182}Hf-^{182}W chronometer, where Hf is lithophile and W is siderophile such that major fractionation occurs during separation of the metallic melt from silicates) or of core crystallization (e.g., ^{187}Re-^{187}Os, ^{107}Pd-^{107}Ag, and ^{53}Mn-^{53}Cr, where the parent-daughter elements are fractionated during crystallization of Fe-Ni phases, sulfides, and phosphates from the metallic melt).

3.4.1. Metal segregation. The W-isotopic compositions of a variety of magmatic and nonmagmatic iron meteorites have been reported within the last decade (e.g., *Lee and Halliday,* 1995, 1996; *Horan et al.,* 1998) and have the lowest $^{182}W/^{184}W$ ratios of any solar system material, with $\varepsilon^{182}W$ values ranging from ~–3 to –5 (with most values between –3.5 and –4.1). As emphasized by *Horan et al.* (1998), these $\varepsilon^{182}W$ values suggest that metal segregation on different planetesimals occurred within a period of ~5 m.y. in the early history of the solar system. As mentioned earlier, however, the lower end of the range of $\varepsilon^{182}W$ values in the iron meteorites have been called into question, since such values have the improbable implication that some iron meteorites predate CAIs and chondrites (e.g., *Kleine et al.,* 2005). *Yin and Jacobsen* (2003) have measured the W-isotopic compositions of several iron meteorites, including some of those previously analyzed by *Lee and Halliday* (1996) and *Horan et al.* (1998), and reported that none had $\varepsilon^{182}W$ values lower than the chondritic initial value of –3.45 ± 0.25. Recently, *Quitté and Birck* (2004) reanalyzed the W-isotopic compositions of two iron meteorites, Duel Hill (IVA) and Tlacotepec (IVB), which were reported by *Horan et al.* (1998) to have among the lowest $\varepsilon^{182}W$ values. These authors reported an $\varepsilon^{182}W$ value of –3.5 ± 0.3 for Duel Hill, significantly higher than that previously reported for this meteorite by *Horan et al.* (1998) (i.e., –5.1 ± 1.1). However, for Tlacotepec, they obtained an $\varepsilon^{182}W$ value of –4.4 ± 0.3, which agrees well with the previous value (–4.5 ± 0.4) (*Horan et al.,* 1998). Therefore, it appears that the W-isotopic composition of the iron meteorites ranges down to at least this value. However, only 4 out of 28 iron meteorites measured by *Lee and Halliday* (1996) and *Horan et al.* (1998) have $\varepsilon^{182}W$ values below the chondritic initial value, outside of uncertainty, and the mean $\varepsilon^{182}W$ value for all these samples is –3.8 ± 0.5 (1σ). Therefore, the results from these studies are consistent with most iron meteorites having $\varepsilon^{182}W$ values that are not resolvably below the initial chondritic value. This is further supported by several recent investigations that have remeasured the W-isotopic compositions of iron meteorites for which previous work had indicated $\varepsilon^{182}W$ values lower than –3.5 (e.g., *Yin and Jacobsen,* 2003; *Scherstén et al.,* 2004). As to the $\varepsilon^{182}W$ values of some few iron meteorites lying resolvably and reproducibly below the initial chondritic value, one possibility is that the W-isotopic compositions of the components (mainly metal and silicates) of the chondrites analyzed so far, and used to defined the initial chondritic $\varepsilon^{182}W$ value, may have been equilibrated to some degree. Another possibility is that exposure to cosmic radiation could result in spallation and neutron capture reactions on W isotopes that could result in apparently lower $\varepsilon^{182}W$ values in some irons with long exposure ages (*Markowski et al.,* 2005; *Leya et al.,* 2000, 2003a,b; *Masarik,* 1997). These possibilities have implications for the initial $^{182}Hf/^{180}Hf$ ratio of the solar system, but remain to be rigorously evaluated.

The timing of metal segregation (core formation) on the HED parent body has been estimated from ^{182}Hf-^{182}W systematics in bulk samples of eucrites and diogenites (*Kleine et al.,* 2004; *Quitté and Birck,* 2004). Assuming that the HED parent body had a bulk chondritic Hf/W composition and that core formation resulted in a $^{180}Hf/^{184}W$ ratio in theHED mantle of ~40 (i.e., close to the highest Hf/W ratio measured in eucrites that are the products of partial melting of this mantle), *Kleine et al.* (2004) estimated that core formation on the HED parent planetesimal occurred 0.9 ± 0.3 m.y. prior to global mantle (silicate) differentiation (Fig 5b).

3.4.2. Core crystallization. Once the metal has segregated into the core of a planetesimal, the timescales involved in the crystallization of this metal may be constrained by isochrons based on bulk samples and mineral phases of magmatic iron meteorites and pallasites. In recent years, precise Re-Os isochrons have been obtained for various groups of the iron meteorites. Most iron meteorite groups give relatively old Re-Os ages that range from 4557 ± 12 Ma (IIIAB magmatic irons) to 4526 ± 27 Ma (IVB magmatic irons) (*Shen et al.,* 1996; *Smoliar et al.,* 1996; *Horan et al.,* 1998; *Cook et al.,* 2004). There is a discrepancy, however, in the Re-Os systematics in the IVA magmatic irons. While *Shen et al.* (1996) report a Re-Os age for the IVA irons that is 60 ± 45 m.y. older than for other iron meteorite groups, the data of *Smoliar et al.* (1996) give an age of 4456 ± 25 Ma, significantly younger than other iron meteorite groups, which these authors attributed to later disturbance of the Re-Os system. *Horan et al.* (1998) subsequently reported the Re-Os-isotopic compositions of several IVA irons and showed that while some lie on the 4456 ± 25 Ma isochron of *Smoliar et al.* (1996), others were consistent with an older age. This suggests that the Re-Os-isotopic systematics in the IVA are indeed disturbed, perhaps by processes such as breakup and reassembly of the parent planetesimal, which have been invoked to explain the range of metallographic cooling rates of the IVA irons (e.g., *Haack et al.,* 1996).

Rhenium-osmium ages reported so far use a ^{187}Re decay constant that was determined by assuming that the IIIAB isochron should give the same age as the U-Pb age of the angrites (*Smoliar et al.,* 1996), so the accuracy of these ages is only as good as the validity of this assumption. Nevertheless, the age range indicated by Re-Os isochrons is independent of the half-life, so the results for various iron meteorite groups, with the exception of the IVA irons, suggest that core crystallization (or more specifically, Re-Os isotopic closure) in planetesimals spanned a period of ~30 m.y. (although the relatively large errors certainly allow this time interval to be significantly narrower than this). Rhenium-osmium systematics in pallasites indicate that they may be younger than iron meteorites by ~60 m.y.; however, this apparently young age may be due to later reequilibration of the Re-Os system (*Chen et al.,* 2002). *Cook et al.* (2004) recently reported the first high-precision ^{190}Pt-^{186}Os isochrons for the IIAB and IIIAB magmatic irons, and estimated ages for these meteorite groups of 4323 ± 80 Ma and 4325 ± 26 Ma, respectively. These ages are somewhat younger than Re-Os ages for iron meteorites, and these authors suggested that this discrepancy could reflect an error in the decay constant for ^{190}Pt.

In contrast to the more leisurely pace of core crystallization suggested by Re-Os systematics (although as indicated above, the errors are rather large and could accommodate a much narrower time interval for this process), evidence for the former presence of live ^{107}Pd in a variety of metal-rich meteorites, including irons and pallasites, indicates that this process occurred rapidly, well within the lifetime of this short-lived radionuclide. For most iron meteorites, closure of the ^{107}Pd-^{107}Ag-isotopic system occurred within a narrow time interval of only ~4 m.y. The highest inferred ^{107}Pd/^{108}Pd ratio of ~2.4×10^{-5} is in samples such as Gibeon (IVA) and Canyon Diablo (IA), for which well-defined ^{107}Pd-^{107}Ag isochrons have been obtained (*Chen and Wasserburg,* 1996; *Carlson and Hauri,* 2001). Pallasites have somewhat younger Pd-Ag ages, most of which indicate isotopic closure ≤10 m.y. after Gibeon and Canyon Diablo. Of the pallasites investigated so far, Brenham has the highest ^{107}Pd/^{108}Pd ratio of $(1.65 \pm 0.05) \times 10^{-5}$, defined by a four-point internal isochron comprised of silicates and metal (*Carlson and Hauri,* 2001). This more precisely defined value is only marginally higher than the ^{107}Pd/^{108}Pd ratio of ~1.1×10^{-5} inferred from a single bulk data point (*Chen and Wasserburg,* 1996; *Chen et al.,* 2002), and indicates that closure of the Pd-Ag system occurred ~3.5 m.y. after Gibeon and Canyon Diablo. The recently acquired high-precision Pd-Ag-isotopic data for the Canyon Diablo (IA) and Grant (IIIB) iron meteorites and the Brenham pallasite have additionally demonstrated that ε-level differences in the initial Ag-isotopic compositions of these meteorites can be resolved, and this has important implications for their formation histories (*Carlson and Hauri,* 2001). The initial Ag-isotopic composition of Brenham indicates that during the 3.5-m.y. time interval between closure of its Pd-Ag system and that of Canyon Diablo, the parent planetesimal of Brenham evolved with a nearly chondritic Pd/Ag ratio, unlike the high Pd/Ag ratio characterizing the bulk composition of this pallasite. This may suggest that metal-silicate separation on the Brenham parent body did not occur until shortly before its crystallization (*Carlson and Hauri,* 2001). This result is distinct from that of the IIIB iron Grant. Grant has a similar inferred initial ^{107}Pd/^{108}Pd as does Brenham, but a more radiogenic initial Ag-isotopic composition, which indicates that the high bulk Pd/Ag of Grant was a characteristic of its source (*Carlson and Hauri,* 2001). With the exception of the disturbed Re-Os systematics in the IVA irons, the Re-Os and Pd-Ag ages provide a broadly similar temporal sequence of formation of the iron meteorites and pallasites.

Manganese-53–chromium-53 systematics in IIIAB magmatic irons and pallasites also indicate early fractionation and closure of the Mn-Cr system in these metal-rich meteorites. The first evidence of live ^{53}Mn in IIIAB irons was provided by ion microprobe (SIMS) studies of phosphates that showed a wide range of ^{53}Mn/^{55}Mn ratios from ~1×10^{-6} to ~2×10^{-5} (*Davis and Olsen,* 1991; *Hutcheon and Olsen,* 1991; *Hutcheon et al.,* 1992). Taken at face value, this range of ^{53}Mn/^{55}Mn ratios translates to a time interval of ~16 m.y., which is inconsistent with the Pd-Ag systematics that indicate that most of these meteorites formed contemporaneously. A more recent SIMS study of ^{53}Mn-^{53}Cr systematics in phosphates in several IIIAB iron meteorites has demonstrated that the Mn-Cr ages based on inferred ^{53}Mn/^{55}Mn ratios are different for the different phosphate minerals, most likely due to the slow cooling rate of the IIIAB iron meteorites combined with the difference in the diffusion behaviors of Mn and Cr in the different phosphates (*Sugiura and Hoshino,* 2003); the apparently wide range of inferred ^{53}Mn/^{55}Mn ratios in the previous studies of IIIAB irons may be similarly explained. *Sugiura and Hoshino* (2003) also showed that one of the phosphates (i.e., the Fe-Mn phosphate sarcopside) consistently recorded the same ^{53}Mn/^{55}Mn ratio of ~3.4×10^{-6} in all the IIIAB iron meteorites they investigated. These authors reasoned that the closure temperature of the Pd-Ag system in iron meteorites may be similar to that of the Mn-Cr system in sarcopside, and that the ^{107}Pd/^{108}Pd ratio of ~2×10^{-5} in these meteorites corresponds to a ^{53}Mn/^{55}Mn ratio of ~3.4×10^{-6}. This would imply that, compared to the ^{53}Mn/^{55}Mn ratio of 1.25×10^{-6} in the LEW angrite (*Lugmair and Shu-*

kolyukov, 1998), isotope closure of the Mn-Cr (and Pd-Ag) systems in the IIIAB irons occurred ~5 m.y. before angrite crystallization (i.e., at ~4563 Ma). Furthermore, it was suggested by these authors that in pallasites, the $^{107}Pd/^{108}Pd$ ratio of ~1×10^{-5} (inferred from Glorieta Mountain and Brenham pallasites) (*Chen and Wasserburg,* 1996; *Carlson and Hauri,* 2001; *Chen et al.,* 2002) corresponded to a $^{53}Mn/^{55}Mn$ ratio of ~1×10^{-6} (based on data for the Omolon pallasite) (*Lugmair and Shukolyukov,* 1998), implying broad consistency between Pd-Ag and Mn-Cr systematics (both reflecting a time interval of ~7 m.y.) between the iron meteorites and pallasite.

However, when all the available data for Pd-Ag and Mn-Cr systematics in various iron meteorites and pallasites are examined in detail, there are evident discrepancies between these two systems. For example, the Eagle Station pallasite has an inferred $^{53}Mn/^{55}Mn$ ratio of ~2.3×10^{-6} (*Birck and Allègre,* 1988), but there is no detectable evidence of live ^{107}Pd in this meteorite (i.e., $^{107}Pd/^{108}Pd <$ ~7×10^{-6}) (*Chen and Wasserburg,* 1996). Eagle Station is an anomalous pallasite not belonging to the "main group" (*Mittlefehldt et al.,* 1998, and references therein) and, as discussed by *Lugmair and Shukolyukov* (1998), has a Cr-isotopic composition with an anomalously high $^{54}Cr/^{52}Cr$ ratio of ~1.5 ε. Therefore, this meteorite may contain an anomalous (most likely of presolar nucleosynthetic origin) Cr component that could complicate Mn-Cr systematics such that the $^{53}Mn/^{55}Mn$ ratio inferred from the measured $^{53}Cr/^{52}Cr$ ratios would not reflect the true value. Among the main-group pallasites, a $^{53}Mn/^{55}Mn$ ratio of $(1.4 \pm 0.4) \times 10^{-5}$ has been inferred from a SIMS study of Springwater (*Hutcheon and Olsen,* 1991); this value was since confirmed by another ion microprobe study (*Hsu et al.,* 1997). However, based on thermal ionization mass spectrometry (TIMS) analysis of a separated olivine fraction from Springwater, *Lugmair and Shukolyukov* (1998) inferred a $^{53}Mn/^{55}Mn$ ratio for this pallasite more than an order of magnitude lower (~1×10^{-6}). While TIMS studies of Eagle Station, Omolon, and Springwater (*Birck and Allègre,* 1988; *Lugmair and Shukolyukov,* 1998) have yielded $^{53}Mn/^{55}Mn$ ratios of ~$1–2 \times 10^{-6}$, the SIMS investigations of Albin, Brenham, and Springwater (*Hutcheon and Olsen,* 1991; *Hsu et al.,* 1997) all give values of ~$1–4 \times 10^{-5}$. The fact that analyses of olivine from the same pallasite sample (i.e., Springwater) by bulk (TIMS) or microanalytical (SIMS) techniques yield such different inferred $^{53}Mn/^{55}Mn$ ratios could be indicative of redistribution of Mn and/or Cr within these olivines. If this is the case, then the bulk technique (which would "average" over any redistribution of the Mn and/or Cr) may be more reliable in terms of yielding the true $^{53}Mn/^{55}Mn$ ratio. As such, given the above discussion, the suggestion of *Sugiura and Hoshino* (2003) that pallasites (at least those belonging to the "main group") are characterized by a $^{53}Mn/^{55}Mn$ ratio of ~1×10^{-6} may be a reasonable one. Nevertheless, some outstanding problems with the consistency of the Mn-Cr and Pd-Ag ages in pallasites remain.

4. SUMMARY AND CONCLUSIONS

Figure 6 summarizes the timescales discussed here for the processes involved in planetesimal differentiation. Based on the constraints available from long- and short-lived chronometers, differentiation of planetesimals in the early history of the solar system occurred very rapidly. Melting was initiated on some of the planetesimals, which were subsequently arrested in this state and did not differentiate fully.

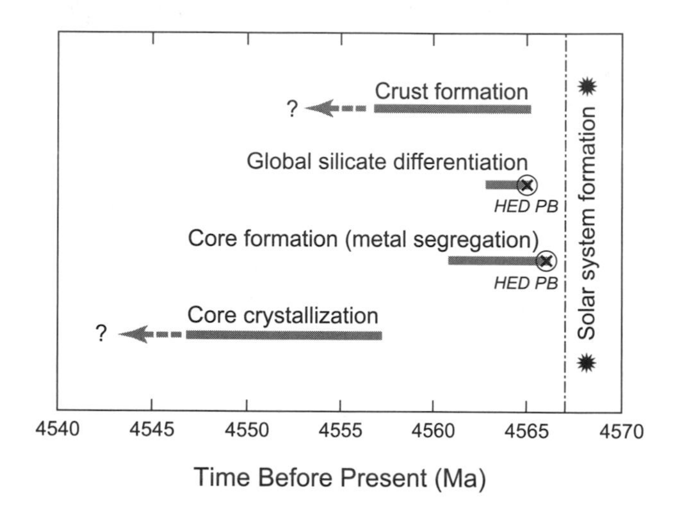

Fig. 6. Timescales of planetesimal differentiation as inferred from chronological investigations of differentiated meteorites (see text for details). The time of formation of the refractory calcium-aluminum-rich inclusions (the earliest solids to form in the protosolar nebula) at 4567 Ma is assumed to be coincident with the formation of the solar system (dotted-dashed line). Based on the ages of noncumulate eucrites and angrites, it is inferred that crust formation on planetesimals began within ~2 m.y. and continued until ~10 m.y. after CAI formation; Pb-Pb ages of cumulate eucrites may imply that igneous activity continued on the howardite-eucrite-diogenite parent body (HED PB) for another ~100–150 m.y. On the HED PB, global silicate (mantle) differentiation occurred within ~2–3 m.y. of CAI formation, and was preceded by core formation by ~1 m.y. (indicated by the two crossed circles). Global silicate (mantle) differentiation in the parent bodies of the eucritic clasts in Vaca Muerta and the aubrite parent body occurred ~2 m.y. after this same process in the HED PB. Metal segregation in the parent bodies of magmatic iron meteorites and pallasites may have occurred over a period of ~5 m.y. (possibly less); assuming that the timing of this process was earliest in the HED PB (inferred to have occurred only ~1–2 m.y. after CAI formation based on Hf-W systematics in the HEDs), this process probably lasted at most until ~4560 Ma. Pd-Ag and Mn-Cr systematics indicate that crystallization of the metallic cores of the extensively differentiated parent bodies of the magmatic irons and pallasites is likely to have occurred over a time interval of ≤10 m.y. (Re-Os systematics suggest that the duration of this process may have been longer, but the large uncertainties in these ages allow for the short time interval inferred from the Pd-Ag and Mn-Cr systematics). This time interval is anchored to the oldest absolute Re-Os age for a magmatic iron meteorite, i.e., 4557 Ma (although, in fact, the uncertainty of ±12 m.y. in this age is comparable to the time interval over which this process is inferred to occur).

On some of these bodies, such incipient melting began as early as ~3–5 m.y. after the formation of the first solids in the solar nebula (i.e., CAIs), but possibly extended to more than ~10 m.y. on others. Other planetesimals underwent more extensive melting and differentiation and acquired the layered structure comprised of a silicate crust and mantle and a metallic core.

Crystallization ages of the basaltic angrites and noncumulate eucrites provide the best means of constraining the timing of crust formation. Chronological evidence for these achondrites suggests that the earliest basalts were emplaced (and formed the crusts of planetesimals) within ~3 m.y. of the formation of CAIs. Primary basalt formation is likely to have continued until ~10 m.y. after CAI formation.

Global silicate differentiation (i.e., the process that established the mantle source reservoirs of the crustal materials) on the parent planetesimals of the angrites and eucrites is likely to have occurred contemporaneously, ~2–3 m.y. after CAI formation. However, this process occurred ~2 m.y. later on the parent body of the HED-like clasts from the Vaca Muerta mesosiderite and on the aubrite parent body. Therefore, the process of global mantle differentiation of planetesimals began early but possibly extended over a time interval of a few million years for different parent bodies.

Metal segregation on different planetesimals, represented by different groups of iron meteorites, is likely to have occurred over a short time interval of ~5 m.y. (and possibly even shorter). In the particular case of the HED parent body, metal segregation may have occurred ~1 m.y. prior to the global mantle differentiation as reflected in the Hf-W systematics of the HEDs. Subsequent core crystallization is thought to have occurred over a somewhat longer period of time. Rhenium-osmium systematics suggest that core crystallization could have continued for ~30 m.y. (although the larger errors in absolute Re-Os ages could accommodate a significantly narrower time interval for this process), but short-lived chronometers (Pd-Ag and Mn-Cr) in most iron meteorites and pallasites indicate that core crystallization on their parent planetesimals spanned a shorter time interval of ≤10 m.y.

The various chronometers (long- and short-lived) applied toward obtaining the timescales discussed here are not always consistent with each other. In some cases, this is to be expected since these chronometers may be dating different events owing to differences in their closure temperatures. However, in other cases, discrepancies may result from disturbance of isotope systematics by secondary events. In the specific case of chronometers based on the short-lived radionuclides (particularly those with half-lives ≤10 m.y.), there is the additional assumption of homogeneous distribution in the meteorite-forming region of the early solar system. This assumption still remains to be rigorously evaluated for several of the short-lived chronometers considered here. Nevertheless, the apparent convergence of the relative timescales indicated by Al-Mg and Mn-Cr with the absolute timescales provided by U-Pb dating suggests that these systems can indeed provide a robust chronology for early solar system events.

The main challenge for future studies seeking to better clarify the reasons for discrepancies between various radiogenic isotope systematics will be to extend chronological investigations to a greater variety of meteoritic materials, such that there are many more instances than there are at present of multiple chronometers being applied to the same objects. Such studies have been inhibited in the past by the limited fractionation in meteoritic materials of various parent-daughter element pairs, such that it has been difficult to apply multiple chronometers to a particular meteorite. These types of investigations should become increasingly feasible with future advancements in analytical capabilities that will allow high-precision isotopic analyses on meteoritic components having relatively low parent/daughter ratios.

Acknowledgments. We are grateful to Q. Yin and an anonymous reviewer for their constructive and thorough reviews that helped to improve the content and to K. McKeegan for the editorial handling of this manuscript. This work was supported NASA grant NAG5-7196 to M.W.

REFERENCES

Allègre C. J., Birck J. L., Fourcade S., and Semet M. P. (1975) Rubidium-87/strontium-87 age of Juvinas basaltic achondrite and early igneous activity in the solar system. *Science, 187,* 436–438.

Amelin Y., Krot A. N., Hutcheon I. D., and Ulyanov A. A. (2002) Lead isotopic ages of chondrules and calcium-aluminum-rich inclusions. *Science, 297,* 1678–1683.

Baker J., Bizzarro M., Wittig M., Connelly J., and Haack H. (2005) Early planetesimal melting from an age of 4.5662 Gyr for differentiated meteorites. *Nature, 436,* 1127–1131.

Becker H. and Walker R. J. (2003a) In search of extant Tc in the early solar system: ^{98}Ru and ^{99}Ru abundances in iron meteorites and chondrites. *Chem. Geol., 196,* 43–56.

Becker H. and Walker R. J. (2003b) Efficient mixing of the solar nebula from uniform Mo isotopic composition of meteorites. *Nature, 425,* 152–155.

Birck J.-L. and Allègre C. J. (1978) Chronology and chemical history of the parent body of the basaltic achondrites studied by the ^{87}Rb-^{87}Sr method. *Earth Planet. Sci. Lett., 39,* 37–51.

Birck J.-L. and Allègre C. J. (1988) Manganese-chromium isotope systematics and development of the early solar system. *Nature, 331,* 579–584.

Bizzarro M., Baker J., and Haack H. (2004) Mg isotope evidence for contemporaneous formation of chondrules and refractory inclusions. *Nature, 431,* 275–278.

Bizzarro M., Baker J., and Haack H. (2005) Timing of crust formation on differentiated asteroids inferred from Al-Mg chronometry (abstract). In *Lunar and Planetary Science XXXVI,* Abstract #1312. Lunar and Planetary Institute, Houston (CD-ROM).

Bogard D. D. and Garrison D. H. (1997) ^{39}Ar-^{40}Ar ages of igneous non-brecciated eucrites (abstract). In *Lunar and Planetary Science XXVIII,* pp. 127–128. Lunar and Planetary Institute, Houston.

Bogard D. D., Nyquist L. E., Johnson P., Wooden J., and Bansal B. (1983) Chronology of Brachina (abstract). *Meteoritics, 18,* 269–270.

Bogdanovski O. and Jagoutz E. (1996) Sm-Nd system in the Divnoe meteorite (abstract). In *Lunar and Planetary Science XXVII,* pp. 129–130. Lunar and Planetary Institute, Houston.

Bogdanovski O., Shukolyukov A., and Lugmair G. W. (1997) ^{53}Mn-^{53}Cr isotope system in the Divnoe meteorite (abstract). *Meteoritics & Planet. Sci., 32,* A16–A17.

Brazzle R. H., Pravdivtseva O., Meshik A. P., and Hohenberg C. M. (1999) Verification and interpretation of the I-Xe chronometer. *Geochim. Cosmochim. Acta, 63,* 739–760.

Cameron A. G. W. (2001) From interstellar gas to the Earth-Moon system. *Meteoritics & Planet. Sci., 36,* 9–22.

Cameron A. G. W., Hoeflich P., Myers P. C., and Clayton D. D. (1995) Massive supernovae, Orion gamma rays, and the formation of the solar system. *Astrophys. J. Lett., 447,* L53.

Carlson R. W. and Hauri E. H. (2001) Extending the ^{107}Pd-^{107}Ag chronometer to low Pd/Ag meteorites with multicollector plasma-ionization mass spectrometry. *Geochim. Cosmochim. Acta, 65,* 1839–1848.

Carlson R. W. and Lugmair G. W. (2000) Timescales of planetesimal formation and differentiation based on extinct and extant radionuclides. In *The Origin of the Earth and Moon* (R. M. Canup and K. Righter, eds.), pp. 25–44. Univ. of Arizona, Tucson.

Clayton R. N. (1993) Oxygen isotopes in meteorites. *Annu. Rev. Earth Planet. Sci., 21,* 115–149.

Clayton R. N. and Mayeda T. K. (1996) Oxygen isotope studies of achondrites. *Geochim. Cosmochim. Acta, 60,* 1999–2017.

Clayton R. N., Grossman L., and Mayeda T. K. (1973) A component of primitive nuclear composition in carbonaceous meteorites. *Science, 182,* 485–488.

Chen J. H. and Wasserburg G. J. (1985) U-Th-Pb isotopic studies on meteorite ALHA 81005 and Ibitira (abstract). In *Lunar and Planetary Science XVI,* pp. 119–120. Lunar and Planetary Institute, Houston.

Chen J. H. and Wasserburg G. J. (1996) Live ^{107}Pd in the early solar system and implications for planetary evolution. In *Earth Processes: Reading the Isotope Code* (A. Basu and S. R. Hart, eds.), pp. 1–20. AGU Monograph 95, American Geophysical Union, Washington, DC.

Chen J. H., Papanastassiou D. A., and Wasserburg G. J. (2002) Re-Os and Pd-Ag systematics in Group IIAB irons and in pallasites. *Geochim. Cosmochim. Acta, 66,* 3793–3810.

Chen J. H., Papanastassiou D. A., and Wasserburg G. J. (2003) Endemic Ru isotopic anomalies in iron meteorites and in Allende (abstract). In *Lunar and Planetary Science XXXIV,* Abstract #1789. Lunar and Planetary Institute, Houston (CD-ROM).

Chen J. H., Papanastassiou D. A., Wasserburg G. J., and Ngo H. H. (2004) Endemic Mo isotopic anomalies in iron and carbonaceous meteorites (abstract). In *Lunar and Planetary Science XXXV,* Abstract #1431. Lunar and Planetary Institute, Houston (CD-ROM).

Cook D. L., Walker R. J., Horan M. F., Wasson J. T., and Morgan J. W. (2004) Pt-Re-Os systematics of group IIAB and IIIAB iron meteorites. *Geochim. Cosmochim. Acta, 68,* 1413–1431.

Crozaz G. and Pellas P. (1983) Where does Brachina come from? (abstract). In *Lunar and Planetary Science XXIV,* pp. 142–143. Lunar and Planetary Institute, Houston.

Dauphas N., Marty B., and Reisberg L. (2002) Molybdenum evidence for inherited planetary scale isotope heterogeneity of the protosolar nebula. *Astrophys. J., 565,* 640–644.

Dauphas N., Davis A. M., Marty B., and Reisberg L. (2004a) The cosmic molybdenum-ruthenium isotope correlation. *Earth Planet Sci. Lett., 226,* 465–475.

Dauphas N., Foley N., Wadhwa M., Davis A. M., Gopel C., Birck J.-L., Janney P. E., and Gallino R. (2004b) Testing the homogeneity of the solar system for iron (54, 56, 57, and 58) and tungsten (182, 183, 184, and 186) isotope abundances (abstract). In *Lunar and Planetary Science XXXV,* Abstract #1498. Lunar and Planetary Institute, Houston (CD-ROM).

Dauphas N., Janney P. E., Mendybaev R. A., Wadhwa M., Richter F. M., Davis A. M., Zuilen M., Hines R., and Foley C. N. (2004c) Chromatographic separation and MC-ICPMS analysis of iron, investigating mass dependent and independent isotope effects. *Anal. Chem., 76,* 5855–5863.

Davis A. M. and Olsen E. J. (1991) Phosphates in pallasite meteorites as probes of mantle processes in small planetary bodies. *Nature, 353,* 637–640.

Desch S. J., Connolly H. C. Jr., and Srinivasan G. (2004) An interstellar origin for the beryllium-10 in calcium-rich, aluminum-rich inclusions. *Astrophys. J., 602,* 528–542.

Eugster O., Michel T., and Niedermann S. (1991) ^{244}Pu-Xe formation and gas retention age, exposure history and terrestrial age of angrites LEW 86010 and LEW 87051. *Geochim. Cosmochim. Acta, 55,* 2957–2964.

Galer S. J. G. and Lugmair G. W. (1996) Lead isotope systematics of noncumulate eucrites (abstract). *Meteoritics & Planet. Sci., 31,* A47–A48.

Gallino R., Busso M., Wasserburg G. J., and Straniero O. (2004) Early solar system radioactivity and AGB stars. *New Astron. Rev., 48,* 133–138.

Ghosh A. and McSween H. Y. Jr. (1998) A thermal model for the differentiation of asteroid 4 Vesta, based on radiogenic heating. *Icarus, 134,* 187–206.

Glavin D. P., Kubny A., Jagoutz E., and Lugmair G. W. (2004) Mn-Cr isotope systematics of the D'Orbigny angrite. *Meteoritics & Planet. Sci., 39,* 693–700.

Goodrich C. A. (1998) Residues from low degrees of melting of a heterogeneous parent body (abstract). *Meteoritics & Planet. Sci., 33,* A60.

Goodrich C. A. and Lugmair G. W. (1995) Stalking the LREE-enriched component in ureilites. *Geochim. Cosmochim. Acta, 59,* 2609–2620.

Goodrich C. A., Hutcheon I. D., and Keil K. (2002) ^{53}Mn-^{53}Cr age of a highly-evolved igneous lithology in polymict ureilite DaG 165 (abstract). *Meteoritics & Planet. Sci., 37,* A54.

Göpel C., Manhès G., and Allègre C. J. (1992) U-Pb study of the Acapulco meteorite (abstract). *Meteoritics, 27,* 226.

Göpel C., Manhès G., and Allègre C. J. (1994) U-Pb systematics of phosphates from equilibrated ordinary chondrites. *Earth Planet. Sci. Lett., 121,* 153–171.

Gounelle M., Shu F. H., Shang H., Glassgold A. E., Rehm E. K., and Lee T. (2001) Extinct radioactivities and protosolar cosmic-rays: Self-shielding and light elements. *Astrophys. J., 548,* 1051–1070.

Gray C. M., Papanastassiou D. A., and Wasserburg G. J. (1973) The identification of early condensates from the solar nebula. *Icarus, 20,* 213–239.

Haack H., Scott E. R. D., Love S. G., Brearley A. J., and McCoy T. J. (1996) Thermal histories of IVA stony-iron meteorites: Evidence for asteroid fragmentation and reaccretion. *Geochim. Cosmochim. Acta, 60,* 3103–3113.

Hohenberg C. M. (1970) Xenon from the Angra dos Reis meteorite. *Geochim. Cosmochim. Acta, 34,* 185–191.

Hohenberg C. M., Bernatowicz T. J., and Podosek F. A. (1991) Comparative xenology of two angrites. *Earth Planet. Sci. Lett., 102,* 167–177.

Horan M. F., Smoliar M. I., and Walker R. J. (1998) ^{182}W and ^{187}Re-^{187}Os systematics of iron meteorites: Chronology for melting, differentiation, and crystallization in asteroids. *Geochim. Cosmochim. Acta, 62,* 545–554.

Hsu W., Huss G. R., and Wasserburg G. J. (1997) Mn-Cr systematics of differentiated meteorites (abstract). In *Lunar and Planetary Science XXVIII,* Abstract #1783. Lunar and Planetary Institute, Houston (CD-ROM).

Hutcheon I. D. and Olsen E. J. (1991) Cr isotopic composition of differentiated meteorites: A search for ^{53}Mn (abstract). In *Lunar and Planetary Science XXII,* pp. 605–606. Lunar and Planetary Institute, Houston.

Hutcheon I. D., Olsen E., Zipfel J., and Wasserburg G. J. (1992) Chromium isotopes in differentiated meteorites: Evidence for ^{53}Mn (abstract). In *Lunar and Planetary Science XXIII,* pp. 565–566. Lunar and Planetary Institute, Houston.

Jacobsen S. B. and Wasserburg G. J. (1984) Sm-Nd isotopic evolution of chondrites and achondrites, II. *Earth Planet. Sci. Lett., 67,* 137–150.

Jagoutz E., Jotter R., Kubny A., Varela M. E, Zartman R., Kurat G. and Lugmair G. W. (2003) Cm?-U-Th-Pb isotopic evolution of the D'Orbigny angrite (abstract). *Meteoritics & Planet. Sci., 38,* Abstract #5048.

Kita N., Ikeda Y., Shimoda H., Morishita Y., and Togashi S. (2003) Timing of basaltic volcanism in ureilite parent body inferred from the ^{26}Al ages of plagioclase-bearing clasts in DaG-319 polymict ureilite (abstract). In *Lunar and Planetary Science XXXIV,* Abstract #1557. Lunar and Planetary Institute, Houston (CD-ROM).

Kleine T., Münker C., Mezger K., and Palme H. (2002) Rapid accretion and early core formation on asteroids and the terrestrial planets from Hf-W chronometry. *Nature, 418,* 952–955.

Kleine T., Mezger K., Münker C., Palme H., and Bischoff A. (2004) ^{182}Hf-^{182}W isotope systematics of chondrites, eucrites, and martian meteorites: Chronology of core formation and early mantle differentiation in Vesta and Mars. *Geochim. Cosmochim. Acta, 68,* 2935–2946.

Kleine T., Mezger K., Palme H., and Scherer E. (2005) Tungsten isotopes provide evidence that core formation in some asteroids predates the accretion of chondrite parent bodies (abstract). In *Lunar and Planetary Science XXXVI,* Abstract #1431. Lunar and Planetary Institute, Houston (CD-ROM).

Kumar A., Gopalan K., and Bhandari N. (1999) ^{147}Sm-^{143}Nd and ^{87}Rb-^{87}Sr ages of the eucrite Piplia Kalan. *Geochim. Cosmochim. Acta, 63,* 3997–4001.

Lee D.-C. and Halliday A. N. (1995) Hafnium-tungsten chronometry and the timing of terrestrial core formation. *Nature, 378,* 771–774.

Lee D.-C. and Halliday A. N. (1996) Hf-W isotopic evidence for rapid accretion and differentiation in the early solar system. *Science, 274,* 1876–1879.

Lee D.-C., Halliday A. N., Singletary S. J., and Grove T. L. (2005) ^{182}Hf-^{182}W chronometry and an early differentiation on the parent body of urelilites (abstract). In *Lunar and Planetary Science XXXVI,* Abstract #1638. Lunar and Planetary Institute, Houston (CD-ROM).

Lee T., Papanastassiou D. A., and Wasserburg G. J. (1976) Demonstration of ^{26}Mg excess in Allende and evidence for ^{26}Al. *Geophys. Res. Lett., 3,* 109–112.

Leya I., Weiler R., and Halliday A. N. (2000) Cosmic-ray production of tungsten isotopes in lunar samples and meteorites and its implications for Hf-W chronochemistry. *Earth Planet. Sci. Lett., 175,* 1–12.

Leya I., Halliday A. N., and Wieler R. (2003a) The predictable collateral consequences of nucleosynthesis by spallation reactions in the early Solar System. *Astrophys. J., 594,* 605–616.

Leya I., Weiler R., and Halliday A. N. (2003b) The influence of cosmic-ray production on extinct radionuclide systems. *Geochim. Cosmochim. Acta, 67,* 529–541.

Loss R. D., Lugmair G. W., Davis A. M., and MacPherson G. J. (1994) Isotopically distinct reservoirs in the solar nebula: Isotopic anomalies in Vigarano meteorite inclusions. *Astrophys. J. Lett., 436,* L193–L196.

Lugmair G. W. (1974) Sm-Nd ages: A new dating method. *Meteoritics, 9,* 369.

Lugmair G. W. and Galer S. J. G. (1992) Age and isotopic relationships among angrites Lewis Cliff 86010 and Angra dos Reis. *Geochim. Cosmochim. Acta, 56,* 1673–1694.

Lugmair G. W. and Marti K. (1977) Sm-Nd-Pu timepieces in the Angra dos Reis meteorite. *Earth Planet. Sci. Lett., 35,* 273–284.

Lugmair G. W. and Shukolyukov A. (1998) Early solar system timescales according to ^{53}Mn-^{53}Cr systematics. *Geochim. Cosmochim. Acta, 62,* 2863–2886.

Lugmair G. W., Scheinin N. B., and Carlson R. W. (1977) Sm-Nd systematics of the Serra de Magé eucrite (abstract). *Meteoritics, 10,* 300–301.

Lugmair G. W., Galer S. J. G., and Carlson R. W. (1991) Isotope systematics of cumulate eucrite EET-87520 (abstract). *Meteoritics, 26,* 368.

Manhès G., Göpel C., and Allègre C. J. (1987) High resolution chronology of the early solar system based on lead isotopes (abstract). *Meteoritics, 22,* 453–454.

Markowski A., Quitté G., Kleine T., and Halliday A. N. (2005) Tungsten isotopic constraints on the formation and evolution of iron meteorite parent bodies (abstract). In *Lunar and Planetary Science XXXVI,* Abstract #1308. Lunar and Planetary Institute, Houston (CD-ROM).

Masarik J. (1997) Contribution of neutron-capture reactions to observed tungsten isotopic ratios. *Earth Planet Sci. Lett., 152,* 181–185.

McCoy T. J., Keil K., Clayton R. N., Mayeda T. K., Bogard D. D., Garrison D. H., Huss G. R., Hutcheon I. D., and Wieler R. (1996) A petrologic, chemical, and isotopic study of Monument Draw and comparison with other acapulcoites: Evidence for formation by incipient partial melting. *Geochim. Cosmochim. Acta, 60,* 2681–2708.

McCoy T. J., Keil K., Clayton R. N., Mayeda T. K., Bogard D. D., Garrison D. H., and Wieler R. (1997a) A petrologic and isotopic study of lodranites: Evidence of early formation as partial melt residues from heterogeneous precursors. *Geochim. Cosmochim. Acta, 61,* 623–637.

McCoy T. J., Keil K., Muenow D. W., and Wilson L. (1997b) Partial melting and melt migration in the acapulcoite-lodranite parent body. *Geochim. Cosmochim. Acta, 61,* 639–650.

McCoy T. J., Mittlefehldt D. W., and Wilson L. (2006) Asteroid differentiation. In *Meteorites in the Early Solar System II* (D. S. Lauretta and H. Y. McSween Jr., eds.), this volume. Univ. of Arizona, Tucson.

Meyer B. S. and Clayton D. D. (2000) Short-lived radioactivities and the birth of the sun. *Space Sci. Rev., 92,* 133–152.

Mittlefehldt D. W., Lindstrom M. M., Bogard D. D., Garrison

D. H., and Field S. W. (1996) Acapulco- and Lodran-like achondrites: Petrology, geochemistry, chronology and origin. *Geochim. Cosmochim. Acta, 60,* 867–882.

Mittlefehldt D. W., McCoy T. J., Goodrich C. A., and Kracher A. (1998) Non-chondritic meteorites from asteroidal bodies. In *Planetary Materials* (J. J. Papike, ed.), pp. 4-1 to 4-195. Reviews in Mineralogy, Vol. 36, Mineralogical Society of America.

Mostefaoui S., Lugmair G. W., and Hoppe P. (2004) In-situ evidence for live iron-60 in the early solar system: A potential heat source for planetary differentiation from a nearby supernova explosion (abstract). In *Lunar and Planetary Science XXXV,* Abstract #1271. Lunar and Planetary Institute, Houston (CD-ROM).

Nehru C. E., Prinz M., Delaney J. S., Dreibus G., Palme H., Spettel B., and Wänke H. (1983) Brachina. A new type of meteorite, not a Chassignite. *Proc. Lunar Planet. Sci. Conf. 13th,* in *J. Geophys. Res., 87,* A365–A373.

Nehru C. E., Prinz M., Weisberg M. K., Ebihara M. E., Clayton R. N., and Mayeda T. K. (1992) Brachinites: A new primitive achondrites group (abstract). *Meteoritics, 27,* 267.

Nehru C. E., Prinz M., Weisberg M. K., Ebihara M. E., Clayton R. N., and Mayeda T. K. (1996) A new brachinites and petrogenesis of the group (abstract). In *Lunar and Planetary Science XXVII,* pp. 943–944. Lunar and Planetary Institute, Houston.

Nichols R. H., Hohenberg C. M., Kehm K., Kim Y., and Marti K. (1994) I-Xe studies of the Acapulco meteorite: Absolute I-Xe ages of individual phosphate grains and the Bjurböle standard. *Geochim. Cosmochim. Acta, 58,* 2553–2561.

Nyquist L. E., Takeda H., Bansal B. M., Shih C.-Y., Wiesmann H., and Wooden J. L. (1986) Rb-Sr and Sm-Nd internal isochron ages of a subophitic basalt clast and a matrix sample from the Y75011 eucrite. *J. Geophys. Res., 91,* 8137–8150.

Nyquist L. E., Bansal B., Wiesmann H., and Shih C.-Y. (1994) Neodymium, strontium and chromium isotopic studies of the LEW 86010 and Angra dos Reis meteorites and the chronology of the angrite parent body. *Meteoritics, 29,* 872–885.

Nyquist L. E., Bogard D., Wiesmann H., Shih C.-Y., Yamaguchi A., and Takeda H. (1996) Early history of the Padvarninkai eucrite. *Meteoritics, 31,* A101–A102.

Nyquist L., Bogard D., Takeda H., Bansal B., Wiesmann H., and Shih C.-Y. (1997a) Crystallization, recrystallization, and impact-metamorphic ages of eucrites Y792510 and Y791186. *Geochim. Cosmochim. Acta, 61,* 2119–2138.

Nyquist L. E., Wiesmann H., Reese Y., Shih C.-Y., and Borg L. E. (1997b) Samarium-neodymium age and manganese-chromite systematics of eucrite Elephant Moraine 90020. *Meteoritics, 32,* A101–A102.

Nyquist L. E., Shih C.-Y., Wiesmann H., and Mikouchi T. (2003a) Fossil ^{26}Al and ^{53}Mn in D'Orbigny and Sahara 99555 and the timescale for angrite magmatism (abstract). In *Lunar and Planetary Science XXXIV,* Abstract #1388. Lunar and Planetary Institute, Houston (CD-ROM).

Nyquist L. E., Reese Y., Wiesmann H., Shih C.-Y., and Takeda H. (2003b) Fossil ^{26}Al and ^{53}Mn in the Asuka 881394 eucrite: Evidence of the earliest crust on asteroid 4 Vesta. *Earth Planet. Sci. Lett., 214,* 11–25.

Ott U., Begemann F., and Löhr H. P. (1987) Noble gases in ALH 84025: Like Brachina, unlike Chassigny (abstract). *Meteoritics, 22,* 476–477.

Papanastassiou D. A. and Wasserburg G. J. (1969) Initial strontium isotopic abundances and the resolution of small time differences in the formation of planetary objects. *Earth Planet. Sci. Lett., 5,* 361–376.

Papanastassiou D. A., Chen J. H., and Wasserburg G. J. (2004) More on Ru endemic isotope anomalies in meteorites (abstract). In *Lunar and Planetary Science XXXV,* Abstract #1828. Lunar and Planetary Institute, Houston (CD-ROM).

Pellas P., Fieni C., Trieloff M., and Jessberger E. K. (1997) The cooling history of the Acapulco meteorite as recorded by the ^{244}Pu and ^{40}Ar-^{39}Ar chronometers. *Geochim. Cosmochim. Acta, 61,* 3477–3501.

Petaev M. I., Baruskova L. D., Lipschutz M. E., Wang M.-S., Arsikan A. A., Clayton R. N., and Mayeda T. K. (1994) The Divnoe meteorite: Petrology, chemistry, oxygen isotopes and origin. *Meteoritics, 29,* 182–199.

Podosek F. A., Zinner E. K., MacPherson G. J., Lundberg L. L., Brannon J. C. and Fahey A. J. (1991) Correlated study of initial ^{87}Sr/^{86}Sr and Al/Mg isotopic systematics and petrologic properties in a suite of refractory inclusions from the Allende meteorite. *Geochim. Cosmochim. Acta, 55,* 1083–1110.

Pravdivtseva O. V., Hohenberg C. M., and Meshik A. P. (2005) I-Xe dating: The time line of chondrule formation and metamorphism in LL chondrites (abstract). In *Lunar and Planetary Science XXXVI,* Abstract #2354. Lunar and Planetary Institute, Houston (CD-ROM).

Prinzhofer A., Papanastassiou D. A., and Wasserburg G. J. (1992) Samarium-neodymium evolution of meteorites. *Geochim. Cosmochim. Acta, 56,* 797–815.

Quitté G. and Birck J. L. (2004) Tungsten isotopes in eucrites revisited and the initial ^{182}Hf/^{180}Hf of the solar system based on iron meteorite data. *Earth Planet. Sci. Lett., 219,* 201–207.

Quitté G., Birck J. L., and Allègre C. J. (2000) ^{182}Hf-^{182}W systematics in eucrites: The puzzle of iron segregation in the early solar system. *Earth Planet. Sci. Lett. 184,* 83–94.

Quitté G., Latkoczy C., Halliday A. N., Schönbächler M., and Günther D. (2005) Iron-60 in the eucrite parent body and the initial ^{60}Fe/^{56}Fe of the solar system (abstract). In *Lunar and Planetary Science XXXVI,* Abstract #1827. Lunar and Planetary Institute, Houston (CD-ROM).

Russell S. S., Hartmann L., Cuzzi J., Krot A. N., Gounelle M., and Weidenschilling S. (2006) Timescales of the solar protoplanetary disk. In *Meteorites in the Early Solar System II* (D. S. Lauretta and H. Y. McSween Jr., eds.), this volume. Univ. of Arizona, Tucson.

Scherstén A., Elliott T., Russell S., and Hawkesworth C. J. (2004) W isotope chronology of asteroidal core formation (abstract). *Geochim. Cosmochim. Acta, 68,* A729.

Shen J. J., Papanastassiou D. A., and Wasserburg G. J. (1996) Precise Re-Os determinations and systematics of iron meteorites. *Geochim. Cosmochim. Acta, 60,* 2887–2900.

Shu F. H., Shang H., and Lee T. (1996) Toward an astrophysical theory of chondrites. *Science, 271,* 1545–1552.

Shukolyukov A. and Lugmair G. W. (1993a) Live iron-60 in the early solar system. *Science, 259,* 1138–1142.

Shukolyukov A. and Lugmair G. W. (1993b) ^{60}Fe in eucrites. *Earth Planet. Sci. Lett., 119,* 159–166.

Shukolyukov A. and Lugmair G. W. (2004) Manganese-chromium isotope systematics of enstatite chondrites. *Geochim. Cosmochim. Acta, 68,* 2875–2888.

Shukolyukov A., Lugmair G. W., and Bogdanovski O. (2003) Manganese-chromium isotope systematics of Ivuna, Kainsaz, and other carbonaceous chondrites (abstract). In *Lunar and Planetary Science XXXIV,* Abstract #1279. Lunar and Planetary Institute, Houston (CD-ROM).

Smoliar M. I. (1993) A survey of Rb-Sr systematics of eucrites. *Meteoritics, 28,* 105–113.

Smoliar M. I., Walker R. J., and Morgan J. W. (1996) Re-Os ages of group IIA, IIIA, IVA, and IVB iron meteorites. *Science, 271,* 1099–1102.

Spivak-Birndorf L., Wadhwa M., and Janney P. E. (2005) ^{26}Al-^{26}Mg chronology of the D'Orbigny and Sahara 99555 angrites (abstract). *Meteoritics & Planet. Sci., 40,* A145.

Srinivasan G. (2002) ^{26}Al-^{26}Mg systematics in eucrites A881394, A87122 and Vissannapeta (abstract). *Meteoritics & Planet. Sci., 37,* A135.

Srinivasan G., Goswami J. N., and Bhandari N. (1999) ^{26}Al in eucrite Piplia Kalan: Plausible heat source and formation chronology. *Science, 284,* 1348–1350.

Sugiura N. and Hoshino H. (2003) Mn-Cr chronology of five IIIAB iron meteorites. *Meteoritics & Planet. Sci., 38,* 117–143.

Swindle T. D., Kring D. A., Burkland M. K., Hill D. H., and Boynton W. V. (1998) Noble gases, bulk chemistry, and petrography of olivine-rich achondrites Eagles Nest and Lewis Cliff 88763: Comparison to brachinites. *Meteoritics & Planet. Sci., 33,* 31–48.

Tachibana S. and Huss G. R. (2003) The initial abundance of ^{60}Fe in the solar system. *Astrophys. J. Lett., 588,* L41–L44.

Tatsumoto M., Knight R. J., and Allègre C. J. (1973) Time differences in the formation of meteorites as determined from the ratio of lead-207 to lead-206. *Science, 180,* 1279–1283.

Tera F., Carlson R. W., and Boctor N. Z. (1997) Radiometric ages of basaltic achondrites and their relation to the early history of the solar system. *Geochim. Cosmochim. Acta, 61,* 1713–1731.

Tonui E. K., Ngo H. H., and Papanastassiou D. A. (2003) Rb-Sr and Sm-Nd study of the D'Orbigny angrite (abstract). In *Lunar and Planetary Science XXXIV,* Abstract #1812. Lunar and Planetary Institute, Houston (CD-ROM).

Torigoye-Kita N., Misawa K., and Tatsumoto M. (1995a) U-Th-Pb and Sm-Nd isotopic systematics of the Goalpara ureilites: Resolution of terrestrial contamination. *Geochim. Cosmochim. Acta, 59,* 381–390.

Torigoye-Kita N., Tatsumoto M., Meeker G. P., and Yanai K. (1995b) The 4.56 Ga age of the MET 78008 ureilite. *Geochim. Cosmochim. Acta, 59,* 2319–2329.

Trinquier A., Birck J. L., and Allègre C. J. (2005a) ^{54}Cr anomalies in the solar system: Their extent and origin (abstract). In *Lunar and Planetary Science XXXVI,* Abstract #1259. Lunar and Planetary Institute, Houston (CD-ROM).

Trinquier A., Birck J. L., and Allègre C. J. (2005b) Reevaluation of the ^{53}Mn-^{53}Cr systematic in the basaltic achondrites (abstract). In *Lunar and Planetary Science XXXVI,* Abstract #1946. Lunar and Planetary Institute, Houston (CD-ROM).

Unruh D. M., Nakamura N., and Tatsumoto M. (1977) History of the Pasamonte achondrite: Relative susceptibility of the Sm-Nd, Rb-Sr, and U-Pb systems to metamorphic events. *Earth Planet. Sci. Lett., 37,* 1–12.

Vanhala H. A. T. and Boss A. P. (2002) Injection of radioactivities into the forming solar system. *Astrophys. J., 575,* 1144–1150.

Wadhwa M. and Lugmair G. W. (1995) Sm-Nd systematics of the eucrite Chervony Kut (abstract). In *Lunar and Planetary Science XXVI,* pp. 1453–1454. Lunar and Planetary Institute, Houston.

Wadhwa M. and Lugmair G. W. (1996) Age of the eucrite "Caldera" from convergence of long-lived and short-lived chronometers. *Geochim. Cosmochim. Acta, 60,* 4889–4893.

Wadhwa M., Shukolyukov A., and Lugmair G. W. (1998) ^{53}Mn-^{53}Cr systematics in Brachina: A record of one of the earliest phases of igneous activity on an asteroid (abstract). *Meteoritics & Planet. Sci., 29,* 1480.

Wadhwa M., Shukolyukov A., Davis A. M., Lugmair G. W., and Mittlefehldt D. W. (2003) Differentiation history of the mesosiderite parent body: Constraints from trace elements and manganese-chromium isotopic systematics of Vaca Muerta silicate clasts. *Geochim. Cosmochim. Acta, 67,* 5047–5069.

Wadhwa M., Amelin Y., Bogdanovski O., Shukolyukov A., Lugmair G. W. and Janney P. E. (2005) High precision relative and absolute ages for Asuka 881394, a unique and ancient basalt (abstract). In *Lunar and Planetary Science XXXVI,* Abstract #2126. Lunar and Planetary Institute, Houston (CD-ROM).

Warren P. H. and Kallemeyn G. W. (1989) Allan Hills 84025: The second brachinites, far more differentiated than Brachina, and an ultramafic achondritic clast from the L chondrite Yamato 75097. *Proc. Lunar Planet. Sci. Conf. 19th,* pp. 475–486.

Wasserburg G. J., Tera F., Papanastassiou D. A., and Huneke J. C. (1977) Isotopic and chemical investigations on Angra dos Reis. *Earth Planet. Sci. Lett., 35,* 294–316.

Weigel A., Eugster O., Koeberl C., and Krähenbühl U. (1996) Primitive differentiated achondrite Divnoe and its relationship to brachinites (abstract). In *Lunar and Planetary Science XXVII,* pp. 1403–1404. Lunar and Planetary Institute, Houston.

Wiechert U. H., Halliday A. N., Palme H., and Rumble D. (2004) Oxygen isotope evidence for rapid mixing of the HED meteorite parent body. *Earth Planet. Sci. Lett., 221,* 373–382.

Yamaguchi A., Taylor G. J., Keil K., Floss C., Crozaz G., Nyquist L. E., Bogard D. D., Garrison D., Reese Y., Wiesman H. and Shih C.-Y. (2001) Post-crystallization reheating and partial melting of eucrite EET 90020 by impact into the hot crust of asteroid 4 Vesta ~4.50 Ga ago. *Geochim. Cosmochim. Acta, 65,* 3577–3599.

Yin Q. and Jacobsen S. B. (2003) The initial ^{182}W/^{183}W and ^{182}Hf/^{180}Hf of the solar system and a consistent chronology with Pb-Pb ages (abstract). In *Lunar and Planetary Science XXXIV,* Abstract #1857. Lunar and Planetary Institute, Houston (CD-ROM).

Yin Q., Jacobsen S. B., and Yamashita K. (2002a) Diverse supernova sources of pre-solar material inferred from molybdenum isotopes in meteorites. *Nature, 415,* 881–883.

Yin Q., Jacobsen S. B., Yamashita K., Blichert-Toft J., Télouk P., and Albarède A. (2002b) A short time scale for terrestrial planet formation from Hf-W chronometry of meteorites. *Nature, 418,* 949–952.

Young E., Simon J., Galy A., Russell S. S., Tonui E., and Lovera O. (2005) Supra-canonical ^{26}Al/^{27}Al and the residence time of CAIs in the solar protoplanetary disk. *Science, 308,* 223–227.

Zinner E. (1996) Presolar material in meteorites: An overview. In *Astrophysical Implications of the Laboratory Study of Presolar Materials* (T. J. Bernatowicz and E. K. Zinner, eds.), pp. 3–26. AIP Conference Proceedings 402, American Institute of Physics, New York.

Zipfel J., Shukolyukov A., and Lugmair G. W. (1996) Manganese-chromium systematics in the Acapulco meteorite (abstract). *Meteoritics & Planet. Sci., 31,* A160.

Zhu X. K., Guo Y., O'Nions R. K., Young E. D., and Ash R. D. (2001) Isotopic homogeneity of iron in the early solar nebula. *Nature, 412,* 311–313.

Asteroid Differentiation

Timothy J. McCoy
Smithsonian Institution

David W. Mittlefehldt
NASA Johnson Space Center

Lionel Wilson
University of Hawai'i at Manoa

Differentiation of silicate bodies was widespread in the inner solar system, producing an array of igneous meteorites. While thc precursor material was certainly chondritic, direct links between chondrites and achondrites are few and tenuous. Heating by short-lived radionuclides (^{26}Al and ^{60}Fe) initiated melting. Formation of a basaltic crust would occur soon after the onset of melting, as would migration of the cotectic Fe,Ni-FeS component, perhaps forming a S-rich core. While some asteroids appear to be arrested at this point, many continued to heat up, with efficient segregation of metal to form large cores. Extensive melting of the silicate mantle occurred, perhaps reaching 40–75%. In the largest asteroids, the original, basaltic crust may have been resorbed into the mantle, producing a magma ocean. The crystallization of that magma ocean, whether by equilibrium or fractional crystallization, could have produced the layered structure typical of large differentiated asteroids like 4 Vesta. Spacecraft exploration of 4 Vesta promises to further extend our knowledge of asteroid differentiation.

1. INTRODUCTION

The transformation of rocky bodies from an agglomerate of nebular particles to fully differentiated, layered worlds consisting of core, mantle, and crust is, without question, one of the most fundamental processes to have occurred during the birth of our solar system. The products of this transformation abound. The basaltic eucrites, iron meteorites, and pallasites all originated on asteroidal bodies that experienced extensive differentiation. Since the publication of *Meteorites and the Early Solar System* (*Kerridge and Matthews,* 1988), primitive achondrites — those meteorites that experienced only partial differentiation — have also been widely recognized, marking one of the major advances in our understanding of asteroidal differentiation. While primitive chondrites may dominate the flux of material encountering Earth, the majority of distinct asteroids that have delivered meteorites to Earth are differentiated (*Burbine et al.,* 2002). All the terrestrial planets and our own Moon experienced igneous differentiation. In short, differentiation appears to have been widespread in the inner solar system.

While planetary differentiation in general appears to have been a dominant process in the inner solar system, the differentiation of asteroids in particular forms an end member in understanding large-scale igneous processes. In comparison to Earth, asteroidal differentiation occurred at much lower gravity (radii $\leq$ 500 km; g < 0.03 terrestrial); was driven by early, intense heat sources; and operated over a much wider range of f_{O_2} conditions [~IW+2 for angrites (*Jurewicz et al.,* 1993) to IW-5 for aubrites (*Keil,* 1989)] and over a broader range of states of hydration. Thus, studies of asteroidal differentiation provide a natural complement to studies of igneous systems under terrestrial conditions.

Despite the obvious interest in asteroid differentiation, a complete understanding of this process has been and remains elusive, in large part owing to the remoteness in time and space of the locations in which this event occurred. While Earth experienced planetwide melting and differentiation, subsequent geologic processes have largely obliterated the early record of its differentiation. Large parts of the differentiated Earth remain inaccessible, and thus analogous materials to asteroidal cores are not available. Finally, all meteorites are samples without geologic context and enormous effort has been expended to link meteorites from common asteroidal parents. Even now, the first detailed examination of a differentiated asteroid and its components remains years in the future.

2. APPROACH

In this chapter, we address the differentiation of an asteroidal body from its primitive, precursor material through heating; the onset of differentiation; the separation of a core, mantle, and crust; to the solidification of the silicate portion of the asteroid. We briefly discuss possible heat sources for differentiation. Core crystallization is addressed by *Chabot and Haack* (2006). As we demonstrate, this is an enormously complex process both physically and chemically. Furthermore, myriad variations on the theme of asteroid differentiation must have occurred to produce the variety of meteorites observed in our collections. Most notable among these variations are differences in the chemical composition of the starting material, the style of melting that might be engendered by different heat sources, the role of f_{O_2} during igneous processes, and the random influence of impacts occurring during igneous differentiation [as opposed to those occurring later and forming breccias; these are addressed by *Bischoff et al.* (2006)]. For our purposes, we outline differentiation of a common precursor material and attempt to trace differentiation of an asteroid throughout that process from the onset of melting through solidification. We discuss excursions from this path only as they are warranted, particularly where differences in condition (e.g., f_{O_2}) produce dramatically different results. Finally, we emphasize the *process* of, rather than the *products* of, differentiation. Specific meteorite types are presented as evidence supporting the broad conceptual framework we outline here, but we do not discuss specific meteorite types and their origin in a systematic manner. The reader is referred to the works of *Mittlefehldt et al.* (1998), *Mittlefehldt* (2004), and *Haack and McCoy* (2004) for treatments of individual meteorite types.

3. STARTING MATERIAL

The nature of the precursor material for a differentiated asteroid can have a profound influence on the process of differentiation. It is therefore disconcerting that, for most differentiated asteroids, solid links to a precursor assemblage simply do not exist. It is widely assumed that differentiated asteroids began their existence as chondritic materials. This in large part is owed to the complete lack of any chemically and petrologically primitive materials that fall outside the broad scope of the term "chondritic" (*Brearley and Jones,* 1998; *Scott and Krot,* 2003) and, to a lesser extent, limited evidence linking chondrites and achondrites. Among the latter is the presence of relict chondrules in some acapulcoites (*McCoy et al.,* 1996), winonaites (*Benedix et al.,* 2003), and silicate-bearing IIE irons (*Bild and Wasson,* 1977), and similarities in O-isotopic composition, mineralogy, and mineral compositions between silicate-bearing IIE irons and H chondrites (*Bogard et al.,* 2000) and enstatite chondrites and aubrites (*Keil,* 1989). Even within these groups, it remains highly disputed whether we have any achondrite groups that can be linked directly to any specific group of chondrites. *Meibom and Clark* (1999) addressed the issue of the abundance of ordinary chondritic composition parent bodies within the subset of asteroids sampled as meteorites. In their assessment of likely precursors for various groups of achondrites, they found that carbonaceous chondrites appeared to be the dominant precursor material. There are, of course, important exceptions, such as the silicate-bearing IIE and IVA irons and the aubrites.

For the purposes of this chapter, we will trace the evolution of an ordinary chondrite protolith through melting and differentiation. There are several reasons for following this course. First, ordinary chondrites are well characterized both chemically and mineralogically. Normative mineralogies provide a reasonable approximation of the modal mineralogy for ordinary chondrites (e.g., *McSween et al.,* 1991) and allow comparison across groups with differing oxidation states. [The normative calculation for ordinary chondrites utilizes wet-chemistry analyses of bulk samples. A standard CIPW norm is calculated for the metal- and sulfide-free portion of the analysis, and then adding the metal and sulfide components and renormalizing to 100%; see *McSween et al.* (1991) for full description of the method.] Ordinary chondrites can be described as subequal mixtures of olivine and pyroxene, with lesser amounts of feldspar, metal, troilite, phosphates, and chromite. Carbonaceous chondrites are more olivine normative, while enstatite chondrites lack olivine in their normative compositions. Second, the melting of ordinary chondrites can be reasonably well understood both theoretically [through application of the Fe-FeS and olivine-anorthite-silica phase diagrams (e.g., *Stolper,* 1977)] and experimentally [through partial melting of ordinary chondrites (e.g., *Kushiro and Mysen,* 1979; *Takahashi,* 1983; *Jurewicz et al.,* 1995; *Feldstein et al.,* 2001)]. This approach allows reasonably robust conclusions about the temperatures for onset of melting, the composition of the melts, and, by extension, the physical properties (e.g., density, viscosity) of those melts. Together, this information proves invaluable in tracing the differentiation of an asteroid-sized body.

4. HEAT SOURCES

The differentiation of an asteroid requires a potent heat source, capable of producing high temperatures [in some cases, in excess of 1773 K (*Taylor et al.,* 1993)] in bodies with high surface-to-volume ratios. The heat source must have also operated quite early in the history of the solar system. Hafnium-182/tungsten-182 and ^{187}Re-^{187}Os systematics in iron meteorites (e.g., *Horan et al.,* 1998) suggest core formation within 5 m.y. of the formation of the first solids in the solar system. Early differentiation is also supported by the presence of excess ^{26}Mg from the decay of extinct ^{26}Al (half-life of 0.73 Ma) in the eucrites Piplia Kalan (*Srinivasan et al.,* 1999) and Asuka 881394 (*Nyquist et al.,* 2001). The identity of the heat source (or sources) remains unresolved. Several heat sources can be ruled out. Heating was not the result of the decay of long-lived radionuclides (e.g., U, K, Th) that contributed to the melting of Earth and other terrestrial planets. Similarly, accretional heating was probably minimal. *Haack and McCoy* (2004) argue that accretional heating would raise the temperature only 6 K on a body 100 km in diameter.

Impacts may have contributed to the heating of asteroids, either through thermal metamorphism or limited melting. Some authors have championed complete melting of asteroids by impact (e.g., *Lodders et al.,* 1993). However, *Keil et al.* (1997) effectively marshal the case against impacts as a global heat source and this view is now widely accepted. Perhaps most compelling among their arguments is that impacts large enough to produce postshock temperature increases of hundreds of degrees would shatter the asteroidal target. There is no question that impact played a major, perhaps even dominant, role in shaping individual asteroids and the asteroid belt as it exists today. It is also possible that these impacts produced local heating. *Rubin* (2004) documented postshock annealing in ordinary chondrites that occurred hundreds of millions of years after the birth of the solar system and must have resulted from impact heating. He argued that localized heating of the walls and floors of craters in rubble-pile asteroids might produce sufficient heat for thermal metamorphism. Impact as a heat source for local melting and differentiation has been championed in the case of the silicate-bearing IAB irons by *Wasson et al.* (1987) and *Choi et al.* (1995). While this idea continues to stimulate debate, it is clear that impact did not produce complete differentiation of asteroids and the wide range of achondrites observed in our collections.

Electrical conduction heating by the T Tauri solar wind from the pre-main-sequence Sun (e.g., *Sonett et al.,* 1970) and short-lived radioactive isotopes remain the most likely heat sources for melting of an asteroidal body. The latter were discussed extensively by *McSween et al.* (2002) in their review of asteroid thermal histories. Among the short-lived radioactive isotopes, ^{26}Al has gained increasing acceptance since the discovery of excess ^{26}Mg in Piplia Kalan (*Srinivasan et al.,* 1999). In feldspar-bearing chondritic materials, Al is evenly distributed and could act as an effective heat source. However, once melting begins, Al is sequestered into early partial melts and could effectively migrate out of the core and mantle, leaving these parts of the parent body with essentially no heat source. *Shukolyukov and Lugmair* (1992, 1996) have argued that ^{60}Fe might play a role as a heat source, although *McSween et al.* (2002) suggest that its concentration would be too low to effectively heat the mantle, producing a reverse thermal gradient. However, more recent analyses of chondritic troilite imply a higher ^{60}Fe/^{56}Fe ratio for the early solar system than previously thought (*Mostefaoui et al.,* 2003; *Tachibana and Huss,* 2003), suggesting that metallic cores might contain a potent heat source. It seems likely that a range of short-lived radioactive isotopes might play a role in heating asteroids and their interplay is likely complex.

5. ONSET OF MELTING

Heating and, ultimately, melting of asteroids is inherently a high-temperature, low-pressure process. Even within the centers of the largest asteroids, static pressures do not exceed 2 kbar. If the protolith is an ordinary chondrite, melting also occurs under very dry conditions. Thus, the first melts generated do not occur until ~1223 K and the first silicate melts at 1323 K. This is in sharp contrast to Earth, where melting under even shallow crustal conditions can occur at ~1073 K for mafic rocks in the presence of water.

The earliest melt produced from an ordinary chondrite protolith occurs at the Fe,Ni-FeS eutectic at ~1223 K and contains ~85 wt% FeS (*Kullerud,* 1963). There are numerous experimental, theoretical, and petrologic data to constrain the processes that occur once this temperature is reached and melting begins. A considerable number of experimental studies (e.g., *Herpfer and Larimer,* 1993; *Ballhaus and Ellis,* 1996; *Minarik et al.,* 1996; *Shannon and Agee,* 1996, 1998; *Gaetani and Grove,* 1999; *Rushmer et al.,* 2000; *Bruhn et al.,* 2000) discuss melt migration and core formation in olivine-dominated systems. Although most of these studies were conducted at pressures considerably greater than those occurring during asteroidal melting, taken as a whole, they suggest that only anion-dominated metallic melts (mostly S in the experiments) exhibit dihedral angles less than 60° and form interconnected networks (Fig. 1). Thus, even the FeS-rich eutectic melt (anion/cation ~0.8) would not migrate to form a core under static conditions.

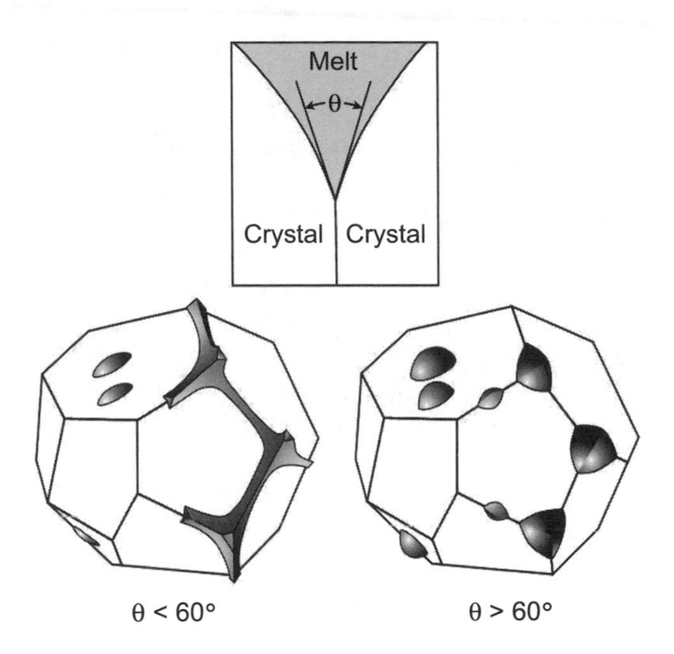

Fig. 1. Illustration of the dihedral angle θ and a depiction of two endmember microstructures for static systems. At $\theta < 60°$, an interconnected network will form and melt migration can occur. If $\theta > 60°$, melt forms isolated pockets. In experimental systems, interconnectedness occurs in static systems only if they are anion-rich.

During Fe,Ni-FeS eutectic melting, dynamic conditions may actually dominate in the mobility of these melts. *Keil and Wilson* (1993) suggested that overpressure created by melting at the eutectic might create veins and these veins might actually rise within the parent body due to their incorporating bubbles of preexisting gases, thus overcoming the gravitational influence on these high-density melts. The effect of this melting and melt migration is clearly seen both petrographically and geochemically in several groups of meteorites, most notably the acapulcoites and lodranites. *McCoy et al.* (1996) documented micrometer-sized veinlets occurring ubiquitously in all acapulcoites and rarer, centimeter-sized veins in the Monument Draw acapulcoite (Fig. 2).

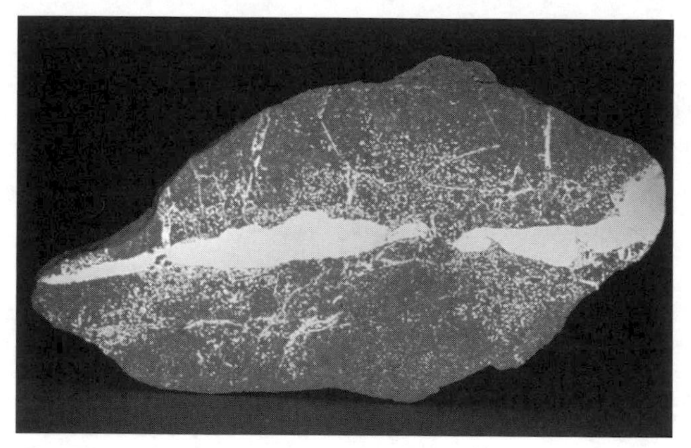

Fig. 2. Centimeter-scale veins of Fe,Ni metal and troilite in the acapulcoite Monument Draw formed during the first melting of an asteroid. In the absence of silicate melting, these veins were unable to migrate substantial distances and thus would not have contributed to core formation. Length of specimen is ~7.5 cm. (Smithsonian specimen USNM 7050; photo by C. Clark.)

These veins cross-cut silicates, suggesting that the silicates were solid at the time of Fe,Ni-FeS melt intrusion and remained solid throughout the history of the rocks. The veins migrated reasonably short distances (hundreds of micrometers to centimeters), leaving essentially unfractionated compositions. Similar veins and veinlets are observed in winonaites (*Benedix et al.,* 1998). Selenium/cobalt ratios (a proxy for sulfide/metal) in acapulcoites are essentially unfractionated relative to H-group ordinary chondrites, suggesting that the sulfide-rich early partial melt was not separated from the metal-rich residue (*Mittlefehldt et al.,* 1996) (Fig. 3).

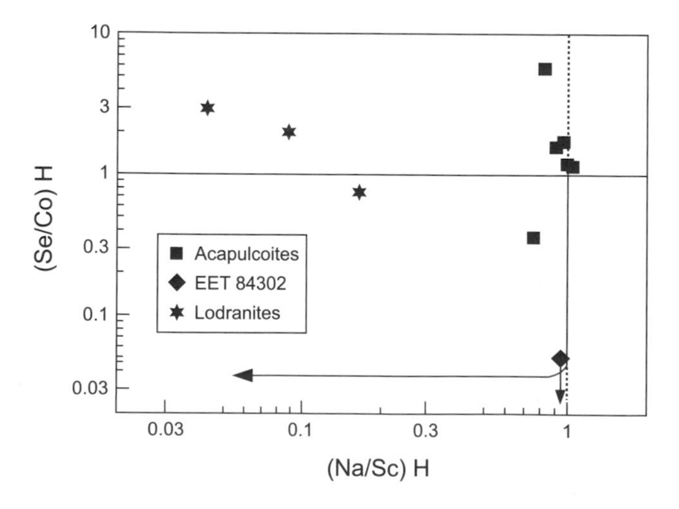

Fig. 3. H-chondrite normalized Se/Co vs. Na/Sc for acapulcoites, lodranites, and the intermediate EET 84302 after *Mittlefehldt et al.* (1996). The ratios serve as proxies for troilite/metal and plagioclase/pyroxene respectively. Partial melting should follow the arrowed curve. While lodranites have clearly lost the plagioclase-rich silicate partial melt, redistribution of the S-rich Fe,Ni-FeS cotectic melt was uneven, with melt probably being both removed from and introduced into the lodranites.

Silicate partial melting of an ordinary chondrite begins at between ~1323 and 1373 K (*Kushiro and Mysen,* 1979; *Takahashi,* 1983). This first melt is broadly basaltic in composition (dominated by pyroxene and plagioclase). In the Fo-An-SiO_2 system, the first melt is at the peritectic and contains ~55% plagioclase (*Morse,* 1980), although the phase relations in an iron-bearing system with albitic feldspar are incompletely understood. Despite these uncertainties, removal of a plagioclase-rich melt (basalt) from an ordinary chondrite protolith that contains ~10 wt% plagioclase (*McSween et al.,* 1991) would produce a residue substantially depleted in feldspar, provided that the melt separates efficiently. This must have occurred in the case of the lodranites, as evidenced by Fig. 3. Ratios of Na/Sc (a proxy for plagioclase/pyroxene) are substantially lower in lodranites relative to acapulcoites or ordinary chondrites, reflecting efficient removal of feldspar (in the form of basaltic melt). It is interesting to note that basalt-depleted lodranites exhibit chondritic to slightly enriched Se/Co ratios, suggesting that some melt of the Fe-Ni-S system might have even been introduced from another region of the parent body. Some intermediate members of the group, such as EET 84302, exhibit petrologic evidence for silicate melting, but have chondritic Na/Sc ratios and highly depleted Se/Co ratios (*Mittlefehldt et al.,* 1996). Thus, it appears that the onset of silicate partial melting first provided the conduit for efficient melt migration and set the stage for the formation of a basaltic crust and metallic core.

6. CRUST FORMATION

There are two linked requirements for the formation of a compositionally distinct crust on an asteroid: a sufficiently large degree of partial melting of the region that becomes the mantle, and efficient spatial separation of melt from the residual mantle. The relevant issues are (1) the nature of the mineral population, which determines which grain-grain contacts are the sites of first melting and also the typical separation between such sites; (2) the contact angles between melt and unmelted solid for each mineral species present, which determine the minimum degree of melting that must be attained before a widespread continuous fluid network exists; (3) the difference between the density of the fluid network and the bulk density of the unmelted matrix, which determines the buoyancy force (which could be positive or negative) acting on the melt; (4) the volume fraction of empty pore space between mineral grains; (5) the nature of, and initial pressure in, any gas species occupying the pore space; and (6) the uniformity of the ambient stress regime in the region of melt formation.

Numerical simulations imply that a few percent partial melting must occur before interconnected networks are formed (*Maaloe,* 2003). Initial basaltic melts have a somewhat smaller density and hence larger volume than the solids from which they form and so will invade available pore space, compressing whatever gas species may be present and causing a pressure increase. Buoyant melt will percolate upward along grain boundaries due to gravity alone, especially if the presence of a significant shear stress due to inhomogeneity in the asteroid structure encourages grain compaction. However, at low pore space volume fractions the large pressure rise in trapped gas may become a significant factor in fracturing the contacts between mineral grains and facilitating formation of buoyant liquid-filled veins that are longer than the typical mineral grain size of the matrix (*Muenow et al.,* 1992). As soon as veins become able to expand by fracturing their tips they will grow and may intersect other veins. A vein network will form, and larger veins will preferentially drain melt from smaller veins (*Sleep,* 1988) and grow at their expense (*Keil and Wilson,* 1993), eventually forming structures large enough to qualify as dikes.

The speed at which melt can move through a fracture, whether it be classified as a vein or a dike, depends on the inherent driving forces of buoyancy and excess pressure and on the shortest dimension of the fracture. Once a sufficiently wide dike forms, it will be able to propagate toward the surface and very efficiently drain much of the melt from the vein network feeding it. Thus two competing mechanisms operate in the mantles of asteroids (and larger bodies): a slow, steady percolation of melt along grain boundaries, and an intermittent but fast drainage of most of the available melt from a laterally and vertically extensive region into a dike that can penetrate a large fraction of the way from the melt source zone to the asteroid surface. There is a very large difference between these two mechanisms in regard to the time that a given batch of melt spends in contact with rocks at depths shallower than its source (*Kelemen et al.,* 1997; *Niu,* 1997), and so the relative importance of these two mechanisms is beginning to be understood from studies of the partitioning of rare Earth elements in, for example, the residual asteroid mantle rocks represented by the ureilites (*Goodrich and Lugmair,* 1995; *Warren and Kallemeyn,* 1992).

The timescale τ for the heating of asteroids due to the decay of ^{26}Al is on the order of 1 m.y. The distance λ that a thermal wave can travel in silicate rocks with thermal diffusivities κ typically on the order of 10^{-6} m^2 s^{-1} is $\lambda = (\kappa\tau)^{1/2} = 5.5$ km. Thus, the ability of heat to be conducted radially within an asteroid during the time for which the most plausible heat source is available is limited to a few kilometers, and so in asteroids with radii larger than a few tens of kilometers and ^{26}Al abundances sufficient to reach the solidus temperature, the melting process must occur simultaneously over almost the entire range of depths, as is found in detailed numerical models of asteroid evolution (*Grimm and McSween,* 1989; *Ghosh and McSween,* 1998). By the same token, there will be a region occupying the outermost ~5 km of all asteroids that is completely unmelted because it can always lose heat to the surface fast enough to compensate for heat flow from the interior (*Ghosh and McSween,* 1998).

The initial structure of this near-surface region at the end of asteroid accretion is critical. If it has a high porosity and low bulk density, melts rising from the mantle will not be able to penetrate it due to buoyancy alone unless they are very volatile-rich and contain a sufficiently large proportion of gas bubbles. Instead they will stall to form intrusions at a depth where they are neutrally buoyant. On large bodies like Earth, Mars, and Venus, tectonic control of major locations of mantle melting ensures that magma injection events are repeated frequently in essentially the same place. Previous magma batches then remain molten until subsequent ones arrive and intersect them, leading to development of shallow magma reservoirs and the growth of volcanic edifices such as shield volcanos on the surface above them. It is by no means clear that this pattern of activity will have dominated on any asteroids, and volcanism may have been characterized by the random distribution in space and time of monogenetic events, each of which produced either an eruption or the intrusion of a dike or sill, the outcome depending on the previous history of the eruption site and the volatile content of the magma. Bearing in mind the fact that initial crustal growth involves eruptions onto and intrusions into the primitive parent material of the asteroid, and that random impacts will be taking place on the surface of the growing crust, it is clear that predicting the pattern of crustal growth is a complex problem. There is probably still much to be learned by studying the proportions of, and detailed morphologies and thermal histories of, groups of meteorites from the outer layers of a common parent body, such as the eucrites, howardites, and diogenites (*Taylor et al.,* 1993; *Wilson and Keil,* 1996a).

A major issue related to the growth of asteroid crusts is the efficiency of volcanic advection of heat. There are two aspects of this. The first relates to the release of heat at the asteroid surface. Clearly even shallow intrusions were less efficient than surface eruptions at releasing heat. When surface eruptions occurred, explosive activity was probably more efficient than lava flow formation, but the situation is not clear cut. The negligible atmospheric pressure caused maximal expansion of released gas and hence high velocities of ejected magma droplets, and the relatively low gravity allowed wide dispersal of pyroclasts (*Wilson and Keil,* 1991). However, the high degree of gas expansion would also have fragmented the erupted magma into very small (submillimeter to tens of micrometers to judge by lunar samples) droplets and there is then a wide range of eruption conditions (*Wilson and Keil,* 1996b) that could have produced optically dense fire fountains in which all droplets except those in the outer envelope of the fountain retained their heat and formed uncooled ponds feeding lava flows.

7. CORE FORMATION

Formation of a metal-sulfide core would occur essentially contemporaneously with the early stages of crust formation. As discussed earlier in this chapter, cotectic Fe,Ni-FeS melts are unlikely to have dihedral angles exceeding 60° and migrate under static conditions. However, dynamic conditions may produce very different results. *Keil and Wil-*

son (1993) explored the loss of the Fe,Ni-FeS melt by explosive volcanism as a mechanism for producing the S depletions (relative to S within the Fe,Ni-FeS component of ordinary chondrites) inferred for several groups of iron meteorites (most notably the IIIAB, IVA, and IVB groups). These authors argued that even the dense cotectic Fe,Ni-FeS melts, which would normally be negatively buoyant, can be driven to the surface and lost by expansion of trapped gases on small asteroidal bodies (radii < 100 km). While this mechanism may set the stage for S loss on small asteroidal bodies, it has equal implications for larger bodies. On those asteroids with radii in excess of 100 km, such melts would likely migrate to the center of the body and form a small, S-rich core. If partial melting were essentially arrested at this step (such as appears to have happened on the parent bodies of many primitive achondrites), this small S-rich core might crystallize, overlain by a shell of unfractionated or weakly fractionated silicates. Several authors (e.g., *Kracher,* 1982; *Benedix et al.,* 2000) have argued that this is the mechanism by which IAB and IIICD irons formed, although they recognize that impact must have played a role in mixing silicate inclusions into these meteorites. An alternative hypothesis, championed most recently by *Wasson and Kallemeyn* (2002), argues that these groups are formed in localized impact-melt pools near the surface of their parent body.

In many asteroids, much more extensive melting must have occurred, leading to formation of a core that incorporated essentially the entire metal-sulfide component present in the asteroid (recognizing that small droplets could have remained entrained in an overlying, convecting silicate shell). *Taylor* (1992) examined the physics of core formation, both at low and high degrees of partial melting. While the conclusions of this work at low degrees of partial melting have been challenged by subsequent experimental work, the calculations for separation of metallic melt blobs at high degrees of partial melting appear sound. *Taylor* (1992) argued that metal droplets might coarsen from the millimeter-size range seen in chondrites to 10 cm by a variety of largely unknown processes (e.g., Ostwald ripening) during early heating and melting of an asteroid. For these droplets to overcome the yield strength of the melt depends on the

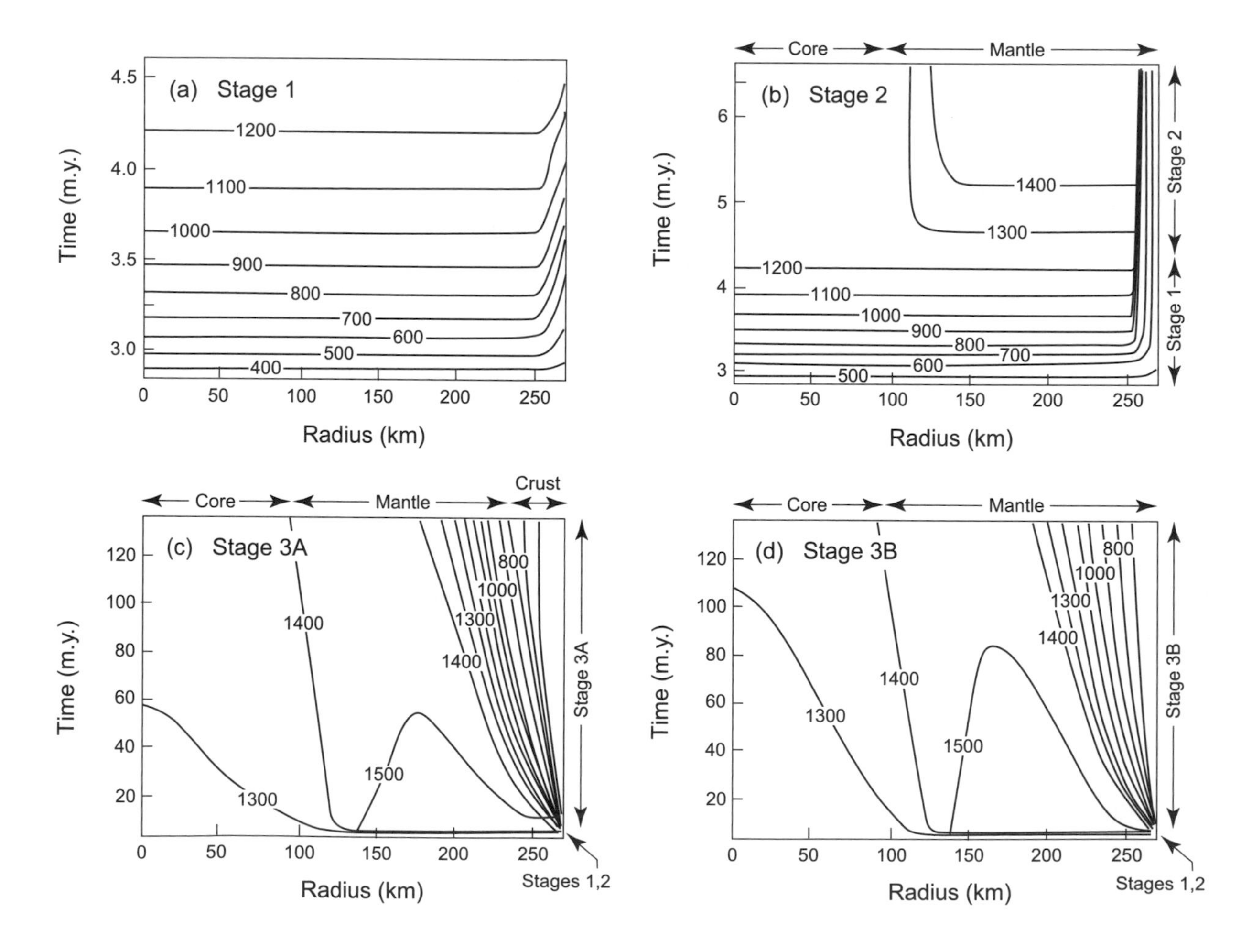

Fig. 4. Temperature contours for 4 Vesta on plots of time elapsed since CAI formation and radial distance from asteroid center after *Ghosh and McSween* (1998). **(a)** Stage 1 is the interval from accretion to core formation. **(b)** In stage 2, core formation redistributes ^{26}Al into the mantle, where temperatures continue to increase. Stages 3A **(c)** and 3B **(d)** compare heating in the cases of complete melt extraction from the mantle and retention of the melt in the mantle respectively. Peak temperatures reach 1500 K in the mantle, corresponding to 25% melting of their model mantle.

gravity and, hence, on the radius of the parent body. The percentage of melting required exceeded 50% in each case, reaching as much as 90% melting for efficient core formation on a 10-km-radius asteroid. Uncertainties in the yield strength propagate to uncertainties of ±10% in the degree of melting. Given that iron meteorite parent bodies were in the range of 10–100 km in radius (*Haack and McCoy,* 2004), high degrees of partial melting must have been reasonably widespread, given the abundance of iron meteorite groups and ungrouped meteorites (see *Weisberg et al.,* 2006).

8. COMPLETE MELTING?

While the presence of iron meteorites implies high degrees of partial melting (≥50%), did differentiated asteroids melt completely? We can examine this issue from a number of theoretical standpoints, as well as by looking at the meteorite record.

Several authors, most notably *Righter and Drake* (1997), have argued that the parent body of the howardite-eucrite-diogenite (HED) meteorites — likely 4 Vesta — had a magma ocean formed by extensive partial melting. They modeled partitioning and concentrations for several siderophile elements for a variety of likely chondritic precursors (H, L, LL, CI, CM, CO, CV) and found their best matches with peak temperatures of 1773–1803 K and degrees of silicate melting between 65% and 77%.

Ghosh and McSween (1998) did modeling of the thermal history of an asteroid of chondritic composition the size of Vesta heated mostly by ^{26}Al (Fig. 4). The modeling was simplified in that instantaneous cold accretion was assumed, and core formation was assumed to occur instantaneously when the "Fe-FeS liquidus temperature" was reached. (The temperature used in the model was 1223 K, too low to explain the low-Ni, high-Ir irons of classic magmatic iron meteorite groups such as IIAB and IIIAB.) Because ^{60}Fe produced only a small fraction of the heat, the core acted as a heat sink in the early history of the asteroid (*Ghosh and McSween,* 1998). More recent estimates of the nebular abundance of ^{60}Fe are 30–40× higher than the value used in the thermal models (*McKeegan and Davis,* 2003). In their models, peak temperatures within the mantle reach ~1500 K, although this peak temperature appears to be constrained by the model parameter of reaching only 25% melting in the mantle.

Somewhat higher temperatures and degrees of partial melting were inferred by *Taylor et al.* (1993) in examining the liquidus temperatures for the Fe,Ni-FeS component of LL (1630 K) and H (1700 K) ordinary chondrites and the S-depleted IVB irons (1770 K). Comparing these temperatures with melting relationships for peridotites (Fig. 5), these authors inferred 35% and 60% melting.

In each case, the inference is that high degrees of partial melting were achieved for many asteroidal parent bodies, but that a substantial fraction of unmelted crystals (25–50%) might remain. *Taylor* (1992) argued that these high degrees of melting were consistent with the olivine-only silicate residues at partial melting in excess of 45% and the absence

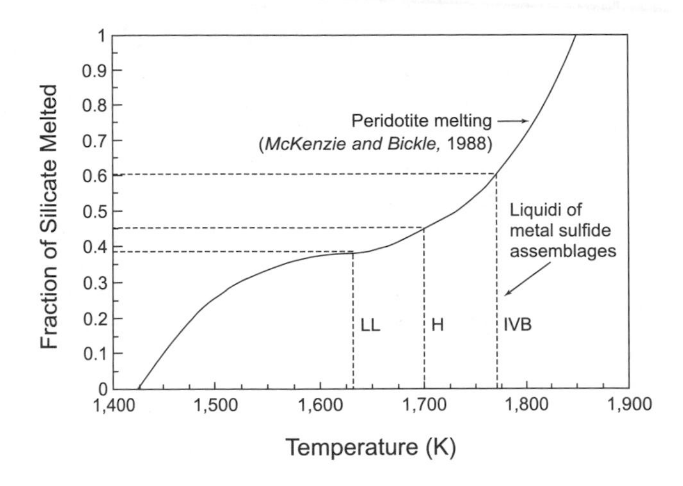

Fig. 5. Fraction of silicate melted vs. temperature for peridotite melting (*McKenzie and Bickel,* 1988). Liquidi are shown for H and LL chondrites and the S-poor IVB irons. Degrees of partial melting between 39% and 60% are indicated, consistent with the presence of magma oceans on the parent asteroids of most iron meteorites and a residual mantle with 40–60% residual crystals. The residual crystals would be dominantly olivine, although a small pyroxene component might be present at the lower ranges of degrees of partial melting, consistent with the discovery of pyroxene pallasites. Figure from *Taylor et al.* (1993).

of significant pyroxene in pallasites. Since the publication of that work, the recognition of pyroxene pallasites might suggest that the degree of partial melting achieved on some asteroidal parent bodies that formed cores was lower than that inferred by *Taylor* (1992).

The enigma that remains is that of whether differentiated asteroids achieved a truly homogeneous magma ocean prior to solidification or whether an early-formed crust overlay a peridotite mantle. As *Ghosh and McSween* (1998) note, the realistic case is probably intermediate between early and complete extrusion of basaltic melts from the mantle and their complete retention in the mantle. In addition, these authors point out that the outer 5 km of an asteroid dissipates heat as quickly as it enters from below, and thus might remain largely unmelted. If an early basaltic crust forms or a relict chondritic layer remains, the two most likely scenarios for incorporation of this material back into the mantle are recycling by impacts and remelting of this crust, primarily at the base.

It is difficult to model the former. *Ghosh and McSween* (1998) cited earlier workers that suggested either minimal heating by impact or localized heating. However, we envision remelting not due to heating as a consequence of impact, but by recycling cold outer crust into the deeper, hotter parts of the mantle.

It is easier to estimate the ability to melt the crust from below and this appears to be particularly dependent on parent-body radius. For example, on a 20-km-radius parent body, the volume of melt in a 10-km-thick mantle at 50% melting would equal only 35% of the volume of the outer 5-km shell. Thus, the ability of this relatively small amount of melt to recycle the entire outer shell would be very limited. In contrast, a 250-km-radius asteroid (e.g., 4 Vesta) with a 50-km-diameter core and 5-km outer shell at 50% melting of the mantle would contain nearly 32× as much melt in the mantle as the volume of cold, outer crust. Thus, it seems quite probable that large, differentiated asteroids may have in fact reincorporated any early-formed crust and achieved a true magma ocean, while smaller asteroids retained the early-formed basaltic crust and perhaps even highly metamorphosed traces of their chondritic precursor.

9. SOLIDIFICATION

In this section, we address the solidification of an initially chemically uniform, ultramafic magma ocean. As a starting point in this discussion, we will assume that the rate of heating is much more rapid than the rate of heat dissipation. Thus, for example, if ^{26}Al is the heat source, Al-rich melt does not migrate out of the interior rapidly enough to shut down heating of the deep interior. Under this assumption, immiscible metal-sulfide melts and refractory metallic solids will segregate downward to form a core, which will be overlain by a molten silicate shell of ultramafic composition.

At present, little detailed modeling has been done to try to understand the physical behavior of such a silicate melt shell, and thus chemical modeling of solidification is somewhat ungrounded. If the heat source is a short-lived lithophile nuclide such as ^{26}Al, then the silicate melt shell will be approximately evenly heated internally, with thermal boundaries at the top and bottom (see *Ghosh and McSween,* 1998). This could act to subdue convection, although calculations using the canonical solar nebula value for $^{26}Al/^{27}Al$ indicate that molten asteroids would be within the convection realm (*Righter and Drake,* 1997; *Taylor et al.,* 1993). On the other hand, if ^{60}Fe is a major heat source, the silicate melt shell would be heated from the base and would contain an evenly distributed heat source, and strong convection would certainly occur. The style of solidification — equilibrium vs. fractional crystallization — could be very different in these two cases.

A compositionally uniform, heated silicate shell maintains a high temperature for an extended period of time. For example, at a radius of 175 km (depth of 100 km) the temperature is above 1400 K for 140 m.y. [For comparison, the silicate solidus for ordinary chondrites is probably <1375 K; *Jurewicz et al.* (1995) found ~13% melt in ordinary chondrite experimental charges run at 1445 K.] The "hot zone" in the asteroid thermal model is located in the mid regions of the silicate shell, roughly at a radius of 200 km at the start, lowering to radius of roughly 150 km as the body cools (*Ghosh and McSween,* 1998). (The core-mantle boundary is at radius 100 km.) If this thermal model is correct, solidification would be going on simultaneously at the base and top of the molten silicate shell.

Tonks and Melosh (1990) discussed the nature of convection in planetary-scale magma oceans. In a terrestrial or lunar magma ocean, the Rayleigh number would be very high, and convection would be in the realm of inertial flow, where turbulent eddies overcome settling forces to keep growing crystals in suspension, although this conclusion has been challenged by subsequent workers (see *Solomatov,* 2000). *Taylor et al.* (1993) and *Righter and Drake* (1997) have estimated the Rayleigh numbers for a molten silicate shell over a metallic core on 100-km-radius and Vesta-sized asteroids. Both studies found that an asteroidal magma ocean would convect in the inertial realm, keeping crystals suspended in a well-mixed melt. Under these conditions, equilibrium crystallization would occur for a substantial fraction of the crystallization sequence (*Righter and Drake,* 1997).

Taylor et al. (1993) calculated that crystals would not grow large enough for settling to occur. However, they based this conclusion on calculations that the magma ocean would cool and solidify on a very short timescale for the 100-km-radius asteroid they considered. Their model did not include continued heating of the asteroid through ^{26}Al decay even while heat is being lost to space. The thermal models of *Ghosh and McSween* (1998) for a Vesta-sized asteroid have a prolonged cooling history, and crystals would reach much larger sizes, possibly allowing settling to occur. The settling velocity of a crystal is proportional to the square of the crystal radius (*Walker et al.,* 1978), so a growing crystal rapidly increases its settling velocity.

All models of the physical setting of asteroid solidification assume uniform low gravity, but this is a simplification. Figure 6a shows the gravity as a function of fractional radius for a Vesta-sized asteroid, assuming core masses of 5% and 15% the mass of the asteroid (cf. *Righter and Drake,* 1997). The core and molten silicate shell are each assumed to have uniform density, with the silicate melt density that of *Righter and Drake* (1997). Gravity reaches a local minimum in the lower portion of the magma shell before increasing to the maximum at the surface. Figure 6b shows calculated settling velocity following *Walker et al.* (1978) for 1-cm-radius crystals in the silicate magma shell. Because of the linear dependence on gravity, the settling velocity similarly reaches minimum in the lower portion of the magma shell. The mean convection velocity scales as the cube-root of gravity (*Tonks and Melosh,* 1990, their equation (17)), so the grain-settling velocity (linear function of gravity) is a stronger function of depth within the magma shell than is convection velocity. There is thus a potential that grains growing in the upper region of the magma sphere could settle, only to become stalled in convecting magma at an intermediate depth.

As outlined above, we have only a rudimentary understanding of the physics of a cooling asteroid magma shell. For this reason, models for the petrologic evolution of the magma are crude approximations. Only two end members have been modeled in detail in the literature: equilibrium crystallization (*Righter and Drake,* 1997) and fractional crystallization (e.g., *Ruzicka et al.,* 1997). Both of these have been done to model the evolution of the howardite-

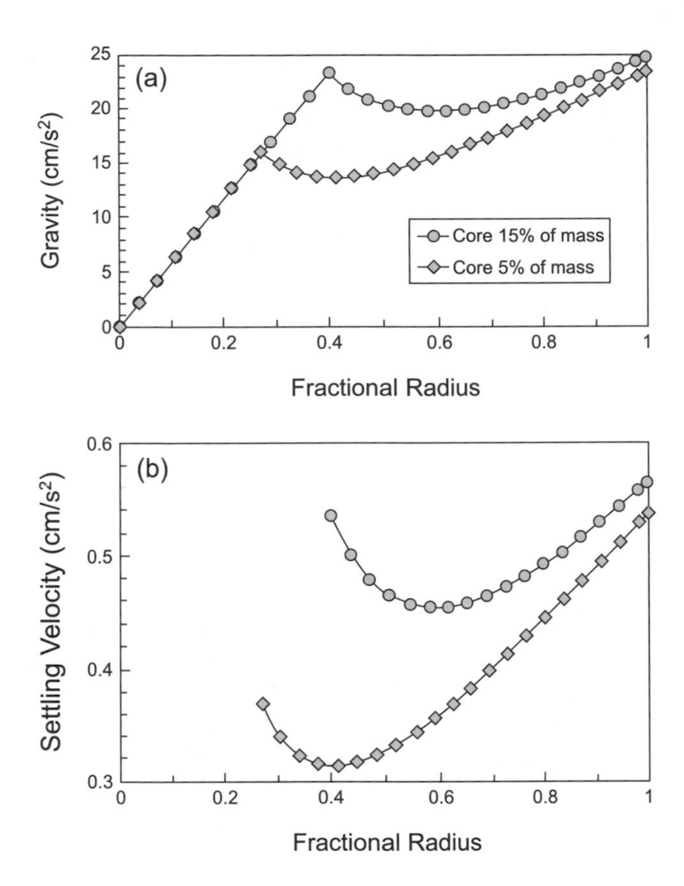

Fig. 6. Calculated **(a)** gravity and **(b)** grain-settling velocity vs. fractional radius for a Vesta-sized molten asteroid. Core density calculated for a mixture of 90% Fe,Ni and 10% FeS; silicate melt density is 2900 kg m^{-3}, following *Righter and Drake* (1997). Settling velocity calculated following *Walker et al.* (1978) assuming 1-cm grains with a density of 3200 kg m^{-3}.

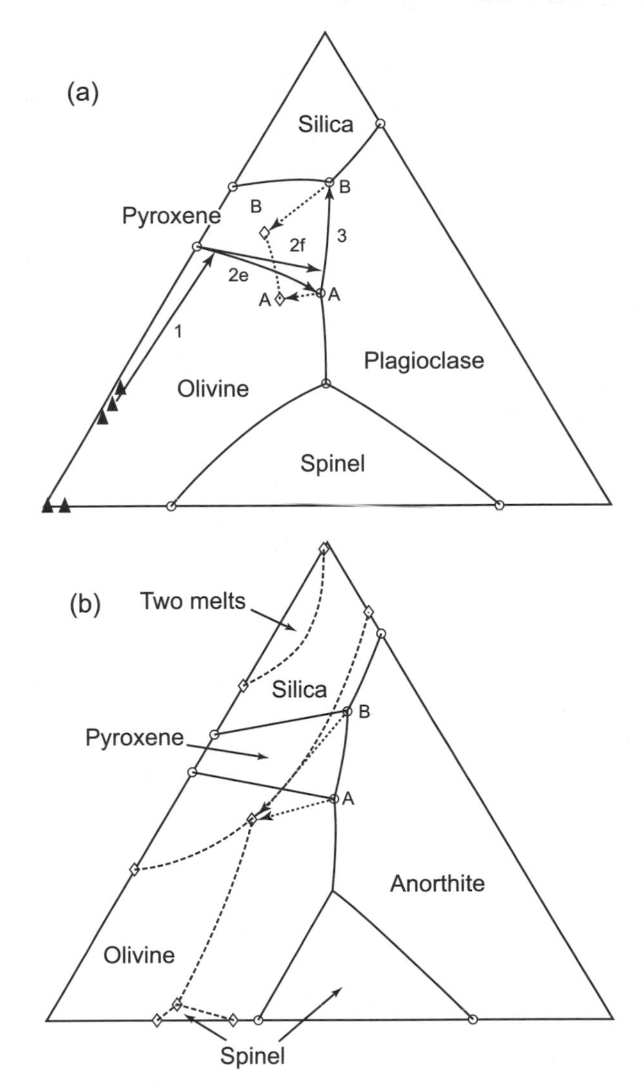

Fig. 7. **(a)** Schematic olivine-plagioclase-silica pseudoternary diagrams showing idealized paths for crystallization of molten chondritic asteroids. Phase boundaries are taken from *Walker et al.* (1973), devised for modeling lunar compositions; mg# of 70 and Na-depleted relative to chondrites. Dashed arrows show the effect of decreasing mg# on the peritectic (A) and eutectic (B) points based on experiments on eucritic compositions (connected by dotted line), with mg# ~40 (taken from *Stolper,* 1977). A schematic equilibrium crystallization path is given by arrows 1 and 2e; crystallization will terminate at the peritectic point (A). A schematic fractional crystallization path is given by arrows 1, 2f, and 3; crystallization will terminate at the eutectic (B). Phase boundaries shift with changes in composition, pressure, and f_{O_2}, and no single diagram can truly represent full crystallization path of a magma. Solid triangles are the locations of the silicate portions of ordinary chondrites (upper three) and CM and CV chondrites, using average fall data from *Jarosewich* (1990). **(b)** The shifting phase boundaries with Fe content are illustrated by comparing the Fo-An-Si (solid lines) and Fa-An-Si (dashed lines) phase diagrams. With increasing Fe content, the anorthite and silica stability fields expand, the peritectic (A) and eutectic (B) points merge, and the stability field of pyroxene disappears. Substituting albite for anorthite (not shown) in the Fo-Ab-Si system greatly decreases the stability field of plagioclase, and points A and B shift toward the Si-Ab join.

eucrite-diogenite parent body, presumed to be 4 Vesta. The compositions investigated are roughly chondritic, but that modeled by *Ruzicka et al.* (1997) is depleted in Na.

9.1. Equilibrium Crystallization

A magma with roughly chondritic relative proportions of lithophile elements will be highly olivine normative (Fig. 7). The maximum temperature reached in the Vesta thermal models is ~1540 K (*Ghosh and McSween,* 1998). *Jurewicz et al.* (1995) found that LL-chondrite charges run at 1600 K contained ~55% residual olivine, while H-chondrite charges run at 1550 K contained ~70% residual olivine. Thus an asteroid magma shell might consist of turbulently convecting melt entraining refractory olivine grains in equilibrium with the melt.

Righter and Drake (1997) modeled the equilibrium crystallization of a Vestan magma shell assuming a composition equivalent to 30% CV–70% L chondrite (chosen to approximate the HED O-isotopic composition) and an initial temperature of 1875 K. They calculated the Rayleigh number for their model magma, and concluded that convection would keep the crystal and melt well-stirred up to about 80% crystallization. Convection lockup would occur at this point, and the remaining 20% solidification would follow a fractional crystallization path. An idealized equilibrium crystallization path is shown in Fig. 7a. Olivine is the sole crystallizing phase for about the first 35% solidification (Fig. 7a, arrow 1). At that point, orthopyroxene and minor Cr-rich spinel join the crystallization sequence at nearly the same time (*Righter and Drake,* 1997) (Fig. 6). These three phases co-crystallize until about 1515 K (Fig. 7a, arrow 2e). [The olivine-orthopyroxene boundary is normally a reaction boundary at low pressure. In Fe-rich mafic melts, it becomes a cotectic boundary. The modeling programs used by *Righter and Drake* (1997) treated the boundary as a cotectic.] Pigeonite replaces orthopyroxene below 1515 K, when the silicate system is ~85% crystallized. Plagioclase joins the crystallization sequence at the end, and some olivine will react with the residual, basaltic melt to form pigeonite. Crystallization ends at the pseudoternary peritectic point, A.

If equilibrium is maintained throughout the crystallization sequence, the asteroid would end up as a metal-sulfide core surrounded by a homogeneous shell of coarse-grained olivine, orthopyroxene, pigeonite, plagioclase, and chromite. This seems unlikely. At some point the convective motion in the magma will cease to entrain the growing crystals, and as discussed by *Righter and Drake* (1997) the residual melt will begin to follow a fractional crystallization path. Should this occur before the melt composition reaches the peritectic point in the pseudoternary system, orthopyroxene will fractionate from the melt, followed by fractionation of pigeonite plus plagioclase. In this case, the asteroid would consist of a metal-sulfide core, a thick shell of harzburgite, overlain by a thin shell of orthopyroxenite, capped by a gabbroic/basaltic crust. *Righter and Drake* (1997) estimate a thickness for the mafic crust on a Vesta-sized asteroid of 10–15 km. The thickness of the orthopyroxenite shell would depend on the point at which the transition from equilibrium to fractional crystallization occurred. If the transition to fractional crystallization occurs only after the peritectic point is reached, than there will be no orthopyroxenite layer. The asteroid mantle would consist of harzburgite.

9.2. Fractional Crystallization

Ruzicka et al. (1997) modeled fractional crystallization of molten asteroids of several chondritic compositions; all Na-poor because the intent was to model the eucrite-diogenite suite. Olivine was the sole crystallizing phase for ~60 vol% crystallization (Fig. 7a, arrow 1). [*Ruzicka et al.* (1997) tabulated model results by volume, rather than by mass.] The olivine-orthopyroxene boundary was treated as a reaction boundary. When the melt composition reached this boundary, orthopyroxene became the sole crystallizing phase, followed by orthopyroxene + chromite, followed by pigeonite (Fig. 7a, arrow 2f). The melt composition reached the pyroxene-plagioclase cotectic, and formation of gabbroic cumulates (plagioclase + pigeonite) began (Fig. 7a, arrow 3). *Ruzicka et al.* (1997) noted that melt viscosity would be substantially higher by this stage, and suggested the further crystallization would approach equilibrium crystallization. Finally, the plagioclase-pyroxene-silica ternary eutectic was reached (point B) and silica plus ferroan olivine joined the crystallization sequence. The resulting asteroid is slightly different from the model produced by equilibrium crystallization: a metal-sulfide core, overlain by a 150-km-thick dunite shell, overlain by roughly 13 km of orthopyroxenite, capped by a 26-km-thick mafic crust (*Ruzicka et al.,* 1997).

The discussions of equilibrium and fractional crystallization paths given above in relation to phase relations (Fig. 7a) are highly schematic because phase boundaries shift as the system evolves. This is qualitatively illustrated in Fig. 7b where the phase boundaries in the forsterite-anorthite-silica (Fo-An-Si) and fayalite-anorthite-silica (Fa-An-Si) are compared. Adding Fe to the system almost completely changes the topology of the diagram. In real silicate systems, the mg# [molar 100*MgO/(MgO + FeO)] decreases as crystallization proceeds, and the position of the pyroxene-plagioclase cotectic shifts toward the evolving melt composition, foreshortening the pyroxene crystallization interval (Fig. 7a, arrow 2f). The olivine-pyroxene boundary also changes from a reaction boundary to a cotectic as mg# decreases, which will alter the crystallization sequence. Finally, the peritectic point crosses the plagioclase-pyroxene join as pressure is increased, and the olivine-pyroxene boundary will become a cotectic by ~100 MPa (*Bartels and Grove,* 1991). This corresponds to a depth of ~190 km in a Vesta-sized asteroid. These issues have been discussed by *Bartels and Grove* (1991), *Boesenberg and Delaney* (1997), and *Warren* (1985), for example.

Figure 7 can be thought of as an asteroidal equivalent to the tholeiitic basalt sequence on Earth — crystallization results in low-Ca pyroxene and silica-normative residua. *Jurewicz et al.* (1993) and *Longhi* (1999) have shown that increasing the f_{O_2} of chondritic systems can change the phase relations, resulting in minimum melts that are critically silica unsaturated, and magmas with angritic affinities are produced. The solidification of such an asteroid would be quite different from that outlined above for HED analog asteroids, but no detailed modeling has been done. Olivine would still be the liquidus phase, and form the bulk of the solidified asteroid. Plagioclase ± spinel would next join olivine in the crystallization sequence, followed by augite. Finally, kirschsteinite ($CaFeSiO_4$) joins the crystallization sequence at the very end.

The modeling by *Ruzicka et al.* (1997) considered Na-poor starting compositions because the object was to model the HED parent body. Although *Righter and Drake* (1997) were also attempting to explain HED petrogenesis, they used true chondritic Na contents. The phase relations will be different at higher, more chondrite-like, Na contents (see *Stolper et al.,* 1979). The stability field of plagioclase greatly decreases, while those of olivine and pyroxene increase as the Na content of the system increases. Pyroxene also becomes more calcic as Na replaces Ca in plagioclase (*Stolper et al.,* 1979). Some of the difference in model results between *Ruzicka et al.* (1997) and *Righter and Drake* (1997) is probably a reflection of the different assumed magma compositions. There have been no attempts to model the solidification of true chondritic composition asteroids (e.g., CV, H, EH, etc.) using a common methodology. Because of this, detailed predictions of the solidification and structure of chondritic composition asteroids are not known.

10. SYNOPSIS

Despite the broad array of chondritic precursors known to exist, the wide range of geologic processes acting during melting of these bodies (e.g., heating, differentiation, impact), and the complex interplay between these parameters, we can define — in the broadest sense — the melting of a typical asteroid. As heating exceeded the solidus, melting would begin first at the Fe,Ni-FeS cotectic and be followed by melting of silicates. These low-degree partial melts would likely not form networks, but would migrate primarily through the formation of veins and dikes driven in part by expansion of the melts and their included gases. Formation of a basaltic crust would occur soon after the onset of melting, perhaps in as little as 1 m.y. At the same time, the Fe,Ni-FeS melts would begin to migrate and, depending on parent-body size, might be erupted either from the asteroid or the sink to form a small S-rich core. The melting of the parent bodies of some meteorites (e.g., ureilites, acapulcoites-lodranites, perhaps IAB irons) were arrested at this stage. These parent bodies produced achondrites variably depleted in basaltic components and melts of the Fe-Ni-S system (ureilites, lodranites). Most differentiated asteroids continued to higher degrees of melting, eventually reaching percentages of partial melting where efficient metal segregation occurred by sinking of metallic melt droplets and metal-dominated cores were formed. The mantles of these asteroids likely did not melt completely, with percentages of melting ranging from ~40% to 75%. In the largest asteroids, impacts and basal melting may have recycled the early-formed basaltic crust (as well as any remaining chondritic material at the surface) back into the mantle, producing a homogeneous magma ocean. In contrast, smaller asteroids might have retained this primary basaltic crust. The crystallization of that magma ocean, whether by equilibrium or fractional crystallization, would produce a core mantled with a thick layer of olivine. Depending on the style of crystallization and the nature of the olivine-pyroxene phase boundary, the dunite layer will be overlain by harzburgite, orthopyroxenite, or some combination of the two. The ultramafic shell will be capped by a 10–25-km-thick basaltic to gabbroic crust. This gross structure should be typical of large differentiated asteroids like 4 Vesta.

11. FUTURE DIRECTIONS

Our understanding of asteroid differentiation is likely to continue unabated for the foreseeable future. Recovery of new meteorites and attendant geochemical and experimental studies of the processes that formed them proceed and shed new light on these processes. In many respects, the most exciting new avenue for research into the next decade will be the first spacecraft exploration of a differentiated asteroid. The DAWN mission is scheduled to orbit 4 Vesta beginning in 2010, collecting a variety of spectral, geochemical, shape, and magnetic data. Existing groundbased rotational spectroscopy (*Binzel et al.,* 1997; *Gaffey,* 1997) suggests geologic units consistent with dunite, basalt, and orthopyroxenite interpreted as the mantle, eucrite, and diogenite components of the HED suite. Given the importance that the study of this suite has in shaping our views of asteroidal differentiation (perhaps even to the exclusion of considering alternative scenarios for smaller and/or less-melted asteroids), these new data could prove revolutionary in shaping our understanding of the process by which primitive, chondritic asteroids ultimately become the layered worlds we know from our meteorite collections.

REFERENCES

Ballhaus C. and Ellis D. J. (1996) Mobility of core melts during Earth's accretion. *Earth Planet. Sci. Lett., 143,* 137–145.

Bartels K. S. and Grove T. L. (1991) High-pressure experiments on magnesian eucrite compositions: Constraints on magmatic processes in the eucrite parent body. *Proc. Lunar Planet. Sci. Conf., Vol. 21,* pp. 351–365.

Benedix G. K., McCoy T. J., Keil K., Bogard D. D., and Garrison D. H. (1998) A petrologic and isotopic study of winonaites: Evidence for early partial melting, brecciation, and metamorphism. *Geochim. Cosmochim. Acta, 62,* 2535–2553.

Benedix G. K., McCoy T. J., Keil K., and Love S. G. (2000) A petrologic study of the IAB iron meteorites: Constraints on the formation of the IAB-winonaite parent body. *Meteoritics & Planet. Sci., 35,* 1127–1141.

Benedix G. K., McCoy T. J., and Lauretta D. S. (2003) Is NWA 1463 the most primitive winonaite? (abstract). *Meteoritics & Planet. Sci., 38,* A70.

Bild R. W. and Wasson J. T. (1977) Netschaevo: A new class of chondritic meteorite. *Science, 197,* 58–62.

Binzel R. P., Gaffey M. J., Thomas P. C., Zellner B. H., Storrs A. D., and Wells E. N. (1997) Geologic mapping of Vesta from 1994 Hubble Space Telescope images. *Icarus, 128,* 95–103.

Bischoff A., Scott E. R. D., Metzler K., and Goodrich C. A. (2006) Origin and formation of meteoritic breccias. In *Meteorites and the Early Solar System II* (D. S. Lauretta and H. Y. McSween Jr., eds.), this volume. Univ. of Arizona, Tucson.

Bosenberg J. S. and Delaney J. S. (1997) A model composition of the basaltic achondrite planetoid. *Geochim. Cosmochim. Acta, 61,* 3205–3225.

Bogard D. D., Garrison D. H., and McCoy T. J. (2000) Chronology and petrology of silicates from IIE iron meteorites: Evidence of a complex parent body evolution. *Geochim. Cosmochim. Acta, 64,* 2133–2154.

Brearley A. J. and Jones R. H. (1998) Chondritic meteorites. In *Planetary Materials* (J. J. Papike, ed.), pp. 3-1 to 3-98. Reviews in Mineralogy, Vol. 36, Mineralogical Society of America.

Bruhn D., Groebner N., and Kohlstedt D. L. (2000) An interconnected network of core-forming melts produced by shear deformation. *Nature, 403,* 883–886.

Burbine T. H., McCoy T. J., Meibom A., Gladman B., and Keil K. (2002) Meteoritic parent bodies: Their number and identification. In *Asteroids III* (W. F. Bottke Jr. et al., eds.), pp. 653–667. Univ. of Arizona, Tucson.

Chabot N. L. and Haack H. (2006) Evolution of asteroidal cores. In *Meteorites and the Early Solar System II* (D. S. Lauretta and H. Y. McSween Jr., eds.), this volume. Univ. of Arizona, Tucson.

Choi B. G., Ouyang X. W., and Wasson J. T. (1995) Classification and origin of IAB and IIICD iron meteorites. *Geochim. Cosmochim. Acta, 59,* 593–612.

Feldstein S. N., Jones R. H., and Papike J. J. (2001) Disequilibrium partial melting experiments on the Leedey L6 chondrite: Textural controls on melting processes. *Meteoritics & Planet. Sci., 36,* 1421–1441.

Gaffey M.J. (1997) Surface lithologic heterogeneity of asteroid 4 Vesta. *Icarus, 127,* 130–157.

Gaetani G. A. and Grove T. L. (1999) Wetting of mantle olivine by sulfide melt: Implications for Re/Os ratios in mantle peridotite and late-stage core formation. *Earth Planet. Sci. Lett., 169,* 147-163.

Ghosh A. and McSween H. Y. Jr. (1998) A thermal model for the differentiation of asteroid 4 Vesta, based on radiogenic heating. *Icarus, 134,* 187–206.

Goodrich C. A. and Lugmair G. W. (1995) Stalking the LREE-enriched component in ureilites. *Geochim. Cosmochim. Acta, 59,* 2609–2620.

Grimm R. E. and McSween H. Y. (1989) Water and the thermal evolution of carbonaceous chondrite parent bodies. *Icarus, 82,* 244–280.

Haack H. and McCoy T. J. (2004) Iron and stony-iron meteorites. In *Treatise on Geochemistry, Vol. 1: Meteorites, Comets and Planets* (A. M. Davis, ed.), pp. 325–345. Elsevier, Oxford.

Herpfer M. A. and Larimer J. W. (1993) Core formation: An experimental study of metallic melt-silicate segregation. *Meteoritics, 28,* 362.

Horan M. F., Smoliar M. I., and Walker R. J. (1998) ^{182}W and ^{187}Re-^{187}Os systematics of iron meteorites: Chronology for melting, differentiation and crystallization in asteroids. *Geochim. Cosmochim. Acta, 62,* 545–554.

Jarosewich E. (1990) Chemical analyses of meteorites: A compilation of stony and iron meteorite analyses. *Meteoritics, 25,* 323–337.

Jurewicz A. J. G., Mittlefehldt D. W., and Jones J. H. (1993) Experimental partial melting of the Allende (CV) and Murchison (CM) chondrites and the origin of asteroidal basalts. *Geochim. Cosmochim. Acta, 57,* 2123–2139.

Jurewicz A. J. G., Mittlefehldt D. W., and Jones J. H. (1995) Experimental partial melting of the St. Severin (LL) and Lost City (H) chondrites. *Geochim. Cosmochim. Acta, 59,* 391–408.

Keil K. (1989) Enstatite meteorites and their parent bodies. *Meteoritics, 24,* 195–208.

Keil K. and Wilson L. (1993) Explosive volcanism and the compositions of cores of differentiated asteroids. *Earth Planet. Sci. Lett., 117,* 111–124.

Keil K., Stöffler D., Love S. G., and Scott E. R. D. (1997) Constraints on the role of impact heating and melting in asteroids. *Meteoritics & Planet. Sci., 32,* 349–363.

Kelemen P. B., Hirth G., Shimizu N., Spiegelman M., and Dick H. J. B. (1997) A review of melt migration processes in the adiabatically upwelling mantle beneath oceanic spreading ridges. *Philos. Trans. R. Soc. Lond., Ser. A, 355,* 283–318.

Kerridge J. F. and Matthews M. S., eds. (1988) *Meteorites and the Early Solar System.* Univ. of Arizona, Tucson. 1269 pp.

Kracher A. (1982) Crystallization of a S-saturated -melt, and the origin of the iron meteorite groups IAB and IIICD. *Geophys. Res. Lett., 9,* 412–415.

Kullerud G. (1963) The Fe-Ni-S system. *Annu. Rep. Geophys. Lab., 1412,* 175–189. Washington, DC.

Kushiro I. and Mysen B. O. (1979) Melting experiments on a Yamato chondrite. *Proc. 4th Symp. Antarctic Met., Mem. Natl. Inst. Polar Res., 15,* 165–170.

Lodders K., Palme H., and Wlotzka F. (1993) Trace elements in mineral separates of the Pena Blanca Spring aubrite: Implications for the evolution of the aubrite parent body. *Meteoritics, 28,* 538–551.

Longhi J. (1999) Phase equilibrium constraints on angrite petrogenesis. *Geochim. Cosmochim. Acta, 63,* 573–585.

Maaloe S. (2003) Melt dynamics of a partially molten mantle with randomly oriented veins. *J. Petrol., 44,* 1193–1210.

McCoy T. J., Keil K., Clayton R. N., Mayeda T. K., Bogard D. D., Garrison D. H., Huss G. R., Hutcheon I. D., and Wieler R. (1996) A petrologic, chemical and isotopic study of Monument Draw and comparison with other acapulcoites: Evidence for formation by incipient partial melting. *Geochim. Cosmochim. Acta, 60,* 2681–2708.

McKeegan K. D. and Davis A. M. (2003) Early solar system chronology. In *Treatise on Geochemistry, Vol. 1: Meteorites, Comets and Planets* (A. M. Davis, eds.), pp. 431–460. Elsevier, Oxford.

McKenzie D. P. and Bickel M. J. (1988) The extraction of magma from the crust and mantle. *Earth Planet. Sci. Lett., 74,* 81–91.

McSween H. Y. Jr., Bennett M. E. III, and Jarosewich E. (1991) The mineralogy of ordinary chondrites and implications for asteroid spectrophotometry. *Icarus, 90,* 10–116.

McSween H. Y. Jr., Ghosh A., Grimm R. E., Wilson L., and Young E. D. (2002) Thermal evolution models of asteroid. In *Asteroids III* (W. F. Bottke Jr. et al., eds.), pp. 559–571. Univ. of Arizona, Tucson.

Meibom A. and Clark B. E. (1999) Evidence for the insignificance of ordinary chondritic material in the asteroid belt. *Meteoritics & Planet Sci., 34,* 7–24.

Minarik W. G., Ryerson F. J., and Watson E. B. (1996) Textural entrapment of core-forming melts. *Science, 272,* 530–533.

Mittlefehldt D. W. (2004) Achondrites. In *Treatise on Geochemistry, Vol. 1: Meteorites, Comets and Planets* (A. M. Davis, eds.), pp. 291–324. Elsevier, Oxford.

Mittlefehldt D. W., Lindstrom M. M., Bogard D. D., Garrison D. H., and Field S. W. (1996) Acapulco- and Lodran-like achondrites: Petrology, geochemistry, chronology and origin. *Geochim. Cosmochim. Acta, 60,* 867–882.

Mittlefehldt D. W., McCoy T. J., Goodrich C. A., and Kracher A. (1998) Non-chondritic meteorites from asteroidal bodies. In *Planetary Materials* (J. J. Papike, ed.), pp. 4-1 to 4-195. Reviews in Mineralogy, Vol. 36, Mineralogical Society of America.

Morse S. A. (1980) *Basalts and Phase Diagrams.* Springer-Verlag, Berlin. 493 pp.

Mostefaoui S., Lugmair G. W., Hoppe P., and El Goresy A. (2003) Evidence for live iron-60 in Semarkona and Chervony Kut: A NanoSIMS study (abstract). In *Lunar Planet. Sci. Conf. XXXIV,* Abstract #1585. Lunar and Planetary Institute, Houston (CD-ROM).

Muenow D. W., Keil K., and Wilson L. (1992) High-temperature mass spectrometric degassing of enstatite chondrites — Implications for pyroclastic volcanism on the aubrite parent body. *Geochim. Cosmochim. Acta, 56,* 4267–4280.

Niu Y. L. (1997) Mantle melting and melt extraction processes beneath ocean ridges: Evidence from abyssal peridotites. *J. Petrol., 38,* 1047–1074.

Nyquist E. L., Reese Y., Wiesmann H., Shih C. Y., and Takeda H. (2001) Dating eucrite formation and metamorphism. *Antarctic Meteorites XXVI,* 113–115. National Institute for Polar Research, Tokyo.

Righter K. and Drake M. J. (1997) A magma ocean on Vesta: Core formation and petrogenesis of eucrites and diogenites. *Meteoritics & Planet. Sci., 32,* 929–944.

Rubin A. E. (2004) Postshock annealing and postannealing shock in equilibrated ordinary chondrites: Implications for the thermal and shock histories of chondritic asteroids. *Geochim. Cosmochim. Acta, 68,* 673–689.

Rushmer T., Minarik W. G., and Taylor G. J. (2000) Physical processes of core formation. In *Origin of the Earth and Moon* (R. M. Canup and K. Righter, eds.), pp. 227–243. Univ. of Arizona, Tucson.

Ruzicka A., Snyder G. A., and Taylor L. A. (1997) Vesta as the howardite, eucrite and diogenite parent body: Implications for the size of a core and for large-scale differentiation. *Meteoritics & Planet. Sci., 32,* 825–840.

Scott E. R. D. and Krot A. N. (2003) Chondrites and their components. In *Treatise on Geochemistry, Vol. 1: Meteorites, Comets and Planets* (A. M. Davis, ed.), pp. 143–200. Elsevier, Oxford.

Shannon M. C. and Agee C. B. (1996) High pressure constraints on percolative core formation. *Geophys. Res. Lett., 23,* 2717–2720.

Shannon M. C. and Agee C. B. (1998) Percolation of core melts at lower mantle conditions. *Science, 280,* 1059–1061.

Shukolyukov A. and Lugmair G. (1992) ^{60}Fe — Light my fire. *Meteoritics, 27,* 289.

Shukolyukov A. and Lugmair G. W. (1996) ^{60}Fe-^{60}Ni systematics in the eucrite caldera (abstract). *Meteoritics, 31,* A129.

Sleep N. (1988) Tapping of melt by veins and dikes. *J. Geophys. Res., 93,* 10255–10272.

Solomatov V. S. (2000) Fluid dynamics of a terrestrial magma ocean. In *Origin of the Earth and Moon* (R. M. Canup and K. Righter, eds.), pp. 323–338. Univ. of Arizona, Tucson.

Sonett C. P., Colburn D. S., Schwartz K., and Keil K. (1970) The melting of asteroidal-sized parent bodies by unipolar dynamo induction from a primitive T Tauri sun. *Astrophys. Space Sci., 7,* 446–488.

Srinivasan G., Goswami J. N., and Bhandari N. (1999) ^{26}Al in eucrite Piplia Kalan: Plausible heat source and formation chronology. *Science, 284,* 1348–1350.

Stolper E. (1977) Experimental petrology of eucrite meteorites. *Geochim. Cosmochim. Acta, 41,* 587–611.

Stolper E., McSween H. Y. Jr., and Hays J. F. (1979) A petrogenetic model of the relationships among achondritic meteorites. *Geochim. Cosmochim. Acta, 43,* 589–602.

Tachibana S. and Huss G. I. (2003) The initial abundance of ^{60}Fe in the solar system. *Astrophys. J. Lett., 588,* L41–L44.

Takahashi E. (1983) Melting of a Yamato L3 chondrite (Y-74191) up to 30 Kbar. *Proc. 8th Symp. Antarctic Met., Mem. Natl. Inst. Polar Res., 30,* 168–180.

Taylor G. J. (1992) Core formation in asteroids. *J. Geophys. Res., 97,* 14717–14726.

Taylor G. J., Keil K., McCoy T., Haack H., and Scott E. R. D. (1993) Asteroid differentiation: Pyroclastic volcanism to magma oceans. *Meteoritics, 28,* 34–52.

Tonks W. B. and Melosh H. J. (1990) The physics of crystal settling and suspension in a turbulent magma ocean. In *Origin of the Earth* (H. E. Newsom and J. H. Jones, eds.), pp. 151–174. Oxford Univ., New York.

Walker D., Longhi J., Grove T. L., Stolper E., and Hays J. F. (1973) Experimental petrology and origin of rocks from the Descartes Highlands. *Proc. Lunar. Sci. Conf. 4th,* pp. 1013–1032.

Walker D., Stolper E. M., and Hays J. F. (1978) A numerical treatment of melt/solid segregation: Size of the eucrite parent body and stability of the terrestrial low-velocity zone. *J. Geophys. Res., 83,* 6005–6013.

Warren P. H. (1985) Origin of howardites, diogenites and eucrites: A mass balance constraint. *Geochim. Cosmochim. Acta, 49,* 577–586.

Warren P. H. and Kallemeyn G. W. (1992) Explosive volcanism and the graphite-oxygen fugacity buffer on the parent asteroid(s) of the ureilite meteorites. *Icarus, 100,* 110–126.

Wasson J. T. and Kallemeyn G. W. (2002) The IAB iron-meteorite complex: A group, five subgroups, numerous grouplets, closely related, mainly formed by crystal segregation in rapidly cooling melts. *Geochim. Cosmochim. Acta, 66,* 2445–2473.

Wasson J. T., Rubin A. E., and Benz W. (1987) Heating of primitive, asteroid-size bodies by large impact (abstract). *Meteoritics, 22,* 525–526.

Weisberg M. K., McCoy T. J., and Krot A. N. (2006) Systematics and evaluation of meteorite classification. In *Meteorites and the Early Solar System II* (D. S. Lauretta and H. Y. McSween Jr., eds.), this volume. Univ. of Arizona, Tucson.

Wilson L. and Keil K. (1991) Consequences of explosive eruptions on small solar system bodies: The case of the missing basalts on the aubrite parent body. *Earth Planet. Sci. Lett., 104,* 505–512.

Wilson L. and Keil K. (1996a) Volcanic eruptions and intrusions on the asteroid 4 Vesta. *J. Geophys. Res., 101,* 18927–18940.

Wilson L. and Keil K. (1996b) Clast sizes of ejecta from explosive eruptions on asteroids: Implications for the fate of the basaltic products of differentiation. *Earth Planet. Sci. Lett., 140,* 191–200.

Evolution of Asteroidal Cores

N. L. Chabot
The Johns Hopkins Applied Physics Laboratory

H. Haack
University of Copenhagen

Magmatic iron meteorites provide the opportunity to study the central metallic cores of asteroid-sized parent bodies. Samples from at least 11, and possibly as many as 60, different cores are currently believed to be present in our meteorite collections. The cores crystallized within 100 m.y. of each other, and the presence of signatures from short-lived isotopes indicates that the crystallization occurred early in the history of the solar system. Cooling rates are generally consistent with a core origin for many of the iron meteorite groups, and the most current cooling rates suggest that cores formed in asteroids with radii of 3–100 km. The physical process of core crystallization in an asteroid-sized body could be quite different than in Earth, with core crystallization probably initiated by dendrites growing deep into the core from the base of the mantle. Utilizing experimental partitioning values, fractional crystallization models have examined possible processes active during the solidification of asteroidal cores, such as dendritic crystallization, assimilation of new material during crystallization, incomplete mixing in the molten core, the onset of liquid immiscibility, and the trapping of melt during crystallization.

1. INTRODUCTION

From Mercury to the moons of the outer solar system, central metallic cores are common in the planetary bodies of our solar system. However, the only samples we have of any of these planetary cores are some iron meteorites, believed to have originated from the cores of up to 60 different asteroids. As our only samples of any planetary cores, these iron meteorites provide a unique opportunity to study planetary cores and how cores evolve. This chapter reviews what we have learned about the evolution of asteroidal cores from studies based on iron meteorites. Section 2 introduces iron meteorites, providing general compositional and structural information and establishing a basis for the following sections. Section 3 presents different timescales that have been determined for the evolution of asteroidal cores. Section 4 discusses the physical process of core crystallization in an asteroid-sized parent body, and section 5 compares recent core crystallization models.

2. OBSERVATIONS FROM "MAGMATIC" IRON METEORITES

2.1. Statistics of Iron Meteorites

All iron meteorites are classified as belonging to one of 13 groups or as being "ungrouped." A group is defined as having a minimum of five members that appear to be genetically related and are consequently believed to have formed on the same parent body. The classification is based largely on the trace element concentrations in the metal of the iron meteorites, although other factors, such as the structure of the metal and the presence of secondary minerals, are also considered (*Scott and Wasson,* 1975). In the first attempt to classify iron meteorites, groups I–IV were defined on the basis of their Ga and Ge concentrations. As more and better data became available, complex names such as the IAB complex and IIIAB evolved. For a recent review of the nomenclature and history of iron meteorite classification, see *Haack and McCoy* (2003). The ungrouped iron meteorites, those that have less than four other iron meteorites to which they seem to be related, may sample up to 50 different asteroids (*Scott,* 1979b; *Wasson,* 1990). The ungrouped irons were previously referred to as "anomalous," but the term "ungrouped" is preferred to avoid suggesting a different history for the ungrouped and grouped meteorites (*Scott,* 1979b; *Kracher et al.,* 1980). Table 1 lists the number of meteorites per group for irons that have been analyzed at UCLA by J. T. Wasson and colleagues, which is estimated to be >90% of all iron meteorites that are available for such analysis (J. T. Wasson, personal communication, 2004). Iron meteorites exhibit a wide range of chemical diversity, as displayed in Fig. 1.

The groups are further subdivided into two main types: those believed to sample central metallic cores and those that do not. Significant differences exist between the two types of groups. "Magmatic" iron meteorite groups exhibit similar elemental fractionation trends within each group that are generally consistent with being formed by fractional crystallization of a large, single melt body, commonly agreed to be the central metallic cores of asteroid-sized parent bodies (*Scott,* 1972, 1979a; *Scott and Wasson,* 1975; *Wasson,* 1985; *Haack and McCoy,* 2003). In contrast, the "nonmagmatic" or "silicate-bearing" groups do not have large frac-

TABLE 1. Classification of iron meteorites.

Group	No. of members*
Magmatic	
IIIAB	217 (31.5%)
IIAB	71 (10.3%)
IVA	59 (8.6%)
IID	14 (2.0%)
IIIE	12 (1.7%)
IVB	11 (1.6%)
IIIF	9 (1.3%)
IC	7 (1.0%)
IIC	7 (1.0%)
IIF	6 (0.9%)
IIG	5 (0.7%)
Silicate-bearing	
IAB complex†	171 (24.8%)
IIE	17 (2.5%)
Ungrouped	83 (12.1%)

*Number in parentheses is the percent relative to all irons. Statistics include all irons that have been analyzed by J. T. Wasson and colleagues at UCLA, which is estimated at >90% of all iron meteorites that are available for such analysis. Statistics furnished by J. T. Wasson, and are the most recent as of March 2004.

†IAB complex discussed in *Wasson and Kallemeyn* (2002).

tionations in many siderophile elements, such as Ir, and have more variable Ga, Ge, and Ni contents as compared to the ranges of these elements within any one magmatic group, as shown in Fig. 1 (*Scott,* 1972; *Scott and Wasson,* 1975). Additionally, the silicate-bearing iron meteorites commonly contain silicate inclusions, often of primitive, chondritic compositions (*Bunch et al.,* 1970; *Bild,* 1977; *Scott and Wasson,* 1975; *Wasson and Wang,* 1986; *Mittlefehldt et al.,* 1998). The silicate-bearing irons are thus not believed to have experienced the same fractional crystallization process as the magmatic irons.

The terms "magmatic" and "nonmagmatic" have been used as labels for the two group types (*Wasson,* 1985; *Wasson and Wang,* 1986), although such terms can be deceiving, as both types of iron meteorites are currently believed to have formed from metal that was molten at one time. The label "silicate-bearing" is consequently preferred to the term "nonmagmatic" by some (*Mittlefehldt et al.,* 1998; *Haack and McCoy,* 2003). Regardless of the label, the distinction between the two types of groups is an important one, as the different types of groups appear to have formed by different processes. Theories for the formation of silicate-bearing iron meteorites are currently debated and include processes such as the crystallization of a S-rich metal (*Kracher,* 1982; *McCoy et al.,* 1993), the breakup and reassembly of an incompletely differentiated parent body (*Benedix et al.,* 2000), the partial separation of silicate and metallic melts in the parent asteroid (*Takeda et al.,* 2000), and the generation of molten metal in impact melt pools on chondritic asteroids (*Choi et al.,* 1995; *Wasson and Kallemeyn,* 2002). The rest of this chapter deals with magmatic iron meteorites and the insights that they have provided about the evolution of asteroidal cores.

2.2. Metallic Structure and Composition

As the name implies, iron meteorites are dominantly composed of Fe-Ni metal. The slow cooling of Fe-Ni metal allows the formation of the Widmanstätten pattern, which develops as an intergrowth of low-Ni kamacite and high-Ni taenite crystals during cooling (see review by *Buchwald,* 1975). The exact structure of the Widmanstätten pattern depends on the composition and the cooling rate of the metal. Iron meteorites that exhibit a Widmanstätten pattern are labeled as octahedrites; octahedrites are further subdivided into five types, ranging from coarsest to finest depending

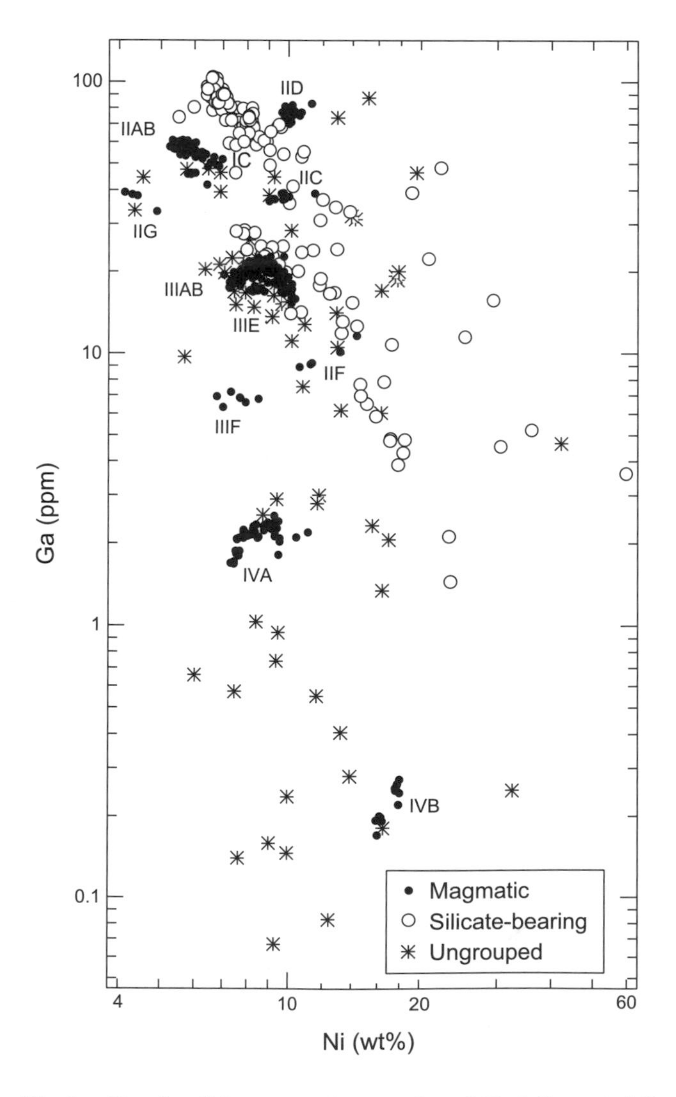

Fig. 1. Data for all iron meteorites are plotted. Each "magmatic" group exhibits limited variations in Ga and Ni, in contrast to the "nonmagmatic" irons. The ungrouped irons have a large range of chemical compositions. Magmatic groups, believed to each represent samples from a different asteroidal core, are labeled in the figure. References for the iron meteorite compositional data are given in the text.

Fig. 2. The IID magmatic iron meteorite Carbo exhibits a well-formed Widmanstätten pattern characteristic of a medium octahedrite. Two troilite inclusions are also shown, the larger of which is 38 mm long. Carbo specimen (#1990.143) was provided by the Geological Museum, University of Copenhagen.

on the width of the kamacite bands. Iron meteorites with high Ni contents (generally ≥15 wt%) do not begin to grow kamacite crystals until lower temperatures and consequently any intergrowth texture is only microscopically visible; these meteorites have been termed ataxites. The metal in low-Ni iron meteorites (<6 wt%) also exhibits no visible Widmanstätten pattern due to the metal consisting almost entirely of kamacite; the term hexahedrites refers to such meteorites. The growth of the Widmanstätten pattern is discussed in detail in section 3.4, as understanding its formation has led to information regarding the cooling rates of asteroidal cores. Figure 2 shows an example of the Widmanstätten pattern in the medium octahedrite IID magmatic iron meteorite Carbo. Numerous images and descriptions of the different metallic structures observed in iron meteorites can be found in *Buchwald* (1975).

The vast majority of compositional information about the metal in iron meteorites comes from an ongoing series of papers by J. T. Wasson and colleagues regarding the chemical classification of iron meteorites (*Wasson,* 1967, 1969, 1970; *Wasson and Kimberlin,* 1967; *Wasson and Schaudy,* 1971; *Schaudy et al.,* 1972; *Scott et al.,* 1973; *Scott and Wasson,* 1976; *Kracher et al.,* 1980; *Malvin et al.,* 1984; *Wasson et al.,* 1989, 1998). Additional compositional studies are listed in the recent review of *Haack and McCoy* (2003). Figure 3 shows some elemental trends for the three largest magmatic iron meteorite groups: IIIAB, IVA, and IIAB. Traditionally, element vs. element diagrams for iron meteorites were plotted against Ni (e.g., *Scott,* 1972), but recent studies have shown that due to the larger variation in Au within iron meteorite groups and the precision of Au measurements, Au is a better choice (*Haack and Scott,* 1993; *Wasson et al.,* 1998; *Wasson,* 1999). Magmatic groups cluster in different areas on Figs. 1 and 3 due to the specific history each group experienced prior to crystallizing in an asteroidal core. Both nebular and planetary processes, such as condensation from the solar nebula, accretion of the asteroid parent body, oxidation, and core formation, may have affected the mean compositions of the groups (*Scott,* 1972; *Haack and McCoy,* 2003). The asteroidal cores sampled by magmatic iron meteorites thus had different initial bulk compositions. However, the similar elemental fractionation trends suggest a similar evolution of the asteroidal cores subsequent to their formation. The trends are attributed to the fractional crystallization of the initially fully molten metallic cores (*Scott,* 1972), and recent crystallization models are discussed in section 5.

2.3. Secondary Minerals

Magmatic iron meteorites contain a number of secondary minerals, the most common of which are troilite (FeS), schreibersite ($(FeNi)_3P$), graphite (C), cohenite ($(FeNi)_3C$), chromite ($FeCr_2O_4$), and daubréelite ($FeCr_2S_4$). These minerals were recently described by *Mittlefehldt et al.* (1998). A comprehensive list of minerals in iron meteorites may also be found in *Buchwald* (1975). *Mittlefehldt et al.* (1998) lists a total of 36 minerals in iron meteorites, where 22 are either rare or occur only in one or a few meteorites. With the exception of lonsdaleite and diamond, which probably are a result of transforming graphite during crater-forming events on Earth, all these minerals are believed to have formed within the core of the parent body. Several of the secondary minerals are common in some groups of iron meteorites but not in others. This is not only useful for classification purposes but also demonstrates that the bulk chemistry differed from group to group. Two main trends are apparent. One is that S- and P-rich minerals, such as troilite and schreibersite, decrease in abundance from the original group I through groups II, III, and IV. The second is that some groups appear to have crystallized from a more-reduced core liquid than others. The best example is group IIAB, where graphite and cohenite are common but phosphates are never observed. The apparently most oxidized group is IIIAB, whose meteorites contain phosphates (e.g., *Olsen et al.,* 1999) but never graphite and only rarely cohenite.

2.4. Evidence from Cape York Irons

The IIIAB Cape York irons, with a recovered mass of 58 tons, form the largest known meteorite shower and are only exceeded in mass by the 60-ton IVB meteorite, Hoba. The Cape York irons deserve special attention since they are the only iron meteorites where chemical gradients of

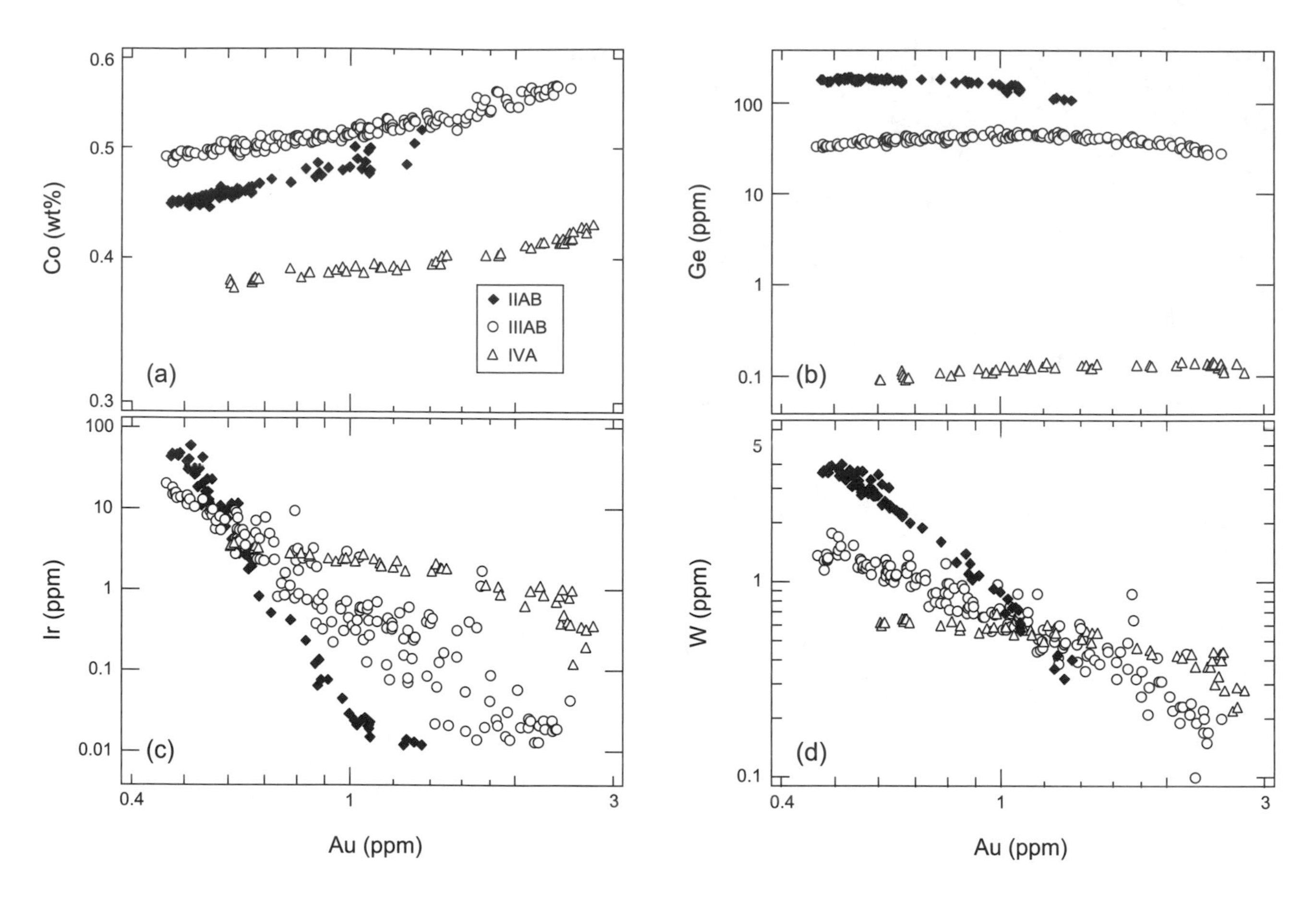

Fig. 3. The concentrations of **(a)** Co, **(b)** Ge, **(c)** Ir, and **(d)** W in the three largest magmatic iron meteorite groups, IIAB, IIIAB, and IVA, are plotted against the concentration of Au. Element vs. element plots, such as these and many others, are used to classify iron meteorites into different groups. The similar fractionation trends observed in the groups are attributed to the crystallization process experienced by all the groups in their parental asteroidal cores. Meteorite data provided by J. T. Wasson.

the parent core have been studied. Chemical analysis shows that the Cape York irons are chemically diverse and form a chemical trend different from that of the other IIIAB iron meteorites (*Esbensen and Buchwald,* 1982; *Esbensen et al.,* 1982). The second largest recovered Cape York mass, the 20-ton Agpalilik, is itself significantly chemically zoned; a 90-cm-long profile across a slice of Agpalilik has an Ir variation of about 12% and a Au variation of about 4%. The magnitude of the chemical gradient observed within the Agpalilik mass suggests that the distance between the chemical end members, Thule and Savik, in the original Cape York meteoroid was on the order of 10 m, meaning that chemical gradients must have existed in the IIIAB core on a much smaller scale than the approximately 10-km radius of the entire core (*Haack et al.,* 1990; *Haack and Scott,* 1993).

For those elements where the Cape York trend deviates significantly from the general IIIAB trend, the scatter from the general trend by other IIIAB irons is also greater, suggesting that the process responsible for the deviating Cape York trends is also potentially responsible for the scatter in the compositions of the other IIIAB irons (*Haack and Scott,* 1992; *Haack and Scott,* 1993). *Wasson* (1999) suggested that the divergent trends observed among Cape York irons represented a mixing line between trapped melt and solid. Additionally, elongated troilite nodules with buoyant phosphate at one end and dense chromite grains at the opposite end seem to indicate the direction of the ancient gravity field (*Buchwald,* 1971; *Buchwald,* 1975). The inferred ancient gravity field is perpendicular to that of the chemical gradient and thus argues against a concentric growth pattern for the core akin to the crystallization of Earth's core. These observations suggest that the structure of the crystallizing solid was complex, at least in the portion of the IIIAB core where the Cape York irons crystallized.

2.5. Relationships to Other Meteorite Classes

For samples of an asteroidal core to reach Earth, the parent body was likely significantly disrupted. Unfortunately, it seems that our meteorite collections contain surprisingly few samples of the corresponding silicate portions of the planetary bodies from which magmatic iron meteorites originated.

Pallasites, composed dominantly of mixtures of olivine and metal, are commonly believed to be samples from the core-mantle boundary of asteroid-sized parent bodies. The trace element composition of the metal in main-group pallasites overlaps with the concentrations measured in more-evolved IIIAB irons (*Scott,* 1977; *Wasson and Choi,* 2003), suggesting a genetic relationship. The O-isotopic compositions of IIIAB irons and main-group pallasites are also consistent with being derived from a single parent body (*Clayton and Mayeda,* 1996).

Oxygen-isotopic abundances for mesosiderites and HED (howardite, eucrite, diogenite) meteorites also fall into the same grouping as main-group pallasites and IIIAB irons (*Clayton and Mayeda,* 1996). Mesosiderites, like pallasites, are mixtures of metal and silicate, but the silicate component is mineralogically more consistent with a crustal rather than a mantle origin (*Mittlefehldt et al.,* 1998). Analysis of the metal in mesosiderites revealed a composition similar to that of IIIAB irons, but the small range of measured Ir concentrations is inconsistent with the source of the metal for mesosiderites being a fractionally crystallized core (*Hassanzadeh et al.,* 1990). The HED meteorites are crustal igneous rocks (*Mittlefehldt et al.,* 1998), suggested to be samples from the large asteroid 4 Vesta (*Consolmagno and Drake,* 1977; *Binzel and Xu,* 1993; *Keil,* 2002). If HED meteorites are from Vesta, it seems that the main-group pallasites and IIIAB irons must be from a different parent body than the HED meteorites; having >200 samples of the central metallic core of an asteroid would seem to require severe disruption of the body. However, if Vesta is not the HED parent body (*Cruikshank et al.,* 1991), a genetic link between HED meteorites, main-group pallasites, and IIIAB irons could be suggested by the O isotopes, although unrelated groups of meteorites with similar O isotopes do exist. The IVA irons have O-isotopic ratios that are distinct from any known achondrite group (*Clayton and Mayeda,* 1996).

Thus, while sampling at least 11 different asteroidal cores, our meteorite collections seem to be lacking the corresponding crustal and mantle materials for the same parent bodies. Intuitively, iron meteorites are more resistant to terrestrial weathering than stony meteorites, resulting in longer survival times on Earth. The strength of Fe-Ni metal would also make it more durable than silicate materials in space, and the longer cosmic-ray exposure ages of irons than chondrites seem to support this statement (*Voshage and Feldman,* 1979; *Wasson,* 1985). *Bottke et al.* (2000), studying the Yarkovsky effect, found that stony meteorites are transferred from the asteroid belt to the inner solar system on much shorter timescales than iron meteorites, consistent with the differences in exposure ages. Whether the Yarkovsky effect, the greater strength in space, and/or the higher resistance to terrestrial weathering is enough of an explanation to account for the overabundance of irons remains to be thoroughly investigated.

3. TIMESCALES OF CORE EVOLUTION

3.1. Formation of Asteroidal Cores

The separation of metal from silicate to form the central metallic cores of asteroid-sized bodies occurred early in the history of the solar system (e.g., *Wadhwa et al.,* 2006). A few Hf-W studies have provided specific information regarding the timing of the formation of metallic cores on the iron meteorite parent bodies. The short-lived ^{182}Hf-^{182}W isotope system, with a half-life of 9 m.y., has been used to date the relative timing of core formation in planetary bodies (*Harper and Jacobsen,* 1996). Due to its lithophile tendency, Hf will remain in the silicate mantle during metal-silicate differentiation; in contrast, some portion of the moderately siderophile W will partition into the forming metallic core. If differentiation occurred when there was still a substantial amount of live ^{182}Hf, the metal would consequently have a deficit of ^{182}W relative to an undifferentiated chondritic value.

All measurements of magmatic iron meteorites show a deficit of ^{182}W (*Lee and Halliday,* 1995, 1996; *Horan et al.,* 1998), indicating that the formation of the parent body cores occurred early in the history of the solar system, such as within the first 30 m.y. (*Yin et al.,* 2002; *Kleine et al.,* 2002). Additionally, no resolvable W-isotopic variations between the groups have been detected. Thus, all the parent bodies underwent core formation within a very narrow time interval of <6 m.y. (*Horan et al.,* 1998). Table 2 lists the relative timing of core formation on four of the larger magmatic iron meteorite parent bodies, which show no resolvable differences between the groups within the given errors. Because early and late crystallizing members for each group show no ^{182}W differences, *Horan et al.* (1998) concluded that less than 2% of late segregating metal could have been added to the core during crystallization without leaving a discernable effect on the W-isotopic ratio. This indicates that metal segregation to form the parent cores of iron meteorites was complete to very nearly complete prior to the start of core crystallization.

3.2. Crystallization of Asteroidal Cores

When metallic cores first formed, the cores were completely molten (e.g., *McCoy et al.,* 2006). As the parent bodies cooled, the cores began to solidify by a fractional crystallization process. Some isotopic systems have provided information about how long it took for the cores to crystallize.

3.2.1. Rhenium-osmium system. The best-understood information about the timing of core crystallization comes from studies of the long-lived Re-Os isotopic system. Rhenium-^{187} decays to ^{187}Os with a half-life of 41.6 G.y. (*Smoliar et al.,* 1996). Both elements are highly siderophile and are consequently concentrated in metallic cores during core formation. However, during core crystallization, Re and Os have slightly different tendencies for partitioning between the growing solid core and the remaining molten portion, creating a fractionation of the two elements that can be dated.

The recent study of *Cook et al.* (2004) compiled and compared all the previous Re-Os work for the IIAB and IIIAB groups, finding good agreement between the results from the previous studies. Using the Mn-Cr age of IIIAB phosphates (discussed in section 3.3), *Cook et al.* (2004)

TABLE 2. Summary of timescales of core evolution and implied parent body sizes.

	IIIAB	IIAB	IVA	IVB
Relative timing of core formation, Δt, relative to IIIAB formation (m.y.)*	0 (+2.2, −1.9)	1.8 (+1.4, −1.3)	−0.6 (+2.5, −2.1)	0 (+1.5, −1.4)
Timing of core crystallization (Ma)	4517 ± 28†	4546 ± 46†	4464‡–4606§	4527 ± 29‡
Core cooling rate (K/m.y.)	49 (+38, −21)¶	2**	200–1500††	4300 (+4300, −2150)‡‡
Parent body radius (km)§§	20–30	100	5–10	2–4
Disruption of the core (Ma)¶¶	675 ± 100		450 ± 100	

* *Horan et al.* (1998), ^{182}Hf-^{182}W chronometer, $\pm 2\sigma$.
† *Cook et al.* (2004), ^{187}Re-^{187}Os chronometer, $\pm 2\sigma$.
‡ *Smoliar et al.* (1996), ^{187}Re-^{187}Os chronometer, $\pm 2\sigma$.
§ *Shen et al.* (1996), ^{187}Re-^{187}Os chronometer.
¶ *Rasmussen* (1989b), metallographic techniques, $\pm 1\sigma$.
** *Randich and Goldstein* (1978), modeling phosphide growth.
†† *Goldstein and Hopfe* (2001) and *Yang and Goldstein* (2005), metallographic techniques.
‡‡ *Rasmussen* (1989a), metallographic techniques, $\pm 1\sigma$.
§§ *Haack et al.* (1990), equation (1).
¶¶ *Voshage and Feldmann* (1979), cosmic-ray exposure ages, ±100 is the general estimated uncertainty of the method.

determined absolute ages of 4546 ± 46 Ma and 4517 ± 28 Ma for crystallization of the IIAB and IIIAB cores respectively. Thus, within the given errors, the timing of core crystallization for the IIAB and IIIAB groups show no resolvable difference, although an earlier study by *Smoliar et al.* (1996) on a more limited set of samples suggested otherwise. The IVB core is also found to have crystallized around the time of the IIAB and IIIAB cores (*Smoliar et al.,* 1996). Table 2 summarizes the crystallization ages.

The ability for multiple meteorites from the same group to define a single Re-Os isochron suggests that the amount of time for the core to evolve from completely molten to mostly solidified was relatively quick and not extended over tens of millions of years (*Shirey and Walker,* 1998). Isochrons determined from early-crystallizing IIAB irons and late-crystallizing IIAB irons were found to be indistinguishable within analytical uncertainties, suggesting crystallization of the IIAB core was completed in less than 50 m.y. (*Cook et al.,* 2004). Similarly, *Cook et al.* (2004) found no evidence for resolvable age differences between the early- and late-crystallizing IIIAB members or for the IIIAB Cape York shower. If resolvable Re-Os differences between members of the same group can be measured in the future, such work has the potential to provide information about the rate of crystallization of the parent body cores.

In contrast to the studies of other iron meteorite groups, the work of *Smoliar et al.* (1996) and *Shen et al.* (1996) are in disagreement about the relative timing of the crystallization of the IVA core. Of the ten IVA iron meteorites measured by *Smoliar et al.* (1996), seven defined an isochron with a relative age about 70 m.y. younger than the IIAB irons; the three IVA meteorites that were inconsistent with the isochron suggested an older age. *Shen et al.* (1996) reported that the IVA irons recorded a crystallization age that was older than the IIAB irons by about 60 m.y. Thus, the relative timing of core crystallization on the IVA parent body is not well established. However, although there are contradictory results between the two studies, both suggest that the IIAB and IVA cores crystallized within 100 m.y. of each other.

3.2.2. Platinum-osmium system. Similar to the Re-Os system, Pt and Os are fractionated from each other due to differing partitioning behaviors during the crystallization of the core. The long-lived isotope ^{190}Pt decays to ^{186}Os with a half-life of about 470 G.y. (*Begemann et al.,* 2001). *Cook et al.* (2004) determined Pt-Os ages of 4323 ± 80 Ma for the IIAB group and 4315 ± 29 Ma for the IIIAB group, ages too young to be consistent with the corresponding Re-Os data. Error in the value of the ^{190}Pt decay constant was suggested as the most likely source for the discrepancy. In addition to the IIAB and IIIAB groups having Pt-Os ages unresolvable from each other, Pt-Os isochrons also showed no evidence for age differences within either group. However, due to larger fractionations between Pt-Os than between Re-Os, future Pt-Os studies may place important constraints on the timing of core crystallization.

3.2.3. Palladium-silver system. In contrast to the Re-Os and Pt-Os systems, Pd-Ag is a short-lived isotopic system: ^{107}Pd decays to ^{107}Ag with a half-life of 6.5 m.y. (*Chen and Wasserburg,* 1996). The Pd-Ag chronometer reflects the time of major chemical fractionation of Pd and Ag. However, it is unclear what event caused the major Pd-Ag fractionation recorded in magmatic iron meteorites, with proposed choices including nebular condensation, planetary body formation, and solidification of the parent body core (*Chen and Wasserburg,* 1996). Regardless of the process, all the measured groups showed evidence for excess ^{107}Ag, indicating the presence of live ^{107}Pd at some stage during their formation and implying that the the Pd-Ag system is recording an early fractionation event (*Chen and Wasserburg,* 1996, *Chen et al.,* 2002). Additionally, *Chen and Wasserburg* (1996) noted that the time differences suggested by their Pd-Ag-isotopic data for the IIAB, IIIAB, IVA, and IVB iron meteorite groups were within 12 m.y. of each other.

Recently, a new technique involving multicollector plasma-ionization mass spectrometry has improved the precision of Ag-isotopic measurements, allowing such measurements to be made on samples with lower Pd/Ag ratios than previously possible (*Carlson and Hauri,* 2001). Currently, the new technique has only been applied to one magmatic iron, the IIIAB iron Grant, with results that were consistent with the previous work of *Chen and Wasserburg* (1996). With this advancement, future studies may yield more precise relative Pd-Ag ages.

3.2.4. Technetium-ruthenium systems. In these short-lived isotopic systems, ^{98}Tc decays to ^{98}Ru with a poorly constrained half-life of 4 to10 m.y., and ^{99}Tc decays to ^{99}Ru with a half-life of just 0.21 m.y. Examining meteorites from the IIAB and IIIAB groups, *Becker and Walker* (2003) found no signature from live ^{98}Tc or ^{99}Tc in any of the irons.

3.3. Formation of Secondary Minerals

The formation processes of secondary minerals in iron meteorites are often not fully understood, and consequently, interpreting isotopic age studies of these inclusions can be difficult; it is not always clear what event the isotopic system is recording. However, such studies can still provide some constraints about the crystallization and post-crystallization evolution of asteroidal cores.

In addition to determining Re-Os isochrons for the metal, *Shen et al.* (1996) also dated a schreibersite-metal pair from a IIIAB iron and determined that the schreibersite became closed to the ^{187}Re-^{187}Os system 0.3–0.5 G.y. after the neighboring metal. Schreibersite grows through solid-state precipitation and diffusion (e.g., *Goldstein and Ogilvie,* 1963; *Randich and Goldstein,* 1975), and thus the Re-Os age likely reflects nucleation and growth following crystallization.

The ^{107}Pd-^{107}Ag system has been used to examine both schreibersite and troilite nodules in iron meteorites (*Chen and Wasserburg,* 1990, 1996). In IIIAB and IIAB irons, excess ^{107}Ag was measured in schreibersite but not in troilite. IVA troilite showed measurable amounts of ^{107}Ag, although in many of the IVA nodules, the isotopic ratios were not consistent with ^{107}Ag due to the *in situ* decay of ^{107}Pd. Cooling is not instantaneous for iron meteorites, and ^{107}Pd is a short-lived isotope. Consequently, *Chen and Wasserburg* (1996) suggested that considerable diffusion could have taken place that may have significantly affected the Pd-Ag systematics, making any interpretation of the data difficult.

The short-lived isotope ^{53}Mn decays to ^{53}Cr with a half-life of 3.7 m.y., and an advantage of the Mn-Cr system is that it has been applied to CAIs in the primitive meteorite Allende as well as other specimens (*Birck and Allegre,* 1988). The Mn-Cr isotopic system thus can provide ages relative to the formation of the first material in the solar system. Evidence for excess ^{53}Cr in Mn-rich phosphates of IIIAB irons has been reported (*Davis and Olsen,* 1990; *Hutcheon and Olsen,* 1991; *Hutcheon et al.,* 1992). In the same IIIAB meteorites, chromites, troilite nodules, and other types of phosphates all had normal Cr-isotopic compositions, likely due to the Mn-poor nature of these phases. The presence of excess ^{53}Cr indicates such phosphates formed early in the solar system, when ^{53}Mn was still alive, and IIIAB Mn-rich phosphates are estimated to have formed 7–27 m.y. after the formation of Allende CAIs (*Davis and Olsen,* 1990; *Hutcheon et al.,* 1992).

3.4. Cooling of Asteroidal Cores

Information about the rate of cooling of metallic cores following crystallization comes from understanding the Widmanstätten pattern in magmatic iron meteorites. The Widmanstätten pattern develops as a two-phase intergrowth of the low-Ni, body-centered cubic α-phase kamacite and the higher-Ni, face-centered cubic γ-phase taenite over the temperature range of about 800° to 200°C.

3.4.1. Determining cooling rates. There are two main methods currently in use for determining metallographic cooling rates. The taenite central Ni content method (*Wood,* 1964) plots the central Ni content of measured meteoritic taenite against the halfwidth of the taenite; this plot is compared to computer calculations, and meteorite data are expected to fall along one of the computed isocooling rate curves. The taenite profile-matching method (*Goldstein and Ogilvie,* 1965) works by matching the calculated Ni compositional profiles in taenite obtained from computer simulations with the Ni profiles measured in meteoritic taenite.

Other methods have also been previously used to determine cooling rates, but subsequent work demonstrated that these methods were less accurate. The kamacite bandwidth width method (*Goldstein and Short,* 1967; *Narayan and Goldstein,* 1985) directly related the kamacite bandwidth with the cooling rate, but the assumption that the diffusion distance between neighboring kamacite plates was infinite was shown to be invalid in most cases (*Saikumar and Goldstein,* 1988). Similar to the methods involving taenite, cooling rates were also determined based on the Ni profile in kamacite or the kamacite central Ni content. However, the sensitivity of the methods based on kamacite rather than taenite are much lower and can be insufficient to determine cooling rates accurately to within even an order of magnitude (*Hopfe and Goldstein,* 2001).

Modeling the formation of the Widmanstätten pattern requires knowledge about the relevant diffusion coefficients and phase diagram. Interdiffusion coefficients for Ni in kamacite and taenite were most recently determined by *Dean and Goldstein* (1986). Over the temperature range of 925°–550°C, the experimentally measured interdiffusion coefficients for Ni ranged from 10^{-12} to 10^{-18} cm^2/s. The latest Fe-Ni phase diagram is shown in Fig. 4 (*Yang et al.,* 1996). On the Fe-Ni phase diagram, the $\alpha + \gamma$ (kamacite + taenite) two-phase field was determined largely from the experimental data of *Romig and Goldstein* (1980), who formed coexisting kamacite and taenite over the tempera-

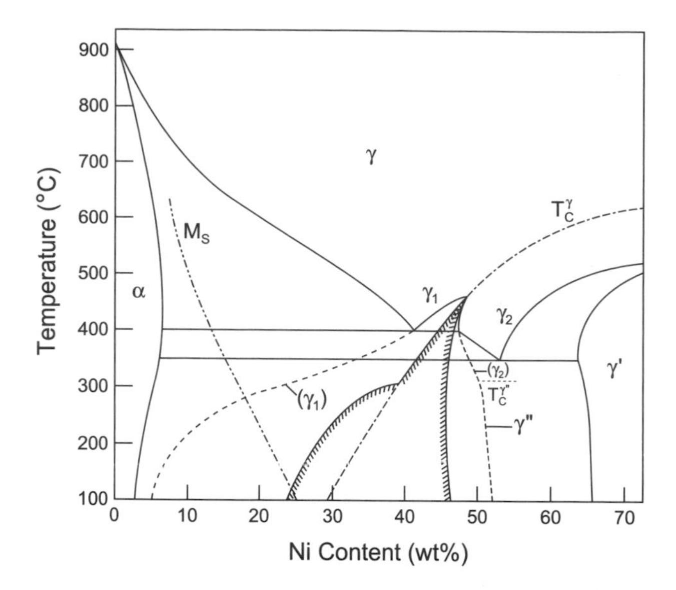

Fig. 4. The subsolidus Fe-Ni phase diagram from *Yang et al.* (1996) is shown. At high temperatures, taenite (γ) is the stable phase. Given the bulk composition of most iron meteorites, as the temperature drops, it is common for iron meteorites to encounter the γ + α (kamacite) field, where the Widmanstätten pattern will begin to form. Many of the phases shown below ~400°C are metastable. The label M_S refers to the low-Ni bcc phase martensite, γ_1 represents a low-Ni paramagnetic fcc phase, γ_2 represents a high-Ni ferromagnetic fcc phase, γ' refers to ordered Ni_3Fe, and γ"represents ordered FeNi. The label T_C refers to the Curie temperature.

ture range of 700°–300°C and measured the compositions of both phases. In contrast, the other phase transformations of taenite below 400°C were determined by examining the metallic regions of meteorites using high-resolution analytical electron microscopy techniques (*Yang et al.,* 1996).

Although composed dominantly of Fe and Ni, magmatic iron meteorites also contain ~0.02–1 wt% P (*Scott,* 1972), which can significantly affect the formation of the Widmanstätten pattern. The presence of P increases the diffusivity of Ni (*Dean and Goldstein,* 1986) and decreases the extent of the α + γ two-phase field (*Romig and Goldstein,* 1980). However, the most important effect P has on the formation of the Widmanstätten pattern could be in determining the nucleation process for the transformation. Experimental studies that began with taenite and were cooled to grow kamacite found that the presence of phosphides was necessary to form intragranular kamacite (*Narayan and Goldstein,* 1984, 1985). In the presence of phosphides, the nucleation of kamacite required little to no undercooling in the experiments. *Narayan and Goldstein* (1985) suggested that intragranular kamacite would not grow until the temperature had dropped to a point where phosphides would form.

In iron meteorites with low P contents, P saturation and the subsequent formation of phosphides will not occur until quite low temperatures, if at all. *Yang and Goldstein* (2005) recently proposed that in low-P meteorites, the growth of kamacite follows the stabilization of the low-Ni bcc martensite phase. They suggested that the growth of kamacite, and hence the formation of the Widmanstätten pattern, will dominantly follow one of two reactions paths, to which they gave numerical labels: In Mechanism II, as the temperature drops, irons with higher P contents will reach P saturation and grow phosphides, leading to the nucleation of kamacite; in Mechanism V, irons with lower P contents will cross the martensite start temperature, forming martensite, which will subsequently transform to kamacite and taenite. *Yang and Goldstein* (2005) indicated that the critical P content that would determine which of the mechanisms was followed was dependent on the exact Ni content of the meteorite but generally ranged from about 0.2–0.3 wt% P.

The taenite in iron meteorites is actually composed of multiple phases, due to a series of complex phase transformations that occur at or below 400°C (*Yang et al.,* 1997a), as shown on Fig. 4. *Yang et al.* (1997b) observed an empirical relationship between the size of the "island phase," found next to the outer taenite rim in meteoritic metal, and the metallographic cooling rate determined by the techniques just discussed. The "island phase" is composed of high-Ni ordered FeNi tetrataenite (γ" on Fig. 4) with a few low-Ni bcc precipitates (martensite on Fig. 4), and *Yang et al.* (1997b) found the width of the island phase varied from 20 to 500 nm as a regular function of the cooling rate. Thus, the size of the island phase can potentially be used to estimate the cooling rate of a meteorite. Although not an independent method because it is fundamentally based on cooling rates determined by other metallographic techniques, such estimates are still valuable for assessing relative rates within groups.

3.4.2. Cooling rates of magmatic irons. A metallic asteroidal core is expected to be essentially isothermal, due to the much higher thermal conductivity of the Fe-Ni metal than that of the overlying mantle. Consequently, if each iron meteorite group samples an asteroidal core, the cooling rates determined for all members of that iron meteorite group should be the same. The most recent estimate for the cooling rate of the IIIAB group is 49 K/m.y. (*Rasmussen,* 1989b). Although there is nearly an order-of-magnitude scatter among the cooling rates of individual IIIAB irons, the lack of a correlation with composition suggests that the results are generally consistent with a single cooling rate and hence with a core origin for the group. Additionally, *Yang et al.* (1997b) found the island phases in four IIIAB iron meteorites to be essentially the same size, also suggesting a common cooling rate.

In IVB irons, due to the high Ni content, the Widmanstätten pattern did not start to form until a lower temperature than in other groups. Applying updated diffusion rate and phase diagram data to the IVB study of *Rasmussen et al.* (1984), *Rasmussen* (1989a) concluded that the IVB data were consistent within a factor of 2 with a cooling rate of 4300 K/m.y. Although *Rasmussen* (1989a) used the kamacite bandwidth method, which is generally considered inaccurate (*Saikumar and Goldstein,* 1988), the result may be correct, as the high Ni content results in large kamacite spacing, potentially making the assumption of infinite diffusion length between kamacite plates valid.

Due to the low Ni content, taenite is only present in some IIAB meteorites, and the metallographic cooling rate methods can only be applied to these irons. *Randich and Goldstein* (1978) developed a technique for estimating the cooling rate of low-Ni IIAB meteorites based on the width of phosphides and suggested that the measured IIAB irons were consistent with a single cooling rate of approximately 2 K/m.y. The higher Ni members of the IIAB group do have a Widmanstätten pattern to which metallographic cooling rate techniques can be applied, although no such study has been undertaken.

In contrast to other magmatic iron meteorite groups, studies have debated whether different IVA meteorites experienced different cooling rates (*Moren and Goldstein,* 1978, 1979; *Willis and Wasson,* 1978a,b; *Rasmussen,* 1982; *Rasmussen et al.,* 1995). Employing the latest *Hopfe and Goldstein* (2001) cooling-rate model, *Goldstein and Hopfe* (2001) found that cooling rates for the IVA group varied from 200 to 1500 K/m.y. and, as found in previous studies, were inversely correlated with the bulk Ni content of the meteorites. Variable cooling rates within the IVA group are inconsistent with a simple core origin, although the chemical trends within the IVA group suggest such an origin. Because silicate inclusions in IVA iron meteorites show evidence of rapid cooling from high temperatures, *Haack et al.* (1996) proposed that the IVA parent body was catastrophically fragmented and subsequently reassembled after the core was nearly crystallized but prior to the formation of the Widmanstätten pattern. Such an event would be consistent with a variation in metallographic cooling rates, but offers no explanation for the apparent correlation between cooling rate and Ni content. In contrast, *Yang et al.* (1997b) found that the size of the island phase in IVA iron meteorites was nearly the same, implying a similar cooling rate for all of the irons at temperatures around 300°C. Citing the constant island phase result, *Wasson and Richardson* (2001) argued that the correlation with Ni content could imply a systematic error in the cooling rate model rather than a true variation in the IVA cooling rates.

Applying the *Yang and Goldstein* (2005) model to the IVA meteorite Bristol, as shown in Fig. 5, a cooling rate of 220 K/m.y. is estimated, which is consistent with the 200–1500 K/m.y. range determined by *Goldstein and Hopfe* (2001). Currently, the *Yang and Goldstein* (2005) model has not been applied to any other IVA irons, so it is not possible to evaluate if the new calculation method will solve the issue of variable cooling rates among IVA irons. Additional work is needed before the cooling rates of IVA meteorites will be fully understood. Table 2 summarizes the most recent estimates for the cooling rates of the IIAB, IIIAB, IVA, and IVB parental cores.

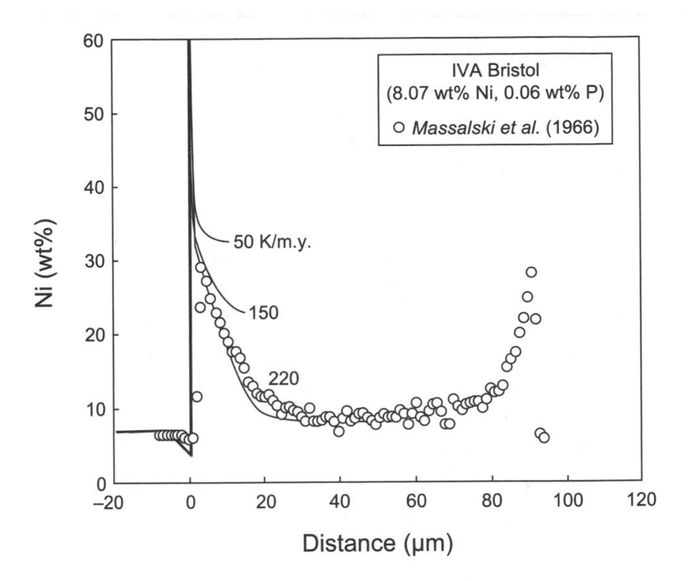

Fig. 5. An application of the taenite profile-matching method for determining the metallographic cooling rate is illustrated for the IVA iron meteorite Bristol. The measurements by *Massalski et al.* (1966) show a characteristic "M"-shaped diffusion profile for Ni across the taenite grain. Because the IVA meteorite Bristol is low in P (0.06 wt%), the model of *Yang and Goldstein* (2005) is used to determine a cooling rate of 220 K/m.y. Model calculations performed by J. Yang and J. I. Goldstein.

3.5. Implications for the Size of the Parent Asteroids

Modeling the thermal evolution of an asteroid-sized body has led to estimates of the sizes of the parent asteroids of different meteorite groups (e.g., *Wood,* 1979). Larger asteroids cool more slowly, resulting in slower cooling rates. Compared to the overlying silicates, the metallic core has an extremely high thermal conductivity. The outer silicate layers thus exclusively control the cooling of the entire asteroid. *Haack et al.* (1990) demonstrated that the presence of an insulating regolith layer on the asteroid would especially influence the cooling rate of the asteroid's interior. Assuming a regolith thickness of 0.3% of the asteroidal radius, *Haack et al.* (1990) introduced an approximate relationship between the radius of the asteroid and the cooling rate of its metallic core

$$R = 149(CR)^{(-0.465)} \qquad (1)$$

The radius of the asteroid, R, is in km, and the cooling rate, CR, is in K/m.y. *Haack et al.* (1990) also modified equation (1) to model the effects of varying mantle conductivities, initial temperatures, concentrations of radioactive elements, temperatures at which the cooling rate is measured, and, most crucial, regolith and/or megaregolith thicknesses. Given that the potential thickness of an asteroidal regolith layer is fairly unconstrained and that a number of factors can influence the calculation, equation (1) provides just a rough estimate for the size of the parent asteroid.

Using equation (1) to estimate the parent body size to just one significant figure, the IIIAB cooling rate suggests a parent asteroid with a radius of 20–30 km while the higher IVB cooling rate implies a smaller radius of just 2–4 km. The limited IIAB cooling rate data suggest a parent body radius of slightly over 100 km. The current cooling rate data for the IVA group are not consistent with a single cooling rate, and consequently equation (1) is not strictly applicable. However, the range in IVA cooling rates suggests a parent body with a radius of about 5–10 km. These estimates are summarized in Table 2. The cooling rates of magmatic iron meteorites are consistent with being cooled in the interiors, not at the surfaces, of asteroids, as would be expected from a core origin, although a large range of parent body sizes is suggested.

3.6. Disruption of the Parent Asteroids

The presence of material from asteroidal cores in our meteorite collections is evidence for the fragmentation and break-up of the parent asteroids. The IIIAB core seems to be compositionally well-sampled by the >200 IIIAB irons, implying that the former asteroidal core has been completely disrupted. While in space, materials are exposed to cosmic rays, and therefore cosmic-ray exposure ages indicate how long a sample has been in space within about 1 m of the surface (e.g., *Eugster et al.,* 2006). Nearly all ^{41}K/^{40}K cosmic-ray exposure ages of IIIAB iron meteorites cluster around 675 ± 100 Ma (*Voshage and Feldmann,* 1979). The similar age implies that the IIIAB irons are from a single parent body that was catastrophically disrupted into meter-sized pieces about 650 m.y. ago by a single event (*Keil et al.,* 1994). Similarly, the large majority of IVA irons have ^{41}K/^{40}K cosmic-ray exposure ages of 450 ± 100 Ma (*Voshage and Feldmann,* 1979), also implying catastrophic fragmentation of the parental core in a single event. Table 2 summarizes the cosmic-ray exposure ages. It is unknown if the IIIAB and IVA parent asteroids still retained any overlying silicate layers at the time of the catastrophic events that disrupted the metallic cores.

In contrast to the IIIAB and IVA groups, the IIAB and IVB irons exhibit a wide range of cosmic-ray exposure ages and hence were likely produced by numerous, separate collisions in the last billion years (*Voshage and Feldman,* 1979). It may thus be expected that remnants of central metallic cores are currently present in the asteroid belt. Due to the relatively featureless spectra of Fe-Ni, iron meteorites have been commonly linked to M-type asteroids, although the M-type spectral signature is also consistent with that of other meteorites; additionally, M asteroids do tend to have higher radar albedos, as would be expected from a metallic composition, but groundbased measurements indicate that the bulk densities may be too low to be Fe-Ni metal without considerable porosity (e.g., *Burbine et al.,* 2002). The parent bodies of magmatic iron meteorites are thus not compellingly linked to any asteroid type.

4. PHYSICS OF CORE CRYSTALLIZATION

4.1. Heat Sources

Although the presence of iron meteorites shows that there was a heat source present in their parent bodies, no evidence of the heat source(s) has been found in these meteorites. Three types of heat sources have been suggested: short-lived radioactive heat sources like ^{26}Al (*Grimm and McSween,* 1993, *Bizzarro et al.,* 2004) and ^{60}Fe (*Tachibana and Huss,* 2003), electromagnetic heating (*Herbert and Sonett,* 1979), and impact heating (*Rubin et al.,* 2001). Long-lived radioactive heat sources, such as ^{235}U, ^{238}U, and ^{232}Th, are not important for asteroid-sized bodies since their heat production within the geological lifetime of asteroids is insignificant (*Haack et al.,* 1990; *Ghosh and McSween,* 1998). The role of impact heating has also been questioned, since impacts large enough to contribute significantly to the heating of a small body were also calculated to be sufficiently energetic to catastrophically disrupt it (*Keil et al.,* 1997). The only heat source that could potentially heat the core directly is ^{60}Fe, although with a half-life of only 1.5 m.y., any live ^{60}Fe had likely decayed prior to the onset of crystallization in the core.

4.2. State of the Core Prior to Crystallization

Under the assumption that asteroidal heat sources were short-lived and became extinct a few million years after accretion, heating times are short compared to the characteristic thermal diffusion time, $t = L^2/D$, for a differentiated asteroid. If the characteristic length scale, L, is 20 km (i.e., the depth of the mantle) and the thermal diffusivity, D, for the mantle is 4.8×10^{-7} m^2/s (*Haack et al.,* 1990), the characteristic thermal diffusion time becomes 25 m.y., which is much longer than the heating time of short-lived heat sources. Since thermal diffusion is a slow process compared to the heating timescale, the temperature distribution of asteroids was controlled by the distribution of heat sources. If ^{26}Al was the dominating heat source, the surface may have been warmer than the interior of the asteroid, due to the concentration of Al in crustal basaltic melts. The core may have been the coldest part of the asteroid since it contained no ^{26}Al and thus may not have been heated, unlike the mantle, following core formation. A partially or fully molten mantle would have been convecting, suppressing any mantle temperature gradients and accelerating cooling of the asteroid. The onset of mantle crystallization would inhibit convection, and the thermal gradient across the mantle would then increase.

Cooling of the core could not begin until a thermal gradient had developed across the mantle in response to cooling at the surface. Depending on the size of the parent body, this would probably take a few tens of millions of years. Although the gravity is a modest 0.1 m/s^2 at the surface of a 50-km-radius asteroid, core convection is still expected to be vigorous at this stage. For a 20-km-radius core, the Rayleigh number is 10^{19}, which is 15 orders of magnitude above the critical value (*Haack and Scott,* 1992). This would lead to thorough stirring of the molten core for a very long time, and it is therefore reasonable to assume that the metallic liquid was homogeneous prior to the onset of crystallization.

With a well-stirred molten core, the core's vertical temperature gradient can be calculated. The temperature gradient can influence if the onset of crystallization is at the center or at the top of the core. The only planetary body for which we can presently study the macroscopic characteristics of ongoing core crystallization is Earth, where the core crystallizes outward from the center. The reason for the outward growth of Earth's core is the increase in pressure across the core from 135.7 GPa at the core-mantle interface to 363.9 GPa at the center of the core (*Dziewonski and Anderson,* 1981). The pressure-induced increase in the liquidus temperature of the metallic liquid is greater than the actual increase in temperature from the core-mantle boundary to Earth's center. In comparison, the pressure at the center of even a 100-km-radius asteroid is less than 0.1 GPa, and the pressure variation across the core is only ~0.025 GPa. Since the liquidus of iron increases by 0.027–0.030 K/MPa (*MacLachlan and Ehlers,* 1971), the total pressure-induced variation in liquidus temperature across an asteroidal core is less than 0.8 K. The actual temperature variation across the core must follow an adiabatic profile if the liquid is well stirred. This gives a temperature gradient of 0.034 K/MPa (*Haack and Scott,* 1992) and a total variation of up to 0.85 K in the core of a 100-km-radius asteroid, which is slightly more than the pressure-induced variation in liquidus temperature.

Given the nearly isothermal state of the core, the first solid to crystallize was probably at the only place where nucleation sites were available: the base of the mantle. The first phase to nucleate for a typical core composition is the γ-taenite phase. Phosphides and other minerals, which could have acted as nucleation sites for the metal, formed as secondary phases (*Doan and Goldstein,* 1970; *Sellamuthu and Goldstein,* 1983). The nucleation of metal at the base of the mantle could have been instantaneous when the liquidus temperature was reached, whereas the absence of nucleation sites in the central parts of the core may have required the metallic liquid to become undercooled before crystallizing.

Another factor that could have controlled the initial crystallization is the shape of the core. The core cavity may not have had time to become perfectly spherical prior to core crystallization, in particular since the driving force, gravity, was small. If core crystallization commenced a few million years after core formation, *Haack and Scott* (1992) found that the core-mantle boundary was likely to have topography on the order of tens of meters. If the mantle was convecting, lateral temperature variations at the base of the mantle could have been coupled with the topographic relief. If such temperature variations were at least a few degrees (*Haack and Scott,* 1992), they could have dominated over the temperature variation in the core.

4.3. The Significance of Sulfur during Core Crystallization

The single most important element in controlling the structure and chemistry of the growing solid in the core is S. Sulfur is almost insoluble in solid metal (*Willis and Goldstein,* 1982) and consequently becomes concentrated in the liquid during crystallization. As a result of their lower densities, S-bearing liquids are buoyant relative to Fe-Ni-rich fluids. The convection pattern in the core following the onset of crystallization was therefore likely controlled by the distribution of S rather than temperature differences. As the liquid S concentration increases from a few weight percent to about 30 wt% at the eutectic composition, the freezing point of the liquid decreases by more than 500 K. Crystallization will thus be enhanced in regions with lower S concentrations relative to areas with higher S contents. Again, this effect is much more significant than any effects from the <1 K temperature variations that were likely present in the core. The distribution of S in the core therefore controls where the crystallization takes place and the convection pattern in the remaining fluid.

Unfortunately, it is difficult to determine the S concentration in the core prior to the onset of crystallization. Due to the low solubility in solid Fe-Ni, S in iron meteorites is almost exclusively found in heterogeneously distributed troilite nodules. This makes it difficult to constrain the concentration of S in the original liquid. Since S can have a significant effect on the partition coefficients between liquid and solid metal, it is possible to indirectly estimate the S content of the liquid by adjusting the S concentration until an acceptable match to the iron meteorite elemental fractionation trends is achieved. This is discussed in section 5. The models suggest that S was present in most cores at the weight percent level, making S one of the three most abundant elements in the cores.

A significant problem in our understanding of the role of S is the large discrepancy between the amount of eutectic liquid predicted by the models and the almost complete absence of samples of the solidified S-rich melt in the form of meteorites. Although S-rich meteorites are probably more fragile and ablate more readily during atmospheric entry (*Kracher and Wasson,* 1982), it remains a mystery why we see so few in our collections.

4.4. Dendritic or Concentric Growth?

Dendritic growth is a consequence of insufficient mixing of the liquid. During crystallization, the boundary layer liquid adjacent to the crystallizing solid becomes enriched in incompatible elements, and if these elements are not removed, crystallization within the boundary layer will be halted. This effect is known as constitutional supercooling. Liquid diffusion may prevent constitutional supercooling if the characteristic diffusion length, $L = (D\ \Delta t)^{1/2}$ (where D is the liquid diffusivity and Δt is a characteristic timescale for crystallization), is shorter than the crystal growth scale in the same time interval. Assuming a D of 10^{-9} cm^2/s (*Geiger et al.,* 1975) and a timescale of 10 m.y. yields a characteristic diffusion length of only 17 cm, which is orders of magnitude less than the characteristic length scale for crystal growth in the same time interval. Liquid diffusion is therefore too slow to prevent the formation of a constitutionally supercooled boundary layer, although it cannot be ruled out that convective mixing may suffice. The boundary layer effect will be more pronounced in areas where the crystallizing front is concave rather than convex. In concave areas, the liquid volume is smaller relative to the surface area of the growing solid and removal of the boundary liquid by convection is less efficient. Only that portion of the solid that extends through the constitutionally supercooled boundary layer, such as the tips, will be able to grow. This will eventually lead to a highly complex solid growth front known as a dendritic structure.

Empirical relationships between dendrite dimensions and cooling rates suggest that kilometer-sized dendrites existed in asteroidal cores (*Narayan and Goldstein,* 1982). In the case of inward growth of the core, the S-enriched boundary layer liquid would concentrate in the upper parts of the core due to its buoyancy, further promoting dendritic growth. Dendrites extending deep into the core would grow in liquid lower in S and consequently may have crystallized faster than solids higher up. The deepest parts of the core-mantle interface would have been the most likely nucleation sites for the earliest dendrites, especially if such locations were associated with cool, mantle downwelling (e.g., Fig. 4 of *Haack and Scott,* 1992). During crystallization, the S-rich liquids produced would rise and accumulate between these protruding parts of the mantle, thereby halting crystallization in the uppermost portions of the core.

4.5. Assimilation of Liquid during Crystallization

The idea of continued addition of metallic liquid to the core during crystallization was originally introduced in order to explain an apparent leveling off of the Ir vs. Ni trend for groups IIAB and IIIAB (*Pernicka and Wasson,* 1987; *Malvin,* 1988). The model of *Malvin* (1988) is discussed in more detail in section 5.5. It is an open question whether the leveling of Ir concentrations in the more evolved iron meteorites of these groups represents a genuine trend representative of the entire core, an artifact of poor sampling, or a local sequence of meteorites similar to the Cape York irons. Physically, a problem with continued assimilation is that during core crystallization, the temperature in most of the mantle had already dropped below the freezing point of the metallic liquid. Release of latent heat from the core during crystallization acts as a buffer for the temperature of the core and results in a steady-state situation where the temperature drops off linearly with decreasing depth in the mantle. Therefore, it does not seem possible for any metallic liquid that accreted to the surface at a late time to subsequently travel all the way to the core. Also, since the mantle is cooling, it seems unlikely that any metal in the mantle that had not already drained down to the core should do so at this stage.

4.6. Trapped Melt

The ubiquitous presence of troilite nodules in even the earliest irons to crystallize seems to require that pockets of melt were trapped (*Esbensen et al.,* 1982; *Haack and Scott,* 1993; *Wasson,* 1999). If the trapped melt had the same composition as the bulk core liquid, it should crystallize just like the rest of the core and thus produce similar fractionation trends. However, if the dimensions of the trapped volume were small enough, solid state diffusion may have erased these trends resulting in a volume of solid metal with the same composition as the original liquid. Assuming a diffusivity of 10^{-4} m^2/s for Ni at 1300°C (*Dean and Goldstein,* 1986), the characteristic diffusion length in a 10-m.y. time interval is 5 m. Therefore, if the growing solid formed isolated pockets with dimensions <10 m, it seems likely that we should see geochemical effects from trapped melt. If the core crystallized as a highly complex dendritic structure, it may be expected that melt was trapped between dendrites, in particular during the later stages of crystallization. These pockets of trapped melt could have spanned a wide range of dimensions, all the way from sub-meter- to kilometer-sized. Eventually, as the dimensions of the residual melt pockets became sufficiently small, all traces of the crystallization process could be erased by solid-state diffusion.

The best examples of geochemical trends caused by trapped melt are probably the Cape York irons. *Esbensen et al.* (1982) and *Wasson* (1999) suggested that the divergent compositional trends, defined by the fragments from the Cape York shower, form a mixing line between coexisting solid and liquid in the crystallizing IIIAB core, the modeling of which is discussed in section 5.8. The observation that fragments plotting closer to the inferred liquid composition are also richer in troilite speaks in favor of the model (*Wasson,* 1999).

4.7. Different Modes of Crystallization

Although we will argue that inward dendritic growth is the most likely mode of crystallization, no alternative can be ruled out conclusively in the absence of any direct observations of asteroidal cores. Given the evidence that iron meteorite parent bodies constituted a highly diverse group in terms of size, peak temperature, bulk chemistry, or impact/collision history (to name just a few parameters for which we may have some constraints), it should also be pointed out that the mode of crystallization may not have been the same for all parent bodies. Two different approaches have

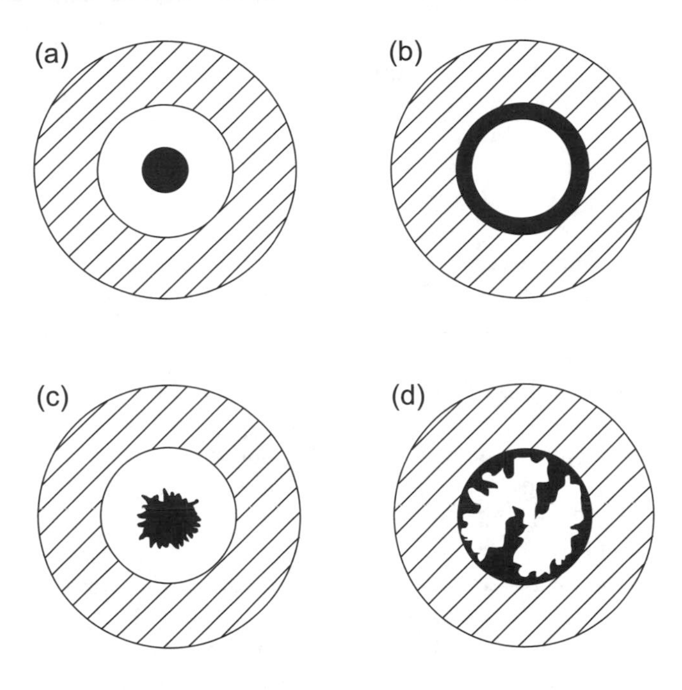

Fig. 6. Cross sections of differentiated asteroids undergoing different modes of core crystallization are depicted: **(a)** Outward concentric growth akin to the ongoing crystallization of the Earth's core (section 4.7.1.). **(b)** Inward concentric growth (section 4.7.3.). **(c)** Outward dendritic growth (section 4.7.2.). **(d)** Inward dendritic growth (section 4.7.4.). Mantle and crust are shown as hatched regions, while liquid metal is white and solid metal is black. Redrawn from *Haack and Scott* (1992).

been used in the past to constrain the crystallization mode of asteroidal cores. One approach is to use the iron meteorite data to construct models that are able to match the observed compositional features. Another approach is to model the physics of asteroidal core crystallization and then evaluate how well the resulting model reproduces the observed properties of iron meteorites. In this section, we take the latter approach and base much of the following discussion on a study of the physics of core crystallization (*Haack and Scott,* 1992). Figure 6 illustrates the different modes of crystallization that we discuss.

4.7.1. Outward concentric growth. Although we have argued that the onset of core crystallization was probably not at the center of the core, we cannot rule out that outward crystallization became important later. This could be the case if the liquid in the outer regions of the core was enriched in S during the course of crystallization. The boundary layer liquid next to the crystallizing solid would be enriched in S and thus rise toward the base of the mantle. In the case of outward concentric growth, the rising S-rich liquid would lead to enhanced convection and very efficient stirring of the liquid. Since the fractional crystallization trends observed within iron meteorite groups indicate that the metallic liquid was fairly well stirred during crystallization, this has been used as an argument for outward crystallization (*Esbensen et al.,* 1982).

There are three arguments against this type of crystallization. Outward concentric growth requires that the solid nucleates at the center of the core where there are no nucleation sites available. Without nucleation sites, this would likely not be possible until the core was significantly undercooled, and central crystallization may therefore not have taken place until significant dendritic growth from the base of the mantle had already occurred. However, it cannot be ruled out that, due to impacts or other means, dendrites or other material may have been dislodged from the ceiling and settled to the center of the core, providing potential nucleation sites. Second, pure outward crystallization could stir the liquid so efficiently that we should expect that the liquid remained well-mixed and thus that the solid crystallized along a well-defined chemical trend. This is inconsistent with the observation that all iron meteorite groups have members that plot far from the average trend. Also, due to efficient mixing, crystallization at the colder, outer regions of the core would become more favorable. This would probably lead to a situation where crystallization was partially in the form of both outward concentric and inward dendritic growth. Third, concentric crystallization is inconsistent with the large gradient of the chemical zoning observed in the Cape York irons. The Cape York chemical gradient is orders of magnitude greater than that expected for concentric crystallization of an asteroidal core; in addition, the chemical gradient is horizontal, not vertical as expected for concentric growth.

4.7.2. Outward dendritic growth. Although it is possible that a fraction of the growth was outward from the center of the core, it seems unlikely that it would be dendritic. The buoyant boundary layer liquid produced at the center of the core would readily rise, which would lead to enhanced convection, thus creating good conditions for concentric growth. It may be argued that at the center of the core, the gravity is zero and the buoyant liquid was thus unable to rise. However, the gravity increases linearly with distance from the center, and the volume with near-zero gravity is insignificant. Also, convection patterns driven by thermal and compositional differences likely incorporated the central liquid as well.

4.7.3. Inward concentric growth. As already argued, the initiation of crystallization was likely at the base of the mantle. At this stage, the base of the mantle probably possessed topography as well as lateral temperature variations related to the convection pattern in the mantle. Therefore, crystallization may not have resulted in a uniform layer of solid metal covering the entire inner surface of the mantle but rather may have resembled isolated patches of metal crystallizing on the coldest parts of the mantle. The onset of crystallization would increase the S concentration of the liquid under the base of the mantle. Since this liquid is buoyant, efficient mixing with the bulk core liquid may have been difficult and inward concentric growth is therefore inconsistent with the observation that all magmatic groups crystallized from initially well-mixed magmas.

4.7.4. Inward dendritic growth. This has been suggested as the most likely mode of crystallization (*Esbensen and Buchwald,* 1982; *Sellamuthu and Goldstein,* 1983; *Haack and Scott,* 1992). As discussed, the initial crystallization was likely at the base of the mantle and resulted in an increase in the S content of the local liquid, which may have been difficult to efficiently remix with the bulk core liquid. Thus, the increasing S concentration would lower the liquidus of the melt, potentially quickly inhibiting further crystallization in the upper parts of the core. Crystallization at the deepest parts of the solid would therefore be enhanced relative to the upper parts, leading to a situation where the growth was dominantly at the tips of dendrites extending deep into the core. The liquid in the upper parts of the core, between the dendrites, would be enriched in S and thus lighter than the central liquid below the deepest dendrite. As the dendrites reach the central parts of the core, liquid enriched in S would rise along the dendrites, forcing core-wide mixing (*Haack and Scott,* 1992). At a later stage, the entire core may have been transformed into a network of dendrites separated by residual liquid. Eventually, these dendrites would inhibit core-wide mixing and create pools of trapped melt.

Inward dendritic growth seems consistent with the iron meteorite data. The observed increasing scatter of the chemical trends as crystallization proceeds is to be expected if a complex network of dendrites develops across the core, inhibiting core-wide mixing and facilitating the trapping of melt in pockets between dendrites. The large horizontal chemical gradients in the Cape York irons are also consistent with inward dendritic growth. The larger surface area of the solid in the case of dendritic growth results in chemical gradients over shorter distances; the entire chemical evolution of the core could have been recorded in a single dendrite cross section if the dendrite had grown slowly since the onset of crystallization. However, the compositional trends recorded across a growing dendrite are not expected to differ significantly from the average trend recorded throughout the crystallizing core, and the deviation of the Cape York compositional trends from those of the average IIIAB group has been suggested to be due to melt trapping during core crystallization (*Wasson,* 1999). Although growing dendrites are not the only way to trap melt, a complex solid structure, such as the one expected from inward dendritic growth, could lead to increased amounts of trapped melt.

If main-group pallasites are samples of the core-mantle boundary of the IIIAB core, their metal composition should allow us to determine when solid metal formed at the base of the mantle. Assuming that the core crystallized inward from the mantle, we would expect the metal in main-group pallasites to be similar to the IIIAB iron meteorites representing early crystallization. In actuality, main-group pallasites contain Ni-rich metal that is consistent with forming after about 80% of the IIIAB iron meteorites had crystallized (*Scott,* 1977; *Haack and Scott,* 1993; *Wasson et al.,* 1999; *Wasson and Choi,* 2003). Possible scenarios that have been suggested to reconcile inward growth with the pallasite compositions include crystallizing the pallasites at a late stage from pockets of S-rich liquid trapped between dendrites, mixing of early crystallized metal with late stage melt, and dragging minor pieces of the mantle into the core by dendrites broken off from the base of the mantle.

4.8. Late-Stage Crystallization: The Role of Liquid Immiscibility

The last stages of crystallization of the core are by far the least understood. The large scatter in the compositional trends suggest that crystallization was quite far from ideal fractional crystallization during this time, we lack S-rich meteorites from this late period, and liquid immiscibility could have played a significant but poorly understood role during this stage. The Fe-S-P phase diagram predicts that at high S and P concentrations, the metallic liquid will enter a two-phase field where the liquid splits into a S-rich liquid and a P-rich liquid (*Raghavan,* 1988). It has been hypothesized that the size of the liquid immiscibility field may be influenced by the oxygen fugacity (*Malvin et al.,* 1986), and experiments have suggested that for conditions relevant to the crystallization of iron meteorites, the size of the liquid immiscibility field may be decreased from that given by the Fe-S-P phase diagram (*Chabot and Drake,* 2000). However,

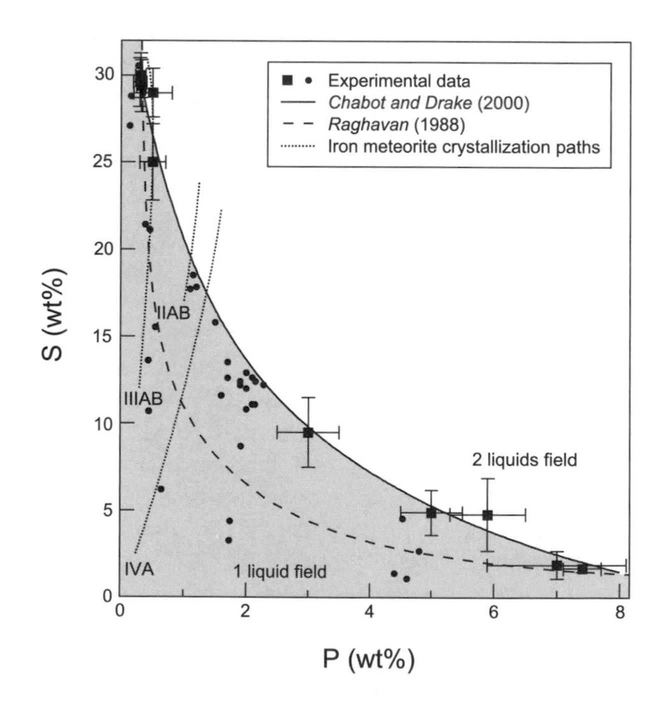

Fig. 7. A large liquid immiscibility field exists in the Fe-Ni-S-P system, shown by the unshaded region. The size of the liquid immiscibility field determined by experiments examining the crystallization of iron meteorites [solid line (*Chabot and Drake,* 2000)] is smaller than the two immiscible liquids field given by the Fe-S-P phase diagram [dashed line (*Raghavan,* 1988)]. However, in either case, the IIAB, IIIAB, and IVA groups may have eventually encountered liquid immiscibility during their crystallization, depending on the S-content estimates used for the modeling of each group. References for the experimental data and iron meteorite crystallization paths are given in *Chabot and Drake* (2000).

as shown in Fig. 7, with either of the two-phase field boundaries, the IIAB, IIIAB, and IVA groups may still be expected to have encountered liquid immiscibility during their evolution, although this is dependent on the S contents of their cores, estimates of which are discussed in section 5. Liquid immiscibility may have eventually occurred in the large majority of asteroidal cores, as P and S are both incompatible and thus become increasingly concentrated in the liquid as crystallization proceeds. The effects from liquid immiscibility were potentially quite important for the IIAB group, which may have been rich in S and P and very reduced.

The resulting effects of liquid immiscibility depend strongly on how well the core liquid remained mixed during continued crystallization. If the P- and S-rich liquids remained in equilibrium with each other, the solid crystallizing from the two liquids would be different from the solid crystallizing in a situation where there was not liquid immiscibility, but the effect would be fairly minor. However, if the two liquids, which have very different densities, became isolated from one another, the solids crystallizing from the two different liquids would follow divergent trends. The effects of liquid immiscibility on the IIIAB trend, as modeled by *Ulff-Møller* (1998), are discussed in section 5.7. Having disequilibrium between the two immiscible liquids may not have been unlikely toward the end of crystallization, especially if the structure of the solid was highly complex and core-wide mixing was inhibited. Lacking good constraints in the form of iron meteorites representing the last stages of crystallization, we seem to be quite far from a good understanding of the last, complex portion of the crystallization process.

5. MODELING CORE CRYSTALLIZATION

5.1. Fractional Crystallization

Asteroidal cores seem to have solidified by fractional crystallization, accounting for the well-defined elemental trends observed within the magmatic iron meteorite groups, such as shown in Fig. 3. The process of fractional crystallization can be modeled using simple mass balance equations

$$C_S(E) = \frac{[D(E)][C_i(E)]}{[1 - f + fD(E)]} \quad (2)$$

$$C_L(E) = \frac{C_S(E)}{D(E)} \quad (3)$$

$C_i(E)$ is the initial concentration of an element E in the molten portion of the core, while $C_S(E)$ and $C_L(E)$ are the concentrations in the crystallizing solid and remaining liquid respectively. The variable f refers to the fraction of solid metal formed, and D(E) is the solid metal-liquid metal partition coefficient. Modeling can be done by setting f to a small fraction and determining the composition of the solid metal (C_S) and the remaining metallic liquid in the core (C_L). The composition of the remaining metallic liquid then becomes the initial liquid composition (C_i) for the next solid metal to form, and the calculations are repeated, resulting in fractional crystallization trends.

5.2. Experimental Partition Coefficients

As is clear from equations (2) and (3), how an element will fractionate during crystallization of an asteroidal core is dependent on the element's partition coefficient, D. The partition coefficient is simply the ratio of the concentration of the element in the solid metal divided by the concentration of that element in the liquid metal. Early modeling of iron meteorite crystallization trends used values for the partition coefficients that were constant throughout the crystallization process (e.g., *Scott,* 1972). However, experimental determinations of partition coefficients demonstrated that the concentrations of S and P in the metallic liquid can dramatically affect the value of D (*Willis and Goldstein,* 1982; *Jones and Drake,* 1983).

Numerous experimental solid metal-liquid metal partitioning studies have been conducted in the last 25 years. All the equilibrium experimental data for Ag, As, Au, Co, Cr, Cu, Ga, Ge, Ir, Ni, Os, P, Pd, Pt, Re, and W were recently compiled in *Chabot et al.* (2003) and *Chabot and Jones* (2003). *Malvin et al.* (1986) demonstrated that some early experiments that employed a temperature gradient to attempt to simulate fractional crystallization (*Sellamutha and Goldstein,* 1983, 1984, 1985a,b) did not produce partitioning results consistent with the equilibrium studies; although consequently excluded from the compilations, these studies also illustrated the important influence of metallic liquid composition. Additionally, experimental values for D(Ru), D(Rh), D(Pb), D(Tl), D(Sb), D(Mo), and D(Sn) have also been determined (*Fleet and Stone,* 1991; *Jones et al.,* 1993; *Jones and Casanova,* 1993; *Liu and Fleet,* 2001).

During the fractional crystallization of an asteroidal core, some P will partition into the solid metal, but S, which is effectively excluded from the crystallizing solid, will be concentrated in the metallic liquid. Consequently, the S content of the molten portion of the core will increase as fractional crystallization proceeds. It is thus necessary to account for the compositional dependency of D when modeling the crystallization of iron meteorites. *Jones and Malvin* (1990) developed a parameterization method for expressing D mathematically as a function of the S and P contents of the metallic liquid, based on the theory that the availability of Fe domains in the metallic liquid is the dominant influence on the partitioning behavior. Recently, this parameterization method was revised slightly by *Chabot and Jones* (2003).

Figure 8 shows how the S content of the metallic liquid affects D(Au), D(Ge), and D(Ir), three elements commonly measured in iron meteorites. The partition coefficient for Ir changes by about 3 orders of magnitude from the S-free system to a composition near the Fe-FeS eutectic. Although

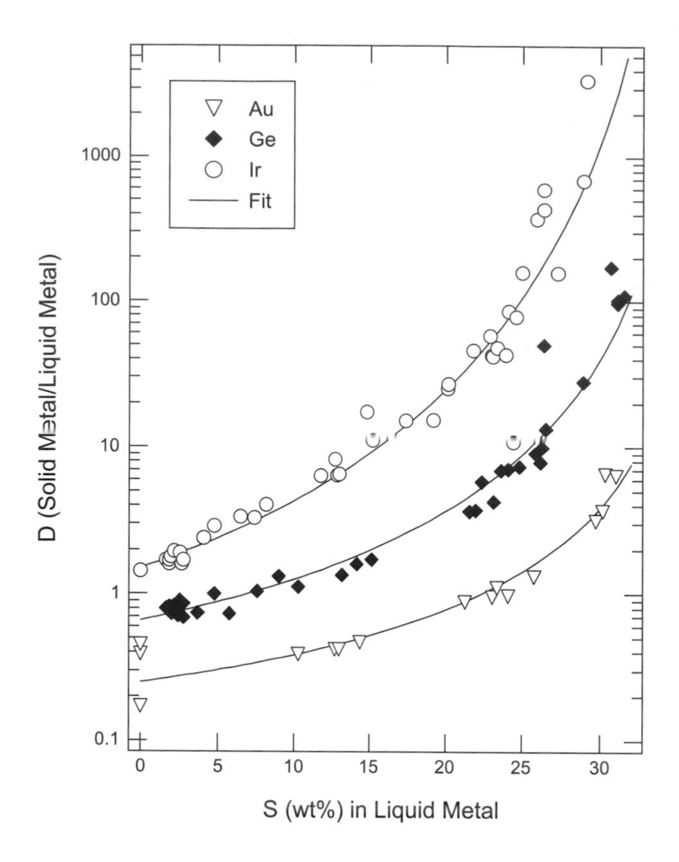

Fig. 8. Experimental solid metal-liquid metal partition coefficients (D) for three siderophile elements commonly measured in iron meteorites, Au, Ge, and Ir, are shown as an example of the significant effects the S content of the metallic liquid can have on the partitioning values. The fitted curves are from the parameterizations of *Chabot and Jones* (2003). Experimental partitioning data are also compiled in *Chabot and Jones* (2003).

the absolute change in D(Ge) is less than that of D(Ir), D(Ge) exhibits the interesting behavior of switching from incompatible (D < 1) to compatible (D > 1) in solid metal with increasing S content of the metallic liquid. The partition coefficients in Fig. 8 all increase with increasing S content, but some elements, such as Ag, Cr, and Cu, are known to be chalcophile and have experimental partition coefficients that decrease with increasing S content.

5.3. Simple Fractional Crystallization Model

With >200 members and a wide range of compositions, the IIIAB group is a good choice for comparing different models of asteroidal core crystallization. Data are available for many elements in iron meteorites, and any successful crystallization model must account for all the elemental trends. However, some elements are not as severely fractionated by core crystallization as others, and a wide range of conditions can explain their trends. Highly siderophile elements, such as Ir, have concentrations that span more than 3 orders of magnitude in the IIIAB group. The elements of Ga and Ge exhibit uniquely curved crystallization trends. Consequently, here, different crystallization models for the IIIAB parent body core are compared by examining trends involving Ir, Ge, and Au. Other elements, such as Pt, Re, and Os, also offer worthwhile challenges for modeling (*Cook et al.,* 2004).

In simple fractional crystallization, the fundamental assumption is that even though the composition of the molten portion of the core changes during crystallization, the metallic liquid quickly restores a single homogeneous composition through vigorous convection. The simple fractional crystallization model has been applied to the IIIAB group by *Willis and Goldstein* (1982), *Jones and Drake* (1983), *Haack and Scott* (1993), *Chabot and Drake* (1999), and *Chabot* (2004), with the difference being simply the amount of experimental partitioning data available at the time of each study. As shown in Fig. 9, the simple fractional crystallization model is able to explain the 3-orders-of-magnitude Ir fractionations and the curved Ge trend. However, the simple fractional crystallization model fails to reproduce the observed Ge vs. Ir IIIAB trend. Specifically, the later crystallizing members of the IIIAB group appear to have higher Ir contents than predicted by the simple fractional crystallization model.

The S content of the parent body core cannot be determined through direct measurements of iron meteorites, since S is excluded from the crystallizing solid. However, the partition coefficients depend strongly on the choice of the S content of the metallic liquid. For the IIIAB group, the best fits are obtained with a S content of about 12 wt% S (*Chabot,* 2004). With an initial S content of 12 wt%, the trends in Fig. 9 are matched by about the first 55% of the core to crystallize. After 61% of the core has solidified, the composition of the remaining metallic liquid is that of the Fe-FeS eutectic. The simple fractional crystallization model thus implies that IIIAB iron meteorites only sample slightly over half of the parent body core. It has been suggested that the weaker nature of any S-rich meteorites would result in a sampling bias in the meteorite collections (*Kracher and Wasson,* 1982), although such an idea is difficult to test.

Simple fractional crystallization cannot explain the observed scatter around the general IIIAB trends in Fig. 9, scatter that is greater than the analytical uncertainty in the measurements (*Wasson,* 1999). The IIIAB elemental trends are generally consistent with being due to fractional crystallization of an asteroidal core, but the crystallization process appears to have been more complex than just simple fractional crystallization.

5.4. Dendritic Crystallization Model

As discussed in section 4, it has been proposed that asteroidal cores solidified through the growth of large dendrites rather than from a planar crystallization front. The dendritic crystallization model of *Haack and Scott* (1993) suggested that light S-rich liquid could accumulate preferentially at

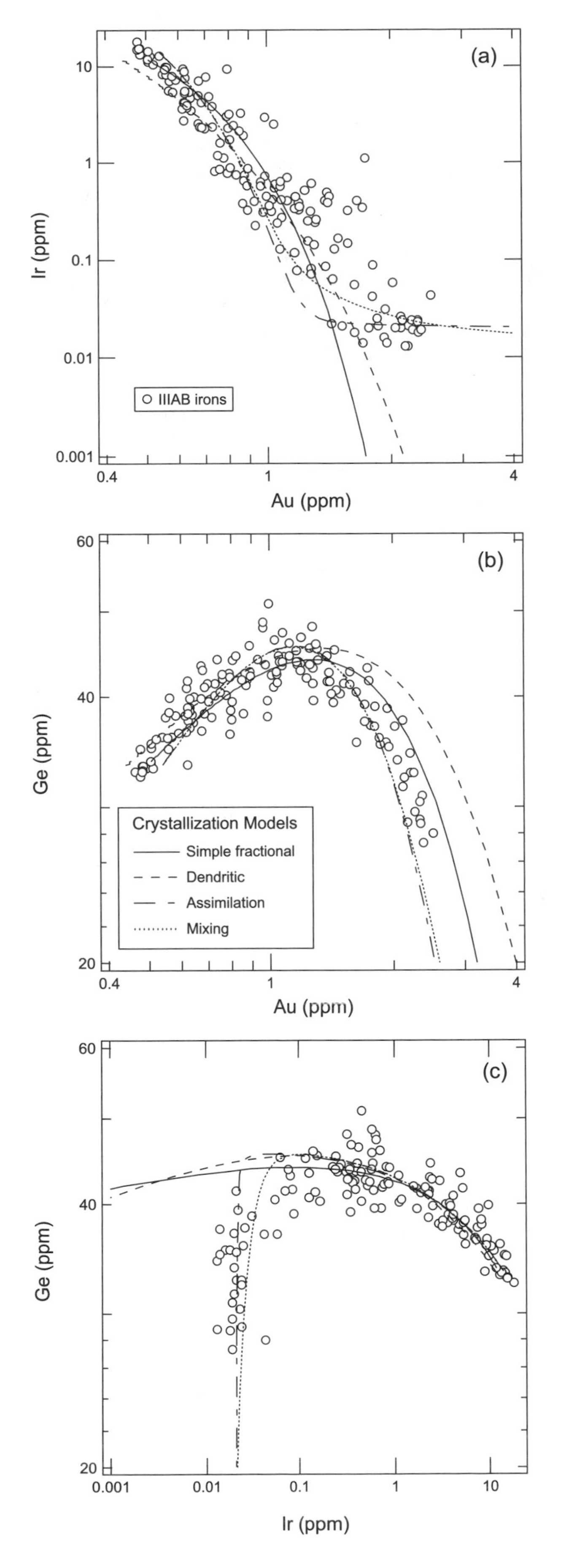

Fig. 9. Four different crystallization models are applied to the IIIAB **(a)** Ir vs. Au, **(b)** Ge vs. Au, and **(c)** Ge vs. Ir trends. All four models can produce the large, 3-orders-of-magnitude variations measured in Ir and the general curved Ge trend. The assimilation fractional crystallization model (*Malvin,* 1988) and the mixing in the molten core model (*Chabot and Drake,* 1999) can also explain the leveling of the late stage Ir values, which the simple fractional crystallization model (*Chabot,* 2004) and the dendritic model (*Haack and Scott,* 1993) cannot. IIIAB iron meteorite data provided by J. T. Wasson.

the top of the core or in structural traps formed by the growing dendrites, effectively removing that liquid from the crystallizing system.

Mathematically, the model of *Haack and Scott* (1993) is identical to simple fractional crystallization except for the value used for D(S). To mimic the loss or trapping of S-rich liquid due to the growing dendrites, *Haack and Scott* (1993) used values for D(S) that ranged from 0.6 to 0.8; a D(S) of ~0.01 is generally used to reflect the very limited solubility of S in Fe-Ni metal. *Haack and Scott* (1993) did not suggest that the solubility of S in metal was larger, but rather, altering D(S) was a simple way to investigate the effects of removing S from the system. Figure 9 shows that the trends produced by the dendritic crystallization model do not differ much from those of the simple fractional crystallization model, especially considering the use of slightly different parameterizations for D(Ir), D(Ge), and D(Au) by the two models. Like the simple fractional crystallization calculations, the dendritic crystallization model also cannot match the IIIAB Ge vs. Ir trend.

However, due to the different D(S) value, *Haack and Scott* (1993) determined an initial S content of 6 wt% for the IIIAB core, half the value determined by the simple fractional crystallization model. Furthermore, with a lower initial S content and by removing S-rich material from the system, the dendritic crystallization model is able to solidify 94% of the core to create the observed IIIAB elemental trends. It is much more satisfying to think that the >200 IIIAB irons sample 94% of the parent body core, as compared to only 55% as implied by the simple fractional crystallization model. Yet, the concentration of troilite in IIIAB irons is too low to account for all the S-rich material removed in the dendritic crystallization model, leading to the conclusion that S-rich material from the IIIAB core is still underrepresented in our meteorite collections.

The growth of large dendrites in the IIIAB core could explain the scatter around the general crystallization trends. Large dendrites could impede thorough mixing in the metallic liquid, leading to local heterogeneities. Although the scatter can possibly be explained by the growth of dendrites in general, the model of *Haack and Scott* (1993) was unable to match the specific elemental trends observed in the Cape York irons.

5.5. Assimilation Fractional Crystallization Model

In the assimilation fractional crystallization model, fresh metal is continually assimilated into the asteroidal core even as the core crystallizes. *Malvin* (1988) suggested that such a scenario could occur if the asteroid parent body was continuing to accrete material during the time the core was crystallizing. The physical plausibility of segregating metallic liquid through a solid mantle can be questioned, as discussed in section 4.5. Additionally, adding liquid to the molten core would be much more difficult if the outermost core was solid Fe-Ni metal due to crystallization by inward dendritic growth. The model calculations of *Malvin* (1988) are the same as for simple fractional crystallization except that during each crystallization step, some amount of original composition liquid is added into the remaining metallic liquid, modifying the composition of the molten core.

Malvin (1988) found that assimilating material into the core at a rate of just 1% of the rate of crystallization gave a better fit to the IIIAB Ir trend than simple fractional crystallization, as shown in Fig. 9. At that low rate, the small amount of new liquid added to the core has little effect on the fractionation trends of most elements. However, if the concentration of an element, such as Ir, becomes severely depleted in the molten core due to its large partitioning value, the continual addition of undepleted metallic liquid, even at a low rate, becomes the primary source for that element in the molten core and the crystallizing solid. Consequently, the crystallization trend of that element will level off at an amount consistent with the amount of new material being supplied to the molten core.

5.6. Mixing in the Molten Core Model

A fundamental assumption of simple fractional crystallization is that the metallic liquid remains well-mixed throughout the crystallization process. *Chabot and Drake* (1999) developed a crystallization model to explore the possible effects that heterogeneities in the molten core could have on the resulting IIIAB crystallization trends. Mathematically, the model is a small variation on the simple fractional crystallization calculations. *Chabot and Drake* (1999) divided the molten core into two zones of liquid, one zone actively involved in the crystallization process and another zone too far from the crystallization front to be immediately involved. In the mixing model, the two zones were allowed to interact by mixing between the two zones and by changing the size of the two zones. If mixing between the two zones was set to a high value, the resulting elemental trends were identical to those from simple fractional crystallization. If, however, mixing was not efficient enough to homogenize the compositions of the two zones, the composition of the crystallizing solid metal could differ from the metal produced by simple fractional crystallization.

Shown in Fig. 9 are the results from a scenario in which the molten core is initially homogeneous but becomes less well-mixed as crystallization proceeds. In the mixing model, having some liquid not actively involved in the crystallization process is able to buffer the Ir concentration in the forming solid metal and consequently match the Ge vs. Ir IIIAB trend. The bulk compositional differences between the two zones are small, but even small heterogeneities can affect the resulting crystallization trends. Based on their mixing model, *Chabot and Drake* (1999) suggested that asteroidal cores crystallized from the core-mantle boundary inward, since the liquid near the crystallization front would be less dense and consequently above the second zone of liquid not actively involved in crystallization. However, it is questionable how possible it is to maintain compositional differences in an essentially isothermal metallic core. Heterogeneities in the molten core could potentially explain the scatter in the IIIAB trend, but *Chabot and Drake* (1999) were not able to match the Cape York trends in any mixing model scenarios. The mixing model of *Chabot and Drake* (1999) also does not propose a unique solution, since multiple scenarios examined in their study resembled the IIIAB trends.

5.7. Liquid Immiscibility Model

As discussed in section 4.8, as an asteroidal core crystallizes, the molten portion of the core becomes enriched in S and P, and its composition may consequently fall into the Fe-Ni-S-P liquid immiscibility field. The liquid immiscibility model of *Ulff-Møller* (1998) is just simple fractional crystallization until the composition of the metallic liquid encounters the liquid immiscibility field on the Fe-S-P phase diagram (*Raghavan,* 1988). *Ulff-Møller* (1998) then allowed the molten core to separate into two immiscible liquids and examined the resulting crystallization trends from three scenarios: crystallization from the P-rich liquid only, crystallization from the S-rich liquid only, and crystallization if the two immiscible liquids remained in intimate equilibrium. *Ulff-Møller* (1998) investigated these three scenarios because they represented the most extreme cases in equilibrium and disequilibrium conditions; any other solution would fall in between these extremes.

The IIIAB Ir vs. Au trend is well matched by the scenario involving equilibrium between the two immiscible liquids, as shown in Fig. 10. Additionally, the Ir vs. Au trend is bounded by the two extreme disequilibrium cases of the model, suggesting the scatter in the trend can be explained by various degrees of local disequilibrium between the immiscible liquids in the molten core. The study of *Ulff-Møller* (1998) does not show the Ge vs. Au trend specifically, but it shows the similar Ga vs. Ni trend and states that the results for Ge were comparable. Like the Ir vs. Au trend, the IIIAB irons that compose the Ga vs. Ni trend are bounded by the two extreme disequilibrium cases, but in contrast, the equilibrium scenario does not match the curved Ga trend. The curved nature of the Ga (and Ge) trend is attributed to the onset of liquid immiscibility and some amount of disequilibrium associated with that event.

As shown in Fig. 10, the liquid immiscibility model cannot explain the IIIAB Ge vs. Ir trend. *Ulff-Møller* (1998) was also unable to produce any trends consistent with those

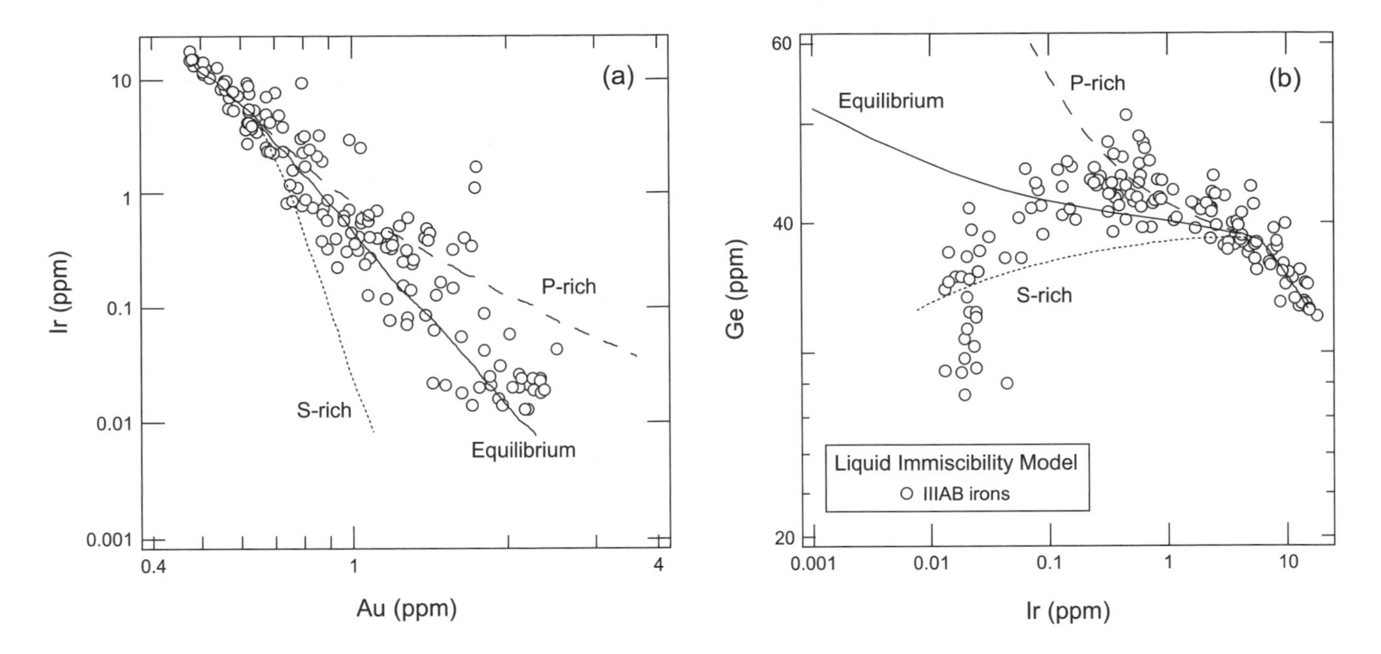

Fig. 10. The effects from the onset of liquid immiscibility in the Fe-Ni-S-P system on the IIIAB **(a)** Ir vs. Au and **(b)** Ge vs. Ir trends are plotted (*Ulff-Møller,* 1998). Three extreme endmember cases are explored: The two immiscible liquids remain in intimate equilibrium during crystallization, crystallization proceeds from the P-rich immiscible liquid only, or crystallization proceeds from the S-rich liquid only. All other cases should fall in between these extremes. IIIAB iron meteorite data provided by J. T. Wasson.

observed in the Cape York irons. However, any shortcomings of the liquid immiscibility model do not mean that liquid immiscibility did not occur in asteroidal cores or that its effects can be neglected while modeling core crystallization.

5.8. Trapped Melt Model

During the crystallization of iron meteorites, some amount of metallic liquid was trapped by the growing solid, as is evidenced by the presence of troilite nodules. *Wasson* (1999) examined the effects that the trapping of liquid during crystallization could have on the observed IIIAB elemental trends, the details of which were later modified slightly by *Wasson and Richardson* (2001) and *Wasson and Choi* (2003). The approach of *Wasson* (1999) in modeling the IIIAB trends differs from the other models discussed in two ways. First, while other models attempted to fit all the IIIAB data with just the crystallizing solid track, *Wasson* (1999) argued that iron meteorites represented mixtures between the solid and liquid compositions. Second, other models all used fitted parameterizations to the experimental partition coefficients and varied the initial S content to fit the IIIAB trends. *Wasson* (1999) based the choice of the initial IIIAB S content on measurements of the amount of troilite in large Cape York specimens and by attributing the troilite to melt trapped during crystallization. By determining the total amount of trapped melt each sample contained, the partition coefficients were varied to produce a match to the IIIAB data, using an initial S content of 2.5 wt% (*Wasson and Richardson,* 2001). Mathematically, the model of *Wasson* (1999) is identical to simple fractional crystallization calculations. The difference is just the choice of partition coefficients. The constraints, or lack thereof, that experimental studies can place on the partition coefficients applicable to iron meteorite crystallization is a subject of debate (*Wasson,* 1999; *Wasson and Richardson,* 2001; *Chabot et al.,* 2003; *Chabot,* 2004).

In the crystallization model of *Wasson* (1999), the scatter in the IIIAB trends is due to different amounts of trapped melt in the IIIAB irons, as illustrated in Fig. 11. In the model, the IIIAB irons are bounded by the compositions of the crystallizing solid metal and the molten core; meteorites that fall to the right of the solid trend on the Ir vs. Au plot do so because they contain some amount of melt that was trapped during crystallization. The solid and liquid paths similarly bound the Ge vs. Au trend and explain the scatter. Thus, the trapped melt model is the first to offer a quantitative explanation for the composition of each individual IIIAB iron. The trapped melt model is also the first to quantitatively explain the Cape York trends; the positions of the Cape York irons are consistent with various degrees of mixing between solid metal and trapped melt when the IIIAB core was about 30% solidified.

The trapped melt model of *Wasson* (1999) is thus a very attractive model for its ability to quantitatively account for the scatter and the Cape York trends in the IIIAB group. Additionally, with a low initial S content of 2.5 wt%, the IIIAB irons are concluded to represent about 80% of the parent body core, and the possibility of liquid immiscibility is greatly diminished. The trapped melt model has not been applied to the Ge vs. Ir trend, but it is possible that the leveling off of the Ir concentrations may be accounted for by a mixing line between the solid metal and coexist-

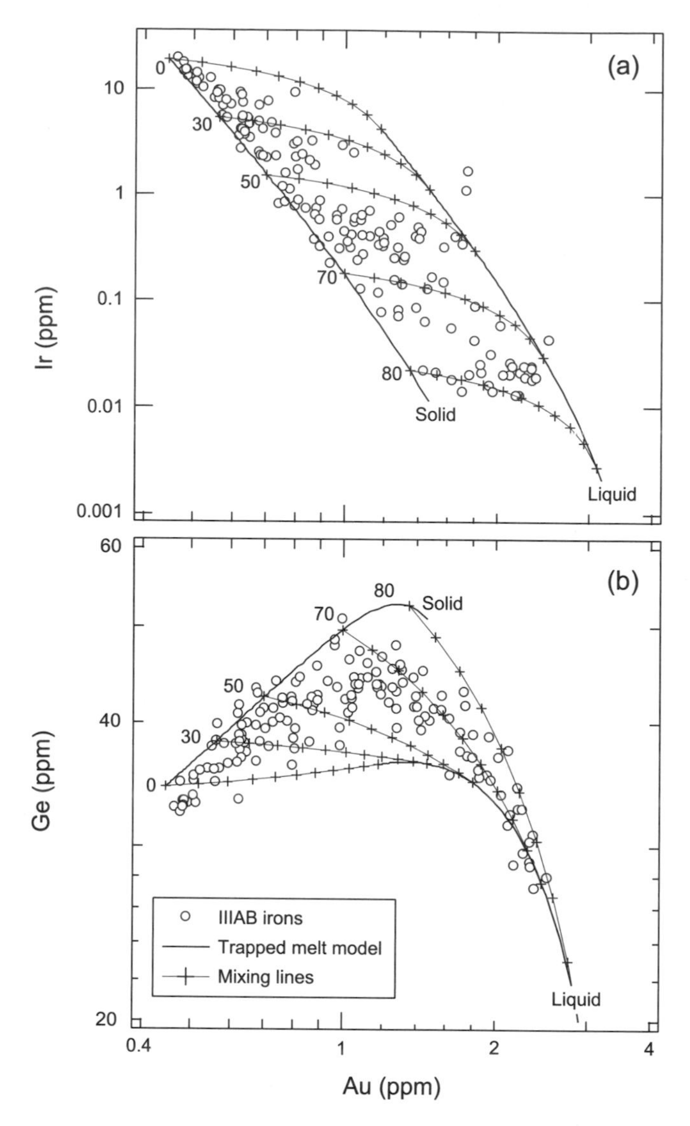

Fig. 11. The trapped melt model (*Wasson,* 1999; *Wasson and Choi,* 2003) is applied to the IIIAB **(a)** Ir vs. Au and **(b)** Ge vs. Au trends. In the model, meteorites that fall on the solid metal path contain no significant amount of trapped melt, while meteorites that fall in between the solid and liquid paths contain varying fractions of melt trapped during crystallization. Mixing lines between coexisting solid and liquid compositions at different degrees of crystallization of the core are shown. IIIAB iron meteorite data were provided by J. T. Wasson.

ing liquid. The largest issue for the trapped melt model is whether all the experimental partitioning data must be used as a constraint, specifically for the choice of D(Ir), which as used in the trapped melt model is inconsistent with the experimental data.

5.9. Model Comparisons

Different crystallization models for the IIIAB parent body core are compared in Table 3. The ideal model would be able to meet all the criteria listed in Table 3. Each model does have its positives and negatives, but at present, none of the models discussed can meet all the criteria.

It is important to keep in mind that some models are just rough attempts at approximating a process mathematically, such as the dendritic crystallization (*Haack and Scott,* 1993) and mixing (*Chabot and Drake,* 1999) models. Any failure of these models does not necessarily mean that the IIIAB core did not solidify by growing large dendrites or that mixing between local heterogeneities did not occur in the molten core. These models were designed to explore the possible effects such processes might have on the crystallization of an asteroidal core. Similarly, the effects from liquid immiscibility must be included in any model of the IIIAB core if the core composition enters the Fe-Ni-S-P two-liquid field, regardless of the success or failure of the liquid immiscibility model of *Ulff-Møller* (1998). Currently, the trapped melt model of *Wasson* (1999) is the only model that attempts to match the composition of each individual IIIAB iron, which a successful model must be able to do. Additionally, the models in Table 3 are not necessarily mutually exclusive. Multiple processes could have affected the crystallizing IIIAB core. Conceptually, the models also have overlap. Incomplete mixing in the molten core may result from the crystallization of large dendrites or the onset of liquid immiscibility. Dendritic crystallization of the core may enhance the amount of melt trapped by the solid. Although the models in Table 3 are different, they all arise from the common conclusion that a process more complex than just simple fractional crystallization led to the solidification of the IIIAB parent body core.

5.10. Modeling Other Groups

Most modeling studies have focused on the IIIAB group, but studies of other iron meteorite groups provide the opportunity to compare how the cores from different asteroidal parent bodies crystallized.

A simple fractional crystallization model is able to match the curved IIAB Ge trend and the large fractionation in Ir concentrations using an initial S content of 17 wt% (*Chabot,* 2004). Such a high S content implies that well over half the IIAB parent body core is unsampled. Additionally, with an initial S content of 17 wt%, the onset of liquid immiscibility is predicted to occur early in the crystallization process (*Chabot and Drake,* 2000). Similar to the IIIAB group, there is evidence for leveling off of the Ir concentrations in the later-crystallizing IIAB irons; the assimilation fractional crystallization (*Malvin,* 1988) and mixing (*Chabot and Drake,* 1999) models have both had success at explaining the IIAB Ir fractionation.

In contrast to the IIIAB and IIAB groups, simple fractional crystallization is not able to explain both the IVA Ge vs. Au and Ir vs. Au trends with a single S content (*Chabot,* 2004). *Scott et al.* (1996) applied the dendritic crystallization model of *Haack and Scott* (1993) to the IVA group but were also unable to match all the IVA trends. *Wasson and Richardson* (2001) applied the trapped melt model to the IVA group and were able to match the Ir vs. Au trend.

TABLE 3. Comparison of crystallization models for the IIIAB iron meteorite group.

Crystallization Models (IIIAB)	Simple Fractional*	Dendritic†	Assimilation‡	Mixing§	Liquid Immiscibility¶	Trapped Melt**
Initial wt% S	12	6	12	12	10.8	2.5
% of core represented	55	94	55	55	60	80
Match Ir vs. Au?	yes	yes	yes	yes	yes	yes
Match Ge vs. Au?	yes	yes	yes	yes	yes	yes
Match Ge vs. Ir?	no	no	yes	yes	no	(n.d.)
Consistent with all experimental data?	yes	yes	yes	yes	yes	no
Explain scatter in IIIAB trend?	no	maybe	no	maybe	yes	yes
Explain Cape York trend?	no	no	no	no	no	yes

* *Chabot* (2004).
† *Haack and Scott* (1993).
‡ *Malvin* (1988).
§ *Chabot and Drake* (1999).
¶ *Ulff-Møller* (1998).
** *Wasson* (1999), *Wasson and Choi* (2003).

n.d. = not determined by the study.

However, *Wasson and Richardson* (2001) did not attempt to fit the curved IVA Ge trend, and currently, no crystallization model has fit both the IVA Ir vs. Au and Ge vs. Au trends simultaneously.

The elemental trends in the IVB group are consistent with simple fractional crystallization of an asteroidal core with 0–2 wt% S (*Chabot,* 2004). With only just over 10 members, crystallization trends of the IVB group are not as well defined as the larger iron meteorite groups, and consequently the IVB group is currently not well suited for more-detailed crystallization studies.

6. CONCLUSIONS

Reviewing our current knowledge of the evolution of asteroidal cores calls attention to the need for additional research and future work. Current isotopic studies indicate that cores formed and evolved early in the history of the solar system, but there are discrepancies regarding the relative timing of core crystallization of the different iron meteorite groups. Future work may not only better resolve the relative timing between groups but may perhaps even determine the relative timing of crystallization within a single group. The model for determining the cooling rates of iron meteorites is still undergoing revision. Ongoing and future research in this area will hopefully clarify if the apparent correlation between cooling rates and Ni concentration among group IVA iron meteorites is real and if the current contradiction between variable IVA cooling rates and a core origin can be resolved. An improved understanding of the formation of the Widmanstätten pattern is not only of interest for studies of the enigmatic group IVA, but also of general interest to all studies that make use of one of the key techniques used to study the thermal evolution of asteroidal cores: metallographic cooling rates. The mode of asteroidal core crystallization, whether by dendritic growth from the core-mantle boundary inward or concentric growth from the center outward, has implications for the chemical evolution of the core, since the physical structure of the crystallizing solid will likely control mixing of the liquid and the possible trapping of melt during crystallization. A better understanding of trace element partitioning behavior and its sensitivity to the metallic liquid composition has led to more complex fractional crystallization models for asteroidal cores. As discussed, each of the current models has its strengths as well as weaknesses. Along with the development and testing of improved models, applying the models to other groups in addition to the IIIAB irons will allow a comparison between the processes involved in core crystallization on different parent bodies. As future research promises further insights into the evolution of asteroidal cores, iron meteorites will remain important and unique as our only samples of any planetary cores.

Acknowledgments. This chapter benefited from thoughtful and thorough reviews by J. I. Goldstein, A. Kracher, R. J. Walker, and J. T. Wasson, and we greatly appreciate their time and effort. Additionally, we thank J. T. Wasson for generating the statistics shown in Table 1 and for the use of his compiled iron meteorite datasets, some of which contained substantial amounts of as yet unpublished data. We also thank J. Yang and J. I. Goldstein for providing Figs. 4 and 5. Support for this work was provided by NASA grants NAG5-12831 to N.L.C. and NAG5-11122 to R. P. Harvey.

REFERENCES

Becker H. and Walker R. J. (2003) In search of extant Tc in the early solar system: ^{98}Ru and ^{99}Ru abundances in iron meteorites and chondrites. *Chem. Geol., 196,* 43–56.

Begemann F., Ludwig K. R., Lugmair G. W., Min K., Nyquist L. E., Patchett P. J., Renne P. R., Shih C.-Y., Villa I. M., and Walker R. J. (2001) Call for an improved set of decay constants for geochronological use. *Geochim. Cosmochim. Acta, 65,* 111–121.

Benedix G. K., McCoy T. J., Keil K., and Love S. G. (2000) A petrologic study of the IAB iron meteorites: Constraints on the formation of the IAB-Winonaite parent body. *Meteoritics & Planet. Sci., 35,* 1127–1141.

Bild R. W. (1977) Silicate inclusions in group IAB irons and a relation to the anomalous stones Winona and Mt. Morris (Wis). *Geochim. Cosmochim. Acta, 41,* 1439–1456.

Binzel R. P. and Xu S. (1993) Chips off of asteroid 4 Vesta: Evidence for the parent body of basaltic achondrite meteorites. *Science, 260,* 186–191.

Birck J.-L. and Allegre C. J. (1988) Manganese-chromium isotope systematics and the development of the early Solar System. *Nature, 331,* 579–584.

Bizzarro M., Baker J. A., and Haack H. (2004) Mg isotope evidence for contemporaneous formation of chondrules and refractory inclusions. *Nature, 431,* 275–278.

Bottke W. F. Jr., Rubincam D. P., and Burns J. A. (2000) Dynamical evolution of main belt meteoroids: Numerical simulations incorporating planetary perturbations and Yarkovsky thermal forces. *Icarus, 145,* 301–331.

Buchwald V. F. (1971) The Cape York shower, a typical group IIIA iron meteorite, formed by directional solidification in a gravity field. *Meteoritics, 6,* 252–253.

Buchwald V. F. (1975) *Handbook of Iron Meteorites, Vols. 1, 2, and 3.* Univ. of California, Berkeley.

Bunch T. E., Keil K., and Olsen E. (1970) Mineralogy and petrology of silicate inclusions in iron meteorites. *Contrib. Mineral. Petrol., 25,* 297–340.

Burbine T. H., McCoy T. J., Meibom A., Gladman B., and Keil K. (2002) Meteoritic parent bodies: Their number and identification. In *Asteroids III* (W. F. Bottke Jr. et al., eds), pp. 653–667. Univ. of Arizona, Tucson.

Carlson R. W. and Hauri E. H. (2001) Extending the ^{107}Pd-^{107}Ag chronometer to low Pd/Ag meteorites with multicollector plasma-ionization mass spectrometry. *Geochim. Cosmochim. Acta, 65,* 1839–1848.

Chabot N. L. (2004) Sulfur contents of the parental metallic cores of magmatic iron meteorites. *Geochim. Cosmochim. Acta, 68,* 3607–3618.

Chabot N. L. and Drake M. J. (1999) Crystallization of magmatic iron meteorites: The role of mixing in the molten core. *Meteoritics & Planet. Sci., 34,* 235–246.

Chabot N. L. and Drake M. J. (2000) Crystallization of magmatic iron meteorites: The effects of phosphorus and liquid immiscibility. *Meteoritics & Planet. Sci., 35,* 807–816.

Chabot N. L. and Jones J. H. (2003) The parameterization of solid metal-liquid metal partitioning of siderophile elements. *Meteoritics & Planet. Sci., 38,* 1425–1436.

Chabot N. L., Campbell A. J., Jones J. H., Humayun M., and Agee C. B. (2003) An experimental test of Henry's Law in solid metal-liquid metal systems with implications for iron meteorites. *Meteoritics & Planet. Sci., 38,* 181–196.

Chen J. H. and Wasserburg G. J. (1990) The isotopic composition of Ag in meteorites and the presence of ^{107}Pd in protoplanets. *Geochim. Cosmochim. Acta, 54,* 1729–1743.

Chen J. H. and Wasserburg G. J. (1996) Live ^{107}Pd in the early solar system and implications for planetary evolution. In *Earth Processes: Reading the Isotopic Code* (A. Basu and S. R. Hart, eds.) pp. 1–20. Geophysical Monograph 95, American Geophysical Union, Washington, DC.

Chen J. H., Papanastassiou D. A., and Wasserburg G. J. (2002) Re-Os and Pd-Ag systematics in Group IIIAB irons and in pallasites. *Geochim. Cosmochim. Acta, 66,* 3793–3810.

Choi B.-G., Ouyang X., and Wasson J. T. (1995) Classification and origin of IAB and IIICD iron meteorites. *Geochim. Cosmochim. Acta, 59,* 593–612.

Clayton R. N. and Mayeda T. K. (1996) Oxygen isotope studies of achondrites. *Geochim. Cosmochim. Acta, 60,* 1999–2017.

Consolmagno G. J. and Drake M. J. (1977) Composition and evolution of the eucrite parent body: Evidence from rare earth elements. *Geochim. Cosmochim. Acta, 41,* 1271–1282.

Cook D. L., Walker R. J., Horan M. F., Wasson J. T., and Morgan J. W. (2004) Pt-Re-Os systematics of group IIAB and IIIAB iron meteorites. *Geochim. Cosmochim. Acta, 68,* 1413–1431.

Cruikshank D. P., Tholen D. J., Hartmann W. K., Bell J. F., and Brown R. H. (1991) Three basaltic Earth-approaching asteroids and the source of the basaltic meteorites. *Icarus, 89,* 1–13.

Davis A. M. and Olsen E. J. (1990) Phosphates in the El Sampal IIIA iron meteorite have excess ^{53}Cr and primordial lead (abstract). In *Lunar and Planetary Science XXI,* pp. 258–259. Lunar and Planetary Institute, Houston.

Dean D. C. and Goldstein J. I. (1986) Determination of the interdiffusion coefficients in the Fe-Ni and Fe-Ni-P systems below 900°C. *Metall. Trans., 17A,* 1131–1138.

Doan A. S. and Goldstein J. I. (1970) The ternary phase diagram Fe-Ni-P. *Metall. Trans., 1,* 1759–1767.

Dziewonski A. M. and Anderson D. L. (1981) Preliminary reference earth model. *Phys. Earth Planet. Inter., 25,* 297–356.

Eugster O., Marti K., Caffee M., and Herzog G. (2006) Irradiation records, cosmic-ray exposure ages, and transfer times of meteorites. In *Meteorites and the Early Solar System II* (D. S. Lauretta and H. Y. McSween Jr., eds.), this volume. Univ. of Arizona, Tucson.

Esbensen K. H. and Buchwald V. F. (1982) Planet(oid) core crystallization and fractionation. Evidence from the Cape York iron meteorite shower. *Phys. Earth Planet. Inter., 29,* 218–232.

Esbensen K. H., Buchwald V. F., Malvin D. J., and Wasson J. T. (1982) Systematic compositional variations in the Cape York iron meteorites. *Geochim. Cosmochim. Acta, 46,* 1913–1920.

Fleet M. E. and Stone W. E. (1991) Partitioning of platinum-group elements in the Fe-Ni-S system and their fractionation in nature. *Geochim. Cosmochim. Acta, 55,* 245–253.

Geiger G. H., Kozakevitch P., Olette M., and Riboud P. V. (1975) Theory of BOF reaction rates. In *BOF Steelmaking, Vol. 2* (J. M. Gaines, ed.), pp. 191–321. American Institute of Mining, Metallurgical, and Petroleum Engineers, New York.

Ghosh A. and McSween H. Y. (1998) A thermal model for the differentiation of asteroid 4 Vesta, based on radiogenic heating. *Icarus, 134,* 187–206.

Goldstein J. I. and Hopfe W. D. (2001) The metallographic cooling rates of IVA iron meteorites. *Meteoritics & Planet. Sci., 36,* A67.

Goldstein J. I. and Ogilvie R. E. (1963) Electron microanalysis of metallic meteorites. Part 1 — Phosphides and sulfides. *Geochim. Cosmochim. Acta, 27,* 623–637.

Goldstein J. I. and Ogilvie R. E. (1965) The growth of the Widmanstätten pattern in metallic meteorites. *Geochim. Cosmochim. Acta, 29,* 893–920.

Goldstein J. I. and Short J. M. (1967) The iron meteorites, their thermal history and parent bodies. *Geochim. Cosmochim. Acta, 31,* 1733–1770.

Grimm R. E. and McSween H. Y. (1993) Heliocentric zoning of the asteroid belt by Al-26 heating. *Science, 259,* 653–655.

Haack H. and McCoy T. J. (2003) Iron and stony-iron meteorites. In *Treatise on Geochemistry, Vol. 1: Meteorites, Comets, and Planets* (A. M. Davis, ed.), pp. 325–346. Elsevier, Oxford.

Haack H. and Scott E. R. D. (1992) Asteroid core crystallization by inward dendritic growth. *J. Geophys. Res.–Planets, 97,* 14727–14734.

Haack H. and Scott E. R. D. (1993) Chemical fractionations in Group IIIAB iron meteorites: Origin by dendritic crystallization of an asteroidal core. *Geochim. Cosmochim. Acta, 57,* 3457–3472.

Haack H., Rasmussen K. L., and Warren P. H. (1990) Effects of regolith/megaregolith insulation on the cooling histories of differentiated asteroids. *J. Geophys. Res., 95,* 5111–5124.

Haack H., Scott E. R. D., Love S. G., Brearley A. J., and McCoy T. J. (1996) Thermal histories of IVA stony-iron and iron meteorites: Evidence for asteroid fragmentation and reaccretion. *Geochim. Cosmochim. Acta, 60,* 3103–3113.

Harper C. L. Jr. and Jacobsen S. B. (1996) Evidence for ^{182}Hf in the early Solar System and constraints on the timescale of terrestrial accretion and core formation. *Geochim. Cosmochim. Acta, 60,* 1131–1153.

Hassanzadeh J., Rubin A. E., and Wasson J. T. (1990) Compositions of large metal nodules in mesosiderites: Links to iron meteorite group IIIAB and the origin of mesosiderite subgroups. *Geochim. Cosmochim. Acta, 54,* 3197–3208.

Herbert F. and Sonett C. P. (1979) Electromagnetic heating of minor planets in the early solar-system. *Icarus, 40,* 484–496.

Hopfe W. D. and Goldstein J. I. (2001) The metallographic cooling rate method revised: Application to iron meteorites and mesosiderites. *Meteoritics & Planet. Sci., 36,* 135–154.

Horan M. F., Smoliar M. I., and Walker R. J. (1998) ^{182}W and ^{187}Re-^{187}Os systematics of iron meteorites: Chronology for melting, differentiation, and crystallization in asteroids. *Geochim. Cosmochim. Acta, 62,* 545–554.

Hutcheon I. D. and Olsen E. (1991) Cr isotopic composition of differentiated meteorites: A search for ^{53}Mn (abstract). In *Lunar and Planetary Science XXII,* pp. 605–606. Lunar and Planetary Institute, Houston.

Hutcheon I. D., Olsen E., Zipfel J., and Wasserburg G. J. (1992) Cr isotopes in differentiated meteorites: Evidence for ^{53}Mn (abstract). In *Lunar and Planetary Science XXIII,* pp. 565–566. Lunar and Planetary Institute, Houston.

Jones J. H. and Casanova I. (1993) Experimental partitioning of As and Sb among metal, troilite, screibersite, barringerite, and metallic liquid. *Meteoritics, 28,* 374–375.

Jones J. H. and Drake M. J. (1983) Experimental investigations of trace element fractionation in iron meteorites, II: The influence of sulfur. *Geochim. Cosmochim. Acta, 47,* 1199–1209.

Jones J. H. and Malvin D. J. (1990) A nonmetal interaction model for the segregation of trace metals during solidification of Fe-Ni-S, Fe-Ni-P, and Fe-Ni-S-P alloys. *Metall. Trans., 21B,* 697–706.

Jones J. H., Hart S. R., and Benjamin T. M. (1993) Experimental partitioning studies near the Fe-FeS eutectic, with an emphasis on elements important to iron meteorite chronologies (Pb, Ag, Pd, Tl). *Geochim. Cosmochim. Acta, 57,* 453–460.

Keil K. (2002) Geological history of asteroid 4 Vesta: The "smallest terrestrial planet." In *Asteroids III* (W. F. Bottke Jr. et al., eds.), pp. 653–667. Univ. of Arizona, Tucson.

Keil K., Haack H., and Scott E. R. D. (1994) Catastrophic fragmentation of asteroids: Evidence from meteorites. *Planet. Space Sci., 42,* 1109–1122.

Keil K., Stöffler D., Love S. G., and Scott E. R. D. (1997) Constraints on the role of impact heating and melting in asteroids. *Meteoritics & Planet. Sci., 32,* 349–363.

Kleine T., Münker C., Mezger K., and Palme H. (2002) Rapid accretion and early core formation on asteroids and the terrestrial planets from Hf-W chronometry. *Nature, 418,* 952–955.

Kracher A. (1982) Crystallization of a S-saturated Fe, Ni melt, and the origin of iron meteorite groups IAB and IIICD. *Geophys. Res. Lett., 9,* 412–415.

Kracher A. and Wasson J. T. (1982) The role of S in the evolution of the parental cores of the iron meteorites. *Geochim. Cosmochim. Acta, 46,* 2419–2426.

Kracher A., Willis J., and Wasson J. T. (1980) Chemical classification of iron meteorites — IX. A new group (IIF), revision of IAB and IIICD, and data on 57 additional irons. *Geochim. Cosmochim. Acta, 44,* 773–787.

Lee D. C. and Halliday A. N. (1995) Hafnium-tungsten chronometry and the timing of terrestrial core formation. *Nature, 378,* 771–774.

Lee D.-C. and Halliday A. N. (1996) Hf-W isotopic evidence for rapid accretion and differentiation in the early solar system. *Science, 274,* 1876–1879.

Liu M. and Fleet M. E. (2001) Partitioning of siderophile elements (W, Mo, As, Ag, Ge, Ga, and Sn) and Si in the Fe-S system and their fractionation in iron meteorites. *Geochim. Cosmochim. Acta, 65,* 671–682.

MacLachlan D. and Ehlers E. G. (1971) Effect of pressure on the melting temperatures of metals. *J. Geophys. Res., 76,* 2780–2789.

Malvin D. J. (1988) Assimilation-fractional crystallization modeling of magmatic iron meteorites (abstract). In *Lunar and Planetary Science XIX,* pp. 720–721. Lunar and Planetary Institute, Houston.

Malvin D. J., Jones J. H., and Drake M. J. (1986) Experimental investigations of trace element fractionation in iron meteorites. III: Elemental partitioning in the system Fe-Ni-S-P. *Geochim. Cosmochim. Acta, 50,* 1221–1231.

Malvin D. J., Wang D., and Wasson J. T. (1984) Chemical classification of iron meteorites — X. Multielement studies of 43 irons, resolution of group IIIE from IIIAB, and evaluation of Cu as a taxonomic parameter. *Geochim. Cosmochim. Acta, 48,* 785–804.

Massalski T. B., Park F. R., and Vassamillet L. F. (1966) Speculations about plessite. *Geochim. Cosmochim. Acta, 30,* 649–662.

McCoy T. J., Keil K., Scott E. R. D., and Haack H. (1993) Genesis of the IIICD iron meteorites: Evidence from silicate-bearing inclusions. *Meteoritics, 28,* 552–560.

McCoy T. J., Mittlefehldt D. W., and Wilson L. (2006) Asteroid differentiation. In *Meteorites and the Early Solar System II* (D. S. Lauretta and H. Y. McSween Jr., eds.), this volume. Univ. of Arizona, Tucson.

Mittlefehldt D. W., McCoy T. J., Goodrich C. A., and Kracher A. (1998) Non-chondritic meteorites from asteroidal bodies. In *Planetary Materials* (J. J. Papike, ed.), pp. 4-1 to 4-195. Reviews in Mineralogy, Vol. 36, Mineralogical Society of America.

Moren A. E. and Goldstein J. I. (1978) Cooling rate variations of group IVA iron meteorites. *Earth Planet. Sci. Lett., 40,* 151–161.

Moren A. E. and Goldstein J. I. (1979) Cooling rates of group IVA iron meteorites determined from a ternary Fe-Ni-P model. *Earth Planet. Sci. Lett., 43,* 182–196.

Narayan C. and Goldstein J. I. (1982) A dendritic solidification model to explain Ge-Ni variations in iron meteorite chemical groups. *Geochim. Cosmochim. Acta, 46,* 259–268.

Narayan C. and Goldstein J. I. (1984) Nucleation of intragranular ferrite in Fe-Ni-P alloys. *Metall. Trans., 15A,* 861–865.

Narayan C. and Goldstein J. I. (1985) A major revision of iron meteorite cooling rates — An experimental study of the growth of the Widmanstätten pattern. *Geochim. Cosmochim. Acta, 49,* 397–410.

Olsen E. J., Kracher A., Davis A. M., Steele I. M., Hutcheon I. D., and Bunch T. E. (1999) The phosphates of IIIAB iron meteorites. *Meteoritics & Planet. Sci., 34,* 285–300.

Pernicka E. and Wasson J. T. (1987) Ru, Re, Os, Pt and Au in iron meteorites. *Geochim. Cosmochim. Acta, 51,* 1717–1726.

Raghavan V. (1988) The Fe-P-S system. In *Phase Diagrams of Ternary Iron Alloys 2,* pp. 209–217. Indian National Scientific Documentation Centre, New Delhi.

Randich E. and Goldstein J. I. (1975) Non-isothermal finite diffusion-controlled growth in ternary systems. *Metall. Trans., 6A,* 1553–1560.

Randich E. and Goldstein J. I. (1978) Cooling rate of seven hexahedrites. *Geochim. Cosmochim. Acta, 48,* 805–813.

Rasmussen K. L. (1982) Determination of the cooling rates and nucleation histories of eight group IVA iron meteorites using local bulk Ni and P variation. *Icarus, 52,* 444–453.

Rasmussen K. L. (1989a) Cooling rates and parent bodies of iron meteorites from group IIICD, IAB, and IVB. *Phys. Scripta, 39,* 410–416.

Rasmussen K. L. (1989b) Cooling rates of IIIAB iron meteorites. *Icarus, 80,* 315–325.

Rasmussen K. L., Malvin D. J., Buchwald V. F., and Wasson J. T. (1984) Compositional trends and cooling rates of group IVB iron meteorites. *Geochim. Cosmochim. Acta, 48,* 805–813.

Rasmussen K. L., Ulff-Møller F., and Haack H. (1995) The thermal evolution of IVA iron meteorites: Evidence from metallographic cooling rates. *Geochim. Cosmochim. Acta, 59,* 3049–3059.

Romig A. D. and Goldstein J. I. (1980) Determination of the Fe-Ni and Fe-Ni-P phase diagrams at low temperatures (700 to 300°C). *Metall. Trans., 11A,* 1151–1159.

Rubin A. E., Ulff-Møller F., Wasson J. T., and Carlson W. D. (2001) The Portales Valley meteorite breccia: Evidence for impact-induced melting and metamorphism of an ordinary chondrite. *Geochim. Cosmochim. Acta, 65,* 323–342.

Saikumar V. and Goldstein J. I. (1988) An evaluation of the methods to determine the cooling rates of iron meteorites. *Geochim. Cosmochim. Acta, 52,* 715–726.

Schaudy R., Wasson J. T., and Buchwald V. F. (1972) The chemical classification of iron meteorites. VI. A reinvestigation of irons with Ge concentrations lower than 1 ppm. *Icarus, 17,* 174–192.

Scott E. R. D. (1972) Chemical fractionation in iron meteorites and its interpretation. *Geochim. Cosmochim. Acta, 36,* 1205–1236.

Scott E. R. D. (1977) Pallasites — Metal composition, classification and relationships with iron meteorites. *Geochim. Cosmochim. Acta, 41,* 349–360.

Scott E. R. D. (1979a) Origin of iron meteorites. In *Asteroids* (T. Gehrels, ed.), pp. 892–925. Univ. of Arizona, Tucson.

Scott E. R. D. (1979b) Origin of anomalous iron meteorites. *Mineral. Mag., 43,* 415–421.

Scott E. R. D. and Wasson J. T. (1975) Classification and properties of iron meteorites. *Rev. Geophys. Space Phys., 13,* 527–546.

Scott E. R. D. and Wasson J. T. (1976) Chemical classification of iron meteorites — VIII. Groups IC, IIE, IIIF and 97 other irons. *Geochim. Cosmochim. Acta, 40,* 103–115.

Scott E. R. D., Haack H., and McCoy T. J. (1996) Core crystallization and silicate-metal mixing in the parent body of the IVA iron and stony-iron meteorites. *Geochim. Cosmochim. Acta, 60,* 1615–1631.

Scott E. R. D., Wasson J. T., and Buchwald V. F. (1973) The chemical classification of iron meteorites — VII. A reinvestigation of iron with Ge concentrations between 25 and 80 ppm. *Geochim. Cosmochim. Acta, 37,* 1957–1983.

Sellamuthu R. and Goldstein J. I. (1983) Experimental study of segregation in plane front solidification and its relevance to iron meteorite solidification. *J. Geophys. Res., 88,* B343–B352.

Sellamuthu R. and Goldstein J. I. (1984) Measurement and analysis of distribution coefficients in Fe-Ni alloys containing S and/or P: Part 1. K_{Ni} and K_P. *Metall. Trans., 15A,* 1677–1685.

Sellamuthu R. and Goldstein J. I. (1985a) Analysis of segregation trends observed in iron meteorites using measured distribution coefficients. *Proc. Lunar Planet. Sci. Conf. 15th,* in *J. Geophys. Res., 90,* C677–C688.

Sellamuthu R. and Goldstein J. I. (1985b) Measurement and analysis of distribution coefficients in Fe-Ni alloys containing S and/or P: Part II. K_{Ir}, K_{Ge}, and K_{Cu}. *Metall. Trans., 16A,* 1871–1878.

Shen J. J., Papanastassiou D. A., and Wasserburg G. J. (1996) Precise Re-Os determinations and systematics of iron meteorites. *Geochim. Cosmochim. Acta, 60,* 2887–2900.

Shirey S. B. and Walker R. J. (1998) The Re-Os isotope system in cosmochemistry and high-temperature geochemistry. *Annu. Rev. Earth Planet. Sci., 26,* 423–500.

Smoliar M. I., Walker R. J., and Morgan J. W. (1996) Re-Os ages of group IIA, IIIA, IVA, and IVB iron meteorites. *Science, 271,* 1099–1102.

Tachibana S. and Huss G. R. (2003) The initial abundance of Fe-60 in the solar system. *Astrophys. J. Lett., 588,* L41–L44.

Takeda H., Bogard D. D., Mittlefehldt D. W., and Garrison D. H. (2000) Mineralogy, petrology, chemistry, and ^{39}Ar-^{40}Ar and exposure ages of the Caddo County IAB iron: Evidence for early partial melt segregation of a gabbro area rich in plagioclase-diopside. *Geochim. Cosmochim. Acta, 64,* 1311–1327.

Ulff-Møller F. (1998) Effects of liquid immiscibility on trace element fractionation in magmatic iron meteorites: A case study of group IIIAB. *Meteoritics & Planet. Sci., 33,* 207–220.

Voshage H. and Feldmann H. (1979) Investigations on cosmic-ray-produced nuclides in iron meteorites, 3. Exposure ages, meteoroid sizes and sample depths determined by mass spectrometric analyses of potassium and rare gases. *Earth Planet. Sci. Lett., 45,* 293–308.

Wadhwa M., Srinivasan G., and Carlson R. W. (2006) Timescales of planetesimal differentiation in the early solar system. In *Meteorites and the Early Solar System II* (D. S. Lauretta and H. Y. McSween Jr., eds.), this volume. Univ. of Arizona, Tucson.

Wasson J. T. (1967) The chemical classification of iron meteorites: I. A study of iron meteorites with low concentrations of gallium and germanium. *Geochim. Cosmochim. Acta, 31,* 161–180.

Wasson J. T. (1969) The chemical classification of iron meteorites — III. Hexahedrites and other irons with germanium concentrations between 80 and 200 ppm. *Geochim. Cosmochim. Acta, 33,* 859–876.

Wasson J. T. (1970) The chemical classification of iron meteorites IV. Irons with Ge concentrations greater than 190 ppm and other meteorites associated with group I. *Icarus, 12,* 407–423.

Wasson J. T. (1985) *Meteorites: Their Record of Early Solar-System History*. W. H. Freeman and Company, New York. 267 pp.

Wasson J. T. (1990) Ungrouped iron meteorites in Antarctica: Origin of anomalously high abundance. *Science, 249,* 900–902.

Wasson J. T. (1999) Trapped melt in IIIAB irons; solid/liquid elemental partitioning during the fractionation of the IIIAB magma. *Geochim. Cosmochim. Acta, 63,* 2875–2889.

Wasson J. T. and Choi B.G. (2003) Main-group pallasites: Chemical composition, relationship to IIIAB irons, and origin. *Geochim. Cosmochim. Acta, 67,* 3079–3096.

Wasson J. T. and Kallemeyn G. W. (2002) The IAB iron-meteorite complex: A group, five subgroups, numerous grouplets, closely related, mainly formed by crystal segregation in rapidly cooling melts. *Geochim. Cosmochim. Acta, 66,* 2445–2473.

Wasson J. T. and Kimberlin J. (1967) The chemical classification of iron meteorites — II. Irons and pallasites with germanium concentrations between 8 and 100 ppm. *Geochim. Cosmochim. Acta, 31,* 2065–2093.

Wasson J. T. and Richardson J. W. (2001) Fractionation trends among IVA iron meteorites: Contrasts with IIIAB trends. *Geochim. Cosmochim. Acta, 65,* 951–970.

Wasson J. T. and Schaudy R. (1971) The chemical classification of iron meteorites — V. Groups IIIC and IIID and other irons with germanium concentrations between 1 and 25 ppm. *Icarus, 14,* 59–70.

Wasson J. T. and Wang J. (1986) A nonmagmatic origin of group-IIE iron meteorites. *Geochim. Cosmochim. Acta, 50,* 725–732.

Wasson J. T., Ouyang X., Wang J., and Jerde E. (1989) Chemical classification of iron meteorites: XI. Multi-element studies of 38 new irons and the high abundance of ungrouped irons from Antarctica. *Geochim. Cosmochim. Acta, 53,* 735–744.

Wasson J. T., Choi B.-G., Jerde E. A., and Ulff-Møller F. (1998) Chemical classification of iron meteorites: XII. New members of the magmatic groups. *Geochim. Cosmochim. Acta, 62,* 715–724.

Wasson J. T., Lange D. E., Francis C. A., and Ulff-Møller F. (1999) Massive chromite in the Brenham pallasite and the fractionation of Cr during the crystallization of asteroidal cores. *Geochim. Cosmochim. Acta, 63,* 1219–1232.

Willis J. and Wasson J. T. (1978a) Cooling rates of group IVA iron meteorites. *Earth Planet. Sci. Lett., 40,* 141–150.

Willis J. and Wasson J. T. (1978b) A core origin for group IVA iron meteorites: A reply to Moren and Goldstein. *Earth Planet. Sci. Lett., 40,* 162–167.

Willis J. and Goldstein J. I. (1982) The effects of C, P, and S on trace element partitioning during solidification in Fe-Ni alloys. *Proc. Lunar Planet. Sci. Conf. 13th,* in *J. Geophys. Res., 87,* A435–A445.

Wood J. A. (1964) The cooling rates and parent planets of several iron meteorites. *Icarus, 3,* 429–459.

Wood J. A. (1979) Review of the metallographic cooling rates of meteorites and a new model for the planetesimals in which they formed. In *Asteroids* (T. Gehrels, eds.), pp. 849–891. Univ. of Arizona, Tucson.

Yang C.-W., Williams D. B., and Goldstein J. I. (1996) A revision of the Fe-Ni phase diagram at low temperatures (<400°C). *J. Phase Equilibria, 17(6),* 522–531.

Yang C.-W., Williams D. B., and Goldstein J. I. (1997a) Low-temperature phase decomposition in metal from iron, stony-iron, and stony meteorites. *Geochim. Cosmochim. Acta, 61,* 2943–2956.

Yang C.-W., Williams D. B., and Goldstein J. I. (1997b) A new empirical cooling rate indicator for meteorites based on the size of the cloudy zone of the metallic phases. *Meteoritics & Planet. Sci., 32,* 434–429.

Yang J. and Goldstein J. I. (2005) The formation mechanism of the Widmanstätten structure in meteorites. *Meteoritics & Planet. Sci., 40,* 239–253.

Yin Q., Jacobsen S. B., Yamashita K., Blichert-Toft J., Télouk P., and Albarède F. (2002) A short timescale for terrestrial planet formation from Hf-W chronometry of meteorites. *Nature, 418,* 949–952.

Part IX:

The Planetary Epoch: Meteorites and the Earth

Meteorites and the Timing, Mechanisms, and Conditions of Terrestrial Planet Accretion and Early Differentiation

Alex N. Halliday
University of Oxford

Thorsten Kleine
Eidgenössische Technische Hochschule (ETH) Zentrum

Isotopic studies of meteorites provide the fundamental data for determining how the terrestrial planets, including Earth, accreted. Over the past few years there have been major advances in our understanding of both the timescales and processes of terrestrial planet accretion, largely as a result of better definition of initial solar system abundances of short-lived nuclides and their daughters, as determined from meteorites. Cosmogenic effects, cross calibrations with other isotopic systems, and decay constant uncertainties are also critical in many instances. Recent improvements to ^{182}Hf-^{182}W chronology in all these areas have been particularly noteworthy. Uncertainty has surrounded the initial Hf- and W-isotopic composition of the solar system and the meaning of the unradiogenic W-isotopic compositions of iron meteorite data, rendered complicated by cosmogenic effects. However, the timescales for the formation of certain iron meteorite parent bodies would appear to be very fast (<1 m.y.). Similarly, the accretion of Mars would appear to have been very fast, consistent with rapid accretion via runaway growth. Precise quantification is difficult, however, because martian meteorites display variable W-isotopic compositions that relate in part to the levels of depletion in siderophile elements. This is as expected from a planet that never achieved a well-mixed silicate reservoir characterized by uniform siderophile-element depletion, as is found on Earth. Therefore, attempts to apply W-isotopic models to martian meteorites need to be treated with caution because of this demonstrable variability in early source Hf/W presumably resulting from partial metal or core segregation. The current best estimates for martian reservoirs as represented by Zagami would imply formation within the first 1 m.y. of the solar system. The modeled timescales for metal segregation from the source of other meteorites, Nakhla, for example, would appear to be more like 10 m.y. These estimates are based on trace-element and isotopic data obtained from different and probably unrepresentative aliquots. Further high-quality combined trace-element and isotopic studies are needed to confirm this. Nevertheless, the chondritic ^{142}Nd abundance for Zagami provides powerful supporting evidence that the W-isotopic effects record extremely rapid (<1 m.y.) accretion and core formation on Mars. The timescales for Earth accretion are significantly more protracted. The last major stage of accretion is thought to be the Moon-forming giant impact, the most recent Hf-W age estimates for which are in the range 40–50 m.y. after the start of the solar system. Applying this to accretion models for Earth provides evidence that some of the accreted metal did not fully equilibrate with silicate reservoirs. This cannot explain the very late apparent accretion ages deduced from other chronometers, in particular U-Pb. Either all the estimates for the Pb-isotopic composition of the bulk silicate Earth are in error or there was some additional late-stage U/Pb fractionation that removed Pb from the silicate Earth. If the latter was the case there was either late segregation of Pb to the core after W removal, or removal of Pb via atmospheric escape following the "giant impact." Changes in the mechanisms and partitioning associated with core formation are indeed predicted from the stability in the mantle of S-rich metal before, and sulfide after, the giant impact. However, losses from Earth also need to be evaluated. Strontium-isotopic data provide evidence of major late (>10 m.y.) losses of moderately volatile elements from the material that formed the Moon and probably Earth. The Earth's nonchondritic Mg/Fe may similarly reflect silicate losses during growth of Earth itself or the protoplanets that accreted to Earth. The budgets for plutonogenic Xe provide evidence that some erosion was extremely late (>100 m.y.), clearly postdating the giant impact and presumably related to irradiation and bombardment during the Hadean.

1. INTRODUCTION

For the past 50 years it has been clear that meteorites provide the crucial evidence that permits determination of the formation times of the terrestrial planets and tests of dynamic models for their growth and early differentiation. The most remarkable example of this was the classic experiment by *Patterson* (1956), in which the age of Earth was determined by measuring the Pb-isotopic compositions of iron meteorites and achondrites. Prior to this time, *Houtermans* (1946) and *Holmes* (1946) had both independently shown that Pb isotopes could determine the age of Earth. However, at that time they had to make their best estimates of the minimum age (roughly 3 Ga) on the basis of the Pb-isotopic compositions of terrestrial samples measured by *Nier et al.* (1941). Although it was Houtermans who in this context first coined the term "isochron," it was Patterson who, by measuring meteorites, got the right points and slope that defined the age of the solar system and Earth.

In this review the latest pertinent information on the various isotopic chronometers as obtained from meteorites is summarized and it is explained how this yields constraints on the age and early evolution of the terrestrial planets. Some of the ways in which various isotopic systems can be jointly utilized are discussed. The chapter finishes with the best current estimates of the timescales for the formation of the terrestrial planets and asteroid belt, closing with some speculations on how the use of these isotopic systems may develop in the future.

2. CONSTRAINTS ON PLANET FORMATION PROCESSES AND TIMESCALES

The timescales and processes associated with planetary accretion can be studied with three complementary approaches.

1. Theoretical calculations and dynamic simulations provide clear indications of what is expected in terms of the timescales and mechanisms of accretion and core formation given certain assumptions about the prevailing conditions (see *Chambers,* 2004, for a recent review). However, the dependence on the amount of gas present, the potential for planet migration since accretion, and the difficulty in deriving a clearly workable mechanism to build objects that are of sufficient size for gravity to play a dominant role leave considerable uncertainty about the degree to which these models represent a real solar system. There is still no understanding of how circumstellar dust sticks and eventually becomes kilometer-scale objects that can accrete with one another gravitationally (*Blum,* 2000; *Benz,* 2000), and/or whether gravitational instability may play a role (*Ward,* 2000).

2. Observations can now be made of circumstellar disks (*McCaughrean and O'Dell,* 1996) as well as extrasolar planets (*Mayor and Queloz,* 1995). However, the extrasolar planets found so far are almost exclusively limited by the detection methods to Jupiter-sized objects orbiting with a very short period. Young stars that are the most interesting from the point of view of planet formation are sufficiently energetic and internally unstable that they are largely unsuitable for detecting the small Doppler shift effects used so far to identify most extrasolar planets.

3. Isotopic measurements have long provided key evidence for the processes, rates, and timing of formation of inner solar system objects. Most of this work has been restricted to studies of planetesimals. However, with the development of multiple-collector inductively coupled plasma mass spectrometry (MC-ICPMS) (*Halliday et al.,* 1995), this field has taken off in several new directions involving short-lived isotopic chronometers, including the determination of accretion rates for the terrestrial planets. It now appears that calculated timescales differ depending on the chronometer. These differences may be providing us with important new information about how planets are built, as discussed below.

3. THE MAIN ISOTOPIC SYSTEMS RELEVANT TO PLANET FORMATION

The kinds of isotopic systems that could conceivably be used for studying how planets form can be crudely categorized as follows:

1. Long-lived radioactive decay systems have mainly been used to provide information on the absolute age (*Patterson,* 1956; *Hanan and Tilton,* 1987; *Wasserburg et al.,* 1977a; *Lugmair and Galer,* 1992; *Carlson and Lugmair,* 2000; *Amelin et al.,* 2002). Such systems also can provide constraints on the time-averaged chemical composition of the materials from which an object formed by defining the radioactive parent element/radiogenic daughter element ratio (*Gray et al.,* 1973; *Wasserburg et al.,* 1977b; *Hanan and Tilton,* 1987; *Halliday and Porcelli,* 2001).

2. Short-lived nuclides have been extremely successful as high-resolution chronometers of events in the early solar system, in particular ^{26}Al (*Lee et al.,* 1977), ^{53}Mn (*Birck and Allègre,* 1988; *Lugmair and Shukolyukov,* 1998), ^{107}Pd (*Chen and Wasserburg,* 1996), ^{129}I (*Reynolds,* 1960), and ^{182}Hf (*Harper et al.,* 1991a; *Lee and Halliday,* 1995). However, they need to be "mapped" onto an absolute timescale. They too can provide time-averaged chemical compositions (*Halliday et al.,* 1996; *Halliday,* 2004; *Jacobsen and Harper,* 1996).

3. Nucleosynthetic isotopic heterogeneity in the disk from which the planets form has been proposed on the basis of O-isotopic data (*Clayton,* 1986, 1993; *Clayton and Mayeda,* 1975, 1996), although the currently favored mechanism for production of mass-independent anomalies in O isotopes is by photochemical means either in the solar nebula or in the precursor molecular cloud (*Clayton,* 2002; *Yurimoto and Kuramoto,* 2004). Whatever their cause, O-isotopic data provide a tool for studying the provenance of early solar system materials and the precision on these

measurements, and hence the ability to resolve small differences in average provenance has moved to new levels (*Wiechert et al.,* 2001, 2004). Mass-independent isotopic effects of plausible nucleosynthetic origin increasingly are being demonstrated as very high precision measurements of a wide range of elements have become possible using MC-ICPMS and thermal ionization mass spectrometry (TIMS) (*Dauphas et al.,* 2002a,b, 2004; *Yin et al.,* 2002a; *Schönbächler et al.,* 2003; *Podosek et al.,* 1997; *Trinquier et al.,* 2005).

4. Isotopic heterogeneity in highly volatile elements like the noble gases provides clues to the mixture of nebular and dust-derived volatile phases (*Wieler et al.,* 1992) and have been crucial for understanding the timescales of volatile accretion and loss (*Porcelli et al.,* 2001; *Porcelli and Pepin,* 2000).

5. Mass-dependent stable-isotope fractionation of light elements such as H, C, N, O, Mg, Si, S, and K has been thoroughly characterized in recent years. Such fractionations or lack thereof provide powerful constraints on the conditions and processes associated with planetary accretion and development (*Young,* 2000; *Humayun and Clayton,* 1995). Evidence for large isotopic fractionations associated with volatile loss in the early solar system has been found in the gases Ne and Xe (*Hunten et al.,* 1987; *Pepin,* 1997, 2000). This field is likely to expand rapidly and become increasingly interesting as a wide range of elements from Li to U become amenable to study at high precision with new MC-ICPMS techniques (*Galy et al.,* 2000; *Poitrasson et al.,* 2004).

6. Cosmic-ray exposure ages do not provide much direct evidence for how the planets formed. They do provide evidence of transit times of material delivered to Earth (e.g., *Heck et al.,* 2004) and it has occasionally been proposed that some isotopic effects in chondrites reflect primordial preexposure on a meteorite parent-body regolith.

7. Local irradiation phenomena in the early Sun also provide a mechanism for generating isotopic heterogeneity and a local source of short-lived nuclides in the early solar system (*Lee et al.,* 1998; *Gounelle et al.,* 2001; *Leya et al.,* 2000, 2003). Studies of some of the isotopic variations predicted should shed new light on how material that formed close to the Sun was scattered across the disk (for example, by X-winds).

This paper mainly focuses on the conclusions one derives from the first two of these: long- and short-lived (extinct) nuclides.

4. CALIBRATION OF CHRONOMETERS FROM THE START OF THE SOLAR SYSTEM

4.1. Long-lived Chronometers

The definition of events in absolute time at the start of the solar system is accomplished using long-lived radionuclides (Table 1). Although several of these have been utilized, the precision and accuracy that can be achieved with $^{238/235}U$-$^{207/206}Pb$ is so much greater that the results of every other kind of chronometry, short- or long-lived, are referenced to the data derived from this system.

The $^{238/235}U$-$^{207/206}Pb$ system is very reliable and well understood. The uncertainties on the decay constants are of potential importance in defining an absolute age (*Schön et al.,* 2004) but are small and insignificant when it comes to resolving time differences at the start of the solar system. There does exist uncertainty concerning the amount of ^{247}Cm in the early solar system (*Stirling et al.,* 2005), and because this decays to ^{235}U with a long characteristic half-life of 12 m.y. it could potentially change early solar system $^{238/235}U$-$^{207/206}Pb$ chronometry. However, the ^{247}Cm abundance is probably very small (*Stirling et al.,* 2005) and so at present there exists no case for a correction.

Other long-lived isotopic systems have, in general terms, been more problematic. The greatest difficulties have been with ^{87}Rb and ^{176}Lu. In both cases it is found that to achieve concordance with isotopic ages obtained by $^{238/235}U$-$^{207/206}Pb$ or ^{147}Sm-^{143}Nd in meteorites, a different decay constantfrom that obtained by counting experiments or by comparing the ages of terrestrial samples is required. The reasons for this are unclear at present and seem to be different in the two cases. One should also take note of the recommendations of *Steiger and Jäger* (1977). Some of the recent measurements relevant to ^{176}Lu are found in *Scherer et al.* (2001), *Blichert-Toft et al.* (2002), *Bizzarro et al.* (2003), and *Söderlund et al.* (2004). A recent compilation and discussion of this very topical issue can be found in *Scherer et al.* (2003).

These decay constant uncertainties of a few percent render as limited the usefulness of the decay schemes for defining a precise absolute timescale. However, such problems are of lesser consequence when it comes to determining small time differences between early objects and events. A decay constant error of 1% produces a huge uncertainty in absolute time at the start of the solar system at 4.5 Ga that renders it of little use given that the solar system was probably completed in <50 m.y. However, this same percentage uncertainty also translates into a similar fractional error on time differences, where it is of negligible importance. The simplest and most relevant example is for long-lived chronometers where the time difference D_t between t_1 and t_2 is $\ll 1/\lambda$ (the inverse of the decay constant or mean life) and can therefore be simplified to

$$\Delta t = \lambda \times (D_{t_1} - D_{t_2})/P$$

where D is the atomic ratio of a radiogenic daughter to a nonradiogenic stable isotope of the same element and P is the ratio of the radioactive parent nuclide to the same stable denominator isotope of the daughter element. It can be seen that the fractional error on λ translates linearly to a fractional error on Δt. As such, Sr-isotopic compositions, for example, can be used to infer time differences of less than a million years at the start of the solar system for which a decay con-

TABLE 1. Main isotopic decay systems in use in studying the origin and early evolution of the terrestrial planets.

Parent	Daughter(s)	Half-Life (yr)	Status*	Principal Applications and Comments
Long-lived				
^{40}K (10.7%)	^{40}Ar	1.25×10^{9}	✓	Chronology, especially lunar bombardment
^{40}K (89.3%)	^{40}Ca	1.25×10^{9}	✓	Not used greatly
^{87}Rb	^{87}Sr	4.75×10^{10}	✓	Chronology, refractory/volatile fractionations
^{138}La (66.4%)	^{138}Ba	1.05×10^{11}	✓	Not used greatly
^{138}La (33.6%)	^{138}Ce	1.05×10^{11}	✓	Not used greatly
^{147}Sm	^{143}Nd	1.06×10^{11}	✓	Chronology and early silicate differentiation
^{176}Lu	^{176}Hf	3.57×10^{10}	✓	Chronology and early silicate differentiation
^{187}Re	^{187}Os	4.23×10^{10}	✓	Chronology of metals
^{190}Pt	^{186}Os	6.5×10^{11}	✓	Not used greatly
^{232}Th	^{208}Pb	1.40×10^{10}	✓	Time-averaged Th/U
^{235}U	^{207}Pb	7.04×10^{8}	✓	Chronology, refractory/volatile fractionations
^{238}U	^{206}Pb	4.47×10^{9}	✓	Chronology, refractory/volatile fractionations
Extinct				
^{7}Be	^{7}Li	1.459×10^{-1}	?	CAI formation
^{10}Be	^{10}B	1.5×10^{6}	✓	CAI formation
^{26}Al	^{26}Mg	7.3×10^{5}	✓	CAIs, chondrules, early silicate melting, heat
^{41}Ca	^{41}K	1.04×10^{5}	✓	CAI formation
^{53}Mn	^{53}Cr	3.7×10^{6}	✓	Early silicate melting and provenance
^{60}Fe	^{60}Ni	1.49×10^{6}	✓	Massive star signature, early metals, heat
^{92}Nb	^{92}Zr	3.6×10^{7}	✓	*p*-process signature, disk heterogeneity
^{93}Zr	^{93}Nb	1.53×10^{6}	?	Daughter is monoisotopic
^{97}Tc	^{97}Mo	2.6×10^{6}	?	*p*-process signature, disk heterogeneity
^{98}Tc	^{98}Mo	4.2×10^{6}	?	*s*-process signature, disk heterogeneity
^{99}Tc	^{99}Ru	2.13×10^{5}	?	*s*-process signature, disk heterogeneity
^{107}Pd	^{107}Ag	6.5×10^{6}	✓	Metal and refractory/volatile chronology
^{126}Sn	^{126}Te	2.35×10^{5}	?	*r*-process signature
^{129}I	^{129}Xe	1.57×10^{7}	✓	CAI, chondrule chronology, degassing
^{135}Cs	^{135}Ba	2.3×10^{6}	?	Refractory/volatile fractionations
^{146}Sm	^{142}Nd	1.03×10^{8}	✓	Early solar system, crustal evolution
^{182}Hf	^{182}W	8.9×10^{6}	✓	*r*-process, metal, accretion chronology
^{205}Pb	^{205}Tl	1.5×10^{7}	✓	*s*-process signature
^{244}Pu	^{136}Xe†	8×10^{7}	✓	CAI, chondrule chronology, degassing
^{247}Cm	^{235}U	1.6×10^{7}	?	Supernova *r*-process

*A check mark means that the nuclide's presence or former presence has been established. Nuclides with a question mark are still not convincingly demonstrated.

† Spontaneous fissionogenic product.

stant error of 1% is completely negligible (*Gray et al.*, 1973; *Papanastassiou and Wasserburg*, 1976; *Wasserburg et al.*, 1977b; *Lugmair and Galer*, 1992; *Halliday and Porcelli*, 2001).

4.2. Uncertainties in Short-lived Systems — Decay Constants

The relative importance of uncertainties for short-lived systems is somewhat different. The errors on the decay constants are generally insignificant but with increasing precision and greater opportunity for cross-calibration these become critical. The ^{182}Hf decay constant is a particularly good (or bad) case in point and has been the most problematic in recent years when it comes to early solar system timescales. Unfortunately, the artificial production of significant ^{182}Hf for activity measurements requires very large neutron fluxes. As a consequence, little work has been done until recently. The half-life of 9 m.y. has long been limited by an uncertainty of more than ±20% (*Wing et al.*, 1961). This stated uncertainty does not affect first-order conclusions regarding the accretion rates of the planets but has started to become significant when resolving time differences in the early solar system and comparing results between chronometers (*Halliday*, 2003, 2004). Of greater concern perhaps is that some half-life determinations have been inaccurate by far outside their stated uncertainties, as was the case with ^{60}Fe until the work of *Kutschera et al.* (1984). The new study by *Vockenhuber et al.* (2004) addresses the ^{182}Hf issue and has produced a new, precise, and highly reliable value of 8.904 ± 0.088 m.y. for the ^{182}Hf decay constant. Therefore, this issue can now be considered closed.

4.3. Uncertainties in Short-lived Systems — Initial Abundances

The uncertainties in the bulk solar system initial (BSSI) isotopic composition of radioactive parent or radiogenic daughter elements introduce further and frequently the largest error in precise calibration of the timescale. All the short-lived nuclide systems suffer from this problem but it has been particularly apparent for ^{26}Al-^{26}Mg (*Bizzarro et al.*, 2004; *Young et al.*, 2005), ^{53}Mn-^{53}Cr (*Lugmair and Shukolyukov*, 1998; *Birck et al.*, 1999; *Nyquist et al.*, 1997, 2001),

^{60}Fe-^{60}Ni (*Shukolyukov and Lugmair,* 1993; *Mostefaoui et al.,* 2004), ^{92}Nb-^{92}Zr (*Harper et al.* 1991b; *Hirata,* 2001; *Münker et al.,* 2000; *Sanloup et al.,* 2000; *Yin et al.,* 2000; *Schönbächler et al.,* 2002, 2003), ^{107}Pd-^{107}Ag (*Chen and Wasserburg,* 1996; *Carlson and Hauri,* 2001; *Woodland et al.,* 2005), ^{182}Hf-^{182}W (*Harper and Jacobsen,* 1996; *Jacobsen and Harper,* 1996; *Lee and Halliday,* 1995, 1996,2000a; *Kleine et al.,* 2002; *Quitté et al.,* 2000; *Quitté and Birck,* 2004; *Schönberg et al.,* 2002; *Yin et al.,* 2002b), and ^{247}Cm-^{235}U (*Arden,* 1977; *Blake and Schramm,* 1973; *Chen and Wasserburg,* 1980; *Tatsumoto and Shimamura,* 1980; *Stirling et al.,* 2005). In the first three cases (^{26}Al-^{26}Mg, ^{53}Mn-^{53}Cr, and ^{60}Fe-^{60}Ni) the uncertainty is largely dominated by studying different archives and issues having to do with the handling and interpretation of the data. In the case of ^{92}Nb-^{92}Zr and ^{107}Pd-^{107}Ag there are apparent discrepancies in the data produced from different laboratories that have not been satisfactorily explained. In the case of ^{247}Cm-^{235}U there is a clear indication that the U-isotopic variations reported in some studies were probably analytical artifacts (*Chen and Wasserburg,* 1980; *Stirling et al.,* 2005). In the case of ^{182}Hf-^{182}W there are both discrepancies in the data and differences in interpretation. However, it is now known that certain early W-isotopic data for carbonaceous chondrites (*Lee and Halliday,* 1995, 1996) are definitely incorrect by about 180–200 ppm. The discrepancies over ^{107}Pd-^{107}Ag and ^{247}Cm-^{235}U are of little relevance to planetary timescales at the present time and will not be discussed further here. In the following the uncertainties over the other systems are described briefly.

4.3.1. Aluminum-26–magnesium-26. The canonical value of $(^{26}Al/^{27}Al)_{BSSI}$ is 5×10^{-5} (*Lee et al.,* 1977). However, new Mg-isotopic data for bulk CAIs and minerals have led *Young et al.* (2005) to propose that this only represents the isotopic abundance at the time of recrystallization of the CAIs. Young et al. present a large amount of new data for CAIs measured by laser ablation MC-ICPMS. From the scatter of data for minerals and bulk CAIs they extract a minimum apparent $(^{26}Al/^{27}Al)_{BSSI} \sim 7 \times 10^{-5}$ using the upper portion of the scatter to the Al-Mg isochron. The difficulty with the approach is that nonsystematic and matrix-sensitive errors in element ratios can be quite significant using laser ablation. These are hard to quantify in fine-grained materials, particularly in the absence of detailed matrix matching and laser artifact studies. As a result the meaning of the upper limit of such a scattered dataset becomes ambiguous. A few bulk CAIs that were analyzed conventionally in the same study lie within the range of the laser ablation measurements but do not extend to such high ^{26}Al/^{27}Al. The Young et al. study appears to be in conflict with the very precise $(^{26}Al/^{27}Al)_{BSSI} = (5.25 \pm 0.10) \times 10^{-5}$ obtained for bulk Allende CAIs by *Bizzarro et al.* (2004). The reasons for the apparent difference are complex. The two groups have analyzed different CAIs, in some cases from different meteorites. However, they also treat the correction for isotopic fractionation differently. The raw data are distorted by both natural and instrument-induced processes following different laws. With only three isotopes, one of which is radiogenic, a judgment has to be made about how best to determine the radiogenic difference. This becomes important in strongly fractionated residues like bulk CAIs and has the potential to leave an apparent residual isotopic anomaly that is significant relative to the small effect being expected from low Al/Mg CAIs. *Davis et al.* (2005) argue that the high ^{26}Al/^{27}Al reported by *Young et al.* (2005) could reflect inaccurate characterization of this natural fractionation. Young et al. claim that on its own, this is insufficient to explain the total amount of the apparent discrepancy with the results reported by *Bizzarro et al.* (2004). However, *Bizzarro et al.* (2005) have now reported an error in their treatment of their data resulting in a corrected $^{26}Al/^{27}Al = (5.83 \pm 0.11) \times 10^{-5}$. This is consistent with the conventional bulk data in *Young et al.* (2005). In fact, regression of all data from Young et al. gives an identical, although less-precise, ^{26}Al/^{27}Al of $5.9 \pm 0.3 \times 10^{-5}$. However, the very high "supracanonical" $(^{26}Al/^{27}Al)_{BSSI}$ value of 7×10^{-5} is determined as an upper limit to the scatter of the data obtained by laser ablation and has not yet been ratified by any conventional measurements.

4.3.2. Manganese-53–chromium-53. The former presence of live ^{53}Mn in the early solar system is now well established (*Birck and Allègre,* 1988; *Lugmair and Shukolyukov,* 1998). However, the study of different meteorite archives has yielded different values for the initial abundance and this has a dramatic effect on the timescales deduced for Mn/Cr fractionation in planetary bodies. The Mn-Cr-isotopic data for CAIs (*Birck and Allègre,* 1988) are consistent with a solar system initial ^{53}Mn/^{55}Mn that is significantly higher than that implied from the Mn-Cr systematics of angrites and eucrites (*Lugmair and Shukolyukov,* 1998, 2001). The data for CAIs could conceivably reflect nucleosynthetic effects that are well established in chondrites (*Podosek et al.,* 1997) and/or disturbance of Mn-Cr-isotopic systematics (*Papanastassiou et al.,* 2005). Manganese-chromium chronometry is left in a somewhat equivocal status until this is clarified. The ramifications for early planetesimal accretion are discussed further below.

4.3.3. Iron-60–nickel-60. The demonstration of formerly live ^{60}Fe in the early solar system (*Shukolyukov and Lugmair,* 1993) was accomplished by measuring the Ni-isotopic compositions of basaltic achondrites and in particular eucrites. Recently other groups have demonstrated very high ^{60}Fe/^{56}Fe in some sulfides from chondrites (e.g., *Mostefaoui et al.,* 2004). The recently reported evidence for very high ^{60}Fe/^{56}Fe (*Moynier et al.,* 2004) in iron meteorites has not been confirmed by more recent measurements (*Cook et al.,* 2005; *Quitté et al.,* 2006). Exactly how high is the value of $(^{60}Fe/^{56}Fe)_{BSSI}$ is now unclear but it could be as high as10^{-6} in which case early solar system events recorded in eucrites are more protracted than previously thought or the Ni-isotopic systems have been partially reset at some late stage.

4.3.4. Niobium-92–zirconium-92. In the case of ^{92}Nb there exist basically two values. The first is $(^{92}Nb/^{92}Nb)_{BSSI} \sim 10^{-3}$ (*Münker et al.,* 2000; *Sanloup et al.,* 2000; *Yin et al.,*

2000). The second is $(^{92}Nb/^{92}Nb)_{BSSI} \sim 10^{-5}$ (*Harper et al.,* 1991b; *Hirata,* 2001; *Schönbächler et al.,* 2002, 2003, 2005). These two very distinct estimates result in radically different constraints on the early differentiation history of the terrestrial planets. If the former is correct the Zr-isotopic compositions for the silicate Earth imply late formation of Earth's core and continental crust (*Münker et al.,* 2000; *Jacobsen and Yin,* 2001). The reasons for the differences are unclear. *Schönbächler et al.* (2003) were unable to reproduce the Zr-isotopic variations reported by *Münker et al.* (2000) and *Sanloup et al.* (2000) when similar materials (the same bulk chondrites and similar refractory low Nb/Zr Allende CAIs) were analyzed. Similarly, the results for iron meteorite rutiles reported by *Yin et al.* (2000) differ from those reported by *Harper et al.* (1991b) and the results for an early (meteorite) zircon with Nb/Zr = 0 reported by *Yin et al.* (2000) are different from the results for similar phases studied by *Hirata* (2001). These apparent discrepancies warrant further studies to settle the issue definitively.

4.3.5. Hafnium-182–tungsten-182. In the case of the value of $(^{182}Hf/^{180}Hf)_{BSSI}$ the initial estimate of *Harper and Jacobsen* (1996) and *Jacobsen and Harper* (1996) of $\sim 10^{-5}$ was low because of a paucity of data (*Ireland,* 1991) and the modeling available from comparisons with other isotopic systems (*Wasserburg et al.,* 1994). *Lee and Halliday* (1995, 1996) presented the first W-isotopic data for carbonaceous chondrites and inferred a $(^{182}Hf/^{180}Hf)_{BSSI} > 10^{-4}$. They then appeared to confirm this with a value of $^{182}Hf/^{180}Hf \sim 2 \times 10^{-4}$ for internal Hf-W isochrons of rather variable quality for the ordinary chondrites Forest Vale and Ste. Marguerite (*Lee and Halliday,* 2000b).

Three groups have independently shown that some of these University of Michigan data are incorrect (*Kleine et al.,* 2002; *Schönberg et al.,* 2002; *Yin et al.,* 2002b). They have instead demonstrated a $(^{182}Hf/^{180}Hf)_{BSSI} \sim 1 \times 10^{-4}$. The most clear-cut error is in the compositions of the carbonaceous chondrites Allende and Murchison. These, and all carbonaceous chondrites measured thus far (*Kleine et al.,* 2002, 2004a), have a $^{182}W/^{184}W$ that is 180–200 ppm lower than that originally determined by *Lee and Halliday* (1995, 1996) (Fig. 1).

The reason for the error in the Michigan measurements is unclear. The MC-ICPMS instrument used was the first built and had relatively poor sensitivity compared with more recent instruments. This should not matter directly unless it resulted in a background problem.

The original internal isochrons for Forest Vale and Ste. Marguerite (*Lee and Halliday,* 2000b) also differ in part from the results presented by *Kleine et al.* (2002). The explanation could again be analytical error. There may also have been differences between the particular phases analyzed. Clearly, the age or time of closure of a particular phase is critical to the $^{182}Hf/^{180}Hf$ determined. However, the data from CAIs and other ordinary chondrite isochrons (*Yin et al.,* 2002b) and the data obtained on bulk meteorites all appear to confirm a $(^{182}Hf/^{180}Hf)_{BSSI}$ of about 1×10^{-4}, rather than 2×10^{-4}.

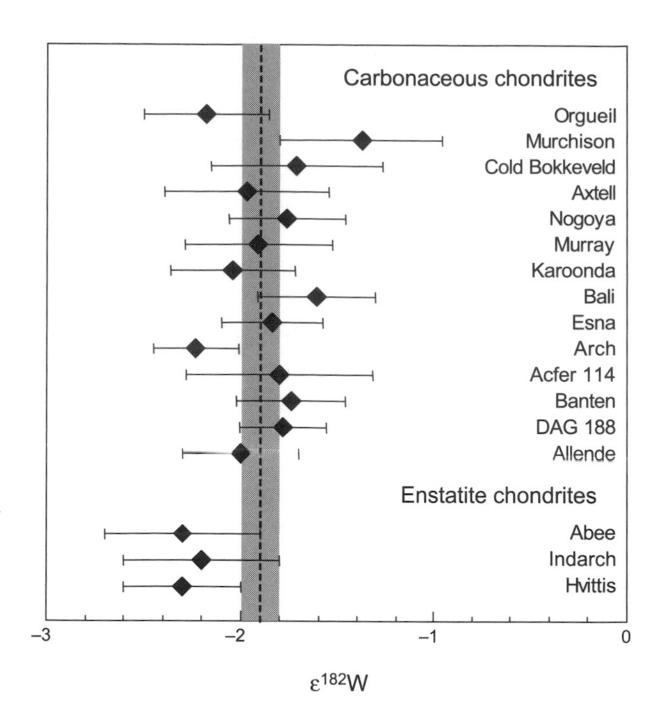

Fig. 1. The W-isotopic compositions of carbonaceous chondrites, as determined by *Kleine et al.* (2002, 2004a), shown here, but also by *Yin et al.* (2002b) and *Schoenberg et al.* (2002), demonstrate that the earlier measurements of Allende and Murchison (*Lee and Halliday,* 1995, 1996) are inaccurate by roughly 180 to 200 ppm. The new values agree within error with the values determined for enstatite chondrites (*Lee and Halliday,* 2000a) shown here but previously considered anomalous and hard to explain. The average for carbonaceous chondrites (vertical line) is $\varepsilon^{182}W = -1.9 \pm 0.1$.

The W-isotopic composition of iron meteorites may, however, provide evidence that the initial abundance lies between these two values. Iron meteorites sample early solar system W without Hf and therefore provide an indication of the W-isotopic composition at the start of the solar system in an analogous fashion to the way in which the initial Pb-isotopic composition at the start of the solar system has been defined by Canyon Diablo troilite. The most recent estimate of $(^{182}Hf/^{180}Hf)_{BSSI}$ is $(1.7 \pm 0.3) \times 10^{-4}$ based on the least-radiogenic W-isotopic composition of solar system materials yet determined, as measured in the iron meteorite Tlacotopec (*Quitté and Birck,* 2004). This particular meteorite has a very long cosmic-ray exposure age and the effects on the W-isotopic composition (*Leya et al.,* 2000, 2003) could be significant. This has been confirmed by more detailed high-precision W-isotopic measurements of iron meteorites (*Markowski et al.,* 2006). Although the differences over the BSSI abundance of Hf and W isotopes would appear to be minor compared with the uncertainties and disputes over other isotopic systems, in particular ^{53}Mn-^{53}Cr, ^{60}Fe-^{60}Ni, and ^{92}Nb-^{92}Zr, they do have a nonnegligible effect on calculated early solar system timescales and their interpretation, and are particularly important because of the wider utility of ^{182}Hf-^{182}W.

TABLE 2. Isotopic ages of early objects and best estimates of early solar system timescales.

Type of Event	Object or Model	Isotopic System	Reference	Age (Ga)	Time (m.y.)
Start of solar system	Efremovka CAIs	$^{235/238}$U-$^{207/206}$Pb	*Amelin et al.* (2002)	4.5672 ± 0.0006	0.0 ± 0.6
Start of solar system	Allende CAIs	$^{235/238}$U-$^{207/206}$Pb	*Göpel et al.* (1991)	4.566 ± 0.002	1 ± 2
Start of solar system	Allende CAIs	^{26}Al-^{26}Mg	*Bizzarro et al.* (2004)	4.567	0.00 ± 0.03
Chondrule formation	Acfer chondrules	$^{235/238}$U-$^{207/206}$Pb	*Amelin et al.* (2002)	4.5647 ± 0.0006	2.5 ± 1.2
Chondrule formation	UOC chondrules	^{26}Al-^{26}Mg	*Russell et al.* (1996)	<4.566 to 4.565	>1 to 2
Chondrule formation	Allende chondrule	^{26}Al-^{26}Mg	*Galy et al.* (2000)	<4.5658 ± 0.0007	>1.4 ± 0.7
Chondrule formation	Allende chondrules	^{26}Al-^{26}Mg	*Bizzarro et al.* (2004)	4.567 to <4.565	0 to ≥1.4
H chondrite parent body metamorphism	Ste. Marguerite phosphate	$^{235/238}$U-$^{207/206}$Pb	*Göpel et al.* (1994)	4.5627 ± 0.0006	4.5 ± 1.2
Asteroidal core formation	Magmatic irons	^{182}Hf-^{182}W	*Markowski et al.* (2006)	>4.566	<1
Vesta differentiation	Silicate-metal	^{182}Hf-^{182}W	*Kleine et al.* (2002)	4.563 ± 0.001	4 ± 1
Vesta accretion	Earliest age	^{87}Rb-^{87}Sr	*Halliday and Porcelli* (2001)	<4.563 ± 0.002	>4 ± 2
Vesta differentiation	Silicate-metal	^{182}Hf-^{182}W	*Lee and Halliday* (1997)	4.56*	10*
Vesta differentiation	Silicate-silicate	^{53}Mn-^{53}Cr	*Lugmair and Shukulyokov* (1998)	4.5648 ± 0.0009	1 ± 2
Vesta differentiation	Silicate-metal	^{182}Hf-^{182}W	*Quitté et al.* (2000)	4.550 ± 0.001*	16 ± 1*
Vesta differentiation	Silicate-metal	^{182}Hf-^{182}W	*Kleine et al.* (2002)	4.563 ± 0.001	4 ± 1
Vesta differentiation	Silicate-metal	^{182}Hf-^{182}W	*Yin et al.* (2002b)	4.564	3
Early eucrites	Asuka 881394	$^{235/238}$U-$^{207/206}$Pb	*Wadhwa et al.* (2005), M. Wadhwa (personal communication, 2005)	4.5665 ± 0.0009	0.7 ± 1.1
Early eucrites	Noncumulate eucrites	^{182}Hf-^{182}W	*Quitté and Birck* (2004)	4.558 ± 0.003	9 ± 3
Early eucrites	Chervony Kut	^{53}Mn-^{53}Cr	*Lugmair and Shukulyokov* (1998)	4.563 ± 0.001	4 ± 1
Angrite formation	Angra dos Reis and LEW 86010	$^{235/238}$U-$^{207/206}$Pb	*Lugmair and Galer* (1992)	4.5578 ± 0.0005	9 ± 1
Mars accretion	Youngest age	^{146}Sm-^{142}Nd	*Harper et al.* (1995)	≥4.54	≤30
Mars accretion	Mean age	^{182}Hf-^{182}W	*Lee and Halliday* (1997)	4.560*	6*
Mars accretion	Youngest age	^{182}Hf-^{182}W	*Lee and Halliday* (1997)	≥4.54*	≤30*
Mars accretion	Youngest age	^{182}Hf-^{182}W	*Halliday et al.* (2001b)	≥4.55*	≤20*
Mars accretion	Youngest age	^{182}Hf-^{182}W	*Kleine et al.* (2002)	≥4.55	<13 ± 2
Mars accretion	Youngest age	^{182}Hf-^{182}W	This study	>4.566	<1
Earth accretion	Mean age	$^{235/238}$U-$^{207/206}$Pb	*Halliday* (2000)	4.527 to 4.562	15 to 40
Earth accretion	Mean age	^{182}Hf-^{182}W	*Yin et al.* (2002b)	4.556 ± 0.001	11 ± 1
Earth accretion	Mean age	$^{235/238}$U-$^{207/206}$Pb	*Halliday* (2004)	4.550 ± 0.003	17 ± 3
Moon formation	Best estimate of age	$^{235/238}$U-$^{207/206}$Pb	*Tera et al.* (1973)	4.47 ± 0.02	100 ± 20
Moon formation	Best estimate of age	$^{235/238}$U-$^{207/206}$Pb and ^{147}Sm-^{143}Nd	*Carlson and Lugmair* (1988)	4.44 to 4.51	60 to 130
Moon formation	Best estimate of age	^{182}Hf-^{182}W	*Halliday et al.* (1996)	4.47 ± 0.04*	100 ± 40*
Moon formation	Best estimate of age	^{182}Hf-^{182}W	*Lee et al.* (1997)	4.51 ± 0.01*	55 ± 10*
Moon formation	Earliest age	^{182}Hf-^{182}W	*Halliday* (2000)	≤4.52*	≥45*
Moon formation	Earliest age	^{87}Rb-^{87}Sr	*Halliday and Porcelli* (2001)	<4.556 ± 0.001	>11±1
Moon formation	Best estimate of age	^{182}Hf-^{182}W	*Lee et al.* (2002)	4.51 ± 0.01*	55 ± 10*
Moon formation	Best estimate of age	^{182}Hf-^{182}W	*Kleine et al.* (2002)	4.54 ± 0.01	30 ± 10
Moon formation	Best estimate of age	^{182}Hf-^{182}W	*Yin et al.* (2002b)	4.546	29
Moon formation	Best estimate of age	^{182}Hf-^{182}W	*Halliday* (2004)	4.52 ± 0.01	45 ± 10
Lunar highlands	Ferroan anorthosite 60025	$^{235/238}$U-$^{207/206}$Pb	*Hanan and Tilton* (1987)	4.50 ± 0.01	70±10
Lunar highlands	Ferroan anorthosite 60025	^{147}Sm-^{143}Nd	*Carlson and Lugmair* (1988)	4.44 ± 0.02	130 ± 20
Lunar highlands	Norite from breccia 15445	^{147}Sm-^{143}Nd	*Shih et al.* (1993)	4.46 ± 0.07	110 ± 70
Lunar highlands	Ferroan noritic anorthosite 67016	^{147}Sm-^{143}Nd	*Alibert et.al.* (1994)	4.56 ± 0.07	10 ± 70
Earliest crust	Jack Hills zircons	$^{235/238}$U-$^{207/206}$Pb	*Wilde et al.* (2001)	4.44 ± 0.01	130 ± 10

*Based on early solar system initial abundances now thought incorrect.

CAIs = calcium-aluminum-rich refractory inclusions. UOC = unequilibrated ordinary chondrite.

4.4. Uncertainties in Short-lived Systems — Conversion to Absolute Time

To convert the results from short-lived decay systems like Hf-W into an absolute timescale one needs cross-calibration points. In principle, cross-calibrations with precise $^{235/238}$U-$^{207/206}$Pb ages should allow one to determine the initial abundances and the decay constants of the short-lived nuclides independently. There now exist at least four of these, all of which can be tied in to absolute timescales with the very precise $^{235/238}$U-$^{207/206}$Pb ages shown in Table 2.

1. The ^{26}Al-^{26}Mg chronometer yields time differences (*Russell et al.,* 1996) that are fully consistent with differences in absolute time deduced from $^{235/238}$U-$^{207/206}$Pb data for CAIs and chondrules (*Amelin et al.,* 2002).

2. The very precise $^{235/238}$U-$^{207/206}$Pb data for angrites (*Lugmair and Galer,* 1992) provides a powerful means of mapping ^{53}Mn-^{53}Cr onto an absolute timescale (*Lugmair and Shukolyukov,* 1998, 2001).

3. Similarly, the $^{235/238}$U-$^{207/206}$Pb data for ordinary chondrites, in particular Ste. Marguerite (*Göpel et al.,* 1994), allow the ^{182}Hf-^{182}W chronometer to be cross-calibrated onto an absolute timescale (*Kleine et al.,* 2002, 2005b) if they date the same event. In fact, the U-Pb closure temperature in phosphates is 400°–500°C whereas Hf-W closure between metal and silicates takes place at temperatures in excess of 600°C (*Kleine et al.,* 2005b). Using the ^{53}Mn-^{53}Cr chronometry (*Polnau and Lugmair,* 2001), which itself is mapped onto an absolute timescale by cross calibration with $^{235/238}$U-$^{207/206}$Pb (*Lugmair and Shukolyukov,* 1998, 2001),

results in better agreement between the Hf-W and Mn-Cr whole-rock ages of the eucrites and also between the Hf-W and $^{235/238}$U-$^{207/206}$Pb age for CAIs (*Kleine et al.,* 2005a).

4. Most recently, the ^{182}Hf-^{182}W chronometer has been cross-calibrated onto an absolute timescale by determining an internal isochron for CAIs (*Kleine et al.,* 2005a). This permits direct comparison with $^{235/238}$U-$^{207/206}$Pb and ^{26}Al-^{26}Mg data.

These calibrations are tremendously important as the field develops in the direction of increasing age resolution. For example, for many years there was a debate about whether ^{26}Al/^{27}Al variations in early solar system objects reflected time differences, variable resetting, or differences in the abundance of ^{26}Al in different settings. The latter model was fueled by more recent models of ^{26}Al production close to the Sun (*Lee et al.,* 1998) and the detection of formerly live ^{10}Be in CAIs (*McKeegan et al.,* 2000). However, since ^{10}Be is clearly spallogenic and excess ^{10}B (from the decay of ^{10}Be) has been detected in CAIs lacking evidence of excess ^{26}Mg, the vast majority of ^{26}Al cannot be spallogenic (*Marhas et al.,* 2002). The close agreement between ^{26}Al-^{26}Mg and $^{235/238}$U-$^{207/206}$Pb chronometry, as well as the very precise initial abundance of $(^{26}Al/^{27}Al)_{BSSI}$ recently determined for bulk CAIs (*Bizzarro et al.,* 2004, 2005), appears to demonstrate that any variation in initial abundance is very small.

Although such issues mandate further effort to achieve clarity, there is no question that we now know the absolute age of certain key objects extremely well and can get a good idea of how quickly they formed.

5. CHONDRITES, ACHONDRITES, IRONS, AND EARLIEST PLANETESIMALS

The most precise and accurate current estimate of the absolute age of the solar system is taken to be the $^{235/238}$U-$^{207/206}$Pb age of Efremovka CAIs of 4.5672 ± 0.0006 Ga (*Amelin et al.,* 2002). The most widely held view is that CAIs did form within the earliest solar system. The reason for believing this is that they are the earliest objects yet dated, have the highest abundances of short-lived nuclides, and have roughly chondritic proportions of highly refractory elements and isotopic compositions that are, broadly speaking, similar to those of average solar system.

When and how CAIs formed relative to the collapse of the portion of the molecular cloud that formed the Sun is unknown. The most precise estimate of the duration of CAI formation is provided by the new high-precision ^{26}Al-^{26}Mg data produced by *Bizzarro et al.* (2004, 2005), who obtained an identical apparent initial ^{26}Al abundance for all CAIs measured, implying a time interval of perhaps as little as 50,000 years for Allende CAI formation. These measurements need to be extended to CAIs from other meteorites but the preliminary data are consistent with a very short duration of CAI production. As discussed above, there are apparent discrepancies between this result and the Mg-isotopic data of *Young et al.* (2005), who found a higher apparent ^{26}Al/^{27}Al for bulk CAIs than the canonical value based on minerals. It is currently unclear whether this difference is real. Either way, the resolution of these chronometers is getting to the point at which incredible detail should be revealed about the earliest solar system over the coming years as these problems are resolved.

How chondrules form is also unknown and highly controversial. The most widely accepted theory is that they form within the dusty circumstellar disk from which the planetesimals formed. A variety of studies have now shown that chondrule formation overlapped with CAI production, but was protracted, extending to roughly 2 m.y. after the start of the solar system (*Russell et al.,* 1996; *Galy et al.,* 2000; *Bizarro et al.,* 2004). Whether all these objects were formed in a similar manner is unclear. Some may have been produced as a result of molten planetesimal impacts (*Zook,* 1981; *Sanders,* 1996). Another "planetary" model for chondrule formation involves bow shocks around eccentric asteroids driven by jovian resonances (*Weidenschilling et al.,* 1998). Similarly, *Lugmair and Shukolyukov* (2001) argue that early planetesimals would melt rapidly and that collisions among them would release melt that ultimately forms chondrules.

The question then arises as to how well we know how quickly the first planetesimals formed. From dynamic considerations it is thought that planetesimals formed within a few times 10^5 yr (e.g., *Kortenkamp et al.,* 2000). Optically dense circumstellar disks appear to form very early and last for about a million years or more. Some dusty disks like HR 4796A and Beta Pictoris are an order of magnitude older (*Schneider et al.,* 1999). The presence of these disks does not imply that planetesimals or planets have not yet formed. It is assumed that they are already formed and are present in the midplane.

Chondrites themselves, while judged to be undifferentiated, cannot have formed very early because they contain a variety of CAIs, chondrules, and presolar grains that must have become mixed together some time after chondrule formation. Certain kinds of presolar grains could not have survived the temperatures and conditions of chondrule formation. Therefore, a scenario for chondrite-parent-body formation must involve mixing of CAIs, chondrules, and presolar grains into a particular region where chondrite parent bodies were able to form by sticking or local gravitational collapse of these small objects. The ^{26}Al/^{27}Al (*Bizzarro et al.,* 2004, 2005; *Young et al.,* 2005) and high ^{60}Fe/^{56}Fe (*Mostefaoui et al.,* 2004) now thought to have characterized the earliest solar system should have produced molten and differentiated planetesimals at an early stage. The very existence of chondrites provides evidence of small-body accretion of undifferentiated material millions of years after the start of the solar system. As such, their existence in the inner solar system is dynamically problematic.

Isotopic data for differentiated meteorites provide clear evidence that some of the earliest planetesimals formed within the first 10 m.y. of the solar system. Exactly how early is a matter of some uncertainty. The primary lines of evidence come from short-lived nuclides and in particular ^{26}Al-^{26}Mg, ^{53}Mn-^{53}Cr, and ^{182}Hf-^{182}W. For example, Chervony Kut is an unbrecciated eucrite with relatively high ^{60}Fe (*Shukolyukov and Lugmair,* 1993), lending confidence to the view that it is a less-disturbed sample well suited for defining early solar system timescales. The ^{53}Mn-^{53}Cr data for Chervony Kut can be compared with the data for the angrites Angra dos Reis and LEW 86010 (Table 2). These objects are very precisely dated by $^{235/238}U$-$^{207/206}Pb$ and, although relatively young, they permit mapping of the Mn-Cr timescale for eucrites to an absolute age. On this basis Chervony Kut formed within 5 m.y. of the start of the solar system, allowing for the uncertainties in Pb-Pb chronometry of the angrites and CAIs (*Lugmair and Shukolyukov,* 1998). The eucrite whole-rock Mn-Cr isochron, thought to reflect planetary differentiation, defines an earlier age no more than 4 m.y. after the start of the solar system (Table 2).

Despite these lines of evidence that differentiated objects formed early, the Sr-isotopic data for eucrites are difficult to explain as the primary planetesimals that supposedly formed within the first few hundred thousand years of the solar system. There is a difference between the initial Sr-isotopic compositions of CAIs (*Gray et al.,* 1973; *Podosek et al.,* 1991) on the one hand and eucrites and angrites (*Wasserburg et al.,* 1977; *Lugmair and Galer,* 1992) on the other. This cannot be readily explained unless the eucrite parent body formed more than 2 m.y. after the start of the solar system (*Halliday and Porcelli,* 2001). The basis for this is as follows. All these objects are strongly depleted in moderately volatile Rb relative to refractory Sr and, as such, there is no possibility of the Sr-isotopic composition of the parent body or its mantle changing significantly by decay of ^{87}Rb. Yet the Sr-isotopic difference requires a significant period of time in a high Rb/Sr environment. The highest Rb/Sr environment possible is the Rb/Sr of the solar nebula (0.3). The minimum time required to generate the difference in Sr-isotopic composition then defines the earliest time these objects could have formed. This time difference is at least 2 m.y. (*Halliday and Porcelli,* 2001).

Basaltic achondrites are relatively rare and it could well be that in the harsh energetic processes associated with accretion and destruction of asteroids the silicate portions of many primary planetesimals have been destroyed. A better sample may be found among the many iron meteorites (*Halliday,* 2003). It has long been known from Pb-isotopic data as well as ^{107}Pd-^{107}Ag chronometry (*Chen and Wasserburg,* 1996) that iron meteorites are early objects. However, the Pb-isotopic data for iron meteorites appear to be offset from the initial compositions defined by CAIs for reasons that are not well understood (*Tera and Carlson,* 1999; *Carlson and Lugmair,* 2000) and the ^{107}Pd-^{107}Ag system is not well calibrated against other chronometers (*Chen and Wasserburg,* 1996). Alternative chronometers such as ^{129}I-^{129}Xe and ^{53}Mn-^{53}Cr probably date cooling rather than the primary formation of these objects.

The most powerful chronometer of the accretion and metal segregation processes associated with iron meteorite formation is ^{182}Hf-^{182}W. The timing of metal segregation is well defined by the W-isotopic composition because Hf and W are both refractory and early solar system parent bodies can be safely assumed to have chondritic Hf/W (*Lee and Halliday,* 1996; *Horan et al.,* 1998) (Fig. 2). As such the parent-body isotopic evolution is predictable and the W-isotopic composition of the iron meteorite can define the time of metal separation to extremely high precision using modern MC-ICPMS techniques (*Halliday,* 2003, *Halliday et al.,* 2003; *Lee and Halliday,* 2003). The W-isotopic data for iron meteorites are strikingly unradiogenic, indicating that they formed within a few million years of each other at the start of the solar system (*Lee and Halliday,* 1996; *Horan*

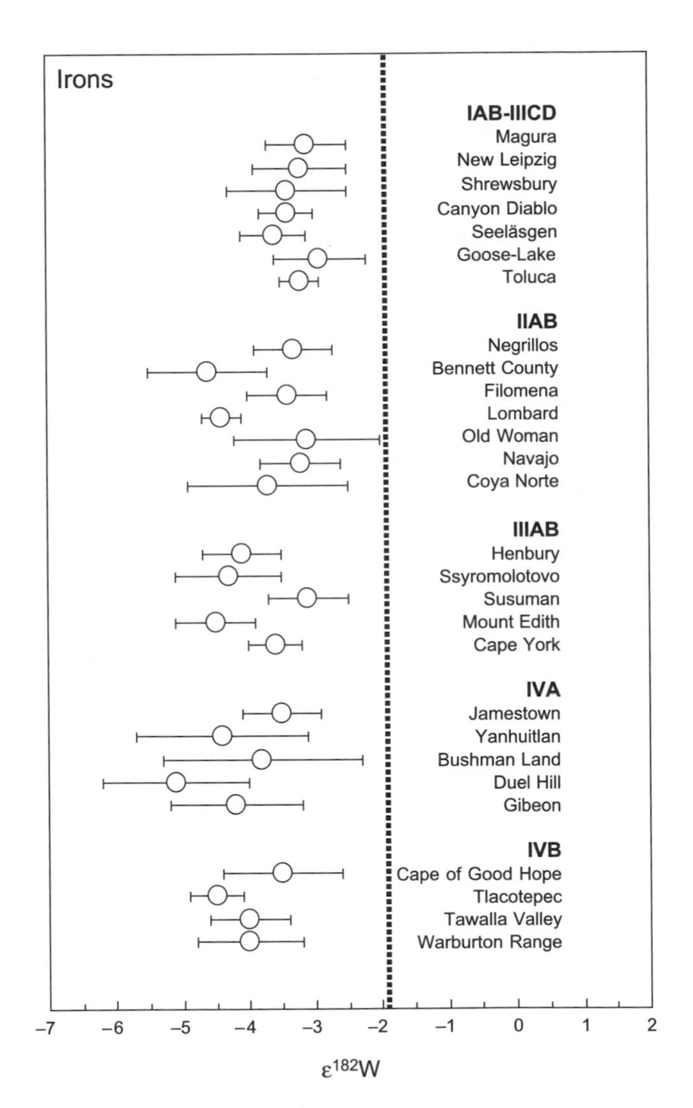

Fig. 2. The W-isotopic compositions of iron meteorites provide unequivocal evidence of early metal segregation. Assuming the parent bodies had chondritic relative proportions of refractory Hf and W, one can calculate a model time difference between the times of formation of different metals. The most comprehensive dataset published so far is that of *Horan et al.* (1998) shown here. The vertical line is the best estimate of average solar system based on data for carbonaceous chondrites (*Kleine et al.,* 2002, 2004a).

et al., 1998). The absolute age can be calculated from cross-calibrations of the Hf-W-isotopic system such as that provided by Ste. Marguerite, discussed above. New studies now show that the W-isotopic compositions of iron meteorites yield resolvable differences at very high precision (*Lee and Halliday*, 2003; *Markowski et al.*, 2004, 2006; *Schersten et al.*, 2004; *Kleine et al.*, 2005a; *Lee,* 2005).

The least-radiogenic $^{182}W/^{184}W$ reported to high precision is $\varepsilon^{182}W = -4.4 \pm 0.2$ (the deviation in parts per 10^4 relative to bulk silicate Earth), obtained for Tlacotopec (*Quitté and Birck,* 2004). However, this low value may in part reflect burnout of W isotopes caused by the interaction with thermal neutrons produced during a prolonged exposure to cosmic rays (*Masarik,* 1997; *Leya et al.*, 2003). Several iron meteorites, including Tlacotopec, have very long cosmic-ray exposure ages of hundreds of millions of years. As such, their W-isotopic compositions are not expected to be pristine. The production rate of thermal neutrons and, hence, the cosmogenic effects on W isotopes vary with depth in a meteorite. Therefore, different W-isotopic data may be obtained from different aliquots of the same meteorite. To obtain really precise estimates of the timescales of metal segregation it may be necessary to measure exposure ages and W-isotopic compositions on the same aliquots and then to apply a correction (*Markowski et al.*, 2006).

Even if the least-radiogenic W in iron meteorites is only as low as $\varepsilon W = -3.7$, other problems are apparent. The Hf-W data for Allende CAIs (*Kleine et al.*, 2005a) define a $(^{182}W/^{184}W)_{BSSI}$ that is higher than $\varepsilon^{182}W = -3.7$, apparently suggesting that some iron meteorites formed at least a few hundred thousand years before CAIs. This is contrary to the reigning paradigm that CAIs are the first solids that formed in the solar nebula. Challenging this paradigm will require a precise quantification of the ^{182}W-burnout effects in iron meteorites (*Markowski et al.*, 2006). It will also require demonstration that the Hf-W systematics of CAIs have not been preferentially reset. Given the presence of W compounds as Fremdlinge in Allende CAIs and the demonstration that these are secondary features (*Armstrong et al.*, 1985, 1987; *Hutcheon et al.*, 1987), one is left with some level of uncertainty as to whether there has been preferential disturbance of the W isotopes. To counter this, it seems unlikely that the CAI isochron is reset for the following reasons [a detailed discussion of this issue can be found in *Kleine et al.* (2005a)]: (1) The isochron is based on mineral separates from one CAI *and* bulk analyses from two other CAIs. (2) Resetting of the Hf-W system would require mobilization of radiogenic W from silicates into metal. The diffusion of W from silicates into metal, however, appears to require temperatures in excess of 600°C (*Kleine et al.*, 2005b), which were not achieved on the Allende parent asteroid. The formation of scheelite and powellite as secondary products from Fremdlinge as well as the mobilization of W from Fremdlinge to surrounding silicates would not cause resetting of the isochron. (3) The ^{182}Hf-^{182}W age derived from the CAI isochron (calculated relative to the Ste. Marguerite H chondrite) is 4568.0 ± 1.7 Ma, which is identical to the U-Pb age for Efremovka CAIs (*Amelin et al.*, 2002).

Recently, very precise W-isotopic data have been obtained for a large range of magmatic and nonmagmatic irons. By correcting for the maximum possible cosmogenic effect, it can be demonstrated that some magmatic irons segregated within <0.5 m.y. of the start of the solar system, as defined by W-isotopic data for Allende CAIs (*Markowski et al.*, 2006).

6. MARTIAN METEORITES AND EARLIEST PLANETARY DIFFERENTIATION PROCESSES

Martian meteorites provide the most powerful constraints on the timescales of formation of another terrestrial planet. This is not just of interest to comparative planetology. In many respects, Mars represents an example of how Earth and other terrestrial planets may have first started. There is evidence that a Mars-sized object sometimes referred to as "Theia" (the Greek goddess who was the mother of Selene, the goddess of the Moon) was responsible for the Moon-forming giant impact (*Canup and Asphaug,* 2001). There also is evidence that certain features of the composition of Theia were similar to those of Mars (*Halliday,* 2004; *Halliday and Porcelli,* 2001).

The constraints from short-lived nuclide data for martian meteorites are dominated by the data for four isotopic systems: ^{129}I-^{129}Xe, ^{146}Sm-^{142}Nd, ^{182}Hf-^{182}W, and ^{244}Pu-^{136}Xe (Table 1). Although all these data provide a similar picture of rapid development of Mars, the comparisons between W and Nd have proved most interesting. The latest compilation of W-isotopic data is shown in Fig. 3. It can be seen that all W-isotopic compositions are now resolvable from that of the silicate Earth, with radiogenic values in the range $\varepsilon^{182}W = 0$–3. In general terms the shergottites tend to be closer to 0 (*Kleine et al.*, 2004a; *Foley et al.*, 2005) whereas the nahklites are closer to 3 (*Lee and Halliday,* 1997). The original W-isotopic data (*Lee and Halliday,* 1997; *Kleine et al.*, 2004a) show a weakly defined relationship with the original $\varepsilon^{142}Nd$ (*Harper et al.*, 1995; *Jagoutz and Jotter,* 2000; *Jagoutz et al.*, 2003) that provided evidence that the early processes that affected each system were somehow correlated. Recently this has been questioned (*Foley et al.*, 2005) on the basis of new data, as discussed in detail below. The processes fractionating Hf/W are partial melting and core formation (*Lee and Halliday,* 1997; *Halliday and Lee,* 1999), whereas Sm/Nd responds only to partial melting because both parent and daughter elements are lithophile. *Lee and Halliday* (1997) proposed that this reflects contemporaneous partial melting and core formation. However, this was based on the assumption that the least-radiogenic W was chondritic, as was the least-radiogenic Nd. Now it is known that chondrites do not have $\varepsilon^{182}W = 0$ (*Lee and Halliday,* 1995, 1996) but rather $\varepsilon^{182}W \sim -2$ (*Lee and Halliday,* 2000a; *Kleine et al.*, 2002; *Schoenberg et al.*, 2002; *Yin et al.*, 2002b). It has been pro-

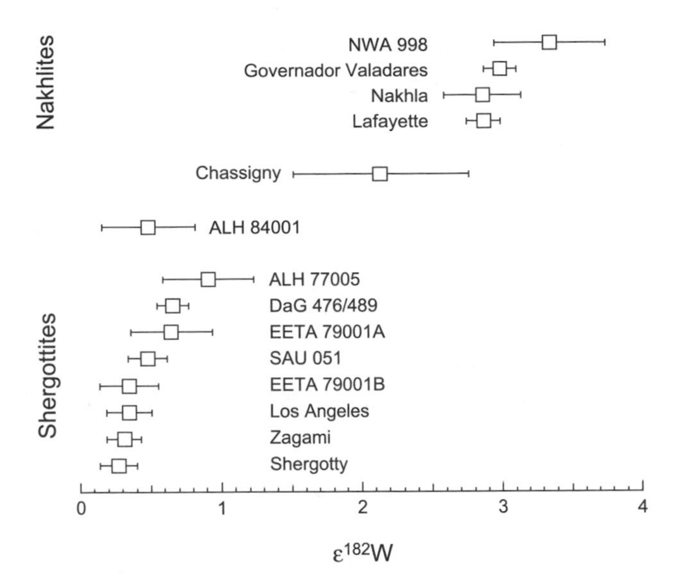

Fig. 3. Tungsten-isotopic data for martian meteorites reveal a significant offset relative to the newly defined average solar system at εW = –1.9. Data from *Lee and Halliday* (1997), *Kleine et al.* (2004a), and *Foley et al.* (2005). In general terms the sequence of papers is accompanied by a dramatic improvement in precision. Where the same meteorite has been analyzed in more than one study, only the most precise measurements are shown here and in the following figures. The three studies show generally good agreement but the imprecise value for EETA 79001 in *Lee and Halliday* (1997) differs significantly from the more precise measurements reported by *Kleine et al.* (2004a) and *Foley et al.* (2005). Assuming that these more precise recently determined values are likely to be correct, the data obtained for shergottites are relatively uniform and distinct from nakhlites, as pointed out by Kleine et al. and Foley et al.

posed that there is a small (~20 ppm) offset in chondritic ε^{142}Nd as well (*Boyet and Carlson,* 2005). However, this is of second-order importance for the comparison with W isotopes. The least-radiogenic W-isotopic composition measured thus far for Mars is significantly offset to higher than chondritic values (Fig. 3). *Kleine et al.* (2002, 2004a) have explained this in terms of a period of core formation prior to silicate melting, whereas the correlation with ε^{142}Nd reflects the effects of silicate melting only.

More precise ^{182}W and ^{142}Nd data on previously studied and additional martian meteorites have now been used to argue that no such overall Nd-W correlation exists (*Foley et al.,* 2005). The analytical uncertainties on ε^{182}W are considerably improved in the Foley et al. study. Most previously published W data agree with the new data within the stated uncertainties except for the early report of ε^{182}W ~ 2 for EETA 79001 (*Lee and Halliday,* 1997), which is not replicated by both of the more recent studies (*Kleine et al.,* 2004; *Foley et al.,* 2005). There are also apparent discrepancies between the ε^{142}Nd of two Saharan martian meteorites, DaG 476 and SAU 051 (*Jagoutz and Jotter,* 2000; *Jagoutz et al.,* 2003; *Foley et al.,* 2005). The most precise Nd- and W-isotopic data available as of August 2005 are plotted in Fig. 4. The nakhlites form a group with relatively uniform and radiogenic Nd and W (ε^{182}W ~ 3). Chassigny is slightly less radiogenic. The shergottites are less radiogenic in W, as already noted. However, with the inclusion of the latest data for the two Saharan shergottites DaG 476 and SAU 051 (*Foley et al.,* 2005), there is a broad spread in ε^{142}Nd from –0.3 to radiogenic values of ~+1.0, like those of the nakhlites, and offset from the main correlation. The addition of the new data for the Saharan meteorites therefore changes the apparent distribution somewhat. In Fig. 4 both the new Foley data and the Jagoutz ε^{142}Nd data are plotted to illustrate the effect. The reason for this apparent discrepancy is not certain although, as discussed by Foley et al., the lower ε^{142}Nd value reported by *Jagoutz et al.* (2003) in the Saharan martian meteorites could result from the presence of a terrestrial weathering component. Until this discrepancy is clarified it is perhaps premature to venture an explanation for the apparent offset of these two Saharan Nd data from

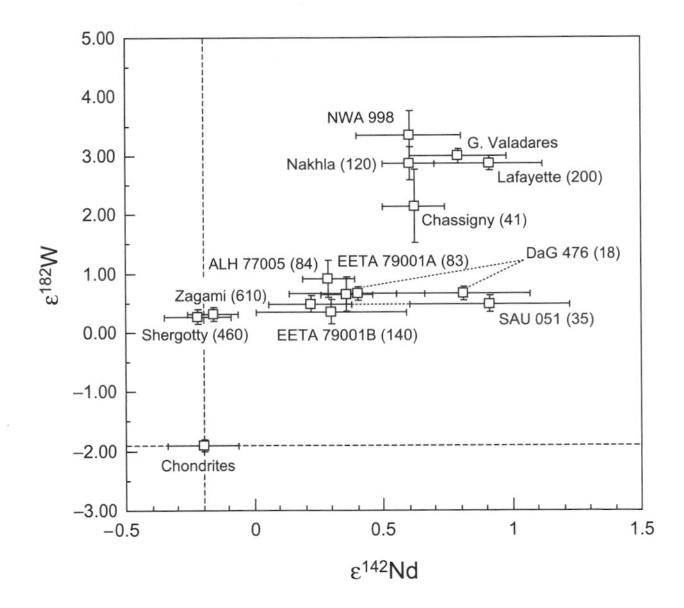

Fig. 4. The W-isotopic compositions of martian meteorites display a broad relationship with their corresponding ε^{142}Nd resulting from decay of formerly live ^{146}Sm. Data from *Harper et al.* (1995), *Jagoutz et al.* (2003), and *Foley et al.* (2005). The latest ^{142}Nd data for the two Saharan meteorites (*Foley et al.,* 2005) differ from those reported by *Jagoutz et al.* (2003). The reasons for these discrepancies are unclear and both sets of values are shown here. *Foley et al.* (2005) point out that inclusion of their new Saharan data seriously reduces the evidence for a relationships between Nd- and W-isotopic effects and the proposal that metal segregation and silicate melting overlapped in time (*Lee and Halliday,* 1997). The fact that some samples have W that is more radiogenic than chondritic, while still possessing nearly chondritic Nd, provides evidence that a component of core formation preceded large-scale silicate differentiation (*Kleine et al.,* 2002, 2004a). Note the new values for chondritic W (*Kleine et al.,* 2004a) and chondritic Nd (*Boyet and Carlson,* 2005). The concentrations of W (in ppb) are shown in parentheses.

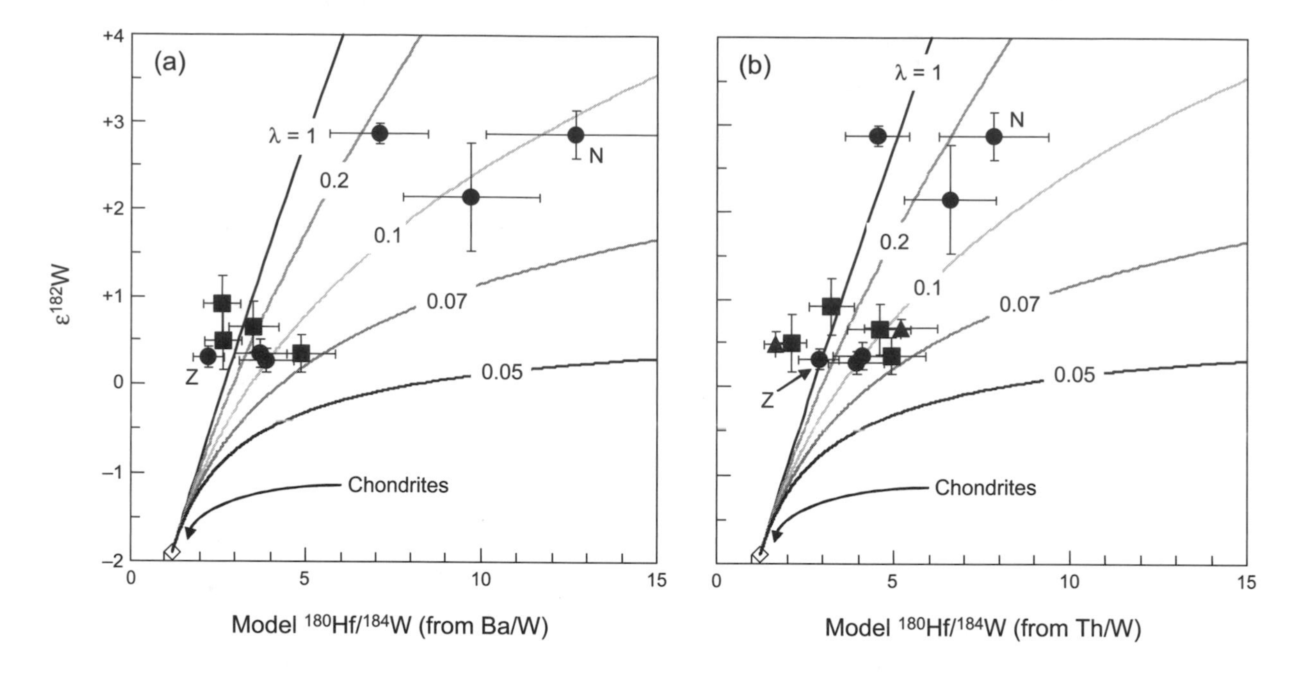

Fig. 5. Tungsten-isotopic compositions of martian meteorites do not correlate with measured Hf/W but do correlate with model source Hf/W determined from the chondritic value of **(a)** Hf/Ba and the sample's Ba/W and **(b)** Hf/Th and the sample's Th/W. Assuming Hf/W is only fractionated by core formation and not silicate partial melting, these data can be modeled in terms of a progressive core formation model. However, it is very well established that W is much more incompatible than Hf during partial melting of the mantle (*Newsom et al.,* 1986, 1996; *Halliday and Lee,* 1999). Note therefore that metal segregation probably only accounts for a fraction of the W-isotopic heterogeneity. The present-day W-isotopic compositions are shown for a range of martian reservoirs that have suffered different degrees of W depletion by core formation over different timescales. Each model curve deploys exponentially increasing Hf/W with time and a time constant (m.y.$^{-1}$) as shown. There is no particular basis for assuming that W depletion did change exponentially with time. Other models show a similar general curvature and the timescales calculated (see Fig. 6) are similar. However, a portion of the apparent rapidity of differentiation could reflect the unknown magnitude of increase in Hf/W resulting from silicate partial melting. See text for details.

the main trend. *Foley et al.* (2005) note that there still exists a correlation between the ε^{182}W and ε^{142}Nd values for the shergottites, as can be seen from Fig. 4, but it has a very different (shallower) slope.

A concern with some of the Saharan samples is with mobility of certain trace elements. Some such samples have anomalous trace-element compositions, thought to reflect partial terrestrial contamination or leaching. As shown in Fig. 4, the meteorites DaG 476 and SAU 051 have extremely depleted W concentrations (*Kleine et al.,* 2004) — among the most depleted yet recorded from shergottites and nahklites (*Lodders,* 1998). The Nd/W ratios are perfectly normal (*Kleine et al.,* 2004; *Dreibus et al.,* 2003). Therefore, whether this depletion reflects leaching/dissolution of incompatible matrix elements or is a primary igneous feature is unclear. There is no question from the bulk compositions that there has been enrichment in some fluid mobile elements (e.g., Ba) and so leaching and exchange of either Nd or W may have also occurred. Whether this has anything to do with the isotopic data distribution or the interlaboratory differences is also unclear at present.

The W-isotopic compositions of martian meteorites are not correlated with chemical indices of magmatic fractionation (*Halliday and Lee,* 1999). This is not unexpected because martian meteorites are mainly young and their sources have experienced several billion years of evolution since the very early processes that produced the W-isotopic effects. However, the first, imprecise, W-isotopic compositions do show a trend with Ba/W (*Halliday et al.,* 2001b). This is less convincing with more precise measurements (Fig. 5a), as discussed below. Barium and W are equally incompatible in mantle melting and their relative proportions in terrestrial basalts are more or less uniform (*Newsom et al.,* 1986, 1996). The Ba/W ratios of terrestrial basalts dominantly reflect the amount of W depletion in the mantle caused by core formation. Assuming this is also true of silicate partitioning on Mars, one can use the variation in Ba/W as a proxy for different degrees of metal segregation. The implication is that the incompatible trace elements in martian basalts carry a record of very early metal segregation processes on the planet. This at first seems quite remarkable. On Earth such effects have largely been homogenized by billions of years of mantle convection. The oldest incompatible-element isotopic heterogeneities in the present-day convecting mantle, as preserved in modern basalt magmas, are <2 b.y. old. Even the subcratonic lithospheric mantle is overwhelmingly dominated by heterogeneity that is ≤3 b.y. Traces of very early (>4.4 Ga) ε^{142}Nd heterogeneity have been found in the ancient sources of early Archean rocks from West Greenland (*Boyet et al.,* 2003; *Caro et al.,* 2003). Therefore, Earth had small mantle isotopic heterogeneities in the early Archean that have since been homogenized. It has recently been proposed that the silicate Earth, as sampled by magmas, is offset from chondrites in ε^{142}Nd by 20 ± 14 ppm (*Boyet and Carlson,* 2005). Whether this reflects a complementary hidden reservoir or a slightly nonchondritic silicate Earth is unclear. None of these effects compare with the large heterogeneities preserved within the martian mantle from the earliest solar system.

The fact that the martian W-isotopic compositions relate to Ba/W provides evidence that the isotopic variations do define differences in the degree of W depletion related to core formation and not just partial melting. The samples carry a chemical as well as isotopic record of variations in the degree of siderophile depletion. One can simply convert the Ba/W to a model source Hf/W that would have been produced as a result of metal-silicate fractionation by multiplying by the chondritic Hf/Ba ratio (0.044). This is illustrated (Fig. 5a) for all the most precise W-isotopic data for which Ba/W ratios are available. Saharan meteorites are strongly altered and as such it is not possible to use Hf/Ba as a refractory lithophile multiplier in such samples. Unfortunately, U, the commonly used alternative to Ba, is also highly mobile in Saharan meteorites. Thorium is probably the best, although also imperfect, alternative and is used here to calculate a second set of model source Hf/W values from the measured Th/W and chondritic Hf/Th (4.0). When the W-isotopic data are compared to these model source Hf/W ratios, a trend is evident, whichever multiplier is deployed (Fig. 5a,b). This cannot be explained by silicate partial melting. It indicates that core formation was a primary factor in generating the variations in W-isotopic composition.

The scatter to the data distribution in Fig. 5 could be caused by one or more of four factors:

1. *Alteration.* The different symbols in Fig. 5 refer to Saharan (diamonds), Antarctic (squares), and other (circles) martian meteorites. The model Hf/W values calculated for the two Saharan meteorites, using Th/W, define the full spread in shergottite values (Fig. 5b) and should therefore be treated with some caution.

2. *Analytical uncertainties and sample heterogeneity issues.* The source Hf/W calculated either way has been assigned a somewhat arbitrary uncertainty of ±20%. However, the trace-element compositions are generally determined on different aliquots from those used for measuring isotopic compositions. Given the variability in trace-element concentrations reported by different workers, this may be an underestimate of the real uncertainties. (See, for example, C. Meyer's *Mars Meteorite Compendium* of available data at *http://www-curator.jsc.nasa.gov/curator/antmet/mmc/mmc.htm.*) Furthermore, some of the apparent variability in Ba/W for the same sample could be related to these elements being predominantly cited in different minor phases that could be distributed heterogeneously on the scale of the typical sample amounts used by various workers.

3. *Partial melting.* A second effect may have fractionated the source Hf/W, resulting in changes in W-isotopic composition. The variability in ε^{142}Nd (Fig. 4) indicates that some silicate partial melting was consanguineous with core formation. This would have produced source variations in Hf/W that are not determinable with the approach utilized here.

4. *Variable siderophile-element depletion in time and space.* Given the preservation of heterogeneity in the W-isotopic composition and degree of W depletion, there is no basis for assuming that metal segregation proceeded at a constant rate in all portions of Mars.

This last feature can be modeled quite easily. This is important because clearly the W-isotopic compositions were not generated from a single primitive mantle reservoir with uniform Hf/W as is commonly deployed to calculate a model age of core formation. The data distributions shown in Fig. 5a,b are plotted along with model curves, each of which defines the predicted present-day W-isotopic compositions of a reservoir that became isotopically heterogeneous by undergoing variable degrees of metal segregation, hence W depletion, with a particular time constant, λ, i.e.,

$$^{180}\text{Hf}/^{184}\text{W} = (^{180}\text{Hf}/^{184}\text{W})_{\text{BSS}} \times e^{\lambda t}$$

where $(^{180}\text{Hf}/^{184}\text{W})_{\text{BSS}}$ is the parent/daughter ratio of the bulk solar system (or chondrites), λ is the time constant, and t is time. The model assumes that Mars was a chondritic object that developed various such heterogeneous reservoirs, which differed in the rate at which they underwent W depletion because of core formation. If the W depletion develops rapidly the Hf/W ratio increases early and the reservoir develops W that is relatively radiogenic for a given Hf/W.

The longevity defined by the locus of each of these different time constant curves is shown in Fig. 6. It can be seen that the timescales that are required to generate martian meteorite compositions like that of Nakhla could in fact be quite long (>10 m.y.) because, despite having radiogenic W, the model Hf/W is also very high. Unless other effects (1 through 3 above) are responsible, the nakhlite source appears to have carried on segregating metal over long (10^7 yr) timescales. The steepest line in Fig. 5 defines a very fast, effectively instantaneous differentiation rate ($\lambda = 1$ m.y.$^{-1}$) and several of the martian meteorites plot along this line or slightly to the left with low Hf/W for a given W-isotopic composition. The differentiation timescales for generating the W-isotopic compositions along this low Hf/W line are incredibly fast — around 1 m.y., as shown for Zagami (Fig. 6).

Coeval silicate partial melting effects mean that these Hf/W values could both over- and underestimate the real source Hf/W values. This could explain why some samples have slightly radiogenic W for a given model Hf/W and plot to the left in Fig. 5, for example. That is, the source was also depleted by partial melting generating higher Hf/W than

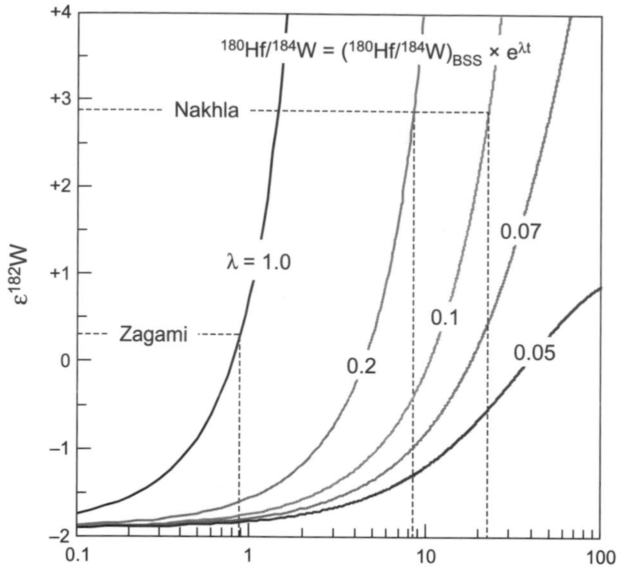

Fig. 6. Timescales implied by the model development of Hf/W with time in the martian mantle assuming all the W-isotopic effects are the result of siderophile depletion associated with core formation. Two extreme examples are highlighted. The W depletion of the Nakhla source would have taken more than 10 m.y. to develop, whereas the time necessary to form the source of Zagami is <1 m.y.

accounted for with the approach adopted here that assumes that all the W-isotopic effects result from core formation. However, this does *not* appear to apply to Zagami and Shergotty (Fig. 4). These meteorites yield chondritic Nd-isotopic compositions; there is no evidence of fractionation as a result of silicate partial melting. Further combined W- and Nd-isotopic data and high-quality trace-element data on the same and representative sample aliquots are needed to evaluate this more generally. Model attempts based on the lithophile behavior of Nd have been utilized by *Kleine et al.* (2004a) and *Foley et al.* (2005). The deduced timescales for martian mantle reservoirs tend to be similar. *Brandon et al.* (2000) have also presented a broadly similar story from Re-Os systematics. *Marty and Marti* (2002) similarly have presented the case for very rapid (<35 m.y.) differentiation based on Xe-isotopic systematics but with a more protracted degassing history for the nakhlite source. All these approaches yield approximately the same view of Mars, that the accretion and primary differentiation were exceedingly rapid.

Nevertheless, the Hf-W data provide unique evidence that core formation proceeded over a range of timescales that in some cases were quite long (10^7 yr). Other meteorites such as Zagami appear to record more rapid timescales (10^6 yr). This in turn implies that accretion of Mars was early and rapid, on the order of 1 m.y. Such timescales are similar to those predicted from runaway growth (*Weidenschilling et al.,* 1997). There is little evidence for late-stage planetary-scale collisions affecting and effecting the growth of Mars.

7. THE AGE OF THE MOON

The discovery of ^{182}W variability in lunar samples has provided the most powerful constraint on the age of the Moon (*Lee et al.,* 1997, 2002; *Leya et al.* 2000; *Kleine et al.,* 2005c). The W-isotopic compositions of lunar samples are offset to radiogenic values relative to chondrites. At first, it appeared these effects were relatively large such that a rather simple model age calculation could be applied. However, the most enhanced ε_W values are now known to be the product of cosmogenic effects on ^{181}Ta, which produces ^{182}Ta, the intermediate decay product of ^{182}Hf decay (*Leya et al.,* 2000; *Lee et al.,* 2002). The isotopic variations appear to be at most 1 or 2 ε units (*Lee et al.,* 2002). However, variations within the lunar mantle, even of this magnitude, restrict the age of the Moon to the first 30 to 55 m.y. of the solar system (*Kleine et al.,* 2005c; *Yin et al.,* 2002b; *Halliday,* 2003, 2004), in stark contrast to the limited constraints based on the oldest dated rocks (*Wasserburg et al.,* 1977b) (Table 2).

The determination of the Hf-W age of the Moon depends on knowing the magnitude of the W-isotopic effect produced by decay of ^{182}Hf within the Moon itself (*Halliday,* 2004). For this it is necessary to know (1) the initial Hf- and W-isotopic compositions of the solar system, (2) the initial W-isotopic composition of the Moon, (3) the present-day W-isotopic composition of the lunar mantle, and (4) the Hf/W of the lunar mantle.

The initial $^{182}Hf/^{180}Hf$ of the solar system is now fairly well established (*Kleine et al.,* 2002, 2005a; *Schoenberg et al.,* 2002; *Yin et al.,* 2002b). The initial $\varepsilon^{182}W$ of the solar system based on the W-isotopic compositions of CAIs and chondrites is established to be –3.5 (*Kleine et al.,* 2002, 2005a; *Yin et al.,* 2002b); nevertheless, as discussed earlier, there is still some degree of uncertainty because of the apparently less-radiogenic W-isotopic compositions of some iron meteorites.

The initial W-isotopic composition of the Moon is at present unknown. The argument has been made (*Halliday,* 2003, 2004) that the large number of samples with W-isotopic compositions close to $\varepsilon^{182}W = 0$ most likely means that this approximates a common composition from which the Moon started. This could well be wrong; however, it is unlikely to be far wrong. Clearly the initial composition of the Moon cannot be higher than its least-radiogenic value. It is unlikely to be much lower either since the current theory for the origin of the Moon involves a giant impact. The Moon is thought to have been derived from the silicate mantles of the proto-Earth and the impacting planet Theia. All such silicate reservoirs studied thus far have $\varepsilon^{182}W \geq -0.5$ (*Quitté et al.,* 2000) and unless the Moon formed very early (<20 m.y.), there is no reason why it would have a lower value.

Most of the present-day W-isotopic compositions of lunar samples corrected for cosmogenic effects lie in the range $\varepsilon^{182}W$ = 0–1. Most cosmogenic corrections are very approximate. Some values are unequivocally higher (*Lee et al.,* 2002). Lunar sample 15555 with no cosmogenic effect yields a value of 1.30 ± 0.39. From the current database it would appear that the average composition of the lunar mantle is unlikely to be greater than $\varepsilon^{182}W$ = 1, but it could conceivably be this high if 15555 is especially representative of the lunar mantle. Therefore, a likely scenario is an average radiogenic increase of about 0.5–1.0 (*Halliday,* 2004). A more conservative estimate of the amount of radiogenic increase would be that it is less than 2.0 $\varepsilon^{182}W$ units.

The fact that there are endemic, indigenous W-isotopic variations at all provides clear evidence that they were produced by radioactive decay within the lunar mantle; they cannot be residual from Earth or Theia given the inevitable mixing associated with the high temperatures of accretion and early convective overturn. The initial W-isotopic composition of the Moon does reflect the mix of components from these parent planets, however, and provides information on their history, as described below.

The Hf/W ratio of the lunar mantle has been estimated most recently and comprehensively by *Jones and Palme* (2000) to be 25.2. (Note that this value changes to 26.5 when using more up-to-date CI abundances; the estimate is based on U/W = 1.93 for the lunar mantle and depends on the Hf/U used for chondrites.) This is similar to previously used estimates (*Halliday,* 2003, 2004).

The age of the Moon can be deduced from these combined estimates as shown in Fig. 7 and would appear to lie in the range 40–50 m.y. (*Halliday,* 2004; *Kleine et al.,* 2004b). The age can be refined more closely once more precise estimates of the amount of radiogenic growth on the Moon are known. This estimate is later than the 29 m.y. proposed by *Kleine et al.* (2002) and the value of 30 m.y. calculated by *Yin et al.* (2002b). These are model ages relative to a chondritic reservoir and are the equivalent of assuming that the Moon formed in a single step from an undifferentiated chondritic reservoir. We now know the W-isotopic composition of such a reservoir very well (*Kleine et al.,* 2002; *Schoenberg et al.,* 2002; *Yin et al.,* 2002b). However, the Moon has a relatively low uncompressed density compared with that of the terrestrial planets, and this has long been taken as evidence of an origin from a silicate-rich precursor, i.e., a planetary mantle. Therefore, current models of lunar formation (e.g., *Canup and Asphaug,* 2001) involve a collision between Earth and a Mars-sized planet that is already differentiated into silicate and metal. As such, the prior W-isotopic evolution of these parent planetary mantles will have contributed to the W-isotopic composition recorded in lunar basalts. A model age relative to chondrites is the equivalent of assuming that none of this had happened. It provides a useful upper limit, that is, the earliest likely time that the Moon can have formed, given its W-isotopic composition.

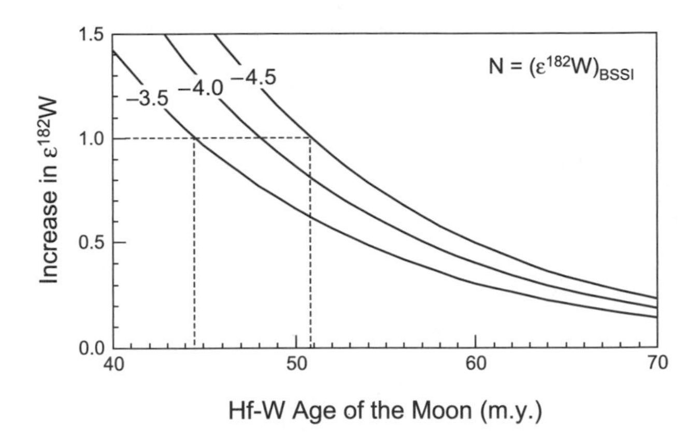

Fig. 7. The age information that can be derived from current W-isotopic data for the Moon at the present time is strongly model dependent. The effects on the calculated age of the Moon of not knowing the average amount of radiogenic, as opposed to inherited or cosmogenic, ^{182}W are shown. The increase in εW within the Moon as a function of decay of primordial ^{182}Hf, as can be deduced from currently available data, is probably ≤1 ε unit. The data are therefore consistent with an age of the Moon (hence giant impact) of about 40 to 50 m.y. after the start of the solar system. The different curves are based on differing values of the $\varepsilon^{182}W_{BSSI}$, which directly affects the calculated $(^{182}Hf/^{180}Hf)_{BSSI}$. All calculations assume $\lambda^{182}Hf = 0.077 \times 10^{-6}\ yr^{-1}$, Hf/W_{MOON} = 22, and $\varepsilon^{182}W_{BSS} = -1.9$.

Recently, *Kleine et al.* (2005c) have reported W-isotopic data for metals separated from lunar samples that provide what appears to be the best constraint on the exact age of the Moon. The significance of these measurements lies in the fact that the metal has very low Hf/W and Ta/W such that the W-isotopic composition is a real initial ratio for the igneous rocks studied whatever their age and exposure. By comparing these W-isotopic compositions with independent estimates of the Hf/W ratio of the magma sources, one can derive a differentiation age for the Moon. On this basis Kleine et al. derive an age for lunar differentiation of 40 ± 10 m.y. This is the most precise estimate yet obtained for the age of crystallization of the lunar magma ocean.

Whatever the exact age of the Moon, there are clear implications for these results:

1. First, there is no question that the Moon formed late relative to other small objects. Therefore, models for its formation must explain this. The fission and late impact theories would seem best suited. Capture and coaccretion do not predict a late formation.

2. A significant time gap can now be identified between the age of crystallization of the lunar magma ocean deduced from ^{182}Hf-^{182}W and the precise ages deduced from $^{235/238}U$-$^{207/206}Pb$ and ^{147}Sm-^{143}Nd chronology for early lunar rocks and lunar differentiation events, previously thought to re-

late to the formation and crystallization of the lunar magma ocean.

3. Assuming the Moon was formed in a giant impact when the proto-Earth was ~90% formed, its age provides the single most important piece of independent evidence that can be used with the meteorite reference frame to calibrate Earth's growth history, as explained next.

8. METEORITES AND THE GROWTH OF EARTH

As explained above, the W-isotopic data for iron meteorites provide evidence of very rapid accretion and core formation of earliest planetesimals in the inner solar system. Similarly, the data for Mars are hard to explain unless accretion and earliest core formation were extremely rapid (<1 m.y.). The isotopic data for Earth, however, provide strong support for protracted accretion over tens of millions of years. This is consistent with dynamic simulations that predict timescales for terrestrial planet accretion that are on the order of a few tens of millions of years (*Safronov,* 1954; *Wetherill,* 1980, 1986; *Agnor et al.,* 1999; *Canup and Agnor,* 2000; *Chambers,* 2001a,b, 2004). These simulations can be extremely sensitive to the amount of nebular gas present (*Agnor and Ward,* 2002; *Kominami and Ida,* 2002). Their accuracy for describing the real Earth needs to be tested rigorously, hence the central importance of isotopic approaches.

The three most powerful and effective techniques for determining the growth rate of Earth are $^{235/238}$U-$^{207/206}$Pb, ^{244}Pu/^{129}I-$^{136/129}$Xe, and ^{182}Hf-^{182}W. These yield differing age estimates, and this in turn may provide insights into the processes that are likely to have accompanied Earth accretion. As the first ^{182}Hf-^{182}W data became available there appeared to be excellent agreement with the protracted timescales for planetary accretion and atmospheric loss as deduced from ^{92}Nb-^{92}Zr (*Münker et al.,* 2000; *Jacobsen and Yin,* 2001), ^{129}I/^{244}Pu-$^{129/136}$Xe (*Porcelli and Pepin,* 2000), and $^{235/238}$U-$^{207/206}$Pb chronometry (*Halliday,* 2000). The timescales deduced from ^{97}Tc-^{97}Mo (*Yin and Jacobsen,* 1998) and ^{107}Pd-^{107}Ag (*Carlson and Hauri,* 2001) chronometry were shorter.

Now it is clear that some of these constraints are not very strong because some of the critical Mo, Zr, and Ag data appear to be incorrect or misinterpreted (*Dauphas et al.,* 2002a,b, 2004; *Chen et al.,* 2004; *Lee and Halliday,* 2003; *Schönbächler et al.,* 2002, 2003; *Woodland et al.,* 2004). For example, the anomalies in Mo are hard to distinguish from nucleosynthetic effects and different groups have reported significantly different results for some of the isotopes even when the same normalization procedure is deployed (*Yin et al.,* 2002a; *Becker and Walker,* 2003; *Dauphas et al.,* 2002a,b, 2004; *Chen et al.,* 2004; *Lee and Halliday,* 2003). The reasons for these discrepancies are unclear. *Dauphas et al.* (2004) reported that his Mo-isotopic data correlated with effects for Ru reported by *Chen et al.* (2003), apparently confirming their validity as nucleosynthetic heterogeneities. The latest results for Mo reported by *Chen et al.* (2004) reveal differences relative to previously published results. They confirm only some of the nucleosynthetic effects and find a decoupling between *p*- and *r*-process components. The results no longer correlate with Ru so well.

The apparent consistency between $^{235/238}$U-$^{207/206}$Pb, ^{244}Pu/^{129}I-$^{136/129}$Xe, and ^{182}Hf-^{182}W chronometry may also be incorrect. The early W-isotopic data for chondrites (*Lee and Halliday,* 1995, 1996) are wrong by about 180–200 ppm (*Kleine et al.,* 2002; *Schönberg et al.,* 2002; *Yin et al.,* 2002b). The new ^{182}Hf-^{182}W timescales are shorter and appear to be inconsistent with $^{235/238}$U-$^{207/206}$Pb timescales (*Halliday,* 2003, 2004; *Wood and Halliday,* 2005). Even the latest ^{244}Pu/^{129}I-$^{136/129}$Xe timescales appear to be too long (*Porcelli et al.,* 2001). In detail, therefore, it now appears as though the chronometers are reflecting different chemical fractionations that took place over distinct timescales, or that the degree to which the real processes are accurately simulated by the isotopic models differ between elements (*Halliday,* 2003, 2004; *Wood and Halliday,* 2005).

The $^{235/238}$U-$^{207/206}$Pb and ^{182}Hf-^{182}W chronometers both rely on determining the timing of fractionation during core formation. Both are usually utilized in one of two ways: (1) A *model age* can be calculated as the time that the silicate Earth was last in total isotopic equilibrium, since the time when its silicate (high U/Pb and Hf/W) and metal (low U/Pb and Hf/W) reservoirs developed distinct compositions. This can be thought of as defining an age of instantaneous core formation in a completely formed planet. Alternatively, it can be thought of as catastrophic reequilibration of silicate and metal reservoirs during some major overturn mixing event. Finally, it can be thought of as simply the age of the planet itself assuming the core formed simultaneously. There is considerable evidence against any of these representing the way Earth formed and developed in its earliest stages. (2) A *mean age* for accretion can be calculated as the inverse of the time constant for exponentially decreasing growth of Earth, assuming that, as it grew, the accreted material isotopically equilibrated with the primitive mantle and continuously segregated further core material in present-day proportion to Earth's mass (*Jacobsen and Harper,* 1996). There seems little question that this is more realistic (*Halliday,* 2004).

The first of these models represents the standard procedure adopted for dating Earth or core formation (*Allègre et al.,* 1995; *Galer and Goldstein,* 1996; *Lee and Halliday,* 1995, 1996; *Halliday and Lee,* 1999). However, the "event" defined probably never occurred as such. The second of these models is more sophisticated and useful and is based on the formalism first proposed by *Jacobsen and Wasserburg* 1979). It was explored for ^{182}Hf-^{182}W by *Harper and Jacobsen* (1996) and *Jacobsen and Harper* (1996) and further explored in a series of models by *Halliday et al.* (1996, 2000), *Halliday and Lee* (1999), and *Halliday* (2000). These all predate the correct determination of the isotopic compositions of chondrites (*Kleine et al.,* 2002, 2004a; *Lee and Halliday,* 2000a; *Schoenberg et al.,* 2002; *Yin et al.,* 2002b). Recent studies have made extensive use of this approach (and the correct reference compositions) to obtain an accurate ^{182}Hf-^{182}W accretion timescale for Earth (*Yin et al.*, 2002b; *Halliday*, 2003, 2004; *Kleine et al.*, 2004b). The same idea was developed and exploited for $^{235/238}U$-$^{207/206}Pb$ by *Halliday* (2000, 2003, 2004).

When the results for ^{182}Hf-^{182}W and $^{235/238}U$-$^{207/206}Pb$ are compared there appears to be a discrepancy (*Halliday*, 2003, 2004; *Wood and Halliday*, 2005). The $^{235/238}U$-$^{207/206}Pb$ accretion rates appear to be more protracted than those determined from ^{182}Hf-^{182}W (Fig. 8). One possible explanation to consider is that it relates to the differences in the rates of metal-silicate equilibration of refractory W vs. volatile Pb during the accretion process itself (*Halliday*, 2003, 2004). The possibility that there had been a lack of equilibration during accretion was discussed by *Halliday* (2001), but at that stage it appeared that the W-isotopic composition of the silicate Earth was identical to that of chondrites such that equilibration had to be the general rule (*Halliday*, 2001). Now that a resolvable difference has been established between the silicate Earth and chondrites, the question arises as to whether this is caused by faster accretion, or lack of isotopic equilibration, or both (*Halliday*, 2004). The only way to test this is with some independent assessment of the age of the Moon, or the giant impact.

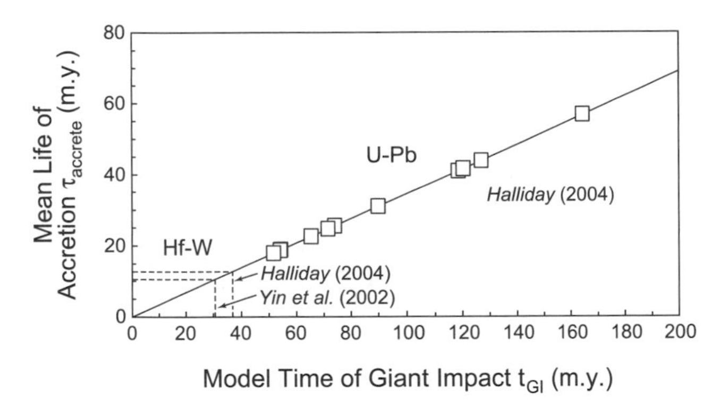

Fig. 8. Calculated values for Earth's mean life of accretion (τ) and time of the giant impact (T_{GI}) given in million years as deduced from the different estimates of the Pb-isotopic composition of the bulk silicate Earth (BSE) (*Doe and Zartman*, 1979; *Davies*, 1984; *Allègre et al.*, 1988; *Zartman and Haines*, 1988; *Allègre and Lewin*, 1989; *Kwon et al.*, 1989; *Galer and Goldstein*, 1991; *Liew et al.*, 1991; *Kramers and Tolstikhin*, 1997; *Kamber and Collerson*, 1999; *Murphy et al.*, 2003) using the type of accretion model shown in Fig. 9 (see *Halliday*, 2004). Note that, while all calculations assume continuous core formation and total equilibration between accreted material and the BSE, the accretion is punctuated, as predicted from the planetesimal theory of planetary accretion. This generates more protracted calculated timescales than those that assumer smooth accretion (*Halliday*, 2004). The $^{238}U/^{204}Pb$ values assumed for Earth = 0.7 (*Allègre et al.*, 1995). Even the Hf-W timescales using the exact same style of model are significantly shorter. The mean life model ages given in Table 2 are for a smooth accretion model that yields less-protracted results (*Halliday*, 2000) and are for the average (*Halliday*, 2004) of the most recent (*Kramers and Tolstikhin*, 1997; *Kamber and Collerson*, 1999; *Murphy et al.*, 2003) of these Pb-isotopic estimates.

The age of the Moon provides the only firm independent constraint on the accretion rate because it is thought to be the byproduct of the last major growth phase of Earth (*Canup and Asphaug*, 2001). The later the age of the Moon, the more disequilibrium is needed to explain the radiogenic W-isotopic composition of the silicate Earth. Using a window of 45 ± 5 m.y. (Fig. 7) as the most realistic for the giant impact (*Halliday*, 2003, 2004; *Kleine et al.*, 2004b, 2005c) allows a model accretion curve to be constructed using larger-sized impacts with time, culminating in the giant impact (Fig. 9). From this accretion curve one can deduce the level of W-isotopic equilibration that would be required assuming a general level of disequilibrium throughout the accretion history or disequilibrium during the giant impact alone (Fig. 10). Assuming the latter case it can be seen that

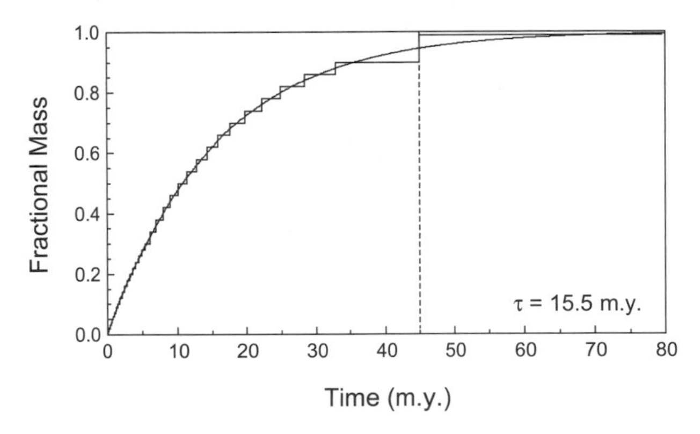

Fig. 9. Change in mass fraction of Earth as a function of time in the model used in *Halliday* (2004) illustrated with an accretion scenario calculated from a giant impact at 50 m.y. after the start of the solar system. The approximated mean life of accretion (τ) is the time taken to achieve 63% growth. Both this and the timing of each increment are calculated from the timing of the giant impact (t_{GI}) to achieve an overall exponentially decreasing rate of growth for Earth broadly consistent with dynamic simulations. The smooth curves show the corresponding exponentially decreasing rates of growth. The step function curves define the growth used in the isotopic calculations. Growth of Earth is modeled as a series of collisions between differentiated objects. The overall rate of accretionary growth of Earth may have decreased in some predictable fashion with time, but the growth events would have become more widely interspersed and larger. Therefore, the model simulates further growth by successive additions of 1% $M_{\oplus}$ objects from 1% to 10% of the current mass, then by 2% objects to 30%, and then by 4% objects to 90%. The Moon-forming giant impact is modeled to take place when Earth was ~90% of its current mass and involved an impactor planet Theia that was ~10% of the (then) mass of Earth. Therefore, in the model the giant impact at 45 m.y. after the start of the solar system contributes a further 9% of the current Earth mass. There is evidence against large amounts of accretion after the giant impact. Although episodic relative to more conventional continuous core formation models (*Jacobsen and Harper*, 1996; *Halliday et al.*, 1996, 2000; *Halliday and Lee*, 1999; *Halliday*, 2000), the model is still smooth relative to accretion simulations (e.g., *Agnor et al.*, 1999; *Chambers*, 2001a,b, 2004).

the amount of equilibration of Theia's core with the silicate Earth required to explain its W-isotopic composition was limited to <60% during a giant impact at 45 m.y. and could have been as little as 40% (cf. *Halliday,* 2004).

This point is made primarily to illustrate the importance of considering equilibration when determining model isotopic age constraints such as the accretionary mean life. If the giant impact involved a larger impactor, for example, a 0.15 instead of 0.10 $M_\oplus$ object, then a greater proportion of Earth would have been accreted at a later time. Therefore, the radiogenic W in the silicate Earth would imply greater amounts of disequilibrium during accretion. The size of the impactor affects the energy that is released. As such, it can be tuned in dynamic simulations to produce an Fe-depleted Moon at the right distance with the correct angular momentum. However, a critical issue is the equation of state, which defines the amount of energy released for a given sized impactor, and which is poorly known for high-pressure materials such as perovskite, even though advances are being made. A closer integration of isotopic modeling with dynamic simulations and experimental and theoretical mineral physics should help in this respect.

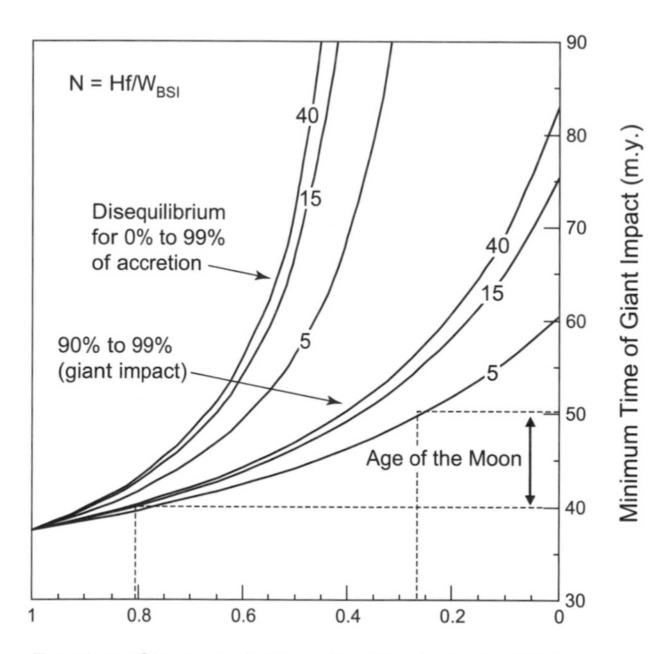

Fig. 10. An illustration of the effect on calculated Hf-W timescales for Earth's formation of incomplete mixing and equilibration of impacting core material. This plot shows the true time of the giant impact that generates $\varepsilon^{182}W_{BSE}$ of zero as a function of various levels of incomplete mixing of the impacting core material with the BSE. The lower curves are for disequilibrium during the giant impact alone. The upper curves correspond to disequilibrium during the entire accretion process up to and including the giant impact. The different curves for different Hf/W_{BSI} (the Hf/W in the silicate portion of the impactor) are also shown. All curves are calculated with $\varepsilon^{182}W_{BSSI} = -3.5$. Assuming an age for the Moon in the range 40–50 m.y. after the start of the solar system, it would seem likely that there was significant isotopic disequilibrium during the giant impact (*Halliday,* 2004).

Another variable that is becoming critical is the exact composition of Earth. We now have excellent W-isotopic data for chondrites (*Kleine et al.,* 2004a), assumed to represent the bulk Earth. This is unlikely to be far wrong. However, it now becomes important to know the exact Hf/W of the total Earth. If Earth has a Mg/Fe ratio that is lower than chondritic (*Halliday et al.,* 2001a; H. Palme, personal communication, 2003), then it probably has a Hf/W ratio that is also subchondritic. The effect on W-isotopic models depends on when this Hf/W depletion occurred. If it is an early feature it requires that the amount of disequilibrium during accretion was greater. Even the chondritic Hf/W ratios are not so well known. New high-precision isotope-dilution data by *Kleine et al.* (2004a) are, on average, slightly lower than the previously utilized "standard" data (*Newsom,* 1995). Although the data agree within error, these uncertainties are of sufficient magnitude that they now significantly limit attempts to produce high-resolution model age and accretion rate calculations. Sample size effects have plagued Lu-Hf and Sm-Nd studies of chondrites and new Hf-W studies of large representative samples would be worthwhile.

A Moon-forming impact at 45 m.y. does little to explain the discrepancy with the rates determined from the Pb-isotopic composition of Earth (Fig. 8). Assuming any one of these 11 estimates is approximately correct, the most likely explanation is that there was a change in U/Pb in the bulk silicate Earth during accretion. This might have been caused by changes in partitioning or volatile loss (*Halliday,* 2004). A recent study by *Wood and Halliday* (2005) proposes a new version of the former explanation. Tungsten and Pb may well segregate into planetary cores at different stages. Tungsten is more siderophile than chalcophile, whereas Pb displays the opposite behavior. During planetary accretion and differentiation a point will be reached where metal segregates as a result of the removal of ferric iron into perovskite. The metallic iron removes W at an early stage. However, Pb remains in the silicate Earth until the upper mantle of Earth becomes oxidized as a result of the giant impact disrupting perovskite in the lower mantle. The S added from the impactor, which appears to have been rich in chalcophile elements (*Yi et al.,* 2000), results in the formation of sulfides as Earth cools. These will strip the silicate Earth of its Pb and, if sufficiently abundant, will accumulate at the base of the magma ocean, eventually sinking to the core (*Wood and Halliday,* 2005). Therefore, there appears to be a logical explanation for why the Pb-isotopic system records a more protracted timescale than the Hf-W system. Cooling of Earth's lower mantle should have been very fast after the giant impact (*Solomatov,* 2000). However, the upper mantle from which the sulfide segregated could have cooled much more slowly (up to 10^8 yr). The earliest calculated timescales for removing Pb in this fashion are about 30 m.y. after a 45 m.y. giant impact based on the most recent estimate for the Pb-isotopic composition of the BSE (*Wood and Halliday,* 2005). Temperatures of roughly 3000 K are required to segregate sulfide in this fashion. The latest models of the giant impact (*Canup,* 2004) raise the temperature of Earth to about 7000 K. Therefore, broadly speaking, Earth cooled at more than 100 K per m.y. after the giant impact. This would correspond to the time required to cool the upper portion of the mantle to the temperature at which sulfide stripped Pb from the mantle.

9. TIME-INTEGRATED CHEMICAL COMPOSITIONS AND THE VOLATILE BUDGETS OF EARLY PLANETESIMALS AND TERRESTRIAL PLANETS

The radiogenic isotopic compositions of early solar system objects can be used to define time-averaged parent/daughter ratios of precursor materials. This can be explored for any isotopic system and provides useful insights into the paleocosmochemistry of the early solar system. There are four systems to which this can be very usefully applied to constrain early chemical evolution during accretion.

9.1. Time-integrated Uranium/Lead

Lead-isotopic compositions provide simultaneous constraints on the age of an object and the parent/daughter ratios ($^{238}U/^{204}Pb$ or primary "μ") of the precursor materials. For objects like Earth and Mars the present-day Pb-isotopic compositions of the silicate reservoirs are overwhelmingly dominated by the time-integrated μ since accretion and core formation. As such, these data say little about any earlier history. However, several lunar rocks formed early and sampled the Pb-isotopic compositions of their precursor reservoirs. Lead is moderately volatile, and most lunar rocks contain so little inherited Pb that the initial composition is hard to resolve (*Tera et al.,* 1973). It has been proposed that this may represent the early silicate Earth (*Galer,* 1993). However, an early high U/Pb silicate Earth is not easy to reconcile with the present Pb-isotopic composition of the silicate Earth (*Halliday,* 2004). Furthermore, some lunar rocks carry initial Pb-isotopic compositions that are relatively unradiogenic (*Meyer et al.,* 1975; *Hanan and Tilton,* 1987; *Torigoye-Kita et al.,* 1995) and provide evidence that the Moon formed from material that was not so depleted in Pb. In fact, the primary μ values are broadly similar to that of the bulk silicate Earth (*Halliday et al.,* 1996). Assuming that the Moon sampled a mixture of material from the proto-Earth and Theia, there is no clear evidence for any great difference in the magnitude of volatile depletion in these precursor materials compared with the present Earth. This raises the issue of how the strong depletion in ^{204}Pb in the Moon was introduced. Presumably it was a product of the giant impact.

9.2. Time-integrated Rubidium/Strontium

The Rb-Sr system also provides an indication of the magnitude of volatile-element depletion (*Halliday and Porcelli,* 2001) of precursor materials. The differences between the Sr-isotopic compositions of early objects like eucrites, angrites, and lunar rocks relative to CAIs provides strong evidence that some process causes loss of Rb during accretion (*Halliday and Porcelli,* 2001). This process cannot fractionate K isotopes (*Humayun and Clayton,* 1995), and partial evaporation would therefore appear unlikely. However, elemental fractionations unaccompanied by isotopic effects can in fact occur as a result of partial evaporation or condensation, as long as environmental conditions change sufficiently slowly to allow for thermodynamic equilibrium (*Richter,* 2003). In the case of early planetesimals like the angrite and eucrite parent bodies, it is thought that there was very early melting, differentiation, and volcanism predating the formation of the rocks sampled as achondrites (*Lugmair and Shukolyukov,* 1998). It is conceivable that such an early global differentiation was accompanied by extensive outgassing and almost complete loss of moderately volatile elements. The volatile budgets would then be added by subsequent accretion of new material. In the case of the Moon, however, it would appear more likely that the major loss (>90%) of Rb was associated with the giant impact (*O'Neill,* 1991a,b).

9.3. Time-integrated Hafnium/Tungsten

The Hf/W of a silicate reservoir is a function of the partitioning of W during core formation but may also be rendered heterogeneous by silicate partial melting (*Halliday and Lee,* 1999). The metal/silicate partition coefficient for W is strongly dependent on oxygen fugacity (*Schmitt et al.,* 1989; *Walter et al.,* 2000). It presumably is also affected by volatile contents and the pressure and temperature of core formation (*Righter and Drake,* 1996, 1999; *Righter et al.,* 1997; *Walter et al.,* 2000). The lack of depletion of W in the martian mantle has long been taken as providing evidence of more oxidizing conditions thought to be linked to the greater abundances of moderately volatile elements (for a review, see *Halliday et al.,* 2001b). The fact that the W-isotopic compositions recorded in martian meteorites are only 200–500 ppm more radiogenic than chondritic despite being generated very early is entirely consistent with this lack of depletion. In the case of the silicate Earth the fact that the W-isotopic composition also is within 200 ppm of chondritic despite a Hf/W that is an order of magnitude greater than chondritic is explained by protracted accretion of chondritic material equilibrating with the silicate Earth. However, the fact that the initial W-isotopic composition of the Moon is close to chondritic is more problematic (*Halliday,* 2004). The Moon has a high Hf/W, and it is thought that it was mainly formed from the silicate portions of Theia, with subordinate contributions from the proto-Earth. If Theia was only a Mars-sized object it should have formed relatively fast like Mars itself, as discussed above. However, with such a high Hf/W it would then be expected to have produced highly radiogenic W by the time of the giant impact. Instead, the Moon started with $\varepsilon W \sim 0$ (*Halliday,* 2004; *Kleine et al.,* 2005c), within ~200 ppm of chondritic values. Assuming the Moon was mainly derived from Theia, there are four possible explanations: (1) The bulk silicate Theia had its composition modified by large impacts shortly before the giant impact (*Halliday et al.,* 2000), which is somewhat untestable but not impossible. (2) Theia formed very slowly, and would need to be accreting at a rate much slower than Earth, Mars, or Vesta to have such unradiogenic W. (3) The material that formed the Moon equilibrated with quite large amounts of metal; to achieve an initial W-isotopic composition of $\varepsilon W \sim 0$ in the Moon the amount of metal required is roughly 10% of the total material, which is well in excess of that predicted in giant impact simulations; (4) The silicate portions of Theia and/or the proto-Earth had a Hf/W that was close to chondritic; W partitioning changes as a function of oxygen fugacity (*Walter et al.,* 2000), and the implication would be that the combination of silicate material that formed the Moon was significantly more oxidizing than the present silicate Earth.

The last of these scenarios seems the most likely. When the time-integrated Hf/W for the source of the Moon's constituents (generally thought to be dominated by the bulk silicate Theia) is compared with the time-integrated Rb/Sr (*Halliday and Porcelli,* 2001), the combined composition is strikingly similar to that of the martian mantle (*Halliday,* 2004), lending support for the idea that Mars represents a type of normal protoplanet that was an essential building block in the formation of larger terrestrial planets. In fact, the composition plots at the extreme end of an array of known compositions for terrestrial planets and differentiated planetesimals (Fig. 11), providing evidence that loss of moderately volatile elements is somehow linked to the metal/silicate partitioning of W.

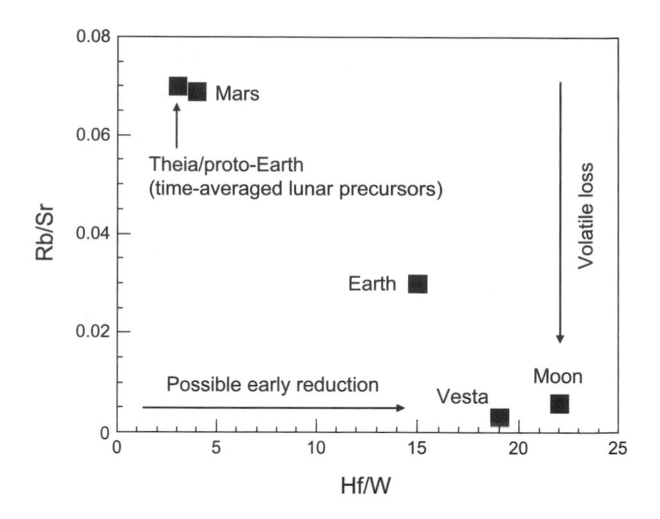

Fig. 11. Hafnium/tungsten appears to be negatively correlated with Rb/Sr in the primitive mantles of inner solar system planetesimals and planets. A possible explanation for this is that the loss of moderately volatile elements was linked to loss of other volatiles during planetary collisions such that the mantles changed from more oxidizing to more reducing. The Moon is an extreme example of this. The fact that the mixture of material from the proto-Earth and Theia that is calculated to have formed the Moon is so like Mars provides evidence that such volatile-rich objects may have been common in the inner solar system during the early stages of planetary accretion. See *Halliday and Porcelli* (2001) and *Halliday* (2004) for further details.

Mars has a higher FeO/Fe ratio than Earth, which means that the average oxidation state of Mars is higher, so this would explain why W was more lithophile (*Walter et al.,* 2000). The Earth's upper mantle has a higher ferric iron content and thus oxygen fugacity than Mars. However, this is thought to be due to the self-oxidation process of Earth's mantle associated with growth of the core. Once Earth had established a certain mass, greater than that of Mars, perovskite would have become stable in the lower mantle. Therefore, during accretion and despite its low FeO/Fe, Earth started producing ferric iron that pushed up the oxygen fugacity (*Wood and Halliday,* 2005). This would not have happened on Mars because perovskite is only just stable at the base of the martian mantle.

The striking differences between the present composition of the Moon and the time-integrated Rb/Sr, U/Pb, and Hf/W of Theia provide evidence that a major compositional change was effected by the giant impact. In some respects this is no surprise because the energy was enormous. However, little has been established about the physical chemistry of such accretion processes and remarkably little is known about the feasibility of losing volatile elements from the debris of a giant impact.

9.4. Time-integrated Manganese/Chromium

The depletion of Earth's and the Moon's inventories of some volatile elements must have occurred earlier than the giant impact and probably predates planetary formation. Models for the collapse of the solar nebula and accretion of a planetary disk predict high temperatures (1500 K) in the inner solar system (*Boss,* 1990). Therefore, it is likely that the inner solar system became depleted in volatile elements before accretion of sizable bodies. The Earth and Moon share similar Cr-isotopic compositions, distinct from those of chondrites (*Lugmair and Shukolyukov,* 1998), consistent with an early depletion in (more volatile) Mn (*Halliday et al.,* 1996; *Cassen and Woolum,* 1997). The magnitude of the difference in Cr-isotopic composition between the BSE and some chondritic meteorites (about 0.5 $\varepsilon^{53}Cr/^{52}Cr$ units) would be consistent with the depletion occurring within about 3.5 m.y. after the formation of the material in Allende, given the differences between the estimated Mn/Cr ratios of the total Earth (~0.18) and chondrites (~0.40) and assuming an initial $^{53}Mn/^{55}Mn$ ratio of 4.4×10^{-5}. It should be emphasized that the interpretation of Mn-Cr data is controversial and complex (*Lugmair and Shukolyukov,* 1998; *Birck et al.,* 1989). The primary hypothesis proposed by *Lugmair and Shukolyukov* (1998) is that the variations in Cr-isotopic composition simply reflect a radial gradient in the ^{53}Mn distribution resulting from incomplete mixing of material injected into the disk from a nearby stellar source that synthesized this and other nuclides.

Isotope geochemistry therefore provides limited evidence to support the widely accepted and almost certainly correct view that some volatile-element depletion in the planets was very early. Other lines of isotopic evidence indicate that some changes in volatile abundance were late and some may have been related to energetic accretion processes like the Moon-forming giant impact.

10. CONCLUSIONS AND OUTLOOK

Isotope geochemistry of meteorites has been central to the determination of the ages, rates, and mechanisms of accretion of the terrestrial planets. Not only do isotopic studies of meteorites provide information on the average solar system and hence a reference for models of accretion and differentiation, meteorites provide important clues about first planetesimals and the formation of the earliest planetary embryos from which the rest of the solar system was built. The development of the Hf-W chronometer has had a bigger effect on this area of science than any other aspect of isotope geochemistry. With the excellent reference dataset now available for carbonaceous chondrites and the newly determined highly precise value for the ^{182}Hf decay constant (*Vockenhuber et al.,* 2004), the precision and accuracy of Hf-W chronometry has taken major steps forward.

At the time of this writing, several important conclusions can be drawn from Hf-W and other isotopic data about how the inner solar system was built (Fig. 12):

1. Iron meteorites appear to provide the best bet for samples of the first planetesimals. The ubiquitous unradiogenic W, sometimes lower in ^{182}W than any other solar system sample measured so far, provides strong evidence that they formed early.

2. The exact timing of metal segregation as represented by iron meteorites requires improved understanding of cosmogenic effects, which tend to decrease ^{182}W and will be present in many iron meteorites with long exposure ages. By applying a maximum correction for cosmogenic effects it appears that some iron meteorites formed within <1 m.y. of the start of the solar system, as defined by the W-isotopic composition of Allende CAIs (*Markowski et al.,* 2006).

3. It is also essential to provide better cross calibration of chronometers. These will be important as improvements in mass spectrometric techniques will provide increasing age resolution.

4. There exist conflicts between the interpretations of different isotopic data for angrites and eucrites. The Sr-isotopic data are hard to explain unless the parent body formed >2 m.y. after the start of the solar system.

5. The W-isotopic data for Mars are best explained if parts of the martian mantle differentiated very rapidly. The current data are consistent with accretion and differentiation of some reservoirs within less than 1 m.y. Other reservoirs appear to have developed over longer timescales (>10 m.y.). However, these timescales assume that other factors such as alteration, analytical and sampling issues, and silicate differentiation have not affected the inferred Hf/W ratios in the source reservoirs of the martian meteotites. Further studies that combine high-quality trace-element determination with new W and Nd measurements are needed to confirm this.

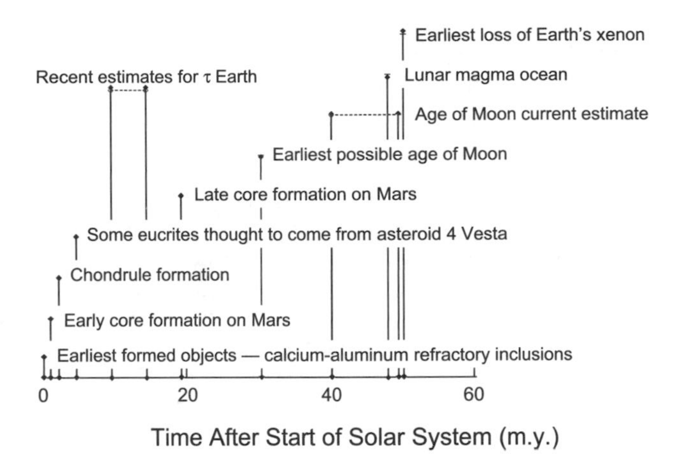

Fig. 12. The current best estimates for the timescales over which very early inner solar system objects and the terrestrial planets formed. The approximated mean life of accretion (τ) is the time taken to achieve 63% growth at exponentially decreasing rates of growth. The dashed lines indicate the mean life for accretion deduced for Earth based on W and Pb isotopes (*Halliday,* 2000, 2003, 2004; *Kleine et al.,* 2002; *Yin et al.,* 2002b). The earliest age of the Moon assumes separation from a reservoir with chondritic Hf/W (*Kleine et al.,* 2002; *Yin et al.,* 2002b). The best estimates are based on the radiogenic ingrowth deduced for the interior of the Moon (*Halliday,* 2003, 2004; *Kleine et al.,* 2005c). See Table 2 for details of other sources.

6. The age of the Moon is best defined by Hf-W data that provide strong evidence of an origin from a giant impact, in the time range 30–50 m.y. after the start of the solar system, the exact age depending on the model deployed.

7. The age of the Moon provides an important, independent, and unique constraint on the accretion rate of Earth assuming it defines the last major phase adding roughly 10% of the present mass of Earth.

8. The W-isotopic composition of the silicate Earth is consistent with this timescale for accretion. The small excess of ^{182}W in the bulk silicate Earth relative to chondrites is explicable if the initial W-isotopic composition of the solar system is $\varepsilon^{182}W < -3.5$, or if there was some level of disequilibrium between incoming metal and the BSE during accretion.

9. The difference between W- and Pb-isotopic estimates for the rates of accretion of Earth cannot be explained by disequilibrium during accretion. Either these estimates for the average Pb-isotopic composition of the bulk silicate Earth are in error or there has been an additional process that has fractionated U/Pb at a relatively late stage during Earth accretion.

10. Time-integrated parent/daughter ratios provide limited evidence that some volatile-element depletion in the planets was very early. Other lines of isotopic evidence indicate that some changes in volatile abundance were late and possibly related to energetic accretion processes like the Moon-forming giant impact. It is essential to understand how planetary chemical compositions are achieved given

the evidence from isotopic compositions that these have changed over time.

Although the development of Hf-W has resulted in major progress in quantifying the accretion of the terrestrial planets, further progress is essential in certain critical areas.

Acknowledgments. We are very grateful to S. Jacobsen, C. Münker, and an anonymous reviewer for their comments on an earlier version of this chapter. M. Wadhwa kindly provided access to the unpublished martian meteorite data of N. Foley and coworkers. B. Wood is thanked for discussions on mantle oxidation and siderophile-element partitioning during core formation. We are deeply indebted to D. Lauretta and M. Wadhwa for editorial advice and for their patience in giving us time to produce and revise this paper in the midst of many complications that intervened.

REFERENCES

Agnor C. B. and Ward W. R. (2002) Damping of terrestrial-planet eccentricities by density-wave interactions with a remnant gas disk. *Astrophys. J., 567,* 579–586.

Agnor C. B., Canup R. M., and Levison H. F. (1999) On the character and consequences of large impacts in the late stage of terrestrial planet formation. *Icarus, 142,* 219–237.

Alibert C., Norman M. D., and McCulloch M. T. (1994) An ancient Sm-Nd age for a ferroan noritic anorthosite clast from lunar breccia 67016. *Geochim. Cosmochim. Acta, 58,* 2921–2926.

Allègre C. J. and Lewin E. (1989) Chemical structure and history of the Earth: Evidence from global non-linear inversion of isotopic data in a three box model. *Earth Planet. Sci. Lett., 96,* 61–88.

Allègre C. J., Lewin E., and Dupré B. (1988) A coherent crust-mantle model for the uranium-thorium-lead isotopic system. *Chem. Geol., 70,* 211–234.

Allègre C. J., Manhès G., and Göpel C. (1995) The age of the Earth. *Geochim. Cosmochim. Acta, 59,* 1445–1456.

Amelin Y., Krot A. N., Hutcheon I. D., and Ulyanov A. A. (2002) Lead isotopic ages of chondrules and calcium-aluminum-rich inclusions. *Science, 297,* 1678–1683.

Arden J. W. (1977) Isotopic composition of uranium in chondritic meteorites. *Nature, 269,* 788–789.

Armstrong J. T., El Goresy A., and Wasserburg G. J. (1985) Willy: A prize noble Ur-Fremdling — Its history and implications for the formation of Fremdlinge and CAI. *Geochim. Cosmochim. Acta, 49,* 1001–1022.

Armstrong J. T., Hutcheon I. D., and Wasserburg G. J. (1987) Zelda and company: Petrogenesis of sulfide-rich Fremdlinge and constraints on solar nebular processes. *Geochim. Cosmochim. Acta, 51,* 3155–3173.

Becker H. and Walker R. J. (2003) Efficient mixing of the solar nebula from uniform Mo isotopic composition of meteorites. *Nature, 425,* 152–155.

Benz W. (2000) Low velocity collisions and the growth of planetesimals. In *From Dust to Terrestrial Planets* (W. Benz et al., eds.), pp. 279–294. Space Sci. Reviews, Vol. 92.

Birck J.-L. and Allègre C. J. (1988). Manganese-chromium isotope systematics and the developement of the early solar system. *Nature, 331,* 579–584.

Birck J. L., Rotaru M., and Allègre C. J. (1999) ^{53}Mn-^{53}Cr evolution of the early solar system. *Geochim. Cosmochim. Acta, 63,* 4111–4117.

Bizzarro M., Baker J. A., Haack H., Ulfbeck D., and Rosing M. (2003) Early history of Earth's crust-mantle system inferred from hafnium isotopes in chondrites. *Nature, 421,* 931–933.

Bizzarro M., Baker J. A., and Haack H. (2004) Mg isotope evidence for contemporaneous formation of chondrules and refractory inclusions. *Nature, 431,* 275–278.

Bizzarro M., Baker J. A., and Haack H. (2005) Mg isotope evidence for contemporaneous formation of chondrules and refractory inclusions — correction. *Nature, 435,* 1280.

Blake J. D. and Schramm D. N. (1973) ^{247}Cm as a short-lived *r*-process chronometer. *Nature, 243,* 138–140.

Blichert-Toft J., Boyet M., Télouk P., and Albarède F. (2002) ^{147}Sm-^{143}Nd and ^{176}Lu-^{176}Hf in eucrites and the differentiation of the HED parent body. *Earth Planet. Sci. Lett., 204,* 167–181.

Blum J. (2000) Laboratory experiments on preplanetary dust aggregation. In *From Dust to Terrestrial Planets* (W. Benz et al., eds.). pp. 265–278. Space Sci. Reviews, Vol. 92.

Boss A. P. (1990) 3D solar nebula models: Implications for Earth origin. In *Origin of the Earth* (H. E. Newsom and J. H. Jones, eds.), pp. 3–15. Oxford Univ., Oxford.

Boyet M. and Carlson R. W. (2005) ^{142}Nd evidence for early (>4.53 Ga) global differentiation of the silicate Earth. *Science, 309,* 576–581.

Boyet M., Blichert-Toft J., Rosing M., Storey M., Telouk P., and Albarede F. (2003) ^{142}Nd evidence for early Earth differentiation. *Earth Planet. Sci. Lett., 214,* 427.

Brandon A. D., Walker R. J., Morgan J., and Goles C. (2000) Re-Os isotopic evidence for early differentiation of the Martian mantle. *Geochim. Cosmochim. Acta, 64,* 4083–4095.

Canup R. M. (2004) Simulation of a late lunar forming impact. *Icarus, 168,* 433–456.

Canup R. M. and Agnor C. B. (2000) Accretion of the terrestrial planets and the Earth-Moon system. In *Origin of the Earth and Moon* (R. M. Canup and K. Righter, eds.), pp. 113–129. Univ. of Arizona, Tucson.

Canup R. M. and Asphaug E. (2001) Origin of the moon in a giant impact near the end of the Earth's formation. *Nature, 412,* 708–712.

Carlson R. W. and Hauri E. H. (2001) Extending the ^{107}Pd-^{107}Ag chronometer to low Pd/Ag meteorites with the MC-ICPMS. *Geochim. Cosmochim. Acta, 65,* 1839–1848.

Carlson R. W. and Lugmair G. W. (1988) The age of ferroan anorthosite 60025: Oldest crust on a young Moon? *Earth Planet. Sci. Lett., 90,* 119–130.

Carlson R. W. and Lugmair G. W. (2000) Timescales of planetesimal formation and differentiation based on extinct and extant radioisotopes. In *Origin of the Earth and Moon* (R. M. Canup and K. Righter, eds.), pp. 25–44. Univ. of Arizona, Tucson.

Caro G., Bourdon B., Birck J. L., and Moorbath S. (2003) Sm-146-Nd-142 evidence from Isua metamorphosed sediments for early differentiation of the Earth's mantle. *Nature, 423,* 428–432.

Cassen P. and Woolum D. S. (1997) Nebular fractionations and Mn-Cr systematics (abstract). In *Lunar and Planetary Science XXVIII,* pp. 211–212. Lunar and Planetary Institute, Houston.

Chambers J. E. (2001a) Comparing planetary accretion in two and three dimensions. *Icarus, 149,* 262–276.

Chambers J. E. (2001b) Making more terrestrial planets. *Icarus, 152,* 205–224.

Chambers J. E. (2004) Planetary accretion in the inner solar system. *Earth Planet. Sci. Lett., 223,* 241–252.

Chen J. H. and Wasserburg G. J. (1996) Live ^{107}Pd in the early solar system and implications for planetary evolution. In *Earth Processes: Reading the Isotopic Code* (A. Basu and S. R. Hart, eds.), pp. 1–20. Geophysics Monograph 95, American Geophysical Union, Washington, DC.

Chen J. H. and Wasserburg G. J. (1980) A search for isotopic anomalies in uranium *Geophys. Res. Lett., 7,* 275–278.

Chen J. H., Papanastassiou D. A., and Wasserburg G. J. (2003) Endemic Ru isotopic anomalies in meteorites (abstract). In *Lunar and Planetary Science XXXIV,* Abstract #1789. Lunar and Planetary Institute, Houston (CD-ROM).

Chen J. H., Papanastassiou D. A., Wasserburg G. J., and Ngo H. H. (2004) Endemic Mo isotopic anomalies in iron and carbonaceous meteorites (abstract). In *Lunar and Planetary Science XXXV,* Abstract #1431. Lunar and Planetary Institute, Houston (CD-ROM).

Clayton R. N. (1986) High temperature isotope effects in the early solar system. In *Stable Isotopes in High Temperature Geological Processes* (J. W. Valley et al., eds.), pp. 129–140. Mineralogical Society of America, Washington, DC.

Clayton R. N. (1993) Oxygen isotopes in meteorites. *Annu. Rev. Earth Planet. Sci., 21,* 115–149.

Clayton R. N. (2002) Self-shielding in the solar nebula. *Nature, 415,* 860–861.

Clayton R. N. and Mayeda T. K. (1975) Genetic relations between the moon and meteorites. *Proc. Lunar Planet. Sci. Conf. 11th,* pp. 1761–1769.

Clayton R. N. and Mayeda T. K. (1996) Oxygen isotope studies of achondrites. *Geochim. Cosmochim. Acta, 60,* 1999–2017.

Clayton R. N., Grossman L., and Mayeda T. K. (1973) A component of primitive nuclear composition in carbonaceous meteorites. *Science, 182,* 485–487.

Cook D. L., Wadhwa M., Clayton R. N., Janney P. E., Dauphas N., and Davis A. M. (2005) Nickel isotopic composition of meteoritic metal: Implications for the initial ^{60}Fe/^{56}Fe ratio in the early solar system (abstract). *Meteoritics & Planet. Sci.,* Abstract #5136.

Dauphas N., Marty B., and Reisberg L. (2002a) Molybdenum evidence for inherited planetary scale isotope heterogeneity of the protosolar nebula. *Astrophys. J., 565,* 640–644.

Dauphas N., Marty B., and Reisberg L. (2002b) Molybdenum nucleosynthetic dichotomy revealed in primitive meteorites. *Astrophys. J. Lett., 569,* L139–L142.

Dauphas N., Marty B., Davis A. M., Reisberg L., and Gallino R. (2004) The cosmic molybdenum-ruthenium isotope correlation. *Earth Planet. Sci. Lett., 226,* 465–476.

Davies G. F. (1984) Geophysical and isotopic constraints on mantle convection: An interim synthesis. *J. Geophys. Res., 89,* 6017–6040.

Davis A. M., Richter F. M., Mendybaev R. A., Janney P. E., Wadhwa M., and McKeegan K. D. (2005) Isotopic mass fractionation laws and the initial solar system ^{26}Al/^{27}Al ratio (abstract). In *Lunar and Planetary Science XXXVI,* Abstract #2334. Lunar and Planetary Institute, Houston (CD-ROM).

Doe B. R. and Zartman R. E. (1979) Plumbotectonics I: The Phanerozoic. In *Geochemistry of Hydrothermal Ore Deposits* (H. L. Barnes, ed.), pp. 22–70. Wiley, New York.

Dreibus G., Haubold R., Huisl W., and Spettel B. (2003). Comparison of the chemistry of Yamato 980459 with DaG 476 and SaU 005 (abstract). In *Intl. Symp. on Evolution of Solar System Materials: A New Perspective from Antarctic Meteorites,* pp. 19–20. National Institute of Polar Research, Tokyo.

Foley C. N., Wadhwa M., Borg L. E., Janney P. E., Hines R., and Grove T. L. (2005) The early differentiation history of Mars from ^{182}W-^{142}Nd isotope systematics in the SNC meteorites. *Geochim. Cosmochim. Acta, 69,* 4557–4571.

Galer S. J. G. (1993) Significance of lead isotope similarities between the early Earth and the Moon. *Eos Trans. AGU, 74,* 655.

Galer S. J. G. and Goldstein S. L. (1991) Depleted mantle Pb isotopic evolution using conformable ore leads. *Terra Abstracts (EUG VI), 3,* 485–486.

Galer S. J. G. and Goldstein S. L. (1996) Influence of accretion on lead in the Earth. In *Isotopic Studies of Crust-Mantle Evolution* (A. R. Basu and S. R. Hart, eds.), pp. 75–98. American Geophysical Union, Washington, DC.

Galy A., Young E. D., Ash R. D., and O'Nions R. K. (2000) High precision magnesium isotopic composition of Allende material: A multiple collector inductively coupled mass spectrometry study (abstract). In *Lunar and Planetary Science XXXI,* Abstract #1193. Lunar and Planetary Institute, Houston (CD-ROM).

Göpel C., Manhès G., and Allègre C. J. (1991) Constraints on the time of accretion and thermal evolution of chondrite parent bodies by precise U-Pb dating of phosphates. *Meteoritics, 26,* 73.

Göpel C., Manhès G., and Allègre C. J. (1994) U-Pb systematics of phosphates from equilibrated ordinary chondrites. *Earth Planet. Sci. Lett., 121,* 153–171.

Gounelle M., Shu F. H., Shang H., Glassgold A. E., Rehm K. E., and Lee T. (2001) Extinct radioactivities and protosolar cosmic-rays: Self-shielding and light elements. *Astrophys. J., 548,* 1051.

Gray C. M., Papanastassiou D. A., and Wasserburg G. J. (1973) The identification of early condensates from the solar nebula. *Icarus, 20,* 213–239.

Halliday A. N. (2000) Terrestrial accretion rates and the origin of the Moon. *Earth Planet. Sci. Lett., 176,* 17–30.

Halliday A. N. (2003) The origin and earliest history of the Earth. In *Treatise on Geochemistry, Vol. 1: Meteorites, Comets and Planets* (A. M. Davis, eds.), pp. 509–557. Elsevier, Oxford.

Halliday A. N. (2004) Mixing, volatile loss and compositional change during impact-driven accretion of the Earth. *Nature, 427,* 505–509.

Halliday A. N. and Lee D-C. (1999) Tungsten isotopes and the early development of the Earth and Moon. *Geochim. Cosmochim. Acta, 63,* 4157–4179.

Halliday A. N. and Porcelli D. (2001) In search of lost planets — The paleocosmochemistry of the inner solar system. *Earth Planet. Sci. Lett., 192,* 545–559.

Halliday A. N., Lee D-C., Christensen J. N., Walder A. J., Freedman P. A., Jones C. E., Hall C. M., Yi W., and Teagle D. (1995) Recent developments in inductively coupled plasma magnetic sector multiple collector mass spectrometry. *Intl. J. Mass Spec. Ion Processes, 146/147,* 21–33.

Halliday A. N., Rehkämper M., Lee D-C., and Yi W. (1996) Early evolution of the Earth and Moon: New constraints from Hf-W isotope geochemistry. *Earth Planet. Sci. Lett., 142,* 75–89.

Halliday A. N., Lee D-C., and Jacobsen S. B. (2000) Tungsten isotopes, the timing of metal-silicate fractionation and the origin of the Earth and Moon. In *Origin of the Earth and Moon* (R. M. Canup and K. Righter, eds.), pp. 45–62. Univ. of Arizona, Tucson.

Halliday A. N., Lee D-C., Porcelli D., Wiechert U., Schönbächler M., and Rehkämper M. (2001a) The rates of accretion, core formation and volatile loss in the early solar system. *Phil. Trans. R. Soc. London, 359,* 2111–2135.

Halliday A. N., Wänke H., Birck J-L., and Clayton R. N. (2001b) The accretion, bulk composition and early differentiation of Mars. *Space Sci. Rev., 96,* 197–230.

Halliday A., Quitté G., and Lee D.-C. (2003) Tungsten isotopes and the time-scales of planetary accretion. *Meteoritics & Planet. Sci., 38,* A133.

Hanan B. B. and Tilton G. R. (1987) 60025: Relict of primitive lunar crust? *Earth Planet. Sci. Lett., 84,* 15–21.

Harper C. L. and Jacobsen S. B. (1996) Evidence for ^{182}Hf in the early solar system and constraints on the timescale of terrestrial core formation. *Geochim. Cosmochim. Acta, 60,* 1131–1153.

Harper C. L., Nyquist L. E., Bansal B., Wiesmann H., and Shih C.-Y. (1995) Rapid accretion and early differentiation of Mars indicated by ^{142}Nd/^{144}Nd in SNC meteorites. *Science, 267,* 213–217.

Harper C. L., Völkening J., Heumann K. G., Shih C.-Y., and Wiesmann H. (1991a) ^{182}Hf-^{182}W: New cosmochronometric constraints on terrestrial accretion, core formation, the astrophysical site of the *r*-process, and the origin of the solar system (abstract). In *Lunar and Planetary Science XXII,* pp. 515–516. Lunar and Planetary Institute, Houston.

Harper C. L., Wiesmann H., and Nyquist L. E., Howard W. M., Meyer B., Yokoyama Y., Rayet M., Arnould M., Palme H., Spettel B., and Jochum K. P. (1991b) ^{92}Nb/^{93}Nb and ^{92}Nb/^{146}Sm ratios of the early solar system: Observations and comparison of *p*-process and spallation models (abstract). In *Lunar and Planetary Science XXII,* pp. 519–520. Lunar and Planetary Institute, Houston.

Heck P. R., Schmitz B., Baur H., Halliday A. N., and Wieler R. (2004) Fast delivery of meteorites from a major asteroid collision. *Nature, 430,* 323–325.

Hirata T. (2001) Determinations of Zr isotopic composition and U-Pb ages for terrestrial and extraterrestrial Zr-bearing minerals using laser ablation-inductively coupled plasma mass spectrometry: Implications for Nb-Zr isotopic systematics. *Chem. Geol., 176,* 323–342.

Holmes A. (1946) An estimate of the age of the Earth. *Nature, 157,* 680–684.

Horan M. F., Smoliar M. I., and Walker R. J. (1998) ^{182}W and ^{187}Re-^{187}Os systematics of iron meteorites: Chronology for melting, differentiation, and crystallization in asteroids. *Geochim. Cosmochim. Acta, 62,* 545–554.

Houtermans F. G. (1946) Die Isotopenhäufigkeitem in natürlichen Blei und das Alter des Urans. *Naturwissenschaften, 33,* 185–186.

Humayun M. and Clayton R. N. (1995) Potassium isotope cosmochemistry: Genetic implications of volatile element depletion. *Geochim. Cosmochim. Acta, 59,* 2131–2151.

Hunten D. M., Pepin R. O., and Walker J. C. G. (1987) Mass fractionation in hydrodynamic escape. *Icarus, 69,* 532–549.

Hutcheon I. D., Armstrong J. T., and Wasserburg G. J. (1987) Isotopic studies of Mg, Fe, Mo, Ru, and W in Fremdlinge from Allende refractory inclusions. *Geochim. Cosmochim. Acta, 51,* 3175–3192.

Ireland T. R. (1991). The abundance of ^{182}Hf in the early solar system (abstract). In *Lunar and Planetary Science XXII,* pp. 609–610. Lunar and Planetary Institute, Houston.

Jacobsen S. B. and Harper C. L. Jr. (1996) Accretion and early differentiation history of the Earth based on extinct radionuclides. In *Earth Processes: Reading the Isotope Code* (A. Basu and S. Hart, eds.), pp. 47–74. American Geophysical Union, Washington, DC.

Jacobsen S. B. and Wasserburg G. J. (1979). The mean age of mantle and crust reservoirs. *J. Geophys. Res., 84,* 7411–7427.

Jacobsen S. B. and Yin Q. Z. (2001) Core formation models and extinct nuclides (abstract). In *Lunar and Planetary Science XXXII,* Abstract #1961. Lunar and Planetary Institute, Houston (CD-ROM).

Jagoutz E. and Jotter R. (2000) New Sm-Nd isotope data on Nakhla minerals (abstract). In *Lunar and Planetary Science XXXI,* p. 1609. Lunar and Planetary Institute, Houston.

Jagoutz E., Dreibus G., and Jotter R. (2003) New ^{142}Nd data on SNC meteorites. *Geochim. Cosmochim. Acta, 67,* A184.

Jones J. H. and Palme H. (2000) Geochemical constraints on the origin of the Earth and Moon. In *Origin of the Earth and Moon* (R. M. Canup and K. Righter, eds.), pp. 197–216. Univ. of Arizona, Tucson.

Kamber B. S. and Collerson K. D. (1999) Origin of ocean-island basalts: A new model based on lead and helium isotope systematics. *J. Geophys. Res., 104,* 25479–25491.

Kleine T., Münker C., Mezger K., and Palme H. (2002) Rapid accretion and early core formation on asteroids and the terrestrial planets from Hf-W chronometry. *Nature, 418,* 952–955.

Kleine T., Mezger K., Munker C., Palme H., and Bischoff A. (2004a) ^{182}Hf-^{182}W isotope systematics of chondrites, eucrites and martian meteorites: Chronology of core formation and early mantle differentiation in Vesta and Mars. *Geochim. Cosmochim. Acta, 68,* 2935–2946.

Kleine T., Mezger K., Palme H., and Münker C. (2004b) The W isotope evolution of the bulk silicate Earth: Constraints on the timing and mechanisms of core formation and accretion. *Earth Planet. Sci. Lett., 228,* 109.

Kleine T., Mezger K., Palme H., Scherer E., and Münker C. (2005a) Early core formation in asteroids and late accretion of chondrite parent bodies: Evidence from ^{182}Hf-^{182}W in CAIs, metal-rich chondrites and iron meteorites. *Geochim. Cosmochim. Acta, 69,* 5805–5818.

Kleine T., Mezger K., Palme H., Scherer E., and Münker C. (2005b) The W isotope composition of eucrites metal: Constraints on the timing and cause of the thermal metamorphism of basaltic eucrites. *Earth Planet. Sci. Lett., 231,* 41.

Kleine T., Palme H., Mezger K., and Halliday A. N. (2005c) Hf-W chronometry of lunar metals and the age and early differentiation of the Moon. *Science, 310,* 1671–1674.

Kominami J. and Ida S. (2002) The effect of tidal interaction with a gas disk on formation of terrestrial planets. *Icarus, 157,* 43–56.

Kortenkamp S. J., Kokubo E., and Weidenschilling S. J. (2000) Formation of planetary embryos. In *Origin of the Earth and Moon* (R. M. Canup and K. Righter, eds.), pp. 85–100. Univ.of Arizona, Tucson.

Kramers J. D. and Tolstikhin I. N. (1997) Two terrestrial lead isotope paradoxes, forward transport modeling, core formation and the history of the continental crust. *Chem. Geol., 139,* 75–110.

Kutschera W. and 10 colleagues (1984) Half-life of Fe-60. *Nucl. Instr. Meth. Phys. Res., B5,* 430–435.

Kwon S.-T., Tilton G. R., and Grünenfelder M. H. (1989) Lead isotope relationships in carbonatites and alkaline complexes: An overview. In *Carbonatites — Genesis and Evolution* (K. Bell, ed.), pp. 360–387. Unwin-Hyman, London.

Lee D-C. (2005) Protracted core formation in asteroids: Evidence from high precision W isotopic data. *Earth Planet. Sci. Lett., 198,* 267–274.

Lee D.-C. and Halliday A. N. (1995) Hafnium-tungsten chronometry and the timing of terrestrial core formation. *Nature, 378,* 771–774.

Lee D.-C. and Halliday A. N. (1996) Hf-W isotopic evidence for rapid accretion and differentiation in the early solar system. *Science, 274,* 1876–1879.

Lee D.-C. and Halliday A. N. (1997) Core formation on Mars and differentiated asteroids. *Nature, 388,* 854–857.

Lee D-C. and Halliday A. N. (2000a) Accretion of primitive planetesimals: Hf-W isotopic evidence from enstatite chondrites. *Science, 288,* 1629–1631.

Lee D-C. and Halliday A. N. (2000b) Hf-W isotopic systematics of ordinary chondrites and the initial $^{182}Hf/^{180}Hf$ of the solar system. *Chem. Geol., 169,* 35–43.

Lee D.-C. and Halliday A. N. (2003) High precision W and Mo isotope compositions for iron meteorites. *Geochim. Cosmochim. Acta, 67,* A246.

Lee D.-C., Halliday A. N., Snyder G. A., and Taylor L. A. (1997) Age and origin of the Moon. *Science, 278,* 1098–1103.

Lee D-C., Halliday A. N., Leya I., Wieler R., and Wiechert U. (2002) Cosmogenic tungsten and the origin and earliest differentiation of the Moon. *Earth Planet. Sci. Lett., 198,* 267–274.

Lee T., Papanastassiou D. A., and Wasserburg G. J. (1977) ^{26}Al in the early solar system: Fossil or fuel? *Astrophys. J. Lett., 211,* L107–L110.

Lee T., Shu F. H., Shang H., Glassgold A. E., and Rehm K. E. (1998) Protostellar cosmic rays and extinct radioactivities in meteorites. *Astrophys. J., 506,* 898–912.

Leya I., Wieler R., and Halliday A. N. (2000) Cosmic-ray production of tungsten isotopes in lunar samples and meteorites and its implications for Hf-W cosmochemistry. *Earth Planet. Sci. Lett., 175,* 1–12.

Leya I., Wieler R., and Halliday A. N. (2003) The influence of cosmic-ray production on extinct nuclide systems. *Geochim. Cosmochim. Acta, 67,* 527–541.

Liew T. C., Milisenda C. C., and Hofmann A. W. (1991) Isotopic constrasts, chronology of element transfers and high-grade metamorphism: The Sri Lanka Highland granulites, and the Lewisian (Scotland) and Nuk (S. W. Greenland) gneisses. *Geol. Rundsch., 80,* 279–288.

Lodders K. (1998) A survey of shergottite, nakhlite and chassigny meteorites whole-rock compositions. *Meteoritics & Planet. Sci., 33,* A183–A190.

Lugmair G. W. and Galer S. J. G. (1992) Age and isotopic relationships between the angrites Lewis Cliff 86010 and Angra dos Reis. *Geochim. Cosmochim. Acta, 56,* 1673–1694.

Lugmair G. W. and Shukolyukov A. (1998) Early solar system timescales according to Mn-53-Cr-53 systematics. *Geochim. Cosmochim. Acta, 62,* 2863–2886.

Lugmair G. W. and Shukolyukov A. (2001) Early solar system events and timescales. *Meteoritics & Planet. Sci., 36,* 1017–1026.

Marhas K. K., Goswami J. N., and Davis A. M. (2002) Short-lived nuclides in hibonite grains from Murchison: Evidence for solar system evolution. *Science, 298,* 2182–2185.

Markowski A., Quitté G., and Halliday A. N. (2004) Systematic differences in $^{182}W/^{184}W$ between iron meteorite groups. *Meteoritics & Planet. Sci., 39,* 5075.

Markowski A., Quitté G., Halliday A. N., and Kleine T. (2006) Tungsten isotopic compositions of iron meteorites: Chronological constraints vs. cosmogenic effects. *Earth Planet. Sci. Lett.,* in press.

Marty B. and Marti K. (2002) Signatures of early differentiation of Mars. *Earth Planet. Sci. Lett., 196,* 251–263.

Masarik J. (1997) Contribution of neutron-capture reactions to observed tungsten isotopic ratios. *Earth Planet. Sci. Lett., 152,* 181–185.

Mayor M. and Queloz D. (1995) A Jupiter-mass companion to a solar-type star. *Nature, 378,* 355–359.

McCaughrean M. J. and O'Dell C. R. (1996) Direct imaging of circumstellar disks in the Orion nebula. *Astronom. J., 111,* 1977–1986.

McKeegan K. D., Chaussidon M., and Robert F. (2000) Incorporation of short-lived Be-10 in a calcium-aluminum-rich inclusion from the Allende meteorite. *Science, 289,* 1334–1337.

Meyer J. C., McKay D. C., Anderson D. H., and Butler P. Jr. (1975) The source of sublimates on the Apollo 15 green and Apollo 17 orange glass samples. *Proc. Lunar Sci. Conf. 6th,* pp. 1637–1699.

Mostefaoui S, Lugmair G. W., Hoppe P., and El Goresy A. (2004) Evidence for live Fe-60 in meteorites. *New Astron. Rev., 48,* 155–159.

Moynier F., Blichert-Toft J., Telouk P., and Albarede F. (2004) The Ni isotope geochemistry of chondrites and iron meteorites. *Geochim. Cosmochim. Acta, 68,* A757–A757.

Münker C., Weyer S., Mezger K., Rehkämper M., Wombacher F., and Bischoff A. (2000) ^{92}Nb-^{92}Zr and the early differentiation history of planetary bodies. *Science, 289,* 1538–1542.

Murphy D. T., Kamber B. S., and Collerson K. D. (2003) A refined solution to the first terrestrial Pb-isotope paradox. *J. Petrol., 44,* 39–53.

Newsom H. E. (1995) Composition of the solar system, planets, meteorites, and major terrestrial reservoirs. In *Global Earth Physics, A Handbook of Physical Constants* (T. J. Ahrens, ed.), pp. 159–189. AGU Reference Shelf 1, American Geophysical Union, Washington, DC.

Newsom H. E., White W. M., Jochum K. P., and Hofmann A. W. (1986) Siderophile and chalcophile element abundances in oceanic basalts, Pb isotope evolution and growth of the earth's core. *Earth Planet. Sci. Lett., 80,* 299–313.

Newsom H. E., Sims K. W. W., Noll P. D. Jr., Jaeger W. L., Maehr S. A., and Bessera T. B. (1996) The depletion of W in the bulk silicate Earth. *Geochim. Cosmochim. Acta, 60,* 1155–1169.

Nier A. O., Thompson R. W., and Murphey B. F. (1941) The isotopic composition of lead and the measurement of geological time. III. *Phys. Rev., 60,* 112–116.

Nyquist L. E., Lindstrom D., Shih C-Y., Weismann H., Mittlfelhdt D., Wentworth S., and Martinez R. (1997) Mn-Cr systematics of chondrules from the Bishunpur and Chainpur meteorites (abstract). In *Lunar and Planetary Science XXVIII,* pp. 1033–1034. Lunar and Planetary Institute, Houston.

Nyquist L. E., Lindstrom D., Mittlefehldt D., Shih C. Y., Wiesmann H., Wentworth S., and Martinez R. (2001) Manganese-chromium formation intervals for chondrules from the Bishunpur and Chainpur meteorites. *Meteoritics & Planet. Sci., 36,* 911.

O'Neill H. St. C. (1991a) The origin of the Moon and the early history of the Earth — A chemical model; Part I: The Moon. *Geochim. Cosmochim. Acta, 55,* 1135–158.

O'Neill H. St. C. (1991b) The origin of the Moon and the early history of the Earth — A chemical model; Part II: The Earth. *Geochim. Cosmochim. Acta, 55,* 1159–1172.

Papanastassiou D. A. and Wasserburg G. J. (1976) Early lunar differentiates and lunar initial $^{87}Sr/^{86}Sr$ (abstract). In *Lunar Sci-*

ence VII, pp. 665–667. Lunar Science Institute, Houston.

Papanastassiou D. A., Wasserburg G. J., and Bogdanovski O. (2005) The ^{53}Mn-^{53}Cr system in CAIs: An update (abstract). In *Lunar and Planetary Science XXXVI,* Abstract #2198. Lunar and Planetary Institute, Houston (CD-ROM).

Patterson C. C. (1956) Age of meteorites and the Earth. *Geochim. Cosmochim. Acta, 10,* 230–237.

Pepin R. O. (1997) Evolution of Earth's noble gases: Consequences of assuming hydrodynamic loss driven by giant impact. *Icarus, 126,* 148–156.

Pepin R. O. (2000) On the isotopic composition of primordial xenon in terrestrial planet atmospheres. *Space Sci. Rev., 92,* 371–395.

Podosek F. A., Zinner E. K., MacPherson G. J., Lundberg L. L., Brannon J. C., and Fahey A. J. (1991) Correlated study of initial Sr-87/Sr-86 and Al-Mg isotopic systematics and petrologic properties in a suite of refractory inclusions from the Allende meteorite. *Geochim. Cosmochim. Acta, 55,* 1083–1110.

Podosek F. A., Ott U., Brannon J. C., Neal C. R., Bernatowicz T. J., Swan P., and Mahan S. E. (1997) Thoroughly anomalous chromium in Orgueil. *Meteoritics & Planet. Sci., 32,* 617–627.

Poitrasson F., Halliday A. N., Lee D-C., Levasseur S., and Teutsch N. (2004) Iron isotope differences between Earth, Moon, Mars and Vesta as possible records of contrasted accretion mechanisms. *Earth Planet. Sci. Lett., 223,* 253–266.

Polnau E. and Lugmair G. W. (2001) Mn-Cr isotope systematics in the two ordinary chondrites Richardton (H5) and Ste. Marguerite (H4) (abstract). In *Lunar and Planetary Science XXXII,* Abstract #1527. Lunar and Planetary Institute, Houston (CD-ROM).

Porcelli D. and Pepin R. O. (2000) Rare gas constraints on early earth history. In *Origin of the Earth and Moon* (R. M. Canup and K. Righter, eds.), pp. 435–458. Univ. of Arizona, Tucson.

Porcelli D., Cassen P., and Woolum D. (2001) Deep Earth rare gases: Initial inventories, capture from the solar nebula and losses during Moon formation. *Earth Planet. Sci. Lett, 193,* 237–251.

Quitté G. and Birck J-L. (2004) Tungsten isotopes in eucrites revisited and the initial $^{182}Hf/^{180}Hf$ of the solar system based on iron meteorite data. *Earth Planet. Sci. Lett., 219,* 201–207.

Quitté G., Birck J-L., and Allègre C. J. (2000) ^{182}Hf-^{182}W systematics in eucrites: The puzzle of iron segregation in the early solar system. *Earth Planet. Sci. Lett., 184,* 83–94.

Quitté G., Meier M., Latkoczy C., Halliday A. N., and Günther D. (2006) Nickel isotopes in iron meteorites — nucleosynthetic anomalies in sulphides with no effects in metals and no trace of ^{60}Fe. *Earth Planet. Sci. Lett.,* in press.

Reynolds J. H. (1960) Determination of the age of the elements. *Phys. Rev. Lett., 4,* 8.

Richter F. M. (2003) Time scales for elemental and isotopic fractionation by evaporation and condensation (abstract). In *Lunar and Planetary Science XXXVI,* Abstract #1231. Lunar and Planetary Institute, Houston (CD-ROM).

Righter K. and Drake M. J. (1996) Core formation in Earth's Moon, Mars, and Vesta. *Icarus, 124,* 513–529.

Righter K. and Drake M. J. (1999) Effect of water on metal-silicate partitioning of siderophile elements: A high pressure and temperature terrestrial magma ocean and core formation. *Earth Planet. Sci. Lett., 171,* 383–399.

Righter K., Drake M. J., and Yaxley G. (1997) Prediction of siderophile element metal/silicate partition coefficients to 20 GPa and 2800°C: The effects of pressure, temperature, oxygen fugacity, and silicate and metallic melt compositions. *Phys. Earth Planet. Inter., 100,* 115–142.

Russell S. S., Srinivasan G., Huss G. R., Wasserburg G. J., and MacPherson G. J. (1996) Evidence for widespread ^{26}Al in the solar nebula and constraints for nebula time scales. *Science, 273,* 757–762.

Safronov V. S. (1954) On the growth of planets in the protoplanetary cloud. *Astron. Zh., 31,* 499–510.

Sanders I. S. (1996) A chondrule-forming scenario involving molten planetesimals. In *Chondrules and the Protoplanetary Disk* (R. Hewins et al., eds.), pp. 327–334. Cambridge Univ., Cambridge.

Sanloup C., Blichert-Toft J., Télouk P., Gillet P., and Albarède F. (2000) Zr isotope anomalies in chondrites and the presence of live ^{92}Nb in the early solar system. *Earth Planet. Sci. Lett., 184,* 75–81.

Scherer E. E., Münker C., and Mezger K. (2001) Calibration of the lutetium-hafnium clock. *Science, 293,* 683–687.

Scherer E. E., Mezger K., and Münker C. (2003) The ^{176}Lu decay constant discrepancy: Terrestrial samples versus meteorites. *Meteoritics & Planet. Sci., 38,* A136.

Scherstén A., Elliott T., Russell S. S., and Hawkesworth C. (2004) W isotope chronology of asteroidal core formation. *Geochim. Cosmochim. Acta, 68,* Abstract 1240.

Schmitt W., Palme H., and Wänke H. (1989) Experimental determination of metal/silicate partition coefficients for P, Co, Ni, Cu, Ga, Ge, Mo, and W and some implications for the early evolution of the Earth. *Geochim. Cosmochim. Acta, 53,* 173–185.

Schneider G., Smith B. A., Becklin E. E., Koerner D. W., Meier R., Hines D. C., Lowrance P. J., Terrile R. I., and Rieke M. (1999) Nicmos imaging of the HR 4796A circumstellar disk. *Astrophys. J. Lett., 513,* L127–L130.

Schön R., Winkler G., and Kutschera W. (2004) A critical review of experimental data for the half-lives of the uranium isotopes ^{238}U and ^{235}U. *Appl. Radiation Isotopes, 60,* 263–273.

Schönbächler M., Rehkämper M., Halliday A. N., Lee D. C., Bourot-Denise M., Zanda B., Hattendorf B., and Günther D. (2002) Niobium-zirconium chronometry and early solar system development. *Science, 295,* 1705–1708.

Schönbächler M., Lee D.-C., Rehkämper M., Halliday A. N., Fehr M. A., Hattendorf B., and Günther D. (2003) Zirconium isotope evidence for incomplete admixing of *r*-process components in the solar nebula. *Earth Planet. Sci. Lett., 216,* 467–481.

Schönbächler M., Lee D.-C., Rehkämper M., Halliday A. N., Hattendorf B., and Günther D. (2005) Nb/Zr fractionation on the Moon and the search for extinct $_{92}Nb$. *Geochim. Cosmochim. Acta, 69,* 775–785.

Schönberg R., Kamber B. S., Collerson K. D., and Eugster O. (2002) New W isotope evidence for rapid terrestrial accretion and very early core formation. *Geochim. Cosmochim. Acta, 66,* 3151–3160.

Shih C.-Y., Nyquist L. E., Dasch E. J., Bogard D. D., Bansal B. M., and Wiesmann H. (1993) Age of pristine noritic clasts from lunar breccias 15445 and 15455. *Geochim. Cosmochim. Acta, 57,* 915–931.

Shukolyukov A. and Lugmair G. W. (1993a) ^{60}Fe in eucrites. *Earth Planet Sci. Lett., 119,* 159–166.

Söderlund U., Patchett J. P., Vervoort J. D., and Isachsen C. E. (2004) The ^{176}Lu decay constant determined by Lu-Hf and U-Pb isotope systematics of Precambrian mafic intrusions. *Earth Planet. Sci. Lett., 219,* 311–324

Solomatov V. S. (2000) Fluid dynamics of a terrestrial magma ocean. In *Origin of the Earth and Moon* (R. M. Canup and K.

Righter, eds.), pp. 323–338. Univ. of Arizona, Tucson.
Steiger R. H. and Jäger E. (1977) Subcommission on geochronology: Convention on the use of decay constants in geo- and cosmochronology. *Earth Planet. Sci. Lett., 36,* 359–362.
Stirling C. H., Halliday A. N., and Porcelli D. (2005) In search of live ^{247}Cm in the early solar system. *Geochim. Cosmochim. Acta, 69,* 1059–1071.
Tatsumoto M. and Shimamura T. (1980) Evidence for live Cm-247 in the early solar system. *Nature, 286,* 118–122.
Tera F. and Carlson R. W. (1999) Assessment of the Pb-Pb and U-Pb chronometry of the early solar system. *Geochim. Cosmochim. Acta, 63,* 1877–1889.
Tera F., Papanastassiou D. A., and Wasserburg G. J. (1973) A lunar cataclysm at ~3.95 AE and the structure of the lunar crust (abstract). In *Lunar Science IV,* pp. 723–725. Lunar Science Institute, Houston.
Torigoye-Kita N., Misawa K., Dalrymple G. B., and Tatsumoto M. (1995) Further evidence for a low U/Pb source in the Moon: U-Th-Pb, Sm-Nd, and Ar-Ar isotopic systematics of lunar meteorite Yamato-793169. *Geochim. Cosmochim. Acta, 59,* 2621–2625.
Trinquier A., Birck J-L., and Allègre C. J. (2005) ^{54}Cr anomalies in the solar system: Their extent and origin (abstract). In *Lunar and Planetary Science XXXVI,* Abstract #1259. Lunar and Planetary Institute, Houston (CD-ROM).
Vockenhuber C., Oberli F., Bichler M., Ahmad I., Quitté G., Meier M., Halliday A. N., Lee D-C., Kutschera W., Steier P., Gehrke R. J., and Helmer R. G. (2004) A new half-life measurement of ^{182}Hf — Sharpening a tool for the chronology of the early solar system. *Phys. Rev. Lett., 93,* Article No. 172501.
Wadhwa M., Amelin Y., Bogdanovski O., Lugmair G. W., and Janney P. E. (2005) High precision relative and absolute ages for Asuka 881394, a unique and ancient basalt (abstract). In *Lunar and Planetary Science XXXVI,* Abstract #2126. Lunar and Planetary Institute, Houston (CD-ROM).
Walter M. J., Newsom H. E., Ertel W., and Holzheid A. (2000) Siderophile elements in the Earth and Moon: Metal/silicate partitioning and implications for core formation. In *Origin of the Earth and Moon* (R. M. Canup and K. Righter, eds.), pp. 265–289. Univ.of Arizona, Tucson.
Ward W. R. (2000) On planetesimal formation: The role of collective behavior. In *Origin of the Earth and Moon* (R. M. Canup and K. Righter, eds.), pp. 75–84. Univ.of Arizona, Tucson.
Wasserburg G. J., Busso M., Gallino R., and Raiteri C. M. (1994) Asymptotic giant branch stars as a source of short-lived radioactive nuclei in the solar nebula. *Astrophys. J., 424,* 412–428.
Wasserburg G. J., Papanastassiou D. A., Tera F., and Huneke J. C. (1977a) Outline of a lunar chronology. *Philos. Trans. R. Soc. London, A285,* 7–22.
Wasserburg G. J., Tera F., Papanastassiou D. A., and Huneke J. C. (1977b) Isotopic and chemical investigations on Angra dos Reis. *Earth Planet. Sci. Lett., 35,* 294–316.
Weidenschilling S. J., Spaute D., Davis D. R., Marzari F., and Ohtsuki K. (1997) Accretional evolution of a planetesimal swarm. 2. The terrestrial zone. *Icarus, 128,* 429–455.
Weidenschilling S. J., Marzari F., and Hood L. L. (1998) The origin of chondrules and jovian resonances. *Science, 279,* 681–684.
Wetherill G. W. (1980) Formation of the terrestrial planets. *Annu. Rev. Astron. Astrophys., 18,* 77–113.
Wetherill G. W. (1986) Accmulation of the terrestrial planets and implications concerning lunar origin. In *Origin of the Moon* (W. K. Hartmann et al., eds.), pp. 519–550. Lunar and Planetary Institute, Houston.
Wiechert U., Halliday A. N., Lee D.-C., Snyder G. A., Taylor L. A., and Rumble D. A. (2001) Oxygen isotopes and the Moon-forming giant impact. *Science, 294,* 345–348.
Wiechert U., Halliday A. N., Palme H., and Rumble D. (2004) Oxygen isotopes and the differentiation of planetary embryos. *Earth Planet. Sci. Lett., 221,* 373–382.
Wieler R., Anders E., Baur H., Lewis R. S., and Signer P. (1992) Characterisation of Q-gases and other noble gas components in the Murchison meteorite. *Geochim. Cosmochim. Acta, 56,* 2907–2921.
Wilde S. A., Valley J. W., Peck W. H., and Graham C. M. (2001) Evidence from detrital zircons for the existence of continental crust and oceans on the Earth 4.4 G.y. ago. *Nature, 409,* 175–178.
Wing J., Schwartz B., and Huizenga J. (1961) New hafnium isotope, ^{182}Hf. *Phys. Rev., 123,* 1354–1355.
Wood B. J. and Halliday A. N. (2005) Kelvin revisited: Cooling and core formation after the giant impact. *Nature, 437,* 1345–1348.
Woodland S. J., Rehkämper M., Halliday A. N., Lee D-C., Hattendorf B., and Günther D. (2005) Accurate measurement of silver isotopic compositions in geological materials including low Pd/Ag meteorites. *Geochim. Cosmochim. Acta, 69,* 2153–2163.
Yi W., Halliday A. N., Alt J. C., Lee D.-C., Rehkämper M., Garcia M., Langmuir C., and Su Y. (2000) Cadmium, indium, tin, tellurium and sulfur in oceanic basalts: Implications for chalcophile element fractionation in the earth. *J. Geophys. Res., 105,* 18927–18948.
Yin Q. Z. and Jacobsen S. B. (1998) The ^{97}Tc-^{97}Mo chronometer and its implications for timing of terrestrial accretion and core formation (abstract). In *Lunar and Planetary Science XXIX,* Abstract #1802. Lunar and Planetary Institute, Houston (CD-ROM).
Yin Q. Z., Jacobsen S. B., McDonough W. F., Horn I., Petaev M. I., and Zipfel J. (2000) Supernova sources and the ^{92}Nb-^{92}Zr *p*-process chronometer. *Astrophys. J. Lett., 535,* L49–L53.
Yin Q., Jacobsen S. B., and Yamashita K. (2002a) Diverse supernova sources of pre-solar material inferred from molybdenum isotopes in meteorites. *Nature, 415,* 881–883.
Yin Q. Z., Jacobsen S. B., Yamashita K., Blichert-Toft J., Télouk P., and Albarède F. (2002b) A short time scale for terrestrial planet formation from Hf-W chronometry of meteorites. *Nature, 418,* 949–952.
Young E. D. (2000) Assessing the implications of K isotope cosmochemistry for evaporation in the preplanetary solar nebula. *Earth Planet. Sci. Lett., 183,* 321–333.
Young E. D., Simon J. I., Galy A., Russell S. S., Tonui E., and Lovera O. (2005) Supra-canonical ^{26}Al/^{27}Al and the residence time of CAIs in the solar protoplanetary disk. *Science, 308,* 223–227.
Yurimoto H. and Kuramoto K. (2004) Molecular cloud origin for the oxygen isotope heterogeneity in the solar system. *Science, 305,* 1763–1766.
Zartman R. E. and Haines S. M. (1988) The plumbotectonic model for Pb isotopic systematics among major terrestrial reservoirs — A case for bi-directional transport. *Geochim. Cosmochim. Acta, 52,* 1327–1339.
Zook H. A. (1981) On a new model for the generation of chondrules (abstract). In *Lunar and Planetary Science XII,* pp.1242–1244. Lunar and Planetary Institute, Houston.

Compositional Relationships Between Meteorites and Terrestrial Planets

Kevin Righter
NASA Johnson Space Center

Michael J. Drake
University of Arizona

Ed Scott
University of Hawai'i

Terrestrial planets exhibit a range of core sizes, mantle, and surficial magma compositions. Determining the bulk compositions of terrestrial planets presents a challenge faced by geochemistry and cosmochemistry. Traditionally, chondritic meteorites have been called upon as suitable bulk compositions. Detailed examination of such building blocks, however, frequently results in a problem for any specific meteorite type. In this contribution, we examine whether the terrestrial planets can be made from (1) known meteorite types; (2) familiar types that have been transformed at higher pressures, temperatures, or different f_{O_2} or f_{H_2O}; or (3) from materials no longer falling to Earth because they have all been swept up in the accretion process. Bulk properties such as Mg-Si-Al, D/H ratios, O, Cr, and Os isotopes, and noble gases provide constraints on the bulk composition. The bulk compositions of Mars and asteroid 4 Vesta can be explained by mixtures of known meteorite types. However, the idea is entertained that Earth and perhaps other terrestrial planets are not made from familiar meteorite types, but instead from a material that is not currently represented in our collection. Our limited understanding of terrestrial planet formation and origin is due primarily to the restriction of samples to Earth and Moon, Mars, and some asteroids. Sample return missions such as Genesis will greatly enhance our understanding of planet formation, as would samples returned from Mercury and Venus.

1. INTRODUCTION

In order to elucidate compositional relationships between meteorites and terrestrial planets, it is necessary to identify the mechanisms by which, and the materials from which, the planets accreted and differentiated into their cores, mantles, crusts, oceans, and atmospheres. The problem we face is that we sample planets approximately 4.5 G.y. after they formed. Rocks have been subjected to various differentiation processes, and we need to find a way to "see through" 4.5 G.y. of solar system history. Fortunately, nature has provided us with appropriate samples of some terrestrial objects.

For Earth we have lavas, typically basanites or kimberlites, that transport samples of the mantle to the surface for our study. Mantle nodules and unaltered basanites are particularly useful for deducing the state of the silicate Earth immediately after primordial differentiation into metallic core and silicate mantle. For the Moon, we have mare basalts and feldspathic highland rocks, but no samples of the lunar mantle. For Mars we have martian meteorites and Mars Pathfinder and Spirit and Opportunity *in situ* analyses of martian rocks but, again, no mantle samples. We have samples of primitive and differentiated meteorites. Unfortunately, we have no samples of Mercury or Venus delivered as meteorites. For Mercury, we must make do with remote sensing measurements of composition (*Strom,* 1997) and gross geophysical measurements of density. For Venus we must rely on chemical analyses from the Soviet Venera and Vega spacecraft.

Evidence for dependence of composition upon heliocentric distance comes from consideration of metallic core masses of terrestrial planets, along with their mantle FeO content (Fig. 1). Mercury has a very large metallic core, but only a small amount of FeO in its mantle based on spectroscopic measurements (e.g., *Emery et al.,* 1998; *Solomon,* 2003). Earth and Venus have intermediate core sizes and mantle FeO contents. And Mars has the smallest core of the terrestrial planets, and is thought to have the highest mantle FeO content (e.g., *Longhi et al.,* 1992; *Bertka and Fei,* 1998a,b). Estimates of the core size in asteroid 4 Vesta [based on HED meteorites (e.g., *Righter and Drake,* 1997b; *Ruzicka et al.,* 1997)] have converged on values between 10 and 20 mass%, and mantle FeO contents as high as 25 wt% (Fig. 1). Other elements and all but one (Cr) isotopic compositions do not vary systematically with distance from the Sun (e.g., *Palme,* 2000; *Scott and Newsom,* 1989; *Taylor,* 1991), and such fundamental observations must be explained by any model for making terrestrial planets.

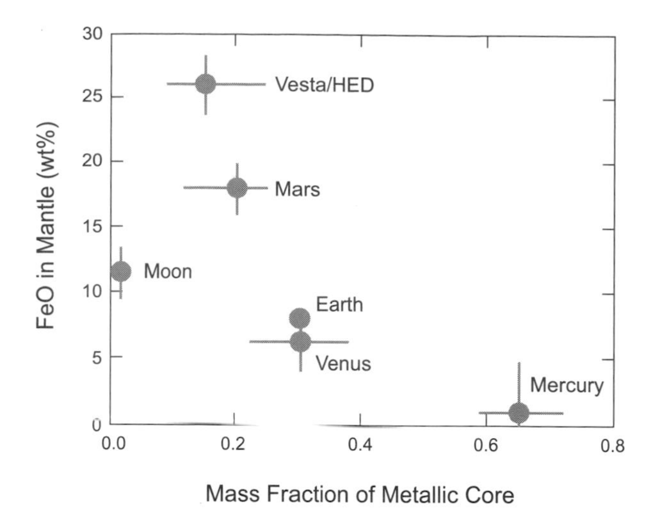

Fig. 1. Core mass fraction and FeO (wt%) in the mantle of the inner solar system planets, asteroid 4 Vesta, and Earth's Moon. 4 Vesta core mass and FeO contents are based on the range of values derived from modeling by *Righter and Drake* (1997b), *Ruzicka et al.* (1997), and *Dreibus and Wänke* (1980). Mars and Mercury estimates are based on reviews presented by *Longhi et al.* (1992) (Mars) and *Strom* (1997) and *Taylor and Scott* (2003) (Mercury). Mantle and core values for Earth's Moon are from *Righter* (2002) and references therein. Terrestrial information is from *Turcotte and Schubert* (1982).

There are three approaches that have been adopted for considering the bulk compositions of terrestrial planets. The formation of the Earth has been debated in terms of heterogeneous vs. homogeneous accretion, and both of these fall into the first class of model that terrestrial planets are made of extant meteoritic material that originated in the asteroid belt. Early theories favored homogeneous accretion of Earth, from materials that are represented in our meteorite collections (e.g., *Ringwood,* 1961). Heterogeneous accretion envisions the material accreting to Earth changing in composition and oxidation state with time. *Eucken* (1944) first proposed this idea, and later work of *Wood* (1962), *Larimer and Anders* (1967), and *Cameron* (1962) was all consistent with the existence of temperature and compositional gradients imposed by conditions in the early solar system (*Turekian and Clark,* 1969). Heterogeneous accretion was subsequently modeled in detail by later workers such as *Wänke* (1981), *Newsom and Sims* (1991), and *O'Neill* (1991). For example, driven by the "stairstep" abundance pattern of siderophile (metal-seeking) elements in the terrestrial upper mantle (Fig. 2), *Wänke* (1981) suggested that the first 80–90% of Earth accreted from very reducing materials. All elements except the refractory lithophile elements such as Sc and the rare earth elements (REE) (Fig. 2) would be quantitatively extracted into the core, and the mantle would be devoid of Fe^{2+}. The next roughly 20–10% of material accreting to Earth would be more oxidizing, and all but the highly siderophile elements (Ir, Os, Au, etc.) would remain stranded in the mantle. The highly siderophile elements were again quantitatively extracted into the core. The last roughly 1% added [the so-called "late veneer" (*Chou,* 1978)] was so oxidizing that metallic Fe did not exist. The "stairstep" pattern of siderophile elements in Fig. 2 is thus explained.

A second class of model has been devised since the 1990s in light of developments in high-temperature and high-pressure experimental data and dynamical and planetary accretion modeling. The realization that high temperatures (>3000 K) are possible during planetary growth stages is a result of the giant impact theory for the origin of the Moon (*Hartmann and Davis,* 1975; *Cameron and Ward,* 1976), and its most recent version (*Canup and Asphaug,* 2001; *Cameron,* 2000). In addition, there is general agreement (e.g., *Drake,* 2000; *Walter et al.,* 2000; *Rubie et al.,* 2003; *Wade and Wood,* 2005) that Earth developed one or more magma oceans late in its accretion, effectively homogenizing any preexisting material. The bulk geochemical properties of Earth were likely established by magmatic processes in a high-pressure and high-temperature magma ocean environment (see solid symbols in Fig. 2). This second class of model allows for the possibility that Earth is made of known meteorites that have been chemically transformed by high temperature and pressure during growth.

A third class of model holds that terrestrial planets are not made of materials known in our meteorite collections, but rather swept-up material available in the region of the inner solar system in which they formed. *Wetherill*'s (1975) pioneering dynamical work indicated that, at any stage during the growth of a planet, the largest objects accreting to it were one-tenth to one-third of the mass of the growing planet (e.g., *Wetherill,* 1975). The modern view is that

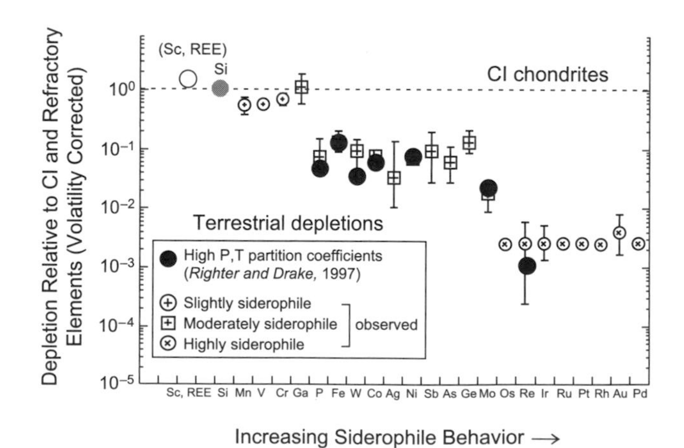

Fig. 2. The abundances of elements in Earth's primitive upper mantle (PUM) after core formation show a stairstep pattern, a fingerprint that contains clues to accretion and planetary differentiation. Comparison of the observed siderophile-element depletions in Earth's upper mantle (open symbols — abundances normalized to CI chondrites and refractory elements), with those calculated using 1-bar partition coefficients (crosses) and high-pressure/high-temperature calculated partition coefficients (solid circles). Calculated depletions using the latter partition coefficients overlap the observed depletions, consistent with metal-silicate equilibrium and homogeneous accretion (*Righter and Drake,* 1997a).

"feeding zones" for planets are relatively narrow, with only late-stage accretion sampling material from significant distances (e.g., *Chambers,* 2001; *Drake and Righter,* 2002). Thus, chemical and isotopic reservoirs should have been preserved in some form as a function of heliocentric distance, and it is unlikely that any planet was made primarily from material currently existing in the asteroid belt.

In this paper we review these three approaches to terrestrial planet formation: known building blocks, known but transformed building blocks, and unknown building blocks. Where necessary, we review aspects of the meteoritic materials that are potential building blocks for terrestrial planets, the processes that were operating in the early solar nebula, and accretion processes (planetesimal formation and giant impacts). We also review the relative effects of such processes that can transform such materials (temperature, f_{O_2}, pressure, and water content) as planets start to grow. Finally, we illustrate gaps in understanding and sampling, and show that an understanding of the volatile contents of terrestrial planets depends upon all these processes.

2. MATERIALS AND CONDITIONS IN THE EARLY SOLAR SYSTEM

2.1. Chondrites and Their Properties

To understand how the terrestrial planets may be related to the chondrites requires considerable insights into the processes that made the components in chondrites and accreted the chondritic components into planetesimals, as well as how planetesimals evolved into planets and asteroids.

2.1.1. Chondrite groups and chondritic components. Fifteen groups of chondrites are defined primarily by their bulk chemical compositions (Figs. 3 and 4), although secondary parameters such as bulk O-isotopic compositions, chondrule size and abundance, and mineralogy are also useful (Table 1). The redox state of Fe is a fundamental variable in the chondrite groups with the two extremes defined by reduced enstatite chondrites containing FeO-free silicates and Fe metal, and oxidized carbonaceous chondrites containing very little metal and FeO-rich silicates (Table 1,

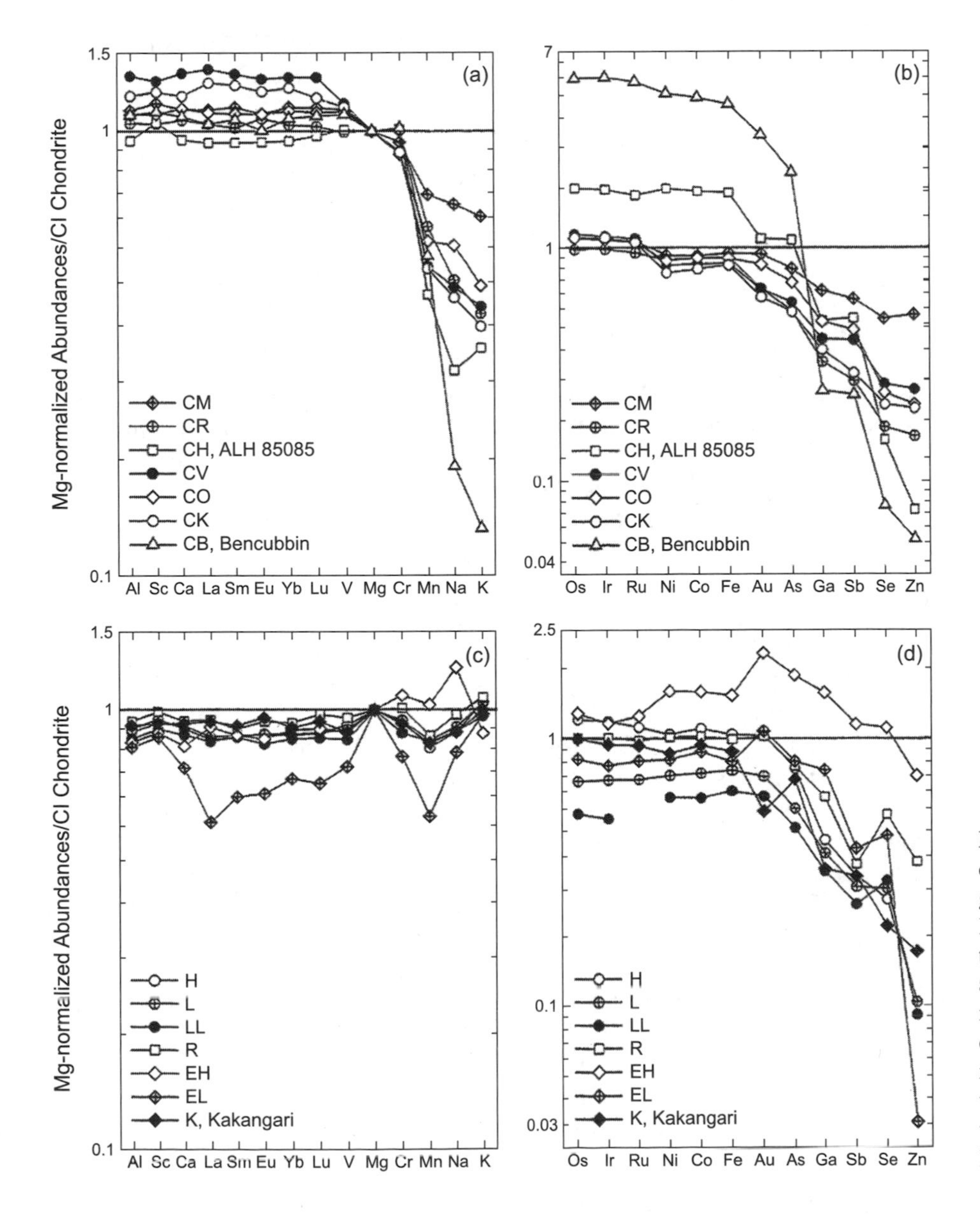

Fig. 3. Compositional variation in carbonaceous chondrites (top diagrams) and ordinary, enstatite, R, and K chondrites (bottom diagrams) for the lithophile elements (left diagrams) and siderophile and chalcophile elements (right diagrams). Note the close correspondence of most chondritic materials for the lithophile elements, with the greatest disparity being for volatile elements such as Mn, Na, and K. Figure after *Krot et al.* (2003).

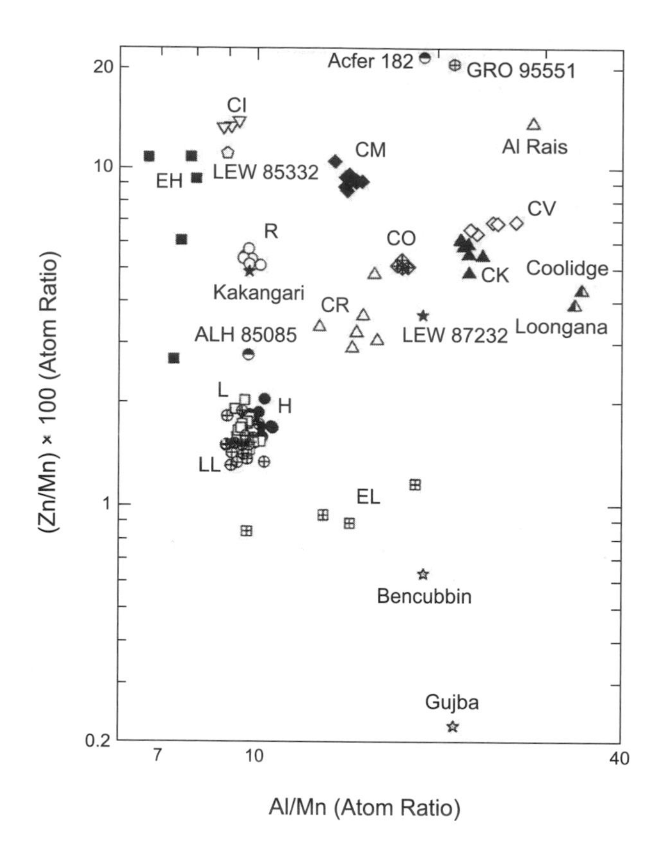

Fig. 4. Compositional variation in chondrite groups for Zn/Mn vs. Al/Mn. Abbreviations are: EL and EH = enstatite chondrite groups; H, L, and LL = ordinary chondrite groups; R = R chondrites; CI, CV, CR, CM, CO, and CK = carbonaceous chondrite groups. Figure after *Krot et al.* (2003).

Fig. 5). A second fundamental variable is the degree of volatile-element (e.g., K, Na, Mn, Zn, Se, Sb) depletion; when compared to CI chondrites, the other carbonaceous groups exhibit varying degrees of depletion from CM through CB (Fig. 3). All but two groups, R and K, can be assigned to the carbonaceous, ordinary, or enstatite chondrite classes. CI chondrites, which contain only traces of chondrules and refractory inclusions, occupy a unique position in the chondrite stable. They consist largely of fine-grained matrix and are closest in composition to the solar photosphere, neglecting H, He, C, N, O, and the noble gases. The chondrites in Table 1 are divided into six petrologic types, which indicate the extent of metamorphism and alteration in their parent asteroids. Types 3–6 reflect increasing thermal metamorphism while types 3–1 reflect increasing degrees of alteration. In some groups, the type 3 chondrites have been subdivided into types 3.0–3.9 (or even further to include types 3.00–3.15) in order to identify the rare chondrites that have suffered the least metamorphism and alteration (e.g., *Scott and Krot,* 2003; *Grossman and Brearley,* 2005).

There are four major constituents of chondrites — chondrules, refractory inclusions, metal, and matrix. Chondrules are millimeter- to centimeter-sized largely spherical objects composed primarily of ferromagnesian silicates that were wholly or partly molten in the solar nebula (or protoplanetary disk) and cooled in minutes to hours (e.g., *Zanda,* 2004). They commonly contain grains of metallic Fe,Ni and, in most but not all groups, troilite. The refractory inclusions, which account for <0.1–10 vol% of the chondrite groups, formed entirely from minerals that were stable above 1400 K in the solar nebula, assuming canonical pressures of 10^{-3}–10^{-4} bar. Two varieties are recognized: Ca-Al–rich inclusions (CAIs), which are composed entirely of refractory Ca-Al-Ti minerals and may be rimmed by forsterite grains, and amoeboid-olivine aggregates (AOAs), which are mixtures of fine-grained forsterite and Ca-Al-Ti minerals (*MacPherson,* 2003). The fine-grained and irregularly shaped refractory inclusions probably formed from grains that condensed in the solar nebula, and include so-called "fluffy" type A CAIs and AOAs. Coarse-grained CAIs with igneous textures and spheroidal shapes are probably melted condensates that cooled during crystallization at rates comparable to those of chondrules.

Matrix material is a mixture of grains with sizes of 10 nm to ~5 μm that coated the chondrules and refractory inclusions before they accreted (see *Scott and Krot,* 2005a). The bulk composition of matrix is approximately CI-like, but the components are derived from diverse locations in the disk and include ~10–50 ppm of presolar silicates and oxides. Because fine-grained material was readily altered in asteroids, only type 3–4 chondrites appear to contain mostly unaltered matrix material and high concentrations of presolar silicates (*Brearley,* 1993; *Greshake,* 1997; *Nguyen and Zinner,* 2004; *Kobayashi et al.,* 2005). Hydrated silicates and Fe-rich olivine, which were once thought to be nebula products, are rare in the least-altered chondrite matrices. These matrices contain instead crystalline Mg-rich silicates, amorphous Fe-rich silicates, plus grains of metallic Fe,Ni, sulfides, and refractory oxides. The minerals in pristine chondrite matrices are surprisingly similar to those in the chondritic, anhydrous interplanetary dust particles (IDPs), which are thought to come from comets, although the latter contain 10× higher concentrations of C and presolar grains (*Bradley,* 2003). (Hydrated, chondritic IDPs are thought to come from asteroids, but some chondritic IDPs contain both anhydrous and hydrated minerals.) Thus the nebular silicate dust that accreted at 2.5 AU was not fundamentally different from the silicate dust at 5–10 AU, and probably also resembled the dust that accreted inside 2 AU.

The major constraints on the time and location of chondrule, CAI, and matrix grain formation come from isotopic data, especially radiometric ages and O-isotopic compositions, that has been reviewed elsewhere (e.g., *Scott and Krot,* 2005b).

2.1.2. Chondrite parent bodies. Chondrites come from diverse parent bodies, and evidence from cosmic-ray exposure ages and chondrite breccias suggest that each group is probably derived from one or a few asteroids (see *Scott and Krot,* 2003). The major clues to the formation locations of the chondrites are those derived from inferred relationships between asteroids and chondrite parent bodies. Carbonaceous chondrites are probably derived from C-type asteroids that are dominant in the belt beyond 2.5 AU (*Bell*

TABLE 1. Characteristic properties of the chondrite groups.

Group	Type	Refractory Inclusions (vol%)	Chondrules (vol%)*	Chondrules Average Diameter (mm)	Fe,Ni Metal (vol%)	Matrix (vol%)†	Fall Frequency (%)‡	Refractory Lithophile/Mg Relative CI§	Examples
Carbonaceous									
CI	1	<0.01	<5	—	<0.01	95	0.5	1.00	Ivuna, Orgueil
CM	1–2	5	20	0.3	0.1	70	1.6	1.15	Murchison
CO	3	13	40	0.15	1–5	30	0.5	1.13	Ornans
CV	2–3	10	45	1.0	0–5	40	0.6	1.35	Vigarano, Allende
CK	3–6	4	15	0.8	<0.01	75	0.2	1.21	Karoonda
CR	1–2	0.5	50–60	0.7	5–8	30–50	0.3	1.03	Renazzo
CH	3	0.1	~70	0.05	20	5	0	1.00	ALH 85085
CB_a	3	<0.1	40	~5	60	<5	0	1.0	Bencubbin
CB_b	3	<0.1	30	~0.5	70	<5	0	1.4	QUE 94411
Ordinary									
H	3–6	0.01–0.2	60–80	0.3	8	10–15	34.4	0.93	Dhajala
L	3–6	<0.1	60–80	0.5	3	10–15	38.1	0.94	Khohar
LL	3–6	<0.1	60–80	0.6	1.5	10–15	7.8	0.90	Semarkona
Enstatite									
EH	3–6	<0.1	60–80	0.2	8	<0.1–10	0.9	0.87	Qingzhen, Abee
EL	3–6	<0.1	60–80	0.6	15	<0.1–10	0.8	0.83	Hvittis
Other									
K	3	<0.1	20–30	0.6	6–9	70	0.1	0.9	Kakangari
R	3–6	<0.1	>40	0.4	<0.1	35	0.1	0.95	Rumuruti

* Includes chondrule fragments and silicates inferred to be fragments of chondrules.
† Includes matrix-rich clasts, which account for all matrix in CH and CB_b chondrites.
‡ Fall frequencies based on 918 falls of differentiated meteorites and classified chondrites (*Grady,* 2000).
§ Mean ratio of refractory lithophiles relative to Mg, normalized to CI chondrites.

See *Scott and Krot* (2003) and sources therein.

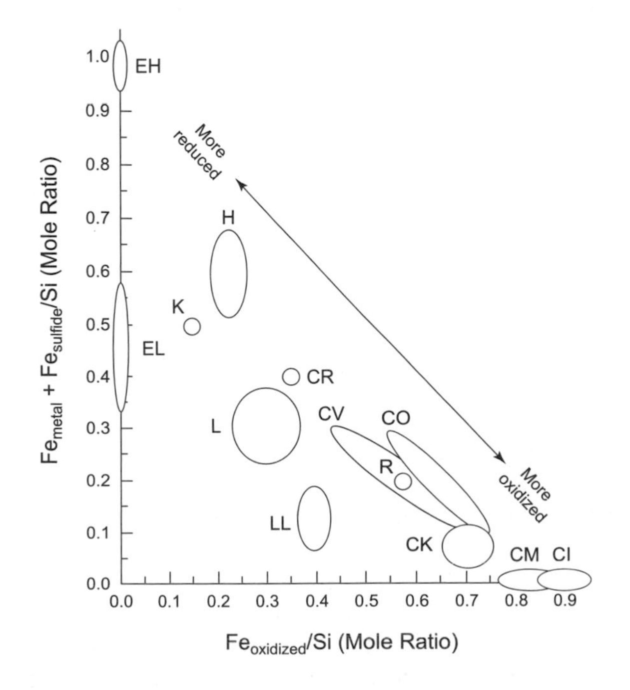

Fig. 5. Urey-Craig diagram, illustrating the proportion of oxidized vs. reduced Fe in chondritic meteorites (from *Brearley and Jones,* 1998). Abbreviations are the same as in Fig. 4.

et al., 1989). The abundance of CM-like clasts in chondrite breccias and micrometeorites suggest that this is the dominant material in the belt, consistent with the overall dominance of C-type asteroids. Ordinary chondrites probably formed at <2.6 AU judging from their close relationship with S-type asteroids (*Chapman,* 2004). Enstatite chondrites may form closer, as enstatite achondrites are probably derived from E-type asteroids that are concentrated at 1.9–2.0 AU.

We do not know the extent to which these chondrites are representative of the chondritic asteroids that formed in the main belt. Since the number of chondrite groups has almost doubled in the past 30 years, one could argue that the known groups are only a tiny fraction of the chondritic planetesimals (e.g., *Scott and Newsom,* 1989). Certainly, the CH, CB, and K chondrites are all very strange groups that do not fit comfortably among the other groups. However, these may simply be rare, atypical types, as they do not match any of the abundant asteroid types. Most of the ungrouped chondrites and grouplets are matrix-rich carbonaceous groups, which have comparable properties to known groups (*Krot et al.,* 2003). Poorly consolidated, carbonaceous asteroids are certainly underrepresented among the chondrite groups. These probably include the volatile-rich D-type asteroids that probably supplied the Tagish Lake chondrite. The chondrites in Table 1 may be reasonably representative of the tougher asteroids. Supporting evidence comes from the Kaidun breccia, which is composed entirely of fragments of seven groups of enstatite, carbonaceous, ordinary, and R chondrites plus some ungrouped carbonaceous samples (excluding CH and CB chondrites) and some achondritic fragments (*Zolensky and Ivanov,* 2003; *Mikouchi et al.,* 2005).

2.2. Solar Nebula Models

Our knowledge of the conditions in the solar nebula is based on observations of young stars as well as thermodynamic models for the condensation of solids and liquids from a gas of solar composition. Historically, variation in conditions within the nebula has been linked to radial distance from the Sun (e.g., *Wasson,* 1985; *Lewis,* 1974). Factors controlling the thermal profile within the nebula have been reviewed by *Boss* (1998) and include mass accretion rates, opacity (dust/gas ratio), and total mass. Although estimates of nebular midplane temperatures have in the past been typically cool [300–600 K (*Woolum and Cassen,* 1999)], developments within this field have led to revision to higher midplane temperatures as much as a factor of 3 higher (*Humayun and Cassen,* 2000). These hotter temperatures most likely led to vaporization of all but the most refractory materials within a few astronomical units of the Sun, but only for a short (10^5 yr) period of time (*Humayun and Cassen,* 2000). Increasing sophistication has led to models combining gas/dust ratios and thermodynamics (*Ebel and Grossman,* 2001), as well as treating specific trace-element groups, such as the volatile elements (*Humayun and Cassen,* 2000) or highly siderophile elements (e.g., *Campbell et al.,* 2001).

Additional evidence for conditions in the solar nebula comes from the meteorites. Virtually all the chondritic components except for the presolar grains and organics formed in the nebula at high temperatures during relatively brief heating events. Refractory inclusions formed in the inner part of the disk where the bulk O-isotopic composition equaled that of the proto-Sun (*Hashizume and Chaussidon,* 2005). Refractory inclusions probably formed in <0.3 m.y. at high ambient temperatures (1300 K) when the proto-Sun was accreting rapidly and were then distributed across the disk with refractory dust by bipolar outflows or nebula turbulence (*Shu et al.,* 1996; *Cuzzi et al.,* 2003; *Wood,* 2004). Chondrules and most matrix silicates formed under more oxidizing conditions at lower ambient temperatures when the proto-Sun was accreting more slowly. Chondrule melts were stabilized by higher total pressures or high dust/gas ratios. Chondrule formation involved melting, evaporation, condensation, and accretion of solid, partly melted, and wholly melted materials.

Chondrules and matrix grains formed after the nebula gas was depleted in ^{16}O, probably as a result of evaporation of meter-sized icy bodies that were enriched in ^{17}O and ^{18}O due to self-shielding during UV photodissociation of CO in the protosolar molecular cloud or in the disk (*Clayton,* 2002; *Lyons and Young,* 2004; *Yurimoto and Kuramoto,* 2004; *Krot et al.,* 2005). Chondrules may have formed soon after CAIs (*Bizzarro et al.,* 2004), but most chondrules appear to have formed 1–3 m.y. after CAI formation (see *Kita et al.,* 2005). Given the unique chemical and isotopic properties of the chondrules in many different groups and the evidence that CAIs and presolar grains were spread throughout the inner solar nebula, it seems likely that chondrules in a single group accreted soon after they formed, i.e., <0.3 m.y.

Accretion timescales in the nebula increased with increasing distance from the Sun as the surface density of solids decreased and the orbital periods increased. As a result, we might expect that the chemical compositions of the chondrites would correlate simply with the heliocentric distance at which they formed. However, chondrites cannot be simply arranged into a single chemical sequence because their concentrations of volatile, refractory, and siderophile elements are not simply correlated. The simplest explanation for the chemical complexity of chondrites is that many accreted close to the snow-line and the tar-line where the density of solids showed major step increases with increasing heliocentric distance. Thus, two (or perhaps three) accretion fronts may have advanced across the belt from opposite directions, requiring us to disentangle time and distance to reconstruct the primordial chemical zoning patterns of planetesimals in the inner disk.

Additional clues to the accretion locations of the chondrites have been derived from a wide variety of chondrite properties (e.g., *Rubin and Wasson,* 1995). The concentrations of water and volatiles in chondrites increase through the E-O-C sequence, supporting the view that this represents a sequence of increasing heliocentric distance. In addition, the concentration of chondrules probably decreased with increasing distance as the C-rich, volatile-rich carbonaceous chondrites like CIs and Tagish Lake have fewer chondrules than enstatite and ordinary chondrites (Table 1). *Wasson* (1988) argued that E chondrites formed <1 AU from the Sun based on chondritic and planetary properties. Evidence for an inner nebula origin (although at >1 AU) comes from the correlation of $^{53}Cr/^{52}Cr$ with heliocentric distance, which places E chondrites between Earth and Mars (*Lugmair and Shukolyukov,* 1998; *Shukyolyukov and Lugmair,* 1999).

2.3. Accretion Models

Planetary accretion is typically split into three distinct epochs. The first epoch comprised accretion of dust from a gas-rich disk. Turbulence and gas drag transported gas dust and solid bodies smaller than 1 km over large radial distances (e.g., *Chambers,* 2006), within 10^5 yr from T_0. The second epoch consisted of accretion of planetesimals from dust-sized particles. That these planetesimals were homogeneous is attributed to homogeneity in both Mg isotopes (*Norman et al.,* 2004; *Zhu et al.,* 2001), and this occurred on a 10^5–10^6-yr timeframe. The third epoch involved accretion of planet-sized bodies from planetesimals and requires on the order of 10^7 yr for completion. These three epochs of growth define the accretion process, and here we include only a brief discussion of the third epoch, with an emphasis on the width of feeding zones in the nebula during this stage, and the possibility that two stages of this epoch be distinguished — an early to mid stage and a late stage.

Early models for accretion of the terrestrial planets included narrow feeding zones for materials that ended up being incorporated into planets (e.g., *Safronov,* 1969; *Lewis,* 1972; *Wetherill,* 1980). Zones as narrow as 0.1 AU were considered to be sources for individual bodies, such as Earth and the Moon. This scenario has held true for the midstages of planet formation when eccentricities and inclinations are low (e.g., *Canup and Agnor,* 2000). For example, planetesimals that accreted inside 2 AU probably differed from the planetesimals that accreted in the asteroid belt as a result of both spatial and temporal variations. The inner planetesimals would presumably have had lower water and matrix abundances, as a result of differences in location, and higher ^{26}Al concentrations as a result of more rapid accretion closer to the proto-Sun. Dynamical modeling by *Bottke et al.* (2005) suggests that the differentiated meteorite parent asteroids themselves may be derived from planetesimals that accreted inside 2 AU and were then scattered into the belt by planetary embryos. Thus the parent asteroids of the differentiated meteorites may give us the best clues to the nature of the innermost planetesimals. Enstatite chondrites remain a likely component of the 1–2-AU region, although their accretion time is not constrained by precise chondrule formation ages.

More recent models of the later stages of accretion have shown that the feeding zones widen substantially (*Wetherill,* 1994). In fact, the results of terrestrial planet simulations in which the starting embryos are "tagged" show that there is extensive mixing of material ending up in the surviving planets, between 0.5 and 3.5 AU (e.g., *Chambers and Wetherill,* 1998; *Chambers,* 2001). In light of these findings, discussions of "feeding zones" must explicitly define a timeframe since the extent of mixing apparently increases with age.

3. KEY ISOTOPIC AND ELEMENTAL TOOLS

The question of whether samples of the "building blocks" of Earth and other planets still exist as meteorites and comets is addressed by examining specific elemental and isotopic properties of the Earth, Mars, Venus, Mercury, comets, and meteorites. The ratios of Mg/Si and Al/Si and the isotopic compositions of O, H, Cr, Os, and noble gases are particularly useful.

3.1. Major Elements in Earth, Moon, Mars, Venus, and Meteorites

Magnesium/silicon vs. Al/Si ratios for a variety of solar system objects illustrate that these ratios vary significantly (Fig. 6). Primitive meteorites have distinct major-element

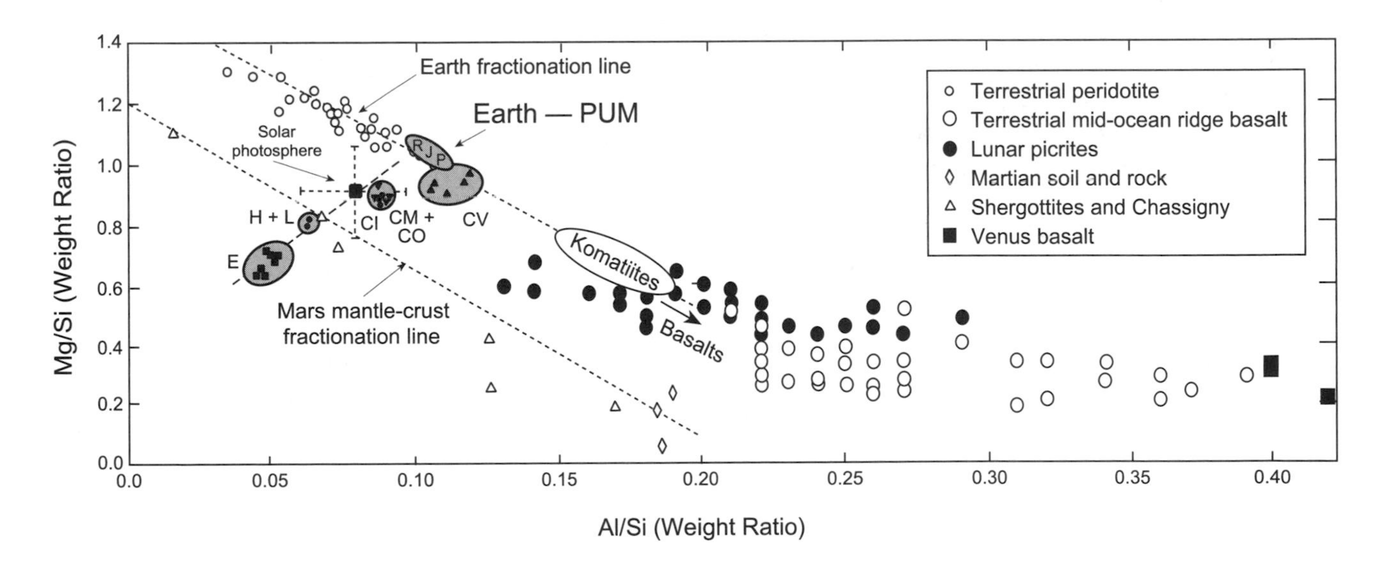

Fig. 6. The major-element composition of primitive material in the inner solar system is *not* of uniform composition, but defines an unexplained trend. Mg/Si vs. Al/Si ratios in chondritic, terrestrial, lunar, martian, and venusian materials (*Jagoutz et al.,* 1979; *Taylor,* 1992; *Dreibus et al.,* 1998; *Kargel et al.,* 1993; *Basaltic Volcanism Study Project,* 1981). Abbreviations are: E = enstatite chondrites; H, L = ordinary chondrites; CI, CM, CO, and CV = carbonaceous chondrites. Small open circles are terrestrial peridotite samples; terrestrial fractionation line is defined by peridotites, komatiites, and basalts. The letters R, J, and P refer to estimates of the bulk silicate Earth composition by *Ringwood* (1979), *Jagoutz et al.* (1979), and *Palme and Nickel* (1986), respectively. Solid circles are lunar picritic glasses and large open circles are MORB (*Basaltic Volcanism Study Project,* 1981). Martian fractionation line is defined by Chassigny, shergottites, and martian soils and rocks from the Viking and Pathfinder mission (*Dreibus et al.,* 1998). Venusian samples are analyses obtained from the Vega and Venera missions (*Kargel et al.,* 1993).

compositions and define a line with a positive slope, sometimes called the "cosmochemical trend" (*Jagoutz et al.,* 1979). The origin of this trend is not well understood but undoubtedly results from condensation processes, differences in modal mineralogy (CAI, chondrules, metal), etc. The bulk Mg/Si and Al/Si ratios for differentiated objects such as Earth and Mars must be inferred from samples available at their surfaces. These ratios can be fractionated by the addition or subtraction of phases in mantle differentiation, such a Mg perovskite, garnet, pyroxene, and olivine (*Jones,* 1996; *Agee and Walker,* 1993). Also, Si in the core can lead to elevated Mg/Si ratios (e.g., *Malavergne et al.,* 2004; *Gessmann et al.,* 2001; *Kilburn and Wood,* 1997).

3.2. Oxygen Isotopes in Earth, Moon, Mars, and Meteorites

Various meteorite groups preserve unique O-isotopic compositions (Fig. 7a). Some meteorite groups define a slope of 1 : 2, which corresponds to control by mass-dependent fractionation processes. However, other bodies and groups define a slope of 1 : 1. The origin of the 1 : 1 slopes is debated and has been attributed to either ^{16}O additions to the nebula (e.g., *Clayton,* 1993), or to photochemical reactions involving CO in the solar nebula (e.g., *Clayton,* 2002). Part of the uncertainty is due to lack of consensus on the position of the solar value on such a diagram, and this is the primary scientific goal of the Genesis mission (e.g., *Burnett et al.,* 2003). Despite the large range of values recorded in meteorites, it appears that distinct O reservoirs are preserved over relatively small annuli of heliocentric distance, such as Earth, Mars, and the HED parent body (Fig. 7b). This makes O isotopes a very useful tool for tracing provenance. However, we still have no constraints on the O-isotopic composition of the inner solar system (Venus and Mercury).

3.3. Initial Osmium-187/Osmium-188 Ratio in Primitive Upper Mantles of Earth, Mars, and Meteorites

Highly siderophile elements (Re, Au, and the platinum group elements) can provide information about the conditions and/or timing of core formation and metal-silicate segregation since they are stripped almost entirely into metallic Fe. Rhenium and Os are two HSEs that are linked by beta decay, ^{187}Re = ^{187}Os + β. The concentrations of highly siderophile elements and ratios of Os isotopes have been used to argue for the late addition of chondritic material to the mantle of Earth, after the main stage of core formation had ended (e.g., *Chou,* 1978; *Morgan et al.,* 1981). This is because the large differences in partition coefficients (both metal/silicate and solid silicate/liquid silicate) for various HSEs indicate that fractionation of these elemental ratios is otherwise easy (e.g., *Righter et al.,* 2000). Although this idea initially seems rigorous, the latest results (Fig. 8) show

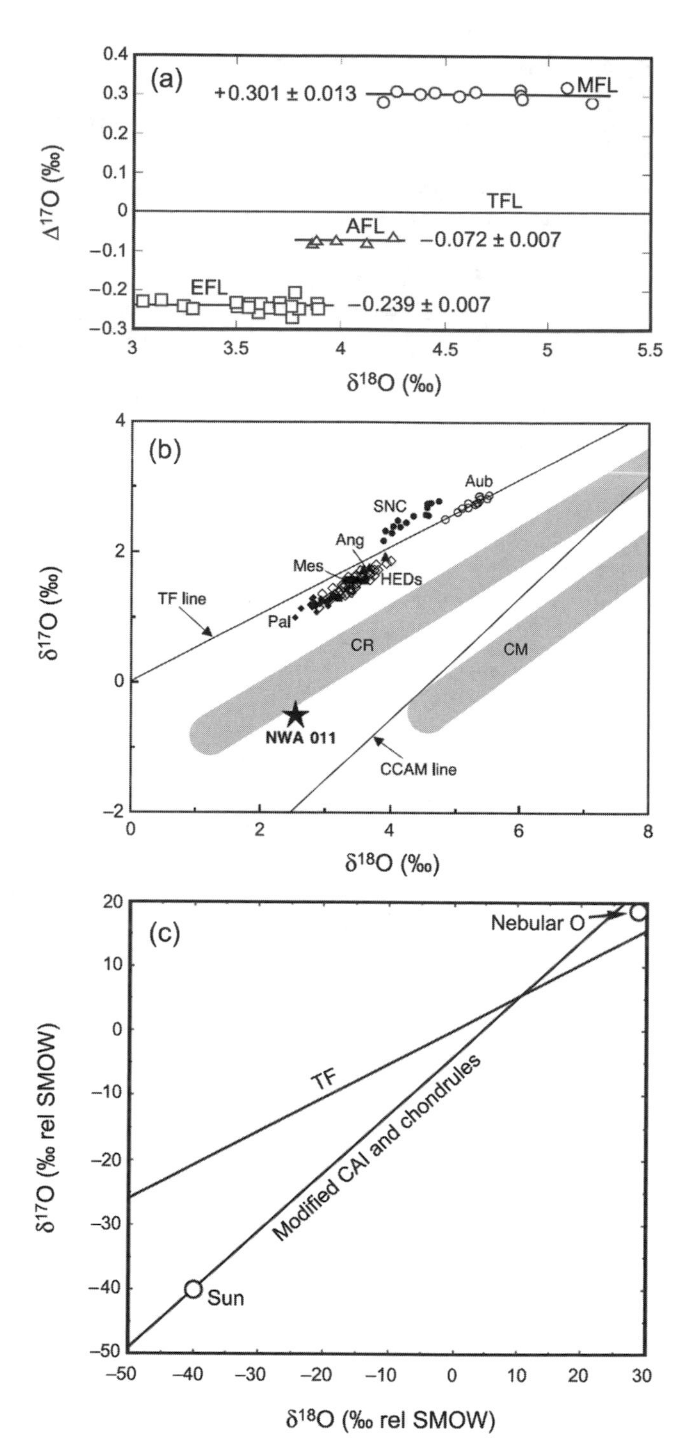

Fig. 7. **(a)** The Mars, terrestrial, angrite, and eucrite fractionation lines (MFL, TFL, AFL, and EFL, respectively) from *Greenwood et al.* (2004), in terms of their Δ^{17}O values. **(b)** Simplified O-isotopic plot of inner solar system material, based on the work of R. N. Clayton and colleagues at the University of Chicago (*Clayton et al.,* 1991; *Clayton,* 1993; *Clayton and Mayeda,* 1999, 2001). Also shown are the aubrites, SNCs, angrites, HEDs, mesosiderites, pallasites, CR and CM chondrites, and the unusual eucrite NWA 011. **(c)** Two possible extreme end members for solar O isotopes taken from *Clayton* (2004). In **(b)** and **(c)**, the terrestrial fractionation line is simply indicated as "TF."

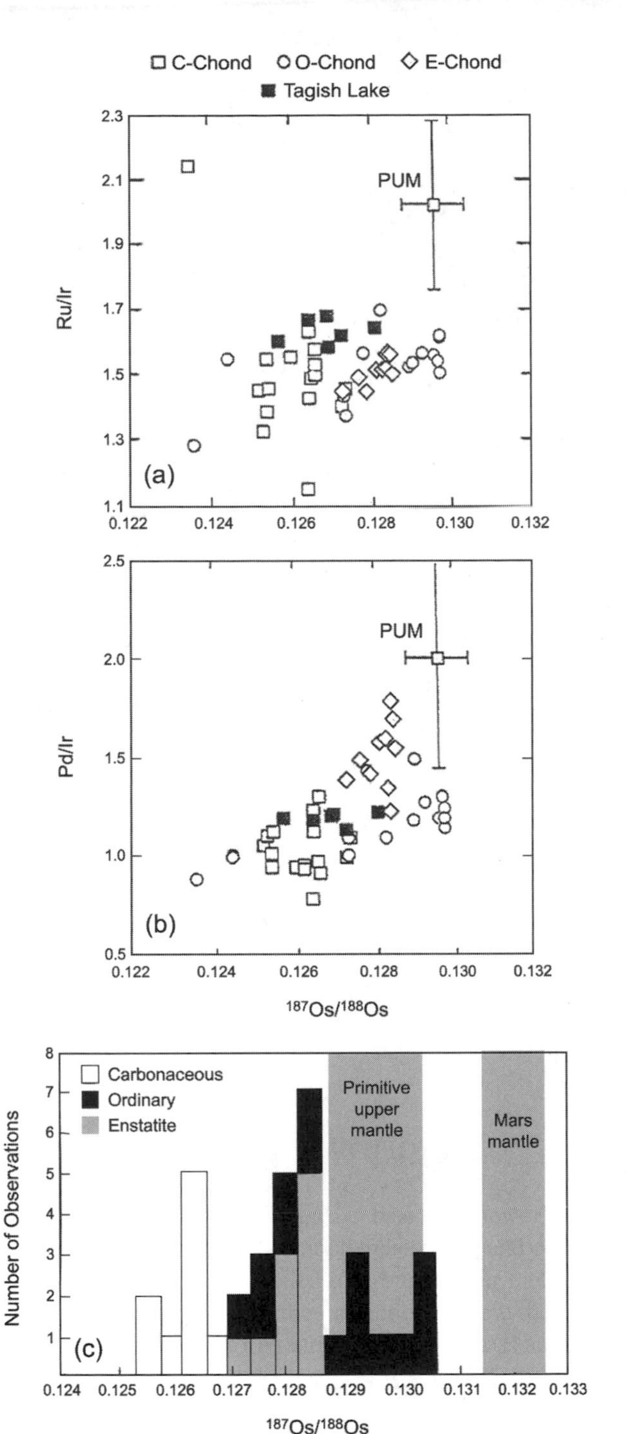

Fig. 8. $^{187}Os/^{188}Os$, Ru/Ir, and Pd/Ir ratios in carbonaceous, ordinary, and enstatite chondrites, Tagish Lake, and Earth's primitive upper mantle (PUM) (*Meisel et al.,* 2001). Also shown is the primitive martian mantle with nonchondritic $^{187}Os/^{188}Os$ (from *Brandon et al.,* 2005).

that none of the chondrite groups match the $^{187}Os/^{188}Os$ isotopic, Ru/Ir, or Pd/Ir ratios of Earth's primitive upper mantle. Carbonaceous chondrites, the only abundant water-bearing meteorites, have a significantly lower $^{187}Os/^{188}Os$ ratio of 0.1265 than Earth's primitive upper mantle (PUM) value of 0.1295, effectively ruling them out as the source of the "late veneer" (Fig. 4). Mars' primitive upper mantle appears to have an even higher $^{187}Os/^{188}Os$ ratio of 0.132 (*Brandon et al.,* 2005). The Earth's mantle $^{187}Os/^{188}Os$ ratio overlaps anhydrous ordinary chondrites and is distinctly higher than anhydrous enstatite chondrites, while the martian mantle is higher than all common meteorite types.

3.4. Deuterium/Hydrogen Ratios in Earth, Mars, Meteorites, and Comets

A fundamental question is whether planets accreted "wet" or "dry," i.e., did their building blocks contain water, or was water delivered from exogenous sources after the planets formed. Meteorites, comets, and the interstellar medium define a large range of D/H ratios, and therefore the origin of planetary water can be investigated by examining D/H ratios in various planetary materials (Fig. 9) (*Geiss and Reeves,* 1972; *Owen and Bar-Nun,* 2000). Although D/H ratios may have varied substantially with heliocentric distance in the early solar system, partial to complete homogenization was likely on the timescale of 10^6 yr (*Robert et al.,* 2000). As a result a range of D/H ratios could have been preserved in ancient materials, all of which could later be supplies for planet making.

3.5. Chromium-Isotopic Ratios in Earth, Moon, Mars, and Meteorites

Manganese-53 decays to ^{53}Cr with a half-life of 3.7 ± 0.4 m.y. Daughter ^{53}Cr will be unmeasurable after 5–7 half-lives, or about 20 m.y. to 26 m.y. after nucleosynthesis of ^{53}Mn. Thus, variations in the $^{53}Cr/^{52}Cr$-isotopic ratio in solar system materials must be established early in solar system history and reflect initial heterogeneities in the Cr-isotopic

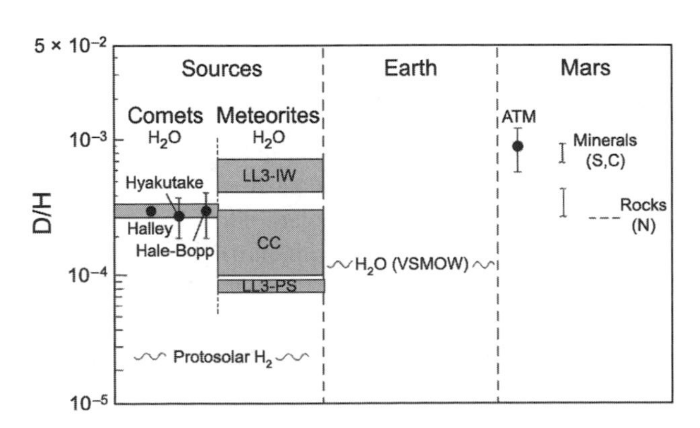

Fig. 9. The D/H ratios in H_2O in three comets, meteorites, Earth (Vienna standard mean ocean water — VSMOW), protosolar H_2, and Mars (*Owen and Bar-Nun,* 2000; *Lecluse and Robert,* 1994; *Deloule and Robert,* 1995; *Robert et al.,* 2000; *Deloule et al.,* 1998; *Robert,* 2001). CC = carbonaceous chondrites, LL3-IW = interstellar water in Semarkona, LL3-PS = protostellar water in Semarkona.

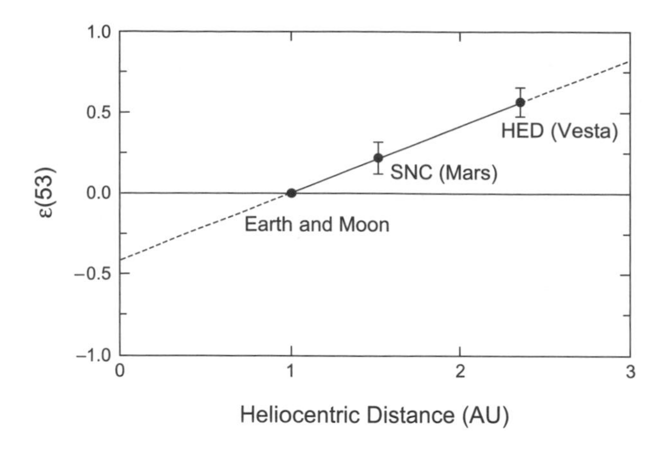

Fig. 10. Variation in Cr-isotopic composition with heliocentric distance. Ordinary chondrites plot with HED meteorites. After *Lugmair and Shukolyukov* (1998). ε(53) is defined as $[(^{53}Cr/^{52}Cr)_{measured}/(^{53}Cr/^{52}Cr)_{standard} - 1] \times 10^4$ (*Lugmair and Shukolyukov,* 1998).

ratio (as with O), variations in the Mn/Cr ratio, or some combination of the two (Fig. 10). One cautionary note is that ^{54}Cr anomalies have been found in carbonaceous chondrites (*Rotaru et al.,* 1992), and these must be considered carefully when discussing Cr isotopes in the context of potential building blocks.

3.6. Noble Gases

The traditional view and interpretation of noble gases is that the solar compositions measured in many planetary atmospheres is the result of planetary accretion and retention (with perhaps later degassing) of solar nebular gases. Meteorites exhibited a range of values, but they were rarely identified as solar (e.g., *Pepin,* 1991). However, measurements of solar noble gases in angrites (*Busemann et al.,* 2004) and enstatite chondrites (*Busemann et al.,* 2003) shows that there might be a significant solar component preserved in meteorites (see also *Wieler et al.,* 2006; *Podosek,* 2003). In addition, ices and outer solar system objects can potentially carry noble gases, and *Owen and Bar-Nun* (2000) discuss this possibility with respect to Earth and Mars (Fig. 11).

4. INFERRING PLANETARY BULK COMPOSITIONS FROM SURFACE ROCKS: "SEEING THROUGH" ACCRETION AND IGNEOUS DIFFERENTIATION

As noted in the introduction, chondritic meteorite groups are commonly considered as building blocks for terrestrial planets. Mixtures of ordinary and carbonaceous chondrites (e.g., *Lodders and Fegley,* 1998), or even enstatite chondrites (*Javoy,* 1995), have been proposed. However, *Drake and Righter* (2002) and *Burbine and O'Brien* (2004) have demonstrated the difficulty of making Earth from any extant meteorite type. Regardless of the nature of the building blocks of the terrestrial planets, it is unlikely that a large planet will retain the exact chemistry of its building blocks, because such materials can be readily transformed by variations in f_{O_2}, temperature, pressure, reaction with water or other volatile species, and also by magmatic or mantle fractionation. Furthermore, large planets might sequester differentiated materials deep in their mantles, making inference of bulk composition from surface rocks problematic (*Boyet and Carlson,* 2005).

4.1. Oxygen Fugacity

Oxygen fugacity is an important parameter in terrestrial planets, as it controls the relative proportions of FeO and Fe_2O_3 and the stability of metallic Fe. On Earth, variations in f_{O_2} can affect magmatic fractionation trends and largely control Fe enrichment or depletion trends in oceanic and continental basalt suites (e.g., *Carmichael, et al.,* 1974). Similarly, a large range of f_{O_2} is defined by meteoritic materials (Fig. 12) (*Rubin et al.,* 1988), and variations in f_{O_2} in meteoritic materials can change chemical features and phase equilibria dramatically.

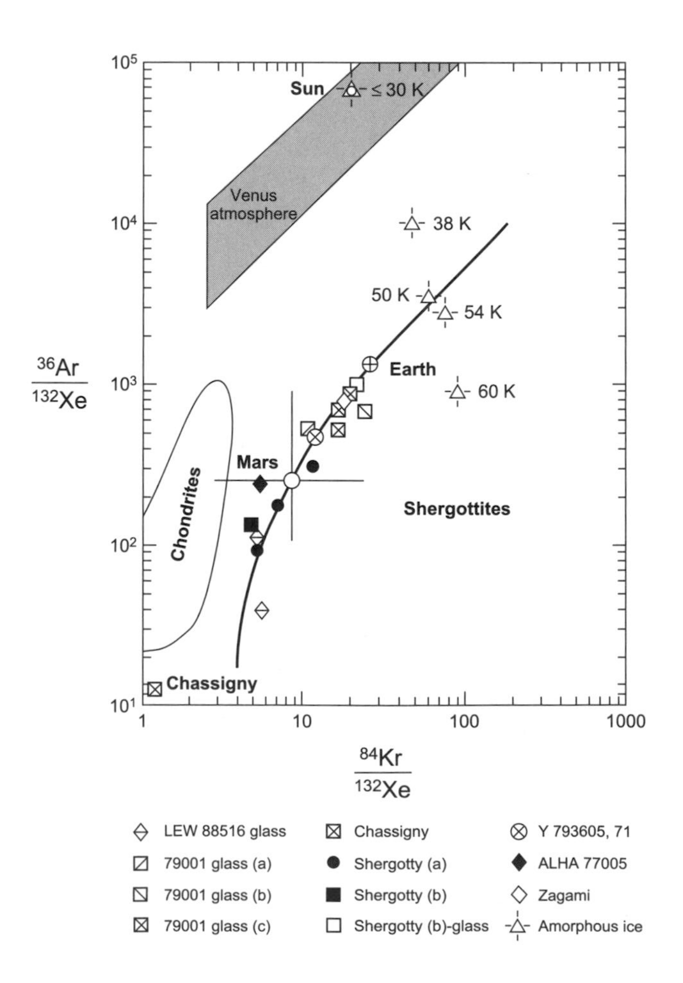

Fig. 11. $^{36}Ar/^{132}Xe$ vs. $^{84}Kr/^{132}Xe$ for Earth, Mars, chondrites, venusian atmosphere, shergottites, Chassigny, ices, and the Sun (after *Owen and Bar-Nun,* 2000).

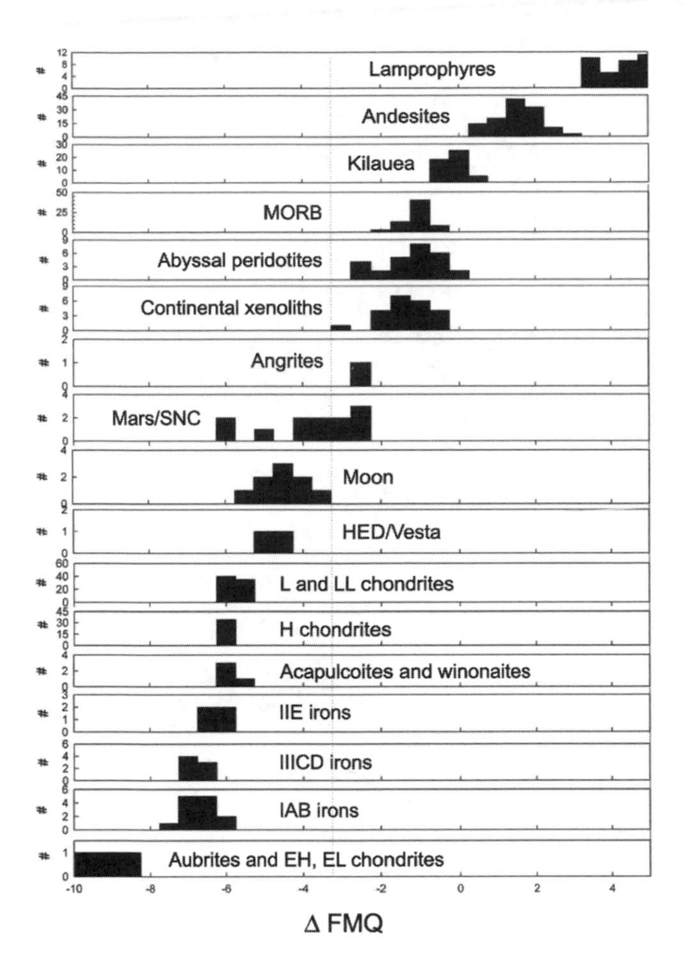

Fig. 12. Histogram of f_{O_2} of solar system materials, including terrestrial samples [lamprophyres and andesites, Kilauea and MORB glasses, and abyssal and continental xenolith peridotites from *Carmichael* (1991)], angrites (from *McKay et al.,* 1994; *Jurewicz et al.,* 1991), martian meteorites (from *Righter and Drake,* 1996), Earth's Moon (from *Collins et al.,* 2005; *Righter et al.,* 2005), HED meteorites (from *Stolper,* 1977; *Righter et al.,* 1990), ordinary chondrites, acapulcoites, lodranites, winonaites, and IIE, IIICD, and IAB irons (all from *Righter and Drake,* 1996), and aubrites and EH and EL chondrites (from *Fogel et al.,* 1989; *Casanova et al.,* 1993). All referenced to the FMQ buffer of *O'Neill* (1987). The iron-wüstite buffer (IW) is approximately 3.5 log units below FMQ.

Equilibrium between graphite, CO, and CO_2 buffers f_{O_2} as discussed by *Sato* (1978). At low pressures, such f_{O_2} are lower than the IW buffer, i.e., in an Fe-metal stability field. However, the volume change of this reaction is quite large, such that at higher pressures, the f_{O_2} defined by C-CO-CO_2 is higher than the stability field of Fe metal. In fact, olivine-pyroxene-graphite-gas equilibria in ureilites allow the equilibrium pressure to be defined, since ureilites contain graphite (e.g., *Goodrich et al.,* 1987; *Walker and Grove,* 1993). At even higher pressures, C-CO-CO_2 define f_{O_2} within the range of the relatively oxidized terrestrial mantle (*LaTourette and Holloway,* 1994). Because graphite is possibly present during the early stages of terrestrial planet formation, this buffer may control f_{O_2} and thus have an effect on the bulk chemistry of chondritic materials involved with planet formation (Fig. 13).

The f_{O_2} range over which Fe metal becomes unstable is relatively narrow (*Arculus et al.,* 1990). Reduction or oxidation of Fe metal in chondrite material can occur over 2 log f_{O_2} units, a range easily covered by pressure effects on the C-CO-CO_2 surface, or even produced by equilibria involving H_2-H_2O (see below). Such oxidation will affect not only the amount of Fe metal stable during the oxidation (or reduction), but will also affect the distribution of siderophile elements such as Ni, Co, and W. Equilibria such as these can control the mass of a planetary core, as well as its overall composition (e.g., O-, H- or C-bearing), due to equilibria such as $Fe + CO_2 = FeO + CO$.

Differences in f_{O_2} may also produce very different differentiated meteorites that are otherwise derived from similar bulk compositions. For example, it has been shown that partial melting of chondritic material at IW–1 can produce eucritic liquids, whereas melting of the same material at IW + 2 results in angritic liquids (*Jurewicz et al.,* 1991).

Finally, increasing temperature also causes reduction. For example, in an FeO-bearing system at QFM, if f_{O_2} is held constant and temperature is increased, FeO will eventually be converted to Fe metal. Similarly, but at higher f_{O_2}, Fe_2O_3 will eventually form Fe_3O_4, or FeO. Equally, a Fe_2O_3bearing silicate melt will eventually become a FeO-bearing silicate melt with heating at a constant f_{O_2} (*Kress and Carmichael,* 1991). This simple relation could possibly be related to the presence of larger metallic cores in the hotter inner part of the solar system (i.e., Mercury with large core) compared to the cooler, outer region (i.e., Mars with small core).

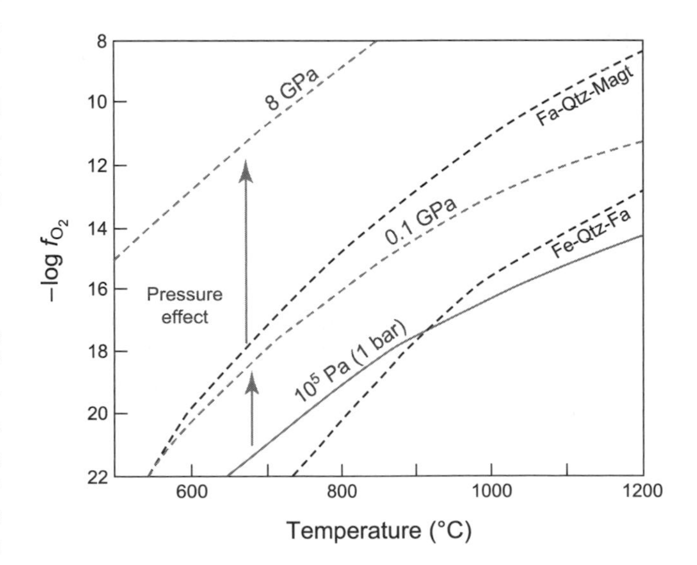

Fig. 13. Change in f_{O_2} of C-CO-CO_2 buffer with pressure relative to the FMQ buffer; gray curves shown for 1 bar, 1 kb (or 0.1 GPa), and 80 kb (or 8 GPa). Shown for reference are the FMQ and IQF buffers (dark dashed lines). Data are from *Arculus et al.* (1990); *O'Neill* (1987); *LaTourrette and Holloway* (1994).

4.2. Pressure and Temperature: Melting and Fractionation of Phases

Pressure can affect phase equilibria of planetary materials, the compositions of phases that have solid solution, and the physical properties of minerals and melts such as density. All these effects can have significant leverage on transforming planetary materials at high pressures.

One of the most important phenomena with respect to planet formation is the density crossover between minerals and melts at interior pressures. For example, plagioclase flotation on the Moon is a now-classical concept of differentiation. Less-dense plagioclase will float on a denser ultrabasic or peridotitic magma ocean, thus leading to the anorthositic lunar highlands with positive Eu anomalies (and a corresponding negative Eu anomaly in mantle-derived lunar basalts). This example has become a textbook lesson on how geophysics and geochemistry are linked, and also how surficial products can be quite different as a result of postaccretional processes (e.g., *Warren,* 2003). Similarly, olivine and garnet will become less dense than their coexisting silicate melt at specific pressures within planetary mantles (Fig. 14). Because these pressures are strongly dependent upon FeO content of both the liquid and mineral, it is not possible to generalize about any given mineral — the crossover points will be a function of composition, which is in turn a function of pressure, temperature, and f_{O_2}, to the extent to which these three variables control composition.

Finally, a topic of continued and ongoing research is the effect of pressure on the composition of metallic liquids that form planetary cores. This topic was honed on Earth because it has been recognized for some time that the density of Earth's outer core is about 10% lower than that expected for a pure FeNi alloy. The identity of the light element in the core remains uncertain, but the list of possibilities includes Si, S, C, H, P, or O (see recent review of this topic by *Hillgren et al.,* 2000). Most of these elements will not have a large effect on the surficial composition of planetary materials. But there are two notable exceptions. First, if a significant (>2%) amount of Si were partitioned into Earth's outer core, the Mg/Si ratio of the mantle would increase. This very mechanism has been called on to explain the superchondritic Mg/Si of the terrestrial upper mantle (*Allegre et al.,* 1995; *Gessmann et al.,* 2001; *Wade and Wood,* 2001). Second, if a significant amount of O was partitioned into the core of Earth (as FeO), the mantle FeO content would be lowered accordingly (*Rubie et al.,* 2004; *Ringwood,* 1979).

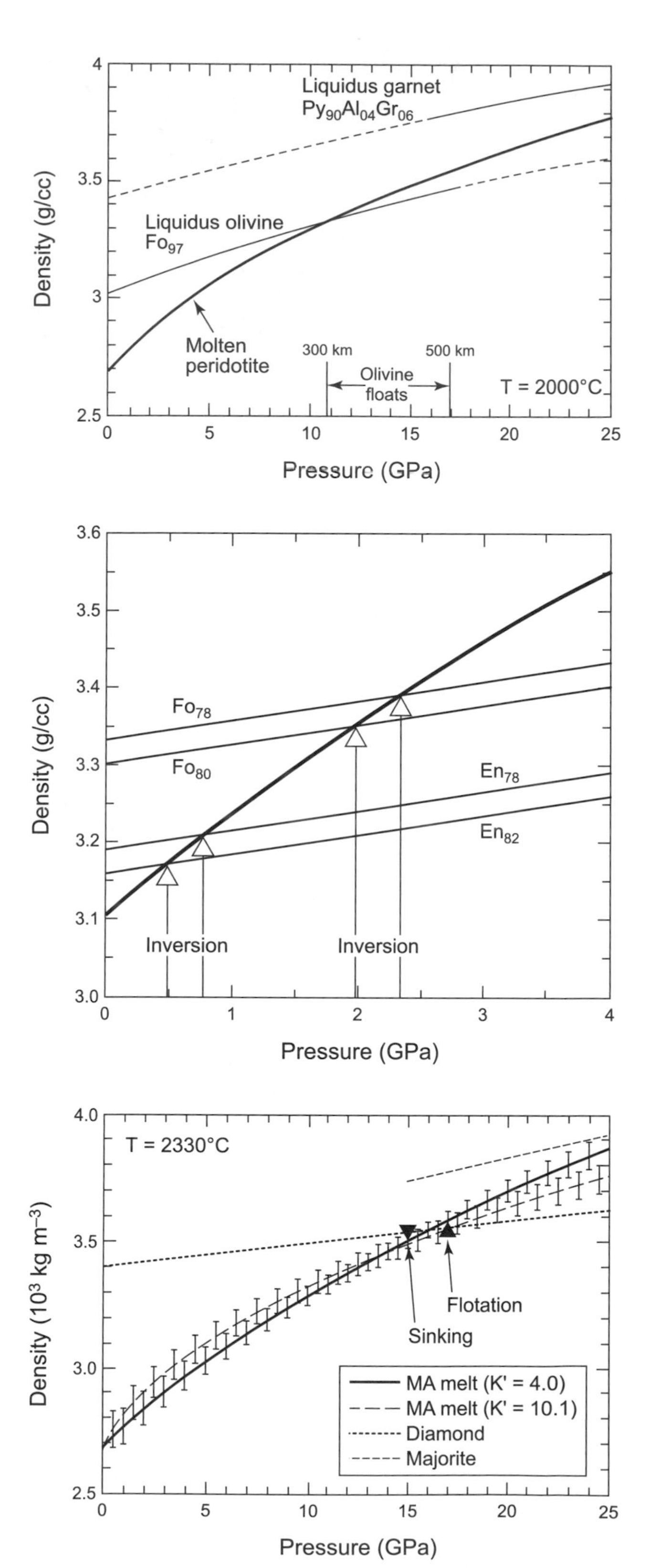

Fig. 14. Density crossovers for terrestrial peridotite liquid and olivine (top) (*Agee,* 1998); Apollo 14 black glass, olivine, and orthopyroxene (middle) (*Circone and Agee,* 1996); and synthetic martian mantle melt, diamond, and majoritic garnet (bottom) (*Ohtani et al.,* 1995).

4.3. Water

It has been known for nearly 40 years that water is very soluble in silicate melts, with as much as 15 wt% water entering a basaltic liquid in a system saturated with a pure H_2O fluid at 1.5 GPa (*Hamilton et al.,* 1964). Water solubility as a function of pressure and temperature is well

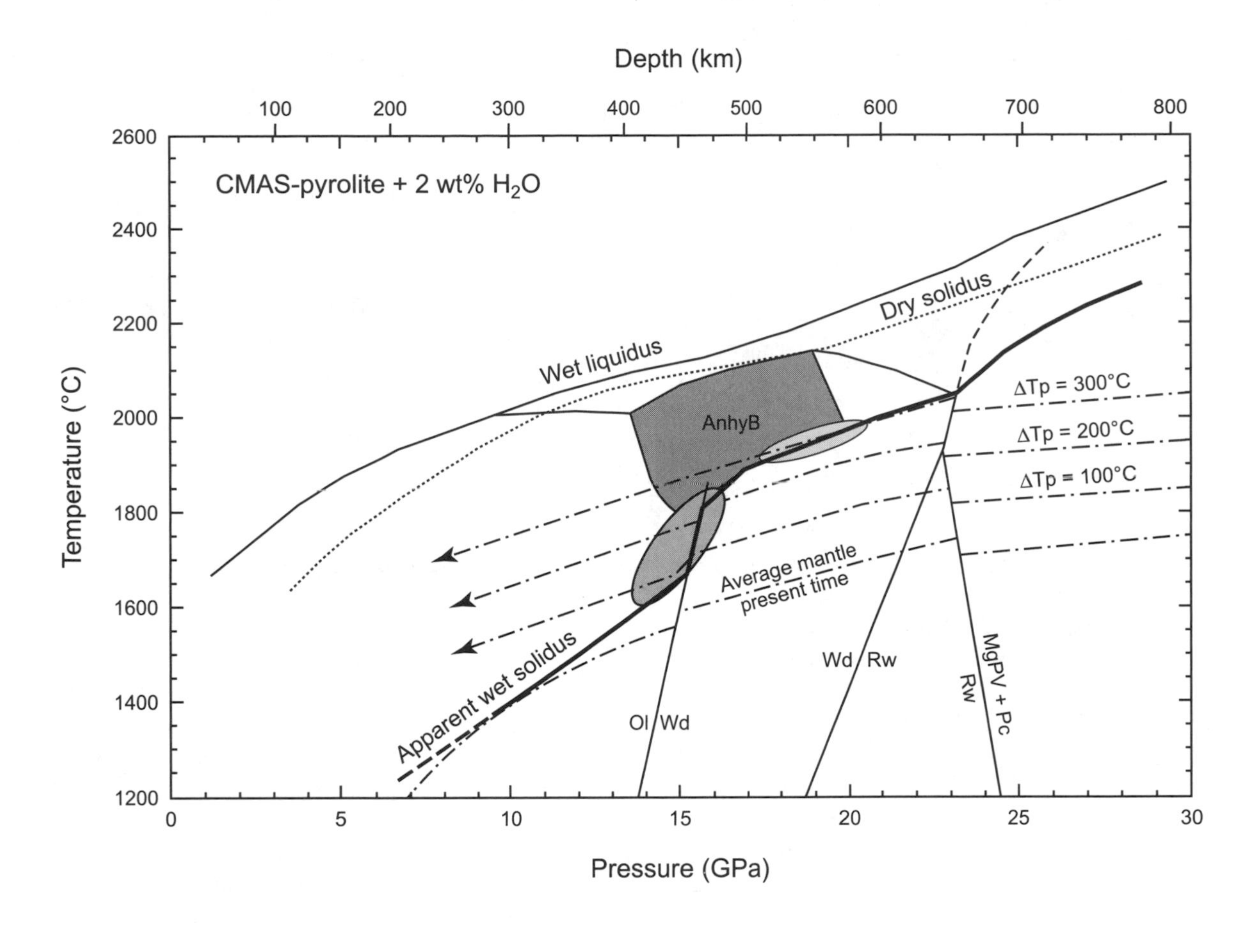

Fig. 15. The effect of water on phase equilibria in the CaO-MgO-Al_2O_3-SiO_2 system up to 25 GPa (from *Litasov and Ohtani,* 2002). Compare dry solidus to wet solidus, indicating >300°C difference up to 20–25 GPa.

known now for a large range of natural liquids, and much research has focused on systems relevant to terrestrial magmatic systems (basalt, andesite, dacite, rhyolite), both in water-saturated and mixed speciation (e.g., CO_2-H_2O) conditions (e.g., *Moore et al.,* 1995, 1998).

The effect of water on phase relations of magmatic systems is significant. The olivine phase field is expanded in basaltic systems by the addition of even a small amount of water (e.g., *Kushiro,* 1969), and plagioclase is generally suppressed by dissolution of water in basic magmas (*Sisson and Grove,* 1993). Large changes have been documented in more basic systems as well. The liquidus in the enstatite system is 600°C lower in the water-saturated system (*Inoue,* 1994) than it is in the anhydrous system (*Gasparik,* 1993). Similarly, the forsterite liquidus has a huge depression at water contents of 20 wt% compared to dry conditions (e.g., *Abe et al.,* 2000). Experiments on hydrous peridotitic materials have shown not only a large reduction in liquidus and solidus temperatures (*Inoue and Sawamoto,* 1992; *Kawamoto and Holloway,* 1997), but also a change in the liquidus phases from olivine in a dry system, to orthopyroxene and garnet in a hydrous system (*Inoue and Sawamoto,* 1992). A direct comparison of the effect of 2 wt% water on the phase relations of the CaO-MgO-Al_2O_3-SiO_2 system up to 25 GPa shows that it will have a very large effect, reducing the solidus and liquidus by hundreds of degrees (Fig. 15).

Finally, water can react with Fe metal to form FeO (dissolved in solid or liquid silicate) or FeH (dissolved in metal). If water is available during accretion, metal-bearing materials could become oxidized

$$Fe + H_2O = FeO + H_2$$

thus changing the bulk composition of the silicate materials that will ultimately become part of the mantle. Alternatively, Fe-metal-bearing materials could react with water to form FeH (e.g., *Okuchi,* 1997; *Abe et al.,* 2000)

$$4Fe + 2H_2O = 4FeH + O_2$$

in which case the metallic core may contain some H as a light element.

With respect to the first reaction, an illustration of various kinds of water sources and how much would be required to raise the FeO content of a mantle is constructed in Fig. 16. It would require 3,000,000 Borrelly-sized com-

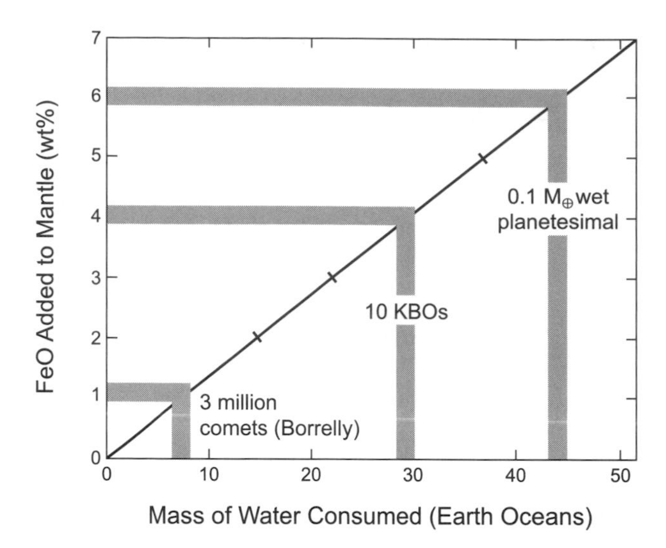

Fig. 16. Mass of water (in Earth oceans) consumed vs. FeO produced in equilibrium Fe + H_2O = FeO + H_2, assuming all water is consumed in the reaction for simplicity (to illustrate the amount of water required to do this).

ets to oxidize Earth's mantle by 1 wt% FeO. It would take 10 large (1000 km) KBOs to oxidize the mantle to 4 wt% FeO higher than its standard value. And finally, it would take one 0.1 $M_\oplus$ wet planetesimal (e.g., *Morbidelli et al.,* 2000) to oxidize the mantle of an Earth-sized planet to 6 wt% higher. Thus water, FeO, the Mg# of the mantle, and the size of a metallic core are all potentially linked in a planet that accretes wet.

4.4. Fractionation

Lava flows can be used as probes of a planet's interior, assuming that they have not fractionated dramatically as they migrated to the surface. It turns out that FeO is not greatly fractionated during partial melting and modest fractional crystallization, although the Fe/Mg ratio varies dramatically because of Mg fractionation. *Robinson and Taylor* (2001) assembled experimental and observational data to show this quantitatively. For example, the average FeO concentration of terrestrial mid-ocean-ridge basalts (MORB) divided by the composition of the primitive terrestrial mantle is only 1.3. Thus, the FeO concentration of a lava is only about 30% higher than the concentration of its mantle source region. This led *Robinson and Taylor* (2001) to conclude that Mercury contained 2–3 wt% FeO in its interior.

Magmatic fractionation can also have a large effect on the surficial properties of planetary materials. For example, it is well known that Ni, Cr, Co, and Mg are all very compatible in olivine and pyroxene or spinel during fractional crystallization of a basaltic liquid. In addition, it is also known that elements such as La, Ce, Sm (LREE), Gd, Ho, Yb (HREE), Hf, Ta, Nb (HFSE), and Th are incompatible in these same phases during fractional crystallization of a basaltic liquid. As a result, a basaltic liquid that has undergone extensive fractional crystallization will have much higher concentrations of incompatible elements such as Sm and Th, and lower Ni and MgO than a primary mantle melt. Thorium and Sm concentration data from orbital gamma-ray spectroscopy should thus be interpreted in this context, and attempts should be made to be correlate such concentrations with MgO to assess whether high values are due to inherent planetary properties or simply to fractionation.

In summary, C-H-O fluids (and specifically f_{O_2} and f_{H_2O}), pressure, temperature, and fractionation all affect the potential building blocks for planets and must be well understood in order to assess from what materials a planet has been constructed.

5. TERRESTRIAL PLANETS: MADE FROM EXTANT MATERIALS?

5.1. Earth

5.1.1. Composed of known materials? The composition of Earth's primitive upper mantle (PUM — the Earth's mantle immediately after core formation) is distinct from that of any kind of extant meteorite, barely overlapping CV3s (Fig. 6). The higher Mg/Si ratio of Earth's mantle relative to chondrites led *Hewins and Herzberg* (1996) to suggest that Earth formed from Mg-rich type I chondrules, which are depleted in moderately volatile elements. Chondrites rich in such chondrules with only minor matrix provide a plausible precursor material, especially as olivine-rich type I chondrules in Semarkona tend to be ~0.5 m.y. older than pyroxene-rich chondrules (*Kita et al.,* 2005). However, as has been shown in the previous section, it is unlikely that such low-pressure and low-temperature (relative to Earth's interior during differentiation) building blocks would survive the many changes in T, P, f_{O_2}, and volatile contents to be recognizable at the end of accretion.

A variety of other studies have proposed that Earth is made of extant meteorites. For example, *Wänke* (1981), *O'Neill* (1991), and *Newsom and Sims* (1991) all argue that Earth is made from a mixture of enstatite chondrites and carbonaceous chondrites (component A, reduced, and component B, oxidized). Then a chondritic late veneer is added (<1%) that accounts for the HSE in the primitive mantle. However, Earth's PUM has a significantly higher $^{187}Os/^{188}Os$ ratio than carbonaceous chondrites, effectively ruling them out as the source of the "late veneer" (Fig. 8). The PUM $^{187}Os/^{188}Os$ ratio overlaps anhydrous ordinary chondrites and is distinctly higher than anhydrous enstatite chondrites. In summary, these models cannot be reconciled with the Os-isotopic data, and also can only be satisfied if the late veneer is a reduced (enstatite or ordinary) meteorite, which is the opposite of the originally proposed oxidized veneer (*Chou,* 1978).

Palme and O'Neill (2004) argue that Earth has many similarities to carbonaceous chondrites, such as refractory lithophile elements and volatile-element depletions. Similarly, *Allegre et al.* (2001) demonstrate that volatile-element depletions in Earth could be controlled by a range of carbonaceous chondrites from CI (highly refractory), to CM (moderately refractory), to CV-like (volatile). Carbonaceous chondrites could also provide water for Earth of the right D/H ratio. The D/H ratio of Vienna standard mean ocean water (VSMOW) on Earth is 150×10^{-6}, similar to the average for carbonaceous chondrites (150×10^{-6}). However, either of these ideas is inconsistent with terrestrial O-isotopic data, Os-isotopic data, and noble gas data, which are distinctly dissimilar to carbonaceous chondrites (Table 2). Earth's noble gases, specifically ^{84}Kr/^{130}Xe ratios in Earth's atmosphere, are distinct from both CI carbonaceous chondrites and enstatite chondrites. These differences may have arisen from some secondary process. Not surprisingly, then, the best explanations for noble gas concentrations and origins in Earth involve two-stage models where the first stage is hydrodynamic escape of a primary hydrogen-rich, isotopically solar atmosphere, followed by degassing of Kr, Ar, and Ne from deep within the planet. Subsequently, it appears unlikely that Earth's Kr/Xe or other noble gas ratios are derived from carbonaceous chondrites (e..g., *Pepin,* 1991). *Pepin*'s (1991) modeling of noble gases coupled with C and N compounds resulted in the conclusion that a late accretional veneer similar in composition to enstatite chondrites could satisfy the C, N, and noble gas in surficial reservoirs. More recent modeling of N and Ne abundances in the terrestrial mantle by *Owen and Bar-Nun* (2000) suggest that icy planetesimals may have contributed some of terrestrial volatiles, including the light N component that has been difficult to explain. On the other hand, the extent to which solar components could have been supplied by meteoritic building blocks has recently received attention, but a full discussion of this possibility is beyond the scope of this paper (*Wieler et al.,* 2006; *Becker et al.,* 2003).

Motivated by O-isotopic and bulk compositional data, *Javoy* (1995) argued for an enstatite chondrite bulk Earth composition. This hypothesis satisfies the O-isotopic constraints, and is consistent with Os-isotopic data for the terrestrial mantle. However, there are a number of problems with this approach — E chondrites contain more Fe than is observed in Earth's core. Oxidation of this Fe to FeO would result in a lower mantle with a higher FeO content than the upper mantle. This is not consistent with the bulk of geophysical constraints for the upper and lower mantle.

TABLE 2. Ratios of ^{84}Kr/^{130}Xe in various solar system objects (*Pepin,* 1991); note that extant meteorites have ^{84}Kr/^{130}Xe ratios that are very different from Earth's atmosphere.

Object	^{84}Kr/^{130}Xe Ratio
Solar system	0.7×10^{-2}
CI carbonaceous chondrites	5.1×10^{-2}
Enstatite chondrites	(5–23)
Earth atmosphere	1.2×10^{-2}

Finally, attempts have been made to explain the source of Earth's water through known materials (some nonmeteoritic), such as comets (e.g., *Delsemme,* 1997). There are caveats to using cometary D/H ratios to limit the delivery of cometary water to Earth. First, we do not know that Oort cloud comets Halley, Hale-Bopp, and Hyakutake are representative of all comets. Certainly they are unlikely to be representative of Kuiper belt objects, the source of Jupiter-family comets, as Oort cloud comets formed in the region of the giant planets and were ejected while Kuiper belt objects have always resided beyond the orbit of Neptune. Second, D/H measurements are not made of the solid nucleus, but of gases emitted during sublimation. Differential diffusion and sublimation of HDO and H_2O may make such measurements unrepresentative of the bulk comet, as the D/H ratio would be expected to rise in diffusion and sublimation, as has been confirmed in preliminary laboratory experiments (*Weirich et al.,* 2004). Intriguing experiments on mixtures of water ice and TiO_2 grains by *Moores et al.* (2005) suggest that D/H ratios could be lowered in sublimates. Lower bulk D/H ratios would increase the allowable amount of cometary water. Third, the D/H ratio of organics and hydrated silicates in comets is unknown, although that situation may be rectified by analysis of samples returned by the Stardust mission. Note, however, that D/H ratios up to 50× VSMOW have been measured in some chondritic porous interplanetary dust particles (CP-IDPs) that may have cometary origins (*Messenger,* 2000), and higher aggregate D/H ratios of comets would decrease the allowable cometary contribution to Earth's water.

5.1.2. Composed of known, but transformed, materials? There are several models that explicitly evaluate the possibility that Earth accreted from known materials, but were then transformed at higher temperature and pressure conditions. For example, *Ringwood* (1979) argued that Earth's mantle has a low FeO content due to dissolution of FeO (or O) into the core at high pressures. Although this has shown to be true to a limited extent, it is difficult to partition enough O into the core the explain Earth's composition by starting with a known chondritic material (e.g., *Rubie et al.,* 2004).

Several suggestions have been made to explain the elevated Mg/Si and Al/Si ratios in Earth's PUM relative to undifferentiated meteorites. The most prominent of these include (1) sequestering Si in the core of Earth, raising the Mg/Si and Al/Si ratios in the silicate mantle (*Wänke,* 1981, *Allegre et al.,* 1995); (2) appealing to the possibility that the lower mantle of Earth has a different composition from the upper mantle, i.e., PUM is not representative of the bulk mantle (*Anderson,* 1989; *Javoy,* 1995); or (3) flotation or sinking of olivine or perovskite, respectively (e.g., *Agee and Walker,* 1988; *McFarlane et al.,* 1994).

While Si can be extracted into metallic cores under very reducing conditions, it is difficult to partition enough Si into core metal at deep mantle conditions, or those at the base of a high-pressure/high-temperature terrestrial magma ocean, to account for the light element in the core, or the superchondritic Mg/Si ratio. At conditions of a deep mid-mantle magma ocean, the metal/silicate partition coefficient for Si is approximately 10^{-3}–10^{-2} (*Wade and Wood,* 2001). Partitioning under very reducing conditions is also inconsistent with the primitive upper mantle abundances of first transition series elements such as V, Cr, and Mn, as they would be depleted far below observed abundances if conditions were sufficiently reducing to allow Si to dissolve in Earth's core (*Drake et al.,* 1989). Only for metal-silicate equilibria at much higher pressures and temperatures, 50 GPa and 3000 K, would the abundances of Si, V, Mn, and V be satisfied (*Wade and Wood,* 2005; *Chabot and Agee,* 2003), but other elements such as Mo, W, and P have not yet been adequately evaluated at these conditions.

The question of whether the composition of the lower mantle is different from the upper mantle is more complicated. If the lower mantle, constituting 68% by volume of the mantle, has a different bulk composition than PUM, then arguments about the bulk composition of the silicate Earth based on PUM are flawed. While we have no direct samples of the lower mantle, geophysical measurements provide indirect clues to its composition (e.g., *Stixrude et al.,* 1992). Although there is no firm consensus, *Bina* (2004) concludes that high-pressure mineral physics constraints, including elasticity and thermal conductivity data, are most consistent with the lower and upper mantle having the same composition (see also discussion in *Drake and Righter,* 2002). Furthermore, *Boyet and Carlson* (2005) argue on the basis of ^{146}Sm-^{142}Nd systematics that most of Earth's mantle has the composition of the MORB source.

Agee and Walker (1988, 1993) proposed making Earth from CV3 chondrites and then floating olivine or removing perovskite into the upper or lower mantle, respectively. Olivine flotation (Fo_{90}) occurs at pressures of approximately 8.0 GPa, and has been proposed as a mechanism for giving the upper mantle of Earth a superchondritic Mg/Si ratio (*Agee and Walker,* 1993). It has also been proposed as a possibility in the early martian mantle (*Righter and Drake,* 1996). Although olivine flotation is most likely inconsistent with our knowledge of the Ni, Co, Ir, Pd, Re, and Os contents of the upper mantle (*McFarlane and Drake,* 1990; *Righter,* 2005), it remains an important concept and could undoubtedly be an important postaccretional process altering planetary surficial compositions.

5.1.3. Composed of unknown materials? Difficulties in explaining the bulk composition of Earth from either a known or transformed meteoritic material have led some to conclude that Earth may have accreted from materials that no longer exist and are unrepresented in the collections (e.g. *Drake,* 1984; *Drake and Righter,* 2002). A cursory examination of the O-isotopic composition of meteorites indicates that only the enstatite meteorites (chondrites and achondrites) share a common O reservoir with Earth and the Moon (Fig. 7). This link is strengthened by similarities of Cr isotopes. There appears to be a systematic variation in the ^{53}Cr/^{52}Cr ratio, usually expressed in units of ε^{53}Cr or simply $\varepsilon(53)$ with heliocentric distance (Fig. 10), unlike any other isotopic system such as O (*Lugmair and Shukolyukov,* 1998). Earth is by definition at $\varepsilon(53) = 0$. The identical value for the Moon implies a close connection in terrestrial and lunar origin, a conclusion supported by O-isotope studies of *Wiechert et al.* (2001). There is a linear trend of increasing $\varepsilon(53)$ through Mars to the asteroid belt. Such systematic variation, unseen in any other isotopic system, most probably reflects a primordial gradient in the Mn/Cr ratio, with more volatile Mn increasing in composition with distance from the Sun. *Birck et al.* (1999) argue that the solar system started with homogenous ^{53}Mn and ^{53}Cr, with differences in $\varepsilon(53)$ reflecting the timing and relative volatility of Mn and Cr during isolation from the solar nebula. A correlation with other volatile-refractory parent-daughter pairs would support this alternative view. It is possible that Earth accreted from a larger reservoir of material that no longer exists.

Extensive computational modeling by *Burbine and O'Brien* (2004) has also demonstrated that there is no known combination of building blocks that can explain Earth's bulk composition if PUM is representative of the bulk silicate Earth and Si has not been sequestered in the core or added to the upper mantle. Additionally, *Zolensky* (1998) and *Zolensky et al.* (2006) have shown that the flux of material to Earth's surface changes over time such that material accreting to Earth today may not be entirely representative of that available in the early solar system. Thus Earth could be made from material no longer extant, i.e., it has all been swept up by Earth (and the Moon-forming impactor — see section 5.2).

5.1.4. Moon. We have no direct samples of the lunar mantle, but our detailed models of its bulk composition are based on basaltic samples from the Apollo, Luna, and meteorite collections (e.g., Fig. 6), as well as geophysical data. The distinguishing features of the Moon are its small metallic core (<2 mass%), high mantle FeO content relative to Earth and low total Fe content, and volatile-element depletion. Some of the aspects of the Moon's composition can be easily explained by an impact origin, such as its small metallic core.

The Moon is very similar to Earth in a number of compositional ways (e.g., *Jones and Palme,* 2000). Since impact modeling predicts that Earth's Moon formed in a giant impact and the Moon is largely derived from the impactor (*Cameron,* 2000; *Canup and Asphaug,* 2001), it is remarkable that the O- and Cr-isotopic compositions of Earth and the Moon are so similar (Δ^{17}O for the Moon is <15% of the Mars value). This could imply that at least two large bodies, proto-Earth and the impactor, had identical O- and Cr-isotopic compositions (*Wiechert et al,* 2001; *Lugmair and Shukolyukov,* 1998). If the Moon is made from the mantle of a differentiated impactor, either both Earth and the impactor formed in the same region of space, or both Earth and the Moon were chemically and isotopically homogenized in the impact event(s). Alternatively, Earth and the Moon were isotopically equilibrated during the impact (*Pahlevan and Stevenson,* 2005).

There has been very limited work on Os-isotopic compositions of lunar materials, because the concentrations of Re and Os in pristine materials are so low that the measurements are very difficult. Analyses of dunite 72415 (*Walker et al.,* 2004) yield results indicating that the Re/Os ratio of the lunar mantle and its corresponding $^{187}Os/^{188}Os$ ratio are most likely superchondritic. This is, however, in contrast to measurements by *Birck and Allegre* (1994) on two lunar basalts showing nearly chondritic present-day $^{187}Os/^{188}Os$ ratios. A near-chondritic lunar mantle source is not favored or ruled out by this limited dataset, and resolution requires additional measurements.

The volatile content of the Moon is low (e.g., *Taylor et al.,* 1995; *Humayun and Clayton,* 1995), making a discussion of its D/H and noble gas composition difficult. However, this highlights the issue of the Moon's volatile element depletion — its origin is not well understood and remains one of the largest outstanding questions in lunar science.

In summary, the Moon may have been derived from the mantle of the proto-Earth, from a differentiated protoplanet [Theia (*Halliday and Porcelli,* 2001)] of unknown origin and composition, or a mixture of the two (*Canup and Asphaug,* 2001). Because of this uncertainty, compositional information provided by the Moon must be interpreted carefully.

5.2. Mars

5.2.1. Composed of known materials? In terms of major elements, Mars is different from Earth in that the array of martian meteorites and Mars Pathfinder analyses fall beneath carbonaceous chondrites and the solar value. Bulk martian Mg/Si and Al/Si ratios may plausibly be near the H and L chondrites (Fig. 6). Later processes acting on these bulk compositions, such as melting and fractionation, may have produced a linear array for martian materials, similar to that for Earth (e.g., *McSween,* 2002; *Dreibus et al.,* 1998). Interestingly, most Mars Spirit and Opportunity rocks do not plot on these "primitive" trends, but plot on trends expected if fractionation of aluminous phases (feldspar, clays) was involved. This similarity to a known meteorite group, however, precludes the need to call upon secondary processing, such as Si in the core or garnet or perovskite fractionation, for the martian mantle. Mars is distinct from the Earth-Moon system and from all chondrite and achondrite parent bodies in that its O-isotopic composition lies above the terrestrial fractionation line by 0.3 $\Delta^{17}O$ units (Fig. 7), and does not fall near any other known chondrite group, but can be satisfied by a mixture of known groups.

Wänke and Dreibus (1988) applied their two-component model (component A, reduced, and component B, oxidized) to Mars and found that many geochemical features of the martian mantle are consistent with a Mars bulk composition of 40% component B (CI-like) and 60% component A (enstatite chondrite-like), with a S-bearing metallic core of ~22 mass% of the planet. However, contrary to their models for Earth, they argued that these components may have accreted simultaneously and equilibrated in a homogeneous accretion scenario (see also *Righter et al.,* 1998; *Treiman et al.,* 1986, 1987). Similarly, but using O isotopes to guide specific chondritic building blocks, *Lodders and Fegley* (1997) defined a martian bulk composition of 85% H chondrite, 11% CV chondrite, and 4% CI chondrite. Although this matches O isotopes and is consistent with acceptable core masses and mantle Mg#s, it fell short of explaining the lower alkali and halogen contents of martian meteorites.

Carrying out wetting experiments in L5 chondrites, *Agee and Draper* (2004) show that the high Ca/Al of shergottites can be explained by high-pressure melting (3–5 GPa). However, their liquids contain between 22 and 30 wt% FeO, much higher than that measured in shergottites (15–29 wt%). A solution to this problem may come from studies with lower FeO compositions that also satisfy density and moment-of-inertia data for Mars (e.g., *Bertka and Fei,* 1998a,b). *Bertka and Fei* (1998a,b) show that the moment-of-inertia data for Mars (Pathfinder) are inconsistent with a CI chondritic martian bulk composition, as proposed by *Wänke and Dreibus* (1988). Even attempts to fit the data with endmember core compositions such as FeH, Fe_3O_4, FeS, and Fe_7C_3 result in mantle Fe/Si ratios that are unlike CI — Mg-Fe-Si and chondrite types. A detailed analysis using this same approach for bulk compositions that are ordinary-chondrite-rich has not yet been done.

The Cr-isotopic composition of Mars is clearly higher from that of Earth, probably reflecting a higher Mn/Cr ratio in the material accreted at 1.5 AU. Higher Mn is consistent with the higher volatile-element content of Mars compared to Earth. This difference, along with the differences in O isotopes and major-element abundances, highlights the argument that the terrestrial planets had narrow feeding zones for the majority of their growth.

Martian volatiles could also have come from a known source. The average D/H for Mars (there is considerable spread and there is no extant water ocean) is about 300 × 10^{-6}, similar to the value for comets — 300 × 10^{-6} (Fig. 9). It is possible that the D/H ratio on Mars has been raised from its original value by differential loss of H, as is suggested by the elevated D/H ratio in the martian atmosphere (Fig. 9). However, Mars has been outgassed less than Earth because its smaller mass would lead to more rapid loss of heat. Thus, the lower martian mineral and rock D/H ratios are most likely representative of the solid undegassed planet (Fig. 9). An alternative interpretation (*Owen and Bar-Nun,* 2000) notes that the lack of exchange between the martian lithosphere and hydrosphere allows for the possibility that the D/H ratio of the unsampled martian interior could be very different from rocks, minerals, and atmosphere. Sampled rocks, minerals, and atmosphere could reflect addition of unrecycled cometary material.

The noble gas distributions on Mars are generally similar to those on Earth and can be explained by a two-stage model that involves early escape of a H-rich solar atmosphere, followed by degassing of Kr, Ar, and Ne from the interior (*Pepin,* 1991; *Becker et al.,* 2003). As for Earth, the extent to which solar components could have been supplied by meteoritic building blocks has recently received attention and a full discussion of this possibility is beyond the scope of this paper (see *Wieler et al.,* 2006). However, as opposed to Earth, Xe-isotopic data indicate that the degassed material from the interior could have been CI-chondrite like, with the additional possibility that radiogenic ^{129}Xe was added by decay of extinct ^{129}I in surficial and crustal processes (*Musselwhite et al.,* 1991; *Musselwhite and Drake,* 2000).

In summary, there is a general consensus that Mars can be made from a mixture of existing meteorite types, with possible input from comets. Remaining issues are whether specific ordinary chondrite groups can provide the right amount of FeO for the mantle and still be consistent with the moment-of-inertia data and core mass.

5.2.2. Composed of known, but transformed, materials? No specific models of this type have been proposed for Mars, most likely because so many of its properties can be explained without appealing to major transformations. For example, pressure at the base of the martian mantle is not high enough to stabilize perovskite, which is often called upon in terrestrial models for promoting mantle differentiation. In addition, the water content of Mars is unlikely to be high enough to have had a major effect on Fe-FeO equilibria. And the f_{O_2} of martian materials exhibit a lower range than those for Earth, thus suggesting f_{O_2} changes have been less important in defining the bulk composition.

5.2.3. Composed of unknown materials? Measurements of Os isotopes on martian meteorites show that the ^{187}Os/^{188}Os of the primitive martian upper mantle (PMUM) is 0.132 (±0.001), outside the range of any known chondritic value [highest known is 0.1305 (*Brandon et al.,* 2005)]. As a result, the martian mantle is nonchondritic in its HSE content, which is also reflected in the concentrations of HSEs in martian meteorites (e.g., *Kong et al.,* 1999; *Warren,* 1999; *Jones et al.,* 2002). If Mars received a late accretional veneer as some have argued for Earth, it may have been provided by a chondrite group that is not currently represented in our meteorite collection.

5.3. Asteroid 4 Vesta

The howardite-eucrite-diogenite (HED) clan of differentiated meteorites is thought to be from asteroid 4 Vesta (e.g., *McCord et al.,* 1970; *Drake,* 2002). They provide the opportunity to learn more about the bulk composition of a large, differentiated asteroid. *Dreibus and Wänke* (1980) proposed a MgO-rich bulk composition based on geochemical modeling of a mixture of eucrite and diogenite materials. Attempts to determine how large a metallic core could have produced the significant siderophile-element depletions in eucrites (e.g., *Newsom and Drake,* 1982) resulted in estimates of core mass of approximately 30–50%. Coupling of core formation with basaltic eucrite liquid formation led *Righter and Drake* (1997b) and *Ruzicka et al.* (1997) to propose that Vesta has a 15–20 mass% core. In particular, *Righter and Drake* (1997b) chose the modeling results of *Boesenberg and Delaney* (1997) to constrain bulk compositions. This combined approach led to a complete model for the origin of HED meteorites by differentiation of a mixture of CM and L chondrites that satisfied O isotopes, Fe/Mn, and siderophile and lithophile trace-element data (*Righter and Drake,* 1997b). This is another case where the possible building blocks are represented in our collection, and the planetary body formed at low pressure, f_{O_2}, and f_{H_2O}. However, one drawback to the model is its inability to explain the volatile element depletions in HEDs if they truly had CM/L bulk composition.

5.4. Mercury

Not much is known empirically about the bulk chemical composition of Mercury. Because we have no samples of Mercury, we have no constraints on its O-, Os-, or Cr-isotopic compositions. We emphasize the three compositional features that are reasonably well determined: a large metallic core, a low FeO content in the mercurian crust, and the presence of at least a little Na. This skimpy list gives us surprisingly useful information.

The mean density of Mercury is 5.43 ± 0.01 g/cm^3 (*Anderson et al.,* 1987), and has been interpreted to indicate a large metallic FeNi core making up ~70% of the planet's mass and ~40% of its volume (assuming a silicate density of about 3 g/cm^3). If this is correct, Mercury has a very high Fe/Si ratio. However, as *Goettel* (1988) points out, we have no direct evidence that the core is composed of Fe and Ni only. If the core is made of pure FeS, it would make up an even larger percentage of the planet.

A variety of spectral observations using Earth-based telescopes show that the surface of Mercury is low in FeO + TiO_2 (*Vilas,* 1988; *Sprague et al.,* 1994; *Jeanloz et al.,* 1995). *Robinson and Lucey* (1997) recalibrated the Mariner 10 data for Mercury and concluded that there were compositional heterogeneities in the crust, although all of it is low in FeO. All these constraints suggest that the crust and perhaps mantle of Mercury contain a very low FeO content, near 3 wt%.

We have no knowledge of noble gases or D/H ratios for Mercury, but there is some information about volatiles from a tenuous atmosphere that contains Na and K (*Potter and Morgan,* 1985, 1986). These elements are derived from the surface by a variety of loss mechanisms, apparently dominated by sputtering (e.g., *Killen and Ip,* 1999). Although it is not possible to determine the Na and K abundances on the surface from their abundances in the atmosphere, their presence shows that some Na and K must be present, which is important in assessing the volatile inventory of the planet. Similarly, *Sprague et al.* (1995) suggested on the basis of midinfrared spectral observations that the surface of Mercury is relatively rich in Fe sulfides. They also suggest that the radar anomalies in polar regions, which are generally attributed to cometary water ice (*Slade et al.,* 1992; *Harmon and Slade,* 1992), might be caused by deposits of elemental S in cold traps.

We do not know anything about the differentiation history of Mercury, ignorance that currently limits our understanding of its interior. Did Mercury have a magma ocean? Or did its ancient crust form by serial magmatism early in its history? It is thus difficult to ascertain whether Mercury is made of known building blocks. We hope that data from

NASA's MESSENGER and ESA's BepiColumbo missions will help to answer these questions.

5.6. Venus

As with Mercury, because we have no samples of Venus, we have no constraints on its O-, Os-, or Cr-isotopic composition. However, we do have data for Venus from the Soviet-landed spacecraft Veneras 13 and 14 and Vega 2. While we have no direct sample of the Venus mantle, Venus crustal basalts fall on the trend of lunar picritic glasses and terrestrial MORB, suggesting that Venus may have the same bulk Mg, Al, and Si abundances as Earth (Fig. 6).

The D/H ratio of the Venus atmosphere has been measured by spacecraft and is about 100× that of Earth's atmosphere, but cannot represent the original venusian D/H ratio because of significant differential loss of H relative to D (e.g., *Hunten and Donahue,* 1976; *Donahue et al.,* 1982). The Venus greenhouse vaporized water, and solar UV at the top of the atmosphere dissociated the water. Hydrogen is lost preferentially to D by Jeans escape to space (*Hunten and Donahue,* 1976).

The venusian atmosphere is much richer in noble gases than Earth's, by as much as a factor of 70 for ^{40}Ar (*Pepin,* 1991). In addition, Xe/Kr-, Ar/Kr-, and Ne-isotopic ratios all have a solar-like signature, albeit with large error bars, suggesting that Venus retained a slightly fractionated version of a primary solar atmosphere. The much-lower-than-solar Ne and N abundances may indicate that icy planetesimals may have played an important role in providing venusian volatiles, as they may have for Earth as well. Furthermore, coupling of noble gas with C and N data indicates that the inventories of these elements are consistent with the addition of enstatite-chondrite-like material (*Pepin,* 1991), late in the hydrodynamic escape period.

6. CONCLUSIONS AND UNRESOLVED ISSUES REGARDING THE ACCRETION OF THE TERRESTRIAL PLANETS

A major deficiency in our understanding of the relationships between meteorites and planets is our inadequate understanding of the elemental and isotopic compositions of Venus, Mercury, and the Sun, as well as the inventory of volatiles available to planets during and after accretion. It is likely that some of these uncertainties will be resolved in the near future, by recent and upcoming missions such as Genesis, MESSENGER, Stardust, and BepiColumbo. Several issues, however, such as O-, Os-, and Cr-isotopic measurements, can only be resolved by sample return, and thus will have to await future mission planning.

6.1. Compositional Uncertainties in Venus, Mercury, and the Sun

6.1.1. Major elements in Venus, Mercury, and the Sun. As shown in Fig. 6, the major elements Mg, Al, and Si appear to be robust discriminators among different differentiated and undifferentiated planetary objects, provided the most "primitive" samples from differentiated objects are used in the comparison and major reservoirs are not sequestered in the interiors of differentiated planets. Primitive objects preserve their original bulk compositions and the most "primitive" samples of differentiated objects define trends with negative slopes that include their bulk compositions, i.e., they are either mantle samples or high-degree melts of the mantle. The reason is that Mg is compatible (remains behind in silicate mantles during melting) while Al is incompatible (is extracted into magmas during melting).

Measurements of the moment of inertia, topography, and gravity field of Mercury by future missions will be able to place strong constraints on the size of the core and the thickness of the crust (e.g., *Spohn et al.,* 2001).

Basaltic rocks from Venus also fall on an aluminous fractionation line in the Mg/Si vs. Al/Si diagram (Fig. 6), based on analyses by Venera and Vega spacecraft summarized in *Kargel et al.* (1993), and overlap the terrestrial MORB and lunar picritic glass data. This similarity to Earth is also seen in U/Th ratios (Fig. 11 of *Kargel et al.,* 1993), as expected of two refractory elements. More importantly, Venus data fall on the terrestrial lines or fields in volatile element/refractory element diagrams such as K/Th, K/U, and MnO/FeO (Figs. 12–14 respectively of *Kargel et al.,* 1993). Venus even falls in the terrestrial fields for FeO/MgO and Al_2O_3/MgO, although these fields have more overlap with other terrestrial planetary bodies. It seems reasonable to conclude that Venus is Earth like within analytical error for major and minor elements, perhaps suggesting mixing between 1 AU and 0.72 AU. It is unknown whether Mercury differs from Earth solely in degree of reduction and mass of the core, or formed from profoundly different material.

6.1.2. Oxygen-isotopic ratios in Venus, Mercury, and the Sun. As noted earlier, only with sample return will the O-isotopic compositions of Venus and Mercury be known with adequate precision. There is an intriguing possibility that we may soon know the O-isotopic composition of the Sun, however, as a result of the Genesis mission, which collected solar wind. Currently there are two conflicting predictions of the isotopic composition of the Sun. *Clayton* (2002) and *Yurimoto and Kuramoto* (2004) predict that the Sun should be rich in ^{16}O, near the ^{16}O-rich end of the CAI line in Fig. 7, based on considerations of photodissociation, and self-shielding in molecular clouds can yield to ^{16}O enrichment in the CO gas. In contrast, *Ireland et al.* (2005) argues that the Sun is ^{16}O-poor, based on measurements of lunar metal grains exposed to the solar wind. Samples returned by the Genesis mission should resolve this disagreement. If Venus and Mercury share the same O-isotopic reservoir as Earth and Moon (and by inference the same Mg, Al, and Si composition), the implication would be that the solar system inward of 1 AU was well-mixed, but remained heterogeneous outward from 1 AU. Such a result would provide significant constraints on the thermal history of the inner solar system.

6.1.3. Chromium-isotopic ratios in Venus, Mercury, and the Sun. It is unknown if Venus and Mercury fall on the extrapolated trend indicated by the dashed line, or whether the solar system inside 1 AU is well mixed, with Venus and Mercury plotting at $\varepsilon(53) = 0$. Clearly sample return missions to Venus and Mercury would reveal much about the chemical and isotopic structure, and hence mixing processes, in the inner solar system.

6.2. Volatile-Element Inventories

Some bodies in the inner solar system are volatile-element-depleted (e.g., the Moon, 4 Vesta, Mercury), and it is not clear if this is due to the materials from which they accreted, or to subsequent modification of those compositions [e.g., loss of lunar volatiles (e.g., *Becker et al.,* 2003)] due to a process such as impact events (e.g., *Genda and Abe,* 2005). These two endmember ideas have formed the basis of evaluations for the Moon.

On the other hand, those planets that have retained volatiles to a certain degree (e.g., Mars, Earth, Venus) could have acquired volatiles from a number of sources and by a number of processes. Many of the potential sources of volatiles are not well characterized, such as Jupiter-family comets, asteroids, and Kuiper belt objects, and better characterization by future spacecraft (and sample return) missions would be hugely beneficial. Further work on the processes affecting volatile retention in planets, such as impacts (e.g., *Pierazzo and Melosh,* 2001) and deep mantle storage (e.g., *Abe et al.,* 2000; *Williams and Hemley,* 2001), will aid our understanding of the ability of small and large planetesimals to acquire and store volatiles over geologic time.

It is clear that integration of all geochemical data will not be simple, and is an enticing challenge around the corner.

Acknowledgments. This manuscript benefited from the reviews of C. J. Allegre and C. B. Agee. Work on this chapter was supported by a NASA RTOP to K.R., and NASA grants NAG5-12795 (M.J.D.) and NAG5-11591 (E.R.D.S.).

REFERENCES

Abe Y., Ohtani E., Okuchi T., Righter K., and Drake M. J. (2000) Water in the early Earth. In *Origin of the Earth and Moon* (R. Canup and K. Righter, eds.), pp. 413–434. Univ. of Arizona, Tucson.

Agee C. B. (1998) Crystal-liquid density inversions in terrestrial and lunar magmas. *Phys. Earth Planet. Inter., 107,* 63–74.

Agee C. B. and Draper D. S. (2004) Experimental constraints on the origin of martian meteorites and the composition of the martian mantle. *Earth Planet. Sci. Lett., 224,* 415–429.

Agee C. B. and Walker D. (1988) Mass balance and phase density constraints on early differentiation of chondritic mantle. *Earth Planet. Sci. Lett., 90,* 144–156.

Agee C. B. and Walker D. (1993) Olivine flotation in mantle melt. *Earth Planet. Sci. Lett., 114,* 315–324.

Allegre C. J., Poirer J.-P., Humler E., and Hofmann A. W. (1995) The chemical composition of the Earth. *Earth Planet. Sci. Lett., 134,* 515–526.

Allegre C. J., Manhes G., and Lewin E. (2001) Chemical composition of the Earth and the volatility control on planetary genetics. *Earth Planet. Sci. Lett., 185,* 49–69.

Anderson D. L. (1989) Composition of the Earth. *Science, 243,* 367–370.

Anderson J. D., Colombo G., Espsitio P. B., Lau E. L., and Trager G. B. (1987) The mass, gravity field, and ephemeris of Mercury. *Icarus, 71,* 337–349.

Arculus R. J., Holmes R. D., Powell R., and Righter K. (1990) Metal-silicate equilibria and core formation. In *Origin of the Earth* (H. E. Newsom and J. H. Jones, eds.), pp. 251–271. Oxford Univ., New York.

Basaltic Volcanism Study Project (1981) *Basaltic Volcanism on the Terrestrial Planets.* Pergamon, New York. 1286 pp.

Becker R. H., Clayton R. N., Galimov E. M., Lammer H., Marty B., Pepin R. O., and Wieler R. (2003) Isotopic signatures of volatiles in terrestrial planets — Working group report. *Space Sci. Rev., 106,* 377–410.

Bell J. F., Davis D. R., Hartmann W. K., and Gaffey M. J. (1989) Asteroids: The big picture. In *Asteroids II* (R. P. Binzel et al., eds.), pp. 921–945. Univ. of Arizona, Tucson.

Bertka C. M. and Fei Y. (1998a) Implications of Mars Pathfinder data for the accretion history of the terrestrial planets. *Science, 281,* 1838–1840.

Bertka C. M. and Fei Y. (1998b) Density profile of an SNC model martian interior and the moment-of-inertia factor of Mars. *Earth Planet. Sci. Lett., 157,* 79–88.

Bina C. R. (2004) Seismological constraints upon mantle composition. In *Treatise on Geochemistry, Vol. 2: Mantle and Core* (R. W. Carlson, ed.), pp. 39–60. Elsevier, Oxford.

Birck J. L. and Allegre C. J. (1994) Contrasting Re/Os magmatic fractionation in planetary basalts. *Earth Planet. Sci. Lett., 124,* 139–148.

Birck J. L., Rotaru M., and Allegre C. J. (1999) ^{53}Mn-^{53}Cr evolution of the early solar system. *Geochim. Cosmochim. Acta, 63,* 4111–4117.

Bizzarro M., Baker J. A., and Haack H. (2004) Mg isotope evidence for contemporaneous formation of chondrules and refractory inclusions. *Nature, 431,* 275–278.

Boesenberg J. S. and Delaney J. S. (1997) A model composition for the basaltic achondrite planetoid. *Geochim. Cosmochim. Acta, 61,* 3205–3225.

Boss A. P. (1998) Temperatures in protoplanetary disks. *Annu. Rev. Earth Planet. Sci., 26,* 53–80.

Bottke W. F., Durda D. D., Nesvorný D., Jedicke R., Morbidelli A., Vokrouhlický D., and Levison H. (2005) The fossilized size distribution of the main asteroid belt. *Icarus, 175,* 111–140.

Boyet M. and Carlson R. W. (2005) ^{142}Nd evidence for early (>4.53 Ga) global differentiation of the silicate Earth. *Science, 309,* 576–581.

Bradley J. P. (2003) Interplanetary dust particles. In *Treatise on Geochemistry, Vol. 1: Meteorites, Comets and Planets* (A. M. Davis, ed.), pp. 689–711. Elsevier, Oxford.

Brandon A. D., Humayun M., Puchtel I. S., and Zolensky M. E. (2005) Re-Os isotopic systematics and platinum group element composition of the Tagish Lake carbonaceous chondrite. *Geochim. Cosmochim. Acta, 69,* 1619–1631.

Brearley A. J. (1993) Matrix and fine-grained rims in the unequilibrated CO3 chondrite, ALHA77307: Origins and evidence for diverse, primitive nebular dust components. *Geochim. Cosmochim. Acta, 57,* 1521–1550.

Brearley A. J. and Jones R. H. (1998) Chondritic meteorites. In *Planetary Materials* (J. J. Papike, ed.), pp. 3-1 to 3-398. Reviews in Mineralogy, Vol. 36, Mineralogical Society of America.

Burbine T. H. and O'Brien K. M. (2004) Determining the possible building blocks of the Earth and Mars. *Meteoritics & Planet. Sci., 37,* 667–680.

Burnett D. S., Barraclough B. L., Bennett R., Neugebauer M., Oldham L. P., Sasaki C. N., Sevilla D., Smith N., Stansbery E., Sweetnam D., and Wiens R. C. (2003) The Genesis Discovery mission: Return of solar matter to Earth. *Space Sci. Rev., 105,* 509–534.

Busemann H., Baur H., and Wieler R. (2003) Solar noble gases in enstatite chondrites and implications for the formation of the terrestrial planets (abstract). In *Lunar and Planetary Science XXXIV,* Abstract #1665. Lunar and Planetary Institute, Houston (CD-ROM).

Busemann H., Lorenzetti S., and Eugster O. (2004) Solar noble gases in the angrite parent body — Evidence from volcanic volatiles trapped in D'Orbigny glass (abstract). In *Lunar and Planetary Science XXXV,* Abstract #1705. Lunar and Planetary Institute, Houston (CD-ROM).

Cameron A. G. W. (1962) The formation of the Sun and planets. *Icarus, 1,* 13–69.

Cameron A. G. W. (2000) High resolution simulations of the giant impact. In *Origin of the Earth and Moon* (R. M. Canup and K. Righter, ed.), pp. 133–144. Univ. of Arizona, Tucson, Arizona.

Cameron A. G. W. and Ward W. R. (1976) The origin of the Moon (abstract). In *Lunar Science VII,* pp. 120–122. Lunar Science Institute, Houston.

Campbell A. J., Humayun M., Meibom A., Krot A. N., and Keil K. (2001) Origin of zoned metal grains in the QUE94411 chondrite. *Geochim. Cosmochim. Acta, 65,* 163–180.

Canup R. M. and Agnor C. B. (2000) Accretion of the terrestrial planets and the Earth-Moon system. In *Origin of the Earth and Moon* (R. M. Canup and K. Righter, eds.), pp. 113–132. Univ. of Arizona, Tucson.

Canup R. M. and Asphaug E. (2001) The Moon-forming impact. *Nature, 412,* 708–712.

Carmichael I. S. E. (1991) The redox state of basic and silicic magmas: A reflection of their source regions? *Contrib. Mineral. Petrol., 106,* 129–141.

Carmichael I. S. E., Turner F. J., and Verhoogen J. (1974) *Igneous Petrology.* McGraw-Hill, New York. 564 pp.

Casanova I., Keil K., and Newsom H. E. (1993) Composition of metal in aubrites: Constraints on core formation. *Geochim. Cosmochim. Acta, 57,* 675–682.

Chabot N. L. and Agee C. (2003) Core formation in the Earth and Moon: New experimental constraints from V, Cr, and Mn. *Geochim. Cosmochim. Acta, 67,* 2077–2091.

Chambers J. (2001) Making more terrestrial planets. *Icarus, 152,* 205–224.

Chambers J. E. (2006) Meteoritic diversity and planetesimal formation. In *Meteorites and the Early Solar System II* (D. S. Lauretta and H. Y. McSween Jr., eds.), this volume. Univ. of Arizona, Tucson.

Chambers J. E. and Wetherill G. W. (1998) Making the terrestrial planets: N-body integrations of planetary embryos in three dimensions. *Icarus, 136,* 304–327.

Chambers J. E. and Wetherill G. W. (2001) Planets in the asteroid belt. *Meteoritics & Planet. Sci., 36,* 381–399.

Chapman C. R. (2004) Space weathering of asteroid surfaces. *Annu. Rev. Earth Planet. Sci., 32,* 539–567.

Chou C.-L. (1978) Fractionation of siderophile elements in the Earth's upper mantle. *Proc. Lunar. Planet. Sci. Conf. 9th,* pp. 219–230.

Circone S. and Agee C. B. (1996) Compressibility of molten high-Ti mare glass: Evidence for crystal-liquid density inversions in the lunar mantle. *Geochim. Cosmochim. Acta, 60,* 2709–2720.

Clayton R. N. (1993) Oxygen isotopes in meteorites. *Annu. Rev. Earth Planet. Sci., 21,* 115–149.

Clayton R. N. (2002) Self-shielding in the solar nebula. *Nature, 415,* 860–861.

Clayton R. N. (2004) The origin of oxygen isotope variations in the early solar nebula (abstract). In *Lunar and Planetary Science XXXV,* Abstract #1682. Lunar and Planetary Institute, Houston (CD-ROM).

Clayton R. N. and Mayeda T. K. (1999) Oxygen isotope studies in carbonaceous chondrites. *Geochim. Cosmochim. Acta, 63,* 2089–2104.

Clayton R. N. and Mayeda T. K. (2001) Oxygen isotope signatures of hydration reactions in solar system materials. In *Eleventh Annual V. M.Goldschmidt Conference,* Abstract #3648. LPI Contribution No. 1088, Lunar and Planetary Institute, Houston (CD-ROM).

Clayton R. N., Mayeda T. K., Goswami J. N., and Olsen E. J. (1991) Oxygen isotope studies in ordinary chondrites. *Geochim. Cosmochim. Acta, 55,* 2317–2337.

Collins S. J., Righter K., and Brandon A. D. (2005) Mineralogy, petrology and oxygen fugacity of the LaPaz Icefield lunar basaltic meteorites and the origin of evolved lunar basalts (abstract). In *Lunar and Planetary Science XXXVI,* Abstract #1141. Lunar and Planetary Institute, Houston (CD-ROM).

Cuzzi J. N., Davis S. S., and Dobrovolskis A. R. (2003) Blowing in the wind. II. Creation and distribution of refractory inclusions in a turbulent protoplanetary nebula. *Icarus, 166,* 385–402.

Deloule E. and Robert F. (1995) Interstellar water in meteorites? *Geochim. Cosmochim. Acta, 59,* 4695–4706.

Deloule E., Robert F., and Doukhan J. C. (1998) Interstellar hydroxyl in meteoritic chondrules: Implications for the origin of water in the inner solar system. *Geochim. Cosmochim. Acta, 62,* 3367–3378.

Delsemme A. (1997) The origin of the atmosphere and the oceans. In *Comets and the Origin and Evolution of Life* (P. J. Thomas et al., eds.), pp. 29–67. Springer-Verlag, New York.

Donahue T. M., Hoffman J. H., Hodges R. R. Jr., and Watson A. J. (1982) Venus was wet: A measurement of the ratio of deuterium to hydrogen. *Science, 216,* 630–633.

Drake M. J. (1984) Geochemical constraints on the early thermal history of the Earth. *Z. Naturforsch., 44a,* 883–890.

Drake M. J. (2000) Accretion and primary differentiation of the Earth: A personal journey. *Geochim. Cosmochim. Acta, 64,* 2363–2370.

Drake M. J. (2002) The eucrite/Vesta story. *Meteoritics & Planet. Sci., 36,* 501–513.

Drake M. J. (2005) Origin of water in the terrestrial planets. *Meteoritics & Planet. Sci., 40,* 519–527.

Drake M. J. and Righter K. (2002) Determining the composition of the Earth. *Nature, 416,* 39–44.

Drake M. J., Newsom H. E., and Capobianco C. J. (1989) V, Cr, and Mn in the Earth, Moon, EPB, and SPB and the origin of the Moon: Experimental studies. *Geochim. Cosmochim. Acta, 53,* 2101–2111.

Dreibus G. and Wänke H. (1980) The bulk composition of the eucrite parent asteroid and its bearing on planetary evolution. *Z. Naturforsch., 35,* 204–216.

Dreibus G., Ryabchikov I., Rieder R., Economou T., Brückner J., McSween H. Y., and Wänke H. (1998) Relationship between rocks and soil at the Pathfinder landing site and the martian meteorites (abstract). In *Lunar and Planetary Science XXIX,* Abstract #1348. Lunar and Planetary Institute, Houston (CD-ROM).

Ebel D. S. and Grossman L. (2001) Condensation in dust-enriched systems. *Geochim. Cosmochim. Acta, 64,* 339–366.

Emery J. P., Sprague A. L., Witteborn F. C., Colwell J. E., Kozlowski R. W. H., and Wooden D. H. (1998) Mercury: Thermal modeling and mid-infrared (5–12 μm) observations. *Icarus, 136,* 104–123.

Eucken A. (1944) Physikalisch-chemische Betrachtungen uber die fruheste Entwicklunsgeschiste der Erde. *Nachr. Akad. Wiss. Gottingen, Math-Phys., 1,* 1–15.

Fogel R. A, Hess P. C., and Rutherford M. J. (1989) Intensive parameters of enstatite chondrite metamorphism. *Geochim. Cosmochim. Acta, 53,* 2735–2746.

Gasparik T. (1993) The role of volatiles in the transition zone. *J. Geophys. Res., 98,* 4287–4300.

Geiss J. and Reeves H. (1972) Cosmic and solar system abundances of deuterium and helium 3. *Astron. Astrophys., 18,* 126–132.

Genda H. and Abe Y. (2005) Enhanced atmospheric loss on protoplanets at the giant impact phase in the presence of oceans. *Nature, 433,* 842–844.

Gessmann C. K., Wood B. J., Rubie D. C., and Kilburn M. R. (2001) Solubility of silicon in liquid metal at high pressure: Implications for the composition of the Earth's core. *Earth Planet. Sci. Lett., 184,* 367–376.

Goettel K. A. (1988) Present bounds on the bulk composition of Mercury — Implications for planetary formation processes. In *Mercury* (F. Vilas et al., eds.), pp. 613–621. Univ. of Arizona, Tucson.

Goodrich C. A., Jones J. H., and Berkley J. L. (1987) Origin and evolution of the ureilite parent magmas: Multi-stage igneous activity on a large planet. *Geochim. Cosmochim. Acta, 51,* 2255–2273.

Grady M. M. (2000) *Catalogue of Meteorites, 5th edition.* Cambridge Univ., Cambridge. 750 pp.

Greenwood R. C., Franchi I. A., Jambon A., and Buchanan P. C. (2004) Widespread magma oceans on asteroidal bodies in the early solar system. *Nature, 435,* 916–918.

Greshake A. (1997) The primitive matrix components of the unique carbonaceous chondrite Acfer 094: A TEM study. *Geochim. Cosmochim. Acta, 61,* 437–452.

Grossman J. N. and Brearley A. J. (2005) The onset of metamorphism in ordinary and carbonaceous chondrites. *Meteoritics & Planet. Sci., 40,* 87–122.

Halliday A. N. and Porcelli D. (2001) In search of lost planets — the paleocosmochemistry of the inner solar nebula. *Earth Planet. Sci. Lett., 192,* 545–559.

Hamilton D. L., Burnham C. W., and Osborn E. F. (1964) The solubility of water and the effects of oxygen fugacity and water content on crystallization in mafic magmas. *J. Petrol., 5,* 21–39.

Harmon J. K. and Slade M. K. (1992) Radar mapping of Mercury — Full-disk images and polar anomalies. *Science, 258,* 640–643.

Hartmann W. K. and Davis D. R. (1975) Satellite-sized planetesimals and lunar origin. *Icarus, 24,* 504–515.

Hashizume K. and Chaussidon M. (2005) A non-terrestrial ^{16}O-rich isotopic composition for the protosolar nebula. *Nature, 434,* 619–621.

Hewins R. H. and Herzberg C. T. (1996) Nebular turbulence, chondrule formation, and the composition of the Earth. *Earth Planet. Sci. Lett., 144,* 1–7.

Hillgren V. J., Gessmann C. K., and Li J. (2000) An experimental perspective on the light element in earth's core. In *Origin of the Earth and Moon* (R. M. Canup and K. Righter, eds.), pp. 245–263. Univ. of Arizona, Tucson.

Humayun M. and Cassen P. (2000) Processes determining the volatile abundances of the meteorites and terrestrial planets. In *Origin of the Earth and Moon* (R. M. Canup and K. Righter, eds.), pp. 3–24. Univ. of Arizona, Tucson.

Humayun M. and Clayton R. N. (1995) Potassium isotope cosmochemistry: Genetic implications of volatile element depletion. *Geochim. Cosmochim. Acta, 59,* 2131–2148.

Hunten D. M. and Donohue T. M. (1976) Hydrogen loss from the terrestrial planets. *Annu. Rev. Earth Planet. Sci., 4,* 265–292.

Inoue T. (1994) The effect of water on phase relations and melt compositions in the system Mg_2SiO_4-$MgSiO_3$-H_2O up to 15 GPa. *Phys. Earth Planet. Inter., 85,* 237–263.

Inoue T. and Sawamoto H. (1992) High pressure melting of pyrolite under hydrous conditions and its geophysical implications. In *High-Pressure Research: Application to Earth and Planetary Sciences* (Y. Syono and M. H. Manghnani, eds.), pp. 323–331. Tokyo/Washington, DC: TERRAPUB/American Geophysical Union, Tokyo/Washington, DC.

Ireland T. R., Holden P., and Norman M. D. (2005) The oxygen isotope composition of the Sun and implications for oxygen processing in molecular clouds, star-forming regions, and the solar nebula (abstract). In *Lunar and Planetary Science XXXVI,* Abstract #1572. Lunar and Planetary Institute, Houston (CD-ROM).

Jagoutz E., Palme H., Baddenhausen H., Blum K., Cendales M., Dreibus G., Spettel B., Lorenz V., and Wänke H. (1979) The abundances of major, minor and trace elements in the Earth's mantle as derived from primitive ultramafic nodules. *Proc. Lunar Planet. Sci. Conf. 10th,* pp. 2031–2050.

Javoy M. (1995) The integral enstatite chondrite model of the Earth. *Geophys. Res. Lett., 22,* 2219–2222.

Jeanloz R., Mitchell D. L., Sprague A. L., and de Pater I. (1995) Evidence for a basalt-free surface on Mercury and implications for internal heat. *Science, 268,* 1455–1457.

Jones J. H. (1996) Chondrite models for the composition of the Earth's mantle and core. *Philos. Trans. R. Soc. Lond., A354,* 1481–1494.

Jones J. H. and Palme H. (2000) Geochemical constraints on the origin of the Earth and Moon. In *Origin of the Earth and Moon* (R. M. Canup and K. Righter, eds.), pp. 197–216. Univ. of Arizona, Tucson.

Jones J. H., Neal C. R., and Ely J. C. (2002) Signatures of the highly siderophile elements in the SNC meteorites and Mars: A review and petrologic synthesis. *Chem. Geol., 196,* 21–41.

Jurewicz A. J. G., Mittlefehldt D. W., and Jones J. H. (1991) Partial melting of the Allende (CV3) meteorite — Implications for origins of basaltic meteorites. *Science, 252,* 695–698.

Kargel J. S., Komatsu G., Baker V. R., and Strom R. G. (1993) The volcanology of Venera and VEGA landing sites and the geochemistry of Venus. *Icarus, 103,* 253–275.

Kawamoto T. and Holloway J. R. (1997) Melting temperature and partial melt chemistry of H_2O-saturated mantle peridotite to 11 gigapascals. *Science, 276,* 240–243.

Kilburn M. R. and Wood B. J. (1997) Metal-silicate partitioning and the incompatibility of S and Si during core formation. *Earth Planet. Sci. Lett., 152,* 139–148.

Killen R. M. and Ip W. H. (1999) The surface bounded atmospheres of Mercury and the Moon. *Rev. Geophys., 37,* 361–406.

Kita N. T., Huss G. R., Tachibana S., Amelin Y., Nyquist L. E., and Hutcheon I. D. (2005) Constraints on the origin of chondrules and CAIs from short-lived and long-lived radionuclides. In *Chondrites and the Protoplanetary Disk* (A. N. Krot et al., eds.), p. 558. ASP Conference Series 341, Astronomical Society of the Pacific, San Francisco.

Kobayashi S., Tonotani A., Sakamoto N., Nagashima K., Krot A. N., and Yurimoto H. (2005) Presolar silicate grains from primitive carbonaceous chondrites Y-81025, ALHA 77307, Adelaide and Acfer 094 (abstract). In *Lunar and Planetary Science XXXVI,* Abstract #1931. Lunar and Planetary Institute, Houston (CD-ROM).

Kong P., Ebihara Y., and Palme H. (1999) Siderophile elements in martian meteorites and implications for core formation in Mars. *Geochim. Cosmochim. Acta, 63,* 1865–1875.

Kress V. C. and Carmichael I. S. E. (1991) The compressibility of silicate liquids containing Fe_2O_3 and the effect of composition, temperature, oxygen fugacity and pressure on their redox states. *Contrib. Mineral. Petrol., 108,* 82–92.

Krot A. N., Keil K., Goodrich C. A., Scott E. R. D., and Weisberg M. K. (2003) Classification of Meteorites, In *Treatise on Geochemistry, Vol. 1: Meteorites, Comets, and Planets* (A. M. Davis, ed.), pp. 83–128. Elsevier, Oxford.

Krot A. N., Hutcheon I. D., Yurimoto H., Cuzzi J. N., McKeegan K. D., Scott E. R. D., Libourel G., Chaussidon M., Aléon J., and Petaev M. I. (2005) Evolution of oxygen isotopic composition in the inner solar nebula. *Astrophys. J., 622,* 1333–1342.

Kushiro I. (1969) Liquidus relations in the system forsterite diopside-silica-H_2O at 20 kb. *Am. J. Sci., 267A,* 269–294.

Larimer J. W. and Anders E. (1967) Chemical fractionations in meteorites, II: Abundance patterns and their interpretation. *Geochim. Cosmochim. Acta, 31,* 1239–1270.

LaTourette T. and Holloway J. R. (1994) Oxygen fugacity of the diamond + C-O fluid assemblage and CO_2 fugacity at 8 GPa. *Earth Planet. Sci. Lett., 128,* 439–451.

Lecluse C. and Robert F. (1994) Hydrogen isotope exchange reaction rates: Origin of water in the inner solar system. *Geochim. Cosmochim. Acta, 58,* 2927–2939.

Lewis J. S. (1972) Metal-silicate fractionation in the solar system. *Earth Planet. Sci. Lett., 15,* 286–190.

Lewis J. S. (1974) The temperature gradient in the solar nebula. *Science, 186,* 440–443.

Litasov K. and Ohtani E. (2002) Phase relations and melt compositions in CMAS-pyrolite-H_2O system up to 25 GPa. *Phys. Earth Planet. Inter., 134,* 105–127.

Lodders K. and Fegley B. (1997) An oxygen isotopic model for the composition of Mars. *Icarus, 126,* 373–394.

Lodders K. and Fegley B. (1998) *The Planetary Scientist's Companion.* Oxford Univ., New York.

Longhi J., Knittle E., Holloway J. R., and Wänke H. (1992) The bulk composition, mineralogy and internal structure of Mars. In *Mars* (H. H. Kieffer et al., eds.), pp. 184–208. Univ. of Arizona, Tucson.

Lugmair G. W. and Shukolyukov A. (1998) Early solar system timescales according to ^{53}Mn-^{53}Cr systematics. *Geochim. Cosmochim. Acta, 62,* 2863–2886.

Lyons J. and Young E. D. (2004) Evolution of oxygen isotopes in the solar nebula (abstract). In *Lunar and Planetary Science XXXV,* Abstract #1970. Lunar and Planetary Institute, Houston (CD-ROM).

MacPherson G. J. (2003) Calcium-aluminum-rich inclusions in chondritic meteorites. In *Treatise on Geochemistry, Vol. 1: Meteorites, Comets, and Planets* (A. M. Davis, ed.), pp. 201–246. Elsevier, Oxford.

Malavergne V., Siebert J., Guyot F., Gautron L., Combes R., Hammouda T., Borensztajn S., Frost D. J., and Martinez I. (2004) Si in the core? New high pressure and high temperature experimental data. *Geochim. Cosmochim. Acta, 68,* 4201–4211.

McCord T. B., Adams J. B., and Johnson T. V. (1970) Asteroid 4 Vesta: Spectral reflectivity and compositional implications. *Science, 168,* 1445.

McFarlane E. A. and Drake M. J. (1990) Element partitioning and the early thermal history of the Earth. In *Origin of the Earth* (H. E. Newsom and J. H. Jones, eds.), pp. 135–150. Oxford Univ., New York.

McFarlane E. A., Drake M. J., and Rubie D. C. (1994) Element partitioning between Mg-perovskite, magnesiowüstite, and silicate melt at conditions of the Earth's mantle. *Geochim. Cosmochim. Acta, 58,* 5161–5172.

McKay G. A., Le L., Wagstaff J., and Crozaz G. (1994) Experimental partitioning of rare earth elements and strontium: Constraints on petrogenesis and redox conditions during crystallization of Antarctic angrite Lewis Cliff 86010. *Geochim. Cosmochim. Acta, 58,* 2911–2919.

McSween H. Y. Jr. (2002) Leonard Medal Address: The rocks of Mars, from far and near. *Meteoritics & Planet. Sci., 37,* 7–25.

Meisel T., Walker R. J., Irving A. J., and Lorand J-P. (2001) Osmium isotopic compositions of mantle xenoliths: A global perspective. *Geochim. Cosmochim. Acta, 65,* 1311–1323.

Messenger S. (2000) Identification of molecular-cloud material in interplanetary dust particles. *Nature, 404,* 968–971.

Mikouchi T., Makishima J., Koizumi E., and Zolensky M. E. (2005) Porphyritic olivine-pyroxene clast in Kaidun: First discovery of an ordinary chondrite clast? (abstract). In *Lunar and Planetary Science XXXVI,* Abstract #1956. Lunar and Planetary Institute, Houston (CD-ROM).

Moore G. M., Righter K., and Carmichael I. S. E. (1995) The effect of dissolved water on the oxidation state of iron in natural silicate liquids. *Contrib. Mineral. Petrol., 120,* 170–179.

Moore G. M., Vennemann T., and Carmichael I. S. E. (1998) An empirical model for the solubility of H_2O in magmas to 3 kilobars. *Am. Mineral., 83,* 36–42.

Moores J. E., Brown R. P., Lauretta D. S., and Smith P. H. (2005) Preliminary results of sublimations fractionation in dusty disaggregated samples (abstract). In *Lunar and Planetary Science XXXVI,* Abstract #1973. Lunar and Planetary Institute, Houston (CD-ROM).

Morbidelli A. J., Chambers J. I., Lunine J. M., Petit F., Robert G. B., Valsecchi G. B., and Cyr K. (2000) Source regions and timescales for delivery of water to the Earth. *Meteoritics & Planet. Sci., 35,* 1309–1320.

Morgan J. W., Wandless G. A., Petrie R. K., and Irving A. J. (1981) Composition of the Earth's upper mantle; I. Siderophile trace elements in ultramafic nodules. *Tectonophys., 75,* 47–67.

Musselwhite D. S. and Drake M. J. (2000) Early outgassing of Mars: Implications from experimentally determined solubilities of iodine in silicate magmas. *Icarus, 148,* 160–175.

Musselwhite D. S., Drake M. J., and Swindle T. D. (1991) Early outgassing of Mars: Inferences from the geochemistry of iodine and xenon. *Nature, 352,* 697–699.

Newsom H. E. and Drake M. J. (1982) The metal content of the eucrite parent body: Constraints from the partitioning behavior of tungsten. *Geochim. Cosmochim. Acta, 46,* 2483–2489.

Newsom H. E. and Sims K. W. W. (1991) Core formation during early accretion of the Earth. *Science, 252,* 926–933.

Nguyen A. N. and Zinner E. (2004) Discovery of ancient silicate stardust in a meteorite. *Science, 303,* 1496–1499.

Norman M., McCulloch M., O'Neill H., and Brandon A. (2004) Magnesium isotopes in the Earth, Moon, Mars, and pallasite parent body: High-precision analysis of olivine by laser-ablation multi-collector ICPMS (abstract). In *Lunar and Planetary Science XXXV,* Abstract #1447. Lunar and Planetary Institute, Houston (CD-ROM).

O'Neill H. St. C. (1987) Quartz-fayalite-iron and quartz-fayalite-magnetite equilibria and the free energy of formation of fayalite (Fe_2SiO_4) and magnetite (Fe_3O_4). *Am. Mineral., 72,* 67–75.

O'Neill H. St. C. (1991) The origin of the Moon and the early history of the Earth — A chemical model 2: The Earth. *Geochim. Cosmochim. Acta, 55,* 1159–1172.

Ohtani E., Nagata Y., Suzuki Y., and Kato T. (1995) Melting relations of peridotite and the density crossover in planetary mantles. *Chem. Geol., 120,* 207–221.

Okuchi T. (1997) Hydrogen partitioning into molten iron at high pressure: Implications for Earth's core. *Science, 278,* 1781–1784.

Owen T. and Bar-Nun A. (2000) Volatile contributions from icy planetesimals. In *Origin of the Earth and Moon* (R. M. Canup and K. Righter, eds.), pp. 459–473. Univ. of Arizona, Tucson.

Pahlevan K. and Stevenson D. J. (2005) The oxygen isotope similarity of the Earth and Moon: Source region or formation process? (abstract). In *Lunar and Planetary Science XXXVI,* Abstract #2382. Lunar and Planetary Institute, Houston (CD-ROM).

Palme H. (2000) Are there chemical gradients in the inner solar system? *Space Sci. Rev., 192,* 237–262.

Palme H. and Nickel K. G. (1986) Ca/Al ratio and composition of the Earth's primitive upper mantle. *Geochim. Cosmochim. Acta, 49,* 2123–2132.

Palme H. and O'Neill H. St. C. (2004) Cosmochemical estimates of mantle composition. In *Treatise on Geochemistry, Vol. 2: Mantle and Core* (R. W. Carlson, ed.), pp. 1–38. Elsevier, Oxford.

Pepin R. O. (1991) On the origin and early evolution of terrestrial planet atmospheres and meteoritic volatiles. *Icarus, 92,* 2–79.

Pierazzo E. and Melosh H. J. (2001) Melt production in oblique impacts. *Icarus, 145,* 252–261.

Podosek F. A. (2003) Noble gases. In *Treatise on Geochemistry, Vol. 1: Meteorites, Comets, and Planets* (A. M. Davis, ed.), pp. 381–405. Elsevier, Oxford.

Potter A. E. and Morgan T. H. (1985) Discovery of sodium in the atmosphere of Mercury. *Science, 229,* 651–653.

Potter A. E. and Morgan T. H. (1986) Potassium in the atmosphere of Mercury. *Icarus, 67,* 336–340.

Righter K. (2002) Does the Moon have a metallic core? Constraints from giant impact modelling and siderophile elements. *Icarus, 158,* 1–13.

Righter K. (2005) Highly siderophile elements: Constraints on Earth accretion and early differentiation. In *Earth's Deep Mantle: Structure, Composition, and Evolution* (J. Bass et al., eds.), pp. 201–218. AGU Monograph Series 160, American Geophysical Union, Washington, DC.

Righter K. and Drake M. J. (1996) Core formation in Earth's Moon, Mars and Vesta. *Icarus, 124,* 512–528.

Righter K. and Drake M. J. (1997a) Metal/silicate equilibrium in a homogeneously accreting Earth: New results for Re. *Earth Planet. Sci. Lett., 146,* 541–554.

Righter K. and Drake M. J. (1997b) A magma ocean on Vesta: Core formation and petrogenesis of eucrites and diogenites. *Meteoritics & Planet. Sci., 32,* 929–944.

Righter K., Arculus R. J., Delano J. W., and Paslick C. (1990) Electrochemical measurements and thermodynamic calculations of redox equilibria in pallasite meteorites — Implications for the eucrite parent body. *Geochim. Cosmochim. Acta, 54,* 1803–1815.

Righter K., Hervig R. L., and Kring D. A. (1998) Accretion and core formation in Mars: Mo contents of melt inclusions in three SNC meteorites. *Geochim. Cosmochim. Acta, 62,* 2167–2177.

Righter K., Walker R. J., and Warren P. H. (2000) Significance of highly siderophile elements and osmium isotopes in the lunar and terrestrial mantles. In *Origin of the Earth and Moon* (R. M. Canup and K. Righter, eds.), pp. 291–322. Univ. of Arizona, Tucson.

Righter K., Collins S. J., and Brandon A. D. (2005) Mineralogy and petrology of the LaPaz Icefield lunar mare basaltic meteorites. *Meteoritics & Planet. Sci., 40,* 1703–1722.

Ringwood A. E. (1961) Chemical evolution of the terrestrial planets. *Geochim. Cosmochim. Acta, 30,* 41–104.

Ringwood A. E. (1979) *Origin of the Earth and Moon.* Springer, New York.

Robert F. (2001) The origin of water on Earth. *Science, 293,* 1056–1058.

Robert F., Gautier D., and Dubrulle B. (2000) The solar system D/H ratio: Observations and theories. *Space Sci. Rev., 92,* 201–224.

Robinson M. S. and Lucey P. G. (1997) Recalibrated Mariner 10 color mosaics: Implications for mercurian volcanism. *Science, 275,* 197–200.

Robinson M. S. and Taylor G. J. (2001) Ferrous oxide in Mercury's crust and mantle. *Meteoritics & Planet. Sci., 36,* 841–847.

Rotaru M., Birck J. L., and Allegre C. J. (1992) Clues to early solar system history from chromium isotopes in carbonaceous chondrites. *Nature, 358,* 465–470.

Rubie D. C., Melosh H. J., Reid J. E., Liebske C., and Righter K. (2003) Mechanisms of metal-silicate equilibration in the terrestrial magma ocean. *Earth Planet. Sci. Lett., 205,* 239–255.

Rubie D. C., Gessmann C. K., and Frost D. J. (2004) Partitioning of oxygen during core formation on the Earth and Mars. *Nature, 429,* 58–61.

Rubin A. E. and J. T. Wasson (1995) Variations of chondrite properties with heliocentric distance (abstract). *Meteoritics & Planet. Sci., 30,* 569.

Rubin A. E., Fegley B., and Brett R. (1988) Oxidation state in chondrites. In *Meteorites and the Early Solar System* (J. F. Kerridge and M. S. Matthews, eds.), pp. 488–511. Univ. of Arizona, Tucson.

Ruzicka A., Snyder G. A., and Taylor L. A. (1997) Vesta as the HED parent body: Implications for the size of a core and for large-scale differentiation. *Meteoritics & Planet. Sci., 32,* 825–840.

Safronov V. S. (1969) *Evolution of the Protoplanetary Cloud and Formation of the Earth and Planets.* Nauka, Moscow. (In Russian. Translated 1972, NASA TT F-677.)

Sato M. (1978) A possible role of carbon in characterizing the oxidation state of a planetary interior and originating a metallic core (abstract). In *Lunar and Planetary Science IX,* pp. 990–992. Lunar and Planetary Institute, Houston.

Scott E. R. D. and Krot A. N. (2003) Chondrites and their components. In *Treatise on Geochemistry, Vol. 1: Meteorites, Comets, and Planets* (A. M. Davis, ed.), pp. 143–200. Elsevier, Oxford.

Scott E. R. D. and Krot A. N. (2005a) Thermal processing of silicate dust in the solar nebula: Clues from primitive chondrite matrices. *Astrophys. J., 623,* 571–578.

Scott E. R. D. and Krot A. N. (2005b) Chondritic meteorites and the high-temperature nebular origins of their components. In *Chondrites and the Protoplanetary Disk* (A. N. Krot et al., eds.), p. 15. ASP Conference Series 341, Astronomical Society of the Pacific, San Francisco.

Scott E. R. D. and Newsom H. (1989) Planetary compositions — Clues from meteorites and asteroids. *Z. Naturforsch., 44a,* 924–934.

Shu F. H., Shang H., and Lee T. (1996) Toward an astrophysical theory of chondrites. *Science, 271,* 1545–1552.

Shukolyukov A. and Lugmair G. W. (1999) The ^{53}Mn-^{53}Cr isotope systematics of the enstatite chondrites (abstract). In *Lunar and Planetary Science XXX,* Abstract #1093. Lunar and Planetary Institute, Houston (CD-ROM).

Sisson T. W. and Grove T. L. (1993) Experimental investigations of the role of water in calc-alkaline differentiation and subduction zone magmatism. *Contrib. Mineral. Petrol., 113-143-166.*

Slade M. A., Butler B. J., and Muhleman D. O. (1992) Mercury radar imaging — Evidence for polar ice. *Science, 258,* 635–640.

Solomon S. C. (2003) Mercury: The enigmatic innermost planet. *Earth Planet. Sci. Lett., 216,* 441–455.

Spohn T., Sohl F., Wieczerkowski K., and Conzelmann V. (2001) The interior structure of Mercury: What we know, what we expect from BepiColombo. *Planet. Space Sci., 49,* 1561–1570.

Sprague A. L., Kozlowski R. W. H., Witteborn F. C., Cruikshank D. P., and Wooden D. H. (1994) Mercury: Evidence for anorthosite and basalt from mid-infrared (7.3–13.5 micrometers) spectroscopy. *Icarus, 109,* 156–167.

Sprague A. L., Hunten D. M., and Lodders K. (1995) Sulfur at Mercury, elemental at the poles and sulfides in the regolith. *Icarus, 118,* 211–215.

Stolper E. (1977) Experimental petrology of eucritic meteorites. *Geochim. Cosmochim. Acta, 41,* 587–611.

Stixrude L., Hemley R. J., Fei Y., and Mao H. K. (1992) Thermoelasticity of silicate perovskite and magnesiowustite and stratification of the earth's mantle. *Science, 257,* 1099–1101.

Strom R. G. (1997) Mercury: An overview. *Adv. Space Res., 19,* 1471–1485.

Taylor G. J. and Scott E. R. D. (2003) Mercury. In *Treatise on Geochemistry, Vol. 1: Meteorites, Comets, and Planets* (A. M. Davis, ed.), pp. 477–485. Elsevier, Oxford.

Taylor L. A., Rossman G. R., and Qi Q. (1995) Where has all the lunar water gone? (abstract). In *Lunar and Planetary Science XXVI,* p. 1399. Lunar and Planetary Institute, Houston.

Taylor S. R. (1991) Accretion in the inner nebula: The relationship between terrestrial planetary compositions and meteorites. *Meteoritics, 26,* 267–277.

Taylor S. R. (1992) *Solar System Evolution: A New Perspective.* Cambridge Univ., Cambridge. 307 pp.

Treiman A. H., Drake M. J., Janssens M-J., Wolf R., and Ebihara M. (1986) Core formation in the Earth and shergottite parent body (SPB): Chemical evidence from basalts. *Geochim. Cosmochim. Acta, 50,* 1071–1091.

Treiman A. H., Jones J. H., and Drake M. J. (1987) Core formation in the shergottite parent body and comparison with the Earth. *Proc. Lunar Planet. Sci. Conf. 17th,* in *J. Geophys. Res., 92,* E627–E632.

Turekian K. K. and Clark S. P. (1969) Inhomogeneous accretion of the Earth from the primitive solar nebula. *Earth Planet. Sci. Lett., 6,* 346–348.

Turcotte D.L. and Schubert G. (1982) *Geodynamics: Applications of Continuum Physics to Geological Problems.* Wiley and Sons, New York. 450 pp.

Vilas F. (1988) Surface composition of Mercury from reflectance spectrophotometry. In *Mercury* (F. Vilas et al., eds.), pp. 59–76. Univ. of Arizona, Tucson.

Wade J. and Wood B. J. (2001) The Earth's "missing" niobium may be in the core. *Nature, 409,* 75–78.

Wade J. and Wood B. J. (2005) Core formation and the oxidation state of the Earth. *Earth Planet. Sci. Lett., 236,* 78–95.

Walker D. and Grove T. L. (1993) Ureilite smelting. *Meteoritics, 28,* 629–636.

Walker R. J., Horan M. F., Shearer C. K., and Papike J. J. (2004) Low abundances of highly siderophile elements in the lunar mantle: Evidence for prolonged late accretion. *Earth Planet. Sci. Lett., 224,* 399–413.

Walter M. J., Newsom H. E., Ertel W., and Holzheid A. (2000) Siderophile elements in the Earth and Moon: Metal/silicate partitioning and implications for core formation. In *Origin of the Earth and Moon* (R. M. Canup and K. Righter, eds.), pp. 265–290. Univ. of Arizona, Tucson.

Wänke H. (1981) Constitution of terrestrial planets. *Philos. Trans. R. Soc. London, A393,* 287–302.

Wänke H. and Dreibus G. (1988) Chemical composition and accretion history of terrestrial planets. *Philos. Trans. R. Soc. London, A325,* 545–557.

Warren P. H. (1999) Origin of planetary cores: Evidence from highly siderophile elements in martian meteorites. *Geochim. Cosmochim. Acta, 63,* 2105–2122.

Warren P. H. (2003) The Moon. In *Treatise on Geochemistry, Vol. 1: Meteorites, Comets, and Planets* (A. M. Davis, ed.), pp. 559–599. Elsevier, Oxford.

Wasson J. T. (1985) *Meteorites: Their Record of Early Solar-System History.* Freeman, New York. 274 pp.

Wasson J. T. (1988) The building stones of planets. In *Mercury* (F. Vilas et al., eds.), pp. 622–650. Univ. of Arizona, Tucson.

Weirich J. R., Brown R. H., and Lauretta D. S. (2004) Cometary D/H fractionation during sublimation. *36th Annual DPS Meeting 36,* Abstract #33.01.

Wetherill G. W. (1975) Occurrence of giant impacts during the growth of the terrestrial planets. *Science, 228,* 877–879.

Wetherill G. W. (1980) Formation of the terrestrial planets. *Annu.*

Rev. Astron. Astrophys., 18, 77–113.

Wetherill G. W. (1990) Formation of the Earth. *Annu. Rev. Earth Planet. Sci., 18,* 205–256.

Wetherill G. W. (1994) Provenance of the terrestrial planets. *Geochim. Cosmochim. Acta, 58,* 4513–4520.

Wiechert U., Halliday A. N., Lee D-C., Snyder G., Taylor L. A., and Rumble D. (2001) Oxygen isotopes and the Moon forming giant impact. *Science, 294,* 345–348.

Wieler R., Busemann H., and Franchi I. A. (2006) Trapping and modification processes of noble gases and nitrogen in meteorites and their parent bodies. In *Meteorites and the Early Solar System II* (D. S. Lauretta and H. Y. McSween Jr.), this volume. Univ. of Arizona, Tucson.

Williams Q. and Hemley R. J. (2001) Hydrogen in the deep Earth. *Annu. Rev. Earth Planet. Sci., 29,* 365–418.

Wood J. A. (1962) Chondrules and the origin of terrestrial planets. *Nature, 194,* 127–130.

Wood J. A. (2004) Formation of chondritic refractory inclusions: The astrophysical setting. *Geochim. Cosmochim. Acta, 68,* 4007–4021.

Woolum D. S. and Cassen P. (1999) Astronomical constraints on nebular temperatures: Implications for planetesimal formation. *Meteoritics & Planet. Sci., 34,* 897–907.

Yurimoto H. and Kuramoto K (2004) Molecular cloud origin for the oxygen isotope heterogeneity in the solar system. *Science, 305,* 1763–1766.

Zanda B. (2004) Chondrules. *Earth Planet. Sci. Lett., 224,* 1–17.

Zhu X. K., Guo Y., O'Nions R. K., Young E. D., and Ash R. D. (2001) Isotopic homogeneity of iron in the early solar nebula. *Nature, 412,* 311–313.

Zolensky M. E. (1998) Flux of extraterrestrial materials to Earth. In *Meteorites: Flux with Time and Impact Effects* (M. M. Grady, ed.), pp. 158–182. Geological Society Special Publication 140.

Zolensky M. E. and Ivanov A. V. (2003) The Kaidun microbreccia meteorite: A harvest from the inner and outer asteroid belt. *Chem. Erde, 63,* 185–246.

Zolensky M., Bland P., Brown P., and Halliday I. (2006) Flux of extraterrestrial materials. In *Meteorites and the Early Solar System II* (D. S. Lauretta and H. Y. McSween Jr.), this volume. Univ. of Arizona, Tucson.

Irradiation Records, Cosmic-Ray Exposure Ages, and Transfer Times of Meteorites

O. Eugster
University of Bern

G. F. Herzog
Rutgers University

K. Marti
University of California, San Diego

M. W. Caffee
Purdue University

During the 4.56-G.y. history of the solar system, every meteorite experienced at least one exposure to cosmic rays as a meter-sized meteoroid. The cosmic-ray exposure (CRE) age of a meteorite measures the integral time of exposure to galactic cosmic rays (GCRs). A fraction of meteoritic material was also irradiated by cosmic rays before ejection from kilometer-sized parent bodies (pre-irradiation); studies of pre-irradiation effects yield information about surface processes on planetary objects. In this review we discuss some methods of calculation for CRE ages of meteorites and present CRE age histograms for asteroidal, martian, and lunar meteorites. Compositional, formation-age, and CRE records indicate that probably the ~18,000 meteorites in our collections come from about 100 different asteroids. Stone meteorites exhibit CRE ages ≤120 m.y., while those of iron meteorites are generally, but not always, longer (up to 1500 m.y.); these CRE ages also show evidence for long-term GCR flux variations in the solar neighborhood. The CRE age differences of the various classes either signal different source regions, variability in the size of the Yarkovsky effect, or different resistance against crushing. Rocks blasted off the Moon and Mars by asteroidal or cometary impacts represent surface areas unlikely to be sampled by manned or automated missions; their CRE ages indicate that they come from some eight different sites on the Moon and also on Mars.

1. INTRODUCTION

The apparent emptiness of the interplanetary space belies its complex nature. The interplanetary medium is populated by a variety of energetic particles originating from both within and outside our solar system. Earth-based ionization measurements first established the presence of galactic cosmic rays (GCRs) and energetic solar-flare particles in 1912 and 1942 respectively (cf. *Pomerantz and Duggal,* 1974). Based on the observation of comet plasma tails, *Biermann* (1951, 1953) deduced the presence of solar-wind (SW) particles in the interplanetary medium. Since then instruments on satellites have measured the primary particle fluxes directly, along with their energy spectra and elemental and isotopic abundances.

The goal of this review is to present some characteristics of the energetic particle environment during the past 0.03–1500 m.y. as inferred from the effects of particle irradiation on meteorites and lunar surface materials. Toward that end we will discuss the basic principles of nuclide production in meteorites, the calculation of exposure times or cosmic-ray exposure (CRE) ages, and the statistical distributions of these ages. We will show that the CRE ages have implications for several interrelated questions. From how many different parent bodies do meteorites come? What was their pre-atmospheric size? How well do meteorites represent the population of the asteroid belt? How many collisions on their parent bodies have created the known meteorites of each type? Is there a time variation in the cosmic-ray (CR) flux? What factors control the CRE age of a meteorite and how do meteoroid (the immediate meter-sized precursor of meteorites) orbits evolve through time?

In this review we summarize the exposure histories of asteroidal, martian, and lunar meteorites. This field of science experienced remarkable development during the past 50 years and some comprehensive reviews were published (e.g., *Anders,* 1964; *Lal,* 1972; *Reedy et al.,* 1983; *Caffee et al.,* 1988; *Vogt et al.,* 1990; *Marti and Graf,* 1992; *Wieler and Graf,* 2001; *Wieler,* 2002a; *Herzog,* 2003; *Eugster,* 2003). In this review we will not be concerned with CR interactions with the lunar surface or the surface exposure ages of lunar rocks and soils from the Apollo and Luna missions; we refer the interested reader to the reviews by *Vogt et al.* (1990) and *Eugster* (2003). We will also not discuss the ways that trapping of solar particles modifies noble gas concentrations in meteorites. *Wieler et al.* (2006) treat this subject.

The energetic particles found in our solar system can be divided into three classes: SW particles, solar energetic particles, and GCRs. Here we mainly focus on the GCR effects in meteorites. For a detailed discussion of the interactions of solar particles we refer to the reviews by *Caffee et al.* (1988) and *Wieler* (2002b).

1.1. Nature of Galactic Cosmic Rays

Galactic cosmic rays, the most energetic particles in the interplanetary medium, originate from outside our solar system. The omnidirectional flux of the GCR nuclei at 1 AU is about 3 nuclei cm^{-2} s^{-1} for particles having kinetic energy >1 GeV/nucleon. Their energy spectrum obeys a power law in energy (Fig. 1). Above ~10 GeV, the number of particles is proportional to $E^{-2.65}$, where E is the particle energy. As noted above, solar modulation influences the number of GCR nuclei penetrating the solar system, this effect being most pronounced for the low-energy (<1 GeV/nucleon) component (cf. *Castagnoli and Lal,* 1980).

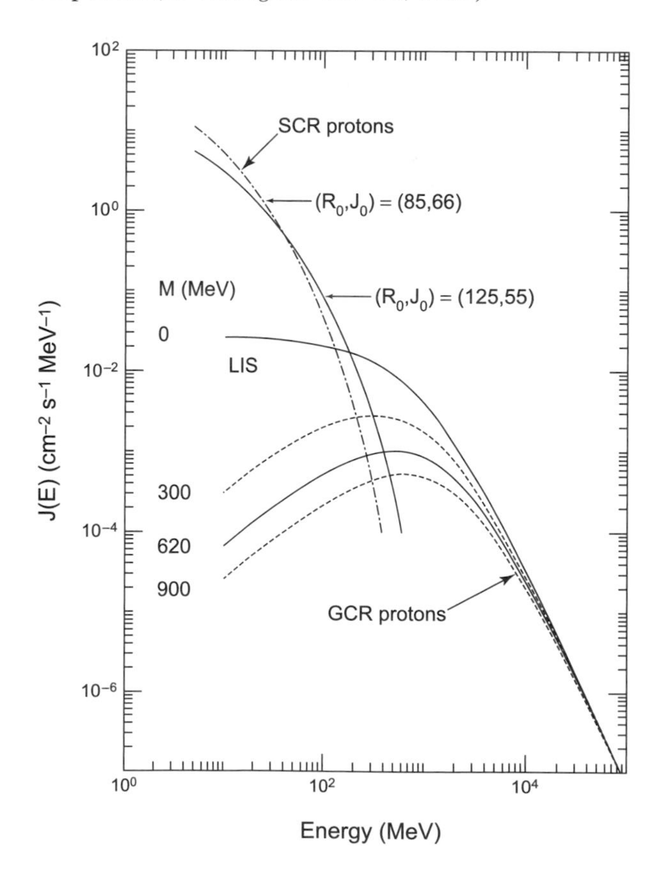

Fig. 1. Spectra of solar (SCR) and galactic (GCR) cosmic-ray protons at 1 AU. The modulation parameter, M, is shown vs. proton energy. GCR spectra are plotted for times of an active (M = 900 MeV) and a quiet (M = 300 MeV) Sun, as well as for the average GCR spectrum during the last 10 m.y. and for the local interstellar spectrum (LIS, M = 0). For original data and details see *Michel et al.* (1996).

Observed elemental abundances of GCRs are similar to solar system abundances. The similarity suggests that CRs and the nuclei in the Sun were synthesized by similar processes. The high energy of the cosmic radiation links it to shock acceleration mechanisms in supernovae environments (*Axford,* 1981; *Lingenfelter et al.,* 2000). Propagation through the interstellar medium results in the production of the light elements Li, Be, and B via nuclear "spallation" interactions, causing them to be overabundant relative to solar system values.

The CR Isotope Spectrometer (*Binns et al.,* 2001) can measure isotopic ratios of several elements in GCRs. The data of these authors support models where GCRs preferentially are from dust or gas of Wolf Rayet stars or supernovae from the central region of the galaxy. The isotopic composition of these elements usually agrees with the solar system abundances. An outstanding exception to this is the $^{20}Ne/^{22}Ne$ ratio. In galactic CRs, ^{22}Ne is enriched relative to either SW-implanted Ne or the trapped Ne component observed in many meteorites (*Binns et al.,* 2001). This observation is interesting in light of the occurrence of essentially pure ^{22}Ne (Ne-E) in several meteorites (*Eberhardt,* 1974; *Eberhardt et al.,* 1979).

1.2. Products of Interactions of Energetic Particles with Extraterrestrial Matter

Energetic particles interact with solid matter in a variety of ways. In order to decipher the exposure history of a meteorite it is necessary to understand these interactions. The dominant modes of interaction are determined by the energy and mass of the incident particle. At the low end of the energy spectrum are the SW nuclei. Their energies are so low that they simply stop; in the language of specialists, they are referred to as "implanted." At the other end of the energy spectrum are mainly protons and α-particles in the GCRs. These particles possess enough energy to induce nuclear reactions in material.

1.2.1. Direct implantation. Particles in the SW with low energy (~1 keV/nucleon) only penetrate solids to depths of ~50 nm, although subsequent diffusion of the implanted nuclei within the solid may result in an altered depth distribution. In stopping, the low-energy SW ions also cause radiation damage on the surface of solids. Because of the high flux of SW ions, mineral grains exposed to the SW can become saturated (that is, reach a limiting equilibrium concentration), especially for lighter elements. For instance, at 1 AU, 200 year is sufficient to reach He saturation. Not surprisingly, many lunar grains are saturated with SW-implanted He.

1.2.2. Energy loss and lattice damage. Energetic very heavy particles with Z > 20 (VH nuclei) slow down, near the end of their range. In many silicate materials they create permanent "latent" damage trails. These trails are observed directly by transmission electron microscopy or are chemically etched and enlarged to produce a conical hole that is visible by optical microscopy. These damage trails or etch holes are termed nuclear tracks. [For a review of this technique see *Fleischer et al.* (1975).] The tracks in silicate grains are produced by nuclei from both solar cosmic ray (SCRs) and GCRs. These two sources of tracks can be distinguished from each other. Galactic-cosmic-ray VH nuclei can penetrate solid matter to depths of several centimeters, whereas SF-VH nuclei have a range of <0.1 cm in solid matter. In this shorter range, SCR-produced tracks are far more prevalent than GCR-produced tracks, as shown in Fig. 1. Thus SCR-produced tracks are usually characterized by high track densities and track-density gradients, reflecting the steeply falling energy spectra of the SCR nuclei. The relationship between the track density, ρ, and the depth, x, can be described by the power law

$$\rho\ (x)\ \text{proportional to}\ x^{\eta} \qquad (1)$$

where η, the power-law exponent, is the slope of the semilogarithmic track-density profile and depends on the size of the object (in space). The relationship between the energy, E, of the incident particle and its range S is given by

$$S\ \text{proportional to}\ E^{\beta} \qquad (2)$$

where β can be determined in experiments at particle accelerators. Since tracks are produced at the end of a particle's range, and the track density as a function of depth x is proportional to the number of track-forming particles stopping at depth x, we have

$$\rho\ \text{proportional to} \left(\frac{dN}{dS}\right)_x = \left(\frac{dN}{dE} \times \frac{dE}{dS}\right)_x \qquad (3)$$

where the decrease in kinetic energy, E, is expressed by

$$\frac{dN}{dE} = \text{constant} \times E^{-\gamma} \qquad (4)$$

The power law exponent, γ, defines the spectral hardness, which, for regions below the surface, also depends on the size of the irradiated object. The term dE/dS can be derived from equation (2). Substituting these derivatives into equation (3) and equating it to equation (1) produces

$$\eta = (1/\beta)(\gamma + \beta - 1) \qquad (5)$$

This equation describes the dependence of the spectral shape of the track-forming heavy nuclei on the power-law exponent, η, of the incident primary nuclei (*Fleischer et al.,* 1967, 1975; *Bhattacharya et al.,* 1973).

1.2.3. Nuclear reactions. The energetic, charged particles that bombard a meteoroid can have any one of three fates: (1) the particles can escape; (2) they can stop without causing nuclear reactions; or (3) they can enter into a nuclear reaction. The relative probabilities depend on initial particle energy and atomic number, and on meteoroid size and composition. At low energies, stopping without reaction is the most likely outcome. The particle energy at which the probability of loss by nuclear reaction (3) is equal to the probability of survival, the crossover point, is ~300 MeV/nucleon. Since the ionization loss is proportional to Z^2, the crossover energy increases for high Z particles.

To induce a nuclear reaction, the bombarding particles must have at least several megaelectron volts of energy. Of the more energetic particles in the solar system, protons and α-particles from the Sun (SCRs) as well as GCRs produce nuclear reactions. Solar-cosmic-ray primaries typically produce reactions with small mass losses, such as $^{56}Fe(p,n)^{56}Co$, where an energetic proton reacts with a ^{56}Fe target nucleus to produce ^{56}Co and a (secondary) neutron. Primary and secondary GCRs have a broad energy spectrum and accordingly induce a large variety of high- and low-energy nuclear reactions. Galactic-cosmic-ray primaries account for most high-energy reactions where nuclear spallation reactions dominate. The secondary GCR particle cascade, especially the neutrons, account for lower-energy reactions (<100 MeV). At the low-energy end of nuclear reactions, epithermal and thermal neutron reactions, such as $^{157}Gd(n,\gamma)^{58}Gd$, produce nuclides by neutron capture (see below).

Nuclear reactions in meteorites are quite rare. Over a period of typical CRE ages of meteorites (~10 m.y.), only about one in every 10^8 nuclei undergoes a nuclear transformation. The "cosmogenic" nuclides are products of GCR- or SCR-produced nuclides, either radioactive or stable nuclides of exceedingly low initial abundance, such as noble gases. Table 1 shows a compilation of the measured cosmogenic nuclides in meteorites.

2. SHIELDING AND PRE-ATMOSPHERIC SIZE

The size of a meteoroid in space cannot be determined directly from the recovered mass of a meteorite because of the unknown degrees of ablation and fragmentation. Size must therefore be inferred from various experimental measurements and model calculations. Because of the imprecision of these models and for convenience, meteoriticists usually assume that meteoroids are spherical and express their "sizes" as radii. The most common units for the radii are either centimeters or grams per square centimeters, where unit (g/cm^2) = radius (cm) × density (g/cm^3). The use of units of grams per square centimeters, which borrows from the practice of nuclear physicists, has the advantage of correlating better with the number of atoms available for nuclear reactions in a meteoroid than does the radius expressed in centimeters.

As we will see below, the rates of nuclear reactions in a spherical meteoroid are strongly influenced by the original "depth" of the sample analyzed in the meteoroid, which is to say the shortest distance from the sample to the pre-atmospheric surface. In a spherical meteoroid, but not necessarily in real ones, the "depth" is a well-defined quantity.

TABLE 1. Cosmogenic nuclides measured in meteorites.

Nuclide	Half-Life (yr)	Main Targets
^{3}H	12.26	O, Mg, Si, Fe
^{3}He, ^{4}He	S*	O, Mg, Si, Fe
^{10}Be	1.51×10^{6}	O, Mg, Si, Fe
^{14}C	5730	O, Mg, Si, Fe
^{20}Ne, ^{21}Ne, ^{22}Ne	S	Mg, Al, Si, Fe
^{22}Na	2.6	Mg, Al, Si, Fe
^{26}Al	7.17×10^{5}	Si, Al, Fe
^{36}Ar, ^{38}Ar	S	Fe, Ca, K
^{36}Cl	3.01×10^{5}	Fe, Ca, K
^{37}Ar	35 d	Fe, Ca, K
^{39}Ar	269	Fe, Ca, K
^{40}K	1.251×10^{9}	Fe, Ni
^{39}K, ^{41}K	S	Fe, Ni
^{41}Ca	1.034×10^{5}	Ca, Fe
^{53}Mn	3.74×10^{6}	Fe, Ni
^{54}Mn	312 d	Fe, Ni
^{59}Ni	7.6×10^{4}	Ni
^{60}Co	5.27	Co, Ni
^{81}Kr	2.29×10^{5}	Rb, Sr, Y, Zr
^{78}Kr, ^{80}Kr, ^{82}Kr, ^{83}Kr	S	Rb, Sr, Y, Zr
^{129}I	1.57×10^{7}	Te, Ba, La, Ce
$^{124-132}Xe$	S	Te, Ba, La, Ce, (I)
^{150}Sm	S	Sm
^{156}Gd, ^{158}Gd	S	Gd

*S denotes that the nuclide is stable.

Other observed nuclides (few measurements): ^{7}Be, ^{44}Ti, ^{46}Sc, ^{55}Fe, ^{56}Co, and ^{60}Fe.

As mentioned above, secondary neutrons produced by CRs induce "low-energy" nuclear reactions. These neutrons are by custom very roughly divided into three groups according to their energy: >1 MeV, 0.1–300 eV (epithermal), and <0.1 eV (thermal). In large meteoroids of radius >200 g cm^{-2} the flux of thermal and epithermal neutrons peaks at shielding depths of 100–300 g cm^{-2} (cf. *Eberhardt et al.,* 1963; *Lingenfelter et al.,* 1972; *Spergel et al.,* 1986). The neutron flux within a meteoroid is thus a function of its size and of the shielding depth of the investigated sample within the meteoroid. Epithermal and thermal neutrons especially engage in what are called neutron-capture reactions that give rise to distinct and readily identifiable products. We show below that shielding depths and pre-atmospheric radii are important parameters for calculating the production rates of cosmogenic nuclides. Furthermore, the determination of the pre-atmospheric size of meteorites based on neutron-produced nuclides is significant for the study of the parent-body ejection dynamics of meteoritic matter. Finally, neutron-capture species may give information about exposure histories of large shielded objects that cannot be obtained from measurements of other types (high-energy spallation) of CR irradiation products alone. Complex exposures (see below) can be revealed from the relation of species produced by neutron capture that are stable with those that have relatively short half-lives.

Table 2 gives the neutron capture reactions that have been observed in meteorites. The references indicate additional information and work in which isotopic anomalies have been observed for the first time.

The determination of sample depth and meteoroid size are closely linked. The three most reliable methods for their determination within stony meteoroids are (1) combined measurements of the cosmogenic ratio $^{22}Ne/^{21}Ne$ and a radionuclide, (2) measurements of the production rates of nuclides produced by thermal neutrons, and (3) nuclear tracks. The $^{22}Ne/^{21}Ne$ ratio taken alone is not a unique function of size and depth. As a ratio, it depends strongly on the shape of the energy spectra of the primary and secondary particles and on the number of nuclear-active particles. To obtain both size and depth one must also determine the concentration of a cosmogenic radionuclide that is more sensitive to the number of active particles. For theoretical calculations showing the relation between $^{22}Ne/^{21}Ne$ and depth we refer to *Leya et al.* (2000) and *Masarik et al.* (2001). The reaction $^{24}Mg(n,\alpha)^{21}Ne$ dominates the production of ^{21}Ne: At average shielding corresponding to about

TABLE 2. Observed CR-produced neutron capture reactions in meteorites.

Reaction	Neutron energy	Reference
$^{24}Mg(n,\alpha)^{21}Ne$	> 5 MeV	[1]
$^{35}Cl(n,\gamma)^{36}Cl(\beta^-)^{36}Ar$	thermal	[2]
$^{40}Ca(n,p)^{40}K$	~6 MeV	[3]
$^{40}Ca(n,\gamma)^{41}Ca$	thermal	[4]
$^{58}Ni(n,\gamma)^{59}Ni$	thermal	[5]
$^{59}Co(n,\gamma)^{60}Co$	thermal	[6]
$^{79,81}Br(n,\gamma)^{80,82}Br(\beta^-)^{80,82}Kr$	epithermal (30–300 eV)	[7]
$^{127}I(n,\gamma)^{128}I(\beta^-)^{128}Xe$	epithermal (30–300 eV)	[7]
$^{130}Ba(n,\gamma)^{131}Ba(\beta^+)^{131}Cs(e\ capt.)^{131}Xe$	epithermal (resonance capture)	[8, 9]
$^{149}Sm(n,\gamma)^{150}Sm$	thermal	[4]
$^{157,155}Gd(n,\gamma)^{158,156}Gd$	thermal	[10]

References: [1] *Eberhardt et al.* (1966); [2] *Spergel et al.* (1986); [3] *Burnett et al.* (1966); [4] *Bogard et al.* (1995); [5] *Schnabel et al.* (1999); [6] *Cressy* (1972); [7] *Marti et al.* (1966); [8] *Eberhardt et al.* (1971); [9] *Kaiser and Rajan* (1973); [10] *Eugster et al.* (1970).

30 g cm^{-2}, about 70% of ^{21}Ne is produced by this reaction (*Michel et al.,* 1991). Neutron production of ^{22}Ne is somewhat less important than that of ^{21}Ne although it also increases with depth up to 200 g cm^{-2}. The strength of the method based on Ne lies in the fact that the analyses of the three stable Ne isotopes alone — ^{20}Ne, ^{21}Ne, and ^{22}Ne (together with the knowledge of the target element concentrations for ^{21}Ne production) — are sufficient to derive a shielding-corrected CRE age. Therefore, the ^{21}Ne–^{22}Ne/^{21}Ne dating method is most widely applied to stone meteorites. A determination of the ^{20}Ne concentration has to be included in the analyses in order to correct for other contributions to meteoritic Ne, such as primordial Ne. All other neutron-capture reactions are less suited for the determination of the shielding corrections as they require the isotopic analysis of an additional element.

The basis for using thermal neutrons for the determination of the pre-atmospheric sizes and sample depths of meteorites was set by the pioneering work of *Eberhardt et al.* (1963). These authors calculated the production of ^{36}Cl, ^{59}Ni, and ^{60}Co by n,γ-processes in spherical chondrites as a function of the meteorite radius. *Marti et al.* (1966) showed that epithermal neutrons in the energy range 30–300 eV must be responsible for an ^{80}Kr, ^{82}Kr, and ^{128}Xe anomaly observed in large meteorites. These authors calculated for the first time pre-atmospheric sizes of meteoroids based on excesses of ^{80}Kr from the reaction ^{79}Br(n,γ)^{80}Br(β$^-$)^{80}Kr and obtained minimal pre-atmospheric masses of 220 kg and 1400 kg for the Abee and the Mezö-Maderas chondrites, respectively. The detailed procedure for the calculation of the $^{80}Kr_n$ production rate, the epithermal neutron flux, the slowing-down density, and the pre-atmospheric size is given for 19 chondrites by *Eugster et al.* (1993) and for 6 martian meteorites by *Eugster et al.* (2002a). *Bogard et al.* (1995) measured significant concentrations of ^{36}Cl, ^{36}Ar, ^{41}Ca, and ^{150}Sm/^{149}Sm excesses from the capture of thermalized neutrons in the large Chico L6 chondrite. The fluence calculated from these nuclides is about 1.2×10^{16} n cm^{-2}. Neutron fluences of 10^{15}–10^{16} n cm^{-2} were also observed for three other chondrites and two achondrites. These authors concluded that the neutron-induced nuclides were produced during CRE. The rarely measured nuclide ^{59}Ni, resulting mainly from thermalized neutron capture of ^{58}Ni, was measured by *Schnabel et al.* (1999) in meteorite fragments and spherules of the Canyon Diablo impactor. These measurements imply that the liquid precursor material for the spheroids came from depths of 1.3–1.6 m beneath the pre-atmospheric surface of the impactor. *Hidaka et al.* (2000) analyzed the Sm- and Gd-isotopic composition in eight chondrites and observed evidence for neutron-capture effects. This enabled them to calculate pre-atmospheric sizes for six L and LL chondrites.

In principle, it is also possible to deduce the pre-atmospheric size and the approximate shape of a meteorite from a study of the track-density contours, since the track density drops rapidly with depth. Mass ablation estimates for 160 meteorites range between 27% and 99% with a weighted mean ablation of 85% (*Bhandari et al.,* 1980). These authors observed a relation between the pre-atmospheric mass of a meteorite, the track-production rate, and the cosmogenic ^{22}Ne/^{21}Ne ratios for shielded samples. This empirical relation can be used to determine the pre-atmospheric masses of meteorites from measurements of track densities and Ne-isotopic abundances of a few samples from interior locations.

3. DATING METHODS

3.1. Assumptions and Equations

Given certain assumptions about the conditions of irradiation and the history of a meteorite, we can calculate for that meteorite a CRE age, T. For this purpose we need measurements of the concentrations of stable and/or radioactive cosmogenic nuclides (Table 1). The assumptions are that (1) the flux of primary cosmic rays was constant in time, (2) the flux of primary cosmic rays was constant in space, (3) the shape of the sample did not change appreciably, (4) the chemical composition of the sample did not change appreciably, (5) any cosmogenic contributions from prior periods of irradiation are known, (6) all non-cosmogenic contributions to the inventory of the nuclide of interest are known, and (7) the sample did not lose nuclides of interest except by known rates of radioactive decay.

Remarkably, many meteorites seem to satisfy this set of requirements (*Leya et al.,* 2001a; *Welten et al.,* 2001a). For those meteorites, we may write for the concentration of a stable cosmogenic nuclide

$$S = P_S T \tag{6}$$

and for the concentration R of a radioactive nuclide

$$R = P_R \lambda^{-1}(1 - e^{-\lambda T}) \tag{7}$$

where P denotes the production rate and λ the decay constant.

3.2. Cosmogenic Nuclide Production Rates

Calibrations of the production rates may not be straightforward, since the production rate depends on the position of a sample within the pre-atmospheric object, which in general is difficult to ascertain. There are several methods for determining production rates: The physical methods (semi-empirical calculations) are based on measurements of cross sections and particle spectra and extensive computer calculations (cf. *Michel et al.,* 1991, 1996). Another method is based on measurements of pairs of radioactive and stable nuclides with similar nuclear-production parameters. This method requires experimental concentration data for both nuclides. The two approaches that we now discuss in more detail are not always cleanly separated.

In order to model a particular cosmogenic-nuclide-production rate (cf. *Graf et al.,* 1990a; *Michel et al.,* 1995; *Honda et al.,* 2002) it is necessary to specify several physical parameters. The production rate of a particular nucleus, P(R,d), assumed to be independent of time, is given by the relation

$$P(R,d) = \Sigma N_j \Sigma \int \sigma_{ij}(E) F_i(E,R,d) dE \quad (8)$$

where N_j is the abundance of the jth target element, $\sigma_{ij}(E)$ is the cross section at energy E for producing the nuclide from particle i reacting with target nucleus j, and $F_i(E,R,d)$ is the flux of particle i at energy E in a meteoroid of radius R and at a depth d (*Arnold et al.,* 1961; *Kohmann and Bender,* 1967; *Reedy,* 1985, 1987). The basic shapes for $F_i(E,R,d)$ are fairly well known, especially for high-energy GCRs. Factors influencing $F_i(E,R,d)$ are the size, shape, and composition of a meteoroid (*Reedy,* 1985). If the radius of a meteoroid is less than the interaction length of the GCRs (~100 g/cm^2), the production rates due to the primary particles do not decrease much from surface to center. The primary particles are only part of the story, however. Galactic cosmic rays also produce secondary particles, which have important effects on production rates in all but the smallest meteorites (radii <10 cm). These effects are obvious in all stony meteorites with radii between 30 and 150 g/cm^2 where production rates actually increase with depth. The energy distribution of the particles influences the production profile. Profiles of nuclides produced by relatively low-energy SCR particles, for example, decrease sharply with depth and are essentially confined to the outermost few centimeters because the SCRs produce few or no nuclear-active secondary particles. The depth profiles of nuclides produced by higher-energy GCRs, in contrast, vary by only about 30% in small- to medium sized meteorites (R ~ 30 cm) because of the ingrowth of secondary particle fluxes. Toward the interior (depths >15 cm) of larger meteorites, fluxes of both primary and secondary particles eventually decrease, typically exponentially with a half thickness of about 40 cm in stones (*Leya et al.,* 2000).

The production rate is proportional to the (energy-dependent) cross section for the production of a particular nuclide. Whenever possible, cross sections are based on direct measurements, but many are not known and must be extrapolated from experimental data or calculated from theoretical considerations. Combining the particle fluxes (both primary and secondary) with excitation functions and target-element chemistry according to equation (8) yields the production profile for a particular nuclide. Figure 2 shows the production profile for ^{21}Ne, a low-energy reaction product, and Fig. 3 shows that for ^{3}He, produced primarily in high-energy reactions.

Experimental approaches have been employed to determine the production profiles of cosmogenic nuclides in meteorites. *Graf et al.* (1990b) measured CR-produced nuclides from high- and low-energy reactions in samples from known positions within the L5 chondrite Knyahinya (R ~ 45 cm; pre-atmospheric mass 1300–1400 kg). Simulations of the GCR bombardment of meteorites were carried out using accelerator techniques, in which thick targets were irradiated with energetic protons (*Honda,* 1962). The thick-target production profiles are then translated into production

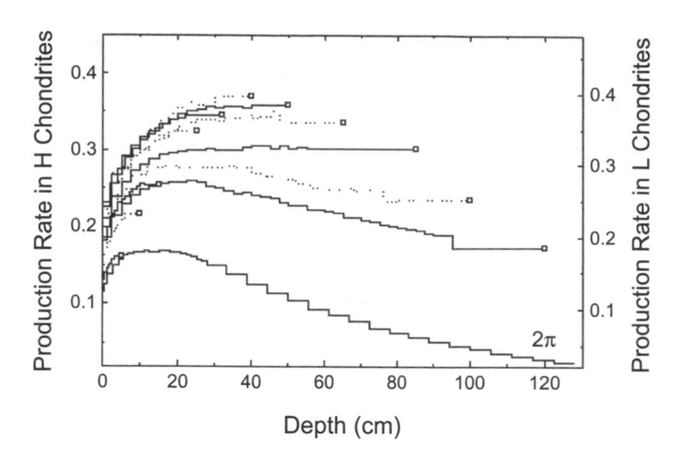

Fig. 2. Calculated production rates (in units of 10^{-8} cm^3 STP/g m.y.$^{-1}$) of ^{21}Ne for ordinary chondrites, shown for several radii vs. depth of an irradiated sample (from *Leya et al.,* 2000).

profiles for a hemispherical object (*Kohman and Bender,* 1967). This approach has been successful in predicting the production of cosmogenic nuclides in large iron meteorites.

Several thick-target experiments for the simulation of irradiations of stony and iron meteoroids with GCR protons were performed during the last two decades (e.g., *Michel et al.,* 1986; *Lüpke,* 1993; *Leya and Michel,* 1998; *Leya et al.,* 2004a). Such irradiations can be used to test and improve model calculations; for the recent examples we refer to *Michel et al.* (1991, 1995, 1996) and *Leya et al.* (2000). It is not practical to conduct simulation experiments for all the possible exposure geometries of meteorites, so theoretically derived production profiles in agreement with simulated irradiations give us confidence that the underlying theory can reliably predict production profiles for meteorites of different sizes.

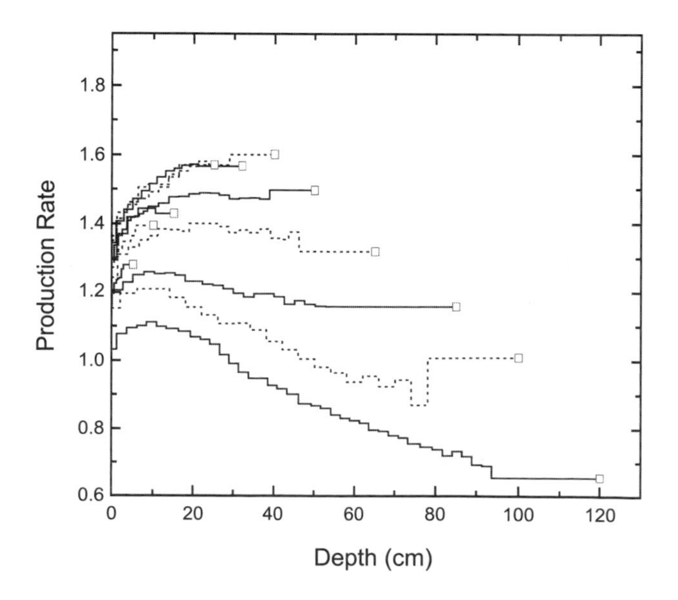

Fig. 3. Calculated production rates (in units of 10^{-8} cm^3 STP/g m.y.$^{-1}$) of ^{3}He as a function of meteorite radius and depth (*Leya et al.,* 2004a).

In the analysis of cosmogenic nuclide abundances, it is convenient to be able to estimate separately and easily the effects of composition (*Stauffer,* 1962) and shielding (*Eberhardt et al.,* 1966) on production rates without having to examine the output of detailed model calculations. So useful are the results of such formulations that since the 1960s meteoriticists have been writing production rates in the general form

$$P = f_1(\text{shielding}) \times f_2(\text{composition}) \times P_{Std} \quad (9)$$

where P_{Std} is a normative production rate under certain "standard" conditions [cosmogenic $^{22}Ne/^{21}Ne$ ratio = 1.11; chondritic composition (see *Eugster,* 1988)]. The shielding factor, f_1, is written in terms of one or two geometry-dependent quantities such as $^{22}Ne/^{21}Ne$ ratios or radionuclide concentrations (*Graf et al.,* 1990a; *Leya et al.,* 2000). Ideally, f_1 varies systematically and uniquely with meteoroid size and sample depth.

The compositional factor, f_2, is normally written as shown in equation (10)

$$\frac{\sum_j^{sa} N_j P_j}{\sum_j^{std} N_j P_j} \quad (10)$$

where sa stands for "sample" and std for "standard." N_j are elemental abundances (assumed uniform throughout the meteoroid) and P_j are the corresponding (nominal) production rates for each element that are taken from model calculations, but can also be obtained from meteoritic analyses (e.g., *Stauffer,* 1962). A key assumption here is that under fixed geometric irradiation conditions the values of P_j are constant and secondary particle fluxes do not depend on composition. However, equation (8) cannot be transformed into equation (9) without (falsely) assuming composition-independent secondary particle fluxes contributing to f values. For the relatively small ranges in composition found in H, L, and LL chondrites, this assumption is appropriate and equation (10) works well. On the other hand, for large differences in composition larger errors arise in the production rates based on equation (9) (*Begemann and Schultz,* 1988). *Masarik and Reedy* (1994) compared the model-production rates in mesosiderites and ordinary chondrites. For samples irradiated at an average depth of 57 g/cm^2 in objects with radii of 100 g/cm^2, they found the production rate of ^{21}Ne from Mg was 80% higher in the mesosiderites. The dependence of the nominal elemental production rates, P_j, on bulk composition is a consequence of what is called the matrix effect. Stated in general terms, the matrix effect refers to how, under fixed geometric conditions, elemental composition influences the secondary flux of nuclear particles and, through those fluxes, production rates.

Production rates may also be determined solely from analyses of meteorites. We consider P_R first. For meteorites with CRE ages long compared to the relevant half-life, $T \gg t_{1/2} = \ln(2)/\lambda$, we have $R = P_R \lambda^{-1}$. The quantity P_R varies from sample to sample as composition and shielding change. Shielding can depress P_R for a given nuclide tenfold or more in a large meteorite, but variations of less than ±30% around the average are much more typical for small- and medium-sized objects. Thus, we can apply equation (7) using an average value of P_R, obtained by averaging measurements of many meteorites with long exposure ages, and a value of R in another meteorite with an unknown age. This procedure yields a first approximation for the CRE age for objects with $T \leq 2 \times t_{1/2}$. Analogously, we can obtain relative CRE ages from equation (6) simply by comparing the measured concentrations of stable nuclides.

To obtain more precise and absolute CRE ages one must account for the effects of shielding sample by sample. The corrections for shielding are made by determining pairs of carefully selected stable and radioactive cosmogenic nuclides and forming the ratio of their concentrations. By combining equations (6) and (7), we have

$$S/R = \lambda P_s P_R^{-1} T (1 - e^{-\lambda T})^{-1} \quad (11)$$

If similar nuclear reactions produce nuclides S and R, then we expect the ratio P_S/P_R to vary little with shielding. Thus, to calculate absolute CRE ages from equation (11), we need to determine P_S/P_R at least once, and for the most precise work, to develop an expression for P_S/P_R that has a known dependence on shielding (see below) and compensates adequately for matrix effects. Examples of reliable methods for calculating CRE ages that incorporate these requirements for stones are based on (1) $^{81}Kr/Kr$ ratios and (2) coupled analyses of ^{21}Ne and the $^{22}Ne/^{21}Ne$ ratio. This latter method is not independent because it is pegged to the $^{81}Kr/Kr$ CRE ages of groups of meteorites (see below). For iron meteorites a frequently used method is based on $^{36}Cl/^{36}Ar$ concentration ratios. These methods will be discussed below. For other methods, such as those based on other stable noble gas species and appropriate production rates, $^{10}Be/^{21}Ne$, $^{26}Al/^{21}Ne$, and $^{40}K/K$ we refer to the reviews by *Herzog* (2003) and *Wieler* (2002a).

3.3. Calculation of Cosmic-Ray Exposure Ages

3.3.1. Krypton-81/krypton-83 cosmic-ray exposure ages. The main target nuclei for the production of Kr isotopes in stony meteorites are those of the trace elements Rb, Sr, Y, and Zr. As the isotopic Kr production rate ratios depend very weakly on the relative concentrations of these elements, their concentrations must be known only for the most precise CRE ages. Adapting equation (11), we have

$$^{83}Kr/^{81}Kr = (\lambda P_{83} T)/[P_{81}(1 - e^{-\lambda T})] \quad (12)$$

where λ is the ^{81}Kr decay constant. By measuring all Kr-isotopic ratios in one mass spectrometer run, one avoids the systematic biases that may affect the determination of CRE ages that rest on concentration measurements made with different instruments (*Marti,* 1967; *Eugster et al.,* 1967).

The production-rate ratio was originally determined from considerations of nuclear reaction systematics and nuclear cross-section measurements on the element Ag. However, it is now updated as

$$P_{81}/P_{83} = 0.92(^{80}Kr + ^{82}Kr)/(2 \times ^{83}Kr) \quad (13)$$

where the factor 0.92 is the isobaric fraction yield of ^{81}Kr determined from proton cross-section measurements on the main meteoritic target elements Rb, Sr, Y, and Zr (*Gilabert et al.,* 2002). Nuclear reactions on Br induced by low-energy secondary neutrons affect the relative production rates of ^{80}Kr and ^{82}Kr. Consequently, variations in the production-rate ratios $^{80}Kr/^{83}Kr$ and $^{82}Kr/^{83}Kr$ may often occur, thereby invalidating the use of equation (13). *Marti and Lugmair* (1971) circumvented the problem by determining P_{81}/P_{83} based on the $^{78}Kr/^{83}Kr$ ratio in lunar rocks. The approach also works for meteorites because low-energy neutrons do not affect the relative concentrations of either ^{78}Kr or ^{83}Kr, even in larger meteorites. For constants in equation (14) see *Eugster et al.* (2002a) and *Leya et al.* (2004b), who calculated slightly different values for chondrites

$$\frac{P_{81}}{P_{83}} = 1.174\left(\frac{^{78}Kr}{^{83}Kr}\right)_{spallogenic} + 0.354 \quad (14)$$

3.3.2. Neon-21 ages with shielding corrections based on neon-22/neon-21 ratios. *Nishiizumi et al.* (1980) and *Eugster* (1988) developed an empirical expression for the production rates of ^{21}Ne in ordinary chondrites, and *Eugster and Michel* (1995) extended this work to achondrites. *Eugster* (1988) used the $^{81}Kr/Kr$ CRE ages, cosmogenic ^{21}Ne contents, and $^{22}Ne/^{21}Ne$ ratios of a suite of ordinary chondrites and determined ^{21}Ne production rates based on $^{81}Kr/Kr$ CRE ages. In previous work *Eberhardt et al.* (1966) had argued that the ^{21}Ne production rate depends on the $^{22}Ne/^{21}Ne$ ratio according to the empirical relation

$$P^{21} = [a(^{22}Ne/^{21}Ne) + b]^{-1} \quad (15)$$

Though not rigorously defensible, this mathematical form fits the available data for many meteorites reasonably well, and yields the following equation for L chondrites, where the production rate results in units of 10^{-8} cm^3 STP/g m.y.$^{-1}$

$$P_{21} = [13.52(^{22}Ne/^{21}Ne) - 12.00]^{-1} \quad (16)$$

Minor adjustments for the other chondritic meteorites are discussed by *Eugster* (1988). However, equation (16) is not suitable for $^{22}Ne/^{21}Ne$ ratios less than about 1.08, i.e., for samples recovered from the interiors of larger meteoroids (*Masarik et al.,* 2001; *Leya et al.,* 2001b).

As developed by *Eugster* (1988) for ordinary chondrites, equation (16) assumes that composition dependence can be expressed by equation (9) with particular values of P_j taken from the work of *Freundel et al.* (1986) on ordinary chondrites. The matrix effect does not allow one to apply equation (16) to meteorites with compositions very different from those of ordinary chondrites. Recognizing this problem, *Eugster and Michel* (1995) recalibrated equation (16) for achondrites. *Albrecht et al.* (2000) found that on average, the equations of *Eugster and Michel* (1995) gave reliable production rates also for mesosiderites and concluded that the $^{22}Ne/^{21}Ne$ ratio has considerable capacity to correct for both shielding and composition. In sum, although the matrix effect can be large and is of theoretical interest, modeling calculations automatically take it into account, and standard calculations of CRE ages compensate adequately for it.

3.3.3. Chlorine-36/argon-36 cosmic-ray exposure ages. The principal target nuclei for the production of ^{36}Cl and ^{36}Ar in iron meteorites and in the metal phase of stony ones are the elements Fe and Ni (e.g., *Begemann et al.,* 1976). Once produced, ^{36}Cl decays with a half-life of 0.30 m.y. in part to ^{36}Ar. Adapting equation (11), we find for T much larger than the half-life of ^{36}Cl

$$^{36}Ar/^{36}Cl = [\lambda P_{^{36}Ar}T]/[P_{^{36}Cl}(1 - e^{-\lambda T})] \quad (17)$$

where the term $P_{^{36}Ar}$ includes the indirect contribution from the decay of ^{36}Cl, about 83.5% of the total. Equation (17) incorporates the approximation that all ^{36}Cl produced has already decayed, mostly to ^{36}Ar. This approximation introduces a negligible error for meteorites with CRE ages >10 m.y., but a slightly more complex equation applies to meteorites with short CRE ages (*Nyquist et al.,* 1973).

Direct cross-section measurements at various energies for the nuclear reactions such as $Fe(p,x)^{36}Ar$ and $Fe(p,x)^{36}Cl$ give fairly good estimates for the production rate ratio in iron meteorites or the metal phases of stony ones (*Lavielle et al.,* 1999).

On conversion to measurement units, *Albrecht et al.* (2000) obtained the equation

$$T = (430 \pm 16) \times (1 - e^{-\lambda_{36}T}) \times \frac{^{36}Ar}{^{36}Cl} \quad (18)$$

where ^{36}Ar is inserted in units of 10^{-8} cm^3 STP/g metal and ^{36}Cl in dpm/kg metal. *Lavielle et al.* (1999) recommend a very similar constant of 427.

3.3.4. Potassium-40/potassium-41 cosmic-ray exposure ages. Although the method based on $^{40}K/^{41}K$ systematics is the most complex, it is important for the study of the long-term constancy of the cosmic radiation. *Voshage* (1967) developed the method (equation (19)) during the period from about 1960 to 1985 and applied it mainly to irons and the metal phases of stony irons with CRE ages greater than about 100 m.y.

$$^{41}K/^{40}K - c^{39}K/^{40}K = (P_{41}/P_{40} - cP_{39}/P_{40})\lambda T/(1 - e^{-\lambda T}) \quad (19)$$

where the constant c is the primordial $^{41}K/^{39}K$ ratio in the meteorite and λ the ^{40}K decay constant.

The first term on the righthand side of equation (19) represents a shielding-dependent production-rate ratio, as discussed by *Voshage and Hintenberger* (1963). The evaluation is not straightforward since we do not know the relevant nuclear cross sections. These authors estimated production-rate ratios under various shielding conditions from theoretical model calculations and using measured $^{4}He/^{21}Ne$ ratios as the shielding parameter, permitting a relation between the K-production-rate ratios and the $^{4}He/^{21}Ne$ ratios. Using a least-squares fitting procedure and considering only data for meteorites with one-stage exposure histories, *Lavielle et al.* (1999) revisited the evaluation of $^{41}K/K$ ages.

The CRE ages obtained from equations (12), (17), and (19) are not sensitive to matrix effects, since these equations involve the ratio of two nuclides produced by similar nuclear reactions. In this context, "similar" means that the two nuclides are produced from the same set of target elements and that the cross sections for the relevant nuclear reactions depend on energy in similar ways. Although in reality variations in the behavior of the cross sections are to be expected, their effects on the exposure ages appear to be smaller than present measurement uncertainties.

3.4. Complex Exposure Histories

The one-stage irradiation model for meteorites has proved quite robust. As we shall demonstrate in section 4, good evidence for it comes from the clustering of CRE ages, as exemplified by the 6–8-m.y. peak in the H-chondrite CRE age distribution. In numerous instances, however, meteoritic inventories of cosmogenic nuclides also include sizeable contributions from earlier periods of irradiation. Such meteorites are said to have had complex exposure histories, with successive stages presumably initiated by collisions that changed the geometric conditions of irradiation and hence production rates. To comply with space limitations, we give only an overview of this subject, moving backward in time.

1. Meteoroids may undergo collisions in space. A collision will usually decrease the size of the irradiated object, and as a result, average GCR production rates will increase in most instances. If the longer-lived and stable nuclides are retained, they record both the earlier and later stages of irradiation. The relative contributions from the two stages depend, of course, on the durations and the shielding conditions. In contrast, the shorter-lived radionuclides produced in the earlier stages have time to decay before Earth arrival. For meteorites that have had a two-stage irradiation, the nominal one-stage CRE ages calculated for longer-lived nuclides will be larger than the CRE age calculated from the shorter lived nuclides. Examples include the meteorites Pitts (*Begemann et al.,* 1970), Bur Gheluai (*Vogt et al.,* 1993a), Torino (*Wieler et al.,* 1996), Shaw (*Herzog et al.,* 1997), Peekskill (*Graf et al.,* 1997), and Kobe (*Caffee et al.,* 2000; *Goswami et al.,* 2001).

Erosion in space can be regarded as special type of complex history characterized by a multitude of small, virtually continuous, size-reducing collisions. Erosion is well established in lunar rocks (*Hörz et al.,* 1991) and is typically estimated to be ~1 mm/m.y. Erosion rates of meteoroids were inferred to be somewhat lower, 0.65 mm/m.y. for stones and 0.022 mm/m.y. for irons (*Schaeffer et al.,* 1981; *Welten et al.,* 2001a).

2. Proto-meteoroids may undergo irradiation as parts of large fragments in parent-body regoliths. Such irradiation is detectable as neutron-capture effects in the meteoritic matter (see also section 2). In particular, the aubrites show evidence for preexposure on their parent body. For the Norton County aubrite, *Eugster et al.* (1970) found anomalies in the Gd-isotopic composition due to neutron capture of ^{157}Gd. Subsequently, these effects in the form of Sm- and Gd-isotopic shifts were observed by *Hidaka et al.* (1999) for five aubrites. These authors concluded that the meteoritic matter was located near the surface of its parent body for several hundred million years. *Lorenzetti et al.* (2003) confirmed the high neutron fluxes experienced by many aubrites: Excesses of neutron-produced ^{80}Kr were measured in Shallowater, Khor Temiki, Cumberland Falls, and Mayo Belwa, the latter two also showing the above-mentioned Sm- and Gd-isotopic shifts. The neutron-exposure time must have been considerably longer than the time indicated by the CRE age, requiring a regolith history for the aubritic matter before compaction into the present meteorite.

Herzog (2003) suggests that an appreciable number of micrometeorites likely belong in this group and retain records of surface exposure on parent bodies in the relatively recent past. *Pepin et al.* (2001) discuss other possible CRE histories for micrometeorites.

For the other end of the size scale, *Welten et al.* (2004) propose that several large meteorites with complex histories were irradiated essentially as boulders on parent-body surfaces. Meteorites in this group include Tsarev (*Herzog et al.,* 1997), Jilin (*Heusser et al.,* 1996), Gold Basin (*Welten et al.,* 2003), and QUE 90201 (*Welten et al.,* 2004). The criteria for distinguishing between irradiation in a boulder on a parent body and a boulder orbiting in space, i.e., between our cases (1) and (2), are not well defined.

3. Small, independent meteorite grains (<1 mm) may undergo irradiation on parent-body surfaces or even in other meteoroids prior to compaction into the matter that becomes the meteorite studied. Many meteorites contain gases implanted by SW and SCRs, which as noted earlier, travel only a few millimeters in matter. Thus the presence of solar gases implies exposure to GCR as well. Good examples are some aubrites that contain solar gases and neutron-produced nuclides, indicating that they have been located near the surface of their parent body (*Hidaka et al.,* 1999; *Lorenzetti et al.,* 2003). *Hohenberg et al.* (1990) showed that small, solar-gas-rich grains in carbonaceous chondrites contain more GCR-produced ^{21}Ne than do samples of the corresponding bulk meteorites. The timing and intensity of the early irradiation is controversial (*Rao et al.,* 1997; *Wieler et al.,* 2000),

although it seems likely that compaction occurred very early in the history of the solar system.

As recognized early on, meteoritic breccias may also contain somewhat larger fragments from various sources, not all of which contain solar noble gases, and each with its own irradiation history. A few examples of meteorites in this category include Weston (*Schultz et al.,* 1972), St. Mesmin (*Schultz and Signer,* 1977), Fayetteville (*Wieler et al.,* 1989; *Padia and Rao,* 1989; *Nishiizumi et al.,* 1993b), and possibly Kaidun (*Kalinina et al.,* 2004).

4. Components of meteorites may have undergone irradiation not in regoliths, but as free-floating bodies in the early solar nebula. *Polnau et al.* (2001) showed that chondrules in several meteorites contain more GCR-produced nuclides than the average bulk meteorites from which they come. They attribute the excesses to a 4π irradiation shortly after chondrule formation and before incorporation in a parent body. Certain theoretical models explain the abundances of now-extinct radionuclides in meteoritic matter (inclusions rich in Ca and Al and chondrules) by postulating an early irradiation close to an active ancient Sun (*Gounelle et al.,* 2001, 2004; *Leya et al.,* 2003; *Chaussidon and Gounelle,* 2006).

5. Extrasolar grains preserved in meteorites must have been irradiated in interstellar space. *Ott and Begemann* (2000) discuss this challenging frontier of CRE ages.

3.5. Time Variations

Evidence of variations of the particle environment, i.e., in the flux of GCRs and of the energetic particles emitted by the Sun, is recorded in meteorites for two endpoints in the evolutionary history of the solar system: the earliest stages of the solar nebula and the last 1 G.y. The variation of the particle environment in the early solar nebula is a separate chapter in this book (*Chaussidon and Gounelle,* 2006). We consider here the more recent changes. Shock acceleration from supernovae explosions is now considered to be a major source of the GCRs (*Reeves,* 1978; *Axford,* 1981; *Ramaty et al.,* 1996; *Ustinova,* 2002). The observed differential stellar motions in active star-forming regions can be used to reconstruct the past locations of supernovae and therefore the particle environment of the solar system (*De Zeeuw et al.,* 1999; *Branham,* 2002).

It has long been recognized that iron meteorites are excellent fossil detectors of cosmic radiation. Their CRE ages are calculated from a substantial database of GCR-produced nuclides that includes the data used for the $^{40}K/^{41}K$ method discussed earlier (*Voshage,* 1967; *Voshage and Feldmann,* 1979; *Voshage et al.,* 1983). Interestingly, several studies indicate systematic differences between ages obtained by stable nuclides and radioactive nuclides of with different half-lives, such as $^{36}Cl/^{36}Ar$, $^{39}Ar/^{38}Ar$, $^{10}Be/^{21}Ne$, and $^{129}I/^{129}Xe$ when compared to the $^{40}K/^{41}K$ CRE ages. These discrepancies have been interpreted as evidence for a time-variation in the GCR flux (*Marti et al.,* 2004). However, alternative interpretations, such as erosion in space and complex exposure histories of iron meteorites (*Voshage,* 1982; *Lavielle et al.,* 1999), need to be considered.

Lavielle et al. (1999) used an approach in the evaluation of the constancy of GCRs that eliminates irons with complex exposure histories. They assume a constant GCR flux over a limited time interval of 150–700 m.y. ago, but then find that meteorites with longer exposure ages are in conflict with this assumption.

To summarize, CRE ages of iron and stony iron meteorites based on the ^{40}K half-life disagree in a systematic way with radionuclides of <5-m.y. half-lives, such as ^{81}Kr, ^{36}Cl, ^{10}Be, and ^{53}Mn (e.g., *Eugster,* 2003). The average production rates of ^{36}Cl or ^{36}Ar over the time interval of 150–700 m.y. inferred for CRE ages based on K isotopes are lower by 28% when compared to production rates commonly used for the recent GCR flux. A recent GCR flux increase offers a most straightforward explanation. A new alternative calibration of interest for recent variations in the GCR flux is proposed based on the 16-m.y. radionuclide ^{129}I, which decays to ^{129}Xe (*Marti,* 1986; *Marti et al.,* 2004). Iodine-129 is efficiently produced by neutron reactions on Te that is mainly located in troilite. *Schnabel et al.* (2004) modeled the production of ^{129}I in stony meteorites based on experimental production rates of ^{129}I from thick-target experiments. Iodine-129 integrates the GCR flux over several half-lives and is thus well suited to study the flux records during the last 100 m.y.

4. COSMIC-RAY EXPOSURE AGES

Most CRE ages are calculated from noble gas data. The latest compilation (*Schultz and Franke,* 2004) gives He-, Ne-, and Ar-isotopic abundances for more than 1600 meteorites. For the calculation of CRE ages <4 m.y., radionuclide data are also used in particular for CM and CI chondrites (*Nishiizumi and Caffee,* 2002) for which large trapped-noble-gas contributions make the quantification of the GCR-produced component difficult.

Dynamical aspects and orbital evolution considerations are required for the understanding of the systematics in CRE age distributions. A decade ago it was thought that transit times to Earth are determined by proximity and injection of the meteoroid into "chaotic resonances in the inner main belt" (e.g., *Bottke et al.,* 2000). Such dynamics may account for longer exposure ages of stony irons and irons and possibly the low ^{3}He concentrations of not only irons, but also H5 chondritic metals (*Graf et al.,* 2001). These authors showed that metal in some H5 chondrites had been heated to temperatures where ^{3}H was diffusing out of the metal, requiring orbits with close approaches to the Sun. The chaotic injection model failed to explain some observed properties of asteroids (*Bottke,* 2002) and especially the long CRE ages for implied orbits with solar heating effects. Calculations by *Gladman* (1997) showed that once a meteoroid's orbit is in a resonance with that of Jupiter, its orbit quickly evolves to an Earth-crossing one. Consideration of the Yarkovsky effect seems to resolve several problems

(*Farinella et al.,* 1998; *Vokrouhlicky et al.,* 2000; *Spitale and Greenberg,* 2002). The Yarkovsky effect, the asymmetric heating of meteoroids by solar radiation, leads to small forces of acceleration that change the principal elements of the meteoroid orbit. The changes occur slowly enough that a meteoroid may spend tens of millions of years or more in the main belt before it reaches a resonance that brings it to the inner solar system, specifically to Earth. Smaller orbital changes by the Yarkovsky effect are expected for irons because of their optical properties, which would allow for longer random-walk passages through the main belt. Although most physical parameters that control CRE age distributions are understood, we should not forget that the ejection of meteoroids begins with random collisional events on parent objects, which account for the general clumpiness of the CRE age distributions.

Several lines of evidence (chemical, mineralogical, textural, and isotopic) indicate that the meteorites in the world's collections came from about 15 unmelted (undifferentiated, chondritic) asteroids and possibly 70 melted (differentiated) asteroids (*Keil et al.,* 1994; *Burbine et al.,* 2003). We discuss the CRE age histograms of several different meteorite classes. These data are used to obtain collisional information, constraints on their origin and parent-body history, break-up events, and ejection of meteorites, as well as on the dynamical systematics of their immediate precursor bodies.

4.1. Chondritic Meteorites

4.1.1. H, L, LL, E, and R chondrites. The CRE ages of the three major "ordinary" (H, L, LL) types of chondrites were reviewed by *Marti and Graf* (1992) and the CRE age histograms of these meteorites are presented in Fig. 4. These authors showed that none of the histograms is consistent with a continuous delivery of asteroidal material to Earth, as the observed CRE data clearly disagree with expected exponential distributions for a variety of orbital lifetimes. They conclude that CRE age histograms are dominated by stochastic events and that the continuous supply of asteroidal material can account for only a minor background. It is difficult therefore to infer collisional or dynamical half-lives for chondrite populations from measured histograms. The H chondrites recorded two major events, ~7 m.y. and ~33 m.y. ago; among the L chondrites, events at ~28 m.y. and ~40 m.y. ago were identified, and the LL chondrites suffered a collision ~15 m.y. ago.

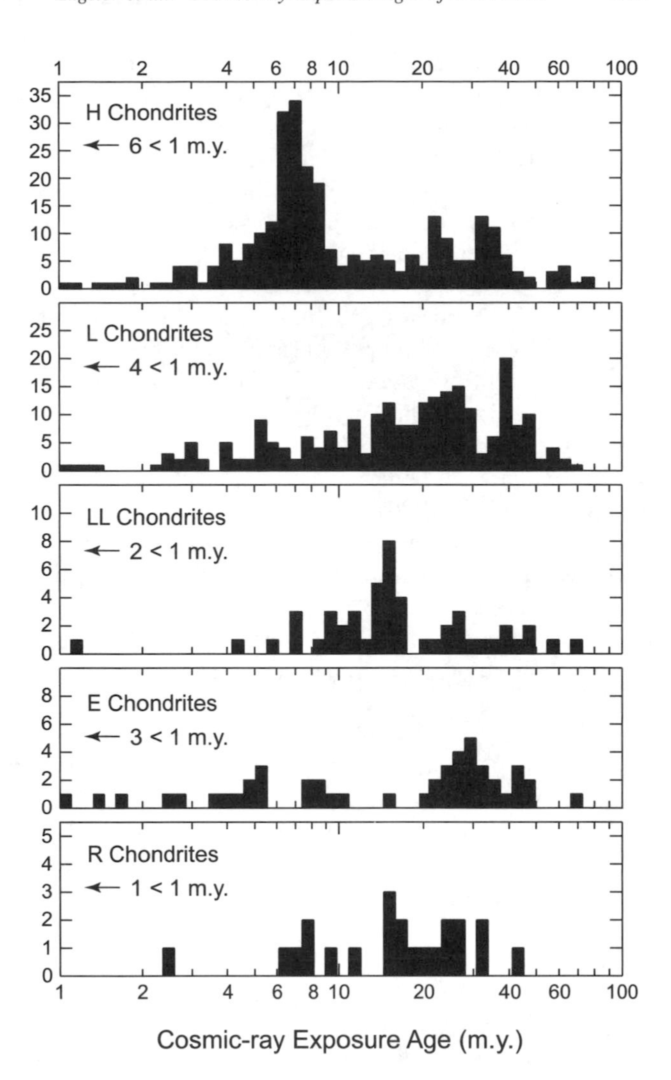

Fig. 4. Cosmic-ray exposure age distributions for H, L, LL, enstatite, and Rumuruti-type chondrites. Ordinary chondrite data are from *Marti and Graf* (1992), *Graf and Marti* (1994, 1995), and updates; the E-chondrite histogram is from *Patzer and Schultz* (2001) and that for the R chondrites from *Schultz et al.* (2005). Off-scale numbers of meteorites are indicated by arrows.

Graf and Marti (1995) conclude that very few H chondrites have exposure ages of less than 1 m.y. The CRE age distribution of H chondrites is consistent with the production of H chondrites by a relatively small number of events involving only a few parent bodies. Orbital information can be obtained from the p.m./total fall ratio among observed falls: H3, 4, and 6 chondrites show the typical afternoon excess of falls, but the H5 subgroup reveals a clear exception with a ratio of p.m./total falls <0.5. This ratio of fall times suggests the possibility of two collisional events, and an evolved orbit for one H5 parent object involved in the 7-m.y. collisional events. These authors further point out that for each petrologic type, there is a similar fraction of gas-rich stones. The implanted solar-type noble gases indicate that the gas-rich material had a regolith history. *Graf and Marti* (1995) conclude that the immediate parent of H chondrites was not an ancient layered object, since in this case the higher petrographic types should be devoid of solar-type noble gases, contrary to observation.

Marti and Graf (1992) reviewed L-chondrite exposure ages and, as in the case of H chondrites, the CRE age histograms (Fig. 4) of L5 and L6 chondrites do not fit a continuum; they also differ for the petrologic types. A strong CRE age peak occurs at about 40 m.y., a less well-defined peak is at 5 m.y., and a broad peak runs from 20 to 30 m.y. with a maximum in the vicinity of 28 m.y. All L3 or L4 CRE ages are <50 m.y. Many L chondrites have lost radiogenic

^{40}Ar and ^{4}He. Numerous authors have searched for relations between these gas losses and CRE ages. The CRE age peaks at 5 m.y. and 28 m.y. noted above stand out much more clearly for L chondrites that exhibit significant losses of ^{40}Ar than for other L chondrites. This difference in turn suggests that these events sampled strongly heated portion(s) of the L-chondrite parent body or bodies.

Okazaki et al. (2000) studied the light noble gases in 11 enstatite (E) chondrites and found CRE ages (based on ^{21}Ne) divided in two groups: <15 m.y. and >40 m.y. *Patzer and Schultz* (2001) reported ^{21}Ne CRE ages of about 60 E chondrites. This distribution (Fig. 4) is somewhat similar to that of the L chondrites, but clearly differs from the CRE age histogram of enstatite achondrites, the aubrites (Fig. 6). *Patzer and Schultz* (2001) identified clusters at about 3.5, 8, and 25 m.y., but caution that confirmation is needed. The CRE ages show no systematic correlation with either petrologic type or iron content (high-iron enstatites, EH chondrites; low-iron enstatites, EL chondrites). Thus, the CRE age distribution of the E chondrites yields no information on the structure of the E-chondrite parent asteroid.

The CRE ages of 23 individual R-chondrite falls (Fig. 4) range from 0.2 to 42 m.y. Six of them cluster at 16.6 ± 1.1 m.y. and may suggest a breakup on their parent body (*Schultz et al.,* 2005). Eleven of these meteorites contain solar gases and are thus regolith breccias. This percentage of almost 50% is higher than that of ordinary chondrites, howardites, and aubrites and may imply that the parent body of the R chondrites has a relatively thick regolith.

4.1.2. Carbonaceous chondrites. Most CI and CM chondrites have very short CRE ages of <2 m.y. (Fig. 5). Only the lunar meteorites have similarly short ages. *Scherer and Schultz* (2000) discuss three possible reasons why meteorites have short exposure ages: (1) the parent body was close to a resonance, (2) the parent body was in an Earth crossing orbit when a collision ejected the meteoroid, (3) their fragile character diminishes the rate of survival in space. The CRE ages of CI and CM chondrites are largely based on data from cosmogenic radionuclides. Based on ^{26}Al and ^{10}Be measurements, *Caffee and Nishiizumi* (1997) report a CRE age peak at 0.2 m.y. for C2 (mostly CM) meteorites. No peak is observed in the distribution of ^{21}Ne ages, but this may reflect the uncertainties in spallation noble gas components in CI and CM chondrites, which are affected by components from earlier irradiations (see section 5).

The average CRE (^{21}Ne) ages for CV, CK, and CO chondrites (not corrected for pairing) are 13 ± 10, 23 ± 14, and 22 ± 18 m.y., respectively. These ages are about one order of magnitude larger than those of the CI and CM chondrites. Cosmic-ray exposure measurements of other clans of carbonaceous chondrites (CR, CH, and CB) are rare, and data are not shown in Fig. 5. Cosmic-ray exposure ages for the CR and CH chondrites range from 1 to 25 m.y. and 1 to 12 m.y., respectively, while those of the CB chondrites Bencubbin (*Begemann et al.,* 1976) and Gujba (*Rubin et al.,* 2003) are about 30 m.y.

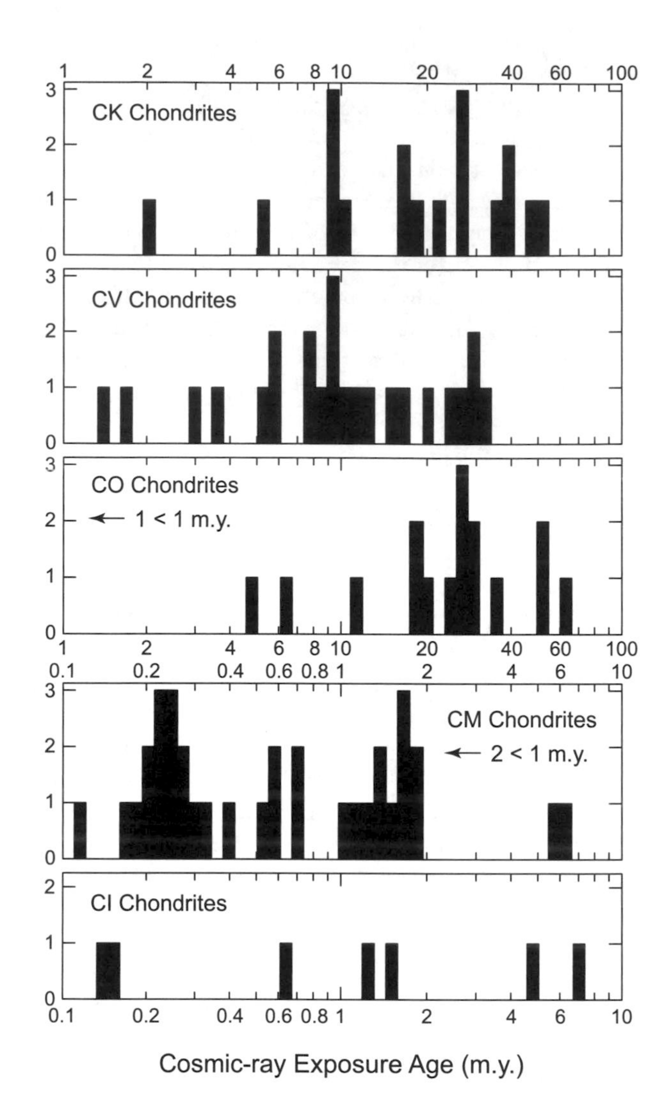

Fig. 5. Cosmic-ray exposure age distributions for carbonaceous chondrites. Data for CK, CV, and CO chondrites are from *Scherer and Schultz* (2000); a few updates are included. Data for CM chondrites from *Nishiizumi et al.* (1993a), *Caffee and Nishiizumi* (1997), *Eugster et al.* (1998), and *Nishiizumi and Caffee* (2002). Data for CI chondrites from *Nishiizumi et al.* (1993a) and K. Nishiizumi (personal communication, 2001).

4.2. Achondritic Meteorites

4.2.1. Howardites, eucrites, and diogenites (HED meteorites). The CRE age histogram of the HED meteorites shown in Fig. 6 reveals the following clusters: All three types of HED meteorites are represented in the 22 ± 3 m.y. age peak and in a cluster at 36 ± 3 m.y., while eucrites cluster also at 12 ± 3 m.y.; two eucrites and one diogenite have CRE ages of 7 ± 1 m.y. Of all 86 dated HEDs, at least 70% fall in these four clusters and more, if only ^{81}Kr-Kr ages are considered. For discussions of the statistical significance of the observed CRE age clusters see the original literature (*Eugster and Michel,* 1995; *Shukolyukov and Begemann,*

1996; *Welten et al.,* 1997). The rather similar histograms for H, E, and D clans suggest collisional breakup processes and ejections from the same parent object(s).

Asteroid Vesta was suggested as the common parent of the HEDs (*Consolmagno and Drake,* 1977), but transfers from Vesta to Earth appear to be dynamically difficult. A possible solution to the dynamical problem was proposed by *Binzel and Xu* (1993): a two-step evolution of orbital elements. These authors discovered several small asteroids with diameters of 5–10 km of basaltic achondritic composition with semimajor axes ranging between those of Vesta and the distance of the 3:1 resonance with Jupiter. They conclude that injection of large (>1 km) fragments into this resonance for subsequent delivery of meteoritic material to the inner solar system might explain the transfer dynamics. In this case the common ejection times of HEDs would imply that multikilometer-sized objects contain eucritic, diogenitic, and howarditic materials. *Bottke et al.* (2000) pointed out that the typical CRE ages of the HEDs (10–40 m.y.) are consistent with estimates obtained based on the Yarkovsky effect (see below) and Vesta as the parent object. In case of a model where HEDs were delivered directly from Vesta to Earth, the collisional events at 22 m.y., 12 m.y., and 36 m.y. must have been large, since they must liberate simultaneously Vesta's surface (eucrites, howardites) and interior (diogenites) layers. Smaller and frequent collisional events can be expected to eject only eucrites and howardites, a scenario consistent with the CRE age clusters.

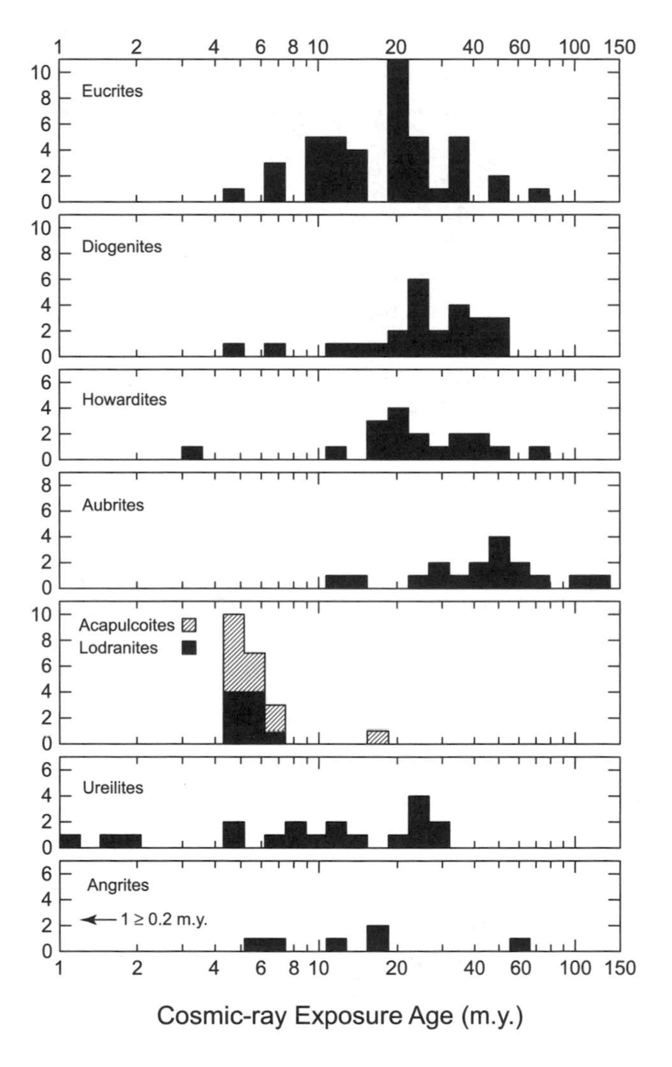

Fig. 6. Cosmic-ray exposure age distributions for achondrites. Data for eucrites, diogenites, and howardites were taken from *Eugster and Michel* (1995), *Miura* (1995), *Shukolyukov and Begemann* (1996), *Welten et al.* (1997), *Miura et al.* (1998), *Caffee and Nishiizumi* (2001), and *Welten et al.* (2001a); for aubrites from *Lorenzetti et al.* (2003); for acapulcoites and lodranties from *Weigel et al.* (1999), *Terribilini et al.* (2000a), and *Patzer et al.* (2003); for ureilites from *Goodrich* (1992) and *Scherer et al.* (1998); for angrites from *Eugster et al.* (2002b).

4.2.2. Aubrites (enstatite achondrites). The CRE ages of aubrites range from 12 to 116 m.y. (Fig. 6) and their distribution is nonuniform. Although a cluster appears around 50 m.y., an implied collisional event is highly doubtful, as seven of the eight aubrites with ages between 40 and 60 m.y. are breccias (*Lorenzetti et al.,* 2003). Six of these show characteristics of precompaction regolith exposure, chondritic inclusions, shock blackening, implanted solar noble gases in the surface layers of regolith grains, or isotopic variations produced by neutron capture (e.g., Sm, Gd, Br; see section 2). The two aubrites with the longest CRE ages of all stone meteorites, Norton County and Mayo Belwa, were not exposed to the solar wind. The relatively long CRE ages of aubrites (given their low physical strength) suggest that orbital elements of their parent body are distinct from those of other classes of stone meteorites. Specifically, the origin of the enstatite chondrites, with nearly half of their CRE ages <10 m.y., must differ from that of the enstatite achondrites, as first pointed out by *Eberhardt et al.* (1965).

4.2.3. Acapulcoites and lodranites. For no other meteorite group do the CRE ages cluster as tightly as for the acapulcoites and lodranites (Fig. 6). This clustering suggests ejection from a parent body in a breakup event about 6 ± 1.5 m.y. ago (*Terribilini et al.,* 2000a). Some fine structure in CRE age peak exists (*Eugster and Lorenzetti,* 2005), and there is an additional CRE age of 15 m.y. (*Patzer et al.,* 2003). Several other lines of evidence indicate that the acapulcoites and lodranites originated from a common parent body (*McCoy et al.,* 1997). *Eugster and Lorenzetti* (2005) proposed a model in which acapulcoites originate from the outer layer where the temperature was never high enough for silicate partial melting, accounting for old ^{39}Ar-^{40}Ar ages of 4503–4556 m.y., while lodranites would represent the inner shell of their parent asteroid, with ^{39}Ar-^{40}Ar ages at the later end of this range.

4.2.4. Ureilites. Ureilites share several properties characteristic of primitive meteorites, e.g., concentrations of heavy noble gases and O-isotopic signatures representing primitive solar system materials (*Scherer et al.,* 1998). The CRE age distribution of ureilites ranges from ~0.1 to about 34 m.y. (Fig. 6) and show no age clusters that would indicate breakup events; however, statistics are still poor. The large spread in CRE ages and the evidence for regolith evolution (some breccias contain solar gases) suggest a relatively large parent asteroid.

4.2.5. Angrites and brachinites. Cosmic-ray exposure ages are available for seven of the nine known angrites (Fig. 6). Statistics for CRE ages are poor, but suggest ejection events 6, 12, 17, and 56 m.y. ago (*Eugster et al.,* 2002b). Approximate CRE ages are available for three of seven known brachinites (not shown in Fig. 6): Brachina, 3 m.y. (*Ott et al.,* 1985); ALH 84025, 10 m.y. (*Ott et al.,* 1987); and Eagles Nest, 25 m.y. (*Swindle et al.,* 1998).

4.3. Stony-Iron Meteorites

Cosmic-ray exposure ages based on noble gases and cosmogenic radionuclides are available for mesosiderites (*Begemann et al.,* 1976; *Terribilini et al.,* 2000b; *Albrecht et al.,* 2000; *Nishiizumi et al.,* 2000; *Welten et al.,* 2001b). Figure 7 shows the distribution of the CRE ages for the mesosiderites and pallasites. The average, ~90 m.y., is intermediate between CRE ages for stones (younger) and irons (older). The age distribution has no obvious clusters, but *Welten et al.* (2001b) note that 3 of 19 mesosiderites have CRE ages close to 70 m.y.

4.4. Iron Meteorites

A long space exposure is evident for most, but not all, iron meteorites that record CR histories during the last ~1 G.y. Many also show multistep exposure histories. Cosmic-ray exposure ages for the four main groups I–IV are shown in Fig. 8.

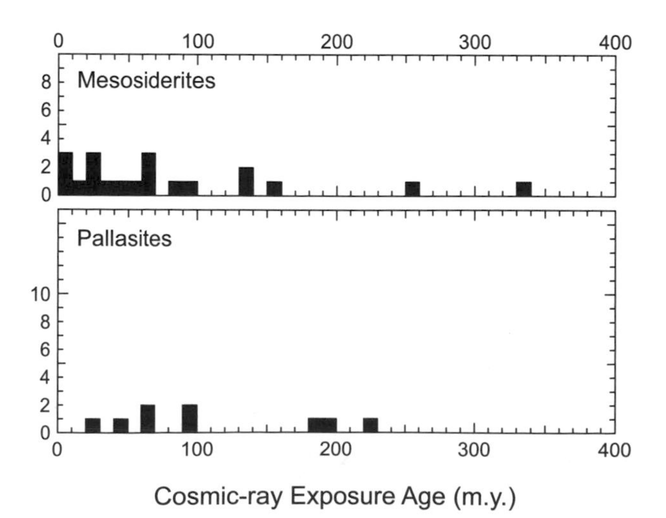

Fig. 7. Cosmic-ray exposure age distributions for stony-iron meteorites. Data for the mesosiderites are from *Begemann et al.* (1976), *Albrecht et al.* (2000), *Terribilini et al.* (2000b), and *Welten et al.* (2001b); for pallasites from *Megrue* (1968), *Voshage and Feldmann* (1979), *Shukolyukov and Petaev* (1992), and *Honda et al.* (1996).

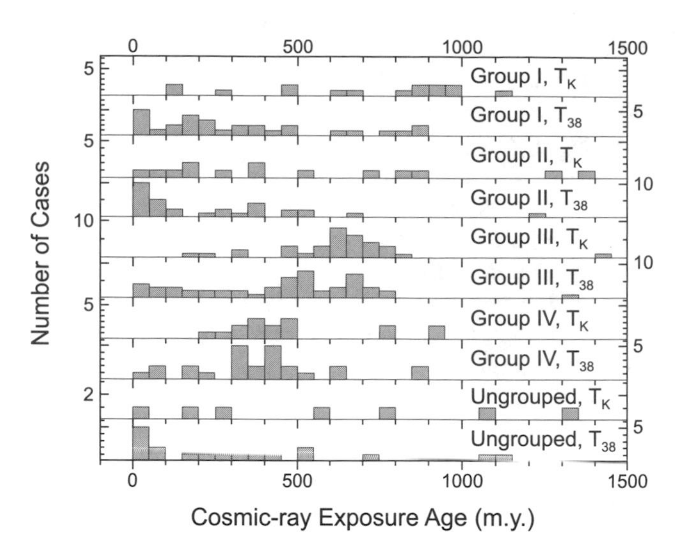

Fig. 8. Cosmic-ray exposure ages of iron meteorites. The $^{40}K/K$ ages (T_K) are from *Voshage* (1967), *Voshage and Feldmann* (1979), and *Voshage et al.* (1983). The ^{38}Ar–$^4He/^{21}Ne$ ages are from *Lavielle et al.* (1985).

As noted above, CRE ages based on Ar data and those based on K disagree in a systematic way (*Schaeffer et al.,* 1981; *Lavielle et al.,* 1999). Potassium-40/potassium-41-based CRE ages are less accurate for ages <200 m.y. because of limitations inherent to the method and to the experimental technique. There are few irons with CRE ages >1 G.y. For group III irons, both the $^{40}K/K$- and the ^{38}Ar-based CRE ages cluster at about 650 m.y.; the ^{38}Ar-based CRE ages suggest a second peak at about 450 m.y. (*Lavielle et al.,* 1985). *Lavielle et al.* (1999) documented a peak in CRE ages of IVA irons based on $^{40}K/^{41}K$ ages and more recently *Lavielle et al.* (2001) resolved this peak into two clusters at 255 ± 15 m.y. and 217 ± 13 m.y., respectively, using data obtained by the $^{36}Ar/^{36}Cl$ method.

4.5. Martian Meteorites

At present about 50 meteorites representing about 30 individual falls of meteorites whose parent body is Mars have been recovered. In a comprehensive review on ages and geological histories of martian meteorites, *Nyquist et al.* (2001) discussed the characteristics of crystallization age and isotopic and elemental composition, as well as the arguments for their martian origin.

The martian meteorites can be divided into seven petrographic types, characterized, e.g., by their ratio of pyroxene to olivine (*Meyer,* 2003): basaltic shergottites (average px/ol = 400), olivine-phyric shergottites (3.3), lherzolites (0.77), clinopyroxenites (nakhlites, 5.5), a dunite (only member Chassigny, 0.05), an orthopyroxenite (only member ALH 84001, ≫100), and EET 79001, consisting of two main lithologies of olivine-phyric and of basaltic composition, respectively. Figure 9 shows that these meteorites were ejected in a number of discrete events. The earliest event is represented by Dhofar 019, an olivine-phyric shergottite, 19.8 ± 2.3 m.y. ago. The only orthopyroxenite, ALH 84001, has an ejection age of 15.0 ± 0.8 m.y. It is the only meteorite from the ancient martian crust (for references see *Nyquist et al.,* 2001). All five clinopyroxenites (nakhlites), consisting primarily of magnesian augite with lesser amounts of Fe-rich olivine, have a common ejection age of 10.6 ± 1.0 m.y. The same impact event may also be responsible for the only dunite (Chassigny) since its T_{ej} is 11.1 ± 1.6 m.y. The six basaltic shergottites yield ejection ages in the range of 2.4–3.0 m.y. They consist predominantly of the clinopyroxenes pigeonite and augite and differ from the four lherzolites that consist mainly of olivine, orthopyroxene, and chromite. Lherzolites were ejected 3.8–4.7 m.y. ago. *Nyquist et al.* (2001) also discuss the possibility that the lherzolites were preexposed on the martian surface and that they were ejected at the same time as the group of basaltic shergottites. However, there is no evidence from radionuclide activities (*Schnabel et al.,* 2001) that supports such a scenario. Five olivine-phyric shergottites were ejected 1.2 ± 0.1 m.y. ago. Finally, the ejection age of EET 79001 is 0.73 ± 0.15 m.y. (*Nyquist et al.,* 2001). We conclude that eight events may account for the ejection of the martian meteorites, if Chassigny was not ejected together with the nakhlites.

Model calculations for the transfer times of rocks ejected from Mars are consistent with the distribution of observed CRE ages for material ejected slightly above escape velocity of 5 km/s. *Gladman et al.* (1996) found that 95% of all ejected rocks reach Earth within 20 m.y. and only about 20% are expected to have CRE ages <1 m.y. In contrast to lunar meteorites (see below), no clear evidence for a complex exposure history has been observed for martian meteorites. If the impacts on Mars occur at few-million-year intervals, as indicated by the CRE ages, the size of the impactors are likely to be on the order of kilometers and the resulting craters tens of kilometers in diameter (*Gladman,* 1997). Thus, most ejected rocks are expected to originate from more than a few meters depth and did not experience a preexposure to cosmic rays on Mars. The size of the ejected rocks from such impacts is estimated to be in a range of 20–200 cm (*Artemieva and Ivanov,* 2002; *Eugster et al.,* 2002a). Consequently, the number of martian meteoroids from a cratering event must be enormous, and source crater pairing of specimens collected on Earth is not surprising.

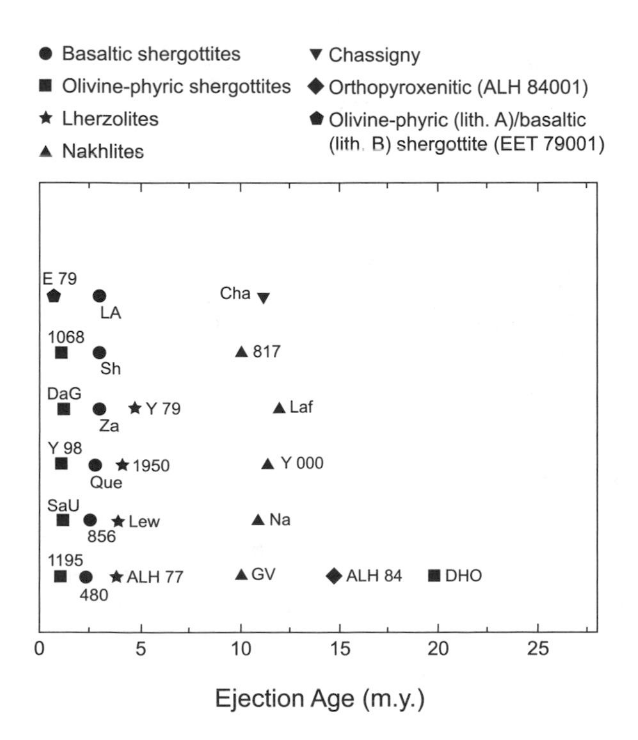

Fig. 9. Ejection ages of martian meteorites. For references of data source and meteorite names see *Meyer* (2003).

4.6. Lunar Meteorites

During the past two decades numerous (more than 40) lunar meteorites were found in several different locations, including Antarctica, North Africa, Oman, and Australia. These stones were blasted off the Moon by impacts and are identical in all relevant characteristics with the specimens returned by the Apollo missions.

Although lunar meteorites fell at different locations at different times, some of them may originate from the same lunar cratering event, i.e., they may be source-crater paired. The relevant parameter for judging source-crater pairing, the ejection age, is the sum of the Moon-Earth transfer time and the terrestrial age. Source-crater pairing is further constrained by the chemical characteristics of the meteorites.

Figure 10 shows measured Ca/Ti elemental ratios vs. ejection ages (*Lorenzetti et al.,* 2005). This ratio is chosen because in highland rocks the Ca concentration is higher and that of Ti lower than those in mare material. Several groupings are identified for meteorites with similar Ca/Ti ratios and ejection ages: the five highland breccias, DaG 400, QUE 93069/94269 (fall paired), DaG 262, ALHA 81005, and Y 791197 all have ejection ages of 0.06 m.y., and source-crater pairing cannot be excluded; MAC 88104/88105 (fall paired) and NWA 482 have T_{ej} = 0.25–0.30 m.y.; Y 82192/82193/86032 (fall paired) have T_{ej} = 8 ± 3 m.y. For the mare basalts and the highland/mare rocks we observe five different ejection ages: Y 793274 and Y 981031 are paired and have T_{ej} < 0.04 m.y.; EET 87521/96008 (fall paired) T_{ej} = 0.05 to 0.11 m.y.; QUE 94281 T_{ej} = 0.35 ± 0.10 m.y.; both Asuka 881757 and Y 793169 have T_{ej} = 1.0 ± 0.2 m.y., and Calcalong Creek 3 ± 1 m.y. From these data we conclude that at least three impact events in highland areas and five events in mare regions ejected the lunar meteorites collected until now.

The escape speed for rocks ejected from the Moon is 2.4 km/s. About 97% of all rocks ejected with ≤3.2 km/s reach the Earth within <1 m.y. (*Arnold,* 1965; *Gladman et al.,* 1995) and only a small fraction of all rocks is launched at speeds >3.2 km/s (*Gladman et al.,* 1996). *Warren* (1994) concluded that most lunar meteorites are ejected by impact

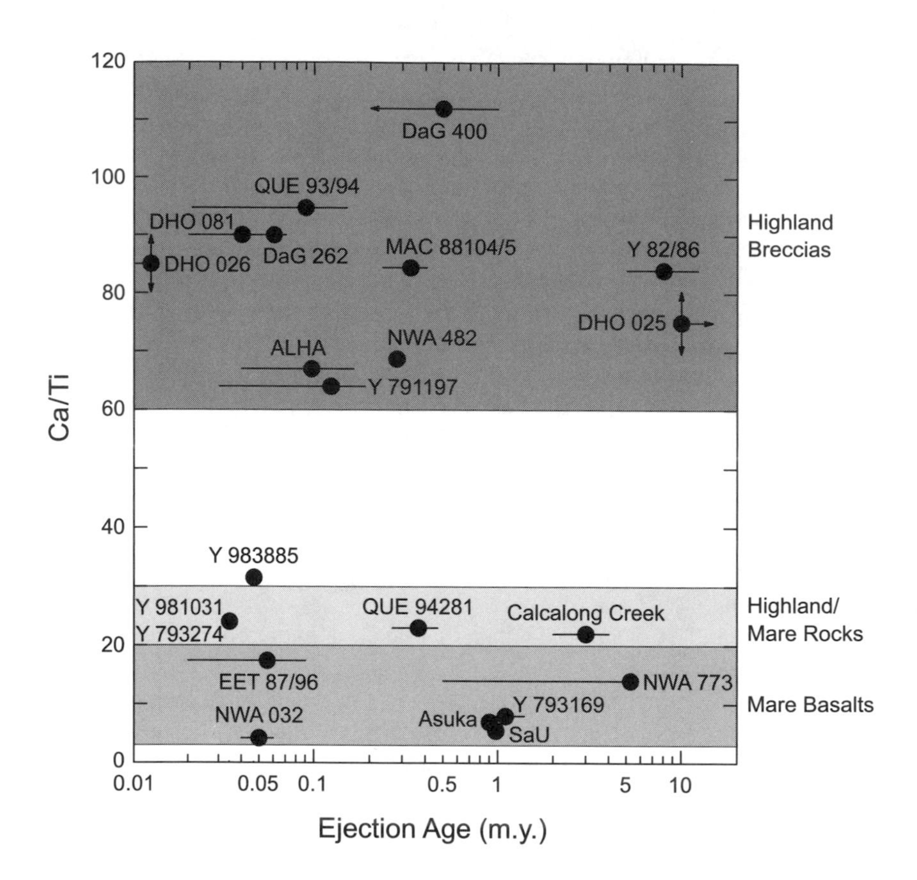

Fig. 10. Calcium/titanium ratios of lunar meteorites vs. times of ejection from the lunar surface. This ratio is diagnostic for the provenance of the lunar meteorites: High ratios are typical for lunar highlands, low ratios for lunar mare regions. For references see *Lorenzetti et al.* (2005).

events, resulting in craters with diameters of much less than 10 km. The smaller size of these craters relative to those responsible for the ejection of martian meteorites and the regolith structure of the lunar surface may explain why lunar meteorites were generally ejected from <1000 g/cm^2 (<5 m) depth (*Vogt et al.,* 1993b) and often show a complex exposure history.

5. RECORDS OF EARLY PARTICLE RADIATION

Processes of light-element synthesis in solar system materials by SCRs from an active early Sun were suggested by *Fowler et al.* (1962), and recently received renewed attention from the evidence for the presence of short-lived nuclides ^{10}Be, ^{26}Al, ^{36}Cl, ^{41}Ca, and ^{60}Fe in CAIs from primitive chondrites, as well as in chondrules. The evaluation of required particle fluxes and of variations in the isotopic compositions of elements due to decay of these short-lived radionuclides are discussed by *Chaussidon and Gounelle* (2006). These authors also summarize evidence for and estimates of the required fluences to explain the excesses of ^{21}Ne and ^{38}Ar in gas-rich meteorites, as well as recently suggested models for producing non-mass-dependent isotopic anomalies by solar UV radiation.

Solar-mass stars during their early evolution go through a phase of increased activity while en route to the main sequence, and this evolution is characterized by strong stellar winds, increased SCR activity, and increased ultraviolet luminosity. During this phase, particle fluxes would be considerably different from contemporary particle fluxes, and increased activity would have resulted in increased SCR production rates. There is considerable uncertainty about the duration of this phase, and it is not clear when this phase occurred with respect to the formation of the meteorites. If the active phase of the Sun ended prior to the formation of meteorites, then the meteorites themselves would contain no direct evidence for the existence of increased stellar activity. Also, solar nebular gases or solids may have partially or totally blocked this radiation at asteroidal distances. As a result, meteoritic matter may or may not have recorded effects from an enhanced particle flux.

Further complicating the picture is our limited knowledge of the early history of the meteorites and their precursor components that may carry records of the energetic-particle environment. As discussed in this chapter, the much better knowledge of the contemporary energetic particle environment permits calculation of CRE ages. However, these inferred production rates may differ significantly from the rates prevalent in the early solar system. In the introduction we mentioned that precursors of the meteorites may have been exposed during accretion and that other components were probably exposed during regolith processes. *Polnau et al.* (2001) presented evidence for excess amounts of light noble gas spallation components in chondrules, particularly in chondrites with short CRE ages, such as the Sena chondrite. These authors inferred that a precompaction irradiation is the likely source of excess concentrations.

Furthermore, fossil records of solar particle radiations in asteroidal regoliths were observed in gas-rich meteorites. The evidence was discussed in detail by *Caffee et al.* (1988). Specific records of particle radiations were recently also studied in a variety of meteoritic regolith breccias (*Ferko et al.,* 2002; *Lorenzetti et al.,* 2003; *Mathew and Marti,* 2003).

6. CONCLUSIONS AND OUTLOOK

The presence of peaks in CRE exposure age distributions, taken in conjunction with spectral observations of asteroids and observations of meteorite falls, shows that a few asteroids have produced most of the rocks that fall to Earth. To illustrate this point we note that about 45% of all H chondrites (Fig. 4) have CRE ages between either 6 and 10 m.y. or 30 and 35 m.y. Thus we infer that a maximum of 4000 of the 7000 H chondrites in collections may have come from collisions other than those leading to the two peaks in the CRE age distribution for H chondrites. Although we cannot yet infer with confidence which spectral type of asteroid produced the H chondrites, we know that one possible class of asteroids, the S-type, comprises about 15% of all asteroids with known spectra. The total number of asteroids with radii greater than 1 km in size is about 1,000,000. Thus, even in the exceedingly unlikely event that each known H chondrite originates on a different asteroid, we would have sampled only a small fraction of the total observed population, a compelling reason to explore the asteroid belt.

As was recognized early on, the short CRE ages (<1500 m.y. for irons and <120 m.y. for stones) relative to the 4560-m.y. age of the solar system permit several conclusions: (1) Ongoing processes (collisions in the asteroid belt, breakup of comets) must replenish the meteoroid population. (2) Meteoroid destruction events (collisions, capture by the Sun and planets, and ejection from the solar system) limit CRE ages of meteoroids, otherwise CRE ages would range up to 4.56 G.y. (3) Matter residing in the top few meters of meteorite parent asteroids either exhibits short lifetimes or constitutes a small fraction of the collisional ejecta. (4) The distribution of collisional events shows that mechanical strength (low for aubrites) cannot be the only important factor in limiting CRE ages. Also, subclasses of iron meteorites (IIE) have short CRE ages (Fig. 8).

Meteoroids populate a size range that astronomers cannot observe directly with present-day optical methods; as a result, the CRE ages and the influx of meteoritic matter have been longstanding problems for dynamicists. Dynamical calculations showed that once a meteoroid's orbit is in resonance with Jupiter, the orbital evolution is chaotic (*Wisdom,* 1985) and the transport to Earth follows a chaotic route and implies short (less than 1 m.y.) CRE ages. As discussed in section 4, the calculations show that the Yarkovsky effect delivers asteroidal meteoroids to resonances on timescales consistent with observed CRE ages. This agreement clearly suggests that the Yarkovsky effect is a major mechanism for orbital evolutions of meteoroids.

Ultimately the attempt should be to tie CRE ages to detailed cratering histories of specific parent bodies. Except for the lunar and martian meteorites, the only meteorites for which a parent body has been assigned are the howardites, eucrites, and diogenites. Their probable parent asteroid is Vesta or multikilometer-sized fragments of Vesta with Vesta-consistent orbital elements. Cosmic-ray exposure ages of this group cluster at 22 m.y. and show minor clusters at 36 and 12 m.y. (Fig. 6). Periodicities in the distribution of CRE ages have been suggested, but the evidence is restricted to possible common clusterings at 6, 13, 25, and 36 m.y. Common cluster statistics were interpreted to reflect enhanced collisional activity in the asteroid belt. For a larger number of meteorites (H chondrites, L chondrites, CM and CI chondrites, acapulcoites, and lodranites), CRE ages inform when, but not on which specific bodies, major collisional events took place.

The available data leave many large gaps. Exposure histories of lunar meteorites, for example, reflect only a blurry picture regarding the "when" of the collisions, i.e., the cratering history of the Moon. The Moon's proximity to Earth suggests that more-frequent, smaller collisions produce recoverable meteorites once escape velocities are achieved. Smaller collisions, in turn, favor ejection of near-surface material with a significant irradiation history on the Moon. Unfortunately, pre-irradiation on the Moon makes it more difficult to resolve the extra cosmogenic nuclide signal accumulated during the transit from Moon to Earth. Thus many of the transit times of lunar meteorites, although clearly short, are known only as upper bounds.

It is desirable to follow the arrival of ejecta from a single collisional event over millions of years in order to test models of orbital evolution. The Antarctic meteorites may supply such a source of information, but detailed results will be a long time coming for two reasons. First, this work requires determinations of terrestrial age and CRE age for each meteorite. Second, there is only a handful of meteorites with terrestrial ages of 1 m.y. or more. In this regard, the recent work of *Heck et al.* (2004) represents an intriguing new approach to tracking meteorite influxes over time.

The one-stage exposure history model breaks down in regolith samples. For this reason, in returned samples from the surfaces of other solar system bodies, the calculation of CRE ages for bulk samples must give way to statistical characterization of exposure. In a sample of lunar soil or a meteoritic breccia containing 1000 grains, for example, each one may have a distinct history of exposure. No experimentalist in the foreseeable future can hope to track these stochastic variations in detail. The relevant extractable parameters are nominal times of exposure at various depths, or perhaps rates of vertical turnover; analyses of cosmogenic nuclides supply little information about horizontal motions in a regolith. Possibly micrometeorites may represent stand-ins for sources of material eroded from asteroidal surfaces.

We emphasize that all discussed CRE ages (except for ^{40}K/K ages and revised ^{38}Ar ages of irons) are based on the assumption of a constant CR flux. We have shown that CR fluxes differed from present values in the early solar system environment and during the period from 150 to 700 m.y. before present. As a goal of future research, the variation of the particle environment and of the GCR flux in the "local" region of the Sun during its galactic rotation needs to be assessed in detail, in particular during its motion in star-forming regions of the galaxy.

Some of today's frontiers of CRE histories lie at the temporal and spatial boundaries of the solar system, in histories of particles of stardust sequestered for 4.5 G.y. in meteorites, and in solar system and extrasolar particles recovered by missions and spacecraft. The study of these and other planetary samples requires improvement in instrumental sensitivity by orders of magnitude.

Acknowledgments. We thank M. Gounelle, R. Michel, K. Welten, and R. Wieler for helpful comments. Technical support was provided by A. Chaoui and K. Bratschi.

REFERENCES

Albrecht A., Schnabel C., Vogt S., Xue S., Herzog G. F., Begemann F., Weber H. W., Middleton R., Fink D., and Klein J. (2000) Light noble gases and cosmogenic radionuclides in Estherville, Budulan, and other mesosiderites: Implications for exposure histories and production rates. *Meteoritics & Planet. Sci., 35,* 975–986.

Anders E. (1964) Origin, age, and composition of meteorites. *Space Sci. Rev., 3,* 583–714.

Arnold J. R. (1965) The origin of meteorites as small bodies. II. The model. *Astrophys. J., 141,* 1536–1547.

Arnold J. R., Honda M., and Lal D. (1961) Record of cosmic-ray intensity in meteorites. *J. Geophys. Res., 66,* 3519–3531.

Artemieva N. A. and Ivanov B. A. (2002) Ejection of martian meteorites — can they fly? (abstract). In *Lunar and Planetary Science XXXIII,* Abstract #1113. Lunar and Planetary Institute, Houston (CD-ROM).

Axford W. I. (1981) Acceleration of cosmic rays by shock waves. *Proceedings of the 17th International Cosmic-Ray Conference, Paris, 12,* 155.

Begemann F. and Schultz L. (1988) The influence of bulk chemical composition on the production rate of cosmogenic nuclides in meteorites (abstract). In *Lunar and Planetary Science XXIX,* pp. 51–52. Lunar and Planetary Institute, Houston.

Begemann F., Vilcsek E., Nyquist L. E., and Signer P. (1970) The exposure history of the Pitts meteorite. *Earth Planet. Sci. Lett., 9,* 317–321.

Begemann F., Weber H. W., Vilcsek E., and Hintenberger H. (1976) Rare gases and ^{36}Cl in stony-iron meteorites: Cosmogenic elemental production rates, exposure ages, diffusion losses, and thermal histories. *Geochim. Cosmochim. Acta, 40,* 353–368.

Bhandari N., Lal D., Rajan R. S., Arnold J. R., Marti K., and Moore C. B. (1980) Atmospheric ablation in meteorites: A study based on cosmic ray tracks and neon isotopes. *Nucl. Tracks, 4,* 213–262.

Bhattacharya S. K., Goswami J. N., Gupta S. K., and Lal D. (1973) Cosmic ray effects induced in a rock exposed on the Moon or in free space: Contrast in patterns for tracks and isotopes. *Moon, 8,* 253–286.

Biermann L. (1951) Kometenschweife und solare Korpuskularstrahlung. *Z. Astrophys., 29,* 274–286.

Biermann L. (1953) Physical processes in comet tails and their relation to solar activity. *Extrait des Mem. Soc. Roy. Sci. Liege Quatr. Ser.,* pp. 291–302.

Binns W. R., Wiedenbeck M. E., Christian E. R., Cummings A. C., George J. S., Israel H. M., Leske R. A., Mewaldt R. A., Stone E. C., Rosenvinge T. T., and Yanasak N. E. (2001) GCR neon isotopic abundances: Comparison with Wolf-Rayet star models and meteorite abundances. In *Solar and Galactic Composition* (R. F. Wimmer-Schweingruber, ed.), pp. 257–262. American Institute of Physics, New York.

Binzel R. P. and Xu S. (1993) Chips off of asteroid 4 Vesta: Evidence for the parent body of basaltic achondrite meteorites. *Science, 260,* 186–191.

Bogard D. D., Nyquist L. E., Bansal B. M., Garrison D. H., Wiesmann H., Herzog G. F., Albrecht A. A., Vogt S., and Klein J. (1995) Neutron-capture ^{36}Cl, ^{41}Ca, ^{36}Ar, and ^{150}Sm in large chondrites: Evidence for high fluences of thermalized neutrons. *J. Geophys. Res., 100,* 9401–9416.

Bottke W. F. (2002) The dynamical evolution of asteroids and meteoroids via Yarkovsky thermal forces (abstract). *Bull. Am. Astron. Soc., 34,* 934.

Bottke W. F., Rubincam D. P., and Burns J. A. (2000) Dynamical evolution of main belt meteoroids: Numerical simulations incorporating planetary perturbations and Yarkovsky thermal forces. *Icarus, 145,* 301–331.

Branham R. L. Jr. (2002) Kinematics of OB stars. *Astrophys. J., 570,* 190–197.

Burbine T. H., McCoy T. J., Meibom A., Gladman B., and Keil K. (2003) Meteoritic parent bodies: Their number and identification. In *Asteroids III* (W. F. Bottke Jr. et al., eds.), pp. 653–667. Univ. of Arizona, Tucson.

Burnett D. S., Lippolt H. J., and Wasserburg G. J. (1966) The relative isotopic abundance of K^{40} in terrestrial and meteoritic samples. *J. Geophys. Res., 71,* 1249–1269.

Caffee M. W. and Nishiizumi K. (1997) Exposure ages of carbonaceous chondrites: II (abstract). *Meteoritics & Planet. Sci., 32,* A26.

Caffee M. W. and Nishiizumi K. (2001) Exposure history of separated phases from the Kapoeta meteorite. *Meteoritics & Planet. Sci., 36,* 429–437.

Caffee M. W., Goswami J. N., Hohenberg C. M., Marti K., and Reedy R. C. (1988) Irradiation records in meteorites. In *Meteorites and the Early Solar System* (J. F. Kerridge and M. S. Matthews, eds), pp. 205–245. Univ. of Arizona, Tucson.

Caffee M. W., Nishiizumi K., Matsumoto Y., Matsuda J., Komura K., and Nakamura N. (2000) Noble gases and cosmogenic radionuclides in Kobe CK meteorite (abstract). *Meteoritics & Planet. Sci., 35,* A37.

Castagnoli G. C. and Lal D. (1980) Solar modulation effects in terrestrial production of carbon 14. *Radiocarbon, 22,* 133–159.

Chapman C. R., Merline W. J., Thomas P. C., Joseph J., Cheng A. F., and Izenberg N. (2002) Impact history of Eros: Craters and boulders. *Icarus, 155,* 104–118.

Chaussidon M. and Gounelle M. (2006) Irradiation processes in the early solar system. In *Meteorites and the Early Solar System II* (D. S. Lauretta and H. Y. McSween Jr., eds.), this volume. Univ. of Arizona, Tucson.

Consolmagno G. J. and Drake M. J. (1977) Composition and evolution of the eucrite parent body: Evidence from rare earth elements. *Geochim. Cosmochim. Acta, 41,* 1271–1282.

Cressy P. J. Jr. (1972) Cosmogenic radionuclides in the Allende and Murchison carbonaceous chondrites. *J. Geophys. Res., 77,* 4905–4911.

De Zeeuw P. T., Hoogerwerf R., De Bruijne J. H. J., Brown A. G. A., and Blaauw A. (1999) A Hipparcos census of the nearby OB associations. *Astronom. J., 117,* 354–399.

Eberhardt P. (1974) A neon-E-rich phase in the Orgueil carbonaceous chondrite. *Earth Planet. Sci. Lett., 24,* 182–187.

Eberhardt P., Geiss J., and Lutz H. (1963) Neutrons in meteorites. In *Earth Science and Meteoritics* (J. Geiss and E. D. Goldberg, eds.), pp. 143–168. North Holland, Amsterdam.

Eberhardt P., Eugster O., and Geiss J. (1965) Radiation ages of aubrites. *J. Geophys. Res., 70,* 4427–4434.

Eberhardt P., Eugster O., Geiss J., and Marti K. (1966) Rare gas measurements in 30 stone meteorites. *Z. Naturforsch., 21A,* 414–426.

Eberhardt P., Geiss J., Graf H., and Schwaller H. (1971) On the origin of excess ^{131}Xe in lunar rocks. *Earth Planet. Sci. Lett., 12,* 260–262.

Eberhardt P., Jungck M. H. A., Meier F. O., and Niederer F. (1979) Presolar grains in Orgueil: Evidence from neon-E. *Astrophys. J. Lett., 234,* L169–L171.

Eugster O. (1988) Cosmic-ray production rates for ^{3}He, ^{21}Ne, ^{38}Ar, ^{83}Kr, and ^{126}Xe in chondrites based on ^{81}Kr-Kr exposure ages. *Geochim. Cosmochim. Acta, 52,* 1649–1662.

Eugster O. (2003) Cosmic-ray exposure ages of meteorites and lunar rocks and their significance. *Chem. Erde/Geochemistry, 63,* 3–30.

Eugster O. and Lorenzetti S. (2005) Cosmic-ray exposure ages of four acapulcoites and two differentiated achondrites and evidence for a two-layer structure of the acapulcoite-lodranite parent asteroid. *Geochim. Cosmochim. Acta, 69,* 2675–2685.

Eugster O. and Michel T. (1995) Common asteroid break-up events of eucrites, diogenites, and howardites and cosmic-ray production rates for noble gases in achondrites. *Geochim. Cosmochim. Acta, 59,* 177–199.

Eugster O., Eberhardt P., and Geiss J. (1967) ^{81}Kr in meteorites and ^{81}Kr radiation ages. *Earth Planet. Sci. Lett., 2,* 77–82.

Eugster O., Tera F., Burnett D. S., and Wasserburg G. J. (1970) Neutron capture effects in Gd from the Norton County meteorite. *Earth Planet. Sci. Lett., 7,* 436–440.

Eugster O., Michel Th., Niedermann S., Wang D., and Yi W. (1993) The record of cosmogenic, radiogenic, fissiogenic, and trapped noble gases in recently recovered Chinese and other chondrites. *Geochim. Cosmochim. Acta, 57,* 1115–1142.

Eugster O., Eberhardt P., Thalmann Ch., and Weigel A. (1998) Neon-E in CM-2 chondrite LEW90500 and collisional history of CM-2 chondrites, Maralinga, and other CK chondrites. *Geochim. Cosmochim. Acta, 62,* 2573–2582.

Eugster O., Busemann H., Lorenzetti S., and Terribilini D. (2002a) Ejection ages from krypton-81–krypton-83 dating and pre-atmospheric sizes of martian meteorites. *Meteoritics & Planet. Sci., 37,* 1345–1360.

Eugster O., Busemann H., Kurat G., Lorenzetti S., and Varela M. E. (2002b) Characterization of the noble gases and the CRE age of the D'Orbigny angrite (abstract). *Meteoritics & Planet. Sci., 37,* A44.

Farinella P., Vokrouhlicky D., and Hartmann W. K. (1998) Meteorite delivery via Yarkovsky orbital drift. *Icarus, 132,* 378–387.

Ferko T. E., Wang M-S., Hillegonds D. J., Lipschutz M. E., Hutchison R., Franke L., Scherer P., Schultz L., Benoit P. H., Sears D. W. G., Singhvi A. K., and Bhandari N. (2002) The irradiation history of the Ghubara (L5) regolith breccia. *Meteoritics & Planet. Sci., 37,* 311–327.

Fleischer R. L., Price P. B., Walker R. M., and Maurette M. (1967) Origin of fossil charged particle tracks in meteorites. *J. Geophys. Res., 72,* 333–353.

Fleischer R. L., Price P. B., and Walker R. M. (1975) *Nuclear Tracks in Solids: Principles and Applications.* Univ. of California, Berkeley. 705 pp.

Fowler W. A., Greenstein J. L., and Hoyle F. (1962) Nucleosynthesis during the early history of the solar system. *Geophys. J. Roy. Astron. Soc., 6,* 148–220.

Freundel M., Schultz L., and Reedy R. C. (1986) ^{81}Kr-Kr ages of Antarctic meteorites. *Geochim. Cosmochim. Acta, 50,* 2663–2673.

Gilabert E., Lavielle B., Michel R., Leya I., Neumann S., and Hepers U. (2002) Production of krypton and xenon isotopes in thick stony and iron targets isotropically irradiated with 1600 MeV protons. *Meteoritics & Planet. Sci., 37,* 951–976.

Gladman B. J. (1997) Destination: Earth. Martian meteorite delivery. *Icarus, 130,* 228–246.

Gladman B. J., Burns J. A., Duncan M. J., and Levison H. F. (1995) The dynamical evolution of lunar impact ejecta. *Icarus, 118,* 302–321.

Gladman B. J., Burns J. A., Duncan M., Lee P., and Levison H. F. (1996) The exchange of impact ejecta between terrestrial planets. *Science, 271,* 1387–1392.

Goodrich C. A. (1992) Ureilites: A critical review. *Meteoritics, 27,* 327–352.

Goswami J. N., Sinha N., Nishiizumi K., Caffee M. W., Komura K., and Nakamura K. (2001) Cosmogenic records in Kobe (CK4) meteorite: Implications of transport of meteorites from the asteroid belt (abstract). *Meteoritics & Planet. Sci., 36,* A70–A71.

Gounelle M., Shu F. H., Shang H., Glassgold A. E., Rehm K. E., and Lee T. (2001) Extinct radioactivities and protosolar cosmic rays: Self shielding and light elements. *Astrophys. J., 548,* 1051–1070.

Gounelle M., Shu F. H., Shang H., Glassgold A. E., Rehm K. E., and Lee T. (2004) The origin of short-lived radionuclides and early solar system irradiation (abstract). In *Lunar and Planetary Science XXXV,* Abstract #1629. Lunar and Planetary Institute, Houston (CD-ROM).

Graf T. and Marti K. (1994) Collisional records in LL-chondrites. *Meteoritics, 29,* 643–648.

Graf T. and Marti K. (1995) Collisional history of H chondrites. *J. Geophys. Res., 100,* 21247–21263.

Graf T., Baur H., and Signer P. (1990a) A model for the production of cosmogenic nuclides in chondrites. *Geochim. Cosmochim. Acta, 54,* 2521–2534.

Graf T., Signer P., Wieler R., Herpers U., Sarafin R., Vogt S., Fieni Ch., Pellas P., Bonani G., Suter M., and Wölfli W. (1990b) Cosmogenic nuclides and nuclear tracks in the chondrite Knyahinya. *Geochim. Cosmochim. Acta, 54,* 2511–2520.

Graf T., Marti K., Xue S., Herzog G. F., Klein J., Middleton R., Metzler K., Herd R., Brown P., Wacker J. F., Jull A. J. T., Masarik J., Koslowsky V. T., Andrews H. R., Cornett R. J. J., Davies W. G., Greiner B. F., Imahori Y., McKay J. W., Milton G. M., and Milton J. C. D. (1997) Exposure history of the Peekskill (H6) meteorite. *Meteoritics & Planet. Sci., 32,* 25–30.

Graf T., Caffee M. W., Marti K., Nishiizumi K., and Ponganis K. V. (2001) Dating collisional events: ^{36}Cl-^{36}Ar exposure ages of H-chondritic metals. *Icarus, 150,* 181–188.

Heck P. R., Schmitz B., Baur H., Halliday A. N., and Wieler R. (2004) Fast delivery of meteorites to Earth after a major asteroid collision. *Nature, 430,* 323–325.

Herzog G. F. (2003) Cosmic-ray exposure ages of meteorites. In *Treatise on Geochemistry, Vol. 1: Meteorites, Comets, and*

Planets (A. M. Davis, ed.), pp. 347–380. Elsevier, Oxford.

Herzog G. F., Vogt S., Albrecht A., Xue S., Fink D., Klein J., Middleton R., Weber H. W., and Schultz L. (1997) Complex exposure histories for meteorites with "short" exposure ages. *Meteoritics & Planet. Sci., 32,* 413–422.

Heusser G., Ouyang Z., Oehm J., and Yi W. (1996) Aluminum-26, sodium-22 and cobalt-60 in two drill cores and some other samples of the Jilin chondrite. *Meteoritics & Planet. Sci., 31,* 657–665.

Hidaka H., Ebihara M., and Yoneda S. (1999) High fluences of neutrons determined from Sm and Gd isotopic compositions in aubrites. *Earth Planet. Sci. Lett., 173,* 41–51.

Hidaka H., Ebihara M., and Yoneda S. (2000) Isotopic study of neutron capture effects on Sm and Gd in chondrites. *Earth Planet. Sci. Lett., 180,* 29–37.

Hohenberg C. M., Nichols R. H., Olinger C. T., and Goswami J. N. (1990) Cosmogenic neon from individual grains of CM meteorites: Extremely long pre-exposure histories or an enhanced early particle flux. *Geochim. Cosmochim. Acta, 54,* 2133–2140.

Honda M. (1962) Spallation products distributed in a thick iron target bombarded by 3 BeV protons. *J. Geophys. Res., 67,* 4847–4858.

Honda M., Nagai H., Nagao K., and Miura Y. N. (1996) Cosmogenic products in metal phase of the Brenham pallasite (abstract). *Antarctic Meteorites, 21,* 48–50.

Honda M., Caffee M. W., Miura Y. N., Nagai H., Nagao K., and Nishiizumi K. (2002) Cosmogenic nuclides in the Brenham pallasite. *Meteoritics & Planet. Sci., 37,* 1711–1728.

Hörz F., Grieve R., Heiken G., Spudis P., and Binder A. (1991) Lunar surface processes. In *Lunar Sourcebook: A User's Guide to the Moon* (G. Heiken et al., eds.), pp. 61–120. Cambridge Univ., Cambridge.

Kaiser W. A. and Rajan R. S. (1973) The variation of cosmogenic Kr and Xe in a core from Esterville mesosiderite: Direct evidence that the lunar ^{131}Xe anomaly is a depth effect. *Earth Planet. Sci. Lett., 20,* 286–294.

Kalinina G. V., Kashkarov L. L., Ivleiv A. I., and Skripnik A. Y. (2004) Radiation and shock thermal history of the Kaidun CR2 chondrite glass inclusions (abstract). In *Lunar and Planetary Science XXXV,* Abstract #1075. Lunar and Planetary Institute, Houston (CD-ROM).

Keil K., Haack H., and Scott E. R. D. (1994) Catastrophic fragmentation of asteroids: Evidence from meteorites. *Planet. Space Sci., 42,* 1109–1122.

Kohman T. P. and Bender M. L. (1967) Nuclide production by cosmic rays in meteorites and on the moon. In *High-Energy Reactions in Astrophysics* (B. S. P. Shen, ed.), pp. 169–245. Benjamin, New York.

Lal D. (1972) Hard rock cosmic ray archaeology. *Space Sci. Rev., 14,* 3–102.

Lavielle B., Marti K., and Regnier S. (1985) Ages d'exposition des meteorites de fer: Histoires multiples et variations d'intensité du rayonnement cosmique. In *Isotopic Ratios in the Solar System,* pp. 15–20. Cepadeus-Editions, Toulouse, France.

Lavielle B., Marti K., Jeannot J.-P., Nishiizumi K., and Caffee M. (1999) The ^{36}Cl-^{36}Ar-^{40}K-^{41}K records and cosmic ray production rates in iron meteorites. *Earth Planet. Sci. Lett., 170,* 93–104.

Lavielle B., Caffee M., Gilabert E., Marti K., Nishiizumi K., and Ponganis K. (2001) Irradiation records in group IVA irons. *Meteoritics & Planet. Sci., 36,* A110–A111.

Leya I. and Michel R. (1998) Determination of neutron cross sections for nuclide production at intermediate energies by deconvolution of thick-target production rates. In *Proceedings of the International Conference on Nuclear Data for Science and Technology* (G. Reffo, ed.), pp. 1463–1467. Trieste, Italy.

Leya I., Lange H.-J., Neumann S., Wieler R., and Michel R. (2000) The production of cosmogenic nuclides in stony meteoroids by galactic cosmic ray particles. *Meteoritics & Planet. Sci., 35,* 259–286.

Leya I., Wieler R., Aggrey K., Herzog G. F., Schnabel C., Metzler K., Hildebrand A. R., Bouchard M., Jull A. J. T., Andrews H. R., Wang M.-S., Ferko T. E., Lipschutz M. E., Wacker J. F., Neumann S., and Michel R. (2001a) Exposure history of the St-Robert (H5) fall. *Meteoritics & Planet. Sci., 36,* 1479–1494.

Leya I., Graf T., Nishiizumi K., and Wieler R. (2001b) Cosmic-ray production rates of He-, Ne-, and Ar-isotopes in H-chondrites based on ^{36}Cl-^{36}Ar ages. *Meteoritics & Planet. Sci., 36,* 963–973.

Leya I., Halliday A., and Wieler R. (2003) The predictable collateral consequences of nucleosynthesis by spallation reactions in the early solar system. *Astrophys. J., 594,* 605–616.

Leya I., Begemann F., Weber H. W., Wieler R., and Michel R. (2004a) Simulation of the interaction of galactic cosmic ray protons with meteoroids: On the production of ^{3}H and light noble gas isotopes in isotropically irradiated thick gabbro and iron targets. *Meteoritics & Planet. Sci., 39,* 367–386.

Leya I., Gilabert E., Lavielle B.,Wiechert U., and Wieler R. (2004b) Production rates for cosmogenic krypton and argon isotopes in H-chondrites with known ^{36}Cl-^{36}Ar ages. *Antarct. Met. Res., 17,* 185–199.

Lingenfelter R. E., Canfield E. H., and Hampel V. E. (1972) The lunar neutron flux revisited. *Earth Planet. Sci. Lett., 16,* 355–369.

Lingenfelter R. E., Higdon J. C., and Ramaty R. (2000) Cosmic ray acceleration in superbubbles and the compositon of cosmic rays. In *Acceleration and Transport of Energetic Particles Observed in the Heliosphere* (R. Mewaldt et al., eds.), pp. 375. American Institute of Physics, New York.

Lorenzetti S., Eugster O., Busemann H., Marti K., Burbine T. H., and McCoy T. (2003) History and origin of aubrites. *Geochim. Cosmochim. Acta, 67,* 557–571.

Lorenzetti S., Busemann H., and Eugster O. (2005) Regolith history of lunar meteorites. *Meteoritics & Planet. Sci., 40,* 315–327.

Lüpke M. (1993) Untersuchungen zur Wechselwirkung galaktischer Protonen mit Meteoroiden-Dicktarget Simulationsexperimenten und Messung von Dünntarget-Wirkungsquerschnitten. Ph.D. thesis, University of Hannover, Germany.

Marti K. (1967) Mass-spectrometric detection of cosmic-ray-produced Kr^{81} in meteorites and the possibility of Kr-Kr dating. *Phys. Rev. Lett., 18,* 264–266.

Marti K. (1986) Live ^{129}I-^{129}Xe dating. In *Workshop on Cosmogenic Nuclides* (R. C. Reedy and P. Englert, eds.), pp. 49–51. LPI Tech. Rpt. 86-06, Lunar and Planetary Institute, Houston.

Marti K. and Graf T. (1992) Cosmic-ray exposure history of ordinary chondrites. *Annu. Rev. Earth Planet. Sci., 20,* 221–243.

Marti K. and Lugmair G. W. (1971) Kr^{81}-Kr and K-Ar^{40} ages, cosmic-ray spallation products, and neutron effects in lunar samples from Oceanus Procellarum. *Proc. Lunar. Sci. Conf. 2nd,* pp. 1591–1605.

Marti K., Eberhardt P., and Geiss J. (1966) Spallation, fission, and neutron capture anomalies in meteoritic krypton and xenon.

Z. Naturforsch., 21a, 398–413.

Marti K., Lingenfelter R., Mathew K. J., and Nishiizumi K. (2004) Cosmic rays in the solar neighborhood (abstract). *Meteoritics & Planet. Sci., 39,* A63.

Masarik J. and Reedy R. C. (1994) Effects of bulk composition on nuclide production processes in meteorites. *Geochim. Cosmochim. Acta, 58,* 5307–5317.

Masarik J., Nishiizumi K., and Reedy R. C. (2001) Production rates of ^{3}He, ^{21}Ne, and ^{22}Ne in ordinary chondrites and the lunar surface. *Meteoritics & Planet. Sci., 36,* 643–650.

Mathew K. J. and Marti K. (2003) Solar wind and other gases in the regoliths of the Pesyanoe parent object and the moon. *Meteoritics & Planet. Sci., 38,* 627–643.

McCoy T. J., Keil K., Clayton R. N., Mayeda T. K., Bogard D. D., Garrison D. H., Wieler R. (1997) A petrologic and isotopic study of lodranites: Evidence for early formation as partial melt residues from heterogeneous precursors. *Geochim. Cosmochim. Acta, 61,* 623–637.

Megrue G. H. (1968) Rare gas chronology of hypersthene achondrites and pallasites. *J. Geophys. Res., 73,* 2027–2033.

Meyer C. (2003) *Mars Meteorite Compendium-2003, Revision B.* JSC Publ. No. 27672, NASA Johnson Space Center, Houston.

Michel R., Dragovitsch P., Englert P., Peiffer F., Stuck R., Theis S., Begemann F., Weber H., Signer P., Wieler R., Filges D., and Cloth P. (1986) On the depth dependence of spallation reactions in a spherical thick diorite target homogeneously irradiated by 600 MeV protons: Simulation of production of cosmogenic nuclides in small meteorites. *Nucl. Instrum. Methods, B16,* 61–82.

Michel R., Dragovitsch P., Cloth P., Dagge G., and Filges D. (1991) On the production of cosmogenic nuclides in meteoroids by galactic protons. *Meteoritics, 26,* 221–242.

Michel R., Lüpke M., Herpers U., Rösel R., Suter M., Dittrich-Hannen B., Kubik P. W., Filges D., and Cloth P. (1995) Simulation and modelling of the interaction of galactic protons with stony meteoroids. *Planet. Space Sci., 43,* 557–572.

Michel R., Leya I., and Borges L. (1996) Production of cosmogenic nuclides in meteoroids — Accelerator experiments and model calculations to decipher the cosmic ray record in extraterrestrial matter. *Nucl. Instrum. Methods Phys. Res., B113,* 434–444.

Miura Y. N. (1995) Studies on differentiated meteorites: Evidence from ^{244}Pu-derived fission Xe, ^{81}Kr, other noble gases and nitrogen. Ph.D. thesis, Univ. of Tokyo. 212 pp.

Miura Y. N., Nagao K., Sugiura N., Fujitani T., and Warren P. H. (1998) Noble gases ^{81}Kr-Kr exposure ages and ^{244}Pu-Xe ages of six eucrites, Béréba, Binda, Camel Donga, Juvinas, Millbillillie, and Stannern. *Geochim. Cosmochim. Acta, 62,* 2369–2387.

Nishiizumi K. and Caffee M. W. (2002) Exposure histories of C2 carbonaceous chondrites (abstract). *Meteoritics & Planet. Sci., 37,* A109.

Nishiizumi K., Regnier S., and Marti K. (1980) Cosmic ray exposure ages of chondrites, pre-irradiation and constancy of cosmic ray flux in the past. *Earth Planet. Sci. Lett., 50,* 150–170.

Nishiizumi K., Arnold J. R., Caffee M. W., Finkel R. C., Southon J. R., Nagai H., Honda M., Imamura M., Kobayashi K., and Sharma P. (1993a) Exposure ages of carbonaceous chondrites (abstract). In *Lunar and Planetary Science XXIV,* pp. 1085–1086. Lunar and Planetary Institute, Houston.

Nishiizumi K., Arnold J. R., and Sharma P. (1993b) Two-stage exposure of the Fayetteville meteorite based on ^{129}I (abstract). *Meteoritics, 28,* 412.

Nishiizumi K., Caffee M. W., Bogard D. D., Garrison D. H., and Kyte F. T. (2000) Noble gases and cosmogenic radionuclides in the Eltanin meteorite (abstract). In *Lunar and Planetary Science XXXI,* Abstract #2070. Lunar and Planetary Institute, Houston (CD-ROM).

Nyquist L. E., Funk H., Schultz L., and Signer P. (1973) He, Ne, and Ar in chondritic Ni-Fe as irradiation hardness sensors. *Geochim. Cosmochim. Acta, 37,* 1655–1685.

Nyquist L. E., Bogard D. D., Shih C.-Y., Greshake A., Stöffler D., and Eugster O. (2001) Ages and geologic histories of martian meteorites. *Space Sci. Rev., 96,* 105–164.

Okazaki R., Takaoka N., Nakamura T., and Nagao K. (2000) Cosmic-ray exposure ages of enstatite chondrites. *Antarct. Meteorite Res., 13,* 153–169.

Ott U. and Begemann F. (2000) Spallation recoil and age of presolar grains in meteorites. *Meteoritics & Planet. Sci., 35,* 53–63.

Ott U., Löhr H.-P., and Begemann F. (1985) Noble gases and the classification of Brachina. *Meteoritics, 20,* 69–78.

Ott U., Löhr H. P., and Begemann F. (1987) Noble gases in ALH 84025: Like Brachina, unlike Chassigny. *Meteoritics, 22,* 476–477.

Padia J. T. and Rao M. N. (1989) Neon isotope studies of Fayetteville and Kapoeta meteorites and clue to ancient solar activity. *Geochim. Cosmochim. Acta, 53,* 1461–1467.

Patzer A. and Schultz L. (2001) Noble gases in enstatite chondrites I: Exposure ages, pairing, and weathering effects. *Meteoritics & Planet. Sci., 36,* 947–961.

Patzer A., Schultz L., and Francke L. (2003) New noble gas data of primitive and differentiated achondrites including NWA 011 and Tafassasset. *Meteoritics & Planet. Sci., 38,* 1485–1498.

Pepin R. O., Palma R. L., and Schlutter D. J. (2001) Noble gases in interplanetary dust particles, II: Excess helium-3 in cluster particles and modeling constraints on interplanetary dust particle exposures to cosmic-ray irradiation. *Meteoritics & Planet. Sci., 36,* 1515–1534.

Polnau E., Eugster O., Burger M., Krähenbühl U., and Marti K. (2001) The precompaction exposure of chondrules and implications. *Geochim. Cosmochim. Acta, 65,* 1849–1866.

Pomerantz M. A. and Duggal S. P. (1974) The Sun and cosmic rays. *Rev. Geophys. Space Phys., 12,* 343–361.

Ramaty R., Kozlovsky B., and Lingenfelter R. E. (1996) Light isotopes, extinct radioisotopes, and gamma-ray lines from low-energy cosmic-ray interactions. *Astrophys. J., 456,* 525–540.

Rao M. N., Garrison D. H., Palma R. L., and Bogard D. D. (1997) Energetic proton irradiation history of the howardite parent body regolith and implications for ancient solar activity. *Meteoritics & Planet. Sci., 32,* 531–543.

Reedy R. C. (1985) A model for GCR-particle fluxes in stony meteorites and production rates of cosmogenic nuclides. *Proc. Lunar Planet. Sci. Conf. 15th,* in *J. Geophys. Res., 90,* C722–C728.

Reedy R. C. (1987) Predicting the production rates of cosmogenic nuclides in extraterrestrial matter. *Nucl. Instrum. Meth., B29,* 251–261.

Reedy R., Arnold J. R., and Lal D. (1983) Cosmic-ray record in solar system matter. *Science, 219,* 127–135.

Reeves H. (1978) Supernovae contamination of the early solar system in an OB stellar association or the "Bing Bang" theory of the origin of the solar system. In *Protostars and Planets* (T. Gehrels, ed.), pp. 399–426. Univ. of Arizona, Tucson.

Rubin A., Kallemeyn G. W., Wasson J. T., Clayton R. N., Mayeda T. K., Grady M., Verchovsky A. B., Eugster O., Lorenzetti S. (2003) Formation of metal and silicate globules in Gujba: A new Bencubbin-like meteorite fall from Nigeria. *Geochim. Cosmochim. Acta, 67,* 3283–3298.

Schaeffer O. A., Nagel K., Fechtig H., and Neukum G. (1981) Space erosion of meteorites and the secular variation of cosmic rays (over 10^9 years). *Planet. Space Sci., 29,* 1109–1118.

Scherer P. and Schultz L. (2000) Noble gas record, collisional history, and paring of CV, CO, CK, and other carbonaceous chondrites. *Meteoritics & Planet. Sci., 35,* 145–153.

Scherer P., Zipfel J., and Schultz L. (1998) Noble gases in two new ureilites from the Saharan desert (abstract). In *Lunar and Planetary Science XXIX,* Abstract #1383. Lunar and Planetary Institute, Houston (CD-ROM).

Schnabel C., Herzog G. F., Pierazzo E., Xu S., Masarik J., Cresswell R. G., di Tada M. L., Liu K., and Fifield L. K. (1999) Shock melting of the Canyon Diablo impactor: Constraints from nickel-59 and numerical modeling. *Science, 285,* 85–88.

Schnabel C., Ma P., Herzog G. F., Faestermann T., Knie K., and Korschinek G. (2001) ^{10}Be, ^{26}Al, and ^{53}Mn in martian meteorites (abstract). In *Lunar and Planetary Science XXXII,* Abstract #1353. Lunar and Planetary Institute, Houston (CD-ROM).

Schnabel C., Leya I., Gloris M., Michel R., Lopez-Gutirérrez J. M., Krähenbühl U., Herpers U., Kuhnhenn J., and Synal H. A. (2004) Production rates and proton-induced production of cross sections of ^{129}I from Te and Ba. An attempt to model the ^{129}I production in stony meteoroids and ^{129}I in a Knyahinya sample. *Meteoritics & Planet. Sci., 39,* 453–466.

Schultz L. and Franke L. (2004) Helium, neon, and argon in meteorites: A data collection. *Meteoritics & Planet. Sci., 39,* 1889–1890.

Schultz L. and Signer P. (1977) Noble gases in the St. Mesmin chondrite: Implications to the irradiation history of a brecciated meteorite. *Earth Planet. Sci. Lett., 36,* 363–371.

Schultz L., Signer P., Lorin J. C., and Pellas P. (1972) Complex irradiation history of the Weston chondrite. *Earth Planet. Sci. Lett., 15,* 403–410.

Schultz L., Weber H. W., and Franke L. (2005) Rumuruti (R)-chondrites: Noble gases, exposure ages, pairing and parent body history. *Meteoritics & Planet. Sci., 40,* 557–571.

Shukolyukov Y. A. and Begemann F. (1996) Cosmogenic and fissiogenic noble gases and ^{81}Kr-Kr exposure age clusters of eucrites. *Meteoritics & Planet. Sci., 31,* 60–72.

Shukolyukov Y. A. and Petaev M. I. (1992) Noble gases in the Omolon pallasite (abstract). In *Lunar and Planetary Science XXIII,* pp. 1297–1298. Lunar and Planetary Institute, Houston.

Spergel M. S., Reedy R. C., Lazareth O. W., Levy P. W., and Slates L. A. (1986) Cosmogenic neutron-capture-produced nuclides in stony meteorites. *Proc. Lunar Planet. Sci. Conf. 16th,* in *J. Geophys. Res., 91,* D483–D494.

Spitale J. and Greenberg R. (2002) Numerical evaluation of the general Yarkovsky effect: Effects on eccentricity and longitude of periapse. *Icarus, 156,* 211–222.

Stauffer H. (1962) On the production ratios of rare gas isotopes in stone meteorites. *J. Geophys. Res., 67,* 2023–2028.

Swindle T. D., Kring D. A., Burkland M. K., Hill D. H., and Boynton W. F. (1998) Noble gases, bulk chemistry, and petrography of olivine-rich achondrites Eagles Nest and Lewis Cliff 88763: Comparison to brachinites. *Meteoritics & Planet. Sci., 33,* 31–48.

Terribilini D., Eugster O., Herzog G. F., and Schnabel C. (2000a) Evidence for common break-up events of the acapulcoites/lodranites and chondrites. *Meteoritics & Planet. Sci., 35,* 1043–1050.

Terribilini D., Eugster O., Mittlefehldt D. W., Diamond L. W., Vogt S., and Wang D. (2000b) Mineralogical and chemical composition and cosmic-ray exposure history of two mesosiderites and two iron meteorites. *Meteoritics & Planet. Sci., 35,* 617–628.

Ustinova G. K. (2002) Mechanisms of isotopic heterogeneity generation by shock waves in the primary matter of the solar system. *Geochem. Intl., 40,* 827–842.

Vogt S., Herzog G. F., and Reedy R. C. (1990) Cosmogenic nuclides in extraterrestrial materials. *Rev. Geophys., 28,* 253–275.

Vogt S., Aylmer D., Herzog G. F., Wieler R., Signer P., Pellas P., Fiéni C., Tuniz C., Jull A. J. T., Fink D., Klein J., and Middleton R. (1993a) Multi-stage exposure history of the H5 chondrite Bur Gheluai. *Meteoritics, 28,* 71–85.

Vogt S. K., Herzog G. F., Eugster O., Michel Th., Niedermann S., Krähenbühl U., Middleton R., Dezfouly-Arjomandy B., Fink D., and Klein J. (1993b) Exposure history of the lunar meteorite Elephant Moraine 87521. *Geochim. Cosmochim. Acta, 57,* 3793–3799.

Voshage H. (1967) Bestrahlungsalter und Herkunft der Eisenmeteorite. *Z. Naturforsch., 22a,* 477–506.

Voshage H. (1982) Investigations of cosmic-ray-produced nuclides in iron meteorites, 4. Identification of noble gas abundance anomalies. *Earth Planet. Sci. Lett., 61,* 32–40.

Voshage H. and Feldmann H. (1979) Investigations of cosmic-ray-produced nuclides in iron meteorites, 3: Exposure ages, meteoroid sizes and sample depths determined by spectrometric analyses of potassium and rare gases. *Earth Planet. Sci. Lett., 45,* 293–308.

Voshage H. and Hintenberger H. (1963) The cosmic-ray exposure ages of iron meteorites as derived from the isotopic composition of potassium and the production rates of cosmogenic nuclides in the past. In *Radioactive Dating,* pp. 367–379. International Atomic Energy Agency, Vienna.

Voshage H., Feldmann H., and Braun O. (1983) Investigations of cosmic-ray-produced nuclides in iron meteorites: 5. More data on the nuclides of potassium and noble gases, on exposure ages and meteoroid sizes. *Z. Naturforsch., 38a,* 273–280.

Vokrouhlicky D., Milani A., and Chesley S. R. (2000) Yarkovsky effect on small near-Earth asteroids: Mathematical formulation and examples. *Icarus, 148,* 118–138.

Warren P. H. (1994) Lunar and Martian meteorite delivery services. *Icarus, 111,* 338–363.

Weigel A., Eugster O., Koeberl C., Michel R., Krähenbühl U., and Neumann S. (1999) Relationships among lodranites and acapulcoites: Noble gas isotopic abundances, chemical composition, cosmic-ray exposure ages, and solar cosmic ray effects. *Geochim. Cosmochim. Acta, 63,* 175–192.

Welten K. C., Lindner L., van der Berg K., Loeken T., Scherer P., and Schultz L. (1997) Cosmic-ray exposure ages of diogenites and the recent collisional history of the howardite, eucrite and diogenite parent body/bodies. *Meteoritics & Planet. Sci., 32,* 891–902.

Welten K. C., Nishiizumi K., Caffee M. W., and Schultz L. (2001a) Update on exposure ages of diogenites: The impact history of the HED parent body and evidence of space erosion and/or collisional disruption of stony meteoroids (abstract). *Meteoritics & Planet. Sci., 36,* A223.

Welten K. C., Bland P. A., Russell S. S., Grady M. M., Caffee

M. W., Masarik J., Jull A. J. T., Weber H. W., and Schultz L. (2001b) Exposure age and pre-atmospheric radius of the Chinguetti-mesosiderite: Not part of a much larger mass. *Meteoritics & Planet. Sci., 36,* 939–946.

Welten K. C., Caffee M. W., Leya I., Masarik J., Nishiizumi K., and Wieler R. (2003) Noble gases and cosmogenic radionuclides in the Gold Basin L4-chondrite shower: Thermal history, exposure history, and pre-atmospheric size. *Meteoritics & Planet. Sci., 38,* 157–174.

Welten K., Nishiizumi K., Caffee M. W., Hillegonds D. J., Leya I., Wieler R., and Masarik J. (2004) The complex exposure history of a very large L/LL5 chondrite shower: Queen Alexandra Range 90201 (abstract). In *Lunar and Planetary Science XXXV,* Abstract #2020. Lunar and Planetary Institute, Houston (CD-ROM).

Wieler R. (2002a) Cosmic-ray-produced noble gases in meteorites. *Rev. Mineral. Geochem., 47,* 125–170.

Wieler R. (2002b) Noble gases in the solar system. *Rev. Mineral. Geochem., 47,* 21–70.

Wieler R. and Graf T. (2001) Cosmic ray exposure history of meteorites. In *Accretion of Extraterrestrial Material Throughout Earth's History* (B. Peucker-Ehrenbrink and B. Schmitz, eds.), pp. 221–240. Kluwer Academic/Plenum, New York.

Wieler R., Graf T., Pedroni A., Signer P., Pellas P., Fieni C., Suter M., Vogt S., Clayton R. N., and Laul J. C. (1989) Exposure history of the regolithic chondrite Fayetteville: II. Solar-gas-free light inclusions. *Geochim. Cosmochim. Acta, 53,* 1449–1459.

Wieler R., Graf T., Signer P., Vogt S., Herzog G. F., Tuniz C., Fink D., Fifield L. K., Klein J., Middleton R., Jull A. J. T., Pellas P., Masarik J., and Dreibus G. (1996) Exposure history of the Torino meteorite. *Meteoritics & Planet. Sci., 31,* 265–272.

Wieler R., Pedroni A., and Leya I. (2000) Cosmogenic neon in mineral separates from Kapoeta: No evidence for an irradiation of its parent body regolith by an early active Sun. *Meteoritics & Planet. Sci., 35,* 251–257.

Wieler R., Busemann H., and Franchi I. A. (2006) Trapping and modification processes of noble gases and nitrogen in meteorites and their parent bodies. In *Meteorites and the Early Solar System II* (D. S. Lauretta and H. Y. McSween Jr., eds.), this volume. Univ. of Arizona, Tucson.

Wisdom J. (1985) A perturbative treatment of motion near 3:1 commensurability. *Icarus, 63,* 272–289.

Weathering of Chondritic Meteorites

P. A. Bland
Imperial College London and Natural History Museum (London)

M. E. Zolensky
NASA Johnson Space Center

G. K. Benedix
Natural History Museum (London)

M. A. Sephton
Imperial College London

Meteorites are frequently not the pristine record of early solar system processes that we would like them to be. Many unique meteorites, samples that can further our understanding of the environment of the early solar system, are finds, so a knowledge of terrestrial weathering becomes a requirement to accurately interpret their preterrestrial history. This chapter outlines the weathering process in chondrites, the mineralogical changes that occur in response to weathering, the variations in weathering in different environments, the effect of weathering on O isotopes and chemistry, and how weathering varies between different chondrite groups. As the most abundant meteorite group, most studies of meteorite weathering have concentrated on ordinary chondrites, although weathering affects all classes of meteorites. In many cases, these data are relevant to other chondrite groups, as well as irons and achondrites, but there are a number of specific issues related to carbonaceous chondrites (degradation of organic C, mobilization of S, reordering of phyllosilicates) that require a separate discussion. In particular, carbonaceous chondrites appear to be uniquely susceptible to terrestrial weathering, and recent studies have shown a number of alteration effects, even in falls curated in museum collections. The terrestrial alteration of carbonaceous chondrites concludes our chapter.

1. INTRODUCTION

Meteorite weathering can be regarded as the alteration of original component phases of the meteorite to phases that are more stable at Earth's surface. On entering Earth's atmosphere, interaction with the terrestrial environment begins. Understanding the nature of this interaction is extremely important, given that modern meteoritics involves the analysis of a much higher proportion of finds, following the recovery of large populations of samples from desert accumulations, than was the norm a few decades ago. Knowing the details of the terrestrial weathering process allows those effects to be deconvolved from any asteroidal alteration that may be present (e.g., *Abreu and Brearley,* 2005). In addition, recent studies have observed a number of parallels between weathering of chondrites in Antarctica and aqueous alteration on asteroids (e.g., *Lee and Bland,* 2004; *Sephton et al.,* 2004). There are parallels, both in the effect of alteration on organic material and in the nature of the weathering products produced. It may be that the Antarctic environment can provide an effective analog in helping us to understand asteroidal alteration.

Terrestrial weathering of a meteorite is often viewed as an obstacle to interpreting the preterrestrial history of a sample, but meteorites have value to terrestrial researchers involved in low-temperature aqueous alteration studies in that they provide a "standard sample" for tracking the effects of rock weathering. Meteorites are unique among geological materials in that they show relatively minor intragroup variations and their terrestrial ages can be established by measuring the decay of cosmogenic radionuclides by accelerator mass spectrometry. They may therefore be a potential "chronometer" of environmental conditions during their terrestrial residency, given the proper analytical approach. The processes and rates of this weathering depend on a similar group of factors to those that control the alteration of terrestrial rocks, namely the chemistry of weathering fluids and the nature of reactions at mineral and alloy surfaces. The microenvironmental conditions within a rock that regulate the rates of chemical reactions are frequently influenced by climate and include such factors as the pH and f_{O_2} of pore- and groundwaters, temperature, drainage, and microbial activity. Meteorites have been recovered from all over the world and therefore have been exposed to all types of climatic regimes. Other factors that control the response to weathering include the composition and crystal structure of the silicate or alloy, the nature of secondary oxidation products, the degree of porosity or fracturing, and the exposure time of the sample (e.g., *Gooding,* 1981; *Velbel et al.,* 1991).

1.1. Falls, Finds, and Meteorite Accumulations

Prior to 1969 the total number of known meteorites was around 2100 (*Grady,* 2000). Many of these had been observed as fireballs as they entered Earth's atmosphere, and were then recovered later ("falls"), but around 60% were chance discoveries or "finds." Currently, meteorite samples available to researchers number in the tens of thousands, following extensive collecting programs in the Antarctic and in hot desert regions such as the Sahara and the Nullarbor of Australia. With only 1000 falls in modern collections (*Grady,* 2000), the ratio of falls to finds has changed dramatically since 1969. Preservation of meteorites depends mainly on climate. In temperate and tropical zones weathering is relatively rapid. However, in persistently dry climates chemical weathering is less vigorous, and this process might be slowed down to the point where meteorites begin to accumulate (*Bland et al.,* 1996a, 1998a). Meteorites may be preserved for thousands, tens of thousands, or, as in Antarctica, millions of years after their fall.

The cold desert of Antarctica was the first place where concentrations of meteorites were recognized (*Yoshida et al.,* 1971). In Antarctica, meteorites survive for longer owing to relatively constant environmental conditions and low chemical weathering, and are physically concentrated by processes of ice movement and ablation. As ice moves from the center of the continent to the coast it encounters obstacles in the form of mountain ranges. While most of the ice flows around the barriers, some remains in front of the rock and is only removed gradually by wind ablation. Any meteorites that may have fallen into the ice upstream of this point will thus gradually be exposed, forming a concentration on this stranding surface (*Cassidy et al.,* 1992). In addition to Antarctica, a number of the world's other arid and semi-arid regions have proved to be prolific sources of meteorites. Large numbers of meteorites have been recovered from "bad lands" subject to wind deflation in Roosevelt County, New Mexico (*Scott et al.,* 1986; *Sipiera et al.,* 1987; *Zolensky et al.,* 1990), and many more meteorites have also been found in the stony deserts of North Africa (*Wlotzka,* 1989; *Jull et al.,* 1990, *Schlüter et al.,* 2002). More than 3.8 million km^2 of Australia is arid or semi-arid land that has also provided conditions suitable for the prolonged preservation of meteorites (*Bevan,* 1992a,b).

1.2. Differences Between Antarctic and Non-Antarctic Populations

Terrestrial age dating of meteorites by cosmogenic radionuclides has revealed that there are large differences in the residence time between Antarctic and hot-desert meteorites. Meteorites from Antarctica show a mean residence time of around 10^5 yr (*Nishiizumi et al.,* 1989; *Nishiizumi,* 1995), while hot desert meteorites have much younger terrestrial ages, typically between 15,000 and 20,000 yr (*Jull et al.,* 1993a, 1995; *Wlotzka et al.,* 1995). This raises the possibility that the two meteorite populations might sample different asteroidal parent bodies, since it has been suggested (*Wetherill,* 1986) that "meteorite streams" may be the source of many of Earth's more common meteorites, and that these streams may vary over the timescale of 10^5–10^6 yr (*Greenberg and Chapman,* 1983). *Dennison et al.* (1986) and *Dennison and Lipschutz* (1987) observed differences in the trace-element composition between Antarctic and non-Antarctic H5 ordinary chondrite populations, and suggested a temporal variation in the source regions from which the meteoroids derive as a cause. *Benoit and Sears* (1992) also observed differences in the metallographic cooling rate between Antarctic and non-Antarctic H5 chondrites. Nonetheless, while it appears likely that there may be generic differences between Antarctic and non-Antarctic meteorite populations, it is clear that many of the variations are a function of climate and its effect on the weathering of meteorites (*Buchwald,* 1989): A better understanding of terrestrial weathering will help in identifying generic differences between meteorite populations and make the identification of extraterrestrial alteration simpler.

1.3. Weathering Scales

In an attempt to quantify the degree of alteration that a sample has experienced, a number of qualitative weathering indices have been applied to cold- and hot-desert populations. The Meteorite Working Group at the NASA Johnson Space Center in Houston uses weathering categories "A," "B," and "C" to denote minor, moderate, and severe rustiness of Antarctic finds, and have a similar scale for degree of fracturing. A scale of weathering effects seen in polished thin sections of meteorites has been proposed by *Jull et al.* (1991) and updated by *Wlotzka* (1993), *Wlotzka et al.* (1995), and *Al-Kathiri et al.* (2005) for meteorites from hot deserts. In this system six categories of weathering are recognized, beginning with minor to complete oxidation of metal and then troilite (categories W1–W4) and continuing with at first minor (W5) and then massive (W6) alteration of mafic silicates. *Bland et al.* (1996a,b, 1998a, 2000a) discussed a quantitative index of alteration for ordinary chondrites, using ^{57}Fe Mössbauer spectroscopy to determine the relative abundance of Fe^{3+} in a sample (unweathered ordinary chondrite falls contain negligible Fe^{3+}).

2. PROCESSES AND EFFECTS OF WEATHERING

2.1. Alteration of Ferromagnesian Silicates

Determining the differences in alteration rate between phases in a meteorite is a rather difficult analytical problem. Mössbauer spectroscopy is a useful tool in this regard. Observing how the spectral area of primary phases varies with the total amount of oxidation allows us to determine which phases are most susceptible to weathering: An increase in a ferric component should be accompanied by a concomitant decrease in an absorption associated with a primary mineral. *Bland et al.* (1998a) compared the spectral area of "opaque" phases (Fe-Ni and troilite) and ferromagnesian silicates to total oxidation, and observed a decrease in primary phases with increasing oxidation, suggesting that all Fe-containing minerals within the meteorite are affected by weathering to some degree. A similar observation was made by *Burns et al.* (1995) on weathered Antarctic meteorites. Interestingly, although ferromagnesian silicates appear to be weathered at a constant rate, there was no detectable difference in the rate of weathering between olivine and pyroxene. Although a similar feature has been observed in studies of weathered Antarctic meteorites (*Fisher and Burns,* 1992), its interpretation remains problematic given the well-documented susceptibility of olivine to dissolution in experimental weathering studies (*Luce et al.,* 1972; *Siever and Woodford,* 1979; *Grandstaff,* 1978). One explanation relates to how the O-isotopic composition of meteorites varies with weathering (discussed in more detail below). *Bland et al.* (2000b) found evidence for topotactic alteration during the incipient weathering of olivine. *Banfield et al.* (1990, 1991) note that olivine weathered in this manner is comparatively resistant to subsequent oxidation: a possible explanation for the similarity between olivine and pyroxene weathering rates in ordinary chondrites.

2.2. Rusting of Meteorites and Corrosion Products

In common with the iron meteorites, the Fe-Ni metal, troilite, and mafic silicates — the main components in ordinary chondrites — render these meteorites susceptible to rusting; ferruginous oxidation products are the most obvious weathering products on meteorites. *Gooding* (1986) recognized two forms of rust: "metallic" (Fe-Ni-S) rust, formed by weathering of primary Fe-Ni metal and troilite, and "sialic" (Fe-Si-Al) rust, which forms by the weathering of mafic silicates. In a model based on the variable ability of Fe-rich minerals in meteorites to nucleate water to ice (and so retard weathering), *Gooding* (1986) found that troilite is less susceptible to weathering than Fe-Ni metal (composed of the two polymorphs taenite and kamacite), taenite is less susceptible than kamacite, and silicates are less susceptible than sulfides and metal. Using scanning electron microscope, electron microprobe, and X-ray powder diffraction, akaganéite [β-FeOOH], goethite [α-FeOOH], lepidocrocite [γ-FeOOH], and maghemite [γ-Fe_2O_3] were identified by *Buchwald and Clarke* (1989) as constituting the bulk of the corrosion products in a suite of Antarctic meteorites. Magnetite [Fe_3O_4] was also observed by *Buchwald* (1989). *Marvin* (1963), using X-ray powder photographs, recognized akaganéite as a corrosion product of iron meteorites along with goethite and magnetite. Hematite was found in very small quantities in only one meteorite. Other studies have also tentatively identified jarosite [$KFe_3(SO_4)_2(OH)_6$] (*Gooding,* 1981, 1986; *Zolensky and Gooding,* 1986) as another component of "metallic" rust. Metallic rust commonly occurs as mantles on Fe-Ni metal or as veins, and may act as a pore-filling cement in a similar way to diagenetic cements in sedimentary rocks (*Gooding,* 1986). The importance of Mössbauer spectroscopy as a technique in characterizing these types of mineral assemblages is evident. The presence of this suite of oxides and oxyhydroxides was confirmed in a ^{57}Fe Mössbauer spectroscopy study of weathered ordinary chondrites by *Bland et al.* (1998a). Only jarosite was not observed.

In an extensive study of weathering in Antarctic meteorites, *Buchwald and Clarke* (1989) found that akaganéite was the mineralogical key to an understanding of the corrosion of meteoritic metal. It has also been shown to be important in the weathering of a modern fall: Tsarev, an L5 chondrite that fell in Russia in 1922 (*Honda et al.,* 1993). β-FeOOH (akaganéite) was described in detail by *Mackay* (1960, 1962); *Waychunas* (1991) describes the mineral as having a tunnel structure similar to hollandite [$KAlSi_3O_8$], with a variable amount of a large anion (Cl^- is common) occupying the alkali site. Evidence suggests that akaganéite is not stable without this extra species (*Feitknecht et al.,* 1973; *Chambaere and DeGrave,* 1984), and as such a new and more detailed formula for akaganéite may be written $[Fe_{15}Ni][O_{12}(OH)_{20}]Cl_2(OH)$ (*Buchwald and Clarke,* 1989), which is a variation of that proposed by *Keller* (1970).

Buchwald and Clarke (1989) postulated a process of corrosion in meteorites involving anodic metal going into solution and Cl^- ions from the terrestrial environment moving to the reaction surface to maintain a charge balance, Cl-containing akaganéite then precipitating at the reaction surface. Over time, these initial corrosion products decompose to intimate intergrowths of goethite and maghemite, releasing Cl^- for further corrosive action. As suggested in this model, the Cl content is found to be highest at the interface with the metal. In temperate zones, Cl could be introduced from soil groundwater; however, in Antarctica the source is probably sea-spray, carried inland as an aerosol or gas (see below). The transition from kamacite to akaganéite appears to take place without significant volume change (the silicates remain undisturbed). The Ni content of the oxide reflects the composition of the phase from which it formed, so Fe/Ni ratios of around 16 result when akaganéite is formed from kamacite, and about 4 when formed from taenite. In a further study involving Antarctic and non-Antarctic meteorites *Buchwald* (1989) noted that in cold, dry environments (<20% relative humidity) akaganéite is stable and long lived, while in meteorites from temperate climates, akaganéite is only seen in the active corrosion zone, most of it having decomposed to the oxides goethite, maghemite, and magnetite. Using conventional half-reactions, the corrosion mechanism proposed by *Buchwald and Clarke* (1989) may be expressed as follows. Fe-Ni metal (Fe^0) is oxidized at the anode

$$Fe^0 \rightarrow Fe^{2+} + 2e$$

and O_2 is reduced at the cathode

$$O_2 + 2H_2O + 4e \rightarrow 4OH^-$$

or alternatively, H is reduced at the cathode

$$2H^+ + 2e \rightarrow H_2$$

Combining the first two equations

$$2Fe^0 + O_2 + 2H_2O \xrightarrow[\text{irreversible}]{} 2Fe(OH) + 2OH^-$$

This is followed under oxidizing conditions by the formation of akaganéite

$$2Fe(OH)^+ + 2OH^- + \tfrac{1}{2}O_2 \xrightarrow[\text{irreversible}]{} \underbrace{2\beta\text{–}Fe^{3+}OOH}_{\text{akaganéite}} + H_2O$$

Using a modified version of the formula proposed by *Keller* (1970), *Buchwald and Clarke* (1989) express the reaction as

$$\underbrace{30Fe^0 + 2Ni^0}_{\text{kamacite}} + 47O + 19H_2O + 4H^+ + 4Cl^- \xrightarrow[\text{irreversible}]{} \underbrace{2[Fe_{15}Ni][O_{12}(OH)_{20}]Cl_2(OH)}_{\text{akaganéite}}$$

This reaction occurs slowly in the Antarctic environment, but more rapidly in temperate climates. Cl^- in akaganéite exchanges with OH^-, and Cl^- is released into solution to move again to the corrosion front or to be flushed from the system. The akaganéite left behind will approach the composition $[Fe_{15}Ni][O_{12}(OH)_{20}](OH)_3$ and, having lost its Cl, becomes unstable. Under conditions of modest heating and/or seasonal changes in temperature and humidity, it decomposes according to a reaction such as

$$2[Fe_{15}Ni][O_{12}(OH)_{20}](OH)_3 \longrightarrow$$

$$\underbrace{7\gamma\text{–}Fe_2O_3 + NiO}_{\text{maghemite}} + \underbrace{16\alpha\text{–}FeOOH + NiO}_{\text{goethite}} + 15H_2O$$

The recognition of a similar suite of corrosion products, including akaganéite, in meteorites recovered from a variety of climatic regimes (*Bland et al.,* 1998a) indicates that this corrosion mechanism may well be acting in the oxidation of all meteoritic metal. This is also substantiated by the presence of the mixed valence ironhydroxychloride, $Fe_2(OH)_3Cl$, in a number of meteorites from temperate climates (*Buchwald,* 1989), a mineral that may represent the initial step in the corrosion process. More recently, the suggestion that akaganéite requires Cl for stability has been questioned by *Rézel and Génin* (1990). These workers used Mössbauer spectroscopy and X-ray analysis to demonstrate that akaganéite retained its crystal structure following substitution of Cl^- ions by OH^-. In addition, *Bland et al.* (1997) have recorded akaganéite in a meteorite weathered in the absence of Cl^-. But whether or not akaganéite requires Cl^- for stability, the ubiquitous nature of Cl in the environment means that natural occurrences of akaganéite contain Cl.

While the formation of the majority of iron oxides and oxyhydroxides found in ordinary chondrites may be explained by the process detailed above, the presence of magnetite as a widespread corrosion product in meteorites from temperate climates (*Buchwald,* 1989; *Bland et al.,* 1998a,b; *Verma and Tripathi,* 2004) suggests that an additional mechanism may be involved. The correlation between spectral areas associated with Fe-Ni metal and magnetite in Mössbauer spectra from an artificially weathered sample and from meteorites recovered from hot and cold desert regions (*Bland et al.,* 1997, 1998a) indicate that magnetite is formed as a direct result of the dissolution of Fe-Ni metal. A possible explanation for the stability of a mixed Fe^{2+}-Fe^{3+} oxide in this environment may be that in a situation of rapid dissolution and oxidation of Fe-Ni metal, porewaters around metal grains become saturated with Fe^{2+} ions. While some proportion are oxidized to Fe^{3+}, others remain in solution, the combination of both species allowing the formation of magnetite. A similar process was suggested by *Faust et al.* (1973) to explain oxidation observed in the Wolf Creek meteorite.

Although *Buchwald and Clarke* (1989) only examined the corrosion of meteoritic metal, similar reactions occur for the dissolution of other Fe-containing meteoritic minerals

(i) troilite (FeS):

$$FeS + H_2O + 3O^{2-} \rightarrow Fe^{2+} + H_2SO_4 \quad (1)$$

(ii) olivine ($Fo_{0.80}Fa_{0.20}$):

$$(Mg_{0.80}Fe_{0.20})_2SiO_4 + 4H^+ \rightarrow 1.60Mg^{2+} + 0.40Fe^{2+} + H_4SiO_4 \quad (2)$$

(iii) pyroxene ($En_{0.80}Fs_{0.20}$):

$$Mg_{0.80}Fe_{0.20}SiO_3 + 2H^+ + H_2O \rightarrow 0.80Mg^{2+} + 0.20Fe^{2+} + H_4SiO_4 \quad (3)$$

Once dissolution of Fe^{2+} ions from primary meteoritic minerals has occurred, oxidation of dissolved Fe^{2+} ions and hydrolysis of Fe^{3+} ions takes place

$$Fe^{2+} + \tfrac{1}{4}O_{2(aq)} + H^+ = Fe^{3+} + \tfrac{1}{2}H_2O \quad (4)$$

$$Fe^{3+} + 2H_2O = FeOOH_{(s)} + 3H^+ \quad (5)$$

where the overall hydrolysis reaction (equation (5)) involves intermediate steps, including

$$Fe^{3+} + H_2O = [FeOH]^{2+}_{(aq)} + H^+ \quad (6)$$

$$[FeO]^{2+} + H_2O = \tfrac{1}{2}[Fe_2(OH)_4]^{2+}_{(aq)} + H^+ \quad (7)$$

$$\tfrac{1}{2}[Fe_2(OH)_4]^{2+} = Fe(OH)_{3(aq)} + H^+ \quad (8)$$

$$Fe(OH) \rightarrow \text{polymer} \rightarrow 5Fe_2O_3.9H_2O \text{ (ferrihydrite)} \rightarrow \gamma\text{-FeO (maghemite)}, \text{ FeOOH (goethite, lepidocrocite, etc.)} \quad (9)$$

In the above reactions, Mg^{2+} and other nonferric cations may be incorporated in carbonates, as traces in some iron oxides/oxyhydroxides, in clay minerals, or removed from the system entirely.

Ferrihydrite is the mineral name identifying poorly crystalline to amorphous hydrous iron oxides with chemical composition near $5Fe_2O_3.9H_2O$. It is formed abundantly as the initial product of the hydrolysis and precipitation of dissolved Fe in a variety of geological systems (*Waychunas,* 1991). It is likely that this is also true for the oxidation of meteorites. In ^{57}Fe Mössbauer spectra of a sample of Allegan weathered in the laboratory, *Bland et al.* (1997) observed ferrihydrite as one of the suite of corrosion products that formed. The artificially accelerated oxidation and dissolution of this sample makes it likely that a poorly crystalline, amorphous iron oxide such as ferrihydrite would be the initial weathering product.

The weathering of ordinary chondrite meteorites therefore appears to involve a combination of a number of processes. The initial product of the weathering of Fe-Ni metal is akaganéite (or magnetite where dissolution is exceptionally rapid). For other primary meteoritic minerals (troilite, olivine, and pyroxene), ferrihydrite may be the first iron oxide mineral following the dissolution of these phases. Over time, the initial corrosion products undergo an aging process that converts them to goethite, maghemite, and lepidocrocite. Magnetite may remain as a metastable reaction product.

2.3. Differences in Weathering Between Cold and Hot Deserts

In Mössbauer studies of weathered meteorites, two classes of alteration phases are readily distinguished: paramagnetic weathering minerals, which include akaganéite, small-particle goethite, lepidocrocite, and phyllosilicates; and magnetically ordered phases, which include magnetite and maghemite. *Bland et al.* (1998a) observed that the ratio of paramagnetic to magnetically ordered ferric species varies widely in weathered ordinary chondrites, with a high abundance of magnetically ordered phases typically associated with more humid environments. It was noted that magnetite was extremely uncommon in Antarctic samples.

In a TEM study of Antarctic and hot desert finds, *Lee and Bland* (2004) observed cronstedtite, a Fe-rich phyllosilicate common in CM chondrites in the heavily weathered Antarctic finds ALHA 78045 and 77002. This is mineral is rare on Earth, and was not observed in hot desert samples. An unidentified hydrous Si-Fe-Ni-Mg mineral or gel had also partially replaced taenite in ALHA 78045. In addition to Fe-rich weathering products, "hot" desert meteorites contain sulfates, Ca-carbonate, and silica, minerals that were largely absent from Antarctic finds. As preterrestrial cronstedtite is abundant in CM2 carbonaceous chondrites, the Antarctic environment may be a powerful analog for aqueous alteration in the asteroidal parent bodies of primitive meteorites.

2.4. Formation of Evaporites and Carbon Content

Weathering of meteorites involves not only the formation of rust and oxidation of metal but also hydrolysis of silicates, hydration, carbonation, and solution. Clay mineraloids have been identified compositionally (*Gooding,* 1986, 1981) but are often insufficiently crystalline to allow identification by XRD (*Velbel and Gooding,* 1988). An integral part of the process of meteorite weathering is the production of a variety of terrestrial evaporitic carbonates and sulfates (*Velbel et al.,* 1991; *Jull et al.,* 1988; *Grady et al.,* 1989a). *Miyamoto* (1989, 1991) used infrared spectroscopy to analyze a suite of ordinary chondrites including modern falls and Antarctic and non-Antarctic finds, and found that hydrated carbonates were ubiquitous in Antarctic samples and common in non-Antarctic meteorites. X-ray diffraction studies of Antarctic chondrites (*Marvin,* 1980; *Gooding et al.,* 1988) have shown that the specific mineral constituents of evaporitic carbonates and sulfates include nesquehonite [$Mg(HCO_3)(OH).2H_2O$], hydromagnesite [$Mg_5(CO_3)_4(OH)_2.4H_2O$], starkcyite [$MgSO_4.4H_2O$], epsomite [$MgSO_4.7H_2O$], and gypsum [$CaSO_4.2H_2O$]. Measurements of $\delta^{13}C$ and $\delta^{18}O$ from nesquehonite in one Antarctic H5 chondrite indicate that it formed at near 0°C by reaction with terrestrial water and CO_2 (*Grady et al.,* 1989a). Carbon-14 dating of this phase showed that it formed in the last 40 years (*Jull et al.,* 1988). An analysis of the stable-isotopic ratios in Antarctic ordinary chondrites (*Grady et al.,* 1989b; *Socki et al.,* 1991) has confirmed that carbonates originate from a reaction between atmospheric CO_2 and the meteorite, while the source of the cations in evaporites (Na, Mg, K, Ca, Rb) is probably meteoritic (*Velbel et al.,* 1991). The primary mineral likely to react to produce a hydrous magnesium carbonate would be olivine (*Velbel et al.,* 1991) via decomposition with water and CO_2.

Total C content, and macromolecular C, may also be affected by weathering. Studies of the Holbrook meteorite, which fell in 1915 in Arizona and has been collected at various times over a period of 60 years, have shown that weathering of the meteorite has been accompanied by an increase in the C content (*Gibson and Bogard,* 1978). A study of the unequilibrated ordinary chondrite, Roosevelt County 075, has shown that this meteorite has a higher C content than any other bulk ordinary chondrite (*Ash and Pillinger,* 1992a). Stepped combustion experiments suggest that about 50% of the C is from terrestrial organic material (*Ash and Pillinger,* 1992a). Further work has shown some correlation between the degree of C contamination and terrestrial residence time in other meteorites from Roosevelt County (*Ash and Pillinger,* 1993).

2.5. Oxygen Isotopes and Weathering

Although the oxidative interaction of terrestrial porewaters with extraterrestrial meteoritic minerals must necessarily involve the incorporation of terrestrial O into the stable crystal lattices of secondary alteration minerals within the meteorite, very little has been done to systematically quantify the effect of weathering on the bulk O-isotopic composition of meteorites. *Clayton et al.* (1984) recognized that a major drawback in an O-isotopic study of Antarctic meteorite finds is the possibility of contamination by O derived from the polar ice. In the case of ice from the Allan Hills region, from which a large number of meteorites have been recovered, *Fireman and Norris* (1982) found a $\delta^{18}O$ of

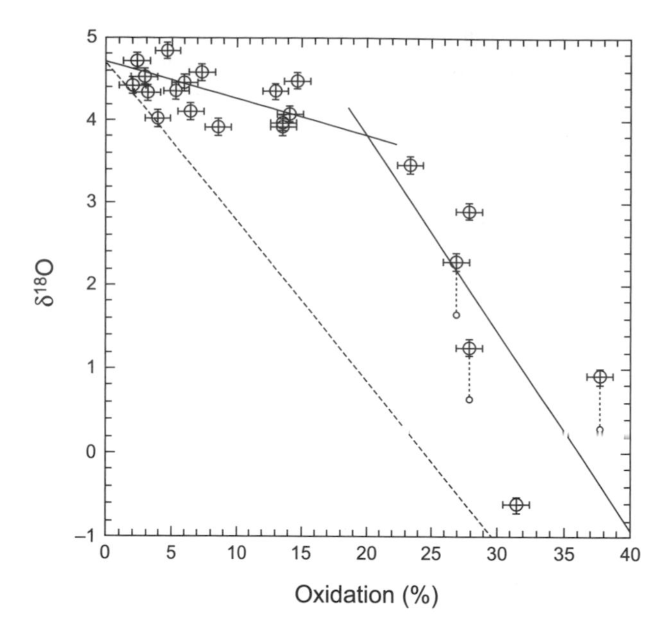

Fig. 1. Ferric oxidation (percentage of total iron now present as ferric iron, derived from ^{57}Fe Mössbauer spectroscopy analyses) plotted against $\delta^{18}O$ in L-chondrite finds from Antarctica. The dashed line shows the expected trend; solid lines are modeled trends assuming surface-mediated alteration initially and solution-mediated alteration after 20% oxidation.

about –40‰ relative to SMOW, i.e., even a small contamination should lead to a large displacement of the isotopic composition of the meteorite to lower $\delta^{18}O$ values (*Clayton et al.,* 1984). What is surprising is that significant amounts of alteration are required before any isotopic effect is observed. *Bland et al.* (2000b) studied how O isotopes varied in L-group ordinary chondrites weathered in Antarctica, using Mössbauer spectroscopy to quantify the degree weathering. The data indicate that alteration is a two-stage process, with an initial phase producing only a negligible isotopic effect. L-group falls contain no Fe^{3+} detectable using Mössbauer spectroscopy. Even though these meteorites begin with a $\delta^{18}O$ value of 4.7‰ (*Clayton et al.,* 1991), and are being altered by water at a $\delta^{18}O$ of –40‰, there is no discernible shift in O away from fall values until >25% of the Fe in the meteorite has been converted to Fe^{3+} (Fig. 1). Although surprising, an explanation may be found in the alteration of terrestrial silicates. Numerous studies report pervasive development of channels a few to a few tens of nanometers wide in the incipient alteration of silicates (*Eggleton,* 1984; *Smith et al.,* 1987; *Banfield et al.,* 1990; *Hochella and Banfield,* 1995). A similar texture is observed in the samples studied by *Bland et al.* (2000b). Alteration involves a restructuring to clay minerals along these narrow channels, in which access of water is restricted. The clay shows a topotactic relationship to the primary grain, suggesting either epitaxial growth of the clay using the silicate as a substrate, or inheritance of the original O structure by the clay. The data presented by *Bland et al.* (2000b) indicate the latter: With extensive inheritance of structural polymers by the weathering product, the bulk O-isotopic composition is comparatively unaffected. This work shows that the relationship between O isotopes and terrestrial alteration is not straightforward. It has significance in the area of asteroidal aqueous alteration in that substantial modification of chondritic materials may occur without a pronounced isotopic effect.

2.6. Elemental Mobilization During Weathering

The literature on this topic is sparse, although studies involving both ordinary chondrites and carbonaceous chondrites have been carried out. A discussion of mobilization of aqueous species during carbonaceous chondrite weathering is included at the end of this chapter.

In the case of ordinary chondrites, *Gibson and Bogard* (1978) observed a decrease in SiO_2 and MgO contents, and no change in the elemental concentrations for Ti, Ca, Al, P, and Mn, in the weathering of Holbrook. *Bland et al.* (1998a) also observed a decrease in Si and Mg (and minor decreases in Na and Ca) with increasing weathering. No significant changes were observed in elemental concentrations for Fe, Ti, Al, Mn, K, or P. These data are consistent with the dissolution products of ferromagnesian silicates (Si and Mg) being flushed from the system as weathering proceeds, while Fe is bound in hydrated ferric oxides, a scenario that is similar to that proposed by *Velbel et al.* (1991) for the mineral alteration and element mobilization involved in the growth of Mg-carbonates on Antarctic meteorites. As noted above, *Velbel et al.* (1991) found that Mg, Na, K, Ca, and Rb were mobilized from within an Antarctic H5 chondrite to form evaporate efflorescence on the surface. More recently, *Al-Kathiri et al.* (2005) have observed depletion in S in meteorites weathered in Oman, and also mobilization of Ni and Co, related to breakdown of Fe-Ni metal. Meteorites were commonly found to be enriched in Sr and Ba, both highly mobile trace elements. In Antarctica, the action of circulating waters in meteorites stranded on the ice has also produced elemental enrichments. *Delisle et al.* (1989) measured an enrichment of U in some meteorites from Frontier Mountain, Antarctica, on the order of 300× the normal value, a feature that was also noted by *Welten and Nishiizumi* (2000). The suggestion is that U salts have precipitated from solution in the meteorite after being leached from nearby granite.

2.7. Iodine Contamination

Although they have been stored in one of the most sterile terrestrial environments, I overabundances appear to be common in both Antarctic meteorites and rocks (*Dreibus and Wänke,* 1983; *Heumann et al.,* 1987; *Ebihara et al.,* 1989; *Langenauer and Krähenbühl,* 1992). Iodine is also seen to vary with depth in sample: *Ebihara et al.* (1989) found 11.2 ppm I in the outermost portion of an L6 chondrite, decreasing to 0.5 ppb at the center, concluding that the I overabundance was a result of interaction of atmospheric I compounds originating from the ocean with the surfaces of Antarctic rocks and meteorites. They postulated that the majority of the sea salt found in meteorites is emitted as small particles by sea spray deposited on the Antarctic ice close to the ice edge. Longer-distance transport may involve oxidation of iodide at the surface of the ocean to elemental I that exists as a gas before being precipitated in snow. From analyses of atmospheric halogens, *Heumann et al.* (1987) concluded that the gaseous compound methyl iodide (CH_3I) is probably responsible for the enrichment of I in Antarctic meteorites. This organoiodine compound is formed by microorganisms in the surface layers of the ocean. Methyl iodide has a relatively long atmospheric lifetime compared with elemental I or other iodides, allowing the compound to be transported long distances in the atmosphere. After deposition it would be converted photochemically into other I species.

2.8. Noble Gases

In a study of the isotopic composition and concentration of noble gases in a suite of ordinary chondrites from the Sahara desert and Antarctica, *Scherer et al.* (1992) found that the concentration of atmospheric noble gases (specifically the heavy noble gases Kr and Xe) appears to vary with weathering. Krypton-84 is higher in chondrites from hot deserts than in modern falls or Antarctic finds. In later work *Scherer et al.* (1995) also showed that hot desert meteorites have excess ^{132}Xe. Xenon-132 was also found to be greater in Antarctic samples than in modern falls. The implication is that atmospheric Kr and Xe are incorporated into weathering products, presumably after being dissolved in ground water, the most highly weathered samples (hot desert meteorites) showing the largest enrichment (*Scherer et al.,* 1992).

2.9. Water-Soluble Minerals in Falls

It has recently become apparent that equilibrated ordinary chondrites may contain primary halides, recording aqueous alteration on the ordinary chondrite parent body (e.g., *Zolensky et al.,* 1998). These minerals are highly water soluble, and have only been observed in meteorites recovered soon after their fall. This raises the possibility that the mineral suite we observe in meteorites may be rapidly modified by the terrestrial environment. This is discussed in more detail in the section on carbonaceous chondrites.

2.10. Relationship Between Weathering and Terrestrial Age

Chondrites in hot deserts rarely have terrestrial ages >50 ka, but in exceptional cases they can survive for longer periods — *Welten et al.* (2004) recorded an H5 chondrite with a terrestrial age of 150 ± 40 ka. In Antarctica finds, terrestrial ages can be much greater. *Scherer et al.* (1997) and *Welten et al.* (1997) have discovered meteorites with terrestrial ages of 2 Ma and 2.35 Ma respectively.

A number of radionuclides, most notably ^{26}Al, ^{36}Cl, ^{81}Kr, and ^{14}C, have been used to estimate the terrestrial age of meteorites. Exposure ages of meteorites in space range from a few million years to 50 m.y. (*Crabb and Schultz,* 1981). During this time, a meteorite is bombarded with cosmic rays, and through the interaction of these with meteoritic minerals accumulates a population of cosmogenic stable nuclides and cosmogenic radionuclides with variable half-lives. Following its fall to Earth, the meteorite is shielded from cosmic-ray bombardment, so a terrestrial age can be calculated using the known level of saturation of radionuclides such as ^{14}C within the sample and the observed concentration of cosmogenic radionuclides (*Jull,* 2006). The half-life of ^{14}C is 5730 yr, giving a maximum measurable terrestrial age of about 40 ka (*Jull et al.,* 1989). This isotope is commonly used to measure terrestrial ages of hot-desert meteorites (*Jull et al.,* 1991). Krypton-81 and ^{36}Cl have longer half-lives (210,000 and 301,000 yr respectively) and are more often used to measure the terrestrial age (or residence time) of Antarctic meteorites (*Freundel et al.,* 1986). The calculation of terrestrial ages from the ^{14}C composition of meteorites involves the heating of samples to >250°–500°C to remove terrestrial contamination. The bulk of the error on these measurements comes from the possibility of so-called "self-shielding," i.e., from a variation in ^{14}C with depth in the original stone. The maximum terrestrial age measurable by ^{14}C dating is around 42 ka years due to *in situ* production of ^{14}C at Earth's surface from cosmic rays that reach the ground.

In a ^{14}C study of terrestrial ages of meteorites from the central and southwestern United States, *Jull et al.* (1993a) estimated a weathering "half-life" for meteorites (or decay time) of between 10,000 and 15,000 yr, 4–5× longer than previous estimates (*Boeckl,* 1972) (the earlier estimate was based on a small sample population and substantially underestimated the average terrestrial age of meteorite finds from this region). Only a weak correlation between terrestrial age and weathering was found (*Jull et al.,* 1993a), although there appears to be a better correlation in the population of meteorites from Roosevelt County, New Mexico (*Jull et al.,* 1991; *Wlotzka et al.,* 1995). In a similar study of 13 meteorites from Western Libya, *Jull et al.* (1990) found a peak in the age distribution at 4–8 ka years and only two samples with older ages. They suggest the reason for this may be related to climate changes in North Africa around this time (*Butzer et al.,* 1972), which saw increased precipitation (>200 mm/year) between 10,000 and 7700 years ago (*Pachur,* 1980). In contrast, the meteorites from Texas and New Mexico show a deficit of "young" samples, so some selection process also appears to be at work in this region (*Jull et al.,* 1993b).

More recently, *Bland et al.* (1996a,b, 1998a,b,c) have used Mössbauer spectroscopy to show that oxidation increases over time, with a rapid initial period followed by more gradual subsequent alteration, and that weathering rate varies substantially between different regions. Correlations between the degree of weathering that a sample exhibits, and its terrestrial age, are observed for meteorites from all hot desert accumulations studied (Fig. 2). In addition [and contrary to earlier studies by *Burns et al.* (1995)], although the data show more scatter than is typical of hot desert

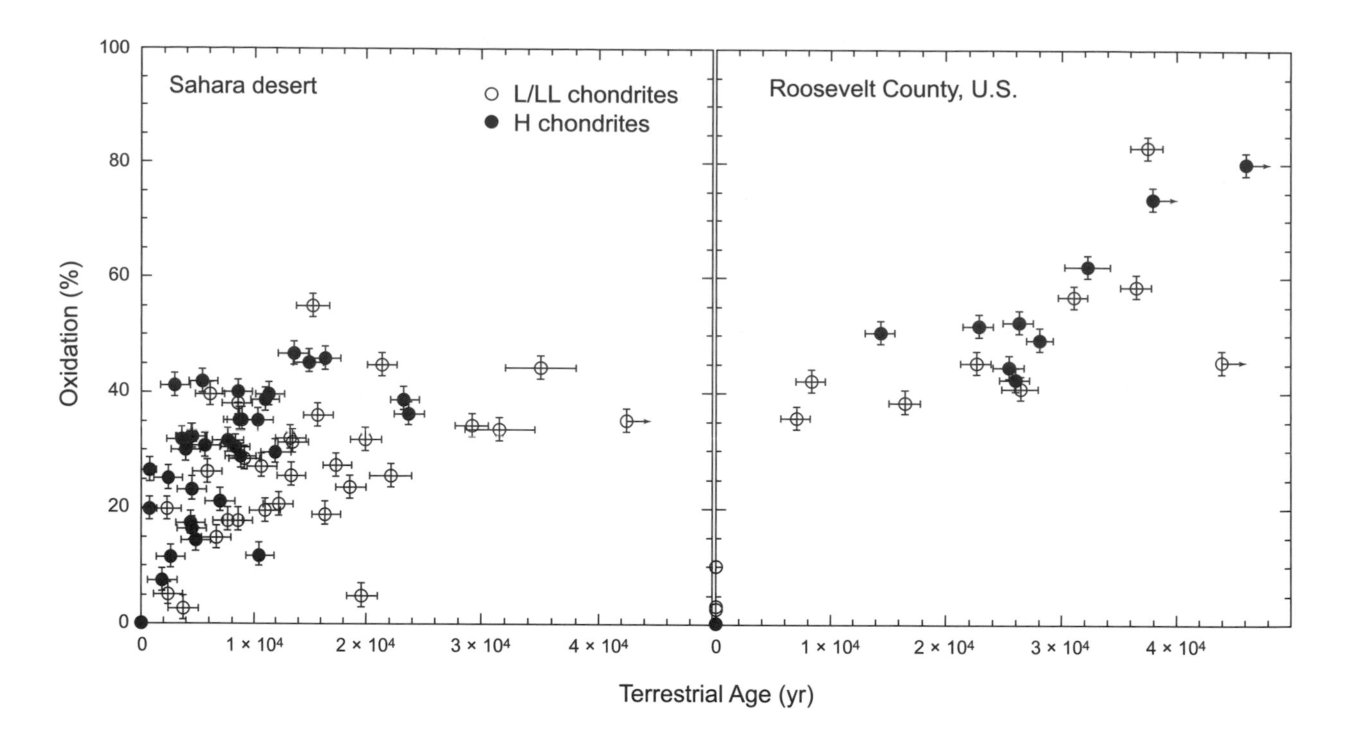

Fig. 2. Scatter of total ferric oxidation (as measured by ^{57}Fe Mössbauer spectroscopy) against terrestrial age for H and L (LL) ordinary chondrites from the Sahara Desert and Roosevelt County. Note the relatively rapid initial weathering phase.

populations, there is a correlation between weathering and terrestrial age for meteorites from Allan Hills, allowing an estimate of the rate of weathering over time for this accumulation (*Bland et al.,* 1998c). Overall, weathering rates in meteorites may be approximated by an appropriate power law (Fig. 3). It is apparent from this work that H chondrites weather significantly faster than L and LL chondrites.

2.11. Passivation of Weathering by Porosity Reduction

Ordinary chondrite falls are quite porous [11–15% (*Wasson,* 1974)]. Oxidation of metal leads to substantial volume expansion (*Buddhue,* 1957), which may reduce porosity, thus reducing the ability of water to penetrate the sample. Since porosity and permeability are among the most important factors influencing weathering rates in silicate rocks (*Noack et al.,* 1993), any variation in porosity (and permeability) in meteorites would affect the rate at which samples are weathered. *Bland et al.* (1998a) compared porosity and oxidation data (as measured by Mössbauer spectroscopy) for 23 ordinary chondrites, weathered in a variety of climatic regimes, to average data for falls. The results indicate an initial rapid decrease in porosity with weathering until after ~20% of the original Fe has been converted to Fe^{3+}, and porosity then stabilizes.

Given that we observe an initial rapid weathering phase, followed by a much-reduced weathering rate, some passivating mechanism is required to explain the survival of meteorites with low levels of weathering over extended timescales. *Bland et al.* (1998a) suggest that the most significant process involved is reduction in porosity, restricting water flow throughout the sample. *Britt and Consolmagno* (2003) reached a similar conclusion following a review of meteorite porosity data. The act of precipitating a diffusion-resistant film around Fe-Ni and troilite grains may also inhibit further oxidation, once some critical thickness is reached.

In an early study of meteorite weathering, *Buddhue* (1957) calculated that (given a specific gravity of pure Fe of 7.85, and meteorite rust of 4.24, and assuming spherical metal grains) oxidation of 10 wt% metal would produce a total volume increase of ~12%. The average metal content of an H-chondrite fall is 16–18 wt% (*Keil,* 1962; *Jarosewich,* 1990; *McSween et al.,* 1991), so oxidation of half the metal in a fresh H chondrite would be sufficient to reduce porosity to zero. As such, the minimum oxidation (measured by Mössbauer spectroscopy) required to passivate weathering may be as low as 8–10%. *Lee and Bland* (2004) modeled the volume changes following alteration of primary minerals for a range of weathering products, and showed that the primary porosity of most meteorites is sufficient to accommodate weathering products. Dilation of primary pores and brecciation, which has been observed in parts of some meteorites, will only occur if the meteorite is especially metal-rich, or has a low primary porosity.

3. WEATHERING OF CARBONACEOUS CHONDRITES

In terms of the alteration of anhydrous silicates, metal, and sulfide, there will clearly be a number of similarities between ordinary and carbonaceous chondrite weathering.

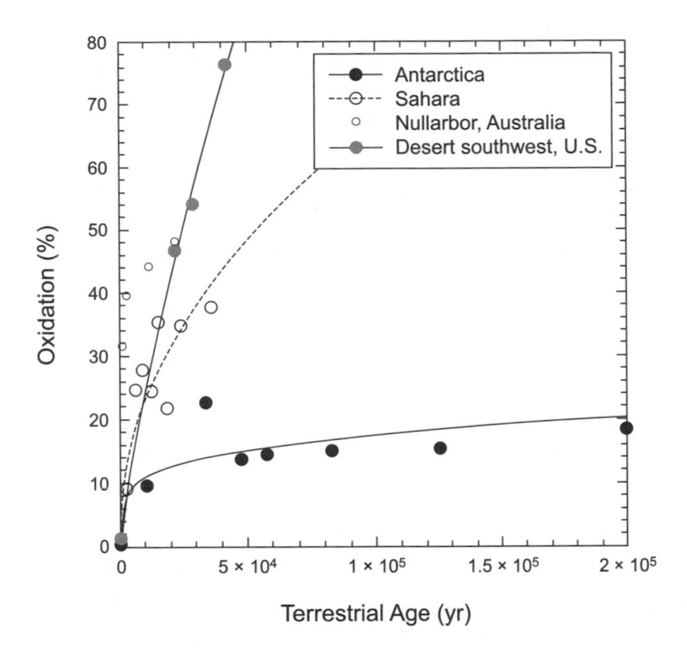

Fig. 3. Terrestrial age vs. oxidation compared between sites. Data are binned by terrestrial age (for each site) so that a similar number of data points fall in each bin. This comparison shows that there is significant variation in weathering rate between sites, presumably as a result of broad differences in regional climate over the lifetime of the accumulation.

However, carbonaceous chondrites differ in a number of ways that are significant in considering how they respond to the terrestrial environment. These include a larger proportion of fine grained matrix (and a correspondingly larger surface area of reactant), enrichment in volatile elements, and the presence of abundant organic compounds. Although meteorite finds will clearly be affected to varying degrees by weathering, some of these factors make carbonaceous chondrite falls susceptible to alteration. It is noteworthy, for instance, that although we have tens of thousands of recovered finds, the only CI1 chondrites in our collections are falls: CIs are clearly not robust enough to last long in the terrestrial environment.

3.1. Alteration of Carbonaceous Chondrite Finds

3.1.1. Elemental mobilization. In a INAA study of CR chondrites, including falls and finds, *Kallemeyn et al.* (1994) observed a large variability in Br abundance in the Antarctic finds — as in the case of I contamination discussed previously, Br can be transported from sea spray (*Heumann et al.,* 1987). Concentrations of Na are also low and highly variable, which *Kallemeyn et al.* (1994) suggests may be related to loss of Na by leaching of sulfates during weathering. Saharan samples also appear to be anomalously high in K (*Kallemeyn et al.,* 1994). In a trace-element study of carbonaceous chondrite matrix, *Bland et al.* (2005) also observed superchondritic abundances of Sr in two Saharan finds, similar to observations of weathered Omani ordinary chondrites (*Al-Kathiri et al.,* 2005).

3.1.2. Weathering of meteoritic organic matter and contamination by terrestrial carbon. Within carbonaceous chondrites the effects of weathering are likely to be most severe for the more labile materials, such as organic matter. The majority of organic matter in meteorites comprises aromatic rings with C-bearing side chains and O-containing functional groups. This material is, at present, our only available record of natural prebiotic chemical evolution and appears to contain some exceedingly ancient materials that are only partially transformed from presolar matter (*Sephton,* 2002, 2004). Hence, it is important to understand just how pristine and uncorrupted the organic matter is in carbonaceous finds.

It is apparent that the organic material in carbonaceous chondrites recovered from hot deserts has been substantially modified. Meteorites from Reg el Açfer (Algeria) show a depletion in bulk C, faster than any gain in C from evaporitic carbonate (*Ash and Pillinger,* 1992b), and it is likely that the macromolecular C these meteorites contained has in most cases been destroyed by extreme weathering conditions. *Newton et al.* (1995) also has found a relationship between the isotopic composition of C in CO chondrites and the recovery site: Fresh falls have the most primitive isotopic composition; meteorites from Allan Hills, Antarctica, trend toward more terrestrial isotopic values; while the most contaminated samples are those from Reg el Açfer, Algeria, and the Yamato Mountains, Antarctica. This is interesting, as Antarctic samples are frequently assumed to have experienced only limited alteration.

The effect of Antarctic weathering on meteoritic organic matter has recently been investigated using a combination of Mössbauer spectroscopy and thermal-extraction/pyrolysis-gas chromatography-mass spectrometry (Py-GC-MS) (*Sephton et al.,* 2004). Preterrestrial aqueous processing and terrestrial weathering of meteorite samples is usually associated with the oxidation of Fe-bearing minerals that can be quantified using Mössbauer spectroscopy, allowing samples to be placed on a scale of relative alteration. Any corresponding alteration of the overall organic constitution of small meteorite samples can be determined by using Py-GC-MS. In an attempt to discriminate between non-Antarctic and Antarctic processes, *Sephton et al.* (2004) studied a series of organic-rich CM falls and CM Antarctic finds.

Interestingly, the inorganic and organic constituents of CM chondrites seem to exhibit responses to both parent-body aqueous alteration and terrestrial weathering. Increased oxidation levels for Fe-bearing minerals within the non-Antarctic chondrites appear to be a response to greater amounts of parent-body aqueous alteration. Moreover, when the Mössbauer responses are used as a guide it is evident that parent-body processing removes O-containing organic entities from organic material and C-bearing side chains from its constituent units.

Mössbauer spectroscopy indicates that the level of alteration (oxidation) displayed by Fe-bearing minerals in the Antarctic samples is generally higher than the non-Antarctic sample set. Organic responses to Antarctic weathering proceed in a similar fashion to parent-body aqueous alteration (removal of side chains and loss of O-containing functional groups) but are more extreme, leading to an enhancement of the organic responses observed in the falls. The record in Antarctic chondrites is clearly not pristine. Instead, these meteorites have been altered in a subtle way that mimics genuine preterrestrial alteration. As was noted in the discussion of cronstedtite recently observed in ordinary chondrites weathered in Antarctica (*Lee and Bland,* 2004), there appear to be a number of similarities between the chemical microenvironments occurring within chondrites undergoing weathering on the Antarctic ice at the present day, and conditions during aqueous alteration on carbonaceous chondrite parent bodies in the early solar system.

3.2. Alteration of Carbonaceous Chondrite Falls

What makes carbonaceous chondrites unusual among the various meteorite groups is that there is significant evidence for terrestrial alteration of falls. This takes a number of forms, including addition of hygroscopic water, formation of sulfate and carbonate veins, mobilization of aqueous trace elements, and possible reordering of phyllosilicates.

3.2.1. Elemental mobilization. *Friedrich et al.* (2002) compared the composition of pristine Tagish Lake to that from samples that had come into contact with icewater. Elements depleted in the altered sample were Sr and Bi, both highly mobile trace elements. Uranium, In, and Cd were enriched in the altered sample, most likely deposited from contaminated meltwater.

3.2.2. Sulfate and carbonate veining. *Gounelle and Zolensky* (2001) showed that the ubiquitous sulfate veins observed in CI1 chondrites were of terrestrial origin. The first description of sulfate veins in Orgueil was made in 1961 by *Dufresne and Anders* (1962), almost 100 years after the meteorite fell. A close reading of the earlier descriptions of CI1 chondrites finds no mention of sulfate veins, despite the fact that these earlier workers were limited to describing the macroscopic properties of a sample. Rather, the extremely friable, loose nature of the material, as well as its high porosity, were regularly noted. An additional property that interested early workers was the extreme avidity of CI1 material for water. *Pisani* (1864) dehydrated a sample of Orgueil, and observed that after a few hours it had regained almost all the hygroscopic water that it had lost (9.15 wt%). It appears that over the period since these meteorites fell, they have continued to acquire water — *Cloëz* (1864) reported 9.06 wt% water in Orgueil a few weeks after the fall, and nearly a century later, *Wiik* (1956) found 19.89 wt%. It is probable that this property of CI1 chondrites has led to the remobilization of any initial fine-grained sulfate, and the oxidation of sulfides, to produce vein sulfate (*Gounelle and Zolensky,* 2001). The final word on a terrestrial origin is provided by the O-isotopic composition of Orgueil sulfate, which is terrestrial (*Aireau et al.,* 2001), and the example of the 2.5-kg sample of Orgueil at Musée d'Histoire Naturelle de Montauban, which shows abundant white sulfate veins growing up through the black fusion crust (*Gounelle and Zolensky,* 2001).

A recent study of the CV3 chondrite Vigarano (*Abreu and Brearley,* 2005) observed a variety of calcite morphologies, including networks of veins, vesicle fillings in the fusion crust, and pseudomorphic replacement of augite. Most veins vary in thickness from <1 μm to 25 μm, and extend for several hundred micrometers. Some veins crosscut the fusion crust and are connected to a carbonate coating on the exterior of the meteorite. The conclusion is that the veins are terrestrial in origin (*Abreu and Brearley,* 2005).

3.2.3. Alteration of CM chondrite falls? It may be that similar processes are also acting to modify CM falls. CM chondrites contain minerals that formed by interaction between liquid water and anhydrous minerals (*McSween,* 1979). How this process progressed on the parent body is important for understanding how to unravel the effects of aqueous alteration in these meteorites. Studies undertaken to qualify (*McSween,* 1979) and quantify (*Browning et al.,* 1996) the extent of alteration found in the CMs resulted in a mineralogic alteration index (MAI) (*Browning et al.,* 1996). The MAI attempted to provide a numerical representation of the Fe^{3+}/Si ratio in matrix phyllosilicates in CM meteorite falls. It has been found to correlate with carbonate (*Benedix et al.,* 2003) and sulfate (*Aireau et al.,* 2001) O-isotopic composition. But as we have seen, these minerals may form in carbonaceous chondrite falls in the terrestrial environment. So a key question is the extent of terrestrial alteration within these meteorites.

Figure 4 is a plot of MAI vs. year of fall (*Benedix and Bland,* 2004) for all the CM chondrites for which an MAI was determined (*Browning et al.,* 1996). The MAI has a

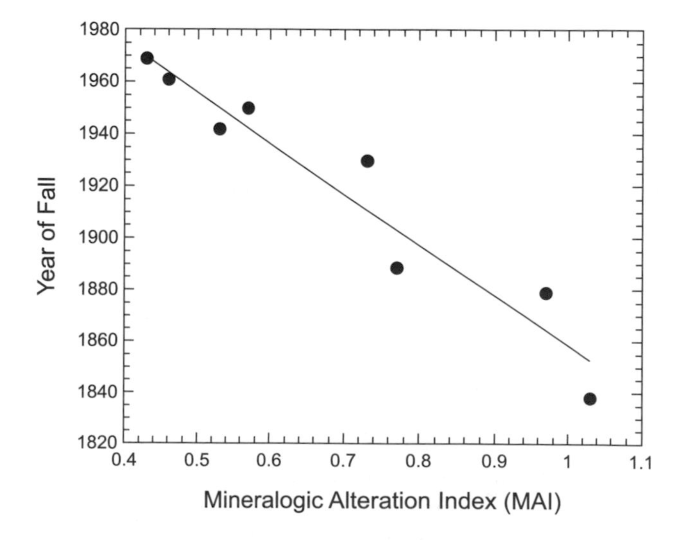

Fig. 4. Mineralogic alteration index (from *Browning et al.,* 1996) vs. year of fall of CM chondrites. The line has a correlation coefficient of 0.93.

strong positive correlation with the year of fall ($R^2 = 0.93$); i.e., the more altered meteorites are older. There is no such systematic correlation between year of fall and matrix or bulk $\Delta^{17}O$, $\delta^{18}O$ (*Clayton and Mayeda,* 1999), water content, or δD (*Eiler and Kitchen,* 2004). Although not as robust, a correlation also exists between the year of fall and $\Delta^{17}O$ in sulfate and carbonate, two minerals that are susceptible to interaction with liquid water. The fact that MAI and carbonate and sulfate $\Delta^{17}O$ track with the year of fall indicates that interaction with the terrestrial atmosphere may have taken place, even though the meteorite was recovered soon after its fall (*Benedix and Bland,* 2004).

If the MAI is an indicator of terrestrial contamination, what does this measure? The procedure for determining an MAI involves an assumption of a generalized stoichiometry for phyllosilicate to derive Fe^{3+} content. But determining the abundance of Fe^{3+} is difficult using probe techniques, even in monomineralic, well-crystalline materials. CM matrix is fine-grained, and each probe analysis includes multiple phases. In addition, phyllosilicates are frequently interlayered and of poor crystallinity. As such, the MAI may be recording a variety of factors in addition to Fe^{3+} abundance. One of these that may be significant for this discussion involves S-bearing phases. The MAI procedure involves the use of multiple screening devices that exclude analyses containing obvious unwanted phases (>50% of probe analyses are excluded in this way): S-rich analyses were excluded, but S is ubiquitous, and so S-bearing inclusions (assumed to be pyrrhotite) were also subtracted from each analysis. It is apparent that S is highly mobile in carbonaceous chondrite falls, and it appears at least possible that this mobilization is influencing the measured MAI.

Whether the MAI is recording terrestrial or preterrestrial alteration remains an open question. What is apparent is that carbonaceous chondrite falls are susceptible to terrestrial weathering: Determining where asteroidal alteration ends and terrestrial alteration begins is clearly a nontrivial task.

4. CONCLUSIONS

Terrestrial weathering can affect many of the properties of meteorites that make them valuable to researchers: chemistry, mineralogy, composition of organic matter, C- and O-isotopic composition, noble gases, etc. In analyzing meteorite finds it is important to be aware of any overprint on a sample's primordial composition imposed by the terrestrial environment. For a number of reasons, carbonaceous chondrites are more susceptible to terrestrial alteration than other meteorite types. It appears that in this group of meteorites, even falls may display the effects of significant terrestrial modification.

Although terrestrial weathering may obscure many of the properties that make a meteorite interesting to researchers, it can also provide an analog to help us understand asteroidal alteration. The weathering environment of Antarctica is similar in a number of ways to the alteration environment in asteroids in the early solar system: The recognition that the same alteration products are produced (*Lee and Bland,* 2004), and that organic material responds in a similar way in both environments, serves to highlight this.

Finally, the fact that meteorites of similar composition are deposited over Earth's entire surface, and that their terrestrial exposure time may be determined quite precisely by measuring their cosmogenic radionuclide abundance, makes them an effective "standard sample" for studying how rocks are affected by the terrestrial environment. Meteorites are one of the few geological materials for which field weathering rates have been determined, allowing the influence of climate on rock weathering rate to be deduced.

Acknowledgments. P.A.B. thanks the Royal Society for their support.

REFERENCES

Abreu N. M. and Brearley A. J. (2005) Carbonates in Vigarano: Terrestrial, preterrestrial, or both? *Meteoritics & Planet. Sci., 40,* 609–625.

Airieau S., Farquhar J., Jackson T. L., Leshin L. A., Thiemens M. H., and Bao H. (2001) Oxygen isotope systematics of CI and CM chondritic sulphate: Implications for evolution and mobility of water in planetesimals (abstract). In *Lunar and Planetary Science XXXII,* Abstract #1744. Lunar and Planetary Institute, Houston (CD-ROM).

Al-Kathiri A., Hofmann B. A., Jull A. J. T., and Gnos E. (2005) Weathering of meteorites from Oman: Correlation of chemical/mineralogical weathering proxies with ^{14}C terrestrial ages and the influence of soil chemistry. *Meteoritics & Planet. Sci., 40,* 1215–1239.

Ash R. D. and Pillinger C. T. (1992a) Carbon and nitrogen in Roosevelt County 075: An unusual organic-rich UOC (abstract). *Meteoritics, 27,* 198–199.

Ash R. D. and Pillinger C. T. (1992b) The effects of Saharan weathering on light element contents of various primitive chondrites (abstract). *Meteoritics, 27,* 199.

Ash R. D. and Pillinger C. T. (1993) Carbon in weathered ordinary chondrites from Roosevelt County (abstract). In *Lunar and Planetary Science XXIV,* p. 43. Lunar and Planetary Institute, Houston.

Banfield J. F., Veblen D. R., and Jones B. F. (1990) Transmission electron microscopy of subsolidus oxidation and weathering of olivine. *Contrib. Mineral. Petrol., 106,* 110–123.

Banfield J. F., Jones B. F., and Veblen D. R. (1991) An AEM-TEM study of weathering and diagenesis, Abert Lake, Oregon: I. Weathering reactions in the volcanics. *Geochim. Cosmochim. Acta, 55,* 2781–2793.

Benedix G. K. and Bland P. A. (2004) What does the CM chondrite mineralogical alteration index really mean? (abstract). *Meteoritics & Planet. Sci., 39,* A14.

Benedix G. K., Leshin L. A., Farquhar J., Jackson T., and Thiemens M. H. (2003) Carbonates in CM2 chondrites: Constraints on alteration conditions from oxygen isotopic compositions and petrographic observations. *Geochim. Cosmochim. Acta, 67,* 1577–1588.

Benoit P. H. and Sears D. W. G. (1992) Metallographic cooling rate differences between Antarctic and non-Antarctic H5 chondrites and some implications (abstract). In *Lunar and Planetary Science XXIII,* pp. 87–88. Lunar and Planetary Institute, Houston.

Bevan A. W. R. (1992a) Australian meteorites. *Rec. Austral. Mus., Suppl., 15,* 1–27.

Bevan A. W. R. (1992b) 1992 WAMET/EUROMET joint expedition to search for meteorites in the Nullarbor Region, Western Australia (abstract). *Meteoritics, 27,* 202–203.

Bland P. A., Smith T. B., Jull A. J. T., Berry F. J., Bevan A. W. R., Cloudt S., and Pillinger C. T. (1996a) The flux of meteorites to the Earth over the last 50,000 years. *Mon. Not. R. Astron. Soc., 283,* 551–565.

Bland P. A., Berry F., Smith T. B., Skinner S., and Pillinger C. T. (1996b) Flux of meteorites to the Earth and weathering in hot desert ordinary chondrite finds. *Geochim. Cosmochim. Acta, 60,* 2053–2059.

Bland P. A., Kelley S. J., Berry F. J., Cadogan J., and Pillinger C. T. (1997) Artificial weathering of an ordinary chondrite fall, Allegan: Implications for the presence of Cl^- as a structural component in akaganéite. *Am. Mineral., 82,* 1187–1197.

Bland P. A., Sexton A., Jull A. J. T., Bevan A. W. R., Berry F. J., Thornley D., Astin T., and Pillinger C. T. (1998a) Climate and rock weathering: A study of terrestrial age dated ordinary chondritic meteorites from hot desert regions. *Geochim. Cosmochim. Acta, 62,* 3169–3184.

Bland P. A., Berry F. J., and Pillinger C. T. (1998b) Rapid weathering in Holbrook: An ^{57}Fe Mössbauer spectroscopy study. *Meteoritics & Planet. Sci., 33,* 127–129.

Bland P. A., Conway A., Smith T. B., Berry F. J., and Pillinger C. T. (1998c) Calculating flux from meteorite decay rates: A discussion of problems encountered in deciphering a 10^5 to 10^6 year integrated meteorite flux at Allan Hills and a new approach to pairing. In *Meteorites: Flux with Time and Impact Effects* (M. M. Grady et al., eds.), pp. 43–58. Geological Society London Special Publication 140.

Bland P. A., Bevan A. W. R., and Jull A. J. T. (2000a) Ancient meteorite finds and the Earth surface environment. *Quarternary Res., 53,* 131–142.

Bland P. A., Lee M. R., Sexton A. S., Franchi I. A., Fallick A. E. T., Miller M. F., Cadogan J. M., Berry F. J., and Pillinger C. T. (2000b) Aqueous alteration without a pronounced oxygen isotope shift: Implications for the asteroidal processing of chondritic materials. *Meteoritics & Planet. Sci., 35,* 1387–1395.

Bland P. A., Alard A., Benedix G. K., Kearsley A. T., Menzies O. N., Watt L. E., and Rogers N. W. (2005) Volatile fractionation in the early solar system and chondrule/matrix complementarity. *Proc. Natl. Acad. Sci., 102,* 13755–13760.

Boeckl R. S. (1972) Terrestrial age of nineteen stony meteorites derived from their radiocarbon content. *Nature, 236,* 25–26.

Britt D. T. and Consolmagno G. J. (2003) Stony meteorite porosities and densities: A review of the data through 2001. *Meteoritics & Planet. Sci., 38,* 1161–1180.

Browning L. B., McSween H. Y. Jr., and Zolensky M. E. (1996) Correlated alteration effects in CM carbonaceous chondrites. *Geochim. Cosmochim. Acta, 60,* 2621–2633.

Buchwald V. F. (1989) On the difference between weathering products on Antarctic and non-Antarctic meteorites. In *Differences Between Antarctic and Non-Antarctic Meteorites,* pp. 24–26. LPI Tech. Rpt. 90-01, Lunar and Planetary Institute, Houston.

Buchwald V. F. and Clarke R. S. Jr. (1989) Corrosion of Fe-Ni alloys by Cl-containing akaganéite (β-FeOOH): The Antarctic meteorite case. *Am. Mineral., 74,* 656–667.

Buddhue J. D. (1957) *The Oxidation and Weathering of Meteorites.* Univ. of New Mexico, Albuquerque.

Burns R. G., Burbine T. H., Fisher D. S., and Binzel R. P. (1995) Weathering in Antarctic H and CR chondrites: Quantitative analysis through Mössbauer spectroscopy. *Meteoritics, 30,* 1–9.

Butzer K. W., Isaac G. L., Richardson J. L., and Washbourn-Kamau C. (1972) Radiocarbon dating of east African lake levels. *Science, 175,* 1069–1076.

Cassidy W. A., Harvey R., Schutt J., Delisle G., and Yanai K. (1992) The meteorite collection sites of Antarctica. *Meteoritics, 27,* 490–525.

Chambaere D. G. and DeGrave E. (1984) A study of the non-stoichiometrical halogen and water content of β-FeOOH. *Phys. Status Sol., 83,* 93–102.

Clayton R. N., Mayeda T. K., and Yanai K. (1984) Oxygen isotopic compositions of some Yamato meteorites. *Proc. Ninth Symposium Antarct. Meteorites,* 267–271.

Clayton R. N., Mayeda T. K., Goswami J. N., and Olsen E. J. (1991) Oxygen isotope studies of ordinary chondrites. *Geochim. Cosmochim. Acta, 55,* 2317–2337.

Clayton R. N. and Mayeda T. (1999) Oxygen isotope studies of carbonaceous chondrites. *Geochim. Cosmochim. Acta, 63,* 2089–2104.

Cloëz S. (1864) Note sur la composition chimique de la pierre météoritique d'Orgueil. *Compt. R. Acad. Sci. Paris, 58,* 986–988.

Crabb J. and Schultz L. (1981) Cosmic-ray exposure ages of the ordinary chondrites and their significance for the parent-body stratigraphy. *Geochim. Cosmochim. Acta, 45,* 2151–2160.

Delisle G., Schultz L., Spettel B., Weber H. W., Wlotzka F., Hofle H. C., Thierbach R., Vogt S., Herpers U., Bonani G., and Suter M. (1989) Meteorite finds near the Frontier Mountain Range in North Victoria Land. *Geol. Jb., 38,* 483–513.

Dennison J. E., Lingner D. W., and Lipschutz M. E. (1986) Antarctic and non-Antarctic meteorites form different populations. *Nature, 319,* 390–393.

Dennison J. E. and Lipschutz M. E. (1987) Chemical studies of H chondrites. II: Weathering effects in the Victoria Land, Antarctic population and comparison of two Antarctic populations with non-Antarctic falls. *Geochim. Cosmochim. Acta, 51,* 741–754.

Dreibus G. and Wänke H. (1983) Halogens in Antarctic meteorites (abstract). *Meteoritics, 18,* 291–292.

DuFresne E. R. and Anders E. (1962) On the chemical evolution of the carbonaceous chondrites. *Geochim. Cosmochim. Acta, 26,* 1085–1114.

Ebihara M., Shinonaga T., Nakahara H., Kondoh A., Miyamoto M., and Kojima H. (1989) Depth-profiles of halogen abundance and integrated intensity of hydration band near 3 μm in ALH 77231, Antarctic L6 chondrite. In *Differences Between Antarctic and Non-Antarctic Meteorites,* pp. 32–37. LPI Tech. Rpt. 90-01, Lunar and Planetary Institute, Houston.

Eggleton R. A. (1984) Formation of iddingsite rims on olivine: A transmission electron microscope study. *Clays Clay Minerals, 32,* 1–11.

Eiler J. M. and Kitchen N. (2004) Hydrogen isotope evidence for the origin and evolution of carbonaceous chondrites. *Geochim. Cosmochim. Acta, 68,* 1395–1411.

Faust G. T., Fahey J. J., Mason B. H., and Dwornik E. J. (1973) *The Disintegration of the Wolf Creek Meteorite, and the For-*

mation of Pecoraite, the Nickel Analog of Clinochrysotile. U.S. Geological Survey Professional Paper No. 384-C.

Feitknecht W., Giovanoli R., Michaelis W., and Muller M. (1973) Uber die Hydrolyse von Eisen (III) Salzlösungen. 1. Die Hydrolyse der Lösungen von Eisen (III) chlorid. *Helv. Chim. Acta, 56,* 2847–2856.

Fireman E. L. and Norris T. L. (1982) Ages and compositions of gas trapped in Allan Hills and Byrd core ice. *Earth Planet. Sci. Lett., 60,* 339–350.

Fisher D. S. and Burns R. G. (1992) Mössbauer spectra of H-5 chondrites from Antarctica (abstract). In *Lunar and Planetary Science XXIII,* pp. 367–368. Lunar and Planetary Institute, Houston.

Freundel M., Schultz L., and Reedy R. C. (1986) Terrestrial ^{81}Kr-Kr ages of Antarctic meteorites. *Geochim. Cosmochim. Acta, 50,* 2663–2673.

Friedrich J. M., Wang M-S., and Lipschutz M. E. (2002) Comparison of the trace element composition of Tagish Lake with other primitive carbonaceous chondrites. *Meteoritics & Planet. Sci., 37,* 677–686.

Gibson E. K. Jr. and Bogard (1978) Chemical alterations of the Holbrook chondrite resulting from terrestrial weathering. *Meteoritics, 13,* 277.

Gooding J. L. (1981) Mineralogical aspects of terrestrial weathering effects in chondrites from Allan Hills, Antarctica. *Proc. Lunar Planet. Sci. 12A,* pp. 1105–1122.

Gooding J. L. (1986) Weathering of stony meteorites in Antarctica. In *International Workshop on Antarctic Meteorites,* pp. 48–54. LPI Tech. Rpt. 86-01, Lunar and Planetary Institute, Houston.

Gooding J. L., Jull A. J. T., Cheng S., and Velbel M. A. (1988) Mg-carbonate weathering products in Antarctic meteorites: Isotopic composition and origin of nesquehonite from LEW 85320 (abstract). In *Lunar and Planetary Science XIX,* pp. 397–398. Lunar and Planetary Institute, Houston.

Gounelle M. and Zolensky M. E. (2001) A terrestrial origin for sulphate veins in CI1 chondrites. *Meteoritics & Planet. Sci., 36,* 1321–1329.

Grady M. M., Gibson E. K., Wright I. P., and Pillinger C. T. (1989a) The formation of weathering products on the LEW 85320 ordinary chondrite: Evidence from carbon and oxygen stable isotope compositions and implications for carbonates in SNC meteorites. *Meteoritics, 24,* 1–7.

Grady M. M., Wright I. P., and Pillinger C. T. (1989b) Comparisons between Antarctic and non-Antarctic meteorites based on carbon stable isotope geochemistry. In *Differences Between Antarctic and Non-Antarctic Meteorites,* pp. 38–42. LPI Tech. Rpt. 90-01, Lunar and Planetary Institute, Houston.

Grady M. M. (2000) *Catalogue of Meteorites.* Natural History Museum, London.

Grandstaff D. E. (1978) Changes in surface area and morphology and the mechanism of forsterite dissolution. *Geochim. Cosmochim. Acta, 42,* 1899–1901.

Greenberg R. and Chapman C. R. (1983) Asteroids and meteorites: Parent bodies and delivered samples. *Icarus, 55,* 455–481.

Heumann K. G., Gall M., and Weiss H. (1987) Geochemical investigations to explain iodine overabundances in Antarctic meteorites. *Geochim. Cosmochim. Acta, 51,* 2541–2547.

Hochella M. F. Jr. and Banfield J. F. (1995) Chemical weathering of silicates in nature: A microscopic perspective with theoretical considerations. In *Chemical Weathering Rates of Silicate Minerals* (A. F. White and S. L. Brantley, eds.), pp. 353–406. Mineralogical Society of America, Washington, DC.

Honda M., Nagai H., Nishiizumi K., and Ebihara M. (1993) Weathering of a chondrite, Tsarev, L5. *Proc. Eighteenth Symp. Antarct. Meteorites,* 176–179.

Jarosewich E. (1990) Chemical analyses of meteorites: A compilation of stony and iron meteorite analyses. *Meteoritics, 25,* 323–337.

Jull A. J. T., Cheng S., Gooding J. L., and Velbel M. A. (1988) Rapid growth of magnesium carbonate weathering products in a stony meteorite from Antarctica. *Science, 242,* 417–419.

Jull A. J. T., Donahue D. J., and Linick T. W. (1989) Carbon-14 activities in recently fallen meteorites and Antarctic meteorites. *Geochim. Cosmochim. Acta, 53,* 2095–2100.

Jull A. J. T., Donahue D. J., Wlotzka F., and Palme H. (1990) Distribution of terrestrial age and petrologic type of meteorites from western Libya. *Geochim. Cosmochim. Acta, 54,* 2895–2898.

Jull A. J. T., Wlotzka F., and Donahue D. J. (1991) Terrestrial ages and petrologic description of Roosevelt County meteorites (abstract). In *Lunar and Planetary Science XXIV,* pp. 667–668. Lunar and Planetary Institute, Houston.

Jull A. J. T., Donahue D. J., Cielaszyk E., and Wlotzka F. (1993a) Carbon-14 terrestrial ages and weathering of 27 meteorites from the Southern High Plains and adjacent areas (USA). *Meteoritics, 28,* 188–195.

Jull A. J. T., Wlotzka F., Bevan A. W. R., Brown S. T., and Donahue D. J. (1993b) ^{14}C terrestrial ages of meteorites from desert regions: Algeria and Australia (abstract). *Meteoritics, 28,* 376–377.

Jull A. J. T., Bevan A. W. R., Cielaszyk E., and Donahue D. J. (1995) Carbon-14 terrestrial ages and weathering of meteorites from the Nullarbor Region, Western Australia. In *Meteorites from Cold and Hot Deserts,* pp. 37–38. LPI Tech. Rpt. 95-02, Lunar and Planetary Institute, Houston.

Jull A. J. T. (2006) Terrestrial ages of meteorites. In *Meteorites and the Early Solar System II* (D. S. Lauretta and H. Y. McSween Jr., eds.), this volume. Univ. of Arizona, Tucson.

Karlsson H. R., Jull A. J. T., Socki R. A., and Gibson E. K. Jr. (1991) Carbonates in Antarctic ordinary chondrites: Evidence for terrestrial origin (abstract). In *Lunar and Planetary Science XXII,* pp. 689–690. Lunar and Planetary Institute, Houston.

Kallemeyn G. W., Rubin A. E., and Wasson J. T. (1994) The compositional classification of chondrites: VI. The CR carbonaceous chondrite group. *Geochim. Cosmochim. Acta, 58,* 2873–2888.

Keil K. (1962) On the phase composition of meteorites. *J. Geophys. Res., 67,* 4055–4061.

Keller P. (1970) Eigenshaften von $(Cl,F,OH)_{<2}Fe_2(O,OH)_{16}$ und Akaganéite. *Neues Jahrbuch für Mineralogie Abhandlungen, 113,* 29–49.

Langenauer M. and Krahenbuhl U. (1992) Depth-profiles of the trace elemental concentration of fluorine, chlorine, bromine and iodine in two Antarctic H5 chondrites (abstract). *Meteoritics, 27,* 247–248.

Lee M. R. and Bland P. A. (2004) Mechanisms of weathering of meteorites recovered from hot and cold deserts and the formation of phyllosilicates. *Geochim. Cosmochim. Acta, 68,* 893–916.

Luce R. W., Bartlett R. W., and Parks G. A. (1972) Dissolution kinetics of magnesium silicates. *Geochim. Cosmochim. Acta, 36,* 35–50.

Mackay A. L. (1960) β-Ferric oxyhydroxide. *Mineral. Mag., 32,* 545–557.

Mackay A. L. (1962) β-Ferric oxyhydroxide-akaganeite. *Mineral. Mag., 33,* 270–280.

Marvin U. B. (1963) Mineralogy of the oxidation products of the Sputnik 4 fragment and of iron meteorites. *J. Geophys. Res., 68(17),* 5059–5068.

Marvin U. B. (1980) Magnesium carbonate and magnesium sulfate deposits on Antarctic meteorites. *Antarctic J., 15,* 54–55.

McSween H. Y. (1979) Alteration in CM carbonaceous chondrites inferred from modal and chemical variations in matrix. *Geochim. Cosmochim. Acta, 43,* 1761–1770.

McSween H. Y. Jr. and Bennett M. E. III (1991) The mineralogy of ordinary chondrites and implications for asteroid spectrophotometry. *Icarus, 90,* 107–116.

Miyamoto M. (1989) Differences in the degree of weathering between Antarctic and non-Antarctic ordinary chondrites: Infrared spectroscopy. In *Differences Between Antarctic and Non-Antarctic Meteorites,* pp. 68–71. LPI Tech. Rpt. 90-01, Lunar and Planetary Institute, Houston.

Miyamoto M. (1991) Differences in the degree of weathering between Antarctic and non-Antarctic meteorites inferred from infrared diffuse reflectance spectra. *Geochim. Cosmochim. Acta, 55,* 89–99.

Nishiizumi K., Elmore D., and Kubik P. W. (1989) Update on terrestrial ages of Antarctic meteorites. *Earth Planet. Sci. Lett., 93,* 299–313.

Nishiizumi K. (1995) Terrestrial ages of meteorites from cold and cold regions. In *Meteorites from Cold and Hot Deserts,* pp. 53–55. LPI Tech. Rpt. 95-02, Lunar and Planetary Institute, Houston.

Newton J., Sephton M. A., and Pillinger C. T. (1995) Contamination differences between CO3 falls and Antarctic and Saharan finds: A carbon and nitrogen isotope study. In *Meteorites from Cold and Hot Deserts,* pp. 51–53. LPI Tech. Rpt. 95-02, Lunar and Planetary Institute, Houston.

Noack Y., Colin F., Nahon D., Delvigne J., and Michaux L. (1993) Secondary-mineral formation during natural weathering of pyroxene: Review and thermodynamic approach. *Am. J. Sci., 293,* 111–134.

Pachur H-J (1980) Climatic history in the Late Quaternary in southern Libya and the western Libyan desert. In *The Geology of Libya* (M. J. Salem and M. T. Busrewil, eds.) pp. 781–788. Academic, London.

Pisani (1864) Étude chimique et analyse de l'aérolithe d'Orgueil. *Compt. R. Acad. Sci. Paris, 59,* 132–135.

Rézel D. and Génin J. M. R. (1990) The substitution of chloride ions to OH^--ions in the akaganéite beta ferric oxyhydroxide studied by Mössbauer effect. *Hyperfine Interactions, 57,* 2067–2076.

Scherer P., Loeken T., and Schultz L. (1992) Differences of terrestrial alteration effects in ordinary chondrites from hot and cold deserts: Petrography and noble gases (abstract). *Meteoritics, 27,* 314–315.

Scherer P., Schultz L., and Loeken T. (1995) Weathering and atmospheric noble gases in chondrites from hot deserts. In *Meteorites from Cold and Hot Deserts,* pp. 58–59. LPI Tech. Rpt. 95-02, Lunar and Planetary Institute, Houston.

Scherer P., Schultz L., Neupert U., Knauer M., Neumann S., Leya I., Michel R., Mokos J., Lipschutz M. E., Metzler K., Suter M. and Kubik P. W. (1997) Allan Hills 88019: An Antarctic H-chondrite with a very long terrestrial age. *Meteoritics & Planet. Sci., 32,* 769–773.

Schlüter J., Schultz L., Thiedig F., Al-Mahdi B. O., and Abu Aghreb A. E. (2002) The Dar al Gani meteorite field (Libyan Sahara): Geological setting, pairing of meteorites, and recovery density. *Meteoritics & Planet. Sci., 37,* 1079–1093.

Scott E. R. D., McKinley S. G., Keil K., and Wilson I. E. (1986) Recovery and classification of thirty new meteorites from Roosevelt County, New Mexico. *Meteoritics, 21,* 303–308.

Sephton M. A. (2002) Organic compounds in carbonaceous meteorites. *Natural Product Rept., 19,* 292–311.

Sephton M. A. (2004) Organic matter in ancient meteorites. *Astron. Geophys., 45,* 2.8–2.14.

Sephton M. A., Bland P. A., Pillinger C. T., and Gilmour I. (2004) The preservation state of organic matter in meteorites from Antarctica. *Meteoritics & Planet. Sci., 39,* 747–754.

Siever R. and Woodford N. (1979) Dissolution kinetics and the weathering of mafic minerals. *Geochim. Cosmochim. Acta, 43,* 717–724.

Sipiera P. P., Becker M. J., and Kawachi Y. (1987) Classification of twenty-six chondrites from Roosevelt County, New Mexico. *Meteoritics, 22,* 151–155.

Socki R. A., Gibson E. K., Jull A. J. T., and Karlsson H. R. (1991) Carbon and oxygen isotope composition of carbonates from an L6 chondrite: Evidence for terrestrial weathering from the Holbrook meteorite (abstract). *Meteoritics, 26,* 396.

Smith K. L., Milnes A. R., and Eggleton R. A. (1987) Weathering of basalt: Formation of iddingsite. *Clays Clay Minerals, 35,* 418–428.

Velbel M. A. and Gooding J. L. (1988) X-ray diffraction evidence for weathering products on Antarctic basaltic achondrites (abstract). *Meteoritics, 23,* 306.

Velbel M. A., Long D. T., and Gooding J. L. (1991) Terrestrial weathering of Antarctic stone meteorites: Formation of Mg-carbonates on ordinary chondrites. *Geochim. Cosmochim. Acta, 55,* 67–77.

Verma H. C. and Tripathi R. P. (2004) Anomalous Mossbauer parameters in the second generation regolith Ghubara meteorite. *Meteoritics & Planet. Sci., 39,* 1755–1759.

Wasson J. T. (1974) *Meteorites: Classification and Properties.* Springer-Verlag, New York.

Waychunas G. A. (1991) Crystal chemistry of oxides and oxyhydroxides. In *Oxide Minerals: Petrologic and Magnetic Significance* (D. H. Lindsley, ed.), pp. 11–68. Mineralogical Society of America, Washington, DC.

Welten K. C. and Nishiizumi K. (2000) Degree of weathering of H-chondrites from Frontier Mountain, Antarctica (abstract). In *Workshop on Extraterrestrial Materials from Cold and Hot Deserts,* p. 83. LPI Contribution No. 997, Lunar and Planetary Institute, Houston.

Welten K. C., Alderliesten C., van der Borg K., Lindner L., Loeken T., and Schultz L. (1997) Lewis Cliff 86360: An Antarctic L-chondrite with a terrestrial age of 2.35 million years. *Meteoritics & Planet. Sci., 32,* 775–780.

Welten K. C., Nishiszumi K., Finkel R. C., Hillegonds D. J., Jull A. J. T., Franke L., and Schultz L. (2004) Exposure history and terrestrial ages of ordinary chondrites from the Dar al Gani region, Libya. *Meteoritics & Planet. Sci., 39,* 481–498.

Wetherill G. W. (1986) Unexpected Antarctic chemistry. *Nature, 319,* 357–358.

Wiik H. B. (1956) The chemical composition of some stony meteorites. *Geochim. Cosmochim. Acta, 9,* 279–289.

Wlotzka F. (1989) Meteoritical Bulletin No. 68. *Meteoritics, 24,* 57.

Wlotzka F. (1993) A weathering scale for the ordinary chondrites (abstract). *Meteoritics, 28,* 460.

Wlotzka F., Jull A. J. T., and Donahue D. J. (1995) Carbon-14 terrestrial ages of meteorites from Acfer, Algeria. In *Meteorites from Cold and Hot Deserts,* pp. 72–73. LPI Tech. Rpt. 95-02, Lunar and Planetary Institute, Houston.

Yoshida M., Ando H., Omoto K., and Naruse R. (1971) Discovery of meteorites near Yamato Mountains, East Antarctica. *Japanese Antarct. Rec., 39,* 62–65.

Zolensky M. E. and Gooding J. L. (1986) Aqueous alteration on carbonaceous-chondrite parent bodies as inferred from weathering of meteorites in Antarctica. *Meteoritics, 21,* 548–549.

Zolensky M. E., Wells G. L., and Rendell H. M. (1990) The accumulation rate of meteorite falls at the Earth's surface: The view from Roosevelt County, New Mexico. *Meteoritics, 25,* 11–17.

Zolensky M. E., Bodnar R. J., Gibson E. K. Jr., Nyquist L. E., Reese Y., Shih C-Y, Wiesmann H. (1998) Asteroidal water within fluid inclusion-bearing halite in an H5 Chondrite, Monahans. *Science, 285,* 1377–1379.

Flux of Extraterrestrial Materials

Michael Zolensky
NASA Johnson Space Center

Phil Bland
Imperial College

Peter Brown
University of Western Ontario

Ian Halliday
Ottawa, Ontario, Canada

---

1. INTRODUCTION

In its annual journey around the Sun, Earth continuously experiences collisions with solid objects derived predominantly from asteroidal, cometary, and planetary debris. This debris ranges in size from large objects capable of forming impact craters; down to boulder-sized chunks, some of which deposit meteorites on the ground; to a steady rain of interplanetary and interstellar dust particles. The flux of these materials has been assessed using space-, atmosphere-, and groundbased techniques, and is observed to vary with time to a significant degree. We will describe the resultant flux of extraterrestrial materials in order of decreasing incoming mass, beginning with the current flux of meteorite-producing fireballs.

2. CURRENT METEORITE FLUX FROM FIREBALLS AND GROUND OBSERVATIONS

2.1. Introduction

In this section we consider objects that produce very bright meteors during their atmospheric passage, frequently called "fireballs" or sometimes "bolides" in more technical literature. The faint limit of their brightness is an apparent peak magnitude near –5 (brighter than Venus) with an upper limit of about –15 to –20, limited only by the extreme scarcity of such bright events. This range is suitable for optical detection, traditionally by photography and more recently with various electronic and satellite techniques. The observational data may be used to derive solar-system orbits and, using a knowledge of meteor and atmospheric physics, estimates of mass upon entry into Earth's atmosphere and of any surviving meteorites that reach the ground. For determining flux values, this approach has the obvious advantage that the number of events in a known time interval is rather easily determined. The greatest challenge is to derive valid values for the masses involved. Efforts to determine meteor composition from optical observations have been attempted, with variable success.

Early estimates of the flux of meteorites (on the ground) were based on recovery rates in regions with high population density. It was clear that this method involved large uncertainties, but with the advent of three photographic camera networks in the 1960s and early 1970s, instrumental data gradually became available. These three networks were the European Network in central Europe, with headquarters at the Ondrejov Observatory in the Czech Republic, the Prairie Network in the central plains of the United States, and the Meteorite Observation and Recovery Project (MORP) in the prairie provinces of western Canada. The first flux estimate was for the mass distribution of all objects at the top of the atmosphere (*McCrosky and Ceplecha,* 1969) based on Prairie Network data. The fraction of events related to actual meteorite falls was estimated, indicating a flux 1 to 2 orders of magnitude higher than the previous estimates. This suggested that either the efficiency of meteorite recognition was much lower than expected or there was some problem with the reduction of the camera data.

2.2. Meteorite Observation and Recovery Project Data on Meteorite Masses

One of the goals of the MORP network was to tackle this problem of the flux of meteorites. The network consisted of 12 camera stations providing effective two-station coverage over an area of 1.26×10^6 km^2. The network was in operation from 1971 to 1985 and detailed records of the clear-sky area covered for each night were maintained for 11 years, from 1974 to 1985. A description of the network and details of the techniques used for the flux project were published in several papers (*Halliday et al.,* 1978, 1984, 1989, 1991, 1996). Only a summary of the results is presented here. The total clear-sky coverage in 11 years was

1.51×10^{10} km^2 h. We quote flux values in units of events per 10^6 km^2 per year; the total coverage is equivalent to 1.72 years over 10^6 km^2. Typical search areas in desert regions might be 10 km^2 so the network observed the equivalent infall for nearly 200,000 years for such an area.

Essentially all stony meteorites, and a large proportion of irons and stony irons, fracture during their atmospheric flight. There is convincing evidence that for a relatively small event such as the Innisfree, Canada, fall of 1977, with an estimated entry mass of about 40 kg (*Halliday et al.,* 1981) and for the massive daytime Moravka, Slovakia, event of 2000, whose preatmospheric mass is estimated at 1500 kg (*Borovička and Kalenda,* 2003; *Borovička et al.,* 2003), there was considerable fragmentation above a height of 50 km. At such heights the pressure loading is small enough that fragmentation must be due to preexisting cracks in a meteoroid of low intrinsic strength. Successive fragmentation events at lower heights lead to numerous fragments spread over a strewn field, typically several kilometers in extent, and frequently an order of magnitude larger. It is thus very important to be precise in defining an estimate of flux. Does it refer to the total distribution of all fragments from many events, to the distribution of total masses for each event, to the distribution of the largest fragment from each event, or to the distribution of entry masses at the top of the atmosphere? Attempts may be made to quantify each of these as a flux in some time interval.

Camera networks such as the MORP system are best suited to estimating the mass of the largest fragment for each event. They can also provide estimates of entry masses, although these are less secure. With some assumptions based on favorable meteorite events, the distribution of the largest fragment masses can be converted to the distribution of total masses on the ground.

Two suitable photographs of a fireball can be reduced to give values of height, velocity, and deceleration along the path. Values for deceleration become reliable late in the trail for those meteoroids that penetrate deep into the atmosphere, which are the best candidates for recoverable meteorite falls. The conventional drag equation can then be used to estimate a "dynamic mass" using observed or known values for velocity, deceleration, and atmospheric density, together with reasonable choices for the meteoroid density, drag coefficient, and shape factor, the latter a measure of the ratio of mass to cross-section area. By choosing the lowest portion of the path for which the deceleration is reliable, further ablation or fragmentation is likely not a problem and the dynamic mass obtained should be close to the actual mass of the largest fragment. Normally the largest fragment is the leading one in the cluster of fragments in the trail and the only one for which good decelerations can be derived. In a favorable case, such as Innisfree (*Halliday et al.,* 1981), dynamic masses were obtained for four separate fragments that could be matched to the recovered meteorites on the ground. The agreement between dynamic and recovered masses was quite satisfactory, providing confidence in the values of dynamic masses.

The flux of the largest fragments can be obtained directly from dynamic masses for those objects that are observed under suitable clear-sky conditions. It is, however, more meaningful to estimate the total mass of meteorites from each fall. From consideration of various falls for which recovery is believed to have been reasonably complete, it appears that the largest fragment constitutes a smaller fraction of the total mass as we consider larger events. The most comprehensive examination of this (*Halliday et al.,* 1984, 1989) found an increase from largest fragment to total mass ranged from a factor of 1.25 at 50 g to 4.0 at 20 kg.

Meteorite flux values are usually quoted as cumulative frequencies of the form

$$\log N = A \log m + B$$

where N is the number of events per unit time and area with mass equal to or larger than m. We typically choose N to be the number per 10^6 km^2 per year and express m in grams. (An area of 10^6 km^2 is close to the size of the country of Egypt.)

Figure 1 shows the cumulative distribution for main masses on the ground, shown as M – M and also the estimate for total masses for each event, marked T – T in the figure. There is a marked change in slope near 600 g for the M – M plot, which corresponds to a mass of 1030 g for total masses T – T. This surprising result was considered in some detail by *Halliday et al.* (1989), where it was shown that the deficiency of smaller masses is due entirely to the slower fireballs with initial velocity less than 16 km s^{-1}. The change cannot be due to observational selection effects and is unlikely to be caused by changes in ablation as a function of mass. The most likely cause appears to be a fragmentation effect, although the mechanism is not understood. The T – T plot predicts that nine events per year will de-

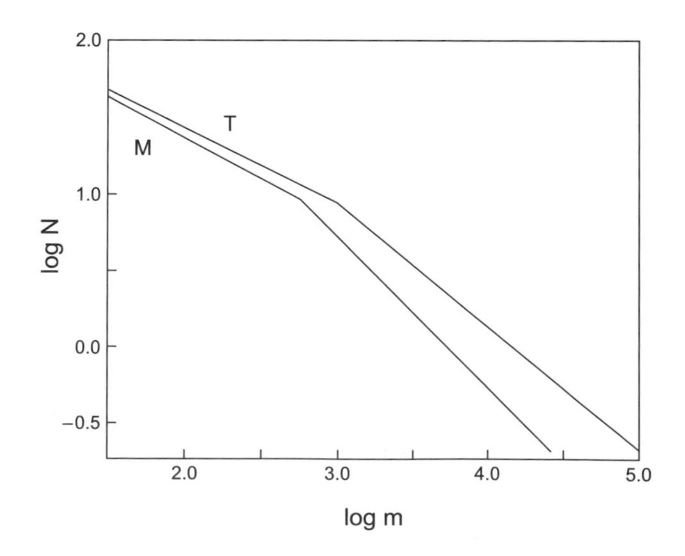

Fig. 1. The cumulative frequencies, N, of number of events per year per 10^6 km^2, as a function of mass, m, in grams, for meteorites on the ground. The line M shows values for the largest fragment of each event while T refers to the total estimated mass for each event.

posit at least 1 kg in an area of 10^6 km^2, while one event in 2 years will deposit 30 kg. The mathematical expressions for Fig. 1 are

for M – M: $\log N = -0.53 \log m + 2.42$ $m < 600$ g
$\log N = -1.00 \log m + 3.72$ $m > 600$ g

for T – T: $\log N = -0.49 \log m + 2.41$ $m < 1030$ g
$\log N = -0.82 \log m + 3.40$ $m > 1030$ g

The nature of the flux problem does not make it appropriate for a formal evaluation of probable errors. Errors in counting the number of events and recording total areas and times of observation are trivial compared to errors in mass calibration. An individual dynamic mass could possibly be in error by a factor of 3 or 4 but it is unlikely that any systematic error greater than a factor of 2 exists in positioning the mass scale in Fig. 1 for the M – M plot. The conversion to total masses on the ground may lead to a somewhat greater uncertainty for the T – T plot.

2.3. Preatmospheric Masses

The MORP observations may also be used to derive the mass influx at the top of the atmosphere. The so-called "photometric masses" are derived from measures of the luminosity along the trail combined with the observed velocity at each point and some knowledge of the "luminous efficiency" with which kinetic energy is converted to luminosity. Much effort has been devoted to the luminous efficiency of meteorites as shown in the discussion by *Ceplecha et al.* (1998). Early estimates of the efficiency were generally considerably less than 1% but values of several percent have been adopted more recently. Some authors have used a simple dependence of the efficiency with velocity and, of course, the values should depend on the wavelength used in a particular set of observations. The MORP reductions used a constant value of 0.04 for the panchromatic wavelength band of the film in use. *Borovička et al.* (2003) determined a value of 0.09 for the larger bolometric range for the Moravka meteorite event. Considerable mass loss may occur during brief flares along a fireball's path where a sudden fragmentation event produces a cloud of small particles that then ablate very rapidly. For the Moravka event, *Borovička and Kalenda* (2003) state that such events were the major mass-loss process. It is thus important to measure the brightness of these flares to determine the mass loss. Moravka, with an initial velocity of 22.5 km s^{-1}, is among the fastest group of meteorite events, so the importance of fragmentation here may also have been greater than normal.

The MORP clear-sky survey observed 754 fireballs from which an unbiased sample of 213 events was studied in detail, and photometric mass estimates were published for each of these (*Halliday et al.*, 1996). This group includes fragile cometary material such as bright members of numerous annual meteor showers. (All potential meteorite events were studied, including those observed under poor sky conditions, but the group of 213 represents a typical sample of bright meteoroids.) This data can be divided into several subgroups. One group of predominantly asteroidal origin was defined by initial velocity less than 25 km s^{-1} and nonmembership in a known meteor shower, while the faster objects and all shower members were called the "cometary" group. From the physical data, primarily density estimates for the meteoroids, it appears that perhaps one-fourth of both groups are misplaced in the wrong group by this rather arbitrary division.

Figure 2 shows some of the mass distributions at the top of the atmosphere. The line A – A represents the total influx of fireballs, B – B the "asteroidal" group defined above, C – C the "cometary" group, and D – D those objects that survive to drop meteorites. For an initial mass of 10 kg our result is a factor of 10 below the early Prairie Network value (*McCrosky and Ceplecha,* 1969), due mainly to the low value for luminous efficiency in use at that time. For entry masses of a few kilograms it appears there are roughly two objects of asteroidal origin for every cometary object. The asteroidal plot B – B has a steeper slope for masses larger than about 2.7 kg, which is also seen in the A – A plot for all objects since asteroidal meteoroids dominate the entire group.

The mathematical expressions for the plots of Fig. 2 are listed below, together with the approximate mass range for each expression.

All objects:
A – A $\log N = -1.06 \log m + 5.26$ 2.4 to 12 kg
A – A $\log N = -0.48 \log m + 3.50$ 0.1 to 2.4 kg

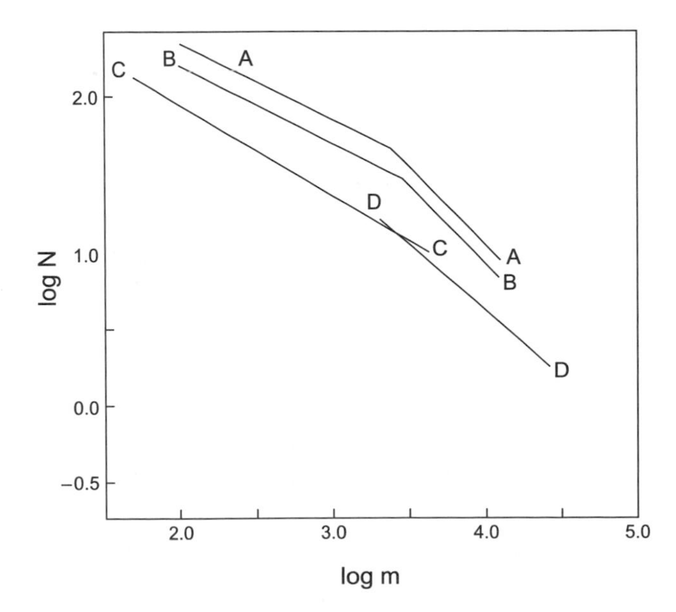

Fig. 2. Cumulative frequencies as in Fig. 1 for preatmospheric masses. A = the total influx of all meteoroids; B = the influx of asteroidal material; C = the influx of cometary meteoroids; D = those objects that deposit meteorites of at least 50 g on the ground.

Asteroidal:
B – B log N = –1.00 log m + 4.92 2.7 to 12 kg
B – B log N = –0.50 log m + 3.20 0.1 to 2.7 kg

Cometary:
C – C log N = –0.60 log m + 3.15 0.05 to 4 kg

Meteorite:
D – D log N = –0.87 log m + 4.09 2.0 to 25 kg

Photometric masses are generally less secure than dynamic masses due to problems in calibrating the luminosity scale, especially for very bright events, and the appropriate value for luminous efficiency is also in some doubt. One might have used a higher value for the efficiency in the MORP reductions, which would lower the mass estimates. We can examine the fraction of the initial mass that survives as meteorites for meteorite events by comparing the preatmospheric mass and the estimate for total masses on the ground. There is a large range in the survival fraction due primarily to velocity differences, but equally important are the slope of the atmospheric path and the severity of fragmentation. The median value for the survival from the MORP data is 0.20, a reasonable value that lends some confidence to the photometric masses.

For comparison, *Ceplecha* (2001) adopts a different value and functional dependence for luminous efficiency than *Halliday et al.* (1996) and makes use of Prairie Network and European Network data for large fireballs and finds a flux ~2× higher than Halliday et al. over the mass interval from 10 g–10 kg (Fig. 3). The physical makeup of these bodies appears to be roughly 50% cometary and 50% asteroidal. Halliday et al. finds that 50–70% of all fireballs in this mass range are asteroidal in physical strength (with the proportion becoming higher as the mass increases), while *Ceplecha* (1994) finds 45–60% are asteroidal over the observed size range from 0.1–1 m, with asteroidal bodies again increasing as mass increases. Here we take asteroidal bodies to be type I and II fireballs in the scheme of *Ceplecha et al.* (1998), which are assumed to correspond to ordinary and carbonaceous chondrites respectively. *Ceplecha* (1994) also estimates that the proportion of cometary-type bodies begins increasing again above 1 m in diameter; however, this is largely an extrapolated result with little constraining observational data. Based on the recent work of *Brown et al.* (2002a) and its agreement with *ReVelle* (1997), it seems probable that the trend toward increasing proportions of solid (asteroidal) bodies continues to much larger sizes (>10 m). Indeed, the best current telescopic estimates from dynamical arguments and spectral measurements have the fraction of the near-Earth asteroid (NEA) population, which might be extinct cometary nuclei, as 10–18% (*Binzel et al.,* 2004), basically consistent with this result (cf. *Weissman et al.,* 2002).

Camera networks provide influx rates for at most a few decades, while collections of meteorites from hot deserts and Antarctica should represent an average over many thousands of years. We have good agreement between the camera results and the analysis of desert meteorite collections, while spacecraft and infrasonic observations extend the coverage to both larger and smaller mass ranges.

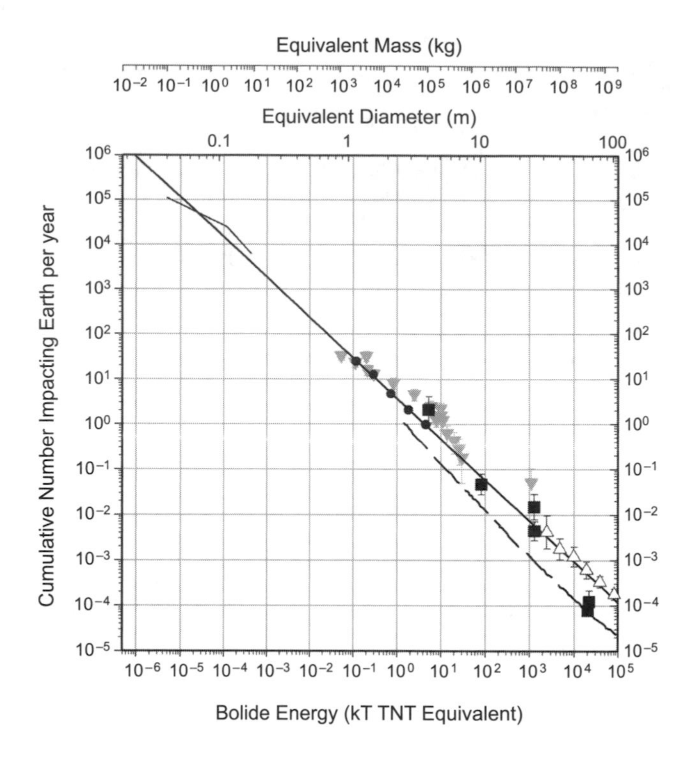

Fig. 3. Fluxes for meteoroids in the 0.01–100-m size range. Data are from groundbased camera observations of bright meteors/fireballs (*Halliday et al.,* 1996) (thin solid line), satellite fireball data (*Brown et al.,* 2002a) (filled circles), infrasound measurements (*ReVelle,* 1997) (gray inverted triangles), telescopic observations [*Rabinowitz et al.* (2000), from Spacewatch data, solid black squares; *Harris* (2002), from LINEAR data, open triangles], and estimates from lunar cratering [*Werner et al.* (2002), lower dashed line]. All flux curves are computed assuming a mass weighted average impact velocity of 20.3 km/s. Diameters are computed assuming bulk densities of 3 g cm^{-3}. The solid line running through the figure is the power law $N = 3.7E^{-0.90}$ fit to the satellite fireball data of *Brown et al.* (2002a). 1 kt TNT equivalent = 4.185×10^{12} J.

2.4. Satellite and Infrasound Observations

Satellite observations of meteoroids entering Earth's atmosphere began with the spectacular August 10, 1972, Earth-grazing fireball over the western portion of North America (*Rawcliffe et al.,* 1974). *Tagliaferri et al.* (1994) summarize the basic properties of the spacebased sensors involved in gathering these data. Regular, automated detection of fireballs with these spacebased sensors began in 1994 and continues to the present time. These sensors detect ~meter-sized and larger bodies impacting Earth, although the actual data measured is integrated optical brightness for a particular fireball in the silicon passband (cf. *Brown et al.,* 1996). To generate meaningful flux values from these optical energies, the conversion efficiency of initial kinetic energy of the body to light in the passband of the sensor is needed, as is an assumed entry velocity. The calibration of this satellite luminous efficiency has been attempted by *Brown et al.* (2002a). They make use of satellite fireball events detected with other instruments or by comparison of satellite energies with events having recovered meteorites to constrain the initial meteoroid size. Implicit in all these determinations is the assumption that fireball radiation follows a 6000 K blackbody; this is at best a gross approximation and at worse incorrect (cf. *Jenniskens et al.,* 2000). Nevertheless, *Brown et al.* (2002a) were able to derive an empirical fit to the luminous efficiency that matched theoretical predictions well (*Nemtchinov et al.,* 1997) and produced flux values compatible with telescopic observations of 10-m+-sized bodies (cf. *Rabinowitz et al.,* 2000). An updated version of their calibration curve is shown in Fig. 4 along with a comparison to theoretical values computed by *Nemtchinov et al* (1997). Using these calibration values for the luminous efficiency, the satellite-determined cumulative annual global flux of meteoroids with energy E (in kilotons of TNT equivalent energy = 4.185×10^{12} J) was found to be $N = (3.70 \pm 0.13)E^{-0.90 \pm 0.03}$ valid in the energy interval from $0.1 < E < 20$ kt. Assuming a mean impact velocity with Earth of 20.3 km s^{-1}, these bodies range in size from 1–7 m in diameter.

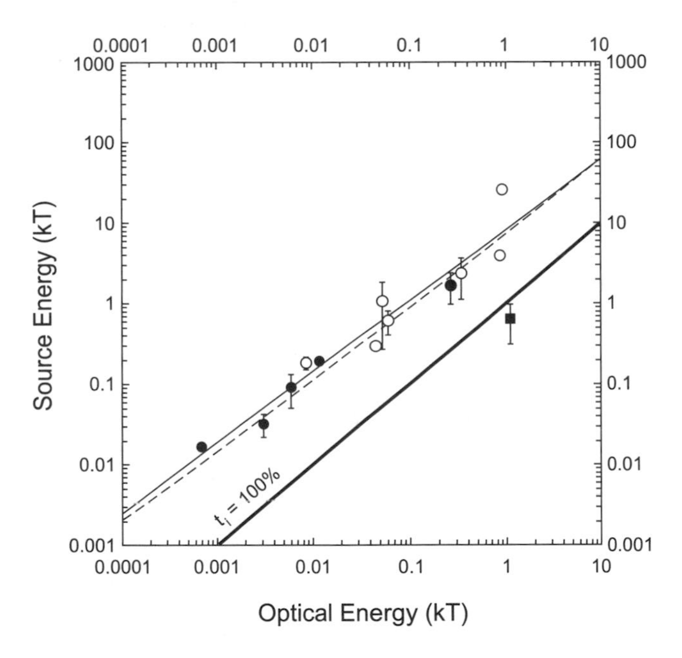

Fig. 4. An empirical fit to the integral luminous efficiency for meter-sized impactors detected by spacebased sensors. The empirical fit (thin solid line) to integral luminous efficiency in this figure is used to compute the total energy (and through an assumed velocity impacting mass) for bolides in the 0.1–10-m range. The filled circles represent individual bolide events that have energy estimates of highest reliability based on groundtruth data (meteorite recoveries, photographic or video data), while open circles are events that have energy measurements based on acoustic data only [following the technique described in *ReVelle* (1997)]. The dashed line is the theoretical integral luminous efficiency expected for bodies with H-chondrite composition (*Nemtchinov et al.,* 1997). The area below the thick solid line shows the region where more than 100% of the initial energy would be required to match the optical brightness, hence this region is unphysical.

Other flux measurements in this size range are available from infrasound records of bolide airbursts (*ReVelle,* 1997), from a small number of NEA observations (*Rabinowitz et al.,* 2000), and from extrapolation of small lunar cratering events (*Werner et al.,* 2002). Werner et al. found an impact rate on Earth for objects >4 m of 0.9 ± 0.1 per year from the lunar cratering record. In contrast, *Rabinowitz et al.* (2000), based on two NEA detections by the Spacewatch telescope, normalized the Spacewatch size-frequency distribution to the calibrated NEAT flux and found a cumulative impact rate of 2.1 ± 1.1 objects >4 m in diameter per year globally. The discrepancy between the lunar result and data based on telescopic observations is likely not significant considering the uncertainties. For example, the lunar flux values represent long-term averages; if the present impactor flux is slightly elevated, this could easily account for the factor of 2–3 disparity. *ReVelle* (1997), based on some 21 infrasonically detected bolide airwave events, derives a value of 2.5 ± 0.8 to the same equivalent energy/size per year globally. The largest single event in the infrasonic record reported by ReVelle is a 1.1-Mt-equivalent fireball that occurred over the South Atlantic on August 3, 1963. If the yield for this event is accurate, it implies an influx rate of 0.05 ± 0.05 objects globally per year >25 m diameter, a value that is much larger than either the *Rabinowitz et al.* (2000) value of $4 \times 10^{-3} \pm 3 \times 10^{-3}$ or the *Werner et al.* (2002) value of 6×10^{-3} for the same size. The small number statistics, combined with uncertainties in the albedo for such small asteroidal bodies (cf. *Binzel et al.,* 2002), means that the flux of bodies in the 10–30-m size range to Earth is the most uncertain portion of the size-frequency spectrum. More recently, *Harris* (2002) has derived impact rates from LINEAR data for objects with diameters >30 m. His results are comparable (within error) to more recent values, also from the LINEAR survey (*Stuart and Harris,* 2004), and represent a first step in filling in the uncertainties in this size range (see Fig. 3 for a summary).

3. METEORITE FLUX ESTIMATES FOR THE PAST BASED ON METEORITE RECOVERY RATES

3.1. Introduction

Initial attempts to determine the flux of meteorites in the mass range $10–10^6$ g used data on the recovery of eyewitnessed falls in densely populated areas (*Brown,* 1960, 1961; *Hawkins,* 1960; *Millard,* 1963; *Buchwald,* 1975; *Hughes,* 1980). Unfortunately, estimates of flux calculated using this technique have varied by more than 4 orders of magnitude, probably as a result of the difficulty of unambiguously constraining a number of parameters inherent in this method, e.g., population density, season, time of day, etc. As described above, direct observations of fireballs using camera networks have provided an estimate of the present flux of meteorites and its mass distribution. In particular, data from the Canadian MORP camera network (*Halliday et al.,* 1989) suggest a present flux of 83 events of mass equal to or greater than 1 g per 10^6 km^2 per year. This estimate is well constrained, but there is an uncertainty in deriving a dynamical mass from a fireball image, since at the time of the study only three imaged meteorites had been recovered to provide a comparison. This uncertainty gave rise to an approximate factor of 2 error in the MORP data (*Halliday et al.,* 1991).

Given that only one well-constrained dataset exists, and considering that flux in the meteorite size range helps to constrain the wide range of estimates for the rate at which small crater-forming events occur on Earth, an independent estimate of flux in the meteorite size range is valuable.

Another method for calculating meteorite flux, and more importantly, potential temporal variations in flux, involves analyses of meteorite accumulation sites. There are several locations on Earth where, due to favorable conditions, meteorites accumulate in large numbers. Among these are numerous hot desert sites, including the Sahara, Arabian Peninsula, Roosevelt County, New Mexico, and the Nullarbor Region of Australia, and the cold desert of Antarctica (*Annexstad et al.,* 1995). Meteorites survive for long periods in these areas because moisture level and weathering rates are low, and they are frequently recognized because of a contrast between "dark" meteorites and "pale" country rock or ice. In theory, these accumulation sites should record the integrated flux throughout the meteorite mass range over the lifetime of the accumulation surface.

Aside from offering an independent means of confirming the present flux, this method also offers the intriguing possibility of observing changes in flux over time. This would potentially elucidate meteoroid delivery mechanisms, the operation of meteoroid streams, and the compositional nature of the main asteroid belt. For example, there is compelling evidence that some portion of the meteoroid complex is in the form of short-lived (10^4–10^5 yr) co-orbital streams (*Jopek et al.,* 2002; *Gladman and Pauls,* 2005), such as groups of asteroid families with similar orbits (*Drummond,* 1991; *Rabinowitz et al.,* 1993), identical fall times of meteorites with similar orbits (*Halliday,* 1987), and differences between "old" Antarctic meteorite populations and more recent falls (*Dennison et al.,* 1986). Some claims for identification of meteorites resulting from streams remain controversial, since the cosmic-ray-exposure histories of some of these meteorites span a large spectrum of ages (*Schultz and Weber,* 1996), while calculations indicate that co-orbital streams would not be stable over the excessive timescales required (*Wetherill,* 1986; *Jopek et al.,* 2002; *Gladman and Pauls,* 2005). Analyses of meteorite populations and modeling their flux (by constraining weathering rates) may help in resolving this controversy.

Finally, although we can derive the flux of small (less than a few kilograms) meteorites from camera network data or analysis of meteorite accumulations on Earth, neither approach is effective in estimating the impact rate of larger (~10–10^5 kg) meteorites. The number of fireball events of this size is too low, and the number of desert finds too small, to allow an accurate estimate. However, as we noted above, there are data for the flux of large meteoroids at the top of Earth's atmosphere. If we can model the effects of the atmosphere on a given bolide, and we perform that modeling for a large number of cases over a large enough mass range, then it should be possible to extrapolate the upper atmosphere flux to a flux at Earth's surface. *Bland and Artemieva* (2003) took this approach to constrain the impact rate of objects from 10^2 to 10^{12} kg at Earth's surface.

Given knowledge of meteorite pairing, weathering, and removal and recovery rates it should be possible to discern changes in flux and mass distribution with time (hot deserts retain meteorites for up to at least 10^5 yr, while meteorites may survive in Antarctica for more than 10^6 yr). The idea behind this approach is that if the number of meteorite samples per unit area on the ground today is known, and the rate at which they are removed by weathering (characterized by a so-called "decay constant," λ) is calculated, then ideally we can work back to derive an estimate of the number that fell to Earth initially. However, several difficult problems with this approach must first be addressed: (1) An accurate age for either the accumulation surface and/or terrestrial ages for the meteorite population is required. (2) The action of chemical and physical weathering in removing samples must be quantified and, in any case, be minimal. (3) No other process can have acted to remove samples, e.g., no streams. Ideally, meteorites must be buried for the majority of their terrestrial history, and only recently have been excavated. (4) All the meteorite samples within a known catchment area are recovered and a study to pair samples successfully undertaken, so that a "population density" of meteorites per unit area is known. (5) The search area is large compared to the typical strewn field size. Failure to take this factor into account may mean that an area samples an unrepresentative number of falls (see *Halliday et al.,* 1991). We discuss aspects of these factors below.

3.2. Weathering Rate

Accurately modeling the weathering decay rate of meteorite samples has significance principally for two reasons. First, it allows us to quantify part of the interaction between Earth-surface environments and rock weathering. Second, in hot deserts, weathering (chemical and mechanical) is likely to be the sole means by which meteorite samples are removed from a population. As such, given a well-constrained decay rate and a measure of the number of meteorites per unit area, it is possible to make an accurate estimate of the flux of meteorites to Earth over the accumulation period. Terrestrial weathering is covered in detail in *Bland et al.* (2006).

3.3. Terrestrial Ages

In order to quantify the weathering decay rate for meteorites, it is necessary to establish an absolute chronology. Fortunately, this is possible based on the measured exposure times of the meteorites since landing on Earth, known as the terrestrial age. Over the last 30 years a variety of methods have been used to determine the terrestrial ages of meteorite finds, most of which analyze the abundance of cosmogenic radionuclides produced while the meteorite is in space. In almost all cases, the time it takes for a sample to migrate from the asteroid belt to an eventual impact with Earth (its cosmic-ray-exposure age) is significantly greater than the time needed for a saturation level of the appropriate cosmogenic radionuclides to build up in the meteorite. After the sample lands on Earth, it is largely shielded from cosmic rays and the abundance of cosmogenic radionuclides declines from a saturation level as the unstable isotopes decay. Depending on their period of terrestrial residence, the decay of several cosmogenically produced radionuclides, such as ^{14}C (halflife 5730 yr), ^{81}Kr (halflife 210,000 yr), ^{36}Cl (halflife 301,000 yr), and ^{26}Al (halflife 740,000 yr), have been used to determine the terrestrial ages of meteorites (e.g., *Honda and Arnold,* 1964; *Chang and Wänke,* 1969; *Fireman,* 1983; *Jull et al.,* 1990; *Nishiizumi,* 1990; *Nishiizumi et al.,* 1989a; *Freundel et al.,* 1986; *Aylmer et al.,* 1988; *Scherer et al.,* 1987; *Welten et al.,* 1997). These workers have reported terrestrial ages for Antarctic meteorites ranging up to ~3 m.y. For stony meteorites from the world's "hot" desert regions that have been dated so far, *Jull et al.* (1993a,b, 1995) and *Wlotzka et al.* (1995a,b) have reported ^{14}C ages >40,000 yr. The ^{14}C-dating technique measures the decay of cosmogenic radiocarbon produced in space and separates ^{14}C produced by cosmic-ray bombardment from ^{14}C accumulated by the meteorite from terrestrial sources such as organic contamination and weathering. The main source of error in this analysis comes from the possibility that some level of shielding occurred if the meteorite was buried at some depth on a large meteoroid while in space. Consideration of this (e.g., *Jull et al.,* 1995) gives rise to much of the comparatively large (~1300 yr) error on an individual ^{14}C terrestrial age.

One means of estimating decay rates for hot desert (*Zolensky et al.,* 1992) and Antarctic (*Freundel et al.,* 1986) meteorite populations has been to use terrestrial ages plotted as a frequency distribution. In a complete population for which reliable pairing data are available, an exponential decrease in frequency over time is expected. The slope of the distribution approximates to a weathering decay rate. This approach is effective provided the terrestrial age dataset is large enough to enable the slope to be well constrained; however, for most populations either sufficient samples are not available or the terrestrial-age dataset is not yet large enough. As described in *Bland et al.* (2006), one can constrain the effect of weathering in actually removing samples from a population. This, combined with the weathering rate, allows an estimate of the decay constant λ to be made (see *Bland et al.,* 1996a,b).

3.4. Number and Mass of Meteorites Arriving on Earth

To obtain an estimate of flux for each desert accumulation site, calculated values for λ are related to the density of paired meteorites over a given mass per unit area. For example, for Roosevelt County meteorites, λ is 0.032 $k.y.^{-1}$, and the density of meteorites per km^2 over 10 g in mass is 39/10.8 (39 meteorites >10 g found in an area of 10.8 km^2), which is 3.6. Given these data, the flux calculation for meteorites over 10 g in mass per 10^6 km^{-2} yr^{-1} is $0.032 \times 3.6 \times 1000$, which yields an estimate of 115 events. This estimate can be extrapolated to larger masses by using the slope of the mass distribution for each site. Generally a power law is chosen for the flux rate of the form $N_o(M) = cM^A$. A is the slope of the mass distribution and has different values over different mass ranges (*Halliday et al.,* 1989). MORP camera network data indicate a value of –0.49 for the mass range ~0.01–1 kg. The mass distribution may also be derived from an analysis of paired sample populations in accumulation sites. *Zolensky et al.* (1990, 1992) found a value of –0.54 for the Roosevelt County population, so to extrapolate to events over 500 g from the initial 10 g one takes $50^{-0.54} \times 115$, which is approximately 14. In this manner, given data on the local decay constant, mass, and aerial distribution for each site, a meteorite flux can be calculated for approximately the 10 g–1 kg mass range. Taking a similar approach for the Nullarbor Region yields 36 meteorites over 10 g in mass per 10^6 km^{-2} yr^{-1}, and for the Sahara, 95 meteorites over 10 g in mass per 10^6 km^2 yr^{-1}. Averaging the three sites, the meteorite accumulation site method yields an estimate of 82 meteorites over 10 g in mass per 10^6 km^2 yr^{-1}, remarkably close to the *Halliday et al.* (1989) estimate of 83 meteorites. Thus there is good agreement between the meteorite flux as calculated from camera networks and desert recovery sites. Given that the surface area of Earth is 5.11×10^8 km^2, these results would suggest ~42,000 meteorites >10 g in mass arriving at Earth per year (or more than 100 per day).

Bland et al. (1996a) also arrived at an independent estimate of the total amount of material in the meteorite size range that arrives at Earth's surface, by integrating over the mass range of interest. *Halliday et al.* (1989) (equations (6) and (7)) give

$$\log N = \begin{cases} 2.41 - (0.49) \log m_T & \text{for } m_T \leq 1030 \text{ g} \\ 3.40 - (0.82) \log m_T & \text{for } m_T \geq 1030 \text{ g} \end{cases}$$

so

$$\log N(m_T) = \begin{cases} 2.41 - (0.49) \log m_T & \text{for } m_T \leq 1030 \text{ g} \\ 3.40 - (0.82) \log m_T & \text{for } m_T \geq 1030 \text{ g} \end{cases}$$

where N is the number of mass $\geq m_T$ grams per 10^6 km^2 yr^{-1}.

Bland et al. (1996a,b) derive a similar relationship for a number of hot desert accumulation sites for masses <1 kg

$\log N = 2.13 - (0.58) \log m_T$ for the NR
$\log N = 2.60 - (0.54) \log m_T$ for RC
$\log N = 2.65 - (0.67) \log m_T$ for the SD ($R^* = 0.46$)
$\log N = 3.31 - (0.67) \log m_T$ for the SD ($R^* = 0.32$)

If N(m) is equal to the number per 10^6 km^2 yr^{-1} of mass greater than m, then the number per 10^6 km^2 yr^{-1} lying between m and m + dm is

$$N(m) - N(m + dm) = -\frac{dN(m)}{dm} dm$$

So the mass on the ground (per 10^6 km^2 yr^{-1}) between m and m + dm is

$$-m\frac{dN(m)}{dm} dm$$

From *Halliday et al.* (1989) we can write

$$N(m) = \begin{cases} c_1 m^{A_1} \text{ for } m \le 1030 \text{ g}; c_1 = 10^{2.41}, A_1 = -0.49 \\ c_2 m^{A_2} \text{ for } m \ge 1030 \text{ g}; c_2 = 10^{3.40}, A_2 = -0.82 \end{cases}$$

This gives

$$-m\frac{dN}{dm} = \begin{cases} -A_1 c_1 m^{A_1} \text{ for } m \le 1030 \text{ g} \\ -A_2 c_2 m^{A_2} \text{ for } m \ge 1030 \text{ g} \end{cases}$$

Considering all masses between m_L and m_U (where m_L = 1030 g and m_U = 1030 g) then the total mass per 10^6 km^2 yr^{-1} is

$$M = \int_{m_L}^{1030} -A_1 c_1 m^{A_1} dm + \int_{1030}^{m_U} -A_2 c_2 m^{A_2} dm$$

$$= -\left(\frac{A_1 c_1}{A_1 + 1}\right)\left[(1030)^{A_1+1} - (m_L)^{A_1+1}\right]$$

$$-\left(\frac{A_2 c_2}{A_2 + 1}\right)\left[(m_U)^{A_2+1} - (1030)^{A_2+1}\right]$$

Taking the above values for A_1, c_1, A_2, and c_2 from *Halliday et al.* (1989), between m_L = 10 g and m_U = 10^6g, *Bland et al.* (1996a) find M = 105.3 kg per 10^6 km^2 yr^{-1}. Given that the surface area of Earth is 5.11×10^8 km^2, this analysis suggests a total mass flux to Earth in the meteorite size range of 5.38×10^4 kg yr^{-1}, consistent with earlier estimates.

3.5. The Flux of Large Meteorites at Earth's Surface

Given the flux of large meteoroids at the top of Earth's atmosphere, and a well-constrained model of the effect of the atmosphere on a given bolide, it is possible to extrapolate the upper atmosphere flux to a flux at Earth's surface. *Bland and Artemieva* (2003) used this method to constrain the impact rate of bodies <1 km diameter at Earth's surface. Although there are gaps in the upper atmosphere data, *Bland and Artemieva* (2003) showed that overall shape of the size-frequency distribution (SFD) is similar to that calculated for impactors on the Moon and Mars, and to the observed population of near-Earth objects (NEOs) and main-belt asteroids. Thus, constraining one portion of the SFD constrains the rest. Based on this calculated SFD for the upper atmosphere, and using a model that calculates motion, aerodynamic loading, and ablation for each individual particle in a fragmented impactor, they were able to extrapolate the upper atmosphere impactor SFD to an impactor SFD for Earth's surface. The so-called "separated fragments" (SF) model allowed them to calculate fragmentation and ablation in Earth's atmosphere for a range of impactor types and masses. The SFD that *Bland and Artemieva* (2003) derived for Earth's surface suggests that a 1-t meteoritic mass arrives on Earth approximately once a year, a 10-t mass every 15 years, and a 100-t mass every 200 years. Their modeling further suggests that >95% of these bodies will be irons.

3.6. Meteorite Flux and Populations Based on Antarctic Finds: A Window into the Meteorite Flux over the Past 3 Million Years

Since preservation and recovery factors are best attained in Antarctica, as shown by 30 difficult years of collection activity, we discuss work on these meteorites as a special case. Meteorite recovery expeditions over the past 30 years by the National Institute of Polar Research, National Science Foundation, and EUROMET have returned more than 30,000 meteorite specimens from Antarctica, probably representing more than 2000 individual falls (*Annexstad et al.,* 1995). The population of recovered Antarctic meteorites thus outnumbers the entire harvest from falls over the remainder of Earth that has collected over the past 13 centuries (some very old falls have been preserved in Japan). Terrestrial age determinations have revealed that meteorites have been preserved for up to 3,000,000 years in Antarctica, although most have a terrestrial residence time of less than 100,000 years (*Nishiizumi et al.,* 1989a; *Nishiizumi,* 1995; *Welten,* 1995; *Scherer et al.,* 1987). Elsewhere on Earth meteorites are generally completely weathered away in 10^4–10^5 yr. The population of Antarctic meteorites contains numerous types not found (or rarely so) elsewhere on Earth, such as lunar meteorites, certain irons, and metamorphosed carbonaceous chondrites. Some workers report compositional variations in the H chondrites with time. These various reports suggest changes in the source asteroids feeding material to Earth over the past several hundred thousand years (*Michlovich et al.,* 1995). However, many of these unusual meteorites are small in mass, and therefore are most easily located on a white, rock-free environment, leading one to speculate that a collection bias is at work (*Huss,* 1991). It is thus desirable to reexamine how the population of meteorites to Antarctica through the ages compares to the modern record. Here we will explain the barriers to this goal and progress that is being made.

According to the general model, as meteorites fall from space onto the Antarctic plateau, they get incorporated into a growing pile of snow and ice within a few years (*Whillans and Cassidy,* 1983, *Cassidy et al.,* 1992; *Delisle and Sievers,* 1991). From this time until they emerge somewhere downslope they are locked into cold ice, subject only to very slow weathering [by the action of "unfrozen water" (see *Zolensky and Paces,* 1986)] and some dynamic stress as the ice flows toward the periphery of the continent, where it drops into the ocean. However, in some areas the ice has to pass over or around major mountain ranges (often buried but still felt by the ice), and in these areas the ice has to slow down, increasing exposure of the ice to the dry katabatic winds and causing increased sublimation. Over thousands of years meteorites can be highly concentrated at the surface on temporarily stranded, dense, blue ice. In some situations (e.g., the Pecora Escarpment) meteorite stranding zones are not directly downflow of glaciers, but rather appear to have moved laterally for poorly understood reasons, in a sort of "end run" around the front of an obstruction (J. Schutt, personal communication, 1991). These stranding surfaces appear to be stable until increased snowfall upslope increases the flow of ice downslope, flushing the meteorites away. It is possible for one stranding surface to form and be flushed downstream only to reemerge later. Each stranding surface contains at least two populations of meteorites. The first are those that traveled in the ice to the stranding site, and therefore represent a snapshot of the meteorite flux from the past (older meteorites have been flushed away, younger ones are upstream still entrained in the ice). The second population is provided by recent direct falls onto the stranding surface. The presence of at least two chronologically distinct meteorite populations considerably complicates any calculation of meteorite flux based upon statistical considerations. In fact, *Huss* (1990a,b) has argued that nearly all meteorites in stranding surfaces fell directly there, that the stranding surface is not in fact moving significantly, and that the accumulation times calculated from the terrestrial ages of the meteorites can therefore be used to date the period over which the ice has been stalled. This model is based upon a careful consideration of the meteorite mass/number populations. The conflicts between Huss' model and that by *Cassidy et al.* (1992) have not been resolved. This is obviously a major impediment to the use of meteorite populations to calculate meteorite flux; nevertheless, we move bravely onward.

3.7. Calculation of Antarctic Meteorite Flux

There are numerous uncertainties that currently impede calculation of meteorite flux from Antarctic collections, most of which cannot be easily circumvented (*Harvey and Cassidy,* 1991). We review these factors below.

3.7.1. Terrestrial age dating of meteorites. In terrestrial age determinations of Antarctic meteorites (principally chondrites) the isotopes ^{36}Cl, ^{81}Kr, ^{10}Be, ^{14}C, and ^{26}Al are most effectively employed (*Nishiizumi et al.,* 1989a; *Wieler et al.,* 1995; *Welten,* 1995). There is a measurement gap between ^{14}C ages (<40 ka) and ^{36}Cl ages (>70 ka). *Nishiizumi* (1995) has suggested that this gap might be closed by future investigations of cosmogenic ^{41}Ca. Unfortunately, it is not currently easy to measure ^{41}Ca in chondrites.

Due principally to the sheer bulk of recovered meteorites, available funding, and finite lifetimes of meteorite investigators, only about 1% of the Antarctic meteorites have had their terrestrial ages determined (*Nishiizumi et al.,* 1989a; *Nishiizumi,* 1995; *Wieler et al.,* 1995; *Welten,* 1995); these are principally by ^{36}Cl. Indirect means must be found to date the remainder.

Benoit, Sears, and coworkers have suggested that natural thermoluminescence (TL) can be used to help address this problem in the interim (*Benoit et al.,* 1992, 1993, 1994). In this technique, one assumes a mean annual temperature for the storage of Antarctic meteorites and a "surface exposure age" as part of the terrestrial age is deduced. According to these TL measurements most of the meteorites spend <50% of their terrestrial history on the surface of the ice in Antarctica, with this conclusion being based upon their low relative TL sensitivity, which in turn is related to their degree of postexposure weathering. While this technique cannot be used to obtain absolute terrestrial ages, the calculated exposure ages could be of value in comparing the histories of the different Antarctic stranding surfaces. However, we note that *Welten et al.* (1995) have sounded a dissenting note on the applicability of TL to this problem.

We show results for approximately 280 dated chondrites in Fig. 5 [data from *Nishiizumi et al.* (1989b), *Wieler et al.* (1995), *Welten* (1995), and *Michlovich et al.* (1995)]; the complex nature of meteorite preservation is well displayed.

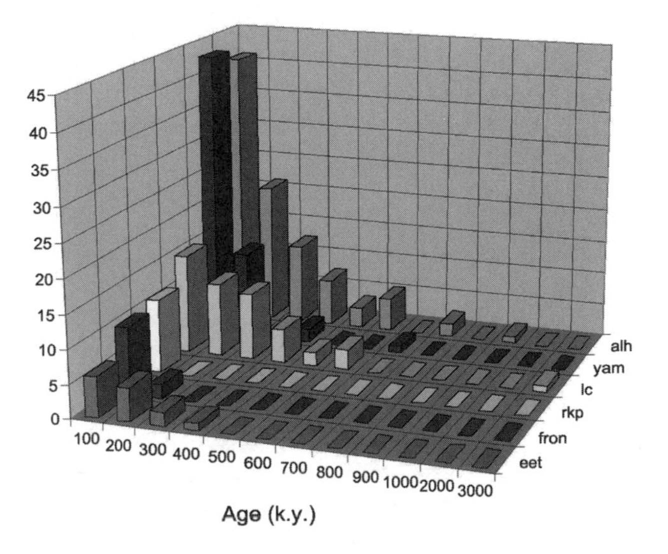

Fig. 5. Distributions of terrestrial ages of meteorites from diverse locations in Antarctica. Alh = Allan Hills, yam = Yamato Mountains, lc = Lewis Cliff, rkp = Reckling Peak, fron = Frontier Mountains, eet = Elephant Moraine.

The Antarctic meteorites display terrestrial ages ranging up to 3 Ma, although chondritic terrestrial ages from different Antarctic ice fields have different distributions. The oldest meteorites are generally found at the main ice field of Allan Hills and Lewis Cliff. In particular, the Lewis Cliff ice field appears to have the best record of meteorites, since the decay of its meteorite population with terrestrial age is least pronounced (Fig. 5). The Lewis Cliff ice tongue appears to be the oldest of the investigated ice field stranding areas (*Welten,* 1995; *Welten et al.,* 1995). At the other end of the scale, terrestrial ages of chondrites recovered from Frontier Mountain are <140 ka, indicating that this is a relatively recent stranding surface.

Various schemes have been proposed for the estimation of terrestrial ages for entire stranding surface meteorite populations. For a given stranding surface one could assume that all meteorites are younger than the oldest measured meteorite, and that meteorite recovery is 100%, but this is certainly not the case, as explained below. Workers have also attempted to characterize the accumulation age of stranding surfaces by examining the mass distributions and weathering lifetimes of meteorites (*Huss,* 1990b). Finally, it would appear that merely measuring the age of a particular blue ice stranding surface and counting up the meteorites presented there would permit the measurement of the flux rate of meteorites to that surface, but we have shown above that at least two distinct meteorite populations are always present. Nevertheless, of all factors to be considered in the derivation of past meteorite flux for Antarctica, terrestrial age is one of the best constrained.

3.7.2. Uncertain meteorite catchment area. In other regions of Earth you can easily measure the area you are searching for meteorites, enabling a flux per unit area to be calculated. This is not directly possible in Antarctica because the meteorites exposed today at a stranding surface have fallen over a far larger glacial tributary system. This situation is further complicated by the fact that the ice movement is irregular, and velocity is not constant. Tributary ice streams can provide additional meteorites with an irregular periodicity. G. Delisle (*Delisle and Sievers,* 1991; *Delisle,* 1993, 1995) has shown that glacial stages result in thicker coastal ice, while interglacials produce thinner coastal ice. The consequences of this are seen in a change of velocity and mass transport of ice with the entrained meteorites. The meteorites at Allan Hills and possibly Frontier Mountain are suspected to have been buried and reemerged, as the interglacial sublimation rates are much higher than the glacial sublimation rates. In this picture the meteorite traps are not stagnant regions, but are continually readjusting to the pattern of glacial and interglacial periods. The uncertain nature of the meteorite catchment area appears to be a severe impediment to the successful calculation of meteorite flux from Antarctic data.

3.7.3. Paleo-stranding surfaces. During periods of increased snowfall upstream, ice thickness will increase, and a downstream stranding surface (with its population of direct falls) can be flushed further downstream, only to be stranded again and acquire a fresh brood of falls (*Cassidy et al.,* 1992; *Delisle and Sievers,* 1991).

3.7.4. Direct infall. Fresh meteorites fall directly onto recovery stranding surfaces, which can cause confusion if they are being added to a recently exposed paleo-stranding surface.

3.7.5. Recovery rate. One might assume that all meteorites from a given stranding surface can be easily recovered. However, considerable experience by the first author (M.E.Z.) indicates that meteorites are often missed due to the wind getting in a searcher's eyes, or burial by a transient snow patch. Meteorite recognition skills also vary considerably from person to person (even among veteran meteoriticists!), and luck plays a role. *Harvey* (1995) has discussed the problem of searching efficiency in some detail, and attempted to quantify a correction factor.

3.7.6. Weathering and differential survival of meteorites. Several factors have been identified that have major effects on the survival of meteorites on the ice. The chemical weathering of meteorites, although minimal in Antarctica by the cold and aridity, is still significant. Even while stored in ice, the action of thin monolayers of "unfrozen water" coating meteorites will inevitably cause chemical weathering (*Zolensky and Paces,* 1986). Meanwhile, ice movement can cause shearing and consequent breakage of the entrained meteorites (physical weathering). Finally, once the meteorites have emerged on the stranding surface, they are subject to accelerated chemical weathering on hot days when ice melts around them (the meteorites soak up solar radiation) and physical abrasion from the katabatic winds.

In "hot" deserts such as the Libyan Desert, Algeria, and Roosevelt County, New Mexico, the mean survival time of chondrites is >10 k.y. (*Boeckl,* 1972; *Jull et al.,* 1995; *Wlotzka et al.,* 1995a), although far older meteorites (~100 ka) are found in some fortuitous regions [e.g., Roosevelt County (see *Zolensky et al.,* 1992)].

Certain easily weathered meteorite types will be preferentially lost from the stranding record. Attempts to correct the Antarctic record for weathering effects (*Huss,* 1990a; *Harvey,* 1995) have had unknown success. *Bland et al.* (1996a,b) suggest that H chondrites weather faster than L chondrites (as might be expected from the former's greater metal content), although *Zolensky et al.* (1995) found just the opposite for meteorites recovered from the Atacama Desert. *Bland et al.* (1996a,b) point out that carbonates and hydrous silicates could, in some situations, armor mafic silicates and thereby provide a certain degree of protection against further alteration. Also, the changing temperature and aridity conditions attending regional climatic change would have further complicated this situation.

The main results of weathering will be threefold: (1) to decrease the population of all meteorites and reduce the mass of the survivors; (2) to increase the number of samples, through physical breakup, and subtly change the composition of the survivors (see below), increasing the importance and difficulty of pairing studies; and (3) removing from the record especially friable meteorites (e.g., CI1, of which no Antarctic specimens have been found). Various workers have proposed correction factors for weathering, but these appear to be site specific (*Boeckl,* 1972; *Zolensky et al.,* 1992; *Harvey,* 1995).

3.7.7. Removal of meteorites by wind. The unrelenting katabatic winds in Antarctica promote rapid ablation [on the order of 5 cm/yr at the Allan Hills (*Schutt et al.,* 1986; *Cassidy et al.,* 1992)], and once meteorites emerge from the sublimating ice they are subject to transport downwind by these same winds (*Schutt et al.,* 1986; *Delisle and Sievers,* 1991; *Harvey,* 1995). This was elegantly demonstrated by an experiment at the Allan Hills main ice field. During the 1984–1985 field season, the NSF Antarctic Search for Meteorites (ANSMET) field party placed surrogate meteorites (rounded basalt samples of varying masses) onto a "starting line," one rock "team" sitting on the ice surface and the second team embedded just below the surface (*Schutt et al.,* 1986). This rock race was visited the following year, and the results were that all but the three largest buried meteorites had emerged. A number of stones were found to have moved downwind from the starting line, and eight had blown so far away that they could not be located.

Eolian effects are more pronounced in Antarctica than in other meteorite recovery regions, and the experiment described above has shown that specimen size inversely affects specimen movement, and that the mass threshold is a function of wind velocity, which relates to a specific area and may not be the same over different search regions. Small meteorites will be removed from the stranding surface by the action of wind alone, which undoubtedly skews the constitution of the recovered meteorite population.

3.7.8. Meteorite pairing. Many (most?) meteorites fall as showers of individuals that are subject to later breakage. In addition, physical weathering breaks meteorites into smaller, more numerous specimens. Flux calculations concern the number of individual falls, not total specimens, necessitating that meteorite pairing be determined. Pairing studies for unusual meteorites are well established; however, few of the ordinary chondrites (i.e., 90% of the collection) have received adequate study (because they are generally unexciting to most meteoriticists). Pairing for ordinary chondrites is most often done using comparative petrological, ^{26}Al, TL, and noble gas analyses (the latter revealing cosmic-ray-exposure histories) (*Schultz et al.,* 1991; *Benoit et al.,* 1993; *Scherer et al.,* 1995).

Unfortunately, as for the determination of terrestrial ages, no workers can analyze all meteorites on a given stranding surface, much less all 15,000 Antarctic specimens. In addition, weathering has been shown to deleteriously affect the noble gas content of chondrites, seriously complicating pairing studies (*Scherer et al.,* 1995). Therefore one must derive and apply a correction factor for the pairing of the ubiquitous ordinary chondrites; these have ranged from 2 to 6 (*Ikeda and Kimura,* 1992; *Lindstrom and Score,* 1995). Lindstrom and Score derived a chondrite pairing value of 5, based on the known pairing incidences of Antarctic meteorites. By applying this simple pairing correction value to the full population of Antarctic chondrites, they were able to address the question of possible differences in the meteorite population through time (see below).

Weathering the foregoing difficulties, workers have attempted to calculate the meteorite flux for Earth over the past several million years based upon the Antarctic meteorite record. As one example, *Huss* (1990a) concluded that that the ages of the Antarctic ice fields, as calculated from meteorite mass distributions, indicate an infall rate only slightly higher than the modern rate, concluding that the flux rate had not changed by more than a factor of 2 over the past million years. However, most agree that it is premature to attempt direct measurement of the past meteorite flux rate from the Antarctic record, despite the potential value in doing so. Additionally, considerable uncertainties exist in attempts to estimate the most critical variables, which continue to prevent accurate estimation of the flux. To advance this effort, the following studies should be undertaken: (1) The pairing of ordinary chondrites needs to be better constrained with real data. (2) We need to settle the nagging issue of the usefulness of TL in estimating meteorite terrestrial ages. (3) Investigation into the potential effects of weathering in the differential destruction of meteorite types, work begun now by Bland as described above, needs to continue. (4) Most critically, we need to better understand ice flow dynamics at each of the oldest meteorite stranding sites (Allan Hill and Lewis Cliff) to permit real meteorite fluxes to be calculated for the past 3 m.y.

Despite the difficulties faced in unraveling the record of past meteorite fluxes based on the record of the Antarctic meteorites, workers have also sought changes in the distribution of meteorite types among these same samples, as we shall see below.

3.8. More Ancient Meteorite Flux from Deposits of Fossil Meteorites

Quite surprisingly, more than 40 meteorites preserved in anoxic marls have lately been recovered from quarries in Scandinavia (*Schmitz et al.,* 1997, 2003). These "fossil" meteorites are recognizable due to pseudomorphed chondrule textures (Fig. 6) and preserved spinel grains with Al_2O_3/MgO and Al_2O_3/TiO_2 ratios similar to those in L and LL chondrites (Fig. 7). Workers have used the results of a systematic meteorite recovery operation from one particular quarry to attempt a calculation of chondrite flux some 420 m.y. ago (in the early Ordovician). As reported by *Schmitz et al.* (1997, 2003) this age is close to the calculated breakup age of the L-chondrite parent asteroid (420 ± 20 Ma). From a consideration of the number of recovered meteorites, the volume of material sampled, and sediment deposition rate, *Schmitz et al.* (1997, 2003) calculate an early Ordovician flux of L chondrites some 2 orders of magnitude greater than the modern rate. They further report, from these same sedimentary rocks, Os-isotopic and Ir contents consistent with an order of magnitude increase in the flux of interplanetary dust, ascribing this to the same L-

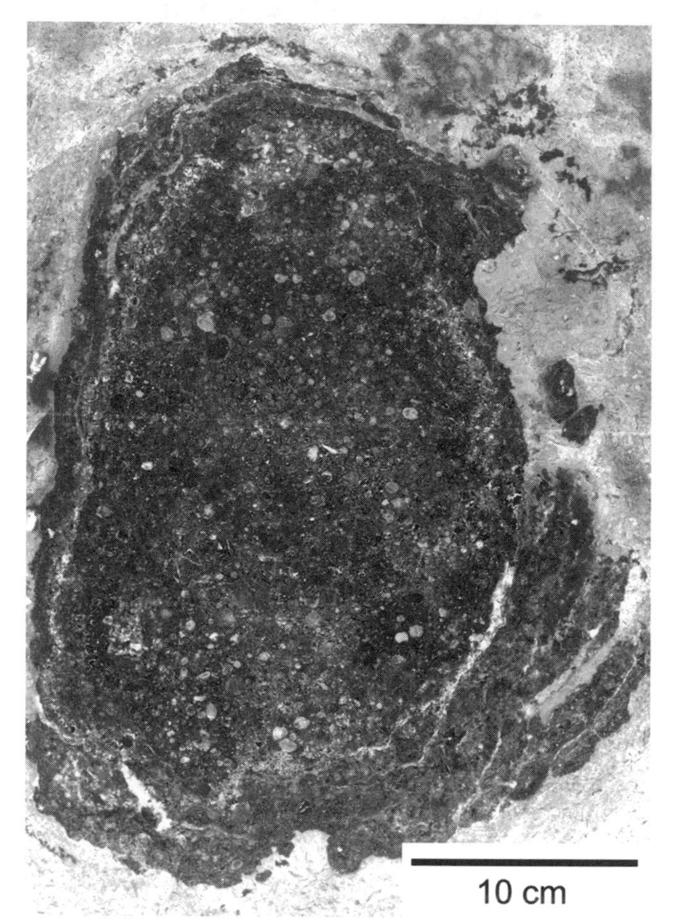

Fig. 6. Chondrule textures preserved in fossilized meteorites found in quarries in Scandinavia. After *Schmitz et al.* (2001). The millimeter- to submillimeter-sized round features apparent in this figure are fossil chondrules.

chondrite parent asteroid. Given their study assumptions, this report requires verification from another early Ordovician site (if this is even possible). However, if correct, this result suggests that directly upon breakup of the L-chondrite parent asteroid, Earth experienced a dramatic increase in L-chondrite falls. To generalize, the breakup of any body near Earth should temporarily increase the flux of extraterrestrial material — something that could and should have happened many times throughout Earth's history.

4. METEORITE TYPE STATISTICS

Where are meteorites coming from? How fast do these meteorite source regions evolve, as reflected in a changing complexion of meteorite types? What is the danger to life and structures from falling meteorites, and is this threat changing (*Beijing Observatory,* 1988)? Some geochronological techniques rely on a constant flux of extraterrestrial material to Earth — but how reasonable is this assumption? To discover the answers to these critical questions we need to determine the meteorite flux to Earth not just for the present day, but back into the past.

4.1. Meteorite Population Determinations from Falls

Without pressing a permanent force of persons into meteorite watch gangs, it is impossible to derive accurate meteorite fluxes from fall observations. But fall information probably does provide the best information on the relative types of meteorites that today pepper Earth (*Wasson,* 1985). Some workers suggest that the modern fall record favors

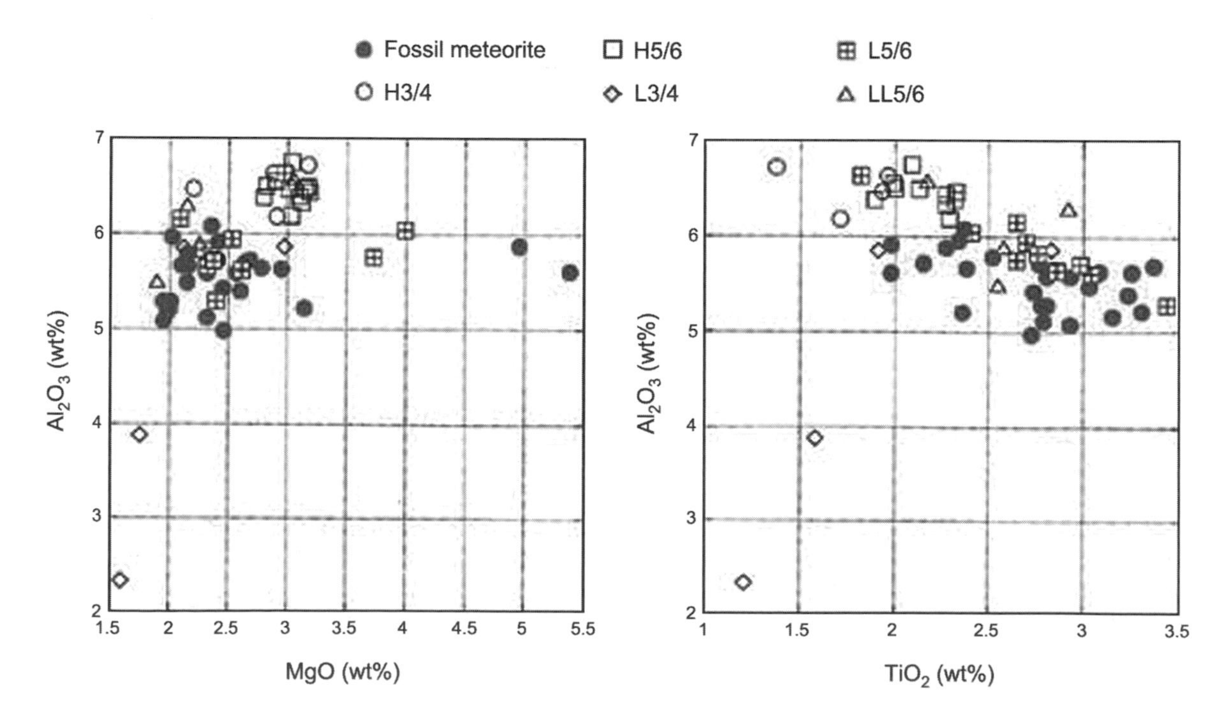

Fig. 7. Comparison of average chemical composition of chromite grains from 26 fossil meteorites and 33 ordinary chondrites. After *Schmitz et al.* (2001).

large showers over small, individual falls, but it is not clear what effect this potential bias would have. In any case, we wish to learn how fast the meteorite population changes, i.e., are we sampling different meteorite parent bodies through time?

4.2. Have Meteorite Populations Changed Through Time?

A very contentious subject has been that of potential compositional and mineralogical differences between younger finds and falls vs. older finds. An entire workshop was devoted to this subject (*Koeberl and Cassidy,* 1990). In particular, it appears that the important issue of possible observed changes in the H-chondrite populations remains unresolved.

In numerous papers, Wolf, Lipschutz, and coworkers argue that there is chemical evidence that Antarctic H4–6 and L4–6 chondrites with terrestrial ages >50 ka (from Victoria Land) are different from those that are younger (Queen Maud Land meteorites and modern falls) (*Dennison,* 1986; *Michlovich et al.,* 1995; *Wolf and Lipschutz,* 1995a,b,c). A difference between these meteorites would imply a change in the sources providing these meteorites over a timescale deemed too rapid by many dynamicists (*Wetherill,* 1986). Critics suggest that weathering may have redistributed the volatile trace elements being used for the difference arguments, but this view is disputed by some recent studies (*Wolf and Lipschutz,* 1995c; *Burns et al.,* 1995). Other critics point out that if it takes a sophisticated statistical examination of fully 10 trace elements in order to see a difference between the putative meteorite groups, then they are actually very similar. In a test of the Lipschutz-Wolf model, *Loeken and Schultz* (1995) performed analyses of noble gases in H chondrites from Antarctica (from Allan Hills and the Yamato Mountains) and modern falls. The authors were careful to select only Antarctic meteorites analyzed by Wolf and Lipschutz. They found no correlation between terrestrial age and either exposure age, radiogenic ^{4}He, or radiogenic ^{40}Ar over the period of the last 200 k.y. One might expect to see such correlations if the population source for Antarctic H-chondrite finds and modern H-chondrite falls indeed differed. Lipschutz and Wolf steadfastly maintained their position that the apparent population differences are real, and the reader will just have to decide the issue for himself.

As described above, *Lindstrom and Score* (1995) have examined the population statistics for the Antarctic meteorites recovered by the ANSMET expeditions. By applying a simple pairing correction value to the full population of Antarctic ordinary chondrites, they find that the relative numbers of Antarctic finds match modern fall statistics, so there is no difference in these populations when they are viewed in this manner (Table 1). Application of this simple pairing statistic suggests that there were approximately 1300 separate known Antarctic falls represented in the ANSMET collection in 1995. They also reported that most rare types of meteorites (with the exception of lunar and SNC meteorites) are small in mass.

TABLE 1. Meteorite populations for modern falls and Antarctic finds.

Meteorite Type	Modern Falls* (%)	Antarctic Finds† (%)
Ordinary chondrites	79.5	79.5
Carbonaceous chondrites	4.2	5.2
Enstatite chondrites	1.6	1.7
Achondrites	8.3	8.5
Stony irons	1.2	0.9
Irons	5.1	4.3
Total meteorites	830	1294

*From *Graham et al.* (1985)

†Numbers are given only for meteorites recovered by the ANSMET program. Data from *Lindstrom and Score* (1995), corrected for pairing (a pairing value of 5 was assumed for ordinary chondrites).

However, it is undeniable that a significant number of unique meteorites are found principally in Antarctica. These include lunar meteorites (all of which have been found in deserts, with the majority having been found in Antarctica), certain irons, CM1 chondrites, chondrite regolith breccias, and metamorphosed carbonaceous chondrites (*Zolensky et al.,* 2005). The majority of these unusual meteorites are small, under 1 kg, and it is undeniable that small meteorites are more likely to be recovered in Antarctica than elsewhere on Earth. This is despite the fact that small meteorites can be removed from stranding areas by the katabatic winds; those that remain are generally easily found, particularly in stranding areas devoid of terrestrial rocks. Nevertheless, it is intriguing that few of these types of meteorites have been found outside Antarctica.

As mentioned above, more than a dozen metamorphosed carbonaceous chondrites have been recognized from the Japanese Antarctic (JARE) collections, but only one has been recognized from the U.S. Antarctic collection (ANSMET), and there are none from modern falls or hot desert finds (*Zolensky et al.,* 2005). Typically the JARE meteorites have terrestrial ages intermediate between modern falls and meteorites recovered by ANSMET. This observation suggests that the source asteroid(s) of these metamorphosed carbonaceous chondrites may have broken up near Earth in the last 100,000 years. Recent modeling of such breakups show that they can signicifantly affect meteorite fluxes for up to 10^5 yr (*Gladman and Pauls,* 2005). These observations all require verification, but are sufficient to show that the flux rate and type of meteorites is not constant even at the scale of 10^5 yr.

4.3. Meteorite Masses

To a large degree the arguments for differences between Antarctic (old terrestrial age) and non-Antarctic (generally younger) meteorites comes down to the point that these unusual meteorites are generally small in size. It is certainly a fact that all Antarctic finds have smaller mean masses than comparable meteorites from the modern fall record (*Huss,* 1990a,b; *Harvey,* 1990, 1995; *Harvey and Cassidy,* 1991).

Four explanations for this latter observation are immediately apparent (others are undoubtedly also possible): (1) Small, unusual meteorites are falling today, but because they are small, are not generally recovered from most locales. Smaller meteorites are relatively easy to locate in Antarctica, due to the generally white to blue background and, locally, the low number of admixed terrestrial rocks. (2) Antarctic meteorites were identical to modern falls in mass, possibly different in type, and have been weathered into a steady-state population of smaller mass bits. (3) Antarctic meteorites are identical to modern falls, but the largest individuals have been preferentially removed from the Antarctic population, leaving only smaller specimens. Note that this would this make the Antarctic find record complementary to the modern fall record (where generally only the largest specimens are recovered). This explanation is, however, totally unsubstantiated by any reasonable mechanism. (4) The ancient meteorite flux was truly different from the modern one, in type and mass. Thus there are subtle indications that the old falls represented by the Antarctic collections have sampled a slightly different meteorite parent-body population. These differences are present at the 1–2% level of the total populations, however.

5. THE FLUX OF MICROMETEOROIDS TO EARTH

5.1. Satellite and Long Duration Exposure Facility Data

Satellite impact observations [such as the Long Duration Exposure Facility (LDEF) or Pegasus] only begin in the 1–100-mg range (*Humes,* 1991; *Naumann et al.,* 1966). For masses in the range of a few kilograms to tens of micrograms, available flux data are primarily camera, video, and radar observations of meteors. These techniques make use of the atmosphere as a detector. The major advantage of these approaches are the large collecting areas that can be utilized — up to 10^4 km^2 and higher for some radar and optical systems (cf. *Brown et al.,* 2002a,b; *Thomas et al.,* 1988). Each technique, however, suffers from different biases that make final flux interpretations somewhat problematic. For photographic/video observations, weather and daylight may bias accessible radiants and sensitivity is a function of angular velocity, making correction to a single limiting magnitude problematic (cf. *Hawkes,* 2002). Furthermore, these optical instruments detect radiation emitted by the meteor in the specific passband of the instrument — transforming to mass or size of the incident meteoroid requires knowledge of the luminous efficiency, a value that is poorly constrained over much of the size/velocity range of interest (cf. *Bronshten,* 1983, for a review). These physical uncertainties are further compounded by the general logistical difficulty of running optical networks for long periods (and reducing the resulting data); the few successful examples are the three large fireball networks (*Halliday et al.,* 1989), of which only the European Network is still in operation (*Spurný,* 1997). Video systems (which access a smaller particle population than the camera networks) operating continuously are just now becoming feasible with faster computers and cheap, mass-storage devices being developed.

Radar systems suffer a greater range of biases than optical instruments, including poorly known ionization efficiencies, fragmentation effects, recombination, and initial radius biases (*Baggaley,* 2002). In addition, most backscatter meteor radar observations reflect off a small segment (1–2 km) of the meteor trail only and so can provide (at best) a limiting value for the mass of any particular incident meteoroid. Historically, underestimation of the effects of the initial trail radius bias have led to radar influx measurements more than an order of magnitude smaller than comparable satellite measurements at the same mass (*Hughes,* 1980, 1994). Low-frequency radar observations (*Thomas et al.,* 1988) and modeling of this effect (*Campbell-Brown and Jones,* 2003) have led to flux estimates more consistent with dust detector values. Computation of the radar collecting area is also more complicated than is the case for optical instruments (*Brown and Jones,* 1995). Technically, radars also tend to be labor intensive to operate and maintain; however, once working and with suitable processing algorithms they are best suited for long-term surveys and tend to produce much larger datasets, in some cases exceeding 10^4 meteor echoes per day (*Baggaley,* 2002). Cross-calibration of optical and radar measurements of meteors is a major outstanding issue to address the flux at masses of micrograms to milligrams.

Examination of Fig. 8 shows the flux measurements available in the 10–10^{-5} g range (meteoroids to micrometeoroids). Most remarkable is the lack of modern measurements over much of this range; no dedicated optical sporadic meteor fluxes have been made using video or CCD instruments, and only LDEF satellite data covers the lower decade in mass. *Hawkins* (1960) used Super-Schmidt camera data to compute meteor fluxes at bright (+2 and brighter) visual magnitudes, although these are based on a relatively small sample and suffer from the uncertainties in panchromatic luminous efficiencies at these sizes (cf. *Ceplecha et al.,* 1998). Among the best modern meteor radar measurements are those from the high-power, low-frequency Jindalee radar (*Thomas and Netherway,* 1989). These data are particularly suited to flux determinations as the low frequencies involved produce negligible initial radius biases, are less affected by fragmentation interference effects, and the collecting area for the system is well-determined. The estimated flux from this source is also shown in Fig. 8.

Ceplecha (2001) has produced a compendium of primary data sources over this mass interval, although the exact reduction technique used to transform to final flux values is not discussed in detail (see Fig. 8). Ceplecha's flux curve is generally below that found from *Hawkins* (1960) values and also below the standard *Grün et al.* (1985) interplanetary dust flux model, which is itself based on earlier Apollo-era satellite and radar meteor data.

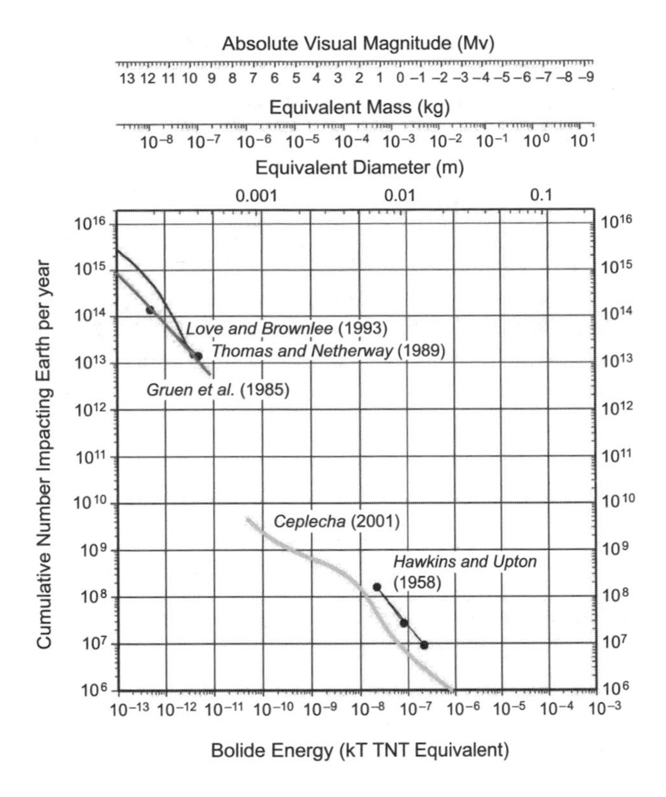

Fig. 8. Flux measurements available in the 10^{-8}–10-kg range. The same velocity and bulk density assumptions are used as in Fig. 9. The equivalent absolute visual magnitude is found following the relation given by *Verniani* (1973). Among the best modern meteor radar measurements are those from the high-power, low-frequency Jindalee radar [*Thomas and Netherway* (1989), triangles in the upper left portion of the figure], which suffers much less from biases that heavily affect higher-frequency meteor radar meteor flux measurements. The standard interplanetary flux curve produced by *Grün et al.* (1985) from a compilation of satellite impact data and other sources is also shown as a thick black line. The flux derived from impact craters measured on LDEF (*Love and Brownlee,* 1993) is given up to the size of the largest equivalent impactor near 10^{-7} kg. The compilation of *Ceplecha* (2001) based on video and photographic observations (thick gray line) is also shown (valid for masses $>10^{-6}$ kg). Measurements of Super-Schmidt meteors by *Hawkins and Upton* (1958) (solid circles) differ from that of *Ceplecha* (2001), in part due to differences in the mass scaling used to convert luminous intensity to mass. This uncertainty in the optical mass scale may also partially explain the apparent gap between the *Ceplecha* (2001) curve and the measurements valid for $m < 10^{-7}$ kg. The *Halliday et al.* (1996) fireball curve shown in Fig. 9 is also given for comparison. *Hawkins* (1960) used Super-Schmidt camera data to compute meteor fluxes at bright (+2 and brighter) visual magnitudes, although these are based on a relatively small sample and suffer from the uncertainties in panchromatic luminous efficiencies at these sizes (cf. *Ceplecha et al.,* 1998).

Numerous datasets from satellite impact detectors and/or returned surfaces have provided micrometeoroid flux estimates in the $<10^{-4}$-g mass range (cf. *Staubach et al.,* 2001, for a review). Results from LDEF, in particular, provide the largest time-area exposure product of any dataset to date. The micrometeoroid flux from LDEF returned surfaces have been determined by counting the cumulative number vs. size for visible impact craters on the space-facing end of the returned satellite surface (*Love and Brownlee,* 1993). Laboratory calibration of impact crater depth-diameter-impactor density-velocity relations permit reconstruction of the original particle mass. The original LDEF micrometeoroid flux curve measured by *Love and Brownlee* (1993) is shown in Fig. 8.

One crucial assumed input parameter in this analysis (and others that use satellite measurements to infer impactor masses) is meteoroid velocity. To date, accurate individual meteoroid velocities have only been measured for meteoroids impacting Earth's atmosphere. Historically, mass-averaged velocities derived from the Harvard Radio Meteor Project (*Sekanina and Southworth,* 1975) or the photographically inferred velocity distribution for larger meteoroids (*Erickson,* 1968) have been used. The *Erickson* (1968) distribution is appropriate to large meteoroids (gram-sized and larger) and is based on less than 300 meteors total. There are less than 65 meteors with velocities in excess of 50 km/s in this survey. The original HRMP velocity distribution has also been found to be in error by 2 orders of magnitude at higher velocities due to a typographic error (*Taylor and Elford,* 1998). The revised HRMP velocity distribution pushes the mass-averaged normal velocities from 12 km/s as used in the LDEF analysis to 18 km/s. This has the effect of shifting the LDEF flux curve downward by a factor of 2 in mass.

More recently, two independent radar surveys, one in the southern hemisphere (*Galligan and Baggaley,* 2004) and one in the northern hemisphere (*Brown et al.,* 2005), have remeasured velocities for faint radar meteors using a technique different from that utilized by the HRMP (see Fig. 9). Both find that the mass-averaged mean velocity is much higher than the original HRMP value of 14 km/s (*Southworth and Sekanina,* 1973). This may represent an unrecognized bias against fragmenting meteors in the original HRMP survey due to the velocity measurement technique. If this finding is confirmed, the effect will be to move the mass flux curve for microgram and smaller measurements downward by close to an order of magnitude at some masses. Remaining differences between the various velocity distributions as shown in Fig. 9 may reflect differences in the choice of bias-correction terms.

There have been several attempts to determine the composition of meteoroids and micrometeoroids of cometary origin by averaging the composition of the radiating gas along the fireball path (e.g., *Ceplecha,* 1964, 1965; *Millman,* 1972; *Harvey,* 1973; *Nagasawa,* 1978; *Borovička,* 1993; *Borovička and Spuný,* 1996; *Borovička and Betlem,* 1997; *Ceplecha et al.,* 1998). The most recent study (*Trigo-Rodriguez et al.,* 2004a,b) suggests that these meteoroids have Si-normalized compositions of most determined elements (Mg, Fe, Ni, Cr, Mn, and Co) in the range of the CM and CI chondrites, but exhibited depletions in Ca and Ti,

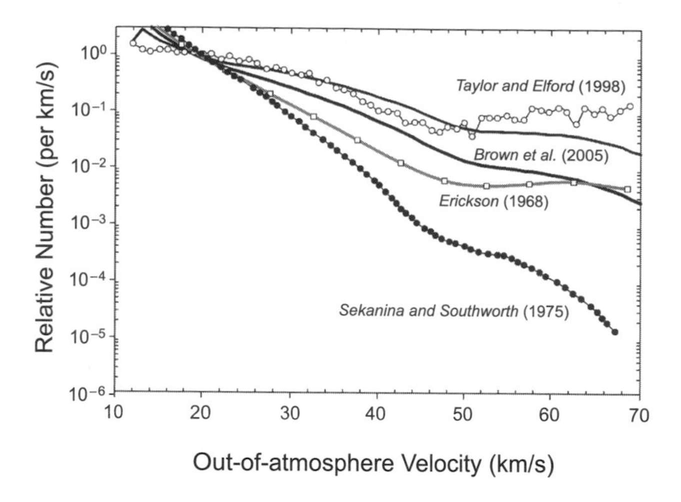

Fig. 9. The velocity distribution of meteoroids at Earth as measured by two radar surveys and from photographic observations. All data are normalized to the number of meteoroids measured in the 20 km/s velocity bin. The solid circles are the measured velocity distribution from the Harvard Radio Meteor Project (HRMP) as published by *Sekanina and Southworth* (1975), which has been used extensively in the literature (e.g., *Love and Brownlee,* 1993). The open circles are the revision of these same data using modern correction factors by *Taylor and Elford* (1998). The solid lines represent the upper and lower error limits of the distribution measured by the Canadian Meteor Orbit Radar (*Brown et al.,* 2005). Photographic estimates of the velocity distribution from Super-Schmidt camera measurements by *Erickson* (1968) are shown for comparison. Both sets of radar data are applicable to meteoroid masses in the 10^{-4}–10^{-5}-g range, while the photographic data represent meteoroids with masses on the order of 1 g.

ascribed to incomplete volatilization of these elements. Surprisingly, the meteoroids appeared to have a factor of 2–3 greater Na than CI or CM chondrites and even chondritic IDPs of probably cometary origin. They suggest that this Na enrichment is real, and that Na has been removed from the IDPs by atmospheric heating. Hopefully, the cometary dust being returned to Earth in January 2006 by the Stardust mission will help settle this issue.

5.2. Past Dust Flux

Since most of the ^{3}He in ocean sediments derives from IDPs, its concentration can be used as a sensor of the flux of IDPs back through time. As for the meteorite record in Earth's geologic column, there is also evidence that the flux of extraterrestrial dust to Earth has not been constant in even the recent geologic past. The ^{3}He concentrations in slowly accumulating pelagic clays suggest considerable changes (up to 5×) in the IDP flux over the past 70 m.y. (*Farley,* 1995) (Fig. 10). The pronounced peak in ^{3}He at approximately 8 Ma appears to correspond to the breakup of the Veritas family progenitor asteroid, which is the most recent of the major asteroid breakup events. It has been estimated that as much as 25% of the IDPs reaching Earth today derive from this family (W. Bottke, personal communication, 2005), but it is clear from the ^{3}He spike that immediately following the impact event that created and dispersed the Veritas family asteroids, dust from this source dominated the flux to Earth. Since the asteroids in this family are type Ch (i.e., C-type asteroids containing appreciable hydrated minerals), this dust would have brought a significant quantity of water to Earth over the past 8 m.y.

In addition, there appears to be an approximately 100-k.y. periodicity in ^{3}He concentration back though at least the past 500 k.y. (*Muller and MacDonald,* 1995; *Farley and Patterson,* 1995) (Fig. 11). The cause of the apparent periodicity remains unexplained.

6. SUMMARY

All measurements of the present fluxes of extraterrestrial materials can be combined with the measured and calculated fluxes of entire asteroids and comets to Earth (crater-forming objects) into a single diagram (Fig. 12). As we

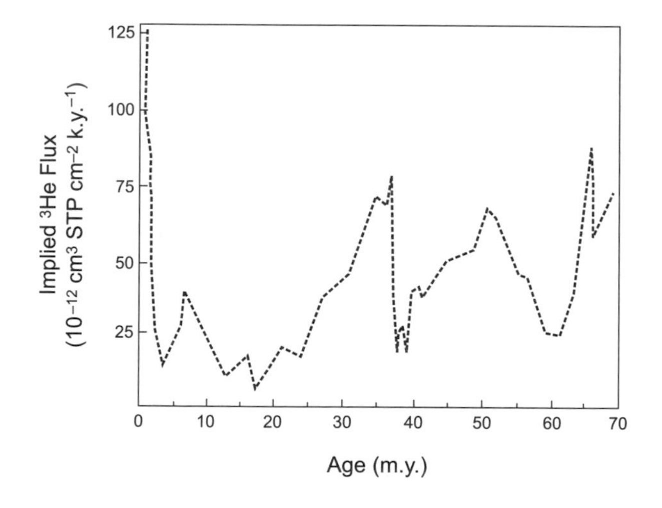

Fig. 10. The ^{3}He concentrations in slowly accumulating pelagic clays suggest considerable changes (up to 5×) in the IDP flux over the past 70 m.y. (after *Farley,* 1995).

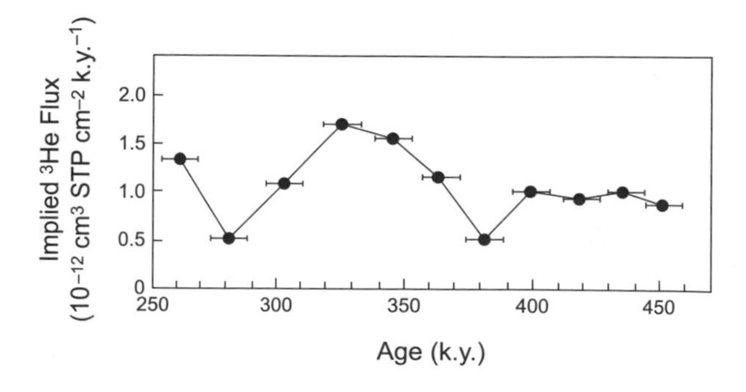

Fig. 11. There appears to be an approximately 100-k.y. periodicity in ^{3}He concentration over at least the past 500 k.y. (after *Farley and Patterson,* 1995).

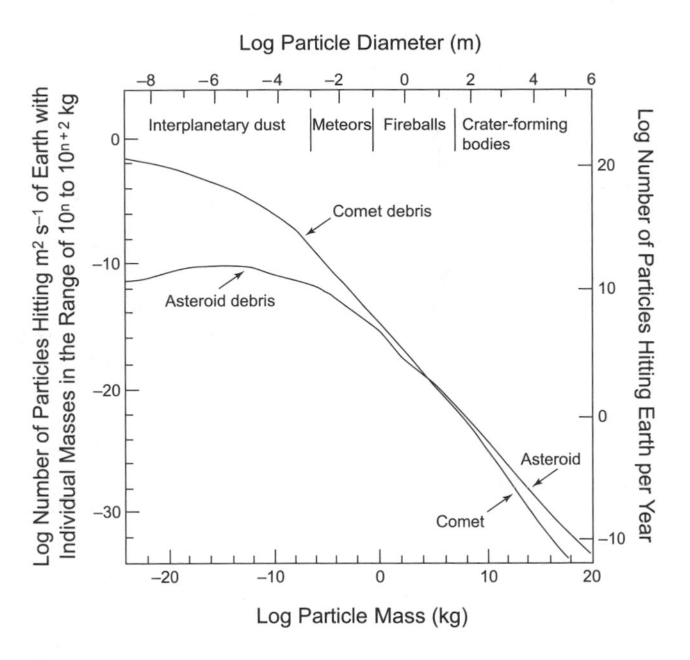

Fig. 12. Histogram showing the annual flux of extraterrestrial material to Earth (after *Hughes,* 1994). The flux is divided into four regions: IDPs and micrometeorites, meteors, fireballs, and crater-forming bodies.

have seen, the flux of meteorite and smaller-sized objects has varied considerably over the recent geologic past, and so the diagram in Fig. 12 is accurate only for the present. Future studies will shed additional light on the types and magnitudes of changes to this flux in the past.

REFERENCES

Annexstad J., Schultz L., and Zolensky M. E., eds. (1995) *Workshop on Meteorites From Cold and Hot Deserts.* LPI Tech. Rpt. 95-02, Lunar and Planetary Institute, Houston. 83 pp.

Aylmer D., Bonanno V., Herzog G. F., Weber H., Klein J., and Middleton R. (1988) Al-26 and Be-10 production in iron meteorites. *Earth Planet. Sci. Lett., 88,* 107–118.

Baggaley W. J. (2002) Radar observations. In *Meteors in the Earth's Atmosphere* (E. Murad and I. P. Williams, eds.), p. 123. Cambridge Univ., Cambridge.

Beijing Observatory, ed. (1988) *Zhongguo Gudai Tianxiang Jilu Zongji* (*A Union Table of Ancient Chinese Records of Celestial Phenomena*), pp. 63–105. Kexue Jishu Chubanshe, Kiangxu.

Benoit P. H., Sears H., and Sears D. W. G. (1992) The natural thermoluminescence of meteorites. 4. Ordinary chondrites at the Lewis Cliff Ice Field. *J. Geophys. Res., 97,* 4629–4647.

Benoit P. H., Sears H., and Sears D. W. G. (1993) The natural thermoluminescence of meteorites. 5. Ordinary chondrites at the Allan Hills Ice Fields. *J. Geophys. Res., 98,* 1875–1888.

Benoit P. H., Roth J., Sears H., and Sears D. W. G. (1994) The natural thermoluminescence of meteorites. 7. Ordinary chondrites from the Elephant Moraine region, Antarctica. *J. Geophys. Res., 99,* 2073–2085.

Binzel R. P., Lupishko D., di Martino M., Whiteley R. J., and Hahn G. J. (2002) Physical properties of near-Earth objects. In *Asteroids III* (W. F. Bottke Jr. et al., eds.), pp. 255–271. Univ. of Arizona, Tucson.

Binzel R P., Rivkin A. S., Stuart J. S., Harris A. W., Bus S. J., and Burbine T. H. (2004) Observed spectral properties of near-Earth objects: Results for population distribution, source regions, and space weathering processes. *Icarus, 170,* 259–294.

Bland P. A. and Artemieva N. A. (2003) Efficient disruption of small asteroids by Earth's atmosphere. *Nature, 424,* 288–291.

Bland P., Berry F. J., Smith T. B., Skinner S. J., and Pillinger C. T. (1996a) The flux of meteorites to the Earth and weathering in hot desert ordinary chondrite finds. *Geochim. Cosmochim. Acta, 60,* 2053–2059.

Bland P., Smith T. B., Jull A. J. T., Berry F. J., Bevan A. W. R, Cloud S., and Pillinger C. T. (1996b) The flux of meteorites to the Earth over the last 50,000 years. *Mon. Not. R. Astron. Soc., 283,* 551–565.

Bland P. A., Zolensky M. E., Benedix G. K., and Sephton M. A. (2006) Weathering of chondritic meteorites. In *Meteorites and the Early Solar System II* (D. S. Lauretta and H. Y. McSween Jr., eds.), this volume. Univ. of Arizona, Tucson.

Boeckl R. (1972) Terrestrial age of nineteen stony meteorites derived from their radiocarbon content. *Nature, 236,* 25–26.

Borovička J. (1993) A fireball spectrum analysis. *Astron Astrophys., 279,* 627–645.

Borovička J. and Betlem H. (1997) Spectral analysis of two Perseid meteors. *Planet. Space Sci., 45,* 563–575.

Borovička J. and Kalenda P. (2003) The Moravka meteorite fall: 4. Meteoroid dynamics and fragmentation in the atmosphere. *Meteoritics & Planet. Sci., 38,* 1023–1043.

Borovička J. and Spurný P. (1996) Radiation study of two very bright terrestrial bolides. *Icarus, 121,* 484–510.

Borovička J., Weber H. W., Jopek T., Jakes P., Randa Z., Brown P. G., ReVelle D. O., Kalenda P., Schultz L., Kucera J., Haloda J., Tycova P., Fryda J., and Brandstatter F. (2003) The Moravka meteorite fall: 3. Meteoroid initial size, history, structure, and composition. *Meteoritics & Planet. Sci., 38,* 1005–1021.

Bronshten V. A. (1983) *Physics of Meteor Phenomena.* Reidel, Dordrecht. 356 pp.

Brown H. (1960) The density and mass distribution of meteoritic bodies in the neighbourhood of the Earth's orbit. *J. Geophys. Res., 65,* 1679–1683.

Brown H. (1961) Addendum: The density and mass distribution of meteoritic bodies in the neighbourhood of the Earth's orbit. *J. Geophys. Res., 66,* 1316–1317.

Brown P. and Jones J. (1995) A determination of the strengths of the sporadic radio-meteor sources. *Earth Moon Planets, 68,* 223–245.

Brown P., Hildebrand A. R., Green D. W. E., Page D., Jacobs C., ReVelle D., Tagliaferri E., Wacker J., and Wetmiller B. (1996) The fall of the St-Robert meteorite. *Meteoritics & Planet. Sci., 31,* 502–517.

Brown P., Spalding R. E., ReVelle D. O., Tagliaferri E., and Worden S. P. (2002a) The flux of small near-Earth objects colliding with the Earth. *Nature, 420,* 294–296.

Brown P., Campbell M. D., Hawkes R. L., Theijsmeijer C., and Jones J. (2002b) Multi-station electro-optical observations of the 1999 Leonid meteor storm. *Planet. Space Sci., 50,* 45–55.

Brown P., Jones J., Weryk W. J., and Campbell-Brown M. D. (2005) The velocity distribution of meteoroids at the earth as measured by the Canadian meteor orbit radar (CMOR). *Earth Moon Planets,* in press.

Buchwald V. F. (1975) *Handbook of Iron Meteorites, Volume 1: Iron Meteorites in General*. Univ. of California, Berkeley.

Burns R. G., Burbine T. H., Fisher D. S., and Binzel R. P. (1995) Weathering in Antarctic H and CR chondrites: Quantitative analysis through Mössbauer spectroscopy. *Meteoritics, 30,* 625–633.

Campbell-Brown M and Jones J. (2003) Determining the initial radius of meteor trains. *Mon. Not. R. Astron. Soc., 343,* 3.

Cassidy W., Harvey R., Schutt J., Delisle G., and Yanai K. (1992) The meteorite collection sites of Antarctica. *Meteoritics, 27,* 490–525.

Ceplecha Z. (1964) Study of a bright meteor flare by mean of emission curve of growth. *Bull. Astron. Inst. Czech., 15,* 102–112.

Ceplecha Z. (1965) Complete data on bright meteor 32281. *Bull. Astron. Inst. Czech., 16,* 88–101.

Ceplecha Z. (1994) Impacts of meteoroids larger than 1 m into the Earth's atmosphere. *Astron. Astrophys., 286,* 967–970.

Ceplecha Z. (2001) The meteoroidal influx to the Earth. In *Collisional Processes in the Solar System* (M. Ya. Marov and H. Rickman, eds.), pp. 35–50. Astrophysics and Space Science Series Vol. 261, Kluwer, Dordrecht.

Ceplecha Z., Borovička J., Elford W. G., ReVelle D. O., Hawkes R. L., Porubcan V., and Simek M. (1998) Meteor phenomena and bodies. *Space Sci. Rev., 84,* 327–471.

Chang C. T. and Wänke H. (1969) Beryllium-10 in iron meteorites, their cosmic ray exposure and terrestrial ages. In *Meteorite Research* (P. M. Millman, ed.), pp. 397–406. Reidel, Dordrecht.

Delisle G. (1993) Global change, Antarctic meteorite traps and the East Antarctic ice sheet. *J. Glaciol., 39,* 397–408.

Delisle G. (1995) Storage of meteorites in Antarctic ice during glacial and interglacial stages. In *Workshop on Meteorites from Cold and Hot Deserts* (L. Schultz et al., eds.), pp. 26–27. LPI Tech. Rpt. 95-02, Lunar and Planetary Institute, Houston.

Delisle G. and Sievers J. (1991) Sub-ice topography and meteorite finds near the Allan Hills and the Near Western Ice Field, Victoria Land, Antarctica. *J. Geophys. Res., 96,* 15577–15587.

Dennison J. E., Ligner D. W., and Lipschutz M. E. (1986) Antarctic and non-Antarctic meteorites form different populations. *Nature, 319,* 390–393.

Drummond J. D. (1991) Earth-approaching asteroid streams. *Icarus, 89,* 14–25.

Erickson J. E. (1968) Velocity distribution of sporadic photographic meteors. *J. Geophys. Res., 73,* 3721–3762

Farley K. A. (1995) Cenozoic variations in the flux of interplanetary dust recorded by ^{3}He in a deep-sea sediment. *Nature, 376,* 153–156.

Farley K. A. and Patterson D. B. (1995) A 100-kyr periodicity in the flux of extraterrestrial ^{3}He to the sea floor. *Nature, 378,* 600–603.

Fireman E. L. (1983) Carbon-14 ages of Allan Hills meteorites and ice (abstract). In *Lunar and Planetary Science XIV,* pp. 195–196. Lunar and Planetary Institute, Houston.

Freundel M., Schultz L., and Reedy R. C. (1986) Terrestrial ^{81}Kr-Kr ages of Antarctic meteorites. *Geochim. Cosmochim. Acta, 50,* 2663–2673.

Galligan D. P. and Baggaley W. J. (2004) The orbital distribution of radar-detected meteoroids of the solar system dust cloud. *Mon. Not. R. Astron. Soc., 353,* 422–446.

Gladman B. and Pauls A. (2005) Decoherence scales for meteoroid streams (abstract). In *Asteroids, Comets and Meteors,* p. 66. IAU Symposium No. 229.

Graham A. L., Bevan A. W. R., and Hutchison R. (1985) *Catalogue of Meteorites*. Univ. of Arizona, Tucson.

Grün E., Zook H., Fechtig H., and Giese R. H. (1985) Collisional balance of the meteoritic complex. *Icarus, 62,* 244–272.

Halliday I. (1987) Detection of a meteorite stream: Observations of a second meteorite fall from the orbit of the Innisfree chondrite. *Icarus, 69,* 550–556.

Halliday I., Blackwell A. T., and Griffin A. A. (1978) The Innisfree meteorite and the Canadian camera network. *J. R. Astron. Soc. Canada, 72,* 15–39.

Halliday I., Griffin A. A., and Blackwell A. T. (1981) The Innisfree meteorite fall: A photographic analysis of fragmentation, dynamics and luminosity. *Meteoritics, 16,* 153–170.

Halliday I., Blackwell A. T., and Griffin A. A. (1984) The frequency of meteorite falls on the Earth. *Science, 223,* 1405–1407.

Halliday I., Blackwell A. T., and Griffin A. A. (1989) The flux of meteorites on the Earth's surface. *Meteoritics, 24,* 173–178.

Halliday I., Blackwell A. T., and Griffin A. A. (1991) The frequency of meteorite falls: Comments on two conflicting solutions to the problem. *Meteoritics, 26,* 243–249.

Halliday I., Griffin A. A., and Blackwell A. T. (1996) Detailed data for 259 fireballs from the Canadian camera network and inferences concerning the influx of large meteoroids. *Meteoritics, 31,* 185–217.

Harvey G. A. (1973) Elemental abundance determinations for meteors by spectroscopy. *J. Geophys. Res., 78,* 3913–3926.

Harvey R. P. (1990) Statistical differences between Antarctic finds and modern falls: Mass frequency distributions and relative abundance by type. In *Workshop on Differences Between Antarctic and Non-Antarctic Meteorites* (C. Koeberl and W. A. Cassidy, eds.), pp. 43–45. LPI Tech. Rept. 90-01, Lunar and Planetary Institute, Houston.

Harvey R. P. (1995) Moving targets: The effect of supply, wind movement, and search losses on Antarctic meteorite size distributions. In *Workshop on Meteorites from Cold and Hot Deserts* (L. Schultz et al., eds.), pp. 34–37. LPI Tech. Rpt. 95-02, Lunar and Planetary Institute, Houston.

Harvey R. P. and Cassidy W. A. (1991) A statistical comparison of Antarctic finds and modern falls: Mass frequency distributions and relative abundance by type. *Meteoritics, 24,* 9–14.

Harris A. W. (2002) A new estimate of the population of small NEAs. *Bull. Am. Astron. Soc., 34,* 835.

Hawkes R. L. (2002) Detection and analysis procedures for visual, photographic and image intensified CCD meteor observations. In *Meteors in the Earth's Atmosphere* (E. Murad and I. Williams, eds.), p. 97. Cambridge Univ., Cambridge.

Hawkins G. S. (1960) Asteroidal fragments. *Astron. J., 65,* 318–322.

Hawkins G. S. and Upton E. K. L. (1958) The influx rate of meteors in the Earth's atmosphere. *Astrophys. J., 128,* 727–735.

Honda M. and Arnold J. R. (1964) Effects of cosmic rays on meteorites. *Science, 143,* 203–212.

Hughes D. W. (1980) On the mass distribution of meteorites and their influx rate. In *Solid Particles in the Solar System* (I. Halliday and B. A. McIntosh, eds.), pp. 207–210. Reidel, Dordrecht.

Hughes D. W. (1994) Comets and asteroids. *Contemporary Phys., 35,* 75–93.

Humes D. (1991) Large craters on the Meteoroid and Space Debris Impact Experiment. In *LDEF, 69 Months in Space: First Post-Retrieval Symposium* (A S. Levine, ed.), p. 399. NASA, Washington, DC.

Huss G. (1990a) Meteorite infall as a function of mass: Implications for the accumulation of meteorites on Antarctic ice. *Meteoritics, 25,* 41–56.

Huss G. (1990b) Meteorite mass distributions and differences between Antarctic and non-Antarctic meteorites. In *Workshop on Differences Between Antarctic and Non-Antarctic Meteorites* (C. Koeberl and W. A. Cassidy, eds.), pp. 49–53. LPI Tech. Rpt. 90-01, Lunar and Planetary Institute, Houston.

Huss G. (1991) Meteorite mass distribution and differences between Antarctic and non-Antarctic meteorites. *Geochim. Cosmochim. Acta, 55,* 105–111.

Ikeda Y. and Kimura M. (1992) Mass distribution of Antarctic ordinary chondrites and the estimation of the fall-to-specimen ratios. *Meteoritics, 27,* 435–441.

Jenniskens P., Wilson M. A., Packan D., Laux C. O., Krüger C. H., Boyd I. D., Popova O. P., and Fonda M. (2000) Meteors: A delivery mechanism of organic matter to the early Earth. *Earth Moon Planets, 82/83,* 57–70.

Jopek T. J., Valsecchi G. B., and Froeschlé C. (2002) Asteroid meteoroid streams. In *Asteroids III* (W. F. Bottke Jr. et al., eds.), pp. 645–652. Univ. of Arizona, Tucson.

Jull A. J. T., Wlotzka F., Palme H. and Donahue D. J. (1990) Distribution of terrestrial age and petrologic type of meteorites from western Libya. *Geochim. Cosmochim. Acta, 54,* 2895–2898.

Jull A. J. T., Donahue D. J., Cielaszyk E., and Wlotzka F. (1993a) Carbon-14 terrestrial ages and weathering of 27 meteorites from the southern high plains and adjacent areas (USA). *Meteoritics, 28,* 188–195.

Jull A. J. T., Wloztka F., Bevan A. W. R., Brown S. T., and Donahue D. J. (1993b) ^{14}C terrestrial ages of meteorites from desert regions: Algeria and Australia. *Meteoritics, 28,* 376–377.

Jull T., Bevan A. W. R., Cielaszyk E., and Donahue D. J. (1995) Carbon-14 terrestrial ages and weathering of meteorites from the Nullarbor region, Western Australia. In *Workshop on Meteorites from Cold and Hot Deserts* (L. Schultz et al., eds.), pp. 37–38. LPI Tech. Rpt. 95-02, Lunar and Planetary Institute, Houston.

Koeberl C. and Cassidy W. A., eds. (1990) *Workshop on Differences Between Antarctic and Non-Antarctic Meteorites.* LPI Tech. Rpt. 90-01, Lunar and Planetary Institute, Houston. 102 pp.

Lindstrom M. and Score R. (1995) Populations, pairing and rare meteorites in the U.S. Antarctic meteorite collection. In *Workshop on Meteorites from Cold and Hot Deserts* (L. Schultz et al., eds.), pp. 43–45. LPI Tech. Rpt. 95-02, Lunar and Planetary Institute, Houston.

Loeken M. and Schultz L. (1995) The noble gas record of H chondrites and terrestrial age: No correlation. In *Workshop on Meteorites from Cold and Hot Deserts* (L. Schultz et al., eds.), pp. 45–47. LPI Tech. Rpt. 95-02, Lunar and Planetary Institute, Houston.

Love S. G. and Brownlee D. E. (1993) A direct measurement of the terrestrial mass accretion rate of cosmic dust. *Science, 262,* 550–553.

McCrosky R. E. and Ceplecha Z. (1969) Photographic networks for fireballs. In *Meteorite Research* (P. M. Millman, ed.), pp. 600–612. Reidel, Dordrecht.

Michlovich E. S., Wolf S. F., Wang M.-S., Vogt S., Elmore D., and Lipschutz M. (1995) Chemical studies of H chondrites. 5. Temporal variations of sources. *J. Geophys. Res., 100,* 3317–3333.

Millard H. T. Jr. (1963) The rate of arrival of meteorites at the surface of the Earth. *J. Geophys. Res., 68,* 4297–4303.

Millman P. M. (1972) Giacobinid meteor spectra. *J. R. Astron. Soc. Canada, 66,* 201–211.

Muller R. A. and MacDonald G. (1995) Glacial cycles and orbital inclination. *Nature, 377,* 107–108.

Nagasawa K. (1978) Analysis of spectra of Leonid meteors. *Ann. Tokyo Astron. Obs. 2nd Series, 16,* 157–187.

Naumann N. R. J. (1966) *The Near Earth Meteoroid Environment.* NASA TND 3717, U.S. Government Printing Office, Washington, DC.

Nemtchinov I. V., Svetsov V. V., Kosarev I. B., Golub A. P., Popova O. P., Shuvalov V. V., Spalding R. E., Jacobs C., and Tagliaferri E. (1997) Assessment of kinetic energy of meteoroids detected by satellite-based light sensors. *Icarus, 130,* 259–274.

Nishiizumi K. (1990) Update on terrestrial ages of Antarctic meteorites. In *Antarctic Meteorite Stranding Surfaces* (W. A. Cassidy and I. M. Whillans, eds.), pp. 49–53. LPI Tech. Rpt. 90-03, Lunar and Planetary Institute, Houston.

Nishiizumi K. (1995) Terrestrial ages of meteorites from cold and cold regions. In *Workshop on Meteorites from Cold and Hot Deserts* (L. Schultz et al., eds.), pp. 53–55. LPI Tech. Rpt. 95-02, Lunar and Planetary Institute, Houston.

Nishiizumi K., Elmore D., and Kubik P. W. (1989a) Update on terrestrial ages of Antarctic meteorites. *Earth Planet. Sci. Lett., 93,* 299–313.

Nishiizumi K., Winter E., Kohl C., Klein J., Middleton R., Lal D., and Arnold J. (1989b) Cosmic-ray production rates of Be-10 and Al-26 in quartz from glacially polished rocks. *J. Geophys. Res.–Solid Earth and Planets, 94,* 17907–17915.

Rabinowitz D. L. et al. (1993) Evidence for a near-Earth asteroid belt. *Nature, 363,* 704–706.

Rabinowitz D., Helin E., Lawrence K., and Pravdo S. (2000) A reduced estimate of the number of kilometer-sized near-Earth asteroids. *Nature, 403,* 165–166.

Rawcliffe R. D., Bartky C. D., Gordon E., and Carta D. (1974) Meteor of August 10, 1972. *Nature, 247,* 284–302.

ReVelle D. O. (1997) Historical detection of atmospheric impacts of large super-bolides using acoustic-gravity waves. *Ann. N. Y. Acad. Sci., 822,* 284–302

Scherer P., Schultz L., Neupert U., Knauer M., Neumann S., Leya I., Michel R., Mokos J., Lipschutz M. E., Metzler K., Suter M., and Kubik P. W. (1987) Allan Hills 88019: An Antarctic H-chondrite with a very long terrestrial age. *Meteoritics & Planet. Sci., 32,* 769–773.

Scherer P., Schultz L., and Loeken Th. (1995) Weathering and atmospheric noble gases in chondrites from hot deserts. In *Workshop on Meteorites from Cold and Hot Deserts* (L. Schultz et al., eds.), pp. 58–60. LPI Tech. Rpt. 95-02, Lunar and Planetary Institute, Houston.

Schmitz B., Peucker-Ehrenbrink B., Lindström M., and Tassinari M. (1997) Accretion rates of meteorites and cosmic dust in the early Ordovician. *Science, 278,* 88–90.

Schmitz B., Tassinari M., and Peucker-Ehrenbrink B. (2001) A rain of ordinary chondrite meteorites in the early Ordovician. *Earth Planet. Sci. Lett., 194,* 1–15.

Schmitz B., Häggström T., and Tassinari M. (2003) Sediment-dispersed extraterrestrial chromite traces a major asteroid disruption event. *Science, 300,* 961–964.

Schultz L. and Weber H. W. (1996) Noble gases and H chondrite meteoroid streams: No confirmation. *J. Geophys. Res., 101,* 21177–21181.

Schultz L., Weber W., and Begemann F. (1991) Noble gases in H chondrites and potential differences between Antarctic and non-Antarctic meteorites. *Geochim. Cosmochim. Acta, 55,* 59–66.

Schutt J., Schultz L., Zinner E., and Zolensky M. (1986) Search for meteorites in the Allan Hills region, 1985–1986. *Antarct. J., 21,* 82–83.

Sekanina Z. and Southworth R. B. (1975) *Physical and Dynamical Studies of Meteors: Meteor Fragmentation and Stream-Distribution Studies.* NASA CR-2615, U.S. Government Printing Office, Washington, DC.

Southworth R. B. and Sekanina Z. (1973) *Physical and Dynamical Studies of Meteors.* NASA CR-2316, U.S. Government Printing Office, Washington, DC.

Spurný P. (1997) Photographic monitoring of fireballs in central Europe. *Proc. SPIE, 3116,* 144–155.

Staubach P., Grün E., and Matney M. J. (2001) Synthesis of observations. In *Interplanetary Dust* (E. Grün et al., eds.), pp. 347–384. Springer-Verlag, Berlin.

Stuart J. S. and Binzel R. P. (2004) Bias-corrected population, size distribution, and impact hazard for the near-Earth objects, *Icarus, 170,* 295–311.

Taylor A. D. and Elford W. G. (1998) Meteoroid orbital element distributions at 1 AU deduced fromo the Harvard Radio Meteor Project observations. *Earth Planets Space, 50,* 569–575.

Tagliaferri E., Spalding R., Jacobs C., Worden S. P., and Erlich A. (1994) Detection of meteoroid impacts by optical sensors in Earth orbit. In *Hazards Due to Comets and Asteroids* (T. Gehrels, ed.), pp. 199–220. Univ. of Arizona, Tucson.

Thomas R. M. and Netherway D. J. (1989) Observations of meteors using an over-the-horizon radar. *Proc. Astron. Soc. Austral., 8,* 88–93.

Thomas P., Whitham P. S., and Wilford W. G. (1988) Response of high frequency radar to meteor backscatter. *J. Atmos. Terrestrial Phys., 50,* 703–724.

Trigo-Rodriguez J. M., Llorca J., Borovička J., and Fabregat J. (2004a) Chemical abundances from meteor spectra: I. Ratios of the main chemical elements. *Meteoritics & Planet. Sci., 38,* 1283–1294.

Trigo-Rodriguez J. M., Llorca J., and Fabregat J. (2004b) Chemical abundances from meteor spectra: II. Evidence for enlarged sodium abundances in meteoroids. *Mon. Not. R. Astron. Soc., 348,* 802–810.

Verniani F. (1973) An analysis of the physical parameters of 5759 faint radio meteors. *J. Geophys. Res., 78,* 8429–8462.

Wasson J. T. (1985) *Meteorites: Their Record of Early Solar-System History.* Freeman, New York.

Weissman P., Bottke W. F., and Levison H. F. (2002) Evolution of comets into asteroids. In *Asteroids III* (W. F. Bottke Jr. et al., eds.), pp. 669–686. Univ. of Arizona, Tucson.

Welten K. C. (1995) Exposure histories and terrestrial ages of Antarctic meteorites. Ph.D. thesis, Univ. of Utrecht. 150 pp.

Welten K. C., Alderliesten C., van der Borg K., and Lindner L. (1995) Cosmogenic beryllium-10 and aluminum-26 in Lewis Cliff meteorites. In *Workshop on Meteorites from Cold and Hot Deserts* (L. Schultz et al., eds.), pp. 65–70. LPI Tech. Rpt. 95-02, Lunar and Planetary Institute, Houston.

Welten K. C., Alderliesten C., van der Borg K., Lindner L., Loeken T. and Schultz L. (1997) Lewis Cliff 86360: An Antarctic L-chondrite with a terrestrial age of 2.35 million years. *Meteoritics & Planet. Sci., 32,* 775–780.

Werner S. C., Harris A. W. Neukum G., and Ivanov B. A. (2002) The near-earth asteroid size-frequency distribution: A snapshot of the lunar impactor size-frequency distribution. *Icarus, 156,* 287–290.

Wetherill G. W. (1986) Unexpected Antarctic chemistry. *Nature, 319,* 357–358.

Whillans I. M. and Cassidy W. A. (1983) Catch a falling star: Meteorites and old ice. *Science, 222,* 55–57.

Wieler R., Caffee M. W., and Nishiizumi K. (1995) Exposure and terrestrial ages of H chondrites from Frontier Mountain. In *Workshop on Meteorites from Cold and Hot Deserts* (L. Schultz et al., eds.), pp. 70–72. LPI Tech. Rpt. 95-02, Lunar and Planetary Institute, Houston.

Wlotzka F., Jull A. J. T., and Donahue D. J. (1995a) Carbon-14 terrestrial ages of meteorites from Acfer, Algeria. In *Workshop on Meteorites from Cold and Hot Deserts* (L. Schultz et al., eds.), pp. 72–73. LPI Tech. Rpt. 95-02, Lunar and Planetary Institute, Houston.

Wlotzka F., Jull A. J. T. and Donahue D. J. (1995b) Carbon-14 terrestrial ages of meteorites from Acfer, Algeria. In *Workshop on Meteorites from Cold and Hot Deserts* (L. Schultz et al., eds.), pp. 37–38. LPI Tech. Rpt. 95-02, Lunar and Planetary Institute, Houston.

Wolf S. F. and Lipschutz M. E. (1995a) Applying the bootstrap to Antarctic and non-Antarctic H-chondrite volatile-trace-element data. In *Workshop on Meteorites from Cold and Hot Deserts* (L. Schultz et al., eds.), pp. 73–75. LPI Tech. Rpt. 95-02, Lunar and Planetary Institute, Houston.

Wolf S. F. and Lipschutz M. E. (1995b) Yes, meteorite populations reaching the Earth change with time. In *Workshop on Meteorites from Cold and Hot Deserts* (L. Schultz et al., eds.), pp. 75–76. LPI Tech. Rpt. 95-02, Lunar and Planetary Institute, Houston.

Wolf S. F. and Lipschutz M. E. (1995c) Chemical studies of H chondrites — 7. Contents of Fe^{3+} and labile trace elements in Antarctic samples. *Meteoritics, 30,* 621–624.

Zolensky M. E. and Paces J. (1986) Alteration of tephra in glacial ice by "unfrozen water." *Geol. Soc. Am. Abstr. with Progr., 18,* 801.

Zolensky M. E., Wells G. L., and Rendell H. M. (1990) The accumulation rate of meteorite falls at the Earth's surface: The view from Roosevelt County, New Mexico. *Meteoritics, 25,* 11–17.

Zolensky M., Rendell H., Wilson I., and Wells G. (1992) The age of the meteorite recovery surfaces of Roosevelt County, New Mexico, USA. *Meteoritics, 27,* 460–462.

Zolensky M. E., Martinez R., and Martinez de los Rios E. (1995) New L chondrites from the Atacama Desert, Chile. *Meteoritics, 30,* 785–787.

Zolensky M. E., Abell P., and Tonui E. (2005) Metamorphosed CM and CI carbonaceous chondrites could be from the breakup of the same Earth-crossing asteroid (abstract). In *Lunar and Planetary Science XXXVI,* Abstract #2048. Lunar and Planetary Institute, Houston (CD-ROM).

Terrestrial Ages of Meteorites

A. J. Timothy Jull
University of Arizona

The terrestrial age, or the terrestrial residence time of a meteorite, together with its exposure history provides us with useful insight into the history of the meteorite. It is easy to observe that stony meteorites can weather quickly in humid environments. However, we find that large numbers of meteorites found in semi-arid and arid environments can survive for much longer times. Meteorites in desert environments can survive for at least 50,000 yr, and there are some meteorites over 250,000 yr old from these locations. The cold and dry conditions of polar regions such as Antarctica are also good for the storage of meteorites. A considerable number of meteorites survive there for hundreds of thousands of years. Some meteorites have been found in Antarctica with ages of up to 2 m.y. In this paper, we discuss the terrestrial residence times or terrestrial ages of these meteorites. We will show the wide range of terrestrial ages from different environments.

1. INTRODUCTION

It is important to determine the terrestrial age, or residence time, of a meteorite on the surface of Earth, as this gives us useful information that can be applied to studies of infall rates, meteorite distributions, weathering of meteorites, and meteorite concentration mechanisms. Most meteorites are recovered as "finds" and not as freshly fallen material. The study of the terrestrial ages of these meteorites gives us useful information concerning the storage and weathering of meteorites, as well as the effect of local geology and climate on meteorite storage. We would expect that weathering of meteorites and their eventual destruction would be a function of the terrestrial age. A direct connection of weathering rates to the terrestrial survival times of meteorites was initially shown by *Jull et al.* (1990) and *Wlotzka et al.* (1995) and later by *Bland et al.* (1996, 1998).

1.1. Infall

It is a general assumption that meteorites fall equally all over the world (*Halliday et al.,* 1989; *Halliday,* 2001). It has been known for many years that the distribution of potential meteoroids in the solar system has an inverse power law dependence on mass, where $n \sim m^{-0.8}$. There have been a number of models and discussions of the expected infall rate onto the surface of Earth. Several models have discussed the ablation of material during fall through the atmosphere. This results in a shift to low fragment sizes and total mass of the infalling material at the surface (*Love and Brownlee,* 1993; *Cziczo et al.,* 2001).

The best description of infall is the analysis of *Halliday et al.* (1989), who formulated an equation to describe the apparent mass dependence of observed meteorite falls, based on fireball observations and other information. This equation was a refinement of earlier studies of infalling material, which showed an approximately $M^{-0.83}$ mass dependence for material in the vicinity of Earth. This is also used to predict the infall rates of objects to the surface of Earth. The infall rate was described as a function of mass where

$$\log N = a \log M + b \quad (1)$$

where N is the number of meteorites that fall per 10^6 km^2 per year, of greater than mass M in grams (see *Halliday et al.,* 1989). *Halliday et al.* (1989) determined the constants a and b to be –0.49 and +2.41 for M < 1030 g, and –0.82 and +3.41 for M > 1030 g, based on observations of meteoroids. This would result in an infall rate of M > 10 g of 83 events per 10^6 km^2/yr, or roughly one event per km^2 in 10,000 yr. Using a different approach, combining weathering rates and recovery statistics, *Bland* (2001) estimated an infall rate of 36–116 events per 10^6 km^2.

Different flux estimates have been derived from other sources. *Zolensky et al.* (1990) overestimated the infall rate, based on recovery of meteorites at Roosevelt County (New Mexico), although this was later corrected by *Zolensky et al.* (1992). *Huss* (1990) also calculated high rates of influx based on statistical analysis of the collections from different Antarctic ice fields. The model of *Huss* (1990) had assumed there was no transport of meteorites in Antarctic ice. We can explain the apparent high infall rates of *Huss* (1990) due to incorrect assumptions. Ice transport is further and the age range of Antarctic meteorites is larger than assumed by *Huss* (1990). Also, some authors have proposed that there are compositional differences in meteorites falling in different parts of the world. For example, *Lipschutz* (1989, 1990) and *Schultz et al.* (1995) discussed whether the Antarctic collection and the non-Antarctic collection sampled the same distribution of meteorites. There now appears to be little disagreement with the infall rate similar to those of *Halliday et al.* (1989), at least as a first-order approximation to the infall rate (*Bland et al.,* 1996; *Bland,* 2001). There is no evidence for fluctuations in the infall rate, at least on timescales of less than tens of millions of years (*Shoemaker,* 1997). *Karner et al.* (2003) found no changes in dust flux over the last 20 k.y., although *Farley and Patterson* (1995) and others have argued for accretionary pulses on the scale of 100 k.y. for dust-sized particles. Over longer time periods, there is possible evidence of an enhanced flux. *Schmitz et al.* (2001) found over 50 "fossil" meteorites in an Ordovician quarry in Sweden that were pseudomorphs of chondritic textures and also contained spinel grains of L-chondrite composition. The material is consistent with an age of ca. 480 Ma and their results suggest the infall rate might have been 1–2 orders of magnitude higher at that time, with the assumptions of the model used by *Schmitz et al.* (2001).

1.2. Find Locations

In warm and humid regions, recovering meteorites of any significant terrestrial age is a difficult task. Meteorites weather rapidly in warm and humid conditions, especially in the presence of Cl^- (*Buchwald,* 1975). However, there are many places where large concentrations of meteorites can be recovered. In such regions, particularly arid environments such as deserts and cold, dry polar regions, we can expect that weathering rates are much reduced and so meteorites can be preserved for longer periods of time. In addition, in polar regions, meteorites may be concentrated by ice flow, as observed in Antarctica (*Cassidy,* 2003).

Great numbers of meteorites have been recovered from many areas of the world (*Koblitz,* 2003). The largest single source is Antarctica (estimated >20,000 individual specimens), but large numbers have been recovered from North African deserts, including Morocco (>2000), Libya (over 1270 meteorites recovered), and Algeria (~500 meteorites), the deserts of Oman (>1100), and the Arabian peninsula (*Franchi et al.,* 1995; *Hofmann et al.,* 2003). In addition, some 300 meteorites have been recovered from the Nullarbor Plain of Australia (*Bevan and Binns,* 1989a,b; *Bevan et al.,* 1998; *Bland et al.,* 2000). Roosevelt County, New Mexico (*Sipiera et al.,* 1987), where 105 individuals have been recovered, was the first area of large-scale recovery in North America. *Kring et al.* (1999, 2001) have discussed the large strewnfield of the Gold Basin meteorite, where thousands of individual fragments have been recovered. *Verish et al.* (1999) have discussed the Lucerne Valley meteorites recovered from a dry lakebed in California. It is now common to hear of organized searches of meteorites, arranged by scientists, meteorite enthusiasts, and collectors, especially to the deserts of North Africa, North America, southern Africa, and the Middle East (Fig. 1). To add to the interest, many rare and valuable achondrites, identified as being from Mars or the Moon, have been recovered from these locations. A total of about 32 meteorites from Mars (representing 24 discrete falls) and the same number from the Moon were known as of 2003 (*Cassidy,* 2003; *Hofmann et al.,* 2003).

Although one could assume the length of time a meteorite has resided on Earth's surface affects its degree of weathering, this correlation is not direct, as has been shown by the study of *Bland et al.* (1996). Desert meteorites can be particularly weathered, due to the extreme conditions of storage *(Welten et al.,* 1999a). Weathering of meteorites received much less attention before the recovery of large numbers of desert meteorites. Although the Roosevelt County meteorites, collected from the semi-arid areas of eastern New Mexico, were very weathered, studies on Antartic meteorites initially assumed that meteorites collected from Antarctica were pristine. This became an issue in studies of carbonates of apparently martian origin in the sher-

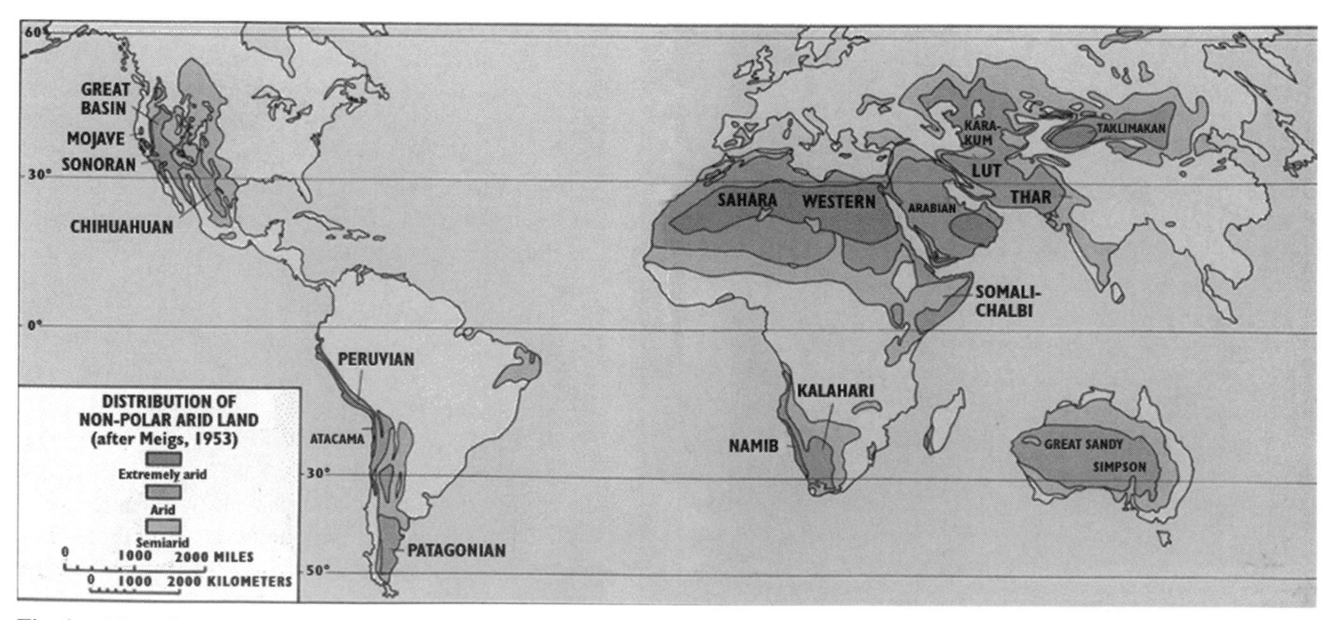

Fig. 1. Map of desert regions of the world based on rates of precipitation (*Fraser,* 1997). Courtesy of the U.S. Geological Survey.

TABLE 1. Radionuclides of interest for terrestrial-age studies.

	Half-life	Phase	Saturated Activity (dpm/kg) L	H	References
^{39}Ar	269 yr	metal	25	21	*Begemann et al.* (1969)
^{14}C	5.73 k.y.	bulk	51	46	*Jull et al.* (1998b); *Jull* (2001)
^{41}Ca	100 k.y.	metal	24		*Fink et al.* (1991); *Herzog et al.* (1997)
^{59}Ni	108 k.y.*	metal	~350		*Schnabel et al.* (1999a)
^{81}Kr	229 k.y.	bulk	0.003–0.0045†		*Freundel et al.* (1986); *Eugster* (1988a)
^{36}Cl	300 k.y.	metal	22.8		*Nishiizumi et al.* (1989a); *Herzog et al.* (1997)
^{26}Al	700 k.y.	bulk	60	56	*Evans et al.* (1982); *Vogt et al.* (1990)
^{60}Fe	1.49 m.y.				*Knie et al.* (1999); *Goel and Honda* (1965)
^{10}Be	1.5 m.y.	bulk	22	20	*Nishiizumi et al.* (1989a); *Nishiizumi* (1995)
^{53}Mn	3.7 m.y.	metal	434		*Nishiizumi et al.* (1977, 1989a)

Data adapted from *Jull* (2001).
* *Ruhm et al.* (1994).
† Based on the average production rates for ^{81}Kr of 6–9 × 10^{-14} cm^3/g/m.y. (*Eugster,* 1988a); depends on composition.

gottite EETA 79001 and later some other martian meteorites. The degree of terrestrial alteration of EETA 79001 carbonates (*Jull et al.,* 1997) was shown to be considerable. However, this discussion became much more important due to a much larger debate over Allan Hills 84001 (ALH 84001), since *McKay et al.* (1996) asserted that they observed fossil-bacterial forms as well as evidence for organic material in this meteorite. This initiated a considerable debate about weathering and contamination effects. As part of this discussion, *Jull et al.* (1998a) and *Bada et al.* (1998) showed that over the terrestrial age of ALH 84001 of some 13,000 yr, there had been weathering and adsorption of organic materials.

1.3. Weathering

The original scheme for collection of meteorites from Antarctica assumed that weathering might be an indication of terrestrial residence time. However, it is well known to many meteoriticists that weathering is dependent on the composition of the meteorites, particularly for irons and that the presence of Cl^-, incorporated into the meteorite from the environment, seems be an important catalyst for the oxidation of meteoritic metal (*Buchwald,* 1975). A common observation is that certain compositions seem to be more susceptible to weathering than others. The weathering criteria used by the U.S. Antarctic Search for Meteorites (ANSMET) uses a simple A-B-C scale for hand specimens, which does not correlate easily with terrestrial age (*Jull et al.,* 1998b). *Wlotzka et al.* (1995) devised a five-step scale, which more broadly follows terrestrial age, at least for meteorites from Roosevelt County, New Mexico.

1.4. How We Can Determine Terrestrial Age

Although some thermoluminescence (TL) measurements have been used to estimate terrestrial ages (*McKeever,* 1982; *Ninagawa et al.,* 1983; *Benoit et al.,* 1993), this method is of limited use because of the number of variables involved, and thus is not easily utilized for terrestrial ages. In one case, *Benoit et al.* (1993) were able to show a fair correlation of TL vs. terrestrial age for some meteorites from the southwestern U.S. However, attempts to apply TL to Antarctic meteorites have not been very successful.

The best approach for measurements of terrestrial age has been to study the amounts of cosmic-ray-produced radionuclides (Table 1). This was originally demonstrated in 1962, when terrestrial ages were measured by *Suess and Wänke* (1962) and about the same time by *Goel and Kohman* (1962) on large meteorite samples (10–100 g), using 5.73-k.y. ^{14}C β-decay counting. A decade later, *Boeckl* (1972) used ^{14}C to estimate terrestrial ages of meteorites from the central and southwestern U.S., using ~10-g samples. In meteorites, the longer-lived nuclide ^{36}Cl, with a half-life of 300 k.y., was also first measured by decay counting (*McCorkell et al.,* 1968; *Chang and Wänke,* 1969). *Chang and Wänke*'s (1969) pioneering work was the first attempt to systematically determine terrestrial ages of iron meteorites using ^{36}Cl; these authors also showed that some meteorites could apparently survive for very long times on the surface of Earth. *Chang and Wänke* (1969) determined that the iron meteorite Tamarugal (Chile) had a terrestrial age of ~2.7 Ma. *McCorkell et al.* (1968) studied six different radionuclides in three iron meteorites, two of unknown age (Hoba and Deelfontein), and estimated their terrestrial age. As we will discuss later, the entire field of measurement of many long-lived radionuclides was revolutionized by the introduction of accelerator mass spectrometry (AMS) in 1977 (see *Tuniz et al.,* 1998; *Fifield,* 1999). In general, ^{14}C is most useful for the terrestrial age of stony meteorites and ^{36}Cl for iron meteorites. The radionuclide ^{81}Kr, with a half-life of 229 k.y., is measured by noble-gas mass spectrometry (*Freundel et al.,* 1986; *Miura et al.,* 1993; *Eugster et al.,* 1997). This is accomplished by comparing the ^{81}Kr-^{83}Kr apparent exposure age to an independent measure of exposure age, such as ^{38}Ar.

1.5. Radionuclide Production

Radionuclides are produced in meteorites by interactions of primary cosmic rays (mostly protons and α particles) and the secondary particle cascade resulting from this irradiation (*Tuniz et al.,* 1998). The two main modes of production of nuclides are by spallation and neutron-capture reactions. Spallation is a result of disintegration of a nucleus as a result of a collision with a high-energy cosmic-ray particle. Secondary particles produced by such interactions include many neutrons, which peak at a depth of about 50 g/cm^2 (an expression of depth in the units of length times density), equivalent to ~17 cm in material of meteoritic composition. These interactions occur at high energies above about 10–20 MeV. A depth profile of a radionuclide produced purely by spallation in an infinitely large object will increase to about 50 g/cm^2 and then decrease by a factor of ~2 every 100 g/cm^2. We refer to this as a "2π irradiation." For smaller object such as meteorites, since the irradiation occurs from all solid angles, the production rate will continue to increase into the center of the object until we approach a certain size of object, which has a radius of about 150 g/cm^2 (i.e., 50 cm). We refer to this situation as a "4π irradiation." When we study larger objects, the production-rate maximum will occur before the center and we will eventually reach a condition where the cosmic rays can only penetrate from one side, that is, we eventually reach a "2π irradiation."

The secondary neutrons will eventually slow to thermal energies with depth in the sample, if they do not react or escape from the object. The nuclear reactions of these thermal neutrons with some target nuclei gives rise to reactions at greater depths, due to neutron capture or other thermal neutron reactions. In this case, production rate for the neutron-produced radionuclides will increase with depth and the size of the meteoroid, up to about a depth of 200 g/cm^2 (2π). Cobalt-60 is an example of a short-lived radionuclide produced by thermal neutrons, and most of the ^{36}Cl and ^{41}Ca produced in silicates is from thermal-neutron reactions, as is ^{59}Ni in meteoritic metal. It is also important that we can estimate the depth of a meteorite sample in its parent meteoroid in space during irradiation, which is usually unknown. We have to estimate this depth, often termed "shielding depth," from ratios of isotopes sensitive to depth, such as ^{22}Ne/^{21}Ne (*Cressy and Bogard,* 1976). Alternatively, shielding-independent age measurements can be obtained by using pairs of different radionuclides of different half-life, but similar production reactions.

After the collection of meteorites from Antarctica, there was a renewed interest in understanding the terrestrial ages of these meteorites. This began with the work of *Fireman* (1978). *Kigoshi and Matsuda* (1986) also made some ^{14}C measurements on Antarctic meteorites using similar β-counting methods. All these early measurements required the use of gas or solid counters and usually required large samples and gave low count rates. Aluminum-26 measurements were performed by counting the annihilation γ-rays from ^{26}Al on whole-rock specimens *(Evans and Rancitelli,* 1979; *Evans et al.,* 1982, 1992), but it was found that determining this nuclide alone could not be easily used to determine terrestrial age.

Few measurements of terrestrial age were available, because of the size of sample needed and the slow nature of the counting process. This was changed radically by the development of accelerator mass spectrometry (AMS), which allowed the use of much smaller samples. *Nishiizumi et al.* (1979) were the first to exploit the usefulness of this new technology for meteoritics. A review of the extensive literature up to 2001 has been discussed by *Jull et al.* (1998b) and *Jull* (2001). In the last several years, the number of meteorites recovered from arid desert regions has continued to increase rapidly and many more measurements on various meteorites have been made. An example of this focus is the recovery of a number of lunar and martian meteorites (*Nishiizumi et al.,* 2002, 2004). Apart from some rare cases of very old achondrites (*Nishiizumi et al.,* 2002), it is still true that for desert meteorites, ^{14}C is the most useful radionuclide for terrestrial-age determinations (*Jull et al.,* 2002; *Welten et al.,* 2003, 2004).

2. ACCELERATOR MASS SPECTROMETRY METHODS

In accelerator mass spectrometry (AMS), the atoms of the radionuclide of interest are counted directly using mass spectrometry combined with nuclear accelerator techniques (*Tuniz et al.,* 1998). The sample size requirements are correspondingly reduced by factors of 1000 to 10,000 compared to decay counting of β particles or γ rays and the counting time is also reduced from days or weeks to a measurement time by AMS of ~20–30 min. At the Arizona AMS laboratory, the detection limit for ^{14}C is ~2×10^{-15} ^{14}C/^{12}C and for ^{10}Be ~5×10^{-15} ^{10}Be/^{9}Be. Detection limits for other radionuclides are similar (*Tuniz et al.,* 1998; *Fifield,* 1999). For typical sample and carrier, this is about 10^5 atoms for both nuclides. In the case of meteorite samples, blanks are typically higher for ^{14}C, ~10^6 atoms, due to chemical processing and high-temperature of extraction of the CO_2 from the sample. Although the chemistry used to extract the nuclide of interest varies considerably, the principles are very similar to other radionuclide methodology. The sample is processed and a carrier of the element of interest is usually added. A concentration step allows the separation of the carrier and the radionuclide and the final product is a solid target consisting of 0.1–1 mg of the element of interest in a suitable compound. In all cases of AMS measurements, a solid target material is pressed into a target holder and negative ions are produced from the sample by Cs sputtering. The negative ions are accelerated, stripped of several electrons, and then further accelerated. These positive ions are then separated by mass and energy, to the final detector (see *Tuniz et al.,* 1998; *Fifield,* 1999).

3. PRODUCTION AND DECAY OF RADIONUCLIDES

3.1. Principles

The buildup and decay of the radionuclides in meteorites are the same as for any other radioactive process. The buildup of the nuclide is controlled by the production rate of the given nuclide in the meteorite, at the location of the sample within the larger meteoroid. In most cases, the meteoroid has been exposed in space for longer than several mean lives of the radionuclide and the nuclide reaches a saturation equilibrium, termed "saturation," where the rate of production equals the rate of decay. We show some typical production rates of radionuclides in ordinary chondrites in Table 1.

We can relate the amount of any radionuclide remaining at time t by integrating the basic radioactive decay equation $dN/dt = -\lambda N$. The equations below are given in number of atoms present, but similar equations can be derived for activity (dN/dt)

$$t = -\frac{1}{\lambda}\ln\left(\frac{N}{N_0}\right) \quad (2)$$

where N_0 is the initial number of atoms, N is the number of atoms remaining, and λ is the decay constant. Except for samples exposed at altitudes above a few kilometers (*Tuniz et al.,* 1998), the exposure to cosmic radiation after the meteorite fell to Earth is negligible and is not used for most calculations of terrestrial ages. However, this radioactivity can become significant in cases where most of the radionuclide produced in space has decayed. This allows us to calculate an age from this equation, defining N_0 as Ps/λ

$$N = \frac{P_s}{\lambda}e^{-\lambda t} + \frac{P_t}{\lambda}(1 - e^{-\lambda t}) \quad (3)$$

Here, P_s is the production rate in space and P_t is the production rate in the meteorite after it fell to Earth. P_s is a function that depends on the position of the sample in the parent meteoroid and the original size of that object (*Graf et al.,* 1990a,b). We can apply this equation to any radionuclide. In some cases, we can get better precision in the age measurement by using the ratio of two radionuclides with different half-lives, since the effects of shielding on both nuclides is minimized (*Nishiizumi et al.,* 1997)

$$\frac{N_1}{N_2} = \frac{P_1\lambda_2}{P_2\lambda_1}e^{(\lambda_2 - \lambda_1)t} \quad (4)$$

where the subscripts 1 and 2 refer to the two radionuclides with different half-lives. Nishiizumi and co-workers have used the combinations of ^{10}Be-$^{36}Cl/^{10}Be$ and ^{36}Cl-$^{41}Ca/^{36}Cl$ to calculate terrestrial ages.

3.2. Production Rates

We can estimate the dependence of different radionuclides in small meteoroids from calculations using computer modeling codes (*Reedy,* 1987, 2000; *Vogt et al.,* 1990). *Reedy* (1985) adopted the original *Reedy and Arnold* (1972) model for use with meteorites. *Graf et al.* (1990a,b) developed models for both the size of the meteoroid and depth dependence of major radionuclides. This work was also utilized by *Wieler et al.* (1996). There are three computer codes currently in use: HERMES (*Leya et al.,* 2000, 2004), the LAHET code system (*Reedy and Masarik,* 1995), and MCNPX (*Kim and Reedy,* 2003).

Graf et al. (1990a,b) compared measured and calculated levels of cosmogenic nuclides in Knyahinya to show the production rates as a function of depth. In the case of ^{14}C, *Jull et al.* (1994), *Jull* (2001), and *Wieler et al.* (1996) have discussed the variation in production rate at different depths in meteorites of different sizes. Recent falls generally show activities of ^{14}C equivalent in a range of production rates from about 38–58 atoms/min/kg, with an average value of 51 ± 1 dpm/kg for L chondrites (*Jull,* 2001). It was estimated by *Wieler et al.* (1996) that in H chondrites with pre-atmospheric radii from 20 to 45 cm, the saturated activity (or production rate) should vary from about 38 to 52 dpm/kg. Smaller objects have lower production rates of ^{14}C. *Jull et al.* (1994) measured samples from different depths in the Knyahinya L chondrite (Fig. 2), which had a preatmospheric radius ~45 cm. These measurements gave values of 37 at the surface to 58 dpm/kg at the center of the meteorite (*Jull et al.,* 1994). In most cases, nearly all ^{14}C is produced from spallation of O, only about 3% produced from Si (*Sisterson*

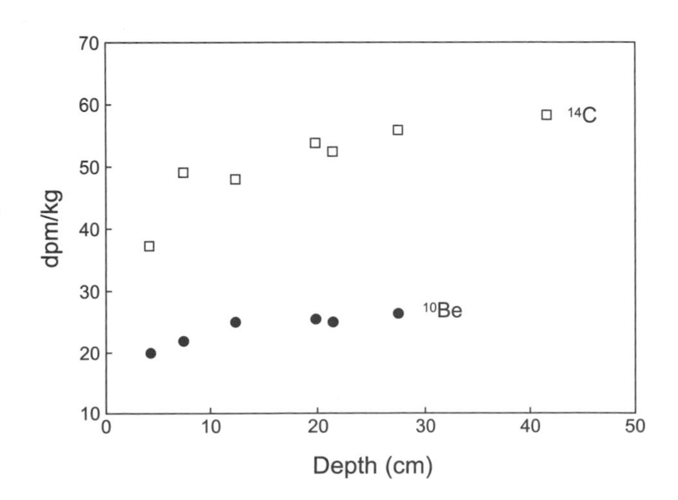

Fig. 2. Depth dependence of ^{14}C and ^{10}Be in samples of Knyahinya (*Jull et al.,* 1994, 2004a). The increase with depth is due to the buildup of secondary-neutron production of the radionuclides.

et al., 1995). Hence, we can use the O content as a scaling parameter. We estimate the saturated activity for a given class of meteorite by normalizing the mean value of the ^{14}C content of Bruderheim (51 ± 1 dpm/kg) to the O content of the meteorite determined from bulk chemistry or from average compositions (*Mason,* 1979).

4. OTHER RADIONUCLIDE TERRESTRIAL-AGE DATING METHODS

4.1. Carbon-14–Beryllium-10 Dating

A limitation of using ^{14}C for terrestrial ages is that shielding corrections may be required if the original meteoroid was very large or small. One way to make these depth estimates is by $^{22}Ne/^{21}Ne$ ratios (*Welten et al.,* 2003). An excellent method is to use the production rate of a more long-lived nuclide such as ^{10}Be to normalize to the ^{14}C production rate. Since both of these radionuclides are produced by spallation reactions on O, we can assume their production ratio in meteorites to be reasonably constant. We assume that the exposure age of the meteorite is sufficient to saturate ^{10}Be and that the $^{14}C/^{10}Be$ production ratio is reasonably constant. In the past, we have taken the production ratio $^{14}C/^{10}Be$ to be 2.5 ± 0.1 (*Kring et al.,* 2001; *Welten et al.,* 2001; *Jull et al.,* 2001), based on a few irradiation experiments, as well as observations of the $^{14}C/^{10}Be$ in recently fallen meteorites. We first observed the stability of the ^{14}C-^{10}Be system for Knyahinya (*Jull et al.,* 1994). Later, we were able to show the utility of the ^{14}C-^{10}Be method for the large fall of the Gold Basin (Arizona) meteorites (*Kring et al.,* 2001) and for Frontier Mountain (Antarctica) meteorites (*Welten et al.,* 2001). In Fig. 3, we show the behavior of the ^{14}C-^{10}Be ratio for average chondrites [where we have taken $^{14}C/^{10}Be$ = 2.5 (*Jull et al.,* 2004a)] and also Knyahinya (L5), which gives a ratio of ~2.1 (*Jull et al.,* 2004a). For comparison, we also plot the results for the Gold Basin meteorite shower (*Kring et al.,* 2001). For Gold Basin, we can observe the much lower ratio due to the decay of ^{14}C over the 15-ka terrestrial age of these meteorite fragments.

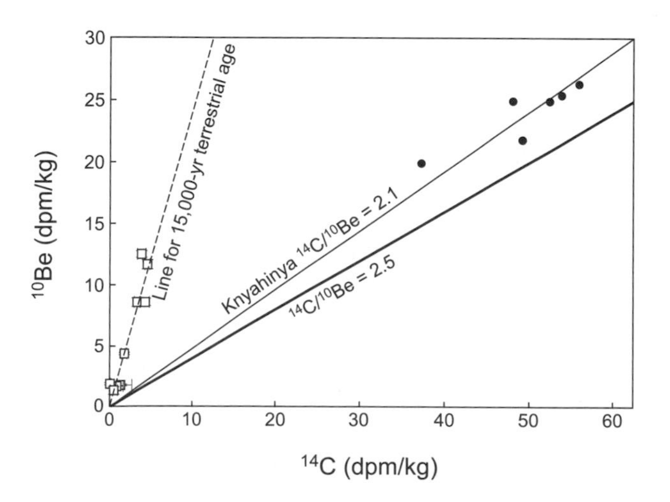

Fig. 3. Behavior of ^{14}C and ^{10}Be for chondrites of average size, with a production rate of ~2.5, Knyahinya ($^{14}C/^{10}Be$ = 2.1) and Gold Basin. For Gold Basin, the low $^{14}C/^{10}Be$ is due to the decay of ^{14}C over its 15-ka terrestrial age.

Western Australian meteorites (*Jull et al.,* 2001) generally show lower shielding and higher ^{10}Be production in the range 13–18 dpm/kg ^{10}Be. Recently, it has become apparent that there may be other meteorites where the $^{14}C/^{10}Be$ is not as well constrained. We have therefore undertaken a new set of modeling studies using the MCNPX code system to model the dependence of ^{14}C and ^{10}Be production as a function of depth and size of the meteoroid (*Jull et al.,* 2004a). The calculated ratios range from ~1.2 at the surface of small meteoroids to ~3.0 for deep samples in larger meteoroids for R = 10–1000 cm. Most of the ratios ranged from 1.8 to 3.0 and include the measured ratios of 2.5 ± 0.1 adopted by *Jull et al.* (2001). However, meteoroids with pre-atmospheric radii of ~25–100 cm, the most common size for chondrites, scatter around 2.5 but with a spread greater than 0.1. Our calculations (*Jull et al.,* 2004a) indicate that using the $^{14}C/^{10}Be$ ratio to get a rate for ^{14}C would yield slightly higher values for cases of low shielding (near surface locations or very small pre-atmospheric size) and slightly lower values for very large radii.

4.2. Chlorine-36–Calcium-41 and Chlorine-36–Beryllium-10 Dating

Welten et al. (2001) have developed a new method of dating using both ^{41}Ca and ^{36}Cl measured in the metal phase of chondrites and small irons. They found good agreement of terrestrial ages derived from these ratios compared to ^{14}C ages. This is possible due to the near-constant production ratio of these two radionuclides in the metal phase (*Nishiizumi and Caffee,* 1998). Also, these nuclides have the potential to cover the range beyond ^{14}C (>50,000 yr) and below ^{36}Cl (<150,000 yr). This method was first used for lunar meteorites, where a range of radionuclides were measured (*Nishiizumi et al.,* 1996; *Nishiizumi and Caffee,* 1998). *Nishiizumi et al.* (1997) have also reported on an alternative method for normalizing shielding corrections for iron meteorites using the combination of ^{36}Cl-^{10}Be. *Nishiizumi et al.* (1999) have claimed that using these two radionuclide combinations can reduce the uncertainty in the terrestrial age to about ±40 ka.

4.3. Other Radionuclides

Two other radionuclides of potential interest are ^{59}Ni and ^{60}Fe. Nickel-59 is a potentially useful nuclide with a half-life of ~100,000 yr, similar to ^{41}Ca. However, this radionuclide has only been used in some very limited cases and cannot be considered a routine measurement. Accelerator mass spectrometry measurements would require the use of very large accelerators. Recently, AMS measurements of ^{59}Ni and ^{60}Fe have become feasible using very large accelerators at Munich and Canberra. *Knie et al.* (1999) have studied ^{60}Fe in two iron meteorites and *Schnabel et al.* (1999a) determined ^{59}Ni in fragments of Canyon Diablo. This nuclide is potentially useful for large iron meteorites, where the production rate of the spallogenic nuclides is least useful. For example, using decay counting, *Kaye* (1963) measured ^{59}Ni in fragments of the Cañon Diablo (Arizona) iron meteorite, from which he derived a terrestrial age of $<4 \times 10^4$ yr. This result can be compared to the age estimate of Meteor Crater of ~50 ka, which is discussed below in section 5.5. In another early counter study, *McCorkell et al.* (1968) estimated the terrestrial age of Hoba to be <80,000 yr. This value was determined from the levels of ^{59}Ni present, 355 ± 100 dpm/kg, which suggested little decay of ^{59}Ni, compared to a known fall (Sikhote-Alin, Russian Far East), which gave 285 ± 70 dpm/kg.

Another radionuclide of potential interest for meteoritics is ^{39}Ar ($t_{1/2}$ 269 yr). *Begemann and Vilcsek* (1969) reported on some measurements using gas counters. Improvements in small gas-counter technology make it feasible that such studies might be done in the future. Perhaps future technical improvements will allow these three radionuclides to be added to the stable of techniques available to us for cosmogenic-nuclide dating.

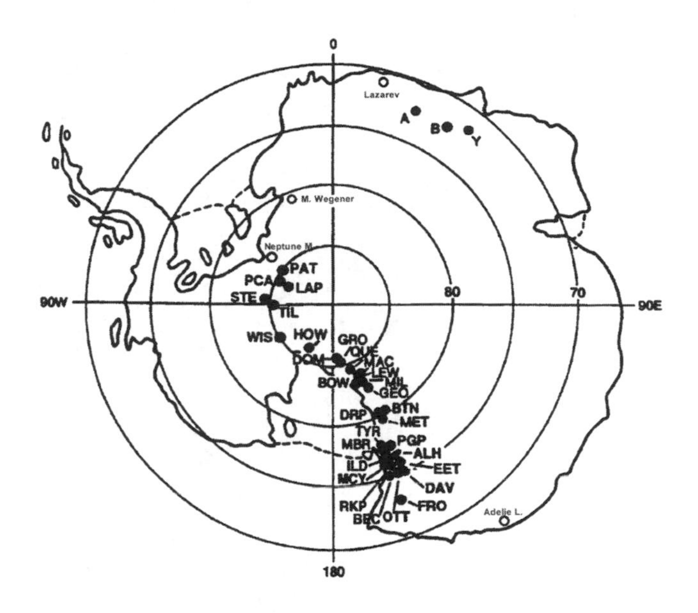

Fig. 4. Map of Antarctica showing the primary recovery locations for meteorites. The sites are encoded as follows: A — Asuka, ALH — Allan Hills, B — Belgica Mountains, BEC — Beckett Mountains, BOW — Bowden Neve, BTN — Bates Nunataks, DAV — David Glacier, DOM — Dominion Range, DRP — Derrick Peak, EET — Elephant Moraine, GEO — Geologists Range, GRO — Grosvenor Mountains, HOW — Mt. Howe, ILD — Inland Forts, LAP — LaPaz Icefield, LEW — Lewis Cliff, MAC — MacAlpine Hills, MBR — Mount Baldr, MCY — MacKay Glacier, MET — Meteorite Hills, MIL — Miller Range, OTT — Outpost Nunatak, QUE — Queen Alexandra Range, PAT — Patuxent Range, PCA — Pecora Escarpment, PGP — Purgatory Peak, RKP — Reckling Peak, STE — Stewart Hills, TIL — Thiel Mountains, TYR — Taylor Glacier, WIS — Wisconsin Range, Y — Yamato Mountains. Adapted from the *Antarctic Meteorite Newsletter,* courtesy of NASA Johnson Space Center.

5. TERRESTRIAL AGES OF METEORITES FROM DIFFERENT REGIONS

Although meteorites are found everywhere, the arid and polar regions of the world appear to be the best locations for storage of meteorites. They can survive for long periods of time in such environments (*Nishio and Annexstad,* 1980; *Nishiizumi et al.,* 1989b, 2002; *Jull et al.,* 1990, 1993a, 1995, 1998b).

5.1. Antarctic Meteorites

Beginning in 1969, Japanese researchers recovered a number of meteorites from Antarctica. They have continued to recover meteorites annually. In 1976, Cassidy and Olsen undertook an expedition to Antarctica to recover meteorites from the Allan Hills blue icefield, located in easy range of the U.S. base at McMurdo (*Cassidy,* 2003). This program has since developed into the U.S. Antarctic Search for Meteorites (ANSMET), and at least 20,000 meteorites have been recovered from Antarctica by various international teams (*Marvin and Mason,* 1984; *Marvin and MacPherson,* 1989, 1992; *Grossman,* 1994; *Cassidy*, 2003). Figure 4 shows a map of the various recovery locations.

The U.S. collection contained 8792 meteorites in 2002 with a total weight of close to 2 metric tons. In principle, we would expect that the cold and dry conditions would allow the storage of meteorites with low rates of weathering and meteorite destruction. The observation of old terrestrial ages, some at the limit of ^{14}C dating, was initially shown by *Fireman* (1978), *Fireman and Norris* (1981), and *Kigoshi and Matsuda* (1986), using β-decay counting. This also confirmed the long terrestrial ages of *Nishizumi et al.* (1979). The change to AMS measurements had already begun to have its effect and Fireman worked with other colleagues (*Brown et al.,* 1984) on the first AMS ^{14}C measurements on meteorites. Subsequently, there is considerable literature on ^{14}C terrestrial age measurements using smaller sample sizes (0.1–0.7 g) and AMS measurements on Antarctic meteorites, as summarized by *Jull et al.* (1998b) and *Jull* (2001). As we have already discussed, the longer-lived isotopes ^{81}Kr and ^{36}Cl can determine longer terrestrial ages, beyond the useful range of ^{14}C of ~40,000 yr. The combination of two nuclide measurements can constrain the age of the meteorite; for example, a meteorite with no ^{14}C, but close to saturated ^{36}Cl, can be constrained to be between >40 and <80 ka. *Nishiizumi et al.* (1989a), *Cresswell et al.* (1993), *Jull et al.* (1993b, 1998b, 1999a), and *Michlovich et al.* (1995) have shown that the age distributions of meteorites at the Allan Hills (Victoria Land) and Yamato (Queen Maud Land) collection sites in Antarctica can be very different, as seen in Fig. 5. *Nishiizumi et al.* (1989a) reported that many meteorites from the Allan Hills Main Icefield have long terrestrial ages, as determined by ^{36}Cl ($t_{1/2}$ =

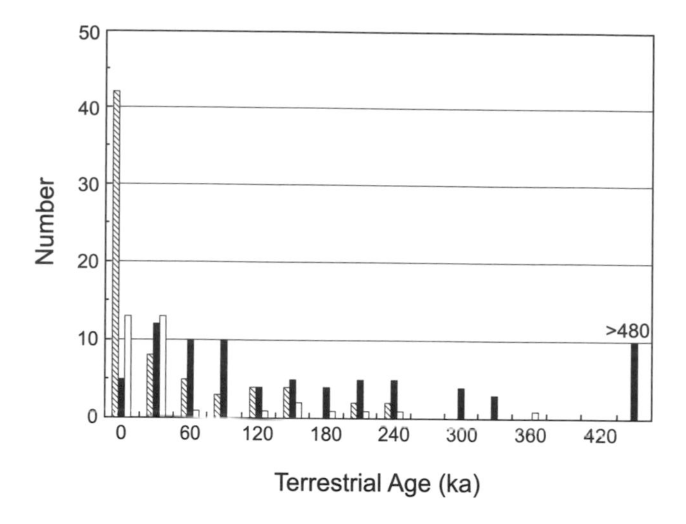

Fig. 5. Terrestrial age distributions of meteorites from the Yamato site — slanted bars (*Jull et al.,* 1993b, 1999a, unpublished data; *Nishiizumi et al.,* 1989a, 1999) — the Allan Hills main icefield (black) and Elephant Moraine (white) taken from *Nishiizumi et al.* (1999), *Michlovich et al.* (1995), *Jull et al.* (1998b), and A. J. T. Jull (unpublished data, 2003).

301,000 yr). Similarly, *Nishiizumi et al.* (1999) summarized the age distribution of Allan Hills Main and Near Western Icefield meteorites and determined that 49 of 107 samples were <70 ka.

Because of low storage temperatures, it makes sense that Antarctic meteorites can be stored for long periods of time in or on the ice. Interestingly, at Allan Hills, an H chondrite, Allan Hills 88019, and an L chondrite, Lewis Cliff 86360, were recovered, which have very long terrestrial ages in excess of 2 Ma (*Scherer et al.,* 1997; *Welten et al.,* 1997).

In Fig. 5, we show a summary of terrestrial ages determined by ^{14}C, ^{36}Cl, and ^{81}Kr for meteorites from Allan Hills, Elephant Moraine (*Welten et al.,* 1999b; *Jull et al.,* 1998b; and A. J. T. Jull et al., unpublished data, 2003), and also the Yamato site (*Jull et al.,* 1993b, 1999a, *Nishiizumi et al.* 1987). The age distribution of samples from the Allan Hills Far Western Icefield and ^{14}C ages from the Yamato site show a generally significantly younger population of meteorites. The Yamato site does show a large range of terrestrial ages, with four eucrites of terrestrial age >200 ka (*Miura et al.,* 1993).

Nishio and Annexstad (1980) developed a model involving long-range transport of meteorites, flowing along with the ice, to explain these long terrestrial ages. At the Allan Hills Far Western Icefield there are few meteorites with less than saturated ^{36}Cl (*Nishiizumi et al.,* 1989a), which would indicate short terrestrial ages consistent with the observation of significant levels of ^{14}C in most samples. In such cases, meteorites cannot have been transported any significant distance in the ice, and most likely fell at the location where they were recovered. Many Antarctic meteorites are clearly fragments of the same fall, which raises the question of "pairing." Two very large falls, one of an H5 chondrite at Lewis Cliffs and another of an LL5 chondrite at Queen Alexandra Range, are well documented in the U.S. Antarctic collection (*Zolensky,* 1998). *Lindstrom and Score* (1994) estimated a "pairing probability" for Antarctic meteorites, which they proposed might reduce the number of discrete falls by a factor of 2–6. If pairing is also taken into account, it may be possible to relate the distribution of meteorite terrestrial ages to ice flow patterns in the Allan Hills region.

5.2. Desert Meteorites

One of the first recognized areas for collections of meteorites from nonpolar regions was Roosevelt County, New Mexico (*Scott et al.,* 1986; *Zolensky et al.,* 1990). The deserts in Australia and the northern Sahara Desert in Africa have also become sources of a large number of meteorites (*Wlotzka et al.,* 1995; *Jull et al.,* 1995; *Bevan et al.,* 1998). Recent searches in Oman (*Hofmann et al.,* 2003), Libya (*Ghadi et al.,* 2003; *Schlüter et al.,* 2002), and other desert areas have been very productive. Smaller-scale searches have been undertaken at the dry lakes in the Mojave Desert, California (*Verish et al.,* 1999), and less-explored areas such as the remote deserts of southern Africa and South America may yield many more meteorites. Many meteorites are found in Northwest Africa and are often purchased from sources with unclear provenance of the material. These meteorites are labeled "Northwest Africa" meteorites.

The collection of meteorites from the Nullarbor Plain in Australia (*Bevan and Binns,* 1989a,b) and from Roosevelt County, New Mexico (*Sipiera et al.,* 1987; *Scott et al.,* 1986) has catalyzed other searches in many desert environments. Searches have been conducted in the Saharan and Saudi deserts, in North and South America, and in central Asian deserts. Where meteorites were found, many of these meteorites were heavily weathered, and thus the actual length of time they have resided on the desert floor or in surface sediments became of great interest. It would appear that achondrites can survive longer than chondrites, based on observations from Antarctica (see *Jull et al.,* 1998b). *Miura et al.* (1993) demonstrated the longevity of some Yamato eucrites. Another example is the long terrestrial ages observed for a martian and a lunar meteorite at Dhofar, Oman, of 340 and 500 Ma, respectively, by *Nishiizumi et al.* (2002).

It is interesting to note that differences in trace-element composition may not be due to differences in meteoritic composition of infall streams. *Al-Kathiri et al.* (2003) showed that uptake of trace elements from the local soil was an important source of changes in trace-element chemistry. This is the likely focus of future studies of contamination of meteorites from desert environments.

In recent years, large numbers of meteorites have been recovered from desert environments and these meteorites are becoming the major source (other than Antarctica) for meteorites. We expect that weathering occurs at different rates depending on sample chemistry and local climatic effects. Measurement of terrestrial ages of desert meteorites has been performed almost exclusively by ^{14}C, due to the problem of recovery of sufficient metal for ^{41}Ca and ^{36}Cl analysis (*Welten et al.,* 1999a).

5.2.1. The Sahara. The first terrestrial-age study of a group of meteorites from one region, outside the Antarctic, was conducted by *Jull et al.* (1990), who reported on the ^{14}C terrestrial ages of meteorites that had been recovered by a private collector near the town of Daraj, in the Hammadah al Hamra (stony desert) region of Libya. We measured the terrestrial ages of 13 meteorites, apparently from 12 discrete falls, which gave terrestrial ages between 3.5 and 35 ka. There also appeared to be a correlation between the storage of the meteorites and changing climatic conditions (*Jull et al.,* 1990). The potential of desert regions as storehouses for meteorites was slow to be recognized, but a workshop in Nördlingen, Germany, in 1994 showed much interest in the potential of these areas (*Schultz et al.,* 1995). Similarly, *Wlotzka et al.* (1995) reported on similar age distributions of 53 meteorites found in Acfer, Algeria. Two of the meteorites from Acfer were close to the limit of ^{14}C measurements, indicating ages of 35 to 42 ka. *Schlüter et al.* (2002) collected 869 meteorites of varying size from the central Sahara, south of Tripoli, from 6 g to 95 kg from Dar al Gani and the Hammadah al Hamra regions. *Welten et al.* (2004) reported on the terrestrial ages on 17 of these meteorites, which ranged from 0–30 ka, except for one that had a terrestrial age (based on ^{41}Ca) of 150 ± 40 ka.

5.2.2. Roosevelt County, New Mexico. *Jull et al.* (1991) had also reported on the terrestrial ages of 17 meteorites from Roosevelt County, New Mexico. This study contradicted higher estimates of the meteorites' age, which had been based on age estimates of the cover sands at Roosevelt County. *Zolensky et al.* (1990) made a much higher estimate, based on recovery of meteorites at Roosevelt County, but they had used a thermoluminescence age of cover sands

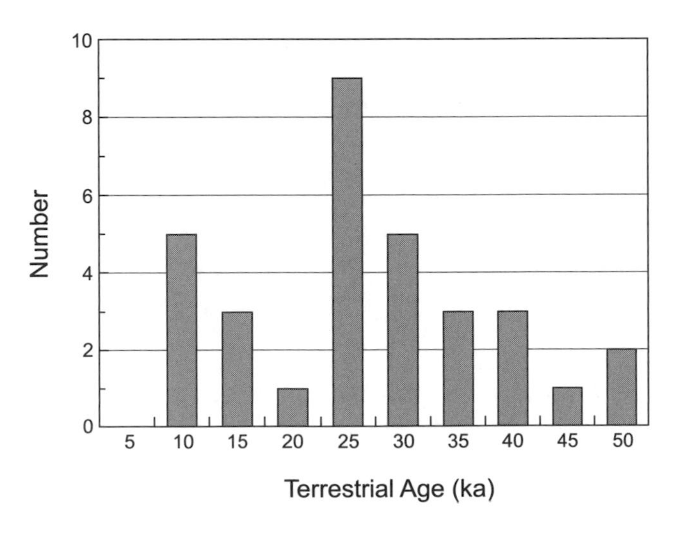

Fig. 6. Distribution of terrestrial ages from Roosevelt County meteorites (*Jull et al.,* 1991, and unpublished data). Note the peak at 20–25 ka, which may be due to pairing of several meteorites.

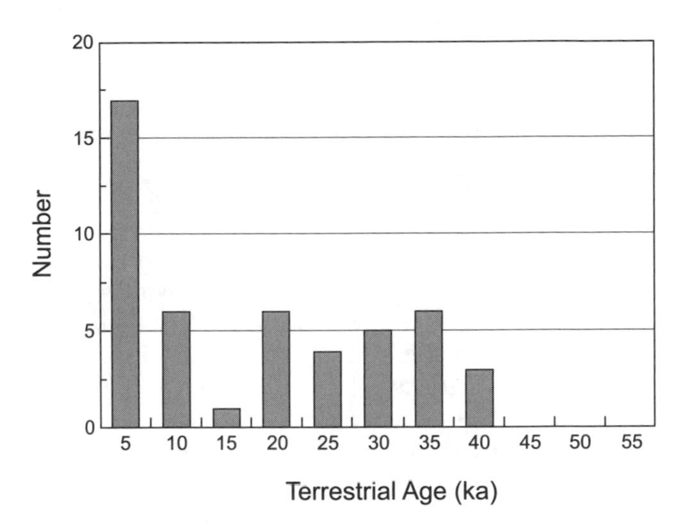

Fig. 7. Distribution of terrestrial ages from Western Australian meteorites (*Jull et al.,* 1995, 2001; *Bland et al.,* 2000).

at the Roosevelt County sites as the estimate of the meteorites' age. Later studies on the age of the cover sands were more consistent with the terrestrial age of the meteorites (*Zolensky et al.,* 1992). The terrestrial ages of the meteorites ranged from 7 to >44 ka and there was a deficiency, with respect to the expected exponential dropoff with age, of meteorites of <20 ka terrestrial age. These results, together with newer measurements on a total of 32 Roosevelt County meteorites (A. J. T. Jull, unpublished data, 2004), are shown in Fig. 6. The deficiency of younger meteorites can be explained by the fact that the Roosevelt County meteorites are found in blowouts in the cover sands of late Quaternary age (Blackwater Draw formation) found in this region.

5.2.3. Western Australia. *Jull et al.* (1995) presented results of terrestrial age determinations from 22 Western Australian meteorites, which showed that these meteorites could survive for up to 30 k.y. Further results were reported by *Bland et al.* (2000). *Jull et al.* (2001) also discussed $^{14}C/^{10}Be$ ages for some of these meteorites. In Fig. 7, the age distribution for the Western Australian meteorites shows an exponential dropoff of number of meteorites with increasing age. This indicates a mean residence time of perhaps 10 k.y. at this location. The difference in the age distribution with that of Roosevelt County (Fig. 6), where the meteorites were protected by late Quaternary cover sands, indicates a marked difference in storage conditions.

5.2.4. The Arabian Peninsula. At the Nördlingen workshop, *Franchi et al.* (1995) discussed the potential of the Arabian peninsula for recovery of meteorites. At that time, only 24 meteorites were known from this region, mostly collected by geologists in the 1930s and 1950s. *Franchi et al.* (1995) had only measured about six ^{14}C ages on meteorites from Oman and Saudi Arabia. All these measurements were made on museum samples. By 2003, several different groups of collectors had recovered thousands of specimens from Oman alone (*Hofmann et al.,* 2003; *Gnos et al.,*

2003). *Gnos et al.* (2003) have found a large strewnfield at the Jiddat al Harasis, consisting of nearly 3000 individual stones ranging in size from 1 g to 54 kg. During three meteorite field search seasons in Oman by a Swiss-Omani team (2001–2003), more than 200 individual meteorites were recovered plus several strewnfields, including that at the Jiddat al Harasis. The measured ^{14}C terrestrial ages of Omani meteorites is in the range of 2.2 to >49 ka, with many between 10 and 40 ka. The older meteorites generally are more weathered. *Gnos et al.* (2003) also observed that meteorites from the same strewnfield may have different weathering grades depending on burial conditions and the size of the meteorite. These authors also studied the chemical analysis of some of the meteorites with mean compositions of H and L chondrites and found that enrichments of Sr and Ba are most prominent. These compositional changes increase with age and weathering grade.

Nishiizumi and Caffee (2001) and *Nishiizumi et al.* (2002) have used the nuclides ^{36}Cl and ^{41}Ca for estimates of the ages of achondrites. In particular, they found terrestrial ages of a lunar and martian meteorite from Dhofar, Oman, to be 500 and 350 ka respectively. Such terrestrial ages had previously only been observed for a few achondrites in Antarctica (*Miura et al.,* 1993).

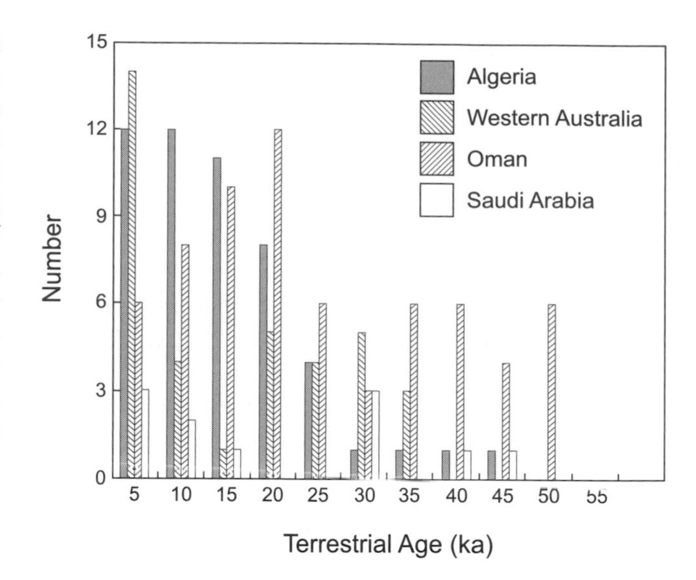

Fig. 8. Comparison of the terrestrial ages of meteorites recovered from arid regions of Algeria, Western Australia, Oman, and Saudi Arabia (*Wlotzka et al.,* 1995; *Bevan et al.,* 1998; *Bland et al.,* 2000; *Jull et al.,* 1990, 1995, 2001; *Stelzner et al.,* 1999; *Hofmann et al.,* 2003).

5.3. Trends in Terrestrial Ages

For the Nullarbor Plain, Western Australia, we observe (see Fig. 7) an approximately exponential dropoff of number of meteorites with increasing terrestrial age (*Bland et al.,* 2000; *Bevan et al.,* 1998; *Jull et al.,* 1995). *Jull et al.* (1990, 1995, 1999b), *Knauer et al.* (1995), *Neupert et al.* (1997), and *Wlotzka et al.* (1995) have reported on terrestrial ages of meteorites from the Sahara in Libya and Algeria, which show similar trends. Different climatic regimes and local geology can affect the distribution of terrestrial ages of meteorites from areas such as the Sahara and Roosevelt County, as weathering occurs at different rates depending on sample chemistry and local climatic effects.

Figure 8 summarizes ^{14}C terrestrial ages observed for meteorites from Algeria, the Nullarbor Plain, New Mexico, and Libya. In general, the Nullarbor and Sahara meteorites show an approximately exponential decrease of number of finds with terrestrial age to at least 30 ka. Since weathering gradually destroys meteorites, we expect that the resulting distribution should show some exponential dependence on age. As an example, consider a collection where meteorites fell continuously directly on the collection area. The meteorites then should eventually disintegrate and reach a steady state where the disintegration rate will match the infall rate. Therefore, the number will decrease with increasing age, and so there should be more young meteorites than older ones. This is similar to a simple first-order model of meteorite accumulation (*Freundel et al.,* 1986; *Jull et al.,* 1993a). However, as *Bland et al.* (1996, 1998, 2000) have shown, weathering of meteorites is a complex process, where an initial rapid weathering phase is followed by slower, longer-term effects. We can show an example of the case of the Western Australian meteorites. Here, few meteorites beyond the range of ^{14}C are observed and this profile shows the simple decay model for meteorite ages. In a new approach to the question of meteorite survival and weathering, *Bland et al.* (1996, 1998, 2000) compared the terrestrial ages of many meteorites with the degrees of weathering, measured by Mössbauer spectroscopy, and they further showed that meteorites of different composition weather at different rates, a fact known qualitatively to many meteoriticists, but difficult to quantify. *Bland et al.* (1998) noted that the degree of oxidation observed for Antarctic finds was 25–30% lower than for desert meteorites, which they presumed was due to the low storage temperatures. The burial of the meteorites in the ice during transport also likely protects the meteorites from weathering. As a result, correlations between degree of oxidation and weathering and terrestrial age are much weaker for Antarctic meteorites than for those from other locations.

However, the situation was much clearer for desert meteorites. *Bland et al.* (1996) showed that H chondrites, which contain more metallic iron, weather at faster rates than L chondrites. We can also conclude that achondrites, since they contain no iron metal, ought to survive the longest, which is in agreement with the observations of *Nishiizumi et al.* (2002) and *Miura et al.* (1993).

5.4. Martian and Lunar Meteorites

Meteorites from the Moon and Mars are of particular interest to many meteoriticists. A total of about 32 meteorites from Mars (representing 24 discrete falls) and 29 from

TABLE 2. Terrestrial and ejection ages of lunar meteorites.

Meteorite	Ejection Depth (g/cm^2)	Ejection Age (m.y.)	Transit Time (m.y.)	Terrestrial Age (ka)	Reference
ALHA 81005	150–180	0.04 ± 0.02	<0.05	18 ± 1	*Nishiizumi et al.* (1991a); *Jull and Donahue* (1992)
Asuka 881757	>1000	0.9 ± 0.1	0.9 ± 0.1	<50	*Nishiizumi et al.* (1992a)
Calcalong Creek	>1000	3 ± 1	3 ± 1	<30	*Nishiizumi et al.* (1992a)
Dar al Gani 262	75–90	0.16 ± 0.01	0.08 ± 0.01	80 ± 10	*Nishiizumi et al.* (1998)
Dar al Gani 400	>1100	0.24 ± 0.02	0.22 ± 0.02	17.3 ± 1.4	*Nishiizumi* (2003)
Dhofar 025	>1000	13–20	13–20	~500	*Nishiizumi et al.* (2002)
Dhofar 026	~1400	0.003	0.003		*Nishiizumi et al.* (2002)
Dhofar 081	200–230	0.04 ± 0.02	<0.01	40 ± 20	*Nishiizumi et al.* (2002)
Dhofar 280, 281	200–230	0.04 ± 0.02	<0.01	40 ± 20	*Nishiizumi et al.* (2004)
Dhofar 489	1100	—	6 ± 2	—	*Nishiizumi et al.* (2004)
EET 87521, 96008	540–600	0.08 ± 0.04	<0.01	80 ± 30	*Nishiizumi et al.* (1999)
MAC 88104, 88105	360–400	0.28 ± 0.02	0.04–0.05	230 ± 20	*Nishiizumi et al.* (1991a)
NWA 482	240	0.29	0.27	8.6 ± 1.3	*Nishiizumi and Caffee* (2001); *Daubar et al.* (2002)
NWA 773	800–1000	—	1-30	17 ± 1	*Nishiizumi et al.* (2004)
NWA 032	>1100	0.047 ± 0.010	0.042 ± 0.005	<10	*Nishiizumi and Caffee* (2001)
QUE 93069/94269	65–80	0.16 ± 0.02	0.15 ± 0.02	10 ± 2	*Nishiizumi et al.* (1996)
QUE 94281	270–320	0.05 ± 0.03	<0.05	23.5 ± 1.8	*Nishiizumi and Caffee,* (1996); A. J. T. Jull, unpublished data, 1996
Sayh al Uhaymir 169	>1000	<0.85	<0.85	9.7 ± 1.3	*Gnos et al.* (2004)
Y 791197	4–8	<0.1	<0.1	30–90	*Nishiizumi et al.* (1991a)
Y 793169	500	1.1 ± 0.2	1.1 ± 0.2	<50	*Nishiizumi et al.* (1992a); *Thalmann et al.* (1996)
Y 793274, 981031	140–180	0.032 ± 0.003	<0.01	20–35	*Nishiizumi et al.* (1991c, 2004); *Eugster et al.* (1992)
Y 793885	~90	—	—	>36	*Jull et al.* (2004b); *Nishiizumi et al.* (2004)
Y 82192, 82193, 86032	>1000	11	11	80	*Nishiizumi et al.* (1988); *Eugster* (1988b)

the Moon were known as of 2003 (*Cassidy,* 2003; *Gnos et al.,* 2004). It is interesting to note that the first lunar meteorite to be identified, ALHA 81005, was only found in 1981. Martian meteorites were not proposed to be a group, having previously been classified as shergottites, nakhlites, and chassignites, until the observation that they formed a discrete grouping in O-isotopic composition (*Clayton and Mayeda,* 1983) and also contained "trapped" gas similar to Mars atmospheric composition (*Bogard and Johnson,* 1983). The terrestrial age and the transit time (the cosmic-ray exposure age of a meteorite in free space) tells us much about its history. By calculating the sum of the terrestrial age and transit time, we can reconstruct the ejection times for meteorites that land on Earth and have an origin on Mars or the Moon. Hence, we can identify meteorites possibly related to the same impact event or common source areas. In Table 2, we give the ejection age, transit time, and terrestrial age for lunar meteorites. Similarly, martian (or SNC) meteorites are shown in Table 3; we can similarly observe that many shergottites have shorter exposure ages than nakhlites. For a detailed discussion of the ejection times of lunar and martian meteorites, the reader is referred to *Zolensky et al.* (2006).

5.5. Dating of Meteorite Craters Using *In Situ*-produced Cosmogenic Radionuclides

In the last 15 years, methods of *in situ* terrestrial cosmogenic nuclide (TCN) dating have developed (see *Tuniz et al.,* 1998). In these methods, the buildup of cosmogenic nuclides at a low level due to cosmic-ray interactions with material at Earth's surface can be used. This method can then be used to estimate the terrestrial age of a very large meteorite that caused an impact crater, by dating the crater itself. *Shoemaker et al.* (1990) considered these methods for the dating of some Australian meteor craters, and obtained estimates of the ages of Wolfe Creek (~300 ka), Boxhole (~30 ka), and Dalgaranga (apparent age ~270 ka) craters. However, the best example is the application to Meteor Crater, Arizona. *Nishiizumi et al.* (1991b) obtained an exposure age of 49.2 ± 1.7 ka reported on this new dating method using ^{10}Be-^{26}Al surface exposure dating, and *Phillips et al.* (1991) reported a similar value, using ^{36}Cl, of 49.7 ± 0.9 ka. Both of these results suggest that the impact age of Meteor Crater, and therefore the Cañon Diablo, Arizona, iron meteorite, is ~50 ka. Intriguingly, *Schnabel*

TABLE 3. Terrestrial and exposure ages of martian meteorites.

Meteorite	Classification	Ejection Age (Ma)	Terrestrial Age (ka)	Reference
ALH 77005	Shergottite	3.32 ± 0.55	190 ± 70	*Eugster et al.* (1997); *Nishiizumi et al.* (1994)
ALH 84001	Orthopyroxenite (unique)	14.4 ± 0.7	13 ± 1	*Eugster et al.* (1997)
Chassigny	Dunite	12 ± 4	Fall	*Eugster et al.* (1997)
Dar al Gani cluster	Shergottite	1.05 ± 0.1	60 ± 20	*Nishiizumi et al.* (2001)
Dhofar 019	Shergottite	18.7 ± 4.3	340 ± 40	*Nishiizumi et al.* (2002)
EET 79001	Shergottite	0.65 ± 0.20	12 ± 1	*Eugster et al.* (1997); *Jull and Donahue* (1988)
Governador Valadares	Nakhlite	10.1 ± 2.2	Fall	*Eugster et al.* (1997)
Lafayette	Nakhlite	11.4 ± 2.1	2.9 ± 1.0	*Eugster et al.* (1997); *Jull et al.* (1999c)
LEW 88516	Shergottite	4.1 ± 0.6	21.5 ± 1.5	*Eugster et al.* (1997); *Jull et al.* (1994); *Nishiizumi et al.* (1992b)
Los Angeles	Shergottite	3.0 ± 0.4	9.9 ± 1.3	*Nishiizumi et al.* (2000); A. J. T. Jull, unpublished data, 2004
Nakhla	Nakhlite	11.6 ± 1.8	Fall	*Eugster et al.* (1997)
NWA 1068/1110	Shergottite	2.2-3	>40	*Nishiizumi et al.* (2004)
NWA 1195	Shergottite	1.1 ± 0.2	>37	*Nishiizumi et al.* (2004)
NWA 1460	Shergottite	2.2-3	—	*Nishiizumi et al.* (2004)
NWA 480	Shergottite	2.4 ± 0.2	—	*Marty et al.* (2001)
NWA 817	Nakhlite	9.7 ± 1.1	—	*Marty et al.* (2001)
NWA 998	Nakhlite	—	6 ± 1	*Nishiizumi et al.* (2004)
QUE 94201	Shergottite	2.6 ± 0.5	290 ± 50	*Eugster et al.* (1997); *Kring et al.* (2003)
Sayh al Uhaymir cluster	Shergottite	1.5 ± 0.3	—	*Nishiizumi et al.* (2001)
Shergotty	Shergottite	2.71 ± 0.45	Fall	*Eugster et al.* (1997)
Y 793604	Shergottite	4.4 ± 1.0	—	*Eugster and Polnau* (1997)
Y 980459	Shergottite	1.1 ± 0.2	—	*Nishiizumi and Hillegonds* (2004)
Y 000593, 000749, 000802	Nakhlite	12.1 ± 0.7	55 ± 20	*Okazaki et al.* (2003); *Jull et al.* (2004b); *Nishiizumi and Hillegonds* (2004)
Zagami	Shergottite	2.81 ± 0.18	Fall	*Eugster et al.* (1996)

et al. (1999b) obtained an approximate age based on ^{41}Ca/^{36}Cl on Cañon Diablo meteorite fragments of 82 ± 20 ka and concluded, when considering the other data, that the crater must be <75 ka.

6. CONCLUSIONS

To determine the terrestrial ages of meteorites, the levels of radionuclides produced by exposure to cosmic radiation in space must be measured. If we compare these levels to the expected activity of a recently fallen meteorite, we can then determine the terrestrial age. From measurements of ^{14}C and ^{36}Cl, we can show that the residence time of most desert meteorites is <50 k.y. Longer residence times are observed for some chondrites and achondrites in Antarctica determined by ^{36}Cl and ^{81}Kr studies. Some meteorites can survive for a very long time in desert environments, with a few achondrites surviving for hundreds of thousands of years (*Nishiizumi et al.,* 2002). There are also some large iron meteorites with a terrestrial age of up to 2.7 Ma (*Aylmer et al.,* 1988; *Chang and Wänke,* 1969), as well as a few Antarctic chondrites that have survived for 2 m.y. (*Scherer et al.,* 1997; *Welten et al.,* 1997). As a result of studies of the terrestrial ages and weathering of meteorites, we conclude that the infall rates agree with those discussed by *Halliday* (2001), and that there is still much to learn about the weathering processes that affect meteorites.

Acknowledgments. The author is grateful for the support of the staff of the NSF Arizona AMS Laboratory. I am also grateful to K. Nishiizumi and P. A. Bland for useful discussions and to O. Eugster and two anonymous reviewers for helpful comments. This work was supported in part by NASA grant NAG5-11979 and NSF grant EAR-0115488.

REFERENCES

Al-Kathiri A., Hofmann B. A., Gnos E., and Jull A. J. T. (2003) Weathering of meteorites from Oman: The influence of soil chemistry and correlation of chemical/mineralogical weathering proxies with ^{14}C terrestrial ages (abstract). *Meteoritics & Planet. Sci., 38,* A55.

Aylmer D., Bonnano V., Herzog G. F, Klein J., and Middleton R. (1988) ^{26}Al and ^{10}Be production in iron meteorites. *Earth Planet. Sci. Lett., 88,* 107–118.

Bada J. L., Glavin D. P., McDonald G. D., and Becker L. (1998) A search for endogenous amino acids in the Martian meteorite,

ALH84001. *Science, 279,* 362–365.

Begemann F. and Vilcsek E. (1969) Chlorine-36 and argon-39 production rates in the metal of stone and stony-iron meteorites. In *Meteorite Research* (P. Millman, ed.), pp. 355–362. Reidel, Dordrecht.

Begemann F., Rieder R., Vilcsek E., and Wänke H (1969) Cosmic-ray produced radionuclides in the Barwell and Saint-Séverin meteorites. In *Meteorite Research* (P. Millman, ed.), pp. 267–274. Reidel, Dordrecht.

Benoit P. H., Jull A. J. T., McKeever S. W. S., and Sears D. W. G. (1993) The natural thermoluminescence of meteorites VI: Carbon-14, thermoluminescence and the terrestrial ages of meteorites. *Meteoritics, 28,* 196–203.

Beukens R. P., Rucklidge J. C., and Miura Y. (1988) ^{14}C ages of Yamato and Allan Hills meteorites. *Proc. NIPR Symp. Antarctic Meteorites, 1,* 224–230.

Bevan A. W. R. and Binns R. A. (1989a) Meteorites from the Nullarbor Region, Western Australia: I. A review of past recoveries and a procedure for naming new finds. *Meteoritics, 24,* 127–133.

Bevan A. W. R and Binns R. A. (1989b) Meteorites from the Nullarbor Region, Western Australia: II. Recovery and classification of 34 new meteorite finds from the Mundrabilla, Forrest, Reid and Deakin areas. *Meteoritics, 24,* 134–141.

Bevan A. W. R., Bland P. A., and Jull A. J. T. (1998) Meteorite flux on the Nullarbor Region, Australia. In *Meteorites: Flux with Time and Impact Effects* (M. M. Grady et al., eds.), pp. 58–73. Geological Society of London Special Publication 140.

Bland P. A. (2001) Quantification of meteorite infall rates from accumulations in deserts and meteorite accumulations on Mars. In *Accretion of Extraterrestrial Matter throughout Earth's History* (B. Peuker-Ehrenbrink and B. Schmitz, eds.), pp. 267–303. Kluwer, New York.

Bland P. A., Smith T. B., Jull A. J. T., Berry F. J., Bevan A. W. R., Cloudt S., and Pillinger C. T. (1996) The flux of meteorites to the Earth over the last 50,000 years. *Mon. Not. R. Astron. Soc., 283,* 551–565.

Bland P. A., Sexton A. S., Jull A. J. T., Bevan A. W. R., Berry F. J., Thornley D. M., Astin T. R., Britt D. T., and Pillinger C. T. (1998) Climate and rock weathering: A study of terrestrial age dated ordinary chondritic meteorites from hot desert regions. *Geochim. Cosmochim. Acta., 62,* 3169–3184.

Bland P. A., Bevan A. W. R., and Jull A. J. T. (2000) Ancient meteorite finds and the Earth's surface environment. *Quaternary Res., 53,* 131–142.

Boeckl R. S. (1972) Terrestrial ages of nineteen stony meteorites derived from their radiocarbon content. *Nature, 236,* 25–26.

Bogard D. D. and Johnson P. (1983) Martian gases in an Antarctic meteorite. *Science, 221,* 651–654.

Brown R. M., Andrews H. R., Ball G. C., Burn N., Imahori Y., Milton J. C. D., and Fireman E. L. (1984) ^{14}C content of ten meteorites measured by tandem accelerator mass spectrometry. *Earth Planet. Sci. Lett., 67,* 1–8.

Buchwald V. F. (1975) *Handbook of Iron Meteorites: Their History, Distribution, Composition, and Structure.* Univ. of California, Berkeley.

Cassidy W. A. (2003) *Meteorites, Ice and Antarctica: A Personal Account.* Cambridge Univ., Cambridge. 349 pp.

Chang C. and Wänke H. (1969) Beryllium-10 in iron meteorites: Their cosmic-ray exposure and terrestrial ages. In *Meteorite Research* (P. Millman, ed.), pp. 397–406. Reidel, Dordrecht.

Clayton R. N. and Mayeda T. K. (1983) Oxygen isotopes in eucrites, shergottites, nakhlites and chassignites. *Earth Planet. Sci. Lett., 62,* 1–6.

Cresswell R. G., Miura Y., Beukens R. P., and Rucklidge J. C. (1993) ^{14}C terrestrial ages of nine Antarctic meteorites using CO and CO_2 temperature extractions. *Proc. NIPR Symp. Antarctic Meteorites, 6,* 381–390.

Cressy P. J. and Bogard D. D. (1976) Calculation of cosmic ray exposure ages of stone meteorites. *Geochim. Cosmochim. Acta, 40,* 749–762.

Cziczo D. J., Thompson D. S., and Murphy D. M. (2001) Ablation, flux and atmospheric implications of meteors inferred from stratospheric aerosol. *Science, 291,* 1772–1775.

Daubar I. J., Kring D. A., Swindle T. D., and Jull A. J. T. (2002) Northwest Africa 482: A crystalline impact-melt breccia from the lunar highlands. *Meteoritics & Planet. Sci., 37,* 1797–1813.

Eugster O. (1988a) Cosmic-ray production rates for He-3, Ne-21, Ar-38, Kr-83 and Xe-126 in chondrites based on ^{81}Kr-Kr exposure ages. *Geochim. Cosmochim. Acta, 52,* 1649–1662.

Eugster O. (1988b) Exposure age and terrestrial age of the paired meteorites Yamato 82192 and 82193 from the Moon. *Proc. NIPR Symp. Antarctic Meteorites, 1,* 135–141.

Eugster O. and Polnau E. (1997) Mars-Earth transfer time of lherzolite Yamato-793605. *Proc. NIPR Symp. Antarctic Meteorites, 10,* 142–148.

Eugster O., Michel Th., and Niedermann S. (1992) Solar wind and cosmic ray exposure history of lunar meteorite Yamato 793274. *Proc. NIPR Symp. Antarctic Meteorites, 5,* 23–35.

Eugster O., Weigel A., and Polnau E. (1996) Two different ejection events for basaltic shergottites QUE94201, Zagami and Shergotty (2.6 Ma ago) and lherzolitic shergottites LEW88516 and ALH77005 (3.5 Ma ago) (abstract). In *Lunar and Planetary Science XXVII,* pp. 345–346. Lunar and Planetary Institute, Houston.

Eugster O., Weigel A., and E. Polnau (1997) Ejection times of Martian meteorites. *Geochim. Cosmochim. Acta, 61,* 2749–2757.

Evans J. C. and Rancitelli L. A. (1979) Terrestrial ages. *Smithson. Contrib. Earth Sci., 23,* 45–46.

Evans J. C., Reeves J. H., and Rancitelli L. A. (1982) Aluminum-26: Survey of Victoria Land meteorites. *Smithson. Contrib. Earth Sci., 24,* 70–74.

Evans J., Wacker J., and Reeves J. (1992) Terrestrial ages of Victoria Land meteorites derived from cosmogenic radionuclides. *Smithson. Contrib. Earth Sci., 30,* 45–56.

Farley K. and Patterson D. B. A. (1995) A 100ka periodicity in the flux of extraterrestrial ^{3}He to the seafloor. *Nature, 378,* 600–603.

Fifield L. K. (1999) Accelerator mass spectrometry and its applications. *Rept. Prog. Phys., 62,* 1223–1274.

Fink D., Klein J., Middleton R., Vogt S., and Herzog G. F. (1991) Ca-41 in iron falls, Grant and Estherville — Production rates and related exposure age calculations. *Earth Planet. Sci. Lett., 107,* 115–128.

Fireman E. L. (1978) Carbon-14 in lunar soil and in meteorites. *Proc. Lunar Planet. Sci. Conf. 9th,* pp. 1647–1654.

Fireman E. L. and Norris T. L. (1981) Carbon-14 ages of Allan Hills meteorites and ice. *Proc. Lunar Planet. Sci. 12A,* pp. 1019–1025.

Franchi I. A., Delisle G., Jull A. J. T., Hutchison R., and Pillinger C. T. (1995) An evaluation of the meteorite potential of the Jiddat al Harasis and the Rub al Khali regions of southern

Arabia. In *Workshop on Meteorites from Cold and Hot Deserts* (L. Schultz et al., eds.), pp. 29–30. LPI Tech. Rpt. 95-02, Lunar and Planetary Institute, Houston.

Freundel M., Schultz L., and Reedy R. C. (1986) Terrestrial ^{81}Kr-Kr ages of Antarctic meteorites. *Geochim. Cosmochim. Acta, 50,* 2663–2673.

Fraser K. (1997) Map of Desert Regions of the World. U.S. Geological Survey. Available on line at http://pubs.usgs.gov/gip/deserts/what/world.html.

Goel P. S. and Kohman T. L. (1962) Cosmogenic carbon-14 in meteorites and terrestrial ages of 'finds' and craters. *Science, 136,* 875–876.

Ghadi A. M., Abu Aghreb A. E., Schlüter J., Schultz L., and Thiedig F. (2003) The Dar al Gani meteorite field in the Libyan Sahara (abstract). *Meteoritics & Planet. Sci., 38,* A49.

Gnos E., Hofmann B. A., Al-Kathari A., Lorenzetti S., Eugster O., and Jull A. J. T. (2003) The newly-discovered Jiddat al Harasis strewnfield in Oman (abstract). *Meteoritics & Planet. Sci., 38,* A38.

Gnos E., Hofmann B. A., Al-Kathiri A., Lorenzetti S., Eugster O., Whitehouse M. J., Villa I. M., Jull A. J. T., Eikenberg J., Spettel B., Krähenbühl U., Franchi I. A., and Greenwood R. C. (2004) Pinpointing the source of a lunar meteorite: Implications for the evolution of the moon. *Science, 305,* 657–659.

Goel P. S. and Honda M. (1965) Cosmic-ray-produced iron-60 in Odessa meteorite. *J. Geophys. Res., 70,* 747.

Graf Th., Baur H., and Signer P. (1990a) A model for the production of cosmogenic nuclides in chondrites. *Geochim. Cosmochim. Acta, 54,* 2521–2534.

Graf Th., Signer P., Wieler R., Herpers U., Sarafin R., Vogt S., Fieni Ch., Bonani G., Suter M., and Wölfli W. (1990b) Cosmogenic nuclides and nuclear tracks in the chondrite Knyahinya. *Geochim. Cosmochim. Acta, 54,* 2511–2520.

Grossman J. (1994) The U.S. Antarctic meteorite collection. *Meteoritical Bulletin, 76,* 1994 January, in *Meteoritics, 29,* 100–143.

Halliday I. (2001) The present-day flux of meteorites to the Earth. In *Accretion of Extraterrestrial Matter Throughout Earth's History* (B. Peuker-Ehrenbrink and B. Schmitz, eds.), pp. 305–318. Kluwer, New York.

Halliday I., Blackwell A. T., and Griffin A. A. (1989) The flux of meteorites on the Earth's surface. *Meteoritics, 24,* 173–178.

Herzog G. F., Vogt S., Albrecht A., Xue S., Fink D., Klein J., Middleton R., Weber H. W., and Schultz L. (1997) Complex exposure histories for meteorites with "short" exposure ages. *Meteoritics & Planet. Sci., 32,* 413–422.

Hofmann B., Gnos E., Al-Kathari A., Al-Azri H., and Al-Murazzi A. (2003) Omani-Swiss meteorite search 2001–2003: Project overview (abstract). *Meteoritics & Planet. Sci., 38,* A54.

Huss G. R. (1990) Meteorite infall as a function of mass: Implications for the accumulation of meteorites on Antarctic ice. *Meteoritics, 25,* 41–56.

Jull A. J. T. (2001) Terrestrial ages of meteorites. In *Accretion of Extraterrestrial Matter Throughout Earth's History* (B. Peuker-Ehrenbrink and B. Schmitz, eds.), pp. 241–266. Kluwer, New York.

Jull A. J. T. and Donahue D. J. (1988) Terrestrial age of the Antarctic shergottite EETA79001. *Geochim. Cosmochim. Acta, 52,* 1295–1297.

Jull A. J. T. and Donahue D. J. (1992) ^{14}C terrestrial ages of two lunar meteorites, ALHA 81005 and EET 87521 (abstract). In *Lunar and Planetary Science XXIII,* pp. 637–638. Lunar and Planetary Institute, Houston.

Jull A. J. T., Wlotzka F., Palme H., and Donahue D. J. (1990) Distribution of terrestrial age and petrologic type of meteorites from western Libya. *Geochim. Cosmochim. Acta, 54,* 2895–2899.

Jull A. J. T., Woltkza F., and Donahue D. J. (1991) Terrestrial ages and petrologic description of Roosevelt County meteorites (abstract). In *Lunar and Planetary Science XXII,* pp. 667–668. Lunar and Planetary Institute, Houston.

Jull A. J. T., Donahue D. J., Cielaszyk E., and Wlotzka F. (1993a) Carbon-14 terrestrial ages and weathering of 27 meteorites from the southern high plains and adjacent areas (USA). *Meteoritics, 28,* 188–195.

Jull A. J. T., Miura Y., Cielaszyk E, Donahue D. J., and Yanai K. (1993b) AMS ^{14}C ages of Yamato achondritic meteorites. *Proc. NIPR Symp. Antarctic Meteorites, 6,* 374–380.

Jull A. J. T., Donahue D. J., Reedy R. C., and Masarik J. (1994) A carbon-14 depth profile in the L5 chondrite Knyahinya. *Meteoritics, 29,* 649–738.

Jull A. J. T., Bevan A. W. R., Cielaszyk E., and Donahue D. J. (1995) Carbon-14 terrestrial ages and weathering of meteorites from the Nullarbor Plain, Western Australia. In *Workshop on Meteorites from Cold and Hot Deserts* (L. Schultz et al., eds.), pp. 37–38. LPI Tech. Rpt. 95-02, Lunar and Planetary Institute, Houston.

Jull A. J. T., Eastoe C. J., and Cloudt S. (1997) Isotopic composition of carbonates in the SNC meteorites, Allan Hills 84001 and Zagami. *J. Geophys. Res., 102,* 1663–1669.

Jull A. J. T., Courtney C., Jeffrey D. A., and Beck J. W. (1998a) Isotopic evidence for a terrestrial source of organic compounds found in Martian meteorites, Allan Hills 84001 and Elephant Moraine 79001. *Science, 279,* 366–368.

Jull A. J. T., Cielaszyk E., and Cloudt S. (1998b) ^{14}C terrestrial ages of meteorites from Victoria Land, Antarctica and the infall rates of meteorites. In *Meteorites: Flux with Time and Impact Effects* (M. M. Grady et al., eds.), pp. 75–91. Geological Society of London Special Publication 140.

Jull A. J. T., Klandrud S. E., Cielaszyk E., and Cloudt S. (1999a) Carbon-14 terrestrial ages of meteorites from the Yamato region, Antarctica. In *Antarctic Meteorites XXIV,* pp. 62–63. National Institute of Polar Research, Tokyo

Jull A. J. T., Bland P. A., Klandrud S. E., McHargue L. R., Bevan A. W. R., Kring D. A., and Wlotzka F. (1999b) Using ^{14}C and ^{14}C-^{10}Be for terrestrial ages of desert meteorites. In *Workshop on Extraterrestrial Materials from Cold and Hot Deserts,* pp. 41–43. LPI Contribution No. 997, Lunar and Planetary Institute, Houston.

Jull A. J. T., Klandrud S. E., Schnabel C., Herzog G. F., Nishiizumi K., and Caffee M. W. (1999c) Cosmogenic radionuclide studies of the nakhlites (abstract). In *Lunar and Planetary Science XXX,* Abstract #1004. Lunar and Planetary Institute, Houston (CD-ROM).

Jull A. J. T., Bland P. A., Bevan A. W. R., Klandrud S. E., and McHargue L. R. (2001) ^{14}C and ^{14}C-^{10}Be terrestrial ages of meteorites from Western Australia (abstract). *Meteoritics & Planet. Sci., 36,* A91.

Jull A. J. T., Koblitz J., Hofmann B., Franchi I. A., McHargue L. R., and Shahab A. (2002) Terrestrial ages of some meteorites from Oman. *Meteoritics & Planet. Sci., 37,* A74.

Jull A. J. T., Kim K. J., Reedy R. C., McHargue L. R., and Johnson J. A. (2004a) Modeling of ^{14}C and ^{10}Be production rates in meteorites and lunar samples (abstract). In *Lunar and Plane-*

tary Science XXXV, Abstract #1191. Lunar and Planetary Institute, Houston (CD-ROM).

Jull A. J. T., McHargue L. R., Johnson J. A., and Nishiizumi K. (2004b) Terrestrial ^{14}C ages of Yamato meteorites. In *Abstracts for the 28th Symposium on Antarctic Meteorites,* pp. 25–26.

Karner D. B., Levine J., Muller R. A., Asaro F., Ram M., and Stolz M. R. (2003) Extraterrestrial accretion from the GISP2 ice core. *Geochim. Cosmochim. Acta, 67,* 751–763.

Kaye J. H. (1963) Cosmogenic X-ray and β emitters in iron meteorites. Ph.D. thesis, Carnegie Institute of Technology, Pittsburgh.

Kigoshi K. and Matsuda E. (1986) Radiocarbon datings of Yamato meteorites. In *International Workshop on Antarctic Meteorites* (J. O. Annexstad et al., eds.), pp. 58–60. LPI Tech. Rpt. 86-01, Lunar and Planetary Institute, Houston.

Kim K. J. and Reedy R. C. (2003) Terrestrial ages using ratios of cosmogenic radionuclides: Numerical simulations. In *Evolution of Solar System Materials: A New Perspective from Antarctic Meteorites,* p. 51. National Institute of Polar Research, Tokyo.

Knauer M., Neupert U., Michel R., Bonani G., Dittrich-Hannen B., Hajdas I., Ivy-Ochs S., Kubik P. W., and Suter M. (1995) Measurement of the long-lived radionuclides beryllium-10, carbon-14 and aluminum-26 in meteorites from hot and cold deserts by accelerator mass spectrometry (AMS). In *Workshop on Meteorites from Cold and Hot Deserts* (L. Schultz et al., eds.), pp. 38–42. LPI Tech. Rpt. 95-02, Lunar and Planetary Institute, Houston.

Knie K., Merchel S., Korschinek G., Faestermann T., Herpers U., Gloris M, and Michel R. (1999) Accelerator mass spectrometer measurements and model calculations of iron-60 production rates in meteorites. *Meteoritics & Planet. Sci., 34,* 729–734.

Koblitz J. (2003) *Metbase Data Retrieval Software.* Metbase, Fischerhude, Germany.

Kring D. A., Jull A. J. T., and Bland P. A. (1999) The Gold Basin strewn field, Mojave Desert, and its survival from the late Pleistocene to the present. In *Workshop on Extraterrestrial Materials from Cold and Hot Deserts,* pp. 44–45. LPI Contribution No. 997, Lunar and Planetary Institute, Houston.

Kring D. A., Jull A. J. T., McHargue L. R., Bland P. A., Hill D. H., and Berry F. J. (2001) Gold Basin meteorite strewn field, Mojave Desert, northwestern Arizona: Relic of a small late Pleistocene impact event. *Meteoritics & Planet. Sci., 36,* 1057–1066.

Kring D. A., Gleason J. D., Swindle T. D., Nishiizumi K., Caffee M. W., Hill D. H., Jull A. J. T., and Boynton W. V. (2003) Composition of the first bulk melt sample from a volcanic region of Mars: Queen Alexandra Range 94201. *Meteoritics & Planet. Sci., 38,* 1833–1848.

Leya I., Lange H.-J., Neumann S., Wieler R., and Michel R. (2000) The production of cosmogenic nuclides in stony meteoroids by galactic cosmic-ray particles. *Meteoritics & Planet. Sci., 35,* 259–286.

Leya I., Begemann F., Weber H. W., Wieler R., and Michel R. (2004) Simulation of the interaction of galactic cosmic ray protons with meteoroids: On the production of ^3H and light noble gas isotopes in isotropically irradiated thick gabbro and iron targets. *Meteoritics & Planet. Sci., 39,* 367–386.

Lindstrom M. M. and Score R. (1994) Populations, pairing, and rare meteorites in the U.S. Antarctic meteorite collection. In *Workshop on Meteorites from Cold and Hot Deserts* (L. Schultz et al., eds.), pp. 43–45. LPI Tech. Rpt. 95-02, Lunar and Planetary Institute, Houston.

Lipschutz M. (1989) Trace-element variations between Antarctic (Victoria Land) and non-Antarctic meteorites. *Smithson. Contrib. Earth Sci., 28,* 99–102.

Lipschutz M. (1990) Differences between Antarctic and non-Antarctic meteorite populations. In *Workshop on Differences Between Antarctic and Non-Antarctic Meteorites* (C. Koeberl and W. A. Cassidy, ed.), pp. 59–61. LPI Tech. Rpt. 90-01, Lunar and Planetary Institute, Houston.

Love S. G. and Brownlee D. E. (1993) A direct measurement of the terrestrial mass accretion rate of cosmic dust. *Science, 262,* 550–553.

Marty B., Marti K., Barrat J., Birck J. L., Blichert-Toft J., Chaussidon M., Deloule E., Gillet P., Göpel C., Jambon A., Manhès G., and Sautter V. (2001) Noble gases in new SNC meteorites NWA817 and 480 (abstract). *Meteoritics & Planet. Sci., 36,* A122.

Marvin U. B. and Mason B. (1984) Field and laboratory investigations of meteorites from Victoria Land, Antarctica. *Smithson. Contrib. Earth Sci., 26,* 1–4.

Marvin U. B. and MacPherson G. J. (1989) Field and laboratory investigations of meteorites from Victoria Land and the Thiel Mountains, Antarctica, 1982–1983 and 1983–1984. *Smithson. Contrib. Earth Sci., 28,* 1–3.

Marvin U. B. and MacPherson G. J. (1992) Field and laboratory investigations of Antarctic meteorites collected by United States expeditions. *Smithson. Contrib. Earth Sci., 30,* 1–3.

Mason B. (1979) Cosmochemistry, part 1. Meteorites. In *Data of Geochemistry, 6th edition* (M. Fleischer, ed.), U.S. Geological Survey Professional Paper 440-B-1.

McCorkell R. H., Fireman E. L., d'Amico J., and Thompson S. O. (1968) Radiometric isotopes in Hoba West and other iron meteorites. *Meteoritics, 4,* 113–122.

McKay D. S., Gibson E. K. Jr., Thomas-Keprta K. L., Vali H., Romanek C. S., Clemett S. J., Chillier X. D. F., Maechling C. R., and Zare R. N. (1996) Search for past life on Mars: Possible relic biogenic activity in Martian meteorite ALH84001. *Science, 273,* 924–930.

McKeever S. W. S. (1982) Dating meteorite falls using thermoluminescence — Application to Antarctic meteorites. *Earth Planet. Sci. Lett., 58,* 419–429.

Meigs P. (1953) World distribution of arid and semi-arid homoclimates. In *Reviews of Research on Arid Zone Hydrology,* pp. 203–209. United National Educational Scientific and Cultural Organization, Paris.

Miura Y., Nagao K., and Fujitani T. (1993) ^{81}Kr terrestrial ages and grouping of Yamato eucrites based on noble-gas and chemical compositions. *Geochim. Cosmochim. Acta, 57,* 1857–1866.

Michlovich E. S., Wolf S. F., Wang M. S., Vogt S., Elmore D., and Lipschutz M. E. (1995) Chemical studies of H chondrites 5. Temporal variations of sources. *J. Geophys. Res., 100,* 3317–3333.

Neupert U., Michel R., Leya I., Neumann S., Schultz L., Scherer P., Bonani G., Hajdas I., Ivy-Ochs S., Kubik P. W., and Suter M. (1997) Ordinary chondrites from the Açfer region: A study of exposure histories. *Meteoritics & Planet. Sci., 32,* A98–99.

Ninagawa K., Miono S., Yoshida M., and Takaoka N. (1983) Measurement of terrestrial age of Antarctic meteorites by thermoluminescence technique. *Mem. Natl. Inst. Polar Res. Special Issue, 30,* 251–258.

Nishio F. and Annexstad J. O. (1980) Studies on the ice flow in the bare ice area near the Allan Hills in Victoria Land, Antarctica. *Mem. Natl. Inst. Polar Res. Special Issue, 17,* pp. 1–13.

Nishiizumi K. (1995) Terrestrial ages of meteorites from cold and hot deserts. In *Workshop on Meteorites from Cold and Hot Deserts* (L. Schultz et al., eds.), pp. 53–55. LPI Tech. Rpt. 95-02, Lunar and Planetary Institute, Houston.

Nishiizumi K. (2003) Exposure histories of lunar meteorites. In *Evolution of Solar System Materials: A New Perspective from Antarctic Meteorites,* p. 104. National Institute of Polar Research, Tokyo.

Nishiizumi K. and Caffee M. W. (1996) Exposure histories of lunar meteorites Queen Alexandra Range 94281 and 94269 (abstract). In *Lunar and Planetary Science XXVII,* pp. 959–960. Lunar and Planetary Institute, Houston.

Nishiizumi K. and Caffee M. W. (1997) Exposure history of shergottite Yamato 793605. In *Antarctic Meteorites XXII,* pp. 149–151. National Institute of Polar Research, Tokyo.

Nishiizumi K. and Caffee M. W. (1998) Measurement of cosmogenic calcium-41 and calcium-41/chlorine-36 terrestrial ages (abstract). *Meteoritics & Planet. Sci., 33,* A117.

Nishiizumi K. and Caffee M. W. (2001) Exposure histories of lunar meteorites Dhofar 025, 026 and Northwest Africa 482 (abstract). *Meteoritics & Planet. Sci., 36,* A148.

Nishiizumi K. and Hillegonds D. (2004) Exposure and terrestrial histories of new Yamato lunar and Martian meteorites. In *Abstracts for the 28th Symposium on Antarctic Meteorites,* pp. 60–61.

Nishiizumi K., Arnold J. R., Elmore D., Ferraro R. D., Gove H. E., Finkel R. C., Beukens R. P., Chang K. H., and Kilius L. R. (1979) Measurements of ^{36}Cl in Antarctic meteorites and Antarctic ice using a van de Graaff accelerator. *Earth Planet. Sci. Lett., 45,* 285–292.

Nishiizumi K., Arnold J. R., Elmore D., Kubik P. W., Klein J., and Middleton R. (1987) Terrestrial ages of Antarctic and Greenland meteorites. *Meteoritics, 22,* 473.

Nishiizumi K., Reedy R. C., and Arnold J. R. (1988) Exposure history of four lunar meteorites. *Meteoritics, 23,* 294–295.

Nishiizumi K., Elmore D., and Kubik P. W. (1989a) Update on terrestrial ages of Antarctic meteorites. *Earth Planet. Sci. Lett., 93,* 299–313.

Nishiizumi K., Jull A. J. T., Bonani G., Suter M., Wölfli W., Elmore D., Kubik P., and Arnold J. R (1989b) Age of Allan Hills 82102, a meteorite found inside the ice. *Nature, 340,* 550–551.

Nishiizumi K., Arnold J. R., Klein J., Fink D., Middleton R., Kubik P. W., Sharma P., Elmore D., and Reedy R. C (1991a) Exposure histories of lunar meteorites: ALHA81005, MAC88104, MAC88105 and Y791197. *Geochim. Cosmochim. Acta, 55,* 3149–3155.

Nishiizumi K., Kohl C. P., Shoemaker E. M., Arnold J. R., Klein J., Fink D., and Middleton R. (1991b) In-situ ^{10}Be-^{26}Al exposure ages at Meteor Crater, Arizona. *Geochim. Cosmochim. Acta, 55,* 2699–2703.

Nishiizumi K., Arnold J. R., Klein J., Fink D., Middleton R., Sharma P., and Kubik P. (1991c) Cosmic ray exposure history of lunar meteorite Yamato 793274. In *Abstracts for the 16th Symposium on Antarctic Meteorites,* pp. 188–190.

Nishiizumi K., Arnold J. R., Caffee M. W., Finkel R. C., and Reedy R. C. (1992a) Cosmic ray exposure histories of lunar meteorites Asuka-881757, Yamato-793169 and Calcalong Creek. In *Proc. 17th NIPR Symp. Antarctic Meteorites,* pp. 129–132.

Nishiizumi K., Arnold J. R., Caffee M. W., Finkel R. C., and Southon J. (1992b) Exposure histories of Calcalong Creek and LEW 88516 meteorites. *Meteoritics, 27,* 270.

Nishiizumi K., Caffee M. W., and Finkel R. C. (1994) Exposure histories of ALH84001 and ALHA77005. *Meteoritics, 29,* 511.

Nishiizumi K., Caffee M. W., Jull A. J. T., and Reedy R. C. (1996) Exposure history of lunar meteorites Queen Alexandra Range 93069 and 94269. *Meteoritics, 31,* 893–896.

Nishiizumi K., Caffee M. W., Jeannot J.-P., and Lavielle B. A. (1997) Systematic study of the cosmic-ray exposure history of iron meteorites: ^{10}Be-^{36}Cl/^{10}Be terrestrial ages (abstract). *Meteoritics & Planet. Sci., 32,* A100.

Nishiizumi K., Caffee M. W., and Jull A. J. T. (1998) Exposure histories of Dar al Gani 262 lunar meteorites (abstract). In *Lunar and Planetary Science XXIX,* Abstract #1957. Lunar and Planetary Institute, Houston (CD-ROM).

Nishiizumi K., Caffee M. W., and Welten K. C. (1999) Terrestrial ages of Antarctic meteorites — Update 1999. In *Workshop on Extraterrestrial Materials from Cold and Hot Deserts,* p. 64. LPI Contribution No. 997, Lunar and Planetary Institute, Houston.

Nishiizumi K., Caffee M. W., and Masarik J. (2000) Cosmogenic radionuclides in the Los Angeles martian meteorite (abstract). *Meteoritics & Planet. Sci., 35,* A120.

Nishiizumi K., Caffee M. W., Jull A. J. T., and Klandrud S. E. (2001) Exposure history of shergottites Dar al Gani 476/489/670/735 and Sayh al Uhaymir 005 (abstract). In *Lunar and Planetary Science XXXII,* Abstract #2117. Lunar and Planetary Institute, Houston (CD-ROM).

Nishiizumi K., Okazaki R., Park J., Nagao K., Masarik J., and Finkel R. C. (2002) Exposure and terrestrial histories of Dhofar 019 Martian meteorite (abstract). In *Lunar and Planetary Science XXXIII,* Abstract #1355. Lunar and Planetary Institute, Houston.

Nishiizumi K., Hillegonds D. J., McHargue L. R., and Jull A. J. T. (2004) Exposure and terrestrial histories of new lunar and Martian meteorites (abstract). In *Lunar and Planetary Science XXXV,* Abstract #1130. Lunar and Planetary Institute, Houston.

Okazaki R., Keisuki N., Imae N., and Kojima H. (2003) Noble gas signatures of Antarctic nakhlites, Yamato (Y) 000593, Y000749 and Y000802. *Antarct. Meteorite Res., 16,* 58–79. National Institute of Polar Research, Tokyo.

Phillips F. M., Zreda M. G., Smith S. S. , Elmore D., Kubik P. W., Dorn R. I., and Roddy D. J. (1991) Age and geomorphic history of Meteor Crater, Arizona, from cosmogenic ^{36}Cl and ^{14}C in rock varnish. *Geochim. Cosmochim. Acta, 55,* 2695–2698.

Reedy R. C. (1985) A model for GCR-particle fluxes in stony meteorites and production rates of cosmogenic nuclides. *Proc. Lunar Planet. Sci. Conf. 15th,* in *J. Geophys. Res., 90,* C722–C728.

Reedy R. C. (1987) Predicting the production rates of cosmogenic nuclides in extraterrestrial matter. *Nucl. Instr. Meth. Phys. Res. Sec. B, 29,* 251–261.

Reedy R. C. (2000) Predicting the production rates of cosmogenic nuclides. *Nucl. Instr. Meth. Phys. Res. Sec. B, 172,* 782–785.

Reedy R. C. and Arnold J. R. (1972) Interaction of solar and galactic cosmic ray particles with the Moon. *J. Geophys. Res., 77,* 537–555.

Reedy R. C. and Masarik J. (1995) Production profiles of nuclides by galactic-cosmic-ray particles in small meteoroids. In *Workshop on Meteorites from Cold and Hot Deserts* (L. Schultz et al., eds.), pp. 55–57. LPI Tech. Rpt. 95-02, Lunar and Planetary Institute, Houston.

Ruhm W., Schneck B., Knie K., Korschinek G., Zerle L., Nolte E., Weselka D., and Vonach H. A. (1994) New half-life determination of Ni-59. *Planet. Space. Sci., 42,* 227–230.

Scherer P., Schultz L., Neupert U., Knauer M., Neumann S., Leya I., Michel R., Mokos J., Lipschutz M. E., Metzler K., Suter M., and Kubik P. W. (1997) Allan Hills 88019: An Antarctic H-chondrite with a very long terrestrial age. *Meteoritics & Planet. Sci., 32,* 769–773.

Schlüter J., Schultz L., Thiedig F., Al-Mahdi B. O., and Abu Aghreb A. E. (2002) The Dar al Gani meteorite field (Libyan Sahara): Geological setting, pairing of meteorites and recovery density. *Meteoritics & Planet. Sci., 37,* 1079–1093.

Schmitz B., Tassinari M., and Peucker-Ehrenbrink B. (2001) A rain of ordinary chondrite meteorites in the early Ordovician. *Earth Planet. Sci. Lett., 194,* 1–15.

Schnabel C., Pierazzo E., Xue S., Herzog G. F., Masarik J., Cresswell R. G., di Tada M. I., Liu K., and Fifield L. K. (1999a) Shock melting of the Canyon Diablo impactor: Constraints from nickel-59 contents and numerical modeling. *Science, 285,* 85–88.

Schnabel C., Ma P., Herzog G. F., di Tada M. L., Hausladen P. A., and Fifield L. K. (1999b) Terrestrial age of Canyon Diablo meteorites (abstract). *Meteoritics & Planet. Sci., 36,* A184.

Schultz L., Annexstad J. O., and Zolensky M. E., eds. (1995) *Workshop on Meteorites from Cold and Hot Deserts.* LPI Tech. Rpt. 95-02, Lunar and Planetary Institute, Houston. 83 pp.

Scott E. R. D., McKinley S. G., Keil K., and Wilson I. E. (1986) Recovery and classification of thirty new meteorites from Roosevelt County, New Mexico. *Meteoritics, 21,* 303–309.

Shoemaker E. M. (1997) Long-term variations in the impact cratering rate on Earth. In *Meteorites: Flux with Time and Impact Effects* (M. M. Grady et al., eds.), pp. 7–10. Geological Society of London Special Publication 140.

Shoemaker E. M., Shoemaker C. S., Nishiizumi K., Kohl C. P., Arnold J. R., Klein J., Fink D., Middleton R., Kubik P. W., and Sharma P. (1990) Age of Australian meteorite craters — A preliminary report. *Meteoritics, 25,* 409.

Sipiera P. P., Becker M. J., and Kawachi Y. (1987) Classification of twenty-six chondrites from Roosevelt County, New Mexico. *Meteoritics, 22,* 151–155.

Sisterson J. M., Schneider R. J., Jull A. J. T., Donahue D. J., Cloudt S., Beverding A., Englert P. A. J., Castaneda C., Vincent J., and Reedy R. C. (1995) Production cross sections for ^{14}C from elements found in lunar rocks: Implications for cosmic ray studies (abstract). In *Lunar and Planetary Science XXVI,* pp. 1309–1310. Lunar and Planetary Institute, Houston.

Stelzner Th., Heide K., Bischoff A., Weber D., Scherer P., Schultz L., Happel M., Schrön W., Neupert U., Michel R., Clayton R. N., Mayeda T. K., Bonani G., Ivy-Ochs S., and Suter M. (1999) An interdisciplinary study of weathering effects in ordinary chondrites from the Acfer region, Algeria. *Meteoritics & Planet. Sci., 34,* 787–794.

Suess H. and Wänke H. (1962) Radiocarbon content and terrestrial age of 12 stony meteorites and one iron meteorite. *Geochim. Cosmochim. Acta, 26,* 475–480.

Thalmann C., Eugster O., Herzog G. F., Klein J., Krähenbühl U., Vogt S., and Xue S. (1996) History of lunar meteorites Queen Alexandra Range 93069, Asuka 881757 and Yamato 793169 based on noble gas isotopic abundances, radionuclide concentrations and chemical composition. *Meteoritics & Planet. Sci., 31,* 857–866.

Tuniz C., Bird J. R., Fink D., and Herzog G. F. (1998) *Accelerator Mass Spectrometry: Ultrasensitive Analysis for Global Science.* CRC Press, Boca Raton, Florida.

Verish R. S., Rubin A. E., Moore C. B., and Oriti R. A. (1999) Deflation and meteorite exposure on playa lakes in the southwestern United States: Unpaired meteorites at Lucerne Dry Lake, California. In *Workshop on Extraterrestrial Materials from Cold and Hot Deserts,* pp. 74–75. LPI Contribution No. 997, Lunar and Planetary Institute, Houston.

Vogt S., Herzog G. F., and Reedy R. C. (1990) Cosmogenic nuclides in extraterrestrial materials. *Rev. Geophys., 28,* 253–275.

Welten K. C., Alderliesten C., van der Borg K., Lindner L., Loeken T., and Schultz L. (1997) Lewis Cliff 86360: An Antarctic L-chondrite with a terrestrial age of 2.35 million years. *Meteoritics & Planet. Sci., 32,* 775–780.

Welten K. C., Nishiizumi K., and Caffee M. W. (1999a) Cosmogenic radionuclides in hot desert chondrites with low 14C activities: A progress report. In *Workshop on Extraterrestrial Materials from Cold and Hot Deserts,* pp. 88–89. LPI Contribution No. 997, Lunar and Planetary Institute, Houston.

Welten K. C., Linder L., Alderliesten C., and van der Borg K (1999b) Terrestrial ages of ordinary chondrites from the Lewis Cliff stranding area, East Antarctica. *Meteoritics & Planet. Sci., 34,* 559–569.

Welten K. C., Nishiizumi K., Masarik J., Caffee M. W., Jull A. J. T., Klandrud S. E., and Wieler R. (2001) Neutron-capture production of chlorine-36 and calcium-41 in two H-chondrite showers from Frontier Mountain, Antarctica. *Meteoritics & Planet. Sci., 36,* 301–317.

Welten K. C., Caffee M. W., Leya I., Masarik J., Nishiizumi K., and Wieler R. (2003) Noble gases and cosmogenic radionuclides in the Gold Basin L4 chondrite shower: Thermal history, exposure history and pre-atmospheric size. *Meteoritics & Planet. Sci., 38,* 157–173.

Welten K. C., Nishiizumi K., Finkel R. C, Hillegonds D. J., Jull A. J. T., and Schultz L. (2004) Exposure history and terrestrial ages of ordinary chondrites from the Dar al Gani region, Libya. *Meteoritics & Planet. Sci., 39,* 481–498.

Wieler R., Graf Th., Signer P., Vogt S., Herzog G. F., Tuniz C., Fink D., Fifield L. K., Klein J., Middleton R., Jull A. J. T., Pellas P., Masarik J., and Dreibus G. (1996) Exposure history of the Torino meteorite. *Meteoritics & Planet. Sci., 31,* 265–272.

Wlotzka F., Jull A. J. T., and Donahue D. J. (1995) Carbon-14 terrestrial ages of meteorites from Acfer, Algeria. Workshop on meteorites from cold and hot deserts. In *Workshop on Meteorites from Cold and Hot Deserts* (L. Schultz et al., eds.), pp. 72–73. LPI Tech. Rpt. 95-02, Lunar and Planetary Institute, Houston.

Zolensky M. E. (1998) The flux of meteorites to Antarctica. In *Meteorites: Flux with Time and Impact Effects* (M. M. Grady et al., eds.), pp. 93–104. Geological Society of London Special Publication 140.

Zolensky M. E., Wells G. L., and Rendell H. M. (1990) The accumulation rate of meteorite falls at the Earth's surface: The view from Roosevelt County, New Mexico. *Meteoritics, 25,* 11–17.

Zolensky M. E., Rendell H., Wilson I., and Wells G. (1992) The age of the meteorite recovery surfaces of Roosevelt County, New Mexico. *Meteoritics, 27,* 460–462.

Zolensky M. E., Bland P., Brown P., and Halliday I. (2006) Flux of extraterrestrial materials. In *Meteorites and the Early Solar System II* (D. S. Lauretta and H. Y. McSween Jr., eds.), this volume. Univ. of Arizona, Tucson.

Glossary

acapulcoite — fine-grained (0.1–0.2 mm), equigranular primitive achondrites with near-chondritic abundances of olivine, pyroxene, plagioclase, metal, and troilite.

AMS — accelerator mass spectrometry.

accretion disk — a structure formed by material falling into a gravitational source. Conservation of angular momentum requires that, as a large cloud of material collapses inward, any small rotation it may have will increase. Centrifugal force causes the rotating cloud to collapse into a disk, and tidal effects will tend to align this disk's rotation with the rotation of the gravitational source in the middle.

accretion — growth by assimilation of material from the outside, e.g., formation of planets and planetesimals by the accumulation of smaller objects in the primordial nebula.

accretion shock — the dynamic compression and heating of gas and dust resulting from collapse into the nebular disk.

achondrite — a stony meteorite of nonsolar composition. Achondrites are formed by the melting and recrystallization on or within meteorite parent bodies and have distinct textures and mineralogies indicative of igneous processes. They are also known as differentiated stony meteorites.

adiabat — a line on a thermodynamic chart relating the pressure and temperature of a substance that is undergoing a transformation in which no heat is exchanged with its environment.

aerolite — historical term for stony meteorites in the classification system developed by Maskelyne.

AGB star — a cool, luminous, and pulsating unstable red giant star on the asymptotic giant branch of the Hertzsprung-Russell (HR) diagram. Such stars gradually develop an intense wind that removes material from the surface at an increasing rate as the end of its lifecyles approaches. AGB stars are the proposed source of many types of presolar grains.

albite — a common Na-rich feldspar that is the "pivot" mineral of two different feldspar series. It is the end member of the plagioclase feldspar solid-solution series. It is also an end member of the alkali feldspar series.

alcohol — a class of organic compounds containing one or more hydroxyl groups (OH).

aliphatic — a class of saturated or unsaturated carbon compounds in which the carbon atoms are joined in open chains, and that contains no aromatic rings.

alkane — an organic molecule that contains only single carbon-carbon bonds.

alkene — an unsaturated hydrocarbon that contains one or more C = C double bonds.

alkyl — any of a series of univalent groups of the general formula C_nH_{2n+1} derived from aliphatic hydrocarbons.

alkyne — hydrocarbons that have at least one triple bond between two carbon atoms.

ALMA — Atacama Large Millimeter Array.

ambipolar diffusion — a process that occurs in a low-ionization gas in which a magnetic field tied only to the ions slips past the neutral gas because of the weak collisional coupling between the low density ions and the neutral gas. It occurs in the initial stage of star formation whereby clumps of a molecular cloud uncouple from the interstellar magnetic field, which would otherwise resist the further gravitational collapse.

amide — a nitrogenous organic compound, generally containing the $-CONH_2$ radical.

amine — a derivative of ammonia in which hydrocarbon radicals have replaced one or more of the H atoms.

amino acid — one of a group of nitrogenous organic compounds that serve as the structural units of proteins.

amoeboid — having a variable irregular shape.

amphiphilic — a molecule that contains both hydrophobic and hydrophilic groups.

angrite — medium- to coarse-grained achondrite meteorites of basaltic composition that are depleted in alkali elements relative to chondritic composition and contain abundant Ca-Al-Ti-rich pyroxene, Ca-rich olivine, and anorthitic plagioclase.

angular momentum — a quantity obtained by multiplying the mass of an orbiting body by its velocity and the radius of its orbit. According to the conservation laws of physics, the angular momentum of any orbiting body must remain constant at all points in the orbit, i.e., it cannot be created or destroyed.

AOA — amoeboid olivine aggregate.

aromatic — one of a series of organic compounds characterized by a cyclic ring structure.

asteroid — a class of small natural bodies that follow independent orbits about the Sun and that appear as unresolved point sources when present in the inner solar system.

ataxite — a class of iron meteorites with little or no structure visible to the naked eye, containing >16 wt% Ni.

atmophile — one of the geochemical classes of elements. Atmophiles are those elements that tend to be present in planetary bodies as gases, e.g., the inert gases and nitrogen.

aubrite — a class of highly reduced achondrite meteorites that may be related to the enstatite chondrites. They are composed predominantly of enstatite. Albitic plagioclase, diopside, and forsterite are present in minor amounts.

Balbus-Hawley instability — a global magnetorotational instability that occurs in differentially rotating gas disks and results in a turbulent outward flux of angular momentum. This instability results from points on a magnetic field line being forced to corotate. By virtue of their connection, the outermost point "tugs" on the inner point, a process that speeds up the former and slows down the latter. Consequently the loss of energy of the inner point causes it to "fall" inward, while the energy gain of the outer point causes it to find a new equilibrium farther out.

basalt — a dark, fine-grained, mafic extrusive igneous rock composed primarily of plagioclase and pyroxene.

bipolar outflow — a stream of matter in two opposing directions from a central object, usually a star. Bipolar outflows represent significant periods of mass loss in a star's life. They tend to occur during the protostar and pre-main-sequence phase and, again, during the red giant phase just before the production of a planetary nebula.

brachinite — medium- to coarse-grained (0.1–3 mm) olivine-rich achondrite meteorites.

breccia — rock composed of fragments derived from previous generations of rocks, cemented together to form a new lithology.

CAI — calcium-aluminum-rich inclusion.

carbonaceous chondrite — chondritic meteorite that has bulk refractory lithophile abundances, normalized to Si, that are equal or greater that those in the solar photosphere. Despite the name, many carbonaceous chondrites are relatively poor in carbon.

carbon star — a red giant star whose atmosphere contains more carbon than oxygen. The two elements combine in the upper layers of the star, forming carbon monoxide and other carbon compounds. The abundance of carbon is thought to be a product of helium fusion within the star, a process that red giants reach toward the end of their lives.

carbonyl — functional group composed of a carbon atom double-bonded to an oxygen atom.

carboxylic acid — organic acids characterized by the presence of a carboxyl (COOH) group.

catalysis — a chemical reaction involving a material that promotes or increases the rate of the reaction without itself undergoing any permanent chemical change.

cathodoluminescence — light induced in a crystal by the action of an electron beam.

C class — a very common asteroid type in the outer part of the main belt; they typically have flat visible wavelength spectra, and may have surface compositions that are related to some carbonaceous chondrites.

chalcophile — one of the geochemical classes of elements that tend to be present in the sulfide phase, e.g., sulfur, selenium, cadmium, and zinc.

Chandrasekhar mass — the mass above which a white dwarf star would formally shrink to zero radius owing to its inability to counter its own gravity.

chassignite — a very rare type of martian meteorite consisting of olivine with minor amounts of pyroxene, plagioclase, chromite, and sulfide.

chirality — the ability of a chemical substance to exist in two mirror-image forms.

chondrite — originally defined as a meteorite that contained chondrules; now also implies a bulk chemical composition, for all but the most volatile elements, that is not far removed from that of the Sun.

chondrule — approximately spherical assemblages, characteristic of most chondrites, that existed independently prior to incorporation in the meteorite and that show evidence for partial or complete melting.

chromosphere — a normally transparent region lying between the photosphere and the corona of a main-sequence star such as the Sun, and of intermediate temperature.

circumstellar disk — a broad ring of material, orbiting around a star, that may be either a protoplanetary nebula or a regenerated dust disk.

clast — an individual lithic unit, e.g., mineral grain or rock fragment, produced by disintegration of a larger rock unit.

clathrate — a chemical substance consisting of a lattice of one type of molecule trapping and containing a second type of molecule.

clinopyroxene — a mineral of the pyroxene group that crystallizes in the monoclinic system.

CMAS — calcium-magnesium-aluminum-silicon chemical system.

comet — a class of small natural bodies that follow independent orbits about the Sun and that develop an extended coma when present in the inner solar system.

commensurability — a condition in which orbital frequencies are in a ratio of small integers.

condensation — transformation from the gaseous to a solid or liquid phase.

cool-bottom processing — a mixing mechanism that transports material from the cool bottom of the stellar convective envelope of an AGB star down to the hydrogen-burning shell, where nuclear burning can alter its composition, then back up to be mixed back into the convective envelope.

corona — the greatly extended and very hot outermost region of a main-sequence star such as the Sun, sometimes used for a similar region of a planetary atmosphere.

cosmic rays — energetic particles of extraterrestrial origin. The composition includes electrons, protons, γ rays, and atomic nuclei from a large region of the periodic table. The kinetic energies of these particles span over fourteen orders of magnitude.

cosmic-ray exposure age — the period of time during which a meteorite was exposed to cosmic rays, commonly the time between its final reduction in size by impact and its arrival on Earth. More generally, it is the time spent within a few meters of the space environment. Nuclear reactions between the radiation and nuclides in the meteorite produce new nuclides, or associated phenomena such as tracks, whose abundances can be used to estimate the exposure age.

cosmochronology — the science of determining the age of meteorites and their components.

cosmogenic — produced by interaction with cosmic radiation, e.g., ^{21}Ne is a cosmogenic nuclide, produced by spallation reactions.

cotectic — the simultaneous crystallization, as a function of temperature, pressure, and composition, of two or more phases from a single liquid.

Coulomb barrier — the repulsive force between charged particles of the same sign that acts to prevent bombardment by charged particles from penetrating the nucleus.

crater — bowl-shaped hole on a surface made by a volcanic explosion or the impact of a body.

cryptocrystalline — a rock texture in which the individual crystallites are too small to be distinguished by conventional optical microscopy.

cumulate — a plutonic igneous rock composed chiefly of crystals that accumulated by sinking or floating from a magma.

CVD — chemical vapor deposition.

daughter nuclide — a nuclide produced by decay of a radioactive parent. The daughter nuclide may be stable or radioactive.

decay constant — the decay constant multiplied by the number of atoms of a radioactive nuclide equals the instantaneous decay rate. Units are in reciprocal time.

dendrite — a crystal that has grown along certain preferred directions resulting in a pattern resembling the twigs and branches of a tree.

desorption — changing from an adsorbed state on a surface to a gaseous or liquid state.

deuterium — heavy stable hydrogen isotope that has a mass of 2 amu.

devitrification — crystallization of a glass in the solid state.

diagenetic — pertaining to the chemical and physical changes brought about in a sediment after it is buried.

diaplectic glass — a disordered, glass-like substance formed by shock metamorphism of a mineral or mineral assemblage without melting.

diastereomers — stereoisomers that are not mirror-image molecules. Diastereomers, unlike enantiomers, have different physical and chemical characteristics.

dicarboxylic acids — a family of organic compounds with a chemical formula of $HOOC(CH_2)_nCOOH$.

diogenite — a class of coarse-grained orthopyroxene cumulate achondrite meteorites.

EDS — energy dispersive spectroscopy.

Eh — oxidation potential; the difference in potential between an atom or ion and the state in which an electron has been removed to an infinite distance from that atom or ion.

ejecta — materials ejected either from a crater by the action of volcanism or a meteoroid impact, or from a stellar object, such as a supernova, by shock waves.

electron microprobe — an analytical instrument that utilizes a finely focused beam of electrons to excite characteristic X-rays in the sample. These are then analyzed using either a crystal spectrometer or a solid-state, energy-dispersive detector.

enantiomers — nonsuperimposable mirror-image molecules.

endogenic — originating within a planetary or planetesimal object.

endothermic reaction — a reaction that consumes energy.

enstatite — the Mg-rich end member of the orthopyroxene solid-solution series.

enstatite chondrite — highly reduced chondritic meteorites that are composed predominantly of enstatite, metal, and sulfide.

e-process — equilibrium process; a complex set of nucleosynthetic reactions in which photodisintegration of previously synthesized nuclides leads to a population of nuclides that approaches local statistical equilibrium. Believed to be largely responsible for the so-called iron peak.

Epstein drag — a law describing the drag of a medium (e.g., a gas) on an object moving through it. The Epstein drag law applies when the object is smaller than the mean free path of the constituent particles (or molecules) of the medium.

eucrite — a broad class of achondritic meteorite. Both basaltic and cumulate eucrites have been recognized. Basaltic eucrites are pigeonite-plagioclase rocks thought to represent either residual liquids or primary partial melts. Cumulate eucrites are coarse-grained gabbros composed of low-calcium pyroxene and plagioclase and are cumulates from a fractionally crystallizing mafic melt.

euhedral — well formed; used to describe a mineral crystal that is completely or mainly bounded by its own regularly developed crystal faces.

europium anomaly — a situation in which the element europium is fractionated relative to the other rare-Earth elements. It is the result of its divalent nature in reducing environments, causing it to partition preferentially into minerals in which the other rare Earths are incompatible, most notably plagioclase.

eutectic — the lowest temperature at which a mixture of two or more elements can be wholly or partly liquid. Crystallization under these conditions frequently leads to a petrologically distinctive texture.

evaporite — deposit caused from the evaporation of water that causes crystals of several minerals to form.

evaporation — the process by which a liquid changes into a gas.

exogenic — originating externally to a planetary or planetesimal object.

Exor outburst — small, short (~1 yr duration) optical outbursts observed in pre-main-sequence stars.

exothermic reaction — a reaction that liberates heat energy.

exsolution — formation of a second crystalline phase by decomposition of a primary crystalline phase at subsolidus temperature.

extinct radionuclide — a radioactive nuclide with a short half-life (compared with the age of the solar system) that was present when a meteorite or meteoritic component formed but that has now decayed below detection limits. Their one-time presence in the material is indicated by their decay products.

extrasolar — originating outside our solar system; e.g., an extrasolar planet is a planet that orbits a star other than the Sun.

face-centered cubic — a type of crystal structure in which each atom has 12 nearest neighbors.

fall — a meteorite that was seen to fall. Such meteorites are usually recovered soon after fall and are relatively free of terrestrial contamination and weathering effects.

fassaite — a variety of augite that has a very low iron content. It is not recognized as an offical, separate mineral.

fayalite — the iron-rich end member of the olivine solid-solution series.

feeding zone — the region of the protoplanetary disk from which a growing protoplanet can easily acquire material.

feldspar — a group of rock-forming minerals having the general formula $MAl(Al,Si)_3O_8$, where M is typically calcium, sodium, or potassium.

ferromagnesian — containing iron and magnesium.

find — a meteorite that was not seen to fall but was found and recognized subsequently.

fireball — a meteor that is brighter than any planet or star, i.e., brighter than magnitude –4.

Fischer-Tropsch synthesis — production of organic molecules by the hydrogenation of carbon monoxide in the presence of a suitable catalyst.

fission — a radioactive decay process in which a heavy nucleus fragments into two or more pieces of roughly equal mass. The process can be spontaneous or induced by particle bombardment.

fractionation — the physical separation of one phase, element, or isotope from another.

fugacity — a measure of the chemical potential of a gaseous species; it is the equivalent for a nonideal gas of the partial pressure of an ideal gas.

fullerene — a form of carbon having a large molecule consisting of an empty cage of 60 or more carbon atoms.

FUN — fractionation and unknown nuclear effects.

FU Orionis star — a pre-main-sequence star that displays an extreme change in magnitude and spectral type. FU Orionis stars appear to be a stage in the development of T Tauri stars. They gradually brighten by about six magnitudes over several months, during which time matter is ejected, then remain almost steady or slowly decline by a magnitude or two over a period of years.

FU Orionis outbursts — an astronomical phenomenon associated with FU Orionis stars. During an outburst, the accretion disk outshines the central star by factors of 100–1000, and a powerful wind emerges. These outbursts are thought to occur when the mass accretion rate through the circumstellar disk of a young star increases by orders of magnitude.

galactic cosmic rays — high-energy particles that flow into our solar system from far away in the galaxy. GCRs are mostly pieces of atoms — protons, electrons, and atomic nuclei — that have had all the surrounding electrons stripped during their high-speed (almost the speed of light) passage through the galaxy.

gardening — reworking and overturning of a regolith, principally by micrometeoroid bombardment.

gas-retention age — the age of a meteorite as calculated from the abundance of gaseous daughter products.

GEMS — glass with embedded metal and sulfide.

genomict — a breccia in which the clasts have the same class but different petrographic properties.

geobarometry — determination of the pressure at which an assemblage (mineral pairs, minerals and a presumed gas or liquid, or different phases of a given mineral) was formed.

granulite — a rock consisting of interlocking grains of roughly uniform size.

graphite — crystalline form of carbon having a laminar structure.

Hα emission — a particular emission line created by hydrogen that has a wavelength of 656.3 nm, is visible in the red part of the electromagnetic spectrum, and is the easiest way for astronomers to trace the ionized hydrogen content of gas clouds.

half-life — the length of time required for half the atoms in a given sample of a radioactive nuclide to decay.

halite — a chloride mineral with the chemical formula of NaCl.

Hayashi stage — a hypothetical, high-luminosity stage early in the life of a star.

HED — the howardite-eucrite-diogenite clan of meteorites.

HED parent body — the parent asteroid of the HED clan of meteorites, presumed to be the asteroid 4 Vesta based on the spectral similarities between this object and the HED meteorites as well as a proven dynamical mechanism to deliver material from Vesta to Earth-crossing orbits.

Herbig-Haro objects — semistellar, emission-line nebulae that are produced by shock waves in the supersonic outflow of material from young stars; also referred to as Herbig-Haro nebulae.

heterocycle — an organic chemical structure containing noncarbon elements.

H II region — region of ionized hydrogen in interstellar space. H II regions occur near stars with high luminosities and high surface temperatures. Ionized hydrogen, having no electrons, does not produce spectral lines; however, occasionally a free electron will be captured by a free proton and the resulting radiation can be studied optically. H II regions have lifetimes of only a few million years.

hopper crystal — a crystal in which the faces have grown more at the edges than at the center.

hot bottom burning — a nucleosynthetic process that occurs in AGB stars with masses greater than 4× that of the Sun. The bottom of the deep convective envelope penetrates the top of the hydrogen-burning shell, initiating hydrogen-burning nucleosynthesis at the bottom of the convective envelope itself.

howardite — polymict brecciated achondrite consisting predominantly of lithic units similar to eucrites and diogenites, although more extreme compositions are also found.

HR diagram — Hertzsprung-Russell diagram; a plot of bolometric magnitude against effective temperature for a population of stars. Related plots are the color-magnitude plot (absolute or apparent visual magnitude against color index) and the spectrum-magnitude plot (visual magnitude vs. spectral type, the original form of the HR diagram).

HRTEM — high-resolution transmission electron microscopy.

Hugoniot — locus of pressure-volume-energy states achieved by shock compression; frequently represented in the pressure-volume plane.

hydroxy acid — any acid that has hydroxyl groups in addition to the hydroxyl group in the acid itself.

IDP — interplanetary dust particle.

igneous — a rock or mineral that solidified from molten or partly molten material.

imidazole — a heterocyclic organic compound containing two nitrogen atoms within a five-membered ring.

impactite — glassy rocks associated with craters, made by the fusion of local target rock by the heat of impact.

INAA — instrumental neutron activation analysis.

incompatible element — an element that tends to be excluded from a growing crystal (and hence concentrated in the residual magmatic liquid), commonly because its ionic radius is too large to fit comfortably in the crystal lattice.

induction — the process by which a time-varying electromagnetic field causes electrical currents to flow in any object with nonzero conductivity.

initial mass function — the distribution in mass of a newly formed stellar population.

IOM — insoluble organic macromolecules.

ionization — the process by which a neutral atom or molecule acquires a positive or negative charge.

ion microprobe — an analytical instrument in which a finely focused beam of ions ionizes atoms in the sample and ejects them into a mass spectrometer where they may be analyzed.

ion-molecule reaction — a reaction between an ionized atom or molecular fragment and a neutral atom or molecule.

IRAS — Infrared Astronomical Satellite.

iron-wüstite buffer — the curve produced on a phase diagram by plotting the oxygen fugacity in equilibrium with metallic iron and wüstite ($Fe_{0.947}O$) in any proportions as a function of temperature.

irradiation — the process by which an item is exposed to radiation.

ISM — interstellar medium.

ISP — isotope-selective photodestruction.

ISO — Infrared Space Observatory.

isochemical — without change in bulk chemical composition.

isochron — refers to a line of equal age for a sample suite when the radiogenic daughter nuclide is plotted against the radioactive parent nuclide. The two nuclides are usually normalized to a nonradiogenic nuclide.

isomer — one of a number of molecules that all have the same elemental composition but that differ from each other in structure.

Jeans escape — escape of the fastest atoms in a Maxwell-Boltzmann distribution when their speed exceeds the escape velocity.

kamacite — Fe,Ni alloy of 7 wt% Ni or less with the body-centered-cubic structure. It occurs as large plates or single crystals in iron meteorites, abundant grains in chondrites, and rare grains in most achondrites.

KBO — Kuiper belt object.

Kepler velocity — the orbital velocity of a gravitationally bound object around the central object, i.e., the velocity that leads to a centrifugal force exactly balancing the gravitational attraction between the two objects.

kerogen — insoluble macromolecular organic matter, operationally defined as the organic residue left after acid demineralization of a rock.

keto acid — an organic acid containing a ketone group along with the acid groups.

kinetic isotope effect — changing of an element's isotopic proportions as a result of differing reaction rates for each isotope.

Kuiper belt — A region in our outer solar system where many short-period comets (possessing orbits of less than 200 yr) originate. This region begins beyond Neptune's orbit and encompasses an estimated distance of between 30 and 100 astronomical units (AU).

lodranite — coarse-grained (0.5–0.7 mm) primitive achondrites that contain abundant olivine and pyroxene and are depleted in troilite and plagioclase, relative to chondrites.

Le Chatelier's principle — if a system in equilibrium is altered by an outside force, the system will tend to minimize the effect of the force by changes in itself.

lightcurve — brightness values plotted as a function of time.

liquidus — the line or surface in a phase diagram above which the system is completely liquid.

lithology — the physical character of a rock.

lithophile — one of the geochemical classes of elements. Lithophile elements are those that tend to concentrate in the silicate phase, e.g., silicon, magnesium, calcium, aluminum, sodium, potassium, and the rare-Earth elements.

lithostatic pressure — pressure due to the weight of overlying rock.

luminosity — a measure of the energy output of a star.

Lyman lines — spectroscopic features due to hydrogen atoms.

mafic — term used to describe a silicate mineral whose cations are predominantly magnesium and/or iron. It is also used for rocks made up principally of such minerals.

magma ocean — a hypothetical stage in the evolution of a planetary object during which virtually the entire surface of the object is covered with molten lava.

magmatic — associated with molten silicate material.

magnetohydrodynamics — the study of the behavior of an electrically conducting fluid in the presence of a magnetic field.

MAI — mineralogic alteration index.

martensite — distorted body-centered cubic structure formed in Fe,Ni-alloys, originally in the γ field, as they cool rapidly.

maskelynite — glass formed from plagioclase feldspar by disordering during intense shock.

mass fractionation — a process that causes the proportions of the isotopes or elements to change in a manner that is dependent on the differences in their mass.

mass spectrometer — an analytical instrument in which the sample is converted into a beam of ions that can be separated from each other on the basis of their mass-to-charge ratio, generally by a magnetic or electrostatic field, so that the relative proportions of entities of different mass, commonly isotopes, can be determined.

matrix — the fine-grained material that occupies the space in a rock, such as a meteorite, between the larger, well-characterized components such as chondrules, inclusions, etc.

M class — a fairly common asteroid type in the main belt with featureless visible wavelength spectra distinguished by moderate albedos. They are presumed to have metallic (nickel-iron) compositions and may be generally related to iron meteorites.

mesosiderite — class of stony-iron meteorite consisting of subequal proportions of silicate material (related to eucrites and diogenites) and iron-nickel metal.

mesostasis — interstitial, generally fine-grained material occupying the space between larger mineral grains in an igneous rock; generally, therefore, the last material to solidify from a melt.

metallicity — the proportion of matter in a star that is made up of chemical elements other than hydrogen and helium.

metamorphism — solid-state modification of a rock, e.g., recrystallization, caused by elevated temperature and pressure.

metasomatism — selected addition and removal of chemical elements by hot, briny fluids that soak rocks during metamorphism.

metastable — said of an energy state, or the material in that state, that is characterized by a potential-energy minimum that is not, however, the ground state of the system.

meteor — the light phenomenon produced by a meteoroid experiencing frictional heating when entering a planetary atmosphere; also used for the glowing meteoroid itself. If particularly large, it is described as a fireball.

meteorite — a natural object of extraterrestrial origin that survives passage through a planetary atmosphere.

meteoroid — a natural, small (subkilometer) object in an independent orbit in the solar system.

micrometeorite — a small extraterrestrial particle that has survived entry into Earth's atmosphere.

Miller-Urey synthesis — formation of organic molecules by the passage of an electric discharge, or energetic radiation, through a mixture of methane and ammonia over refluxing water. It is often used for analogous experiments.

model age — a radiometric age determination based on parameters at least one of which is assumed rather than measured.

molecular cloud — a cold, dense interstellar cloud that contains a high fraction of molecules. It is widely believed that the relatively high density of dust particles in these clouds plays an important role in the formation and protection of the molecules.

monoclinic — a crystal system characterized by either a single twofold symmetry axis, a single symmetry plane, or a combination of the two.

Monte Carlo simulation — a computational procedure in which random numbers are used to approximate the solution to otherwise intractable mathematical or physical problems.

monomict — description of a breccia in which the matrix and clasts are of the same class and type.

MRI — magnetorotational instability.

nakhlite — a type of martian meteorite consisting of calcic pyroxene (augite) and olivine.

Neumann lines — lines in kamacite, visible upon etching a polished face with mild acid, caused by mechanical twinning following mild shock.

neutron star — a star whose core is composed primarily of neutrons, as is expected to occur when the mean density is in the range 10^{13}–10^{15} g/cm^3. Under current theories pulsars are thought to be rotating magnetic neutron stars.

nicotinic acid — also known as niacin, a water-soluble vitamin whose derivatives such as NADH play essential roles in energy metabolism in the living cell and DNA repair.

noble gas — the gases helium, argon, krypton, neon, xenon, and radon that rarely undergo chemical reactions, also known as inert gases and rare gases.

norite — a type of igneous rock containing plagioclase and in which the pyroxene is mainly of the orthorhombic form, rather than monoclinic.

nova — a star that exhibits a sudden surge of energy, temporarily increasing its luminosity by as much as 17 magnitudes or more (although 12 to 14 magnitudes is typical). Novae are old disk-population stars, and are all close binaries with one component, a main-sequence star filling its Roche lobe, and the other component a white dwarf. Unlike supernovae, novae retain their stellar form and most of their substance after the outburst. Since 1925, novae have been given variable-star designations.

NSE — nuclear statistical equilibrium.

nucleosynthesis — the process of creating new atomic nuclei either by nuclear fusion or nuclear fission.
obliquity — the angle between an object's axis of rotation and the pole of its orbit.
Occam's Razor — an axiom enunciated by William of Ockham (?–1348) of which one translation is "It is vain to do with more what may be done with less."
octahedrite — a type of iron meteorite consisting of a mixture of taenite and kamacite in a characteristic Widmanstätten pattern, revealed by mild etching, that reflects the octahedral symmetry of the taenite.
olefin — alternate term for alkene.
olivine — a magnesium iron silicate with the formula $(Mg,Fe)_2SiO_4$ in which the ratio of magnesium and iron varies between the two end members of the series: forsterite (magnesium-rich) and fayalite (iron-rich). It gives its name to the group of minerals with a related structure (the olivine group) that includes monticellite and kirschsteinite.
onion-shell — a structural model in which varying layers in a spherical body are arranged concentrically.
Oort cloud — a spherical cloud of comets having semimajor axes> 20,000 AU found by J. H. Oort in his empirical study of the orbits of long-period comets. Comets in this shell can be sufficiently perturbed by passing stars or giant molecular clouds so that a fraction of them acquire orbits that take them within the orbits of Jupiter and Saturn.
ophitic — a texture found in many igneous rocks, characterized by laths of plagioclase partially or completely included with crystals of pyroxene.
orbital resonance — occurs when two orbiting bodies have periods of revolution that are in a simple integer ratio so that they exert a regular gravitational influence on each other.
ordinary chondrite — collective name for the most common variety of chondritic meteorite, subdivided into H, L, and LL groups on the basis of iron content and distribution.
orthopyroxene — a mineral of the pyroxene group that crystallizes in the orthorhombic form.
orthorhombic — a crystal system characterized by three mutually perpendicular twofold symmetry axes.
oxidation — the addition of oxygen, removal of hydrogen, or removal of electrons from an element or compound.
oxidation state — the charge on an element if it is assumed that it is behaving as an ion.
oxygen fugacity — a function expressing the molar free energy of oxygen in a manner analogous to the way pressure measures free energy of an ideal gas. In practice, oxygen fugacity is equivalent to the partial pressure of oxygen.
PAH — polycyclic aromatic hydrocarbon.
paired meteorites — specimens originally recovered some distance apart and hence given separate names, but later recognized as fragments of a single parent mass, on the basis of classification, cosmic-ray or gas-retention age, texture, or other diagnostic features.
pallasite — class of stony-iron meteorites in which the iron-nickel metal forms a continuous framework enclosing nodules of olivine.
parent body — the object on or in which a given meteorite or class of meteorites was located prior to ejection as approximately meter-sized objects.
parent nuclide — an unstable nuclide that changes spontaneously into another (daughter) nuclide.
parsec — measure of distance used in astronomy. One parsec is the distance at which 1 AU subtends an angle of 1 arcsecond, equivalent to 3.26 light years.
PCP — poorly characterized phase, the term initially applied to an ill-defined mineral found in certain carbonaceous chondrites. It is now known to be either tochilinite or a tochilinite-phyllosilicate intergrowth.
PDB — Pee Dee belemnite, fossil carbonate from the Cretaceous Pee Dee formation in South Carolina. The isotopic composition of its carbon is used as the international standard for reporting $^{13}C/^{12}C$ ratios.
PDF — planar deformation feature.
peridotite — a coarse-grained, olivine-rich plutonic rock.
perihelion — the point in the orbit of an object gravitationally bound to the Sun at which it is nearest to the Sun.
peritectic — a situation in which a phase crystallizes from a mixture of a liquid plus a previously crystallized solid phase, with the newly forming solid growing at the expense of the earlier one.
phenocryst — a relatively large, and therefore conspicuous, grain in a porphyritic rock.
photochemistry — the study of the chemical and physical changes occurring when a molecule or atom absorbs light.
photolysis — chemical decomposition brought about by the action of light.
phosphonic acid — an oxoacid of phosphorus. More generally, an organic group bonded to phosphorus.
photosphere — the normally visible region of a main-sequence star such as the Sun. Most solar spectroscopic abundance data have been obtained for this region.
phyllosilicate — one of a family of silicate minerals characterized by a structure that consists of sheets or layers, invariably hydrated.
plagioclase — an important group of igneous rock forming tectosilicate minerals. Plagioclase comprises a solid-solution series: $NaAlSi_3O_8$ (albite)–$CaAl_2Si_2O_8$ (anorthite).
planetary nebula — an expanding envelope of rarefied ionized gas surrounding a hot white dwarf. The envelope receives ultraviolet radiation from the central star and reemits it as visible light by the process of fluorescence. During the core contraction that terminates the red giant stage, a shell of material is ejected and becomes separated from the core.
planetesimal — small rocky or icy body formed in the primordial solar nebula.
plasmasphere — a region in the upper atmosphere of a

planet in which most of the constituents are ionized.

plessite — a fine-grained intergrowth of kamacite and taenite.

PMD — percent mean deviation, a statistical term commonly used as a measure of how far the silicate minerals in a chondrite are from chemical equilibrium.

poikilitic — a rock texture in which many small euhdral mineral grains are contained within a single, generally anhedral, mineral grain.

poikiloblastic — a rock texture in which relatively large recrystallized grains surround relicts of the original minerals.

polymer — a large molecule made by polymerization.

polymerization — joining molecules together to form a larger, macromolecular complex. Originally the joined units (monomers) were identical to each other; now the term commonly includes heteromolecular complexes.

polymict — description of a breccia in which the clasts and/or matrix have differing composition.

population I stars — young stars with relatively high abundances of metals that occur in the disk of a galaxy, especially the spiral arms, in dense regions of interstellar gas.

population II stars — older stars with low abundances of metals that are typically found in the nuclear bulge of a galaxy or in globular clusters.

porphyritic — a texture found in many igneous rocks, characterized by relatively large crystals, known as phenocrysts, set in a fine-grained or glassy matrix.

Poynting-Robertson effect — an effect of radiation on a small particle orbiting the Sun that causes it to spiral slowly toward the Sun. It occurs because the orbiting particle absorbs energy and momentum streaming radially outward from the Sun, but reradiates energy isotropically in its own frame of reference.

ppm — parts per million, generally by weight.

polyol — an alcohol containing more than two hydroxyl groups.

p-process — the name of the hypothetical nucleosynthetic process thought to be responsible for the synthesis of the rare heavy proton-rich nuclei that are bypassed by the *r*- and *s*-processes. It is manifestly less efficient (and therefore rarer) than the *s*- or *r*-process because the protons must overcome the Coulomb barrier, and may in fact work as a secondary process on the *r*- and *s*-process nuclei. It seems to involve primarily (p, γ) reactions at masses above cerium (where neutron separation energies are low).

Prairie Network — a network of 16 cameras that provided photometric and astrometric data for bright meteors, ranging in mass from a few grams up to hundreds of kilograms. Because of their spectacular appearance, these objects are termed "fireballs." This network, operated by the Smithsonian Astrophysical Observatory, surveyed the fireball flux over an area of 10^6 km^2 in the central U.S. for a period of about 10 yr. Several thousand fireballs were photographed, and reduced data are available for about 335 objects.

prebiotic — the state of matter before life existed, but conducive to the formation of life.

presolar — said of objects that are assumed to be older than the solar system.

protostar — the period in the evolution of a star after the cloud of hydrogen, helium, and dust has started contraction and before the star has reached the main sequence on the HR diagram.

protostellar disk — an accretion disk surrounding a protostar.

purine — a heterocyclic compound containing fused pyrimidine and imidazole rings.

pyrimidine — a heterocyclic organic compound containing two nitrogen atoms within the six-membered ring.

pyrolysis — heating and decomposition under controlled conditions, typically in a vacuum or an inert atmosphere.

pyroxene — a group of rock-forming silicate minerals that share a common structure comprised of single chains of silica tetrahedra. They crystallize in the monoclinic and orthorhombic system.

Q class — a rare asteroid classification denoted by moderate albedos and spectra with a strong absorption features near 1 and 2 μm, indicative of olivine and pyroxene compositions. Their spectra are interpreted as being similar to ordinary chondrites.

Q-gases — a noble-gas component in carbonaceous chondrites that is heavily fractionated relative to solar, favoring the heavy elements.

QSE — quasistatistical equilibrium.

quantum tunelling — penetration by a particle into a potential energy region that is classically forbidden.

racemic — consisting of equal molar proportions of both left- and right-handed optically active forms.

radiogenic — that which is made by radioactive decay.

radiogenic nuclide — a nuclide produced by decay of a radioactive parent nuclide, e.g., ^{206}Pb produced from the decay of ^{238}U.

radiometric — referring to any dating technique based on the measurement of radioactive decay.

radionuclide — a nuclide that is unstable against radioactive decay.

Raman spectroscopy — analysis of the characteristic spectrum produced when monochromatic light is scattered by a transparent substance.

Rankine-Hugoniot equations — a set of equations that govern the behavior of shock waves normal to the oncoming flow and considering the conservation of mass, momentum, and energy.

Rayleigh criterion — a criterion for the stability of inviscid circular flow between concentric cylinders. It has been summarized as "the varying centrifugal force of the different layers of fluid plays the part of gravity and the resulting condition for stability is that the square of the circulation must increase continuously in passing from the inner to the outer cylinder, just as the density of a fluid under gravity must decrease continuously with height in order that it may be in stable equilibrium."

Rayleigh fractionation — an open-system process that involves the progressive removal of a fractional increment of a trace substance or isotope from a larger reservoir. A consistent relationship, such as a distribution coefficient, equilibrium constant, or fractionation factor, is maintained between the reservoir and each increment at the instant of its formation, but, once formed, each increment is thereafter removed or otherwise isolated from the system.

Rayleigh-Taylor instability — a type of fluid instability that occurs any time a dense, heavy fluid is being accelerated by light fluid. Rayleigh-Taylor instabilities are common in supernova remnants, such as the Crab Nebula, in which hot gas from the explosion is ramming into the surrounding interstellar medium, and they give rise to the familiar clumpy appearance of material in these objects.

reaction cross section — proportionality constant that relates the abundance of a target nucleus to the rate at which a given nuclear reaction occurs.

red giant — a late-type (K or M) high-luminosity (brighter than $M_v = 0$) star of very large radius that occupies the upper-right portion of the HR diagram. Red giants are post-main-sequence stars that have exhausted the nuclear fuel in their cores, and whose luminosity is supported by energy production in a hydrogen-burning shell. Within the lifetime of the galaxy, only main-sequence stars of type F and earlier have had time to evolve to the red giant phase (or beyond). The red giant phase corresponds to the establishment of a deep convective envelope.

reduction — process involving the addition of hydrogen, the removal of oxygen, or decreasing the valence of an element; the opposite of oxidation.

REE — rare-Earth elements, the lanthanide series in the periodic table.

refractory — term describing the high-temperature stability of an element or phase, the opposite of volatile.

regolith — the fragmented layer found on the surface of many planetary or subplanetary objects. It is created from the local competent lithologies by meteoroid impact and subsequently comminuted and turned over by such impacts.

remainder ratio — the ratio of the true abundance of a radioactive nuclide to the abundance it would have if it were stable.

resonance — a dynamical configuration of bodies in which Newtonian gravitation maintains commensurability.

Reynolds number — the ratio of inertial forces to viscous forces that is used for determining whether a flow will be laminar or turbulent. It is the most important dimensionless number in fluid dynamics and provides a criterion for determining dynamic similitude.

RNAA — acronym for radiochemical neutron activation analysis, an analytical technique for determining trace and ultratrace elements in meteorites.

Rossby waves — large-scale motions in a fluid, such as an ocean or atmosphere, whose restoring force is the variation in the Coriolis effect with latitude. Rossby waves have been proposed to occur in the solar nebula as a result of Keplerian rotation coupled with a source of vortices.

r-process — the capture of neutrons on a very rapid timescale (i.e., one in which a nucleus can absorb neutrons in rapid succession, so that regions of great nuclear instability are bridged), a theory advanced to account for the existence of all elements heavier than bismuth as well as the neutron-rich isotopes heavier than iron. The essential feature of the *r*-process is the production and consumption of great numbers of neutrons in a very short time (<100 s). The presumed source for such a large flux of neutrons is a supernova, at the boundary between the neutron star and the ejected material.

rubble-pile — a hypothetical structural model for planetesimals, asteroids, and comets in which different, solid components are arranged in random fashion.

S class — a very common asteroid class in the inner main belt with moderate albedos and reddish-sloped spectra with moderate absorption features at 1 and 2 μm, indicative of olivine and pyroxene compositions. The association of this common asteroid class to ordinary chondrite meteorites has been long debated.

SCR — solar cosmic ray.

SEM — scanning electron microscope.

SETI — the search for extraterrestrial intelligence.

shergottite — a type of martian meteorite, consisting of pyroxene (pigeonite) and maskelynite.

shock — momentary excursion to high pressures and temperatures caused by impact, accretion, or some other dynamic event.

shock lithification — welding of regolith by shock-induced intergranular melting.

shock metamorphism — alteration of a rock by shock-induced mechanical deformation or phase transformation above or below the solidus.

shock wave — discontinuity in temperature and pressure propagating in a solid, liquid, or gas with supersonic velocity, caused by impact or explosion.

siderites — historical term for iron meteorites in the classification system developed by Maskelyne.

siderolites — historical term for stony-iron meteorites in the classification system developed by Maskelyne.

siderophile — one of the geochemical classes of elements. Siderophile elements are those that tend to go into the metal phase, e.g., nickel, cobalt, gold, arsenic, germanium, gallium, iridium, osmium, and rhenium.

SIMS — secondary ion mass spectrometry.

SMA — Submillimeter Array.

SMOW — standard mean ocean water, the reference standard for the measurement of oxygen isotopes.

SOHO — Solar and Heliospheric Observatory.

solar flare — sudden outburst of energy on the solar photosphere.

solar nebula — the disk of gas and dust around the proto-Sun from which the solar system formed.

solar wind — expansion of the solar corona to form a stream of ions away from the Sun.

solid solution — a substance in which two or more components, such as the atoms of two or more elements, are randomly mixed on so fine a scale that the resulting solid is homogeneous.

solidus — the line or surface in a phase diagram below which the system is completely solid.

solvus — the curved line in the phase diagram of a binary system separating a field of homogeneous solid solution from a field of two or more phases.

space weathering — a term describing possible processes through which spectral properties of asteroid surfaces may be modified in a manner that may wholly or partially disguise their meteorite association. These processes include micrometeorite impact and reworking, implantation of solar wind and flare particles, radiation damage and chemical effects from solar particles and cosmic rays, and sputtering erosion and deposition.

spallation — a nuclear reaction in which a bombarded nucleus breaks up into many particles. This process accounts for the relatively high abundance of beryllium, lithium, and boron in cosmic rays.

spallogenic — produced by the violent partial disintegration of an atomic nucleus. In a meteoritic context, this is generally due to cosmic-ray bombardment.

s-process — a process in which heavy, stable, neutron-rich nuclei are synthesized from iron-peak elements by successive captures of free neutrons in a weak neutron flux, so that there is time for β decay before another neutron is captured (cf. *r*-process). This is a process of nucleosynthesis that is believed to take place in intershell regions during the red giant phase of evolution.

sputtering — expulsion of atoms or ions from a solid, caused by impact of energetic particles.

STEM — scanning transmission electron microscopy.

Stokes drag — viscous drag law stating that the force that retards a body moving through a fluid is directly proportional to the velocity, the radius of the body, and the viscosity of the fluid. Stokes drag applies for low Reynolds number and if the mean free path of fluid molecules is small compared with the body's size (*see* Epstein drag).

Strecker-cyanohydrin synthesis — the synthesis of amino acids from aldehydes, HCN, and NH_3 in the presence of an aqueous fluid.

sublimation — the transition of a substance from the solid phase directly to the vapor phase, or vice versa, without passing through an intermediate liquid phase.

suevite — a fragmental breccia containing melt particles of the same composition formed in a single impact event.

sulfonic acid — an organic compound containing the functional group $-SO_2OH$, which consists of a sulfur atom, S, bonded to a carbon atom that may be part of a large aliphatic or aromatic hydrocarbon, and also bonded to three oxygen atoms, O, one of which has a hydrogen atom, H, attached to it.

supernova, Type I — in a star about to become a Type I supernova, the star's hydrogen is exhausted, and the star's gravity pulling inward overcomes the forces of the thermonuclear reactions pushing the material outward. As the core begins to contract, the remaining hydrogen ignites in a shell, swelling the star into a giant and beginning the process of helium burning. Eventually the star is left with a still-contracting core of carbon and oxygen. If the star, now a white dwarf, has a nearby stellar companion, it will begin to pull matter from the companion. In many stars the excess matter is blown off periodically as a nova; if it is not, the star continues to get more and more massive until the matter in the core begins to contract again. When the star gets so massive that it passes Chandrasekhar's limit, it collapses very quickly and all its matter explodes, forming a Type I supernova.

supernova, Type II — Type II supernovas involve massive stars that burn their gases out within a few million years. If the star is massive enough, it will continue to undergo nucleosynthesis after the core has turned to helium and then to carbon. Heavier elements such as phosphorus, aluminum, and sulfur are created in shorter and shorter periods of time until silicon results. It takes less than a day for the silicon to fuse into iron; the iron core gets hotter and hotter and in less than a second the core collapses. Electrons are forced into the nuclei of their atoms, forming neutrons and neutrinos, and the star explodes, throwing as much as 90% of its material into space at speeds exceeding 30,000 km/s. After the supernova explosion, there remains a small, hot neutron star, possibly visible as a pulsar, surrounded by an expanding cloud.

taenite — iron with a face-centered cubic structure that is stable at high temperatures and/or when alloyed with a suitable proportion of a face-centered metal such as nickel.

TDU — third dredge-up.

TEM — transmission electron microscope.

tektite — glass objects, up to a few centimeters in size, that are formed by the impact of large meteorites on Earth's surface. They are totally melted into spherical, elongated, teardrop, or twisted shapes and cooled before returning to Earth.

terrestrial age — the period of time since the fall of a meteorite.

thermoluminescence (TL) — an emission of light brought about by heating certain natural materials.

T Tauri stars — a very young, low-mass star, less than 10 m.y. old and under 3 $M_\odot$, that is still undergoing gravitational contraction. T Tauri stars represent an intermediate stage between a protostar and a low-mass main-sequence star like the Sun. The prototype for this class of stars is T Tau.

turbulence — a flow regime characterized by low-momentum diffusion, high-momentum convection, and rapid variation of pressure and velocity in space and time. Flow that is not turbulent is called laminar flow. The Reynolds number characterizes whether flow conditions lead to laminar or turbulent flow.

turbostratic — partially ordered polymer-like structure.

ultramafic — an igneous rock consisting predominantly of mafic silicate minerals.

ureilite — olivine-pyroxene-bearing primitive achondrite meteorites with interstitial carbon (in the form of graphite and microdiamonds) mixed with minor metal, sulfides, and silicates.

van der Waals force — an attractive force between two atoms, due to a fluctuating dipole moment in one molecule inducing a dipole moment in the other molecule, that then interact.

V class — a rare asteroid classification exemplified by 4 Vesta. Spectra show very strong 1- and 2-μm absorption bands interpreted to be due to pyroxene-rich mineralogies. The spectra are interpreted as being similar to HED meteorites.

vesicle — bubble-shaped cavity in a mineral or rock, generally produced by expansion of gas in a magma.

viscosity — the measure of a material's resistance to flow. Viscosity is a result of the internal friction of the material's molecules. Materials with a high viscosity do not flow readily; materials with a low viscosity are more fluid.

volatile — an element that condenses from a gas or evaporates from a solid at a relatively low temperature.

VSMOW — Vienna standard mean ocean water.

vug — a vesicle.

white dwarf — a dim, dense, planet-sized star that marks the evolutionary endpoint for all but the most massive stars. White dwarfs form from the collapse of stellar cores in which nuclear fusion has stopped, and are exposed to space following the loss of the star's outer envelope, typically as a planetary nebula. White dwarfs consist of electron degenerate matter, which provides the pressure needed to prevent further collapse, provided that the mass of the dwarf does not exceed the Chandrasekhar mass.

Widmanstätten pattern — structure in iron meteorites, revealed by etching with mild acid, in which large plates of octahedrally arranged kamacite enclose small fields of taenite.

winonaite — fine- to medium-grained equigranular, primitive achondrite meteorites. Winonaites have roughly chondritic mineralogy and chemical composition but achondritic, recrystallized textures.

Wolf-Rayet star — a hot (25,000 to 50,000 K), massive (more than 25 $M_{\odot}$), luminous star in an advanced stage of evolution that is losing mass in the form of a powerful stellar wind. Wolf-Rayets are believed to be oxygen stars that have lost their hydrogen envelopes, leaving their helium cores exposed.

xenolith — fragment in a rock or meteorite that is foreign to its host.

XRF — X-ray fluorescence.

X-wind — a theoretical construct in which coronal winds from a central star and the inner edge of its accompanying accretion disk join together to form magnetized jets of gas capable of launching material above the disk at velocities of hundreds of kilometers per second.

Yarkovsky effect — a nongravitational thrust produced when small bodies absorb sunlight, heat up, and then reradiate the energy after a short delay produced by thermal inertia that produces a slow but steady drift in the semimajor axis of an object's orbit.

zodiacal light — a faint glow that extends away from the Sun in the ecliptic plane of the sky, visible to the naked eye in the western sky shortly after sunset or in the eastern sky shortly before sunrise. The zodiacal light contributes about a third of the total light in the sky on a moonless night.

Color Section

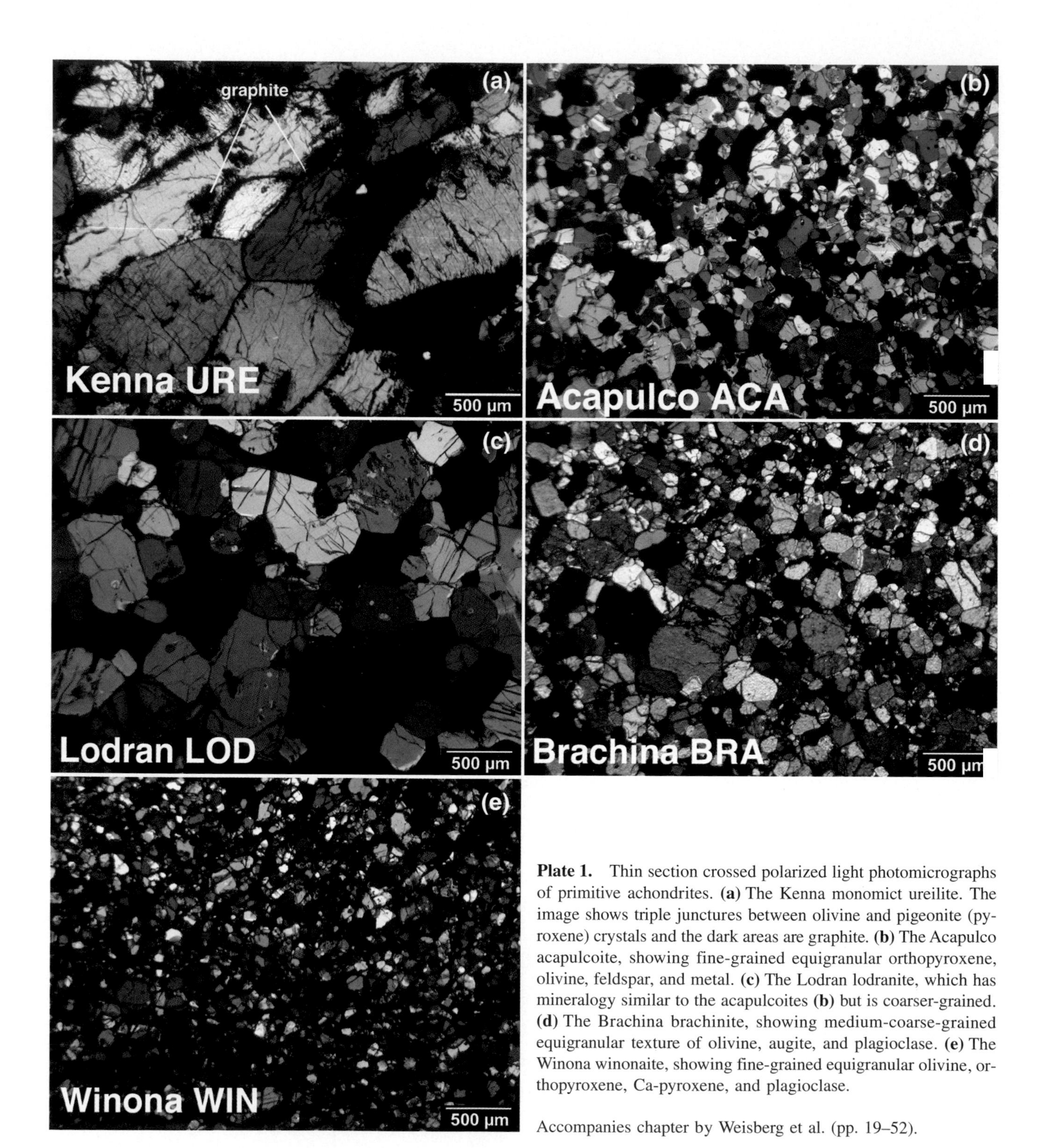

Plate 1. Thin section crossed polarized light photomicrographs of primitive achondrites. **(a)** The Kenna monomict ureilite. The image shows triple junctures between olivine and pigeonite (pyroxene) crystals and the dark areas are graphite. **(b)** The Acapulco acapulcoite, showing fine-grained equigranular orthopyroxene, olivine, feldspar, and metal. **(c)** The Lodran lodranite, which has mineralogy similar to the acapulcoites **(b)** but is coarser-grained. **(d)** The Brachina brachinite, showing medium-coarse-grained equigranular texture of olivine, augite, and plagioclase. **(e)** The Winona winonaite, showing fine-grained equigranular olivine, orthopyroxene, Ca-pyroxene, and plagioclase.

Accompanies chapter by Weisberg et al. (pp. 19–52).

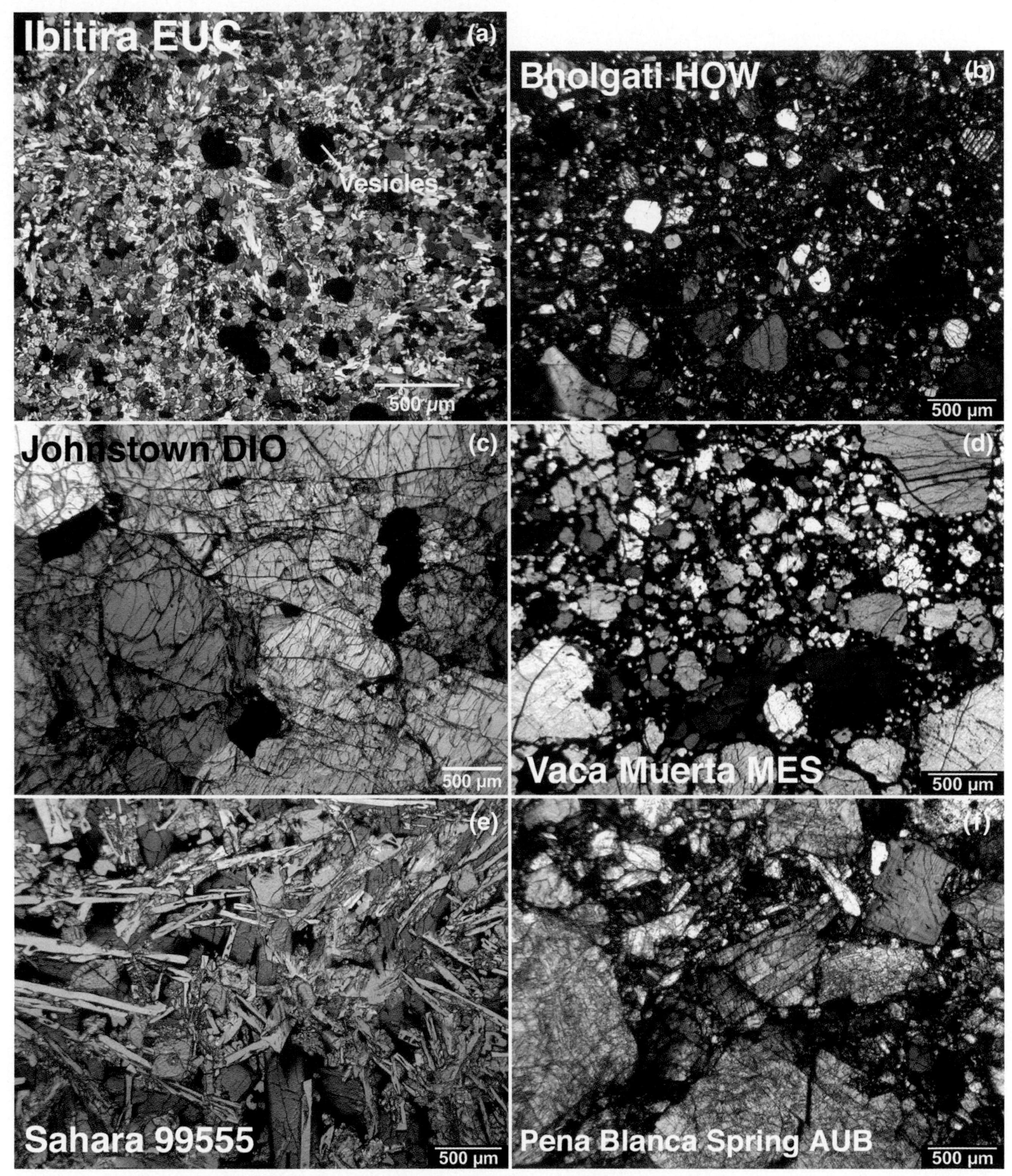

Plate 2. Thin section crossed polarized light photomicrographs of achondrites **(a)** Ibitira showing a eucrite, which is a vesicular basalt; the large dark areas in the image are vesicles; **(b)** the Bholgati howardite, which is a breccia of eucritic and diogenitic clasts; **(c)** the Johnstown diogenite, which consists mostly of orthopyroxene; **(d)** a region from the Vaca Muerta mesosiderite, showing a mixture of silicate and metal; **(e)** the Sahara 99555 angrite (in plane polarized light), showing an intergrowth of zoned Ca-Ti-rich pyroxene, Ca-rich olivine, and anorthite; **(f)** the Pena Blanca Spring aubrite, which is a breccia composed mostly of enstatite and enstatite-rich clasts.

Accompanies chapter by Weisberg et al. (pp. 19–52).

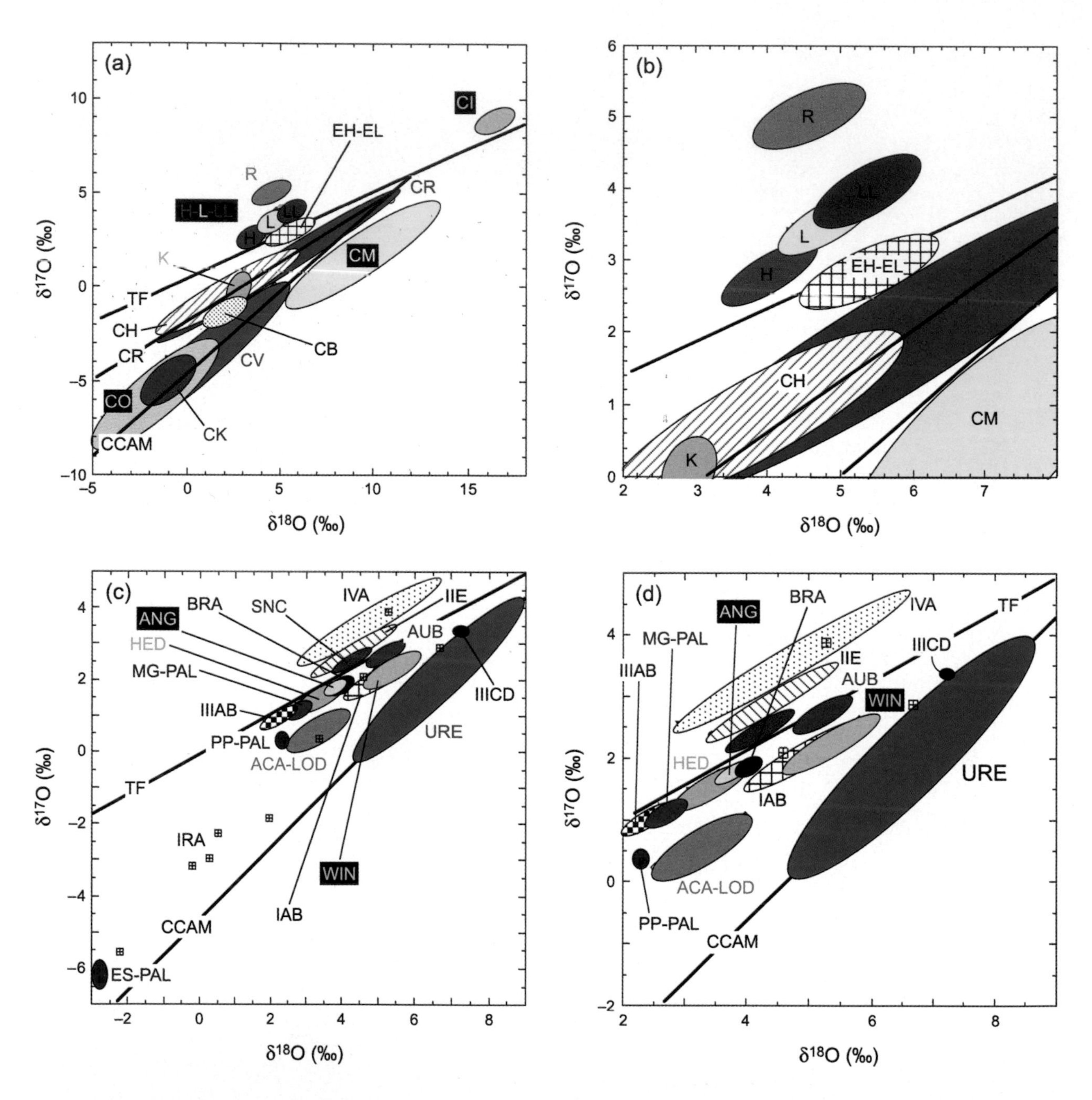

Plate 3. **(a)** Oxygen three-isotope diagram showing the O-isotopic differences among the chondrite clans and groups and the fields in the diagram occupied by each group. **(b)** A portion of the diagram shown in **(a)** enlarged to show the region in which O, E, and R chondrites occupy on the diagram. **(c)** Oxygen diagram showing the differences among and the regions of the diagram occupied by the achondrite and primitive achondrite clans and groups. **(d)** A portion of the diagram shown in **(c)** enlarged to show the region the ureilites and other achondrites occupy on the diagram. The ureilites are primitive achondrites that plot along the slope-1 CCAM mixing line. Because meteorite groups occupy specific regions on the diagram, O isotopes have become a powerful tool for classifying meteorites. Additionally, they can help reveal relationships among meteorite groups, such the CR, CB, and CH chondrites, which form the CR clan. TF — terrestrial fractionation line; CCAM — carbonaceous chondrite anhydrous mineral mixing line; CR — CR mixing line. Symbols for the achondrite groups are the same as in Fig. 1. IAB, IIIAB, IVA, IICD, and IIE are iron meteorite groups. IRA are ungrouped irons. Data are from *Clayton et al.* (1984) and *Clayton and Mayeda* (1996, 1999) and references therein.

Accompanies chapter by Weisberg et al. (pp. 19–52).

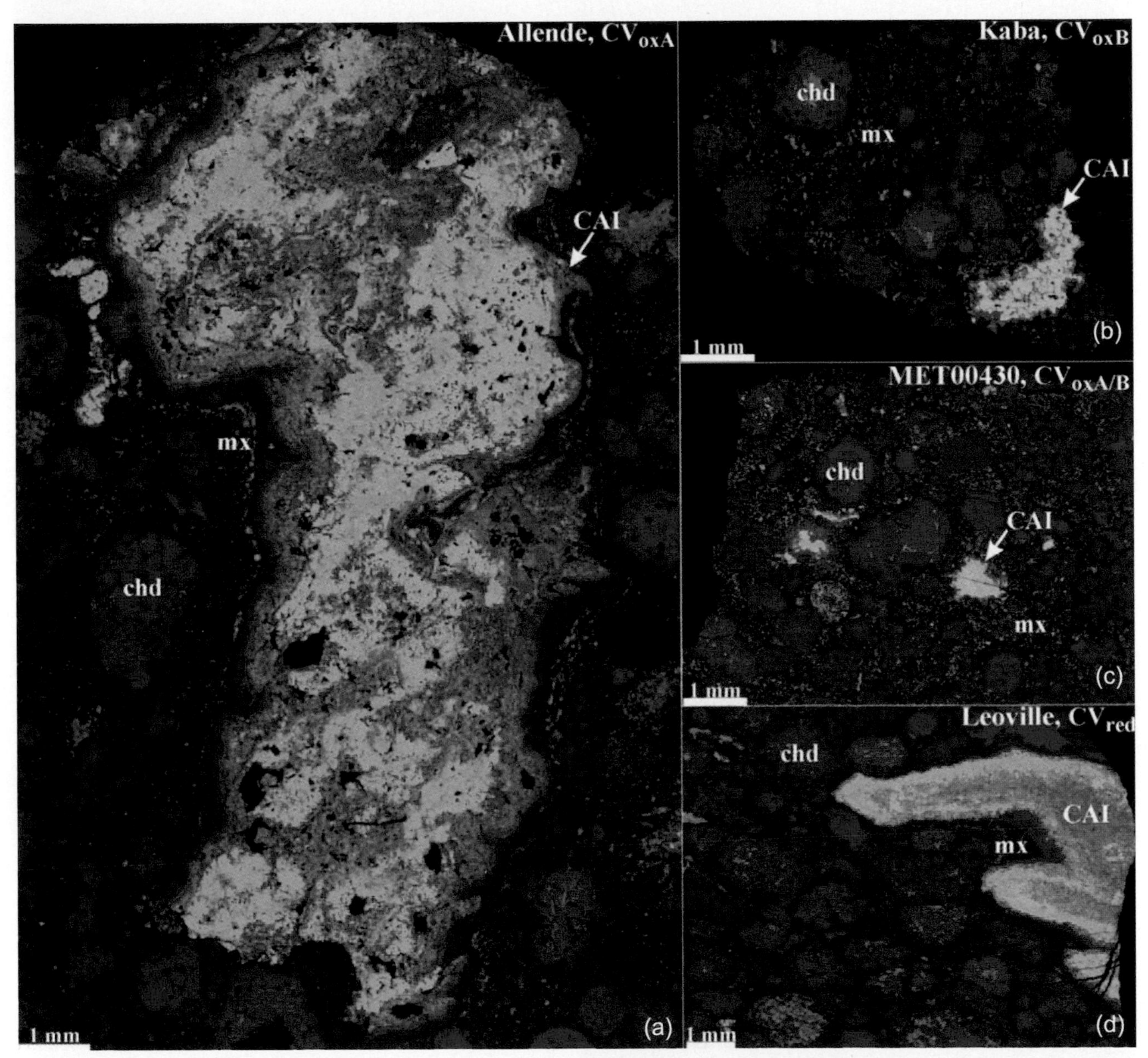

Plate 4. Combined elemental maps in Mg (red), Ca (green), and Al (blue) Kα X-rays of the CV carbonaceous chondrites **(a)** Allende (CV_{oxA}), **(b)** Kaba (CV_{oxB}), **(c)** MET 00430 ($CV_{oxA/B}$), and **(d)** Leoville (CV_{red}). The CV chondrites contain large Ca,Al-rich inclusions (CAIs), mostly magnesian, porphyritic chondrules (chd), and fine-grained matrix (mx), and show large variations in chondrule/matrix ratios (0.5–1.2). Matrices in the oxidized CVs contain higher abundance of Ca,Fe-pyroxene ± andradite nodules (green spots) than the Leoville matrix.

Accompanies chapter by Weisberg et al. (pp. 19–52).

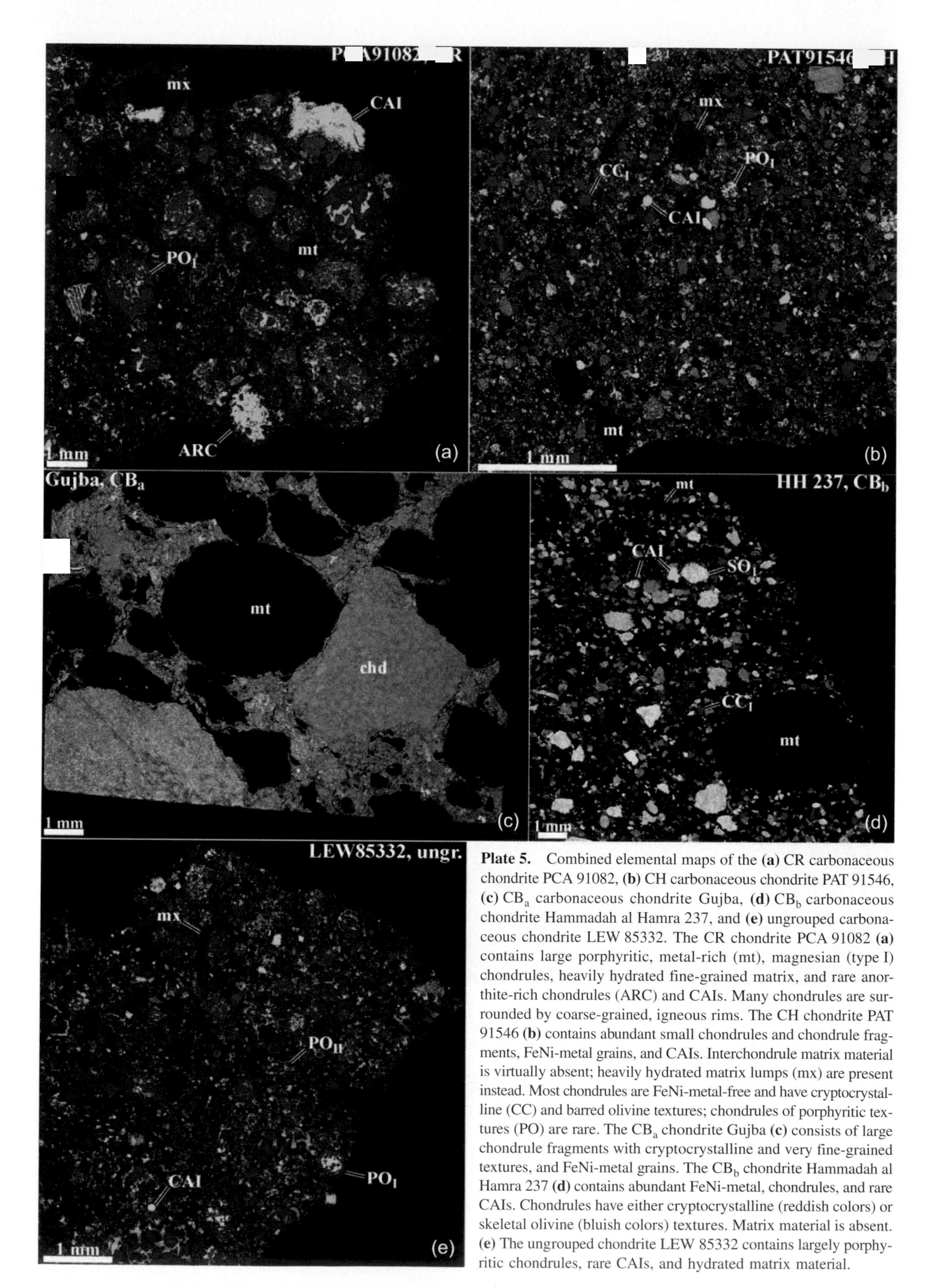

Plate 5. Combined elemental maps of the **(a)** CR carbonaceous chondrite PCA 91082, **(b)** CH carbonaceous chondrite PAT 91546, **(c)** CB_a carbonaceous chondrite Gujba, **(d)** CB_b carbonaceous chondrite Hammadah al Hamra 237, and **(e)** ungrouped carbonaceous chondrite LEW 85332. The CR chondrite PCA 91082 **(a)** contains large porphyritic, metal-rich (mt), magnesian (type I) chondrules, heavily hydrated fine-grained matrix, and rare anorthite-rich chondrules (ARC) and CAIs. Many chondrules are surrounded by coarse-grained, igneous rims. The CH chondrite PAT 91546 **(b)** contains abundant small chondrules and chondrule fragments, FeNi-metal grains, and CAIs. Interchondrule matrix material is virtually absent; heavily hydrated matrix lumps (mx) are present instead. Most chondrules are FeNi-metal-free and have cryptocrystalline (CC) and barred olivine textures; chondrules of porphyritic textures (PO) are rare. The CB_a chondrite Gujba **(c)** consists of large chondrule fragments with cryptocrystalline and very fine-grained textures, and FeNi-metal grains. The CB_b chondrite Hammadah al Hamra 237 **(d)** contains abundant FeNi-metal, chondrules, and rare CAIs. Chondrules have either cryptocrystalline (reddish colors) or skeletal olivine (bluish colors) textures. Matrix material is absent. **(e)** The ungrouped chondrite LEW 85332 contains largely porphyritic chondrules, rare CAIs, and hydrated matrix material.

Accompanies chapter by Weisberg et al. (pp. 19–52).

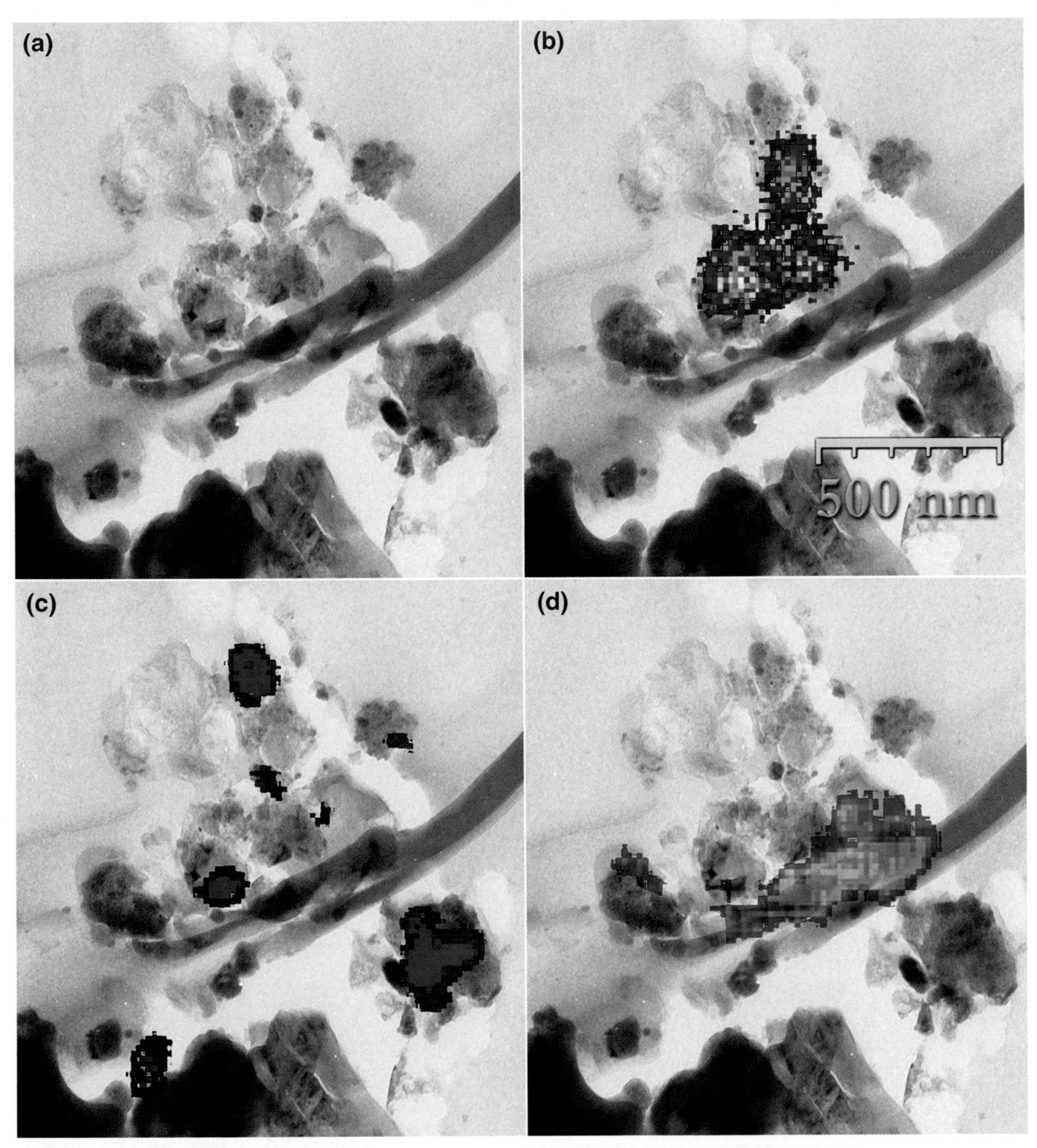

Plate 6. **(a)** TEM brightfield image of IDP L2011 B10A containing forsterite, enstatite, GEMS grains, and carbonaceous matter. **(b)** Overlay of ^{18}O-rich region identifying a supernova olivine grain. **(c)** ^{32}S hotspots show the locations of FeS and GEMS grains. **(d)** ^{15}N-rich carbonaceous matter associated with presolar olivine grain. From *Messenger et al.* (2005).

Accompanies chapter by Messenger et al. (pp. 187–208).

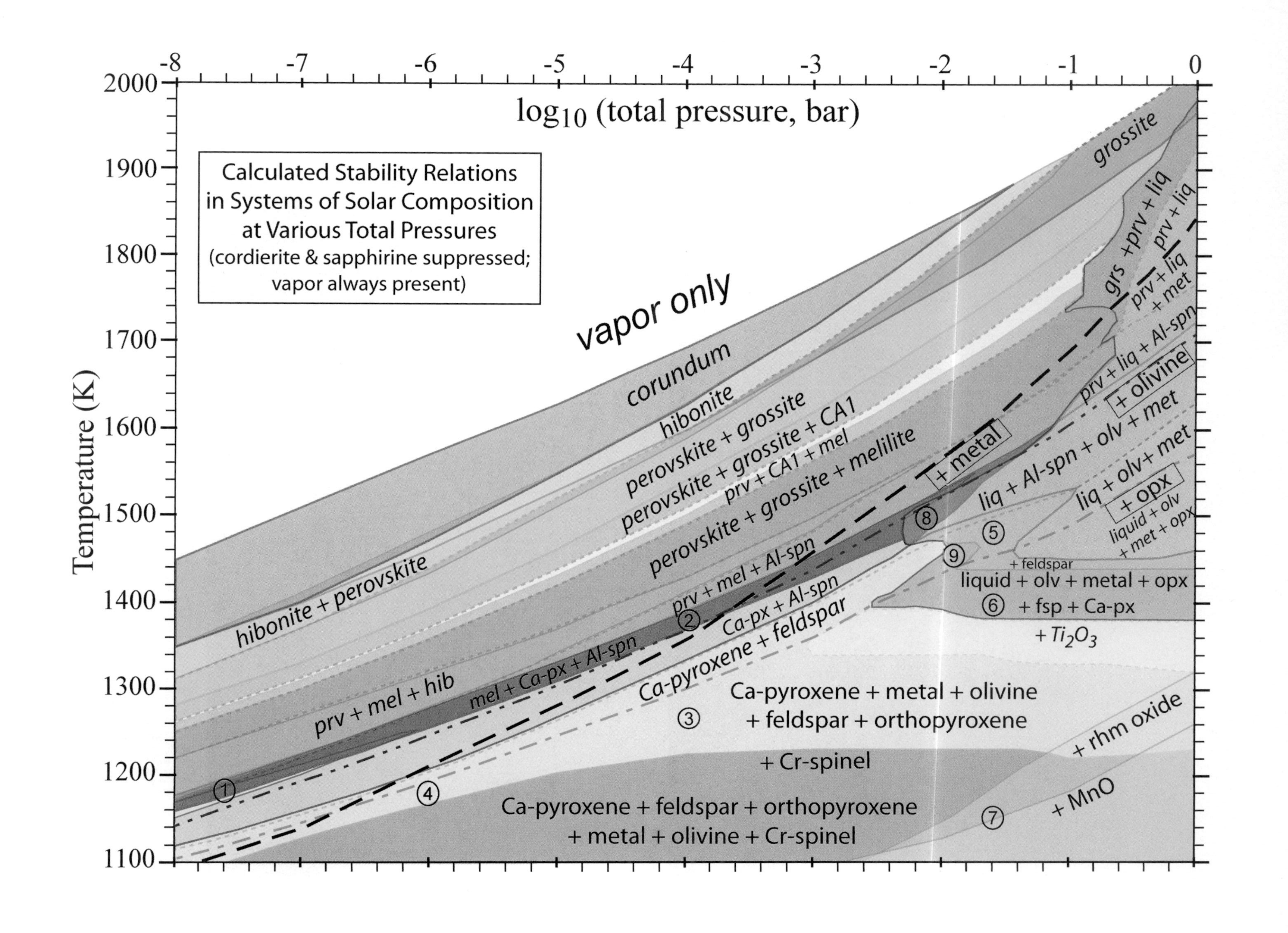

log10 (total pressure, bar)
Temperature (K)
Calculated Stability Relations
in Systems of Solar Composition
at Various Total Pressures
(cordierite & sapphirine suppressed;
vapor always present)
vapor only
corundum
hibonite
grossite
perovskite + grossite
perovskite + grossite + CA1
prv + CA1 + mel
perovskite + grossite + melilite
hibonite + perovskite
prv + mel + hib
mel + Ca-px + Al-spn
prv + mel + Al-spn
Ca-px + Al-spn
Ca-pyroxene + feldspar
Ca-pyroxene + metal + olivine
+ feldspar + orthopyroxene
+ Cr-spinel
Ca-pyroxene + feldspar + orthopyroxene
+ metal + olivine + Cr-spinel
+ rhm oxide
+ MnO
+ metal
grs + prv + liq
prv + liq
prv + liq + met
prv + liq + Al-spn
+ olivine
liq + Al-spn + olv + met
liq + olv + met
+ opx
liquid + olv + met + opx
+ feldspar
liquid + olv + metal + opx
+ fsp + Ca-px
+ Ti2O3

Plate 7. Stability of condensed phases with temperature and pressure, in a vapor of solar composition. The temperatures below which metal, olivine, and orthopyroxene are stable are shown as a function of P^{tot} by dashed, dash-dot-dot, and dash-dot lines, respectively. At $P^{tot} > \sim 10^{-2.5}$ bar, liquids, primarily of CMAS composition, become stable. The feldspar here is nearly pure anorthite (alkali-free). Even at high P^{tot}, silicates contain almost no FeO-bearing component, in contrast to Mg-silicates at high dust enrichments in Plate 10. Circled numerals illustrate features of this result: At ①, Ca-feldspar is stable at $P^{tot} < \sim 10^{-7.6}$ bar over a small T-range above its persistent T, which is about 80° lower (see Plate 8). At ②, melilite, Ca-pyroxene, and Al-spinel are stable, and melilite will disappear, and olivine will become stable, about 15° above the appearance of metal, as the system cools. The assemblage at ② is commonly observed in melted CAIs. Upon cooling of the system at ②, large quantities of olivine and metal will condense, with Ca-pyroxene and Al-spinel. At the same P^{tot}, feldspar will replace Al-spinel at lower temperatures, then orthopyroxene becomes stable, yielding the assemblage at ③, which is the same assemblage as at ④. Phases stable at ⑤ are liquid + olivine + metal + feldspar, with orthopyroxene and then Ca-pyroxene condensing at successively lower T to yield assemblage ⑥, in which all liquid crystallizes to solids at ~1380 K. Olivine + melilite + Al-spinel + liquid + metal are stable at ⑧, and metal + olivine + feldspar + Ca-pyroxene + liquid at ⑨. Assemblages ③–⑥ are relevant to FeO-poor chondrules (e.g., *Jones and Scott,* 1989). The oxide Ti_2O_3 is predicted to form when liquid crystallizes; however, this Ti should, in real systems, dissolve into silicates or oxides, as should the MnO predicted to condense at ⑦ at $P^{tot} \sim 10^{-1.6}$. These "shoulds" stem from the extreme difficulty of accurately calibrating solid-solution models of minor components.

Accompanies chapter by Ebel (pp. 253–277).

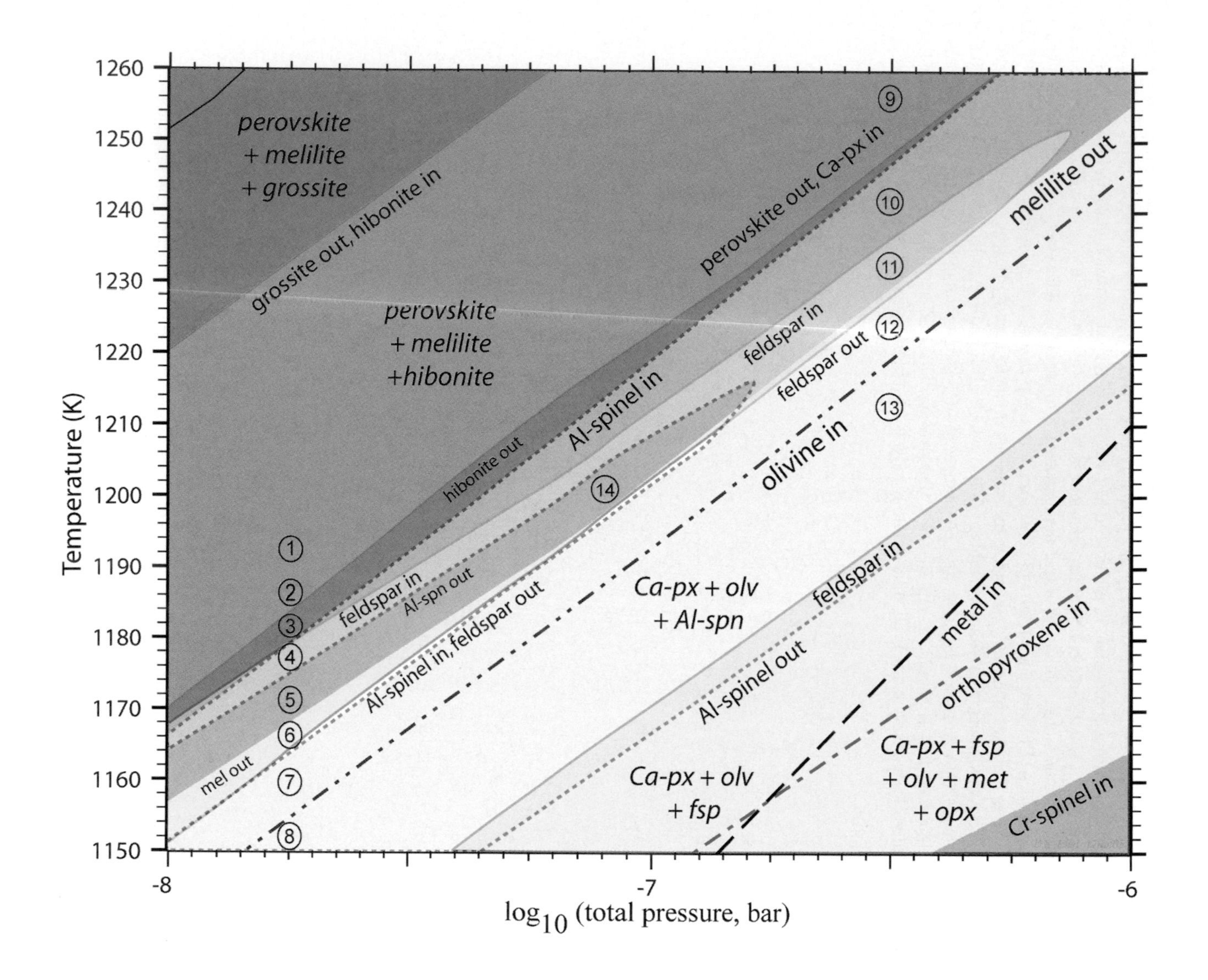

Plate 8. Stability of feldspar at low total pressure, in a vapor of solar composition. This is a detail of Plate 7. The stable solids at $P^{tot} = 10^{-7.75}$, as Si and Mg condense from vapor as the system cools, are ① perovskite + melilite + hibonite; ② melilite + hibonite + Ca-pyroxene; ③ melilite + Ca-pyroxene; ④ melilite + Ca-pyroxene + Al-spinel + feldspar; ⑤ melilite + Ca-pyroxene + feldspar (as at ⑭); ⑥ Ca-pyroxene + feldspar; ⑦ Ca-pyroxene + Al-spinel; ⑧ Ca-pyroxene + Al-spinel + olivine. At $P^{tot} = 10^{-6.5}$, as the system cools, assemblages are ⑨ perovskite + melilite + hibonite; ⑩ melilite + Ca-pyroxene + Al-spinel; ⑪ melilite + Ca-pyroxene + Al-spinel + feldspar; ⑫ Ca-pyroxene + Al-spinel + feldspar; ⑬ Ca-pyroxene + Al-spinel + feldspar + olivine.

Accompanies chapter by Ebel (pp. 253–277).

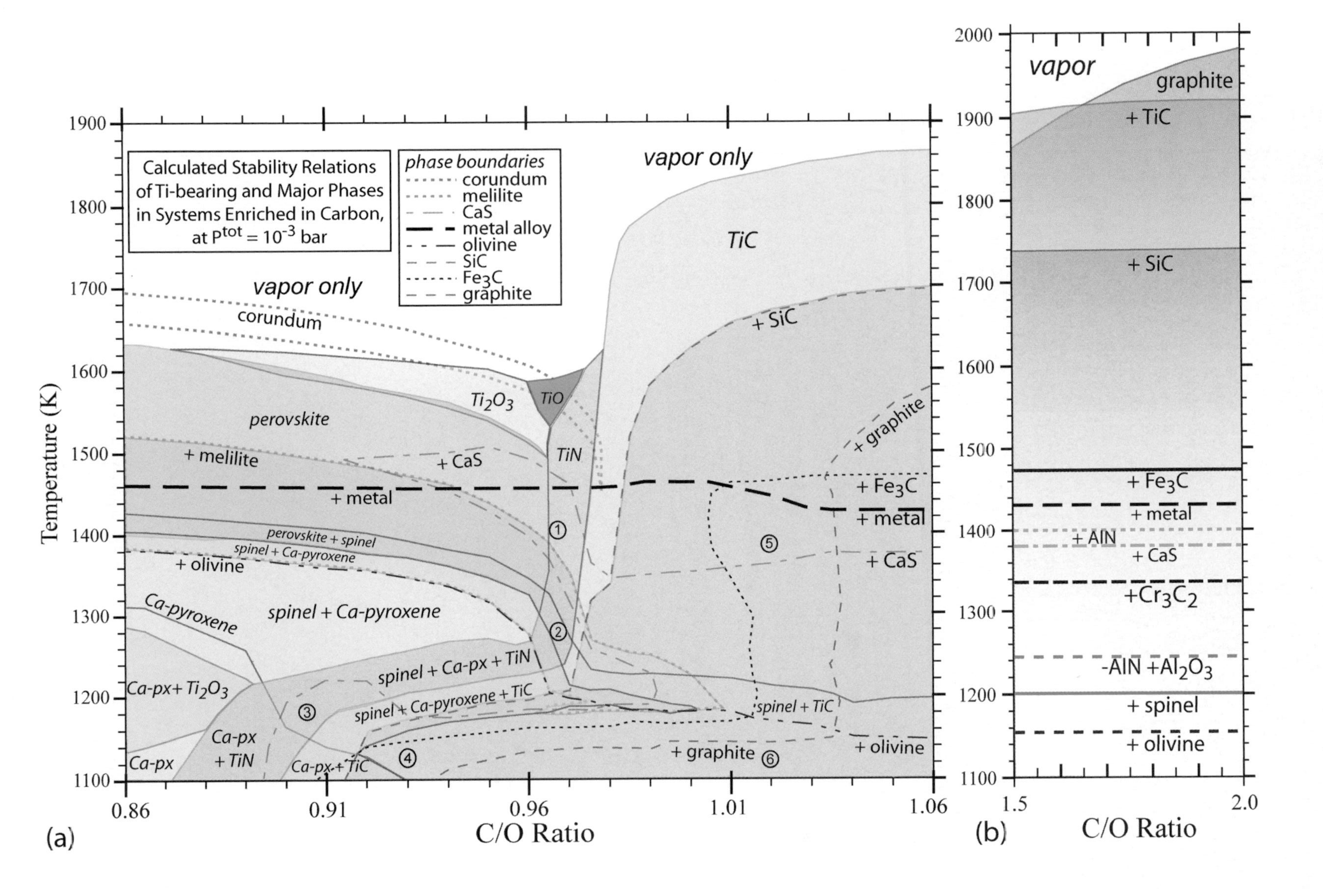

Plate 9. **(a)** Stability of Ti-bearing phases, and selected other phases, in a vapor of solar composition with C added to increase the C/O ratio. **(b)** Detail of phase stability at high C/O ratio, as excess C condenses as graphite. Circled numbers in **(a)** are: At ①, C/O ~ 0.967, 1400 K, CaS (oldhamite) is stable, with TiN (osbornite), as well as hibonite and metal alloy. At ②, 120° lower, Ti is in spinel solid solution, TiN, and perovskite, and Ca is in melilite. Melilite is stable over a successively smaller T region up to C/O ~ 1.01. At ③, C/O = 0.905, 1180 K, Ti is in spinel, Ca-pyroxene, and TiN, coexisting with olivine, metal, and CaS. At ④, and ⑥, Ti is in spinel and TiC, while at ⑤ all the Ti is in TiC. Metal coexists with Fe_3C (cohenite) at all three points. At ④, olivine and CaS are also stable; at ⑤, SiC and AlN; and at ⑥, olivine, spinel, SiC, CaS, graphite, and Cr_3C_2. Field boundaries for AlN and Cr_3C_2 are not shown.

Accompanies chapter by Ebel (pp. 253–277).

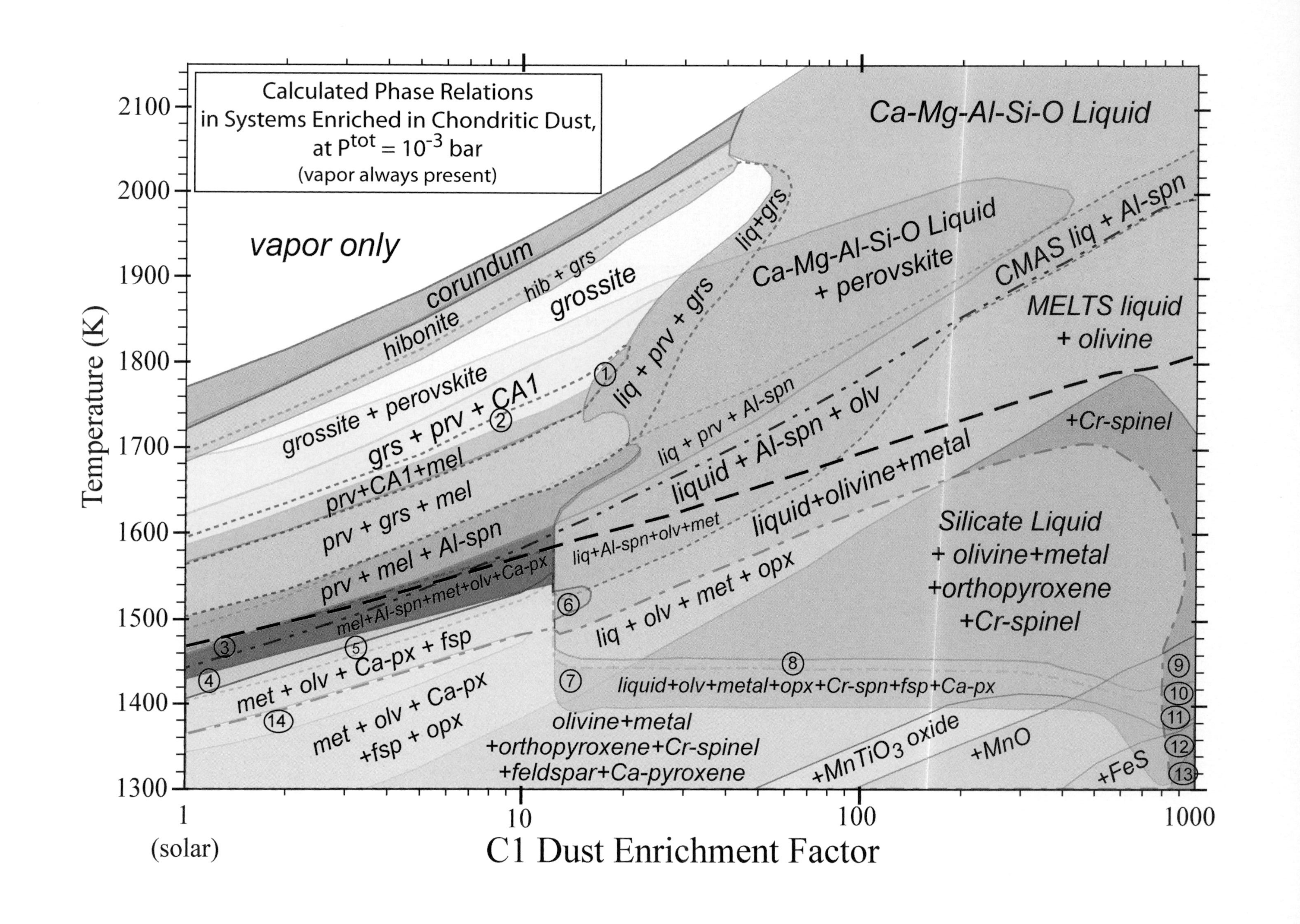

Calculated Phase Relations
in Systems Enriched in Chondritic Dust,
at $P^{tot} = 10^{-3}$ bar
(vapor always present)
vapor only
corundum
hibonite
hib + grs
grossite
grossite + perovskite
grs + prv + CA1
prv+CA1+mel
prv + grs + mel
prv + mel + Al-spn
mel+Al-spn+met+olv+Ca-px
met + olv + Ca-px + fsp
met + olv + Ca-px
+fsp + opx
liq + prv + grs
liq+grs
Ca-Mg-Al-Si-O Liquid
Ca-Mg-Al-Si-O Liquid
+ perovskite
CMAS liq + Al-spn
MELTS liquid
+ olivine
liq + prv + Al-spn
liquid + Al-spn + olv
liq+Al-spn+olv+met
liquid+olivine+metal
liq + olv + met + opx
+Cr-spinel
Silicate Liquid
+ olivine+metal
+orthopyroxene
+Cr-spinel
liquid+olv+metal+opx+Cr-spn+fsp+Ca-px
olivine+metal
+orthopyroxene+Cr-spinel
+feldspar+Ca-pyroxene
+MnTiO3 oxide
+MnO
+FeS
1
2
3
4
5
6
7
8
9
10
11
12
13
14
Temperature (K)
1300
1400
1500
1600
1700
1800
1900
2000
2100
C1 Dust Enrichment Factor
1
(solar)
10
100
1000

Plate 10. Calculated phase relations in systems enriched in a CI-chondrite-like dust at a total pressure (P^{tot}) of 10^{-3} bar, calculated as in *Ebel and Grossman* (2000). See Table 1 for formulae and abbreviations. Circled numbers are assemblages: ① perovskite + Ca-monoaluminate + liquid, ② perovskite + Ca-monoaluminate, ③ melilite + Al-spinel + metal + Ca-pyroxene, ④ Al-spinel + metal + olivine + Ca-pyroxene, ⑤ Al-spinel + metal + olivine + Ca-pyroxene + feldspar, ⑥ liquid + olivine + metal + feldspar, ⑦ liquid + olivine + metal + orthopyroxene + feldspar + Ca-pyroxene, ⑧ liquid + olivine + metal + orthopyroxene + Cr-spinel + feldspar, ⑨ liquid + olivine + metal + Cr-spinel + MnO, ⑩ liquid + olivine + metal + Cr-spinel + feldspar + MnO, ⑪ = ⑩ + Ca-pyroxene, ⑫ = ⑪ + sulfide + $MnTiO_3$, ⑬ = ⑫ + whitlockite, ⑭ Ti_3O_5 is briefly stable in this field. Manganese is calculated to condense as MnO, as well as $MnTiO_3$ oxide, that would both dissolve in real coexisting silicates (e.g., olivine, pyroxene).

Accompanies chapter by Ebel (pp. 253–277).

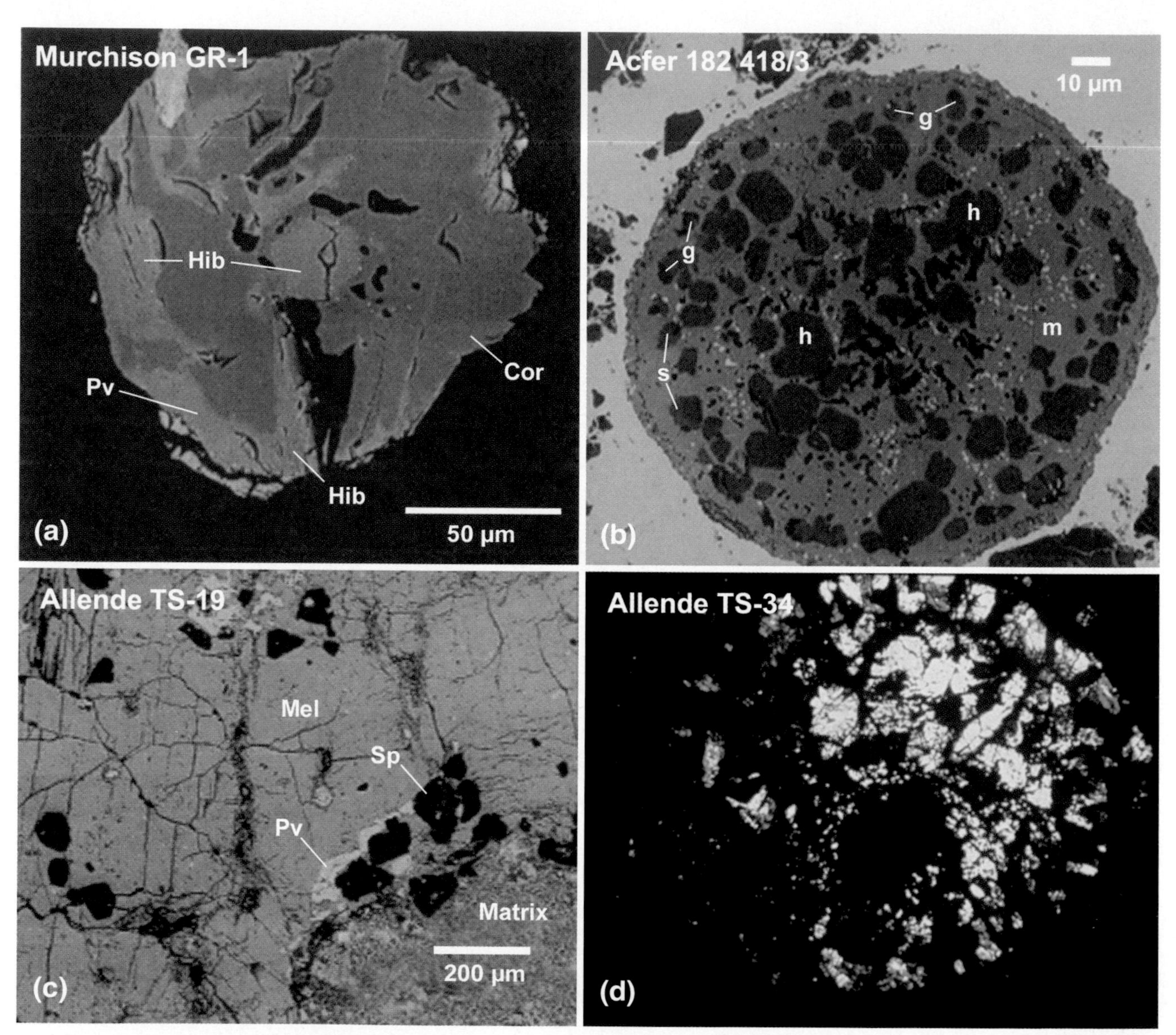

Plate 11. Examples of meteoritic inclusions thought to have crystallized from a melt. **(a)** Backscattered electron (BSE) photomicrograph of GR-1, a Murchison (CM2) inclusion from *MacPherson et al.* (1984a). Phases indicated are Cor: corundum; Hib: hibonite; Pv: perovskite. **(b)** BSE photomicrograph of inclusion 418/3 from the CH chondrite Acfer 182. The small white blebs are perovskite. Other phases include h: hibonite, g: grossite, s: spinel, and m: melilite. From *Weber and Bischoff* (1994). **(c)** BSE photomicrograph of Allende (CV3) compact type A inclusion TS-19. Matrix refers to material outside the inclusion. Mel: melilite; Pv: perovskite; Sp: spinel. From *Simon et al.* (1999). **(d)** Crossed polars view of an Allende type B1 inclusion, TS-34. A melilite-rich mantle (blue to yellow) surrounds a core containing spinel (isotropic, hence black in the image), fassaite, and melilite. Field of view, ~2.5 cm. Photo courtesy of G. J. MacPherson.

Accompanies chapter by Beckett et al. (pp. 399–429).

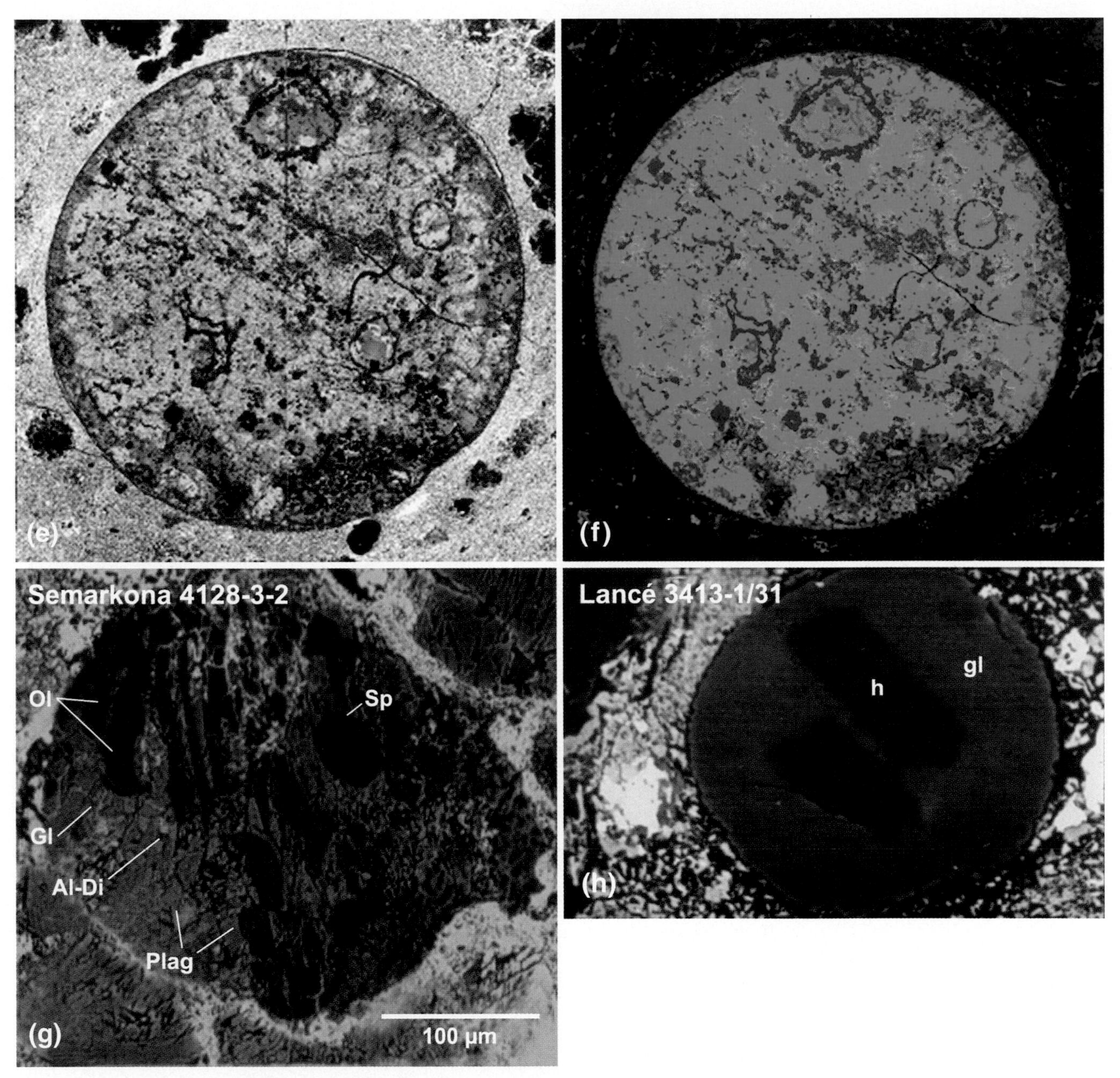

Plate 11. (continued). **(e)** BSE photomicrograph of an Allende type B2 inclusion from USNM 4947-1. The vertical line bisecting the inclusion is an artifact of figure preparation. Field of view, 6138 µm. **(f)** False-color mosaic of the type B2 inclusion depicted in **(e)**. This panel was produced through an overlay of three elemental X-ray maps onto the BSE image with Ti in red, Ca in green, and Al in blue, each scaled to generate a combined map that allows major phases to be distinguished. In the overlay, spinel is blue, greens are melilite (light) and anorthite (darker and much less abundant), and reddish-orange regions show fassaite. **(g)** BSE photomicrograph of an Al-rich chondrule 4128-3-2 from the LL3.0 ordinary chondrite Semarkona (*MacPherson and Huss,* 2005). Al-Di: aluminous diopside; Gl: glass; Ol: olivine; Plag: plagioclase; Sp: spinel. **(h)** BSE photomicrograph of a hibonite-glass spherule LA 3413-1/31 from the CO3 chondrite Lancé (*Ireland et al.,* 1991). h: hibonite; gl: glass. The spherule is ~50 µm in diameter.

Accompanies chapter by Beckett et al. (pp. 399–429).

Index

Note: Material in tables is indicated by the letter *T* after the page number. Figures are indicated by the letter *F* after the page number.